AF814303

The Most Accepted
Crash Course Programme

JEE Main in 40 DAYS

Chemistry

The Most Accepted

Crash Course Programme

JEE Main in 40 DAYS

Chemistry

Arvind Kumar
Ayush Kumar

arihant
SUCCESS AND NOTHING LESS

arihant
SUCCESS AND NOTHING LESS

ARIHANT PRAKASHAN (Series), MEERUT

卐 **Administrative & Production Offices**

Regd. Office

'Ramchhaya' 4577/15, Agarwal Road, Darya Ganj, New Delhi -110002
Tele: 011- 47630600, 43518550

卐 **Head Office**

Kalindi, TP Nagar, Meerut (UP) - 250002
Tel: 0121-7156203, 7156204

卐 **Sales & Support Offices**

Agra, Ahmedabad, Bengaluru, Bareilly, Chennai, Delhi, Guwahati, Hyderabad, Jaipur, Jhansi, Kolkata, Lucknow, Nagpur & Pune.

卐 **PRICE** ₹525.00

卐 **PO No :** TXT-XX-XXXXXXX-X-XX

Published by Arihant Publications (India) Ltd.

For further information about the books published by Arihant, log on to www.arihantbooks.com or e-mail at info@arihantbooks.com

Follow us on

PREFACE

It is a fact that nearly 10 lacs students would be in the race with you in JEE Main, the gateway to some of the prestigious engineering and technology institutions in the country, requires that you take it seriously and head-on. A slight underestimation or wrong guidance will ruin all your prospects. You have to earmark the topics in the syllabus and have to master them in concept-driven-problem-solving ways, considering the thrust of the questions being asked in JEE Main.

The book **40 Days JEE Main Chemistry** serves the above cited purpose in perfect manner. At whatever level of preparation you are before the exam, this book gives you an accelerated way to master the whole JEE Main Chemistry Syllabus. It has been conceived keeping in mind the latest trend of questions, and the level of different types of students.

The whole syllabus of Chemistry has been divided into day-wise-learning modules with clear groundings into concepts and sufficient practice with solved and unsolved questions on that day. After every few days you get a Unit Test based upon the topics covered before that day. On last three days you get three full-length Mock Tests, making you ready to face the test. It is not necessary that you start working with this book in 40 days just before the exam. You may start and finish your preparation of JEE Main much in advance before the exam date. This will only keep you in good frame of mind and relaxed, vital for success at this level.

Salient Features

- Concepts discussed clearly and directly without being superfluous. **Only the required material** for JEE Main being described comprehensively to keep the students focussed.
- Exercises for each day give you the collection of **only the Best Questions** of the concept, giving you the perfect practice in less time.
- Each day has two Exercises; **Foundation Questions Exercise** having Topically Arranged Questions & **Progressive Question Exercise** having higher Difficulty Level Questions.
- **All types of Objective Questions** included in Daily Exercises (Single Option Correct, Assertion & Reason, etc).
- Along with Daywise Exercises, there above also the **Unit Tests** & **Full Length Mock Tests**.
- At the end, there are all Online Solved Papers of JEE Main 2023; Jan & April attempts.

We are sure that 40 Days Chemistry for JEE Main will give you a fast way to prepare for Chemistry without any other support or guidance.

Publisher

CONTENTS

SYLLABUS

CHEMISTRY

SECTION- A (Physical Chemistry)

UNIT 1
Some Basic Concepts in Chemistry

Matter and its nature, Dalton's atomic theory; Concept of atom, molecule, element and compound; Physical quantities and their measurements in Chemistry, precision and accuracy, significant figures, S.I. Units, dimensional analysis; Laws of chemical combination; Atomic and molecular masses, mole concept, molar mass, percentage composition, empirical and molecular formulae; Chemical equations and stoichiometry.

UNIT 2 States of Matter

Classification of matter into solid, liquid and gaseous states.

Gaseous State Measurable properties of gases; Gas laws - Boyle's law, Charle's law, Graham's law of diffusion, Avogadro's law, Dalton's law of partial pressure; Concept of Absolute scale of temperature; Ideal gas equation, Kinetic theory of gases (only postulates); Concept of average, root mean square and most probable velocities; Real gases, deviation from Ideal behaviour, compressibility factor, van der Waals'

Equation, liquefaction of gases, critical constants.

Liquid State Properties of liquids - vapour pressure, viscosity and surface tension and effect of temperature on them (qualitative treatment only).

Solid State Classification of solids: molecular, ionic, covalent and metallic solids, amorphous and crystalline solids (elementary idea); Bragg's Law and its applications, Unit cell and lattices, packing in solids (fcc, bcc and hcp lattices), voids, calculations involving unit cell parameters, imperfection in solids; electrical, magnetic and dielectric properties.

UNIT 3 Atomic Structure

Discovery of sub-atomic particles (electron, proton and neutron); Thomson and Rutherford atomic models and their limitations; Nature of electromagnetic radiation, photoelectric effect; spectrum of hydrogen atom, Bohr model of hydrogen atom - its postulates, derivation of the relations for energy of the electron and radii of the different orbits, limitations of Bohr's model; dual nature of matter, de-Broglie's relationship, Heisenberg uncertainty principle.

Elementary ideas of quantum mechanics, quantum mechanical model of atom, its important features, ψ and ψ^2, concept of atomic orbitals as one electron wave functions; Variation of ψ and ψ^2 with r for 1*s* and 2*s* orbitals; various quantum numbers (principal, angular momentum and magnetic quantum numbers) and their significance; shapes of *s*, *p* and *d* - orbitals, electron spin and spin quantum number; rules for filling electrons in orbitals – aufbau principle, Pauli's exclusion principle and Hund's rule, electronic configuration of elements, extra stability of half-filled and completely filled orbitals.

UNIT 4 Chemical Bonding and Molecular Structure

Kossel Lewis approach to chemical bond formation, concept of ionic and covalent bonds.

Ionic Bonding Formation of ionic bonds, factors affecting the formation of ionic bonds; calculation of lattice enthalpy.

Covalent Bonding Concept of electronegativity, Fajan's rule, dipole moment; Valence Shell Electron Pair Repulsion (VSEPR) theory and shapes of simple molecules.

Quantum mechanical approach to covalent bonding Valence bond theory - Its important features, concept of **hybridization** involving s, p and d orbitals; Resonance.

Molecular Orbital Theory Its important features, LCAOs, types of molecular orbitals (bonding, antibonding), sigma and pi-bonds, molecular orbital electronic configurations of homonuclear diatomic molecules, concept of bond order, bond length and bond energy.

Elementary idea of metallic bonding. Hydrogen bonding and its applications.

UNIT 5 Chemical Thermodynamics

Fundamentals of thermodynamics System and surroundings, extensive and intensive properties, state functions, types of processes.

First law of thermodynamics Concept of work, heat internal energy and enthalpy, heat capacity, molar heat capacity, Hess's law of constant heat summation; Enthalpies of bond dissociation, combustion, formation, atomization, sublimation, phase transition, hydration, ionization and solution.

Second law of thermodynamics Spontaneity of processes; ΔS of the universe and ΔG of the system as criteria for spontaneity, ΔG° (Standard Gibb's energy change) and equilibrium constant.

UNIT 6 Solutions

Different methods for expressing concentration of solution - molality, molarity, mole fraction, percentage (by volume and mass both), vapour pressure of solutions and Raoult's Law - Ideal and non-ideal solutions, vapour pressure - composition plots for ideal and non-ideal solutions.

Colligative properties of dilute solutions - relative lowering of vapour pressure, depression of freezing point, elevation of boiling point and osmotic pressure; Determination of molecular mass using colligative properties; Abnormal value of molar mass, van't Hoff factor and its significance.

UNIT 7 Equilibrium

Meaning of equilibrium, concept of dynamic equilibrium.

Equilibria involving physical processes
Solid -liquid, liquid - gas and solid - gas equilibria, Henry's law, general characteristics of equilibrium involving physical processes.

Equilibria involving chemical processes Law of chemical equilibrium, equilibrium constants (K and K) and their significance, significance of ΔG and ΔG° in chemical equilibria, factors affecting equilibrium concentration, pressure, temperature, effect of catalyst; Le -Chatelier's principle.

Ionic equilibrium Weak and strong electrolytes, ionization of electrolytes, various concepts of acids and bases (Arrhenius, Bronsted - Lowry and Lewis) and their ionization, acid-base equilibria (including multistage ionization) and ionization constants, ionization of water, pH scale, common ion effect, hydrolysis of salts and pH of their solutions, solubility of sparingly soluble salts and solubility products, buffer solutions.

UNIT 8
Redox Reactions and Electrochemistry

Electronic concepts of oxidation and reduction, redox reactions, oxidation number, rules for assigning oxidation **number,** balancing of redox reactions.

Eectrolytic and metallic conduction, conductance in electrolytic solutions, specific and molar conductivities and their variation with concentration: Kohlrausch's **law and its applications.**

Electrochemical cells - Electrolytic and Galvanic cells, different types of electrodes, electrode potentials including standard electrode potential, half - cell and cell reactions, emf of a Galvanic cell and its measurement; Nernst equation and its applications; Relationship between cell potential and Gibbs' energy change; Dry cell and lead accumulator; Fuel cells; Corrosion and its prevention.

UNIT 9 Chemical Kinetics

Rate of a chemical reaction, factors affecting the rate of reactions concentration, temperature, pressure and catalyst; elementary and complex reactions, order and molecularity of reactions, rate law, rate constant and its units, differential and integral forms of zero and first order reactions, their characteristics and half - lives, effect of temperature on rate of reactions - Arrhenius theory, activation energy and its calculation, collision theory of bimolecular gaseous reactions (no derivation).

UNIT 10 Surface Chemistry

Adsorption Physisorption and chemisorption and their characteristics, factors affecting adsorption of gases on solids- Freundlich and Langmuir adsorption isotherms, adsorption from solutions.

Catalysis Homogeneous and heterogeneous, activity and selectivity of solid catalysts, enzyme catalysis and its mechanism.

Colloidal state distinction among true solutions, colloids and suspensions, classification of colloids: - lyophilic, lyophobic; multi molecular, macromole-cular and associated colloids (micelles), preparation and properties of colloids Tyndall effect, Brownian movement, electrophoresis, dialysis, coagulation and flocculation; Emulsions and their characteristics.

SECTION- B (Inorganic Chemistry)

UNIT 11

Classification of Elements and Periodicity in Properties

Periodic Law and Present Form of the Periodic Table, *s, p, d* and *f* Block Elements, Periodic Trends in Properties of Elementsatomic and Ionic Radii, Ionization Enthalpy, Electron Gain Enthalpy, Valence, Oxidation States and Chemical Reactivity.

UNIT 12

General Principles and Processes of Isolation of Metals

Modes of occurrence of elements in nature, minerals, ores; steps involved in the extraction of metals - concentration, reduction (chemical and electrolytic methods) and refining with special reference to the extraction of Al, Cu, Zn and Fe; Thermodynamic and electrochemical principles involved in the extraction of metals.

UNIT 13 Hydrogen

Position of hydrogen in periodic table, isotopes, preparation, properties and uses of hydrogen; physical and chemical properties of water and heavy water; Structure, preparation, reactions and uses of hydrogen peroxide; Classification of hydrides ionic, covalent and interstitial; Hydrogen as a fuel.

UNIT 14 *s* - Block Elements

(Alkali and Alkaline Earth Metals)

Group 1 and 2 Elements

General introduction, electronic configuration and general trends in physical and chemical properties of elements, anomalous properties of the first element of each group, diagonal relationships.

Preparation and properties of some important compounds - sodium carbonate, sodium chloride, sodium hydroxide and sodium hydrogen carbonate; Industrial uses of lime, limestone, Plaster of Paris and cement; Biological significance of Na, K, Mg and Ca.

UNIT 15 *p* - Block Elements

Group 13 to Group 18 Elements

General Introduction Electronic configuration and general trends in physical and chemical properties of elements across the periods and down the groups; unique behaviour of the first element in each group.

Group wise study of the *p* – block elements

Group 13 Preparation, properties and uses of boron and aluminium; structure, properties and uses of borax, boric acid, diborane, boron trifluoride, aluminium chloride and alums.

Group 14 Tendency for catenation; Structure, properties and uses of allotropes and oxides of carbon, silicon tetrachloride, silicates, zeolites and silicones.

Group 15 Properties and uses of nitrogen and phosphorus; Allotrophic forms of phosphorus; Preparation, properties, structure and uses of ammonia nitric acid, phosphine and phosphorus halides,(PCl_3, PCl_5); Structures of oxides and oxoacids of nitrogen and phosphorus.

Group 16 Preparation, properties, structures and uses of dioxygen and ozone; Allotropic forms of sulphur; Preparation, properties, structures and uses of sulphur dioxide, sulphuric acid (including its industrial preparation); Structures of oxoacids of sulphur.

Group 17 Preparation, properties and uses of chlorine and hydrochloric acid; Trends in the acidic nature of hydrogen halides; Structures of Interhalogen compounds and oxides and oxoacids of halogens.

Group 18 Occurrence and uses of noble gases; Structures of fluorides and oxides of xenon.

UNIT 16 *d*-and *f*-Block Elements

Transition Elements General introduction, electronic configuration, occurrence and characteristics, general trends in properties of the first row transition elements - physical properties, ionization enthalpy, oxidation states,

atomic radii, colour, catalytic behaviour, magnetic properties, complex formation, interstitial compounds, alloy formation; Preparation, properties and uses of $K_2Cr_2O_7$ and $KMnO_4$.

Inner Transition Elements Lanthanoids Electronic configuration, oxidation states, chemical reactivity and lanthanoid contraction.

Actinoids Electronic configuration and oxidation states.

UNIT 17 Coordination Compounds

Introduction to coordination compounds, Werner's theory; ligands, coordination number, denticity, chelation; IUPAC nomenclature of mononuclear coordination compounds, isomerism; Bonding Valence bond approach and basic ideas of Crystal field theory, colour and magnetic properties; importance of coordination compounds (in qualitative analysis, extraction of metals and in biological systems).

Unit 18 Environmental Chemistry

Environmental pollution Atmospheric, water and soil. Atmospheric pollution Tropospheric and stratospheric.

Tropospheric pollutants Gaseous pollutants Oxides of carbon, nitrogen and sulphur, hydrocarbons; their sources, harmful effects and prevention; Green house effect and Global warming; Acid rain;

Particulate pollutants Smoke, dust, smog, fumes, mist; their sources, harmful effects and prevention.

Stratospheric pollution Formation and breakdown of ozone, depletion of ozone layer - its mechanism and effects.

Water pollution Major pollutants such as, pathogens, organic wastes and chemical pollutants their harmful effects and prevention.

Soil pollution Major pollutants such as: Pesticides (insecticides, herbicides and fungicides), their harmful effects and prevention.

Strategies to control environmental pollution.

SECTION- C (Organic Chemistry)

UNIT 19 Purification & Characterisation of Organic Compounds

Purification Crystallization, sublimation, distillation, differential extraction and chromatography principles and their applications.

Qualitative analysis Detection of nitrogen, sulphur, phosphorus and halogens.

Quantitative analysis (basic principles only) Estimation of carbon, hydrogen, nitrogen, halogens, sulphur, phosphorus.

Calculations of empirical formulae and molecular formulae; Numerical problems in organic quantitative analysis.

UNIT 20 Some Basic Principles of Organic Chemistry

Tetravalency of carbon; Shapes of simple molecules hybridization (s and p); Classification of organic compounds based on functional groups: —C=C—,—C=C— and those containing halogens, oxygen, nitrogen and sulphur, Homologous series; Isomerism - structural and stereoisomerism.

Nomenclature (Trivial and IUPAC)

Covalent bond fission Homolytic and heterolytic free radicals, carbocations and carbanions; stability of carbocations and free radicals, electrophiles and nucleophiles.

Electronic displacement in a covalent bond Inductive effect, electromeric effect, resonance and hyperconjugation.

Common types of organic reactions Substitution, addition, elimination and rearrangement.

UNIT 21 Hydrocarbons

Classification, isomerism, IUPAC nomenclature, general methods of preparation, properties and reactions.

Alkanes Conformations: Sawhorse and Newman projections (of ethane); Mechanism of halogenation of alkanes.

Alkenes Geometrical isomerism; Mechanism of electrophilic addition: addition of hydrogen, halogens, water, hydrogen halides (Markownikoff's and peroxide effect); Ozonolysis, oxidation, and polymerization.

Alkynes acidic character; addition of hydrogen, halogens, water and hydrogen halides; polymerization.

Aromatic hydrocarbons Nomenclature, benzene structure and aromaticity; Mechanism of electrophilic substitution: halogenation, nitration, Friedel – Craft's alkylation and acylation, directive influence of functional group in mono-substituted benzene.

UNIT 22 Organic Compounds Containing Halogens

General methods of preparation, properties and reactions; Nature of C—*X* bond; Mechanisms of substitution reactions. Uses/environmental effects of chloroform, iodoform

UNIT 23 Organic Compounds Containing Oxygen

General methods of preparation, properties, reactions and uses.

Alcohols, Phenols and Ethers

Alcohols Identification of primary, secondary and tertiary alcohols; mechanism of dehydration.

Phenols Acidic nature, electrophilic substitution reactions: halogenation, nitration and sulphonation, Reimer - Tiemann reaction.

Ethers Structure.

Aldehyde and Ketones Nature of carbonyl group; Nucleophilic addition to \C=O group, relative reactivities of aldehydes and ketones; Important reactions such as - Nucleophilic addition reactions (addition of HCN, NH_3 and its derivatives), Grignard reagent; oxidation; reduction (Wolff Kishner and Clemmensen) acidity of α-hydrogen, aldol condensation, Cannizzaro reaction, Haloform reaction; Chemical tests to distinguish between aldehydes and Ketones.

Carboxylic Acids Acidic strength and factors affecting it.

UNIT 24 Organic Compounds Containing Nitrogen

General methods of preparation, properties, reactions and uses.

Amines Nomenclature, classification, structure basic character and identification of primary, secondary and tertiary amines and their basic character.

Diazonium Salts Importance in synthetic organic chemistry.

UNIT 25 Polymers

General introduction and classification of polymers, general methods of polymerization-addition and condensation, copolymerization; Natural and synthetic rubber and vulcanization; some important polymers with emphasis on their monomers and uses - polythene, nylon, polyester and bakelite.

UNIT 26 Biomolecules

General introduction and importance of biomolecules.

Carbohydrates Classification: aldoses and ketoses; monosaccharides (glucose and fructose), constituent monosaccharides of oligosaccharides (sucrose, lactose, maltose) and polysaccharides (starch, cellulose, glycogen).

Proteins Elementary Idea of α-amino acids, peptide bond, . polypeptides; proteins: primary, secondary, tertiary and quaternary structure (qualitative idea only), denaturation of proteins, enzymes.

Vitamins Classification and functions.

Nucleic Acids Chemical constitution of DNA and RNA. Biological functions of Nucleic acids.

UNIT 27 Chemistry in Everyday Life

Chemicals in medicines Analgesics, tranquilizers, antiseptics, disinfectants, antimicrobials, antifertility drugs, antibiotics, antacids, antihistamins - their meaning and common examples.

Chemicals in food Preservatives, artificial sweetening agents - common examples.

Cleansing agents Soaps and detergents, cleansing action.

UNIT 28 Principles Related to Practical Chemistry

- Detection of extra elements (N, S, halogens) in organic compounds; Detection of the following functional groups: hydroxyl (alcoholic and phenolic), carbonyl (aldehyde and ketone), carboxyl and amino groups in organic compounds.
- Chemistry involved in the preparation of the following
- Inorganic compounds Mohr's salt, potash alum.
- Organic compounds Acetanilide, *p*-nitroacetan ilide, aniline yellow, iodoform.
- Chemistry involved in the titrimetric excercises - Acids bases and the use of indicators, oxalic acid vs $KMnO_4$, Mohr's salt vs $KMnO_4$.
- Chemical principles involved in the qualitative salt analysis
- Cations — Pb^{2+}, Cu^{2+}, Al^{3+}, Fe^{3+}, Zn^{2+}, Ni^{2+}, Ca^{2+}, Ba^{2+}, Mg^{2+} NH^{4+}. Anions –CO_3^{2-}, S^{2-}, SO_4^{2-}, NO_2, NO_3, Cl^-, Br^-, I^- (Insoluble salts excluded).
- Chemical principles involved in the following experiments

1. Enthalpy of solution of $CuSO_4$
2. Enthalpy of neutralization of strong acid and strong base.
3. Preparation of lyophilic and lyophobic sols.
4. Kinetic study of reaction of iodide ion with hydrogen peroxide at room temperature.

HOW THIS BOOK IS
USEFUL FOR YOU ?

As the name suggests, this is the perfect book for your recapitulation of the whole syllabus, as it provides you a capsule course on the subject covering the syllabi of JEE Main, with the smartest possible tactics as outlined below:

1. REVISION PLAN

The book provides you with a practical and sound revision plan. The chapters of the book have been designed day-wise to guide the students in a planned manner through day-by-day, during those precious 35-40 days. Every day you complete a chapter/a topic, also take an exercise on the chapter so that you can check & correct your mistakes, answers with the provided hints & solutions. By 37th day from the date you start using this book, entire syllabus gets revisited.

Again, as per your convenience/preparation strategy, you can also divide the available 30-35 days into two time frames, first time slot of 3 weeks and last slot of 1 & 1/2 week. Utilize first time slot for studies and last one for revising the formulas and important points. Now fill the time slots with subjects/topics and set key milestones. Keep all the formulas, key points on a couple of A4 size sheets as ready-reckoner on your table and go over them time to time. If you are done with notes, prepare more detailed inside notes and go over them once again. Study all the 3 subjects every day. Concentrate on the topics that have more weightage in the exam that you are targeting.

2. MOCK TESTS

Once you finish your revision on 37th day, the book provides you the full length mock tests for day 38th, 39th, & 40th, thereby ensures your total & full proof preparation for the final show.

The importance of solving previous years' papers and 10-15 mock tests cannot be overemphasized. Identify your weaknesses and strengths. Work towards your strengths i.e., devote more time to your strengths to be 100% sure and confident. In the last time frame of 1 & 1/2 week, don't take-up anything new, just revise what you have studied before. Be exam-ready with quality mock tests in between to implement your winning strategy.

3. FOCUS TOPICS

Based on past years question paper trends, there are few topics in each subject which have more questions in exam than others. So far Chemistry is concerned it may be summed up as below:

Chemical Bonding, Alkyl Halide, Alcohol and Ether are such topics from which more than 80% of questions are normally asked.

However, be prepared to find a completely changed pattern for the exam then noted above as examiners keep trying to weed out 'learn by rot practice'. One should not panic by witnessing a new pattern , rather should be tension free as no one will have any upper hand in the exam.

4. IMPROVES STRIKE RATE AND ACCURACY

The book even helps to improve your strike rate & accuracy. When solving practice tests or mock tests, try to analyze where you are making mistakes-where are you wasting your time; which section you are doing best. Whatever mistakes you make in the first mock test, try to improve that in second. In this way, you can make the optimum use of the book for giving perfection to your preparation.

What most students do is that they revise whole of the syllabus but never attempt a mock and thus they always make mistake in main exam and lose the track.

5. LOG OF LESSONS

During your preparations, make a log of Lesson's Learnt. It is specific to each individual as to where the person is being most efficient and least efficient. Three things are important - what is working, what's not working and how would you like to do in your next mock test.

6. TIME MANAGEMENT

Most candidates who don't make it to good medical colleges are not good in one area- Time Management. And, probably here lies the most important value addition that's the book provides in an aspirant's preparation. Once the students go through the content of the book precisely as given/directed, he/she learns the tactics of time management in the exam.

Realization and strengthening of what you are good at is very helpful, rather than what one doesn't know. Your greatest motto in the exam should be, how to maximize your scoring with the given level of preparation. You have to get about 200 plus marks out of a total of about 400 marks for admission to a good NIT (though for a good branch one needs to do much better than that). Remember that one would be doomed if s/he tries to score 400 in about 3 hours.

7. ART OF PROBLEM SOLVING

The book also let you to master the art of problem solving. The key to problem solving does not lie in understanding the solution to the problem but to find out what clues in the problem leads you to the right solution. And, that's the reason Hints & Solutions are provided with the exercises after each chapter of the book. Try to find out the reason by analyzing the level of problem & practice similar kind of problems so that you can master the tricks involved. Remember that directly going though the solutions is not going to help you at all.

8. POSITIVE PERCEPTION

The book put forth for its readers a 'Simple and Straightforward' concept of studies, which is the best possible, time-tested perception for 11th hour revision / preparation.

The content of the book has been presented in such a lucid way so that you can enjoy what you are reading, keeping a note of your already stressed mind & time span.

Cracking JEE Main is not a matter of life and death. Do not allow panic and pressure to create confusion. Do some yoga and prayers. Enjoy this time with studies as it will never come back.

DAY ONE

Some Basic Concepts of Chemistry

Learning & Revision for the Day

- Matter and its Nature
- Physical Quantities and their Measurements
- Laws of Chemical combinations
- Dalton's Atomic Theory
- Equivalent Weight
- Mole Concept and Molar Mass
- Chemical Equations and Stoichiometry
- Various Concentration Terms

Chemistry is the branch of science which deals with the composition, properties and interaction of all kinds of matter such as air, water, rocks, plants, earth etc.

Matter and Its Nature

Anything that occupies space and possesses mass is called **matter**. On the basis of physical state of substance, matter is divided into three types:

1. **Solids** have definite volume and definite shape.
2. **Liquid** have definite volume but not the definite shape.
3. **Gases** have neither definite volume nor definite shape.

On the basis of chemical composition of substance. It is of three types:

(i) **Elements** These are the substances that cannot be decomposed into simpler substances by chemical change.

(ii) **Compounds** These can be decomposed into simpler substances by chemical changes. Compound is always homogeneous.

(iii) **Mixtures** These have variable composition and variable properties due to the fact that components retain their characteristic properties. Components of a mixture can be separated by applying physical methods.

Every substance has unique property and these can be measured qualitatively and quantitatively.

Physical Quantities and Their Measurements

- Mass, length, time and temperature are physical quantities. These are expressed in numerals with suitable units. Units may be basic (fundamental) or derived.
- The **SI system** has seven base units. These units pertain to the seven fundamental scientific quantities. The units of mass (kg), length (m), time (s), electric current (A),

PREP MIRROR
Your Personal Preparation Indicator

- No. of Questions in Exercises (x)—
- No. of Questions Attempted (y)—
- No. of Correct Questions (z)—
  (Without referring Explanations)

- Accuracy Level $(z/y \times 100)$—
- Prep Level $(z/x \times 100)$—

In order to expect good rank in JEE, your Accuracy Level should be above 85 & Prep Level should be above 75.

temperature (K), luminous intensity (cd), and amount of substance (mol) are fundamental units.

- The lowest temperature permitted in nature is −273.15°C (0 K). This temperature is known as **absolute zero**.
- Relationship between Celsius and Kelvin scale is $K = °C + 273.15$.
- Relationship between the Celsius and Fahrenheit scales are related as $°C = \frac{5}{9}(°F - 32)$
- A number of quantities must be derived from measured value of the SI base quantities. These are called **derived units**. e.g. Units of density (kg m^{-3}) is derived from the units of mass (kg) and volume (m^3).

NOTE
- The term **precision** refers for the closeness of the set of values obtained from identical measurements of a quantity. Precision is simply a measure of reproducibility of an experiment.
- **Accuracy**, a related term, refers to the closeness of a single measurement to its true value.

Significant Figures

Significant figures are meaningful digits which are known with certainty. These are the total number of digits in a number including last digit whose value is uncertain. The uncertainty is indicated by writing the certain digits and the last uncertain digit.

Certain rules for determining the number of significant figures are as follows

- Read the number from left to right and count all the digits, starting with the first digit that is non-zero.
- In addition or subtraction, the number of decimal places in the answer should not exceed the number of decimal places in either of the numbers.
- In multiplication and division, the result should be reported to the same number of significant figures as that in the quantity with least number of significant figures.
- When a number is rounded off, the number of significant figures is reduced. The last digit retained is increased by 1 only if the following digit is ≥ 5 and is left as such if the following digit is ≤ 4.

Dimensional Analysis

In calculations, many of the times it become necessary to convert units from one system to another. This is achieved by factor label method or unit factor method or dimensional analysis. The dimensions of a derived quantity are the powers to which the basic quantities have to be raised in a product defining the quantity. Dimensional analysis involves calculations based on the fact that if two quantities have to be equated, they must have the same dimensions or the same units.

Laws of Chemical Combinations

The combination of elements to form compounds is governed by the following basic laws:

1. **Law of conservation of mass** (Lavoisier, 1789) Total mass of reactants = total mass of products.
2. **Law of constant composition/Definite proportions** (Proust, 1799) For the same compound, obtained by different methods, the percentage of each element should be same in each case.
3. **Law of multiple proportions** (Dalton, 1803) An element may form more than one compound with another element. For a given mass of an element, the masses of other elements (in two or more compounds) are in the ratio of small whole numbers. For example, in NH_3 and N_2H_4, fixed mass of nitrogen requires hydrogen in the ratio 3 : 2.
4. **Law of equivalent/reciprocal proportions** (Ritcher, 1794) When two different elements combines with a fixed weight of a third element, the ratio of their combination will either be same or multiple of the ratio in which they combine with each other. e.g. CH_4, CO_2 and H_2O.
5. **Law of combining volumes** (Gay-Lussac, 1808) It states that when gases combine or are produced in a chemical reaction they do so in a simple ratio by volume provided all gases are at same temperature and pressure.
6. **Avogadro's Law** It states that equal volume of all gases at same temperature and pressure should contain equal number of molecules.

Dalton's Atomic Theory

John Dalton developed his famous theory of atoms in 1808. The main postulates of this theory were:

- Atom was considered as hard, dense and smallest indivisible particle of matter.
- Atom is indestructible i.e. it cannot be destroyed or created in a chemical reaction.
- Atom is the smallest portion of matter which takes part in chemical combination.
- Atoms combine with each other, to form compound (or molecules) in simple whole number ratio.
- Atoms of same elements are identical in mass and chemical properties.
- Chemical reactions involve reorganisation of atoms. These are neither created nor destroyed in a chemical reaction.

Atomic, Molecular and Formula Masses

1. **Atomic Mass** It is defined as the number which indicates how many times the mass of one atom of the element is heavier as compared to $\frac{1}{12}$th part of the mass of one atom of C-12.

2. The **gram atomic mass** of an element should not be mass of their atoms. e.g. Gram atomic mass of H-element is 1.008 g but mass of H-atoms is 1 μ [1.67×10^{-24} g].

   The approximate atomic mass of solid elements except Be, B, C and Si, is related to specific heat as

   $$\text{Average atomic mass} = \frac{6.4}{\text{specific heat}}$$

   (from Dulong and Petit's law for metals)

   Exact atomic mass = Equivalent mass × valency

   As most of the elements have isotopes, so their actual atomic mass is the average of atomic masses of all the isotopes.

   Average atomic mass is calculated as

   $$M_{av} = \frac{m_1 \times r_1 + m_2 \times r_2 + m_3 \times r_3}{r_1 + r_2 + r_3}$$

   where, r_1, r_2 and r_3 = relative abundances of the isotopes.

3. **Molecular Mass** It is the sum of atomic masses of the elements present in a molecule. It is obtained by multiplying the atomic mass of each element by the number of its atoms and adding them together.

4. **Formula Mass** It is the sum of the atomic masses of all atoms in the formula unit of the compound. It is normally calculated for ionic compounds. Formula mass of NaCl is 23 + 35.5 = 58.5 amu or 58.5u.

Equivalent Weight

It is the weight of an element or of a compound, which would combine with or displace (by weight) 1 part of hydrogen or 8 parts of oxygen or 35.5 parts of chlorine.

Equivalent weight. (Eq. wt.)

$$= \frac{\text{atomic wt. or molecular wt.}}{\text{'}n\text{' factor}}$$

'n' factor for various compounds can be obtained as :

1. **'n' factor for acids, i.e. basicity** is the number of ionisable H^+ per molecule is the basicity of acids. e.g. basicity of HCl = 1
2. **'n' factor for bases, i.e. acidity** is the number of ionisable OH^- per molecule is the acidity of bases. e.g. Acidity of NaOH = 1
3. **'n' factor for salt** is total positive or negative charge of ions. e.g. $Na_2CO_3 \longrightarrow 2Na^+ + CO_3^{2-}$
4. **'n' factor for ion** is equal to charge of that ion.

   e.g. $$E_{Cl^-} = \frac{35.5}{1} = 35.5$$

5. **In redox titration,** n factor for reducing agent is number of electrons lost by the molecule and for oxidising agent is number of electron gained by the molecule.

   e.g. $FeSO_4 \Rightarrow$ As reducing agent

   $$Fe^{2+} \longrightarrow Fe^{3+} + e^- \qquad \text{'}n\text{' factor} = 1$$

   $\Rightarrow$ As an oxidising agent

   $$Fe^{2+} + 2e^- \longrightarrow Fe(s) \qquad \text{'}n\text{' factor} = 2$$

- Equivalent mass of oxidising / reducing agent

$$= \frac{\text{molecular mass of oxidising / reducing agent}}{\text{number of electrons gained or lost by one molecule}}$$

Mole Concept and Molar Mass

- Mole is the amount of substance which contains 6.022×10^{23} (Avogadro's number) particles and has mass equal to gram-atomic mass or gram-molecular mass.
- Mole is related to the mass of the substance (in grams), volume of gaseous substance and number of particles. Therefore, number of moles

$$= \frac{\text{Mass of substance (g)}}{\text{Molar mass (g mol}^{-1}\text{)}}$$

$$= \frac{\text{Volume of gas at STP (L)}}{22.4\text{(L)}}$$

$$= \frac{\text{Number of particles at STP}}{N_A}$$

- Moles of atoms/molecules/ions/ electrons

  $= N_A \times$ Number of atoms/molecules/ions/electrons
- Total charge present on an ion

  $= \text{mole} \times N_A \times \text{charge on one ion} \times 1.6 \times 10^{-19}$ C
- **Molar mass** of an element is defined as mass of 1 mole of a substance in grams, i.e. mass of 6.023×10^{23} entities or particles of that element.

Percentage Composition, Empirical and Molecular Formulae

- The percentage of any element or constituent in a compound is the number of parts by mass of that element or constituent present in two parts, by mass of the compound.

  Mass % of an element

$$= \frac{\text{mass of element in the compound} \times 100}{\text{molar mass of compound}}$$

- An **empirical formula** represents the simplest whole number ratio of various atoms present in a molecule of the compound, whereas the **molecular formula** shows the exact number of different types of atoms present in a molecule of a compound.

$$n = \frac{\text{molecular mass}}{\text{empirical formula mass}}$$

  here, n is any integer such 1, 2, 3, ... etc.

  Molecular formula = $n \times$ empirical formula

  Molecular/molar mass = 2 × vapour density

Chemical Equations and Stoichiometry

- A **balanced chemical equation** with suitable stoichiometric coefficients represent the ratio of number of moles of reactants and products.
- The chemical equation provides qualitative and quantitative information about a chemical change in a simple manner.
- Stoichiometry deals with calculation of masses of the reactants and the products involved in a chemical reaction.
- The numerals used to balance a chemical equation is called **stoichiometric coefficients**.
- For the stoichiometric calculations, the mole relationships between different reactants and products are required. The mass-mass, mass-volume and volume- volume relationship can be obtained between different reactants and products.

Limiting Reagent

- The substance which is completely consumed in a reaction is called limiting reagent. It determines the amount of product.

$$\text{Reaction yield} = \frac{\text{actual yield} \times 100}{\text{theoretical yield}}$$

- In stoichiometry, if the quantities of two or more reactants are given, the amount of products formed depend upon the limiting reactant (the reactant which consumed first in the reaction).

Various Concentration Terms

Different concentration terms are given below :

1. **Molarity** (M) It is defined as the number of moles of solute per litre of solution or as the number of mg-molecules per millilitre of solution. The molarity is usually designated by M. It is dependent upon the temperature, as it depends on volume which changes with temperature.
   e.g. if the molarity of H_3PO_4 is 0.18, it means a concentration corresponding to 0.18 mole of H_3PO_4 per litre of solution.
   Thus, molarity is given as

$$\text{Molarity, } (M) = \frac{\text{moles of solute}}{\text{volume of solution (in L)}}$$

   If specific gravity is given,

$$\text{molarity} = \frac{\text{specific gravity} \times \%\ \text{strength} \times 10}{\text{molecular weight}}$$

   If molarity and volume of solution are changed from m_1, V_1 to m_2, V_2 then

$$m_1V_1 = m_2V_2 \qquad \text{(molarity equation)}$$

   If two solutions of the same solute are mixed the molarity of resulting solution

$$m_3 = \frac{m_1V_1 + m_2V_2}{V_1 + V_2}$$

   Strength of solution (S)
   Amount of solute present in 1 L solution

$$\text{Strength, } (S) = \frac{\text{weight of solute}}{\text{volume of solution (in L)}}$$

2. **Molality** (m) It is defined as the number of moles of solute dissolved in 1000 g of the solvent. It is designated by m. Molality is independent of temperature, as it depends only upon the mass which does not vary with temperature.

$$\text{Molality, } (m) = \frac{\text{moles of solute}}{\text{weight of solvent (in g)}} \times 1000$$

3. **Normality** (N) It is defined as the number of g-equivalents of solute per litre of solution or as the number of mg-equivalents of a substance per millilitre of solution. e.g. 0.12 N H_2SO_4 means a solution which contains 0.12 g-equivalent of H_2SO_4 per litre of solution.
   This also means that each millilitre of this solution can react, for example, with 0.12 mg-eq. of CaO or with 0.12 mg-eq. of Na_2CO_3. Thus,

$$\text{Normality } (N) = \frac{\text{gram-equivalent of solute}}{\text{volume of solution (in L)}}$$

   or

$$= \frac{\text{gram-equivalent of solute}}{\text{volume of solution (in mL)}} \times 1000$$

   If specific gravity is known, normality is calculated as

$$\text{Normality} = \frac{\text{specific gravity} \times \%\ \text{strength} \times 10}{\text{equivalent weight}}$$

   If normality and volume of solution are changed from N_1, V_1 to N_2, V_2 then $N_1V_1 = N_2V_2$ if two solution of same solute are mixed then normality or resulting solution

$$N_3 = \frac{N_1V_1 + N_2V_2}{V_1 + V_2}$$

4. **Mole fraction** (χ) It is the ratio of number of moles of a particular component to the total number of moles of the solution.
   Thus, mole fraction is given as

$$\text{Mole fraction } (\chi) = \frac{\text{number of moles of component}}{\text{number of moles of solution}}$$

The sum of mole fractions of the component is equal to 1.

DAY PRACTICE SESSION 1

FOUNDATION QUESTIONS EXERCISE

1 The correctly reported answer of the addition of 4.523, 2.3 and 6.24 will have significant figures
(a) two (b) three
(c) four (d) five

2 A student performs a titration with different burettes and finds titre values of 25.2 mL, 25.25 mL and 25.0 mL. The number of significant figures in the average titre value is → AIEEE 2010
(a) 1 (b) 2
(c) 3 (d) 4

3 A metal oxide contains 53% metal and carbon dioxide contains 27% carbon. Assuming the law of reciprocal proportions, the percentage of metal in the metal carbide is
(a) 75% (b) 25%
(c) 37% (d) 66%

4 Two oxides of metal were found to contain 31.6% and 48% of oxygen respectively. If the formula of first is represented by M_2O_3, then formula of second is
(a) MO_3 (b) MO_2
(c) M_2O (d) M_2O_2

5 The equivalent weight of H_3PO_2, when it disproportionates into PH_3 and H_3PO_4, is
(a) 82 (b) 61.5
(c) 33 (d) 20.5

6 3g of an oxide of a metal is converted to chloride completely and it yielded 5 g of chloride. The equivalent weight of the metal is
(a) 33.25 (b) 3.325
(c) 12 (d) 20

7 Sea water contains 65×10^{-3} g L^{-1} of bromide ions. If all the bromide ions are converted to produce Br_2, how much sea water is needed to prepare 1 kg Br_2 ?
(a) 15.38 L (b) 15.38×10^3 L
(c) 7.69×10^3 L (d) 76.9 L

8 0.376 g of Al reacted with an acid to displace 0.468 L of H_2 measured in standard conditions. Equivalent volume of H_2 formed is (equivalent mass of Al is 9 g equiv^{-1})
(a) 22.4 L (b) 5.6 L
(c) 11.2 L (d) 2.24 L

9 The same amount of a metal combines with 0.200 g of oxygen and with 3.17 g of a halogen. Hence, equivalent mass of halogen is
(a) 127 g (b) 80 g
(c) 35.5 g (d) 9 g

10 If we consider that 1/6 in place of 1/12, mass of carbon atom is taken to be the relative atomic mass unit, the mass of one mole of a substance will
(a) to be a function of the molecular mass of the substance
(b) remain unchanged
(c) increase two fold
(d) decrease twice

11 How many moles of magnesium phosphate, $Mg_3(PO_4)_2$ will contain 0.25 mole of oxygen atoms?
(a) 0.02 (b) 3.125×10^{-2}
(c) 1.25×10^{-2} (d) 2.5×10^{-2}

12 Rearrange the following (I to IV) in the order of increasing masses and choose the correct answer
(atomic mass; O = 16, Cu = 63, N = 14)
I. 1 molecule of oxygen.
II. 1 atom of nitrogen.
III. 1×10^{-10} g molecular weight of oxygen.
IV. 1×10^{-10} g atomic weight of copper.
(a) II < I < III < IV (b) IV < III < II < I
(c) II < III < I < IV (d) III < IV < I < II

13 Number of atoms in the following samples of substances is largest in → JEE Main (Online) 2013
(a) 4.0 g of hydrogen (b) 70.0 g of chlorine
(c) 127.0 g of iodine (d) 48.0 g of magnesium

14 The total number of electrons present in 18 mL of water (density of water is 1 gmL^{-1}) is
(a) 6.02×10^{23} (b) 6.02×10^{23}
(c) 6.02×10^{24} (d) 6.02×10^{25}

15 The weight of 1×10^{22} molecules of $CuSO_4 \cdot 5H_2O$ is
(a) 41.59 g (b) 415.9 g
(c) 4.159 g (d) None of these

16 The number of moles of $(NH_4)_2SO_4Fe_2(SO_4)_3 \cdot 24H_2O$ formed from a sample containing 0.0056 g of Fe is
(a) 10^{-4} mol (b) 0.5×10^{-4} mol
(c) 2×10^{-4} mol (d) 0.33×10^{-4} mol

17 If 10^{21} molecules are removed from 200 mg of CO_2, the number of moles of CO_2 left are
(a) 2.88×10^{-3} (b) 28.8×10^{-3}
(c) 0.288×10^{-3} (d) 1.66×10^{-2}

18 Medical experts generally consider a lead level of 30 μg Pb per dL of blood to pose a significant health risk (1 dL = 0.1 L). Express this lead level as the number of Pb atoms per cm^3 blood (Pb = 207).
(a) 8.72×10^{14} (b) 8.72×10^{15}
(c) 8.72×10^{13} (d) 8.72×10^{16}

19 Which of the following pairs of gases contains the same number of molecules ?

(a) 16 g of O_2 and 14 g of N_2
(b) 8 g of O_2 and 22 g of CO_2
(c) 28 g of N_2 and 22 g of CO_2
(d) 32 g of O_2 and 32 g of N_2

20 The number of H-atoms present in 25.6 g of sucrose which has a molar mass of 342.3 g is

(a) 22×10^{23} (b) 9.91×10^{23} (c) 11×10^{23} (d) 44×10^{23}

21 The most abundant elements by mass in the body of a healthy human adult are oxygen (61.4%), carbon (22.9%), hydrogen (10.0%) and nitrogen (2.6%). The weight which a 75 kg person would gain if all H atoms are replaced by ^2H atoms is

(a) 15 kg (b) 37.5 kg (c) 7.5 kg (d) 10 kg

22 A gaseous hydrocarbon gives upon combustion 0.72 g of water and 3.08 g of CO_2. The empirical formula of the hydrocarbon is → JEE Main (Online) 2013

(a) C_2H_4 (b) C_3H_4 (c) C_6H_5 (d) C_7H_8

23 A 0.2075 g sample of an oxide of cobalt on analysis was found to contain 0.1475 g cobalt. The empirical formula of the oxide is

(a) CoO_2 (b) Co_2O_3
(c) CoO (d) Co_4O_6

24 At 300 K and 1 atom, 15 mL of a gaseous hydrocarbon requires 375 mL air containing 20% O_2 by volume for complete combustion. After combustion, the gases occupy 330 mL. Assuming that the water, formed is in liquid form and the volumes were measured at the same temperature and pressure, the formula of the hydrocarbon is → JEE Main 2016

(a) C_3H_8 (b) C_4H_8
(c) C_4H_{10} (d) C_3H_6

25 The ratio of mass per cent of C and H of an organic compound ($C_xH_yO_z$) is 6 : 1. If one molecule of the above compound ($C_xH_yO_z$) contains half as much oxygen as required to burn one molecule of compound C_xH_y completely to CO_2 and H_2O. The empirical formula of compound $C_xH_yO_z$ is → JEE Main 2018

(a) $C_3H_6O_3$ (b) C_2H_4O
(c) $C_3H_4O_2$ (d) $C_2H_4O_3$

26 A mixture of FeO and Fe_3O_4 when heated in air to constant weight, gains 5% in its weight. What is the percentage of Fe_3O_4 in mixture?

(a) 73.87% (b) 26.13%
(c) 79.75% (d) 20.25%

27 1.00×10^{-3} moles of Ag^+ and 1.00×10^{-3} moles of CrO_4^{2-} react together to form solid Ag_2CrO_4. Calculate the amount of Ag_2CrO_4 formed (Ag_2CrO_4 = 331.73 g mol^{-1}).

(a) 0.268 g (b) 0.166 g
(c) 0.212 g (d) 1.66 g

28 Mixture X = 0.02 mole of $[Co(NH_3)_5SO_4]$ Br and 0.02 mole of $[Co(NH_3)_5Br]$ SO_4 was prepared in 2 L of solution.

1 L of mixture X + excess $AgNO_3 \longrightarrow Y$
1 L of mixture X + excess $BaCl_2 \longrightarrow Z$

Number of moles of Y and Z are

(a) 0.01, 0.01 (b) 0.02, 0.01
(c) 0.01, 0.02 (d) 0.02, 0.02

29 How much AgCl will be formed by adding 200 mL of 5 N HCl to a solution containing 1.7 g $AgNO_3$ (Ag = 108)?

(a) 0.1435 g (b) 1.435 g
(c) 14.35 g (d) 143.5 g

30 If 0.5 mole of $BaCl_2$ are mixed with 0.2 mole of Na_3PO_4, the maximum number of moles of $Ba_3(PO_4)_2$ that can be formed, is

(a) 0.7 (b) 0.5
(c) 0.30 (d) 0.10

31 The reaction between yttrium metal and dil. HCl produces H_2S and y^{3+} ions. The molar ratio of yttrium to that hydrogen produced is

(a) 1:2 (b) 2:1
(c) 2:3 (d) 5:2

32 3 g of activated charcoal was added to 50 mL of acetic acid solution (0.06 N) in a flask. After an hour it was filtered and the strength of the filtrate was found to be 0.042 N. The amount of acetic acid adsorbed (per gram of charcoal) is → JEE Main 2015

(a) 18 mg (b) 36 mg
(c) 42 mg (d) 54 mg

33 The molecular formula of a commercial resin used for exchanging ions in water softening is $C_8H_7SO_3Na$ (Mol. wt. = 206). What would be the maximum uptake of Ca^{2+} ions by the resin when expressed in mole per gram resin? → JEE Main 2015

(a) 1/103 (b) 1/206 (c) 2/309 (d) 1/412

34 The mass of potassium dichromate crystals required to oxidise 750 cm^3 of 0.6 M Mohr's salt solution is (Given, molar mass : Potassium dichromate = 294, Mohr's salt = 392) → AIEEE 2011

(a) 0.49 g (b) 0.45 g
(c) 22.05 g (d) 2.2 g

35 5 mL of N HCl, 20 mL of $N/2$ H_2SO_4 and 30 mL of $N/3$ HNO_3 are mixed together and volume made to 1 L. The normality of resulting solution is

(a) 0.45 (b) 0.025 (c) 0.9 (d) 0.05

36 Two solutions of a substance (non-electrolyte) are mixed in the following manner : 480 mL of 1.5 M of first solution with 520 mL of 1.2 M of second solution. The molarity of final solution is

(a) 1.20 M (b) 1.50 M (c) 1.344 M (d) 2.70 M

Direction (Q.Nos. 37-38) *In the following questions assertion followed by a reason is given. Choose the correct answer out of the following choices.*

(a) Both A and R are true and R is correct explanation of A
(b) Both A and R are true but R is not correct explanation of A
(c) A is true but R is false
(d) Both A and R are false

37 **Assertion** (A) On changing volume of the solution by 20%, molarity of solution also changes by 20%.

Reason (R) Molar concentration or molarity of solution is not affected on dilution.

38 **Assertion** (A) If 30 mL of H_2 combines with 20 mL of O_2 to form water, 5 mL of H_2 left after the reaction.

Reason (R) O_2 is the limiting reagent.

DAY PRACTICE SESSION 2

PROGRESSIVE QUESTIONS EXERCISE

1 The reaction, $N_2 + 3H_2 \longrightarrow 2NH_3$ is used to produce ammonia. When 450 g of hydrogen was reacted with nitrogen, 1575 g of ammonia were produced. What is the per cent yield of this reaction?

(a) 61.8% (b) 41.5%
(c) 30.8% (d) 20.7%

2 A sample of ammonium phosphate, $(NH_4)_3PO_4$ contains 3.18 moles of hydrogen atoms. The number of moles of oxygen atoms in the sample is

(a) 0.265 (b) 0.795
(c) 1.06 (d) 3.18

3 1 mole of potassium chlorate is thermally decomposed and excess of aluminium is burnt in the gaseous product. How many moles of aluminium oxide are formed?

(a) 1 (b) 2 (c) 1.5 (d) 3

4 Amount of A $FeSO_4(NH_4)_2SO_4 \cdot 6H_2O$ (molar mass $= 392\,g\,mol^{-1}$) must be dissolved and diluted to 250 mL to prepare an aqueous solution of density 1.00 g mL^{-1}, i.e. 1.00 ppm Fe^{2+} by weight is

(a) 3.50×10^{-3} g (b) 1.75×10^{-3} g
(c) 7.00×10^{-3} g (d) 0.35×10^{-3} g

5 Excess of carbon dioxide is passed through 50 mL of 0.5 m calcium hydroxide solution. After completion of the reaction, the solution was evaporated to dryness. The solid calcium carbonate was completely neutralised with 0.1 N hydrochloric acid. The volume of hydrochloric acid required is (atomic mass of calcium = 40).

(a) 300 cm^3 (b) 200 cm^3
(c) 500 cm^3 (d) 400 cm^3

6 Sodium nitrate on reduction with Zn in the presence of NaOH solution produces NH_3. Mass of sodium nitrate absorbing 1 mole of electron will be

(a) 7.750 (b) 10.625
(c) 8.000 (d) 9.875

7 1 mole of N_2 and 4 moles of H_2 are allowed to react in a vessel and after reaction, H_2O is added. Aqueous solution required 1 mole of HCl. Mole fraction of H_2 in the gaseous mixture after reaction is

(a) $\frac{1}{6}$ (b) $\frac{5}{6}$
(c) $\frac{1}{3}$ (d) None of these

8 Potassium selenate is isomorphous with potassium sulphate and contains 50.0% of Se. Find the atomic weight of Se.

(a) 47.33 (b) 71
(c) 142 (d) 284

9 Acidified $KMnO_4$ oxidises oxalic acid to CO_2. What is the volume (in litres) of 10^{-4} M $KMnO_4$ required to completely oxidise 0.5 L of 10^{-2} M oxalic acid in acidic medium?

(a) 125 (b) 1250
(c) 200 (d) 20

10 Haemoglobin contains 0.33% of iron by weight. The molecular weight of haemoglobin is approximately 67200. The number of iron atoms (at. wt. of Fe is 56) present in one molecule of haemoglobin are

(a) 1 (b) 6
(c) 4 (d) 2

11 What volume of hydrogen gas at 273 K and 1 atm pressure will be consumed in obtaining 21.6 g of elemental boron (atomic mass = 10.8) from the reduction of boron trichloride by hydrogen ?

(a) 89.6 L (b) 67.2 L
(c) 44.8 L (d) 22.4 L

12 Density of 2.05 M solution of acetic acid in water is 1.02 g/mL. The molality of same solution is

(a) 1.14 mol kg^{-1} (b) 3.28 mol kg^{-1}
(c) 2.28 mol kg^{-1} (d) 0.44 mol kg^{-1}

13 A 15.00 mL sample of a solution is 0.04 M in Sn^{2+} and x M in Fe^{2+}. Both are easily oxidised by $Cr_2O_7^{2-}$ in acidic solution to Sn^{4+} and Fe^{3+} and itself reduced to Cr^{3+}. 18.0 mL of 0.125 M $Cr_2O_7^{2-}$ is required. Thus, x is

(a) 0.410 (b) 0.205
(c) 0.820 (d) 1.640

14 A mixture contains $Na_2C_2O_4$ and KHC_2O_4 in 1:1 molar ratio. Mixture is neutralised by 100 mL of 0.01 M KOH. What volume of 0.01 M $KMnO_4$?

Thus, the same mixture is oxidised by

(a) 200 mL (b) 100 mL
(c) 90 mL (d) 80 mL

15 Weight of 1 L milk is 1.032 kg. It contains butter fat (density 865 kg m^{-3}) to the extent of 4% by volume/volume. The density of the fat free skimmed milk will be

(a) 1038.5 kg m^{-3} (b) 1032.2 kg m^{-3}
(c) 997 kg m^{-3} (d) 1000.5 kg m^{-3}

ANSWERS

SESSION 1

1 (b)	**2** (c)	**3** (a)	**4** (a)	**5** (c)	**6** (a)	**7** (b)	**8** (c)	**9** (a)	**10** (b)
11 (b)	**12** (a)	**13** (a)	**14** (c)	**15** (c)	**16** (b)	**17** (a)	**18** (a)	**19** (a)	**20** (b)
21 (c)	**22** (d)	**23** (b)	**24** (d)	**25** (d)	**26** (c)	**27** (b)	**28** (a)	**29** (b)	**30** (d)
31 (c)	**32** (a)	**33** (d)	**34** (c)	**35** (a)	**36** (c)	**37** (d)	**38** (d)		

SESSION 2

1 (a)	**2** (c)	**3** (a)	**4** (b)	**5** (c)	**6** (b)	**7** (b)	**8** (c)	**9** (d)	**10** (c)
11 (b)	**12** (c)	**13** (c)	**14** (d)	**15** (a)					

Hints and Explanations

SESSION 1

1 4.523 + 2.3 + 6.24 = 13.063. As 2.3 has least number of decimal places, i.e. one, therefore sum should be reported to one decimal place only. After rounding off, reported sum = 13.1 which has three significant figures.

2 Average value $= \frac{25.2 + 25.25 + 25.0}{3} = \frac{75.45}{3} = 25.15 = 25.2$ mol

Number of significant figure is 3.

3 In metal oxide, metal = 53%, O = 47%

In CO_2, C = 27%, O = 73%

$\because$ 73 parts of oxygen combines with 27 parts of carbon.

$\therefore$ 47 parts of oxygen will combine $= \frac{27}{73} \times 47 = 17.38$ parts of C.

Thus, metal and carbon will be present in the ratio of 53 : 17.38. Hence, % of metal

$$= \frac{53}{53 + 17.38} \times 100 = 75.3\% \approx 75\%$$

4 First oxide is M_2O_3. Here, O is 31.6% and M is 68.4%.

Let second oxide be M_2O_x. Here, O is 48% and M is 52%.

So, 68.4g of M in first oxide

= 2 atom of M and 52 g of M contains

$= \frac{2 \times 52}{68.4} = 1.5$ atom of M and 52 g of M contains

31.6 g of oxygen in first oxide contains 3 atom of oxygen.

48 g of oxygen contains $= \frac{3 \times 48}{31.6} = 4.5$ atoms of oxygen

Ratio of M:O = 1.5 : 4.5 = 1 : 3

Thus, the formula is MO_3

5 H_3PO_2 disproportionates as

$$2H_3\overset{+1}{P}O_2 \longrightarrow \overset{-3}{P}H_3 + H_3\overset{+5}{P}O_4$$

Molecular wt. of $H_3PO_2 = 3 + 31 + 32 = 66$

$\therefore$ Eq. wt. $= \frac{66}{4} + \frac{66}{4} = 33$

6 $\frac{\text{Wt. of metal oxide}}{\text{Wt. of metal chloride}} = \frac{\text{Eq. wt. of metal + eq. wt. of oxide}}{\text{Eq. wt. of metal + eq. wt. of chloride}}$

$$\frac{3}{5} = \frac{E + 8}{E + 35.5}$$

where, [E = Eq. wt. of metal]

or $5E + 40 = 3E + 106.5$ or $2E = 66.5$

$\therefore$ $E = 33.25$

7 $2Br^- \longrightarrow Br_2$

Equivalents of Br^- = Equivalents of Br_2

$$\frac{w}{80} = \frac{10^3}{160/2}$$

$$w_{Br^-} = 10^3 \text{ g}$$

$\therefore$ Sea water needed $= \frac{10^3}{65 \times 10^{-3}} = 15.38 \times 10^3$ L

8 0.376 g Al = 0.468 L H_2

$\frac{0.376}{9}$ equivalent of Al = 0.468 L H_2

∴ 1 equivalent of Al = 11.2 L H_2

9 0.20 g oxygen ≡ 3.17 g halogen

∴ Equivalent mass halogen

$\equiv \frac{3.17}{0.20} \times 8 = 126.8 \text{ g} \simeq 127 \text{ g}$

10 Mass of the given amount of a substance is a constant quantity.

11 In $Mg_3(PO_4)_2$; 1 moles of O-atoms are present in 1 mole of $Mg_3(PO_4)_2$.

Hence, 0.25 mole of O-atom are contained

$= \frac{1}{8} \times 0.25$

$= 3.125 \times 10^{-2}$

12 I. 1 molecule of $O_2 = \frac{32}{6.022 \times 10^{23}}$ g

$= 5.3 \times 10^{-23}$ g

II. 1 atom of N $= \frac{14}{6.022 \times 10^{23}}$g

$= 2.3 \times 10^{-23}$ g

III. 10^{-10} g mol. wt. of oxygen

$= 10^{-10} \times 32 = 3.2 \times 10^{-9}$g

IV. 10^{-10}g atomic weight of copper

$= 10^{-10} \times 63.5 = 6.35 \times 10^{-9}$ g

∴ Order of increasing mass is

II < I < III < IV.

13 Number of atoms

$= \frac{\text{weight}}{\text{atomic weight}} \times N_A \times \text{species}$

∴ In 4 g of hydrogen,

Number of atoms $= \frac{4}{2} \times N_A \times 2 = 4N_A$

[Here, species = 2, because hydrogen is present as H_2]

In 71 g of chlorine,

Number of atoms $= \frac{71}{71} \times N_A \times 2 = 2\,N_A$

In 127 g of iodine,

Number of atoms $= \frac{127}{127} \times N_A \times 2 = 2N_A$

In 48 g of magnesium,

Number of atoms $= \frac{48}{24} \times N_A \times 1 = 2N_A$

[Here, Mg is present as Mg so species = 1] Thus, the number of atoms are largest in 4 g of hydrogen.

14 18 mL H_2O = 18 g H_2O = 1 mol

$= 6.02 \times 10^{23}$ molecules/atoms

In 1 atom of water 10 electrons are present.

∴ electrons in 1 mole H_2O

$= (2 + 8) \times 6.023 \times 10^{23}$ electrons

$= 10 \times 6.02 \times 10^{23}$

$= 6.02 \times 10^{24}$ electrons

15 ∵ 6.02×10^{23} molecules of $CuSO_4 \cdot 5H_2O$

= 63.5 + 32 + 64 + 90 = 249.5 g

∴ 1×10^{22} molecules of $CuSO_4 \cdot 5H_2O$

$= \frac{249.5}{6.02 \times 10^{23}} \times 10^{22} = 4.15$ g

16 Number of moles of Fe

$= \frac{0.0056}{56} = 10^{-4}$ mol

2 moles of Fe is present in 1 mole of $(NH_4)_2SO_4\,Fe_2(SO_4)_3$.

Therefore, 10^{-4} mole of Fe is present in

$= \frac{10^{-4} \times 1}{2}$ mol

$= 0.5 \times 10^{-4}$ mol

17. 200 mg CO_2 = 0.2 g $= \frac{0.2}{44}$ mol

$= 0.00454$ mol $= 4.54 \times 10^{-3}$mol

10^{21} molecules of $CO_2 = \frac{10^{21}}{6.02 \times 10^{23}}$

$= 1.66 \times 10^{-3}$ mol

∴ Number of moles left

$= (4.54 - 1.66)\ 10^{-3}$

$= 2.88 \times 10^{-3}$

18 0.1 L = 100 mL has Pb = 30 mg

$= 30 \times 10^{-6}$ g

$= \frac{30 \times 10^{-6}}{207}$ mole of Pb

$= \frac{30 \times 10^{-6}}{207} \times 6.02 \times 10^{23}$ Pb atoms

Number of atoms per cm^3 blood

$= \frac{30 \times 10^{-6} \times 6.02 \times 10^{23}}{207 \times 100} = 8.72 \times 10^{14}$

19 16 g of $O_2 = \frac{16}{32} = 0.5$ mol

$= \frac{N_A}{2}$ molecules

14 g of $N_2 = \frac{14}{28} = 0.5$ mol

$= \frac{N_A}{2}$ molecules

20 Moles of sucrose $[C_{12}H_{22}O_{11}]$

$= \frac{25.6}{342.3} = 0.0747$

Number of H-atoms in 1 mole of sucrose

$= 22 \times 6.023 \times 10^{23}$

Number of H-atoms in 0.0747 mole of sucrose

$= 22 \times 6.023 \times 10^{23} \times 0.0747$

$= 9.9 \times 10^{23}$

21 Given, abundance of elements by mass oxygen = 61.4%, carbon = 22.9%, hydrogen = 10%

Total weight of person = 75 kg

Mass due to $^1H = \frac{75 \times 10 \times 1}{100} = 7.5$ kg

^1H atoms are replaced by ^2H atoms.

Mass due to $^2H = (7.5 \times 2)$ kg

∴ Mass gain by person = 7.5 kg

22 18 g H_2O contains 2 g of H

∴ 0.72 g H_2O contains 0.08 g of H.

44 g CO_2 contains 12 g of C

∴ 3.08 gCO_2 contains 0.84 g of C

∴ $C : H = \frac{0.84}{12} : \frac{0.08}{1} = 0.07 : 0.08 = 7 : 8$

∴ Empirical formula = C_7H_8

23 Weight of oxygen in sample

$= 0.2075 - 0.1475 = 0.06$ g

Moles of cobalt $= \frac{0.1475}{59} = 0.0025$

Moles of oxygen $= \frac{0.06}{16} = 0.0037$

Simplest ratio of Co $= \frac{0.0025}{0.0025} = 1.0$

Simplest ratio of O $= \frac{0.0037}{0.0025} = 1.48 \approx 1.5$

Ratio of Co:O = 1:1.5 = 2:3

So, the formula is Co_2O_3.

24 $\underset{15}{C_xH_y} + \underset{15(x+y)_4}{(x + y)_4\,O_2} \longrightarrow \underset{15x}{xCO_2(g)} + y_2H_2O(l)$

Before Combustion

O_2 used = 20% of 375 = 75 mL

After Combustion

Inert part of air = 80% of 375 = 300 mL

Total volume of gases = CO_2 + Inert part of air

$330 = 15x + 300 \Rightarrow x = 2$

$\frac{x + (y/4)}{1} = \frac{75}{15} \Rightarrow x + \frac{y}{4} = 5$

$\Rightarrow x = 2,\ y = 12 \Rightarrow C_2H_{12}$ or C_3H_6.

Thus empirical formula of compound is C_3H_6

25 We can calculate the simplest whole number ratio for C and H from the data given as:

Element	Relative mass	Molar mass	Relative mole	Simplest whole number ratio
C	6	12	$\frac{6}{12} = 0.5$	$\frac{0.5}{0.5} = 1$
H	1	1	$\frac{1}{1} = 1$	$\frac{1}{0.5} = 2$

Now, after calculating this ratio look for condition 2 given in the question, i.e. quantity of oxygen is half of the quantity required to burn one molecule of compound C_xH_y completely to CO_2 and H_2O. We can calculate number of oxygen atoms from this as consider the equation.

$$C_xH_y + \left[x + \frac{y}{4}\right]O_2 \longrightarrow xCO_2 + \frac{y}{2}H_2O$$

Number of oxygen atoms required

$$= 2 \times \left[x + \frac{y}{4}\right] = \left[2x + \frac{y}{2}\right]$$

Now given, $z = \frac{1}{2}\left[2x + \frac{y}{2}\right] = \left[2x + \frac{2}{4}\right]$

Here, we consider x and y as simplest ratios for C and H so, now putting the values of x and y in the above equation.

$$z = \left[x + \frac{y}{4}\right] = \left[1 + \frac{2}{4}\right] = 1.5$$

Thus, the simplest ratio figures for x, y and z are $x = 1$, $y = 2$ and $z = 1.5$.

Now, put these values in the formula given, i.e. $C_xH_yO_z = C_1H_2O_{1.5}$

So,empirical formula will be

$$[C_1H_2O_{1.5}] \times 2 = C_2H_4O_3$$

26 Let wt. of FeO = a g and wt. of $Fe_3O_4 = b$ g

$$2FeO + \frac{1}{2}O_2 \longrightarrow Fe_2O_3$$

$$2Fe_3O_4 + \frac{1}{2}O_2 \longrightarrow 3Fe_2O_3$$

$\because$ 144 g of FeO gives 160 g Fe_2O_3.

$\therefore$ a g FeO will give $= \frac{160 \times a}{144}$ g Fe_2O_3

Similarly, weight of Fe_2O_3 formed by b g

$$Fe_3O_4 = \frac{160 \times 3 \times b}{464}$$

Now, if $a + b = 100$ ...(i)

Then, $\frac{160 \times a}{144} + \frac{160 \times 3 \times b}{464} = 105$ ...(ii)

From Eqs. (i) and (ii), $a = 20.25$ g

and $b = 79.75$ g

$\therefore$ Percentage of $Fe_3O_4 = 79.75\%$

27 The reaction is

$$2Ag^+ + CrO_4^{2-} \longrightarrow Ag_2CrO_4.$$

Using the limiting reagent concept, number of moles of Ag_2CrO_4

$$= 0.5 \times 10^{-3}$$

Amount of Ag_2CrO_4 formed

$$= 0.5 \times 10^{-3} \times 331.73$$
$$= 0.166 \text{ g}$$

28 Mixture X will contain 0.02 mole of Br^- ions and 0.02 mole of SO_4^{2-} ions in 2 L solution. Hence, 1 L of mixture X will contain 0.01 mole of Br^- and 0.01 SO_4^{2-} ions. With excess of $AgNO_3$, 0.01 moles of AgBr, i.e. Y is formed and with excess of $BaCl_2$, 0.01 moles of $BaSO_4$, i.e. Z is formed.

29 200 mL of 5 N HCl

$$= 200 \times 5 \text{ milliequivalents}$$
$$= 1000 \text{ millimoles} = 1 \text{ mol HCl}$$

1.7g of $AgNO_3$ = 0.01 mol

$$\underset{\substack{1\text{ mol}\\0.01\text{ mol}}}{AgNO_3} + \underset{\substack{1\text{ mol}\\1\text{ mol}}}{HCl} \longrightarrow \underset{\substack{1\text{ mol}\\0.01\text{ mol}}}{AgCl} + HNO_3$$

$AgNO_3$ is the limiting reagent.

Thus, AgCl formed = 0.01 mol

$$= 0.01 \times 143.5 = 1.435 \text{ g}$$

30 $3BaCl_2 + 2Na_3PO_4 \longrightarrow Ba_3(PO_4)_2 + 6NaCl$

Here, limiting reactant is Na_3PO_4.

0.2 mole of Na_3PO_4 will give $Ba_3(PO_4)_2$

$$= \frac{1}{2} \times 0.2 = 0.1 \text{ mol}$$

31 $2y \longrightarrow 2y^{3+} + 6e^-$ $[y \to y^{3+} + 3e^-]$

$6H^+ + 6e^- \longrightarrow 3H_2$ $[2H^+ + 2e^- \to H_2]$

$2y + 6H^+ \longrightarrow 2y^{3+} + 3H_2$

The above individual equations suggest that,

$$\underset{(n = 3)}{1 \text{ eq. of } y} = \underset{(n = 2)}{1 \text{ eq. of } H_2}$$

$\Rightarrow$ $\frac{1}{3}$ mol $y = \frac{1}{2}$ mol H_2

Thus, $H_2 : y = 2 : 3$

32 Initial strength of acetic acid = 0.06 N

Final strength = 0.042 N

Given volume = 50 mL

$\therefore$ Initial millimoles of CH_3COOH

$$= 0.06 \times 50 = 3$$

Final millimoles of CH_3COOH

$$= 0.042 \times 50 = 2.1$$

$\therefore$ Millimoles of CH_3COOH adsorbed

$$= 3 - 2.1 = 0.9 \text{ mmol}$$

Hence, mass of CH_3COOH adsorbed per gram of charcoal $= \frac{0.9 \times 60}{3}$ [molar mass of $CH_3COOH = 60 \text{g mol}^{-1}$]

$$= \frac{54}{3} = 18 \text{ mg}$$

33 Molecular weight of $C_8H_7SO_3Na$

$= (12 \times 8) + (1 \times 7) + 32 + (3 \times 16) + 23$

= 206u

Number of moles in 206 g of $C_8H_7SO_3$ Na resin $= \frac{1}{206}$ mol

Now, reaction would be

$$2C_8H_7SO_3Na + Ca^{2+} \longrightarrow (C_8H_7SO_3)_2Ca + 2Na^+$$

$\therefore$ 2 moles of $C_8H_7SO_3Na$ combines with 1 mole of Ca^{2+}.

$\therefore$ 1 mole of $C_8H_7SO_3Na$ combines with $\frac{1}{2}$ mole of Ca^{2+}.

$\therefore \frac{1}{206}$ mole of $C_8H_7SO_3Na$ will combine with

$\frac{1}{2} \times \frac{1}{206}$ mole of Ca^{2+}

$$= \frac{1}{412} \text{ mole of } Ca^{2+}$$

34 Mohr's salt is $FeSO_4 \cdot (NH_4)_2SO_4 \cdot 6H_2O$

Only oxidisable part is Fe^{2+}.

$$[Fe^{2+} \longrightarrow Fe^{3+} + e^-] \times 6$$
$$Cr_2O_7^{2-} + 14H^+ + 6e^- \longrightarrow 2Cr^{3+} + 7H_2O$$
$$6Fe^{2+} + Cr_2O_7^{2-} + 14H^+ \longrightarrow 6Fe^{3+} + 2Cr^{3+} + 7H_2O$$

Millimoles of $Fe^{2+} = 750 \times 0.6$

$$= 450 \text{ mmol}$$

Moles of $Fe^{2+} = \frac{450}{1000} = 0.450$ mol

6 moles of $Fe^{2+} \equiv 1$ mole of $Cr_2O_7^{2-}$

$\therefore$ 0.450 mole of $Fe^{2+} \equiv \frac{0.450}{6}$

$$= 0.0075 \text{ mole of } Cr_2O_7^{2-}$$

Mass of $K_2Cr_2O_7$ required

$$= 0.075 \times 294 \text{ g} = 22.05 \text{ g}$$

35 Normality equation is,

$$N_1V_1 + N_2V_2 + N_3V_3 = N_4(V_1 + V_2 + V_3)$$

or $1 \times 5 + 20 \times \frac{1}{2} + 30 \times \frac{1}{3} = N_4(5 + 20 + 30)$

$\therefore$ Resulting normality $(N_4) = \frac{25}{55} = 0.45$ N

36 For I solution : millimoles

$$= MV = 480 \times 1.5 = 720$$

For II solution : millimoles

$$= MV = 520 \times 1.2 = 624$$

Total millimoles = 720 + 624 = 1344

$\therefore$ Molarity $= \frac{\text{Moles of solute}}{\text{Total volume of solution (L)}}$

$$= \frac{1344}{480 + 520}$$
$$= 1.344 \text{ M}$$

37 Assertion and Reason both are false. As volume of solution changes by 20%, so it becomes $= 1 + \frac{20}{100} = 1.2$ L

$\therefore$ Molarity of resulting solution

$$= \frac{\text{Moles of solute}}{\text{Total volume of solution (L)}}$$
$$= 0.8 \text{ M}$$

and change in molarity = 1 − 0.8 = 0.2 M

$\therefore$ % change in molarity

$$= \frac{0.2}{1.2} \times 100 = 16.66\%$$

38 Both Assertion and Reason are false.

$$H_2 + \frac{1}{2}O_2 \rightarrow H_2O$$
30 mL 15 mL

Volume of O_2 left = 20 − 15 mL = 5 mL

Therefore, no H_2 left after the reaction hence, H_2 is the limiting reagent.

SESSION 2

1 $N_2 + 3H_2 \longrightarrow 2NH_3$

6 g → 2 × 17 g

450 g → $\frac{34}{6} \times 450 = 2550$ g

Actual = 1575 g

$$\% \text{ yield} = \frac{1575 \times 100}{2550} = 61.76\%$$

2 $(NH_4)_3PO_4$ has 12 H atoms and 4 O atoms; H:O = 3:1

Hence, O relative to 3.18 mol H
= 1.06 mol

3 $2KClO_3 \longrightarrow 2KCl + 3O_2$

$4Al + 3O_2 \longrightarrow 2Al_2O_3$

2 mol of $KClO_3 \equiv$ 2 mol Al_2O_3

∴ 1 mol $KClO_3 \equiv$ 1 mol Al_2O_3

4 10^6 g solution = 1 g Fe^{2+}

250 mL = 250 g solution

$$= \frac{1}{10^6} \times 250 \text{ g } Fe^{2+}$$

$$= \frac{250 \times 392}{10^6 \times 56} \text{ g } (A)$$

$$= 1.75 \times 10^{-3}$$

5 Number of millimoles of $Ca(OH)_2$
= 50 × 0.5 = 25

Number of millimoles of $CaCO_3$ = 25

Number of milliequivalents of $CaCO_3$ = 50

∴ Volume of 0.1 N HCl

$$= \frac{50}{0.1} = 500 \text{ cm}^3$$

6 Required equation is given below:

$$Zn + 2OH^- \longrightarrow ZnO_2^{2-} + 2H^+ + 2e^-$$

$$NO_3^- + 8H^+ + 8e^- \longrightarrow OH^- + 2H_2O + NH_3$$

From the above equation,

∵ 8 mole of electrons are absorbed by 85 g of $NaNO_3$

∴ 1 mole of electron will be absorbed by $\frac{85}{8}$ g of $NaNO_3$ = 10.625g

7 $N_2 + 3H_2 \longrightarrow 2NH_3$

1 mol 4 mol

$NH_3 + HCl \longrightarrow NH_4Cl$

1 1

NH_3 formed = 1 mol

N_2 reacted = 0.5 mol

∴ H_2 reacted = 1.5 mol

N_2 unreacted = 1 − 0.5 = 0.5 mol

H_2 unreacted = 4 − 1.5 = 2.5 mol

NH_3 formed = 1 mol

But NH_3 gets dissolved in H_2O leaving N_2 and H_2

$$\therefore \quad x_{H_2} = \frac{2.5}{3.0} = \frac{5}{6}$$

8 K_2SO_4 is isomorphous with K_2SeO_4.

The molar weight of K_2SeO_4 is given by

$(2 \times 39) + x + (4 \times 16)$

[∵ x = atomic wt. of Se]

⇒ $(142 + x)$g

If $(142 + x)$g of K_2SeO_4 contains x g of K_2SO_4. So, therefore,

$$100 \text{ g of } K_2SeO_4 = \frac{x}{(142 + x)} \times 100$$

But K_2SeO_4 contains 50% of Se, thus

$$50 = \frac{x}{(142 + x)} \times 100$$

or $x = 142$ g

Hence, the atomic wt. of Se is 142 g.

9 $KMnO_4$ reacts with oxalic acid according to the following equation.

$$2MnO_4^- + 5C_2O_4^{2-} + 16H^+ \longrightarrow 2Mn^{2+} + 10CO_2 + 8H_2O$$

Equivalent mass of $KMnO_4$

$$= \frac{\text{molecular mass}}{(7-2)}$$

$$N_{KMnO_4} = 5 \times \text{molarity} = 5 \times 10^{-4}$$

Equivalent mass of

$$C_2O_4^{2-} = \frac{\text{molecular mass}}{2(4-3)}$$

$$C_2O_4^{2-} = \frac{\text{molecular mass}}{2}$$

$$N_{C_2O_4^{2-}} = 2 \times \text{molarity} = 2 \times 10^{-2}$$

According to normality equation,

$$N_1V_1 = N_2V_2$$

$$5 \times 10^{-4} \times V_1 = 2 \times 10^{-2} \times 0.5$$

$$V_1 = \frac{2 \times 10^{-2} \times 0.5}{5 \times 10^{-4}} = 20 \text{ L}$$

10 100 g haemoglobin contains 0.33 g Fe.

∴ 67200 g haemoglobin contains

$$= \frac{0.33 \times 67200}{100} \text{ g Fe} = 221.76 \text{ g Fe}$$

$$\therefore \text{Number of Fe-ato} = \frac{221.76}{56}$$

$$= 3.96 \approx 4$$

11 $2BCl_3 + \frac{3}{2}H_2 \longrightarrow B + 3HCl$

No. of moles of elemental boron

$$= \frac{21.6}{10.8} = 2 \text{ mol}$$

∴ No. of moles of H_2 consumed

$$= \frac{3}{2} \times 2 = 3 \text{ moles}$$

Volume of H_2 at NTP = (22.4 × 3) L
= 67.2 L

12 Molarity of acetic acid = 2.05 M

Mass of CH_3COOH in 1 L solution
= 2.05 × 60 = 123 g

Mass of 1 L solution = 1000 × 1.02 = 1020 g (since density = 1.02 g/mL)

Mass of water in solution
= 1020 − 123 = 897 g

$$\therefore \text{Molality} = \frac{2.05}{897 \times 10^{-3}} = 2.28 \text{ mol kg}^{-1}$$

13

$Sn^{2+} \longrightarrow Sn^{4+} + 2e^-$		2 units
$Fe^{2+} \longrightarrow Fe^{3+} + e^-$		1 unit
$Cr_2O_7^{2-} + 6e^- \longrightarrow 2Cr^{3+}$		6 units

Milliequivalent of Sn^{2+} = 15 × 0.04 × 2 = 1.2

Milliequivalent of Fe^{2+} = $15x \times 1 = 15x$

Milliequivalent of $Cr_2O_7^{2-}$ = 18 × 0.125 × 6
= 13.5

∴ $1.2 + 15x = 13.5 \Rightarrow x = 0.82$

14 $KHC_2O_4 \equiv KOH$

100 mL of 0.01 M KOH
≡ 100 mL of 0.01 M KHC_2O_4
= 1 millimol KHC_2O_4

∴ $Na_2C_2O_4$ = 1 millimol

Total $C_2O_4^{2-}$ = 2 millimol
= 4 milliequiv

Let volume of $KMnO_4$ = V mL

(MnO_4^- reduced to Mn^{2+})

∴ $V \times 0.01 \times 5 = 4$

⇒ V = 80 mL

15 Let 100 m^3 milk contains 4 m^3 fat.

∴ Weight of butter fat in 1 m^3 milk

$$= \frac{4}{100} \times 865 = 35 \text{ kg}$$

Weight of 10^3 cm^3 (1 L) milk = 1.032 kg

∴ Wt. of 10^6 cm^3 (1 m^3) milk = 1032 kg

∴ Weight of skimmed milk = 1032 − 35
= 997 kg

and volume of skimmed milk = 1 − 0.04
= 0.96 m^3

∴ Density of fat free skimmed milk

$$= \frac{997}{0.96} = 1038.5 \text{ kg m}^{-3}$$

DAY TWO

States of Matter

Learning & Revision for the Day

- Classification of Matter
- Gaseous State
- Kinetic Theory of Gases
- Liquid State
- Solid State
- Bragg's Law and Its Applications
- Unit Cell and Lattices

Classification of Matter

Anything which has mass and occupies space is called matter. Three states of matter are as follows

1. **Solids** have definite volume, a definite shape and are rigid.
2. **Liquids** have definite volume, but not definite shape and are non-rigid.
3. **Gases** have neither definite shape, nor definite volume and are non-rigid. It is the simplest state and shows great uniformity in behaviour.

- Besides the three states of matter, two more states of matter have been found to exist. These are **plasma** and **Bose-Einstein condensate** (predicted by Albert Einstein and Indian physicist Satyendra Nath Bose).
- If plasma are super-hot and super-excited atoms (a mixture of electrons and positively charged ions formed by superheating, e.g. in the sun), the Bose-Einstein condensate are total opposite, i.e. super-cold and super- unexcited atoms formed by super-cooling to such an extent that the atoms lose their individual identity and condensate to form a single super-atom.

Following two factors determine the physical state of matter:

1. **Thermal agitation** due to kinetic energy of the particles, atoms [or molecules] of a matter are in a state of continuous vibration and agitation. This increases with increase in temperature due to increase in their kinetic energy.
2. **Cohesive forces** (intermolecular forces) may be attractive or repulsive. Attractive intermolecular forces are called van der Waals' forces.

   The nature of intermolecular forces, molecular interactions and effect of thermal energy on the motion of particles helps to determine the state of a substance.

   Attractive intermolecular forces are known as van der Waals' forces. The different types of van der Waals' forces are as follows:

   (i) **Dispersion or London forces** are the forces that exists between non-polar molecules possessing temporary dipole moments. The interaction energy of these forces is proportional to $1/r^6$. [where, r is the distance between two particles].

   (ii) **Dipole-dipole forces** are forces that exists between dipole ends of polar molecules and are strongest of all van der Waals' forces.

PREP MIRROR
Your Personal Preparation Indicator

- No. of Questions in Exercises (x)—
- No. of Questions Attempted (y)—
- No. of Correct Questions (z)—
  (Without referring Explanations)

- Accuracy Level ($z/y \times 100$)—
- Prep Level ($z/x \times 100$)—

In order to expect good rank in JEE, your Accuracy Level should be above 85 & Prep Level should be above 75.

Dipole-dipole interaction energy between stationary polar molecules is proportional to $1/r^3$ and that between rotating polar molecules is proportional to $1/r^6$.

(iii) **Dipole-induced dipole forces** are forces that exists between the polar molecules having permanent dipole and the molecules lacking permanent dipole. Permanent dipole of the polar molecule induces dipole on the electrically neutral molecule by deforming its electron cloud.

In dipole-induced dipole interactions, the interaction energy is proportional to $1/r^6$.

(iv) **Ion-dipole and ion-Induced dipole forces** are similar to dipole-dipole and induced-dipole interactions but involves only ions, instead of only polar and non-polar molecules. These forces are stronger than dipole-dipole interactions because the charge of any ion is much greater than the charge of a dipole moment.

(v) **Hydrogen bond** it is a special case of dipole-dipole interaction.

(a) In a solid, the cohesive forces predominate the effect of thermal agitation. Consequently, the particles are held together in rigid, highly-oriented and close-packed structure.

(b) In a liquid, the cohesive forces are no longer strong enough, however, these are still sufficient, so that particles cannot escape each others environment, they have sufficient mobility.

(c) In a gas, the thermal agitation dominates the effect of cohesive forces, thus the gas molecules acquire the unrestricted and independent mobility of the vapour state.

Gaseous State

It is the most disordered state of matter. In this state, matter neither have fixed volume nor fixed shape. It takes the shape and volume of the container in which it is placed.

Gases exhibit the following characteristic properties:

- Gases **expand indefinitely**.
- Gases are **highly compressible**.
- Gases intermix freely with one another or move from one place to other without any difficulty. This is known as **diffusion**.
- Pure gases or their mixtures are all **homogeneous in composition**.
- Gases possess very **low density**. The density of gas when compared to that of hydrogen is termed as relative density.
- Due to collision of molecules on the walls of container, **gases exert pressure** which obviously increases due to increase in temperature.
- Gaseous molecules move very rapidly in all directions in a random manner, i.e. gases have **highest kinetic energy**.

Measurable Properties of Gases

The characteristics of gases are described fully in terms of four parameters, i.e. pressure (p), volume (V), temperature (T) and mass or mole, which are termed as **measurable properties**.

(i) **Mass** is expressed in gram or kilogram and is equal to the difference in masses of empty vessel and vessel containing gas.

(ii) **Volume** is equal to the volume of the container and is expressed in terms of litre (L), millilitre (mL), cubic centimetre (cm^3), cubic metre (m^3) or cubic decimetre (dm^3).

$1\ \text{L} = 1000\ \text{mL} = 1000\ \text{cm}^3 = 1\ \text{dm}^3$

$1\ \text{m}^3 = 10^3\text{dm}^3 = 10^6\text{cm}^3 = 10^6\text{mL} = 10^3\ \text{L}$

(iii) **Pressure** is defined as force per unit area. The greater the force acting on a given area, the greater is the pressure. The origin of the force exerted by a gas is the continuous collisions of the molecules against the walls of the container.

The units of pressure are atm, mm Hg, torr etc.

$1\ \text{atm} = 76\ \text{cm of Hg} = 760\ \text{mm of Hg} = 760\ \text{torr}$

$1\ \text{atm} = 101.325\ \text{kPa} = 101325\ \text{Pa} = 101.325\ \text{Nm}^{-2}$

$= 1.01325\ \text{bar}$

$1\ \text{bar} = 10^5\ \text{Pa} = 0.987\ \text{atm}$

(iv) **Temperature** is the measurement of hotness or coldness of an object. It is measured in Celsius scale or absolute scale (Kelvin scale). Celsius scale was earlier known as the centigrade scale.

Gas Laws

Various types of gas laws are as follows:

1. Boyle's Law (1662)

At constant temperature, the pressure of a fixed amount of gas varies inversely with its volume.

At constant temperature (T), $p \propto \frac{1}{V}$

$$pV = \text{constant (for given moles and } T)$$

or $$p_1V_1 = p_2V_2 = \text{constant}$$

The two conventional ways of graphically representing Boyle's law are as follows:

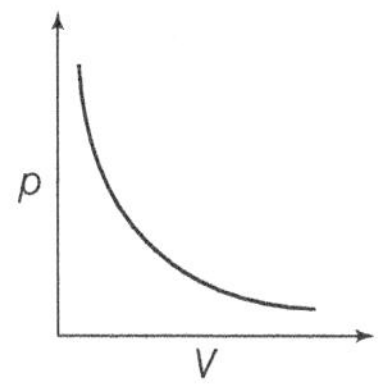

p vs V plot

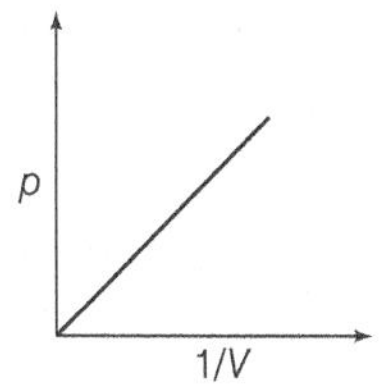

p vs 1/V plot

2. Charles' Law (1787)

At constant pressure, the volume of a fixed mass of a gas is directly proportional to its absolute temperature.

At constant pressure (p), $V \propto T$

$$\frac{V}{T} = \text{constant (for given } n \text{ and } p)$$

or

$$\frac{V_1}{T_1} = \frac{V_2}{T_2}$$

- Charle's found that for all gases at given pressure, graph of V vs T is a straight line.

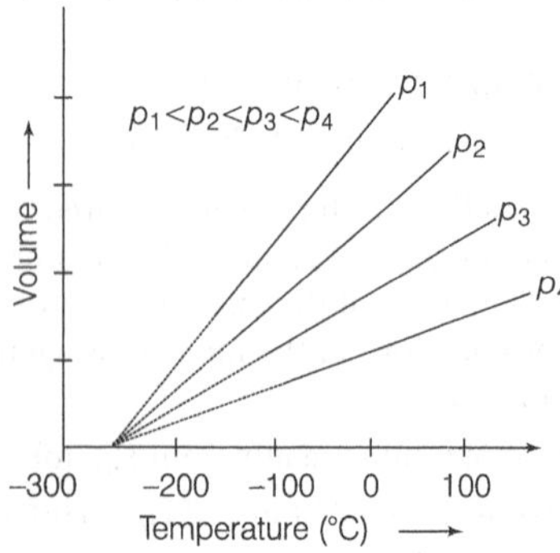

***V vs T* graph at different constant pressures**

- Charles' obtained experimentally that for 1°C change in temperature of a gas, the volume get changed by a fraction of $\frac{1}{273}$. Hence, if V_0 be the volume of a gas at 0°C and it is recorded to be V at t°C then

$$V = V_0 + \frac{1}{273.15} t\, V_0$$

$$= V_0\left(1 + \frac{t}{273.15}\right) = \frac{V_0}{273.15} T$$

If temperature is lowered, the volume decreases at constant pressure and become zero at -273.15°C.

Further lowering is not possible thus, it is called the lowest possible temperature or absolute zero of temperature. Scale of temperature based on this choice of zero is called **absolute scale of temperature** or thermodynamic scale of temperature.

$$T(\text{K}) = t\,(°\text{C}) + 273.15$$

3. Gay-Lussac's Law

At constant volume, pressure of a fixed amount of a gas varies directly with the temperature.

At constant volume (V), $p \propto T$

$$\frac{p}{T} = \text{constant}$$

(for given n and V)

or

$$\frac{p_1}{T_1} = \frac{p_2}{T_2}$$

4. Avogadro's Law

It states that equal volumes of all gases at same pressure and temperature contain equal number of molecules.

$$V \propto n \qquad \text{(at given } T \text{ and } p)$$

or

$$\frac{V_1}{n_1} = \frac{V_2}{n_2}$$

Ideal Gas Equation

The Boyle's law, Charle's law and Avogadro's law can be combined together in a single equation known as **ideal gas equation**.

$$V \propto \frac{nT}{p}$$

$$pV = nRT \quad \text{or} \quad pV = \frac{w}{M} RT$$

$$\therefore \left[\text{number of moles } n = \frac{w}{M}\right]$$

or $pM = dRT$

where, w = mass of gas,

M = molecular weight of gas

d = density of gas,

p = pressure of gas

T = temperature, V = volume

R = universal gas constant.

Values of $R = 0.0821\ \text{L atm K}^{-1}\ \text{mol}^{-1}$

$= 8.314\ \text{J K}^{-1}\ \text{mol}^{-1}$

$= 1.987\ \text{cal K}^{-1}\ \text{mol}^{-1}$

$= 8.314\ \text{kPa dm}^3\ \text{K}^{-1}\ \text{mol}^{-1}$

$= 0.083\ \text{bar dm}^3\ \text{K}^{-1}\ \text{mol}^{-1}$

NOTE Standard temperature and pressure (**STP** or **NTP**) means 273.15 K (0°C) temperature and 1 bar (i.e. exactly 10^5 pascal) pressure. At STP, molar volume of an ideal gas is 22.71098 L mol^{-1}.

Dalton's Law of Partial Pressures

- Total pressure exerted by the mixture of non-reacting gases is equal to the sum of the partial pressures of individual gases.

  $p_{\text{Total}} = p_1 + p_2 + p_3 + p_4 + \ldots$ (at constant T, V)

- Partial pressure of a gas (p) = mole fraction (χ) × total pressure exerted by the mixture (p_{total})
- Percentage of a gas in the mixture

$$= \frac{\text{partial pressure of gas}}{\text{total pressure}} \times 100$$

$$= \text{mole fraction of that gas} \times 100$$

- When a gas is collected over the water, the pressure of the gas $p_{\text{observed or moist}}$ = pressure of dry gas + aqueous tension or water vapour pressure.
- Relative humidity (R_H) is given by R_H = partial pressure of water in air/vapour pressure of water.

Graham's Law of Diffusion

According to this law "if a gas is allowed to escape from its container through a small hole into vacuum, the process is called effusion. On the other hand, diffusion is the passage of gas through a porous partition.

- Rate of diffusion or effusion $\propto \dfrac{1}{\sqrt{\text{molecular weight}}}$

  [$\because$ Molecular weight = 2 × vapour density $\Rightarrow M \propto d$]

  Rate of diffusion or effusion $\propto \dfrac{1}{\sqrt{\text{density}}}$

- Rate of diffusion or effusion (r)

  $$= \frac{\text{volume diffused or effused}}{\text{time taken}}$$

  Hence, $\dfrac{r_1}{r_2} = \dfrac{t_2}{t_1} = \sqrt{\dfrac{d_2}{d_1}} = \sqrt{\dfrac{M_2}{M_1}}$

- If p is not constant $r \propto p$.

  $\therefore \quad \dfrac{r_1}{r_2} = \dfrac{p_1}{p_2}\sqrt{\dfrac{M_2}{M_1}}$

Kinetic Molecular Theory of Gases

It is based on the following postulates:

- A gas consists of extremely small discrete molecules. These molecules are so small and so far apart that actual volume of the molecules is negligible as compared to the total volume of the gas.
- Gas molecules are in constant random motion with high velocities that moves in a straight line and changes their directions on collision with other molecules or walls of container.
- The intermolecular forces of attraction are negligible and the effect of gravity on them is also negligible.
- The collisions are perfectly elastic, therefore, there is no loss of kinetic energy during collision. However, there may be redistribution of energy during such a collision.
- The pressure of a gas is caused by the bombardment of moving molecules against the walls of container.
- In a gas, different molecules have different kinetic energies but the average kinetic energy of molecules is proportional to absolute temperature of the gas.
- On the basis of these assumptions, the following mathematical expression for kinetic gas equation is derived.

$$pV = \frac{1}{3} mNu^2$$

or

$$pV = \frac{1}{3} Mu^2$$

($\because M = m \times N$, for one mole $m = 1$)

where, u = rms velocity.

Different kind of velocities of molecules are as follows:

(i) **Root mean square velocity**

$$(u_{\text{rms}}) = \sqrt{\frac{3RT}{M}} = \sqrt{\frac{3p}{d}} = \sqrt{\frac{3KT}{M}}$$

For the same gas at two different temperatures, the ratio of RMS velocities will be

$$\frac{u_1}{u_2} = \sqrt{\frac{T_1}{T_2}}$$

For two different gases, the ratio of the RMS velocities will be $\dfrac{u_1}{u_2} = \sqrt{\dfrac{M_2}{M_1}}$

(ii) **Average velocity** $(u_{\text{av}}) = \sqrt{\dfrac{8RT}{\pi M}} = \sqrt{\dfrac{8KT}{\pi M}}$

(iii) **Most probable velocity** $(u_{\text{mp}}) = \sqrt{\dfrac{2RT}{M}} = \sqrt{\dfrac{2pV}{M}} = \sqrt{\dfrac{2p}{d}}$

Relation between different molecular velocities are as follows:

(i) $u_{\text{mp}} : u_{\text{av}} : u_{\text{rms}} :: \sqrt{2} : \sqrt{\dfrac{8}{\pi}} : \sqrt{3} = 1 : 1.128 : 1.224$

(ii) $u_{\text{av}} = u_{\text{rms}} \times 0.9213$

(iii) Kinetic energy of molecules

KE (per molecule) $= \dfrac{3}{2}KT$

$\left[\text{where, } K = \dfrac{R}{N_A}\right]$, K is Boltzmann constant

$\dfrac{R}{N_0} = 1.38 \times 10^{-23}$ J K^{-1} molecule

KE (per mol) $= \dfrac{3}{2}pV = \dfrac{3}{2}RT$

At absolute zero, the kinetic energy of gas molecules become zero.

Real Gases

The gas which does not obey gas laws at all conditions of temperature and pressure is called real gas. Real gases obey the gas laws under moderate conditions of temperature and pressure.

Deviation from Ideal Behaviour

At very low temperature and high pressure, real gases show deviation from the ideal gas behaviour.

The causes of deviation are:

(i) At low temperature and high pressure, volume of a real gas is larger than that predicted for an ideal gas as the molecules of gas have their own volume.

(ii) Intermolecular forces are not negligible.

The deviations from ideal gas behaviour are expressed in terms of **compressibility factor** (Z).

Compressibility factor $(Z) = \frac{pV}{nRT} = 1$ (for ideal gas)

Different conditions of compressibility factor (Z) for real gases are as follows:

- $Z < 1$ (as for CH_4, NH_3, CO_2) can be attributed to predominance of attractive forces among the molecules of these gases at the temperature of experimentation.
- $Z > 1$ (for H_2, He) can be attributed to the dominance of strong repulsive forces among the molecules. **Boyle temperature** is the characteristic temperature of the gas at which it obeys ideal gas nature. Below this temperature, the value of Z at first decreases, approaches a minimum and then increases as the pressure increases continuously.

van der Waals' Equation

To rectify the errors caused by ignoring the intermolecular forces of attraction and the volume occupied by molecules, Johannes van der Waals' proposed following equation, which is valid for real gases upto a large range of temperature and pressure.

$$\left[p + \frac{n^2 a}{V^2}\right](V - nb) = nRT$$

where, n = number of moles, a = constant (measure of magnitude of attractive forces between the molecules)
b = constant (due to occupied volume of gaseous molecules = co-volume or excluded volume).

Units of 'a' are atm $L^2 mol^{-2}$ or atm m^6 mol^{-2} or $Nm^4 mol^{-2}$ and that of 'b' are $L mol^{-1}$ or dm^3 mol^{-1}.

van der Waals' equation get reduced at different conditions of pressures and temperatures. These are shown below:

1. **When Pressure is Low** At low pressure, V is very large, thus, $(V - b) \approx V$ as b is neglected in comparison to V, therefore equation becomes:

$$\left(p + \frac{a}{V^2}\right) V = RT \text{ or } pV + \frac{a}{V} = RT$$

$$\Rightarrow \quad \frac{pV}{RT} = 1 - \frac{a}{VRT} \quad \Rightarrow \quad Z = \frac{pV}{RT} = 1 - \frac{a}{VRT}$$

At extremely low pressures, $\frac{a}{V}$ is very small as V is very large. Thus, it can be neglected.

$$\therefore \quad pV = RT$$

2. **When Pressure is High** At high pressure, V is quite small, thus, the term $(V - b)$ cannot be neglected while the term $\frac{a}{V^2}$ is quite large but in comparison to high p, $\frac{a}{V^2}$ can be neglected.

$$\therefore \quad p(V - b) = RT$$

$$\Rightarrow \quad pV = RT + pb$$

$$\frac{pV}{RT} = 1 + \frac{pb}{RT} \quad \text{or} \quad Z = 1 + \frac{pb}{RT}$$

3. **When Temperature is High** At high temperature, V is very large, the terms $\frac{a}{V^2}$ and b can be neglected thus, $pV = RT$.

4. **When Temperature is Low** At low temperature, both p and V are small thus, both pressure and volume corrections are appreciable.

The values of van der Waals' constant 'a' and 'b' do not remain constant over the entire range of temperature and pressure. Hence, van der Waals' equation is valid only over specific range of temperature and pressure.

Liquefaction of Gases and Critical Constants

- The phenomenon of converting a gas into liquid is known as **liquefaction** at particular temperature, pressure and volume. When a gas changes into liquid, these conditions are known as **critical conditions** e.g. T_c, V_c, p_c etc.
- **Critical Temperature (T_c)** is the maximum temperature at which a gas can be liquefied. **Critical pressure (p_c)** is the minimum pressure which is required to liquefy a gas. **Critical volume** (V_c) is the volume of one mole of a gas which is occupied at critical temperature.

Critical temperature $(T_c) = \frac{8a}{27\,Rb}$

Critical pressure $(p_c) = \frac{a}{27 b^2}$

Critical volume $(V_c) = 3\,b$

NOTE
- Relation between critical p, V and T is $p_c V_c = \frac{3}{8} RT_c$ or $\frac{p_c V_c}{RT_c} = \frac{3}{8}$ $Z < 1$ at critical point.
- At critical point, there is no distinction between gaseous state and liquid state as densities of a substance in gaseous and liquid states are same.

Liquid State

It is the state of matter in which the molecules are held close to each other and execute random motion through intervening spaces.

Liquids have definite volume. They can move past one another freely. They can flow, can be poured and can assume the shape of the container in which these are stored.

Physical Properties of Liquid

1 Vapour Pressure

- Inside a closed vessel, the liquid and its vapours are in dynamic equilibrium. The pressure exerted by the vapours on the walls of the container increases. After sometime, it becomes constant, an equilibrium is established between liquid phase and vapour phase. Vapour pressure at this stage is known as **equilibrium vapour pressure**.

- Since, the vapour pressure is a kinetic phenomena, it is independent of the amount of a liquid. It only depends on the temperature.
- When a liquid is heated in an open vessel, the liquid vapourises from the surface. The temperature at which vapour pressure of the liquid becomes equal to the external pressure, vaporisation can occur throughout the bulk of the liquid is called **boiling point** and the condition is called **boiling**.

2 Viscosity

- The ease with which a fluid can flow is determined by its property, called **viscosity**. It arises due to the internal friction between layers of fluids.
- Viscosity of liquid decreases with increase in temperature because rise in temperature increases the kinetic energy of molecules, this results in decrease in inter-layer friction.

3 Surface Tension

- The force acting along the surface of a liquid at right angle to any line per unit length is called **surface tension**. The unit of surface tension is **Nm^{-1}**. Surface tension decreases with rise in temperature. It is different for different liquids due to different intermolecular forces.
- Due to surface tension, the small drops of liquids are spherical as the liquid tends to have a minimum surface area. Due to surface tension, liquids rise or fall in capillary tube.

Solid State

It is defined as that form of matter which has rigidity and due to which it possesses a definite volume and a definite shape. These characteristics are due to the existence of very strong forces of attraction among the constituent species of the solid.

On the basis of the arrangement of their atoms or ions or molecules, solids are broadly classified into two groups:

1. **Crystalline Solids** They have systematic and regular arrangement of particles and sharp melting point.Crystalline solids are anisotropic, i.e. their physical properties have different values in different directions. On the basis of types of bond present, they are further divided into four groups.
   (a) **Molecular crystal**s or van der Waals' crystals, e.g. $I_2(s)$, $H_2O\ (s)$ etc.
   (b) **Atomic crystals** or (covalent crystals), e.g. graphite, diamond etc.
   (c) **Metallic crystals**, e.g. all metallic solids etc.
   (d) **Ionic crystals**, e.g. NaCl, CsCl, ZnS etc.
2. **Amorphous Solids** They are unsystematic and irregular arrangement of particles. These are isotropic, i.e. their physical properties have same values in different directions.

Bragg's Law and Its Applications

The fundamental relation among the wavelength of the X-rays, the inter-planar distance in the crystal and the angle of reflection is known as the Bragg's law.

$$2d \sin \theta = n\lambda$$

where,

λ = wavelength of X-ray used

θ = angle between incident X-rays and the plane of the crystal

d = distance between planes of the constituent particle in a crystal

n = an integer (1, 2, 3, 4, ...) which represents the serial order of diffracted beams.

Various applications of Bragg's law are as follows:

- Structure of solids are determined by X-ray diffraction methods.
- Structures of complex substances such as proteins and nucleic acids are also determined by this method.

Unit Cell and Lattices

- An array of lattice points showing arrangement of constituent particles in different position in three dimensional space, is known as **space lattice** or **crystal lattice**.
- **Unit cell** is the smallest group of lattice points which when repeated in all directions develop the entire lattice.
- A unit cell is characterised by its dimensions (a, b and c) and angles (α, β, γ). These parameters of a typical unit cell are tabulated as follows:

Seven crystal system and their unit cell parameters

Crystal systems and Bravais lattices	Unit cell parameters
Cubic [primitive, face centred, body centred] e.g. NaCl, KCl, diamond, copper, zinc blende etc.	$a = b = c, \alpha = \beta = \gamma = 90°$
Rhombohedral or trigonal [primitive] e.g. $NaNO_3$, $CaCO_3$, HgS, etc.	$a = b = c, \alpha = \beta = \gamma \neq 90°$
Orthorhombic [primitive, face centred, body centred, end centred] e.g. Rhombic sulphur, $BaSO_4$ (baryta), $PbCO_3$ etc.	$a \neq b \neq c, \alpha = \beta = \gamma = 90°$
Triclinic (most asymmetric) [primitive] e.g. $K_2Cr_2O_7$, H_3BO_3, $CuSO_4 \cdot 5H_2O$ etc.	$a \neq b \neq c, \alpha \neq \beta \neq \gamma \neq 90°$
Monoclinic [primitive, end centred] e.g. $Na_2SO_4 \cdot 10H_2O$, $Na_2B_2O_7 \cdot 10H_2O$, monoclinic S etc.	$a \neq b \neq c$, $\alpha = \gamma = 90°, \beta \neq 90°$
Tetragonal [primitive, body centred] e.g. Cassiterite, white tin, rutile etc.	$a = b \neq c, \alpha = \beta = \gamma = 90°$
Hexagonal [primitive] e.g. Mg, beryl, ZnO (zincite), PbI_2, graphite etc.	$a = b \neq c$, $\alpha = \beta = 90°, \gamma = 120°$

Various types of unit cells are as follows:

1. **Simple unit cell or primitive unit cell** in which particles are present only at corners.

   ∴ Number of atoms per unit cell $= \frac{1}{8} \times 8 = 1$

2. **Face centred cell** in which the particles are present at the corners as well as at the centre of each face.

   ∴ Number of atoms per unit cell

   $= \frac{1}{8} \times 8 + \frac{1}{2} \times 6 = 4$

3. **Body centred unit cell** in which the particles are present at the corners of the cube as well as one particle is present at the centre within the body.

   ∴ Number of atoms per unit cell $= \frac{1}{8} \times 8 + 1 = 2$

Coordination Number

The number of atoms in a crystal which surrounds particular atoms as its nearest atoms in its neighbour is called its coordination number. Each coordination number corresponds to a fixed radius ratio (the ratio of radius of cation to that of anion).

Radius ratio $\left(\frac{r_+}{r_-}\right)$	Coordination number	Structure
< 0.155	2	Linear
0.155 – 0.225	3	Planar triangle
0.225 – 0.414	4	Tetrahedral
0.414 – 0.732	6	Octahedral
0.732 – 1	8	Cubic

Packing in Solids

The packing of spheres of equal size in solids takes place as follows:

1. One Dimensional Packing

When the spheres are placed in horizontal row, touching each other, an edge of the crystal is formed.

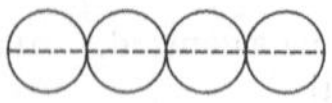

Close packing of spheres in one dimension

2. Two Dimensional Packing

It is done in two different types:

(i) **Square Close Packing** (SCP) The particles, when placed in the adjacent rows show a horizontal as well as vertical alignment and form squares.

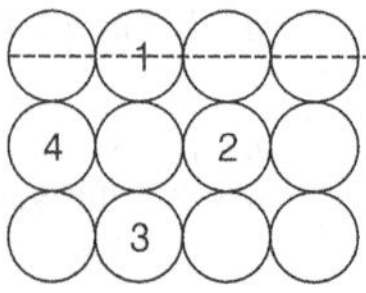

Square close packing

(ii) **Hexagonal Close Packing** (HCP) The particles in every next row are placed in the depression between the particles of the first row. The particles in the third row will be vertically aligned with those in the first row.

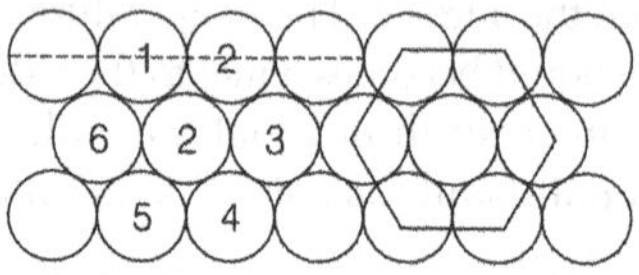

Coordination number is 6 (HCP)

Hexagonal close packing

3. Three Dimensional Packing

It can be obtained by stacking two dimensional layers one above the other.

It is of two types:

(i) **Hexagonal Closed Packing** (HCP) The first layer as layer *A* and second layer as layer *B*, the arrangement is called *AB AB* ... pattern or hexagonal closed packing (hcp).

(ii) **Cubic Close Packing** (CCP) When the third layer is placed over the second layer in such a way that spheres cover the octahedral voids, a layer different from layers *A* and *B* is produced. Let us call it as layer *C*. This pattern of stacking spheres is called *ABC ABC* ... pattern or cubic close packing (ccp). It is similar to face centred cubic (fcc) packing.

Voids

In closely packed structures, the empty space is called interstitial site or void. The void can be a simple triangular space in the case of two dimensional packing and called the **trigonal void**.

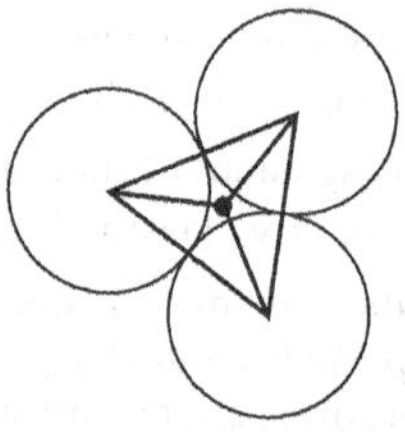

Trigonal void

In three dimensional close packing patterns, the voids can be of two types:

1. **Tetrahedral Void** It is a simple triangular space surrounded by four spheres.

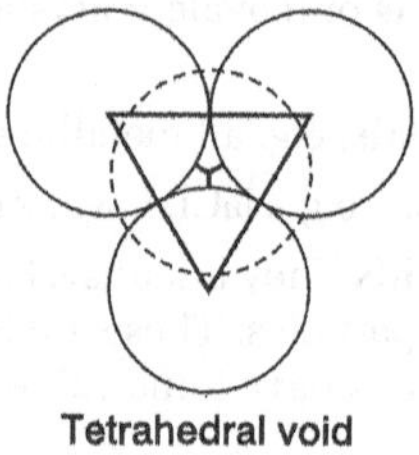

Tetrahedral void

2. **Octahedral Void** It is a double triangular void surrounded by six spheres.

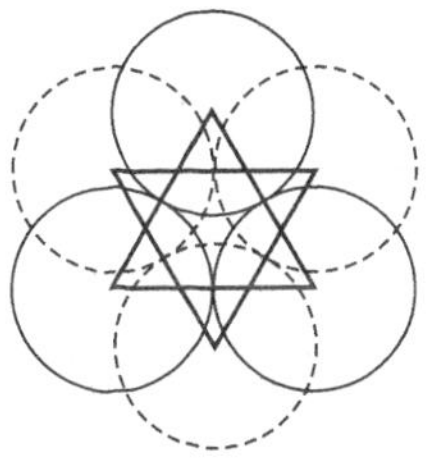

Octahedral void

The sizes of above voids can be seen in the increasing order as: Trigonal < tetrahedral < octahedral are N sphere (atoms or ions) in the packing then.

Number of octahedral voids = N

Number of tetrahedral voids = $2N$

Calculations Involving Unit Cell Parameters

1. **Relation between Edge Length and Radius of Sphere**

(i) **Simple cubic**, $r = a/2$

Percentage of packing fraction = 52 %

(ii) **Face centred cubic** for ccp and hcp structures $r = \frac{a}{2\sqrt{2}}$

Percentage of packing fraction = 74%

(iii) **Body centred cubic** $r = \sqrt{\frac{3}{4}}\, a$

Percentage of packing fraction = 68%

2. **Density of unit cell** $d = \frac{M \times Z}{N_A \times a^3}$

where, d = density

M = molecular weight

Z = number of atoms per unit cell

N_A = Avogadro number

a = edge length of unit cell

Structures of Some Ionic Solids

1. **Rock salt** (NaCl) **type** It has fcc arrangement and coordination number $Na^+ = 6, Cl^- = 6$, e.g. LiCl, AgF, AgCl, NH_4Br etc.
2. **CsCl type** It has the bcc arrangement with Cs^+ at the body centre and Cl^- ion at corners of a cube or *vice-versa*. Coordination number $Cs^+ = 8, Cl^- = 8$, e.g. CsCN, TiCl, TiCN etc.
3. **Zinc blende** (ZnS) **type** It has ccp or fcc arrangement. Coordination number $Zn^{2+} = 4, S^{2-} = 4$. Structure of ZnS is identical to diamond, e.g. CuCl, AgI etc.
4. **Fluorite** (CaF_2) **type** It has ccp arrangement. F^- ions occupy all the tetrahedral voids. Coordination number $Ca^{2+} = 8, F^- = 4$, e.g. CdF_2, PbF_2, $BaCl_2$, $SrCl_2$.
5. **Anti fluorite** (A_2B) **type** It has ccp arrangement. The cation occupy half of the tetrahedral voids. Coordination number $Na^+ = 4, O^{2-} = 8$, e.g. Na_2O.
6. On applying pressure, NaCl structure (6:6 coordination) changes into CsCl (8 : 8 coordination) and reverse of this occurs at high temperature.

Imperfections in Solids or Defects

Any departure from perfectly ordered arrangement of atoms in a crystal is called **imperfection or defect.**

The defects are of two types:

(i) **Line defects** are the irregularities or deviations from ideal arrangement in entire rows of lattice points.

(ii) **Point defects** are the irregularities or deviations from ideal arrangement around a point or an atom in crystalline substance.

Points defects are classified into three types. These are as follows:

1. **Stoichiometric Defects** are the point defects that do not disturb the stoichiometry of the solid.
   - These are of two types: Vacancy and interstitial defects.
   - Schottky and Frenkel defects are vacancy and interstitial defects respectively.

Schottky Defect	Frenkel Defect
It is mainly a vacancy defect arises due to equal number of cations and anions missing from the lattice sites.	It is mainly an interstitial defect arises due to the missing of ions (usually cations) from the lattice sites and they occupy the interstitial sites.
This results in the decrease in density of crystal.	It has no effect on the density of crystal.
This type of defect is found in highly ionic compounds with high coordination number, e.g. NaCl, CsCl, AgBr etc.	This type of defect is found in crystals, where the difference in the size of cations and anions is very large, e.g. AgCl, AgBr, ZnS etc.

2. **Non-Stoichiometric Defects** contain the constituent elements is non-stoichiometric ratio due to defects in their crystal structures. These defects result in either **excess of metal atom** or **deficiency of metal atom.**

   In **metal excess defect** due to anionic vacancies, a negative ion may be missing from its lattice site, leaving a hole which is occupied by an electron, thereby maintaining the electrical balance. The electrons, thus trapped in the anion vacancies are called ***F*-centres** because they are responsible for imparting colour to the crystals.

3. **Impurity Defects** is a common method of introducing defects in ionic solids. It is done by adding impurity ions having different charge than host ion.

   These foreign atoms are present at lattice site in substitutional solids and at vacant interstitial sites in interstitial solids.

Electrical Properties of Solids

Solids are classified into three groups on the basis their conductivity. These are as follows:

(i) **Conductors** are the solids with conductivity range of the order of 10^4 to $10^7 \Omega^{-1} m^{-1}$, e.g. metals.

(ii) **Semiconductors** are the solids with conductivity range $10^{-6} - 10^4 \ \Omega^{-1} \ m^{-1}$, e.g. semi-metals.

(iii) **Insulators** are the solids with conductivity range $10^{-20} - 10^{-10} \ (\Omega \ m)^{-1}$, e.g. non-metals.

1. In **conductors**, the conduction band is almost overlapping with the valence band, i.e. there is no energy gap present between these two bands or valence band is not completely filled. Electrons can flow easily under the influence of electric field, in both the cases.

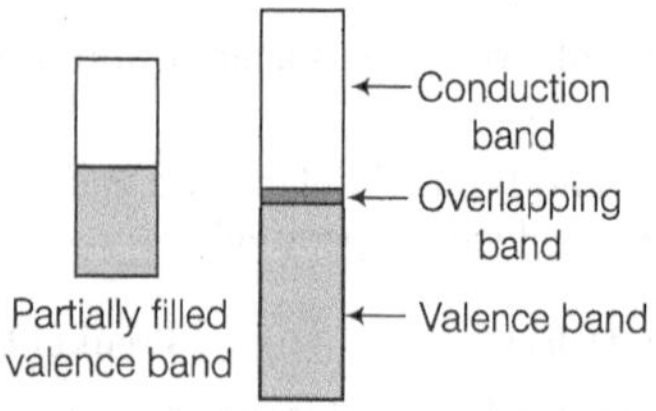

Conductors

2. In **semiconductors**, the gap between valence band and conduction band is small and therefore, some of the electrons may jump from valence band to conduction band and some conductivity is observed.

   The conductivity here increases with increase in temperature due to lowering of energy gap as shown in the figure.

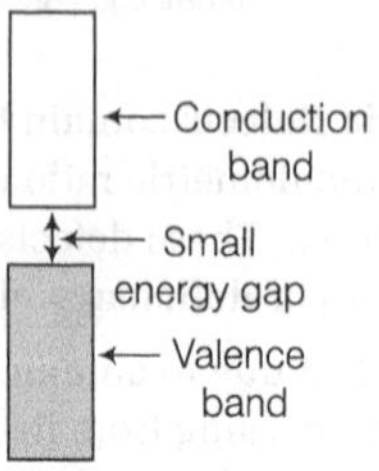

Semiconductors

- Electrical conductivity of semiconductors increases with rise in temperature, since more electrons can jump to the conduction band. Substances like silicon and germanium show this type of behaviour and are called **intrinsic semiconductor (undoped)**.

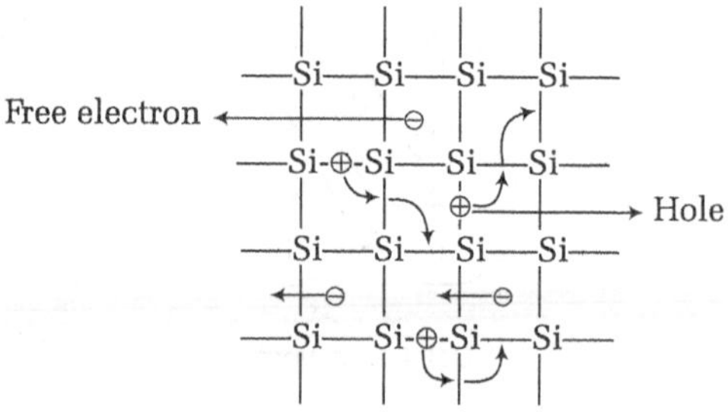

Electrical conductance

- The conductivity of intrinsic semiconductors is too low. So, their conductivity is increased by adding an appropriate amount of suitable impurity. This process is called **doping**.
- Introduction of impurities in the form of electrons of 13 and 15 group of periodic table in perfect crystal. Such conductors is called **extrinsic conductors.** Depending upon the nature of the elements used for doping, the extrinsic semiconductor can be of two types, *viz*, *n*-type semiconductors and *p*-type semiconductors.

  (i) ***n*-type semiconductors** The group 15 elements have 1 electron excess to Si or Ge, after forming 4 covalent bonds with group 14 member (Si or Ge). This excess free electron is responsible for electrical conductivity in them. Group 14 elements when doped with group 15 elements are called ***n*-type semiconductors**. Here, *n* specifies that negative charge flows in them.

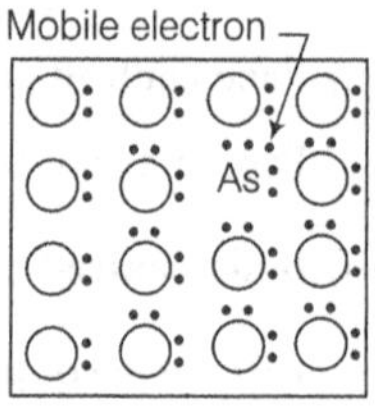

***n*-type semiconductor**

  (ii) ***p*-type semiconductors** The group 13 elements have 1 electron short to group 14 elements, thus giving rise to electron deficient band or a hole. Here, such holes are responsible for electrical conductivity.

  Thus, group 14 elements when doped with elements of group 13 are called ***p*-type semiconductors.**

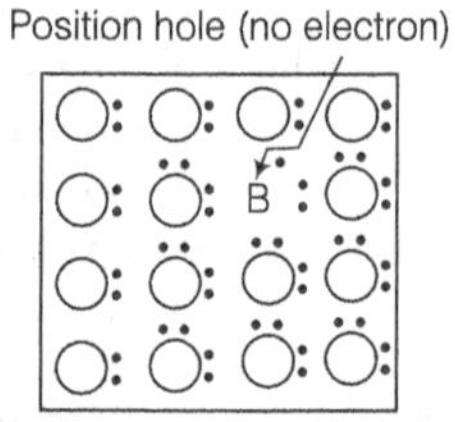

***p*-type semiconductor**

3. In case of **insulators**, the energy gap between valence band and conduction band is so large that it cannot even covered up by supplying energy in the form of heat.

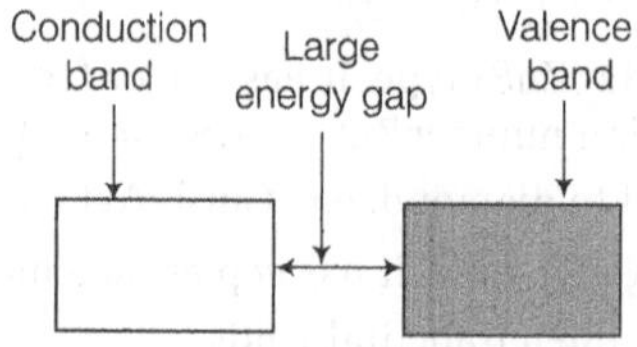

Insulator

Magnetic Properties of Solids

- The origin of magnetic properties lies in the electrons. Each electron in an atom behaves like a tiny magnet. Its magnetic moment originates from two type of motions (i) its orbital motion around the nucleus and (ii) its spin around its own axis.
- Electron being a charged particle and undergoing these motions can be considered as a small loop of current which possesses a magnetic moment.
- Each electron has a permanent spin and an orbital magnetic moment associated with it whose magnitude is very small and is measured in the unit called Bohr magneton, μ_B. It is equal to 9.27×10^{-24} Am^2.

Solids can be divided into different classes depending on their response to magnetic field. These are as follows :

1 **Diamagnetic substances** are the substances that are weakly repelled by the magnetic field and do not have any unpaired electron. e.g. TiO_2, V_2O_5, C_6H_6, NaCl etc.

2 **Paramagnetic substances** are the substances that are attracted by the magnetic field and have unpaired electrons. They lose magnetism in the absence of magnetic field. e.g. O_2, Cu^{2+}, Fe^{3+} etc.

3 **Ferromagnetic substances** are attracted by the magnetic field and show permanent magnetism even in the absence of magnetic field. e.g. Fe, Co, Ni and CrO_2. Ferromagnetic property decreases from iron to nickel because number of unpaired electrons decreases.

4 **Antiferromagnetic substances** have net magnetic moment zero due to compensatory alignment of magnetic moments. e.g. MnO, MnO_2, FeO, NiO, Cr_2O_3 etc.

5. **Ferrimagnetic substances** have a net dipole moment due to unequal parallel and anti-parallel alignment of magnetic moments. e.g. Fe_3O_4, ferrites $MgFe_2O_4$ and Zn Fe_2O_4 etc.

Dielectric Properties of Solids

These are seen in insulators. Insulators show generation of dipoles in them, when they are placed in an electric field. These dipoles are given below:

(i) May align themselves in an order manner, so that there is a net dipole moment in the crystal.

(ii) May align themselves in such a manner that dipole moments may cancel each other.

(iii) It is also possible that there are no dipoles in the crystal but only ions are present.

Some other important dielectric properties of solids are:

1. **Piezoelectricity** It is produced on applying mechanical stress on polar crystals.
2. **Pyroelectricity** It is produced on heating some polar crystals.
3. **Ferroelectricity** Piezoelectric crystals with permanent dipoles are said to have ferroelectricity. e.g. $BaTiO_3$, KH_2PO_4 etc. All ferroelectric solids are piezoelectric but all piezoelectric solids are not ferroelectric, e.g. quartz is piezoelectric but not ferroelectric.
4. **Antiferroelectricity** In antiferroelectric crystals, the dipoles in alternate polyhedra point up and down in such a way that the crystal does not possess any net dipole moment.

DAY PRACTICE SESSION 1

FOUNDATION QUESTIONS EXERCISE

1 The intermolecular interaction that is dependent on the inverse cube of distance between the molecules is → JEE Main 2015

(a) ion-ion interaction (b) ion- dipole interaction
(c) London force (d) hydrogen bond

2 A certain mass of the oxygen gas occupies 7 L volume under a pressure of 380 mm Hg. The volume of the same mass of the gas at standard pressure with temperature remaining constant shall be

(a) 26.60 L (b) 54.28 L (c) 3.5 L (d) 7 L

3 A bubble of gas released at the bottom of a lake increases to eight times its original volume when it reaches the surface.Assuming that atmospheric pressure is equivalent to the pressure exerted by a column of water 10 m high, what is the depth of the lake?

(a) 90 m (b) 10 m (c) 70 m (d) 80 m

4 Equal weights of ethane and hydrogen are mixed in an empty container at 25°C. The fraction of the total pressure exerted by hydrogen is

(a) 1 : 2 (b) 1 : 1 (c) 1 : 16 (d) 15 : 16

5 Through the two ends of a glass tube of length 200 cm HCl gas and NH_3 are allowed to enter. At what distance, the white ring of ammonium chloride will first appear?

(a) 81.1 cm from HCl end (b) 81.1 cm from NH_3 end
(c) 118.9 cm from HCl end (d) The ring will not form

6 At STP, 0.48 g of O_2 diffused through a porous partition in 1200 s. What volume of CO_2 will diffuse in the same time and under the same conditions?

(a) 286.6 mL (b) 346.7 mL
(c) 112.2 mL (d) 224.8 mL

7 What are the relative rates of diffusion for $^{235}UF_6$ and $^{238}UF_6$?

(a) 1.0043 (b) 2.540 (c) 1.503 (d) 1.119

8 Assuming the same pressure in each case, what is the mass of hydrogen required to inflate a balloon to a certain volume V at 100°C, if 3.5 g helium is required to inflate the balloon to half the volume V at 25°C?

(a) 3.8 g (b) 2.8 g (c) 4.2 g (d) 4.0 g

9 Helium atom is two times heavier than a hydrogen molecule at 298 K. The average kinetic energy of helium is

(a) two times that of hydrogen molecule
(b) same as that of hydrogen molecule
(c) four times that of hydrogen molecule
(d) half that of hydrogen molecule

10 For a monoatomic gas, kinetic energy = E. The relation with rms velocity is

(a) $u = \left(\frac{2E}{M}\right)^{\frac{1}{2}}$ (b) $u = \left(\frac{3E}{2M}\right)^{\frac{1}{2}}$

(c) $u = \left(\frac{E}{2M}\right)^{\frac{1}{2}}$ (d) $u = \left(\frac{E}{3M}\right)^{\frac{1}{2}}$

11 At what temperature, the rms velocity of SO_2 be same as that of O_2 at 303 K?

(a) 273 K (b) 606 K (c) 303 K (d) 403 K

12 The ratio between the root mean square velocity of H_2 at 50 K and that of O_2 at 800 K is

(a) 4 (b) 2 (c) 1 (d) 1/4

13 As the temperature is raised from 20°C to 40°C, the average kinetic energy of neon atoms changes by a factor of which of the following?

(a) $\frac{1}{2}$ (b) $\sqrt{\frac{313}{293}}$ (c) $\frac{313}{293}$ (d) 2

14 To an evacuated vessel with movable piston under external pressure of 1 atm, 0.1 mole of He and 1.0 mole of an unknown compound (vapour pressure 0.68 atm at 0°C) are introduced.
Considering the ideal gas behaviour, the total volume (in litre) of the gases at 0°C is close to → AIEEE 2011

(a) 3 (b) 5 (c) 7 (d) 9

15 If 10^{-4} dm^3 of water is introduced into a 1.0 dm^3 flask at 300 K, how many moles of water are in the vapour phase when equilibrium is established?
(Given, vapour pressure of H_2O at 300 K is 3170 Pa; $R = 8.314$ JK^{-1} mol^{-1}) → AIEEE 2010

(a) 5.56×10^{-3} mol (b) 1.53×10^{-2} mol
(c) 4.46×10^{-2} mol (d) 1.27×10^{-3} mol

16 The compressibility factor for a real gas at high pressure is → AIEEE 2012

(a) $1 + \frac{RT}{pb}$ (b) 1
(c) $1 + \frac{pb}{RT}$ (d) $1 - \frac{pb}{RT}$

17 The compressibility of a gas is less than unity at STP. Therefore,

(a) $V_m > 22.4$ L (b) $V_m < 22.4$ L
(c) $V_m = 22.4$ L (d) $V_m = 44.8$ L

18 If Z is a compressibility factor, van der Waals' equation at low pressure can be written as → JEE Main 2014

(a) $Z = 1 + \frac{RT}{pb}$ (b) $Z = 1 - \frac{a}{VRT}$
(c) $Z = 1 - \frac{pb}{RT}$ (d) $Z = 1 + \frac{pb}{RT}$

19 In van der Waals' equation of state for a non-ideal gas, the term that accounts for intermolecular force is

(a) $(V - b)$ (b) $(RT)^{-1}$
(c) $\left(p + \frac{a}{V^2}\right)$ (d) RT

20 For one mole of a van der Waals' gas when $b = 0$ and $T = 300$ K, the pV vs $\frac{1}{V}$ plot is shown below. The value of the van der Waals' constant a (atm L mol^{-2}) is → AIEEE 2012

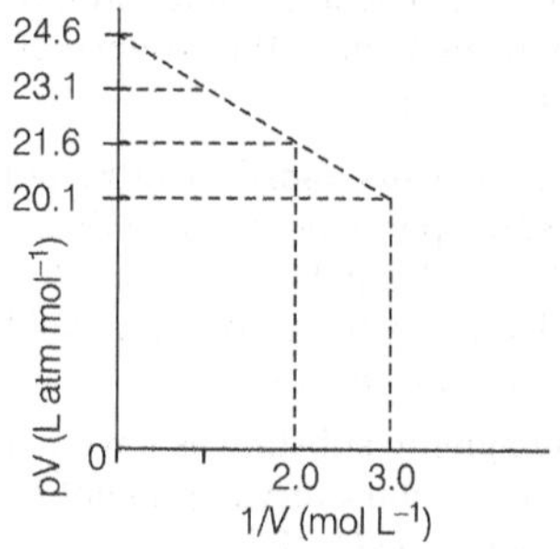

(a) 1.0 (b) 4.5
(c) 1.5 (d) 3.0

21 a and b are van der Waals' constants for gases. Chlorine is more easily liquefied than ethane because → AIEEE 2011

(a) a and b for $Cl_2 >$ a and b for C_2H_6
(b) a and b for $Cl_2 <$ a and b for C_2H_6
(c) a for $Cl_2 <$ a for C_2H_6 but b for $Cl_2 >$ b for C_2H_6
(d) a for $Cl_2 >$ a for C_2H_6 but b for $Cl_2 <$ b for C_2H_6

22 The value of compression factor, Z for critical constants is

(a) $\frac{1}{2}$ (b) $\frac{3}{4}$
(c) $\frac{2}{3}$ (d) $\frac{3}{8}$

23 Which of the following exists as crystals in the solid state? → JEE Main 2013

(a) Iodine (b) Silicon
(c) Sulphur (d) Phosphorus

24 The number of hexagonal faces that are present in truncated octahedron is → AIEEE 2011

(a) 2 (b) 4 (c) 6 (d) 8

25 The first order diffraction of X-rays from a certain set of crystal planes occur at an angle of 11.8° from the planes. If the planes are 0.281 nm apart, what is the wavelength of X-rays?

(a) 0.281 nm (b) 0.2044 nm
(c) 0.1149 nm (d) 0.1180 nm

26 In diamond, the coordination number of carbon is

(a) four and its unit cell has eight carbon atoms
(b) four and its unit cell has six carbon atoms
(c) six and its unit cell has four carbon atoms
(d) four and its unit cell has four carbon atoms

27 Sodium metal crystallises in a body centred cubic lattice with a unit cell edge of 4.29 Å. The radius of sodium atom is approximately → JEE Main 2015

(a) 1.86 Å (b) 3.22 Å
(c) 5.72 Å (d) 0.93 Å

28 Lithium forms body centred cubic structure. The length of the side of its unit cell is 351 pm. Atomic radius of the lithium will be → AIEEE 2012

(a) 75 pm (b) 300 pm (c) 240 pm (d) 152 pm

29 CsCl crystallises in body centred cubic lattice. If '*a*' its edge length, then which of the following expressions is correct? → JEE Main 2014

(a) $r_{Cs^+} + r_{Cl^-} = 3a$ (b) $r_{Cs^+} + r_{Cl^-} = \frac{3a}{2}$

(c) $r_{Cs^+} + r_{Cl^-} = \frac{\sqrt{3}}{2}a$ (d) $r_{Cs^+} + r_{Cl^-} = \sqrt{3a}$

30 In a face centred cubic lattice, atom *A* occupies the corner positions and atom *B* occupies the face centre positions. If one atom of *B* is missing from one of the face centred points, the formula of the compound is → AIEEE 2010

(a) A_2B (b) AB_2 (c) A_2B_2 (d) A_2B_5

31 An element having an atomic radius of 0.14 nm crystallises in an fcc unit cell. What is the length of a side of the cell? → JEE Main 2013

(a) 0.56 nm (b) 0.24 nm (c) 0.96 nm (d) 0.4 nm

32 If the distance between Na^+ and Cl^- ion in NaCl crystal is '*a*' pm. What is the length of the cell edge? → JEE Main 2013

(a) 4 *a* pm (b) $\frac{a}{4}$ pm
(c) 2 *a* pm (d) $\frac{a}{2}$ pm

33 A metal crystallises in a face centred cubic structure. If the edge length of its unit cell is '*a*' the closest approach between two atoms in metallic crystal will be → JEE Main 2017

(a) $2a$ (b) $2\sqrt{2}a$ (c) $\sqrt{2}a$ (d) $\frac{a}{\sqrt{2}}$

34 How many unit cells are present in a cube shaped ideal crystal of NaCl of mass 1.00 g?

(a) 2.57×10^{21} (b) 5.14×10^{21}
(c) 1.28×10^{21} (d) 1.71×10^{21}

35 Iron crystallises in a bcc system with a lattice parameter of 2.861 Å. Calculate the density of iron in the bcc system (atomic weight of Fe = 56, $N_A = 6.02 \times 10^{23}\ mol^{-1}$)

(a) $7.94\ g\ mL^{-1}$ (b) $8.96\ g\ mL\ g^{-1}$
(c) $2.78\ g\ mL^{-1}$ (d) $6.72\ g\ mL^{-1}$

36 Calcium crystallises in a face centred cubic unit cell with $a = 0.556$ nm. Calculate the density, if it contained 0.1% Schottky defects.

(a) $1.5463\ g/cm^3$ (b) $1.4962\ g/cm^3$
(c) $1.5448\ g/cm^3$ (d) $1.5943\ g/cm^3$

37 To get *n*-type semiconductor from silicon, it should be doped with a substance with valence

(a) 1 (b) 2
(c) 3 (d) 5

38 Which type of 'defect' has the presence of cations in the interstitial sites? → JEE Main 2018

(a) Schottky defect (b) Vacancy defect
(c) Frenkel defect (d) Metal deficiency defect

39 Which of the following compounds is metallic and ferromagnetic? → JEE Main 2016

(a) CrO_2 (b) VO_2
(c) MnO_2 (d) TiO_2

Direction (Q. Nos. 40-43) *In the following questions Assertion (A) followed by a Reason (R) is given. Choose the correct answer out of the following choices.*

(a) Both A and R are true and R is correct explanation of A
(b) Both A and R are true but R is not correct explanation of A
(c) A is true but R is false
(d) Both A and R are false

40 **Assertion** (A) Average speed of molecules, if a gas in a container moving only in one direction, will be zero.

Reason (R) The molecules of gas are not collected in one direction.

41 **Assertion** (A) The temperature at which vapour pressure of a liquid is equal to the external pressure is called boiling temperature

Reason (R) At high altitude atmospheric pressure is high.

42 **Assertion** (A) Liquids tends to have maximum number of molecules at their surface.

Reason (R) Small liquid drops have spherical shape.

43 **Assertion** (A) Solids containing *F*-centres are paramagnetic.

Reason (R) *F*-centres solids possess holes occupied by unpaired electrons.

DAY PRACTICE SESSION 2

PROGRESSIVE QUESTIONS EXERCISE

1 A mixture of $NH_3(g)$ and $N_2H_4(g)$ is placed in a sealed container at 300 K. The total pressure is 0.5 atm. When the container is heated to 1200 K, both substances decompose completely according to the equations :

$$2NH_3(g) \longrightarrow N_2(g) + 3H_2(g)$$

and $$N_2H_4(g) \longrightarrow N_2(g) + 2H_2(g)$$

After decomposition, the total pressure at 1200 K is found to be 4.5 atm. What is the per cent of $N_2H_4(g)$ in the original mixture? (assuming ideal behaviour)

(a) 35% (b) 40% (c) 75% (d) 25%

2 For gaseous state, if most probable speed is denoted by C^*, average speed by $\bar{C}$ and mean square speed by C, then for a large number of molecules, the ratio of these speeds are :

(a) $C^*: \bar{C}: C = 1.128 : 1.225 : 1$ (b) $C^*:\bar{C}:C = 1 : 1.128 : 1.225$
(c) $C^*:\bar{C}:C = 1 : 1.125 : 1.128$ (d) $C^*:\bar{C}:C = 1.225 : 1.128 : 1$

3 When 3.2 g sulphur is vaporised at 450°C and 723 mm Hg pressure, the vapours occupy a volume of 780 mL. What is the molecular formula of S vapours?

(a) S_2 (b) S_4 (c) S_6 (d) S_8

4 What is the difference in the density of dry air at 1 atm and 25°C and moist air with 50% relative humidity under the same condition? The vapour pressure of water at 25°C is 23.7 torr and dry air has 75.5% N_2 and 24.5% O_2.

(a) 0.005 kg m^{-3} (b) 0.007 kg m^{-3}
(c) 0.01 kg m^{-3} (d) 0.05 kg m^{-3}

5 In a spinel structure, oxide ions are cubical closed packed whereas 1/8th of tetrahedral voids are occupied by *A* ions and 1/2 of octahedral voids are occupied by *B* ions. The charge present on *A* and *B* ions will be

(a) A^{2+}, B^{3+} (b) A^{3+}, B^{2+} (c) A^{4+}, B^{2-} (d) A^{-1}, B^{2-}

6 The given graph represents the variation of Z (compressibility factor $= \frac{pV}{nRT}$) *versus* p, for three real gases *A*, *B* and *C*. Identify the only incorrect statement.

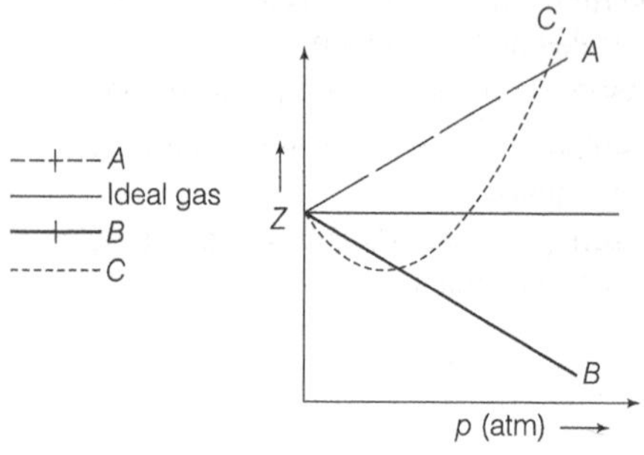

(a) For the gas A, $a = 0$ and its dependence on p is linear at all pressures
(b) For the gas B, $b = 0$ and its dependence on p is linear at all temperatures
(c) For the gas C, which is typical real gas for which neither $a = 0$ nor $b = 0$, by knowing the minima and the point of intersection, with $Z = 1$, a and b can be calculated
(d) At high pressure, the slope is positive for all real gases

7 At 100°C and 1 atm, if the density of liquid water is 1.0 g cm^{-3} and that of water vapour is 0.0006 g cm^{-3}, the volume occupied by water molecules in 1 L of steam at that temperature is

(a) 6 cm^3 (b) 60 cm^3 (c) 0.6 cm^3 (d) 0.06 cm^3

8 Experimentally, it was found that a metal oxide has formula $M_{0.98}O$. Metal M is present as M^{2+} and M^{3+}. It oxide fraction of the metal which exists as M^{3+} would be

(a) 4.08% (b) 6.05% (c) 5.08% (d) 7.01%

9 Which of the following order of root mean square speed of different gases at the same temperature is true?

(a) $(\mu_{rms})H_2 > (\mu_{rms})CH_4 > (\mu_{rms})NH_3 > (\mu_{rms})CO_2$
(b) $(\mu_{rms})H_2 < (\mu_{rms})CH_4 < (\mu_{rms})NH_3 < (\mu_{rms})CO_2$
(c) $(\mu_{rms})H_2 < (\mu_{rms})CH_4 > (\mu_{rms})NH_3 > (\mu_{rms})CO_2$
(d) $(\mu_{rms})H_2 > (\mu_{rms})CH_4 < (\mu_{rms})NH_3 < (\mu_{rms})CO_2$

10 The behaviour of a real gas is usually depicted by plotting compression factor (Z) *versus* p at a constant temperature. At low temperature and low pressure, Z is usually less than one. This fact can be explained by van der Waals' equation when

(a) the constant a is negligible and not b
(b) the constant b is negligible and not a
(c) both the constant a and b are negligible
(d) both the constant a and b are not negligible

11 When a capillary tube of diameter 0.4 mm is dipped in a liquid having density 800 kg m^{-3}, then the height of liquid in the capillary tube rises to 4 cm. The surface tension of liquid is ($g = 9.8$ m / s^2).

(a) 1.6×10^{-2} Nm^{-1} (b) 5.6×10^{-2} Nm^{-1}
(c) 6.3×10^{-2} Nm^{-1} (d) 7.3×10^{-2} Nm^{-1}

12 When a sample of gas is compressed at constant temperature from 15 atm to 6 atm, its volume changes from 76 cm^3 to 20.5 cm^3. Which of the following statements are possible explanations of this behaviour?

I. The gas behaves non-ideally.
II. The gas dimerises.
III. The gas is absorbed into the vessel walls.

(a) I, II and III (b) I and II only
(c) II and III (d) Only I

13 For a crystal, the angle of diffraction (2θ) is 90° and the second order line has a d value of 2.28 Å. The wavelength (in Å) of X-rays used for Bragg's diffraction is

(a) 1.612 (b) 2.00
(c) 2.28 (d) 4.00

14 How many unit cells are present in a cube shaped ideal crystal of NaCl of mass 1.00 g? (Atomic mass of Na = 23, Cl = 35.5)

(a) 2.57×10^{21} (b) 5.14×10^{21}
(c) 1.28×10^{21} (d) 1.71×10^{21}

15 Schottky defects occur mainly in electrovalent compounds where

(a) positive ions and negative ions are of different size
(b) positive ions and negative ions are of same size
(c) positive ions are small and negative ions are big
(d) positive ions are and negative ions are small

ANSWERS

SESSION 1	**1** (d)	**2** (c)	**3** (c)	**4** (d)	**5** (a)	**6** (a)	**7** (a)	**8** (b)	**9** (b)	**10** (a)
	11 (b)	**12** (c)	**13** (c)	**14** (c)	**15** (d)	**16** (c)	**17** (b)	**18** (b)	**19** (c)	**20** (c)
	21 (d)	**22** (d)	**23** (b)	**24** (d)	**25** (c)	**26** (a)	**27** (a)	**28** (d)	**29** (c)	**30** (d)
	31 (d)	**32** (c)	**33** (d)	**34** (a)	**35** (a)	**36** (c)	**37** (d)	**38** (c)	**39** (a)	**40** (a)
	41 (c)	**42** (d)	**43** (a)							
SESSION 2	**1** (d)	**2** (b)	**3** (d)	**4** (b)	**5** (a)	**6** (c)	**7** (c)	**8** (a)	**9** (a)	**10** (b)
	11 (c)	**12** (d)	**13** (a)	**14** (a)	**15** (b)					

Hints and Explanations

SESSION 1

1 Ion-ion interaction is dependent on the square of distance, i.e. ion-ion interaction $\propto \frac{1}{r^2}$.

Similarly, ion-dipole interaction $\propto \frac{1}{r^2}$.

London forces $\propto \frac{1}{r^6}$ and dipole-dipole interactions [H-bonds] $\propto \frac{1}{r^3}$. Thus, H-bonds being dipole-dipole interactions is dependent on the inverse cube of distance between the molecules.

2 At constant temperature,

$$p_1V_1 = p_2V_2 \text{ (Boyle's law)}$$
$$380 \times 7 = 760 \times V_2$$
$$V_2 = \frac{380 \times 7}{760} = 3.5 \text{ L}$$

3 By using Boyle's law,

$$p \times 8\,V = p_1 \times V;\ 8p = p_1$$

p_1 = atmosphere pressure + pressure due to water in lake

∴ Pressure of water in lake = $7p$

Since, p = pressure exerted by 10 m of water, the depth of the lake = 70 m.

4 $n_{C_2H_6} = \frac{w}{30}, n_{(H_2)} = \frac{w}{2}$

$$p_{H_2} = \frac{w/2}{\frac{w}{2} + \frac{w}{30}} \cdot p = \frac{1}{1 + \frac{1}{15}} = \frac{15}{16} \cdot p$$

Hence, (d) option is correct.

5 Let a cm from HCl end, white fumes of NH_4Cl are noticed.

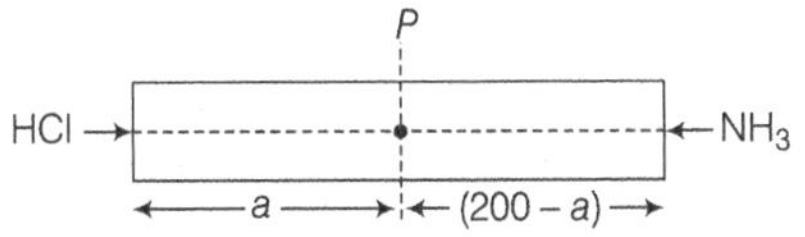

From Graham's law; $\frac{r_{HCl}}{r_{NH_3}} = \sqrt{\frac{M_{NH_3}}{M_{HCl}}}$

or $$\frac{a}{t} \times \frac{t}{(200 - a)} = \sqrt{\frac{17}{36.5}}$$

[∵ Time of diffusion in tube for both is same.]

∴ $a = 81.1$ cm from HCl end.

6 Volume of O_2 diffused $= \frac{22400 \times 0.48}{32} = 336$ mL

Let the volume of CO_2 diffused be x mL.

Rate of diffusion of $O_2 = \frac{336}{1200}$ mL s^{-1}

Rate of diffusion of $CO_2 = \frac{x}{1200}$ mL s^{-1}

$$\frac{r_{O_2}}{r_{CO_2}} = \frac{V_{O_2}/t}{V_{CO_2}/t} = \sqrt{\frac{M_{CO_2}}{M_{O_2}}}$$

or $$\frac{\frac{336}{1200}}{\frac{x}{1200}} = \sqrt{\frac{44}{32}}$$

∴ $x = 286.6$ mL

7 According to Graham's law of diffusion,

$$\frac{r_1}{r_2} = \sqrt{\frac{M_2}{M_1}}$$

$$\therefore \quad \frac{r_{235_{UF_6}}}{r_{238_{UF_6}}} = \sqrt{\frac{352}{349}} = 1.0043$$

8 For H_2 : $p \times V = \frac{w}{2} \times R \times 373$

For He : $p \times \frac{V}{2} = \frac{3.5}{4} \times R \times 298$

Hence, $w_{H_2} = 2.8$ g

9 Average kinetic energy depends only on temperature and does not depend upon the nature of gas.

10 $u_{rms} = \sqrt{\frac{3pV}{M}}$

$pV = nKT$ (K = Boltzmann's constant)

For a molecule, $n = 1$

$pV = KT \Rightarrow u_{rms} = \sqrt{\frac{3KT}{M}}$

Kinetic energy $(E) = \frac{3}{2}KT$ or $KT = \frac{2}{3}E$

$$u_{rms} = \sqrt{\frac{3 \times \frac{2}{3}E}{M}} = \sqrt{\frac{2E}{M}}$$

11 $\frac{u_{rms}\ (SO_2)}{u_{rms}\ (O_2)} = \sqrt{\frac{T\ (SO_2)}{M\ (SO_2)} \times \frac{M\ (O_2)}{T\ (O_2)}}$

i.e. $1 = \sqrt{\frac{T\ (SO_2)}{64} \times \frac{32}{303}}$ or,

$T(SO_2) = 606$ K

12 $u_{rms} = \sqrt{\frac{3RT}{M}} \therefore \frac{u_1}{u_2} = \sqrt{\frac{3RT_1}{M_1} \times \frac{M_2}{3RT_2}}$

$$= \sqrt{\frac{T_1 M_2}{T_2 M_1}} = \sqrt{\frac{50 \times 32}{800 \times 2}} = 1$$

13 Average KE $= \frac{3}{2}RT/N_0$

$$\text{KE} \propto T \Rightarrow \frac{(\text{KE})_{313}}{(\text{KE})_{293}} = \frac{313}{293}$$

14 Since, the external pressure is 1.0 atm, the gas pressure is also 1.0 atm as piston is movable. Out of this 1.0 atm, partial pressure due to known compound is 0.68 atm. Therefore, partial pressure of He

$= 1.00 - 0.68 = 0.32$ atm

$$\text{Volume} = \frac{n_{(He)}\ RT}{p_{(He)}}$$

$$= \frac{0.1 \times 0.082 \times 273}{0.32} = 7 \text{ L}$$

15 $n = \frac{pV}{RT} = \frac{3170 \times 10^{-3}}{8.314 \times 300}$

$= 1.27 \times 10^{-3}$ mol

16 van der Waals' equation for one mole of real gas is $\left(p + \frac{a}{V^2}\right)(V - b) = RT$

When pressure is high $p >> \frac{a}{V^2}$ such that $\left(p + \frac{a}{V^2}\right) = p$

Thus, $p(V - b) = RT \Rightarrow pV = RT + pb$

$\therefore$ Compressibility factor, $Z = \frac{pV}{RT}$

$$= \left(1 + \frac{pb}{RT}\right)$$

17 $Z = \frac{pV}{nRT}$, i.e. $\frac{pV}{nRT} < 1$ or $pV < nRT$

(p = 1 atm at STP)

$V < 1 \times 0.0821 \times 273$ or $V < 22.4$ L

18 To solve this problem, the stepwise approach is required, i.e.

(i) Write the van der Waals' equation, then apply the condition that at low pressure, volume become high, i.e. $V - b \approx V$

(ii) Now, calculate the value of compressibility factor (Z).

$[Z = pV/RT]$

According to van der Waals' equation, $\left(p + \frac{a}{V^2}\right)(V - b) = RT$

At low pressure,

$$\left(p + \frac{a}{V^2}\right)V = RT \Rightarrow pV + \frac{a}{V} = RT$$

$$pV = RT - \frac{a}{V}$$

Divide both side by RT,

$$\frac{pV}{RT} = 1 - \frac{a}{RTV}$$

19 Intermolecular forces are given by the term $\left(p + \frac{a}{V^2}\right)$.

20 $\left(p + \frac{a}{V^2}\right)V = RT \Rightarrow pV = RT - \frac{a}{V}$ ...(i)

Eq. (i) is a straight line equation between pV and $\frac{1}{V}$ where slope is $-a$.

Equating with slope of the straight line given in the graph

$$-a = \frac{20.1 - 21.6}{3 - 2} = -1.5, a = 1.5$$

21 van der Waas' constant a is due to force of attraction and b is due to finite size of molecules. Thus, greater the value of a and smaller the value of b, larger the liquefaction.

Thus, $a\ (Cl_2) > a\ (C_2H_6)$ and $b\ (Cl_2) < b\ (C_2H_6)$.

22 We know that, $p_c = \frac{a}{27b^2}$, $V_c = 3b$

and $T_c = \frac{8a}{27Rb}$

$$= \frac{p_c V_c}{T_c} = \frac{a \times 3b \times 27Rb}{27b^2 \times 8a} = \frac{3}{8}R$$

Compression factor, $Z = \frac{p_c V_c}{RT_c} = \frac{3}{8}$

23 Silicon exists as covalent crystal in solid state. (Network like structure, like diamond).

24 The truncated octahedron is the 14 faced Archimedean solid, with 14 total faces : 6 squares and 8 regular hexagonals. The truncated octahedron is formed by removing the six right square pyramids one from each point of a regular octahedron.

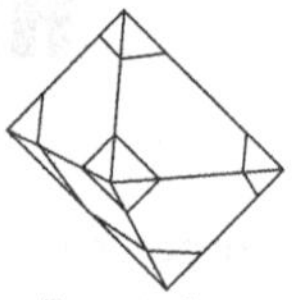

Truncated octahedron

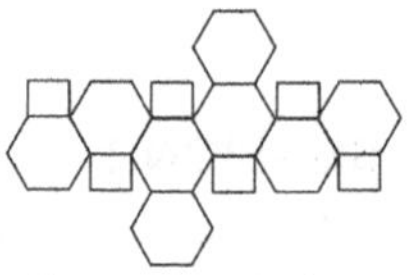

Truncated octahedron unfolded in two dimensions

25 Given, $n = 1, d = 0.281 \times 10^{-9}$ m,

$\theta = 11.8°$

$$\therefore \quad \lambda = \frac{2 \times 0.281 \times 10^{-9} \times \sin 11.8}{1}$$

$= 2 \times 0.281 \times 10^{-9} \times 0.2044$

$= 0.1149$ nm

26 The space lattice of diamond is fcc. The primitive basis has two identical atoms (0,0,0), $\left(\frac{1}{4}, \frac{1}{4}, \frac{1}{4}\right)$ associated with each point of the fcc lattice. Thus, the conventional unit cube contains eight atoms. Also, diamond is tetrahedral.

27

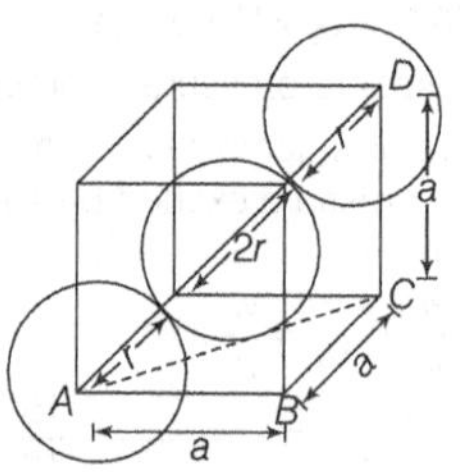

From this figure,

$(AC)^2 = (AB)^2 + (BC)^2 = a^2 + a^2 = 2a^2$

Also, $(AD)^2 = (AC)^2 + (DC)^2$

$(4r)^2 = 2a^2 + a^2, 16r^2 = 3a^2, r = \frac{\sqrt{3}}{4}a$

Now, when Na metal crystallises in bcc unit cell with unit cell edge length $a = 4.29 Å$

$\therefore \quad r = \frac{\sqrt{3}}{4} \times 4.29 Å = 1.86 Å$

28 Atomic radius of lithium $= \frac{\sqrt{3}}{4}a$

$= \frac{1.73}{4} \times 351 = 151.8 \text{ pm} \approx 152 \text{ pm}$

29 CsCl crystal has bcc structure.

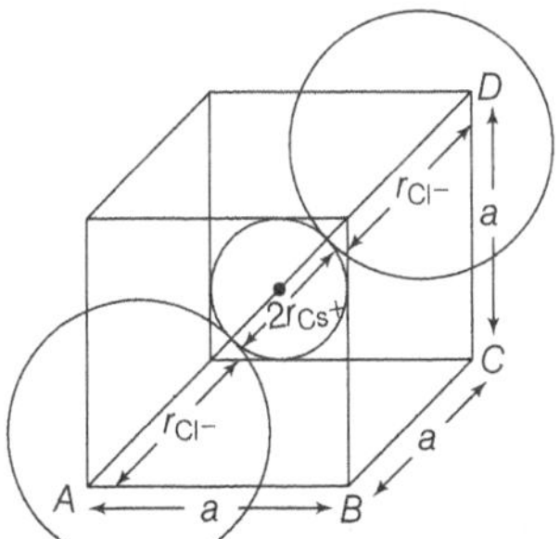

$AD^2 = AC^2 + CD^2 = (a\sqrt{2})^2 + a^2 = 3a^2$

$\Rightarrow AD = a\sqrt{3}$

Also, $AD = r_{Cl^-} + 2r_{Cs^+} + r_{Cl^-} = a\sqrt{3}$

$\Rightarrow \quad r_{Cs^+} + r_{Cl^-} = \frac{\sqrt{3}}{2}a$

30 Number of atoms (*A*) per unit cell

$= 8 \times \frac{1}{8} = 1$

Number of atoms (*B*) per unit cell

$= (6 - 1) \times \frac{1}{2} = \frac{5}{2}$

($\because$ One *B* atom is missing.)

Thus, formula is $A_1B_{5/2}$ or A_2B_5.

31 For fcc unit cell,

$r = \frac{\sqrt{2}}{4}a$

$\Rightarrow 0.14 = \frac{\sqrt{2}}{4}a$

$a = \frac{4 \times 0.14}{\sqrt{2}} = 0.396 \text{ nm} = 0.4 \text{ nm}$

32 Length of the edge of NaCl unit cell
$= 2 \times$ distance between Na^+ and Cl^-
$= 2 \times a = 2a$ pm

33 For fcc arrangement, $4r = \sqrt{2}a$
where, r = radius and a = edge length

$\therefore$ Closest distance $= 2r = \frac{\sqrt{2}a}{2} = \frac{a}{\sqrt{2}}$

34 Mass of one unit cell $= V \times d = a^3 \times d$,

Density $= \frac{Z \cdot M}{a^3 \cdot N_0}$

Mass of 1 unit cell $= \frac{a^3 \times Z \times M}{a^3 \times N_0} = \frac{ZM}{N_0}$

$= \frac{4 \times 58.5}{6.02 \times 10^{23}} = 38.87 \times 10^{-23}$ g

Number of unit cells in 1g

$= \frac{1}{38.87 \times 10^{-23}} = 2.57 \times 10^{21}$

35 $d = \frac{ZM}{N_A a^3}$ (for bcc, $Z = 2$)

$d_{Fe} = \frac{(2) \times 56.0 \text{ g mol}^{-1}}{(6.02 \times 10^{23} \text{mol}^{-1})(2.861 \times 10^{-8})^3 \text{ cm}^3}$

$= 7.94 \text{ g cm}^{-3}$

36 $d = \frac{ZM}{N_0 a^3} \Rightarrow Z = 4 \times \left(1 - \frac{0.1}{100}\right) = 3.996$

(because it contains 0.1% Schottky defect)

$d = \frac{3.996 \times 40}{6.02 \times 10^{23} \times (0.556 \times 10^{-7})^3}$

$= 1.5448 \text{ g/cm}^3$

37 For the preparation of *n*-type semiconductor from silicon, it should be doped with a substance of 5 valence electrons.

38 It is the "Frenkel defect" in which cations leave their original site and occupy interstitial site as shown below:

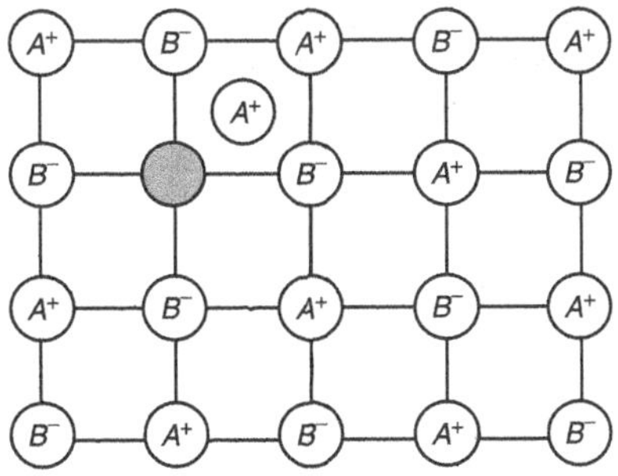

39 Only three elements iron (Fe), cobalt (Co) and nickel (Ni) show ferromagnetism at room temperature. CrO_2 is also a metallic and ferromagnetic compound which is used to make magnetic tapes for cassette recorders.

40 Both Assertion and Reason are true, Reason is a correct explanation for Assertion. Because motion of gas in one direction gives zero average speed.

41 Assertion is true but Reason is false. Because at high altitude, atmospheric pressure is low.

42 Both Assertion and Reason are false. Liquids tends to possess minimum number of molecules on the surface due to surface tension.

43 Both Assertion and Reason are true. Reason is a correct explanation for Assertion. Because *F*-centre contains unpaired electron and show paramagnetic character.

SESSION 2

1 Let the moles of NH_3 and N_2H_4 in original mixture are n_1 and n_2 respectively.

From given decomposition reactions :

2 moles of NH_3 produce total 4 moles of N_2 and H_2.

1 mole of N_2H_4 produces total 3 moles of N_2 and H_2.

Hence, total moles after decomposition $= 2n_1 + 3n_2$

Ideal gas equation : $pV = nRT$

Now, (0.5 atm) $V = (n_1 + n_2) R$ (300 K) ...(i)

and (4.5 atm)

$V = (2n_1 + 3n_2) R$ (1200 K) ...(ii)

On dividing the two equations and rearranging

$\frac{2n_1 + 3n_2}{n_1 + n_2} = \frac{9}{4}$ or $\frac{2(n_1 + n_2) + n_2}{n_1 + n_2} = \frac{9}{4}$

or $\frac{n_2}{n_1 + n_2} = \frac{9}{4} - 2 = \frac{1}{4}$

$\therefore$ % of N_2H_4 is $\frac{n_2}{n_1 + n_2} \times 100 = 25\%$

or $\frac{1}{4} \times 100 = 25\%$

2 $C^* = \sqrt{\frac{2RT}{M}}, \ \bar{C} = \sqrt{\frac{8RT}{\pi M}}, \ C = \sqrt{\frac{3RT}{\pi M}}$

$\therefore \quad C^* : \bar{C} : C = 1 : 1.128 : 1.225$

3 Given, $p = (723/760)$ atm, $w = 3.2$ g

$T = 273 + 450 = 723$ K

$\Rightarrow \quad V = (780/1000)$ L

$\because \quad pV = \frac{w}{M}RT$

$\therefore M = \frac{3.2 \times 0.0821 \times 760 \times 1000 \times 723}{723 \times 780}$

$= 256$

$\because$ 32 g sulphur has 1 atom of S.

$\therefore$ 256 g S has 256/32 = 8 atoms of S

$\therefore$ Molecular formula $= S_8$

4 $M_{(\text{dry air})}$

$$= \frac{M_1(O_2)\, x_1(\%\text{ of } O_2) + M_2\,(N_2)\, x_2\,(\%\text{ of } N_2)}{x_1 + x_2}$$

$$= \frac{32 \times 24.5 + 28 \times 75.5}{100} = 28.98 \text{ g mol}^{-1}$$

$$d_{(\text{dry air})} = \frac{pM_{(\text{air})}}{RT} = \frac{1 \times 28.98}{0.0821 \times 298.15}$$

$$= 1.184 \text{ g L}^{-1} = 1.184 \text{ kgm}^{-3}$$

Relative humidity (50 %)

$$= \frac{\text{partial pressure of } H_2O \text{ in air}}{\text{vapour pressure of } H_2O}$$

$$p\,(H_2O) = 0.50 \times 23.7 \text{ torr}$$

$$= 11.85 \text{ torr} = \frac{11.85}{760} \text{ atm}$$

$$= 0.0156 \text{ atm}$$

% of H_2O vapour in air

$$= \frac{0.0156 \times 100}{1} = 1.56\%$$

% of N_2 and O_2 in air

$$= 100 - 1.56 = 98.44\%$$

$M_{(\text{wet air})}$

$$= \frac{28.98 \times 98.44 \text{ (air)} + 18 \times 1.56 \text{ (water vapour)}}{100}$$

$$= 28.81 \text{ g mol}^{-1}$$

$$\therefore d_{(\text{wet air})} = \frac{pM_{(\text{wet air})}}{RT}$$

$$= \frac{1 \times 28.81}{0.0821 \times 298.15}$$

$$= 1.177 \text{ g L}^{-1}$$

$$= 1.177 \text{ kg m}^{-3}$$

Difference

$$= 1.184 - 1.177$$

$$= 0.007 \text{ kg m}^{-3}$$

5 In cubic closed packing, there will be four oxide ions;

$8 \times \frac{1}{8} =$ one A ion and $4 \times \frac{1}{2} = 2B$ ions.

Hence, formula will be $\overset{+2}{A}\,\overset{+3}{B_2}\,\overset{-2}{O_4}$.

6 van der Waals' equation is

$$\left(p + \frac{a}{V^2}\right)(V - b) = RT$$

When $a = 0$, $p\,(V - b) = RT$

or $pV = RT + pb$

or $\frac{pV}{RT}$, i.e. $Z = 1 + \frac{pb}{RT}$.

Thus, $Z > 1$ and increases with increase of pressure. Hence, (a) is correct.

When $b = 0$, $\left(p + \frac{a}{V^2}\right) V = RT$

or $pV + \frac{a}{V} = RT$ or $\frac{pV}{RT}$, i.e. $Z = 1 - \frac{a}{VRT}$.

Thus, $Z < 1$ and decreases with increase of pressure. Hence, (b) is correct. (d) is also correct and only wrong statement is (c).

7 Mass of 1L steam

$= 1000 \times 0.0006 = 0.6$ g

$=$ mass of liquid water

Volume of water $= 0.6$ cc

$(d = 1 \text{ g cm}^{-3})$.

Same volume will be occupied by water molecules in steam.

8 Given, metal oxide $= M_{0.98}O$

Let x ions are of M in $+3$ states,

Therefore, $3x + (0.98 - x) \times 2 = 2$,

$x = 0.04$

$$\% \text{ of } M \text{ in } +3 \text{ states} = \frac{0.04}{0.98} \times 100$$

$$= 4.08\%$$

9 The correct order of root mean square speed of different gases at the same temperature is

$$(\mu_{rms})_{H_2} > (\mu_{rms})_{CH_4} > (\mu_{rms})_{NH_3} > (\mu_{rms})_{CO_2}$$

We know that,

$$\mu_{rms} \propto \frac{1}{M}$$

More the mass of the molecule, less is the μ_{rms}.

10 At low temperature and low pressure, Z is usually less than one. This fact can be explained by van der Waals' equation when the constant b is negligible and not a.

11 $T = \frac{rhdg}{2} = \frac{(4 \times 10^{-4}) \times (0.04) \times 800 \times 9.8}{2}$

$(r = 0.4 \text{ mm} = 4 \times 10^{-4} \text{ m})$

$$= 4 \times 10^{-4} \times 0.04 \times 400 \times 9.8$$

$$= 4 \times 4 \times 4 \times 98 \times 10^{-5}$$

$$= 6.3 \times 10^{-2} \text{ Nm}^{-1}$$

12 When a sample of gas is compressed at constant temperature from 15 atm to 6 atm, its volume changes from 76 cm^3 to 20.5 cm^3. This behaviour shows that gas behaves non-ideally.

13 Given, angle of diffraction $(2\theta) = 90°$,

$\theta = 45°$

Distance between two planes,

$d = 2.28$ Å, $n = 2$

Bragg's equation is $n\lambda = 2d \sin \theta$

$$2 \times \lambda = 2 \times 2.28 \times \sin 45°$$

$$\lambda = 1.612$$

14 Mass of one unit-cell (m) = volume $\times$ density

$$= a^3 \times d$$

$$= a^3 \times \frac{MZ}{N_0 a^3}$$

$$\frac{Mz}{N_0} \times m = \frac{58.5 \times 4}{6.02 \times 10^{23}} \text{ g}$$

$\therefore$ Number of unit cell in 1 g

$$= \frac{1}{m} = \frac{6.02 \times 10^{23}}{584 \times 4} = 2.57 \times 10^{21}$$

15 Schottky defects occur mainly in electrovalent compounds where positive ions and negative ions are of same size.

DAY THREE

Atomic Structure

Learning & Revision for the Day

- Subatomic Particles
- Atomic Models (Thomson and Rutherford)
- Developments Leading to the Bohr's Model of an Atom
- Bohr's Model of Hydrogen Atom
- Dual Behaviour of Matter
- Elementary Ideas of Quantum Mechanics
- Rules for Filling Electrons in Orbitals
- Quantum Numbers and their Significance

Subatomic Particles

A large number of subatomic particles have been discovered but among them only electron, proton and neutron are of great importance and hence, these three are called **fundamental particles.**

Various experiments that lead to the **discovery of fundamental particles** are as follows:

Discovery of Electron

Cathode rays were discovered by Sir J.J. Thomson. Cathode rays were a stream of fast moving negatively charged particles, called **electrons.** The specific charge is the ratio of charge to mass of an electron, i.e. e/m ratio of electron was found to be same for all gases. Its value is $= 1.758 \times 10^{11}$ C/kg.

Positive Rays or Canal Rays (Discovery of Proton)

Positive rays (canal rays) were discovered by Goldstein. These rays consist of positively charged particles, called **protons.** Unlike cathode rays, their e/m value depends upon the nature of gas taken in the tube. The e/m value is maximum for hydrogen gas.

Discovery of Neutrons

Neutrons are neutral particles and discovered by Chadwick. These are the heaviest particles of the atom.

Masses of electron, proton and neutron respectively, are 9.1×10^{-31} kg, 1.672×10^{-27} kg and 1.674×10^{-27} kg and their charges are 1.6×10^{-19}C, 1.6×10^{-19} C and zero respectively.

Different Types of Atomic Species

1. **Isotopes** Isotopes are the species with same atomic number but different mass numbers. e.g. ${}_1H^1$, ${}_1H^2$
2. **Isobars** Isobars are the species with same mass number but different atomic numbers. e.g. ${}_{18}Ar^{40}$, ${}_{19}K^{40}$
3. **Isotones** Isotones are the species having same number of neutrons. e.g. ${}_1H^3$, ${}_2He^4$

PREP MIRROR
Your Personal Preparation Indicator

- No. of Questions in Exercises (x)—
- No. of Questions Attempted (y)—
- No. of Correct Questions (z)—
  (Without referring Explanations)
- Accuracy Level ($z/y \times 100$)—
- Prep Level ($z/x \times 100$)—

In order to expect good rank in JEE, your Accuracy Level should be above 85 & Prep Level should be above 75.

4. **Isodiaphers** Isodiaphers are the species with same isotopic number. e.g. ${}_{19}K^{39}$, ${}_{9}F^{19}$

   Isotopic number = mass number − 2 × atomic number

Atomic Models
(Thomson and Rutherford)

The atomic model which describe the atomic structure are given below

1. Thomson Model of an Atom

Thomson model of an atom proposed that an atom is a sphere of positive charges uniformly distributed, with the electrons scattered as points throughout the sphere. This was also known as **plum pudding model**.

Limitations of Thomson Model

Thomson's model was able to explain the overall neutrality of the atom, but it could not satisfactorily explain the results of scattering.

2. Rutherford Atomic Model

Rutherford bombarded a thin foil of gold with high speed positively charged α-particles and made the following observations and conclusions :

- Most of the α-particles passed through the foil undeflected, it means that most of the space in atom is empty.
- Some of them were deflected, but only at small angles. This shows that there is something positively charged at the centre.
- A very few α-particles were deflected by nearly 180°, it means that the positively charged solid thing (called nucleus) is very small.
- The electrons present in empty space, around the nucleus, revolve around with a very high speed.

Lmitations of Rutherford Model

Rutherford's model cannot explain the stability of an atom. When a charge is subjected to acceleration around an opposite charge, it emits radiation continuously.

Developments Leading to the Bohr's Model of an Atom

Nature of electromagnetic radiations and spectrum play an important role in the development of Bohr's model.

Wave Nature of Electromagnetic Radiation

- James Maxwell explains the interaction between the charged bodies and the behaviour of electrical and magnetic fields on macroscopic level.
- He suggested that when electrically charged particle moves under acceleration, alternating electrical and magnetic fields are produced and transmitted.

These fields are transmitted in the form of waves called **electromagnetic waves.**

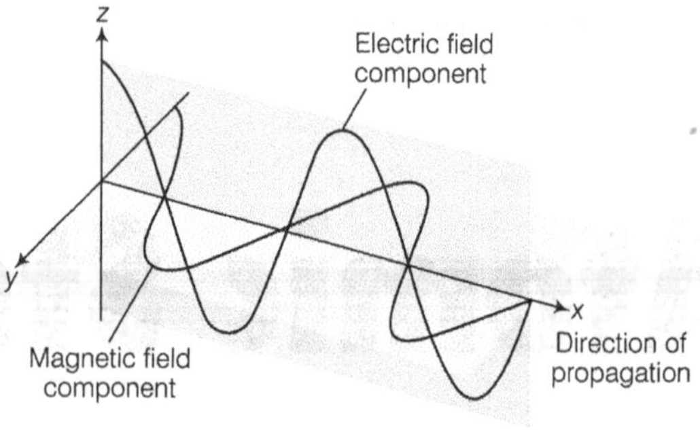

Simplified picture of electromagnetic waves

- A wave is a periodic disturbance in space or in a medium that involves elastic displacement of material particles or a periodic change in some physical quantities such as temperature, pressure, electric potential, electromagnetic field. Thus, wave motion represents propagation of a periodic disturbance carrying energy.

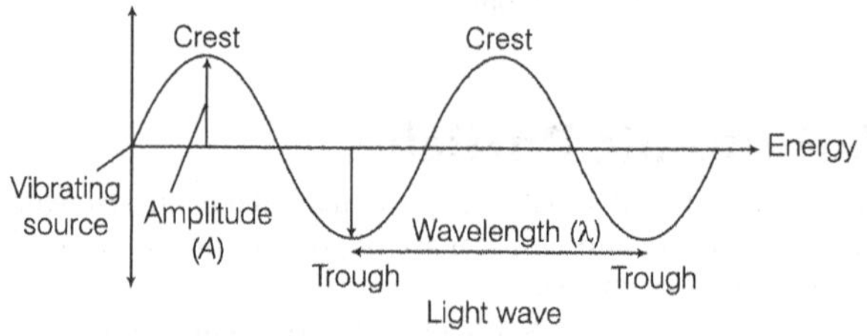

Wave motion

Characteristics of Wave

The waves are characterised in terms of their wavelength (λ), frequency (ν), velocity (c), amplitude and wave number ($\bar{\nu}$). These characteristics are related as

$$c = \lambda \times \nu$$

or $$\nu = \frac{c}{\lambda} \text{ and } \frac{1}{\lambda} = \bar{\nu}$$

∴ $$\nu = c\bar{\nu}$$

Particle Nature of Electromagnetic Radiation (Planck's Quantum Theory)

The radiant energy which is emitted or absorbed in the form of small discrete packets of energy known as **quantum** and in case of light, the quantum of energy is called **photon.**

$$E = h\nu \qquad (c = \nu\lambda)$$

$$E = \frac{hc}{\lambda}$$

where, h = Planck's constant = 6.63×10^{-34} Js

E = energy of photon or quantum

If n is the number of quanta of a particular frequency and E_T be the total energy, then $E_T = nh\nu$.

The energy possessed by one mole of quanta (or photon), i.e. Avogadro's number (N_0) of quanta, is called one Einstein of energy, i.e. 1 Einstein of energy.

$$E = N_0 h\nu = N_0 \frac{hc}{\lambda}$$

Photoelectric Effect

The phenomenon of ejection of electrons from a metal surface when a light of certain frequency strikes on its surface is called **photoelectric effect.**

- The minimum energy required to eject electrons from a surface is called **work function** (W_0)

$$\text{Work function } (W_0) = h\nu_0 = \frac{hc}{\lambda_0}$$

Here ν_0 is called **threshold frequency**, the minimum frequency of incident radiation to eject electrons from the metal surface.

λ_0 is **threshold wavelength**, the maximum wavelength in incident radiation below which, electrons can be emitted from metal surface.

- If a photon of energy E is incident of a metallic surface, then it is absorbed by the surface and electrons are emitted. The energy (E) is consumed in two ways; in during work function and providng kinetic energy to emitted electrons.

$$\therefore \quad E = W_0 + \text{KE}$$
$$h\nu = h\nu_0 + \text{KE}$$
$$\therefore \quad \text{KE} = h(\nu - \nu_0)$$

If velocity of ejected electrons is v, then $\frac{1}{2}mv^2 = h(\nu - \nu_0)$

Spectrum of Hydrogen Atom

The pictorial representation of arrangement of various types of EMR in their increasing order of wavelength (or decreasing order of frequency) is known as **spectrum**. The order of radiation would be:

γ-rays < X-rays < UV-rays < Visible < IR < Microwave < Radiowave

In vacuum, all types of electromagnetic radiation regardless of their wavelength, travel at the same speed, i.e. 3×10^8 ms^{-1}.

The two most common spectra are as follows:

(i) The arrangement of the radiation emitted by an atom of an element on the absorption of energy in the increasing order of the wavelengths or decreasing frequencies is called **emission spectrum**.

(ii) The arrangement of absorbed radiation by an atom of element on the emission of energy in the increasing wavelength is **absorption spectrum**.

If $\Delta E = E_{n_2} - E_{n_1}$; $n_2 > n_1 \Rightarrow$ emission spectra
If $\Delta E = E_{n_2} - E_{n_1}$; $n_2 < n_1 \Rightarrow$ absorption spectra
Here, n = energy levels

- When an electric discharge is passed through gaseous hydrogen, the H_2 molecules dissociate and the energetically excited hydrogen atoms produced, emit electromagnetic radiation of discrete frequencies.
- The line spectra of hydrogen lies in three regions of electromagnetic spectrum *viz*, infrared, visible and UV region. The set of lines in the visible region is known as **Balmer series**, that in ultraviolet region as **Lyman series** and there are three sets of lines in infrared region namely **Paschen, Brackett** and **Pfund series.**
- **Wave number** ($\bar{\nu}$) is defined as reciprocal of the wavelength, i.e. $\bar{\nu} = \frac{1}{\lambda} = RZ^2\left(\frac{1}{n_1^2} - \frac{1}{n_2^2}\right)$

where, λ = wavelength, R = Rydberg constant = 109677 cm^{-1}
For Lyman series $n_1 = 1$ and $n_2 = 2, 3, 4, \ldots$.
For Balmer series $n_1 = 2$, $n_2 = 3, 4, 5, \ldots$
For Paschen series $n_1 = 3$, $n_2 = 4, 5, 6, \ldots$ and so on.

- Wave number ($\bar{\nu}$) and frequency (ν) of radiations emitted when electron drops from n_2 to n_1 are obtained by the following expressions

$$\bar{\nu} = 109678\,[\text{cm}^{-1}] \times Z^2 \left[\frac{1}{n_1^2} - \frac{1}{n_2^2}\right] \text{ (For hydrogen, } Z = 1)$$

$$\nu = 3.289 \times 10^{15}\,(\text{s}^{-1}) \times Z^2 \left[\frac{1}{n_1^2} - \frac{1}{n_2^2}\right]$$

- The number of spectral lines in the spectrum when the electron comes from nth level to the ground level will be $\frac{n(n-1)}{2}$. The intensities of spectral lines decreases with increase in the value of n. e.g. the intensity of first Lyman series ($2 \rightarrow 1$) is greater than second line ($3 \rightarrow 1$) and so on.

Bohr's Model of Hydrogen Atom

Bohr's model is applicable only for one electron system like H, He^+, Li^{2+}etc.

Postulates of Bohr's Atomic Model

Its main postulates are :

- Electrons revolve around the nucleus only in stationary path, called energy levels or **orbits** or shells. Each energy level has a definite energy associated with it. As one moves away from the nucleus, the energy of the states increases.
- Angular momentum of an electron is an integral multiple of $\frac{h}{2\pi}$, i.e. $mvr = n\frac{h}{2\pi}$

where, m = mass of the electron, v = velocity of the electron, r = radius of the orbit, n = number of orbit

- Energy is emitted or absorbed, only when an electron jumps from higher energy level to lower energy level and *vice-versa*. $\Delta E = E_2 - E_1 = h\nu = \frac{hc}{\lambda}$

- The frequency of radiation absorbed or emitted when transition occurs between two stationary states that differ in energy by ΔE, $\nu = \frac{\Delta E}{\lambda}$

Derivation of the Relations for Energy of the Electron and Radii of Different Orbitals

From Bohr's model, energy and radius of an electron in n^{th} orbit are calculated as:

1. **Radius of *n*th Bohr Orbit**

$$r_n = \frac{n^2h^2}{4\pi^2me^2Z} = \frac{r_0n^2}{Z}$$

where, $r_0 = 52.9$ pm

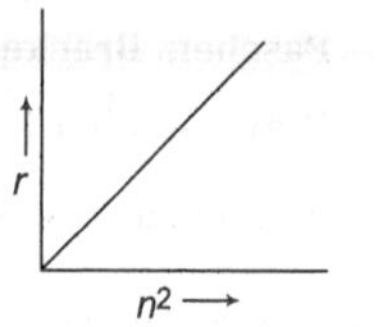

Variation of *r* with n^2

The above relation suggests that with an increase of energy levels (n), the radius of atom also increases.

Thus, $\frac{r_1}{r_2} = \frac{n_1^2}{n_2^2}$

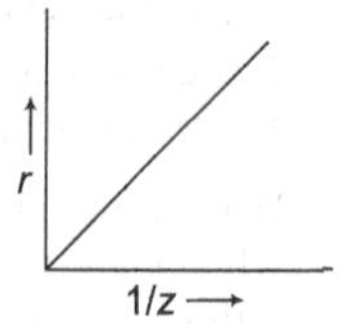

Plot of *r* vs 1/Z

This was derived by equating electrostatic force of attraction $\left(F = \frac{kZe^2}{r^2}\right)$ and centrifugal force $\left(\frac{mv^2}{r}\right)$ when both of them are equal and opposing to each other.

2. **Energy of an Electron in *n*th Orbit**

$$E_n = -\frac{2\pi^2mZ^2e^4}{n^2h^2}k^2 = -\frac{21.78 \times 10^{-19}Z^2}{n^2} \text{ J/atom}$$

$$\Delta E = -2.178 \times 10^{-18}\left(\frac{1}{n_1^2} - \frac{1}{n_2^2}\right)Z^2 \text{ J/atom}$$

where, n = number of shell, Z = atomic number

Thus was derived by considering sum of kinetic energy $\left(\frac{1}{2}mv^2\right)$ and potential energy $\left(\frac{-kZe^2}{r}\right)$ and then substituting in equation of radius.

NOTE • Velocity of an Electron in nth Bohr Orbit

$$v_n = \frac{2\pi e^2Z}{nh} = v_0 \times \frac{Z}{n}$$

where, $v_0 = 2.188 \times 10^8$ cm s^{-1}

Limitations of Bohr's Model

Bohr left the following facts unexplained.

- Fine structure of atom.
- Spectrum of multielectron system.
- Zeeman effect and Stark effect (i.e. splitting of spectral lines under the influence of magnetic and electric field respectively).
- Three dimensional existence of atom.
- Dual nature of electron.

Dual Behaviour of Matter

Concept of movement of an electron in an orbit was replaced by the concept of probability of finding electron in an orbital due to **de-Broglie concept** of dual nature of electron and **Heisenberg uncertainty principle.**

de-Broglie Principle (Dual Nature of Matter)

According to de-Broglie, moving particles behave like wave and matter both i.e. show dual behaviour. The wavelength of wave associated with moving particle is given by

$$\lambda = \frac{h}{mv} = \frac{h}{P}$$

where, λ = wavelength, v = velocity of particle
m = mass of particle, P = momentum

Heisenberg's Uncertainty Principle

It states that it is impossible to determine at any given instant, both the momentum and the position of subatomic particles like electron, simultaneously.

$$\Delta x \cdot \Delta p \geq \frac{h}{4\pi}$$

Since, $p = mv$

Hence $\Delta p = m\Delta v$

So, $(m \cdot \Delta v)\,\Delta x \geq \frac{h}{4\pi}$

where, Δx = uncertainty in position,
ΔP = uncertainty in momentum.

The effect of Heisenberg uncertainity principle is significant only for motion of microscopic objects and is negligible for that of macroscopic objects.

Elementary Ideas of Quantum Mechanics

On the basis of dual nature of matter and Heisenberg's uncertainty principle, Erwin Schrodinger developed a new branch of science, called quantum mechanics.

Quantum Mechanical Model of Atom

Quantum mechanical model of atom is the picture of the structure of atom, which energies from the application of the Schrodinger equation to atoms.

Schrödinger Wave Equation

- Schrödinger derived an equation for an electron which describes the wave motion of an electron along any of three axis, i.e. x, y and z.

This Schrödinger wave equation is given as

$$\frac{\partial^2\psi}{\partial x^2} + \frac{\partial^2\psi}{\partial y^2} + \frac{\partial^2\psi}{\partial z^2} + \frac{8\pi^2m}{h^2}(E - U)\psi = 0$$

where, ψ = wave function, m = mass of electron,
E = permissible total energy of electron,
U = potential energy of electron $= -\frac{Ze^2}{r}$,
h = Planck's constant.

- For H-atom, the equation is solved on the basis of $\hat{H}\psi = E\psi$, where, $\hat{H}$ is the total energy operator, called Hamiltonian; the sum of kinetic energy operator $(\hat{T})$ and potential energy operator $(\hat{V})$, is the total energy, E of the system,

$$\therefore \quad \hat{H} = \hat{T} + \hat{V}$$
$$(\hat{T} + \hat{V})\psi = E\psi$$

Important Features of the Quantum Mechanical Model of Atom

- The energy of electrons in atoms is quantised.
- The existence of quantised electronic energy levels is a direct consequence of the wave like properties of electrons and allowed solutions of Schrondinger wave equation.

Concept of Atomic Orbitals as One Electron Wave Function

The atomic orbitals or orbital wave functions can be represented by the product of two wave functions, radial and angular wave function.

Significance of ψ and ψ^2

- The orbital wave function ψ has no significance but ψ^2 measures the electron probability density at a point in an atom.
- Variation of ψ and ψ^2 with distance from nucleus for $1s$ and $2s$-orbitals.
- Wave function ψ and ψ^2 can be plotted against distance 'r' from the nucelus as:

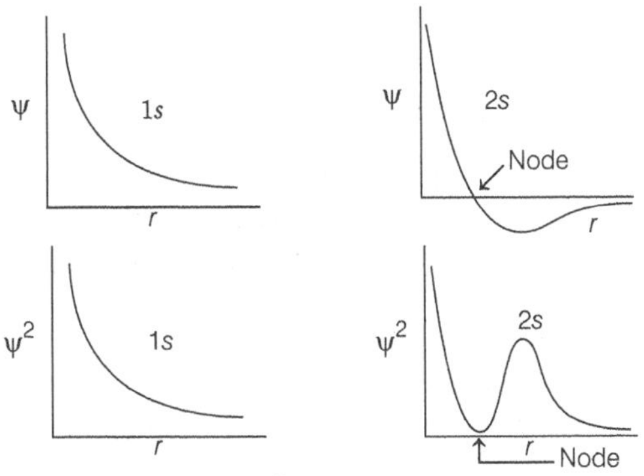

Plot of ψ and ψ^2 for 1s and 2s orbitals

- A node is a region of space, where probability of finding an electron is zero. Different nodes can be calculated as
  (i) $(n - l - 1)$ = radial nodes
  (ii) l = angular nodes
  (iii) $(n - 1)$ = total number of nodes

Rules for Filling Electrons in Orbitals

There are some rules for filling the electrons in orbitals are given below:

1. **Aufbau Principle** It states that "electrons are filled to the various orbitals, in their order of increasing energy starting with the orbital of lowest energy".

   As a working rule, a new electron enters in an empty orbital for which the value of $(n + l)$ is minimum, if the value of $(n + l)$ is same for two or more orbitals, the new electron enters in an orbital having lower value of 'n'.

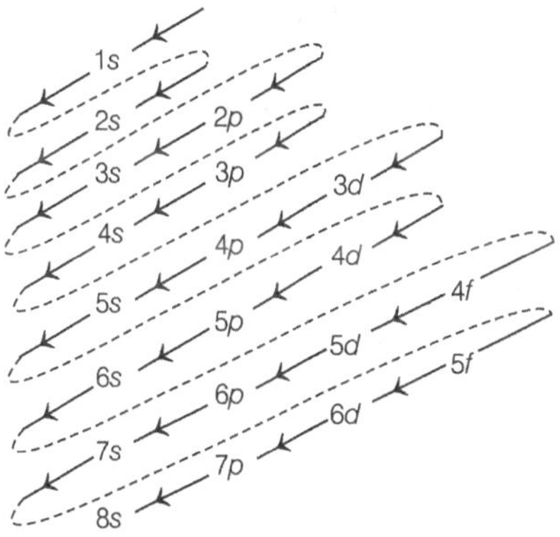

Order of filling of orbitals

   The energy of atomic orbitals for H-atom varies as :
   $1s < 2s = 2p < 3s = 3p = 3d < 4s = 4p = 4d = 4f$

2. **Hund's Rule of Maximum Multiplicity** It states that "Electrons never pair up until each orbital of a given sub-shell contains one electron or is singly occupied.

Electronic Configuration of Elements

The arrangement of electrons in various shells, sub-shells and orbitals in an atom is termed as **electronic configuration**. It is written as nl^x, where n = order of shell, l = sub-shell and x = number of electrons present.

e.g. $C(6) = 1s^2, 2s^2, 2p^2$
$Na(11) = 1s^2, 2s^2, 2p^6, 2s^1$

Extra Stability of Completely Filled one Half-Filled Orbitals

When a set of equivalent orbitals (degenerate orbitals) is either **fully filled** or **half-filled**, i.e. each containing one or a pair of electrons, the atom gain more stability. This effect is more dominant in d and f-sub-shells. Therefore, the outer shell configuration for Cr is $3d^5\,4s^1$ and for Cu is $3d^{10}\,4s^1$.

Quantum Numbers and their Significance

The term quantum number is used to identify the various energy and levels that are available to an electron.

There are four types of quantum numbers which are given below:

1. **Principal Quantum Number** (n) It gives an idea about the position and energy of an orbital. Principal quantum number also identifies the shell number.
2. **Azimuthal Quantum Number** (l) Represents the sub-shells (s, p, d, f) present in the main shell and angular momentum of the electron. It tells about the three dimensional shape of sub-shells. It is also called the orbital angular momentum.

   For values n, l can have n values ranging from 0 to $n-1$.

   Value for l: 0 1 2 3 4......

   s p d f g......
3. **Magnetic Quantum Number** (m_l) It gives information about the spatial orientation of the orbital with respect to standard set of coordinate axis.

   $$m = -l \text{ to } +l$$

   Number of sub-shells $= 2l + 1$

   Number of orbitals $= n^2$

   s-sub-shell contains only one orbital. p-sub-shell has three orbitals p_x, p_y, p_z (axis perpendicular to each other). d-sub-shell has 5 orbitals, i.e. $d_{xy}, d_{yz}, d_{xz}, d_{x^2-y^2}, d_{z^2}$. In f-sub-shell, there are seven orbitals.

   Maximum number of electrons in a sub-shell is equal to $2(2l + 1)$.
4. **Spin Quantum number** (m_s) These are distinguished by the two orientations of an electron in an orbital which can take values of $+\frac{1}{2}$ and $-\frac{1}{2}$. These are called two spin states of the electron and are generally represented by two arrows ↿(spin up) and ⇂(spin down) respectively. Maximum number of electrons in a main energy level $= 2n^2$ and total spin $= \pm\left(\frac{1}{2} \times n\right)$.

NOTE
- **Pauli's Exclusion Principle** states that "No two electrons in an atom can have the same set of all the four quantum numbers."
- The pairing of electrons will start in that p, d nd f orbitals with the entry of 4th, 6th and 8th electron.
- Number of sub-shells in main energy level $= n$
- Orbital angular momentum of electron in an orbital $= \left(\frac{h}{2\pi}\right)\sqrt{[l(l+1)]}$
- Spin angular momentum $= \left(\frac{h}{2\pi}\right)\sqrt{[s(s+1)]}$
- Shapes of s, p, d and f-orbitals
  - s-orbital — spherical
  - p-orbital — dumble
  - d-orbital — double-dumble
  - f-orbital — complicated

DAY PRACTICE SESSION 1

FOUNDATION QUESTIONS EXERCISE

1 The radii of an atom and atomic nucleus is of the order of
(a) 10^{-12} m and 10^{-10} m
(b) 10^{-10} cm and 10^{-8} cm
(c) 10^{-10} m and 10^{-15} m
(d) 10^{-15} m and 10^{-10} m

2 Atom consists of electrons, protons and neutrons. If the mass attributed to neutron is halved and that attributed to the electrons is doubled, the atomic mass of $_6C^{12}$ would be approximately
(a) same (b) doubled
(c) halved (d) reduced by 25

3 Many elements have non-integral atomic masses because
(a) their isotopes have same number of neutrons
(b) their isotopes have non-integral masses
(c) they exist as isotopes
(d) their constituents neutrons, protons and electrons combine to give fractional masses

4 Rutherford's α-particle scattering experiment eventually concluded that
(a) mass and energy are related
(b) neutrons are buried deep in the nucleus
(c) electrons occupy space around the nucleus
(d) the point of impact with matter can be precisely determined

5 A photon of light of wavelength 6000 Å has energy E. What will be the wavelength of photon of a light which has energy of photon $4E$?
(a) 1500 Å (b) 6000 Å (c) 2000 Å (d) 750 Å

6 Calculate the wavelength of a helium atom whose speed is equal to root mean square speed at 293 K.
(rms $= \sqrt{\frac{3RT}{M}}$)
(a) 6.51×10^{-5} nm (b) 7.38×10^{-5} nm
(c) 7.96×10^{-5} nm (d) None of these

7 The work function (ϕ) of some metals is listed below. The number of metals which will show photoelectric effect when light of 300 nm wavelength falls on the metals is

→ AIEEE 2011

Metal	ϕ (eV)
Li	2.4
Na	2.3
K	2.2
Mg	3.7
Cu	4.8
Ag	4.3
Fe	4.7
Pt	6.3
W	4.75

(a) 2 (b) 4 (c) 6 (d) 8

8 A 25 Watt bulb emits monochromatic yellow light of wavelength of 0.57 μm. What will be the rate of emission of quanta per second?

(a) 7.17×10^{19} (b) 0.717×10^{-19}
(c) 71.7×10^{19} (d) 7.17×10^{-19}

9 The critical wavelength for producing the photoelectric effect in tungsten metal is 2600 Å. What wavelength would be necessary to produce photoelectrons from tungsten having twice the kinetic energy of those produced at 2200 Å?

(a) 1800 Å (b) 1907 Å (c) 1926 Å (d) 2015 Å

10 Energy required to stop the ejection of electrons from Cu plate is 0.24 eV. If radiation of $\lambda = 253.7$ nm strikes the plate, the work function is

(a) 4.65 eV (b) 4.89 eV
(c) 4.24 eV (d) 3.0 eV

11 When an electron is excited from ground level to 5th orbit, the number of spectral lines obtained in Bohr spectrum of H-atom is

(a) 5 (b) 8 (c) 10 (d) 15

12 How many times does the electron go round the first Bohr's orbit of hydrogen in one second ?

(a) 0.657×10^{15} (b) 6.57×10^{15}
(c) 6.57×10^{10} (d) 65.7×10^{12}

13 Which hydrogen like species will have same radius as that of Bohr orbit of hydrogen atom ?

(a) $n = 2$, Li^{2+} (b) $n = 2$, Be^{3+} (c) $n = 2$, He^+ (d) $n = 3$, Li^{2+}

14 In Bohr's series of lines of hydrogen spectrum, the third line from the red end corresponds to which one of the following inter-orbit jumps of the electron for Bohr's orbit in an atom of hydrogen?

(a) $5 \rightarrow 2$ (b) $4 \rightarrow 1$ (c) $2 \rightarrow 5$ (d) $3 \rightarrow 2$

15 When the electron of a hydrogen atom jumps, from the $n = 4$ to the $n = 1$ state, the number of spectral lines emitted is

(a) 15 (b) 6 (c) 3 (d) 4

16 The wave number of first line of Balmer series of hydrogen is 15200 cm^{-1}. The wave number of first Balmer line of Li^{2+} ion is

(a) 15200 cm^{-1} (b) 60800 cm^{-1}
(c) 76000 cm^{-1} (d) 136800 cm^{-1}

17 The wave number of the first emission line in the Balmer series of H-spectrum is (R = Rydberg constant)

→ JEE Main (Online) 2013

(a) $\frac{5}{36}R$ (b) $\frac{9}{400}R$
(c) $\frac{7}{6}R$ (d) $\frac{3}{4}R$

18 If the shortest wavelength of H-atom in Lyman series is X, the longest wavelength in Balmer series of He^+ is

(a) $\frac{9X}{5}$ (b) $\frac{36X}{5}$
(c) $\frac{X}{4}$ (d) $\frac{5X}{9}$

19 If the speed of an electron in the Bohr's first orbit of hydrogen atom be x, speed of the electron in 3rd orbit is

(a) $x/9$ (b) $x/3$ (c) $3x$ (d) $9x$

20 The ionisation enthalpy of hydrogen atom is 1.312×10^6 J mol^{-1}. The energy require to excite the electrons in an atom from $n = 1$ to $n = 2$ is

(a) 7.56×10^5 J mol^{-1} (b) 9.84×10^5 J mol^{-1}
(c) 8.51×10^5 J mol^{-1} (d) 6.56×10^5 J mol^{-1}

21 Ionisation energy of He^+ is 19.6×10^{-18} J $atom^{-1}$. The energy of the first stationary state ($n = 1$) of Li^{2+} is

→ AIEEE 2010

(a) 4.41×10^{-16} J $atom^{-1}$ (b) -4.41×10^{-17} J $atom^{-1}$
(c) -2.2×10^{-15} J $atom^{-1}$ (d) 8.82×10^{-17} J $atom^{-1}$

22 The wavelength of a neutron with a translatory kinetic energy equal to kT at 300 K is

(a) 17.8 pm (b) 20.0 pm
(c) 200 pm (d) 178 pm

23 Energy of an electron is given by $E = -2.178 \times 10^{-18} \times \frac{Z^2}{n^2}$. Wavelength of light required to excite an electron in the hydrogen atom from level $n = 1$ to $n = 2$ will be

→ JEE Main (Online) 2013

(a) 2.816×10^{-7} m (b) 6.500×10^{-7} m
(c) 8.500×10^{-7} m (d) 1.214×10^{-7} m

24 A gas absorbs photon of 355 nm and emits at two wavelengths. If one of the emission is at 680 nm, the other is at

→ AIEEE 2011

(a) 1035 nm (b) 325 nm (c) 743 nm (d) 518 nm

25 The energy required to break one mole of Cl—Cl bonds in Cl_2 is 242 kJ mol^{-1}. The longest wavelength of light capable of breaking a single Cl—Cl bond is

→ AIEEE 2010

(a) 594 nm (b) 640 nm
(c) 700 nm (d) 494 nm

26 A photon of 300 nm is absorbed by a gas which then re-emits two photons. One re-emitted photon has wavelength of 496 nm. Wavelength of other re-emitted photon is

(a) 300 nm (b) 496 nm
(c) 759 nm (d) 550 nm

27 Which of the following is the energy of a possible excited state of hydrogen? → JEE Main 2015

(a) +13.6 eV (b) −6.8 eV
(c) −3.4 eV (d) +6.8 eV

28 The radius of the second Bohr orbit for hydrogen atom is (Planck's constant $(h) = 6.6262 \times 10^{-34}$ Js; mass of electron $= 9.1091 \times 10^{-31}$ kg ; charge of electron $(e) = 1.60210 \times 10^{-19}$ C; permitivity of vacuum $(\epsilon_0) = 8.854185 \times 10^{-12}\, kg^{-1} m^{-3} A^2$) → JEE Main 2017

(a) 1.65 Å (b) 4.76 Å
(c) 0.529 Å (d) 2.12 Å

29 The kinetic energy of an electron in the second Bohr orbit of a hydrogen atom is [a_0 is Bohr radius] → AIEEE 2011

(a) $\frac{h^2}{4\pi^2 m a_0^2}$ (b) $\frac{h^2}{16\pi^2 m a_0^2}$
(c) $\frac{h^2}{32\pi^2 m a_0^2}$ (d) $\frac{h^2}{64\pi^2 m a_0^2}$

30 The de-Broglie wavelength of a car of mass 1000 kg and velocity 36 km/h is ($h = 6.63 \times 10^{-34}$ Js) → JEE Main 2013

(a) 6.626×10^{-34} m (b) 6.626×10^{-38} m
(c) 6.626×10^{-31} m (d) 6.626×10^{-30} m

31 A stream of electrons from a heated filament was passed between two charged plates kept at a potential difference V esu. If e and m are charge and mass of an electron, respectively, then the value of h/λ (where, λ is wavelength associated with electron wave) is given by → JEE Main 2016

(a) $2\,meV$ (b) $\sqrt{meV}$
(c) $\sqrt{2meV}$ (d) meV

32 The uncertainty in the position of an electron moving with a velocity of 300 ms^{-1} alongwith an accuracy of 0.001% is

(a) 3.84×10^{-2} m (b) 5.76×10^{-2} m
(c) 1.93×10^{-2} m (d) 19.3×10^{-2} m

33 A dust particle has mass equal to 10^{-11} g, diameter 10^{-4} cm and velocity 10^{-4} cm s^{-1}. The error in measurement of velocity is 0.1%. What will be the uncertainty in its position ?

(a) 0.527×10^{-10} m (b) 5.27×10^{-9} cm
(c) 0.527×10^{-15} cm (d) 0.527×10^{-9} cm

34 The uncertainty involved in the measurement of velocity of an electron within a distance of 0.1Å is

(a) 5.79×10^8 ms^{-1}
(b) 5.79×10^5 ms^{-1}
(c) 5.79×10^6 ms^{-1}
(d) 5.79×10^7 ms^{-1}

35 In an atom, an electron is moving with a speed of 600 m/s with an accuracy of 0.005%. Certainty with which the position of the electron can be located is ($h = 6.6 \times 10^{-34}$ kg $m^2 s^{-1}$, mass of electron, $e_m = 9.1 \times 10^{-31}$ kg

(a) 1.52×10^{-4} m (b) 5.10×10^{-3} m
(c) 1.92×10^{-3} m (d) 3.84×10^{-3} m

36 Amongst the following set of quantum numbers, the impossible set is

	n	l	m	s		n	l	m	s
(a)	3	2	−3	1/2	(b)	4	0	0	1/2
(c)	5	3	0	−1/2	(d)	3	2	−2	1/2

37 In an atom how many orbital(s) will have the quantum numbers $n = 3, l = 2$ and $m_l = +2$? → JEE Main (Online) 2013

(a) 5 (b) 3 (c) 1 (d) 7

38 The correct set of four quantum numbers for the valence electrons of rubidium atom ($Z = 37$) is → JEE Main 2014

(a) $5, 0, 0, +\frac{1}{2}$ (b) $5, 1, 0, +\frac{1}{2}$
(c) $5, 1, 1, +\frac{1}{2}$ (d) $5, 0, 1, +\frac{1}{2}$

39 Which of the following is not possible for $4p$ or $3d$ electrons?

(a) $n = 3, l = 2, m = +1, s = +1/2$
(b) $n = 4, l = 1, m = 0, s = +1/2$
(c) $n = 3, l = 3, m = +3, s = +1/2$
(d) $n = 4, l = 1, m = -1, s = +1/2$

40 Given, (i) $n = 5, m_l = +1$

(ii) $n = 2, l = 1, m_l = -1, m_s = -1/2$

The maximum number of electron(s) in an atom that can have the quantum numbers as given in (i) and (ii) respectively are → JEE Main 2013

(a) 25 and 1 (b) 8 and 1 (c) 2 and 4 (d) 4 and 1

41 The electrons identified by quantum numbers n and l

(I) $n = 4, l = 1$ (II) $n = 4, l = 0$
(III) $n = 3, l = 2$ (IV) $n = 3, l = 1$

can be placed in the order of increasing energy as → AIEEE 2012

(a) (III) < (IV) < (II) < (I) (b) (IV) < (II) < (III) < (I)
(c) (II) < (IV) < (I) < (III) (d) (I) < (III) < (II) < (IV)

42 The orbital diagram in which Aufbau principle is violated is

(a)  (b) 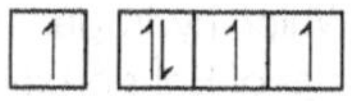

(c)  (d) 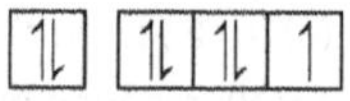

Direction (Q Nos. 43-45) *In the following questions assertion* (A) *followed by reason* (R) *is given. Choose the correct answer out of the following choices.*

(a) Both A and R are true and R is the correct explanation of A.
(b) Both A and R are true but R is not the correct explanation of A.
(c) A is true but R is false.
(d) Both A and R are false.

43 Assertion (A) All isotopes of a given element show the same type of chemical behaviour.

Reason (R) The chemical properties of an atom are controlled by the number of electrons in the atom.

44 Assertion (A) The path of an electron in an atom is clearly defined.

Reason (R) It is impossible to determine the exact position and exact momentum of an electron simultaneously.

45 Assertion (A) On heating a solid for a longer time, radiations become white and then blue as the temperature becomes very high.

Reason (R) Radiations emitted go from a lower frequency to higher frequency as the temperature increases.

DAY PRACTICE SESSION 2

PROGRESSIVE QUESTIONS EXERCISE

1 The atomic numbers of elements x, y and z are 19, 21 and 25 respectively. The number of electrons present in M-shell of these elements follow the order

(a) $z > x > y$ (b) $x > y > z$ (c) $z > y > x$ (d) $y > z > x$

2 The uncertainities in the velocities of two particles A and B are 0.05 and 0.02 m s^{-1} respectively. The mass of B is five times to that of mass A. What the ratio of uncertainities $\left(\frac{\Delta x_A}{\Delta x_B}\right)$ in their positions.

(a) 2 (b) 0.25 (c) 4 (d) 1

3 I_2 molecule dissociates into atoms after absorbing light of 4500 Å. If one quantum of energy is absorbed by each molecule, the KE of iodine atoms will be (BE of I_2 = 240 kJ / mol)

(a) 240×10^{-19} J (b) 0.216×10^{-19} J
(c) 2.16×10^{-19} J (d) 2.40×10^{-19} J

4 In a Bohr's model of atom when an electron in H-atom jumps from $n = 1$ to $n = 3$, how much energy will be emitted or absorbed?

(a) 2.15×10^{-11} erg/atom (b) 1.936×10^{-11} erg/atom
(c) 2.389×10^{-12} erg/atom (d) 0.239×10^{-10} erg/atom

5 The total mass of neutrons in 7 mg of C^{14} (assume mass of a neutron $= 1.675 \times 10^{-27}$kg) is

(a) 1.25×10^{-9} (b) 2.40×10^{-8}
(c) 4.03×10^{-6} (d) 5.36×10^{-7}

6 The frequency of light emitted for the transition $n = 4$ to $n = 2$ of He^+ is equal to the transition in H atom corresponding to which of the following?

(a) $n = 3$ to $n = 1$ (b) $n = 2$ to $n = 1$
(c) $n = 3$ to $n = 2$ (d) $n = 4$ to $n = 3$

7 To move an electron in one H-atom from the ground state to the second excited state, 12.084 eV are needed. How much energy is needed to cause 1 mole of H-atoms to undergo this transition ?

(a) 728 kJ/mol (b) 984 kJ/mol
(c) 1036 kJ/mol (d) 1164 kJ/mol

8 Which of the following statements in relation to the hydrogen atom is correct?

(a) 3s, 3p and 3d orbitals all have the same energy
(b) 3s and 3p orbitals are of lower energy than 3d-orbital
(c) 3p-orbital is lower in energy than 3d-orbital
(d)3s-orbital is lower in energy than 3p-orbital

9 The radiation is emitted when a hydrogen atom goes from a higher energy state to a lower energy state. The wavelength of one line in visible region of atomic spectrum of hydrogen is 6.63×10^{-7}m. Energy difference between the two states is

(a) 3.0×10^{-19} J (b) 1.0×10^{-18} J
(c) 5.0×10^{-10} J (d) 6.5×10^{-7} J

10 Given the set of quantum numbers for a multi-electron atom 2, 0, 0, $\frac{1}{2}$ and 2, 0, 0, $-\frac{1}{2}$. What is the next higher allowed set of n and l quantum numbers for this atom in its ground state?

(a) $n = 2, l = 0$ (b) $n = 2, l = 1$
(c) $n = 3, l = 0$ (d) $n = 3, l = 1$

11 In a multielectron atom, which of the following orbitals described by the three quantum numbers wil have the same energy in the absence of magnetic and electric fields?

(A) $n = 1, l = 0, m = 0$ (B) $n = 2, l = 0, m = 0$
(C) $n = 2, l = 1, m = 1$ (D) $n = 3, l = 2, m = 1$
(E) $n = 3, l = 2, m = 0$

(a) (D) and (E) (b) (C) and (D)
(c) (B) and (C) (d) (A) and (B)

12. Consider the following sets of quantum numbers.

	n	l	m	s
I.	3	0	0	+1/2
II.	2	2	1	+1/2
III.	4	3	−2	−1/2
IV.	1	0	−1	−1/2
V.	2	3	3	+1/2

Which of the following sets of the quantum number is not possible?

(a) II, III and IV (b) I, II, III and IV
(c) II, IV and V (d) I and III

13 For the electrons of oxygen atom, which of the following statements is correct?

(a) Z_{eff} for an electron in a 2*s*-orbital is the same as Z_{eff} for an electron in a 2*p*-orbital
(b) An electron in the 2*s*-orbital has the same energy as an electron in the 2*p*-orbital
(c) Z_{eff} for an electron in 1*s*-orbital is the same as Z_{eff} for an electron in a 2*s*-orbital
(d) The two electrons present in the 2*s*-orbital have spin quantum numbers m_s but of opposite sign

14 Which of the following statements about electromagnetic spectrum is not correct?

(a) Infrared radiations have larger wavelength than cosmic rays
(b) The frequency of microwaves is less than that of ultraviolet rays
(c) X-rays have larger wave number than microwaves
(d) The velocity of X-rays is more than that of microwaves

15 Consider the following statements.

I. $|\psi|^2$ is a measure of electron density at a point in an atom.
II. Radial probability function ($= 4\pi r^2 R^2 \psi^2$) gives the probability of finding the electron at a distance *r* (atomic radius) from the nucleus regardless of direction.
III. The shape of an orbital is defined as a surface of constant probability density that encloses some large fractions of the probability of finding the electron.

Select the correct statements.

(a) Both I and II (b) Both II and III
(c) Both I and III (d) All of these

ANSWERS

SESSION 1	1 (c)	2 (d)	3 (c)	4 (c)	5 (a)	6 (b)	7 (b)	8 (a)	9 (b)	10 (a)
	11 (c)	12 (b)	13 (b)	14 (a)	15 (b)	16 (d)	17 (a)	18 (a)	19 (b)	20 (b)
	21 (b)	22 (d)	23 (d)	24 (c)	25 (d)	26 (c)	27 (c)	28 (d)	29 (c)	30 (b)
	31 (c)	32 (c)	33 (d)	34 (c)	35 (c)	36 (a)	37 (c)	38 (a)	39 (c)	40 (b)
	41 (b)	42 (b)	43 (a)	44 (a)	45 (a)					
SESSION 2	1 (c)	2 (a)	3 (b)	4 (b)	5 (c)	6 (b)	7 (d)	8 (a)	9 (a)	10 (b)
	11 (a)	12 (c)	13 (d)	14 (b)	15 (d)					

Hints and Explanations

SESSION 1

1 The radii of an atom and atomic nucleus is in the order of 10^{-10} m and 10^{-15} m.

2 No change by doubling mass of electron, however by reducing mass of neutron to half, total atomic mass becomes 6 + 3 instead of 6 + 6.

Now, atomic mass $= \frac{9}{12} \times 100 = 75\%$

∴ Reduction in atomic mass = 25%

3 Many elements have several isotopes, for such elements atomic mass is average of the atomic masses of different isotopes, which is usually non-integral.

4 According to Rutherford, extra nuclear part is present around the nucleus in which electrons are contained.

5 $E = \frac{hc}{\lambda}$ or $E \propto \frac{1}{\lambda}$

∴ $E_1 \propto \frac{1}{6000}$ and $E_2 \propto \frac{1}{\lambda_2}$

∴ $\frac{E_1}{E_2} = \frac{\lambda_2}{6000}$

∵ $E_1 = E; E_2 = 4E$

∴ $\lambda_2 = 1500$ Å

6 Mass of one He-atom $= \frac{4}{N_0} = \frac{4}{6.02 \times 0^{23}}$

$$v = \sqrt{\frac{3RT}{M}} = \sqrt{\frac{3 \times 8.3143 \times 293}{4 \times 10^{-3}}}\ \text{ms}^{-1}$$

$= 1351.69\ \text{ms}^{-1}$

$$\lambda = \frac{h}{mv} = \frac{6.626 \times 10^{-34}}{\frac{4}{6.02 \times 10^{23}} \times 1351.69}$$

$\lambda = 7.38 \times 10^{-5}$ nm

7 Energy of photon $= \frac{hc}{\lambda}$ J $= \frac{hc}{e\lambda}$ eV

$$= \frac{6.626 \times 10^{-34} \times 3 \times 10^8}{300 \times 10^{-9} \times 1.602 \times 10^{-19}}$$

= 4.14 eV

For photoelectric effect to occur, energy of incident photons must be greater than work function of metals. Hence, only Li, Na, K and Mg have work function less than 4.14 eV.

8 $E_{photon} = \frac{hc}{\lambda} = \frac{6.626 \times 10^{-34} \times 3.0 \times 10^8}{0.57 \times 10^{-6}}$

$= 34.86 \times 10^{-20}$ J

Watt = power
=number of photons emitted/s $\times E_{photon}$

25 = number of photons emitted/s $\times 34.86 \times 10^{-20}$

$\therefore$ Number of photons emitted/s
$$= \frac{25}{34.86 \times 10^{-20}} = 7.17 \times 10^{19}/s$$

9 Critical wavelength corresponds to work function.

$\therefore$ Work function, $h\nu_0 = \dfrac{hc}{\lambda_0}$

Also, KE at λ_1 = 2 (KE) at 2200 Å

$$\left(\frac{hc}{\lambda_1} - \frac{hc}{\lambda_0}\right) = 2\left(\frac{hc}{\lambda_2} - \frac{hc}{\lambda_0}\right)$$

$$\left(\frac{1}{\lambda_1} - \frac{1}{2600}\right) = 2\left(\frac{1}{2200} - \frac{1}{2600}\right)$$

$$\frac{1}{\lambda_1} = \frac{1}{1100} - \frac{1}{2600}$$

$$\frac{1}{\lambda_1} = \frac{1500}{1100 \times 2600}$$

$$\lambda_1 = \frac{2600 \times 11}{15} = 1907 \text{ Å}$$

10 Energy of photon (E_{photon})

= work function + kinetic energy absorbed

$= W_0 + eV_0$

where, e = electric charge,

V_0 = stopping potential and eV_0 = KE, i.e. energy required to stop the ejection of electrons.

$$E_{photon} = \frac{hc}{\lambda} = \frac{6.626 \times 10^{-34} \times 3.0 \times 10^8}{253.7 \times 10^{-9}}$$

$= 7.835 \times 10^{-19}$ J

$$= \frac{7.835 \times 10^{-19}}{1.602 \times 10^{-19}} \text{ eV}$$

= 4.89 eV

$\because \quad E_{photon} = W_0 + 0.24$ eV

$\therefore \quad W_0 = 4.65$ eV

11 Number of spectral lines from ground state is

$$\frac{n(n-1)}{2} = \frac{5 \times 4}{2} = 10$$

12 For H-atom, $v_1 = \dfrac{2\pi e^2}{h}, \quad r_1 = \dfrac{h^2}{4\pi^2 me^2}$

$\therefore$ Circumference of 1st Bohr orbit $= 2\pi r_1$

$$= \frac{2\pi h^2}{4\pi^2 me^2} = \frac{h^2}{2\pi me^2}$$

$\therefore$ Number of rounds by electron in 1st orbit

$$= \frac{v_1}{2\pi r_1} = \frac{2\pi e^2 \times 2\pi me^2}{h \times h^2} = \frac{4\pi^2 me^4}{h^3}$$

$$= \frac{4 \times (3.14)^2 \times 9.108 \times 10^{-28} \times (4.8 \times 10^{-10})^4}{(6.626 \times 10^{-27})^3}$$

$= 6.57 \times 10^{15}$ round/s

13 Radius of orbit $(r) = \dfrac{n^2h^2}{4\pi^2 me^2} \times \dfrac{1}{Z}$

$$= \frac{0.529}{Z} n^2 \text{Å}$$

For H atom. $r_H = \dfrac{0.528n^2}{1}$ Å

$rBe^{3+} = 0.529n^2/4$Å and $r_n = rBe^{3+}$

$\therefore \quad n = 2$

14 As red colour is sun, so the line lies in visible region and obviously only Balmer series corresponds to the visible region. For Balmer series $n_1 = 2$ and red end means low energy or third end from red end means $n_2 = 5$. So, jump is involved from $5 \rightarrow 2$.

15 N = Number of lines emitted $= \dfrac{1}{2}n(n-1)$

$$= \frac{1}{2} \times 4(4-1) = 6$$

16 For H, $\bar{\nu} = R\left(\dfrac{1}{n_1^2} - \dfrac{1}{n_2^2}\right)$

For other atoms/ions, $\bar{\nu}$

$$= R\left(\frac{1}{n_1^2} - \frac{1}{n_2^2}\right)Z^2$$

$\therefore$ For Li^{2+},

$\bar{\nu} = 15200 \times 3^2 = 136{,}800 \text{ cm}^{-1}$

17 For Balmer series, $n_1 = 2$ and

$n_2 = (n_1 + 1)$ for first emission line
$= (2 + 1) = 3$

We know that,

Wave number, $\bar{\nu} = R_H\left(\dfrac{1}{n_1^2} - \dfrac{1}{n_2^2}\right)$

$$= R\left(\frac{1}{(2)^2} - \frac{1}{(3)^2}\right)$$

$$= R\left(\frac{1}{4} - \frac{1}{9}\right) = \frac{5R}{36}$$

18 For the shortest wavelength in Lyman series of H-atom, $n_1 = 1, n_2 = \infty$

$$\therefore \quad \frac{1}{\lambda_{min}} = R_H \times 1^2\left(\frac{1}{1^2} - \frac{1}{\infty^2}\right)$$

or $\quad \dfrac{1}{X} = R_H \quad \ldots$(i)

For the longest wavelength in Balmer series; $n_1 = 2, n_2 = 3$ and $Z = 2$ for He^+ ion.

$$\frac{1}{\lambda_{max}} = R_H \times 2^2\left(\frac{1}{2^2} - \frac{1}{3^2}\right) = R_H \cdot \frac{5}{9} \quad \ldots\text{(ii)}$$

For He^+, $\dfrac{1}{\lambda_{max}} = \dfrac{1}{X} \cdot \dfrac{5}{9}$

or $\quad \lambda_{max} = \dfrac{9}{5}X$

19 $\quad v_n = k\dfrac{Z}{n}$

i.e. $\quad v_n \propto \dfrac{1}{n}$

$\therefore \quad \dfrac{v_1}{v_3} = \dfrac{3}{1}$

As $\quad v_1 = x, \quad \therefore \quad v_3 = \dfrac{x}{3}$

20 IE $= E_\infty - E_1 = 0 - E_1 = -E_1$

$E_1 = -1.312 \times 10^6$ J mol^{-1}

$$= \frac{-1.312 \times 10^6}{1^2} \text{ J mol}^{-1}$$

$$E_2 = \frac{-1.312 \times 10^6}{2^2} \text{ J mol}^{-1}$$

$$\therefore E_2 - E_1 = -1.312 \times 10^6 \left[\frac{1}{2^2} - \frac{1}{1^2}\right] \text{J mol}^{-1}$$

$= 1.312 \times 10^6 \times \dfrac{3}{4}$ J mol^{-1}

$= 9.84 \times 10^5$ J mol^{-1}

21 IE $= E_1$

E_1 for $He^+ = -19.6 \times 10^{-18}$ J atom^{-1}

$$\frac{(E_1)_{He^+}}{(E_1)_{Li^{2+}}} = \frac{(Z_{He^+})^2}{(Z_{Li^{2+}})^2}$$

$$\frac{-19.6 \times 10^{-18}}{(E_1)_{Li^{2+}}} = \frac{4}{9}$$

or $\quad E_{1(Li^{2+})} = \dfrac{-19.6 \times 9 \times 10^{-18}}{4}$

$= -4.41 \times 10^{-17}$ J atom^{-1}

22 KE $= kT = \dfrac{RT}{N_0}$

$$= \frac{8.3143 \text{ J mol}^{-1} \text{ K}^{-1} \times 300 \text{ K}}{6.02 \times 10^{23} \text{ mol}^{-1}}$$

$= 4.143 \times 10^{-21}$ J

m (neutron) $= 1.67495 \times 10^{-27}$ kg

$$\lambda = \frac{h}{mv} = \frac{h}{\sqrt{2m(\text{KE})}}$$

$$= \frac{6.626 \times 10^{-34}}{\sqrt{2 \times 1.67495 \times 10^{-27} \times 4.143 \times 10^{-21}}}$$

$= 1.778 \times 10^{-10}$ m

$= 177.8$ pm ≈ 178 pm

23 $E = \frac{hc}{\lambda} = 2.178 \times 10^{-18} Z^2 \left[\frac{1}{n_1^2} - \frac{1}{n_2^2}\right]$

$$\frac{6.6 \times 10^{-34} \times 3 \times 10^8}{\lambda} = 2.178 \times 10^{-18} \times (1)^2 \times \left[\frac{1}{(1)^2} - \frac{1}{(2)^2}\right]$$

$$\lambda = 1.214 \times 10^{-7} \text{ m}$$

24 Energy values are always additive.

$$E_{total} = E_1 + E_2$$

$$\frac{hc}{\lambda} = \frac{hc}{\lambda_1} + \frac{hc}{\lambda_2}$$

$$\frac{1}{355} = \frac{1}{680} + \frac{1}{\lambda_2}$$

$$\lambda_2 = 742.94 \text{ nm} \approx 743 \text{ nm}$$

25 Energy required for 1 Cl_2 molecule

$$= \frac{242 \times 10^3}{N_A} \text{ J}$$

$$E = \frac{hc}{\lambda} \text{ or } \lambda = \frac{hc}{E}$$

$$= \frac{6.626 \times 10^{-34} \times 3 \times 10^8 \times 6.02 \times 10^{23}}{242 \times 10^3}$$

$$= 494 \times 10^{-9} \text{ m} = 494 \text{ nm}$$

26 λ of photon absorbed = 300 nm

λ of I photon re-emitted out = 496 nm

Let λ of II photon re-emitted out = λ_H

∵ Total energy absorbed = Total energy re-emitted out

$$\therefore \frac{hc}{300 \times 10^{-9}} = \frac{hc}{496 \times 10^{-9}} + \frac{hc}{\lambda_H}$$

$$\therefore \lambda_H = 759 \times 10^{-9} \text{ m} = 759 \text{ nm}$$

27

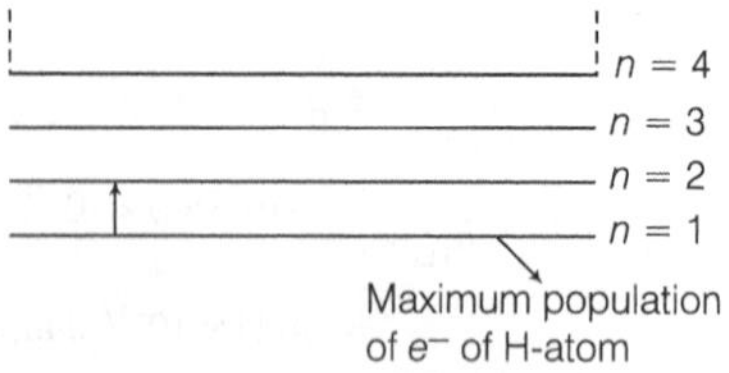

Since, at $n = 1$, the population of electrons is maximum, i.e. at ground state. So, maximum excitation will take place from $n = 1$ to $n = 2$.

Hence, $n = 2$ is the possible excited state.

Now, we have the formula for energy of H-atom $(E_n)_H = -13.6 \frac{Z^2}{n^2}$ eV

where, Z = atomic number, Z for H-atom = 1

$$\therefore (E_n)_H = -13.6 \times \frac{1}{2^2} \text{eV}$$

$$= -\frac{13.6}{4} \text{eV} = -3.4 \text{eV}$$

28 Bohr radius $(r_n) = \epsilon_0 n^2 h^2$

$$r_n = \frac{n^2 h^2}{4\pi^2 me^2 kZ}$$

$$k = \frac{1}{4\pi \epsilon_0}$$

$$\therefore r_n = \frac{n^2 h^2 \epsilon_0}{\pi me^2 Z} = n^2 \frac{a_0}{Z}$$

where, m = mass of electron

e = charge of electron

h = Planck's constant

k = Coulomb constant

$$r_n = \frac{n^2 \times 0.53}{Z} \text{Å}$$

Radius of nth Bohr orbit for H-atom

$= 0.53 \, n^2$ Å $[Z = 1 \text{ for H-atom}]$

∴ Radius of 2nd Bohr orbit for H-atom

$= 0.53 \times (2)^2 = 2.12$ Å

29 According to Bohr's model,

$$mvr = \frac{nh}{2\pi}$$

$$(mv)^2 = \frac{n^2 h^2}{4\pi^2 r^2}$$

$$\text{KE} = \frac{1}{2} mv^2 = \frac{n^2 h^2}{8 \pi^2 r^2 m} \quad \dots \text{(i)}$$

Radius of the orbit $r_n = \frac{a_0 \times n^2}{Z}$

For H-atom $Z = 1$

Substituting the value of r in Eq. (i) gives

$$\text{KE} = \frac{h^2}{8 \pi^2 n^2 a_0^2 m}$$

When $n = 2$, $\text{KE} = \frac{h^2}{8 \pi^2 (2)^2 a_0^2 m}$

$$= \frac{h^2}{32 \pi^2 a_0^2 m}$$

30 de-Broglie equation is $\lambda = \frac{h}{mv}$

Given, $m = 1000$ kg and $v = 36$ km/h

$$= \frac{36 \times 1000}{60 \times 60} = 10 \text{ m/s}$$

On putting values, $\lambda = \frac{6.626 \times 10^{-34}}{1000 \times 10}$

$$= 6.626 \times 10^{-38} \text{ m}$$

31 As you can see in options, energy term is mentioned hence, we have to find out relation between $\frac{h}{\lambda}$ and energy. For this, we shall use de-Broglie wavelength and kinetic energy term in eV.

de-Broglie wavelength for an electron

$$(\lambda) = \frac{h}{p}$$

$$\Rightarrow p = \frac{h}{\lambda} \quad \dots \text{(i)}$$

Kinetic energy of an electron = eV

As we know that, $\text{KE} = \frac{p^2}{2m}$

$$\therefore \text{eV} = \frac{p^2}{2m} \text{ or } p = \sqrt{2meV} \quad \dots \text{(ii)}$$

From Eqs. (i) and (ii), we get

$$\frac{h}{\lambda} = \sqrt{2meV}$$

32 $\Delta x \cdot \Delta v \geq \frac{h}{4\pi m} \Rightarrow \Delta x = \frac{h}{4\pi m \times \Delta v}$

$$= \frac{6.626 \times 10^{-34}}{4 \times \frac{22}{7} \times 9.1 \times 10^{-31} \times 300 \times 0.001 \times 10^{-2}}$$

$$= 1.93 \times 10^{-2} \text{ m}$$

33 $v = 10^{-4} \text{ cm}^{-1} \text{ s}^{-1}$

$$\therefore \Delta v = \frac{0.1 \times 10^{-4}}{100} = 1 \times 10^{-7} \text{ cm s}^{-1}$$

Now, $\Delta v \cdot \Delta x = \frac{h}{4\pi m}$

$$\therefore \Delta x = \frac{6.626 \times 10^{-27}}{4 \times 3.14 \times 10^{-11} \times 10^{-7}}$$

$$= 0.527 \times 10^{-9} \text{ cm}$$

34 $\Delta v \cdot \Delta x = \frac{h}{4\pi m}$

$$\therefore \Delta v = \frac{h}{4\pi m \cdot \Delta x}$$

$$= \frac{6.626 \times 10^{-34}}{4 \times 3.14 \times 9.11 \times 10^{-31} \times 0.1 \times 10^{-10}}$$

$$= 5.79 \times 10^6 \text{ ms}^{-1}$$

35 By Heisenberg's uncertainty principle

$$\Delta m \Delta v = \frac{h}{4\pi} \Rightarrow \Delta v = 0.005\%$$

or $600 \text{ m/s} = \frac{600 \times 0.005}{100} = 0.03$

$$\Delta x \times 9.1 \times 10^{-31} \times 0.03 = \frac{6.6 \times 10^{-34}}{4 \times 3.14}$$

Hence, $\Delta x = \frac{6.6 \times 10^{-34}}{4 \times 3.14 \times 0.03 \times 9.1 \times 10^{-31}}$

$$= 1.92 \times 10^{-3} \text{ m.}$$

36 For each value of l, m is $-l$ to $+l$. For $l = 2$, $m = -3$ is not possible.

37 Quantum numbers $n = 3$, $l = 2$, $m_l = +2$ represents only one orbital.

38 Given, atomic number of Rb, $Z = 37$

Thus, its electronic configuration is [Kr] $5s^1$. Since, the last electron or valence electron enter in 5s-sub-shell.

So, the quantum numbers are
$n = 5, l = 0$, (for s-orbital)

$m = 0 (\because m = +l \text{ to } -l), s = +1/2 \text{ or } -1/2$

39 For 4p orbital, $n = 4$ and $l = 1$

and for 3d orbital, $n = 3$ and $l = 2$

So, option (c) doesn't show 4p and 3d orbital, it indicates 3f-orbital which is not possible.

40 (i) When $n = 5, l = 0, 1, 2, 3, 4$

When $l = 0, m_l = 0$

When $l = 1, m_l = 0, +1, -1$ (two e^-)

When $l = 2, m_l = 0, \pm 1, \pm 2$ (two e^-)

When $l = 3, m_l = 0, \pm 1, \pm 2, \pm 3$ (two e^-)

When $l = 4, m_l = 0, \pm 1, \pm 2, \pm 3, \pm 4$ (two e^-)

(Because each orbital can accommodate a maximum of $2e^-$.)

$\therefore$ Total electrons having $n = 5$

and $m_l = \pm 1 = 2 + 2 + 2 + 2 = 8$

(ii) Only one electron is associated with $n = 2, l = 1, m_l = -1$

and $m_s = -1/2$ because Pauli's principle states that no two electrons can have same value for all the four quantum numbers.

41

n	l	nl (sub-orbit)	$n + 1$
4	1	4p	5
4	0	4s	4
3	2	3d	5
3	1	3p	4

Higher the value of $(n + l)$, higher the energy. If $(n + l)$ are same, sub-orbit with lower value of n has lower energy.

Thus, $3p < 4s < 3d < 4p$

Hence correct order is (iv) < (ii) < (iii) < (i).

42 Aufbau principle states that electrons are filled in the increasing order of energy. Hence, 2s-orbital should be filled first before filling 2p-orbitals.

43 Isotopes have the same atomic number i.e. same number of electrons which are responsible for their chemical behaviour. Hence, these exhibit similar chemical properties.

44 According to Heisenberg's uncertainty principle, the exact position and exact momentum of an electron cannot be determined simultaneously. Thus, the path of electron in an atom is not clearly defined. Hence (a) option is correct.

45 On heating a solid for a longer time, radiations become white and then blue as the temperature becomes very high because radiations emitted go from a lower frequency to higher frequency as the temperature increases.

SESSION 2

1 $x = 19 \Rightarrow$ The electronic configuration is
$1s^2 2s^2 2p^6 \underbrace{3s^2 3p^6}_{M \text{ shell}} 4s^1$.

$y = 21 \Rightarrow$ The electronic configuration is
$1s^2 2s^2 2p^6 \underbrace{3s^2 3p^6 3d^1}_{M \text{ shell}} 4s^2$.

$z = 25 \Rightarrow$ The electronic configuration is
$1s^2 2s^2 2p^6 \underbrace{3s^2 3p^6 3d^5}_{M \text{ shell}} 4s^2$.

$\therefore$ The number of electrons present in M shell of these element follow the order

$$z > y > x.$$

2 From Heisenberg uncertainty principle

$$= \Delta x \cdot \Delta p = \frac{h}{4\pi} = \Delta x \cdot m\Delta v = \frac{h}{4\pi}$$

For two particle A and B the ratio can be written as:

$$= \frac{\Delta x_A}{\Delta x_B} \frac{\Delta v_A}{\Delta v_B} = \frac{m_B}{m_A}.$$

$$\frac{\Delta x_A \times 0.05}{\Delta x_B \times 0.02} = \frac{5}{1}$$

$$\frac{\Delta x_A}{\Delta x_B} = \frac{5 \times 0.02}{1 \times 0.05} = \frac{2}{1}$$

3 Energy given to I_2 molecule $= \frac{hc}{\lambda}$

$$= \frac{6.626 \times 10^{-34} \times 3 \times 10^8}{4500 \times 10^{-10}}$$

$$= 4.4 \times 10^{-19} \text{ J}$$

Energy of I_2 molecule $= 240 \times 10^3$ J/mol

$$= \frac{240 \times 10^3}{6.023 \times 10^{23}} \text{ J/atom}$$

$$= 3.98 \times 10^{-19} \text{ J/atom}$$

$$\text{KE} = (4.4 \times 10^{-19}) - (3.98 \times 10^{-19})$$

$$= 0.42 \times 10^{-19} \text{ J/atom}$$

KE of I atoms

$$= \frac{0.42}{2} \times 10^{-19} = 0.21 \times 10^{-19}$$

4 The amount of energy required by an electron to excite or jump from $n = 1$ (ground state) to $n = 3$ (excited state)

$$\Delta E = E_3 - E_1$$

According to Bohr's model

$$E = \frac{-13.6}{n^2} \text{ eV/atom}$$

$$\therefore \quad \Delta E = \frac{-13.6}{3^2} - \left(\frac{-13.6}{1^2}\right)$$

$$= \frac{-13.6}{3^2} + \frac{13.6}{1}$$

$$= \frac{-13.6 + 122.4}{9}$$

$$= \frac{108.8}{9} \text{ eV/atom}$$

$$= 12.08 \text{ eV/atom}$$

$$= 12.08 \times 1.602 \times 10^{-12}$$

$$(1 \text{ eV} = 1.602 \times 10^{-12} \text{ erg})$$

$$= 19.36 \times 10^{-12}$$

or 1.936×10^{-11}

5 1 mole of C-14

$= 14 \text{ g} = 6.022 \times 10^{23}$ carbon atoms

Number of neutrons in 1 carbon atom

= mass number − atomic number

$= 14 - 6 = 8$ neutrons

6.023×10^{23} carbon atoms will contain $6.023 \times 10^{23} \times 8$ neutrons

$\therefore$ 14 g C-14 have $6.023 \times 10^{23} \times 8$ neutrons

Hence, 7×10^{-3} g C-14 will have

$$= \frac{7 \times 10^{-3} \times 6.023 \times 10^{23} \times 8}{14} \text{ neutrons}$$

$= 24.092 \times 10^{20}$ neutrons

$= 2.4092 \times 10^{21}$ neutrons

Mass of 1 neutron $= 1.675 \times 10^{-27}$ kg

$\therefore$ Mass of 2.4092×10^{21} neutrons

$= 2.4092 \times 10^{21} \times 1.675 \times 10^{-27}$ kg

$= 4.0354 \times 10^{-6}$ kg

6 $\nu = RZ^2\left(\frac{1}{n_1^2} - \frac{1}{n_2^2}\right)$

For He^+, $Z = 2, n_1 = 2, n_2 = 4$

$$\nu = R(2)^2\left[\frac{1}{2^2} - \frac{1}{4^2}\right]$$

$$= 4R\left[\frac{3}{16}\right] \quad \ldots(i)$$

$$= \frac{3R}{4}$$

For H, $Z = 1$, as the frequency of light emitted for the transition $n = 4$ to

$n = 2$ of He^+ is equal to the transition in H-atom corresponds to $n = 2$ to $n = 1$.

$$\nu = R(1)^2\left[\frac{1}{1^2} - \frac{1}{2^2}\right]$$

$$= R\left[\frac{3}{4}\right] = \frac{3R}{4}.$$

7 $\Delta E = 12.084$ eV

$= 12.084 \times 1.6 \times 10^{-19}$ J/atom

$= 12.084 \times 1.6 \times 10^{-19} \times 6.02 \times 10^{23}$ J mol^{-1}

$= 1164$ kJ

8 For any atom the order of energy for the orbitals can be given as:

$$1s < 2s < 2p < 3s < 3p < 3d$$

Hence, statement (a) is an incorrect statement.

9 $\Delta E = \frac{hc}{\lambda}$

$$= \frac{6.626 \times 10^{-34} \times 3 \times 10^8}{6.63 \times 10^{-7}}$$

$= 2.9 \times 10^{-19}$ J $\simeq 3 \times 10^{-19}$ J

10 For any orbital set of quantum number possible are,

$l = n - 1$

$m = -l$ to l

$s = \frac{-1}{2}, \frac{1}{2}$

For given set of quantum numbers for a multi-electron atom 2, 0, 0, $\frac{1}{2}$ and 2, 0, 0, $\frac{-1}{2}$. The next higher allowed set of n and l quantum number for this atom in its ground state is $n = 2, l = 1$.

11 Orbitals having quantum number $n = 3$, $l = 2, m = 1$ and $n = 3, l = 2, m = 0$ represents $3d$-orbitals which are degenerate in the absence of magnetic and electric field.

12 For any orbital, set of quantum number possible are

$l = n - 1$

$m = -l$ to l

$s = \frac{-1}{2}, \frac{1}{2}$.

Thus, in case (II) $n = l = 2$ hence this set of quantum number is not possible.

In case (IV) $n = 1$ to $l = 0, m = -1$ but it should be 0 hence, this is not possible.

In case (V) $n = 2, l = 3$ which is greater than n thus this is also not possible.

13 (a) Electrons in $2s$ and $2p$-orbitals have different screening effect. Hence, their Z_{eff} is different.

Z_{eff} of $2s$- orbital $> Z_{eff}$ of $2p$-orbital

Therefore, it is not correct.

(b) Energy of $2s$-orbital < energy of $2p$-orbital.
Hence, it is not correct.

(c) Z_{eff} of $1s$-orbital $\neq Z_{eff}$ of $2s$-orbital
Hence, it is incorrect.

(d) For the two electrons of $2s$-orbital, the value of m_s is $+\frac{1}{2}$ and $-\frac{1}{2}$.
Hence, it is correct.

14 The order of frequency of radiation is given as γ-rays < X-rays < UV rays > Visible < IR < Microwave < Radiowave

Thus, statement (b) is incorrect.

15 I. $|\psi|^2$ is a measure of electron density at a point in an atom. Thus, the statement is correct.

II. Radial probability function $4\pi^2R^2\psi^2$ gives the probability of finding the electron at a distance r from the nucleus regardless of direction. Thus, the statement is correct.

III. The shape of an orbital is defined as a surface of constant probability density that encloses some large fractions of the probability of finding the electron. Thus, the statement is correct.

DAY FOUR

Chemical Bonding & Molecular Structure

Learning & Revision for the Day

- Ionic or Electrovalent Bond
- Covalent Bond
- Bond Parameters
- Resonance
- Coordinate Bonding
- Valence Bond Theory
- Concept of Hybridisation Involving *s*, *p* and *d*-orbital
- VSEPR Theory
- Molecular orbital Theory
- Hydrogen Bonding
- Metallic Bonding

A **chemical bond** is the attractive force which holds the various constituents (atoms, ions etc.) together in different chemical species.

The different types of chemical bonds formed between atoms are as follows:

(i) Ionic bond (ii) Covalent bond (iii) Coordinate bond

Kossel-Lewis Approach (Octet Rule)

- According to this approach, the atoms of different elements take part in chemical combination in order to complete their octet (to have eight electrons in the outermost valence shell) or duplet (to have two valence electrons) in some cases such as H, Li, Be etc., or to attain the nearest noble gas configuration. This is known as octet rule.
- In Lewis symbol, the number of dots around the symbol represents the number of valence electrons. The number of electrons helps to calculate the common valency of the element.

Limitations of Octet Rule

In accordance to this rule, the shape of the molecule cannot be predicted. The relative stability of molecule cannot be known by this rule.

However, the octet rule is violated in a significant number of cases. These are:

(i) Electron deficient compounds : $BeCl_2$, BF_3, $AlCl_3$ etc.

(ii) Hypervalent compounds : PCl_5, SF_6, IF_7, H_2SO_4 etc.

(iii) Compounds of noble gases : XeF_2, XeF_6, XeF_4, KrF_2 etc.

(iv) Odd electron molecules : NO, NO_2, O_2^-, O_3 etc.

H_2^+, He_2^+, O_2, NO, NO_2, ClO_2 are some of the examples of stable molecules having odd electron bonds (bonds formed by sharing of usually one or three electrons).

PREP MIRROR
Your Personal Preparation Indicator

- No. of Questions in Exercises (x)—
- No. of Questions Attempted (y)—
- No. of Correct Questions (z)—
  (Without referring Explanations)

- Accuracy Level ($z/y \times 100$)—
- Prep Level ($z/x \times 100$)—

In order to expect good rank in JEE, your Accuracy Level should be above 85 & Prep Level should be above 75.

Ionic or Electrovalent Bond

The attractive forces of ionic bond (i.e. electrostatic force of attraction) are developed between an electropositive atom and an electronegative atom due to complete transfer of electrons from former to later. It is generally formed between the atoms having large difference in their electronegativity.

Formation of Ionic Bond

- Ionic bond is formed by the complete transfer of electron(s) from one atom to the other.

  e.g. $_{11}Na$ (2, 8, 1) $_{17}Cl$ (2, 8, 7)

  $$\dot{Na} + :\ddot{Cl}: \longrightarrow \overset{+}{Na}[:\ddot{Cl}:]^-$$

  $$Na^+ + Cl^- \longrightarrow NaCl$$

- Ionic bonds are non-directional and also known as electrovalent bonds or polar bonds.

Factors Affecting the Formation of Ionic Bonds

The formation of an ionic bond is related to cation and anions which depends upon the following factors:

(i) Low ionisation energy of the electropositive element.
(ii) High electron affinity of electronegative element.
(iii) High lattice enthalpy

$$\left(\text{Lattice enthalpy} \propto \frac{\text{charge on ions}}{\text{size of ion}}\right)$$

NOTE
- Elements of group 1 and group 2 on combining with halogens, oxygen and sulphur generally form ionic bonds.
- Bonding in compounds of transition metals (in lower oxidation state) is ionic with partial covalent character.

Lattice Enthalpy and Its Calculation

The lattice enthalpy of an ionic solid is defined as the energy required to completely separate one mole of a solid ionic compound into gaseous constituent ions. This energy is calculated by **Born-Haber cycle.**

Born-Haber cycle includes

(i) vaporisation of reactants into gaseous state
(ii) conversion of gaseous atoms into ions
(iii) combination of gaseous ions to form ionic lattice of molecules

e.g. $M(s) + \frac{1}{2}X_2(g) \longrightarrow MX(s), \Delta H = Q$

Here, ΔH = enthalpy change of the reaction
Q = heat of the reaction.

The above reaction includes the various steps:

(a) $M(s) \longrightarrow M(g);\ \Delta H_1 = S$ (S = sublimation energy)
(b) $M(g) \longrightarrow M^+(g) + e^-;\ \Delta H_2 = I$ (I = ionisation energy)
(c) $\frac{1}{2}X_2(g) \longrightarrow X(g);\ \Delta H_3 = \frac{D}{2}$ (D = dissociation energy)
(d) $X(g) + e^- \longrightarrow X^-(g);\ \Delta H_4 = -E$ (E = electron affinity)
(e) $M^+(g) + X^-(g) \longrightarrow MX(s);\ \Delta H_5 = -U$ (U = lattice enthalpy)

Overall enthalpy change of the reaction is given by

$$\Delta H = \Delta H_1 + \Delta H_2 + \Delta H_3 + \Delta H_4 + \Delta H_5$$

or alternatively, the above equation can be written as:

$$\Delta H = S + I + \frac{D}{2} - E - U$$

or

$$U = S + I + \frac{D}{2} - E - \Delta H$$

NOTE
- A number of ionic solids are almost insoluble in water because hydration energy is smaller than their lattice energy. Examples of water insoluble salts are AgCl, AgBr, AgI, Ag_2CrO_4, $PbSO_4$, $BaSO_4$, $CaCO_3$.
- Both lattice energy and hydration energy decreases with increase in ionic size.
- If both anion and cation are of comparable size, the cationic radius will influence the lattice energy.

Covalent Bond

The chemical bonds that are formed by sharing of electrons between the elements of almost same electronegativity or between the elements having less difference in electronegativity are called covalent bonds.

e.g. Formation of O_2 molecule.

$$_8O = 2, 6$$

O=O Or 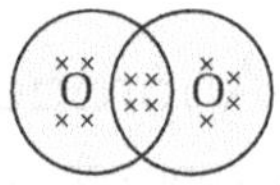

The covalent bond can be of the following two types:

1. **Non-polar Covalent Bond** If the covalent bond is formed between two homonuclear atoms, i.e. between atoms of exactly equal electronegativity, the electron pair is equally shared between them. e.g. H_2, Cl_2, F_2, Br_2 etc.
2. **Polar Covalent Bond** If the bond forming entities are dissimilar, i.e. heteronuclear or with different electronegativity, the bond formed has partial ionic character as the electron pair is attracted by more electronegative entity.

$$\overset{\delta+}{H}\cdots\overset{\delta-}{O}\cdots\overset{\delta+}{H} \quad \overset{\delta+}{H}\cdots\overset{\delta-}{Cl}$$

The greater the difference in electronegativity, higher is the polar nature. The relative order of electronegativity of some important elements is

F	> O	> Cl ≈ N	> Br	> S ≈ C ≈ I	> H
4.0	3.5	3.0	2.8	2.5	2.1

Calculation of Percentage Ionic Character

The ionic character in polar bond can be calculated by the following methods:

1. **Pauling equation**
   Percentage ionic character = $18\,(\chi_A - \chi_B)^{1.4}$
2. **Hannay and Smith equation**
   Percentage ionic character = $16\,(\chi_A - \chi_B) + 3.5\,(\chi_A - \chi_B)^2$
   where, $(\chi_A - \chi_B)$ = electronegativity difference.

Dipole Moment (μ)

- It is defined as the product of the magnitude of charge (q) and distance (d) separating the centres of positive and negative charges.
- Its direction is from positive end to negative end,

$$\mu = q \times d$$

- Its unit in CGS system is debye (D).
- A molecule is said to be polar if the net dipole moment of the molecule is not equal to zero. e.g. NF_3 has lower dipole moment than NH_3 because resulting vector is towards the lone pair in NH_3 but in NF_3 it is opposite of lone pair, which cancels the resultant moment.

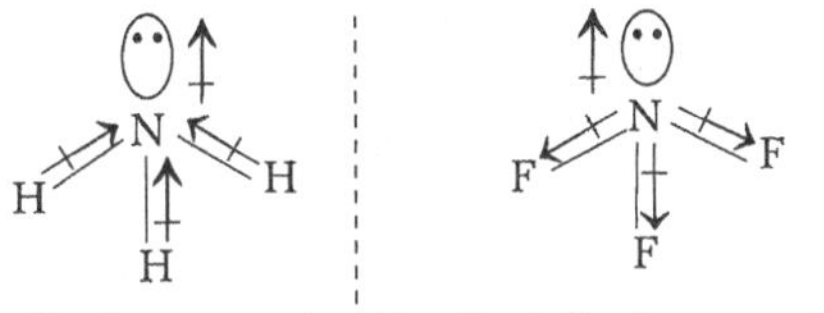

Resultant dipole moment in $NH_3 = 4.90 \times 10^{-30}$ cm | Resultant dipole moment in $NF_3 = 0.80 \times 10^{-30}$ cm

A comparison dipole moments of NH_2 and NF_3

- Percentage ionic character of any molecule can be calculated by dipole moment (μ).

$$\text{Percentage ionic character} = \frac{\mu_{\text{observed}}}{\mu_{\text{ionic}}} \times 100$$

where, $\mu_{\text{ionic}} = q \times d$ $\quad [q = 4.8 \times 10^{-10} \text{ esu}]$

Partial Ionic Character of Covalent Bond

On the basis of electronegativity difference, partial ionic character of covalent bond can be summarised as :

(i) Electronegativity difference between combining atoms = 1.7, then bond is 50% ionic and 50% covalent.

(ii) Electronegativity difference > 1.7, ionic character in bond is more than 50%.

(iii) Electronegativity difference < 1.7, ionic character is less than 50%.

Fajan's Rule

- It states that the magnitude of covalent character in an ionic bond depends upon the polarising power. Higher the polarising power, more will be the covalent character.

In general,

$$\text{Polarising power} \propto \frac{1}{\text{size of cation}} \propto \text{size of anion}$$

$$\propto \text{charge on ions}$$

e.g. In between NaI and NaCl, NaI due to larger size of I^- ion has more covalent character. $FeCl_2$ is less covalent than $FeCl_3$ because polarising power of Fe^{2+} is less than that of Fe^{3+} ion having smaller size and higher oxidation state.

- Cation with pseudo noble gas configuration has greater polarising power than the other noble gas configuration cation.

Formal Charge

It is defined as the difference between the number of valence electrons of that atom in an isolated or free state and the number of electrons assigned in the Lewis structure.

Formal charge (FC) on an atom in a Lewis structure

$$= \begin{bmatrix}\text{Total number of}\\ \text{valence electrons}\\ \text{in the free atom}\end{bmatrix} - \begin{bmatrix}\text{Total number of}\\ \text{non-bonding (lone pair)}\\ \text{electrons}\end{bmatrix} - \frac{1}{2}\begin{bmatrix}\text{Total number of}\\ \text{bonding (shared) electrons}\end{bmatrix}$$

Bond Parameters

The covalent bonds are characterised by the some parameters which are as follows:

1. **Bond Length** In general the average distance between the centre of nuclei of the two bonded atoms in a molecule is known as bond length. It depends upon the size of atoms, hybridisation, steric effect, resonance etc. Usually bond length of polar bond is smaller as compared to a non-polar bond. Bond length increases as the size of atom or orbital increases.
2. **Bond Enthalpy** It is the amount of energy required to break one mole of bonds of a particular type between two atoms in gaseous state.

   $$\text{Bond enthalpy} \propto \text{electronegativity}$$
   $$\propto \frac{1}{\text{size of atoms}}$$
   $$\propto \frac{1}{\text{number of lone pair of electrons}}$$

3. **Bond Order** Bond order is just like number of bond(s) between two atoms in a molecule.

   $$\text{Bond order} \propto \frac{1}{\text{bond length}} \propto \text{bond stability}$$

   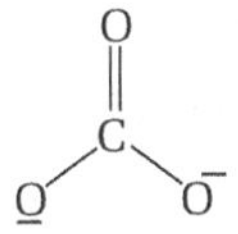

   Bond order $\propto$ bond strength (stronger the bond, larger will be the bond dissociation energy and bond enthalpy) e.g. CO_3^{2-}.

$$\text{Bond order} = 1 + \frac{\text{Number of double bonds}}{\text{Number of delocalisation position}}$$

For O_3, $\left(1+\frac{1}{2}\right) = 1.5$, SO_4^{2-}, $\left(1+\frac{2}{4}\right) = 1.5$

4. **Bond Angle** It is defined as the angle between the orbitals containing bonding electron pairs around the central atom in a molecule or complex ion.

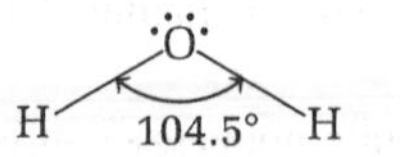

Resonance

There are certain molecules (benzene, ozone, nitric acid and many more), whose all the properties cannot be explained by a single structure, then two or more structures, called **resonating structures** are required to explain all the properties and the actual structure is intermediate of these structures. e.g. CO_3^{2-}, O_3 etc.

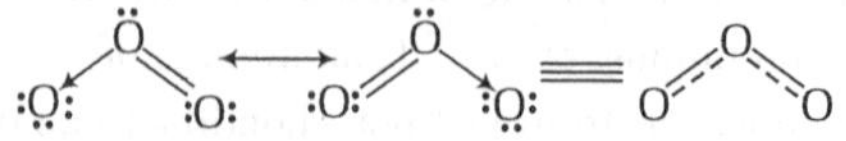

- Resonance is shown by only those molecules which possess conjugate single and multiple bonds.
- It imparts stability to the molecule and hence, decreases its reactivity (due to resonance).
- Since, the electrons are not localised between any particular atoms and are uniformly distributed in the resonance hybrid, all the bonds are similar and are of equal bond lengths.
- Resonance averages the bond characteristic as a whole.
- The difference in the energy of a resonance hybrid and most stable structure (with least energy) is called **resonance energy**.

Coordinate Bonding

The bonding in which one atom furnishes a pair of electrons to the other atom, but shared by both the atoms in such a manner that both atoms achieve stability, is called coordinate bonding or dative bonding. e.g.

$$H^+ + \ddot{N}H_3 \longrightarrow \overset{+}{N}H_4$$

Acceptor Donor

$$\ddot{N}H_3 + BF_3 \longrightarrow [H_3N \longrightarrow BF_3]$$

Donor Acceptor Complex

Valence Bond Theory (VBT)
(Given by Heitler and London)

- It explains bond formation in terms of overlapping of orbitals, e.g. the formation of H_2 molecule from two hydrogen atoms involves the overlap of 1s-orbital of the two H-atoms which are singly occupied.
- Because of orbital overlap the electron density between the nuclei increases which helps in bringing them closer.

The overlapping of orbitals may results in two types of bonds given below :

1. **Sigma (σ)-Bond** It is the result of end to end overlapping or axial overlapping between *s*-*s*, *p*-*p*, *s*-*p* orbitals. Single bond is always σ-bond. The electron density accumulates between the centre of the atoms being bonded.

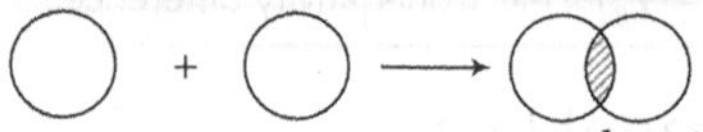

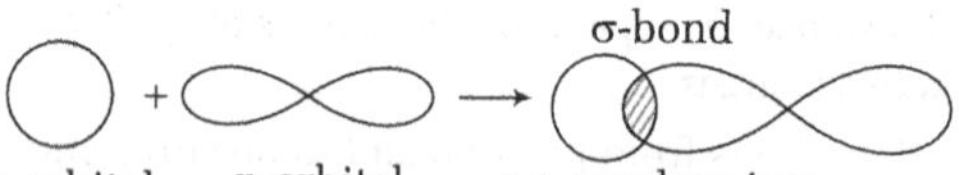

2. **Pi (π)-Bond** It is formed by the incomplete overlapping of orbitals or in other words sidewise or parallel overlapping of *p*-*p* orbitals results in the π-bond formation.

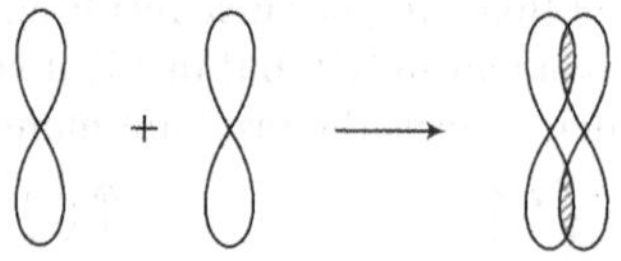

Double bond has one σ-bond and one π-bond. Triple bond has two π-bonds and one σ-bond.

Concept of Hybridisation Involving *s*, *p* and *d*-orbital

- Pauling introduced the concept of hybridisation. It is defined as intermixing of atomic orbitals of nearly the same energy and resulting in the formation of new atomic orbitals same in number and identical in all respects (shape, energy and size).
- sp, sp^2, sp^3 hybridisations of atomic orbitals of Be, B, C, N and O are used to explain the formation and geometrical shapes of molecules like $BeCl_2$, BCl_3, CH_4, NH_3 and H_2O.

NOTE (i) The *d*-orbital taking part in dsp^2-hybridisation is $d_{x^2-y^2}$ while in sp^3d it is d_{z^2} and in sp^3d^2, the two *d*-orbitals are d_{z^2} and $d_{x^2-y^2}$.

(ii) In dsp^3, if d_{z^2} is used, the shape is trigonal bipyramidal. If in dsp^3, $d_{x^2-y^2}$ is used, the shape is square pyramidal.

- In case of 3rd period elements, the energy of 3*d*-orbitals are comparable to the energy of 3*s* and 3*p*-orbitals as well as to the energy of 4*s* and 4*p* orbitals.
- As a result of this, hybridisation involves either 3*s*, 3*p* and 3*d* or 3*d*, 4*s* and 4*p* is possible.

Important hybridisation schemes involving *s*, *p* and *d* orbitals are as follows:

Shape of molecules/ions	Hybridisation type	Atomic orbitals	Examples
Square planar	dsp^2	$d+s+p(2)$	$[Ni(CN)_4]^{2-}$, $[Pt(Cl)_4]^{2-}$
Trigonal bipyramidal	sp^3d	$s+p(3)+d$	PF_5, PCl_5
Square pyramidal	sp^3d^2 with one lone pair	$s+p(3)+d(2)$	BrF_5, IF_5
Octahedral	sp^3d^2, d^2sp^3	$s+p(3)+d(2)$ $d(2)+s+p(3)$	SF_6, $[CrF_6]^{3-}$, $[Co(NH_3)_6]^{3+}$

Some important examples are :

1. **sp^3d-hybridisation** e.g. PCl_5 molecule. In PCl_5 the two axial bonds are slightly elongated as the axial bond pairs suffer more repulsive interaction from equivalent bond pairs.

   Hence, axial bonds found to be slightly longer and weaker than equatorial bonds. Thus, PCl_5 is more reactive.

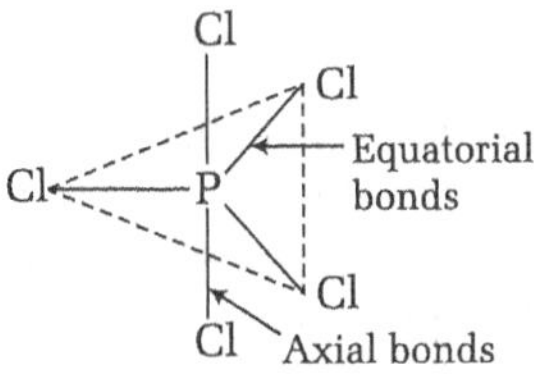

2. **sp^3d^2-hybridisation** e.g. SF_6 molecule. In SF_6, four S—F bonds are in same plane at right angles to one another and are directed towards the corner of a square. The other two F-atom lie at right angle above and below the plane of F-atoms.

SF_6 molecule is a symmetrical molecule and therefore is stable and less reactive.

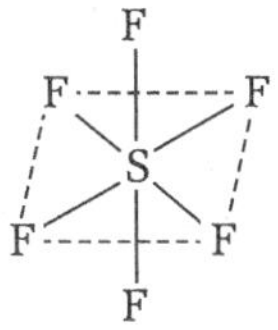

The formula for predicting hybridisation of central atom and the number of hybrid orbitals (X) is given below :

$$X = \frac{1}{2}[V + M + A - C]$$

where, V = Number of valence electrons of central atom

M = Number of monovalent atoms attached

C = Cation and A = Anion

While determining the type of hybridisation on the atom, π-bonds are never taken into account, but lone pairs are always considered.

VSEPR Theory

(Given by Gillespie and Nyholm)

- VSEPR stands for Valence Shell Electron Pair Repulsion. According to this theory, all valence shell electron pairs, surrounding, the central atom arrange themselves in such a manner, so that they are as far away from each other as possible.
- There are two types of electron pairs around the central atom; bonding electron pair (*bp*) and non-bonding electron pairs (*lp*). The strength of repulsion between the electron pairs varies as:

$$lp - lp > lp - bp > bp - bp$$

Hybrid orbitals and molecular shapes involving *s*, *p* and *d*-orbitals

Number of electron pairs	Geometry	Hybridisation	Examples
2 *bp*	Linear	sp	BeF_2
3 *bp*	Trigonal planar	sp^2	BF_3
2*bp* + 1*lp*	Bent	sp^2	SO_2, O_3
4 *bp*	Tetrahedral	sp^3	CH_4
3*bp*+1 *lp*	Pyramidal	sp^3	$\ddot{N}H_3$
2*bp*+2 *lp*	Angular or V-shape	sp^3	$H_2\ddot{O}$:
5 *bp*	Trigonal bipyramidal	sp^3d	PF_5, PCl_5
4*bp* + 1*lp*	*See-saw*	sp^3d	$:SF_4$
3*bp* + 2*lp*	T-shaped	sp^3d	$:\ddot{C}lF_3$
6*bp*	Octahedral	sp^3d^2	SF_6
5*bp* + 1*lp*	Square pyramidal	sp^3d^2	BrF_5
4*bp*+2*lp*	Square planar	sp^3d^2	$:XeF_4$
7*bp*	Pentagonal bipyramidal	sp^3d^3	IF_7

Molecular Orbital Theory
(Given by Hund and Mulliken)

This theory is a method for determining molecular structure in which electrons are not assigned to individual bonds between atoms, but are treated as moving under the influence of the nuclei in the whole molecule. Formation of molecular orbitals involve linear combination of atomic orbitals.

Molecular orbitals (MO) formed are of two types :

(i) **Bonding MO** which are of lower energy, are represented as σ or π. These makes bond stronger.

(ii) **Antibonding MO** which are of higher energy, are represented as $\overset{*}{\sigma}$ or $\overset{*}{\pi}$. These reduces the stability of molecules.

The electronic configuration of molecular orbitals (MO) are written in the following manner:

(a) For species like O_2, F_2 etc.

$$\sigma 1s,\ \overset{*}{\sigma} 1s,\ \sigma 2s,\ \overset{*}{\sigma} 2s,\ \sigma 2p_z, (\pi\ 2p_x = \pi 2p_y),$$
$$(\overset{*}{\pi} 2p_x = \overset{*}{\pi} 2p_y),\ \overset{*}{\sigma} 2p_z$$

(b) For species like $Li_2, Be_2, B_2,\ C_2,\ N_2$ etc.

$$\sigma 1s,\ \overset{*}{\sigma} 1s,\ \sigma 2s,\ \overset{*}{\sigma} 2s, (\pi\ 2p_x = \pi 2p_y), \sigma 2p_z,$$
$$(\overset{*}{\pi} 2p_x = \overset{*}{\pi} 2p_y), \overset{*}{\sigma} 2p_z$$

Bond Order (B.O.) for diatomic molecule or ions,

$$\text{B.O.} = \frac{N_b - N_a}{2}$$

N_b = number of bonding electrons

N_a = number of antibonding electrons

- The molecule is stable if $N_b > N_a$, i.e. if bond order is positive. The molecule is unstable if $N_b < N_a$ or $N_b = N_a$, i.e. if the bond order is negative or zero.
- Magnetic behaviour of a molecule can also be conveyed from its electronic configuration. If any unpaired electron is present in electronic configuration, the molecule is paramagnetic and in case of paired electrons, molecule is diamagnetic.

  Magnetic moment $= \sqrt{n(n+2)}$ BM

  (BM = Bohr Magneton)

  where, n = number of unpaired electrons

Hydrogen Bonding

Hydrogen bond can be defined as the attractive force, which binds hydrogen atom of one molecule with the electronegative atom (F,O or N) of another molecule. Cl has same electronegativity as nitrogen but it does not form strong H-bonds due to its large size. Strongest H-bond exist in KHF_2.

There are two types of H-bonds:

1. **Intermolecular hydrogen bonding** is a type of H-bond that is formed between the different molecules of same substance or different substance.

   e.g. $\overset{+\delta}{H}-\overset{-\delta}{F}\cdots\overset{+\delta}{H}-\overset{-\delta}{F}\cdots\overset{+\delta}{H}-\overset{-\delta}{F}$

   Intermolecular H-bonding decreases the volatility and increases the boiling point, viscosity and surface tension of a substance.

2. **Intramolecular hydrogen bonding** is a type of H-bond that is formed within the same molecule. Intramolecular H-bonding increases the volatility, decreases the boiling point of the compound and also decreases its solubility in water.

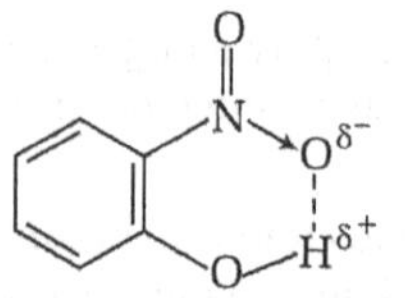

Intramolecular H-bonding in *o*-nitrophenol

- Boiling point of H_2O is more than that of HF because number of H-bonds formed by H_2O is more than that by HF.
- Hydrogen bonding is strongest when the bonded structure is stabilised through resonance.

Effect of H-Bonding

Various effects arise due to H-bonding in molecules are as follows:

- Due to polar nature of H_2O, there is association of water molecules giving a **liquid state of abnormally high boiling point.**

  $$\overset{+\delta}{H}-\underset{\underset{H_{+\delta}}{|}}{\overset{-\delta}{O}}\cdots\overset{+\delta}{H}-\underset{\underset{H_{+\delta}}{|}}{\overset{-\delta}{O}}\cdots\cdots\overset{+\delta}{H}-\underset{\underset{H_{+\delta}}{|}}{\overset{-\delta}{O}}\cdots\overset{+\delta}{H}-\underset{\underset{H_{+\delta}}{|}}{\overset{-\delta}{O}}$$

  Actually, in water, one water molecule is joined to four water molecules (two with H-atom and other two with O-atoms.)

  Thus, coordination number of water molecule in water is four.
- When ice is formed from liquid water, some air gap is formed (in tetrahedral packing of water molecules). Due to this, volume of ice is greater than liquid water and thus, ice is lighter than water. In another words, we can say that when ice melts, density increases but only upto 4°C, after this intermolecular H-bonding between water molecules breaks, hence volume increases and hence, density decreases. Thus, water has maximum density at 4°C.

- In the gaseous state, several polymeric forms of HF molecules exist in which the monomers are held together through H-bonding. A pentagonal arrangement of H—F molecules is shown below:

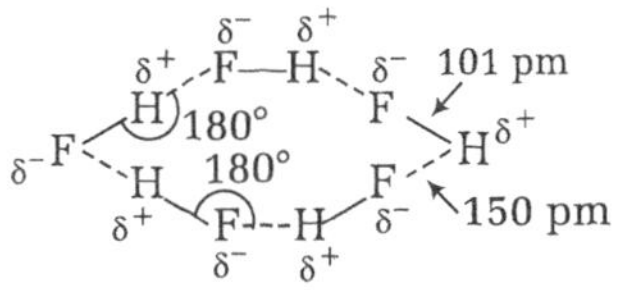

- Carboxylic acid dimerises in gaseous state due to H-bonding. The dimerisation of carboxylic acids is given below.

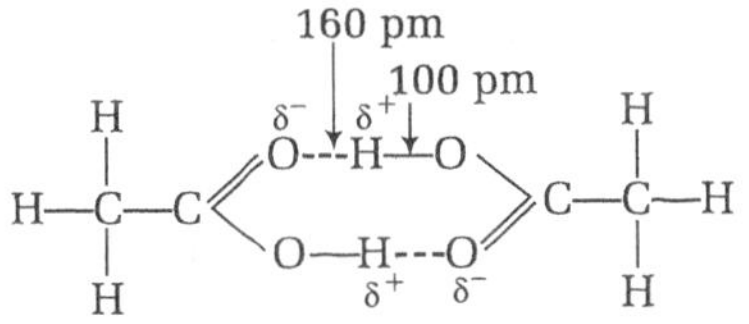

Metallic Bonding

The attractive force which hold together the constituent particles in a metal is known as **metallic bonding**.

The two models which explain metallic bonding are :

(i) **According to electron sea model**, metallic crystal consists of positive kernels packed together as closely as possible in a regular geometric pattern and immersed in a sea of mobile electrons.

(ii) **According to band model**, atomic orbitals of atoms with same energy and same symmetry overlap to form energy bands. The highest occupied energy band is valence band and lowest unoccupied energy band is conduction band. The gap between these two bands is called energy gap. In insulators, energy gap is very large while in semiconductors it is very small. On increasing temperature, electrical conductivity of semiconductors increases because some electrons move from valence band to conduction band.

DAY PRACTICE SESSION 1

FOUNDATION QUESTIONS EXERCISE

1 Among the following, electron deficient molecule is
(a) CCl_4 (b) PCl_5
(c) BF_3 (d) SF_6

2 Sodium chloride is soluble in water but not in benzene because
(a) $\Delta H_{\text{Hydration}} < \Delta H_{\text{Lattice energy in water}}$
$\Delta H_{\text{Hydration}} > \Delta H_{\text{Lattice energy in benzene}}$
(b) $\Delta H_{\text{Hydration}} > \Delta H_{\text{Lattice energy in water}}$
$\Delta H_{\text{Hydration}} < \Delta H_{\text{Lattice energy in benzene}}$
(c) $\Delta H_{\text{Hydration}} = \Delta H_{\text{Lattice energy in water}}$
$\Delta H_{\text{Hydration}} < \Delta H_{\text{Lattice energy in benzene}}$
(d) $\Delta H_{\text{Hydration}} < \Delta H_{\text{Lattice energy in water}}$
$\Delta H_{\text{Hydration}} = \Delta H_{\text{Lattice energy in benzene}}$

3 Which of the following compounds contain(s) no covalent bond(s)? → JEE Main 2018
KCl, PH_3, O_2, B_2H_6, H_2SO_4
(a) KCl, B_2H_6, PH_3 (b) KCl, H_2SO_4
(c) KCl (d) KCl, B_2H_6

4 The % ionic character in Cs—Cl bond present in CsCl molecule will be, if the electronegativity values for Cs and Cl are 0.8 and 3.0 respectively
(a) 62.9% (b) 60%
(c) 75% (d) 52.14%

5 Which one of the following molecule is polar? → JEE Main (Online) 2013
(a) XeF_4 (b) IF_5
(c) SbF_5 (d) CF_4

6 Which of the following pairs has zero dipole moment?
(a) CH_2Cl_2 and NF_3 (b) SiF_4 and BF_3
(c) PCl_3 and ClF (d) BF_3 and NF_3

7 Bond distance in HF is 9.17×10^{-11} m. Dipole moment of HF is 6.104×10^{-30} cm. The per cent ionic character in HF will be (electron charge $= 1.60 \times 10^{-19}$ C) → JEE Main (Online) 2013
(a) 61.0% (b) 38.0%
(c) 35.5% (d) 41.6%

8 Among the following the maximum covalent character is shown by the compound → AIEEE 2011
(a) $FeCl_2$ (b) $SnCl_2$ (c) $AlCl_3$ (d) $MgCl_2$

9 The correct statement for the molecule, CsI_3 is → JEE Main 2014
(a) it is a covalent molecule
(b) it contains Cs^+ and I_3^-
(c) It contains Cs^{3+} and I^- ions
(d) it contains Cs^+, I^- and lattice I_2 molecule

10 Bond order normally gives idea of stability of a molecular species. All the molecules *viz.* H_2, Li_2 and B_2 have the same bond order yet they are not equally stable. Their stability order is → JEE Main (Online) 2013
(a) $H_2 > B_2 > Li_2$ (b) $H_2 > Li_2 > B_2$
(c) $Li_2 > B_2 > H_2$ (d) $B_2 > H_2 > Li_2$

11 Which of the following compounds has the smallest bond angle in its molecule?
(a) H_2O (b) H_2S (c) NH_3 (d) SO_2

12 Arrange the following molecules in the increasing order of bond angle.

H_2O H_2S H_2Se H_2Te
I II III IV

(a) I < II < III < IV (b) IV < III < II < I
(c) I < III < II < IV (d) IV < II < III < I

13 Consider the following iodides,

PI_3—102°, AsI_3—100.2, SbI_3—99°

The bond angle is maximum in PI_3 which is
(a) due to small size of phosphorus
(b) due to more *bp-bp* repulsion in PI_3
(c) due to less electronegativity of phosphorus
(d) None of the above

14 Bond energy of H—H, F—F and H—F bonds are 104, 38 and 135 kcal mol^{-1} respectively. The resonance energy in the H—F molecule will be
(a) 142 kcal mol^{-1} (b) 66 kcal mol^{-1}
(c) 72.14 kcal mol^{-1} (d) 79.26 kcal mol^{-1}

15 The species in which the N-atom is in a state of *sp*- hybridisation is → JEE Main 2016
(a) NO_2^- (b) NO_3^- (c) NO_2 (d) NO_2^+

16 The *d*-orbital involved in sp^3d-hybridisation is
(a) d_{xy} (b) d_{zx} (c) d_{z^2} (d) $d_{x^2-y^2}$

17 The hybridisation of orbitals of N-atom in NO_3^-, NO_2^+ and NH_4^+ respectively are → AIEEE 2011
(a) sp, sp^2, sp^3 (b) sp^2, sp, sp^3
(c) sp, sp^3, sp^2 (d) sp^2, sp^3, sp

18 The states of hybridisation of boron and oxygen atoms in boric acid (H_3BO_3) are respectively
(a) sp^2 and sp^3
(b) sp^2 and sp^3
(c) sp^3 and sp^2
(d) sp^3 and sp^3

19 Match the following and choose the correct option.

	Column I		Column II
A.	SF_4	1.	sp^3d^2
B.	IF_5	2.	sp^3
C.	NO_2^+	3.	sp
D.	NH_4^+	4.	sp^3d

Codes

	A	B	C	D		A	B	C	D
(a)	4	1	3	2	(b)	1	3	2	4
(c)	3	2	4	1	(d)	3	1	2	4

20 Isostructural species are those which have the same shape and hybridisation. Among the given species identify the isostructural pairs.
(a) NF_3 and BF_3 (b) BF_4^- and NH_4^+
(c) BCl_3 and $BrCl_3$ (d) NH_3 and NO_3^-

21 In which of the following pairs the two species are not isostructural? → AIEEE 2012
(a) CO_3^{2-} and NO_3^- (b) PCl_4^+ and $SiCl_4$
(c) PF_5 and BrF_5 (d) AlF_6^{3-} and SF_6

22 Total number of lone pair of electron in I_3^- ion is
(a) 3 (b) 6 (c) 9 (d) 2

23 In XeF_2, XeF_4 and XeF_6, the number of lone pairs of Xe respectively are
(a) 2, 3, 1 (b) 1, 2, 3
(c) 4, 1, 2 (d) 3, 2, 1

24 The structure of IF_7 is → AIEEE 2010
(a) square pyramidal (b) trigonal bipyramidal
(c) octahedral (d) pentagonal bipyramidal

25 Which has trigonal bipyramidal shape? → JEE Main (Online) 2013
(a) $XeOF_4$ (b) XeO_3
(c) XeO_3F_2 (d) $XeOF_2$

26 The shape of IF_6^- is → JEE Main (Online) 2013
(a) trigonally distorted octahedron
(b) pyramidal
(c) octahedral
(d) square antiprism

27 The molecular shapes of SF_4, CF_4 and XeF_4 are
(a) different with 1, 0 and 2 lone pairs of electrons on the central atoms, respectively
(b) different with 0, 1 and 2 lone pairs of electrons on the central atoms, respectively
(c) the same with 1, 1 and 1 lone pairs of electrons on the central atoms, respectively
(d) the same with 2, 0 and 1 lone pairs of electrons on the central atoms, respectively

28 The structure of which of the following chloro species can be explained on the basis of dsp^2-hybridisation ? → JEE Main (Online) 2013
(a) $PdCl_4^{2-}$ (b) $FeCl_4^{2-}$
(c) $CoCl_4^{2-}$ (d) $NiCl_4^{2-}$

29 XeO_4 molecule is tetrahedral having → JEE Main (Online) 2013
(a) two $p\pi - d\pi$ bonds (b) one $p\pi - d\pi$ bond
(c) four $p\pi - d\pi$ bonds (d) three $p\pi - d\pi$ bonds

30 Which of the following species is not paramagnetic? → JEE Main 2017
(a) NO (b) CO (c) O_2 (d) B_2

31 Which of the following order of energies of molecular orbitals of N_2 is correct?
(a) $\sigma 2s^2 < \sigma^* 2s^2 < \pi 2p_x^2 \approx \pi 2p_y^2 < \sigma 2p_z^2$
(b) $\sigma 2s^2 < \sigma^* 2s^2 < \sigma 2p_z^2 < \pi 2p_x^2 \approx \pi 2p_y^2$
(c) $\sigma^* 2s^2 < \sigma 2s^2 < \sigma 2p_z^2 < \pi 2p_x^2 < \pi 2p_y^2$
(d) $\sigma 2s^2 < \sigma 2p_z^2 < \sigma 2s^2 < \pi 2p_x^2 < \pi 2p_y^2$

32 Which one of the following molecules is/are expected to exhibited diamagnetic behaviour? → JEE Main 2013

(a) N_2 (b) O_2
(c) S_2 (d) C_2

33 Which of the following statement(s) is/are correct?

(a) In the formation of dioxygen from oxygen atoms 10 molecular orbitals will be formed.
(b) All the molecular orbitals in the dioxygen will be completely filled
(c) Total number of bonding molecular orbitals will not be same as total number of anti-bonding orbitals in dioxygen
(d) Number of filled bonding orbitals will be same as number of filled anti-bonding orbitals

34 In accordance to molecular theory,

(a) O_2^+ is diamagnetic and bond order is more than O_2
(b) O_2^+ is diamagnetic and bond order is less than O_2
(c) O_2^+ is paramagnetic and bond order is more than O_2
(d) O_2^+ is paramagnetic and bond order is less than O_2

35 The stability of the species Li_2, Li_2^- and Li_2^+ increases in the order of → JEE Main 2013

(a) $Li_2^- < Li_2^+ < Li_2$
(b) $Li_2 < Li_2^- < Li_2^+$
(c) $Li_2^- < Li_2 < Li_2^+$
(d) $Li_2 < Li_2^+ < Li_2^-$

36 Which of the following pairs of molecules/ions are not likely to exist? → JEE Main 2013

(a) H_2^+, He_2^{2-}
(b) H_2^-, He_2^{2-}
(c) H_2^{2+}, He_2
(d) H_2^-, He_2^{2+}

37 *Ortho*-nitrophenol is less soluble in water than *p*-and *m*-nitrophenols because → AIEEE 2012

(a) *o*-nitrophenol is more steam volatile than those of *m*- and *p*-isomers
(b) *o*-nitrophenol shows intramolecular H-bonding
(c) *o*-nitrophenol shows intermolecular H-bonding
(d)melting point of *o*-nitrophenol is lower than those of *m*- and *p*-isomers

38 Consider the following compounds,

I. 1, 2-hydroxybenzene
II. 1, 3-dihydroxybenzene
III. 1, 4-dihydroxybenzene
IV. hydroxybenzene

The increasing order of their boiling points is

(a I < II < III < IV
(b) IV < I < II < III
(c) IV < II < I < III
(d) I < II < IV < III

39 Given,

Reaction	Energy change (in kJ)
$Li(s) \longrightarrow Li(g)$	161
$Li(g) \longrightarrow Li^+(g)$	520
$\frac{1}{2}F_2(g) \longrightarrow F(g)$	77
$F(g) + e^- \longrightarrow F^-(g)$	(Electron gain enthalpy)
$Li(g) + F^-(g) \longrightarrow LiF(s)$	−1047
$Li(s) + \frac{1}{2}F_2(g) \longrightarrow LiF(s)$	− 617

Based on data provided, the value of electron gain enthalpy of fluorine would be → JEE Main 2013

(a) $-300\ kJ\ mol^{-1}$ (b) $-350\ kJ\ mol^{-1}$
(c) $-328\ kJ\ mol^{-1}$ (d) $-228\ kJ\ mol^{-1}$

Direction (Q. No 40-43) *In the following question, a statement of Assertion (A) followed by a statement of Reason (R) is given. Choose the correct answer out of the following choices.*

(a) Both A and R are true and R is the correct explanation of A
(b) Both A and R are true but R is not the correct explanation of A
(c) A is true but R is false
(d) Both A and R are false

40 **Assertion** (A) The bond angles in molecules depend upon electronegativity of central atom, number of lone pairs, odd electrons and multiplicity of bonds.

Reason (R) NO_2 and NO_2^- have angles 134° and 115° respectively.

41 **Assertion** (A) Among the two O—H bonds in H_2O molecule, the energy required to break the first O—H bond and the other O—H bonds is the same.

Reason (R) This is because the electronic environment around oxygen is the same even after breakage of one O—H bond.

42 **Assertion** (A) SF_4 has a lone pair of electrons at equatorial position in preference to axial position in the overall trigonal bipyramidal geometry.

Reason (R) If lone pair of electrons is at equatorial position, repulsion will be minimum.

43 **Assertion** (A) *p*-dimethoxy benzene is a polar molecule.

Reason (R) The two methoxy groups at *para* position are located as

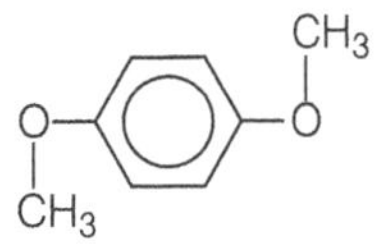

DAY PRACTICE SESSION 2

PROGRESSIVE QUESTIONS EXERCISE

1 Which of the following statement(s) is/are true?
(a) HF is less polar than HBr
(b) Absolutely pure water does not contain any ions
(c) Chemical bond formation takes place when forces of attraction overcome the forces of repulsion
(d) In covalency, transference of electrons take place

2 Consider the Born-Haber cycle for the formation of ionic compound given below and identify the compound formed.

$$\left[\begin{array}{l} M(s) \xrightarrow{\Delta H_1} M(g) \xrightarrow{\Delta H_2} M^+(g) \\ \frac{1}{2}X_2(g) \xrightarrow{\Delta H_3} X(g) \xrightarrow{\Delta H_4} X^-(g) \end{array}\right] \xrightarrow{\Delta H_5} Z$$

(a) M^+X^-
(b) $M^+X^-(s)$
(c) MX
(d) $M^+X^-(g)$

3 What is the geometry of nitrogen atom in NH_3, $N(CH_3)_3$ and $N[Si(CH_3)_3]_3$ molecules?
I. Trigonal planar
II. Trigonal pyramidal
III. Square planar
Choose the correct option.
(a) II, II and I
(b) III, III and I
(c) I, II and III
(d) I, III and II

4 O_2 molecule is paramagnetic due to the presence of
(a) two unpaired electrons in anti-bonding MO
(b) one unpaired electron in anti-bonding MO
(c) two unpaired electrons in bonding MO
(d) one unpaired electron in bonding MO

5 The BCl_3 is a planar molecule whereas NCl_3 is pyramidal because
(a) B—Cl bond is more polar than N—Cl bond
(b) N—Cl bond is more covalent that B—Cl bond
(c) nitrogen atom is smaller than boron atom
(d) BCl_3 has no lone pair electrons but NCl_3 has one lone pair of electrons

6 H_2O is dipolar whereas BeF_2 is not. It is because
(a) the electronegativity of F is greater than that of O
(b) H_2O involves hydrogen bonding whereas BeF_2 is angular discrete molecule
(c) H_2O is linear and BeF_2 is angular
(d) H_2O is angular and BeF_2 is linear

7 In terms of polar character, which one of the following order is correct?
(a) $NH_3 < H_2O < HF < H_2S$
(b) $H_2S < NH_3 < H_2O < HF$
(c) $H_2O < NH_3 < H_2S < HF$
(d) $HF < H_2O < NH_3 < H_2S$

8 What will be the lattice enthalpy of $CaCl_2$, if the enthalpy of
(i) sublimation energy of Ca is 121 kJ mol^{-1}
(ii) dissociation energy of Cl_2 to Cl is 242.8 kJ mol^{-1}
(iii) ionisation energy Ca to Ca^{2+} is 2422 kJ mol^{-1}
(iv) electron gain enthalpy for Cl to Cl^- is −355 kJ mol^{-1}
(v) $\Delta_f H$ overall is −795 kJ mol^{-1}
(a) −3225.8 kJ mol^{-1}
(b) −980.8 kJ mol^{-1}
(c) −2870.8 kJ mol^{-1}
(d) −2628 kJ mol^{-1}

9 Select the correct statement(s).
(a) Both lattice energy and hydration energy decrease with the increase in ionic size
(b) Lattice energy can be calculated using Born-Haber cycle
(c) If the anion is larger than the cation, the lattice energy will remain almost constant within a particular group
(d) All of the above statements are correct

10 The bond dissociation energy of B—F in BF_3 is 646 kJ mol^{-1}, whereas that of C—F in CF_4 is 515 kJ mol^{-1}. The correct reason for higher B—F bond dissociation energy as compared to that of C—F bond is
(a) smaller size of B-atom as compared to that of C-atom
(b) stronger σ bond between B and F in BF_3 as compared to that between C and F is CF_4
(c) significant $p\pi$-$p\pi$ interaction between B and F in BF_3 whereas there is no possibility of such interaction between C and F in CF_4
(d) lower degree of $p\pi$-$p\pi$ interaction betwen B and F in BF_3 than that between C and F in CF_4

11 If H—X bond length is 2Å and H–X bond has dipole moment 5.12×10^{-30} cm, the percentage of ionic characters in the molecule will be
(a) 10%
(b) 16%
(c) 18%
(d) 20%

12 The geometrical shapes of XeF_5^+, XeF_6 and XeF_8^{2-} respectively are
(a) trigonal bipyramidal, octahedral and square planar
(b) square pyramidal, distorted octahedral and square antiprismatic
(c) planar pentagonal, octahedral and square antiprismatic
(d) square pyramidal, distorted octahedral and octahedral

13 Suppose the observed value of dipole moment of H_2O molecule is 1.83 D. What will be the H—O—H bond angle in H_2O molecule? (Given bond moment of O—H bond is 1.5 D.)
(a) 104° 20″
(b) 105°
(c) 105°20″
(d) 104°

ANSWERS

SESSION 1	**1** (c)	**2** (b)	**3** (c)	**4** (d)	**5** (b)	**6** (b)	**7** (d)	**8** (c)	**9** (b)	**10** (b)
	11 (d)	**12** (b)	**13** (b)	**14** (c)	**15** (d)	**16** (c)	**17** (b)	**18** (b)	**19** (a)	**20** (b)
	21 (c)	**22** (c)	**23** (d)	**24** (b)	**25** (d)	**26** (a)	**27** (a)	**28** (a)	**29** (c)	**30** (b)
	31 (a)	**32** (a,d)	**33** (a)	**34** (c)	**35** (a)	**36** (c)	**37** (b)	**38** (b)	**39** (c)	**40** (a)
	41 (d)	**42** (a)	**43** (a)							
SESSION 2	**1** (c)	**2** (b)	**3** (a)	**4** (a)	**5** (d)	**6** (d)	**7** (b)	**8** (c)	**9** (d)	**10** (c)
	11 (b)	**12** (b)	**13** (a)							

Hints and Explanations

SESSION 1

1 The compounds in which central atom has less than 8 electrons in their valence shell, are electron deficient.

(a) CCl_4,

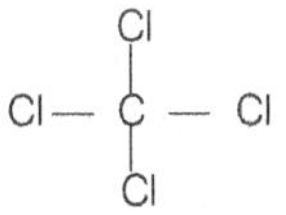

Carbon (central atom) has 8 electrons in the valence shell, 4 from carbon and 1 electron each from four chlorine atoms.

(b) PCl_5,

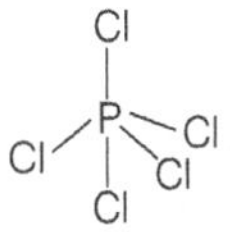

Phosphorus (central atom) has 10 electrons in valence shell, 5 from P and one each from five chlorine atoms.

(c) BF_3,

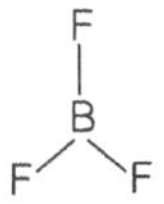

Boron (central atom) has 6 electrons in valence shell, 3 from boron and one each from three flourine atoms.

Boron in BF_3 has less than 8 electrons. So, it is electron deficient molecule.

(d) SF_6

Sulphur (central atom) has 12 electrons in valence shell, 6 from sulphur and one each from six fluorine atoms.

2 For a compound to be soluble, the hydration energy must be greater than lattice energy. Since, NaCl is soluble in water but insoluble in benzence due to,

$\Delta H_{\text{Hydration}} > \Delta H_{\text{Lattice energy in water}}$ and

$\Delta H_{\text{Hydration}} < \Delta H_{\text{Lattice energy in benzene}}$

3 KCl is the only ionic compound. The covalent bonds in PH_3, O_2, B_2H_6 and H_2SO_4 are shown below:

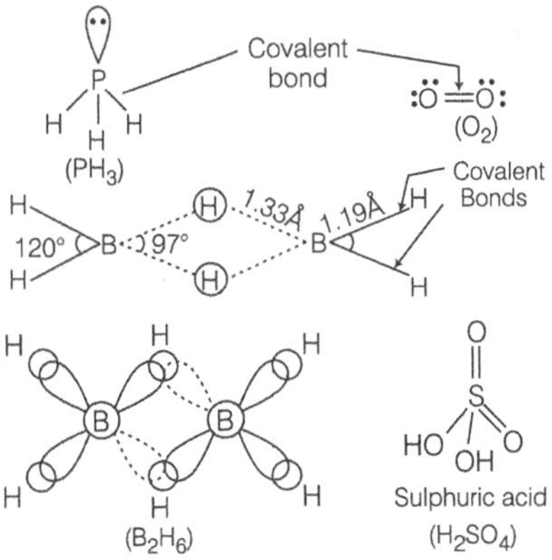

In H_2SO_4 all bond between S and O atom are covalent bonds.

4 % ionic character

$= [16\,(X_{Cl} - X_{Cs}) + 3.5\,(X_{Cl} - X_{Cs})^2]$

$= [16\,(3.0 - 0.8) + 3.5\,(3.0 - 0.8)^2]$

$= [16 \times 2.2 + 3.5 \times (2.2)^2]$

$= [35.2 + 16.94] = 52.14\%$

5 Only IF_5 is polar because of its unsymmetrical structure. Rest of the molecules have zero dipole moment as they are symmetrical in nature.

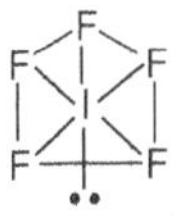

Structure of IF_5

6 SiF_4 is a symmetrical tetrahedral molecule and BF_3 is a triangular planar (symmetrical) structure and hence, have zero dipole moment.

7 $\%$ ionic character $= \dfrac{\mu_{\text{observed}}}{\mu_{\text{calculated}}} \times 100$

$\mu_{\text{calculated}} = e \times d$

$= 1.6 \times 10^{-19} \text{C} \times 9.17 \times 10^{-11} \text{m}$

$= 1.467 \times 10^{-29}$ cm

$\therefore$ % ionic character

$= \dfrac{6.104 \times 10^{-30}}{1.467 \times 10^{-29}} \times 100$

$= 41.6\%$

8 Covalent character can be determined by applying Fajan's rule.

In all the given compounds, anion is same (Cl^-), hence polarising power is decided by size and charge of cation. Al^{3+} with maximum charge and smallest size has maximum polarising power hence, $AlCl_3$ has maximum covalent character.

9 I_3^- is an ion made up of I_2 and I^- which has linear shape. While Cs^+ is an alkali metal cation.

10 The MO configuration of the given species can be written as :

$H_2\,(1 + 1 = 2) = \sigma 1s^2$

$Li_2\,(3 + 3 = 6) = \sigma 1s^2, \sigma^* 1s^2, \sigma 2s^2$

$B_2 (5 + 5 = 10) = \sigma 1s^2, \sigma^* 1s^2, \sigma 2s^2, \sigma^* 2s^2, \pi 2p_x^1 \approx \pi 2p_y^1$

As the number of antibonding electrons (σ^* or π^*) increases, energy increases and stability decreases.

Thus, the correct order of stability is $H_2 > Li_2 > B_2$

11 Bond angle of H_2S is smallest because S-atom is larger in size and has low electronegativity.

12 *lp-bp* repulsion is maximum in H_2Te due to least electronegativity of Te and minimum in H_2O due to high electronegativity of O. So, the correct order is IV < III < II < I.

13 P is most electronegative among the given options due to this there is more *bp-bp* repulsion.

14 Resonance energy;

$$\Delta_{H-F} = (BE)_{H-F} - \sqrt{(BE)_{H_2} \times (BE)_{F_2}}$$
$$= 135 - \sqrt{104 \times 38}$$
$$= 135 - 62.86$$
$$= 72.14 \text{ kcal mol}^{-1}$$

15

Species	Hybridisation
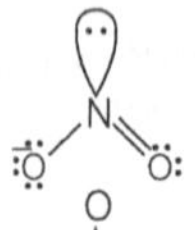	sp^2
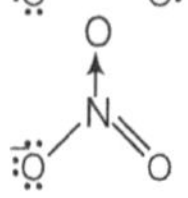	sp^2
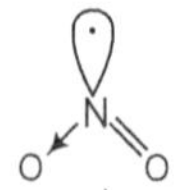	sp^2
$O=\overset{+}{N}=O$	sp

16 The d_{z^2} orbital is involved in sp^3d-hybridisation. In sp^3d hybridisation, 1 *s* and 2 *p*-orbitals combine to form three planar triangular hybrid orbitals. The remaining *p*-orbital may combine with d_{z^2} orbital to form two axial orbitals.

17 Count σ-bond, lone pairs and unpaired electron or count number of atoms directly attached, lone pairs and unpaired electrons.

NO_3^- (I) 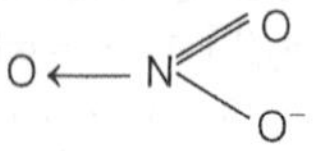

NO_2^+ (II) $O=\overset{+}{N}=O$

NH_4^+ (III) 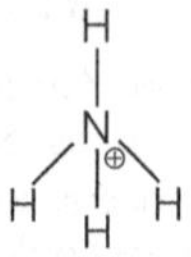

S.No.	σ-bond	Lone pair	Unpaired electron	Total
I.	3	×	×	3 (sp^2)
II.	2	×	×	2 (sp)
III.	4	×	×	4 (sp^3)

18 H_3BO_3 has structure

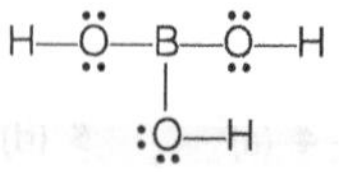

Boron has three bonds, thus sp^2 hybridised. Each oxygen atom has two bonds and two lone pairs, hence sp^3 hybridised.

19 (a) $SF_4 = 4bp + 1lp = sp^3d$-hybridisation

(b) $IF_5 = 5bp + 1lp = sp^3d^2$-hybridisation

(c) $NO_2^+ = 2bp + 0lp = sp$-hybridisation

(d) $NH_4^+ = 4bp + 0lp = sp^3$-hybridisation

20 The species BF_4^- and NH_4^+ both 4*bps* and 0*lp*, thus they are sp^3-hybridised, therefore, both these species have similar shape and hybridisation. i.e. They are isostructural species.

21 (a) CO_3^{2-}, NO_3^- — triangular planar

(b) PCl_4^+, $SiCl_4$ — tetrahedral

(c) PF_5, BrF_5 — trigonal bipyramidal, square pyramidal

(d) AlF_6^{3-}, SF_6 — octahedral

22 The structure of I_3^- ion is

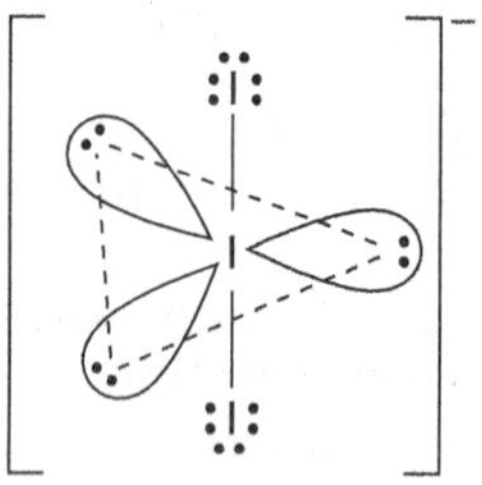

Hence, 9 is the correct answer.

23 Xe-atom has 8 electrons in its outermost shell. In case of XeF_2 out of these 8 electrons 2 are used for bond formation while 3 pairs remain as such, i.e. it has 3 lone pairs.

In case of XeF_4, 4 electrons of Xe are used for bonding, therefore number of lone pairs (non-bonding electrons) is 2.

In case of XeF_6, 6 electrons are involved for bond formation, thus, number of lone pair is only 1.

24 IF_7

In ground state

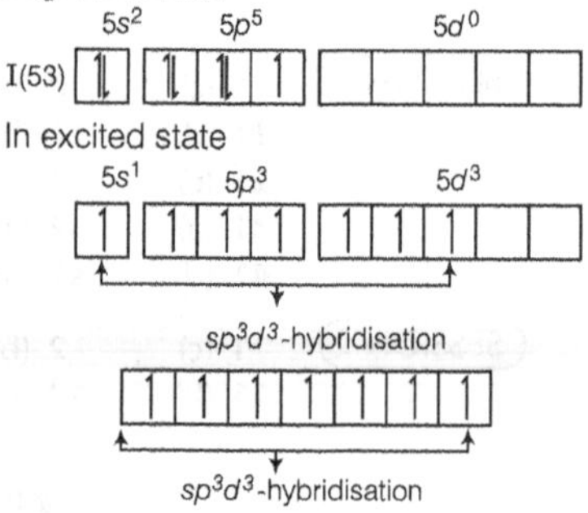

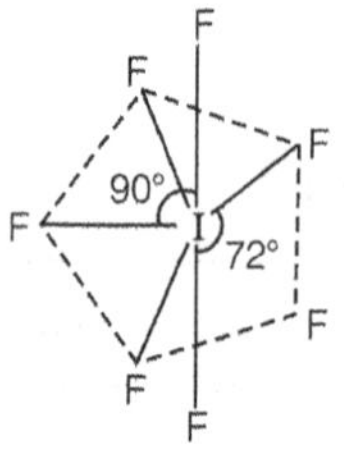

The structure of IF_7 is pentagonal-bipyramidal.

25 $XeOF_4 = 5bp + 1lp$ = square-pyramidal shape

$XeO_3 = 3bp + 1lp$ = pyramidal shape

$XeO_3F_2 = 5bp + 0lp$ = trigonal-bipyramidal shape

$XeOF_2 = 3bp + 2lp$ = T-shape

The shape is trigonal bipyramidal, if a compound have five bond pairs and zero lone pairs.

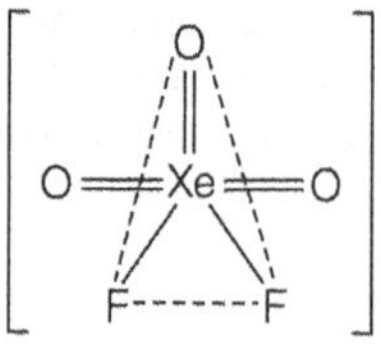

Trigonal bipyramidal

26 Number of hybrid orbital,

$$H = \frac{1}{2}[V - C + A + M]$$

[Here V = valence e^- of central atom, C and A = positive and negative charge respectively, M = monovalent atoms]

In IF_6^-, $H = \frac{1}{2}[7 - 0 + 1 + 6] = 7$

So, the hybridisation is sp^3d^3 and structure should be pentagonal bipyramidal. But it contains one lone pair and 6 bond pairs. Hence, its actual geometry is trigonally distorted octahedron.

27 SF_4 (sp^3d), CF_4 (sp^3) and XeF_4 (sp^3d^2) contain 1,0 and 2 lone pairs respectively. Therefore, their shapes are also different.

28 Although Cl^- is a weak field ligand but in case of $[PdCl_4]^{2-}$, it pair up the electrons of Pd and results in dsp^2-hybridisation because of the large size of Pd^{2+}. In all other options, pairing is not possible, so hybridisation is sp^3 with tetrahedral structure.

29 The structure of XeO_4 molecule is shown below :

It contains four $p\pi$- $d\pi$ bonds.

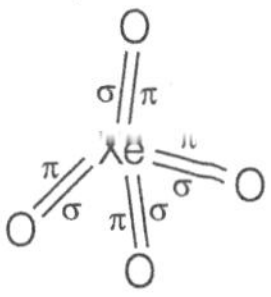

30 To identify the magnetic nature we need to check the molecular orbital configuration. If all orbitals are fully occupied, species is diamagnetic while when one or more molecular orbitals is/are singly occupied, species is paramagnetic.

(a) NO (7 + 8 = 15)

$\sigma 1s^2, \sigma^* 1s^2, \sigma 2s^2, \sigma^* 2s^2,$

$\pi 2p_x^2 = \pi 2p_y^2, \pi 2p_z^2, \pi^* 2p_x^1 = \pi^* 2p_y^0$

One unpaired electron is present. Hence, it is paramagnetic.

(b) CO (6 + 8 = 14) -

$\sigma 1s^2, \sigma^* 1s^2, \sigma 2s^2, \sigma^* 2s^2,$

$\pi 2p_x^2 = \pi 2p_y^2, \sigma 2p_z^2$

No unpaired electron is present. Hence, it is diamagnetic.

(c) O_2 (8 + 8 = 16) -

$\sigma 1s^2, \sigma^* 1s^2, \sigma 2s^2, \sigma^* 2s^2, \sigma 2p_z^2,$

$\pi 2p_x^2 = \pi 2p_y^2, \pi^* 2p_x^1 = \pi^* 2p_y^1$

Two unpaired electrons are present. Hence, it is paramagnetic.

(d) B_2(5 + 5) -

$\sigma 1s^2, \sigma^* 1s^2, \sigma 2s^2, \sigma^* 2s^2, \pi 2p_x^1$

$= \pi 2p_y^1$

Two unpaired electrons are present.

Hence, it is paramagnetic.

31 The correct increasing order of energies of molecular orbitals of N_2 is given below

$$\sigma 1s^2 < \overset{*}{\sigma} 1s^2 < \sigma 2s^2 < \overset{*}{\sigma} 2s^2 < (\pi 2p_x^2 \approx \pi 2p_y^2) < \sigma 2p_z^2$$

32 $C_2 = (6 + 6 = 12) =$

$\sigma 1s^2\ \sigma^* 1s^2\ \sigma 2s^2\ \sigma^* 2s^2\ \pi 2p_x^2 = \pi 2p_y^2$

$N_2 = (7 + 7 = 14)$

$\sigma 1s^2, \sigma^* 1s^2, \sigma 2s^2, \sigma^* 2s^2, \pi 2p_x^2,$

$= \pi 2p_y^2, \sigma 2p_z^2$

As all the electrons are paired, thus both C_2 and N_2 are diamagnetic.

33 In the formation of dioxygen from oxygen atoms, ten molecular orbitals will be formed.

$O_2 = \sigma 1s^2, \overset{*}{\sigma} 1s^2, \sigma 2s^2, \overset{*}{\sigma} 2s^2, \sigma 2p_z^2, \pi 2p_x^2$

$= \pi 2p_y^2,\ \overset{*}{\pi} 2p_x^1 = \overset{*}{\pi} 2p_y^1, \overset{*}{\sigma} 2p_z^0$

34 O_2^+ contains one unpaired electron, so it is paramagnetic and therefore it has bond order of 2.5 while O_2 contains only 2 unpaired electrons. So, it posses bond order of 2.

35 Stability of a molecule $\propto$ bond order

$Li_2(6) = \sigma 1s^2, \sigma^* 1s^2, \sigma 2s^2$

$$\text{Bond order} = \frac{4-2}{2} = 1$$

$Li_2^-(7) = \sigma 1s^2, \sigma^* 1s^2\ \sigma 2s^2, \sigma^* 2s^1$

$$\text{Bond order} = \frac{4-3}{2} = 0.5$$

$Li_2^+(5) = \sigma 1s^2, \sigma^* 2s^2, \sigma 2s^1$

$$\text{Bond order} = \frac{3-2}{2} = 0.5$$

As both Li_2^- and Li_2^+ has 0.5 bond order but Li_2^- is less stable because its valence electron is present in antiorbital. The stability order is $Li_2^- < Li_2^+ < Li_2$

36 Species having zero or negative bond order do not exist.

$H_2^{2+}(1 + 1 - 2 = 0) = \sigma 1s^0$

Bond order = 0

$He_2(2 + 2 = 4) = \sigma 1s^2, \overset{*}{\sigma} 1s^2$

$$\text{Bond order} = \frac{N_b - N_a}{2} = \frac{2-2}{2} = 0$$

So, both H_2^{2+} and He_2 do not exist.

37 There is intramolecular H-bonding in *o*-nitrophenol and thus, solubility in water is decreased.

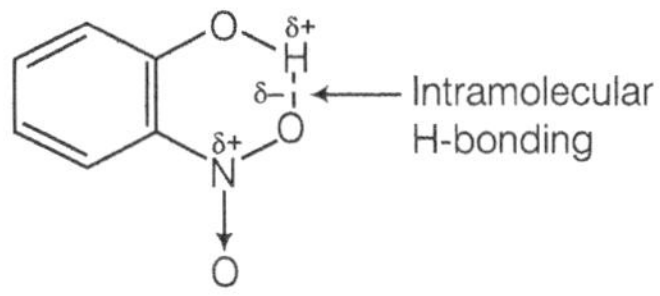

38

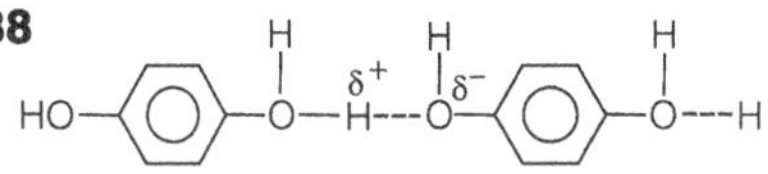

1,4-dihydroxybenzene

1, 4-dihydroxybenzene and isomer of dihydroxybenzene shows highest boiling point due to intermolecular H-bonding followed by *meta* and *ortho* isomer which shows intramolecular H-bonding. Hence, correct option is (b).

39 From Born-Haber cycle,

$$Q = S + I + D + EA + U$$

$$-617 = 161 + 520 + 77 + EA - 1047$$

[∵ Here, S = sublimation energy, I = ionisation energy,

D = dissociation energy, EA = electron gain enthalpy

and U = lattice energy]

$\therefore EA = 289 - 617 = -328$ kJ mol^{-1}

40 Reason is the correct explanation for Assertion.

(NO_2) (NO_2^-)

41 The bond enthalpies of the two O—H bonds in H—O—H are not equal. This is because electronic environment around O is not same after breakage of one O—H bond.

42 Both Assertion and Reason are correct.

43 *p*-dimethoxy benzene is polar due to unsymmetrical orientation of CH_3 group as shown below:

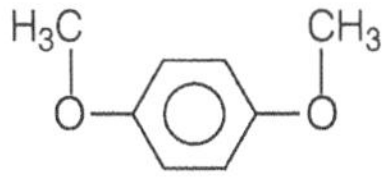

SESSION 2

1 A chemical bond is formed when forces of attraction are greater than the forces of repulsion.

2 The Born-Haber cycle takes place as follows :

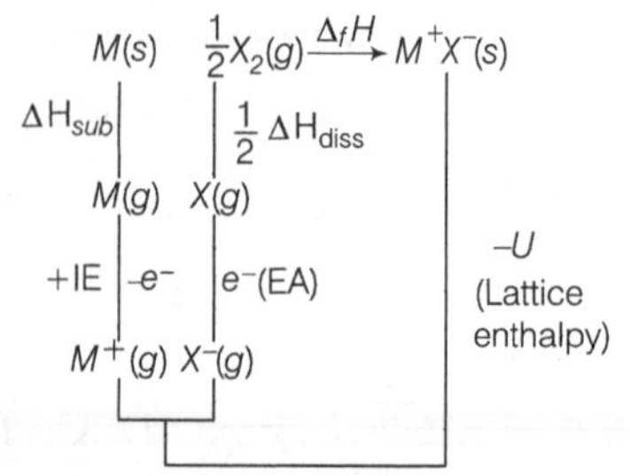

Hence, Z is $M^+ X^-(s)$

3 In case of NH_3 and $N(CH_3)_3$: Trigonal pyramidal due to *lp-bp* repulsion.

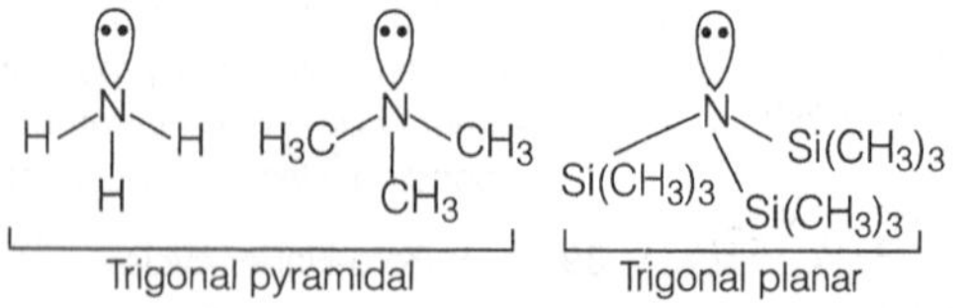

In case of $N[Si(CH_3)_3]$ it is planar. It is due to the fact that lone-pair on N-atom is transferred to the empty *d*-orbitals of silicon ($p\pi$-$d\pi$ overlapping) and its Lewis base character is lost.

4 O_2 $(8+8=16) = \sigma 1s^2, \sigma^* 2s^2, \sigma 2s^2, \sigma^* 2s^2, \sigma 2p_z^2,$

$$\pi 2p_x^2 = \pi 2p_y^2, \pi^* 2p_x^1 = \pi^* 2p_y^1$$

According to MO theory as there are two unpaired electron present in antibonding MO hence it is paramagnetic.

5

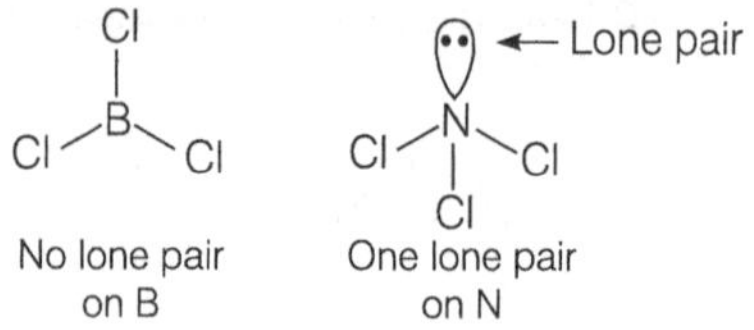

6 The structure of H_2O is angular V-shape and has sp^3 hybridisation and bond angle is 105°. Its dipole moment value is positive or more than zero.

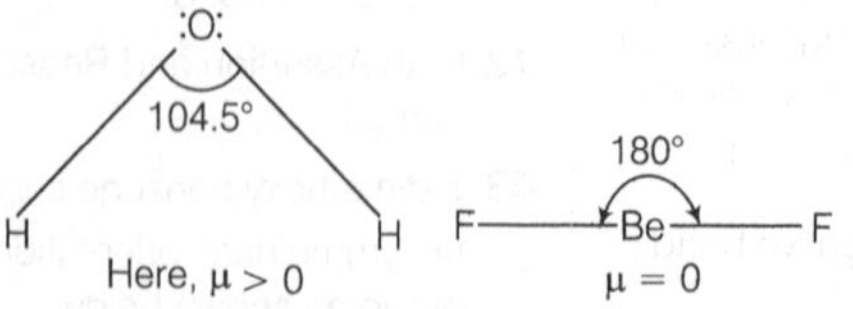

But in BeF_2, structure is linear due to *sp*-hybridisation ($\mu = 0$) Thus, due to $\mu > 0$, H_2O is dipolar and due to $\mu = 0$, BeF_2 is non-polar.

7 Polarity is decided by (EN) difference between two adjacent atoms. The correct order of polarity is :

$$H_2S < NH_3 < H_2O < HF$$

8 Given, $\Delta_{sub} H_{Ca} = 121\ \text{kJ mol}^{-1}$

$$\Delta H_{Cl-Cl} = 242.8\ \text{kJ mol}^{-1}$$

$$\Delta_i H_{Ca^{2+}} = 2422\ \text{kJ mol}^{-1}$$

$$\Delta_{eg} H_{Cl} = -355\ \text{kJ mol}^{-1}$$

$$\Delta_f H_{CaCl_2} = -795\ \text{kJ mol}^{-1}$$

$$\Delta_f H_{CaCl_2} = \Delta_{sub} H_{Ca} + \Delta H_{Cl-Cl} + \Delta_i H_{Ca^{2+}} + 2 \times \Delta_{eg} H_{Cl} + \Delta_l H_{CaCl_2}$$

or $\quad -795 = 121 + 242.8 + 2422 - (355 \times 2) + \Delta_l H_{CaCl_2}$

$\therefore \quad \Delta_l H_{CaCl_2} = -2870.80\ \text{kJ mol}^{-1}$

9 All given statements are correct.

10 In BF_3 there is significant $p\pi$-$p\pi$ interaction between unshared *p*-orbital (having no electron) of boron and the lone pair of electron of fluorine in 2*p*-orbital.

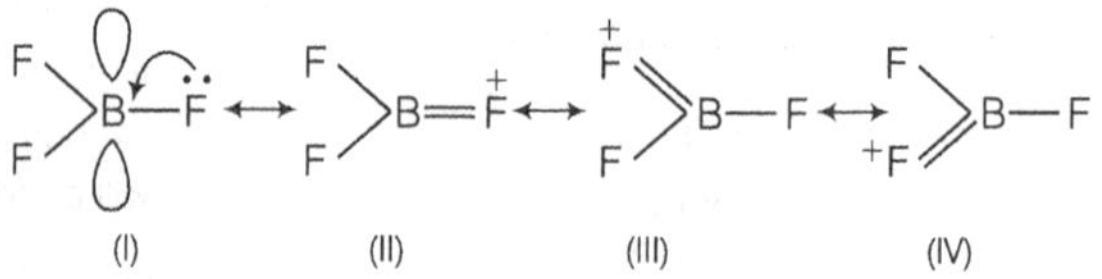

11 $\mu = q \times d$

$\mu_{cal} = 2.0 \times 10^{-10}\ \text{m} \times 1.6 \times 10^{-19}\ \text{C}$

$= 3.2 \times 10^{-29}$ C-m.

$$\text{Percentage of ionic character} = \frac{\mu_{exp}}{\mu_{cal}} \times 100$$

$$= \frac{5.12 \times 10^{-29}}{3.2 \times 10^{-29}} \times 100 = 16\%$$

12 $XeF_5^+ \Rightarrow 5\ bp + 1\ lp \Rightarrow$ (square pyramidal)

$XeF_6 \Rightarrow 6\ bp + 1\ lp \Rightarrow$ (distorted octahedral)

$XeF_8^{2-} \Rightarrow 8\ bp + 1\ lp \Rightarrow$ (square antiprismatic)

13 As dipole moment of H_2O is the resultant of the two vectors (O—H bonds), therefore if α is the angle between the two vectors, then

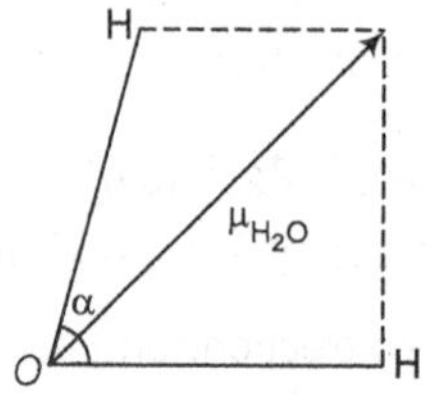

$$\mu_{H_2O} = \sqrt{\mu_{O-H}^2 + \mu_{O-H}^2 + 2\mu_{O-H}^2 \cos\alpha}$$

$$1.84 = \sqrt{(1.5)^2 + (1.5)^2 + 2(1.5)^2 \cos\alpha}$$

or $\quad 3.3856 = 2.25 + 2.25 + 4.50 \cos\alpha$

or $\quad \cos\alpha = -0.2476$

or $\quad \alpha = 104°20''$

[$\cos(180 - \theta) = -\cos\theta$

i.e., $\cos(180 - \alpha) = 0.2476$]

DAY FIVE

Unit Test 1

(General Chemistry)

1 A gaseous mixture contains oxygen and nitrogen in the ratio of 1 : 4 by weight. Therefore, the ratio of their number of molecules is

(a) 1 : 4
(b) 1 : 8
(c) 7 : 32
(d) 3 : 16

2 The iron (III) oxide, Fe_2O_3 reacts with CO to give iron and CO_2

$$Fe_2O_3 + 3CO \longrightarrow 2Fe + 3CO_2$$

The reaction of 480.0 g Fe_2O_3 with excess of CO yields 252.0 g iron. Thus, percentage yield is (Fe = 56)

(a) 100.0%
(b) 91.2%
(c) 75.0%
(d) 50.0%

3 When same amount of zinc is treated separately with excess of sulphuric acid and excess of sodium hydroxide, the ratio of volumes of hydrogen evolved is

(a) 1 : 1
(b) 1 : 2
(c) 2 : 1
(d) 9 : 4

4 The equivalent weight of $MnSO_4$ is half of its molecular weight, when it is converted to

(a) Mn_2O_3
(b) MnO_2
(c) MnO_4^-
(d) MnO_4^{2-}

5 The number of moles of $KMnO_4$ that will be needed to react with one mole of sulphite ion in acidic solution is

(a) $\frac{2}{5}$
(b) $\frac{3}{5}$
(c) $\frac{4}{5}$
(d) 1

6 Match the following and choose the correct option.

	Column I		Column II
A.	Mole fraction	1.	Candela
B.	Luminous intensity	2.	mol kg^{-1}
C.	Molality	3.	Unit less
D.	Molarity	4.	mol L^{-1}

Codes

	A	B	C	D		A	B	C	D
(a)	4	1	2	3	(b)	3	4	2	1
(c)	3	1	2	4	(d)	3	2	4	1

7 If the equivalent weight of an element is 32, percentage of oxygen in its oxide is

(a) 16
(b) 40
(c) 32
(d) 20

8 According to kinetic theory of gases for a diatomic molecule

(a) the pressure exerted by the gas is proportional to mean velocity of the molecule
(b) the pressure exerted by the gas is proportional to the root mean velocity of the molecule
(c) the root mean square velocity of the molecule is inversely proportional to the temperature
(d) the average translational kinetic energy of the molecule is proportional to the absolute temperature

9 X mL of H_2 gas effuses through a hole in container in 5 s. The time taken for the effusion of the same volume of the gas specified below under identical conditions is

(a) 10 s : He
(b) 20 s : O_2
(c) 25 s : CO
(d) 55 s : CO_2

10 The rms velocity of hydrogen is $\sqrt{7}$ times the rms velocity of nitrogen. If T is the temperature of the gas,

(a) $T(H_2) = T(N_2)$
(b) $T(H_2) > T(N_2)$
(c) $T(H_2) < T(N_2)$
(d) $T(H_2) = \sqrt{7}\,T(N_2)$

11 The ratio of rate of diffusion of helium and methane under identical conditions of pressure and temperature will be

(a) 4
(b) 2
(c) 1
(d) 0.5

12 The root mean square velocity of an ideal gas at constant pressure varies with density (d) as

(a) d^2 (b) d
(c) $\sqrt{d}$ (d) $1/\sqrt{d}$

13 The compound with longest bond length is → NCERT Exemplar

(a) NO^- (b) NO^+ (c) CN^- (d) CN

14 The numbers of σ and $d\pi - p\pi$ bonds in pyrophosphoric acid, ($H_4P_2O_7$) are → NCERT Exemplar

(a) 6 and 2 (b) 8 and 2 (c) 12 and 0 (d) 12 and 2

15 Match the Column I (quantum numbers) with the Column II (information provided) and choose the correct option.

	Column I (Quantum number)		Column II (Information provided)
A.	Principal quantum number	1.	Orientation of the orbital
B.	Azimuthal quantum number	2.	Energy and size of orbital
C.	Magnetic quantum number	3.	Spin of electron
D.	Spin quantum number	4.	Shape of the orbital

Codes

	A	B	C	D		A	B	C	D
(a)	1	3	4	2	(b)	2	4	1	3
(c)	2	3	4	1	(d)	3	1	2	4

16 The ratio of energy of a photon of 2000Å wavelength radiation to that of 4000Å radiation is

(a) 1/4 (b) 4
(c) 1/2 (d) 2

17 For a d-electron, the orbital angular momentum is

(a) $\sqrt{6}\left(\frac{h}{2\pi}\right)$ (b) $\sqrt{2}\left(\frac{h}{2\pi}\right)$
(c) $\left(\frac{h}{2\pi}\right)$ (d) $2\left(\frac{h}{2\pi}\right)$

18 The wavelength associated with a golf ball weighing 200 g and moving at a speed of 5 m/h is of the order

(a) 10^{-10}m (b) 10^{-20}m
(c) 10^{-30}m (d) 10^{-40}m

19 Which of the following emissions will generate an isotope?

(a) Neutron particle emission (b) Positron particle emission
(c) α-particle emission (d) β-particle emission

20 In which of the following molecules ion of all the bonds are not equal? → NCERT Exemplar

(a) XeF_4 (b) BF_4^- (c) C_2H_4 (d) SiF_4

21 Which of the following statements in relation to the hydrogen atom is correct?

(a) 3s, 3p and 3d orbitals all have the same energy
(b) 3s and 3p orbitals are of lower energy than 3d orbital
(c) 3p orbital is lower in energy than 3d orbital
(d) 3s orbital is lower in energy than 3p orbital

22 The potential energy of an electron in the first Bohr orbit in the He^+ ion is

(a) – 13.6 eV (b) – 27.2 eV
(c) – 54.4 eV (d) – 108.8 eV

23 **Assertion** (A) Significant figures for 0.200 is 3 whereas for 200 it is 1.

Reason (R) Zero at the end or right of a number are significant provided they are not on the right side of the decimal point. → [NCERT Exemplar]

(a) Both A and R are true and R is the correct explanation of A
(b) Both A and R are true but R is not the correct explanation of A
(c) A is true but R is false
(d) Both A and R are false

24 What electronic transition in a hydrogen atom starting from the orbit $n = 7$ will produce infrared light of wavelength 2170 nm? ($\bar{R}_H = 1.09677 \times 10^7\ m^{-1}$)

(a) $n = 7$ to $n = 5$ (b) $n = 7$ to $n = 4$
(c) $n = 7$ to $n = 3$ (d) $n = 7$ to $n = 2$

25 Photoelectrons are liberated by ultraviolet light of wavelength 3000 Å from a metallic surface for which the photoelectric threshold is 4000 Å. The de-Broglie wavelength of electron emitted with maximum kinetic energy is

(a) 1.2×10^{-9} m (b) 5.49×10^{-25} m
(c) 7.28×10^{-7} m (d) 1.65×10^{-19} m

26 Carbon suboxide (C_3O_2) has recently been shown as a component of the atmosphere of Venus. Which of the following formulation represents the correct ground state Lewis structure for carbon suboxide?

(a) :O: C:: C : C: O: (b) : O :: C:: C: C :: O:
(c) :$\ddot{O}$:: C:: C:: C ::$\ddot{O}$: (d) : O: C: C: C: O:

27 The atomic weight of an element is double of its atomic number. If there are four electrons in 2p orbital, the element is isotonic with

(a) $_{20}Ca^{40}$ (b) $_7N^{15}$
(c) $_7N^{14}$ (d) $_8O^{17}$

28 In PO_4^{3-} ion, the formal charge on the oxygen atom of P—O bond is → NCERT Exemplar

(a) +1 (b) –1 (c) –0.75 (d) + 0.75

29 The compound which contains both ionic and covalent bonds is

(a) CH_4 (b) H_2 (c) KCN (d) KCl

30 The total number of electrons that take part in forming the bond in N_2 is

(a) 2 (b) 4 (c) 6 (d) 10

31 Carbon tetrachloride has no net dipole moment because of

(a) its planar structure
(b) its regular tetrahedral structure
(c) similar sizes of carbon and chlorine atoms
(d) similar electron affinities of carbon and chlorine atoms

32 On hybridisation of one *s*-and one *p*-orbital, we get
(a) two mutually perpendicular orbitals
(b) two orbitals at 180°
(c) four orbitals directed tetrahedrally
(d) three orbitals in a plane

33 The molecule having one unpaired electron, is
(a) NO (b) CO (c) CN^- (d) O_2

34 The species in which the central atom uses sp^2 hybrid orbitals in its bonding is
(a) PH_3 (b) NH_3 (c) CH_3^+ (d) SbH_3

35 Among the following species, identify the isostructural pairs.
$NF_3, NO_3^-, BF_3, H_3O^+, N_3H$
(a) $[NF_3, NO_3^-]$ and $[BF_3, H_3O^+]$
(b) $[NF_3, N_3H]$ and $[NO_3^-, BF_3]$
(c) $[NF_3, H_3O^+]$ and $[NO_3^-, BF_3]$
(d) $[NF_3, H_3O^+]$ and $[N_3H, BF_3]$

36 10^{21} molecules are removed from 200 mg of CO_2. Thus, number of moles left is
(a) 1.73×10^{21} (b) 2.87×10^{-3}
(c) 0.287×10^{-4} (d) 2.73×10^{21}

37 What per cent of a sample of nitrogen must be allowed to escape if its temperature, pressure and volume are to be changed as given?

273°C, 3.00 atm, 1.65 L	→	0°C, 0.75 atm, 0.55 L

(a) 16.67% (b) 83.64% (c) 75.00% (d) 25.00%

38 If E_e, E_α and E_p represent the kinetic energies of an electron, alpha particle and a proton respectively, each moving with same de-Broglie wavelength then
(a) $E_e = E_\alpha = E_p$ (b) $E_e > E_\alpha > E_p$
(c) $E_\alpha > E_p > E_e$ (d) $E_e > E_p > E_\alpha$

39 Surface tension of water is 73 dyne cm^{-1} at 20°C. If surface area is increased by 0.10 m^2, workdone is
(a) 7.3 erg (b) 7.3×10^4 erg
(c) 73 J (d) 0.73 J

40 Phosphorus pentachloride dissociates in the vapour phase according to the equation

$$PCl_5 \rightleftharpoons PCl_4^+ + Cl^-$$

The hybridisation of phosphorus in PCl_5 is sp^3d. What is the hybridisation of P in PCl_4^+?
(a) sp^3d (b) sp^3d^2 (c) sp^3 (d) dsp^3

41 Among the species: CO_2, CH_3COO^-, CO, CO_3^{2-}, HCHO which has the weakest C—O bond?
(a) CO (b) CO_2, (c) CO_3^{2-} (d) CH_3COO^-,

42 A gas bulb of 1 mL capacity contains 2.0×10^{21} molecules of nitrogen exerting a pressure of 7.57×10^3 Nm^{-2}. The root mean square speed of the gas molecules is
(a) 274 ms^{-1} (b) 494 ms^{-1}
(c) 690 ms^{-1} (d) 988 ms^{-1}

43 A vessel contains 1.6 g of dioxygen at STP (273.15 K, 1 atm pressure). The gas is now transferred to another vessel at constant temperature, where pressure becomes half of the original pressure. The volume of the new vessel and number of molecules of dioxygen are respectively. → **NCERT Exemplar**
(a) 2.24, 6.02×10^{23} (b) 1.12, 3.01×10^{22}
(c) 2.24, 3.01×10^{21} (d) 2.24, 3.01×10^{22}

44 1.020 g of metallic oxide contains 0.540 g of the metal. If the specific heat of the metal, *M* is 0.216 cal $deg^{-1}g^{-1}$, the molecular formula of its oxide is
(a) *MO* (b) M_2O_3
(c) M_2O_4 (d) M_2O

45 A liquid of density 850 kg / m^3 having a surface tension of 0.055 N/m will rise how far in a glass capillary of 1.40 mm inside diameter in N/m
(a) 18.86 (b) 14.34
(c) 17.89 (d) 15.38

46 Assuming that Hund's rule is violated, the bond order and magnetic nature of the diatomic molecule B_2 is
(a) 1 and diamagnetic
(b) 0 and diamagnetic
(c) 1 and paramagnetic
(d) 0 and paramagnetic

Direction (Q. Nos. 47-50) *Each of these questions contains two statements : A (Assertion) and R (Reason). Each of these questions also has four alternative choices, only one of which is the correct answer. You have to select one of the codes (a), (b), (c) and (d) given below.*
(a) Both A and R are true and R is a correct explanation for Statement I
(b) Both A and R are true. But R is not a correct explanation for Statement I
(c) A is true; R is false
(d) Both A and R are false

47 **Assertion** (A) The value of van der Waals' constant *'a'* is larger for ammonia than for nitrogen.
Reason (R) Hydrogen bonding is present in ammonia.

48 **Assertion** (A) Cathode rays consist of negatively charged particles, called electrons.
Reason (R) In the presence of electrical/magnetic field, the behaviour of cathode rays is similar to the negatively charged particles.

49 **Assertion** (A) In H_2O molecule, O-atom has two lone pairs of electrons.
Reason (B) The geometry of H_2O is tetrahedral due to two sp^3-hybrid orbitals.

50 **Assertion** (A) Sulphur forms many compounds in which the octet rule is obeyed.
Reason (R) Due to the absence of *d*-orbitals in sulphur, it follows the octet rule.

ANSWERS

1 (c)	**2** (c)	**3** (a)	**4** (b)	**5** (a)	**6** (c)	**7** (d)	**8** (d)	**9** (b)	**10** (c)
11 (b)	**12** (d)	**13** (a)	**14** (d)	**15** (b)	**16** (d)	**17** (a)	**18** (c)	**19** (a)	**20** (c)
21 (a)	**22** (d)	**23** (c)	**24** (b)	**25** (a)	**26** (c)	**27** (b)	**28** (c)	**29** (c)	**30** (c)
31 (b)	**32** (b)	**33** (a)	**34** (c)	**35** (c)	**36** (b)	**37** (b)	**38** (d)	**39** (b)	**40** (c)
41 (c)	**42** (b)	**43** (d)	**44** (b)	**45** (a)	**46** (a)	**47** (a)	**48** (a)	**49** (c)	**50** (c)

Hints and Explanations

1 Let the weight of oxygen = x

$\therefore$ The weight of nitrogen = $4x$

Number of molecules of oxygen

$$= \frac{x \times N_0}{32}$$

Number of molecules of nitrogen

$$= \frac{4x}{28} \times N_0$$

$$\frac{\text{Number of molecules of oxygen}}{\text{Number of molecules of nitrogen}} = \frac{x \times 28}{32 \times 4x} = \frac{7}{32}$$

2 $Fe_2O_3 + 3CO \longrightarrow 2Fe + 3CO_2$

160 g Fe_2O_3 give = 112 g Fe

480 g Fe_2O_3 give = $\frac{112}{160} \times 480$ g Fe

= 336 g Fe (theoretical)

But actual yield is = 252.0 g

Thus, % yield

$$= \frac{\text{actual yield}}{\text{theoretical yield}} \times 100$$

$$= \frac{252}{336} \times 100 = 75\%$$

3 $Zn + H_2SO_4 \longrightarrow ZnSO_4 + \underset{1\text{ mol}}{H_2} \uparrow$

$Zn + NaOH \longrightarrow Na_2ZnO_2 + \underset{1\text{ mol}}{H_2} \uparrow$

Hence, the ratio of volumes of hydrogen evolved = 1 : 1

4 Equivalent weight

$$= \frac{\text{molecular weight}}{\text{valency factor}}$$

When $\overset{+2}{Mn}SO_4$ is converted into $\overset{+4}{Mn}O_2$, the valency factor is 2 and the equivalent weight of $MnSO_4$ will be half of its molecular weight.

5 $2MnO_4^- + 6H^+ + 5SO_3^{2-} \longrightarrow 2Mn^{2+} + 5SO_4^{2-} + 3H_2O$

$\because$ 5 moles of SO_3^{2-} required = 2 moles of MnO_4^-

$\therefore$ 1 mole of SO_3^{2-} will require = $\frac{2}{5}$ moles of MnO_4^-

6 A $\to$ 3, B $\to$ 1, C $\to$ 2, D $\to$ 4

A. Mole fraction = Unitless

B. Luminous Intensity = Candela

C. Molality = Concentration in mole per kg solvent

D. Molarity = Concentration in mol L^{-1}.

7 Equivalent weight of element = 32 g and that of oxygen = 8 g

Thus, one equivalent of oxide = 40 g

Percentage of oxygen in oxide

$$= 8 \times \frac{100}{40} = 20\%$$

8 According to kinetic theory of gases, for a diatomic molecule, the average translational kinetic energy of the molecule is proportional to the absolute temperature.

9 $r_{H_2} = \frac{\text{volume of } H_2}{\text{used time}} = \frac{X}{5}$ mL/s

$$\frac{r_{H_2}}{r_{O_2}} = \sqrt{\frac{M_{O_2}}{M_{H_2}}} = \sqrt{\frac{32}{2}} = 4$$

$\therefore$ $r_{H_2} = 4 \times r_{O_2}$

$$= \frac{X}{4 \times 5} = \frac{X}{20} \text{ mL/s}$$

10 $U_{rms}H_2 = \sqrt{7} \times U_{rms}N_2$

or $\sqrt{\frac{3RT_{H_2}}{2}} = \sqrt{7} \times \sqrt{\frac{3RT_{N_2}}{28}}$

$$\left(\because U_{rms} = \sqrt{\frac{3RT}{M}}\right)$$

$$\frac{3RT_{H_2}}{2} = 7 \times \frac{3RT_{N_2}}{28}$$

$$T_{N_2} = 2T_{H_2}$$

$\therefore$ $T_{N_2} > T_{H_2}$

11 By Graham's diffusion law,

$$\frac{r_{He}}{r_{CH_4}} = \sqrt{\frac{M_{CH_4}}{M_{He}}}$$

$$\Rightarrow \frac{r_{He}}{r_{CH_4}} = \sqrt{\frac{16}{4}} = \sqrt{\frac{4}{1}} = 2$$

Hence, ratio of rate of diffusion of He and CH_4 is 2.

12 $U_{rms} = \sqrt{\frac{3pV}{M}} = \sqrt{\frac{3p}{M/V}} = \sqrt{\frac{3p}{d}}$

(where, d = density)

Hence, at constant pressure,

$$U_{rms} \propto \frac{1}{\sqrt{d}}.$$

13 Bond length is inversely proportional to their bond order. The bond order for NO^-, NO^+, CN^- and CN are 2, 3, 3, 2.5. Therefore, NO^- has highest bond length.

14 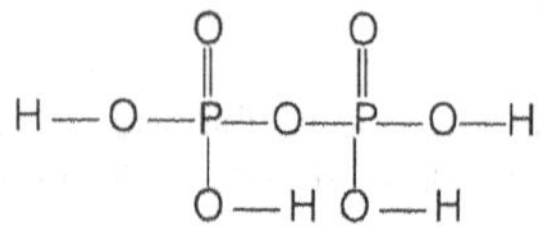

$\Rightarrow$ 12σ and $2d\pi - p\pi$ bonds.

15 A $\to$ 2, B $\to$ 4, C $\to$ 1, D $\to$ 3

16 $E_1 = \frac{hc}{\lambda_1}$ and $E_2 = \frac{hc}{\lambda_2}$

$$\frac{E_1}{E_2} = \frac{hc}{\lambda_1} \times \frac{\lambda_2}{hc}$$

$$\Rightarrow \frac{E_1}{E_2} = \frac{\lambda_2}{\lambda_1} = \frac{4000}{2000} = 2$$

17 For d-electron, $l = 2$

Angular momentum

$$= \frac{h}{2\pi}\sqrt{l(l+1)} = \frac{h}{2\pi}\sqrt{2(2+1)} = \frac{h}{2\pi}\sqrt{6}$$

18 $\lambda = \dfrac{h}{p}$ or $\dfrac{h}{mv}$

$$= \frac{6.626 \times 10^{-34} \text{ Js}}{200 \times 10^{-3} \times \text{kg} \times \dfrac{5}{60 \times 60} \text{ m/s}}$$

$$= \frac{6.626 \times 10^{-34} \times 60 \times 60}{200 \times 10^{-3} \times 5}$$

$$= 2.385 \times 10^{-30} \text{m}$$

19 ${}^{A}_{Z}X \longrightarrow {}^{A-1}_{Z}X + {}^{1}_{0}n$

Isotopes are the species having same number of protons but different number of neutrons, i.e. different mass number.

20 $XeF_4 = 4bp + 2lp \Rightarrow$ square planar $\Rightarrow$ all bonds are equal

$BF_4^- = 4bp + 0lp \Rightarrow$ tetrahedral $\Rightarrow$ all bonds are equal

C_2H_4 = 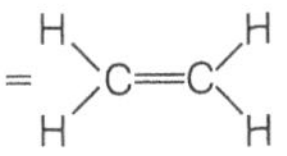

$\Rightarrow$ C=C bond is not equal to C—H bond

$SiF_4 = 4bp + 0lp \Rightarrow$ tetrahedral $\Rightarrow$ all bonds are equal.

21 Hydrogen atom has $1s^1$ configuration and these 3s, 3p and 3d-orbitals will have same energy with respect to 1s-orbital.

22 PE = −2 × total energy

$$= -2 \times \left(\frac{13.6Z^2}{n^2}\right)$$

For helium, $Z = 2$, and $n = 1$ in first Bohr orbit.

Thus, $\quad PE = -2 \times \dfrac{13.6 \times (2)^2}{1^2}$

$$= -108.8 \text{ eV}$$

23 0.200 contains 3 while 200 contains only one significant figure because zero at the end or right of a number are significant provided they are on the right side of the decimal point.

24 $\bar{\nu} = \dfrac{1}{\lambda} = \bar{R}_H Z^2 \left[\dfrac{1}{n_1^2} - \dfrac{1}{n_2^2}\right]$

$\lambda = 2170$ nm

For H-atom, $Z = 1$

and $n_2 = 7$ (given)

$$\therefore \frac{1}{2170 \times 10^{-9} \text{ m}} = 1.09677 \times 10^7 \text{m}^{-1} \left[\frac{1}{n_1^2} - \frac{1}{7^2}\right]$$

$$\therefore \frac{1}{n_1^2} = \frac{1}{49} + \frac{1}{\left[\begin{array}{l}2170 \times 10^{-9} \text{ m} \\ \times 1.09677 \times 10^7 \text{m}^{-1}\end{array}\right]}$$

$$= \frac{1}{49} + 0.042$$

$$\frac{1}{n_1^2} = 0.0624$$

$$n_1^2 = 16.02$$

$$n_1 = 4$$

25 KE = quantum energy − threshold energy

$$= \frac{6.626 \times 10^{-34} \times 3 \times 10^8}{3000 \times 10^{-10}} - \frac{6.626 \times 10^{-34} \times 3 \times 10^8}{4000 \times 10^{-10}}$$

$$= 6.626 \times 10^{-19} - 4.9695 \times 10^{-19}$$

$$= 1.6565 \times 10^{-19} \text{ J}$$

$$KE = 1.6565 \times 10^{-19} = \frac{1}{2} mv^2$$

$$\therefore m^2v^2 = 2 \times 1.6565 \times 10^{-19} \times 9.1 \times 10^{-31}$$

$$mv = 5.49 \times 10^{-25}$$

de-Broglie wavelength,

$$\lambda = \frac{h}{mv} = \frac{6.626 \times 10^{-34}}{5.49 \times 10^{-25}}$$

$$= 1.2 \times 10^{-9} \text{ m}$$

26 In structure (c), all the atoms have complete octet. Thus, it is the correct representation of carbon suboxide.

27 Since, there are four electrons in 2p orbital, the outer configuration of the element is $2s^2, 2p^4$. The complete configuration of the element is $1s^2, 2s^2, 2p^4$.

The element contains total 8 electrons, so its atomic number is 8. Then, the number of neutrons

$$= 16 - 8 = 8$$

It is isotonic with ${}_7N^{15}$ (as it also contains 8 neutrons).

28 In PO_4^{3-} ion, formal charge on each O-atom of P—O bond

$$= \frac{\text{Total charge}}{\text{Number of O-atom}} = \frac{-3}{4} = -0.75$$

29 KCN molecule contains both ionic and covalent bonds as follows

$K^+ [C{\equiv}N]^-$
↑
covalent bond

30 In nitrogen molecule, triple bond is present between two nitrogen atoms, i.e. N≡N.

So, number of electrons take part in forming bond in $N_2 = 3 \times 2 = 6$.

31 Carbon tetrachloride has no net dipole moment because of its regular tetrahedral structure which is symmetrical.

32 Bond angle between two sp hybrid orbitals is 180°.

33 NO (15) = $\sigma 1s^2, \overset{*}{\sigma} 1s^2, \sigma 2s^2, \overset{*}{\sigma} 2s^2, \sigma 2p_z^2, \pi 2p_x^2, \approx \pi 2p_y^2, \overset{*}{\pi} 2p_x^1 \approx \overset{*}{\pi} 2p_y^0$

34 In CH_3^+, carbon atom is in sp^2-hybridised state.

35 Both $[NF_3, H_3O^+]$ are pyramidal and sp^3-hybridised and $[NO_3^-, BF_3]$ are triangular planar.

36 200 mg or 0.2 g of CO_2 contains molecules

$$= \frac{6.023 \times 10^{23}}{44} \times 0.2 = 2.73 \times 10^{21}$$

10^{21} molecules are removed, hence molecules left

$$= 2.73 \times 10^{21} - 10^{21} = 1.73 \times 10^{21}$$

The moles of CO_2 left $= \dfrac{1.73 \times 10^{21}}{6.023 \times 10^{23}}$

$$= 2.87 \times 10^{-3} \text{ mol}$$

37 Number of moles of nitrogen gas in 1.65 L at 273°C and 3 atm,

$$n_1 = \frac{pV}{RT} = \frac{3 \times 1.65}{0.0821 \times 546} = 0.11 \text{mol}$$

Number of moles of nitrogen gas in 0.55 L at 0°C and 0.75 atm,

$$n_2 = \frac{pV}{RT} = \frac{0.75 \times 0.55}{0.0821 \times 273}$$

$$= 0.018 \text{ mol}$$

Hence, % of nitrogen escaped, when temperature, pressure and volume is converted.

$$= \frac{(0.11 - 0.018)}{0.11} \times 100$$

$$= 83.64\%$$

38 de-Broglie wavelength,

$$\lambda = \frac{h}{mv} \quad \text{...(i)}$$

$$KE = \frac{1}{2} mv^2 \quad \text{...(ii)}$$

Now, putting the value of v from Eq. (i), we get,

$$KE = \frac{1}{2} m \times \left[\frac{h}{m \times \lambda}\right]^2 = \frac{1}{2} \times \frac{h^2}{m\lambda^2}$$

i.e. $KE \propto \frac{1}{m}$ [$\because \lambda$ are same.]

$\therefore \quad E_e > E_p > E_\alpha$

39 Workdone = Surface tension × increase in area

$= 73 \text{ dyne cm}^{-1} \times 0.10 \text{ m}^2$

$= 73 \text{ dyne cm}^{-1} \times 0.10 \times 10^4 \text{ cm}^2$

$= 7.3 \times 10^4$ erg

40 PCl_4^+ is isostructural with NH_4^+. Hence, the central atom in both have the same type of hybridisation, i.e. sp^3.

41 Among the given species, the bond dissociation energy of C—O bond is minimum in case of CO_3^{2-} by which C—O bond become more weaker in CO_3^{2-} or the bond order of CO_3^{2-}(1.33) is minimum so, the bond become weaker.

42 Number of moles of the gas

$= \frac{2.0 \times 10^{21}}{6.023 \times 10^{23}}$ mol

$= 3.32 \times 10^{-3}$ mol

From, $pV = nRT$

$T = \frac{pV}{nR}$

$= \frac{7.57 \times 10^3 \times 10^{-3}}{3.32 \times 10^{-3} \times 8.314}$

= 274.25 K

$\because$ Root mean square speed,

$v_{rms} = \sqrt{\frac{3RT}{M}}$

$\therefore \quad v_{rms} = \sqrt{\frac{3 \times 8.314 \times 274.25}{28 \times 10^{-3}}}$

$= 494.26 \text{ ms}^{-1}$

43 (i) $p_1 = 1$ atm $\quad T_1 = 273\text{K}, \; V_1 = ?$

32 g of oxygen occupies 22.4 L of volume at STP.

Hence, 1.6 g of oxygen will occupy,

$1.6 \text{ g oxygen} \times \frac{22.4 \text{ L}}{32 \text{ g oxygen}}$

= 1.12 L volume

$V_1 = 1.12\text{L}$

$p_2 = \frac{p_1}{2} = \frac{1}{2} = 0.5$ atm

$V_2 = ?$

According to Boyle's law,

$p_1V_1 = p_2V_2$

$\therefore \quad V_2 = \frac{p_1 \times V_1}{p_2}$

$= \frac{1 \text{ atm} \times 1.12 \text{ L}}{0.5 \text{ atm}}$

= 2.24L

(ii) Number of molecules of oxygen in the vessel

$= \frac{6.022 \times 10^{23} \times 1.6}{32}$

$= 3.011 \times 10^{22}$

44 Mass of oxygen in the oxide

$= (1.020 - 0.540) = 0.480$ g

Equivalent mass of the metal

$= \frac{0.540}{0.480} \times 8 = 9.0$

According to Dulong and Petit's law,

Approx, atomic mass

$= \frac{6.4}{\text{sp. heat}} = \frac{6.4}{0.216}$

= 29.63

Valency of the metal

$= \frac{\text{at. mass}}{\text{eq. mass}}$

$= \frac{29.63}{9.0} \approx 3$

Hence, the formula of the oxide is M_2O_3.

45 $\gamma = \frac{rh\rho g}{2}, \; \gamma = 0.055$ N/m

$r = \frac{1.40}{2} = 0.70 \text{ mm} = 0.70 \times 10^{-3}$ m,

$\rho = 850 \text{ kg/m}^3$,

$g = 9.80 \text{ m/s}^2, h = ?$

$h = \frac{2\gamma}{r\rho g} = \frac{2 \times 0.055}{0.70 \times 10^{-3} \times 850 \times 9.8}$

= 0.01886 m = 18.86 mm

46 B_2 : Total electrons = 10

Configuration:

$\sigma 1s^2, \overset{*}{\sigma}2s^2, \sigma 2s^2, \overset{*}{\sigma}2s^2, \pi 2p_x^1 \approx \pi 2p_y^1$

If Hund's rule is violated, then

$\sigma 1s^2, \overset{*}{\sigma}1s^2, \sigma 2s^2, \overset{*}{\sigma}2s^2, \pi 2p_x^2 \approx \pi 2p_y^0$

So, diamagnetic,

bond order $= \frac{6-4}{2} = 1$

47 The value of van der Waals' constant '*a*' is larger for ammonia than for nitrogen due to presence of hydrogen bonding. This is because value of '*a*' is a measure of intermolecular interaction.

48 Most probably the cathode rays consist of negatively charged particles called electrons because in the presence of electrical/magnetic field, the behaviour of cathode rays are similar to the negatively charged particles.

49 The H_2O molecule has sp^3-hybridisation with two lone pair of electrons, due to which its geometry is bent.

50 Sulphur has vacant *d*-orbital so it can expand its octet. In general, it forms many compounds which obey octet rule.

DAY SIX

Chemical Thermodynamics

Learning & Revision for the Day

- Fundamental of Thermodynamics
- Types of Processes
- Concept of Work
- Heat
- First Law of Thermodynamics
- Entropy Change (ΔS)
- Second Law of Thermodynamics
- Third Law of Thermodynamics
- Gibb's Free Energy (ΔG)

The laws of thermodynamics deal with the energy changes of macroscopic systems involving a large number of molecules rather than microscopic systems containing a few molecules.

Fundamentals of Thermodynamics

Some basic terms and concepts commonly used in thermodynamics are briefly explained below:

System and Surroundings

System is a part of universe under observation and the part of the universe except system is called **surroundings**. System basically a specific portion which is considered under thermodynamic studies.

Thus, system and the surrounding together constitute the universe, i.e.

$$\text{universe} = \text{system} + \text{surroundings}$$

On the basis of exchange of mass and energy, systems are of three types:

1. **Open System** In which energy and matter both can be exchanged with surroundings.
2. **Closed System** In which only energy can be exchanged with surroundings.
3. **Isolated System** In which neither matter nor energy can be exchanged with surroundings.

Extensive and Intensive Properties

Properties used to define the state of a system are extensive and intensive properties.

1. **Intensive Properties** Those properties that depend on nature of matter but do not depend on quantity of the matter, e.g. pressure, temperature, specific heat, melting point etc.

PREP MIRROR
Your Personal Preparation Indicator

- No. of Questions in Exercises (x)—
- No. of Questions Attempted (y)—
- No. of Correct Questions (z)—
  (Without referring Explanations)

- Accuracy Level ($z/y \times 100$)—
- Prep Level ($z/x \times 100$)—

In order to expect good rank in JEE, your Accuracy Level should be above 85 & Prep Level should be above 75.

2. **Extensive Properties** Those properties that depend on quantity of the matter present in the system, e.g. internal energy, heat, total moles, volume, enthalpy, entropy, free energy etc.

   Force is extensive property but pressure is intensive property.

State and Path Functions

1. **State Functions** Those functions which depend only on the state of the system and not on how it is reached. e.g. pressure, volume, temperature, ΔH, ΔE etc.
2. **Path Functions** The functions that depend upon the path followed are called path functions, e.g. work done, heat etc.

Types of Processes

The state of a variable can be changed by means of a thermodynamic process. These are of the following types :

1. **Adiabatic Process** In which system does not exchange heat with its surrounding, i.e. $dQ = 0$.
2. **Isothermal Process** In which temperature remains fixed, i.e. $dT = 0$.
3. **Isobaric Process** In which change of state is brought about at constant pressure, i.e. $dp = 0$.
4. **Isochoric Process** In which volume of the system remains constant, i.e. $dV = 0$.
5. **Cyclic Process** This is the process in which a system undergoes a number of different states and finally returns to its initial state. For such a process, change in internal energy and enthalpy is zero, i.e. $dE = 0$ and $dH = 0$.
6. **Reversible Process** In this process, (quasistatic system), change taken place is infinitesimally slow and their direction at any point can be reversed by infinitesimal change in the state of the system. Reversible process is an ideal process and here, every intermediate state is in equilibrium with others, if any.
7. **Irreversible Process** This process, is the one which cannot be reversed. In this process, amount of energy increases. All natural processes are irreversible in nature.

NOTE

- If macroscopic properties like temperature, pressure, etc., do not change with the time, the system is said to be in **thermodynamic equilibrium**.
- If the two systems are in thermal equilibrium with a third system, they must be in thermal equilibrium with each other. This law is known as **zeroth law of thermodynamics**.

  The law defines the temperature as the property which determines, whether the body is in thermal equilibrium or not.

Concept of Work

- The work is said to be done when gas expands or contracts against the external pressure. Work done is a path function not a state function as depends upon the path followed.

$$W = -p_{ext} \times \Delta V$$

For expansion $W = -p_{ext}(V_2 - V_1)$ [Here, $V_2 > V_1$]

For compression $W = -p_{ext}(V_2 - V_1)$ [Here, $V_2 < V_1$]

- Maximum work done for reversible isothermal process

$$W_{rev} = -2.303\, nRT \log \frac{V_2}{V_1}$$

where, V_2 = final volume, V_1 = initial volume

Also, $W_{rev} = -2.303 nRT \log \frac{p_1}{p_2}$

Maximum work done for irreversible isothermal expansion

(i) **Intermediate expansion**

$$p_{ext} < p_{gas},\ W_{irr} = \int_{V_1}^{V_2} p_{ext}\, dV$$

$$W = -p_{ext}(V_2 - V_1)$$

(ii) **Free expansion** If gas expands in vacuum, $p_{ext} = 0$, therefore, $W = 0$

Sign Convention

(i) If W is positive — work done on the system

(ii) If W is negative — work done by the system

Heat

It is defined as the quantity of energy, which flows between system and surroundings on account of temperature difference. It is also a path function, i.e. depends upon the path followed.

Sign Convention

(i) If q is positive — heat is supplied to the system

(ii) If q is negative — heat is lost by the system

Internal Energy (*E* or *U*)

- It is the total energy within the substance. It is equal to the sum of translational energy, vibrational energy, potential energy etc. We can only determine the change in internal energy.

i.e. $$\Delta U = U_2 - U_1$$

- At constant temperature, internal energy change (ΔE) will be zero. Internal energy depends on temperature, pressure, volume and quantity of matter.

First Law of Thermodynamics

According to this law, energy can neither be created nor destroyed although it may be converted from one form to another.

- Mathematical form of first law of thermodynamics,
$$\Delta E = q + W$$
- Both W and q are not state functions but quantity $W + q$ is a state function.

Enthalpy

- The total heat content of a system at constant pressure is called the enthalpy of the system. It is the sum of internal energy and the product of pressure-volume work.
- It is an extensive quantity, state function and represented by the symbol H.

  The equation is
$$H = E + pV$$
$$\Delta H = \Delta E + p\Delta V$$
$$\Delta H = \Delta E + \Delta n_g RT$$
  where, ΔH = enthalpy change

  Δn_g = gaseous moles of products − gaseous moles of reactants
- If $\Delta n_g = 0$, then $\Delta H = \Delta U$; If $\Delta n_g > 0$ then $\Delta H > \Delta U$ and if $\Delta n_g < 0$ then, $\Delta H < \Delta U$.
- For reaction involving solids and liquids only, $\Delta H = \Delta E$. Enthalpy also changes, when a substance undergoes phase transition.

Heat Capacity

Heat capacity (C) of a system is defined as the amount of heat required to raise the temperature of the system by 1°C.

$$C = \frac{q}{\Delta T}$$

- If the system consists of a single substance or a solution and weighs 1 mol, the heat capacity of the system is referred as **molar heat capacity** (C_m).
- If the system consists of a single substance or a solution and weights 1 g, the heat capacity of the system is referred as **specific heat of the system** (C_s).
$$q = C \times m \times \Delta T$$
  where, m = mass of substance,

  C = specific heat capacity

  and ΔT = temperature difference.

There are two types of heat capacities based on the conditions such as volume and pressure.

1. **Heat Capacity at Constant Pressure** (C_p) The heat capacity of a substance measured when the gaseous system is at constant pressure.
$$C_p = \left(\frac{q}{\Delta T}\right)_p \quad \text{or} \quad C_p = \frac{dH}{dT}$$
2. **Heat Capacity at Constant Volume** (C_V) The heat capacity of a substance measured when the gaseous system is at constant volume.
$$C_V = \left(\frac{q}{\Delta T}\right)_V \quad \text{or} \quad C_V = \frac{dV}{dT}$$

- Relationship between C_p and C_V is given as $C_p - C_V = R$ (R is the molar gas constant.)
$$C_p - C_V = \frac{R}{M}$$
- For monoatomic gas the molar heat capacity at constant volume, $C_V = \left(\frac{3}{2}\right)R$
- For monoatomic gas molar heat capacity at constant pressure,
$$C_p = \left(\frac{3}{2}\right)R + R = \left(\frac{5}{2}\right)R$$
- Poisson's ratio, $\gamma = \frac{C_p}{C_V} = \left(\frac{5}{3}\right) = 1.66$

  $\gamma = 1.66$ for monoatomic gases (like He, Ar)

  $\gamma = 1.40$ for diatomic gases (like H_2, O_2, CO)

  $\gamma = 1.33$ for triatomic gases (like H_2O, O_3)
- Work done in reversible adiabatic expansion

  Work done $= C_V \Delta T = C_V \times (T_2 - T_1)$

NOTE The molar heat capacity for any process is given by
$$C = C_V + \frac{R}{1-\gamma}$$
when, pV^{γ} = constant

Kirchhoff's Equation

According to Kirchhoff's equation, the partial derivatives of the change of enthalpy (or of internal energy) during a reaction, with respect to temperature at constant pressure (or volume) equals to the change in heat capacity at constant pressure (or volume).

$$\Delta C_p = \frac{\Delta H_2 - \Delta H_1}{T_2 - T_1} \quad \text{and} \quad \Delta C_V = \frac{\Delta E_2 - \Delta E_1}{T_2 - T_1}$$

where, $\Delta C_p = \Sigma C_p$ of products $- \Sigma C_p$ of reactants

$\Delta C_V = \Sigma C_V$ of products $- \Sigma C_V$ of reactants

Entropy Change (ΔS)

Entropy is the measurement of randomness or disorder of the molecules. A process which proceeds at its own without any outside help is termed as spontaneous process. The total entropy change (ΔS_{total}) for the system and surrounding of a spontaneous process is always positive,

$$\Delta S_{total} = \Delta S_{system} + \Delta S_{surrounding} > 0.$$

- Entropy is a state function and depends only on initial and final states of the system.

  i.e. $\Delta S = S_{final} - S_{initial}$
- Unit of entropy is joule per Kelvin per mole.
- For a reversible change at constant temperature

  $$\Delta S = \frac{q_{rev}}{T} = S_{final} - S_{initial}$$

  q_{rev} = heat absorbed or evolved at absolute temperature T. If $\Delta S > 0$, heat is absorbed and if $\Delta S < 0$, heat is evolved.
- For a spontaneous isothermal expansion

  $$\Delta S = 2.303\, nR \log \frac{V_2}{V_1}$$
- The change of matter from one state to another is called **phase transition.**

  The entropy changes at the time of phase transition are as follows:

  (i) $\Delta S_{melting} = \frac{\Delta H_{fusion}}{T_m}$; ($T_m$ = melting point of substance)

  (ii) $\Delta S_{vaporisation} = \frac{\Delta H_{vaporisation}}{T_b}$; ($T_b$ = boiling point of substance)

  (iii) $\Delta S_{sublimation} = \frac{\Delta H_{sublimation}}{T_{sub}}$; ($T_{sub}$ = sublimation temperature)

Second Law of Thermodynamics

The second law of thermodynamics tells us whether a given process can occur spontaneously and to what extent, it also helps us to calculate maximum fraction of heat that can be converted into work in a given process. The second law states that the entropy of the universe is continuously increasing and tends to a maximum.

For an irreversible process,

$$dS > \frac{dq}{T}$$

$$\therefore \quad dS \geq \frac{dq}{T}$$

This is the mathematical statement of the second law of thermodynamics.

Carnot Engine

Carnot in 1824 gave an imaginary reversible cycle which demonstrates the maximum conversion of heat into work. He actually proposed a theoretical heat engine to show that its efficiency was based upon the temperatures, between which it operated.

A Carnot cycle comprises four operations or processes:

(i) Isothermal reversible expansion
(ii) Adiabatic reversible expansion
(iii) Isothermal reversible compression
(iv) Adiabatic reversible compression

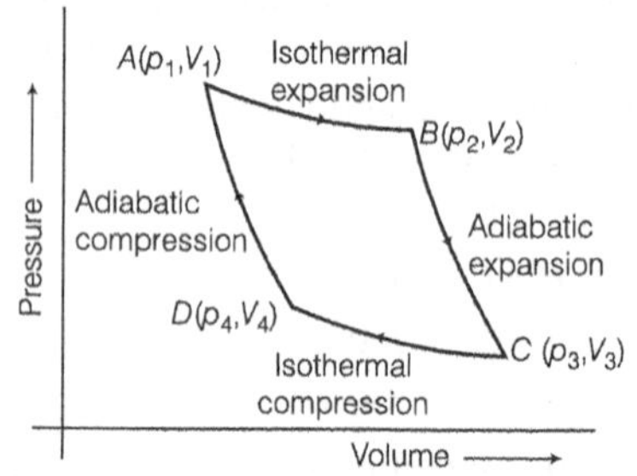

Carnot cycle

- Net work done in 1 cycle is

  $$W = RT_2 \ln \frac{V_2}{V_1} + RT_1 \ln \frac{V_4}{V_3}$$
- Net heat absorbed in the whole cycle is

  $$q = R(T_2 - T_1) \ln \frac{V_2}{V_1}$$
- Efficiency of a heat engine (denoted by η) in a Carnot cycle is given

  $$\eta = \frac{W}{q_2} = \frac{T_2 - T_1}{T_2} = 1 - \frac{T_1}{T_2}$$
- The above relation was stated in the form of Carnot theorem by Carnot, i.e. "Every perfect engine working reversibly between the same temperature limits has the same efficiency, whatever be the working substance."

Third Law of Thermodynamics

- This law was proposed by German chemist Walther Nernst. According to this law, "The entropy of a perfectly crystalline substance approaches zero as the absolute zero of temperature is approached".
- It forms the basis from which entropies at other temperatures can be measured, $\lim_{T \to 0} S = 0$

Gibbs Free Energy (ΔG)

- The maximum amount of energy available to a system, during a process that can be converted into useful work is called free energy or Gibbs free energy.
- Under standard conditions of temperature (i.e. at 298 K) and pressure (i.e. 1 atm), the free energy change for a process in which reactants in their standard state is converted into products in their standard state is called standard free energy change (i.e. ΔG°).
- The changes in the Gibbs energy of a system as a function of temperature can be calculated by the equation known as Gibbs Helmholtz Equation.

$$\Delta G = \Delta H - T\Delta S$$

where, ΔG = Gibbs free energy (measurement of useful work)

- The following cases are considered for ΔG.
  (i) $\Delta G > 0$, for non-spontaneous process
  (ii) $\Delta G < 0$, for spontaneous process
  (iii) $\Delta G = 0$, at equilibrium

Criteria for Spontaneity

- In the determination of spontaneity, Gibbs energy criteria is better than entropy criteria because Gibbs energy refers to the system only while entropy refers to both system and surroundings.
- For negative value of ΔG several conditions exist which plays an important role in deciding the spontaneity of the process.

ΔS	ΔH	ΔG	**Spontaneity of a Process**
+	–	Negative at all temperatures.	Spontaneous at all temperatures.
–	–	Negative at all temperatures.	Spontaneous at low temperature but non-spontaneous at high temperature.
+	+	Positive at low temperature and negative at high temperature.	Spontaneous at high temperature.
+	+	Positive at all temperature.	Non-spontaneous at all temperature.

Relation between Gibbs Energy Change and Equilibrium Constant

Relation between Gibbs free energy with reaction quotient and equilibrium constant are as follows :

- $\Delta G = \Delta G^\circ + 2.303\, RT \log Q$
  where, Q = reaction quotient
- At equilibrium, $\Delta G = 0$
  $\therefore \quad \Delta G^\circ = -2.303\, RT \log K$
  K = equilibrium constant
- $\Delta G^\circ = -n E^\circ_{cell} F$
  where, n = number of electrons lose or gained,
  E°_{cell} = standard electrode potential,
  1F = 96500 C
- Efficiency of the fuel cell $= \Delta G / \Delta H \times 100$

DAY PRACTICE SESSION 1

FOUNDATION QUESTIONS EXERCISE

1 Which of the following statement is correct?
(a) The presence of reacting species in a covered beaker is an example of open system
(b) There is an exchange of energy as well as matter between the system and the surroundings in a closed system
(c) The presence of reactants in a closed vessel made up of copper is an example of a closed system
(d) The presence of reactants in a thermos flask or any other closed insulated vessel is an example of a closed system

2 The pressure-volume work for an ideal gas can be calculated by using the expression $W = -\int_{V_i}^{V_f} p_{ex} dV$. The work can also be calculated from the pV – plot by using the area under the curve within the specified limits. When an ideal gas is compressed (a) reversibly or (b) irreversibly from volume V_i to V_f. Choose the correct option.
(a) $W_{(reversible)} = W_{(irreversible)}$
(b) $W_{(reversible)} < W_{(irreversible)}$
(c) $W_{(reversible)} > W_{(irreversible)}$
(d) $W_{(reversible)} = W_{(irreversible)} + p_{ex}\Delta V$

3 An ideal gas expands in volume from 1×10^{-3} m^3 to 1×10^{-2} m^3 at 300 K against a constant pressure of 1×10^5 N/m^2. The work done is

(a) −900 J (b) −900 kJ
(c) 270 kJ (d) 900 kJ

4 When 1 mole of a gas is heated at constant volume, temperature is raised from 298 K to 308 K. Heat supplied to the gas is 500 J. Then, which statement is correct?

(a) $q = -W = 500$ J, $\Delta E = 0$
(b) $q = W = 500$ J, $\Delta E = 0$
(c) $q = \Delta E = 500$ J, $W = 0$
(d) $\Delta E = 0$, $q = W = -500$ J

5 In an adiabatic process, no transfer of heat takes place between system and surroundings. Choose the correct option for free expansion of an ideal gas under adiabatic condition from the following.

(a) $q = 0, \Delta T \neq 0, W = 0$
(b) $q \neq 0, \Delta T = 0, W = 0$
(c) $q = 0, \Delta T = 0, W = 0$
(d) $q = 0, \Delta T < 0, W \neq 0$

6 1 mole of an ideal gas at 300 K is expanded isothermally from an initial volume of 1 L to 10 L. The ΔE for this process is ($R = 2$ cal K^{-1} mol^{-1})

(a) 270 cal (b) zero (c) 10 L atm (d) 181.7 cal

7 ΔU is equal to → JEE Main 2017

(a) isochoric work (b) isobaric work
(c) adiabatic work (d) isothermal work

8 A piston filled with 0.04 mole of an ideal gas expands reversibly from 50.0 mL to 375 mL at a constant temperature of 37.0°C. As it does so, it absorbs 208 J of heat. The values of q and W for the process will be [$R = 8.314$ J/mol K) (ln 7.5 = 2.01) → JEE Main 2013

(a) $q = -208$ J, $W = -208$ J
(b) $q = -208$ J, $W = +208$ J
(c) $q = +208$ J, $W = +208$ J
(d) $q = +208$ J, $W = -208$ J

9 Latent heat of vaporisation of a liquid at 500 K and 1 atm pressure is 10.0 kcal/mol. What will be the change in internal energy (ΔE) of 3 moles of liquid at same temperature?

(a) 30 kcal (b) −54 kcal (c) 27.0 kcal (d) 50 kcal

10 ($\Delta H - \Delta E$) for the formation of carbon monoxide (CO) from its elements at 298 K is ($R = 8.314$ JK^{-1} mol^{-1})

(a) −2477.57 J mol^{-1} (b) 2477.57 J mol^{-1}
(c) −1238.78 J mol^{-1} (d) 1238.78 J mol^{-1}

11 Consider the reaction,
$N_2 + 3H_2 \longrightarrow 2NH_3$ carried out at constant temperature and pressure. If ΔH and ΔE are the enthalpy and internal energy changes for the reaction, which of the following expressions is true?

(a) $\Delta H > \Delta E$ (b) $\Delta H < \Delta E$
(c) $\Delta H = \Delta E$ (d) $\Delta H = 0$

12 The combustion of benzene (l) gives $CO_2(g)$ and $H_2O(l)$. Given that heat of combustion of benzene at constant volume is −3263.9 kJ mol^{-1} at 25° C; heat of combustion (in kJ mol^{-1}) of benzene at constant pressure will be ($R = 8.314$ JK^{-1} mol^{-1}) → JEE Main 2018

(a) 4152.6 (b) −452.46
(c) 3260 (d) −3267.6

13 Heat required to raise the temperature of 1 mole of a substance by 1°C is called

(a) specific heat (b) molar heat capacity
(c) water equivalent (d) specific gravity

14 The molar heat capacity of water at constant pressure is 75 J K^{-1}mol^{-1}. When 1.0 kJ of heat is supplied to 100 g of water, which is free to expand, the increase in temperature of water is

(a) 1.2 K (b) 2.4 K
(c) 4.8 K (d) 6.8 K

15 For two moles of an ideal gas

(a) $(C_p - C_V) = 2R$ (b) $(C_p - C_V) = 0$
(c) $(C_p - C_V) = R$ (d) $(C_p - C_V) = R/2$

16 The heat of sublimation of iodine is 24 cal g^{-1} at 50°C. If specific heat of solid iodine and its vapours are 0.055 and 0.031 cal g^{-1} respectively, the heat of sublimation of iodine at 100°C is

(a) 22.8 cal g^{-1} (b) 25.2 cal g^{-1}
(c) −22.8 cal g^{-1} (d) −25.2 cal g^{-1}

17 What will be the change of entropy $\Delta_r S°$ at 298 K for the reaction, in which urea is formed from NH_3 and CO_2?

$$2NH_3(g) + CO_2(g) \longrightarrow NH_2CONH_2(aq) + H_2O(l)$$

[Given, the standard entropy of $NH_2CONH_2(aq)$, $CO_2(g)$, $NH_3(g)$ and $H_2O(l)$ are 174.0, 213.7, 192.3 and 69.9 JK mol^{-1} respectively]

(a) 200 JK^{-1} mol^{-1} (b) −35.44 JK^{-1} mol^{-1}
(c) −354.4 JK^{-1} mol^{-1} (d) 425.2 JK^{-1} mol^{-1}

18 2 moles of an ideal gas at 27°C are expanded reversibly from 2 L to 20 L. Find entropy change (in cal/mol K). ($R = 2$ cal/mol K)

(a) 0 (b) 4
(c) 9.2 (d) 92.0

19 For an isolated system, $\Delta E = 0$, then

(a) $\Delta S = 0$
(b) $\Delta S < 0$
(c) $\Delta S > 0$
(d) the value of ΔS cannot be predicted

20 When 1.8 g of steam at the normal boiling point of water is converted into water, at the same temperature, enthalpy and entropy changes respectively will be [Given, ΔH_{vap} for water = 40.8 kJ mol^{-1}]

(a) −8.12 kJ, 11.89 JK^{-1} (b) 10.25 kJ, 12.95 JK^{-1}
(c) −4.08 kJ, − 10.93 JK^{-1} (d) 10.93 kJ, −4.08 JK^{-1}

21 Which of the following does not have zero entropy even at absolute zero?

CO, CO_2, NaCl, NO

(a) CO, CO_2 (b) CO, NO
(c) CO_2, NaCl (d) NaCl

22. The enthalpy of vaporisation of liquid diethyl ether $(C_2H_5)_2O$, is 26.0 kJ mol^{-1} at its boiling point (35°C). What will be the ΔS for conversion of liquid to vapour and vapour to liquid respectively?

(a) + 84.41 and − 84.41 JK^{-1} mol^{-1}
(b) + 80.90 and − 68.83 JK^{-1} mol^{-1}
(c) − 84.41 and + 90.63 JK^{-1} mol^{-1}
(d) + 68.83 and − 84.41 JK^{-1} mol^{-1}

23 The direct conversion of A to B is difficult, hence it is carried out by the following shown path

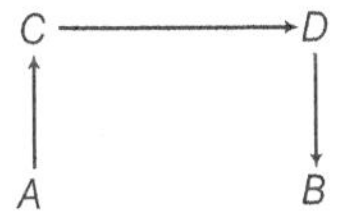

Given, $\Delta S_{(A \to C)} = 50$ eu

$\Delta S_{(C \to D)} = 30$ eu

$\Delta S_{(B \to D)} = 20$ eu

where, eu is entropy unit, then $\Delta S_{(A \to B)}$ is

(a) + 100 eu (b) + 60 eu
(c) − 100 eu (d) − 60 eu

24 Match the following and choose the correct option.

	Column I		Column II
A.	Entropy of vaporisation	1.	decreases
B.	K for spontaneous process	2.	is always +ve
C.	Crystalline solid state	3.	lowest entropy
D.	ΔE in adiabatic expansion of ideal gas	4.	$\frac{\Delta H_{vap}}{T_b}$

Codes

	A	B	C	D		A	B	C	D
(a)	2,4	2	3	1	(b)	3	1,2	4	2
(c)	2	4	1	3	(d)	3,4	2	1	2

25 The entropy change involved in the isothermal reversible expansion of 2 moles of an ideal gas from a volume of 10 dm^3 to a volume of 100 dm^3 at 27°C is → AIEEE 2011

(a) 38.3 J mol^{-1} K^{-1}
(b) 35.8 J mol^{-1} K^{-1}
(c) 32.3 J mol^{-1} K^{-1}
(d) 42.3 J mol^{-1} K^{-1}

26 A Carnot engine operates between temperature T and 400 K (T > 400 K). If efficiency of engine is 25%, the temperature T is

(a) 666.0 K (b) 498.5 K
(c) 533.3 K (d) 500.0 K

27 In view of the signs of $\Delta_r G°$ for the following reactions

$PbO_2 + Pb \longrightarrow 2PbO, \ \Delta_r G° < 0$

$SnO_2 + Sn \longrightarrow 2SnO, \ \Delta_r G° > 0$

(a) For lead + 4, for tin + 2 (b) For lead + 2 , for tin + 2
(c) For lead + 4, for tin + 4 (d) For lead + 2, for tin + 4

28 When the heat of a reaction at constant pressure is -2.5×10^3 cal and entropy change for the reaction is 7.4 cal deg^{-1}, it is predicted that the reaction at 25°C is

(a) reversible (b) spontaneous
(c) non-spontaneous (d) irreversible

29 At 1 atm pressure, $\Delta S = 75$ J K^{-1} mol^{-1}; $\Delta H = 30$ kJ mol^{-1}. The temperature of the reaction at equilibrium is

(a) 400 K (b) 330 K (c) 200 K (d) 110 K

30 For a particular reversible reaction at temperature T, ΔH and ΔS were found to be both positive. If T_e is the temperature at equilibrium, the reaction would be spontaneous when → AIEEE 2010

(a) $T_e > T$ (b) $T > T_e$
(c) T_e is 5 times T (d) $T = T_e$

31 For the process,

$H_2O(l)$ (1 bar, 273 K) $\to H_2O(g)$ (1 bar, 373 K),

The correct set of thermodynamic parameters is

(a) $\Delta G = 0, \Delta S = +ve$ (b) $\Delta G = 0, \ \Delta S = -ve$
(c) $\Delta G = +ve, \Delta S = 0$ (d) $\Delta G = -ve, \Delta S = +ve$

32 Considering entropy (S) as a thermodynamic parameter, the criterion for the spontaneity of any process is

(a) $\Delta S_{system} + \Delta S_{surrounding}$ be +ve
(b) ΔS_{system} be zero
(c) $\Delta S_{system} - \Delta S_{surrounding}$ be + ve
(d) $\Delta S_{surrounding}$ be zero

33 For a reaction, $M_2O(s) \longrightarrow 2M(s) + \frac{1}{2}O_2(g)$,

$\Delta H = 30$ kJ mol^{-1} and $\Delta S = 0.07$ kJ K^{-1} mol^{-1} at 1 atm. The temperature upto which the reaction would not be spontaneous is

(a) $T < 400.08$ K (b) $T < 273.15$ K
(c) $T < 428.57$ K (d) $T < 473.50$ K

34 For the process, $H_2O(l) \longrightarrow H_2O(g)$ at T= 100 °C and 1 atmosphere pressure, the correct choice is → JEE Main 2014

(a) $\Delta S_{system} > 0$ and $\Delta S_{Surrounding} > 0$
(b) $\Delta S_{system} > 0$ and $\Delta S_{Surrounding} < 0$
(c) $\Delta S_{system} < 0$ and $\Delta S_{Surrounding} > 0$
(d) $\Delta S_{system} < 0$ and $\Delta S_{Surrounding} < 0$

35 The incorrect expression among the following is → AIEEE 2012

(a) $\frac{\Delta G_{system}}{\Delta S_{total}} = -T$

(b) In isothermal process, $W_{reversible} = -nRT \ln \frac{V_f}{V_i}$

(c) $\ln K = \frac{\Delta H° - T\Delta S°}{RT}$

(d) $K = e^{-\Delta G°/RT}$

36 The value of $\log_{10} K_C$ for a reaction $A \rightleftharpoons B$ is
[Given, $\Delta_r H^\circ_{298\,K} = -54.07\ \text{kJ mol}^{-1}$,

$\Delta_r S^\circ_{298\,K} = 10\ \text{JK}^{-1}\ \text{mol}^{-1}$ and $R = 8.314\ \text{JK}^{-1}\ \text{mol}^{-1}$
and $2.303 \times 8.314 \times 298 = 5705$]

(a) 5 (b) 10
(c) 95 (d) 100

37 For a spontaneous reaction, the ΔG, equilibrium constant (K) and E°_{cell} will be respectively

(a) – ve, > 1, – ve (b) – ve, < 1, – ve
(c) + ve, > 1, – ve (d) – ve, > 1, + ve

38 In a fuel cell, methanol is used as fuel and oxygen gas is used as an oxidiser. The reaction is

$$CH_3OH(l) + \frac{3}{2}O_2(g) \longrightarrow CO_2(g) + 2H_2O(l)$$

At 298 K standard Gibb's energies of formation for $CH_3OH(l)$, $H_2O(l)$ and $CO_2\ (g)$ are –166.2, –237.2 and –394.4 kJ mol^{-1}, respectively. If standard enthalpy of combustion of methanol is –726 kJ mol^{-1}, efficiency of the fuel cell will be → AIEEE 2009

(a) 80% (b) 87%
(c) 90% (d) 97%

Direction (Q.Nos. 39 and 40) *In the following questions assertion followed by a reason is given. Choose the correct answer out of the following choices.*

(a) Both Assertion and Reason are correct statements and Reason is correct explanation of Assertion.
(b) Both Assertion and Reason are correct statements but Reason is not correct explanation of Assertion.
(c) Assertion is correct statements but Reason is wrong statement.
(d) Both Assertion and Reason are wrong.

39 Assertion (A) Spontaneous process is an irreversible process and may only be reversed by some external agency.

Reason (R) Decrease in enthalpy is a contributory factor for spontaneity.

40 Assertion (A) The thermodynamic function which determines the spontaneity of a process is the free energy. For a process to be spontaneous, the change in free energy must be negative.

Reason (R) The change in free energy is related to the change in enthalpy and change in entropy. The change in entropy for a process must be always positive if it is spontaneous.

DAY PRACTICE SESSION 2

PROGRESSIVE QUESTIONS EXERCISE

1. An ideal gas is allowed to expand both reversibly and irreversibly in an isolated system. If T_i is the initial temperature and T_f is the final temperature, which of the following statement is correct?

(a) $(T_f)_{irrev} > (T_f)_{rev}$
(b) $T_f > T_i$ for reversible process, but $T_f = T_i$ for irreversible process
(c) $(T_f)_{rev} > (T_f)_{irrev}$
(d) $T_f = T_i$ for both reversible and irreversible processes

2. When one mole of monoatomic ideal gas at T K undergoes adiabatic change under a constant external pressure of 1 atm, changes volume from 1 L to 2 L. The final temperature in Kelvin would be

(a) $\frac{T}{2^{2/3}} Z$ (b) $T + \frac{2}{3 \times 0.0821}$
(c) T (d) $T - \frac{2}{3 \times 0.0821}$

3. Water is brought to boil under a pressure of 1.0 atm. When an electric current of 0.50 A from a 12 V supply is passed for 300 s through a resistance in thermal contact with it, it is found that 0.798 g of water is vaporised. Calculate the molar internal energy change at boiling point (373.15 K).

(a) 37.5 kJ mol^{-1} (b) 3.75 kJ mol^{-1}
(c) 42.6 kJ mol^{-1} (d) 4.26 kJ mol^{-1}

4. 1 mole of CO_2 gas at 300 K is expanded under adiabatic conditions such that its volume becomes 27 times. What is work done? ($\gamma = 1.33$ and $C_V = 6$ cal mol^{-1} for CO_2)

(a) 900 cal (b) 1000 cal (c) 1200 cal (d) 1400 cal

5. A monoatomic ideal gas undergoes a process in which the ratio of p to V at any instant is constant and equals to 1. What is the molar heat capacity of the gas?

(a) $4R/2$ (b) $3R/2$
(c) $5R/2$ (d) Zero

6. $CaCO_3$ exists in two forms calcite and aragonite. The conversion of 1 mole of calcite to aragonite is accompanied by internal energy change equal to + 0.21 kJ. Given that the densities of calcite and aragonite are 2.71 cm^{-3} and 2.73 g cm^{-3} respectively. The enthalpy change at the pressure of 1.0 bar will be

(a) 200.72 J mol^{-1} (b) 309.72 J mol^{-1}
(c) 209.72 J mol^{-1} (d) 315.00 J mol^{-1}

7 Identify the correct statement regarding a spontaneous process.

(a) For a spontaneous process in an isolated system, the change in entropy is positive
(b) Endothermic processes are never spontaneous
(c) Exothermic processes are always spontaneous
(d) Lowering of energy in the reaction process is the only criteria for spontaneity

8 The reversible expansion of an ideal gas under adiabatic and isothermal conditions is shown in the figure. Which of the following statement is incorrect?

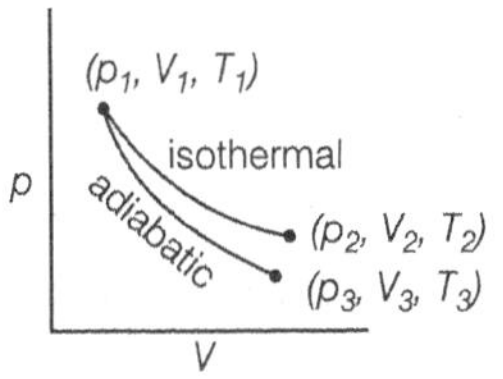

(a) $T_1 = T_2$ (b) $T_3 > T_1$
(c) $W_{(isothermal)} > W_{(adiabatic)}$ (d) $\Delta E_{(isothermal)} > \Delta E_{(adiabatic)}$

9 A gas present in a cylinder, fitted with a frictionless piston, expands against a constant pressure of 1 atm from a volume of 2 litre to a volume of 6 litre. In doing so, it absorbs 800 J heat from surroundings. The increase in internal energy of process is

(a) 305.85 J (b) 394.95 J
(c) 405.83 J (d) –463.28 J

10 In an irreversible process taking place at constant T and p and in which only pressure volume work is being done, the change in Gibbs free energy (ΔG) and change in entropy (ΔS), satisfy the criteria

(a) $(\Delta S)_{V,E} < 0, (\Delta G)_{T,p} < 0$ (b) $(\Delta S)_{V,E} > 0, (\Delta G)_{T,p} < 0$
(c) $(\Delta S)_{V,E} = 0, (\Delta G)_{T,p} = 0$ (d) $(\Delta S)_{V,E} = 0, (\Delta G)_{T,p} > 0$

11 A reaction is non-spontaneous at the freezing point of water but is spontaneous at the boiling point of water, then

	ΔH	ΔS
(a)	+ ve	+ ve
(b)	– ve	– ve
(c)	– ve	+ ve
(d)	+ ve	– ve

12 One mole of a liquid (1 bar, 100 mL) is taken in an adiabatic container and the pressure increases steeply to 100 bar. Then at constant pressure of 100 bar, volume decreases by 1 mL. ΔU and ΔH will be respectively.

(a) 10 J, 10 J
(b) 10 J, 0
(c) 0, 10 J
(d) 5 J, 10 J

13 Standard entropy of X_2, Y_2 and XY_3 are 60, 40 and 50 $JK^{-1} mol^{-1}$ respectively. For the reaction,

$$\frac{1}{2}X_2 + \frac{3}{2}Y_2 \longrightarrow XY_3, \ \Delta H = -30 \text{ kJ}$$

to be at equilibrium, the temperature will be

(a) 750 K (b) 1000 K
(c) 1250 K (d) 500 K

14 For an isomerisation reaction $A \rightleftharpoons B$, the temperature dependence of equilibrium constant is given by

$$\log_e K = 4.0 - \frac{2000}{T}$$

The value of $\Delta S°$ at 300 K is therefore,

(a) $4R$ (b) $5R$
(c) $400R$ (d) $2000R$

15 1 kg block of ice at 0°C is placed into a perfectly insulated, sealed container that has 2 kg of water also at 0°C. The water and ice completely fill the container is flexible. After sometime one can except that

(a) the water will freeze so that the mass of the ice will increase
(b) the ice will melt so that the mass of ice will decrease
(c) both the amount of water and the amount of ice will remain constant
(d) both the amount of water and the amount of ice will decrease

ANSWERS

SESSION 1	1 (c)	2 (b)	3 (a)	4 (c)	5 (c)	6 (b)	7 (c)	8 (d)	9 (c)	10 (d)
	11 (b)	12 (d)	13 (b)	14 (b)	15 (a)	16 (a)	17 (c)	18 (c)	19 (c)	20 (c)
	21 (b)	22 (a)	23 (b)	24 (a)	25 (a)	26 (c)	27 (d)	28 (b)	29 (a)	30 (b)
	31 (a)	32 (a)	33 (c)	34 (b)	35 (c)	36 (b)	37 (d)	38 (d)	39 (b)	40 (a)
SESSION 2	1 (a)	2 (d)	3 (a)	4 (c)	5 (a)	6 (c)	7 (a)	8 (b)	9 (b)	10 (b)
	11 (a)	12 (b)	13 (a)	14 (a)	15 (b)					

Hints and Explanations

SESSION 1

1 For a closed vessel made up of copper, no matter can exchange between the system and the surrounding but energy exchange can occur through its walls.

Presence of reaction species in a covered beaker-closed system and exchange of matter as well as energy-open system. Presence of reactant in a closed vessel is a closed system and presence of reactant flask is an isolated system.

2

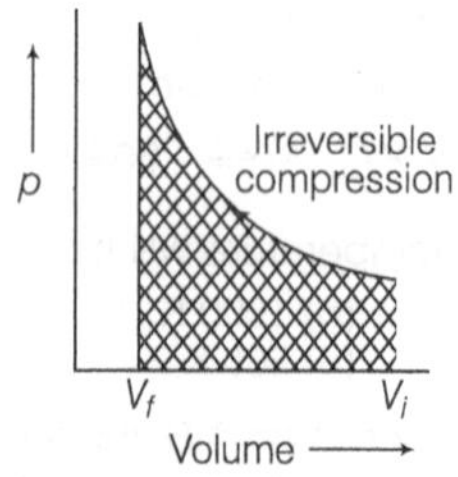

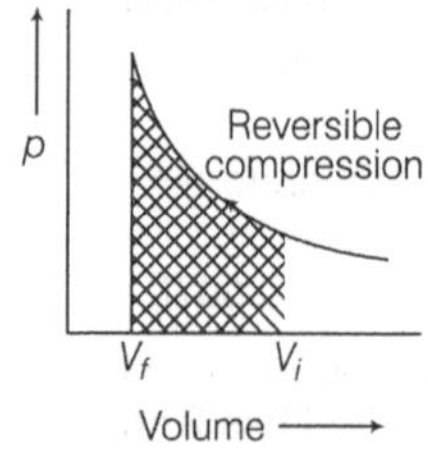

As the area under the curve is more in irreversible compression than the area under curve of reversible compression. Thus, $W_{\text{irreversible}} > W_{\text{reversible}}$.

3 Work done due to change in volume against constant pressure is

$$W = -p(V_2 - V_1)$$
$$= -1\times10^5\,\text{N/m}^2\,(1\times10^{-2} - 1\times10^{-3})\,\text{m}^3$$
$$= -900\,\text{Nm} = -900\,\text{J}$$

4 At constant volume, $p\Delta V = 0$, i.e. $W = 0$

From first law of thermodynamics,

$$\Delta E = q + W$$
$$\therefore \quad q = \Delta E$$

5 In free expansion, $W = 0$ while in adiabatic process, $q = 0$.

$$\Delta U = q + W = 0$$

This suggests that internal energy remains constant. Therefore, $\Delta T = 0$. Expansion of an ideal gas under adiabatic conditions in a vacuum leads to no absorption/evolution of heat.

Thus, no external work is done for the separation of gaseous molecules.

6 For isothermal process, $\Delta T = 0$, then

$$\Delta E = 0$$

7 According to first law of thermodynamics,

$$\Delta U = q + W = q - p\Delta V$$

In isochoric process ($\Delta V = 0$),

$$\Delta U = q$$

In isobaric process ($\Delta p = 0$),

$$\Delta U = q$$

In adiabatic process ($q = 0$)

$$\Delta U = W$$

In isothermal process $\Delta T = 0$ and $\Delta U = 0$

$\therefore$ ΔU is equal to adiabatic work.

8 As the process is carrying out at constant temperature, therefore this type of expansion is called **isothermal reversible expansion**, for which, $\Delta U = 0$. Hence, $q = -W$

i.e. heat absorbed by the system is equal to the work done by the system.

Thus, $\quad q = +208\text{ J}$

$\quad W = -208\text{ J}$

9 $\Delta H = \Delta E + \Delta n_g RT$

Given, $\Delta H = 30$ kcal for 3 mol

$$\Delta n_g = 3$$

because, liquid $\rightleftharpoons$ vapour

$$30 = \Delta E + 3\times2\times500\times10^{-3}$$
$$\Delta E = 27\text{ kcal}$$

10 $C(s) + \frac{1}{2}O_2(g) \longrightarrow CO(g)$

$$\Delta n_g = 1 - \frac{1}{2} = \frac{1}{2}$$
$$\Delta H - \Delta E = \Delta n_g RT$$
$$= \frac{1}{2}\times 8.314\text{ JK}^{-1}\text{ mol}^{-1}\times 298\text{ K}$$
$$= 1238.78\text{ J mol}^{-1}$$

11 According to relationship of ΔH and ΔE,

$$\Delta H = \Delta E + \Delta n_g RT$$

ΔH = enthalpy change (at constant pressure)

ΔE = internal energy change (at constant volume)

(given reaction is exothermic)

(Δn_g = moles of gaseous products − moles of gaseous reactants)

For the reaction,

$$N_2(g) + 3H_2(g) \longrightarrow 2NH_3(g)$$
$$\Delta n_g = 2 - 4 = -2$$

Thus, $\quad \Delta H = \Delta E - 2RT$

$\Rightarrow \quad \Delta H < \Delta E$

Numerical value of $\Delta H < \Delta E$ in exothermic reaction and when $\Delta n_g < 0$.

12 Calculate the heat of combustion with the help of following formula,

$$\Delta H_p = \Delta U + \Delta n_g RT$$

where, ΔH_p = Heat of combustion at constant pressure

ΔU = Heat at constant volume (It is also called ΔE)

Δn_g = Change in number of moles (In gaseous state

R = Gas constant

T = Temperature

From the equation,

$$C_6H_6(l) + \frac{15}{2}O_2(g) \longrightarrow 6CO_2(g) + 3H_2O(l)$$

Change in the number of gaseous moles, i.e.

$$\Delta n_g = 6 - \frac{15}{2} = -\frac{3}{2}\text{ or } -1.5$$

Now, we have Δn_g and other values given in the question are

$$\Delta U = -3263.9\text{ kJ/mol}$$
$$T = 25°\text{C}$$
$$= 273 + 25 = 298\text{ K}$$
$$R = 8.314\text{ JK}^{-1}\text{mol}^{-1}$$

So, $\Delta H_p = (-3263.9) + (-1.5)\times 8.314 \times 10^{-3} \times 298$

$$= -3267.6\text{ kJ mol}^{-1}$$

13 The amount of heat required to raise the temperature of one mole of substance through 1°C is called molar heat capacity.

$$C = \frac{q}{T_2 - T_1}$$

14

$$Q = mC\Delta T$$
$$1000 = \frac{100}{18}\times75\times\Delta T$$
$$\Delta T = 2.4\text{ K}$$

15 The equation, $C_p - C_V = nR$

2 moles of an ideal gas is equal to

$$C_p - C_V = 2R$$

16 Using Kirchhoff's equation,

$$\Delta C_p = \frac{\Delta H_2 - \Delta H_1}{T_2 - T_1}$$

$$\Delta H_2 - \Delta H_1 = \Delta C_p (T_2 - T_1)$$

$$\Delta H_2 - 24 = (0.031 - 0.055)(100 - 50)$$

$$\Delta H_2 = 22.8 \text{ cal g}^{-1}$$

17 For the given change,

$$\Delta_r S^\circ = \Sigma n_P S_P^\circ - \Sigma n_R S_R^\circ$$

$$= S^\circ_{NH_2CONH_2} + S^\circ_{H_2O} - [2 \times S^\circ_{NH_3} + S^\circ_{CO_2}]$$

$$= 174.0 + 69.9 - [2 \times 192.3 + 213.7]$$

$$= -354.4 \text{ JK}^{-1} \text{ mol}^{-1}$$

18 $\Delta S = 2.303 nR \log \frac{V_2}{V_1}$

$$= 2.303 \times 2 \times 2 \log \frac{20}{2} = 9.2$$

19 For an isolated system,

$\Delta H = \Delta E + p\Delta V$ or $T\Delta S = \Delta E + p\Delta V$

But $\Delta E = 0$

$\therefore$ $T\Delta S = p\Delta V$

$$\Delta S = \frac{p\Delta V}{T} > 0$$

20 $\Delta H_{\text{condensation}}$ for 1.8 g of steam

$$= (-40.8) \times \frac{1.8}{18} = -4.08 \text{ kJ}$$

$$\Delta S = \frac{\Delta H}{T_b} = \frac{-4.08 \times 10^3}{373.15}$$

$$= -10.93 \text{ JK}^{-1}$$

21 CO and NO molecules in solid states at 0 K adopt a nearly random arrangement indicating a positive value of entropy. It is due to their dipole moment which results in disorder.

22 $\Delta S_{\text{vap}} = \frac{\Delta H_{\text{vap}}}{T} = \frac{26 \times 10^3}{308}$

$$= +84.41 \text{ JK}^{-1} \text{ mol}^{-1}$$

$\Delta S_{\text{cond}} = \frac{\Delta H_{\text{cond}}}{T}$ $[\because \Delta H_{\text{cond}} = -26 \text{ kJ}]$

$$= -\frac{26 \times 10^3}{308} = -84.41 \text{ JK}^{-1}\text{mol}^{-1}$$

23

		ΔS
$A \to C$	=	50 eu
$C \to D$	=	30 eu
$D \to B$	=	−20 eu
$A \to B$		60 eu

24 A → (2, 4), B → (2), C → (3), D → (1)

$A \to$ Entropy of vaporisation is always positive, i.e. $\frac{\Delta H_{\text{vap}}}{T_b}$

$B \to \Delta G^\circ = -RT\ln K$

If K is +ve, then, $\Delta G^\circ = $ −ve, i.e. spontaneous process.

$C \to$ Crystalline state is most orderly arranged state of particles, thus lowest entropy.

$D \to$ During adiabatic expansion of an ideal gas, $q = 0$ thus, $\Delta E = W$ i.e.

Work done at the cost of internal energy which decreases when $q = 0$.

25 Entropy change for n moles of isothermal expansion of an ideal gas from volume V_1 to volume V_2 is

$$\Delta S = 2.303 \, nR \log \frac{V_2}{V_1}$$

$$= 2.303 \times 2 \times 8.3143 \log \frac{100}{10}$$

$$= 38.296 \text{ J mol}^{-1} \text{ K}^{-1}$$

26 $\eta \text{ (efficiency)} = \frac{T_2 - T_1}{T_2}$

$\Rightarrow$ $0.25 = \frac{T - 400}{T}$

$\therefore$ $T = 533.3 \text{ K}$

27 $PbO_2 + Pb \longrightarrow 2PbO$

Oxidation state +4 0 +2

Since, $\Delta G^\circ < 0$, i.e. it is negative. Therefore, the reaction is spontaneous in the forward direction. This suggest that Pb^{2+} is more stable than Pb^{4+}.

$SnO_2 + Sn \longrightarrow 2SnO$

Oxidation state +4 0 +2

Since, $\Delta_r G^\circ > 0$, i.e. it is positive, therefore, the reaction is non-spontaneous in the forward direction. But it will be spontaneous in the backward direction. This suggests that Sn^{4+} will be more stable than Sn^{2+}. These facts are also supported by the inert pair effect down the group.

28 Heat at constant pressure means enthalpy, i.e.

$$\Delta H = -2.5 \times 10^3 \text{ cal}$$

$$\Delta S = 7.4 \text{ cal deg}^{-1}$$

$$T = 298 \text{ K}$$

$$\Delta G = \Delta H - T\Delta S = -2.5 \times 10^3 - 298 \times 7.4$$

$$= -4705 \text{ cal}$$

As, $\Delta G = $ − ve

Hence, the process is spontaneous.

29 $\Delta G = \Delta H - T\Delta S$

At equilibrium, $\Delta G = 0$

$$T = \frac{\Delta H}{\Delta S} = \frac{30 \times 10^3}{75} = 400 \text{ K}$$

30 $\Delta G = \Delta H - T\Delta S$

At equilibrium, $\Delta G = 0$

$\therefore$ $T_e = \Delta H / \Delta S$

For a reaction to be spontaneous ΔG should be negative, so this implies $\Delta H < T\Delta S$.

$$\frac{\Delta H}{\Delta S} < T; T_e < T$$

Therefore, T should be greater than T_e.

31 $H_2O(l)$ and $H_2O(g)$ both exist together at same temperature and pressure,

$$H_2O(l) \rightleftharpoons H_2O(g)$$

In the state of equilibrium, $\Delta G = 0$ and conversion of liquid into gas increases the disorderness.

Hence, entropy $\Delta S = $ +ve

32 For a spontaneous process,

$$\Delta S_{\text{total}} = \Delta S_{\text{system}} + \Delta S_{\text{surroundings}}$$

be positive.

33 For a non-spontaneous reaction,

$$\Delta G = + \text{ve}$$

$$\Delta G = \Delta H - T\Delta S$$

$\therefore$ $\Delta H - T\Delta S$ should be +ve or $\Delta H > T\Delta S$

which is possible if $T < \frac{\Delta H}{\Delta S}$

Given, $\Delta H = 30 \times 10^3 \text{ J mol}^{-1}$

$\Delta S = 70 \text{ JK}^{-1} \text{ mol}^{-1}$

$$T < \frac{30 \times 10^3}{70}$$

$$T < 428.57 \text{ K}$$

34 Total entropy change of universe is zero. At 100°C and 1 atmosphere pressure,

$$H_2O(l) \rightleftharpoons H_2O(g)$$

is at equilibrium.

For equilibrium, $\Delta S_{\text{total}} = 0$

and $\Delta S_{\text{system}} + \Delta S_{\text{surrounding}} = 0$

As we know, during conversion of liquid to gas, entropy of system increases, in a similar manner entropy of surroundings decreases.

$\therefore$ $\Delta S_{\text{system}} > 0$

and $\Delta S_{\text{surrounding}} < 0$

Hence, (b) is the correct choice.

35 (a) $\Delta G = \Delta H - T\Delta S$

For a system, total entropy change

$$= \Delta S_{total}$$

$$\Delta H_{total} = 0$$

$$\therefore \quad \Delta G_{system} = -T\Delta S_{total}$$

$$\therefore \quad \frac{\Delta G_{system}}{\Delta S_{total}} = -T$$

(b) For isothermal reversible process,

$$\Delta E = 0$$

By first law of thermodynamics

$$\Delta E = q + W$$

$$\therefore \; W_{reversible} = -q = -\int_{V_i}^{V_f} p\,dV$$

$$W_{reversible} = -nRT \ln \frac{V_f}{V_i}$$

(c) $\Delta G = \Delta H^\circ - T\Delta S^\circ = -RT\log K$

$$\log K = -\left(\frac{\Delta H^\circ - T\Delta S^\circ}{RT}\right)$$

(d) $\Delta G^\circ = -RT \log K$

$$\therefore \quad \log K = -\frac{\Delta G^\circ}{RT}$$

$$\therefore \quad K = e^{-\Delta G^\circ/RT}$$

36 $\Delta G^\circ = \Delta H^\circ - T\Delta S^\circ = -2.303\, RT \log_{10} K$

$\therefore \; -2.303\, RT \log_{10} K = \Delta H^\circ - T\Delta S^\circ$

i.e. $\quad -2.303 \times 8.314 \times 298 \times \log_{10} K$

$$= [-54.07 \times 1000] - [298 \times 10]$$

$$-5705 \log_{10} K = -54070 - 2980$$

$$-5705 \log_{10} K = -57050$$

or $\quad \log_{10} K = 10$

37 The standard free energy related to equilibrium constant, K as

$$\Delta G^\circ = -2.303\, RT \log K_{eq}$$

$$\Delta G^\circ = -nFE^\circ_{cell}$$

If a cell reaction is spontaneous (proceeding in forward side), it means

$$K_{eq} > 1 \text{ and } E^\circ_{cell} = +\text{ ve}$$

Thus, $\quad \Delta G^\circ = -\text{ ve}$

38 Percentage efficiency of the fuel cell

$$= \frac{\Delta G}{\Delta H} \times 100$$

The concerned reaction is

$$CH_3OH(l) + \frac{3}{2} O_2(g) \longrightarrow CO_2(g) + 2H_2O(l);$$

$$\Delta G_r = [\Delta G_f(CO_2, g) + 2\Delta G_f(H_2O, l)] - [-\Delta G_f(CH_3OH, l) - \frac{3}{2}\Delta G_f(O_2, g)]$$

$$= -394.4 + 2(-237.2) - (-166.2) - 0$$

$$= -394.4 - 474.4 + 166.2$$

$$= -702.6 \text{ kJ mol}^{-1}$$

Percentage efficiency

$$= \frac{702.6}{726} \times 100 = 96.78\% \approx 97\%$$

39 Spontaneous processes are accompanied by decrease in energy and increase in randomness.

40 Reason is the correct explanation for Assertion.

SESSION 2

1. The pV diagram for the reversible and irreversible expansion of an ideal gas is shown below :

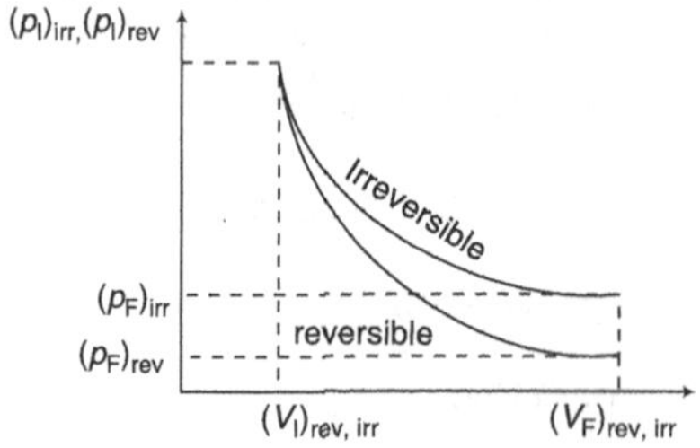

From ideal gas equation

$$p = \frac{nRT}{V}$$

For final condition of temperature, pressure and volume

$$p_F = \frac{nRT_f}{V_f}$$

From the curve it is seen that

$$(p_F)_{irr} > (p_F)_{rev} \text{ or } (T_F)_{irr} > (T_f)_{rev}$$

2 For adiabatic change, $\Delta E = \Delta W$

$$\Delta E = nC_V(T_2 - T)$$

$$\Delta W = -p(V_2 - V_1)$$

$$nC_V(T_2 - T) = -p(V_2 - V_1) = -1(2 - 1)$$

$$n \times \frac{R}{(\gamma - 1)} \times (T_2 - T) = -1$$

For monoatomic gas $n = 1, \gamma = \frac{5}{3}$,

$R = 0.0821$ L atm K^{-1} mol^{-1}

$$1 \times \frac{0.0821}{(5/3) - 1} \times (T_2 - T) = -1$$

$$T_2 = T - \frac{2}{3 \times 0.0821}$$

3 $\Delta H =$ work done $= i \times V \times t$

$$= 0.50 \text{ A} \times 12 \text{ V} \times 300 \text{ s}$$

$$= 1800 \text{ J} = +1.8 \text{ kJ}$$

Molar enthalpy of vaporisation,

$$\Delta H_m = \frac{\Delta H}{\text{moles of } H_2O} = \frac{\Delta H}{n_{H_2O}}$$

$$= \frac{1.8 \text{ kJ}}{\frac{0.798}{18}} = 40.6 \text{ kJ mol}^{-1}$$

$$\Delta H_m = \Delta E_m + \Delta n_g RT$$

$$\Delta H_m = \Delta E_m + RT$$

$[\Delta n_g = 1 \text{ for } H_2O(l) \rightleftharpoons H_2O(g)]$

Molar internal energy change,

$$\Delta E_m = \Delta H_m - RT$$

$$= 40.6 - 8.314 \times 10^{-3} \times 373.15$$

$$= 37.5 \text{ kJ mol}^{-1}$$

4 For adiabatic condition, $\frac{T_2}{T_1} = \left(\frac{V_1}{V_2}\right)^{\gamma - 1}$

$$\frac{T_2}{T_1} = \left(\frac{1}{27}\right)^{1.33 - 1} \Rightarrow \left(\frac{1}{27}\right)^{0.33}$$

$$= \left(\frac{1}{27}\right)^{1/3} = \frac{1}{3}$$

$$T_2 = 300 \times \frac{1}{3} = 100 \text{ K}$$

Thus, $T_2 < T_1$, hence cooling takes place due to expansion under adiabatic condition.

$$\Delta E = q + W = W$$

($\because q = 0$ for adiabatic change.)

Sign of W is negative because the gas expands.

$$W = -\Delta E = -C_V(T_2 - T_1)$$

$$= -6(100 - 300) = 1200 \text{ cal}$$

5. The molar heat capacity for any process is given by following expression,

$C = C_V + \frac{R}{1 - \gamma}$ when $pV^\gamma =$ constant

and $C_p / C_V = \gamma$

Here, $\quad \frac{p}{V} = 1$, i.e. pV^{-1} constant

$$C = \frac{3}{2}R + \frac{R}{1 - (-1)} = \frac{3}{2}R + \frac{R}{2} = \frac{4}{2}R$$

6 $\Delta H = \Delta U + p\Delta V$

Here, $\Delta U = +0.21$ kJ mol^{-1}

$$= 0.21 \times 10^3 \text{ J mol}^{-1}$$

$$= 210 \text{ J mol}^{-1}$$

$$p = 1 \text{ bar} = 10^5 \text{ Pa}$$

$\Delta V =$ Molar volume of aragonite − Molar volume of calcite

$$= \frac{100}{2.93} - \frac{100}{2.71} \text{ cm}^3 \text{ mol}^{-1}$$

($\because$ Molar mass of $CaCO_3 = 100$ g mol^{-1})

$$= 100\left(\frac{1}{2.93} - \frac{1}{2.71}\right)$$

$$= 100 \times \left(\frac{0.22}{2.93 \times 2.71}\right)$$

$= 2.77 \text{ cm}^3 \text{ mol}^{-1}$

$= -2.77 \times 10^{-6} \text{ m}^3$

$\therefore \Delta H = 210 \times 10^5 (-2.77 \times 10^{-6})$

$= 210 - 0.277$ J

$= 209.72 \text{ J mol}^{-1}$

7 For spontaneous process

$$\Delta S_{total} = \Delta S_{sys} + S_{surr} > 0$$

For an isolated system $\Delta S_{surr} = 0$

Hence, $\Delta S_{sys} > 0$.

8 (a) Since, change of state (p_1, V_1, T_1) to (p_2, V_2, T_2) is isothermal therefore, $T_1 = T_2$ (correct statement)

(b) Since, change of state (p_1, V_1, T_1) to (p_3, V_3, T_3) is an adiabatic expansion, it brings about cooling of gas, therefore, $T_3 < T_1$. Thus, it is incorrect.

(c) Work done is given by the area under the curve of p-V diagram. As obvious from the given diagram, magnitude of area under the isothermal curve is greater than that under adiabatic curve, hence $W_{isothermal} > W_{adiabatic}$ (correct statement).

(d) $\Delta E = nC_V \Delta T$

In isothermal process, $\Delta E = 0$ as $\Delta T = 0$

In adiabatic process,

$$\Delta E = nC_v (T_3 - T_1) < 0 \text{ as } T_3 < T_1$$

$\Delta E_{isothermal} > \Delta E_{adiabatic}$

(correct statement)

9 Work is done against constant pressure and thus, irreversible.

Given, $\Delta V = 6 - 2 = 4$ L and $p = 1$ atm

$W = -p \times dV$

$= -1 \times 4$ L atm

$$= -\frac{1 \times 4 \times 1.987}{0.0821} \text{ cal}$$

($\because$ 0.0821 L atm = 1.987 cal)

$= -96.81$ cal

$= -96.81 \times 4.184 \text{ J} = -405.04$ J

From first law of thermodynamics,

$q = \Delta E - W$

$\Delta E = q + W = 800 + (-405.04)$

$= 394.95$ J

10 For an irreversible process to be spontaneous,

$(\Delta S)_{V,E}$ (change in entropy)

$= +\text{ve} > 0$

$\Rightarrow (\Delta G)_{T,p}$ (change in Gibbs free energy)

$= -\text{ve} < 0$

11 For spontaneity, $\Delta G = -$ ve

$$\Delta G = \Delta H - T\Delta S$$

ΔH, for endothermic process + ve

At lower temperature,

$T\Delta S = +$ve

At lower temperature, $T\Delta S$ will be less than ΔH.

Hence, $\Delta G = +$ve

But at high temperature, $T\Delta S$ will be greater than ΔH.

Hence, $\Delta G = -$ ve, spontaneous.

12 As the process is carried out under adiabatic conditions, $q_p = 0$. But by definition $q_p = \Delta H$.

Hence, $\Delta H = 0$

Further, at constant pressure of 100 bar, volume has decreased by 1 mL, therefore, work of contraction

$= p\Delta V = 100 \text{ bar} \times 1 \text{ mL}$

$= (100 \times 10^5 \text{ Nm}^{-2})(10^{-6} \text{ m}^3)$

$= 10$ J

During adiabatic process, work done

= change in internal energy

Hence, $\Delta U = 10$ J

13 $\Delta_r S° = S°_{XY_3} - \left[\frac{1}{2} S°_{X_2} + \frac{3}{2} S°_{Y_2}\right]$

$$= 50 - \left[\frac{1}{2} \times 60 + \frac{3}{2} \times 40\right]$$

$= 50 - (30 + 60) = -40 \text{ JK}^{-1} \text{ mol}^{-1}$

$\Delta_r G° = \Delta_r H° - T\Delta_r S°$

(At equilibrium, $\Delta_r G° = 0$)

$$\therefore \quad T = \frac{\Delta_r H°}{\Delta_r S°} = \frac{-30{,}000 \text{ J mol}^{-1}}{-40 \text{ JK}^{-1} \text{ mol}^{-1}} = 750 \text{ K}$$

14 $\Delta G = \Delta H - T\Delta S$

Also, $\Delta G = -2.303 RT \log K$

Variation of K with temperature is given by $\log K = \frac{\Delta S°}{R} - \frac{\Delta H°}{RT}$

Given, $\log K = 4.0 - \frac{2000}{T}$

On comparing, $\frac{\Delta S°}{R} = 4$ or

$\Delta S° = 4R$

15 In accordance to second law of thermodynamics, if an irreversible process occurs in a closed system, the entropy of that system always increases, it never decreases. As the ice is more ordered then water, i.e. it has less entropy. Therefore, the ice will melt so that the mass of the ice will decrease.

DAY SEVEN

Thermochemistry

Learning & Revision for the Day

- Thermochemical Standard State
- Heat or Enthalpy of Reaction
- Hess's Law of Constant Heat Summation
- Bond Energy or Enthalpy
- Calorimetry

Thermochemistry is a branch of chemistry which deals with energy exchange between a chemical system and its surrounding. Heat is generally evolved or absorbed whenever a chemical reaction takes place, or change in the state of matter (vaporisation, fusion, phase transition) occurs.

- Thermochemistry focuses on these energy changes, particularly on the system's energy exchange with its surroundings.
- The heat change accompanying chemical reactions or physical changes are measured experimentally with the help of calorimeter.
- **Chemical Reactions** are invariably associated with transfer of energy and most frequently, energy transfer in chemical reactions takes place in the form of heat. Reactions may be exothermic or endothermic.
  (i) **Exothermic** reactions transfer heat to the surroundings.
  $$4Al(s) + 3O_2(g) \longrightarrow 2Al_2O_3(s);\ \Delta H = -1676\,kJ$$
  (ii) **Endothermic** reactions transfer heat from the surroundings.
  $$N_2(g) + O_2(g) \longrightarrow 2NO(g);\ \Delta H = +90.4\,kJ$$

Thermochemical Standard State

- A thermochemical standard state of a substance is its most stable state under 1 atm pressure (standard pressure) and 298 K (standard temperature). Under these conditions, any parameter is designated the superscript θ or 0.
- Purest form for liquids and solids, most stable states are considered with 1 bar pressure condition and with 298 K temperature.
- For a gas, the standard state is considered at a pressure of one atmosphere. In a mixture of gases, its partial pressure must be one atmosphere.
- For a substance in solution, the standard state refers to one molar concentration.

Heat or Enthalpy of Reaction

- It is the amount of heat absorbed or evolved at constant pressure, when the quantities of substance indicated by thermochemical equation have completely reacted. It is denoted by ΔH_r, e.g. $CH_4(g) + 2O_2(g) \longrightarrow CO_2(g) + 2H_2O(g);\ \Delta H_r = -890.3\ kJ$
- ΔH_r° = (sum of enthalpies of products) − (sum of enthalpies of reactants), i.e.

$$\Delta H_r^\circ = \Sigma H_{\text{products}} - \Sigma H_{\text{reactants}}$$

PREP MIRROR
Your Personal Preparation Indicator

- No. of Questions in Exercises (x)—
- No. of Questions Attempted (y)—
- No. of Correct Questions (z)—
  (Without referring Explanations)

- Accuracy Level ($z/y \times 100$)—
- Prep Level ($z/x \times 100$)—

In order to expect good rank in JEE, your Accuracy Level should be above 85 & Prep Level should be above 75.

Standard Enthalpy When reaction is carried out at standard conditions (i.e. at 1 atm pressure and 298 K) It is denoted as by $\Delta_r H^\circ$.

(i) If $H_{\text{products}} = H_{\text{reactants}}$; $\Delta H = 0$

(ii) If $H_{\text{products}} > H_{\text{reactants}}$; $\Delta H = +\text{ve}$, reaction is said to be endothermic.

(iii) If $H_{\text{products}} < H_{\text{reactants}}$; $\Delta H = -\text{ve}$, reaction is said to be exothermic.

Factors Influencing Enthalpy of Reaction

Various factors that affect the enthalpy of reaction are :

(i) Physical state of reactants and products
(ii) Allotropic forms of elements involved
(iii) Chemical composition state of reactants and products
(iv) Amount of reactants
(v) Temperature
(vi) Reaction conditions

Types of Standard Enthalpy of Reaction

Different types of standard enthalpy of reactions are given below:

1. Standard Enthalpy of Formation (ΔH_f°)

- It is the standard enthalpy change for the formation of one mole of a compound from its elements in their most stable state of aggregation. e.g.

$$C_{(\text{graphite})} + 2H_2(g) \longrightarrow CH_4(g);\ \Delta H_f^\circ = -74.8\ \text{kJ mol}^{-1}$$

- Enthalpy of formation of an element at standard state by convention is taken as zero. e.g. enthalpy of formation of Mg, Al, Na, H_2, O_2 etc., is taken as zero.
- The standard enthalpy of the chemical reaction is given by

$$\Delta_r H^\circ = \Delta_f H_P^\circ - \Delta_f H_R^\circ$$

Here, $\Delta_f H^\circ_P$ = standard enthalpy of formation for products

$\Delta_f H_R^\circ$ = standard enthalpy of formation for reactants

2. Standard Enthalpy of Combustion (ΔH_c°)

- It is the standard enthalpy change per mole of a substance, when it undergoes complete combustion.

e.g. $CH_4(g) + 2O_2(g) \longrightarrow CO_2(g) + 2H_2O(l)$;

$$\Delta H_c^\circ = -192\ \text{kcal mol}^{-1}$$

- ΔH combustion is always negative but for certain reactions it is positive. For example,

$$N_2 + O_2 \longrightarrow 2NO;\ \Delta H = \text{positive}$$

$$F_2 + \frac{1}{2}O_2 \longrightarrow OF_2;\ \Delta H = \text{positive}$$

3. Enthalpy of Neutralisation (ΔH_n°)

- It is the amount of heat liberated when 1g equivalent of an acid is completely neutralised by 1g equivalent of a base. ΔH_n is constant for strong acid and strong base,

i.e. $\Delta H_n = -13.7\ \text{kcal mol}^{-1}$ or $-57.27\ \text{kJ mol}^{-1}$

$$HCl(aq) + NaOH(aq) \longrightarrow NaCl(aq) + H_2O(l);\quad \Delta H_n = -57.3\ \text{kJ mol}^{-1}$$

- For a weak acid against a strong base or weak base, the numerical value of ΔH_n is always less than 13.7 kcal due to the fact that here the heat is used up in ionisation of weak acid or weak base.
- The absolute value of heat of neutralization of HF is more than 57.3 kJ. This is due to very high heat of hydration of fluoride ion.

4. Standard Enthalpy of Atomisation (ΔH_a°)

It is the enthalpy change on breaking one mole of bonds completely to obtain atoms in the gas phase. In case of diatomic molecules (X_2), the enthalpy of atomisation, bond dissociation enthalpy and bond enthalpy are same thing.

5. Lattice Enthalpy ($\Delta H_{\text{lattice}}$)

It is the enthalpy change, which occurs when one mole of an ionic compound dissociates into its ions in gaseous state.

$$NaCl(s) \longrightarrow Na^+(g) + Cl^-(g);\ \Delta H_{\text{lattice}}^\circ = -787\ \text{kJ mol}^{-1}$$

6. Enthalpy of Hydration (ΔH_{hyd})

It is the enthalpy change, when one mole of anhydrous or partially hydrated salt combines with required number of moles of water to form a specific hydrate.

$$CuSO_4(s) + 5H_2O(l) \longrightarrow CuSO_4 \cdot 5H_2O(s);\quad \Delta H_{\text{hydration}} = -78.21\ \text{kJ mol}^{-1}$$

During dissolution, physical state of the compound changes while during hydration, there is no change in the physical state of compound.

7. Standard Enthalpy of Solution ($\Delta H_{\text{sol}}^\circ$)

It is the standard enthalpy change, when one mole of substance dissolves in a specified amount of solvent.

$$\Delta H_{\text{sol}}^\circ = \Delta H_{\text{lattice}} - \Delta H_{\text{hydration}}$$

Integral heat of solution is the enthalpy change when 1 mole of solute is dissolved in a pure solvent to form a solution of desired concentration.

- Differential heat of solution is the enthalpy change when 1 mole of solute is dissolved in such a large volume of solution so that no enthalpy change occurs on further dilution.
- If the solubility of a substance is known at two different temperatures, the mean molar enthalpy of solution over this temperature range can be calculated by applying an equation similar to van't Hoff equation;

$$\frac{\log S_1}{\log S_2} = \frac{\Delta H}{2.303\,R}\left(\frac{1}{T_1} - \frac{1}{T_2}\right)$$

where, S_1 and S_2 are solubilities at T_1 and T_2 temperatures respectively.

8. Standard Enthalpy of Hydrogenation ($\Delta H^{\circ}_{\text{hydrogenation}}$)

It is the amount of enthalpy change that takes place when one mole of unsaturated organic compound is completely hydrogenated.

9. Standard Enthalpy of Dilution ($\Delta H^{\circ}_{\text{dil.}}$)

The standard enthalpy change, when 1 mole of a substance is diluted to such an extent that on further dilution no heat is evolved or absorbed is termed as standard enthalpy of dilution.

10. Standard Enthalpy of Transition (ΔH°_{t})

It is the enthalpy change when one mole of the substance undergoes transition from one allotropic form to another.

S(rhombic) $\rightarrow$ S(monoclinic); $\Delta H^{\circ}_{t} = -13.14$ kJ

11. Standard Enthalpy of Fusion ($\Delta H^{\circ}_{\text{fus}}$)

It is the enthalpy change that accompanies melting of one mole of a solid substance.

$$\Delta H^{\circ}_{\text{fus}} = -\Delta H^{\circ}_{\text{freez}}$$

12. Standard Enthalpy of Vaporisation ($\Delta H^{\circ}_{\text{vap}}$)

It is the amount of heat required to convert one mole of liquid into its vapour state.

$$\Delta H^{\circ}_{\text{vap}} = -\Delta H^{\circ}_{\text{cond}}$$

13. Standard Enthalpy of Sublimation ($\Delta H^{\circ}_{\text{sub}}$)

At standard conditions, change in enthalpy, when one mole of a solid substance sublimes is called the standard enthalpy of sublimation.

$$\Delta_{\text{sub}}H^{\circ} = \Delta_{\text{fus}}H^{\circ} + \Delta_{\text{vap}}H^{\circ}$$

14. Enthalpy of Ionisation (ΔH_{ions})

It is the amount of enthalpy change that take place when one mole of a covalent compound on dissolution in water splits to produce ions in the solution.

$$CH_3COOH(aq) \rightleftharpoons CH_3COO^-(aq) + H^+(aq);$$

$$\Delta H_{\text{ions}} = -55.43 \text{ kJ mol}^{-1}$$

Hess's Law of Constant Heat Summation

The enthalpy change in a particular reaction is always constant and is independent of the path by which the reaction takes place.

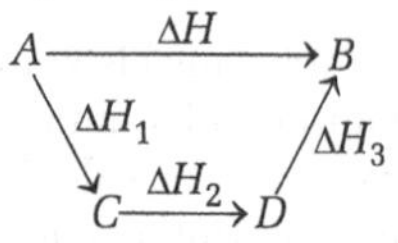

In other words, the total heat change (ΔH) accompanying a chemical reaction is the same whether, the reaction takes place in one step or in more steps.

According to Hess's law heat of summation is given as:

$$\Delta H = \Delta H_1 + \Delta H_2 + \Delta H_3$$

Some applications of Hess's law are :

- In determination of heat of formation.
- In determination of heat of transition.
- In determination of bond enthalpy
- In determination of resonance energy

Resonance energy = experimental heat of formation – calculated heat of formation

Bond Energy or Enthalpy

- When a bond is formed between two atoms in gaseous state to form a molecule, some heat is always evolved which is called bond energy or bond formation energy.
- **Bond dissociation energy** is the amount of energy required to break/dissociate bond of a particular type present in one molecule of the compound.
- Bond energy is taken as the average value of dissociation energies of same type of bonds present in one mole.
- Since, a chemical reaction involves the breaking of old bonds in reactants and formation of new bonds in products, the enthalpy change of a reaction,

  ΔH_r = Sum of BE of reactants – sum of BE of products
- In a polyatomic molecule containing two or more covalent bonds between same atoms (e.g. CH_4), the term **average bond energy** is preferred in place of bond dissociation energy.

  For CH_4, average BE of C—H bond

  $$= \frac{\text{Bond dissociation energy of } CH_4}{4}$$
- Bond dissociation enthalpy values are negative, if bond formation occurs whereas bond energy values are positive, if bond dissociation occurs.

Factor Affecting Bond Enthalpy

There are various factors affecting bond enthalpy which are as follows:

1. Size of Atoms

- Smaller the size of atom, more closer the atoms to each other during bonding hence, larger is the bond enthalpy. e.g. order of bond enthalpy of halogens is F—F < Cl—Cl > Br—Br > I—I.
- Bond enthalpy of fluorine is smaller than chlorine because of the high degree of lone pair repulsions in F_2 due to its smaller size.

2. Electronegativity

Larger the electronegativity difference between two atoms, more is the polarity in bond and thus, more is the bond strength as well as bond enthalpy.

e.g. F—H > O—H > N—H (Bond enthalpy decreases)

3. Bond Length

Shorter the bond length, more is the bond dissociation enthalpy.

4. Number of Bonding Electrons

As the number of electrons involved in bond increases, strength of the bond increases. This increases the bond enthalpy.

e.g. $C \equiv C > C = C > C - C$ (Bond enthalpy decreases)

Calorimetry

- The experimental measurement of the heat change of reaction or enthalpy change is known as calorimetry. In laboratory, heat changes in physical and chemical processes are measured with a calorimeter which is an insulated container

$$q = mc\Delta t = C\,\Delta t \qquad \text{(Heat capacity, } C = mc)$$
$$\Delta t = t_{final} - t_{initial}$$

where, m is the mass of the substance in grams, c is the specific heat and C is the heat capacity.

- For endothermic change, q is positive and for exothermic change, q is negative.

Constant Volume Calorimetry (Bomb Calorimeter)

- Heat of combustion is measured by placing a known mass of a compound in a **constant volume bomb calorimeter** which is filled with oxygen at about 30 atm pressure.
- On ignition of the sample electrically, there is evolution of heat which can be calculated by recording the rise in temperature of water.
- Heat lost by the sample = Heat gained by the water

$$q_{combustion} = -[q_{water} + q_{bomb}]$$
$$= -[m_{water} \times c_{water} + m_{bomb} \times c_{bomb}] \times \Delta t$$
$$q_{combustion} = \Delta E_{combustion}$$

(combustion in bomb calorimeter at constant V)

Therefore, $\Delta H_{combustion} = \Delta E + \Delta n_g RT$

- The calorific value of a fuel or food is the amount of heat in calories or joules produced from the complete combustion of one gram of the fuel or the food.

$$\text{Calorific value} = \frac{\Delta H_{comb}}{\text{Molecular mass}}$$

DAY PRACTICE SESSION 1

FOUNDATION QUESTIONS EXERCISE

1 $\Delta H^\circ_{f(NO_2)} = 33.84$ kJ mol^{-1} and $\Delta H^\circ_{f(N_2O_4)} = 9.66$ kJ mol^{-1}, hence, dimerisation of NO_2 forming N_2O_4 is

(a) endothermic (b) isothermic
(c) exothermic (d) isochoric

2 In the reaction, $y \to z, \Delta H = +100$ kcal/mol and for the reaction, $z \to x, \Delta H = -80$ kcal/mol and $y \to x$, $\Delta H = 20$ kcal
Rank the enthalpies of formation of x, y and z in increasing order

(a) y, x, z (b) y, z, x (c) x, z, y (d) x, y, z

3 2.1 g of Fe combines with S evolving 3.77 kJ. The heat of formation of FeS in kJ/mol is

(a) −1.79 (b) −100.5
(c) −3.77 (d) None of these

4 Which of the following is (are) exothermic reactions.

I. Combustion of methane
II. Decomposition of water
III. Dehydrogenation of ethane to ethylene
IV. Conversion of graphite to diamond

(a) (I) and (II) (b) (II) and (III)
(c) (III) and (IV) (d) (II), (III) and (IV)

5 The ΔH_f° for $CO_2(g)$, $CO(g)$ and $H_2O(g)$ are −393.5, −110.5 and −241.8 kJ mol^{-1} respectively, the standard enthalpy change (in kJ) for the reaction,

$$CO_2(g) + H_2(g) \longrightarrow CO(g) + H_2O(g), \text{ is}$$

(a) 524.1 (b) 41.2 (c) −262.5 (d) −41.2

6 Consider the following,

$$C + O_2 \longrightarrow CO_2;\ \Delta H = x$$
$$CO + \frac{1}{2}O_2 \longrightarrow CO_2;\ \Delta H = y$$

Then, the heat of formation of CO is

(a) $x - y$ (b) $y - 2x$ (c) $x + y$ (d) $2x - y$

7 On the basis of the following thermochemical data $(\Delta_f H^\circ_{H^+(aq)} = 0)$

$$H_2O(l) \longrightarrow H^+(aq) + OH^-(aq);\ \Delta H = 57.32 \text{ kJ}$$
$$H_2(g) + \frac{1}{2}O_2(g) \longrightarrow H_2O(l);\ \Delta H = -286.20 \text{ kJ}$$

Also $H^\circ(H^+, aq) = 0$
The value of enthalpy of formation of OH^- ion at 25°C is

→ JEE Main 2009

(a) − 22.88 kJ (b) − 228.88 kJ
(c) + 228.88 kJ (d) − 343.52 kJ

8 The heat of combustion of carbon and carbon monoxide are – 393.5 and – 283.5 kJ mol^{-1}, respectively. The heat of formation (in kJ) of carbon monoxide per mole is → JEE Main 2016
(a) 676.5 (b) –676.5 (c) –110.5 (d) 110.5

9 If enthalpies of formation of $C_2H_4(g)$, $CO_2(g)$ and $H_2O(l)$ at 250°C and 1 atm pressure be 52, –394 and –286 kJ mol^{-1} respectively, the enthalpy of combustion of $C_2H_4(g)$ will be
(a) 1412 kJ mol^{-1} (b) –1412 kJ mol^{-1}
(c) +141. 2 kJ mol^{-1} (d) –141.2 kJ mol^{-1}

10 Heat of combustion of $H_2(g) = -241.8$ kJmol^{-1}
$C(s) = -393.5$ kJmol^{-1}, $C_2H_5OH\,(l) = -1234.7$ kJmol^{-1}
Hence, heat of formation of $C_2H_5OH(l)$ is
(a) – 2747.1 kJ mol^{-1} (b) –277.7 kJ mol^{-1}
(c) 277.7 kJ mol^{-1} (d) 2747.1 kJ mol^{-1}

11 Combustion of glucose takes place according to the equation,
$C_6H_{12}O_6 + 6O_2 \longrightarrow 6CO_2 + 6H_2O$; $\Delta H = -72$ kcal
How much energy will be required for the production of 1.6 g of glucose (molecular mass of glucose = 180) ?
(a) 0.064 kcal (b) 0.64 kcal (c) 6.4 kcal (d) 64 kcal

12 The heat evolved in the combustion of methane is given by the following equation,
$$CH_4(g) + 2O_2(g) \longrightarrow CO_2(g) + 2H_2O(l); \quad \Delta H = -\,890.3\text{ kJ}$$
How many grams of methane would be required to produce 445.15 kJ of heat of combustion?
(a) 4 g (b) 8 g (c) 12 g (d) 16 g

13 Equal volumes of methanoic acid and sodium hydroxide are mixed, if x is the heat of formation of water, heat evolved in neutralisation is
(a) more than x (b) equal to x
(c) twice of x (d) less than x

14 When 1 mole of oxalic acid is treated with excess of NaOH in dilute aqueous solution, 106 kJ of heat is liberated. The enthalpy of ionisation of the acid is
(a) 4.3 kJ mol^{-1} (b) – 4.3 kJ mol^{-1}
(c) – 8.6 kJ mol^{-1} (d) 8.6 kJ mol^{-1}

15 Consider the reaction,
$$4NO_2(g) + O_2(g) \longrightarrow 2N_2O_5(g); \quad \Delta_r H = -111\text{ kJ}.$$
If $N_2O_5(s)$ is formed instead of $N_2O_5(g)$ in the above reaction, the $\Delta_r H$ value will be
(Given, ΔH of sublimation for N_2O_5 is 54 kJ mol^{-1})
→ AIEEE 2011
(a) – 165 kJ (b) +54 kJ (c) +219 kJ (d) – 219 kJ

16 The amount of heat absorbed by 70.09 of water for their complete vaporisation is
(a) 23,352 J (b) 7000 J (c) 15,813 J (d) 158,200 J

17 The H_f° of O_3 , CO_2, NH_3 and HI are 142.2, – 393.3, – 46.2 and + 25.9 kJ per mol respectively. The order of their increasing stabilities will be
(a) O_3, CO_2, NH_3, HI (b) CO_2, NH_3, HI, O_3
(c) O_3, HI, NH_3, CO_2 (d) NH_3, HI, CO_2, O_3

18 ΔH°_f of atom B is 134.5 kcal / mol and ΔH°_f of atom F is 118.86 kcal / mol. ΔH°_f of $BF_3(g)$ is –271.75 kcal / mol. Average B—F bond energy would be
(a) 97.7 kcal / mol (b) 116.6 kcal / mol
(c) 135.4 kcal / mol (d) 254.3 kcal / mol

19 In a calorimeter, the temperature of the calorimeter increases by 6.12 K, the heat capacity of the system is 1.23 kJ/g /deg. What is the molar heat of decomposition for the ammonium nitrate?
(a) –7.53 kJ / mol (b) –398.1 kJ / mol
(c) –16.1 kJ/mol (d) –602 kJ/mol

20 The bond dissociation energies of gaseous H_2, Cl_2 and HCl are 104, 58 and 103 kcal respectively. The enthalpy of formation of HCl gas would be
(a) –44 kcal (b) 44 kcal (c) –22 kcal (d) 22 kcal

21 Given that, $C(g) + 4H(g) \longrightarrow CH_4(g)$; $\Delta H = -166$ kJ
The bond energy of C—H will be
(a) –415 kJ / mol (b) –41.5 kJ / mol
(c) 832 kJ / mol (d) None of these

22 If at 298 K, the bond energies of C—H, C—C, C=C and H—H bonds are 414, 347, 615 and 435 kJ mol^{-1} respectively,
The value of enthalpy change for the reaction,
$H_2C{=}\,CH_2(g) + H_2(g) \longrightarrow H_3C{-}CH_3(g)$ at 298 K will be
(a) + 250 kJ mol^{-1} (b) – 250 kJ mol^{-1}
(c) + 125 kJ mol^{-1} (d) – 125 kJ mol^{-1}

23 Given that,
I. $\Delta_f H^\circ$ of N_2O is 82 kJ mol^{-1}
II. Bond energies of N≡N, N=N, O=O and N=O are 946, 418, 498 and 607 kJ mol^{-1} respectively. The resonance energy of N_2O is
→ JEE Main (Online) 2013
(a) – 88 kJmol^{-1} (b) – 66 kJmol^{-1}
(c) – 62 kJmol^{-1} (d) – 44 kJmol^{-1}

24 Using the data provided, calculate the multiple bond energy (kJ mol^{-1}) of a C≡C bond in C_2H_2.
The energy is (take the bond energy of a C—H bond as 350 kJ mol^{-1}) → AIEEE 2012
$2C(s) + H_2(g) \longrightarrow C_2H_2(g)$; $\Delta H = 225$ kJ mol^{-1}
$2C(s) \longrightarrow 2C(g)$; $\Delta H = 1410$ kJ mol^{-1}
$H_2(g) \longrightarrow 2H(g)$; $\Delta H = 330$ kJ mol^{-1}
(a) 1165 kJ mol^{-1} (b) 837 kJ mol^{-1}
(c) 865 kJ mol^{-1} (d) 815 kJ mol^{-1}

25 The standard enthalpy of formation of NH_3 is -46.0 kJ mol^{-1}. If the enthalpy of formation of H_2 from its atoms is -436 kJ mol^{-1} and that of N_2 is -712 kJ mol^{-1}, the average bond enthalpy of N—H bond in NH_3 is →AIEEE 2010

(a) -964 kJ mol^{-1} (b) -352 kJ mol^{-1}
(c) $+1056$ kJ mol^{-1} (d) -1102 kJ mol^{-1}

26 The heat of combustion of benzoic acid is -2546kJ mol^{-1} at 25°C and 1 atm. The ΔE is

(a) -1555.6 kJ mol^{-1} (b) -2244.7 kJ mol^{-1}
(c) -2544.76 kJ mol^{-1} (d) -2868.66 kJ mol^{-1}

27 ΔH for combustion of ethane and ethyne are -341.1 and -310.0 kcal respectively. What will be the ratio of calorific values of ethane and ethyne respectively?

(a) 1 : 0.95 (b) 0.65 : 2
(c) 0.95 : 1 (d) 0.002 : 1

28 **Assertion** (A) Combustion of all organic compounds is an exothermic reaction.

Reason (R) The enthalpies of all elements in their standard state are zero.

(a) Assertion and Reason both are correct statements and Reason is the correct explanation of the Assertion
(b) Assertion and Reason both are correct statements but Reason is not the correct explanation of the Assertion
(c) Assertion is correct incorrect and Reason is incorrect
(d) Both Assertion and Reason are incorrect

29 **Statement I** The enthalpy of formation of $H_2O(l)$ is greater than that of $H_2O(g)$.

Statement II The enthalpy change for condensation reaction, i.e. $H_2O(g) \longrightarrow H_2O(l)$ is negative.

(a) Statement I is true, Statement II is true; Statement II is a correct explanation for Statement I
(b) Statement I is true, Statement II is true; Statement II is not a correct explanation for Statement I
(c) Statement I is true; Statement II is false
(d) Both the Statements I and II are false

DAY PRACTICE SESSION 2

PROGRESSIVE QUESTIONS EXERCISE

1 The heat of combustion of carbon to CO_2 is -393.5 kJ/mol. The heat released upon formation of 35.2 g of CO_2 from carbon and oxygen gas is

(a) + 315 kJ (b) − 31.5 kJ
(c) − 315 kJ (d) + 31.5 kJ

2 Which of the reactions defines, ΔH_f°?

(a) $C_{(diamond)} + O_2(g) \longrightarrow CO_2(g)$

(b) $\frac{1}{2}H_2(g) + \frac{1}{2}F_2(g) \longrightarrow HF(g)$

(c) $N_2(l) + 3H_2(g) \longrightarrow 2NH_3(g)$

(d) $CO(g) + \frac{1}{2}O_2(g) \longrightarrow CO_2(g)$

3 The enthalpy of combustion of H_2, cyclohexene and cyclohexane are -241, -3800 and -3920 kJ mol^{-1} respectively. Heat of hydrogenation of cyclohexene is

(a) 121 kJ mol^{-1} (b) -121 kJ mol^{-1}
(c) $+242$ kJ mol^{-1} (d) -242 kJ mol^{-1}

4 The enthalpy changes for the following processes are listed below :

$$Cl_2(g) \longrightarrow 2Cl(g),\ 242.3 \text{ kJ mol}^{-1}$$
$$I_2(g) \longrightarrow 2I(g),\ 151.0 \text{ kJ mol}^{-1}$$
$$ICl(g) \longrightarrow I(g) + Cl(g),\ 211.3 \text{ kJ mol}^{-1}$$
$$I_2(s) \longrightarrow I_2(g),\ 62.76 \text{ kJ mol}^{-1}$$

Given that the standard states for iodine and chlorine are $I_2(s)$ and $Cl_2(g)$, the standard enthalpy of formation of $ICl(g)$ is

(a) -14.6 kJ mol^{-1} (b) -16.8 kJ mol^{-1}
(c) $+16.8$ kJ mol^{-1} (d) $+244.8$ kJ mol^{-1}

5 On burning 4.0 g of iron to ferric oxide at constant pressure, the heat evolved is 29.28 kJ. The enthalpy of formation of ferric oxide is (At. wt. of Fe = 56)

(a) $+81.98$ kJ mol^{-1} (b) -81.98 kJ mol^{-1}
(c) $+819.8$ kJ mol^{-1} (d) -819.8 kJ mol^{-1}

6 A cooking gas cylinder is assumed to contain 11.2 kg *iso*-butane. The combustion of *iso*-butane is given by

$$C_4H_{10}(g) + \frac{13}{2}O_2(g) \longrightarrow 4CO_2(g) + 5H_2O(l);$$
$$\Delta H = -2658 \text{ kJ}$$

If a family needs 15,000 kJ of energy per day for cooking, how long would the cylinder last?

[Assuming that 30% of the gas is wasted due to incomplete combustion.]

(a) 34 days (b) 30 days (c) 31 days (d) 24 days

7 If the heat of neutralisation for a strong acid-base reaction is -57.1 kJ, what would be the heat released when 350 cm^3 of 0.20 M of a dibasic strong acid is mixed with 650 cm^3 of 0.10 M monoacidic base?

(a) 57.1 kJ (b) 3.71 kJ (c) -57.1 kJ (d) 0.317 kJ

8 Consider the reactions given below. On the basis of these reactions find out which of the algebraic relationship given in options (a) to (d) is correct?

I. $C(g) + 4H(g) \longrightarrow CH_4(g)$; $\Delta_r H = x$ kJmol^{-1}

II. $C(\text{graphite}) + 2H_2(g) \longrightarrow CH_4(g)$; $\Delta_r H = y$ kJmol^{-1}

(a) $x = y$ (b) $x = 2y$ (c) $x > y$ (d) $x < y$

9 The dissolution of $CaCl_2 \cdot 6H_2O$ in large volume of water is endothermic to the extent of 3.5 kcal mol^{-1}. For the reaction,

$$CaCl_2(s) + 6H_2O(l) \longrightarrow CaCl_2 \cdot 6H_2O(s)$$

$\Delta H = -23.2$ kcal. Hence, heat of solution of $CaCl_2$ (anhydrous) in a large volume of water is

(a) 26.7 kcal (b) −26.7 kcal
(c) 19.7 kcal (d) −19.7 kcal

10 The temperature of a bomb calorimeter rises by 1.6 K when a current of 3.2 A is passed for 27s from a 12 V source. Which of the following statements is true?

(a) The calorimeter constant is 648 JK^{-1}
(b) This calorimeter constant will be same if the calorimeter is open
(c) The information is insufficient for calculating calorimeter constant
(d) The calorimeter constant is independent of calorimeter content

11 For the complete combustion of ethanol, $C_2H_5OH(l) + 3O_2(g) \longrightarrow 2CO_2(g) + 3H_2O(l)$, the amount of heat produced as measured in bomb calorimeter, is 1364.47 kJ mol^{-1} at 25°C Assuming ideality the enthalpy of combustion, $\Delta_c H$, for the reaction will be ($R = 8.314$ JK^{-1}mol^{-1}) → JEE Main 2015

(a) −1366.95 kJ mol^{-1} (b) −1361.95 kJ mol^{-1}
(c) −1460.50 kJ mol^{-1} (d) −1350.50 kJ mol^{-1}

12 The enthalpy of dissolution of $BaCl_2(s)$ and $BaCl_2 \cdot 2H_2O(s)$ are −20.6 and 8.8 kJ per mol respectively. The enthalpy of hydration for,

$$BaCl_2(s) + 2H_2O \longrightarrow BaCl_2 \cdot 2H_2O(s), \text{ is}$$

(a) 29.4 kJ (b) −29.4 kJ (c) −11.8 kJ (d) 38.2 kJ

13 The standard enthalpy of formation (ΔH_f°) at 298 K for methane, $CH_4(g)$ is − 74.8 kJ mol^{-1}. The addition information required to determine the average energy for C—H bond formation would be

(a) the dissociation energy of H_2 and enthalpy of sublimation of carbon
(b) latent heat of vaporisation of methane
(c) the first four ionisation energies of carbon and electron gain enthalpy of hydrogen
(d) the dissociation energy of hydrogen molecule, H_2

14 What will be the enthalpy change for the combustion of cyclopropane at 298 K? The enthalpies of formation of $CO_2(g)$, $H_2O(l)$ and propene (g) are − 393.5, −285.8 and 20.42 kJ mol^{-1} respectively. The enthalpy of isomerisation of cyclopropane to propene is − 33.0 kJ mol^{-1}.

(a) 844.63 kJ
(b) − 844.63 kJ
(c) − 2090.342 kJ
(d) 1893.44 kJ

15 If the bond dissociation energies of XY, X_2 and Y_2 (all diatomic molecules) are in the ratio of 1:1: 0.5 and ΔH_f for the formation of XY is −200 kJ mol^{-1}. The bond dissociation energy of X_2 will be

(a) 400 kJ mol^{-1}
(b) 300 kJ mol^{-1}
(c) 200 kJ mol^{-1}
(d) None of the above

16 1.0 L sample of mixture of CH_4 and O_2 measured at 25°C and 740 torr, was allowed to react at constant pressure in a calorimeter, together with its contents had a heat capacity of 1260 cal K^{-1}. The complete combustion of CH_4 to CO_2 and water caused a temperature rise in calorimeter of 0.667 K. What will be the mole % of CH_4 in the original mixture?

[Heat of combustion of CH_4 is − 215 kcal mol]

(a) 25% (b) 15%
(c) 40% (d) 10%

ANSWERS

SESSION 1	1 (c)	2 (a)	3 (b)	4 (d)	5 (b)	6 (a)	7 (b)	8 (c)	9 (b)	10 (b)
	11 (b)	12 (b)	13 (d)	14 (d)	15 (a)	16 (d)	17 (b)	18 (d)	19 (d)	20 (c)
	21 (b)	22 (d)	23 (a)	24 (d)	25 (b)	26 (c)	27 (c)	28 (b)	29 (d)	
SESSION 2	1 (c)	2 (b)	3 (b)	4 (c)	5 (d)	6 (d)	7 (b)	8 (c)	9 (c)	10 (a)
	11 (a)	12 (b)	13 (a)	14 (c)	15 (d)	16 (d)				

Hints and Explanations

SESSION 1

1 $2NO_2 \longrightarrow N_2O_4$

$\Delta H^\circ = \Delta H^\circ_{f\,(N_2O_4)} - 2 \times \Delta H^\circ_{f\,(NO_2)}$

$\Delta H^\circ = 9.66 - 2 \times 33.84 = -58.02$ kJ

Since, $\Delta H = -$ve thus, reaction is exothermic.

2 $\Delta H_{reaction} = \Sigma\Delta H_{f\,(Products)} - \Sigma\Delta H_{f\,(reactants)}$

$\Rightarrow \quad \Delta H^\circ_{f(z)} - \Delta H^\circ_{f(y)} = 100$ kcal

i.e. $\quad \Delta H^\circ_{f(z)} > \Delta H^\circ_{f(y)}$

also, $\Delta H^\circ_{f(x)} - \Delta H^\circ_{f(z)} = -80$ kcal

i.e. $\quad \Delta H^\circ_{f(z)} > \Delta H^\circ_{f(x)}$

Also $\quad y \rightarrow x, \Delta H = 20$ kcal

i.e. $\quad \Delta H^\circ_{f(x)} > \Delta H^\circ_{f(y)}$

$\Rightarrow$ Overall order is $y < x < z$.

3 When 2.1 g Fe combines with S, heat evolved = 3.77 kJ

When 56 g (atomic mass) combines with S, then heat evolved.

ΔH / mole of FeS $= \dfrac{3.77 \times 56}{2.1} = 100.5$

Since, the heat is evolved in the formation of FeS, thus, the heat of formation of FeS is -100.5 kJ mol^{-1}.

4 $CH_4(g) + O_2 \xrightarrow{\Delta} 2CO_2(g) + H_2O(l)$

$H_2O(l) \longrightarrow H_2(g) + \frac{1}{2}O_2(g) + \text{Heat}$

$C_2H_6 \longrightarrow CH_2{=}CH_2 + H_2 + \text{Heat}$

$C(\text{graphite}) + \frac{1}{2}O_2(g) \longrightarrow CO(g) + \text{Heat}$

(II), (III) and (IV) are exothermic reactions because they proceeds by the evolution of heat.

5. $C + O_2 \longrightarrow CO_2(g)$;

$\Delta H_f^\circ = -393.5$ kJ mol^{-1} ...(i)

$C + \frac{1}{2}O_2 \longrightarrow CO(g)$;

$\Delta H_f^\circ = -110.5$ kJ mol^{-1} ...(ii)

$H_2 + \frac{1}{2}O_2 \longrightarrow H_2O(g)$;

$\Delta H_f^\circ = -241.8$ kJ mol^{-1} ...(iii)

Eq. (ii) + Eq. (iii) – Eq. (i)

$CO_2(g) + H_2(g) \longrightarrow CO(g) + H_2O(g)$

$\Delta H_r = \Delta H^\circ_{f(CO)} + \Delta H^\circ_{f(H_2O)} - \Delta H^\circ_{f(CO_2)}$

$= -110.5 - 241.8 - (-393.5)$

$= -352.3 + 393.5 = 41.2$ kJ

6 $C + O_2 \longrightarrow CO_2$; $\Delta H = x$

$CO + \frac{1}{2}O_2 \longrightarrow CO_2$; $\Delta H = y$

$C + \frac{1}{2}O_2 \longrightarrow CO$; $\Delta H = x - y$

7 Consider the heat of formation of H_2O.

$H_2(g) + \frac{1}{2}O_2(g) \longrightarrow H_2O(l)$;

$\Delta H = -286.20$ kJ

i.e. $\Delta H_{f(H_2O,\,l)} = -286.20$ kJ

Now, consider the ionisation of H_2O.

$H_2O(l) \longrightarrow H^+(aq) + OH^-(aq)$;

$\Delta H = 57.32$ kJ

$\Delta H_r = \Delta H_{f(H^+,\,aq)} + \Delta H_{f(OH^-,\,aq)} - \Delta H_{f(H_2O,\,l)}$

$57.32 = 0 + \Delta H_{f(OH^-,\,aq)} - (-286.20)$

Thus, $\Delta H_{f(OH^-,\,aq)} = 57.32 - 286.20$

$= -228.88$ kJ

8 $C(s) + O_2(g) \longrightarrow CO_2(g)$;

$\Delta H = -393.5$ kJ mol^{-1} ...(i)

$CO + \frac{1}{2}O_2 \longrightarrow CO_2(g)$;

$\Delta H = -283.5$ kJ mol^{-1} ...(ii)

On subtracting Eq. (ii) from Eq. (i), we get

$C(s) + \frac{1}{2}O_2(g) \longrightarrow CO(g)$;

$\Delta H = (-393.5 + 283.5)$ kJ mol^{-1}

$= -110$ kJ mol^{-1}

9. $C_2H_4 + 3O_2 \longrightarrow 2CO_2 + 2H_2O$

$\Delta H_{reaction} = [2 \times \Delta H^\circ_{f\,(CO_2)} + 2 \times \Delta H^\circ_{f\,(H_2O)}]$

$- [\Delta H^\circ_{f\,(C_2H_4)} + 3 \times \Delta H^\circ_{f\,(O_2)}]$

$= [2(-394) + 2(-286)] - [52 + 0]$

$= -1412$ kJ mol^{-1}

10 $H_2(g) + \frac{1}{2}O_2(g) \longrightarrow H_2O(l)$;

$\Delta H^\circ = -241.8$ kJ mol^{-1}

$\therefore \quad \Delta H^\circ_{f\,(H_2O)} = -241.8$ kJ mol^{-1}

$C(s) + O_2(g) \longrightarrow CO_2(g)$,

$\Delta H^\circ = -393.5$ kJ mol^{-1}

$\therefore \quad \Delta H^\circ_{f\,(CO_2)} = -393.5$ kJ mol^{-1}

$C_2H_5OH(l) + 3O_2(g) \longrightarrow 2CO_2(g) + 3H_2O(l)$;

$\Delta H^\circ = -1234.7$ kJ mol^{-1}

$\Delta H^\circ = 2\Delta H^\circ_{f(CO_2)} + 3\Delta H^\circ_{f\,(H_2O)} - \Delta H^\circ_{f\,(C_2H_5OH)}$

$-1234.7 = -2 \times 393.5 - 3 \times 241.8 - \Delta H^\circ_{f\,(C_2H_5OH)}$

$\therefore \quad \Delta H^\circ_{f\,(C_2H_5OH)} = -277.7$ kJ mol^{-1}

11 ΔH per 1.6 g of glucose

$= \dfrac{72 \times 1.6}{180} = 0.64$ kcal

12 CH_4 required $= \dfrac{445.15 \times 16}{890.3} = 8$ g

13 As methanoic acid is a weak acid, heat of neutralisation is less than x.

14 $H_2C_2O_4(aq) \longrightarrow 2H^+(aq) + C_2O_4^{2-}(aq)$;

$\Delta_{ion}H = x$ kJ

$[H^+(aq) + OH^-(aq) \longrightarrow H_2O(l)$;

$\Delta_{heat}H = -57.3$ kJ$] \times 2$

$H_2C_2O_4 + 2OH^- \longrightarrow 2H_2O(l) + C_2O_4^{2-}$;

$\Delta H = x - 114.6$ kJ

Also, $\Delta H_i = -106$ kJ (given) heat is liberated

But $\quad x - 114.6 = -106$

$\therefore \quad x = 8.6$ kJ mol^{-1}

15 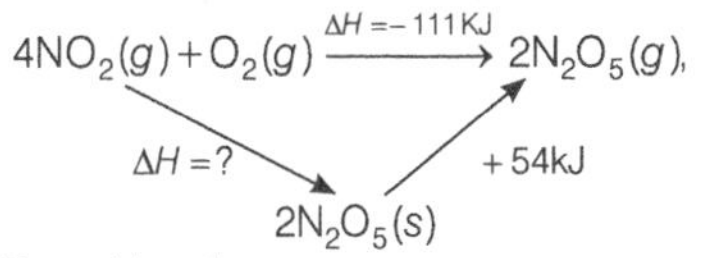

From Hess law,

$\Delta H_f + \Delta H_{sub} = \Delta H_{reaction}$

$\Delta H_f = \Delta H_{reaction} - \Delta H_{sub}$

$= -111$ kJ $- (54$ kJ$)$

$= -111 - 54$ kJ $= -165$ kJ

Thus, the enthalpy of formation, ΔH_f for N_2O_5 (s) is -165 kJ.

16 The heat absorbed, Q is given by

Q = mass × latent heat of vaporisation (L_v)

mass = 70.09 g = 0.07 kg

$L_v = 2260$ kJ

$\therefore Q = 0.07 \times 2260 = 158.2$ kJ $= 158{,}200$ J

17 Energy absorbed

$\propto \dfrac{1}{\text{stability of compound}}$

Energy released $\propto$ stability of compound.

Thus, the correct order of increasing stabilities will be

$CO_2 < NH_3 < HI < O_3$

18 The dissociation of BF_3 *(g)* can be written as

$BF_3(g) \longrightarrow B(g) + 3F(g)$;

$\Delta H_f^\circ = -271.75$ kcal/mol

$\Delta H = [\Delta H_B^\circ + 3 \times \Delta H^\circ_F] - [\Delta H^\circ_{BF_3}]$

$= -271.75$ kcal/mol

$\Delta H = 134.5 + 3 \times 118.86 - (-271.75)$

$= 762.83$ kcal

$= 3 \times$ Avg B—F bond energy

B—F bond energy $= 254.3$ kcal/mol

19 Molecular weight of $NH_4NO_3 = 80$

Heat evolved $= 1.23 \times 6.12$

Molar heat of decomposition

$= 1.23 \times 6.12 \times 80$

$= -602$ kJ mol^{-1} (as heat is evolved)

20 $\frac{1}{2}H_2 + \frac{1}{2}Cl_2 \longrightarrow HCl$

$\Delta H = \Sigma BE_{reactants} - \Sigma BE_{products}$

$= \left[\frac{1}{2}BE(H_2) + \frac{1}{2}BE(Cl_2)\right] - BE(HCl)$

$= \left[\left(\frac{1}{2} \times 104\right) + \left(\frac{1}{2} \times 58\right)\right] - 103$

$= (52 + 29) - 103 = -22$ kcal

21 The bond energy C—H $= -166/4$

$= -41.5$ kJ mol^{-1}

22 $CH_2{=}CH_2 + H_2 \longrightarrow CH_3{-}CH_3$

$\Delta H = (BE)_{reactants} - (BE)_{products}$

$= 4\,(BE)_{C-H} + (BE)_{C=C} + (BE)_{H-H}$

$-[6(BE)_{C-H} + (BE)_{C-C}]$

$= -125$ kJmol^{-1}

23 $N_2 + \frac{1}{2}O_2 \longrightarrow N_2O$

or $N{\equiv}N + \frac{1}{2}O{=}O \longrightarrow N{=}N{=}O$

$\therefore\ \Delta H = BE_{N\equiv N} + \frac{1}{2}BE_{O=O}$

$-[BE_{N=N} + BE_{N=O}]$

$= 946 + \frac{1}{2} \times 498 - [418 + 607]$

$= 1195 - 1025 = +170$

Resonance energy = Experimental value − calculated value

$= 82 - 170 = -88$ kJ

24 For calculation of C≡C bond energy, we must first calculate dissociation energy of C_2H_2 as

$C_2H_2(g) \longrightarrow 2C\,(g) + 2H(g)$ ...(i)

By using the given bond energies and enthalpies,

$C_2H_2(g) \longrightarrow 2C(g) + H_2(g)$;

$\Delta H = -225$ kJ mol^{-1} ...(ii)

$2C\,(s) \longrightarrow 2C(g)$; $\Delta H = 1410$ kJ mol^{-1} ...(iii)

$H_2(g) \longrightarrow 2H(g)$; $\Delta H = 330$ kJ mol^{-1} ...(iv)

On adding Eqs. (ii), (iii) and (iv) gives Eq. (i) $C_2H_2(g) \longrightarrow 2C(g) + 2H(g)$;

$\Delta H = 1515$ kJ mol^{-1}

1515 kJ mol$^{-1} = 2 \times$ (C—H) BE + (C≡C) BE

$= 2 \times 350 +$ (C≡C) BE

(C≡C) BE $= 1515 - 700 = 815$ kJ mol^{-1}

25 Given, $\frac{1}{2}N_2(g) + \frac{3}{2}H_2(g) \longrightarrow NH_3(g)$;

$\Delta H_f^\circ = -46.0$ kJ mol^{-1}

∴ Bond enthalpy of N—H = 46 kJ mol^{-1} (+ve sign is taken because energy is supplied to break N—H bonds).

$2H(g) \longrightarrow H_2(g)$;

$\Delta H_f^\circ = -436$ kJ mol^{-1}

$BE_{(H-H)} = 436$ kJ mol^{-1}

$2N(g) \longrightarrow N_2(g)$;

$\Delta H_f^\circ = -712$ kJ mol^{-1}

$BE_{(N-N)} = 712$ kJ mol^{-1}

$\Delta H_r = \frac{1}{2}BE_{N_2} + \frac{3}{2}BE_{H_2} - 3BE_{N-H}$

$\Rightarrow -46 = \left[\frac{1}{2} \times 712 + \frac{3}{2} \times 436\right] - 3BE_{(N-H)}$

$3BE_{(N-H)} = 1010 + 46$

$3E_{(N-H)} = 1056$

$BE_{(N-H)} = \frac{1056}{3} = 352$ kJ mol^{-1}

26 $C_6H_5COOH(s) + 7.5O_2(l)$

$\longrightarrow 7CO_2(g) + 3H_2O(l)$

$\Delta n_g = -0.5$

$\Delta H = \Delta E + \Delta n_g RT$

$-2546 = \Delta E - 0.5 \times 8.314 \times 10^{-3} \times 298$

$\Delta E = -2544.76$ kJ mol^{-1}

27 Calorific value of heat produced by 1 g of fuel.

Calorific value of ethane

$= \frac{-341.1}{30} = -11.37$ kcal / g

Calorific value for ethyne

$= \frac{-310.0}{26} = -11.92$ kcal / g

∴ Ratio of calorific values of ethane and ethyne = 11.37 : 11.92 = 0.95 : 1

28 In a combustion reaction, sum of enthalpies of reactants is greater than the sum of the enthalpies of products. Therefore, combustion of all organic compounds is an exothermic reaction. Also, enthalpies of all elements in their standard state are zero.

So, both A and R are true but R is not the correct explanation of A.

29 Both the Statements I and II are correct and Statement II is the correct explanation for Statement I.

SESSION 2

1 $C + O_2 \longrightarrow CO_2$; $\Delta H = -393.5$ kJ/mol

∵ Heat released during the formation of 44 g of $CO_2 = -393.5$ kJ

Heat released during the formation of 1 g of $CO_2 = \frac{-393.5}{44}$ kJ

Heat released during the formation of 35.2 g (given) of CO_2

$= -\frac{393.5 \times 35.2}{44} = -315$ kJ

2 ΔH_f°, standard heat of formation is the amount of heat evolved or absorbed when one gram mole of a substance is formed from its constituent elements. For the formation of a compound the constituent must be in elemental form therefore formation of HF represents the standard heat of formation of HF.

3 $H_2 + \frac{1}{2}O_2 \longrightarrow H_2O$;

$\Delta H = -241$ kJ mol^{-1} ...(i)

$C_6H_{10} + \frac{17}{2}O_2 \longrightarrow 6CO_2 + 5H_2O$;

$\Delta H = -3800$ kJ mol^{-1} ...(ii)

$C_6H_{12} + 9O_2 \longrightarrow 6CO_2 + 6H_2O$;

$\Delta H = -3920$ kJ mol^{-1} ...(iii)

For the reaction,

$C_6H_{10} + H_2 \longrightarrow C_6H_{12}$

Eq. (i) + Eq. (ii) − Eq. (iii)

$\Delta H = -241 - 3800 - (-3920)$

$= -121$ kJ mol^{-1}

4 $\frac{1}{2}I_2(s) + \frac{1}{2}Cl_2(g) \longrightarrow ICl(g)$

$\Delta H = \left[\frac{1}{2}\Delta H_{s \to g(I_2)} + \frac{1}{2}\Delta H_{diss(Cl_2)} + \frac{1}{2}\Delta H_{diss(I_2)}\right] - \Delta H_{ICl}$

$$= \left(\frac{1}{2} \times 62.76 + \frac{1}{2} \times 242.3 + \frac{1}{2} \times 151.0\right) - 211.3$$

$= 228.03 - 211.3$

$= 16.73 \approx 16.8$ kJ/mol

5 $\underset{2\,mol}{2Fe(s)} + \frac{3}{2}O_2(g) \longrightarrow \underset{1\,mol}{Fe_2O_3(s)}$

Moles of Fe $= \frac{4}{56} = \frac{1}{14}$ mol

Moles of Fe_2O_3 formed $= \frac{1}{28}$ mol

Heat released = 29.28 kJ

$\therefore \Delta_f H°(Fe_2O_3) = \frac{-29.28}{1/28}$ kJ mol^{-1}

$= -819.84$ kJ mol^{-1}

6 $\because$ 58 g *iso*-butane provides energy

$= 2658$ kJ

11.2×10^3 g *iso*-butane will provide energy

$= \frac{2658 \times 11.2 \times 10^3}{58} = 513268.9$ kJ

The daily requirement of energy

$= 15000$ kJ

$\because$ Loss of energy = 30%

$\therefore$ Total energy used for work

$= \frac{513268.9 \times 70}{100}$

$\therefore$ Cylinder will last

$= \frac{513268.9 \times 70}{100 \times 15000} = 24$ days

7 Millimoles of dibasic strong acid

$= M \times V = 0.20 \times 350 = 70$ mmol

$\therefore$ Amount of H^+ ions in the acid

$= 2 \times 70 = 140$ mmol

Similarly, amount of OH^- ions in monobasic strong base

$= 0.10 \times 650 \times 1 = 65$ mmol

[Here, OH^- is the limiting reactant].

$\because$ 1 mole of OH^- ions produces

$= 57.1$ kJ heat 65×10^{-3} moles of OH^- ions will produce

$= 57.1 \times 65 \times 10^{-3} = 3.71$ kJ

8 Same bonds are formed in reaction (1) and (2) but no bonds are broken in reaction (1) whereas, bonds in the reactant molecules are broken in reaction (2).

As energy is absorbed when bonds are broken, energy released in reaction (1) is greater than that in reaction (2) hence, $x > y$.

9 $CaCl_2(s) + 6H_2O \longrightarrow CaCl_2 \cdot 6H_2O(s)$;

$\Delta H = -23.2$ kcal

$CaCl_2 \cdot 6H_2O(s) + (aq) \longrightarrow CaCl_2 \cdot 6H_2O$;

$\Delta H = +3.5$ kcal

Thus, $CaCl_2(s) + (aq)$ (excess)

$\longrightarrow CaCl_2 \cdot 6H_2O(aq)$,

$\Delta H = -23.2 + 3.5 = 19.7$ kcal

10 $Q = It = 3.2 \times 27$

Energy produced

$= QV = 3.2 \times 27 \times 12 = 1036.8$ J

Calorimeter constant

$= \frac{1036.8}{1.6} = 648$ JK^{-1}

11 $C_2H_5OH(l) + 3O_2(g) \longrightarrow 2CO_2(g) + 3H_2O(l)$

$\Delta U = -1364.47$ kJ/mol

$\Delta H = \Delta U + \Delta n_g RT$

$\Delta n_g = -1$

$\Delta H = -1364.47 + \frac{-1 \times 8.314 \times 298}{1000}$

[Here, value of R is unit of J must be converted into kJ]

$= -1364.47 - 2.4776$

$= -1366.95$ kJ/mol

12 $BaCl_2(s) + 2H_2O(l)$

$\longrightarrow BaCl_2 \cdot 2H_2O(aq)$;

$\Delta H = H_1 = -20.6$ kJ ...(i)

$BaCl_2(s) + 2H_2O(aq)$

$\longrightarrow BaCl_2 \cdot 2H_2O(aq)$;

$\Delta H = H_2 = +8.8$ kJ ...(ii)

For the reaction,

$BaCl_2(s) + 2H_2O \longrightarrow BaCl_2 \cdot 2H_2O(s)$

Eq (i) – Eq (ii)

$\Delta H = \Delta H_1 - \Delta H_2 = -20.6 - 8.8$

$= -29.4$ kJ

13 Carbon is found in solid state. The state of substance affects the enthalpy change. For formation of CH_4, the equation of can be written as

$C(s) + H_2(g) \longrightarrow CH_4(g)$

$C(s) \longrightarrow C(g)$ sublimation

and $H_2(g) \longrightarrow 2H(g)$ dissociation are required for the average energy of C—H formation

14 Given,

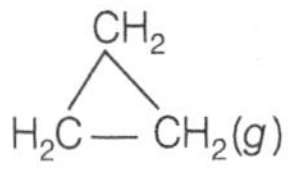
$(g) \longrightarrow CH_3CH = CH_2(g)$;

$\Delta H = -33.0$ kJ ...(i)

$C + O_2 \longrightarrow CO_2(g)$;

$\Delta H = -393.5$ kJ ...(ii)

$H_2 + \frac{1}{2}O_2 \longrightarrow H_2O(l)$;

$\Delta H = -285.8$ kJ ...(iii)

$3C + 3H_2 \longrightarrow CH_3—CH{=}CH_2(g)$;

$\Delta H = 20.42$ kJ ...(iv)

Eq. (iv) – Eq. (i) yields

$3C + 3H_2 \longrightarrow$ 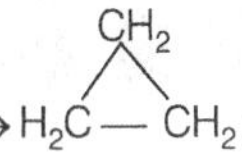 ;

$\Delta H = 53.42$ kJ ...(v)

3 Eq (ii) + 3 Eq (iii) – Eq (v) yields

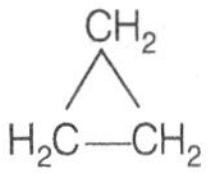 $+ \frac{9}{2}O_2 \longrightarrow 3CO_2 + 3H_2O$

$\Delta_r H = [3 \times (-393.5)] + [3 \times (-285.5)] - 53.42 = -2090.342$ kJ

15 $X_2 + Y_2 \longrightarrow 2XY$

$\Delta H = (BE)_{X—X} + (BE)_{Y—Y} - 2(BE)_{X—Y}$

If BE of $X—Y = a$,

(BE) of $(X—X) = a$ and (BE) of $(Y—Y) = a/2$

and BE of $X - Y = a$

$\Delta H_{f(X—Y)} = 200$ kJ

$-400 = a + \frac{a}{2} - 2a$

$-400 = -\frac{a}{2}$ $a = +800$ kJ

The bond dissociation energy of $X_2 = 800$ kJmol^{-1}

16 Heat generated $= C \times \Delta T$

$= 1260 \times 0.667 = 840.4$ cal

Molar mass of $CH_4 = 16$ g/mol

Enthalpy of combustion for 16 g $CH_4 = -215$ kcal/mol

$= -215000$ cal/mol

Thus, enthalpy equal to -840.42 cal

$= \frac{16}{215000} \times 840.42 = 0.0625$ g

Number. of moles of CH_4 in the reaction

$= \frac{0.0625}{16} = 3.91 \times 10^{-3}$ mol

$pV = nRT = 0.973 \times 1$

$= n \times 0.082 \times 298$

$n = 3.92 \times 10^{-2}$ mol

$\therefore$ mol% of $CH_4 = \frac{3.91 \times 10^{-3}}{3.92 \times 10^{-2}} \times 100$

$= 9.97 \approx 10\%$

DAY EIGHT

Solutions

Learning & Revision for the Day

- Concentration of Solution
- Solubility
- Vapour Pressure
- Raoult's Law
- Ideal Solutions
- Non-ideal Solutions
- Henry's Law
- Colligative Properties
- Abnormal Molar Masses

Solution is a homogeneous mixture of two or more substances on molecular level. The substances forming the solution are called **components** (solute, solvent) of the solutions.

Concentration of Solution

Various methods for expressing concentration of solutions are as follows:

1. Molality (m)

It is the number of moles of solute dissolved in 1000 g of solvent.

$$\text{Molality} = \frac{\text{Number of moles of solute}}{\text{Weight of solvent(in kg)}}$$

2. Molarity (M)

It is the number of solute present in one litre of the solution.

$$\text{Molarity} = \frac{\text{Number of moles of solute}}{\text{Volume of solution (in L)}}$$

- Relation between molality (m) and molarity (M),

$$\text{Molarity} = \frac{m \times d}{1 + \dfrac{m \times M_2}{1000}}$$

where, M_2 = molar mass of solute, d = density of the solution, m = molality of solution

- Relation between molarity and density

$$\text{Molarity} = 10 \times \text{density} \times \% \text{ by wt. of solute/mol. wt. of solute}$$

3. Mole Fraction (χ)

It is the ratio of the number of moles of one component (i.e. solute or solvent) to the total number of moles of solute and solvent (i.e. moles of solution).

Mole fraction of solute in the solution, $\chi_{\text{solute}} = \dfrac{\text{Moles of solute } (n_{\text{solute}})}{\text{Total moles of solution } (n_{\text{solute}} + n_{\text{solvent}})}$

PREP MIRROR
Your Personal Preparation Indicator

- No. of Questions in Exercises (x)—
- No. of Questions Attempted (y)—
- No. of Correct Questions (z)—
  (Without referring Explanations)

- Accuracy Level ($z/y \times 100$)—
- Prep Level ($z/x \times 100$)—

In order to expect good rank in JEE, your Accuracy Level should be above 85 & Prep Level should be above 75.

Mole fraction of solvent in the solution

$$\chi_{solvent} = \frac{\text{Moles of solvent } (n_{solvent})}{\text{Total moles of solution } (n_{solute} + n_{solvent})}$$

Sum of mole fractions is always equal to one, i.e. $\chi_{solute} + \chi_{solvent} = 1$

- Relation between molarity (M) and mole fraction (χ)

$$\chi = \frac{MM_1}{M(M_1 - M_2) + d}$$

Here, M_1 = molar mass of solvent
M_2 = molar mass of solute
d = density of the solution

4. Percentage by Weight or Mass Per cent

It is the amount of solute present in 100 g of the solution.

$$\text{Percentage by weight} = \frac{\text{Weight of solute} \times 100}{\text{Weight of solution}}$$

5. Percentage by Volume

The volume of solute (in mL) present in 100 mL of the solution is called voume per cent

$$\text{Percentage by volume} = \frac{\text{Volume of solute (mL)}}{\text{Volume of solution (mL)}} \times 100$$

6. Mass by Volume Percentage

It is the mass of solute dissolved in 100 mL of the solution.

$$\text{Mass/volume percentage} = \frac{\text{Mass of solute}}{\text{Total volume of solution (mL)}} \times 100$$

7. Parts Per Million (ppm)

When the amount of solute (in g) is present in one million mL (1 million = 10^6) per mL of solution, the concentration of the solution is expressed in ppm.

$$\text{ppm concentration} = \frac{10^6 \times \text{mass of solute (g)}}{\text{Mass of (solute + solvent) (mL)}}$$

8. Normality

It is the number of gram-equivalents of substance dissolved per litre of the solution.

$$\text{Normality} = \frac{\text{Number of gram-equivalent of solute}}{\text{Volume of solution (in L)}}$$

$$\text{where, gram-equivalents} = \frac{\text{Weight of solute}(w)}{\text{Equivalent weight of solute}(E)}$$

Solubility

The maximum amount of a solute that can be dissolved in a given amount of solvent at a given temperature is termed as **solubility** at that temperature. Solubility of a gas in liquid depends upon temperature and pressure.

The quantitative relation between pressure and solubility of a gas in a solvent is given by Henry.

Vapour Pressure

The pressure exerted by the vapour molecules above liquid surface which are in equilibrium with it at a given temperature is called **vapour pressure**.

Factor Affecting Vapour Pressure

Vapour pressure gets affected by the following factors:

1. **Purity of the Liquid** Pure liquid always has a vapour pressure higher than its solution.
2. **Nature of the Liquid** Liquids which have weak intermolecular forces are volatile and have greater vapour pressure.
3. **Temperature** The vapour pressure of a liquid increases with increase in temperature. This is because on increasing the temperature, the kinetic energy of molecules increases that results into the fact that more molecules of the liquid can go into vapour phase.
4. **Effect of Adding Solute** When a liquid contains a solute, some of the solvent molecules are replaced by the solute particles on the liquid surface and therefore, the available surface area for the escape of solvent molecule decreases.

   Due to the less available area on the surface of liquid for escape, rate of evaporation and the rate of condensation both decreases.

- The vapour pressure of liquid in solution is known as its **partial vapour pressure** and is less than the vapour pressure of the pure liquid at the same temperature.
- If $p°$ be the vapour pressure of pure liquid and p_s be that of liquid in solution then lowering of vapour pressure of the liquid = $p° - p_s$

$\therefore$ Relative lowering in vapour pressure $= \dfrac{p° - p_s}{p°}$

Raoult's Law

Raoult's law for the solutions of liquids in liquids states "The partial vapour pressure of each component of the solution is directly proportional to the mole fraction of that component present in solution.

$$p_A = \chi_A p_A^\circ \quad \text{and} \quad p_B = \chi_B p_B^\circ$$

where, A and B are volatile solute and solvent respectively and p_A° and p_B° are the vapour pressures of A and B in pure state.

- If χ_A and χ_B are the mole fraction of the components A and B respectively in the vapour phase then,

$$p_A = \chi_A \cdot p_{total}$$
$$p_B = \chi_B \cdot p_{total}$$

- Raoult's law for solutions of solids in liquids, i.e. for non-volatile solutes,

$$p_{solution} = \chi_{solvent}\, p^\circ_{solvent}$$

Limitations of Raoult's Law are as follows :

(i) It is applicable only to very dilute solutions.

(ii) It is applicable only to solutions containing non-volatile and non-electrolytic solutes which exist as a single molecule.

(iii) It is not applicable to solutes which dissociate or associate in the particular solution.

Ideal Solutions

- The solutions, which obey Raoult's law over all ranges of concentration and temperature are called **ideal solutions**.

  For ideal solutions, $\Delta H_{mix} = 0$, $\Delta V_{mix} = 0$

  where, ΔH_{mix} = enthalpy of mixing to form the solution

  ΔV_{mix} = volume of mixing to the solution

- For such solutions, solute-solute and solvent-solvent interactions ≈ solute-solvent interactions.
- Solution of *n*-hexane and *n*-heptane, benzene and toluene etc., fall in this category.

  Practically, no solution is ideal. The graphical representation of ideal solutions is given below:

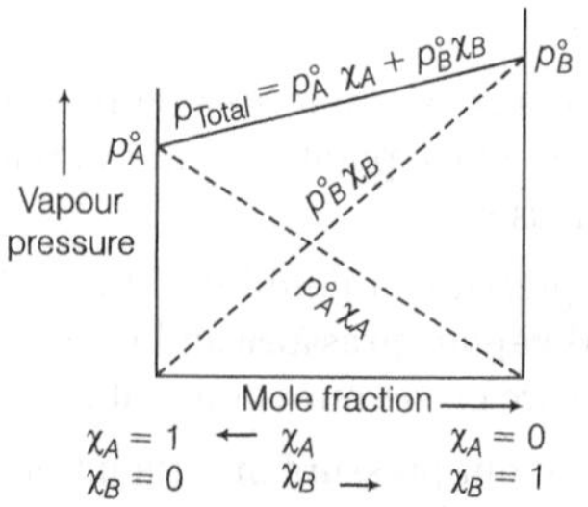

Graphical representation of the ideal solution

Non-ideal Solutions

The solutions which shows deviation from Raoult's law are called **non-ideal solution**.

For such solutions, $\Delta H_{mix} \neq 0$, $\Delta V_{mix} \neq 0$

Non-ideal Solutions Showing Positive Deviation

- When the observed partial vapour pressure is more than that expected by Raoult's law, then positive deviation is observed.
- For such a deviation,

$$p_A > p_A^o \chi_A$$

$$p_B > p_B^o \chi_B$$

$$p_{Total} > p_A^o \chi_A + p_B^o \chi_B$$

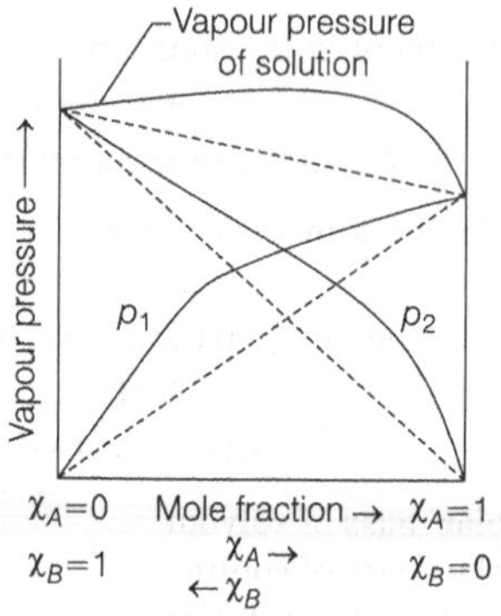

Graphical representation of the solution showing positive deviation

- For such solutions, $\Delta H_{mix} > 0$, $\Delta V_{mix} > 0$. These are usually obtained by mixing of polar liquids with non-polar ones. e.g. cyclohexane and ethanol, H_2O and C_2H_5OH.

Non-ideal Solutions Showing Negative Deviation

- When the observed partial vapour pressure is less than that expected by Raoult's law, then negative deviation is observed.
- For such solutions,

  $\Delta H_{mix} < 0$, $\Delta V_{mix} < 0$, e.g. chloroform and acetone.

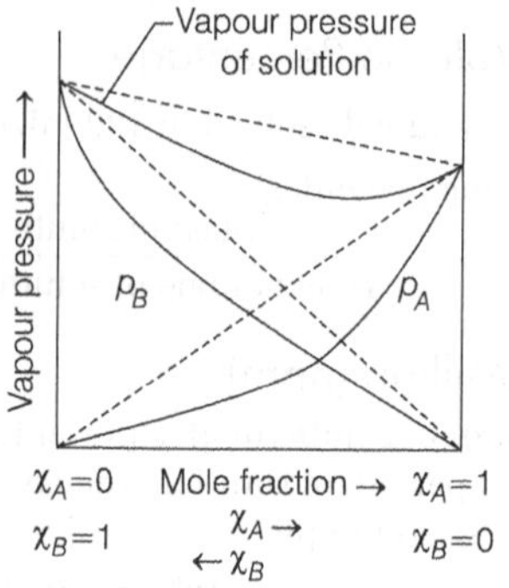

Graphical representation of the solution showing negative deviation

For a solution showing negative deviation,

$$p_A < p_A^o \times \chi_A \text{ or } p_B < p_B^o \times \chi_B$$

$$p_{total} < p_A^o \chi_A + p_B^o \chi_B$$

Azeotropes

Mixture of liquids that boils at a constant temperature and distills over without any change in composition are called **azeotropes**.

Azeotropes are of two types :

1. **Minimum Boiling Azeotropes** These are formed by those liquid pairs which show positive deviation from ideal behaviour. Such azeotropes have boiling points lower than either of the components. e.g. ethanol + water, ether-acetone, water-methanol.

2. **Maximum Boiling Azeotropes** These are formed by those liquid pairs which show negative deviation from ideal behaviour. Such azeotropes have boiling points higher than either of the components. e.g. water (20.22% by mass) + hydrochloric acid, acetone + chloroform, water + nitric acid.

Henry's law

It states that, at constant temperature, the solubility of a gas in a liquid is directly proportional to the partial pressure of the gas present above the surface of liquids or solution. In other words, "The partial pressure of the gas in vapour phase (p) is proportional to the mole fraction of the gas (χ) in the solution". Thus,

$$p \propto \chi \Rightarrow p = K_H \chi$$

where, K_H is Henry's law constant.

Higher the value of K_H at a given pressure, the lower is the solubility of the gas in the liquid.

Raoult's Law as a Special Case of Henry's Law

- According to Raoult's law, the vapour pressure of a volatile component in a given solution is $p_i = \chi_i\, p_i^\circ$. In the solution of a gas in a liquid, one of the components is so volatile that it exists as a gas and its partial pressure is directly proportional to its mole fraction in the solution and this is stated by Henry's law as $p = K_H \chi$.
- If we compare the equations for Raoult's law and Henry's law, it can be seen that the partial pressure of the volatile component or gas is directly proportional to its mole fraction in the solution. Only the proportionality constant K_H differs from p_i°, Thus, Raoult's law becomes a special case of Henry's law in which K_H becomes equal to p_i°.

Colligative Properties

The properties which depend only on the number of solute particles irrespective of their nature relative to the total number of particles present in the solution as **colligative properties**, e.g. relative lowering of vapour pressure, depression in freezing point, elevation in boiling point, osmotic pressure etc.

- For different solutions of same molar concentration of different non-electrolyte solutes, the colligative properties have the same value for all.
- For different molar concentrations of the same solute, the colligative property has greater value for more concentrated solution.
- For solutions of different solutes having same percent strength, the colligative property has greater value for the solute having least molecular weight.

There are four types of colligative properties as given below:

1. Relative Lowering of Vapour Pressure

Addition of non-volatile solute leads to the lowering of vapour pressure.

$$\frac{p^\circ - p}{p^\circ} = \chi_{\text{solute}} \text{ or } \frac{p^\circ - p}{p^\circ} = \frac{n_B}{n_A}$$

[For dilute solution $n_B << n_A$]

where, $\frac{p^\circ - p}{p^\circ}$ = relative lowering of vapour pressure

n_B = moles of solute, n_A = moles of solvent

p° = vapour pressure of pure solvent

Determination of Molecular Mass from Relative Lowering of Vapour Pressure

Relationship between molecular mass and relative lowering of vapour pressure can be given as:

$$M_B = \frac{w_B \cdot M_A}{w_A \cdot \left(\frac{p^\circ - p}{p^\circ}\right)}$$

where, w_B and w_A = mass of solute and solvent respectively and M_B and M_A = molecular weight of solute and solvent respectively.

2. Elevation in Boiling Point (ΔT_b)

As the vapour pressure of a solution is decreased due to presence of a non-volatile solute, the solution boils at a higher temperature as compared to the pure solvent. Thus, the boiling point of the solution is higher than the boiling point of the pure solvent.

For dilute solutions, $\Delta T_b = K_b \times \text{molality}$

$$\Delta T_b = T_b - T_b^\circ$$

where, T_b = boiling point of solution

T_b° = boiling point of pure solvent.

$$K_b = \frac{M_A R (T_b^\circ)^2}{1000\, \Delta H_V}$$

where, R = gas constant and ΔH_V = latent heat of vaporisation.

K_b is ebullioscopic constant or molal elevation constant (K_b depends only on solvent).

Determination of Molecular Mass from Elevation in Boiling Point

The relationship between molar mass and elevation in boiling point can be given as :

$$M_B = \frac{K_b \times w_B \times 1000}{\Delta T_b \times w_A}$$

3. Depression in Freezing Point (ΔT_f)

When a non-volatile solute is added to a solvent, the freezing point of the solution is always lower than that of pure solvent.

For dilute solutions,

$$\Delta T_f = K_f \times \text{molality}\ ,\ \Delta T_f = T_f^\circ - T_f$$

where, T_f° = freezing point of pure solvent

T_f = freezing point of solution

K_f = cryoscopic constant or molal depression constant

$$K_f = \frac{MR(T_f^\circ)^2}{1000\ \Delta H_f}$$

where, ΔH_f = latent heat of fusion

Determination of Molecular Mass from Depression in Freezing Point

The relationship between molar mass and depression in freezing point given as:

$$M_B = \frac{K_f \times w_B \times 1000}{\Delta T_f \times w_A},$$

4. Osmosis and Osmotic Pressure

Spontaneous flow of solvent molecules through a semipermeable membrane from a pure solvent to the solution (or from a dilute solution to concentrated solution) is termed as **osmosis**. The external pressure which must be applied on the solution in order to stop the flow of solvent into the solution through semi-permeable membrane is termed as **osmotic pressure**.

$$\text{Osmotic pressure } (\pi) = \frac{n}{V} RT = CRT$$

where,

n = moles of solute, C = molar concentration

V = volume of solution (in litre), R = gas constant

T = temperature in Kelvin (K).

Determination of Molecular Mass from Osmotic Pressure

Relationship between osmotic pressure and molar mass can be given as:

$$M_B = \frac{w_B RT}{\pi V}$$

Two solutions having same osmotic pressures at same temperature are termed as **isotonic solutions.** When two solutions are being compared, the solution with higher osmotic pressure is termed as **hypertonic** and the solution with lower osmotic pressure is termed as **hypotonic.**

Abnormal Molar Masses

Molar masses that are either lower or higher than the expected or normal values are called **abnormal molar masses**.

Association of molecules leads to decreases in the number of particles in solution resulting in decrease in the value of colligative properties while in case of dissociation, the number of particles increases in solution (when substance is an electrolyte) and the colligative properties increases accordingly.

van't Hoff Factor (i)

In 1880 van't Hoff introduced a factor i, known as the vant Hoff factor, to account for the extent of dissociation or association. This factor i is defined as

(i) $i = \dfrac{\text{number of particles after association / dissociation}}{\text{number of particles before association / dissociation}}$

(ii) $i = \dfrac{\text{normal molecular mass}}{\text{abnormal molecular mass}}$

(iii) $i = \dfrac{\text{(abnormal) observed value of colligative property}}{\text{(normal) calculated value of colligative property}}$

Significance of van't Hoff Factor

Quantitatively the extent to which a solute is being associated or dissociated can be expressed by van't Hoff factor i.

- For dissociation, $i > 1$ while for association $i < 1$.
- **Degree of dissociation,** $\alpha = \dfrac{i-1}{n-1}$

  where, n = number of particles after dissociation.
- **Degree of association**

$$\alpha = \frac{i-1}{\frac{1}{n}-1}$$

  Here, n = number of particles after association.

Modified Expressions of Colligative Properties

Modified expressions of colligative properties are as follows:

- Relative lowering of vapour pressure

$$\frac{p^\circ - p}{p^\circ} = i\,\frac{n_B}{n_B + n_A};\ \begin{pmatrix} n_B = n \\ n_B + n_A \approx N \end{pmatrix}$$

- Elevation in boiling point, $\Delta T_b = i \cdot K_b \cdot m$

  Here, m = molality
- Depression in freezing point, $\Delta T_f = i \cdot K_f \cdot m$ (m = molality)
- Osmotic pressure $(\pi) = i \cdot CRT$

DAY PRACTICE SESSION 1

FOUNDATION QUESTIONS EXERCISE

1 A one molal solution of sodium chloride has a density of 1.21 g mL^{-1}. The molarity of this solution is
(a) 4.15 M (b) 1.143 M
(c) 2.95 M (d) 3.15 M

2 The density (in g mL^{-1}) of a 3.60 M sulphuric acid solution having 29% H_2SO_4 (molar mass = 98 g mol^{-1}) by mass, will be
(a) 1.64 (b) 1.88 (c) 1.22 (d) 1.45

3 The molarity of a solution obtained by mixing 750 mL of 0.5 M HCl with 250 mL of 2 M HCl will be → JEE Main 2013
(a) 1.00 M (b) 1.75 M
(c) 0.975 M (d) 0.875 M

4 10 mL of 2 M NaOH solution is added to 200 mL of 0.5 M of NaOH solution. What is the final concentration? → JEE Main (Online) 2013
(a) 0.57 M (b) 5.7 M (c) 11.4 M (d) 1.14 M

5 The density of a solution prepared by dissolving 120 g of urea (mol. mass = 60 u) in 1000 g of water is 1.15 g/mL. The molarity of this solution is → AIEEE 2012
(a) 0.50 M (b) 1.78 M
(c) 1.02 M (d) 2.05 M

6 The density of 3 M solution of sodium chloride is 1.252 g mL^{-1}. The molality of the solution will be (molar mass, NaCl = 58.5 g mol^{-1}) → JEE Main 2013
(a) 2.60 m (b) 2.18 m
(c) 2.79 m (d) 3.00 m

7 A 5.2 molal aqueous solution of methyl alcohol, CH_3OH is supplied. What is the mole fraction of methyl alcohol in the solution? → AIEEE 2011
(a) 0.100 M (b) 0.190 M
(c) 0.086 M (d) 0.050 M

8 25 mL of a solution of barium hydroxide on titration with 0.1 molar solution of hydrochloric acid gave a titre value of 35 mL. The molarity of barium hydroxide solution was
(a) 0.07 M (b) 0.14 M
(c) 0.28 M (d) 0.35 M

9 At 100°C, benzene and toluene have vapour pressure of 1375 torr and 558 torr, respectively. Assuming, these two form an ideal binary solution, then calculate the mole fraction of benzene in vapour phase at 1 atm and 100°C?
(a) 0.247 (b) 0.753
(c) 0.447 (d) 0.553

10 An aqueous solution of 2% (wt/wt) non-volatile solute exerts a pressure of 1.004 bar at the boiling point of the solvent. What is the molecular mass of the solute?
(a) 0.3655 (b) 36.55 (c) 41.37 (d) 40.16

11 Vapour pressure of pure benzene is 119 torr and that of toluene is 37.0 torr at the same temperature. Mole fraction of toluene in vapour phase, which is in equilibrium with a solution of benzene and toluene having a mole fraction of 0.50 will be → JEE Main (Online) 2013
(a) 0.137 (b) 0.237
(c) 0.435 (d) 0.205

12 On mixing, heptane and octane form an ideal solution at 373 K, the vapour pressures of the two liquid components (heptane and octane) are 105 kPa and 45 kPa respectively. Vapour pressure of the solution obtained by mixing 25 g of heptane and 35 g of octane will be (molar mass of heptane = 100 g mol^{-1} and of octane = 114 g mol^{-1}). → AIEEE 2010
(a) 72.0 kPa (b) 36.1 kPa
(c) 96.2 kPa (d) 144.5 kPa

13 A binary liquid solution is prepared by mixing *n*-heptane and ethanol. Which one of the following statements is correct regarding the behaviour of the solution? → AIEEE 2009
(a) The solution formed is an ideal solution
(b) The solution is non-ideal, showing positive deviation from Raoult's law
(c) The solution is non-ideal, showing negative deviation from Raoult's law
(d) *n*-heptane shows positive deviation while ethanol show negative deviation from Raoult's law

14 If two liquids *A* and *B* form minimum boiling azeotrope at some specific composition then ....
(a) *A*—*B* interactions are stronger than those between *A*—*A* or *B*—*B*
(b) vapour pressure of solution increases because more number of molecules of liquids *A* and *B* can escape from the solution
(c) vapour pressure of solution decreases because less number of molecules of only one of the liquids escape from the solution
(d) *A*—*B* interactions are weaker than those between *A*—*B* or *B*—*B*

15 An unknown compound is immiscible with water. It is steam distilled at 98.0°C. At 98.0°C, p and $p^\circ_{H_2O}$ are respectively 737 and 707 torr. This distillate was 75% by weight of water. The molecular weight of the unknown compound will be
(a) 318.15 g mol^{-1} (b) 300 g mol^{-1}
(c) 306.76 g mol^{-1} (d) None of these

16 The relative lowering of vapour pressure of an aqueous solution containing a non-volatile solute is 0.0125. The molality of the solution is

(a) 0.69 m (b) 0.50 m
(c) 0.80 m (d) 0.40 m

17 The vapour pressure of water at 20°C is 17.5 mm Hg. If 18 g glucose ($C_6H_{12}O_6$) is added to 178.2 g of water at 20°C, the vapour pressure of the resulting solution will be

(a) 16.500 mm Hg (b) 17.325 mm Hg
(c) 17.675 mm Hg (d) 15.750 mm Hg

18 *A* and *B* are ideal gases. The molecular weight of *A* and *B* are in the ratio of 1 : 4 . The pressure of a gas mixture containing equal weight of *A* and *B* is *p* atm. What is the partial pressure (in atm) of *B* in the mixture?

(a) *p*/5 (b) *p*/2 (c) *p*/2.5 (d) 3*p*/4

19 18 g of glucose ($C_6H_{12}O_6$) is added to 178.2 g water. The vapour pressure of water (in torr) for this aqueous solution is → JEE Main 2016

(a) 76.0 (b) 752.4 (c) 759.0 (d) 7.6

20 The vapour pressure of acetone at 20°C is 185 torr. When 1.2 g of a non-volatile substance was dissolved in 100 g of acetone at 20 °C, its vapour pressure was 183 torr. The molar mass of the substance is → JEE Main 2015

(a) 32 (b) 64 (c) 128 (d) 488

21 12 g of a non-volatile solute dissolved in 108g of water produces the relative lowering of vapour pressure of 0.1. The molecular mass of the solute (in g mol^{-1}) is → JEE Main (Online) 2013

(a) 80 (b) 60 (c) 20 (d) 40

22 The Henry's law constant for the solubility of N_2 gas in water at 298 K is 1.0×10^5 atm. The mole fraction of N_2 in air is 0.8. The number of moles of N_2 from air dissolved in 10 moles of water at 298 K and 5 atm pressure is

(a) 4.0×10^{-4} (b) 4.0×10^{-5}
(c) 5.0×10^{-4} (d) 4.0×10^{-6}

23 For a dilute solution containing 2.5 g of non-volatile, non-electrolyte solute in 100 g of water, the elevation in boiling point at 1 atm pressure is 2°C. Assuming concentration of solute is much lower than the concentration of solvent, the vapour pressure of solution is ($K_b = 0.76$ K kg mol^{-1}) → AIEEE 2012

(a) 724 (b) 740 (c) 736 (d) 718

24 An aqueous solution freezes at −0.186° C ($K_f = 1.86$ K kg mol^{-1}, $K_b = 0.512$ K kg mol^{-1}). What is the elevation in boiling point?

(a) 0.186°C (b) 0.512°C
(c) 0.86°C (d) 0.0512°C

25 When mercuric iodide is added to the aqueous solution of potassium iodide, the

(a) freezing point is raised
(b) freezing point is lowered
(c) freezing point does not change
(d) boiling point does not change

26 The molecular weight of benzoic acid in benzene is determined by depression in freezing point method corresponds to

(a) ionisation of benzoic acid
(b) dimerisation of benzoic acid
(c) trimerisation of benzoic acid
(d) solvation of benzoic acid

27 The depression in freezing point of 0.01 M aqueous solution of urea, sodium chloride and sodium sulphate is in the ratio of

(a) 1 : 1 : 1 (b) 1 : 2 : 3 (c) 1 : 2 : 4 (d) 2 : 2 : 3

28 During depression in freezing point of a solution, the following are in equilibrium

(a) liquid solvent, solid solvent
(b) liquid solvent, solid solute
(c) liquid solute, solid solute
(d) liquid solute, solid solvent

29 How many grams of methyl alcohol should be added to 10 L tank of water to prevent its freezing at 268 K? (K_f for water is 1.86 K kg mol^{-1}) → JEE Main (Online) 2013

(a) 880.07 g (b) 899.04 g
(c) 886.02 g (d) 868.06 g

30 K_f for water is 1.86 K kg mol^{-1}. If your automobile radiator holds 1.0 kg of water, how many grams of ethylene glycol ($C_2H_6O_2$) you must add to get the freezing point of the solution lowered to − 2.8° C? → AIEEE 2012

(a) 72 g (b) 93 g (c) 39 g (d) 27 g

31 Ethylene glycol is used as an antifreeze in cold climate. Mass of ethylene glycol which should be added to 4 kg of water to prevent it from freezing at − 6°C will be (K_f for water = 1.86 K kg mol^{-1} and molar mass of ethylene glycol = 62 g mol^{-1}). → AIEEE 2011

(a) 800.00 g (b) 204.30 g (c) 400.00 g (d) 304.60 g

32 The molar mass of a solute χ in g mol^{-1}. If its 1% solution is isotonic with 95% solution of cane sugar (molar mass = 342 g mol^{-1}) is

(a) 68.4 (b) 34.2 (c) 136.2 (d) 171.2

33 Match the following and choose the correct option.

	Column I		Column II
A.	Raoult's law	1.	$p = K_H \chi$
B.	Elevation of boiling point	2.	$\pi = CRT$
C.	Henry's law	3.	$p = \chi_1 p_1^\circ + \chi_2 p_2^\circ$
D.	Osmotic pressure	4.	$\Delta T_b = K_b m$

	A	B	C	D
(a)	3	4	1	2
(b)	2	1	4	3
(c)	4	1	2	3
(d)	1	3	2	4

34 When 20 g of naphthoic acid ($C_{11}H_8O_2$) is dissolved in 50 g of benzene ($K_f = 1.72$ K kg mol^{-1}), a freezing point depression of 2 K is observed. The van't Hoff factor (*i*) is

(a) 0.5 (b) 1 (c) 2 (d) 3

35 We have 3 aqueous solution of NaCl labelled as '*A*', '*B*' and '*C*' with concentrations 0.1 M, 0.01 M and 0.001 M, respectively. The value of van't Hoff factor for these solution will be in the order of

(a) $i_A < i_B < i_C$ (b) $i_A > i_B > i_C$
(c) $i_A = i_B = i_C$ (d) $i_A < i_B > i_C$

36 Which one of the following aqueous solutions will exhibit highest boiling point?

(a) 0.01 M Na_2SO_4 (b) 0.01 M KNO_3
(c) 0.015 M urea (d) 0.015 M glucose

37 A compound *X* undergoes tetramerisation in a given organic solvent. The van't Hoff factor is

(a) 4.0 (b) 0.25
(c) 0.125 (d) 2.0

38 The degree of dissociation (α) of a weak electrolyte, A_xB_y is related to van't Hoff factor (i) by the expression → AIEEE 2011

(a) $\alpha = \dfrac{i-1}{(x+y-1)}$ (b) $\alpha = \dfrac{i-1}{x+y+1}$
(c) $\alpha = \dfrac{x+y-1}{i-1}$ (d) $\alpha = \dfrac{x+y+1}{i-1}$

39 Which of the following aqueous solutions should have the highest boiling point?

(a) 1.0 M NaOH (b) 1.0 M Na_2SO_4
(c) 1.0 M NH_4NO_3 (d) 1.0 M KNO_3

40 In a 0.2 molal aqueous solution of a weak acid H*X*, the degree of ionisation is 0.3. Taking K_f for water as 1.85 K kg mol^{-1}, the freezing point of the solution will be nearest to

(a) −0.360°C (b) −0.260°C
(c) +0.480°C (d) −0.481°C

41 In comparison to a 0.01M solution of glucose, the depression in freezing point of a 0.01M $MgCl_2$ solution is ....

(a) the same (b) about twice
(c) about three times (d) about six times

42 The freezing point (in °C) of a solution containing 0.1 g of $K_3[Fe(CN)_6]$ (mol. wt. 329 gmol^{-1}) in 100 g of water (K_f =1.86 K kg mol^{-1}) is

(a) -2.3×10^{-2} (b) -5.7×10^{-2}
(c) -5.7×10^{-3} (d) -1.2×10^{-2}

43 For 1 molal aqueous solution of the following compounds, which one will show the highest freezing point? → JEE Main 2018

(a) $[Co(H_2O)_6]Cl_3$ (b) $[Co(H_2O)_5Cl]Cl_2 \cdot H_2O$
(c) $[Co(H_2O)_4Cl_2]Cl \cdot 2H_2O$ (d) $[Co(H_2O)_3Cl_3] \cdot 3H_2O$

44 The freezing point of benzene decreases by 0.45°C when 0.2 g of acetic acid is added to 20 g of benzene. If acetic acid associates to form a dimer in benzene, percentage association of acetic acid in benzene will be (K_f for benzene = 5.12 K kg mol^{-1}) → JEE Main 2017

(a) 64.6 % (b) 80.4 %
(c) 74.6 % (d) 94.6 %

45 If sodium sulphate is considered to be completely dissociated into cations and anions in aqueous solution, the change in freezing point of water (ΔT_f), when 0.01 mole of sodium sulphate is dissolved in 1 kg of water, is (K_f =1.86 K kg mol^{-1}) → AIEEE 2010

(a) 0.0372 K (b) 0.0558 K
(c) 0.0744 K (d) 0.0186 K

46 Consider separate solution of 0.500 M C_2H_5OH (*aq*), 0.100M $Mg_3(PO_4)_2$(*aq*), 0.250M KBr(*aq*) and 0.125M Na_3PO_4(*aq*) at 25° C. Which statement is true about these solution, assuming all salts to be strong electrolytes? → JEE Main 2014

(a) They all have the same osmotic pressure
(b) 0.100 M $Mg_3(PO_4)_2$ (*aq*) has the highest osmotic pressure
(c) 0.125 M Na_3PO_4 (*aq*) has the highest osmotic pressure
(d) 0.500 M C_2H_2OH(*aq*) has the highest osmotic pressure

Direction (Q. Nos. 47-50) *In the following question, Assertion (A) followed by a Reason (R) is given. Choose the correct option.*

(a) Both A and R are true and R is correct explanation of A
(b) Both A and R are true but R is not correct explanation of A
(c) A is true but R is false
(d) Both A and R are false

47 **Assertion** (A) When methyl alcohol is added to water, boiling point of water increases.

Reason (R) When a volatile solute is added to a volatile solvent, elevation in boiling point is observed.

48 **Assertion** (A) Molarity of a solution in liquid state changes with temperature.

Reason (R) The volume of a solution changes with change in temperature.

49 **Assertion** (A) Osmotic pressure of 1 M glucose is lesser than 1 M NaCl (*aq*) but vapour pressure of 1 M glucose is higher than 1 M NaCl.

Reason (R) Osmotic pressure is a colligative property but vapour pressure is not a colligative property, however relative lowering in vapour pressure is a colligative property.

50 **Assertion** (A) Ebullioscopy or cryoscopy cannot be used for the determination of molecular weight of polymers.

Reason (R) High molecular weight solute leads to very low value of ΔT_b or ΔT_f.

DAY PRACTICE SESSION 2

PROGRESSIVE QUESTIONS EXERCISE

1 The elevation in boiling point of a solution of 13.44 g of $CuCl_2$ in 1 kg of water using the following information will be (molecular weight of $CuCl_2 = 134.4$ and $K_b = 0.52 \text{ K mol}^{-1}$)

(a) 0.16 (b) 0.05 (c) 0.1 (d) 0.2

2 How many mL of 0.1 M HCl are required to react completely with 1g mixture of Na_2CO_3 and $NaHCO_3$ containing equimolar amount of these two?

(a) 157.8 mL (b) 0.1578 mL (c) 210.4 mL (d) 105.2 mL

3 Which of the following liquid pairs shows a positive deviation from Raoult's law?

(a) Water — Hydrochloric acid
(b) Benzene — Methanol
(c) Water — Nitric acid
(d) Acetone — Chloroform

4 To neutralise completely 20 mL of 0.1 M aqueous solution of phosphorus acid (H_3PO_3), the volume of 0.1 M aqueous KOH solution required is

(a) 10 mL (b) 20 mL
(c) 40 mL (d) 60 mL

5 At 10°C, the osmotic pressure of urea solution is 500 mm Hg. The solution is diluted and the temperature is raised to 25°C. The osmotic pressure of dilute solution is 105.3 mm Hg at 25°C. The extent of dilution can be shown as

(a) $V_{\text{final}} = 5\, V_{\text{initial}}$ (b) $V_{\text{initial}} > V_{\text{final}}$
(c) $V_{\text{final}} = 4\, V_{\text{initial}}$ (d) $V_{\text{final}} = 6\, V_{\text{initial}}$

6 Plot of $\frac{1}{\chi_A}$ vs $\frac{1}{\chi_A}$ (χ_A = mole fraction of A in liquid and χ_A in vapour) is linear whose slope and intercept respectively are given

(a) $\frac{p_B^\circ}{p_A^\circ}, \frac{p_B^\circ - p_A^\circ}{p_B^\circ}$ (b) $p_A^\circ - p_B^\circ, \frac{p_A^\circ - p_B^\circ}{p_B^\circ}$
(c) $\frac{p_A^\circ}{p_B^\circ}, \frac{p_B^\circ - p_A^\circ}{p_B^\circ}$ (d) $\frac{p_B^\circ}{p_A^\circ}, \frac{p_A^\circ - p_B^\circ}{p_B^\circ}$

7 Two liquids X and Y form an ideal solution at 300 K, vapour pressure of the solution containing 1 mole of X and 3 moles of Y is 550 mm Hg. At the same temperature, if 1 mole of Y is further added to this solution, vapour pressure of the solution increases by 10 mm Hg. Vapour pressure (in mm Hg) of X and Y in their pure states will be, respectively

(a) 200 and 300
(b) 300 and 400
(c) 400 and 600
(d) 500 and 600

8 58.5 g of NaCl and 180 g of glucose were separately dissolved in 1000 mL of water. Identify the correct statement regarding the elevation of boiling point (b.p.) of the resulting solutions.

(a) NaCl solution will show higher elevation of boiling point
(b) Glucose solution will show higher elevation of boiling point
(c) Both the solutions will show equal elevation of boiling point
(d) The boiling point elevation will be shown by neither of the solution

9 The vapour pressure of benzene at a certain temperature is 640 mm Hg. A non-volatile, non-electrolyte solute weighing 2.175 g, is added to 39.0 g of benzene. The vapour pressure of the solution is 600 mm Hg. What is the molecular weight of the solid substance?

(a) 6.96
(b) 65.3
(c) 63.8
(d) None of the above

10 The vapour pressure of a solvent decreases by 10 mm of mercury, when a non-volatile solute was added to the solvent. The mole fraction of the solute in the solution is 0.2. What should be the mole fraction of the solvent, if the decrease in vapour pressure is to be 20 mm of mercury?

(a) 0.8 (b) 0.6
(c) 0.4 (d) 0.7

11 At 310 K, the vapour pressure of an ideal solution containing 2 moles of A and 3 moles of B is 550 mm of Hg. At the same pressure if one mole of B is added to this solution, the vapour pressure of solution increased by 10 mm of Hg. What is the vapour pressure of A in its pure state?

(a) 460 mm (b) 610 mm
(c) 360 mm (d) 750 mm

12 Which one of the following statements is false?

(a) Raoult's law states that the vapour pressure of a component over a solution is proportional to its mole fraction
(b) The osmotic pressure (π) of a solution is given by the equation $\pi = MRT$, where M is the molarity of the solution
(c) The correct order of osmotic pressure for 0.01 M aqueous solution of each compound is $BaCl_2 > KCl > CH_3COOH >$ sucrose
(d) Two sucrose solutions of same molality prepared in different solvents will have the same freezing point depression

13 Two beakers of capacity 500 mL were taken. One of these beakers, labelled as *A*, was filled with 400 mL water whereas, the beaker labelled *B* was filled with 400 mL of 2 M solution of NaCl. At the same temperature, both the beakers were placed in closed containers of same material and same capacity as shown in the figure.

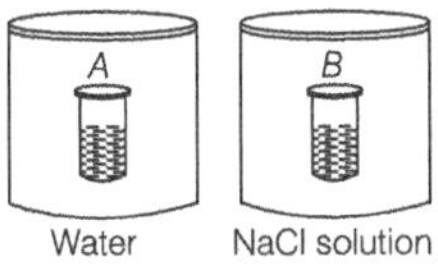

At a given temperature, which of the following statement is correct about the vapour pressure of pure water and that of NaCl solution?

(a) Vapour pressure in container *A* is more than that in container *B*
(b) Vapour pressure in container *A* is less than that in container *B*
(c) Vapour pressure is equal in both the containers
(d) Vapour pressure in container *B* is twice the vapour pressure in container *A*

14 A 0.001 molal solution of $[Pt(NH_3)_4Cl_4]$ in water had a freezing point depression of 0.0054° C. If K_f for water is 1.80, the correct formulation of the above molecule is

(a) $[Pt(NH_3)_4Cl_3]Cl$ (b) $[Pt(NH_3)_4Cl_2]Cl_2$
(c) $[Pt(NH_3)_4Cl]Cl_3$ (d) $[Pt(NH_3)_4Cl_4]$

15 Sea water is found to contain 5.85% NaCl and 9.5% $MgCl_2$ by weight of solution. Calculate its normal boiling point assuming 70% ionisation for NaCl and 50% ionisation of $MgCl_2$ $[K_b\,(H_2O) = 0.51 \text{ K kg mol}^{-1}]$

(a) 101.4°C (b) 102.29°C
(c) 103.27°C (d) 99.46°C

ANSWERS

SESSION 1	**1** (b)	**2** (c)	**3** (d)	**4** (a)	**5** (d)	**6** (c)	**7** (c)	**8** (b)	**9** (c)	**10** (c)
	11 (b)	**12** (a)	**13** (b)	**14** (b)	**15** (a)	**16** (a)	**17** (b)	**18** (a)	**19** (b)	**20** (b)
	21 (c)	**22** (a)	**23** (a)	**24** (d)	**25** (a)	**26** (b)	**27** (b)	**28** (a)	**29** (c)	**30** (b)
	31 (a)	**32** (a)	**33** (a)	**34** (a)	**35** (b)	**36** (a)	**37** (b)	**38** (a)	**39** (b)	**40** (d)
	41 (c)	**42** (a)	**43** (d)	**44** (d)	**45** (b)	**46** (a)	**47** (d)	**48** (a)	**49** (a)	**50** (a)
SESSION 2	**1** (a)	**2** (a)	**3** (b)	**4** (c)	**5** (a)	**6** (c)	**7** (c)	**8** (a)	**9** (b)	**10** (b)
	11 (a)	**12** (d)	**13** (a)	**14** (b)	**15** (b)					

Hints and Explanations

SESSION 1

1 Molar mass of NaCl $= M_2 = 58.5\text{ g}$

$$M = \frac{m \times d}{1 + \dfrac{m \times M_2}{1000}}$$

$$M = \frac{1 \times 1.21}{1 + \dfrac{1 \times 58.5}{1000}} = 1.143\text{ M}$$

2 Molarity

$$= \frac{10 \times \text{density} \times \%\text{ by weight of solute}}{\text{molecular weight of the solute}}$$

$$\text{Density} = \frac{3.60 \times 98}{10 \times 29} = 1.216\text{ g/mL}$$

$$\approx 1.22\text{ g/mL}$$

3 $M_1V_1 + M_2V_2 = MV$ (Total moles)

$$M = \frac{M_1V_1 + M_2V_2}{V_1 + V_2}$$

$$M = \frac{0.5 \times 750 + 2 \times 250}{1000}$$

$$M = 0.875\text{ M}$$

4 Final concentration, $M = \dfrac{M_1V_1 + M_2V_2}{V_1 + V_2}$

$$= \frac{10 \times 2 + 200 \times 0.5}{200 + 10}$$

$$= \frac{20 + 100}{210} = \frac{120}{210} = 0.57\text{ M}$$

5 Total mass of solution

= 1000 g water + 120 g urea

= 1120 g

Density of solution = 1.15 g/mL

Volume of solution

$$= \frac{\text{Mass}}{\text{Density}}$$

$$= \frac{1120\text{ g}}{1.15\text{ g/mL}}$$

$$= 973.91\text{ mL} = 0.974\text{ L}$$

$$\text{Moles of solute} = \frac{120}{60} = 2$$

$$\text{Molarity} = \frac{\text{Moles of solute}}{\text{Volume (L) of solution}}$$

$$= \frac{2}{0.974} = 2.05\text{ mol/L}$$

6 3M solution means 3 moles of solute (NaCl) are present in 1000 L of solution.

Mass of solution = volume of solution × density

= 1000 × 1.252 = 1252 g

Mass of solute = Number of moles × molar mass of NaCl

= 3 × 58.5 g

= 175.5 g

Mass of solvent = (1252 − 175.5) g

= 1076.5 g

= 1.076 kg

$$\text{Molality} = \frac{\text{Moles of solute}}{\text{Mass of solvent (in kg)}}$$

$$= \frac{3}{1.076} = 2.79\text{ m}$$

7 Molality $= \frac{\text{Moles of solute}}{\text{Mass of solvent (in kg)}}$

$= \frac{5.2 \text{ mol } CH_3OH}{1 \text{ kg (1000 g) } H_2O}$

$n_1\ (CH_3OH) = 5.2$

$n_2\ (H_2O) = \frac{1000}{18} = 55.56$

$\therefore\ n_1 + n_2 = 5.20 + 55.56 = 60.76$ mol

$\therefore$ Mole fraction of CH_3OH, χ_{CH_3OH}

$= \frac{n_1}{n_1 + n_2} = \frac{5.2}{60.76} = 0.086$

8 Let, molarity of $Ba(OH)_2 = M_1$

Let the molarity of $HCl = M_2$

$\therefore\ M_1V_1 = M_2V_2$

$M_1 \times 25 = 0.1 \times 35$

$M_1 = \frac{3.5}{25}\text{M} = 0.14\text{ M}$

9 Let χ_1 be the mole fraction of benzene in solution and $(1 - x)$ be the mole fraction of toluene.

Applying Raoult's law,

$pr = \chi_1 p^\circ_1 + (1 - \chi_1)p^\circ_2$

or $760 = \chi_1(1375) + (1 - \chi_1)(558)$

or $\chi_1 = \frac{760 - 558}{1375 - 558} = 0.247$ torr

Mole fraction of benzene in vapour phase at 1 atm

$\chi_1 = \frac{\chi_1 p^\circ_1}{p} = \frac{0.247 \times 1375}{760} = 0.447$

10 $p^\circ = 1.013$ bar $= 1$ atm (at boiling point)

$w_{solvent} = 100 - 2 = 98$ g,

According to Raoult's law,

$\frac{p^\circ - p_1}{p^\circ} = \frac{w_{solute}}{M_{solute}} \times \frac{M_{solvent}}{w_{solvent}}$

$\frac{1.013 - 1.004}{1.013} = \frac{2 \times 18}{M_{solute} \times 98}$

$\therefore\ M_{solute} = 41.37 \text{ gmol}^{-1}$

11 From Raoult's law, for ideal solution,

$p_{total} = p^\circ_B\chi_B + p^\circ_T\chi_T$

$= 119 \times 0.5 + 37 \times 0.5$ ($\because \chi_B = 1 - \chi_T$)

$= 59.5 + 18.5 = 78$ torr

Mole fraction of toluene in vapour phase

$(\chi_T)_v = \frac{p^\circ_T\chi_T}{p} = \frac{18.5}{78} = 0.237$

12 $p_T = \chi_H \cdot p^\circ_H + \chi_O \cdot p^\circ_O$

$\chi_H = \frac{25/100}{\frac{25}{100} + \frac{35}{114}} = 0.45$

$\therefore\ \chi_O = 1 - 0.45 = 0.55$

$p_T = 0.45 \times 105 + 0.55 \times 45$

$= 72$ kPa

13 *n*- heptane and ethanol forms non-ideal solution. As a result, *n*-heptane-ethanol molecular interaction is very poor in comparison to ethanol-ethanol or *n*-heptane-*n*-heptane interactions. So, the resulting solution gives positive deviation from Raoult's law.

14 Minimum boiling azeotrope shows positive deviations from Raoult's law due to the stronger solute-solute interactions.

15 Since, unknown compound is immiscible with water, hence vapour pressure $\propto$ moles.

Given, $p^\circ_{total} = 737$ torr

$p^\circ_{H_2O} = 707$ torr,

$\therefore\ p^\circ_{unknown} = 737 - 707 = 30$ torr

$\frac{p^\circ_{unknown}}{p^\circ_{water}} = \frac{n_{unknown}}{n_{H_2O}}$

$= \frac{W_{unknown} \times m_{H_2O}}{W_{H_2O} \times m_{unknown}}$

or $\frac{30}{707} = \frac{75.0 \times 18}{100 \times m_{unknown}}$

or $m_{unknown} = 318.15 \text{ g mol}^{-1}$

16 $\frac{p^\circ - p_1}{p^\circ} = x_2 = \frac{mM_1}{1000 + mM_1}$

(For dilute solutions, $1000 >> mM_1$)

$\frac{p^\circ - p_1}{p^\circ} = \frac{m \times M}{1000}$

$0.0125 = \frac{m \times 18}{1000}$

$m = \frac{0.0125 \times 1000}{18} = 0.69$ m

17 $\frac{p^\circ - p_1}{p^\circ} = \frac{n_2}{n_1} = \frac{w_2/M_2}{w_1/M_1 + w_2/M_2}$

or $\frac{17.5 - p_1}{p^\circ} = \frac{18/180}{178.2/18 + 18/180}$

or $\frac{17.5 - p_1}{17.5} = \frac{1/10}{9.9 + 0.1}$

or $p_1 = 17.325$ mm Hg.

18 Molecular weight ratio of *A* and *B* = 1:4

$\therefore$ Mole ratio of *A* and *B*, if equal weight of *A* and *B* are taken = 4:1

Partial pressure of $B = \frac{1}{5} \times p = \frac{p}{5}$

19 (b) Vapour pressure of water (p°) = 760 torr

Number of moles of glucose

$= \frac{\text{Mass (in g)}}{\text{Molecular mass (g mol}^{-1})}$

$= \frac{18 \text{ g}}{180 \text{ g mol}^{-1}} = 0.1$ mol

Molar mass of water = 18 g/mol

Mass of water (given) = 178.2g

Number of moles of water

$= \frac{\text{Mass of water}}{\text{Molar mass of water}} = \frac{178.2\text{g}}{18\text{g/mol}}$

$= 6.6$ mol

Total number of moles = (0.1 + 9.9) moles = 10 mol

Now, mole fraction of glucose in solution = Change in pressure with respect to initial pressure,

i.e. $\frac{\Delta p}{p^\circ} = \frac{0.1}{10}$

or $\Delta p = 0.01p^\circ = 0.01 \times 760 = 7.6$ torr

$\therefore$ Vapour pressure of solution

$= (760 - 7.6)$ torr $= 752.4$ torr

20 Given, $p^\circ = 185$ torr at 20 °C,

$p_1 = 183$ torr at 20°C

Mass of non-volatile substance,

$m = 1.2$ g

Mass of acetone taken = 100 g

As, we have $\frac{p^\circ - p_1}{p_1} = \frac{n_B}{n_A}$

$\therefore\ \frac{185 - 183}{183} = \frac{\frac{1.2}{M}}{\frac{100}{58}} \Rightarrow \frac{2}{183} = \frac{1.2 \times 58}{100 \times M}$

$\therefore\ M = \frac{183 \times 1.2 \times 58}{2 \times 100}$,

$M = 63.684 \approx 64$ g/mol

21 The relationship between molar mass of solute and relative lowering in vapour pressure is given as:

$\Delta p = \frac{p^\circ - p}{p^\circ} = \frac{n_B}{n_A} = \frac{w}{m} \times \frac{M}{W}$

$w = 12$ g; $W = 108$ g, $m = ?$

$M = 18$ g, $\Delta p = 0.1$

$\Delta p = \frac{w}{m} \times \frac{M}{W}$, $0.1 = \frac{12}{m} \times \frac{18}{108}$

$m = \frac{12 \times 18}{10.8} = 20$

22 According to Henry's law, $p = K_H \chi$

where, χ = mole fraction

$n_{N_2} = n_{H_2O} \times \frac{p_{N_2}}{K_H} \Rightarrow p_{N_2} = \chi_{N_2} \times p$

$p_{N_2} = 0.8 \times 5 = 4$ atm

$\therefore\ n_{N_2} = 10 \times \frac{4}{1 \times 10^5} = 4 \times 10^{-4}$ moles

23 The elevation in boiling point is

$\Delta T_b = K_b \cdot m \left(\text{where, } m = \frac{n_2 \times 1000}{w_1}\right)$

$\Delta T_b = 2 = \frac{0.76 \times n_2 \times 1000}{100}$ or $n_2 = \frac{5}{19}$

From Raoult's law of lowering of vapour pressure

$$\frac{\Delta p}{p^\circ} = \chi_2 = \frac{n_2}{n_1 + n_2} \approx \frac{n_2}{n_1} \quad (\because n_1 >> n_2)$$

$$\Delta p = 760 \times \frac{5 \times 18}{19 \times 100} = 36 \text{ mm of Hg}$$

$$p = 760 - 36 = 724 \text{ mm of Hg}$$

24 $\Delta T_f = T_f^\circ - T_f = 0 - (-0.186) = 0.186°C$

$\Delta T_f = K_f \times m \Rightarrow 0.186 = 1.86 \times m$

or $m = 0.1$,

$\Delta T_b = K_b \times m = 0.512 \times 0.1 = 0.0512°C$

25 $2KI + HgI_2 \longrightarrow K_2HgI_4$

As a result of this reaction, number of ions decreases. So, the lowering in freezing point is less or the actual freezing point is raised.

26 Benzoic acid undergoes dimerisation in benzene.

27 Concentrations of particles of 0.01 M urea, NaCl and Na_2SO_4 are 0.01 M, 0.02 M, 0.03 M respectively, i.e. they are in the ratio of 1:2:3. Hence, depression in freezing point will be in the same ratio.

28 When freezing starts, liquid solvent is in equilibrium with the solid solvent (both have the same vapour pressure).

29 Normal freezing point of water = 273.15K.

In order to prevent freezing at 268 K, let the amount of methanol added be x g.

$\therefore$ Molality, $m = \dfrac{x}{32 \times 10} = \dfrac{x}{320}$

[$\because$ Molar mass of CH_3OH = 32 g mol^{-1} and mass of H_2O = V of H_2O because density of water $\simeq 1\text{ gL}^{-1}$]

Lowering, in freezing point = $K_f \cdot m$

$$273.15 - 268 = 1.86 \times \frac{x}{320}$$

$$5.15 = \frac{1.86x}{320}$$

or $$x = \frac{5.15 \times 320}{1.86} = 886.02\text{ g}$$

30 Coolant is glycol ($C_2H_6O_2$), which is non-electrolyte.

$$\Delta T_f = 2.8°\text{ C}$$

$$\Delta T_f = \frac{1000\, K_f\, w_2}{M_1 \times w_1}$$

$$2.8 = \frac{1000 \times 1.86 \times w_1}{62 \times 1000}$$

$\therefore$ $w_1 = 93.33\text{ g} \approx 93\text{ g}$

31 ΔT_f = Freezing point of H_2O – freezing point of ethylene glycol solution

= 0 – (– 6) = 6°C

K_f = 1.86 K kg mol^{-1}

w_2 = Mass of ethylene glycol in grams

w_1 = Mass of solvent (H_2O) in grams

= 4000 g

m_2 = Molar mass of ethylene glycol

= 62 g mol^{-1}

i = van't Hoff factor = 1

($\because$ ethylene glycol is non-electrolyte)

From, $$\Delta T_f = \frac{i \times 1000\, K_f\, w_2}{m_2\, w_1}$$

$\therefore$ $$6 = \frac{1000 \times 1.86 \times w_2 \times 1}{62 \times 4000}$$

$$w_2 = 800\text{ g}$$

32 For isotonic solutions, $\pi_1 = \pi_2$

$$\frac{n_1RT}{V_1} = \frac{n_2RT}{V_2} \Rightarrow \frac{n_1}{V_1} = \frac{n_2}{V_2}$$

Thus, $$\frac{w_1}{V_1m_1} = \frac{w_2}{V_2m_2}, \frac{w_1}{V_1} = 1, \frac{w_2}{V_2} = 5$$

$$m_1 = ?,\ m_2 = 342, \frac{1}{m_1} = \frac{5}{342}$$

$\therefore$ $$m_1 = \frac{342}{5} = 68.4 = x$$

i.e. molar mass of solute x is 68.4 g mol^{-1}.

33 A → 3, B → 4, C → 1, D → 2

34 Actual molecular weight of naphthoic acid, ($C_{11}H_8O_2$) = 172

Molecular mass (calculated)

$$= \frac{1000 \times K_f \times w_{\text{solute}}}{w_{\text{solvent}} \times \Delta T_f}$$

$$= \frac{1000 \times 1.72 \times 20}{50 \times 2} = 344$$

van't Hoff factor (i)

$$= \frac{\text{Actual molecular weight}}{\text{Calculated molecular weight}}$$

$$= \frac{172}{344} = 0.5$$

35 Greater the dilution, greater is the dissociation into ions and hence, greater is the van't Hoff factor.

The value of van't Hoff factor for these solution will be in the order of, $i_A > i_B > i_C$

36 Boiling point of the solution is directly proportional to van't Hoff factor. Thus, for

(a) Na_2SO_4

$Na_2SO_4 \rightleftharpoons 2Na^+ + SO_4^{2-}$ [3 ions]

$\therefore$ $i = 3$

(b) KNO_3

$KNO_3 \rightleftharpoons K^+ + NO_3^-$ [2 ions]

$\therefore$ $i = 2$

Urea [(NH_2CONH_2) and glucose are poor electrolytes, therefore i for these two solutes are equal to zero.

Hence, 0.01 M Na_2SO_4 possess highest boiling point.

37 $4A \rightleftharpoons (A)_4$,

	$4A$	$(A)_4$
Initially	1	0
After association	$1-\alpha$	$\alpha/4$

$$i = \frac{1 - \alpha + \frac{\alpha}{4}}{1} = 1 - \frac{3}{4}\alpha$$

α = degree of dissociation = 1 or 100%

$\therefore$ $i = 1 - \dfrac{3}{4} = 0.25$

38 $A_xB_y \rightleftharpoons xA^{y+} + yB^{x-}$

	A_xB_y	xA^{y+}	yB^{x-}
Initially	1	0	0
After dissociation	$(1-\alpha)$	$x\alpha$	$y\alpha$

$$i = n(A_xB_y) + n(A^{y+}) + n(B^{x-})$$

$$= 1 - \alpha + x\alpha + y\alpha$$

$$= 1 + \alpha(x + y - 1)$$

$\therefore$ $$\alpha = \frac{i - 1}{(x + y - 1)}$$

39 Boiling point of the solution is directly proportional to van't Hoff factor. Na_2SO_4 possess larger value of i than other given solutions. Hence, 1.0 M Na_2SO_4 solution would have highest boiling point.

40 $HX \rightleftharpoons H^+ + X^-$

	HX	H^+	X^-
Initially	1 mol	0	0
After dissociation	1 – 0.3	0.3	0.3

Total moles = 1 – 0.3 + 0.3 + 0.3 = 1.3,

$$i = \frac{1.3}{1} = 1.3$$

$\Delta T_f = iK_fm = 1.3 \times 1.85 \times 0.2 = 0.418\text{ K}$

$\therefore T_f = 273 - 0.418\text{ K} = 272.519\text{ K}$

or – 0.418°C

41 Being a non-electrolyte glucose does not undergo ionisation when dissolved in water whereas, $MgCl_2$ releases 3 ions [$Mg^+ + 2Cl^-$] when dissolved in water. Thus, i for $MgCl_2$ is 3, while that of glucose is 0. Therefore, depression of freezing point of 0.01 M. $MgCl_2$ solution is about three times than 0.01 M glucose solution.

42 van't Hoff factor (i) = 4

$\{\because K_3[Fe(CN)_6] \rightleftharpoons 3K^+ + [Fe(CN)_6]^{3-}\}$

$$\text{Molality} = \frac{0.1 \times 1000}{329 \times 100} = \frac{1}{329}$$

$$-T_f = iK_fm$$

$$\Rightarrow -4 \times 1.86 \times \frac{1}{329} = -2.3 \times 10^{-2}$$

(As freezing point of water is 0°C.)

43 "Addition of solute particles to a pure solvent results into depression in its freezing point."

All the compounds given in question are ionic in nature. So, consider their van't Hoff factor (i) to reach at final conclusion.

The solution with maximum freezing point must have minimum number of solute particles. This generalisation can be done with the help of van't Hoff factor (i), i.e. number of solute particles $\propto$ van't Hoff factor (i).

Thus, we can say directly solution with maximum freezing point will be the one in which solute with minimum van't Hoff factor is present

Now, for

$Co(H_2O)_6]Cl_3 \rightleftharpoons [Co(H_2O)_6]^{3+} + 3Cl^-$

van't Hoff factor (i) is 4.

Similarly for, $[Co(H_2O)_5Cl]Cl_2 \cdot H_2O \rightleftharpoons$

$[Co(H_2O)_5Cl]^{2+} + 2Cl^-$ 'i' is 3

$[Co(H_2O)_4Cl_2]Cl \cdot 2H_2O \rightleftharpoons$

$[Co(H_2O)_4Cl_2]^+ + Cl^-$ 'i' is 2

and for $[Co(H_2O)_3Cl_3] \cdot 3H_2O$, '$i$' is 1 as it does not show ionisation. Hence, $[Co(H_2O)_3Cl_3] \cdot 3H_2O$ have minimum number of particles in the solution.

So, freezing point of its solution will be maximum.

44 Let the degree of association of acetic acid (CH_3COOH) in benzene is α, then

$$2CH_3COOH \rightleftharpoons (CH_3COOH)_2$$

Initial moles	1	0
Moles at equili.	$1-\alpha$	$\frac{\alpha}{2}$

$\therefore$ Total moles $= 1 - \alpha + \frac{\alpha}{2} = 1 - \frac{\alpha}{2}$

or $\quad i = 1 - \frac{\alpha}{2}$

Now, depression in freezing point (ΔT_f) is given as

$$\Delta T_f = i\,K_f m \quad \ldots \text{(i)}$$

where, K_f = molal depression constant or cryoscopic constant.

m = molality

$$\text{Molality} = \frac{\text{Number of moles of solute}}{\text{Weight of solvent (in kg)}}$$

$$= \frac{0.2}{60} \times \frac{1000}{20}$$

Putting the values in Eq. (i),

$$\therefore \quad 0.45 = \left[1 - \frac{\alpha}{2}\right](5.12)\left[\frac{0.2}{60} \times \frac{1000}{20}\right]$$

$$1 - \frac{\alpha}{2} = \frac{0.45 \times 60 \times 20}{5.12 \times 0.2 \times 1000}$$

$$\Rightarrow 1 - \frac{\alpha}{2} = 0.527 \Rightarrow \frac{\alpha}{2} = 1 - 0.527$$

$\therefore \quad \alpha = 0.946$

Thus, percentage of association = 94.6%

45 $Na_2SO_4 \longrightarrow 2Na^+ + SO_4^{2-}$

$\therefore$ van't Hoff factor (i) for $Na_2SO_4 = 3$

$\Delta T_f = i \times K_f \times m$

$= 3 \times 1.86 \times 0.01 = 0.0558$ K

$$\left[\because m = \frac{0.01}{1} = 0.01\right]$$

46 Effective molarity

= van't Hoff factor × molarity

0.5 M $C_2H_5OH(aq)$ $\quad i = 1$

Effective molarity = 0.5

0.25 M KBr (aq), $\quad i = 2$

Effective molarity = 0.5 M

0.1 M $Mg_3(PO_4)_2(aq)$, $i = 5$

Effective molarity = 0.5 M

0.125M $Na_3PO_4(aq)$, $i = 4$

Effective molarity = 0.5 M

Hence, all solutions have some osmotic pressure.

47 **Assertion** When methyl alcohol is added to water, boiling point of water decreases.

Reason When a more volatile solute is added to volatile solvent, vapour pressure of solvent increases and hence boiling point decreases but for non-volatile solute it is *vice-versa*.

48 Assertion and Reason both are correct statements and Reson is the correct explanation of Assertion.

Volume of solutions is a function of temperature which varies with temperature. Hence, molarity of solution in liquid state changes with temperature.

$$\text{Molarity} = \frac{\text{Moles of solute}}{\text{Volume of solution in litre}}$$

49 Both Assertion and Reason are correct and Reason is the correct explanation of the Assertion. Osmotic pressure is a colligative property while vapour pressure is not a colligative property.

50 The changes observed in these properties are very small, (e.g. 0.00001 K for substances having molar masses of the order of 10^6 g mol^{-1}). A little error in measurement of ΔT_b or ΔT_f will cause abnormal values of molecular weight.

SESSION 2

1

	$CuCl_2$	$\rightleftharpoons Cu^{2+}$	$+ 2Cl^-$
Initially	1 mol	0	0
After ionisation	$(1-\alpha)$ mol	α mol	2α mol

Thus, number of particles after ionisation

$= 1 - \alpha + \alpha + 2\alpha = 1 + 2\alpha$

van't Hoff factor (i)

$$= \frac{\text{Number of particles after ionisation}}{\text{Number of particles before ionisation}}$$

$$i = \frac{1 + 2\alpha}{1} \text{ (on 100\% ionisation, } \alpha = 1)$$

$$= \frac{1 + 2 \times 1}{1} = 3$$

$\because \Delta T_b = iK_b m$

$$\therefore \Delta T_b = \frac{3 \times 0.52 \times 13.44}{134.4 \times 1}$$

$= 0.156 \approx 0.16$ °C

2 Let, amount of Na_2CO_3 be 'x' g and $NaHCO_3$ will be $(1 - x)$ g.

$\because n_{Na_2CO_3} = n_{NaHCO_3}$

$$\therefore \quad \frac{x}{106} = \frac{1-x}{84} \Rightarrow x = \frac{106}{190} = 0.5578 \text{ g},$$

$1 - x = 0.4422$ g

$n_{Na_2CO_3} = n_{NaHCO_3}$

$$= \frac{0.5578}{106} = 0.00526 \text{ mol}$$

$Na_2CO_3 + 2HCl \longrightarrow 2NaCl + H_2O + CO_2$

and $NaHCO_3 + HCl \longrightarrow NaCl + H_2O + CO_2$

$M_1V_1 = M_2V_2 + M_3V_3$

(HCl) $\quad (Na_2CO_3) \quad (NaHCO_3)$

$0.1 \times V_1 = 2 \times 0.00526 + 0.00526$

$V_1 = 157.8$ mL

3 Water and hydrochloric acid and water and nitric acid form miscible solutions. They show negative deviation.

In case of CH_3COCH_3 and $CHCl_3$, there is interaction between them, thus force of attraction between $CH_3COCH_3 \ldots CHCl_3$ is larger than between $CHCl_3 \ldots CHCl_3$ or $CH_3COCH_3 \ldots CH_3COCH_3$ and thus, vapour pressure is less than expected—a negative deviation.

In case of CH_3OH, there is association by intermolecular H-bonding. When benzene is added to CH_3OH, H-bonding breaks and thus, force of attraction between CH_3OH and benzene molecule is smaller than between CH_3OH or benzene molecules (in pure state). Vapour pressure of mixture is greater than expected—a positive deviation.

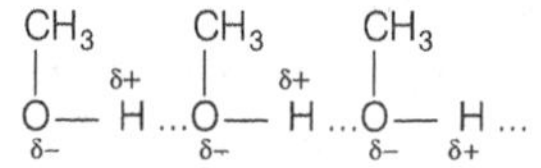

4 H_3PO_3 is a dibasic acid

(i.e. contains two ionisable protons attached directly to O).

$H_3PO_3 \rightleftharpoons 2H^+ + HPO_4^{2-}$

$\therefore$ 0.1 M H_3PO_3 = 0.2 N H_3PO_3

and 0.1 M KOH = 0.1 N KOH

$N_1V_1 = N_2V_2$

(KOH) $\quad (H_3PO_3)$

$0.1\,V_1 = 0.2 \times 20$

$V_1 = 40$ mL

5 For initial solution,

$$\because \pi = \frac{500}{760} \text{ atm}, T = 283 \text{ K}$$

$$\therefore \quad \frac{500}{760} \times V_{\text{initial}} = n \times R \times 283 \quad \ldots \text{(i)}$$

$$\text{After dilution, } \pi = \frac{105.3}{760} \text{ atm}, T = 298 \text{ K}$$

$\therefore \frac{105}{760} \times V_{final} = n \times R \times 298$ ...(ii)

From Eqs. (i) and (ii), we get

$\frac{V_{initial}}{V_{final}} = \frac{1}{5}$,

i.e. solution was diluted to 5 times.

6 $\chi_A = \frac{\chi_A p^\circ_A}{p}$ and $\chi_B = \frac{\chi_B p^\circ_B}{p}$

$\therefore \frac{\chi_B p^\circ_B}{\chi_A p^\circ_A} = \frac{\chi_B}{\chi_A}$

or $\frac{p^\circ_B (1-\chi_A)}{p^\circ_A \chi_A} = \frac{1-\chi_A}{\chi_A}$

or $\frac{p^\circ_B}{p^\circ_A}\left(\frac{1}{\chi_A} - 1\right) = \left(\frac{1}{\chi_A} - 1\right)$

or $\frac{1}{\chi_A} = \frac{p^\circ_A}{p^\circ_B} \times \frac{1}{\chi_A} + \left[1 - \frac{p^\circ_A}{p^\circ_B}\right]$

or $\frac{1}{\chi_A} = \frac{p^\circ_A}{p^\circ_B} \times \frac{1}{\chi_A} + \frac{p^\circ_B - p^\circ_A}{p^\circ_B}$

This is the equation of straight line, where, $\frac{p^\circ_A}{p^\circ_B}$ is slope and $\frac{p^\circ_B - p^\circ_A}{p^\circ_B}$ is intercept.

7 According to Dalton's law,

$p_T = p^\circ_A \chi_A + p^\circ_B \chi_B$

$550 = p^\circ_A \times \frac{1}{4} + p^\circ_B \times \frac{3}{4}$

Thus, $p^\circ_A + 3p^\circ_B = 2200$ ...(i)

When, 1 mole of y is further added to the solution

$560 = p^\circ_A + \frac{1}{5} + p^\circ_B \times \frac{4}{5}$

Thus, $p^\circ_A + 4p^\circ_B = 2800$ ...(ii)

On subtracting, Eq. (ii) and Eq (i)

$p^\circ_B = 2800 - 2200 = 600$ mm Hg

Putting the value of p°_B in Eq. (i)

$p^\circ_A + 3 \times 600 = 2200$

$p^\circ_A = 2200 - 1800 = 400$ mm Hg

8 Elevation in boiling point,

$\Delta T_b = i \times K_b \times m$

Molality of NaCl solution

$= \frac{n}{w} \times 1000 = \frac{\frac{58.5}{58.5}}{w_{H_2O}} \times 1000 = \frac{1000}{w_{H_2O}}$

Molality of $C_6H_{12}O_6$ (glucose) solution

$= \frac{\frac{180}{180} \times 1000}{w_{H_2O}} = \frac{1000}{w_{H_2O}}$

Both the solutions have same molality, but the value of i for NaCl and glucose are 2 and 1 respectively.

$\therefore \Delta T_{b(NaCl)} = 2 \times \Delta T_{b(C_6H_{12}O_6)}$

Hence, NaCl will show higher elevation boiling point.

9 $\frac{p^\circ - p_1}{p^\circ} = \frac{w_2/M_2}{w_2/M_2 + w_1/M_1}$

$\frac{640 - 600}{640} = \frac{2.175/M_2}{2.175/M_2 + 39/78}$

or $0.0625\left[\frac{2.175}{M_2} + 0.5\right] = \frac{2.175}{M_2}$

or $0.9375 \times \frac{2.175}{M_2} = 0.0625 \times 0.5$

or $M_2 = 65.3$

10 $\frac{\Delta p}{p^\circ} = \chi_2$

$\therefore \frac{10}{p^\circ} = 0.2$ or $p^\circ = 50$ mm

Similarly, $\frac{20}{50} = \chi_2$, i.e. $\chi_2 = 0.4$

$\therefore \chi_1 = 1 - 0.4 = 0.6$

11 According to Dalton's law,

$p = \chi_A p^\circ_A + \chi_B p^\circ_B$

or $550 = p^\circ_A\left(\frac{2}{5}\right) + p^\circ_B\left(\frac{3}{5}\right)$

or $2p^\circ_A + 3p^\circ_B = 2750$ ...(i)

$\Rightarrow$ When 1 mole of B is added to it,

$560 = p^\circ_A\left(\frac{2}{6}\right) + p^\circ_B\left(\frac{4}{6}\right)$

or $2p^\circ_A + 4p^\circ_B = 3360$ ...(ii)

From Eqs. (i) and (ii), we get

$p^\circ_A = 460$ mm

12 (a) $p_A = X_A p^\circ_A$ true

(b) $\pi = iMRT = MRT$

true [If van't Hoff factor, $i = 1$]

(c) $\pi \propto i$

Greater the value of i, larger is the value of π.

i, for $BaCl_2$ (strong electrolyte) is 3.

$[BaCl_2 \rightleftharpoons Ba^{2+} + 2Cl^-$ (3 ions)]

i for KCl is 2. $[KCl \rightleftharpoons K^+ + Cl^-]$

i for CH_3COOH is less than 2

$[CH_3COOH \rightleftharpoons CH_3COO^- + H^+]$

But, CH_3COOH is weak electrolyte than KCl.

i for sucrose is 1, as it is a non-electrocyte.

Thus, i (for $BaCl_2$) > KCl > CH_3COOH < sucrose

Thus, (c) is also true.

(d) $\Delta T_f = K_f m$

K_f is dependent on solvent.

Thus, freezing point $= [T$ (solvent) $- \Delta T_f]$ are different.

Thus, (d) is false.

13 Due to the presence of non-volatile solute, NaCl in the beaker B, the vapour pressure of beaker B decreases and becomes less than that of A.

14 Suppose,

$[Pt(NH_3)_4Cl_4] \xrightarrow{\text{Dissociation}} n$ moles of product ions

$\therefore i = n.$

But from the given data

$\Delta T_f = iK_f m$

$\Rightarrow 0.0054 = n \times 1.80 \times 0.001$

$\therefore n = 3$

Hence, the formula must be the one which gives 3 ions of products, i.e. $[Pt(NH_3)_4Cl_2]Cl_2$

15 100 g solution contains

= 5.85 g NaCl

= 0.1 mole of NaCl

100 g solution contains

$= 9.50$ g $MgCl_2$

$= 0.1$ mole of $MgCl_2$

Hence, weight of solvent (H_2O)

$= 100 - (5.85 + 9.50)$

$= 84.65$ g

NaCl ionises 80%,

$NaCl \rightleftharpoons Na^+ + Cl^-$

$i = 1 + (y-1)x = (1+x)$

$= 1 + 0.8 = 1.8$

Here, y is the number of ions per mole of solute and x is the degree of ionisation.

Hence, number of moles of NaCl from 0.1 mole due to ionisation

$= 1.8 \times 0.1 = 0.18$ mol

$MgCl_2$ ionises 50%,

$MgCl_2 \rightleftharpoons Mg^{2+} + 2Cl^-$

$\therefore i = 1 + (y-1)x = 1 + 2x$

$= 1 + 2 \times 0.5 = 2$

Hence, number of moles of $MgCl_2$ from 0.1 mole

$= 2 \times 0.1 = 0.20$

Total moles of NaCl and $MgCl_2$ in solution $= 0.18 + 0.20 = 0.38$

$(n_1 + n_2)i = 0.38$

Elevation in boiling point

$(\Delta T_b) = \frac{1000 K_b (n_1 + n_2) i}{w_2}$

$= \frac{1000 \times 0.51 \times 0.38}{84.65} = 2.29°C$

Hence, boiling point of solution

$= 100 + 2.29 = 102.29$ °C

DAY NINE

Physical and Chemical Equilibrium

Learning & Revision for the Day

- Equilibrium State
- Equilibria Involving Physical Processes
- Equilibria Involving Chemical Processes
- Factor Affecting Equilibrium
- Equilibrium constant (K_p and K_c) and their Significance

Equilibrium is a general term which applies not only to chemical reactions but applies to physical changes as well, e.g. ice and water are in equilibrium at 0°C and atmospheric pressure.

Equilibrium State

In a reversible reaction, the point at which there is no further change in concentration of reactants and products, is called **equilibrium state**. It gives rise to a constant vapour pressure because of an equilibrium in which the number of molecules leaving the liquid equals the number returning to liquid from the vapour.

Some important features of equilibrium state are

(i) Equilibrium is attained in a closed container.

(ii) Rate of forward reaction = Rate of backward reaction.

(iii) At equilibrium, concentration of reactants and products becomes constant.

Concept of Dynamic Equilibrium

Equilibrium is always dynamic in nature, i.e. the reaction does not stop but goes on forward and backward directions with equal speed. In other words, the equilibrium state is a dynamic balance between the forward and backward reaction.

e.g. solid $\rightleftharpoons$ liquid (physical equilibria)

$H_2O(s) \rightleftharpoons H_2O(l)$ (273 K, 1 atm)

$N_2O_4(g) \rightleftharpoons 2NO_2(g)$ (chemical equilibria)

Catalyst helps in attaining the state of equilibrium quickly without changing the state of equilibrium.

PREP MIRROR
Your Personal Preparation Indicator

- No. of Questions in Exercises (x)—
- No. of Questions Attempted (y)—
- No. of Correct Questions (z)—
  (Without referring Explanations)

- Accuracy Level ($z/y \times 100$)—
- Prep Level ($z/x \times 100$)—

In order to expect good rank in JEE, your Accuracy Level should be above 85 & Prep Level should be above 75.

Equilibria Involving Physical Processes

The characteristic of system at equilibrium are better understood by physical processes.

This may be attained by the following ways

1. **Solid-liquid equilibrium**

$$\text{Ice} \rightleftharpoons \text{Water}$$

Rate of melting of ice = rate of freezing of water

2. **Liquid-gas equilibrium**

$$\text{Water} \rightleftharpoons \text{Water vapours}$$

Rate of evaporation of water = rate of condensation of water vapours

3. **Solid-gas equilibrium**

Certain solids on heating directly change from solid into vapour state (sublimation).

$$\underset{\text{(solid)}}{\text{Naphthalene}} \rightleftharpoons \underset{\text{(vapour)}}{\text{Naphthalene}}$$

Henry's Law

The equilibrium between the molecules in the gaseous state and the molecules dissolved in the liquid under pressure is governed by **Henry's law** which states that the mass of a gas dissolved in a given mass of a solvent at any temperature is proportional to the pressure of the gas above the solvent.

General Characteristics of Equilibrium Involving Physical Processes

General characteristics of equilibrium involving a physical process are as follows:

(i) At equilibrium, all the measurable properties of the system remain constant.

(ii) Equilibrium is attained only in a closed system at a given temperature.

(iii) Equilibrium is dynamic in nature, i.e. the opposing processes do not stop but take place simultaneously and with the same rate.

(iv) The physical equilibrium is chracterised by constant value of one of its parameter (such as melting point) at a given temperature.

(v) The magnitude of these parameters at any stage shows the extent of physical process upto which it has proceeded before reaching equilibrium.

Equilibria Involving Chemical Processes

The rate of reversible reactions at which the concentration of the reactants and products do not change with time is known as chemical equilibrium.

Active mass = molar concentration (mol/L)

e.g.

$$H_2 + I_2 \rightleftharpoons 2HI$$

$$PCl_5 \rightleftharpoons PCl_3 + Cl_2$$

Types of Chemical Equilibrium

On the basis of physical state of reactants and products, equilibrium may be of the following two types:

1. **Homogeneous equilibria**, in which reactants and products are in same phase, e.g.

$$N_2(g) + 3H_2(g) \longrightarrow 2NH_3(g)$$

$$CH_3COOH(l) + C_2H_5OH(l) \rightleftharpoons CH_3COOC_2H_5(l) + H_2O(l)$$

2. **Heterogeneous equilibria**, in which reactants and products are in different phase, e.g.

$$CaCO_3(s) \rightleftharpoons CaO(s) + CO_2(g)$$

$$NH_4HS(s) \rightleftharpoons NH_3(g) + H_2S(g)$$

Law of Chemical Equilibrium

At a given temperature, the product of concentrations of the reaction products raised to the respective stoichiometric coefficient in the balanced chemical equation divided by the product of concentrations of the reactants raised to their individual stoichiometric coefficients has a constant value. This is known as **"law of chemical equilibrium."**

$$aA + bB \rightleftharpoons cC + dD$$

Rate of forward reaction $(R_f) \propto [A]^a\,[B]^b$

$$\therefore \quad R_f = k_f\,[A]^a\,[B]^b$$

Rate of backward reaction $(R_b) \propto [C]^c\,[D]^d$

$$\therefore \quad R_b = k_b\,[C]^c\,[D]^d$$

At equilibrium, $k_f\,[A]^a[B]^b = k_b\,[C]^c[D]^d$

$$k = \frac{k_f}{k_b} = \frac{[C]^c[D]^d}{[A]^a[B]^b}$$

where, k_f = rate constant for forward reaction

k_b = rate constant for backward reaction

The above expression is also known as law of chemical equilibrium.

For a gaseous reaction, $K_p = \dfrac{p_C^c \cdot p_D^d}{p_A^a \cdot p_B^b}$

where, K_p = equilibrium constant in terms of partial pressure.

Equilibrium Constants (K_p and K_C) and their Significance

The ratio of the concentrations of the products raised to the power of their stoichiometric coefficients to the concentrations of reactants raised to the power of their stoichiometric coefficients at equilibrium state is called the **equilibrium constant** (K).

The equilibrium constant (K) can be expressed in three ways:

1. **Equilibrium constant in terms of molar concentration (K_C).**

$$K_C = \frac{[C]^c \cdot [D]^d}{[A]^a \cdot [B]^b}$$

2. **Equilibrium constant in terms of partial pressure (K_p).**

$$K_p = \frac{p_C^c p_D^d}{p_A^a p_B^b}$$

3. **Equilibrium constant in terms of mole fraction (K_χ).**

$$K_\chi = \frac{\chi_C^c \cdot \chi_D^d}{\chi_A^a \cdot \chi_B^b}$$

Significance of equilibrium constants are as follows :

(i) If $K_C > 10^3$, products predominate over reactants. In other words, if K_C is very large, the reaction proceeds almost in all the way to completion.

(ii) If $K_C < 10^{-3}$, reactants predominate over products. In other words, if K_C is very small, the reaction proceeds hardly at all.

(iii) If K_C is in the range 10^{-3} to 10^3, appreciable concentration of both reactants and products are present.

Equilibrium constant helps in predicting the direction in which given reaction will proceed at any stage. For this purpose, we find out **reaction quotient**.

It is defined as the ratio of the molar concentration or reaction quotient partial pressure of the product species to that of reactant species at any stage in the reaction.

For a general reaction, $aA + bB \rightleftharpoons cC + dD$

$$Q_C = \frac{[C]^c [D]^d}{[A]^a [B]^b}$$

and

$$Q_p = \frac{p_C^c \cdot p_D^d}{p_A^a \cdot p_B^b}$$

At any stage of the reaction,

(i) if $Q_C > K_C$, the reaction will proceed in the direction of reactants (reverse reaction).

(ii) if $Q_C < K_C$, the reaction will proceed in the direction of the products.

(iii) if $Q_C = K_C$, reaction mixture is already at equilibrium.

Relation between K_p, K_C and K_χ

$$K_p = K_C (RT)^{\Delta n_g}$$

$$K_p = K_\chi (p)^{\Delta n_g}$$

R = gas constant, T = temperature in Kelvin

Δn_g = gaseous moles of products – gaseous moles of reactants

Hence,

(i) If $\Delta n_g = 0$, K_p then $= K_C$ (no units for both K_C and K_p)

(ii) If $\Delta n_g = +$ ve, then $K_p > K_C$ (unit of K_C is $(\text{mol L}^{-1})^{\Delta n_g}$ and that of K_p is $(\text{atm})^{\Delta n_g}$)

(iii) If $\Delta n_g = -$ve then $K_p < K_C$ (unit of K_C is $(\text{L mol}^{-1})^{\Delta n_g}$ and that of K_p is $(\text{atm})^{\Delta n_g}$)

Equilibrium constant in terms of activities, K_a

$$K_a = \frac{a_C^c \times a_D^d}{a_A^a \times a_B^b}$$

Activity = activity coefficient × molality (or molarity)

i.e. $a = \gamma \times m$

Characteristics of Equilibrium Constant

(i) Equilibrium constant (K_p or K_C) does not depend on pressure, volume, concentration and catalyst but depends only upon temperature.

(ii) Equilibrium constant for a given reaction is independent of the reaction mechanism.

(iii) Equilibrium constant depends on stoichiometric coefficient, e.g.

- $H_2(g) + I_2(g) \overset{K_{C_1}}{\rightleftharpoons} 2HI(g);\ K_{C_1}$
- $\frac{1}{2}H_2(g) + \frac{1}{2}I_2(g) \overset{K_{C_2}}{\rightleftharpoons} HI(g);\ K_{C_2} = \sqrt{K_{C_1}}$
- $HI(g) \overset{K_{C_3}}{\rightleftharpoons} \frac{1}{2}H_2(g) + \frac{1}{2}I_2(g);\ K_{C_3} = \frac{1}{K_{C_2}} = \frac{1}{\sqrt{K_{C_1}}}$

(iv) If a reaction is multiplied by n, the rate constant, K_C becomes $(K_C)^n$. n can be fractional also (+ve only).

(v) If K_1 be equilibrium constant for $P \rightleftharpoons Q$ and K_2 be equilibrium constant for $R \rightleftharpoons S$, equilibrium constant for $P + R \rightleftharpoons Q + S$ is K_1K_2.

(vi) For exothermic reactions, K_C decreases with increase in temperature. For endothermic reactions, K_C increases with increase in temperature. For reactions having zero heat energy, temperature has no effect.

Significance of ΔG and $\Delta G°$ in Chemical Equilibria

- ΔG for a reactions under any set of conditions is related to its value under standard conditions, i.e. $\Delta G°$ by the equation,

$$\Delta G = \Delta G° + 2.303\ RT \log Q$$

- Under equilibrium condition, for same number of moles of reactants and products.

$$Q = K_p = K_C = K \text{ and } \Delta G = 0$$

$$\therefore \quad \Delta G° = -2.303\ RT \log K$$

Significance of $\Delta G°$ are given below:

(i) If $\Delta G° < 0$, $\log K > 1 \Rightarrow K > 1$
Therefore, forward reaction is spontaneous.

(ii) If $\Delta G° > 0$, $\log K < 1 \Rightarrow K < 1$
Therefore, backward reaction is spontaneous.

(iii) If $\Delta G° = 0$, $\log K = 0$, $\Rightarrow K = 1$
Therefore, reaction is at equilibrium.

NOTE **Relation between Degree of Dissociation and Density**
Degree of dissociation (α) of a gaseous compound is related to its vapour density by

$$\alpha = \frac{D-d}{d(y-1)}$$

Here, D = molar density before dissociation / initial density
d = density after dissociation/density of the gaseous mixture
y = number of moles of products

$$\text{Density of gas} = \frac{\text{molecular weight}}{2}$$

Molecular weight = Density of gas × 2

$$M_O = D \times 2 \quad \text{or} \quad M_C = d \times 2$$

Here,
M_O = Observed molecular weight [Abnormal]
M_C = Calculated molecular weight [Theoretical]

Factors Affecting Equilibrium

(Le-Chatelier's Principle)

The principle states that change in any of the factors that determine the equilibrium conditions of a system, will cause the system to change in such a manner, so as to reduce or to counteract the effect of the change.

Different factors affecting equilibrium are discussed below:

1. Effect of Concentration Change

The concentration stress of an added reactant or product is relieved by net reaction in the direction that consumes the added substance. e.g.

$$A + B \rightleftharpoons C$$

(i) If we increase the concentration of either A or B (reactants), the equilibrium goes in the direction that consumes A or B, i.e. forward direction.

(ii) If we increase the concentration of C (product), the equilibrium goes in the direction that consumes C, i.e. backward direction.

(iii) If we remove C (product), the equilibrium goes in the direction in which its concentration increases, i.e. forward direction.

(iv) If any of the species is in solid or liquid state, its addition does not alter the original equilibrium.

Sweet substances cause tooth decay because on fermentation these produce H^+ ions which combine with OH^- ions and shift the equilibrium in forward direction.

$$Ca_5(PO_4)_3OH(s) \underset{\text{Remineralisation}}{\overset{\text{Demineralisation}}{\rightleftharpoons}} 5Ca^{2+}(aq) + 3PO_4^{3-}(aq) + OH^-(aq)$$

2. Effect of Pressure

At high pressure, reaction goes from higher number of moles to lower number of moles or from higher volume to lower volume and *vice-versa*.

(i) If $\Delta n_g = 0$, no effect on equilibrium due to pressure change.

(ii) If $\Delta n_g > 0$, the increase in pressure favours backward reaction.

(iii) If $\Delta n_g < 0$, the increase in pressure favours forward reaction. (Δn_g = number of moles of gaseous products − number of moles of gaseous reactants).

3. Effect of Temperature

At high temperature, reaction goes to **endothermic direction** while at low temperature reaction goes to **exothermic direction.** The equilibrium constant for an endothermic reaction (positive ΔH) increases as the temperature increases.

$K \propto T$; if $\Delta H° = +$ ve (endothermic)

$K \propto \frac{1}{T}$; if $\Delta H° = -$ ve (exothermic)

4. Effect of Catalyst

A catalyst increases the rate of forward reaction as well as the rate of backward reaction, so it does not affect the equilibrium and equilibrium constant.

5. Effect of Inert Gas

At constant volume, there is no effect of addition of inert gas. At constant pressure, when inert gas is added, reaction goes from lower number of moles to higher number of moles.

Application of Le-Chatelier Principle

The Le-Chatelier's principle is applicable to physical as well as chemical equilibria. Some of its applications are mentioned below:

1. To Chemical Equilibria

Haber's processes of synthesis of ammonia

$$N_2(g) + 3H_2(g) \rightleftharpoons 2NH_3(g); \quad \Delta H = -\text{ve}$$

To get better yield of ammonia, required conditions are:

- High pressure as $\Delta n_g = -$ ve
- Low temperature as reaction is exothermic
- High concentration of reactants
- Removal of NH_3

2. To Physical Equilibria

Some of the application for physical equilibria is as follows:

(i) Effect of pressure on melting point

(ii) Ice-water equilibrium,

$$\text{Ice } (s) \rightleftharpoons \text{water } (l)$$

An increase in pressure favours the melting of ice into water because $V_{ice} > V_{water}$. Therefore, on increasing pressure, more and more ice will melt.

(ii) **Effect of pressure on solubility of gases**

Gas + solvent $\rightleftharpoons$ solution

An increase in pressure always favours the dissolution of gas in the solvent. Therefore, solubility of gas increases with increase in pressure.

The amount of gas dissolved per unit volume of solvent is directly proportional to its pressure (Henry's law), i.e. $m \propto p$.

(iii) **Effect of temperature on solubility of solids**

Solute + solvent $\rightleftharpoons$ solution ; $\Delta H = +$ ve

An increase in temperature always favours endothermic reaction, therefore solutes having endothermic dissolution show an increase in their solubility with temperature and solutes having exothermic dissolution (dissolution of lime in water) show a decrease in their solubility with temperature.

DAY PRACTICE SESSION 1

FOUNDATION QUESTIONS EXERCISE

1 In a chemical reaction, the rate constant for the backward reaction is 7.5×10^{-4} and the equilibrium constant is 1.5. The rate constant for the forward reaction is

(a) 5×10^{-4}
(b) 2×10^{-3}
(c) 1.125×10^{-3}
(d) 9.0×10^{-4}

2 56 g of N_2 and 6 g of H_2 were kept at 400°C in 1 L vessel. The equilibrium mixture contained 27.54 g of NH_3. The approximate value of K_C for the above reaction in $mol^{-2} L^2$ is

(a) 0.128 (b) 0.118
(c) 0.148 (d) 0.008

3 The concentration of CO_2 which will be in equilibrium with 2.5×10^{-2} mol L^{-1} of CO at 100°C. For the reaction $FeO(s) + CO(g) \rightleftharpoons Fe(s) + CO_2(g)$; $K_C = 5.0$ will be

(a) 0.5×10^{-1} mol L^{-1} (b) 1.25×10^{-1} mol L^{-1}
(c) 2×10^{-2} mol L^{-1} (d) None of these

4 1.1 moles of A are mixed with 2.2 moles of B and the mixture is kept in a 1 L flask till the equilibrium, $A + 2B \rightleftharpoons 2C + D$ is reached. At equilibrium, 0.2 mole of C is formed. The equilibrium constant of the above reaction is

(a) 0.0002 (b) 0.004
(c) 0.001 (d) 0.003

5 In aqueous solution, the ionisation constants for carbonic acid are $K_1 = 4.2 \times 10^{-7}$ and $K_2 = 4.8 \times 10^{-11}$. Select the correct statement for a saturated 0.034 M solution of the carbonic acid. → AIEEE 2010

(a) The concentration of CO_3^{2-} is 0.034 M
(b) The concentration of CO_3^{2-} is greater than that of HCO_3^-
(c) The concentration of H^+ and HCO_3^- are approximately equal
(d) The concentration of H^+ is double that of CO_3^{2-}

6 The standard Gibbs energy change at 300 K for the reaction, $2A \rightleftharpoons B + C$ is 2494. 2 J.
At a given time, the composition of the reaction mixture is $[A] = \frac{1}{2}$, $[B] = 2$ and $[C] = \frac{1}{2}$. The reaction proceeds in the
[$R = 8.314$ JK/mol, $e = 2.718$] → JEE Main 2015

(a) forward direction because $Q > K_C$
(b) reverse direction because $Q > K_C$
(c) forward direction because $Q < K_C$
(d) reverse direction because $Q < K_C$

7 For the reaction,

$$2NO_2(g) \rightleftharpoons 2NO(g) + O_2(g)$$

[$K_C = 1.8 \times 10^{-6}$ at 184°C, $R = 0.00831$ kJ/(mol K)]
when K_p and K_C are compared at 184°C it is found that

(a) whether K_p is greater than, less than or equal to K_C depends upon the total gas pressure
(b) $K_p = K_C$
(c) K_p is less than K_C
(d) K_p is greater than K_C

8 For the reaction,
$CO(g) + Cl_2(g) \rightleftharpoons COCl_2(g)$, the K_p/K_C is equal to

(a) $1/RT$ (b) RT
(c) $\sqrt{RT}$ (d) 1.0

9 For the following reaction in gaseous phase

$$CO + \frac{1}{2}O_2 \longrightarrow CO_2 . K_C/K_p \text{ is}$$

(a) $(RT)^{1/2}$ (b) $(RT)^{-1/2}$
(c) (RT) (d) $(RT)^{-1}$

10 For the reaction,

$$SO_2(g) + \frac{1}{2}O_2(g) \rightleftharpoons SO_3(g)$$

if $K_p = K_C(RT)^x$ where, the symbols have their usual meaning then the value of x is (assuming ideality) → JEE Main 2014

(a) -1 (b) $-\frac{1}{2}$ (c) $\frac{1}{2}$ (d) 1

11 The ratio $\frac{K_p}{K_C}$ for the reaction,

$$CO(g) + \frac{1}{2}O_2(g) \rightleftharpoons CO_2(g) \text{ is}$$

→ JEE Main (Online) 2013

(a) $\frac{1}{\sqrt{RT}}$ (b) $(RT)^{1/2}$ (c) RT (d) 1

12 A reversible chemical reaction is having two reactants in equilibrium. If the concentration of the reactants are doubled then the equilibrium constant will

(a) be doubled (b) become one fourth
(c) be halved (d) remain the same

13 Consider the equilibrium reactions,

$$H_3PO_4 \overset{K_1}{\rightleftharpoons} H^+ + H_2PO_4^-$$
$$H_2PO_4^- \overset{K_2}{\rightleftharpoons} H^+ + HPO_4^{2-}$$
$$HPO_4^{2-} \overset{K_3}{\rightleftharpoons} H^+ + PO_4^{3-}$$

The equilibrium constant, K for the following dissociation

$$H_3PO_4 \rightleftharpoons 3H^+ + PO_4^{3-}, \text{ is}$$

(a) $K_1 / K_2 \cdot K_3$ (b) $K_1 \cdot K_2 \cdot K_3$
(c) $K_2 / K_1 \cdot K_3$ (d) $K_1 + K_2 + K_3$

14 If K_1 and K_2 are the equilibrium constants of the equilibria (I) and (II) respectively, what is the relationship between the two constants ?

I. $SO_2(g) + \frac{1}{2}O_2(g) \overset{K_1}{\rightleftharpoons} SO_3(g)$

II. $2SO_3(g) \overset{K_2}{\rightleftharpoons} 2SO_2(g) + O_2(g)$

(a) $(K_1)^2 = \frac{1}{K_2}$ (b) $K_2 = (K_1)^2$
(c) $K_1 = \frac{1}{K_2}$ (d) $K_1 = K_2$

15 For the following three reactions (I), (II) and (III), equilibrium constants are given

I. $CO(g) + H_2O(g) \rightleftharpoons CO_2(g) + H_2(g); K_1$
II. $CH_4(g) + H_2O(g) \rightleftharpoons CO(g) + 3H_2(g); K_2$
III. $CH_4(g) + 2H_2O(g) \rightleftharpoons CO_2(g) + 4H_2(g); K_3$

Which of the following relation is correct?

(a) $K_3 = K_1K_2$ (b) $K_3K_2^3 = K_1^2$
(c) $K_1\sqrt{K_2} = K_3$ (d) $K_2K_3 = K_1$

16 The equilibrium constant, K_C for the reaction,

$$SO_3(g) \rightleftharpoons SO_2(g) + \frac{1}{2}O_2(g)$$

is 4.9×10^{-2}. The value of K_C for the reaction,

$$2SO_2(g) + O_2(g) \rightleftharpoons 2SO_3(g)$$

will be approximately equal to

(a) 416 (b) 2.40×10^{-3}
(c) 9.8×10^{-2} (d) 4.9×10^{-2}

17 An aqueous solution contains 0.10 M H_2S and 0.20 M HCl. If the equilibrium constants for the formation of HS^- from H_2S is 1.0×10^{-7} and that of S^{2-} from HS^- ions is 1.2×10^{-13} then the concentration of S^{2-} ions in aqueous solution is → JEE Main 2018

(a) 5×10^{-8} (b) 3×10^{-20}
(c) 6×10^{-21} (d) 5×10^{-19}

18 $N_2(g) + 3H_2(g) \rightleftharpoons 2NH_3(g); K_1$ ...(i)

$N_2(g) + O_2(g) \rightleftharpoons 2NO(g); K_2$ ...(ii)

$H_2(g) + \frac{1}{2}O_2(g) \rightleftharpoons H_2O(g); K_3$ ...(iii)

The equation for the equilibrium constant of the reaction

$2NH_3(g) + \frac{5}{2}O_2(g) \rightleftharpoons 2NO(g) + 3H_2O(g), (K_4)$ in terms of K_1, K_2 and K_3 is → JEE Main (Online) 2013

(a) $\frac{K_1K_2}{K_3}$ (b) $\frac{K_1K_3^2}{K_2}$ (c) $K_1K_2K_3$ (d) $\frac{K_2K_3^3}{K_1}$

19 The equilibrium constant (K_C) for the reaction $N_2(g) + O_2(g) \rightleftharpoons 2NO(g)$ at temperature T is 4×10^{-4}. The value of K_C for the reaction.

$NO(g) \rightleftharpoons \frac{1}{2}N_2(g) + \frac{1}{2}O_2(g)$ at the same temperature is

→ AIEEE 2012

(a) 0.02 (b) 2.5×10^2 (c) 4×10^{-4} (d) 50.0

20 For the following equilibrium, $N_2O_4 \rightleftharpoons 2NO_2$ in gaseous phase, NO_2 is 50% of the total volume when equilibrium is set up. Hence, per cent of dissociation of N_2O_4 is

(a) 50% (b) 25%
(c) 66.66% (d) 33.33%

21 At certain temperature and a total pressure of 10^5 Pa, iodine vapours contains 40% by volume of iodine atoms. K_p for the equilibrium $I_2(g) \rightleftharpoons 2I(g)$ will be

(a) 0.67 (b) 1.5
(c) 2.67×10^4 (d) 9.0×10^4

22 For the reaction, $N_2O_4(g) \rightleftharpoons 2NO_2(g)$ the degree of dissociation at equilibrium is 0.2 at 1 atm pressure. The equilibrium constant, K_p will be

(a) 1/2 (b) 1/4 (c) 1/6 (d) 1/8

23 At 30°C, K_p for the dissociation reaction

$$SO_2Cl_2(g) \rightleftharpoons SO_2(g) + Cl_2$$

is 2.9×10^{-2} atm. If the total pressure is 1 atm, the degree of dissociation of SO_2Cl_2 is (assume, $1 - \alpha^2 = 1$)

(a) 87% (b) 13%
(c) 17% (d) 29%

24 The equilibrium constant K_p for the reaction,

$$PCl_5 \rightleftharpoons PCl_3 + Cl_2 \text{ is 1.6 at 200°C.}$$

The pressure at which PCl_5 will be 50% dissociated at 200°C is

(a) 3.2 atm (b) 4.8 atm
(c) 2.4 atm (d) 6.4 atm

25 XY_2 dissociates as $XY_2(g) \rightleftharpoons XY(g) + Y(g)$.
When the initial pressure of XY_2 is 600 mm Hg, the total equilibrium pressure is 800 mm Hg. Calculate K for the reaction assuming that the volume of the system remains unchanged.
(a) 50 (b) 100 (c) 166.6 (d) 400.0

26 A vessel at 1000 K contains CO_2 with a pressure of 0.5 atm. Some of the CO_2 is converted into CO on the addition of graphite. If the total pressure at equilibrium is 0.8 atm, the value of K_p is → AIEEE 2011
(a) 1.8 atm (b) 3 atm (c) 0.3 atm (d) 0.18 atm

27 Ammonium carbamate decomposes as
$$NH_2COONH_4(s) \rightleftharpoons 2NH_3(g) + CO_2(g)$$
For this reaction, $K_p = 2.9 \times 10^{-5} atm^3$. If we start with 1 mole of the compound, the total pressure at equilibrium would be
(a) 0.0766 atm (b) 0.0581 atm
(c) 0.0388 atm (d) 0.0194 atm

28 In reaction $A + 2B \rightleftharpoons 2C + D$, initial concentration of B was 1.5 times of [A], but at equilibrium the concentrations of A and B became equal. The equilibrium constant for the reaction is → JEE Main (Online) 2013
(a) 8 (b) 4 (c) 12 (d) 6

29 The equilibrium constant at 298K for a reaction, $A + B \rightleftharpoons C + D$ is 100. If the initial concentrations of all the four species were 1M each, then equilibrium concentration of D (in mol L^{-1}) will be → JEE Main 2016
(a) 0.818 (b) 1.818
(c) 1.182 (d) 0.182

30 Match standard free energy of the reaction with the corresponding equilibrium constant.

Column I		Column II
A. $\Delta G^{\ominus} > 0$	1.	$K > 0$
B. $\Delta G^{\ominus} < 0$	2.	$K = 1$
C. $\Delta G^{\ominus} = 0$	3.	$K = 0$
	4.	$K < 1$

Codes

	A	B	C		A	B	C
(a)	2	1	3	(b)	3	4	1
(c)	4	1	2	(d)	3	1	4

31 On increasing the pressure, in which direction will the gas phase reaction proceed to re-establish equilibrium, is predicted by applying the Le-Chatelier's principle. Consider the reaction,
$$N_2(g) + 3H_2(g) \rightleftharpoons 2NH_3(g)$$
Which of the following is correct, if the total pressure at which the equilibrium is established, is increased without changing the temperature?
(a) K will remain same
(b) K will decrease
(c) K will increase
(d) K will increase initially and decrease when pressure is very high

32 In the given reaction,
$$2X(g) + Y(g) \rightleftharpoons Z(g) + 80 \text{ kcal}$$
which combination of pressure and temperature gives the highest yield of Z at equilibrium?
(a) 1000 atm and 500°C (b) 500 atm and 500°C
(c) 1000 atm and 100°C (d) 500 atm and 100°C

33 The following reactions are known to occur in the body
$$CO_2 + H_2O \longrightarrow H_2CO_3 \rightleftharpoons H^+ + HCO_3^-$$
If CO_2 escapes from the system
(a) pH will decreases
(b) hydrogen ion concentration will diminish
(c) H_2CO_3 concentration will be altered
(d) the forward reaction will be promoted

34 Consider the following equilibrium in a closed container
$$N_2O_4(g) \rightleftharpoons 2NO_2(g)$$
At a fixed temperature, the volume of the reaction container is halved. For this change, which of the following statements holds true regarding the equilibrium constant K_p and degree of dissociation (α)?
(a) Neither K_p nor α changes
(b) Both K_p and α changes
(c) K_p changes but α does not
(d) K_p does not change but α changes

35 The equilibrium constant for the reaction, $H_2(g) + I_2(g) \rightleftharpoons 2HI(g)$ is 64. If the volume of the container is reduced to half of the original volume, the value of the equilibrium constant will be
(a) 16 (b) 32 (c) 64 (d) 128

36 The exothermic formation of $ClF_3(g)$ is represented by the equation,
$$Cl_2(g) + 3F_2(g) \rightleftharpoons 2ClF_3(g);\ \Delta H_r = -329 \text{ kJ}$$
Which of the following will increase the quantity of ClF_3 in an equilibrium mixture of Cl_2, F_2 and ClF_3?
(a) Adding F_2
(b) Increasing the volume of the container
(c) Removing Cl_2
(d) Increasing the temperature

37 Consider the reaction equilibrium,
$$2SO_2(g) + O_2(g) \rightleftharpoons 2SO_3(g);\quad \Delta H° = -198 \text{ kJ}$$
On the basis of Le-Chatelier's principle, the condition favourable for the forward reaction is
(a) lowering of temperature as well as pressure
(b) increasing of temperature as well as pressure
(c) lowering the temperature and increasing the pressure
(d) any value of temperature and pressure

Direction (Q. Nos. 38-41) *In the following questions Assertion (A) followed by Reason (R) is given. Choose the correct optoin out of the choices given below.*

(a) Both A and R are true and R is the true explanation of A
(b) Both A and R are true but R is not the true explanation of A
(c) A is true but R is true
(d) Both A and R are false

38 Assertion (A) In the dissociation of PCl_5 at constant pressure and temperature, addition of helium at equilibrium increases the dissociation of PCl_5.

Reason (R) Helium removes Cl_2 from the field of action.

39 Assertion (A) The reaction,

$$2NO\,(g) + O_2\,(g) \rightleftharpoons 2NO_2(g)$$

is favoured in the forward direction with increase of pressure.

Reason (R) The reaction is exothermic.

40 Assertion (A) The equilibrium constant for a reaction having positive $\Delta H°$ increases with increase of temperature.

Reason (R) The temperature dependence of the equilibrium constant is related to $\Delta S°$ and $\Delta H°$ for the reaction.

41 Assertion (A) Haber's synthesis of NH_3 is carried out in the presence of a catalyst.

Reason (R) The catalyst shifts the position of the equilibrium of reaction,

$$N_2(g) + 3H_2(g) \rightleftharpoons 2NH_3(g)$$

to the product side.

DAY PRACTICE SESSION 2

PROGRESSIVE QUESTIONS EXERCISE

1 Formaldehyde polymerises to form glucose according to the reaction, $6HCHO \rightleftharpoons C_6H_{12}O_6$

The theoretically computed equilibrium constant for this reaction is found to be 6×10^{22}. If 1M solution of glucose dissociates according to the above equilibrium, the concentration of formaldehyde in the solution will be

(a) 1.6×10^{-2} M (b) 1.6×10^{-4} M
(c) 1.6×10^{-6} M (d) 1.6×10^{-8} M

2 1% CO_2 in air is sufficient to prevent any loss in weight when M_2CO_3 is heated at 120°C.

$$M_2CO_3(s) \rightleftharpoons M_2O(s) + CO_2(g),$$

$K_p = 0.0095$ atm at 120°C. How much would the partial pressure of CO_2 have to be to promote this reaction at 120°C ?

(a) $p'_{CO_2} < 0.0095$ (b) $p'_{CO_2} = 0.0095$
(c) $p'_{CO_2} > 0.0095$ (d) Cannot be predicted

3 Starting with one mole of O_2 and two moles of SO_2, the equilibrium for the formation of SO_3 was established at a certain temperature. If V is the volume of the vessel and $2x$ is the number of moles of SO_3 present, the equilibrium constant will be

(a) $\dfrac{x^2V}{(1-x)^3}$ (b) $\dfrac{4x^2}{(2-x)(1-x)}$
(c) $\dfrac{(1-x)^3}{2V}$ (d) $\dfrac{x^2}{(2-x)(1-x)}$

4 For $N_2 + 3H_2 \rightleftharpoons 2NH_3$, 1 mole of N_2 and 3 moles of H_2 are at 4 atm. Equilibrium pressure is found to be 3 atm. Hence, K_p is

(a) $\dfrac{1}{(0.5)\,(0.15)^3}$ (b) $\dfrac{1}{(0.5)\,(1.5)^3}$
(c) $\dfrac{3\times3}{(0.5)\,(0.5)^3}$ (d) None of these

5 Reaction between nitrogen and oxygen takes place as following

$$2N_2(g) + O_2(g) \rightleftharpoons 2N_2O\,(g)$$

If a mixture of 0.482 mole of N_2 and 0.933 mole of O_2 is placed in a reaction vessel of volume 10 L and allowed to form N_2O at a temperature for which $K_C = 2.0 \times 10^{-37}$, the equilibrium concentration of $[N_2O]$ will be

(a) 7.06×10^{-20} mol L^{-1} (b) 6.58×10^{-21} mol L^{-1}
(c) 4.82×10^{-4} mol L^{-1} (d) 9.36×10^{-7} mol L^{-1}

6 The values of K_{p1} and K_{p2} for the reactions,

$$X \rightleftharpoons Y + Z \quad ...(i)$$
$$A \rightleftharpoons 2B \quad ...(ii)$$

are in the ratio of 9 : 1. If degree of dissociation of X and A be equal, then total pressure at equilibrium of reactions (I) and (II) are in the ratio.

(a) 3 : 1 (b) 1 : 9 (c) 36 : 1 (d) 1 : 1

7 The following equilibrium constants are given,

$$N_2 + 3H_2 \rightleftharpoons 2NH_3;\ K_1$$
$$N_2 + O_2 \rightleftharpoons 2NO;\ K_2$$
$$H_2 + \frac{1}{2}O_2 \rightleftharpoons H_2O;\ K_3$$

The equilibrium constant for the oxidation of NH_3 by oxygen to given NO is

(a) $\dfrac{K_2K_3^3}{K_1}$ (b) $\dfrac{K_2K_3^2}{K_1}$ (c) $\dfrac{K_2^2K_3^6}{K_1^2}$ (d) $\dfrac{K_1K_2}{K_3}$

8 At 600°C, K_p for the following reaction is 1atm,

$$X(g) \rightleftharpoons Y(g) + Z(g)$$

at equilibrium, 50%, of $X(g)$ is dissociated. The total pressure of the equilibrium system is p atm. What is the partial pressure (in atm) of $X(g)$ at equilibrium?

(a) 1 (b) 4 (c) 2 (d) 0.5

9 N_2 and H_2 in 1 : 3 molar ratio are heated in a closed container having a catalyst. When the following equilibrium, $N_2(g) + 3H_2(g) \rightleftharpoons 2NH_3(g)$ is attained, the total pressure is 10 atm and mole fraction of NH_3 is 0.60. The equilibrium constant K_p for dissociation of NH_3 is

(a) 1.333 atm^{-2} (b) 0.75 atm^2
(c) 0.75 atm^{-2} (d) 1.333 atm^2

10 At 77°C and one atmospheric pressure, N_2O_4 is 70% dissociated into NO_2. What will be the volume occupied by the mixture under these conditions if we start with 10g of N_2O_4?

(a) 6.32 L (b) 5.32 L
(c) 6.38 L (d) 5.38 L

11 Volume of the flask in which species are transferred is double of the earlier flask. In which of the following cases, equilibrium constant is affected?

I. $N_2(g) + 3H_2(g) \rightleftharpoons 2NH_3(g)$
II. $N_2(g) + O_2(g) \rightleftharpoons 2NO(g)$
III. $PCl_5(g) \rightleftharpoons PCl_3(g) + Cl_2(g)$
IV. $2NO(g) \rightleftharpoons N_2(g) + O_2(g)$

(a) Both I and II (b) Both II and III
(c) Both I and III (d) Both III and IV

12 The rate of the elementary reaction, $H_2(g) + I_2(g) \longrightarrow 2HI(g)$ at 25°C is given by

$$\text{rate} = 1.7 \times 10^{-18} [H_2][I_2]$$

The rate of decomposition of gaseous HI to $H_2(g)$ and $I_2(g)$ at 25°C is rate $= 2.4 \times 10^{-21} [HI]^2$. Equilibrium constant for the formation of one mole of gaseous HI from the $H_2(g)$ and $I_2(g)$ is

(a) 708 (b) 354 (c) 0.0014 (d) 26.6

13 In the reaction, $AB(g) \rightleftharpoons A(g) + B(g)$ at 30°C, K_p for the dissociation pressure at equilibrium is 2.56×10^{-2} atm. If the total equilibrium is 1 atm, then the percentage dissociation of AB is

(a) 43.5% (b) 13% (c) 87% (d) 16%

14 The rate of the elementary gaseous phase reaction,

$2NO + O_2 \rightleftharpoons 2NO_2$ at 380°C is given by

$$\text{rate} = 2.6 \times 10^3 [NO_2]^2[O_2]$$

The rate of the reverse reaction at 380°C is given by

$$\text{rate} = 4.1 [NO_2]^2$$

Hence, equilibrium constant for the formation of NO_2 by chemical equation, $NO + \frac{1}{2} O_2 \rightleftharpoons NO_2$ is

(a) 6.34×10^2 (b) 1.577×10^{-3}
(c) 25.18 (d) 3.97×10^{-2}

15 When NH_3 is heated in a 0.50 L flask at 700 K and 100 atm pressure, it decomposes into N_2 and H_2 and their equilibrium moles are given below:

$$\underset{0.30}{2NH_3(g)} \rightleftharpoons \underset{0.30}{N_2(g)} + \underset{0.90}{3H_2(g)}$$

Which of the pairs given represents correct value?

	Initial moles of NH_3	**K_c for NH_3 formation**
(a)	0.80	9.72
(b)	0.90	9.72
(c)	0.80	0.103
(d)	0.90	0.103

ANSWERS

SESSION 1

1 (c)	**2** (c)	**3** (b)	**4** (c)	**5** (c)	**6** (b)	**7** (d)	**8** (a)	**9** (a)	**10** (b)
11 (a)	**12** (d)	**13** (b)	**14** (a)	**15** (a)	**16** (a)	**17** (b)	**18** (d)	**19** (d)	**20** (d)
21 (c)	**22** (c)	**23** (c)	**24** (b)	**25** (b)	**26** (a)	**27** (b)	**28** (b)	**29** (b)	**30** (c)
31 (a)	**32** (c)	**33** (b)	**34** (d)	**35** (c)	**36** (a)	**37** (c)	**38** (c)	**39** (b)	**40** (b)
41 (c)									

SESSION 2

1 (b)	**2** (c)	**3** (a)	**4** (b)	**5** (b)	**6** (c)	**7** (c)	**8** (a)	**9** (b)	**10** (b)
11 (c)	**12** (d)	**13** (d)	**14** (c)	**15** (d)					

Hints and Explanations

SESSION 1

1 $K = k_f / k_b$, $1.5 = k_f/(7.5\times10^{-4})$

or $k_f = 1.125\times10^{-3}$

2 $56\,g\ N_2 = \frac{56}{28} = 2$ mol

$6\,g\ H_2 = \frac{6}{2} = 3$ mol

$27.54\,g\ NH_3 = \frac{27.54}{17} = 1.62$ mol

At equilibrium $N_2 = \frac{1.62}{2} = 0.81$

At equilibrium $H_2 = \frac{1.62}{3} = 0.54$

	N_2	$+\ 3H_2 \rightleftharpoons$	$2NH_3$
Initial moles	2	3	0
At equil.	2 − 0.81 = 1.19	3 − 0.54 = 2.46	1.62

$$K_C = \frac{(1.62)^2}{(1.19)(2.46)^3} = 0.148$$

3 $K_C = \frac{[CO_2]}{[CO]}$ or $5 = \frac{[CO_2]}{2.5\times10^{-2}}$

or $[CO_2] = 1.25\times10^{-1}$ mol L^{-1}

4

	A	$+\ 2B \rightleftharpoons$	$2C$	$+D$
Initial moles	1.1	2.2	0	0
Moles at eq.	1.0	2.0	0.2	0.1

$$\therefore \quad K = \frac{(0.2/V)^2(0.1/V)}{(1/V)\,(2/V)^2} = 0.001$$

5 $H_2CO_3 \rightleftharpoons H^+ + HCO_3^-$; $K_1 = 4.2\times10^{-7}$

$HCO_3^- \rightleftharpoons H^+ + CO_3^{2-}$;

$K_2 = 4.8\times10^{-11}$; Thus, $K_1 >> K_2$

$\therefore \quad [H^+] = [HCO_3^-]$

$$K_2 = \frac{[H^+][CO_3^{2-}]}{[HCO_3^-]}$$

So, $[CO_3^{2-}] = K_2 \Rightarrow 4.8\times10^{-11}$

6 Given, $\Delta G° = 2494.2$ J

For the reaction,

$2A = B + C$

$$Q = \frac{[B][C]}{[A]^2} = \frac{2\times(1/2)}{(1/2)^2} = 4$$

$\therefore$ We know, $\Delta G = \Delta G° + RT\ln Q$

$= 2494.2 + 8.314\times300\ln 4$

$= 5951.90 =$ positive value

Also, we have $\Delta G = RT\ln\frac{Q}{K_C}$

If ΔG is positive, $Q > K_C$.

Therefore, reaction is spontaneous in reverse direction.

7 $2NO_2(g) \rightleftharpoons 2NO(g) + O_2(g)$

$K_C = 1.8\times10^{-6}$ at 184°C (= 457 K)

$R = 0.00831$ kJ mol^{-1} K^{-1}

$K_p = K_C\,(RT)^{\Delta n_g}$

$\Delta n_g = 3 - 2 = 1$

$\therefore \quad K_p = 1.8\times10^{-6}\times0.00831\times457$

$= 7.0637\times10^{-6} > 1.8\times10^{-6}$

Thus, $K_p > K_C$

8 $K_p = K_C(RT)^{\Delta n_g}$

$CO(g) + Cl_2(g) \longrightarrow COCl_2(g)$

$\therefore \quad \Delta n_g = 1 - 2 = -1$

$\therefore \quad K_p = K_C\times(RT)^{-1}$

$$\frac{K_p}{K_C} = (RT)^{-1} = \frac{1}{RT}$$

9 $K_p = K_C(RT)^{\Delta n_g} \Rightarrow \Delta n_g = 1 - 1.5 = -0.5$

$K_p = K_C(RT)^{-1/2}$

$$\therefore \quad \frac{K_C}{K_p} = (RT)^{1/2}$$

10 For the given reaction, $\Delta n_g = n_p - n_R$

where, n_p = number of moles of products

n_R = number of moles of reactants

$K_p = K_C(RT)^{\Delta n_g}$

$$\Delta n_g = 1 - 1.5 = -\frac{1}{2}$$

11 We know that, $K_p = K_C(RT)^{\Delta n_g}$

For the given reaction,

$$\Delta n_g = 1 - \left(1 + \frac{1}{2}\right) = -0.5$$

On putting the value of Δn_g,

$$K_p = K_C(RT)^{-0.5} \Rightarrow K_p = \frac{K_C}{\sqrt{RT}}$$

$$\frac{K_p}{K_C} = \frac{1}{\sqrt{RT}}$$

12 Equilibrium constant depends only upon the temperature, not upon the concentration of reactant. Thus, on doubling concentration, equilibrium constant remains the same.

13 For the reaction,

$H_3PO_4 \rightleftharpoons 3H^+ + PO_4^{3-}$

$K = K_1\cdot K_2\cdot K_3$

14 Reaction (ii) is double and reverse of (i).

Hence, $K_2 = \frac{1}{K_1^2}$ or $K_1^2 = \frac{1}{K_2}$

15 $K_1 = \frac{[CO_2][H_2]}{[CO][H_2O]} \Rightarrow K_2 = \frac{[CO][H_2]^3}{[CH_4][H_2O]}$

$$K_3 = \frac{[CO_2][H_2]^4}{[CH_4][H_2O]^2}$$

Thus, $K_1\times K_2 = K_3$

16 Equilibrium constant for the reaction,

$SO_2(g) + \frac{1}{2}O_2(g) \rightleftharpoons SO_3(g)$

$$K_C = \frac{1}{4.9\times10^{-2}}$$

and for $2SO_2(g) + O_2(g) \rightleftharpoons 2SO_3(g)$

$$K_C = \left(\frac{1}{4.9\times10^{-2}}\right)^2 = \frac{10^4}{(4.9)^2} = 416.493$$

≈ 416

17 Given, $[H_2S] = 0.10$ M

$[HCl] = 0.20$ M So, $[H^+] = 0.20$ M

$H_2S \rightleftharpoons H^+ + HS^-$; $K_1 = 1.0\times10^{-7}$

$HS^- \rightleftharpoons H^+ + S^{2-}$; $K_2 = 1.2\times10^{-13}$

It means for,

$H_2S \rightleftharpoons 2H^+ + S^{2-}$

$K = K_1\times K_2 = 1.0\times10^{-7}\times1.2\times10^{-13}$

$= 1.2\times10^{-20}$

Now, $[S^{2-}] = \frac{K\times[H_2S]}{[H^+]^2}$

[according to the final equation]

$$= \frac{1.2\times10^{-20}\times0.1}{(0.2)^2}$$

$$= \frac{12\times10^{-20}\times1\times10^{-1}\text{ M}}{4\times10^{-2}\text{M}}$$

$= 3\times10^{-20}$M

18 In the required equation, NH_3 is on LHS, so invert the equation (i)

$2NH_3(g) \rightleftharpoons N_2(g) + 3H_2(g)$;

$K_5 = \frac{1}{K_1}$ ...(iv)

Moreover, there are three moles of H_2O, so multiply Eq. (iii) by 3.

$3H_2(g) + \frac{3}{2}O_2(g) \rightleftharpoons 3H_2O(g)$;

$K_6 = K_3^3$ ...(v)

(because when a reaction is multiplied by n, K becomes K^n).

$N_2(g) + O_2(g) \rightleftharpoons 2NO(g)$, K_2

On adding Eqs. (iv), (ii) and (v) we get

$$2NH_3(g) + \frac{5}{2}O_2(g) \rightleftharpoons 2NO(g) + 3H_2O(g);$$

$$K_4 = K_2 \times K_5 \times K_6$$

On putting the values of K_5 and K_6, we get

$$K_4 = K_2 \cdot \frac{1}{K_1} \cdot K_3^3 = \frac{K_2 K_3^3}{K_1}$$

19 $N_2(g) + O_2(g) \rightleftharpoons 2NO(g)$

$$K_C = \frac{[NO]^2}{[N_2][O_2]} = 4 \times 10^{-4}$$

$$2NO(g) \rightleftharpoons N_2(g) + O_2(g)$$

$$K_C' = \frac{1}{K_C} = \frac{[N_2][O_2]}{[NO]^2} = \frac{1}{4 \times 10^{-4}} = \frac{10^4}{4}$$

$$NO(g) \rightleftharpoons \frac{1}{2}N_2(g) + \frac{1}{2}O_2(g)$$

$$K_C'' = \frac{[N_2]^{1/2}[O_2]^{1/2}}{[NO]} = \sqrt{K_C'} = \sqrt{\frac{10^4}{4}} = \frac{100}{2} = 50$$

20

	$N_2O_4 \rightleftharpoons$	$2NO_2$
Initial at moles	1	0
Moles at equili.	$(1-x)$	$2x$

Total moles $= 1 - x + 2x = 1 + x$

$\therefore$ % of NO_2 by volume $= \frac{2x}{1+x} \times 100 = 50$

or $x = \frac{1}{3} = 0.33$

Hence, per cent of dissociation of N_2O_4 33.33%.

21 $I_2(g) \rightleftharpoons 2I(g)$

Partial pressure of iodine atoms (p_I)

$$= \frac{40}{100} \times 10^5 = 0.40 \times 10^5 \text{ Pa}$$

Partial pressure of $I_2 (p_{I_2})$

$$= \frac{60}{100} \times 10^5 \text{ Pa} = 0.60 \times 10^5 \text{ Pa}$$

$$\therefore K_p = \frac{p_I^2}{p_{I_2}} = \frac{(0.40 \times 10^5)^2}{0.60 \times 10^5} = 2.67 \times 10^4$$

22

	$N_2O_4 \rightleftharpoons$	$2NO_2$
Initial moles	a	0
Moles at equili.	$a - 0.2a = 0.8a$	$0.4a$

Total moles $= 1.2a$

$$p_{N_2O_4} = \frac{0.8a}{1.2a} \times 1 \text{ atm} = 2/3 \text{ atm},$$

$$p_{NO_2} = \frac{0.4a}{1.2a} \times 1 \text{ atm} = 1/3 \text{ atm}$$

$$K_p = \frac{(1/3)^2}{(2/3)} = 1/6$$

23

	$SO_2Cl_2 \rightleftharpoons$	$SO_2 +$	Cl_2
Initial	1	0	0
At equil.	$1-\alpha$	α	α

Total moles $= 1 + \alpha$

$$p_{SO_2Cl_2} = \frac{1-\alpha}{1+\alpha}, \quad p_{SO_2} = \frac{\alpha}{1+\alpha},$$

$$p_{Cl_2} = \frac{\alpha}{1+\alpha}$$

$$K_p = \left(\frac{\alpha}{1+\alpha}\right)^2 \bigg/ \left(\frac{1-\alpha}{1+\alpha}\right) = \frac{\alpha^2}{1-\alpha^2}$$

$$= \alpha^2; (1 - \alpha^2 = 1)$$

$$\alpha = \sqrt{K_p} = \sqrt{2.9 \times 10^{-2}} = 0.17$$

$\therefore$ Degree of dissociation = 17%

24

	$PCl_5 \rightleftharpoons$	$PCl_3 +$	Cl_2
Initial moles	1	0	0
Moles at equil.	0.5	0.5	0.5

Total moles = 1.5

Partial pressure

$$p_{PCl_5} = \frac{0.5}{1.5}p; \quad p_{PCl_3} = \frac{0.5}{1.5}p; \quad p_{Cl_2} = \frac{0.5}{1.5}p$$

(where, p is the total pressure)

$$K_p = \left(\frac{0.5}{1.5}p\right)^2 \bigg/ \left(\frac{0.5}{1.5}p\right) = \frac{1}{3}p$$

$$k_p = 1.6 \text{ (Given)}$$

$$\therefore \quad p = 4.8 \text{ atm}$$

25

	$XY_2 \rightleftharpoons$	$XY +$	Y
Initial	600 mm	0	0
At equili.	$600 - p$	p	p

Total pressure

$$= 600 - p + p + p = 600 + p$$

or $600 + p = 800$ mm

$\Rightarrow$ $p = 200$ mm

$$XY_2 = 400 \text{ mm}$$

$$K = \frac{200 \times 200}{400} = 100$$

26

	$CO_2(g) + C(s) \rightleftharpoons$	$2CO(g)$
Initial	0.5 atm	0
At equil .	$(0.5 - p)$	$2p$ atm

This is a case of heterogeneous equilibrium.

C(s) being solid is not considered.

$$p_{CO_2} + p_{CO} = p_{total}$$

$$0.5 - p + 2p = 0.8$$

$$p = 0.3 \text{ atm}$$

$$\therefore \quad p_{CO_2} = 0.5 - 0.3 = 0.2 \text{ atm}$$

$$p_{CO} = 2p = 0.6 \text{ atm}$$

$$K_p = \frac{p_{CO}^2}{p_{CO_2}} = \frac{0.6 \times 0.6}{0.2} = 1.8 \text{ atm}$$

27 At equilibrium, if partial pressure of $CO_2 = p$, and that of $NH_3 = 2p$

$$K_p = p^2_{NH_3} \times p_{CO_2} = (2p)^2 \times p = 4p^3$$

$$K_p = 2.9 \times 10^{-5} = 4p^3$$

$$p^3 = 7.25 \times 10^{-6}$$

$$p = 1.935 \times 10^{-2}$$

Hence, total pressure

$$= 3p \Rightarrow 3 \times 1.935 \times 10^{-2}$$

$$= 5.81 \times 10^{-2} = 0.0581 \text{ atm}$$

28 Let the degree of dissociation $= x$

	$A +$	$2B \rightleftharpoons$	$2C +$	D
Initial conc.	1	1.5	0	0
At equili.	$(1-x)$	$(1.5-2x)$	$2x$	x

Given, $(1 - x) = (1.5 - 2x)$

$$1 = 1.5 - 2x + x$$

$$1 = 1.5 - x$$

$$x = 1.5 - 1 = 0.5$$

Equilibrium constant for the reaction

$$K_C = \frac{[C]^2[D]}{[A][B]^2} = \frac{(2x)^2(x)}{(1-x)(1.5-2x)^2} = 4$$

$(\because x = 0.5)$

29

	$A +$	$B \rightleftharpoons$	$C +$	D
Initially at $t = 0$	1	1	1	1
At equili.	$1-x$	$1-x$	$1+x$	$1+x$

$$K_{eq} = \frac{[C][D]}{[A][B]} = \frac{(1+x)(1+x)}{(1-x)(1-x)} = \frac{(1+x)^2}{(1-x)^2}$$

or $100 = \left(\frac{1+x}{1-x}\right)^2$ or $10 = \frac{1+x}{1-x}$

or $10 - 10x = 1 + x$

$$10 - 1 = x + 10x$$

$$9 = 11x$$

$$x = \frac{9}{11} = 0.818$$

$\therefore [D] = 1 + x = 1 + 0.818 = 1.818$

30 A $\rightarrow$ 4; B $\rightarrow$ 1; C $\rightarrow$ 2

31 In the reaction,

$$N_2(g) + 3H_2(g) \rightleftharpoons 2NH_3(g)$$

If the total pressure at which the equilibrium is established, is increased without changing the temperature, K will remain same. K changes only with change in temperature.

32 Since, $n_p < n_R$ and the reaction is exothermic. So, high pressure and low temperature favour forward reaction. To favour forward reaction, pressure of 1000 atm and temperature of 100°C is required.

33 If CO_2 escapes, equilibrium will shift in the backward direction, so that $[H^+]$ will diminish.

34 K_p is constant at constant temperature. As volume is halved, pressure will be doubled. Hence, equilibrium will shift in the backward direction, i.e. degree of dissociation (α) decreases.

35 $H_2(g)+I_2(g) \rightleftharpoons 2HI(g)$

For this reaction, $\Delta n_g = 0$

$\therefore$ The reaction and its equilibrium constant is not affected by change in volume. Moreover, equilibrium constant depends only on temperature.

36 Reaction is exothermic.By Le-Chatelier's principle, a reaction is spontaneous in forward direction (in the direction of formation of more ClF_3, if F_2 is added, temperature is lowered and ClF_3 is removed.

37 $\Delta n_g = -$ ve ; Reaction takes place with decrease in number of moles or pressure; hence increase in pressure shifts the equilibrium in forward side.

$\Delta H° = -$ ve ; Reaction takes place with evolution of heat or increase in temperature, hence decrease in temperature shifts the equilibrium in forward side.

38 In dissociation of $PCl_5(g) \rightleftharpoons PCl_3(g)+Cl_2(g)$, the addition of inert gast (He) at constant pressure and temperature results increase in the dissociation of PCl_5 in accordance with Le-Chatelier's principle.

39 With increase of pressure, equilibrium shifts in that direction in which lesser number of gaseous moles are produced.

40 Increased temperature always shifts the equilibrium towards endothermic reaction.

41 Catalyst does not affect equilibrium. It increases the rate of chemical reactions by providing a new low energy pathway for the forward and reverse reactions by exactly the same amount.

SESSION 2

1 A very high value of K for the given equilibrium shows that dissociation of glucose to form HCHO is very small. Hence, at equilibrium, we can take $C_6H_{12}O_6 = 1M$

$$K = \frac{C_6H_{12}O_6}{[HCHO]^6}$$

i.e. $$6\times10^{22} = \frac{1}{[HCHO]^6}$$

or $$[HCHO] = \left(\frac{1}{6\times10^{22}}\right)^{\frac{1}{6}} = 1.6\times10^{-4} M$$

2 $M_2CO_3(s) \rightleftharpoons M_2O(s)+CO_2(g)$

$\therefore$ $K_p = p'_{CO_2} = 0.0095$ atm

$\because$ CO_2 is 1% in air,

$$\therefore p'_{CO_2} = \frac{1}{100}\times p_{air} = \frac{1}{100}\times 1 = 0.01$$

3 $$2SO_2 + O_2 \rightleftharpoons 2SO_3$$

	$2SO_2$	O_2	$2SO_3$
Initial	2 mol	1 mol	0 mol
At equil.	$(2-2x)$ $=2(1-x)$	$(1-x)$	$(2x)$
Molar conc.	$2(1-x)/V$	$(1-x)/V$	$2x/V$

$$K_C = \frac{(2x/V)^2}{[2(1-x)/V]^2(1-x)/V} = \frac{x^2V}{(1-x)^3}$$

4 $$N_2 + 3H_2 \rightleftharpoons 2NH_3$$

	N_2	$3H_2$	$2NH_3$
Initial	1	3	0
Equili.	$(1-x)$	$(3-3x)$	$2x$

Initial moles, $n_1 = 4$

Pressure, $p_1 = 4$ atm

At equilibrium moles, $n_2 = 4-2x$

At equilibrium pressure, $p_2 = 3$ atm

$$\frac{n_2}{n_1} = \frac{p_2}{p_1} \Rightarrow \frac{4-2x}{4} = \frac{3}{4}$$

$$4-2x = 3 \text{ or } x = 0.5$$

$$\chi_{N_2} = \frac{1-x}{4-2x} = \left(\frac{0.5}{3}\right)$$

$$\chi_{H_2} = \frac{3-3x}{4-2x} = \frac{1.5}{3}$$

$$\chi_{NH_3} = \frac{2x}{4-2x} = \frac{1}{3}$$

$$p_{N_2} = 3\times\frac{0.5}{3} = 0.5 \text{ atm}$$

$$p_{H_2} = 0.5\times3 = 1.5 \text{ atm}$$

and $$p_{NH_3} = \frac{1}{3}\times3 = 1 \text{ atm}$$

$$K_p = \frac{(p_{NH_3})^2}{p_{N_2}(p_{H_2})^3} = \frac{1}{(0.5)(1.5)^3}$$

5 $$2N_2(g) + O_2(g) \rightleftharpoons 2N_2O(g)$$

	$2N_2(g)$	$O_2(g)$	$2N_2O(g)$
Initial	0.482	0.933	0
At equil.	$(0.482-2x)$	$0.933-x$	$2x$
Active mass	$\left(\frac{0.482-2x}{10}\right)$	$\left(\frac{0.933x}{10}\right)$	$\left(\frac{2x}{10}\right)$

$$K_C = 2.0\times10^{-37}$$

Magnitude of K_C is very very small such that at equilibrium

$$\frac{0.482-2x}{10} \approx \frac{0.482}{10} = 0.0482$$

$\therefore$ $[N_2]_{eq} = 0.0482$ mol L^{-1}

$[O_2]_{eq} = 0.0933$ mol L^{-1},

$$K_C = \frac{[N_2O]^2}{[N_2]^2[O_2]}$$

$$2.0\times10^{-37} = \frac{\frac{4x^2}{100}}{(0.0482)^2\ (0.0933)}$$

$$x = 3.29\times10^{-20}$$

$\therefore$ $$[N_2O] = \frac{2x}{10}$$

$\therefore$ $$[N_2O] = \frac{2\times3.292\times10^{-20}}{10}$$

$$= 6.58\times10^{-21} \text{ mol } L^{-1}$$

6 Suppose total pressure at equilibrium for reactions (i) and (ii) are p_1 and p_2 respectively, then

$$X \rightleftharpoons Y + Z$$

	X	Y	Z
Initial moles	1 mol	0	0
Moles at equil.	$1-\alpha$	α	α

Total moles $= 1-\alpha+2\alpha = 1+\alpha$

$$p_X = \frac{1-\alpha}{1+\alpha}\times p_1$$

$$p_Y = \frac{\alpha}{1+\alpha}\times p_1$$

$$p_Z = \frac{\alpha}{1+\alpha}\times p_1$$

$$K_{p1} = \frac{\left(\frac{\alpha}{1+\alpha}p_1\right)^2}{\frac{1-\alpha}{1+\alpha}\times p_1}$$

$$= \frac{\alpha^2 p_1}{1-\alpha^2} \approx \alpha^2 p_1$$

$$A \rightleftharpoons 2B$$

	A	$2B$
Initially	1 mol	0
At equil.	$1-\alpha$	2α

Total moles $= 1+\alpha$

$$p_A = \frac{1-\alpha}{1+\alpha}\times p_2,$$

$$p_B = \frac{2\alpha}{1+\alpha}p_2$$

$$K_{p2} = \frac{\left(\frac{2\alpha}{1+\alpha} p_2\right)^2}{\frac{1-\alpha}{1+\alpha} p_2} = \frac{4\alpha^2}{1-\alpha^2} p_2 = 4\alpha^2 p_2$$

$$\frac{K_{p1}}{K_{p2}} = \frac{\alpha^2 p_1}{4\alpha^2 p_2} = \frac{p_1}{4p_2} = \frac{9}{1} \text{ (given)}$$

or $\frac{p_1}{p_2} = \frac{36}{1} = 36:1$

7 The required equation for the oxidation of NH_3 by oxygen to given NO is

$$4NH_3 + 5O_2 \xrightarrow[800°C]{\text{Pt (gauze)}} 4NO + 6H_2O$$

For this, $K = \frac{[NO]^4[H_2O]^6}{[NH_3]^4[O_2]^5}$

For the Eq. (i), $K = \frac{[NH_3]^2}{[N_2][H_2]^3}$

For the Eq. (ii), $K_2 = \frac{[NO]^2}{[N_2][O_2]^{1/2}}$

For the eq. (iii) $K_3 = \frac{[H_2O]}{[H_2][O_2]^{1/2}}$

For getting the K, we must do

$$K_1^2 = \frac{[NH_3]^4}{[N_2]^2[H_2]^6},$$

$$K_2^2 = \frac{[NO]^4}{[N_2]^2[O_2]^2},$$

$$K_3^6 = \frac{[H_2O]^6}{[H_2]^6[O_2]^{\frac{6}{2}=3}},$$

$$K = \frac{K_2^2 \times K_3^6}{K_1^2}$$

On substituting the values, we get

$$K = \frac{[NO]^4[H_2O]^6}{[NH_3]^4[O_2]^5}$$

8 $X(g) \rightleftharpoons Y(g) + Z(g)$

	$X(g)$	$Y(g)$	$Z(g)$
Initial	1	0	0
At equil.	0.5	0.5	0.5
Total pressure = 1.5			
∴ Partial pressure	$\frac{0.5p}{1.5}$	$\frac{0.5p}{1.5}$	$\frac{0.5p}{1.5}$

$$K_p = \frac{p_Y \cdot p_Z}{p_X}$$

$$1 = \frac{\frac{p}{3} \times \frac{p}{3}}{\frac{p}{3}} \Rightarrow p = 3 \text{ atm}$$

Partial pressure of $X = \frac{p}{3} = 1$ atm

9 $N_2(g) + 3H_2(g) \rightleftharpoons 2NH_3(g)$

Mole fraction of NH_3 $(\chi_{NH_3}) = 0.6$

Mole fraction of N_2 and $H_2 = 1 - 0.6 = 0.4$

Total number of moles of N_2 and $H_2 = 1 + 3 = 4$

Then, $\chi_{N_2} = \frac{1}{4} \times 0.4 = 0.1$,

$\chi_{H_2} = \frac{3}{4} \times 0.4 = 0.3$

Partial pressure of

$N_2 (p_{N_2}) = 0.1 \times 10 = 1$ atm

Partial pressure of

$H_2 (p_{H_2}) = 0.3 \times 10 = 3$ atm

Partial pressure of

$NH_3 (p_{NH_3}) = 0.6 \times 10 = 6$ atm

Dissociation of ammonia

$$2NH_3 \rightleftharpoons N_2 + 3H_2$$

$$K_p = \frac{p_{N_2} \times p_{H_2}^3}{p_{NH_3}^2} = \frac{1 \text{ atm} \times (3 \text{ atm})^3}{(6 \text{ atm})^2}$$

$$= \frac{27}{36} = 0.75 \text{ atm}^2$$

10 Molar mass of

$N_2O_4 = 28 + 64 \Rightarrow 92$ g mol^{-1}

	N_2O_4	$\rightleftharpoons$ $2NO_2$
Initial moles	$\frac{10}{92}$	0
After dissociation	$\frac{10}{92} - \frac{70}{100} \times \frac{10}{92}$	2×0.076
	$= 0.109 - 0.076$	$= 0.152$
	$= 0.033$	

∴ Total moles after dissociation

$= 0.033 + 0.152 = 0.185$

$T = 77°C = 77 + 273$ K $= 350$ K

$pV = nRT$ or $V = \frac{nRT}{p}$

$$= \frac{0.185 \text{ mol} \times 0.0821 \text{ L atm K}^{-1} \text{mol}^{-1} \times 350 \text{ K}}{1 \text{ atm}}$$

$V = 5.32$ L

11 In case I and III $\Delta n_g \neq 0$, thus, equilibrium constant is affected.

12 $H_2 + I_2 \longrightarrow 2HI$

$$\left(\frac{dx}{dt}\right)_f = K_f[H_2][I_2]$$

$$= 1.7 \times 10^{-18} [H_2][I_2]$$

$$2HI \longrightarrow H_2 + I_2$$

$$\left(\frac{dx}{dt}\right)_b = K_b[HI]^2$$

$$= 2.4 \times 10^{-21} [HI]^2$$

∴ $H_2 + I_2 \rightleftharpoons 2HI$

$$K_C = \frac{K_f}{K_b} = \frac{1.7 \times 10^{-18}}{2.4 \times 10^{-21}}$$

∴ $\frac{1}{2}H_2 + \frac{1}{2}I_2 \rightleftharpoons HI$

$K_C = \sqrt{K_C} = 26.6$

13 $AB(g) \rightleftharpoons A(g) + B(g)$

∴ $K_p = p\frac{\alpha^2}{(1-\alpha^2)} = 2.56 \times 10^{-2}$ atm

$\alpha^2 \times \frac{1 \text{ atm}}{(1-\alpha^2)} = 2.56 \times 10^{-2}$ atm

Assuming $1 - \alpha \approx 1$

$\Rightarrow$ $\alpha = 0.16\%$ or 16%

14 $(\text{Rate})_f = K_f[NO]^2[O_2]$;

$(\text{Rate})_b = K_b[NO_2]^2$

∴ At equilibrium

$$\frac{K_f}{K_b} = \frac{[NO_2]^2}{[NO]^2[O_2]}$$

$$= \frac{2.6 \times 10^3}{4.1} = K_C$$

∴ For $NO + \frac{1}{2}O_2 \rightleftharpoons NO_2$

$K_C' = \sqrt{K_C} = 25.18$

15 $2NH_3 \rightleftharpoons N_2(g) + 3H_2(g)$

	$2NH_3$	$N_2(g)$	$3H_2(g)$
Initial	a	0	0
Equil.	$(a - 2x)$	x	$3x$

$x = 0.30$

∴ $a - 2x = 0.30$

∴ $a = 0.90$

Thus, initial moles of $NH_3 = 0.90$ mol

$$K_C = \frac{[N_2][H_2]^3}{[NH_3]^2}$$

$$= \frac{\left(\frac{0.30}{0.50}\right)\left(\frac{0.90}{0.50}\right)^3}{\left(\frac{0.30}{0.50}\right)^2} = 9.72$$

Thus, K_C for NH_3 formation

$$= \frac{1}{9.72} = 0.103$$

DAY TEN

Ionic Equilibrium

Learning & Revision for the Day

- Weak and Strong Electrolytes
- Degree/Dissociation
- Various concepts of Acids and Bases and their Ionisation
- Ionisation of Water
- pH Scale
- Solubility of Sparingly Soluble Salts
- Buffer Solution

The equilibrium concept also extends to ionic reactions. There it is known as **ionic equilibrium** and applied as equilibrium between dissociated and undissociated form of an ionic reaction of weak acid, weak base, hydrolysis reaction etc.

Weak and Strong Electrolytes

1. **Weak electrolytes** dissociate partially in the solutions and such solutions are poor conductor of electricity. e.g. CH_3COOH, H_3PO_4, H_3BO_3, NH_4OH, HCN etc.
2. **Strong electrolytes** dissociate completely into their ions in solution and such solutions are very good conductor of electricity. e.g. HCl, H_2SO_4, NaOH, KOH, NaCl, KCl etc.

Ionisation of Electrolytes

Separation of an electrolyte into their ions either on fusion or dissolution is called **ionisation** or **dissociation.**

$$NaCl(aq) \longrightarrow Na^+(aq) + Cl^-(aq)$$

(Usually the term dissociation is used for weak electrolyte and ionisation for strong electrolyte).

The solution of weak electrolytes contain ions, which are in equilibrium with unionised molecules.

$$CH_3COOH \rightleftharpoons CH_3COO^- + H^+$$

$$NH_4OH \rightleftharpoons NH_4^+ + OH^-$$

This equilibrium is known as **ionic equilibrium** and is dynamic in nature.

PREP MIRROR
Your Personal Preparation Indicator

- No. of Questions in Exercises (x)—
- No. of Questions Attempted (y)—
- No. of Correct Questions (z)—
  (Without referring Explanations)

- Accuracy Level ($z/y \times 100$)—
- Prep Level ($z/x \times 100$)—

In order to expect good rank in JEE, your Accuracy Level should be above 85 & Prep Level should be above 75.

Degree/Dissociation

- The fraction of total number of moles undergoing ionisation is called degree of ionisation or dissociation (α). Alternately, the fraction of the amount of an electrolyte present in the solution as free ions is called degree of ionisation (α).

$$\alpha = \frac{\text{number of moles of electrolyte dissociated ionized as ions}}{\text{total number of moles of electrolyte dissolved}}$$

- The extent of ionisation depends upon the strength of bond and extent of solution of ions obtained.

  Various factors influencing degree of ionisation/ dissociation are as follows:

  (i) For strong electrolyte, $\alpha = 1$ at normal dilution while for most of the polar covalent compounds, i.e. weak electrolytes, $\alpha <<< 1$.

  (ii) Degree of ionisation of an electrolyte increases with polarity of the solvent.

  (iii) The degree of ionisation of an electrolyte decreases with increase in concentration of the electrolyte.

  (iv) The degree of ionisation rises with raise in temperature of the solution.

  (v) The addition of species possessing a common ion to that of weak electrolyte causes a decrease in the degree of dissociation of weak electrolyte.

Ostwald's Dilution Law

It states that degree of dissociation of weak electrolyte is inversely proportional to the square root of concentration or directly proportional to the square root of volume containing one mole of the solute.

$$\alpha \propto \frac{1}{\sqrt{C}}, \quad \alpha = \sqrt{\frac{K}{C}} = \sqrt{KV}$$

where, α = degree of dissociation
V = volume containing 1 mole of weak electrolyte
K = dissociation constant

NOTE Ostwald's dilution law is applicable only to weak electrolytes because for strong electrolytes,

$$\alpha \simeq 1, \text{ i.e. } K_a \rightarrow \infty$$

Various applications of Ostwald's dilution law are as follows:

- To calculate the value of dissociation constant (K) of the weak acids and weak bases, by determining the degree of dissociation (α) from conductance measurement $\left(\frac{\lambda_V}{\lambda_\infty}\right)$ at any concentration C.
- To calculate the degree of dissociation, α of weak acids and bases by knowing the value of K.
  For weak acid, put $[A^+] = [H^+]$ and $K = K_a$
  Similarly for weak bases,
  $$[B^-] = [OH^-] \quad \text{and} \quad K = K_b$$

- This law holds good only in case of weak electrolytes. The value of K can be calculated only in dilute solutions of weak electrolytes.

Various Concept of Acids and Bases and their Ionisation

An acid is that whose aqueous solution tastes sour, turns blue litmus red, neutralises bases and so on. On the other hand, the aqueous solution of a base tastes bitter, turns red litmus blue, neutralises acid and so on.

1. Arrhenius Theory

According to Arrhenius theory, acids are substances that dissociates in water to give hydrogen ions $H^+(aq)$ and bases are substances that produce hydroxyl ions OH^- (aq).

Limitations of Arrhenius theory are as follows:

(i) Free H^+ and OH^- ions do not exist in water.

(ii) The concept is limited to aqueous solutions only.

(iii) It cannot explain the acidic character of certain substances which don't contain free H^+ ions like, $AlCl_3$, BF_3 and basic character of substances which don't contain free OH^- ions. NH_3, PH_3.

2. Bronsted-Lowry Concept

According to Bronsted-Lowry concept, acids are proton donors while bases are proton acceptors. A conjugate pair of acid and a base differs by a proton.

Conjugate base

$$\underset{\text{Acid}}{HCl} + \underset{\text{Base}}{H_2O} \longrightarrow H_3O^+ + Cl^-$$

Conjugate acid

The substances which behave both as an acid and a base are called **amphiprotic**. e.g. H_2O, HSO_3^-, HS^- etc.

Conjugate base of weak acid is strong or *vice-versa*.

NOTE
- The strength of acid depends upon the nature of solvent. e.g. H_2SO_4, $HClO_4$, HNO_3 and HCl, all have same strength in water due to levelling effect of water.
- In acetic acid solvent, the order of their acidic strength is $HClO_4 > H_2SO_4 > HCl > HNO_3$.

Limitations of Bronsted-Lowry concept are as follows:

(i) The protonic definition cannot be used to explain the reactions occurring in the non-protonic solvents such as $COCl_2, SO_2, N_2O_4$ etc.

(ii) This concept cannot explain the reactions between some acidic oxides (such as CO_2, SO_2, SO_3) and basic oxides (such as CaO, BaO, MgO) which take place even in the absence of the solvent, e.g. $CaO + SO_3 \longrightarrow CaSO_4$

(iii) BF_3, $AlCl_3$ etc., do not have any hydrogen and hence, cannot give a proton but are known to behave as acids.

3. Lewis Concept

According to Lewis concept acids are the substances which accept a pair of electrons to form a coordinate bond and bases are the substances which donate a pair of electrons to form coordinate bond.

Following species can act as Lewis acids are

(i) Molecules in which central atom has incomplete octet e.g. BF_3, $AlCl_3$, $FeCl_3$ etc.

(ii) Molecules in which the central atom is either non-metal cation or metal cation with empty *d*-orbital. (*d*-block elements)
e.g. SiX_4, GeX_4, PX_3, $TiCl_4$, H^+, Ag^+etc.

Lewis bases should satisfy following conditions are

(a) Octet should be complete and central atom should be more electronegative.

(b) Lone pair/pairs should be present, e.g. $\ddot{N}H_3$, $H_2\ddot{\underset{\cdot\cdot}{O}}$, $R—\ddot{\underset{\cdot\cdot}{O}}H$, $R—\ddot{\underset{\cdot\cdot}{O}}R$ etc.

(c) Negatively charged species, e.g. CN^-, OH^-, Cl^- etc.

Limitations of Lewis concept are as follows:

(i) The strength of Lewis acids and bases is found to depend on the type of reaction, it is not possible to arrange them in any order of their relative strength.

(ii) It does not explain the behaviour of protonic acids which do not form coordinate bond such as HNO_3, HCl, H_2SO_4 etc.

(iii) Catalytic activity of Lewis acid cannot be explained because the catalytic activity of many acids is due to their tendency to furnish H^+. Lewis acid does not do so.

Relative Strength of Mono Acidic Bases

Relative strength of mono acidic bases (or mono basic acids) of equimolar concentrations can be given as

$$\frac{\text{Strength of base }(BOH)_1}{\text{Strength of base }(BOH)_2} = \sqrt{\frac{K_{b_1}}{K_{b_2}}} = \frac{\alpha_1}{\alpha_2}$$

$$\frac{\text{Strength of acid }(HA)_1}{\text{Strength of acid }(HA)_2} = \sqrt{\frac{K_{a_1}}{K_{a_2}}} = \frac{\alpha_1}{\alpha_2}$$

where, K_a and K_b are the dissociation constants of acid and base respectively.

Ionisation of Water

- Pure water is a weak electrolyte and is ionised according to following equation.

$$H_2O(l) + H_2O(l) \rightleftharpoons H_3O^+(aq) + OH^-(aq)$$

At 25°C, for pure water $[H_3O^+] = [OH^-] = 10^{-7}$ mol/L

$$K_w = [H_3O^+][OH^-] = 10^{-14}$$

Here, K_w is called the **ionic product** of water and is defined as the product of molar concentration of H^+ ions and OH^- ions

$$K = \frac{K_w}{55.55}$$

where, K = ionisation constant

- Value of K_w depends upon temperature as it is equilibrium constant. If temperature increases, value of K_w also increases.
- For dissociation of weak acids in water, K_a is called **acid ionisation constant.**

$$CH_3COOH + H_2O \rightleftharpoons CH_3COO^- + H_3O^+ \;;$$

$$K_a = \frac{[CH_3COO^-]\,[H_3O^+]}{(CH_3COOH)}$$

- Similarly, for a weak base, K_b is called base ionisation constant.

$$NH_4OH \rightleftharpoons NH_4^+ + OH^-;\; K_b = \frac{[NH_4^+][OH^-]}{(NH_4OH)}$$

Relation between K_a and K_b

- Relation between acid dissociation (K_a) and base dissociation (K_b) constants is

$$K_a \times K_b = K_w$$

where, $pK_a = -\log K_a$, $pK_b = -\log K_b$ and $pK_w = -\log K_w$

- Many acids are capable of furnishing more than one protons in water. Such acids are called **polybasic** or **polyprotic** acids, e.g. H_2SO_4, H_3PO_4 etc.
- The first ionisation constant (K_{a_1}) is always greater than the second ionisation constant (K_{a_1}).

pH Scale

It is used to express and compare the acidic and basic strength of a solution. pH is defined as the negative logarithm of H_3O^+ ion concentration (in moles per litre) present in it.

Thus, $\quad pH = -\log[H_3O^+]$

Similarly, $\quad pOH = -\log[OH^-]$

$$pH + pOH = 14$$

Relation between pK_a and pK_b is given as:

$$pK_a + pK_b = 14 = pK_w$$

pH scale range is 0 to 14 and it depends upon the value of K_w. As temperature increases, value of pH decreases at 25°C.

Following observations are seen with solutions:

(i) pH of very dilute ($\sim 10^{-8}$ M or lower) acids or bases is nearly 7 but not 7 (i.e. not simply – log [acid or base]) due to ionisation of water.

(ii) pH of strong acids with concentration > 1 M is never negative, it is zero only.

NOTE Greater the value of K_a or K_b smaller is the value of pK_a and pK_b and stronger is acid or base.

pH of Mixtures of Acid and Bases

The rules for determining the pH of mixtures of acids and bases are as follows:

(i) If strong acid or strong base remains unused, calculate the concentration or molarity of H^+ ions and OH^- ions left in the solution and then calculate the pH or pOH accordingly.

(ii) If weak acid or weak base is left behind or remains unused, a buffer (acidic or basic) is formed. Calculate the concentration of salt formed (millimoles of salt formed/volume of solution) and the concentration of weak acid or weak base left behind. Use the buffer equation to calculate the pH of the solution.

(iii) If acids or bases are completely neutralised, then salt is formed. Calculate the concentration of the salt formed and use the hydrolysis equation to calculate the pH of the solution.

NOTE
- pH value of a solution decreases on heating because ionisation of water is an endothermic process. pH of boiling water is 6.5625, although it is neutral.
- When pH decreases by one unit, H^+ ion concentration increases by a factor of 10. Similarly, when pH decreases by two units, H^+ ion concentration increases by a factor of 100.

Hydrolysis of Salts and pH of their Solutions

- The process of salt hydrolysis is actually the reverse process of neutralisation. The reaction of an anion or cation of the salt with water accompanied to produce acidic and basic solution is called **salt hydrolysis**. Salt hydrolysis affects the pH of the solution.
- **Salts of strong acids and strong bases** (i.e. neutral salts) do not undergo hydrolysis are called **neutral salt**, e.g. NaCl, $CaSO_4$ etc. If such salt is dissolved in water, pH of the solution remains 7.
- **Salt of a strong acid and weak base**, e.g. NH_4Cl are called **acidic salts**. Such salts undergo cationic hydrolysis. pH of acidic salt solution will be less than 7.

  For such salts, $[H_3O^+] = \sqrt{K_h \times C}$

  $$K_h = \frac{K_w}{K_b} \quad \text{or} \quad pH = 7 - \frac{1}{2}[pK_b + \log C] \text{ at } 25°C$$

  where, K_h = hydrolysis constant
  K_b = ionisation constant for weak base
  C = molar concentration of salt.
- **Salt of strong base and weak acid**, e.g. $NaNO_2$, NaCN, CH_3COONa are termed as **basic salts**. Such salts undergo anionic hydrolysis. pH of basic salt solution will be more than 7.

  For basic salts, $[OH^-] = \sqrt{K_h \times C}$ or $K_h = \frac{K_w}{K_a}$

  $$pH = 7 + \frac{1}{2}[pK_a + \log C] \text{ at } 25°C$$
- The salts other than halides, sulphates, nitrates of metals fall into salts of weak acid and base category, e.g. CH_3COONH_4 etc.

  For such salts, $K_h = \frac{K_w}{K_a \times K_b}$, $pH = 7 + \frac{1}{2}[pK_a - pK_b]$

NOTE There is no effect of dilution on the hydrolysis of salts of weak acid and weak base because pH and K_h are all independent of concentration, C.

Solubility of Sparingly Soluble Salts

Lattice enthalpy and solvation enthalpy play an important role in deviding the solubility of the salts in particular solvent. For a salt to be able to dissolve in a particular solvent, the solvation enthalpy must be higher than its lattice enthalpy.

Solubility Product

It is defined as the product of molar concentration of its ions in a saturated solution, each concentration terms raised to the power equal to the number of ions produced on dissociation of one molecule of electrolyte.

$$A_xB_y \rightleftharpoons xA^+ + yB^-$$

$$\therefore \quad K_{sp} = [A^+]^x [B^-]^y$$

e.g. $$A_2X_3 \longrightarrow \underset{2S}{2A^{3+}} + \underset{3S}{3X^{2-}}$$

$$K_{sp} = [A^{3+}]^2[X^{2-}]^3$$

$$K_{sp} = (2S)^2(3S)^3$$

$$\therefore \quad K_{sp} = 108S^5$$

$$BaSO_4(s) \xrightleftharpoons[\text{in water}]{\text{Saturated solution}} \underset{S}{Ba^{2+}}(aq) + \underset{S}{SO_4^-}(aq)$$

$$K_{sp} = [S][S]$$

$$K_{sp} = S^2$$

where, S = molar solubility, K_{sp} = solubility product

NOTE If the ionic product exceeds the value of the solubility product of a sparingly soluble salt then precipitation will occur.

Common Ion Effect

- It states that if to the solution of a weak electrolyte, a solution of strong electrolyte is added which furnishes an ion common to that furnished by the weak electrolyte, the ionisation of the weak electrolyte is suppressed.

  e.g. $NH_4OH \rightleftharpoons NH_4^+ + OH^-$
- If NH_4Cl or NaOH is added to NH_4OH solution, the above equilibrium will shift to the left due to high concentration of common ion and therefore, the ionisation of NH_4OH is further suppressed.
- In IInd group of qualitative analysis, H_2S is passed in the presence of HCl. This is due to the fact that HCl suppresses the ionisation of weakly dissociated H_2S. Due to this only

sulphides of II group radicals are precipitated. Sulphides of III, IV etc., groups are not precipitated because of their high solubility product.

- The common ion effect is also used for almost complete precipitation of a particular ion as its sparingly soluble salt, with very low value of solubility product for gravimetric estimation.

Buffer Solution

The solution, which maintains its pH constant or reserve acidic or basic nature even upon addition of small amounts of acid or base is called **buffer solution**. The ability of buffer solution to resist changes in pH on addition of acid or base is called **buffer action**.

A buffer solution should exhibit following characteristics:

(i) Buffer solutions possess a definite pH value.

(ii) Their pH value remains constant on keeping long or dilution.

(iii) The pH value is not changed on the addition of a strong acid in acidic buffer and a strong base in basic buffer.

Two types of buffer are as follows:

(a) A buffer solution pH of which is less than 7 is called **acidic buffer**. Weak acid with its salt of strong base gives acidic buffer.
e.g. $CH_3COOH + CH_3COONa$; HCN + NaCN

(b) A buffer solution having pH more than 7 is called **basic buffer**. Weak base with its salt of strong acid gives basic buffer.
e.g. $NH_4OH + NH_4Cl$, $C_6H_5NH_2 + C_6H_5NH_3^+Cl^-$

Henderson's Equation for Buffer Solution

For acidic buffer; $\text{pH} = \text{p}K_a + \log \frac{[\text{salt}]}{[\text{acid}]}$

For basic buffer; $\text{pOH} = \text{p}K_b + \log \frac{[\text{salt}]}{[\text{base}]}$

Buffer Capacity

Buffer capacity is quantitatively defined as the number of moles of acid or base added to 1 L in of buffer solution to change the pH by unity.

$$\text{Buffer capacity} = \frac{\text{number of moles of acid / base added to 1 L of buffer}}{\text{change in pH}}$$

Buffer capacity is maximum when

(a) $[\text{salt}] = [\text{acid}]$, $\text{pH} = \text{p}K_a$ for acidic buffer

(b) $[\text{salt}] = [\text{base}]$, $\text{pH} = \text{p}K_b$ for basic buffer

Greater the buffer capacity, larger is its capacity to resist the change in pH value.

DAY PRACTICE SESSION 1

FOUNDATION QUESTIONS EXERCISE

1 Of the given anions, the strongest Bronsted base is
(a) ClO^- (b) ClO_3^- (c) ClO_2^- (d) ClO_4^-

2 Four species are listed below:
I. HCO^{3-} II. H_3O^+ III. HSO_4^- IV. HSO_3F
Which one of the following is the correct sequence of their acidic strength?
(a) IV<II<III<I (b) II<III<I<IV
(c) I<III< II<IV (d) III<I<IV<II

3 Three reactions involving $H_2PO_4^-$ are given below:
I. $H_3PO_4 + H_2O \longrightarrow H_3O^+ + H_2PO_4^-$
II. $H_2PO_4^- + H_2O \longrightarrow HPO_4^{2-} + H_3O^+$
III. $H_2PO_4^- + OH^- \longrightarrow H_3PO_4 + O^{2-}$
In which of the above does $H_2PO_4^-$ act as an acid?
→ AIEEE 2010
(a) Only II (b) Both I and II
(c) Only III (d) Only I

4 The pK_w of a neutral solution at 50°C is 13.36, what would be the pH of the solution at this temperature?
(a) 7.63 (b) 7.00 (c) 6.68 (d) 7.96

5 The ionic product of water at 310 K is 2.7×10^{-14}. What is the pH of neutral water at this temperature?
(a) 7.0 (b) 5.98 (c) 6.78 (d) 4.58

6 Dissociation constant of a weak acid is 1×10^{-4}. Equilibrium constant of its reaction with strong base is
(a) 1×10^{-4} (b) 1×10^{10} (c) 1×10^{-10} (d) 1×10^{4}

7 The pH of the neutralisation point of 0.1 N ammonium hydroxide with 0.1 N HCl is
(a) 8 (b) 6 (c) 7 (d) 9

8 NaOH is a strong base. What will be pH of 5.0×10^{-2}M NaOH solution ? (log 2 = 0.3) → JEE Main (Online) 2013
(a) 14.00 (b) 13.70 (c) 13.00 (d) 12.70

9 What is the pH of a 10^{-4}M OH^- solution at 330 K, if K_w at 330 K is $10^{-13.6}$? → JEE Main (Online) 2013
(a) 4 (b) 9.0 (c) 10 (d) 9.6

10 How many litres of water must be added to 1 L of an aqueous solution of HCl with a pH of 1 to create an aqueous solution with pH of 2? → JEE Main (Online) 2013
(a) 0.1 L (b) 0.9 L (c) 2.0 L (d) 9.0 L

11 The pH of a 0.1 molar solution of the acid HQ is 3. The value of the ionisation constant, K_a of the acid is → AIEEE 2012

(a) 3×10^{-1} (b) 1×10^{-3} (c) 1×10^{-5} (d) 1×10^{-7}

12 Values of dissociation constant, K_a are given as follows:

Acid	K_a
HCN	6.2×10^{-10}
HF	7.2×10^{-4}
HNO_2	4.0×10^{-4}

Correct order of increasing base strength of the base CN^-, F^- and NO_2^- will be → JEE Main (Online) 2013

(a) $F^- < CN^- < NO_2^-$ (b) $NO_2^- < CN^- < F^-$
(c) $F^- < NO_2^- < CN^-$ (d) $NO_2^- < F^- < CN^-$

13 An acid HA ionises as $HA \rightleftharpoons H^+ + A^-$

The pH of 1.0 M solution is 5. Its dissociation constant would be → AIEEE 2011

(a) 1×10^{-10} (b) 5.0 (c) 5×10^{-8} (d) 1×10^{-5}

14 What will be the degree of ionisation of 0.05 M acetic acid if its pK_a value is 4.74 ?

(a) 0.019% (b) 1.9% (c) 3.0% (d) 4.74%

15 HX is a weak acid ($K_a = 10^{-5}$). It forms a salt NaX(0.1 M) on reacting with caustic soda. The degree of hydrolysis of NaX is

(a) 0.01% (b) 0.0001% (c) 0.1% (d) 0.5%

16 pK_a of a weak acid (HA) and pK_b of a weak base (BOH) are 3.2 and 3.4 respectively. The pH of their salt (AB) solution is → JEE Main 2017

(a) 7.2 (b) 6.9 (c) 7.0 (d) 1.0

17 Number of $(OH)^-$ in 1 mL solution of pH = 13 is

(a) 1×10^{-13} (b) 6.00×10^{7}
(c) 6.00×10^{13} (d) 6.02×10^{19}

18 pH of 0.005 M calcium acetate (pK_a of CH_3COOH = 4.7) is

(a) 7.04 (b) 9.37 (c) 9.26 (d) 8.37

19 The pK_a of a weak acid, HA is 4.80. The pK_b of a weak base BOH is 4.78. The pH of an aqueous solution of the corresponding salt BA will be

(a) 7.01 (b) 9.22 (c) 9.58 (d) 4.79

20. The solubility of a sparingly soluble salt, AB_2 in water is 1.0×10^{-5} mol L^{-1}. Its solubility product will be

(a) 4×10^{-10} (b) 1×10^{-15}
(c) 1×10^{-10} (d) 4×10^{-15}

21 The solubility of $PbCl_2$ is

(a) $\sqrt{K_{sp}}$ (b) $(K_{sp})^{1/3}$
(c) $(K_{sp}/4)^{1/3}$ (d) $(8\,K_{sp})^{1/2}$

22 A solution which is 10^{-3} M each in Mn^{2+}, Fe^{2+}, Zn^{2+} and Hg^{2+} is treated with 10^{-16} M sulphide ion. If K_{sp} of MnS, FeS, ZnS and HgS are 10^{-15}, 10^{-23}, 10^{-20} and 10^{-54} respectively, which one will precipitate first?

(a) FeS (b) MgS (c) HgS (d) ZnS

23 Solubility product constant (K_{sp}) of salts of types MX, MX_2 and M_3X at temperature T are 4.0×10^{-8}, 3.2×10^{-14} and 2.7×10^{-15} respectively. Solubilities (mol dm^{-3}) of the salts at temperature 'T' are in the order

(a) $MX > MX_2 > M_3X$ (b) $M_3X > MX_2 > MX$
(c) $MX_2 > M_3X > MX$ (d) $MX > M_3X > MX_2$

24 The solubility product of $BaSO_4$ is 1.5×10^{-9}. The precipitation in a 0.01 M Ba^{2+} solution will start on adding H_2SO_4 of concentration.

(a) 10^{-9} M (b) 10^{-8} M (c) 10^{-7} M (d) 10^{-6} M

25 Solid $Ba(NO_3)_2$ is gradually dissolved in a 1.0×10^{-4} M Na_2CO_3 solution. At which concentration of Ba^{2+}, precipitate of $BaCO_3$ begins to form?
(K_{sp} for $BaCO_3 = 5.1 \times 10^{-9}$) → AIEEE 2012

(a) 5.1×10^{-5} M (b) 7.1×10^{-5} M
(c) 4.1×10^{-5} M (d) 8.1×10^{-2} M

26 The molar solubility (in mol L^{-1}) of a sparingly soluble salt MX_4 is S. The corresponding solubility product is K_{sp}. S is given in terms of K_{sp} by the relation.

(a) $S = (K_{sp}/128)^{1/4}$ (b) $S = (128\,K_{sp})^{1/4}$
(c) $S = (256K_{sp})^{1/5}$ (d) $S = (K_{sp}/256)^{1/5}$

27 Solubility product of silver bromide is 5.0×10^{-13}. The quantity of potassium bromide (molar mass taken as 120 g mol^{-1}) to be added to 1 L of 0.05 M solution of silver nitrate to start the precipitation of AgBr is → AIEEE 2010

(a) 1.2×10^{-10} g (b) 1.2×10^{-9} g
(c) 6.2×10^{-5} g (d) 5.0×10^{-8} g

28 At 25°C, the solubility product of $Mg(OH)_2$ is 1.0×10^{-11}. At which pH, will Mg^{2+} ions start precipitating in the form of $Mg(OH)_2$ from a solution of 0.001 M Mg^{2+} ions? → AIEEE 2010

(a) 9 (b) 10 (c) 11 (d) 8

29 The precipitate of CaF_2 ($K_{sp} = 1.7 \times 10^{-10}$) is obtained when equal volumes of the following are mixed

(a) 10^{-4} M Ca^{2+} + 10^{-4} M F^- (b) 10^{-2} M Ca^{2+} + 10^{-3} M F^-
(c) Both (a) and (b) (d) None of these

30 The K_{sp} for $Cr(OH)_3$ is 1.6×10^{-30}. The molar solubility of this compound in water is → AIEEE 2012

(a) $\sqrt[2]{1.6 \times 10^{-30}}$ (b) $\sqrt[4]{1.6 \times 10^{-30}}$
(c) $\sqrt[4]{1.6 \times 10^{-30}/27}$ (d) $1.6 \times 10^{-30}/27$

31 Which one of the following arrangements represents the correct order of solubilities of sparingly soluble salts Hg_2Cl_2, $Cr_2(SO_4)_3$, $BaSO_4$ and $CrCl_3$ respectively? → JEE Main (Online) 2013

(a) $\left(\frac{K_{sp}}{108}\right)^{1/5}, \left(\frac{K_{sp}}{27}\right)^{1/4}, (K_{sp})^{1/2}, \left(\frac{K_{sp}}{4}\right)^{1/3}$

(b) $(K_{sp})^{1/2}, \left(\frac{K_{sp}}{4}\right)^{1/3}, \left(\frac{K_{sp}}{27}\right)^{1/4}, \left(\frac{K_{sp}}{108}\right)^{1/5}$

(c) $(K_{sp})^{1/2}, \left(\frac{K_{sp}}{108}\right)^{1/5}, \left(\frac{K_{sp}}{27}\right)^{1/4}, \left(\frac{K_{sp}}{4}\right)^{1/3}$

(d) $\left(\frac{K_{sp}}{4}\right)^{1/3}, \left(\frac{K_{sp}}{108}\right)^{1/5}, (K_{sp})^{1/2}, \left(\frac{K_{sp}}{27}\right)^{1/4}$

32 The solubility of $AgBrO_3$ in an aqueous solution of $NaBrO_3$ (as compared to that in water) is
(a) the same
(b) more
(c) less
(d) unpredictable due to a new chemical reaction

33 The solubility of AgI in NaI solution is less than that in pure water because
(a) AgI forms complex with NaI
(b) of common ion effect
(c) solubility product of AgI is less
(d) the temperature of the solution decreases

34 How do we differentiate between Fe^{3+} and Cr^{3+} in group III?
(a) By taking excess of NH_4OH
(b) By increasing NH_4^+ ion concentration
(c) By decreasing OH^- ion concentration
(d) Both (b) and (c)

35 0.1 mole of CH_3NH_2 ($K_b = 5 \times 10^{-4}$) is mixed with 0.08 mole of HCl and diluted to one litre. What will be the H^+ concentration in the solution?
(a) 8×10^{-2} M (b) 8×10^{-11} M
(c) 1.6×10^{-11} M (d) 8×10^{-5} M

36 Which of the following will produce a buffer solution when mixed in equal volumes?
(a) 0.1 mol dm^{-3} NH_4OH and 0.1mol dm^{-3} HCl
(b) 0.05 mol dm^{-3} NH_4OH and 0.1 mol dm^{-3} HCl
(c) 0.1 mol dm^{-3} NH_4OH and 0.05 mol dm^{-3} HCl
(d) 0.1 mol dm^{-3} CH_4COONa and 0.1 mol dm^{-3} NaOH

37 What would be the pH of a solution obtained by mixing 5g of acetic acid and 7.5 g of sodium acetate and making the volume equal to 500 mL?
($K_a = 1.75 \times 10^{-5}, pK_a = 4.76$) → JEE Main (Online) 2013
(a) pH = 4.70
(b) pH < 4.70
(c) pH of solution will be equal to pH of acetic acid
(d) 4.76 < pH < 5.0

38 The pK_a of a weak acid (HA) is 4.5. The pOH of an aqueous buffered solution of HA in which 50% of the acid ionised is
(a) 4.5 (b) 2.5 (c) 9.5 (d) 7.0

39 An aqueous solution of 1 M NaCl and 1 M HCl is
(a) not a buffer but pH < 7 (b) not a buffer but pH > 7
(c) a buffer with pH < 7 (d) a buffer with pH > 7

Direction (Q. Nos. 40-44) *In the following questions assertion followed reason is given. Choose the correct answer out of the following choices.*
(a) Both Assertion and Reason are true and Reason is the correct explanation of Assertion.
(b) Both Assertion and Reason are true but Reason is not the correct explanation of Assertion.
(c) Assertion is true but Reason is false.
(d) Assertion and Reason both are wrong.

40 **Assertion** (A) When HCl gas passes through saturated solution of NaCl, a solid NaCl separates out from the solution.
Reason (R) HCl decreases the solubility product of NaCl.

41 **Assertion** (A) The addition of Ag^+ ions to a mixture of an aqueous solution of NaCl and NaBr causes precipitation of AgBr, instead of AgCl.
Reason (R) Solubility product of AgCl is less than that of AgBr.

42 **Assertion** (A) The ionisation of hydrogen sulphide in water is low in the presence of hydrochloric acid.
Reason (R) Hydrogen sulphide is a weak acid.

43 **Assertion** (A) According to the principle of common ion effect, the solubility of HgI_2 is expected to be less in aqueous solution of KI than in water. But HgI_2 dissolves in an aqueous solution of KI to form a clear solution.
Reason (R) Iodide ion (I^-) is highly polarisable.

44 **Assertion** (A) A solution containing a mixture of acetic acid and sodium acetate maintains a constant value of pH on addition of small amounts of acid or alkali.
Reason (R) A solution containing a mixture of acetic acid and sodium acetate acts as buffer solution around pH 4.75.

DAY PRACTICE SESSION 2

PROGRESSIVE QUESTIONS EXERCISE

1. The pH values of 0.1 M solution of HCl, CH_3COOH, NH_4Cl and CH_3COONa will have the order
(a) $HCl < CH_3COOH < \overset{-}{N}H_4 \overset{+}{C}l < CH_3CO\overset{-}{O}\overset{+}{N}a$
(b) $CH_3COONa < NH_4Cl < CH_3COOH < HCl$
(c) $NH_4Cl < CH_3COONa < CH_3COOH < HCl$
(d) All will have same pH value

2 Which one of the following statements is not true?
(a) The conjugate base of $H_2PO_4^-$ is HPO_4^{2-}
(b) pH + pOH = 14 for all aqueous solutions
(c) The pH of 1×10^{-8} M HCl is 8
(d) $\alpha = \sqrt{K_v}$ for a weak electrolyte

3 What will be the value of pH of 0.01 mol dm^{-3} CH_3COOH? ($K_a = 1.74 \times 10^{-5}$) [log 4.17 = 0.6201]
(a) 3.4 (b) 3.6 (c) 3.9 (d) 3.0

4 A weak acid (*HA*) is 4% ionised at 1 M. Per cent ionisation is 1% in the presence of A^-. The concentration of A^- is
(a) 0.04 M (b) 0.16 M (c) 0.02 M (d) 0.10 M

5 An aqueous solution contains an unknown concentration of Ba^{2+}. When 50 mL of a 1 M solution of Na_2SO_4 is added, $BaSO_4$ just begins to precipitate. The final volume is 500 mL. The solubility product of $BaSO_4$ is 1×10^{-10}. What is the original concentration of Ba^{2+}? → JEE Main 2018
(a) 5×10^{-9} M (b) 2×10^{-9} M
(c) 1.1×10^{-9} M (d) 1.0×10^{-10} M

6 Hydroxyl ion concentration $[OH^-]$ in the case of sodium acetate can be expressed as (where, K_a is dissociation constant of CH_3COOH and *C* is the concentration of sodium acetate)
(a) $[OH^-] = (CK_w\, K_a)^{1/2}$ (b) $[OH^-] = C \cdot K_w \sqrt{K_a}$
(c) $[OH^-] = \left(\frac{C \cdot K_w}{K_a}\right)^{1/2}$ (d) $[OH^-] = C \cdot K_a \cdot K_w$

7 The silver ion in a solution is precipitated by addition of chloride ion. The final volume of solution is 500 mL. What should be the concentration of Cl^- ion if not more than 0.10 mg of Ag^+ ion remains unprecipitated? ($K_{sp}(AgCl) = 1.0 \times 10^{-10}$)
(a) 5.4×10^{-5} M (b) 1.9×10^{-6} M
(c) 5.4×10^{-4} M (d) 1.9×10^{-7} M

8 The degree of ionisation of 1.0 M weak acid, *HA* is 0.5%. If 2 mL of 1.0 M *HA* solution is diluted to 32 mL, the degree of ionisation of the acid and H_3O^+ ion concentration in the resulting solution will be respectively.
(a) 0.02 and 3.125×10^{-4} (b) 0.02 and 1.25×10^{-3}
(c) 1.25×10^{-3} and 0.02 (d) 0.02 and 8.0×10^{-12}

9 1 dm^3 solution containing 10^{-5} moles each of Cl^- ions and CrO_4^{2-} ions is treated with 10^{-4} moles of silver nitrate. Which one of the following observation is made? ($K_{sp} Ag_2CrO_4 = 4 \times 10^{-12}$), ($K_{sp}$ AgCl $= 1 \times 10^{-10}$)
(a) Precipitation does not occur
(b) Silver chromate gets precipitated first
(c) Silver chloride gets precipitated first
(d) Both silver chromate and silver chloride start precipitating simultaneously

10 In a saturated solution of the sparingly soluble strong electrolyte $AgIO_3$ (molecular mass = 283), the equilibrium sets as $AgIO_3(s) \rightleftharpoons Ag^+(aq) + IO_3^-(aq)$. If the solubility product constant, K_{sp} of $AgIO_3$ at a given temperature is 1.0×10^{-8}, what is the mass of $AgIO_3$ contained in 100 mL of its saturated solution?
(a) 28.3×10^{-2} g (b) 2.83×10^{-3} g
(c) 1.0×10^{-7} g (d) 1.0×10^{-4} g

11 A solution of monoprotic weak acid has ionisation constant K_a. What is the minimum concentration *C* in terms of K_a, such that the concentration of the undissociated acid can be equated to *C* within a 10% limit of error. [Assume that activity coefficient correction are negligible.]
(a) $45K_a$ (b) $10K_a$
(c) $90K_a$ (d) $80K_a$

12 What will be the pH of a buffer solution prepared by dissolving 30 g of Na_2CO_3 in 500 mL of an aqueous solution containing 150 mL of 1 M HCl?
[K_a for $HCO_3^- = 5.63 \times 10^{-11}$]
(a) 10.197 (b) 8.089
(c) 9.858 (d) 6.400

13 2.5 mL of 2/5 M weak monoacidic base ($K_b = 1 \times 10^{-12}$ at 25°C) is titrated with 2/15 M HCl in water at 25°C. The concentration of H^+ at equivalence point is ($K_w = 1 \times 10^{-14}$ at 25°C)
(a) 3.7×10^{-13}M (b) 3.2×10^{-7}M
(c) 3.2×10^{-2}M (d) 2.7×10^{-2}M

14 1 M NH_4OH and 1M HCl are mixed to make total volume of 300 mL. If pH of the mixture is 9.26 and pK_a $[NH_4^+] = 9.26$ then volume ratio of NH_4OH and HCl will be
(a) 1 : 1 (b) 1 : 2
(c) 2 : 1 (d) 3 : 1

ANSWERS

SESSION 1	1 (a)	2 (c)	3 (a)	4 (c)	5 (c)	6 (b)	7 (b)	8 (d)	9 (d)	10 (d)
	11 (c)	12 (c)	13 (a)	14 (b)	15 (a)	16 (b)	17 (d)	18 (d)	19 (a)	20 (d)
	21 (c)	22 (c)	23 (d)	24 (d)	25 (a)	26 (d)	27 (b)	28 (b)	29 (b)	30 (c)
	31 (d)	32 (c)	33 (b)	34 (d)	35 (b)	36 (c)	37 (d)	38 (c)	39 (a)	40 (b)
	41 (c)	42 (b)	43 (b)	44 (a)						
SESSION 2	1 (a)	2 (c)	3 (a)	4 (b)	5 (c)	6 (c)	7 (a)	8 (b)	9 (c)	10 (b)
	11 (c)	12 (a)	13 (c)	14 (c)						

Hints and Explanations

SESSION 1

1 HClO is the weakest acid, hence its conjugate base, i.e. ClO^- is the strongest Bronsted base.

2 Acidity order is as follows:

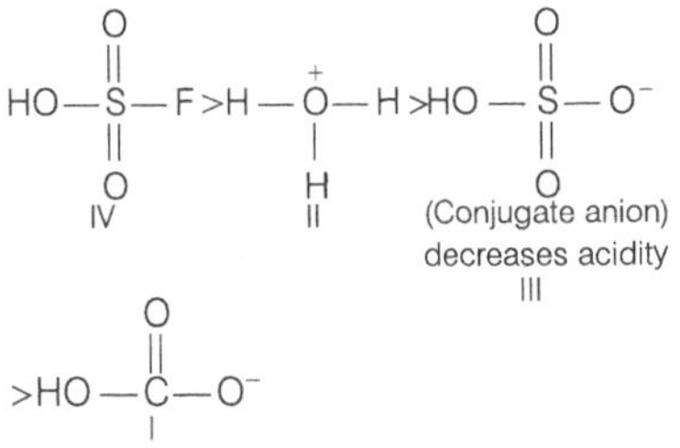

3 Only in reaction (ii) $H_2PO_4^-$, gives H^+ to H_2O thus, behaves as an acid. In other two reactions it accept H^+ and thus behaves as a base.

4 $pK_w = pH + pOH$

But $[H^+] = [OH^-] \Rightarrow pH = pOH$,

$\therefore \quad pK_w = 2\,pH$

$$pH = \frac{pK_w}{2} = \frac{13.36}{2} = 6.68$$

5 $K_w = [H^+][OH^-] = 2.7 \times 10^{-14}$

But $[H^+] = [OH^-]$, therefore,

$$K_w = [H^+]^2$$

$$[H^+] = \sqrt{K_w} = \sqrt{2.7 \times 10^{-14}}$$

$$= 1.643 \times 10^{-7}\ M$$

$$pH = -\log[H^+]$$

$$= -\log[1.643 \times 10^{-7}] = 6.78$$

6 $HA \rightleftharpoons H^+ + A^-$

$$K_a = \frac{[H^+][A^-]}{[HA]} \quad ...(i)$$

$$HA + NaOH \rightleftharpoons ANa + H_2O$$

or $\quad HA + OH^- \rightleftharpoons A^- + H_2O$

$$K_C = \frac{[A^-][H_2O]}{[HA][OH^-]} \quad ...(ii)$$

Also, $K_w = [H^+][OH^-] = 10^{-14}$ ...(iii)

From equations (i), (ii) and (iii)

$$K_C = \frac{K_a}{K_w} = \frac{10^{-4}}{10^{-14}} = 1 \times 10^{10}$$

7 NH_4OH (weak base) + HCl (strong acid) forms NH_4Cl, which gives acidic solution with pH< 7.

8 5.0×10^{-2} M NaOH $\equiv [OH^-]$

$$= 5 \times 10^{-2} M$$

$$K_w = [H^+][OH^-] = 1 \times 10^{-14}$$

$$[H^+] \cdot 5 \times 10^{-2} = 1 \times 10^{-14}$$

$$[H^+] = \frac{1 \times 10^{-14}}{5 \times 10^{-2}}$$

$$= 2 \times 10^{-13}$$

$$pH = -\log[H^+]$$

$$= -\log(2 \times 10^{-13})$$

$$= 12.69 \approx 12.70$$

9 At 330 K,

$$[H^+][OH^-] = K_w = 1 \times 10^{-13.6}$$

$$[H^+](10^{-4}\ M) = 1 \times 10^{-13.6}$$

$$[H^+] = \frac{1 \times 10^{-13.6}}{10^{-4}} = 10^{-9.6}$$

$$pH = -\log[H^+]$$

$$= -\log(10^{-9.6}) = 9.6$$

10 $pH = 1 \ \therefore [H^+] = 10^{-1} = 0.1M$

$pH = 2 \ \therefore [H^+] = 10^{-2} = 0.01M$

For dilution of HCl, $M_1V_1 = M_2V_2$

$$0.1 \times 1 = 0.01 \times V_2$$

$$V_2 = 10\,L$$

Volume of water to be added

$$= 10 - 1 = 9\,L$$

11 $pH = 3 \Rightarrow H^+ = 10^{-3}\ M = \alpha C$

$$\alpha = \frac{10^{-3}}{C} = \frac{10^{-3}}{0.1} = 0.01 << 1$$

According to Ostwald dilution law

$\therefore \quad K_a = \alpha^2 C$

$$= (0.01)^2 \times 0.1$$

$$= 1 \times 10^{-5}$$

12 We know that acidic strength $\propto K_a$ value.

Thus, on the basis of K_a value, order of acidic strength is

$$HCN < HNO_2 < HF.$$

Conjugate base of a strong acid is weak.

Therefore, the order of base strength of conjugate base is

$$F^- < NO_2^- < CN^-.$$

13

	HA	$\rightleftharpoons$	H^+	+	A^-
At $t=0$	1		0		0
At $t=t$	$(1-10^{-5})$		(10^{-5})		(10^{-5})

$$[A^-] = [H^+]$$

$$\Rightarrow \quad K_a = \frac{[H^+][A^-]}{[HA]} = \frac{[10^{-5}][10^{-5}]}{[1-10^{-5}]}$$

Since, $1 >>> 10^{-5}$, therefore, $(1-10^{-5}) \approx 1$.

$\therefore \quad K_a = 1 \times 10^{-10}$

14 $CH_3COOH \rightleftharpoons CH_3COO^- + H^+$

	CH_3COOH	CH_3COO^-	H^+
Initially	1	0	0
After ionisation	$1-\alpha$	α	α

$$pK_a = -\log K_a = 4.74$$

$\therefore \quad K_a = 1.82 \times 10^{-5}$

From, $K_a = \dfrac{C\alpha^2}{(1-\alpha)} = C\alpha^2 \quad (\because 1-\alpha \approx 1)$

$$\alpha = \sqrt{\frac{K_a}{C}} = \sqrt{\frac{1.82 \times 10^{-5}}{0.05}} = 0.019 \text{ or } 1.9\%$$

15 $NaX + H_2O \rightleftharpoons NaOH + HX$

HX is a weak acid, so NaX is a salt of weak acid and strong base.

Hydrolysis constant of NaX,

$$K_h = \frac{K_w}{K_a} = \frac{1 \times 10^{-14}}{10^{-5}} = 1 \times 10^{-9}$$

Again, $K_h = \dfrac{h^2}{V} = Ch^2$

(where, h = degree of hydrolysis)

$$1 \times 10^{-9} = 0.1 \times h^2$$

$$h^2 = \frac{1 \times 10^{-9}}{0.1} = 1 \times 10^{-8} \Rightarrow h = 1 \times 10^{-4}$$

% of degree of hydrolysis of NaX salt

$$= 1 \times 10^{-4} \times 100 = 1 \times 10^{-2} = 0.01\%$$

16 For a salt of weak acid and weak base,

$$pH = 7 + \frac{1}{2}pK_a - \frac{1}{2}pK_b$$

Given, $pK_a\ (HA) = 3.2$, $pK_a\ (BOH) = 3.4$

$$\therefore \quad pH = 7 + \frac{1}{2}(3.2) - \frac{1}{2}(3.4)$$

$$= 7 + 1.6 - 1.7 = 6.9$$

17 $pH = 13, pOH = 1$

$[OH^-] = 10^{-1}$ M

mole of $(OH)^- = \dfrac{1 \times 0.1}{1000} = 1 \times 10^{-4}$ mol

$= 1 \times 10^{-4} \times 6.02 \times 10^{23}\ OH^-$ ions per mL

$= 6.02 \times 10^{19}$

18 $(CH_3COO)_2Ca \rightarrow Ca^{2+} + 2CH_3COO^-$

(0.005 M) $\quad (2 \times 0.005 = 0.01)$

$[CH_3COO^-] = 0.01$ M

$$CH_3COO^- + H_2O \rightleftharpoons CH_3COOH + OH^- \text{ (Alkaline)}$$

$$pH = 7 + \frac{pK_a}{2} + \frac{\log C}{2}$$

$$= 7 + 2.37 + \frac{\log 0.01}{2}$$

$$= 7 + 2.37 - 1 = 8.37$$

19 For salt of a weak acid and weak base,

$$pH = 7 + \frac{1}{2}[pK_a - pK_b]$$

$$= 7 + \frac{1}{2}[4.80 - 4.78]$$

$$= 7 + \frac{1}{2}(0.02)$$

$$= 7.01$$

20 $AB_2 \longrightarrow A^{2+} + 2B^-$

$$K_{sp} = [A^{2+}][2B^-]^2$$

$$= (1.0 \times 10^{-5})(2 \times 1.0 \times 10^{-5})^2$$

$$= 4 \times 10^{-15}$$

21 Let the solubility of $PbCl_2 = x$ mol

$$\underset{x \text{ mol/L}}{PbCl_2} \rightleftharpoons \underset{x}{Pb^{2+}} + \underset{2x}{2Cl^-}$$

$$K_{sp} = [Pb^{2+}][Cl^-]^2$$

or $= (x) \times (2x)^2$ or $= 4x^3$

$$\therefore \quad x = \sqrt[3]{\frac{K_{sp}}{4}} = \left(\frac{K_{sp}}{4}\right)^{1/3}$$

22 The one with lowest value of K_{sp}, i.e. HgS will precipitate out first.

23

Salt	Solubility product	Solubility
MX	$S_1^2 = 4.0 \times 10^{-8}$	$S_1 = 2 \times 10^{-4}$
MX_2	$4S_2^3 = 3.2 \times 10^{-14}$	$S_2 = 2 \times 10^{-5}$
M_3X	$27S_3^4 = 2.7 \times 10^{-15}$	$S_3 = 1 \times 10^{-4}$

Thus, order of solubilities

$$= MX > M_3X > MX_2$$

24 K_{sp} of $BaSO_4 = 1.5 \times 10^{-9}$,

$$Ba^{2+} = 0.01 \text{ M}$$

$$SO_4^{2-} > \frac{1.5 \times 10^{-9}}{0.01} > 1.5 \times 10^{-7}$$

i.e. $SO_4^{2-} > 10^{-6}$ M

25 Concentration of CO_3^{2-} ions

$$= 1.0 \times 10^{-4} \text{ M}$$

For precipitation $K_{sp} \leq [Ba^{2+}][CO_3^{2-}]$

Given, $K_{sp} = 5.1 \times 10^{-9}$

Hence, minimum concentration of Ba^{2+} ions should be

$$= \frac{K_{sp}}{[CO_3^{2-}]} = \frac{5.1 \times 10^{-9}}{1.0 \times 10^{-4}} = 5.1 \times 10^{-5} \text{ M}$$

26 For the solute, $A_xB_y \rightleftharpoons xA + yB$

$$K_{sp} = x^x y^y (s)^{x+y}$$

Thus, $Mx_4 \rightleftharpoons M^{4+} + 4x^-$, $x = 1$, $y = 4$

$$K_{sp} = (1)^1 (4)^4 (S)^{(1+4)} = 256 S^5$$

$$S = \left(\frac{K_{sp}}{256}\right)^{1/5}$$

27 $K_{sp} = [Ag^+][Br^-] = 5.0 \times 10^{-13}$

$$[Ag^+] = 0.05 \text{ M}$$

$$Br^- = \frac{5 \times 10^{-13}}{0.05} = 1 \times 10^{-11} \text{M}$$

Moles of KBr $= 1 \times 10^{-11} \times 1 = 1 \times 10^{-11}$

Weight of KBr $= 1 \times 10^{-11} \times 120$

$$= 1.2 \times 10^{-9} \text{g}$$

28 $Mg(OH)_2 \rightleftharpoons Mg^{2+} + 2OH^-$

$$K_{sp} = [Mg^{2+}][OH^-]^2$$

$$[OH^-] = \sqrt{\frac{K_{sp}}{[Mg^{2+}]}} = 10^{-4}$$

pOH = 4 and pH = 10.

29 When ionic product is greater than K_{sp}, then precipitation occurs.

$\therefore K_{sp} <$ Ionic product.

This happens in option (b) where

$$1.7 \times 10^{-10} < (10^{-2})(10^{-3})$$

$$1.7 \times 10^{-10} < 10^{-6}$$

30 $$\underset{S}{Ca(OH)_3(s)} \rightleftharpoons \underset{S}{Cr^{3+}(aq)} + \underset{3S}{3OH^-(aq)}$$

$$K_{sp} = [Cr^{3+}][OH^-]^3$$

$$1.6 \times 10^{-30} = (S)(3S)^3$$

$$1.6 \times 10^{-30} = 27(S)^4$$

$$\therefore \quad S^4 = \frac{1.6 \times 10^{-30}}{27}$$

$$S = \sqrt[4]{\frac{1.6 \times 10^{-30}}{27}}$$

31 (i) $$\underset{S_1 \text{ mol L}^{-1}}{Hg_2Cl_2} \rightleftharpoons \underset{S_1}{Hg_2^{2+}} + \underset{2S_1}{2Cl^-}$$

$$K_{sp} = [Hg_2^{2+}][Cl^-]^2 = (S_1)(2S_1)^2$$

$$= S_1 \cdot 4S_1^2 = 4S_1^3$$

$$\therefore \text{ Solubility, } S_1 = \left(\frac{K_{sp}}{4}\right)^{1/3}$$

(ii) $$\underset{S_2 \text{ mol L}^{-1}}{Cr_2(SO_4)_3} \rightleftharpoons \underset{2S_2}{2Cr^{3+}} + \underset{3S_2}{3SO_4^{2-}}$$

$$K_{sp} = [Cr^{3+}]^2 [SO_4^{2-}]^3$$

$$= (2S_2)^2(3S_2)^3$$

$$= 4S_2^2 \times 27S_2^3 = 108 S_2^5$$

$$\therefore \text{ Solubility, } S_2 = \left(\frac{K_{sp}}{108}\right)^{1/5}$$

(iii) $$\underset{S_3 \text{ mol L}^{-1}}{BaSO_4} \rightleftharpoons \underset{S_3}{Ba^{2+}} + \underset{S_3}{SO_4^{2-}}$$

$$K_{sp} = [Ba^{2+}][SO_4^{2-}] = (S_3)^2$$

$$\therefore \text{ Solubility, } S_3 = (K_{sp})^{1/2}$$

(iv) $$\underset{S_4 \text{ mol L}^{-1}}{CrCl_3} \rightleftharpoons \underset{S_4}{Cr^{3+}} + \underset{3S_4}{3Cl^-}$$

$$K_{sp} = [Cr^{3+}][Cl^-]^3$$

$$= (S_4)(3S_4)^3 = 27S_4^4$$

$$\therefore \text{ Solubility, } S_4 = \left(\frac{K_{sp}}{27}\right)^{1/4}$$

32 $NaBrO_3 \rightleftharpoons Na^+ + BrO_3^-$

$AgBrO_3 \rightleftharpoons Ag^+ + BrO_3^-$

Due to common ion effect (BrO_3^-), the solubility of $AgBrO_3$ in an aqueous solution of $NaBrO_3$ is suppressed.

33 Solubility is decreased due to common ion effect.

$$AgI \rightleftharpoons Ag^+ + I^-, \; NaI \rightleftharpoons Na^+ + I^-$$

34 In group III of analysis, addition of NH_4Cl increases NH_4^+ ion and decreases OH^- ion concentration produced from NH_4OH due to common ion effect.

35 CH_3NH_2 (base) on reaction with HCl (acid) give a salt of weak base and strong acid as $CH_3NH_3^+Cl^-$.

	CH_3NH_2	+ HCl	$\longrightarrow CH_3NH_3^+Cl^-$
At $t = 0$	0.1 mol	0.08 mol	0
After reaction	(0.1 − 0.08) = 0.02 mol	0	0.08 mol

So, it acts as basic buffer solution due to the presence of weak base and its salt in solution of one litre.

$$pOH = -\log K_b \times \log \frac{[\text{salt}]}{[\text{base}]}$$

$$pOH = -\log K_b \times \log \frac{[CH_3NH_3^+Cl^-]}{[CH_3NH_2]}$$

$$= -\log (5 \times 10^{-4}) + \log \frac{[0.08]}{[0.02]}$$

$$= -\log 5 + 4 \log 10 + \log 4$$

$$= -0.699 + 4 + 0.602 = 3.903$$

$$pH = 14 - pOH = 14 - 3.903$$

$$= 10.097 = -\log [H^+]$$

$$[H^+] = 8.0 \times 10^{-11}$$

36 When the concentration of NH_4OH (weak base) is higher than the strong acid (HCl), a mixture of weak base and its conjugate acid is obtained, which acts as basic buffer.

$$NH_4OH + HCl \longrightarrow NH_4Cl + H_2O$$

	NH_4OH	HCl	NH_4Cl
Initially	0.1M	0.05M	0
After reaction	0.05M	0	0.05M

37 Concentration $= \dfrac{\text{mass}}{\text{molar mass} \times V\,(\text{mL})}$

$$\therefore\ [CH_3COOH] = \frac{5 \times 1000}{60 \times 500} = 0.166\ M$$

Similarly, $[CH_3COONa] = \dfrac{7.5 \times 1000}{82 \times 500}$

$$= 0.183\ M$$

For the buffer solution,

$$CH_3COOH + NaOH \rightleftharpoons CH_3CO\overset{-}{O}\overset{+}{Na} + H_2O$$

$$pH = pK_a + \log\frac{[CH_3COONa]}{[CH_3COOH]}$$

$$= 4.76 + \log\left[\frac{0.183}{0.166}\right]$$

$$= 4.76 + \log(1.10)$$

$$= 4.76 + 0.042 = 4.80$$

$$\therefore\ 4.76 < pH < 5.0$$

38 From the aqueous buffered solution of *HA*, 50% *HA* is ionised.

$$[HA] = [A^-]$$

Buffer solution of weak acid $HA \longrightarrow$ acidic buffer

$$pH = pK_a + \log\frac{[A^-]}{[HA]}$$

$$= pK_a + \log 1$$

$$pH = pK_a = 4.5$$

$$pOH = pK_w - pH$$

$$pOH = 14 - 4.5 = 9.5$$

39 NaCl is the salt of strong acid and strong base. It is not a buffer as aqueous solution of NaCl is itself an exact neutral solution.

40 The ionisation of NaCl is suppressed due to the common ion effect of Cl^- which results in the precipitation of NaCl.

41 The value of solubility product of AgCl is greater than that of AgBr. Since, compounds with lower value of solubility product is precipitated first, therefore AgBr precipitates out more easily than AgCl.

42 HCl gives the common H^+ ions and hence, ionisation equilibrium $H_2S \rightleftharpoons 2H^+ + S^{2-}$ is suppressed.

43 HgI_2 combines with KI to form the soluble complex K_2HgI_4 and thus the solubility of HgI_2 is expected to be less in KI due to common ion effect.

44 A solution containing a mixture of acetic acid and the sodium acetate acts as a buffer solution as it maintains a constant value of pH (= 4.75) and its pH is not affected on addition of small amounts of acid or alkali.

SESSION 2

1 Strong acids have low pH value while strong bases have high pH value. Hence, the order of pH is

$$\underset{\text{Strong acid}}{HCl} < \underset{\text{Weak acid}}{CH_3COOH} < \underset{\text{Salt of weak base and strong acid}}{NH_4Cl} < \underset{\text{Salt of weak acid and strong base}}{CH_3CO\overset{-}{O}\overset{+}{Na}}$$

2 In 1×10^{-8} M HCl solution, H_2O is also present and undergoes self ionisation.

$$H_2O \rightleftharpoons H^+ + OH^-$$

$$[H^+] = 10^{-7}\ M \text{ at } 25°C$$

H^+ from HCl decreases self ionisation which decreases $[H^+]$ concentration, hence net concentration must be smaller than 10^{-7} M.

Thus, the pH of 1×10^{-8} M HCl is less than 8.

3 Given that, $K_a = 1.74 \times 10^{-5}$

Concentration of CH_3COOH

$$= 0.01\ \text{mol dm}^{-3}$$

$$[H^+] = \alpha C$$

According to Ostwald dilution law,

$$K_a = \alpha^2 C$$

$$\alpha = \sqrt{\frac{K_a}{C}}$$

$$\therefore\ H^+ = \sqrt{\frac{K_a}{C}} \times C$$

$$[H^+] = \sqrt{K_a \cdot C}$$

$$= \sqrt{1.74 \times 10^{-5} \times 0.01}$$

$$= 4.17 \times 10^{-4}$$

$$pH = -\log[H^+]$$

$$= -\log(4.17 \times 10^{-4}) = 3.4$$

4 $HA \rightleftharpoons H^+ + A^-$

$$K_a = C\alpha^2 = 1 \times \left(\frac{4}{100}\right)^2 = 1.6 \times 10^{-3}$$

$$K_a = \frac{[H^+][A^-]}{[HA]},\ [H^+] = C\alpha = 1 \times 0.01$$

$$= 0.01\ M$$

$$\therefore\ 1.6 \times 10^{-3} = \frac{0.01\,[A^-]}{1};\quad [A^-] = 0.16\ M$$

5 Its given that the final volume is 500 mL and this final volume was arrived when 50 mL of 1 M Na_2SO_4 was added to unknown Ba^{2+} solution. So, we can interpret the volume of unknown Ba^{2+} solution as 450 mL i.e.

$$\underset{Ba^{2+}\ \text{solution}}{450\text{mL}} + \underset{Na_2SO_4\ \text{solution}}{50\text{mL}} \longrightarrow \underset{BaSO_4\ \text{solution}}{500\text{mL}}$$

From this we can calculate the concentration of SO_4^{2-} ion in the solution *via*

$$M_1V_1 = M_2V_2$$

$$1 \times 50 = M_2 \times 500$$

(as 1M Na_2SO_4 is taken into consideration)

$$M_2 = \frac{1}{10} = 0.1\ M$$

Now for just precipitation,

Ionic product = Solubility product (K_{sp})

i.e. $[Ba^{2+}][SO_4^{2-}] = K_{sp}$ of $BaSO_4$

Given, K_{sp} of $BaSO_4 = 1 \times 10^{-10}$

So, $[Ba^{2+}][0.1] = 1 \times 10^{-10}$

or $[Ba^{2+}] = 1 \times 10^{-9}$ M

This is the concentration of Ba^{2+} ions in final solution. Hence, for calculating the $[Ba^{2+}]$ in original solution we have to use

$$M_1V_1 = M_2V_2$$

as $M_1 \times 450 = 10^{-9} \times 500$

so, $M_1 = 1.1 \times 10^{-9}$M

6 CH_3COO^- acts as buffer

$$CH_3COO^- + H_2O \rightleftharpoons CH_3COOH + OH^-$$

$$\therefore\ OH^- = Cx = C\sqrt{\frac{K_b}{C}} = \sqrt{C K_b}$$

But $K_a \cdot K_b = K_w$

$$\therefore\ K_b = \frac{K_w}{K_a}$$

$$\therefore\ OH^- = \sqrt{\frac{CK_w}{K_a}}$$

7 $[Ag^+] = 0.1$ mg

$$= (0.1 \times 10^{-3}\ g) \text{ in } 500\ mL\ H_2O$$

$= \frac{0.1\times10^{-3}}{108}$ mol in 500 mL H_2O

$= \frac{0.1\times10^{-3}\times2}{108}$ mol in 1 L H_2O

$= 1.85\times10^{-6}$ M

$[Ag^+][Cl^-] = 1.0\times10^{-10}$

$\therefore \quad [Cl^-] = \frac{1.0\times10^{-10}}{1.85\times10^{-6}} = 5.4\times10^{-5}$ M

8 $\alpha_1 = 0.005 = \sqrt{K_a}$ ($\because C_1 = 1$ mol L^{-1})

Molarity of the diluted solution,

$C_2 = \frac{2}{32} = \frac{1}{16}$ mol L^{-1}

According to Ostwald dilution law,

$\alpha_2 = \sqrt{\frac{K_a}{C_2}} = 0.005\sqrt{16} = 0.02$

$[H_3O^+] = C_2\,\alpha_2$

$= \frac{1\times0.02}{16} = 1.25\times10^{-3}$M

9 For precipitation,

Ionic product > solubility product (K_{sp})

For Ag_2CrO_4, Ionic product

$= [Ag^+]^2[CrO_4^-]$

$= (10^{-4})^2(10^{-5}) = 10^{-13}$

K_{sp} of $Ag_2CrO_4 = 4\times10^{-12}$

Here, K_{sp} > Ionic product

Thus, no precipitate is obtained.

For AgCl, ionic product

$= [Ag^+][Cl^-] = (10^{-4})(10^{-5})$

$= 10^{-9}$

K_{sp} (AgCl) $= 1\times10^{-10}$

Here, Ionic product > K_{sp}

So, precipitate will form. Thus, silver chloride gets precipitated first.

10 $AgIO_3(s) \rightleftharpoons Ag^+(aq) + IO_3^-(aq)$

$K_{sp} = [Ag^+][IO_3^-]$

$1.0\times10^{-8} = (S)(S)$

$1.0\times10^{-8} = S^2$

$S = \sqrt{1.0\times10^{-8}}$

$= 1.0\times10^{-4}$ mol/L

In 1000 mL, moles of $AgIO_3$ dissolved

$= 1\times10^{-4}$ mol

In 100 mL, moles of $AgIO_3$ dissolved

$= 1\times10^{-5}$ mol

Mass of $AgIO_3$ in 100 mL

$= 1\times10^{-5}\times283 = 2.83\times10^{-3}$ g

11 $\underset{C}{HA} + H_2O \longrightarrow \underset{0}{H_3^+O} + \underset{0}{A^-}$

$C(1-\alpha) \qquad C\alpha \qquad C\alpha$

$K_a = \frac{[H_3^+O][A^-]}{[HA]} = \frac{C^2\alpha^2}{C(1-\alpha)}$

With in an error of 10%,

$C(1-\alpha)$ or $C - C\alpha = 0.90\,C$

or $C\alpha = 0.10\,C$

$\because \quad K_a = \frac{(0.10\,C)^2}{0.90\,C} = \frac{C}{90}$

or $C = 90K_a$

12 $Na_2CO_3 + HCl \rightarrow NaCl + NaHCO_3$

	Na_2CO_3	HCl	NaCl	$NaHCO_3$
Meq. before reaction	$\frac{30}{106}\times1000 = 283$	$150\times1 = 150$	0	0
Meq after reaction	133	0	150	150

The solution contains Na_2CO_3 and HCO_3^- and thus, acts as buffer.

$\therefore$ pH $= -\log K_a + \log\frac{[CO_3^{2-}]}{[HCO_3^-]}$

pH $= -\log K_a + \log(133/150)$

pH $= -\log(5.63\times10^{-11}) + \log(133/150)$

$= 10.249 - 0.052$

pH $= 10.197$

13 $\underset{\text{(Base)}}{N_1V_1} = \underset{\text{(Acid)}}{N_2V_2}$

$2.5\times\frac{2}{5} = \frac{2}{15}\times V_2$ or $V_2 = \frac{15}{2} = 7.5$ mL

$BOH + HCl \longrightarrow BCl + H_2O$

2.5 mL of $\frac{2}{5}$ M base contains, base

$= 2.5\times\frac{2}{5} = 1$ mmol

$\therefore$ Salt BCl formed = 1m mol

Volume of solution = 2.5 mL + 7.5 mL = 10 mL

$\therefore$ Concentration of salt [BCl] in the solution $= \frac{1}{10}$ M = 0.1 M

For salt of weak base and strong acid

$[H^+] = \sqrt{\frac{K_wC}{K_b}} = \sqrt{\frac{10^{-14}\times0.1}{10^{-12}}}$

$= 3.2\times10^{-2}$ M

14 pH = 9.26 indicates $[NH_4OH] > [HCl]$ and thus mixture is a basic buffer since, HCl will react with equivalent amount of NH_4OH forming NH_4Cl.

Let HCl $= x$ mL $= x$ millimol

$NH_4OH = (300 - x)$ mL

$= (300 - x)$ millimol

NH_4Cl formed $= x$ millimol

NH_4OH unreacted

$= 300 - x - x = (300 - 2x)$ millimol

pOH = 14 − 9.26 = 4.74

$pK_b = 14 - 9.26 = 4.74$

pOH $= pK_b + \log\frac{[NH_4^+]}{[NH_4OH]}$

or $4.74 = 4.74 + \log\frac{x}{300 - 2x}$

$\Rightarrow \frac{x}{300 - 2x} = 1$

$x = 100$ mL = volume of HCl

$(300 - x) = 200$ mL = volume of NH_4OH

Hence, volume ratio of NH_4OH and HCl = 2 : 1.

DAY ELEVEN

Unit Test 2
(Physical Chemistry I)

1 The free energy for the following reactions are as follows:

$$C_2H_2(g) + \frac{5}{2}O_2(g) \longrightarrow 2CO_2(g) + H_2O(l); \Delta G° = -1234 \text{ kJ}$$

$$C(s) + O_2(g) \longrightarrow CO_2(g); \Delta G° = -394 \text{ kJ}$$

$$H_2(g) + \frac{1}{2}O_2(g) \longrightarrow H_2O(l); \Delta G° = -237 \text{ kJ}$$

The standard free energy change for the following reaction is → NCERT Exemplar

$$H_2(g) + 2C(s) \longrightarrow C_2H_2(g)$$

(a) −209 kJ
(b) −2259 kJ
(c) +209 kJ
(d) +2259 kJ

2 The enthalpies of elements in their standard states are taken as zero. The enthalpy of formation of a compound → NCERT Exemplar

(a) is always negative
(b) is always positive
(c) may be positive or negative
(d) is never negative

3 The internal energy change when a system goes from state A to B is 40 kJ mol^{-1}. If the system goes from A to B by a reversible path and returns to state A by an irreversible path, what would be the net change in internal energy?

(a) 40 kJ (b) > 40 kJ
(c) < 40 kJ (d) Zero

4 Standard entropy of X_2, Y_2 and XY_3 are 60, 40 and 50 JK^{-1} mol^{-1}, respectively. For the reaction,
$\frac{1}{2}X_2 + \frac{3}{2}Y_2 \longrightarrow XY_3; \Delta H = -30$ kJ, to be at equilibrium, the temperature will be

(a) 1250 K (b) 500 K
(c) 750 K (d) 1000 K

5 An ideal gas heat engine operates in Carnot cycle between 227°C and 127°C. It absorbs 6×10^4 cal of heat at high temperature. Amount of heat converted to work is

(a) 1.2×10^4 cal (b) 4.8×10^4 cal
(c) 6×10^4 cal (d) 2.4×10^4 cal

6 When one mole of anhydrous $FeSO_4$ is dissolved in excess of water, there in evolution of 58.2 kJ of heat. But when one mole of $FeSO_4 \cdot 5H_2O$ is dissolved in water, the heat change is + 8.6 kJ. Calculate the enthalpy of hydration of anhyd. $FeSO_4$.

(a) − 49.6 kJ (b) − 66.8 kJ (c) + 49.6 kJ (d) + 66.8 kJ

7 $A \longrightarrow B$, $\Delta H = +$ve. Graph between $\log_{10} p$ and $\frac{1}{T}$ is a straight line of slope $\frac{1}{4.606}$. Hence, ΔH is

(a) 1 (b) 2 (c) 4 (d) −1

8 The heat of atomisation of $PH_3(g)$ is 228 kcal mol^{-1} and that of $P_2H_4(g)$ is 355 kcal mol^{-1}. The energy of the P—P bonds (in kcal mol^{-1}) is

(a) 102 (b) 51 (c) 26 (d) 204

9 Match the following and choose the correct option.

	Column I		Column II
A.	$Q = K$	1.	Reaction is nearer to completion.
B.	$Q < K$	2.	Reaction is not at equilibrium.
C.	$Q > K$	3.	Reaction is fast in forward direction.
D.	$K >>> 1$	4.	Reaction at equilibrium.
		5.	Reaction proceeds in backward direction.

	A	B	C	D
(a)	4	2,3	2,5	1
(b)	1	2,3	4	2
(c)	2	3	1,4	3,4
(d)	1	2	3	2,4

10 1 mole of non-ideal gas undergoes a change of state (2.0 atm, 3.0 L , 95K) $\longrightarrow$ (4.0 atm, 5.0 L, 245 K) with a change in internal energy, $\Delta E = 30.0$ L atm. The change in enthalpy (ΔH) of the process in L atm is

(a) 40.0 (b) 42.3 (c) 44.0
(d) not defined because pressure is not constant

11 **Assertion** (A) Aqueous solution of ammonium carbonate is basic.

Assertion (R) Acidic/basic nature of a salt solution of a salt of weak acid and weak base depends on K_a and K_b value of the acid and the base forming it.

→ [NCERT Exemplar]

(a) Assertion and Reason both are correct statements and Reason is the correct explanation of the Assertion
(b) Assertion and Reason both are correct statements but Reason is not the correct explanation of the Assertion
(c) Assertion is correct and Reason is incorrect
(d) Both Assertion and Reason are incorrect

12 The polymerisation of ethene to linear polythene is represented by the reaction.

$$n(CH_2{=}CH_2) \longrightarrow -\!\!\!\left(CH_2{-}CH_2\right)\!\!\!-_n$$

Given that the average enthalpies of bond dissociation for C=C and C—C at 298 K are + 590 and + 331 kJ mol^{-1} respectively. The enthalpy of polymerisation per mole of ethene at 298 K, is

(a) 72 kJ mol^{-1} (b) 27 kJ mol^{-1}
(c) 1144 kJ mol^{-1} (d) 172 kJ mol^{-1}

13 When 1 mole of an ideal gas is compressed to half its initial volume and simultaneously heated to twice its initial temperature, the change in entropy (ΔS) is

(a) C_V ln 2 (b) C_p ln 2 (c) R ln 2 (d) $(C_V - R)$ ln2

14 The enthalpy changes of formation of the gaseous oxides of nitrogen (N_2O and NO) are positive because of

(a) the high bond energy of the nitrogen molecule
(b) the high electron affinity of oxygen atoms
(c) the high electron affinity of nitrogen atoms
(d) the tendency of oxygen to form O^{2-} ion

15 1.0 g of pure calcium carbonate was found to require 50 mL of dilute HCl for complete reaction. The strength of the HCl solution is given by

(a) 0.2 N (b) 0.4 N (c) 2.0 N (d) 4.0 N

16 What is the molarity of H_2SO_4 solution that has a density 1.84 g/cc at 35°C and contains solute 98% by weight ?

(a) 4.18 M (b) 1.84 M (c) 8.41 M (d) 18.4 M

17 The lowering in vapour pressure caused by the addition of 100 g of sucrose (molecular mass = 342) to 1000 g of lowering in water, if the vapour pressure of pure water at 25°C is 23.8 mm Hg, is

(a) 0.012 mm Hg (b) 0.125 mm Hg
(c) 1.15 mm Hg (d) 1.25 mm Hg

18 Two solutions of KNO_3 and CH_3COOH are prepared separately. Molarity of both is 0.1 M and their osmotic pressures are p_1 and p_2 respectively. The correct relationship between the osmotic pressures is

(a) $p_1 = p_2$ (b) $p_1 > p_2$
(c) $p_2 > p_1$ (d) $\frac{p_1}{p_1 + p_2} + \frac{p_2}{p_1 + p_2}$

19 The osmotic pressure of a 5% (w/V) solution of cane sugar at 150°C is

(a) 3.078 atm (b) 4.078 atm
(c) 5.071 atm (d) 2.45 atm

20 Which will show maximum depression in freezing point when concentration is 0.1 M?

(a) Urea (b) $BaCl_2$
(c) $Al_2(SO_4)_3$ (d) KBr

21 A 5% solution of sugarcane (mol wt. = 342) is isotonic with 1% solution of x. The mol. wt. of x is

(a) 34.2 (b) 68.4 (c) 136.2 (d) 171.2

22 Match the following and choose the correct option.

	Column I		Column II
A.	ΔG	1.	$\Delta U + nRT$
B.	ΔH	2.	$\Delta H - T\Delta S$
C.	ΔU	3.	$nC\,vdT$
D.	ΔS	4.	$2.303nR\log 10\left(\frac{V_2}{V_1}\right)$

	A	B	C	D		A	B	C	D
(a)	2	1	3	4	(b)	1	2	3	4
(c)	3	2	4	1	(d)	4	3	2	1

23 An amount of solid NH_4HS is placed in a flask already containing ammonia gas at a certain temperature and 0.5 atm pressure. Ammonium hydrogen sulphate decomposes to yield NH_3 and H_2S gases in the flask. When the decomposition reaction reaches at equilibrium, the total pressure in the flask rises to 0.84 atm.
The equilibrium constant for the decomposition of NH_4HS at this temperature, is

(a) 0.11 (b) 0.17
(c) 0.18 (d) 0.30

24 For the reaction,

$2NO_2(g) \rightleftharpoons 2NO(g) + O_2(g)$; $K_C = 1.8 \times 10^{-6}$ at 184°C
$(R = 0.00831$ kJ mol^{-1} K$^{-1})$

When K_p and K_C are compared at 184°C, it is found that

(a) whether K_p is greater than, less than or equal to K_C depends upon the total gas pressure
(b) $K_p = K_C$
(c) K_p is less than K_C
(d) K_p is greater than K_C

25 For the equilibrium,

$$Ag^+ + 2NH_3 \rightleftharpoons Ag(NH_3)_2^+; K_1 = 1.8 \times 10^7$$

and $Ag^+ + Cl^- \rightleftharpoons AgCl; K_2 = 5.6 \times 10^9$

Hence, for the equilibrium,

$$AgCl + 2NH_3 \rightleftharpoons Ag(NH_3)_2^+ + Cl^-$$

equilibrium constant is

(a) 0.32×10^{-2} (b) 0.31×10^{-21}
(c) 1.01×10^{17} (d) 1.01×10^{-17}

26 Given pH of a solution *A* is 3 and it is mixed with another solution having pH 2. If both are mixed, resultant pH of the solution will be

(a) 3.21 (b) 1.96 (c) 3.42 (d) 3.58

27 For the reaction,

$$CuSO_4 \cdot 5H_2O(s) \rightleftharpoons CuSO_4 \cdot 3H_2O(s) + 2H_2O(g)$$

K_p at 298 K is 1.086×10^{-4} atm^2 and vapour pressure of water is 23.8 torr. The salt $CuSO_4 \cdot 5H_2O$ will be efforescent when the relative humidity is

(a) 80% (b) 60%
(c) 50% (d) less than 33.1%

28 50.0 mL of 0.3 M HCl is mixed with 50 mL of 0.4 M NH_3 solution. If pK_a of NH_4^+ is 9.26, pH of the mixture is

(a) 5.22 (b) 1.30 (c) 8.78 (d) 12.70

29 pK_a of acetyl salicylic acid (aspirin) is 3.5. The pH of gastric juice in human stomach is about 2–3 and the pH in the small intestine is about 8. Aspirin will be

(a) unionised in the small intestine and in the stomach
(b) completely ionised in the small intestine and in the stomach
(c) ionised in the stomach and almost decrease in the small intestine
(d) ionised in the small intestine and almost unionised in the stomach

30 A weak acid HX has the dissociation constant 1×10^{-5} M. It forms a salt NaX on reaction with alkali. The degree of hydrolysis of 0.1 M solution of NaX is

(a) 0.0001% (b) 0.01%
(c) 0.1% (d) 0.15%

31 A sample of $Na_2CO_3 \cdot H_2O$ weighing 0.62 g is added to 100 mL of 0.1 N $(NH_4)_2SO_4$ solution. What will be the resulting solution?

(a) Acidic (b) Neutral
(c) Basic (d) None of these

32 At infinite dilution, the percentage ionisation for both strong and weak electrolyte is

(a) 1% (b) 20%
(c) 50% (d) 100%

33 A litre of solution is saturated with AgCl. To this solution if 1.0×10^{-4} moles of solid NaCl are added, what will be the $[Ag^+]$ assuming no volume change?

(a) More (b) Less (c) Equal (d) Zero

34 Which hydroxide will have lowest value of solubility product at normal temperature (25°C)?

(a) $Mg(OH)_2$ (b) $Ca(OH)_2$
(c) $Ba(OH)_2$ (d) $Be(OH)_2$

35 When solid potassium cyanide is added in water, the

(a) pH will increase
(b) pH will decrease
(c) pH will remain same
(d) electrical conductivity will not change

36 Heat obtained due to expansion of 1 mole of H_2 gas at 1000 K from 10 L to 100 L under isothermal reversible condition is absorbed by an engine having a sink at 300 K. Useful work obtained is

(a) – 1382 cal (b) – 3224 cal
(c) 1382 cal (d) 3224 cal

37 The enthalpy of hydrogenation of 1-pentene is + 126 kJ mol^{-1}. The enthalpy of hydrogenation of 1,3-pentadiene is + 230 kJ mol^{-1}. Hence, resonance (delocalisation) energy of 1,3-pentadiene is

(a) 22 kJ (b) 104 kJ
(c) 252 kJ (d) 11 kJ

38 An aqueous solution of liquid '*X*' [mol. weight 56] 28% by weight has a vapour pressure 150 mm. Find the vapour pressure of '*X*' if vapour pressure of water is 155 mm of Hg.

(a) 110 mm (b) 150 mm
(c) 220 mm (d) 125 mm

39 A monoprotonic weak acid[H*A*] is ionised 5% in 0.1 M aqueous solution. What is the equilibrium constant for its ionisation?

$$HA(aq) + H_2O(l) \rightleftharpoons H_3O^+(aq) + A^-(aq)$$

(a) 9.5×10^{-2} (b) 2.63×10^{-4}
(c) 2.303×10^{-3} (d) 5×10^{-3}

Direction (Q. Nos. 40-42) *Aqueous calcium chloride solution is mixed with sodium oxalate and precipitate of calcium oxalate formed is filtered and dried. Its saturated solution was prepared and 250 mL of this solution was titrated with 0.001 M $KMnO_4$ solution, when 6.0 mL of this was required.*

40 100 mL of pH = 6 (acidic) is diluted to 1000 mL by H_2O. pH will increase approximately by

(a) 9 unit (b) 1 unit (c) 0.7 unit (d) – 0.7 unit

41 Consider the following reaction,

$$N_2(g) + 3H_2 \longrightarrow 2NH_3(g);$$

$$\Delta H = -95.4 \text{ kJ and } \Delta S = -198.3 \text{ JK}^{-1}$$

Find the temperatures at which Gibbs energy change (ΔG) is zero and predict if the reaction below this temperature would be spontaneous or non-spontaneous.

(a) 481 K ; non-spontaneous (b) 281 K ; spontaneous
(c) 281 K ; non-spontaneous (d) 481 K ; spontaneous

42 pK_b of NH_3 is 4.74 and pK_b of A^-, B^- and C^- are 4, 5 and 6, respectively. Aqueous solution of 0.01 M has pH in the increasing order

(a) $NH_4A < NH_4B < NH_4C$
(b) $NH_4C < NH_4B < NH_4A$
(c) $NH_4C < NH_4A < NH_4B = 7$
(d) All have equal pH being salt of weak acid and weak base

43 Given that $dE = TdS - pdV$ and $H = E + pV$. Which one of the following relations is true?

(a) $dH = TdS + Vdp$
(b) $dH = SdT + Vdp$
(c) $dH = -SdT - Vdp$
(d) $dH = dE - pdV$

44 1.0 mole of a monoatomic idea gas is expanded from state (1) to state (2) as shown in the figure. Calculate the work done for the expansion of gas from state (1) to state (2) at 298 K → NCERT Exemplar

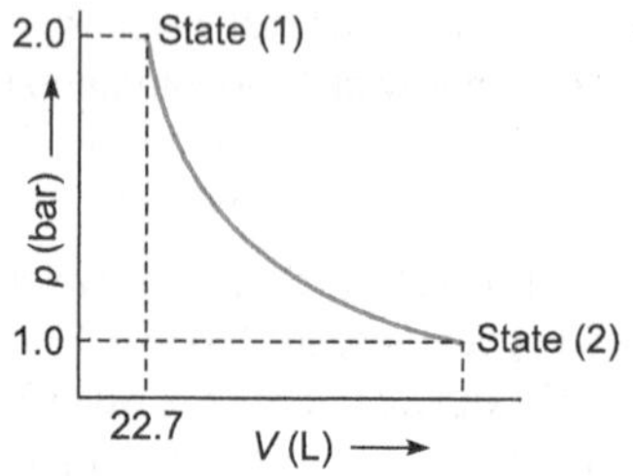

(a) –1717.46 J
(b) +1717.46 J
(c) –1908.2 J
(d) +1908.2 J

Direction (Q. Nos. 45-50) *Each of these questions contains two statements: Assertion and* Reason. *Each of these questions also has four alternative choices, only one of which is the correct answer. You have to select one of the codes (a), (b), (c) and (d) given below.*

(a) Both A and R are true and R is correct explanation of A
(b) Both A and R are true but R is not correct explanation of A
(c) A is true but R is false
(d) Both A and R are false

45 **Assertion** (A) An aqueous solution of ammonium acetate can act as a buffer.

Assertion (R) Acetic acid is a weak acid and NH_4OH is a weak base. → NCERT Exemplar

46 **Assertion** (A) On mixing equal volumes of 1 M HCl and 2M CH_3COONa, an acidic buffer solution is formed.

Reason (R) Resultant mixture contains CH_3COOH and CH_3COONa which are parts of acidic buffer.

47 **Assertion** (A) A liquid crystallises into a solid and is accompanied by decrease in entropy.

Reason (R) In crystals, molecules organise in an ordered manner.

48 **Assertion** (A) The pK_a of a weak acid becomes equal to pH of the solution at the mid point of its titration.

Reason (R) The molar concentrations of proton acceptor and proton donor become equal at the mid point of titration of a weak acid.

49 **Assertion** (A) Heat of neutralisation is always less than zero.

Reason (R) Neutralisation involves reaction between an acid and a base.

50 **Assertion** (A) The molality of the solution does not change with change in temperature.

Reason (R) The molality is expressed in units of moles per 1000 g of solvent.

ANSWERS

1. (a)	**2.** (c)	**3.** (d)	**4.** (c)	**5.** (a)	**6.** (b)	**7.** (a)	**8.** (b)	**9.** (a)	**10.** (b)
11. (a)	**12.** (a)	**13.** (d)	**14.** (a)	**15.** (b)	**16.** (d)	**17.** (b)	**18.** (b)	**19.** (c)	**20.** (c)
21. (b)	**22.** (a)	**23.** (a)	**24.** (d)	**25.** (a)	**26.** (b)	**27.** (d)	**28.** (c)	**29.** (d)	**30.** (b)
31. (a)	**32.** (d)	**33.** (b)	**34.** (d)	**35.** (a)	**36.** (d)	**37.** (a)	**38.** (a)	**39.** (b)	**40.** (c)
41. (d)	**42.** (b)	**43.** (a)	**44.** (a)	**45.** (b)	**46.** (a)	**47.** (a)	**48.** (a)	**49.** (b)	**50.** (a)

Hints and Explanations

1 $C_2H_2(g) + \frac{5}{2}O_2(g) \rightarrow 2CO_2(g) + H_2O(l);$

$$\Delta G° = -1234 \text{ kJ}$$

(b) $[C(s) + O_2(g) \longrightarrow CO_2(g);$

$$\Delta G° = -394 \text{ kJ}] \times 2$$

(c) $H_2(g) + \frac{1}{2}O_2(g) \longrightarrow H_2O(l);$

$$\Delta G° - 237 \text{ kJ}$$

Add Eq. (b) + (c) and then subtract from (a), to get,

$2C(s) + H_2(g) \longrightarrow C_2H_2(g);$

$$\Delta G° = -209 \text{ kJ}$$

2 Combustion of elements to form a compound can be exothermic or endothermic e.g. $C + O_2 \longrightarrow CO_2$ is exothermic.

whereas, $C + 2S \longrightarrow CS_2$ is endothermic

Hence,enthalpy of formation can be positive or negative.

3 In case of cyclic process, $\Delta E = 0$.

4 $\Delta S = S(XY_3) - \frac{1}{2}S(X_2) - \frac{3}{2}S(Y_2)$

$= 50 - 30 - 60$

$= -40 \text{ J mol}^{-1} \text{ K}^{-1}$

$\Delta H = -30 \text{ kJ} = -30000 \text{ J}$

$\Delta G = \Delta H - T\Delta S$

At equilibrium, $\Delta G = 0$

$$T = \frac{\Delta H}{\Delta S} = \frac{-30000}{-40} = 750 \text{ K}$$

5 $\eta = \frac{T_2 - T_1}{T_2} = \frac{500 - 400}{500} = \frac{1}{5}$

$$W = \eta \times Q = \frac{1}{5} \times 6 \times 10^4 = 1.2 \times 10^4$$

6 The hydration of $FeSO_4$ in depicted as

$FeSO_4 + 5H_2O \longrightarrow FeSO_4 \cdot 5H_2O;$

$$\Delta_{hyd}H^- = ?$$

We have,

$FeSO_4(\text{anhyd}) + H_2O(\text{excess})$

$$\longrightarrow FeSO_4(aq);$$

$$\Delta H_1 = -58.2 \text{ kJ}$$

$FeSO_4 \cdot 5H_2O + H_2O(\text{excess})$

$$\longrightarrow FeSO_4(aq);$$

$$\Delta H_2 = +8.6 \text{ kJ}$$

Thus, we can calculate,

$$\Delta_{hyd}H = \Delta H_1 - \Delta H_2$$

$$= -58.2 - 8.6 = -66.8 \text{ kJ}$$

7 By Clapeyron – Clausius equation

$$\log p = -\frac{\Delta H}{2.303\,RT} + \text{constant}$$

$$\frac{\Delta H}{2.303\,R} = \frac{1}{4.606}$$

$$\Delta H = \frac{2.303 \times 2}{4.606} = 1 \text{ cal}$$

8 Bond dissociation energy of PH_3 = 228 kcal mol^{-1}

P—H bond energy

$$= \frac{228}{3} = 76 \text{ kcal mol}^{-1}$$

Structure of P_2H_4 is 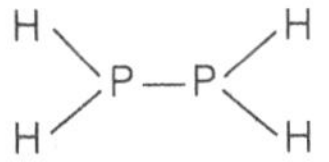

$\therefore$ Bond dissociation energy of P_2H_4 can be given as

$4(P—H) + (P—P) = 355 \text{ kcal mol}^{-1}$

$4 \times 76 + (P—P) \approx 355 \text{ kcal mol}^{-1}$

$\therefore$ P—P bond energy = 51 kcal mol^{-1}

9 A $\rightarrow$ 4, B $\rightarrow$ 2,3 ,C $\rightarrow$ 2, 5 D $\rightarrow$ 1

10 $\Delta H = \Delta E + \Delta n_g RT, \quad \Delta n_g = 1 \text{ mol},$

$\Delta T = 245 - 95 = 150 \text{ K},$

$R = 0.0821 \text{ L atm K}^{-1} \text{ mol}^{-1}$

$\Delta E = 30.0 \text{ L atm}$

$\Delta H = 30 + 1 \times 0.0821 \times 150 = 42.3 \text{ L atm}$

11 $(NH_4)_2CO_3 \longrightarrow 2\overset{+}{N}H_4 + CO_3^{2-}$

$$2H_2O \longrightarrow \underset{\substack{NH_4OH \\ \text{Weak base}}}{2OH^-} + \underset{\substack{H_2CO_3 \\ \text{Weak acid}}}{2H^+}$$

If K_b of $NH_4OH > K_a$ of H_2CO_3. The solution is basic

or, if K_a of $H_2CO_3 > K_b$ of NH_4OH; the solution is acidic.

12 In this polymerisation reaction, every molecule of ethene involves breaking of one C ═ C (double bond) and formation of two C—C single bonds.

The amount of energy required to break one mole of >C ═ C< (double bond) into >C — C< (single bond) = 590 kJ.

The energy released in the formation of two moles of >C—C< single bond

$= 2 \times 331 = 662 \text{ kJ}$

Net energy released per mole of ethene

$= 662 - 590 = 72 \text{ kJ}$

Enthalpy of polymerisation per mole of ethene at 298 K,

$$\Delta H = 72 \text{ kJ mol}^{-1}$$

13 When there is simultaneously change in temperature and volume (or pressure)

$$\Delta S = nC_V \ln\left(\frac{T_2}{T_1}\right) + nR\ln\left(\frac{V_2}{V_1}\right)$$

$$= C_V \ln\left(\frac{2}{1}\right) + R\ln\frac{1}{2}$$

$$= C_V \ln 2 - R\ln 2 = (C_V - R)\ln 2$$

14 The enthalpy changes of formation of the gaseous oxides of nitrogen are positive due to high bond energy of the nitrogen molecule.

15 Meq. of HCl = M eq. of $CaCO_3$

$$N \times 50 = \frac{1}{50} \times 1000$$

$$N = \frac{1 \times 1000}{50 \times 50} = 0.4\text{N}$$

16 98% H_2SO_4 means 98 g H_2SO_4 in 100 g solution.

Volume of solution $= \frac{100}{1.84}$ cc = 54.3 cc

Number of moles of solute $= \frac{98}{98} = 1$ mol

Molarity $= \frac{1}{54.3} \times 1000 = 18.4$ M

17 Molecular mass of sucrose = 342

Moles of sucrose $(x) = \frac{100}{342} = 0.292$ mol

Moles of water $(N) = \frac{1000}{18} = 55.5$ mol

Vapour pressure of pure water

$p° = 23.8$ mm Hg

$$\frac{\Delta p}{p°} = \frac{n}{n+N}$$

$$\frac{\Delta p}{23.8} = \frac{0.292}{0.292 + 55.5}$$

$$\Delta p = \frac{23.8 \times 0.292}{55.792}$$

$$= 0.125 \text{ mm Hg}$$

18 KNO_3 dissociates completely while CH_3COOH dissociates to a smaller extent, hence $p_1 > p_2$.

19 $C = \frac{5 \times 1000}{342 \times 100} = \frac{50}{342} \text{mol L}^{-1}$

$\pi = \frac{50}{342} \times 0.082 \times 423 = 5.07$ atm

20 $KBr = K^+ + Br^- = 2$ ions

$BaCl_2 = Ba^{2+} + 2Cl^- = 3$ ions

$Al_2(SO_4)_3 = 2Al^{3+} + 3SO_4^{2-} = 5$ ions

Urea does not get ionised.

Hence, $Al_2(SO_4)_3$ shows maximum depression in freezing point.

21 For isotonic solutions,

$\frac{w_1}{m_1} = \frac{w_2}{m_2}$

$\frac{5}{342} = \frac{1}{m_2}$

$m_2 = 68.4$

22 A → 2, B → 1, C → 3, D → 4

23 $NH_4HS(s) \rightleftharpoons NH_3(g) + H_2S(g)$

	$NH_4HS(s)$	$NH_3(g)$	$H_2S(g)$
Initially	1	0.5	0
At equilibrium	$(1-x)$	$(0.5+x)$	x

Total pressure at equilibrium

$= p_{NH_3} + p_{H_2S}$

$= 0.5 + x + x = 0.84$

$x = 0.17$ atm

$p_{NH_3} = 0.50 + 0.17 = 0.67$ atm

$p_{H_2S} = 0.17$ atm

$K_p = p_{NH_3} \cdot p_{H_2S}$

$= 0.67 \times 0.17$

$= 0.114 \text{ atm} \approx 0.11$ atm

24 $2NO_2(g) \rightleftharpoons 2NO(g) + O_2(g)$

$K_C = 1.8 \times 10^{-6}$ at 184°C (= 457 K)

$R = 0.00831 \text{ kJ mol}^{-1} \text{ K}^{-1}$

$K_p = K_C (RT)^{\Delta n_g}$

where, Δn_g = gaseous products − gaseous reactants = 3 − 2 = 1

$K_p = 1.8 \times 10^{-6} \times 0.00831 \times 457$

$= 6.836 \times 10^{-6}$

Thus, $K_p > K_C$

25 $Ag^+ + 2NH_3 \rightleftharpoons Ag(NH_3)_2^+$;

$K_1 = 1.8 \times 10^7$

$AgCl \rightleftharpoons Ag^+ + Cl^-$;

$K_2' = \frac{1}{K_2} = \frac{1}{5.6 \times 10^9}$

$\therefore AgCl + 2NH_3 \rightleftharpoons Ag(NH_3)_2^+ + Cl^-$

$K = K_1 K_2' = \frac{K_1}{K_2}$

$= \frac{1.8 \times 10^7}{5.6 \times 10^9}$

$= 0.32 \times 10^{-2}$

26 pH of the solution A = 3

$[H^+]_A = 10^{-3}$ M

pH of the solution B = 2

$[H^+]_B = 10^{-2}$ M

$[H^+] = 10^{-3} + 10^{-2} = 11 \times 10^{-3}$

$pH = -\log (11 \times 10^{-3})$

$= 3 - \log 11 = 3 - 1.04 = 1.96$

27 An efflorescent salt is one that loses water to the atmosphere. This will occur if in the equilibrium water vapour pressure, with the salt is greater than the water vapour pressure in the atmosphere. For the given hydrated salt, equilibrium is

$CuSO_4 \cdot 5H_2O(s) \rightleftharpoons CuSO_4 \cdot 3H_2O(s) + 2H_2O(g)$

$K_p = p^2_{H_2O} = 1.086 \times 10^{-4} \text{ atm}^2$

$\therefore p_{H_2O} = 1.042 \times 10^{-2} \text{ atm} = 7.89$ torr

Since, p_{H_2O} is less than the vapour pressure of water in air at the same temperature, $CuSO_4 \cdot 5H_2O$ will not always efflorescence. It will efflorescence only on a dry day, when the partial pressure of moisture in the air is less than 7.89 torr.

Relative humidity

$= \frac{7.89}{23.8} = 0.331 = 33.1\%$

Thus, this salt will efflorescence when the relative humidity is less than 33.1%.

28. (c) $pK_a(NH_4^+) = 9.26$

$\therefore pK_b(NH_3) = 14 - 9.26 = 4.74$

50 mL of 0.3 M HCl = 15 millimol

50 mL of 0.4 M NH_3 = 20 millimol

$NH_3 + HCl \rightleftharpoons NH_4Cl$

	NH_3	HCl	NH_4Cl
Initial millimol	20	15	0
At equilibrium	5	0	15

Mixture is a buffer containing 5 millimol of NH_3 (base) and 15 millimol of NH_4^+ (conjugate acid)

$\therefore pOH = pK_b + \log \frac{[NH_4^+]}{[NH_3]}$

$= 4.74 + \log 3$

$\therefore pH = 14 - pOH = 9.26 - \log 3$

$= 9.26 - 0.48 = 8.78$

29. Aspirin is a weak acid. Due to common ion effect, it is unionised in acidic medium i.e. stomach but completely ionised in alkaline medium i.e. small intestine.

30. $X^- + H_2O \rightleftharpoons HX + OH^-$

$K_h = \frac{10^{-14}}{10^{-5}}$

So, $x = \sqrt{\frac{K_h}{C}} = \sqrt{\frac{10^{-9}}{10^{-1}}} = 10^{-4}$

$= 100 \times 10^{-4} = 10^{-2} = 0.01$

So, degree of hydrolysis = 0.01%.

31. Gram equivalent of

$(NH_4)_2SO_4 = \frac{100}{1000} \times \frac{1}{10} \times 66 = 0.66$

Gram equivalent of

$Na_2CO_3 \cdot H_2O = \frac{0.62}{62} = 0.01$

Left $(NH_4)_2SO_4 = 0.66 - 0.01 = 0.65$

Since, $(NH_4)_2SO_4$ is a salt of strong acid and weak base therefore solution will be acidic in nature.

32. According to Ostwald's dilution law, Degree of ionisation ∝ dilution

∴ At infinite dilution, strong and weak both electrolytes will be 100% ionised.

33 $AgCl \longrightarrow Ag^+ + Cl^-$

	Ag^+	Cl^-
	x	x
After adding NaCl	x	$x + 1 \times 10^{-4}$

$[Ag^+]$ decreases due to common ion effect.

34 $Be(OH)_2$ has lowest solubility and hence, lowest solubility product.

35 $KCN + H_2O \rightleftharpoons KOH + HCN$; KOH is a strong base and HCN is a weak acid, due to which solution will be basic in nature. Therefore, pH of the solution will increase.

36 $-W = q = 2.303 nRT \log \frac{V_2}{V_1}$

$= 2.303 \times 1 \times 2 \times 1000 \log \frac{100}{10}$

$= 4606$ cal

$\eta = \frac{T_2 - T_1}{T_2} = \frac{q_2 - q_1}{q_2}$

or $1 - \frac{T_1}{T_2} = 1 - \frac{q_1}{q_2}$

or $\frac{T_1}{T_2} = \frac{q_1}{q_2}$ or $\frac{300}{1000} = \frac{q_1}{4606}$,

$q_1 = 1381.8 = 1382$ cal

$W = q_2 - q_1$

$= 4606 - 1382 = 3224$ cal

37 $CH_2 = CHCH_2CH_2CH_3 + H_2 \longrightarrow CH_3CH_2CH_2CH_2CH_3$;

$\Delta H = 126 \text{ kJ mol}^{-1}$

$$H_2C{=}CH{-}CH_2{-}CH{=}CH_2 + 2H_2 \longrightarrow CH_3CH_2CH_2CH_2CH_3; \quad \Delta H = 230\ \text{kJ mol}^{-1}$$

Theoretical value of hydrogenation of two [C=C] bonds = 252 kJ mol^{-1}. Thus,

resonance energy = 252 − 230 = 22 kJ

38 According to Raoult's law for liquid mixtures,

$$p_t = p_A + p_B$$

$$\therefore \quad p_t = p^\circ_A \times \left[\frac{\frac{w_A}{M_A}}{\frac{w_A}{M_A}+\frac{w_B}{M_B}}\right] + p^\circ_B \times \left[\frac{\frac{w_B}{M_B}}{\frac{w_A}{M_A}+\frac{w_B}{M_B}}\right]$$

Given that, $w_A = 28\ g$, $w_{H_2O} = 72\ g$

$p^\circ_A = ?$

$p^\circ_{H_2O} = 155$, $M_A = 56\ g$, $M_{H_2O} = 18g$

and $p_t = 150$ mm

$$\therefore \quad 150 = p^\circ_A \times \left[\frac{\frac{28}{56}}{\frac{28}{56}+\frac{72}{18}}\right] + 155 \times \left[\frac{\frac{72}{18}}{\frac{28}{56}+\frac{72}{18}}\right]$$

$$150 = p^\circ_A \times \frac{1}{2} \times \frac{2}{9} + 155 \times 4 \times \frac{2}{9}$$

$$\therefore \quad p^\circ_A = 110\ \text{mm}$$

39 $H_2O(l)$ is taken as pure liquid, hence is not included in equilibrium.

$$HA(aq) + H_2O(l) \rightleftharpoons H_3O^+(aq) + A^-(aq)$$

	$HA(aq)$	$H_3O^+(aq)$	$A^-(aq)$
Initial	0.1 M	0	0
Equilibrium	$0.1 - \frac{0.1 \times 5}{100}$	$+\frac{0.1 \times 5}{100}$	$+\frac{0.1 \times 5}{100}$
Conc.	0.095 M	0.005 M	0.005 M

$$K_C = \frac{[H_3O^+][A^-]}{[HA]} = \frac{0.005 \times 0.005}{0.095} = 2.63 \times 10^{-4}$$

40 On dilution, $N_1V_1 = N_2V_2$

$$10^{-6} \times 100 = N_2 \times 1000$$

$$N_2 = 10^{-7}\ \text{N}$$

Since, acid solution has pH < 7, it will remain acidic even after dilution.

Hence, total $[H^+] = [H^+]_{Acid} + [H^+]_{Water}$

$$= 10^{-7} + 10^{-7} = 2 \times 10^{-7}\ \text{N}$$

$$\therefore \quad pH = -\log [H^+] = -\log (2 \times 10^{-7})$$

$$= 7 - \log 2 = 6.7$$

Thus, change in pH = 0.7 unit

41

$$\Delta G = \Delta H - T\Delta S$$

$$\Delta G = 0$$

$$\Delta H - T\Delta S = 0 \quad \text{or} \quad \Delta H = T\Delta S$$

$$T = \frac{\Delta H}{\Delta S} = \frac{-95.4 \times 1000\ \text{J}}{-198.3\ \text{JK}^{-1}} = 481\ \text{K}$$

At this temperature, the reaction would be in equilibrium and with the increase in temperature the opposing factor $T\Delta S$ would become more and hence, ΔG would become positive and the reaction would become non-spontaneous. The reaction would be spontaneous at the temperature below 481 K.

42 Salts (NH_4A, NH_4B, NH_4C) are of weak acid and weak base.

$$pH = 7 + \frac{pK_a}{2} - \frac{pK_b}{2}$$

Thus, greater the value of pK_a of HA, greater the pH.

$$pK_b(A^-, B^-, C^-) = \underset{4}{A^-} < \underset{5}{B^-} < \underset{6}{C^-}$$

$$pK_a(HA, HB, HC) = \underset{8}{HC} < \underset{9}{HB} < \underset{10}{HA}$$

Greater the pK_a, greater is the value of pH

Thus, increasing order of pH

$$NH_4C < NH_4B < NH_4A$$

43 Given, $dE = TdS - pdV$ ...(i)

$$H = E + pV \quad \text{...(ii)}$$

Differentiating Eq. (ii)

$$dH = dE + pdV + Vdp \quad \text{...(iii)}$$

From Eq. (i) and (iii), we get

$$dH = TdS + Vdp$$

44 It is clear from the figure that the process has been carried out in infinite steps, hence it is isothermal reversible expansion.

$$W = -2.303nRT\log\frac{V_2}{V_1}$$

But, $p_1V_1 = p_2V_2$

$$\Rightarrow \quad \frac{V_2}{V_1} = \frac{p_1}{p_2} = \frac{2}{1} = 2$$

$$\therefore \quad W = -2.303\ nRT\log\frac{p_1}{p_2}$$

$$= -2.303 \times 1\ \text{mol} \times 8.314\ \text{Jmol}^{-1}\text{K}^{-1} \times 298\text{K} \times \log 2$$

$$= -2.303 \times 8.314 \times 298 \times 0.3010\ \text{J}$$

$$= -1717.46\ \text{J}$$

45 Ammonium acetate is a salt of weak acid (CH_3COOH) and weak base (NH_4OH). Hence, Both *A* and *R* are true but *R* is not the correct explanation of *A*.

46 A buffer solution containing mixture of CH_3COOH and CH_3COONa is known as acidic buffer. When a equal volume of 1 M HCl is added, the pH of the solution almost remained unchanged and acidic buffer is formed.

47 When liquid crystalises, entropy decreases because in crystalline form the molecules are more ordered as campared to the liquid.

48 Both Assertion and Reason are correct and Reason is the correct explanation of the assertion.

49 Heat of neutralisation refers to the amount of heat liberated in the combination of H^+ and OH^- ions in the solution to form one mole of water. Hence, Reason is not the correct explanation of Assertion.

50 The molality of solution is expresed in terms of moles per 1000 g of solvent, i.e. depends only on masses and hence. remains unaffected by the temperature.

DAY TWELVE

Redox Reactions

Learning & Revision for the Day

- Electronic Concept of Oxidation and Reduction
- Oxidation Number
- Redox Reactions
- Balancing of Redox Reactions

Redox reactions involve both oxidation and reduction simultaneously. Originally, the term oxidation was used to describe the addition of oxygen to an element and reduction was used to describe the removal of oxygen from an element.

Electronic Concept of Oxidation and Reduction

Loss of electron by an atom is called oxidation or de-electronation, while gain of electron by an atom is called reduction or electronation.

1. **Oxidants or oxidising agents** are the substances which oxidise other and itself, get reduced by gaining electrons (i.e. their oxidation number decreases during a reaction).
2. **Reductants or reducing agents** are the substances which reduce other and itself by get oxidised losing electrons (i.e. their oxidation number increases during a reaction).

Some Important Oxidants

Some important oxidants used in oxidation are given below:

- Molecules of most electronegative elements such as O_2, O_3, halogens.
- Oxides of metals and non-metals such as MgO, CaO, CrO_3, H_2O, CO_2 etc.
- The compounds having either of an element in their highest oxidation state such as $K_2Cr_2O_7$, $KMnO_4$, $HClO_4$, H_2SO_4, HNO_3, $FeCl_3$, $HgCl_2$, $KClO_3$ etc.
- **Permanganate ion** acts as strong oxidising agent and in acidic medium, it always produces 5 electrons per formula unit irrespective of the reducing agent.

Some Important Reductants

Some important reductants used in reduction are given below:

- All metals such as Na, Al, Zn etc. and some non-metals, e.g. C, S, P, H_2 etc.
- Metallic hydrides such as NaH, LiH, KH, CaH_2 and halogen acids such as HCl, HBr, HI.
- The compounds having either of an element in their lowest oxidation state such as $H_2C_2O_4$, $FeSO_4$, Hg_2Cl_2, Cu_2O, $SnCl_2$ etc.

PREP MIRROR
Your Personal Preparation Indicator

- No. of Questions in Exercises (x)—
- No. of Questions Attempted (y)—
- No. of Correct Questions (z)—
  (Without referring Explanations)

- Accuracy Level ($z/y \times 100$)—
- Prep Level ($z/x \times 100$)—

In order to expect good rank in JEE, your Accuracy Level should be above 85 & Prep Level should be above 75.

Equivalent Weights of Oxidising Agent (OA) or Reducing Agent (RA)

$$E_{OA/RA} = \frac{\text{Molar mass of OA / RA agent}}{\text{Number of electrons lost or gained per formula unit of RA / OA}}$$

H_2O_2 is both oxidising and reducing agent but its equivalent weight as either oxidising or reducing agents are the same, i.e. 17.

Oxidation Number

The real or imaginary charge, which an atom appears to have in its combined state is called oxidation number of that atom.

Rules for Assigning Oxidation Number

- The oxidation number of an element in its elementary state and a compound is zero. e.g. H in H_2, S in S_8.
- Oxidation number of an ion is equal to the electrical charge present on it.
- Oxidation number of fluorine is always −1 in all of its compounds.
- The oxidation number of alkali metals is always +1 and those of alkaline earth metals is +2.
- Oxidation number of hydrogen is +1 except in ionic hydrides, where it is −1.
- Two oxidation numbers of N are −3 and +3, when it is bonded with less electronegative and more electronegative atoms, respectively.
- Oxidation number of oxygen is −2 except in $OF_2(+2)$, $O_2F_2(+1)$, peroxides (−1) and superoxides (−1/2).
- The oxidation number of halogens is always −1 in metal halides.
- In interhalogen compounds, the more electronegative element out of the two halogens gets the oxidation number of −1.
- Oxidation number of metals in amalgams and carbonyls, e.g. $[Fe(CO)_5]$ is zero.

NOTE Valency of an element is always a whole number. It can neither be zero nor fractional. While oxidation number may be positive or negative. It can be zero or fractional.

Redox Reactions

The reaction, which involves oxidation and reduction as its two half reactions is called redox reaction. A redox change occurs simultaneously.

Redox reactions are of following three types :

1. **Intermolecular redox reactions** are the reactions that involve the reaction between two substances, one of them is oxidant and other is reductant.

   e.g. $\underset{\text{Reductant}}{10FeSO_4} + \underset{\text{Oxidant}}{2KMnO_4} + 8H_2SO_4 \longrightarrow 2\,MnSO_4 + 5Fe_2(SO_4)_3 + K_2SO_4 + 8H_2O$

   These are further divided into two types :

   (i) **Combination reactions** in which two atoms or molecules (in their zero oxidation state) combine together and one gets oxidised while the other gets reduced.

   $$\underset{\text{Reductant}}{\overset{0}{C}} + \underset{\text{Oxidant}}{\overset{0}{O_2}} \longrightarrow \overset{+4\,-2}{CO_2}$$

   (ii) **Displacement reactions** in which an atom or ion in a compound is replaced by other atom or ion. These are further divided in two types.

   (a) **Metal displacement reactions** in which metal with low reactivity is displaced by metal with high reactivity.

   $$\underset{\text{Oxidant}}{CuSO_4} + \underset{\text{Reductant}}{Zn} \longrightarrow Cu + ZnSO_4$$

   (b) **Non-metal displacement reactions** in which non-metal is displaced.

   $$\underset{\text{Reductant}}{2\,Na} + \underset{\text{Oxidant}}{2H_2O} \longrightarrow 2NaOH + H_2$$

2. **Intramolecular redox reactions** are the reactions that involve oxidation of one element of a compound as well as reduction of other element of the same compound. Decomposition reactions are also called intramolecular redox reactions, but to be a redox reaction, it is essential that one of the products of decomposition must be in the elemental state. e.g.

   $$(NH_4)_2Cr_2O_7 \xrightarrow{\Delta} N_2 + Cr_2O_3 + 4H_2O$$

3. **Auto-redox or disproportionation reactions** are the reactions that involves oxidation and reduction of the same element.

   e.g. $Cl_2 + 2OH^- \longrightarrow ClO^- + Cl^- + H_2O$

Balancing of Redox Reactions

Following two methods are used to balance the redox reactions:

1. Ion Electron Method

This method involves the following steps :

- Write redox reaction in ionic form.
- Split redox reaction into oxidation half and reduction half reactions.
- Balance atoms of each half reactions by using simple multiples.

- For balancing H and O, if the reaction is carried out in acidic medium add H^+ ion and H_2O to the appropriate sides, similarly for basic medium add OH^- and H_2O to the appropriate sides.
- Balance the charge on both the sides and multiply one or both half reactions by suitable number to equalise number of electrons in both equations.
- Add the two balance half-reactions and cancel common terms.

2. Oxidation Number Method

The method involves the following steps :

- Assign oxidation number to the atoms in the equation and write separate equations for atoms undergoing oxidation and reduction.
- Find the change in oxidation number in each equation and make the change equal in both the equations by multiplying with suitable integers.
- After adding both the equations complete the balancing by balancing H and O.

DAY PRACTICE SESSION 1

FOUNDATION QUESTIONS EXERCISE

1 Which of the following is not a reducing agent?

(a) SO_2 (b) H_2O_2 (c) CO_2 (d) NO_2

2 In which of the following reaction, nitric oxide acts as a reducing agent?

(a) $4NH_3 + 5O_2 \longrightarrow 2NO + 6H_2O$
(b) $2NO + 3I_2 + 4H_2O \longrightarrow 2NO_3^- + 6I^- + 8H^+$
(c) $2NO + H_2SO_3 \longrightarrow N_2O + H_2SO_4$
(d) $2NO + H_2S \longrightarrow N_2O + S + H_2O$

3 Which one of the following cannot function as an oxidising agent? → JEE Main (Online) 2013

(a) I^- (b) S(s) (c) $NO_3^-(aq)$ (d) $Cr_2O_7^{2-}$

4 In which of the following reactions, hydrogen is acting as an oxidising agent ?

(a) With iodine to give hydrogen iodide
(b) With lithium to give lithium hydride
(c) With nitrogen to give ammonia
(d) With sulphur to give hydrogen sulphide

5 The compound that can work both as an oxidising as well as reducing agent is

(a) $KMnO_4$ (b) H_2O_2 (c) $Fe_2(SO_4)_3$ (d) $K_2Cr_2O_7$

6 MnO_4^- is a good oxidising agent in different medium changing to

$MnO_4^- \longrightarrow Mn^{2+} \longrightarrow MnO_4^{2-} \longrightarrow MnO_2 \longrightarrow Mn_2O_3$

Change in oxidation number respectively, are

(a) 1, 3, 4, 5 (b) 5, 4, 3, 2
(c) 5, 1, 3, 4 (d) 2, 6, 4, 3

7. The oxidation numbers of phosphorus in $Ba(H_2PO_2)_2$ and xenon in Na_4XeO_6 respectively are

(a) +3 and +4 (b) +2 and +6 (c) +1 and +8 (d) –1 and –6

8 In which of the following pairs, there is greatest difference in the oxidation number of the underlined elements?

(a) $\underline{N}O_2$ and $\underline{N}_2O_4$ (b) $\underline{P}_2O_5$ and $\underline{P}_4O_{10}$
(c) $\underline{N}_2O$ and $\underline{N}O$ (d) $\underline{S}O_2$ and $\underline{S}O_3$

9 The oxidation number of an element in a compound is evaluated on the basis of certain rules. Which of the following rules is not correct in this respect?

(a) The oxidation number of hydrogen is always +1
(b) The algebraic sum of all the oxidation numbers in a compound is zero
(c) An element in the free or the uncombined state bears oxidation number zero
(d) In all its compounds, the oxidation number of fluorine is –1

10 The oxidation number of sulphur in S_8, S_2F_2 and H_2S respectively are

(a) 0, +1 and –2 (b) +2, +1 and –2
(c) 0, + 1 and + 2 (d) –2, + 1 and – 2

11 In which of the following, the oxidation number of oxygen has been arranged in increasing order?

(a) $BaO_2 < KO_2 < O_3 < OF_2$
(b) $OF_2 < KO_2 < BaO_2 < O_3$
(c) $BaO_2 < O_3 < OF_2 < KO_2$
(d) $KO_2 < OF_2 < O_3 < BaO_2$

12 In the reaction,

$$3Br_2 + 6CO_3^{2-} + 3H_2O \longrightarrow 5Br^- + BrO_3^- + 6HCO_3^-$$

(a) bromine is oxidised and the carbonate radical is reduced
(b) bromine is reduced and the carbonate radical is oxidised
(c) bromine is neither reduced nor oxidised
(d) bromine is both reduced and oxidised

13 The largest oxidation number exhibited by an element depends on its outer electronic configuration. With which of the following outer electronic configurations, the element will exhibit largest oxidation number?

(a) $3d^1 4s^2$ (b) $3d^3 4s^2$ (c) $3d^5 4s^1$ (d) $3d^5 4s^2$

14 In which of the following compounds, an element exhibits two different oxidation states?

(a) NH_2OH (b) NH_4NO_3
(c) N_2H_4 (d) N_3H

15 When SO_2 is passed through an acidified solution of potassium dichromate, the oxidation state of S changes from

(a) + 4 to 0 (b) + 4 to +2
(c) + 4 to +6 (d) + 6 to +4

16 In which of the compounds does manganese exhibit highest oxidation number?

(a) MnO_2 (b) Mn_3O_4
(c) K_2MnO_4 (d) $MnSO_4$

17 Which order of compounds is according to the decreasing order of the oxidation state of nitrogen?

(a) HNO_3, NO, NH_4Cl, N_2 (b) HNO_3, NO, N_2, NH_4Cl
(c) HNO_3, NH_4Cl, NO, N_2 (d) NO, HNO_3, NH_4Cl, N_2

18 The difference in the oxidation numbers of the two types of sulphur atoms in $Na_2S_4O_6$ is

(a) 4 (b) 5 (c) 6 (d) 7

19 Oxidation states of the metal in the minerals haematite and magnetite, respectively are

(a) II, III in haematite and III in magnetite
(b) II, III in haematite and II in magnetite
(c) II in haematite and II, III in magnetite
(d) III in haematite and II, III in magnetite

20 The oxidation state of chromium in the final product formed by the reaction between KI and acidified potassium dichromate solution is

(a) +3 (b) +2 (c) +6 (d) +4

21 Oxidation state of sulphur in anions SO_3^{2-}, $S_2O_4^{2-}$ and $S_2O_6^{2-}$ increases in the order → JEE Main (Online) 2013

(a) $S_2O_6^{2-} < S_2O_4^{2-} < SO_3^{2-}$ (b) $SO_3^{2-} < S_2O_4^{2-} < S_2O_6^{2-}$
(c) $S_2O_4^{2-} < SO_3^{2-} < S_2O_6^{2-}$ (d) $S_2O_4^{2-} < S_2O_6^{2-} < SO_3^{2-}$

22 Six moles of Cl_2 undergo a loss and gain of 10 moles of electrons to form two oxidation states of Cl in an autoredox change. What are the two oxidation states of Cl in this change?

(a) +5, −1 (b) +7, −1
(c) +3, 0 (d) +3, −1

23 Among the properties *A* reducing, *B* oxidising and *C* complexing, the set of properties shown by CN^- ion towards metal species is

(a) *A*, *B* (b) *B*, *C*
(c) *C*, *A* (d) *A*, *B*, *C*

24 Which of the following is a redox reaction?

(a) Formation of glucose from CO_2 and water
(b) Reaction of potassium cyanide with silver cyanide
(c) Hydration of rubidium
(d) Reaction of barium chloride with sulphuric acid

25 Which of the following is a redox reaction?

(a) $NaCl + KNO_3 \longrightarrow NaNO_3 + KCl$
(b) $CaC_2O_4 + 2HCl \longrightarrow CaCl_2 + H_2C_2O_4$
(c) $Ca(OH)_2 + 2NH_4Cl \longrightarrow CaCl_2 + 2NH_3 + 2H_2O$
(d) $2K[Ag(CN)_2] + Zn \longrightarrow 2Ag + K_2[Zn(CN)_4]$

26 For the redox reaction,

$$MnO_4^- + C_2O_4^{2-} + H^+ \longrightarrow Mn^{2+} + CO_2 + H_2O$$

the correct coefficients of the reactants for the balanced reaction are

	MnO_4^-	$C_2O_4^{2-}$	H^+
(a)	2	5	16
(b)	16	5	2
(c)	5	16	2
(d)	2	16	5

27 $C_2H_6(g) + nO_2 \longrightarrow CO_2(g) + H_2O(l)$

In this equation, ratio of the coefficients of CO_2 and H_2O is

(a) 1 : 1 (b) 2 : 3
(c) 3 : 2 (d) 1 : 3

28 Given, $x\,Na_2HAsO_3 + y\,NaBrO_3 + z\,HCl \longrightarrow NaBr + H_3AsO_4 + NaCl$

The value of *x*, *y* and *z* in the above redox reaction respectively are → JEE Main (Online) 2013

(a) 2, 1, 2 (b) 2, 1, 3
(c) 3, 1, 6 (d) 3, 1, 4

29 Consider the following reaction,

$$x\,MnO_4^- + y\,C_2O_4^{2-} + z\,H^+ \longrightarrow x\,Mn^{2+} + 2y\,CO_2 + \frac{z}{2}H_2O.$$

The values of *x*, *y* and *z* in the reaction respectively are → JEE Main 2013

(a) 2, 5 and 8 (b) 2, 5 and 16
(c) 5, 2 and 8 (d) 5, 2 and 16

30 **Assertion** (A) Oxidation number of chromium in CrO_5 is + 6.

Reason (R) Oxidation number of each oxygen atom is −1.

(a) Assertion and Reason both are correct statements and Reason is the correct explanation of the Assertion
(b) Assertion and Reason both are correct statements but Reason is not the correct explanation of the Assertion
(c) Assertion is correct and Reason is incorrect
(d) Both Assertion and Reason are incorrect

DAY PRACTICE SESSION 2

PROGRESSIVE QUESTIONS EXERCISE

1 Which of the following statements is not correct?

(a) The oxidation number of S in $(NH_4)_2S_2O_8$ is +6

(b) The oxidation number of Os in OsO_4 is +8

(c) The oxidation number of S in H_2SO_5 is +8

(d) The oxidation number of O in KO_2 is $-\frac{1}{2}$

2 In a reaction, 4 moles of electrons are transferred to 1 mole of HNO_3. The possible product obtained due to reduction is

(a) 0.5 mole of N_2 (b) 0.5 mole of N_2O

(c) 1 mole of NO_2 (d) 1 mole of NH_3

3 When the following half-reaction is balanced

$$CN^- \longrightarrow CNO^-$$

Which of the following statements is true regarding the balance half-reaction?

(a) Carbon is losing two electrons per atom

(b) Oxidation number of carbon increases from +1 to +3

(c) Oxidation number of nitrogen remains constant

(d) Statements (a) and (c) both are true

4 Photographic paper is developed with alkaline hydroquinone

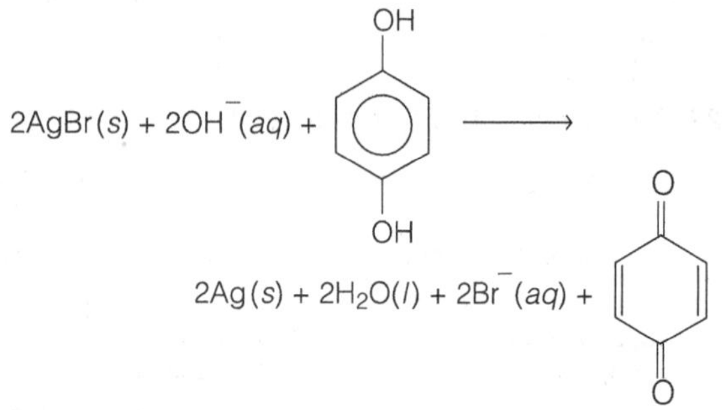

Select the correct statement.

(a) Hydroquinone is the oxidant

(b) Ag^+ is the oxidant

(c) Br^- is the oxidant

(d) Ag^+ is the reductant

5 For the following reaction, consider the following statements

$$2Cr(OH)_3 + 3H_2O_2 + 4OH^- \longrightarrow 2CrO_4^{2-} + 8H_2O$$

I. there is colour change from green precipitate to yellow coloured solution.

II. oxidation number of Cr changes from +3 to +6.

III. oxidation number of O in H_2O_2 changes from −2 to −1.

Select the correct statement(s).

(a) Only I (b) Both I and II

(c) Only II (d) Both I and III

6 Oxidation states of X, Y, Z are +2 +5 and −2 respectively. Formula of the compound formed by these will be

(a) X_2YZ_6 (b) XY_2Z_6

(c) XY_5 (d) X_3YZ_4

7 I^- reduces IO_3^- to I_2 and itself oxidised to I_2 in acidic medium. Final reaction is

(a) $I^- + IO_3^- + 6H^+ \longrightarrow I_2 + 3H_2O$

(b) $I^- + IO_3^- \longrightarrow I_2 + O_3$

(c) $5I^- + IO_3^- + 6H^+ \longrightarrow 3I_2 + 3H_2O$

(d) None of the above

8 Thiosulphate reacts differently with iodine and bromine in the reactions given below:

$$2S_2O_3^{2-} + I_2 \longrightarrow S_4O_6^{2-} + 2I^-$$

$$S_2O_3^{2-} + 2Br_2 + 5H_2O \longrightarrow 2SO_4^{2-} + 2Br^- + 10H^+$$

Which of the following statements justifies the above dual behaviour of thiosulphate?

(a) Bromine is a stronger oxidant than iodine

(b) Bromine is a weaker oxidant than iodine

(c) Thiosulphate undergoes oxidation by bromine and reduction by iodine in these reactions

(d) Bromine undergoes oxidation and iodine undergoes reduction in these reactions

9 Which has maximum number of equivalent per mole of the oxidant?

(a) $Zn(s) + VO^{2+}(aq) \longrightarrow Zn^{2+}(aq) + V^{3+}(aq)$

(b) $Ag(s) + NO_3^-(aq) \longrightarrow Ag^+(aq) + NO_2(g)$

(c) $Mg(s) + VO_4^{3-}(aq) \longrightarrow Mg^{2+}(aq) + V^{2+}(aq)$

(d) $I^-(aq) + IO_3^-(aq) \longrightarrow I_3^-(aq)$

10 For the reaction between MnO_4^- and $C_2O_4^{2-}$ in basic solution, the unbalanced equation is

$$MnO_4^- + C_2O_4^{2-} \longrightarrow MnO_2 + CO_2$$

In a balanced equation, the number of OH^- ions is

(a) 0 (b) 8 on the right

(c) 4 on the left (d) 2 on the left

ANSWERS

SESSION 1	1 (c)	2 (b)	3 (a)	4 (b)	5 (b)	6 (c)	7 (c)	8 (d)	9 (a)	10 (a)
	11 (a)	12 (d)	13 (d)	14 (b)	15 (c)	16 (c)	17 (b)	18 (b)	19 (d)	20 (a)
	21 (c)	22 (a)	23 (c)	24 (a)	25 (d)	26 (a)	27 (b)	28 (c)	29 (b)	30 (c)
SESSION 2	1 (c)	2 (b)	3 (d)	4 (b)	5 (b)	6 (b)	7 (c)	8 (a)	9 (d)	10 (b)

Hints and Explanations

SESSION 1

1 In CO_2, the oxidation number of C is +4 and is already the maximum. It cannot increase its oxidation number, hence, it does not act as a reducing agent.

2 $2\overset{+2}{N}O + 3I_2 + 4H_2O \longrightarrow 2\overset{+5}{N}O_3^- + 6I^- + 8H^+$

Hence, NO acts as a reducing agent and reduces I_2 to I^- since, the oxidation number of nitrogen changes from +2 in NO to +5 in NO_3^-.

3 In I^-, iodine is present in its lowest possible oxidation state. Further, reduction in oxidation state is not possible. Hence, it cannot function as oxidising agent.

4 $2\overset{0}{Li} + \overset{0}{H_2} \longrightarrow 2\overset{+1}{Li}\overset{-1}{H}$

5 The oxidation number of O in H_2O_2 is −1. It can either increases to zero in O_2 or decreases to −2 in H_2O. Therefore, H_2O_2 can act both as an oxidising as well as reducing agent.

6 $\overset{+7}{Mn}O_4^- \longrightarrow Mn^{2+}$ change in oxidation number 5

$\longrightarrow \overset{+6}{Mn}O_4^{2-}$ change in oxidation number 1

$\longrightarrow \overset{+4}{Mn}O_2$ change in oxidation number 3

$\longrightarrow \overset{+3}{Mn_2}O_3$ change in oxidation number 4

7 $\overset{+2}{Ba}(\overset{+1}{H_2}\overset{x}{P}\overset{-2}{O_2})_2$

$\therefore \quad 2 + 2\,[2 \times (+1) + x + 2 \times (-2)] = 0$

or $\quad 2 + 4 + 2x - 8 = 0$ or $x = +1$

and $\overset{+1}{Na_4}\overset{x}{Xe}\overset{-2}{O_6}$

$\therefore 4 \times 1 + x + 6 \times (-2) = 0$ or $x = +8$

8 (a) $\overset{+4}{N}O_2$ and $\overset{+4}{N_2}O_4$; Difference = 0

(b) $\overset{+5}{P_2}O_5$ and $\overset{+5}{P_4}O_{10}$; Difference = 0

(c) $\overset{+1}{N_2}O$ and $\overset{+2}{N}O$; Difference $= +2 - 1 = +1$

(d) $\overset{+4}{S}O_2$ and $\overset{+6}{S}O_3$; Difference $= +6 - 4 = +2$

9 Oxidation number of hydrogen is +1 except in ionic hydrides, where it is −1.

10 $\overset{0}{S_8}, \overset{+1}{S_2}\overset{-1}{F_2}, \overset{+1}{H_2}\overset{-2}{S}$

$\therefore$ Oxidation number of S in S_8, S_2F_2 and H_2S respectively are 0, +1, −2.

11 $Ba\overset{-1}{O_2} < K\overset{-0.5}{O_2} < \overset{0}{O_3} < \overset{+2}{O}F_2$

12 $3\overset{0}{Br_2} + 6CO_3^{2-} + 3H_2O \longrightarrow 5\overset{-1}{Br^-} + \overset{+5}{Br}O_3^- + 6HCO_3^-$

Br_2 is reduced to Br^- (oxidation number decreases from zero to −1) and Br_2 is oxidised to BrO_3^- (oxidation number increases from zero to +5).

13 $3d^5 4s^2 = +7$ oxidation number

14 NH_4NO_3 is an ionic compound exist in the form of NH_4^+, NO_3^-
In NH_4^+ ; $x + 4 = +1$ or $x = -3$
In NO_3^- ; $x - 6 = -1$ or $x = +5$

15 $K_2Cr_2O_7 + 4H_2SO_4 + 3SO_2 \longrightarrow K_2SO_4 + Cr_2(SO_4)_3 + 3\overset{+6}{S}O_3 + 4H_2O$

Oxidation number of S changes from +4 in SO_2 to +6 in SO_3.

16 Highest oxidation number of Mn in K_2MnO_4 is, $2 + x - 8 = 0$ or $x = +6$, while in all other compounds oxidation number of Mn is lower than 6.

17 Let the oxidation number of N be x

In $HNO_3 \rightarrow +1 + x + 3(-2) = 0$ or $x = +5$

In $NO \rightarrow \quad x - 2 = 0$ or $x = +2$

In $N_2 \rightarrow \quad x = 0$

In $NH_4Cl \rightarrow x + 4(1) - 1 = 0$ or $x = -3$

Thus, the order of oxidation states of nitrogen is

$HNO_3 > NO > N_2 > NH_4Cl$.

18 $Na_2S_4O_6$ is a salt of $H_2S_4O_6$, which has the following structure.

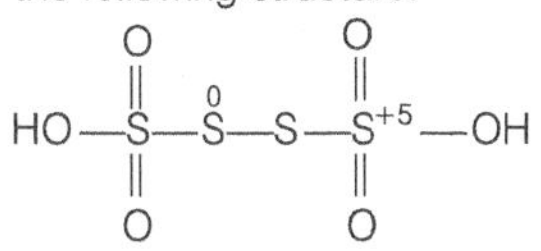

$\therefore$ The difference in oxidation number is 5.

19 Haematite is Fe_2O_3, in which oxidation number of iron is (III). Magnetite is Fe_3O_4, which is infact a mixed oxide $(FeO \cdot Fe_2O_3)$, hence, iron is present in both (II) and (III) oxidation states.

20 $Cr_2O_7^{2-} + 14H^+ + 6I^- \longrightarrow 2Cr^{3+} + 7H_2O + 3I_2$

$Cr_2O_7^{2-}$ is reduced to Cr^{3+}. Thus, final state of Cr is + 3.

21 Let the oxidation state of S be x.

$SO_3^{2-} \quad x + (-2) \times 3 = -2$

$x - 6 = -2$

$\therefore \quad x = +4$

$S_2O_4^{2-} \quad 2 \times x + (-2)4 = -2$

$2x - 8 = -2$

$2x = 6$ or $x = +3$

$S_2O_6^{2-} \quad 2x + (-2)6 = -2$

$2x = 10 \Rightarrow x = +5$

$\therefore$ The increasing order of oxidation states is $S_2O_4^{2-} < SO_3^{2-} < S_2O_6^{2-}$.

22 **Oxidation** $Cl_2 \longrightarrow 2\overset{5+}{Cl} + 10e^-$

Reduction $5Cl_2 + 10e^- \longrightarrow 10Cl^-$

Overall redox reaction,

$6Cl_2 \longrightarrow 2\overset{5+}{Cl} + 10Cl^-$

23 CN^- is a better complexing agent (*C*) as well as a reducing agent (*A*)

Thus, properties (*A*) and (*C*) are shown.

Property (*C*): $Ni^{2+} + 4CN^- \longrightarrow [Ni(CN)_4]^{2-}$

Property (*A*):

$$\overset{+2}{Cu}Cl_2 + 5KCN \longrightarrow K_3[\overset{+1}{Cu}(CN)_4] + \frac{1}{2}(CN)_2 + 2KCl$$

(CN^- reduces Cu^{2+} to Cu^+)

24. $$\overset{+4}{6CO_2} + 6H_2\overset{-2}{O} \xrightarrow{\text{Sunlight}} \overset{0}{C_6}H_{12}O_6 + 6\overset{0}{O_2}$$

(Oxidation: O from −2 to 0; Reduction: C from +4 to 0)

Since, oxidation and reduction both occurs simultaneously in the above equation, so it is a redox reaction.

25 (a) $\overset{+1}{Na}\overset{-1}{Cl} + \overset{+1}{K}\overset{-1}{NO_3} \longrightarrow \overset{+1}{Na}\overset{-1}{NO_3} + \overset{+1}{K}\overset{-1}{Cl}$

(b) $\overset{+2}{Ca}\overset{-2}{C_2O_4} + 2\overset{+1}{H}\overset{-1}{Cl} \longrightarrow \overset{+2}{Ca}\overset{-1}{Cl_2} + \overset{+1}{H_2}\overset{-2}{C_2O_4}$

(c) $\overset{+2}{Ca}\overset{-1}{(OH)_2} + 2\overset{-3}{N}\overset{+1}{H_4}\overset{-1}{Cl} \longrightarrow \overset{+2}{Ca}\overset{-1}{Cl_2} + 2\overset{-3}{N}\overset{+1}{H_3} + 2\overset{+1}{H_2}\overset{-2}{O}$

In all these (i.e. a, b, c) cases during reaction, there is no change in oxidation state of ion or molecule or constituent atom.

∴ These are simply ionic reactions.

(d) $2K[Ag(CN)_2] + Zn \longrightarrow 2Ag + K_2[Zn(CN)_4]$

$Ag^+ \longrightarrow Ag$; gain of e^-, (reduction)

$Zn \longrightarrow Zn^{2+}$; loss of e^-, (oxidation)

26 $[MnO_4^- + 8H^+ + 5e^- \longrightarrow Mn^{2+} + 4H_2O] \times 2$

$[C_2O_4^{2-} \longrightarrow 2CO_2 + 2e^-] \times 5$

$2MnO_4^- + 5C_2O_4^{2-} + 16H^+ \longrightarrow 2Mn^{2+} + 10CO_2 + 8H_2O$

Thus, the coefficients of MnO_4^-, $C_2O_4^{2-}$ and H^+ in the above balanced equation respectively are 2, 5, 16.

27 The balanced equation is

$$2C_2H_6 + 5O_2 \longrightarrow 4CO_2 + 6H_2O$$

Ratio of the coefficient of CO_2 and H_2O is 4 : 6 or 2 : 3.

28 $3Na_2HAsO_3 + NaBrO_3 + 6HCl \longrightarrow NaBr + 3H_3AsO_4 + 6NaCl$

29 $2MnO_4^- + 5C_2O_4^{2-} + 16H^+ \longrightarrow 2Mn^{2+} + 10\,CO_2 + 8H_2O$

$x = 2, y = 5, z = 16$

30 In CrO_5, four oxygen atoms are in −1 oxidation state.

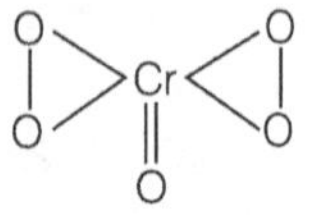

$$x + 1(-2) + 4(-1) = 0 \Rightarrow x = +6$$

Thus, Assertion is true but Reason is false.

SESSION 2

1 (a) $(NH_4)_2S_2O_8^{2-}$

∴ Oxidation state of S = + 6

(Since, $S_2O_8^{2-}$ has one peroxide bond)

(b) Oxidation state of Os = + 8

(c) Oxidation state of S in H_2SO_5 = + 6

(Since, it has one peroxide bond)

(d) $K^+O_2^-$ oxidation state of O $= -\frac{1}{2}$

2 $\underset{+5}{HNO_3} + 4e^- \longrightarrow$ Product has oxidation number = 1

In N_2O oxidation number of N = +1. Thus, 0.5 mole of N_2O are formed by reduction of 1 mole of HNO_3 by 4 moles of electrons.

3 In the given redox reaction, $C^- \equiv N \longrightarrow O^- — C \equiv N$

Oxidation number of nitrogen remains unchanged as − 3. Oxidation number of carbon is increasing from + 2 to + 4. Hence, (a) and (c) are the only correct response. Therefore, option (d) is correct.

4 Ag^+ is reduced hence, it is an oxidant. Hydroquinone is oxidised hence, it is a reductant.

5 (I) $Cr(OH)_3$ (green precipitate) $\longrightarrow CrO_4^{2-}$ (yellow)

(II) Cr(+3) $\longrightarrow$ Cr(+6)

(III) oxidation number of O in H_2O_2 changes from −1 to −2.

Thus, (I) and (II) are true.

6 The oxidation state of *X*, *Y* and *Z* are +2, +5 and −2 respectively.

I. In $X_2YZ_6 = 2 \times 2 + 5 + 6(-2) \neq 0$

II. In $XY_2Z_6 = 2 + 5 \times 2 + 6(-2) = 0$

III. In $XY_5 = 2 + 5 \times 2 \neq 0$

IV. In $X_3YZ_4 = 3 \times 2 + 5 + 4(-2) \neq 0$

Hence, formula of compound is XY_2Z_6.

7 $5I^- + IO_3^- + 6H^+ \longrightarrow 3I_2 + 3H_2O$

8 $2\overset{+2}{S_2}\overset{-2}{O_3^{2-}}(aq) + \overset{0}{I_2}(s) \longrightarrow \overset{2.5}{S_4}O_6^{2-}(aq) + 2I^-(aq)$

$\overset{+2}{S_2}\overset{-2}{O_3^{2-}}(aq) + 2\overset{0}{Br_2}(l) + 5H_2O(l) \longrightarrow 2\overset{+6}{S}\overset{2-}{O_4^{2-}}(aq) + 4Br^-(aq) + 10H^+(aq)$

Bromine being stronger oxidising agent than I_2, oxidises S of $S_2O_3^{2-}$ to SO_4^{2-}, whereas I_2 oxidises it only into $S_4O_6^{2-}$ ion.

9

	Reaction	Change in oxidation number	Number of equivalent
(a)	$VO^{2+}(aq) \longrightarrow V^{3+}(aq)$; $x - 2 = +2$, $x = +4$ → +3	1 unit	1
(b)	$NO_3^-(aq) \longrightarrow NO_2(g)$; +5 → +4	1 unit	1
(c)	$VO_4^{3-}(aq) \longrightarrow V^{2+}(aq)$; +5 → +2	3 units	3
(d)	$IO_3^-(aq) \longrightarrow I_3^-(aq)$; +5 → $-\frac{1}{3}$	$\frac{16}{3}$	$\frac{16}{3}$

10 $2MnO_4^- + 3C_2O_4^{2-} + 4H_2O \longrightarrow 2MnO_2 + 6CO_2 + 8OH^-$

∴ The number of OH^- ions is 8 on the right side.

DAY THIRTEEN

Electrochemistry

Learning & Revision for the Day

- Metallic and Electrolytic Conduction
- Electrolysis
- Kohlrausch's Law
- Electrochemical Cell
- Electrochemical Series
- Nernst Equation and its Applications
- Batteries
- Fuel Cells
- Corrosion and its Prevention

The study of the chemical reactions which take place in a solution at the interface of an electron conductor and an ionic conductor is considered under the branch of chemistry namely as electrochemistry.

Metallic and Electrolytic Conduction

The substance which can conduct electricity are called conductors. On the basis of species that conduct electricity (current), conductors are of two types :

(i) **Metallic or electronic conductors** can conduct current by transfer of free electrons. This process is known as **metallic conduction**.

(ii) **Electrolytic conductors** can conduct current by the mobility of ions. This process is known as **electrolytic conduction**.

Conductance in Electrolytic Solutions

- The power of an electrolyte to conduct electric current is called conductance or **conductivity**. Just like metallic conductors, electrolytic solutions also obey Ohm's law.
- Reciprocal of resistance is called conductance, G.

$$G = \frac{1}{\text{resistance}} = \frac{1}{R}$$. Its unit is mho or (Ω^{-1}) or Siemens (S).

Conductivity or Specific Conductance (κ)

- The resistance of any conductor varies directly as its length (l) and inversely as its cross-sectional area (a), i.e. $R \propto \frac{l}{a}$ or $R = \rho \frac{l}{a}$

  where, ρ is called the **resistivity** or **specific resistance**.
- If $l = 1$ cm and $a = 1\text{ cm}^2$ then $R = \rho$

$$\therefore \quad \kappa = \frac{1}{\rho} = \frac{l}{a} \times \text{conductance } (G)$$

PREP MIRROR
Your Personal Preparation Indicator

- No. of Questions in Exercises (x)—
- No. of Questions Attempted (y)—
- No. of Correct Questions (z)—
  (Without referring Explanations)

- Accuracy Level ($z/y \times 100$)—
- Prep Level ($z/x \times 100$)—

In order to expect good rank in JEE, your Accuracy Level should be above 85 & Prep Level should be above 75.

where, $\frac{l}{a}$ = cell constant, it is determined with the help of conductivity bridge, where a standard solution of KCl is used.

- The unit of specific conductance (κ) is $ohm^{-1}\ cm^{-1}$ or $S\ cm^{-1}$.

Molar Conductivity (Λ_m)

It is the conducting power of all the ions produced by one gram mole of an electrolyte in a given solution.

Thus, $\Lambda_m = \frac{\kappa \times 1000}{\text{molarity}}$

The units for molar conductivity

$= ohm^{-1} cm^2 mol^{-1}$ or $S\ cm^2\ mol^{-1}$

Equivalent Conductivity (Λ_{eq})

It is the conducting power of all the ions produced by one gram equivalent of an electrolyte in a given solution.

Thus, $\Lambda_{eq} = \frac{\kappa \times 1000}{\text{normality}}$. The units for equivalent conductivity

$= ohm^{-1}\ cm^2\ (g\text{-}eq)^{-1}$ or $S\ cm^2\ (g\text{-}eq)^{-1}$

Equivalent as well as molar conductivity $\propto$ dilution and specific conductivity $\propto \frac{1}{\text{dilution}}$

Variation of Molar and Equivalent Conductivities with Concentration

Strong electrolytes, like KCl, have high value of conductance even at low concentration and there is no rapid increase in their equivalent or molar conductance on dilution and can be represented by Debye-Huckel-Onsager equation,

$$\Lambda_m = \Lambda_m^\circ - AC^{\frac{1}{2}}$$

where, Λ_m = molar conductivity

Λ_m° = limiting molar conductivity

A = Debye-Huckel constant

C = concentration

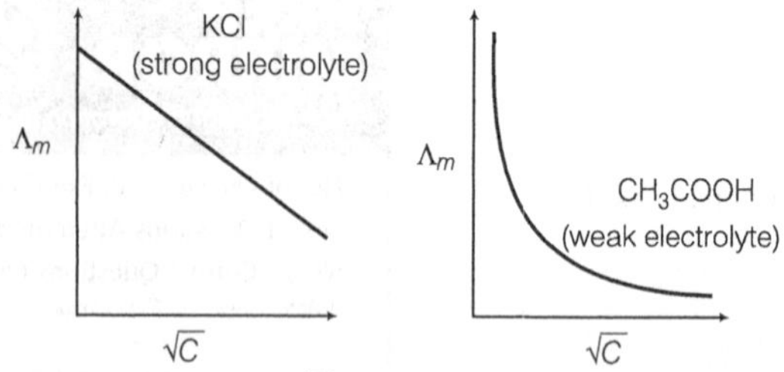

Variation of Λ_m with $\sqrt{C}$ for strong and weak electrolytes

- In case of weak electrolytes, like acetic acid, have a low value of conductance at high concentration and there is a rapid increase in the value of equivalent conductance (molar conductance) with dilution.
- Limiting molar conductivity or infinite conductivity (Λ_m° or Λ_m^∞) is defined as the molar conductivity of electrolyte when concentration of electrolyte approaches zero (i.e. at infinite dilution).

Electrolysis

- Electrolysis is a process of decomposition of an electrolyte by the passage of electricity through its aqueous solution or molten state. As a result cations move towards cathode while anions towards anode.
- The reduction and oxidation changes occurs at respective electrodes during electrolysis. Oxidation occurs at anode while reduction occurs at cathode.
- An electrolytic solution usually consists of more than two ions. When electrolysis is done then all the ions are not discharged at electrodes simultaneously. Certain ions are liberated at the electrodes in prefer to other.
For cations, which is stronger oxidising agent is discharged at cathode.

$$\xrightarrow[\text{Increasing order of decomposition}]{K^+, Ca^{2+}, Na^+, Mg^{2+}, Al^{3+}, Zn^{2+}, Fe^{2+}, H^+, Cu^{2+}, Ag^+, Au^{3+}}$$

For anions, which is stronger reducing agent is liberated at anode.

$$\xrightarrow[\text{Increasing order of discharge}]{SO_4^{2}, NO_3^{2-}, OH^-, Cl^-, Br^-, I^-}$$

Faraday's Laws of Electrolysis

1. **According to First Law of Faraday** The amount of chemical change produced is proportional to the quantity of electric charge passing through an electrolytic cell.

$$w \propto Q, w = ZQ. \quad w = Zit$$

where, w = mass of the metal deposited/gas liberated (in grams);

Q = charge (in Coulomb)

i = current (in amperes), t = time (in second)

Z = electrochemical equivalent.

$$= \frac{\text{equivalent wt.}}{96500\ C} \left[\text{Equivalent wt.} = \frac{\text{atomic wt.}}{\text{valency}}\right]$$

- 1 Faraday = charge of one mole of electrons
$1F = 6.022 \times 10^{23} \times 1.6 \times 10^{-19} = 96500\ C$ (approx.)
- Number of gram equivalents = Number of Faradays of electricity, other forms of Faraday first law expression are $w = ZQ = \frac{E}{F} \times Q = \frac{Q}{F} \times \frac{M}{Z}$
$= \frac{it}{F} \times \frac{M}{Z}$
- One Faraday or 96500 C or 1 mole of electron cause the reduction of 1 mole of monovalent cation or 1/2 mole of divalent cation.

2. **According to Second Law of Faraday** The number of equivalents of any substance produced by a given quantity of electricity during electrolysis are same.

$$\frac{w_A}{w_B}=\frac{E_A}{E_B}$$

where, w_A = deposited mass of substance A,
E_A = equivalent weight of A,
w_B = deposited mass of substance B,
E_B = equivalent weight of B.

Kohlrausch's Law

It states that molar conductance at infinite dilution for any electrolyte is the sum of contribution of its constituent ions, i.e. anions and cations.

$$\overset{\circ}{\Lambda}_m = \overset{\circ}{\Lambda}_a + \overset{\circ}{\Lambda}_c$$

e.g. $\overset{\circ}{\Lambda}_m CH_3COOH \longrightarrow \overset{\circ}{\Lambda}_m CH_3COO^- + \overset{\circ}{\Lambda}_m H^+$

$\overset{\circ}{\Lambda}_m Al_2(SO_4)_3 \longrightarrow 2\times\overset{\circ}{\Lambda}_m Al^{3+} + 3\times\overset{\circ}{\Lambda}_m SO_4^{2-}$

Applications of Kohlrausch's Law

- For the determination of equivalent/molar conductivity at infinite dilution.
- For the determination of degree of dissociation.

  Degree of dissociation (α)

  $$=\frac{\text{molar conductance at a given concentration}}{\text{molar conductance at infinite dilution}}$$

  $$\alpha=\frac{\Lambda_m^c}{\Lambda_m^\infty}$$

- For the calculation of dissociation constant of a weak electrolyte.

  $$K_a=\frac{C\alpha^2}{1-\alpha}=C\alpha^2 \quad [\because \text{For weak electrolyte, } \alpha <<< 1]$$

  Here, K_a = equilibrium dissociation constant
  C = molar concentration of weak electrolyte

- For the determination of solubility of sparingly soluble salt.

  $$\text{Solubility}=\frac{\kappa\times 1000}{\overset{\circ}{\Lambda}_m}$$

Electrochemical Cell

A cell is an arrangement in which two electrodes are fitted in the same electrolyte or in which two different electrolytes are joined by a salt-bridge.

Cells are of the following two types:

1. Electrolytic Cells

It is a device in which electrolysis (chemical reaction involving oxidation and reduction) is carried out by using electricity or in which conversion of electrical energy into chemical energy is done.

$$H_2SO_4\,(l) \underset{\text{(Acidified/dil.)}}{\overset{H^+}{\rightleftharpoons}} H_2O(g) + O_2(g) + \text{dil. } H_2SO_4$$

2. Galvanic or Voltaic Cells

It is a device in which a redox reaction used to convert chemical energy into electrical energy.

$$Cu(s) + 2Ag^+(aq) \rightleftharpoons Cu^{2+}(aq) + 2Ag(s)$$

Electrode and Half-Cell

A strip of metal, M called electrode is immersed in a solution containing the metal ion M^{n+}, this combination of the metal electrode and solution is called a half-cell.

Three kinds of interactions are possible between metal atom on the electrode and metal ion in solution.

- A metal ion M^{n+} may collide with the electrode and undergo no change.
- A metal ion M^{n+} may collide with the electrode, gain n electrons and be converted to a metal atom M. The ion is reduced.

  $$M^{n+}(aq) + ne^- \xrightarrow{\text{Reduction}} M(s)$$

- A metal atom M on the electrode may lose n electrons and enter the solution as ion M^{n+}. The metal atom is oxidised.

  $$M\,(s) \xrightarrow{\text{Oxidation}} M^{n+}\,(aq) + ne^-$$

	Electrolytic cell		Voltaic or Galvanic cell	
	Anode	**Cathode**	**Anode**	**Cathode**
Sign	+	–	–	+
Electron flow	out	in	out	in
Half-reaction	oxidation	reduction	oxidation	reduction

Some different types of electrodes are as follows :

1. **Normal hydrogen electrode** is the primary reference electrode and is used normally to know the electrode potential of a half-cell in galvanic all.
2. **Standard hydrogen electrode** is the hydrogen electrode in which the pressure of H_2 gas is maintained at 1 atm and concentration of H^+ ions is 1M.
3. **Calomel electrode** is an another generally used reference electrode but it is an secondary reference electrode. Calomel is the common name for mercury (I) chloride.

$$Hg + Cl^- \longrightarrow \frac{1}{2}Hg_2Cl_2 + e^-$$

Salt-Bridge

It is a U-shaped tube containing a gel permeated with a solution of an inert electrolyte such as NH_4NO_3. The ions of the inert electrolyte do not react with the other ions in the solutions and they are not oxidised or reduced at the electrodes. The salt-bridge is necessary to complete the electrical circuit and to maintain electrical neutrality in both compartments (by flow of ions).

Difference between Electrolytic and Electrochemical Cells

The main points of difference between these two cells are :

- The device in which electrolysis (chemical reaction involving oxidation and reduction) is carried out by using electricity or in which conversion of electrical energy into chemical energy is done is called **electrolytic cell** while **galvanic** or **voltaic** cell is a device in which **redox** reaction is used to convert chemical energy to electrical energy.
- In 'electrolytic cell', anode is positive electrode, while cathode is negative electrode. On the other hand, in 'galvanic cell' anode is negative electrode and cathode is positive electrode. In both the cells, anode is always the site of oxidation and cathode is of reduction.
- In electrolytic cell, ions are discharged at both the electrodes, while in galvanic cell ions are discharged only at cathode.
- In electrolytic cell, both the electrodes are fitted in same compartment, while in galvanic cell both the electrodes are fitted in different compartments.

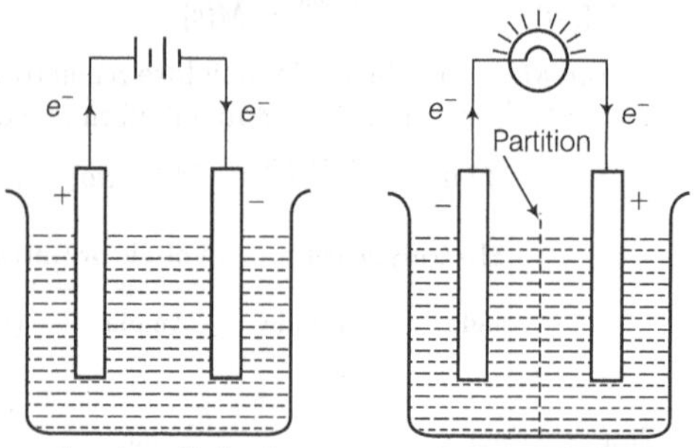

Anode Cathode
(a) Electrolytic cell
(emf is applied to cell)

Anode Cathode
(b) Voltaic or galvanic cell
(emf is generated by cell)

Representation of a Cell

Consider the Daniell cell with following cell reaction:

$$Zn(s) + Cu^{2+}(aq) \longrightarrow Zn^{2+}(aq) + Cu(s)$$

In writing a cell diagram, following points are considered. We divide the cell into two half-cells.

	The anode	The cathode
Reaction	Oxidation takes place $Zn(s) \longrightarrow Zn^{2+}(aq) + 2e^-$	Reduction takes place $Cu^{2+}(aq) + 2e^- \longrightarrow Cu(s)$
Terminal	Negative	Positive
Side	LHS	RHS
Diagram	$Zn(s) / Zn^{2+}(aq)$	$Cu^{2+}(aq) / Cu(s)$

Complete cell diagram may be represented as follows:

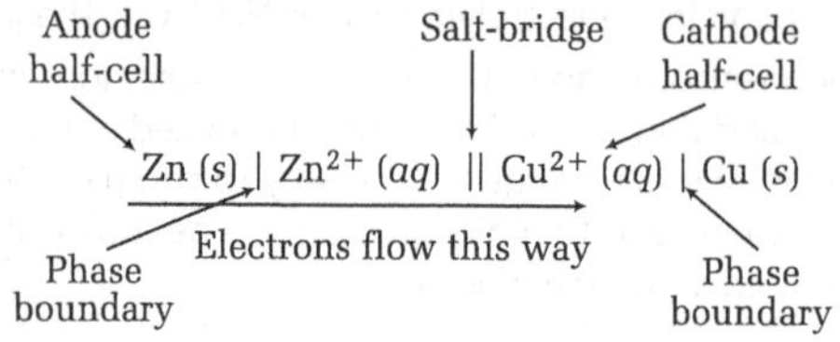

EMF of a Galvanic Cell

In electrochemical cell, the electrodes in different half-cells have different reduction potential. This difference in electrode potential of electrodes due to which electrons flow from anode to cathode is called **electromotive force** or **cell potential of a cell**. This is the driving force for all cell reactions.

Electrode Potential and Standard Electrode Potential

- If we connect two different electrodes, electrons will flow from the electrode of higher negative electric charge density to the electrode with a lower negative electric charge density. A property closely related to the density of negative electric charge is called the **electrode potential.** Potential difference between the metal and the metal ion in which electrode is dipped is called electrode potential denoted as E.
- In the standard state, when pressure is 1 atm, (latest IUPAC correction use 1 bar) temperature is 298 K and concentration is 1M, electrode potential is called **standard electrode potential** denoted as E°.
- According to international convention, if the standard electrode potential of a half-cell is 0.34 V, it means it is the standard reduction potential $E^\circ_{M^{n+}/M}$, of the half-cell M^{n+}/M with half-cell reaction

$$M^{n+} + ne^- \longrightarrow M.$$

- If $E^\circ_{ox} = x\text{V}$, then $E^\circ_{red} = -x\text{V}$
- E°_{cell} or E_{cell} is the potential difference between the two half-cells. Since, the potential difference is the driving force for electrons, it is also called the electromotive force (emf) of the cell or the cell potential or the cell voltage.
- The SI unit of cell potential is the volt (V) and the potential of a galvanic cell is defined as the positive quantity.

$$E^\circ_{cell} = E^\circ_{ox} + E^\circ_{red},$$
$$E_{cell} = E_{ox} + E_{red}$$

- Standard Hydrogen Electrode, SHE used as a reference electrode. Its reduction potential is taken as zero. It is represented as Pt, H_2(1 atm) | H^+(1 M) or H^+ (1M)|H_2 (1 atm), Pt depending upon whether it acts as anode or cathode respectively.
- The difference between the potential required for the evolution of gas and the standard electrode potential of that gas is called over voltage/over potential.

Electrochemical Series

The arrangement of metals in decreasing order of tendency to lose electrons (e^-) is called electrochemical series or it is the series in which the elements are arranged on the basis of the values of their standard reduction potential at 25°C.

Element	Electrode reaction (Reduction)	Standard electrode reduction potential E° (Volt)
Li	$Li^+ + e^- \longrightarrow Li$	− 3.05
K	$K^+ + e^- \longrightarrow K$	− 2.925
Ca	$Ca^{2+} + 2e^- \longrightarrow Ca$	− 2.87
Na	$Na^+ + e^- \longrightarrow Na$	− 2.714
Mg	$Mg^{2+} + 2e^- \longrightarrow Mg$	− 2.37
Al	$Al^{3+} + 3e^- \longrightarrow Al$	− 1.66
Zn	$Zn^{2+} + 2e^- \longrightarrow Zn$	− 0.7628
Cr	$Cr^{3+} + 3e^- \longrightarrow Cr$	− 0.74
Fe	$Fe^{2+} + 2e^- \longrightarrow Fe$	− 0.44
Cd	$Cd^+ + e^- \longrightarrow Cd$	− 0.403
Ni	$Ni^{2+} + 2e^- \longrightarrow Ni$	− 0.25
Sn	$Sn^{2+} + 2e^- \longrightarrow Sn$	− 0.14
H_2	$2H^+ + e^- \longrightarrow H_2$	0.00
Cu	$Cu^{2+} + 2e^- \longrightarrow Cu$	+ 0.337
I_2	$I_2 + 2e^- \longrightarrow 2I^-$	+ 0.535
Ag	$Ag^+ + e^- \longrightarrow Ag$	+ 0.799
Hg	$Hg^{2+} + 2e^- \longrightarrow Hg$	+ 0.885
Br_2	$Br_2 + 2e^- \longrightarrow 2Br^-$	+ 1.08
Cr_2	$Cl_2 + 2e^- \longrightarrow 2Cl^-$	+ 1.36
Au	$Au^{3+} + 3e^- \longrightarrow Au$	+ 1.50
F_2	$F_2 + 2e^- \longrightarrow 2F^-$	+ 2.87

Increasing tendency for reduction ↓ / Increasing strength of oxidising agent ↓ / Increasing strength of reducing agent ↑ / Increasing tendency for oxidation ↑

Nernst Equation and its Applications

The relationship between electrode potential and concentration of solution is called Nernst equation.

$$E_{cell} \text{ or EMF} = E^\circ_{cell} - \frac{2.303RT}{nF} \log Q$$

where, n = total number of electrons lost or gained

1F = 96500 C, T = 298 K, R = 8.314 J/K/mol

Q = reaction coefficient of overall reaction $= \frac{[O.S.]}{[R.S.]}$

where, [O.S.] = concentration of oxidised state

[R.S.] = concentration of reduced state

At T = 298 K, the Nerst equation becomes,

$$E_{cell} = E^\circ_{cell} - \frac{0.0591}{n} \log Q$$

The emf of a standard cell does not change with temperature. Weston cell is a common example of standard cell.

There are two important applications of Nernst equation are:

1. **To find equilibrium constant**

   At equilibrium, E_{cell} = zero

   $$\therefore \quad E^\circ_{cell} = \frac{0.0591}{n} \log K$$

   where, K = equilibrium constant.

2. **To find Gibb's free energy change**

   $$\Delta G^\circ = -n E^\circ_{cell} F$$

   ΔG° = standard Gibbs free energy change

Relationship between Cell Potential and Gibbs Energy Change (ΔG)

- In an electrochemical cell, maximum work done (ΔG) is given by $\Delta G = -nF \times E_{cell}$

  where, F = Faraday's constant

  n = number of moles of electrons transferred.

- A galvanic cell does electrical work by transferring electrical charge through an external circuit. When small amount of current is drawn from the cell then

  $$-\Delta G = W_{elec.}$$

  For spontaneous reactions,

  $$\Delta G = -ve$$

  $$\therefore \quad E_{cell} = +ve$$

  (i) If ΔG of the system is positive then the process would be possible only if the surrounding do electrical work on the system (as in electrolysis).

  (ii) If ΔG of the system is negative then the system does electric work on the surroundings (as in galvanic cells). For a voltaic cell, the work is done on the surroundings, thus given a negative sign. Hence,

  $$\Delta G = W_{max} = -nFE_{cell}$$

  and the standard free energy (ΔG°) is given by,

  $$\Delta G^\circ = -nFE^\circ_{cell}$$

- The relation between cell potential (E°_{cell}), free energy (ΔG°) and equilibrium constant (K_C) is given by,

  $$\Delta G^\circ = -2.303RT \log K_C$$

  as $\quad \Delta G^\circ = nFE^\circ_{cell}$

  and $\quad E^\circ_{cell} = \frac{2.303RT}{nF} \log K_C$

- If two half-reactions having potentials E_1° and E_2° are combined to give a third half-reaction having a potential E_3°, then

  $$\Delta G_3^\circ = \Delta G_1^\circ + \Delta G_2^\circ$$

  or $\quad -n_3FE_3^\circ = -n_1FE_1^\circ - n_2FE_2^\circ$

  or $\quad n_3E_3^\circ = n_1E_1^\circ + n_2E_2^\circ$

  or $\quad E_3^\circ = \frac{n_1E_1^\circ + n_2E_2^\circ}{n_3}$

Batteries

A cell or a battery (arrangement of 1 or more cells connected in series) is basically a galvanic cell and used where the chemical energy of redox reaction is converted into electrical energy.

There are two types of batteries.

1. The Primary Batteries

The battery in which the cell reaction occurs only once and the battery becomes dead after use over a period of time and cannot be reused again, e.g. dry cells like Leclanche cell, mercury cell etc. Hence, primary batteries are not rechargeable. e.g. dry cell, mercury cell, etc.

(i) Dry Cell or Leclanche Cell

It is also called primary voltaic cell. The electrode reactions involved in this cell are :

At anode $Zn(s) \longrightarrow Zn^{2+} + 2e^-$

At cathode $MnO_2 + NH_4^+ + e^- \longrightarrow MnO(OH) + NH_3$

The cell potential is 1.6 V.

(ii) Mercury Cell

- Mercury cell, suitable for low current devices like hearing aids, watches.
- The electrode reactions for the cell are :

At anode $Zn(Hg) + 2OH^- \longrightarrow ZnO(s) + H_2O + 2e^-$

At cathode $HgO + H_2O + 2e^- \longrightarrow Hg(l) + 2OH^-$

The overall cell reaction is

$$Zn(Hg) + HgO(s) \longrightarrow ZnO(s) + Hg(l)$$

2. Secondary Batteries

These are also called **reversible galvanic** or **voltaic cell**. They are rechargeable because on charging reaction becomes reverse, e.g. lead accumulator (lead storage) battery, nickel cadmium cell etc.

Lead Accumulator (Lead-Storage Battery)

- In lead accumulator the electrodes are made of lead and the electrolyte consists of dilute sulphuric acid. The electrodes are usually cast from a lead alloy containing 7–12% of antimony (to give increased hardness and corrosion resistance) and a small amount of tin (for better casting properties).
- The electrodes are coated with a paste of lead (II) oxide (PbO) and finely divided lead; after insertion into the electrolyte a 'forming' current is passed through the cell to convert the PbO on the negative plate into a sponge of finely divided lead. On the positive plate the PbO is converted to lead (IV) oxide (PbO_2).
- The half-cell reactions, when the battery is being used up are:

At anode $Pb(s) + SO_4^{2-} \longrightarrow PbSO_4(s) + 2e^-$

At cathode $PbO_2(s) + SO_4^{2-} + 4H^+ + 2e^- \longrightarrow PbSO_4(s) + 2H_2O$

Overall reaction $Pb(s) + PbO_2(s) + 4H^+(aq) + 2SO_4^{2-}(aq) \longrightarrow 2PbSO_4(s) + 2H_2O$

The reaction is reversed during charging.

During charging following reactions occur :

At anode $PbSO_4(s) + 2H_2O \longrightarrow PbO_2(s) + SO_4^{2-} + 4H^+ + 2e^-$

At cathode $PbSO_4(s) + 2e^- \longrightarrow Pb(s) + SO_4^{2-}$

Overall reaction $2PbSO_4(s) + 2H_2O \longrightarrow Pb(s) + PbO_2(s) + 4H^+ + 2SO_4^{2-}$

- Each cell gives an emf of about 2-V and in motor vehicles a 12-V battery of six cells is usually used. The lead-acid battery produces 80–120 kJ per kilogram.

Fuel Cells

- Fuel cells are another means by which chemical energy may be converted into electrical energy. Energy can be obtained indefinitely from a fuel cell as long as outside supply of fuel is maintained, e.g. $H_2—O_2$ fuel cell.
- This cell was used as a primary source of electrical energy on the moon flights. The overall cell reaction produces water, which was used for drinking by the astronauts.

  The half-reactions are

  At anode $2H_2(g) + 4OH^-(aq) \longrightarrow 4H_2O(l) + 4e^-$

  At cathode $O_2(g) + 2H_2O(l) + 4e^- \longrightarrow 4OH^-(aq)$

  Overall cell reaction $2H_2(g) + O_2(g) \longrightarrow 2H_2O(l)$

- The efficiency of the fuel cell is the ratio of change in Gibbs energy ΔG to the heat of combustion ΔH and mathematically can be given as,

$$\eta = \frac{\Delta G}{\Delta H} \times 100$$

where, η = thermodynamic efficiency of a fuel cell

ΔH = heat of combustion

ΔG = work done = $-nFE^\circ_{cell}$

Corrosion and its Prevention

- Corrosion is basically an electrochemical phenomenon. A metal is oxidised by loss of electrons to oxygen and form metal oxide, e.g. conversion of iron to rust $[Fe_2O_3 \cdot xH_2O]$, the tarnishing of silver (due to formation of Ag_2S), development of a green coating on copper and bronze.
- Corrosion of iron, known as rusting, occurs in the presence of water and oxygen.

At anode $2Fe(s) \longrightarrow 2Fe^{2+}(aq) + 4e^-$

At cathode $O_2(g) + 4H^+(aq) + 4e^- \longrightarrow 2H_2O(l)$

The overall reaction

$$2Fe(s) + O_2(g) + 4H^+(aq) \longrightarrow 2Fe^{2+}(aq) + 2H_2O(l)$$

- Fe^{2+} ions further oxidised by atmospheric oxygen to Fe^{3+} ions and form hydrated ferric oxide $[Fe_2O_3 \cdot xH_2O]$.

$$4Fe^{2+} + O_2 + 4H_2O \longrightarrow 2Fe_2O_3(s) + 8H^+(aq)$$

$$Fe_2O_3 + xH_2O \longrightarrow \underset{\text{Hydrated ferric oxide (rust)}}{Fe_2O_3 \cdot xH_2O}$$

Rusting of iron can be prevented by the following methods.

(i) Barrier protection through coating of paints or electroplating.

(ii) Galvanisation or coating of surface with tin metal.

(iii) By the use of antirust solutions.

DAY PRACTICE SESSION 1

FOUNDATION QUESTIONS EXERCISE

1 Conductance (unit Siemen's) is directly proportional to area of the vessel and the concentration of the solution in it and is inversely proportional to the length of the vessel, then the unit of constant of proportionality is

(a) $S\ m\ mol^{-1}$ (b) $S\ m^2\ mol^{-1}$
(c) $S^{-2}\ m^2\ mol$ (d) $S^2\ m^2\ mol^{-2}$

2 Resistance of 0.2 M solution of an electrolyte is 50 Ω. The specific conductance of the solution is $1.3\ S\ m^{-1}$. If resistance of the 0.4 M solution of the same electrolyte is 260 Ω, its molar conductivity is → AIEEE 2011

(a) $6250\ Sm^2\ mol^{-1}$ (b) $6.25 \times 10^{-4}\ Sm^2\ mol^{-1}$
(c) $625 \times 10^{-4}\ Sm^2\ mol^{-1}$ (d) $62.5\ Scm^2\ mol^{-1}$

3 Resistance of 0.2 M solution of an electrolyte is 50Ω. The specific conductance of the solution of 0.5 M solution of same electrolyte is $1.4\ S\ m^{-1}$ and resistance of same solution of the same electrolyte is 280Ω. The molar conductivity of 0.5M solution of the electrolyte in $S\ m^2\ mol^{-1}$ is → JEE Main 2014

(a) 5×10^{-4} (b) 5×10^{-3} (c) 5×10^3 (d) 5×10^2

4 The equivalent conductance of NaCl at concentration (*C*) and at infinite dilution are λ_C and λ_∞, respectively. The correct relationship between λ_C and λ_∞ is given as (Where, the constant *B* is positive) → JEE Main 2014

(a) $\lambda_C = \lambda_\infty + (B)C$ (b) $\lambda_C = \lambda_\infty - (B)C$
(c) $\lambda_C = \lambda_\infty - (B)\sqrt{C}$ (d) $\lambda_C = \lambda_\infty + (B)\sqrt{C}$

5 How many Faraday's are required to reduce 1 mole of BrO_3^- to Br^-?

(a) 3 (b) 5 (c) 6 (d) 4

6 Two Faraday of electricity is passed through a solution of $CuSO_4$. The mass of copper deposited at the cathode is (atomic mass Cu = 63.5u) → JEE Main 2015

(a) 0 g (b) 63.5 g (c) 2 g (d) 127 g

7 A solution of copper sulphate ($CuSO_4$) is electrolysed for 10 min with a current of 1.5 A. The mass of copper deposited at the cathode (Atomic mass of Cu = 63u) is → JEE Main (Online) 2013

(a) 0.3892 g (b) 0.2938 g (c) 0.2398 g (d) 0.3928 g

8 *X* g of silver is plated out on a serving tray by electrolysis of a solution containing silver in +1 oxidation state for a period of 8.0 h at a current of 8.46 A. What is the area of the tray if the thickness of silver plating is 0.00254 cm? [Given, density of silver = $10.5\ g\ cm^{-3}$]

(a) $10.7 \times 10^4\ cm^2$ (b) $1.02 \times 10^4\ cm^2$
(c) $4.1 \times 10^3\ cm^2$ (d) $10.2 \times 10^4\ cm^2$

9 A current of 2 A is passed for 5 h through a molten tin salt to deposit 22.2 g tin. What is the oxidation state of tin in salt ? [At. wt. of Sn = 118.69 g]

(a) +2 (b) +5 (c) +3 (d) +4

10 In the electrolytic cell, flow of electrons is from

(a) cathode to anode in solution
(b) cathode to anode through external supply
(c) cathode to anode through internal supply
(d) anode to cathode through internal supply

11 When the sample of copper with the zinc impurity is to be purified by electrolysis, the appropriate electrodes are

	Cathode	Anode
(a)	pure zinc	pure copper
(b)	impure zinc	pure copper
(c)	impure zinc	impure sample
(d)	pure copper	impure sample

12 $\Lambda^\circ_{m\,(NH_4OH)}$ is equal to ....

(a) $\Lambda^\circ_{m\,(NH_4OH)} + \Lambda^\circ_{m\,(NH_4Cl)} - \Lambda^\circ_{(HCl)}$
(b) $\Lambda^\circ_{m\,(NH_4Cl)} + \Lambda^\circ_{m\,(NaOH)} - \Lambda^\circ_{(NaCl)}$
(c) $\Lambda^\circ_{m\,(NH_4Cl)} + \Lambda^\circ_{m\,(NaCl)} - \Lambda^\circ_{(NaOH)}$
(d) $\Lambda^\circ_{m\,(NaOH)} + \Lambda^\circ_{m\,(NaCl)} - \Lambda^\circ_{(NH_4Cl)}$

13 For the reaction,

$$\frac{1}{2}H_2(g) + AgCl(s) \longrightarrow H^+(aq) + Cl^-(aq) + Ag(s)$$

occurs in the galvanic cell

(a) Ag| AgCl(*s*) | KCl(*aq*) || $AgNO_3$(*aq*)|Ag(*s*)
(b) Pt | H_2(*g*) | HCl (*aq*) || $AgNO_3$(*aq*) | Ag(*s*)
(c) Pt | H_2(*g*) | HCl(*aq*) || AgCl(*s*)|Ag(*s*)
(d) Pt | H_2(*g*) | KCl(*aq*) || AgCl(*s*)|Ag(*s*)

14 Which of the following is not true?
(a) Cathode is negative terminal in an electrolytic cell
(b) Anode is positive terminal in a galvanic cell
(c) Reduction occurs at cathode in either of cells
(d) Oxidation occurs at anode

15 The standard reduction potential for Fe^{2+}/Fe and Sn^{2+}/Sn electrodes are – 0.44 and – 0.14 V respectively. For the cell reaction; $Fe^{2+} + Sn \longrightarrow Fe + Sn^{2+}$, the standard emf is
(a) +0.30 V (b) – 0.58 V
(c) +0.58 V (d) – 0.30 V

16 Of the following metals that cannot be obtained by electrolysis of the aqueous solution of their salts are
(a) Ag and Mg (b) Ag and Al
(c) Mg and Al (d) Cu and Cr

17 The standard reduction potential values of the three metallic cations X, Y and Z are 0.52, –3.03 and –1.18 V respectively. The order of reducing power of the corresponding metals is
(a) $Y > Z > X$ (b) $X > Y > Z$
(c) $Z > Y > X$ (d) $Z > X > Y$

18 Given, $E^\circ_{Cl/Cl^-} = 1.36$ V; $E^\circ_{Cr^{3+}/Cr} = -0.74$ V
$E^\circ_{Cr_2O_7^{2-}/Cr^{3+}} = 1.33$ V; $E^\circ_{MnO_4^-/Mn^{2+}} = 1.51$ V
Among the following, the strongest reducing agent is
(a) Cr (b) Mn^{2+}
(c) Cr^{3+} (d) Cl–

19 The metal that cannot be obtained by electrolysis of an aqueous solution of its salts is → JEE Main 2014
(a) Ag (b) Ca (c) Cu (d) Cr

20 Given, $E^\circ_{\frac{1}{2}Cl_2/Cl} = 1.36$ V, $E^\circ_{Cr^{3+}/Cr} = -0.74$ V,
$E^\circ_{Cr_2O_7^{2-}/Cr^{3+}} = 1.33$ V, $E^\circ_{MnO_4^-/Mn^{2+}} = 1.51$ V
The correct order of reducing power of the species (Cr, Cr^{3+}, Mn^{2+} and Cl^-) will be → JEE Main (Online) 2013
(a) $Mn^{2+} < Cl^- < Cr^{3+} < Cr$ (b) $Mn^{2+} < Cr^{3+} < Cl^- < Cr$
(c) $Cr^{3+} < Cl^- < Mn^{2+} < Cr$ (d) $Cr^{3+} < Cl^- < Cr < Mn^{2+}$

21 Given, $E^\circ_{Cr^{3+}/Cr} = -0.74$ V; $E^\circ_{MnO_4^-/Mn^{2+}} = 1.51$ V
$E_{Cr_2O_7^{2-}/Cr^{3+}} = 1.33$ V; $E^\circ_{Cl/Cl^-} = 1.36$ V
Based on the data given above, strongest oxidising agent will be → JEE Main 2013
(a) Cr^{3+} (b) Mn^{2+} (c) MnO_4^- (d) Cl^-

22 Electrode potentials (E°) are given below :
$Cu^+/Cu = +\ 0.52$ V, $Fe^{3+}/Fe^{2+} = +\ 0.77$ V
$\frac{1}{2}I_2(s)/I^- = +\ 0.54$ V, $Ag^+/Ag = +\ 0.88$ V
Based on the above potentials strongest oxidising agent will be → JEE Main (Online) 2013
(a) Cu (b) Fe^{3+} (c) Ag^+ (d) I_2

23 In the cell reaction,
$Cu(s) + 2Ag^+(aq) \longrightarrow Cu^{2+}(aq) + 2Ag(s), E^\circ_{cell} = 0.46$ V
by doubling the concentration of Cu^{2+}, E_{cell} is
(a) doubled
(b) halved
(c) increased but less than double
(d) decreased by a small fraction

24 For the reaction,
$$Zn(s) + Cu^{2+}(0.1M) \longrightarrow Zn^{2+}(1M) + Cu(s)$$
taking place in a cell, E°_{cell} is 1.10 V. E_{cell} for the cell will be $\left(2.303\frac{RT}{F} = 0.0591\right)$
(a) 1.80 V (b) 1.07 V (c) 0.82 V (d) 2.14 V

25 Given, $E^\circ_{Cr^{3+}/Cr} = -0.72$ V; $E^\circ_{Fe^{2+}/Fe} = -\ 0.42$ V
The potential for the cell
$Cr|Cr^{3+}(0.1M)||Fe^{2+}(0.01M)|Fe$ is
(a) – 0.339 V (b) – 0.26 V (c) 0.26 V (d) 0.339 V

26 What will be the emf for the given cell?
$$Pt\,|\,H_2(p_1)\,|\,H^+(aq)\,|\,H_2(p_2)\,|\,Pt$$
(a) $\frac{RT}{F}\ln\frac{p_1}{p_2}$ (b) $\frac{RT}{2F}\ln\frac{p_1}{p_2}$
(c) $\frac{RT}{2F}\ln\frac{p_2}{p_1}$ (d) None of these

27 The value of the reaction quotient, Q for the cell
$Zn(s)|Zn^{2+}(0.01M)||Ag^+(1.25\ M)|Ag(s)$ is
(a) 156 (b) 125 (c) 1.25×10^{-2} (d) 6.4×10^{-3}

28 The reduction potential of hydrogen half-cell will be negative if → AIEEE 2011
(a) $p(H_2) = 1$ atm and $|H^+| = 2.0$ M
(b) $p(H_2) = 1$ atm and $|H^+| = 1.0$ M
(c) $p(H_2) = 2$ atm and $|H^+| = 1.0$ M
(d) $p(H_2) = 2$ atm and $|H^+| = 2.0$ M

29 Which of the following statement is correct?
(a) E_{cell} and $\Delta_r G$ of cell reaction both are extensive properties
(b) E_{cell} and $\Delta_r G$ of cell reaction both are intensive properties
(c) E_{cell} is an intensive property while $\Delta_r G$ of cell reaction is an extensive property
(d) E_{cell} is an extensive property while $\Delta_r G$ of cell reaction is an intensive property

30 The standard emf of a cell, involving one electron change is found to be 0.591V at 25°C. The equilibrium constant of the reaction is(1F = 96500 C mol^{-1})
(a) 1.0×10^1 (b) 1.0×10^5 (c) 1.0×10^{10} (d) 1.0×10^{30}

31 Given below are the half-cell reactions
$$Mn^{2+} + 2e^- \longrightarrow Mn; E^\circ = -1.18\,eV$$
$$2(Mn^{3+} + e^- \longrightarrow Mn^{2+});\ E^\circ = +\ 1.51\ eV$$
The E° for $3Mn^{2+} \longrightarrow Mn + 2Mn^{3+}$ will be
→ JEE Main 2014

(a) −2.69V; the reaction will not occur
(b) −2.69V; the reaction will occur
(c) −0.33V; the reaction will not occur
(d) −0.33V; the reaction will occur

32 The standard reduction potentials for Zn^{2+}/Zn, Ni^{2+}/Ni and Fe^{2+}/Fe are − 0.76, − 0.23 and − 0.44 V respectively. The reaction $X + Y^{2+} \longrightarrow X^{2+} + Y$ will be spontaneous when → AIEEE 2012

(a) $X = Ni, Y = Fe$ (b) $X = Ni, Y = Zn$
(c) $X = Fe, Y = Zn$ (d) $X = Zn, Y = Ni$

33 Given, $E^\circ_{Fe^{3+}/Fe} = -0.036$ V, $E^\circ_{Fe^{2+}/Fe} = -0.439$ V
The value of standard electrode potential for the charge, $Fe^{3+}(aq) + e^- \longrightarrow Fe^{2+}(aq)$ will be → AIEEE 2009

(a) − 0.072 V (b) 0.385 V (c) 0.770 V (d) − 0.270 V

34 Galvanisation is applying a coating of → JEE Main 2016

(a) Cr (b) Cu (c) Zn (d) Pb

35 Match the following and choose the correct option.

	Column I		Column II
A.	Lead storage battery	1.	Maximum efficiency
B.	Mercury cell	2.	Prevented by galvanisation
C.	Fuel cell	3.	Gives steady potential
D.	Rusting	4.	Pb is anode, PbO_2 is cathode

Codes

	A	B	C	D		A	B	C	D
(a)	4	3	1	2	(b)	2	1	3	4
(c)	1	4	3	2	(d)	4	1	3	2

Direction (Q. Nos. 36 and 37) *In the following questions, Assertion (A) followed by Reason (R) is given. Choose the correct answer out of the following choices.*

(a) Assertion and Reason both are correct statements and Reason is the correct explanation of the Assertion
(b) Assertion and Reason both are correct statements but Reason is not the correct explanation of the Assertion
(c) Assertion is correct and Reason is incorrect
(d) Both Assertion and Reason are incorrect

36 Assertion (A) E_{cell} should have a positive value for the cell to function.

Reason (R) $E_{cathode} < E_{anode}$

37 Assertion (A) $E_{Ag^+/Ag}$ increases with increase in concentration of Ag^+ ions.

Reason (R) $E_{Ag^+/Ag}$ has a positive value.

Direction (Q. Nos. 38 and 39) *Each of these questions contains two statements : Statement I (Assertion) and Statement II (Reason). Each of these questions also has four alternative choices, only one of which is the correct answer. You have to select one of the codes (a), (b), (c) and (d) given below :*

(a) Statement I is true; Statement II is true; Statement II is a correct explanation for Statement I.
(b) Statement I is true, Statement II is true; Statement II is not a correct explanation for Statement I.
(c) Statement I is true; Statement II is false.
(d) Statement I is false; Statement II is true.

38 Statement I For the Daniell cell, $Zn|Zn^{2+}||Cu^{2+}|Cu$ with $E_{cell} = 1.1V$, the application of opposite potential greater than1.1V results into flow of electron from cathode to anode.

Statement II Zinc is deposited at anode and Cu is deposited at cathode.

39 Statement I Galvanised iron does not rust.

Statement II Zinc has more negative electrode potential than iron.

DAY PRACTICE SESSION 2

PROGRESSIVE QUESTIONS EXERCISE

1 Aluminium oxide may be electrolysed at 1000°C to furnish aluminium metal
(Atomic mass = 27u; 1F = 96500 C).
The cathode reaction is $Al^{3+} + 3e^- \longrightarrow Al$.
To prepare 5.12 kg of aluminium metal by this method required electricity will be

(a) $5.49 \times 10^1 C$ (b) $5.49 \times 10^4 C$
(c) $1.83 \times 10^7 C$ (d) $5.49 \times 10^7 C$

2 For a saturated solution of AgCl at 25°C, specific conductance is 3.41×10^{-6} omh^{-1} cm^{-1} and that water used for preparing the solution was 1.60×10^{-6} ohm^{-1} cm^{-1}. The solubility product of AgCl is [$\Lambda^\infty_{eq}(AgCl) = 138.3$ ohm^{-1} cm^{-1} equiv^{-1}]

(a) 1.31×10^{-5} (b) 1.74×10^{-8}
(c) 1.72×10^{-10} (d) 3.61×10^{-3}

3 The Edison storage cell is represented as

$$Fe(s)|FeO(s)|KOH(aq)|Ni_2O_3(s)|NiO(s)|Ni(s)$$

the half-cell reactions are

$$Ni_2O_3(s) + H_2O(l) + 2e^- \rightleftharpoons 2NiO(s) + 2OH^-; \quad E^\circ = +0.40\text{ V}$$

$$FeO(s) + H_2O(l) + 2e^- \rightleftharpoons Fe(s) + 2OH^-; E^\circ = -0.87\text{V}$$

What is the maximum amount of electrical energy that can be obtained from one mole of Ni_2O_3?

(a) 127 kJ (b) 245.11 kJ (c) 90.71 kJ (d) 122.55 kJ

4 An inaccurate ammeter and silver coulometer is connected in series in an electric circuit through which a constant direct current flows. If ammeter reads 0.6 ampere throughout one hour, the silver deposited on coulometer was found to be 2.16 g. What per cent error is in the reading of ammeter ? [Assume 100% current efficiency].

(a) 1% (b) 0.54% (c) 0.06 % (d) 10%

5 An aqueous solution of NaCl on electrolysis gives $H_2(g)$, $Cl_2(g)$ and NaOH. According to reaction,

$2Cl^-(aq) + 2H_2O \longrightarrow 2OH^-(aq) + H_2(g) + Cl_2(g)$

A direct current of 25 A with a current efficiency of 62% is passed through 20 L of NaCl solution (20% by weight). How long it take to produce 1 kg of Cl_2?

(a) 30.20 h (b) 12.17 h (c) 48.71 h (d) 14.61 h

6 $Zn|Zn^{2+}(a = 0.1\ M)||Fe^{2+}(a = 0.01\ M)|Fe$.

The emf of the above cell is 0.2905 V.

Equilibrium constant for the cell reaction is

(a) $10^{0.32/0.0591}$ (b) $10^{0.32/0.0295}$
(c) $10^{0.26/0.0295}$ (d) $10^{0.32/0.295}$

7 The standard reduction potentials at 298 K for the following half reactions are given against each.

$$Zn^{2+}(aq) + 2e^- \rightleftharpoons Zn(s);\ -0.762\ V$$
$$Cr^{3+}(aq) + 3e^- \rightleftharpoons Cr(s);\ -0.740\ V$$
$$2H^+(aq) + 2e^- \rightleftharpoons H_2(g);\ 0.00\ V$$
$$Fe^{3+}(aq) + e^- \rightleftharpoons Fe^{2+}(aq);\ 0.770\ V$$

Which is the strongest reducing agent?

(a) Zn (*s*) (b) Cs (*s*) (c) $H_2(g)$ (d) $Fe^{3+}(aq)$

8 Same quantity of charge is being used to liberate iodine (at anode) and a metal *M* (at cathode). The mass of metal *M* liberated is 0.617 g and the liberated iodine is completely reduced by 46.3 mL of 0.124 M sodium thiosulphate. What is the total time to bring this change if 10 A current is passed through solution of metal iodide?

(a) 55.4 s (b) 25.2 s (c) 5.54 s (d) 16.8 s

9 If λ^∞_{eq} (NaCl), λ^∞_{eq} (KCl) and $\lambda^\infty_{eq}\ K_2SO_4$ are 123.7, 147.0 and 152.1 $\Omega^{-1}\ cm^2\ eq^{-1}$ then $\lambda^\infty_{eq}(Na_2SO_4)$ would be

(a) $128.8\ \Omega^{-1}\ cm^2 eq^{-1}$ (b) $257.6\ \Omega^{-1}\ cm^2 eq^{-1}$
(c) $105.5\ \Omega^{-1}\ cm^2 eq^{-1}$ (d) $118\ \Omega^{-1}\ cm^2 eq^{-1}$

10 A gas *X* at 1 atm is bubbled through a solution containing a mixture of 1 M Y^- and 1 M Z^- at 25°C. If the order of reduction potentials is $Z > Y > X$, then

(a) *Y* will oxidise *X* and not *Z* (b) *Y* will oxidise *Z* and not *X*
(c) *Y* will oxidise both *X* and *Y* (d) *Y* will reduce both *X* and *Y*

11 The equivalent conductances of two strong electrolytes at infinite dilution in H_2O (where ions move freely through a solution) at 25°C are given below

$$\Lambda^\circ_{CH_3COONa} = 91.0\ S\ cm^2/equiv$$
$$\Lambda^\circ_{HCl} = 426.2\ S\ cm^2/equiv$$

What additional information/quantity one needs to calculate Λ° of an aqueous solution of acetic acid ?

(a) Λ° of NaCl
(b) Λ° of CH_3COOK
(c) The limiting equivalent conductance of H^+ $[\lambda^\circ_{H^+}]$
(d) Λ° of chloroacetic acid ($ClCH_2COOH$)

12 The rusting of iron takes place as follows:

$$2H^+ + 2e^- + \frac{1}{2}O_2 \longrightarrow H_2O(l); E^\circ = +1.23\ V$$
$$Fe^{2+} + 2e^- \longrightarrow Fe(s); E^\circ = -0.44\ V$$

Calculate ΔG° for the net process.

(a) $-322\ kJ\ mol^{-1}$ (b) $-161\ kJ\ mol^{-1}$
(c) $-152\ kJ\ mol^{-1}$ (d) $-76\ kJ\ mol^{-1}$

13 Standard electrode potential data are useful for understanding the suitability of an oxidant in a redox titration. Some half-cell reactions and their standard potentials are given below:

$$MnO_4^-(aq) + 8H^+(aq) + 5e^- \longrightarrow Mn^{2+}(aq) + 4H_2O(l);\quad E^\circ = 1.51\ V$$
$$Cr_2O_7^{2-}(aq) + 14H^+(aq) + 6e^- \longrightarrow 2Cr^{3+}(aq) + 7H_2O(l);\quad E^\circ = 1.38\ V$$
$$Fe^{3+}(aq) + e^- \longrightarrow Fe^{2+}(aq); E^\circ = 0.77\ V$$
$$Cl_2(g) + 2e^- \longrightarrow 2Cl^-(aq);\ E^\circ = 1.40\ V$$

Identify the incorrect statement regarding the quantitative estimation of gaseous $Fe(NO_3)_2$.

(a) MnO_4^- can be used in aqueous HCl
(b) $Cr_2O_7^{2-}$ can be used in aqueous HCl
(c) MnO_4^- can be used in aqueous H_2SO_4
(d) $Cr_2O_7^{2-}$ can be used in aqueous H_2SO_4

14 The cell, $Zn|Zn^{2+}(1M)||Cu^{2+}(1M)|Cu$ ($E^\circ_{cell} = 1.10$ V), was allowed to be completely discharged at 298 K. The relative concentration of Zn^{2+} to Cu^{2+} $\left(\frac{[Zn^{2+}]}{[Cu^{2+}]}\right)$ is

(a) antilog (24.08) (b) 37.2
(c) $10^{37.2}$ (d) 9.65×10^4

15 How long (approximate) should water be electrolysed by passing through 100 A current so that the oxygen released can completely burn 27.66 g of diborane? (Atomic weight of $B = 10.8\ \mu$) → JEE Main 2018

(a) 6.4 h (b) 0.8 h (c) 3.2 h (d) 1.6 h

16 Given the data at 25°C,

$$Ag + I^- \longrightarrow AgI + e^-;\ E^\circ = 0.152\ V$$
$$Ag \longrightarrow Ag^+ + e^-;\ E^\circ = -0.800\ V$$

What is the value of $\log K_{sp}$ for AgI ?

$$\left(2.303\frac{RT}{F} = 0.059\ V\right)$$

(a) −8.12 (b) +8.612 (c) −37.83 (d) −16.14

ANSWERS

SESSION 1	1 (b)	2 (b)	3 (a)	4 (c)	5 (c)	6 (b)	7 (b)	8 (b)	9 (a)	10 (d)
	11 (d)	12 (b)	13 (c)	14 (b)	15 (d)	16 (c)	17 (a)	18 (a)	19 (b)	20 (a)
	21 (c)	22 (c)	23 (d)	24 (b)	25 (c)	26 (b)	27 (d)	28 (c)	29 (c)	30 (c)
	31 (a)	32 (d)	33 (c)	34 (c)	35 (a)	36 (c)	37 (b)	38 (a)	39 (a)	
SESSION 2	1 (d)	2 (c)	3 (b)	4 (d)	5 (c)	6 (b)	7 (a)	8 (a)	9 (a)	10 (a)
	11 (a)	12 (a)	13 (a)	14 (c)	15 (c)	16 (d)				

Hints and Explanations

SESSION 1

1 Conductance $= \kappa \times \dfrac{\text{area} \times \text{conc.}}{\text{length}}$

or $\kappa = \dfrac{\text{Conductance} \times \text{length}}{\text{area} \times \text{conc.}}$

$= \dfrac{S \times m}{m^2 \times \text{mol } m^{-3}} = S\, m^2 \text{mol}^{-1}$

2 Specific conductance

= conductance × cell constant

$1.3 = \dfrac{1}{50} \times$ cell constant

$\therefore$ Cell constant $= 1.3 \times 50\ m^{-1}$

$= 65\ m^{-1} = \left(\dfrac{65}{100}\right) cm^{-1}$

Molar conductivity

$= \dfrac{1000 \times \text{conductance} \times \text{cell constant}}{\text{molarity}}$

$= \dfrac{1000}{0.4} \times \dfrac{1}{260} \times \dfrac{65}{100}$

$= 6.25\ S\, cm^2\ mol^{-1}$

$= 6.25 \times 10^{-4}\ S\, m^2\ mol^{-1}$

3 For first solution,

$k = 1.4\ Sm^{-1}, R = 50\,\Omega, M = 0.2$

Specific conductance $(\kappa) = \dfrac{1}{R} \times \dfrac{l}{A}$

$\Rightarrow\ 1.4\ Sm^{-1} = \dfrac{1}{50} \times \dfrac{l}{A}$

$\dfrac{l}{A} = 50 \times 1.4\ m^{-1}$

For second solution,

$R = 280, \dfrac{l}{A} = 50 \times 1.4\ m^{-1}$

$\kappa = \dfrac{1}{280} \times 1.4 \times 50 = \dfrac{1}{4}$

Now, molar conductivity

$\lambda_m = \dfrac{\kappa}{1000 \times M}$

$= \dfrac{1/4}{1000 \times 0.5} = \dfrac{1}{2000}$

$= 5 \times 10^{-4}\ S\, m^2\, mol^{-1}$

4 The correct relationship between λ_C and λ_∞ is given by to Debye-Huckel Onsager equation,

$\lambda_C = \lambda_\infty - B\sqrt{C}$

where, λ_C = limiting equivalent conductivity at concentration C

λ_∞ = limiting equivalent conductivity at infinite dilution

C = concentration

5 $BrO_3^- + 6H^+ + 6e^- \longrightarrow Br^- + 3H_2O$

$\therefore$ Number of Faraday required = 6

6 Given, $Q = 2F$

Atomic mass of Cu = 63.5 u

Valency of the metal $Z = 2$

$W = ZQ = \dfrac{E}{F} \cdot 2F = 2E$

$= \dfrac{2 \times 63.5}{2} = 63.5\ g$

7 From Faraday's first law,

Mass of Cu deposited, $w = Zit$

$= \dfrac{E}{96500} \cdot i \cdot t$

$= \dfrac{63}{2 \times 96500} \times 1.5\ A \times 10 \times 60\ s$

$= 0.2938\ g$

8 $\dfrac{w_{Ag}}{\text{eq.wt.}_{Ag}} = \dfrac{i \cdot t}{96500}$

or $w_{Ag} = \dfrac{107.8 \times 8.46 \times 8 \times 60 \times 60}{96500}$

or $w_{Ag} = 272.18$ g.

Volume of Ag $= \dfrac{272.18}{10.5} = 25.92\ cm^3$

$\therefore$ Surface area

$= \dfrac{25.92}{0.00254} = 1.02 \times 10^4\ cm^2$

9 Equivalents of tin $= \dfrac{i \cdot t}{96500}$

or $\dfrac{22.2}{\text{eq.wt.}} = \dfrac{2 \times 5 \times 60 \times 60}{96500}$

Eq. wt. = 59.5

$\therefore$ Valency of tin

$= \dfrac{\text{at. wt.}}{\text{eq. wt.}} = \dfrac{118.69}{59.5} \approx 2$ (an integer)

Therefore, $Sn^{2+} + 2e^- \longrightarrow Sn$

Thus, oxidation state of tin (Sn) in salt is +2.

10 In electrolytic cell, flow of electron is possible from anode to cathode through internal supply.

11 Impure sample is made the anode and pure copper acts as the cathode.

12 According to Kohlrausch's law,

$\Lambda^\circ_{m(NH_4Cl)} + \Lambda^\circ_{m(NaOH)}$

$= \Lambda^\circ_{m\,(NH_4OH)} + \Lambda^\circ_{m(NaCl)}$

or $\Lambda^\circ_{m(NH_4OH)} = \Lambda^\circ_{m\,(NH_4Cl)} + \Lambda^\circ_{m(NaOH)} - \Lambda^\circ_{m(NaCl)}$

13 H_2 undergoes oxidation and AgCl (Ag^+) undergoes reduction. Therefore, cell may be represent, as

Pt | $H_2(g)$ | HCl(aq) || AgCl(s) | Ag(s).

14 In galvanic cell, cathode is positive terminal whereas anode is negative terminal.

15 Given, $E^\circ_{Fe^{2+}/Fe} = -0.44\ V$

$E^\circ_{Sn/Sn^{2+}} = 0.14\ V$

$E^\circ_{cell} = E^\circ_{oxi}(Sn) + E^\circ_{red}(Fe)$

$= +0.14 + (-0.44) = -0.30\ V$

16 Mg and Al have lower reduction potentials than H_2O. Hence, H_2O is reduced more easily to give H_2 gas at the cathode.

17 Greater the reduction potential, less is the reducing power.

The decreasing order of reduction potential is $X > Z > Y$.

Therefore, order of reducing power of corresponding metals is $Y > Z > X$.

18 The substances which have lower reduction potentials are stronger reducing agents. Therefore, Cr($E^\circ_{Cr^{3+}/Cr}$ −0.74V) is the strongest reducing agent among all the other given options.

19 Higher the position of element in the electrochemical series more difficult is the reduction of its cations.

If $Ca^{2+}(aq)$ is electrolysed, water is reduced in preference to it. Hence, it cannot be reduced electrolytically from their aqueous solution.

$$Ca^{2+}(aq) + H_2O \longrightarrow Ca^{2+} + OH^- + H_2\uparrow$$

20 More the negative value of reduction potential, higher the reducing power. Thus, the correct order of reducing tendency is

$$\underset{(1.51\text{ V})}{Mn^{2+}} < \underset{(1.36\text{ V})}{Cl^-} < \underset{(1.33\text{ V})}{Cr^{3+}} < \underset{(-0.74\text{ V})}{Cr}$$

21 As per data mentioned MnO_4^- is strongest oxidising agent as it has maximum SRP (Standard Reduction Potential) value.

22 More positive value of electrode potential represents that Ag^+ is the strongest oxidising agent.

23 Applying Nernst equation,

$$E_{cell} = E^\circ_{cell} - \frac{RT}{nF}\ln\frac{[Cu^{2+}]}{[Ag^+]^2}$$

Doubling (Cu^{2+}) decreases the emf by a small fraction.

24 $$E_{cell} = E^\circ_{cell} - \frac{0.0591}{n}\log\frac{[Zn^{2+}]}{[Cu^{2+}]}$$

$$= 1.10 - \frac{0.0591}{2}\log\frac{1}{0.1} = 1.07\text{ V}$$

25 $E^\circ_{cell} = -0.42 - (-0.72) = 0.30$ V

The cell reaction is

$2Cr + 3Fe^{2+} \longrightarrow 2Cr^{3+} + 3Fe, n = 6$

Using Nernst equation,

$$E_{cell} = E^\circ_{cell} - \frac{0.0591}{6}\log\frac{[Cr^{3+}]^2}{[Fe^{2+}]^3}$$

$$= 0.30 - \frac{0.0591}{6}\log\frac{(0.1)^2}{(0.01)^3}$$

$$= 0.30 - \frac{0.0591}{6} \times 4$$

$$= 0.30 - 0.04 = 0.26\text{ V}$$

26 RHS, $2H^+ + 2e^- \rightleftharpoons H_2(p_2)$

LHS, $H_2(p_1) \rightleftharpoons 2H^+ + 2e^-$

Overall reaction, $H_2(p_1) \rightleftharpoons H_2(p_2)$

$$E = E^\circ - \frac{RT}{nF}\ln\frac{p_2}{p_1}$$

$$= 0 - \frac{RT}{2F}\ln\frac{p_2}{p_1} = \frac{RT}{2F}\ln\frac{p_1}{p_2}$$

27 The cell reaction is

$$Zn(s) \longrightarrow Zn^{2+}(0.01M) + 2e^-$$

$$[Ag^+(1.25\text{ M}) + e^- \longrightarrow Ag(s)] \times 2$$

$$Zn(s) + 2Ag^+(1.25M) \longrightarrow Zn^{2+}(0.01M) + 2Ag(s)$$

$$Q = \frac{[Zn^{2+}]}{[Ag^+]^2} = \frac{0.01}{(1.25)^2} = 6.4 \times 10^{-3}$$

28 Reduction hydrogen half-cell is

$$H^+(xM) \mid Pt(H_2)$$

Half-cell reaction is

$$2H^+(aq) + 2e^- \longrightarrow H_2(g)$$

Reaction quotient $= Q = p_{H_2}/[H^+]^2$,

$n = 2$

$$E_{red} = E^\circ_{red} - \frac{0.0591}{n}\log Q$$

$$= 0 - \frac{0.0591}{2}\log Q$$

	p_{H_2}	$[H^+]$	$Q = \frac{p_{H_2}}{[H^+]^2}$	E_{red}
(a)	1 atm	2.0 M	0.25	+ve
(b)	1 atm	1.0 M	1.0	0
(c)	2 atm	1.0 M	2.0	−ve
(d)	2 atm	2.0 M	0.50	+ve

29 E_{cell} is an intensive property because it does not depend on the system size or the amount of material in the system while $\Delta_r G$ of a cell reaction is an extensive property because it depends on the amount of material in the system.

30 Relation between K_{eq} and E°_{cell} is

$$E^\circ_{cell} = \frac{2.303RT}{nF}\log K_{eq}$$

$$E^\circ_{cell} = \frac{0.0591}{n}\log K_{eq} \quad \text{(at 298 K)}$$

$$0.591 = \frac{0.0591}{1}\log K_{eq}$$

$\therefore \log K_{eq} = 10$

$\therefore K_{eq} = 1 \times 10^{10}$

31 $Mn^{2+} \longrightarrow Mn + Mn^{3+}$

O.N. +2   0   +3 (+2 → 0: Reduction; +2 → +3: Oxidation)

Mn^{2+} is oxidised as well as reduced. Thus, it is a disproportionation reaction.

$$2 \times [Mn^{2+} \longrightarrow Mn^{3+} + e^-] \quad E^\circ = -1.51\text{ V}$$

$$Mn^{2+} + 2e^- \longrightarrow Mn, \quad E^\circ = -1.18\text{ V}$$

$$\therefore \quad 3Mn^{2+} \longrightarrow 2Mn^{3+} + Mn \quad E^\circ = -2.69\text{V}$$

Since, $E^\circ < 0$, hence reaction will not occur.

32 A cell reaction is spontaneous, if $\Delta G^\circ < 0$.

Since, $\Delta G^\circ = -nFE^\circ_{cell}$

Thus, $E^\circ_{cell} > 0$

$E^\circ_{cell} = E^\circ_{ox} + E^\circ_{red}$

(a) If X = Ni, Y = Fe

$$Ni + Fe^{2+} \longrightarrow Ni^{2+} + Fe$$

$E^\circ_{Ni/Ni^{2+}} = 0.23$V

$E^\circ_{Fe^{2+}/Fe} = -0.44$V

Thus, $E^\circ_{cell} = E^\circ_{Ni/Ni^{2+}} + E^\circ_{Fe^{2+}/Fe}$

$= +0.23\text{ V} - 0.44\text{V}$

$= -0.21$V

$E^\circ_{cell} < 0$

(b) If X = Ni, Y = Zn

$$Ni + Zn^{2+} \longrightarrow Ni^{2+} + Zn$$

$E^\circ_{Ni/Ni^{2+}} = 0.23$ V

$E^\circ_{Zn^{2+}/Zn} = -0.76$ V

$\therefore \quad E^\circ_{cell} = -0.53$ V

$E^\circ_{cell} < 0$

(c) X = Fe, Y = Zn

$$Fe + Zn^{2+} \longrightarrow Fe^{2+} + Zn$$

$E^\circ_{Fe/Fe^{2+}} = 0.44$ V

$E^\circ_{Zn^{2+}/Zn} = -0.76$ V

$E^\circ_{cell} = E^\circ_{Fe/Fe^{2+}} + E^\circ_{Zn^{2+}/Zn}$

$= 0.44 - 0.76 = -0.32$ V

$E^\circ_{cell} < 0$

(d) X = Zn, Y = Ni

$$Zn + Ni^{2+} \longrightarrow Zn^{2+} + Ni$$

$E^\circ_{Zn/Zn^{2+}} = 0.76$ V

$E^\circ_{Ni^{2+}/Ni} = -0.23$

$E^\circ_{cell} = E^\circ_{Zn/Zn^{2+}} + E^\circ_{Ni^{2+}/Ni}$
$= 0.76 - 0.23 = 0.53$ V
$E^\circ_{cell} > 0$ (spontaneous)

33 Given,
$Fe^{3+} + 3e^- \longrightarrow Fe;\ E^\circ_1 = -0.036$ V ...(i)
$Fe^{2+} + 2e^- \longrightarrow Fe;\ E^\circ_2 = -0.439$ V ...(ii)
We need to calculate
$Fe^{3+} + e^- \longrightarrow Fe^{2+};\ E^\circ_3 = ?$ ...(iii)
We can obtain (iii) by subtracting (ii) from (i) but we cannot obtain E°_3 that way because electrode potential is intensive property. That's why we determine E°_3 calculating
$\Delta G_3 = \Delta G_1 - \Delta G_2$
(ΔG is an extensive property)
$\Delta G_3 = 3 \times 0.036 - 2 \times 0.439$
$-nFE^\circ_3 = (0.108 - 0.878)$ F
$\Rightarrow -1 \times F \times E^\circ_3 = -0.770$
$E^\circ_3 = 0.770$ V

34 Zinc metal is the most stable metal to cover iron surfaces. The process of coating the iron surface by zinc is called galvanisation.

35 A→4, B→3, C→1, D→2
In lead storage, Pb is anode, PbO_2 is cathode. As, mercury cell is primary battery it gives steady potential. Fuel has maximum efficiency which is given as,
$$\eta = \frac{\Delta G}{\Delta H} \times 100$$
Rusting of iron can be prevented by the process of galvanisation.

36 As for spontaneous reaction,
$\Delta G = -ve$
$\therefore\ E_{cell} = +ve$
Correct reason :
$E_{cell} = E_{cathode} - E_{anode}$. E_{cell} is +ve when, $E_{cathode} > E_{anode}$.

37 Correct explanation: $Ag^+ + e^- \longrightarrow Ag$,
$$E_{Ag^+/Ag} = E^\circ_{Ag^+/Ag} - \frac{RT}{nF}\log\frac{1}{[Ag^+]}$$

38 In a Daniell cell, $Zn|Zn^{2+}||Cu^{2+}|Cu$
$E_{cell} = 1.1$ V
The reaction of oxidation half-cell is
$Zn \longrightarrow Zn^{2+} + 2e^-$
The reaction of reduction half-cell is
$Cu^{2+} + 2e^- \longrightarrow Cu$
So, $Zn + Cu^{2+} \longrightarrow Zn^{2+} + Cu$
because Zn is oxidised, it is deposited at anode and Cu is reduced, so it is deposited at cathode. If the opposite potential is greater than 1.1 V, then the electrons flow from cathode to anode. So, both the statements are true and correct explanation.

39 Galvanised iron, i.e. iron coated with zinc does not rust easily as zinc has more negative electrode potential (– 0.76 V) than iron.

SESSION 2

1 $Al^{3+} + 3e^- \longrightarrow Al$
From Faraday's first law,
$w = ZQ$
where, w = amount of metal
$= 5.12$ kg $= 5.12 \times 10^3$ g
Q = charge
Z = electrochemical equivalent
$$= \frac{\text{equivalent weight}}{96500} = \frac{\text{atomic mass}}{\text{number of electrons} \times 96500} = \frac{27}{3 \times 96500}$$
$$5.12 \times 10^3 = \frac{27}{3 \times 96500} \times Q$$
$$Q = \frac{5.12 \times 10^3 \times 3 \times 96500}{27}\text{C}$$
$= 5.49 \times 10^7$ C

2 For AgCl, molarity = normality
Actual specific conductance = (specific conductance of AgCl
– specific conductance of water)
$= (3.41 - 1.60) \times 10^{-6}$
$= 1.81 \times 10^{-6}\ ohm^{-1}\ cm^{-1}$
For saturated solution of sparingly soluble salt,
$\Lambda_{eq} = \Lambda^\infty_{eq}$,
and solubility = concentration
$$\Lambda^\infty_{eq} = \frac{1000 \times \text{specific conductance}}{\text{solubility}}$$
$$\therefore\ 138.3 = \frac{1000 \times 1.81 \times 10^{-6}}{S}$$
$$S\ (\text{mol L}^{-1}) = \frac{1000 \times 1.81 \times 10^{-6}}{138.3}$$
$= 1.31 \times 10^{-5}$ mol L^{-1}
$AgCl(s) \rightleftharpoons Ag^+(aq) + Cl^-(aq)$
$\therefore\ K_{sp} = [Ag^+][Cl^-] = S^2$
$= (1.31 \times 10^{-5})^2$
$= 1.72 \times 10^{-10}$ mol^2 L^{-2}

3 Given, $E^\circ_{FeO/Fe} = -0.87$ V
and $E^\circ_{Ni_2O_3/NiO} = 0.40$ V
$\therefore\ E^\circ_{Fe/FeO} = +0.87$ V
At anode
$Fe + 2OH^- \longrightarrow FeO(s) + H_2O(l) + 2e^-$
At cathode
$Ni_2O_3(s) + H_2O(l) + 2e^- \longrightarrow 2NiO(s) + 2OH^-$
Overall reaction
$Fe(s) + Ni_2O_3(s) \longrightarrow FeO(s) + 2NiO(s)$
Hence,
$E^\circ_{cell} = 0.40 + 0.87 = 1.27$ V
and $\Delta G^\circ = -nFE^\circ_{cell}$
$= 2 \times 1.27 \times 96500$
$= -245110$ J $= -245.11$ kJ
$\Delta G^\circ = -W_{max}$
$W_{max} = 245.11$ kJ

4 $$\frac{\text{wt.}}{\text{eq. wt.}} = \frac{i \cdot t}{96500}$$
or $$\frac{2.16}{108} = \frac{i \times 60 \times 60}{96500}$$
or $i = 0.54$ A
Error in reading of ammeter
$= 0.60 - 0.54 = 0.06$ A
$$\therefore\ \%\ \text{error} = \frac{0.06}{0.60} \times 100 = 10\%$$

5 The redox changes are :
At anode $2Cl^- \longrightarrow Cl_2 + 2e^-$
At cathode
$2e^- + 2H_2O \longrightarrow 2OH^- + H_2$
Now, $$\frac{w}{\text{eq. wt.}} = \frac{i \cdot t}{96500}$$
Given, $w_{Cl_2} = 10^3$ g, eq. wt.$_{Cl_2} = 35.5$
and $$i = \frac{25 \times 62}{100} = 15.5\text{ A}$$
Hence, $$t = \frac{10^3 \times 96500}{35.5 \times 15.5}$$
$= 175374.83$ s
$= 48.71$ h

6 For cell,
$Zn|Zn^{2+}(a = 0.1M)||Fe^{2+}(a = 0.01M)|Fe$
The half-cell reactions are:
(i) $Zn(s) \longrightarrow Zn^{2+}(aq) + 2e^-$
(ii) $Fe^{2+}(aq) + 2e^- \longrightarrow Fe(s)$
$Zn(s) + Fe^{2+}(aq) \longrightarrow Zn^{2+}(aq) + Fe(s)$
On applying Nernst equation,
$$E_{cell} = E^\circ_{cell} - \frac{0.0591}{n}\log_{10}\frac{[Zn^{2+}]}{[Fe^{2+}]}$$

$$0.2905 = E^\circ_{cell} - \frac{0.0591}{2}\log_{10}\frac{0.1}{0.01}$$

$$0.2905 = E^\circ_{cell} - 0.0295 \times 1$$

$\therefore$ $E^\circ_{cell} = 0.2905 + 0.0295 = 0.32\text{ V}$

At equilibrium, ($E_{cell} = 0$)

$$E_{cell} = E^\circ_{cell} - \frac{0.0591}{n}\log_{10}K_C$$

or $E^\circ_{cell} = \frac{0.0591}{2}\log_{10}K_C$

$$0.32 = \frac{0.0591}{2}\log_{10}K_C$$

or $K_C = 10^{0.32/0.0295}$

7 Less the reduction potential, weaker is the oxidising agent or stronger is the reducing agent.
Therefore, Zn is the strongest reducing agent.

8 Eq. of metal = Eq. of iodine
= Eq. of hypo

$I_2 + 2e^- \longrightarrow 2I^-$

$2S_2O_3^{2-} \longrightarrow S_4O_6^{2-} + 2e^-$

$\therefore$ $\frac{0.617}{\text{eq. wt.}} = \frac{46.3}{1000} \times 0.124 \times 1$

Hence, $\frac{0.617}{107.47} = \frac{10 \times t}{96500}$

or $t = 55.4\text{ s}$

9 According to Kohlrausch's law,

$\lambda^\infty_{eq}\ Na_2SO_4 = \lambda^\infty_{eq}\ (NaCl) + \lambda^\infty_{eq}\ K_2SO_4 - \lambda^\infty_{eq}\ KCl$

$= (123.7 + 152.1 - 147.0)\ \Omega^{-1}cm^2\,eq^{-1}$

$= 128.8\ \Omega^{-1}cm^2\,eq^{-1}$

10 Greater the reduction potential, stronger is the oxidising agent. Hence, *Y* is stronger oxidising agent than *X* but weaker than *Z*.

11 Molar conductivity of aqueous CH_3COOH solution can be calculated from Kohlrausch law as

$\lambda^\circ_{CH_3COOH} = \lambda^\circ_{CH_3COONa} + \lambda^\circ_{HCl} - \lambda^\circ_{NaCl}$

Hence, for determining the molar conductivity of CH_3COOH solution, λ°_{NaCl} is required.

12 Reactions,

(i) $Fe(s) \longrightarrow Fe^{2+} + 2e^-$;

$E^\circ = +0.44\text{V}$

and $\Delta G_1^\circ = -nE^\circ F = -2 \times 0.44 \times F$

(ii) $2H^+ + 2e^- + \frac{1}{2}O_2 \longrightarrow H_2O(l)$;

$E^\circ = +1.23\text{V}$

and $\Delta G_2^\circ = -2 \times (+1.23) \times F$

Net reaction,

$Fe(s) + 2H^+ + \frac{1}{2}O_2 \longrightarrow Fe^{2+} + H_2O(l)$;

$\Delta G_3^\circ = \Delta G_1^\circ + \Delta G_2^\circ$

$= [-2 \times (+0.44) + (-2 \times 1.23)]$

$= [-0.88 - 2.46]$

$= -3.34\text{ F} = -3.34 \times 96500\text{ J}$

$= -322.31\text{ kJ/mol} \approx -322\text{ kJmol}^{-1}$

13 The reaction between MnO_4^- and HCl may be represented as follows :

$2MnO_4^-(aq) + 16H^+ + 10Cl^- \longrightarrow 2Mn^{2+}(aq) + 8H_2O(l) + 5Cl_2(g)$

Thus, on the basis of this reaction following electrochemical cell will be represented

Pt $Cl^-(aq)|Cl_2\ (g)(1\text{ atm})||$

$MnO_4^-(aq)|\ Mn^{2+}(aq)$

Since, $E^\circ_{cell} = E^\circ_{cathode} - E^\circ_{anode}$

From given data,

$E^\circ_{cell} = 1.51 - 1.40 = 0.11\text{ V}$

E°_{cell} is positive, hence ΔG° is negative.
Thus, above cell reaction is feasible but MnO_4^- ion can oxidise, Fe^{2+} to Fe^{3+} and Cl^- to Cl_2 in aqueous medium also.
Therefore, for quantitative estimation of aqueous $Fe(NO_3)_2$ it is not a suitable reagent.

14 Cell is completely discharged. It means equilibrium gets established, $E_{cell} = 0$

$Zn|Zn^{2+}\ (1\text{ M})\ ||\ Cu^{2+}\ (1\text{ M})\ |\ Cu$

Cell reaction :

$Zn + Cu^{2+} \rightleftharpoons Zn^{2+} + Cu$

$$K_{eq} = \frac{[Zn^{2+}]}{[Cu^{2+}]}$$

We know that,

$$E_{cell} = E^\circ_{cell} - \frac{0.0591}{n}\log K_{eq}$$

$$E^\circ_{cell} = \frac{0.0591}{n}\log K_{eq}$$

or $1.10 = \frac{0.0591}{2}\log K_{eq}$

$$K_{eq} = \frac{[Zn^{2+}]}{[Cu^{2+}]} = \text{antilog}\frac{2.20}{0.0591}$$

$= \text{antilog } 37.2$ or $10^{37.2}$

15 Given that, $i = 100$ amp. also, 27.66 g of diborane (B_2H_6)

Molecular mass of B_2H_6

$= 10.8 \times 2 + 6 = 27.6$

Number of moles of B_2H_6 in 27.66 g

$= \frac{\text{Given mass}}{\text{Molar mass}} = \frac{27.66}{27.6} = 1$

Now, consider the question,

$B_2H_6 + 3O_2 \longrightarrow B_2O_3 + 3H_2O$

From the equation we can interpret that 3 moles of oxygen is required to burn 1 mole (i.e. 27.6 g) B_2H_6 completely.

Also consider the electrolysis reaction of water, i.e.

$H_2O \rightleftharpoons 2H^+ + O^{--}$

$2H^+ \xrightarrow[\text{Cathode}]{+2e^-} 2H \longrightarrow H_2\uparrow$

$O^{--} \xrightarrow[-2e^-]{\text{Anode}} O \xrightarrow[\text{atoms}]{\text{2 such}} O_2\uparrow$

From the above equation it can be easily interpreted that in electrolysis of water for the production of 1 mole of oxygen from 1 mole of H_2O at anode 4 moles electrons are required.

Likewise for the production of 3 moles of O_2 12(3×4) moles of electrons will be needed. So, the total amount of charge required to produce 3 moles of oxygen will be $12 \times F$ or 12×96500

We know $Q = it$

So, $12 \times 96500 = 100 \times t$ in seconds

or $\frac{12 \times 96500}{100 \times 3600} = t$ in hours = 3.2 hours

16. $AgI(s) + e^- \rightleftharpoons Ag(s) + I^-$;

$E_1^\circ = -0.152$

$Ag(s) \longrightarrow Ag^+ + e^-$; $E_2^\circ = -0.8$

$AgI(s) \longrightarrow Ag^+ + I^-$; $E^\circ = -0.952$

$\left[\because \Delta G^\circ = \Delta G_1^\circ + \Delta G_2^\circ \text{ or } E^\circ = E_1^\circ + E_2^\circ\right]$

$$E^\circ_{cell} = \frac{0.059}{n}\log K_{sp}$$

$$-0.952 = \frac{0.059}{1}\log K_{sp}$$

$$\log K_{sp} = \frac{-0.952}{0.059}$$

$= -16.14$

DAY FOURTEEN

Chemical Kinetics

Learning & Revision for the Day

- Rate of a Reaction
- Rate Law and Rate Constant
- Order of Reaction
- Molecularity
- Differential and Integral forms of Different Order Reactions
- Effect of Temperature on Rate of Reaction
- Collision Theory

Chemical kinetics is the branch of chemistry which addresses the rate of chemical reaction. It includes the investigations of how different experimental conditions can influence the speed of a chemical reaction and yield information about the reaction mechanism and transition rates.

Rate of a Reaction

The rate or speed or velocity of a reaction is the rate of change of concentration of reactants or products in unit time.

$$A \rightarrow B$$

Rate of reaction can be defined in two ways

1. Average rate of reaction, $r_{av} = -\frac{\Delta[A]}{\Delta t} = \frac{\Delta[B]}{\Delta t}$

   The units of rate of a reaction are mol L^{-1} s^{-1} or mol L^{-1} min^{-1}.

2. Instantaneous rate of reaction (r_{inst}) can be calculated from r_{av} in the limit $\Delta t \rightarrow 0$ and is represented as $r_{inst} = -\frac{d[A]}{dt} = \frac{d[B]}{dt}$

   In general for a reaction, $aA + bB \longrightarrow cC + dD$

$$\text{Rate} = -\frac{1}{a}\frac{d[A]}{dt} = -\frac{1}{b}\frac{d[B]}{dt} = +\frac{1}{c}\frac{d[C]}{dt} = +\frac{1}{d}\frac{d[D]}{dt}$$

Factors Affecting the Rate of Reaction

The rate of chemical reaction depends upon a number of factors such as

(i) concentration
(ii) nature of reactants
(iii) temperature
(iv) presence of catalyst
(v) exposure to radiations

PREP MIRROR
Your Personal Preparation Indicator

- No. of Questions in Exercises (x)—
- No. of Questions Attempted (y)—
- No. of Correct Questions (z)—
  (Without referring Explanations)

- Accuracy Level ($z/y \times 100$)—
- Prep Level ($z/x \times 100$)—

In order to expect good rank in JEE, your Accuracy Level should be above 85 & Prep Level should be above 75.

Rate Law and Rate Constant

According to law of mass action, "the rate of a chemical reaction is directly proportional to the product of effective concentrations of reacting species, each raised to a suitable power which may or may not be equal to the respective stoichiometric coefficients."

For a general reaction,

$$aA + bB \longrightarrow \text{Products; Rate} = k[A]^a[B]^b$$

where, k is rate constant or velocity constant or specific reaction rate. The above expression is called **rate law**.

- Rate law cannot be deduced from balanced equation. It is obtained experimentally.
- **Rate constant of reaction** (k) is defined as the rate of reaction when the concentration of each reactant is taken as unity.
- Unit of k in terms of concentration,

$$\text{Rate} = k[A]^n$$

$$\frac{\text{mol}}{\text{L}\cdot\text{time}} = k\left[\frac{\text{mol}}{\text{L}}\right]^n$$

$\therefore$ Unit of $k = (\text{mol})^{1-n}\cdot \text{L}^{n-1}\,\text{s}^{-1}$

where, n = order of reaction

Order of Reaction

The sum of the powers of the reacting species that are involved in the rate law expression for the reaction represents the **order of the reaction**.

For the reaction, $aA + bB \longrightarrow cC + dD$

According to law of mass action, $\frac{dx}{dt} = k[A]^a[B]^b$

According to rate law (rate equation), $\frac{dx}{dt} = k[A]^\alpha[B]^\beta$

Overall order of reaction = $\alpha + \beta$

The order of reaction can only be determined by experiments.

Molecularity

- Molecularity is defined as the number of ions or molecules or atoms taking part in an elementary process of the reaction mechanism.
- In case of simple reactions, known as elementary reactions, the molecularity is simply the sum of molecules of different reactants as represented by balanced equation, e.g.
  - (i) **Unimolecular reaction** $N_2O_4(g) \longrightarrow 2NO_2(g)$
  - (ii) **Bimolecular reaction** $2HI(g) \longrightarrow H_2(g) + I_2(g)$
  - (iii) **Trimolecular reaction**

    $$2NO(g) + O_2(g) \longrightarrow 2NO_2(g)$$
- In case of complex reactions, i.e. the reactions involving more than one step the rate determining step is the slowest step. The atoms, molecules or ions participating in this step decide the molecularity, e.g.

$$HBrO_3 + 6HI \longrightarrow HBr + 3H_2O + 3I_2$$

This reaction takes place through following steps:

$$HBrO_3 + HI \xrightarrow[\text{(Rate determining step)}]{\text{Slow}} HBrO_2 + HIO$$

$$HBrO_2 + 4HI \xrightarrow{\text{Fast}} HBr + 2H_2O + 2I_2$$

$$HIO + HI \xrightarrow{\text{Fast}} H_2O + I_2$$

Hence, two molecules participate in slowest step, the reaction is bimolecular type.

NOTE
- **Order of reaction** may change with the condition such as pressure, temperature etc., whereas molecularity does not change.
- **Molecularity** is always a whole number. It can never be zero, negative or fraction whereas, order of a reaction may be whole number, fractional, zero or even negative.

Differential and Integral forms of Different Order Reactions

Zero Order Reaction

Reactions in which the concentration of reactants do not change with time and the concentration, rates remain constant throughout, are said to be zero order reactions.

$$R \longrightarrow P$$

$$\text{Rate} = \frac{-d[R]}{dt} = k[R]^0$$

The integrated rate equation is given as:

$$k = -\frac{[C]_t - [C]_0}{t}$$

where, C_0 = initial concentration of the reactant

C_t = concentration of the reactant at any time t

- Half-life of a reaction is the time required to reduce the concentration of the reactant to half of its initial value.
- Half-life for zero order is given as:

$$t_{1/2} = \frac{[C]_0}{2k}$$

where, k = rate constant, $t_{1/2}$ = half-life period

- For a zero order reaction, half-life period is directly proportional to the initial concentration of reactant and units of rate constant and rate of reaction are same, i.e. mol $L^{-1}\,s^{-1}$.

First Order Reaction

The first order reaction is defined as "the reaction in which the rate is determined by the change of one concentration term of the reactant only".

For first order reaction,

$$R \longrightarrow P$$

$$\text{Rate} = \frac{-d[R]}{dt} = k[R]$$

Integrated rate equation is given as:

$$t = \frac{2.303}{k} \log \frac{a}{(a-x)}$$

where, t = time, a = initial concentration
k = rate constant,
$(a-x)$ = concentration of a after time 't'

- Half-life of first order reaction is given as:

$$t_{1/2} = \frac{0.693}{k}, \quad t_{1/2} = \text{half-life period}$$

- The unit of rate constant for first order reaction is s^{-1}.
- For any reaction, half-life period, $t_{1/2} \propto \frac{1}{a^{n-1}}$

  where, a = initial concentration, n = order of reaction
- Rate equation for first order gas phase reactions

$$k = \frac{2.303}{t} \log \frac{p_0}{p_0 - p_t}$$

where, p_0 = initial pressure, p_t = pressure after time t

- In a sequence of reactions, $A \xrightarrow{k_A} B \xrightarrow{k_B} C$. $[B]$ is maximum when,

$$t = \frac{2.303}{(k_B - k_A)} \log_{10}\left(\frac{k_B}{k_A}\right)$$

Second Order Reaction

The reaction is said to be of **second order** if its reaction rate is determined by the variation of two concentration terms of reactants.

Its rate equation can be represented as $k = \frac{1}{t}\left[\frac{x}{a(a-x)}\right]$, when both the reactants are same.

The unit of rate constant for second order reaction is $mol^{-1}\ L\ s^{-1}$.

- Reactions of third and higher orders are rare. This is due to the fact that the chances of three molecules of coming to a single point simultaneously, i.e. probability of trimolecular collisions is much less as compared to unimolecular or bimolecular collisions.
- Sometimes the rate of reaction decreases as the concentration of one of the constituent is increased, such reactions are known as **negative order reactions**. e.g. transformation of ozone into oxygen, i.e.

$$2O_3 \longrightarrow 3O_2$$

$$\text{rate} = -\frac{1}{2}\frac{d}{dt}[O_3] = k\frac{[O_3]^2}{[O_2]} = k[O_3]^2[O_2]^{-1}$$

- When one of the reactants is present in large excess, the second order reaction confirms to the first order and is called **pseudo unimolecular reaction**, e.g. hydrolysis of ester in acidic medium.

$$CH_3COOC_2H_5 \xrightarrow{H^+/H_2O} CH_3COOH + C_2H_5OH$$

  This is also known as pseudo first order reaction.
  Rate = $k[CH_3COOC_2H_5]$

Effect of Temperature on Rate of Reaction

- It has been found that for a chemical reaction with rise in temperature by 10°, the rate constant is nearly doubled.
- The effect of temperature is usually expressed in terms of temperature coefficient.
- The value of temperature coefficient for most of the reactions lies between 2 to 3.

Arrhenius Theory

The temperature dependence of the rate of a chemical reaction can be accurately explained by **Arrhenius equation**.

$$k = Ae^{-E_a/RT} \quad \text{...(i)}$$

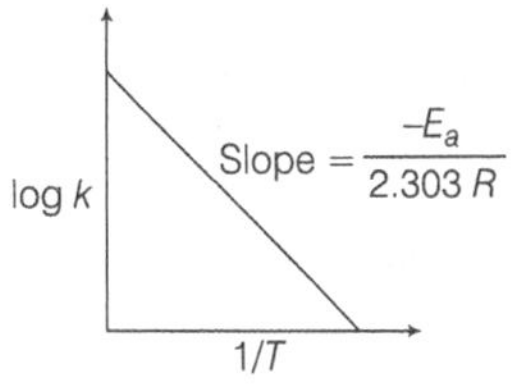

A plot between log K and $1/T$

Here, k = rate constant
A = Arrhenius constant or frequency factor
E_a = activation energy

- It explains that the products are formed through the intermediates or activated complex.
- **Arrhenius equation** is based on two theories of reaction rates, collision theory and transition state theory.

  Since, $k = Ae^{-E_a/RT}$ and E_a is always positive, thus, k always increases with temperature.

  Due to this rate always increases with temperature whether the reaction is exothermic or endothermic.
- When, $T \longrightarrow \infty$, $k = Ae^0$ or rate constant becomes equal to Arrhenius or frequency factor. (For free radical reaction $E_a = 0$, thus $k = A$ and k becomes independent of temperature).

Activation Energy

- The minimum amount of energy required for a chemical reaction to occur is called **activation energy**. It is represented by E_a.

- Activation energy does not depend on temperature, volume, pressure etc., but gets affected by the presence of catalyst.
- Catalyst increases the rate of reaction by providing alternative path of lower activation energy to the reactants.

$A \longrightarrow B \ \Delta H < 0$, i.e. exothermic reaction

- Enzyme catalysed reaction are faster than the metal catalysed reaction because they have lower activation energy. Threshold energy is independent of temperature, but E_a shows temperature dependence.

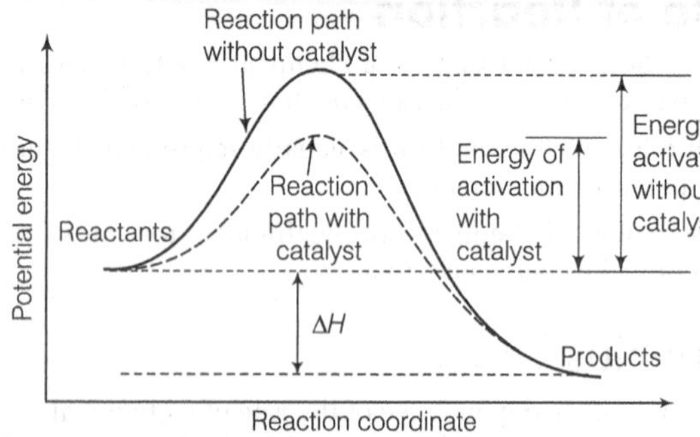

Effect of catalyst on activation energy

Calculation of Activation Energy

Activation energy can be calculated by knowing the rate constants at two different temperatures, assuming that E_a and A remains constant.

Taking log of both the sides in Arrhenius equation, we get

$$\ln k = \ln A - \frac{E_a}{RT}$$

Now, if the values of rate constant at temperatures T_1 and T_2 are k_1 and k_2 respectively then

$$\ln k_1 = \ln A - \frac{E_a}{RT_1} \quad \ldots \text{(i)}$$

$$\ln k_2 = \ln A - \frac{E_a}{RT_2} \quad \ldots \text{(ii)}$$

Subtracting the Eq. (i) from (ii), we get

$$\ln k_2 - \ln k_1 = -\frac{E_a}{RT_2} - \left(-\frac{E_a}{RT_1}\right) = \frac{E_a}{RT_1} - \frac{E_a}{RT_2}$$

$$\ln \frac{k_2}{k_1} = \frac{E_a}{R}\left[\frac{1}{T_1} - \frac{1}{T_2}\right] = \frac{E_a}{R}\left[\frac{T_2 - T_1}{T_1T_2}\right]$$

$$\log \frac{k_2}{k_1} = \frac{E_a}{2.303R}\left(\frac{T_2 - T_1}{T_1T_2}\right) \quad [\text{Here, } T_2 > T_1]$$

- This relation is used when rate constants at two different temperatures are given.

- This reaction is also written as $\frac{d \log k}{dT} = \frac{E}{RT^2}$
- When $\log k$ is plotted against $1/T$, we get a straight line which is represented by Arrhenius equation as

$$\log k = \log A - \frac{E_a}{2.303\,RT}$$

- The intercept of this line is equal to $\log A$ and slope is equal to $-\frac{E_a}{2.303\,R}$. Therefore, $E_a = -2.303\,R \times \text{slope}$

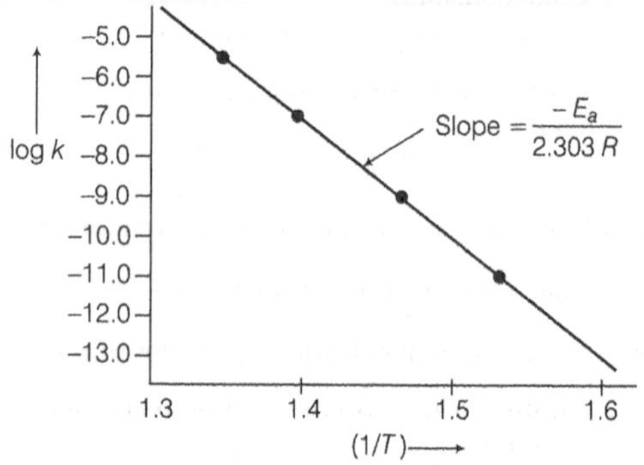

Plot of log *k* vs 1/*T*

Collision Theory

According to **collision theory**, the reactant molecule are assumed to be hard spheres and reaction is postulated to occur when molecules collide with each other.

- The collisions among the reacting species which results in the products are known as the effective collisions.
- The species taking part in the chemical reaction must have a certain minimum energy known as threshold energy ($E°$).
- Threshold energy = activation energy (E_a) + average energy of molecules.
- If we assume that the rate of reaction is equal to rate of collisions per unit volume or per unit time multiplied by fraction of collisions then the rate constant is given by the expression.

$$k = Z_{AB}\, e^{-E_a/RT}$$

where, Z_{AB} = collision frequency factor of reactants A and B

- When steric factor is considered the rate of reaction is given as:

$$\text{Rate} = PZ_{AB}\, e^{-E_a/RT}$$

where, P is probability or steric factor.

DAY PRACTICE SESSION 1

FOUNDATION QUESTIONS EXERCISE

1 A drop of a solution (volume = 0.05 mL) contains 6×10^{-7} mole of H^+. If the rate of disappearance of H^+ is 6.0×10^5 mol $L^{-1}s^{-1}$, how long will it take for the H^+ in the drop to disappear?

(a) 8.0×10^{-8} s (b) 2.0×10^{-8} s
(c) 6.0×10^{-6} s (d) 2.0×10^{-2} s

2 The instantaneous rate of disappearance of MnO_4^- ion in the following reaction is 4.56×10^{-3} M s^{-1}

$2MnO_4^- + 10I^- + 16H^+ \longrightarrow 2Mn^{2+} + 5I_2 + 8H_2O$

The rate of appearance I_2 is → JEE Main (Online) 2013

(a) 4.56×10^{-4} M s^{-1}
(b) 1.14×10^{-2} M s^{-1}
(c) 1.14×10^{-3} M s^{-1}
(d) 5.7×10^{-3} M s^{-1}

3 The rate law for the reaction,

$RCl + NaOH(aq) \longrightarrow ROH + NaCl$ is given by, rate = $k[RCl]$. The rate of the reaction will be

(a) doubled on doubling the concentration of sodium hydroxide
(b) halved on reducing the concentration of alkyl halide to one half
(c) decreased on increasing the temperature of the reaction
(d) unaffected by increasing the temperature of the reaction

4 The units of second order rate constant is

(a) mol dm^{-3}s^{-1} (b) s^{-1}
(c) dm^3 mol^{-1}s^{-1} (d) None of these

5 For the reaction system,

$$2NO(g) + O_2(g) \longrightarrow 2NO_2(g)$$

volume is suddenly reduced to half its value by increasing the pressure on it. If the reaction is of first order with respect to O_2 and second order with respect to NO; the rate of reaction will

(a) diminish to one fourth of its initial value
(b) diminish to one eighth of its initial value
(c) increase to eight times of its initial value
(d) increase to four times of its initial value

6 For the non-stoichiometric reaction, $2A + B \longrightarrow C + D$, the following kinetic data were obtained in three separate experiments, all at 298 K.

	Initial concentration (A)	Initial concentration (B)	Initial rate of formation of (C) (mol L^{-1}s^{-1})
I	0.1 M	0.1 M	1.2×10^{-3}
II	0.1 M	0.2 M	1.2×10^{-3}
III	0.2 M	0.1 M	2.4×10^{-9}

The rate law for the formation of C is → JEE Main 2014

(a) $\frac{dC}{dt} = k[A][B]$ (b) $\frac{dC}{dt} = k[A]^2[B]$
(c) $\frac{dC}{dt} = k[A][B]^2$ (d) $\frac{dC}{dt} = k[A]$

7 Consider a reaction, $aG + bH \longrightarrow$ products, when concentration of both the reactants G and H is doubled, the rate increases by eight times. However, when concentration of G is doubled keeping the concentration of H fixed, the rate is doubled. The overall order of the reaction is

(a) 0 (b) 1 (c) 2 (d) 3

8 At 518°C, the rate of decomposition of a sample of gases acetaldehyde, initially at a pressure of 363 torr, was 1.00 torr s^{-1} when 5% had reacted and 0.5 torr s^{-1}, when 33% had reacted. The order of the reaction is

(a) 2 (b) 3 (c) 1 (d) 0

9 The rate constant for a zero order reaction is

(a) $k = \frac{C_0}{2t}$ (b) $k = \frac{C_0 - C_t}{t}$
(c) $k = \ln\frac{C_0 - C_t}{t}$ (d) $k = \frac{C_0}{C_t}$

10 The rate for the decomposition of NH_3 on platinum surface is zero order. What are the rate of production of N_2 and H_2 respectively, if $k = 2.5 \times 10^{-4}$ mol L^{-1}s^{-1}?

(a) 1.25×10^{-4} mol L^{-1} s^{-1}; 3.75×10^{-4}mol L^{-1}s^{-1}
(b) 3.75×10^{-4} mol L^{-1} s^{-1}; 1.25×10^{-4}mol L^{-1}s^{-1}
(c) 2.5×10^{-4} mol L^{-1} s^{-1}; 3.75×10^{-4}mol L^{-1}s^{-1}
(d) 1.25×10^{-4} mol L^{-1} s^{-1}; 2.5×10^{-4}mol L^{-1}s^{-1}

11 The rate constant of a zero order reaction is 2.0×10^{-2} mol L^{-1} s^{-1}. If the concentration of the reactant after 25 s is 0.5 M, what is the initial concentration?

→ JEE Main (Online) 2013

(a) 0.5 M (b) 1.25 M (c) 12.5 M (d) 1.0 M

12 The time for half-life period of a certain reaction, $A \rightarrow$ products is 1 h. When the initial concentration of the reactant 'A', is 2.0 mol L^{-1}, how much time does it take for its concentration to come from 0.50 to 0.25 mol L^{-1}, if it is a zero order reaction? → AIEEE 2010

(a) 4 h (b) 0.5 h (c) 0.25 h (d) 1 h

13 For a first order reaction, the time required for 99.9% of the reaction to take place is nearly

(a) 10 times that required for half of the reaction
(b) 100 times that required for half of the reaction
(c) 10 times that required for one fourth of the reaction
(d) 20 times that required for half of the reaction

14 A first order reaction is half-completed in 45 min. How long does it need for 99.9% of the reaction to be completed?

(a) 20 h (b) 10 h

(c) $7\frac{1}{2}$ h (d) 5 h

15 For a first order reaction, $A \longrightarrow$ Products, the concentration of A changes from 0.1 M to 0.025 M in 40 min. The rate of reaction when the concentration of A is 0.01 M is →AIEEE 2012

(a) 1.73×10^{-5} M/min
(b) 3.47×10^{-4} M/min
(c) 3.47×10^{-5} M/min
(d) 1.73×10^{-4} M/min

16 Under the same reaction conditions, initial concentration of 1.386 mol dm^{-3} of a substance becomes half in 40 s and 20 s through first order and zero order kinetics respectively. Ratio (k_1/k_0) of the rate constants for first order (k_1) and zero order (k_0) of the reaction is

(a) 0.5 $mol^{-1}dm^3$ (b) 1.0 mol dm^{-3}
(c) 1.5 mol dm^{-3} (d) 2.0 $mol^{-1}dm^3$

17 The half-life period of a first order chemical reaction is 6.93 min. The time required for the completion of 99% of the chemical reaction will be (log 2 = 0.301)

(a) 230.3 min (b) 23.03 min
(c) 46.06 min (d) 460.6 min

18 An organic compound undergoes first order decomposition. The time taken for its decomposition to 1/8 and 1/10 of its initial concentration are $t_{1/8}$ and $t_{1/10}$ respectively.

What is the value of $\frac{[t_{1/8}]}{[t_{1/10}]} \times 10$? ($\log_{10} 2 = 0.3$) →AIEEE 2012

(a) 2 (b) 3
(c) 6 (d) 9

19 Half-life of a hypothetical reaction is found to be inversely proportional to the cube of initial concentration. The order of reaction is

(a) 4 (b) 3
(c) 5 (d) 2

20 The half-life of a second order reaction, $A \longrightarrow B$ is given as

(a) $t_{1/2} = \frac{0.693}{k}$ (b) $t_{1/2} = \frac{k}{[A]_0}$

(c) $t_{1/2} = [A]_0/k$ (d) $t_{1/2} = 1/k[A]_0$

21 The following mechanism has been proposed for the reaction of NO with Br_2 to form NOBr

$$NO(g) + Br_2(g) \rightleftharpoons NOBr_2(g)$$
$$NOBr_2(g) + NO(g) \longrightarrow 2NOBr(g)$$

If the second step is the rate determining step, the order of the reaction with respect to NO(g) is

(a) 1 (b) 0
(c) 3 (d) 2

22 The hydrolysis of methyl formate in acid solution has rate expression : rate $= k\,[HCOOCH_3][H^+]$

The balanced equation being

$$HCOOCH_3 + H_2O \longrightarrow HCOOH + CH_3OH$$

The rate law contains $[H^+]$, though the balanced equation does not contain $[H^+]$ because

(a) H^+ ion is a catalyst
(b) H^+ is an important constituent of any reaction
(c) more for convenience to express the rate law
(d) all acids contain H^+ ions

23 Higher order (> 3) reactions are rare due to →JEE Main 2015

(a) low probability of simultaneous collision of all the reacting species
(b) increase in entropy and activation energy as more molecules are involved
(c) shifting of equilibrium towards reactants due to elastic coillisions
(d) loss of active species on collision

24 The rate of a chemical reaction doubles for every 10°C rise of temperature. If the temperature is raised by 50°C, the rate of the reaction increases by about →AIEEE 2011

(a) 10 times
(b) 24 times
(c) 32 times
(d) 64 times

25 The rate of reaction is doubled for every 10°C rise in temperature. The increase in the reaction rate as a result of temperature rise from 10°C to 100°C is

(a) equal to the energy of activation of products
(b) 112 times
(c) 512 times
(d) 614 times

26 The role of a catalyst is to change ....

(a) Gibbs energy of reaction
(b) enthalpy of reaction
(c) activation energy of reaction
(d) equilibrium constant

27 For an endothermic reaction, where ΔH represents the enthalpy of the reaction in kJ/mol, the minimum value for the energy of activation will be

(a) less than ΔH
(b) zero
(c) more than ΔH
(d) equal to ΔH

28 Which of the following graph represents exothermic reaction?

I
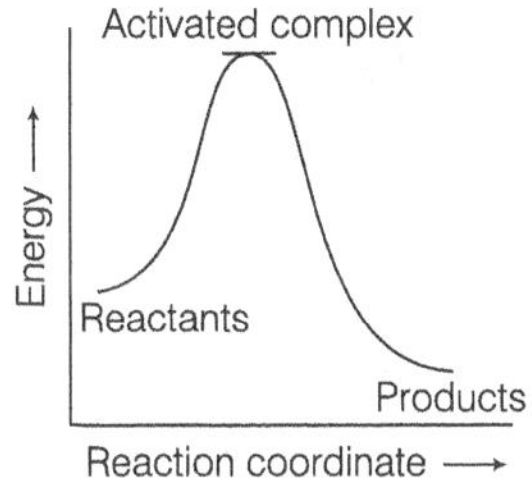

II
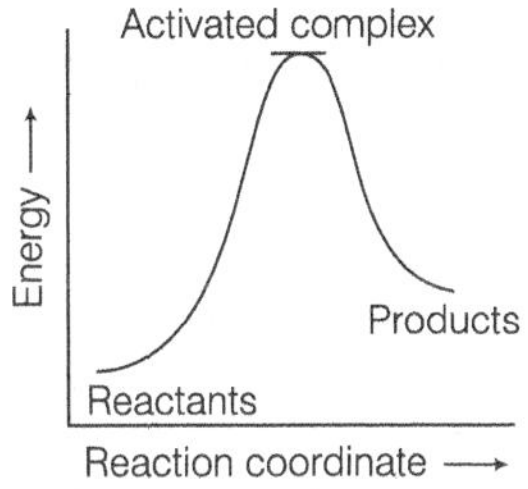

III
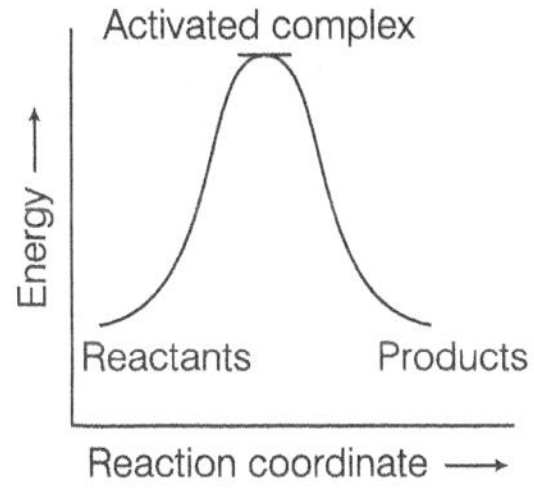

(a) Only (I) (b) Only (II)
(c) Only (III) (d) Both (I) and (II)

29 The reaction, $X \longrightarrow Y$ is an exothermic reaction. Activation energy of the reaction for conversion of X into Y is 150 kJ mol^{-1}. Enthalpy is 135 kJ mol^{-1}. The activation energy for the reverse reaction, $Y \longrightarrow X$ will be

→ JEE Main (Online) 2013

(a) 280 kJ mol^{-1} (b) 285 kJ mol^{-1}
(c) 270 kJ mol^{-1} (d) 15 kJ mol^{-1}

30 H_2O and O-atom react in upper atmosphere bimolecularly to form two OH radicals. ΔH for the reaction is 72 kJ mol^{-1} at 500 K and energy of activation is 77 kJ mol^{-1}. E_a for bimolecular recombination of two OH radicals to form H_2O and O-atom will be

(a) 5 kJ mol^{-1} (b) 72 kJ mol^{-1}
(c) 77 kJ mol^{-1} (d) 149 kJ mol^{-1}

31 The plot of log k *versus* $\frac{1}{T}$ is linear with a slope of

(a) $-E_a/R$ (b) E_a/R
(c) $-E_a/2.303\,R$ (d) $E_a/2.303\,R$

32 The rate of a reaction doubles when its temperature changes from 300 K to 310 K. Activation energy of such a reaction will be
[R = 8.314 JK^{-1} mol^{-1}, log 2 = 0.301] → JEE Main 2013

(a) 48.6 kJ mol^{-1} (b) 58.5 kJ mol^{-1}
(c) 60.5 kJ mol^{-1} (d) 53.6 kJ mol^{-1}

33 A reactant (A) forms two products

$A \xrightarrow{k_1} B$, Activation energy E_{a_1}

$A \xrightarrow{k_2} B$, Activiation energy E_{a_2}

If $E_{a_2} = 2E_{a_1}$, then k_1 and k_2 are related as → AIEEE 2011

(a) $k_1 = 2k_2 e^{E_{a_2}/RT}$ (b) $k_1 = k_2 e^{E_{a_1}/RT}$
(c) $k_2 = 2k_1 e^{E_{a_2}/RT}$ (d) $k_1 = Ak_2 e^{E_{a_1}/RT}$

Direction (Q. Nos. 34 and 35) *In the following questions, Assertion* (A) *followed by a Reason* (R) *is given. Choose the correct answer out of the following choices.*

(a) Both A and R are true and R is correct explanation of A
(b) Both A and R are true but R is not correct explanation of A
(c) A is true but R is false
(d) Both A and R are false

34. **Assertion** (A) Order and molecularity are not same.
Reason (R) Order is determined experimentally and molecularity is the sum of the stoichiometric coefficients of rate determining elementary step.

35. **Assertion** (A) All collisions of reactant molecules do not lead to product formation.
Reason (R) Only those collisions in which molecules have correct orientation and sufficient kinetic energy lead to compound formation.

Direction (Q. Nos. 36 and 37) *Each of these questions contains two statements : Statement* I *(Assertion) and Statement* II *(Reason). Each of these questions also has four alternative choices, only one of which is the correct answer. You have to select one of the codes (a), (b), (c) and (d) given below.*

(a) Statement I is true; Statement II is true; Statement II is a correct explanation for Statement I
(b) Statement I is true, Statement II is true; Statement II is not a correct explanation for Statement I
(c) Statement I is true; Statement II is false
(d) Statement I is false; Statement II is true

36. **Statement I** Order of the reaction can be zero or fractional.
Statement II We cannot determine order from balanced chemical equation.

37. **Statement I** Every collision of reactant molecule is not successful.
Statement II Every collision of reactant molecule with proper orientation is successful one.

DAY PRACTICE SESSION 2

PROGRESSIVE QUESTIONS EXERCISE

1 Consider the reaction,

$Cl_2(aq) + H_2S\,(aq) \longrightarrow S\,(s) + 2H^+(aq) + 2Cl^-(aq)$

The rate equation for this reaction is,

$$\text{rate} = k\,[Cl_2]\,[H_2S]$$

Which of these mechanisms is/are consistent with this rate equation?

I. $Cl_2 + H_2S \longrightarrow H^+ + Cl^- + Cl^+ + HS^-$ (slow)

$Cl^+ + HS^- \longrightarrow H^+ + Cl^- + S$ (fast)

II. $H_2S \rightleftharpoons H^+ + HS^-$ (fast equilibrium)

$Cl_2 + HS^- \longrightarrow 2Cl^- + H^+ + S$ (slow)

(a) Only II (b) Both I and II
(c) Neither I nor II (d) Only I

2 The gas phase decomposition of dimethyl ether follows first order kinetics

$$CH_3OCH_3(g) \longrightarrow CH_4(g) + H_2(g) + CO(g)$$

The reaction is carried out in constant volume container at 500°C and has a half-life of 14.5 min. Initially, only dimethyl ether is present at a pressure of 0.4 atmosphere. What is the total pressure of the system after 12 min ?

(a) 0.564 atm (b) 0.1744 atm
(c) 0.693 atm (d) 0.249 atm

3 The half-life period for first order reaction having activation energy 39.3 kcal mol^{-1} at 300°C and frequency constant $1.11 \times 10^{11} s^{-1}$ will be

(a) 1 h (b) 1.68 h
(c) 1.28 h (d) 1.11h

4 During nuclear explosion, one of the product is ^{90}Sr with half-life of 28.1 yr. If 1 μg of ^{90}Sr was absorbed in the bones of a newly born baby instead of calcium, how much of it will remain after 10 yrs and 60 yrs if it is not lost metabolically ?

(Nuclear explosions follow first order kinetics).

(a) 0.7814 μg and 0.227 μg
(b) 0.227 μg and 0.7814 μg
(c) 0.9338 μg and 0.3323 μg
(d) 0.500 μg and 0.300 μg

5 Consider the Arrhenius equation given below and mark the correct option.

$$k = Ae^{-E_a/RT}$$

(a) Rate constant increases exponentially with increasing activation energy and decreasing temperature
(b) Rate constant decreases exponentially with increasing activation energy
(c) Rate constant increases exponentially with decreasing activation energy and decreasing temperature
(d) Rate constant increases exponentially with decreasing activation energy and increasing temperature

6 The half-life for the reaction, $N_2O_5 \rightarrow 2NO_2 + \frac{1}{2}O_2$ is 24 h at 30°C. Starting with 10 g of N_2O_5, how many grams of N_2O_5 will remain after a period of 96 h ?

(a) 1.25 g (b) 0.63 g
(c) 1.77 g (d) 0.5 g

7 Rate constant k varies with temperature as given by equation

$$\log k\,(\text{min}^{-1}) = 5 - \frac{2000\text{ K}}{T}$$

Consider the following about this equation

I. Pre-exponential factor is 10^5

II. E_a is 9.212 kcal

III. Variation of log k with $\frac{1}{T}$ is linear

Select the correct statement.

(a) I, II, III (b) I, III
(c) II, III (d) I, II

8 Consider the interconversion of the 'boat' and 'chair conformations of cyclohexane.

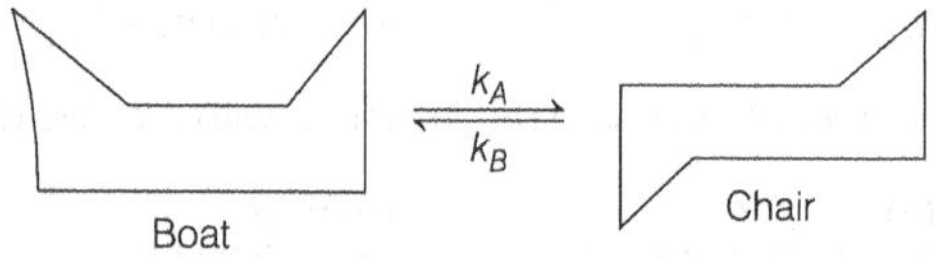

The reaction is first order in each direction with an equilibrium constant of 10^4. The activation energy for the conversion of the chair conformer to the boat conformer is 42 kJ mol^{-1}. Assuming an Arrhenius pre-exponential factor of $10^{12}\,s^{-1}$, what is the expected observed reaction rate constant at 298 K if one were to initiate this reaction starting with only the boat conformer?

(a) $8.01 \times 10^5\,s^{-1}$ (b) $4.34 \times 10^8\,s^{-1}$
(c) $2.56 \times 10^7 s^{-1}$ (d) $3.63 \times 10^7\,s^{-1}$

9 For the reaction following data is given

$$A \longrightarrow B;\ k_1 = 10^{15}\exp^{\left(\frac{-2000}{T}\right)}$$

$$C \longrightarrow D;\ k_2 = 10^{14}\exp^{\left(\frac{-1000}{T}\right)}$$

At what temperature, k_1 and k_2 will be same?

(a) 434.22 K (b) 868.43 K
(c) 217.10 K (d) 130.26 K

10 Select the correct statement.

(a) A collision between reactant molecules must occur with a certain minimum energy before it is effective in yielding product molecule
(b) $\frac{d(\log k)}{dt} = \frac{E_a}{2.303RT^2}$ is called Arrhenius equation
(c) Both (a) and (b)
(d) None of the above

11 Calculate the half-life of the first order reaction,

$$C_2H_4O(g) \longrightarrow CH_4(g) + CO(g)$$

if the initial pressure of C_2H_4O (g) is 80 mm and the total pressure at the end of 20 min is 120 mm.

(a) 40 min (b) 120 min (c) 20 min (d) 80 min

12 The energies of activation for forward and backward reactions for $A_2 + B_2 \rightleftharpoons 2AB$ are 180 kJ mol^{-1} and 200 kJ mol^{-1} respectively. The presence of a catalyst lowers the activation energy of both (forward and backward) reactions by 100 kJ mol^{-1}. The enthalpy change of the reaction ($A_2 + B_2 \longrightarrow 2AB$) in the presence of catalyst will be (in kJ mol^{-1})

(a) 300 (b) −120
(c) 280 (d) − 20

13 A hydrogenation reaction is carried out at 500 K. If the same reaction is carried out in the presence of a catalyst at the same rate, the temperature required is 400 K. What will be the activation energy of the reaction if the catalyst lowers the activation energy barrier by 20 kJ mol^{-1}?

(a) 120 kJ mol^{-1} (b) 80 kJ mol^{-1}
(c) 100 kJ mol^{-1} (d) 20 kJ mol^{-1}

14 Two reactions R_1 and R_2 have identical pre-exponential factors. Activation energy of R_1 exceeds that of R_2 by 20 kJ mol^{-1}. If k_1 and k_2 are rate constant for reactions R_1 and R_2 respectively at 300 K, then $\ln\left(\frac{k_2}{k_1}\right)$ is equal to [R = 8.314 J mol^{-1} K^{-1}]. → JEE Main 2017

(a) 8 (b) 12
(c) 6 (d) 4

15 The chemical reaction, $2O_3 \longrightarrow 3O_2$, proceeds as follows:

$$O_3 \rightleftharpoons O_2 + O \quad \text{(fast)}$$

$$O + O_3 \longrightarrow 2O_2 \quad \text{(slow)}$$

The rate law expression should be

(a) $r = k'[O_3]^2$ (b) $r = k'[O_3]^2[O_2]^{-1}$
(c) $r = k'[O_3][O_2]$ (d) Unpredictable

16 The hydrolysis of sucrose into glucose and fructose,

$$C_{12}H_{22}O_{11} + H_2O \longrightarrow C_6H_{12}O_6 + C_6H_{12}O_6$$

follows first order kinetics. In a neutral solution, if at 27°C, rate constant is 2.1×10^{-11} s^{-1} and at 37°C, rate constant is 8.5×10^{-11} s^{-1}. The rate constant at 47°C will be

(a) 4.04 (b) 3.163×10^{-10}
(c) 8.5×10^{-11} (d) 1.785×10^{-21}

17 Decomposition of H_2O_2 follows a first order reaction. In 50 min, the concentration of H_2O_2 decreases for 0.5 to 0.125 M in one such decomposition. When the concentration of H_2O_2 reaches 0.05 M, the rate of formation of O_2 will be → JEE Main 2016

(a) 6.93×10^{-4} mol min^{-1} (b) 2.66 L min^{-1} at STP
(c) 1.34×10^{-2} mol min^{-1} (d) 6.93×10^{-2} mol min^{-1}

ANSWERS

SESSION 1

1 (b)	2 (b)	3 (b)	4 (c)	5 (c)	6 (d)	7 (d)	8 (a)	9 (b)	10 (a)
11 (d)	12 (c)	13 (b)	14 (c)	15 (b)	16 (a)	17 (c)	18 (d)	19 (a)	20 (d)
21 (d)	22 (a)	23 (a)	24 (c)	25 (c)	26 (c)	27 (c)	28 (a)	29 (b)	30 (a)
31 (c)	32 (d)	33 (b)	34 (a)	35 (a)	36 (b)	37 (c)			

SESSION 2

1 (d)	2 (b)	3 (b)	4 (a)	5 (d)	6 (b)	7 (a)	8 (b)	9 (a)	10 (c)
11 (c)	12 (d)	13 (c)	14 (d)	15 (b)	16 (b)	17 (a)			

Hints and Explanations

SESSION 1

1 $[H^+] = \frac{6\times10^{-7}\text{ mol}}{0.05\times10^{-3}\text{L}} = 1.2\times10^{-2}$ M

or $r = \frac{\Delta x}{\Delta t}$ or $\Delta t = \frac{\Delta x}{r} = \frac{1.2\times10^{-2}\text{ M}}{6\times10^{5}\text{ Ms}^{-1}}$

$\therefore \quad t = 2\times10^{-8}$ s

2 Rate of reaction $= \frac{-1}{2}[MnO_4] = \frac{1}{5}[I_2]$

$\therefore$ Given, rate of disappearance of

$MnO_4^- = d\frac{[MnO_4^-]}{dt} = 4.56\times10^{-3}$

$\because$ Rate of appearance of I_2

$= \frac{5}{2}\times4.56\times10^{-3}$

$= 11.4\times10^{-3} = 1.14\times10^{-2}$ Ms^{-1}

3 As rate $= k[RCl]$, on decreasing the concentration of RCl to half, the rate will also be halved.

4 For second order, $\frac{dx}{dt} = k(\text{conc.})^2$

Hence, $k = \frac{\text{conc.}}{\text{time}}\times\frac{1}{(\text{conc.})^2}$

$= (\text{conc.})^{-1}\cdot\text{time}^{-1}$

$= \text{dm}^3\text{ mol}^{-1}\text{s}^{-1}$

5 $\left(\frac{dx}{dt}\right) = k[NO]^2[O_2] = k\left(\frac{n_{NO}}{V}\right)^2\left(\frac{n_{O_2}}{V}\right)$

$\left(\frac{dx}{dt}\right) = \frac{k}{V^3}(n_{NO})^2(n_{O_2})$

$\left(\frac{dx}{dt}\right)' = \frac{k(n_{NO})^2(n_{O_2})}{\left(\frac{V}{2}\right)^3} = 8\left(\frac{dx}{dt}\right)$

6 $r = \frac{dC}{dt} = k[A]^x[B]^y$

$1.2\times10^{-3} = k(0.1)^x(0.1)^y$ ...(i)

$1.2\times10^{-3} = k(0.1)^x(0.2)^y$ ...(ii)

$2.4\times10^{-9} = k(0.2)^x(0.1)^y$ ...(iii)

As shown above, rate of reaction remains constant as the concentration of reactant B changes from 0.1M to 0.2 M and becomes double when concentration of A changes from 0.1 to 0.2 (i.e. doubled).

$\therefore \quad r = k[A]^1[B]^0$

7 $aG + bH \longrightarrow$ products

Let the rate of reaction be

$r = k[G]^a[H]^b$ ...(i)

When concentration of both reactants G and H is doubled, rate increases by eight times,

$\therefore \quad 8r_1 = k[2G]^a[2H]^b$ ...(ii)

When concentration of G is doubled, keeping the concentration of H fixed, rate is doubled,

$\therefore \quad 2r_2 = k[2G]^a[H]^b$ ...(iii)

On solving Eq (ii) and (iii), we have

$b = 2$

On solving Eq (i) and (iii), we have

$a = 1$

$\therefore$ Overall order of the reaction

$= a + b = 1 + 2 = 3$

8 For the reaction,

$CH_3CHO(g) \xrightarrow{\text{Decomposes}} CH_4 + CO$

Let order of reaction with respect to CH_3CHO is m. Its given, $r_1 = 1$ torr/sec when CH_3CHO is 5% reacted, i.e. 95% unreacted. Similarly, $r_2 = 0.5$ torr/sec when CH_3CHO is 33% reacted, i.e. 67% unreacted.

Use the formula,

$r \propto (a-x)^m$

where, $(a-x)$ = amount unreacted

so, $\frac{r_1}{r_2} = \frac{(a-x_1)^m}{(a-x_2)^m}$ or $\frac{r_1}{r_2} = \left[\frac{a-x_1}{a-x_2}\right]^m$

Now, putting the given values

$\frac{1}{0.5} = \left(\frac{0.95}{0.67}\right)$

$\Rightarrow \quad 2 = (1.41)^m$ or $m = 2$

Therefore, the order of reaction is 2.

9 For a zero order reaction,

$-\frac{d[A]}{dt} = k$

$\int_{C_0}^{C_t} d[A] = k\int_{t=0}^{t=t} dt$

$C_0 - C_t = kt \Rightarrow k = \frac{C_0 - C_t}{t}$

10 $2NH_3 \longrightarrow N_2 + 3H_2$

Rate $= -\frac{1}{2}\frac{d[NH_3]}{dt} = \frac{d[N_2]}{dt} = \frac{1}{3}\frac{d[H_2]}{dt}$

Rate $= k[NH_3]^0$

or $\frac{d[NH_3]}{dt} = 2.5\times10^{-4}$ mol L^{-1} s^{-1}

$\frac{d[N_2]}{dt} = \frac{1}{2}\times2.5\times10^{-4}$

$\frac{d[N_2]}{dt} = 1.25\times10^{-4}$ mol L^{-1} s^{-1}

$\frac{d[H_2]}{dt} = \frac{3}{2}\times2.5\times10^{-4}$

$= 3.75\times10^{-4}$mol L^{-1} s^{-1}

11 For zero order reaction,

rate $= \frac{dx}{dt} = k$

$= \frac{x_1 - x_2}{dt} = 2.0\times10^{-2}$

$\frac{x_1 - 0.5}{25} = 2.0\times10^{-2}$

(Here, x_1 = initial concentration)

$x_1 - 0.5 = 0.5 = 0.5 + 0.5 = 1.0$ M

12 For a zero order reaction, $k_0 = \frac{[A]_0}{2t_{1/2}}$

Since, $[A]_0 = 2$ M, $t_{1/2} = 1$ h

So, $\quad k_0 = 1$ and $k_0 = \frac{\Delta x}{t}$

or $\quad t = \frac{0.50 - 0.25}{1} = 0.25$ h

13 $k = \frac{1}{t}\ln\frac{100}{100-99.9} = \frac{1}{t}\ln\frac{100}{0.1}$

or $\quad \frac{\ln 2}{t_{1/2}} = \frac{1}{t}\ln 10^3$

or $\quad \frac{\log 2}{t_{1/2}} = \frac{1}{t}\times\log 10^3 = \frac{3}{t}$

$t_{1/2} = \frac{\log 2}{3}\times t = \frac{0.30103}{3}\times t \approx 0.10t$

14 $k = \frac{0.693}{45}\text{min}^{-1} = \frac{2.303}{t_{99.9\%}}\log\frac{a}{a-0.999a}$

$[\because a = 1]$

or $\quad t_{99.9\%} = \frac{2.303\times45}{0.693}\log10^3$

$= 448\text{ min} \approx 7\frac{1}{2}$h

15 By first order kinetic, rate constant

$k = \frac{2.303}{t}\log\left(\frac{a}{a-x}\right)$

$a = 0.1$ M

$(a-x) = 0.025$ M

$t = 40$ min

$\therefore \quad k = \frac{2.303}{40}\log\frac{0.1\text{M}}{0.025\text{ M}}$

$= 0.0347$ min^{-1}

Rate $= \left(\frac{dx}{dt}\right) = k[A]^1 = 0.0347\times0.01$

$= 3.47\times10^{-4}$ M min^{-1}

16 First order kinetics,

$$k_1 = \frac{0.693}{t_{1/2}} = \frac{0.693}{40} s^{-1}$$

Zero order kinetics, $k_0 = \frac{C_0}{2t_{1/2}} = \frac{1.386}{2 \times 20}$

Hence, $\frac{k_1}{k_0} = \frac{0.693}{1.386} = 0.5$

17 Half-life, $t_{1/2} = 6.93$ min, $k_1 = \frac{0.693}{6.93} = 0.1$

$$k_1 = \frac{2.303}{t} \log\left(\frac{100}{100-99}\right)$$

$$0.1 = \frac{2.303}{t} \times \log\frac{100}{1} = \frac{2.303}{t} \log 10^2$$

$$t = \frac{2.303 \times 2}{0.1} = 46.06 \text{ min}$$

18 For a first order process, $kt = \ln\frac{[A]_0}{[A]}$

where, $[A]_0$ = initial concentration.
$[A]$ = concentration of reactant remaining at time t.

$$kt_{1/8} = \ln\frac{[A]_0}{[A]_0/8} = \ln 8 \qquad \ldots(i)$$

$$kt_{1/10} = \ln\frac{[A]_0}{[A]_0/10} = \ln 10 \qquad \ldots(ii)$$

Therefore, $\frac{t_{1/8}}{t_{1/10}} = \frac{\ln 8}{\ln 10} = \log 8 = 3\log 2$

$$\frac{t_{1/8}}{t_{1/10}} = 3 \times 0.3 = 0.9$$

$$\frac{t_{1/8}}{t_{1/10}} \times 10 = 0.9 \times 10 = 9.0$$

19 Since, as we know that, $t_{1/2} \propto \frac{1}{a^{n-1}}$

Hence, $t_{1/2} \propto a^{1/3}$, only when $n = 4$

20 For any reaction, halt-life $(t_{1/2}) \propto \frac{1}{a^{n-1}}$

For 2nd order $n = 2$

$\therefore \quad t_{1/2} \propto \frac{1}{[A]_0^{2-1}} \Rightarrow t_{1/2} = \frac{1}{k[A]_0}$

21 As 2nd step is rate determining step.

Rate $= k[NOBr_2][NO] \qquad \ldots(i)$

But, $NOBr_2$ is in equilibrium

$$K_{eq} = \frac{[NOBr_2]}{[NO][Br_2]}$$

$[NOBr_2] = K_{eq}[NO][Br_2] \qquad \ldots(ii)$

Putting the $[NOBr_2]$ in Eq (i)

Rate $= k \cdot K_{eq}[NO][Br_2][NO]$

Hence, Rate $= k \cdot K_{eq}[NO]^2[Br_2]$

Rate $= k'[NO]^2[Br_2] \qquad [k' = k \cdot K_{eq}]$

Hence, order of reaction with respect to NO is two.

22 H^+ ions act as catalyst for the given reaction.

$$HCOOCH_3 + H_2O \longrightarrow HCOOH + CH_3OH$$

23 The main conditions for the occurrence of a reaction is proper orientation and effective collision of the reactants.

Since, the chances of simultaneous collision with proper orientation between more than 3 species is very rare, so reaction with order greater than 3 are rare.

24 For every 10°C rise of temperature, rate is doubled. Thus, temperature coefficient of the reaction = 2.

When temperature is increased by 50°, rate becomes $= 2^{\left(\frac{50}{10}\right)} = 2^5$ times = 32 times

25 Increase in steps of 10°C has been made 9 times, hence rate of reaction should increase 2^9 times, i.e. 512 times.

26 A catalyst lowers the activation energy of reaction.

27 The minimum value for the energy of activation will be more than ΔH.

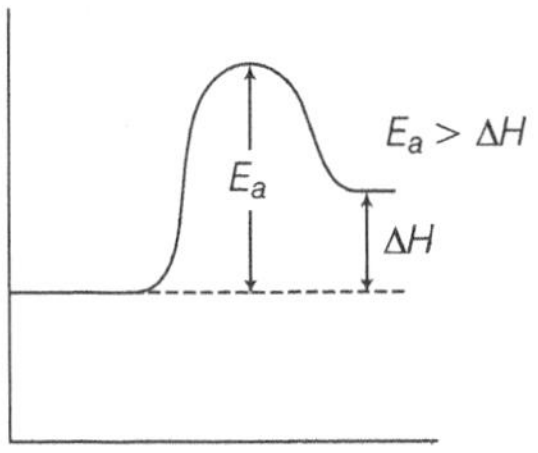

28 Graph (I) only represents exothermic reaction because products have lesser energy than reactants ($\Delta H < 0$).

29 An exothermic reaction can be represented as :

$$X \longrightarrow Y + \text{energy}$$

i.e. energy of $Y << X$

So, the energy profile diagram is

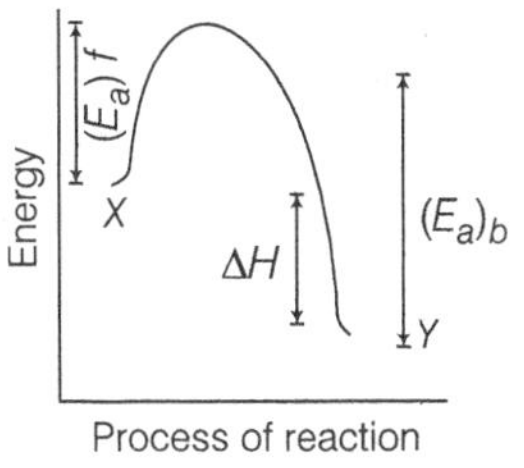

$\therefore \qquad (E_a)_f + \Delta H = (E_a)_b$

$150 + 135 = (E_a)_b$

$(E_a)_b = 285$ kJ mol^{-1}

30 $H_2O + O \xrightarrow{E_a} 2OH; \Delta H = 72 \text{ kJmol}^{-1}$

$2OH \xrightarrow{E_b} H_2O + O; \Delta H = -72 \text{ kJmol}^{-1}$

Also $\qquad E_a - E_b = \Delta H$

or $\qquad 77 - E_b = 72$

$\therefore \qquad E_b = 5$ kJ mol^{-1}

31 According to Arrhenius equation,

$$\log k = \log A - \frac{E_a}{2.303RT}$$

When $\log k$ is plotted against $\frac{1}{T}$, we get a straight line, where intercept of this line is equal to $\log A$ and slope is $\frac{-E_a}{2.303R}$.

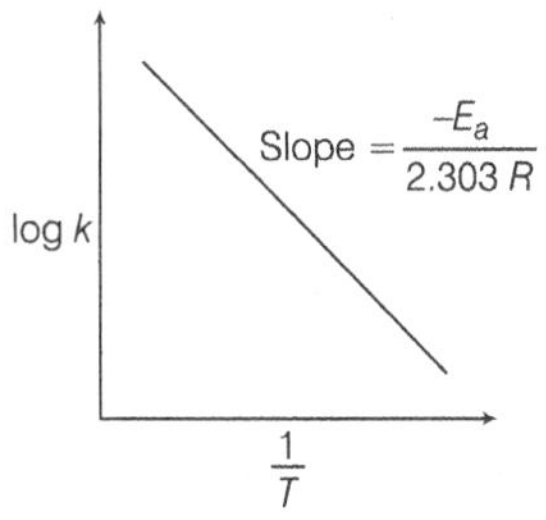

32 As per Arrhenius equation,

$$\ln\frac{k_2}{k_1} = \frac{-E_a}{R}\left[\frac{1}{T_2} - \frac{1}{T_1}\right]$$

$$2.303 \log 2 = \frac{-E_a}{8.314}\left[\frac{1}{310} - \frac{1}{300}\right]$$

$\Rightarrow \qquad E_a = 53.6$ kJ/mol

33 $A \xrightarrow{k_1} B, A \xrightarrow{k_2} C$

By Arrhenius equation,

$$k_1 = A'e^{-E_{a_1}/RT}$$

and $\quad k_2 = A'e^{-E_{a_2}/RT}$

[A' is Arrhenius constant]

$\because \qquad E_{a_2} = 2E_{a_1}$

$\therefore \qquad k_2 = A'e^{-2E_{a_1}/RT}$

$$\frac{k_1}{k_2} = \frac{A'e^{-E_{a_1}/RT}}{A'e^{-2E_{a_1}/RT}} = e^{E_{a_1}/RT}$$

$\therefore \qquad k_1 = k_2 e^{E_{a_1}/RT}$

34 Order may or may not be equal to molecularity. It is determined experimentally and molecularity is the sum of the stoichiometric coefficients of rate determining elementary step.

35 All collisions of reactant molecules do not lead to product formation. Only effective collisions having proper orientation and sufficient kinetic energy lead to compound formation.

36 The order of reaction can only be determined by experiments. Hence, option (b) is correct.

37 A successful collision leads to a chemical reaction, when the reactant molecules collide with proper orientation and attain threshold energy level. Hence, option (c) is correct.

SESSION 2

1 Slowest step is the rate determining step. Thus, in case (*A*), rate law is given as rate $= k[Cl_2][H_2S]$

While for the reaction given in case (*B*), rate law is given as rate $= k[H_2S][Cl_2][H^+]^{-1}$.

Hence, only mechanism (*A*) is consistent with the given rate law.

2 $CH_3OCH_3(g) \rightarrow CH_4(g) + H_2(g) + CO(g)$

At, $t = 0$	p_0	0	0	0
At. t	$p_0 - p$	p	p	p

Total pressure at $t = p_0 + 2p$

Rate expression for first order kinetics :

$$\log \frac{p_0 - p}{p_0} = -\left(\frac{k}{2.303}\right)t = -\left(\frac{0.693/t_{1/2}}{2.303}\right)t$$

$$= -\frac{0.693 / 14.5 \text{ min}}{2.303} \times 12 \text{ min}$$

or $\log \frac{p_0 - p}{p_0} = -0.249$

or $\frac{p_0 - p}{p_0} = 0.564$

or $p = p_0 - 0.564 p_0 = p_0(1 - 0.564) = (0.40 \text{ atm})(0.436) = 0.1744 \text{ atm}$

3 Given, $A = 1.11 \times 10^{11} s^{-1}$; $E_a = 39.3 \times 10^3$ cal mol^{-1}

$R = 1.987$ cal; $T = 573$ K

$\because \log_{10} k = \log_{10} A - \frac{E_a}{2.303RT}$

or $\log_{10} k = \log_{10} 1.11 \times 10^{11} - \left\{\frac{39.3 \times 10^3}{2.303 \times 1.987 \times 573}\right\}$

or $k = 1.14 \times 10^{-4}$ s^{-1}

$$t_{1/2} = \frac{0.693}{k} = \frac{0.693}{1.14 \times 10^{-4}} = 6078 \text{ s} = 1.69 \text{ h}$$

4 Since, nuclear explosion follows the first order kinetics

$\because$ $t_{1/2}$ of $^{90}Sr = 28.1$ yr

$\therefore$ $t_{1/2} = \frac{0.693}{k}$

$\therefore$ $k = \frac{0.693}{28.1} yr^{-1}$

Now, $t = \frac{2.303}{k} \log \frac{a}{(a-x)}$

[Here, $a = 1\,\mu g$]

At $t = 10$ yr;

$$10 = \frac{2.303 \times 28.1}{0.693} \log \frac{1}{(a-x)}$$

$\therefore$ Amount of Sr left $= (a - x) = 0.7814\,\mu g$

At $t = 60$ yr;

$$60 = \frac{2.303 \times 28.1}{0.693} \log \frac{1}{(a-x)}$$

$\therefore (a - x) = 0.227\,\mu g$

5 Rate constant increases exponentially with decreasing activation energy and increasing temperature.

6 $k = \frac{0.693}{24} h^{-1} = \frac{2.303}{96} \log \frac{10}{a-x}$

or $\log \frac{10}{a-x} = 1.2036$

or $1 - \log(a - x) = 1.2036$

or $\log(a - x) = -0.2036 = +1.7964 - 2$

or $(a - x) = \text{antilog}(1.7964 - 2) = 0.6258 = 0.63$ g

7 $\log k \,(\text{min}^{-1}) = 5 - \frac{2000\text{ K}}{T}$

$$\log_{10} k = \log_{10} A - \frac{E_a}{2.303RT}$$

I. $\log A = 5$, $A = 10^5$ is true

II. $\frac{E_a}{2.303R} = 2000$

$E_a = 2000 \times 2.303 \times 0.002$ kcal $= 9.212$ kcal, true

III. Equation represents straight line, hence, true.

8 $k_B = Ae^{-E_a/RT} = 10^{12} e^{-42000/8.314 \times 298} = 4.34 \times 10^4$ s^{-1}

Also, $K_C = 10^4$ and

$$K_C = \frac{\text{rate constant of forward reaction}}{\text{rate constant of backward reaction}}$$

$$10^4 = \frac{k_A}{k_B}$$

$\therefore$ $k_A = 10^4 k_B = 4.34 \times 10^8$ s^{-1}

Expected observed reaction rate constant $= k_A + k_B = 4.34 \times 10^8$ s^{-1}

9 $k_1 = k_2$

$$10^{15} e^{\left(\frac{-2000}{T}\right)} = 10^{14} e^{\left(\frac{-1000}{T}\right)}$$

or $15 - \frac{2000}{T \times 2.303} = 14 - \frac{1000}{T \times 2.303}$

or $\frac{1000}{T \times 2.303} = 1$

$\therefore$ $T = 434.22$ K

10 According to Arrhenius equation,

$\ln k = \ln A - E_a/RT$

or $\log k = \log A - \frac{E_a}{2.303RT}$

Differentiating both sides, we get

$$\frac{d\log k}{dt} = \frac{E_a}{2.303RT^2}$$

Both statements (a) and (b) are correct.

11 $C_2H_4O(g) \longrightarrow CH_4(g) + CO(g)$

At t_0	p_0	0	0
At t_{20}	$p_0 - p$	p	p

Initial pressure, $p_0 = 80$ mm

Total pressure after 20 min,

$p_{total} = p_0 - p + p + p = p_0 + p$

$120 = 80 + p$

or $p = 120 - 80 = 40$ mm

$$k = \frac{2.303}{20} \log \frac{80}{40} = \frac{2.303}{20} \log 2$$

[$\because \log 2 = 0.3010$]

$$k = \frac{2.303 \times 0.3010}{20} = \frac{0.693}{20}$$

We know that, $t_{1/2} = \frac{0.693}{k}$

By putting the value of k in this equation we get,

$$t_{1/2} = \frac{0.693 \times 20}{0.693} = 20 \text{ min}$$

12 $A_2 + B_2 \rightleftharpoons 2AB$

E_a (forward) $= 180$ kJ mol^{-1}

E_a (backward) $= 200$ kJ mol^{-1}

In the presence of catalyst,

E_a (forward) $= 180 - 100 = 80$ kJ mol^{-1}

E_a (backward) $= 200 - 100 = 100$ kJ mol^{-1}

$\Delta H = E_a$ (forward) $- E_a$ (backward) $= 80 - 100 = -20$ kJ mol^{-1}

13 Let E_a = activation energy in the presence of catalyst

E_a' = activation energy in the absence of catalyst

Then, $k = Ae^{-E_a/RT}$

In the presence of catalyst :

$$k_1 = Ae^{-E_a/(R\times 500)}$$

In the absence of catalyst :

$$k_2 = Ae^{-E'_a/(R\times 400)}$$

Given, the rates are same, i.e. $r_1 = r_2$

$\therefore\ e^{-E_a/(R\times 500)} = e^{-E_a'/(R\times 400)}$

or $\dfrac{E_a}{R\times 500} = \dfrac{E_a'}{R\times 400}$

or $\dfrac{E_a}{500} = \dfrac{E_a - 20}{400}$ $[\because E_a - E_a' = 20]$

or $E_a = 100\ \text{kJ mol}^{-1}$

14 According to Arrhenius equation,

$$k = Ae^{-E_a/RT}$$

where, A = frequency factor or pre-exponential factor.

R = gas constant

T = absolute temperature

E_a = energy of activation

For reaction R_1, $k_1 = Ae^{-E_{a_1}/RT}$ ...(i)

For reaction R_2, $k_2 = Ae^{-E_{a_2}/RT}$ ...(ii)

On dividing Eq. (ii) by Eq. (i), we get

$$\frac{k_2}{k_1} = e^{-\frac{(E_{a_2} - E_{a_1})}{RT}} \quad \text{...(iii)}$$

[$\because$ Pre-exponential factor 'A' is same for both reactions]

Taking ln on both sides of Eq. (iii), we get

$$\ln\left(\frac{k_2}{k_1}\right) = \frac{E_{a_1} - E_{a_2}}{RT}$$

Given, $E_{a_1} = E_{a_2} + 10\ \text{kJ mol}^{-1}$

$= E_{a_2} + 10{,}000\ \text{J mol}^{-1}$

$\therefore\ \ln\dfrac{k_2}{k_1} = \dfrac{10{,}000\ \text{J mol}^{-1}}{8.314\ \text{J mol}^{-1}\text{K}^{-1} \times 300\ \text{K}}$

$= 4$

15 From the step (ii), Rate, $r = k[O][O_3]$ ...(i)

For fast reaction, $K_C = \dfrac{[O_2][O]}{[O_3]}$

or $[O] = \dfrac{K_C[O_3]}{[O_2]}$

On putting the value of [O] in Eq. (i), we get

Rate, $r = k\cdot\dfrac{K_C[O_3]}{[O_2]}[O_3]$

$= k\cdot K_C[O_3]^2[O_2]^{-1} = k'[O_3]^2[O_2]^{-1}$

where, $k' = k\cdot K_C$

16 $\log\dfrac{k_2}{k_1} = \dfrac{E_a}{2.303R}\left(\dfrac{T_2 - T_1}{T_1T_2}\right)$

$\log\dfrac{8.5\times 10^{-11}}{2.1\times 10^{-11}}$

$= \dfrac{E_a}{2.303\times 8.314}\left[\dfrac{10}{300\times 310}\right]$

$\therefore\ E_a = 108.1\ \text{kJ}$

$\log k = \log A - \dfrac{E_a}{2.303RT_1}$

$\log 2.1\times 10^{-11}$

$= \log A - \dfrac{108.1\times 10^3}{2.303\times 8.314\times 300}$

$\therefore A = 13.87\times 10^7$

$\log (k)_{47°C} = \log A - \dfrac{E_a}{2.303\times R\times T}$

$= \log 13.87\times 10^7 - \dfrac{108.1\times 10^3}{2.303\times 8.314\times 320}$

$= 3.163\times 10^{-10}\ \text{s}^{-1}$

17 For first order reaction,

$$k = \frac{2.303}{t}\log\frac{a}{a-x}$$

Given, $t = 50$ min, $a = 0.5$ M, $a - x = 0.125$ M

$\therefore\ k = \dfrac{2.303}{50}\log\dfrac{0.5}{0.125} = 0.0277\ \text{min}^{-1}$

Now, as per reaction

$$2H_2O_2 \longrightarrow 2H_2O + O_2$$

$$-\frac{1}{2}\frac{d[H_2O_2]}{dt} = \frac{1}{2}\frac{d[H_2O]}{dt} = \frac{d[O_2]}{dt}$$

Rate of reaction, $-\dfrac{d[H_2O_2]}{dt} = k[H_2O_2]$

$\therefore\ \dfrac{d[O_2]}{dt} = -\dfrac{1}{2}\dfrac{d[H_2O_2]}{dt} = \dfrac{1}{2}k[H_2O_2]$ ...(i)

When the concentration of H_2O_2 reaches 0.05 M,

$\dfrac{d[O_2]}{dt} = \dfrac{1}{2}\times 0.0277\times 0.05$ [from Eq. (i)]

or $\dfrac{d[O_2]}{dt} = 6.93\times 10^{-4}\ \text{mol min}^{-1}$

Alternative Method

In fifty minutes, the concentration of H_2O_2 decreases from 0.5 to 0.125 M or in one half-life, concentration of H_2O_2 decreases from 0.5 to 0.25 M. In two half-lives, concentration of H_2O_2 decreases from 0.5 to 0.125 M or $2t_{1/2} = 50$ min, $t_{1/2} = 25$ min

$\therefore\ k = \left(\dfrac{0.693}{25}\right)\text{min}^{-1}$

or $\dfrac{d[O_2]}{dt} = -\dfrac{1}{2}\dfrac{d[H_2O_2]}{dt}$

$= \dfrac{k[H_2O_2]}{2}$

$= 6.93\times 10^{-4}\ \text{mol min}^{-1}$

DAY FIFTEEN

Adsorption and Catalysis

Learning & Revision for the Day

- Concept of Adsorption
- Freundlich Adsorption Isotherm
- Langmuir Adsorption Isotherm
- Catalysis

The surface of a solid has the tendency to attract and retain the molecules of the phase with which it comes into contact. These molecules remain only at the surface and do not go deeper into the bulk. The accumulation of molecular species at the surface rather than in the bulk of a solid or liquid is called **adsorption.**

Concept of Adsorption

- The solid substances on the surface of which adsorption occurs is called adsorbent. The molecular species that get adsorbed on the solid surface due to intermolecular attractions are called **adsorbate**.
- Rate of adsorption decreases with time whereas rate of absorption remains constant with time.
- If the concentration of an adsorbate at the surface of adsorbent is more than in the bulk of the adjoining phases, it is called **positive adsorption**.
- If concentration of an adsorbate at the surface of adsorbent is less than in the bulk of the adjoining phases, it is called **negative adsorption**.
- The removal of the adsorbed substance from a surface is called **desorption**.
- **Sorption** is a process in which adsorption and absorption take place simultaneously and it is difficult to determine the relative extent of adsorption and absorption, e.g. dyes get adsorbed as well as absorbed on the cotton fibres.

Physisorption and Chemisorption

In **physisorption**, the particles of the adsorbate are held to the surface of the adsorbent by physical forces such as van der Waals' forces while in **chemisorption**, the molecules of the adsorbate are held to the surface of the adsorbate by chemical bonds.

PREP MIRROR
Your Personal Preparation Indicator

- No. of Questions in Exercises (x)—
- No. of Questions Attempted (y)—
- No. of Correct Questions (z)—
  (Without referring Explanations)

- Accuracy Level ($z/y \times 100$)—
- Prep Level ($z/x \times 100$)—

In order to expect good rank in JEE, your Accuracy Level should be above 85 & Prep Level should be above 75.

Comparison between Physisorption and Chemisorption

Physisorption	Chemisorption
Weak van der Waals' forces are present.	Strong chemical bonds are present.
It decreases with increase in temperature.	It first increases with increase in temperature and then decreases.
Heat of adsorption is low and it is in the range of 20-40 kJ/mol.	Heat of adsorption is high and it is in the range of 40-400 kJ/mol.
It is reversible process.	It is an irreversible process.
It is an instantaneous process.	It may be rapid or slow.
ΔS is always negative.	ΔS is positive for endothermic process.
In this, multilayer adsorption occurs and thus, adsorbed layer is several molecules thick.	In this, single layer adsorption occurs. Thus, adsorbed layer is only unimolecular in thickness.

NOTE As the particles of the adsorbate are held on surface, ΔH and ΔS both are negative and $\Delta H > T\Delta S$. This is true in beginning. As adsorption proceeds, ΔH decreases and $T\Delta S$ increases and ultimately, $\Delta H = T\Delta S$ and $\Delta G = 0$. This state is called **adsorption equilibrium**.

Factors Affecting Adsorption of Gases on Solids

Adsorption depends upon the following factors:

1. **Nature of Adsorbent** A gas is adsorbed in different amounts on different adsorbents. Hydrogen is strongly adsorbed on nickel surface while it is weakly adsorbed on alumina surface under identical conditions.
2. **Nature of Adsorbate** Generally, the more liquefiable a gas, the more readily will it be adsorbed. Easily liquefiable gases such as NH_3, HCl, Cl_2, SO_2, etc., are readily adsorbed than permanent gases, such as O_2, H_2, N_2 etc.
3. **Specific Area of Adsorbent** It is the surface area of adsorbent available for adsorption per gram of the adsorbate. Greater the surface area of the solid, larger would be its adsorbing capacity.
4. **Temperature** Low temperature value favours physical adsorption but at high temperature, it decreases. In chemisorption, extent of adsorption first increases and then decreases.
5. **Pressure of Gas** At a given temperature, the extent of adsorption will increase with the increase of pressure of the gas.

Freundlich Adsorption Isotherm

- The variation in the amount of gas adsorbed by the adsorbent with pressure at constant temperature can be expressed by curve termed as adsorption isotherm.
- Freundlich gave an empirical relationship between the quantity of gas adsorbed by unit mass of solid adsorbent and pressure at a particular temperature.
- The extent of adsorption is measured as x/m, where m is the mass of adsorbent and x is the mass of adsorbate. At low pressure, x/m varies linearly with p.

$$\frac{x}{m} = kp^{1/n} \quad \text{or} \quad \log \frac{x}{m} = \log k + \frac{1}{n} \log p$$

At low pressure, $\frac{x}{m} = kp$

At high pressure, $\frac{x}{m} = kp^\circ$

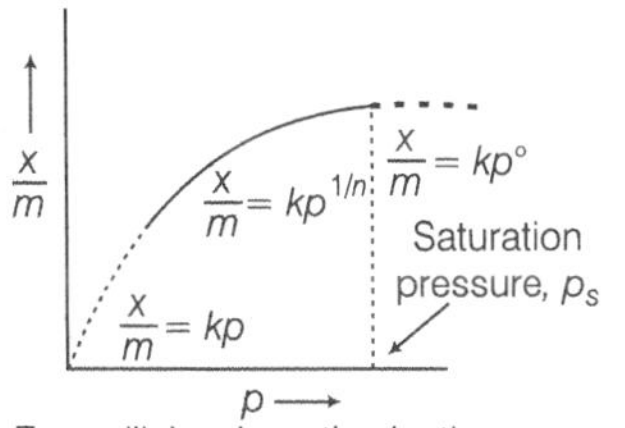

Freundlich adsorption isotherm

This is called **Freundlich adsorption isotherm** at constant temperature.

- Freundlich isotherm fails at high pressure and is only for physical adsorption.

Langmuir Adsorption Isotherm

To overcome the limitation of Freundlich adsorption isotherm, Langmiur developed a new isotherm. It is represented as :

$$\frac{x}{m} = \frac{ap}{1 + bp} \quad (a \text{ and } b \text{ are constants})$$

At very high pressure, $(bp >> 1)\ \frac{x}{m} = \frac{a}{b}$

At very low pressure, $(bp << 1)\ \frac{x}{m} = ap$

- In case of chemisorption, x/m initially increases with temperature and then decreases.
- At low temperature, x/m is small. As temperature is increased, the molecules of the adsorbate gain energy and become equal to activation energy. Therefore, initially, x/m increases with rise in temperature.

  The graph between extent of adsorption (x/m) and temperature T is called **adsorption isobar**.

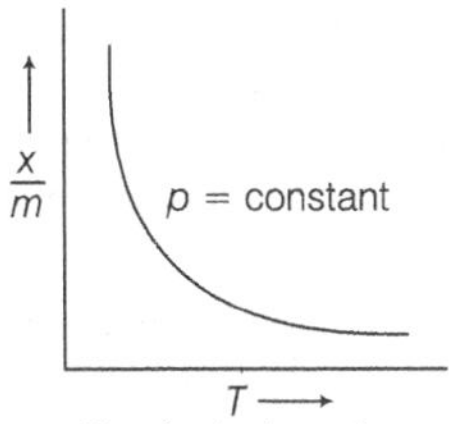

Physical adsorption

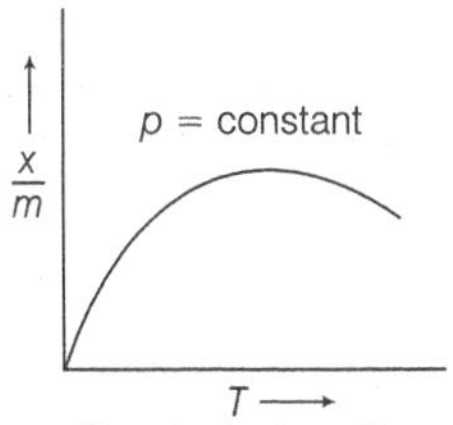

Chemical adsorption

Adsorption from Solutions

- Solids adsorbs dissolved substances from solutions. For example, adsorption of colour impurities by activated carbon in decolourising of solutions.
- Freundlich adsorption isotherm and Langmuir adsorption isotherm are also applicable to adsorption from solutions.

  Freundlich adsorption isotherm, $\frac{x}{m} = kC^{1/n} \quad (n > 1)$

  Langmuir adsorption isotherm, $\frac{x}{m} = \frac{aC}{(1+bC)}$

  Langmuir adsorption isotherm is applicable only to chemisorption.

Characteristics of Adsorption

- Adsorption is **specific and selective** phenomenon which strictly refers to the existence of a higher concentration of any particular component at the surface of a liquid or a solid phase.
- Adsorption is accompanied with decrease in free energy of the system. In adsorption, there is a decrease in entropy of the system.

  As, $\Delta G < 0$ and $\Delta S < 0$,

  Therefore, $\Delta H < 0 \; (\Delta H = \Delta G + T\Delta S)$

  Thus, adsorption is always an exothermic process.
- The atoms or molecules of a solid surface have unbalanced or **residual attractive forces on the surface**, which can hold adsorbate particles together. Thus, the adsorbed atoms or molecules can be held on the surface of a metal by physical van der Waals' forces or chemical forces due to residual valence bonds.
- Adsorption is not very pronounced unless an adsorbent possesses a large surface area for a given mass. Various types of charcoals, silica gels, metal, etc., are used as adsorbents. Appreciable adsorption also takes place on some smooth surfaces, such as those of platinum and glass.
- Charcoal and other solids increased power of adsorption on heating at low pressure with various gases or in air or vacuum at temperature varying from 350-1000°C. Such treated charcoal is called **activated charcoal** and the process is called **activation.**

Catalysis

- The catalyst changes the rate of reaction by providing an alternate path of different activation energy. They themselves remain chemically and quantitatively unchanged after the reaction. The phenomenon is known as catalysis.
- **Promoters** are the substances which can increase the efficiency of a catalyst. These are also known as co-enzymes or activators.
- **Inhibitors** are the substances which can make the catalyst inactive, e.g. Mg^{2+} acts as an activator for many enzymes while $C_2O_4^{2-}$ and F^- act as inhibitors as they form complex with Mg^{2+}.

NOTE
- The catalytic poisoning is specific in nature.
- In former case, catalyst surface is regenerated by scratching the surface while in later, catalyst surface is regenerated by chemical treatment.

Homogeneous and Heterogeneous Catalysis

(i) In **homogeneous catalysis**, reactants and catalyst should be in same phase.

e.g. $$2SO_2(g) + O_2(g) \xrightarrow{NO(g)} 2SO_3(g)$$

(ii) In **heterogeneous catalysis**, reactants and catalyst are in different phase.

$$N_2(g) + 3H_2(g) \xrightarrow[\text{divided Fe}(s)]{\text{Finely}} 2NH_3(g)$$

Activity and Selectivity of Solid Catalysts

1. **Activity** of catalyst depends upon the strength of chemisorption to a large extent.
2. **Selectivity** of a catalyst is its ability to direct a reaction to yield a particular product. e.g.

(i) $CO(g) + 3H_2(g) \xrightarrow{Ni} CH_4(g) + H_2O(g)$

(ii) $CO(g) + H_2(g) \xrightarrow{Cu} HCHO(g)$

(iii) $CO(g) + 2H_2(g) \xrightarrow{Cu/ZnO-Cr_2O_3} CH_3OH(g)$

NOTE The catalytic reaction that depends upon the pore structure of the catalyst and the size of the reactant and product molecules is called **shape-selective catalysis**.

Characteristics of Catalysts

- They become temporarily involved in a reaction providing an alternative reaction path of lower activation energy than that for the uncatalysed reaction.
- They catalyse both forward and backward reactions to the same extent and thus have no effect on the equilibrium constant.
- The catalyst remains unchanged in amount and chemical composition at the end of the reaction. It may undergo some physical change.
- In certain reactions, the rate of the reaction is dependent on the concentration of the catalyst, e.g. rate of inversion of cane sugar is dependent on the concentration of H^+ used as catalyst.

- Rate of the reaction in certain heterogeneous reaction varies with surface area of the catalyst. Hence, finely divided metals are preferred in the form of catalyst.
- The catalyst does not initiate the reaction and are specific in their action, e.g. starting from H_2 and CO, three different products are possible using different catalysts as

$$CO(g) + 3H_2(g) \xrightarrow{Ni} CH_4(g) + H_2O(g)$$

$$CO(g) + 2H_2(g) \xrightarrow{Cu/ZnO\text{-}Cr_2O_3} CH_3OH(g)$$

$$CO(g) + H_2(g) \xrightarrow[250^\circ C]{Cu} HCHO$$

Enzyme Catalysis and its Mechanism

Enzymes are biochemical catalysts. They are proteins and extremely specific in nature. For example,

$$NH_2CONH_2 + H_2O \xrightarrow{Urease} 2NH_3 + CO_2$$

Mechanism for all enzyme catalysed reactions are

$$E + S \rightleftharpoons ES \text{ (fast, reversible)}$$

$$ES \longrightarrow E + P \text{ (slow, rate determining)}$$

1. **Autocatalysis** is the phenomenon in which one of the products formed during the reaction acts as catalyst for the reaction. For example,

$$2KMnO_4 + 5H_2C_2O_4 + 3H_2SO_4 \longrightarrow K_2SO_4 + 2MnSO_4 + 8H_2O + 10CO_2$$

In this reaction, Mn^{2+} ions act as autocatalyst.

2. **Induced catalysis** is the type of catalysis, one reaction influences the rate of other reaction which does not occur under ordinary condition. For example, the reduction of $HgCl_2$ by oxalic acid is slow but becomes faster if reduction is made in mixture of $KMnO_4$ and $HgCl_2$, where both are reduced. Reduction of $KMnO_4$ thus, induces the reduction of $HgCl_2$.

DAY PRACTICE SESSION 1

FOUNDATION QUESTIONS EXERCISE

1 Adsorption is a surface phenomenon because
(a) adsorbent has large surface area
(b) chemical compounds formed are not on the surface of adsorbent
(c) only the surfaces of the adsorbent have unutilised valencies
(d) None of the above

2 Which one of the following characteristics is not correct for physical adsorption?
(a) Adsorption on solids is reversible
(b) Adsorption increases with increase in temperature
(c) Adsorption is spontaneous
(d) Both enthalpy and entropy of adsorption are negative

3 Which of the following statements is incorrect regarding physisorption?
(a) It occurs because of van der Waals' forces
(b) More easily liquefiable gases are adsorbed readily
(c) Under high pressure, it results into multimolecular layer on adsorbent surface
(d) Enthalpy of adsorption ($\Delta H_{adsorption}$) is slow and positive

4 Amount of gas adsorbed per gram of adsorbent increases with pressure, but after certain limit is reached, adsorption becomes constant. It is due to
(a) multilayers are formed
(b) desorption takes place
(c) temperature is increased
(d) absorption also starts

5 Which of the following is the variation of physical adsorption with temperature?

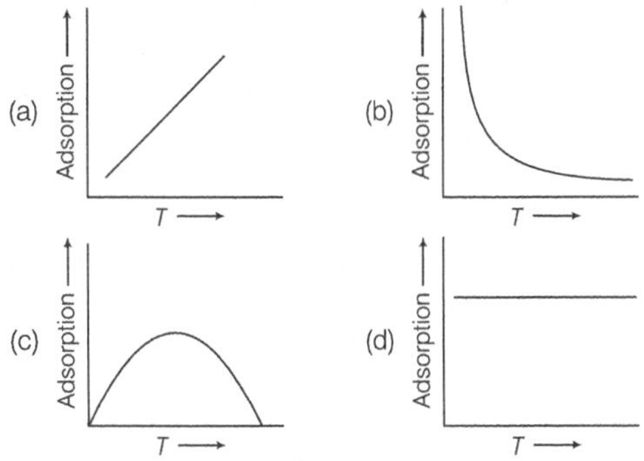

6 In the Freundlich adsorption equation $x/m = kp^{1/n}$, the value of n is
(a) always greater than one
(b) always smaller than one
(c) always equal to one
(d) greater than one at low temperature and smaller than one at high temperature

7 For a linear plot of $\log(x/m)$ *versus* $\log p$ in a Freundlich adsorption isotherm, which of the following statements is correct? (k and n are constants) → JEE Main 2016
(a) $1/n$ appears as the intercept
(b) Only $1/n$ appears as the slope
(c) $\log\left(\frac{1}{n}\right)$ appears as the intercept
(d) Both k and $1/n$ appear in the slope term

8 Which of the following is an adsorption isotherm?

(a) 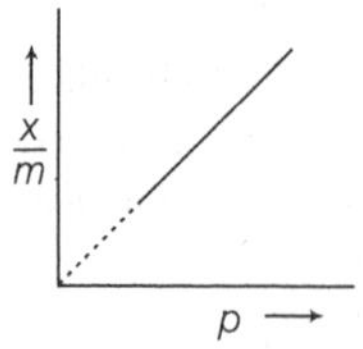

(b) 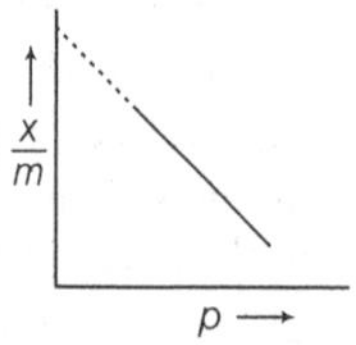

(c) 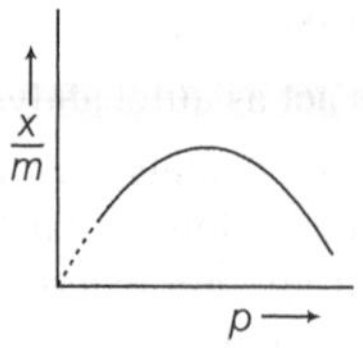

(d) 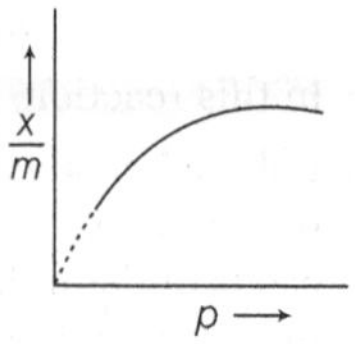

9 The Langmuir adsorption isotherm is deduced using which of the following assumptions?

(a) The adsorption takes place in multilayer
(b) The adsorbed molecules interact with each other
(c) The adsorption sites are equivalent in their ability to absorb the particles
(d) The heat of adsorption varies with coverage

10 Which of the following assumptions is not correct about Langmuir adsorption isotherm?

(a) The solid surface is homogeneous and has a fixed number of adsorption sites
(b) Every adsorption site is equivalent
(c) The adsorbed layer is not uniform all over the adsorbent
(d) The adsorbed gas behaves ideally in the vapour phase

11 In Langmuir's model of adsorption of a gas on a solid surface,

(a) the rate of dissociation of adsorbed molecules from the surface does not depend on the surface covered
(b) the adsorption at a single site on the surface may involve multiple molecules at the same time
(c) the mass of gas striking a given area of surface is proportional to the pressure of the gas
(d) the mass of gas striking a given area of surface is independent of the pressure of the gas

12 Surface area per gram of the adsorbent is called

(a) molar surface area (b) normal surface area
(c) specific surface area (d) equivalent surface area

13 The volume of the gases H_2, CH_4, CO_2 and NH_3 adsorbed by one of activated charcoal at 298 K are in order

(a) $H_2 > CO_2 > CH_4 > NH_3$
(b) $H_2 > CH_4 > CO_2 > NH_3$
(c) $NH_3 > CO_2 > CH_4 > H_2$
(d) $NH_3 > CH_4 > CO_2 > H_2$

14 The decomposition of H_2O_2 may be checked by adding small quantity of phosphoric acid. This is an example of

(a) neutralisation (b) negative catalysis
(c) positive catalysis (d) catalytic poisoning

15 $KClO_3$ on heating decomposes into KCl and O_2. If some MnO_2 is added, the reaction goes much faster because

(a) MnO_2 decomposes to give oxygen
(b) MnO_2 acts as catalyst
(c) MnO_2 provides heat by reacting
(d) MnO_2 provides better contact

16 Arrange the following diagrams in correct sequence of steps involved in the mechanism of catalysis in accordance with modern adsorption theory.

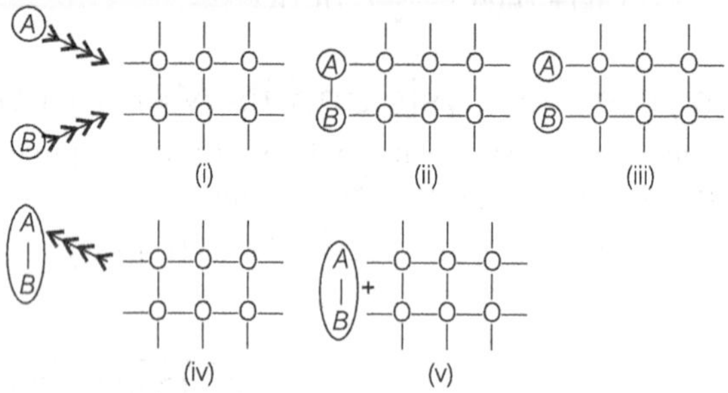

(a) (i) → (ii) → (iii) → (iv) → (v)
(b) (i) → (iii) → (ii) → (iv) → (v)
(c) (i) → (iii) → (ii) → (v) → (iv)
(d) (i) → (ii) → (iii) → (v) → (iv)

17 In which of the following reactions, heterogeneous catalysis is involved?

(i) $2SO_2(g) + O_2(g) \xrightarrow{NO(g)} 2SO_3(g)$

(ii) $2SO_2(g) \xrightarrow{Pt(s)} 2SO_3(g)$

(iii) $N_2(g) + 3H_2(g) \xrightarrow{Fe(s)} 2NH_3(g)$

(iv) $CH_3COOCH_3(l) + H_2O(l) \xrightarrow{HCl(l)} CH_3COOH(aq) + CH_3OH(aq)$

(a) (ii) and (iii) (b) (ii), (iii) and (iv)
(c) (i), (ii) and (iii) (d) Only (iv)

18 Which one of the following is an example of homogeneous catalysis?

(a) Acid hydrolysis of methyl acetate
(b) Catalytic conversion of water gas to methanol
(c) Catalytic conversion of SO_2 to SO_3 in contact process
(d) Haber's process of synthesis of ammonia

19 It is instructed that automobiles with catalytic converter must use unleaded gasolines because

(a) leaded gasolines may give more fumes
(b) surface of the catalyst is rendered ineffective by adsorption of lead
(c) automobiles with catalytic converter cannot run on leaded gasolines
(d) unleaded gasoline is cheaper

20 The void space in zeolites forms more than 50% of the total volume which is occupied by

(a) silicates (b) aluminates
(c) water molecules (d) Na^+, Mg^{2+} and Ca^{2+} ions

21 Which of the following statements is incorrect?

(a) Catalysts only accelerate the rate of chemical equation
(b) Catalysts cannot start a chemical reaction
(c) Catalysts can retard the rate of a chemical reaction
(d) Catalysts can expedite and retard the rate of a chemical reaction

22 Which of the following small sized elements can replace the silicon and aluminium in the frame work at zeolites?

(a) Boron (b) Magnesium
(c) Phosphorus (d) All of these

23 The catalyst which is a zeolite can convert alcohol to various types of gasoline (petrol) by shape selective catalysis is

(a) erionite (b) ZSM-5
(c) gemelinite (d) fanzasite

24 The efficiency of an enzyme in catalysing a reaction is due to its capacity

(a) to form an enzyme substrate complex
(b) to decrease the bond energies of the substrate molecule
(c) to change the shape of the substrate molecule
(d) None of the above

25. Match the catalysts to the correct processes.

	Catalyst		Process
(A)	$TiCl_3$	(i)	Wacker process
(B)	$PdCl_2$	(ii)	*Ziegler- Natta* polymerisation
(C)	$CuCl_2$	(iii)	Contact process
(D)	V_2O_5	(iv)	Deacon's process

(a) (A)- (iii), (B) - (ii), (C) - (iv), (D) - (i) → JEE Main 2015
(b) (A)- (ii), (B) - (i), (C) - (iv), (D) - (iii)
(c) (A)- (ii), (B) - (iii), (C) - (iv), (D) - (i)
(d) (A)- (iii), (B) - (i), (C) - (ii), (D) - (iv)

26 Match the following and choose the correct option.

	Column I		Column II
A.	Activated charcoal	1.	A device to adsorb poisonous gases
B.	$x/m = kp^{1/n}$	2.	One of the adsorbents
C.	For humidity control	3.	Silica gel
D.	Gas masks	4.	Freundlich adsorption isotherm

	A	B	C	D
(a)	2	4	3	1
(b)	4	3	1	2
(c)	2	1	3	4
(d)	2	4	3	2

Direction (Q. Nos. 27-29) *In the following question, an Assertion (A) followed by Reason (R) is given. Choose the correct option out of the following choices.*

(a) Both A and R are true and R is correct explanation of A
(b) Both A and R are true but R is not correct explanation of A
(c) A is true but R is false
(d) Both A and R are false

27 **Assertion** (A) In chemisorption, adsorption keeps on increasing with temperature.

Reason (R) Heat keeps on providing more and more activation energy.

28 **Assertion** (A) Fruit formation process shows increase in the rate with passage of time.

Reason (R) Hydrolysis of ester is homogeneous autocatalytic reaction.

29 **Assertion** (A) Zeolites are water softener as well as catalyst.

Reason (R) The catalytic action of zeolites is based upon their shape selectivity.

DAY PRACTICE SESSION 2

PROGRESSIVE QUESTIONS EXERCISE

1 50 mL of 1 M oxalic acid (molar mass = 126) is shaken with 0.5 g of wood charcoal. The final concentration of the solution after adsorption is 0.5 M. What is the amount of oxalic acid adsorbed per gram of carbon?

(a) 3.15 (b) 1.575
(c) 6.30 (d) 12.60

2 Which of the following statements is false?

(a) Adsorption may be monolayer or multilayer
(b) Increase of pressure increases the amount of adsorption
(c) Increase of temperature may decrease the amount of adsorption
(d) Particle size of adsorbent will not affect the amount of adsorption

3 1g of charcoal adsorbs 100 mL of 0.5 M CH_3COOH to form a monolayer, and thereby the molarity of CH_3COOH reduces to 0.49 M. What is the surface area of charcoal used by each molecule of acetic acid?

[Surface area of charcoal = $3.01 \times 10^2\ m^2/g$]

(a) $6.02 \times 10^{-20}\ m^2$ (b) $5.00 \times 10^{-19}\ m^2$
(c) $3.01 \times 10^{-2}\ m^2$ (d) $2.00 \times 10^{-19}\ m^2$

4 On the basis of data given below, predict which of the following gases shows least adsorption on a definite amount of charcoal?

Gas	CO_2	SO_2	CH_4	H_2
Critical temp. (K)	304	630	190	33

(a) CO_2 (b) SO_2 (c) CH_4 (d) H_2

5 Select the incorrect statement.
   (a) The rate of enzyme catalysed reaction also depends upon enzyme concentration
   (b) The rate of enzyme catalysed reaction depends upon ionic strength
   (c) The rate of enzyme catalysed reaction first increases with temperature and then decreases after attaining optimum temperature
   (d) The increase in activity of protein as enzyme is due to denaturation

6 During the adsorption of krypton on activated charcoal at low temperature,
   (a) $\Delta H > 0$ and $\Delta S > 0$ (b) $\Delta H < 0$ and $\Delta S < 0$
   (c) $\Delta H > 0$ and $\Delta S > 0$ (d) $\Delta H < 0$ and $\Delta S > 0$

7 In the Freundlich adsorption isotherm equation, $\log \frac{x}{m} = \log k + \frac{1}{n} \log p$, the value of n is
   (a) any value from 0 to 1
   (b) a negative integer
   (c) a positive integer
   (d) a positive or negative integer

8 Which one of the following reactions is an example of auto-catalysis?
   (a) $2AsH_3(s) \longrightarrow 2As(s) + 3H_2(g)$
   (b) $N_2(g) + 3H_2(g) \xrightarrow{Fe(s)} 2NH_3(g)$
   (c) $2SO_2(g) + O_2(g) \xrightarrow{No(g)} 2SO_3(g)$
   (d) $C_{12}H_{22}O_{11}(l) + H_2O(l) \xrightarrow{H^+(l)} C_6H_{12}O_6(l) + C_6H_{12}O_6(l)$

9 Given below, catalyst and corresponding process/reaction are matched. The mismatch is
   (a) $[RhCl(PPh_3)_2]$: Hydrogenation
   (b) $TiCl_4 + Al(C_2H_5)_3$: Polymerisation
   (c) V_2O_5 : Haber : Bosch process
   (d) Nickel : Hydrogenation

10 According to Freundlich adsorption isotherm, which of the following is correct? → AIEEE 2012
   (a) $\frac{x}{m} \propto p^0$ (b) $\frac{x}{m} \propto p^1$
   (c) $\frac{x}{m} \propto p^{1/n}$ (d) All of these

ANSWERS

SESSION 1	1 (c)	2 (b)	3 (d)	4 (a)	5 (b)	6 (a)	7 (b)	8 (d)	9 (c)	10 (c)
	11 (c)	12 (c)	13 (c)	14 (b)	15 (b)	16 (b)	17 (a)	18 (a)	19 (b)	20 (c)
	21 (a)	22 (d)	23 (b)	24 (a)	25 (b)	26 (a)	27 (d)	28 (a)	29 (a)	
SESSION 2	1 (c)	2 (d)	3 (b)	4 (d)	5 (d)	6 (b)	7 (c)	8 (a)	9 (c)	10 (d)

Hints and Explanations

SESSION 1

1 Adsorption is a surface phenomenon because unutilised valencies are present only at the surface.

2 For physical adsorption, as temperature increases, adsorption decreases.

$$\text{Adsorbent} + \text{Adsorbate} \rightleftharpoons \text{Adsorbed state} + \Delta E$$

Adsorption is an exothermic process (forward direction), desorption is endothermic process (backward direction).
According to Le-Chatelier's principle, increase in temperature favours endothermic process.

3 Adsorption is an exothermic process, i.e. energy is released against van der Waals' force of attraction (physisorption). Hence, ΔH is always negative.

4 Langmuir showed that at low pressure, the physically adsorbed gas forms only one molecule thick layer. However, above a certain pressure, multimolecular thick layer is formed.

5 Adsorption of gases decreases with increase in temperature.

6 In Freundlich adsorption isotherm, $n > 1$.

7 According to Freundlich adsorption isotherm,

$$\frac{x}{m} = kp^{1/n}$$

On taking logarithm of both sides, we get

$$\log \frac{x}{m} = \log k + \log p^{1/n} \quad \text{or} \quad \log \frac{x}{m} = \log k + \frac{1}{n} \log p$$

$$y = c + mx$$

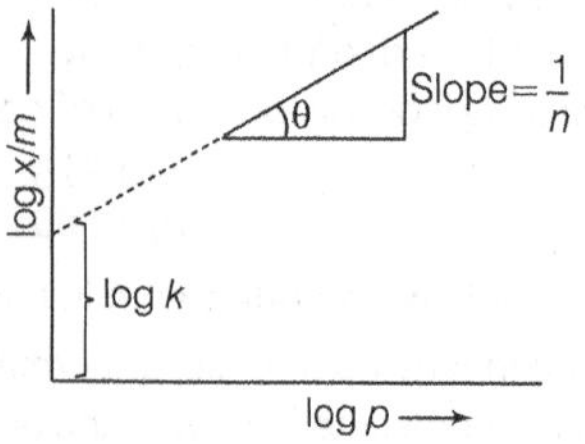

$y = \log \frac{x}{m}$, c = intercept = $\log k$,

m = slope = $\frac{1}{n}$ and $x = \log p$

8 At low pressure, $\frac{x}{m} = kp$ and

at high pressure, $\frac{x}{m} = k$ (i.e.constant)

Hence, option (d) is correct.

9 Langmuir adsorption isotherm is based on various assumptions. These are as follows:

- All sites are equivalent on a surface.
- The layer is uniformly distributed all over i.e. unimolecular.
- At equilibrium, rate of adsorption = rate of desorption.

10 Langmuir adsorption isotherm is based upon the fact that every adsorption site is equivalent and the binding ability of particles does not depend on nearby sites.

11 The adsorption of a gas is directly proportional to the pressure of the gas.

12 Specific surface area is the surface area per gram of the adsorbent.

13 The correct order of gases adsorbed by activated charcoal is

$$NH_2 > CO_2 > CH_4 > H_2$$

Higher the critical temperature, more is the ease of liquefaction of a gas and greater is the amount of gas adsorbed.

14 If the addition of a substance decreases the rate of reaction, the process is called negative catalysis.

15 MnO_2 increases the rate of decomposition of $KClO_3$ and act as catalyst.

16 (i) $\rightarrow$ (iii) $\rightarrow$ (ii) $\rightarrow$ (iv) $\rightarrow$ (v)

17 In heterogeneous catalysis, reactants and catalysts are in different phase. Only (ii) and (iii) reactions have different phase.

18 In acid hydrolysis of methyl acetate, the reactants as well as the catalyst are in same phase, i.e. aqueous. So, it is an example of homogeneous catalysis.

19 Surface of the catalyst is rendered ineffective by adsorption of lead.

20 More than 50% of the total volume of the void space in zeolite is occupied by water molecules.

21 Catalysts can accelerate as well as retard the rate of a chemical reaction. However, they never initiate a chemical reaction.

22 B, Mg and P can replace the Si and Al in the frame work at zeolites.

23 ZSM-5 is a shape selective catalyst that catalyse the conversion of alcohols to various gasoline.

24 Enzyme + substrate $\longrightarrow$ [Complex intermediate] $\longrightarrow$ Product + Enzyme

25 (a) $TiCl_3$ is used as *Ziegler-Natta* catalyst for the polymerisation of ethene.
(b) $PdCl_2$ is used in Wacker process, in which alkene changed into aldehyde *via* catalytic cyclic process initiated by $PdCl_2$.
(c) $CuCl_2$ is used in Deacon's process. (for Cl_2)
(d) V_2O_5 is used in contact process for manufacturing sulphuric acid.

26 Gas mask works on the principle of adsorption.

- Activated charcoal act as adsorbent.
- $x/m = kp^{1/n}$ is the equation for freundlich adsorption isotherm.
- Silica gel is used to control humidity.

Hence, A $\rightarrow$ 2, B $\rightarrow$ 4, C $\rightarrow$ 3, D $\rightarrow$ 1

27 In chemisorption, adsorption first increases and then decreases with change in temperature.

28 Fruit formation process shows increase in the rate with passage of time. It is the homogeneous catalytic reaction.

29 Zeolites are water softener as well as shape selective catalyst.

SESSION 2

1 50 mL of 1 M oxalic acid
$= 50$ millimol $= 0.050$ mol
$= 0.050 \times 126$ g $= 6.3$ g
50 mL of 0.5 M oxalic acid = 3.15 g
Adsorbed oxalic acid
$= 6.30 - 3.15$
$= 3.15$ g on 0.5 g charcoal
Amount of oxalic acid adsorbed per gram of charcoal $= \frac{3.15}{0.5} = 6.3$

2 Physisorption is multilayer while chemisorption is monolayer. Extent of adsorption is affected by pressure, temperature and particle size of adsorbent.

3 100 mL of 0.5 M CH_3COOH contains
$= 0.05$ mol
After adsorption, CH_3COOH remained
$= 0.049$ mol
$\therefore$ Acetic acid adsorbed $= 0.001$ mol
$= 6.02 \times 10^{20}$ molecules

$\therefore$ Surface area of charcoal adsorbed by each molecule

$$= \frac{3.01 \times 10^2 m^2}{6.02 \times 10^{20}} = 5 \times 10^{-19} m^2$$

4 Higher the critical temperature, greater is the adsorption. H_2 has lowest critical temperature and hence, is least adsorbed.

5 Denaturation means biologically inactive protein. Thus activity of enzyme decreases.

6 Adsorption is an exothermic process, thus ΔH is negative (i.e. $\Delta H < 0$). Moreover, adsorption results in more ordered arrangements of molecules, thus entropy decreases (i.e. $\Delta S < 0$).

$$\Delta G = \Delta H - T\Delta S$$

Hence, low temperature favours the reaction.

7 n is a positive integer, and it is a constant that depend upon the nature of adsorbate and adsorbent at a particular temperature. The factor $\frac{1}{n}$ has values between 0 and 1.

8 The reactions which are catalysed by one of the product formed are called autocatalysis reactions and the process is called autocataysis e.g.

$$2AsH_3(s) \longrightarrow 2As(s) + 3H_2(g)$$

9 In Haber-Bosch process, i.e. for the manufacture of NH_3, finely divided iron + molybdenum as promotor is used as catalyst.

$$N_2 + 3H_2 \rightleftharpoons 2NH_3$$

10 By Freundlich adsorption isotherm,

$$\frac{x}{m} = kp^{1/n}$$

(in between low and high pressure)

When $n = 1$, $\frac{x}{m} \propto p^1$

(in lower pressure range)

when n is large, $\frac{x}{m} = k$

(independent of pressure)

Thus, $\frac{x}{m} \propto p^0$

(in high pressure range when saturation point is reached)

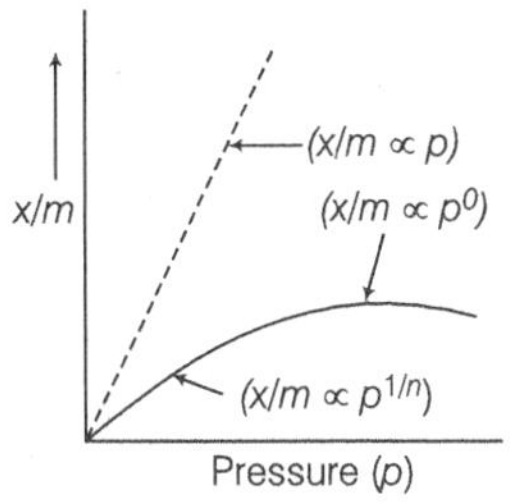

DAY SIXTEEN

Colloidal State

Learning & Revision for the Day

- Distinction Among True Solutions, Colloids and Suspension
- Classification of Colloids
- Preparation of Colloids
- Purification of Colloidal Solutions
- Properties of Colloidal Solutions
- Emulsions

A **colloid** is a heterogeneous system in which one substance is dispersed as very fine particles in another substance called **dispersion medium**. Colloidal particles are larger than simple molecules but small enough to remain suspended.

Distinction Among True Solutions, Colloids and Suspension

- Depending upon the size of dispersed particles, there are three different types of solutions *viz*. **true solutions, suspensions** and **colloidal solutions.** In **true solutions,** the dispersed particles are present as single molecule or ion. The size of the dispersed particles is less than 1 nm. True solutions are homogeneous.
- In **colloidal solution,** colloidal particles have an anormous surface area per unit mass as a result of their small size. It is heterogeneous. In **suspension**, the size of dispersed particles is more than 1000 nm. It is also heterogeneous.

Classification of Colloids

Depending upon whether the dispersed phase and the dispersion medium are solids, liquids or gases, eight types of colloidal systems are possible. These are as follows:

Colloidal system	Dispersed phase	Dispersion medium	Examples
Foam	Gas	Liquid	Soda water, froth, shaving cream
Solid sol	Gas	Solid	Foam, rubber, cork, pumice stone
Aerosol	Liquid	Gas	Fog, mist, cloud
Emulsion	Liquid	Liquid	Milk, hair cream
Solid emulsion (gel)	Liquid	Solid	Butter, cheese
Aerosol of solids	Solid	Gas	Dust in air, smoke
Sol	Solid	Liquid	Paint, ink, colloidal gold
Solid sol	Solid	Solid	Ruby glass, gemstones, alloys, rock salt

PREP MIRROR
Your Personal Preparation Indicator

- No. of Questions in Exercises (x)—
- No. of Questions Attempted (y)—
- No. of Correct Questions (z)—
  (Without referring Explanations)

- Accuracy Level ($z/y \times 100$)—
- Prep Level ($z/x \times 100$)—

In order to expect good rank in JEE, your Accuracy Level should be above 85 & Prep Level should be above 75.

- On the basis of **appearance**, colloids are classified as follows :
  (i) The colloidal systems, in which the dispersion medium is a liquid or gas, are called sols. They are called hydrosol or aquasol, if the dispersion medium is water. When the dispersion medium is alcohol or benzene, they are accordingly called **alcosol** or **benzosol**.
  (ii) The colloidal systems, in which the dispersion medium is gas are called **aerosols.**
  The colloids can be classified into **positive** or **negative colloids** according to the charge present on the dispersed phase particles.
- On the basis of **affinity of phase**, colloids are classified as follows :
  (i) **Lyophilic colloids** represent such colloidal systems in which the particles of dispersed phase have great affinity for the dispersion medium.
  These are reversible colloids, e.g. gum, gelatin, rubber, proteins etc.
  (ii) **Lyophobic colloids** represent such colloidal systems in which particles of the dispersed phase have no affinity for the dispersion medium.
  These are irreversible e.g. sols of metals and their insoluble compounds like sulphides and oxides.
  If water is the dispersion medium, the terms used are **hydrophilic** and **hydrophobic colloids**.
  Lyophobic colloids are less stable due to the presence of electric charge on their particles.
- On the basis of **molecular size**, colloids are classified as follows:
  (i) **Multimolecular colloids** are the colloids in which colloidal particles consist of aggregate of atoms or small molecules with diameter less than 10^{-9} m or 1 nm, e.g. a sol of gold, a sol of sulphur.
  (ii) **Macromolecular colloids** are the colloids in which colloidal particles themselves are large molecules, e.g. starch, proteins, enzymes, nylon etc.
  (iii) **Associated colloids** or **micelles** are the substances which behave as normal electrolytes at low concentration but as colloids at higher concentration. This is because at higher concentration, they form associated particles called micelles, e.g. soap and synthetic detergents.
  The concentration above which micelle formation occurs is called **CMC** (Critical Micelle Concentration) and the temperature above which micelle formation occurs is called **Kraft temperature**.

NOTE
- In **polydisperse colloids**, the colloidal particles are of different sizes while in monodisperse colloids, all the colloidal particles are more or less of identical size.
- The colligative properties of colloidal systems are low due to aggregation. Hence, all colloidal solutions exhibit very low osmotic pressure, very small elevation in boiling point and depression in freezing point.

Preparation of Colloids

Lyophilic colloids may be prepared by simply warming the solid with the liquid dispersion medium. On the other hand, lyophobic colloids have to be prepared by special methods.

Substances are converted into colloidal solutions by the following two methods:

1. Dispersion Methods

These methods involve the breaking of bigger particles to the size of colloidal particles.

The various dispersion methods are:

(i) **Electrodisintegration method** (Bredig's arc method), this process involves dispersion as well as condensation. By this method, colloidal solutions of metals like gold, silver etc., are obtained.

(ii) **Peptisation** involves the conversion of freshly prepared precipitate into colloidal solution by adding suitable electrolyte. The suitable electrolyte is known as peptising agent. e.g.

$$\underset{\text{Fresh ppt.}}{Fe(OH)_3} + \underset{\text{Electrolyte}}{FeCl_3} \longrightarrow \underset{\text{Colloid}}{Fe(OH)_3Fe^{3+}} + 3Cl^-$$

2. Condensation or Chemical Methods

These methods involve the growing of size of the dispersed phase or the size of colloidal particles.

(i) **By double decomposition**

$$As_2O_3 + 3H_2S \longrightarrow \underset{\text{Colloidal solution}}{As_2S_3} + 3H_2O$$

(ii) **By hydrolysis**

$$FeCl_3 + 3H_2O \longrightarrow \underset{\text{Colloidal solution}}{Fe(OH)_3} + 3HCl$$

(iii) **By oxidation**

$$H_2S + 2HNO_3 \longrightarrow 2H_2O + 2NO_2 + \underset{\text{Colloid}}{S}$$

In periodic precipitation, these reactions are carried out in gel medium. As a result of this, rings or layers of precipitates are formed at definite intervals.

Purification of Colloidal Solutions

The following methods are commonly used to purify the colloids:

(i) **Dialysis** It is the process of removing small molecules or ions from a sol by diffusion through a semipermeable membrane. In this process, impure colloidal solution is placed in a bag of semipermeable membrane, dipping in water, the ions diffuse through membrane. Ferric hydroxide sol can be purified by this method.

(ii) **Electrodialysis** If dialysis is carried out under the influence of electric field, it is called electrodialysis. This speeds up the migration of ions to the opposite electrodes.

(iii) **Ultrafiltration** Separation of sol particles from the liquid medium and electrolytes by filtration through an ultrafilter is called ultrafiltration.

(iv) **Ultra-centrifugation** In ultra-centrifuge, the colloidal particles settle down at the bottom and impurities remain in the solution.

Properties of Colloidal Solutions

The important properties of the colloidal solutions are given below:

(i) **Brownian Movement** Colloidal particles are always in a state of rapid random motion (*zig-zag* movement) which is termed as Brownian movement.

(ii) **Tyndall Effect** When a strong and converging beam of light is passed through a colloidal solution, its path becomes visible due to scattering of light by particles. It is called Tyndall effect.

(iii) **Electrophoresis** The phenomenon, involving the migration of colloidal particles under the influence of electric field towards the oppositely charged electrode, is called electrophoresis.

- **Sedimentation potential** or **Dome effect** is the reverse of electrophoresis. It is set up when a particle is forced to move in a resting liquid.
- **Electrophoretic mobility** of colloidal particles is the distance travelled by particles in one second under a potential gradient in one volt per cm. Since, different colloids have different mobilities, so this method is used for separation of proteins, nucleic acids, polysaccharides etc.

(iv) **Coagulation or Flocculation** The precipitation of particles of the dispersed phase in a sol is known as coagulation. The minimum amount of an electrolyte required to cause precipitation of one litre of a colloidal solution is called coagulation value or flocculation value. The reciprocal of coagulation value is regarded as the coagulating power.

(v) **Hardy-Schulze Rule** Higher the valency of the active ion, the greater will be its power to precipitate the sol.

e.g. order of coagulating power is

$$Al^{3+} > Ba^{2+} > Na^{+}$$

$$PO_4^{3-} > SO_4^{2-} > Cl^{-}$$

Protective Action of Lyophilic Colloids

As lyophobic sols are unstable (e.g. Au, Ag) and get easily precipitated, the addition of lyophilic colloids like gums, soaps etc., makes it difficult to precipitate. The process is known as **protection** and the lyophilic colloids are termed as protective colloids.

NOTE **Gold Number** It is the minimum weight (in mg) which must be added to 10 mL of standard red gold sol so that no coagulation of it takes place when 1 mL of 10% NaCl solution is rapidly added to it.

Emulsions

These are the colloidal solutions of two immiscible liquids or partially immiscible liquid in which the liquids act as the dispersed phase as well as the dispersion medium.

There are two types of emulsions:

(a) **Oil in water type**, e.g. milk in which tiny droplets of liquid fat are dispersed in water.

(b) **Water in oil type**, e.g. stiff greases, in which water being dispersed in lubricating oil.

During the preparation of emulsion, a small amount of some substances such as soap, gum, agar and protein etc., are added to stabilise the emulsion. These substances are known as **emulsifying agents.**

Charateristics of Emulsion

- Emulsions can be broken into constituent liquids by heating, freezing, centrifuging or chemical destruction of emulsifying agent.
- They show Brownian movement and Tyndall effect.
- These can be diluted with any amount of the dispersion medium.

NOTE
- **Surfactants** are the substances which get preferentially absorbed at the air-water, oil-water and solid-water interfaces.
- $C_{15}H_{31}COO^{-}Na^{+}$ is an anionic surfactant, $C_{18}H_{37}NH_3^{+}\ Cl^{-}$ is cationic surfactant and $C_nH_{2n+1}(OCH_2CH_2)_xOH$ is **non-ionic surfactant.**

DAY PRACTICE SESSION 1

FOUNDATION QUESTIONS EXERCISE

1 Fog is a colloidal solution of
(a) liquid particles dispersed in a gas
(b) gaseous particles dispersed in a liquid
(c) solid particles dispersed in a liquid
(d) solid particles dispersed in a gas

2 A gel is
(a) a liquid mass of a lyophilic sol in which all the dispersion medium has penetrated into the sol particles
(b) like an emulsion which is stabilised by adding emulsifying agent
(c) a semirigid mass of a lyophobic sol in which all the dispersion medium has penetrated into the sol particles
(d) a semirigid mass of a lyophilic sol in which the dispersion medium has penetrated into the sol particles

3 Smoke is an example of → JEE Main (Online) 2013
(a) solid dispersed in solid (b) gas dispersed in liquid
(c) solid dispersed in gas (d) gas dispersed in solid

4 Which one of the following is correctly matched?
(a) Emulsion - smoke (b) Gel - butter
(c) Aerosol - hair cream (d) Sol - whipped cream

5 Addition of lyophilic solution to the emulsion, forms
(a) a protective film around the dispersed phase
(b) a protective film around the dispersion medium
(c) an aerosol
(d) true solution

6 Lyophilic solutions are more stable than lyophobic solutions because
(a) the colloidal particles have positive charge
(b) the colloidal particles have negative charge
(c) the colloidal particles are solvated
(d) there is strong electrostatic repulsions between the negatively charged colloidal particles

7 Lyophilic sols are
(a) irreversible sols
(b) prepared from inorganic compounds
(c) coagulated by adding electrolytes
(d) self-stabilising

8 Which of the following is not the property of hydrophilic solutions?
(a) High concentration of dispersed phase can be easily obtained
(b) Coagulation is reversible
(c) Viscosity and surface tension are nearly the same as that of water
(d) The charge of the particles depend on the pH of the medium and it may be positive, negative or zero

9 In multimolecular colloidal sols, atoms or molecules are held together by
(a) H-bonding (b) van der Waals' forces
(c) ionic bonding (d) polar covalent bonding

10 Which type of molecules form micelles?
(a) Non-polar molecules
(b) Polar molecules
(c) Surfactant molecules
(d) Salt of weak acid and weak base

11 The critical micelle concentration (CMC) is
(a) the concentration at which micellisation starts
(b) the concentration at which the true solution is formed
(c) the concentration at which one molar electrolyte is present per 1000 g of the solution
(d) the concentration at which $\Delta H = 0$

12 During the micelle formation,
(a) $\Delta H = +\text{ve}, \Delta S = +\text{ve}$ (b) $\Delta H = -\text{ve}, \Delta S = -\text{ve}$
(c) $\Delta H = -\text{ve}, \Delta S = +\text{ve}$ (d) $\Delta H = +\text{ve}, \Delta S = -\text{ve}$

13 Micelles form only
(a) below the critical micelle concentration (CMC) and below the Kraft temperature (T_k)
(b) above CMC and below the T_k
(c) above the CMC and above the T_k
(d) below the CMC and above the T_k

14 Among the following, the surfactant that will form micelles in aqueous solution at the lowest molar concentration at ambient conditions, is
(a) $CH_3(CH_2)_{15}N^+(CH_3)_3Br^-$ (b) $CH_3(CH_2)_{11}OSO_3^-Na^+$
(c) $CH_3(CH_2)_6COO^-Na^+$ (d) $CH_3(CH_2)_{11}N^+(CH_3)_3Br^-$

15. A freshly prepared $Fe(OH)_3$ precipitate is peptised by adding $FeCl_3$ solution. The charge on the colloidal particles is due to preferential adsorption of
(a) Br^- ion (b) Fe^{3+} ion
(c) OH^- ion (d) Ba^{2+} ion

16 Peptisation involves
(a) precipitation of colloidal particles
(b) disintegration of colloidal aggregates
(c) evaporation of dispersion medium
(d) impact of molecules of the dispersion medium on the colloidal particles

17 A particle of radius 1 cm is broken to form colloidal particles of radius 1000 Å. The number of colloidal particles produced are
(a) 10^{15} (b) 6.023×10^{23}
(c) 10^{12} (d) 10^{10}

18 The colloidal sol formed from SnO_2 in acidic and basic medium respectively are
(a) $[SnO_2]SnO_3^{2-}$: 2 Na^+ and $[SnO_2]Sn^{4+}$: $4Cl^-$
(b) $[SnO_2]Sn^{4+}$ and $[SnO_2]SnO_3^{2-}$: $2Na^+$
(c) positively and negatively charged
(d) $[SnO_2]SnO_3^{2-}$: $2Na^+$ in both the media

19 Separation of colloidal particles from those of molecular dimensions by means of electric current is known as
(a) electroosmosis (b) electrophoresis
(c) electrodialysis (d) electrolysis

20 The Brownian movement is due to
(a) enthalpy change during the formation of colloids
(b) attractive force between the colloidal particles and the molecules of dispersion medium
(c) the impact of molecules of the dispersion medium on the colloidal particles
(d) the movement of positively charged colloidal particles to negatively charged particle

21 The migration of dispersion medium under the influence of an electric potential is called ➜ JEE Main (Online) 2013
(a) cataphoresis (b) electroosmosis
(c) electrophoresis (d) gas dispersed in solid

22 Which of the following statements is correct for Tyndall effect?
(a) Scattering and polarising of light by small suspended particles is called Tyndall effect
(b) Tyndall effect of colloidal particles is due to dispersion of light
(c) Tyndall effect is due to refraction of light
(d) *zig-zag* motion of suspended particles

23 Which of the following process is responsible for the formation of delta at a place where rivers meet the sea?
(a) Emulsification (b) Colloidal formation
(c) Coagulation (d) Peptisation

24 Which of the following ions have minimum value of flocculating power?
(a) PO_4^{3-} (b) SO_4^{2-} (c) SO_3^{2-} (d) NO_3^-

25 Match the following and choose the correct option.

Column I	Column II
A. Protective colloid	1. $FeCl_3$ + NaOH
B. Liquid-liquid colloid	2. Lyophilic colloids
C. Positively charged colloid	3. Emulsion
D. Negatively charged colloid	4. $FeCl_3$ + hot water

Codes

	A	B	C	D		A	B	C	D
(a)	1	4	3	2	(b)	2	4	3	1
(c)	2	3	4	1	(d)	1	4	2	3

26 Which of the following has maximum coagulation power with ferric hydroxide sol?
(a) Cryolite (b) $K_2C_2O_4$
(c) $K_3[Fe(CN)_6]$ (d) $K_4[Fe(CN)_6]$

27 The coagulating power of electrolytes having ions Na^+, Al^{3+} and Ba^{2+} for arsenic sulphide solution increases in the order ➜ JEE Main 2013
(a) $Na^+ < Ba^{2+} < Al^{3+}$ (b) $Ba^{2+} < Na^+ < Al^{3+}$
(c) $Al^{3+} < Na^+ < Ba^{2+}$ (d) $Al^{3+} < Ba^{2+} < Na^+$

28 The coagulation values of $AlCl_3$ and NaCl are 0.093 and 52 respectively. The ratio of coagulating powers of both will be
(a) 0.093 : 1 (b) 52 : 1
(c) 559 : 1 (d) 1.788 : 1

29 Which of the following will show Tyndall effect?
(a) Aqueous solution of soap below critical micelle concentration
(b) Aqueous solution of soap above critical micelle concentration
(c) Aqueous solution of sodium chloride
(d) Aqueous solution of sugar

30 Gold number is the index for
(a) protective power of lyophilic colloid
(b) purity of gold
(c) metallic gold
(d) electroplated gold

31 Hardy-Schulze law states that
(a) larger the charge on the coagulating ions, greater is its coagulating power, having opposite sign of solution
(b) solution must have zero gold number
(c) dispersed phase and dispersion medium must be of the same sign
(d) micelles coagulate in the presence of surfactants

32 When dilute aqueous solution of $AgNO_3$ (excess) is added to KI solution, positively charged sol particles of AgI are formed due to adsorption of ion
(a) K^+ (b) Ag^+ (c) I^- (d) NO_3^-

33 Which of the following statements is incorrect?
(a) Emulsions are prepared by shaking two liquid components, say oil and water and adding some emulsifying agent
(b) Water in oil emulsions are formed when the emulsifying agent at the interface is chiefly in the water phase
(c) Water in oil emulsions are formed when the emulsifying agent at the interface is chiefly in the oil phase
(d) Gems and gels mixed together to give emulsions

Direction (Q. Nos. 34-38) *In the following questions, Assertion* (A) *followed by a Reason* (R) *is given. Choose the correct option out of the following choices.*
(a) Both A and R are true and R is correct explanation of A
(b) Both A and R are true but R is not correct explanation of A
(c) A is true but R is false
(d) Both A and R are false

34 **Assertion** (A) An ordinary filter paper impregnated with colloidal solution stops the flow of colloidal particles.
Reason (R) Pore size of the filter paper becomes more than the size of colloidal particle.

35. **Assertion** (A) The Brownian movement has a stirring affect that does not permit the particles to settle.

Reason (R) Brownian motion is responsible for stability of sols.

36. **Assertion** (A) Micelles are formed by surfactant molecules above the Critical Micelle Concentration (CMC).

Reason (R) The conductivity of a solution having surfactant molecules decreases sharply at the CMC.

37. **Assertion** (A) A colloid gets coagulated by addition of an electrolyte.

Reason (R) The rate of coagulation depends on the magnitude and sign of the charge of the coagulant ion.

38. **Assertion** (A) Colloidal AgI is prepared by adding $AgNO_3$ in slight excess to KI solution. When subjected to an electric field, the colloidal particles migrate to the anode.

Reason (R) Colloidal particles adsorb ions and thus, become electrically charged.

DAY PRACTICE SESSION 2

PROGRESSIVE QUESTIONS EXERCISE

1 When 9.0 mL of arsenious sulphide sol and 1.0 mL of 0.1 M $BaCl_2$ are mixed, turbidity due to precipitation just appears after 2 h. The effective ion and its coagulation value respectively are

(a) Cl^-, 10 millimol L^{-1} (b) Cl^-, 20 millimol L^{-1}
(c) Ba^{2+}, 10 millimol L^{-1} (d) Ba^{2+}, 20 millimol L^{-1}

2 The dispersed phase in colloidal iron (III) hydroxide and colloidal gold is positively and negatively charged, respectively. Which of the following statements is not correct?

(a) Coagulation in both sols can be brought about by electrophoresis
(b) Mixing the sols has no effect
(c) Sodium sulphate solution causes coagulation in both sols
(d) Magnesium chloride solution coagulates the gold sol more readily than the iron (III) hydroxide sol

3 The values of colligative properties of colloidal solution are of small order in comparison to those shown by true solutions of same concentration because colloidal particles

(a) exhibit enormous surface area
(b) remain suspended in the dispersion medium
(c) form lyophilic colloids
(d) are comparatively less in number

4 In a coagulating experiment, 5 mL of As_2S_3 is mixed with distilled water and 0.2 M solution of an electrolyte *AB*, so that the total volume is 20 mL. All solutions containing 5.4 mL of *AB* coagulate within 2 min. The flocculation value of *AB* (in millimole) is

(a) 5 (b) 50
(c) 54 (d) None of these

5 Which one of the following does not involve coagulation?

(a) Clotting of blood by the use of ferric chloride
(b) Formation of delta region
(c) Treatment of drinking water by potash alum
(d) Peptisation

6 When 6.0×10^{-5} g of a protective colloid was added to 20 mL of a standard gold sol, the precipitation of latter was just prevented on addition of 1 mL of 10% NaCl solution. The gold number of the protective colloid is

(a) 3 (b) 3×10^{-5}
(c) 0.06 (d) 0.03

7 Gold numbers of protective colloids *A*, *B*, *C* and *D* are 0.50, 0.01, 0.10 and 0.005, respectively. The correct order of their protective powers is

(a) $D < A < C < B$ (b) $C < B < D < A$
(c) $A < C < B < D$ (d) $B < D < A < C$

8 Potassium stearate is obtained by the saponification of an oil or fat. It has formula $CH_3—(CH_2)_{16}—COOK$. The molecule has a lyophobic terminal CH_3 and a lyophilic terminal—COOK. Potassium stearate is an example of

(a) lyophilic colloid
(b) lyophobic colloid
(c) macromolecular colloid
(d) micelle or associated colloid

9 The gold number of some colloidal solutions are given below:

Colloidal solution	Gold number
A	0.01
B	2.5
C	20

The protective nature of these colloidal solutions follow the order

(a) $C > B > A$ (b) $A > B > C$
(c) $A = B = C$ (d) $B > A > C$

10 The disperse phase, dispersion medium and nature of colloidal solution (lyophilic or lyophobic) of 'gold sol' respectively are

(a) solid, solid, lyophobic
(b) liquid, liquid, lyophobic
(c) solid, liquid, lyophobic
(d) solid, liquid, lyophilic

ANSWERS

SESSION 1

1 (a)	**2** (d)	**3** (c)	**4** (b)	**5** (a)	**6** (c)	**7** (d)	**8** (c)	**9** (b)	**10** (c)
11 (a)	**12** (c)	**13** (c)	**14** (a)	**15** (b)	**16** (b)	**17** (a)	**18** (c)	**19** (b)	**20** (c)
21 (b)	**22** (a)	**23** (c)	**24** (a)	**25** (c)	**26** (d)	**27** (a)	**28** (c)	**29** (b)	**30** (a)
31 (a)	**32** (b)	**33** (d)	**34** (c)	**35** (d)	**36** (b)	**37** (b)	**38** (a)		

SESSION 2

1 (c)	**2** (b)	**3** (d)	**4** (c)	**5** (d)	**6** (d)	**7** (c)	**8** (d)	**9** (b)	**10** (c)

Hints and Explanations

SESSION 1

1 Fog is liquid dispersed in gas, a class of colloidal system.

2 A gel is liquid dispersed in solid, e.g. cheese. It is a semirigid mass of a lyophilic sol in which the dispersion medium has penetrated into sol particles.

3 Smoke is a colloid in which dispersed phase is solid and dispersion medium is gas or in other words, in smoke, a solid is dispersed in gas.

4 Gel-butter
Butter is an example for gel.

5 Addition of lyophilic solution to the emulsion forms a protective film around the dispersed phase. Surface phase reaction.

6 Lyophilic colloids are more stable due to solvation.

7 Lyophilic sols are self-stabilising because these sols are reversible and are highly hydrated in the solution.

8 Viscosity is higher than that of dispersion medium (water) while surface tension is usually lower than that of dispersion medium (water).

9 In multimolecular colloidal sols, atoms or molecules are held together by van der Waals' forces.

10 Surfactant molecules form micelles.

11 It is the minimum concentration at which surfactant molecules undergo aggregation.

12 ΔH = negative, ΔS = positive

13 Micelle formation takes place above a particular temperature, called Kraft temperature (T_k) and above a particular concentration, called Critical Micelle Concentration (CMC).

14 Critical concentration for micelle formation decreases as the molecular weight of hydrocarbon chain of surfactant grows. Hexadecyltrimethyl ammonium bromide (CTAB) i.e. $CH_3(CH_2)_{15}N^+(CH_3)_3Br^{\ominus}$, will form micelles in aqueous solution at lowest molar concentration.

15 Solution particle adsorbs common ion present in the medium. So, charge on colloidal particles is due to preferential adsorption of Fe^{3+} ion.

16 Peptisation involves disintegration of colloidal aggregates.

17 Volume of a particle with radius 1 cm

$$= \frac{4}{3}\pi r^3 = \frac{4}{3}\pi\ (1)^3\ cm^3$$

The particle is broken into colloidal particle of radius 1000 Å, i.e. 10^{-5} cm.

Hence, volume of the particle

$$= \frac{4}{3}\pi r^3 = \frac{4}{3} \times \pi \times (10^{-5})^3$$

$$= \frac{4}{3}\pi \times 10^{-15}\ cm^3$$

Hence, number of colloidal particles

$$= \frac{\frac{4}{3}\pi\ cm^3}{\frac{4}{3}\pi \times 10^{-15} cm^3} = 10^{15}$$

18 In acidic medium, Sn^{4+} ion is formed which is preferentially adsorbed on SnO_2 giving positively charged colloidal sol.

$$SnO_2 + Sn^{4+} \longrightarrow \underset{\text{Positively charged}}{[SnO_2]Sn^{4+} : 4Cl^-}$$

In basic medium, SnO_3^{2-} is formed which is preferentially adsorbed on SnO_2, giving negatively charged colloidal sol.

$$SnO_2 + SnO_3^{2-} \longrightarrow \underset{\text{Negatively charged}}{[SnO_2]SnO_3^{2-} : 2Na^+}$$

19 In electrophoresis, charged sol particles move towards opposite electrodes.

20 Brownian movement is *zig-zag* motion of suspended particles. It is due to the impact of molecules of the dispersion medium on the colloidal particles.

21 Electroosmotic flow is caused by the Coulomb force induced by an electric field on net mobile electric charge in a solution. When an electric field is applied to the fluid, the net charge in the electrical double layer (a layer of mobile ions forms in the region near the interface) is induced to move by the resulting Coulomb force.

22 When a beam of light is passed through a colloidal solution, its path becomes visible by light scattering and polarising of light by small suspended particles. This is known as Tyndall effect.

23 The process is responsible for the formation of delta at a place where rivers meet the sea is coagulation.

24 PO_4^{3-} have minimum value of flocculating power.

25 $A \longrightarrow 2;\ B \longrightarrow 3;\ C \longrightarrow 4;\ D \longrightarrow 1$

26 Ferric hydroxide is a positive solution. Thus, a negative ion cause the precipitation of positively charged sol and *vice-versa*. According to Hardy-Schulze rule, greater is the valency of the coagulating or flocculating ion, greater is its power to bring about coagulation.

Thus, in coagulation of a positive solution ferric hydroxide, the flocculating power is maximum for $[Fe(CN)_6]^{4-}$ ion. Thus, $K_4[Fe(CN)_6]$ possesses the maximum coagulating power form $Fe(OH)_3$ solution.

Hence, the correct increasing order is $Na^+ < Ba^{2+} < Al^{3+}$

27 As_2S_3 is an anionic sol (negative sol) hence, coagulation will depend upon coagulating power of cation which is directly proportional to the valency of cation (Hardy-Schulze rule).

28 As coagulating power is inversely proportional to coagulation value, the ratio of their coagulating powers will be

$$\frac{\text{Coagulating power of } AlCl_3}{\text{Coagulating power of NaCl}} = \frac{\text{Coagulation value of NaCl}}{\text{Coagulation value of } AlCl_3} = \frac{52}{0.093} = 559 : 1$$

29 Aqueous solution of soap above critical micelle concentration show Tyndall effect.

30 Protective power of various lyophilic substances is expressed in terms of gold number.

31 Hardy-Schulze law states that larger the charge on the coagulating ions, greater is its coagulating power, having opposite sign of solution.

32 Sol particles possess the tendency to adsorb preferentially the common ion present in solution. Therefore, AgI are formed due to adsorption of Ag^+ ion.

33 Gems are solid and gels are semisolid. This mixture will not form emulsion.

34 **Correct Reason** Pore size of the filter paper becomes smaller than the size of the colloidal particles.

35 The Brownian movement has a stirring affect which does not permit the particles to settle and thus, in responsible for the stability of sols.

36 At a certain concentration, surfactant molecules start to aggregate and form micelle, the concentration is called Critical Micelle Concentration (CMC).

Aggregation of surfactant molecules (ions; $RCOO^-$), i.e. micelle formation cause effective fall in number of free ions to conduct electricity, thus conductivity decreases at CMC.

37 Coagulation of sols by addition of electrolytes is explained by Hardy-Schulze rule. For positive sol, negative ion of electrolyte is the effective ion while for negative sol, positive ion of electrolyte is the effective ion. Greater is the valency of effective ion, more is its coagulating power and lesser is the coagulation value.

38 Colloidal particles are electrically charged, hence they move towards oppositely charged electrode on applying electric field.

SESSION 2

1 As_2S_3 sol is negatively charged due to preferential adsorption of S^{2-} ions by As_2S_3 particles. Hence, cation of largest valency (i.e. Ba^{2+}) would be most effective in causing coagulation of the sol.

$$\text{Moles of } Ba^{2+} = \frac{1.0 \times 0.1}{1000} = 1.0 \times 10^{-4} \text{mol}$$

10 mL (9.0 mL of As_2S_3 and 1.0 mL $BaCl_2$) of As_2S_3 requires $BaCl_2$ for coagulation = 1.0×10^{-4} mol

1000 mL of As_2S_3 requires $BaCl_2$ for coagulation (i.e. coagulation value)

$$= \frac{1.0 \times 10^{-4} \times 1000}{10} = 1.0 \times 10^{-2} \text{ mol L}^{-1} = 10 \text{ millimol L}^{-1}$$

2 Mixing the sols together can cause coagulation, since the charges are neutralised.

3 The values of colligative properties of colloidal solution are of small order in comparison to true solutions of same concentration because colloidal particles are comparatively less in number.

4 Flocculation value is the amount of electrolyte, in millimoles required to bring about complete coagulation of one litre of colloidal solution.

Amount of *AB* in 5.4 mL of 0.2 M *AB*

$$= \frac{5.4 \times 0.2}{1000} = 1.08 \times 10^{-3} \text{ mol}$$

20 mL of As_2S_3 requires *AB* for coagulation $= 1.08 \times 10^{-3}$ mol

1000 mL of As_2S_3 requires *AB* (i.e. flocculation value)

$$= \frac{1.08 \times 10^{-3}}{20} \times 1000 = 0.054 \text{ mol} = 54 \text{ millimol}$$

5 Peptisation is the process of converting a precipitate into colloidal sol by shaking it with dispersion medium in the presence of a small amount of electrolyte. It is the method of preparation of colloids.

6 The protective colloid added to 20 mL of gold sol to completely prevent coagulation by 1 mL of 10 % NaCl solution = 6.0×10^{-5}g = 6.0×10^{-2} mg.

Hence, the protective colloid required to be added to 10 mL of gold sol to completely prevent coagulation by 1 mL of 10% solution $= \frac{6.0 \times 10^{-2} \times 10}{20} = 0.03$ mg

Hence, gold number = 0.03

7 Higher the gold number, lesser will be the protective power of colloid. So, the correct order of their protective power is $A < C < B < D$

8 Potassium stearate is an example of micelle or associated colloid.

9 Smaller the value of gold number, greater will be the protecting power of the protective colloid. Hence, protective nature of *A*, *B* and *C* is as

Colloidal solution : $A > B > C$

Gold number : 0.01 2.5 20

10 Colloidal solution of gold is obtained when dispersed phase is solid and disperstion medium is liquid.

Substances like metals cannot be brought into the colloidal state simply by bringing them in contact with water and therefore, special methods are devised for the purpose.

Hence, they are known as hydrophobic or lyophobic colloids.

DAY SEVENTEEN

Unit Test 3
(Physical Chemistry II)

1 The cell, $Zn|Zn^{2+}(1M)||Cu^{2+}(1M)|\ Cu(E°_{cell} = 1.10\ V)$, was allowed to be completely discharged at 298 K. The relative concentration of Zn^{2+} to $Cu^{2+}\left(\frac{[Zn^{2+}]}{[Cu^{2+}]}\right)$ is

(a) antilog (24.08) (b) antilog 37.3
(c) $10^{37.3}$ (d) 9.65×10^4

2 The molar conductivities Λ°_{NaOAc} and Λ°_{HCl} at infinite dilution in water at 25°C are 91.0 and 426.2 S cm^2/mol respectively. To calculate Λ°_{HOAc}, the additional value required is

(a) $\Lambda^\circ_{H_2O}$ (b) Λ°_{KCl} (c) Λ°_{HOAc} (d) Λ°_{NaCl}

3 Given the data at 25°C,

$$Ag + I^- \longrightarrow AgI + e^-;\ E° = 0.152\ V$$
$$Ag \longrightarrow Ag^+ + e^-;\ E° = -0.800\ V$$

What is the value of $\log K_{sp}$ for AgI ?

$$\left(2.303\frac{RT}{F} = 0.059\ V\right)$$

(a) – 8.12 (b) –8.612 (c) –37.83 (d) –16.14

4 The highest electrical conductivity from the following aqueous solution is of

(a) 0.1 M difluoroacetic acid
(b) 0.1 M fluoroacetic acid
(c) 0.1 M chloroacetic acid
(d) 0.1 M acetic acid

5 When a certain conductivity cell was filled with 0.01 M solution of KCl, it had a resistance of 160 Ω at 25°C and when filled with 0.005 M NaOH, it had a resistance of 190 Ω. If specific resistance of KCl solution is, 700 Ω-cm, specific conductance ($\Omega^{-1}\ cm^{-1}$) of NaOH solution is

(a) 0.00120 (b) 0.00170 (c) 0.00180 (d) 0.00190

6 In a hydrogen-oxygen fuel cell, combustion of hydrogen occurs to

(a) generate heat
(b) create potential difference between the two electrodes
(c) produce water of high purity
(d) remove adsorbed oxygen from electrode surfaces

7 The $E^\circ_{M^{3+}/M^{2+}}$, values for Cr, Mn, Fe and Co are – 0.41, + 1.57, + 0.77 and + 1.97 V respectively. For which one of these metals the change in oxidation state from + 2 to + 3 is easiest ?

(a) Cr (b) Mn
(c) Fe (d) Co

8 Consider the following $E°$ values

$$E^\circ_{Fe^{3+}/Fe^{2+}} = +0.77\ V$$
$$E^\circ_{Sn^{2+}/Sn} = -0.14\ V$$

Under standard conditions the potential for the reaction

$$Sn(s) + 2Fe^{3+}(aq) \longrightarrow 2Fe^{2+}(aq) + Sn^{2+}(aq)\text{ is}$$

(a) 1.68 V (b) 1.40 V
(c) 0.91 V (d) 0.63 V

9 In a cell that utilises the reaction,

$$Zn(s) + 2H^+(aq) \longrightarrow Zn^{2+}(aq) + H_2(g)$$

addition of H_2SO_4 to cathode compartment will

(a) lower the E and shift equilibrium to the left
(b) lower the E and shift the equilibrium to the right
(c) increase the E and shift the equilibrium to right
(d) increase the E and shift the equilibrium to the left

10 Consider the following reactions at 1100°C,

I. $2C + O_2 \longrightarrow 2CO, \Delta G° = -460\ kJ\ mol^{-1}$
II. $2Zn + O_2 \longrightarrow 2ZnO, \Delta G° = -360\ kJ\ mol^{-1}$

Based on these, select the correct alternate.

(a) Zinc can be oxidised by CO
(b) Zinc oxide can be reduced by carbon
(c) Both (a) and (b)
(d) None of the above

11 Consider the reaction, $2A + B \longrightarrow$ products, when concentration of B alone was doubled, the half-life did not change. When the concentration of A alone was doubled, the rate increased by two times. The unit of rate constant for this reaction is

(a) L mol^{-1} s^{-1} (b) no unit
(c) mol $L^{-1}s^{-1}$ (d) s^{-1}

12 A reaction was found to be second order with respect to the concentration of carbon monoxide. If the concentration of carbon monoxide is doubled, with everything else kept the same, the rate of reaction will

(a) remain unchanged (b) triple
(c) increase by a factor of 4 (d) double

13 $t_{1/4}$ can be taken as the time taken for the concentration of a reactant to drop to 3 / 4 of its initial value. If the rate constant for a first order reaction is k, the $t_{1/4}$ can be written as

(a) $0.75/k$ (b) $0.69/k$ (c) $0.29/k$ (d) $0.10/k$

14 Consider an endothermic reaction, $X \longrightarrow Y$ with the activation energies E_b and E_f for the backward and forward reactions respectively. In general,

(a) there is no definite relation between E_b and E_f
(b) $E_b = E_f$
(c) $E_b > E_f$
(d) $E_b < E_f$

15 In a first order reaction, the concentration of the reactant decreases from 0.8 M to 0.4 M in 15 min. The time taken for the concentration to change from 0.1 M to 0.025 M is

(a) 30 min (b) 15 min
(c) 7.5 min (d) 60 min

16 Graph between $\log k$ and $\frac{1}{T}$ (k is rate constant in s^{-1} and T is the temperature in K is a straight line. If $OX = 5$ and slope of the line $= -\frac{1}{2.303}$ then E_a is

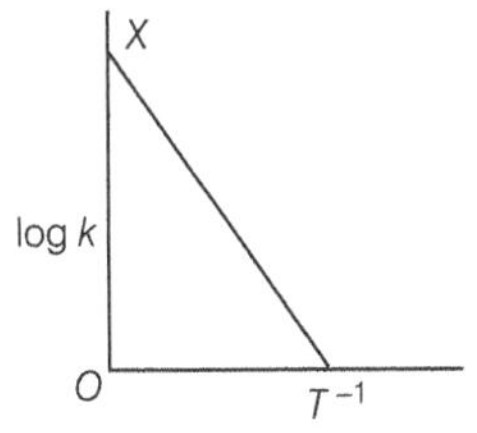

(a) 2.303×2 cal (b) $\frac{2}{2.303}$ cal
(c) 2 cal (d) None of these

17 The rate law for a reaction between the substances A and B is given by rate $= k[A]^n[B]^m$. On doubling the concentration of A and halving the concentration of B, the ratio of the new rate to the earlier rate of the reaction will be

(a) $\frac{1}{2^{m+n}}$ (b) $(m+n)$
(c) $(n-m)$ (d) $2^{(n-m)}$

18 Match the reactions in Column I with the number of electrons lost or gained in Column II.

Column I	Column II
A. $Mn(OH)_2 + H_2O_2 \rightarrow MnO_2$	1. 8
B. $AlCl_3 + 3K \rightarrow Al + 3KCl$	2. 2
C. $3Fe + 4H_2O \rightarrow Fe_3O_4 + 4H_2$	3. 3
D. $H_2S + NO_3^{\ominus} \rightarrow S + NO$	4. 6

Codes

	A	B	C	D		A	B	C	D
(a)	2	4	1	3	(b)	4	1	3	2
(c)	2	3	1	4	(d)	2	1	3	4

19 The average molecular weight of colloidal particles is determined by

(a) Tyndall effect
(b) osmotic pressure measurement
(c) Victor Meyer's method
(d) None of the above

20 Oxidation number of Cr in the following compounds are x, y, z and w. (O_2 given is peroxy linkage)

$\underset{x}{K[CrO(O_2)(OH)]}, \underset{y}{K_3[Cr(O_2)_4]}\underset{z}{(NH_3)_3[Cr(O_2)_2]}, \underset{w}{CrO_2Cl_2}$

These values are

(a) 6, 13, 5, 6 (b) 4, 4, 4, 6
(c) 4, 5, 4, 6 (d) 4, 5, 6, 6

21 Match the following and choose the correct option.

Column I	Column II
A. Placing silica gel in water vapour	1. Occlusion
B. Placing anhydrous $CaCl_2$ in water vapour	2. Adsorption
C. Placing finely divided nickel in a closed vessel containing H_2 gas	3. Absorption
D. Shaking dilute KCl solution with blood charcoal	4. Negative adsorption

Codes

	A	B	C	D		A	B	C	D
(a)	3	1	4	2	(b)	2	3	1	4
(c)	1	4	2	3	(d)	4	2	3	1

22 When a sulphur sol is evaporated, sulphur is obtained. On mixing with water sulphur sol is not formed. The sol is

(a) lyophilic (b) reversible
(c) hydrophobic (d) hydrophilic

23 1 mole of N_2H_4 loses 10 moles of electrons to form a new compound Y. Assuming that, all the nitrogen appear in the new compound, what is the oxidation state of nitrogen in Y? (No change in the oxidation state of H)

(a) −1 (b) −3
(c) +3 (d) +5

24 The process of removing dissolved impurities from a colloidal system, by means of diffusion through suitable membrane under the influence of an electric field is called

(a) electroosmosis (b) electrodialysis
(c) electrophoresis (d) peptisation

Direction (Q. Nos. 25-26) *In the following questions, Assertion (A) followed by Reason (R) is given. Choose the correct option out of the following choices.*

(a) Both (A) and (R) are correct and (R) is correct explanation of (A)
(b) Both (A) and (R) are correct but (R) is not correct explanation of (A)
(c) (A) is correct but (R) is incorrect
(d) Both (A) is and (R) are incorrect

25 **Assertion** (A) Rate constants determined from Arrhenius equation are fairly accurate for simple as well as complex molecules.

Reason (R) Reactant molecules undergo chemical change irrespective of their orientation during collision.

→ NCERT Exemplar

26 **Assertion** (A) Mercury cell does not give steady potential.

Reason (R) In the cell reaction, ions are involved in solution.

27 If dilute HCl is added to a precipitate of stannic oxide,

(a) $SnCl_2$ and $SnCl_4$ are formed
(b) $Sn(OH)_2$ is formed
(c) a stable sol of stannic oxide is formed
(d) Sn_2O_3 is formed

28 Select the incorrect statement.

(a) Physical adsorption is reversible while chemical is irreversible
(b) High pressure favours physical adsorption while low pressure favours chemical adsorption
(c) Physical adsorption is not specific while chemical is highly specific
(d) High activation energy is involved in chemical adsorption.

29 Hemidialysis is used to

(a) separate colloidal sol from water
(b) separate charged and uncharged particles
(c) clean the blood of patients whose kidneys have malfunctioned
(d) None of the above

30 In multimolecular colloidal sols, atoms or molecules are held together by

(a) H-bonding (b) van der Waals' forces
(c) ionic bonding (d) polar covalent bonding

31 SnO_2 is taken in basic medium and current is passed. Colloidal sol migrates toward

(a) anode (b) cathode
(c) Both (a) and (b) (d) None of these

32 $[AgI]I^-$ colloidal sol can be coagulated by the addition of a suitable cation. 1 mole of $[AgI]I^-$ requires mole of $AgNO_3$, $Pb(NO_3)_2$ and $Fe(NO_3)_3$ as

(a) 1, 1, 1 (b) 1, 2, 3 (c) $1, \frac{1}{2}, \frac{1}{3}$ (d) 6, 3, 2

33 Equivalent conductance of $BaCl_2$, H_2SO_4 and HCl are x_1, x_2 and x_3 S cm^2 equiv^{-1} at infinite dilution. If specific conductance of saturated $BaSO_4$ solution is of y S cm^{-1}, then K_{sp} of $BaSO_4$ is

(a) $\frac{10^3 y}{2(x_1 + x_2 - 2x_3)}$ (b) $\frac{10^6 y^2}{(x_1 + x_2 - x_3)^2}$
(c) $\frac{10^6 y^2}{4(x_1 + x_2 - x_3)^2}$ (d) $\frac{x_1 + x_2 - 2x_3}{10^6 y^2}$

34 Select the correct statements.

(a) Emulsifiers stabilise the emulsion
(b) Soaps, detergents, long chain sulphonic acids and lyophilic colloids are emulsifiers
(c) Cleansing action of soap is due to the formation of emulsions
(d) All of the above are correct statements

35 Milk is an emulsion of fat dispersed in water. It is stabilised by

(a) casein—a lyophilic colloidal sol
(b) casein—a lyophobic colloidal sol
(c) lactose—a lyophilic colloidal sol
(d) lactose—a lyophobic colloidal sol

36 $KMnO_4$ reacts with oxalic acid as

$$MnO_4^- + C_2O_4^{2-} + H^+ \longrightarrow Mn^{2+} + CO_2 + H_2O$$

Hence, 50 mL of 0.04 $KMnO_4$ in acidic medium is chemically equivalent to

(a) 100 mL of 0.1 M $H_2C_2O_4$ (b) 50 mL of 0.2 M $H_2C_2O_4$
(c) 50 mL of 0.1 M $H_2C_2O_4$ (d) 25 mL of 0.1 M $H_2C_2O_4$

37 Given that $E^\circ_{Cu^{2+}/Cu} = 0.34$ V; $E^\circ_{Ag^+/Ag} = 0.80$ V;

$E^\circ_{Mg^{2+}/Mg} = 2.37$ V and $E^\circ_{Al^{3+}/Al} = -1.66$ V, in which of the following cells the standard free energy decrease is maximum ?

(a) $Mg|Mg^{2+}(1M)||Cu^{2+}(1\ M)|Cu$
(b) $Mg\ |Mg^{2+}(1\ M)||Ag^+\ (1\ M)|Ag$
(c) $Ag\ |Ag^+(1\ M)||Al^{3+}\ (1\ M)|Al$
(d) $Cu\ |Cu^{2+}(1\ M)||Ag^+\ (1\ M)|Ag$

38 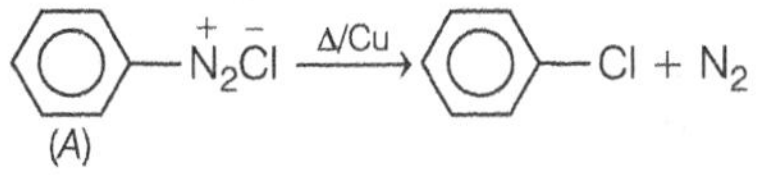

Half-life is independent of concentration of *A*. After 10 min, volume of N_2 gas is 10 L and after complete reaction 50 L. Hence, rate constant is

(a) $\frac{2.303}{10}\log 5\ \text{min}^{-1}$ (b) $\frac{2.303}{10}\log 1.25\ \text{min}^{-1}$
(c) $\frac{2.303}{10}\log 2\ \text{min}^{-1}$ (d) $\frac{2.303}{10}\log 4\ \text{min}^{-1}$

39 Gold number of haemoglobin is 0.03. Hence, 100 mL of gold sol will require haemoglobin so that gold is not coagulated by 10 mL of 10% NaCl sol is

(a) 0.03 mg (b) 30 mg
(c) 0.30 mg (d) 3 mg

40 Graph between $\log\left(\frac{x}{m}\right)$ and $\log p$ is a straight line at angle 45° with intercept *OA* as shown. Hence, $\left(\frac{x}{m}\right)$ at a pressure of 0.2 atm is

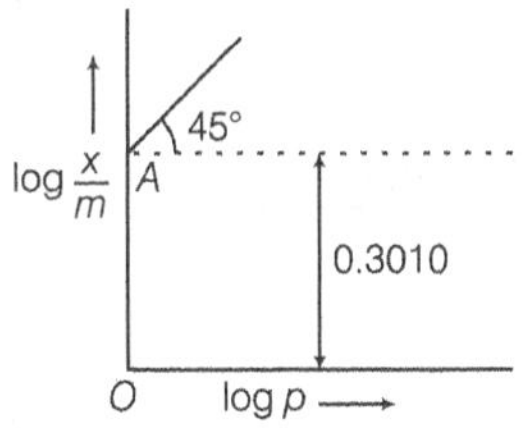

(a) 0.2 (b) 0.4 (c) 0.6 (d) 0.8

41 For the reaction,

$$3BrO^- \longrightarrow BrO_3^- + 2Br^-$$

in alkaline aqueous solution, the value of the second order (in BrO^-) rate constant at 80°C in the rate law for $-\Delta[BrO^-]/\Delta t$ was found to be 0.056 L mol^{-1}s^{-1}.

Thus, select correct alternate

(a) rate constant is 0.019 L mol^{-1} s^{-1} when rate law is $\frac{\Delta[BrO_3^-]}{\Delta t}$

(b) rate constant is 0.037 L mol^{-1} s^{-1} when rate law is $\frac{\Delta[Br^-]}{\Delta t}$

(c) rate constant is 0.037 L mol^{-1}s^{-1} when rate law is $\frac{\Delta[BrO_3^-]}{\Delta t}$

(d) Both (a) and (b) are correct

42 3g of activated charcoal was added to 50 mL of 0.06 N acetic acid solution in a flask.
After an hour, it was filtered and strength of the filtrate was found to be 0.042 N. The amount of acetic acid adsorbed per gram of charcoal is

(a) 18 mg (b) 36 mg
(c) 42 mg (d) 54 mg

43. For the given reaction,

$$A \longrightarrow \text{Product}$$

$$\frac{dx}{dt} = k[A]^2$$

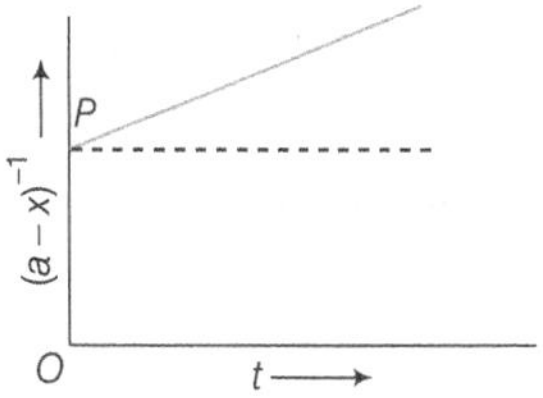

If $k = 1.0 \times 10^{-2}$ L mol^{-1} min^{-1} and $OP = 0.10$ L mol^{-1} then rate at the start of the reaction is

(a) 1.0×10^{-2} mol L^{-1} min^{-1}
(b) 1.0 mol L^{-1} min^{-1}
(c) 1.0×10^{2} mol L^{-1} min^{-1}
(d) 1.0×10^{-4} mol L^{-1} min^{-1}

44. Which one of the following statements is correct ?

(a) Brownian movement is more pronounced for smaller particles than for bigger ones
(b) Sols of metal sulphides are lyophilic
(c) Hardy-Schulze law states, the bigger the size of the ion, the greater is its coagulating power
(d) One would expect charcoal to adsorb chlorine more strongly than hydrogen sulphide

45. The coagulation values of the electrolytes $AlCl_3$ and NaCl for As_2S_3 sol are 0.093 and 52 respectively. How many times has $AlCl_3$ greater coagulating power than NaCl?

(a) 100 (b) 559 (c) 993 (d) 852

46. A colloidal sol is formed by mixing $AgNO_3$ and KI in 2 : 1 molar ratio. This colloidal sol is to be coagulated by

I. NaI II. $MgSO_4$ III. $AlPO_4$

Molar ratio of these required to coagulate 1 mole colloidal sol is

(a) 1 : 1 : 1
(b) 1 : 2 : 3
(c) 3 : 2 : 1
(d) 6 : 3 : 2

Direction (Q. Nos. 47-50) *In the following questions, Assertion* (A) *followed by Reason* (R) *is given. Choose the correct option out of the following choices :*

(a) Both (A) and (R) are correct and (R) is not correct explanation of (A)
(b) Both (A) and (R) are correct but (R) is not correct explanation of (A)
(c) (A) is correct but (R) is incorrect
(d) Both (A) is and (R) are incorrect

47. Assertion (A) Reaction of white phosphorus with NaOH (*aq*) gives PH_3.

Reason (R) The reaction is disproportionation of P in alkaline medium.

48. Assertion (A) An ordinary filter paper impregnated with colloidion solution stops the flow of colloidal particle.

Reason (R) Pore size of the filter paper becomes more than the size of colloidal particle.

49. Assertion (A) For measuring resistance of an ionic solution, an AC source is used.

Reason (R) Concentration of ionic solution will change if DC source is used.

50. Assertion (A) The conversion of a fresh precipitate into a colloidal state by the action of solute or solvent is called peptisation.

Reason (R) Peptisation is a property of colloidal state.

ANSWERS

1 (b)	2 (d)	3 (d)	4 (a)	5 (a)	6 (b)	7 (a)	8 (c)	9 (c)	10 (b)
11 (a)	12 (c)	13 (c)	14 (d)	15 (a)	16 (c)	17 (d)	18 (c)	19 (b)	20 (c)
21 (b)	22 (c)	23 (c)	24 (b)	25 (c)	26 (d)	27 (c)	28 (b)	29 (c)	30 (b)
31 (a)	32 (c)	33 (c)	34 (d)	35 (a)	36 (c)	37 (b)	38 (b)	39 (c)	40 (b)
41 (d)	42 (a)	43 (b)	44 (a)	45 (b)	46 (d)	47 (a)	48 (c)	49 (a)	50 (c)

Hints and Explanations

1 Cell is completely discharged, it means equilibrium gets established, $E_{cell} = 0$

$$Zn \mid Zn^{2+}(1\text{ M}) \parallel Cu^{2+}(1\text{ M}) \mid Cu$$

Cell reaction :

$$Zn + Cu^{2+} \rightleftharpoons Zn^{2+} + Cu$$

$$K_{eq} = \frac{[Zn^{2+}]}{[Cu^{2+}]}$$

We know,

$$E_{cell} = E^\circ_{cell} - \frac{0.0591}{n}\log K_{eq}$$

$$E^\circ_{cell} = \frac{0.0591}{2}\log K_{eq}$$

or $$1.10 = \frac{0.0591}{2}\log K_{eq}$$

$$K_{eq} = \frac{[Zn^{2+}]}{[Cu^{2+}]} = \text{antilog}\,\frac{2.20}{0.0591}$$

$$= \text{antilog } 37.3$$

2 According to Kohlrausch's law

$$[NaOAc = CH_3\overset{\overset{\displaystyle O}{\|}}{C}—O^-Na^+]$$

$$\Lambda^\circ_{CH_3COOH} = \lambda^\circ_{CH_3COO^-} + \lambda^\circ_{H^+} \quad ...(i)$$

$$\Lambda^\circ_{HCl} = \lambda^\circ_{H^+} + \lambda^\circ_{Cl^-} \quad ...(ii)$$

$$\Lambda^\circ_{CH_3COONa} = \lambda^\circ_{CH_3COO^-} + \lambda^\circ_{Na^+} \quad ...(iii)$$

Thus, after adding Eqs. (ii) and (iii) if $\lambda^\circ_{Na^+}$ and $\lambda^\circ_{Cl^-}$ are subtracted, we can obtained the value of Λ°_{HOAc}. Thus, additional value required is Λ°_{NaCl}.

3 $$AgI(s) + e^- \rightleftharpoons Ag(s) + I^-; E^\circ = -0.152$$

$$Ag(s) \longrightarrow Ag^+ + e^-; E^\circ = -0.8$$

$$AgI(s) \longrightarrow Ag^+ + I^-; E^\circ = -0.952$$

$$E^\circ_{cell} = \frac{0.059}{n}\log K_{sp}$$

$$-0.952 = \frac{0.059}{1}\log K_{sp}$$

$$\log K_{sp} = -\frac{0.952}{0.059}$$

$$= -16.135 \approx -16.14$$

4 Fluoro group causes negative inductive effect increasing ionisation, thus 0.1 M difluoroacetic acid has highest electrical conductivity.

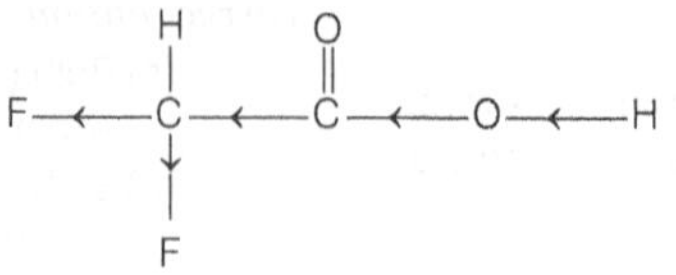

5 Resistance = specific resistance × cell constant

$$R_{(KCl)} = (\rho)_{KCl} \times \text{cell constant}$$

$$R_{(NaOH)} = (\rho)_{NaOH} \times \text{cell constant}$$

$$\frac{(R)_{KCl}}{(R)_{NaOH}} = \frac{(\rho)_{KCl}}{(\rho)_{NaOH}}$$

$$\frac{160}{190} = \frac{700}{(\rho)_{NaOH}}$$

$$(\rho)_{NaOH} = 831.25\ \Omega\text{ cm}$$

Specific conductance, $\kappa_{(NaOH)}$

$$= \frac{1}{831.25}$$

$$= 1.20 \times 10^{-3}\,\Omega^{-1}\text{cm}^{-1} \text{ or } 0.00120$$

6 Any cell (like fuel cell), works when potential difference is developed.

7 $$E^\circ_{Cr^{3+}/Cr^{2+}} = -0.41\text{ V}$$

$$E^\circ_{Mn^{3+}/Mn^{2+}} = +1.57\text{ V}$$

$$E^\circ_{Fe^{3+}/Fe^{2+}} = +0.77\text{ V}$$

$$E^\circ_{Co^{3+}/Co^{2+}} = +1.97\text{ V}$$

More negative value of E°_{red} indicates better reducing agent, thus easily oxidised. Therefore, oxidation of Cr^{2+} to Cr^{3+} is easiest.

8 $$Sn(s) + 2Fe^{3+}(aq) \longrightarrow 2Fe^{2+}(aq) + Sn^{2+}(aq)$$

$$E^\circ_{cell} = E^\circ_{ox} + E^\circ_{red}$$

$$= E^\circ_{Sn/Sn^{2+}} + E^\circ_{Fe^{3+}/Fe^{2+}}$$

Given, $E^\circ_{Sn^{2+}/Sn} = -0.14$ V

$\therefore \quad E^\circ_{Sn/Sn^{2+}} = +0.14$ V

$E^\circ_{Fe^{3+}/Fe^{2+}} = 0.77$ V

$\therefore \quad E^\circ_{cell} = 0.14 + 0.77 = 0.91$ V

9 $Zn(s) + 2H^+(aq) \longrightarrow Zn^{2+}(aq) + H_2(g)$

Reaction quotient, $Q = \dfrac{[Zn^{2+}]}{[H^+]^2}$

Corresponding cell is

$Zn \mid Zn^{2+}(C_1) \| H^+(aq) \mid Pt(H_2)$
Anode Cathode

and $E_{cell} = E^\circ_{cell} - \dfrac{0.0591}{2}\log K$

$= E^\circ_{cell} - \dfrac{0.0591}{2}\log\dfrac{[Zn^{2+}]}{[H^+]^2}$

$= E^\circ_{cell} + \dfrac{0.0591}{2}\log\dfrac{[H^+]^2}{[Zn^{2+}]}$

If H_2SO_4 is added to cathodic compartment (towards reactant side), Q decreases (due to increase in H^+). Hence, equilibrium is displaced towards right and E_{cell} increases.

10 $ZnO + C \longrightarrow Zn + CO, \Delta G^\circ = -ve$.

Hence, this is spontaneous.

11 $2A + B \longrightarrow$ products

$[B]$ is doubled, half-life did not change. Half-life is independent of change in concentration of reactant for first order reactions, i.e. first order w.r.t to B.

When $[A]$ is doubled, rate increases by two times

$\Rightarrow$ First order w.r.t A

Hence, net order of reaction = 1 + 1 = 2

Unit for the rate constant = $\text{conc.}^{(1-n)} t^{-1}$

$= (\text{mol L}^{-1})^{-1} \cdot s^{-1} = \text{L} \cdot \text{mol}^{-1} s^{-1}$

12 Given, $r \propto [CO]^2$

$r' \propto [2CO]^2$

$r' \propto 4[CO]^2$

Hence, $\dfrac{r'}{r} = \dfrac{4[CO]^2}{[CO]^2}; r' = 4r$

13

	$A \longrightarrow$	Product
Initially	a	0
After time t	$(a-x)$	x
After $t_{1/4}$	$\left(a - \dfrac{a}{4}\right)$	$\dfrac{a}{4}$

For first order kinetics,

$k = \dfrac{2.303}{t}\log\left(\dfrac{a}{a-x}\right)$

$\therefore \quad k = \dfrac{2.303}{t_{1/4}}\log\left(\dfrac{a}{3a/4}\right)$ and

$t_{1/4} = \dfrac{2.303 \log \frac{4}{3}}{k} = \dfrac{0.29}{k}$

14 $X \to Y$ is an endothermic reaction, $\Delta H = +ve$

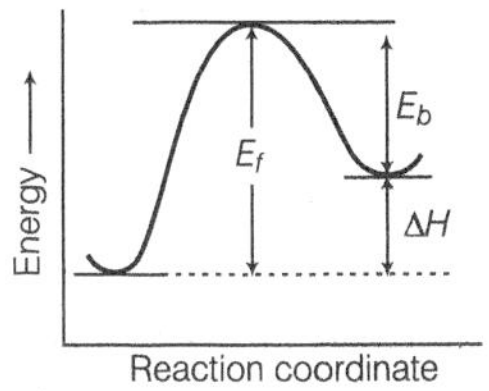

E_b = energy of activation of backward reaction

E_f = energy of activation of forward reaction

ΔH = heat of reaction

Thus, $E_f = E_b + \Delta H$

$\therefore \quad E_f > E_b$

15 Order = 1

Concentration changes from 0.8 M to 0.4 M (50%) in 15 min, thus half-life = 15 min = t_{50}

A change from 0.1 M to 0.025 M is 75% and for first order reaction

$t_{75} = 2 \times t_{50} = 2 \times 15 = 30$ min

or $\quad t_{50} = 15$ min

$k = \dfrac{2.303 \log 2}{t_{50}} = \dfrac{2.303 \log 2}{15}$

$a = 0.1$ M

$(a - x) = 0.025$ M

For first order,

$k = \dfrac{2.303}{t}\log\left(\dfrac{a}{a-x}\right)$

$\dfrac{2.303 \log 2}{15} = \dfrac{2.303}{t}\log\dfrac{0.1}{0.025}$

$= \dfrac{2.303}{t}\log 4$

$\therefore \dfrac{2.303 \log 2}{15} = \dfrac{2 \times 2.303 \log 2}{t};$

$t = 30$ min

16 $\log_{10} K = \log_{10} A - \dfrac{E_a}{2.303RT}$

Slope $= -\dfrac{E_a}{2.303R} = -\dfrac{1}{2.303}$

$\therefore \quad E_a = R = 2$ cal

17 Rate becomes x^y times, if concentration is made x times of a reactant giving y^{th} order reaction.

Rate $= k[A]^n[B]^m$

Concentration of A is doubled, hence $x = 2, y = n$ and rate becomes $= 2^n$ times.

Concentration of B is halved, hence $x = 1/2, y = m$ and rate becomes $= \left(\dfrac{1}{2}\right)^m$ times.

Net rate becomes $= (2)^n\left(\dfrac{1}{2}\right)^m$ times $= (2)^{n-m}$ times.

18 A. $Mn^{2+} \longrightarrow Mn^{4+} + 2e^-$ (Oxidation)

$H_2O_2 + 2e^- \longrightarrow 2H_2O$ (Reduction)

B. $3K \longrightarrow 3K^+ + 3e^-$ (Oxidation)

$Al^{3+} + 3e^- \longrightarrow Al$ (Reduction)

C. $3Fe \longrightarrow Fe_3O_4 + 8e^-$ (Oxidation)

$3x = 0 \quad 3x - 8 = 0$

$3x = 8$

$4H_2O + 8e^- \longrightarrow 4H_2$ (Reduction)

$8x - 8 = 0 \quad 8x = 0$

$8x = 8$

D. $3H_2S \longrightarrow 3S + 6e^-$ (Oxidation)

$2 + X = 0 \quad x = 0$

$X = -2$

$2NO_3^- + 6e^- \longrightarrow 2NO$ (Reduction)

$x - 6 = -1 \quad x - 2 = 0$

$x = 5 \quad x = 2$

Hence, (c) option is correct.

19 The osmotic pressure measurement gives the molecular mass of aggregated molecule.

20 $K\,[\,Cr\ O\ (O_2)\ (OH)]$

$+1 \ +x \ -2 \ -2 \ -1 = 0$

$x = 4$

$K_3[Cr\ (O_2)_4]$

$3 \ +y - 8 = 0, y = 5$

$(NH_3)_3[Cr\ (O_2)_2]$

$0 \ +z - 4 = 0$ thus, $z = 4$

$CrO_2\ Cl_2$

$w - 4 - 2 = 0, w = 6$

21 In adsorption, molecules of substances are accumulated in bulk also. Thus, the correct option is (b).

22 The sol is hydrophobic and hydrophobic sols are irreversible in nature.

23 $N_2H_4 \longrightarrow N_2(...) + 10e^-$

Oxidation number of two N-atoms = − 4 and oxidation number of two N-atoms in oxidised species = −4 + 10 = +6

Thus, oxidation state of N in new compound = + 3.

24 The process of separation of soluble impurities from a colloidal solution by means of a parchment paper (a suitable membrane) is called dialysis. When the process is done under the influence of electric field, it is called **electrodialysis**.

25 **Correct Reason** Arrhenius equation is applicable to collisions between simple as well as complex molecules.

26 Mercury cell gives a steady potential. This is because in the cell reaction, the ions are not involved in the solution whose concentration changes during its life time.

27 Precipitate of stannic oxide is converted to a stable sol of stannic oxide by the addition of dilute HCl (peptisation).

28 Physical as well as chemical adsorptions are favoured by high pressure. However, the difference is that decrease in pressure causes desorption in case of physical adsorption but not in the case of chemical adsorption.

29 Hemidialysis is used to clean the blood of patients having malfunctioned kidneys.

30 Atoms or molecules in multimolecular colloidal sols are held together by van der Waals' forces.

31 In alkaline medium SnO_2 is converted into negatively charged colloidal sol.

$$SnO_2 + 2OH^- \longrightarrow SnO_3^{2-} + H_2O$$

$$SnO_2 + SnO_3^{2-} \longrightarrow \underbrace{[SnO_2]\,SnO_3^{2-}}_{\text{Anion migrate to anode}}$$

32

$$\underset{1\text{ mol}}{[AgI]I^-} + \underset{1\text{ mol}}{AgNO_3} \longrightarrow 2AgI + NO_3^-$$

$$\underset{\substack{2\text{ mol}\\1\text{ mol}}}{2[AgI]I^-} + \underset{\substack{1\text{ mol}\\1/2\text{ mol}}}{Pb(NO_3)_2} \longrightarrow 2AgI\downarrow + PbI_2\downarrow + 2NO_3^-$$

$$\underset{\substack{3\text{ mol}\\1\text{ mol}}}{3[AgI]I^-} + \underset{\substack{1\text{ mol}\\1/3\text{ mol}}}{Fe^{3+}} \longrightarrow 3AgI + FeI_3$$

33 (c) $\Lambda^\circ_{BaSO_4} = \Lambda^\circ_{BaCl_2} + \Lambda^\circ_{H_2SO_4} - \Lambda^\circ_{HCl}$

$$= (x_1 + x_2 - x_3)$$

Equivalent conductance of $BaSO_4$

$$\Lambda^\circ_{BaSO_4} = \frac{1000 \times \text{specific conductance}}{\text{solubility (in saturated solution)}}$$

$$(x_1 + x_2 - x_3) = \frac{1000y}{\text{Solubility}}$$

$$\text{Solubility y }(BaSO_4) = \frac{1000\,y}{(x_1 + x_2 - x_3)}\text{ N}$$

$$= \frac{1000\,y}{2(x_1 + x_2 - x_3)}\text{ M}$$

$$BaSO_4 \rightleftharpoons Ba^{2+} + SO_4^{2-}$$

$$K_{sp}(BaSO_4) = [Ba^{2+}][SO_4^{2-}]\text{ M}^2$$

$$= \frac{10^6 y^2}{4(x_1 + x_2 - x_3)^2}$$

34 All statements are correct.

35 It is stabilised by casein which is a lyophilic colloidal sol.

36 Eq. mass of $MnO_4^- = \frac{\text{molar mass}}{7-2} = \frac{\text{molar mass}}{5}$

$$\text{Eq. mass of } C_2O_4^{2-} = \frac{\text{molar mass}}{2[4-3]} = \frac{\text{molar mass}}{2}$$

Meq. of $KMnO_4 = 50 \times 5 \times 0.04 = 10$

Meq. of $H_2C_2O_4 = 50 \times 2 \times 0.1 = 10$

37 $\because -\Delta G^\circ = nFE^\circ_{cell}$

Thus, the standard free energy decreases is maximum in cell

$$Mg\,|\,Mg^{2+}(1\text{ M})\,||\,Ag^+(1\text{ M})\,|\,Ag$$

38 t_{50} is independent of concentration of A. Hence, the reaction is of first order.

	$A \longrightarrow$	$N_2(g)$
At $t = 0$,	a	0
At time t,	$(a - x)$	$x = 10\text{L}$ [after 10 min]

In the given reaction, $a = 50$ L

$\therefore \quad (a - x) = 40$ L

$$\therefore \quad k = \frac{2.303}{10}\log\frac{50}{40} = \frac{2.303}{10}\log 1.25\text{ min}^{-1}$$

39 Gold number of haemoglobin = 0.03

10mL of gold sol requires haemoglobin to prevent coagulation = 0.03 mg

100 mL of gold sol will require haemoglobin to prevent coagulation = 0.3 mg

40 By Freundlich adsorption isotherm,

$$\left[\frac{x}{m}\right] = k(p)^{1/n}$$

$$\log\left[\frac{x}{m}\right] = \frac{1}{n}\log p + \log k$$

Compare with $y = mx + c$

$\therefore \quad$ Slope $= \tan\theta = \tan 45^\circ = 1$

$\therefore \quad \frac{1}{n} = 1$

$\therefore \quad n = 1$ and $\log k = 0.3010 = \log 2$

$$\log\left[\frac{x}{m}\right] = \log(0.2) + \log 2 = \log 0.4$$

$$\therefore \quad \left[\frac{x}{m}\right] = 0.4$$

41 $-\frac{1}{3}\frac{\Delta[BrO^-]}{\Delta t} = \frac{\Delta[BrO_3^-]}{\Delta t} = \frac{1}{2}\frac{\Delta[Br^-]}{\Delta t}$

$$\therefore \text{ Rate constant of } \frac{\Delta[BrO_3^-]}{\Delta t} = \frac{0.056}{3} = 0.019\text{ L mol}^{-1}\text{ s}^{-1}$$

and rate constant of

$$\frac{\Delta[Br^-]}{\Delta t} = \frac{2}{3} \times 0.056 = 0.037\text{ L mol}^{-1}\text{ s}^{-1}$$

42 CH_3COOH adsorbed

$= (0.06 - 0.042) \times 50 \times 60$ mg

$= 54$ mg

$\therefore$ CH_3COOH adsorbed per gram of charcoal

$$= \frac{54}{3} = 18\text{ mg}$$

43 $\left(\frac{dx}{dt}\right) = k[A]^2$

It represents second order reaction

$$k = \frac{1}{t}\left[\frac{1}{a-x} - \frac{1}{a}\right]$$

$$\frac{1}{(a-x)} = kt + \frac{1}{a}$$

Thus, graph between $(a - x)^{-1}$ and time t is a straight line ($y = mx + c$)

$\therefore \quad OP = \frac{1}{a} = 0.10\text{ L mol}^{-1}$

$\Rightarrow \quad a = 10\text{mol L}^{-1}$

$$\therefore \left(\frac{dx}{dt}\right) = 1 \times 10^{-2}\text{L mol}^{-1}\text{ min}^{-1} \times (10\text{ mol L}^{-1})^2 = 1.0\text{ mol L}^{-1}\text{ min}^{-1}$$

44 Brownian movement is more pronounced for smaller particles than for bigger ones.

45 (b) Coagulating power of the electrolyte is inversely proportional to its coagulating value.

Thus,

$$\frac{\text{coagulating power of } AlCl_3}{\text{coagulating power of NaCl}} = \frac{52}{0.093} = 559$$

Thus, coagulating power of $AlCl_3$ is 559 times more than that of NaCl.

46 $AgNO_3 + KI \longrightarrow \underset{\text{(excess)}}{AgI} \xrightarrow{AgNO_3} \underset{\text{Colloidal sol}}{[AgI]Ag^+}$

Sol is coagulated by anion

$\underset{1}{[AgI]Ag^+} + \underset{1}{NaI} \longrightarrow$

$\underset{1}{[AgI]\,Ag^+} + \frac{1}{2}MgSO_4 \longrightarrow$

$[AgI]Ag^+ + \frac{1}{3}AlPO_4 \longrightarrow$

Thus, ratio $\quad 1 : \frac{1}{2} : \frac{1}{3}$

$6 : 3 : 2$

47 $4\overset{0}{P} + 3NaOH + 3H_2O \longrightarrow 3NaH_2\overset{+1}{P}O_2 + \overset{-3}{P}H_3$

(disproportionation reaction)

48 An ordinary filter paper impregnated with colloidions solution stops the flow of colloidal particle. This is because the pore size of filter paper is less than the size of colloidal particles.

49 Both Assertion and Reason are correct and Reason is the correct explanation of Assertion.

Concentration of ionic solution changes on using DC current as a source of energy while on passing AC current, concentration does not change. Hence, AC source is used for measuring resistance.

50 Peptisation is a method to prepare gold sol.

DAY EIGHTEEN

Classification and Periodicity of Elements

Learning & Revision for the Day

- Earlier Attempts to Classify the Elements
- Periodic Law and Present form of Periodic Table
- Classification of Elements
- Periodic Trends in Properties of Elements

Earlier Attempts to Classify the Elements

Many attempts were made to classify the known elements from time to time. These are Prout's hypothesis, Dobereiner triads, Newland's law of octaves, Lothar Meyer's atomic volume curve, Mendeleef's periodic law (1869). The properties of elements are periodic functions of their atomic weight's and modern periodic law.

Modern Periodic Law and Present form of Periodic Table

- Henry Moseley showed that the atomic number is a more fundamental property of an element than its atomic mass and formulated modern periodic law.
- Modern periodic law can be stated as the physical and chemical properties of the elements. These are periodic functions of their atomic number.
- Modern periodic table is also called Bohr's periodic table and it is just graphical representation of Aufbau principle.
- In this table, elements are arranged in increasing order of atomic number (Z). The isotopes. These are all grouped together as they have the same atomic number.
- The table contains 7 periods (representing 7 orbits) and 18 groups (1-18). The concept of subgroup A and B is removed and groups are given number 1 to 18. The name of zero group is changed to group 18.

PREP MIRROR
Your Personal Preparation Indicator

- No. of Questions in Exercises (x)—
- No. of Questions Attempted (y)—
- No. of Correct Questions (z)—
  (Without referring Explanations)

- Accuracy Level ($z/y \times 100$)—
- Prep Level ($z/x \times 100$)—

In order to expect good rank in JEE, your Accuracy Level should be above 85 & Prep Level should be above 75.

s-Block Elements ← → | d-Block Elements | p-Block Elements | f-Block Elements

Group → 1 ← New Notation → 17, 18
IA ← CAS Version → VIIB, 0 (zero)

STATE: Gas [G], Liquid [L], Solid [S], Not found in nature [X]

Key to Chart: [G] 8 — Atomic number; O — Symbol; Oxygen — Name; 15.9994 — Atomic mass

Period ↓	1 IA	2 IIA	3 IIIA	4 IVA	5 VA	6 VIA	7 VIIA	8	9 VIII	10	11 IB	12 IIB	13 IIIB	14 IVB	15 VB	16 VIB	17 VIB	18 0 (zero)
1	1 [G] H Hydrogen 1.008																	2 [G] He Helium 4.003
2	3 [S] Li Lithium 6.941	4 [S] Be Beryllium 9.0121											5 [S] B Boron 10.811	6 [S] C Carbon 12.011	7 [G] N Nitrogen 14.007	8 [G] O Oxygen 15.999	9 [G] F Fluorine 18.998	10 [G] Ne Neon 20.180
3	11 [S] Na Sodium 22.999	12 [S] Mg Magnesium 24.305											13 [S] Al Aluminium 26.982	14 [S] Si Silicon 28.086	15 [S] P Phosphorus 30.914	16 [S] S Sulphur 32.066	17 [G] Cl Chlorine 35.453	18 [G] Ar Argon 39.948
4	19 [S] K Potassium 39.098	20 [S] Ca Calcium 40.078	21 [S] Sc Scandium 44.956	22 [S] Ti Titanium 47.867	23 [S] V Vanadium 50.942	24 [S] Cr Chromium 51.996	25 [S] Mn Manganese 54.938	26 [S] Fe Iron 55.845	27 [S] Co Cobalt 58.933	28 [S] Ni Nickel 58.693	29 [S] Cu Copper 63.546	30 [S] Zn Zinc 65.39	31 [S] Ga Gallium 69.723	32 [S] Ge Germanium 72.61	33 [S] As Arsenic 74.922	34 [S] Se Selenium 78.96	35 [L] Br Bromine 79.904	36 [G] Kr Krypton 83.30
5	37 [S] Rb Rubidium 85.468	38 [S] Sr Strontium 87. 62	39 [S] Y Yttrium 89.906	40 [S] Zr Zirconium 91.224	41 [S] Nb Niobium 92.906	42 [S] Mo Molybdenum 95.94	43 [X] Tc Technetium (98)	44 [S] Ru Ruthenium 101.07	45 [S] Rh Rhodium 102.906	46 [S] Pd Palladium 106.42	47 [S] Ag Silver 107.868	48 [S] Cd Cadmium 112.411	49 [S] In Indium 114.818	50 [S] Sn Tin 118.710	51 [S] Sb Antimony 121.760	52 [S] Te Tellurium 177.60	53 [S] I Iodine 126.904	54 [G] Xe Xenon 131.29
6	55 [S] Cs Cesium 132.905	56 [S] Ba Barium 137.327	57 [S] La Lanthanum 138.906	72 [S] Hf Hafnium 178.49	73 [S] Ta Tantalum 180.948	74 [S] W Tungsten 183.84	75 [S] Re Rhenium 186.207	76 [S] Os Osmium 190.23	77 [S] Ir Iridium 192.217	78 [S] Pt Platinum 195.078	79 [S] Au Gold 196.967	80 [L] Hg Mercury 200.59	81 [S] Tl Thallium 204.383	82 [S] Pb Lead 207.2	83 [S] Bi Bismuth 208.980	84 [S] Po Polonium (209)	85 [S] At Astatine (210)	86 [G] Rn Radon (222)
7	87 [S] Fr Francium (223)	88 [S] Ra Radium (226)	89 [S] Ac Actinium (227)	104 [X] Rf Rutherfordium (261)	105 [X] Db Dubnium (262)	106 [X] Sg Seaborgium (263)	107 [X] Bh Bohrium (262)	108 [X] Hs Hassium (265)	109 [X] Mt Meitnerium (266)	110 [X] Ds Darmstadtium (269)	111 [X] Rg Rontgenium (272)	112 [X] Cn Copernicium (227)	113 [X] Uut Ununtrium (284)	114 [X] Fl Flerovium (289)	115 [X] Uup Ununpentium (228)	116 [X] Lv Livermorium (292)	117 Uus Ununseptium (292)	118 [X] Uuo Ununoctium (294)

*Lanthanides	58 [S] Ce Cerium 140.116	59 [S] Pr Praseodymium 140.908	60 [S] Nd Neodymium 144.908	61 [X] Pm Promethium (145)	62 [S] Sm Samarium 150.36	63 [S] Eu Europium 151.964	64 [S] Gd Gadolinium 157.25	65 [S] Tb Terbium 158.925	66 [S] Dy Dysprosium 162.50	67 [S] Ho Holmium 164.930	68 [S] Er Erbium 167.26	69 [S] Tm Thulium 168.934	70 [S] Yb Ytterbium 173.04	71 [S] Lu Lutetium 174.967
**Actinides	90 [S] Th Thorium 232.038	91 [S] Pa Protactinium 231.036	92 [S] U Uranium 238.029	93 [S] Np Neptunium (237)	94 [S] Pu Plutonium (244)	95 [X] Am Americium (243)	96 [X] Cm Curium (247)	97 [X] Bk Berkelium (247)	98 [X] Cf Californium (251)	99 [X] Es Einsteinium (252)	100 [X] Fm Fermium (257)	101 [X] Md Mendelevium (258)	102 [X] No Nobelium (259)	103 [X] Lr Lawrencium (262)

☐ Metals
☐ Metalloids
☐ Non-metals

Periodic Table

Classification of Elements

Depending upon the orbital in which last electron enters, the elements are classified as follows:

1. s-Block Elements

- Group I (alkali metals) and group II (alkaline earth metals) elements belong to this block, their last electron enters in s-orbital.
- General electronic configuration of s-block elements is ns^{1-2}.

2. p-Block Elements

- Groups 13th to 18th excluding He, belong to this block. Their last electron enters in p-block.
- General electronic configuration of p-block elements is ns^2, np^{1-6}.
- s and p-block elements are known as representative elements or main group elements.

Diagonal Relationship

The first member of each group of s and p-block elements shows different characteristics from the rest of the members of the same group.

However, the first three elements of second period (Li, Be, B) shows diagonal similarity with elements (Mg, Al, Si) of third period placed on the right hand side. This is called diagonal relationship.

2nd period	Li	Be	B	C
	↘	↘	↘	
3rd period	Na	Mg	Al	Si

3. d-Block or Transition Elements

- Elements of group 3rd to 12th in periodic table belong to d-block. Their last electron enters in d-block.
- Two outermost shells of d-block elements are incomplete.
- General electronic configuration of d-block elements is $(n-1)d^{1-10}\ ns^{0-2}$

4. f-Block or Inner-Transition Elements

- General configuration $(n-2)f^{1-14}(n-1)d^{0-1}ns^2$.
- Last electron enters in f-orbital.
- Two series $4f$ (lanthanoids) and $5f$ (actinoids).
- Also known as rare earth elements.

IUPAC Nomenclature of Elements with Atomic Number >100

The names are derived by using roots for the three digits in the atomic number of the elements followed by adding '–ium' at the end. The roots for the numbers are as:

S.No.	Digit	Name	Abbreviation
1.	0	nil	n
2.	1	un	u
3.	2	bi	b
4.	3	tri	t
5.	4	quad	q
6.	5	pent	p
7.	6	hex	h
8.	7	sept	s
9.	8	oct	o
10.	9	enn	e

e.g.

Atomic number	Name
101	un-nilunium
	1 0 1

Periodic Trends in Properties of Elements

The elements in the periodic table are arranged in order of increasing atomic number. All of these elements display several other trends and we can use the periodic law and table formation to predict their chemical, physical and atomic properties.

Some of the physical and chemical properties of the elements are discussed below :

1. Atomic Radius

Atomic radius is defined as the distance from the centre of the nucleus to the outermost shell containing electrons. It is referred to both covalent or metallic radius depending on whether the element is a non-metal or a metal. On moving left to right in a period atomic radii decreases; on moving down the group atomic radii increases. Atomic radii are of three types

(i) **Covalent radius**

(a) In homodiatomic molecule, $(A-A)$

$$r = \frac{d_{A-A}}{2}$$

d_{A-A} = bond length or distance between two covalently bonded atoms.

(b) In heterodiatomic molecule $(A-B)$:

When $(\chi_A - \chi_B)$ is very small :

Bond length $= d_{A-B} = r_A + r_B$

r_A and r_B = covalent radii of A and B respectively

χ_A and χ_B = electronegativities of A and B

When $(\chi_A - \chi_B)$ is considerable.

$$d_{A-B} = r_A + r_B - 0.09\,(\chi_A - \chi_B)$$

(ii) **Metallic radius,** $r' = \frac{d}{2}$, where $r' >$ actual size of atom

where, d = distance between the adjacent metal ions in the metallic lattice

(iii) **van der Waals' radius,** $r_1 = \frac{d}{2}$

where, d = distance between the nuclei of identical two non-bonded isolated atoms.

$r_1 >>>$ actual size of atom

Covalent radius < Metallic radius < van der Waals' radius

At the end of the period, atomic radii of inert gases are exceptionally large because they do not form molecule and their radii are simply van der Waals' radii.

2. Ionic Radius

- Distance of the outermost shell of an ion from its nucleus is called ionic radius.
- Cationic radii is always smaller than its neutral atom while an ionic radii is always greater than its neutral atom.
- Greater the negative charge on ion,larger is its radii, e.g. $O^{2-} > O^- > O$.
- For isoelectronic species [possess same number of electrons] ionic radii $\propto \frac{1}{\text{magnitude of nuclear charge}}$.

3. Ionisation Enthalpy

It is the energy required when an electron is removed from the outermost orbit from an isolated gaseous atom, so as to convert it into gaseous cation.

Generally left to right in period IE increases; down the group it decreases but half-filled orbitals and fully-filled orbitals are highly stable and thus, have high IE.

Various factors with which ionisation energy varies are:

(i) Atomic size : varies inversely

(ii) Screening effect : varies inversely

(iii) Nuclear charge : varies directly

- Helium has the highest IE_1 while Cs has the lowest.
- IE values of inert gases are exceptionally higher due to stable configuration and smaller size while alkali metals have lowest IE values due to their large size.
- IE_1 of group 2 elements is greater than corresponding elements of group 13 due to their fully-filled s-orbitals.
- The effective nuclear charge experienced by valence electron in an atom will be less than the actual charge on the nucleus because of "shielding" or "screening" electrons of the valence electron from the nucleus by intervening electrons.
- Shielding (or **screening effect**) is the repulsion of valence electrons by the electrons in penultimate shell to reduce effective nuclear charge.

$$Z_{\text{eff}} = Z - \sigma$$

where, Z = atomic number, σ = shielding constant

σ = [0.35 × number of electrons in n^{th} shell excluding last (valence) electron] + [0.85 × number of electrons in $(n-1)^{\text{th}}$ shell] + [1.0 × number of electrons in inner shell]

4. Electron Gain Enthalpy

- It is the energy released when an electron is added to an isolated gaseous atom to form gaseous anion.
- Electron gain enthalpy increases [or becomes more negative] across the period while it decreases down the group.
- Chlorine has the highest electron affinity but oxidising power of fluorine is larger than chlorine.
- Various factors affecting electron gain enthalpy are:
  (i) Atomic size : varies inversely
  (ii) Nuclear charge : varies directly
  (iii) Configuration : Half-filled orbitals and fully-filled orbitals are highly stable, therefore electron gain enthalpy will be low.
- F and O-atoms have small size and high charge density, therefore have lower electron gain enthalpy.
- Electron gain enthalpies of elements of group 2 and 15 is low because of their stable half-filled configuration but for group 18 it is positive because of complete shell.

5. Electronegativity

- It is the tendency of an atom to attract the shared pair of electrons towards itself in a covalent bond. F is the most electronegative element while Cs is the least.
- Decreasing order of electronegativity

$$F > O > Cl \approx N > Br > S \approx C > I > H$$

In periods : left to right electronegativity increases.

In groups : down the group electronegativity decreases.

Measurement of Electronegativity

Some important scales to measure electronegativity of elements are

(i) Pauling Scale

Difference in the electronegativities of two atoms (A and B), $\chi_A - \chi_B = 0.102\sqrt{\Delta}$

where, Δ = actual bond energy – energy for 100% covalent bond.

$\therefore \quad \Delta = E_{A-B} - \sqrt{E_{A-A} \times E_{B-B}}$

Here, E_{A-B} = dissociation enthalpy of $A - B$ (kcal mol^{-1})

E_{A-A} = dissociation enthalpy of $A - A$ (kcal mol^{-1})

E_{B-B} = dissociation enthalpy of $B - B$ (kcal mol^{-1})

(ii) Mulliken's Scale

- Representation of Mulliken's scale is

$$\chi_M = \frac{\text{IE} + \text{EA}}{2}$$

- Mulliken values are ≈ 2.8 times greater than Pauling values

$$\chi_p = \frac{\text{IE} + \text{EA}}{5.6} \quad \text{or} \quad \chi_p = 0.336\,[\chi_M - 0.615]$$

- Unit of electronegativity = eV
- For noble gases, its value is taken as zero.

NOTE *sp* hybridised carbon is more electronegative than a sp^2 hybridised carbon, which in turn is more electronegative than sp^3 hybridised carbon.

6. Valency

- The valency of an element is related to the electronic configuration of its atom and usually determined by electrons present in valence shell (outer shell).
- In a period, valency of the elements with respect to hydrogen increases from 1 to 4 upto 14th group and then decreases to 1. e.g. NaH, CaH_2, AlH_3, SiH_4, PH_3, H_2S, HCl.

  But valency of the elements with respect to oxygen increases from one to seven along a period.
  e.g. Na_2O, CaO, Al_2O_3, SiO_2, P_4O_{10}, SO_3, Cl_2O_7
- All elements in a group have same valency as they have same number of electrons in their outer shell.
- *p*-block elements show variable valency on account of inert pair effect.
- Term oxidation state is also being used for covalence, which indicates the actual charge on atom in that particular molecule. It follows the same trend along period or group as valence.

NOTE Tin has maximum number of isotopes (10 in number) and Os has the highest oxidation state or valence (+8).

- Transition metals and inner-transition metals show variable valence of 1, 2 or 3 as they use electrons from outer as well as penultimate or inner-penultimate shell.

7. Chemical Reactivity

- Reactivity of metals increases with decrease in ionisation energy, electronegativity and increase with atomic radii and electropositive character. Reactivity of non-metals increases with increase in electronegativity and electron gain enthalpy and decreases with increase in atomic radii.
- On moving down the group, reactivity of metals increases while for non-metals it decreases.
- Reactivity of metals decreases while reactivity of non-metals increases across the period.
- The reducing character of the elements increases down the group. The oxidising character of the element decreases down the group. Alkali metals are most reactive, strongest reducing agent and have lowest ionisation enthalpy. Halogen are most reactive strongest oxidising agent and have higher ionisation enthalpy.
- Metals form basic oxides while non-metals form acidic oxides. Basic strength of the oxides or hydroxides increases down the group. Acidic nature of oxides increases along a period. Almost all metallic oxides are basic but ZnO and Al_2O_3 are amphoteric. Similarly non-metallic oxides are acidic but CO, N_2O and NO are neutral. Cl_2O_7 is the most acidic oxide.

DAY PRACTICE SESSION 1

FOUNDATION QUESTIONS EXERCISE

1 The period number in the long form of the periodic table is equal to

(a) magnetic quantum number of any element of the period
(b) atomic number of any element of the period
(c) maximum principal quantum number of any element of the period
(d) maximum azimuthal quantum number of any element of the period

2 Which pair of atomic number of elements has same chemical properties?

(a) 13, 22 (b) 3, 11 (c) 4, 24 (d) 2, 4

3 Both lithium and magnesium display several similar properties due to the diagonal relationship: however, the one which is incorrect is → JEE Main 2018

(a) both form basic carbonates
(b) both form soluble bicarbonates
(c) both form nitrides
(d) nitrates of both Li and Mg yield NO_2 and O_2 on heating

4 Atomic radii of fluorine and neon (in Å) are respectively given by

(a) 0.72, 1.60 (b) 1.60, 1.60
(c) 0.72, 0.72 (d) None of these

5 Which one of the following ions has the highest value of ionic radius?
(a) Li^+ (b) B^{3+} (c) F^- (d) O^{2-}

6 Which of the following series of elements have nearly the same atomic radii?
(a) F, Cl, Br, I (b) Na, K, Rb, Cs
(c) Li, Be, B, C (d) Fe, Co, Ni, Cu

7 Chloride ion and potassium ion are isoelectronic, then
(a) their sizes are same
(b) Cl^- ion is bigger than K^+ ion
(c) K^+ ion is relatively bigger
(d) their sizes depend on other cation and anion

8 The ions O^{2-}, F^-, Na^+, Mg^{2+} and Al^{3+} are isoelectronic. Their ionic radii show
(a) an increase from O^{2-} to F^- and then decrease from Na^+ to Al^{3+}
(b) a decrease from O^{2-} to F^- and then increase from Na^+ to Al^{3+}
(c) a significant increase from O^{2-} to Al^{3+}
(d) a significant decrease from O^{2-} to Al^{3+}

9 The ionic radii (in Å) of N^{3-}, O^{2-} and F^- respectively are → JEE Main 2015
(a) 1.36, 1.40 and 1.71 (b) 1.36, 1.71 and 1.40
(b) 1.71, 1.40 and 1.36 (d) 1.71, 1.36 and 1.40

10 The increasing order of the ionic radii of the given isoelectronic species is → AIEEE 2012
(a) Cl^-, Ca^{2+}, K^+, S^{2-} (b) S^{2-}, Cl^-, Ca^{2+}, K^+
(c) Ca^{2+}, K^+, Cl^-, S^{2-} (d) K^+, S^{2-}, Ca^{2+}, Cl^-

11 The correct sequence which shows decreasing order of the ionic radii of the elements is → AIEEE 2010
(a) $Al^{3+} > Mg^{2+} > Na^+ > F^- > O^{2-}$
(b) $Na^+ > Mg^{2+} > Al^{3+} > O^{2-} > F^-$
(c) $Na^+ > F^- > Mg^{2+} > O^{2-} > Al^{3+}$
(d) $O^{2-} > F^- > Na^+ > Mg^{2+} > Al^{3+}$

12 Which of the following properties shows a clear periodic variation along a group?
(a) First ionisation energy
(b) Molar mass of the element
(c) Number of isotopes of the atom
(d) All of the above

13 Which of the following atoms has the highest first ionisation energy? → JEE Main 2016
(a) Na (b) K (c) Sc (d) Rb

14 Generally the first ionisation energy increases along a period. But there are some exceptions. One which is not an exception is
(a) N and O (b) Na and Mg (c) Mg and Al (d) Be and B

15 A sudden jump between the values of second and third ionisation energy will be associated with the electronic configuration
(a) $1s^2, 2s^2\ 2p^6, 3s^1$ (b) $1s^2, 2s^2\ 2p^6, 3s^2 3p^1$
(c) $1s^2, 2s^2\ 2p^6, 3s^2\ 3p^6$ (d) $1s^2, 2s^2\ 2p^6, 3s^2$

16 The first ionisation potential of Na, Mg and Si are respectively 496, 737 and 786 kJ mol^{-1}. The ionisation potential of Al will be closer to
(a) 760 kJ mol^{-1} (b) 575 kJ mol^{-1}
(c) 801 kJ mol^{-1} (d) 419 kJ mol^{-1}

17 Which of the following represents the correct order of increasing first ionisation enthalpy for Ca, Ba, S, Se and Ar? → JEE Main 2013
(a) S < Se < Ca < Ba < Ar (b) Ba < Ca < Se < S < Ar
(c) Ca < Ba < S < Se < Ar (d) Ca < S < Ba < Se < Ar

18 The first ionisation potential of Na is 5.1 eV. The value of electron gain enthalpy of Na^+ will be → JEE Main 2013
(a) – 5.1 eV (b) – 10.2 eV (c) + 2.55 eV (d) – 2.55 eV

19 The correct order of first ionisation potential of carbon, nitrogen, oxygen and fluorine is
(a) C > N > O > F (b) O > N > F > C
(c) O > F > N > C (d) F > N > O > C

20 The electronic configuration of four elements are given below. Arrange these elements in the correct order of magnitude (without sign) of their electron affinity.
I. $2s^2\ 2p^5$ II. $3s^2\ 3p^5$
III. $2s^2 2p^4$ IV. $3s^2\ 3p^4$
Select the correct answer using the codes given below :
(a) I < II < IV < III (b) II < I < IV < III
(c) I < III < IV < II (d) III < IV < I < II

21 Which is the correct order of second ionisation potential of C, N, O and F in the following? → JEE Main (Online) 2013
(a) O > N > F > C (b) O > F > N > C
(c) F > O > N > C (d) C > N > O > F

22 The successive ionisation energy values for an element are given below:
I. 1st ionisation energy = 410 kJ mol^{-1}
II. 2nd ionisation energy = 820 kJ mol^{-1}
III. 3rd ionisation energy = 1100 kJ mol^{-1}
IV. 4th ionisation energy = 1500 kJ mol^{-1}
V. 5th ionisation energy = 3200 kJ mol^{-1}
Find out the number of valence electron for the atom *x*.
(a) 4 (b) 3 (c) 5 (d) 2

23 The incorrect statements among the following is
(a) the first ionisation potential of Al is less than the first ionisation potential of Mg
(b) the second ionisation potential of Mg is greater than the second ionisation potential of Na
(c) the first ionisation potential of Na is less than the first ionisation potential of Mg
(d) the third ionisation potential of Mg is greater than the third ionisation potential of Al

24 Two elements whose electronegativities are 1.2 and 3.0, the bond formed between them would be
(a) ionic (b) covalent
(c) coordinate (d) metallic

25 *AB* is predominantly ionic as A^+B^- if (IE stands for ionisation energy, EA for electron affinity and EN for electronegativity). Then,

(a) $(IP)_A < (IP)_B$ (b) $(EA)_A < (EA)_B$
(c) $(EN)_A < (EN)_B$ (d) $(IP)_B < (IP)_A$

26 What will be the electronegativity of carbon at Pauling scale?

Given that $E_{H-H} = 104.2$ kcal mol^{-1},
$E_{C-C} = 83.1$ kcal mol^{-1}
$E_{C-H} = 98.8$ kcal mol^{-1}
Electronegativity of hydrogen = 2.1

(a) 0.498 (b) 0.598
(c) 2.134 (d) 2.598

27 The statement that is not correct for the periodic classification of elements, is

(a) the properties of the elements are the periodic function of their atomic number
(b) non-metallic elements are lesser in number than metallic elements
(c) the first ionisation energies of elements along a period do not vary in a regular manner with increase in atomic number
(d) for transition elements the *d*-subshells are filled with electrons monotonically with increase in atomic numbers

28 Match the element (in Column I) with its unique properties (in Column II).

	Column I		Column II
A.	F	1.	Maximum ionisation energy
B.	Cl	2.	Maximum electronegativity
C.	Fe	3.	Maximum electron affinity
D.	He	4.	Recently named by IUPAC
E.	Ds	5.	Variable valence

Codes

	A	B	C	D	E
(a)	5	4	1	2	3
(b)	3	4	2	1	5
(c)	2	3	5	1	4
(d)	3	1	4	2	5

29 Valence electrons in the element *A* are 3 and that in element *B* are 6. Most probable compound formed from *A* and *B* is

(a) A_2B (b) AB_2
(c) A_6B_3 (d) A_2B_3

30 The oxide of an element whose electronic configuration is $1s^2 2s^2 2p^6 3s^1$ is

(a) neutral (b) amphoteric
(c) basic (d) acidic

31 Which among the following factors is the most important in making fluorine the strongest oxidising agent ?

(a) Electron affinity (b) Ionisation enthalpy
(c) Hydration enthalpy (d) Bond dissociation energy

32 The element with which of the following outer electron configuration may exhibit the largest number of oxidation states in its compounds → JEE Main (Online) 2013

(a) $3d^5 4s^2$ (b) $3d^8 4s^2$ (c) $3d^7 4s^2$ (d) $3d^6 4s^2$

33 Which one of the following orders present the correct sequence of the increasing basic nature of the given oxides? → AIEEE 2011

(a) $Al_2O_3 < MgO < Na_2O < K_2O$
(b) $MgO < K_2O < Al_2O_3 < Na_2O$
(c) $Na_2O < K_2O < MgO < Al_2O_3$
(d) $K_2O < Na_2O < Al_2O_3 < MgO$

34 Considering the elements F, Cl, O and N, the correct order of their chemical reactivity in terms of oxidising property is

(a) F > Cl > O > N (b) F > O > Cl > N
(c) Cl > F > O > N (d) O > F > N > Cl

Direction (Q. Nos. 35-39) *In the following questions, Assertion (A) followed by a Reason (R) is given. Choose the correct answer out of the following choices.*

(a) Assertion and Reason both are correct statements and Reason is the correct explanation of the Assertion
(b) Assertion and Reason both are correct statements but Reason is not the correct explanation of the Assertion
(c) Assertion is correct incorrect and Reason is incorrect
(d) Both Assertion and Reason are incorrect

35 **Assertion** (A) Boron has a smaller first ionisation enthalpy than beryllium.

Reason (R) The penetration of 2*s* electron to the nucleus is more than the 2*p* electron hence 2*p* electron is more shielded by the inner core of electrons than the 2*s* electrons.

36 **Assertion** (A) Electron gain enthalpy becomes less negative as we go down a group.

Reason (R) Size of the atom increases on going down the group and the added electron would be farther from the nucleus.

37 **Assertion** (A) Cesium and fluorine both reacts violently.

Reason (R) Cesium is most electropositive and fluorine is most electronegative.

38 **Assertion** (A) The atomic radii of the elements of the oxygen family are smaller than the atomic radii of the corresponding elements of the nitrogen family.

Reason (R) The members of the oxygen family are more electronegative and thus, have lower values of nuclear charge than those of the nitrogen family.

39 **Assertion** (A) Fluorine has a less negative electron affinity than chlorine.

Reason (R) There is relatively greater effectiveness of 2*p* electrons in the small fluorine atom to repel the additional electron entering the atom than to 3*p* electrons in the larger Cl atom.

DAY PRACTICE SESSION 2

PROGRESSIVE QUESTIONS EXERCISE

1 The formation of the oxide ion, O^{2-} (*g*) from oxygen atom requires first an exothermic and then an endothermic step as shown below:

$$O(g) + e^- \rightarrow O^-(g);\ \Delta H^\ominus = -141\ \text{kJ mol}^{-1}$$
$$O^-(g) + e^- \rightarrow O^{2-}(g);\ \Delta H^\ominus = +780\ \text{kJ mol}^{-1}$$

Thus, process of formation of O^{2-} in gas phase is unfavourable even though O^{2-} is isoelectronic with neon. It is due to the fact that,

(a) oxygen is more electronegative
(b) addition of electron in oxygen results in larger size of the ion
(c) electron repulsion outweights the stability gained by achieving noble gas configuration
(d) O^- ion has comparatively smaller size than oxygen atom

2 The increasing order of the first ionisation enthalpies of the elements B, P, S, and F is

(a) F < S < P < B (b) P < S < B < F
(c) B < P < S < F (d) B < S < P < F

3 *A*, *B*, *C* are elements in the third short period. Oxide of *A* is ionic, that of *B* is amphoteric and of *C* a giant molecule. *A*, *B* and *C* will have atomic number in the order

(a) $A < B < C$ (b) $C < B < A$
(c) $A < C < B$ (d) $B < A < C$

4 Consider the isoelectronic species, Na^+, Mg^{2+}, F^- and O^{2-}. The correct order of increasing length of their radii is?

(a) $F^- < O^{2-} < Mg^{2+} < Na^+$ (b) $Mg^{2+} < Na^+ < F^- < O^{2-}$
(c) $O^{2-} < F^- < Na^+ < Mg^{2+}$ (d) $O^{2-} < F^- < Mg^{2+} < Na^+$

5 Mixture containing aqueous Li^+, Na^+, K^+ ions are electrolysed. Cations are discharged at cathode in the order

$$M^+(aq) + e^- \longrightarrow M$$

(a) Li^+, Na^+, K^+ (b) K^+, Na^+, Li^+
(c) Li^+, K^+, Na^+ (d) Na^+, K^+, Li^+

6 The electronic configuration of an element is $1s^2 2s^2 2p^6$, $3s^2 3p^3$. The atomic number and the group number of the element '*X*' which is just below the above element in the periodic table are respectively

(a) 23 and 5 (b) 23 and 15 (c) 33 and 15 (d) 33 and 5

7 The atomic number of elements *A*, *B*, *C* and *D* are $Z-1$, Z, $Z+1$ and $Z+2$ respectively. If *B* is a noble gas, choose the correct option from the following options.

I. *A* has highest electron affinity
II. *C* exists in +2 oxidation state
III. *D* is an alkaline earth metal

(a) I and II (b) II and III (c) I and III (d) I,II and III

8 Ionisation potential and electron affinity of fluorine are 17.42 and 3.45 eV respectively. What will be the ratio of electronegativity of fluorine on Mulliken scale and Pauling scale ?

(a) 2.80 : 1 (b) 1.80 : 1 (c) 2.40 : 1 (d) 4.20 : 1

9 An extra electron is added at the periphery of nitrogen atom in the formation of anion. What will be the ratio of effective nuclear charge of nitrogen atom and nitrogen anion ?

(a) 2.09 : 1 (b) 0.09 : 1 (c) 3.09 : 1 (d) 1.09 : 1

10 The atomic numbers of vanadium (V), chromium (Cr), manganese (Mn) and iron (Fe) are respectively 23, 24, 25 and 26. Which one of these may be expected to have the highest second ionisation enthalpy ?

(a) V (b) Cr (c) Mn (d) Fe

11 The IE values of $Al(g) \longrightarrow Al^+ + e^-$ is 577.5 kJ mol^{-1} and ΔH for $Al(g) \longrightarrow Al^{3+}(g) + 3e^-$ is 5140 kJ mol^{-1}.
If the ratio of second and third IE is 2 : 3, the values of IE_2 and IE_3 are respectively

(a) 1825 and 2737.5 kJ mol^{-1}
(b) 182.5 and 273.75 kJ mol^{-1}
(c) 1825 and 2700 J mol^{-1}
(d) 2737.5 and 1825 kJ mol^{-1}

12 Following transition elements $(IE)_1$ drops abruptly (Ga, In and Tl). This is due to

(a) decrease in effective nuclear charge
(b) increase in atomic radius
(c) removal of an electron from the singly occupied *np*-orbitals of higher energy than the *ns*-orbitals of Zn, Cd and Hg
(d) None of the above is correct

13 Following statements regarding the periodic trends of chemical reactivity to the alkali metals and the halogens are given. Which of these statements give the correct picture?

(a) The reactivity decreases in the alkali metals but increases in the halogens with increase in atomic number down the group
(b) In both the alkali metals and the halogens the chemical reactivity decreases with increase in atomic number down the group
(c) Chemical reactivity increases with increase in atomic number down the group in both the alkali metals and halogens
(d) In alkali metals the reactivity increases but in the halogens it decreases with increase in atomic number down the group

14 In which of the following arrangements the order is not according to the property indicated against it ?
(a) Li < Na < K < Rb : increasing metallic radius
(b) I < Br < F < Cl : increasing electron gain enthalpy (with negative sign)
(c) B < C < N < O : increasing first ionisation enthalpy
(d) $Al^{3+} < Mg^{2+} < Na^{+} < F^{-}$: increasing ionic size

15 Consider the following statements.
I. Cs^+ is highly hydrated than the other alkali metal ions.
II. Among the alkali metals Li, Na, K and Rb, lithium has the highest melting point.
III. Among the alkali metals only lithium forms a stable nitride by direct combination.

Of these statements
(a) I, II and III are correct
(b) I and II are correct
(c) I and III are correct
(d) II and III are correct

ANSWERS

SESSION 1	**1** (c)	**2** (b)	**3** (a)	**4** (a)	**5** (d)	**6** (d)	**7** (b)	**8** (d)	**9** (b)	**10** (c)
	11 (d)	**12** (a)	**13** (c)	**14** (b)	**15** (d)	**16** (b)	**17** (b)	**18** (a)	**19** (d)	**20** (d)
	21 (b)	**22** (a)	**23** (b)	**24** (a)	**25** (a)	**26** (d)	**27** (d)	**28** (c)	**29** (d)	**30** (c)
	31 (c)	**32** (a)	**33** (a)	**34** (a)	**35** (a)	**36** (a)	**37** (a)	**38** (c)	**39** (a)	
SESSION 2	**1** (c)	**2** (d)	**3** (a)	**4** (b)	**5** (b)	**6** (c)	**7** (c)	**8** (a)	**9** (d)	**10** (b)
	11 (a)	**12** (c)	**13** (d)	**14** (c)	**15** (d)					

Hints and Explanations

SESSION 1

1 Since, each period starts with the filling of electrons in a new principal quantum number, therefore, the period number in the long form of the periodic table refers to the maximum principal quantum number of any element in the period.

Period number = maximum n of any element

[where, n = principal quantum number]

2 The pair which belongs to same group, i.e. in which both the elements have same outer electronic configuration has same chemical properties.

$$3 \Rightarrow 1s^2, 2s^1$$

$$11 \Rightarrow 1s^2, 2s^2\, 2p^6, 3s^1$$

3 Mg can form basic carbonate, while Li cannot.

$$5Mg + 6CO_3^{2-} + 7H_2O \longrightarrow 4MgCO_3 \cdot Mg(OH)_2 \cdot 5H_2O + 2HCO_3^-$$

4 Atomic radii of neon (1.60) being van der Waals' radius is higher than that of fluorine (0.72) which is infact covalent radius. Also Ne is an inert gas all the orbital are completely filled and hence the interelectronic repulsion is maximum.

5 The ionic radii of cation is smaller than its parent atom while the size of anion is larger than its parent atom. Greater the negative charge on the ion, larger is its radii. Hence, O^{2-} has the highest value of ionic radius.

6 The atomic and ionic radii of transition elements in a given series show a decreasing trend for first five elements and then becomes almost constant for next five elements. i.e. Fe, Co, Ni, Cu

7 In isoelectronic species, anion is larger in size than the cation. As the magnitude of nuclear charge increases, force of attraction by the nucleus on the electrons increases and hence, ionic radius decreases. Therefore, Cl^- ion is bigger than K^+ ion.

8 Ionic radii $\propto \frac{1}{Z}$ for isoelectronic ion.

Thus, $O^{2-} > F^- > Na^+ > Mg^{2+} > Al^{3+}$
(Z) 8 9 11 12 13

9 N^{3-}, O^{2-} and F^- are isoelectronic species with 10 electrons. For isoelectronic species,

$$\text{Ionic radii} \propto \frac{1}{\text{nuclear charge}}.$$

Thus, the correct order of their radii would be

$$\underset{(1.36)}{F^-} < \underset{(1.40)}{O^{2-}} < \underset{(1.71)}{N^{3-}}$$

10 For isoelectronic species, $r_n \propto \frac{1}{Z}$

Species	Z	Electrons
Cl^-	17	18
Ca^{2+}	20	18
K^+	19	18
S^{2-}	16	18

Thus, ionic size is in order

$$Ca^{2+} < K^+ < Cl^- < S^{2-}.$$

11 $O^{2-}, F^-, Na^+, Mg^{2+}$ and Al^{3+} are isoelectronic species and higher the nuclear charge, smaller the radii of isoelectronic species.

12 First ionisation energy shows a clear periodic variation along a group. It decreases down the group.

13 As, all the given elements belong to IA group except Sc, their ionisation energy decreases down the group. Order of first ionisation energy is
Sc > Na > K > Rb.
Due to poor shielding effect, removal of one electron form 4*s* orbital is difficult as compared to 3*s* orbital.

14 Na and Mg belongs to 2nd period, IA and IIA group. On moving from left to right, first IE increases, therefore IE of Mg is greater than that of Na.

15 As the sudden jump in ionisation is observed between IE_2 and IE_3 thus it has 2 valence electrons, so its electronic configuration must be $1s^2\ 2s^2\ 2p^6\ 3s^2$.

16 Ionisation potential (I.P) increases from left to right while in case of elements possessing fully filled or half-filled orbitals, their ionisation potential is more as compared to the element having unstable configuration.
Since, Na, Mg, Al and Si all belongs to the same period thus, the order of their IP values would be Na<Mg<Al<Si. But Mg possesses fully filled $3s^2$ orbitals, thus, its ionisation potential value would be larger than that of Al. Thus, the correct order would be : Na<Al<Mg<Si. Hence, ionisation potential value of Al would lie between Na and Mg, i.e. 496 and 737. Thus, ionisation potential of Al should be 575 $kJ\,mol^{-1}$.

17 Ionisation energy increases along a period from left to right and decreases down a group. Therefore, increasing order of first ionisation enthalpy is,
Ba < Ca < Se < S < Ar.

18 $Na \underset{\Delta H = -5.1\,eV}{\overset{\Delta H = +5.1\,eV}{\rightleftharpoons}} Na^+ + e^-$, here the backward reaction releases same amount of energy as that of forward reaction and known as electron gain enthalpy.

19 As the first ionisation potential increases down the group, the observed order for increasing IP should be C <N <O <F, but the correct order is C <O <N <F. This is because of the half-filled stable configuration of nitrogen ($1s^2 2s^2 2p^3$), which has more IP than that of oxygen.

20 In general electron affinity increases down the group. Electron affinity for the elements of third period are higher than those of second period due to their small size and high electron density and thus added electron, experience large inter electronic repulsion resulting in release of lower amount of energy. The increasing order of electron affinity is
$$\underset{III}{2s^2 2p^4} < \underset{IV}{3s^2 3p^4} < \underset{I}{2s^2 2p^5} < \underset{II}{3s^2 3p^5}.$$

21 Second ionisation potential represents the amount of energy required to remove an electron from a unipositive ion. The electronic configuration of unipositive ions of given atoms is
$$C^+ (6-1=5) = 1s^2, 2s^2, 2p^1$$
$$N^+ (7-1=6) = 1s^2, 2s^2, 2p^2$$
$$O^+ (8-1=7) = 1s^2, 2s^2, 2p^3$$
$$F^+ (9-1=8) = 1s^2, 2s^2, 2p^4$$
From the electronic configuration, it is clear that the second ionisation potential of O is highest because here the electron which is to be removed belongs to half-filled (so more stable) *p*-orbital. Moreover, along a period, as the number of electrons increases, size decreases, effective nuclear charge increases and hence, ionisation potential increases. Thus, the correct order of second ionisation potential is O>F>N>C.

22 In the given values, there is a biggest jump between IE_4 and IE_5. Hence, there are 4 valence electrons for the atom *X*.

23 IE_2 of Mg is lower than that of Na because in case of Mg^{2+}, 3*s* electron has to be removed while in case of Na^+, an electron from the stable inert gas configuration (neon) has to be removed.

24 When the electronegativity difference is more than 1.7, the bond formed will be ionic in nature.

25 $A \longrightarrow A^+ + e^-$, if $(IE)_A < (IE)_B$
$B + e^- \longrightarrow B^-$, if $(EA)_B > (EA)_A$
$$(EN) = \frac{(IE) + (EA)}{2}$$
A^+B^- is formed if $(EN)_A < (EN)_B$

26 $\chi_C - \chi_H = 0.208\sqrt{\Delta}$
where, $\Delta = E_{C—H} - \sqrt{E_{C—C} \times E_{H—H}}$
$\Delta = 98.8 - \sqrt{83.1 \times 104.2}$
$\therefore$ $\Delta = 5.75$
$\chi_C - 2.1 = 0.208\sqrt{5.75}$
$\chi_C - 2.1 = 0.497$
$\chi_C = 2.598$

27 For transition elements, *d*-subshells are not filled with electrons monotonically (one at a time) because of extra stability of d^5 and d^{10} configuration. Besides these, there are many other exceptions, e.g. Nb, Mo, Ru, Rh, Pb etc.

28 A → 2; B → 3 : C → 5; D → 1; E → 4

29 Valence electrons in *A* = 3
Valence electrons in *B* = 6
Thus, *A* is electropositive and *B* is electronegative; *A* can lose three electrons and *B* can gain two electrons to attain stable configuration.
Thus, *A* exists as A^{3+} and *B* as B^{2-}
Thus, compound is A_2B_3.

30 Na (11) → $1s^2 2s^2 2p^6 3s^1$. It is an alkali metal whose oxide are basic in nature.

31 Due to small size of fluorine, it has high hydration energy which makes it strongest oxidising agent.

32 Due to the presence of maximum number of unpaired electrons, element having $3d^5, 4s^2$, configuration may exhibit the largest number of oxidation states in its compounds.

33 Oxides when dissolved in water form hydroxides.
$$Na_2O + H_2O \longrightarrow 2\,NaOH$$
$$MgO + H_2O \longrightarrow Mg(OH)_2$$
$$Al_2O_3 + 3H_2O \longrightarrow 2Al(OH)_3$$
Smaller the size of cation and larger the charge, greater the polarising power of cation, hence greater the covalent character and lesser the basic nature.

Oxides Basic: Decreases in a period (←); Increases in a group (↓)

Charge $Al^{3+} > Mg^{2+} > Na^+ = K^+$
Size $Al^{3+} < Mg^{2+} < Na^+ < K^+$

Polarising power

$Al^{3+} > Mg^{2+} > Na^{+} > K^{+}$

Basic nature of oxides

$Al_2O_3 < MgO < Na_2O < K_2O$

34 Oxidising property depends on the easiness with which the element is reduced, i.e. it gain electrons which in turn is its electronegativities. Electronegativity F > Cl > O > N. Thus, correct option is (a).

35 Boron has a smaller first ionisation enthalpy than beryllium because the penetration of $2s$ electron to the nucleus is more than the $2p$ electron. Hence, $2p$ electron is more shielded by the inner core of electron than the $2s$ electron.

36 Electron gain enthalpy becomes less negative as the size of an atom increases down the group. This is because within a group screening effect increases on going down in a group and the added electron would be farther away from the nucleus.

37 Cesium and fluorine both reacts violently because cesium is most electropositive and fluorine is most electronegative.

38 The atomic radii of the elements of oxygen family are smaller than atomic radii of the corresponding elements of the nitrogen family because of increase in effective nuclear charge the results in the increased attraction of electrons to the nucleus.

39 Fluorine has a less negative electron affinity than chlorine. There is relatively greater effectiveness of $2p$ electrons in the small fluorine atom to repel the additional electron entering the atom than to $3p$ electrons in the larger Cl atom.

SESSION 2

1 The process of formation of O^{2-} in gas phase is unfavourable even though O^{2-} is isoelectronic with neon because electron repulsion outweights the stability gained by achieving noble gas configuration.

2 Examine the positions in periodic table?

B C N O F

P S

The ionisation enthalpy increases across the period but decreases down the group. Phosphorus has stable half-filled configuration. Hence, order is

B < S < P < F.

3 The correct order for atomic number is

$A < B < C$

4 For isoelectronic,

$$\text{Ionic radii} \propto \frac{1}{\text{Magnitude of nuclear charge}}$$

The correct order for radii is

$Mg^{2+} < Na^{+} < F^{-} < O^{2-}$.

5 Greater the hydration, smaller the ionic mobility of cation to move towards cathode, Li, is heavily hydrated, thus has least ionic mobility and hence, discharged at the end.

K^{+} is least hydrated, due to its large size, has maximum ionic mobility and is discharged first.

6 Atomic number of the given element = 15

Group number = 10 + 5 (valence electron) = 15

Period = 3

Atomic number of the element just below = 33

7 Atomic number of element $B = Z$

[B is noble gas, so it belongs to zero group]

Atomic number of element, $A = Z - 1$ [i.e. halogens]

Atomic number of element, $C = Z + 1$ [i.e. group IA]

Atomic number of element $D = Z + 2$ [i.e. group IIA]

Thus, element A must be a halogen i.e. it has highest electron affinity and element C must be an alkali metal and exist in +1 state and element D must be an alkaline earth metal with + 2 state.

8 $(\chi_F)_M = \dfrac{IP + EA}{2}$

$(\chi_F)_M = \dfrac{17.42 + 3.45}{2} = 10.435$

$(\chi_F)_P = \dfrac{10.435}{2.8} = 3.726$

$(\chi_F)_M : (\chi_F)_P = 2.80 : 1$

9 $Z_{eff} = Z - \sigma$; σ = [0.35 × electrons in nth shell excluding last electron] + [0.85 × electrons in $(n-1)$ shell]

(i) For N atom = $1s^2, 2s^2 2p^3$

$\therefore \quad \sigma = [4 \times 0.35] + [2 \times 0.85] = 3.10$

$\therefore \; Z_{eff} = 7 - 3.10 = 3.90$

(ii) For N^- ion $1s^2, 2s^2, 2p^4$

$\sigma = [5 \times 0.35] + [2 \times 0.85] = 3.45$

$\therefore Z_{eff} = 7 - 3.45$

$= 3.55$

Z_{eff} (N atom) : Z_{eff} (N^- anion)

1.09 : 1

10

$Fe^{+}(25) = [Ar]\, 3d^6, 4s^1$

$Mn^{+}(24) = [Ar]\, 3d^5, 4s^1$

$V^{+}(22) = [Ar]\, 3d^3, 4s^1$

$Cr^{+}(23) = [Ar]\, 3d^5, 4s^0$

By first ionisation potential, the electronic configuration of Cr^{+} is most stable, hence formation of Cr^{2+} by second ionisation potential requires maximum enthalpy.

11 Given, IE_1 of Al = 577.5 kJ mol^{-1} ...(i)

$[IE_1 + IE_2 + IE_3]$ of Al = 5140 kJ mol^{-1} ...(ii)

Given, $IE_2 = (2/3)\, IE_3$ ...(iii)

Solving (i), (ii) and (iii), we get

$IE_2 = 1825$ kJmol^{-1}

$IE_3 = 2737.5$ kJmol^{-1}

12 It is due to the removal of an electron from the singly occupied np-orbitals of higher energy than ns-orbitals of Zn, Cd and Hg.

13 In alkali metals reactivity increases down the group as electropositivity increases but for halogens reactivity decreases down the group as molecular stability of elements increases.

14
- Metallic radii increase in a group from top to bottom. Thus, Li < Na < K < Rb is true.
- Electron gain enthalpy of Cl > F and decreases along a group. Thus, I < Br < F < Cl is true.
- Ionisation enthalpy increases along a period from left to right but due to presence of stable half-filled orbitals in N, ionisation enthalpy of N > O. Thus, B < C < N < O is incorrect.

15 **Correct Statement I** Cs^{+} because of its large size is least hydrated.

DAY NINETEEN

General Principles and Processes of Isolation of Metals

Learning & Revision for the Day

- Occurrence of Elements in Nature
- Minerals and Ores
- Metallurgy
- Extraction of Metals
- Thermodynamics and Electrochemical Principles of Metallurgy
- Refining of Crude Metal
- Occurrence and Extraction of Some Important Metals

About 118 elements are known till today. Among them, few elements occur in free state while others in combined form. Some general principles are adopted for extraction and isolation of an element from its combined form.

Occurrence of Elements in Nature

- Earth crust is the source of many elements. Aluminium is the most abundant metal of earth crust and iron comes second. The percentage of different elements in earth crust is O-49%, Si-26%, Al-7.5%, Fe-4.2%, Ca-3.2%, Na-2.4%, K-2.3%, Mg-2.3%, H-1%.
- Metals occur in two forms in nature (i) in native state and (ii) in combined state.

Metal	Ores	Composition
Aluminium	Bauxite	$AlO_x(OH)_{3-2x}$ [where, $0 < x < 1$]
	Kaolinite (a form of clay)	$[Al_2(OH)_4Si_2O_5]$
Iron	Haematite	Fe_2O_3
	Magnetite	Fe_3O_4
	Siderite	$FeCO_3$
	Iron pyrites	FeS_2
Copper	Copper pyrites	$CuFeS_2$
	Malachite	$CuCO_3 \cdot Cu(OH)_2$
	Cuprite	Cu_2O
	Copper glance	Cu_2S
Zinc	Zinc blende or Sphalerite	ZnS
	Calamine	$ZnCO_3$
	Zincite	ZnO

Minerals and Ores

The substance (or compound) in the form of which metal is found in nature is called a mineral and the mineral from which extraction of metal is beneficial and cheap is called an **ore**.

PREP MIRROR
Your Personal Preparation Indicator

- No. of Questions in Exercises (x)—
- No. of Questions Attempted (y)—
- No. of Correct Questions (z)—
  (Without referring Explanations)

- Accuracy Level ($z/y \times 100$)—
- Prep Level ($z/x \times 100$)—

In order to expect good rank in JEE, your Accuracy Level should be above 85 & Prep Level should be above 75.

The impurities associated with the ore are called **gangue** or **matrix**.

Depending upon the nature of associated group or atom, ores are of following types:

(i) **Oxide ores,** e.g. haematite (Fe_2O_3), zincite (ZnO).

(ii) **Sulphide ores**, e.g. galena (PbS), cinnabar (HgS), argentite (Ag_2S), ruby silver ($Ag_2S \cdot Sb_2S_3$).

(iii) **Carbonate ores**, e.g. magnesite ($MgCO_3$), siderite ($FeCO_3$) etc.

(iv) **Sulphate ores**, e.g. gypsum ($CaSO_4 \cdot 2H_2O$), glauber's salt ($Na_2SO_4 \cdot 10H_2O$) etc.

(v) **Silicate ores**, e.g. willemite (Zn_2SiO_4), feldspar ($NaAlSi_3O_8$).

NOTE **Nitrate ores** are rare because all nitrates are water soluble and at higher temperature, they decompose into oxides or metal.

Metallurgy

Extraction of a metal from its ores is known as metallurgy. Metallurgy of a metal includes several metallurgical operations depending upon the nature of metal, its ore and impurities.

All metallurgical processes may be divided into three processes:

1. **Pyrometallurgical Process** It involves extraction of metals at very high temperature. Cu, Fe, Zn, Sn, Pb, Ni, Cr, Hg are extracted by pyrometallurgical process.
2. **Hydrometallurgical Process** It involves extraction of metal by the use of their aqueous solution. Ag and Au are extracted by this process.

NOTE In **hydrometallurgy**, metals like Fe cannot be used because it is not easy to remove excess iron from precious metal such as Ag, Au; while excess zinc can be easily removed as it is volatile.

3. **Electrometallurgical Process** Sodium, potassium, lithium, calcium, magnesium and aluminium are extracted from their molten salt solutions through electrolytic method.

Extraction of Metals

Various steps/processes involved in the extraction of a metal are as follows:

1. Concentration of Ores

It is the method that involves removal of unwanted materials from the ores. This method is also known as **ore dressing.**

Various methods adopted for the ore dressing are as follows:

(i) **Hydraulic Washing** (Levigation) It is based on the differences in gravities of the ore and gangue particles.

- In this process, an upward stream of running water is used to wash the powdered ore. The lighter gangue particles are washed away and the heavier ores are left behind.
- The oxide ores of iron (Fe_3O_4 and Fe_2O_3) are concentrated by this method.

(ii) **Magnetic-Separation Method** It is based on differences in magnetic properties of the minerals. If either the ore or the gangue is capable of being attracted by a magnetic field, then this process is used.

In electrostatic separation, electrically charged surfaces are used to separate metallic particles from non-metallic particles of ore.

(iii) **Froth-Floatation Method** It is used for the concentration of sulphide ores. In this process, a suspension of the powdered ore is made with water. To it, collectors and froth stabilisers are added.

NOTE **Collectors** (e.g. pine oil, fatty acids, xanthates etc.) enhance non-wettability of the mineral particles and **froth stabilisers** (e.g. cresols, aniline) stabilise the froth.

(iv) **Leaching Method** It is often used if the ore is soluble in some suitable solvent (i.e. acids, bases or other chemicals) but not in the impurities. e.g.

(a) **Leaching of Alumina From Bauxite** (Baeyer's process)

$$Al_2O_3(s) + 2NaOH\ (aq) + 3H_2O\ (l) \xrightarrow{473\text{-}523K} \underset{\text{Sodium meta aluminate}}{2Na\ [Al(OH)_4](aq)}$$

$$SiO_2(s) + 2NaOH\ (aq) \xrightarrow{473\text{-}523\ K} \underset{\text{Sodium silicate}}{Na_2SiO_3(aq)} + H_2O$$

The resulting solution is filtered, cooled and pH is adjusted by neutralisation with CO_2 which causes precipitation of aluminium hydroxide.

$$2Na\ [Al(OH)_4](aq) + CO_2(g) \longrightarrow Al_2O_3 \cdot xH_2O\ (s)\downarrow + 2NaHCO_3(aq)$$

The sodium silicate remains in the solution and hydrated alumina is filtered, dried and heated to give back pure Al_2O_3.

$$Al_2O_3 \cdot xH_2O(s) \xrightarrow{1470K} Al_2O_3(s) + xH_2O(g)$$

(b) **In the Metallurgy of Silver and Gold** The respective metal is leached with dilute solution of NaCN or KCN in the presence of air (from O_2).

$$4M(s) + 8CN^-(aq) + 2\ H_2O(aq) + O_2(g) \longrightarrow 4[M(CN)_2]^-(aq) + 4OH^-(aq)$$

(where, M = Ag or Au))

$$2[M(CN)_2]^-(aq) + Zn(s) \longrightarrow [Zn(CN)_4]^{2-}(aq) + 2\ M(s)$$

2. Extraction of Crude Metal from Concentrated Ore

It involves two major steps, i.e. conversion of concentrated ore to oxide and reduction of the oxide to metal.

In conversion of concentrated ore to oxide following steps are considered:

(i) **Calcination** Strong heating of ore in absence of air. e.g.

$$Fe_2O_3 \cdot xH_2O(s) \xrightarrow{\Delta} Fe_2O_3(s) + xH_2O(g)$$

$$ZnCO_3(s) \xrightarrow{\Delta} ZnO(s) + CO_2(g)$$

$$CaCO_3 \cdot MgCO_3(s) \xrightarrow{\Delta} MgO(s) + CaO(s) + 2CO_2(g)$$

Usually **carbonates and hydroxides are converted** into oxides by this method. Volatile impurities of S, As and P are removed as their volatile oxides.

(ii) **Roasting** Strong heating of ore in presence of air. e.g.

$$2ZnS + 3O_2 \xrightarrow{\Delta} 2ZnO + 2SO_2$$

$$2Cu_2S + 3O_2 \xrightarrow{\Delta} 2Cu_2O + 2SO_2$$

The sulphide ores of copper are heated in reverberatory furnace. If the ore contains iron, it is mixed with silica before heating. Iron oxide slags off as iron silicate and copper is produced in the form of copper matte which contains Cu_2S and FeS.

$$FeO + SiO_2 \longrightarrow FeSiO_3 \text{ (Slag)}$$

SO_2 produced is utilised for manufacturing H_2SO_4.

- **Reduction of the metal oxide to metal** takes place when heated with reducing agents such as C (coke,) or CO or even another metal.

  The process is known as **smelting.**

$$M_xO_y + yC \longrightarrow xM + yCO$$

NOTE
- In chloridising roasting, the ore containing As, S or Sb as impurity is heated with common salt in presence of silver ores.
- In sulphating roasting, sulphide ores are oxidised into sulphate, e.g. ZnS is oxidised to $ZnSO_4$.
- Some metals like Fe dissolve the reducing agent used (carbon) in their extraction. This can be removed by heating the impure metal with more of the ore.
- **Misch metal** is used as a reducing agent for extraction of pure vanadium. Hydrogen is used as a reducing agent in extraction of pure tungsten and vanadium.

Thermodynamics and Electrochemical Principles of Metallurgy

To understand the variation of temperature requirement for thermal reactions and suitable reducing agent for a given metal oxide (M_xO_y), Gibbs energy interpretations are made.

Gibbs equation,

$$\Delta G = \Delta H - T\Delta S$$

where, ΔH = enthalpy change, ΔG = Gibbs free energy,
T = temperature, ΔS = entropy change

$$\Delta G^\circ = -2.303\, RT \log K$$

K = equilibrium constant

- If $\Delta G = -$ ve, process is spontaneous,
  $\Delta G = +$ ve, process is non-spontaneous,
  ΔG = zero, process is at equilibrium.
- A reaction with positive ΔG can still be made to occur by coupling it with another reaction having large negative ΔG. Such coupling is easily understood through **Ellingham diagram**. This diagram help us in predicting the feasibility of thermal reduction of an ore.
- Ellingham diagram plots Gibbs free energy change (ΔG) values for formation of oxides against temperature.
- From Ellingham diagram, it is evident that metals which have more negative $\Delta_f G^\circ$ of their oxides can reduce those metal oxides for which $\Delta_f G^\circ$ is less negative.
- Reduction can also be done by using Al, H_2 etc. By using Al, oxides of Cr, Fe and Mn can be reduced. A mixture of Fe_2O_3 and Al is called **thermite mixture.**
- Highly electropositive metals like Na, K, Al are reduced by the electrolysis of their fused salts.

Refining of Crude Metal

The metal obtained from the above processes is not 100% pure, hence called **crude metal.** From the impure metal, the metal of high purity is obtained by refining.

Several techniques are used in refining depending upon the differences in properties of the metal.

These are as follows :

(i) **Distillation** This method is very useful for low boiling metals like zinc, cadmium and mercury. The impure metal is evaporated to obtain the pure metal as distillate.

(ii) **Liquation** This method, involves low melting metals like tin, bismuth and lead that can be made to flow on a sloping surface of a reverberatory furnace and heated above their melting point. In this way, it is separated from higher melting impurities.

(iii) **Electrolytic Refining** (Electrolysis) Various elements such as Cu, Au, Ag, Pb, Zn and Al can be purified by this method. Here, crude metal is made anode whereas the thin sheet of pure metal is made cathode.

Anode $\underset{\text{(Impure)}}{Cu(s)} \longrightarrow Cu^{2+}(aq) + 2e^-$

Cathode $Cu^{2+}(aq) + 2e^- \longrightarrow \underset{\text{(Pure)}}{Cu(s)}$

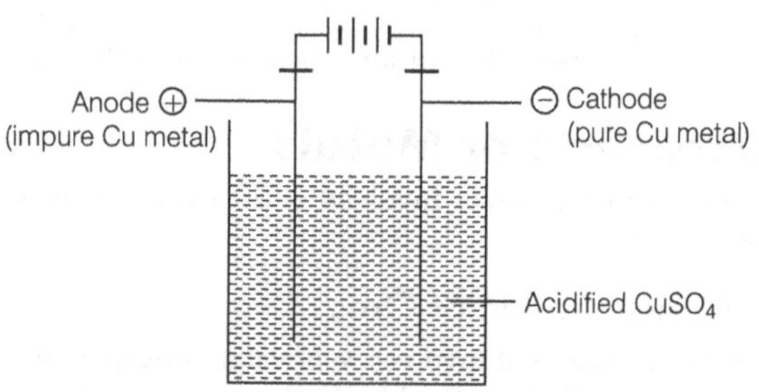

Electro refining

(iv) **Zone-Refining** This method is based on the principle that the impurities are more soluble in the melt than in the solid state of the metal. This method is very useful for producing semiconductors and other metals of very high purity, e.g. germanium, silicon, boron, gallium and indium.

(v) **Vapour Phase Refining** This method, involves the conversion of metal into its volatile compound. It is then decomposed to give pure metal.

e.g.

- **Mond process** for refining nickel.

$$\underset{\text{(Impure)}}{Ni} + 4CO \xrightarrow{330\text{-}350\ K} \underset{\text{Volatile}}{Ni(CO)_4}$$

$$Ni(CO)_4 \xrightarrow{450\text{-}470\ K} \underset{\text{(Pure)}}{Ni} + 4CO$$

- **van-Arkel method** for zirconium or titanium, vanadium or thorium

$$\underset{\text{(Impure)}}{Zr + 2I_2} \longrightarrow \underset{\text{(Volatile)}}{ZrI_4}$$

$$ZrI_4 \xrightarrow{1800\ K} \underset{\text{(Pure)}}{Zr} + 2I_2$$

- **Chromatographic method** This method is based on the principle that different components of a mixture are adsorbed differently on an adsorbent.

Occurrence and Extraction of Some Important Metals

1. Aluminium

Its chief ore is **bauxite** ($Al_2O_3 \cdot 2H_2O$). The outline of its extraction from its ore is shown below:

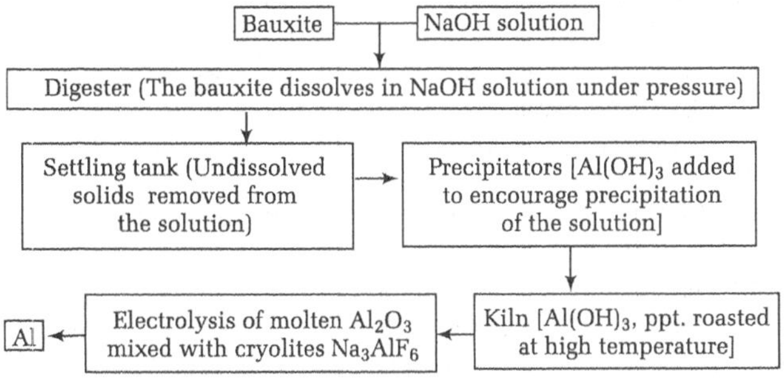

2. Copper

It does not occur abundantly in nature. Its chief ore are **copper pyrites** ($CuFeS_2$), **malachite** ($CuCO_3 \cdot Cu(OH)_2$), **cuperite** (Cu_2O), **copper glance** (Cu_2S). The outline of its extraction from its ore looks like

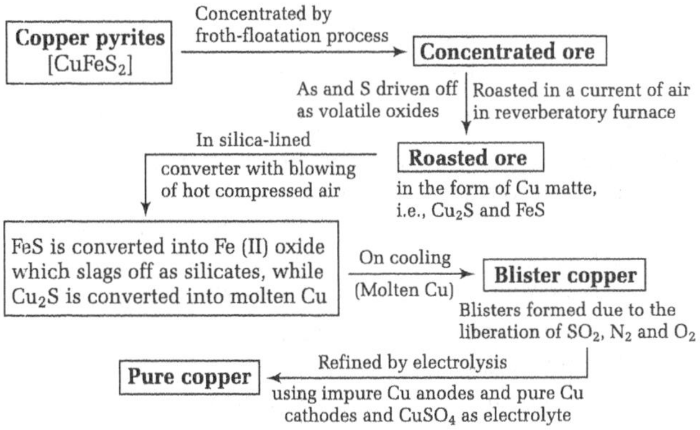

3. Zinc

Its chief ore is **zinc blende** (ZnS) and other ores are **calamine** ($ZnCO_3$), **zincite** (ZnO), **willemite** (Zn_2SiO_4) etc. The outline of its extraction looks like

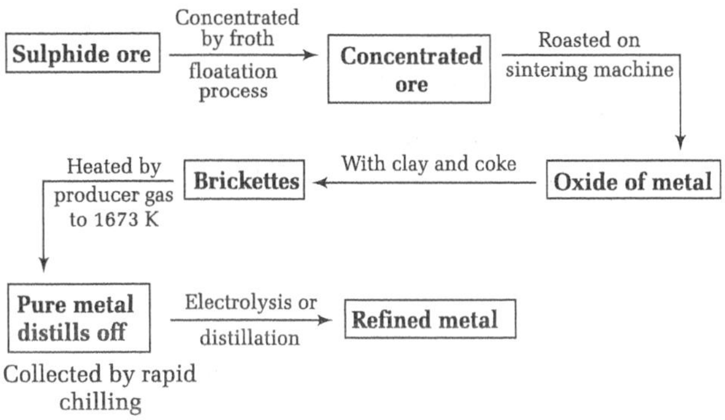

4. Iron

Iron is the second most abundant metal after Al in the earth's crust. Its most important ores are **haematite** (Fe_2O_3), **magnetite** (Fe_3O_4), **siderite** ($FeCO_3$) and **iron pyrites** (FeS_2). The outline of its extraction from its ore looks like

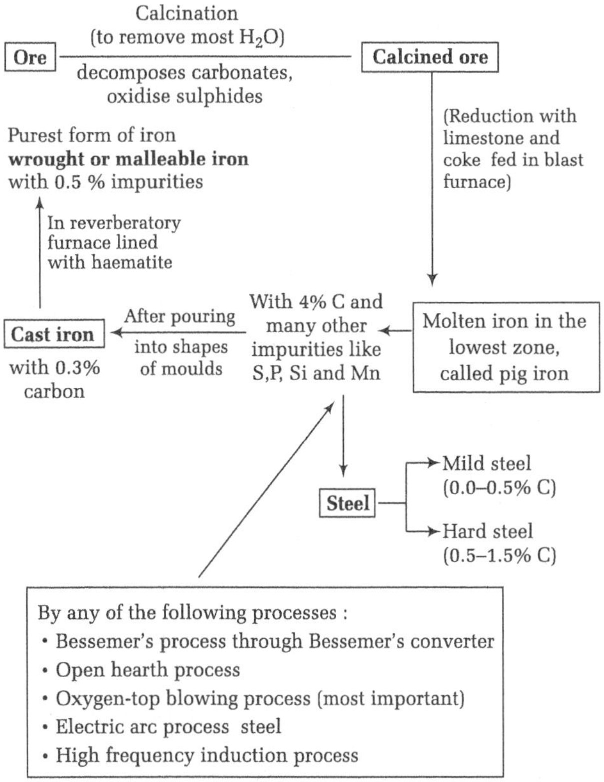

DAY PRACTICE SESSION 1

FOUNDATION QUESTIONS EXERCISE

1 Among the following statement, the incorrect one is
(a) calamine and siderite are carbonates
(b) argentite and cuprite are oxides
(c) zinc blende and pyrites are sulphides
(d) malachite and azurite are ores of copper

2 Which ore contains both iron and copper?
(a) Cuprite (b) Chalcocite
(c) Chalcopyrite (d) Malachite

3 Cassiterite is concentrated by
(a) levigation (b) electromagnetic separation
(c) floatation (d) liquefaction

4 Which one of the following ores is best concentrated by froth floatation method? → JEE Main 2016
(a) Siderite (b) Galena
(c) Malachite (d) Magnetite

5 Which one of the following benefaction processes is used for the mineral, $Al_2O_3 \cdot 2H_2O$?
(a) Froth floatation (b) Leaching
(c) Liquation (d) Magnetic separation

6 In the process of extraction of gold,

$$\text{Roasted gold ore} + CN^- + H_2O \xrightarrow{O_2} [X] + OH^-$$
$$[X] + Zn \longrightarrow [Y] + Au$$

Identify the complexes $[X]$ and $[Y]$.
(a) $X = [Au(CN)_2]^-, Y = [Zn(CN)_4]^{2-}$
(b) $X = [Au(CN)_4]^{3-}, Y = [Zn(CN)_4]^{2-}$
(c) $X = [Au(CN)_2]^-, Y = [Zn(CN)_6]^{4-}$
(d) $X = [Au(CN)_4]^-, Y = [Zn(CN)_4]^{2-}$

7 Extraction of gold and silver involves leaching the metal with CN^- ion. The metal is recovered by
(a) displacement of metal by some other metal from the complex ion
(b) roasting of metal complex
(c) calcination followed by roasting
(d) thermal decomposition of metal complex

8 The process of converting hydrated alumina into anhydrous alumina is called
(a) roasting
(b) smelting
(c) dressing
(d) calcination

9 Calcination is the process in which → Online JEE (Main) 2013
(a) removal of water takes place
(b) decomposition of carbonates takes place
(c) oxidation of sulphides takes place
(d) All of the above

10 Which one of the following reactions is an example of auto-reduction?
(a) $Fe_3O_4 + 4C \longrightarrow 3Fe + 4CO_2$
(b) $Cu_2O + C \longrightarrow 2Cu + CO$
(c) $Cu^{2+}(aq) + Fe(s) \longrightarrow Cu(s) + Fe^{2+}(aq)$
(d) $Cu_2O + \frac{1}{2}Cu_2S \longrightarrow 3Cu + \frac{1}{2}SO_2$

11 The methods chiefly used for the extraction of lead and tin from their ores are respectively
(a) self reduction and carbon reduction
(b) self reduction and electrolytic reduction
(c) carbon reduction and self reduction
(d) cyanide process and carbon reduction

12 Which of the following metals is obtained by electrolytic reduction process?
(a) Fe (b) Cu (c) Ag (d) Al

13 Which of the following statements about the advantage of roasting of sulphide ore before reduction is not true?
(a) $\Delta_f G°$ of the sulphide is greater than those of CS_2 and H_2S
(b) $\Delta_f G°$ is negative for roasting of sulphide ore to oxide
(c) Roasting of sulphide to oxide is thermodynamically feasible
(d) Carbon and hydrogen are suitable reducing agents for metal sulphides

14 The value of $\Delta_f G°$ for the formation of Cr_2O_3 is -540 kJ mol^{-1} and that of Al_2O_3 is -827 kJ mol^{-1}. Is the reduction of Cr_2O_3 with Al is feasible reaction?
(a) The data is incomplete
(b) The reaction is feasible
(c) The reaction is not feasible
(d) The reaction may or may not be feasible

15 Use the relationship, $\Delta G° = -nFE°_{cell}$ to estimate the minimum voltage required to electrolyse Al_2O_3 in the Hall-Heroult process

$$\Delta_f G°(Al_2O_3) = -1520 \text{ kJ mol}^{-1}$$
$$\Delta_f G°(CO_2) = -394 \text{ kJ mol}^{-1}$$

(a) 0.8 V (b) 1.60 V (c) 2.8 V (d) 3.0 V

16 In Goldschmidt aluminothermic process which of the following reducing agent is used? → Online JEE (Main) 2013
(a) Calcium (b) Coke
(c) Al powder (d) Sodium

17 Electrolytic refining is used to purify, which of the following metals?
(a) Cu and Zn (b) Ge and Si
(c) Zr and Ti (d) Zn and Hg

18 Zone-refining is based on the principle that
- (a) impurities of low boiling metals can be separated by distillation
- (b) impurities are more soluble in molten metal than in solid metal
- (c) different components of a mixture are differently adsorbed on an adsorbent
- (d) vapours of volatile compound can be decomposed into pure metal

19 Which method of purification is represented by the following equation?

$$Ti(s) + 2I_2(g) \xrightarrow{523\text{ K}} TiI_4(g) \xrightarrow{1700\text{ K}} Ti(s) + 2I_2(g)$$

(a) Zone-refining (b) Cupellation (c) Polling (d) van-Arkel

20 Which of the following does not represent correct method?
- (a) $TiCl_2 + 2Mg \longrightarrow Ti + 2MgCl_2$: Kroll
- (b) $Ni(CO)_2 \longrightarrow Ni + 4CO$: Mond
- (c) $Ag_2CO_3 \longrightarrow 2Ag + CO_2 + \frac{1}{2}O_2$: van-Arkel
- (d) $ZnI_4 \longrightarrow Zn + 2I_2$: van-Arkel

21 During the process of electrolytic refining of copper, some metals present as impurity settle as 'anode mud'. These are

(a) Fe and Ni (b) Ag and Au (c) Pb and Zn (d) Se and Ag

22 Refining of impure copper with zinc impurity is to be done by electrolysis using electrodes as

	Cathode	Anode
(a)	Pure copper	Pure zinc
(b)	Pure zinc	Pure copper
(c)	Pure copper	Impure copper
(d)	Pure zinc	Impure zinc

23 Which metal can't be obtained from electrolysis?

(a) Ca (b) Mg (c) Cr (d) Al

24 In the context of the Hall-Heroult process for the extraction of Al, which of the following statement is false? → JEE Main 2015
- (a) CO and CO_2 are produced in this process
- (b) Al_2O_3 is mixed with CaF_2 which lowers the melting point of the mixture and brings conductivity
- (c) Al^{3+} is reduced at the cathode to form Al
- (d) Na_3AlF_6 serves as the electrolyte

25 When copper ore is mixed with silica in a reverberatory furnace, copper matte is produced. The copper matte contains
- (a) sulphides of copper (II) and iron (II)
- (b) sulphides of copper (II) and iron (III)
- (c) sulphides of copper (I) and iron (II)
- (d) sulphides of copper (I) and iron (III)

26 Extraction of zinc from zinc blende is achieved by
- (a) electrolytic reduction
- (b) roasting followed by reduction with carbon
- (c) roasting followed by reduction with another metal
- (d) roasting followed by self reduction

27 In blast furnace, maximum temperature is in
- (a) zone of fusion
- (b) zone of combustion
- (c) zone of slag combustion
- (d) zone of reduction

28 Which series of reactions correctly represents chemical relations related to iron and its compound? → JEE Main 2014
- (a) $Fe \xrightarrow{\text{Dil.}H_2SO_4} FeSO_4 \xrightarrow{H_2SO_4, O_2} Fe_2(SO_4)_3 \xrightarrow{\text{Heat}} Fe$
- (b) $Fe \xrightarrow{O_2,\text{Heat}} FeO \xrightarrow{\text{Dil.}H_2SO_4} FeSO_4 \xrightarrow{\text{Heat}} Fe$
- (c) $Fe \xrightarrow{Cl_2,\text{Heat}} FeCl_3 \xrightarrow{\text{Heat, air}} FeCl_2 \xrightarrow{Zn} Fe$
- (d) $Fe \xrightarrow{O_2,\text{Heat}} Fe_3O_4 \xrightarrow{CO,\ 600^\circ C} FeO \xrightarrow{CO,\ 700^\circ C} Fe$

29 Match the following and choose the correct options.

Column I	Column II
A. Blister Cu	1. Aluminium
B. Blast furnace	2. $2Cu_2O + Cu_2S \longrightarrow 6Cu + SO_2$
C. Reverberatory furnace	3. Iron
D. Hall-Heroult process	4. $FeO + SiO_2 \longrightarrow FeSiO_3$

Codes

	A	B	C	D		A	B	C	D
(a)	4	3	1	2	(b)	2	3	4	1
(c)	4	1	3	2	(d)	2	4	3	1

Direction (Q. Nos. 30-33) *In the following questions assertion followed reason is given. Choose the correct answer out of the following choices.*
- (a) Both A and R are true and R is correct explanation of A
- (b) Both A and R are true but R is not correct explanation of A
- (c) A is true but R is false
- (d) Both A and R are false

30 **Assertion** (A) Hydrometallurgy involves dissolving the ore in a suitable reagent followed by precipitation by a more electropositive metal.

Reason (R) Copper is extracted by hydrometallurgy.

31 **Assertion** (A) Nickel can be purified by Mond process.

Reason (R) Ni $(CO)_4$ is a volatile compound which decomposes at 460 K to give pure Ni.

32 **Assertion** (A) Zone refining method is very useful for producing semiconductors.

Reason (R) Semiconductors are of high purity.

33 **Assertion** (A) van-Arkel method is used to prepare ultra pure sample of some metals.

Reason (R) It involves reaction of CO with metals to form volatile carbonyls which decompose on heating to give pure metal.

DAY PRACTICE SESSION 2

PROGRESSIVE QUESTIONS EXERCISE

1 When 1.164 g of a certain metal sulphide was roasted in air, 0.972 g of the metal oxide was formed. If the oxidation number of the metal is +2, calculate the molar mass of the metal.

(a) 25.67 g (b) 31.56 g
(c) 47.35 g (d) 65 g

2 The major role of fluorspar (CaF_2) which is added in small quantities in the electrolytic reduction of alumina dissolved in fused cryolite (Na_3AlF_6) is

(a) that of a catalyst
(b) to make the fused mixture very conducting
(c) to lower the temperature of the melt
(d) to decrease the rate of oxidation of carbon at anode

3 From the Ellingham graph on carbon, which of the following statements is false?

(a) CO reduces Fe_2O_3 to Fe at temperature less than 983 K
(b) CO is less stable than CO_2 at temperature more than 983 K
(c) CO reduces Fe_2O_3 to Fe in the reduction zone of blast furnace.
(d) CO_2 is more stable than CO at temperature less than 983 K

4 In the electrolytic method for obtaining aluminium from purified bauxite, cryolite is added to

(a) minimise the heat loss due to radiation
(b) protect aluminium produced from oxygen
(c) dissolve bauxite and increases its conductivity
(d) lower the melting point of bauxite

5 Bauxite ore is made up of $Al_2O_3 + SiO_2 + TiO_2 + Fe_2O_3$. This ore is treated with conc. NaOH solution at 500 K and 35 bar pressure for few hours and filtered hot. In the filterate, the species present are

(a) $NaAl(OH)_4$ (b) $Na_2Ti(OH)_6$
(c) $Na[Al(OH)_4]$ and Na_2SiO_3 (d) Na_2SiO_3

6 Consider the following reactions:

$$Ag_2S + NaCN \longrightarrow (A)$$
$$(A) + Zn \longrightarrow (B)$$

(B) is a metal. Hence (A) and (B) are

(a) $[Na_2Zn(CN)_4]$, Zn (b) $Na[Ag(CN)_2]$, Ag
(c) $Na_2[Ag(CN)_4]$, Ag (d) $Na_3[Ag(CN)_4]$, Ag

7 When copper pyrites is roasted in excess of air, a mixture of CuO + FeO is formed. FeO is present as impurities. This can be removed as slag during reduction of CuO. The flux added to form slag is

(a) SiO_2, which is an acidic flux
(b) limestone, which is a basic flux
(c) SiO_2, which is a basic flux
(d) CaO, which is a basic flux

8 Free energies of formation ($\Delta_f G$) of MgO (s) and CO (g) at 1273 K and 2273 K are given below:

$\Delta_f G$ [MgO(s)] = –941 kJ / mol at 1273 K

$\Delta_f G$ [MgO(s)] = –314 kJ / mol at 2273 K

and $\Delta_r G = \Delta_f G$ (products) – $\Delta_f G$ (reactants)

$$\Delta_f G[CO(g)] - \Delta_f G[MgO(s)]$$

$$\Rightarrow (-439) - (-941) = +502 \text{ kJ mol}^{-1}$$

$$\Delta_f G[CO(g)] = -439 \text{ kJ/mol at 1273 K}$$

$$\Delta_f G[CO(g)] = -628 \text{ kJ/mol at 2273 K}$$

On the basis of above data, predict the temperature at which carbon can be used as a reducing agent for MgO(s).

(a) 2273
(b) 1273
(c) Both (a) and (b)
(d) None of the above

9 In the extraction of nickel by Mond's process, the metal is obtained by

(a) electrochemical reduction
(b) thermal decomposition
(c) chemical reduction by aluminium
(d) reduction by carbon

10. In the Baeyer's process,

(a) Al_2O_3 goes into solution as soluble $Al(OH)_4^-$, while other basic oxides as TiO_2 and Fe_2O_3 remain insoluble.
(b) Al_2O_3 changes to AlN which in turn decomposed by H_2O
(c) Al_2O_3 changes to $Al_2(CO_3)_3$ which changes to $AlCl_3$
(d) None of the above is correct

11 In the metallurgy of sodium by electrolysis, excess of calcium chloride is mixed with sodium chloride to

(a) make the latter a good conductor
(b) make the latter soft
(c) generate more energy for the electrolytic cell
(d) assist liquefaction of the later much lower temperature

12 Froth-floatation process is used for the concentration of sulphide ore. Which of the following processes is correct?

(a) It is based on the difference in wettability of different minerals
(b) In this process sodium ethyl xanthate, $C_2H_5OCS_2Na$ is used as collector
(c) In this process NaCN is used as depressant in the mixture of ZnS and PbS when ZnS forms soluble complex and PbS forms froth
(d) All of the above

13 An impure metal is allowed to react with carbon monoxide at 50°C and the volatile gas thus, formed is collected and heated further to about 200°C. This process gives the metal of 99.99% purely. What is the metal?

(a) Fe (b) Cr
(c) Co (d) Ni

14 Of the following reduction processes,

I. $Fe_2O_3 + 3C(s) \longrightarrow Fe + 3CO$
II. $ZnO + C(s) \longrightarrow Zn + CO$
III. $Ca_3(PO_4)_2 + C(s) \longrightarrow P$
IV. $PbO + C(s) \longrightarrow Pb + CO$

Correct processes are

(a) All of these (b) All but III
(c) All but IV (d) Both II and IV

ANSWERS

SESSION 1

1 (b)	2 (c)	3 (b)	4 (b)	5 (a)	6 (a)	7 (a)	8 (d)	9 (b)	10 (d)
11 (a)	12 (d)	13 (d)	14 (b)	15 (b)	16 (c)	17 (a)	18 (b)	19 (d)	20 (c)
21 (b)	22 (c)	23 (c)	24 (d)	25 (a)	26 (b)	27 (b)	28 (d)	29 (b)	30 (c)
31 (a)	32 (c)	33 (c)							

SESSION 2

1 (d)	2 (c)	3 (b)	4 (c)	5 (c)	6 (b)	7 (a)	8 (a)	9 (b)	10 (a)
11 (d)	12 (d)	13 (d)	14 (a)						

Hints and Explanations

SESSION 1

1 Argentite (Ag_2S) is a sulphide ore and cuprite (Cu_2O) is an oxide ore.

2 Cuprite — Cu_2O
Chalcocite — Cu_2S
Chalcopyrite — $CuFeS_2$
Malachite — $Cu(OH)_2 \cdot CuCO_3$

3 Cassiterite is tin oxide (SnO_2) which is non-magnetic and contains wolframite, $FeWO_4$ (magnetic) impurities. These are separated by electromagnetic separation.

4 Galena (PbS) is sulphide ore, hence concentrated by froth-floatation process.

5 $Al_2O_3 \cdot 2H_2O$ (bauxite) is concentrated by leaching with NaOH, Al_2O_3 dissolves while other impurities remains undissolved.

6
$$\underset{\text{From gold ore}}{Au} + 4CN^- + H_2O + \frac{1}{2}O_2 \longrightarrow \underset{(X)}{2[Au(CN)_2]^-} + 2OH^-$$
$$\underset{(X)}{2[Au(CN)_2]^-} + Zn \longrightarrow \underset{(Y)}{[Zn(CN)_4]^{2-}} + 2Au$$

7 $2[Ag(CN)_2]^- + Zn \longrightarrow 2Ag + [Zn(CN)_4]^{2-}$

$2[Au(CN)_2]^- + Zn \longrightarrow 2Au + [Zn(CN)_4]^{2-}$

Thus, the metal is recovered by displacement of metal by some other metal from the complex ion.

8 In calcination, moisture or hydrated water is removed on heating.

$$Al_2O_3 \cdot 2H_2O \longrightarrow Al_2O_3 + 2H_2O\uparrow$$

9 Calcination is a thermal treatment in the presence of air to ores and other solid materials to bring about a thermal decomposition, phase transition or removal of volatile fractions.

$$M_2CO_3 \xrightarrow{\Delta} M_2O + CO_2$$

10 Among the given options, example of auto-reduction is

$$Cu_2O + \frac{1}{2}Cu_2S \longrightarrow 3Cu + \frac{1}{2}SO_2$$

11 The process in which the anions associated with the metal help in the reduction is called self-reduction. It is mainly used in case of sulphide ores and the process of extracting a metal by fusion of the oxide ore with carbon is known as carbon reduction.

12 Metals like Na, K, Mg, Ca, Al etc. are reduced by electrolytic reduction.

13 Both C and H_2 are not suitable for reducing sulphide ore because $\Delta_f G^\circ$ of metal sulphide is more than that of CS_2 and H_2S which will be formed as a result of reduction.

14. $\frac{4}{3}Cr\,(s) + O_2(g) \longrightarrow \frac{2}{3}Cr_2O_3(s);$

$\Delta_f G^\circ = -540\ kJ\ mol^{-1}$ ...(i)

$\frac{4}{3}Al\,(s) + O_2(g) \longrightarrow \frac{2}{3}Al_2O_3(s);$

$\Delta_f G^\circ = -827\ kJ\ mol^{-1}$ ...(ii)

Subtracting Eq. (ii) from Eq. (i)

$$\frac{2}{3}Cr_2O_3\,(s) + \frac{4}{3}Al\,(s) \longrightarrow \frac{2}{3}Al_2O_3(s) + \frac{4}{3}Cr\,(s),$$

$\Delta_f G^\circ = -287\ kJ$

Since, $\Delta G°$ comes out to be negative, the reaction is feasible.

15 $2Al_2O_3 + 3C \longrightarrow 4Al + 3CO_2$

$\Delta_r G° = 3\Delta G°(CO_2) - 2\Delta G°(Al_2O_3)$
$= 3 \times (-394) - 2 \times (-1520)$
$= 1858 kJ = 1858 \times 10^3 J$

$\therefore nFE° = 1858 \times 10^3$ ($n = 12$ electrons)

$\therefore E° = \frac{1858 \times 10^3}{12 \times 96500} = 1.60 V$

16 In Goldschmidt aluminothermic process, oxides of Cr, Fe etc., are treated with Al powder which reduces these oxides into crude metal. Thus, Al acts as reducing agent here.

$Fe_2O_3 + \underset{\text{Reducing agent}}{2Al} \longrightarrow Al_2O_3 + 2Fe$

17 Electrolytic refining is used to purify Cu and Zn.

18 Zone-refining is based upon the principle of fractional crystallisation. It is based on the fact that impurities are more soluble in the melt than in the pure metals.

19 van-Arkel process is based on the thermal decomposition of a volatile compound like an iodide, which is first formed by direct combination of metal to be purified and iodine. Metal formed is in purest form. Titanium and zirconium are purified by this method.

20 van-Arkel process is used for Zr or Ti, not for Ag.

21 During electrolysis, noble metals (inert metals) like Ag, Au and Pt are not affected and separate as anode mud from the impure anode.

22 In electrolytic refining of copper, impure copper acts as anode while pure copper acts as cathode. The impurities of iron, nickel, zinc and cobalt present in blister copper being more electropositive pass into solution as soluble sulphates.

23 Cr is less electropositive and can be obtained by the reduction of its oxide by aluminium.

24 (a) In Hall-Heroult process for extraction of Al, carbon anode is oxidised to CO and CO_2.
(b) When Al_2O_3 is mixed with CaF_2, it lowers the melting point of the mixture and brings conductivity.
(c) Al^{3+} is reduced at cathode to get Al.
(d) Here, Al_2O_3 is an electrolyte undergoing the redox process. Na_3AlF_6 although is an electrolyte but serves as a solvent, not electrolyte.

25 Copper matte consists of Cu_2S and FeS.

26 Extraction of Zn from ZnS is done by roasting ZnS and then treated with coke for reduction.

$2ZnS + 3O_2 \xrightarrow{\Delta} 2ZnO + 2SO_2 \uparrow$

$ZnO + C \xrightarrow{\Delta} Zn + CO \uparrow$

27 In blast furnace, zone of combustion (1500-1600C°) has maximum temperature.

28 The correct reaction are as follows:

(a) $Fe + \text{dil. } H_2SO_4 \longrightarrow FeSO_4 + H_2$

$H_2SO_4 + 2FeSO_4 + \frac{1}{4}O_2 \longrightarrow Fe_2(SO_4)_3 + H_2O$

$Fe_2(SO_4)_3 \xrightarrow{\Delta} Fe_2O_3(s) + 3SO_3 \uparrow$

(b) $Fe \xrightarrow[\Delta]{O_2} FeO$

[It could also be Fe_2O_3 or Fe_3O_4]

$FeO + H_2SO_4 \longrightarrow FeSO_4 + H_2O$

$2FeSO_4 \xrightarrow{\Delta} Fe_2O_3 + SO_2 + SO_3$

(c) $Fe \xrightarrow[Cl_2]{\Delta} FeCl_3 \xrightarrow[\text{Air}]{\Delta}$ No reaction

[It cannot give $FeCl_2$]

(d) $Fe \xrightarrow[\Delta]{O_2} Fe_3O_4 \xrightarrow[600°C]{CO} FeO \xrightarrow[700°C]{CO} Fe$

This is correct reaction.

29 A $\longrightarrow$ 2, B $\longrightarrow$ 3, C $\longrightarrow$ 4, D $\longrightarrow$ 1

30 Hydrometallurgy is used for the extraction of low grade copper, while pyrometallurgy is used for the extraction of bulk quantities of copper.

31 Nickel can be purified by Mond's process.

$\underset{\text{Impure}}{Ni} + 4CO \xrightarrow{350-350 K} Ni(CO)_4 \xrightarrow{450-470K} \underset{\text{Pure}}{Ni} + 4CO$

32 **Correct explanation** In the zone refining method, the melt is allowed to crystallise. Since, impurities are more soluble in the melt than in the pure state of the metal therefore, semiconductor metals are obtained after repeated crystallisation.

33 van-Arkel method involve the use of I_2 to form volatile iodide of metals which on decomposition gives pure metal.

SESSION 2

1 Formula of sulphide is *MS* because oxidation number of metal ion being +2.

$$\underset{\substack{(M+32)\\1.164}}{MS} + \frac{3}{2}O_2 \longrightarrow \underset{\substack{(M+16)\\0.972}}{MO} + SO_2$$

$$\frac{M+32}{1.164} = \frac{M+16}{0.972}$$

or $M = 65 g$

2 The major role of fluorspar (CaF_2) which is added in small quantities in the electrolytic reduction of alumina dissolved in fused cryolite is to lower the temperature of the melt.

3 The stability of carbon oxides is decided with the help of Ellingham diagram.

When carbon changes to CO_2, entropy (ΔS) is very small and ΔG hardly shows any change with increasing temperature but when carbon changes to CO, ΔS is positive and ΔG becomes more negative with increasing temperature. Below 983 K. temperature, formation of CO_2 is favoured whereas, above this temperature, formation of CO is preferred, CO is more stable than CO_2 at more than 983 K temperature.

4 Cryolite (Na_3AlF_6) is added to alumina (purified bauxite, i.e. Al_2O_3) for its electrolysis to decrease its melting point and also increase its conductivity.

5 The principal ore of aluminium is bauxite, ($Al_2O_3 \cdot 2H_2O$) usually contains SiO_2, iron oxides, titanium oxide (TiO_2) as impurities. Concentration of ore is carried out by treating the powdered ore with a concentrated solution of NaOH at 473-523 K and 35-36 bar pressure. Al_2O_3 is leached out as sodium aluminate and SiO_2 as sodium silicate leaving the impurities behind.

$Al_2O_3 \cdot 2H_2O + 2NaOH + H_2O \longrightarrow \underset{\text{Sodium aluminate}}{2Na[Al(OH)_4]}$

$SiO_2 + 2NaOH \longrightarrow \underset{\text{Sodium silicate}}{Na_2SiO_3} + H_2O$

Hence, the solution obtained after filtration contains sodium aluminate and sodium silicate.

6 $Ag_2S + NaCN \longrightarrow \underset{(A)}{Na[Ag(CN)_2]}$

$\underset{(A)}{Na[Ag(CN)_2]} + Zn \longrightarrow Na[Zn(CN)_2] + \underset{(B)}{Ag}$

7 The compounds which combine with impurities present in ore (at high temperature) and remove them as a fusible substance (slag), are known as flux. When basic impurities like FeO are present, an acidic flux like SiO_2 is used and *vice-versa*.

$$\underset{\text{Basic impurity}}{FeO} + \underset{\text{Acidic flux}}{SiO_2} \longrightarrow \underset{\text{Slag}}{FeSiO_3}$$

Thus, during reduction of CuO, SiO_2 being an acidic flux is added to remove FeO as $FeSiO_3$.

8 At 1273 K,

(i) $Mg(s) + \frac{1}{2}O_2(g) \longrightarrow MgO(s)$; $\Delta_f G = -941 \text{ kJ mol}^{-1}$

(ii) $C(s) + \frac{1}{2}O_2(g) \longrightarrow CO(g)$; $\Delta_f G = -439 \text{ kJ mol}^{-1}$

The redox equation for the reduction of MgO to Mg by C can be obtained by subtracting Eq (i) from Eq (ii). Thus,

$MgO(s) + C(s) \longrightarrow Mg(s) + CO(g)$, $\Delta_f G = +502 \text{ kJ mol}^{-1}$

Since, $\Delta_r G$ of the above reduction reaction is +ve, the reduction of MgO by C is not feasible at 1273 K.

At 2273 K,

(iii) $Mg(s) + \frac{1}{2}O_2(g) \longrightarrow MgO(s)$; $\Delta_f G = -314 \text{ kJ mol}^{-1}$

(iv) $C(s) + \frac{1}{2}O_2(g) \longrightarrow CO(g)$; $\Delta_f G = -628 \text{ kJ mol}^{-1}$

Subtracting Eq. (iii) from Eq (iv), the redox equation is

$MgO(s) + C(s) \longrightarrow Mg(s) + CO(g)$;

$\Delta_r G = \Delta_f G \text{ (products)} - \Delta_f G \text{(reactants)}$

$= \Delta_f G[CO(g)] - \Delta_f G[MgO(s)]$

$= (-628) - (-314) = -314 \text{ kJ mol}^{-1}$

Since, $\Delta_r G$ for the above reaction is –ve, the reduction of MgO by carbon at 2273 K is feasible.

9 $\underset{\text{Impure}}{Ni} + 4CO \longrightarrow Ni(CO)_4$

$Ni(CO)_4 \xrightarrow[\text{Thermal decomposition}]{\Delta} \underset{\text{Pure}}{Ni} + 4CO$

(Mond's process)

10 In the Baeyer's process, Al_2O_3 goes into the solution as soluble $Al(OH_4)^-$, while that of basic oxides as TiO_2 and Fe_2O_3 remain insoluble.

11 In the metallurgy of sodium by electrolysis, excess of calcium chloride is mixed with NaCl to assist liquefaction of the later at much lower temperature.

12 All the given statements are correct.

13 Ni is the impure metal that react with CO at 50°C and the volatile gas formed is collected and heated to about 200°C.

14 All chemical processes are correct. In these reactions, metal oxide involves heating it with C (acts as reducing agent).

DAY TWENTY

Hydrogen

Learning & Revision for the Day

- Position of Hydrogen in Periodic Table
- Isotopes of Hydrogen
- Dihydrogen
- Hydrides
- Water (H_2O)
- Hard and Soft water
- Hydrogen Peroxide (H_2O_2)

Dihydrogen (H_2) is the most abundant element in the universe (70% of the total mass of the universe) and is the principal element in the solar atmosphere. The giant planets Jupiter and Saturn consist of mainly hydrogen. It is the lightest element known.

Position of Hydrogen in Periodic Table

The position of hydrogen in periodic table is uncertain as it shows resemblance with alkali metals as well as with halogens. However, on the basis of electronic configuration ($1s^1$), it is placed above lithium in the periodic table but still, it is not considered as the member of that group. On the other hand, like halogens (with ns^2, np^5 configuration), it is short by one electron to the corresponding noble gas configuration but it is still not considered as the member of seventeenth group. Thus, it has unique behaviour and is therefore best placed separately in the periodic table.

Isotopes of Hydrogen

- Hydrogen has three isotopes, i.e. protium (^1_1H), deuterium or heavy hydrogen (^2_1H or D) and tritium (^3_1H or T). Out of these isotopes, only tritium is radioactive and emits low energy β^- particles.
- The three isotopes have different masses and different enthalpy of bond dissociation hence, their rates of reaction and equilibrium constants are different. This is known as **isotopic effect.** They have same electronic configuration and chemical properties.

NOTE **Allotropes of hydrogen** are *ortho*-hydrogen (o-H_2) (in which proton spin is in same direction) and *para*-hydrogen (p-H_2) (in which proton spin is in opposite direction). Similarly, deuterium and tritium also exhibit spin isomerism and exist in *ortho*-and *para*-forms.

PREP MIRROR
Your Personal Preparation Indicator

- No. of Questions in Exercises (x)—
- No. of Questions Attempted (y)—
- No. of Correct Questions (z)—
  (Without referring Explanations)

- Accuracy Level ($z/y \times 100$)—
- Prep Level ($z/x \times 100$)—

In order to expect good rank in JEE, your Accuracy Level should be above 85 & Prep Level should be above 75.

Atomic Hydrogen

Hydrogen gas dissociates into atoms when it is subjected to an electric discharge under low pressure.

$$\underset{\text{(Atomic hydrogen)}}{H_2} \xrightarrow{\text{Electric discharge}} \underset{\text{(Nascent hydrogen)}}{2H}$$

Atomic hydrogen is very reactive and can be used as a reducing agent as well as oxidising agent.

Reducing power of atomic hydrogen is more than that of nascent hydrogen.

Dihydrogen

It is the principal element in the solar atmosphere. In the combined form, it occurs in plant and animal tissues.

Preparation of Dihydrogen

Laboratory preparation of dihydrogen are as follows :

(i) It is prepared by the action of dilute hydrochloric acid on metals that are more reactive than hydrogen granulated zinc, iron, magnesium, etc.

$$Zn + H_2SO_4 \longrightarrow ZnSO_4 + H_2 \uparrow$$

(ii) Certain metals like Zn, Sn, Al etc., react with NaOH to liberate H_2

$$Zn + 2NaOH \longrightarrow \underset{\text{Sodium zincate}}{Na_2ZnO_2} + H_2 \uparrow$$

Commercially used processes for the preparation of dihydrogen are as:

- Electrolysis of acidified water using platinum electrodes give dihydrogen.

$$\underset{\substack{\text{Containing small}\\ \text{amount of acid or alkali}}}{2H_2O(l)} \xrightarrow{\text{Traces of acid/ base}} \underset{\text{(At cathode)}}{2H_2(g)} + \underset{\text{(At anode)}}{O_2(g)}$$

- Dihydrogen of high purity (> 99.95%) is obtained by electrolysing warm aqueous barium hydroxide solution between nickel electrodes.
- The mixture of CO and H_2 is called **water gas.** As this mixture is used for the synthesis of methanol and a number of hydrocarbons, it is also called **synthesis gas** or ***syn* gas.** Now-a-days, '*syn* gas' is produced from sewage, scrap wood, newspapers etc. The process of producing '*syn gas*' from coal is called coal gasification. (Bosch process)

$$C(s) + H_2O(g) \xrightarrow{1270\,K} \underset{\text{Water gas}}{CO(g) + H_2(g)}$$

$$CO(g) + H_2O(g) \xrightarrow[\substack{\text{Iron chromate}\\ \text{(Catalyst)}}]{673\,K} CO_2(g) + H_2(g)$$

CO_2 is removed by passing the mixture through water at very high pressure. (CO_2 is dissolved completely in water). This reaction is known as water-gas shift reaction (WGSR)

- **Lane's process** involves decomposition of steam over heated Fe (550° – 800°C). This reaction is known as **gassing reaction**.

$$3Fe + 4H_2O \longrightarrow Fe_3O_4 + 4H_2 \uparrow \; ; \; \Delta H = -160.7\text{ kJ}$$

By the action of water gas, Fe_3O_4 is again reduced to iron. This process is known as **vivifaction.**

$$Fe_3O_4 + 4CO \longrightarrow 3Fe + 4CO_2$$

$$Fe_3O_4 + 4H_2 \longrightarrow 3Fe + 4H_2O$$

- Dihydrogen is obtained as a byproduct of brine electrolysis process for the manufacturing of NaOH, petroleum cracking plants and many electrolytic reactions.

Properties of Hydrogen

- Hydrogen is colourless, tasteless, odourless gas. It is lightest and slightly soluble in water.
- It is a neutral and highly combustible gas, so in the presence of air, it burns with pale blue flame to form water.
- **Reducing action** Hydrogen reduces the oxides of less electropositive elements but cannot reduce the oxides of alkali metals and alkaline earth metals.

$$ZnO\,(s) + H_2(g) \longrightarrow Zn + H_2O$$

$$Fe_3O_4(s) + 4H_2(g) \longrightarrow 3Fe + 4H_2O$$

- Hydrogenation

$$\text{Vegetable oil} + H_2 \xrightarrow{Ni/473K} \text{Vanaspati ghee}$$

$$CH_2 = CH_2 + H_2 \xrightarrow{\text{Ni or Pt}/\Delta} CH_3 - CH_3$$

- Reaction with CO

$$CO\,(g) + 2H_2(g) \xrightarrow[\Delta]{ZnO/Cr_2O_3} CH_3OH(l)$$

- With dinitrogen, it forms ammonia (Haber's process).

$$N_2(g) + 3H_2(g) \xrightarrow[\text{Fe, Mo}]{673\text{ K, }200\text{ atm}} 2NH_3(g)$$

Uses of Hydrogen

- In hydrogenation of oils.
- In preparation of synthetic petrol.
- In oxy-hydrogen flame.
- In hydrogen-oxygen fuel cells to produce electricity.
- In a fuel cell, electrical energy is generated by the reaction of H_2 and O_2 without evolution of heat (cold combustion).
- Liquid hydrogen is used as rocket fuel.
- As a reducing agent in extraction of metals.

Hydrides

Dihydrogen, under certain reaction conditions combine with almost all elements except noble gases to form binary compounds, called hydrides.

Classification of Hydrides

The hydrides are classified into four categories.

1. Ionic or Saline or Salt-like Hydrides

- These hydrides are stoichiometric compounds of dihydrogen formed with most of the *s*-block elements which are highly electropositive in character, e.g. LiH, BeH_2 and MgH_2.
- In fact, BeH_2 and MgH_2 are polymeric in structure. Due to their high reactivity with water, ionic hydrides are used to remove traces of water from organic solvents.

2. Covalent or Molecular Hydrides

- Dihydrogen forms molecular compounds with most of the *p*-block elements. Most familiar examples are CH_4, NH_3, H_2O and HF etc.
- Molecular hydrides are further classified, according to the relative number of electrons and bonds in their Lewis structure into following types:
  (i) **Electron deficient hydrides** have incomplete octet, so behave as Lewis acids, i.e. are electron acceptors. e.g. B_2H_6. These hydrides are formed by the elements of group 13.
  (ii) **Electron precise hydrides** have required number of electrons to write their Lewis structure. These are obtained from elements of group 14 (e.g. CH_4) which are tetrahedral in geometry.
  (iii) **Electron rich hydrides** have excess electrons, which are present as lone pairs. Elements of group 15-17 form such type of hydrides. e.g. $\ddot{N}H_3$ (1 lone pair), $H_2\ddot{\underset{\cdot\cdot}{O}}$ (2 lone pairs) and $H\ddot{\underset{\cdot\cdot}{F}}\colon$ (3 lone pairs).

3. Metallic or Non-stoichiometric or Interstitial Hydrides

- These are formed by many *d*-block and *f*-block elements, however the metals of group 7, 8 and 9 do not form hydride (hydride gap).
- These hydrides are mainly formed by
  (i) transition metals of group 3, 4, 5 of *d*-block
  (ii) Cr metal from group 6
  (iii) *f*-block elements, e.g. $LaH_{2.87}$, $YbH_{2.55}$ etc.
- Metallic hydrides are non-stoichiometric. They have metallic lattice and hydrogen is present at the interstitial sites. These conduct heat and electricity just like metals except hydrides of Eu and Yb. (EuH_2 and YbH_2 are ionic and stoichiometric)

4. Polymeric Hydrides and Complex Hydrides

- Polymeric hydrides are formed by elements having electronegativity in the range 1.4 to 2.0, e.g. $(BeH_2)_n$, $(AlH_3)_n$ etc.
- In complex hydrides, H^- acts as ligand and is attached to central metal atom, e.g. $LiAlH_4$, $LiBH_4$ etc.

Water (H_2O)

The water molecules contains one oxygen and two hydrogen atoms connected by covalent bonds. The physical and chemical properties of water are given below:

Physical Properties

- Water (H_2O) is polar in nature. It exists in liquid state at room temperature due to intermolecular hydrogen bonding.
- HOH bond angle is 104.5° and O—H bond length is 95.7 pm. H_2O (ice) has four hydrogen bonds per molecule and hence, has a highly ordered three dimensional cage-like structure.
- Ice has low density than H_2O (liquid) but H_2O has maximum density at 3.98°C.

Chemical Properties

(i) Water is amphoteric in nature.

$$\underset{\text{Base}}{H_2O(l)} + \underset{\text{Acid}}{HCl(aq)} \rightleftharpoons \underset{\text{Acid}}{H_3O^+(aq)} + \underset{\text{Base}}{Cl^-(aq)}$$

$$\underset{\text{Acid}}{H_2O(l)} + \underset{\text{Base}}{NH_3(aq)} \rightleftharpoons \underset{\text{Acid}}{NH_4^+(aq)} + \underset{\text{Base}}{OH^-(aq)}$$

(ii) Water react with metals and non-metals both.

$$2Na(s) + 2H_2O(l) \longrightarrow 2NaOH(aq) + H_2(g)$$

$$2F_2(g) + 2H_2O(l) \longrightarrow 4H^+(aq) + 4F^-(aq) + O_2(g)$$

(iii) In **hydrated salts**, water may remain in five types such as coordinated water, hydrogen bonded water, lattice water, clathrate water and zeolitic water.
- **Zeolites** are the hydrated sodium aluminium silicate that contains cavities or certain channels in which H_2O molecules get trapped.
- **Clathrate** contains host molecules that crystallises with an open structure containing water molecules.

(iv) A number of compounds such as calcium hydride, calcium phosphide etc., undergo hydrolysis with water. The hydrolysis of hydrides with H_2O is highly exothermic and may be explosive as H_2 catches fire. CO_2 is reduced by hot metal hydride, so it cannot use to extinguish such fire.

Hard and Soft Water

The water which lathers with soap is soft, if not, is hard. Hardness of water is of two types:

1. **Temporary Hardness of Water** is due to the presence of magnesium and calcium hydrogen carbonates. It can be removed either by **boiling,** through which the soluble $Mg(HCO_3)_2$ is converted into insoluble $Mg(OH)_2$ and $Ca(HCO_3)_2$ is changed into insoluble $CaCO_3$. These precipitates can be removed by filtration
2. **Permanent Hardness of Water** is due to the presence of soluble salts of magnesium and calcium in the form of chlorides and sulphates.

Permanent hardness can be removed by the following methods:

(i) **Calgon's Method** Sodium hexametaphosphate ($Na_6P_6O_{18}$), commercially called calgon, when added to hard water, the calcium and magnesium ions present in hard water react with calgon and form soluble complex salts which does not cause any hinderance in other information.

$$2M^{2+} + Na_2[Na_4(PO_3)_6] \longrightarrow \underset{\text{Soluble}}{[Na_2M(PO_3)_6]^{2-}} + 4\,Na^+$$

$$[M^{2+} = Ca^{2+}, Mg^{2+}]$$

(ii) **Ion-exchange Method or Permutit or Zeolite process** is also used to remove permanent hardness.

$$2NaZ(s) + M^{2+}(aq) \longrightarrow MZ_2(s) + 2Na^+(aq) \quad (M = Mg, Ca)$$

where, NaZ is sodium aluminium silicate ($NaAlSiO_4$). This is also known as sodium zeolite. It is regenerated for further use by treating with an aqueous NaCl solution (brine).

$$MZ_2(s) + 2NaCl(aq) \longrightarrow 2NaZ(s) + MCl_2(aq)$$

(iii) **Synthetic Resins Method** Now a days hard water is softened by using synthetic cation exchangers. Cation exchange resins contain large organic molecules with —SO_3H group and are water insoluble. Ion-exchange resin (RSO_3H) is changed to RNa by treating it with NaCl.

$$2RNa + M^{2+}(aq) \longrightarrow R_2M(s) + 2Na^+(aq) \quad (M^{2+} = Ca^{2+}, Mg^{2+})$$

The resin can be regenerated by adding aqueous NaCl solution.

Heavy Water (D_2O)

- It has quite similar physical and chemical properties to those of H_2O. Dielectric constant of D_2O is lower than that of H_2O and rate of reactions are much slower than H_2O.
- It is used as a moderator in nuclear reactions, as trace compound for studying reaction mechanism for the preparation of deuterium.

For e.g.

$$CaC_2 + 2D_2O \longrightarrow C_2D_2 + Ca(OD)_2$$
$$SO_3 + D_2O \longrightarrow D_2SO_4$$
$$Al_4C_3 + 12D_2O \longrightarrow 3CD_4 + 4Al(OD)_3$$

Hydrogen Peroxide (H_2O_2)

- Hydrogen peroxide is a compound with an oxygen-oxygen single bond. It is also a strong oxidiser.
- Hydrogen peroxide has a non-polar structure. The molecular dimensions in solid and gas phase are as follows:

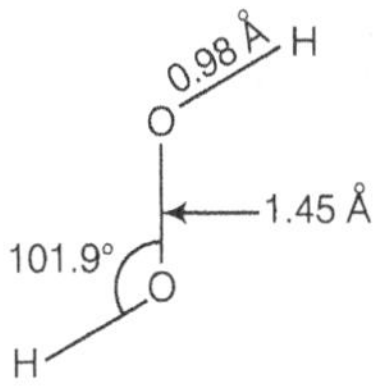

Solid phase

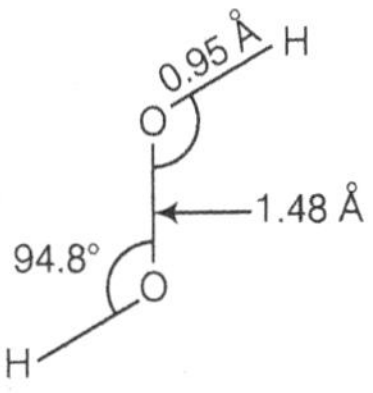

Gas phase

Preparation of H_2O_2

- It can be prepared by treating BaO_2 in presence of acid at high temperature.

$$BaO_2 \cdot 8H_2O(s) + H_2SO_4(aq) \longrightarrow BaSO_4(s) + H_2O_2(aq) + 8H_2O(l)$$

- Industrially, it is prepared by the auto-oxidation of 2-alkylanthraquinols.

Chemical Properties

- It acts as an oxidising as well as reducing agent in both acidic and alkaline media.
- Oxidising action in acidic medium,

$$2Fe^{2+}(aq) + 2H^+(aq) + H_2O_2(aq) \longrightarrow 2Fe^{3+}(aq) + 2H_2O(l)$$
$$PbS(s) + 4H_2O_2 \longrightarrow PbSO_4(s) + 4H_2O(l)$$

- Oxidising action in basic medium,

$$Mn^{2+} + H_2O_2 \longrightarrow Mn^{4+} + 2OH^-$$

- Reducing action in acidic medium,

$$2MnO_4^- + 6H^+ + 5H_2O_2 \longrightarrow 2Mn^{2+} + 8H_2O + 5O_2$$

- Reducing action in basic medium,

$$I_2 + H_2O_2 + 2OH^- \longrightarrow 2I^- + 2H_2O + O_2$$

- Many reactions of H_2O_2 are radical reactions, therefore a mixture of H_2O_2 and Fe (II) is a source of hydroxyl radicals for organic reactions.

Uses of H_2O_2

- Aqueous solution of H_2O_2 is used as germicide, antiseptic, preservative for milk and wine, bleaching agent for soft materials. 30% H_2O_2 is called perhydrol. Its volume strength is 100 and molarity is 8.8.
- It is used as an antichlor and in refreshing old oil paintings which became black.

NOTE H_2O_2 decomposes slowly on exposure to light,

$$2H_2O_2(l) \longrightarrow 2H_2O(l) + O_2(g)$$

In the presence of metal surfaces or traces of alkali (present in glass containers), the above reaction is catalysed. It is therefore, stored in wax lined glass or plastic vessels in dark. Urea can be added as a stabiliser.

DAY PRACTICE SESSION 1

FOUNDATION QUESTIONS EXERCISE

1 *Ortho* and *para* hydrogen differ in the
(a) number of protons
(b) molecular weight
(c) nature of spins of protons
(d) nature of spins of electrons

2 The metal, which gives hydrogen on treatment with acid as well as sodium hydroxide is
(a) Fe (b) Zn
(c) Cu (d) None of these

3 Which of the following reactions increases production of dihydrogen from synthesis gas?
(a) $CH_4(g) + H_2O(g) \xrightarrow[Ni]{1270\,K} CO(g) + 3H_2(g)$
(b) $C(s) + H_2O(g) \xrightarrow{1270\,K} CO(g) + H_2(g)$
(c) $CO(g) + H_2O(g) \xrightarrow[Catalyst]{673K} CO_2(g) + H_2(g)$
(d) $C_2H_6 + 2H_2O \xrightarrow[Ni]{1270K} 2CO + 5H_2$

4 In context with the industrial preparation of hydrogen from water gas $(CO + H_2)$, which of the following is the correct statement?
(a) CO is oxidised to CO_2 with steam in the presence of catalyst followed by absorption of CO_2 in alkali
(b) CO and H_2 are fractionally separated using differences in their densities
(c) CO is removed by absorption in aqueous Cu_2Cl_2 solution
(d) H_2 is removed through occlusion with Pd

5 H_2 gas is liberated at anode by electrolysis of which of the following aqueous solution?
(a) NaH (b) $HCOO^-Na^+$
(c) NaCl (d) LiH

6 Very pure hydrogen (99.9) can be made by which of the following processes?
(a) Reaction of methane with steam
(b) Mixing natural hydrocarbons of high molecular weight
(c) Electrolysis of water
(d) Reaction of salts like hydrides with water

7 In which of the following reactions does dihydrogen act as oxidising agent?
(a) $Ca + H_2 \longrightarrow$ (b) $H_2 + O_2 \longrightarrow$
(c) $H_2 + F_2 \longrightarrow$ (d) $CuO + H_2 \longrightarrow$

8 The maximum possible number of hydrogen bonds in a water molecule can form in ice is
(a) 1 (b) 2 (c) 3 (d) 4

9. Which one is used as propellants for rockets?
(a) Liq. H_2 + Liq. O_2 (b) Liq. N_2 + Liq. O_2
(c) Liq. H_2 + Liq. N_2 (d) Liq. O_2 + Liq. air

10 Hydrogen exists in atomic state in which of the following compounds?
(a) Metallic hydrides (b) Ionic hydrides
(c) Molecular hydrides (d) H_2O

11 Which of the following is electron precise hydride?
(a) B_2H_6 (b) NH_3 (c) H_2O (d) CH_4

12 The hydrides of the first elements in groups 15-17 namely NH_3, H_2O and HF respectively show abnormally high values for melting and boiling points. This is due to
(a) small size of N, O and F
(b) the ability to form extensive intermolecular H-bonding
(c) the ability to form extensive intramolecular H-bonding
(d) effective van der Waals' interaction

13 Ice floats on water because
(a) its density is less than that of water
(b) crystal structure of ice has empty space
(c) Both of the above
(d) None of the above

14 When zeolite, which is hydrated sodium aluminium silicate is treated with hard water, the sodium ions are exchanged with
(a) H^+ ion (b) Ca^{2+} ion
(c) SO_4^{2-} ion (d) OH^- ion

15 Which one of the following statements about water is false? → JEE Main 2016
(a) Water can act both as an acid and as a base
(b) There is extensive intramolecular hydrogen bonding in the condensed phase
(c) Ice formed by heavy water sinks in normal water
(d) Water is oxidised to oxygen during photosynthesis

16 Which one of the following processes will produce hard water?
(a) Saturation of water with $CaSO_4$
(b) Saturation of water with $CaCO_3$
(c) Addition of Na_2SO_4 to water
(d) Saturation of water with $MgCO_3$

17. Which of the following is responsible for the permanent hardness of water?
(a) Calcium bicarbonate
(b) Sodium chloride
(c) Magnesium bicarbonate
(d) Calcium sulphate

18. Permutit is chemically
(a) hydrated sodium aluminium silicate
(b) sodium hexaphosphate
(c) sodium bicarbonate
(d) calcium hydroxide

19. The hardness of water sample containing 0.002 mole of magnesium sulphate dissolved in a litre of water is expressed as

(a) 20 ppm (b) 200 ppm
(c) 2000 ppm (d) 120 ppm

20 Hydrogen peroxide is used as

(a) an oxidant only (b) a reductant only
(c) an acid only (d) All of these

21 Decomposition of H_2O_2 is favoured by

(a) traces of acids (b) alcohol
(c) acetanilide (d) MnO

22 In aqueous solution, hydrogen peroxide oxidises H_2S to

(a) sulphur (b) sulphuric acid
(c) Caro's acid (d) Marshall's acid

23 30 volume hydrogen peroxide means

(a) 30% of H_2O_2 solution
(b) $30 cm^3$ of the solution contains 1g of H_2O_2
(c) $1 cm^3$ of the solution liberates $30 cm^3$ of O_2 at STP
(d) $30 cm^3$ of the solution contains 1 mole of H_2O_2

24 Which of the following statements is not correct ?

(a) H_2O_2 oxidises Fe (II) to Fe (III)
(b) H_2O_2 can be obtained by electrolysis of dil. H_2SO_4
(c) H_2O_2 reduces Mn (VII) to Mn (II)
(d) H_2O_2 is a weak base

25 Acidified solution of chromic acid on treatment with H_2O_2 yields

(a) $CrO_3 + H_2O + O_2$
(b) $Cr_2O_3 + H_2O + O_2$
(c) $CrO_5 + H_2O + K_2SO_4$
(d) $H_2Cr_2O_7 + H_2O + O_2$

26 Hydrogen peroxide when added to a solution of potassium permanganate acidified with H_2SO_4,.

(a) forms water only
(b) acts as an oxidising agent
(c) acts as a reducing agent
(d) reduces sulphuric acid

27 Consider the following reactions :

I. $H_2O_2 + O_3 \longrightarrow H_2O + 2O_2$
II. $H_2O_2 + Ag_2O \longrightarrow 2Ag + H_2O + O_2$

Role of hydrogen peroxide in the above reactions is respectively

(a) oxidising in I and reducing in II
(b) reducing in I and oxidising in II
(c) reducing in I and II
(d) oxidising in I and II

28. Hydrogen peroxide oxidises $[Fe(CN)_6]^{4-}$ to $[Fe(CN)_6]^{3-}$ in acidic medium but reduces $[Fe(CN)_6]^{3-}$ to $[Fe(CN)_6]^{4-}$ in alkaline medium. The other products formed are, respectively. → JEE Main 2015

(a) $(H_2O + O_2)$ and H_2O
(b) $(H_2O + O_2)$ and $(H_2O + OH^-)$
(c) H_2O and $(H_2O + O_2)$
(d) H_2O and $(H_2O + OH^-)$

29. From the following statements regarding H_2O_2, choose the incorrect statement. → JEE Main 2018

(a) It can act only as an oxidising agent
(b) It decomposed on exposure to light
(c) It has to be stored in plastic or wax lined glass bottles in dark
(d) It has to be kept away from dust

Direction (Q. Nos. 30-32) *In the following questions, Assertion* (A) *followed by a Reason* (R) *is given. Choose the correct answer out of the following choices.*

(a) Assertion and Reason both are correct and Reason is the correct explanation of the Assertion
(b) Assertion and Reason both are correct but Reason is not the correct explanation of the Assertion
(c) Assertion is correct and Reason is incorrect
(d) Both Assertion and Reason are incorrect

30. Assertion Density of ionic hydrides such as LiH is greater than that of metals from which they are formed.

Reason Small H^- (hydride ions) occupy holes in the lattice of metal without distorting metal lattice.

31. Assertion Melting point of D_2O is higher than that of H_2O.

Reason Heavy water is more viscous than ordinary water, H_2O.

32. Assertion H_2O_2 acts as an oxidising agent as well as reducing agent.

Reason In H_2O_2, oxygen has −1 oxidation state and it can increase or decrease its oxidation state.

DAY PRACTICE SESSION 2

PROGRESSIVE QUESTIONS EXERCISE

1 Out of LiH, MgH_2 and CuH
(a) all are ionic hydrides
(b) LiH, MgH_2 are ionic and CuH covalent hydride
(c) all are covalent hydrides
(d) LiH is ionic, MgH_2 and CuH are intermediate hydrides

2 The hydride ion H^- is stronger base than its hydroxide ion OH^-. Which of the following reactions will occur if sodium hydride (NaH) is dissolved in water?
(a) $2H^+(aq) \longrightarrow H_2 + 2e^-$
(b) $H^-(aq) + H_2O(l) \longrightarrow OH^- + H_2$
(c) $H^- + H_2O(l) \longrightarrow$ No reaction
(d) None of the above

3 The compound sodium polymeta phosphate $Na_2[Na_4(PO_3)_6]$ is called Calgon because
(a) it was developed by the scientist
(b) it was developed first in California
(c) it refers to calcium gone
(d) it is based on the name of the company which developed

4 Which of the following statements concerning protium, deuterium and tritium is not true?
(a) They are isotopes of each other
(b) They have similar electronic configurations
(c) They exists in the nature in the ratio of 1 : 2 : 3 respectively
(d) Their mass numbers are in the ratio 1 : 2 : 3 respectively

5 Select the incorrect statement.
(a) *Ortho* and *para* hydrogen are different due to difference in their nuclear spins
(b) *Ortho* and *para* hydrogen are different due to different in their electron spins
(c) *Para* hydrogen has a lower internal energy than that of *ortho* hydrogen
(d) *Para* hydrogen is more stable at lower temperature

6 Which one of the following statements is incorrect with regard to *ortho* and *para* dihydrogen?
(a) They are nuclear spin isomers
(b) The *ortho* isomer has zero nuclear spin whereas the *para* isomer has one nuclear spin
(c) The *para* isomer is favoured at low temperatures
(d) The thermal conductivity of the *para* isomer is 50% greater than that of the *ortho* isomer

7 H_2 can be obtained from
(a) water gas ($CO + H_2$) by liquefaction of CO at low temperature under pressure
(b) water gas by oxidation of CO into CO_2 (by steam) which can be easily removed by dissolving in H_2O
(c) electrolysis of water
(d) All the above methods can be used to obtain H_2

8 Which one of the following pairs of substances will not produce hydrogen when reacted together?
(a) Copper and conc. nitric acid
(b) Ethanol and metallic sodium
(c) Magnesium and steam
(d) Phenol and metallic sodium

9 Metal hydrides are ionic, covalent or molecular in nature. Among LiH, NaH, KH, RbH, CsH, the correct order of increasing ionic character is
(a) LiH > NaH > CsH > KH > RbH
(b) LiH < NaH < KH < RbH < CsH
(c) RbH > CsH > NaH > KH > LiH
(d) NaH > CsH > RbH > LiH > KH

10 Which of the following equations depicts the oxidising nature of H_2O_2?
(a) $2MnO_4^- + 6H^+ + 5H_2O_2 \longrightarrow 2Mn^{2+} + 8H_2O + 5O_2$
(b) $2Fe^{3+} + 2H^+ + H_2O_2 \longrightarrow 2Fe^{2+} + 2H_2O + O_2$
(c) $2I^- + 2H^+ + H_2O_2 \longrightarrow I_2 + 2H_2O$
(d) $KIO_4 + H_2O_2 \longrightarrow KIO_3 + H_2O + O_2$

11 Which of the following equation depicts reducing nature of H_2O_2?
(a) $2[Fe(CN)_6]^{4-} + 2H^+ + H_2O_2 \longrightarrow 2[Fe(CN)_6]^{3-} + 2H_2O$
(b) $I_2 + H_2O_2 + 2OH^- \longrightarrow 2I^- + 2H_2O + O_2$
(c) $Mn^{2+} + H_2O_2 \longrightarrow Mn^{4+} + 2OH^-$
(d) $PbS + 4H_2O_2 \longrightarrow PbSO_4 + 4H_2O$

12 Which one of the following statements is correct for H_2O_2?
(a) It reduces ferricyanide to ferrocyanide in acidic medium
(b) It oxidises lead monoxide to lead dioxide
(c) It acts as a reducing agent in the decolourisation of acidified $KMnO_4$
(d) It oxidises hydrides of chlorine and bromine to their diatomic gases

13 Hydrogen resembles halogens in many respects for which several factors are responsible. Of the following factors which one is most important in this respect?
(a) Its tendency to lose an electron to form a cation
(b) Its tendency to gain a single electron in its valence shell to attain stable electronic configuration
(c) Its low negative electron enthalpy value
(d) Its small size

ANSWERS

SESSION 1	**1** (c)	**2** (b)	**3** (c)	**4** (a)	**5** (a)	**6** (d)	**7** (a)	**8** (d)	**9** (a)	**10** (a)
	11 (d)	**12** (b)	**13** (c)	**14** (b)	**15** (b)	**16** (a)	**17** (d)	**18** (a)	**19** (b)	**20** (d)
	21 (d)	**22** (a)	**23** (c)	**24** (d)	**25** (c)	**26** (c)	**27** (a)	**28** (c)	**29** (a)	**30** (a)
	31 (b)	**32** (a)								
SESSION 2	**1** (d)	**2** (b)	**3** (c)	**4** (c)	**5** (b)	**6** (b)	**7** (d)	**8** (a)	**9** (b)	**10** (c)
	11 (b)	**12** (c)	**13** (b)							

Hints and Explanations

SESSION 1

1 *Ortho* and *para*-hydrogens differ in the nature of spins of protons.

2 Zn is amphoteric metal and reacts with acids as well as alkalies to give hydrogen gas.

3 $CO\,(g) + H_2O\,(g) \xrightarrow[\text{Catalyst}]{673\text{ K}} CO_2(g) + H_2\,(g)$

4 In this reaction, CO is oxidised to CO_2 with steam in the pressure of catalyst followed by absorption of CO_2 in alkali.

$$\underset{\text{Water gas}}{CO + H_2} + H_2O \xrightarrow{FeCrO_4} CO_2 + 2H_2$$
$$\xrightarrow{KOH} K_2CO_3$$

5 H_2 gas is liberated at anode by electrolysis of NaH aqueous solution.

$$NaH + H_2O \longrightarrow \underset{\text{Near cathode}}{NaOH} + \underset{\text{Anode}}{H_2}$$

6 Hydrides are instant sources of hydrogen of higher purity. They react with H_2O forming H_2 gas.

$$CaH_2 + 2H_2O \longrightarrow Ca(OH)_2 + 2H_2 \uparrow$$

7 Calcium when reacts with hydrogen forms calcium hydride. Here, dihydrogen act as oxidising agent.

$$Ca + H_2 \longrightarrow 2CaH$$

8 In ice, each H_2O molecule is surrounded tetrahedrally by four other water molecules through H-bonds.

9 Liquid H_2 and liquid O_2 are used as rocket propellants.

10 Hydrogen exists in atomic state in metallic hydrides.

11 The hydride which contains exact number of electrons to form normal covalent bonds are called electron precise hydrides, e.g. CH_4.

12 Hydrides of N, O and F because of the small size and high electronegativity of elements have ability to form extensive intermolecular, (i.e. between two molecules) hydrogen bonding. Thus, a large amount of energy is required to break these bonds, the melting and boiling points of hydrides of these elements are abnormally high.

13 Ice floats on water because its density is less than that of water which in turn is due to empty space in its crystal structure. Crystal structure of ice is regular hexagon with empty spaces at the centre.

14 $$\underset{\text{Sodium zeolite}}{Na_2Z} + \underset{\text{From hard water}}{CaCl_2} \longrightarrow \underset{\text{Calcium zeolite}}{CaZ} + 2NaCl$$

[where, $Z = Al_2Si_2O_8 \cdot x\,H_2O$]

15 Water consists of extensive intermolecular H-bonding in the condensed phase.

16 Permanent hardness is due to the presence of chlorides and sulphates of calcium, magnesium and iron in water.

17 Chlorides and sulphates of calcium and magnesium are responsible for the permanent hardness of water. Hence, $CaSO_4$ is responsible for the permanent hardness of water.

18 Permutit is chemically hydrated sodium aluminium silicate.

19 The hardness of water sample containing 0.002 mole of $MgSO_4$ dissolved in 1 L of water.

Hardness is measured in terms of ppm of $CaCO_3$.

$$MgSO_4 \equiv CaCO_3 \equiv 0.002 \text{ mol L}^{-1}$$
$$CaCO_3 = 0.002 \times 100 \text{ gL}^{-1}$$
$$= 0.002 \times 100 \text{ g in } 10^3 \text{ mL}$$
$$= \frac{0.002 \times 100 \times 10^6}{10^3} \text{ g in } 10^6 \text{ mL(ppm)}$$
$$= 200 \text{ ppm}$$

20 H_2O_2 acts as an oxidant, reductant and an acid (weak).

21 MnO catalysis the decomposition of H_2O_2.

22 In aqueous solution, hydrogen peroxide oxidises H_2S to S.

$$H_2S + H_2O_2 \longrightarrow 2H_2O + S$$

23 30 Volume hydrogen peroxide means 1 cm^3 of this solution produces 30 cm^3 O_2 at NTP.

$$2H_2O_2 \longrightarrow 2H_2O + O_2$$

24 H_2O_2 is not a weak base but a weak acid.

25 $$K_2Cr_2O_7 + H_2SO_4 \longrightarrow K_2SO_4 + H_2Cr_2O_7$$
$$H_2Cr_2O_7 + 4H_2O_2 \longrightarrow 2CrO_5 + 5H_2O$$

26 $$2K\overset{+7}{Mn}O_4 + 3H_2SO_4 + 5H_2O_2 \longrightarrow K_2SO_4 + 2\overset{+2}{Mn}SO_4 + 8H_2O + 5O_2$$

$\therefore$ H_2O_2 is acting as a reducing agent. It reduces $KMnO_4$ to Mn^{2+} ion.

27 In the reaction,

Oxidation

$$\overset{+1\ -1}{H_2O_2} + O_3 \longrightarrow \overset{+1\ -2}{H_2O} + 2O_2$$

Reduction

The oxidation state of oxygen in H_2O_2 changes from −1 to −2 H_2O_2 oxidises O_3 into O_2, thus it behaves as an oxidising agent

For the reaction,

$$\overset{+1\ -1}{H_2O_2} + \overset{+1\ -2}{Ag_2O} \longrightarrow \overset{0}{2Ag} + \overset{+1\ -2}{H_2O} + \overset{0}{O_2}$$

Reduction Oxidation

In the above reaction, H_2O_2 reduces Ag_2O to Ag and the oxidation state of 0 changes from −1 to 0, hence it acts has a reducing agent.

28 Both reactions in their complete format are written below

(i) In acidic medium,

$$[Fe^{2+}(CN)_6]^{4-} + H_2\overset{-1}{O_2} + 2H^+ \longrightarrow [Fe^{3+}(CN)_6]^{3-} + 2H_2\overset{-2}{O}$$

(ii) In alkaline medium,

$$[Fe^{3+}(CN)_6]^{3-} + H_2\overset{-1}{O_2} + 2OH^- \longrightarrow [Fe^{2+}(CN)_6]^{4-} + O_2 + 2H_2O$$

Hence, H_2O (for reaction (i)) and $O_2 + H_2O$ (for reaction (ii)) are produced as by product.

29 H_2O_2 acts as an oxidising as well as reducing agent, because oxidation number of oxygen in H_2O_2 is −1. So, it can be oxidised to oxidation state 0 or reduced to oxidation state −2.

H_2O_2 decomposes on exposure to light. So, it has to be stored in plastic or wax lined glass bottles in dark for the prevention of exposure. It also has to be kept away from dust.

30 Ionic hydrides are solid with high density because small H^- ion occupy holes in the lattice of metal without distorting metal lattice.

31 D_2O shows more stronger H-bonding than H_2O and thus, shows higher melting and boiling points. Also, heavy water is more viscous than ordinary water.

32 H_2O_2 acts as an oxidising as well as reducing agent because oxygen has −10 oxidation state, so it can be oxidised to oxidation state 0 or reduced to oxidation state −2.

$$2e^- + \underset{\text{Oxidant}}{O_2^{1-}} \longrightarrow 2\,O^{2-}$$

$$\underset{\text{Reductant}}{O_2^{1-}} \longrightarrow O_2^0 + 2e^-$$

SESSION 2

1 LiH is ionic, MgH_2 and CuH are intermediate hydrides.

2 Following reactions take place when NaH is dissolved in water.

$$2NaH(s) + H_2O(l) \longrightarrow 2NaOH(aq) + H_2(g)$$

3 The compound sodium polymetaphosphate $Na_2[Na_4(PO_3)_6]$ is called calgon because it refers to calcium gone.

4 Protium, deuterium and tritium exist in nature in the ratio of 3 : 2 : 1 respectively.

5 *Ortho* and *para* hydrogen are different due to difference in proton spin and not electron spin.

6 On the basis of spinning of protons, hydrogen has two allotropes, called *ortho* hydrogen and *para* hydrogen. In *ortho* hydrogen, spins of protons are in same direction, so the nuclear spin $= \frac{1}{2} + \frac{1}{2} = 1$. While in *para* hydrogen, spins of protons are in opposite direction, so the nuclear spin

$$= \frac{1}{2} + \left(-\frac{1}{2}\right) = 0$$

All other statements about *ortho* and *para* hydrogens are true.

7 Hydrogen can be obtained from water gas, i.e. by liquefaction of CO and by oxidation of CO into CO_2. It can also be obtained from electrolysis of water.

Reaction involved :

(i) Electrolysis of H_2O

$$2H_2O(l) \xrightarrow[\text{of acid/base}]{\text{Process}} 2H_2(g) + O_2(g)$$

(ii) $$CO(g) + H_2O(g) \xrightarrow[\text{Catalyst}]{673\text{ K}} CO_2(g) + H_2(g)$$

8 Copper, being less reactive, when reacts with conc. HNO_3, gives NO_2, but not H_2.

$$Cu + 4HNO_3(\text{conc.}) \longrightarrow Cu(NO_3)_2 + 2NO_2 + 2H_2O$$

While Mg and NaH being very reactive give hydrogen gas with steam and water respectively.

$$Mg + 2H_2O \longrightarrow MgO + H_2$$

$$NaH + H_2O \longrightarrow NaOH + H_2$$

Ethanol and phenol due to the presence of acidic hydrogen also produce hydrogen with metallic sodium.

$$C_2H_5OH + Na \longrightarrow C_2H_5O^-\ Na^+ + H_2$$

$$C_6H_5OH + Na \longrightarrow C_6H_5O^-Na^+ + H_2$$

9 Metal hydrides are ionic, covalent or molecular in nature. Ionic character increases as the size of the atom increases or the electronegativity of the atom decreases. The correct order of increasing ionic character is

LiH < NaH < KH < RbH < CsH

10 The reaction in which H_2O_2 is reduced, i.e. oxidation state of oxygen decreases from −1 to −2 depicts the oxidising nature of H_2O_2. e.g.

Reduction

$$2I^- + 2H^+ + \underset{\text{(Oxidising agent)}}{\overset{(-1)}{H_2O_2}} \longrightarrow \overset{0}{I_2} + \overset{(-2)}{2H_2O}$$

Increase in ON (oxidation)

11 $$\overset{0}{I_2} + H_2\overset{-1}{O_2} + 2OH^- \longrightarrow 2\overset{0}{I}^- + 2H_2\overset{0}{O} + O_2$$

depicts reducing nature of H_2O_2

12 H_2O_2 reduces acidified $KMnO_4$ into colourless manganous salt and gets oxidised into O_2.

$$5H_2O_2 + 2KMnO_4 + 3H_2SO_4 \longrightarrow K_2SO_4 + \underset{\text{Colourless}}{2MnSO_4} + 8H_2O + 5O_2$$

13 Hydrogen resemble halogens in many respects for which several factors are responsible. The most important is hydrogen (like halogens) accepts an electron readily to achieve nearest inert gas configuration.

DAY TWENTY ONE

s-Block Elements

Learning & Revision for the Day

- Group 1 Elements (Alkali Metals)
- Compounds of Sodium
- Biological Significance of Na and K
- Group 2 Elements (Alkaline Earth Metals)
- Compounds of Calcium
- Biological Significance of Mg and Ca

Group 1 Elements (Alkali Metals)

- Lithium ($Z = 3$), sodium ($Z = 11$), potassium ($Z = 19$), rubidium ($Z = 37$), caesium ($Z = 55$) and francium ($Z = 87$) are the elements of IA (or 1) group of the periodic table. These elements because of the highly alkaline nature of their water soluble hydroxides are known as **alkali metals.**
- These elements are known as *s*-block elements as their last or valence electron enters in the *s*-orbitals. These all are soft metals.

  Electronic Configuration All the alkali metals have one valence electron ns^1 outside the noble gas core. Their general electronic configuration is [noble gas] ns^1. They readily lose electron to give monovalent M^+ ions. Hence, they are never found in free state in nature.

General Trends in Physical Properties

- Alkali metals are most electropositive elements and on moving down the group the electropositive character increases. Therefore, they readily lose electron to give monovalent, M^+ ions.
- The alkali metal atoms have the largest size and on moving down the group, the size of atom increases. Low densities of alkali metals are due to their large size.
- Alkali metals give flame test. The colour of the flame depends upon the wavelength of radiation emitted. Caesium and potassium are used as electrodes in photoelectric cells due to their low ionisation enthalpy.
- Alkali metals are strong reducing agent and down the group, their reducing nature increases.
- Alkali metal compounds are most ionic in nature and down the group, ionic character increases.
- Alkali metal oxides are most basic in nature and down the group, basic character increases.

PREP MIRROR
Your Personal Preparation Indicator

- No. of Questions in Exercises (x)—
- No. of Questions Attempted (y)—
- No. of Correct Questions (z)—
  (Without referring Explanations)

- Accuracy Level ($z/y \times 100$)—
- Prep Level ($z/x \times 100$)—

In order to expect good rank in JEE, your Accuracy Level should be above 85 & Prep Level should be above 75.

NOTE The ionisation enthalpies of alkali metals are considerably low and decreases down the group from Li to Cs.

- Hydration enthalpy of alkali metal ions decreases with increase in ionic sizes $Li^+ > Na^+ > K^+ > Rb^+ > Cs^+$. Therefore, Li^+ has maximum degree of hydration and due to this reason lithium salts are mostly hydrated, e.g. $LiCl \cdot 2H_2O$.
- Mobility of their cations in aqueous medium is directly proportional to the size of cation due to poor hydration enthalpy

$$\xrightarrow[\text{Mobility}]{Li^+ < Na^+ < K^+ < Rb^+ < Cs^+}$$

General Trends in Chemical Properties

The alkali metals are highly reactive due to their large size and low ionisation enthalpy. The reactivity of these metals increases down the group.

1. **Reactivity Towards Air**

$$4Li + O_2 \longrightarrow 2Li_2O \text{ (oxide)}$$
$$2Na + O_2 \longrightarrow Na_2O_2 \text{(peroxide)}$$
$$M + O_2 \longrightarrow MO_2 \text{ (superoxide) } (M = K, Rb, Cs)$$

- Metal peroxides are diamagnetic and sodium peroxide is widely used as an oxidising agent in inorganic chemistry.
- The superoxides are paramagnetic in nature because of the presence of three electron bonds with one unpaired electron.
- Normal oxides of Li and Na are colourless and diamagnetic, but K_2O is pale yellow, Pb_2O is bright yellow and Cs_2O is orange. Na_2O_2 is yellow due to the presence of small amount of superoxide because superoxide ion, O_2^- has three electron bonds with one unpaired electron.
- KO_2 is used in space capsules, sub-marines and breathing masks as it produces O_2 and removes CO_2 and CO.

2. **Solutions in Liquid Ammonia** The alkali metals dissolve in liquid ammonia giving deep blue solutions which are conducting in nature.

$$M + (x + y)NH_3 \longrightarrow [M(NH_3)_x]^+ + [e(NH_3)_y]^-$$

The blue colour of the solution is due to the ammoniated electrons. The solutions are paramagnetic and on standing, slowly liberate hydrogen resulting in the formation of amide.

$$M^+ + e^- + NH_3(l) \longrightarrow MNH_2 + \frac{1}{2}H_2(g)$$

3. **Reducing Nature** Reducing power of alkali metals in aqueous solution follows the trend: Li > Na > K > Rb > Cs. The standard electrode potential ($E°$) which measures the reducing power represents the overall change.

$M(s) \longrightarrow M(g)$ (sublimation energy)

$M(g) \longrightarrow M^+(g) + e^-$ (ionisation energy)

$H^+(g) + H_2O \longrightarrow M^+(aq)$ (hydration enthalpy)

4. **Carbonates** Alkali metals on exposure to air and moisture get converted into oxides, hydroxides and finally to carbonates,

$$4M + O_2 \longrightarrow 2M_2O \xrightarrow{H_2O} 2MOH \xrightarrow{CO_2} M_2CO_3$$

NOTE (i) Alkali metals are normally kept in kerosene oil because of their high reactivity towards air and water.

(ii) Carbonates of alkali metals are thermally stable except Li_2CO_3 which decomposes readily to evolve CO_2 on heating. Due to the strong polarising action of small cation, Li^+ on the large carbonate ion facilitates their decomposition.

$$Li_2CO_3 \longrightarrow Li_2O + CO_2$$

Anomalous Properties of Lithium

- The anomalous behaviour of Li is due to exceptionally small size (like Mg) and high polarising power (i.e. charge/radius ratio) (like Mg) that exhibits some properties which are different from those of the other members of first group but similar to that of magnesium (present diagonally in the following group, i.e. II group).
- The property of showing similarity in properties with the element present diagonally opposite in the periodic table is called **diagonal relationship**.

These properties are:

(i) Alkali metals do not react with nitrogen except Li.

$$6Li + N_2 \xrightarrow{\text{Heat}} \underset{\text{Lithium nitride}}{2Li_3N}$$

(Li metal is used as scavenger in metallurgy to remove O_2 and N_2 gases.)

$$3Mg + N_2 \longrightarrow \underset{\text{Magnesium nitride}}{Mg_3N_2}$$

(ii) Alkali metal carbonates, nitrates and hydroxides do not decompose on heating into their oxides except lithium.

$$NaOH \xrightarrow{\text{Heat}} \text{No reaction}$$
$$2LiOH \xrightarrow{\text{Heat}} Li_2O + H_2O$$
$$Mg(OH)_2 \xrightarrow{\text{Heat}} MgO + H_2O$$
$$Na_2CO_3 \xrightarrow{\text{Heat}} \text{No reaction}$$
$$Li_2CO_3 \xrightarrow{\text{Heat}} Li_2O + CO_2$$
$$MgCO_3 \xrightarrow{\text{Heat}} MgO + CO_2$$
$$4LiNO_3 \xrightarrow{\text{Heat}} 2Li_2O + 4NO_2 + O_2$$
$$2Mg(NO_3)_2 \xrightarrow{\text{Heat}} 2MgO + 4NO_2 + O_2$$
$$2NaNO_3 \longrightarrow 2NaNO_2 + O_2$$

(iii) $MgCl_2$ and LiCl are deliquescent and crystallises as their hydrates, $LiCl \cdot 2H_2O$ and $MgCl_2 \cdot 8H_2O$.

Compounds of Sodium

The compounds of sodium are industrially important therefore, some of them are discussed below:

1. Sodium Carbonate (Washing Soda, $Na_2CO_3 \cdot 10H_2O$)

Preparation

- Sodium carbonate is generally prepared by Solvay process.

$$2NH_3 + H_2O + CO_2 \longrightarrow (NH_4)_2CO_3$$
$$(NH_4)_2CO_3 + H_2O + CO_2 \longrightarrow 2NH_4HCO_3$$
$$NH_4HCO_3 + NaCl \longrightarrow NH_4Cl + NaHCO_3$$

- Sodium hydrogen carbonate crystals gets separated out. These are heated to obtain sodium carbonate.

$$2NaHCO_3 \xrightarrow{150°C} Na_2CO_3 + CO_2\uparrow + H_2O$$

- In this process, NH_3 is recovered when the solution containing NH_4Cl is treated with $Ca(OH)_2$. Solvay process cannot be extended to the manufacture of potassium carbonate because potassium hydrogen carbonate is too soluble to be precipitated to a saturated solution of potassium chloride.

Properties

- Sodium carbonate is soluble in water with the evolution of considerable amount of heat. This solution is alkaline in nature due to hydrolysis.

$$Na_2CO_3 + 2H_2O \rightleftharpoons 2NaOH + H_2CO_3$$

- On passing CO_2 through the concentrated solution of sodium carbonate, sodium bicarbonate gets precipitated

$$Na_2CO_3 + H_2O + CO_2 \longrightarrow 2NaHCO_3$$

- When solution of sodium carbonate and slaked lime is heated, sodium hydroxide is obtained.

$$Na_2CO_3 + Ca(OH)_2 \longrightarrow 2NaOH + CaCO_3$$

2. Sodium Hydroxide (Caustic Soda, NaOH)

Preparation

- Sodium hydroxide is prepared commercially by the electrolysis of sodium chloride in **Castner-Kellner cell** (or mercury cathode cell).

At cathode $Na^+ + e^- \xrightarrow{Hg} Na\text{-}Hg$ (Amalgam)

At anode $Cl^- \longrightarrow \frac{1}{2}Cl_2 + e^-$ (By product)

- The amalgam is treated with water to give sodium hydroxide and hydrogen gas.

$$2Na\text{-}Hg + 2H_2O \longrightarrow 2NaOH + 2Hg + H_2$$

Properties

- Sodium hydroxide is called **caustic soda** because it breaks down proteins of skin to a pasty mass. On exposure to atmosphere it absorbs moisture and CO_2.

$$2\,NaOH + CO_2 \longrightarrow Na_2CO_3 + H_2O$$

- It is a strong alkali so reacts with acids, acidic oxides and amphoteric oxides to form their corresponding salts. Non-metals like halogens, phosphorus, sulphur, silicon, boron also get attacked by NaOH. It reacts with metals like Zn, Al, Sn and Pb and evolve H_2 gas e.g.

$$Zn + NaOH \longrightarrow Na_2ZnO_2 + H_2$$

- It also reacts with metallic salts to form hydroxides. It is used in soap manufacture, purification of bauxite, manufacture of rayon etc.

3. Sodium Hydrogen Carbonate (Baking Soda, $NaHCO_3$)

Preparation

- Sodium hydrogen carbonate is prepared by saturating a solution of Na_2CO_3 (cold) with CO_2.

$$Na_2CO_3 + CO_2 + H_2O \longrightarrow \underset{\text{Sparingly soluble}}{2NaHCO_3}$$

Properties

- Its aqueous solution is alkaline in nature due to hydrolysis.

$$NaHCO_3 + H_2O \rightleftharpoons NaOH + H_2CO_3$$

- If heated, it decomposes to give Na_2CO_3 and CO_2 gas.

$$2NaHCO_3 \xrightarrow{Heat} Na_2CO_3 + H_2O + CO_2$$

- It is used as a medicine to neutralise acidity in stomach.
- It is used as a constituent of baking powder and in fire extinguisher.

Biological Significance of Na and K

- Na^+ ions are found mainly in extracellular region (outside the cell) and play an important role in the transmission of nerve signals. They also regulate the flow of water across cell membranes and in transport of sugars and amino acid into the cells.
- Prolonged sweating results in sodium ion loss in sweat thus, it is important that Na^+ ions are replaced through proper diet.
- Potassium ions are the most abundant cations within cell fluids where they activate many enzymes, which participate in oxidation of glucose to produce ATP (adenosine triphosphate).
- Potassium ions along with sodium ions are responsible for transmission of nerve signals.
  The functional features of nerve cells depend upon the sodium-potassium ion gradient that is established in the cell. Their ionic gradients are maintained by sodium-potassium pumps that operate across the cell membranes.

Group 2 Elements
(Alkaline Earth Metals)

Be, Mg, Ca, Sr, Ba and Ra are also belong to *s*-block and are called alkaline earth metals. These are called 'alkaline earth metals' (except Be) due to the following facts:

(i) Their hydroxides form alkaline aqueous solutions.

(ii) Their oxides are earthen, i.e. their oxides are found in earth crust.

Electronic Configuration

These elements have two electrons in the *s*-orbital of the valence shell. Their general electronic configuration is represented as [noble gas]ns^2.

General Trends in Physical Properties

- These are less metallic as compared to alkali metals. Radium is the radioactive element of this group.
- The alkaline earth metals also have low ionisation enthalpies but these are higher than that of alkali metals.
- The melting and boiling points of these metals are higher than the corresponding alkali metals due to smaller size.
- Except Be and Mg (due to their small size and high IE), all other alkaline earth metals give flame test like calcium, strontium and barium and impart characteristic brick red, crimson and apple green colours, respectively to the flame.

NOTE Like alkali metal ions, the hydration enthalpies of alkaline metal ions decrease with increase in ionic size on moving down the group.

$$Be^{2+} > Mg^{2+} > Ca^{2+} > Sr^{2+} > Ba^{2+}$$

The hydration enthalpies of alkaline earth metal ions are larger than those of alkali metal ions.

General Trends in Chemical Properties

Alkaline earth metals give all the usual reactions of alkali metals. However, these are less reactive as compared to alkali metals in their chemical reactions.

1. **Reaction with Water and Acid** Except Be and Mg, all other alkaline earth metals give hydroxide with cold water while Be forms hydroxide with steam and Mg with hot water.

$$Ca + 2H_2O \longrightarrow Ca(OH)_2$$

- Except Mg, other members of the group do not give H_2 with HNO_3.

$$Mg + 2HNO_3\ (5\%) \longrightarrow Mg(NO_3)_2 + H_2$$

- With HNO_3, Be becomes passive due to the formation of oxide layer.
- Alkaline earth metal chlorides, nitrates and hydroxides exhibit an increase in their solubilities on moving down the group.

$$\underrightarrow{MgCl_2 < CaCl_2 < SrCl_2 < BaCl_2}$$

$$Mg(OH)_2 < Ca(OH)_2 < Sr(OH)_2 < Ba(OH)_2$$

2. **Carbonates** Carbonates of alkaline earth metals are insoluble in water. The solubility of carbonates in water decreases as the atomic number of metal ion increases.
   - Beryllium carbonate is unstable and can be kept only in the atmosphere of CO_2. The thermal stability increases with increase in cationic size.
   - Bicarbonates of alkali metals are found in solid state whereas, bicarbonates of alkaline earth metals exist only in solution state.
3. **Sulphates** The sulphates of alkaline earth metals are all white solids and stable to heat. $BeSO_4$ and $MgSO_4$ are readily soluble in water. The solubility decreases from $CaSO_4$ to $BaSO_4$.

   The insolubility of $BaSO_4$ is used for detecting obstruction in the digestive system by the technique commonly known as barium meal.
4. **Nitrides** All the alkaline earth metals form nitrides, however among the alkali metals, only Li forms nitrides.

$$3Ca + N_2 \longrightarrow Ca_3N_2$$

5. **Halides** Halides of Be are essentially covalent, hygroscopic and fumes in air while halides of other alkaline earth metals are ionic and ionic character increases down the group.

   In solid state $BeCl_2$ exists in polymeric chain while in vapour state it tends to form a chloro-bridged dimer which dissociates into the linear monomer at high temperature (1200 K).
6. **Reducing Nature** Reducing character or reactivity of alkaline earth metals increases from Be to Ba, i.e. $Be < Mg < Ca < Sr < Ba$.
   - $Mg(ClO_4)_2$ is used as drying agent under the name anhydrone. It is a strong oxidising agent. So, it is not used with organic material.

Anomalous Behaviour of Beryllium

Beryllium due to its exceptionally small size (like Al) and high polarising power (like Al) shows similarities with aluminium, in its properties.

These properties are:

- BeO, like Al_2O_3, is amphoteric and covalent while oxides of other alkaline earth elements are ionic and basic in nature.
- Both $BeCl_2$ and $AlCl_3$ are soluble in organic solvents because of covalent nature and both have a bridged polymeric structure.
- Beryllium hydroxide dissolves in excess of alkali to give a beryllate ion, $[Be(OH)_4]^{2-}$ just as aluminium hydroxide gives aluminate ion, $[Al(OH)_4]^-$.

- Beryllium and aluminium ions have strong tendency to form complexes, BeF_4^{2-}, AlF_6^{3-}.
- Carbides of Be are covalent and react with water to produce methane gas whereas carbides of other members are ionic and produce acetylene with water.

Compounds of Calcium

All the alkaline earth metals form several compounds. Some industrially important compounds of calcium are given below:

1. Lime (Quicklime), CaO

Preparation

$$\underset{\text{Limestone}}{CaCO_3} \underset{1000°C}{\overset{\Delta}{\rightleftharpoons}} CaO + CO_2$$

Uses

- It is used in the manufacture of $CaOCl_2$, $Ca(OH)_2$, CaC_2, glass and cement etc.
- It is used as basic lining in the furnaces.
- It is used in the purification of sugar and in water softening.

2. Limestone, $CaCO_3$

Preparation

$$\underset{\text{Slaked lime}}{Ca(OH)_2} + CO_2 \longrightarrow \underset{\text{Milky}}{CaCO_3} + H_2O$$

Uses

- It is used in the manufacture of quick lime, slaked lime, cement, glass and washing soda etc.
- It is used as a flux in smelting of iron and lead ores.

3. Plaster of Paris, $CaSO_4 \cdot \frac{1}{2}H_2O$

Preparation

$$\underset{\substack{\text{Calcium sulphate}\\\text{dihydrate or gypsum}}}{CaSO_4 \cdot 2H_2O} \xrightarrow[393K]{\Delta} \underset{\substack{\text{Calcium sulphate}\\\text{hemihydrate or}\\\text{plaster of Paris}}}{CaSO_4 \cdot \frac{1}{2}H_2O} + 1\frac{1}{2}H_2O$$

Above 393K, no water of crystallisation is left and $CaSO_4$ (anhydrous) is formed.This is called 'dead burnt plaster'.

Uses

- It is used in making toys, decorative materials and casts of statues.
- It is used in medical applications for setting fractured bones in the right position and in dentistry.

Cement

Preparation

- Cement is an important building material. It is obtained by combining a material such as clay which contains silica, SiO_2 along with the oxides of aluminium, iron and magnesium.
- The raw materials for the manufacture of clement are limestone and clay.

Uses

- It is used in plostering and in the construction of bridges, dams and buildings.
- It is used for making concrete and reinforced concrete.

NOTE The average composition of portland cement is CaO — 50-60%; SiO_2 — 20-25%; Al_2O_3 — 5-10%; MgO — 2-3%; Fe_2O_3 — 1-2% and SO_3 — 1-2%.

Biological Significance of Mg and Ca

- Mg^{2+} ions are present inside the animal cells while Ca^{2+} ions are in the body fluids, i.e. in the extracellular region, the same way as K^+ ions are inside the cell and Na^+ ions outside the cell. All enzymes that utilise ATP in phosphate transfer require Mg^{2+} as cofactor. In green plants, magnesium is present in chlorophyll.
- Ca^{2+} and Mg^{2+} are also essential for the transmission of impulses along nerve fibres.
- In bones and teeth, Ca is present as apatite, $Ca_3(PO_4)_2$ and in enamel on teeth as fluorapatite, $2Ca_3(PO_4)_2 \cdot CaF_2$. Ca^{2+} ions play an important role in blood clotting and are required to trigger the contraction of muscles.
- Ca^{2+} ions also regulate the heart beats.

Comparison between the Properties of Alkaline Earth Metals and Alkali Metals

Unlike the members of group I A, the chemistry of II A group elements is not completely dominated by the chemistry of cations.

When compared with alkali metals, it is found that :

- They are less reactive than alkali metals.
- They are less electropositive, i.e. less metallic than alkali metals.
- Their reducing power is much less than those of alkali metals.
- They are less basic than alkali metals.
- The gradation in properties of these elements is not as regular as in the case of alkali metals, because of different structures of their crystal lattices.

The difference in properties with alkali metals is due to:

(i) Smaller size of atoms and ions.

(ii) Stronger metallic bonds (resulting to more density and hardness).

(iii) Higher melting and boiling points.

DAY PRACTICE SESSION 1

FOUNDATION QUESTIONS EXERCISE

1 The electronic configuration of metal M is $1s^2, 2s^2 2p^6, 3s^1$. The formula of its oxide would be

(a) MO (b) M_2O (c) M_2O_3 (d) MO_2

2 The characteristic not related to alkali metal is

(a) high ionisation energy
(b) their ions are isoelectronic with noble gases
(c) low melting point
(d) low electronegativity

3 Which of the following reacts with water at a high rate?

(a) Li (b) K (c) Na (d) Rb

4 Among Na^+, Na, Mg and Mg^{2+} the largest particle is

(a) Mg^{2+} (b) Mg (c) Na (d) Na^+

5 The ionic conductance of following cation in a given concentration are in the order

(a) $Li^+ < Na^+ < K^+ < Rb^+$ (b) $Li^+ > Na^+ > K^+ > Rb^+$
(c) $Li^+ < Na^+ > K^+ > Rb^+$ (d) $Li^+ = Na^+ < K^+ < Rb^+$

6 The reducing power of a metal depends on various factors. Suggest the factor which makes Li, the strongest reducing agent in aqueous solution.

(a) Sublimation enthalpy (b) Ionisation enthalpy
(c) Hydration enthalpy (d) Electron gain enthalpy

7 The stability of the following alkali metal chlorides follows the order

(a) LiCl > KCl > NaCl > CsCl
(b) CsCl > KCl > NaCl > LiCl
(c) NaCl > KCl > LiCl > CsCl
(d) KCl > CsCl > NaCl > LiCl

8 The main oxides formed on combustion of Li, Na and K in excess of air respectively are → JEE Main 2016

(a) LiO_2, Na_2O_2 and K_2O
(b) Li_2O_2, Na_2O_2 and KO_2
(c) Li_2O, Na_2O_2 and KO_2
(d) Li_2O, Na_2O and KO_2

9 The metallic sodium dissolves in liquid ammonia to form a deep blue coloured solution. The deep blue colour is due to the formation of

(a) solvated electron, $e^-(NH_3)_x$
(b) solvated atomic sodium, $Na(NH_3)_4$
(c) $(Na^+ + Na^-)$
(d) $NaNH_2 + H_2$

10 The solubilities of carbonates decrease down the group due to a decrease in

(a) lattice energies of solid
(b) hydration energies of cations
(c) interionic attraction
(d) entropy of solution formation

11 Consider the following abbreviations for hydrated alkali ions.

$X = [Li(H_2O)_n]^+; Y = [K(H_2O)_n]^+; Z = [Cs(H_2O)_n]^+$

What is the correct order of size of these hydrated alkali ions?

(a) $X > Y > Z$ (b) $Z > Y > X$
(c) $X = Y = Z$ (d) $Z > X > Y$

12 The gas evolved on heating Na_2CO_3 is

(a) CO_2 (b) water vapour
(c) CO (d) No gas is evolved

13 Sodium is heated in air at 350°C to form A. Compound A when reacts with carbon dioxide forms sodium carbonate and Y. Here, Y is

(a) hydrogen peroxide (b) hydrogen
(c) ozone (d) oxygen

14 When CO is passed over solid NaOH heated to 200°C, it forms

(a) Na_2CO_3 (b) H_2CO_3
(c) HCOONa (d) All of these

15 Which one of the following is true?

(a) NaOH is used in the concentration of bauxite ore
(b) NaOH is a primary standard in volumetric analysis
(c) Managanous hydroxide is soluble in excess of NaOH solution
(d) NaOH solution does not react with Cl

16 A sodium salt on treatment with $MgCl_2$ gives white precipitate only on heating. The anion of sodium salt is

(a) HCO_3^- (b) CO_3^{2-}
(c) NO_3^- (d) SO_4^{2-}

17 Solubility of the alkaline earth's metal sulphates in water decreases in the sequence → CBSE-AIPMT 2015

(a) Mg > Ca > Sr > Ba (b) Ca > Sr > Ba > Mg
(c) Sr > Ca > Mg > Ba (d) Ca > Mg > Sr > Ca

18 Solubility of alkaline earth metal sulphates decreases down the group 2 because

(a) they become more ionic
(b) lattice energy of sulphates does not vary significantly
(c) hydration energy decreases rapidly from Be^{2+} to Ba^{2+}
(d) lattice energy plays more predominant role than hydration energy

19 The correct order of increasing ionic character is

(a) $BeCl_2 < MgCl_2 < CaCl_2 < BaCl_2$
(b) $BeCl_2 < MgCl_2 < BaCl_2 < CaCl_2$
(c) $BeCl_2 < BaCl_2 < MgCl_2 < CaCl_2$
(d) $BaCl_2 < CaCl_2 < MgCl_2 < BaCl_2$

20 Which pair of the following chlorides does not impart colour to the flame?
(a) $BeCl_2$ and $SrCl_2$
(b) $BeCl_2$ and $MgCl_2$
(c) $CaCl_2$ and $BaCl_2$
(d) $BaCl_2$ and $SrCl_2$

21 In context with beryllium, which one of the following statements is incorrect?
(a) It is rendered passive by nitric acid
(b) It forms Be_2C
(c) Its salts are rarely hydrolysed
(d) It hydride is electron-deficient and polymeric

22 Beryllium and aluminium exhibit many properties which are similar but the two elements differ in
(a) exhibiting maximum covalency in compounds
(b) forming polymeric hydrides
(c) tendency of forming chelate type complexes.
(d) exhibiting amphoteric nature in their oxides

23 Metal carbonates decompose on heating to give metal oxide and carbon dioxide. Which of the metal carbonates is most stable thermally?
(a) $MgCO_3$ (b) $CaCO_3$
(c) $SrCO_3$ (d) $BaCO_3$

24 Amphoteric hydroxides react with both alkalies and acids.Which of the following group 2 metal hydroxides is soluble in sodium hydroxide?
(a) $Be(OH)_2$ (b) $Mg(OH)_2$
(c) $Ca(OH)_2$ (d) $Ba(OH)_2$

25 Which of the following statements are true about $Ca(OH)_2$?
(a) It is used in the preparation of bleaching powder
(b) It is a light blue solid
(c) It does not possess disinfectant property
(d) It is used in the manufacture of cement

26 The solubilities of carbonates decreases down the magnesium group due to decrease in
(a) lattice energies of solids
(b) hydration energies of cations
(c) interionic attraction
(d) entropy of solution formation

27 Which one of the following alkaline earth metal sulphates has its hydration enthalpy greater than its lattice enthalpy? → JEE Main 2015
(a) $CaSO_4$ (b) $BeSO_4$
(c) $BaSO_4$ (d) $SrSO_4$

28 Among the following, which is insoluble in water?
(a) Sodium fluoride (b) Potassium fluoride
(c) Beryllium fluoride (d) Magnesium fluoride

29 Which of the following on thermal decomposition yields a basic as well as acidic oxide? → AIEEE 2012
(a) $NaNO_3$ (b) $KClO_3$
(c) $CaCO_3$ (d) NH_4NO_3

30 Match the following and choose the correct option.

	Column I		Column II
A.	Li	1.	Insoluble oxalate
B.	Na	2.	Strongest monoacidic base
C.	Ca	3.	Most negative $E^{\ominus}$ value among alkali metals
D.	Ba	4.	$6s^2$outer electronic configuration

Codes

	A	B	C	D		A	B	C	D
(a)	4	2	1	3	(b)	3	1	2	4
(c)	3	2	1	4	(d)	4	1	2	3

31 Which of the following statements(s) is/are correct?
(a) Group 2 metals are denser and harder than group 1 metals
(b) Group 1 metals are less electropositive than group 2 metals
(c) The first ionisation energy of group 1 metals is more than that of group 2 metals
(d) All of the above statements are incorrect

Direction (Q. Nos. 32-35) *In the following questions Assertion (A) followed by reason (R) is given. Choose the correct option out of the choices given bleow:*
(a) Assertion and Reason both are correct statements and Reason is the correct explanation of the Assertion
(b) Assertion and Reason both are correct statements but Reason is not the correct explanation of the Assertion
(c) Assertion is correct and Reason is incorrect
(d) Both Assertion and Reason are incorrect

32 **Assertion** (A) The carbonate of lithium decomposes easily on heating to form lithium oxide and CO_2.
Reason (R) Lithium being very small in size polarises large carbonate ion leading to the formation of more stable Li_2O and CO_2.

33 **Assertion** (A) Beryllium carbonate is kept in the atmosphere of carbon dioxide.
Reason (R) Beryllium carbonate is unstable and decomposes to give beryllium oxide and carbon dioxide.

34 **Assertion** (A) Among the alkali metals, lithium salts exhibit the least electrical conductance in aqueous solutions.
Reason (R) Smaller the radius of the hydrated cation, lower is the electrical conductance in aqueous solutions.

35 **Assertion** (A) $BaCO_3$ is more soluble in HNO_3 than in water.
Reason (R) Carbonate is a weak base and reacts with H^+ from the strong acid, causing the barium salt to dissociate.

DAY PRACTICE SESSION 2

PROGRESSIVE QUESTIONS EXERCISE

1 A mixture contains two moles of Na_2CO_3 and 1 mole of Li_2CO_3. What will be the volume of CO_2 formed on heating this mixture and the data is converted to STP?

(a) 22.4 L (b) 44.8 L (c) 50.2 L (d) 11.2 L

2 A metal *M* readily forms water soluble MSO_4. It also forms oxide *MO* which becomes inert on heating. Hydroxide $M(OH)_2$ is insoluble in water but soluble in NaOH solution. What is *M*?

(a) Mg (b) Ba (c) Ca (d) Be

3 Mg has favourbale standard reduction electrode potential ($E^\circ_{Mg^{2+}/Mg} = -2.37$ V) but it does not react readily. It is because of

(a) its resemblance to Li
(b) formation of protective oxide layer
(c) its oxidising nature
(d) its reaction with acids

4 In group 2, the electrons are more firmly held to the nucleus and hence

(a) ionisation energy of group 2 is greater than group 1
(b) atoms of group 2 are bigger than group 1
(c) reactivity of group 2 is greater than group 1
(d) group 2 metals are less abundant in nature

5 Element *A* burns in nitrogen to give an ionic compound *B*. Compound *B* reacts with water to give *C* and *D*. The solution of *C* becomes milky on bubbling carbon dioxide. The element *A* is

(a) Li (b) Mg
(c) Ca (d) Be

6 On dissolving moderate amount of sodium metal in liquid NH_3 at low temperature, which one of the following does not occur ?

(a) Blue coloured solution is obtained
(b) Na^+ ions are formed in the solution
(c) Liquid NH_3 solution becomes good conductor of electricity
(d) Liquid NH_3 solution remains diamagnetic

7 The atomic radius for Li and Li^+ are 1.23 Å and 0.76 Å respectively. Assuming that the difference in ionic radii relates to the space occupied by 2*s*-electrons. What will be the % of volume of Li atom occupied by single valence electron?

(a) 70.0% (b) 30.0% (c) 24.75% (d) 76.34%

8 A chemical *A* is used for the preparation of washing soda to recover ammonia. When CO_2 is bubbled through an aqueous solution of *A*, the solution turns milky. It is used in white washing due to disinfectant nature. What is the chemical formula of *A*?

(a) $Ca(HCO_3)_2$ (b) CaO (c) $Ca(OH)_2$ (d) $CaCO_3$

9 Some of the group 2 metal halides are covalent and soluble in organic solvents. Among the following metal halides, the one which is soluble in ethanol is

(a) $BeCl_2$ (b) $MgCl_2$ (c) $CaCl_2$ (d) $SrCl_2$

10 A binary salt of potassium '*A*' gives compound '*B*' when heated with S. Compound '*B*' forms white precipitate '*C*' with barium chloride which is insoluble in conc.HCl. 7.1 g of '*A*' gave 11.15 g of white precipitate '*C*'. The compound '*A*' is

(a) KO_2 (b) K_2SO_4 (c) KOH (d) K_2CO_3

11 Select the correct statements(s).

(a) $CaCO_3$ is more soluble in a solution of CO_2 than in H_2O
(b) Na_2CO_3 is converted to Na_2O and CO_2 on heating
(c) Li_2CO_3 is thermally stable
(d) Presence of $CaCl_2$ or $CaSO_4$ in water causes temporary hardness

12 The charge/size ratio of a cation determines its polarising power. Which one of the following sequences represents the increasing order of the polarising power of cationic species; K^+, Ca^{2+}, Mg^{2+}, Be^{2+}?

(a) $Mg^{2+} < Be^{2+} < K^+ < Ca^{2+}$ (b) $Be^{2+} < K^+ < Ca^{2+} < Mg^{2+}$
(c) $K^+ < Ca^{2+} < Mg^{2+} < Be^{2+}$ (d) $Ca^{2+} < Mg^{2+} < Be^{2+} < K^+$

13 0.2 g of magnesium ribbon was placed in a crucible and heated with a lid on until the magnesium began to burn rapidly. At the end of the experiment there was 0.3 g of a white powder left. This result does not agree with the equation; $2Mg(s) + O_2(g) \longrightarrow 2MgO(s)$. It can be due, to

(a) some MgO may have escaped as vapours
(b) some Mg may not have reacted
(c) some Mg_3N_2 can be formed
(d) All of the above

14 Choose the incorrect statement.

(a) $BeCO_3$ is kept in the atmosphere of CO_2 since, it is least thermally stable
(b) Be dissolves in alkali forming $[Be(OH)_4]^{2-}$
(c) BeF_2 forms complex ion with NaF in which Be goes with cation
(d) BeF_2 forms complex ion with NaF in which Be goes with anion

15 Dehydration of hydrates of halides of calcium, barium and strontium, i.e. $CaCl_2 \cdot 6H_2O$, $BaCl_2 \cdot 2H_2O$, $SrCl_2 \cdot 2H_2O$, can be achieved by heating.
These become wet on keeping in air. Which of the following statements are correct about these halides?

(a) Act as dehydrating agent
(b) Can absorb moisture from air
(c) Tendency to form hydrate decreases from to barium
(d) All of the above

ANSWERS

SESSION 1	1 (b)	2 (a)	3 (d)	4 (c)	5 (a)	6 (c)	7 (d)	8 (c)	9 (a)	10 (b)
	11 (a)	12 (d)	13 (d)	14 (c)	15 (a)	16 (a)	17 (a)	18 (c)	19 (a)	20 (b)
	21 (c)	22 (a)	23 (d)	24 (a)	25 (a)	26 (b)	27 (b)	28 (d)	29 (c)	30 (c)
	31 (a)	32 (a)	33 (a)	34 (c)	35 (a)					
SESSION 2	1 (a)	2 (d)	3 (b)	4 (a)	5 (c)	6 (d)	7 (d)	8 (c)	9 (a)	10 (a)
	11 (a)	12 (c)	13 (d)	14 (c)	15 (d)					

Hints and Explanations

SESSION 1

1 Electronic configuration indicates that the metal is univalent (alkali metal). Its oxide will be M_2O.

2 Alkali metals have low ionisation energy. They possess minimum value of ionisation energy in their period due to their large sizes.

3 Rb being most electropositive has the highest reactivity towards water.

4 Atomic radii of alkali metals are largest in the corresponding period. Moreover, size of monovalent alkali metal ion is always smaller than its parent atom. Thus, Na is the largest particle among the given metals.

5 Conductance of cation in aqueous solution is dependent upon its hydrates size as ionic conductance $\propto \dfrac{1}{\text{ionic size of hydrate ion}}$. Thus, ionic conductance in aqueous solution increases in the order $Li^+ < Na^+ < K^+ < Rb^+$. This is because the size of hydrated ion is minimum for Rb^+ among given hydrates alkali metal ions.

6 Hydration enthalpy of Li^+ is highest among all other alkali metal due to its smaller size. Therefore, Li^+ acts as a strong reducing agent.

7 Stability of a compound depends upon its enthalpy of formation ΔH_f. The more negative value of ΔH_f shows more stability of a compound.

ΔH_f for LiCl $= -408.8 \text{ kJ mol}^{-1}$,

ΔH_f for NaCl $= -412.5 \text{ kJ mol}^{-1}$

ΔH_f for CsCl $= -433 \text{ kJ mol}^{-1}$,

ΔH_f for KCl $= -436 \text{ kJ mol}^{-1}$

Thus, KCl is more stable and LiCl is least stable.

Hence, (d) option is correct.

8 $$2Li + \frac{1}{2}O_2(g) \longrightarrow Li_2O$$ (Excess)

$$2Na + O_2(g) \longrightarrow Na_2O_2$$ (Excess)

$$K + O_2(g) \longrightarrow KO_2$$ (Excess)

9 $$Na \rightleftharpoons Na^+(\text{in liq. } NH_3) + e^-$$

$$e^- + xNH_3 \longrightarrow \underset{\text{Ammonia solvated electron}}{e(NH_3)_x^-}$$

10 Solubilities of carbonates decrease down the group because lattice energy decrease is almost constant while decrease in hydration energy happens sharply.

Finally difference of hydration energy and lattice energy decreases, thus solubility decreases.

11 All alkali metal salts are soluble in water. The degree of hydration depends upon the size of the cation. Smaller the size of cation, greater is its charge density and hence, greater is its tendency to withdraw electrons from molecules which are thus polarised. Li^+ ion being smallest in size among alkali metal ions is the most extensively hydrated while Cs^+ ion being the largest alkali metal ion is the least hydrated. The size of hydrated alkali ions is as

$$Li^+ > Na^+ > K^+ > Rb^+ > Cs^+.$$

Hence, the correct order is $X > Y > Z$.

12 On heating Na_2CO_3, no gas is evolved.

13 $$2Na + O_2 \xrightarrow{350^\circ C} \underset{A}{Na_2O_2}$$

$$2Na_2O_2 + 2CO_2 \longrightarrow 2Na_2CO_3 + \underset{Y}{O_2}$$

14 $$\underset{\text{Solid}}{NaOH} + CO \xrightarrow{200^\circ C} \underset{\text{Sodium formate}}{HCOONa}$$

15 NaOH is used in concentration of bauxite ore. Impure bauxite is treated with NaOH with which it forms water soluble sodium meta aluminate complex.

$$Al_2O_3 \cdot 2H_2O + 2NaOH \xrightarrow{420\text{-}425K} \underset{\text{Sodium meta aluminate}}{2NaAlO_2} + 3H_2O$$

Impurities such as Fe_2O_3, TiO_2 and silica are left behind.

Pure alumina is recovered from solution.

$$NaAlO_2 + 2H_2O \longrightarrow Al(OH)_3 + NaOH$$

$$2Al(OH)_3 \xrightarrow{\Delta} Al_2O_3 + 2H_2O$$

16 $$2NaHCO_3 + MgCl_2 \xrightarrow{\Delta} \underset{\text{White ppt.}}{Mg(HCO_3)_2} + 2NaCl$$

17 Mg > Ca > Sr > Ba

This is because down the group the hydration energies of cations decreases.

18 Due to very large size of sulphate ions, the magnitude of lattice energy of alkaline earth metal sulphates remains almost constant. While the hydration energy decreases rapidly down the group. Thus, their solubility is governed by only hydration energy which decreases from Be^{2+} to Ba^{2+}.

19 Larger the size of the cation, smaller the polarising power and thus larger the ionic nature.

Size (cation) $Be^{2+} < Mg^{2+} < Ca^{+} < Ba^{2+}$

Ionic nature

$BeCl_2 < MgCl_2 < CaCl_2 < BaCl_2$

20 Among the alkaline earth metals, the size of beryllium and magnesium metals is very small. Therefore, the electrons in these metals are bounded more strongly and are not excited by the energy of flame to higher energy states.

Hence, these metals or their salts do not impart any colour to the flame.

21 Due to small size and high hydration energy of Be^{2+} ions, its salts are easily hydrolysed.

22 Be ($Z = 4$) has maximum covalency of four while Al ($Z = 13$) has maximum covalency of 6.

23 $BaCO_3$ is thermally most stable because of the small size of resulting oxide ion. With the increase in atomic number, the size of the metal ion, the stability of the metal ion decreases and, hence that of carbonate increases, i.e. from Be to Ba.

Therefore, the increasing size of cation destabilises the oxides and hence, does not favour the decomposition of heavier alkaline earth metal carbonates like $BaCO_3$.

24 The solubility of hydroxides of alkaline earth metals increases from Be to Ba, in water.

Due to high hydration enthalpy and high lattice energy, $Be(OH)_2$ is not soluble in water.

25 Calcium hydroxide is used in the manufacturing of bleaching powder.

$$\underset{\text{Slaked lime}}{2Ca(OH)_2} + 2Cl_2 \longrightarrow CaCl_2 + \underset{\text{Bleaching powder}}{Ca(OCl)_2} + 2H_2O$$

26 The solubilities of carbonates decreases down the magnesium group due to decrease in hydration energy because hydration energy inversely proportional to ionic size. As the size increases, hydration enthalpy of metal ions decreases from Mg^{2+} to Ba^{2+}.

27 As we move down the group, size of metal increases. Be has lower size while SO_4^{2-} has bigger size, that's why $BeSO_4$ breaks easily and lattice energy becomes smaller but due to smaller size of Be, water molecules are gathered around and hence, hydration energy increases. On the other hand, rest of the metals, i.e. Ca, Ba, Sr have bigger size and that's why lattice energy is greater than hydration energy.

28 The fluorides of alkaline earth metals except beryllium fluoride are insoluble in water because of high hydration energy of small Be^{2+} ion. Hence, magnesium fluoride is insoluble in water.

29 (a) $2NaNO_3 \xrightarrow{\Delta} 2NaNO_2 + O_2$

(b) $2KClO_3 \xrightarrow{\Delta} 2KCl + 3O_2$

(c) $CaCO_3 \xrightarrow{\Delta} \underset{\text{Basic}}{CaO} + \underset{\text{Acidic}}{CO_2}$

(d) $NH_4NO_3 \xrightarrow{\Delta} \underset{\text{Neutral}}{N_2O} + \underset{\text{Neutral}}{H_2O}$

30 $A \rightarrow 3, B \rightarrow 2, C \rightarrow 1, D \rightarrow 4$

31 (a) Group 2 metals are denser and harder than group 1 metals. Thus statement is correct.

(b) Group 1 metals are more electropositive than group 2 metals. Thus statement is incorrect.

(c) The first ionisation energy of group 2 metals is more than that of group 1 metals. Thus, statement is incorrect.

32 The thermal stability of carbonates increases down the group. Hence, Li_2CO_3 is least stable.

Due to small size of Li^+, strong polarising power distorts the electron cloud of CO_3^{2-} ion. High lattice energy of Li_2O than Li_2CO_3 also favours the decomposition of Li_2CO_3.

33 BeO is more stable than $BeCO_3$ due to small size and high polarising power of Be^{2+}.

As $BeCO_3$ is unstable and BeO is more stable thus, when $BeCO_3$ is kept in an atmosphere of CO_2, a reversible process takes place and stability of $BeCO_3$ increases.

$$BeCO_3 \rightleftharpoons BeO + CO_2$$

34 Li^+ ion being largest in size becomes highly hydrated, hence moves slowly under the influence of electric field (poorest conductor).

35 Both Assertion and Reason are correct and Reason is the correct explanation of the Assertion. $BaCO_3$ is more soluble in HNO_3 than in water as CO_3^{2-} reacts with H^+ to form HCO_3^- that causes $BaCO_3$ to dissociate more than it would do in water.

SESSION 2

1 Decomposition of Li_2CO_3

$$\underset{\text{1 mol}}{Li_2CO_3} \xrightarrow{\Delta} Li_2O + \underset{\text{22.4 L}}{CO_2}$$

Na_2CO_3 is thermally stable, i.e. it doesn't give CO_2 gas on heating.

Hence, volume of CO_2 formed on heating the mixture at STP is 22.4 L.

2 On moving down, water solubility of alkaline earth metal decreases. Oxides of Be is amphoteric hence soluble in NaOH.

$$BeO \xrightarrow{H_2O} \underset{\text{Insoluble}}{Be(OH)_2} \xrightarrow{NaOH} \underset{\text{Soluble}}{Na_2[Be(OH)_4]}$$

3 It is because of protective oxide (MgO) layer, which makes it unreactive.

4 Since the electrons are more tightly held to the nucleus, because of its small size, thus it is difficult to remove the electrons and hence more energy is required.

5 Carbon dioxide turns only lime water milky. Thus, the compound *C* must be $Ca(OH)_2$ and the element *A* must be Ca. The reactions are as follows:

$$\underset{A}{3Ca} + N_2 \longrightarrow \underset{B}{Ca_3N_2}$$

$$\underset{B}{Ca_3N_2} + 6H_2O \longrightarrow \underset{C}{3Ca(OH)_2} + \underset{D}{2NH_3}$$

$$\underset{C}{Ca(OH)_2} + CO_2 \longrightarrow \underset{\text{Milky}}{CaCO_3} + H_2O$$

6 The alkali metals dissolve in liquid ammonia giving deep blue solution which are conducting in nature.

$$M + (x + 2y)NH_3 \longrightarrow \underset{\text{(Deep blue)}}{[M(NH_3)_x]^{2+}} + 2[e^-(NH_3)_y]^-$$

7 Volume occupied by Li^+

$$= \frac{4}{3}\pi \times (0.76 \times 10^{-8})^3$$

Volume occupied by Li

$$= \frac{4}{3}\pi \times (1.23 \times 10^{-8})^3$$

∴ Volume occupied by 2s-electron

$$= \frac{4}{3}\pi \times [(1.23)^3 - (0.76)^3] \times (10^{-8})^3$$

∴ % volume occupied by 2s-electron of Li- atom

$$= \frac{\frac{4}{3}\pi \times [\{(1.23)^3\} - (0.76)^3\} \times (10^{-8})]^3}{\frac{4}{3}\pi \times [1.23 \times 10^{-8}]^3} \times 100$$

$$= \frac{1.42}{(1.23)^3} \times 100 = 76.34\%$$

8 For the recovery of ammonia, $Ca(OH)_2$ is used during the preparation of washing soda.

$$\underset{A}{Ca(OH)_2} + CO_2 \longrightarrow \underset{\text{Milky}}{CaCO_3} + H_2O$$

9 Ethanol is an organic compound, i.e. of covalent character. As we know "like dissolves like". So, to dissolve in ethanol the compound should have more covalent character. Beryllium halides have covalent character due to small size and high effective nuclear charge. Hence, $BeCl_2$ is most covalent among all other chlorides.

10 $\underset{\substack{142\text{ g}\\A}}{2KO_2} + S \xrightarrow{\Delta} \underset{B}{K_2SO_4}$, $\underset{B}{K_2SO_4} + BaCl_2 \longrightarrow \underset{\substack{223\text{ g}\\C}}{BaSO_4\downarrow} + 2KCl$

∵ 142 g KO_2 gives = 223 g $BaSO_4$

∴ 7.1 g KO_2 will give $= \frac{223}{142} \times 7.1 = 11.15$ g $BaSO_4$

11 (a) $CaCO_3 + H_2O + CO_2 \longrightarrow \underset{\text{Soluble}}{Ca(HCO_3)_2}$

(b) $Na_2CO_3 \xrightarrow{\Delta}$ No effect

(c) $Li_2CO_3 \xrightarrow{\Delta} Li_2O + CO_2$ (even at room temperature)

(d) Temporary hardness is due to soluble carbonates and bicarbonates.

12 Smaller the size of cations, larger the charge, larger the polarising power of the cations.

	Be^{2+}	Mg^{2+}	Ca^{2+}	K^+
Z (atomic number)	4	12	20	19
Charge	+2	+2	+2	+1

Size $Be^{2+} < Mg^{2+} < Ca^{2+} < K^+$

Polarising power $Be^{2+} > Mg^{2+} > Ca^{2+} > K^+$

13 $\underset{2\times 24\text{ g}}{2Mg(s)} + O_2(g) \longrightarrow \underset{2\times 40\text{ g}}{2MgO}$

48 g of Mg gives = 80 g of MgO

0.2 g of Mg gives $= \frac{80}{48} \times 0.2 = 0.33$ g of MgO

But residue left = 0.30. It is due to all of the given options.

14 $BeF_2 + 2NaF \longrightarrow \underbrace{2Na^+}_{\text{Cation}} + \underbrace{BeF_4^{2-}}_{\text{Anion}}$

Thus, (c) is incorrect (Be goes with anion).

15 Chlorides of alkaline earth metals are hydrated salts. Due to their hygroscopic nature, they can be used as a dehydrating agent, to absorb moisture from air.
Extent of hydration decreases from Mg to Ba, i.e.
$MgCl_2 \cdot 6H_2O$, $CaCl_2 \cdot 6H_2O$, $SrCl_2 \cdot 2H_2O$, $BaCl_2 \cdot 2H_2O$

DAY TWENTY TWO

p-Block Elements
(Group-13 to Group-18)

Learning & Revision for the Day

- Periodicity in Properties of *p*-Block Elements
- Group-13 Elements : Boron Family
- Group-14 Elements : Carbon Family
- Group-15 Elements : Nitrogen Family
- Group-16 Elements : Oxygen Family
- Group-17 Elements : Halogens Family
- Group-18 Elements : Noble Gases

In *p*-block elements, the last electron enters in the outermost *p*-orbital. There are six groups of *p*-block elements in the periodic table, numbering from 13 to 18. Their valence shell electronic configuration is ns^2np^{1-6} (except for He). Effective nuclear charge from first to third period decreases and from fourth period and onwards increases, due to poor shielding effect. Therefore, ns^2 electrons bound closely to the nucleus. Therefore, the oxidation state two unit less than the group oxidation state becomes progressively more stable in each group. This is also known as **inert pair effect.**

Periodicity in Properties of p-Block Elements

- Non-metals and metalloids exist only in the *p*-block of the periodic table. On moving down a group metallic character increases and non-metallic character decreases. Between metals and non-metals a few metalloids are also present.
- Second period elements have a tendency to form **$p\pi$- $p\pi$ multiple bond** (double or triple bond) due to their smaller atomic size and high bond energy.
  e.g. $C \equiv C, C = C, N \equiv N, O = O, O = C = O$ etc.
- Elements of second period do not expand their **covalency** due to non-availability of *d*-orbitals. Therefore, halides of such elements do not hydrolysed easily.
  e.g. BCl_3, CCl_4 (halides) etc.
- While elements of third period and onwards only form single bond and can expand their covalency due to the presence of vacant *d*-orbitals.
  e.g. P_4, S_8, $[AlF_6]^{3-}$ ion etc.
- Non-metal oxides are acidic or neutral and metal oxides are basic in nature. Generally, metalloid oxides are amphoteric. In a period, from left to right acidic character increases and down a group, basic character increases. Generally, oxides with higher oxidation state are more acidic in nature.

PREP MIRROR
Your Personal Preparation Indicator

- No. of Questions in Exercises (x)—
- No. of Questions Attempted (y)—
- No. of Correct Questions (z)—
  (Without referring Explanations)

- Accuracy Level ($z/y \times 100$)—
- Prep Level ($z/x \times 100$)—

In order to expect good rank in JEE, your Accuracy Level should be above 85 & Prep Level should be above 75.

Properties of Hydrides

1. **Basic character** of *p*-block hydrides decreases on moving down a group and **acidic character** increases. Generally, left to right in a period, acidic character increases. e.g.

   $NH_3 > PH_3 > AsH_3$
   (down the group basic character decreases)

   $NH_3 > H_2O > HF$
   (left to right basic character decreases)

2. **Thermal stability** of hydrides on moving down a group decreases due to increase in bond length.

   Thermal stability of the hydrides decreases down the group due to which their tendency to act as reducing agent increases down the group.

3. **Bond angle** of hydrides on moving down a group, decreases due to increase in the size of central atom. Charge density and repulsion interaction also decreases on moving down a group, e.g.

   $H_2O > H_2S > H_2Se$
   Bond angle

   If central atoms belong to the same period and have the same hybridisation, then as the number of lone pair increases, bond angle decreases. e.g.

   $CH_4 > \ddot{N}H_3 > H_2\ddot{\underset{\cdot\cdot}{O}}$
   Bond angle

4. **Boiling point** of hydrides increases on moving down a group because van der Waals' forces increases with increase in molecular mass except HF, H_2O, NH_3 etc. HF, NH_3 and H_2O have intermolecular hydrogen bonding as a result of this they have high boiling point.

   Unique behaviour of the first element in each group It is due to the (i) small size (ii) high electronegativity and (iii) non-availability of *d*-orbitals in the valence shell of first element in each group.

Group-13 Elements : Boron Family

Boron (B), aluminium (Al), gallium (Ga), indium (In) and thallium (Tl) are the members of group-13. Boron is a non-metal while rest of the members are metals.

Important Properties

1. Electronic Configuration (ns^2, np^1)

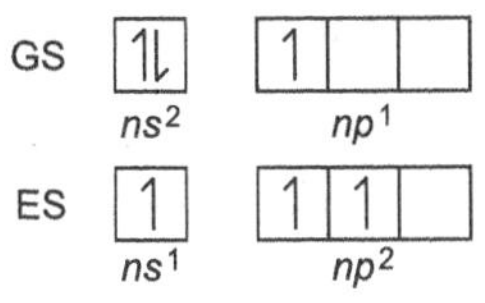

2. Oxidation State (+3, +1)

Stability of +3 oxidation state decreases down the group and that of +1 oxidation state increases down the group due to inert pair effect.

Fourth period and onwards, effective nuclear charge increases due to poor shielding effect, therefore the radius of gallium (135 pm) is less than that of aluminium (143 pm) and on moving down the group inert pair effect increases.

3. Hydrides

Boron hydrides exist in dimeric or polymeric form. The simplest boron hydride is diborane, B_2H_6. Aluminium forms only one colourless, solid polymeric hydride $(AlH_3)_n$ known as alane.

Hydrides of Ga and In are not very much stable. B, Al and Ga form complex anionic hydrides, e.g. $NaBH_4$, $LiAlH_4$ and $LiGaH_4$. All are strong reducing agents.

4. Halides

Halides have incomplete octet, therefore, have a high tendency to accept electrons and behave as Lewis acid.

$BI_3 > BBr_3 > BCl_3 > BF_3$ (Acidic character)

BF_3 is a colourless gas, BCl_3 is a colourless fuming liquid while BI_3 is a white fusible solid. Due to backbonding and resonance, the B—F bond of BF_3 gets a bond order of 1.33.

5. Oxides

On moving down the group, their basic character increases. e.g. B_2O_3 is weakly acidic, Al_2O_3 is amphoteric and other oxides are basic in nature.

Preparation, Properties and Uses of Boron

- The important minerals of boron are borax ($Na_2B_4O_7 \cdot 10H_2O$), orthoboric acid (H_3BO_3) and kernite ($Na_2B_4O_7 \cdot 4H_2O$).
- **Preparation** Now a days, boron is obtained by electrolysis of a fused mixture containing boric anhydride, magnesium oxide and magnesium fluoride at 1100°C.

  $$2MgO \longrightarrow 2Mg + O_2$$
  $$B_2O_3 + 3Mg \longrightarrow 2B + 3MgO$$

  Crystalline boron is obtained by the reduction of B_2O_3 with aluminium powder.

  $$B_2O_3 + 2Al \longrightarrow 2B + Al_2O_3$$

  Amorphous boron of low purity is called **Moissan boron.** It is black in colour.
- **Physical Properties** Boron exists in two allotropic forms, e.g. amorphous and crystalline. Crystalline boron is chemically inert while amorphous boron is chemically active. It is a bad conductor of heat and electricity.

- **Chemical Properties** Amorphous boron on heating with air at 700°C forms a mixture of oxide and nitride.

$$4B + 3O_2 \longrightarrow 2B_2O_3, \; 2B + N_2 \longrightarrow 2BN$$

Boron is attacked by oxidising acids like conc. H_2SO_4 and HNO_3.

$$B + 3HNO_3 \xrightarrow{H_2SO_4} H_3BO_3 + 3NO_2$$

It dissolves in fused alkalies and liberates hydrogen.

$$2B + 6NaOH \longrightarrow 2Na_3BO_3 + 3H_2$$

It reacts with strong electropositive metals at high temperature and forms boride such as Mg_3B_2.

$$3Mg + B_2 \longrightarrow Mg_3B_2$$

It is a powerful reducing agent.

$$3CO_2 + 4B \longrightarrow 2B_2O_3 + 3C$$

$$3SiO_2 + 4B \longrightarrow 2B_2O_3 + 3Si$$

- **Uses** It is used as control rods in atomic reactors and as a deoxidiser.

Preparation, Properties and Uses of Aluminium

The important ores of aluminium are bauxite $AlO_x(OH)_{3-2x}$ (where, $0 < x < 1$) and kaolinite $[Al_2(OH)_4Si_2O_5]$.

- **Preparation** Bauxite contains SiO_2, iron oxides and titanium oxide (TiO_2) as impurities. Leaching of bauxite ore is carried out with conc. NaOH solution. Aluminium is obtained by the electrolysis of Al_2O_3 mixed with Na_3AlF_6 or CaF_2 (Hall-Haroult process). Overall reaction is as follows:

$$2Al_2O_3 \longrightarrow 4Al + 3O_2$$

- **Physical Properties** It is bluish white lustrous metal and loses its lustre due to the formation of protective oxide film Al_2O_3. It is light, malleable, ductile, good conductor of electricity and heat.
- **Chemical Properties** It burns in oxygen producing brilliant light.

$$4Al + 3O_2 \longrightarrow 2Al_2O_3; \Delta H = -\text{ve}$$

This reaction is used in thermite process for the reduction of oxides of Cr, Mn, Fe etc.
It decomposes boiling water evolving hydrogen.

$$2Al + 6H_2O \longrightarrow 2Al(OH)_3 + 3H_2 \uparrow$$

It is rendered passive by nitric acid due to the formation of thin oxide film. It reacts with non-metals and displaces less reactive metals such as copper, zinc and lead from their salt solutions.

- **Uses** It is used for making household utensils, frames, bodies of aircraft automobiles etc. It is used in making paints, as a mordant in dyeing and calico printing.

Ammonal (mixture of Al powder and ammonium nitrate) is used as explosive. Magnalium, duralumin and alnico are important alloys of aluminium.

Some Important Compounds

Some important compounds of group-13 elements or boron family are given below:

1. Diborane (B_2H_6)

- It is a colourless, highly toxic gas with a boiling point of 180 K. Diborane catches fire spontaneously upon exposure to air.

$$B_2H_6 + 3O_2 \longrightarrow B_2O_3 + 3H_2O$$

- With ammonia, diborane gives borazine.

$$3B_2H_6 + 6NH_3 \longrightarrow 3[BH_2(NH_3)_2]^+[BH_4]^- \xrightarrow{\text{Heat}} \underset{\text{Borazine}}{2B_3N_3H_6} + 12H_2$$

- Borazine ($B_3N_3H_6$) is known as **"inorganic benzene"** in view of its ring structure with alternative BH and NH groups.
- However, at high temperature, inorganic graphite $(BN)_x$ is obtained instead of borazine.

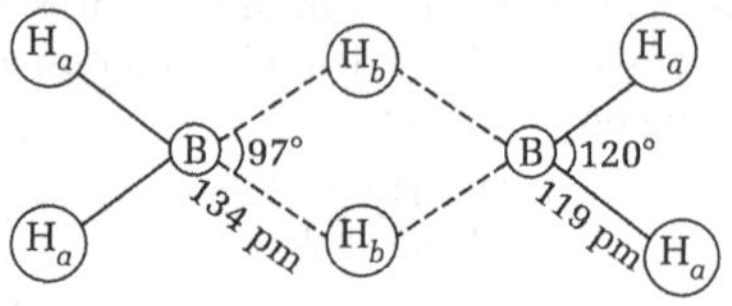

Structure of diborane

- Terminal hydrogens (H_a) in diborane are bonded by $2c-2e$ (two centred two electrons) bonding and bridge hydrogens (H_b) are bonded by $3c-2e$ (three centred held by two electrons) bonding. This type of bonding is also known as **banana bonding.** Orbital structure of B_2H_6 is

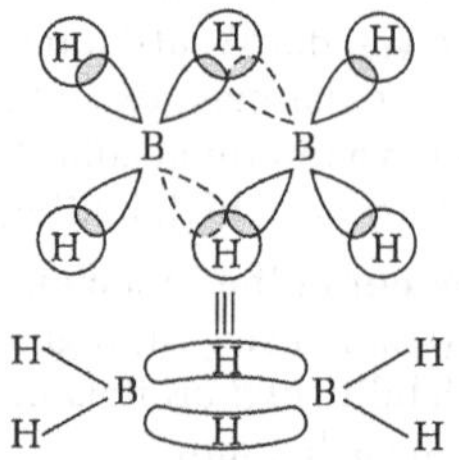

Banana bond

- Due to banana bonding, B_2H_6 complete its electron deficiency and obtain ethane like structure and hybridisation of B_2H_6 becomes sp^3.

$$2MH + B_2H_6 \longrightarrow 2M^+[BH_4]^- \quad (M = \text{Li or Na})$$

It is used as a catalyst in polymerisation process, for welding torch, as reducing agent in organic reactions etc.

2. Borax ($Na_2B_4O_7 \cdot 10H_2O$)

[Sodium Tetra Borate Decahydrate]

- It is the most important compound of boron. It is white crystalline solid.
- Borax dissolves in water to give an alkaline solution.

$$Na_2B_4O_7 + 7H_2O \longrightarrow 2NaOH + \underset{\substack{\text{Orthoboric acid} \\ \text{(Weak acid)}}}{4H_3BO_3}$$

- On heating, it loses their water of crystallisation and swells up. On further heating, it melts into a liquid which then solidifies to form a glassy mass like bead.

$$Na_2B_4O_7 \cdot 10H_2O \xrightarrow[-10H_2O]{\Delta} \underset{\substack{\text{Swell up} \\ \text{(Sodium metaborate)}}}{Na_2B_4O_7} \xrightarrow{\Delta} \underbrace{2NaBO_2 + \underset{\text{Boric anhydride}}{B_2O_3}}_{\text{Glassy mass}}$$

- With acids, it gives orthoboric acid, a weak acid.

$$Na_2B_4O_7 + 2HCl + 5H_2O \longrightarrow 2NaCl + 4B(OH)_3 \text{ or } \underset{\text{Orthoboric acid}}{H_3BO_3}$$

- Borax contains the tetrahedral units, i.e. $[B_4O_5(OH)_4]^{2-}$.

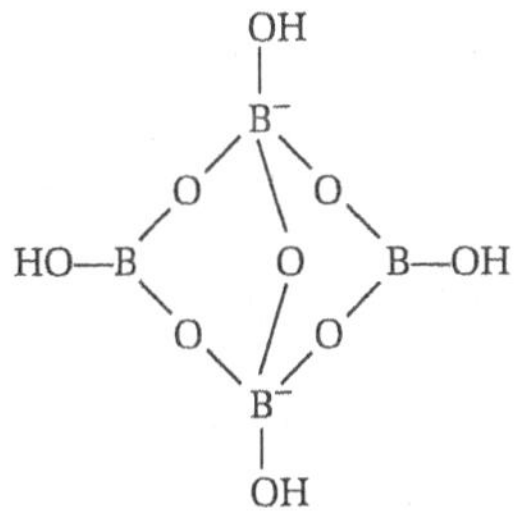

Structure of $[B_4O_5(OH)_4]^{2-}$ ion

3. Boric Acid (H_3BO_3)

- Boric acid is soft white crystalline solid, soapy to touch and less soluble in cold water.
- Boric acid is a weak monobasic acid. It is not a protonic acid but acts as a **Lewis acid**.
- On heating it above 370 K, it forms metaboric acid, HBO_2 which on heating yield B_2O_3.
- $H_3BO_3 \xrightarrow{\text{Heat}} \underset{\text{Metaboric acid}}{HBO_2} \xrightarrow{\text{Heat}} \underset{\text{Boron trioxide}}{B_2O_3}$
- **Structure of boric acid** shows that it has a layer structure in which planar BO_3 units are joined by H-bonds.

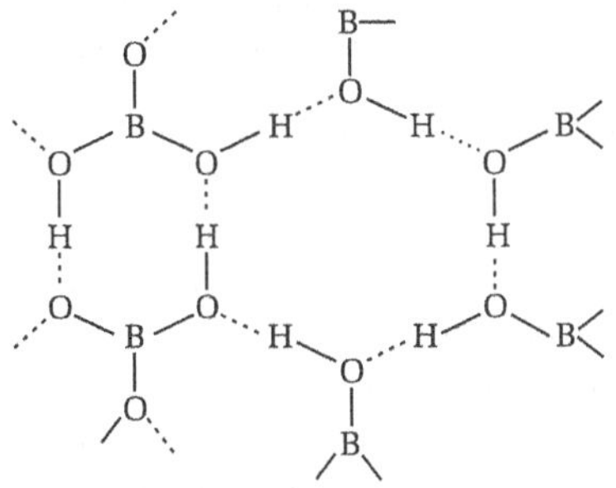

- It is used as an antiseptic, eye wash, as food preservative, in glass industry and in pottery.

4. Boron Trifluoride

- BF_3 is a colourless pungent gas (melting point is –127.1°C and boiling point is – 99.9°C) which strongly fumes in moist air and possesses a pungent smell. It is exceedingly soluble in water that's why it is collected over Hg.
- BF_3 being an electron deficient compound can accept a lone pair of electrons, thus behaves as Lewis acid, e.g. with NH_3, H_2S and F^-, it can form complexes as shown below:

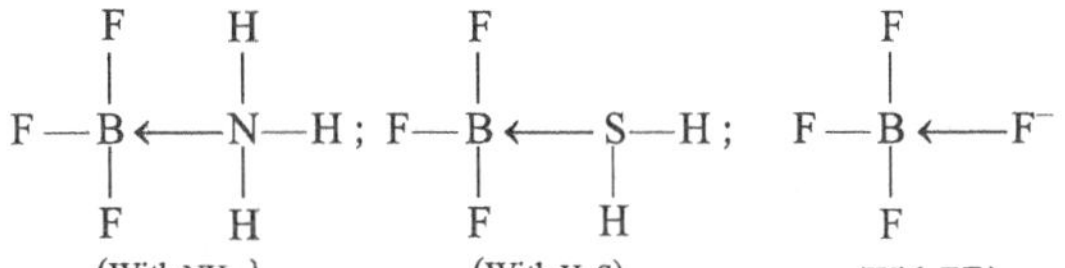

Due to its electron accepting tendency, it is used as a catalyst in a number of organic reactions.

- BF_3 combines with water forming two hydrates, i.e. $BF_3 \cdot H_2O$ (melting point 10.18°C) and $BF_3 \cdot 2H_2O$ (melting point 6.36°C). It gets hydrolysed in aqueous solutions yielding boric acid and hydrofluoroboric acid.

$$4BF_3 + 3H_2O \longrightarrow H_3BO_3 + 3HBF_4$$

The bond length in BF_3 are 1.30 Å and are shorter than the sum of the covalent radii (B = 0.80Å and F = 0.72Å). The bond energy of BF_3 is very high (646 kJ mol^{-1}) and is higher than for any single bond. Based on modern explanation, the double bond is delocalised.

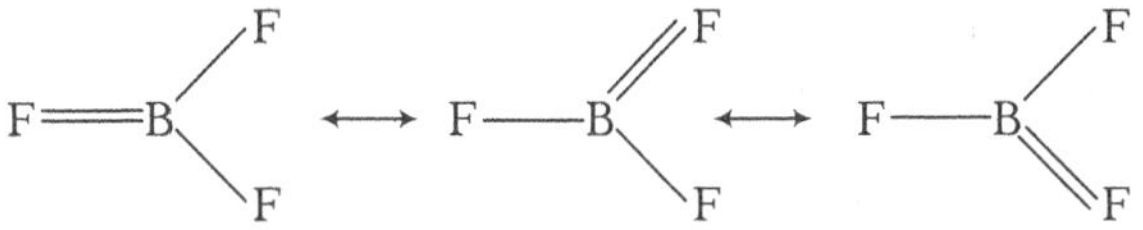

- Various resonating structures of BF_3 molecules involving $p\pi - p\pi$ back bonding.
- It is used as building block for production of other boron compounds.

5. Aluminium Chloride ($AlCl_3$)

- Anhydrous aluminium chloride is a white deliquescent solid which fumes in air. Its vapour density corresponds to the formula Al_2Cl_6.
- It is covalent and soluble in organic solvents.

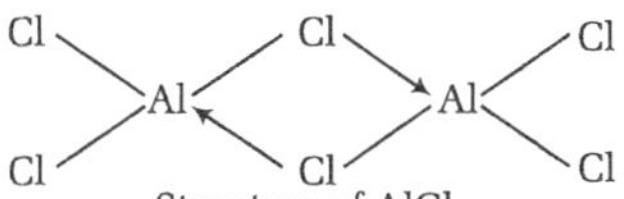

Structure of $AlCl_3$

- Due to the formation of HCl, anhydrous aluminium chloride fumes in moist air.

$$Al_2Cl_6(\text{anhy.}) + 6H_2O \longrightarrow 2Al(OH)_3 + 6HCl$$

$$Al_2Cl_6 + 12\,H_2O \longrightarrow \underset{\text{Hydrated aluminium chloride (ionic)}}{2AlCl_3 \cdot 6H_2O}$$

- With ammonia, it gives addition product.

$$Al_2Cl_6\,(\text{anhy.}) + 12\,NH_3 \longrightarrow \underset{\text{Addition product}}{2[AlCl_3 \cdot 6NH_3]}$$

- With NaOH, sodium meta-aluminate is obtained.

$$AlCl_3 + 3NaOH \longrightarrow Al(OH)_3 \downarrow + 3NaCl$$

$$Al(OH)_3 + NaOH \longrightarrow \underset{\text{Sodium meta-aluminate}}{NaAlO_2} + 2\,H_2O$$

- Anhydrous $AlCl_3$ is used.
  (a) As a reagent in Friedel-Crafts reaction.
  (b) In manufacture of drugs, dyes, perfumes.
  (c) In cracking of high boiling fractions of petroleum to form gasoline.

6. Alums

- These are double sulphates having composition $M_2SO_4 \cdot M_2'(SO_4)_3 \cdot 24\,H_2O$ (M = monovalent basic radicals such as Na^+, K^+ etc., and M' = trivalent basic radicals such as Al^{3+}, Cr^{3+} etc.)

  Potash alums; $K_2SO_4 \cdot Al_2(SO_4)_3 \cdot 24H_2O$

  Sodium alums; $Na_2SO_4 \cdot Al_2(SO_4)_3 \cdot 24H_2O$

  Ferric alums; $(NH_4)_2SO_4 \cdot Fe_2(SO_4)_3 \cdot 24H_2O$
- Alums are isomorphous solids and soluble in hot water. Their aqueous solutions are acidic due to hydrolysis.
- On heating, alum swells up leaving behind a porous mass, called burnt alum.

$$K_2SO_4 \cdot Al_2(SO_4)_3 \cdot 24H_2O \xrightarrow{200°C} K_2SO_4 \cdot Al_2(SO_4)_3 + 24H_2O$$

$$K_2SO_4 \cdot Al_2(SO_4)_3 \xrightarrow{\text{Red hot}} K_2SO_4 + Al_2O_3 + 3SO_3$$

- It is used to purify water, stop bleeding from cuts, as an antiseptic, in leather tanning and as a mordant in dyeing and calico printing.

Group-14 Elements : Carbon Family

Carbon (C), silicon (Si), germanium (Ge), tin (Sn), and lead (Pb) are the members of group-14. Carbon is the seventeenth most abundant element by mass in the earth's crust. C is non-metal, Si, Ge are metalloids and Sn, Pb are metals.

Important Properties

Some of the important properties of group-14 elements or carbon family elements are given below:

1. Electronic Configuration (ns^2np^2)

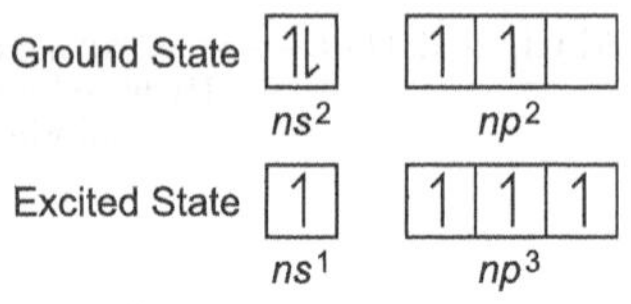

2. Catenation

The tendency for catenation is maximum in carbon and it decreases down the group due to steady decrease in M—M bond strength.

$$C >> Si > Ge \approx Sn >> Pb$$

3. Allotropy

- All the members of group 14 elements (except Pb) show allotropy.
- Carbon exhibits many allotropic forms both crystalline as well as amorphous. Diamond, graphite and fullerene are crystalline forms while coal, charcoal and lamp black are amorphous forms.
- **Diamond** is the purest, hardest form of carbon with high refractive index and density. In it each carbon atom (sp^3hybridised) is tetrahedrally surrounded by four other carbon atoms. It is three dimensional polymer has very high density and very high refractive index. It does not conduct electricity as it has no free electrons. It is used in cutting, grinding and drilling instruments and in making jewellery.
- **Graphite** has two dimensional structure. sp^2 hybridised carbon atom forms three covalent bonds with three other carbon atoms in the same plane and the 4th electron of each carbon remains free and is responsible for electrical conductivity of graphite. The planar hexagonal rings get fused together to form sheets of one atom thickness. These sheets are held together by weak van der Waals' forces.
- **Fullerenes** are the only pure form of carbon. C_{60} molecule contains 12 five membered rings and 20 six-membered rings. The five membered rings are connected to six membered rings while six membered rings are connected to both five and six membered rings. These are used in microscopic ball bearings, light weight batteries, in synthesis of new plastics and new drugs.

4. Oxidation States

The common oxidation states are +4 and +2. Carbon also exhibits negative oxidation states, i.e. –4. Down the group, stability of +4 oxidation state decreases and of +2 oxidation state increases due to inert pair effect.

5. Oxides

CO_2, SiO_2 and GeO_2 are acidic, whereas SnO_2 and PbO_2 are amphoteric in nature.

(i) Among monoxides, CO is neutral, GeO is distinctly acidic whereas SnO and PbO are amphoteric.

(ii) Monomeric form of CO_2 is stable due to non-availability of d-orbitals. Carbon has tendency to form a multiple bond ($O{=}C{=}O$) but SiO_2 exists in three dimensional polymeric form and has high melting point and is solid at room temperature.

6. Reactivity Towards Water

Carbon, silicon and germanium are not affected by water. Tin (Sn) decomposes with steam to form dioxide and dihydrogen gas.

$$Sn + 2H_2O \xrightarrow{\text{Heat}} SnO_2 + 2H_2$$

Lead remains unaffected by water, probably because of a protective oxide film formation.

7. Halides

GeX_4 is more stable than GeX_2, whereas PbX_2 is more stable than PbX_4 and $SnCl_2$ is more stable than $SnCl_4$. $SnCl_4$ is liquid at room temperature due to covalent character and $SnCl_2$ is solid at room temperature.

(i) Except CCl_4, other halides are easily hydrolysed due to availability of vacant d-orbitals.

e.g. $SiCl_4 + 4H_2O \longrightarrow$ 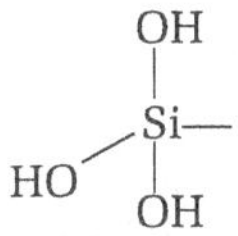 or $H_4SiO_4 + 4HCl$

Silicic acid

(ii) Si has vacant d-orbitals, so can expand its covalency and form higher halides $[SiF_6]^{2-}$ but not $[SiCl_6]^{2-}$ because larger size of chloride ions cannot be accommodated around Si^{4+} due to limitation of its size.

Some Important Compounds

Some of the important compounds of group 14 elements or carbon family elements are given below:

1. Oxides of Carbon

Two important oxides of carbon are carbon monoxide (CO) and carbon dioxide CO_2.

Carbon monoxide can be prepared by the following reactions:

$$C(s) + H_2O(g) \xrightarrow{473\text{ K-}1273\text{ K}} \underbrace{CO(g) + H_2(g)}_{\text{Water gas}}$$

$$2C(s) + O_2(g) + 4N_2(g) \xrightarrow{1273\text{ K}} \underbrace{2CO(g) + 4N_2(g)}_{\text{Producer gas}}$$

Water gas and producer gas are very important industrial fuels.

- Carbon dioxide can be prepared by complete combustion of carbon and carbon containing fuels in excess of air.
- Carbon dioxide can be obtained as a solid in the form of dry ice by allowing the liquified CO_2 to expand rapidly. Dry ice is used as a refrigerant for ice-cream and frozen food.

2. Silicon Tetrachloride ($SiCl_4$)

- It is tetrahedral and essentially covalent. It is readily hydrolysed by water. It fumes in moist air liberating hydrogen chloride.

$$SiCl_4 + 4H_2O \longrightarrow Si(OH)_4 + 4HCl$$

- The hydrolysis of $SiCl_4$ occurs due to coordination of OH with empty 3d-orbitals in Si-atom of $SiCl_4$ molecule.

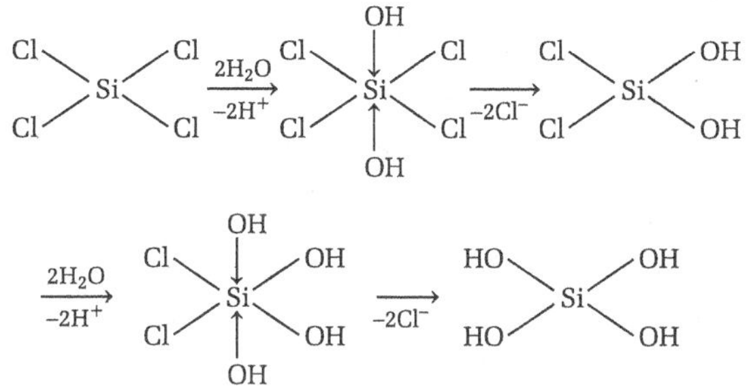

3. Silicones

- These are synthetic organosilicon compounds which has repeated unit of R_2SiO.
- These are prepared from alkyl halides.

$$2CH_3Cl + Si \xrightarrow[570\text{ K}]{\text{Cu powder}} (CH_3)_2SiCl_2 \xrightarrow[-2HCl]{+2H_2O} (CH_3)_2Si(OH)_2$$

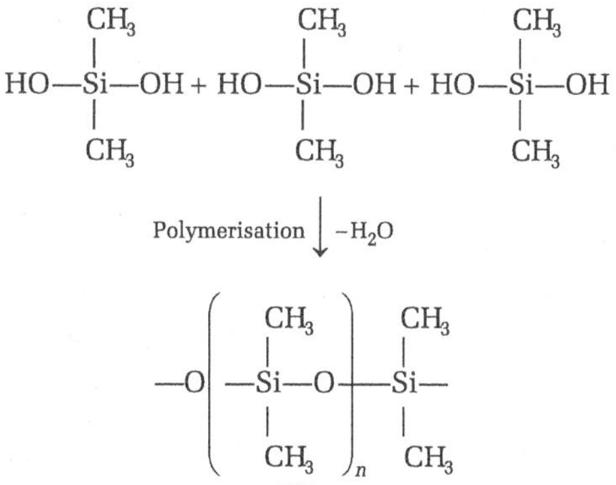

Silicone

Highly cross-linked silicone polymer is obtained by the hydrolysis of $RSiCl_3$.

- They are used as sealant, greases, electrical insulators and for water proofing of fabrics. Being biocompatible, they are also used in surgical and cosmetic plants.

4. Silicates

- A large number of silicate minerals exist in nature. Some of the examples are feldspar, zeolites, mica and asbestos. The basic structural unit of silicates is SiO_4^{4-}.

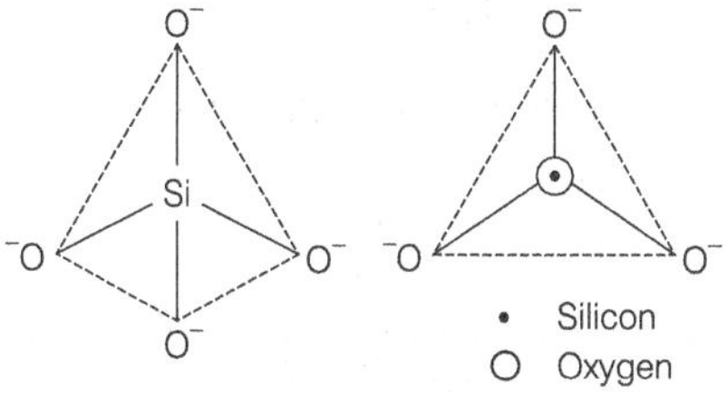

- The SiO_4^{4-} unit is neutralised by positively charged metal ions, if all the four corners are shared with other tetrahedral units.

Generally, the silicates have complex structures but they mainly differ in

(a) Number of oxygen atoms shared between SiO_4^{4-} tetrahedra.

(b) Geometric arrangement of tetrahedra.

(c) The number, type and arrangement of metallic cations.

Different type of silicates are as follows:

(i) **Orthosilicates** Simple silicates containing SiO_4^{4-} tetrahedra.

(ii) **Pyrosilicate** Two tetrahedral units share one O-atom to obtain $Si_2O_7^{6-}$ anion.

(iii) **Cyclic silicates** Two tetrahedral units share two oxygen atoms and form $(SiO_3^{2-})_n$ or $(SiO_3)_n^{2n-}$ anion.

(iv) **Chain silicates** Share two oxygen atoms, $(SiO_3^{2-})_n$ or $(SiO_3)_n^{2n-}$ are obtained.

(v) **Sheet silicates** Involve sharing of three O-atoms per tetrahedron to form $(Si_2O_5^{2-})_n$.

(vi) **Three dimensional silicates** All the four corners (O-atoms) of SiO_4^{4-} tetrahedra are shared with other.

(iv) Two important man-made silicates are glass and cement.

5. Zeolites

- If aluminium atoms replace few silicon atoms in three dimensional network of silicon dioxide, the obtained overall structure is known as alumino silicate and acquires a negative charge. Cations such as Na^+, K^+ or Ca^{2+} balance the negative charge. Their examples include feldspar and zeolites.
- Zeolites are used as a catalyst in petrochemical industries for cracking of hydrocarbons. ZSM-5 (a type of zeolite) is used to convert alcohols directly into gasoline. Hydrated zeolites are used as ion exchangers in softening of hard water.

Group-15 Elements : Nitrogen Family

Group 15 elements are nitrogen (N), phosphorus (P), arsenic (As), antimony (Sb) and bismuth (Bi). Nitrogen and phosphorus are non-metals, arsenic and antimony are metalloids and bismuth is a typical metal.

Important Properties

Electronic Configuration (ns^2np^3)

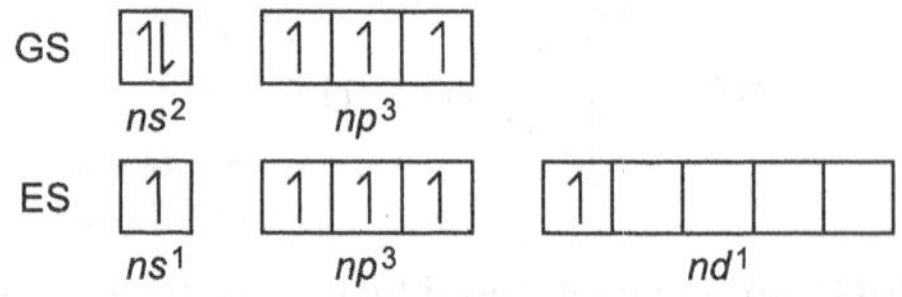

In case of nitrogen, ground state and excited state will be same due to the absence of *d*-orbitals.

Oxidation State

The common **oxidation states** of these elements are –3, +3 and +5. The tendency to exhibit –3 oxidation state decreases down the group due to increase in size and metallic character. The stability of + 5 oxidation state decreases down the group and that of +3 oxidation state increases down the group due to inert pair effect. Nitrogen with oxygen exhibits +1, +2, +4 oxidation states. Oxidation states from +1 to +4 tend to disproportionates in acid solution.

e.g. $$3H\overset{+3}{N}O_2 \longrightarrow H\overset{+5}{N}O_3 + H_2O + 2\overset{+2}{N}O$$

Physical Appearance

Only nitrogen has a tendency to form $p\pi$-$p\pi$ multiple bond ($N \equiv N$) due to the absence of *d*-orbitals.

Chemical Reactivity

N_2 has high bond dissociation energy, therefore reactivity of free N_2 is very less. Phosphorus has tendency to form a single bond due to the presence of *d*-orbitals and thus, exists in polyatomic form (P_4).

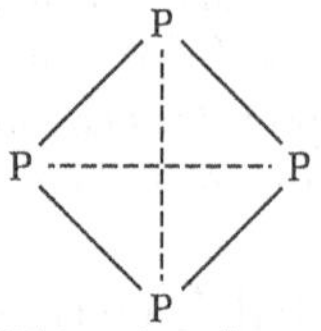

(White phosphorus)

P_4 has cyclic structure, therefore it is highly reactive and can expand its covalency as in PF_6^-.

Preparation, Properties and Uses of Dinitrogen

Preparation

- Dinitrogen is produced commercially by liquefaction and fractional distillation of air. Other methods of preparation are as follows:

$$NH_4Cl(aq) + NaNO_2(aq) \longrightarrow N_2(g) + 2H_2O(l) + NaCl(aq)$$

$$(NH_4)_2Cr_2O_7 \xrightarrow{\text{Heat}} N_2 + 4H_2O + Cr_2O_3$$

$$Ba(N_3)_2 \longrightarrow Ba + 3N_2$$

Properties

- Dinitrogen does not react with alkali metals except Li but reacts with alkaline earth metals to give metal nitride.

$$6Li + N_2 \xrightarrow{\text{Heat}} 2Li_3N, \quad 3Mg + N_2 \xrightarrow{\text{Heat}} Mg_3N_2$$

- It reacts with dioxygen only at high temperature to form nitric oxide.

$$N_2(g) + O_2(g) \xrightarrow{2000\text{ K}} 2NO(g)$$

Uses

- Liquid N_2 is used as refrigerant.
- It is used in the manufacture of HNO_3, NH_3, $CaCN_2$ (calcium cyanamide) and other nitrogenous compounds.
- It is used for filling electric bulbs.

Preparation, Allotropic forms and Uses of Phosphorus

Preparation

1. **Retort process**

$$\underset{\text{Phosphorite}}{Ca_3(PO_4)_2} + 3H_2SO_4 \longrightarrow \underset{\text{Orthophosphoric acid}}{2H_3PO_4} + 3CaSO_4$$

$$H_3PO_4 \longrightarrow \underset{\text{Metaphosphoric acid}}{HPO_3} + H_2O$$

$$4HPO_3 + 10C \longrightarrow P_4 + 10CO + 2H_2O$$

2. **Electrothermal process** Phosphorus is obtained from direct reduction of mineral phosphorite by carbon in the presence of silica.

$$2Ca_3(PO_4)_2 + 6SiO_2 + 10C \xrightarrow{1400\text{-}1500°C} 6CaSiO_3 + P_4 + 10CO$$

Allotropic Forms

Phosphorus exists in three allotropic forms, i.e. white or yellow, red and black (α and β) phosphorus. These forms are interconvertible.

$$\text{Black phosphorus} \xleftarrow[\text{under pressure}]{470\text{ K}} \underset{\text{White phosphorus}}{P_4} \xrightarrow{560\text{ K, inert atm.}} \text{Red phosphorus}$$

1. **White phosphorus** is transparent, soft, poisonous, waxy solid, shows chemiluminescence and chemically more reactive.

$$P_4 + 5O_2 \longrightarrow P_4O_{10} \text{ or } 2P_2O_5$$

$$P_4 + 3NaOH + 3H_2O \longrightarrow \underset{\text{Sodium hypo phosphite}}{3NaH_2PO_2} + \underset{\text{Phosphine}}{PH_3}$$

$$6Mg + P_4 \longrightarrow \underset{\text{Magnesium phosphide}}{2Mg_3P_2}$$

It acts as strong reducing agent.

$$P_4 + 10H_2SO_4 \longrightarrow 4H_3PO_4 + 10SO_2 + 4H_2O$$

2. **Red phosphorus** is the stable form of phosphorus. It is odourless, non-poisonous and less reactive.

$$2P + 5Cl_2 \xrightarrow{\text{Heat}} 2PCl_5$$

$$2P + 3S \xrightarrow{\text{Heat}} P_2S_3$$

$$P + 3Na \longrightarrow Na_3P$$

It is a polymer consists of chains of P_4 tetrahedra linked together.

3. **Black phosphorus** is obtained from white phosphorus.

$$\text{White P} \xrightarrow[\text{high pressure}]{200°C} \text{black phosphorus}$$

It may occur in orthorhombic, rhombohedral and cubic form.

Uses

- Red phosphorus is used in match box industry.
- Radioactive phosphorus is used in treatment of leukaemia and other blood disorders.
- Yellow phosphorus and zinc phosphide are used as a rat poison.

Some Important Compounds

Some of the important compounds of group 15 or nitrogen family elements are given below:

1. Ammonia (NH_3)

Preparation

- It is prepared by following methods.

$$\underset{\text{Urea}}{NH_2CONH_2} + 2H_2O \longrightarrow (NH_4)_2CO_3 \cdot 2NH_3 + H_2O + CO_2$$

$$2NH_4Cl + Ca(OH)_2 \longrightarrow 2NH_3 + 2H_2O + CaCl_2$$

$$(NH_4)_2SO_4 + 2NaOH \longrightarrow 2NH_3 + 2H_2O + Na_2SO_4$$

- On large scale, ammonia is manufactured by **Haber's process.**

$$N_2(g) + 3H_2(g) \underset{723\text{ K; Catalyst}}{\overset{200\text{ atm}}{\rightleftharpoons}} 2NH_3\,;\ \Delta H_f^\circ = -46.1\text{ kJ/mol}$$

Catalyst iron oxide with small amounts of K_2O and Al_2O_3 is used to increase the rate of attainment of equilibrium.

Properties

- Ammonia is covalent N-atom in NH_3 is sp^3-hybridised. Due to the presence of lone pair of electrons, it acquires pyramidal shape.
- NH_3 is basic in nature, has tendency to form hydrogen bond, therefore soluble in water and form NH_4OH or $NH_3(aq)$.
- Due to basic nature NH_3 is a good complexing agent and reducing agent. e.g.

$$2FeCl_3(aq) + 3NH_4OH(aq) \longrightarrow \underset{\text{Brown ppt}}{Fe_2O_3 \cdot xH_2O(s)} + 3NH_4Cl$$

$$ZnSO_4(aq) + 2NH_4OH(aq) \longrightarrow \underset{\text{White ppt}}{Zn(OH)_2(s)} + (NH_4)_2SO_4$$

$$\underset{\text{Blue}}{Cu^{2+}(aq)} + 4NH_3(aq) \rightleftharpoons \underset{\text{Deep blue}}{[Cu(NH_3)_4]^{2+}(aq)}$$

$$\underset{\text{Colourless}}{Ag^+(aq)} + Cl^-(aq) \longrightarrow \underset{\text{White ppt}}{AgCl(s)}$$

$$\underset{\text{White ppt.}}{AgCl(s)} + 2NH_3(aq) \longrightarrow \underset{\text{Colourless}}{[Ag(NH_3)_2Cl(aq)}$$

Structure

NH_3 is a covalent molecule in which nitrogen possess sp^3-hybridisation. But due to the presence of one lone pair it acquires pyramidal structure. The bond angle is 107° due to *lp-bp* repulsion.

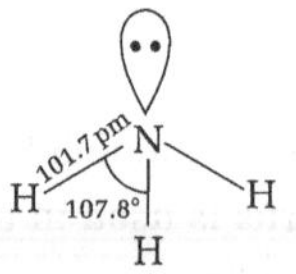

Uses

NH_3 is used in refrigeration due to its large heat of evaporation and in manufacture of HNO_3, $NaHCO_3$, ammonium compounds and nitrogenous fertilizers.

2. Nitric Acid (HNO_3)

Preparation

It was earlier known as aqua fortis. It is prepared by the following processes.

- **Laboratory preparation**

$$NaNO_3(s) + H_2SO_4(aq) \longrightarrow NaHSO_4(aq) + HNO_3(aq)$$

- **Ostwald process**

$$4NH_3(g) + 5O_2(g) \xrightarrow[1100\text{ K}]{\text{Pt gauge}} 4NO(g) + 6H_2O(l)\ ;\ \Delta H = -\text{ve}$$

$$2NO(g) + O_2(g) \longrightarrow 2NO_2(g)$$

$$4NO_2(g) + O_2(g) + 2H_2O(l) \longrightarrow 4HNO_3(aq)$$

Properties

- Anhydrous HNO_3 is a colourless fuming pungent smelling liquid. It acquires yellow colour due to its decomposition.

$$4HNO_3 \xrightarrow{\text{Sunlight}} 4NO_2 + 2H_2O + O_2$$

- It has corrosive action on skin and causes painful sores.
- It is very strong acid and form salts on reaction with basic oxides, carbonates, hydroxides etc.

$$Na_2CO_3 + 2HNO_3 \longrightarrow 2NaNO_3 + H_2O + CO_2$$

- It acts as a strong oxidising agent.
- Various reactions given by conc. HNO_3 that shows chemical properties of it are as follows:
- Non-metals such as C, S, P and I_2 are oxidised to carbonic acid, sulphuric acid, orthophosphoric acid and iodic acid respectively. e.g.

$$C + 4HNO_3 \longrightarrow H_2CO_3 + 4NO_2 + H_2O$$
$$S + 6HNO_3 \longrightarrow H_2SO_4 + 6NO_2 + 2H_2O$$

- Metalloids such as Sb is oxidised to antimonic acid, as to arsenic acid and Sn to stannic acid.

$$Sb + 5HNO_3 \longrightarrow H_3SbO_4 + 5NO_2 + H_2O$$
$$As + 5HNO_3 \longrightarrow H_3AsO_4 + 5NO_2 + H_2O$$
$$Sn + 4HNO_3 \longrightarrow H_2SnO_3 + 4NO_2 + H_2O$$

- **Brown ring test for nitrates**

$$6FeSO_4 + 3H_2SO_4 + 2HNO_3 \longrightarrow 3Fe_2(SO_4)_3 + 2NO + 4H_2O$$

- NO is absorbed by $FeSO_4$ and a dark brown ring of nitroso-ferrous sulphate is formed (ring test for nitrates).
- $C_{12}H_{22}O_{11} + 36HNO_3 \longrightarrow \underset{\text{Oxalic acid}}{6(COOH)_2} + 36NO_2 + 23H_2O$
- Metals like iron, cobalt, nickel, chromium, aluminium become passive in conc. HNO_3 due to the formation of a thin protective film of oxide on the surface of the metal.

Structure

- Nitrogen present in nitric acid is sp^2-hybridised and has a trigonal planar geometry. Structure of HNO_3 can be represented as

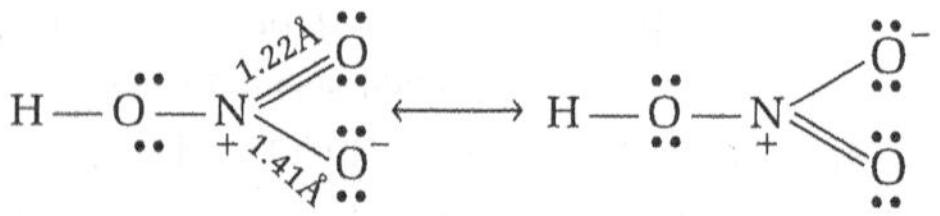

Uses

HNO_3 is used in the manufacture of explosives such as TNT, picric acid, nitroglycerine, dynamite etc., fertilizers, such as NH_4NO_3, basic $Ca(NO_3)_2$ etc., artificial silk, dyes, drugs.

Oxides of Nitrogen

- $N_2O(g)$ is neutral, colourless gas, with *sp*-hybridisation and linear geometry (it is also called laughing gas).

$$NH_4NO_3 \xrightarrow{\text{Heat}} N_2O + H_2O$$

- NO(g) is colourless, neutral gas. NO(g) also contains odd number of electrons (paramagnetic) but in solid or liquid state NO exists in dimeric form and have paired electrons (diamagnetic). It is very reactive and harmful to health.

$$2NaNO_2 + 2FeSO_4 + 3H_2SO_4 \longrightarrow Fe_2(SO_4)_3 + 2NaHSO_4 + 2H_2O + 2NO$$

- N_2O_3 is blue solid. It is acidic and planar with sp^2 hybridisation.

$$2NO + N_2O_4 \xrightarrow{250\text{ K}} 2N_2O_3$$

- $NO_2(g)$ is brown, acidic gas with angular shape and sp^2 hybridisation. NO_2 contains odd number of valence electrons (paramagnetic). On dimerisation, it gets converted to stable N_2O_4 molecule with even number of electrons (diamagnetic).

$$2Pb(NO_3)_2 \xrightarrow{673\text{ K}} 4NO_2 + 2PbO + O_2$$

- N_2O_4 is colourless solid/liquid. It is acidic with planar geometry and sp^2 hybridisation. Its covalency is four (total number of bonds with central atom).

$$2NO_2 \underset{\text{Heat}}{\overset{\text{Cool}}{\rightleftharpoons}} N_2O_4$$

- N_2O_5 is colourless, acidic solid with planar shape and sp^2-hybridisation and four covalency.

$$4HNO_3 + P_4O_{10} \longrightarrow 4HNO_3 + 2N_2O_5$$

Oxides of Phosphorus

Phosphorus trioxide and phosphorus pentoxide are the two oxides of phosphorus. Oxides of phosphorus exist in dimeric forms.

1. Phosphorus trioxide (P_4O_6)

It is prepared from white phosphorus.

$$P_4 + 3O_2 \xrightarrow[\text{(limited supply of air)}]{\text{Burning}} P_4O_6$$

- It is a poisonous waxy solid with garlic odour.
- $4P_4O_6 \xrightarrow{\text{Heat}} \underset{\text{Phosphorus tetroxide}}{3P_4O_8} + 4P$
- $P_4O_6 + 4Cl_2 \longrightarrow \underset{\text{Phosphorus oxychloride}}{2POCl_3} + \underset{\text{Metaphosphorus oxychloride}}{2PO_2Cl}$
- $P_4O_6 + 6H_2O(\text{hot}) \longrightarrow 3H_3PO_4 + PH_3$

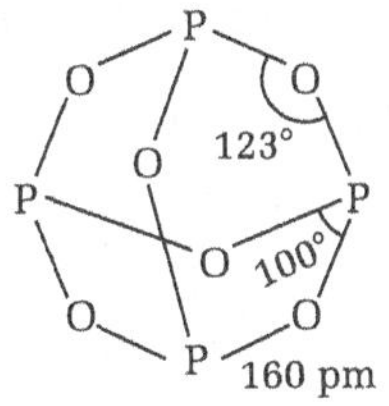

Structure of P_4O_6

2. Phosphorus Pentoxide (P_4O_{10})

It is obtained by burning of white phosphorus in free supply of air.

$$P_4 + 5O_2 \xrightarrow[\text{free supply of air}]{\text{Burning}} P_4O_{10}$$

- It is white crystalline, odourless solid which sublimes on heating.

$$P_4O_{10} \xrightarrow{2H_2O} \underset{\text{Metaphosphoric acid}}{4HPO_3} \xrightarrow{2H_2O} 2H_4P_2O_7$$

$$\underset{\text{Pyrophosphoric acid}}{2H_4P_2O_7} \xrightarrow{2H_2O} \underset{\text{Orthophosphoric acid}}{4H_3PO_4}$$

- It is used as most effective dehydrating agent below 100°C.

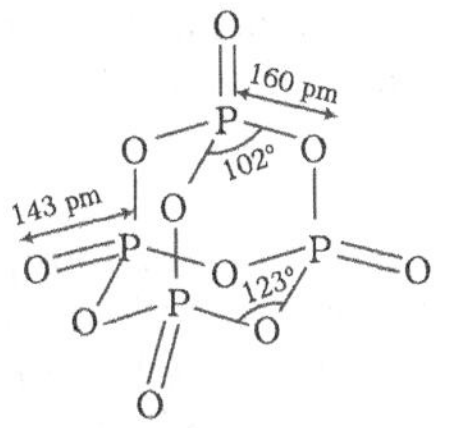

Structure of P_4O_{10}

3. Phosphine (PH_3)

Preparation

It is prepared by following methods:

(a) $Ca_3P_2 + 6H_2O \longrightarrow 3Ca(OH)_2 + 2PH_3$

(b) $Ca_3P_2 + 6HCl \longrightarrow 3CaCl_2 + 2PH_3$

(c) $P_4 + 3NaOH + 3H_2O \longrightarrow PH_3 + 3NaH_2PO_2$

(d) $PH_4I + KOH \longrightarrow KI + H_2O + PH_3$

Properties

- It is a colourless gas with rotten fish like smell and is highly poisonous. It explodes in contact with traces of oxidising agents like HNO_3, Cl_2 and Br_2 vapours.
- It forms phosphides when passed through the solutions of $CuSO_4$, $AgNO_3$ or $HgCl_2$.

$$3CuSO_4 + 2PH_3 \longrightarrow Cu_3P_2 + 3H_2SO_4$$
$$3HgCl_2 + 2PH_3 \longrightarrow Hg_3P_2 + 6HCl$$

- Phosphine is weakly basic $PH_3 + HBr \rightarrow PH_4^+ Br^-$.

Structure

Phosphine is a covalent molecule having pyramidal structure like ammonia. The bond angle H— P— H is 93°. It can be represented as

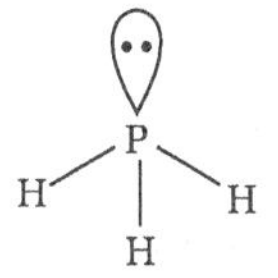

Structure of phosphine

Uses

It is used for making smoke screens, metallic phosphides, holme signals due to spontaneous combustion of PH_3.

4. Phosphorus Halides (PCl_3 and PCl_5)

Preparation

Following methods are used to prepare phosphorus halides.

(a) $P_4 + 6Cl_2 \longrightarrow 4PCl_3$

(b) $P_4 + 10Cl_2 \longrightarrow 4PCl_5$

(c) $P_4 + 8SOCl_2 \longrightarrow 4PCl_3 + 4SO_2 + 2S_2Cl_2$

(d) $P_4 + 10SO_2Cl_2 \longrightarrow 4PCl_5 + 10SO_2$

Properties

- PCl_5 is a yellowish white powder and in moist air it gets hydrolysed to $POCl_3$ and finally gets converted to phosphoric acid.

$$PCl_5 + H_2O \longrightarrow POCl_3 + 2HCl$$
$$POCl_3 + 3H_2O \longrightarrow H_3PO_4 + 3HCl$$

Structure

- PCl_5 in gaseous and liquid phases has sp^3d-hybridisation and its shape is trigonal bipyramidal.

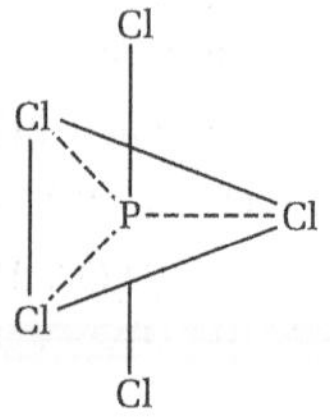

The three equatorial P—Cl bonds are equivalent while the two axial bonds are longer than equatorial bonds.
- In solid state, PCl_5 exists as an ionic solid, $[PCl_4]^+[PCl_6]^-$ in which, the cation, $[PCl_4]^+$ is tetrahedral and the anion $[PCl_6]^-$ is octahedral.
- **Structure of PCl_3** is similar to ammonia. P-atom is sp^3-hybridised.

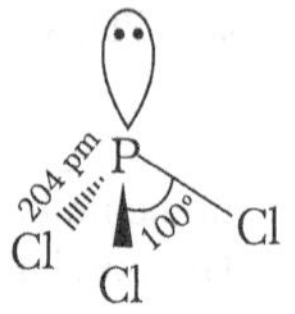

Structure of PCl_3

Uses

- PCl_3 is used in the manufacture of $POCl_3$ and in preparation of organic compounds.
- PCl_5 is used as a chlorinating agent in organic chemistry.

Oxoacids of Phosphorus and Nitrogen

- $H_2N_2O_2$ (hyponitrous acid); HNO_2 (nitrous acid) and HNO_3 (nitric acid) are the oxoacids of nitrogen. Out of these HNO_3 is the most important. HNO_2 (conc.) is strong oxidising agent. e.g.

$$I_2 + 10HNO_3 \longrightarrow 2HIO_3 + 10NO_2 + 4H_2O$$
$$P_4 + 20HNO_3\ (\text{conc.}) \longrightarrow 4H_3PO_3 + 20NO_2 + 4H_2O$$

- The structure of oxoacids of phosphorus are given below :

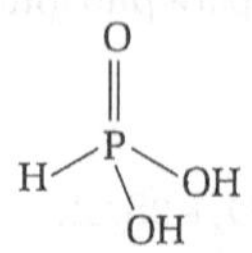

H_3PO_3
(Orthophosphorus acid)
sp^3, diprotic

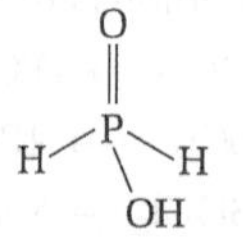

H_3PO_2
(Hypophosphorus acid)
monoprotic, sp^3

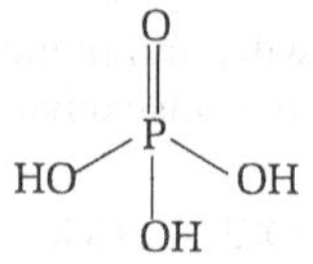

H_3PO_4
(Orthophosphoric acid)
triprotic, sp^3

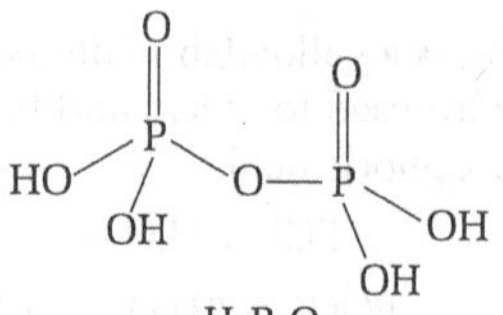

$H_4P_2O_7$
(Pyrophosphoric acid)
sp^3

Pyrophosphorus acid = $H_4P_2O_5$

Hypophosphoric acid = $H_4P_2O_6$

- H_3PO_3 on heating, disproportionates to give orthophosphoric acid and phosphine.

$$4H_3PO_3 \longrightarrow 3H_3PO_4 + PH_3$$

- The acids which contain P—H bond, have strong reducing properties, thus hypophosphorus acid is a good reducing agent as it contains two P—H bonds. e.g.
$4AgNO_3 + 2H_2O + H_3PO_2 \longrightarrow 4Ag + 4HNO_3 + H_3PO_4$

Group-16 Elements : Oxygen Family

Group 16 elements are sulphur (S), selenium (Se), tellurium (Te) and polonium (Po). This is sometimes known as group of **chalcogens** (due to ore forming nature). Oxygen is the most abundant of all the elements on earth. Oxygen and sulphur are non-metals, selenium and tellurium are metalloids and polonium is radioactive metal.

Important Properties

The group-16 elements exihibit the properties may be given as:

Electronic Configuration (ns^2np^4)

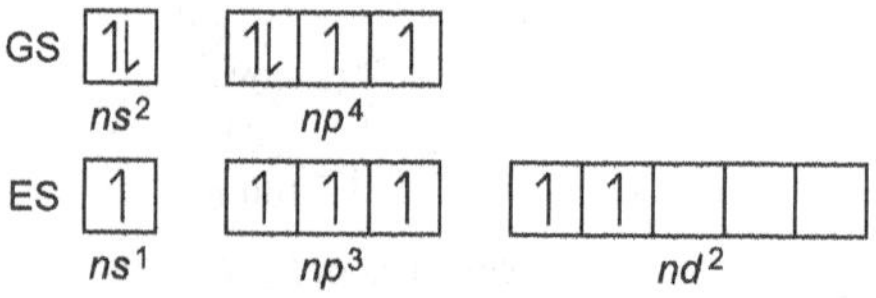

Oxidation States

Oxidation states exhibited by 16th group elements are −2, +2, +4, +6 but +4 and +6 are more common. On moving down the group, stability of +6 oxidation state decreases and that of +4 oxidation state increases due to inert pair effect. The oxidation state of O is −1 and $-\frac{1}{2}$ in peroxides and superoxides respectively. In case of OF_2, O_2F_2, oxidation states of oxygen are +2, +1 respectively.

Physical Appearance

Oxygen has a tendency to form **$p\pi$-$p\pi$ multiple bond** (O═O) due to the absence of *d*-orbitals and high bond dissociation enthalpy. Sulphur has tendency to form single bond due to the presence of *d*-orbitals and exist in polyatomic (S_8) form.

Allotropy

Oxygen exists in two molecular allotropic forms; the diatomic molecule, O_2 (most stable) and the triatomic molecule, O_3 (ozone comparatively less stable). Sulphur forms numerous allotropes, out of which the yellow rhombic (α-sulphur) and monoclinic (β-sulphur) forms are the most important.

$$\text{Rhombic sulphur} \xrightarrow{\text{369 K above}} \text{Monoclinic sulphur}$$

At elevated temperatures (1000 K), S_2 is the dominant species and is paramagnetic like O_2.

Hydrides

Down the group acidic character increases from H_2O to H_2Se. All the hydrides except water possess reducing property and this character increases from H_2S to H_2Te.

Oxides

All these elements (except O_2) form MO_2 type **oxides**. SO_2 is gas while SeO_2 is solid. Reducing property of dioxide decreases from SO_2 to TeO_2. SO_2 is reducing while TeO_2 is an oxidising agent.

Halides

The stability of the **halides** decreases in the order

$$F^- > Cl^- > Br^- > I^-.$$

Amongst hexahalides, hexafluorides are the only stable halides. All hexafluorides are gaseous in nature. SF_6 is exceptionally stable for steric reasons. SF_4 is a gas, SeF_4 is a liquid and TeF_4 is a solid. These fluorides have sp^3d hybridisation and *see-saw* geometry. The well known mono halides are dimeric in nature. e.g. $S_2F_2, S_2Cl_2, S_2Br_2, Se_2Cl_2$ and Se_2Br_2. These dimeric halides undergo disproportionation as given below:

$$2Se_2Cl_2 \longrightarrow SeCl_4 + 3Se$$

Some Important Compounds

Some of the important compounds considered under group 16 are given below:

1. Dioxygen

Preparation

- **Laboratory method**

$$2\,KClO_3 \xrightarrow[\Delta]{MnO_2} 2KCl + 3O_2$$

- It is manufactured from liquid air. Air is liquefied by making use of Joule Thomson effect. Liquid air is a mixture of liquid nitrogen and liquid oxygen. The difference in their boiling points is about 12.8°C, hence they are easily separated by fractional evaporation.
- **By electrolysis of water** (acidified water)

$$H_2SO_4 \rightleftharpoons 2H^+ + SO_4^{2-}$$

At cathode $\quad 2H^+ + 2e^- \longrightarrow H_2$

At anode $\quad 2OH^- \longrightarrow H_2O + \frac{1}{2}O_2 + 2e^-$

Properties

- Colourless, odourless, tasteless gas which is slightly soluble in water. Liquid oxygen exhibits paramagnetism.
- It is non-inflammable but a supporter of combustion.
- O=O, bond dissociation energy is high, therefore it reacts with metals or non-metals after external heating to start the reaction.

$$N_2 + O_2 \xrightarrow{3300\text{ K}} 2\,NO$$

$$4Na(s) + O_2(g) \xrightarrow{\text{Room temperature}} 2Na_2O(s)$$

$$4Al(s) + 3O_2(g) \longrightarrow 2Al_2O_3(s)$$

Uses

- In oxyacetylene and oxyhydrogen flames.
- Liquid O_2 is used as a rocket fuel.
- For life support systems, e.g. in hospitals and in water diving for divers, for miners and mountaineers.
- As an oxidising and bleaching agent.

2. Ozone

It is formed in the upper layer of atmosphere by the action of UV rays from sun on oxygen. It prevents the UV rays from entering the earth's atmosphere. CFCs, common refrigerants deplete the ozone layer.

Preparation

It is prepared in ozoniser by subjecting dry and cold dioxygen to the action of silent electric discharge.

$$3O_2(g) \xrightleftharpoons{\text{Silent electric discharge}} 2\,O_3(g)$$

Properties

- It is pale blue gas with pungent odour. It is diamagnetic and poisonous.
- Ozone is unstable; $2O_3 \xrightarrow{\text{Heat}} 3O_2$
- It acts as a powerful oxidising agent. It liberates iodine from neutral KI solution and the liberated I_2 turns starch paper blue.

$$2\,KI + H_2O + O_3 \longrightarrow 2KOH + I_2 + O_2$$

$$I_2 + \text{starch} \longrightarrow \text{Blue colour}$$

- Alkaline KI is oxidised to potassium iodate and periodate.

$$KI + 3O_3 \longrightarrow \underset{\text{Potassium iodate}}{KIO_3} + 3O_2$$

$$KI + 4O_3 \longrightarrow \underset{\text{Potassium periodate}}{KIO_4} + 4O_2$$

$$H_2S + O_3 \longrightarrow H_2O + S + O_2$$

- Mercury loses its meniscus in contact with ozone (tailing of mercury).

$$2\,Hg + O_3 \longrightarrow Hg_2O + O_2$$

Uses

- As a germicide and disinfectant for sterilising water.
- As a bleaching agent for oils, ivory wax and delicate fibre.
- For detecting the position of double bond in unsaturated compounds.
- In destroying odours coming from cold storage room, slaughter houses and kitchen of hotels.

Allotropic forms of Sulphur

It exists in several allotropic forms. These are as follows:

1. **Rhombic sulphur** (α-sulphur) is common crystalline form of sulphur, yellow in colour, melting point 114.5°C and specific gravity 2.06. Its crystals are prepared by evaporating sulphur solution in CS_2. It is insoluble in water but readily soluble in CS_2.
2. **Monoclinic sulphur** (β-sulphur) is stable above 95.6°C. Its crystals are amber yellow in colour, melting point 119°C, specific gravity 1.98. It is insoluble in CS_2. It is prepared by melting rhombic sulphur in a dish followed by cooling till crust is formed.

$$\text{Rhombic sulphur} \xrightleftharpoons{95.6^\circ C} \text{Monoclinic sulphur.}$$

   At 95.6°C both the forms are stable. This temperature is called **transition temperature.** S_8 rings in both the forms is puckered and has a crown shape.

3. **Plastic sulphur** is obtained by pouring boiling sulphur into cold water. It is amber brown in colour, specific gravity 1.95 and insoluble in CS_2.
4. **Milk of sulphur** is obtained by boiling milk of lime with sulphur and decomposing the products formed with HCl. It is used in medicines.
5. **Colloidal sulphur** is prepared by passing H_2S through a solution of an oxidising agent such as nitric acid etc.

$$2HNO_3 + H_2S \longrightarrow 2NO_2 + 2H_2O + S$$

Colloidal sulphur changes into ordinary form on heating.

Sulphuric Acid (H_2SO_4)

Preparation

It is manufactured by contact process which involves three steps :

(i) Burning of sulphur ores in air to generate SO_2.

(ii) Conversion of SO_2 to SO_3 by the reaction with oxygen in the presence of a catalyst (V_2O_5).

$$2\,SO_2(g) + O_2(g) \xrightarrow[\text{2 bar, 720 K}]{V_2O_5} 2\,SO_3\,(g)$$

(iii) Absorption of SO_3 in H_2SO_4 to give oleum ($H_2S_2O_7$)

$$SO_3 + H_2SO_4 \longrightarrow H_2S_2O_7 \text{ (oleum)}$$

Properties

- Concentrated sulphuric acid is a strong dehydrating agent.

$$C_{12}H_{22}O_{11} \xrightarrow{H_2SO_4} 12C + 11H_2O$$

- Hot concentrated sulphuric acid is a moderately strong oxidising agent. In this respect, it is intermediate between phosphoric and nitric acids.

$$S + 2\,H_2SO_4\text{ (conc.)} \longrightarrow 3SO_2(g) + 2\,H_2O$$
$$C + 2\,H_2SO_4\text{(conc.)} \longrightarrow CO_2 + 2\,SO_2 + 2H_2O$$

Uses

- In the manufacture of fertilisers
- Petroleum refining
- In manufacture of pigments, paints and dyestuff
- Detergent industry
- Metallurgical application
- Storage batteries
- As a laboratory reagent

NOTE H_2SO_3 acts as bleaching agent but it is temporary and the bleached material regains its colour due to oxidation.

Oxoacids of Sulphur

The structures of oxoacids of sulphur are as follows:

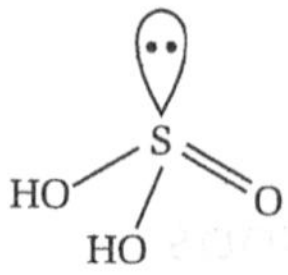

Sulphurous acid (H_2SO_3)

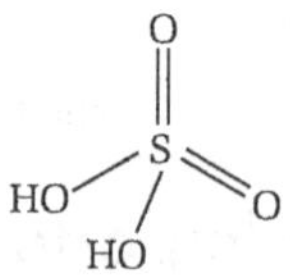

Sulphuric acid (H_2SO_4)

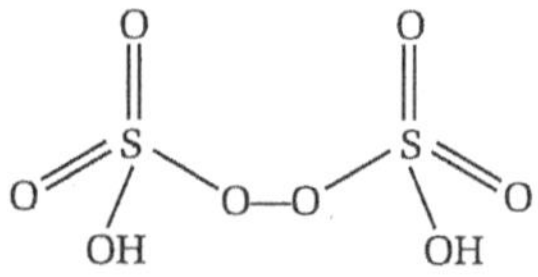

Peroxodisulphuric acid ($H_2S_2O_8$)

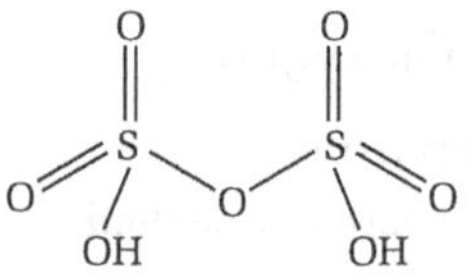

Pyrosulphuric acid (oleum) ($H_2S_2O_7$)

Group-17 Elements :
Halogens Family

- Group-17 members are fluorine (F), chlorine (Cl), bromine (Br), iodine (I) and astatine (At). Astatine is a radioactive element.
- The halogens are highly reactive non-metallic elements. Fluorine is most electronegative atom and strong oxidising agent.

Important Properties

Electronic Configuration

All elements of group have seven electron in their outermost shell $ns^2\ np^2$.

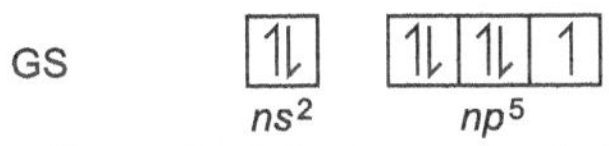

(1 unpaired electron accounts for –1 or +1 oxidation state)

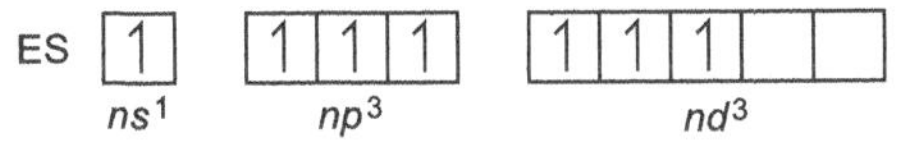

Oxidation State

Ground state and excited state in fluorine will be same due to absence of d-orbitals and oxidation state of fluorine is always –1.

Chemical Reactivity

All the halogens are highly reactive. Halogens are strong oxidising agent and their oxidising power decreases down the group.

$$F_2 > O_2 > Cl_2 > Br_2 > I_2$$

Oxidising power decreases

e.g. $F_2 + 2X^- \longrightarrow 2F^- + X_2 \quad (X = \text{Cl, Br and I})$

$$2F_2(g) + 2H_2O(l) \longrightarrow 4HF(aq) + O_2(g)$$

$$X_2(g) + H_2O(l) \longrightarrow HX(aq) + HOX(aq)$$

(where, X = Cl or Br)

The reaction of I_2 with water is non-spontaneous. However, I^- can be oxidised by oxygen in acidic medium.

$$4I^-(aq) + 4H^+(aq) + O_2(g) \longrightarrow 2I_2(s) + 2H_2O(l)$$

- The order of bond dissociation enthalpy is
$$Cl—F > Cl_2 > Br_2 > F_2 > I_2.$$
- Here, fluorine shows anomalous behaviour because of its very small size and high electronegativity. F_2 has less bond enthalpy than Cl_2 and Br_2 and absence of d-orbitals in valence shell.
- Hydrogen halides dissolve in water to form hydrohalic acids. The acidic strength of these acids varies as
$$HF < FCl < HBr < HI.$$
- HF is a liquid while rest of hydrogen halides are gases. It reacts with glass, so it is stored in waxed glass bottles. This property of HF to act on glass is utilised in etching of glass.

Fluorine forms two oxides, OF_2 and O_2F_2, but only OF_2 is thermally stable at 298 K. O_2F_2 oxidises plutonium to PuF_6 and the reaction is used for removing plutonium as PuF_6 from spent nuclear fuel.

(i) Chlorine forms a number of oxides such as, Cl_2O, Cl_2O_3, Cl_2O_5, Cl_2O_7, ClO_2. ClO_2 is used as a bleaching agent for paper pulp, textiles and in water treatment.

(ii) Br_2O, BrO_2, BrO_3 are the least stable bromine oxides and exist only at low temperatures. They are very powerful oxidising agents.

(iii) The iodine oxides, i.e. I_2O_4, I_2O_5, I_2O_7 are insoluble solids and decompose on heating. I_2O_5 is a very good oxidising agent and is used in the estimation of carbon monoxide.

Some Important Compounds

Some of the important compounds of group 17 elements.

1. Chlorine (Cl_2)

Preparation

It can be prepared by the following methods:

(a) $MnO_2 + 4HCl \longrightarrow MnCl_2 + Cl_2 + 2H_2O$ (Lab method)

(b) $4HCl + O_2 \xrightarrow{CuCl_2} 2Cl_2 + 2H_2O$ (Deacon's process)

Chlorine is obtained by the electrolysis of brine (concentrated NaCl solution). Chlorine is liberated at anode (electrolytic process). It is also obtained as a by product in many chemical industries.

Properties

- It is a greenish yellow gas with pungent and suffocating odour. It is soluble in water.
- $8NH_3 + 3Cl_2 \longrightarrow 6NH_4Cl + N_2$
- $\underset{\text{}}{NH_3} + \underset{\text{Excess}}{3Cl_2} \longrightarrow \underset{\text{Explosive}}{NCl_3} + 3HCl$
- $\underset{\text{Cold and dil.}}{2NaOH} + Cl_2 \longrightarrow NaCl + NaOCl + H_2O$
- $\underset{\text{Hot and conc.}}{6NaOH} + 3Cl_2 \longrightarrow 5NaCl + NaClO_3 + 3H_2O$
- With dry slaked lime, it gives bleaching powder.
$$Ca(OH)_2 + Cl_2 \longrightarrow CaOCl_2 + H_2O$$

Uses

- For bleaching woodpulp, bleaching cotton and textiles.
- In extraction of gold and platinum.
- In manufacture of dyes, drugs and organic compounds.

2. Hydrogen Chloride (HCl)

Preparation

In laboratory, it is prepared by heating sodium chloride with concentrated sulphuric acid.

$$NaCl + H_2SO_4 \xrightarrow{420\text{ K}} NaHSO_4 + HCl$$

$$NaHSO_4 + NaCl \xrightarrow{823\text{ K}} Na_2SO_4 + HCl$$

HCl gas can be dried by passing through concentrated sulphuric acid.

Properties

It is a colourless and pungent smelling gas. Its aqueous solution is called hydrochloric acid.

When three parts of concentrated HCl and one part of concentrated HNO_3 are mixed, *aqua-regia* is formed which is used for dissolving noble metals, e.g. gold, platinum.

Uses

- In manufacture of chlorine, NH_4Cl and glucose.
- For extracting glue from bones.
- In medicine and as laboratory agent.

Oxoacids of Halogens

Higher oxoacids of fluorine such as HFO_2, HFO_3 do not exist because fluorine is most electronegative and has absence of *d*-orbitals. +3 oxidation state of bromine and iodine is unstable due to inert pair effect, therefore $HBrO_2$ and HIO_2 do not exist.

Decreasing order of acidic nature

HOF (Hypofluorous acid) > HOCl (Hypochlorous acid) > HOBr (Hypobromous acid) > HOI (Hypoiodous acid)

$\longrightarrow$ $HOClO_2$ (Chloric acid) > $HOBrO_2$ (Bromic acid) > $HOIO_2$ (Iodic acid)

$\longrightarrow HOClO_3 > HOClO_2 > HOClO > HOCl$

$\longrightarrow HClO_4 > HBrO_4 > HIO_4$

Structures of various oxoacids of chlorine are as follows:

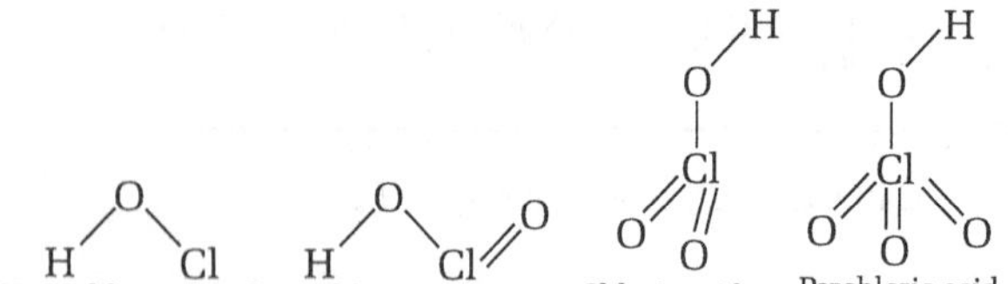

Hypochlorous acid   Chlorous acid   Chloric acid   Perchloric acid

- Pseudo halide ions are stronger ligands than halide ions and these can function as ambidentate ligands as they are made up of two hetero atoms.

Interhalogen Compounds

- When two different halogens react with each other, interhalogen compounds are formed. These compounds are covalent and diamagnetic in nature.
- They are volatile solids or liquids except ClF which is a gas at 298 K. Interhalogen compounds are more reactive than halogens (except fluorine).

Some Properties of Interhalogen Compounds

Type	Formula	Physical state and colour	Structure
XX'	ClF	Colourless gas	Linear
	BrF	Pale brown gas	Linear
	IF (very unstable)	Detected spectroscopically	Linear
	BrCl (Pure solid is known at r.t.)	Ruby red solid (α-form)	Linear
	ICl	Brown red solid (β-form)	Linear
	IBr	Black solid	Linear
XX'_3	ClF_3	Colourless gas	Bent T-shaped
	BrF_3	Yellow green liquid	Bent T-shaped
	IF_3	Yellow power	Bent T-shaped
	ICl_3 (dimerises as Cl-bridged dimer, I_2Cl_6)	Orange solid	Bent T-shaped
XX'_5	IF_5	Colourless gas but solid below 77K	Square pyramidal
	BrF_5	Colourless liquid	Square pyramidal
	ClF_5	Colourless liquid	Square pyramidal
XX'_7	IF_7	Colourless gas	Pentagonal bipyramidal

Group-18 Elements : Noble Gases

- Elements of group 18 are helium (He), neon (Ne), argon (Ar), krypton (Kr), xenon (Xe) and radon (Rn) (radioactive).
- All of these are gases and chemically unreactive. Down the group their ionisation enthalpy decreases, therefore Xe reacts with oxygen and fluorine and forms different compounds. Xenon and radon are the rarest elements of the group. First prepared noble gas compound by Neil Bartlett in 1962 is $XePtF_6$.

Occurrence

On account of their inert nature, the noble gases always occur in the free state. Argon is the most abundant noble gas in the atmosphre while radon is not present in atmosphere. He, Ar and Ne are also found as constituents of dissolved gas of certain spring water.

Important Properties

- **Electronic configuration** is (ns^2np^6) (except helium).

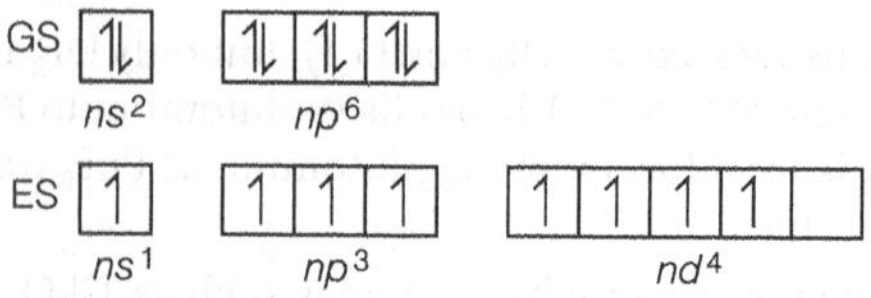

Chemical Reactivity

- $XeF_4 + O_2F_2 \longrightarrow XeF_6 + O_2$
- $6XeF_4 + 12H_2O \longrightarrow 4Xe + 2XeO_3 + 24HF + 3O_2$
- $XeF_6 + 3H_2O \longrightarrow XeO_3 + 6HF$
- Partial hydrolysis of XeF_6 gives oxyfluorides, $XeOF_4$ and XeO_2F_2.
- XeF_2, XeF_4 and XeF_6 are colourless crystalline solids and sublime readily at 298 K. They are powerful fluorinating agents.
- XeF_6 is extremely reactive. It cannot be stored in glass or quartz vessels as it readily reacts with SiO_2 present in glass.

Summary of Stable Compounds of Xe

Compound	Oxidation state of Xe	Hybridisation	No. of lone pair(s) of electrons	Shape
XeF_2	+2	sp^3d	3	Linear
XeF_4	+4	sp^3d^2	2	Square planar
XeF_6	+6	sp^3d^3	1	Distorted octahedral
$XeOF_2$	+4	sp^3d	2	T-shaped
XeO_2F_2	+6	sp^3d	1	Distorted trigonal bipyramidal
$XeOF_4$	+6	sp^3d^2	1	Square pyramidal
XeO_3	+6	sp^3	1	Pyramidal

- **Molecular Structures of Xenon Compounds**

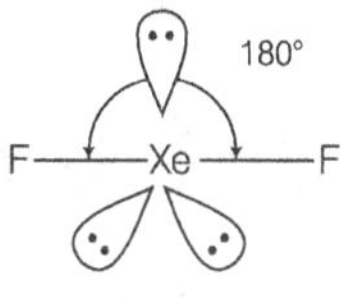

XeF_2

(Xe atom in 1st excited state)

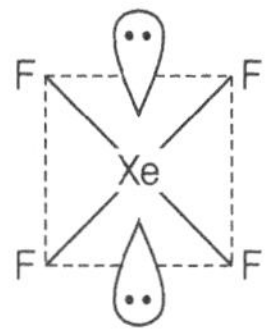

XeF_4

(Xe atom in 2nd excited state)

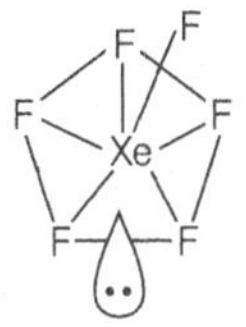

XeF_6

(Xe atom in 3rd excited state)

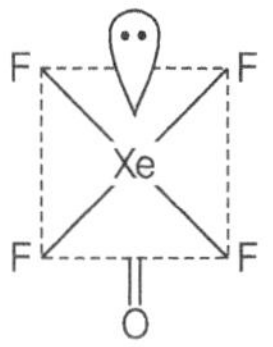

$XeOF_4$

(Xe in 3rd excited state)

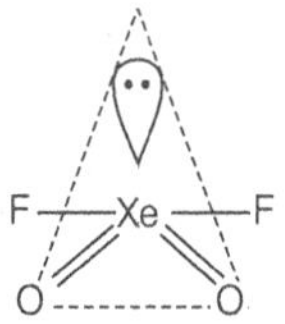

XeO_2F_2

(Xe in 1st excited state)

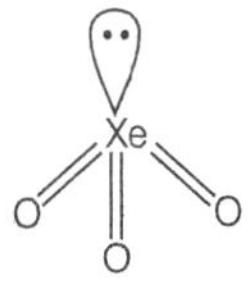

XeO_3

(Xe in 1st excited state)

Uses

- He is used in filling balloons for meteorological observations. It is used in gas cooled reactor to produce powerful super-conducting magnets and as a diluent for oxygen in diving apparatus.
- Ne is used in discharge tubes and fluorescent bulbs.
- Ar is used in filling bulbs and to produce inert atmosphere in various metallurgical operations.

DAY PRACTICE SESSION 1

FOUNDATION QUESTIONS EXERCISE

1 Boric acid is called acid because its molecule
(a) contains replaceable H^+ ion
(b) gives up a proton
(c) accepts OH^- from water releasing proton
(d) combines with proton from water molecule

2 Heating of an aqueous solution of aluminium chloride to dryness will give
(a) $Al(OH)Cl_2$ (b) Al_2O_3 (c) Al_2Cl_6 (d) $AlCl_3$

3 Boron cannot form which one of the following anions? → AIEEE 2011
(a) BF_6^{3-} (b) BH_4^-
(c) $B(OH)_4^-$ (d) BO_2^-

4 Mark the oxide which is amphoteric in character.
(a) CO_2 (b) SiO_2 (c) SnO_2 (d) CaO

5 Graphite is a soft, solid lubricant, extremely difficult to melt. The reason for this anomalous behaviour is that graphite
(a) is an allotropic form of diamond
(b) has molecules of variable molecular masses like polymers
(c) has carbon atoms arranged in large plates of rings of strongly bound carbon atoms with weak interplate bonds
(d) is a non-crystalline substance

6 Carborundum is obtained when silica is heated at high temperature with
(a) carbon (b) carbon monoxide
(c) carbon dioxide (d) calcium carbonate

7 Among the following substituted silanes the one which will give rise to cross linked silicon polymer on hydrolysis, is
(a) R_3SiCl (b) R_4Si
(c) $RSiCl_3$ (d) R_2SiCl_2

8 Name of the structure of silicates in which three oxygen atoms of $[SiO_4]^{4-}$ are shared is
(a) pyrosilicate (b) sheet silicate
(c) linear chain silicate (d) three dimensional silicate

9 In curing cement plasters water is sprinkled from time to time. This helps in
(a) keeping it cool
(b) developing interlocking needle like crystals of hydrated silicates
(c) hydrating sand and gravel mixed with cement
(d) converting sand into silicic acid

10 The catenation tendency of C, Si and Ge is in the order Ge < Si < C. The bond energies (in kJ mol^{-1}) of C—C, Si—Si and Ge—Ge bonds are respectively → JEE Main (Online) 2013
(a) 348, 297, 260 (b) 297, 348, 260
(c) 348, 260, 297 (d) 260, 297, 348

11 Strong reducing behaviour of H_3PO_2 is due to
(a) Low oxidation state of phosphorus
(b) presence of two — OH groups and one P— H bond
(c) Presence of one — OH group and two P— H bonds
(d) high electron gain enthalpy of phosphorus

12 Nitrogen shows different oxidation states in the range
(a) 0 to +5 (b) −3 to +5
(c) −5 to +3 (d) −3 to+3

13 A hydride of nitrogen which is acidic is
(a) NH_3 (b) N_2H_4
(c) N_2H_2 (d) N_3H

14 The compound that does not produce nitrogen gas by the thermal decomposition is → JEE Main 2018
(a) $Ba(N_3)_2$ (b) $(NH_4)_2Cr_2O_7$
(c) NH_4NO_2 (d) $(NH_4)_2SO_4$

15 Extra pure N_2 can be obtained by heating
(a) NH_3 with CuO (b) NH_4NO_3
(c) $(NH_4)_2Cr_2O_7$ (d) $Ba(N_3)_2$

16 The number of σ bonds in P_4O_{10} is
(a) 6 (b) 16
(c) 20 (d) 7

17 PH_3, the hydride of phosphorus is
(a) metallic (b) ionic
(c) non-metallic (d) covalent

18 When concentrated HNO_3 is heated with P_2O_5 it forms
(a) N_2O (b) NO
(c) NO_2 (d) N_2O_5

19 Among the following, the number of compounds that can react with PCl_5 to give $POCl_3$ is

O_2, CO_2, SO_2, H_2O, H_2SO_4, P_4O_{10}.

(a) 1 (b) 2
(c) 3 (d) 4

20 When conc. H_2SO_4 is heated with P_2O_5, the acid is converted to
(a) sulphur trioxide
(b) sulphur dioxide
(c) sulphur
(d) a mixture of sulphur dioxide and sulphur trioxide

21 The pair in which phosphorus atoms have a formal oxidation state of +3 is → JEE Main 2016
(a) pyrophosphorus and hypophosphoric acids
(b) orthophosphorus and hypophophoric acids
(c) pyrophosphorus and pyrophosphoric acids
(d) orthophosphorus and pyrophosphorus acids

22 White phosphorus on reaction with NaOH gives PH_3 as one of the products. This is a
(a) dimerisation reaction (b) disproportionation reaction
(c) condensation reaction (d) precipitation reaction

23 The molecule having smallest bond angle is → AIEEE 2012
(a) NCl_3 (b) $AsCl_3$ (c) $SbCl_3$ (d) PCl_3

24 Which of the following statements is incorrect? → AIEEE 2011
(a) The stability of hydrides increases from NH_3 to BiH_3 in group 15 of the periodic table
(b) Nitrogen cannot form $d\pi$-$p\pi$ bond
(c) Single N—N bond is weaker than the single P—P bond
(d) N_2O_4 has two resonance structures

25 Which one of the following compounds has the smallest bond angle in its molecule?
(a) OH_2 (b) SH_2 (c) NH_3 (d) SO_2

26 The number of S—S bonds in sulphur trioxide trimer (S_3O_9) is
(a) three (b) two (c) one (d) zero

27 Which of the following are peroxoacids of sulphur?
(a) H_2SO_5 and $H_2S_2O_8$ (b) H_2SO_5 and $H_2S_2O_7$
(c) $H_2S_2O_7$ and $H_2S_2O_8$ (d) $H_2S_2O_6$ and $H_2S_2O_7$

28 Which of the following is the wrong statement? → JEE Main (Online) 2013
(a) O_3 molecule is linear
(b) Ozone is violet black in solid state
(c) Ozone is diamagnetic gas
(d) ONCl and ONO^- are not isoelectronic

29 Which of the following statements regarding sulphur is incorrect? → AIEEE 2011
(a) S_2 molecule is paramagnetic
(b) The vapour at 200°C consists mostly of S_8 rings
(c) At 600°C, the gas mainly consists of S_2 molecules
(d) The oxidation state of sulphur is never less than +4 in its compounds

30 Which among the following is the most reactive? → JEE Main 2015
(a) Cl_2 (b) Br_2
(c) I_2 (d) ICl

31 The correct order of the thermal stability of hydrogen halides (H—X) is
(a) HI > HCl > HF > HBr (b) HCl < HF > HBr < HI
(c) HF > HCl > HBr > HI (d) HI > HBr > HCl >HF

32 Among the following oxoacids, the correct decreasing order of acid strength is → JEE Main 2014
(a) $HOCl > HClO_2 > HClO_3 > HClO_4$
(b) $HClO_4 > HOCl > HClO_2 > HClO_3$
(c) $HClO_4 > HClO_3 > HClO_2 > HOCl$
(d) $HClO_2 > HClO_4 > HClO_3 > HOCl$

33 The solubility order of alkali metal fluoride in water is → JEE Main (Online) 2013
(a) LiF < RbF < KF < NaF (b) RbF < KF < NaF < LiF
(c) LiF > NaF > KF > RbF (d) LiF < NaF < KF < RbF

34 Which one has the highest boiling point? → JEE Main 2015
(a) He (b) Ne (c) Kr (d) Xe

35 Argon is used in arc welding because of its
(a) low reactivity with metal
(b) ability to lower the melting point of metal
(c) flammability
(d) high calorific value

36 The structure of XeO_3 is
(a) linear (b) planar
(c) pyramidal (d) T-shaped

37 XeF_4 and XeF_6 are expected to be
(a) oxidising (b) reducing
(c) unreactive (d) strongly basic

38 Which of the following acids possesses oxidising, reducing and complex forming properties?
(a) HCl (b) HNO_2 (c) H_2SO_4 (d) HNO_3

39 Match the species given in Column I with properties given in Column II.

	Column I		Column II
A.	Diborane	1.	Used as a flux for soldering metals
B.	Gallium	2.	Crystalline form of silica
C.	Borax	3.	Banana bonds
D.	Aluminosilicate	4.	Low melting, high boiling,useful for measuring high temperatures
E.	Quartz	5.	Used as catalyst in petrochemical industries

Codes

	A	B	C	D	E
(a)	3	4	1	5	2
(b)	4	3	1	5	2
(c)	4	3	5	2	1
(d)	1	2	5	4	3

Directions (Q. Nos. 40-42) *Each of these questions contains two statements : Statement* I *and Statement* II. *Each of these questions also has four alternative choices, only one of which is the correct answer. You have to select one of the codes (a), (b), (c), (d) given below:*
(a) Statement I is true, Statement II is true; Statement II is a correct explanation for Statement I
(b) Statement I is true, Statement II is true; Statement II is not a correct explanation for Statement I
(c) Statement I is true; Statement II is false
(d) Both the Statement I and II are false

40 **Statement I** Boron always forms covalent bond.

Statement II The small size of B^{3+} favours formation of covalent bond.

41 Statement I In water, orthoboric acid behaves as a weak monobasic acid.

Statement II In water, orthoboric acid acts as a proton donor.

42 Statement I Between $SiCl_4$ and CCl_4, only $SiCl_4$ reacts with water.

Statement II $SiCl_4$ is ionic and CCl_4 is covalent.

Direction (Q. Nos. 43 and 44) *In the following questions, Assertion (A) followed by Reason (R) is given. Choose the correct answer out of the following choices.*

(a) Assertion and Reason both are correct statements and Reason is the correct explanation of the Assertion
(b) Assertion and Reason both are correct statements but Reason is not the correct explanation of the Assertion
(c) Assertion is correct and Reason is incorrect
(d) Both Assertion and Reason are incorrect

43 Assertion (A) Silicones are water repelling in nature.

Reason (R) Silicones are organosilicon polymers, which have (—R_2SiO—) as repeating unit.

44 Assertion (A) Nitrogen and oxygen are the main components in the atmosphere but these do not react to form oxides of nitrogen.

Reason (R) The reaction between nitrogen and oxygen requires high temperature.

DAY PRACTICE SESSION 2

PROGRESSIVE QUESTIONS EXERCISE

1 Identify the incorrect statement among the following.
(a) Ozone reacts with SO_2 to give SO_3
(b) Silicon reacts with NaOH (*aq*) in the presence of air to give Na_2SiO_3 and H_2O
(c) Cl_2 reacts with excess of NH_3 to give N_2 and NH_4Cl
(d) Br_2 reacts with hot and strong NaOH solution to give NaBr, $NaBrO_4$ and H_2O

2 Reduction potentials of some ions are given below. Arrange them in the decreasing order of oxidising power.

Ion	ClO_4^-	IO_4^-	BrO_4^-
Reduction potential $E^\ominus/V$	$E^\ominus = 1.19$ V	$E^\ominus = 1.65$ V	$E^\ominus = 1.74$ V

(a) $ClO_4^- > IO_4^- > BrO_4^-$ (b) $IO_4^- > BrO_4^- > ClO_4^-$
(c) $BrO_4^- > IO_4^- > ClO_4^-$ (d) None of these

3 Which of the following statements is true?
(a) H_3PO_3 is a stronger acid than H_2SO_3
(b) In aqueous medium, HF is a stronger acid than HCl
(c) $HClO_4$ is weaker acid than $HClO_3$
(d) HNO_3 is a stronger acid than HNO_2

4 Which one of the following reaction of xenon compounds is not feasible?
(a) $XeO_3 + 6HF \longrightarrow XeF_6 + 3H_2O$
(b) $3XeF_4 + 6H_2O \longrightarrow 2Xe + XeO_3 + 12HF + 1.5O_2$
(c) $2XeF_2 + 2H_2O \longrightarrow 2Xe + 4HF + O_2$
(d) $XeF_6 + RbF \longrightarrow Rb[XeF_7]$

5 Among the following, the correct statement is :
(a) Between NH_3 and PH_3, NH_3 is a better electron donor because the lone pair of electrons occupies spherical '*s*' orbital and is less directional
(b) Between NH_3 and PH_3, PH_3 is a better electron donor because the lone pair of electrons occupies sp^3 orbital and is more directional
(c) Between NH_3 and PH_3, NH_3 is a better electron donor because the lone pair of electrons occupies sp^3 orbital and is more directional
(d) Between NH_3 and PH_3, PH_3 is a better electron donor because the lone pair of electrons occupies spherical '*s*' orbital and is less directional

6 The formation of $O_2^+[PtF_6]^-$ is the basis for the formation of xenon fluorides. This is because
(a) O_2 and Xe have comparable sizes
(b) both O_2 and Xe are gases
(c) O_2 and Xe have comparable I.E.
(d) Both (a) and (c)

7 The hydrides of the first elements in groups 15-17 namely NH_3, H_2O and HF respectively shows abnormally high values for melting and boiling points. This is due to
(a) small size of N, O and F
(b) the ability to form extensive intermolecular H-bonding
(c) the ability to form extensive intramolecular H-bonding
(d) effective forces of interaction

8 An inorganic compound '*X*', made of two most occurring elements in the earth's crust and used in building construction, when reacts with carbon forms a diatomic molecule, which is poisonous in nature. Compound '*X*' may be
(a) SiO_2 (b) Al_2O_3
(c) CaO (d) CO_2

9 Which one of the following reactions does not occur ?
(a) $F_2 + Cl^- \longrightarrow 2F^- + Cl_2$ (b) $Cl_2 + 2F^- \longrightarrow 2Cl^- + F_2$
(c) $Br_2 + 2I^- \longrightarrow 2Br^- + I_2$ (d) $Cl_2 + 2Br^- \longrightarrow 2Cl^- + Br_2$

10 $B(OH)_3 + NaOH \rightleftharpoons NaBO_2 + Na[B(OH)_4)] + H_2O$
How can this reaction is made to proceed in forward direction?
(a) Addition of *cis* 1, 2-diol (b) Addition of borax
(c) Addition of *trans* 1, 2-diol (d) Addition of Na_2HPO_4

11 An inorganic compound containing ($3c$-$2e^-$) and ($2c$-$2e^-$) bonds when reacts with NH_3, at a certain temperature, gives a compound '*X*' which is isostructural with benzene and when reacts at high temperature forms a substance '*Y*'. The substance '*Y*' is

(a) B_2H_6 (b) $B_3N_3H_6$
(c) inorganic graphite (d) $B_2H_6 \cdot 2NH_3$

12 In which of the following arrangements, the sequence is not strictly according to the property written against it?

(a) $CO_2 < SiO_2 < SnO_2 < PbO_2$: increasing oxidising power
(b) HF < HCl < HBr < HI : increasing acid strength
(c) $NH_3 > PH_3 < AsH_3 < SbH_3$: increasing basic strength
(d) B < C < O < N : increasing first ionisation enthalpy

13 The correct order of pseudohalide, polyhalide and interhalogen are

(a) BrI_2^-, OCN^-, IF_5 (b) IF_5, BrI_2^-, OCN^-
(c) OCN^-, IF_5, BrI_2^- (d) OCN^-, BrI_2, IF_5

14 Consider the following statements.

I. NCl_5 does not exist while PCl_5 does
II. Both O_2^+ and NO are paramagnetic
III. Three C — O bonds are not equal in carbonate ion.
IV. Lead prefers to form tetravalent compound.

Which of the above statements are incorrect?

(a) I and III (b) I, III and IV
(c) II and III (d) III and IV

15. The molecule BF_3 and NF_3 both are covalent compounds, but BF_3 is non-polar and NF_3 is polar. The reason is that

(a) boron is a metal and nitrogen is a gas in uncombined state
(b) BF_3 bonds have no dipole moment whereas NF_3 bond have dipole moment
(c) atomic size of boron is smaller than that of N_2
(d) BF_3 is symmetrical molecule whereas NF_3 is unsymmetrical

ANSWERS

SESSION 1	**1** (c)	**2** (b)	**3** (a)	**4** (c)	**5** (c)	**6** (a)	**7** (a)	**8** (b)	**9** (b)	**10** (a)
	11 (c)	**12** (b)	**13** (d)	**14** (d)	**15** (d)	**16** (b)	**17** (d)	**18** (d)	**19** (d)	**20** (a)
	21 (d)	**22** (b)	**23** (c)	**24** (a)	**25** (a)	**26** (d)	**27** (a)	**28** (a)	**29** (d)	**30** (d)
	31 (c)	**32** (c)	**33** (d)	**34** (d)	**35** (a)	**36** (c)	**37** (a)	**38** (b)	**39** (a)	**40** (b)
	41 (a)	**42** (c)	**43** (a)	**44** (a)						
SESSION 2	**1** (d)	**2** (c)	**3** (d)	**4** (a)	**5** (c)	**6** (d)	**7** (b)	**8** (a)	**9** (b)	**10** (a)
	11 (a)	**12** (c)	**13** (d)	**14** (d)	**15** (d)					

Hints and Explanations

SESSION 1

1 Due to the small size of boron atom and presence of six electrons. Boric acid accepts a pair of electron from OH^- ion of H_2O thereby releasing a proton.

2 Aqueous solution of $AlCl_3$ is acidic due to hydrolysis.

$$AlCl_3 + 3H_2O \rightleftharpoons Al(OH)_3 + 3HCl$$

On strong heating, $Al(OH)_3$ is converted into Al_2O_3.

$$2Al(OH)_3 \xrightarrow{\Delta} Al_2O_3 + 3H_2O$$

3 $_5B$

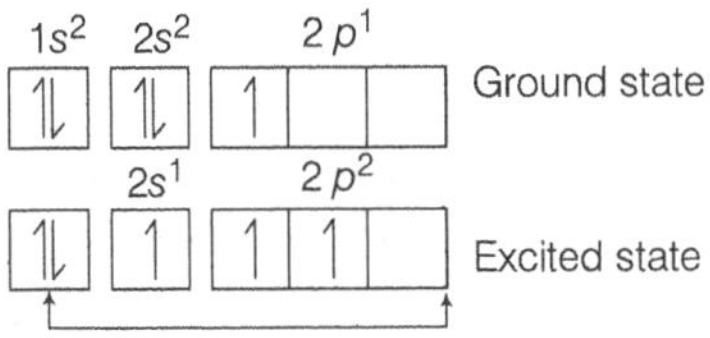

Fourth lone pair is accomodated in this empty orbital
Maximum covalency = 4

Due to the absence of $2d$-orbital, maximum covalency is four. Thus, BF_6^{3-} is not formed.

BH_4^- ($BH_3 + H^-$); $B(OH)_4^-$ ($B(OH)_3 + OH^-$) and BO_2^- are formed.

4 SnO_2 is amphoteric. It dissolves in acids as well as in alkalies e.g.

$$SnO_2 + 2H_2SO_4 \longrightarrow Sn(SO_4)_2 + 2H_2O$$
$$SnO_2 + 2NaOH \longrightarrow Na_2SnO_3 + H_2O$$

5 Graphite has carbon atoms arranged in large hexagonal layers with weak van der Waals' interactions between the layers.

6 $$SiO_2 + 3C \xrightarrow[2300\ K]{\text{High temp.}} \underset{\text{Carborundum}}{SiC} + 2CO$$

7 $RSiCl_3$, gives cross-linked silicon polymer on hydrolysis.

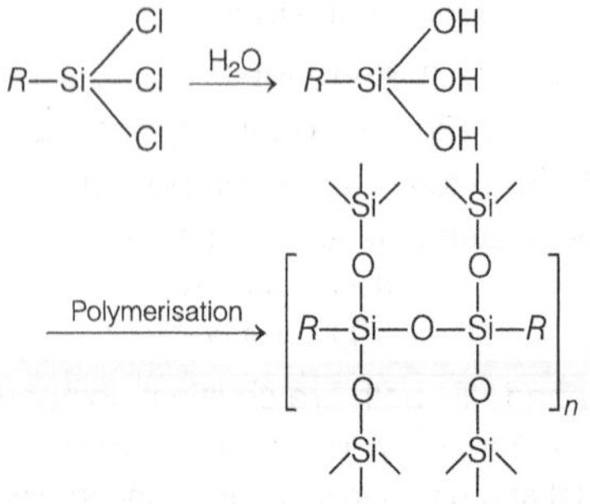

8 The structure of silicates has been found with the help of X-ray diffraction techniques. Sheet silicates are formed when three oxygen atoms (bridging O-atoms) of each $(SiO_4)^{4-}$ unit are shared. Hence, the general formula of sheet silicates is $(Si_2O_5)^{2n-}$.

9 The hydration of calcium aluminates and silicates involved in the setting of cement is an exothermic process, hence cement structures have to be cooled during setting by sprinkling water.

10 Higher the catenation tendency, higher is the bond energy.

$\therefore$ The bond energies of C—C, Si—Si and Ge—Ge bonds are respectively 348, 297 and 260 kJ mol^{-1}.

11 Due to the presence of one —OH group and two P—H bonds, H_3PO_2 act as reducing agent.

12 Nitrogen shows all the oxidation states ranging from −3 to + 5 in its compounds.

13 N_3H is acidic. It shows following resonating structures.

$[N \equiv N^+ - N^-]^- \longrightarrow [N^- - N^+ \equiv N]^-$

14 The thermal decomposition of given compounds is shown below:

$(NH_4)_2Cr_2O_7 \xrightarrow{\Delta} N_2 + 4H_2O + Cr_2O_3$

$NH_4NO_2 \xrightarrow{\Delta} N_2 + 2H_2O$

$(NH_4)_2SO_4 \xrightarrow{\Delta} 2NH_3 + H_2SO_4$

$BO(N_3)_2 \longrightarrow Ba + 3N_2$

Thus, only $(NH_4)_2SO_4$ does not gives N_2 on heating while rest of the given compounds give N_2 on their thermal decomposition.

15 $Ba(N_3)_2 \xrightarrow{\text{Heat}} Ba(s) + 3N_2(g)$

Azide salt of barium can be used to obtain purest form of N_2. The decomposition product contain solid Ba as by product along with gaseous nitrogen.

16 The number of σ-bonds in P_4O_{10} is 16. There are also four π-bonds present in P_4O_{10} molecule.

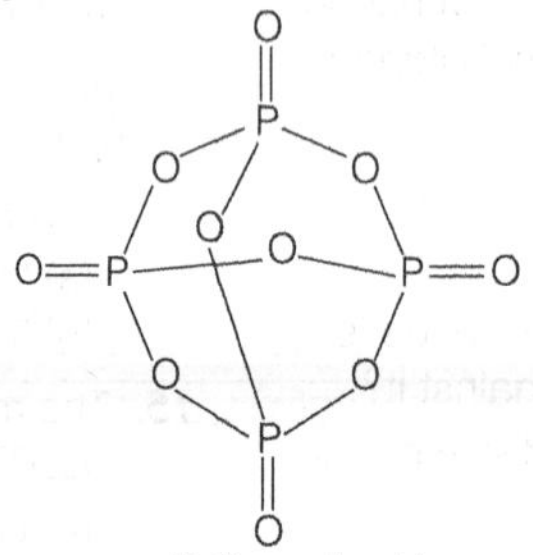

(P_4O_{10} molecule)

17 PH_3 is a covalent hydride.

18 $4HNO_3 + 2P_2O_5 \xrightarrow{\Delta} 2N_2O_5 + 4HPO_3$

19 PCl_5 produces $POCl_3$ with the following reagents.

$2PCl_5 + SO_2 \longrightarrow POCl_3 + SOCl_2$

$PCl_5 + H_2O \longrightarrow POCl_3 + 2HCl$

$2PCl_5 + H_2SO_4 \longrightarrow SO_2Cl_2 + 2POCl_3 + 2HCl$

$6PCl_5 + P_4O_{10} \longrightarrow 10POCl_3$

20 When conc. H_2SO_4 is heated with P_2O_5, the acid is converted into sulphur trioxide.

$2H_2SO_4 + 2P_2O_5 \longrightarrow 2SO_3 + 4HPO_3$

21 Orthophosphorus acid, H_3PO_3

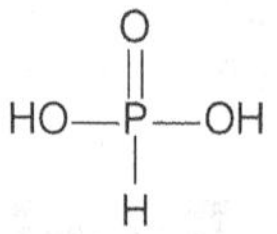

$H_3\overset{x}{P}O_3 = 3 + x + 3(-2) = 0$ or $x = +3$

Pyrophosphorus acid, $H_4P_2O_5$:

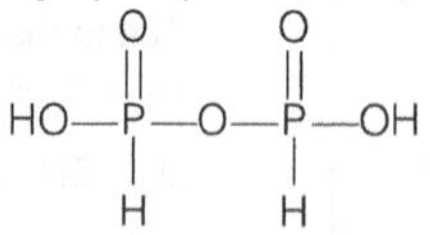

$H_4\overset{x}{P_2}O_5 = 4 + 2x + 5(-2) = 0$

$4 + 2x - 10 = 0 \Rightarrow x = +3$

22 White phosphorus on reaction with NaOH gives PH_3 as one of the product in disproportionation reaction.

$P_4 + 3NaOH + 3H_2O \longrightarrow 3NaH_2PO_2 + PH_3$

23 N, P, As, Sb — Group = 15; Valence electrons = 5

MCl_3 has sp^3 hybridised *M*-element with one lone pair.

Lone pair and bond pair repulsion decreases bond angle. However, the bond-pairs of electrons are much farther away from the central atom in $SbCl_3$ than they are in NCl_3. Thus, lone pair causes even greater distortion in PCl_3, $AsCl_3$ and $SbCl_3$. Hence, bond angle decreases from NCl_3 (maximum) to $SbCl_3$ (minimum).

24 (a) Thermal stability of the hydrides decrease as we go down the group in periodic table for group 15 (N-family).

	BiH_3	< SbH_3	< AsH_3	< PH_3	< NH_3
	Least stable				Most stable
M–H bond energy (kJ mol^{-1})	–	255	247	322	391

(b) Due to the absence of *d*-orbital, nitrogen cannot from *dπ-pπ* bond, thus it is correct.

(c) The N—N bond (BE 160 kJ mol^{-1}) is weaker than P—P bond (BE 209 kJ mol^{-1}) thus, it is correct.

(d) N_2O_4 can form two resonance structures.

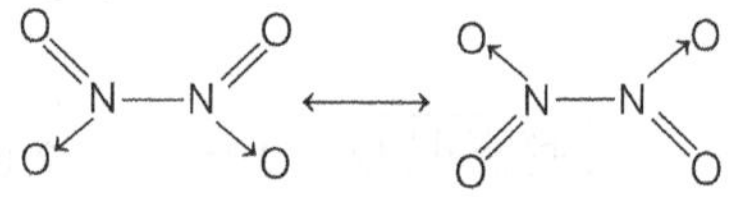

Thus, it is correct.

25 Bond angles are OH_2 (104°), H_2S (92°), NH_3 (107°) and SO_2 (119.5°).

26 S_3O_9 is as

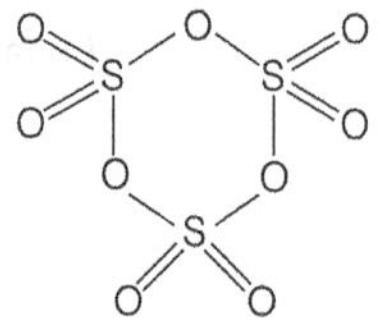

Hence, number of S—S bond = 0.

27

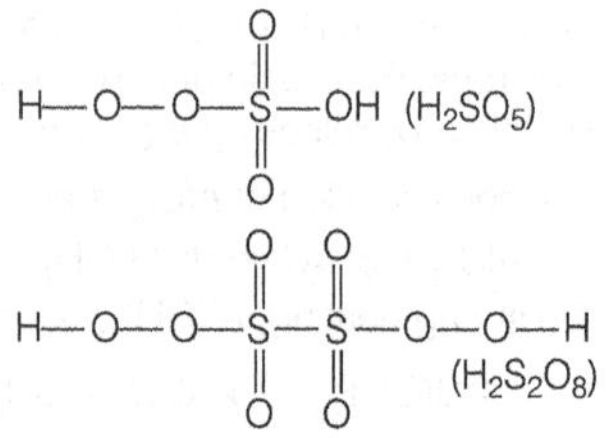

28 O_3 molecule is sp^3-hybridised and has bent shape.

29 (a) S_2 molecule is paramagnetic due to unpaired electrons in MO and is blue-coloured compound, thus true.

(b) The vapour at 200°C consists mostly of S_8 rings, thus correct.

(c) At 600°C, the gas mainly consists of S_2 molecules, thus correct.

(d) Oxidation states of sulphur are:

− 2	in	H_2S
0	in	S_8
+ 2	in	$S_2O_3^{2-}$
+ 4	in	SO_2
+ 6	in	SO_3

Thus, incorrect oxidation state can be less than 4.

30 Cl_2,Br_2 and I_2 are homonuclear diatomic molecule in which electronegativity of the combining atoms is same,so they are more stable and less reactive whereas, I and Cl have different electronegativities and bond between them are polarised and hence, reactive. Therefore, interhalogen compounds are more reactive.

Time Saving Technique In this type of question of halogen, only go through the polarity of the molecule. As we know, diatomic molecule does not have polarity but molecules with dissimilar sizes have polarity resulting in more reactivity.

31 As the size of the halogen atom increases from F to I, the bond length of H — X increases. This increase in bond length decreases the bond strength and thus the thermal stability. Therefore, the order is

$$HF > HCl > HBr > HI.$$

32 Decreasing order of strength of oxoacids

$$HClO_4 > HClO_3 > HClO_2 > HOCl$$

Reason Negative charge is more delocalised on ClO_4^- due to resonance, hence, ClO_4^- is more stable (and less basic).

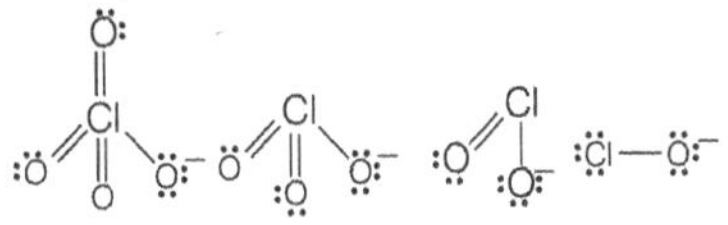

Hence, we can say as the number of oxygen atom (s) around Cl-atom increases as oxidation number of Cl-atom increases and thus, the ability of loose the H^+ increases.

33 For a substance to dissolve, the hydration energy must be greater than lattice energy. Due to the small size of Li^+ ion, the hydration energy of LiF is considerably high, but it has low solubility in water because of its higher lattice energy.

On moving down the group lattice energy decreases which results in increased solubility. Thus, the order of solubility for alkali metal fluorides in water is

$$LiF < NaF < KF < RbF.$$

34 As we move down the group of noble gases, molecular mass increases by which dipole produced for a moment and hence London forces increases from He to Xe. Therefore, more amount of energy is required to break these forces, thus boiling point also increases form He to Xe.

35 Argon is used in arc welding to cause the inert atmosphere around, i.e. without interaction with metal.

36 In XeO_3, total electron pairs

$$= 4\,(3bp + 1lp),$$

i.e. sp^3-hybridisation ⇒ pyramidal shape.

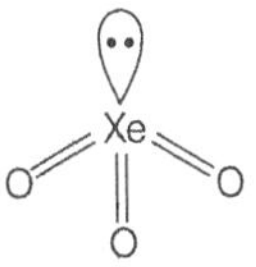

37 XeF_4 oxidises potassium iodide.

$$XeF_4 + 4I^- \longrightarrow 2I_2 + 4F^- + Xe$$

XeF_6 oxidises hydrogen like other xenon fluorides.

38 HNO_2 posseses oxidising, reducing and complex forming properties as in it oxidation number of nitrogen is +3 (i.e. in between −3 to +5).

39 A → 3, B → 4, C → 1, D → 5, E → 2

A. BH_3 is unstable froms diborane B_2H_6 by (3c —$2e^-$) electron bond show banana bond.

B. Gallium with low melting point and high boiling points makes it useful to measure high temperatures.

C. Borax is used as a flux for soldering metals for heat,scratch resistant coating in earthernwares.

D. Alumino silicate used as catalyst in petrochemical industries.

E. Quartz is a crystalline form of silica.

40 Boron always form covalent bond because boron requires very high energy to form B^{3+} and again B^{3+} due to its very small size, have high polarising power, thus cause greater polarisation and eventually significant covalent characteristics (Fajan's rule.)

41 In water, orthoboric acid behave as a weak monobasic Lewis acid.

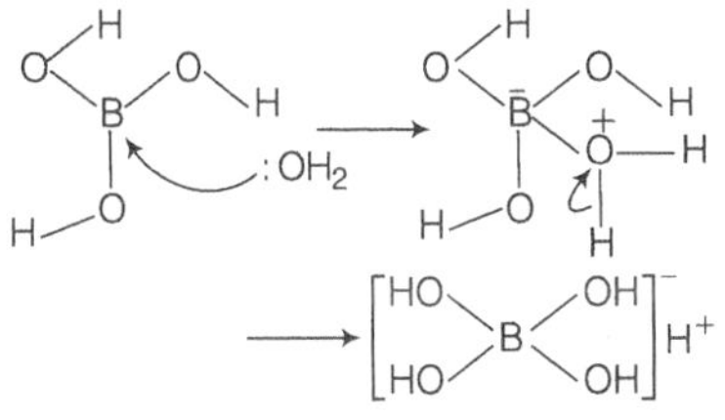

42 Statement I is true but Statement II is false ($SiCl_4$ is also covalent bonded).

43 Silicones are a group of organosilicon polymers which have (R_2SiO) as a repeating unit.

This suggests that silicones are surrounded by non-polar alkyl groups that are water repelling in nature. They have wide applications it is used for water proofing of fabrics.

44 Nitrogen is an inert gas because of the presence of strong bond. That's why although there is 78% N_2 in the atmosphere but nitrogen oxide in not formed under ordinary conditions.

But when temperature is high enough i.e. = 2000 K, it reacts with oxygen to form nitrogen oxide.

$$N_2 + O_2 \xrightarrow{\approx 2000K} 2NO$$

Thus, Assertion and Reason are true and Reason is the correct explanation of the Assertion.

SESSION 2

1 Br_2 reacts with hot and strong NaOH to give NaBr, $NaBrO_3$ and H_2O.

2 Higher the reduction potential stronger the oxidising agent, i.e. option (c) is correct.

3 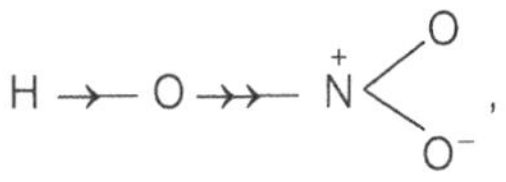 ,

H→—O→—N=O

Polarity along O—H in HNO_3 is more in comparison to —O—H in HNO_2.

4 XeF_6 has much tendency to hydrolyse. The reverse reaction is more spontaneous.

$$XeF_6 + 3H_2O \longrightarrow XeO_3 + 6HF$$

5 NH_3 is better electron donor because the lone pair of electrons occupies sp^3-orbital and is more directional.

6 The first ionisation energy of xenon is quite close to that of oxygen and the molecular diameter of xenon and oxygen are almost identical. Based on the above facts it is suggested that since, oxygen combines with PtF_6, so xenon should also form similar compound with PtF_6.

7 Hydrides like NH_3, H_2O and HF have extensive intermolecular hydrogen bonding due to which they show high melting and boiling point.

8 The two most abundant elements in the earth's crust are Si and oxygen and the compound made by them is SiO_2. This compound is used in building construction.

$$SiO_2 + 2C \longrightarrow Si + \underset{\text{Poisonous}}{2CO}$$

9 With progressive increase in atomic number, the reduction potential of halogens decreases, thus oxidising power also decreases.
Hence, a halogen with lower atomic number will oxidise the halide ion of higher atomic number and therefore, will liberate them from their salt solution. Hence, the reaction,

$$Cl_2 + 2F^- \longrightarrow 2Cl^- + F_2$$

is not possible.

10 $B(OH)_3 + NaOH \rightleftharpoons NaBO_2 + Na^+[B(OH)_4]^- + H_2O$
This reaction is reversible reaction because sodium metaborate, $Na^+[B(OH)_4]^-$ formed by the reaction between $B(OH)_3$ and NaOH gets hydrolysed to regenerate $B(OH)_3$ and NaOH.

$$Na^+[B(OH)_4]^- \xrightarrow{\text{Hydrolysis}} NaOH + B(OH)_3$$

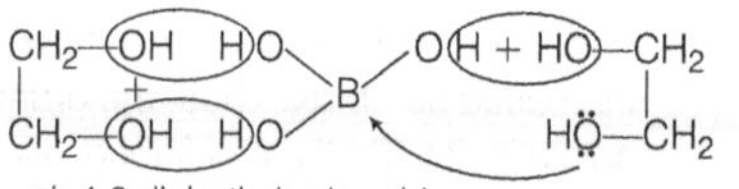

cis 1,2-diol orthoboric acid

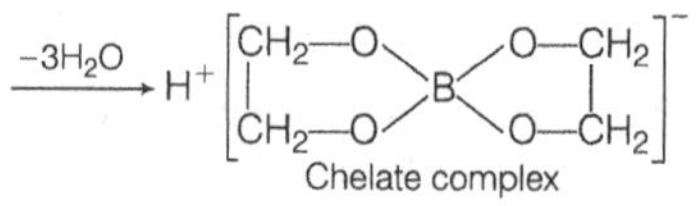

Chelate complex

If some quantity of polyhydroxy compounds like *cis* 1,2-diol, catechol, glycerol etc., is added to the reaction mixture, then the $B(OH)_3$ combines with such polyhydroxy compounds to give chelated complex compound. Due to complex compound formation, stability increases and due to higher stability of complex, reaction moves in forward direction.

11 B_2H_6 contains (3c-2e^-) and (2c-2e^-) bonds.

$$B_2H_6 + 2NH_3 \xrightarrow{\text{Low temp.}} B_2H_6 \cdot 2NH_3 \xrightarrow{200°C} \underset{\text{Inorganic benzene}}{\underset{'X'}{B_3N_3H_6}}$$

$$B_2H_6 + NH_3 \xrightarrow{\text{High temp.}} \underset{\text{Inorganic graphite}}{\underset{'Y'}{(BN)_x}}$$

12 The correct increasing basic strength

$$SbH_3 < AsH_3 < PH_3 < NH_3$$

NH_3 is the most basic because of its small size, the electron density of electron pair is concentrated over small region. As the size increases, the electron density gets diffused over a large surface area and hence, the ability to donate the electron pair (basicity) decrease.

13 Pseudohalide are the combination of more than one electronegative atoms that have one unit negative charge. e.g. OCN^-, CN^-.
Polyhalide ions The complex ions which are formed by reaction of halogens among themselves are called polyhalide ions, e.g. I_3^-, BrI_2^-.

14 I. In nitrogen, *d*-orbitals are absent hence, it does not form NCl_5. Thus, NCl_5 does not exist but PCl_5 does.

II. O_2^+ and NO are isoelectronic and contains one unpaired electron each, thus both are paramagnetic.

III. In carbonate ion CO_3^{2-} all three C—O bonds are identical due to resonance.

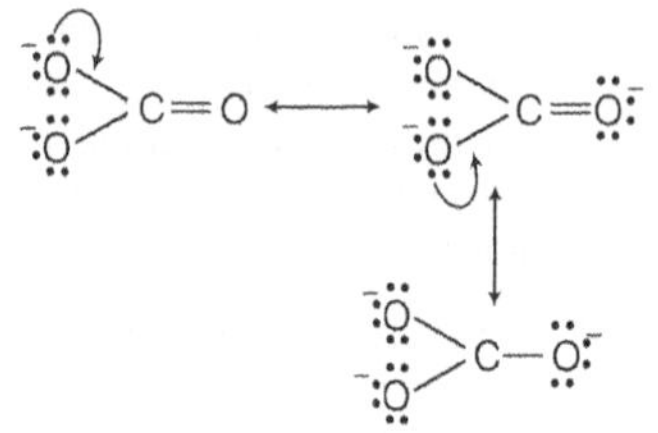

IV. Pb^{2+} is more stable than Pb^{4+} due to inert pair effect, hence, prefers to form divalent compounds.

Thus, the incorrect statements are III and IV.

15 BF_3 is symmetrical planar, although it has polar bonds but resultant dipole moment is zero. In NF_3 lone pair cause distortion hence polarity arises.

DAY TWENTY THREE

The d- and f- Block Elements

Learning & Revision for the Day

- Transition Elements
- Some Important Compounds of Transition Elements
- Inner Transition Elements (f-Block Elements)

The *d*-block elements of the periodic table contains the elements of the group 3-12 in which the *d*-orbitals are progressively filled in each of the four long periods.

The elements constituting the *f*-block are those in which the 4*f* and 5*f*-orbitals are progressively filled in the latter two long periods. These elements are formal members of 3rd group from which they have been taken out to develop a separate *f*-block of the periodic table.

Transition Elements

- *d*-block elements are known as transition elements because their position in the periodic table is between the *s*-block and *p*-block elements.
- Transition elements have incomplete filled *d*-orbital in their ground state or in any of its oxidation state.
- There are mainly three series of the transition metals, (i) 3*d*-series (Sc to Zn), (ii) 4*d*-series (Y to Cd) and (iii) 5*d*-series (La to Hg), omitting (Ce to Lu).

Electronic Configuration

- Electronic configuration of the transition elements is $(n-1)d^{1-10}ns^{1-2}$.
- Zn, Cd, Hg have electronic configuration $[(n-1)d^{10}ns^2]$, are *d*-block elements, but not transition metals because they have completely filled *d*-orbitals.

Transition Metals or *d*-Block Elements

3rd group	4th group	5th group	6th group	7th group
$ns^2(n-1)d^1$	$ns^2(n-1)d^2$	$ns^2(n-1)d^3$	$ns^1(n-1)d^4$	$ns^2(n-1)d^5$
Sc	Ti	V	Cr	Mn
Y	Zr	Nb	Mo	Tc
La	Hf	Ta	W	Re
8th group	**9th group**	**10th group**	**11th group**	**12th group**
$ns^2(n-1)d^6$	$ns^2(n-1)d^7$	$ns^2(n-1)d^8$	$ns^2(n-1)d^9$	$ns^2(n-1)d^{10}$
Fe	Co	Ni	Cu	Zn
Ru	Rh	Pd	Ag	Cd
Os	Ir	Pt	Au	Hg

PREP MIRROR
Your Personal Preparation Indicator

- No. of Questions in Exercises (x)—
- No. of Questions Attempted (y)—
- No. of Correct Questions (z)—
  (Without referring Explanations)

- Accuracy Level $(z/y \times 100)$—
- Prep Level $(z/x \times 100)$—

In order to expect good rank in JEE, your Accuracy Level should be above 85 & Prep Level should be above 75.

- Some of the exceptions are observed in electronic configuration. e.g., $Cr : 3d^5 4s^1$ instead of $3d^4 4s^2$, $Cr : 3d^{10} 4s^1$ instead of $3d^9 4s^2$

General Trends in Properties of the Transition Elements

The general properties of *d*-block elements and their trends are discussed below:

Physical Properties

- All the transition elements exhibit typical metallic properties such as high tensile strength, ductility etc.
- Their melting and boiling points are high which are due to the involvement of $(n-1)\,d$ electrons resulting into strong interatomic bonding.

1. Atomic and Ionic Radii

- In transition metals, on moving from left to right net nuclear charge increases due to poor shielding effect. Due to this, the atomic and ionic radii for transition elements for a given series shows a decreasing trend for first five elements and then becomes almost constant for next five elements of the series. At the end of the period, there is a slight increase in the atomic radii.
- The atomic and ionic radii of the elements of 4*d*-series are higher than corresponding element of 3*d*-series due to increase in number of shells.
- However, the elements of 4*d* and 5*d*-series have almost similar radii and very similar physical and chemical properties.

2. Enthalpy of Atomisation

Transition elements exhibit higher enthalpy of atomisation because of large number of unpaired electrons in their atoms. They have stronger interatomic interaction and hence form stronger bond. Thus, transition elements have very high melting and boiling points.

3. Ionisation Enthalpies

- The first ionisation enthalpy of transition elements lies in between *s*-block and *p*-block elements. In a series from left to right, ionisation enthalpy increases due to increase in nuclear charge.
- The irregular trend in the first ionisation enthalpy of the 3*d*-orbitals, can be accounted by considering that the removal of one electron alters the relative energies of 4*s* and 3*d*-orbitals.
- Ionisation enthalpy values for 5*d*-series are higher than that of 3*d* and 4*d*-series due to weak shielding effect of 4*f*-electrons present in 5*d*-series transition elements.
- Ionisation enthalpy values of Zn, Cd and Hg are abnormally higher on account of greater stability of completely filled *d*-orbital (d^{10}).

4. Oxidation States

Transition metals show variable oxidation states due to the fact that the difference in the energy of $(n-1)d$ electron and *ns* electrons is low which implies that electrons from both energy level can take part in bonding.

Stable oxidation states of the first row transition metals are :

$$Sc(+3), Ti(+4), V(+5), Cr(+3, +6), Mn(+2, +7),$$
$$Fe(+2, +3), Co(+2, +3), Ni(+2), Cu(+2), Zn(+2)$$

- In each period, the highest oxidation state increases with increase in atomic number, attains a maximum value in the middle and then decreases, for example, Mn (3*d*-series), Ru (4*d*-series) and Os (5*d*-series) has maximum value for oxidation state as + 7, + 8, + 8 respectively.
- The transition elements in their lower oxidation states (+2 and + 3) usually forms ionic compounds. In higher oxidation state compounds are normally covalent. Ni and Fe in $Ni(CO)_4$ and $Fe(CO)_5$ show zero oxidation state.

5. Trends in the Standard Electrode Potentials

- The potential difference in transformation of the solid metal atoms to M^{2+} ions in solution is their standard electrode potentials. If sum of the first and second ionisation enthalpies is greater than hydration enthalpy, standard potential ($E^\circ_{M^{2+}/M}$) will be positive and reactivity will be lower and *vice-versa*.
- In general along a period, there is a decrease in electropositive character.

6. Trends in Stability of Higher Oxidation States

- The highest oxidation numbers are achieved in TiX_4, VF_5 and CrF_6. The +7 oxidation state for Mn is not represented in simple halides but MnO_3F is known and beyond Mn, no metal has a trihalide except FeX_3 and CoF_3. The increasing order of oxidising power in the series is
$$VO_2^+ < Cr_2O_7^{2-} < MnO_4^-$$
- The highest oxidation states of transition metals are found in their compounds with fluorine and oxygen only because of their higher electronegativity and small atomic size.
- In low oxidation states, fluorine is unstable. However, Cu(II) halides are known including CuF_2 except iodide.
$$2Cu^{2+} + 4I^- \longrightarrow Cu_2I_2 + I_2$$
Most of the Cu(I) compounds are unstable in aqueous solution and undergo disproportionation.
$$2Cu^+(aq) \longrightarrow Cu^{2+} + Cu(s)$$

7. Magnetic Properties

- The magnetic properties of a *d*-block elements are due to spin of unpaired electrons.
- The magnetic moment is determined by the number of unpaired electrons (n) which is given by
  $\mu = \sqrt{n(n+2)}$ B.M (Bohr Magneton).
- Paramagnetism is due to the presence of unpaired electrons, each such electron having a magnetic moment associated with its spin angular momentum. If all electrons are paired, substance will be diamagnetic and magnetic moment will be zero.
- Substances which are attracted very strongly by applied field and their domains are aligned in one direction are known as ferromagnetic, e.g. Fe, Co, Ni.

8. Formation of Coloured Ions

- The *d*-orbitals are non-degenerated in presence of ligands. When an electron from a lower energy *d*-orbital is excited to a higher energy *d*-orbital, the energy of required wavelength is absorbed and rest light is transmitted out. Therefore, the colour observed corresponds to the complementary colour of the light absorbed.
- In V_2O_5, V is in +5 oxidation state. Inspite of d^0, it is coloured compound. This colour arises due to defects in the crystal lattice.
- Some transition elements are coloured due to charge transfer spectra.

9. Formation of Complex Compounds

Transition metals have small size and high nuclear charge which facilitate the acceptance of lone pair of electrons from ligands. They have vacant *d*-orbitals of appropriate energy in order to accommodate the lone pair of electrons and thus form large number of complex compounds.

10. Catalytic Properties

- Transition metals have the ability to adopt multiple oxidation states and to form complexes, therefore used as a catalyst.
- Transition metals also provide larger surface area for the reactant to be adsorbed.

11. Formation of Interstitial Compounds

Small atoms of non-metals (H, C, N) fit into the voids of crystalline solid of transition metals and form interstitial compounds.

The physical and chemical characteristics of these compounds are as follows:

- They have high melting points, higher than those of pure metals.
- They are very hard, some borides approach diamond in hardness.
- They retain metallic conductivity.
- They are chemically inert.

12. Alloy Formation

Alloy is the homogeneous solid solution of two or more metals. Transition metals have approximate same size, therefore in molten form they can fit to each other's crystalline structure and form alloy. e.g. brass (copper-zinc) and bronze (copper-tin) etc.

Some Important Compounds of Transition Elements

Some important compounds of transition metals are potassium dichromate and potassium permanganate which are discussed below :

1. Potassium Dichromate ($K_2Cr_2O_7$)

Ore : Ferrochrome or chromite ($FeO \cdot Cr_2O_3$) or ($FeCr_2O_4$)

Preparation

$$4FeO \cdot Cr_2O_3 + 8Na_2CO_3 + 7O_2 \longrightarrow \underset{\text{Yellow}}{8Na_2CrO_4} + 2Fe_2O_3 + 8CO_2$$

$$\underset{\text{Yellow}}{2Na_2CrO_4} + 2H^+ \longrightarrow \underset{\text{Orange}}{Na_2Cr_2O_7} + 2Na^+ + H_2O$$

$$Na_2Cr_2O_7 + 2KCl \longrightarrow K_2Cr_2O_7 \downarrow + 2NaCl$$

Sodium dichromate is more soluble than potassium dichromate. Chromates and dichromates are interconvertible in aqueous solution depending upon pH of the solutions.

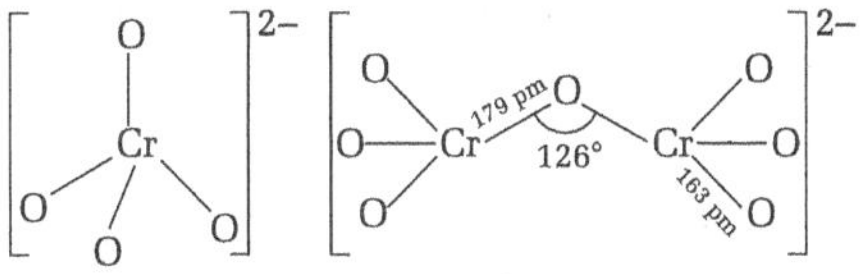

Chromate ion Dichromate ion

Properties

Sodium and potassium dichromates are strong oxidising agents, thus, acidified $K_2Cr_2O_7$ will oxidise iodides to iodine, sulphides to sulphur, tin (II) to tin (IV) and iron (II) salts to iron (III).

$$Cr_2O_7^{2-} + 14H^+ + 6e^- \longrightarrow 2Cr^{3+} + 7H_2O$$

$$Cr_2O_7^{2-} + 3H_2S + 8H^+ \longrightarrow 2Cr^{3+} + 3S + 7H_2O$$

$$Cr_2O_7^{2-} + 14H^+ + 3Sn^{2+} \longrightarrow 3Sn^{4+} + 2Cr^{3+} + 7H_2O$$

$$Cr_2O_7^{2-} + 7H^+ + 6I \longrightarrow 2Cr^{3+} + 3I_2 + 7H_2O$$

$$Cr_2O_7^{2-} + 6Fe^{2+} \longrightarrow 2Cr^{3+} + 6Fe^{3+} + 7H_2O$$

Uses of $K_2Cr_2O_7$

(i) $K_2Cr_2O_7$ is used as oxidising agent in volumetric analysis.
(ii) It is used in mordant dyes, leather industry, photography (for hardening of film).
(iii) It is used in chromyl chloride test.
(iv) It is used in cleaning glassware.
(v) It is used in the preparation of chrome alum $K_2SO_4 \cdot Cr_2(SO_4)_3 \cdot 24H_2O$.

2. Potassium Permanganate ($KMnO_4$)

Ore : Pyrolusite (MnO_2)

Preparation

$$2MnO_2 + 4KOH + O_2 \longrightarrow \underset{\text{Green}}{2K_2MnO_4} + 2H_2O$$

$$3MnO_4^{2-} + 4H^+ \longrightarrow 2MnO_4^- + MnO_2 + 2H_2O$$

Commercial Preparation

$$MnO_2 \xrightarrow[\text{oxidised with air or } KNO_3]{\text{Fused with KOH,}} \underset{\text{Manganate ion (green)}}{MnO_4^{2-}}$$

$$MnO_4^{2-} \xrightarrow[\text{Alkaline medium}]{\text{Electrolytic oxidation}} \underset{\text{Permanganate ion (Purple)}}{MnO_4^-}$$

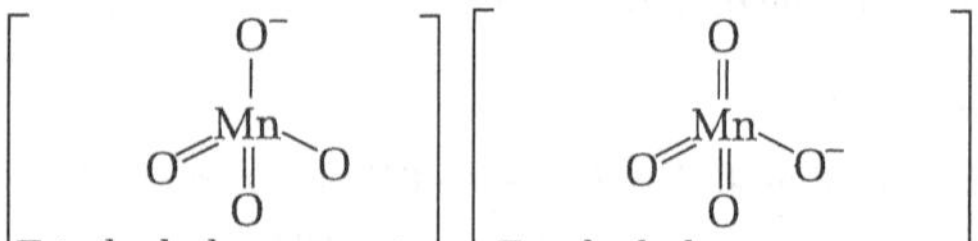

Tetrahedral manganate ion (green) Tetrahedral permanganate ion (purple)

Properties

It acts as strong oxidising agent. The actual oxidising action of $KMnO_4$ depends upon its use in neutral, alkaline or acidic medium.

(i) In presence of dilute H_2SO_4, $KMnO_4$ is reduced to manganous salt (Mn^{2+})

$$MnO_4^- + 8H^+ + 5e^- \longrightarrow Mn^{2+} + 4H_2O$$

Acidified $KMnO_4$ solution oxidises oxalates to CO_2, iron (II) to iron (III), nitrites to nitrates and iodides to iodine. The half-reactions of reductants are

$$5C_2O_4^{2-} \longrightarrow 10CO_2 + 10e^-$$

$$5Fe^{2+} \longrightarrow 5Fe^{3+} + 5e^-$$

$$5NO_2^- + 5H_2O \longrightarrow 5NO_3^- + 10H^+ + 10e^-$$

$$10I^- \longrightarrow 5I_2 + 10e^-$$

To acidify $KMnO_4$, only H_2SO_4 is used and not HCl or HNO_3 because HCl reacts with $KMnO_4$ and produce Cl_2, while HNO_3, itself acts as oxidising agent.

(ii) In alkaline medium, $KMnO_4$ is reduced to MnO_2.

$$MnO_4^- + 3e^- + 2H_2O \longrightarrow MnO_2 + 4OH^-$$

Alkaline or neutral $KMnO_4$ solution oxidises I^- to IO_3^-, $S_2O_3^{2-}$ to SO_4^{2-}, Mn^{2+} to MnO_2 etc.

Uses of $KMnO_4$

(i) In laboratory preparation of Cl_2.

(ii) As an oxidising agent, disinfectant.

(iii) In making Baeyer's reagent.

Inner Transition Elements (*f*-Block Elements)

The *f*-block elements also known as inner transition elements consist of the two series, lanthanoids and actinoids. Lanthanoids are known as rare earth metals (Ce to Lu) and actinoids are known as radioactive elements (Th to Lr).

1. Lanthanoids

- Fourteen elements of first row (4*f*-series) from Ce to Lu which follow La in the periodic table are called lanthanoids.
- General configuration of lanthanoids are $[Xe]\,4f^{1-14}, 5d^{0-1}, 6s^2$.

  Only Ce, Gd, Lu have electron in 5*d*-orbitals as well

  $$Ce = [Xe]\,4f^1\,5d^1 6s^2$$

  $$Gd = [Xe]\,4f^7\,5d^1 6s^2$$

  $$Lu = [Xe]\,4f^{14}\,5d^1 6s^2$$

Properties of Lanthanoids

Some of the properties of lanthanoids are as follows :

(i) Atomic and ionic size of lanthanoids decreases from left to right, due to increase in nuclear charge. This is known as **lanthanoid contraction.**

The consequences of lanthanoid contraction is as follows :

(a) The basic character of hydroxides decreases from $La(OH)_3$ to $Lu(OH)_3$.

(b) The normal increase in size from Sc ⟶ Y ⟶ La disappear after the lanthanides and pair of elements such as Zr and Hf, Nb and Ya, Mo and W have nearly same size and properties.

(ii) All the lanthanoids are silvery white soft metals and tarnish rapidly in air.

(iii) Many trivalent lanthanoid ions are coloured both in the solid state and in aqueous solutions. Neither La^{3+} nor Lu^{3+} ion shows any colour. Lanthanoid ions with *xf*-electrons have similar colour to those with $(14 - x)$ *f*-electrons.

(iv) The lanthanoid ions other than the f^0 type (La^{3+} and Ce^{4+}) and the f^{14} type (Yb^{2+} and Lu^{3+}) are all paramagnetic. Neodymium shows maximum paramagnetism.

(v) **Oxidation States** The most stable oxidation states of lanthanoids is + 3 but some of them also show +2 and +4 oxidation states. Ce^{4+} (in some elements) is favoured by its noble gas configuration, but it is a strong oxidant reverting to the common +3 state.

- The E°_{value} for Ce^{4+} / Ce^{3+} is +1.74 V, the reaction rate is very slow and hence, Ce(IV) is a good analytical reagent.

- Pr, Nd, Tb and Dy also exhibit +4 state but only in oxides. Eu^{2+} is formed by losing the two *s*-electrons and its f^7 configuration accounts for the formation of this ion.
- However, Eu^{2+} is a strong reducing agent changing to the common +3 state. Similarly, Yb^{2+} which has f^{14} configuration is a reductant, Tb^{4+} has half-filled *f*-orbitals and is an oxidant.

(vi) **Reactivity** Lanthanoid are very reactive metals like alkaline earth metals. The generalised reactions of lanthanoids are summarised below:

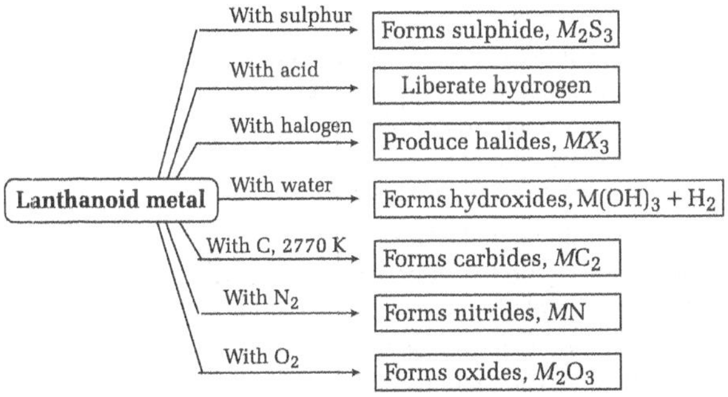

NOTE Misch metals, contain lanthanoids about 90–95%; Ce = 40.5%, lanthanum and neodymium = 44%, iron = 4.5%, calcium, carbon and silicon about 10.5%, are used in cigarette and gas lighters, toys, tank and tracer bullets. It is a pyrophoric alloy, i.e. takes up fire easily.

2. Actinoids

- Fourteen elements of 5*f*-series from Th to Lr which follows actinium in the 7th period of periodic table is known as **actinoids**.
- **General configuration of actinoids** are [Rn] $5f^{1-14}, 6d^{0-2}, 7s^2$.

Properties of Actinoids

(i) Actinoids exhibit a range of **oxidation states** due to comparable energies of 5*f*, 6*d* and 7*s*-orbitals. The general oxidation state of actinoids is +3.

(ii) All the actinoids are strong reducing agents and very reactive.

(iii) Actinoids also react with oxygen, halogen, hydrogen and sulphur, etc., like lanthanoids.

(iv) Actinoids are radioactive in nature and therefore, it is difficult to study their chemical nature.

(v) The atomic and ionic radii of actinoids decreases regularly across the series due to poor shielding effect of 5*f*-electrons. This is known as **actinoid contraction.**

(vi) The allocation of electrons to the 5*f* or 6*d*-orbitals in the actinoid atoms is more uncertain than the allocation to the 4*f* or 5*d*-orbitals in the lanthanoid atoms, because the external shielding of the 5*f*-electrons is less than that of the 4*f*-electrons.

NOTE
- The magnetic properties of *f*-block elements are due to both the orbital motion as well as spin magnetic moment for *d*-block elements.
- Magnetic moment for *f*-block elements, $\mu = \sqrt{4s(s+1) + l(l+1)}$ where, *s* is the sum of spin quantum numbers and *l* is the angular momentum quantum numbers.
- Thorium is used in the atomic reactors and in the treatment of cancer, uranium is used as nuclear fuel.

DAY PRACTICE SESSION 1

GRADED QUESTIONS EXERCISE

1 Electronic configuration of a transition element *X* in +3 oxidation state is [Ar]$3d^5$. What is its atomic number?

(a) 25 (b) 26 (c) 27 (d) 24

2 Metallic radii of some transition elements are given below. Which of these elements will have highest density?

Element	Fe	Co	Ni	Cu
Metallic radii/pm	126	125	125	128

(a) Fe (b) Ni (b) Co (d) Cu

3 The atomic numbers of V, Cr, Mn and Fe respectively are 23, 24, 25 and 26. Which one of these may be expected to have the highest second ionisation enthalpy?

(a) Cr (b) Mn (c) Fe (d) V

4 Highest oxidation state of manganese in fluoride is +4 (MnF_4) but highest oxidation state in oxides is +7(Mn_2O_7) because

(a) fluorine is more electronegative than oxygen

(b) fluorine does not possess *d*-orbitals

(c) fluorine stabilises lower oxidation state

(d) in covalent compounds, fluorine can form single bond only while oxygen forms double bond

5 The element with which of the following outer electron configuration may exhibit the largest number of oxidation states in its compounds. → JEE Main (Online) 2013

(a) $3d^5 4s^2$

(b) $3d^8 4s^2$

(c) $3d^7 4s^2$

(d) $3d^6 4s^2$

6 Iron exhibits + 2 and+3 oxidation states. Which of the following statements about iron is incorrect? → AIEEE 2012

(a) Ferrous oxide is more basic in nature than the ferric oxide
(b) Ferrous compounds are relatively more ionic than the corresponding ferric compounds
(c) Ferrous compounds are less volatile than the corresponding ferric compounds
(d) Ferrous compounds are more easily hydrolysed than the corresponding ferric compounds

7 In context with the transition elements, which of the following statements is incorrect? → JEE Main 2009

(a) In addition to the normal oxidation state, the zero oxidation state is also shown by these elements in complexes
(b) In the highest oxidation states, the transition metal shows basic character and form cationic complexes
(c) In the highest oxidation states of the first five transition elements (Sc to Mn), all the 4*s* and 3*d*-electrons are used for bonding
(d) Once the d^5 configuration is exceeded, the tendency to involve all the 3*d*-electrons in bonding decreases

8 Four successive members of the first row transition elements are listed below with atomic numbers. Which one of them is expected to have the highest $E^\circ_{M^{3+}/M^{2+}}$ value? → JEE Main (Online) 2013

(a) Mn ($Z = 25$) (b) Fe ($Z = 26$)
(c) Co ($Z = 27$) (d) Cr ($Z = 24$)

9 The correct order of $E^\circ_{M^{2+}/M}$ values with negative sign for the four successive elements Cr, Mn, Fe and Co is → AIEEE 2010

(a) Mn > Cr > Fe > Co (b) Cr > Fe > Mn > Co
(c) Fe > Mn > Cr > Co (d) Cr > Mn > Fe> Co

10 Atomic number of an element is 26. The element shows

(a) ferromagnetism (b) diamagnetism
(c) paramagnetism (d) None of these

11 The magnetic moment of a transition metal of 3*d*-series is 6.92 BM. Its electronic configuration would be

(a) $3d^4\,4s^2$ (b) $3d^5\,4s^1$ (c) $3d^8\,4s^1$ (d) $3d^5\,4s^0$

12 The magnetic moment is associated with its spin angular momentum and orbital angular momentum. Spin only magnetic moment value of Cr^{3+} ion is

(a) 2.87 BM (b) 3.87 BM
(c) 3.47 BM (d) 3.57 BM

13 The pair having the same magnetic moment is [at. no. Cr = 24, Mn = 25, Fe = 26 and Co = 27] → JEE Main 2016

(a) $[Cr(H_2O)_6]^{2+}$ and $[Fe(H_2O)_6]^{2+}$
(b) $[Mn(H_2O)_6]^{2+}$ and $[Cr(H_2O)_6]^{2+}$
(c) $[CoCl_4]^{2-}$ and $[Fe(H_2O)_6]^{2+}$
(d) $[Cr(H_2O)_6]^{2+}$ and $[CoCl_4]^{2-}$

14 Which of the following compounds is not coloured?

(a) Na_2CuCl_4 (b) Na_2CdCl_4
(c) $FeSO_4$ (d) VI_3

15 Among the following, the coloured compound is

(a) CuCl (b) $K_3[Cu(CN)_4]$
(c) CuF_2 (d) $[Cu(CH_3CN)_4]BF_4$

16 Which pair of compounds is expected to show similar colour in aqueous medium?

(a) $FeCl_3$ and $CuCl_2$ (b) $VOCl_2$ and $CuCl_2$
(c) $VOCl_2$ and $FeCl_2$ (d) $FeCl_2$ and $MnCl_2$

17 Identify the incorrect statement. → JEE Main (Online) 2013

(a) Cu_2O is colourless
(b) Copper (I) compounds are colourless except where colour results from charge transfer.
(c) Copper (I) compounds are diamagnetic
(d) Cu_2S is black

18 The ability of d-block elements to form complexes is due to

(a) small and highly charged ions
(b) vacant low energy orbitals to accept lone pair of electrons from ligands
(c) Both (a) and (b) are correct
(d) None of the above is correct

19 Which one of the following characteristics of the transition metals is associated with their catalytic activity?

(a) Colour of hydrated ions
(b) Variable oxidation states
(c) High enthalpy of atomisation
(d) Paramagnetic behaviour

20 Which of the following arrangements does not represent the correct order of the property stated against it? → JEE Main (Online) 2013

(a) $V^{2+} < Cr^{2+} < Mn^{2+} < Fe^{2+}$: paramagnetic behaviour
(b) $Ni^{2+} < Co^{2+} < Fe^{2+} < Mn^{2+}$: ionic size
(c) $Co^{3+} < Fe^{3+} < Cr^{3+} < Sc^{3+}$: stability in aqueous solution
(d) Sc < Ti < Cr < Mn : number of oxidation states

21 Which of the following ions will finally give a black precipitate with Ag^+ ion?

(a) SO_3^{2-} (b) Br^- (c) CrO_4^{2-} (d) $S_2O_3^{2-}$

22 Compound *A* on strong heating gives two oxides of sulphur. If aqueous NaOH solution is added to the aqueous solution of *A*, a dirty green precipitate is formed which starts turning brown on exposure to air. Compound *A* is

(a) ferrous hydroxide (b) ferric sulphate
(c) ferrous sulphate (d) ferric oxide

23 For the given aqueous reaction which of the statement is not true?

$$KI + K_3[Fe(CN)_6] \xrightarrow{\text{Dilute } H_2SO_4} \text{Brownish yellow solution}$$
$$\downarrow ZnSO_4$$
$$\text{White precipitate + Brownish yellow filtrate}$$
$$\downarrow Na_2S_2O_3$$
$$\text{Colourless solution.}$$

→ AIEEE 2012

(a) The first reaction is a redox reaction
(b) White ppt. is $Zn_3[Fe(CN)_6]_2$
(c) Addition of filtrate to starch solution gives blue colour
(d) White ppt. is soluble in NaOH solution

24 $(NH_4)_2Cr_2O_7$ on heating gives a gas which is also given by
(a) heating NH_4NO_2 (b) heating NH_4NO_3
(c) $Mg_3N_2 + H_2O$ (d) $Na + H_2O_2$

25 In alkaline H_2O_2, $Cr_2O_7^{2-}$ changes to tetraperoxo species ... having oxidation number of Cr as ...
(a) CrO_4^{2-}, 6 (b) CrO_5, 6
(c) CrO_8^{3-}, 5 (d) CrO_8^{3-}, 11

26 Potassium dichromate when heated with concentrated sulphuric acid and a soluble chloride, gives brown-red vapours of → JEE Main (Online) 2013
(a) CrO_3 (b) $CrCl_3$
(c) CrO_2Cl_2 (d) Cr_2O_3

27 An explosion takes place when conc. H_2SO_4 is added to $KMnO_4$. Which of the following is formed?
(a) Mn_2O_7 (b) MnO_2
(c) $MnSO_4$ (d) Mn_2O_3

28 One of the products formed due to the reaction between $KMnO_4$ and HCl is
(a) red liquid (b) MnO_2
(c) greenish yellow gas (d) $HClO_4$

29 Which one of the following compounds does not decolourise an acidified aqueous solution of $KMnO_4$?
(a) Sulphur dioxide (b) Ferric chloride
(c) Hydrogen peroxide (d) Ferrous sulphate

30 MnO_4^- on reduction in acidic medium forms
(a) MnO_2 (b) Mn^{2+} (c) MnO_4^{2-} (d) Mn

31 When MnO_2 is fused with KOH, a coloured compound is formed, the product and its colour is
(a) K_2MnO_4, purple colour (b) $KMnO_4$, purple
(c) Mn_2O_3, brown (d) Mn_3O_4, black

32 The colour of $KMnO_4$ is due to → JEE Main 2015
(a) $M \rightarrow L$ charge transfer transition
(b) $d - d$ transition
(c) $L \rightarrow M$ charge transfer transition
(d) $\sigma - \sigma$ transition

33 When a small amount of $KMnO_4$ is added to concentrated H_2SO_4, a green oily compound is obtained which is highly explosive in nature. Compound may be → JEE Main (Online) 2013
(a) $MnSO_4$ (b) Mn_2O_7
(c) MnO_2 (d) Mn_2O_3

34 The electronic configuration of Eu (at. no. 63) and Tb (at. no. 65) are
(a) $[Xe]4f^6 5d^1 6s^2$ and $[Xe]4f^9 6s^2$
(b) $[Xe]4f^6 5d^1 6s^2$ and $[Xe]4f^8 5d^1 6s^2$
(c) $[Xe]4f^7 6s^2$ and $[Xe]4f^9 6s^2$
(d) $[Xe]4f^7 5s^2$ and $[Xe]4f^8 5d^1 6s^2$

35 The outer electron configuration of Gd (atomic number 64) is → AIEEE 2011
(a) $4f^3 5d^5 6s^2$ (b) $4f^8 5d^0 6s^2$
(c) $4f^4 5d^4 6s^2$ (d) $4f^7 5d^1 6s^2$

36 Arrange Ce^{3+}, La^{3+}, Pm^{3+} and Yb^{3+} in the increasing order of their ionic radii.
(a) $Yb^{3+} < Pm^{3+} < Ce^{3+} < La^{3+}$
(b) $Ce^{3+} < Yb^{3+} < Pm^{3+} < La^{3+}$
(c) $Yb^{3+} < Pm^{3+} < La^{3+} < Ce^{3+}$
(d) $Pm^{3+} < La^{3+} < Ce^{3+} < Yb^{3+}$

37 Cerium ($Z = 58$) is an important member of the lanthanides. Which of the following statements about cerium is incorrect?
(a) The common oxidation states of cerium are + 3 and +4
(b) The +3 oxidation state of cerium is more stable than + 4 oxidation state
(c) The +4 oxidation state of cerium is not known in solutions
(d) Cerium (IV) acts as an oxidising agent

38 In context of the lanthanoids, which of the following statements is not correct? → AIEEE 2011
(a) There is a gradual decrease in the radii of the members with increasing atomic number in the series
(b) All the members exhibit + 3 oxidation state
(c) Because of similar properties the separation of lanthanoids is not easy
(d) Availability of 4*f*-electrons results in the formation of compounds in + 4 state for all the members of the series

39 Knowing that the chemistry of lanthanoids (Ln) is dominated by its +3 oxidation state, which of the following statements is incorrect? → JEE Main 2009
(a) Because of the large size of the Ln (III) ions the bonding in its compounds is predominantly ionic in character
(b) The ionic sizes of Ln (III) decrease in general with increasing atomic number
(c) Ln (III) compounds are generally colourless
(d) Ln (III) hydroxide are mainly basic in character

40 Identify the incorrect statment among the following.
(a) *d*-block elements show irregular and erratic chemical properties among themselves
(b) La and Lu have partially filled *d*-orbitals and no other partially filled orbitals
(c) The chemistry of various lanthanoids is similar
(d) 4*f* and 5*f*-orbitals are equally shielded

41 Match the properties in Column I with the metals given in Column II.

	Column I		Column II
A.	A element which can show +8 oxidation state	1.	Mn
B.	3*d*-block element that can show upto +7 oxidation state	2.	Cr
C.	3*d*-block element with highest melting point	3.	Os
D.	Radioactive lanthanoid	4.	Pm

Codes

	A	B	C	D		A	B	C	D
(a)	3	1	2	4	(b)	1	2	4	3
(c)	2	4	3	1	(d)	2	3	4	1

Direction (Q. No 42-45) *In the following question, a statement of Assertion (A) followed by a statement of Reason (R) is given. Choose the correct answer out of the following choices.*

(a) Both A and R are true and R is correct explanation of A
(b) Both A and R are true but R is not correct explanation of A
(c) A is true but R is false
(d) Both A and R are false

42 **Assertion** (A) Separation of Zr and Hf is difficult.
Reason (R) Because Zr and Hf lie in the same group of the periodic table.

43 **Assertion** (A) Actinoids form relatively more stable complexes as compared to lanthanoids.
Reason (R) Actinoids can utilise their 5*f*-orbitals along with 6*d*-orbitals in bonding but lanthanoids do not use their 4*f*-orbital for bonding.

44 **Assertion** (A) To a solution of potassium chromate, if a strong acid is added, it changes its colour from yellow to orange.
Reason (R) The colour change is due to the change in oxidation state of potassium chromate.

45. **Assertion** (A) Out of Mn^{2+}, Fe^{2+} and Cr^{2+}, Mn^{2+} is readily oxidised.
Reason (R) Mn^{2+} is more stable than Fe^{2+} and Cr^{2+} due to completely filled *d*-orbitals.

DAY PRACTICE SESSION 2

PROGRESSIVE QUESTIONS EXERCISE

1 Among $FeSO_4 \cdot 7H_2O$ (*A*), $CuSO_4 \cdot 5H_2O$ (*B*), $ZnSO_4 \cdot 7H_2O$ (*C*), $MnSO_4 \cdot 4H_2O$ (*D*) isomorphous salts are
(a) *A* and *C* (b) *A* and *D*
(c) *C* and *B* (d) *A* and *B*

2 Choose the correctly paired gaseous cation and its magnetic (spin only) moment in (B.M).
(a) Ti^{2+}, 3.87 BM (b) Cr^{2+}, 4.90 BM
(c) Co^{3+}, 3.87 BM (d) Mn^{2+}, 4.90 BM

3 Among the following compounds, the one which is both paramagnetic and coloured is
(a) $K_2Cr_2O_7$
(b) $(NH_4)_2[TiCl_2]$
(c) $VOSO_4$
(d) $K_3[Cu(CN)_4]$

4 MnO_4^- is of intense pink colour, though Mn is in (+7) oxidation state. It is due to
(a) oxygen gives colour to it
(b) charge transfer where oxygen gives its electron to Mn making it Mn (VI) hence, coloured
(c) charge transfer when Mn gives its electron to oxygen
(d) None of the above

5 Which one of the following does not correctly represent the correct order of the property indicated against it?
(a) $Ti < V < Cr < Mn$ increasing number of oxidation states
(b) $Ti^{3+} < V^{3+} < Cr^{3+} < Mn^{3+}$ increasing magnetic moment
(c) $Ti < V < Cr < Mn$ increasing melting point
(d) $Ti < V < Mn < Cr$ increasing 2nd ionisation enthalpy

6 When a certain compound *X* (used in laboratory for analysis) is added to copper sulphate solution a yellow-brown precipitate is obtained which turns white on addition of excess of $Na_2S_2O_3$ solution. On addition to Ag^+ ion solution, a yellow curdy precipitate is obtained which is insoluble in NH_4OH. The *X* is
(a) H_2S (b) HCl (c) KI (d) NaCN

7 Amongst the following the lowest degree of paramagnetism per mole of the compound at 298 K will be shown by
(a) $MnSO_4 \cdot 4H_2O$ (b) $CuSO_4 \cdot 5H_2O$
(c) $FeSO_4 \cdot 6H_2O$ (d) $NiSO_4 \cdot 6H_2O$

8 Lanthanoid contraction is caused due to
(a) the appreciable shielding on outer electrons by 4*f*-electrons from the nucleus
(b) the appreciable shielding on outer electrons by 5*d*-electrons from the nucleus
(c) the same effective nuclear charge from Ce to Lu
(d) the imperfect shielding on outer electrons by 4*f*-electrons from the nucleus

9 Larger number of oxidation states are exhibited by the actinoids than those by the lanthanoids, the main reason being
(a) 4*f*-orbitals are more diffused than the 5*f*-orbitals
(b) lesser energy difference between 5*f* and 6*d* than between 4*f* and 5*d*-orbitals
(c) more energy difference between 5*f* and 6*d* than between 4*f* and 5*d*-orbitals
(d) more reactive nature of the actinoids than the lanthanoids

10 A red solid is insoluble in water however, it becomes soluble, if some KI is added to water. Heating the red solid in a test tube results in liberation of some violet coloured fumes and droplets of a metal appear on the cooler parts of the test tube. The red solid is
(a) $(NH_4)_2Cr_2O_7$ (b) HgI_2
(c) Hg (d) Pb_3O_4

11 Copper exhibits only +2 oxidation state in its stable compounds. Why?
(a) Copper is transition metal in +2 state
(b) +2 state compounds of copper are formed by exothermic reactions
(c) Electron configuration of copper in +2 state is $[Ar]\,3d^9, 4s^0$
(d) Copper gives coloured compounds in +2 state

12 $KMnO_4$ can be prepared from K_2MnO_4 as per the reaction, $3MnO_4^{2-} + 2H_2O \longrightarrow 2MnO_4^- + MnO_2 + 4OH^-$
The reaction can go completion by removing OH^- ions by adding
(a) HCl (b) KOH (c) CO_2 (d) SO_2

13 A 0.081 g sample of pyrolusite ore (impure MnO_2) is treated with 1.651 g of oxalic acid ($H_2C_2O_4 \cdot H_2O$) in an acidic medium. Following this reaction the excess oxalic acid is titrated with 30.6 mL of 0.1 M $KMnO_4$, the percentage of MnO_2 in the ore will be (Mn = 55)

$$H_2C_2O_4 + MnO_2 + 2H^+ \longrightarrow Mn^{2+} + 2H_2O + 2CO_2$$

$$5H_2C_2O_4 + 2MnO_4^- + 6H^+ \longrightarrow 2Mn^{2+} + 8H_2O + 10CO_2$$

(a) 25% (b) 50%
(c) 70% (d) 82%

14 Which of the following statements is not true?
(a) On passing H_2S through acidified $K_2Cr_2O_7$ solution, a milky colour is observed
(b) $Na_2Cr_2O_7$ is preferred over $K_2Cr_2O_7$ in volumetric analysis
(c) $K_2Cr_2O_7$ solution in acidic medium is orange
(d) $K_2Cr_2O_7$ solution becomes yellow on increasing the pH beyond 7

15 Cuprous ion is colourless while cupric ion is coloured because
(a) both have half-filled *p* and *d*-orbitals
(b) cuprous ion has incomplete *d*-orbital and cupric ion has a complete *d*-orbital
(c) both have unpaired electrons in the *d*-orbitals
(d) cuprous ion has a completely filled *d*-orbital and cupric ion has an incompletely filled *d*-orbital

ANSWERS

SESSION 1	1 (b)	2 (d)	3 (a)	4 (d)	5 (a)	6 (d)	7 (b)	8 (c)	9 (a)	10 (a)
	11 (b)	12 (b)	13 (a)	14 (b)	15 (c)	16 (b)	17 (a)	18 (c)	19 (b)	20 (a)
	21 (d)	22 (c)	23 (b)	24 (a)	25 (c)	26 (c)	27 (a)	28 (c)	29 (b)	30 (b)
	31 (a)	32 (c)	33 (b)	34 (c)	35 (d)	36 (a)	37 (a)	38 (d)	39 (c)	40 (d)
	41 (a)	42 (b)	43 (a)	44 (c)	45 (d)					
SESSION 2	1 (a)	2 (b)	3 (c)	4 (b)	5 (c)	6 (c)	7 (b)	8 (d)	9 (b)	10 (b)
	11 (b)	12 (c)	13 (d)	14 (b)	15 (d)					

Hints and Explanations

SESSION 1

1 Electronic configuration of transition element *X* in + 3 oxidation state $= [Ar]3d^5$
$\Rightarrow$ It means, it looses $3e^-$ to attain + 3 oxidation state.
$\therefore$ Electronic configuration of element *X* in ground state,

$$= [Ar]3d^6\,4s^2$$

This configuration is of iron which has atomic number = 26

2 Density = mass/volume. As we move from Fe to Cu, mass increases and volume decreases.Hence, density increases [Fe(7.87), Co (8.90), Ni(8.91), Cu (8.95)].Increase in mass in case of Cu dominates over small increases in volume.

3 The electronic configuration of Cr = $3d^5\,4s^1$; After removing $4s^1$-electron, the next electron is to be removed from relatively more stable half-filled $3d^5$-subshell, which will require more energy for its ionisation.

4 Highest oxidation state of manganese in fluoride is +4 (MnF_4) but highest oxidation state in oxides in + 7 (Mn_2O_7). The reason is that in covalent compounds fluorine can form single bond while oxygen forms double bond.

5 Due to the presence of maximum number of unpaired electrons, element having $3d^5$, $4s^2$ (Mn), configuration may exhibit the largest number of oxidation states in its compounds.

6 (a) $FeO > Fe_2O_3$ (basic character). Thus, the statement is correct.
(b) $FeCl_2 > FeCl_3$ (ionic nature), larger the charge greater the polarising power thus, greater the covalent nature. Thus, the statement is correct.
(c) Fe^{2+} salts are more ionic thus, less volatile than Fe^{3+} salts. Thus, the statement is correct.
(d) Greater the covalent nature, more easily they are hydrolysed. Thus, $FeCl_3$ is more hydrolysed than $FeCl_2$. Thus, the statement is incorrect.

7 As oxidation state increases, electronegativity increases thus, acidic character increases and not the basic character.

8 $E^{\circ}_{M^{3+}/M^{2+}}$ value is highest for Co^{3+}/Co^{2+} ion, because of the extra stability of $Co^{2+}(d^6)$ ion than $Co^{3+}(d^7)$.

9 Usually across the first transition series, the negative values for $E^{\circ}_{M^{2+}/M}$ decrease except for Mn due to stable d^5 configuration. So, the correct order is Mn > Cr > Fe > Co.

10 Ferromagnetic substances are those substances which are strongly attracted towards magnetic field and their domain aligned in the same direction. Iron with atomic number = 26 shows ferromagnetism.

11 Magnetic moment $= \sqrt{n(n+2)}$

or $6.92 = \sqrt{n(n+2)}$

[Here, n = number of unpaired electrons]

$$n = 6$$

Hence, metal = Cr = $3d^5 4s^1$

12 The electronic configuration of Cr^{3+} ion is $3d^3$. Hence, spin only magnetic moment $(\mu) = \sqrt{n(n+2)}$ BM where, n = number of unpaired electrons.

$$= \sqrt{3(3+2)} = \sqrt{15} = 3.87 \text{ BM}$$

13 The pair that have same number of unpaired electron will have same magnetic moment.

Complex ion	Electronic configuration of metal ion	Number of unpaired electrons (n)
$[Cr(H_2O)_6]^{2+}$	Cr^{2+}; [Ar] $3d^4$	↑ ↑ ↑ ↑ □ ; 4
$[Fe(H_2O)_6]^{2+}$	Fe^{2+}; [Ar] $3d^6$	↑↓ ↑ ↑ ↑ ↑ ; 4
$[Mn(H_2O)_6]^{2+}$	Mn^{2+}; [Ar] $3d^5$	↑ ↑ ↑ ↑ ↑ ; 5
$[CoCl_4]^{2-}$	Co^{2+}; [Ar] $3d^7$	↑↓ ↑↓ ↑ ↑ ↑ ; 3

14 Those compounds which have unpaired electrons will be coloured. Na_2CdCl_4 does not have unpaired electrons. Hence, it will be colourless.

15 In the crystalline form, CuF_2 is blue coloured as in CuF_2, Cu^{2+} ions exist, having d^9 configuration. The unpaired electron causes colour (d-d) transition.

16 In $VOCl_2$, V is in +4 state ($3d^1$) and in $CuCl_2$, Cu is in +2 state ($3d^9$). Both have one unpaired electron hence, show similar colour.

17 Cu_2O is coloured because of the presence of one unpaired electron, which makes d-d transition possible.

$Cu^{2+} = [Ar]3d^9$

↑↓ ↑↓ ↑↓ ↑↓ ↑

18 Transition metals have small size and high nuclear charge. They also have vacant d-orbitals of appropriate energy which facilitate the acceptance of lone pair of electron from ligands. Thus, they have ability to form complexes.

19 Transition metals have ability to adopt variable oxidation state and to form complexes, therefore they are used as a catalyst.

20 (a) $V^{2+} = 3d^3 4s^0$ –3 unpaired electrons

$Cr^{2+} = 3d^4\, 4s^0$ – 4 unpaired electrons

$Mn^{2+} = 3d^5\, 4s^0$ – 5 unpaired electrons

$Fe^{2+} = 3d^6 4s^0$ – 4 unpaired electrons

Hence, the order of paramagnetic behaviour should be

$$V^{2+} < Cr^{2+} = Fe^{2+} < Mn^{2+}$$

(b) Ionic size decreases from left to right in the same period.

(c) $Co^{3+}/Co^{2+} = 1.97$;

$Fe^{3+}/Fe^{2+} = 0.77$;

$Cr^{3+}/Cr^{2+} = -0.41$

Sc^{3+} is highly stable as it has $3d^0 4s^0$ electronic configuration (it does not show + 2).

(d) The oxidation states increases as we go Sc to Mn in $3d$ transition series.

21 $2Ag + S_2O_3^{2-} \longrightarrow Ag_2S_2O_3$,white ppt which readily changes to yellow, orange, brown and finally black due to formation of silver sulphide.

$$Ag_2S_2O_3 + H_2O \longrightarrow H_2SO_4 + \underset{\text{(Black)}}{Ag_2S}$$

22 $\underset{A}{2FeSO_4} \xrightarrow{\Delta} Fe_2O_3 + SO_2 + SO_3$

$\underset{A}{FeSO_4} + 2NaOH \longrightarrow \underset{\text{Green}}{Fe(OH)_2} + Na_2SO_4$

$Fe(OH)_2 \xrightarrow{[O]} \underset{\text{Brown}}{Fe(OH)_3}$

23 $K_3[Fe^{3+}(CN)_6] + KI(\text{excess}) \xrightarrow{\text{dil. } H_2SO_4} K_4[Fe^{2+}(CN)_6] + \underset{\text{Brownish yellow solution}}{KI_3(\text{redox})}$

$K_4[Fe(CN)_6] + ZnSO_4 \longrightarrow \underset{\text{White ppt.}}{K_2Zn_3[Fe(CN)_6]_2 \text{ or } K_2Zn[Fe(CN)_6]}$

$\underset{\text{Brownish yellow filtrate}}{I_3^-} + Na_2S_2O_3 \longrightarrow \underset{\text{Clear solution}}{Na_2S_4O_6} + 2NaI + \underset{\text{Turns starch solution blue}}{I_2}$

$K_2Zn[Fe(CN)_6]$ reacts with NaOH as follows

$K_2Zn[Fe(CN)_6] + NaOH \longrightarrow [Zn(OH)_4]^{2-} + [Fe(CN)_6]^{4-}$

24 $(NH_4)_2Cr_2O_7 \xrightarrow{\Delta} N_2 + Cr_2O_3 + 4H_2O$

$NH_4NO_2 \xrightarrow{\Delta} 2H_2O + N_2 \uparrow$

25 Tetra peroxo species $[Cr(O_2)_4]^{3-}$

Oxidation number of Cr $= x - 8 = -3$

$$x = +5$$

26 $K_2Cr_2O_7 + 4NaCl + 3H_2SO_4 \xrightarrow{\Delta} \underset{\substack{\text{Chromyl chloride} \\ \text{(Brown red vapours)}}}{2CrO_2Cl_2} + K_2SO_4 + 2Na_2SO_4 + 3H_2$

27 $2KMnO_4 + \underset{\text{Conc.}}{H_2SO_4} \longrightarrow Mn_2O_7 + K_2SO_4 + H_2O$

28 $2KMnO_4 + 3H_2SO_4 \longrightarrow K_2SO_4 + 2MnSO_4 + 3H_2O + 5O$

$2HCl + O \longrightarrow H_2O + \underset{\text{Greenish yellow gas}}{Cl_2 \uparrow}$

29 $FeCl_3$ is not a reducing agent, it will not react with oxidising agent $KMnO_4$ and hence, no colour is discharged.

30 MnO_4^- on reduction in acidic medium forms Mn^{2+} ion .

$MnO_4^- + 8H^+ + 5e^- \longrightarrow Mn^{2+} + 4H_2O$

31 $2MnO_2 + 4KOH + O_2 \longrightarrow 2K_2MnO_4 + 2H_2O$
(Purple colour under K_2MnO_4)

32 $KMnO_4 \longrightarrow K^+ + MnO_4^-$

$\therefore$ In MnO_4^- has +7 oxidation state ($3d^0 4s^0$) having no electron in *d*-orbitals. It is considered that higher the oxidation state of metal, greater is the tendency to occur $L \rightarrow M$ charge transfer becasue ligand is able to donate the electrons into the vacant *d*-orbital of metal. Since, charge transfer is Laporte as well as spin allowed therefore, it shows colour.

33 When a small amount of $KMnO_4$ is added to concentrated H_2SO_4, manganese heptoxide is obtained, which on warming decomposes with a mild explosion into MnO_2 and O_2.

$$2KMnO_4 + H_2SO_4 \longrightarrow \underset{\text{Green liquid}}{Mn_2O_7} + K_2SO_4 + H_2O$$

$$2Mn_2O_7 \longrightarrow 4MnO_2 + 3O_2$$

34 The electronic configuration of Eu (at. no. 63) = [Xe] $4f^7, 6s^2$.

and the electronic configuration of Tb (at. no 65) = [Xe] $4f^9, 6s^2$.

35 Gd (64)

$[Xe]_{54}$ $4f^7$: ↿ ↿ ↿ ↿ ↿ ↿ ↿ $5d^1$: ↿ □ □ □ □ $6s^2$: ⥮

Due to extra stability of half-filled 4*f*-subshell, Gd (64) has EC as $[Xe]_{54}\, 4f^7\, 5d^1\, 6s^2$ instead of $[Xe]_{54}\, 4f^8\, 6s^2$.

36 Ionic size decreases from La^{3+} to Lu^{3+} due to lanthanide contraction. Thus, the correct order of increasing ionic radii is

$$Yb^{3+} < Pm^{3+} < Ce^{3+} < La^{3+}.$$

37 +3 and +4 states are shown by Ce in aqueous solution.

Ce^{4+} is favoured by its noble gas configuration but it is a strong oxidant reverting to common + 3 state.

38 (a) There is a gradual decrease in the radii of the lanthanoids with increasing atomic number which is case of lanthanide contraction, thus statement is true.

(b) Ionisation potential for the formation of Lu^{3+} is comparatively low, hence, + 3 state is favourable, thus statement is true.

(c) Due to lanthanide contraction Zr and Hf; Nb and Ta; Mo and W have the same size and thus similar properties and hence separation is not easy, therefore statement is true.

(d) Formation of + 4 state requires very high energy, thus statement is incorrect.

39 The most common oxidation state of lanthanoid is +3. Lanthanoids in +3 oxidation state usually have unpaired electron in *f*-subshell and impart characteristic colour in solid as well as in solution state due to *f*-*f* transition. (Except La and Lu).

40 The external shielding of the 5*f*-orbital is less than of 4*f*-orbital.

41 A→3, B→1, C→2, D→4

42 **Correct explanation** Zr and Hf have nearly the same size because of lanthanoid contraction.

Hence, Zr and Hf cannot be separated easily.

43 Both Assertion and Reason are correct and Reason is the correct explanation of the Assertion. Actinides form relatively more stable complexes as compared to lanthanide because actinide can utilize their 5*f*-orbitals along with 6*d*-orbitals in bonding but lanthanoids is not use their 4*f* orbitals.

44 $2K_2CrO_4 + H_2SO_4 \longrightarrow K_2Cr_2O_7 + K_2SO_4 + H_2O$

$$\underset{\text{Yellow}}{2\overset{+6}{Cr}O_4^{2-}} + SO_4^{2-} + 2H^+ \longrightarrow \underset{\text{Orange}}{\overset{+6}{Cr_2}O_7^{2-}} + SO_4^{2-} + H_2O$$

No change in oxidation state

Thus, Reason (R) is incorrect.

45 Mn^{2+} cannot be oxidised easily as, it has stable half-filled $3d^5$ electronic configuration. Whereas, Cr^{2+} and Fe^{2+} can be readily oxidised.

SESSION 2

1 $FeSO_4 \cdot 7H_2O$ and $ZnSO_4 \cdot 7H_2O$ have similar crystalline form and chemical composition, i.e. these are isomorphous in nature.

2 Spin only magnetic moment, $\mu = \sqrt{n(n+2)}$ BM

where, n = number of unpaired electrons

Ion	Outer electronic configuration	μ
Ti^{2+}	$3d^2$	2.84
Cr^{2+}	$3d^4$	4.90
Co^{3+}	$3d^6$	4.90
Mn^{2+}	$3d^5$	5.92

3 In $K_2Cr_2O_7$, Cr is +6 oxidation state, the EC of Cr^{6+} is $3d^0$.

In $[(NH_4)_2TiCl_6]$, Ti is + 4 oxidation state. The EC of Ti^{4+} is $3d^0$. In $K_3[Cu(CN_4)_4]$, Cu is in + 1 oxidation state. The EC of Cu^{4+} is $3d^0$. In $VOSO_4$, V is + 4 and it is EC is $3d^1$ electron, due to presence of one unpaired, it is paramagnetic and coloured.

4 Pink colour of MnO_4^- is due to the ligand to metal charge transfer where oxygen gives its electron to Mn making it Mn (+VI).

5 The correct order of melting point is Ti < V < Cr > Mn. The melting point of transition elements first increases to a maximum extent and then fall as atomic number increases. Mn, however, has abnormally low melting point, due to stable configuration and weak intermetallic bonding.

6 *X* gives yellow ppt. insoluble in NH_4OH with Ag^+ solution, so contains I^- ions. It suggests *X* to be KI.

$$2CuSO_2 + \underset{X}{4KI} \longrightarrow 2CuI_2 + 2K_2SO_4$$

$$2CuI_2 \longrightarrow \underset{\text{Yellow brown ppt.}}{Cu_2I_2 \downarrow} + I_2$$

$$I_2 + 2Na_2S_2O_3 \longrightarrow \underset{\text{White}}{Na_2S_4O_6} + 2NaI$$

$$Ag^+ + \underset{X}{KI} \longrightarrow \underset{\substack{\text{Yellow ppt.} \\ \text{(Insoluble in } NH_4OH)}}{AgI \downarrow} + K^+$$

7 The transition metal that has less number of unpaired electron will show low degree of paramagnetism. Mn^{2+} in $MnSO_4 \cdot 4H_2O$ has d^5 configuration (five unpaired electrons); Cu^{2+} in $CuSO_4 \cdot 5H_2O$ has d^9 configuration (one unpaired electron); Fe^{2+} in $FeSO_4 \cdot 6H_2O$ has d^6 configuration (four unpaired electrons); Ni^{2+} in $NiSO_4 \cdot 6H_2O$ has d^2 configuration (two unpaired electrons).

Thus, $CuSO_4 \cdot 5H_2O$ has lowest degree of paramagnetism.

8 Lanthanoid contraction is due to imperfect shielding of outer electrons by 4*f*-electrons from the nucleus.

9 Lanthanoids $[Xe]4f^{1-14}\,5d^{0-1}\,6s^2$

Actinoids $[Rn]\,5f^{1-14}\,6d^{0-1}\,7s^2$

Lanthanoids and actinoids use core *d* and *f*-orbitals to show higher oxidation state. As actinoids have comparatively low energy difference between 5*f* and 6*d*-orbitals because external shielding of 5*f* electrons is less than that of the 4*f* electrons they show more oxidation states.

10 $\underset{\text{Red solid (insoluble in water)}}{HgI_2} + 2KI \longrightarrow \underset{\text{Soluble}}{K_2HgI_4}$

$$HgI_2 \xrightarrow{\Delta} Hg + I_2$$

11 Copper exhibits only +2 oxidation state in its stable compounds because +2 state compounds of copper are formed by exothermic reaction.

12 $3MnO_4^{2-} + 2H_2O \longrightarrow 2MnO_4^- + MnO_2 + 4OH^-$

The reaction can go completion by removing OH^- ions by adding CO_2 as it reacts with OH^- ions to form CO_3^{2-} ions.

$$4OH^- + 2CO_2 \longrightarrow 2CO_3^{2-} + 2H_2O$$

13 MnO_4^- used $= 30.6 \times 0.1$ millimol

$= 3.06 \times 10^{-3}$ mol

2 mole of MnO_4^-

$= 5$ mole of $H_2C_2O_4 \cdot H_2O$

Thus, 3.06×10^{-3} mole of MnO_4^-

$= 7.65 \times 10^{-3}$ mol unreacted

$(H_2C_2O_4 \cdot H_2O)$ taken

$$\frac{1.651}{108} = 1.53 \times 10^{-2} \text{ mol}$$

Used $H_2C_2O_4 \cdot H_2O$ by MnO_2

$= (1.53 \times 10^{-2} - 7.65 \times 10^{-3})$

$= 7.650 \times 10^{-3}$ mol

1 mole of $H_2C_2O_4 \cdot H_2O \equiv 1$ mole of MnO_2

$\therefore$ 7.65×10^{-3} mole of $H_2C_2O_4 \cdot H_2O$

$= 7.65 \times 10^{-3}$ mole of MnO_2

$= 7.65 \times 10^{-3} \times 87$ g pure MnO_2

$= 0.6655$ g in 0.81 g impure MnO_2

$\therefore$ Percentage of pure

$$MnO_2 = \frac{0.6655}{0.81} \times 100 = 82.16\% \approx 82\%$$

14 Being hygroscopic or deliquescent sodium dichromate, $Na_2Cr_2O_7$ cannot be used in volumetric analysis. The rest of the statements are true.

15 Cuprous ion is colourless while cupric ion is coloured because cuprous ion has a completely filled *d*-orbital, i.e. $3d^{10}$, $4s^0$ configuration and cupric ion has an incompletely *d*-orbitals due to $3d^9$, $4s^0$ configuration. So, this one unpaired electron make *d*-*d* transition possible and impart colour.

DAY TWENTY FOUR

Coordination Compounds

Learning & Revision for the Day

- Werner's Theory
- IUPAC Nomenclature of Coordination Compounds
- Isomerism in Coordination Compounds
- Bonding in Coordination Compounds
- Importance and Applications of Coordination Compounds
- Organometallic compounds

Coordination compounds are those molecular compounds which retain their identity in solid as well as in aqueous solution.

In these compounds metal atoms are bound to a number of anions or neutral molecules by coordinate bonds. A part of these compounds is not dissociated in solution and its behaviour is different than its constituents.

e.g. $\underset{\text{Ionisable species}}{K_4}\ \underset{\text{Non-ionisable coordination sphere}}{[Fe(CN)_6]} + H_2O \longrightarrow 4K^+ + [Fe(CN)_6]^{4-}$

Werner's Theory

Alfred Werner a swiss chemist was the first to study the bonding in coordination compounds in 1982.

The main postulates of this theory are:

(i) In complex compounds, metal atom exhibit two types of valencies—primary valency and secondary valency.

(ii) Primary valencies are satisfied by anions only while secondary valencies are satisfied by anions or neutral molecules (ligands). Primary valency depends upon oxidation number of central metal atom while secondary valency represents the coordination number of central metal atom.

(iii) Primary valencies are ionisable and are non-directional while secondary valencies are non-ionisable and directional. Therefore, geometry of complex is decided by secondary valencies. e.g. $[Cr(H_2O)_6]Cl_3$ has primary valency = 3 (OS of Cr) and secondary valency = 6 (CN of Cr)

PREP MIRROR
Your Personal Preparation Indicator

- No. of Questions in Exercises (x)—
- No. of Questions Attempted (y)—
- No. of Correct Questions (z)—
(Without referring Explanations)

- Accuracy Level ($z/y \times 100$)—
- Prep Level ($z/x \times 100$)—

In order to expect good rank in JEE, your Accuracy Level should be above 85 & Prep Level should be above 75.

Definitions of some important terms pertaining to coordination compounds are as follows :

1. Complex Ion

It is an electrically charged species in which central metal atom or ion is surrounded by number of ions or molecules. There are three types of complex ions.

(i) **Anionic Complex** It is the complex ion which carries negative charge.

$$K_3[Fe(C_2O_4)_3] \longrightarrow 3K^+ + \underset{\text{Anionic complex}}{[Fe(C_2O_4)_3]^{3-}}$$

(ii) **Cationic Complex** It is the complex ion which carries positive charge.

$$[CoCl_2(en)_2]Cl \longrightarrow \underset{\text{Cationic complex}}{[CoCl_2(en)_2]^+} + Cl^-$$

(iii) **Neutral Complex** It is the complex ion which does not carry any charge.

$$\underset{\text{Neutral complex}}{[Ni(CO)_4]}$$

2. Ligands

The ions or molecules bound to the central atom/ion in the coordination entity are called ligands.

The number of donor groups in a single ligand that bind to a central atom in a coordination complex is known as **denticity**.

There are different types of ligands which are discussed below :

(i) **Unidentate Ligand** It is bound to a metal ion through a single donor atom. e.g. $H_2O, NH_3, CO, Cl^-, NH_2^-$ etc.

(ii) **Didentate Ligand** It is bound to a metal ion through two donor atoms.

e.g. COO^-—COO^- (Oxalate ion); $CH_2—\ddot{N}H_2$—$CH_2—\ddot{N}H_2$ (Ethylene diamine)

(iii) **Polydentate ligand** It is bound to a metal ion through several donor atoms. e.g. ethylenediamine tetraacetate ion $[EDTA]^{4-}$.

(iv) **Ambidentate Ligand** These are unidentate ligands which can ligate through more than one coordinating atoms.

e.g. $—NO_2$, —ONO, —SCN etc.

(v) **Chelate Ligands** It may be di or polydentate ligands which form closed ring with central metal ion. Closed ring is known as chelate ring and this process is called chelation. Chelate complexes tend to be more stable than similar complexes containing unidentate ligands.

3. Coordination Number

The number of ligand donor atoms to which the metal is directly bonded is called the coordination number of that metal ion in a complex.

- In case of monodentate ligand, it is equal to the number of ligands while in bidentate ligand it is twice the number of ligands present in the complex.
- For example in the complex ion $[Ag(CN)_2]^-$ and $[Fe(C_2O_4)_3]^{2-}$, the coordination numbers of Ag and Fe are 2 and 6 respectively.

4. Coordination Sphere

The central ion and the ligands which are directly attached to it are enclosed in square bracket is collectively known as coordination sphere. The ionisable group written outside the bracket is known as **counter ions.**

For example in coordination complex, $[Cu(NH_3)_4]SO_4$ the complex ion $[Cu(NH_3)_4]^{2+}$ forms coordination sphere and SO_4^{2-} ions are the counter ions.

5. Coordination Polyhedron

The spatial arrangement of the ligands which are directly attached to the central atom or ion, is called **coordination polyhedron around the central atom or ion.**

6. Oxidation Number of Cental Atom

The oxidation number of a central atom is defined as the charge that it carries as calculated by assigning appropriate charges to the ligands and equating the sum of the charges on the central atom and ligands equal to the charge on the coordination sphere.

7. Homoleptic and Heteroleptic Complexes

Complexes in which a metal is bound to only one kind of donor groups, e.g. $[Co(NH_3)_6]^{3+}$ are known as homoleptic complexes.

Complexes in which a metal is bound to more than one kind of donor groups, e.g. $[Co(NH_3)_4Cl_2]^+$, are known as heteroleptic complexes.

IUPAC Nomenclature of Coordination Compounds

The following rules are used while naming coordination compounds.

- The cation is named first in both positively and negatively charged coordination entities.
- The ligands are named in an alphabetical order before the name of the central atom.
- Names of the anionic ligands end in —O, those of neutral and cationic ligands are the same except aqua for H_2O, ammine for NH_3, carbonyl for CO and nitrosyl for NO.

- Prefixes mono, di, tri etc, are used to indicate the number of the individual ligands in the coordination entity. When the name of the ligands include a numerical prefix, then the terms *bis, tris, tetrakis* are used.
- Oxidation state of the metal in cation, anion or neutral coordination entity is indicated by Roman numerical in parenthesis.
- If the complex ion is cation, the metal is named same as the element. If the complex ion is an anion, the name of the metal ends with suffix-ate.
- The neutral complex molecule is named similar to that of the complex cation.

Isomerism in Coordination Compounds

It is a phenomenon, in which compounds have the same molecular formula but different physical and chemical properties on account of different structures. These compounds are called isomers. There are mainly two types of isomerism shown by coordination compounds which are discussed below :

1. Stereoisomerism

Stereoisomerism occurs due to different arrangement of ligands around central metal atom. It is of two types, geometrical isomerism and optical isomerism.

(i) **Geometrical isomerism** It arises in heteroleptic complexes due to different possible geometric arrangement of the ligands. Important examples of this behaviour are found in square planar and octahedral complexes (discussed below), but tetrahedral complexes do not show geometrical isomerism.

(a) **Square planar complex of the type** $[MX_2L_2]$ (X and L are unidentate), the two X ligands may be arranged adjacent to each other in a *cis*-isomer or opposite to each other in a *trans*-isomer, e.g. $[Pt(NH_3)_2Cl_2]$.

(b) **Square planar complex of the type** $[MABXL]$ (where, A, B, X, L are unidentate) shows three isomers, two *cis* and one *trans*. Such isomerism is not possible for tetrahedral geometry, e.g. $[Pt(NH_3)(Br)(Cl)(Py)]$.

(c) **Square planar complex of the type** $M(XL)_2$, here, XL is unsymmetrical didentate ligand, shows two geometrical isomers, i.e. *cis* and *trans* form, e.g. $[Pt(gly)_2]$.

(d) **Octahedral complexes of formula** $[MX_2L_4]$ in which the two ligands X may be oriented *cis* or *trans* to each other, e.g. $[Co(NH_3)_4Cl_2]^+$.

(e) **Octahedral complexes of formula** $[MX_2A_2]$ (where, X are unidentate ligands and A are didentate ligands) form *cis* and *trans*-isomers, e.g. $[CoCl_2(en)_2]$.

(f) **Octahedral coordination entities of the type** $[Ma_3b_3]$ like $[Co(NH_3)_3(NO_2)_3]$ exist in two geometrical isomers. If three donor atoms of the same ligands occupy adjacent positions at the corners of an octahedral face, we have the ***facial*** (*fac*) **isomer**. When the positions are around the meridian of the octahedron, we get the ***meridional*** (*mer*) **isomer**.

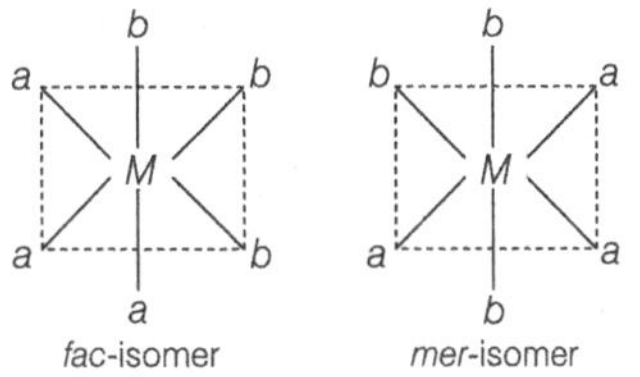

***fac*- and *mer* isomers of Ma_3b_3**

Octahedral complexes of type $[M(AA)_3]$, $[MA_6]$ and $[MA_5B]$ do not show geometrical isomerism.

(ii) **Optical Isomerism** It arises when mirror images cannot be superimposed on one another. These mirror images are called **enantiomers**. The two forms are called *dextro* (*d*) and *laevo* (*l*).

Optical isomerism is common in octahedral complexes having atleast one didentate ligand. Complexes of type $[M(AA)_3]$, $[M(AA)_2B_2]$, $M[(AA)_2BC]$, $M[(AA)B_2C_2]$ show optical isomerism. e.g. $[Co(en)_3]^{3+}$, $[PtCl_2(en)_2]^{2+}$ etc.

NOTE Octahedral complexes of type $MA_2X_2Y_2$ shows both optical and geometrical isomerism.

2. Structural Isomerism

In structural isomerism, isomers have different bonding pattern.

Different types of structural isomerism are as follows:

(i) **Linkage isomerism** arises in a coordination compound containing ambidentate ligand.
e.g. $[Co(NH_3)_5(NO_2)]Cl_2$ and $[Co(NH_3)_5(ONO)]Cl_2$

(ii) **Coordination isomerism** arises from the interchange of ligands between cationic and anionic entities of different metal ions present in a complex.
e.g. $[Co(NH_3)_6][Cr(CN)_6]$ and $[Cr[NH_3)_6][Co(CN)_6]$

(iii) **Ionisation isomerism** arises when the ionisable anion exchange with anion ligand.
e.g. $[Co(NH_3)_5SO_4]Br$ and $[Co(NH_3)_5Br]SO_4$

(iv) **Solvate isomerism** is also known as "hydrate isomerism". In this case water is involved as a solvent. This is similar to ionisation isomerism.
e.g. $[Cr(H_2O)_6]Cl_3$, $[Cr(H_2O)_5Cl]Cl_2 \cdot H_2O$, $[Cr(H_2O)_4Cl_2]Cl \cdot 2H_2O$

Bonding in Coordination Compounds

The bond formation in coordination compounds can be explained by using the two approaches which are given below:

Valence Bond Theory

- According to this theory, the metal atom or ion under the influence of ligands form inner orbital and outer orbital complex. These are hybridised orbitals which are allowed to overlap with ligand orbitals that can donate electron pairs for bonding.
- Octahedral, square planar and tetrahedral complexes are formed as a result of d^2sp^3, dsp^2 and sp^3 hybridisation respectively of the central atom.
- If the metal atom has coordination number six and has d^2sp^3 or sp^3d^2 hybridisation, then the geometry of the complex is octahedral.

Such complexes are of the following two types:

(i) **Inner orbital complexes** (hyperligated complexes) which are formed due to strong field ligands or low spin ligands, has hybridisation d^2sp^3, i.e. involves inner $(n-1)d$ orbital (where two orbitals are of 3d, one orbital of 4s and three orbitals of 4p) and shape of complex will be octahedral. e.g. $[Co(NH_3)_6]^{3+}$

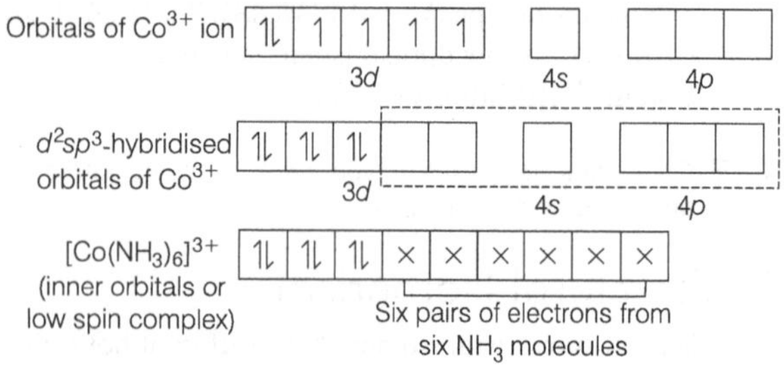

All electrons are paired, therefore complex will be diamagnetic in nature.

(ii) **Outer orbital complexes** (hypoligated complexes) which are formed due to weak field ligands or high spin ligands, has hybridisation sp^3d^2, i.e. uses outer nd orbitals (where one orbital is of 4s, three orbitals of 4p and two orbitals of 4d). Generally halides (F^-, Cl^-, Br^-, I^-), SCN^-, S^{2-} form outer orbital complexes and other ligands form inner orbital complexes. e.g. $[CoF_6]^{3-}$

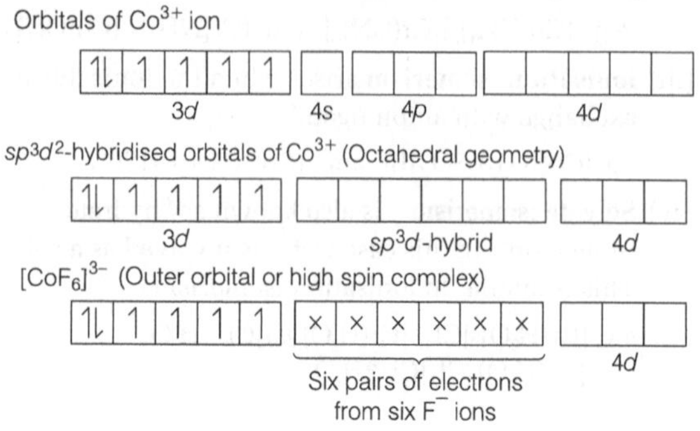

Complex has unpaired electrons, therefore it will be paramagnetic in nature.

- If the metal has coordination number four and has dsp^2 or sp^3-hybridisation, then the geometry of the complex will be square planar or tetrahedral respectively.
- Square planar complexes form inner orbital complexes and tetrahedral complexes form outer orbital complexes, characteristics of which are shown below:

Difference between inner and orbital complexes
(with coordination number four)

Inner orbital complexes	Outer orbital complexes
Strong field or low spin ligands.	Weak field or high spin ligands.
Hybridisation is dsp^2 (where one orbital is of 3d, one orbital of 4s and two orbitals of 4p).	Hybridisation is sp^3 (where one orbital is of 4s and three orbitals of 4p).
Square planar shape.	Tetrahedral shape.

- Generally, halide (F^-, Cl^-, Br^-, I^-) ligands, $[Ni(CO)_4]$, $[Co(CO)_4]$, $[Zn(NH_3)_4]^{2+}$ complexes form outer orbital complexes and other form inner orbital complexes. e.g.

(i) **Inner Orbital Complex** $[Ni(CN)_4]^{2-}$

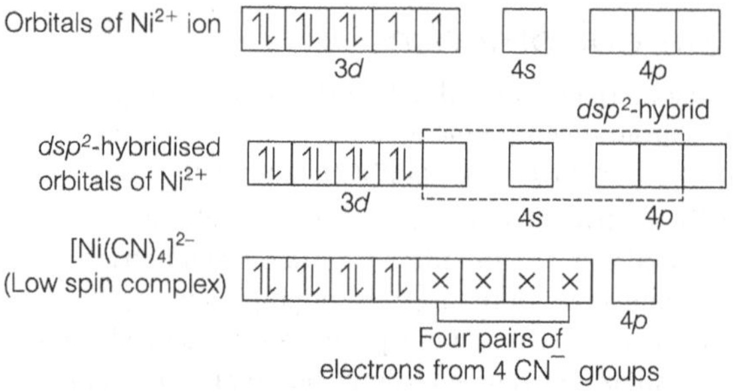

All electrons are paired, thus complex will be diamagnetic in nature.

(ii) **Outer Orbital Complex** $[CoCl_4]^-$

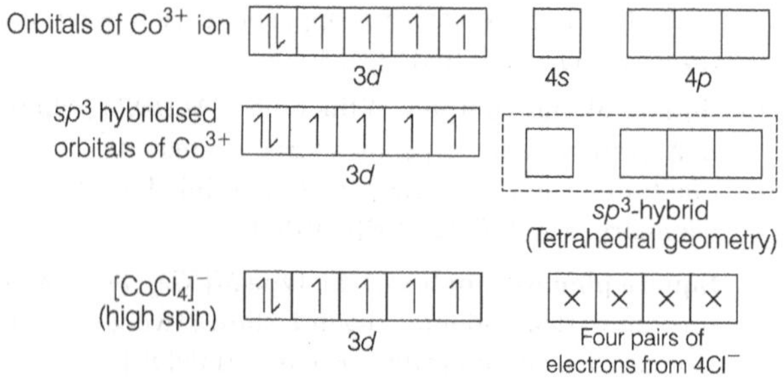

Complex has unpaired electrons, so it will be paramagnetic in nature.

Magnetic Properties

The complex compound is **paramagnetic** if one or more unpaired electrons are present in the d-subshell. If the complex does not contain any unpaired electrons, it is **diamagnetic**.

Magnetic moment, $\mu = \sqrt{n(n+2)}$ BM

Limitation of VBT

Although this theory described the formation, structure and magnetic behaviour of complexes successfully but it suffers from the following short comings.

- It involves a number of assumptions.
- It describes bonding in coordination compounds only qualitatively. It does not offer any explanation for the optical absorption spectra of complexes.
- It does not describe the detailed magnetic properties of coordination compounds.
- It does not give a quantitative interpretation of the thermodynamic or kinetic stabilities of coordination compounds.
- It does not make exact predictions regarding the tetrahedral and square planar structures of 4-coordinate complexes.
- It does not distinguish between weak and strong ligands.

Crystal Field Theory (CFT)

- The splitting of five *d*-orbitals of a metal ion into lower and higher energy levels due to approach of ligands, is explained by crystal field theory.
- The five *d*-orbitals in a gaseous metal atom/ion have same energy, i.e. degenerate.
- However, when the negative field due to ligands surrounds the metal atom, the degeneracy of *d*-orbitals get split depending upon the nature of the crystal field.

Crystal Field Splitting in Octahedral Coordination Entities

- Energy separation of *d*-orbitals is denoted by Δ_o (the subscript *o* is for octahedral). This is also known as crystal field splitting energy (CFSE).
- The energy of the two e_g orbitals (higher energy orbitals) will increase by (3/5) Δ_o and that of the three t_{2g} (lower energy orbitals) will decrease by (2/5) Δ_o.

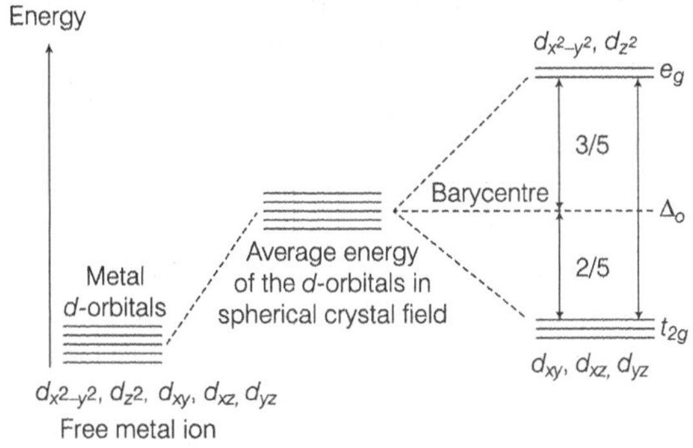

Splitting of *d*-orbitals in octahedral crystal field

- The value of Δ_o is usually compared with the energy required for electron pairing in a single orbital (pairing energy, *p*).
- If $\Delta_o < P$, the fourth electron enters one of the e_g orbitals giving the configuration $t_{2g}^3 e_g^1$. Ligands for which $\Delta_o < P$ are known as weak field ligands and form high spin complexes.
- If $\Delta_o > P$, it becomes more energetically favourable for the fourth electron to occupy a t_{2g} orbital with configuration $t_{2g}^4 e_g^0$. Ligands which produce this effect are known as strong field ligands and form low spin complexes.

Crystal Field Splitting in Tetrahedral Coordination Entities

- In tetrahedral coordination entity formation, *d*-orbital splitting pattern is reverse of splitting pattern in octahedral complexes and $\Delta_t = (4/9)\Delta_o$.
- The orbital splitting energies are not sufficiently large for forcing pairing and therefore, low spin configurations are rarely observed.
- Due to less crystal field stabilisation energy, it is not possible to pair electrons and so all the tetrahedral complexes are high spin.
- An arrangement of ligands in order of increasing crystal field strength is known as **spectrochemical series**.

$$I^- < Br^- < Cl^- < F^- < C_2O_4^{2-} < H_2O < NH_3 < en < NO_2^- < CN^- < CO$$

Colour in Coordination Compounds

- In complex compounds, *d*-orbitals split in two sets t_{2g} and e_g. These have different energies. The difference in energies lies in visible region and electron jump from ground state t_{2g} level to higher state e_g level.
- This is known as $d-d$ transition and is responsible for colour of coordination compounds. $d-d$ transition takes place in d^1 to d^9 ions, so the ions having d^1 to d^9 configuration are coloured.
- On the other hand, the ions with d^0 and d^{10} configuration do not show $d-d$ transition.

NOTE Some coordination complexes have colour due to charge transfer.

Stability Constant and Stability of Complex

- **Stability of a complex** can be expressed in terms of stability constant, *k*. If the complex is ML_n and β_n is the overall formation constant, then

$$M + nL \rightleftharpoons ML_n$$

$$\beta_n = \frac{[ML_n]}{[M][L]^n} = k_1 \times k_2 \times k_3 \ldots k_n$$

- $k_1, k_2, \ldots k_n$ are called stepwise formation constants. Alternatively, $1/k$ is known as instability constant.

The stability of a complex ion depends upon the following factors :

(i) **Higher charge** of the central metal ion, i.e. greater ionic potential $\left(\frac{\text{ionic charge}}{\text{ionic radius}}\right)$ and greater is the stability.

(ii) **Greater base strength** of the ligand, greater will be the stability.

(iii) **Ring formation** (chelation) in structure of the complexes is the chief factor, which increases the stability of the complexes in solution.

(iv) If a multidentate ligand happens to be cyclic without any steric effects, a further increase in stability occurs. This is called **macrocyclic effect.**

The stability of complex can be determined by EAN rule

Effective atomic number EAN of a metal in a complex = atomic number of nearest inert gas.

or EAN = atomic number of metal ± valency + 2 × CN
where, CN = coordination number.

Importance and Applications of Coordination Compounds

- Hardness of water is estimated by simple titration with Na_2EDTA. The Ca^{2+} and Mg^{2+} ions form stable complexes with EDTA.
- Some important extraction processes of metals, like those of silver and gold make use of complex formation.
- Similarly, purification of metals can be achieved through formation and subsequent decomposition of their coordination compounds, e.g. impure nickel is converted to $[Ni(CO)_4]$, which is decomposed to yield pure nickel (Mond's process).
- Coordination compounds are used as catalysts for many industrial processes. e.g. rhodium complex, $[(Ph_3P)_3RhCl]$, (Wilkinson catalyst) is used for the hydrogenation of alkenes.
- Metals present in toxic proportions in animals and plants are removed by chelate therapy, e.g. Cu and Fe are removed by D-penicillamine and desferrioxime-B.

Organometallic Compounds

- Organometallic compounds contain atleast one metal carbon bond. These are of three types, *viz*, σ-bond (e.g. $(C_2H_5)_2Zn$, $(CH_3)_3Al]$, π-bonded (Zeisse's salt, ferrocene) and mixed or σ-and π-bonded (e.g. $Fe(CO)_5$, $[Ni(CO)_4]$.
- π-acid ligands have lone pair of electrons as well as π or π* molecular orbitals. They form σ coordinate bond through lone pair and also form π bond by accepting an appreciable amount of π electron density from metal atom into empty π or π* orbital. e.g. CO is a good π-acceptor (Lewis acid).
- In Zeise's salt, ethylene acts as ligand but it does not have a lone pair of electrons.
- Oxidation state of Fe in ferrocene is Fe^{2+}. In this compound there are two cyclopentadienyl anion $(C_5H_5^-)$ and iron is sandwiched between two aromatic rings.

DAY PRACTICE SESSION 1

FOUNDATION QUESTIONS EXERCISE

1 If excess of $AgNO_3$ is added to 100 mL of a 0.024 M solution of dichlorobis (ethylene diamine) cobalt (III) chloride how many moles of AgCl be precipitated?

(a) 0.0012 (b) 0.0016
(c) 0.0024 (d) 0.0048

2 Consider the following complexes,

1. K_2PtCl_6 2. $PtCl_4 \cdot 2NH_3$
3. $PtCl_4 \cdot 3NH_3$ 4. $PtCl_4 \cdot 5NH_3$

Their respective electrical conductances in aqueous solutions are

(a) 256, 0, 97, 404
(b) 404, 0, 97, 256
(c) 256, 97, 0, 404
(d) 404, 97, 256, 0

3 A solution containing 2.675 g of $CoCl_3 \cdot 6NH_3$ (molar mass = 267.5 g mol^{-1}) is passed through a cation exchanger. The chloride ions obtained in solution were treated with excess of $AgNO_3$ to give 4.78 g of AgCl (molar mass = 143.5 g mol^{-1}). The formula of the complex is (Atomic mass of Ag = 108 u) → AIEEE 2010

(a) $[Co(NH_3)_6]Cl_3$ (b) $[CoCl_2(NH_3)_4]Cl$
(c) $[CoCl_3(NH_3)_3]$ (d) $[CoCl(NH_3)_5]Cl_2$

4 On treatment of 100 mL of 0.1 M solution of $CoCl_3.6H_2O$ with excess of $AgNO_3$; 1.2×10^{22} ions are precipitated. The complex is → JEE Main 2018

(a) $[Co(H_2O)_4Cl_2]Cl\ H_2O$ (b) $[Co(H_2O)_3Cl_3].3H_2O$
(c) $[Co(H_2O)_6]Cl_3$ (d) $[Co(H_2O)_5Cl]Cl_2.H_2O$

5 The coordination number and the oxidation state of the element E in the complex $[E(en)_2(C_2O_4)]NO_2$ are respectively, [where, en is ethylene diamine]
(a) 6 and 2 (b) 4 and 2 (c) 4 and 3 (d) 6 and 3

6 The oxidation states of Cr, in $[Cr(H_2O)_6]Cl_3$, $[Cr(C_6H_6)_2]$, and $K_2[Cr(CN)_2(O)_2(O_2)(NH_3)]$ respectively are → JEE Main 2018
(a) +3, +4 and +6 (b) +3, +2 and +4
(c) +3, 0 and +6 (d) +3, 0 and +4

7 IUPAC name of $[Pt(NH_3)_2Cl(NO_2)]$ is
(a) platinum diamminechloronitrite
(b) chloronitrito-N-ammineplatinum (II)
(c) diamminechloridonitrito-N-platinum (II)
(d) diamminechloronitrito-N-platinate (II)

8 Which among the following will be named as dibromidobis-(ethylenediamine) chromium (III) bromide? → AIEEE 2012
(a) $[Cr(en)_3]Br_3$ (b) $[Cr(en)_2Br_2]Br$
(c) $[Cr(en)Br_4]^-$ (d) $[Cr(en)Br_2]Br$

9 The number of geometrical isomers of the complex $[Co(NO_2)_3(NH_3)_3]$ is
(a) 2 (b) 4 (c) 3 (d) 0

10 The complex, $[Pt(py)(NH_3)BrCl]$ will have how many geometrical isomers? → CBSE-AIPMT 2011
(a) 2 (b) 3 (c) 4 (d) 0

11 The number of geometrical isomers that can exist for square planar $[Pt(Cl)(py)(NH_3)(NH_2OH)]^+$ is (py = pyridine) → JEE Main 2015
(a) 2 (b) 3 (c) 4 (d) 6

12 Consider the following reaction and statements :
$[Co(NH_3)_4Br_2]^+ + Br^- \longrightarrow [Co(NH_3)_3Br_3] + NH_3$

I. Two isomers are produced if the reactant complex ion is a *cis*-isomer.
II. Two isomers are produced if the reactant complex ion is a *trans*-isomer.
III. Only one isomer is produced if the reactant complex ion is a *trans*-isomer.
IV. Only one isomer is produced if the reactant complex ion is a *cis*-isomer. → JEE Main 2018

The correct statements are
(a) (I) and (II) (b) (I) and (III)
(c) (III) and (IV) (d) (II) and (IV)

13 Which one of the following complexes shows optical isomerism? → JEE Main 2016
(a) *cis* $[Co(en)_2Cl_2]Cl$ (b) *trans* $[Co(en)_2Cl_2]Cl$
(c) $[Co(NH_3)_4Cl_2]Cl$ (d) $[Co(NH_3)_3Cl_3]$

14 Which of the following complex species is not expected to exhibit optical isomerism? → JEE Main 2013
(a) $[Co(en)_2Cl_2]^+$ (b) $[Co(NH_3)_3Cl_3]$
(c) $[Co(CN)(NH_3)_2Cl_2]^+$ (d) $[Co(en)_3]^{3+}$

15 Which one of the following has an optical isomer? (en = ethylenediamine) → AIEEE 2010
(a) $[Zn(en)(NH_3)_2]^{2+}$ (b) $[Co(en)_3]^{3+}$
(c) $[Co(H_2O)_4(en)]^{3+}$ (d) $[Zn(en)_2]^{2+}$

16 Type of isomerism which exists between $[Pd(C_6H_5)_2(SCN)_2]$ and $[Pd(C_6H_5)_2(NCS)_2]$ is → JEE Main (Online) 2013
(a) linkage isomerism (b) coordination isomerism
(c) ionisation isomerism (d) solvate isomerism

17 Which of the following pairs represents linkage isomers? → JEE Main 2009
(a) $[Cu(NH_3)_4][PtCl_4]$ and $[Pt(NH_3)_4][CuCl_4]$
(b) $[Pd(PPh_3)_2(NCS)_2]$ and $[Pd(PPh_3)_2(SCN)_2]$
(c) $[Co(NH_3)_5NO_3]SO_4$ and $[Co(NH_3)_5SO_4]NO_3$
(d) $[PtCl_2(NH_3)_4]Br_2$ and $[PtBr_2(NH_3)_4]Cl_2$

18 Match the following and assign the correct code.

	Column I (Complex species)		Column II (Isomerism)
A.	$[Co(NH_3)_4Cl_2]^+$	1.	Optical
B.	*cis* $[Co(en)_2Cl_2]^+$	2.	Ionisation
C.	$[Co(NH_3)_5(NO_2)]Cl_2$	3.	Coordination
D.	$[Co(NH_3)_6][Cr(CN)_6]$	4.	Geometrical
		5.	Linkage

Codes

	A	B	C	D
(a)	1	2	4	5
(b)	4	3	2	1
(c)	4	1	5	3
(d)	4	1	2	3

19 The structure of which of the following chloro species can be explained on the basis of dsp^2 hybridisation? → JEE Main 2013
(a) $PdCl_4^{2-}$ (b) $FeCl_4^{2-}$ (c) $CoCl_4^{2-}$ (d) $NiCl_4^{2-}$

20 Both $[Ni(CO)_4]$ and $[Ni(CN)_4]^{2-}$ are diamagnetic. The hybridisations of nickel in these complexes are respectively,
(a) sp^3, sp^3 (b) sp^3, dsp^2
(c) dsp^2, sp^3 (d) dsp^2, dsp^2

21 Which of the following facts about the complex $[Cr(NH_3)_6]Cl_3$ is wrong? → AIEEE 2011
(a) The complex involves d^2sp^3 hybridisation and is octahedral in shape
(b) The complex is paramagnetic
(c) The complex is an outer orbital complex
(d) The complex gives white precipitate with silver nitrate solution

22 Which of the following is diamagnetic ? → JEE Main (Online) 2013
(a) $[Fe(CN)_6]^{3-}$ (b) $[Co(ox)_3]^{3-}$
(c) $[FeF_6]^{3-}$ (d) $[CoF_6]^{3-}$

23 $NiCl_2[P(C_2H_5)_2(C_2H_5)_2]$ exhibit temperature dependent magnetic behaviour (paramagnetic/diamagnetic). The coordination geometries of Ni^{2+} in the paramagnetic and diamagnetic states are respectively → AIEEE 2012

(a) tetrahedral and tetrahedral
(b) square planar and square planar
(c) tetrahedral and square planar
(d) square planar and tetrahedral

24 The magnetic moment (spin only) of $[NiCl_4]^{2-}$ is → AIEEE 2011

(a) 1.82 BM (b) 5.46 BM (c) 2.82 BM (d) 1.41 BM

25 The magnetic moment of the complex anion $[Cr(NO)(NH_3)(CN)_4]^{2-}$ is → JEE Main (Online) 2013

(a) 5.91 BM (b) 3.87 BM (c) 1.73 BM (d) 2.82 BM

26 In which of the following complexes the crystal field splitting will be least?

(a) $[Fe(H_2O)_6]^{3-}$ (b) $[Cr(NH_3)_6]^{3+}$
(c) $[Co(C_2O_4)_3]^{3-}$ (d) $Ni(CO)_4$

27 In which of the following octahedral complex species the magnitude of Δ_o will be maximum? → JEE Main (Online) 2013

(a) $[Co(H_2O)_6]^{2+}$ (b) $[Co(CN)_6]^{3-}$
(c) $[Co(C_2O_4)_3]^{3-}$ (d) $[Co(NH_3)_6]^{3+}$

28 Among the ligands NH_3, en, CN^- and CO, the correct order of their increasing field strength, is → AIEEE 2011

(a) $CO < NH_3 < en < CN^-$ (b) $NH_3 < en < CN^- < CO$
(c) $CN^- < NH_3 < CO < en$ (d) $en < CN^- < NH_3 < CO$

29 The colour of the coordination compounds depend on the crystal field splitting. What will be the correct order of absorption of wavelength of light in the visible region, for the complexes, $[Co(CN)_6]^{3-}$, $[Co(CN)_6]^{3-}$, $[Co(H_2O)_6]^{3+}$

(a) $[Co(CN)_6]^{3-} > [Co(NH_3)_6]^{3+} > [Co(CN)_6]^{3+}$
(b) $[Co(NH_3)_6]^{3+} > [Co(H_2O)_6]^{3+} > [Co(CN)_6]^{3-}$
(c) $[Co(H_2O)_6]^{3+} > [Co(NH_3)_6]^{3+} > [Co(CN)_6]^{3-}$
(d) $[Co(CN)_6]^{3-} > [Co(NH_3)_6]^{3+} > [Co(H_2O)_6]^{3+}$

30 Which of the following compounds is not yellow coloured? → JEE Main 2015

(a) $Zn_2[Fe(CN)_2]$ (b) $K_3[Co(NO_2)_6]$
(c) $(NH_4)_3[As(Mo_3O_{10})_4]$ (d) $BaCrO_4$

31 The octahedral complex of a metal ion M^{3+} with four monodentate ligands L_1, L_2, L_3 and L_4 absorb wavelengths in the region of red, green, yellow and blue, respectively. The increasing order of ligand strength of the four ligands is → JEE Main 2014

(a) $L_4 < L_3, L_2 < L_1$ (b) $L_1 < L_3 < L_2 < L_4$
(c) $L_3 < L_2 < L_4 < L_1$ (d) $L_1 < L_2 < L_4 < L_3$

32 According to EAN rule, how many CO groups should be attached to Fe?

(a) 4 (b) 5 (c) 6 (d) 8

33 Among the following which is not the π-bonded organometallic compound?

(a) $K[PtCl_3(\eta^2-C_2H_4)]$ (b) $Fe(\eta^5-C_5H_5)_2$
(c) $(CH_3)_4Sn$ (d) $Cr(\eta^6-C_6H_6)_2$

34 Which of the following is an organometallic compound?

(a) $C_2H_5-Zn-C_2H_5$ (b) $H_3C-B(OCH_3)_2$
(c) $H_3C-O-B(O-CH_3)_2$ (d) None of these

35 Ferrocene is an example of

(a) sandwiched complex
(b) pi-bonded complex
(c) a complex in which all the five carbon atoms of cyclopentadiene anion are bonded to the metal
(d) All of the above

36 An aqueous solution of an inorganic salt (X), when added to an aqueous solution of barium chloride, a precipitate insoluble in dil. HCl is obtained. Addition of excess of KI to X gives a brown precipitate which turns white on addition of excess of hypo. With an aqueous solution of potassium ferrocyanide, a chocolate coloured precipitate is formed. X is

(a) $Cu(NH_3)_4SO_4$ (b) $CuSO_4 \cdot 5H_2O$
(c) $ZnSO_4 \cdot 5H_2O$ (d) $AgNO_3$

37 The equation which is balanced and represents the correct product(s) is → JEE Main 2014

(a) $Li_2O + 2KCl \longrightarrow 2LiCl + K_2O$
(b) $[CoCl(NH_3)_5]^+ + 5H^+ \longrightarrow Co^{2+} + 5NH_4^+ + Cl^-$
(c) $[Mg(H_2O_6]^{2+} + (EDTA)^{4-} \xrightarrow{\text{Excess NaOH}} [Mg(EDTA)]^{2+} + 6H_2O$
(d) $CuSO_4 + 4KCN \longrightarrow K_2[Cu(CN)_4] + K_2SO_4$

38 Which of the following statements is incorrect? → JEE Main (Online) 2013

(a) Fe^{3+} ion also gives blood red colour with SCN^- ion
(b) Fe^{3+} ion gives red colour with SCN^- ion
(c) On passing H_2S into Na_2ZnO_2 solution, a white ppt of ZnS is formed
(d) Cupric ion reacts with excess of ammonia solution to give deep blue colour of $[Cu(NH_3)_4]^{2+}$ ion

39 **Assertion** (A) $[Cr(H_2O)_6]Cl_2$ and $[Fe(H_2O)_6]Cl_2$ are reducing in nature.

Reason (R) Chelate complexes tend to be more stable.

(a) Assertion and Reason both are correct statements and Reason is the correct explanation of the Assertion
(b) Assertion and Reason both are correct statements but Reason is not the correct explanation of the Assertion
(c) Assertion is correct incorrect and Reason is incorrect
(d) Both Assertion and Reason are incorrect

Direction (Q. Nos. 40 and 41) *Each of these questions contains two statements : Statement I (Assertion) and Statement II (Reason). Each of these questions also has four alternative choices, only one of which is the correct answer. You have to select one of the codes (a), (b), (c) and (d) given below :*

(a) Statement I is true, Statement II is true; Statement II is a correct explanation for Statement I
(b) Statement I is true, Statement II is true; Statement II is not a correct explanation for Statement I
(c) Statement I is true; Statement II is false
(d) Statement I is false; Statement II is true

40 **Statement I** $[Fe(H_2O)_5NO]SO_4$ is paramagnetic.
Statement II The Fe in $[Fe(H_2O)_5NO]SO_4$ has three unpaired electrons.

41 **Statement I** The geometrical isomers of the complex $[M(NH_3)_4Cl_2]$ are optically inactive.
Statement II Both geometrical isomers of the complex $[M(NH_3)_4Cl_2]$ possess axis of symmetry.

DAY PRACTICE SESSION 2

PROGRESSIVE QUESTIONS EXERCISE

1 The stabilisation of coordination compounds due to chelation is called the chelate effect. Which of the following is the most stable complex species?
(a) $[Fe(CO)_5]$ (b) $[Fe(CN)_6]^{3-}$
(c) $[Fe(C_2O_4)_3]^{3-}$ (d) $[Fe(H_2O)_6]^{3+}$

2 The value of CFSE for complex ion $[CoCl_6]^{4-}$ is 18000 cm^{-1}. The CFSE for $[CoCl_4]^{2-}$ complex ion will be
(a) 18000 cm^{-1} (b) 16000 cm^{-1} (c) 8000 cm^{-1} (d) 2000 cm^{-1}

3 For an octahedral complex, which of the following *d*- electron configuration will give maximum CFSE ?
(a) High spin d^6 (b) Low spin d^4
(c) Low spin d^5 (d) High spin d^7

4 The oxidation state of cobalt in

$$\left[(NH_3)_4Co\left\langle\begin{matrix}NH\\NO_2\end{matrix}\right\rangle Co(NH_3)_4\right](NO_3)_4$$ is

(a) 2 (b) 3
(c) 4 (d) 6

5 IUPAC name for

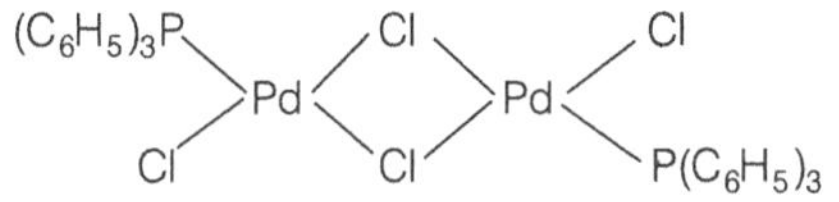

(a) chlorotriphenylphosphinepalladium (II)- μ-dichloro chlorotriphenylphosphinepalladium (II)
(b) chlorotriphenylphosphine palladium (III)- μ-dichlorochlorotriphenyl phosphine palladium (II)
(c) triphenylphosphinechloro palladium (II)- μ-dichlorido triphenylphosphinechloro palladium (III)
(d) triphenylphosphinechloro palladium (III)- μ-dichlorotriphenylphosphinechloro palladium (III)

6 Among the following metal carbonyls, the C—O bond order is lowest in
(a) $[Mn(CO)_6]^+$ (b) $[Fe(CO)_5]^+$
(c) $[Cr(CO)_6]^+$ (d) $[V(CO)_6]^-$

7 Which of the following compounds shows optical isomerism?
(a) $[Co(CN)_6]^{3-}$ (b) $[Cr(C_2O_4)_3]^{3+}$
(c) $[ZnCl_4]^{2-}$ (d) $[Cu(NH_3)_4]^{2+}$

8 Which can exist both as diastereomer and enantiomer?
(a) $[Pt(en)_3]^{4+}$ (b) $[Pt(en)_2ClBr]^{2+}$
(c) $[Ru(NH_3)_4Cl_2]^0$ (d) $[PtCl_2Br_2]^0$

9 The correct order for the wavelength of absorption in the visible region is
(a) $[Ni(NO_2)_6]^{4-} < [Ni(NH_3)_6]^{2+} < [Ni(H_2O)_6]^{2+}$
(b) $[Ni(NO_2)_6]^{4-} < [Ni(H_2O)_6]^{2+} < [Ni(NH_3)_6]^{2+}$
(c) $[Ni(H_2O)_6]^{2+} < [Ni(NH_3)_6]^{2+} < [Ni(NO_2)_6]^{4-}$
(d) $[Ni(NH_3)_6]^{2+} < [Ni(H_2O)_6]^{2+} < [Ni(NO_2)_6]^{4-}$

10 A complex is prepared by mixing $CoCl_3$ and NH_3 in the molar ratio 1 : 4. 0.1 M solution of this complex was found to freeze at −0.372° C. What is the formula of the complex ? [K_f(water) = 1.86° C / m]
(a) $[Co(NH_3)_4Cl_2]Cl$ (b) $[Co(NH_3)_5Cl]Cl_2$
(c) $[Co(NH_3)_3Cl_3]$ (d) $[Co(NH_3)_6]Cl_3$

11 Ammonia forms the complex ion $[Cu(NH_3)_4]^{2+}$ with copper ions in the alkaline solutions but not in acidic solutions. What is the reason for it ?
(a) In acidic solutions hydration protects copper ions
(b) In acidic solutions protons coordinate with ammonia molecules forming NH_4^+ ions and NH_3 molecules are not available
(c) In alkaline solutions insoluble $Cu(OH)_2$ is precipitated which is soluble in excess of any alkali
(d) Copper hydroxide is an amphoteric substance

12 What is the ratio of uncomplexed to complexed Zn^{2+} ion in a solution that is 10 M in NH_3, if the stability constant of $[Zn(NH_3)_4]^{2+}$ is 3×10^9?
(a) 3.3×10^{-9} (b) 3.3×10^{-11}
(c) 3.3×10^{-14} (d) 3×10^{-13}

13 How many moles of AgCl would be obtained, when 100 mL of 0.1 M $Co(NH_3)_5Cl_3$ is treated with excess of $AgNO_3$?

(a) 0.01 (b) 0.02
(c) 0.03 (d) None of these

14 The complex $[Fe(H_2O)_5NO]^{2+}$ is formed in the brown ring test for nitrates when freshly prepared $FeSO_4$ solution is added to aqueous solution of NO_3^- followed by addition of conc. H_2SO_4. Select the correct statement about this complex.

(a) Colour change is due to charge transfer
(b) It has iron in +1 oxidation state and nitrosyl as NO^+
(c) It has magnetic moment of 3.87 BM confirming three unpaired electrons in Fe
(d) All the above are correct statements

ANSWERS

SESSION 1

1 (c)	**2** (a)	**3** (a)	**4** (d)	**5** (d)	**6** (c)	**7** (c)	**8** (b)	**9** (a)	**10** (b)
11 (b)	**12** (b)	**13** (a)	**14** (b)	**15** (b)	**16** (a)	**17** (b)	**18** (d)	**19** (a)	**20** (b)
21 (c)	**22** (b)	**23** (c)	**24** (c)	**25** (c)	**26** (a)	**27** (b)	**28** (b)	**29** (c)	**30** (a)
31 (d)	**32** (b)	**33** (c)	**34** (a)	**35** (d)	**36** (b)	**37** (b)	**38** (a)	**39** (b)	**40** (a)
41 (a)									

SESSION 2

1 (c)	**2** (c)	**3** (c)	**4** (b)	**5** (a)	**6** (b)	**7** (b)	**8** (b)	**9** (a)	**10** (a)
11 (b)	**12** (c)	**13** (b)	**14** (d)						

Hints and Explanations

SESSION 1

1 100 mL of 0.024 M = 0.0024 mole of the complex.
Complex is $[Co(en)_2Cl_2]Cl$ and thus, one Cl^- is formed per mole of the complex which gives 1 mole of AgCl.
Hence, moles of AgCl precipitated are = 0.0024

2 Greater the number of ions produced greater, the conductance.

1. $K_2[PtCl_6] \rightleftharpoons 2K^+ + [PtCl_6]^{2-}$ (3 ions)
2. $PtCl_4 \cdot 2NH_3 \rightleftharpoons [Pt(NH_3)_2Cl_4] \longrightarrow$ no ions (least)
3. $[Pt(NH_3)_3Cl_3]Cl \rightleftharpoons [Pt(NH_3)_3Cl_3]^+ + Cl^-$ (2 ions)
4. $[Pt(NH_3)_5Cl]Cl_3 \rightleftharpoons [Pt(NH_3)_5Cl] + 3Cl^-$ (4 ions maximum)

3 Mole of $CoCl_3 \cdot 6NH_3 = \frac{2.675}{267.5} = 0.01$ mol

$AgNO_3(aq) + Cl^-(aq) \longrightarrow AgCl\downarrow \text{(white)} + NO_3^-(aq)$

Moles of $AgCl = \frac{4.78}{143.5} = 0.03$ mol

0.01 mol $CoCl_3 \cdot 6NH_3$ gives = 0.03 mol AgCl

$\therefore$ 1 mol $CoCl_3 \cdot 6NH_3$ ionises to give = 3 mol Cl^-

Hence, the formula of compound is $[Co(NH_3)_6]Cl_3$.

4 Molarity $(M) = \frac{\text{Number of moles of solute}}{\text{Volume of solution (in L)}}$

$\therefore$ Number of moles of complex

$$= \frac{\text{Molarity} \times \text{volume (in mL)}}{1000} = \frac{0.1 \times 100}{1000} = 0.01 \text{ mol}$$

Number of moles of ions precipitate $= \frac{12 \times 10^{22}}{6.02 \times 10^{23}} = 0.02$ mol

$\therefore$ Number of Cl^- present in ionisation sphere

$$= \frac{\text{Number of moles of ions precipitated}}{\text{Number of moles of complex}} = \frac{0.02}{0.01} = 2$$

$\therefore$ $2Cl^-$ are present outside the square brackets, i.e. in ionisation sphere. Thus, the formula of complex is $[Co(H_2O)_5Cl]Cl_2 \cdot H_2O$.

5 $\begin{matrix} CH_2—NH_2 \\ | \\ CH_2—NH_2 \end{matrix}$ is a bidentate ligand. $C_2O_4^{2-}$ is also a bidentate ligand. Hence, coordination number = 6

Complex can be ionised as

$$[E(en)_2(C_2O_4)]NO_2 \longrightarrow [E(en)_2(C_2O_4)]^+ + NO_2^-$$

Oxidation number $= x + 0 + (-2) = 1$

$\therefore \quad x = 3$

6 Let the oxidation state of Cr in all cases is 'x'.

(i) Oxidation state of Cr in $[Cr(H_2O)_6]Cl_3$

$x + (0 \times 6) + (-1 \times 3) = 0$

or $x + 0 - 3 = 0$ or $x = +3$

(ii) Oxidation state of Cr in $[Cr(C_6H_6)_2]$

$x + (2 \times 0) = 0$ or $x = 0$

(iii) Oxidation state of Cr in $K_2[Cr(CN)_2(O)_2(O_2)(NH_3)]$

$1 \times 2 + x + (-1 \times 2) + (-2 \times 2) + (-2) + 0 = 0$

or $2 + x - 2 - 4 - 2 = 0$ or $x - 6 = 0$

Hence, $x = +6$

Thus, +3, 0 and +6 is the answer.

7 Diamminechloridonitrito-N-platinum (II).

8 Two Br, two (en) and one Cr are parts of complex. Charge on the complex is

$$\left.\begin{aligned}2\ (\text{Br}) &= -2\\ 2\ (\text{en}) &= 0\\ 1\ (\text{Cr}) &= +3\end{aligned}\right\} = +1$$

Thus, complex ion is $[Cr(en)_2Br_2]^+$.

Since, anion is bromide thus, complex is $[Cr(en)_2Br_2]$ Br.

9 Complex $[Co(NO_2)_3(NH_3)_3]$ is of type MA_3B_3 which have 2 geometrical isomers that is *fac* and *mer* isomers.

10 *M(ABCD)* type complex have three geometrical isomers as:

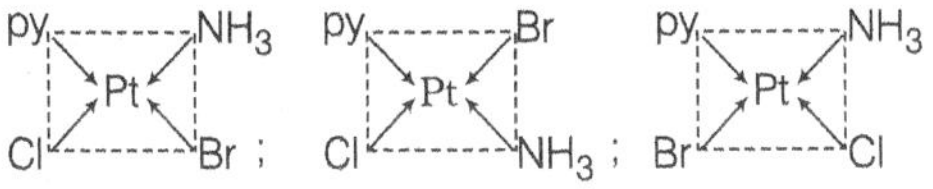

11 $[Pt(Cl)(py)(NH_3)(NH_2OH)]^+$ is square planar complex. The structures are formed by fixing a group and then arranging all the groups.

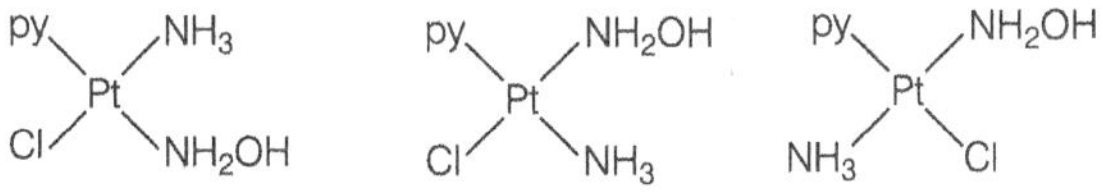

Hence, this complex has 3 geometrical isomers.

12 If the reactant is *cis* isomer than following reaction takes place.

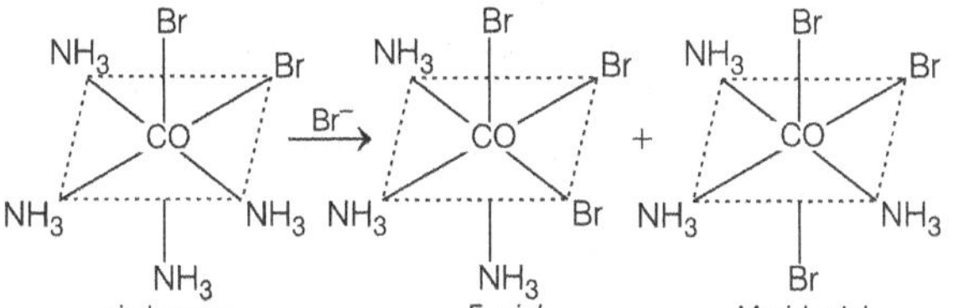

cis-isomer *Facial* *Meridonial*

i.e. two isomers are produced.

If the reactant is *trans* isomer than following reaction takes place.

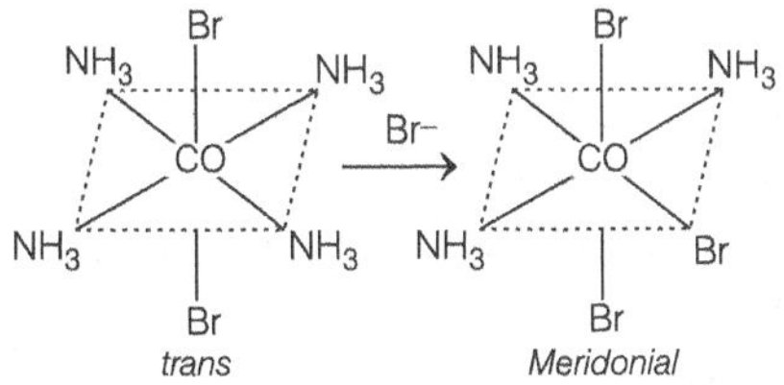

trans *Meridonial*

i.e. only 1 isomer is produced. Thus, statement (I) and (III) are correct resulting to option (b) as the correct answer.

13 $[Co(en)_2Cl_2]Cl$ is of type $M(AA)_2B_2$, where only *cis* isomer shows optical isomerism.

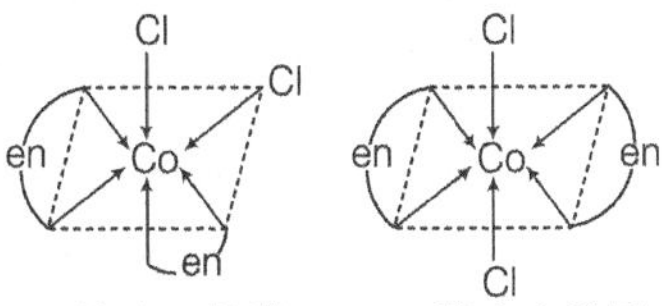

cis-$[Co(en)_2Cl_2]Cl$ (optically active) *trans*-$[Co(en)_2Cl_2]Cl$ (optically inactive due to plane of symmetry)

$[Co(NH_3)_4Cl_2]Cl$ can exist in both *cis* and *trans* forms that are shown below:

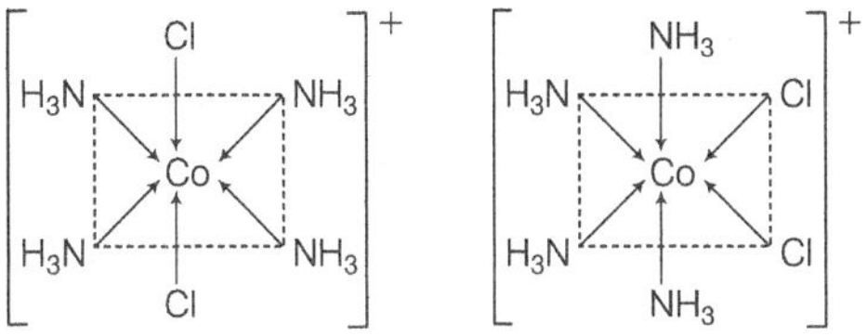

trans-$[Co(NH_3)_4Cl_2]Cl$ (optically inactive) *cis*-$[Co(NH_3)_4Cl_2]Cl$ (optically inactive)

$[Co(NH_3)_3Cl_3]$ exists in *fac* and *mer*-isomeric forms and both are optically inactive.

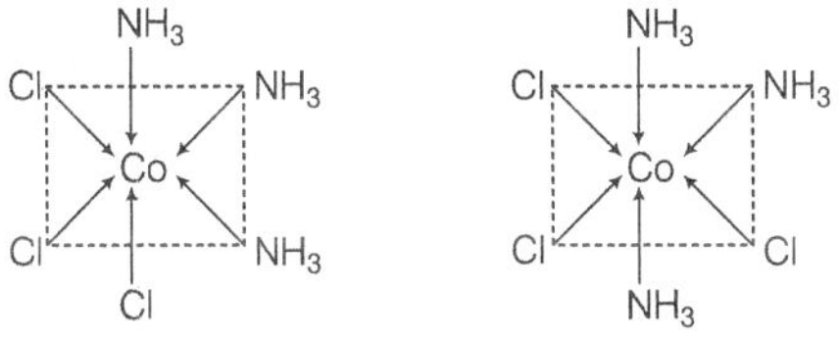

fac-isomer (optically inactive) *mer*-isomer (optically inactive)

14 Optical isomerism is exhibited by only those complexes which lacks elements of symmetry. $[Co(NH_3)_3Cl_3]$ is of type MA_3B_3 which shows *facial* as well as *meridional* isomerism. But both the forms contain plane of symmetry. Thus, this complex does not exhibit optical isomerism.

15 Complex $[Co(en)_3]^{3+}$ is of type MA_3 has no plane of symmetry and centre of symmetry that's why it is optically active.

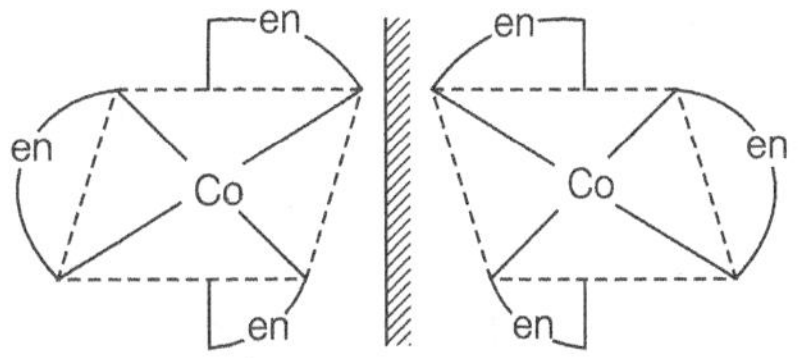

Mirror images (Non-superimposable)

16 Given compound shows linkage isomerism because SCN is ambidentate ligand.

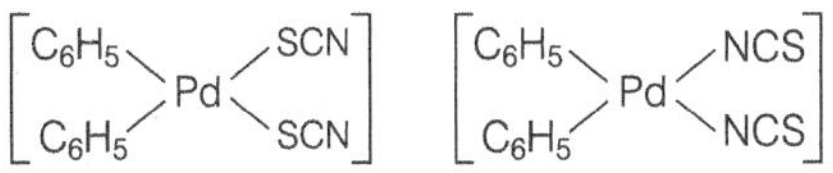

17 Linkage isomers are formed due to the presence of ambidentate ligands. $[Pd(PPh_3)_2(NCS)_2]$ and $[Pd(PPh_3)_2(SCN)_2]$ are linkage isomers due to SCN, ambidentate ligand.

18 A → 4, B → 1, C → 2, D → 3.

19 In $[PdCl_4]^{2-}$, the oxidation state of Pd is +2.

Pd = $[kr]4d^85s^0$ =

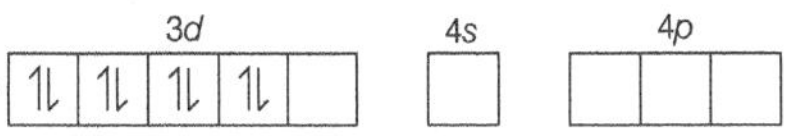

Although Cl^- is a weak field ligand but in case of $[PdCl_4]^{2-}$, the electrons of Pd are paired up because of its large size and results in dsp^2-hybridisation. In all other cases pairing is not possible because of weak ligand Cl^-, so hybridisation is sp^3 with tetrahedral structure.

20 In $[Ni(CO)_4]$, the oxidation state of Ni is zero.

Ni(28) = $[Ar]3d^8, 4s^2$ 3d: | ⇅ | ⇅ | ⇅ | ↑ | ↑ | 4s: | ⇅ | 4p: | | | |

CO is strong field ligand, causes pairing, thus

| ⇅ | ⇅ | ⇅ | ⇅ | ⇅ | | × | × | × | × | (sp^3-hybridisation)

In $[Ni(CN)_4]^{2-}$, the oxidation state of Ni is +2.

$Ni^{2+} = [Ar]3d^8, 4s^0$ 3d: | ⇅ | ⇅ | ⇅ | ↑ | ↑ | 4s: | | 4p: | | | |

CN^- is strong field ligand causes pairing, thus

| ⇅ | ⇅ | ⇅ | ⇅ | × | × | × | × | (dsp^2 hybridisation)

21 $\underbrace{[Cr(NH_3)_6]}_{\text{Coordinate sphere}} \cdot \underbrace{Cl_3}_{\text{Ionisable}}$

$[Cr(NH_3)_6]^{3+} + 3Cl^- \xrightarrow{AgNO_3} \underset{\text{White precipitate}}{AgCl\downarrow}$

$Cr^{3+} = [Ar]3d^3\ 4s^0$

Cr^{3+} = [Ar] 3d: | ↑ | ↑ | ↑ | | | 4s: | | 4p: | | | |

$[Cr(NH_3)_6]^{3+}$ = [Ar] 3d: | ↑ | ↑ | ↑ | × | × | × | × | × | × | (d^2sp^3 hybridised)

Indicates that lone pair of NH_3 are donated to Cr.

(a) d^2sp^3 hybridisation, octahedral. Thus, statement is correct.

(b) There are three unpaired electrons, hence paramagnetic. Thus, statement is correct.

(c) d^2sp^3 inner orbital complex. thus statement is incorrect.

(d) Due to ionisable Cl^- ions, white precipitate with $AgNO_3$ thus, statement is correct.

22 Fe^{3+} and Co^{3+} both the ions have unpaired electrons in their ground state, but strong field ligands, like CN^-, ox if present, pair up the unpaired electrons resulting in diamagnetism.

In $[Fe(CN^-)_6]^{3-}$, $Fe^{3+} = 3d^5 4s^0$

$3d^5$: | ↑ | ↑ | ↑ | ↑ | ↑ |

In $[Co(ox)_3]^{3-}$, $Co^{3+} = 3d^6 4s^0$

$3d^6$: | ⇅ | ⇅ | ⇅ | | | (All electrons are paired)

Also after pairing there is one unpaired electron in $[Fe(CN)_6]^{3-}$. Thus, $[Co(ox)_3]^{3-}$ is a diamagnetic species as all the electrons are paired in it.

23 In the given complex, $NiCl_2\{P(C_2H_5)_2(C_2H_5)_2\}$, Ni is in + 2 oxidation state.

Ni^{2+} = $3d^8$: | ⇅ | ⇅ | ⇅ | ↑ | ↑ | | | | | | (sp^3 hybridised orbitals)

For the given four-coordinated complex to be paramagnetic, it must possess unpaired electrons in the valence shell. The geometry of paramagnetic complex must be tetrahedral. On the other hand, for complex to be diamagnetic, there should not be any unpaired electrons in the valence shell. This condition can be fullfilled by pairing electrons of 3d- orbitals. The geometry of diamagnetic complex is square planar.

Ni^{2+} = 3d: | ⇅ | ⇅ | ⇅ | ↑ | ↑ | | | | | | (sp^3 hybridised)

24 $[NiCl_4]^{2-}$; oxidation number of Ni, $x - 4 = -2 \quad \therefore \quad x = +2$

$Ni_{(28)} = [Ar]\ 3d^8, 4s^2$

Ni^{2+}= [Ar] $3d^8$: | ⇅ | ⇅ | ⇅ | ↑ | ↑ | $4s^0$: | | $4p^0$: | | | |

$[NiCl_4]^{2-}$ $3d^8$: | ⇅ | ⇅ | ⇅ | ↑ | ↑ | | × | × | × | × | (Four sp^3 hybrid orbitals (tetrahedral geometry))

Cl^- is a weak field ligand and thus, unpaired electrons are not paired. Lone pairs from $4Cl^-$ are accommodated in four sp^3 hybrid orbitals and thus $[NiCl_4]^{2-}$ has 2 unpaired electrons.

Magnetic moment (spin only)

$= \sqrt{n(n+2)}$ BM, n = unpaired electrons = 2,

$= \sqrt{2(2+2)}$

$= 2.828 \text{ BM} \approx 2.82 \text{BM}$

25 In $[Cr(NO)(NH_3)(CN)_4]^{2-}$, let the oxidation state of Cr be x.

$\therefore x + (+1) + (0) + (-1)4 = -2, \ x - 3 = -2$ or $x = 1$

$Cr^+ = [Ar]3d^5 4s^0$ = | ↑ | ↑ | ↑ | ↑ | ↑ |

But CN^- and NO being strong field ligands pair up the unpaired electrons of Cr^+.

$[Cr(NO)(NH_3)(CN)_4]^{2-}$ = | ⇅ | ⇅ | ↑ | | | (One unpaired electron)

$\therefore \qquad n = 1$

Magnetic moment, $\mu = \sqrt{n(n+2)} = \sqrt{1(1+2)} = \sqrt{3} = 1.73$ BM

26 Δ_o depends on the strength of negative field ligand. Spectrochemically, it has been found that the strength of splitting is as follows :

$CO > \underline{CN^-} > \underline{NO_2^-} > en > \underline{NH_3} > py > NCS^- > \underline{H_2O} > \underline{O^{2-}} >$

$\underline{ox^{2-}} > OH^- > F^- > Cl^- > SCN^- > S^{2-} > Br^- > I^-$

As H_2O is a weak field ligand among the other given ligands, so it will have Δ_o value least.

27 Δ_o magnitude depends upon the charge of central metal atom and strength of ligand. Among the given, CN^- is the strongest field ligand. So, complex $[Co(CN)_6]^{3-}$ will exhibit the maximum value of Δ_o.

28 Based on spectrochemical series, ligands arranged in increasing order of crystal field strength are as follows:

$$NH_3 < en < CN^- < CO$$

29 The CFSE of the ligands in the order $H_2O < NH_3 < CN^-$.

$$\text{As CFSE} \propto \frac{1}{\lambda}$$

Therefore, wavelengths absorbed will be in the opposite order as $[Co(H_2O)_6]^{3+} > [Co(NH_3)_6]^{3+} > [Co(CN)_6]^{3-}$.

30 $Zn_2[Fe(CN)_6]$, $K_3[Co(NO_2)_6]$ and $[(NH_4)_3As(Mo_3O_{10})_4]$ show colour due to d-d transition while $BaCrO_4$ is coloured due to charge transfer phenomenon.
Further, according to spectro-chemical series the strong ligand possessing complex has higher energy and hence lower wavelength. Therefore, complexes containing NO_2, NH_4^+, O^{2-} etc., ligands show yellow colour while CN^- forces the complex to impart white colour.

31 Ligand field strength $\propto$ Energy of light absorbed

$$\propto \frac{1}{\text{Wavelength of light absorbed}}$$

λ	L_1	L_2	L_3	L_4
Absorbed light	Red	Green	Yellow	Blue

Wavelength of absorbed light decreases.

$\therefore$ Lesser is the wavelength of light absorbed, greater the extent of crystal-field splitting, hence higher is the field strength of the ligand.

The increasing order of wavelength is blue < yellow < green < red. There increasing order of ligand strength $L_1 < L_2 < L_3 < L_4$.

32 Atomic number of Fe = 26

Atomic number of next noble gas = 36

$\therefore$ Number of electrons to be provided by ligands = 36 − 26 = 10

$\because$ Each CO group provide 2 electrons,

$\therefore$ Number of CO groups attached to Fe = 5

33 The organometallic compounds having π-bond between carbon and metal are known as π-bonded organometallic compound. $(CH_3)_4Sn$ is not a π-bonded organometallic compound.

34 An organometallic compound is one, which contains atleast one metal-carbon bond. In $B(OCH_3)_3$ and $CH_3B(OCH_3)_2$ there is no carbon atom that linked directly to a metal. Thus, $C_2H_5—Zn—C_2H_5$ is an organometallic compound.

35 Ferrocene is a sandwich complex compound in which all the five carbon atoms of ferropentadiene anions are linked directly to the metal Fe (π-bonds).

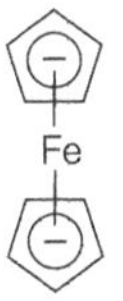

Ferrocene, Fe (η^5–C_5H_5)

36 $\underset{(X)}{CuSO_4 \cdot 5H_2O} + BaCl_2 \longrightarrow \underset{\text{Insoluble in HCl}}{BaSO_4} + CuCl_2$

$$2CuSO_4 + KI\ (\text{excess}) \longrightarrow \underset{\text{Brown ppt}}{Cu_2I_2} + 2K_2SO_4 + I_2$$

$$I_2 + 2Na_2S_2O_3 \longrightarrow 2NaI + Na_2S_4O_6$$

$$2CuSO_4 + K_4[Fe(CN)_6] \longrightarrow \underset{\text{Chocolate ppt.}}{Cu_2[Fe(CN)_6]} + 2K_2SO_4$$

37 (a) $\underset{\text{Weak base}}{Li_2O} + 2KCl \longrightarrow 2LiCl + \underset{\text{Strong base}}{K_2O}$

This is wrong equation, since a stronger base K_2O cannot be generated by a weaker base Li_2O.

(b) $[CoCl(NH_3)_5]^+ + 5H^+ \longrightarrow Co^{2+}(aq) + 5NH_4^+ + Cl^-$

This is correct. All amine complexes can be destroyed by adding $H^{\oplus}$. Hence, on adding acid to $[CoCl(NH_3)_5]$. It gets converted to $Co^{2+}(aq)$, NH_4^+ and Cl^-.

(c) $[Mg(H_2O)_6]^{2+} + EDTA^{4-} \xrightarrow[\text{excess}]{OH^-} [Mg(EDTA)]^{2+} + 6H_2O$

This is wrong equation, since the formula of complex must be $[Mg(EDTA)]^{2-}$.

(d) The 4th reaction is incorrect. It can be correctly represented as:

$$2CuSO_4 + 10KCN \longrightarrow 2K_3[Cu(CN)_4] + 2K_2SO_4 + (CN)_2\uparrow$$

38 $Fe^{3+} + SCN^- \longrightarrow \underset{\text{Red colour, not blood red colour}}{[Fe(SCN)]^{2+}}$

All other given statements are correct.

39 **Correct Explanation** Cr in the given complex $[Cr(H_2O)_6]Cl_2$ is present as Cr^{2+} which undergo oxidation to Cr^{3+} and Fe in the complex $[Fe(H_2O)_6]Cl_2$ is present as Fe^{2+} which can undergo oxidation to Fe^{3+} (H_2O is a weak ligand in both).

40 In $[Fe(H_2O)_5NO]SO_4$; Let oxidation state of Fe be x.

$$x + 5 \times (0) + 1 + (-2) = 0,\ x = +1,\ Fe^+ = [Ar]\,3d^6 4s^1$$

As NO^+ is a strong ligand causes pairing of 4s electron. Thus, the configuration is $3d^7$ and number of unpaired electrons = 3.

41 Both *cis* and *trans* forms of complexes $[M(NH_3)_4Cl_2]$ are optically inactive. There are plane of symmetry in addition to that there is alternate axis of symmetry which makes them opticaly inactive.

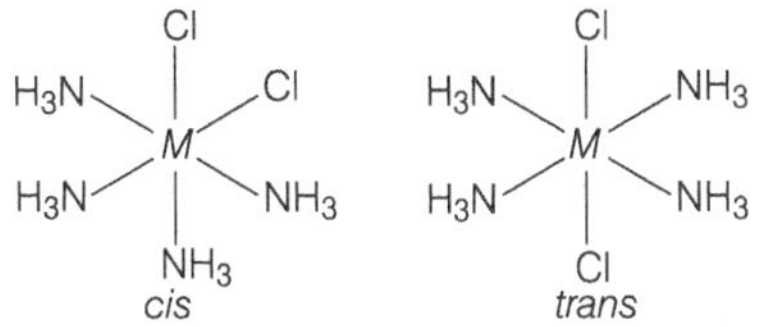

cis *trans*

SESSION 2

1 $C_2O_4^{2-}$ is bidentate ligand that forms the chelate and hence $[Fe(C_2O_4)_3]^{3-}$ is the most stable complex.

2 Δ_o value of $[CoCl_6]^{4-} = 18000\ cm^{-1}$

$$\Delta_t = \frac{4}{9} \times \Delta_o = \frac{4}{9} \times 18000 = 8000\ cm^{-1}$$

3 $\Delta_o = -0.4 \times n_{t_{2g}} + 0.6 \times n_{e_g}$ [where, n = number of electrons]

(a) For high spin d^6, Δ_o (CFSE) $= -0.4 \times 4 + 0.6 \times 2 = -0.4$

(b) For low spin d^4, Δ_o (CFSE) $= -0.4 \times 4 + 0.6 \times 0 = -1.6$

(c) For low spin d^5, Δ_o (CFSE) $= -0.4 \times 5 + 0.6 \times 0 = -2.0$

(d) For high spin d^7, Δ_o (CFSE) $= -0.4 \times 5 + 0.6 \times 2 = -0.8$

The calculations reveal that maximum CFSE is for low spin d^5 configuration.

4 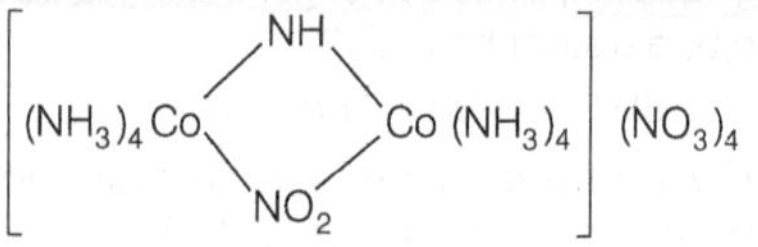

$\because$ $(4 \times 0) + x + (1 \times -1) + (1 \times -1) \times x + (4 \times 0) + (4 \times -1) = 0$

$\therefore$ $x = +3$

5 The IUPAC name of the compound is

chlorotriphenylphosphine palladium (II)-μ- dichloro chlorotriphenylphosphinepalladium (II).

6 If the metal in the metal carbonyl has more number of electrons then back donation will be more, more will the bond order of M—C bond and less is the bond order of C—O bond.

The electrons present in the metals are as follows:

In $[Mn(CO)_6]^+$, $Mn^+ = 3d^5 4s^1$

In $[Fe(CO)_5]^+$, $Fe^+ = 3d^6 4s^1$

In $[Cr(CO)_6]^+$, $Cr = 3d^5 4s^0$

In $[V(CO)_6]^-$, $V^- = 3d^4 4s^0$

As Fe in $[Fe(CO)_5]^+$ has maximum d-electrons therefore bond order of C—O will be least.

7 $[Cr(C_2O_4)_3]^{3+}$ complex is of type $[M(AA)_3]$ shows optical isomerism because its mirror image is not superimposable and do not possess plane of symmetry.

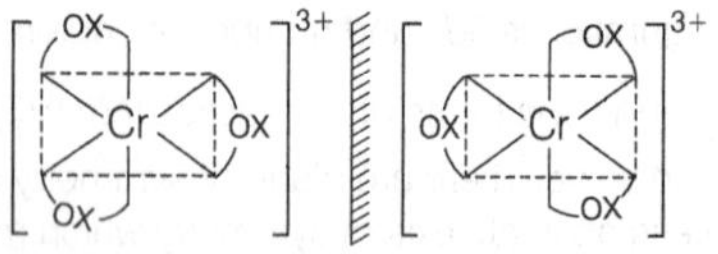

8 $[Pt(en)_2Cl\,Br]^{2+}$ will exist both as diastereoimer and enantiomer.

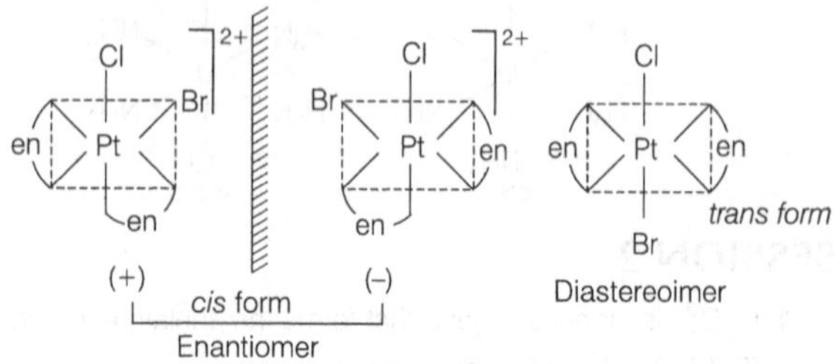

9 The absorption of light in a complex depends upon the charge of the metal ion and on the nature of the ligands. As the complexes given in the options have same charge on the metal ion (Ni^{2+}), therefore absorption is only dependent on the nature of ligand. Stronger field ligand absorbs maximum light, i.e. of minimum wavelength as $\Delta E \propto \frac{1}{\lambda}$.

Since, the Increasing order of nature of ligand given in the options is $H_2O < NH_3 < NO_2^-$.
Therefore, the correct order of wavelength of absorption will be $[Ni(NO_2)_6]^{4-} < [Ni(NH_3)_6]^{2+} < [Ni(H_2O)_6]^{2+}$.

10 $\Delta T_f = K_f \times m = 1.86 \times 0.1 = 0.186\,°C$
Observed value of $\Delta T_f = 0.372\,°C$
As observed value of ΔT_f is double the theoretical value, this shows that each molecule of the complex dissociates to form two ions. Hence, formula is $[Co(NH_3)_4Cl_2]Cl$.

11 Ammonia forms complex ion $[Cu(NH_3)_4]^{2+}$ with Cu^{2+} ions only in the alkaline medium as in acidic medium H^+ coordinates with ammonia molecule forming NH_4^+ ions and the electron pair present on NH_4^+ will not be available for donation.

12 $Zn^{2+} + 4NH_3 \rightleftharpoons [Zn(NH_3)_4]^{2+}$,

$$K_f = \frac{[Zn(NH_3)_4]^{2+}}{[Zn^{2+}]\,[NH_3]^4} = 3 \times 10^9$$

$$\therefore \frac{[Zn(NH_3)_4]^{2+}}{[Zn^{2+}]} = [NH_3]^4 \times 3 \times 10^9 = [10]^4 \times 3 \times 10^9 = 3 \times 10^{13}$$

$$\therefore \frac{[Zn^{2+}]}{[Zn(NH_3)_4]^{2+}} = 3.3 \times 10^{-14}$$

13 $Co(NH_3)_5Cl_3$ is an octahedral complex ionising in aqueous solution as :

$$[Co\,(NH_3)_5Cl]\,Cl_2 \longrightarrow [Co\,(NH_3)_5Cl]^{2+} + 2Cl^-$$

Moles of $[Co\,(NH_3)_5Cl]\,Cl_2 = \frac{0.1}{1000} \times 100 = 0.01$ mol

Moles of AgCl formed = 2 × moles of Cl^- in complex
= 0.01 × 2 = 0.02 mol

14 Fe^{2+} = $3d^6$ [⇅ | ↑ | ↑ | ↑ | ↑]

NO ⟶ $NO^+ + e^-$ (Electron transfer from NO.)
$Fe^{2+} + e^- \longrightarrow Fe^+$
] Charge transfer

(c) Fe^+ = $3d^7$ [⇅ | ⇅ | ↑ | ↑ | ↑]

$\mu = \sqrt{n(n+2)} = \sqrt{3(3+2)} = 3.87$ BM

$\therefore$ Thus, all the given options are correct.

DAY TWENTY FIVE

Unit Test 4

(Inorganic Chemistry)

1 Consider the following reactions.

$$Ag_2S + NaCN \longrightarrow (A)$$
$$(A) + Zn \longrightarrow (B)$$

(*B*) is a metal. Hence, (*A*) and (*B*) are

(a) $Na_2[Zn(CN)_4]$, Zn (b) $Na[Ag(CN)_2]$, Ag
(c) $Na_2[Ag(CN)_4]$, Ag (d) $Na_3[Ag(CN)_4]$, Ag

2 The process of isolation of metals by dissolving the ore in a suitable chemical reagent followed by precipitation of the metal by a more electropositive metal is called → NCERT Exemplar

(a) hydrometallurgy (b) zone refining
(c) electrorefining (d) electrometallurgy

3 When a metal is to be extracted from its ore and the gangue associated with the ore is silica, then

(a) a basic flux is needed
(b) an acidic flux is needed
(c) both basic and acidic flux are needed
(d) neither of them is needed

4 The final step for the extraction of copper from copper pyrite in Bessemers converter involves the reaction

(a) $Cu_2S + 2Cu_2O \longrightarrow 6Cu + SO_2$
(b) $4Cu_2O + FeS \longrightarrow 8Cu + FeSO_4$
(c) $2Cu_2O + FeS \longrightarrow 4Cu + Fe + SO_2$
(d) $Cu_2S + 2FeO \longrightarrow 2Cu + 2Fe + SO_2$

5 van-Arkel method of purification of metals involves converting the metal to a

(a) volatile stable compound
(b) non-volatile stable compound
(c) volatile unstable compound
(d) None of the above

6 Wolframite ore is separated from tin stone ore by the process of

(a) calcination (b) electromagnetic process
(c) roasting (d) smelting

7 Slow acting nitrogenous fertilizer among the following is

(a) $CaCN_2$ (b) NH_2CONH_2
(c) KNO_3 (d) NH_4NO_3

8 Which of the following is correct for hydrogen?

(a) It has same electronegativity as halogen
(b) It has a very high ionisation potential
(c) It is always collected at cathode
(d) It can form bonds in +1 as well as −1 oxidation state

9 Which one of the following equation depicts reducing nature of H_2O_2? → NCERT Exemplar

(a) $2[Fe(CN)_6]^{4-} + 2H^+ + H_2O_2 \longrightarrow 2Fe(CN)_6]^{3-} + 2H_2O$
(b) $I_2 + H_2O_2 + 2OH^- \longrightarrow 2I^- + 2H_2O + O_2$
(c) $Mn^{2+} + H_2O_2 \longrightarrow Mn^{4+} + 2OH^-$
(d) $PbS + 4H_2O_2 \longrightarrow PbSO_4 + 4H_2O$

10 Setting of plaster of Paris is

(a) dehydration
(b) oxidation with atmospheric oxygen
(c) combination with atmospheric CO_2
(d) hydration to yield another hydrate

11 The most stable compound is → NCERT Exemplar

(a) LiF (b) LiCl
(c) LiI (d) LiBr

12 In which of the following processes fused sodium hydroxide is electrolysed at 330°C temperature for extraction of sodium?

(a) Castner's process (b) Cyanide process
(c) Down's process (d) Both (b) and (c)

13 Match the items of Column I with the items of Column II and assign the correct code.

	Column I		Column II
A.	Cyanide process	1.	Ultrapure Ge
B.	Froth floatation process	2.	Dressing of ZnS
C.	Electrolytic reduction	3.	Extraction of Al
D.	Zone refining	4.	Extraction of Au
		5.	Purification of Ni

Codes

	A	B	C	D		A	B	C	D
(a)	4	2	3	1	(b)	2	3	1	5
(c)	1	2	3	4	(d)	3	4	5	1

14 Which of the following is the poorest reducing agent?

(a) Atomic hydrogen
(b) Nascent hydrogen
(c) Dihydrogen
(d) All have same reducing strength

15 **Assertion** (A) The highest oxidation state of osmium is + 8.

Reason (R) Osmium is a 5*d*-block element.

→ NCERT Exemplar

(a) Both A and R are correct and R is correct explanation of A
(b) Both A and R are correct but R is not correct explanation of A
(c) A is correct but R is incorrect
(d) Both A and R are incorrect

16 In the metallurgy of aluminium → NCERT Exemplar

(a) Al^{3+} is oxidised to Al (*s*)
(b) graphite anode is oxidised to carbon monoxide and carbon dioxide
(c) oxidation state of oxygen changes in the reaction at anode
(d) oxidation state of oxygen changes in the overall reaction involved in the process

17 Action of conc. HNO_3 on metallic tin produces

(a) stannic nitrate (b) stannous nitrite
(c) metastannic acid (d) stannous nitrate

18 NH_3 on reaction with hypochlorite anion, can form

(a) NO (b) N_2H_4 (c) NH_4Cl (d) HNO_2

19 The pentavalency in phosphorus is more stable as compared to that of nitrogen even though they belong to the same group. It is due to

(a) inert nature of nitrogen (b) reactivity of phosphorus
(c) larger size of phosphorus atom
(d) dissimilar electronic configuration

20 Oxygen is not evolved on reaction of ozone with

(a) H_2O_2 (b) Hg (c) SO_2 (d) KI

21 Conc. HNO_3 reacts with I_2 to give

(a) HI (b) HOI (c) HIO_3 (d) HIO_4

22 Alkali metals react with water vigorously to form hydroxides and dihydrogen. Which of the following alkali metals reacts with water least vigorously? → NCERT Exemplar

(a) Li (b) Na (c) K (d) Cs

23 Which one of the following oxides is ionic?

(a) MnO (b) CrO_3 (c) P_2O_5 (d) Mn_2O_7

24 What is the magnetic moment of $[FeF_6]^{3-}$?

(a) 4 BM (b) 5.49 BM (c) 2.32 BM (d) 5.92 BM

25 Spiegeleisen is an alloy of

(a) Fe, Co and Cr (b) Fe, Co and Mg
(c) Fe, Mg and C (d) Fe, C and Mn

26 The number of moles of $KMnO_4$ that will be needed to react completely with one mole of ferrous oxalate, $Fe(C_2O_4)$ in acidic solution is

(a) 1 (b) 2/5 (c) 3/5 (d) 4/5

27 Match the following and choose the correct option.

	Column I (Complex Ion)		Column II (Hybridisation, number of unpaired electrons)
A.	$[Cr(H_2O)_6]^{3+}$	1.	dsp^2, 1
B.	$[Co(CN)_4]^{2-}$	2.	sp^3d^2, 5
C.	$[Ni(NH_3)_6]^{2+}$	3.	d^2sp^3, 3
D.	$[MnF_6]^{4-}$	4.	sp^3d^2, 2

Codes

	A	B	C	D
(a)	3	1	4	2
(b)	4	1	2	3
(c)	3	2	4	1
(d)	4	1	2	3

28 The electronic configuration of Cu (II) is $3d^9$ while that of Cu (I) is $3d^{10}$. Which of the following statement is correct?

→ NCERT Exemplar

(a) Cu (II) is more stable
(b) Cu (II) is less stable
(c) Cu (I) and Cu (II) are equally stable
(d) Stability of Cu (I) and Cu (II) depends on the nature of Cu salts

29 Which of the following is not true for ligand metal complex?

(a) Highly charged ligand forms strong bonds
(b) Greater the ionisation potential of central metal, the stronger is the bond
(c) Larger the permanent dipole moment of ligand, the more stable is the bond
(d) Larger the ligand, the more stable is the metal ligand bond

30 The number of ions formed, when cuprammonium sulphate is dissolved in water, is

(a) zero (b) 1 (c) 2 (d) 4

31 Given molecular formula of the hexa coordinated complexes (*A*) $CoCl_3 \cdot 6NH_3$ (*B*) $CoCl_3 \cdot 5NH_3$ (*C*) $CoCl_3 \cdot 4NH_3$. If the number of coordinated NH_3 molecules in *A*, *B* and *C* respectively are 6, 5 and 4, primary valency in *A*, *B* and *C* are

(a) 0, 1, 2 (b) 3, 2, 1 (c) 6, 5, 4 (d) 3, 3, 3

32 1 mole each of H_3PO_2, H_3PO_3 and H_3PO_4 will neutralise *x* moles of NaOH, *y* moles of $Ca(OH)_2$ and *z* moles of $Al(OH)_3$ (assuming all as strong electrolytes). *x*, *y*, *z* are in the ratio of

(a) 3 : 1.5 : 1 (b) 1 : 2 : 3
(c) 3 : 2 : 1 (d) 1 : 1 : 1

33 Finely divided iron combines with CO to give

(a) $Fe(CO)_5$ (b) $Fe_2(CO)_6$
(c) $Fe(CO)_{12}$ (d) $Fe_2(CO)_9$

34 The IUPAC name of $K_2[Cr(CN)_2O_2(O)_2(NH_3)]$ is

(a) potassiumamminedicyanodioxoperoxo chromate (VI)
(b) potassiumamminecyanoperoxodioxo chromate (IV)
(c) potassiumamminecyanoperoxodioxo chromium (VI)
(d) potassiumamminecyanoperoxodioxo chromium (VI)

35 NH_3 cannot be obtained by

(a) heating of NH_4NO_3 or NH_4NO_2
(b) heating of NH_4Cl or $(NH_4)_2CO_3$
(c) heating of NH_4NO_3 with NaOH
(d) reaction of AlN or Mg_3N_2 or $CaCN_2$ with H_2O

36 What mass of CaO will be required to remove the hardness of 1000 L of water containing 1.62 g of calcium bicarbonate per litre?

(a) 0.56 g (b) 560 g (c) 162 g (d) 56 g

37 In which of the following arrangements the order is not according to the property indicated against it?

(a) Li < Mg < Ca (increasing reactivity with N_2)
(b) Be < Mg < Ca< Sr (increasing basic nature)
(c) BeO < MgO < CaO (increasing refractory properties)
(d) $BBr_3 > BCl_3 > BF_3$ (decreasing acidic nature)

38 The optically active compound of boron is

(a) borosalicylic acid (b) borax
(c) borazole (d) boron nitride

39 A black compound (*A*) in solid state is fused with KOH and $KClO_3$. The aqueous extract is green colour solution (*B*). On passing CO_2 gas through it, pink colour of (*C*) is noticed along with some black insoluble mass of *A*. The pink coloured solution is decolourised by Fe^{2+} in acidic medium. Identify *A*.

(a) MnO_2 (b) Fe_2O_3 (c) PbS (d) $ZnSO_4$

40 The dissociation of a complex may be expressed as $[ML_x]^{y+} \rightleftharpoons M^{y+} + xL$ and equilibrium constant of this dissociation is known as instability constant, which is a measure of stability. Hence, identify which of the following complexes is most stable?

(a) $[Cu(CN)_2]^-$, $K_s = 1 \times 10^{-16}$
(b) $[Fe(CN)_6]^{4-}$, $K_s = 1 \times 10^{-37}$
(c) $[Fe(CN)_6]^{3-}$, $K_s = 1 \times 10^{-44}$
(d) $[Ag(CN)_2]^-$, $K_s = 1 \times 10^{-20}$

41 Cu^{2+} has a stronger polarising power than that of Ca^{2+} because

(a) Cu^{2+} ion is smaller than Ca^{2+} ion
(b) Ca^{2+} ion has inert gas configuration whereas Cu^{2+} ion does not
(c) copper shows variable valency, calcium does not
(d) Cu^{2+} ion is smaller than Ca^{2+} ion and the *d*-electrons in Cu^{2+} ion shield the nucleus poorly

42 Magnetic moment of the complexes is zero in

(a) $[Ni(CN)_4]^{2-}$, $[Ni(CO)_4]$
(b) $[Ni(H_2O)_4]^{2+}$, $[Ni(CN)_4]^{2-}$
(c) $[Ni(H_2O)_4]^{2+}$, $[Ni(CO)_4]$
(d) $[Fe(CN)_6]^{4-}$, $[Fe(CN)_6]^{3-}$

43 Which complex gives three chloride ions per formula unit?

(a) $CrCl_3 \cdot 6H_2O$ (b) $CrCl_3 \cdot 5H_2O$
(c) $CrCl_3 \cdot 4H_2O$ (d) All of these

44 Alkaline $KMnO_4$ (Baeyer's reagent) can be used to test unsaturation in (*A*).

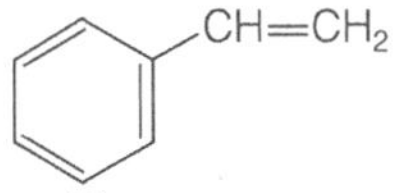

(*A*)

In this case

(a) unsaturation in side-chain is affected
(b) unsaturation in benzene nucleus is affected
(c) unsaturation in both is affected
(d) Baeyer's reagent cannot be used

45 The CFSE for octahedral $[CoCl_6]^{4-}$ is 18000 cm^{-1}. The CFSE for tetrahedral $[CoCl_4]^{2-}$ will be ➔ **NCERT Exemplar**

(a) 18000 cm^{-1} (b) 16000 cm^{-1}
(c) 8000 cm^{-1} (d) 20000 cm^{-1}

46 One mole of the complex $CoCl_3 \cdot 6H_2O$ on reaction with excess of $AgNO_3$ gives two moles of white precipitate. Thus, complex is

(a) $[Co(H_2O)_6]Cl_3$ (b) $[Co(H_2O)_5Cl]Cl_2 \cdot H_2O$
(c) $[Co(H_2O)_4Cl_2]Cl \cdot 2H_2O$ (d) $[Co(H_2O)_3Cl_3] \cdot 3H_2O$

Direction (Q. Nos. 47-50) *In the following questions Assertion* (A) *followed by Reason* (R) *is given. Choose the correct option out the following choices.*

(a) Both (A) and (R) are correct and (R) is correct explanation of (A)
(b) Both (A) and (R) are correct but (R) is not correct
(c) (A) is correct but (R) is incorrect
(d) Both (A) is and (R) are incorrect

47 **Assertion** (A) $E°$ for Mn^{3+}/Mn^{2+} is more positive than Cr^{3+}/Cr^{2+}.

Reason (R) The third ionisation energy of Mn is larger than that of Cr.

48 **Assertion** (A) $K_2Cr_2O_7$ is used as a primary standard in volumetric analysis.

Reason (R) $K_2Cr_2O_7$ has a good solubility in water.

49 **Assertion** (A) Silicones are hydrophobic in nature.

Reason (R) Si — O — Si linkages are moisture sensitive.

50 **Assertion** (A) Potassium ferrocyanide and potassium ferricyanide both are diamagnetic.

Reason (R) The former does not have any unpaired electron.

ANSWERS

1 (b)	2 (a)	3 (a)	4 (a)	5 (a)	6 (b)	7 (a)	8 (d)	9 (b)	10 (d)
11 (a)	12 (a)	13 (a)	14 (c)	15 (b)	16 (b)	17 (c)	18 (b)	19 (c)	20 (c)
21 (c)	22 (a)	23 (a)	24 (d)	25 (d)	26 (c)	27 (a)	28 (a)	29 (d)	30 (c)
31 (b)	32 (d)	33 (a)	34 (a)	35 (a)	36 (b)	37 (c)	38 (a)	39 (a)	40 (c)
41 (a)	42 (a)	43 (a)	44 (a)	45 (c)	46 (b)	47 (b)	48 (d)	49 (c)	50 (d)

Hints and Explanations

1 $Ag_2S + NaCN \longrightarrow \underset{(A)}{Na[Ag(CN)_2]}$

$\underset{(A)}{Na[Ag(CN)_2]} + Zn \longrightarrow \underset{(B)}{Ag}$

Compound (*A*) and (*B*) are $Na[Ag(CN)_2]$ and Ag.

2 Hydrometallurgy is the technique used for the isolation of metals by dissolving the ore in a suitable reagent followed by the precipitation of the metal.

3 To extract a metal from its ore when a gangue associated with are is silica then a basic flux is needed.

$$\underset{\text{Acidic impurity}}{SiO_2} + \underset{\text{Basic flux}}{CaO} \longrightarrow \underset{\text{Slag}}{CaSiO_3}$$

4 In Bessemer's converter, copper sulphide is partially oxidised to cuprous oxide which further reacts with remaining copper sulphide to form copper and SO_2.

$$Cu_2S + 2Cu_2O \longrightarrow 6Cu + SO_2$$

5 $Ti + 2I_2 \xrightarrow{500\,K} \underset{\text{Volatile stable compound}}{TiI_4} \xrightarrow{1700\,K} \underset{\text{Pure metal}}{Ti} + 2I_2$

6 Wolframite ore $[FeWO_4]$ is present in tin stone as impurity and it has same mass per unit volume as that of tin stone, so it is separated by electromagnetic separation because wolframite is magnetic in nature, hence it gets attracted by magnet while tin stone does not.

7 CaNCN decomposes very slowly.

$$CaNCN + 2H_2O \longrightarrow CaCO_3 + NH_2CONH_2$$

$$NH_2CONH_2 + H_2O \longrightarrow CO_2 + 2NH_3$$

$$NH_3 \xrightarrow[\text{bacteria}]{\text{Nitrifying}} \text{Soluble nitrates} \longrightarrow \text{Plants}$$

8 Hydrogen forms bond in +1 and −1 oxidation states.

9 H_2O_2 acts as an oxidising as well as reducing agent in alkaline media. Following reaction shows the reducing action in basic medium.

$$I_2 + H_2O_2 + 2OH^- \longrightarrow 2I^- + 2H_2O + O_2$$

10 Setting of plaster of Paris is hydration process to yield another hydrate. It is an exothermic process.

$$CaSO_4 \cdot \frac{1}{2}H_2O \xrightarrow{H_2O} \underset{\text{Orthorhombic}}{CaSO_4 \cdot 2H_2O} \xrightarrow{\text{Hardening}} \underset{\text{Monoclinic (gypsum)}}{CaSO_4 \cdot 2H_2O}$$

11 Due to the small size of Li and F, LiF has highest lattice enthalpy and hence, most stable compound.

12 Castner's process is used to obtain Na by electrolysis of fused sodium hydroxide.

13 A→4, B→2, C→3, D→1

14 Dihydrogen is the poorest reducing agent as it is least reactive.

15 **Correct Explanation** Osmium has the electronic configuration $5d^6 6s^2$. As 5*d* and 6*s* are close in energy, all the 8 electrons can participate in bonding.

16 In the metallurgy of aluminium, the graphite anode is oxidised to CO and CO_2 by the released O_2.

17 $Sn + \text{Conc. } 4HNO_3 \longrightarrow \underset{\text{Metastannic acid}}{H_2SnO_3} + 4NO_2$

18 $\underset{\text{(Excess)}}{2NH_3} + OCl^- \longrightarrow NH_2{-}NH_2 + NH_4Cl + OH^-$

19 Pentavalency in P is more stable than N due to larger size of phosphorus atom.

20 $3SO_2 + O_3 \longrightarrow 3SO_3$

Whole oxygen of ozone is used up for oxidation.

21 $I_2 + 10HNO_3 \longrightarrow 2HIO_3 + 10NO_2 + 4H_2O$

22 Li has most negative standard reduction potential due to very high enthalpy of hydration, Thus, reaction of Li with water will be most exothermic, but surprisingly Li reacts with water gently, whereas Na and K vigorously.

The explanation is in kinetics and not in thermodynamics of the reaction. No doubt, maximum energy is evolved with Li but its fusion, vaporisation and ionisation consume more energy. As a result reaction proceeds slowly.

Na or K have low melting points and molten metal spreads over water exposing a large surface to water, making the reaction more vigorous.

23 MnO is ionic due to lower oxidation state of Mn i.e. +2.

24 Fe^{3+} ion has five unpaired electrons.

$$\mu = \sqrt{5(5+2)}\text{ BM} = \sqrt{35} = 5.916\text{ BM}$$

25 Spiegeleisen is an alloy of Fe, C and Mn.

26 $3MnO_4^- + 5[Fe^{2+}\ C_2O_4^{2-}] + 24H^+ \longrightarrow 3Mn^{2+} + 5Fe^{3+} + 10CO_2 + 12H_2O$

Thus, 5 moles of FeC_2O_4 are oxidised by 3 moles of $KMnO_4$, therefore 1 mole of FeC_2O_4 is oxidised by 3/5 mole of $KMnO_4$.

27 A → 3, B → 1, C → 4, D → 2

28 Though Cu(I) possess $3d^{10}$ electronic configuration while that of Cu(II) has $3d^9$ configuration yet Cu (II) is more stable than Cu (I) due to greater effective nuclear charge of Cu (II).

29 Large ligand does not destabilise the metal-ligand bond. Higher the charge and smaller the size of ligand, more stable is the complex formed.

30 Cuprammonium salt is $[Cu(NH_3)_4]SO_4$.

$$[Cu(NH_3)_4]SO_4 \rightleftharpoons [Cu(NH_3)_4]^{2+} + SO_4^{2-}$$

31 The complexes can be written as follows:

$[Co(NH_3)_6]Cl_3$, $[Co(NH_3)_5Cl]Cl_2$, $[Co(NH_3)_4Cl_2]Cl$

So, the primary valency in *A*, *B* and *C* respectively are 3, 2, 1.

32. H_3PO_2 is monobasic acid.

$$\underset{1\text{ mol}}{H_3PO_2} + \underset{1\text{ mol}}{NaOH} \longrightarrow NaH_2PO_2 + H_2O$$

$\therefore \quad x = 1$

H_3PO_3 is dibasic acid

$$\underset{1\text{ mol}}{H_3PO_3} + \underset{1\text{ mol}}{Ca(OH)_2} \longrightarrow CaHPO_3 + 2H_2O$$

$y = 1$

H_3PO_4 is tribasic acid

$$\underset{1\text{ mol}}{H_3PO_4} + \underset{1\text{ mol}}{Al(OH)_3} \longrightarrow AlPO_4 + 3H_2O$$

$z = 1$

Thus, $x : y : z :: 1 : 1 : 1$

33. $Fe + 5CO \xrightarrow[\text{pressure}]{\text{Heat}} \underset{\text{Iron penta carbonyl}}{[Fe(CO)_5]}$

34. IUPAC name is potassiumamminedicyano dioxoperoxo chromate (VI).

35. (a) $NH_4NO_2 \xrightarrow{\Delta} N_2O + 2H_2O$

$NH_4NO_2 \xrightarrow{\Delta} N_2 + 2H_2O$

(b) $NH_4Cl \xrightarrow{\Delta} NH_3 + HCl$

(c) $(NH_4)_2CO_3 \xrightarrow{\Delta} 2NH_3 + CO_2 + H_2O$

(d) $NH_4NO_3 + NaOH \xrightarrow{\Delta} NH_3 + NaNO_3 + H_2O$

(e) $\left.\begin{matrix} AlN \\ Mg_3N_2 \\ CaCN_2 \end{matrix}\right] \xrightarrow{H_2O} NH_3$

36. $CaO + Ca(HCO_3)_2 \longrightarrow 2CaCO_3 + H_2O$

For one litre water,

Meq. of CaO = Meq. of $Ca(HCO_3)_2$

$$\frac{w \times 1000}{56/2} = \frac{1.62 \times 1000}{162/2}$$

$\therefore \quad w_{CaO} = 0.56$ g

Thus, CaO required for 10^3 L H_2O

$= 0.56 \times 10^3 = 560$ g

37. Refractory property of alkaline earth metal oxides decreases on moving down the group. Thus, the correct order is BeO > MgO > CaO.

38. Borosalicylic acid is an optically active compound of boron.

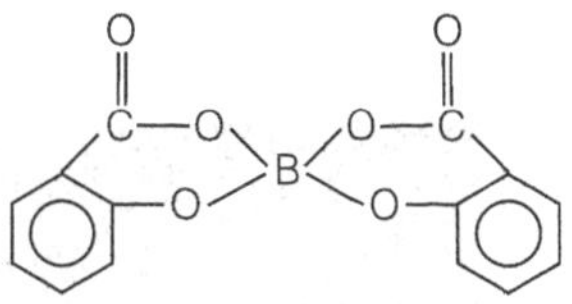

39. $\underset{(A)\ (\text{Black})}{3MnO_2} + 6KOH + KClO_3 \xrightarrow{\Delta} \underset{(B)\ (\text{Green})}{3K_2MnO_4} + KCl + 3H_2O$

In the presence of CO_2, the medium becomes acidic.

$\underset{(B)\ (\text{Green})}{3MnO_4^{2-}} + 4H^+ \longrightarrow \underset{(A)}{MnO_2} + \underset{(C)}{2MnO_4^-} + 2H_2O$

$\underset{(C)\ (\text{Pink})}{MnO_4^-} + 5Fe^{2+} + 8H^+ \xrightarrow{\Delta} Mn^{2+} + 5Fe^{3+} + 4H_2O$

40. As dissociation constants of complexes are inversely proportional to their stability constants, their stability constants are given as:

$[Cu(CN)_2]^- \quad K_S = \frac{1}{K_d} = 1 \times 10^{16}$

$[Fe(CN)_6]^{4-} \quad K_S = 1 \times 10^{37}$

$[Fe(CN)_6]^{3-} \quad K_S = 1 \times 10^{44}$

$[Ag(CN)_2]^- \quad K_S = 1 \times 10^{20}$

41. It is due to screening effect based on Fajan's rule. It states that smaller the cationic size more is the polar using power ($Cu^{2+} < Ca^{2+}$).

42.

Complex	Hybridisation	Unpaired electrons	Effect of ligand
$[Ni(CN)_4]^{2-}$	dsp^2	zero	CN^- is a strong ligand
$[Ni(H_2O)_4]^{2+}$	sp^3	4	H_2O is a weak ligand
$[Ni(CO)_4]$	sp^3	zero	CO is a strong ligand
$[Fe(CN)_6]^{4-}$	d^2sp^3	zero	CN^- is a strong ligand
$[Fe(CN)_6]^{3-}$	d^2sp^3	1	CN^- is a strong ligand

43. $(Cr(H_2O)_6)Cl_3 \rightleftharpoons (Cr(H_2O)_6)^{3+} + 3Cl^-$

C.N. of Cr = 6 in only (a)

44. $KMnO_4$ is decolourised by unsaturated compounds in side chain.

45. CFSE for octahedral and tetrahedral complexes are closely related to each other by formula $\Delta_t = \frac{-4}{9}\Delta_o$.

where, Δ_o = CFSE for octahedral complex, Δ_t = CFSE for tetrahedral complex ($\Delta E = h\nu$, i.e. $\Delta E \propto \nu$)

According to question,

$\Delta_o = 18000\ cm^{-1}$

$\therefore \quad \Delta_t = \frac{4}{9}\Delta_o = \frac{4}{9} \times 18000\ cm^{-1}$

$= 4 \times 2000\ cm^{-1} = 8000\ cm^{-1}$

46. $\underset{1\ \text{mol}}{CoCl_3 \cdot 6H_2O} + AgNO_3 \rightarrow 2$ mol AgCl

Thus, two Cl atoms are outside the coordinate sphere.

C.N. of Co = 6

Thus, $[Co(H_2O)_5Cl]Cl_2 \cdot H_2O$.

47. Mn is present below Cr in electrochemical series. So, its reduction potential is less than that of Cr. The third ionisation energy for Mn is 3258 kJ mol^{-1} and for Cr is 2994 kJ mol^{-1}.

48. Potassium dichromate is used in volumetric analysis mostly in oxidation titration because it is not deliquescent. It forms compound with other elements and precipitate out. It is water soluble.

49. Silicones are insoluble in water and does not react with it. So, they are hydrophobic in nature. The crystal structure of silica has Si—O—Si linkage and it is extremely stable and considerable energy is required to break the silicon oxygen bonds in it. So, silicones are hard and have high melting point.

50. Potassium ferrocyanide is $K_4[Fe(CN)_6]$.

In, this complex, Fe is present as Fe^{2+}.

$Fe^{2+} = [Ar]\ 3d^6$

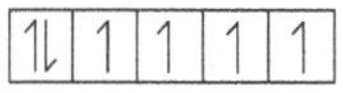

In $[Fe(CN)_6]^{4-}$

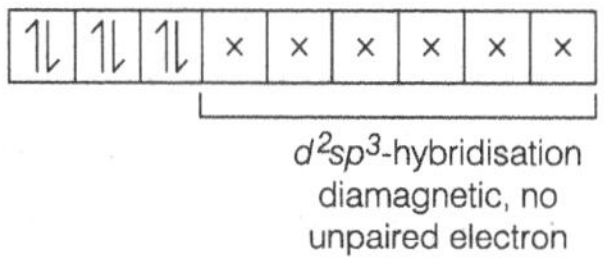

Potassium ferricyanide is $K_3[Fe(CN)_6]$.

Fe is present as $Fe^{3+} = [Ar]\ 3d^5$

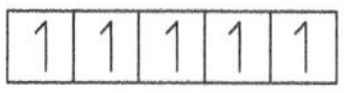

In $[Fe(CN)_6]^{3-}$

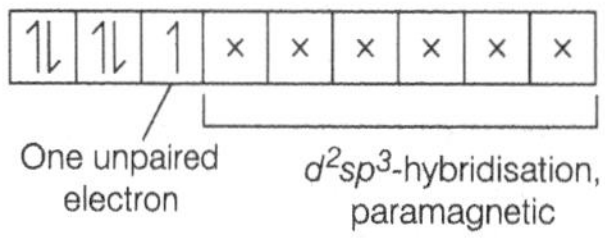

DAY TWENTY SIX

Environmental Chemistry

Learning & Revision for the Day

- Pollutants
- Pollution
- Atmospheric Pollution
- Water Pollution
- Soil and Land Pollution
- Strategies to Control Environmental Pollution

Environmental chemistry deals with the study of origin, transport, reactions, effects and fates of chemical species in the environment.

Pollutants

Any substance which pollutes the environment is known as **pollutant**. A substance becomes a pollutant when it is present in larger concentrations which is harmful to the natural environment. It can be solid, liquid or gaseous.

The pollutants are classified into following categories:

1. **Primary pollutants** are the pollutants persisting in the environment in the form they are produced, e.g. carbon monoxide.
2. **Secondary pollutants** are formed by the combination of primary pollutants present in the environment, e.g. two primary pollutants, nitrogen oxides and hydrocarbons, react together in the presence of sunlight to form secondary pollutant, peroxyacetyl nitrate (PAN).
3. **Biodegradable pollutants** are those pollutants which are decomposed by microorganisms either by itself or by suitable treatment, e.g. sewage, various oxides of nitrogen and sulphur etc.
4. **Non-biodegradable pollutants** are those pollutants which are not decomposed naturally and are not recycled. They are harmful to environment even in low concentrations, e.g. DDT, nuclear waste, lead components etc.

PREP MIRROR
Your Personal Preparation Indicator

- No. of Questions in Exercises (x)—
- No. of Questions Attempted (y)—
- No. of Correct Questions (z)—
  (Without referring Explanations)

- Accuracy Level ($z/y \times 100$)—
- Prep Level ($z/x \times 100$)—

In order to expect good rank in JEE, your Accuracy Level should be above 85 & Prep Level should be above 75.

Pollution

Contamination of the environment (i.e. our surroundings such as air, water, soil etc.) with harmful wastes arising mainly from certain human activities is called environmental pollution.

In the process of environmental pollution, pollutants originate from a source and get transported by air, water or dumped into the soil by human being.

- On the basis of pollutants, the pollution can be classified as:
  (i) Thermal pollution (ii) Noise pollution
  (iii) Chemical pollution (iv) Metal pollution
  (v) Smog pollution (vi) Oil pollution

On the basis of part of environment polluted, the pollution can be classified as:
(i) Atmospheric pollution (ii) Water pollution
(iii) Soil pollution

Atmospheric Pollution

Atmospheric pollution occurs when the concentration of a normal component of the air or a new chemical substance added or formed in air, build up to undesirable proportions causing harm to humans, animals, vegetation and materials.

Structure of Atmosphere

- The lowest region of atmosphere in which the human beings along with other organisms live is called the **troposphere**. It extends upto the height of ~10 km from sea level. Above troposphere, between 10 and 50 km above sea level, lies the **stratosphere**. Troposphere contains about 80% of the total mass of air and water vapours while stratosphere contains nitrogen, oxygen and ozone.
- **Mesosphere** extends 50-85 km from earth's surface. N_2 and O_2 are present in low concentration in this region.
- **Thermosphere** extends between 85-500 km from earth's surface and in it temperature rises to 1200°C.
- The outermost part of atmosphere is **exosphere** and unbounded area beyond exosphere is known as **inter-stellar space**.
- Mesosphere and thermosphere are collectively known as **ionosphere**.
- Sources of air pollution are as follows:
  (i) Burning of fossil fuels such as coal, wood and oil.
  (ii) Exhaust gases emitted by internal combustion engines of vehicles.
  (iii) Chemical industries and their released products.
- Atmospheric pollution is generally studied as tropospheric and stratospheric pollution.

Tropospheric Pollution

It is caused by gaseous pollutants and particulate matter.

Gaseous Air Pollutants

Some gaseous pollutants are as follows :

1. Oxides of Sulphur

Two main oxides of sulphur that act as pollutants are SO_2 and SO_3.

Sources

(i) Burning of sulphur containing fossil fuels, roasting and smelting of sulphide ore.
(ii) Particulate matter in the air oxidises SO_2 to SO_3.

Harmful Effects

- SO_2 in lower concentration irritates the upper respiratory tract, causes cough etc.
- SO_3 is more harmful than SO_2 because it combines with water to form H_2SO_4 and causes acid rain.

Prevention They can be removed by non-regenerative process

$$CaCO_3 \longrightarrow CaO + CO_2$$
$$CaO + SO_2 \longrightarrow CaSO_3$$
$$CaO + SO_3 \longrightarrow CaSO_4$$

2. Oxides of Nitrogen

Nitric oxide (NO) and nitrogen oxide (NO_2) act as pollutant.

Sources

N_2 and O_2 are the main constituents of air but are unreactive at normal temperature. They form oxides when lightening occurs at high altitudes.

$$N_2 + O_2 \xrightarrow[\text{or } 1210°C]{\text{Lightening}} 2NO, \quad 2NO + O_2 \xrightarrow[\text{or } 1100°C]{\text{Lightening}} 2NO_2$$

Harmful Effects

- These oxides can cause pulmonary odema, dilation of arteries, eye irritation, damage to liver and kidney.
- They also retard the rate of photosynthesis.

Prevention

NO_x is controlled by using catalytic converters in automobile exhausts which convert NO_x to free N_2 or to a small amount of NH_3.

3. Oxides of Carbon

The two main oxides of carbon that causes pollution are carbon monoxide (CO) and carbon dioxide (CO_2).

Sources

(i) CO is the product of incomplete combustion of a carbonaceous fuel. Petroleum fuel contribute about 60% of the total CO produced by human activities.
(ii) Main sources of CO_2 production are agricultural burning, metallurgical processes using coke as a reducing agent etc.

Harmful Effects

- CO enters the respiratory system along with O_2 and combines with haemoglobin to form carboxy-haemoglobin, which results in giddiness, headache, decreased vision etc.
- CO_2 is non-toxic upto a certain level, but causes climate changes like greenhouse effect, global warming etc.

Prevention

- Fuel substitution LPG and CNG in place of oil, electric or solar power in place of fossil fuels.
- Use of catalytic converter in automobile exhaust pipe.

4. Hydrocarbons

Hydrocarbons like methane (CH_4) mixed in air act as pollutants.

Sources

These are produced by an aerobic decay of organic matter (like vegetables, dead animal bodies etc), incomplete combustion of fuels and various chemical industries.

Harmful effect

These are carcinogenic and causes suffocation at higher concentration. They also harm plants by causing ageing, breakdown of tissues and shedding of leaves, flowers and twigs.

Prevention

- Adsorption of hydrocarbons on activated charcoal.
- Burning hydrocarbons or catalytic oxidation to CO_2 and H_2O.

NOTE **Leakage of methyl isocyanate** (MIC) gas caused death of approximately 3200 persons (Bhopal gas tragedy).

Greenhouse Effect and Global Warming

- The phenomenon in which atmosphere of earth traps the heat coming from the sun and prevents it from escaping into the outer space is called **greenhouse effect.**
- Certain gases, called greenhouse gases [carbon dioxide, methane, ozone, chlorofluorocarbon compounds (CFCs) and water vapour] in the atmosphere absorb the heat given by earth and radiate back it to the surface of the earth.
- Thus, warming of the earth led to the warming of air due to greenhouse gases, which is called **global warming.**

Some consequences of greenhouse effect are as follows:

(i) The greenhouse gases are useful in keeping the earth warm with an average temperature of about 15° to 20°C.

(ii) There may be less rainfall in this temperature zone and more rainfall in the dried areas of the world.

(iii) Increase in the concentration of CO_2 in the atmosphere leads to increase in the temperature of the earth's surface. As a result, evaporation of surface water will increase which further help in the rise of temperature and results in the melting of glaciers and polar ice-caps and hence, level of sea water may rise.

Acid Rain

- The pH of normal rain water is 5.6 due to the dissolution of CO_2 from atmosphere.

$$H_2O + CO_2 \rightleftharpoons H_2CO_3 \text{ (carbonic acid)}$$

$$H_2CO_3 \rightleftharpoons H^+ + HCO_3^-$$

- When the pH of rain water drops below 5 ppm, it is called acid rain (by Robert Augus). Oxides of N and S are responsible for making rain water acidic.
- Much of the NO_x and SO_x entering in the atmosphere are converted into HNO_3 and H_2SO_4 respectively. The detailed photochemical reactions occurring in the atmosphere are given as:

$$NO + O_3 \longrightarrow NO_2 + O_2$$

$$NO_2 + O_2 \longrightarrow NO_3 + O_2$$

$$NO_2 + NO_3 \longrightarrow N_2O_5$$

$$N_2O_5 + H_2O \longrightarrow 2HNO_3$$

- HNO_3 is removed as a precipitate or as particulate nitrate after reaction with bases (like NH_3, particulate lime etc).

$$SO_2 + \frac{1}{2}O_2 + H_2O \xrightarrow[\text{Soot particles}]{\text{(Hydrocarbon, } NO_x)} H_2SO_4$$

- The presence of hydrocarbons and NO_x step up the oxidation rate of the reaction.
- Soot particles are also known to be strongly involved in catalysing the oxidation of SO_2.
- Acid rain is harmful for agriculture, trees and plants. It causes respiratory ailments in human beings and animals.
- It also causes extensive damage to buildings and sculptural materials of marble, limestone, slate, mortar etc. e.g. acid rain reacts with marble, $CaCO_3$ of Taj Mahal causing damage to this wonderful monument.

$$CaCO_3 + H_2SO_4 \longrightarrow CaSO_4 + CO_2 + H_2O$$

Particulate Pollutants

- Particulate pollutants are the tiny pieces of solid or liquid matter associated with the earth's atmosphere.
- Particulates in atmosphere may be viable or non-viable.
- The **viable particulates** are the minute living organisms such as bacteria, fungi that are dispersed in atmosphere. Human beings are allergic to some of the fungi found in air.
- Mist, smoke, fumes and dust are non-viable particulates in atmosphere.

(i) **Smoke** particulates consist of solid or mixture of solid and liquid particles formed during combustion of organic matter, e.g. cigarette smoke, smoke from garbage, fossil fuel etc.

(ii) **Dust** is composed of fine solid particles produced during crushing, grinding and attribution of solid materials, e.g. fly ash from factories, dust storms etc.

(iii) **Mists** are produced by particles of spray liquids and by condensation of vapours in air e.g. sulphuric acid mist and herbicides and insecticides that miss their target and travel through air and form mists.

(iv) **Fumes** are generally obtained by condensation of vapours during sublimation, boiling and several other chemical reactions. Generally organic solvents, metals and metallic oxides form fume particles.

Harmful Effects The effect of particulate pollutants are larger as they are dependent on the particle size. Thus, particulate particles are dangerous for human health lead is used to be a major air pollutant emitted by vehicle. Leaded petrol used as the primary source for air borne lead mission in Indian cities. It interferes with the development and maturation of red blood cells.

Smog

Smog is the combination of smoke particles with tiny droplets of fog.

Classical Smog (London Smog)

SO_2 and particulate matter are main components of London smog. It is mostly observed in cool humid climate. It is chemically reducing in nature.

Photochemical Smog (Los Angeles Smog)

Photochemical smog contains a mixture of primary pollutants such as nitrogen oxides, carbon monoxide and secondary pollutants such as O_3 and HCHO. It occurs in warm, dry and sunny climate and are caused by the action of sunlight on nitrogen oxides and hydrocarbons. It is oxidising in nature.

Its formation can be shown as follows:

$$N_2 + O_2 \longrightarrow 2NO$$

(In automobile engines or power plant)

$$2NO + O_2 \xrightarrow{\text{Sunlight}} 2NO_2 \quad \text{(Oxidation in air)}$$

$$NO_2 \xrightarrow{h\nu} NO + \underset{\text{Atomic oxygen}}{[O]}$$

$$[O] + O_2 \rightleftharpoons O_3$$

$$NO + O_3 \longrightarrow NO_2 + O_2$$

Both NO_2 and O_3 are strong oxidising agents.

$$3CH_4 + 2O_3 \longrightarrow 3HCHO + 3H_2O$$

$$\text{Hydrocarbons} + O_3, O_2, O, NO, NO_2 \longrightarrow \underbrace{\underset{\text{PAN}}{R\text{—}\overset{\overset{O}{\|}}{C}\text{—O—O—}NO_2} + HCHO + \underset{\text{Acrolein}}{CH_2{=}CH\text{—}CHO}}_{\text{Photochemical smog}}$$

Preventions

Industries should purify the smoke to a certain extent before releasing into air. They should use chimneys. Planting more and more trees is also a method to maintain the oxygen-carbon dioxide balance.

Stratospheric Pollution

Stratospheric pollution is mainly concerned with ozone layer depletion. In stratosphere, there is a region of high concentration of ozone (10 ppm), at a height of 23 km, called ozone layer. This layer does not allow the UV rays coming from the sun to reach on the earth. Thus, protects us from harmful effects of UV-rays.

Ozone Depletion

- Chlorofluorocarbons present in aerosols, air conditioning and refrigeration devices destroy ozone layer and reduce our protection against UV rays from the sun.

$$CF_2Cl_2 + h\nu \longrightarrow \dot{C}l + \dot{C}F_2Cl$$

$$\dot{C}l + O_3 \longrightarrow \dot{C}lO + O_2$$

$$\dot{C}lO + O \longrightarrow \dot{C}l + O_2$$

- One molecule of CFC can destroy more than thousand molecules of O_3. This leads to the formation of ozone hole.
- CFCs are stable in lower atmosphere but when they reach the stratosphere, they split and become unstable by sunlight.

Ozone Depletion in Antarctica

- A large scale depletion in the concentration of O_3 observed over Antarctica is called **ozone hole.**
- In stratosphere, NO_2 and CH_4 act as scavangers for $Cl\dot{O}$ and $\dot{C}l$.

$$Cl\dot{O} + NO_2 \longrightarrow \underset{\text{Chlorine nitrate}}{ClONO_2}$$

$$\dot{C}l + CH_4 \longrightarrow \dot{C}H_3 + HCl$$

- Polar stratospheric clouds (PSCs) are formed over Antarctica in winter. These are of two types: Type I clouds (contain solidified $HNO_3 \cdot 3H_2O$) and type II clouds (contain some ice).

$$ClONO_2 + H_2O \longrightarrow HOCl + HNO_3$$

$$ClONO_2 + HCl \longrightarrow Cl_2 + HNO_3$$

- The ozone depletion over Antarctica occurs during spring but gets replenished after spring is over.

$$HOCl + h\nu \longrightarrow \dot{O}H + \dot{C}l$$

$$Cl_2 + h\nu \longrightarrow 2\dot{C}l$$

- Stable wind patterns in stratosphere are called polar vortex. It does not allow the O_3 rich air to fill up the gap.

Effects of Depletion of the Ozone Layer

Depletion of ozone layer allows UV rays to reach earth's surface. Hence, causes the following problems.

(i) It increases transpiration, therefore decreases soil moisture.

(ii) It results in increased causes of skin cancer in human along with adverse effect on plants and crops.

Water Pollution

Water pollution is the degradation of quality of water due to addition of inorganic, organic, biological or radiological substances, factors (e.g. heat) and deprivation that makes it health hazard, unfit for human use and growth of aquatic biota.

The various pollutants of water pollution are:

(i) **Pathogens** These include bacteria and other organisms that enter water from domestic sewage and animal excreta.

(ii) **Organic Wastes** Organic matter such as leaves, grass, trash etc. pollute water as a consequence of sun off. These wastes are biodegradable.

(iii) **Chemical Pollutants** Water soluble inorganic chemicals that include heavy metals such as Cd, Hg, Ni etc. constitute an important class of pollutants. All these metals are dangerous to humans because our body can't excrete them. They can damage kidneys, CNS, liver etc. The degree of water pollution is measured in terms of BOD and COD.

(a) **Biochemical Oxygen Demand** (BOD) The amount of oxygen consumed by microorganisms in decomposing the waste present in a certain volume of sample of water is called BOD.

- $\text{BOD} = \dfrac{\text{number of milligrams of } O_2 \text{ needed}}{\text{number of litres of the sample}}$
- To determine BOD, water sample is first saturated with oxygen and then incubated at constant temperature for five days.

(b) **Chemical Oxygen Demand** (COD) In COD determination, a known quantity of water sample is oxidised by acidified $K_2Cr_2O_7$. The unused amount of dichromate is determined by back titration. The amount of oxygen used in oxidation is calculated from consumed concentration of $K_2Cr_2O_7$.

Harmful Effects of Water Pollution

- High concentration of fluoride are poisonous and are harmful to bones and teeth at levels over 10 ppm.
- Excessive sulphate (>500 ppm) have a laxative effect.
- Excess nitrate in drinking water can lead to blue baby syndrome (methemoglobinemia).

Prevention of Water Pollution

- Removal of large solids from waste water by filtration (solids are disposed of in landfill sites).
- Settlement of the filtered waste water to remove suspended solids, oily and greasy materials which float on the surface can be skimmed off.
- Degradation of organic content of waste water by microbial oxidation.
- Removal of phosphates, coagulation, filtration and disinfection using chlorine for improving the quality of waste water.

Soil or Land Pollution

Soil pollution is the addition of such chemical substances which decreases its productivity, quality of plants and ground water to the soil system. The polluted soil produces inferior quality of crop.

The major pollutant of soil pollution are :

(i) **Pesticides** like insecticides (e.g. DDT, BHC), herbicides (e.g. $NaClO_3$, Na_2AsO_3), fungicides (e.g. organo mercury compounds), fertilisers and soil conditioners (e.g. compounds of As, Hg, Pb etc).

Harmful Effects These pollutants enter into food or drinking water which adversely affect the health of human beings.

(ii) Dumping of waste such as garbage, industrial wastes, ash, sludge, broken cans and bottles etc.

Harmful Effects They affect the fertility, pH level, microbial population and humidity of soil.

Prevention of Land Pollution

- Forestation to check the spread of desert.
- Pesticides and herbicides should be used only when necessary.

Strategies to Control Environmental Pollution

1. **Waste management and green chemistry** are used to control environmental pollution. Waste management is done by recycling, digestion, incineration, dumping and sewage treatment.
2. **Recycling**, i.e. conversion of waste into useful products. It saves raw materials and reduces the cost of waste disposal, e.g. recycling glass bottles, scrap metal in the manufacture of steel, generation of energy by burning combustible wastes.
3. **Incineration**, i.e. reduction of many combustible wastes from households, hospitals etc., to ash by burning it at very high temperature (> 1000°C) in excess of oxygen. This is one of the best methods for disposal of polychlorinated biphenyls (PCBs) as high temperature breaks C—Cl bonds. The chief disadvantage of incineration is that it leads to air pollution.
4. **Green fuel**, the plastic waste is being converted into fuel which has high octane number and does not contain any lead.
5. **Digestion**, i.e. conversion of the organic material (C, H, O) into carbon dioxide and methane by microorganisms (anaerobic digestion).

Green Chemistry

- It is an alternative tool for reducing pollution. It refers to the production of substances of daily use by chemical reactions which neither employ toxic chemicals nor release the same to atmosphere.
- Green chemistry includes concepts such as waste minimisation, solvent selection, atom utilisation, intensive processing and alternative synthetic routes from sustainable resources.

DAY PRACTICE SESSION 1

FOUNDATION QUESTIONS EXERCISE

1 Environmental pollution affects
(a) biotic components
(b) human beings only
(c) plants only
(d) biotic and abiotic components of environment

2 Which of the following is not a pollutant?
(a) Carbon monoxide (b) Sulphur trioxide
(c) Sulphur dioxide (d) Nitrogen peroxide

3 Which of the following is not considered to be a pollutant?
(a) NO_2 (b) CO_2 (c) O_3 (d) SO_3

4 The lowest layer of earth's atmosphere is
(a) troposphere (b) mesosphere
(c) stratosphere (d) ionosphere

5 Ultraviolet radiation is absorbed by
(a) exosphere (b) ionosphere
(c) mesosphere (d) stratosphere

6 Pollutants released from iron and steel industry are
(a) CO_2, NO_2, H_2S (b) CO, CO_2, SO_2
(c) CO_2, SO_3, NO_2 (d) CO_2, NO, SO_3

7 SO_2 causes
(a) respiratory and lung disease
(b) acid rain
(c) corrosion of building materials
(d) All of the above

8 Main pollutant from automobile exhaust is
(a) CO (b) CO_2
(c) NO (d) hydrocarbons

9 Carbon monoxide, emitted by automobiles, prevents transport of oxygen in the body due to
(a) combining with oxygen to form carbon dioxide
(b) destruction of haemoglobin
(c) preventing reaction between oxygen and haemoglobin
(d) formation of stable compound with haemoglobin

10 Which of the following pollutants is not emitted during volcanic eruptions?
(a) SO_2 (b) H_2S
(c) Hydrocarbons (d) CO

11 The gas leaked from a storage tank of the Union Carbide Plant in Bhopal gas tragedy was → JEE Main 2013
(a) methylamine (b) phosgene
(c) ammonia (d) methyl isocyanate

12 Lead pollution is mainly caused by
(a) sewage (b) insecticide
(c) coal gasoline (d) None of these

13 Which of the following is not a greenhouse gas?
(a) Methane (b) Carbon dioxide
(c) CFCs (d) CH_3Br vapour

14 Acid rains are produced by
(a) excess NO_2 and SO_2 from burning fossil fuels
(b) excess production of NH_3 by industry and coal gas
(c) excess release of carbon monoxide by incomplete combustion
(d) excess formation of CO_2 by combustion and animal respiration

15 Which of the following statements is false?
(a) London smog is oxidising in nature
(b) London smog causes bronchitis
(c) London smog is formed in winter
(d) London smog contains H_2SO_4 droplets

16 The smog is essentially caused by the presence of
(a) O_2 and O_3
(b) O_2 and N_2
(c) oxides of sulphur and nitrogen
(d) O_3 and N_2

17 Which one of the following is not a common component of photochemical smog?
(a) Ozone (b) Acrolein
(c) Peroxyacetylnitrate (d) Chlorofluorocarbons

18 Persons working in cement plants and limestone quarries are more prone to disease like
(a) asthma (b) cancer
(c) silicosis (d) pneumoconiosis

19 Which of the following protects life on earth from harmful effects of UV radiations from sun?
(a) N_2 (b) CO_2
(c) O_2 (d) O_3

20 Ozone has the ability to absorb
(a) UV radiations (b) electromagnetic radiations
(c) CFC (d) greenhouse gases

21 Ozone depletion in the stratosphere is mainly caused by
(a) SO_2 (b) NO_2
(c) NO (d) chlorofluorocarbons

22 Increased UV radiations due to hole in ozone layer
(a) will cause increase in cases of skin diseases
(b) will cause more ice to melt
(c) will cause summer to be more warmer
(d) will cause more rain

23 Ozone hole is maximum over
(a) India (b) Pakistan
(c) Australia (d) Antarctica

24 In Antarctica, ozone depletion is due to the formation of
(a) chlorine nitrate (b) peroxyacetyl nitrate
(c) SO_2 (d) SO_3

25 Identify the incorrect statement from the following. → AIEEE 2011

(a) Oxides of nitrogen in the atmosphere can cause the depletion of ozone layer
(b) Ozone absorbs the intense ultraviolet radiation of the sun
(c) Depletion of ozone layer is because of its chemical reactions with chlorofluoroalkanes
(d) Ozone absorbs infrared radiations

26 Eutrophication of a lake means it

(a) is low in nutrients
(b) is high in nutrients
(c) has excess amount of organic matter
(d) has a high temperature

27 Which of the following does not cause water pollution?

(a) Heavy metals such as Cd, Pb, Hg
(b) Detergents
(c) Polychlorobiphenyls
(d) Freons

28 Fish die in water-bodies polluted by sewage is due to

(a) pathogens (b) reduction in oxygen
(c) foul smell (d) None of these

29 BOD is

(a) waste decomposed in 5 days
(b) oxygen used in 5 days
(c) microorganisms killed in 5 days
(d) dissolved oxygen left after 5 days

30 Mottling of teeth is due to the presence of which of the following element in drinking water?

(a) Mercury (b) Fluorine (c) Boron (d) Chlorine

31 Water is treated with chlorine to

(a) increase oxygen content
(b) increase taste
(c) remove suspended particles
(d) kill germs

32 Sewage water is purfied by

(a) microorganism (b) fishes
(c) aquatic plants (d) All of these

33 Match the pollutants given in Column I with their effects in Column II.

Column I	Column II
A Unsaturated hydrocarbons	1. BOD level of water increases
B Methane in air	2. Acid rain
C. Synthetic detergents in water	3. Global warming
D. Nitrogen oxides in air	4. Photochemical smog

Codes

	A	B	C	D		A	B	C	D
(a)	4	3	1	2	(b)	4	1	3	2
(c)	1	4	2	3	(d)	4	3	2	1

34 Drawback of DDT as pesticides is that

(a) it is less effective than others
(b) it becomes ineffective after some time
(c) it is a non-biodegradable substance
(d) it is very costly

35 What is DDT among the following?

(a) Greenhouse gas
(b) A fertiliser
(c) Biodegradable pollutant
(d) Non-biodegradable pollutant → AIEEE 2012

36 Green chemistry involves

(a) production of chemicals of our daily use from green house gases
(b) such chemical processes in which green plants are used
(c) those reactions which are of biological origin
(d) use of non-toxic reagents and solvents to produce environment friendly products

Direction (Q. Nos. 37-38) *In the following questions, Assertion (A) followed by Reason (R) is given.Choose the correct option out of the choices given below:*

(a) Both A and R are true and R is correct explanation of A
(b) Both A and R are true but R is not correct explanation of A
(c) A is true but R is false
(d) Both A and R are false

37 **Assertion** (A) Greenhouse effect was observed in houses used to grow plants and these are made of green glass.

Reason (R) Greenhouse name has been given because glass houses are made of green glass.

38 **Assertion** (A) Excessive use of chlorinated synthetic pesticides causes soil and water pollution.

Reason (R) Chlorination synthesis pesticides are non-biodegradable.

Direction (Q. Nos. 39 and 40) *Each of these questions contains two statements : Statement* I *and Statement* II. *Each of these questions also has four alternative choices, only one of which is the correct answer. You have to select one of the codes (a), (b), (c) and (d) given below:*

(a) Statement I is true, Statement II is true; Statement II is the correct explanation for Statement I
(b) Statement I is true, Statement II is true; Statement II is not the correct explanation for Statement I
(c) Statement I is true; Statement II is false
(d) Statement I is false; Statement II is true

39 **Statement I** Deforestation is one main factor contributing to global warming.

Statement II Besides CO_2, two other gases methane and CFCs are also included under greenhouse gases.

40 **Statement I** Photochemical smog is produced by nitrogen oxides.

Statement II Vehicular pollution is a major source of nitrogen oxides.

DAY PRACTICE SESSION 2

PROGRESSIVE QUESTIONS EXERCISE

1 The pollutants which come directly in the air from sources are called primary pollutants. Primary pollutants are sometimes converted into secondary pollutants. Which of the following belongs to secondary air pollutants?

(a) CO (b) Hydrocarbon
(c) Peroxyacetyl nitrate (d) NO

2 Sewage containing organic waste should not be disposed in water-bodies because it causes major water pollution. Fishes in such a polluted water die because of

(a) large number of mosquitoes
(b) increase in the amount of dissolved oxygen
(c) decrease in the amount of dissolved oxygen in water
(d) clogging of gills by mud

3. Which of the following statements is wrong?

(a) Polar stratospheric clouds (PSCs) are clouds formed over Antarctica
(b) Acid rain dissolves heavy metals such as Cu, Pb, Hg and Al from soil, rocks and sediments
(c) H_2SO_4 is major contributor to acid rain, HNO_3 ranks second and HCl third in this respect
(d) Fishes grow in warm as well as in cold water

4 Which of the following practices will not come under green chemistry?

(a) If possible, making use of soap made of vegetable oils instead of using synthetic detergents
(b) Using H_2O_2 for bleaching purpose instead of using chlorine based bleaching agents
(c) Using bicycle for travelling small distances instead of using petrol/diesel based vehicles
(d) Using plastic cans for neatly storing substances

5 Identify the incorrect statement in the following.

(a) Chlorofluorocarbons are responsible for ozone layer depletion
(b) Greenhouse effect is responsible for global warming
(c) Ozone layer does not permit infrared radiation from the sun to reach the earth
(d) Acid rain is mostly because of oxides of nitrogen and sulphur

6 The ozone layer forms naturally by

(a) the interaction of CFC with oxygen
(b) the interaction of UV radiation with oxygen
(c) the interaction of IR radiation with oxygen
(d) the interaction of oxygen and water vapour

7 The basic component of smog is

(a) PAN (b) PBN
(c) NO_2 (d) All of these

8 Which of the following statements is false?

(a) The main reason for river water pollution is industrial and domestic sewage discharge
(b) Surface water contains a lot of organic matter, mineral nutrients and radioactive materials
(c) Oil spill in sea water causes heavy damage to fishery
(d) Oil slick in a sea water increases DO value

9. Which of the following statements about photochemical smog is wrong?

(a) It has high concentration of oxidising agents
(b) It has low concentration of oxidising agents
(c) It can be controlled by controlling the release of NO_2, hydrocarbons, ozone etc
(d) Plantation of some plants like pinus helps in controlling photochemical smog

10. Oxidation of sulphur dioxide into sulphur trioxide in the absence of a catalyst is a slow process but this oxidation occurs easily in the atmosphere. Which substance here catalyse the reaction?

(a) Oxygen
(b) Particulate
(c) UV rays
(d) IR rays

11. Negative soil pollution is

(a) reduction in soil productivity due to erosion and over use
(b) reduction in soil productivity due to addition of pesticides and industrial wastes
(c) converting fertile land into barren land by dumping ash, sludge and garbage
(d) None of the above

12 When rain is accompanied by a thunderstorm, the collected rain water will have a pH value

(a) slightly lower than that of rain water without thunderstorm
(b) slightly higher than that of rain water when the thunderstorm is not there
(c) uninfluenced by occurrence of thunderstorm
(d) which depends on the amount of dust in air

13 Which of the following statements about polar stratospheric clouds (PSCs) is not correct?

(a) Type I clouds are formed at about −77°C and contain solid $HNO_3 \cdot 3H_2O$
(b) Type II clouds are formed at about −85°C and contains some ice
(c) A tight whirlpool of wind called polar vortex is formed which surrounds Antarctica
(d) PSCs do not react with chlorine nitrate and HCl

14 For dry cleaning, in the place of tetrachloroethene, liquefied carbon dioxide with suitable detergent is an alternative solvent. What type of harm to the environment will be prevented by stopping use of tetrachloroethene?

(a) It results in tropospheric pollution
(b) It causes depletion of ozone layer
(c) It causes particulate pollution
(d) Both (a) and (b)

15 Average ozone concentration in Jakarta, Indonesia have been reported to be 0.015 mg m^{-3} and those in Tokyo, Japan are 20 ppmV. What is the approximate ratio of these two values, when expressed in the same unit?

(a) 1:0.3 (b) 1: 3
(c) 1: 33 (d) 1: 2

ANSWERS

SESSION 1

1 (d)	2 (d)	3 (c)	4 (a)	5 (d)	6 (b)	7 (d)	8 (a)	9 (d)	10 (c)
11 (d)	12 (c)	13 (d)	14 (a)	15 (a)	16 (c)	17 (d)	18 (c)	19 (d)	20 (a)
21 (d)	22 (a)	23 (d)	24 (a)	25 (d)	26 (b)	27 (d)	28 (b)	29 (b)	30 (b)
31 (d)	32 (a)	33 (a)	34 (c)	35 (d)	36 (d)	37 (a)	38 (a)	39 (b)	40 (b)

SESSION 2

1 (c)	2 (c)	3 (d)	4 (d)	5 (c)	6 (b)	7 (d)	8 (d)	9 (b)	10 (b)
11 (a)	12 (a)	13 (d)	14 (a)	15 (b)					

Hints and Explanations

SESSION 1

1 Environmental pollution affects biotic and abiotic components of environment.

2 Nitrogen peroxide is not a pollutant.

3 NO_2, CO_2 and SO_3 are gaseous pollutants.

4 The lowest layer of earth's atmosphere is troposphere.

5 Ultraviolet radiations are absorbed by the ozone layer present in stratosphere.

6 Major pollutant released from steel industry are CO, CO_2 and SO_2.

7 SO_2 causes eye irritation, damages respiratory tract, produces asthma and bronchitis. It also causes acid rain and destroys building material.

8 CO is the main pollutant from automobile exhaust.

9 Carbon monoxide is highly toxic to living being because it has an ability to form more stable carboxyhaemoglobin complex with haemoglobin due to which the delivery of oxygen to the organs and tissues is blocked.

10 During volcanic eruptions, pollutants like SO_2, H_2S and CO are emitted.

11 The gas leaked from a storage tank of the union carbide plant in Bhopal gas tragedy was methyl isocyanate.

12 Lead pollution is mainly caused by coal gasoline.

13 Methane (CH_4), carbon dioxide (CO_2) CFCs are green house gases.

14 NO_2 and SO_2 released during burning of fossil fuels are responsible for acid rain.

15 London smog is reducing in nature.

16 The smog is caused by the presence of oxides of sulphur and nitrogen.

17 Among the given, CFCs are responsible for ozone depletion. It is not a component of photochemical smog.

18 Persons working in cement plant and limestone quarries are more prone to diseases like silicosis.

19 Ozone protects life on earth by absorbing UV radiations from sun.

20 Ozone layer in the stratosphere (light bluish gas) shields the earth from the harmful ultraviolet radiations of the sun.

21 Chlorofluorocarbons destroy ozone. When CFCs reach the stratosphere, they split and produce reactive free radicals.

22 UV radiations possess high energy and are harmful to human life. They cause skin cancer, swelling of skin, sunburns, burning sensation on the skin. Undesirable mutation may cause more severe problems.

23 Ozone hole is maximum over Antarctica.

24 $\dot{C}lO + NO_2 \longrightarrow \underset{\text{Chlorine nitrate}}{ClONO_2}$

$$ClONO_2 + H_2O \longrightarrow HOCl + HNO_3$$

$$HOCl \xrightarrow{h\nu} \dot{O}H + \dot{C}l$$

The HOCl so formed can get converted into chlorine radicals thus, facilitating ozone depletion.

25 Ozone layer is depleted by oxides of nitrogen and by freons (chlorofluorocarbons). It absorbs harmful UV radiations coming from the sun but it does not absorb infrared radiations.

26 Eutrophication means high concentration of phosphates and nitrates from fertilizers and detergents in aquatic ecosystem.

27 Freons do not cause water pollution, they are responsible for air pollution.

28 As sewage decreases the concentration of dissolved oxygen (concentration) in water and hence fish die in water bodies.

29 To determine BOD (Biochemical Oxygen Demand), water sample is first saturated with oxygen and then it is incubated at constant temperature (20°C) for 5 days.

30 Mottling of teeth is due to the presence of fluorine in drinking water.

31 Chlorine kills germs present in water.

32 Organic matter in sewage water is decomposed by microorganism.

33 $A \rightarrow 4$, $B \rightarrow 3$, $C \rightarrow 1$, $D \rightarrow 2$

34 DDT is a pesticide. It is a non-biodegradable substance which causes soil pollution.

35 DDT is a non-degradable pollutant which causes soil pollution.

36 Green chemistry involves uses of non toxic reagent and solvents to produce environment friendly products.

37 In cold countries, sunlight required to grow plants is less. Hence, plants are kept in a house made of glass, placed in such a manner, so that sunlight enters the greenhouse, heat up the soil and plants.

38 Insecticides, pesticides and herbicides cause soil and water pollution.They are non-biodegradable.

39 If CO_2 concentration increases in the atmosphere, the CO_2 layer also becomes thick. This prevents the heat from being re-radiated back into the outer space. This results in heating up of the earth's surface. CO_2 contributes 57% part in greenhouse effect. Besides it, CFCs 15%, methane 12%, nitrogen oxides 6% and 5% of water contribute to greenhouse effect.

40 It is correct that photochemical smog is produced by oxides of nitrogen and it is also a fact that vehicular pollution is a major source of nitrogen oxides but it is not the correct explanation.

SESSION 2

1 Peroxyacetyl nitrate (PAN) is known as secondary pollutant.

2 Fishes in polluted water die because of the decrease in the amount of dissolved oxygen in water.

3 Fishes do not grow in warm as well as in cold water because warm water contains less amount of dissolved oxygen.

4 Using plastic cans for neatly storing substances will not come under green chemistry. Water in lakes and rivers have been polluted by the use of plastic materials. The plastic materials are non-biodegradable.

5 Ozone layer permits the infrared radiation to pass through it but does not permit the higher range ultraviolet radiation to pass.

6 The ozone layer forms naturally by the interaction of UV radiation with oxygen.

$$O_2 \xrightarrow{\text{UV rays}} O + O$$

$$O_2 + O \longrightarrow \underset{\text{Ozone}}{O_3}$$

7 The basic component of smog is NO_2 which interact with light and ozone to form PAN and PBN.

Formation of smog can be represented as follows:

$$NO_2 \xrightarrow{h\nu} NO + O$$

$$O_3 + NO \longrightarrow NO_2 + O_2$$

$$RCO_3^{\bullet} + NO_2 \longrightarrow RCO_3NO_2$$

If $R = CH_3$, it is called PAN.

If $R = C_6H_5$, it is called PBN.

8 Oil slick causes water pollution, thus it decreases DO value (dissolved oxygen value) of sea water.

9 Photochemical smog has high concentration of oxidising agents and it can be controlled by controlling the release of NO_2, hydrocarbons, ozone (O_3) etc. Plantation of some plants like pinus helps in controlling photochemical smog.

10 The presence of particulate matter in polluted air catalyses the oxidation of SO_2 to SO_3.

$$2SO_2 + O_2 \xrightarrow{\text{Particulates}} 2SO_3$$

11 Negative soil pollution is the reduction in soil productivity due to erosion and over use.

12 During thunderstorm there is formation of NO which changes to NO_2 and ultimately to HNO_3 (acid-rain).

$$N_2 + O_2 \longrightarrow NO \xrightarrow{O_2} NO_2$$

$$\longrightarrow N_2O_5 \xrightarrow{H_2O} HNO_3 \ (pH < 7)$$

13 PSCs (Polar Stratospheric Clouds) of type II provide a surface for the conversion of chlorine nitrate ($ClONO_2$) and HCl into HOCl and Cl_2.

$$ClONO_2 + H_2O \xrightarrow{\text{PSCs}} HOCl + HNO_3$$

$$ClONO_2 + HCl \longrightarrow Cl_2 + HNO_3$$

14 Tetrachloroethene, $Cl_2C = CCl_2$ is suspected to be carcinogenic and also contaminates the ground water. This harmful effect will be prevented using liquefied CO_2 along with suitable detergent.

Use of liquefied CO_2 along with detergent will not be completely safe because most of the detergents are non-biodegradable and they cause water pollution. Moreover, liquefied CO_2 will ultimately enter into the atmosphere and contribute to the green house effect.

15 Ozone concentration in Jakarta, Indonesia

$= 0.015 \text{ mg m}^{-3}$

$= 0.015 \text{ mg L}^{-1}$

Ozone concentration in Tokyo, Japan is 20 ppmV (i.e. by volume). This means that there are 20 μmol of ozone for every 1.0 mol of the components of air.

∴22.4 L (1 mol) mixture of gases contains = 20 μmol of ozone

1 L mixture of gases will contain

$= \frac{20 \times 10^{-6}}{22.4}$ mol of ozone

$= 0.892 \times 10^{-6}$ mol

$= 0.892 \times 10^{-6} \times 48$ g ozone

$= 42.82 \times 10^{-3}$ mg ozone

$= 0.043 \text{ mg L}^{-1}$ ozone

Hence,

$$\frac{\text{concentration of ozone in Jakarta}}{\text{concentration of ozone in Tokyo}} = \frac{0.015}{0.043} = \frac{1}{3}$$

DAY TWENTY SEVEN

General Organic Chemistry

Learning & Revision for the Day

- Bonding in Organic Compounds
- Classification of Organic Compounds
- Functional Group
- Homologous Series
- IUPAC System
- Isomerism
- Fission of a Covalent Bond
- Electronic Displacement Effect in a Covalent Bond
- Types of Organic Reactions
- Qualitative Analysis
- Quantitative Analysis

The hydrides of carbon (hydrocarbons) and their derivatives are called organic compounds. The branch of chemistry which deals with these compounds is called organic chemistry.

Bonding in Organic Compounds

- Carbon is the essential element of all organic molecules and its electronic configuration is 2,4. It always form covalent bonds. The tetravalency of carbon can be explained by the excited state concept, i.e. by moving one of the paired $2s$ electron to empty $2p$ orbital by gaining energy from the system (according to classical concept of bonding) however 4 equivalent C—H bonds in CH_4 is explained on the basis of concept of hybridisation (according to modern concept of bonding).
- Carbon has a unique property that it can form both σ and π-bonds. A single bond contains only one σ-bond. Whereas, there are one σ and one π-bond in case of a double bond and one σ and two π-bonds in case of a triple bond.

Tetravalency of Carbon
(Shapes of Simple Molecules and Hybridisation)

- Hybridisation is the intermixing of orbitals of almost similar energy to form the same number of orbitals of exactly similar energy.
- The carbon atoms show three types of hybridisations depending upon the number of σ-bonds formed in its compounds.
- If there are four σ-bonds, the hybridisation is sp^3 with tetrahedral shape, if three σ-bonds, the hybridisation is sp^2 with trigonal planar shape and if only two σ-bonds, the hybridisation is sp with linear shape.

  Usually saturated hydrocarbons are sp^3-hybridised while unsaturated hydrocarbons may be sp^2 or sp-hybridised.

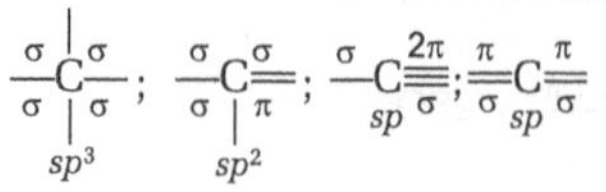

PREP MIRROR
Your Personal Preparation Indicator

- No. of Questions in Exercises (x)—
- No. of Questions Attempted (y)—
- No. of Correct Questions (z)—
  (Without referring Explanations)

- Accuracy Level ($z/y \times 100$)—
- Prep Level ($z/x \times 100$)—

In order to expect good rank in JEE, your Accuracy Level should be above 85 & Prep Level should be above 75.

e.g. But-1-en-3-yne (C_4H_4) consists of 7σ-bonds and 3π-bonds. π-bonds give no contribution to hybridisation.

An organic compound can be represented by the following ways :

(i) In **structural formula**, all the bonds present between any two atoms are shown clearly.

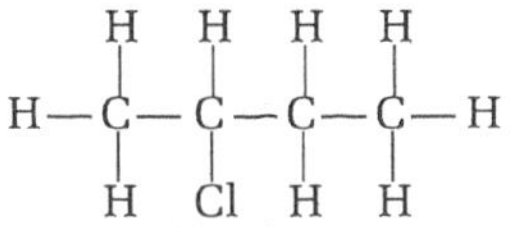

2-chlorobutane

(ii) In **condensed formula**, all the bonds are not shown clearly.

$CH_3CHCH_2CH_3$ (with Cl on the second carbon) or $CH_3CH(Cl)CH_2CH_3$

(iii) In **bond line formula**, every fold and free terminal represents a carbon and lines represent the bond.

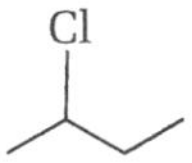

Classification of Organic Compounds

The organic compounds have been classified **on the basis of carbon skeleton (structure)** or functional groups or the concept of homology.

On this basis the organic compounds are classified as :

Organic Compound
- Open-chain, acyclic or aliphatic compounds
- Closed-chain or cyclic compounds
  - Homocyclic compounds
    - Alicyclic compounds
    - Aromatic compounds
      - Benzenoid aromatic compounds
      - Non-benzenoid aromatic compounds
  - Heterocyclic compounds

e.g.

Cyclopropane
(Alicyclic compound)

Benzene

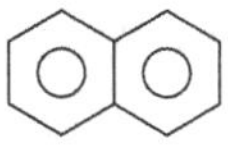

Naphthalene

(Benzenoid aromatic compounds)

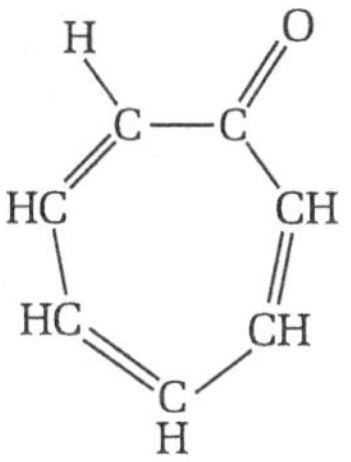

Tropolone

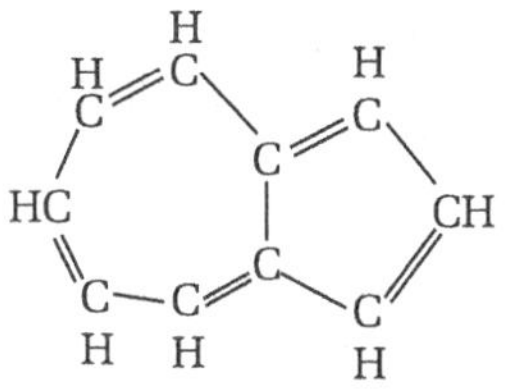

Azulene

(Non-benzenoid aromatic compounds)

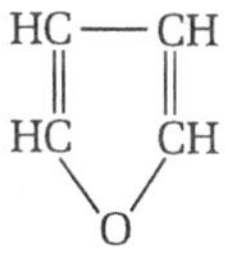

Furan

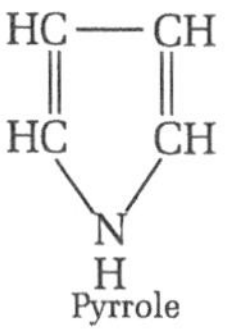

Pyrrole

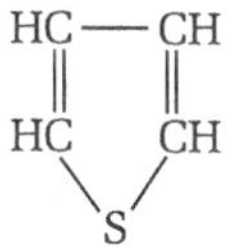

Thiophene

(Heterocyclic compounds)

Compounds having hetero atom in the ring

Functional Groups

An atom or group of atoms joined in a specific manner which is responsible for the characteristic chemical properties of the organic compounds, e.g. carboxylic acid group (—COOH), aldehydic group (—CHO) etc.

- **On the basis of functional groups**, organic compounds are classified into following families or homologous series.

Organic Compounds
- Double/triple bond containing compounds
  - Alkene $>C=C<$
  - Alkyne $-C\equiv C-$
- Halogen derivatives
  - Haloalkane $R-X$
  - Haloarene C_6H_5X
  - (X = F, Cl, Br, I)
- Oxygen containing compounds
  - Alcohol (R—OH)
  - Ether (R OR)
  - Carbonyl compounds
    - Aldehyde (RCHO)
    - Ketone (RCOR)
  - Carboxylic acids (RCOOH)
  - Carboxylic acid derivatives
    - Ester (RCOOR)
    - Amide (RCONH_2)
    - Anhydride $(RCO)_2O$
    - Acid halide (RCOCl)
- Nitrogen containing compounds
  - Amines
    - 1° amine (RNH_2)
    - 2° amine (R_2NH)
    - 3° amine (R_3N)
    - aromatic amine
  - Nitro compounds (RNO_2)
  - Cyanide (RCN)
  - Isocyanides (RNC)

Homologous Series

A group of compounds in which the members have similar structural features and similar chemical properties and the successive members differ in their molecular formula by CH_2 group. The general characteristics of a homologous series are:

(i) All compounds in the series can be represented by one general formula, e.g. the homologous series of monohydric alcohols can be represented by the general formula, $C_nH_{2n+1}OH$.

The formula of various homologous can be written by giving the values 1, 2, 3, ... to *n*.

n	$C_nH_{2n+1}OH$	
1	CH_3OH	
		CH_2
2	C_2H_5OH	
		CH_2
3	C_3H_7OH	
		CH_2
4	C_4H_9OH	

The molecular formula of each member differs from the members above and below by one CH_2 group.

(ii) The molecular mass of every two adjacent members differ by 14 ($CH_2 = 12 + 2 \times 1 = 14$).

(iii) All compounds in the series have similar chemical properties because of the presence of same functional group.

(iv) The members of the series show a gradual gradation in their physical properties like solubility, density, melting and boiling points. The physical properties generally increase as the molecular mass increases.

IUPAC System

In trivial system, name was assigned at the wish of discoverer and had no system. These names are also called **common names.** In 1957, the International Union of Pure and Applied Chemistry (IUPAC) evolved a scheme for giving systematic name to organic compounds on the basis of their structure. This is known as the IUPAC system of nomenclature. This system has set of rules for naming organic molecules from their structure.

An IUPAC name consist of prefixes and suffixes to indicate the substituent or functional group present in compound.

Some functional groups are always indicated by the prefixes instead of secondary suffixes.

Functional Group	Prefix	Family	IUPAC Name
$—NO_2$	Nitro	$R — NO_2$	Nitroalkane
—*OR*	Alkoxy	$R — O — R'$	Alkoxyalkane
—Cl	Chloro	*R*—Cl	Chloroalkane
—Br	Bromo	*R*—Br	Bromoalkane
—I	Iodo	*R* – I	Iodoalkane
—F	Fluoro	*R*—F	Fluoroalkane
—NO	Nitroso	*R*—NO	Nitrosoalkane

Secondary suffix is used to indicate the functional group in the organic compounds.

Family of Compound	Suffix	IUPAC Name of the Family
Alcohols (—OH)	—ol	Alkanol
Amines ($—NH_2$)	—amine	Alkanamine
Aldehydes (—CHO)	—al	Alkanal
Ketones ($>C=O$)	— one	Alkanone
Carboxylic acids (—COOH)	— oic acid	Alkanoic acid
Amides ($—CONH_2$)	—amide	Alkanamide
Acid halides (—COX)	—oyl halide	Alkanoyl halide
Esters (—COO*R*)	—oate	Alkyl alkanoate
Nitriles ($—C\equiv N$)	— nitrile	Alkane nitrile

Classification of Carbon and Hydrogen Atoms

- **Primary Carbon Atom** When carbon atom is attached with only one other carbon atom.
- **Secondary Carbon Atom** When carbon atom is attached with two other carbon atoms.
- **Tertiary Carbon Atom** When carbon atom is attached with three other carbon atoms.
- **Quarternary Carbon Atom** When carbon atom is attached with four other carbon atoms.

Reactivity order of carbon atoms is as follows:

$$3° > 2° > 1°$$

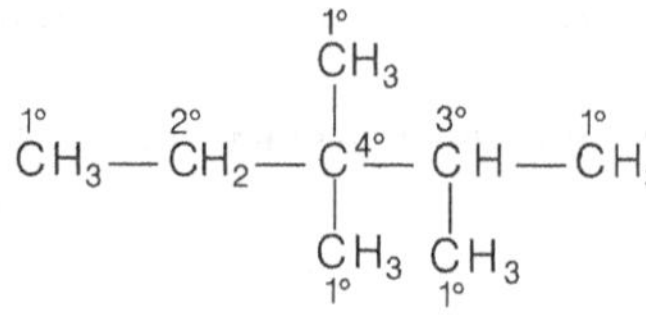

- **1°-hydrogen** (primary) ⟶ attached to 1°-carbon.
- **2°-hydrogen** (secondary) ⟶ attached to 2°-carbon.
- **3°-hydrogen** (tertiary) ⟶ attached to 3°-carbon.
- **α-carbon** Carbon which is directly attached to the functional group.
- **β-carbon** Carbon which is directly attached to the α-carbon.
- **α-hydrogen** (s) Hydrogens which are attached to α-carbon atom.
- **β-hydrogen** (s) Hydrogens which are attached to β-carbon atom.

e.g. $\overset{\beta}{C}H_3—\overset{\alpha}{C}H_2—Cl$, $\overset{\beta}{C}H_3—\overset{\alpha}{C}H_2—COOH$

$\overset{\beta}{C}H_3—\overset{\alpha}{C}H_2—CHO$

Rules for IUPAC Nomenclature

Rules for IUPAC nomenclature are as follows :

- First of all, the longest carbon chain in the molecule is identified.
- Numbering is started from the terminal carbon from which branching is nearest.
- If the two substituents are found at equivalent positions, the numbering is done alphabetically.
- If there are two chains of equal size then that chain is to be selected which contains more number of side chains.
- The longest chain of carbon atoms containing the functional group is numbered in such a way that the functional group is attached at the carbon atom possessing lowest possible number in the chain.
- In polyfunctional compounds, one of the functional group is chosen (priority wise) as the principal functional group and the compound is then named on that basis.

The order of decreasing priority for some functional groups are as follows:

$—COOH, —SO_3H, —COOR, —COCl, —CONH_2, —CN,$

$—CHO, >C{=}O, —OH, —NH_2, >C{=}C<, —C{\equiv}C—.$

Some examples are described below:

(i) $\overset{5}{C}H_3—\overset{4}{C}H(OH)—\overset{3}{C}H_2—\overset{2}{C}H(CH_2CH_3)—\overset{1}{C}H_2—OH$

2-ethylpenta-1, 4-diol (IUPAC name)

Maximum number of functional group should be present in longest carbon chain.

(ii) $\overset{5}{C}H_3—\overset{4}{C}(=O)—\overset{3}{C}H_2—\overset{2}{C}H_2—\overset{1}{C}OOH$

4-ketopentanoic acid (IUPAC name)

(Numbering according to principal functional group).

(iii) 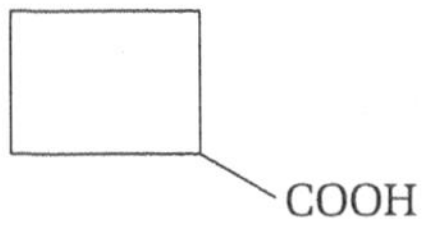

Cyclobutane carboxylic acid (IUPAC name)

(Carbon containing functional group is not considered with cyclic ring).

(iv) 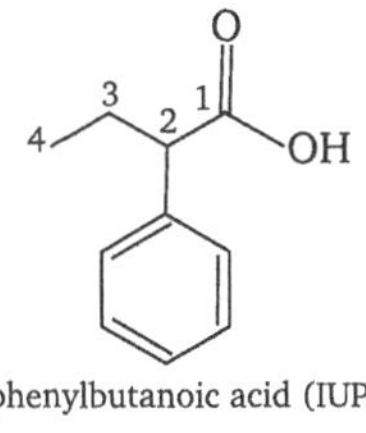

2-phenylbutanoic acid (IUPAC name)

If functional group is not directly attached to benzene ring, the ring is considered as branching and known as phenyl.

(v) In spirocyclic compounds, the common atom is designated as spiro atom. Spiro prefix is written before the alkane.

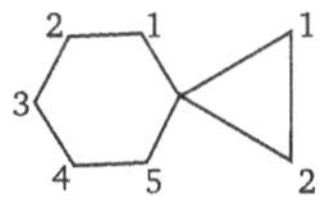

Spiro [2,5] octane

(vi) In bicyclo compounds, the bicyclic systems having two or more atoms in common, are named by prefixing 'bicyclo' to the name of parent hydrocarbon.

Isomerism

The compounds having same molecular formula but differ in properties are known as isomers and the phenomenon is called isomerism. Isomerism can be of the following two types :

(i) Structural Isomerism

(ii) Stereoisomerism

1. Structural Isomerism

In this type of isomerism, compounds have same molecular formula but different structures. Different types of structural isomerism are as follows :

(i) When two or more compounds have similar molecular formula but different carbon skeletons, then these are referred to as chain isomers and phenomenon is termed as **chain isomerism**, e.g. C_5H_{12}.

$CH_3—CH_2—CH_2—CH_2—CH_3$
n-pentane

$CH_3—CH_2—CH(CH_3)—CH_3$, $H_3C—C(CH_3)_2—CH_3$
iso-pentane, *neo*-pentane

(ii) When two or more compounds have same molecular formula but different position of functional groups or substituents then they are called position isomers and phenomenon is called **position isomerism,** e.g. C_3H_8O.

$CH_3—CH_2—CH_2—OH$, $CH_3—CH(OH)—CH_3$
Propan-1-ol, Propan-2-ol

(iii) When two or more compounds have the same molecular formula but different functional group then they are called functional isomers and phenomenon is called **functional isomerism**, e.g. C_3H_6O represents an aldehyde and a ketone as

$CH_3—CH_2—CHO$, $CH_3—C(=O)—CH_3$
Aldehyde, Ketone

C_2H_6O represents an alcohol and an ether

$$\underset{\text{Alcohol}}{CH_3-CH_2-\ddot{\underset{..}{O}}H}, \quad \underset{\text{Ether}}{CH_3-\ddot{\underset{..}{O}}-CH_3} \text{ etc.}$$

(iv) Metamerism arises due to different alkyl chains on either side of the same functional group in the molecule. e.g. $C_4H_{10}O$

$$CH_3-\ddot{\underset{..}{O}}-C_3H_7, C_2H_5-\ddot{\underset{..}{O}}-C_2H_5$$

NOTE **Tautomerism** is a special type of functional isomerism arises in carbonyl compounds containing α–H atom. e.g.

$$\text{(a)}\quad \underset{\substack{\text{Keto form}\\\text{(Acetaldehyde)}}}{CH_3-\overset{\overset{O}{\|}}{C}-H} \rightleftharpoons \underset{\text{Enol form}}{CH_2=\overset{\overset{OH}{|}}{C}-H}$$

(b) α-hydrogen

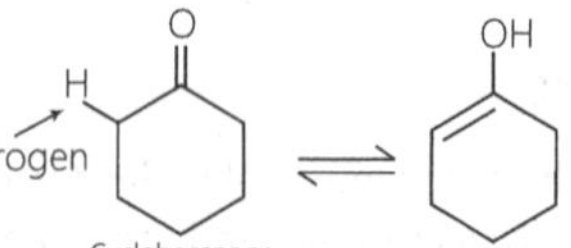

Cyclohexanone (Keto form) Enol form

2. Stereoisomerism

The compounds having same molecular formula but different spatial arrangement of atoms or groups are called stereoisomers and the phenomenon is called stereoisomerism. Stereoisomerism is of two types:

(i) Compounds having similar physical and chemical properties but differ only in behaviour towards plane polarised light are called optical isomers and property is known as **optical isomerism**. e.g.

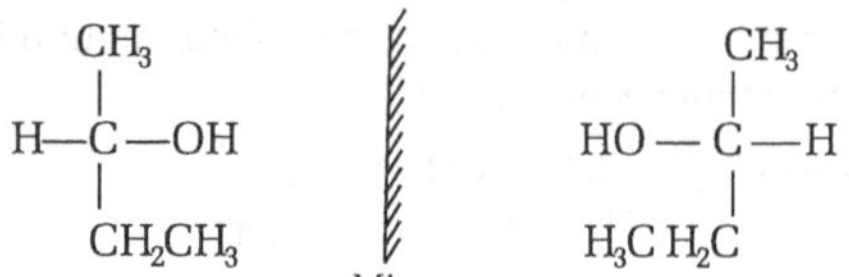

Mirror

Butan-2-ol

(ii) The isomers having same molecular formula but different spatial arrangement of atoms about the double bond are known as geometrical isomers and this phenomenon is called **geometrical isomerism**. e.g.

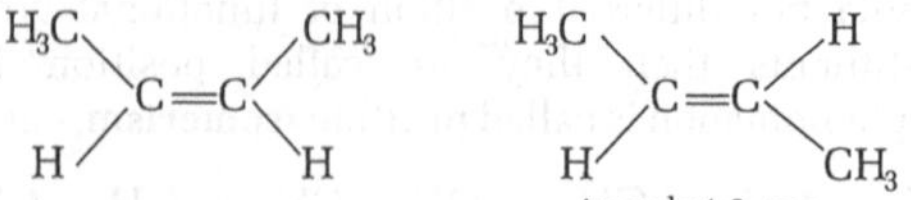

cis-but-2-ene *trans*-but-2-ene

NOTE
- The compound that consists of at least one asymmetric C-atom are capable of showing the phenomenon of optical isomerism.
- Conformational isomerism is not a type of stereoisomerism. These are called conformations that have spatial arrangement of atoms which can be converted into one another by rotation around a C — C single bond. These are non-separable.

Fission of a Covalent Bond

The organic reactions began with the breakage of covalent bond and this breakage or fission is of two types

1. In **homolytic fission**, one of the electrons of the shared pair in a covalent bond goes with each of the bonded atoms. Generally, homolytic fission takes place in non-polar covalent molecules in the presence of sunlight or high temperature.

$$A-B \xrightarrow{\text{Sunlight}} \underbrace{A^{\bullet} + B^{\bullet}}_{\text{Free radicals}}$$

e.g. $$Cl_2 \xrightarrow{\text{Sunlight}} 2Cl^{\bullet}$$

2. In **heterolytic fission**, the bond breaks in such a fashion that the shared pair of electrons goes with one of the fragments.

$$A-B \longrightarrow \underset{\text{Electrophile}}{A^+} + \underset{\text{Nucleophile}}{B^-}$$

It generally takes place in polar covalent molecules but in non-polar molecules, it takes place in the presence of catalyst like $AlCl_3$ (anhyd.), $FeCl_3$ (anhyd.) etc. It requires larger energy than homolytic fission. Heterolytic fission is favoured in acid or base catalysed reaction while homolytic fission is favoured in gas phase reactions.

Electrophiles (Electron Deficient Species)

- All non-metal cations and metal cations which have vacant *d*-orbitals. e.g. Cl^+, NO_2^+, CH_3CO^+ etc.
- Lewis acids (incomplete octet) e.g. BF_3, $ZnCl_2$ (anhydrous), $FeCl_3$ (anhydrous), $AlCl_3$ (anhydrous), $:CH_2$ etc.
- Non-metal acidic oxides, e.g. CO_2, SO_2 etc.
- The electrophile attacks at nucleophilic centre and receives electron pair from nucleophile when the two undergoes bonding interaction.
- In case of same nucleophilic site, nucleophilicity parallels basicity, i.e. as the basicity of nucleophile increases, its strength also increases.
- If attacking atoms are different, nucleophilicity varies inversely with electronegativity.

Nucleophiles (Electron Rich Species)

- All anions behaves as nucleophiles, e.g. Cl^-, NH_2^-, OH^- etc.
- Lewis bases, e.g. $:NH_3$, H_2O, $R-\ddot{\underset{..}{O}}-R$, $-\ddot{\underset{..}{O}}H$ etc.
- Compounds containing C — C multiple bonds behave as nucleophiles. e.g. Benzene, alkenes etc.
- During a polar organic reaction, a nucleophile attacks an electrophilic centre of the substrate which is that part of the electrophile that is electron deficient.
- Nucleophilicity order is $H^- > CH_3^- > NH_2^- > RO^- > OH^-$.

- Organic compounds such as HCHO and CH_3CN having a multiple bond between carbon and more electronegative atom act as nucleophile as well as electrophile.

Reaction Intermediates

Most of the chemical reactions require certain chemical species to occur. These species are generally short lived, highly reactive and are called reactive intermediates. e.g. free radicals, carbocations, carbanions, carbenes, nitrenes etc. These are discussed as follows :

1. **Free radicals** are highly reactive, neutral and electron deficient species. The order of the stability of free radicals is

$$(C_6H_5)_3\dot{C} > (C_6H_5)_2\dot{C}H > C_6H_5\dot{C}H_2 > CH_2{=}CH{-}\dot{C}H_2 > 3° > 2° > 1° > CH_2{=}\dot{C}H$$

2. **Carbocation** Carbon containing chemical species bearing a positive charge on carbon and carry 6 electrons in its valence shell are called **carbocation**. The carbocations follow the following order of stability

$$(C_6H_5)_3\overset{+}{C} > (C_6H_5)_2\overset{+}{C}H > (CH_3)_3\overset{+}{C} > C_6H_5\overset{+}{C}H_2 > 2° > \text{allyl} > 1° > \overset{+}{C_6}H_5 > \text{vinyl}$$

Less stable carbocations (1°/2°) can be converted into more stable carbocations (2°/3°) through either 1, 2-hydride shift or 1, 2-methyl shift.

3. **Carbanion** Carbon containing chemical species bearing a negative charge on carbon atom and carrying 8 electrons in its valence shell is called **carbanion**.

The order of stability of carbanions is as

$$(C_6H_5)_3C^- > (C_6H_5)_2\bar{C}H > C_6H_5\bar{C}H_2 > \text{allyl} > \bar{C}H_3$$
$$(1° > 2° > 3° \text{ carbanions})$$

4. **Carbines** The neutral divalent carbon species in which two non-bonding electrons are present along with 2 bonding pairs are called **carbenes**. Carbenes act as electrophiles as these are electron deficient species. Carbenes are of two types:

(i) **Single carbene**

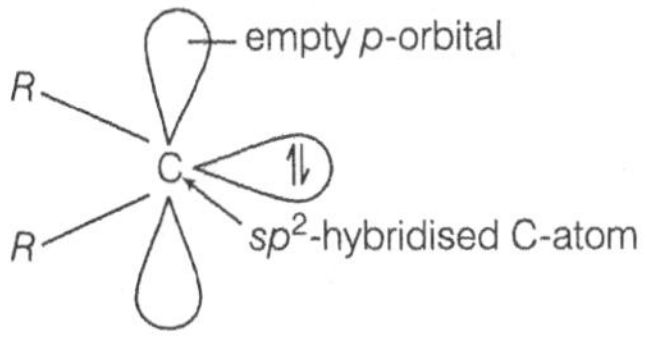

(ii) **Triplet carbene**

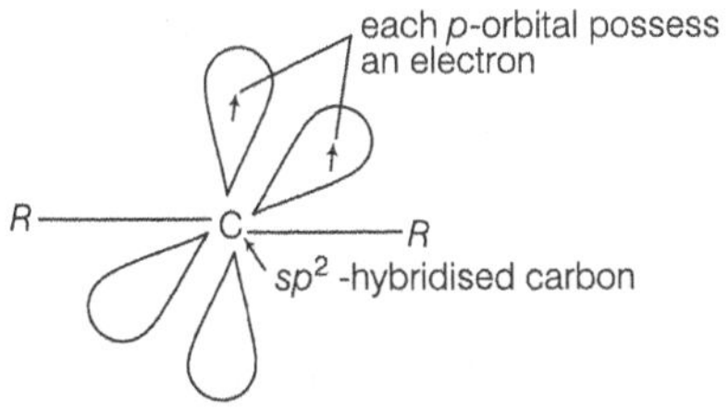

Linear structure

Triplet carbene acts as diradical and it is always more stable than a singlet carbene.

5. **Nitrenes** Neutral monovalent nitrogen species in which nitrogen has two unshared pair of electrons with a monovalent atom or group attached is called **nitrenes,** i.e.

$$\underline{\sigma}\ddot{N}:$$

Nitrenes are produced by thermolysis of azides and are very reactive.

Electronic Displacement in a Covalent Bond

Presence of some atom or group in a molecule or presence of attacking reagent may lead to electronic displacement in a covalent bond. The factors that create the centres of different electron densities are discussed below.

Inductive Effect

- Inductive effect is just like shifting of shared pair of electrons in polar covalent molecules. If shared pair is more shifted towards the more electronegative atom, the less electronegative atom acquire slight positive charge and more electronegative atom acquire partial negative charge,

e.g. $\overset{+\delta}{CH_3} \rightarrow \overset{-\delta}{Cl}$

This is a permanent effect and propagates through carbon chain. The effect decreases rapidly as the number of intervening bonds increases and becomes vanishingly small after three bonds, e.g. $\overset{+\delta\delta\delta}{CH_3} \rightarrow \overset{+\delta\delta}{CH_2} \rightarrow \overset{+\delta}{CH_2} \rightarrow \overset{-\delta}{Cl}$

Here, Cl has $-I$ effect and alkyl group has $+I$ effect.

- Greater the number of C-atoms in alkyl groups, greater would be its $+I$ effect. Therefore, 3°– alkyl halide will be most reactive due to more $+I$ effect.
- Electron donating groups (EDG) such as alkyl groups like $-CH_3$, $-C_2H_5$ etc., produce $+I$ effect.
- Electron withdrawing groups (EWG) such as $-NO_2$, $-CN$, $-COOH$, $-COOR$, $-OC_6H_5$ etc., produce $-I$-effect.

Resonance Effect

- It involves interaction of two π-bonds or a π-bond and lone pair of electron. This effect may be of $+R$ type or $-R$ type.
- Electron donating group with respect to conjugate system shows $+R$ effect. Central atom of functional group should be more electronegative than the surrounding atoms or groups to show $+R$ effect.

e.g. halogens, $-OH$, $-OR$, $-OCOR$, $-NH_2$, $-NHCOR$ etc. This effect in aniline is shown as

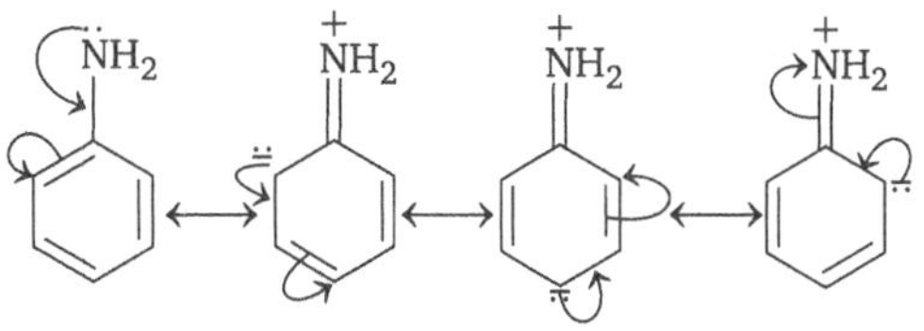

- Electron donating groups producing $+R$ effect are *ortho* and *para* directing. They activate the benzene ring towards the electrophilic substitution reactions except halogens. Halogens slightly deactivate the benzene ring towards the electrophilic substitution reaction. More the EDG, more is the basic nature.
- Electron withdrawing groups with respect to conjugate system shows $-R$ effect. Central atom of functional group should be less electronegative than surrounding atoms or groups to show $-R$ effect.

  e.g. —COOH, —COOR, —CHO, —CN, —NO_2 etc.

  In benzaldehyde, this electron displacement can be depicted as:

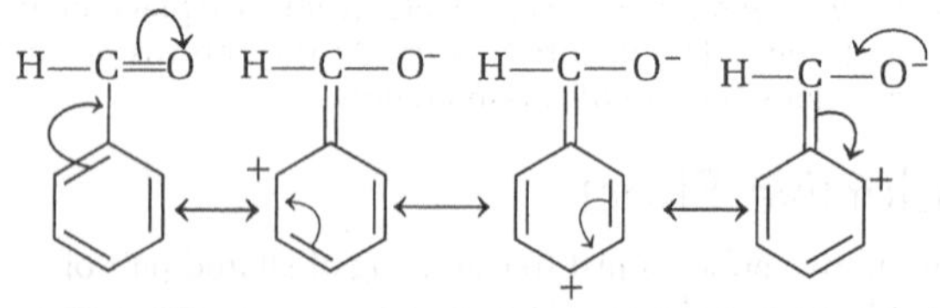

- Electron withdrawing group (EWG) producing $-R$ effect are *meta* directing. They deactivate the benzene ring towards the electrophilic substitution reaction. More the EWG, more is the acidic nature.

Relation between Resonance and Bond Order

In compounds exhibiting resonance, bond order can be given by the formula

$$\text{Bond order} = \frac{\text{Total number of bonds between two atoms}}{\text{Total number of resonating structures}}$$

NOTE Resonance energy is the energy lost when a molecule acquires resonance hybrid formula.

(i) Number of π bonds $\propto$ contributing structures $\propto$ resonance energy $\propto$ stability.

(ii) In benzene, resonance energy is 36 kcal/mol.

Electromeric Effect

- It is defined as the polarity produced in a multiple bonded compound as a reagent approaches it. In the presence of attacking reagent, the two π electrons are completely transferred to any of the one atom. This effect is temporary.
- This may be of $+E$ type (when displacement of electron pair is away from the atom or group) or of $-E$ type (when the displacement is towards the atom or group), e.g.

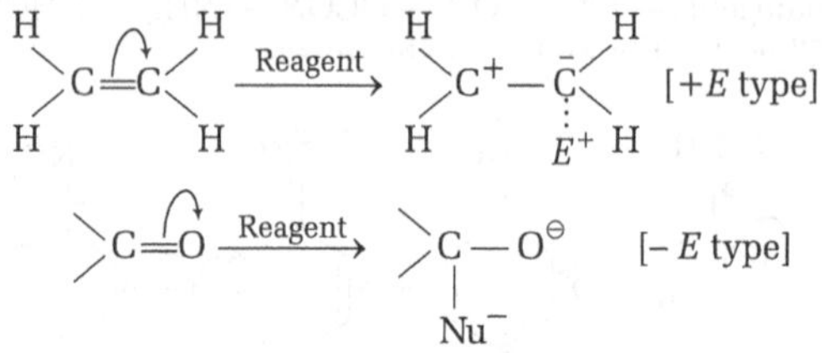

Hyperconjugation

It involves delocalisation of σ electrons of a C—H bond of an alkyl group attached directly to an atom of unsaturated system or to an atom with an unshared p-orbital. e.g.

$$\underset{\text{Prop-1-ene}}{CH_3-CH=CH_2} \longleftrightarrow \overset{H^+}{CH_2=CH-\bar{C}H_2}$$

Requirements for structure of hyperconjugation are as follows:

(a) Compound should have at least one sp^2 hybrid carbon of either alkene, alkyl carbocation or alkyl free radical.

(b) α-carbon with respect to sp^2 hybrid carbon should have at least one hydrogen. More the number of H—C bonds attached to the unsaturated system, more stable will be the alkene.

The following are the important applications of hyperconjugation.

(i) **Stability of Alkenes** More the number of α-hydrogen atoms, more stable is the alkene.

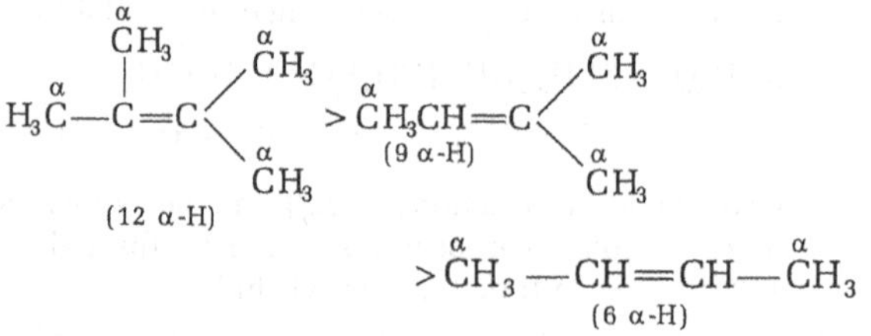

(ii) **Stability of carbocation**: Greater the number of alkyl groups attached to a positively charged carbon atom, the greater is the stability.

$$(CH_3)_3C^+ > (CH_3)_2\overset{+}{C}H > CH_3-\overset{+}{C}H_2 > \overset{+}{C}H_3$$

Types of Organic Reactions

Based on the nature of reactants, products and byproducts as well as the mechanism, organic reactions are classified into four major categories, substitution, addition, elimination and rearrangement reactions.

1. **Substitution reaction** is a type of reaction in which one atom, or a group of atoms, from the reagent replaces or substitutes itself for one atom or a group of atoms on the substrate.

   In nucleophilic substitution, the reagent is a nucleophile (base) whereas in an electrophilic substitution, the reagent is an electrophile (acid). e.g.

$$C_2H_5OH + \underset{\text{Nucleophile}}{Cl^-} \longrightarrow C_2H_5Cl + OH^-$$

(Nucleophilic substitution)

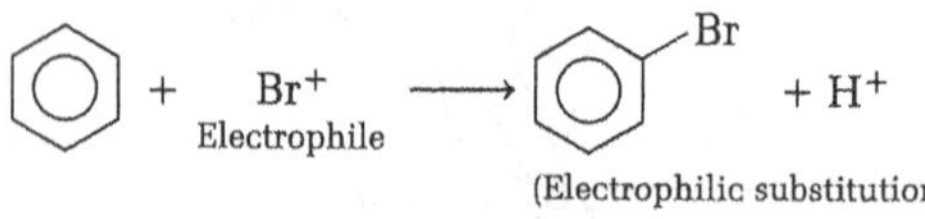

(Electrophilic substitution)

2. **Addition reaction** is a type of reaction in which atoms or group of atoms from the reagent add to the substrate generally without losing any atoms from the substrate.

   e.g. $CH_2=CH_2 + HBr \longrightarrow CH_3CH_2Br$

3. **Elimination reaction** removes two atoms or two group of atoms from the substrate, giving the product a multiple bond or, in some cases, causing it to form a ring.

e.g. $(CH_3)_3COH \xrightarrow[\text{Heat}]{\text{(i) } H_3O^+} \underset{\text{2-methylprop-1-ene}}{(CH_3)_2C{=}CH_2} + H_2O$

4. **Rearrangement reactions** involve the migration of an atom or a group from one atom to another within the same molecule. These reactions usually occur in the presence of a catalyst or under special thermodynamic conditions. e.g.

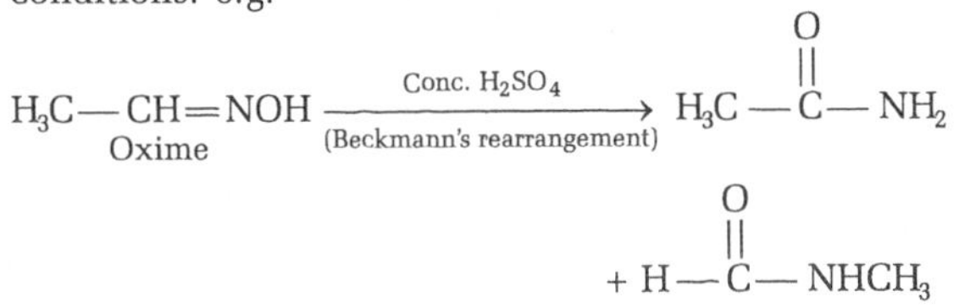

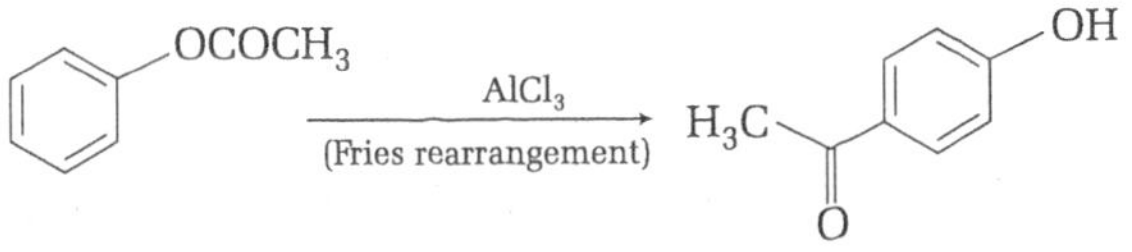

Methods for Purification of Organic Compounds

The common techniques used for purification are as follows:

1. **Sublimation process** is employed for those solid substances that directly changes into vapour phase. This method is used to separate sublimable compounds from non-sublimable impurities.
2. **Crystallisation** is one of the most commonly used techniques for the purification of solid organic compounds. It is based on the difference in the solubilities of the compound and the impurities in a suitable solvent.
3. **Distillation** is used to separate
   (a) volatile liquids from non-volatile impurities.
   (b) the liquids having sufficient difference in their boiling points. Liquids having different boiling points vaporise at different temperatures.

   Several distillation methods employed to separate mixture are as follows :

   (i) **Distillation under reduced pressure** This method is used to purify liquids having very high boiling points and those, which decompose at or below their boiling points. Glycerol can be separated from spentlye in soap industry by this method.

   (ii) **Fractional distillation** The separation of volatile components of different boiling points in a mixture by the gradual increase of temperature and the separate collection of each component.

   (iii) **Steam distillation** This method is used when substances are steam volatile and immiscible with water.

   - In steam distillation, steam from a steam generator is passed through a heated flask containing the liquid to be distilled. The mixture of steam and the volatile organic compound so obtained is condensed and collected. The compound is later separated from water using a separating funnel.
   - Aniline, nitrobenzene, sandal wood oil, bromobenzene, *o*-nitrophenol, *o*-hydroxy acetophenone are obtained or purified by steam distillation method.

4. **Differential extraction** is used to separate organic compound present in aqueous medium, it is separated by shaking it with an organic solvent, in which it is more soluble than in water. The organic solvent and the aqueous solution should be immiscible with each other so that they form two distinct layers which can be separated by separating funnel.
5. **Chromatography** is an important technique extensively used to separate mixtures into their components. In this technique, the mixture of substances is applied onto a stationary phase (solid or liquid). A pure solvent, a mixture of solvents or a gas is allowed to move slowly over the stationary phase.

   Based on the principle involved, chromatography is classified into different categories. Two of these are:
   (a) Adsorption chromatography
   (b) Partition chromatography.
6. **Adsorption chromatography** is based on the fact that different compounds are adsorbed on an adsorbent to different degrees. When a mobile phase is allowed to move over a stationary phase (adsorbent), the components of the mixture move by varying distances over the stationary phase.
   - Following are two main types of chromatographic techniques based on the principle of differential adsorption.
     (a) Column chromatography
     (b) Thin layer chromatography
   - In **partition chromatography** the stationary phase is liquid.

Qualitative Analysis

The qualitative analysis of an organic compound implies the detection of all the major elements which can be present in it with the help of suitable chemical tests.

1. **Detection of Carbon and Hydrogen** It is done with the help of CuO test. Here, carbon changes into CO_2 (tested with lime water which develops turbidity) and hydrogen to H_2O (tested with anhydrous copper sulphate which turns blue).
2. Nitrogen, sulphur, halogens and phosphorus present in an organic compound are detected by Lassaigne's test. In Lassaigne's test, the elements present in the organic compound are converted from covalent form into the ionic form by fusing the compound with sodium metal.

$$Na + C + N \xrightarrow{\Delta} NaCN$$

$$2Na + S \xrightarrow{\Delta} Na_2S$$

$$Na + X \xrightarrow{\Delta} NaX \quad (X = Cl, Br, I)$$

2. **Detection of Nitrogen** To detect the presence of nitrogen, freshly prepared $FeSO_4$ solution is added to the Lassaigne's extract and contents are warmed. To this, drops of $FeCl_3$ solutions are then added and solution is acidified with conc. HCl. Appearance of Prussian blue colour confirms presence of nitrogen.

$$6CN^- + Fe^{2+} \longrightarrow [Fe(CN)_6]^{4-}$$

$$3[Fe(CN)_6]^{4-} + 4Fe^{3+} \xrightarrow{xH_2O} \underset{\text{Prussian blue}}{Fe_4[Fe(CN)_6]_3 \cdot xH_2O}$$

3. **Detection of Sulphur** To detect the presence of sulphur, sodium nitroprusside solution are added to a part of Lassaigne's extract. The appearance of purple colouration confirms its presence.

$$Na_2S + Na_2[Fe(CN)_5NO] \longrightarrow \underset{\text{Purple colour}}{Na_4[Fe(CN)_5NOS]}$$

Lead acetate test is also used to test its presence. Formation of black precipitate is the final observation.

$$Na_2S + (CH_3COO)_2Pb \longrightarrow PbS\downarrow + 2CH_3COO^-Na^+$$

4. **Detection of Halogen** To detect the presence of halogens, sodium fusion extract is acidified with nitric acid and then treated with silver nitrate.

$$X^- + Ag^+ \longrightarrow AgX\downarrow + NaNO_3$$

X represents a halogen — Cl, Br or I.

AgCl-white ppt, AgBr-dull yellow ppt, AgI-bright yellow ppt.

5. **Detection of Phosphorus** To detect the presence of phosphorus compound, compound is heated with an oxidising agent (like sodium peroxide), then the phosphorus present in the compound is oxidised to phosphate. The solution is boiled with nitric acid and then treated with ammonium molybdate. A yellow colouration or precipitate indicates the presence of phosphorus.

Quantitative Analysis

The quantitative analysis is carried out to determine the proportions in which different elements are present in an organic compound.

1. **Estimation of Carbon and Hydrogen** The **carbon and hydrogen** are estimated by **Leibig method.**

Percentage of carbon

$$= \frac{12 \times \text{mass of } CO_2}{44 \times \text{mass of organic compound}} \times 100$$

Percentage of hydrogen

$$= \frac{2}{18} \times \frac{\text{mass of } H_2O}{\text{mass of organic compound}} \times 100$$

2. **Estimation of Nitrogen** Nitrogen can be estimated by the following two methods.

(a) **Dumas' Method** Let the mass of organic compound $= m$g, volume of nitrogen collected $= V_1$ mL, Room temperature $= T_1$ K

Volume of nitrogen at STP $= \dfrac{p_1V_1 \times 273}{760 \times T_1}$ (Let it be V mL)

where, p_1 and V_1 are the pressure and volume of nitrogen.

p_1 = atmospheric pressure − aqueous tension

22400 mL N_2 at STP weighs 28 g

V mL N_2 at STP weighs $= \dfrac{28 \times V}{22400}$ g

Percentage of nitrogen $= \dfrac{28 \times V \times 100}{22400 \times m}$

(b) **Kjeldahl's Method** Let the mass of organic compound taken $= m$ g, volume of H_2SO_4 of molarity, M taken $= V$ mL and volume of NaOH of molarity $M = V_1$mL

Percentage of N $= \dfrac{1.4 \times M \times 2\,(V - V_1/2)}{m}$

Kjeldahl's method is not applicable to compounds containing nitrogen as nitro and azo groups.

3. **Estimation of Halogens** Halogens can be estimated by **carius method**

Let the mass of organic compound $= w$ g

Mass of AgX formed $= w_1$ g

Percentage of halogen

$$= \frac{\text{atomic mass of } X \times \text{mass of Ag}X}{\text{molecular mass of Ag}X \times m} \times 100$$

4. **Estimation of Sulphur** Sulphur be estimated by **Carius method**.

Percentage of sulphur

$$= \frac{32 \times \text{mass of } BaSO_4 \times 100}{233 \times \text{mass of organic compound}}$$

5. **Estimation of Phosphorus** For the estimation of phosphorus weighed amount of organic is heated with conc. HNO_3 in a Carius tube.

Percentage of phosphorus

$$= \frac{62 \times \text{mass of } Mg_2P_2O_7}{222 \times \text{weight of organic compound}} \times 100$$

Percentage of phosphorus

$$= \frac{31 \times \text{mass of } (NH_4)_3PO_4 \cdot 12MoO_3}{1877 \times \text{mass of organic compound}} \times 100$$

Calculation of Empirical and Molecular Formulae

Empirical formula expresses the relative number of atoms present in the molecule. It is calculated from percentage composition of the compound.

Molar mass = (empirical formula mass) × n

$$n = \frac{\text{Molar mass}}{\text{Empirical formula mass}}$$

Eudiometry is an excellent method to determine the molecular formula of a gaseous hydrocarbon.

DAY PRACTICE SESSION 1

FOUNDATION QUESTIONS EXERCISE

1 Electronegativity of carbon atoms depend upon their state of hybridisation. In which of the following compounds, the carbon marked with asterisk is most electronegative?

(a) $CH_3-CH_2-{}^*CH_2-CH_3$
(b) $CH_3-{}^*CH=CH-CH_3$
(c) $CH_3-CH_2-C\equiv{}^*CH$
(d) $CH_3-CH_2-CH={}^*CH_2$

2 Which of the following compounds contain only primary hydrogen atoms?

(a) Isobutene (b) 2, 3 - dimethylbut - 2 - ene
(c) Cyclohexane (d) Propyne

3 The IUPAC name of neopentane is → AIEEE 2009

(a) 2-methylbutane
(b) 2, 2- dimethylpropane
(c) 2-methylropane
(d) 2, 2-dimethylbutane

4 The IUPAC name of the compound shown below is

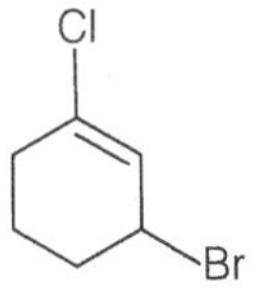

(a) 2-bromo-6-chlorocyclohex-1-ene
(b) 6-bromo-2-chlorocyclohexene
(c) 3-bromo-1-chlorocyclohexene
(d) 1-bromo-3-chlorocyclohexene

5 Which of the following compounds represents 2,2,3 - trimethyl hexane?

(a) $CH_3C(CH_3)_2\ CH_2CH_2CH(CH_3)_2$
(b) $CH_3C(CH_3)_2\ CH_2CH(CH_3)_2CH_2CH_3$
(c) $CH_3C(CH_3)_2\ CH(CH_3)CH_2\ CH_2CH_3$
(d) $CH_3C(CH_3)_2\ CH_2C(CH_3)_2CH_3$

6 The structural formula of 2-oxo-3-methyl-(N-bromo) butanamide is

(a) $CH_3-CH_2-CO-CO-NH-Br$
(b) $CH_3-\underset{}{\overset{CH_3}{\overset{|}{CH}}}-CO-CO-NH-Br$
(c) $CH_3CH-CO-CO-\overset{CH_3}{\overset{|}{N}}OBr$
(d) $(CH_3)_3C-CO-CO-NHBr$

7 The maximum number of isomer for an alkene with the molecular formula C_4H_8 is

(a) two (b) three
(c) four (d) five

8 In which of the following functional group isomerism is not possible?

(a) Alcohols (b) Aldehydes
(c) Alkyl halides (d) Cyanides

9 Identify the compound that exhibits tautomerism

(a) 2-butene (b) lactic acid
(c) 2-pentanone (d) ethane

10 The number of stereoisomers possible for a compound of the molecular formula

$CH_3-CH=CH-CH(OH)\ Me$ is

(a) 3 (b) 2 (c) 4 (d) 6

11 3-methylpent-2-ene on reaction with HBr in presence of peroxide forms an addition product. The number of possible stereoisomers for the product is → JEE Main 2017

(a) six (b) zero
(c) two (d) four

12 Which of the following compound will exhibit geometrical isomerism?

(a) 1- phenyl-2-butene (b) 3 - phenyl-1-butene
(c) 2 - phenyl-1-butene (d) 1,1- diphenyl-1-propane

13 Which of the following compounds will show metamerism?

(a) $CH_3COOC_2H_5$ (b) $C_2H_5-S-C_2H_5$
(c) $C_2H_5-O-C_2H_5$ (d) $CH_3-O-C_2H_5$

14 Nitroethane can exhibit one of the following kind of isomerism.

(a) Metamerism (b) Optical activity
(c) Tautomerism (d) Position isomerism

15 The correct order of nucleophilicity is

(a) $I^- > Br^- > Cl^- > F^-$ (b) $Cl^- > F^- > Br^- > I^-$
(c) $F^- > Cl^- > Br^- > I^-$ (d) $I^- > Cl^- > Br^- > F^-$

16 How many chiral compounds are possible on monochlorination of 2-methylbutane? → AIEEE 2012

(a) 8 (b) 2
(c) 4 (d) 6

17 Which one of the following is most stable?

→ JEE Main (Online) 2013

(a) 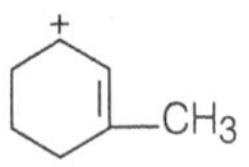

(b) 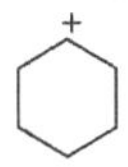

(c) 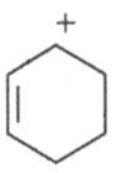

(d) 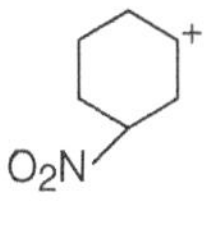

18 The order of stability of the following carbocations → JEE Main (Online) 2013

$CH_2{=}CH{-}\overset{\oplus}{C}H_2$; $CH_3{-}CH_2{-}\overset{\oplus}{C}H_2$; 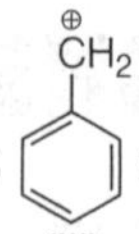 is
(I) (II) (III)

(a) III > II > I (b) II > III > I (c) I > II > III (d) III > I > II

19 Which among the following free radical is most stable?

(a) 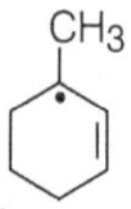

(b) 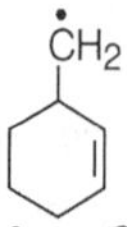

(c) $\dot{C}H_3$

(d) 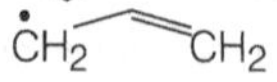

20 Which is the most stable carbocation?

(a) 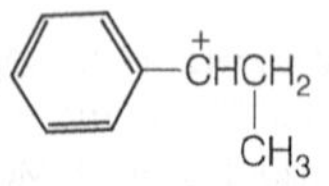

(b) 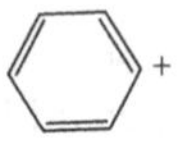

(c) 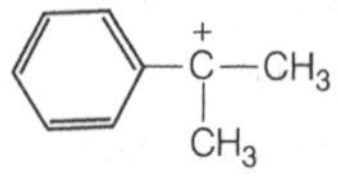

(d) 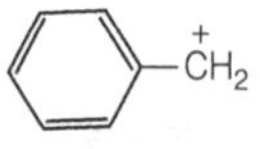

21 The most stable carbanion is

(a) $Ph\overset{\ominus}{C}H_2$

(b) 

(c) 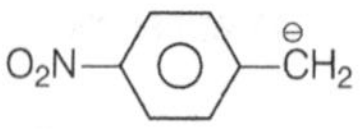

(d) $PhCH_2\overset{\ominus}{C}H_2$

22 The most stable free radical is

(a) $PhCH_2\dot{C}H_2$ (b) $Me\dot{C}H_2$ (c) $Me_2\dot{C}H$ (d) $Ph\dot{C}HMe$

23 Correct order of nucleophilicity is

(a) $CH_3^- < NH_2^- < OH^- < F^-$ (b) $F^- < OH^- < CH_3^- < NH_2^-$
(c) $OH^- < NH_2^- < F^- < CH_3^-$ (d) $F^- < OH^- < NH_2^- < CH_3^-$

24 The correct order of increasing basicity of the given conjugate bases ($R = CH_3$) is → AIEEE 2010

(a) $RCO\bar{O} < HC{\equiv}\bar{C} < \bar{R} < \bar{N}H_2$

(b) $\bar{R} < HC{\equiv}\bar{C} < RCO\bar{O} < \bar{N}H_2$

(c) $RCO\bar{O} < \bar{N}H_2 < HC{\equiv}\bar{C} < \bar{R}$

(d) $RCO\bar{O} < HC{\equiv}\bar{C} < \bar{N}H_2 < \bar{R}$

25 Among the following, the least stable resonance structure is

(a) 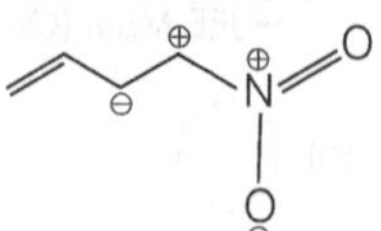

(b) 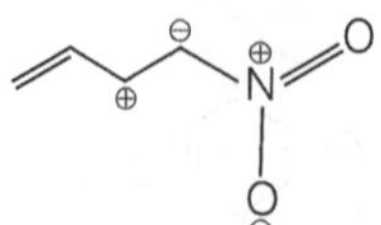

(c) 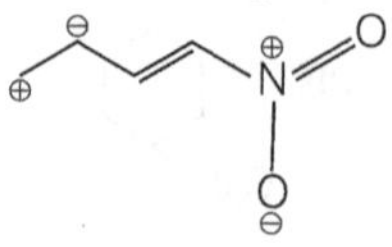

(d) 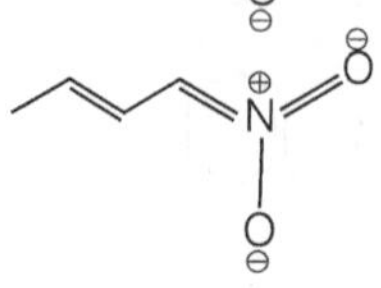

26 Which of the following molecules is least resonance stabilised? → JEE Main 2017

(a) 

(b) 

(c) 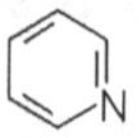

(d) 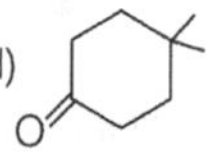

27 The addition of HCl to an alkene proceeds in two steps. The first step is the attack of H^+ ion to portion which can be shown as

(a) 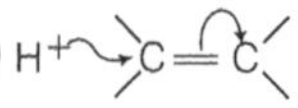

(b) 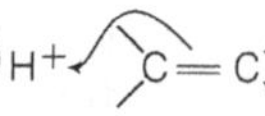

(c) 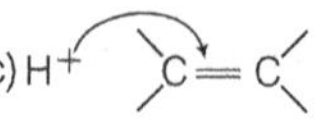

(d) All of these

28 The most suitable method for separation of 1 : 1 mixture of *ortho* and *para*-nitrophenols is

(a) sublimation (b) chromatography
(c) crystallisation (d) steam distillation

29 Two solids *A* and *B* have appreciably different solubilities in water but their melting points are very close. The mixture of *A* and *B* can be separated by

(a) sublimation (b) fractional crystallisation
(c) distillation (d) Both (b) and (c)

30 The distillation technique most suited for separating glycerol from spent lye in the soap industry is → JEE Main 2016

(a) fractional distillation
(b) steam distillation
(c) distillation under reduced pressure
(d) simple distillation

31 The compound that does not give a blue colour in Lassaigne's test is

(a) aniline (b) glycine (c) hydrazine (d) urea

32 The Lassaigne's extract is boiled with dil. HNO_3 before testing for halogens because

(a) Ag_2S is insoluble in HNO_3
(b) AgCN is soluble in HNO_3
(c) Na_2S and NaCN are decomposed by HNO_3
(d) silver halides are soluble in HNO_3

33 In Carius method of estimation of halogens 250 mg of an organic compound gave 141mg of AgBr. The percentage of bromine in the compound is (at. mass Ag = 108, Br = 80) → JEE Main 2015

(a) 24 (b) 36 (c) 48 (d) 60

34 Which of the following compounds will be suitable for Kjeldahl's method for nitrogen estimation? → JEE Main 2018

(a) 

(b) 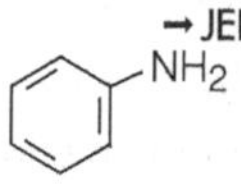

(c) 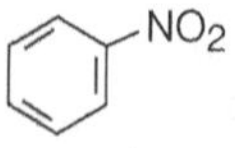

(d) 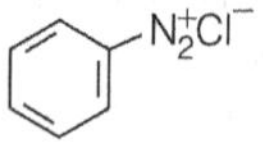

35 Match the following and choose the correct option.

	Column I		Column II
A.	Dumas method	1.	$AgNO_3$
B.	Kjeldahl's method	2.	Silica gel
C.	Carius method	3.	Nitrogen gel
D.	Chromatography	4.	Ammonium Sulphate

Codes

	A	B	C	D		A	B	C	D
(a)	2	4	1	3	(b)	3	4	1	2
(c)	4	2	3	1	(d)	4	1	3	2

36 For the estimation of nitrogen, 1.4 g of an organic compound was digested by Kjeldahl's method and the evolved ammonia was absorbed in 60 mL of $\frac{M}{10}$ sulphuric acid. The unreacted acid required 20 mL of $\frac{M}{10}$ sodium hydroxide for complete neutralisation. The precentage of nitrogen in the compound is → JEE Main 2014

(a) 6% (b) 10% (c) 3% (d) 5%

37 29.5 mg of an organic compound containing nitrogen was digested according to Kjeldahl's method and the evolved ammonia was absorbed in 20 mL of 0.1 M HCl solution. The excess of the acid required 15 mL of 0.1 M NaOH solution for complete neutralisation. The percentage of nitrogen in the compound is → AIEEE 2010

(a) 59.0 (b) 47.4 (c) 23.7 (d) 29.5

38 A gaseous hydrocarbon gives upon combustion 0.72 g of water and 3.08 g of CO_2. The empirical formula of the hydrocarbon is → JEE Main (Online) 2013

(a) C_2H_4 (b) C_3H_4 (c) C_6H_5 (d) C_7H_8

39. 60 g of organic compound on analysis gave following results C = 24 g; H = 4 g and O = 32 g.

The compound can be

(a) CH_2O_2 (b) C_2H_2O (c) $C_2H_2O_4$ (d) CH_2O

Direction (Q. Nos. 40-41) *In the following questions Assertion (A) followed by Reason (R) is given. Choose the correct option out of the choices given below.*

(a) Assertion and Reason both are correct statements and Reason is the correct explanation of the Assertion
(b) Assertion and Reason both are correct statements but Reason is not the correct explanation of the Assertion
(c) Assertion is correct incorrect and Reason is incorrect
(d) Both Assertion and Reason are incorrect

40 Assertion (A) Simple distillation can help in separating a mixture of propan-1-ol (boiling point 97°C) and propanone (boiling point 56°C).

Reason (R) Liquids with a difference of more than 20°C in their boiling points can be separated by simple distillation.

41 Assertion (A) Sulphur present in an organic compound can be estimated quantitatively by Carius method.

Reason (R) Sulphur is separated easily from other atoms in the molecule and gets precipitated as light yellow solid.

42 Assertion (A) Cyclopentadienyl anion is much more stable than allyl anion.

Reason (R) Cyclopentadienyl anion is aromatic in nature.

43 Assertion (A) Phenol is more reactive than benzene towards electrophilic substitution reaction.

Reason (R) In case of phenol, the intermediate carbocation is more resonance stabilised.

DAY PRACTICE SESSION 2

PROGRESSIVE QUESTIONS EXERCISE

1 The bond between carbon atom (1) and carbon atom (2) in compound $N \equiv \overset{1}{C} - \overset{2}{C}H = CH_2$ involves respectively the hybrid orbitals

(a) sp^2 and sp^2 (b) sp^3 and sp
(c) sp and sp^2 (d) sp and sp

2 The name of $ClH_2C - \underset{Br}{\underset{|}{C}} = \underset{Br}{\underset{|}{C}} - CH_2Cl$ according to IUPAC nomenclature system is

(a) dichloro dibromobutene
(b) dichloro dibromobutane
(c) 1,4- dichloro- 2,3- dibromobut- 2- ene
(d) 2,3- dibromo -1,4- dichlorobut- 2- ene

3 Arrange the carbanions, $(CH_3)_3\bar{C}$, $\bar{C}Cl_3$, $(CH_3)_2\bar{C}H$, $C_6H_5\bar{C}H_2$, in order of their decreasing stability.

(a) $C_6H_5\bar{C}H_2 > \bar{C}Cl_3 > (CH_3)_3\bar{C} > (CH_3)_2\bar{C}H$
(b) $(CH_3)_2\bar{C}H > \bar{C}Cl_3 > C_6H_5\bar{C}H_2 > (CH_3)_3\bar{C}$
(c) $\bar{C}Cl_3 > C_6H_5\bar{C}H_2 > (CH_3)_2\bar{C}H > (CH_3)_3\bar{C}$
(d) $(CH_3)_3\bar{C} > (CH_3)_2\bar{C}H > C_6H_5\bar{C}H_2 > \bar{C}Cl_3$

4 Arrange the following free radicals in order of decreasing stability.

Methyl (I), Vinyl (II), Allyl (III), Benzyl (IV)

Codes

(a) I > II > III > IV (b) III > II > I > IV
(c) II > I > IV > III (d) IV > III > I > II

5 On the basis of hybridisation and resonance effect, the correct order of bond length of the following is

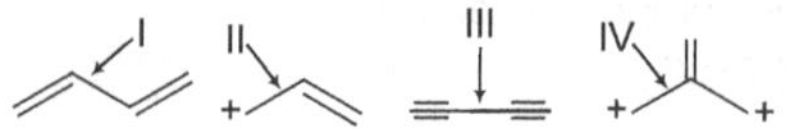

(a) II > IV > III > I (b) I > III > IV > II
(c) II > IV > I > III (d) III > I > IV > II

6 Which of the following order is true regarding the acidic nature of COOH?
(a) Formic acid > acetic acid > propanoic acid
(b) Formic acid > acetic acid < propanoic acid
(c) Formic acid < acetic acid > propanoic acid
(d) Formic acid < acetic acid < propanoic acid

7 Out of the following the alkene that exhibits optical isomerism is
(a) 3-methyl-2-pentene (b) 4-methyl-1-pentene
(c) 3-methyl-1-pentene (d) 2-methyl-2-pentene

8 The correct stability order for the following species is

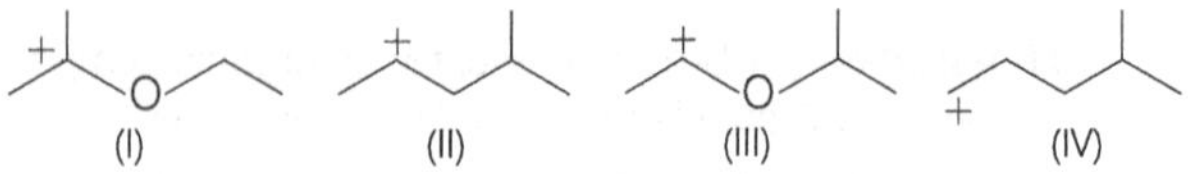

(I) (II) (III) (IV)

(a) (II) > (IV) > (I) > (III) (b) (I) > (II) > (III) > (IV)
(c) (II) > (I) > (IV) > (III) (d) (I) > (III) > (II) > (IV)

9 Which of the following resonating structures of 1-methoxy-1, 3-butadiene is least stable?
(a) $\overset{\ominus}{C}H_2-CH=CH-CH=\overset{\oplus}{O}-CH_3$
(b) $CH_2=CH-\overset{\ominus}{C}H-CH=\overset{\oplus}{O}-CH_3$
(c) $\overset{\ominus}{C}H_2-\overset{\oplus}{C}H-CH=CH-O-CH_3$
(d) $CH_2=CH-\overset{\ominus}{C}H-\overset{\oplus}{C}H-O-CH_3$

10 Consider the following reaction :

$CH_3CH(CH_3)CH=CH_2 + HBr \longrightarrow A$; *A* (predominant) is

(a) $CH_3CH(CH_3)CH(Br)CH_3$ (b) $CH_3CH(CH_3)CH_2CH_2Br$
(c) $CH_3C(Br)(CH_3)CH_2CH_3$ (d) None of these

11 Which of the following compounds is wrongly named?
(a) $CH_3CH_2CH_2CH(Cl)-COOH$; 2–chloropentanoic acid
(b) $CH_3CH=CCH_2-CH(CH_3)COOH$; 2–methylhex–3–enoic acid
(c) $CH_3CH_2CH=CHCOCH_3$; Hex-3-en-2-one
(d) $CH_3CH(CH_3)CH_2CH_2CHO$; 4-methylpentanal

12

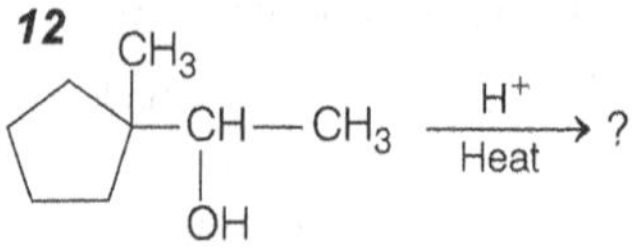

Consider the above written reaction. The product is

(a) 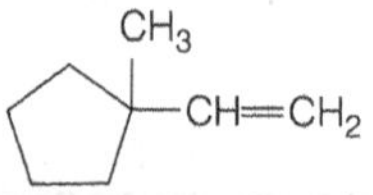 (b) 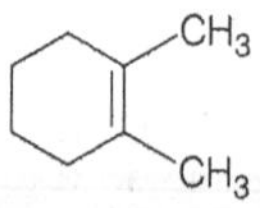

(c) 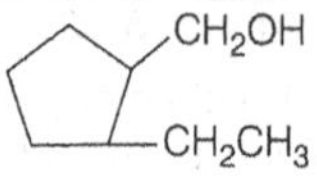 (d) 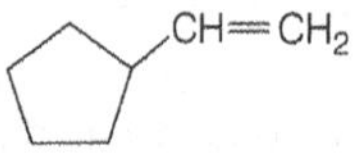

13 Indicate the wrongly named compound.
(a) $CH_3-CH(CH_3)-CH_2-CH_2-CHO$
(4-methyl-1-pentanal)
(b) $CH_3-CH(CH_3)-C\equiv C-COOH$
(4-methyl-2-pentyn-1-oic acid)
(c) $CH_3CH_2CH_2-CH(CH_3)-COOH$
(2-methyl-1-pentanoic acid)
(d) $CH_3CH=CH-CH_2-C(=O)-CH_3$
(3-hexen-2-one)

14 Identify the binary mixtures that can be separated into individual compounds, by differential extraction, as shown in the given scheme.

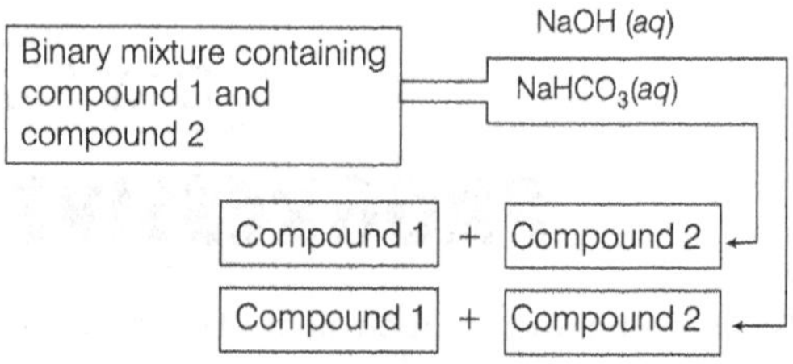

(a) C_6H_5OH and C_6H_5COOH
(b) C_6H_5COOH and $C_6H_5CH_2OH$
(c) $C_6H_5CH_2OH$ and C_6H_5OH
(d) $C_6H_5CH_2OH$, C_6H_5CHO

15 Which of the following statement is wrong?
(a) Using Lassaigne's test nitrogen or sulphur present in an organic compound can be tested
(b) Using Beilstein's test the presence of halogen in a compound can be tested
(c) Lassaigne's test fails to identify nitrogen in diazo compound
(d) In the estimation of carbon an organic compound is heated with CuO in a combustion tube

16 Which of the following orders is not correct regarding $-I$ -effect of the substitutents?

(a) —I<—Cl<—Br<—F (b) $-\overset{+}{N}R_3 < -\overset{+}{O}R_2$

(c) $-\overset{+}{N}R_2 < -OR < -F$ (d) $-SR < -OR < -\overset{+}{O}R_2$

17 Consider the following carbocations.

I. $C_6H_5\overset{+}{C}H_2$ II. $C_6H_5CH_2\overset{+}{C}H_2$

III. $C_6H_5\overset{+}{C}HCH_3$ IV. $C_6H_5\overset{+}{C}(CH_3)_2$

The correct sequence of the stability of these carbocation is

(a) II<I<III<IV (b) II<III<I< IV

(c) III<I<II<IV (d) IV<III<I<II

18 In Kjeldahl's method, the gas evolved from 1.325 g sample of fertiliser is passed into 50.0 mL of 0.2030 N H_2SO_4. 25.32 mL of 0.1980 N NaOH are required for the titration of unused acid. What will be the percentage of nitrogen in fertiliser?

(a) 2.50 % (b) 5.43 %

(c) 6.48 % (d) 12.02 %

19. Among the following four structure I and IV

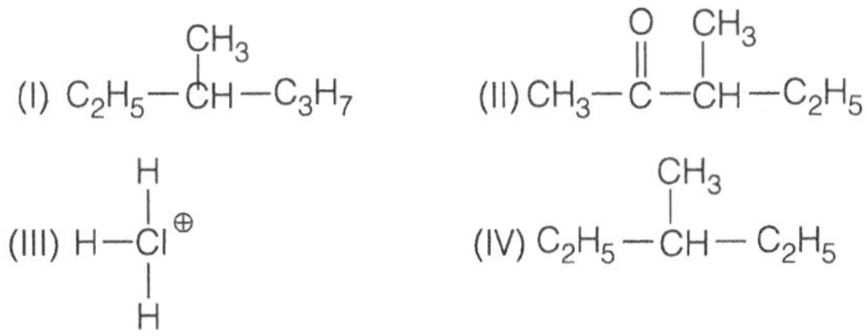

It is true that

(a) all four are chiral compounds

(b) only I and II are chiral compounds

(c) only III is a chiral compound

(d) only II and IV are chiral compounds

ANSWERS

SESSION 1	**1** (c)	**2** (b)	**3** (b)	**4** (c)	**5** (c)	**6** (b)	**7** (c)	**8** (c)	**9** (c)	**10** (c)
	11 (d)	**12** (a)	**13** (a)	**14** (c)	**15** (a)	**16** (b)	**17** (a)	**18** (d)	**19** (a)	**20** (c)
	21 (c)	**22** (d)	**23** (d)	**24** (d)	**25** (a)	**26** (d)	**27** (b)	**28** (d)	**29** (b)	**30** (c)
	31 (c)	**32** (c)	**33** (a)	**34** (b)	**35** (b)	**36** (b)	**37** (c)	**38** (d)	**39** (d)	**40** (a)
	41 (c)	**42** (a)	**43** (a)							

SESSION 2	**1** (c)	**2** (d)	**3** (c)	**4** (d)	**5** (a)	**6** (a)	**7** (c)	**8** (d)	**9** (c)	**10** (c)
	11 (b)	**12** (b)	**13** (d)	**14** (b)	**15** (d)	**16** (a)	**17** (a)	**18** (b)	**19** (b)	

Hints and Explanations

SESSION 1

1 Electronegativity of carbon atom depends on their state of hybridisation.More the s-character, more is the electronegativity.

$$sp^2 < sp^2 < sp$$

s-character 25% 33% 50%

Thus, sp- carbon has the highest electronegativity, i.e. option (c) ($CH_3-CH_2-C\equiv{}^*CH$) is correct.

2 2,3-dimethylbut-2-ene contain only primary hydrogen atoms

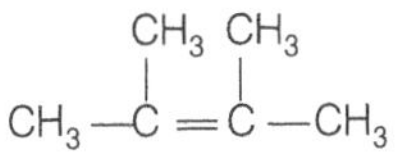

3 The neopentane

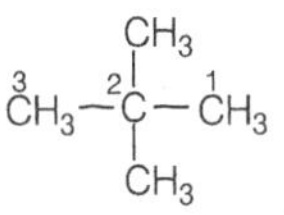

IUPAC name : 2, 2-dimethylpropane

4

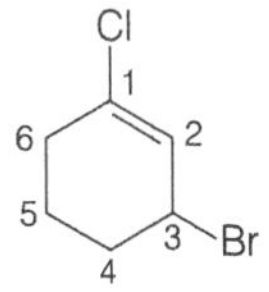

Unsaturation, i.e. double bond is at priority over halogen. So, the correct IUPAC name is 3-bromo-1-chlorocyclohexene.

5 2,2,3-trimethylhexane is represented by

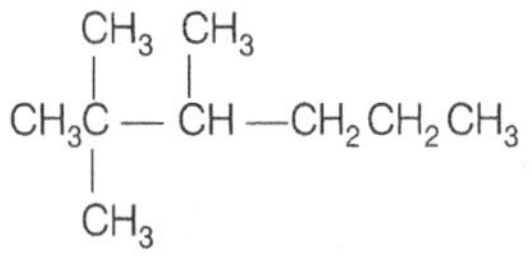

6 2-oxo-3-methyl-(N-bromo) butanamide is

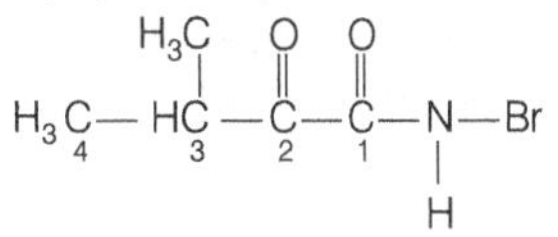

7 The maximum number of isomer for an alkene with molecular formula C_4H_8 is four.

(i) $CH_3CH_2CH=CH_2$ (But-1-ene)

(ii) 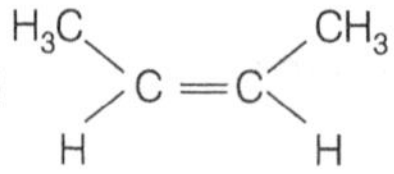 (*cis* but-2-ene)

(iii) 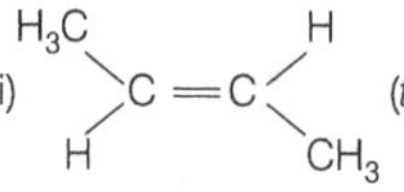 (*trans* but-2-ene)

(iv) 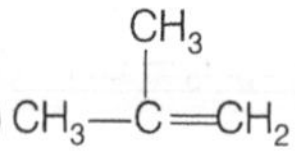 (2-methylpropene)

8 Alkyl halides do not show functional group isomerism.

9 Tautomerism is due to spontaneous interconversion of two isomeric forms with different functional groups into each other. The term tautomer means constitutional isomers that undergo such rapid interconversion so that they cannot be independently isolated.

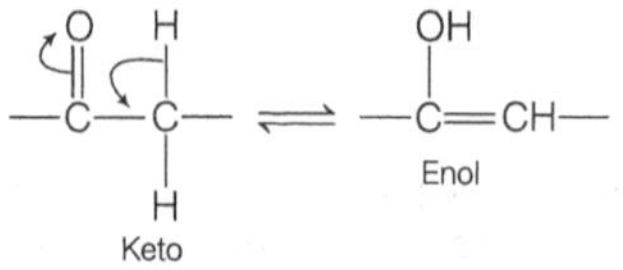

Thus, (c) 2-pentanone exhibit tautomerism.

10 For the given compound ($CH_3-CH=CH-CH(OH)$ Me, the total number of stereoisomers are four.

11 The number of stereoisomers in molecules which are not divisible into two equal halves and have *n* number of asymmetric C-atoms = 2^n.

3-methyl pent-2-ene on reaction with HBr in presence of peroxide forms an addition product, i.e. 2-bromo-3-methyl pentane. It has two chiral centres. Therefore, 4 stereoisomers are possible

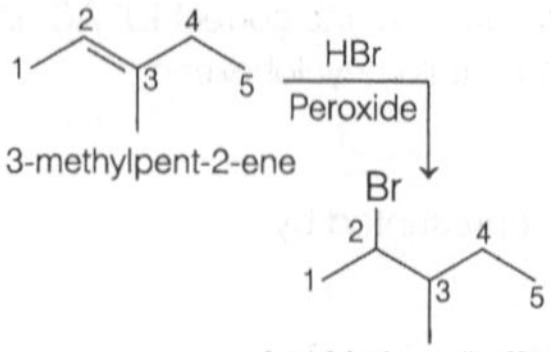

3-methylpent-2-ene

Anti-Markownikoff's addition
Four stereoisomers are possible
(As molecule has two chiral centres and is asymmetric).

12 Alkene in which different groups are attached with the double bonded carbon atoms exhibit geometrical isomerism.

1-phenyl- 2-butene 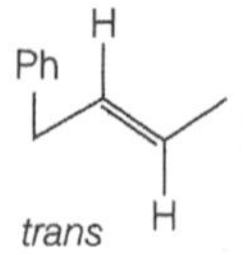,

trans

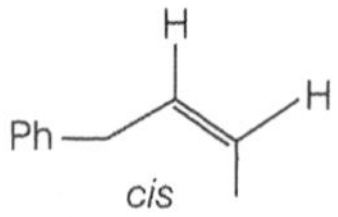

cis

It will show geometrical isomerism.

3-phenyl-1-butene
Same group, does not show geometrical isomerism.

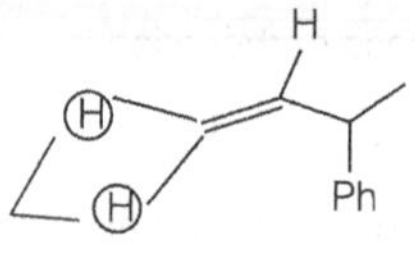

2-phenyl-1-butene
Same group, does not show geometrical isomerism.

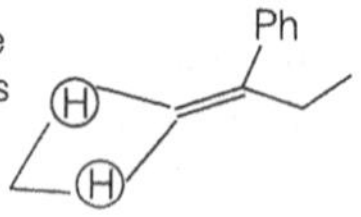

1,1-diphenyl -1-propane being an alkane (saturated compound does show geometrical isomerism.)

13 Metamerism is possible in ethers, thioethers, secondary amines, esters etc. The metamers belong to the same family but differ in the nature of alkyl groups, e.g.

So, $CH_3COOC_2H_5$ ethyl acetate will not show metamerism.

14 Nitroethane exhibits tautomerism

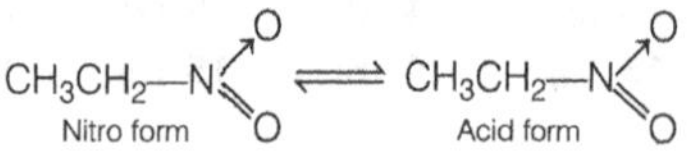

15 The correct order of nucleophilicity is
$I^- > Br^- > Cl^- > F^-$

16 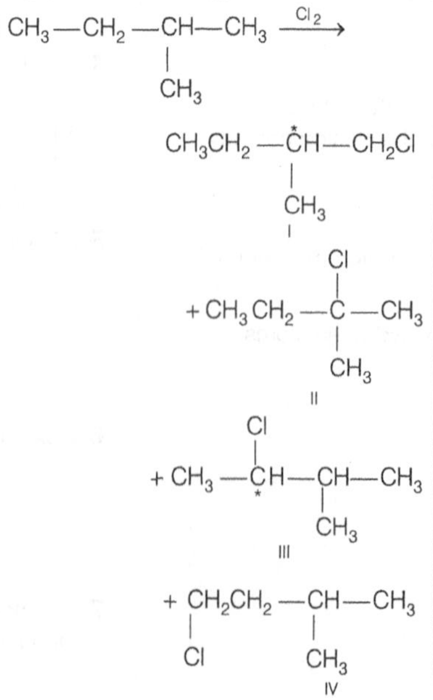

I and III compound are chiral. Therefore, two chiral compounds are possible on monochlorination of 2-methylbutane.

17 Due to the presence of electron releasing group ($-CH_3$) and resonance effect,

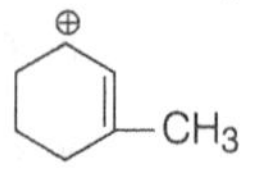 is most stable.

18 The order of stability of carbocation is

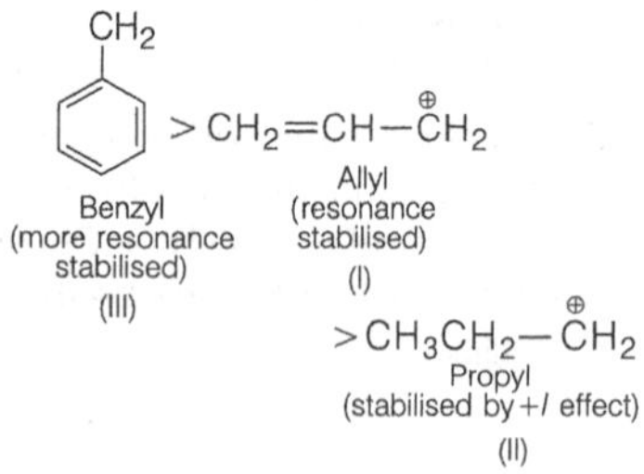

19 Among the given free radicals, allyl free radical 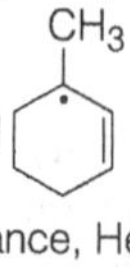 is stabilised by resonance, Hence, is more stable.

20 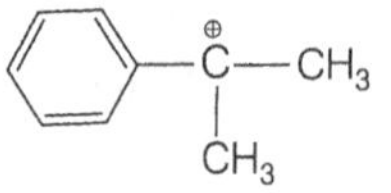

3° carbocation is more stable.

21 EWG stabilises carbanion, $-NO_2$ is EWG both by $-I$ and $-R$. Hence, (c) is the most stable.

22 The most stable free radical is 2° benzyl radical. (PhCHMe) due to resonance.

23 Nucleophilicity

$$\propto \frac{1}{\text{electronegativity of atom bearing negative charge}}$$

So, the correct order is
$F^- < OH^- < NH_2^- < CH_3^-$.

24 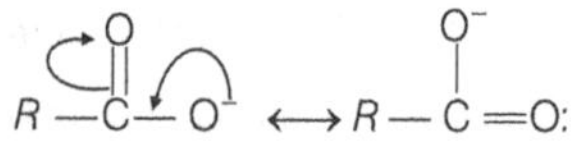

In carboxylate ion, the negative charge is present on oxygen, a most electronegative element here, thus it is resonance stabilised. In $HC\equiv C^-$, carbon is *sp*-hybridised so its electronegativity is much higher than nitrogen.

In $\bar{N}H_2$, nitrogen is more electronegative than sp^3-hybridised C-atom. Thus, the order of the stability of conjugated bases is

$$RCOO^- > HC \equiv C^- > \bar{N}H_2 > R^-$$

and higher is the stability of conjugated bases, lower will be basic character. Hence, the order of basic character is as

$$RCOO^- < HC \equiv C^- < \bar{N}H_2 < R^-$$

25 The least stable resonance structure is (a). Here, two positive charges present at the adjacent place, elevates the energy, thus lowers the stability.

26 Aromatic compounds are stable due to resonance while non-aromatics are not. According to Huckel's rule (or $4n + 2$ rule), "For a planar, cyclic compound to be aromatic, its π cloud must contain $(4n + 2)\pi$ electrons, where, n is any whole number." Thus,

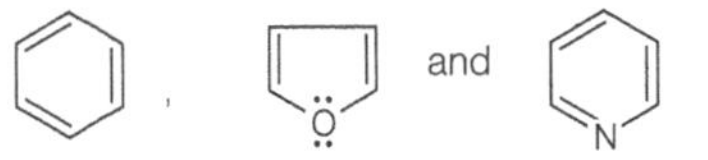

($6\pi e^-$ system) ($4\pi e^- + 2e^-$ system) ($6\pi e^-$ system)

Above given compounds are aromatic and stabilised by resonance. They follow Huckel's rule.

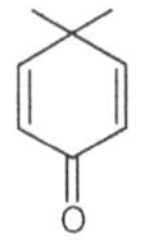

is non-aromatic, hence, least stabilised by resonance.

27 The first step in addition of HCl to an alkene proceeds as

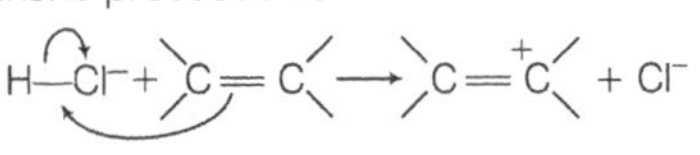

28 *o*-nitrophenol exists as monomer because of chelation and hence steam volatile. Thus, steam distillation method is used for separation of 1 : 1 mixture.

29 Fractional crystallisation is used to separate two solids *A* and *B* having appreciable different solubilities in water but their melting points are very close.

30 Glycerol with high boiling point (290°C) can be separated from spent lye by distillation under reduced pressure. This process is used to purify liquids having very high boiling points. By this process, liquid is made to boil at lower temperature than its boiling point by lowering the pressure on its surface.

31 Hydrazine does not give a blue colour in Lassaigne's, since it does not contain carbon.

32 In order to decompose unreacted sulphide and cyanide, HNO_3 is added.

33 Weight of organic compound = 250 mg

Weight of AgBr = 141 mg

∴ According to formula of % of bromine by Carius method

$$\% \text{ of Br} = \frac{\text{Atomic mass of Br}}{\text{Molecular mass of AgBr}} \times \frac{\text{mass of AgBr}}{\text{mass of organic bromide}} \times 100$$

$$\therefore \quad \% \text{ of Br} = \frac{80}{188} \times \frac{141}{250} \times 100 = \frac{1128000}{47000} = 24\%$$

34 Estimation of nitrogen through Kjeldahl's method is not suitable for organic compounds containing nitrogen in ring or nitrogen in nitro or azo groups. It is because of the fact that nitrogen of these compounds does not show conversion to ammonium sulphate $(NH_4)_2SO_4$ during the process. Hence, among the given compounds only aniline can be used suitably for estimation of nitrogen by Kjeldahl's method.

35 A ⟶ 3; B ⟶ 4 ; C ⟶ 1; D ⟶ 2

Dumas' method - Nitrogen gel

Kjeldahl's method - Ammonium sulphate

Carius method - $AgNO_3$

Chromatography- Silica gel

36 Mass of organic compound = 1.4g

Let it contain x mmole of N-atom.

$$\text{Organic compound} \longrightarrow \underset{x\,\text{mmol}}{NH_3}$$

$$2NH_3 + \underset{\substack{6\text{mmol}\\ \text{initially}\\ \text{taken}}}{H_2SO_4} \longrightarrow (NH_4)_2SO_4 \quad \ldots\text{(i)}$$

$$H_2SO_4 + 2NaOH \longrightarrow Na_2SO_4 + 2H_2O \quad \ldots\text{(ii)}$$

2 mmoles of NaOH reacted.

Hence, mmoles of H_2SO_4 reacted in Eq. (ii) = 1

⇒ mmoles of H_2SO_4 reacted from Eq. (i) = 6 − 1 = 5 mmol

⇒ mmoles of NH_3 in Eq. (i)

$= 2 \times 5 = 10$ mmol

⇒ mmoles of N-atom in the organic compound = 10 mmol

⇒ Mass of N $= 10 \times 10^{-3} \times 14$

$= 0.14$g % of N

$$= \frac{\text{Mass of N present in organic compound}}{\text{Mass of organic compound}} \times 100$$

$$\Rightarrow \% \text{ of N} = \frac{0.14}{1.4} \times 100 = 10\%$$

37 Weight of organic compound = 29.5 mg

$$NH_3 + HCl \longrightarrow NH_4Cl$$

$$\underset{\text{(Remaining)}}{HCl} + \underset{\substack{15 \times 0.1\ M\\ = 1.5\ \text{mmol}}}{NaOH} \longrightarrow NaCl + H_2O$$

Total millimole of HCl = 2

Millimole used by $NH_3 = 2 - 1.5 = 0.5$

Weight of $NH_3 = 0.5 \times 17$ mg = 8.5 mg

Weight of nitrogen

$= \frac{14}{17} \times 8.5$ mg = 7 mg

$$\% \text{ Nitrogen} = \frac{7}{29.5} \times 100 = 23.7\%$$

38 18 g H_2O contains 2 g hydrogen

∴ 0.72 g H_2O contains 0.08 g H. 44 g CO_2 contains 12 g C

∴ 3.08 g CO_2 contains 0.84 g C

$$\therefore C : H = \frac{0.84}{12} : \frac{0.08}{1} = 0.07 : 0.08$$

$= 7 : 8$

∴ Empirical formula = C_7H_8

39 Moles of C is 24 g $= \frac{24\text{g}}{12\text{g}} = 2$

Moles of H is 4 g $= \frac{4\text{g}}{16\text{g}} = 4$

Moles of O is 32 g $= \frac{32\text{g}}{16\text{g}} = 2$

So, molecular formula of compound is $C_2H_4O_2$, i.e. CH_2O

40 Simple distillation can be used to separate a mixture of two liquids which do not react and have boiling point difference of more than 20°C.Hence, a mixture of propan-1-ol and propanone can be separated.

41 S (sulphur) present in an organic compound can be estimated quantitatively by Carius method.In this method, a known weight of organic compound is heated with fuming HNO_3, S present in it gets converted into H_2SO_4. On adding $BaCl_2$,H_2SO_4 gets precipitated as $BaSO_4$ which may be of light yellow or white in colour. If light yellow colour is obtained, it means some impurities are present. It is then filtered, washed, purified and then dried and finally pure $BaSO_4$ of white colour is obtained.

42 Cyclopentadienyl anion is more stable than allyl anion as the former is aromatic and is stabilised by resonance.

43 —OH group is electron donating group, thus activates the benzene ring for electrophilic substitution reaction. Due to more resonance stabilisation of phenoxide ion, it gives more stable carbocation (intermediate).

SESSION 2

1 The bond between carbon atom (1) and carbon atom (2) in compound $N \equiv \overset{1}{C}-\overset{2}{C}H = CH_2$ involves *sp* and sp^2 hybrid orbitals.

2 The IUPAC name of

$$Cl\overset{4}{H_2C}-\underset{Br}{\overset{3}{C}}=\underset{Br}{\overset{2}{C}}-\overset{1}{CH_2}Cl \text{ is}$$

2, 3-dibromo-1,4-dichlorobut-2-ene

3 Presence of electron withdrawing groups stabilises the carbanions. Thus, the order of stability is

$$\bar{C}Cl_3 > C_6H_5\bar{C}H_2 > (CH_3)_2\bar{C}H > (CH_3)_3\bar{C}.$$

4 On the basis of conjugative and hyperconjugative structure. The correct order for the stability of free radical is benzyl (IV) > allyl (III) > methyl (I) > vinyl (II).

5 As the *p*-character increases, bond length also increases. Resonance effect also cause a change in bond length. Thus, the order of bond length is II > IV > III > I.

6 The correct order for the acidic nature of —COOH group is formic acid > acetic acid > propanoic acid

7 $CH_3-CH_2-\underset{CH_3}{\overset{H}{C^*}}-CH=CH_2$

3-methyl-1-pentene

(It has one chiral centre)

8 Carbocation (I) is stabilised by + *R* effect, (effect of O) as well as + *I* effect of two CH_3 groups. Carbocation (II) is stabilised by + *I* effect of CH_3 and $CH_2CH(CH_3)_2$ groups.

Carbocation (III) is stabilised by + *R* effect of O and +*I* effect of one CH_3 group. Carbocation (IV) is stabilised by +*I* effect of $CH_3CH_2CH(CH_3)_2$ group.

Thus, decreasing order of stability of these carbocations is I > III > II > IV

9 The least stable resonating structure of 1-methoxy-1,3-butadiene is $\overset{\ominus}{C}H_2-\overset{\oplus}{C}H-CH=CHOCH_3$. The octet of all atoms are complete in structures (a) and (b). The molecule in which all the atoms have complete octet is more stable than the molecules in which atom have incomplete octet.

In structure (d), the electron deficiency of positive charged carbon is duly compensated by lone pair of electrons of adjacent oxygen atoms while such neighbour group support is not available in structure (c). Hence, structure (c) is least stable in comparison to structure (d).

10 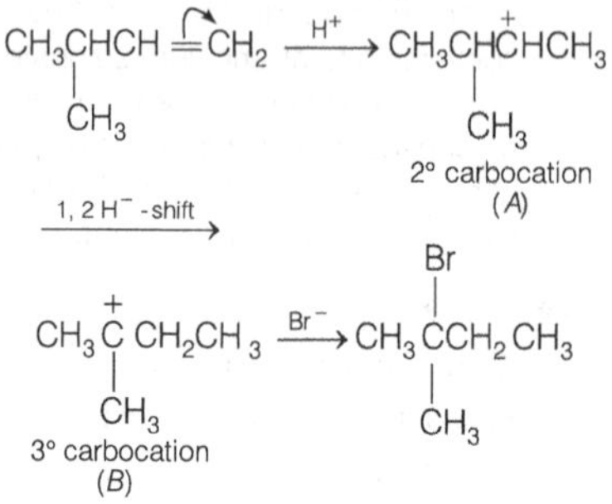

11 The correct IUPAC name of the given structure is 2-methyl hex-2-en-1-oic acid.

$$CH_3CH=\overset{H}{C}CH_2-\overset{CH_3}{C}H-COOH$$

12 It is the elimination reaction with ring expansion.

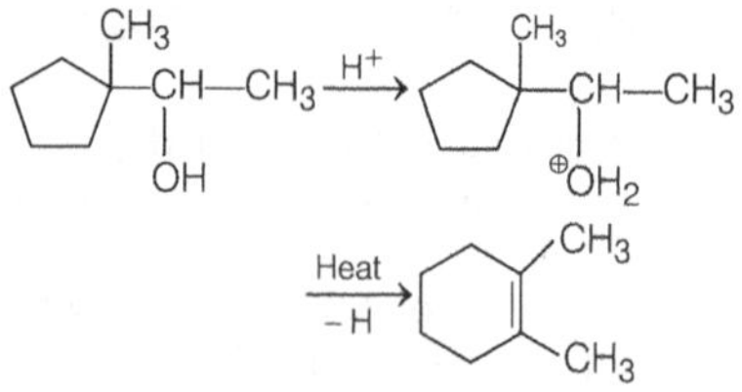

13 The correct IUPAC name is

$$\overset{6}{C}H_3\overset{5}{C}H=\overset{4}{C}H-\overset{3}{C}H_2-\overset{2}{\underset{}{C}}(=O)-\overset{1}{C}H_3$$

(hex-4-en-2-one)

14 For separation of mixture compound III and IV are notchiral due to presence of same group on carbon atom by differential extraction one of the component must form salt with the given base so that the salt will be extracted in aqueous layer leaving other component in organic layer.

(a) Both phenol and benzoic acid forms salt with NaOH but not with $NaHCO_3$ hence, this mixture cannot be separated.

(b) Benzoic acid forms salt with NaOH and $NaHCO_3$ both while benzyl alcohol does not hence, the mixture can be separated using NaOH or $NaHCO_3$.

(c) Neither benzyl alcohol nor phenol form salt with $NaHCO_3$, so mixture cannot be separated by using $NaHCO_3$.

(d) Both $C_6H_5CH_2OH$ and C_6H_5CHO cannot form salt.

15 In estimation of C, organic compound is burnt in the presence of excess of oxygen and CuO.

$$2CuO + C \longrightarrow 2Cu + CO_2$$

16 The correct order regarding —I effect of the substituents is –I < –Br < –Cl < –F

These groups have the ability to withdraw electron density toward themselves.

17 The correct sequence of carbocation is II < I < III < IV. Electron releasing stabilises the carbocation by dispersal of positive charge. More the number of alkyl groups, the greater the dispersal of positive charge and the more stable is the carbocation.

18 Meq. of acid taken

= 0.2030 × 50 = 10.15

Meq. of acid used by NaOH

= 0.1980 × 25. 32 = 5.013

∴ Meq. of acid used for NH_3

= 10.15 – 5.013 = 5.137

or Meq. of NH_3 (i.e.) *N V* = 5.137

$$\therefore \quad \% \ N = \frac{1.4 \times 5.137}{1.325} = 5.43\%$$

19 Only (I) and (II) are chiral compounds as all the four attached substituents are different.

$$C_2H_5-\overset{CH_3}{C}H-C_3H_7$$

(Chiral)

$$H_3C-\overset{O}{\overset{\|}{C}}-\overset{CH_3}{C}H-C_2H_5$$

(Chiral)

DAY TWENTY EIGHT

Hydrocarbons

Learning & Revision for the Day

- Alkanes
- Alkenes
- Alkynes
- Aromatic Hydrocarbons
- Directive Influence of Functional Group in Monosubstituted Benzene

Hydrocarbons means compounds containing carbon and hydrogen only. These may contain only single bonds (saturated hydrocarbons) or double/triple bonds (unsaturated hydrocarbons) alongwith single bond. Hydrocarbons are mainly obtained from petroleum which is the major source of commercial energy.

Hydrocarbons are of different types and classified into three main categories (i) saturated hydrocarbons (alkanes) (ii) unsaturated hydrocarbons (alkenes and alkynes) (iii) aromatic hydrocarbons.

Alkanes

Alkanes are the simplest organic compounds made up of carbon and hydrogen only. Chemically, these are almost inert that's why also called **Paraffins** (Latin, *Parum affinis*=Little affinity). The general formula for this series is C_nH_{2n+2}.

- **Nomenclature** The names of such compounds are based on their chain structure and end with suffix-'ane' and carry a prefix indicating the number of carbon atoms present in the chain.
- **Isomerism** Alkanes shows structural isomerism. Structural isomers which differ in chain of carbon atoms are known as chain isomers. e.g. pentane, 2-methyl butane and 2, 2-dimethyl propane.

General Methods of Preparation

General methods of preparation of alkanes are as follows:

1. **Hydrogenation** Hydrogenation of alkenes and alkynes in the presence of finely divided catalysts like platinum, palladium or nickel to form alkanes.

$$CH_2{=}CH_2 + H_2 \xrightarrow{Pt/Pd/Ni} CH_3{-}CH_3$$

$$CH_3{-}C{\equiv}C{-}H + 2H_2 \xrightarrow{Pt/Pd/Ni} CH_3{-}CH_2{-}CH_3$$

This reaction is known as **Sabatier-Senderens reduction.**

2. **From Alkyl Halides** Alkyl halides on reduction with zinc and dil. HCl produces alkanes.

$$\underset{\text{Chloromethane}}{CH_3Cl} + H_2 \xrightarrow[\text{Reduction}]{Zn,\ H^+} \underset{\text{Methane}}{CH_4} + HCl$$

3. **Wurtz Reaction** Alkyl halides when treated with sodium metal in the presence of dry ether produces higher alkanes. This reaction is known as **Wurtz reaction**.

$$\underset{\text{Ethyl bromide}}{2C_2H_5Br} + 2Na \xrightarrow{\text{Dry ether}} \underset{n\text{-butane}}{C_2H_5{-}C_2H_5} + 2NaBr$$

PREP MIRROR
Your Personal Preparation Indicator

- No. of Questions in Exercises (x)—
- No. of Questions Attempted (y)—
- No. of Correct Questions (z)—
  (Without referring Explanations)

- Accuracy Level ($z/y \times 100$)—
- Prep Level ($z/x \times 100$)—

In order to expect good rank in JEE, your Accuracy Level should be above 85 & Prep Level should be above 75.

4. **Decarboxylation** Sodium salt of carboxylic acid ($R\overset{-}{CO}\overset{+}{ONa}$) when heated with hot sodalime ($CaO + NaOH$) gives alkane.

$$CH_3CO\overset{-}{O}\overset{+}{Na} + NaOH \xrightarrow[\Delta]{CaO} CH_4 + Na_2CO_3$$

This process of elimination of carbon dioxide from a carboxylic acid is known as **decarboxylation.**

5. **Kolbe's Electrolytic Method** Sodium or potassium salt of carboxylic acid on electrolysis gives alkane (R—R) containing even number of carbon atoms at anode.

$$2CH_3CO\overset{-}{O}\overset{+}{Na} + 2H_2O \xrightarrow{\text{Electrolysis}} \underbrace{C_2H_6 + 2CO_2\uparrow}_{\text{Anode}} + \underbrace{H_2 + 2NaOH}_{\text{Cathode}}$$

6. **From Grignard's Reagent** Grignard reagent ($RMgX$) when reacts with H_2O or ROH, gives alkane.

$$CH_3—\overset{-}{CH_2}\boxed{—\overset{+}{Mg}Br + HO}H \longrightarrow \underset{\text{Ethane}}{CH_3—CH_3} + Mg\begin{matrix}OH\\Br\end{matrix}$$

7. **Other Methods of Preparation** Action of water on aluminium carbide or beryllium carbide gives methane.

$$Al_4C_3 + 12H_2O \longrightarrow 4Al(OH)_3 + 3CH_4$$
$$Be_2C + 4H_2O \longrightarrow CH_4 + 2Be(OH)_2$$

These reactions produce methane, hence called methanides.

Physical Properties

Alkanes show following physical properties:

1. **Boiling point** increases with increase in molecular mass. If molecular mass is same and the contact surface area increases, boiling point increases. More branched chain has lesser boiling point.
2. **Melting point** increases with increase in molecular mass. But symmetric molecules have higher melting point than asymmetric ones.

   Therefore, alkanes with even number of carbon atoms have higher melting point than the successive alkanes with odd number of carbon atoms.
3. Alkanes are non-polar, therefore soluble in non-polar solvents such as petroleum, benzene etc.
4. The **densities** of alkanes increase with increase in the molecular masses till the limiting value of about $0.8\ g/cm^3$ is reached.

Chemical Properties

Various chemical properties of alkanes are as follow:

1. **Halogenation of Alkanes** It is a free radical substitution reaction which occurs in the presence of UV light.
   - Rate of replacement of H of alkanes $3° > 2° > 1°$.

   The steps involved in chlorination of methane is

$$CH_4 + Cl_2 \xrightarrow{h\nu} CH_3Cl \xrightarrow{Cl_2/h\nu} CH_2Cl_2 \xrightarrow{Cl_2/h\nu} CHCl_3 \xrightarrow{Cl_2/h\nu} CCl_4$$

   - Bromination is same as chlorination but occurs at a slightly slower rate. Iodination is reversible but fluorination of alkanes is too vigorous.

The **mechanism of halogenation** can be discussed in the following three steps:

(i) **Initiation** The reaction is initiated by homolysis of chlorine molecule in the presence of light or heat.

$$Cl—Cl \xrightarrow[\text{Homolysis}]{h\nu} \underset{\text{Chlorine free radicals}}{\dot{Cl} + \dot{Cl}}$$

(ii) **Propagation** Chlorine free radical reacts with methane molecule and generates methyl free radical.

$$CH_4 + \dot{Cl} \xrightarrow{h\nu} \dot{CH_3} + H—Cl$$
$$\dot{CH_3} + Cl—Cl \xrightarrow{h\nu} CH_3—Cl + \dot{Cl}$$
$$CH_3Cl + \dot{Cl} \longrightarrow \dot{CH_2}Cl + HCl$$
$$\dot{CH_2}Cl + Cl—Cl \longrightarrow CH_2Cl_2 + \dot{Cl}$$

(iii) **Termination** The possible chain terminating steps are

$$\dot{Cl} + \dot{Cl} \longrightarrow Cl—Cl$$
$$H_3\dot{C} + \dot{CH_3} \longrightarrow CH_3—CH_3$$
$$H_3\dot{C} + \dot{Cl} \longrightarrow CH_3—Cl$$

Due to consumption of free radicals, chain is terminated.

2. **Combustion** Combustion of alkanes give CO_2 and H_2O.

$$CH_4(g) + 2O_2(g) \longrightarrow CO_2(g) + 2H_2O(l)$$

   During incomplete combustion alkanes give carbon black which is used in the manufacture of ink, printer ink, black pigments and as filters.

3. **Controlled Oxidation** On controlled oxidation (at high pressure and in presence of suitable catalyst) alkanes give different products as

$$2CH_4 + O_2 \xrightarrow{Cu/523K/100\ atm} 2CH_3OH$$
$$CH_4 + O_2 \xrightarrow{Mo_2O_3} HCHO + H_2O$$
$$2CH_3—CH_3 + 3O_2 \xrightarrow[\Delta]{(CH_3COO)_2Mn} 2CH_3COOH + 2H_2O$$

   Ordinarily alkanes resist oxidation but *iso*-alkanes (3°H) on oxidation with $KMnO_4$ gives tertiary alcohol.

$$(CH_3)_3CH \xrightarrow[\text{Oxidation}]{KMnO_4} \underset{3°\text{alcohol}}{(CH_3)_3C—OH}$$

4. **Isomerisation** On heating in presence of $AlCl_3$ (anhydrous) and $HCl(g)$ n-alkane isomerises to give branched chain alkanes.

$$\underset{n\text{-hexane}}{CH_3—(CH_2)_4—CH_3} \xrightarrow{AlCl_3\,(anhy) + HCl}$$

$$CH_3—\underset{\substack{|\\CH_3}}{CH}—CH_2CH_2CH_3 + CH_3CH_2—\underset{\substack{|\\CH_3}}{CH}—CH_2CH_3$$

5. **Aromatisation** *n*-alkanes having six or more carbon atoms on heating to 773 K at 10-20 atmospheric pressure in the presence of oxides of vanadium, Mo or Cr supported over alumina get dehydrogenated and cyclised to benzene. This reaction is known as **aromatisation**.

$$n\text{-hexane} \xrightarrow[773K,\ 10\text{-}20\ atm]{Cr_2O_3\ or\ V_2O_5}$$ 

Benzene

6. **Reaction With steam** Alkane (such as methane) react with steam at 1273 K in presence of Ni catalyst to form CO and H_2.

$$CH_4 + H_2O(g) \xrightarrow[1273\ K]{Ni} CO + 3H_2$$

This method is used for industrial preparation of H_2 gas.

7. **Pyrolysis or Cracking** Higher alkanes on heating to higher temperature decompose into lower alkanes. This reaction is called **pyrolysis** or **cracking.**

$$C_6H_{14} \xrightarrow{773K} \begin{cases} C_6H_{12} + H_2 \\ C_4H_8 + C_2H_6 \\ C_3H_6 + C_2H_4 + CH_4 \end{cases}$$

Conformations

Spatial arrangements of atoms which can be converted into one another by rotation around a C—C single bond are called conformations or conformers or rotamers. Ethane has three stable conformations. These are follows :

1. **Eclipsed conformation** in which the hydrogen atoms attached to two carbons are as closed together as possible. This conformation is highly unstable and has maximum energy.
2. **Staggered conformation** in which hydrogens are as far apart as possible. It is highly stable and has minimum energy.
3. **Skew conformation** is intermediate conformation of the above two conformations.

Eclipsed and staggered conformations are represented by Sawhorse and Newman projections.

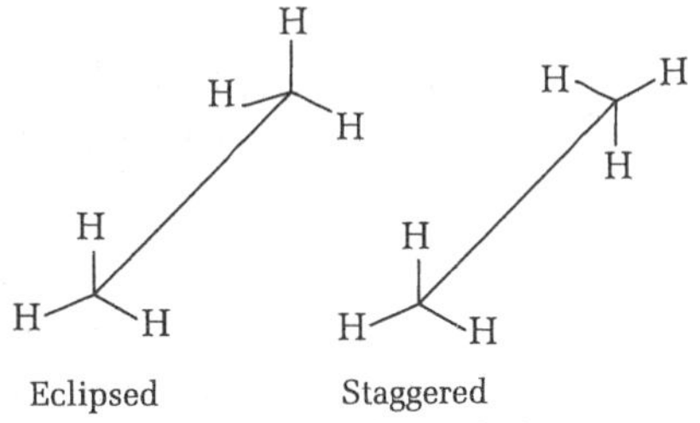

Sawhorse projections of ethane

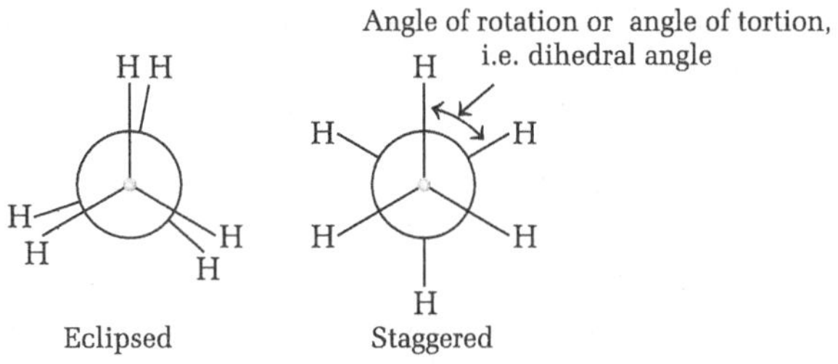

Newman projections of ethane

The energy difference between two extreme forms is of the order of 12.5 kJ/mol which is very small, therefore eclipsed and staggered form cannot be isolated.

Alkenes

Alkenes are unsaturated hydrocarbons containing at least one double bond. Double bonded carbon has sp^2 hybridisation.

General formula for alkenes is C_nH_{2n}.

- **Nomenclature** In IUPAC system of alkenes, numbering of the chain is done from the end which is nearer to the double bond. The suffix 'ene' replaces 'ane' of alkanes.
- **Isomerism** Alkenes show **structural** as well as **geometrical isomerism**. For exhibiting geometrical isomerism two different atoms or groups should be attached to double bonded carbon atoms. If same atoms or groups lie on the same side of the double bond, it is known as ***cis* form** and if same atoms or groups lie on the opposite sides of the double bond, it is known as **trans form**. e.g.

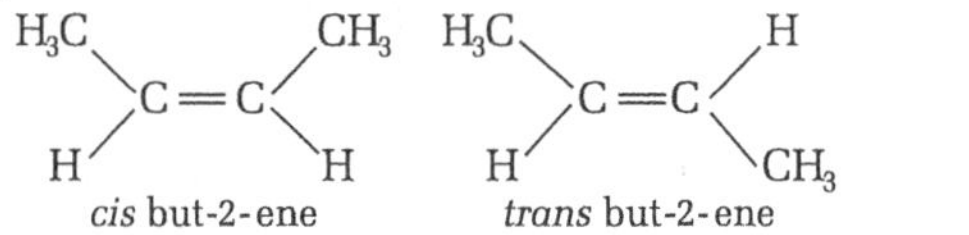

cis but-2-ene *trans* but-2-ene

- *cis*-form of alkene is found to be more polar (i.e. has more dipole moment) than the *trans*.
- *cis*-form has higher boiling point than *trans*-form and *trans*-isomer has higher melting point than that of *cis* form.

General Methods of Preparation

General methods of preparation of alkenes are as follows:

1. **Controlled Hydrogenation** Alkynes when treated with Lindlar's catalyst [Pd/C/H_2(g)] give *cis*-alkenes while on reduction with Na/liquid NH_3 give *trans*-alkenes **(Birch reduction).**

e.g. $\underset{\text{Alkyne}}{R—C\equiv C—R'} + H_2 \xrightarrow{Pd/C}$ 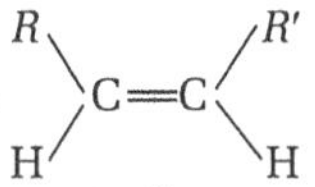

cis-alkene

$\underset{\text{Alkyne}}{R—C\equiv C—R'} + H_2 \xrightarrow[\text{(Birch reduction)}]{Na/Liquid\ NH_3}$ 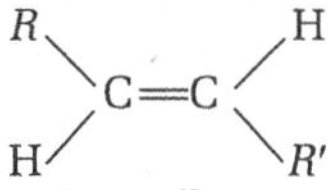

trans-alkene

2. **Dehydrohalogenation** Alkyl halides on heating with alcoholic potash undergoes dehydrohalogenation (**β-elimination reaction**) to form alkenes.

$$\overset{}{CH_3}-\overset{\beta}{CH_2}-\overset{\alpha}{CH_2}Br \xrightarrow[\text{or } KOH + C_2H_5OH]{KOH(alc)/\Delta} CH_3CH{=}CH_2$$

Nature of halogen atom and the alkyl group determine rate of reaction. For halogen, the rate is iodine > bromine > chlorine while for alkyl group, rate is 3° > 2° > 1°.

3. **Dehydration of Alcohols** Alcohols on acidic dehydration give alkenes. This reaction is an example of β-elimination.

$$\underset{\text{3-methylbutanol}}{CH_3-\underset{CH_3}{\underset{|}{CH}}-CH_2-CH_2-OH} \xrightarrow[443\ K]{Conc.\ H_2SO_4} \underset{\text{3-methylbut-2-ene}}{CH_3-\underset{CH_3}{\underset{|}{C}}{=}CH-CH_3} + H_2O$$

In this reaction, intermediate carbocation is formed and rearrange, therefore double bond is formed at maximum degree of carbon.

NOTE Dehydration and dehydrohalogenation both are governed by **Saytzeff's rule**, i.e. the greater is the number of alkyl groups attached to double bonded 'C' atoms, the more stable is the alkene. If dehydrohalogenation of alkyl halide takes place in the presence of potassium *t*-butoxide, less substituted alkene is obtained due to steric hindrance.

4. **Dehalogenation** *Vicinal dihalides* when treated with zinc metal lose a molecule of ZnX_2 to lose a molecule of ZnX_2 to give alkenes. This reaction is known as **dehalogenation**.

$$\underset{\text{1, 2-dibromoethane}}{\underset{Br}{\underset{|}{CH_2}}-\underset{Br}{\underset{|}{CH_2}}} + Zn \xrightarrow{Heat} \underset{\text{Ethene}}{CH_2{=}CH_2} + ZnBr_2$$

Physical Properties

Various physical properties alkenes are as follows:

- The melting point of alkenes do not show regular trend. Alkenes have higher melting point than corresponding alkanes.
- The boiling point of alkenes vary in the same manner as in alkanes. Their boiling points are higher than that of the corresponding alkanes.
- Alkenes are quite soluble in non-polar solvents such as benzene, CCl_4 etc.

Chemical Properties

Alkenes generally undergo **electrophilic addition reactions** because of the presence of a double bond (a strong σ and a weaker π-bond), π electrons of which being loosely held are easily polarisable.

The mechanism involved in electrophilic addition reaction are as follows:

***Step* I** Ionisation of reagent to generate electrophile and nucleophile

$$E-Nu \longrightarrow E^+ + :Nu^-$$

***Step* II** Attack of electrophile on alkene.

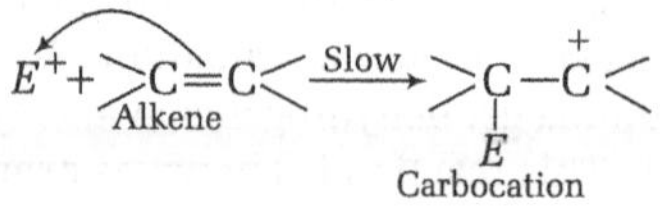

***Step* III** Attack of nucleophile to give addition product.

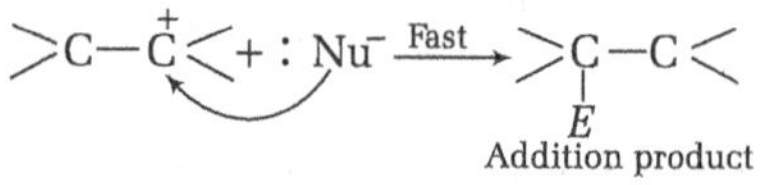

A brief description of different reactions of alkenes is as follows :

1. **Addition of hydrogen** on alkenes occur in the presence of finely divided nickel, palladium or platinum to form alkanes (discussed earlier).
2. **Addition of halogens** like bromine or chlorine when reacted wiith alkene results in the formation of *vicinal* dihalides. However, iodine does not show addition reaction under normal conditions.
3. **Addition of hydrogen halide when** alkenes reacts with hydrogen halides (HCl, HBr, HI) alkyl halides are obtained. According to **Markownikoff's rule** which states that negative part of the adding molecule gets attached to that carbon atom which possesses lesser number of hydrogen atoms.

e.g. $CH_3-CH{=}CH_2 + \underset{(X = Cl,\ Br,\ I)}{HX} \longrightarrow$

$$\underset{\text{(Major product)}}{CH_3-\underset{X}{\underset{|}{CH}}-CH_3} + \underset{\text{(Minor product)}}{CH_3CH_2CH_2-X}$$

- Steps involved in the **mechanism** are as follows :

(i) HBr provides an electrophile, H^+ which attacks the double bond to form carbocation.

$CH_3-CH{=}CH_2 + H-Br$

$\longrightarrow \underset{(1°\ \text{carbocation})}{CH_3-CH_2-\overset{+}{C}H_2} + Br^-$ (less stable)

$\longrightarrow \underset{(2°\ \text{carbocation})}{CH_3-\overset{+}{C}H-CH_3} + Br^-$ (more stable)

(ii) 2° carbocation is formed at a faster rate and it is attacked by Br^- ion to form major product.

$$\underset{(2°\ \text{carbocation})}{CH_3-\overset{+}{C}H-CH_3} + Br^- \longrightarrow \underset{\text{2-bromopropane (Major product)}}{CH_3-\underset{Br}{\underset{|}{CH}}-CH_3}$$

- Alkenes with HBr only in presence of peroxide shows **free radical addition reaction** contrary to the Markownikoff's rule. This is also known as **Kharasch effect**.

$$CH_3—CH=CH_2 + HBr \xrightarrow{\text{Benzoyl peroxide}} \underset{\text{1-bromopropane}}{CH_3—CH_2—CH_2—Br}$$

- Peroxide effect proceeds *via* free radical chain mechanism. The steps involved in the process are as follows :

(i) $\underset{\text{Benzoyl peroxide}}{C_6H_5—\overset{O}{\overset{||}{C}}—O—O—\overset{O}{\overset{||}{C}}—C_6H_5} \xrightarrow{\text{Homolysis}}$

$$2C_6H_5—\overset{O}{\overset{||}{C}}—\dot{\ddot{O}}: \longrightarrow 2\dot{C}_6H_5 + 2CO_2$$

(ii) $\dot{C}_6H_5 + H—Br \xrightarrow{\text{Homolysis}} C_6H_6 + \dot{B}r$

(iii)

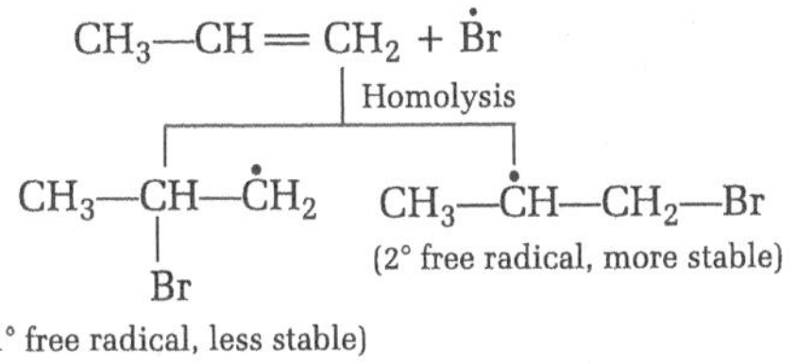

(iv) $CH_3—\dot{C}H—CH_2Br + HBr \xrightarrow{\text{Homolysis}} \underset{\text{Major product}}{CH_3CH_2CH_2Br} + \dot{B}r$

(v) $CH_3—\underset{Br}{CH}—\dot{C}H_2 + HBr \xrightarrow{\text{Homolysis}} \underset{\text{Minor product}}{CH_3—\underset{Br}{CH}—CH_3} + \dot{B}r$

4. **Addition of Sulphuric Acid** Alkene when treated with cold conc. H_2SO_4 gives alkyl hydrogen sulphate by electrophilic addition reaction.

$$CH_3—CH—CH=CH_2 + H_2O \xrightarrow{H^+} CH_3CH_2—\overset{OSO_2OH}{\overset{|}{CH}}—CH_3$$

In this reaction intermediate carbocation is formed, which rearranges to form an alcohol.

5. **Addition of Water** Alkenes react with water in the presence of few drops of conc. H_2SO_4 in accordance with Markownikoff's rule.

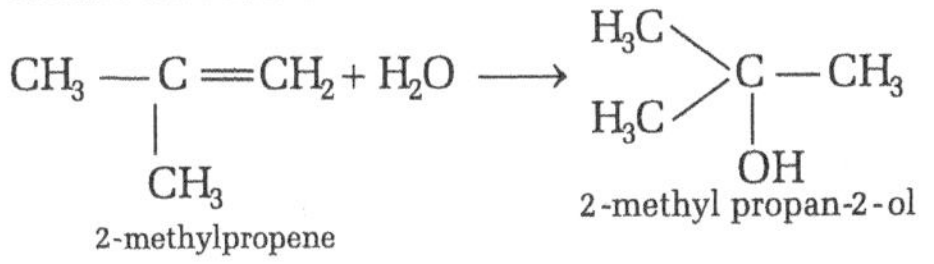

6. **Addition of Hypohalous acid** Alkenes react with hypohalous acid to form halohydrins. These compounds contain —OH and —X groups at successive C-atoms.

$$CH_2=CH_2 + \overset{-}{HO}\overset{+}{X} \longrightarrow \underset{OH}{CH_2}—\underset{Cl}{CH_2}$$

7. **Oxidation** Cold dilute $KMnO_4$ **(Baeyer's reagent)** when treated with alkene undergoes oxidation to give glycol (*vicinal* dihydric alcohol).

$$CH_2=CH_2 + H_2O + O \xrightarrow[273\text{ K}]{\text{Dil. }KMnO_4} \underset{\substack{\textit{cis}\text{-ethane-1, 2-diol} \\ \text{(Ethylene glycol)}}}{\underset{OH}{CH_2}—\underset{OH}{CH_2}}$$

This reaction is used as a test unsaturation as purple colour of $KMnO_4$ disappears during the reaction and a brown precipitate of MnO_2 is obtained.

Acidic $KMnO_4$ or acidic $K_2Cr_2O_7$ oxidises alkenes to ketones and /or acids depending upon the nature of the alkene.

$$(CH_3)_2C=CH_2 \xrightarrow{KMnO_4/H^+} (CH_3)_2C=O + CO_2 + H_2O$$

8. **Ozonolysis of Alkenes** It involves the addition of ozone molecule to alkene to form ozonide and then cleavage of the oxonide by Zn-H_2O to smaller molecules. This reaction is used to locate the position of double bond in an alkene.

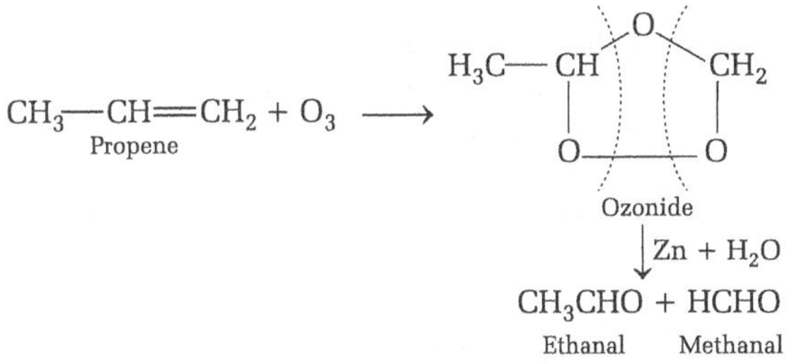

9. **Polymerisation** Alkenes undergo polymerisation and form polymers. Polymers are used for manufacture of plastic bags, radio, T.V. cabinets, buckets etc.

$$\underset{\text{Ethene}}{n(CH_2=CH_2)} \xrightarrow[\text{catalyst}]{\text{High temperature and pressure}} \underset{\text{Polythene}}{-(CH_2—CH_2)_n-}$$

10. **Combustion of Alkene** It involves the formation of carbon dioxide and water. The generalised combustion reaction can be seen as :

$$C_nH_{2n} + \frac{3n}{2}O_2 \longrightarrow nCO_2 + nH_2O + \text{Heat}$$

Alkynes

Alkynes are also unsaturated hydrocarbons. They contain atleast one triple bond between two carbon atoms. Their general formula is C_nH_{2n-2}.

- **Nomenclature** In IUPAC system of alkynes, the suffix 'yne' replaces the 'ane' of alkanes.
- **Isomerism** Alkynes show positional and chain isomerism. The two compounds differ in their structures due to the position of triple bond. e.g. pent-1-yne, pent-2-yne, 3-methylbut-1-yne.

General Methods of Preparation

The various methods of preparation of alkynes are as follows:

1. **By Hydrolysis of Metal Carbides** On industrial scale, ethyne is prepared by treating calcium carbide with water. The steps involved in the process are as follows :

$$CaCO_3 \xrightarrow{\Delta} CaO + CO_2$$

$$CaO + 3C \longrightarrow \underset{\text{Calcium carbide}}{CaC_2} + CO$$

$$CaC_2 \text{ or } \left(C \equiv C\right) \boxed{Ca + 2HO\,H} \longrightarrow \underset{\text{Ethyne}}{CH \equiv CH} + Ca(OH)_2$$

2. **From Vicinal Dihalides** *Vicinal* dihalides on treatment with alc. KOH undergoes dehydrohalogenation.

$$\underset{Br\quad\;\; Br}{CH_2-CH_2} \xrightarrow[\text{KOH + } C_2H_5OH]{\text{KOH (alc.) or}} \underset{\text{1-bromoethene}}{CH_2=CH-Br}$$

$$\xrightarrow{Na\overset{+}{}NH_2^{-}} \underset{\text{Ethyne}}{CH \equiv CH} + NaBr + NH_3$$

3. **From Sodium Acetylide** Higher alkynes can be prepared from lower alkynes by reaction of sodium acetylide with alkyl halide.

$$R-C \equiv CH \xrightarrow{NaNH_2,\ \text{liq}\cdot NH_3} R-C \equiv \bar{C}-\overset{+}{Na}$$

$$\xrightarrow{R'X} R-C \equiv C-R'$$

Physical Properties

Various physical properties of alkynes are as follows:

1. All alkynes are odourless, however acetylene has garlic smell due to the impurity of phosphine.
2. The melting and boiling points of alkynes are slightly higher than those of the corresponding alkenes and alkanes.

Chemical Properties

The various chemical properties of alkynes are as follows:

1. Electrophilic Addition Reaction

Alkynes contain a triple bond, so they add up, two molecules of dihydrogen, halogen, hydrogen halides etc. A few addition reactions are as follows :

(i) **Addition of Dihydrogen**

$$HC \equiv CH + H_2 \xrightarrow{Pt/Pd/Ni} [H_2C=CH_2] \xrightarrow{H_2} CH_3-CH_3$$

(ii) **Addition of Halogen**

$$CH \equiv CH \xrightarrow{Br_2/CCl_4} \underset{}{\overset{Br\quad Br}{HC=CH}} \xrightarrow{Br_2/CCl_4} H-\overset{Br\;Br}{\underset{Br\;Br}{C-C}}-H$$

(iii) **Addition of Hydrogen Halides**

$$\underset{(X = Cl, Br, I)}{CH \equiv CH + HX} \longrightarrow [CH_2=CH-X] \xrightarrow{HX} CH_3CHX_2$$

(iv) **Addition of Water**

$$CH \equiv CH + H_2O \xrightarrow{HgSO_4/H_2SO_4}$$

$$[CH_2=CH-OH] \xrightarrow{\text{Tautomerises}} \underset{\text{Ethanal}}{CH_3CHO}$$

The reaction, called **hydration of alkynes** and is used commercially for the preparation of aldehydes and ketones.

2. Reactions Showing Acidic Character of Alkynes

Due to the presence of acidic H-atom, alkynes (terminal) exhibit following reactions:

(i) **Reaction with Na in liq. NH_3** Terminal alkynes have acidic H because C—H bond has ionic character due to resonance.

$$2CH \equiv CH + 2Na \xrightarrow{475K} 2CH \equiv \bar{C}\overset{+}{Na} + H_2$$

$$CH \equiv CH + \underset{\text{Tollen's reagent}}{2[Ag(NH_3)_2]^+OH^-} \longrightarrow \underset{\substack{\text{Silver acetylide}\\ \text{(white ppt}\cdot)}}{AgC \equiv CAg \downarrow} + 2H_2O + 4NH_3$$

(ii) **Oxidation** Alkyne on oxidation with hot $KMnO_4$ produces carboxylic acids.

$$R-C \equiv C-R' \xrightarrow[\text{Heat}]{KMnO_4/H^+} RCOOH + R'COOH$$

(iii) **Ozonolysis** Alkynes on ozonolysis gives diketones while oxidative ozonolysis of acetylene gives a mixture of glyoxal and formic acid while reductive ozonolysis of acetylene gives glyoxal.

$$R-C \equiv C-R' \xrightarrow{O_3/Zn,\ H_2O} R-\overset{O}{\overset{\|}{C}}-\overset{O}{\overset{\|}{C}}-R'$$

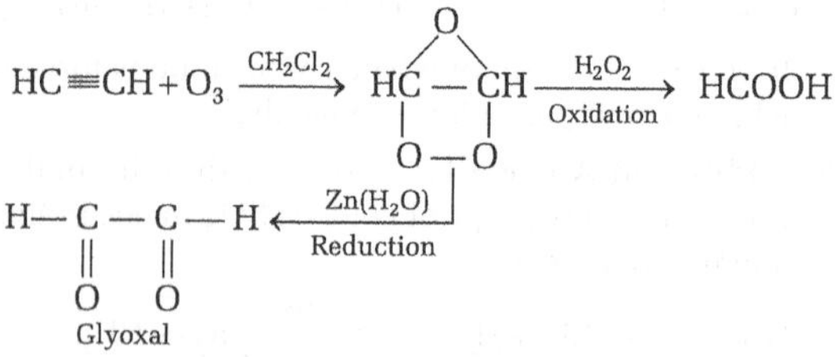

(iv) **Polymerisation** Ethyne undergoes linear polymerisation under suitable conditions to produce polyacetylene or polyethyne,

$$-\!\!\left(CH=CH-CH=CH\right)\!\!-_n$$

while ethyne on passing through red hot iron tube at 873 K undergoes cyclic polymerisation.

$$\underset{\text{Ethyne}}{3CH \equiv CH} \xrightarrow[\text{tube, 873 K}]{\text{Red hot iron}}$$

Benzene

Aromatic Hydrocarbons

Aromatic hydrocarbons are specialised cyclic compounds which are also known as arenes and since, most of them possess pleasant odour [Aroma = fragrance] these are known as **aromatic compounds**.

- **Nomenclature** For IUPAC nomenclature of substituted benzene compounds, the substituents is placed as prefix to the word benzene.
- **Aromaticity** Ring systems whether having benzene ring or not with the following characteristics are said to be **aromatic**.
  - (i) Planarity.
  - (ii) Complete delocalisation of the π-electrons in the ring.
  - (iii) Presence of $(4n + 2)$ π-electrons in the ring, where n is an integer ($n = 0, 1, 2 \ldots$).This is known as **Huckel's rule.**

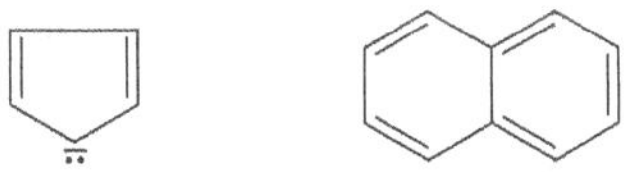

Naphthalene

$n = 1$ $\quad$ $(4n + 2) = 6\pi$-electrons

$n = 2$ $\quad$ $(4n + 2) = 10\pi$-electrons

- Certain compounds such as pyrrole, furan etc., although do not contain benzene ring, but behaves like benzene. These compounds are called **non-benzenoid aromatic compounds**.

Structure of Benzene

- According to Kekule, benzene is said to be resonance hybrid of the following forms.

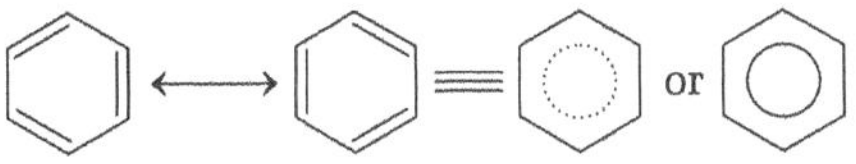

Kekule structure $\qquad$ Delocalised π-bond

- It is aromatic, $(4n+2)\,\pi = 6\pi$. Here, n = number of π-electrons delocalised (Huckel's rule is followed).
- Benzene yields only one substitution product and three isomeric substitution product.
- All C—C bond distance in benzene are equal (139 pm).
- A compound with delocalised electrons is more stable and benzene has six delocalised π-electrons.

General Methods of Preparation

Various methods of preparation of benzene are as follows:

1. **Polymerisation** Polymerisation of ethyne to produce benzene in presence of red hot iron tube (discussed earlier).
2. **Decarboxylation** Sodium salt of benzoic acid on heating with sodalime gives benzene.

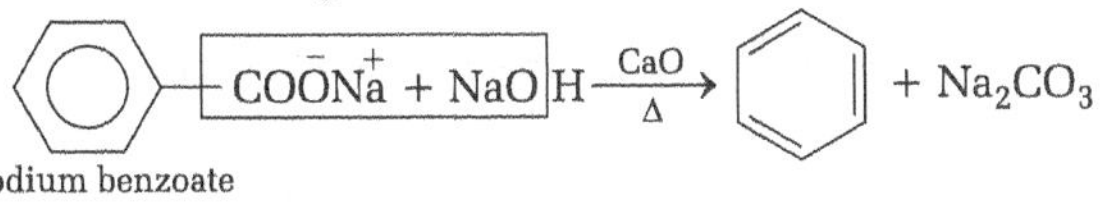

Sodium benzoate

3. **Reduction of benzene diazonium chloride** Diazonium salts are reduced to benzene by passing H_3PO_2/H_2O.

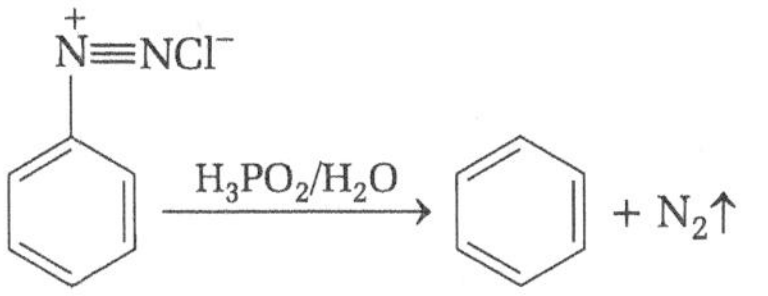

4. **Dehydroxylation** Phenol is reduced to benzene by passing its vapours over heated zinc dust.

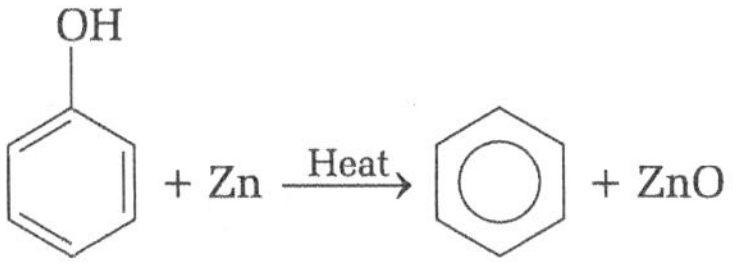

Phenol $\qquad$ Benzene

Physical Properties

Various physical properties of benzene are as follows:

- Aromatic hydrocarbons are non-polar molecules and are usually colourless liquids or solids with a characteristic aroma.
- They are immiscible with water but are readily miscible with organic solvents.
- They burn with sooty flame.

Chemical Properties

Arenes are characterised by electrophilic substitution reactions but under special conditions they can also undergo addition and oxidation reaction.

1. **Electrophilic Substitution Reactions** Various types of electrophilic substitution reactions are as follows :

(i) **Nitration**

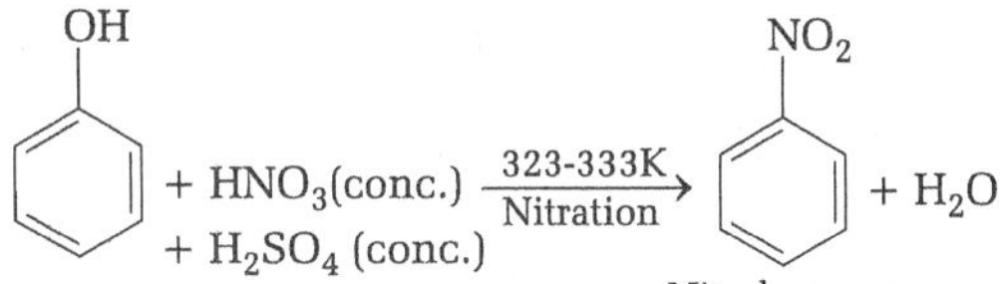

Nitrobenzene

Attacking reagent (electrophile) is NO_2^+.

(ii) **Halogenation**

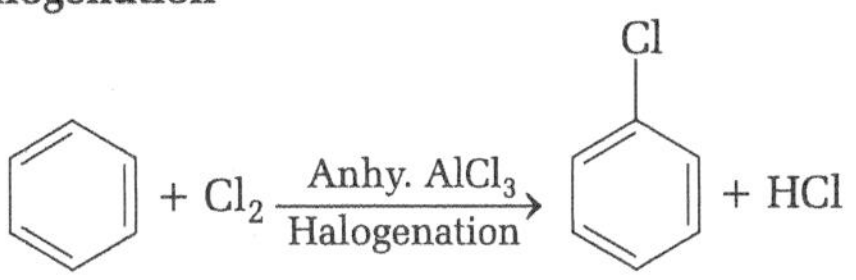

Chlorobenzene

Electrophile is Cl^+.

(iii) **Sulphonation**

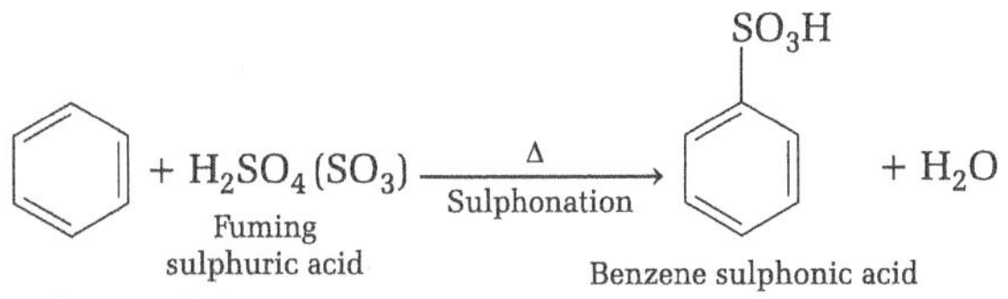

Benzene sulphonic acid

Electrophile is SO_3.

(iv) **Friedel Craft Alkylation**

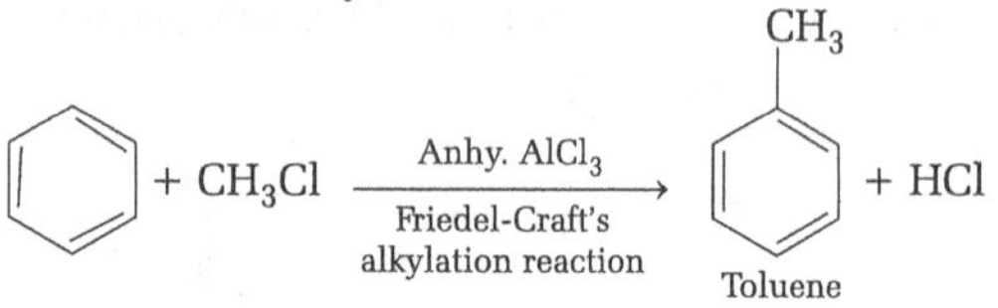

Electrophile is $\overset{+}{C}H_3$.

(v) **Friedel-craft Acylation**

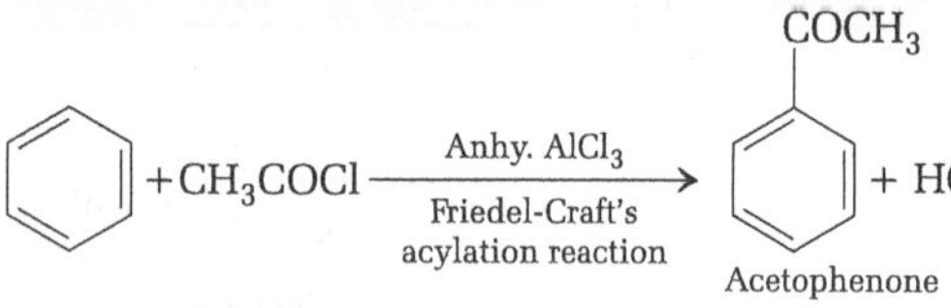

- If excess of electrophilic reagent is used, further substitution reaction may take place in which other hydrogen atoms of benzene ring may also be successively replaced by the electrophile.

Mechanism of electrophilic substitution involves following steps :

***Step* I Formation of σ-complex**

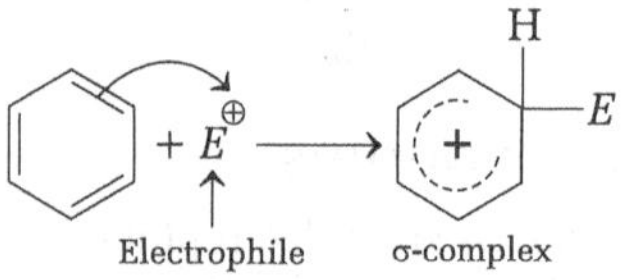

***Step* II Elimination of Proton**

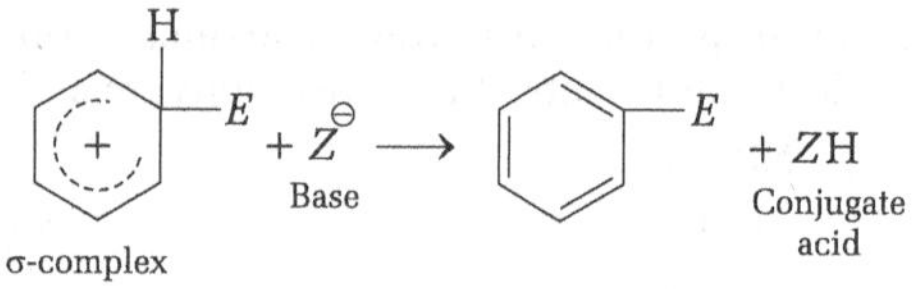

$Z^- = FeX_4^-$ $(X = Cl, Br)$ in halogenation; HSO_4^- in nitration; H_2O in sulphonation, AlX_4^- $(X = Cl, Br)$ in Friedel-Crafts alkylation/ acylation.

2. **Addition Reactions** Under vigorous conditions, i.e. at high temperature or pressure in the presence of Ni catalyst, hydrogenation of benzene gives cyclohexane.

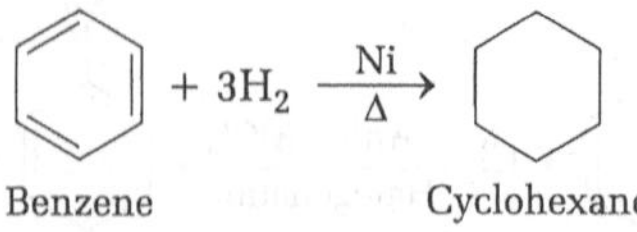

Under UV light, three chlorine molecules add to benzene to produce benzene hexachloride $C_6H_6Cl_3$, also called gammaxane.

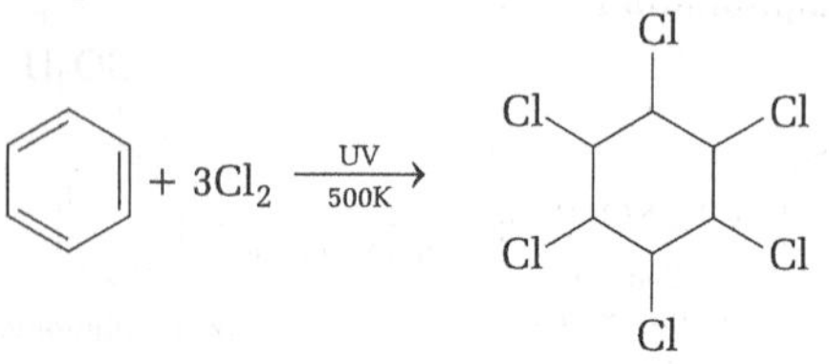

Benzene hexachloride (BHC),

Directive Influence of Functional Group in Monosubstituted Benzene

- The substituent already present in the benzene ring directs the incoming group either to *ortho*- and *para*- or to *meta*-position. *o*-, *p*-directing groups increase the electron density on the ring and thus, increase the reaction rate while *m*-directing groups decrease the electron density on the benzene ring and thus, decrease the reaction rate.
- Example for *ortho* and *para* directing groups (activating).

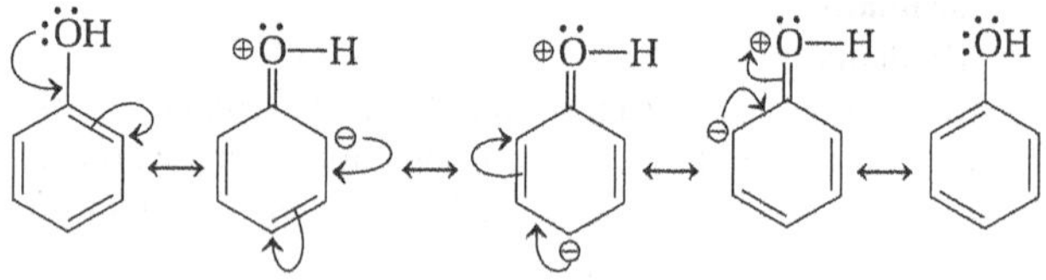

Resonating structures of phenol

In case of phenol, electron density is more concentrated on *o*- and *p*-positions.

Therefore, substitution is mainly in these positions. Other activating groups are —OH, —NH_2, —NHR, —$NHCOCH_3$, —OCH_3, —CH_3, —C_2H_5 etc.

- Example for *meta* directing group (deactivating) :

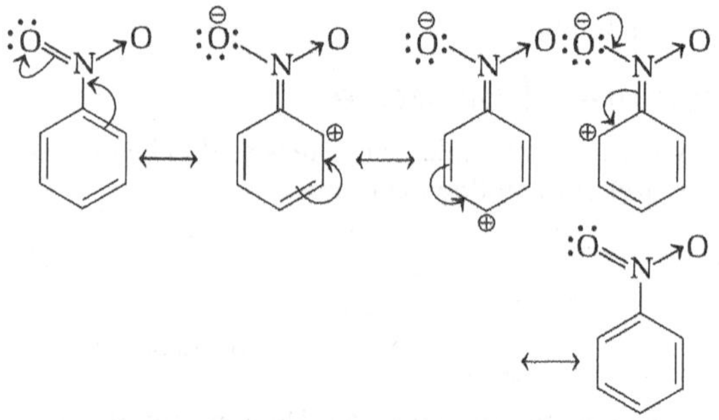

Resonating structures of nitrobenzene

- Other deactivating groups are —NO_2, —CN, —CHO, —COR, — COOH, — COOR, — SO_3H are *meta*-directing groups.
- In case of haloarenes, halogens are moderately deactivating because of –*I*- effect overall electron density on benzene ring decreases. Due to resonance, electron density on *o*- and *p*-position increases, hence halogens are *o*-, *p*-directing groups.

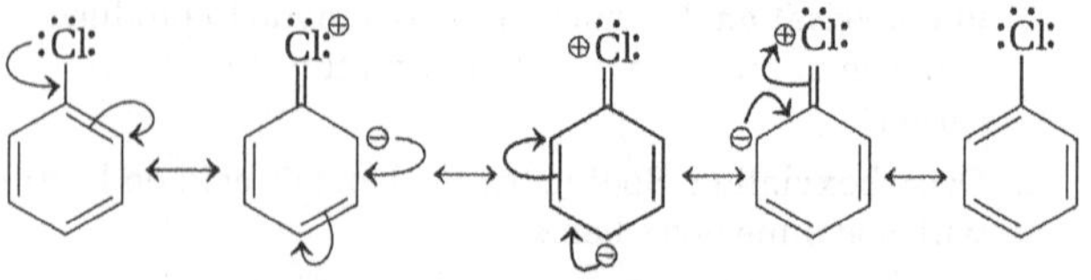

Resonating structures of chlorobenzene

DAY PRACTICE SESSION 1

FOUNDATION QUESTIONS EXERCISE

1 Butene-1 may be converted to butane by reaction with → AIEEE 2003

(a) Zn-HCl (b) Sn-HCl (c) Zn-Hg (d) Pd/H_2

2 The increasing order of reduction of alkyl halides with zinc and dilute HCl is

(a) R— Cl<R — I<R — Br (b) R— Cl<R— Br<R— I
(c) R— I<R— Br<R— Cl (d) R— Br<R— I<R— Cl

3 A mixture of ethyl iodide and *n*-propyl iodide is subjected to Wurtz reaction. The hydrocarbon which will not be formed is

(a) butane (b) propane (c) octane (d) hexane

4 Pure methane can be produced by

(a) Wurtz reaction
(b) Kolbe's electrolytic method
(c) sodalime decarboxylation
(d) reduction with H_2

5 Which one of the following has the minimum boiling point?

(a) *n*-butane (b) 1-butyne (c) 1-butene (d) *Iso* -butene

6 Arrange the following in the decreasing order of their boiling points.

A. *n*-butane B. 2-methylbutane
C. *n*-pentane D. 2, 2-dimethylpropane

(a) $A > B > C > D$ (b) $B > C > D > A$
(c) $D > C > B > A$ (d) $C > B > D > A$

7 The percentage of 1-chloro-2-methylpropane obtained in the chlorination of *iso*-butane is

(a) 38 % (b) 64 % (c) 79 % (d) 36 %

8 On mixing a certain alkane with chlorine and irradiating it with ultraviolet light, it forms only one monochloroalkane. This alkane would be

(a) propane (b) pentane
(c) *iso*-pentane (d) *neo* -pentane

9 Arrange in the correct order of stability (decreasing order) for the following molecules → JEE Main (Online) 2013

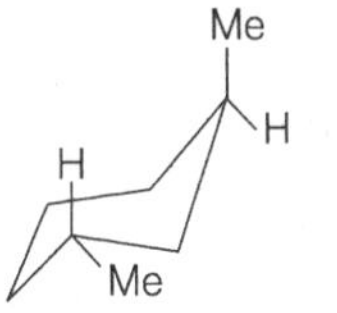

I

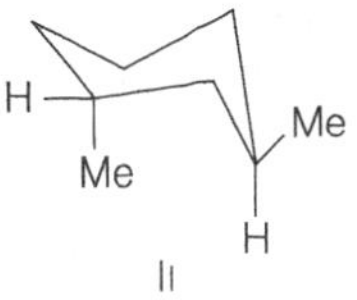

II

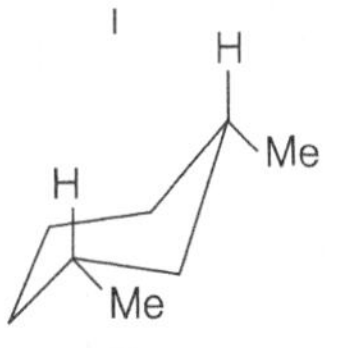

III

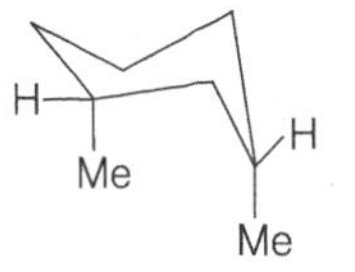

IV

(a) (I) > (II) > (III) > (IV) (b) (IV) > (III) > (II) ≈ (I)
(c) (I) > (II) ≈ (III) > (IV) (d) (III) > (I) ≈ (II) > (IV)

10 Which of the following is the most stable conformation of *n*-butane?

(a)
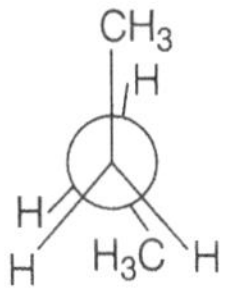

(b)
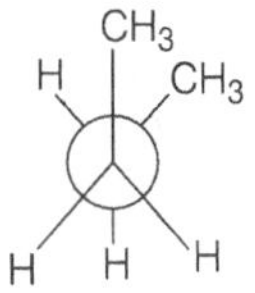

(c)
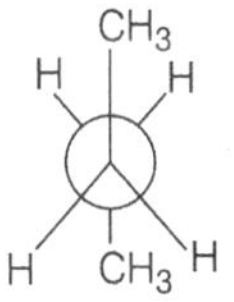

(d)
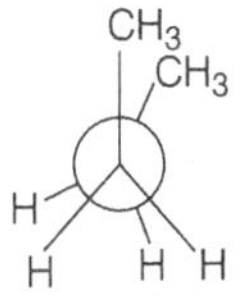

11 The *trans*-alkenes are formed by the reduction of alkynes with → JEE Main 2018

(a) H_2-Pd/C, $BaSO_4$ (b) $NaBH_4$
(c) Na/liq. NH_3 (d) Sn-HCl

12 2-phenylpropene on acidic hydration gives

(a) 2-phenyl propan-2-ol
(b) 2-phenyl propan-1-ol
(c) 3-phenyl propan-1-ol
(d) 1-phenyl propan-2-ol

13 When subjected to acid catalysed hydration, the order of reactivity of the alkenes is

I. $(CH_3)_2C{=}CH_2$ II. $CH_3CH{=}CH_2$
III. $CH_2{=}CH_2$

(a) III > II > I (b) I > III > II
(c) I > II > III (d) II > I > III

14 The intermediate formed during the addition of HCl to propene in the presence of peroxide is

(a) $CH_3\dot{C}HCH_2Cl$ (b) $CH_3\overset{+}{C}HCH_3$
(c) $CH_3CH_2\dot{C}H_2$ (d) $CH_3CH_2\overset{+}{C}H_2$

15 Which of the following is the predominant product in the reaction of HOBr with propene? → JEE Main 2016

(a) 2-bromo-1-propanol (b) 3-bromo-1-propanol
(c) 2-bromo-2-propanol (d) 1-bromo-2-propanol

16 The reaction of propene with HOCl ($Cl_2 + H_2O$) proceeds through the intermediate

(a) $CH_3—\overset{+}{C}H—CH_2—Cl$ (b) $CH_3—CH(OH)—\overset{+}{C}H_2$
(c) $CH_3—CHCl—\overset{+}{C}H_2$ (d) $CH_3—\overset{+}{C}H—CH_2—OH_2$

17 The number of optically active products obtained from the complete ozonolysis of the given compound is

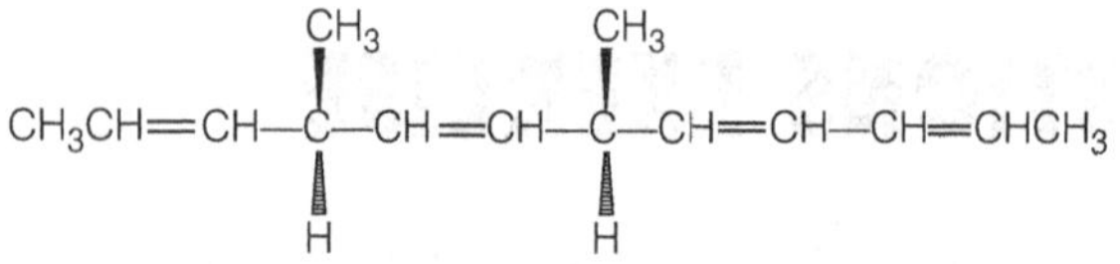

(a) 0 (b) 1 (c) 2 (d) 4

18 Ozonolysis of an organic compound gives formaldehyde as one of the products. This confirms the presence of ➙ AIEEE 2011

(a) two ethylenic double bonds
(b) a vinyl group
(c) an *iso* propyl group
(d) an acetylenic triple bond

19 One mole of a symmetrical alkene on ozonolysis gives two moles of an aldehyde having a molecular mass of 44 u. The alkene is ➙ AIEEE 2010

(a) propene (b) 1-butene (c) 2-butene (d) ethene

20 Ozonolysis of 2,3-dimethyl but-1-ene followed by reduction with zinc and water gives

(a) methanoic acid and 3-methyl butan-2-one
(b) methanal and 2-methyl butan-2-one
(c) methanal and 3-methyl butan-2-one
(d) methanoic acid and 2-methyl butan-2-one

21 Which compound would give 5-keto-2-methyl hexanal upon ozonolysis? ➙ JEE Main 2015

(a) 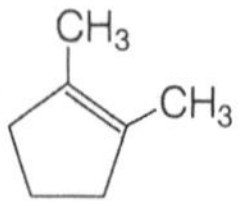
(b) 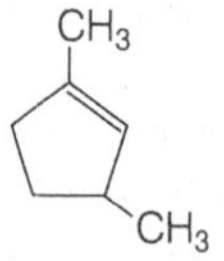
(c) 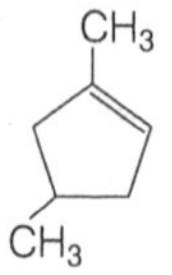
(d) 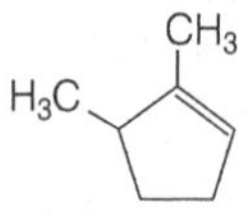

22 6 L of an alkene require 27 L of oxygen at constant temperature and pressure for complete combustion. The alkene is ➙ JEE Main (Online) 2013

(a) ethene (b) propene (c) 1-butene (d) 2-butene

23 Acetylene does not react with ➙ AIEEE 2002

(a) Na (b) ammonical $AgNO_3$
(c) HCl (d) NaOH

24 2-hexyne gives *trans*-2-hexene on treatment with ➙ AIEEE 2012

(a) Pt/H_2 (b) Li/NH_3 (c) $Pd/BaSO_4$ (d) $LiAlH_4$

25 What is the best way to carry out the following transformation?

1-pentyne ⟶ pentanal

(a) $HgSO_4/H_2SO_4$
(b) H_2/Lindlar's catalyst; O_3;Zn-H_2O
(c) HIO_4/H_2O
(d) BH_3; H_2O_2/NaOH

26 1-butyne on oxidation with hot alkaline $KMnO_4$ would yield

(a) $CH_3CH_2CH_2COOH$
(b) CH_3CH_2COOH
(c) $CH_3CH_2COOH + CO_2 + H_2O$
(d) $CH_3CH_2COOH + HCOOH$

27 Match the following and choose the correct option.

	Reactions	Reaction type
A.	$CH_2 = CH_2 + H_2O \xrightarrow{H^+} CH_3CH_2OH$	1. Hydrogenation
B.	$CH_2 = CH_2 + H_2 \xrightarrow{Pd} CH_3—CH_3$	2. Halogenation
C.	$CH_2 = CH_2 + Cl_2 \longrightarrow Cl—CH_2—Cl$	3. Polymerisation
D.	$3CH \equiv CH \xrightarrow[\text{Heat}]{\text{Cu tube}} C_6H_6$	4. Hydration

Codes

	A	B	C	D
(a)	4	2	1	3
(b)	4	1	2	3
(c)	1	3	4	2
(d)	2	1	3	4

28 Which of the following molecules, in pure form, is (are) unstable at room temperature? ➙ IIT JEE 2012

(a) 
(b) 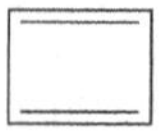
(c) 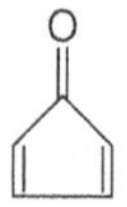
(d) 

29 Which one of the following is the most reactive towards ring nitration?

(a) Benzene (b) Mesitylene
(c) Toluene (d) *m*-xylene

30 Which one is the most reactive towards electrophilic reagent?

(a) 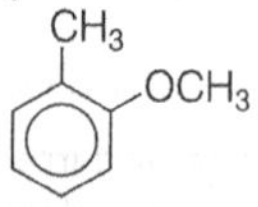

(b) 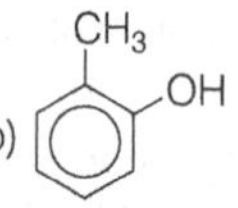

(c) 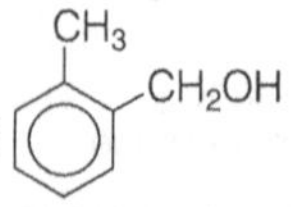

(d) 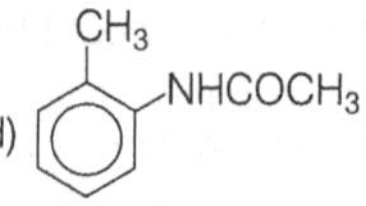

31 The correct sequence of reactions to be performed to convert benzene into *m*-bromoaniline is

(a) nitration, reduction, bromination
(b) bromination, nitration, reduction
(c) nitration, bromination, reduction
(d) reduction, nitration, bromination

32 The electrophile E^+ attacks the benzene ring to generate the intermediate σ-complex. Of the following, which σ-complex is of lowest energy ?

(a) 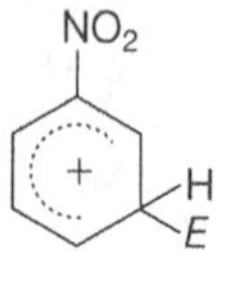
(b) 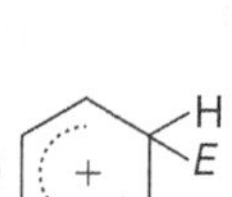
(c) 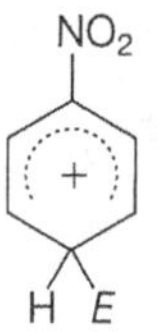
(d) 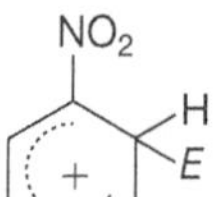

33 Given,

I.  II. 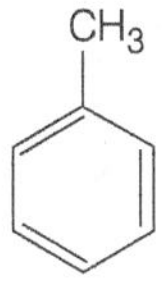

III. 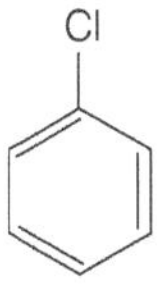 IV. 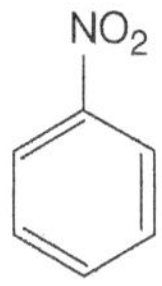

In the above compounds correct order of reactivity in electrophilic substitution reactions will be

(a) II > I > III > IV → JEE Main (Online) 2013
(b) IV > III > II > I
(c) I > II > III > IV
(d) II > III > I > IV

34 Give the major product of the following reaction

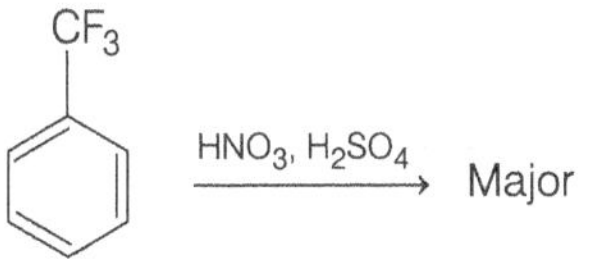

(a) 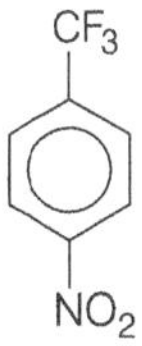 (b) 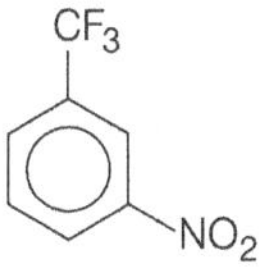

(c) 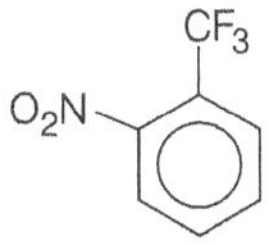 (d) Cannot say

Direction (Q. Nos. 35-36) *In the following questions Assertion (A) followed by a Reason (R) is given. Choose the correct option out of the choices given below.*

(a) Assertion and Reason both are correct statements and Reason is the correct explanation of the Assertion
(b) Assertion and Reason both are correct statements but Reason is not the correct explanation of the Assertion
(c) Assertion is correct and Reason is incorrect
(d) Both Assertion and Reason are incorrect

35 **Assertion** (A) The compound cyclooctatetraene has the following structural formula.

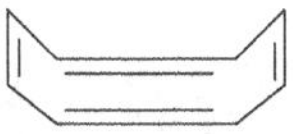

It is cyclic and has conjugated 8π-electron system but it is not an aromatic compound.

Reason (R) $(4n+2)\pi$-electrons rule does not hold good and ring is not planar.

36 **Assertion** (A) Nitration of benzene with nitric acid requires the use of concentrated sulphuric acid.

Reason (R) The mixture of concentrated sulphuric acid and concentrated nitric acid produces the electrophile, NO_2^+.

Direction (Q. Nos. 37-40) *Each of these questions contains two statements : Statement I (Assertion) and Statement II (Reason). Each of these questions also has four alternative choices, only one of which is the correct answer. You have to select one of the codes (a), (b), (c) and (d) given below.*

(a) Statement I is true, Statement II is true; Statement II is a correct explanation for Statement I.
(b) Statement I is true, Statement II is true; Statement II is not a correct explanation for Statement I.
(c) Statement I is true; Statement II is false.
(d) Statement I is false; Statement II is true.

37 **Statement I** Addition of Br_2 to 1-butene gives two optical isomers.

Statement II The product contains one asymmetric carbon atom.

38 **Statement I** 1-butene on reaction with HBr in the presence of a peroxide produces 1-bromobutane.

Statement II It involves the formation of a primary radical.

39 **Statement I** Acetylene on reacting with sodamide gives sodium acetylide and ammonia.

Statement II *sp* hybridised carbon atoms of acetylene are considerably electronegative.

40 **Statement I** Both toluene and *iso*-propyl benzene give the same product on oxidation with $KMnO_4$.

Statement II $KMnO_4$ oxidises side aliphatic chain of arenes to —COOH group.

DAY PRACTICE SESSION 2

PROGRESSIVE QUESTIONS EXERCISE

1 The addition of HBr to 1-butene gives a mixture of products *A*, *B* and *C*

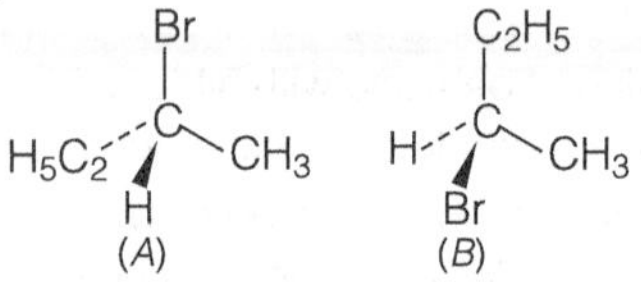

$CH_3CH_2CH_2CH_2Br$
(*C*)

The mixture consists of

(a) *A* and *B* as major and *C* as minor products
(b) *B* as major, *A* and *C* as minor products
(c) *B* as minor, *A* and *C* as major products
(d) *A* and *B* as minor and *C* as major products

2 The most stable conformer of 2, 4-hexadiene is

(a) 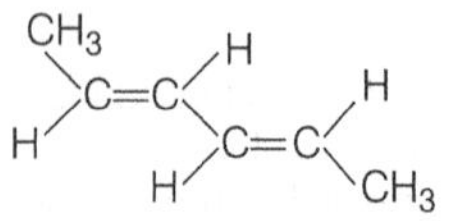

(b) 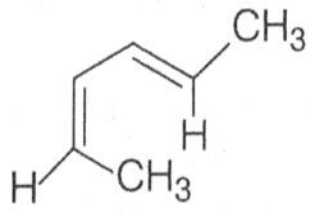

(c) 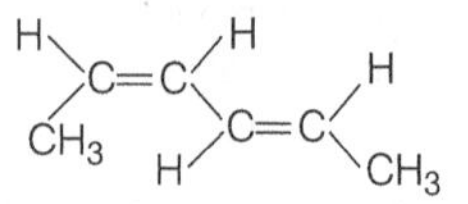

(d) 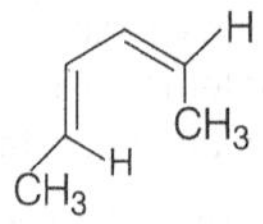

3 When propyne is treated with aqueous H_2SO_4 in presence of $HgSO_4$, the major product is

(a) propanal
(b) *n*-propyl hydrogen sulphate
(c) acetone
(d) propanol

4 Which of the following reactions will yield 2,2-dibromopropane?

(a) $CH_3C\equiv CH + 2HBr \rightarrow$
(b) $CH_3CH{=}CHBr + HBr \rightarrow$
(c) $CH\equiv CH + 2HBr \rightarrow$
(d) $CH_3CH{=}CH_2 + HBr \rightarrow$

5 896 mL vapour of a hydrocarbon '*A*' having carbon 87.80% and hydrogen 12.19% weighs 3.28 g at STP. Hydrogenation of '*A*' gives 2-methylpentane. Also '*A*' on hydration in the presence of H_2SO_4 and $HgSO_4$ gives a ketone '*B*' having molecular formula $C_6H_{12}O$. The ketone '*B*' gives a positive iodoform test. Find the structure of '*A*'.

(a) $(CH_3)_3CC\equiv C-CH_2CH_3$
(b) $(CH_3)_2CHCH_2C\equiv CH$
(c) $CH_3CH_2C\equiv CCH_2CH_2CH_2CH_3$
(d) $(CH_3)_2CHCH_2CH_2C\equiv CH$

6 Cyclohexene on ozonolysis followed by reaction with zinc dust and water gives compound *E*. Compound *E* on further treatment with aqueous KOH yields compound *F*. Compound *F* is

(a) 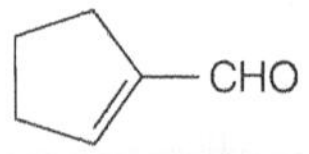

(b) 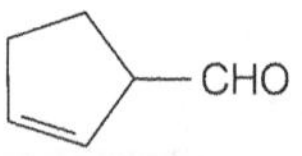

(c) 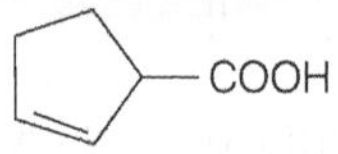

(d) 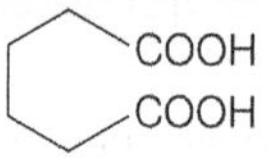

7 Select the correct statement(s).

(a) Addition of Br^- in the following leads to the loss of resonance energy that is associated with the aromatic ring.

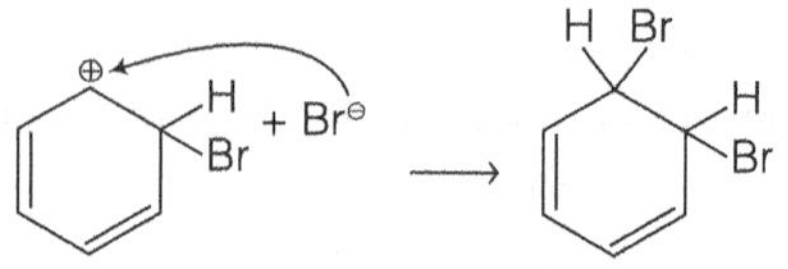

(b) The π-system of the benzene ring is the nucleophile
(c) The sulphur atom in SO_3 is electrophilic by virtue of the formal positive charge in each of the resonance structures
(d) All of the above statements are correct

8 An organic compound, C_9H_{12} *A* can be oxidised with $KMnO_4$ to *B* having molecular formula $C_8H_6O_4$. *B* is a dicarboxylic acid but does not form inner anhydride on heating. *B* on treating with Br in presence of iron gives only one monoderivative *C*, $C_8H_5BrO_4$. What is the structure of compound *C* ?

(a) 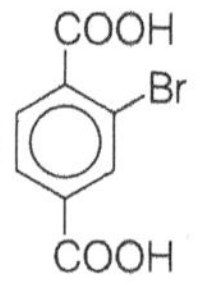

(b) 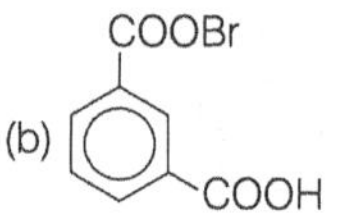

(c) 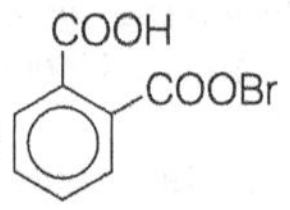

(d) 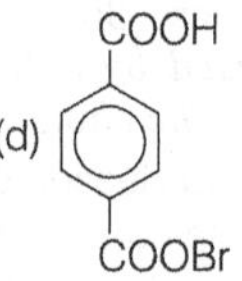

9 Which of the following compounds is expected to give the highest ratio of *ortho*/*para*-isomer (relatively more *ortho*) when reacted with $Cl_2 / FeCl_3$?

(a) 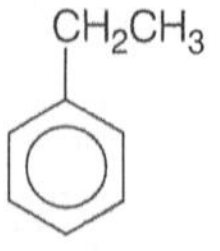

(b) 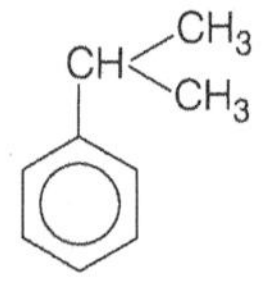

(c) 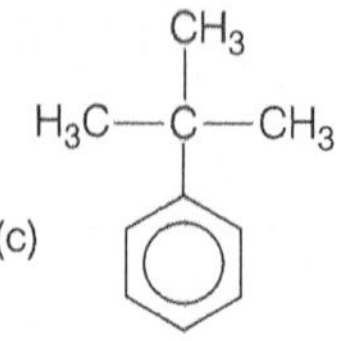

(d) They would all give the *meta*-isomer

10 When 1, 1-dichloropropane and 2, 2-dichloropropane are reacted separately with *aq*. KOH solution, compounds '*A*' and '*B*' are formed. Both '*A*' and '*B*' gave the same product '*C*' on reduction using amalgamated zinc and HCl. Identify *C*.

(a) Propyl alcohol (b) Isopropyl alcohol
(c) Propyl chloride (d) Propane

11 A hydrocarbon '*A*' on ozonolysis gives two isomeric forms '*B*' and '*C*'. *B* on oxidation gives '*D*', silver salt of *D* contains 59.6% Ag. Structure of hydrocarbon is

(a) $CH_3—CH{=}C(CH_3)—CH_3$

(b) $CH_3—CH_2—CH{=}C(CH_3)—CH_3$

(c) $CH_3—CH_2—CH_2—CH{=}C(CH_3)—CH_2—CH_3$

(d) $CH_3—CH(CH_3)—CH{=}C(CH_3)—CH_2—CH_3$

12 Two gases *P* and *Q* decolourise aqueous bromine but only one of them gives a white precipitate with aqueous ammoniacal silver nitrate solution. *P* and *Q* are likely to be

(a) ethane and ethyne
(b) but-1-yne and but-2-yne
(c) ethane and but-2-yne
(d) ethyne and propyne

ANSWERS

SESSION 1

1 (d)	2 (b)	3 (b)	4 (c)	5 (d)	6 (d)	7 (b)	8 (d)	9 (d)	10 (c)
11 (c)	12 (a)	13 (c)	14 (b)	15 (d)	16 (a)	17 (a)	18 (b)	19 (b)	20 (c)
21 (b)	22 (b)	23 (d)	24 (b)	25 (d)	26 (c)	27 (b)	28 (b)	29 (b)	30 (b)
31 (c)	32 (b)	33 (a)	34 (b)	35 (a)	36 (a)	37 (a)	38 (c)	39 (a)	40 (a)

SESSION 2

1 (a)	2 (a)	3 (c)	4 (a)	5 (b)	6 (a)	7 (d)	8 (a)	9 (a)	10 (d)
11 (b)	12 (b)								

Hints and Explanations

SESSION 1

1 Using Pd/H_2, but-1-ene can be converted to butane.

2 The reactivity of reduction of alkyl halides with Zn/HCl increases with the decrease in the strength of C—*X* bond. Therefore, reactivity order will be *R*—Cl < *R*—Br < *R*—I.

3 Propane will not formed as in Wurtz reaction two molecules of each reacting species combine with the removal of Na*X*.

4 Pure methane can be produced from sodalime decarboxylation.

$$CH_3CO\overset{-}{O}\overset{+}{Na} + NaOH \xrightarrow[\Delta]{\text{Sodalime}} CH_4 + Na_2CO_3$$

5 *Iso*-butene $(H_3C)_2C{=}CH_2$ has minimum force of attraction due to steric hindrance. Thus, minimum boiling point.

6 As the number of carbon atom increases, boiling point increases. Moreover, boiling point decreases with branching as shown below:

$H_3C—C(CH_3)_2—CH_3$
(Boiling point = 282.5 K)
2,2- dimethylpropane

$H_3C—CH_2—CH_2—CH_2—CH_3$
(Boiling point = 309.1K)
n-pentane

$H_3C—H_2C—HC(CH_3)—CH_3$
Boiling point = 301 K
2- methylbutane

$H_3C—H_2C—CH_2—CH_3$
Boiling point = 273K
n-butane
(4 carbon atoms with no branching)

Thus, the correct order of boiling point is *C* > *B* > *D* > *A*.

7 *Iso*-butane is $CH_3—CH(CH_3)—CH_3$.

Its chlorination would give two isomers.

$$\frac{1° \text{ isomer}}{3° \text{ isomer}} = \frac{\text{number of } 1° \text{ H}}{\text{number of } 3° \text{ H}} \times \frac{\text{reactivity of } 1° \text{ H}}{\text{reactivity of } 3° \text{ H}} = \frac{9}{1} \times \frac{1}{5} = \frac{9}{5}$$

[∵ The relative reactivities of 3°, 2° and 1° H-atoms are 5.0 : 3.8 : 1] (percentage of 1° isomer)

$$= \frac{9}{9+5} \times 100 = 64.3\% \simeq 64\%$$

8 In *neo*-pentane, all H are equivalent. Thus, it forms only one monochloroalkene

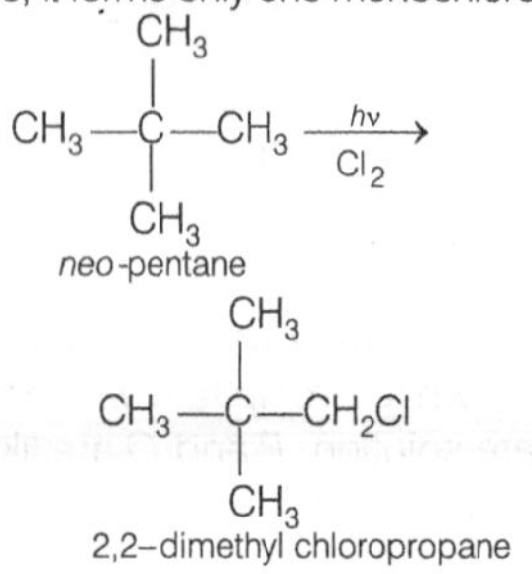

9 The structure in which bulkier groups (which is Me group here) occupy equatorial positions is more stable than the one in which bulkier group are at axial positions. Here, the order of stability is III > I ≈ II > IV.

10 Anti-staggered conformation is most stable. Hence, option (c) is correct.

11 Sodium metal in liquid ammonia reduces alkynes with anti stereochemistry to give *trans* alkenes. The reduction is selectively *anti* since the vinyl radical formed during reduction is more stable in *trans* configuration.

Mechanism

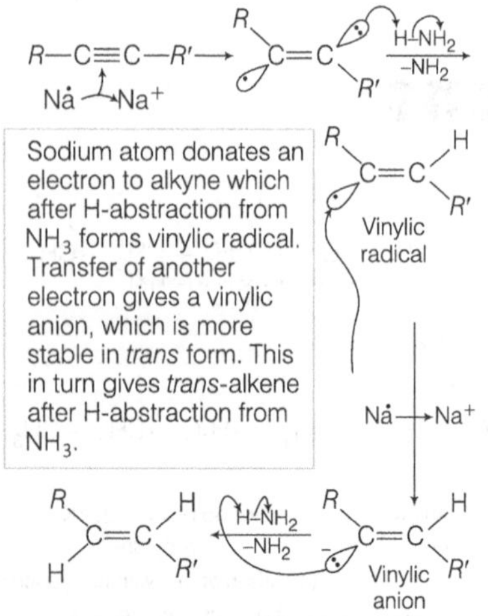

12 Acidic hydration of 2-phenylpropene follows electrophilic reaction mechanism forming an intermediate 3° carbocation (more stable), thereby forming 2-phenyl propan-2-ol.

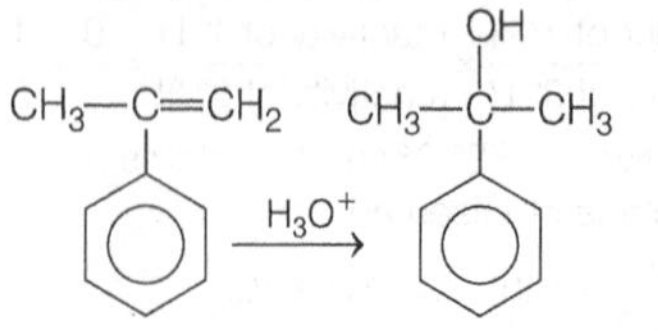

2-phenylpropene 2-phenyl propan-2-ol

13 Acid catalysed hydration of alkenes involve formation of carbocation intermediate. Higher the stability of carbocation intermediate, more is the reactivity towards acid catalysed hydration. The order of stability of carbocation is

$$(CH_3)_2\overset{+}{C}CH_3 > CH_3\overset{+}{C}HCH_3 > \overset{+}{C}H_2CH_3$$

∴ The order of reactivity towards acid catalysed hydration is

$$\underset{I}{(CH_3)_2C=CH_2} > \underset{II}{CH_3CH=CH_2}. > \underset{IIII}{CH_2=CH_2}.$$

14 Peroxide effect (or free radical addition) occurs only in the case of HBr. The addition of HCl in the presence or absence of peroxides occur by an ionic mechanism.

15 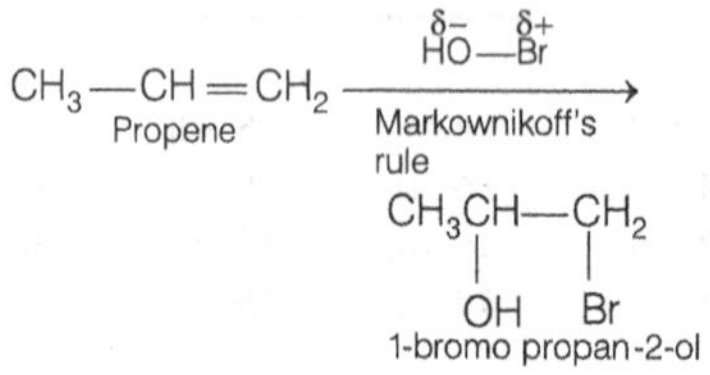

16 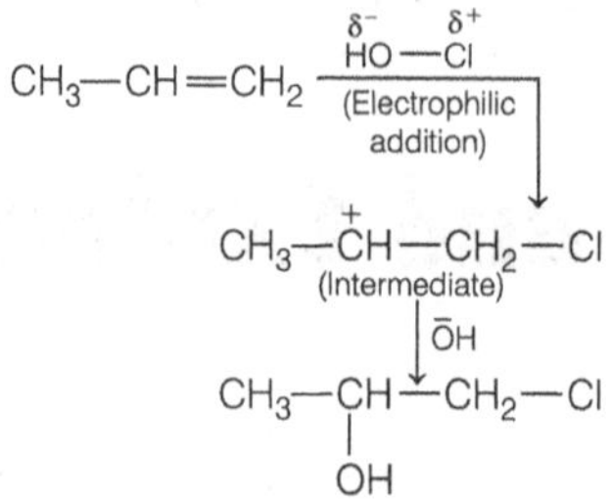

17 Ozonolysis of the given triene occurs as follows :

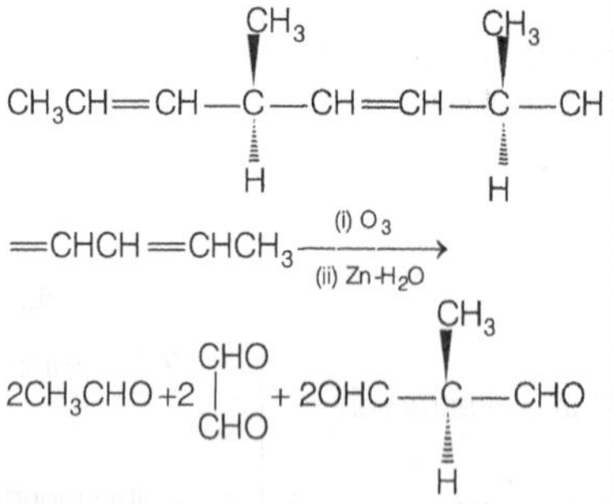

Since, none of the above dial is chiral, no optically active product is obtained.

18 Alkenes give carbonyl compounds on ozonolysis.

$$\text{Alkene} \xrightarrow[\text{Vinyl group}]{O_3/Zn/H_2O} \underset{A}{HCHO} + \text{other}$$

carbonyl compound

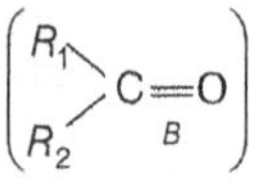

To determine alkene, place carbonyl compounds with their O-atom face to face. Replace O-atom by a double bond.

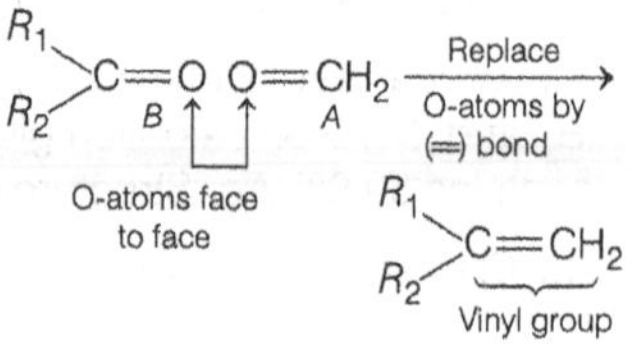

Thus, formation of HCHO confirms the presence of vinyl group.

19 $C_nH_{2n}O = 44 \quad C_nH_{2n} = 44 - 16 = 28$

∴ $n = 2$

So, $CH_3—CH=CH—CH_3$ is the symmetrical alkene.

$$\text{Thus, } \underset{\text{But-2-ene}}{CH_3—CH=CH—CH_3} \xrightarrow[\text{ii. Zn/H}_2\text{O}]{\text{i. O}_3} \underset{\text{Acetaldehyde}}{2CH_3—CH=O}$$

20 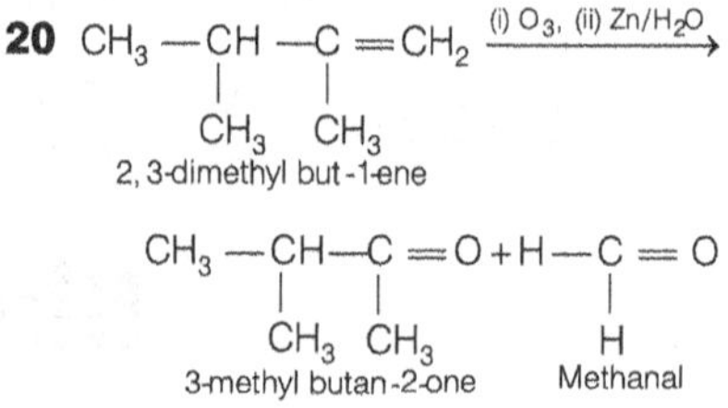

21 (a) 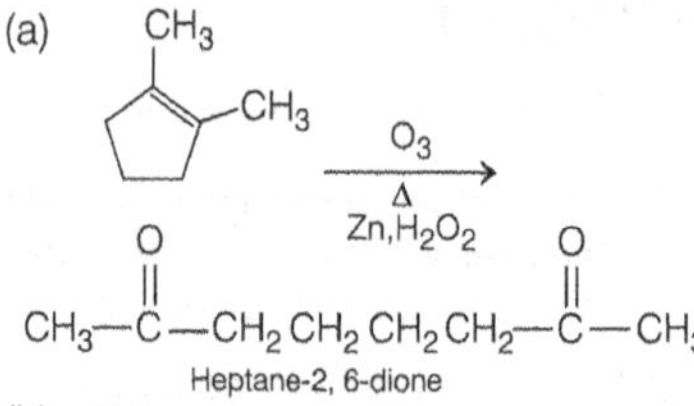

(b) 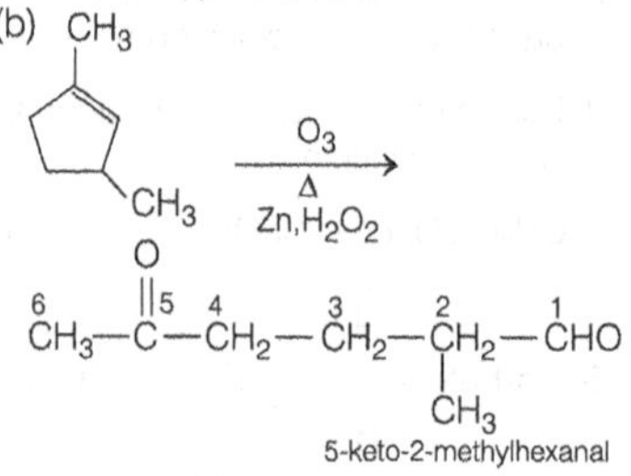

(c) 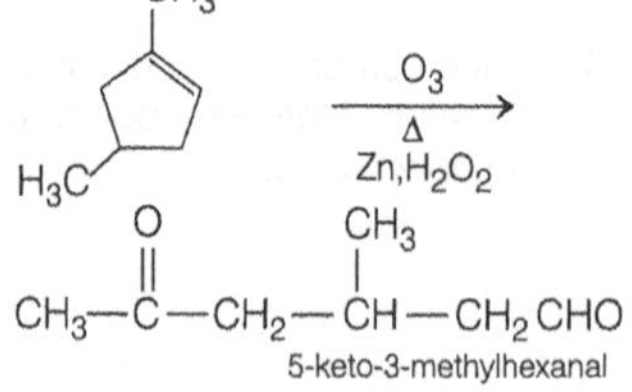

(d)

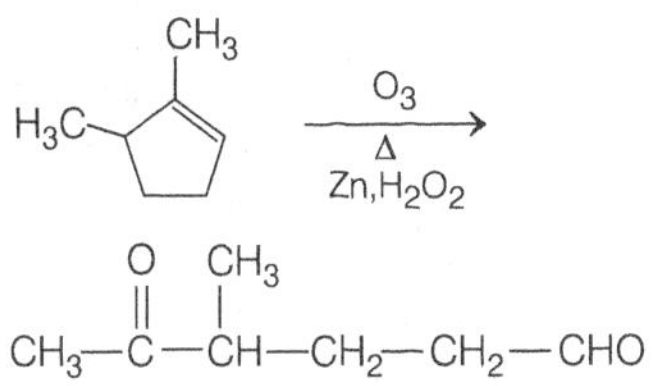

5-keto-4-methyl hexanal

22 General reaction for the combustion of alkene is

$$C_nH_{2n} + \frac{3n}{2}O_2 \longrightarrow nCO_2 + nH_2O$$

Given, 6L alkene require $O_2 = 27\,L$

$\therefore$ 1 L alkene will require $O_2 = \frac{27}{6} L = 4.5$ or $\frac{9}{2} L$

On comparing the amount of O_2,

$$\frac{3n}{2} = \frac{9}{2}$$

$\therefore \quad 3n = 9$

or $\quad n = 3$

On putting the value of n in general formula of alkene, we get $C_3H_{2\times3} = C_3H_6$. Hence, the alkene is propene $(CH_3CH{=}CH_2)$.

23 Acetylene does not react with NaOH. It reacts with other given reagents as follows:

$$CH{\equiv}CH + Na \longrightarrow CH{\equiv}C^{\ominus}Na^{\oplus} + \frac{1}{2}H_2$$

$$\underset{\text{(Ammoniacal)}}{CH{\equiv}CH + AgNO_3} \longrightarrow \underset{\text{White ppt.}}{CH{\equiv}C^{\ominus}Ag^{\oplus}}$$

$$CH{\equiv}CH + HCl \longrightarrow CH_2{=}CH{-}Cl$$

24 $\underset{\text{2-hexyne}}{CH_3CH_2CH_2{-}C{\equiv}C{-}CH_3} \xrightarrow{Li/NH_3}$

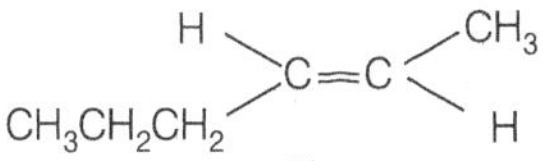

trans-2-hexene

2-hexyne $\xrightarrow{\text{(a) Pt/H}_2}$ *n*-hexane

$\xrightarrow{\text{(c) Pd/BaSO}_4}$ *cis*-2-hexene

$\xrightarrow{\text{(d) LiAlH}_4}$ no reaction

25

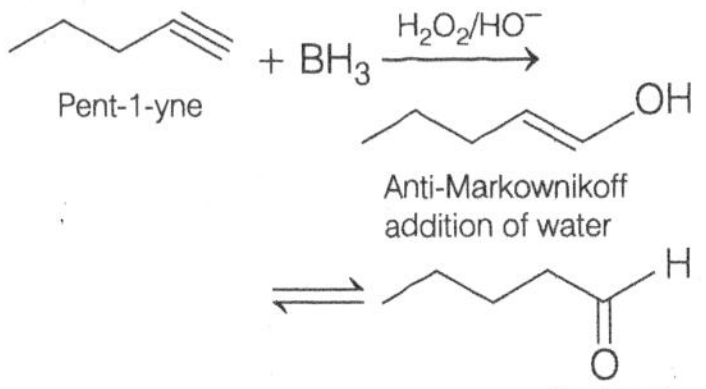

Pentanal

26 $\underset{\text{1-butyne}}{CH_3CH_2C{\equiv}CH} \xrightarrow{\text{Hot KMnO}_4} CH_3CH_2COOH + HCOOH$

$$HCOOH \longrightarrow CO_2 + H_2O$$

27 A $\longrightarrow$ 4, B $\longrightarrow$ 1, C $\longrightarrow$ 2, D $\longrightarrow$ 3

28 According to Huckel rule, the compounds which have $4n$ (n = 0, 1, 2, 3,) delocalised π-electrons in a close-loop are anti aromatic and characteristically unstable. Compound '*B*' satisfy the criteria of anti-aromaticity as

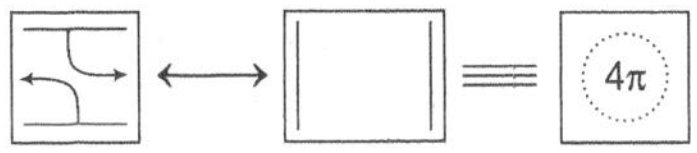

Compound *C* is anti-aromatic in its resonance form

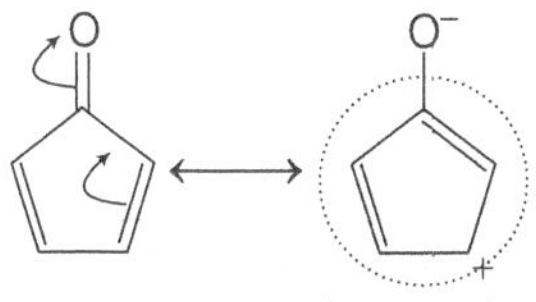

Antiaromatic

Compound '*A*' has 4π-electrons which are also delocalised but do not constitute close loop, hence, non-aromatic. Compound '*D*' is aromatic, characteristically stable.

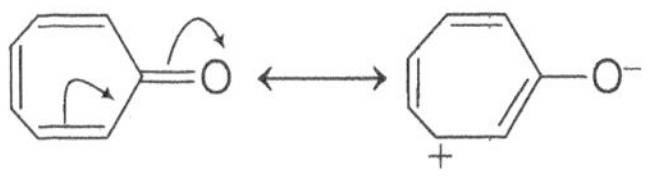

Tropium ion (Aromatic)

29 Mesitylene (1, 3, 5-trimethylbenzene) has three *m*-xylene (1, 3-dimethylbenzene) has two while toluene has only one electron donating —CH_3 group. Therefore electron density is maximum in mesitylene and hence, it is most reactive towards ring nitration.

30 Due to greater electron releasing effect of methyl group in case of *o*-methylphenol, it is most reactive towards electrophilic reagent.

31 Conversion of benzene into *m*- bromoaniline can be done by successive nitration, bromination and reduction using following reagent.

Nitration $\longrightarrow$ conc.HNO_3 / conc.H_2SO_4

Bromination $\longrightarrow Br_2 / H_2O$

Reduction $\longrightarrow$ Sn / HCl

32 Since, NO_2 group is a powerful electron withdrawing group therefore, carbocations given in options (a), (c) and (d) are destabilised. However, in carbocation (b), there is no —NO_2 group. Thus, σ-complex (b) is of lowest energy.

33 Presence of electron releasing groups, (like —*R*) increases the electron density over benzene nucleus and makes it more reactive towards electrophile. Electron withdrawing groups (like Cl^-, NO_2^-; NO_2^- is more electron withdrawing than Cl^-) make the benzene nucleus electrons deficient and hence, decrease its reactivity towards electrophile.

Hence, the correct order of reactivity towards electrophilic substitution reaction will be II > I > III > IV.

34

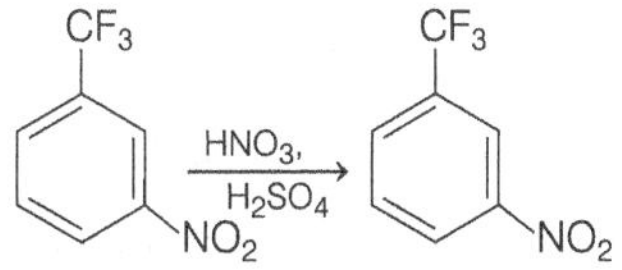

Since, CF_3 is deactivating and *meta*-directing group.

35 Cyclooctatetraene has a tub like structure. It loses planarity. Number of πe^- delocalised = 8. Hence, according to Huckel rule of aromaticity it is a non-aromatic compound.

36 In nitration of benzene with conc. nitric acid, conc. sulphuric acid acts as a calatyst. It helps in the formation of electrophile, i.e. nitronium ion NO_2^+.

$$HNO_3 + H_2SO_4 \longrightarrow NO_2^+ + 2HSO_4^- + H_3O^+$$

37

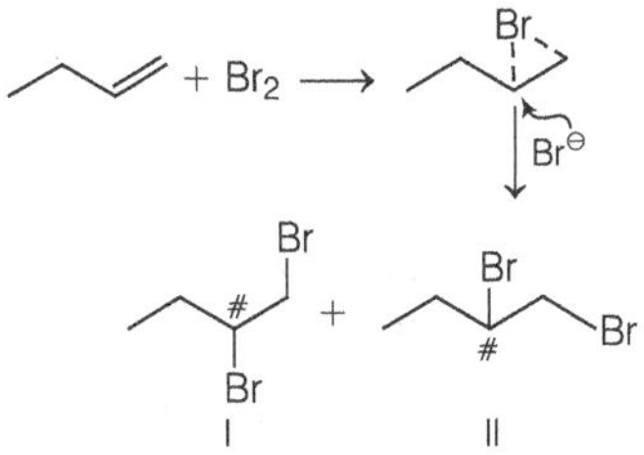

I and II are optical isomers.
denote chiral carbon.

38 1-butene on reaction with HBr in presence of peroxide produces 1-bromobutane. In this reaction, the intermediate product is secondary free radical.

(i) (a) $(C_6H_5CO)_2O_2 \longrightarrow 2C_6H_5COO^{\bullet}$

(b) $C_6H_5COO^{\bullet} + HBr \longrightarrow C_6H_5COOH + Br^{\bullet}$

(ii) $CH_3CH_2CH{=}CH_2 + Br^{\bullet} \longrightarrow CH_3CH_2\dot{C}HCH_2Br$

(iii) $CH_3CH_2\dot{C}HCH_2Br + HBr \longrightarrow CH_3CH_2CH_2CH_2Br + Br^{\bullet}$

39 Acetylene is acidic due to the presence of *sp* hybridised carbon which makes it more electronegative hence, with strong base it undergo elimination of H^+.

40 Alkyl substituent (having H-atom at α-carbon) at phenyl ring undergo oxidation to produce benzoic acid always.

SESSION 2

1 $CH_3CH_2CH{=}CH_2 + H^+ \longrightarrow CH_3CH_2\overset{+}{C}H-CH_3$ 2° (planar) (more stable major)

$+ CH_3CH_2CH_2-\overset{+}{C}H_2$ 1° (less stable)

$CH_3CH_2CH_2\overset{+}{C}H_2 + Br^- \longrightarrow CH_3CH_2CH_2CH_2Br$ (C)

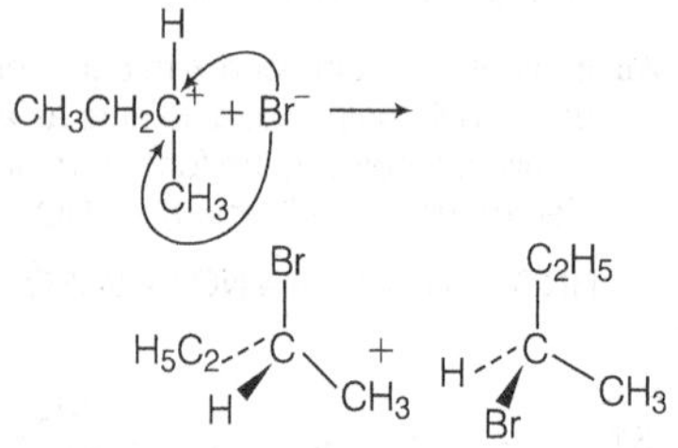

A *B*
(major products)

2 *trans-trans* conformer, i.e. (a) is most stable due to minimum steric hindrance.

3 $CH_3C{\equiv}CH + H_2O \xrightarrow[HgSO_4]{H_2SO_4} [CH_3C(OH){=}CH_2] \xrightarrow{\text{Tautomerisation}} CH_3COCH_3$ (Acetone)

4 $CH_3-C{\equiv}CH + HBr \longrightarrow CH_3-C(Br){=}CH_2$

$\xrightarrow{H^+} CH_3-CH(Br)-\overset{+}{C}H_2$

$\xrightarrow{\text{Rearrangement}} CH_3-\overset{+}{C}(Br)-CH_3$
More stable carbocation

$\xrightarrow{Br^-} H_3C-C(Br)_2-CH_3$
2, 2- dibromopropane

$CH_3-CH{=}CHBr \xrightarrow{HBr} CH_3-CH_2-CHBr_2$

$CH{\equiv}CH \xrightarrow{2HBr} CH_3CHBr_2$

$CH_3-CH{=}CH_2 \xrightarrow{HBr} CH_3-CHBr-CH_3$

7. (i) Determination of empirical formula of hydrocarbon *A*

Element	Percentage	Atomic mass	Relative ratio	Relative number of atoms	Simplest ratio
C	87.8	12	$\frac{87.8}{12} = 7.31$	$\frac{7.31}{7.31} = 1$	$1 \times 3 = 3$
H	12.19	1	$\frac{12.19}{1} = 12.19$	$\frac{12.19}{7.31} = 1.66$	$1.66 \times 3 = 5$

∴ Empirical formula of hydrocarbon $(A) = C_3H_5$

Empirical formula mass $= 12 \times 3 + 5 \times 1 = 41\,u$

(ii) Determination of molecular mass of hydrocarbon (*A*)

896 mL vapours of hydrocarbon (*A*) weigh at STP = 3.28 g

∴ 22400 mL vapours of *A* will weigh at STP

$= \frac{3.28 \times 22400}{896}\ g\ mol^{-1} = 82\ g$

∴ Molecular mass of hydrocarbon $(A) = 82\ g\ mol^{-1}$

(iii) Determination of molecular formula of hydrocarbon *A*.

$$n = \frac{\text{Molecular mass}}{\text{Empirical formula mass}} = \frac{82}{41} = 2$$

Therefore, molecular formula of hydrocarbon

$(A) = n \times$ Empirical formula

$= 2 \times C_3H_5 = C_6H_{10}$

Since, hydrogenation of *A* gives 2-methyl pentane therefore compound '*A*' have 5 carbon atoms in straight chain. It should be an alkyne because compound '*A*' adds a molecule of H_2O in the presence of Hg^{2+} and H^+ to give a ketone '*B*'. Ketone '*B*' gives iodoform test so it should be a methyl ketone. Therefore, the structures of compound *A* and compound *B* are as follows

$CH_3-CH(CH_3)-CH_2-C{\equiv}CH \xrightarrow[Hg^{2+}]{H_2O/H^+}$
4-methylpent-1-yne (*A*)

$CH_3-CH(CH_3)-CH_2-\overset{O}{\overset{\|}{C}}-CH_3$
4-methylpentan-2-one (*B*)

6 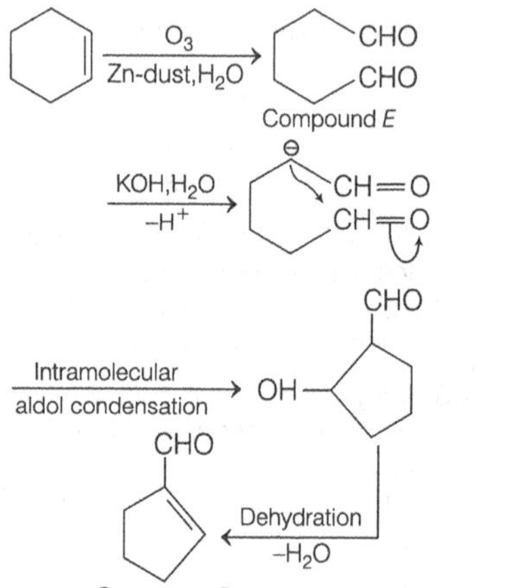

7 (a) Aromatic character in the product is lost due to addition, thus (a) is true.

(b) Sextet of benzene ring is electron rich, thus act as nucleophile. Thus, (b) is true.

(c) Sulphur carries positive charge due to resonance, thus option (d) is true.

8 (i) *A* on oxidation yields dicarboxylic acid, thus it should have two side chains on benzene ring.

(ii) *B* does not form inner anhydride, thus *B* is not an *ortho* carboxylic acid.

(iii) *B* on bromination gives only one product and thus, '*B*' should be *para*-derivative. Thus, *B* is *p*-benzene dicarboxylic acid.

(iv) The above facts and molecular formula of '*A*' suggest that *A* is

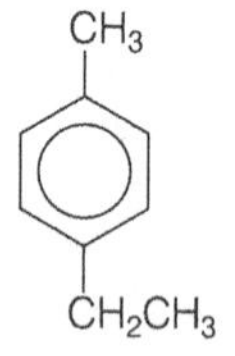

(v) The reactions are

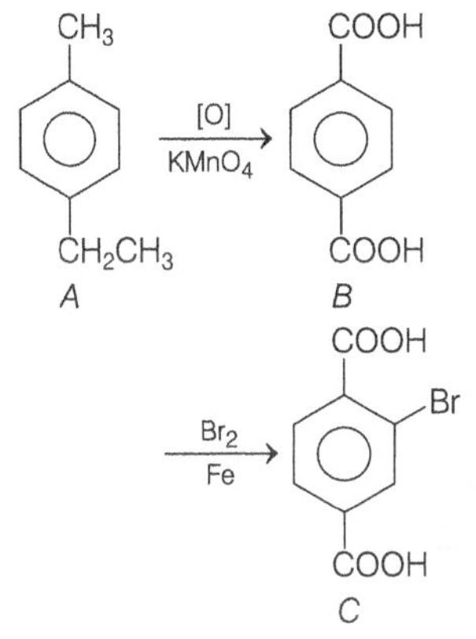

9 Bulky group decreases the proportion of *ortho*-isomer.

Thus, 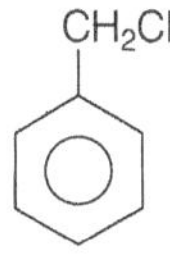 give the highest ratio of *o*/*p* isomer when reacted with $Cl_2/FeCl_3$.

10 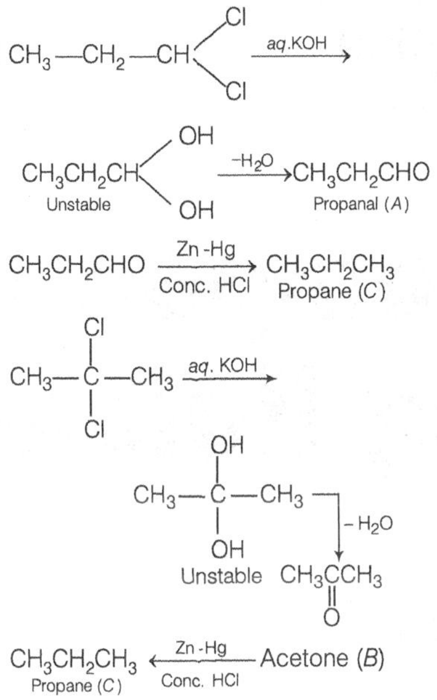

11 Molecular mass of *R*COOAg

$$= \frac{108}{59.6} \times 100 = 181$$

$$RCOOH = RCOOAg - 108 + 1 = 74$$

Thus, acid *R*COOH is CH_3-CH_2-COOH (*D*).

Since, *D* is CH_3-CH_2-COOH, '*B*' will be $CH_3-CH_2-CH=O$ and hence, '*C*' will be

$$(CH_3)_2-C=O$$

On the basis of ozonolysis product *B* and *C*, we can say that hydrocarbon '*A*' is $CH_3-\underset{\underset{CH_3}{|}}{C}=CH-CH_2-CH_3$.

12 Since, both gases *P* and *Q* decolourise aqueous Br_2 solution, therefore both are unsaturated. Since, one of them gives a white ppt with ammoniacal $AgNO_3$ solution while the other does not, one of them is a terminal alkyne (but-1-yne), while the other is a non-terminal alkyne (but-2-yne).

DAY TWENTY NINE

Organic Compounds Containing Halogens

Learning & Revision for the Day

- Alkyl and Aryl Halides
- Polyhalogen Compounds

Alkyl and Aryl Halides

These are formed by the replacement of hydrogen atom(s) from a hydrocarbon (aliphatic or aromatic) by halogen atom(s). These may be classified as mono, di or tri halo compounds depending upon the number of halogen atoms.

1. **Alkyl halides** contain halogen atom(s) attached to sp^3-hybridised carbon atoms of an alkyl group. e.g.

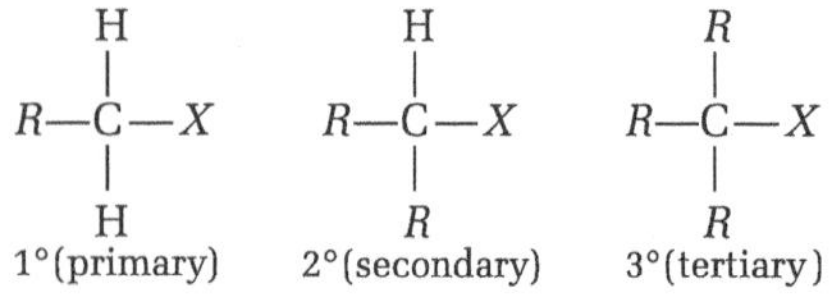

1° (primary) 2° (secondary) 3° (tertiary)

2. **Aryl halides** contain halogen atom(s) attached to sp^2 hybridised carbon atom(s) of an aryl group.

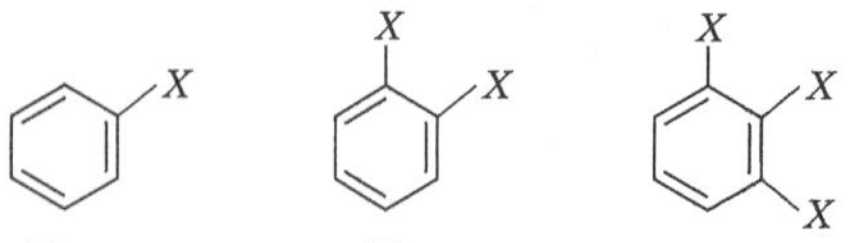

Mono Di Tri

Preparation of Alkyl Halides (Haloalkanes)

Alkyl halides can be prepared by using the following reactions:

- Haloalkanes can be obtained from alcohols by the reaction with halogen acids or phosphorus halides (PX_5 or PX_3) or thionyl chloride.

e.g. $$R\text{—OH} + \text{dry HCl} \xrightarrow{\text{Anhy. ZnCl}_2} R\text{Cl} + H_2O$$

$$R\text{OH} + \overset{+}{\text{Na}}\overset{-}{\text{Br}} \xrightarrow{(aq.)\,H_2SO_4} R\text{Br} + H_2O + \text{NaHSO}_4$$

PREP MIRROR
Your Personal Preparation Indicator

- No. of Questions in Exercises (x)—
- No. of Questions Attempted (y)—
- No. of Correct Questions (z)—
(Without referring Explanations)

- Accuracy Level ($z/y \times 100$)—
- Prep Level ($z/x \times 100$)—

In order to expect good rank in JEE, your Accuracy Level should be above 85 & Prep Level should be above 75.

$$3R\text{—OH} + PX_3 \longrightarrow 3RX + H_3PO_3$$

$$ROH + PCl_5 \longrightarrow RCl + POCl_3 + HCl \quad (X = \text{Cl, Br, I})$$

$$R\text{—OH} \xrightarrow{\text{P(red)}/X_2} RX \; (X_2 = Br_2, I_2)$$

$$R\text{—OH} + SOCl_2 \xrightarrow{\text{Pyridine}} RCl + SO_2\uparrow + HCl\uparrow$$

- Alkanes in the presence of sunlight shows free radical halogenation and gives a complex mixture of isomeric mono and polyhaloalkanes which is difficult to separate as pure compounds. e.g.

$$CH_3\text{—}CH_2\text{—}CH_2\text{—}CH_3 \xrightarrow[\text{or heat}]{Cl_2/\text{UV light}}$$

$$CH_3\text{—}CH_2\text{—}CH_2\text{—}CH_2\text{—}Cl + CH_3\text{—}CH_2\text{—}\underset{\underset{Cl}{|}}{CH}\text{—}CH_3$$

- An alkene is converted to corresponding alkyl halide by reaction with hydrogen chloride, hydrogen bromide or hydrogen iodide.

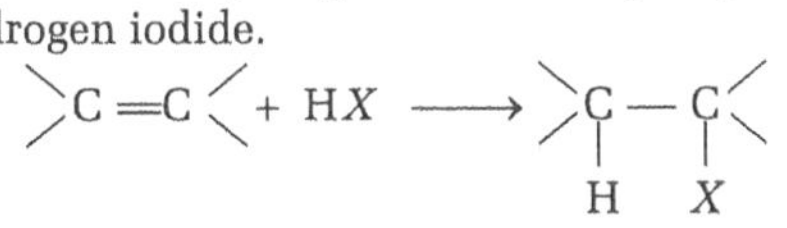

In unsymmetrical alkenes (such as propene), only one product predominates as per Markownikoff's rule

$$CH_3CH{=}CH_2 + HI \longrightarrow \underset{\text{(Minor)}}{CH_3CH_2CH_2I} + \underset{\text{(Major)}}{CH_3\,CH(I)CH_3}$$

- Addition of bromine in CCl_4 to an alkene results in the synthesis of *vic*-dibromides. This reaction discharge the reddish brown colour of bromine and is used in the detection of double bond in a molecule.

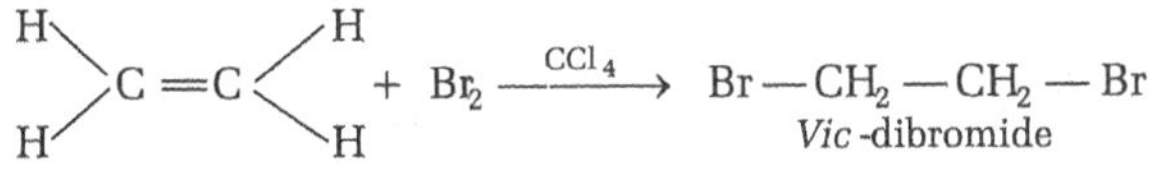

- **Halide transfer method** is used to prepare alkyl fluoride and alkyl iodide. This includes Finkelstein reaction and Swarts reaction.

  These are as follows:

  (i) **Finkelstein reaction** It is used for the conversion of chloroalkane or bromoalkane to iodoalkane.

  $$RX + NaI \xrightarrow{\text{Acetone}} RI + NaX \qquad (X = \text{Br, Cl})$$

  (ii) **Swarts reaction** (Preparation of alkyl fluorides) Alkyl chloride/ bromide when heated with metallic fluoride such as AgF, Hg_2F_2, CoF_2 or SbF_3 gives alkyl fluoride.

  $$CH_3\text{— Br} + AgF \longrightarrow CH_3F + AgBr$$

- **Borodine Hunsdiecker reaction** (used to produce alkyl bromide/chloride) proceeds through free radical mechanism and is used to reduce the length of carbon chain. The reaction is mainly used for the preparation of alkyl bromide

$$RCO_2Ag + Br_2 \longrightarrow RBr + CO_2\uparrow + AgBr$$

Preparation of Aryl Halides (Haloarenes)

Aryl halides can be prepared by the following reactions.

- Aryl chlorides and bromides can be easily prepared by electrophilic substitution of arenes with chlorine and bromine respectively in the presence of a Lewis acids like $FeCl_3$, $FeBr_3$ in dark and at low temperature.

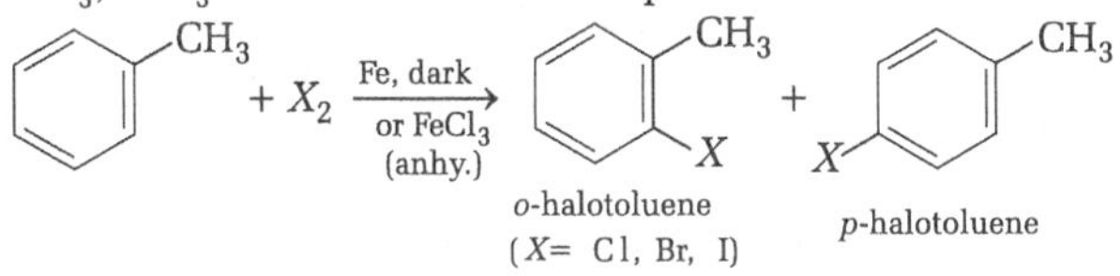

- Aryl halide can be prepared by mixing the solution of freshly prepared diazonium salt with cuprous chloride or cuprous bromide results in the replacement of the diazonium group by-Cl or Br. This reaction is called **Sandmeyer's reaction**

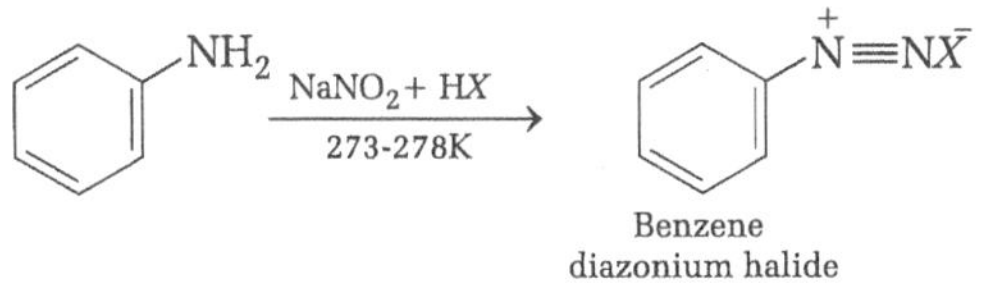

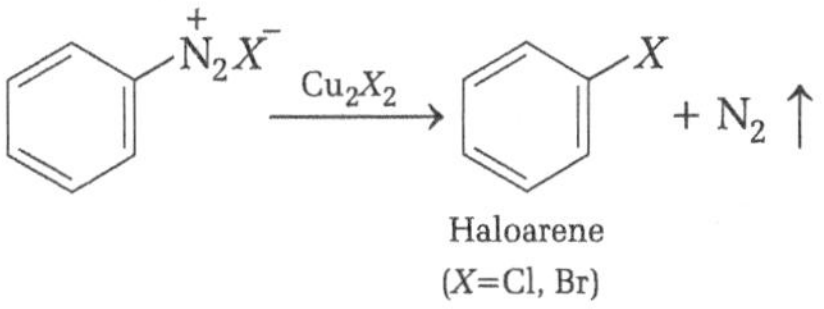

- Iodobenzene is obtained by warming diazonium salt with KI and does not require the presence of cuprous halide.

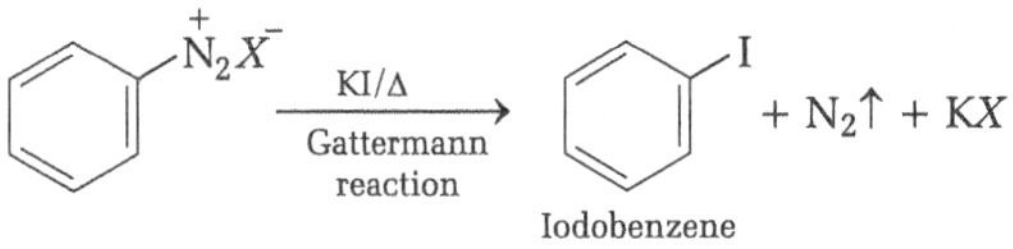

Fluorobenzene is obtained by using fluoroboric acid, HBF_4.

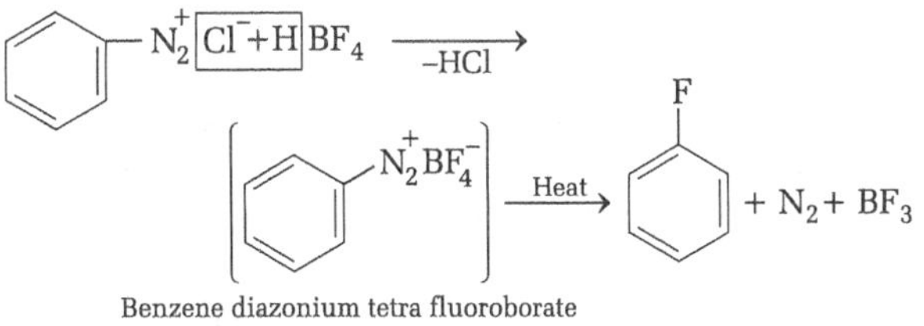

This reaction is known as **Balz-Schiemann reaction.**

Physical Properties of Haloalkanes and Haloarenes

The physical properties of alkyl and aryl halides are as follows.

- Haloalkanes and haloarenes are polar compounds and have higher boiling point due to dipole-dipole interaction. They have higher boiling points than hydrocarbons of comparable molecular mass.

Decreasing order of boiling point of alkyl halides is as follows:

$$RI > RBr > RCl > RF$$

This is because with the increase in size and mass of halogen atom, the magnitude of van der Waals' forces increases.

- Straight chain alkyl halides have higher boiling point as compared with branched chain alkyl halide of similar molecular weight, hence boiling points of isomeric haloalkanes decrease with increase in branching due to decrease in surface area. e.g.

$CH_3—CH_2—CH_2—CH_2—Br$

B.P./K 375

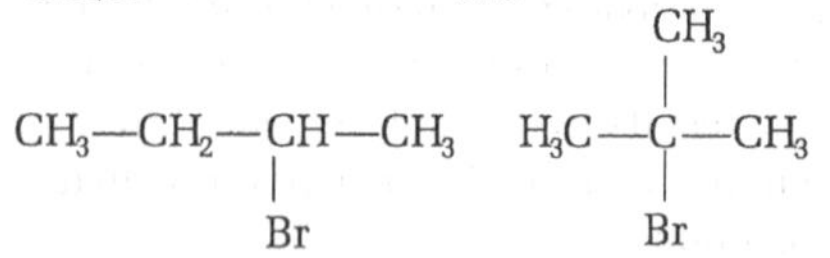

B.P./K 364 346

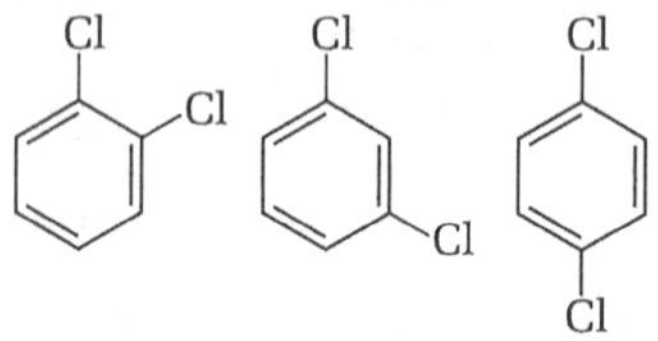

B.P./K 453 446 448

M.P./K 256 249 323

- Bromo and iodo derivatives of hydrocarbons are heavier than water. Lower haloalkanes are very slightly soluble in water but others are insoluble in water because haloalkanes/ haloarenes do not form hydrogen bond with water.

Nature of C—X Bond

- Electronegativity of halogen atom is greater than that of carbon atom due to which the shared pair of electrons in C—X bond closer to the halogen atom.
- On moving down the group, the carbon - halogen bond length increases.
- In haloarenes, sp^2 hybridised carbon of benzene is bonded to halogen and C—X bond is also polar as in haloalkanes due to high electronegativity of halogen.

Chemical Properties of Haloalkanes

Haloalkanes are highly reactive compounds due to the presence of polar $R—X$ bond. The reactions of haloalkanes may be of three types. These are nucleophilic substitution reactions, elimination reactions and reaction with metals. These are as follows.

1. Nucleophilic Substitution Reaction

In nucleophilic substitution reactions, a nucleophile reacts with haloalkane having a partial positive charge on the carbon atom bonded to halogen. These reactions proceed either by S_N1 or by S_N2 type of mechanism.

Main features of substitution nucleophilic unimolecular reaction (S_N1) are :

(i) Tertiary alkyl group and polar solvent favours S_N1.
(ii) It follows first order kinetics, rate = k [substrate].
(iii) Carbocation is formed and gets rearranged, if possible.
(iv) Reaction is completed in two steps.
(v) Nucleophile can attack from front and backside. Therefore, racemic mixture can be formed.

e.g.

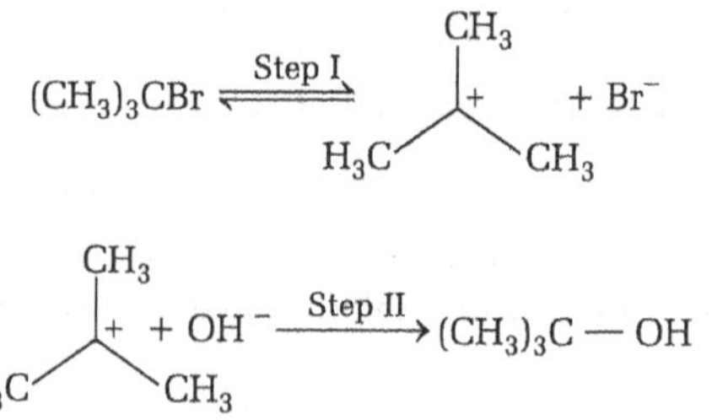

Main features of substitution nucleophilic bimolecular reaction (S_N2) are:

(i) Primary alkyl group and non-polar solvent favour S_N2.
(ii) It follows second order kinetics, rate = k [substrate] [nucleophile]
(iii) Carbocation is not formed.
(iv) Reaction is completed in one step.
(v) It involves complete inversion in configuration as the attack of the nucleophile occurs from the backside of reactant.

e.g.

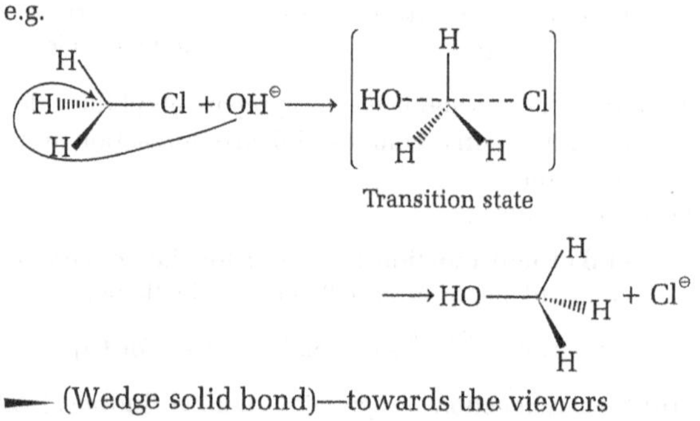

(Wedge solid bond)—towards the viewers

(Dash bond)—away from the viewers

- Reactivity of halides toward S_N2 mechanism is $CH_3X > 1° > 2° > 3°$
- Reactivity of halides towards S_N1 mechanism is $3° > 2° > 1° > CH_3X$.
- In S_N2 mechanism, rate of reactions depends upon the strength of attacking nucleophile.
The strength of different nucleophiles is
$CN^- > I^- > OR^- > OH^- > CH_3COO^- > H_2O > F^-$.

Stereochemical Aspects of Nucleophilic Substitution Reaction

Some basic stereochemical principles and notations are discussed below :

- Certain compounds rotate the plane polarised light when it is passed through solutions. Such compounds are called **optically active** compounds. The angle by which the plane polarised light rotated is measured by an instrument called **polarimeter**.

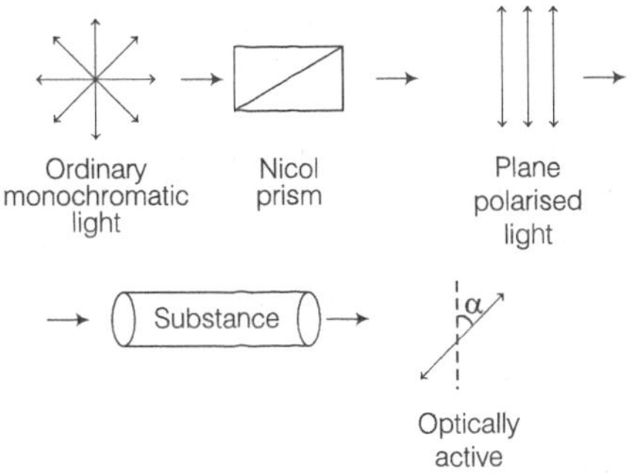

  If plane polarised light rotates in clockwise direction, when passed through a substance, the substance is known as **dextrorotatory** or (+) or (*d*). If it rotates in anticlockwise, the substance is known as **laevorotatory** or *l* or (–).

- Chiral carbon has four different groups or atoms attached to it.
- Chiral molecule (Optically active molecule) possesses at least 1-chiral carbon atom.
- Racemic mixture is the equimolar mixture of enantiomers (*dextro* and *laevo*). Racemic mixture is optically inactive due to external compensation. This phenomenon is known as **racemisation.**
- The backside attack by a nucleophile in S_N2 reactions gives rise to a product whose configuration is opposite to the reactant. Therefore, during S_N2 reaction, **inversion of configuration** occurs. This 100% inversion of configuration is known as **Walden inversion**.

2. Elimination Reactions

- When haloalkane having β-hydrogen atom is heated with alc. KOH, β-elimination takes place. e.g.

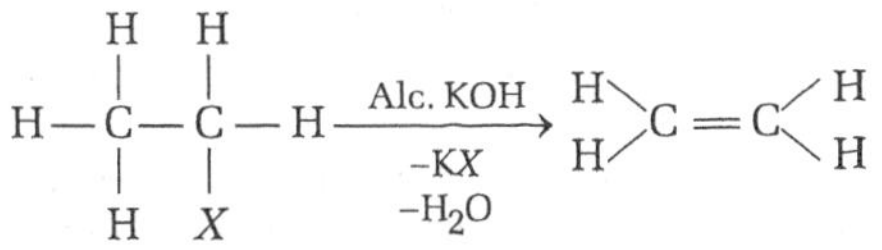

- If there is possibility of formation of more than one alkene due to availability of more than one α-H-atoms, the elimination of HX takes place according to **Saytzeff rule**. It states, "in dehydrohalogenation reactions, the preferred product is that alkene which has the greater number of alkyl groups attached to the doubly bonded carbon atoms, e.g.

$$CH_3CH_2CH_2CH(Br)—CH_3 \text{ (2-bromopentane)} \longrightarrow \begin{cases} \xrightarrow[\text{(Alc.)}]{KOH} CH_3CH_2CH{=}CHCH_3 \text{ Pent-2-ene (81\%)} \\ \xrightarrow[\text{(Alc.)}]{KOH} CH_3CH_2CH_2CH{=}CH_2 \text{ Pent-1-ene (19\%)} \end{cases}$$

3. Reaction with Metals

Grignard reagents are obtained by the reaction of haloalkanes with magnesium metal in dry ether.

$$CH_3CH_2Br + Mg \xrightarrow{\text{Dry ether}} \underset{\text{Grignard reagent}}{CH_3CH_2MgBr}$$

NOTE Some Important Reactions of Haloalkanes:

$$R—X \xrightarrow{AgNO_2} R—NO_2 + AgX$$

$$RX \xrightarrow{NaNO_2 \text{ or } KNO_2} R—ONO + KX \text{ or } NaX$$

$$RX \xrightarrow{KCN \text{ or } NaCN} RCN + KX \text{ or } NaX$$

$$RX \xrightarrow{AgCN} RNC + AgX$$

Hofmann ammonolysis reaction:

$$RX + NH_3 \longrightarrow RNH_2 + R—NH—R + R_3N + [R_4N^+]Cl^-$$

Chemical Properties of Haloarenes

Haloarenes give three types of reactions:

1. Nucleophilic substitution reactions

Haloarenes are less reactive than haloalkanes towards nucleophilic substitution reactions due to resonance effect, difference in hybridisation of carbon atom in C – *X* bond, instability of phenyl cation and repulsion between electron rich attacking nucleophile and electron rich arenes. Various nucleophilic substitution reactions are as follows

(i) 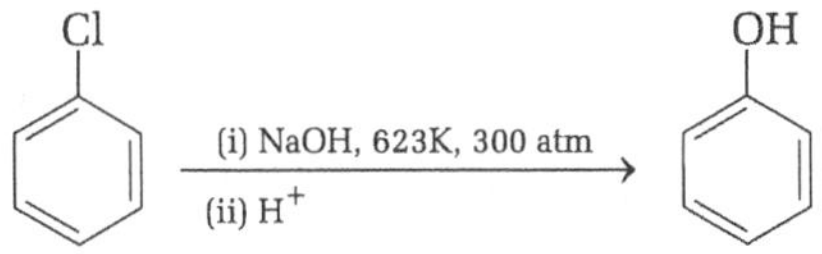

Chlorobenzene Phenol

This is known as **Dow's process.**

(ii) 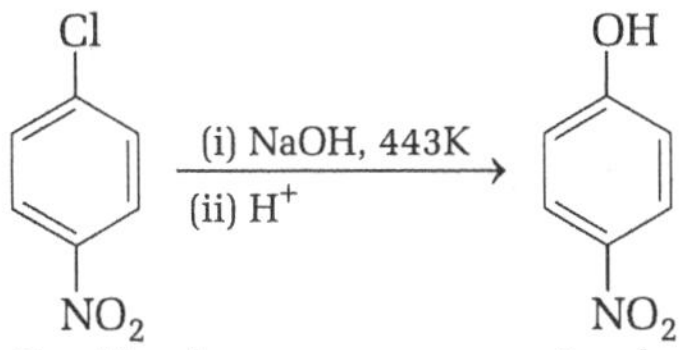

p-nitrochlorobenzene *p*-nitrophenol

(iii) 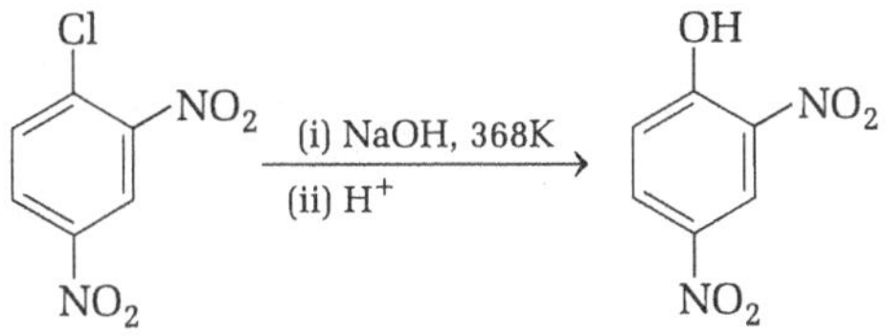

2,4-dinitrochlorobenzene 2,4-dinitrophenol

(iv) 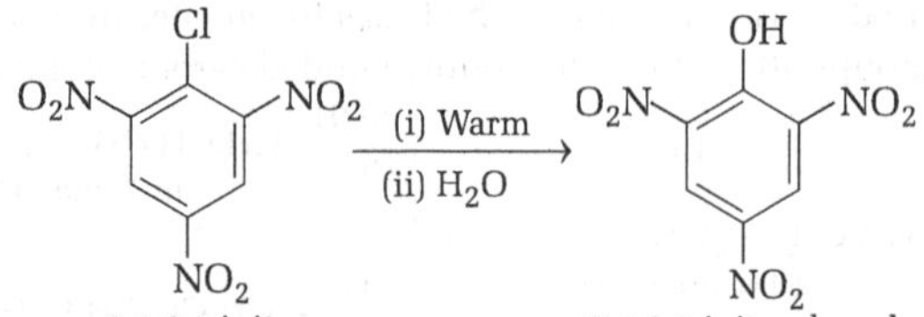

2,4,6-trinitro chlorobenzene → 2,4,6-trinitrophenol or picric acid

2. Electrophilic Substitution Reactions

Haloarenes undergo **electrophilic substitution reactions** of benzene nucleus such as halogenation, nitration, sulphonation and Friedel-Crafts alkylation and acylation.

These are as follows :

(i) **Halogenation**

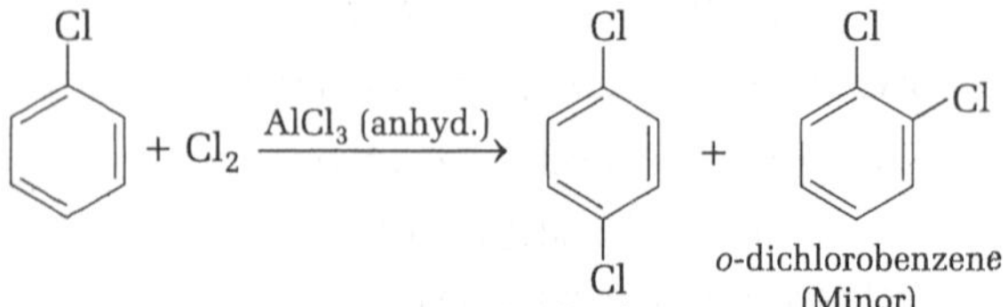

p-dichlorobenzene (Major) + *o*-dichlorobenzene (Minor)

(ii) **Nitration**

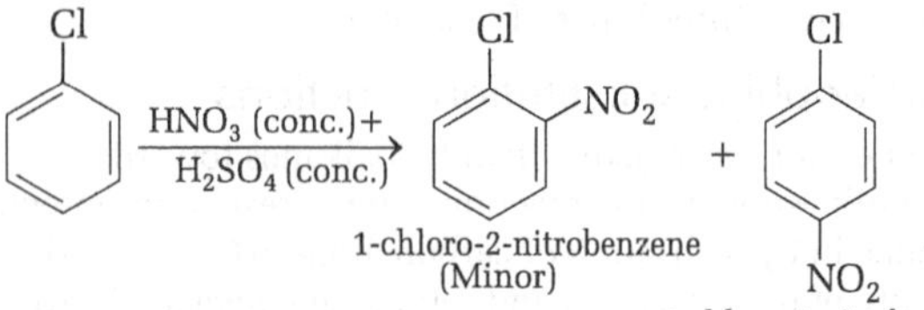

1-chloro-2-nitrobenzene (Minor) + 1-chloro-4-nitrobenzene (Major)

(iii) **Sulphonation**

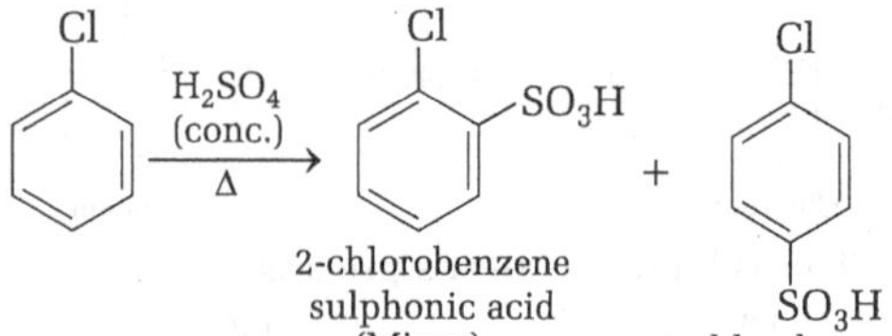

2-chlorobenzene sulphonic acid (Minor) + 4-chlorobenzene sulphonic acid (Major)

(iv) **Friedel-Craft Alkylation**

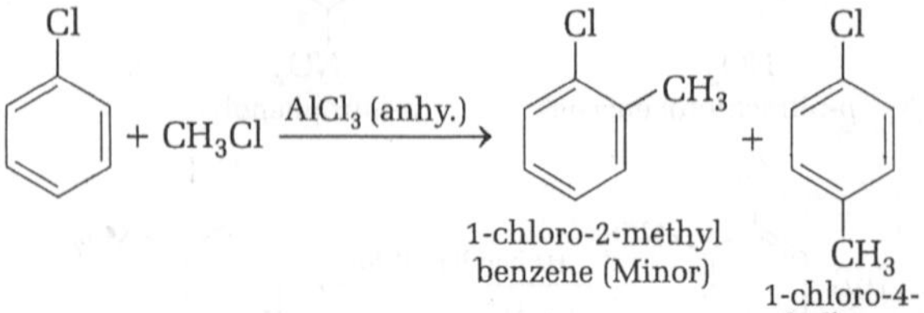

1-chloro-2-methyl benzene (Minor) + 1-chloro-4-methylbenzene (Major)

(v) **Friedel-Craft Acylation**

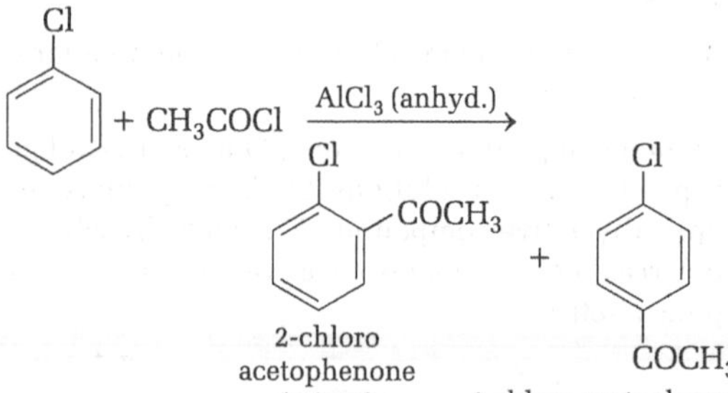

2-chloro acetophenone (minor) + 4-chloro acetophenone (major)

Nuclear halogenation takes place by electrophilic substitution mechanism whereas side chain halogenation takes place by free radical mechanism.

3. Haloarenes React with Metals to Produce Higher Alkanes

These reaction are as follows:

(i) **Wurtz-Fittig Reaction** A mixture of haloarene and haloalkane, when treated with Na in dry ether gives alkyl arene. e.g.

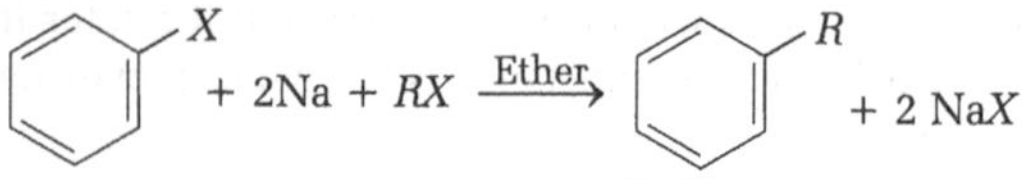

Alkyl benzene

(ii) **Fittig Reaction** Haloarenes on treatment with Na in dry ether give analogous compounds. e.g.

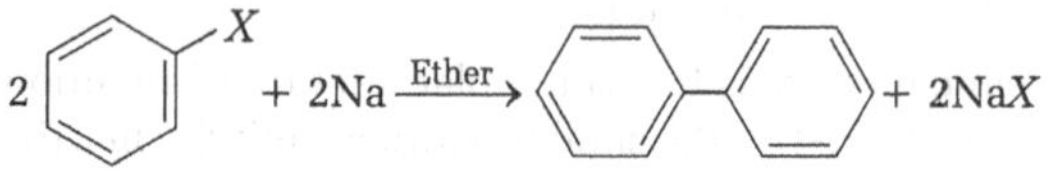

Diphenyl

Polyhalogen Compounds

The polyhalogenated compounds (PHCs) are any compounds with multiple substitutions of halogens. Some of polyhalogen compounds are given below:

1. Trichloromethane (Chloroform; $CHCl_3$)

- It is employed as a solvent for fats, alkaloids, iodine and other substances.
- It was used as a general anaesthetic but now a days it is replaced by less toxic and safe anaesthetic ether.
- The major use of chloroform today is in the production of the refrigerant, freon.
- It is stored in closed dark coloured bottles, filled completely so that air is kept out because it is oxidised by air in the presence of light.

$$2CHCl_3 + \underset{(Air)}{O_2} \xrightarrow{Light} \underset{\substack{\text{Phosgene}\\\text{(Poisonous gas)}}}{2COCl_2} + 2HCl$$

2. Tri-iodomethane (Iodoform; CHI_3)

It is prepared by heating ethanol or acetone with sodium hydroxide and iodine or Na_2CO_3 and I_2 in water.

$$CH_3CH_2OH + 6NaOH + 4I_2 \xrightarrow{\text{Heat}} CHI_3 \downarrow + 5\,NaI + HCOONa + 5\,H_2O$$

$$CH_3COCH_3 + 4NaOH + 3\,I_2 \xrightarrow{\text{Heat}} CHI_3 \downarrow + 3NaI + CH_3COONa + 3H_2O$$

This reaction is called **iodoform reaction.**

Some of the important characteristic of iodoform are as follows:

(i) It is yellow solid having melting point 119°C.

(ii) It is insoluble in water but soluble in ethanol and ether.

(iii) It is used as antiseptic because it liberates iodine but due to obnoxious (unpleasant) smell, it has been replaced by other medicines.

(iv) Its chemical properties are similar to $CHCl_3$.

3. Freons (Chlorofluorocarbon; CCl_2F_2)

They are extremely stable, unreactive non-toxic, non-corrosive and easily liquefiable gases.

- It is manufactured from tetrachloromethane by Swarts reaction and used for aerosol propellants, refrigeration and air conditioning purposes.
- Freons deplete the protective ozone layer surrounding our planet, so their use has been banned in many countries.

4. DDT (p,p′-Dichlorodiphenyltrichloroethane)

- It is a powerful insecticide but is not easily biodegradable. Therefore, its long term effects could be potentially dangerous and its use is banned in many countries.
- IUPAC name of DDT is 2,2-*bis* (4-chloro phenyl)– 1, 1, 1-trichloroethane. It is non-biodegradable and slowly changes to another compound DDE by the loss of a molecule of HCl.

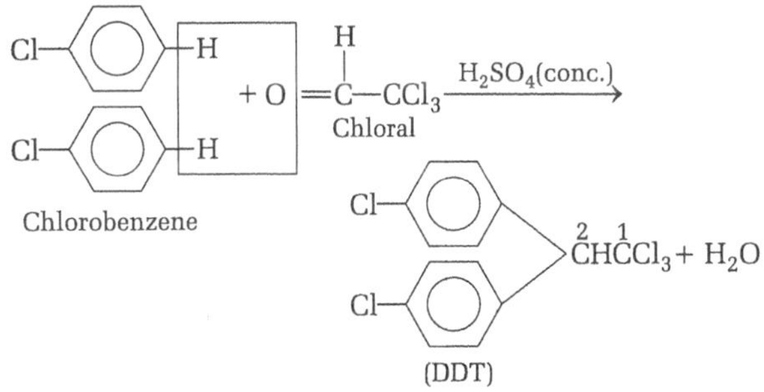

(DDT)
p,p′-dichlorodiphenyltrichloroethane

DAY PRACTICE SESSION 1

FOUNDATION QUESTIONS EXERCISE

1 Preparation of alkyl halides in laboratory is least preferred by
(a) halide exchange
(b) treatment of alcohols
(c) addition of hydrogen halides to alkenes
(d) direct halogenation of alkanes

2 The synthesis of alkyl fluorides is best accomplished by → JEE Main 2015
(a) free radical fluorination (b) Sandmeyer's reaction
(c) Finkelstein reaction (d) Swarts reaction

3 Which branched chain isomer of the hydrocarbon with molecular mass 72 u gives only one isomer of mono substituted alkyl halide? → AIEEE 2012
(a) Tertiary butyl chloride (b) *Neo* -pentane
(c) *Iso* -hexane (d) *Neo* -hexane

4 Given, → JEE Main (Online) 2013

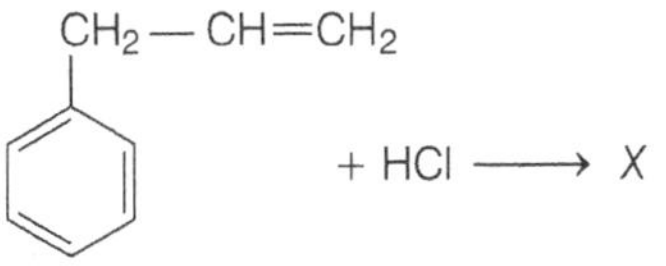

X is

(a) 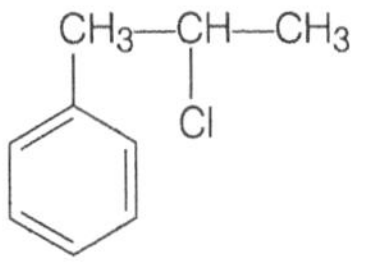

(b) 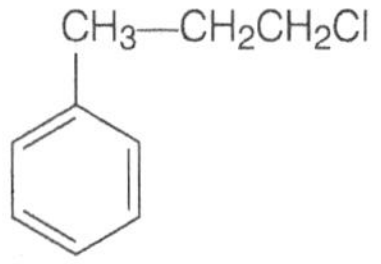

(c) 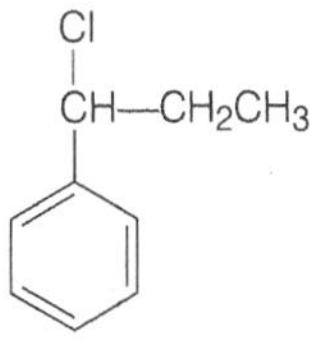

(d) 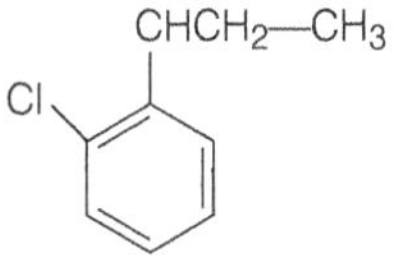

5 Which of the following has highest melting point?
(a) Chlorobenzene
(b) *o*-dichlorobenzene
(c) *m*-dichlorobenzene
(d) *p*-dichlorobenzene

6 Arrange the following compounds, in the increasing order of their boiling points.

(i) $(CH_3)_2CH—CH_2Br$ (ii) $CH_3CH_2CH_2CH_2Br$

(iii) $CH_3—C(CH_3)(Br)—CH_3$

(a) (ii) < (i) < (iii) (b) (i) < (ii) < (iii)
(c) (iii) < (i) < (ii) (d) (iii) < (ii) < (i)

7 Arrange the following compounds in the increasing order of their densities

(i)  (ii) 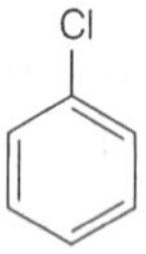 (iii) 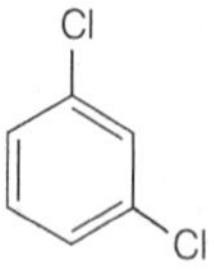 (iv) 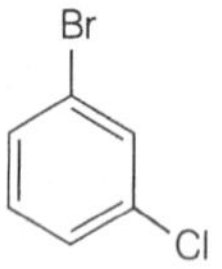

(a) (i) < (ii) < (iii) < (iv) (b) (i) < (iii) < (iv) < (ii)
(c) (iv) < (iii) < (ii) < (i) (d) (ii) < (iv) < (iii) < (i)

8 C—Cl bond is stronger than C—I bond because

(a) C—Cl bond is more ionic than C—I
(b) C—Cl bond is polar covalent bond
(c) C—Cl bond is more covalent than C—Cl
(d) C—Cl bond length is longer than C—I

9 Which of the following alkyl halide is used as a methylating agent ?

(a) C_2H_5Br (b) C_6H_5Cl (c) CH_3Cl (d) C_2H_5Cl

10 The addition of HI in the presence of peroxide catalyst does not follow anti-Markownikoff addition because → JEE Main (Online) 2013

(a) HI is a strong reducing agent
(b) HI bond is too strong to be broken homolytically
(c) I atom combines with H-atom to give back HI
(d) Iodine atom is not reactive enough to add across a double bond

11 Consider the following bromides,

(A)  (B) 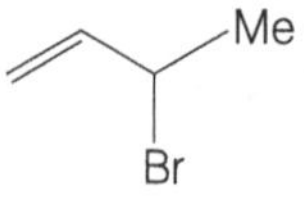

(C) 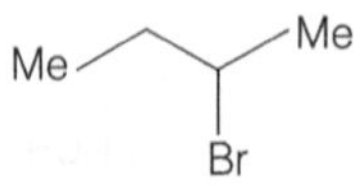

The correct order of S_N1 reactivity is → AIEEE 2010

(a) $B > C > A$ (b) $B > A > C$ (c) $C > B > A$ (d) $A > B > C$

12 The increasing order of reactivity of the following bromides in S_N1 reaction is

I. 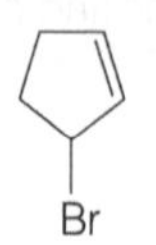 II. 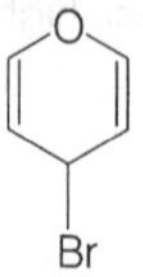 III. 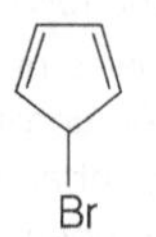 IV. 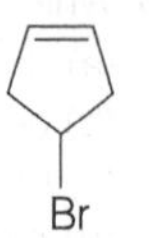

(a) III > I > II > IV (b) III > II > I > IV
(c) II > III > I > IV (d) II > I > IV > III

13 The organic chloro compound which shows complete stereochemical inversion during a S_N2 reaction, is

(a) CH_3Cl (b) $(C_2H_5)_2CHCl$
(c) $(CH_3)_3CCl$ (d) $(CH_3)_2CHCl$

14 The increasing order of reactivity of the following halides for the S_N1 reaction is → JEE Main 2017

I. $CH_3CH(Cl)CH_2CH_3$ II. $CH_3CH_2CH_2Cl$
III. $p-H_3CO—C_6H_4—CH_2Cl$

(a) (III)<(II)<(I) (b) (II)<(I)<(III)
(c) (I)<(III)<(II) (d) (II)<(III)<(I)

15 In S_N2 reactions, the correct order of reactivity for the following compounds CH_3Cl, CH_3CH_2Cl, $(CH_3)_2CHCl$ and $(CH_3)_3CCl$ is → JEE Main 2014

(a) $CH_3Cl > (CH_3)_2CHCl > CH_3CH_2Cl > (CH_3)_3CCl$
(b) $CH_3Cl > CH_3CH_2Cl > (CH_3)_2CHCl > (CH_3)_3CCl$
(c) $CH_3CH_2Cl > CH_3Cl > (CH_3)_2CHCl > (CH_3)_3CCl$
(d) $(CH_3)_2CHCl > CH_3CH_2Cl > CH_3Cl > (CH_3)_3CCl$

16 Choose the incorrect statement.

(a) An S_N1 reaction proceeds with inversion of configuration
(b) An S_N2 reaction proceeds with stereochemical inversion
(c) An S_N2 reaction follows second order kinetics
(d) The reaction of *tert*-butyl bromide with OH^- follows first order kinetics

17 $CH_3Br + Nu^- \longrightarrow CH_3—Nu + Br^-$

The decreasing order of the rate of the above reaction with nucleophile (Nu^-) A to D is

[Nu^- = (A) PhO^-, (B) AcO^-, (C) HO^-, (D) CH_3O^-]

(a) $D > C > A > B$ (b) $D > C > B > A$
(c) $A > B > C > D$ (d) $B > D > C > A$

18 In nucleophilic substitution reaction, order of halogens as incoming (attacking) nucleophile is $I^- > Br^- > Cl^-$. The order of halogens as departing nucleophile should be → JEE Main (Online) 2013

(a) $Br^- > I^- > Cl^-$ (b) $I^- > Br^- > Cl^-$
(c) $Cl^- > Br^- > I^-$ (d) $Cl^- > I^- > Br^-$

19 Match the following and choose the correct option.

	Column I		Column II
A.	S_N1 reaction	1.	*vic*-dibromides
B.	Bromination of alkenes	2.	*gem*-dihalides
C.	Alkylidene halides	3.	Racemisation
D.	Elimination of HX from alkyl halide	4.	Saytzeff rule

Codes

	A	B	C	D
(a)	3	1	2	4
(b)	1	3	2	4
(c)	3	2	1	4
(d)	2	4	3	1

20 In alkaline hydrolysis of a tertiary halide by aqueous alkali, if concentration of alkali is doubled, then the reaction rate

(a) will be doubled (b) will be halved
(c) will remain constant (d) cannot say anything

21 Which halide will be least reactive in respect to hydrolysis ?

(a) Vinyl chloride (b) Allyl chloride
(c) Ethyl chloride (d) *t*-butyl chloride

22 A dihaloalkane *X*, having formula $C_3H_6Cl_2$, on hydrolysis gives a compound, that can reduce Tollen's reagent.The compound '*X*' is

(a) 1, 2- dichloropropane (b) 1, 1- dichloropropane
(c) 1, 3- dichloropropane (d) 2, 2- dichloropropane

23 2-chloro - 2-methylpentane on reaction with sodium methoxide in methanol yields → JEE Main 2016

I. $C_2H_5CH_2C(CH_3)_2—OCH_3$ II. $C_2H_5CH_2C(CH_3)=CH_2$

III. $C_2H_5CH=C(CH_3)—CH_3$

(a) Both I and III (b) Only II
(c) Both I and II (d) All of these

24 Which of the following compounds will give racemic mixture on nucleophilic substitution by OH^- ion?

I. $CH_3—CH(C_2H_5)—Br$ II. $CH_3—C(Br)(C_2H_5)—CH_3$

III. $CH_3—CH(C_2H_5)—CH_2Br$

(a) I (b) I, II, III
(c) II, III (d) I, III

25 A solution of (−) -1-chloro -1-phenylethane in toluene racemises slowly in the presence of a small amount of $SbCl_5$, due to the formation of → JEE Main (Online) 2013

(a) carbene (b) carbocation
(c) free radical (d) carbanion

26 The major product of the following reaction is → JEE Main 2018

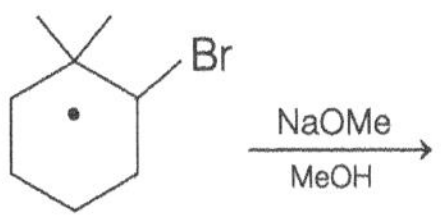

(a) 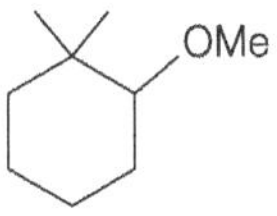

(b) 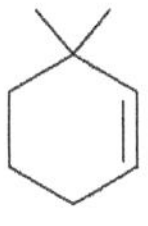

(c) 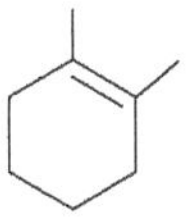
(d) 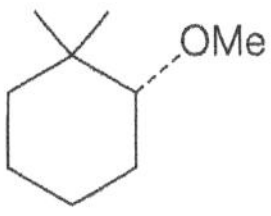

27 How many structures of F are possible?

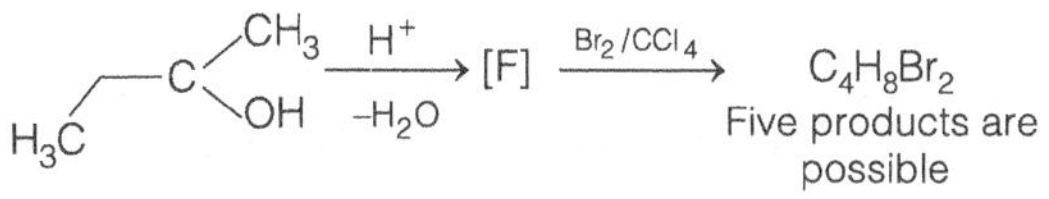

(a) 2 (b) 5 (c) 6 (d) 3

28 The major organic compound formed by the reaction of 1,1,1-trichloroethane with silver powder is → JEE Main 2014

(a) acetylene (b) ethene
(c) 2-butyne (d) 2-butene

29 Aryl halides are less reactive than alkyl halides towards nucleophile due to

(a) resonance (b) stability of carbonium ion
(c) high boiling point (d) None of these

30 Predict the main product in the given reaction :

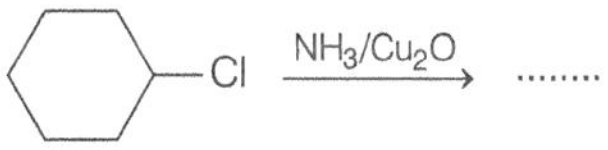

(a) Phenyl cyanide (b) Nitrophenol
(c) Aniline (d) Hydroxylamine

31 Chlorobenzene can be prepared by reacting aniline with:

(a) hydrochloric acid
(b) cuprous chloride
(c) chlorine in the presence of anhydrous aluminium chloride
(d) nitrous acid followed by heating with cuprous chloride

32 The chlorine atom in chlorobenzene is *ortho* and *para* directing because

(a) resonance effect predominates over inductive effect
(b) inductive effect predominates over resonance effect
(c) both inductive and resonance effects are evenly matched
(d) only resonance effect and not inductive effect is operating

33 The reaction of toluene with Cl_2 in the presence of $FeCl_3$ gives predominantly

(a) benzoyl chloride (b) benzyl chloride
(c) *o* - and *p*-chlorotoluene (d) *m*-chlorotoluene

34 Toluene reacts with excess of Cl_2 in the presence of sunlight to give a product which on hydrolysis followed by reaction with NaOH gives

(a) 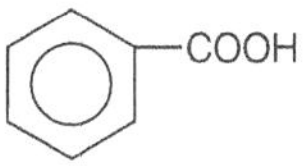

(b) 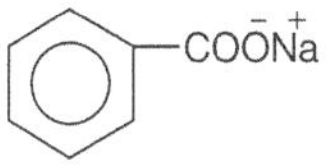

(c) 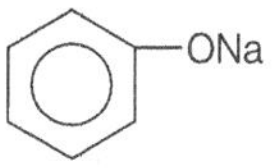

(d) None of these

35 The Wurtz-Fittig reaction involves condensation of → JEE Main (Online) 2013

(a) two molecules of aryl halides
(b) one molecule of each of aryl halide and alkyl-halide
(c) one molecule of each aryl halide and phenol
(d) two molecules of a alkyl halides

36 Compound A, C_8H_9Br gives a white precipitate when warmed with alcoholic $AgNO_3$. Oxidation of A gives an acid B, $C_8H_6O_4$. B easily forms anhydride on heating. Identify the compound A. → JEE Main (Online) 2013

(a) 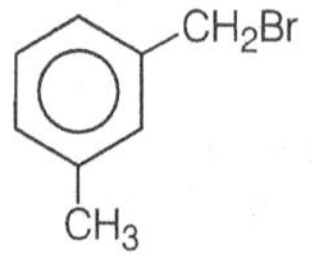

(b) 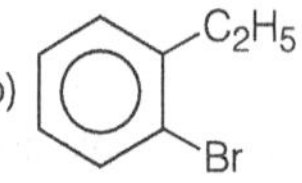

(c) 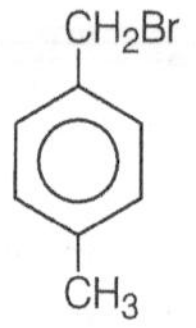

(d) 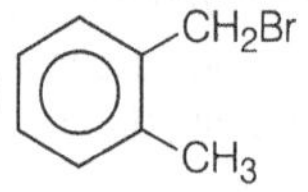

37 CCl_4 is a well known fire extinguisher. However, after using it to extinguish fire, the room should be well ventilated. This is because

(a) it is flammable at higher temperature
(b) it is toxic
(c) it produces phosgene by reaction with water vapour at higher temperature
(d) It is corrosive

38 The reaction of toluene with Cl_2 in the presence of $FeCl_3$ gives X and reaction in presence of light gives Y. Thus, X and Y are

(a) X = benzyl chloride, Y = m- chlorotoluene
(b) X = benzyl chloride, Y = q -chlorotoluene
(c) X = m -chlorotoluene, Y = p- chlorotoluene
(d) X = o and p- chlorotoluene, Y = trichloromethyl benzene

39 Which is not the correct statement?

(a) Chloretone is an insecticide
(b) $COCl_2$ is called phosgene
(c) Chloropicrin is used as an insecticide
(d) CCl_4 is used as fire extinguisher under the name pyrene

40 The compound formed on heating chlorobenzene with chloral in the presence of concentrated sulphuric acid is

(a) gammexane (b) DDT
(c) freon (d) hexachloroethane

Direction (Q. Nos. 41-46) *In the following questions Assertion followed by Reason is given. Choose the correct answer out of the following choices.*

(a) Assertion and Reason both are correct statements and Reason is the correct explanation of the Assertion
(b) Assertion and Reason both are correct statements but Reason is not the correct explanation of the Assertion
(c) Assertion is correct and Reason is incorrect
(d) Both Assertion and Reason are incorrect

41 **Assertion** (A) Phosphorus chlorides (tri and penta) are preferred over thionyl chloride for the preparation of alkyl chloride from alcohols.

Reason (R) Phosphorus chlorides give pure alkyl halides.

42 **Assertion** (A) *Tert*-butyl bromide undergoes Wurtz reaction to give 2, 2, 3, 3-tetramethylbutane.

Reason (R) In Wurtz reaction, alkyl halides react with sodium in dry sodium in dry ether to give hydrocarbon containing double the number of carbon atoms present in the halide.

43 **Assertion** (A) Alkyl halides are not soluble in water.

Reason (R) Although polar in nature, yet alkyl halides are not able to form H-bonds with water molecules.

44 **Assertion** (A) Tertiary alkyl halides are more reactive than 1° alkyl halides towards elimination.

Reason (R) Positive inductive effect of alkyl groups weakens carbon-halogen bond in 3° alkyl halides.

45 **Assertion** (A) Aryl halides undergo nucleophilic substitution with ease.

Reason (R) The carbon halogen bond in aryl halides has partial double bond character.

46 **Assertion** (A) Bromobenzene upon reaction with Br_2 / Fe gives 1,4 -dibromobenzene as the major product.

Reason (R) In bromobenzene, the inductive effect of the bromo group is more dominant than the mesomeric effect in directing the incoming electrophile.

DAY PRACTICE SESSION 2

PROGRESSIVE QUESTIONS EXERCISE

1 Which of the following sequence of reagent is best suited for the reaction shown below?

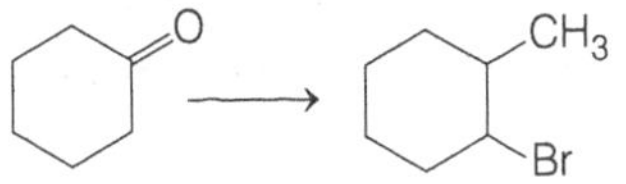

(a) (i) CH_3MgBr, H_3O^+; (ii) H^+ / Δ; (iii) HBr/H_2O_2
(b) (i) CH_3MgBr, H_3O^+; (ii) H^+ / Δ; (iii) HBr
(c) (i) CH_3MgBr, H_3O^+; (ii) HBr;
(d) (i) $HBr/ROOR$; (ii) CH_3MgBr, H_3O^+

2

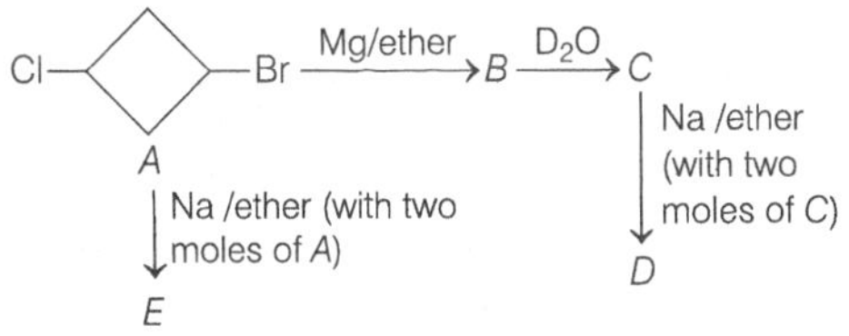

D and *E* respectively are

(a) 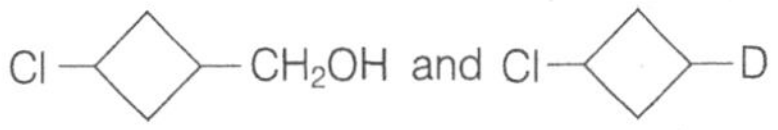

(b) 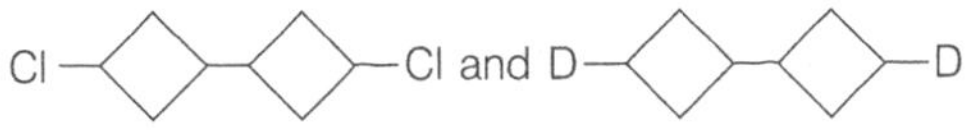

(c) 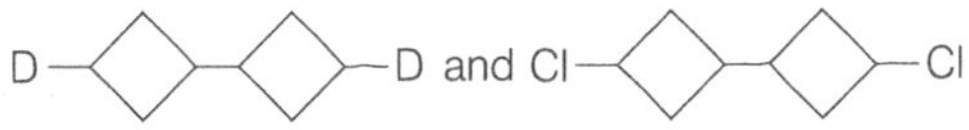

(d) None of the above

3 Which among the following reaction seems to be incorrect?

(a) $Me_3CCl \xrightarrow{NH_3} Me_3CNH_2$
(b) $Me_3CCl \xrightarrow{NH_3} Me_2C{=}CH_2$
(c) 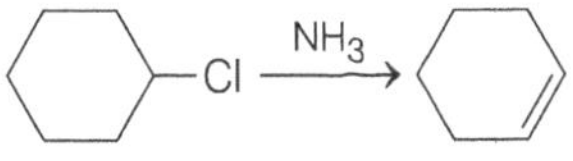

(d) 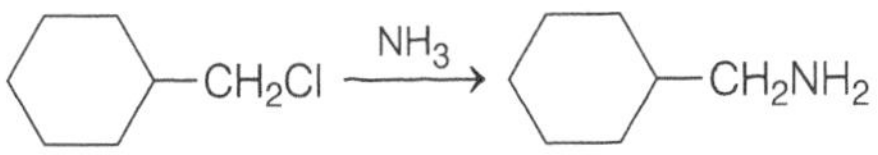

4 An organic compound C_5H_9Br (*A*) which readily decolourises bromine water and cold alkaline $KMnO_4$ solution and gives $C_5H_{11}Br$ (*B*) on catalytic hydrogenation. The reaction of *A* with alcoholic KOH first and then with $NaNH_2$ produces *C* with evolution of NH_3. *C* reacts with Lindlar's catalyst to give *D* and on reaction with Na in liquid NH_3 produces *E*. *D* and *E* are isomeric. The compound *A* is

(a) $CH_3CH_2CH{=}C(Br){-}CH_3$ (b) $CH_3CH_2{-}C(Br){=}CHCH_3$
(c) Both (a) and (b) (d) Neither (a) nor (b)

5 The product of which of the following reaction on dehydration gives 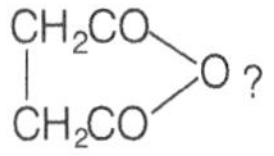 ?

(a) 1, 2-dibromoethane $\xrightarrow{KCN} \xrightarrow{H_3O^+}$
(b) 1, 1-dibromoethane $\xrightarrow{KCN} \xrightarrow{H_3O^+}$
(c) 1, 1, 1-trichloroethane $\xrightarrow{\text{Alkaline hydrolysis}}$
(d) None of the above

6 Product on monobromination of this compound is

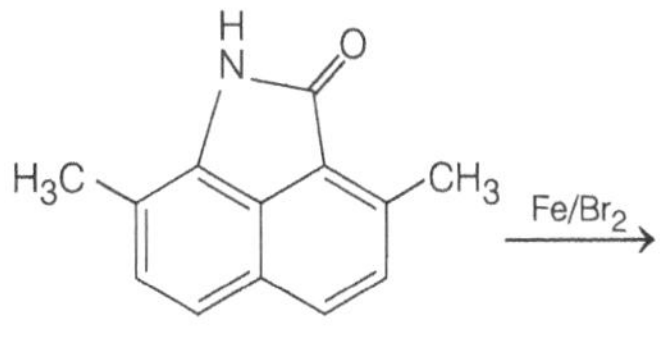

(a) 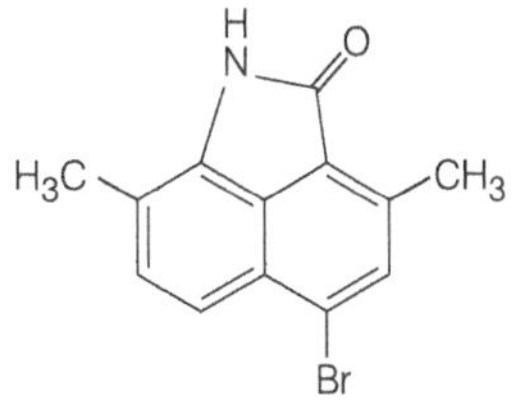 (b) 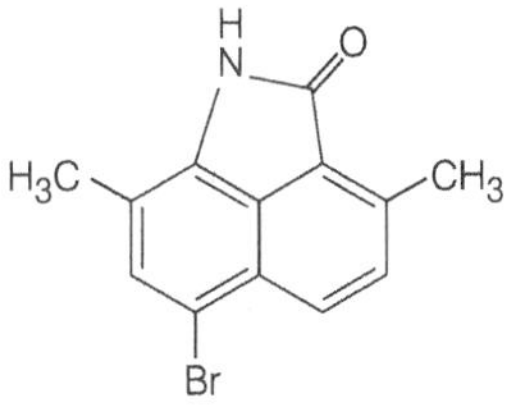

(c) 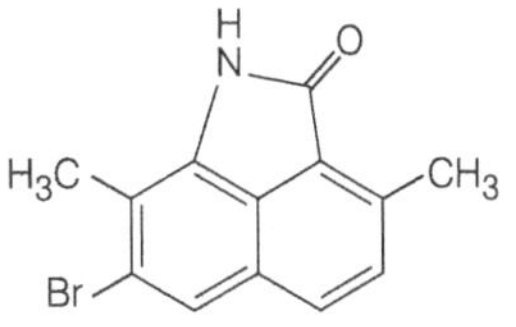 (d) 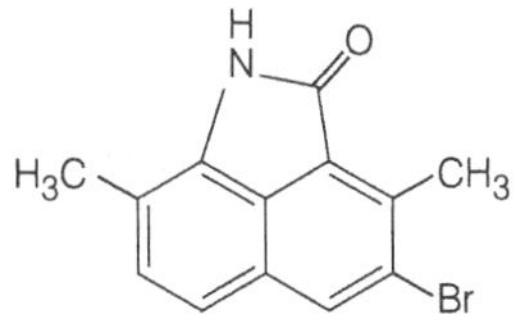

7 Iodine from iodoethane can be substituted by cyanide group in a number of ways. Which of the following is true statement?

(a) With AgCN, EtNC while with KCN, EtCN is formed as major product
(b) With AgCN, EtCN and with KCN, EtNC is formed
(c) With either KCN or AgCN, EtNC is formed as major product
(d) With either KCN or AgCN, EtCN is formed as major product

8 A hydrocarbon has molecular mass = 72. It gives a single monochloride and two dichlorides on further photochlorination. The two dichlorides are

(a) $CH_3CCl_2(CH_2)_2CH_3$; $CH_3(CHCl)_2CH_2CH_3$
(b) $(CH_3)_2CCl_2CH_2CH_3$; $(CH_3)_2(CHCl)_2CH_3$
(c) $CHCl_2C(CH_3)_3$; $(CH_3)_2C(CH_2Cl)_2$
(d) $CHCl_2CH_2CH(CH_3)_2$ $CH_3CHClCH_2CHClCH_3$

9 One mole of hydrocarbon (A) reacts with 1 mole of bromine giving a dibromo compound $C_5H_{10}Br_2$. (A) on treatment with cold dilute alkaline $KMnO_4$ solution forms a compound $C_5H_{12}O_2$. On ozonolysis, (A) gives equimolar quantities of propanone and ethanal. The compound (A) is

(a) $(CH_3)_2C{=}CHCH_3$
(b) $(CH_3)_2CH-CH_2CH_3$
(c) $CH_3CH_2CH{=}CHCH_3$
(d) $CH_3-\underset{\displaystyle CH_3}{\underset{|}{CH}}-CH{=}CH_2$

10 The major product of the following reaction is

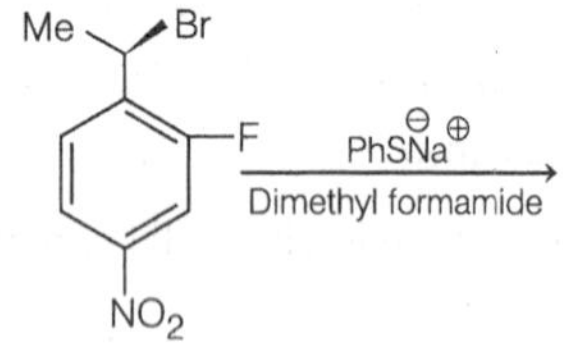

(a)
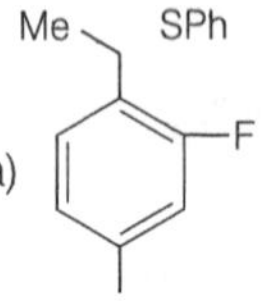

(c)
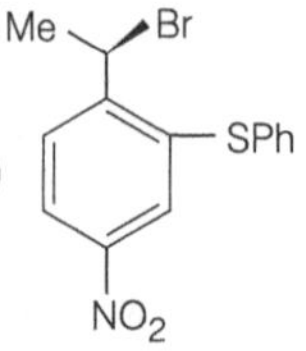

(d)
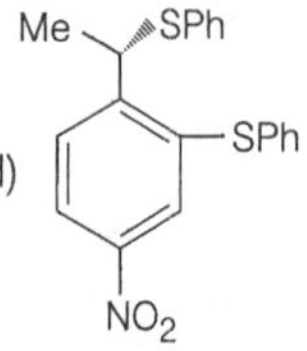

11 Bottles containing C_6H_5I and $C_6H_5CH_2I$ lost their original lables. They were labelled A and B for testing. A and B were separately taken in test tubes and boiled with NaOH solution. The end solution in each tube was made acidic with dilute HNO_3 and then some $AgNO_3$ solution was added. Substance B gave a yellow precipitate. Which one of the following statements is true for this experiment?

(a) A was C_6H_5I
(b) A was $C_6H_5CH_2I$
(c) B was C_6H_5I
(d) Addition of HNO_3 was unnecessary

12 Identify the set of reagents/reaction conditions x and y in the following set of transformations.

$$CH_3-CH_2-CH_2Br \xrightarrow{X} \text{Product} \xrightarrow{Y} CH_3-\underset{\displaystyle Br}{\underset{|}{CH}}-CH_3$$

(a) X = dilute aqueous NaOH, 20° C; Y = HBr/acetic acid, 20°C
(b) X = conc. alcoholic. NaOH, 80°C; Y = HBr/ acetic acid, 20°C
(c) X = dilute aqueous NaOH, 20°C; Y = $Br_2/CHCl_3$, 0°C
(d) X = conc. alcoholic NaOH, 80°C; Y = $Br_2/CHCl_3$, 0°C

13 The reaction condition leading to the best yield of C_2H_5Cl are:

(a) C_2H_6 (excess) + $Cl_2 \xrightarrow{\text{UV light}}$
(b) $C_2H_6 + Cl_2$ (excess) $\xrightarrow[\text{room temp}]{\text{Dark}}$
(c) $C_2H_6 + Cl_2$ (excess) $\xrightarrow{\text{UV light}}$
(d) $C_2H_6 + Cl_2 \xrightarrow{\text{UV light}}$

14 (X) on treatment with sodium hydroxide followed by the addition of silver nitrate give white precipitate at room temperature which is soluble in NH_4OH. (X) can be

(a) chlorobenzene
(b) ethyl bromide
(c) benzyl chloride
(d) vinyl chloride

ANSWERS

SESSION 1	**1** (d)	**2** (d)	**3** (b)	**4** (c)	**5** (d)	**6** (c)	**7** (a)	**8** (a)	**9** (c)	**10** (d)
	11 (a)	**12** (d)	**13** (a)	**14** (b)	**15** (b)	**16** (a)	**17** (a)	**18** (b)	**19** (a)	**20** (c)
	21 (a)	**22** (b)	**23** (d)	**24** (a)	**25** (b)	**26** (b)	**27** (d)	**28** (c)	**29** (a)	**30** (c)
	31 (d)	**32** (a)	**33** (c)	**34** (b)	**35** (b)	**36** (d)	**37** (c)	**38** (d)	**39** (a)	**40** (b)
	41 (d)	**42** (a)	**43** (a)	**44** (a)	**45** (d)	**46** (a)				
SESSION 2	**1** (a)	**2** (c)	**3** (a)	**4** (c)	**5** (a)	**6** (b)	**7** (a)	**8** (c)	**9** (a)	**10** (a)
	11 (a)	**12** (b)	**13** (a)	**14** (c)						

Hints and Explanations

SESSION 1

1 By direct halogenation of alkanes a mixture of monohalogen, dihalogen, trihalogen and tetrahalogen substituted product is obtained by this method. These are difficult to separate.

$$CH_4 \xrightarrow[h\nu]{Cl_2} CH_3Cl \xrightarrow[h\nu]{Cl_2} CH_2Cl_2$$

$$\xrightarrow{Cl_2} CHCl_3 \xrightarrow[h\nu]{Cl_2} CCl_4$$

2 Alkyl fluorides can be prepared by action of metallic fluorides such as AgF, Hg_2F_2 or SbF_3 on corresponding alkyl halide.This reaction is known as Swarts reaction.

$$CH_3Br + AgF \longrightarrow \underset{\text{Methyl fluoride}}{CH_3F} + AgBr$$

When action of NaI/acetone takes place on alkyl chloride or bromide, alkyl iodide forms. This reaction is called Finkelstein reaction.

$$C_2H_5Cl \xrightarrow[\text{acetone}]{NaI} C_2H_5I + NaCl$$

Free radical fluorination is highly explosive reaction, so not preferred for the preparation of fluoride.
Sandmeyer's reaction is used for the preparation of aryl halides.

3 Molar mass $= 72 = C_nH_{2n+2} = 12n + (2n+2)$

$$n = 5$$

Thus, hydrocarbon is C_5H_{12}.
Since, it gives only single $C_5H_{11}Cl$, thus C_5H_{12} is symmetrical. It is *neo* -pentane.

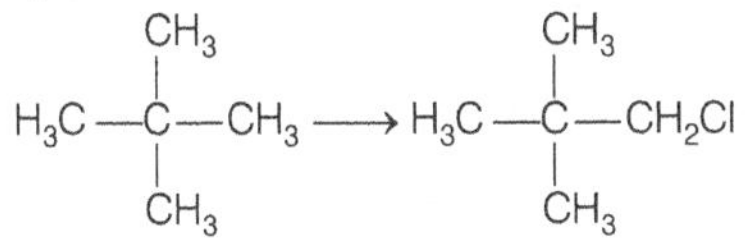

4

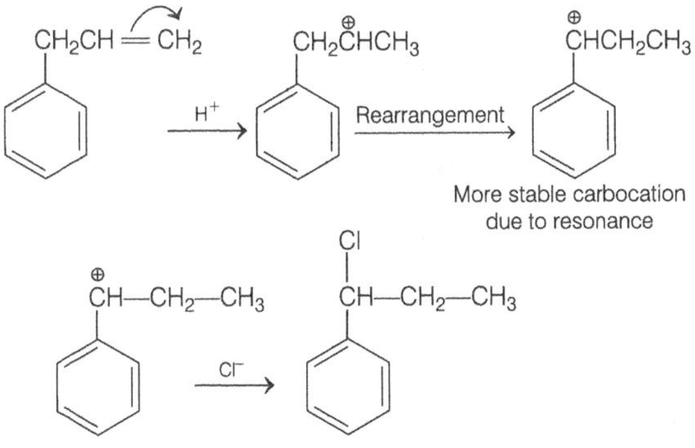

5 *p*-dichlorobenzene has highest melting point because it has symmetrical structure and due to which it can fit well in its crystal lattice. The intermolecular forces of attraction is also strong.

6 Boiling point decreases with branching.
Thus, the correct order is (iii) < (ii) < (i).

7 Densities of alkyl halides decrease in the following order iodide >bromide >chloride >fluoride.
Thus, the correct option is 'a'

8 μ_D for C—Cl bond = 1.56 D
μ_D for C—I bond = 1.29 D
C—Cl bond is more ionic than C—I bond. Among halides, C—X bond length increases as the size of halogen increases. Therefore, C—Cl bond is stronger than C—I bond.

9 CH_3Cl is used as methylating agent. The methylating agent is one which is used for the introduction of methyl groups.

10 Addition of HI in the presence of peroxide catalyst does not follow anti-Markownikoff's rule because iodine radicals have more tendency to form I_2 molecule as compared to form free radical of carbon. Thus, iodine atom is not reactive enough to add across a double bond.

$$I^{\bullet} + I^{\bullet} \longrightarrow I_2$$

11 Higher the stability of carbocation, faster is the reaction because S_N1 reactions involve the formation of carbocation intermediate.

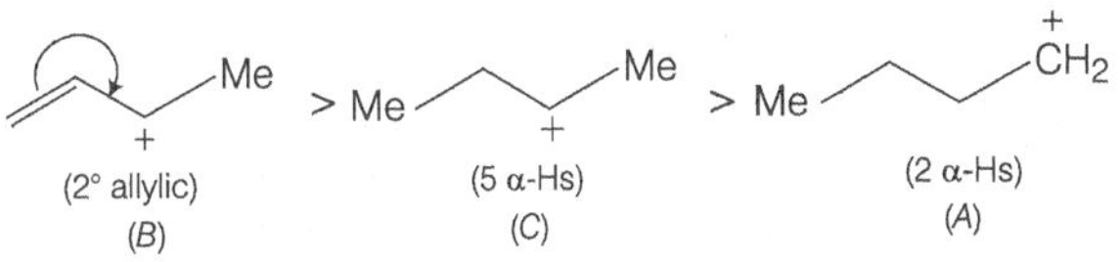

12 II is most reactive as it produces an aromatic carbocation while IV is less reactive as it produces a non-resonance stabilised carbocation.III is least reactive than I as former involve an anti-aromatic carbocation.

13 S_N2 reactions are accomplished by complete stereochemical inversion. Since, 1° alkyl halides are most prone to S_N2 reactions, therefore CH_3Cl undergoes complete stereochemical inversion.

14 (i) The rate of S_N1 reaction depends only upon the concentration of the alkyl halide.

(ii) S_N1 reaction proceeds through the formation of carbocation.
The reactivity is decided by ease of dissociation of alkyl halide.

$$R — X \rightleftharpoons R^{\oplus} + X^{\ominus}$$

Higher the stability of R^+ (carbocation), higher would be the reactivity towards S_N1 reaction.
$p\text{-}H_3CO—C_6H_4—\overset{\oplus}{C}H_2$ is the most stable carbocation due to resonance and then $CH_3\overset{\oplus}{C}HCH_2CH_3$ (2° carbocation) while $CH_3CH_2\overset{\oplus}{C}H_2$ (1°) is least stable.

Thus, the correct increasing order of the reactivity of the given halides towards the S_N1 reaction is

$$\underset{(II)}{CH_3CH_2CH_2Cl} < \underset{(I)}{CH_3\underset{\underset{Cl}{|}}{C}HCH_2CH_3} < \underset{(III)}{p\text{-}H_3CO-C_6H_4-CH_2Cl}$$

15 Since, rate of S_N2 reaction $\propto \dfrac{1}{\text{Steric crowding of C}}$

Thus, the correct order is

$$\underset{\text{(Less crowded)}}{CH_3Cl} > \underset{1^\circ}{CH_3CH_2Cl} > \underset{2^\circ}{CH_3-\underset{\underset{CH_3}{|}}{CH}-Cl} > \underset{\text{(More crowded)}}{CH_3-\underset{\underset{CH_3}{|}}{\overset{\overset{CH_3}{|}}{C}H}-Cl}$$

16 S_N1 reactions are accompanied by racemisation. The attack of nucleophile can occur from both the faces with almost equal ease, thereby giving a 50 : 50 mixture of two enantiomers, i.e. racemic mixture.

17 Nucleophilicity order is

$CH_3O^- > HO^- >$ 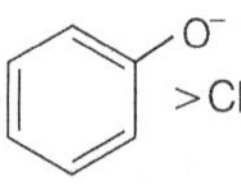 $> CH_3COO^-$

18 The order of departing nature of halides is $I^- > Br^- > Cl^-$. This is because of large size, low electronegativity and nucleophilicity of I^-.

19 A $\rightarrow$ 3, B $\rightarrow$ 1, C $\rightarrow$ 2, D $\rightarrow$ 4

20 Reaction of *tert*-alkyl halides with aqueous alkali gives S_N1 reaction and rate of S_N1 reaction is not based on the concentration of nucleophile (alkali). Hence, reaction rate remains constant.

21 Vinyl chloride is least reactive towards hydrolysis because of the non-reactivity of chlorine atom due to resonance stabilisation. The lone pair of chlorine can participate in delocalisation (resonance) to give two canonical structures.

$$CH_2=CH-\ddot{\underset{..}{Cl}}: \longleftrightarrow \bar{C}H_2-CH=\overset{+}{\underset{..}{Cl}}:$$

22 The obtained compound reduces Tollen's reagent, therefore, it must be an aldehyde. Thus, it is obvious that both the' —Cl atoms are present at C. Hence, the compound 'X' is 1,1-dichloropropane and the reactions are as follows:

$$\underset{\text{1,1-dichloropropane}}{CH_3CH_2CHCl_2} \xrightarrow{\text{Hydrolysis}} \underset{\text{Unstable}}{CH_3CH_2CH(OH)_2} \xrightarrow{-H_2O}$$

$$\underset{\text{Propanaldehyde}}{CH_3CH_2CHO} \xrightarrow[\text{reagent}]{\text{Tollen's}} \underset{\text{Propanoic acid}}{CH_3CH_2COOH} + Ag\downarrow$$

23 Strong nucleophile (OMe) in polar solvent (MeOH) gives elimination products over substitution products but all products are possible in different yields.

$$H_3C-\underset{\underset{CH_3}{|}}{\overset{\overset{Cl}{|}}{C}}-CH_2CH_2CH_3 \xrightarrow[\text{MeOH}]{Me^-ONa^+} \underset{\text{(Less yield)}}{CH_3-\underset{\underset{CH_3}{|}}{\overset{\overset{OCH_3}{|}}{C}}-CH_2CH_2CH_3}$$

$$\underbrace{+\,CH_3-\underset{\underset{CH_3}{|}}{C}=CH-CH_2-CH_3 + CH_2=\underset{\underset{CH_3}{|}}{C}-CH_2-CH_2CH_3}_{\text{(More yield)}}$$

24 $CH_3-\underset{\underset{C_2H_5}{|}}{\overset{*}{C}}-Br$ contains a chiral carbon and gives a racemic product if it undergoes S_N1 reaction.

25

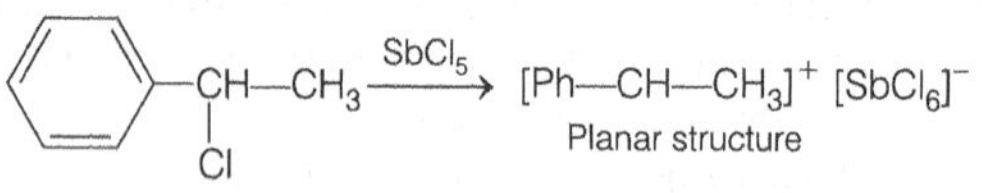

26 Complete reaction can be represented as

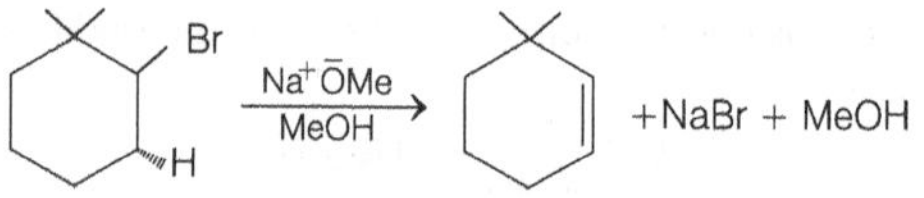

Thus, the given reaction is dehydrohalogenation which is a β-elimination proceeding through E_2 mechanism.

Mechanism The reaction proceeds through the formation of following transition state with simultaneous removal of Br and H-atoms.

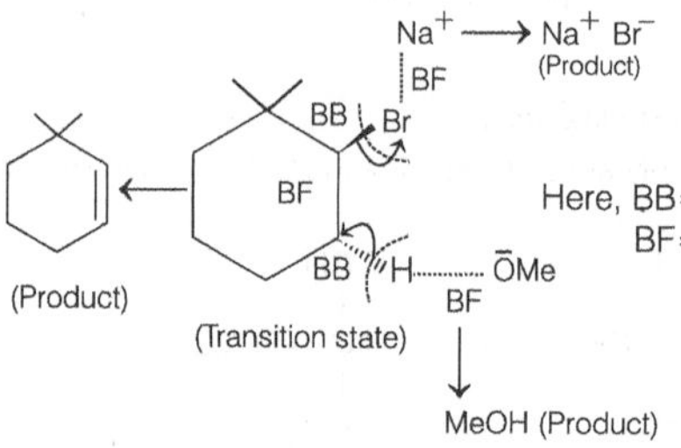

Here, BB= Bond breakage
BF= Bond formation

27

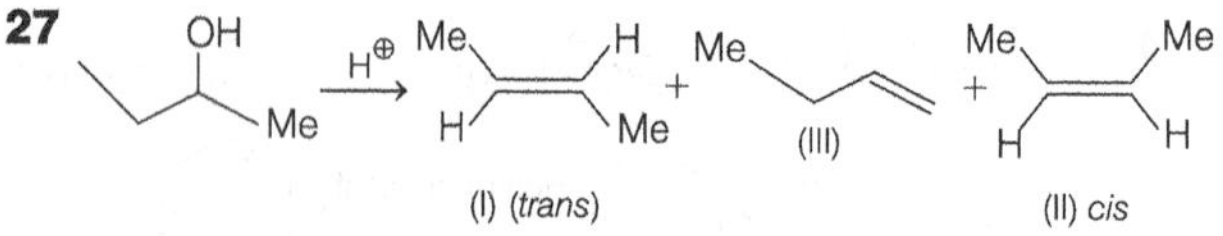

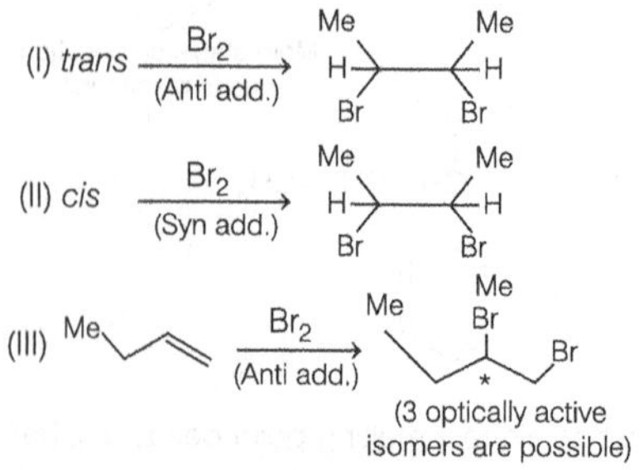

28 The reaction is

$$CH_3-CCl_3 \xrightarrow[\Delta]{6Ag} \underset{\text{But-2-yne}}{CH_3-C\equiv C-CH_3} + 6AgCl$$

29. Aryl halides are less reactive towards nucleophile than alkyl halides due to resonance.

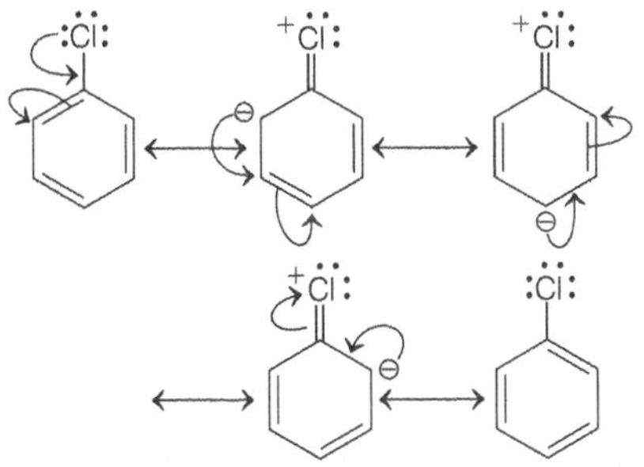

30

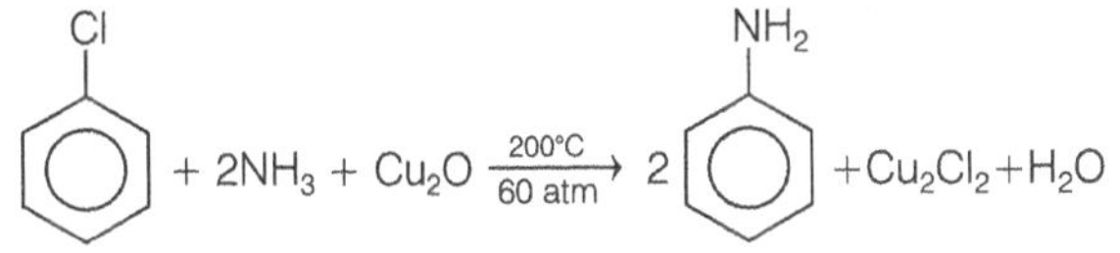

31.

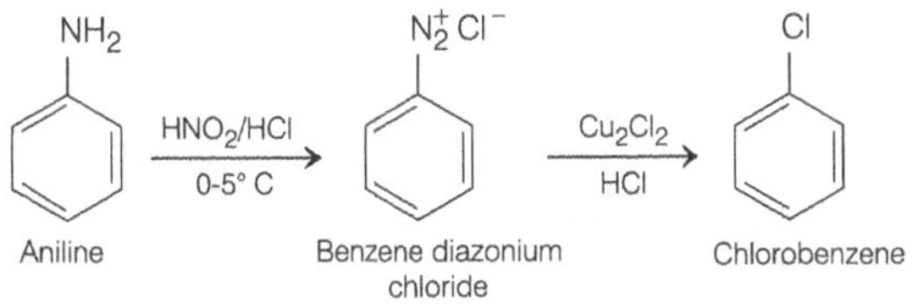

32 Chlorine in chlorobenzene is *ortho*/*para* directing because resonance effect predominates over inductive effect.

33

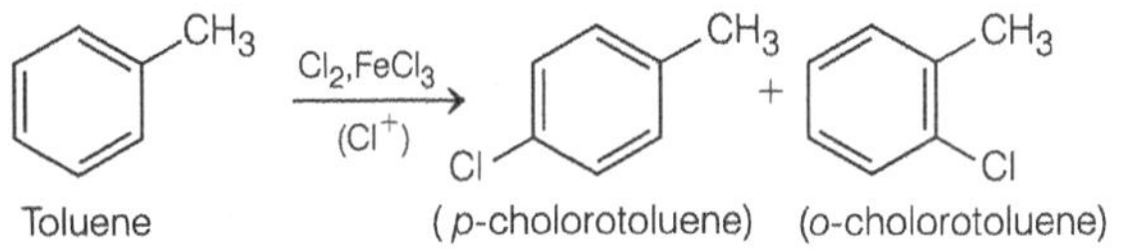

34

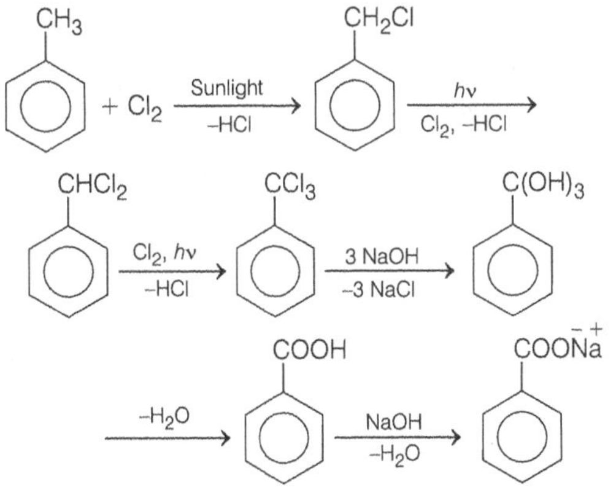

35 When one molecule of alkyl halide undergoes condensation with one molecule of aryl halide in the presence of Na in dry ether, the reaction is called Wurtz–Fittig reaction. e.g.

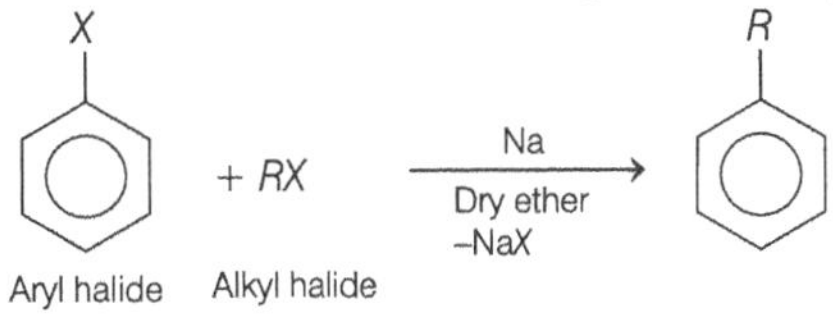

36 Compound *A* gives a precipitate with alcoholic $AgNO_3$ (here white is misprinting because the colour of precipitate is light yellow), so it must contain Br in side chain. On oxidation, it gives $C_8H_6O_4$, which shows the presence of two alkyl chains attached directly with the benzene nucleus. Since, compound *B* gives anhydride on heating, the two alkyl substituent must occupy adjacent (1, 2) positions. Thus, *A* must be

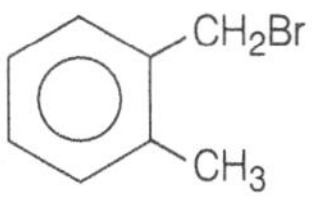

and the reactions are as follows:

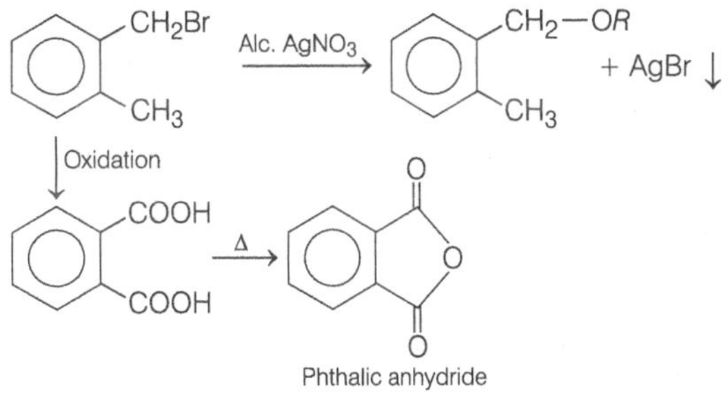

37 CCl_4 at high temperature reacts with water to give phosgene.

$$CCl_4 + H_2O \xrightarrow{500^\circ C} \underset{\text{Phosgene}}{COCl_2} + 2HCl$$

Phosgene is highly poisonous, so after the use of pyrene, CCl_4 to extinguish fire, the room should be well ventilated.

38 Cl_2 in presence of $FeCl_3$ ⟶ ring substitution

Cl_2 in presence of light ⟶ side chain substitution

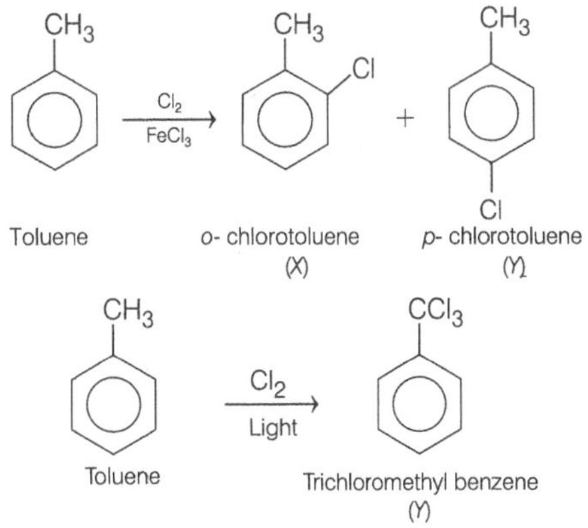

39 Chloretone is a hypnotic drug.

$$CHCl_3 + CH_3COCH_3 \xrightarrow{KOH} CH_3-\underset{CCl_3}{\overset{OH}{C}}-CH_3$$

Chloretone

40 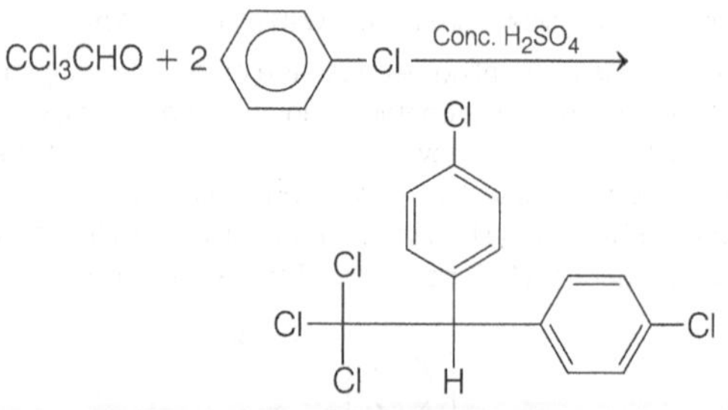

DDT

41 Correct Assertion Thionyl chloride is preferred over phosphorus chlorides (tri and penta) for the preparation of alkyl chlorides from alcohols.
Correct Reason Thionyl chloride gives pure alkyl chlorides.

42 Tertiary butyl bromide undergoes Wurtz reaction to give 2, 2, 3, 3- tetramethylbutane. In Wurtz reaction, product contain double the number of C-atoms.

43 Alkyl halides are not soluble in water as these are not able to form H-bonds with water molecules.

44 Tertiary alkyl halides are more prone to elimination as positive inductive effect of *R*-group weakens C—*X* bond.

45 Aryl halides do not undergo nucleophilic substitution reaction under ordinary conditions. Thus, Assertion I is incorrect. In aryl halides, the carbon-halogen bond has partial double bond character, so it becomes shorter and stronger and cannot be easily replaced by nucleophile.

46 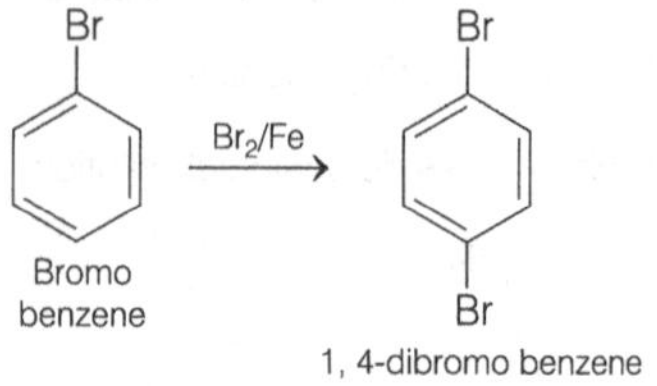

Bromine orientation is controlled by its mesomeric effect, the stabilisation of arenium ion by $+M$ activity. In bromobenzene, the inductive effect of the bromo group is more dominant than the mesomeric effect in the activity of ring.

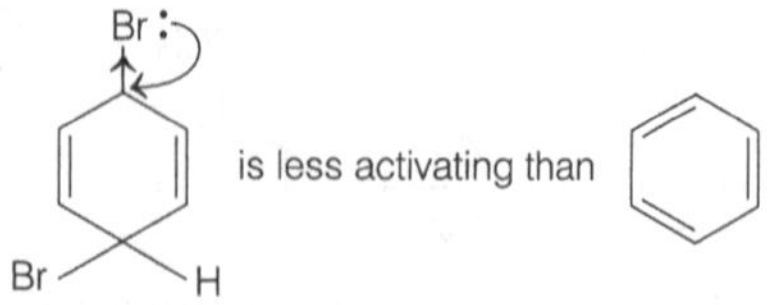

SESSION 2

1 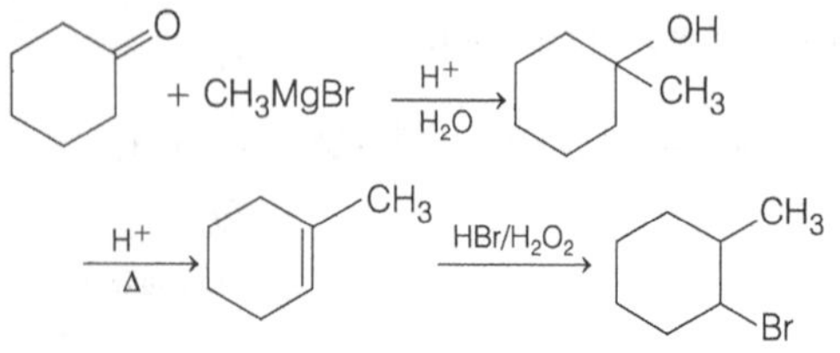

2 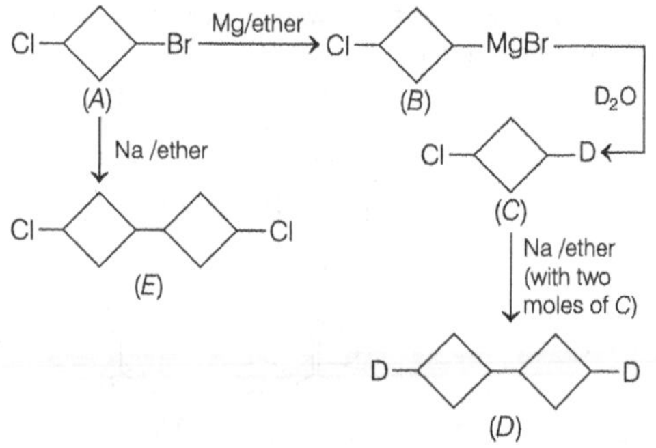

3 Ammonia, being a strong base, favours elimination reaction, rather substitution, when the alkyl halide is 2°or 3°.

4 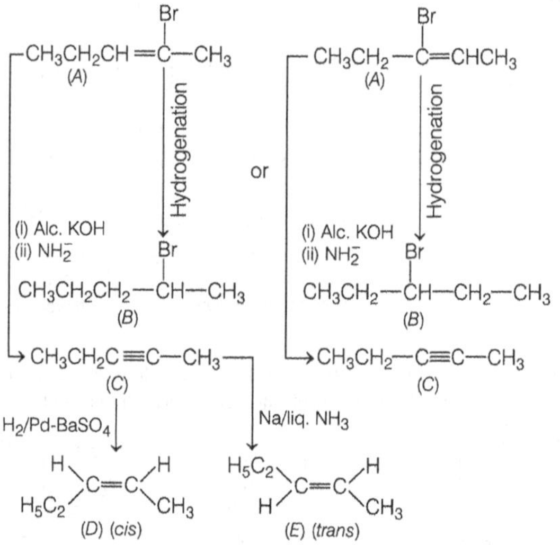

5 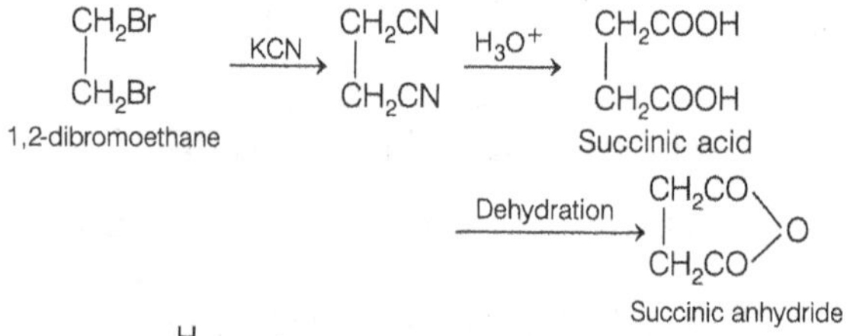

6 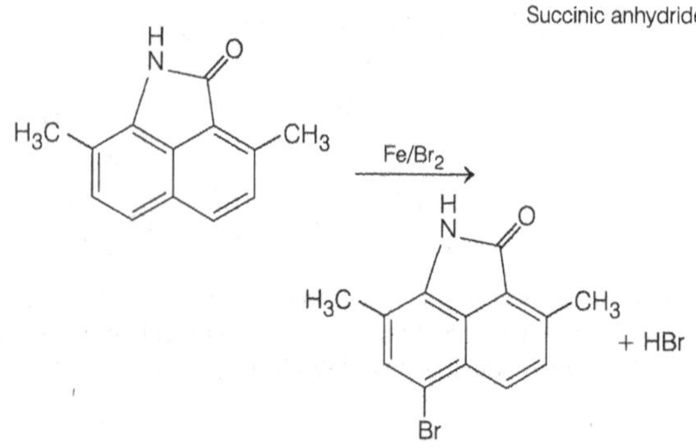

Major product

It is an electrophilic substitution reaction so, electrophile must be attacked on *o*/*p*-positions due to higher electron density on these position.

In this ring, the attached —NH-group will have high electron density due to resonance and *ortho* position is blocked, so electrophile is attacked on *para* position.

7 With AgCN, due to greater covalency of Ag—C bond, N is the donor atom. In KCN, CN^- exists as free ion and better carbon donor forms the bond.

8 Let the hydrocarbon be C_nH_{2n+2},

i.e. $12n + 2n + 2 = 72$

or $n = 5$

The hydrocarbon is C_5H_{12}. As it forms a single mono chloride, all the hydrogen atoms are equivalent. The compound is $(CH_3)_4C$, i.e. *neo* -pentane.

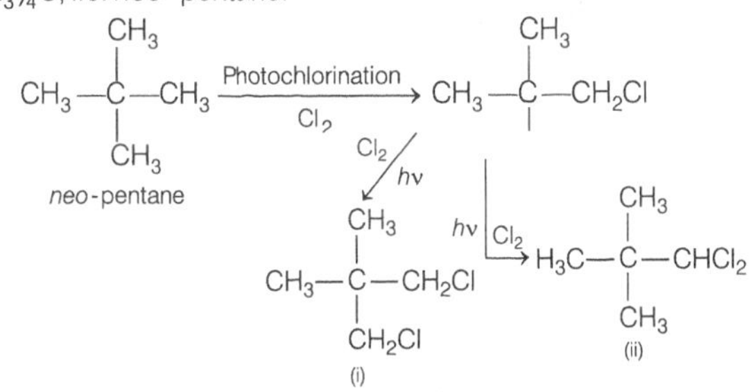

9 (*A*) adds one mole of Br_2 and thus act as an alkene.
(*A*) on ozonolysis forms CH_3COCH_3 and CH_3CHO, thus (*A*) is $(CH_3)_2C{=}CHCH_3$ and the reactions are as follows :

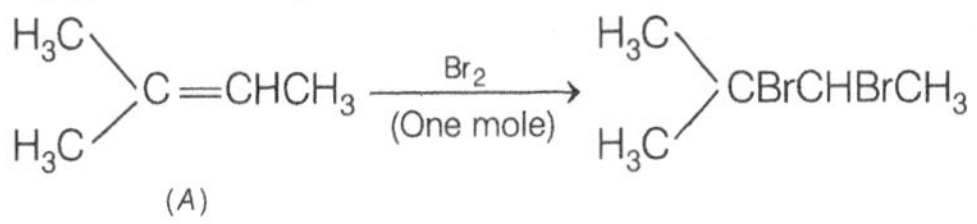

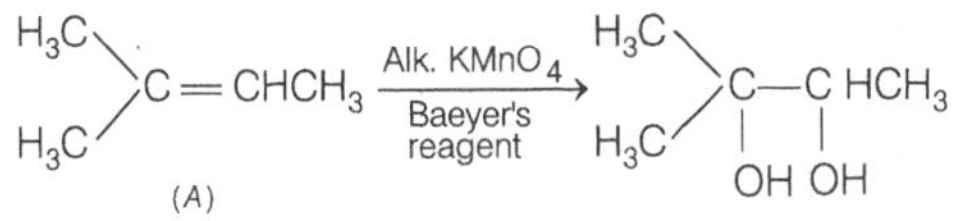

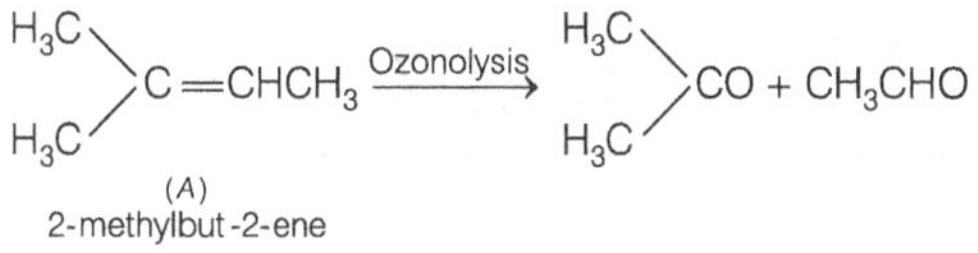

10

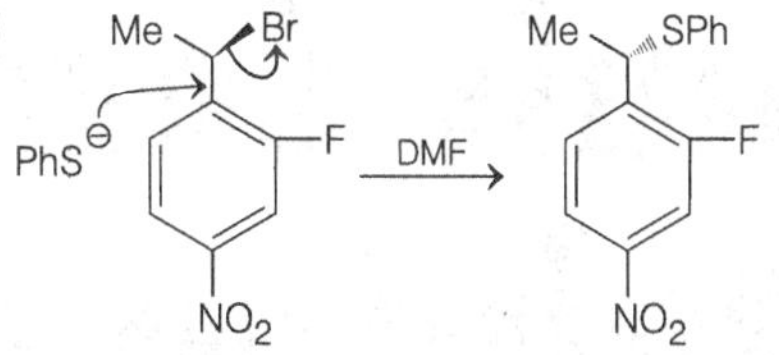

PhS^- is a strong nucleophile and dimethyl formamide (DMF) is a highly polar aprotic solvent.
Condition indicates that bimolecular nucleophilic substitution (S_N2) takes place at 2° benzylic place. Stereochemically, it involves inversion of configuration.

11 In $C_6H_5CH_2$—I, C—I bond is reactive, so it gives yellow precipitate of AgI. On the other hand, C — I bond in C_6H_5I is stable, so it does not give yellow precipitate with $AgNO_3$.

12

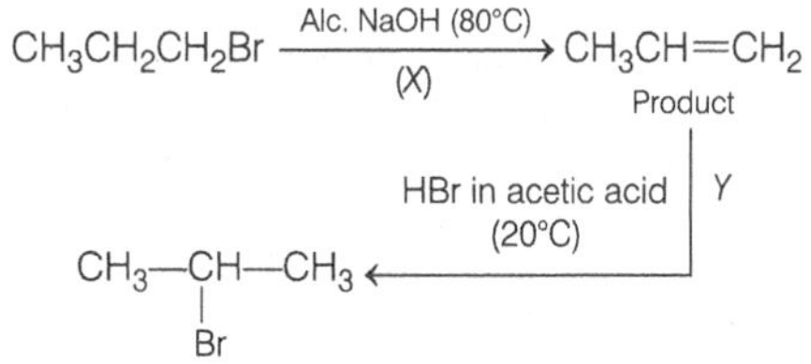

13 The best reaction condition leading to the best yield of C_2H_5Cl is (a)

$$C_2H_6 \text{ (excess)} + Cl_2 \xrightarrow{\text{UV-light}} C_2H_5Cl$$

14 $PHCH_2Cl$ forms stable $Ph\overset{\oplus}{C}H_2$ (benzyl C^+) ion. So, the reaction is feasible.

DAY THIRTY

Organic Compounds Containing Oxygen

Learning & Revision for the Day

- Alcohols
- Phenol
- Ethers
- Aldehydes and Ketones
- Carboxylic Acids

Alcohols and phenols are formed when a hydrogen atom from a hydrocarbon, aliphatic and aromatic respectively, is replaced by —OH group. The substitution of a hydrogen atom from a hydrocarbon by an alkoxy or aryloxy group (*R*—O / Ar—O) gives **ethers**. In **carboxylic acids** and their derivatives (esters, anhydrides), the carbonyl group is bonded to oxygen. In compounds such as amides and acyl halides, carbon is attached to —$CONH_2$ and —*OX* groups respectively.

Alcohols

An alcohol contains one or more hydroxyl group(s) (—OH) directly attached to carbon atom(s) of an aliphatic system (like CH_3OH).

Preparation

These can be prepared by the following methods:

1. From Alkenes

- Alkenes react with water in the presence of acid as catalyst to form alcohols by **acid catalysed hydration.**

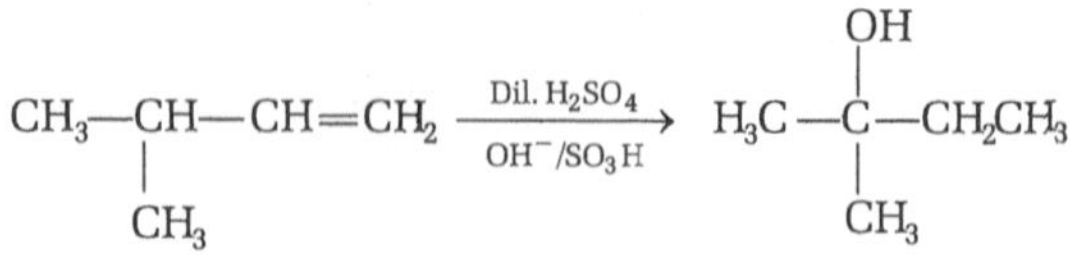

  In this reaction, intermediate carbocation is formed and rearrange, therefore —OH gets attached at maximum degree of carbon.

- **In oxymercuration demercuration reaction**, mercury (II) acetate reacts with alkene in presence of water to give β-hydroxy alkyl mercury (II) acetate which is treated directly with $NaBH_4$ to alcohol.

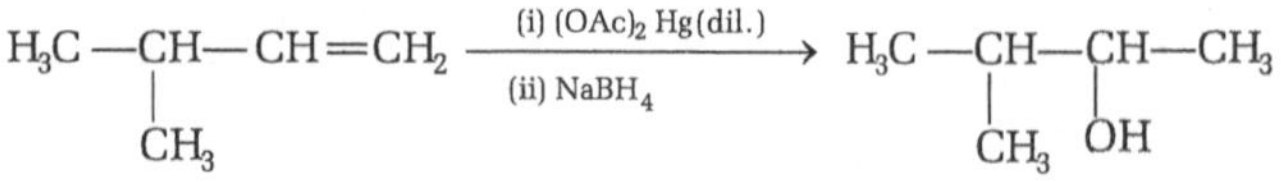

  Alcohol is formed according to Markownikoff's rule.

PREP MIRROR
Your Personal Preparation Indicator

- No. of Questions in Exercises (x)—
- No. of Questions Attempted (y)—
- No. of Correct Questions (z)—
  (Without referring Explanations)

- Accuracy Level (*z*/*y* ×100)—
- Prep Level (*z*/*x* ×100)—

In order to expect good rank in JEE, your Accuracy Level should be above 85 & Prep Level should be above 75.

- Diborane reacts with alkenes to give trialkyl boranes as addition product. This is oxidised to alcohol by hydrogen peroxide in the presence of aqueous sodium hydroxide. This reaction is called **hydroboration oxidation reaction.**

$$H_3C-\underset{\displaystyle CH_3}{\underset{|}{CH}}-CH{=}CH_2 \xrightarrow[\text{(ii) } H_2O_2, \bar{O}H]{\text{(i)} B_2H_6},$$

$$H_3C-\underset{\displaystyle CH_3}{\underset{|}{CH}}-CH_2CH_2OH + B(OH)_3$$

The alcohol so formed looks as if it has been formed by the addition of water to the alkene in a way opposite to the Markownikoff's rule.

2. By Reduction of Carbonyl Compounds

Aldehydes on reduction give primary alcohols and ketones give secondary alcohols in the presence of weak reducing agent ($NaBH_4$) by addition of hydrogen in the presence of catalysts such as platinum, palladium or nickel.

$$RCHO + H_2 \xrightarrow{Pd} RCH_2OH$$

$$RCOR' \xrightarrow{NaBH_4} R-\underset{\displaystyle OH}{\underset{|}{CH}}-R'$$

3. By Reduction of Acid and Esters

Carboxylic acids and esters on reduction, in the presence of strong reducing agent ($LiAlH_4$), give primary alcohols.

$$RCOOH \xrightarrow[\text{(ii) } H_2O]{\text{(i) } LiAlH_4} RCH_2-OH$$

$$CH_3COOCH_2-CH_2-CH_3 \xrightarrow[+4H]{LiAlH_4} CH_3CH_2OH + CH_3CH_2CH_2OH$$

4. Hydrolysis of Haloalkanes

With aqueous sodium or potassium hydroxide, or carbonate or moist silver oxide results in the formation of alcohols.

$$CH_3-CH_2Br + NaOH(aq) \longrightarrow CH_3-CH_2-OH + NaBr$$

5. Aliphatic Primary Amines

Aliphatic primary amines react with nitrous acid to give primary alcohol.

$$NaNO_2 + HCl \xrightarrow{0-5^\circ C} HNO_2 + NaCl$$

$$R-NH_2 + HNO_2 \xrightarrow{0-5^\circ C} R-OH + N_2\uparrow + H_2O$$

6. From Grignard Reagent

Grignard reagent ($R'MgX$) on reaction with aldehydes/ketones followed by hydrolysis gives alcohols.

If the reacting aldehyde is formaldehyde, primary alcohol ($-CH_2OH$) is obtained while other aldehydes give secondary alcohols ($-CHOH$) with Grignard reagent. Ketones give tertiary alcohols with Grignard reagent.

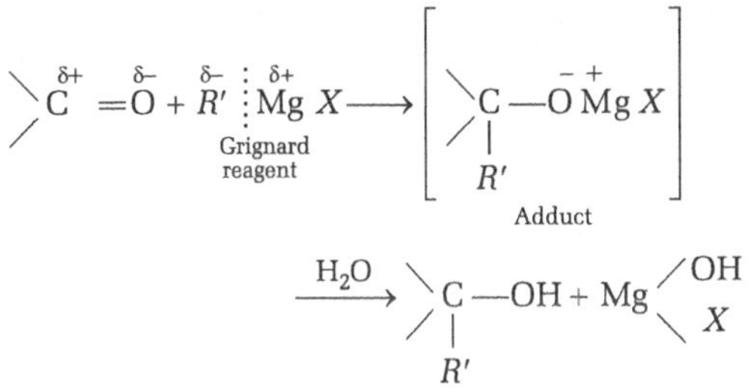

Physical Properties

- Alcohols have higher boiling point than haloalkanes of comparable molecular mass because alcohols form intermolecular hydrogen bonding.

  As the number of carbon atoms increases, boiling point increases. The boiling point decreases with increase of branching in carbon chain.
- Alcohols are soluble in water due to their ability to form hydrogen bonds with water molecules. As the number of carbon atoms increases, solubility decreases.

Chemical Properties

Alcohols react both as nucleophiles and electrophiles. As a result, the bond between O—H and C—O get broken. These are as follows:

1. Reactions Involving Cleavage of O—H Bond

- Alcohols react with active metals such as sodium, potassium and aluminium to yield corresponding alkoxide and hydrogen.

$$2R-OH + 2Na \longrightarrow 2R-\overset{-}{O}-\overset{+}{Na} + H_2$$

- Acidity of alcohols in decreasing order

$$1^\circ > 2^\circ > 3^\circ$$

  Electron releasing group decreases the polarity of —OH bond. This decreases the acidic strength.
- Alcohols when react with carboxylic acids, acid chlorides and acid anhydrides, form esters. This reaction is called **esterification.**

$$RO\,\vdots\,H + R'CO\,\vdots\,OH \overset{H^+}{\rightleftharpoons} R'COOR + H_2O$$

- Alcohols when react with Grignard reagent give alkanes.

$$R'\overset{-}{O}\,\vdots\,\overset{+}{H} + \overset{-}{R}\,\vdots\,\overset{+}{M}gBr \longrightarrow RH + Mg\begin{smallmatrix} OR' \\ Br \end{smallmatrix}$$

2. Reactions Involving Cleavage of Carbon Oxygen (C—OH) Bond

- Alcohols react with hydrogen halides to form alkyl halides.

$$ROH + HX \longrightarrow R-X + H_2O$$

  Reactivity of alcohol in decreasing order $3^\circ > 2^\circ > 1^\circ$.

- **Dehydration of alcohol place** in the presence of protic acids like conc. H_2SO_4 or H_3PO_4 or in the presence of catalysts such as anhy. $ZnCl_2$ or Al_2O_3.

$$\underset{1^\circ\text{alcohol}}{C_2H_5OH} \xrightarrow[443\,K]{H_2SO_4} \underset{\text{Ethene}}{CH_2{=}CH_2} + H_2O$$

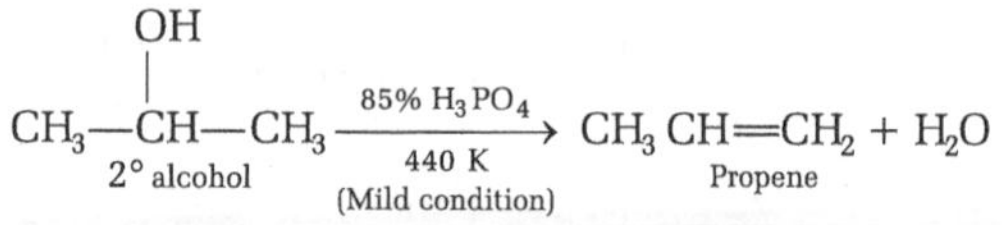

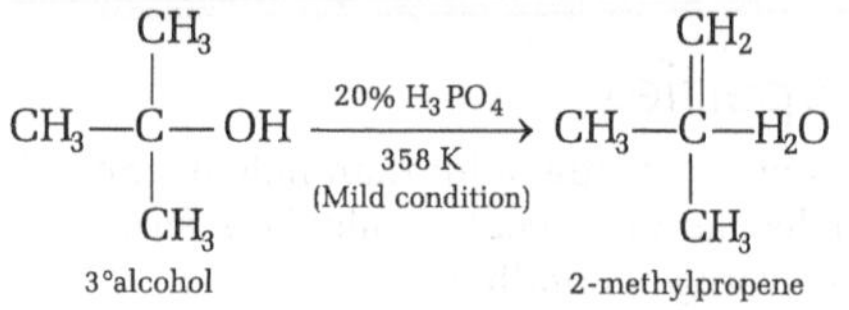

Relative ease of dehydration of alcohols 3° > 2° > 1°.

- Mechanism of **dehydration of ethanol** involves following steps.
  (i) Formation of protonated alcohol.
  (ii) Formation of carbocation
  (iii) Formation of ethene by elimination of proton.

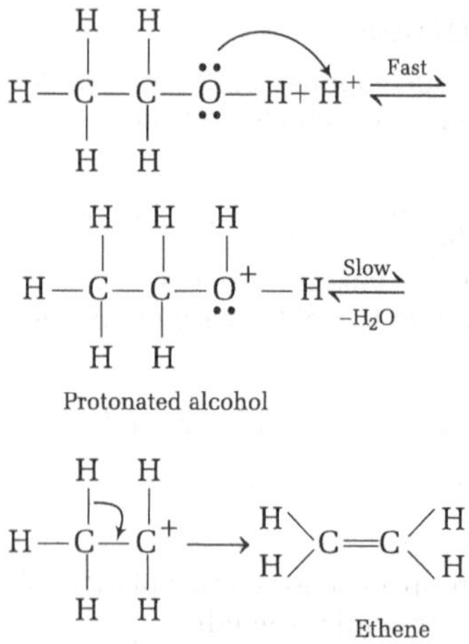

3. Oxidation of Alcohols

It involves the formation of a C—O double bond with cleavage of an O—H and C—H bonds.

- $RCH_2OH \xrightarrow[\text{Acidified }KMnO_4]{[O]} R{-}CHO \xrightarrow{[O]} RCOOH$
- $RCH_2OH \xrightarrow{CrO_3} RCHO$
- $CH_3{-}CH{=}CH{-}CH_2OH \xrightarrow{PCC} CH_3{-}CH{=}CH{-}CHO$

(PCC = pyridinium chloro chromate is a better reagent to convert 1° alcohol to aldehyde)

- 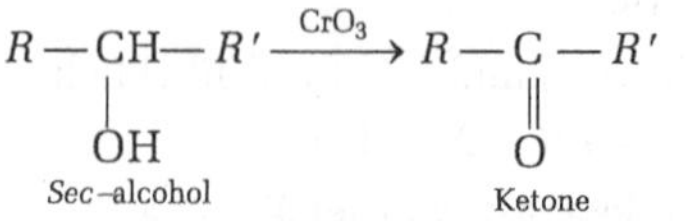

Tertiary alcohols do not undergo oxidation reaction. In presence of strong oxidising agent ($KMnO_4$) and at high temperature, cleavage of C—C bond takes place and a mixture of carboxylic acids containing lesser number of carbon atoms is formed.

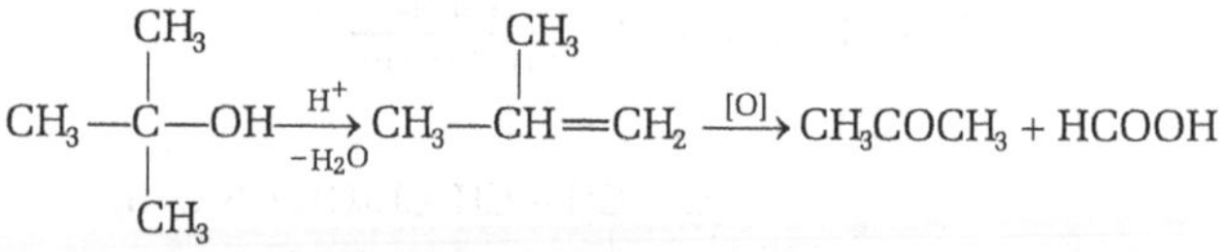

$$HCOOH \xrightarrow{[O]} H_2O + CO_2$$

4. Action of Heated Copper

$$\underset{1^\circ\text{alcohol}}{RCH_2OH} \xrightarrow[573\,K]{Cu} \underset{\text{Aldehyde}}{R{-}CHO} + H_2$$

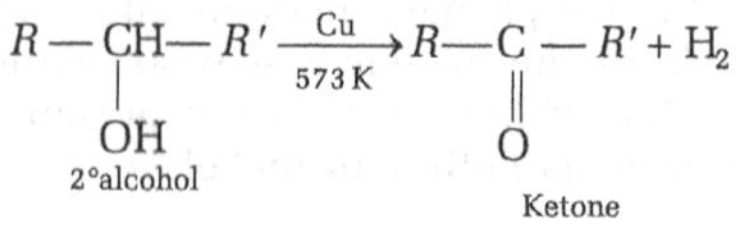

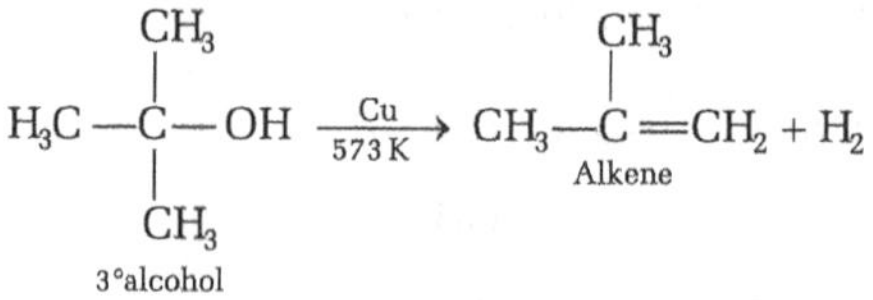

Identification of Primary, Secondary and Tertiary Alcohols

1. With Lucas reagent (conc. HCl and $ZnCl_2$)

- Tertiary alcohols give turbidity immediately.
- Secondary alcohols give turbidity with in five minutes.
- Primary alcohols do not produce turbidity at room temperature.

2. In Victor Meyer's test

- Blood red colour indicates 1° alcohols.
- Blue colour indicates 2° alcohols.
- Colourless solution indicates 3° alcohols.

Uses of Alcohols

- Methanol and ethanol are two commercially important alcohols. Methanol is used as a solvent in paints, varnishes and mainly for making formaldehyde.
- It is highly poisonous in nature. Ingestion of even small quantities of methanol can cause blindness and large quantities cause even death.
- Ethanol is used as a solvent in paint industry and in preparation of a number of carbon compounds.
- Commercial alcohol is made unfit for drinking by mixing $CuSO_4$ and pyridine **(denaturation of alcohol).**
- Ethanol is mainly used in alcoholic beverages.

Phenol (C_6H_5OH)

A phenol contains —OH group(s) directly attached to carbon atom(s) of an aromatic system (like C_6H_5OH).

Preparation

Some general and important **methods of preparation of phenol** are discussed below :

1. From Haloarenes (Dow's Process)

Chlorobenzene is fused with NaOH at 623K and 320 atmospheric pressure. Phenol is obtained by acidification of sodium phenoxide so produced.

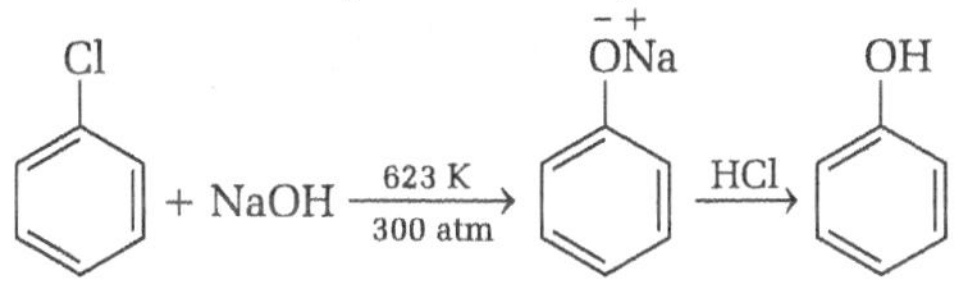

2. From Benzene Sulphonic Acid

Benzene is sulphonated with oleum and benzene sulphonic acid so formed is converted to sodium phenoxide on heating with molten NaOH. Acidification of sodium salt gives phenol.

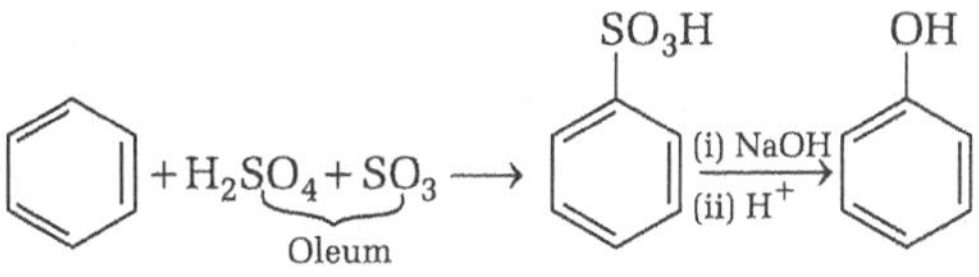

3. From Diazonium Salts

Diazonium salts are hydrolysed to phenols by warming with water or by treating with dilute acids.

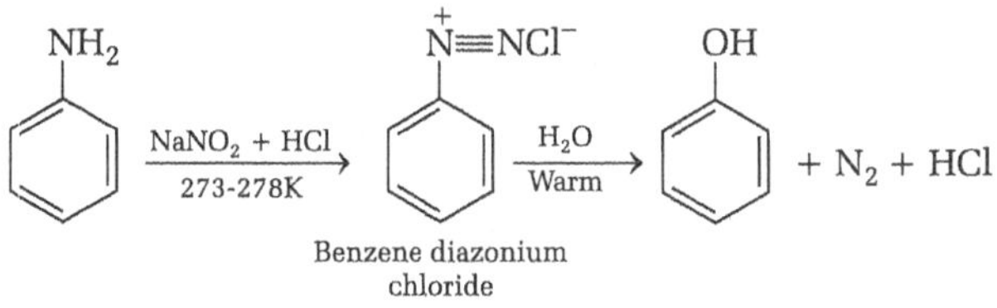

4. From Cumene

Phenol is manufactured from the hydrocarbon, cumene. Cumene is oxidised in presence of air to cumene hydroperoxide. It is converted to phenol and acetone by treating with dilute acid.

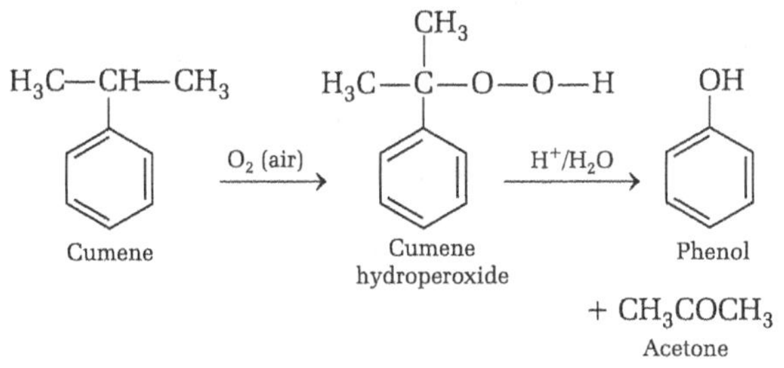

Physical Properties of Phenol

- Phenol has higher boiling point and soluble in water due to its ability to form hydrogen bonding with water.
- **Phenol is acidic in nature**, even more acidic than alcohols due to conjugation between lone pair of oxygen and benzene nucleus, i.e.

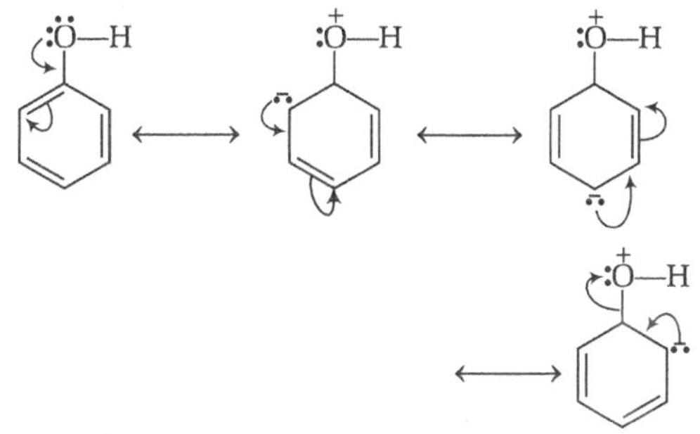

Presence of electron releasing groups like $—CH_3$, $—C_2H_5$ over benzene nucleus destabilises the phenoxide ion, thus, decreases the acidity of phenol whereas presence of electron withdrawing groups like $—NO_2$, —CN, etc., stabilises the phenoxide ion and thus, increases the acidity of phenol.

Chemical Properties

Various chemical properties of phenols are as follows:

1. Electrophilic Substitution Reactions

- In phenol, the —OH group shows $+R$ and $-I$-effect and hence, highly activates the benzene ring towards electrophilic substitution reaction. It is *ortho* and *para* directing group.

Nitration

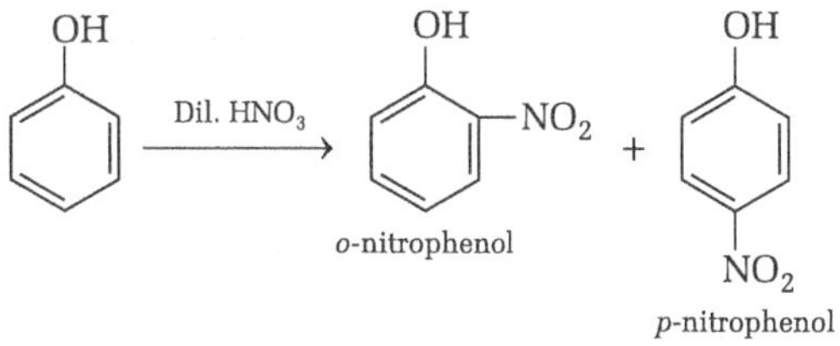

The *ortho* and *para* isomers can be separated by steam distillation.

- With conc. HNO_3, phenol is converted to 2, 4, 6-trinitrophenol

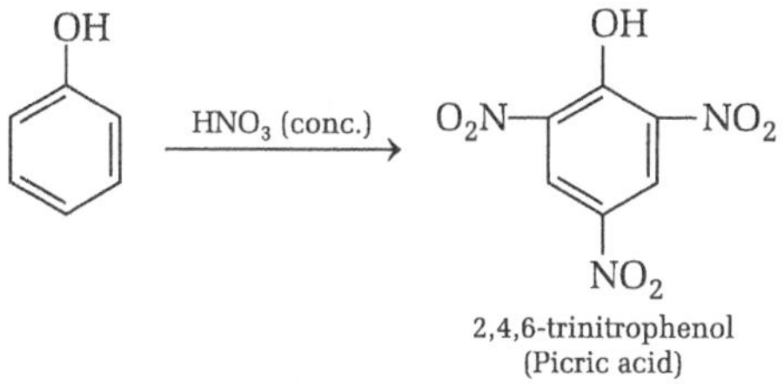

Halogenation

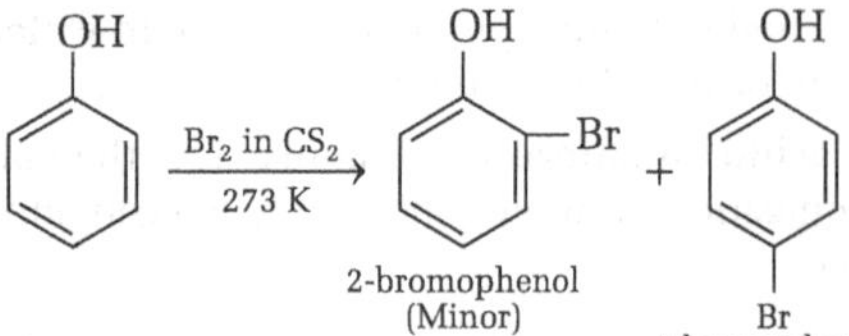

2-bromophenol (Minor)

4-bromophenol (Major)

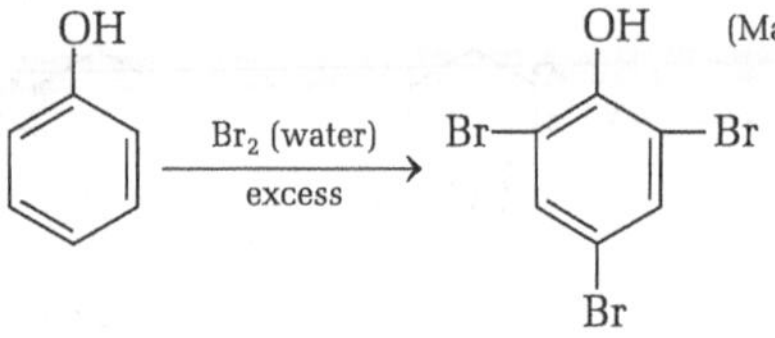

2,4,6-tribromophenol

Kolbe's Reaction

Phenoxide ion undergoes electrophilic substitution with carbon dioxide, a weak electrophile. Here, *ortho*-hydroxybenzoic acid is formed as main product.

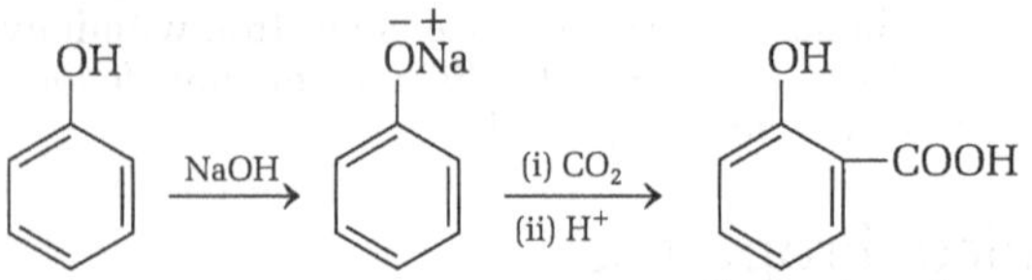

2-hydroxybenzoic acid (Salicylic acid)

2. Reimer-Tiemann Reaction

Phenol on reaction with chloroform in presence of NaOH, produces a — CHO group at *ortho*-position of benzere ring.

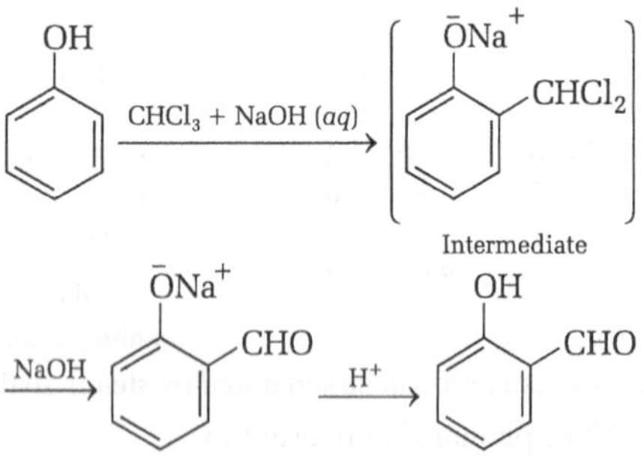

Intermediate

Salicylaldehyde

NOTE Electrophile is CCl_2 (dichlorocarbene)

3. Reaction of Phenol with Zinc Dust

Phenol is connected to benzene on heating with zinc dust.

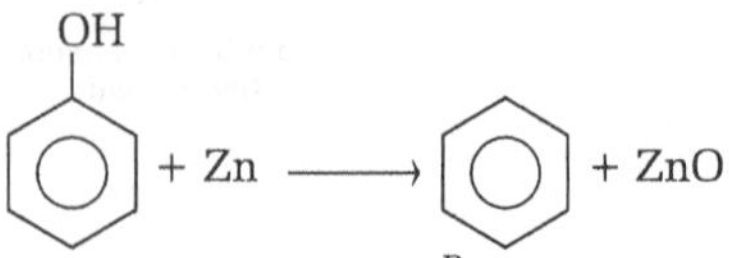

Benzene

4. Oxidation

Oxidation of phenol with chromic acid produces a conjugated diketone known as benzoquinone

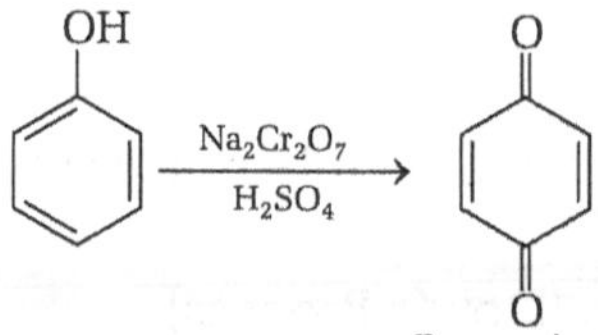

Benzoquinone

Uses of Phenols

- Phenol is used in the manufacture of bakelite, soaps, lotions etc.
- It is used in the manufacture of drugs like aspirin, salol, phenacetin etc.

Ethers

Organic compounds having —O— functional group are called ethers.

- In other words, ethers are the derivatives of water as these are obtained when both the H-atoms of H_2O are replaced by —*R* groups. The —*R* groups may be same or different. When both the —*R* groups (alkyl groups) are same, the ethers are called **simple** or **symmetrical ether** and when both the —*R* groups are different, the ethers are called **mixed** or **unsymmetrical ethers**, e.g.

$CH_3—O—CH_3$ (Simple or symmetrical ethers) and $CH_3OC_2H_5$ (Mixed ether)

- In ethers, two bond pairs and two lone pairs of electrons on oxygen are arranged approximately in tetrahedral manner. The bond angle is slightly greater than tetrahedral angle due to the repulsive forces between the two bulky alkyl groups.

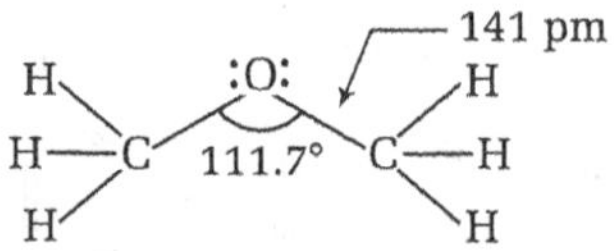

Structure of methoxymethane

Preparation

General methods used to synthesis ethers are as follows:

1. By the Dehydration of Alchols

Alcohols undergo **dehydration** in the presence of protic acids. The formation of the reaction product, alkene or ether depends on the reaction conditions.

$$CH_3CH_2OH \longrightarrow \begin{cases} \xrightarrow[443K]{H_2SO_4} CH_2{=}CH_2 \\ \xrightarrow[413K]{\text{Conc. } H_2SO_4} C_2H_5OC_2H_5 \end{cases}$$

2. Williamson's Synthesis

Only primary alkyl halides when react with sodium alkoxide give ether while tertiary alkyl halide give alkene due to steric hindrance.

$$CH_3CH_2Br + CH_3CH_2\overset{-}{O}\overset{+}{N}a \longrightarrow CH_3CH_2OCH_2CH_3 + NaBr$$

If a tertiary alkyl halide is used, and alkene is the only reaction product.

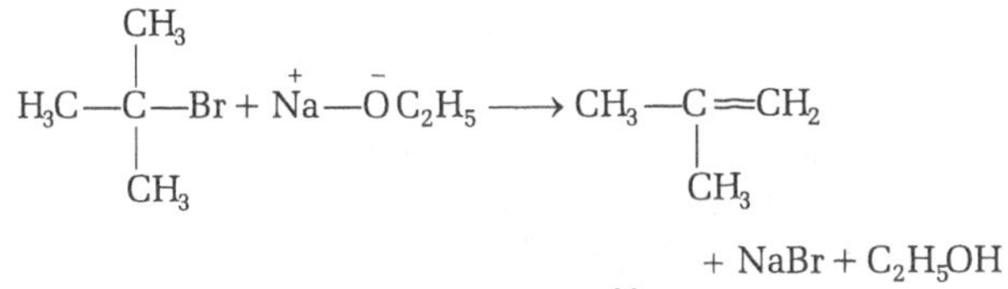

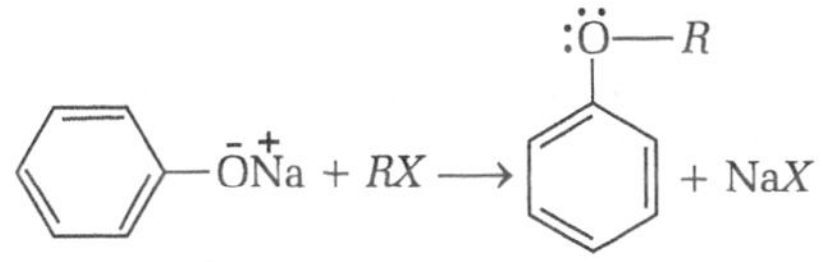

Alkoxy benzene

Physical Properties

- Ethers are polar but insoluble in H_2O and have low boiling point than alcohols (having comparable molecular mass) because ethers do not form hydrogen bond with water.
- Low ethers (upto three carbon atoms) are soluble in water and their miscibility with water resembles those of alcohols of the same molecular mass.

Chemical Properties

The **reactions of ethers** are mainly due to lone pair of ethereal O, cleavage of C—O bond and —*R* group:

1. **Reactions Involving Cleavage of C—O Bonds** Ethers are least reactive of the functional groups. The cleavage of C—O bond in ethers takes place under drastic conditions.

$$CH_3-CH_2-\ddot{\underset{\cdot\cdot}{O}}-CH_3 + HI \xrightarrow{S_N2\text{ mechanism}} CH_3-CH_2-OH + CH_3I$$

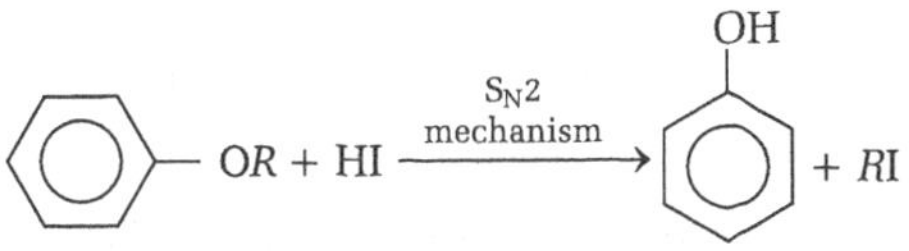

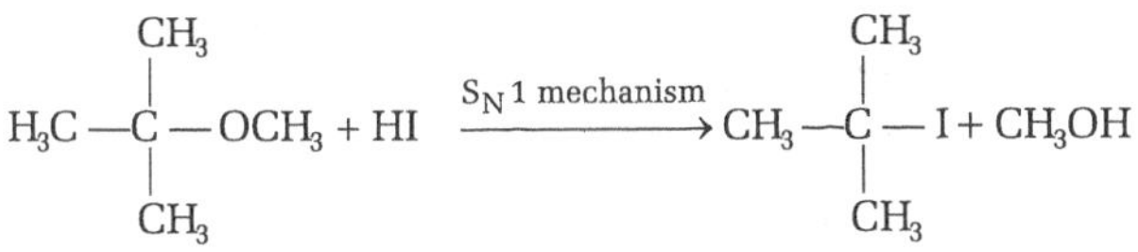

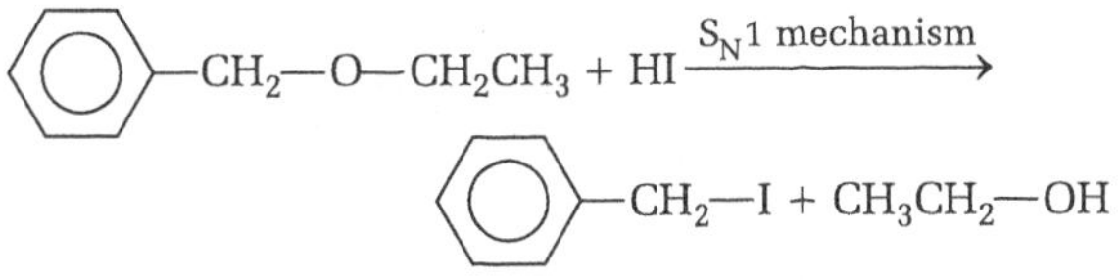

2. Some important **addition reactions** given by ethers are as follows:

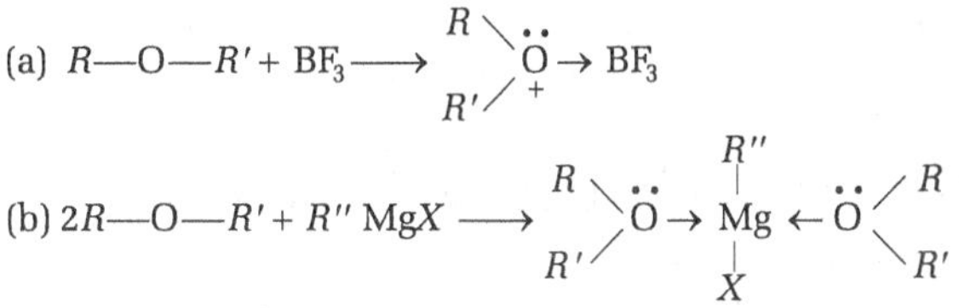

(c) $R—O—R' \xrightarrow[425\,K]{CO/BF_3} RCOOR'$ (Ester)

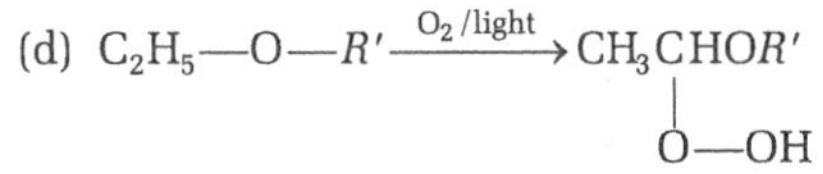

Peroxide

3. **Electrophilic Substitution Reactions** —*OR* is *ortho, para* directing group and activates the aromatic ring towards electrophilic substitution reaction.

(a) **Halogenation**

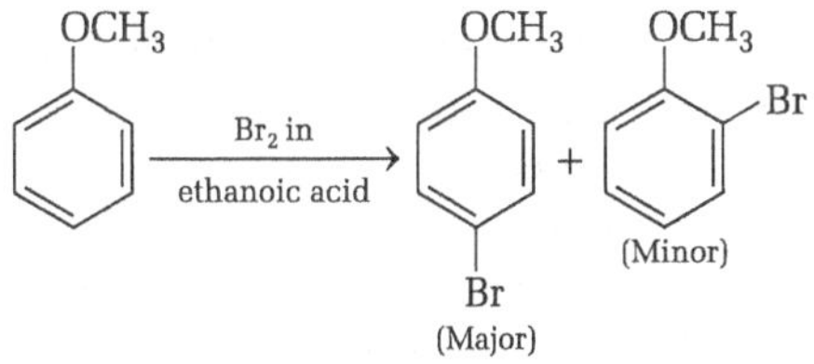

(b) **Friedel-Crafts' Alkylation**

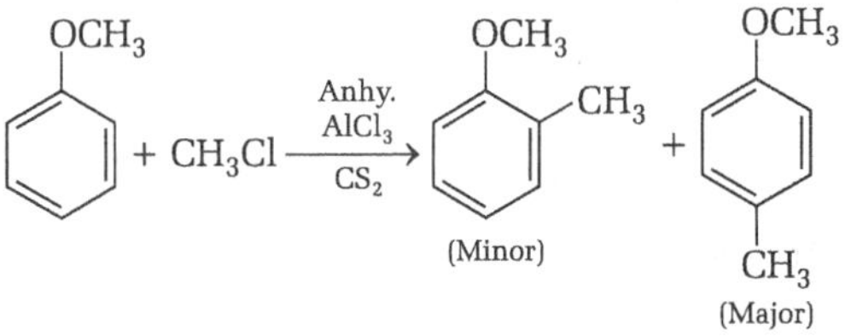

(c) **Friedel-Crafts' Acylation**

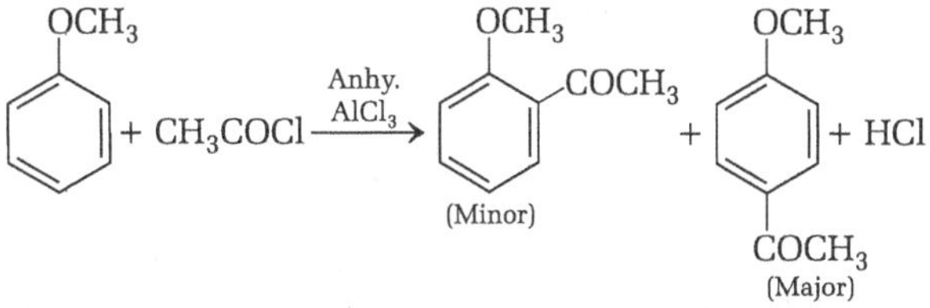

(d) **Nitration**

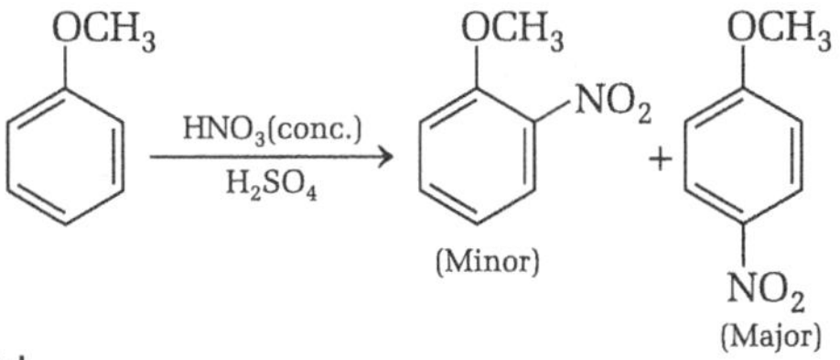

Uses of Ethers

(i) Ethers are used as a solvent for oils, fats and Grignard reagent, etc.

(ii) It is used as general anaesthetic. It provides inert and moisture free medium for various reactions.

Aldehydes and Ketones

The organic compounds containing carbon-oxygen double bond, i.e. $>C=O$ group are called **carbonyl compounds.**

Nature of Carbopnyl Group

In carbonyl group, π-electron cloud is displaced towards more electronegative oxygen atom thus, causing polarisation of the bond so that carbon acquires partially positive and oxygen acquires partially negative charge.

$$>C \overset{\leftarrow\!\!+}{=} O \quad \text{or} \quad >\overset{\delta+}{C}-\overset{\delta-}{O}$$

In resonance terms, electron delocalisation in the carbonyl group is represented by contributions from two principal resonance forms.

$$>C=\ddot{O}: \;\longleftrightarrow\; >\overset{+}{C}-\ddot{O}:^{-}$$

Preparation of Aldehydes and Ketones

There are several method from which aldehydes as well as ketones can be synthesised.

These methods are as follows:

1. **Ozonolysis of alkenes** followed by reaction with zinc dust and water gives aldehyde, ketones and a mixture of both.

$$CH_2 = CH_2 \xrightarrow[\text{(ii) } Zn + H_2O]{\text{(i) } O_3/CH_2Cl_2} 2HCHO$$

$$\underset{H}{\overset{H_3C}{>}}C=C\underset{H}{\overset{CH_3}{<}} + O_3 \xrightarrow[\text{(ii) Zn + } H_2O]{\text{(i) } CH_2Cl_2/O_3} \underset{H}{\overset{H_3C}{>}}C=O$$

2. Acetylene on hydration in presence of H_2SO_4 and $HgSO_4$ gives acetaldehyde while other **alkynes on hydration** gives ketones.

$$\underset{\text{Acetylene}}{CH\equiv CH} + H_2O \xrightarrow[H_2SO_4]{HgSO_4} CH_3CHO$$

$$R-C\equiv CH + H_2O \xrightarrow[H_2SO_4]{HgSO_4} R-\overset{O}{\overset{\|}{C}}-CH_3$$

3. HCN when reacts with **Grignard reagent** and then subjected to hydrolysis, gives aldehydes while RCN gives ketone.

- $R-MgX + H-C\equiv N \xrightarrow{H^+/H_2O} R-CHO + Mg\begin{smallmatrix}X\\OH\end{smallmatrix} + NH_3$

- $RMgX + R'-C\equiv N \xrightarrow{H^+/H_2O} R-\overset{O}{\overset{\|}{C}}-R' + Mg\begin{smallmatrix}X\\OH\end{smallmatrix} + NH_3$
  (R = alkyl group or C_6H_5)

Preparation of Aldehyde

The methods given below can be employed for the synthesis of only aldehydes.

1. **From Acyl Chloride** Acyl chloride is hydrogenated over catalyst, palladium on barium sulphate.

$$R-\overset{O}{\overset{\|}{C}}-X \xrightarrow[Pd\text{-}BaSO_4]{H_2} RCHO + HX$$

(R-alkyl group or C_6H_5)

This reaction is called **Rosenmund reduction.**

Formaldehyde cannot be obtained by this reaction because formyl chloride is unstable at room temperature.

2. **From Nitriles and Esters** Stephen reaction involves the reduction of nitriles to corresponding imine with stannous chloride in presence of HCl, which on hydrolysis gives corresponding aldehyde.

$$R-CN + SnCl_2 + HCl \longrightarrow \underset{\text{Iminochloride}}{RCH=NH\cdot HCl}$$

$$\xrightarrow{H_3O^+} RCHO + NH_4Cl + H_2O$$

Nitriles are selectively reduced by diisobutyl aluminium hydride (DIBAL-H) and gives aldehydes.

$$CH_3CH=CHCH_2CH_2CN \xrightarrow[\text{(ii) } H_2O]{\text{(i) } AlH(i\text{-}Bu)_2} CH_3CH=CHCH_2CH_2CHO$$

$$CH_3(CH_2)_9-\overset{O}{\overset{\|}{C}}-OC_2H_5 \xrightarrow[\text{(ii) } H_2O]{\text{(i) DIBAL-H}} CH_3(CH_2)_9-\overset{O}{\overset{\|}{C}}-H + C_2H_5OH$$

3. **By Side Chain Chlorination followed by Hydrolysis** Toluene gives benzalchloride which on hydrolysis gives benzaldehyde.

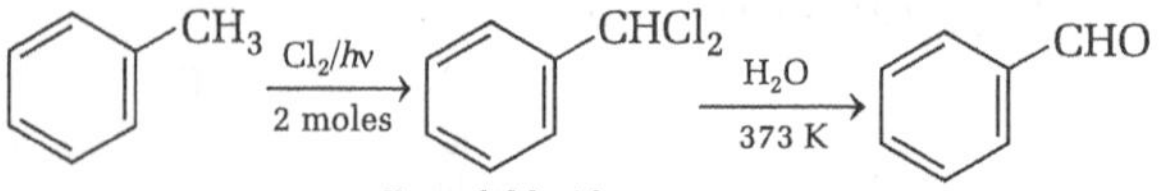

Benzalchloride Benzaldehyde

4. **Use of Chromyl Chloride (CrO_2Cl_2)** Chromyl chloride oxidises methyl group to a chromium complex, which on hydrolysis gives benzaldehyde. This reaction is called **Etard reaction.**

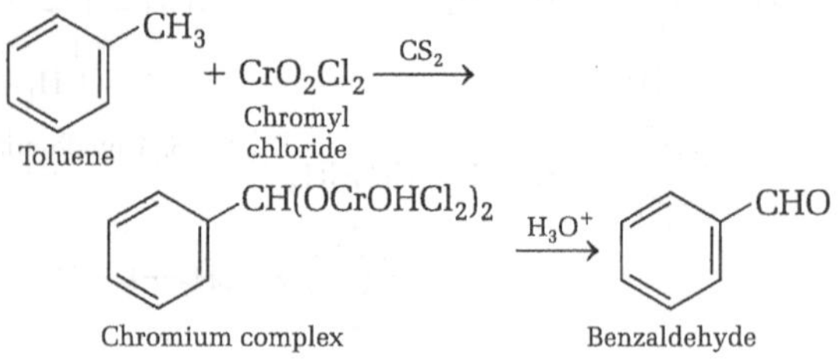

Chromium complex Benzaldehyde

5. **Use of Chromic Oxide (CrO_3)** Toluene or substituted toluene is converted to benzylidene diacetate on treating with chromic oxide in acetic anhydride. The benzylidene diacetate can be hydrolysed to corresponding benzaldehyde with aqueous acid.

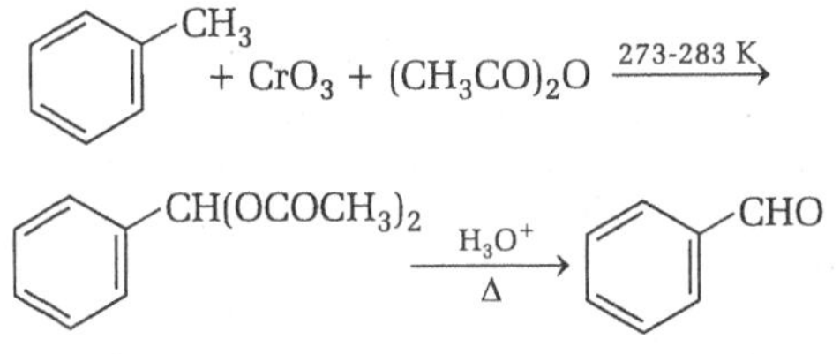

6. **By Gattermann-Koch Reaction** In Gattermann-Koch reaction benzene or its derivative is treated with carbon monoxide and hydrogen chloride in the presence of anhydrous aluminium chloride or cuprous chloride, it gives benzaldehyde or substituted benzaldehyde.

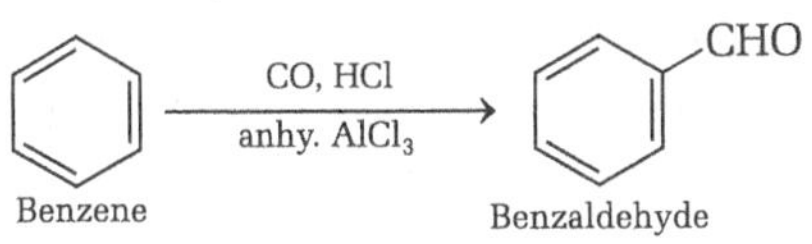

Preparation of Ketones

- The following methods are used to synthesise only ketones.

1. From Acyl Chlorides

Treatment of acyl chlorides with dialkyl cadmium prepared by the reaction of cadmium chloride with Grignard reagent, gives ketones.

$$2R\,MgX + CdCl_2 \longrightarrow R_2Cd + 2Mg(X)Cl$$

$$2R'—\overset{\overset{\displaystyle O}{\|}}{C}{}^{+}\,\vdots\,\overset{-}{Cl} + R_2Cd \longrightarrow 2R'—\overset{\overset{\displaystyle O}{\|}}{C}—R + CdCl_2$$

2. From Benzene or Substituted Benzene

In **Friedel-Crafts acylation**, acid is treated with acid chloride in the presence of anhyd. $AlCl_3$.

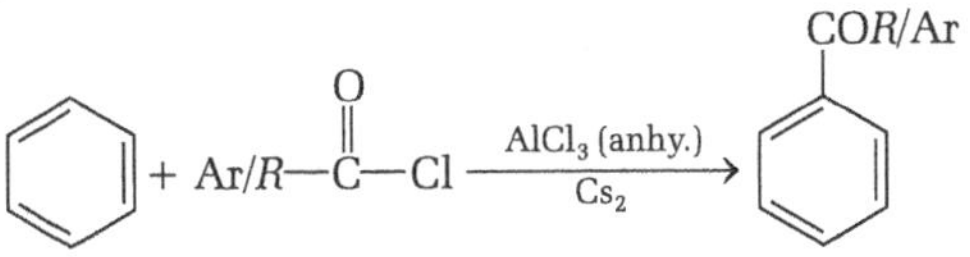

Physical Properties of Aldehydes and Ketones

Physical properties of aldehydes and ketones are as follows:

- Methanal is a gas at room temperature. Ethanal is a volatile liquid. Other aldehydes and ketones are liquid or solid at room temperature.
- The boiling points of aldehydes and ketones are higher than ethers of comparable molecular masses.
- The lower members of aldehydes and ketones such as methanal, ethanal and propanone are miscible with water in all proportions because they form hydrogen bond with water. However, the solubility of aldehydes and ketones decreases rapidly on increasing the length of alkyl chain.

Chemical Properties of Aldehydes and Ketones

The aldehydes and ketones because of the presence of polar carbonyl group exhibit the following **chemical reaction.**

1. Nucleophilic Addition Reactions

A nucleophile attacks at the electrophilic carbon atom of the polar carbonyl group. $\gt C=O$

sp^2, carbonyl group

- As the number of carbon atoms/number of alkyl groups increases, reactivity decreases due to steric hindrance. Hence, the order of reactivity is $HCHO > CH_3CHO > CH_3COCH_3 > C_2H_5COC_2H_5$.
- Various nucleophilic addition reactions are as follows:

(a) **Addition of Hydrogen Cyanide**

$$\gt C=O + HCN \longrightarrow \gt C\begin{matrix} OH \\ CN \end{matrix}$$

Cyanohydrin

(b) **Addition of Sodium Hydrogen Sulphite**

$$\gt C=O + NaHSO_3 \rightleftharpoons \gt C\begin{matrix} OSO_2Na \\ OH \end{matrix}$$

White crystalline solid

This reaction is used for the separation and purification of aldehydes and ketones. The reason for this is that the hydrogen sulphite addition compound formed is water soluble and can be converted back to the original carbonyl compound by treating with dilute mineral acid or alkali.

(c) **Addition of Lower Alcohols**

$$\underset{\underset{\displaystyle H}{|}}{R—C}=O \xrightleftharpoons{R'OH,\ HCl\ gas} \left[R—CH\begin{matrix} OR' \\ OH \end{matrix}\right]$$

Hemiacetal

$$\xrightleftharpoons[H^+]{R'OH} R—CH\begin{matrix} OR' \\ OR' \end{matrix} + H_2O$$

Acetal

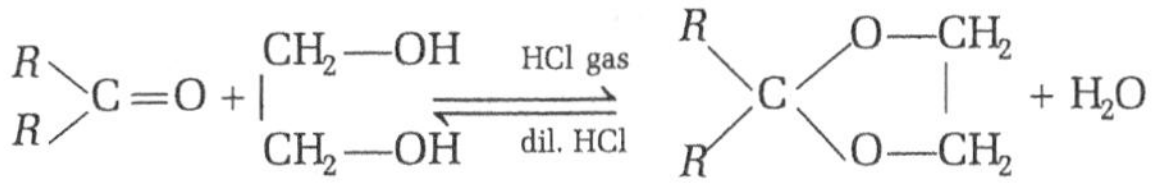

Ethylene glycol ketal

(d) **Addition of Grignard reagent** is used to distingiush aldehyde and ketone. The reactions are as follows:

$$\text{(i) } HCHO + RMgX \longrightarrow \underset{R}{\overset{H}{\gt}}C\begin{matrix} \overset{-\ 2+\ -}{OMgX} \\ H \end{matrix}$$

$$\xrightarrow{Hydrolysis} R—CH_2—OH + Mg(OH)X$$

(1°alcohol other than CH_3OH)

(ii)

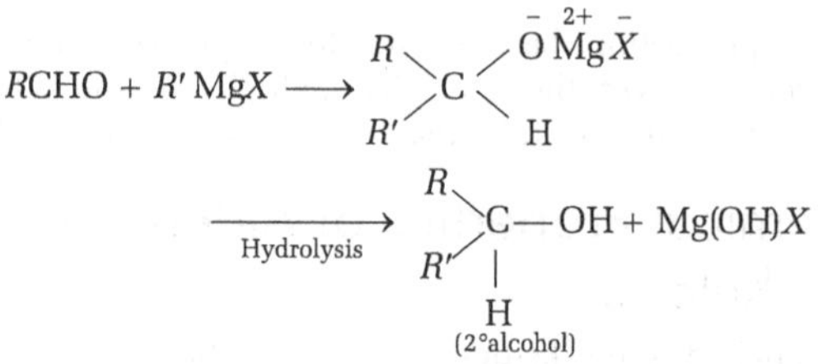

(iii)

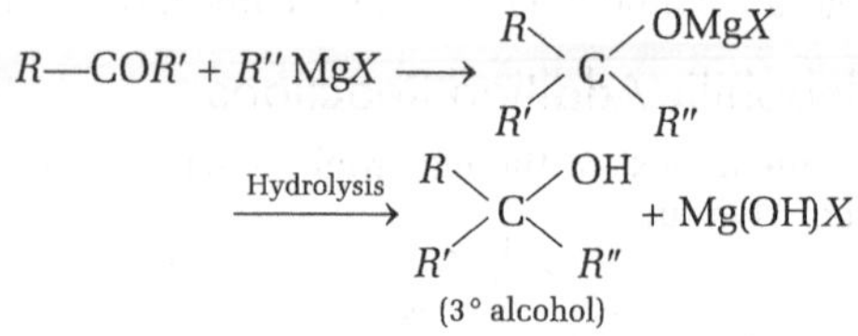

Nucleophilic Addition Elimination Reaction

These are two step reactions involving addition followed by elimination of some small molecules.

Addition Reactions with Derivatives of Ammonia

It is reversible reaction and catalysed by acid (pH ≈ 4). The equilibrium favours the product formation due to rapid dehydration of the intermediate to form $>C=N-Z$.

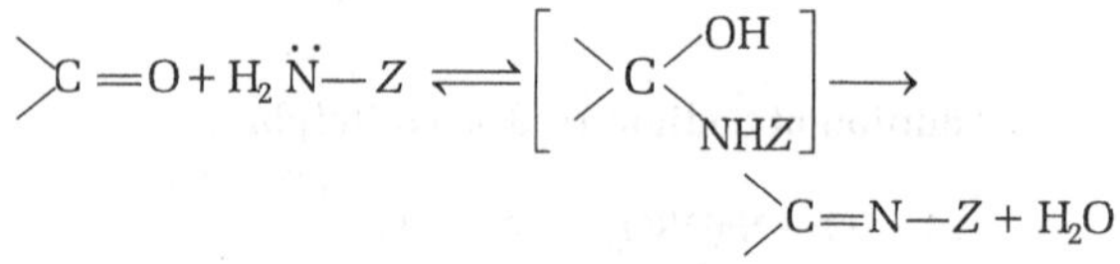

where, Z = alkyl, aryl, —OH, —NH_2, C_6H_5NH—, —$NHCONH_2$ etc.

Formaldehyde reacts with ammonia to form urotropine while acetaldehyde reacts with ammonia to form acetaldimine. While acetone react with ammonia to form diacetonamine.

Reduction

Aldehydes and ketones are reduced to alkanes.

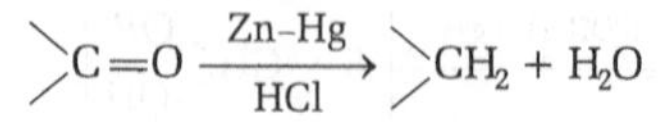

(Clemmensen reduction)

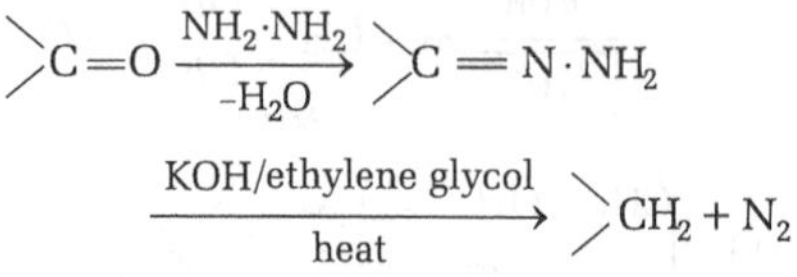

(Wolff-Kishner reduction)

Oxidation

Aldehydes are easily **oxidised** to carboxylic acids on treatment with strong oxidising agents (HNO_3, $K_2Cr_2O_7$, $KMnO_4$, etc.). Even mild oxidising agents, mainly Tollen's reagent and Fehling's reagent also oxidise aldehydes.

$$RCHO + 2[Ag(NH_3)_2]^+ + 3OH^- \longrightarrow$$

(R-alkyl or C_6H_5 group) Tollen's reagent

$$RCOO^- + 2Ag\downarrow + 2H_2O + 4NH_3$$

Silver mirror

$$R—CHO + 2Cu^{2+} + 5OH^- \longrightarrow RCOO^- + Cu_2O + 3H_2O$$

Fehling solution; Cu_2O: Red brown ppt

This reaction is not given by aromatic aldehydes.

- **Ketones cannot be oxidised by weak oxidising agents** such as Tollen's reagent and Fehling's solution. They are oxidised only in the presence of strong oxidising agents and at elevated temperatures. Their oxidation involves carbon-carbon bond cleavage to give a mixture of carboxylic acids having lesser number of carbon atoms than the parent ketone. e.g.

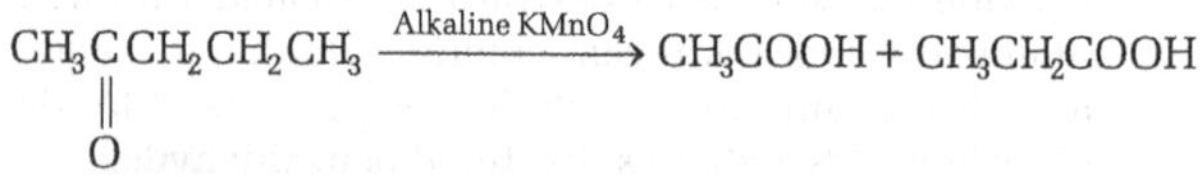

- **Methyl ketones** (—$COCH_3$) are also oxidised by **haloform reaction** in which they are treated with halogen in the presence of alkali or hypohalite salt.

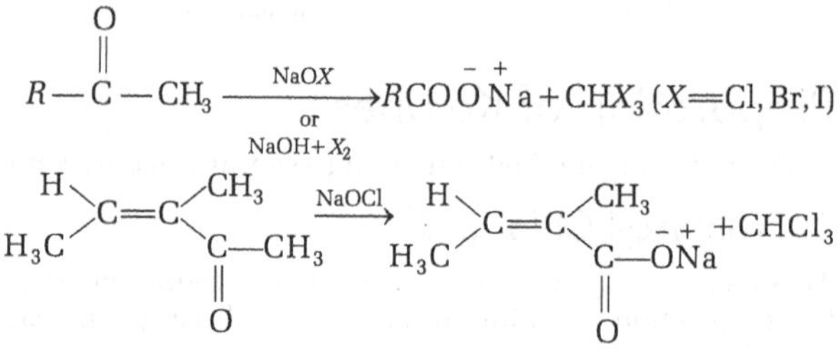

CH_3 group of $COCH_3$ is converted into haloform as it contains acidic hydrogen atoms. Acid salt is obtained corresponding to total number of carbon atoms apart from —CH_3 of $RCOCH_3$.

Reactions due to Acidic α-Hydrogen

- Aldehydes or ketones having atleast one α-hydrogen atom undergo a reaction in the presence of dilute alkali as catalyst to form β–hydroxy aldehydes (aldol) or β-hydroxy ketones (ketol) respectively. This is known as **aldol condensation.**

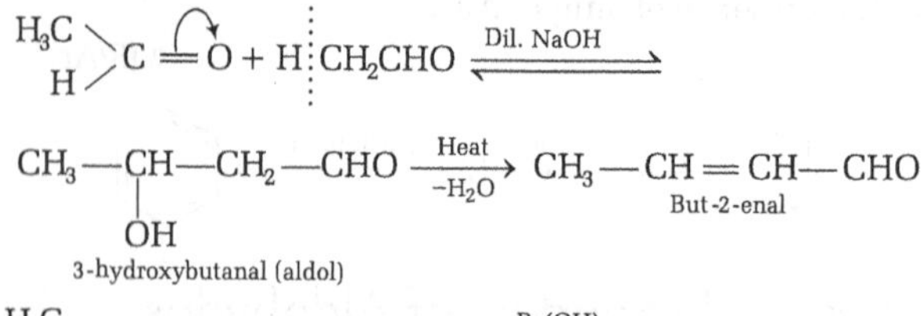

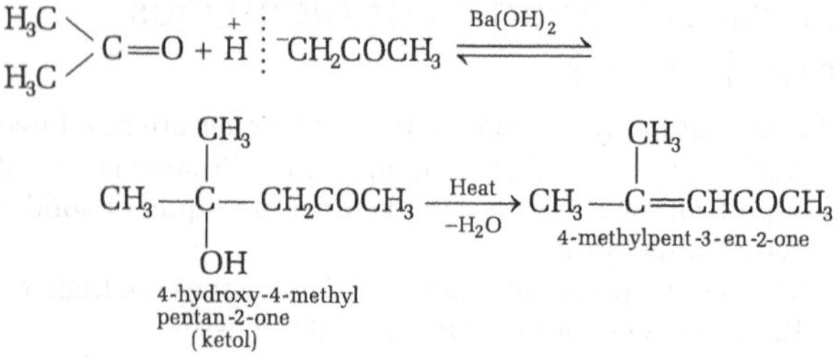

- When aldol condensation is carried out between two different aldehydes and /or ketones containing α-hydrogen, it is called **cross aldol condensation**, e.g.

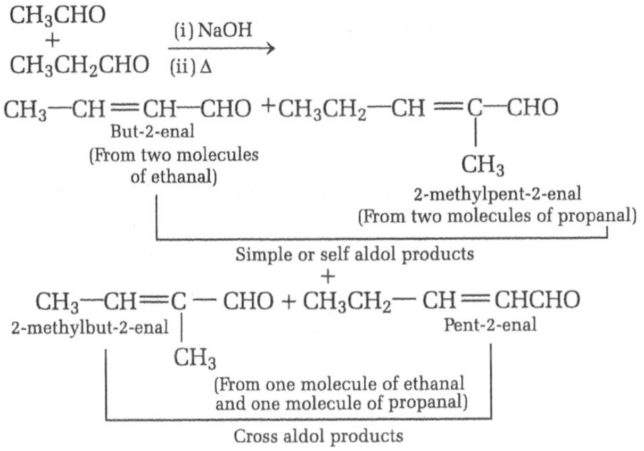

- This reaction is of great synthetic use even if one of the carbonyl group do not possess α-H atom.

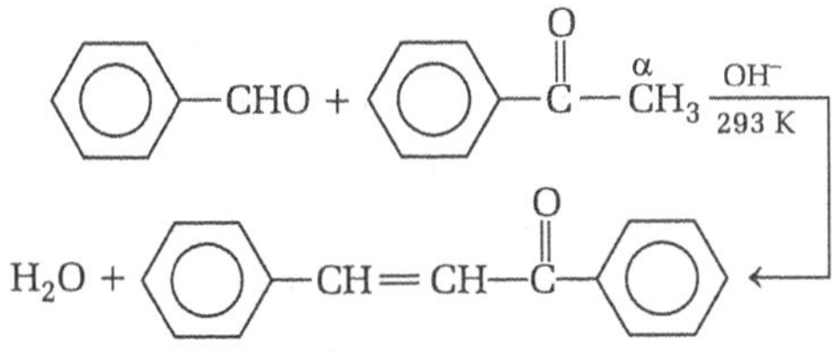

1,3-diphenylprop-2-en-1-one
(Benzalacetophenone)
major product

- Aldehydes which do not have α-hydrogen atom, undergo self oxidation and reduction (disproportionation) reaction on treatment with concentrated alkali. This reaction is known as **Cannizzaro reaction.**

$$2HCHO \xrightarrow[\Delta]{KOH\,(conc.)} \underset{\text{Methanol}}{CH_3OH} + \underset{\text{Potassium formate}}{H{-}\overset{O}{\overset{\|}{C}}{-}\overset{-}{O}\overset{+}{K}}$$

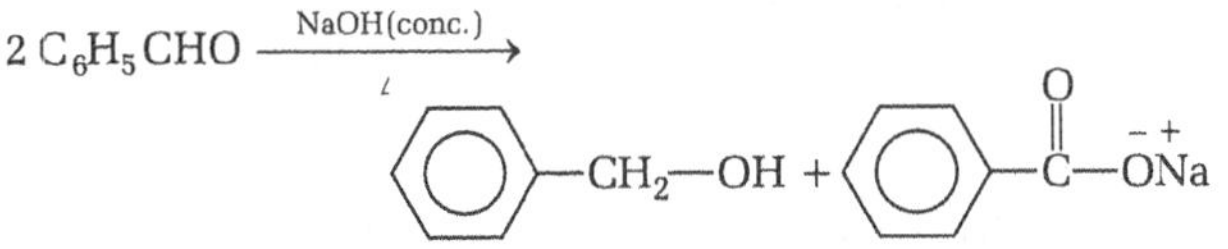

Benzyl alcohol Sodium benzoate

Electrophilic Substitution Reaction

Aromatic aldehydes and ketones undergo electrophilic substitution. Carbonyl group are electron withdrawing in nature, therefore acts as a deactivating and *meta* directing group.

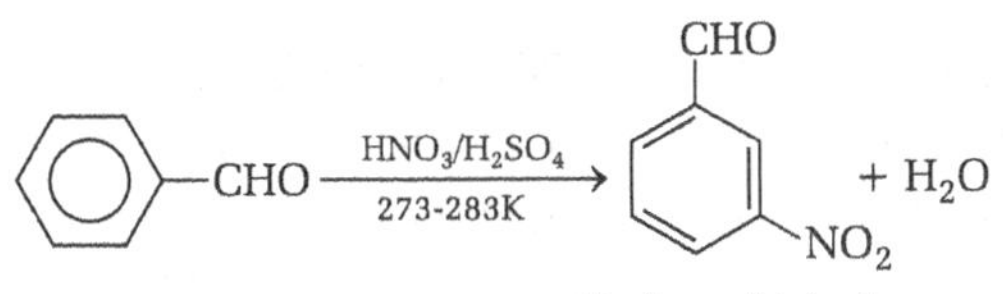

m-nitrobenzaldehyde

Chemical Test to Distinguish Between Aldehydes and Ketones

Aldehydes and ketones can be distinguished by several tests which are tabulated below:

Dishtinguish between Aldehydes and Ketones

Name of test	Reagent	Aldehydes	Ketones
Tollen's test	Ammoniacal silver nitrate	Silver mirror (deposition of Ag on surface of test tube) formed	No change
Fehling's test	Alkaline solution of $CuSO_4$ containing Rochelle salt (sodium potassium tartarate.)	Red precipitate of Cu_2O. (only aliphatic aldehydes give this test)	No change
Benedict test	Alkaline solution of copper acetate and sodium citrate	Red precipitate of Cu_2O. (only aliphatic aldehydes give this test)	No change
Schiff's test	*p*-rosaniline hydrochloride	Pink colour	No change

Uses of Aldehydes and Ketones

- In chemical industry, aldehydes and ketones are used as solvents, starting materials and reagents for the synthesis of other products.
- Formaldehyde, under the name formalin, (40% solution) is used to preserve biological specimens.
- Formaldehyde is also used to prepare bakelite, urea-formaldehyde glues and other polymeric products.

Carboxylic Acids

Carbon compounds containing a carboxylic functional group, —COOH are called carboxylic acids. Some higher members of aliphatic carboxylic acids (C_{12}-C_{18}) are known as fatty acids, occur in natural fats as esters of glycerol.

Preparation

Preparation of carboxylic acids are as follows:

1. From Acid Derivatives

All acid derivatives like amides ($RCONH_2$), acid halides ($RCOCl$), esters ($RCOOR'$), acid anhydrides ($RCO-O-COR$) ($RCO-O-COR$) on hydrolysis give carboxylic acids.

$$RCOZ \xrightarrow[\text{or Dil. NaOH}]{\text{Dil. HCl}} RCOOH$$

$$Z = -NH_2, -X\ (X = Cl, Br, I), OR', RCOO- \text{ etc.}$$

2. From Alkyl Benzene

Alkyl benzene when treated with strong oxidising agent like H_2CrO_4 (chromic acid), acidic or alkaline $KMnO_4$ gives benzoic acid.

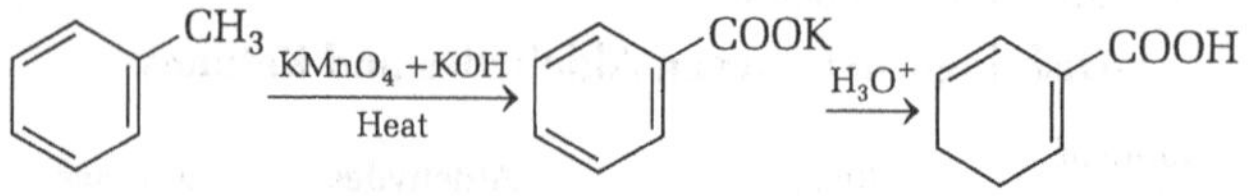

3. From Nitriles

Nitriles are hydrolysed to amides and then to acids in the presence of H^+ or $\bar{O}H$ as catalyst. Mild reaction conditions are used to stop the reaction at milder stage.

$$R-CN \xrightarrow[\Delta]{H^+ \text{ or } ^-OH} R-\overset{\overset{O}{\|}}{C}-NH_2 \xrightarrow[\Delta]{H^+ \text{ or } ^-OH} RCOOH$$

4. From Grignard Reagent

Grignard reagents react with carbon dioxide (dry ice) to form salts of carboxylic acid which give corresponding carboxylic acid after acidification with universal acid.

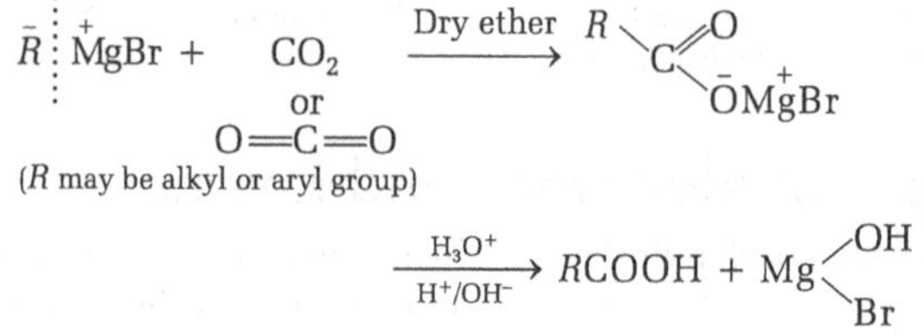

(*R* may be alkyl or aryl group)

Physical Properties

- Carboxylic acids exist in the form of dimer due to hydrogen bonding.
- As compared to aldehydes, ketones and hydrocarbons, carboxylic acids have higher boiling point due to intermolecular hydrogen bonding.

Chemical Properties

The reactions of carboxylic acids are classified as follows :

- Carboxylic acids evolve hydrogen gas with electropositive metals and form salts with alkalies.

$$2\,RCOOH + 2Na \longrightarrow 2RCO\overset{-}{O}\overset{+}{N}a + H_2$$

$$RCOOH + NaOH \longrightarrow RCO\overset{-}{O}\overset{+}{N}a + H_2O$$

$$RCOOH + NaHCO_3 \longrightarrow RCO\overset{-}{O}\overset{+}{N}a + H_2O + CO_2$$

- Carboxylic acids dissociate in water to give resonance stabilised carboxylate anions and hydronium ion.

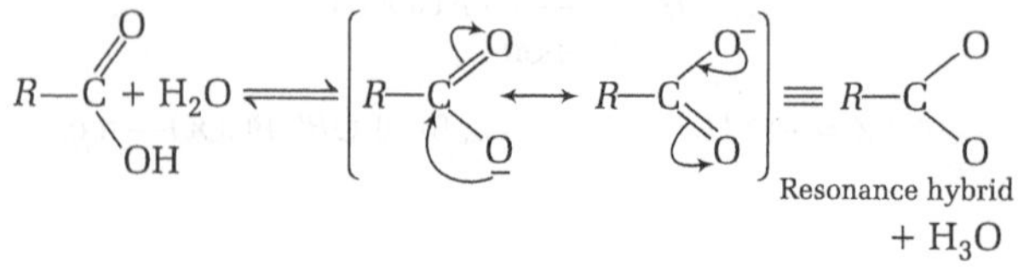

- Substituents may affect the stability of the conjugate base and also affect the acidity of carboxylic acids.
  (a) Electron withdrawing group (E.W.G.) stabilises the carboxylate anion and strengthens the acid.

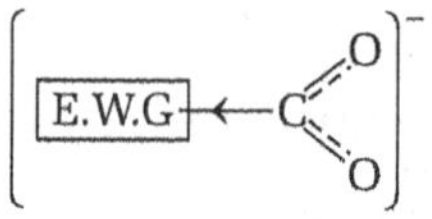

  (b) Electron donating group (E.D.G.) destabilises the carboxylate anion and weakens the acid. Strong acids have higher value of K_a and hence lower value of pK_a.

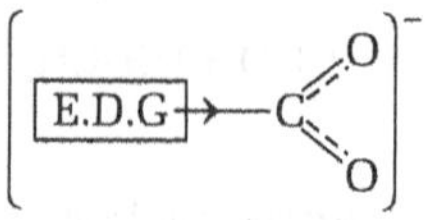

- Following acids are arranged in order of decreasing acidity.

$CF_3COOH > CCl_3COOH > CHCl_2COOH$
$NO_2CH_2COOH > NC-CH_2COOH$
$> FCH_2COOH > ClCH_2COOH > BrCH_2COOH >$
$HCOOH > ClCH_2CH_2COOH >$
$C_6H_5COOH > C_6H_5CH_2COOH > CH_3COOH > CH_3CH_2COOH$

- In case of aromatic carboxylic acids, more the $-R$ effect, more is the acidic nature.

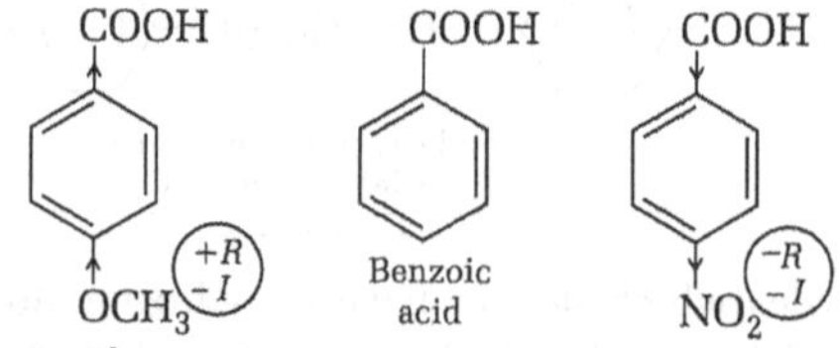

4-methoxy benzoic acid (pK_a = 4.46) | Benzoic acid (pK_a = 4.19) | 4-nitrobenzoic acid (pK_a = 3.41)

Reactions involving cleavage of C—OH bond

1. Formation of Anhydride

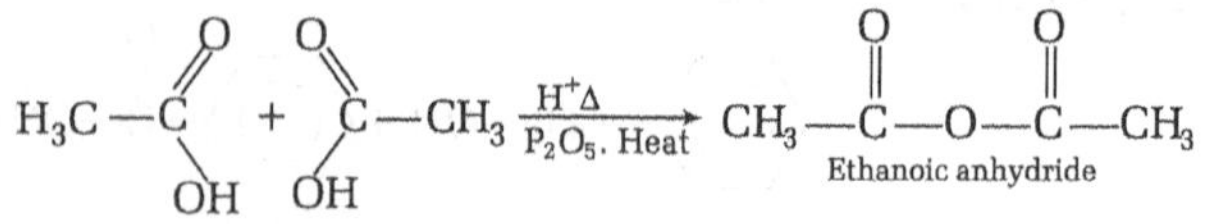

Ethanoic anhydride

2. Esterification

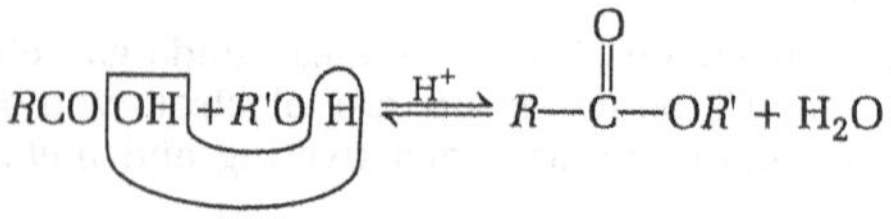

3. Reaction with PCl_5, PCl_3 and $SOCl_2$

- $RCOOH + PCl_5 \longrightarrow RCOCl + POCl_3 + HCl$
- $3RCOOH + PCl_3 \longrightarrow 3RCOCl + H_3PO_3$
- $RCOOH + SOCl_2 \longrightarrow RCOCl + SO_2\uparrow + HCl\uparrow$

4. Reaction with ammonia and its derivatives

- $CH_3COOH + NH_3 \rightleftharpoons \underset{\text{Ammonium acetate}}{CH_3COO^-NH_4^+} \xrightarrow{\text{Heat}} \underset{\text{Acetamide}}{CH_3CONH_2}$

- 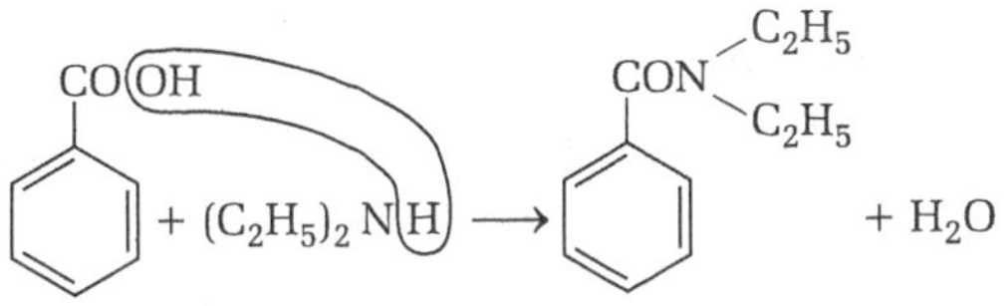

- 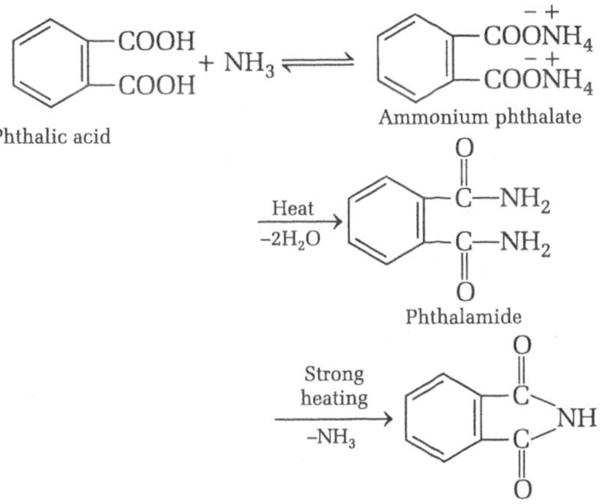

Phthalic acid, Ammonium phthalate, Phthalamide, Phthalimide

Reactions Involving —COOH Group

1. **Reduction** Carboxylic acids are reduced to primary alcohols by $LiAlH_4$ or better with B_2H_6. B_2H_6 does not easily reduce functional groups such as ester, nitro, halo, etc. $NaBH_4$ does not reduce the carboxyl group.

$$RCOOH \xrightarrow[\text{(ii) } H_3O^+]{\text{(i) } LiAlH_4/\text{Ether or } B_2H_6} RCH_2OH + H_2O$$

2. **Decarboxylation** Carboxylic acids lose carbon dioxide to form hydrocarbons when their sodium salts are heated with sodalime.

$$RCO\overset{-}{O}\overset{+}{Na} + NaOH \xrightarrow[\text{Heat}]{CaO} RH + Na_2CO_3$$

Substitution Reactions

1. **Halogenation** Carboxylic acids having an $\alpha-H$ are halogenated at the $\alpha-$ position on treatment with chlorine or bromine in the presence of small amount of red phosphorus to give α-halo carboxylic acids.

$$R—CH_2COOH \xrightarrow[\text{(ii) } H_2O]{\text{(i) } X_2/P \text{ (red)}} R—\underset{\underset{X}{|}}{C}HCOOH$$

$(X = Cl, Br)$

The reaction is known as **Hell Volhard-Zelinsky** reaction.

2. **Ring Substitution** Aromatic carboxylic acids undergo **electrophilic substitution reaction.** —COOH group shows $-R$-effect, therefore acts as a deactivating and *meta*-directing group. Carboxylic acids do not undergo Friedel-Crafts' reaction because the carboxylic group is deactivating and the catalyst $AlCl_3$ (anhyd.) gets bonded to the carboxylic group.

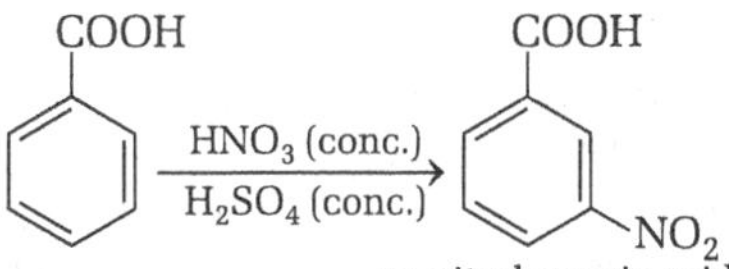

m-nitrobenzoic acid

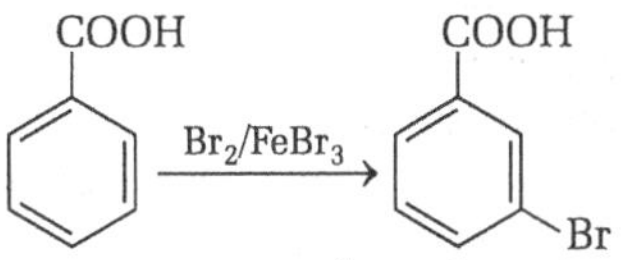

m-bromobenzoic acid

Uses of Carboxylic Acid

- Formic acid is used in leather tanning, textile, dyeing and finishing.
- Acetic acid is used in the manufacture of rayon and in plastics, rubber and silk industries and in cooking.
- Benzoic acid and its salts are used as urinary antiseptics.

DAY PRACTICE SESSION 1

FOUNDATION QUESTIONS EXERCISE

1 Arrange the following compounds in increasing order of boiling point. Propan-1-ol, butan-1-ol, butan-2-ol, pentan-1-ol

(a) Propan-1-ol, butan-2-ol, butan-1-ol, pentan-1-ol
(b) Propan-1-ol, butan-1-ol, butan-2-ol, pentan-1-ol
(c) Pentan-1-ol, butan-2-ol, butan-1-ol, propan-1-ol
(d) Pentan-1-ol, butan-1-ol, butan-2-ol, propan-1-ol

2 From amongst the following alcohols, the one that would react fastest with conc. HCl and anhydrous $ZnCl_2$, is → AIEEE 2010

(a) 2-butanol
(b) 2-methylpropan-2-ol
(c) 2-methylpropanol
(d) 1-butanol

3 Amongst the following alcohols which would react fastest with conc. HCl and $ZnCl_2$? → JEE Main (Online) 2013

(a) Pentanol
(b) 2-methylbutanol
(c) 2-pentanol
(d) 2-methylbutan-2-ol

4 An unknown alcohol is treated with the "Lucas reagent" to determine whether the alcohol is primary, secondary or tertiary. Which alcohol reacts fastest and by what mechanism? → JEE Main (Online) 2013

(a) Secondary alcohol by S_N1
(b) Tertiary alcohol by S_N1
(c) Secondary alcohol by S_N2
(d) Tertiary alcohol by S_N2

5 The product of the reaction given below is → JEE Main 2016

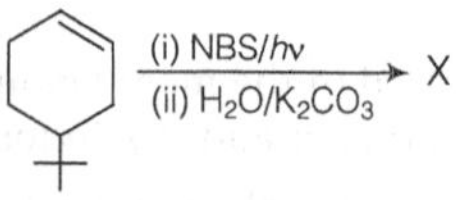

(a) 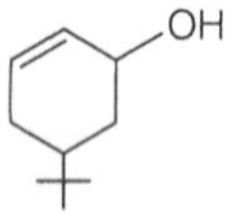

(b) 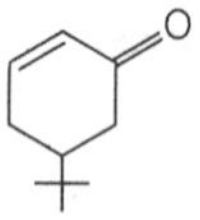

(c) 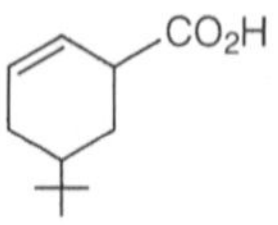

(d) 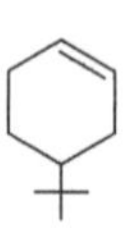

6 Rate of dehydration of alcohols follows the order → JEE Main (Online) 2013

(a) 2°>1°<CH_3OH>3°
(b) 3°>2°>1°>CH_3OH
(c) 2°>3°>1°>CH_3OH
(d) CH_3OH>1°>2°>3°

7 The main product of the following reaction is

$C_6H_5CH_2CH(OH)CH(CH_3)_2 \xrightarrow{\text{Conc. } H_2SO_4} ?$ → AIEEE 2010

(a) 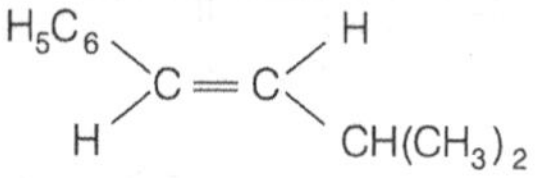

(b) 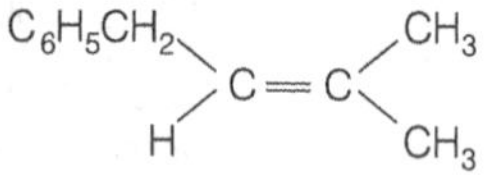

(c) 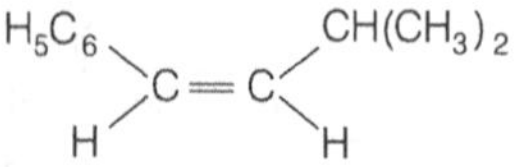

(d) 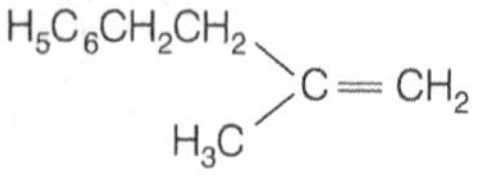

8 For which of the following molecule, significantly $\mu \neq 0$?

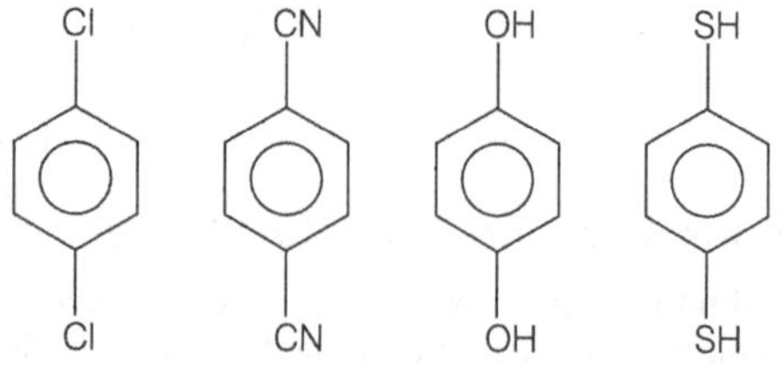

I II III IV

(a) Only I
(b) I and II → JEE Main 2014
(c) Only III
(d) III and IV

9 Arrange the following compounds in the order of decreasing acidity → JEE Main (Online) 2013

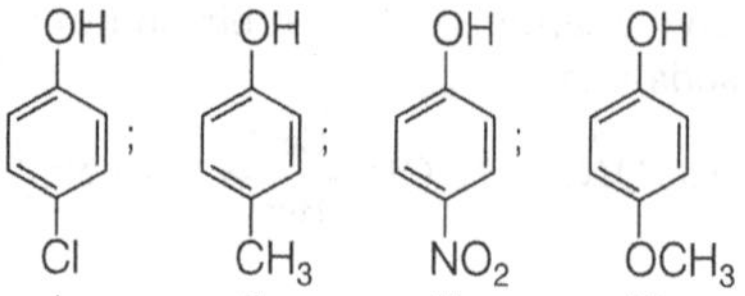

I II III IV

(a) II > IV > I > III
(b) I > II > III > IV
(c) III > I > II > IV
(d) IV > III > I > II

10 The major product obtained on interaction of phenol with sodium hydroxide and carbon dioxide is

(a) benzoic acid
(b) salicylaldehyde
(c) salicylic acid
(d) phthalic acid

11 Sodium phenoxide reacts with CO_2 at 400K and 4.7 atm pressure to give

(a) catechol
(b) salicylaldehyde
(c) sodium salicylate
(d) benzoic acid

12 Phenol on heating with $CHCl_3$ and NaOH gives salicylaldehyde. The reaction is called → JEE Main (Online) 2013

(a) Reimer-Tiemann reaction
(b) Claisen reaction
(c) Cannizzaro's reaction
(d) Hell-Volhard-Zelinsky reaction

13 Phenol is heated with a solution of mixture of KBr and $KBrO_3$. The major product obtained in the above reaction is → AIEEE 2011

(a) 2-bromophenol (b) 3-bromophenol
(c) 4-bromophenol (d) 2, 4, 6-tribromophenol

14 Phenol is heated with phthalic anhydride in presence of conc. H_2SO_4, the product gives pink colour with alkali. The product is

(a) phenolphthalein (b) bakelite
(c) salicylic acid (d) fluorescein

15 The structure of the compound that gives a tribromo derivative on treatment with bromine water is

(a) 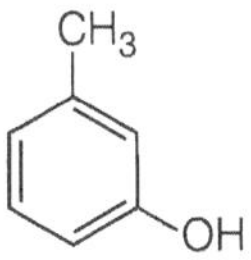

(b) 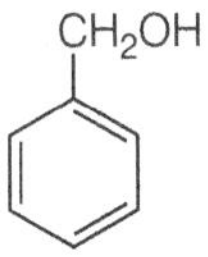

(c) 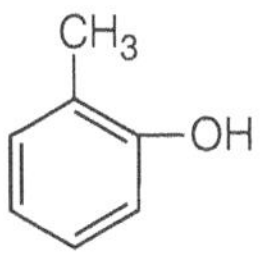

(d) 

16 When benzene sulphonic acid and *p*-nitrophenol are treated with $NaHCO_3$, the gases released respectively, are

(a) SO_2, NO_2 (b) SO_2, NO
(c) SO_2, CO_2 (d) CO_2, CO_2

17 The reaction of phenol with benzoyl chloride to give phenyl benzoate is known as → JEE Main (Online) 2013

(a) Claisen reaction (b) Schotten-Baumann reaction
(c) Reimer-Tiemann reaction (d) Gattermann-Koch reaction

18 Match the following and choose the correct option.

	Column I		Column II
A.	Kolbe's reaction	1.	Conversion of phenol to *o*-hydroxysalicylic acid
B.	Reimer-Tiemann reaction	2.	Conversion of phenol to salicyaldehyde
C.	Conversion of 2°alcohol to ketone	3.	Heated copper at 573 K
D.	Williamson's synthesis	4.	Reaction of alkyl halide with sodium alkoxide

Codes

	A	B	C	D		A	B	C	D
(a)	1	2	3	4	(b)	1	4	3	2
(c)	4	3	2	1	(d)	3	4	1	2

19 The major product in the following reaction. → JEE Main (Online) 2013

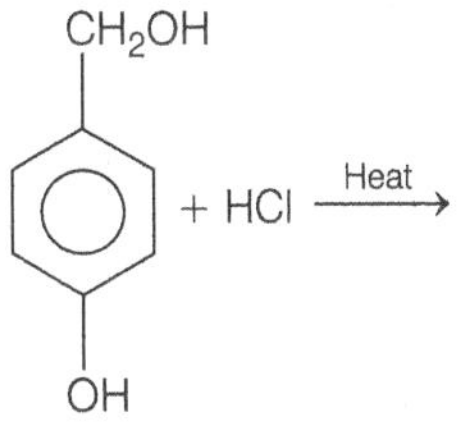

(a) 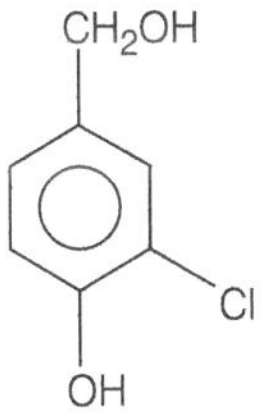

(b) 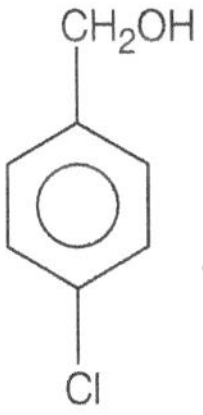

(c) 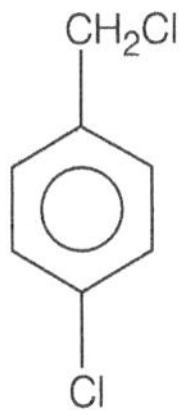

(d) 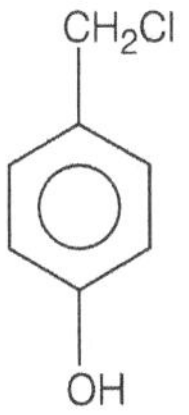

20 What is the structure of the major product when phenol is treated with bromine water? → JEE Main (Online) 2013

(a) 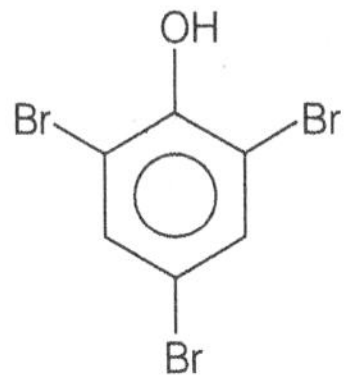

(b) 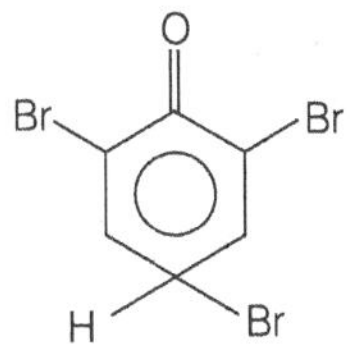

(c) 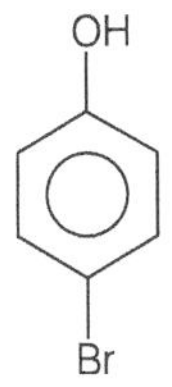

(d) 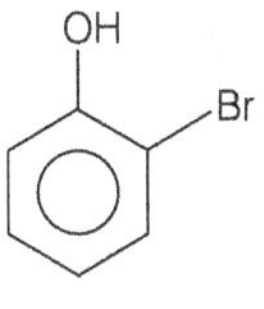

21 An ether (*A*), $C_5H_{12}O$ when heated with excess of hot concentrated HI produced two alkyl halides which when treated with NaOH yielded compounds (*B*) and (*C*). Oxidation of (*B*) and (*C*) gave a propanone and an ethanoic acid respectively. The IUPAC name of the ether (*A*) is → JEE Main (Online) 2013

(a) 2-ethoxypropane (b) ethoxypropane
(c) methoxybutane (d) 2-methoxybutane

22 Phenol on treatment with CO_2 in the presence of NaOH followed by acidification produces compound X as the major product. X on treatment with $(CH_3CO)_2O$ in the presence of catalytic amount of H_2SO_4 produces: → JEE Main 2018

(a) 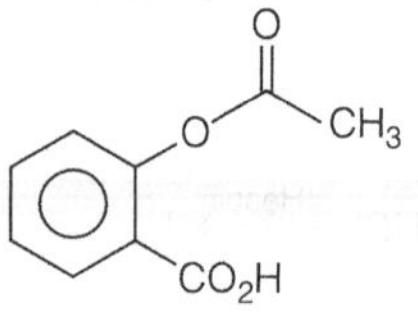

(b) 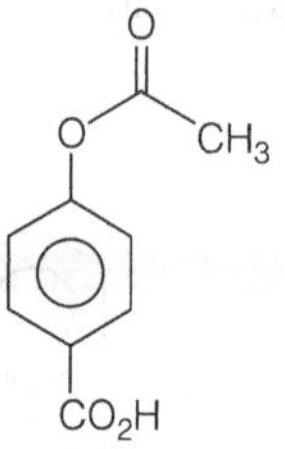

(c) 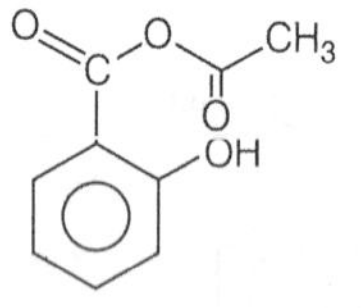

(d) 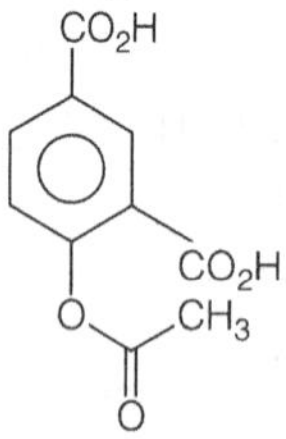

23 Compound Ph—O—C(=O)—Ph can be prepared by the reaction of
(a) phenol and benzoic acid in the presence of NaOH
(b) phenol and benzoyl chloride in the presence of pyridine
(c) phenol and benzoyl chloride in the presence of $ZnCl_2$
(d) phenol and benzaldehyde in the presence of palladium

24 Which is the suitable catalyst for bringing out the transformation given below.

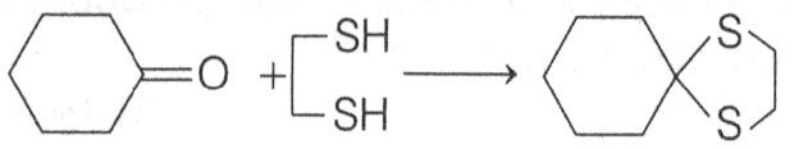

(a) BF_3, Et_2O (b) NaOEt
(c) Tungsten lamp (d) dibenzoyl peroxide

25 Which compound would give 5-keto-2-methyl hexanal upon ozonolysis? → JEE Main 2015

(a) 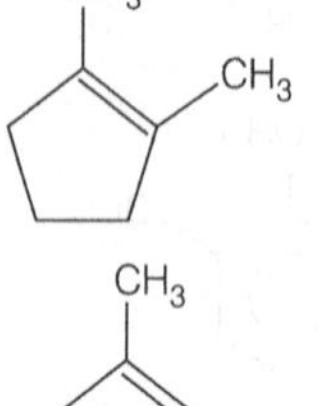

(b) 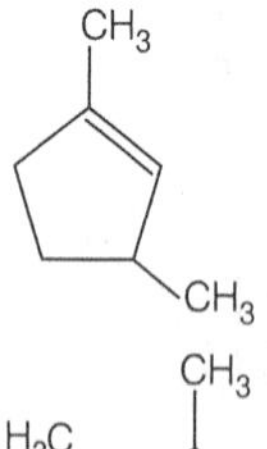

(c) 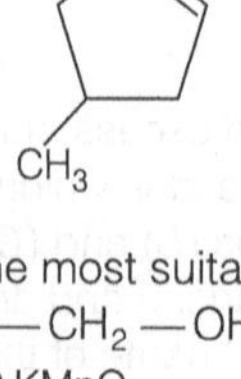

(d) 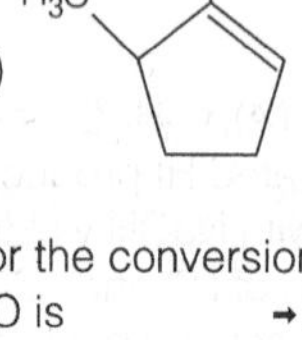

26 The most suitable reagent for the conversion of $R—CH_2—OH \Rightarrow R—CHO$ is → JEE Main 2014
(a) $KMnO_4$
(b) $K_2Cr_2O_7$
(c) CrO_3
(d) PCC (pyridinium chlorochromate)

27 In Rosenmund reaction,

$$R\,COCl + H_2 \xrightarrow{Pd/BaSO_4} RCHO + HCl$$

$BaSO_4$ here,
(a) promotes catalytic activity of Pd
(b) removes the HCl formed in the reaction
(c) deactivates palladium
(d) activates palladium

28 The formation of aldehyde from alkyl cyanide is related with the name
(a) Stephen reaction
(b) Rosenmund reaction
(c) Wurtz reaction
(d) HVZ reaction

29 The product of acid hydrolysis of P and Q can be distinguished by

$P = H_2C{=}C(OCOCH_3)(CH_3)$, $Q = CH_3CH{=}CHOCOCH_3$

(a) Lucas reagent (b) 2, 4-DNP
(c) Fehling's solution (d) $NaHSO_3$

30 Silver mirror test is given by which one of the following compounds? → AIEEE 2011
(a) Acetaldehyde (b) Acetone
(c) Formaldehyde (d) Benzophenone

31 Clemmensen reduction of a ketone is carried out in the presence of → JEE Main (Online) 2013
(a) $LiAlH_4$
(b) Zn-Hg with HCl
(c) glycol with KOH
(d) H_2 with Pt as catalyst

32 Which of the following carbonyl compounds on condensation gives an aromatic compound?
(a) CH_3CHO (b) HCHO
(c) CH_3COCH_3 (d) CH_3CH_2CHO

33 α,β-unsaturated aldehyde is formed in the sequence
(a) $HCHO \xrightarrow{KOH(aq)}$
(b) $CH_3CHO \xrightarrow{\text{Dil. KOH}} A \xrightarrow{\Delta} B$
(c) $CCl_3CHO \xrightarrow{KOH\,(aq)}$
(d) $CH_3—\overset{O}{\overset{\|}{C}}—OC_2H_5 \xrightarrow{KOH\,(aq)}$

34 The number of aldol reactions that occurs in the given transformation is → IIT JEE 2012

$CH_3CHO + 4HCHO \xrightarrow{\text{Conc. } aq.\text{ NaOH}}$ 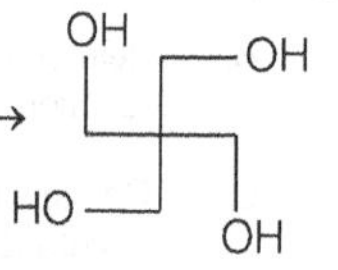

(a) 1 (b) 2
(c) 3 (d) 4

35 Which of the following is the product of aldol condensation? → JEE Main (Online) 2013

(a) 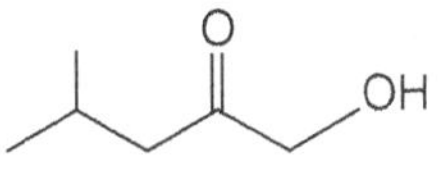 (b) 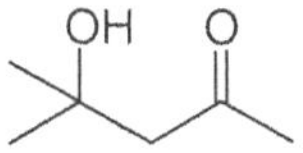

(c) 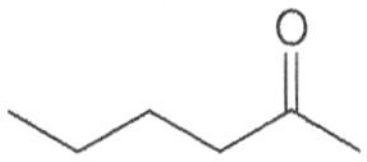 (d) 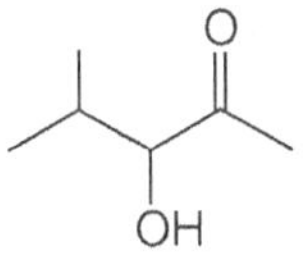

36 In Cannizzaro reaction given below:

$$2PhCHO \xrightarrow{OH^{\ominus}} PhCH_2OH + PhCO_2^{\ominus}$$

The slowest step is
(a) the attack of $OH^{\ominus}$
(b) the transfer of hydride to the carbonyl group
(c) the abstraction of proton from the carboxylic group
(d) the deprotonation of $PhCH_2OH$

37 Cannizzaro's reaction is not given by → JEE Main (Online) 2013

(a) 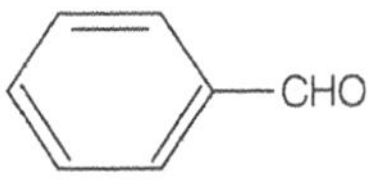

(b) 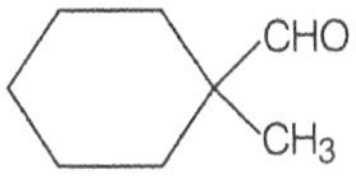

(c) CH_3CHO
(d) HCHO

38 Dry distillation of calcium formate and subsequent treatment with conc. KOH gives the mixture of
(a) CH_3OH, HCOOK (b) CH_3CHO, HCOOK
(c) HCHO, HCOOK (d) None of the above

39 The order of reactivity of phenyl magnesium bromide with the following compounds is

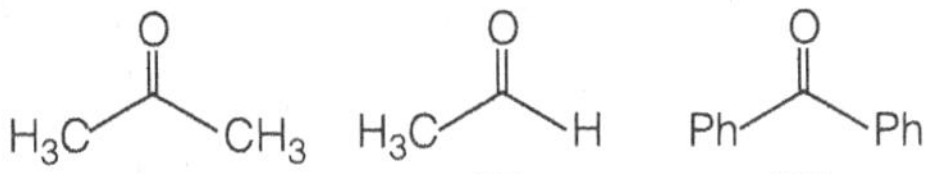

(I) (II) (III)

(a) II > III > I (b) I > III > II
(c) II > I > III (d) All react with the same rate

40 Iodoform can be prepared from all except → AIEEE 2012
(a) ethyl methyl ketone (b) isopropyl alcohol
(c) 3-methyl-2-butanone (d) isobutyl alcohol

41 Both HCHO and CH_3CHO give similar reactions with all the reagents except
(a) Schiff's reagent (b) Fehling solution
(c) ammoniacal $AgNO_3$ (d) ammonia

42 The reagent which does not react with both, acetone and benzaldehyde.
(a) Sodium hydrogen sulphite
(b) Phenyl hydrazine
(c) Fehling's solution
(d) Grignard reagent

43 $2D_2C{=}O + OH^- \xrightarrow{\text{Cannizzaro}} X$ and Y (Y is alcohol, D is deuterium). X and Y will have structure
(a) $D{-}C(=O){-}O^-$, $D{-}CD_2{-}OH$ (b) $D{-}C(=O){-}O^-$, $D{-}CDH{-}OH$
(c) $H{-}C(=O){-}O^-$, $D{-}CD_2{-}OH$ (d) None of these

44

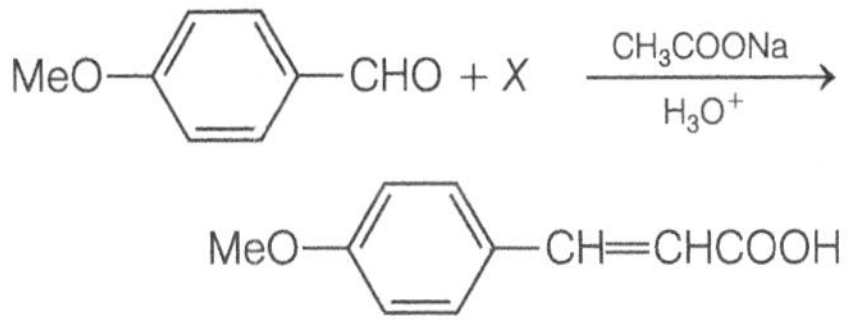

The compound X is
(a) CH_3COOH (b) $BrCH_2-COOH$
(c) $(CH_3CO)_2O$ (d) CHO — COOH

45 Which of the following compounds neither forms semicarbazone nor oxime?
(a) HCHO (b) $CH_3CONHCH_3$
(c) CH_3COCH_2Cl (d) $CH_3CH(CH_3)CHO$

46 In the reaction,

$$C_6H_5CHO + (CH_3CO)_2O \xrightarrow{CH_3COONa} A,$$

product A is
(a) acetaldehyde (b) cinnammic acid
(c) β-naphthol (d) phenol

47 Formaldehyde can be distinguished from acetaldehyde by the use of → JEE Main (Online) 2013
(a) Schiff's reagent (b) Tollen's reagent
(c) I_2, alkali (d) Fehling's solution

48 In the given transformation, which of the following is the most appropriate reagent? → AIEEE 2012

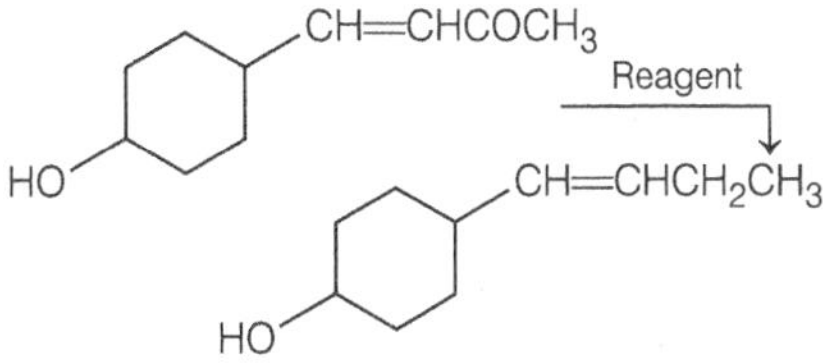

(a) NH_2NH_2, OH^- (b) Zn-Hg /HCl
(c) Na, liq. NH_3 (d) $NaBH_4$

49 Trichloroacetaldehyde was subjected to Cannizzaro's reaction by using NaOH . The mixture of the products contains sodium trichloroacetate ion and another compound. The other compound is → AIEEE 2011
(a) 2, 2, 2-trichloroethanol (b) trichloromethanol
(c) 2, 2, 2-trichloropropanol (d) chloroform

50 The correct sequence of reagents for following conversion will be → JEE Main 2017

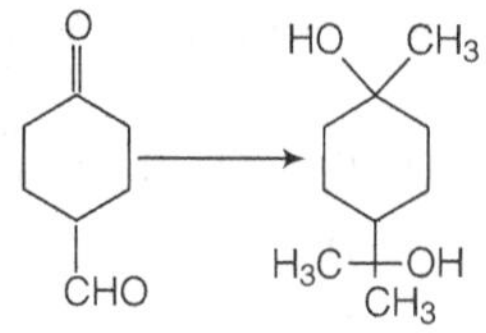

(a) $[Ag(NH_3)_2]^+OH^-H^+/CH_3OH$, CH_3MgBr
(b) CH_3MgBr, H^+/CH_3OH, $[Ag(NH_3)]^+OH^-$
(c) CH_3MgBr, $[Ag(NH_3)^2]^+OH^-$, H^+/CH_3OH
(d) $[Ag(NH_3)_2]^+OH^-$, CH_3MgBr, H^+/CH_3OH

51 Identify the correct order of boiling points of the following compounds

$CH_3CH_2CH_2CH_2OH$ (A), $CH_3CH_2CH_2CHO$ (B), $CH_3CH_2CH_2COOH$ (C)

(a) $A > B > C$ (b) $C > A > B$
(c) $A > C > B$ (d) $C > B > A$

52 Monocarboxylic acids are functional isomers of → JEE Main (Online) 2013

(a) ethers (b) amines
(c) esters (d) alcohols

53 A liquid was mixed with ethanol and a drop of concentrated H_2SO_4 was added. A compound with a fruity smell was formed. The liquid was

(a) CH_3OH (b) HCHO
(c) CH_3COCH_3 (d) CH_3COOH

54 $C_8H_6O_4 \xrightarrow{\Delta} X \xrightarrow{NH_3} Y$

The compound X is
(a) o-xylene (b) phthalic acid
(c) phthalic anhydride (d) salicylic acid

55 In the following sequence of reaction,

$$\text{Toluene} \xrightarrow{KMnO_4} A \xrightarrow{SOCl_2} B \xrightarrow[BaSO_4]{H_2/Pd} C$$

The product C is → JEE Main 2015
(a) C_6H_5COOH (b) $C_6H_5CH_3$
(c) $C_6H_5CH_2OH$ (d) C_6H_5CHO

56

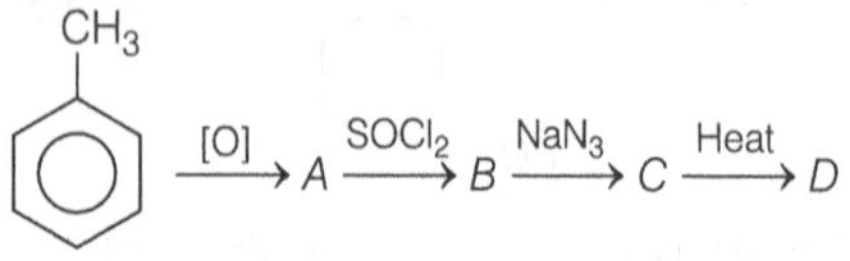

What is D in the above sequence?
(a) An amide
(b) Primary amine
(c) Phenyl isocyanate
(d) None of the above

57 Sodium phenoxide when heated with CO_2 under pressure at 125°C yields a product which on acetylation produces C.

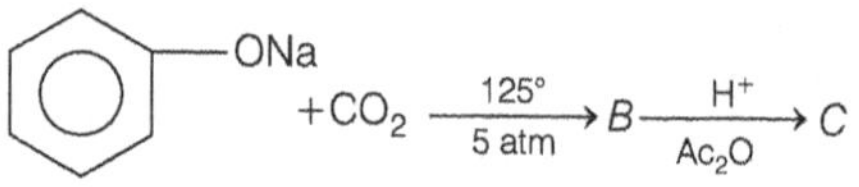

The major product C would be → JEE Main 2014

(a) 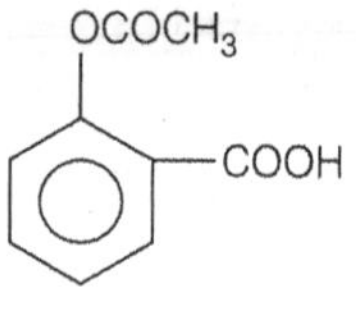 (b) 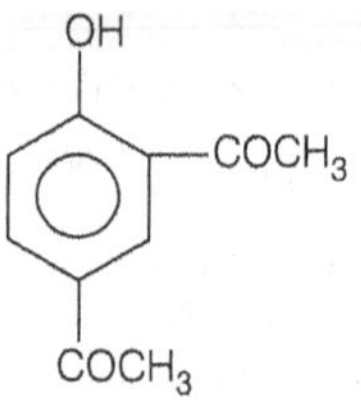

(c) 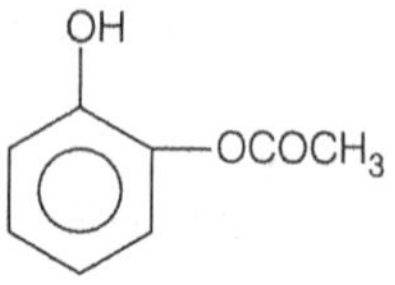 (d) 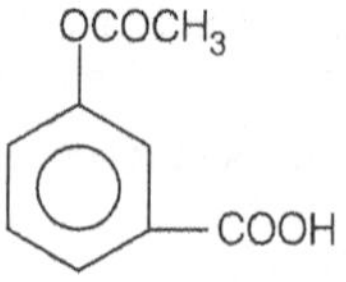

58 In the reaction,

$$CH_3COOH \xrightarrow{LiAlH_4} A \xrightarrow{PCl_5} B \xrightarrow{Alc.KOH} C$$

The product C is → JEE Main 2014
(a) acetaldehyde (b) acetylene
(c) ethylene (d) acetyl chloride

59 Sodium ethoxide has reacted with ethanoyl chloride. The compound that is produced in the above reaction is → AIEEE 2011

(a) diethyl ether (b) 2-butanone
(c) ethyl chloride (d) ethyl ethanoate

60 An organic compound A upon reacting with NH_3 gives B. On heating B, gives C. C in presence of KOH reacts with Br_2 to give $CH_3CH_2NH_2$. A is → IIT JEE Main 2013

(a) $CH_3CH_2CH_2COOH$ (b) $H_3C-\underset{\displaystyle CH_3}{\underset{|}{CH}}-COOH$
(c) CH_3CH_2COOH (d) CH_3COOH

61 The compound that undergoes decarboxylation most readily under mild condition is [IIT JEE 2012]

(a) 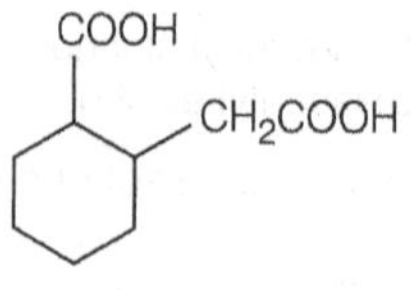 (b) 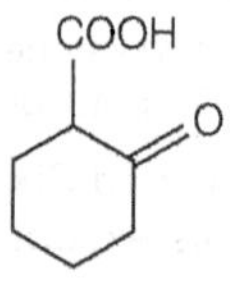

(c) 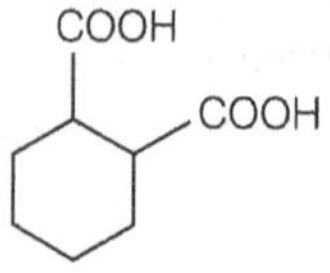 (d) 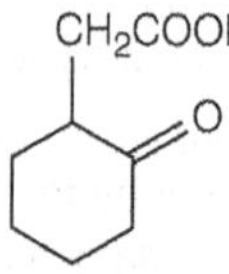

62 The major product obtained in the following reaction is

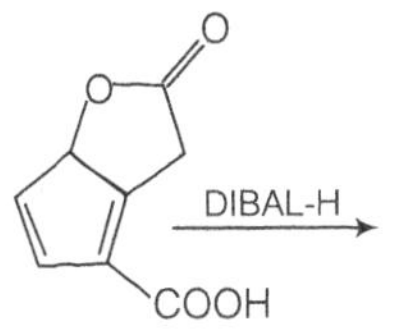

→ JEE Main 2017

(a) 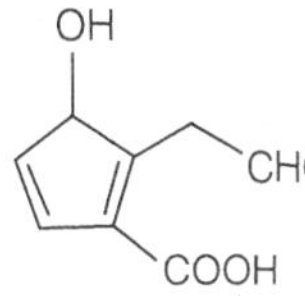

(b) 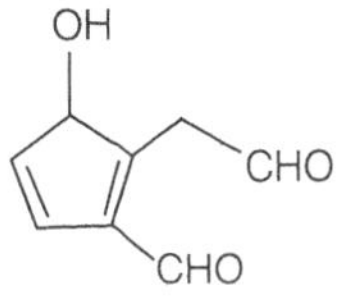

(c) 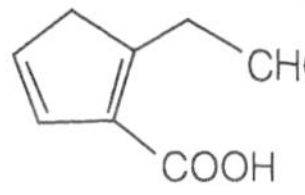

(d) 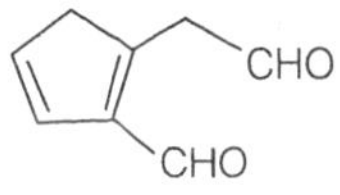

63. Which of the following reagents may be used to distinguish between phenol and benzoic acid?

(a) Aqueous NaOH (b) Tollen's reagent
(c) Molisch reagent (d) Neutral $FeCl_3$

Direction (Q. Nos. 64-69) *In the following questions, Assertion (A) followed by a Reason (R) is given. Choose the correct answer out of the following choices.*

(a) Both A and R are true and R is correct explanation of A
(b) Both A and R are true but R is not correct explanation of A
(c) A is false but R is true
(d) Both A and R are false

64 **Assertion** (A) Formaldehyde is a planar molecule.
Reason (R) It contains sp^2-hybridised carbon atom.

65 **Assertion** (A) Like bromination of benzene, bromination of phenol is also carried out in the presence of Lewis acid.
Reason (R) Lewis acid polarises the bromine molecule.

66 **Assertion** (A) Solubility of *n*-alcohol in water decreases with increase in molecular weight.
Reason (R) The reactive proportion of the hydrocarbon part in alcohols increases with increasing molecular weight which permits enhanced hydrogen bonding with water.

67 **Assertion** (A) *p*-nitrophenol is a stronger acid than *o*-nitrophenol.
Reason (R) Intramolecular hydrogen bonding makes the *o*-isomer weaker than the *p*-isomer.

68 **Assertion** (A) Acetic acid does not undergo haloform reaction.
Reason (R) Acetic acid has no alpha hydrogen.

69 **Assertion** (A) *p*-hydroxy benzoic acid has lower boiling point than *o*-hydroxy benzoic acid.
Reason (R) *o*-hydroxy benzoic acid has intramolecular hydrogen bonding.

DAY PRACTICE SESSION 2

PROGRESSIVE QUESTIONS EXERCISE

1 I. 1, 2– dihydroxy benzene
II. 1, 3– dihydroxy benzene
III. 1, 4 – dihydroxy benzene
IV. Hydroxy benzene

The increasing order of boiling points of above mentioned alcohols is

(a) I < II < III < IV
(b) I < II < IV < III
(c) IV < I < II < III
(d) IV < II < I < III

2 *p*-cresol reacts with chloroform in alkaline medium to give the compound *A* which adds hydrogen cyanide to form the compound *B*. The latter, on acidic hydrolysis gives chiral carboxylic acid. The structure of the carboxylic acid is

(a) 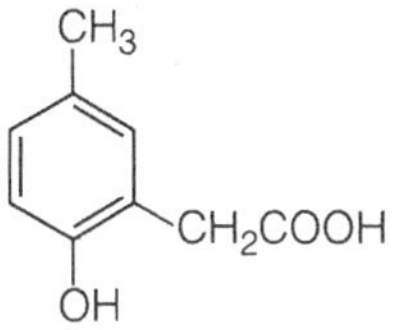

(b) 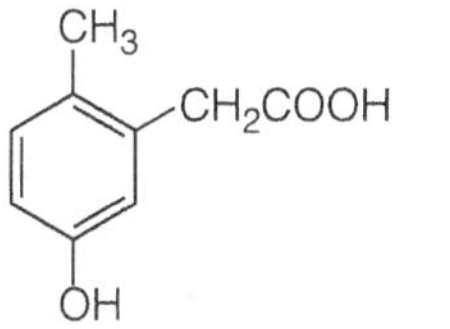

(c) 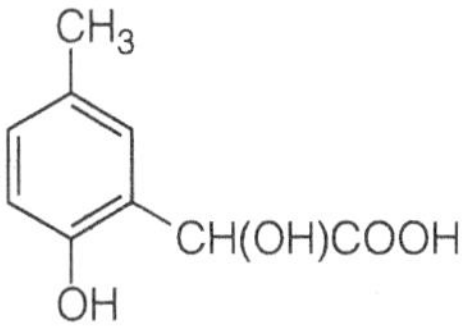

(d) 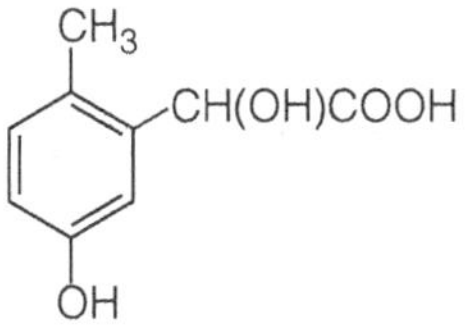

3 Phenol reacts with methyl chloroformate in the presence of NaOH to form product *A*. *A* reacts with Br_2 to form product *B*. *A* and *B* are respectively → JEE Main 2018

(a) 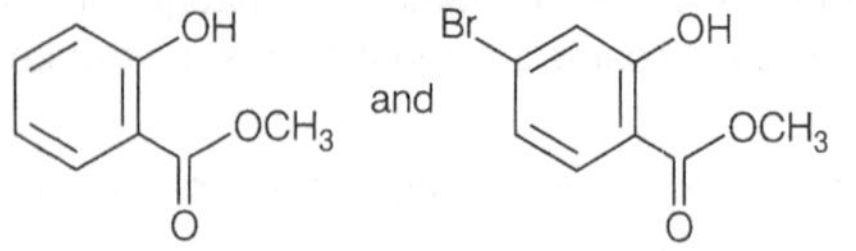

(b) 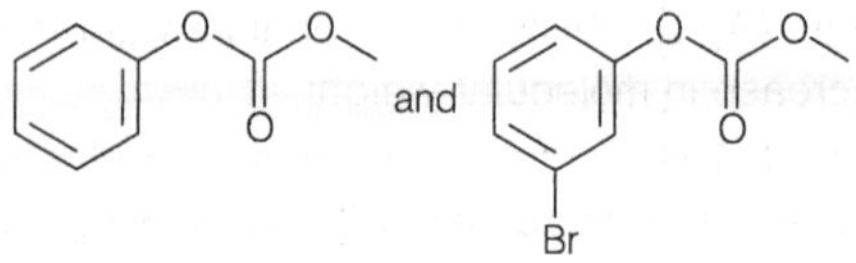

(c) 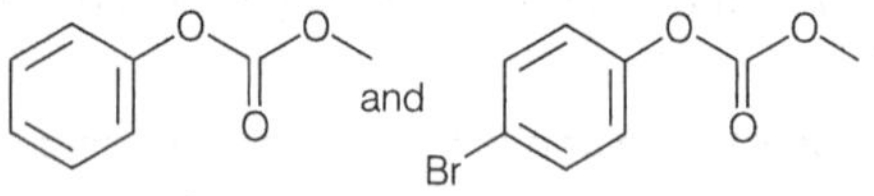

(d) 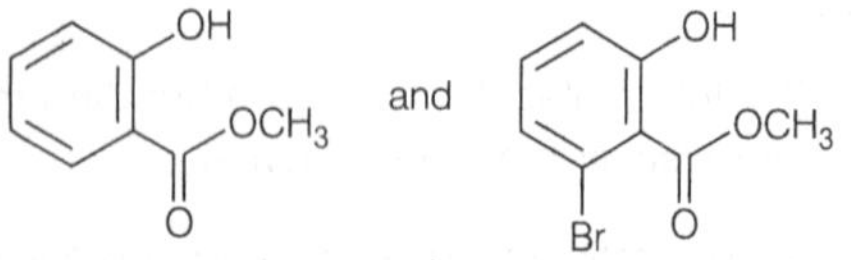

4 An unknown compound (*A*), $C_9H_{12}O$ have four isomeric structures. All the compounds react with sodium to give odourless and colourless gas except one. The three out of the four pre oxidised by hot $KMnO_4$ / OH^- to produce benzoic acid. One another isomer does not respond iodoform test. But one isomer out of these four fulfills all the conditions. The isomer is

(a) 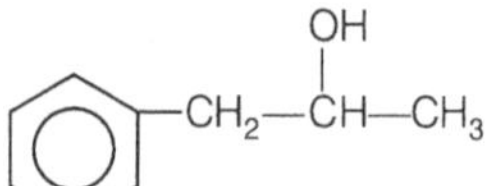 (b) 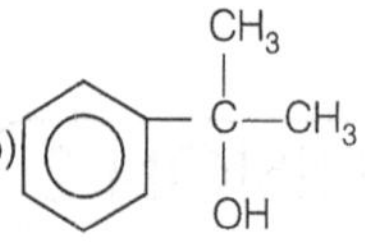

(c) 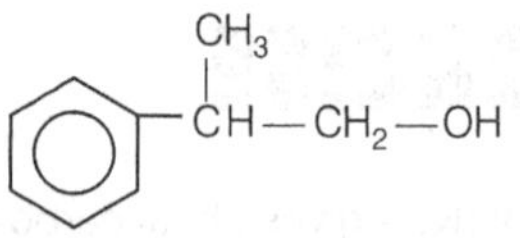 (d) 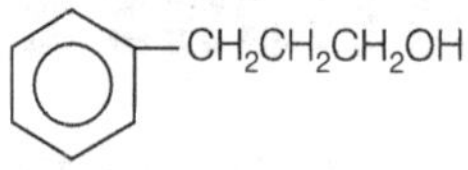

5 An alcohol '*A*' when heated with copper gives a product '*B*' not having oxygen atom. '*B*' on ozonolysis gives two isomeric product '*C*' and '*D*'. '*C*' on oxidation gives a monobasic acid '*E*', silver salt of which contains 59.6% Ag. The structure of '*A*' is

(a) $CH_3—CH(CH_3)—CH(OH)—CH_2—CH_3$

(b) $CH_3—C(OH)(CH_3)—CH_2—CH_2—CH_3$

(c) $CH_3—CH(CH_3)—CH_2—CH_2—CH_2OH$

(d) $CH_3—CH(CH_3)—CH_2—O—CH_2—CH_2OH$

6 With reference to the scheme given, which of the given statement about *T*, *U*, *V* and *W* is incorrect?

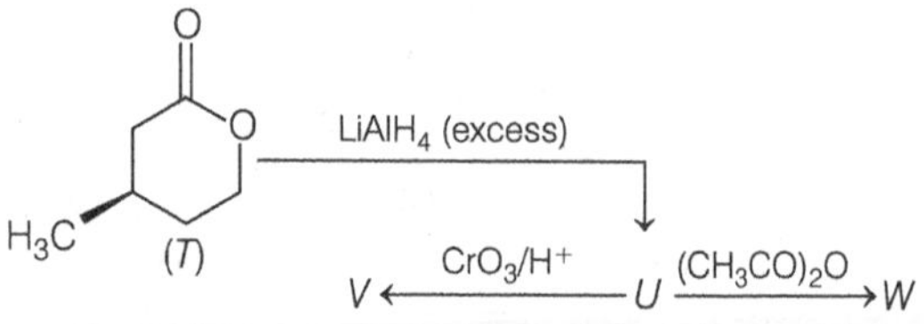

(a) *T* is soluble in hot aqueous NaOH

(b) *V* is optically active

(c) Molecular formula of *W* is $C_{10}H_{18}O_4$

(d) *V* gives effervescence on treatment with aqueous $NaHCO_3$

7 In the following reaction sequence, the correct structures of *E*, *F* and *G* are,

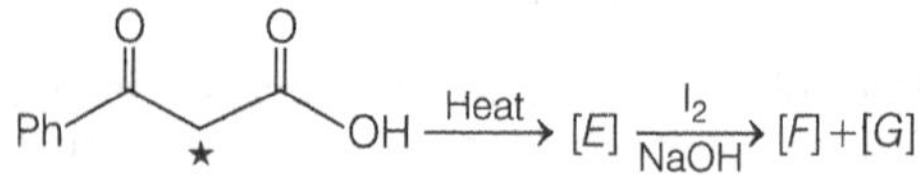

(* implies ^{13}C labelled carbon)

(a) 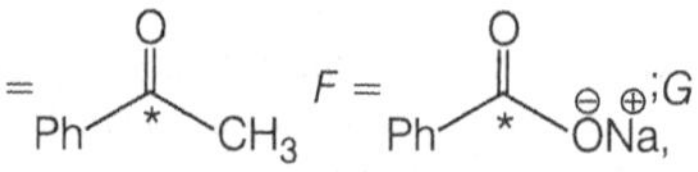 ; $G = CHI_3$

(b) 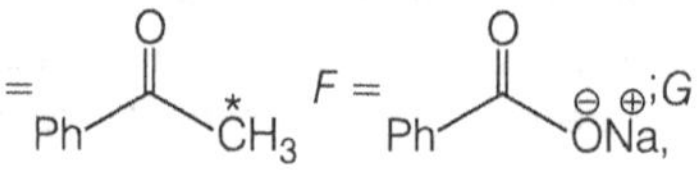 ; $G = CHI_3$

(c) 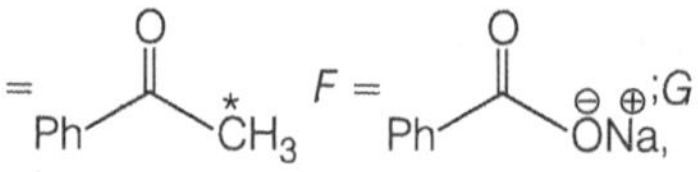 ; $G = \overset{*}{C}HI_3$

(d) 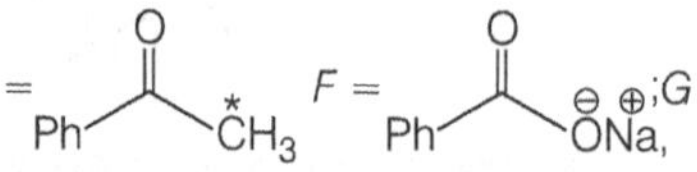 ; $G = \overset{*}{C}H_3I$

8 An organic compound containing C, H and O gives red colouration with sodium nitroprusside solution but does not reduce Tollen's reagent and yields chloroform on treating with NaOH and Cl_2. The compound is

(a) CH_3CH_2OH

(b) $CH_3CH(OH)CH_3$

(c) CH_3COCH_3

(d) $(CH_3)_3CH—CHO$

9 An organic compound (*A*), $C_4H_8O_2$, on treatment with an equivalent of C_2H_5MgBr gives (*B*), which do not respond to iodoform test. (*B*) on treatment with $LiAlH_4$ gives (*C*) which respond to Lucas test in 5-10 minutes. (*C*) upon treatment with H^+/HCl gives (*D*), which on reaction with KOH (*aq*) gave (*E*) which is oxidised with CrO_3 to form (*F*). (*E*) and (*F*) can respond positively to iodoform test. (*F*) can also be obtained by treating butyl methanoate with one equivalent of *n*-prMgBr, H_2O / H^+. Based on this information, (*F*) is

(a) $CH_3-\overset{O}{\overset{\|}{C}}-CH_2CH_2CH_3$ (b) $CH_3-\overset{OH}{\overset{|}{C}H}-CH_2CH_2CH_3$

(c) $CH_3-\underset{OH}{\underset{|}{\overset{CH_3}{\overset{|}{C}}}}-CH_3$ (d) $CH_3-CH\begin{matrix}OH \\ CH_2CHO\end{matrix}$

10 The freezing point constant of C_6H_6 is 4.90 K mol^{-1} and its melting point 5.51°C. A solution of 0.816 g of compound '*A*' when dissolved in 7.5 g of benzene, freezes at 1.59°C. The compound '*A*' has C = 70.58%, H = 5.88% and the rest is oxygen. Compound '*A*' on heating with sodalime gives another compound '*B*' which on oxidation and subsequent acidification gives an acid '*C*' of equivalent weight 122. '*C*' on heating with sodalime gives benzene. The compound '*A*' is

(a) $C_6H_4\begin{matrix}CH_3 \\ COOH\end{matrix}$ (b) $C_6H_5CH_2COOH$

(c) Both (a) and (b) (d) None of these

11 The major product formed in the following reaction is → JEE Main 2018

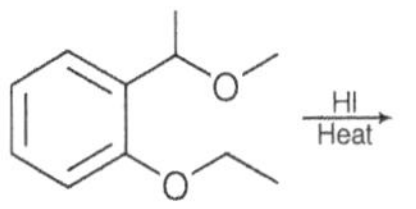

(a) 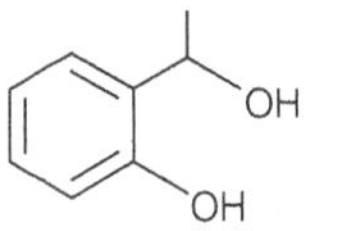 (b) 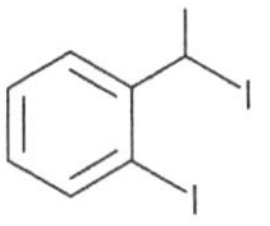

(c) 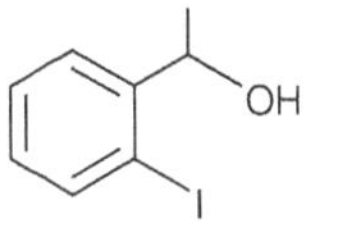 (d) 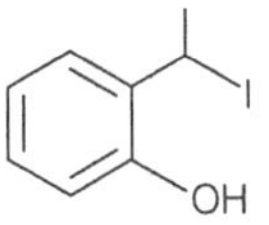

12 Which is the correct combination of reagent which can carry out following conversions?

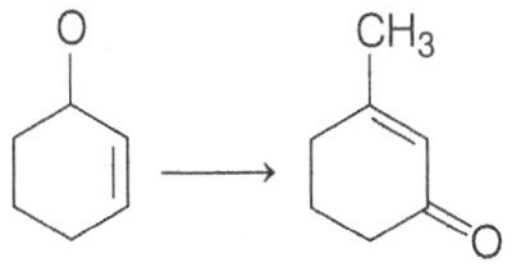

(a) (i) CH_3-MgBr then H^+ (ii) H_2SO_4 / Δ (iii) NH_2-NH_2/KOH

(b) (i) $(CH_3)_2$ CuLi then H^+ (ii) $NaBH_4$/EtOH (iii) H_2SO_4 / Δ

(c) CH_3-Li then H^+ (ii) PCC/Δ

(d) $NaBH_4$. $CeCl_3$ then H^+ (ii) MnO_2 (iii) CH_3-Li

13 An organic compound (*A*) with molecular formula $C_9H_{10}O$ forms an orange-red precipitate with 2,4-DNP reagent and gives yellow precipitate on heating with iodine and NaOH. It does not reduce Tollen's reagent or Fehling's solution nor it decolourises bromine water as Baeyer's reagent. On drastic oxidation with chromic acid, it gives a carboxylic acid having molecular formula $C_7H_6O_2$. Identify the compound (*A*).

(a) 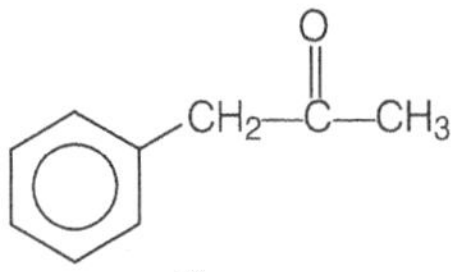

(b) 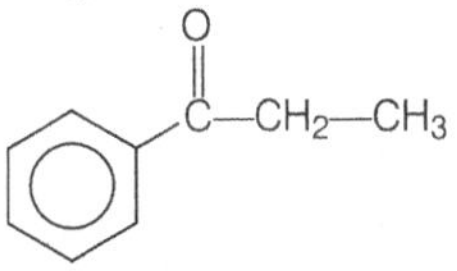

(c) 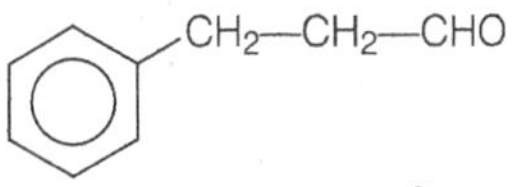

(d) 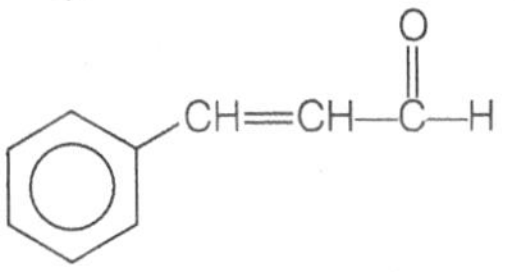

14 Give the order of ease of decarboxylation of the following acids.

$\underset{I}{CH_3COOH}$; $\underset{II}{CH_2=CH-CH_2COOH}$; $\underset{III}{CH_2(COOH)_2}$

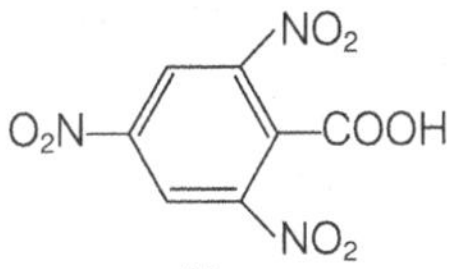

IV

(a) I > II > III > IV (b) III > IV > II > I

(c) IV > III > II > I (d) I > III > II > IV

15 Reaction of cyclohexanone with dimethylamine in the presence of catalytic amount of an acid forms a compound. Water during the reaction is continuously removed. The compound formed is generally known as

(a) an amine (b) an imine

(c) an enamine (d) a Schiff's base

16 A product obtained by the reaction of *X* with hydroxylamine and on further reduction gives

$C_2H_5-\overset{H\ \ NH_2}{\overset{\backslash\ /}{C}}-C(CH_3)_3$. Hence, the compound *X* can be

(a) 2,2-dimethyl-3-pentanone

(b) 3,3-dimethyl-3-butanone

(c) 1-methyl-3-pentanone

(d) diethyl ketone

17 An ester (*A*) with molecular formula $C_9H_{10}O_2$ was treated with excess of CH_3MgBr and the complex so formed was treated with H_2SO_4 to give an olefin (*B*). Ozonolysis of (*B*) gave a ketone with molecular formula C_8H_8O which shows positive iodoform test. The structure of (*A*) is

(a) $C_6H_5COOC_2H_5$ (b) $C_6H_5COOC_6H_5$

(c) $C_6H_5COOCH_3$ (d) p-$H_3CO-C_6H_4COCH_3$

ANSWERS

SESSION 1

1 (a)	**2** (b)	**3** (d)	**4** (b)	**5** (a)	**6** (b)	**7** (a)	**8** (d)	**9** (c)	**10** (c)
11 (c)	**12** (a)	**13** (d)	**14** (a)	**15** (a)	**16** (d)	**17** (b)	**18** (a)	**19** (d)	**20** (a)
21 (a)	**22** (a)	**23** (b)	**24** (a)	**25** (b)	**26** (d)	**27** (c)	**28** (a)	**29** (c)	**30** (a)
31 (b)	**32** (c)	**33** (b)	**34** (d)	**35** (b)	**36** (b)	**37** (c)	**38** (a)	**39** (c)	**40** (d)
41 (d)	**42** (c)	**43** (a)	**44** (c)	**45** (b)	**46** (b)	**47** (c)	**48** (a)	**49** (a)	**50** (a)
51 (b)	**52** (c)	**53** (d)	**54** (c)	**55** (d)	**56** (c)	**57** (a)	**58** (c)	**59** (d)	**60** (c)
61 (b)	**62** (a)	**63** (d)	**64** (a)	**65** (d)	**66** (c)	**67** (a)	**68** (c)	**69** (d)	

SESSION 2

1 (c)	**2** (c)	**3** (c)	**4** (a)	**5** (b)	**6** (b)	**7** (c)	**8** (c)	**9** (a)	**10** (c)
11 (d)	**12** (c)	**13** (a)	**14** (c)	**15** (c)	**16** (a)	**17** (a)			

Hints and Explanations

SESSION 1

1 With the increase in molecular mass, the boiling points also increases. Amongst isomeric alcohols, the boiling points decrease with branching due to decrease in surface area. Thus, the correct increasing order of boiling point is

propan-1-ol < butan-2-ol < butan-1-ol < pentan-1-ol

2 The reaction of alcohol with conc. HCl and anhydrous $ZnCl_2$ follows S_N1 pathway, so greater the stability of carbocation formed, faster is the reaction. 2-methylpropan-2-ol forms 3° carbocation.

Hence, it reacts rapidly with conc. HCl and anhyd. $ZnCl_2$ (Lucas reagent).

3 Conc. HCl and $ZnCl_2$ is Lucas reagent and 3° alcohols (alcohols having —C—OH group) react fastest with it.

The structures of the given alcohols are as

$CH_3CH_2CH_2CH_2CH_2OH$
Pentanol (1° alcohol)

$CH_3CH_2CH(CH_3)CH_2OH$
2-methylbutanol
(1° alcohol)

$CH_3CH_2CH_2CH(OH)CH_3$
2-pentanol
(2° alcohol)

$CH_3—C(CH_3)(OH)—CH_2CH_3$
2-methylbutan-2-ol
(3° alcohol)

Hence, 2-methylbutan-2-ol reacts fastest with the given reagent.

4 The reaction of alcohol with Lucas reagent occur through S_N1 mechanism and the rate of reaction is directly proportional to the stability of carbocation formed in the reaction. Since, 3° alcohol forms 3° carbocation (most stable) hence, it will react fastest.

5

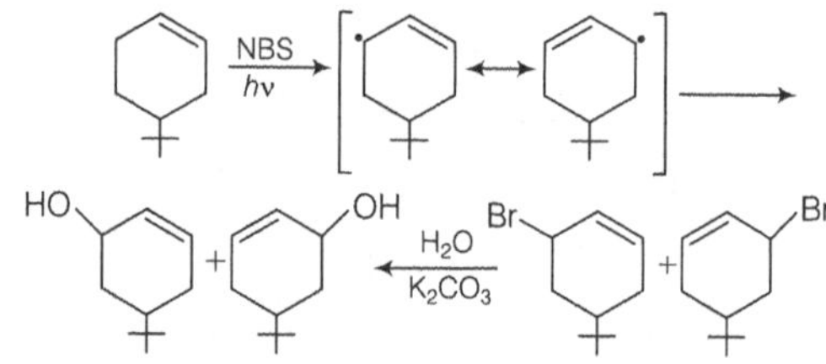

6 Dehydration of alcohols takes place according to Saytzeff rule. In acid catalysed dehydration, formation of most stable carbocation takes place.

Thus, the order of dehydration of alcohols is

$$3° > 2° > 1° > CH_3OH.$$

7

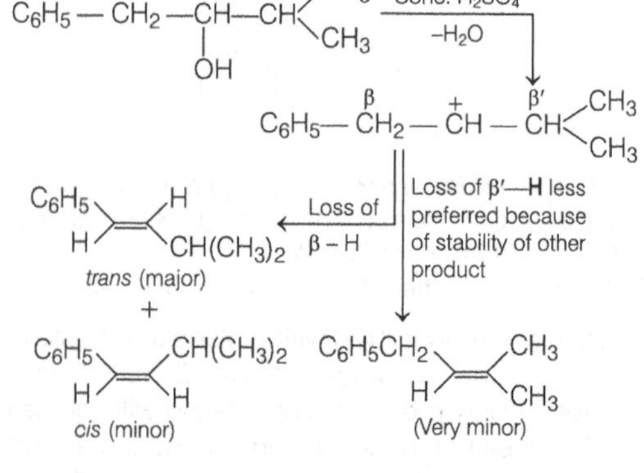

8 In the quinol  and thioquinol 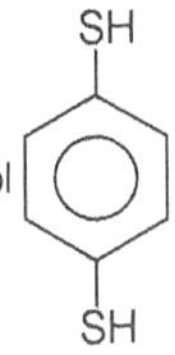

—OH groups and —SH groups do not cancel their dipole moment as they exist in different confirmation.

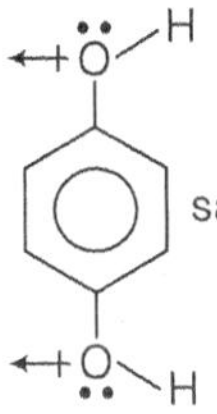

same as in thioquinol.

9 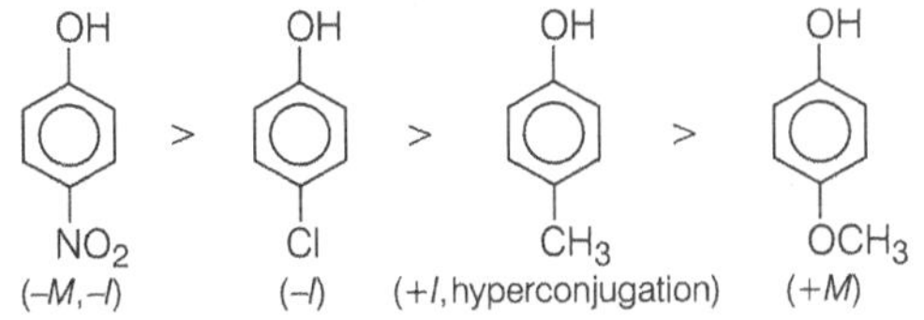

Electron releasing group decreases while electron withdrawing group increases acidic strength by destabilising and stabilising the phenoxide ion formed respectively.

10 Salicylic acid is obtained as a major product by the interaction of phenol with NaOH and CO_2 at 400 K and 4-7 atm. This reaction is called Kolbe's reaction.

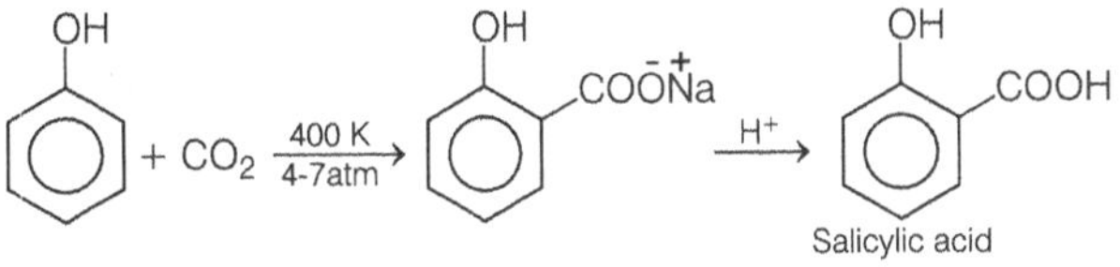

11 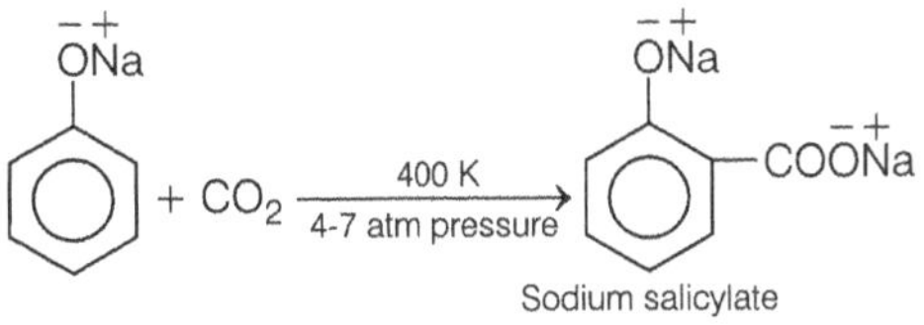

12 This reaction is known as Reimer-Tiemann reaction.

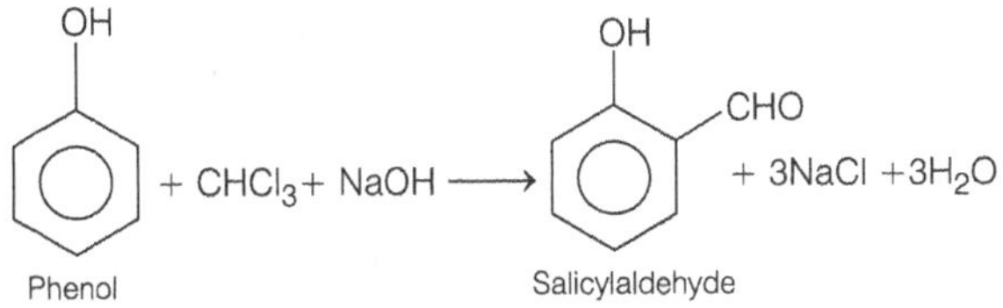

13 Br_2 is formed by a redox reaction,

$$5Br^- + BrO_3^- + 6H^+ \longrightarrow 3\,Br_2 + 3H_2O$$

—OH group is the activating group and there is S_E at *o*-and *p*-positions giving yellowish white precipitate of 2, 4, 6-tribromophenol.

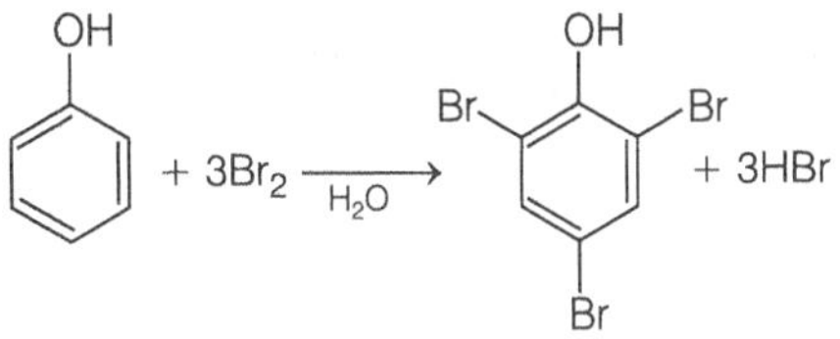

14 Phenol is heated with phthalic anhydride in presence of conc. H_2SO_4 to give phenolphthalein which gives pink colour with alkali.

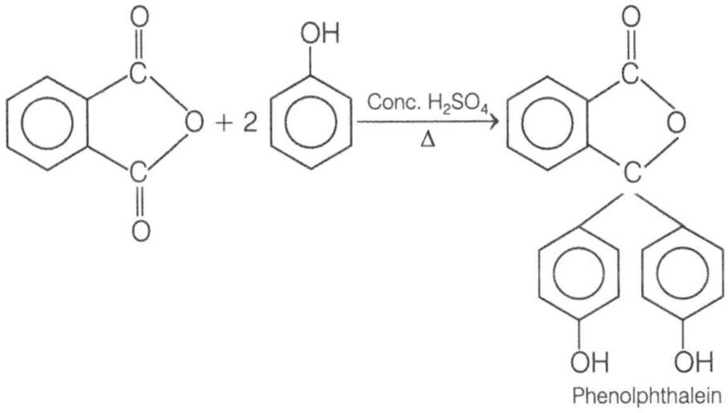

15 *m*-cresol, due to phenoxide ion in H_2O solvent, gives tribromo derivative at all *ortho*-and *para*-positions.

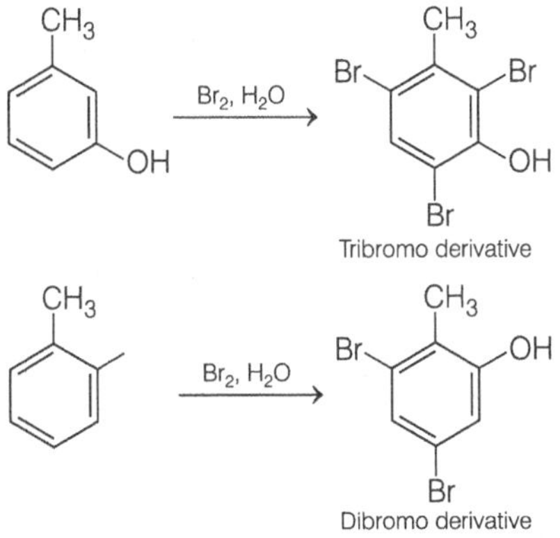

16 Benzene sulphonic acid and *p*-nitrophenol both are stronger acids than $H_2CO_3(H_2O + CO_2)$, so they are capable to evolve CO_2 when reacts with $NaHCO_3$.

$$HCO_3^- + \underset{\text{Acid}}{H^+} \longrightarrow H_2O + CO_2 \uparrow$$

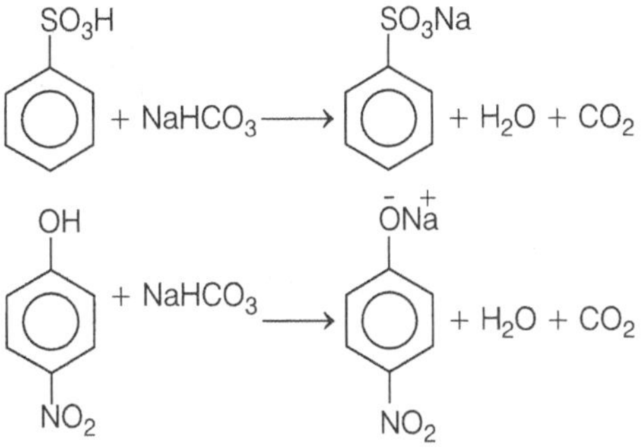

17 The reaction of phenol with benzoyl chloride to give phenyl benzoate is also known as Schotten-Baumann reaction.

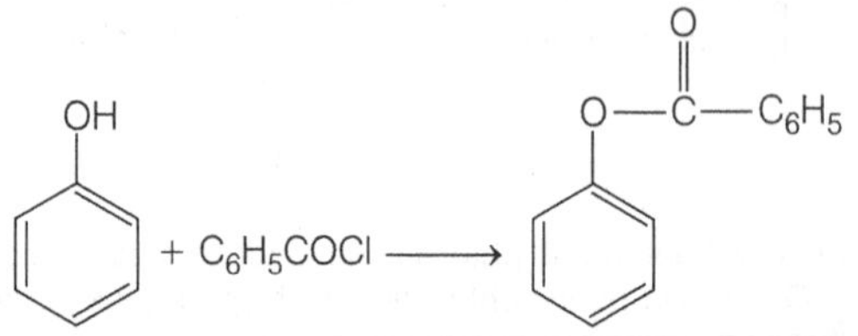

Phenyl benzoate

18 A ⟶ 1, B ⟶ 2, C ⟶ 3, D ⟶ 4

19

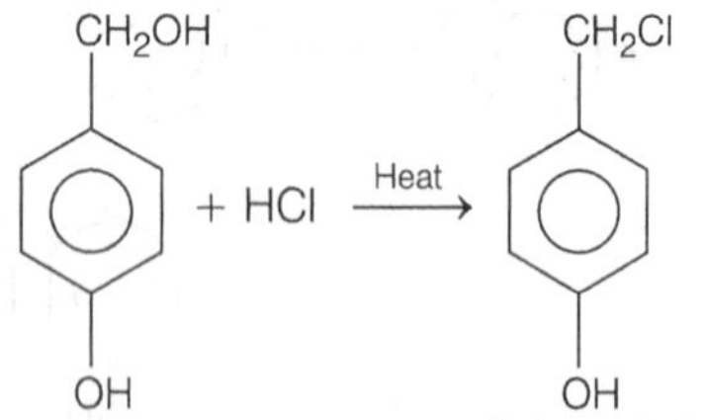

Major product

Only aliphatic —OH is substituted by Cl^-. This is because in phenol the C—O bond is stabilised by resonance.

20

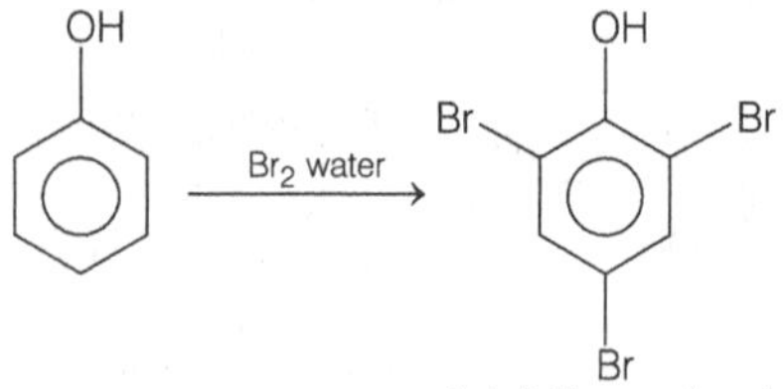

2, 4, 6-tribromophenol

21 Ether (*A*) is 2-ethoxypropane

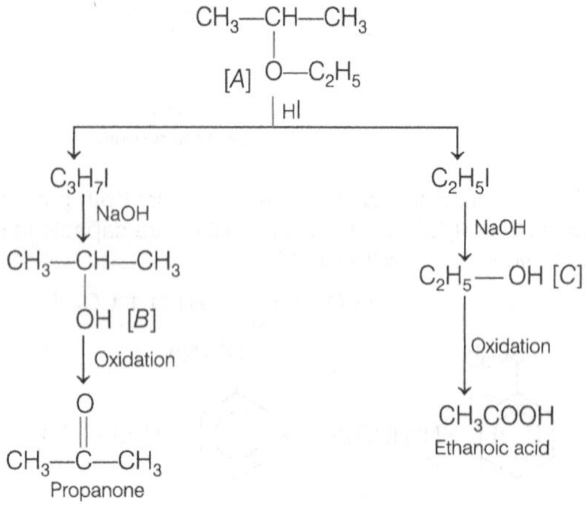

22

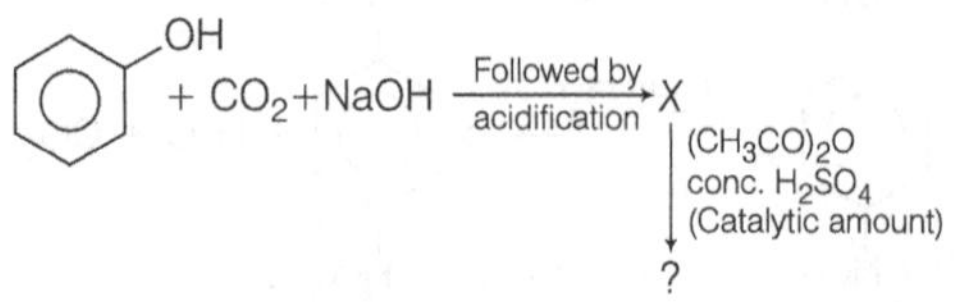

The very first reaction in the above road map looks like Kolbe's reaction which results to salicylic acid as :

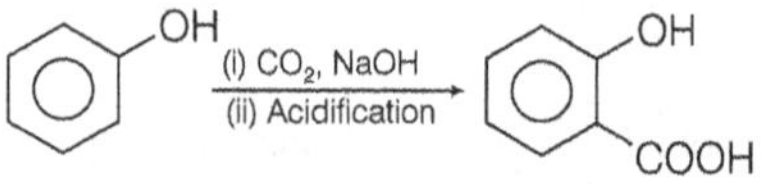

Salicylic acid
(*X*)

The salicylic acid with acetic anhydride [$(CH_3CO)_2O$] in the presence of catalytic amount of conc. H_2SO_4 undergoes acylation to produce aspirin as :

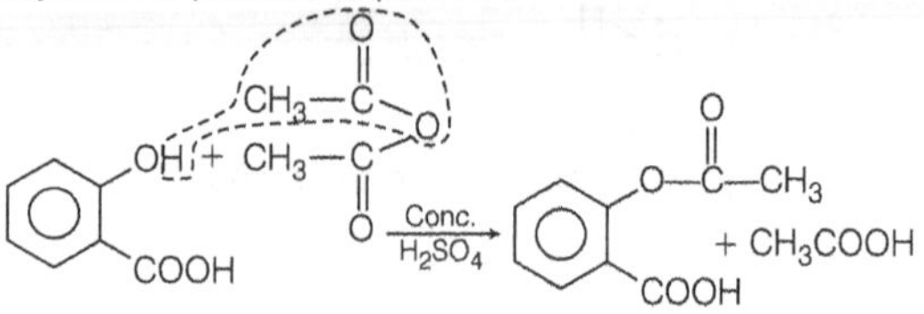

Acetyl salicylic acid (Aspirin)

Aspirin is a non-narcotic analgesic (Cpain killer).

23 For the reaction of 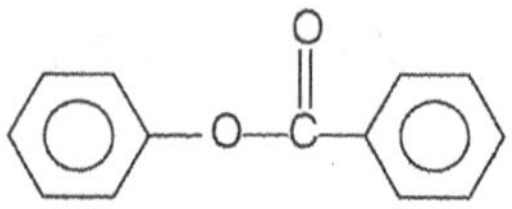 following reaction takes place.

24 BF_3 increases the electrophilicity of carbonyl carbon due to which sulphur attacks on carbonyl carbon.

25

(a)

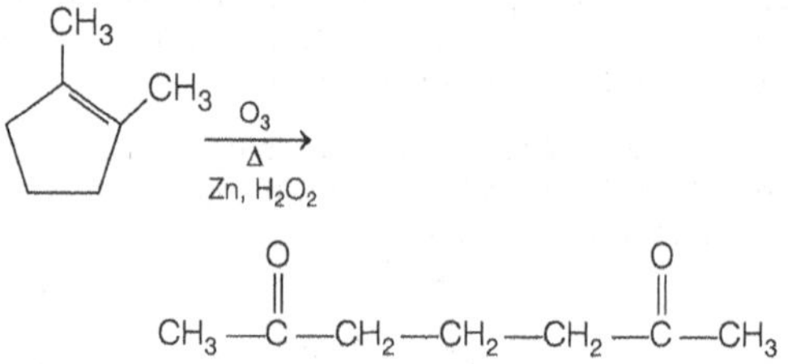

Heptane-2,6-diketone

(b)

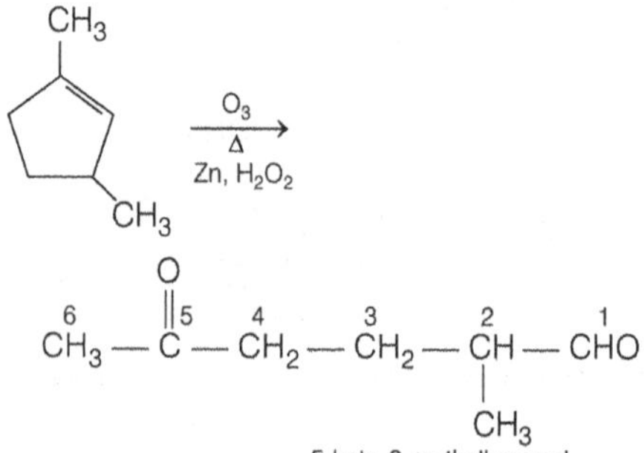

5-keto-2-methylhexanal

(c)

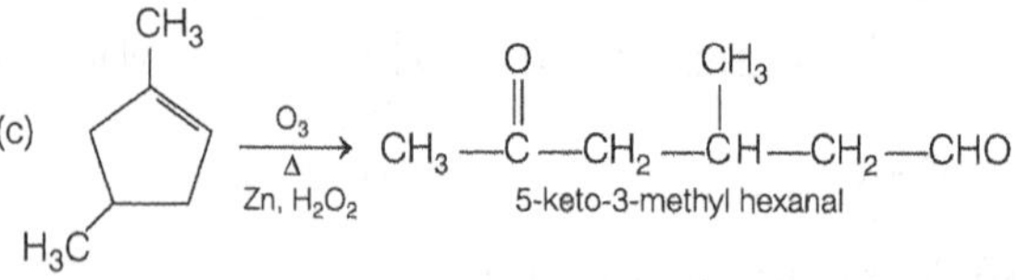

5-keto-3-methyl hexanal

(d)

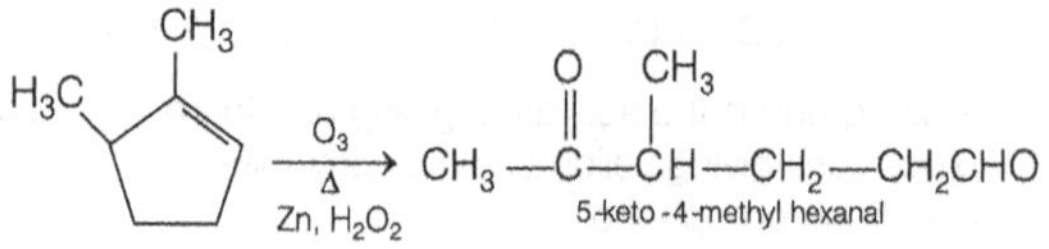

5-keto-4-methyl hexanal

26 $R-CH_2OH \xrightarrow{PCC} R-CH=O$

Pyridinium chlorochromate is the mild oxidising agent which causes conversion of alcohol to aldehyde stage. While others causes conversion of alcohol to acid.

27 $BaSO_4$ decreases the activity of Pd.

$$RCOCl + H_2 \xrightarrow{Pd/BaSO_4} RCHO + HCl$$

28 Formation of aldehyde from alkyl cyanide takes place by Stephen's reaction.

$$\underset{\text{Alkyl cyanide}}{RCN} + 2H \xrightarrow{SnCl_2+HCl} \underset{\text{Iminochloride}}{R\,CH=NH\cdot HCl}$$

$$\xrightarrow{H_2O} \underset{\text{Aldehyde}}{RCHO} + NH_4Cl + H_2O$$

29

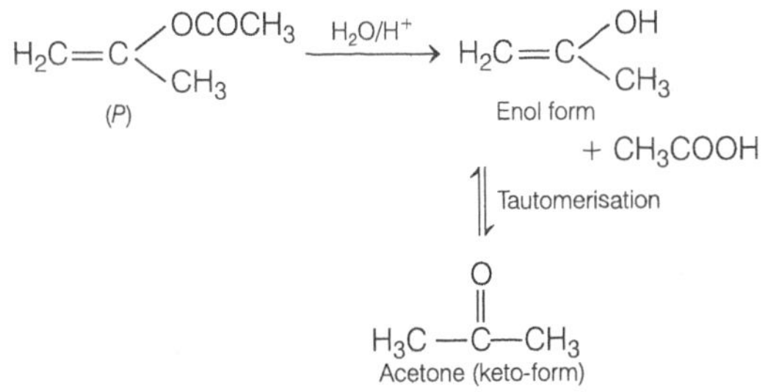

Fehling's solution is not reduced with acetone.

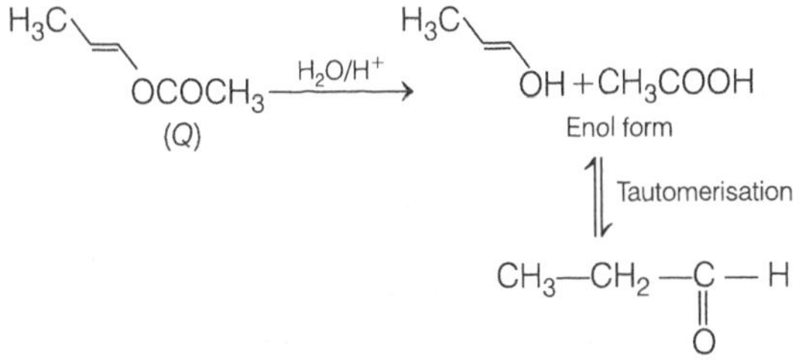

Fehling solution is reduced with aldehyde, i.e. CH_3CH_2CHO. Hence, the product of acid hydrolysis of *P* (ketone) and *Q* (aldehyde) can be distinguished by Fehling's solution.

30 All aldehydes including reducing sugar (as glucose, fructose) give silver-mirror test (with Tollen's reagent).

$$RCHO + 2[Ag(NH_3)_2]^+ + 3OH^- \xrightarrow{\Delta}$$

$$RCOO^- + \underset{\text{Silver mirror}}{2Ag\downarrow} + 4NH_3 + 2H_2O \; (R = H, CH_3)$$

Thus, (a) acetaldehyde and (c) formaldehyde give silver-mirror test.

31 Zn-Hg with HCl is called Clemmensen reagent. It reduces the $>C=O$ group of a ketone into $>CH_2$ group. Thus, the reaction is called Clemmensen reduction.

$$>C=O \xrightarrow[\text{Conc. HCl}]{\text{Zn-Hg}} >CH_2$$

32 $2CH_3COCH_3 \xrightarrow[-3H_2O]{\text{Conc. } H_2SO_4}$

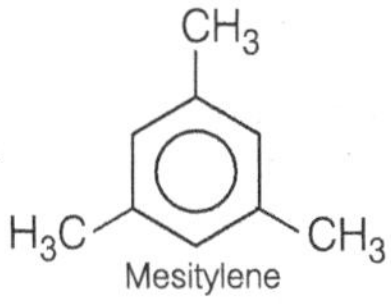

Mesitylene

33 $2CH_3CHO \xrightarrow[\text{Aldol condensation}]{\text{Dil. NaOH}}$

$$\underset{(A)}{CH_3-CH(OH)CH_2CHO} \xrightarrow[-H_2O]{\Delta} \underset{(B)}{CH_3-CH=CH-CHO}$$

34 The given reaction is an example of repeated aldol condensation followed by Cannizzaro reaction.

***Step* I** $CH_3CHO + OH^- \longrightarrow \bar{C}H_2-CHO + H_2O$

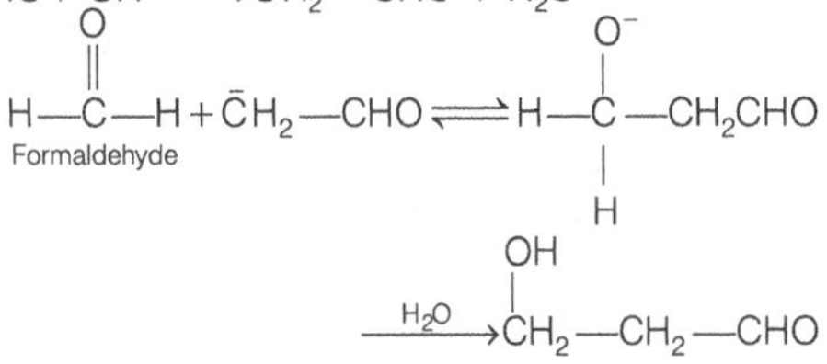

***Step* II** $HOCH_2CH_2CHO + HO^- \rightleftharpoons HOCH_2-\bar{C}HCHO + H_2O$

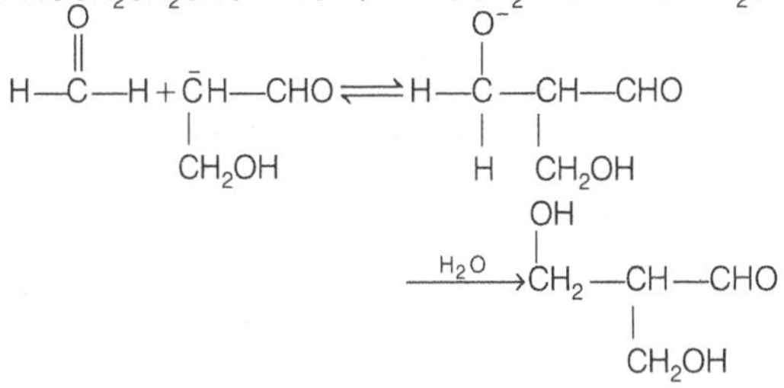

***Step* III** $HOCH_2-CH(CH_2OH)-CHO + HO^- \rightleftharpoons$

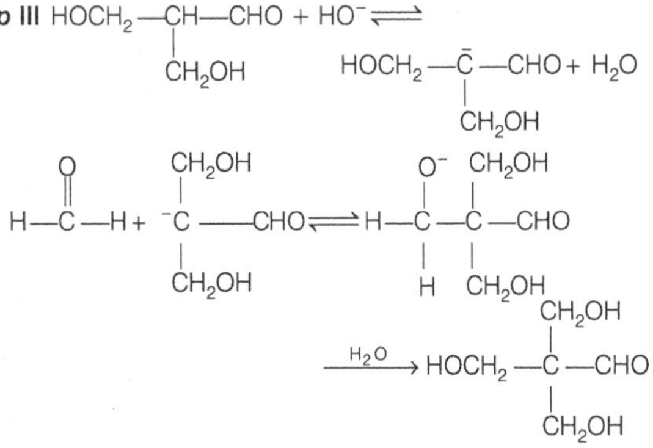

***Step* IV**

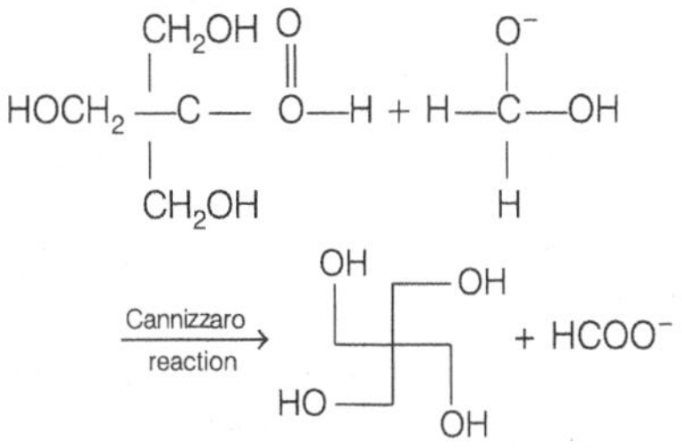

In the last step, formaldehyde is oxidised and the other aldehyde is reduced giving the desired product.

35 When two aldehydes or ketones (or carbonyl compounds) having α-hydrogen atoms react in the presence of a base, they result in the formation of β-hydroxy aldehyde or ketone called the aldol or ketol and the reaction is called aldol condensation.

Thus, 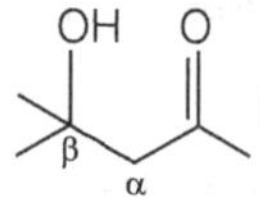 is the product of aldol condensation.

36 In Cannizzaro reaction, the transfer of H^- to another carbonyl group is difficult.

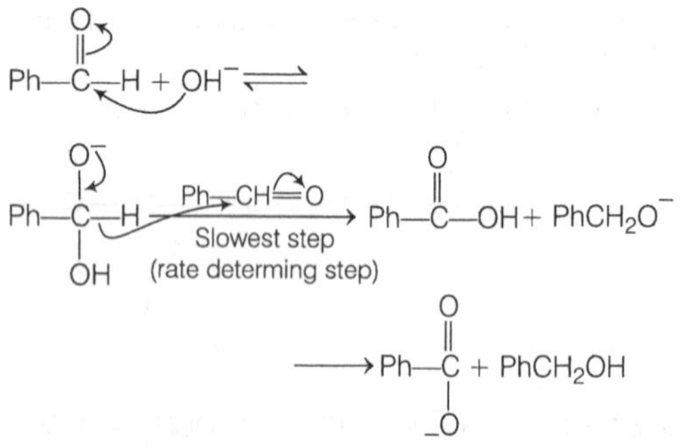

37 Only those aldehydes undergo Cannizzaro reaction in the presence of concentrated base, which have lack of α-H atom. CH_3CHO contains three α-H atoms, that's why it does not undergo Cannizzaro's reaction, rather it undergoes aldol condensation.

38 $(HCOO)_2Ca \xrightarrow{\text{Dry distillation}} HCHO \xrightarrow[\text{Cannizzaro reaction}]{\text{Conc. KOH}} CH_3OH + HCO\overset{-}{O}\overset{+}{K}$

39 In phenyl magnesium bromide, phenyl is attached with that C-atom of carbonyl group which has low electron density (higher electropositive charge). In carbonyl compounds, aldehydes are more reactive, towards nucleophile in nucleophilic addition reactions because in ketones alkyl groups is present on both sides (due to $+I$ effect) which decreases the electropositive charge of carbon of carbonyl group.

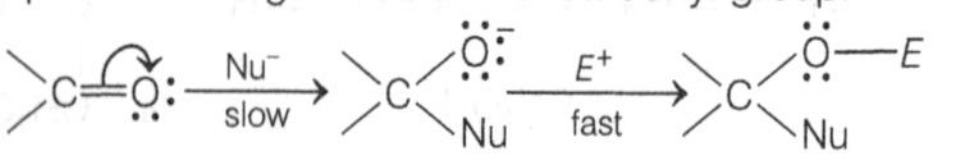

Hence, attraction of nucleophile decreases. Moreover in the tetrahedral intermediate, aldehydes have less steric repulsion than ketones and also the aldehyde increases the negative charge on oxygen less in comparison of ketones.

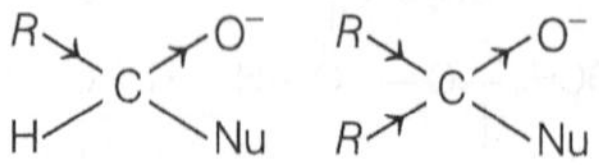

Therefore, the order of reactivity of acetone (I), acetaldehyde (II) and benzophenone (III) with PhMgBr is (II) > (I) > (III).

40 Iodoform reaction shown by the following compounds.

(i) CH_3CH_2OH (ii) CH_3CHO

(iii) All carbonyl compounds of the type $R-\overset{O}{\overset{\|}{C}}-CH_3$.

(iv) 2° alcohol $R-\underset{CH_3}{\underset{|}{C}}HOH$

Therefore, iodoform is not prepared from $CH_3\underset{CH_3}{\underset{|}{C}}HCH_2OH$.

41 HCHO and CH_3CHO give different reactions with NH_3.

$$6HCHO + 4NH_3 \longrightarrow \underset{\text{Urotropine}}{(CH_2)_6N_4} + 6H_2O$$

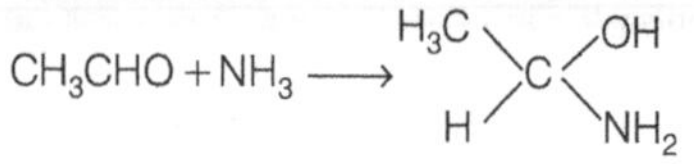

Acetaldehyde ammonia

42 With Fehling's solution, both acetone and benzaldehyde do not react. Only aliphatic aldehydes react with Fehling solution.

43 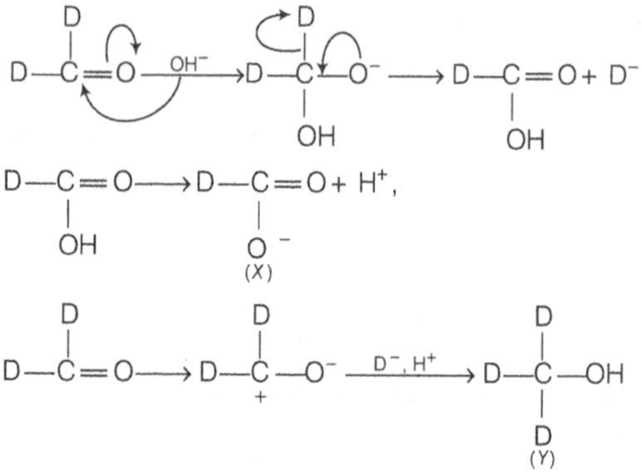

Thus, products are $D-\underset{O^-}{\underset{|}{C}}=O$ and CD_3OH.

44 This reaction is an example of Perkin's reaction because in it, α,β-unsaturated acid is obtained from aromatic aldehydes. Therefore, X is acetic anhydride, i.e.$(CH_3CO)_2O$.

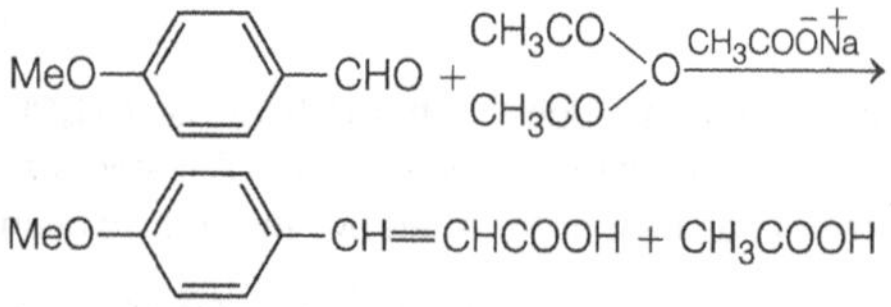

45 $CH_3CONHCH_3$ neither forms semicarbazone nor oxime because it is a substituted amide. While other compounds have carbonyl group, hence they form semicarbazone or oxime.

46 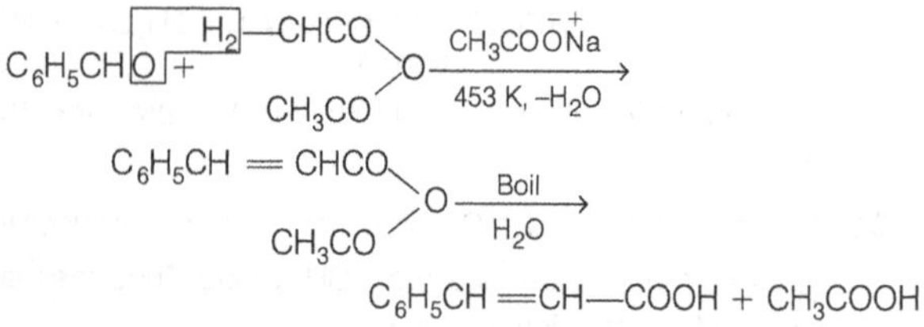

Cinnamic acid

47 $CH_3-\overset{O}{\overset{\|}{C}}-H \xrightarrow{I_2,\ NaOH} H-\overset{O}{\overset{\|}{C}}-O^- + \underset{\substack{\text{Iodoform} \\ \text{(yellow ppt)}}}{CHI_3 \downarrow}$

Formaldehyde does not form iodoform.

48

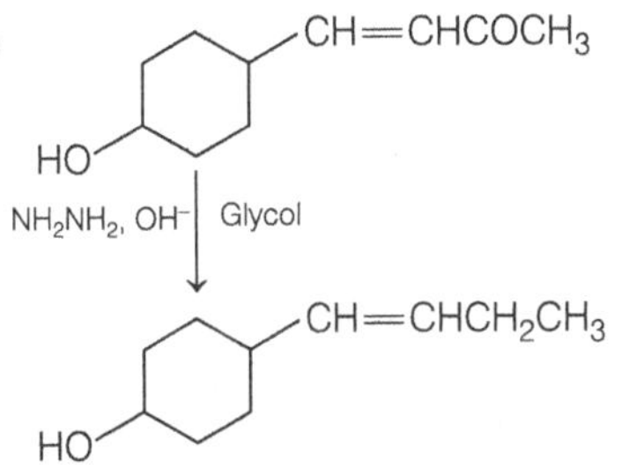

49 Cannizzaro's reaction is given by aldehydes ($RCHO$) lacking H at α-carbon or lacking α-carbon (as in HCHO). With NaOH, there is formation of acid salt ($RCOO^-$) by oxidation and alcohol (RCH_2OH) by reduction.

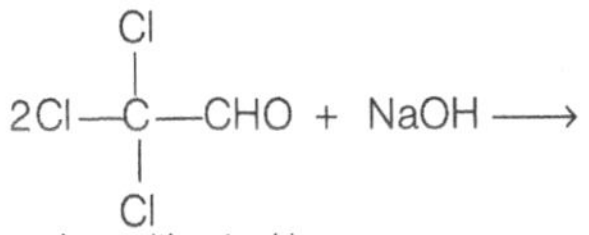

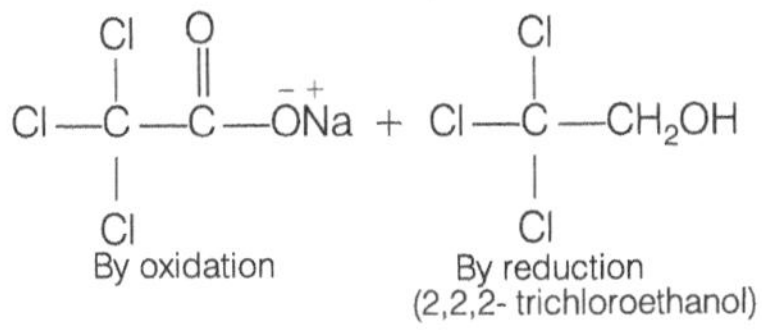

50

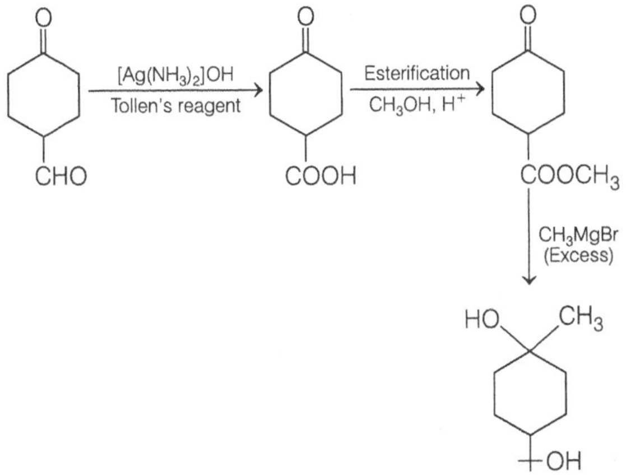

Before final product is formed, intermediate is

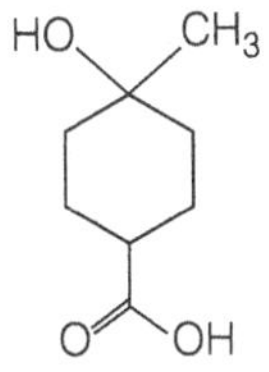

51 Compound *C* forms dimer due to hydrogen bonding and compound *A* also forms hydrogen bonds. Hence, in these, stronger hydrogen bonds are present which are more in compound *C* than compound *A*. Compound *B* does not form hydrogen bond. Hence, they follow the following order of molecular weight $C > A > B$. So, they have same order, in their boiling points.

52 Monocarboxylic acids and esters have the same general formula but different functional groups, so these are called functional isomers. e.g. $C_3H_6O_2$ can be CH_3CH_2COOH or CH_3COOCH_3.

53 A liquid + $C_2H_5OH \xrightarrow{\text{Conc. } H_2SO_4}$ Compound (Fruity smell)

Fruity smell is the characteristic property of ester, thus reaction can be considered as follows :

$$CH_3COOH + C_2H_5OH \xrightarrow{\text{Conc. } H_2SO_4} \underset{\substack{\text{Ethyl acetate} \\ \text{(fruity smell)}}}{CH_3—\overset{\overset{O}{\|}}{C}—OC_2H_5} + H_2O$$

54

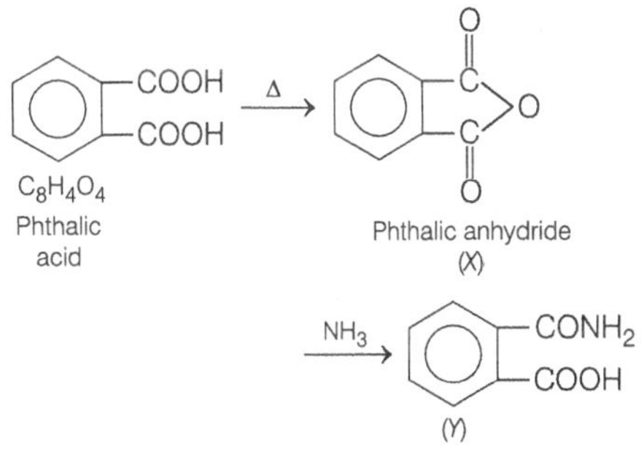

55 The reaction takes place as follows:

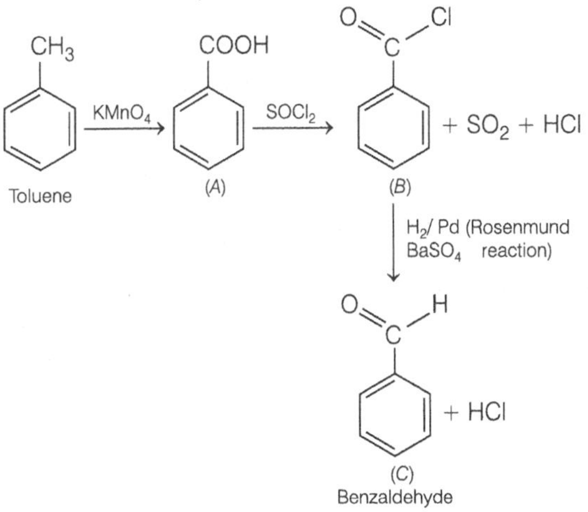

56

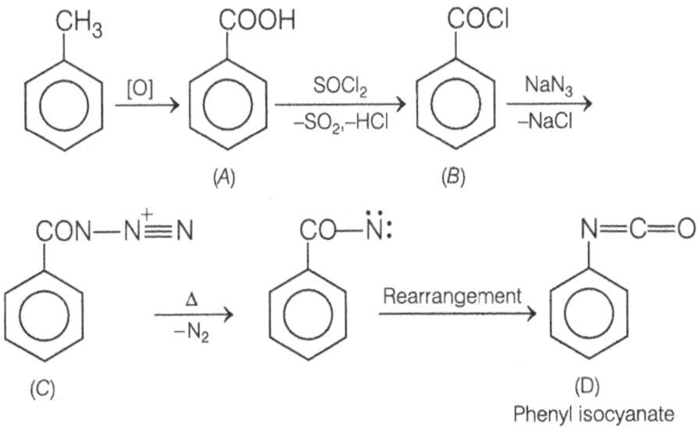

57 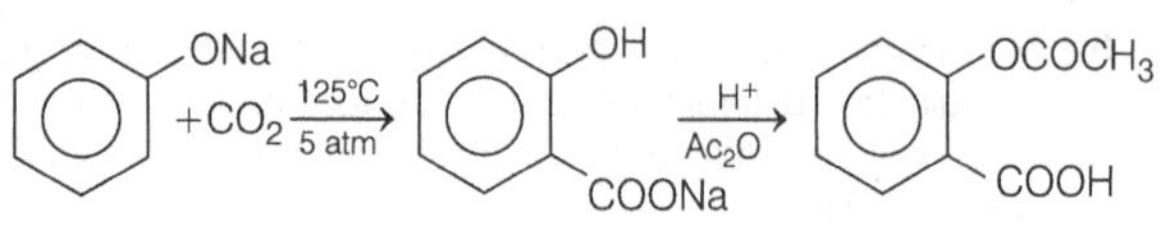

This reaction is known as Kolbe's synthesis.

58 $CH_3COOH \xrightarrow{LiAlH_4} \underset{(A)}{CH_3CH_2OH} \xrightarrow{PCl_5} \underset{(B)}{CH_3CH_2Cl} \xrightarrow[-HCl]{Alc.\,KOH} \underset{\text{(C) Ethylene}}{CH_2=CH_2}$

59 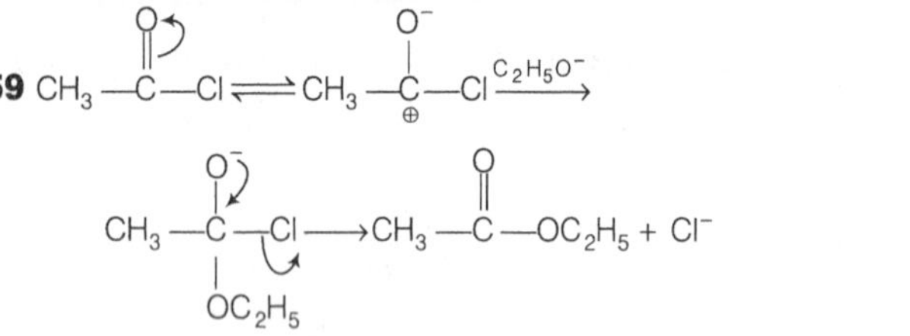

This is by S_N reaction. Cl^- is a better leaving group than $C_2H_5O^-$ and the ethyl ethanoate is formed.

60 $\underset{(A)}{CH_3CH_2-\overset{O}{\overset{\|}{C}}-OH} + NH_3 \longrightarrow \underset{(B)}{CH_3CH_2-\overset{O}{\overset{\|}{C}}-\overset{\ominus\oplus}{O}NH_4}$

$\xrightarrow[-H_2O]{\Delta} \underset{(C)}{CH_3CH_2-\overset{O}{\overset{\|}{C}}-NH_2} \xrightarrow{Br_2/KOH} CH_3CH_2NH_2$

61 It is a β -keto acid which undergoes decarboxylation in very mild condition, i.e. on simple heating. Ordinary carboxylic acid requires sodalime catalyst for decarboxylation (for decarboxylation β-keto acid must contain an α-H atom). Thus, 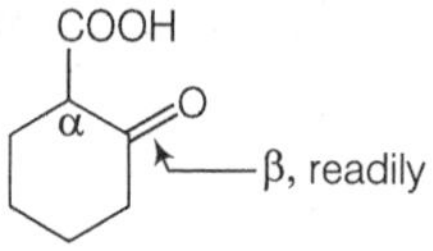 β, readily undergoes decarboxylation.

62 DIBAL-H (Di-isobutyl aluminium hydride) is a reducing agent with formula. This is generally used for the preparation of aldehydes. Using DIBAL —H, lactones are reduced directly to aldehydes.

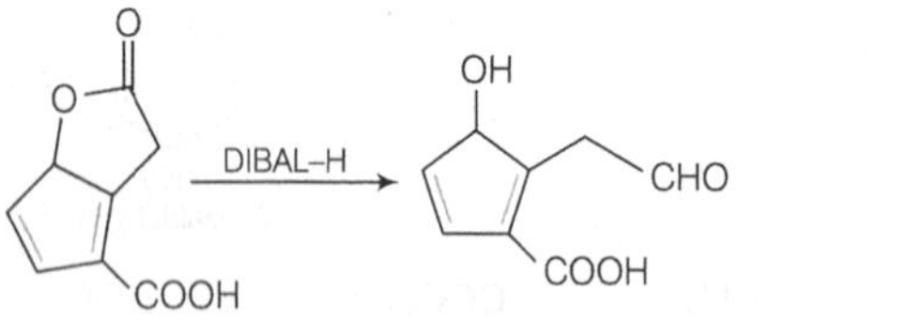

63

	Reagent	Phenol	Benzoic acid	Conclusion
(a)	Aqueous NaOH	Salt formation	Salt formation	No specific colour change
(b)	Tollen's reagent	No effect	No effect	
(c)	Molisch reagent	No effect	No effect	
(d)	Neutral $FeCl_3$	Violet colour	Buff-coloured precipitate	Thus, $FeCl_3$ can be used to make distinction.

64 Formaldehyde is a planar molecule as it contains sp^2 hybrid molecule.

65 **Correct Assertion** Bromination of benzene but not of phenol is carried out in presence of a Lewis acid.

66 Solubility of alcohols $\propto \dfrac{1}{\text{molecular weight}}$

As the molecular weight increases, the hydrophobic part (alkyl part) of alcohol increases. This reduces the solubility of alcohols.

67 The compound which loses a proton more readily is more acidic. *o*-nitrophenol due to intramolecular hydrogen bonding becomes more stable and does not lose a proton, so it is weaker acid than *p*-isomer.

68 Compounds having CH_3CO- or $CH_3-\underset{OH}{\underset{|}{C}}H-$ group undergo haloform reaction (*aq.* $NaOH + I_2$) but haloform reaction is given by only aldehydes, ketones and alcohols, so acetic acid does not undergo haloform reaction. In acetic acid α-hydrogen atoms are present.

69 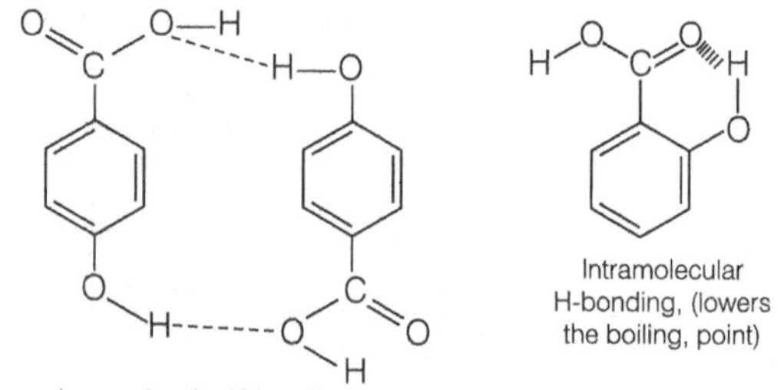

Intermolecular H-bonding, (higher boiling point)

Intramolecular H-bonding, (lowers the boiling, point)

SESSION 2

1 1, 4-dihydroxy benzene shows the highest boiling point among the given compounds because it forms strong intermolecular hydrogen bond

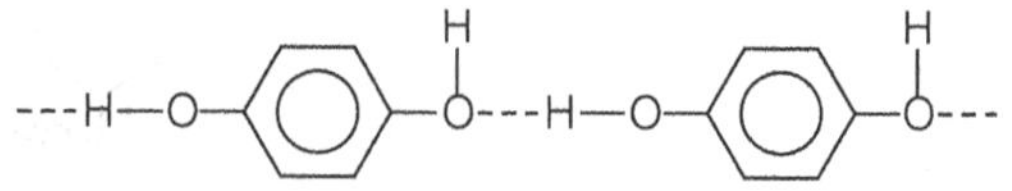

Intermolecular H-bonding

Order of H-bonding in *o*, *m* and *p*-isomers of a compound is given below

Intermolecular H-bonding $o < m < p$-isomers

Intramolecular H-bonding $o > m > p$-isomers

Hydroxy benzene do not form a chain of H-bonding. As the intermolecular H-bond is stronger than intramolecular H-bond, so the stability of 1, 4-dihydroxy benzene is highest. Hence, its boiling point is highest. The increasing order of the boiling points of the given compounds is

IV < I < II < III.

2 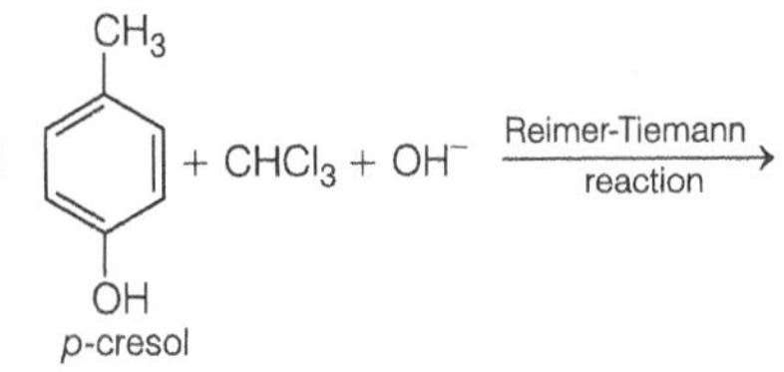

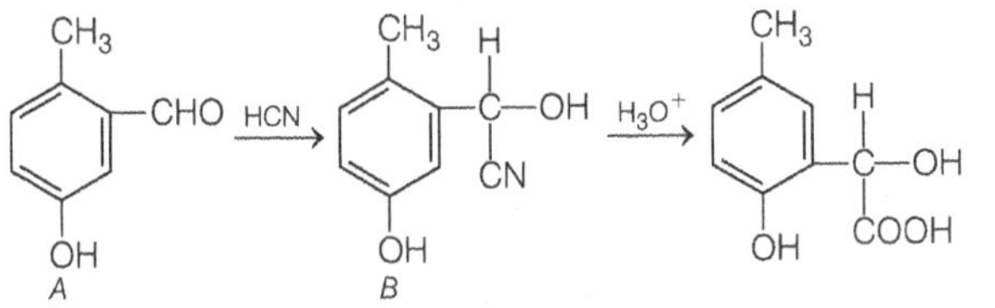

(—OH is more activating than —CH_3 in *o*, *p*-position, thus —CHO goes to *ortho* with respect to —OH.)

3 Given,

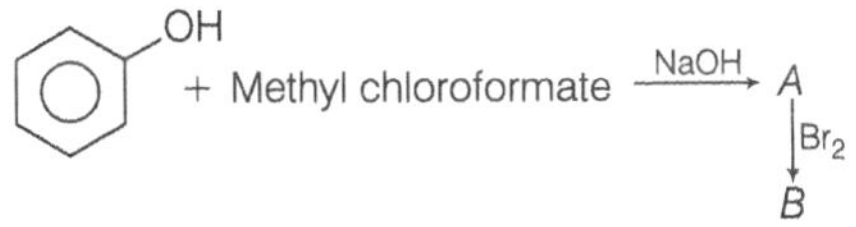

In the above road map, first reaction appears as acid base reaction followed by S_NAE (Nucleophilic substitution through Addition and Elimination). Both the steps are shown below

(i) Acid base reaction

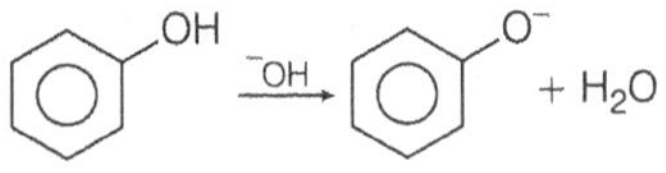

(ii) **S_NAE**

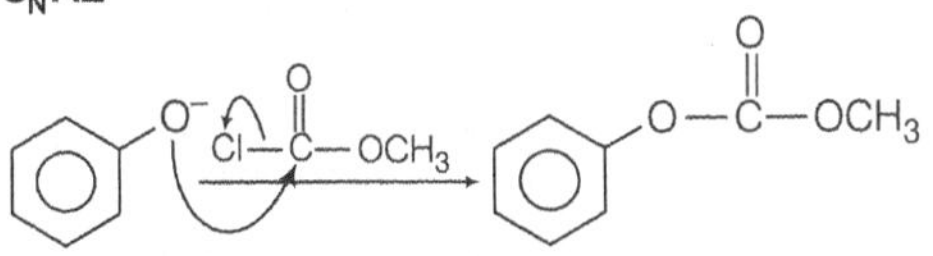

In the product of S_NAE, the attached group is *ortho* and *para*-directing due to following cross conjugation.

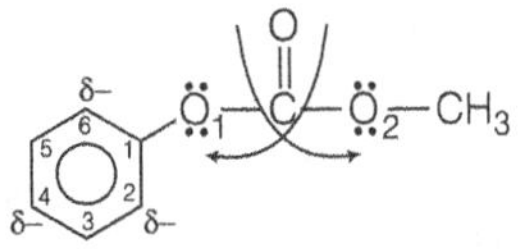

Cross conjugation due to which lone pair of oxygen 1 will be easily available to ring resulting to higher electron density at 2, 4, 6-position with respect to group. However from the stability point of view *ortho* positions are not preferred by substituents as group —O—C(=O)—O—CH_3 is bulky.

Hence, on further bromination of S_NAE product *para* bromo derivative will be the preferred product, i.e.

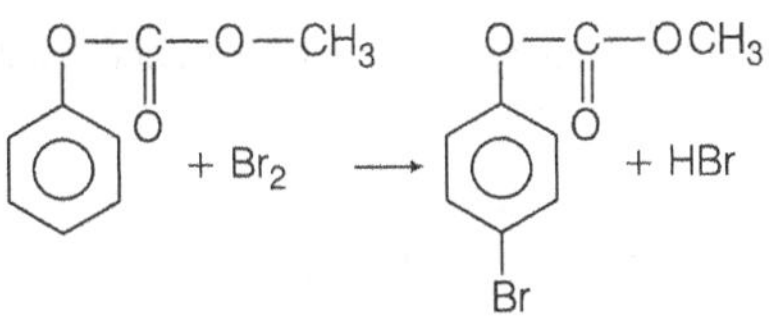

4 Molecular formula ($C_9H_{11}2O$) shows that it is a benzyl alcohol.

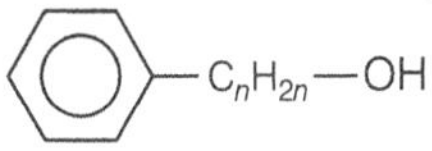

Hence, 4 isomers are

(i) 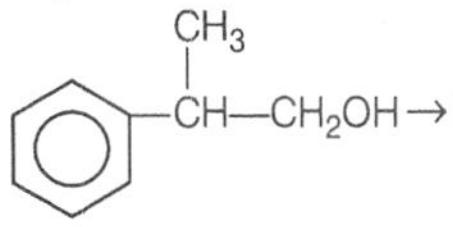

→ Does not respond to iodoform test

(ii) 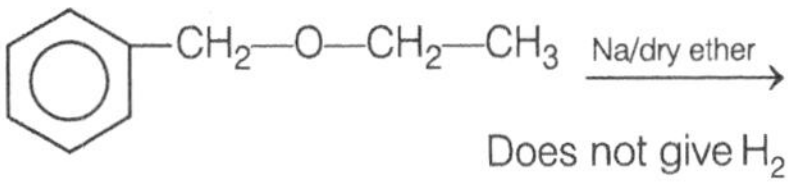

Does not give H_2

(iii) 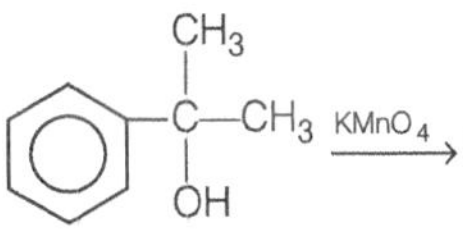

→ Does not give iodoform test

(iv) 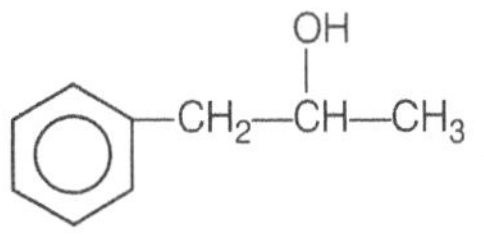

⟶ Fulfills all the conditions

Thus, the correct option is 'a'.

5 Molecular mass of silver salt of $E = \frac{108}{59.6} \times 100 = 181$

$RCOOAg = 181$

$R = 181 - (108 + 12 + 32) = 181 - 152 = 29$

Thus, $R = —CH_2—CH_3$ and hence, acid salt is $CH_3—CH_2—COOAg$.

Corresponding to this salt, acid is $CH_3—CH_2—COOH$, which is obtained by oxidation of (*C*), an ozonolysis product. Thus, (*C*) is $CH_3—CH_2—CH{=}O$. Since, (*C*) and (*D*) are isomers, *D* is

$CH_3—C(CH_3){=}O$. On the basis of (*C*) and (*D*), we can get product (*B*) and finally (*A*).

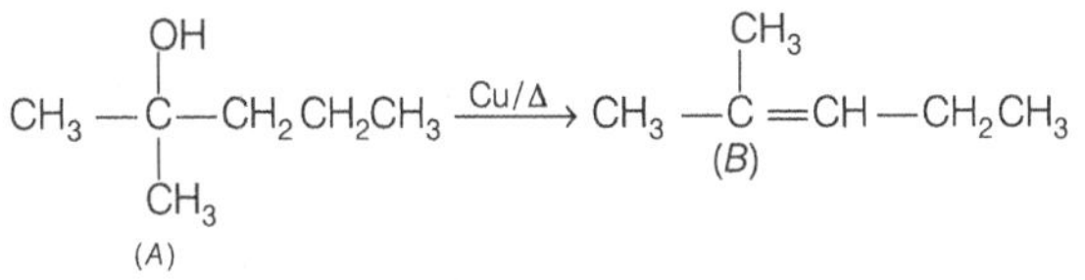

6 (a) *T* is an ester, so get hydrolysed.

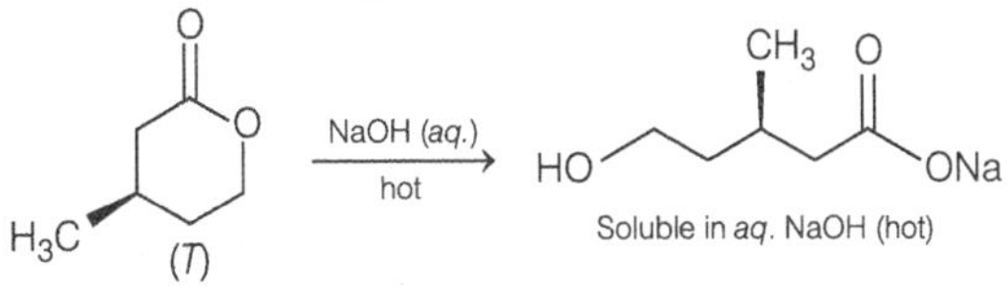

(b) (*T*) + $LiAlH_4$ ⟶ 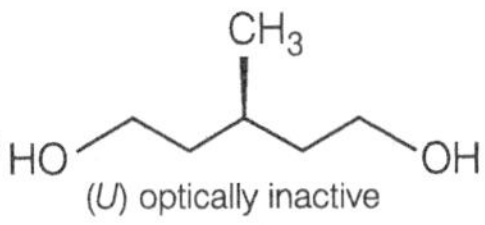

+ CrO_3/H^+ ⟶ 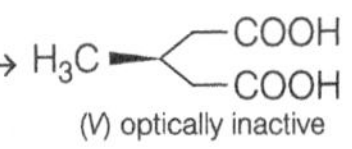

(c) $(U) + (CH_3CO)_2O \longrightarrow$ 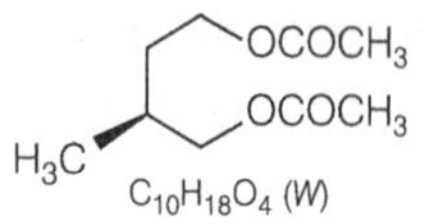

$C_{10}H_{18}O_4$ (W)

(U) On treatment with excess $(CH_3CO)_2O$ forms a diester (W).

(d) $(U) + CrO_3 \xrightarrow{H^+}$ 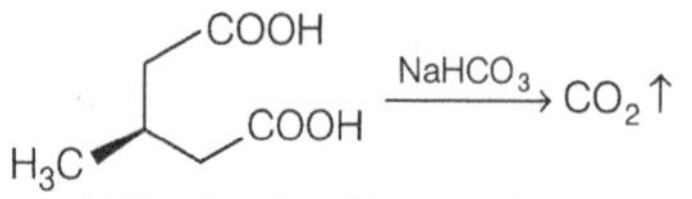

(V) Dicarboxylic acid

7 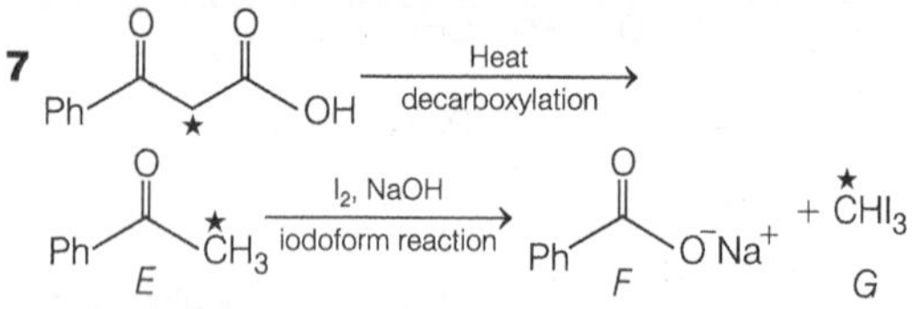

8 CH_3COCH_3 gives red colour with sodium nitroprusside solution but does not reduce Tollen's reagent. Acetone yields chloroform with NaOH/Cl_2. The chemical reaction followed as :

$$\underset{\text{Acetone}}{CH_3COCH_3} + Cl_2 \longrightarrow Cl_3C-COCH_3 \xrightarrow{NaOH} \underset{\text{Chloroform}}{CHCl_3}$$

9 $\underset{(A)}{CH_3-CH_2-\overset{O}{\overset{\|}{C}}-OCH_3}$

$\xrightarrow[\text{(1 equiv)}]{C_2H_5MgBr} \underset{(B)}{CH_3-CH_2-\overset{O}{\overset{\|}{C}}-CH_2CH_3}$

$\xrightarrow{LiAlH_4} \underset{(C)}{CH_3-CH_2-\overset{OH}{\overset{|}{C}H}-CH_2-CH_3}$ 2° alcohol

$\xrightarrow{H^+/HCl} \underset{(D)}{CH_3-\overset{Cl}{\overset{|}{C}H}-CH_2-CH_2-CH_3}$ (Rearranged product)

$\xrightarrow{Aq.\ KOH} \underset{(E)}{CH_3-\overset{OH}{\overset{|}{C}H}-CH_2-CH_2-CH_3}$

$\xrightarrow[H_2SO_4]{CrO_3} \underset{(F)}{CH_3-\overset{O}{\overset{\|}{C}}-CH_2-CH_2CH_3}$

10 (i) Molecular weight of (A), $M = \dfrac{1000\,K_f \times w}{\Delta T_f \times W}$

$\because w = 0.816\,g, W = 7.5\,g, \Delta T_f = 5.51 - 1.59 = 3.92,$

$\therefore \quad M = \dfrac{1000 \times 4.9 \times 0.816}{3.92 \times 7.5} = 136$

(ii) For 'A'

Element	%	Relative number of atoms	Simplest ratio
C	70.58	5.88	4
H	5.88	5.88	4
O	23.54	1.47	1

$\therefore$ Empirical formula of (A) = C_4H_4O

Empirical formula weight = 68

$\therefore$ Molecular formula of (A) is $(C_4H_4O)_2 = C_8H_8O_2$

(iii) Also given,

$$\underset{(A)}{C_8H_8O_2} \xrightarrow[-CO_2]{\text{Sodalime}} \underset{(B)}{C_7H_8} \xrightarrow[H^+]{[O]} \underset{(C)}{R-COOH} \xrightarrow{\text{Sodalime}} \underset{(D)}{\text{Benzene}}$$

(equiv. wt = 122)

Thus, (A) should be

$C_6H_4\begin{smallmatrix}CH_3\\COOH\end{smallmatrix}$ or $C_6H_5CH_2COOH$

o, m, p-toluic acid — Phenyl acetic acid

$\therefore$ Reactions are as follows:

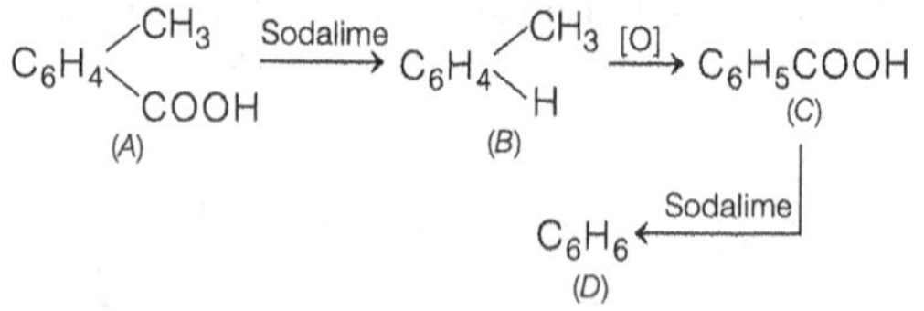

$$\underset{(A)}{C_6H_5CH_2COOH} \xrightarrow{\text{Sodalime}} \underset{(B)}{C_6H_5CH_2CH_3} \xrightarrow{[O]} \underset{(C)}{C_6H_5CH_2COOH} \xrightarrow{\text{Sodalime}} \underset{(D)}{C_6H_6}$$

11 The reaction given is a nucleophilic substitution reaction in which cleavage at C—O bond is visible. The product formation can be visualised with the help of following analysis.

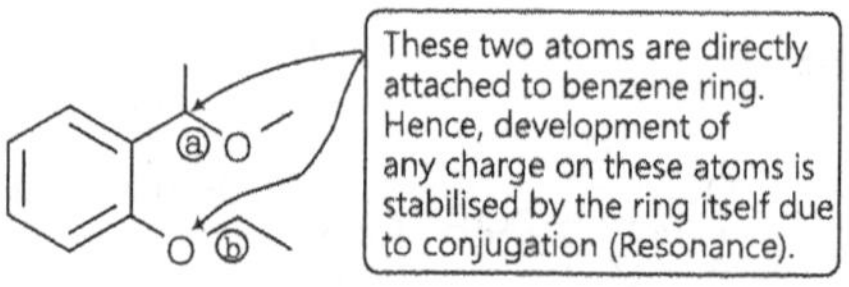

If any one properly visualise the fact written with figure above, than a conclusion can be made that C—O bonds marked (a) and (b) in the figure will undergo heterolysis during the reaction. The reaction can be represented as :

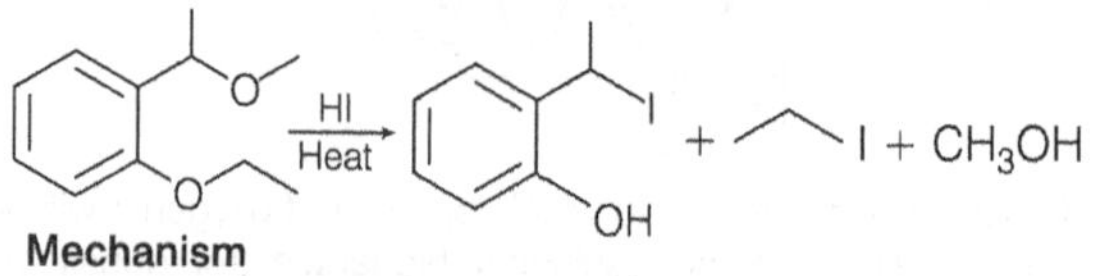

Mechanism

Step I The reaction begins with the attack of H^+ of HI on oxygen to form oxonium ion as:

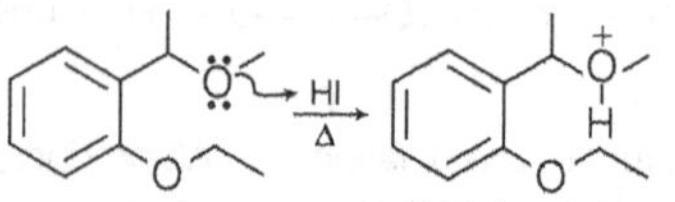

Oxonium ion

Step II This oxonium ion undergoes lysis and addition of I^- to form two products as:

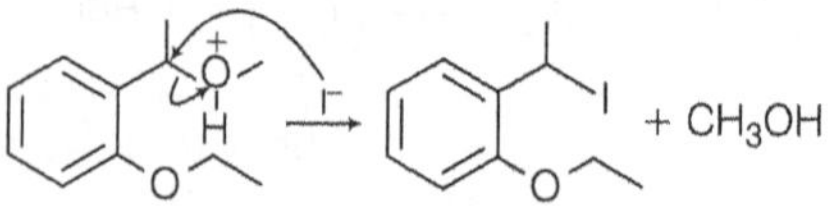

***Step* III** Similar pathway is followed at the other oxygen atom, which can be visualised as:

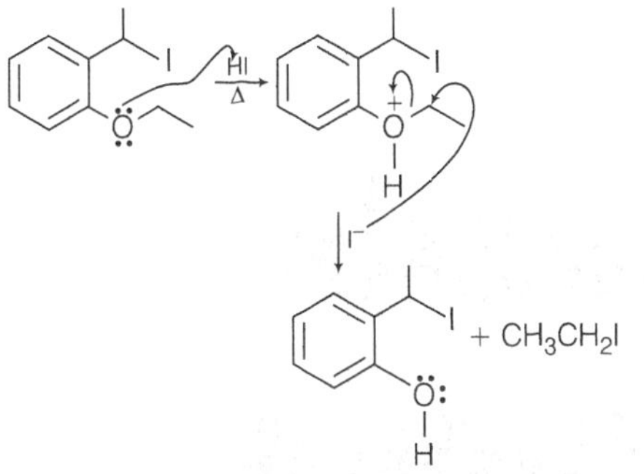

12. Here CH_3 will attack at C = O and dehydration will take place and again water attack at meta position as a nuleophile at the ring. PCC will convert —OH into =O.

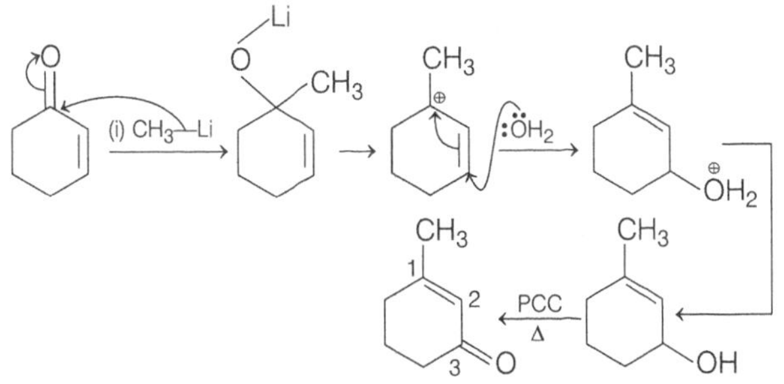

13.

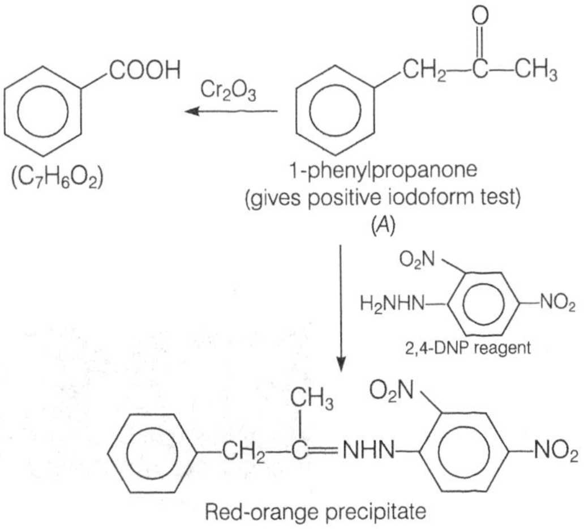

14. In 2,4,6-trinitrobenzoic acid, the decarboxylation takes place most easily, because of −*I* effect of nitro group, whereas in the dicarboxylic acid with one carbon atom having two carboxylic group it is also easier to remove CO_2. Hence, the order of ease of decarboxylation is

IV > III > II > I.

15

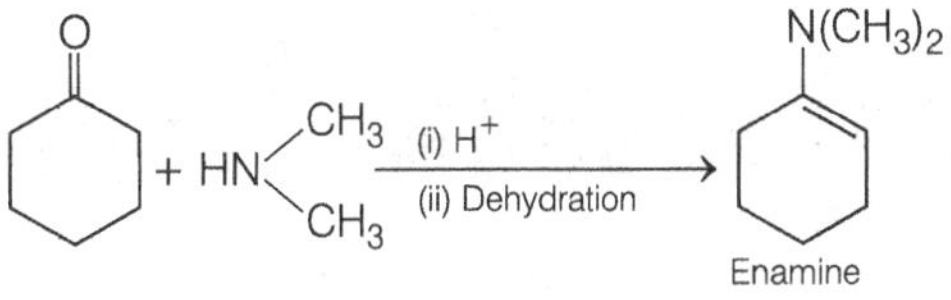

16

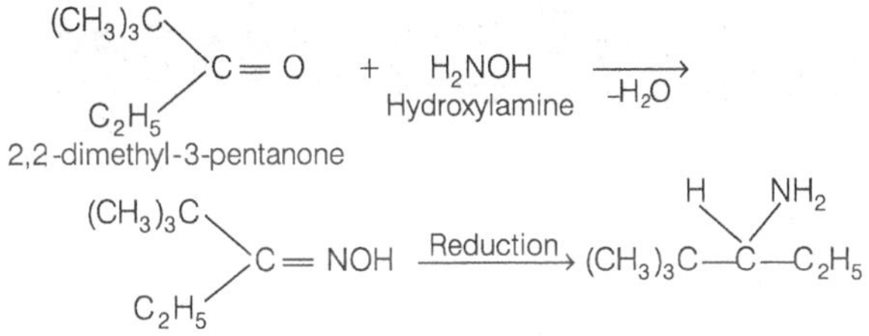

17

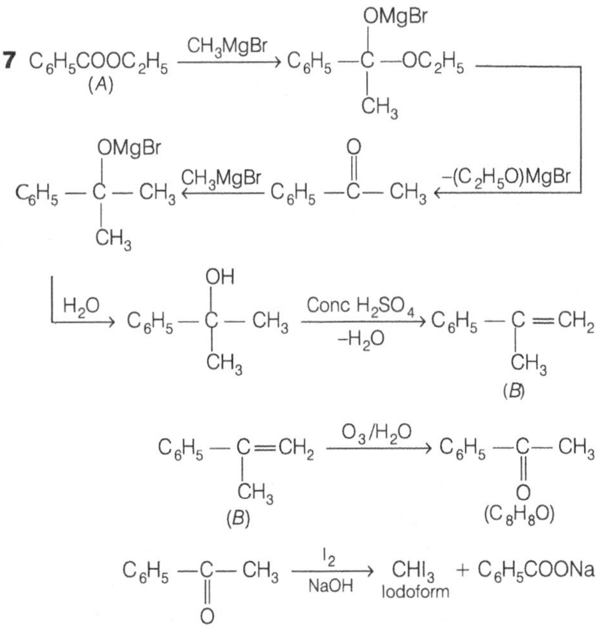

DAY THIRTY ONE

Organic Compounds Containing Nitrogen

Learning & Revision for the Day

- Amines
- Diazonium Salts

Amines

These are the derivatives of ammonia in which one, two or all the three hydrogen atoms are replaced by alkyl or aryl groups. Amines are classified as primary, secondary and tertiary according to the replacement of one, two or three hydrogen atoms from ammonia respectively.

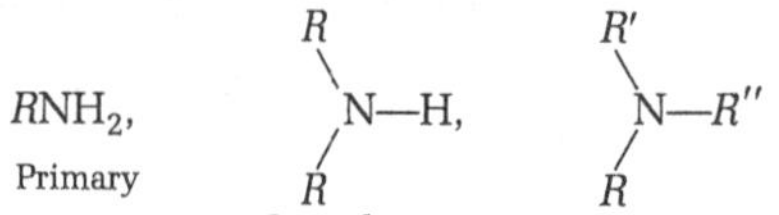

Primary, Secondary, Tertiary

- Nitrogen, in amines, contains four sp^3-hybridised orbitals. Out of these four, the three sp^3-hybridised orbitals of nitrogen overlap either with s-orbitals of hydrogen or sp^3- orbitals carbon depending upon the composition of amines. The fourth hybridised orbital of nitrogen in all amines has unshared pair of electrons.
  Due to the presence of lone pair of electrons, lone pair-bond pair repulsions increases due to which the bond angle C—N—E of nomenclature (where, E is C or H) 109.5° and shape is pyramidal.

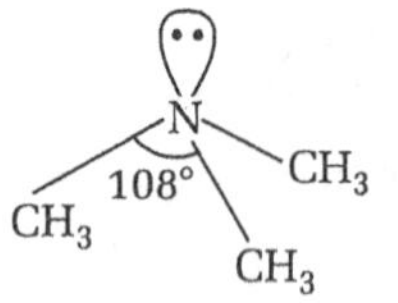

Pyramidal shape of trimethylamine

- In common system of nomenclature, an aliphatic amine is named by prefixing alkyl group to amine, i.e. alkylamine. In **IUPAC system**, amines are named as alkanamines, derived by replacement of 'e' of alkane by the word amine. While naming arylamine suffix 'e' of arene is replaced by 'amine'.

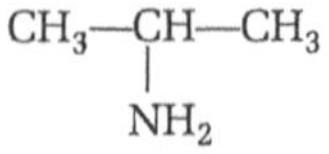

Propan-2-amine

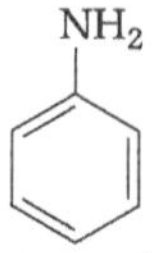

Benzenamine

PREP MIRROR
Your Personal Preparation Indicator

- No. of Questions in Exercises (x)—
- No. of Questions Attempted (y)—
- No. of Correct Questions (z)—
  (Without referring Explanations)

- Accuracy Level ($z/y \times 100$)—
- Prep Level ($z/x \times 100$)—

In order to expect good rank in JEE, your Accuracy Level should be above 85 & Prep Level should be above 75.

Preparation of Amines

Several methods can be used to prepare aliphatic amines, out of which some are specific for 1°, 2°, 3° amines while some gives mixtures of all the three. These are as follows:

(i) Alkyl nitriles, nitroalkanes, oximes of aldehydes, ketones and amides, all on reduction give primary amines. Reduction of nitriles to primary amines by using Na and alcohol is called **Mendius reaction.**

(a) 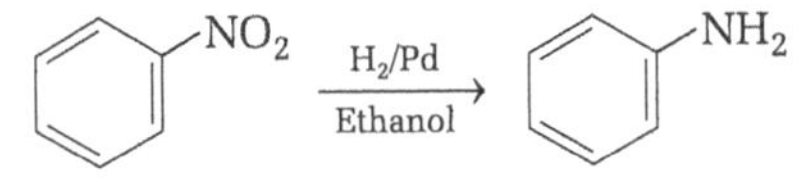

Nitrobenzene Aniline

(b) $R—NO_2 \xrightarrow[\text{or Fe + HCl}]{\text{Sn + HCl}} \underset{\text{Primary amine}}{R—NH_2}$

(c) $R—C\equiv N \xrightarrow[\text{Na(Hg) / } C_2H_5OH]{H_2/\text{Ni}} R—CH_2—NH_2$

(d) $\underset{\text{Ketoxime}}{R_2C{=}NOH} \xrightarrow[\text{or Na/}C_2H_5OH]{\text{LiAlH}_4\text{/ether}} R_2CHNH_2$

(e) $R—\overset{O}{\overset{\|}{C}}—NH_2 \xrightarrow[\text{(ii) } H_2O]{\text{(i) LiAlH}_4} R—CH_2—NH_2$ (R may be alkyl or aryl group)

NOTE $\underset{\text{Isocyanide}}{R—N\equiv C} \xrightarrow[\text{or Na (Hg) + } C_2H_5OH]{H_2/\text{Ni}} \underset{(2°\text{ amine})}{R—NH—CH_3}$

(ii) When alkyl halide is heated with alcoholic solution of ammonia in a sealed tube at about 393 K, a mixture of amines is obtained. This reaction is called **Hofmann's ammonolysis method.**

$$\ddot{N}H_3 + R—X \longrightarrow R—\overset{+}{N}H_3\overset{-}{X}$$

$$\underset{1°}{RNH_2} \xrightarrow{RX} \underset{2°}{R_2NH} \xrightarrow{RX} \underset{3°}{R_3N} \xrightarrow{RX} \underset{\text{Quaternary ammonium salt}}{R_4\overset{+}{N}\overset{-}{X}}$$

The free amine can be obtained from ammonium salt by treatment with a strong base.

$$R\overset{+}{N}H_3X^- + NaOH \longrightarrow R—NH_2 + H_2O + Na^+X^-$$

(iii) **Gabriel phthalimide synthesis** is used for the preparation of 1° amine. In this reaction, phthalimide on treatment with ethanolic KOH forms potassium salt of phthalimide which on heating with RX followed by either alkaline hydrolysis or hydrazinolysis with hydrazine ($H_2N \cdot NH_2$) produces the corresponding 1° amine. Primary aromatic amine (except those containing electron withdrawing group at *o*- and *p*-positions, e.g. —NO_2 group) cannot be synthesised by this method because ArX do not undergo S_N reaction with anion formed by phthalimide.

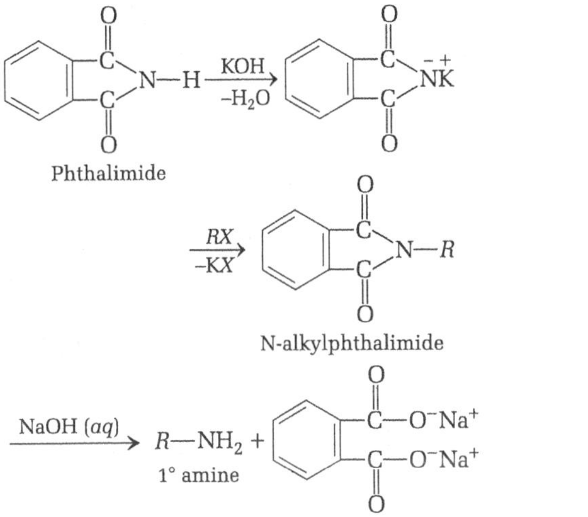

(iv) Amides on reaction with Br_2 in an aqueous or ethanolic solution of NaOH give 1° amine with one C-atom less than that present in the amide. This reaction is called **Hofmann-Bromamide degradation reaction.**

$$R—\overset{O}{\overset{\|}{C}}—NH_2 \xrightarrow[\text{or NaOBr}]{4\,NaOH + Br_2} R—NH_2 + Na_2CO_3 + 2\,NaBr + 2\,H_2O$$

(R = alkyl or C_6H_5)

(v) Carboxylic acid, when warmed with hydrazoic acid (N_3H) in the presence of concentrated sulphuric acid, gives a high yield of primary amines. This reaction is called **Schmidt reaction.**

$$R—\overset{O}{\overset{\|}{C}}—OH + N_3H \xrightarrow[\text{Conc. } H_2SO_4]{\Delta} \underset{1°\text{ amine}}{R—NH_2} + CO_2\uparrow + N_2\uparrow$$

Instead of N_3H and conc. H_2SO_4, a mixture of NaN_3 and conc. H_2SO_4 can also be used.

Physical Properties

Some important physical properties of amines are discussed below :

- The lower aliphatic amines are gases with fishy odour. Aniline and other arylamines are usually colourless but develop colour on keeping it in air for a long time due to atmospheric oxidation.
- Boiling points and solubility of isomeric amines decrease with branching, i.e.

$$\underset{\substack{1°\text{ amine}\\(2\text{H-bonding})}}{CH_3—CH_2—CH_2—NH_2} > \underset{\substack{2°\text{ amine}\\(1\text{ H-bonding})}}{CH_3—NH—CH_2—CH_3} > \underset{\substack{3°\text{ amine}\\(\text{No H-bonding})}}{(CH_3)_3N}$$

- Amines have lower boiling point than corresponding alcohols.

Chemical Properties

Amines are reactive and behave as nucleophiles due to the presence of one unshared pair of electrons. Some important chemical properties of amines are discussed below:

1. Basic Nature of Amines

- Amines are basic in nature, so react with acids and form salt. Larger the value of K_b or smaller the value of pK_b, stronger is the base.
  In $R\,NH_2$, $RN{=}CHR$ and RCN, nitrogen is sp^3, sp^2 and sp-hybridised respectively. More the s-character of hybridised orbital containing lone pair, lesser is the basic character. Hence, RNH_2 is most basic and RCN is least basic among these.
- The order of basic strength in case of methyl substituted amines and ethyl substituted amines in aqueous solution is as follows:

$$(C_2H_5)_2NH > (C_2H_5)_3N > C_2H_5NH_2 > NH_3$$
$$(CH_3)_2NH > CH_3NH_2 > (CH_3)_3N > NH_3$$

- Electron donating substituent such as $—CH_3$, $—OCH_3$ increases the basicity while electron withdrawing group such as $—NO_2$, $—COOH$ decreases the basicity.

NOTE In gaseous phase, the order of basicity would be $3°$ amine > $2°$ amine > $1°$ amine > NH_3.

2. Acylation

Aliphatic and aromatic primary and secondary amines react with acid chlorides, anhydride and esters by nucleophilic substitution reaction. This reaction is known as acylation. In this reaction, 1° amines give N-substituted amide, while 2° amines give N,N-disubstituted amides.

$$\underset{}{C_2H_5—NH_2} + \underset{\text{(Acyl chloride)}}{CH_3COCl} \xrightarrow{OH^-} \underset{\text{N-ethylethanamide}}{C_2H_5—NHCOCH_3} + HCl$$

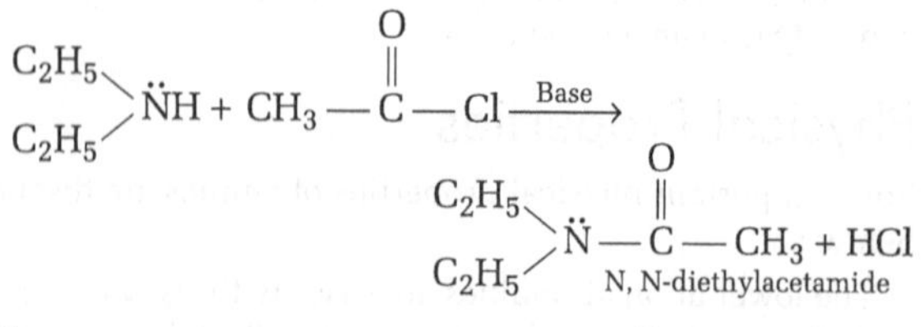

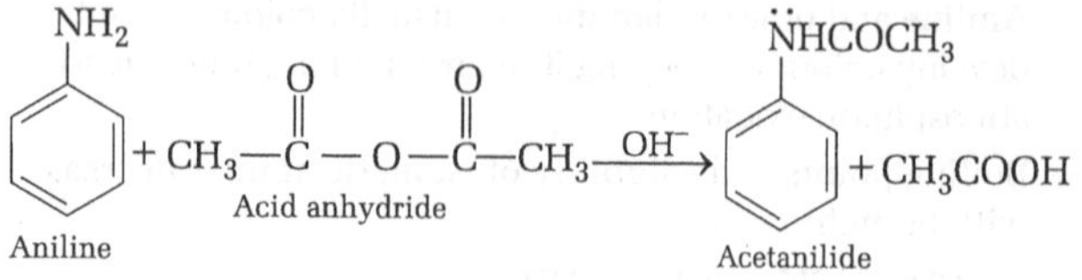

3. Carbylamine Reaction

Aliphatic and aromatic primary amines on heating with $CHCl_3$ and KOH(alc.) form isocyanide or carbylamine which is a foul smelling substance. Secondary ($R—NH—R$) and tertiary amines (R_3N) do not show this reaction. This reaction is known as carbylamine reaction and it is used as a test for primary amines.

$$R—NH_2 + CHCl_3 + 3KOH(alc.) \xrightarrow{\Delta} \underset{\text{Carbylamine}}{R—NC} + 3KCl + 3H_2O$$

4. Reaction with Nitrous Acid

Primary, secondary and tertiary amines react differently with nitrous acid which is prepared *in situ* from a mineral acid and sodium nitrite.

(i) Aliphatic primary amines when react with HNO_2, gives alcohol.

$$R—NH_2 \xrightarrow[NaNO_2 + HCl]{(HNO_2)} [R\overset{+}{N}_2\overset{-}{Cl}] \xrightarrow{H_2O} ROH + N_2\uparrow + HCl$$

(ii) Aromatic primary amines when react with HNO_2 at low temperatures, give diazonium salts.

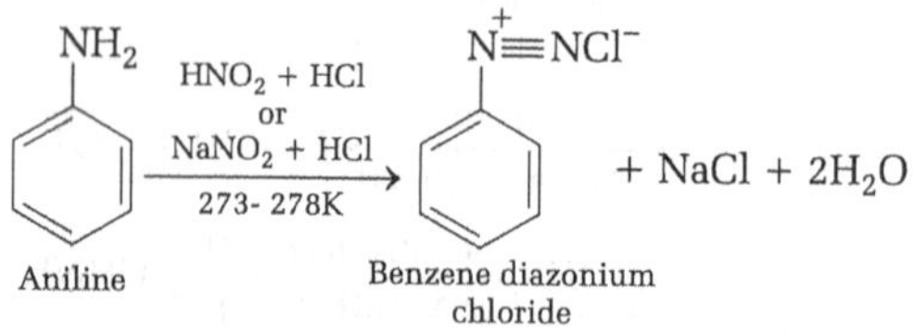

(iii) Secondary amines both aliphatic and aromatic on reaction with HNO_2 give an oily nitroso compound.

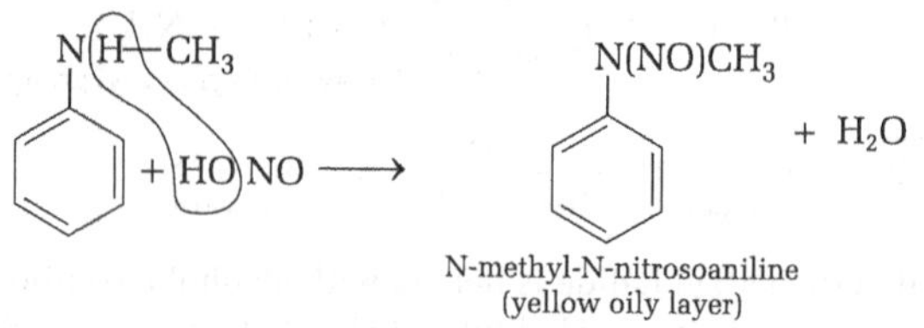

(iv) Tertiary aliphatic amines dissolve in a cold solution of nitrous acid to form water soluble nitrite salts which decompose on warming to give nitrosoamine and alcohol.

$$\underset{3°\text{ amine}}{(C_2H_5)_3N} + HONO \xrightarrow{\text{Cold}} \underset{\text{Trimethylammonium nitrite salt}}{[(C_2H_5)_3N^+H]\,NO_2^-}$$
$$\xrightarrow{\text{Warm}} (C_2H_5)_2N—N{=}O + C_2H_5OH$$

(v) Aromatic amines react with nitrous acid to form aliphatic diazonium salts at low temperatures (273-278 K).

$$C_6H_5—NH_2 \xrightarrow[273–278\,K]{NaNO_2 + 2HCl} C_6H_5—\overset{+}{N}_2\overset{-}{Cl} + NaCl + 2H_2O$$ This reaction is used to distinguish 1°, 2° and 3° amines.

5. Electrophilic Substitution Reaction

In electrophilic substitution reaction, an atom that is attached to an aromatic system is replaced by an electrophile. In case of aniline, $—NH_2$ group is *ortho* and *para* directing and a powerful activating group. These positions become the centre of high electron density.

Some of the most important electrophilic aromatic substitution, i.e. bromination, nitration, sulphonation etc., are given below:

(i) Due to strong activating effect of the amino group, halogenation (or bromination) of amines occur very fast.

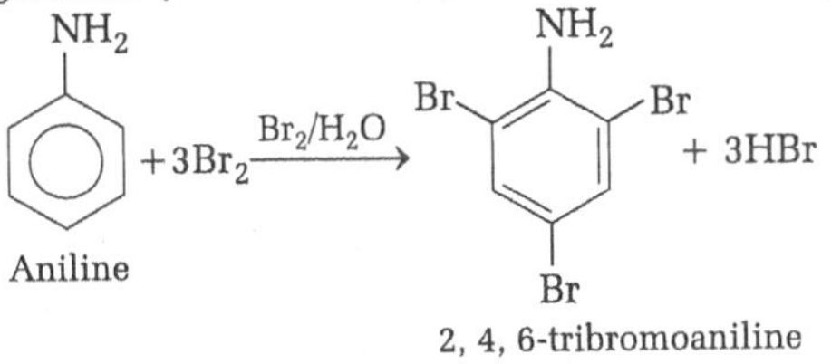

The main problem encountered during electrophilic substitution reactions of aromatic amines is that these have very high reactivity.

A monohalogenated (or monobrominated) product can be obtained by selective bromination. This can be done by protecting the —NH_2 group by acetylation with acetic anhydride then carrying out the desired substitution followed by hydrolysis of amide.

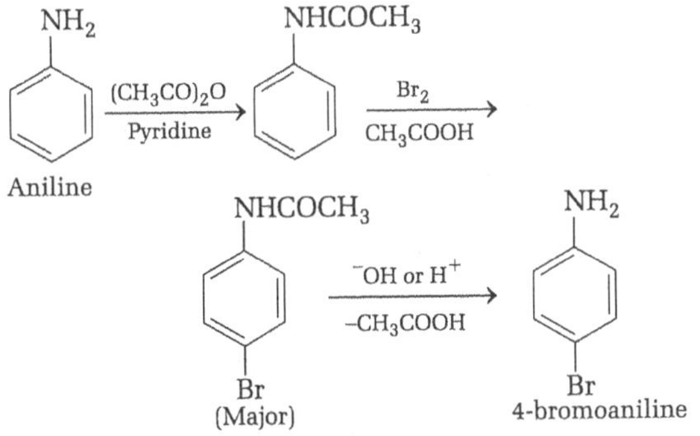

(ii) Aniline can be treated with nitric acid and sulphuric acid to give nitrobenzene *via* nitration process.

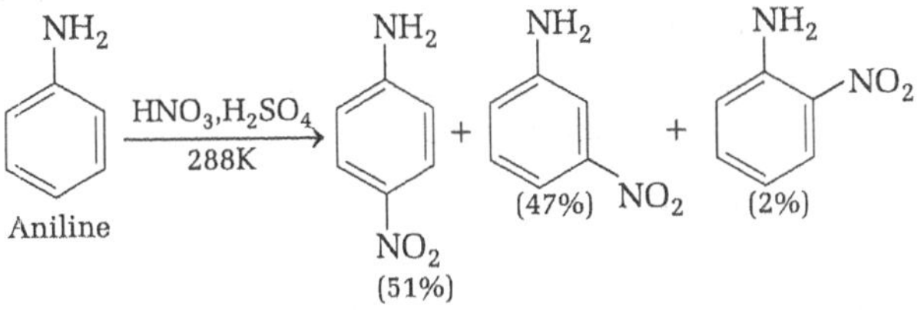

This is due to the fact that aniline gets protonated to form anilinium ion which is *meta*-directing.

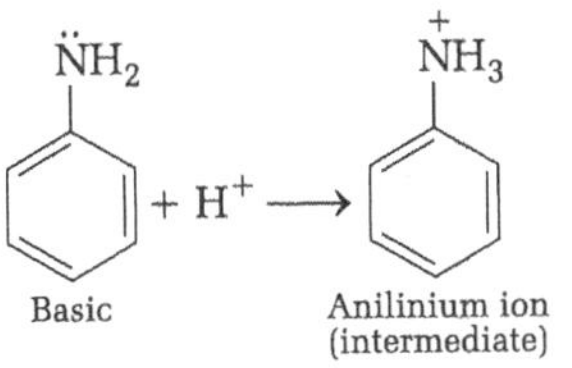

By protecting the —NH_2 group through acetylation with acetic anhydride, the nitration can be controlled and *p*-nitroaniline is obtained as a major product.

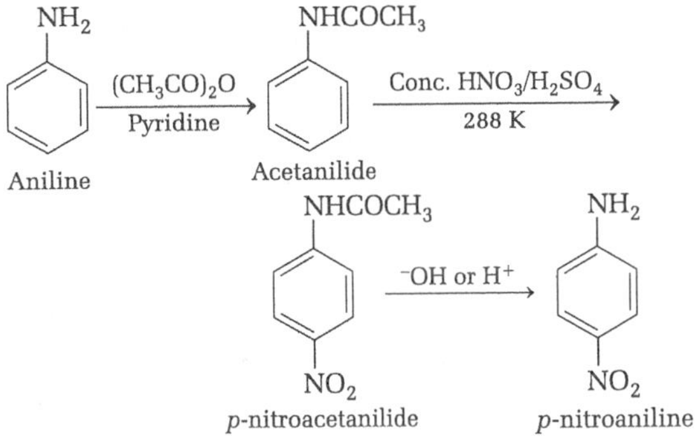

Aniline does not undergo Friedel-Craft's reaction (alkylation and acetylation) due to salt formation with $AlCl_3$ (anhy.), a Lewis acid, which is used as a catalyst.

(iii) Aniline can be treated with a conc. H_2SO_4 to give anilinium hydrogen sulphate which on heating with sulphuric acid at 453-473 K produces sulphanilic acid in a reversible reaction *via* sulphonation.

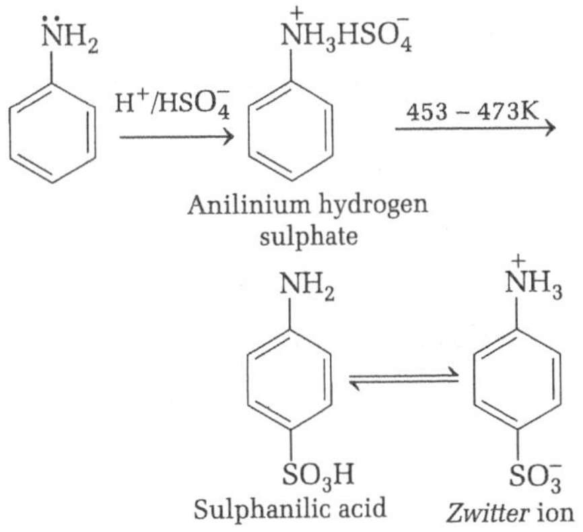

Identification of 1°, 2° and 3° Amines

Several reactions can be used to identify 1°, 2° and 3° amines. Some important test are as follows:

Hinsberg's Test

Distinction between primary, secondary and tertiary amines is made by $C_6H_5SO_2Cl$, which is known as Hinsberg's reagent.

(i) **Primary amines** give an insoluble mass which is soluble in alkali.

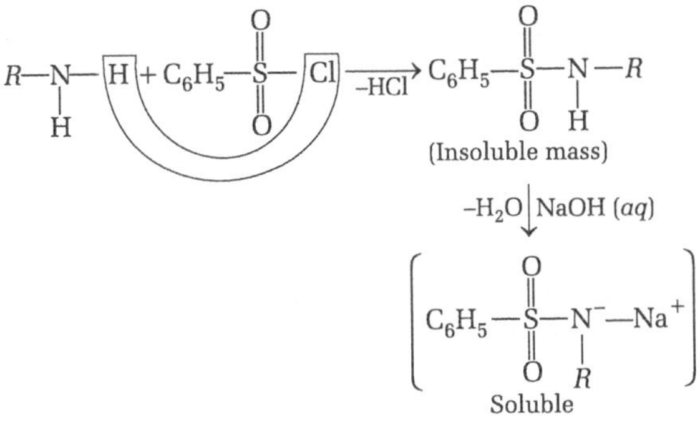

(ii) **Secondary amines** give an insoluble substance, which remains unaffected by alkali.

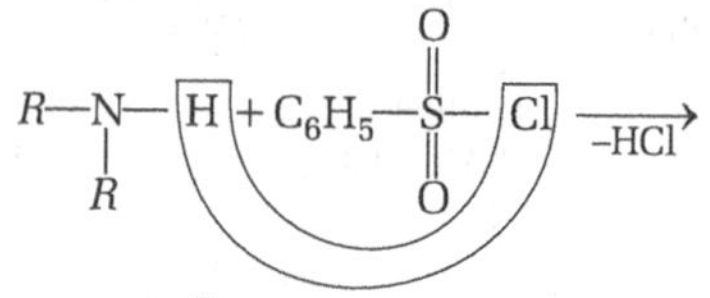

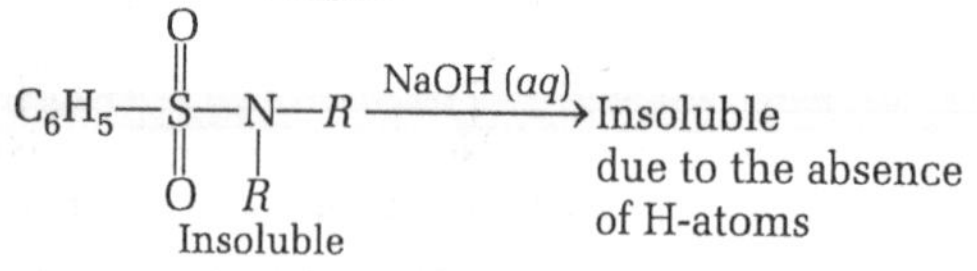

(iii) **Tertiary amines** do not react with benzene sulphonyl chloride because of the absence of replaceable H-atoms on N-atom.

Carbylamine reaction and reaction with nitrous acid can also be used to distinguish between 1°, 2° and 3° amines as discussed before in this chapter.

Uses of Amines

Amines are used as intermediates in drug manufacturing and as reagents in organic synthesis. Aromatic amines are used for manufacture of polymers, dyes and as intermediates for additives in rubber industry. Quaternary ammonium salts of long chain aliphatic tertiary amines are used as detergents.

Diazonium Salts

- Diazonium salts are obtained when primary aromatic amines react with nitrous acid. These salts are used in the preparation of azo dyes and a number of useful halogen substituted arenes.
- Benzene diazonium chloride is prepared by the reaction of aniline with nitrous acid at 273-278 K. The conversion of primary aromatic amines into diazonium salt is known as diazotisation.

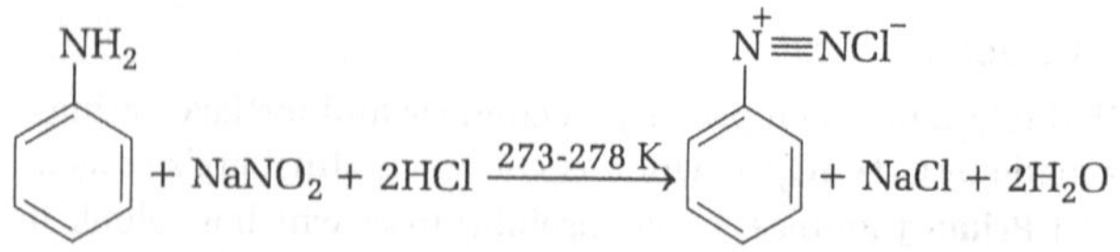

- Benzene diazonium chloride is a colourless crystalline solid. It is readily soluble in water. It is stable in cold but reacts with water when warmed.

Diazonium salts mainly show two types of reactions. These are as follows:

1. Reactions Involving Displacement of Diazo Group

The following reactions are involved in the displacement of diazo group in the benzene can be given as :

(i) Benzene diazonium chloride heated with cuprous chloride or bromide respectively dissolved in HCl or HBr yield chlorobenzene or bromobenzene, respectively. This reaction is called **Sandmeyer's reaction.**

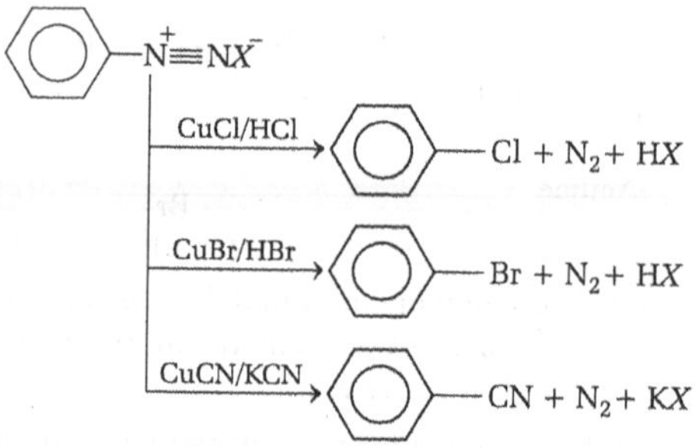

(ii) In **Gattermann reaction**, benzene diazonium chloride is warmed with copper powder and HCl, HBr and KCN to produce chlorobenzene, bromobenzene and cyanobenzene respectively.

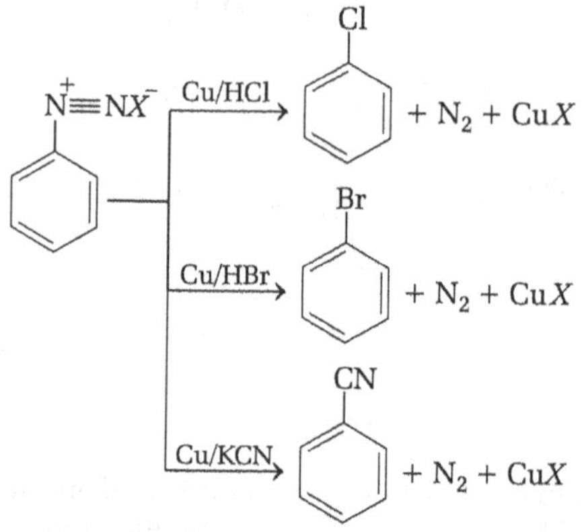

The yield in Sandmeyer's reaction is found to be greater than the Gattermann reaction.

(iii) Some other reactions of diazonium salt involving displacement of diazo group.

(a)

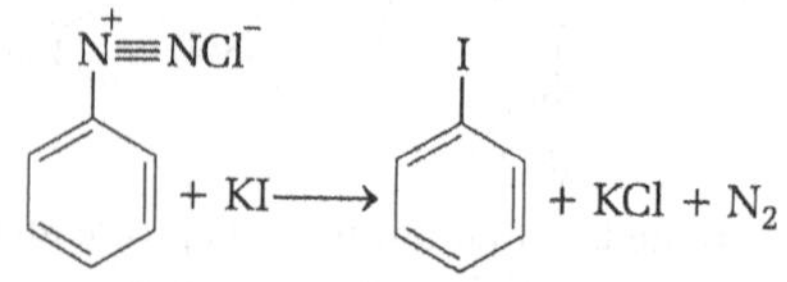

(b)

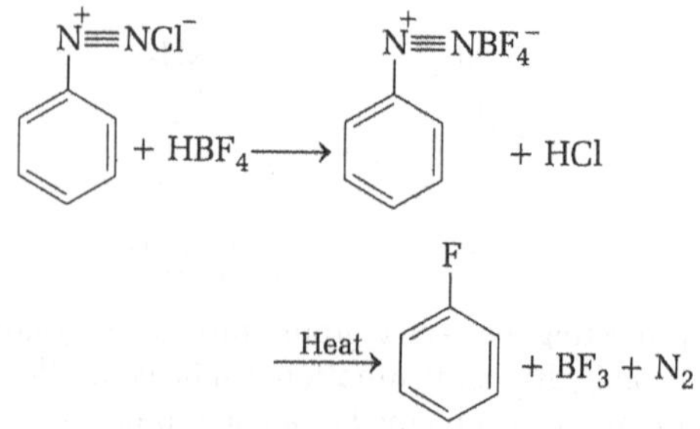

(c) 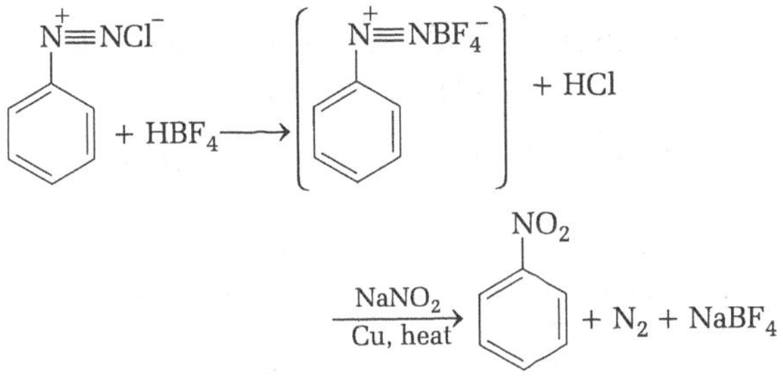

(d) 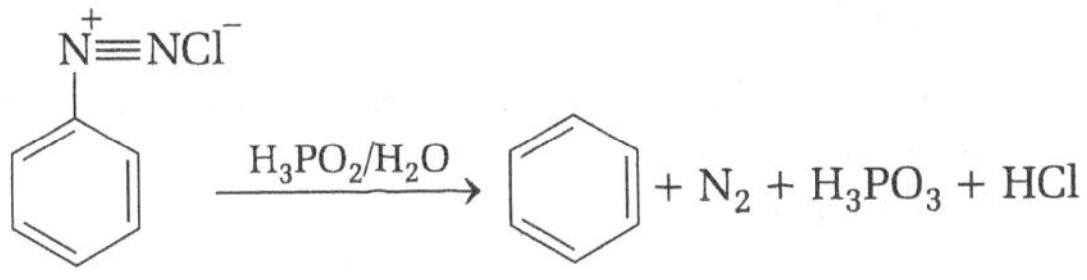

(e) 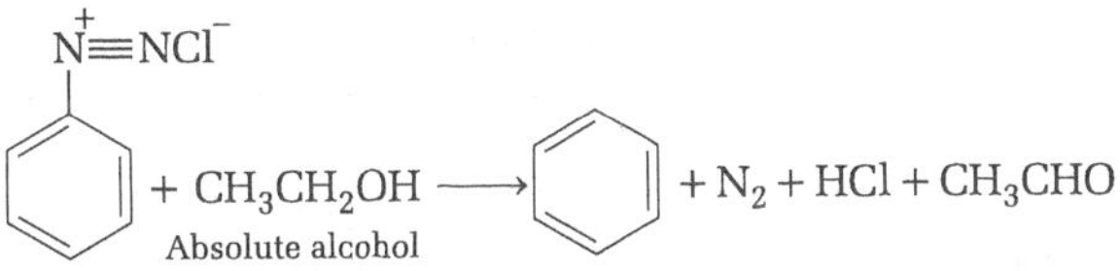

(f) 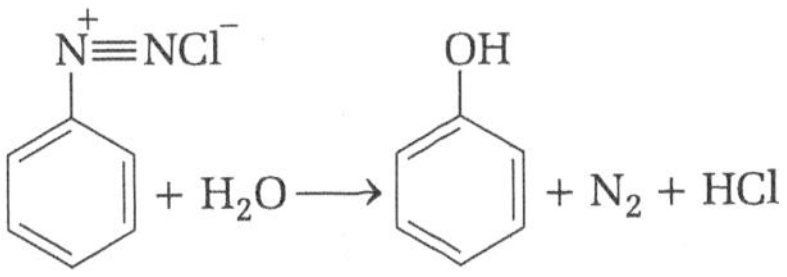

2. Reactions Involving Retention of Diazo Group

Coupling reactions are the example of electrophilic substitution reaction.

The azo products obtained have an extended conjugate system having both the aromatic rings joined through the —N=N— bond.

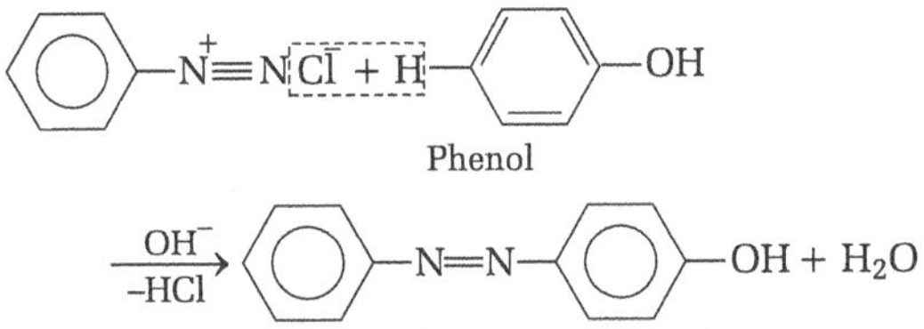

p-hydroxyazobenzene (Orange dye)

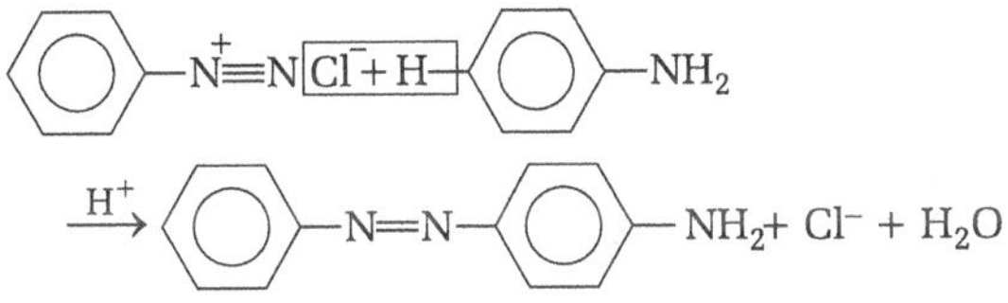

p-aminoazobenzene (Yellow dye)

Coupling with phenols is carried out in weakly alkaline medium (pH 9 to 10) because phenolate ion produced is coupled with diazonium salt more readily. Above pH 10, diazonium salt reacts with hydroxide ion to form diazotate ion which does not take part in coupling reaction.

Uses of Diazonium Salts

Diazonium salts are used for the preparation of a large number of aromatic compounds *viz* substitution and coupling compounds.

DAY PRACTICE SESSION 1

FOUNDATION QUESTIONS EXERCISE

1 Reduction of aromatic nitro compounds using Fe and HCl gives
(a) aromatic oxime (b) aromatic hydrocarbon
(c) aromatic primary amine (d) aromatic amide

2 In which of the following reaction, tertiary amine is obtained?
(a) Aniline $\xrightarrow{CH_3I}\xrightarrow{CH_3I}$ (b) Aniline $\xrightarrow{CH_3I}$
(c) Nitrobenzene $\xrightarrow{Sn/HCl}$ (d) None of these

3 The major product of the following reaction is

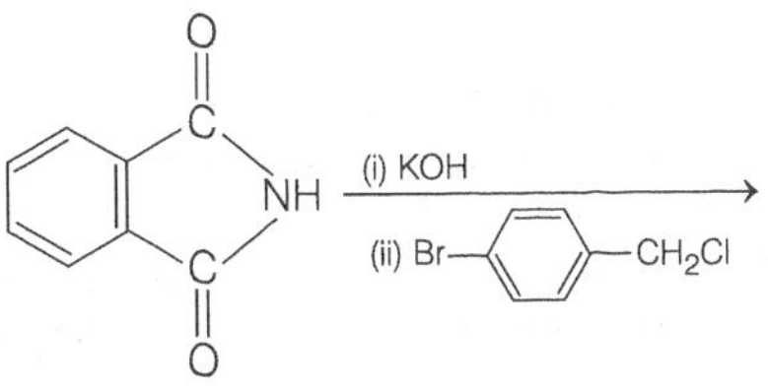

(a) 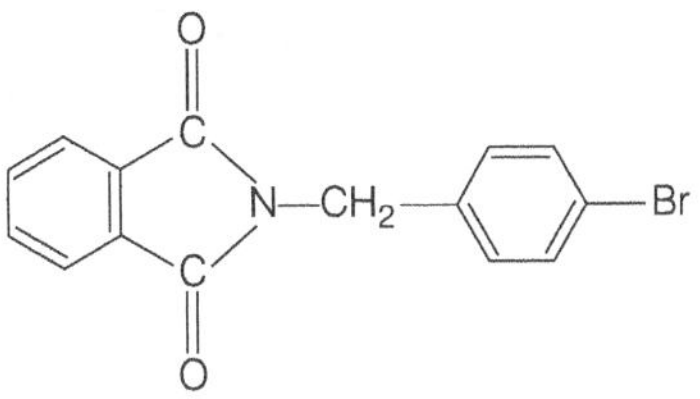

(b) 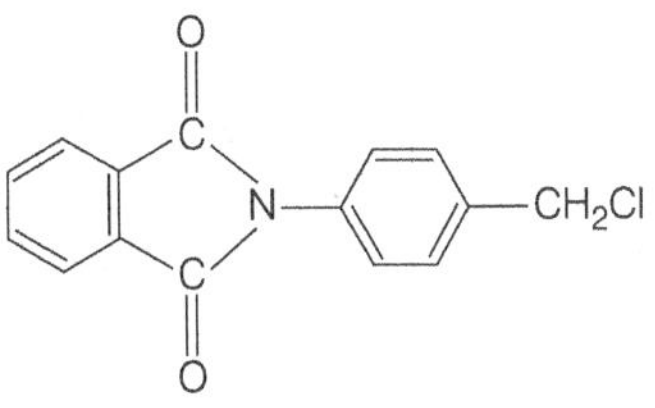

(c) 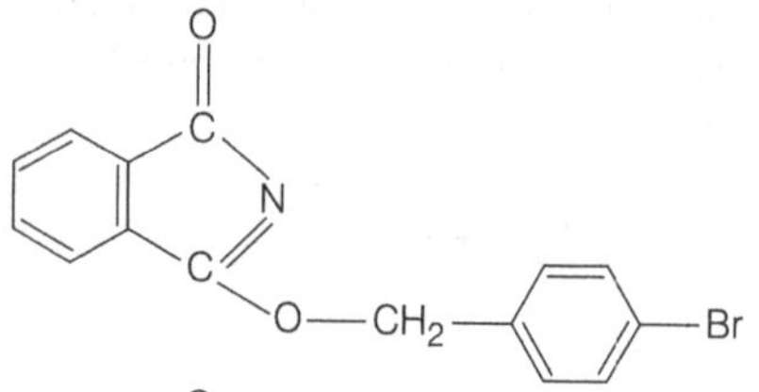

(d) 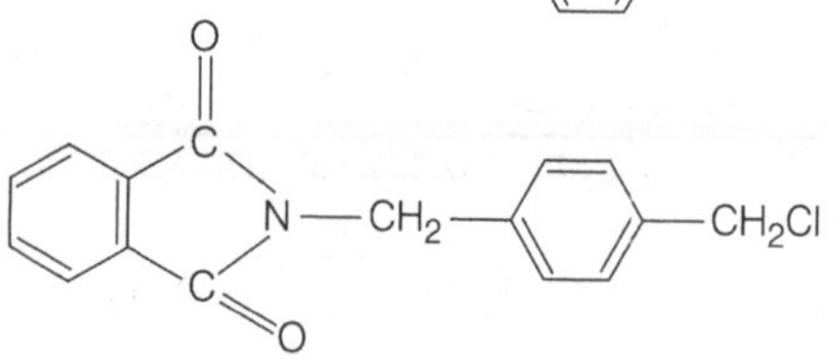

4 The best reagent for converting 2-phenylpropanamide into 2-phenylpropanamine is

(a) excess H_2
(b) Br_2 in aqueous NaOH
(c) iodine in the presence of phosphorus
(d) $LiAlH_4$ in ether

5 Acetamide is treated separately with the following reagents. Which one of these would give methylamine?

(a) PCl_5
(b) NaOH + Br_2
(c) Soda lime
(d) Hot conc. H_2SO_4

6 In the Hofmann-bromamide degradation reaction, the number of moles of NaOH and Br_2 used per mole of amine produced are → JEE Main 2016

(a) four moles of NaOH and two moles of Br_2
(b) two moles of NaOH and two moles of Br_2
(c) four moles of NaOH and one mole of Br_2
(d) one mole of NaOH and two moles of Br_2

7 The major product of the reaction between *m*-dinitrobenzene and NH_4SH is

(a) 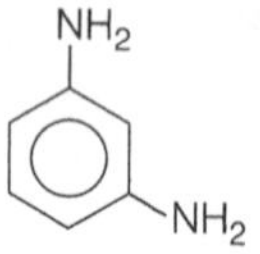
(b) 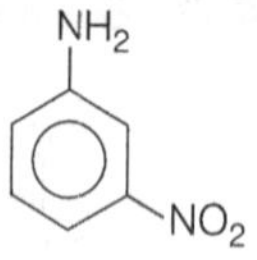
(c) 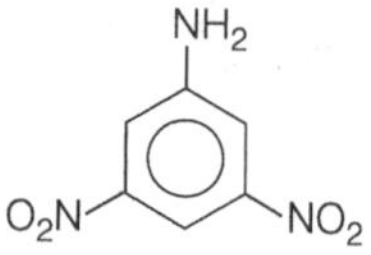
(d) 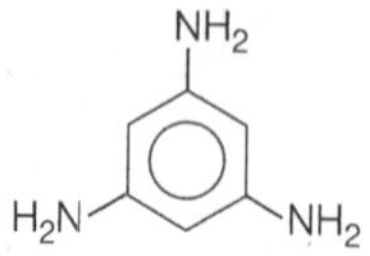

8 An organic compound *A* on reacting with NH_3 gives *B*. On heating, *B* gives *C*. *C* in the presence of KOH reacts with Br_2 to give $CH_3CH_2NH_2$. *A* is → Online JEE Main 2013

(a) CH_3COOH
(b) $CH_3CH_2CH_2COOH$
(c) $CH_3—CH(CH_3)—COOH$
(d) CH_3CH_2COOH

9 Which of the following should be most volatile?

I. $CH_3CH_2CH_2NH_2$ II. $(CH_3)_3N$
III. $\begin{matrix} CH_3CH_2 \\ CH_3 \end{matrix}\!\!>NH$ IV. $CH_3CH_2CH_3$

(a) II (b) IV (c) I (d) III

10 Amongst the following, the strongest base in aqueous medium is

(a) CH_3NH_2 (b) $NCCH_2NH_2$
(c) $(CH_3)_2NH$ (d) $C_6H_5NHCH_3$

11. Which one of the following is the strongest base in aqueous solution?

(a) Trimethyl amine (b) Aniline
(c) Dimethyl amine (d) Methyl amine

12. The order of basicity of amines in gaseous state is → JEE Main (Online) 2013

(a) $1° > 2° > 3° > NH_3$ (b) $3° > 2° > NH_3 > 1°$
(c) $3° > 2° > 1° > NH_3$ (d) $NH_3 > 1° > 2° > 3°$

13. Which of the following is the strongest base?

(a) 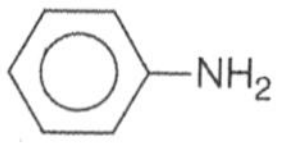
(b) 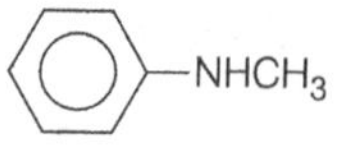
(c) 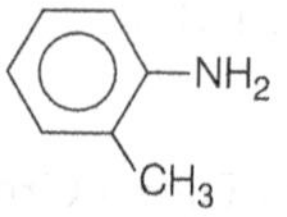
(d) 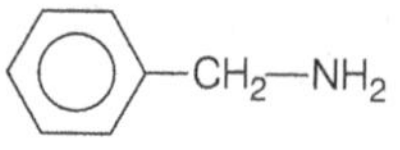

14. Considering the basic strength of amines in aqueous solution, which one has the smallest pK_b value? → JEE Main 2014

(a) $(CH_3)_2NH$ (b) CH_3NH_2
(c) $(CH_3)_3N$ (d) $C_6H_5NH_2$

15. The increasing order of basicity of the following compounds is

I. 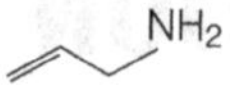
II. 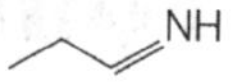
III. 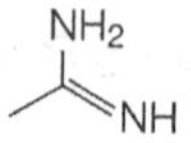
IV. 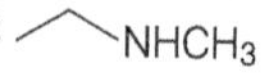

(a) I < II < III < IV (b) II < I < III < IV
(c) II < I < IV < III (d) IV < II < I < III

16 A compound with molecular mass 180 is acylated with CH_3COCl to get a compound with molecular mass 390. The number of amino groups present per molecule of the former compound is → JEE Main (Online) 2013

(a) 2 (b) 5
(c) 4 (d) 6

17 On heating an aliphatic primary amine with chloroform and ethanolic potassium hydroxide, the organic compound formed is → JEE Main 2014

(a) an alkanol (b) an alkanediol
(c) an alkyl cyanide (d) an alkyl isocyanide

18 Carbylamine forms from aliphatic or aromatic primary amine *via* which of the following intermediates?
→ JEE Main (Online) 2013

(a) Carbanion (b) Carbene
(c) Carbocation (d) Carbon radical

19 Which of the following is formed when RNH_2 reacts with $RCHO$?

(a) Hemiacetals (b) Acetals
(c) Ketals (d) Imines

20 The compound, which on reaction with aqueous nitrous acid at low temperature, produces an oily nitrosoamine, is

(a) diethylamine (b) ethylamine
(c) aniline (d) methyl amine

21 RNH_2 reacts with $C_6H_5SO_2Cl$ in aqueous KOH to give a clear solution. On acidification a precipitate is obtained which is due to the formation of

(a) $R—\overset{H}{\underset{H}{N^+}}—SO_2C_6H_5OH^-$ (b) $R—N^-SO_2C_6H_5K^+$
(c) $C_6H_5SO_2NH_2$ (d) $R—NH—SO_2—C_6H_5$

22 Which one of the following methods is neither meant for the synthesis nor for separation of amines?

(a) Curtius reaction (b) Wurtz reaction
(c) Hofmann method (d) Hinsberg method

23 Match the following and choose the correct option.

Column I	Column II
A. Ammonolysis	1. Amine with lesser number of carbon atoms
B. Gabriel phthalimide synthesis	2. Detection test for primary amines
C. Hofmann bromamide reaction	3. Reaction of phthalimide with KOH and $R—X$
D. Carbylamine reaction	4. Reaction of alkyl halides with NH_3

Codes

	A	B	C	D		A	B	C	D
(a)	2	1	3	4	(b)	4	3	1	2
(c)	3	2	4	1	(d)	4	1	3	2

24 On heating aniline with fuming sulphuric acid at 180°C, the compound formed will be

(a) aniline disulphate (b) aniline-2,4,6-trisulphonic acid
(c) sulphanilic acid (d) None of these

25 Nitration of aniline also gives *m*-nitroaniline in strong acidic medium because

(a) in electrophilic substitution reaction amino group is *meta*-directive
(b) inspite of substituents nitro group always goes to *m*-position
(c) in acidic (strong) medium aniline is present as anilinium ion
(d) None of the above

26 Which of the following will give only one monosubstituted product?

(a) *o*-dinitrobenzene (b) *m*-dinitrobenzene
(c) *p*-dinitrobenzene (d) None of these

27 Acetanilide on nitration followed by alkaline hydrolysis mainly gives

(a) *o*-nitroacetanilide (b) *p*-nitroaniline
(c) *m*-nitroaniline (d) 2, 4, 6-trinitroaniline

28 Which of the following compounds will give significant amount of *meta-product* during mononitration reaction?
→ JEE Main 2017

(a) 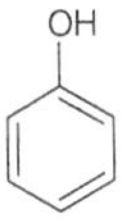

(b) 

(c) 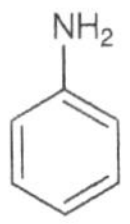

(d) 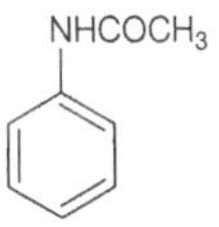

29 Benzene diazonium chloride reacts with hypophosphorous acid to produce

(a) benzene (b) phenol
(c) cyanobenzene (d) chlorobenzene

30 Fluorobenzene (C_6H_5F) can be synthesised in the laboratory

(a) by heating phenol with HF and KF
(b) from aniline by diazotisation followed by heating the diazonium salt with HBF_4
(c) by direct fluorination of benzene with F_2 gas
(d) by reacting bromobenzene with NaF solution

31 In the reaction,

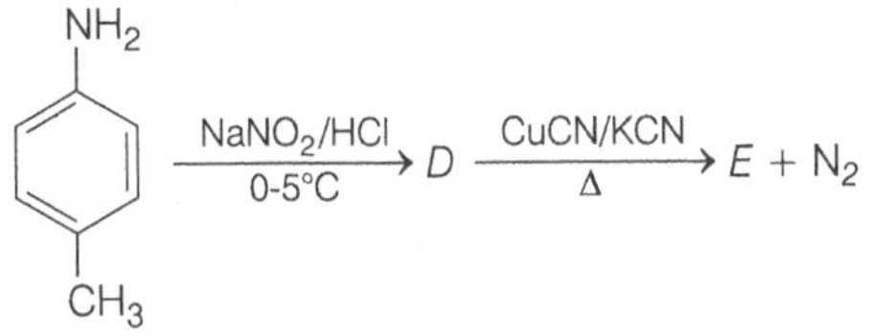

The product *E* is → JEE Main 2015

(a) 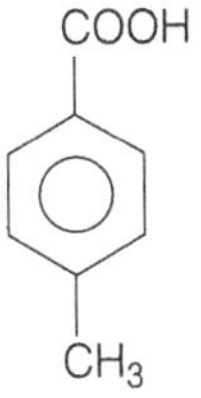

(b) 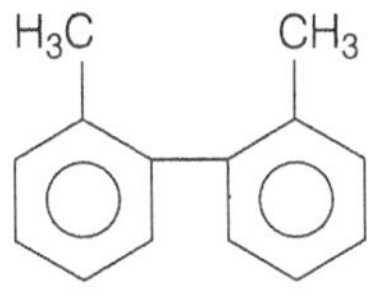

(c) 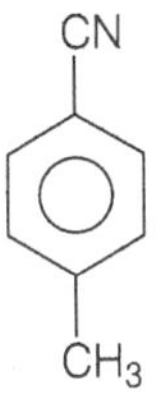

(d) 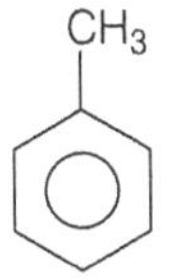

32 In the chemical reaction,

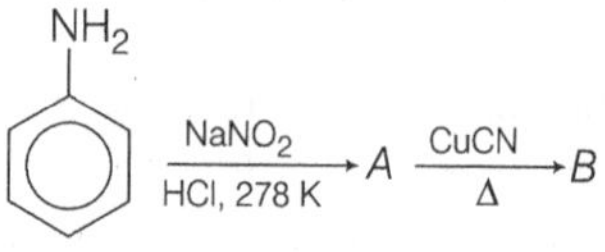

Compounds *A* and *B* respectively are

(a) fluorobenzene and phenol
(b) benzene diazonium chloride and benzonitrile
(c) nitrobenzene and chlorobenzene
(d) phenol and bromobenzene

33 In the chemical reactions,

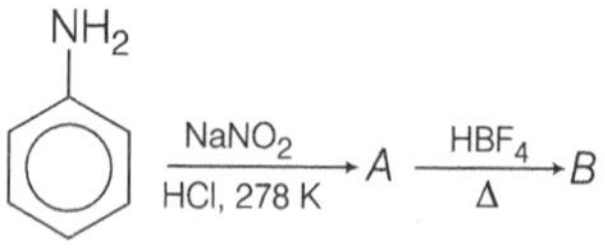

the compounds '*A*' and '*B*' respectively are → AIEEE 2010

(a) nitrobenzene and fluorobenzene
(b) phenol and benzene
(c) benzene diazonium chloride and fluorobenzene
(d) nitrobenzene and chlorobenzene

34 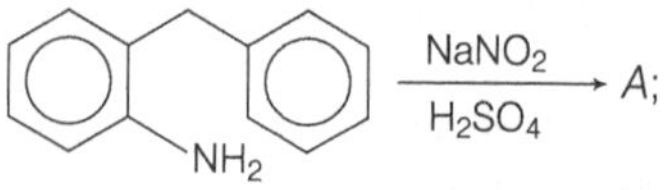

Product of this reaction is

(a) 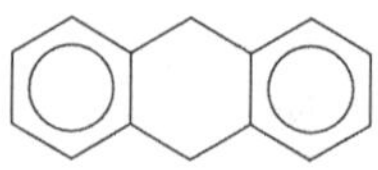 (b) 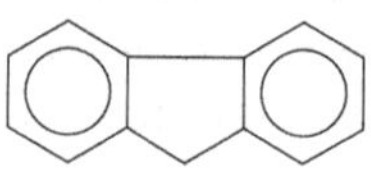

(c) 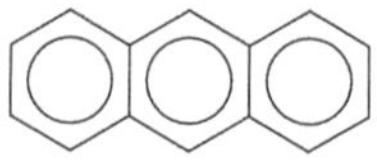 (d) 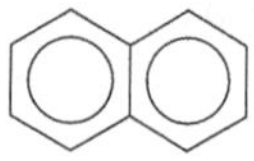

Direction (Q. Nos. 35-36) *In the following questions, Assertion (A) followed by a Reason (R) is given. Choose the correct answer out of the following choices.*

(a) Assertion and Reason both are correct statements and Reason is the correct explanation of the Assertion
(b) Assertion and Reason both are correct statements but Reason is not the correct explanation of the Assertion
(c) Assertion is correct incorrect and Reason is incorrect
(d) Both Assertion and Reason are incorrect

35 Assertion (A) Hofmann's bromamide reaction is given by primary amides.

Reason (R) Primary amines in Hofmann's bromamide reaction follows acidic hydrolysis.

36 Assertion (A) Acylation of amine gives a monosubstituted product whereas alkylation of amines gives polysubstituted product

Reason (R) Acyl group sterically hinders the approach of further acyl groups

Direction (Q. Nos. 37-40) *Each of these questions contains two statements :* Statement I (Assertion) *and* Statement II (Reason). *Each of these questions also has four alternative choices, only one of which is the correct answer. You have to select one of the codes (a), (b), (c) and (d) given below :*

(a) Statement I is true, Statement II is true; Statement II is a correct explanation for Statement I
(b) Statement I is true, Statement II is true; Statement II is not a correct explanation for Statement I
(c) Statement I is true; Statement II is false
(d) Statement I is false; Statement II is true

37 Statement I Amines are pyramidal in shape.

Statement II N-atom is sp^3-hybridised.

38 Statement I Aromatic amines are generally less basic than alkyl amines.

Statement II π-electrons in the ring decreases basic character.

39 Statement I In strongly acidic solutions, aniline becomes more reactive towards electrophilic reagents.

Statement II The amino group being completely protonated in strongly acidic solution, the lone pair of electrons on the nitrogen is no longer available for resonance.

40 Statement I Aniline on reaction with $NaNO_2/HCl$ at 0°C followed by coupling with β-naphthol gives a dark blue coloured precipitate.

Statement II The colour of the compound formed in the reaction of aniline with $NaNO_2/HCl$ at 0°C followed by coupling with β-naphthol is due to the extended conjugation.

DAY PRACTICE SESSION 2

PROGRESSIVE QUESTIONS EXERCISE

1 Compound '*A*' (C_3H_9N) reacts with benzene sulphonyl chloride to form a solid insoluble in alkali. The structure of compound '*A*' is

(a) $CH_3—N(CH_3)—CH_3$

(b) $CH_3—CH_2—NH—CH_3$

(c) $CH_3—CH_2—CH_2—NH_2$

(d) All of these

2 The main product formed in the reaction is

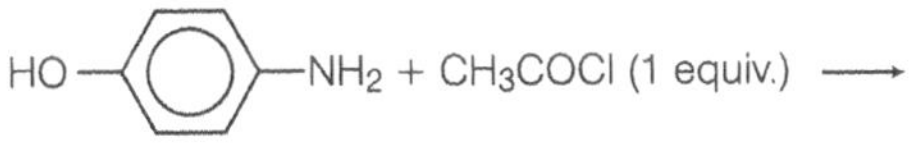

(a) 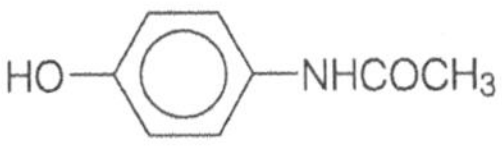

(b) 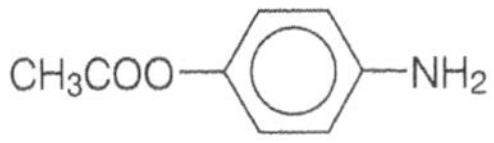

(c) 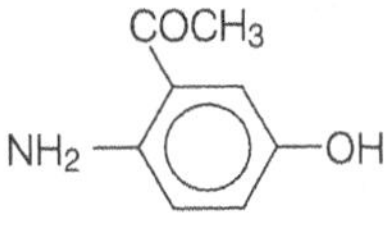

(d) 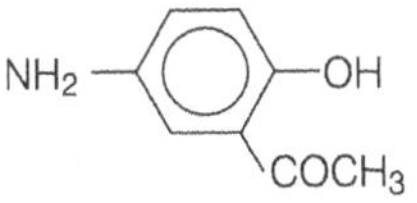

3 Consider the following reaction sequence

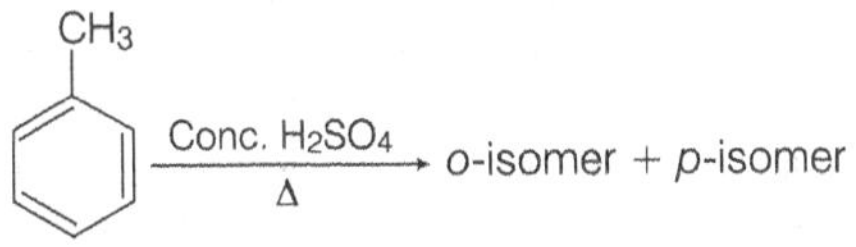

o-isomer $\xrightarrow{PCl_5}$ $\xrightarrow{NH_3}$ $\xrightarrow[35°C]{aq.\ KMnO_4}$ $\xrightarrow{\Delta}$ A

The structure of compound '*A*' is

(a) 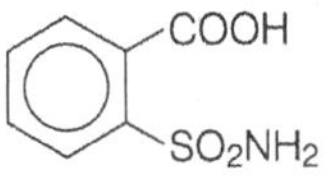

(b) 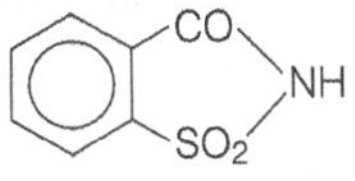

(c) 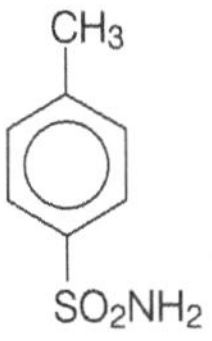

(d) 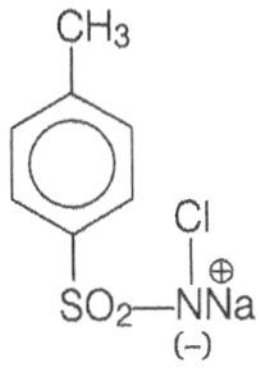

4 A major component of Borsch reagent is obtained by reacting hydrazine hydrate with which of the following?

→ JEE Main (Online) 2013

(a) 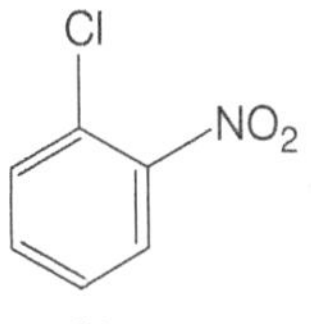

(b) 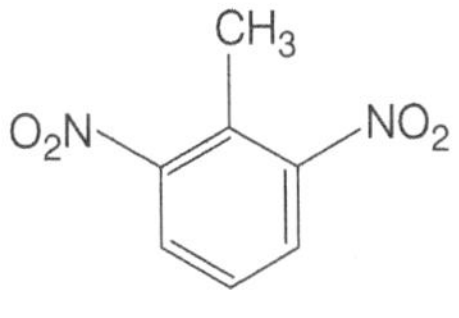

(c) 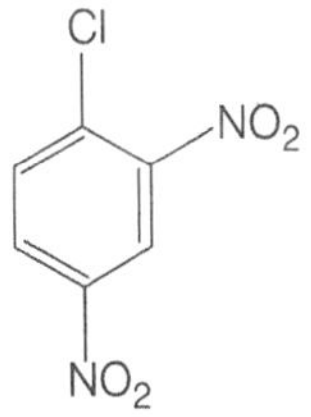

(d) 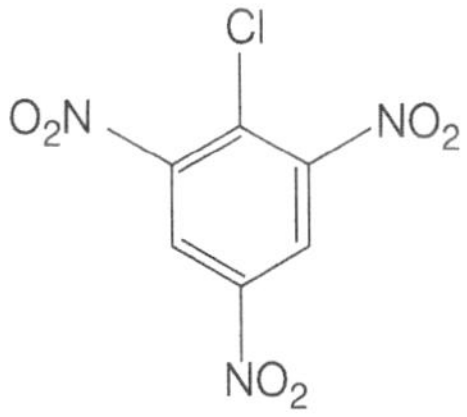

5 An organic compound containing C, H and N only was found to contain C = 61.03, N = 23.71. Its vapour density is 29.5. On treatment with nitrous acid, it gave nitrogen. The compound is

(a) $CH_3—CH(NH_2)—CH_3$

(b) $CH_3—CH_2—NH—CH_3$

(c) $CH_3—N(CH_3)—CH_3$

(d) $CH_3—HC{=}CH—NH_2$

6 Reactants of reaction I are CH_3CONH_2, KOH, Br_2

Reactants of reaction II are CH_3NH_2, $CHCl_3$, KOH

The intermediate species of reaction I and reaction II are respectively

(a) carbonium ion, carbene

(b) carbene, nitrene

(c) nitrene, carbene

(d) carbocation, nitrene

7 Which of the following reactions belong to electrophilic aromatic substitution?

(a) Bromination of acetanilide

(b) Coupling reaction of aryldiazonium salts

(c) Diazotisation of aniline

(d) Both (a) and (b)

8 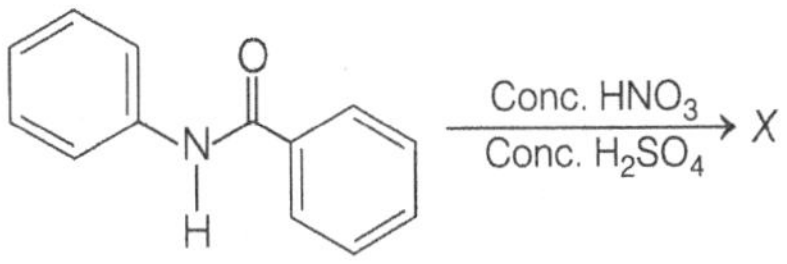

The structure of the major product X is

(a) 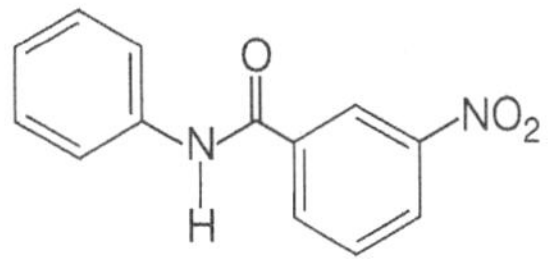

(b) 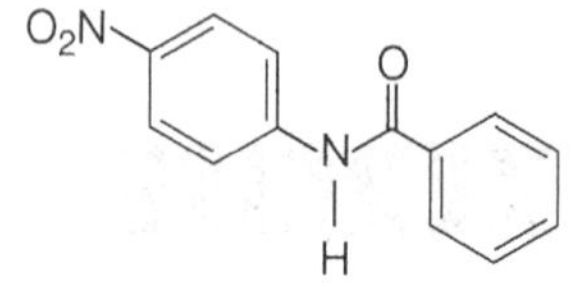

(c) 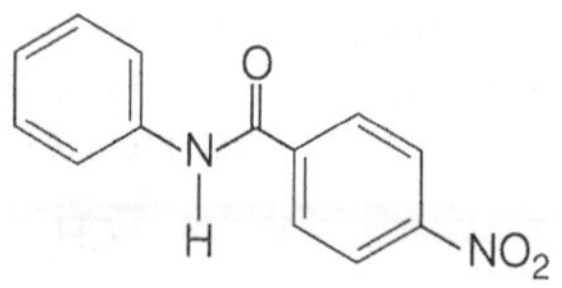

(d) 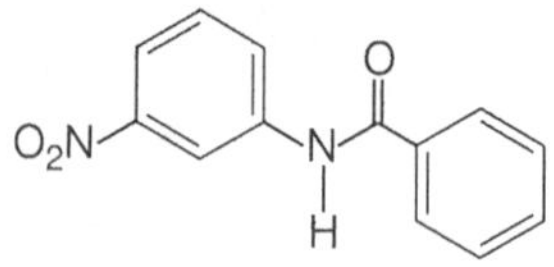

9 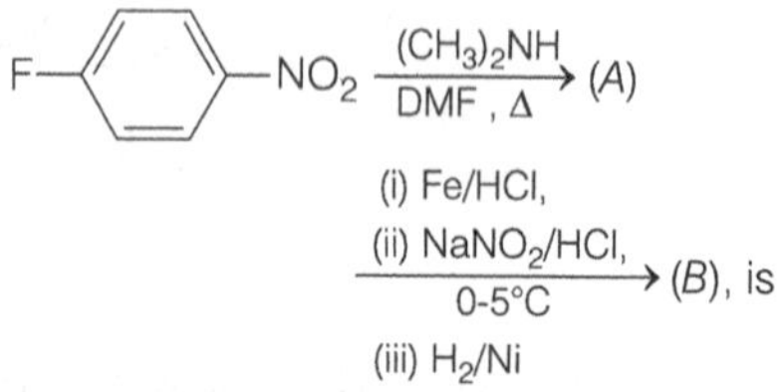

(a) 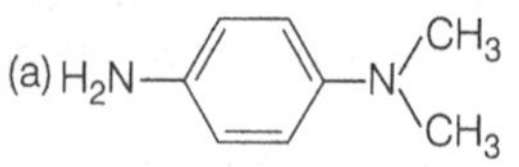 (b) 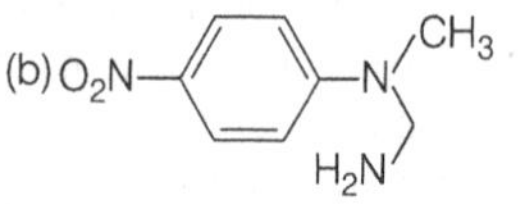

(c) 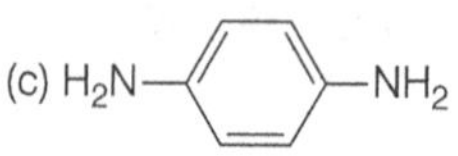 (d) 

10. The best method to synthesise *m*-dibromobenzene is by using the reaction

(a) Benzene $\xrightarrow{Br_2/FeBr_3/heat}$

(b) Aniline $\xrightarrow{Br_2, H_2O}$ [ ] $\xrightarrow[\text{(ii) CuBr}]{\text{(i) HONO}}$

(c) Nitrobenzene $\xrightarrow[H_2SO_4, \Delta]{\text{Fuming } HNO_3}$ [ ] $\xrightarrow[C_2H_5OH, \text{ heat}]{Fe/HCl}$ [ ] $\xrightarrow[\text{(ii) CuBr}]{\text{(i) HONO}}$

(d) Bromobenzene $\xrightarrow[H_2SO_4]{HNO_3}$ [ ] $\xrightarrow[C_2H_5OH, \text{ heat}]{Fe/HCl}$ [ ] $\xrightarrow[\text{(ii) CuBr}]{\text{(i) HONO}}$

13 Toluene is nitrated and the resulting product is reduced with tin and hydrochloric acid. The product so obtained is diazotised and then heated with cuprous bromide solution. The reaction mixture so obtained contains

(a) mixture of *o*-and *p*-bromotoluenes
(b) mixture of *o*-and *p*-dibromobenzenes
(c) mixture of *o*-and *p*-bromoanilines
(d) mixture of *o*-and *m*-bromotoluenes

12 Consider the following reaction sequence
The final product of this reaction sequence is

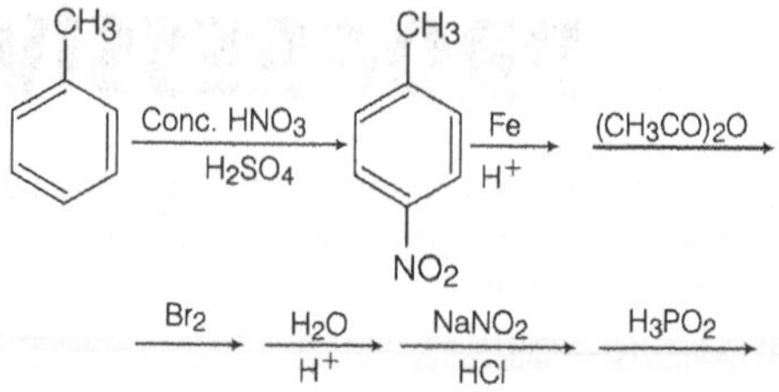

(a) 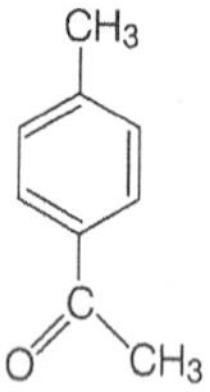 (b) 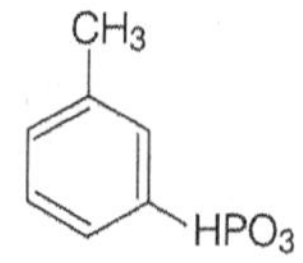

(c) 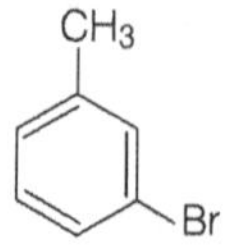 (d) 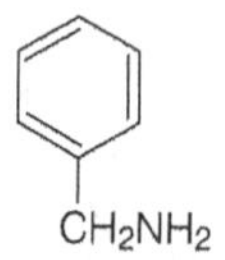

13 The major product of the reaction is

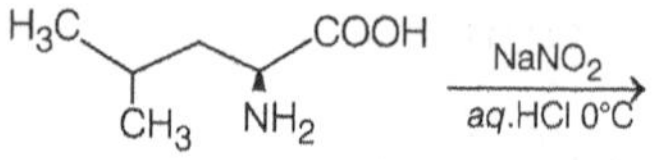

(a) 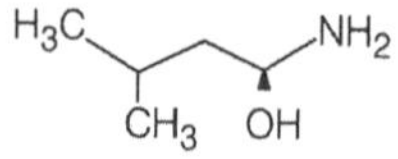 (b) 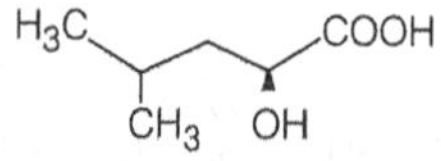

(c) 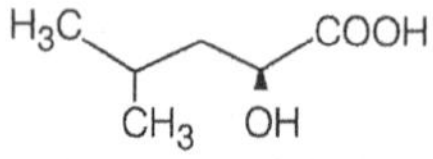 (d) 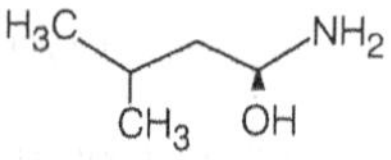

14 A compound (*X*) has the molecular formula C_7H_7NO. On treatment with Br_2 and KOH. (*X*) gives an amine (*Y*); (*Y*) gives carbylamine test. (*Y*) upon diazotisation and coupling with phenol gives an azodye (*Z*). (*X*) is

(a) $PhCONH_2$ (b) $PhCONHCOCH_3$
(c) $PhNO_2$ (d) $PhCOONH_4$

15 (*A*) is subjected to reduction with Zn-Hg/HCl and the product formed is N-methylmethanamine (*A*) can be

(a) ethane nitrile (b) Nitroethane
(c) carbylamino ethane (d) carbylamino methane

16 The correct order of basic strength of the following are

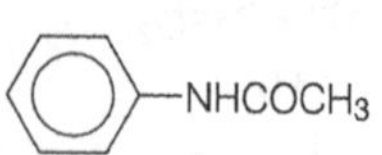
I

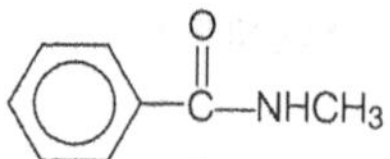
II

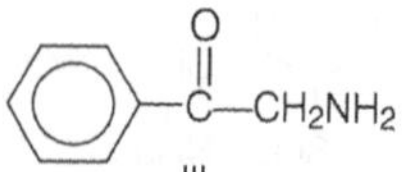
III

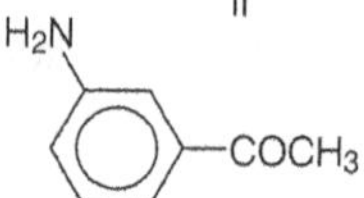
IV

(a) I > II > III > IV (b) IV > II > III > I
(c) III > IV > II > I (d) III > II > IV > I

ANSWERS

SESSION 1	**1** (c)	**2** (a)	**3** (a)	**4** (d)	**5** (b)	**6** (c)	**7** (b)	**8** (d)	**9** (b)	**10** (c)
	11 (c)	**12** (c)	**13** (d)	**14** (a)	**15** (c)	**16** (b)	**17** (d)	**18** (b)	**19** (d)	**20** (a)
	21 (b)	**22** (b)	**23** (b)	**24** (c)	**25** (c)	**26** (c)	**27** (b)	**28** (c)	**29** (a)	**30** (b)
	31 (c)	**32** (b)	**33** (c)	**34** (b)	**35** (c)	**36** (c)	**37** (b)	**38** (c)	**39** (d)	**40** (a)

SESSION 2	**1** (b)	**2** (a)	**3** (b)	**4** (c)	**5** (a)	**6** (c)	**7** (d)	**8** (b)	**9** (a)	**10** (c)
	11 (a)	**12** (c)	**13** (d)	**14** (a)	**15** (c)	**16** (d)				

Hints and Explanations

SESSION 1

1 $C_6H_5-NO_2 \xrightarrow[\text{Reduction}]{Fe/HCl} \underset{\text{Aromatic primary (1°) amine}}{C_6H_5-NH_2}$

2 $C_6H_5NH_2 \xrightarrow{CH_3I} \underset{\text{Secondary amine}}{C_6H_5NHCH_3} \xrightarrow{CH_3I} \underset{\text{Tertiary amine}}{C_6H_5-N(CH_3)_2}$

3 It is the first step of Gabriel's phthalimide synthesis, the hydrogen bonded to nitrogen is sufficiently acidic due to two α-carbonyls.

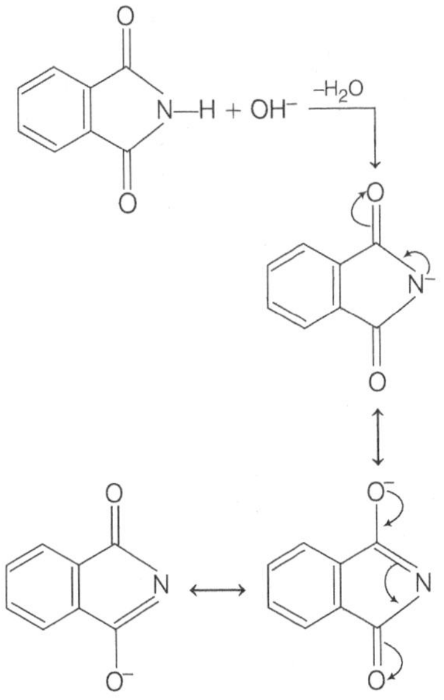

The conjugate base formed above act as nucleophile in the subsequent step of reaction. Nitrogen act as donor as it is better donor than oxygen.

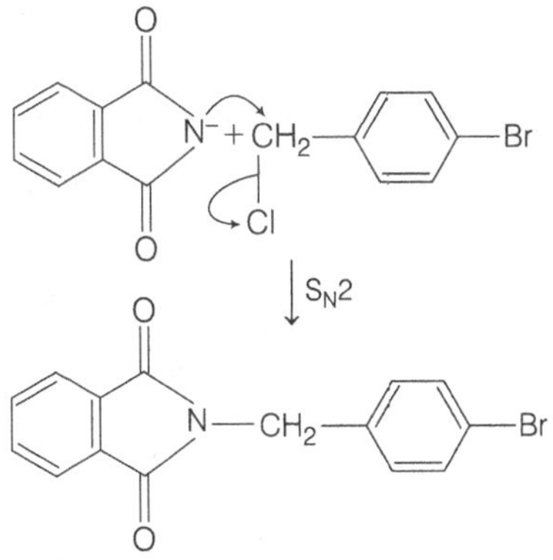

Bromine is not substituted in the above reaction as it is in resonance with benzene ring giving partial double bond character to C—Br bond, hence difficult to break.

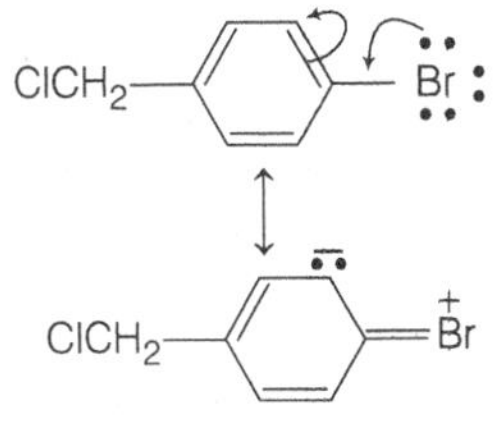

4

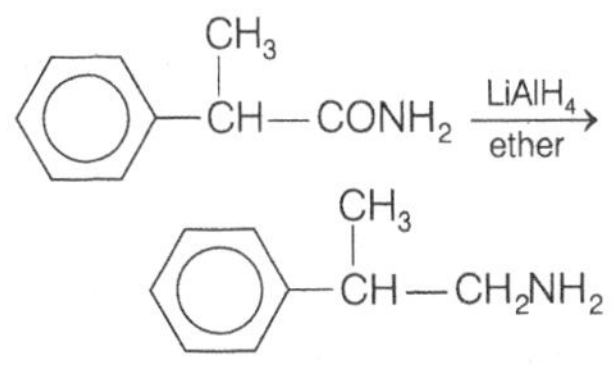

5 $\underset{\text{Acetamide}}{CH_3CONH_2} \xrightarrow{NaOH + Br_2} \underset{\text{Methylamine}}{CH_3NH_2}$

6 Hofmann-bromamide degradation reaction is given as:

$$RCONH_2 + 4NaOH + Br_2 \longrightarrow RNH_2 + NaCO_3 + 2NaBr + 2H_2O$$

Hence, four moles of NaOH and one mole of Br_2 are used.

7

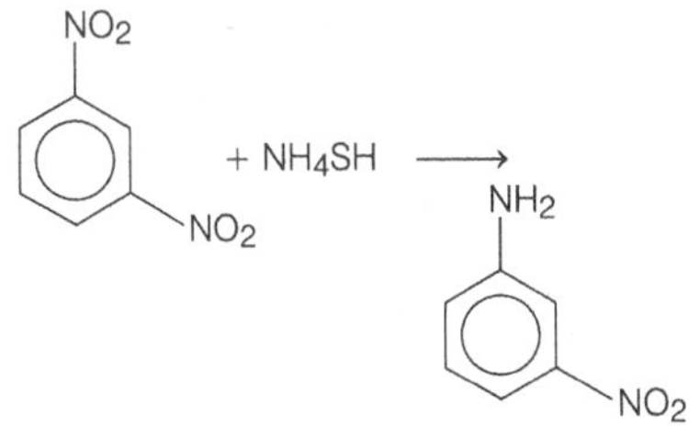

8 $\underset{(A)}{CH_3CH_2COOH} \xrightarrow{NH_3} \underset{(B)}{CH_3CH_2CO\overset{-}{O}\overset{+}{N}H_4} \xrightarrow{\Delta} \underset{(C)}{CH_3CH_2-\overset{O}{\overset{\|}{C}}-NH_2} \xrightarrow[\text{Hofmann-bromamide reaction}]{Br_2, KOH} CH_3CH_2NH_2$

9 1° and 2° amines due to intermolecular H-bonding have higher boiling points (and hence less volatile) than 3° amines and hydrocarbons of comparable molecular mass. Further, due to polar C—N bonds, 3° amines are more polar than hydrocarbons which are almost non-polar. Therefore, due to weak dipole-dipole interactions, 3° amines have higher boiling point (i.e. less volatile) than hydrocarbons. In other words, $CH_3CH_2CH_3$ has the least boiling point and hence, is most volatile.

10 $(CH_3)_2$ NH acts as strong base, due to presence of two methyl groups with +*I*-effect, which intensifies negative charge on N-atom.

11 Dimethyl amine is the strongest base in aqueous solution due to solvation and inductive effect.

12 In gaseous state, as the number of alkyl group increases, electron density also increases and hence, the availability of electrons for donation increases, thus, basicity increases.
Hence, the correct order of basicity of amines in gaseous state is
$3° > 2° > 1° > NH_3$.

13 $—CH_3$ (+*I*-effect) increases electron density at N-atom, hence basic nature is increased. C_6H_5 decreases electron density at N-atom, thus basic nature is decreased.

(a) 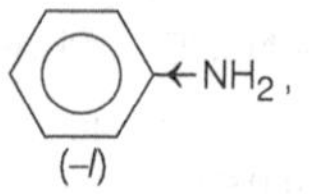

(b) 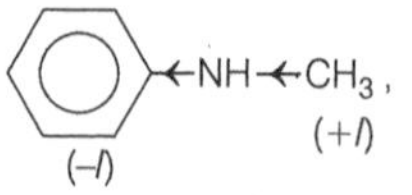

(c) 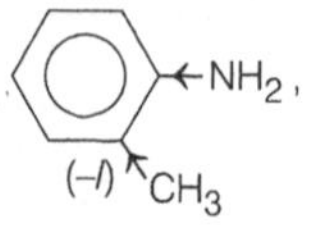

(d) 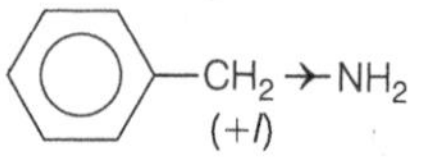

Thus, benzylamine (d) is the strongest base.

14 Order of basic strength of aliphatic amine in aqueous solution is as follows (order of K_b)

$$(CH_3)_2\ddot{N}H > CH_3\ddot{N}H_2 > (CH_3)_3\ddot{N} > C_6H_5\ddot{N}H_2$$

As we know, $pK_b = -\log K_b$. So, $(CH_3)_2\ddot{N}H$ will have smallest pK_b value. In case of phenyl amine, N is attached to sp^2-hybridised carbon, hence it has highest pK_b and least basic strength.

15 Among the given compounds the basic nature depends upon their tendency to donate electron pair.

Since 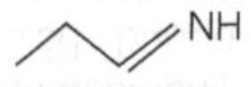 is sp^2-hybridised.

This marginally increases the electron negativity of nitrogen which in turn decreases the electron donating tendency of nitrogen. Thus making compound least basic.

Among the rest, 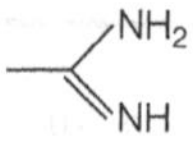 is totally different from others as in this compound lone pair of one nitrogen are in conjugation with π-bond i.e. as a result of this conjugation the cation formed after protonation become resonance stabilised.

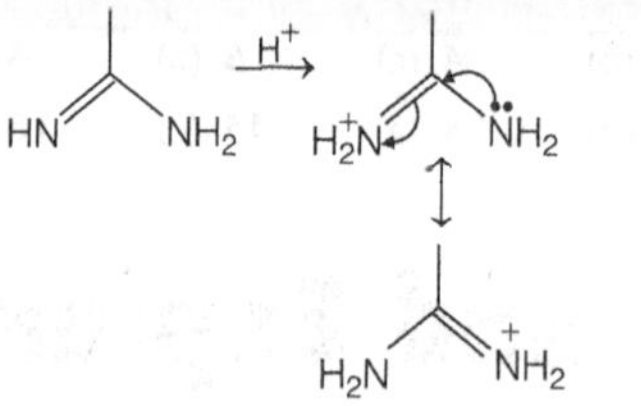

Equivalent resonance

This equivalent resonance in cation makes 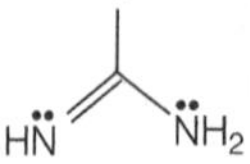 most basic among all.

Categorisation is very simple between rest two as  (primary amine) is less basic than 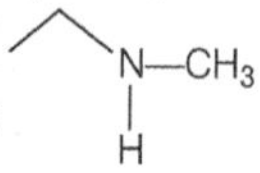 (secondary amine).

Hence, the correct order is II < I < IV < III, i.e. option (c) is correct.

16 $$R—NH_2 + CH_3—\overset{O}{\overset{\|}{C}}—Cl \xrightarrow[(-HCl)]{} R—NH—\overset{O}{\overset{\|}{C}}—CH_3$$

Since, each $—COCH_3$ group displace one H atom in the reaction of one mole of $CH_3—\overset{O}{\overset{\|}{C}}—Cl$ with one $—NH_2$ group, the molecular mass increases with 42 unit. Since, the mass increases by $(390 - 180) = 210$ hence the number of $—NH_2$ group is $\frac{210}{42} = 5$.

17 This reaction is an example of carbylamine reaction which includes conversion of amine to isocyanide.

$$\underset{1° amine}{R—\ddot{N}H_2} + CHCl_3 \xrightarrow[KOH]{C_2H_5OH} \underset{\text{Alkyl isocyanide}}{R—\overset{+}{N}\equiv\overset{-}{C}:}$$

18 The mechanism of carbylamine reaction is given below :

$$CHCl_3 \xrightarrow[\alpha\text{-elimination}]{OH^-} \underset{\text{Dichlorocarbene}}{:CCl_2}$$

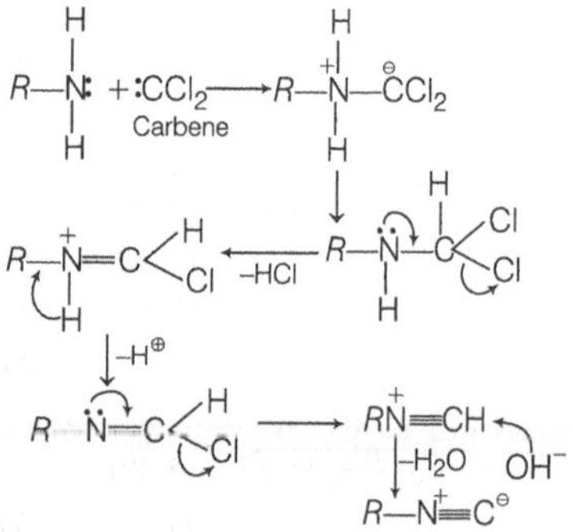

19 When amines reacts with an aldehyde then imines are formed.

$$R—NH_2 + R\overset{O}{\overset{\|}{C}}—H \longrightarrow R—CH=N—R \text{(Imine)}$$

20 $$(C_2H_5)_2NH + HNO_2 \xrightarrow[-H_2O]{} \underset{\text{Nitrosoamine}}{(C_2H_5)_2N \cdot N=O}$$

21 $$C_6H_5SO_2Cl + RNH_2 \longrightarrow \underset{\substack{\text{N- alkyl benzene}\\\text{sulphonamide}\\\text{(Soluble in KOH)}}}{RNHSO_2C_6H_5} + HCl$$

$$\xrightarrow{KOH} [R—\overset{-}{N}SO_2C_6H_5]K^+ + H_2O$$

22 Wurtz reaction is used to prepare alkanes from alkyl halides.

$$2R—X + 2Na \xrightarrow{\text{Dry ether}} R—R + 2NaX$$

23 A ⟶ 4; B ⟶ 3; C ⟶ 1; D ⟶ 2

24. 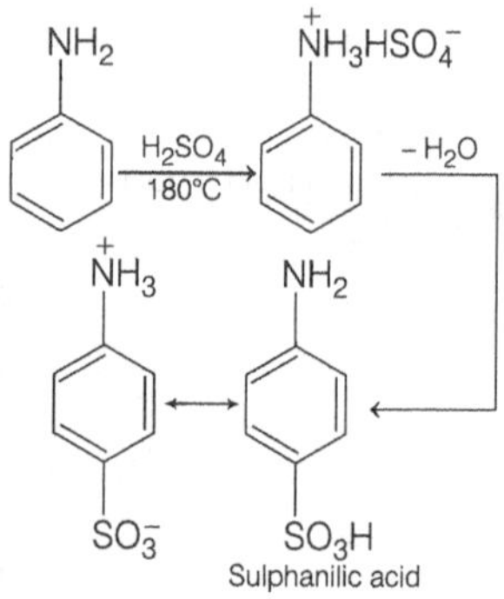

25. 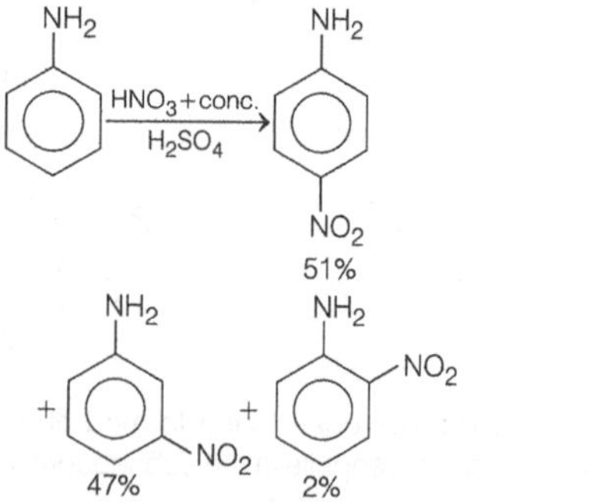

Nitration of aniline also gives *m*-nitroaniline in strong acidic medium because in strong acidic conditions, protonation of $—NH_2$ group gives anilinium ion (NH_3^+) which is deactivating and *m*-directive in nature.

26 The disubstituted benzene in which all the four hydrogen atoms are identical, will give only one monosubstituted product. i.e. *p*-dinitrobenzene.

27 *p*-nitroaniline is obtained as a major product by the nitration of acetanilide followed by alkaline hydrolysis.

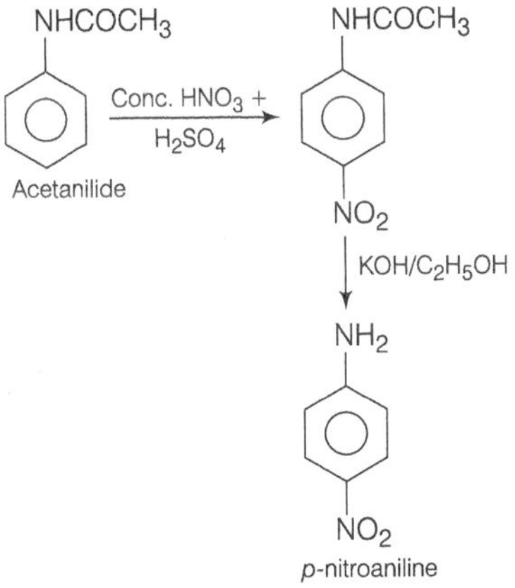

28 Aniline in presence of nitrating mixture (conc. HNO_3 + conc. H_2SO_4) gives significant amount(≈ 47%) of *meta*-product because in presence of H_2SO_4 its protonation takes place and anilinium ion is formed.

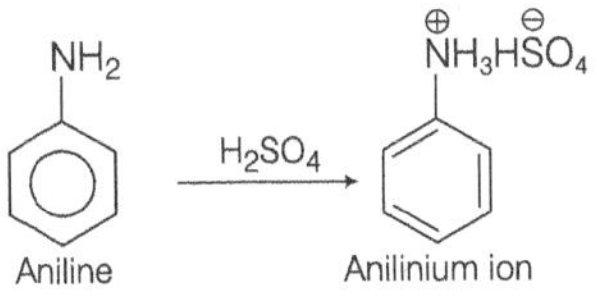

Here, anilinium ion is strongly deactivating group and *meta*-directing in nature. So, it gives *meta*-nitration product.

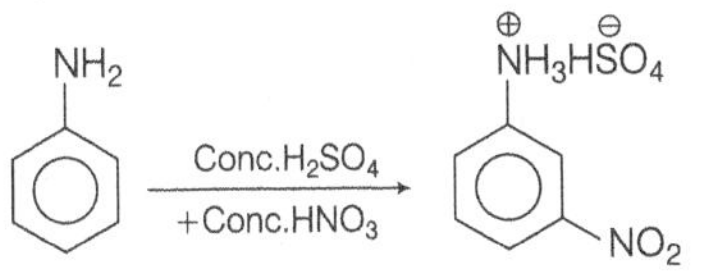

29 Benzene diazonium chloride undergoes deamination upon reaction with hypophosphorus acid and gives benzene.

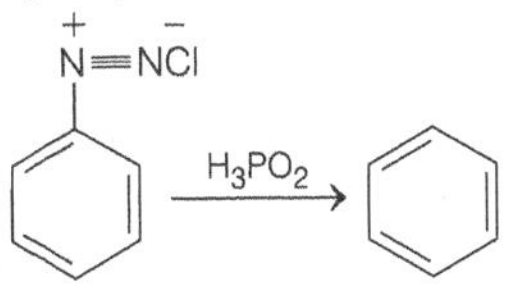

30

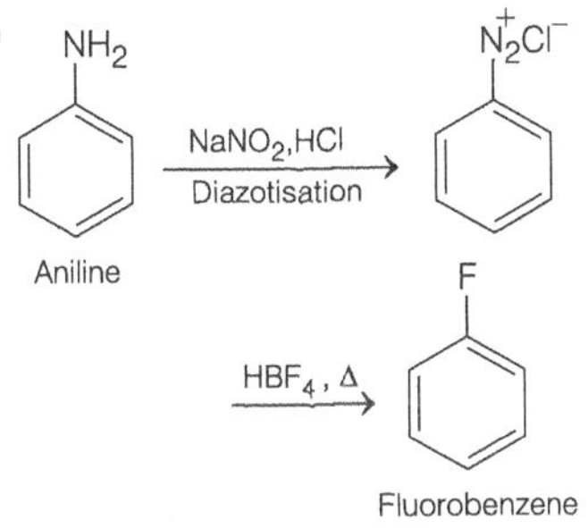

31

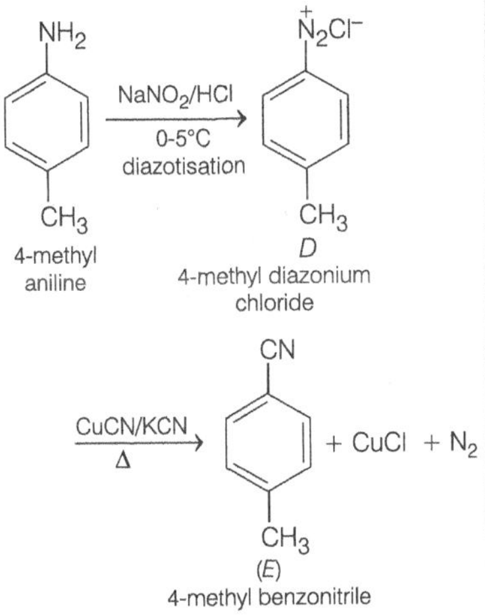

32 Formation of (*A*) is by diazotisation and formation of *B* from *A* is by S_N reaction.

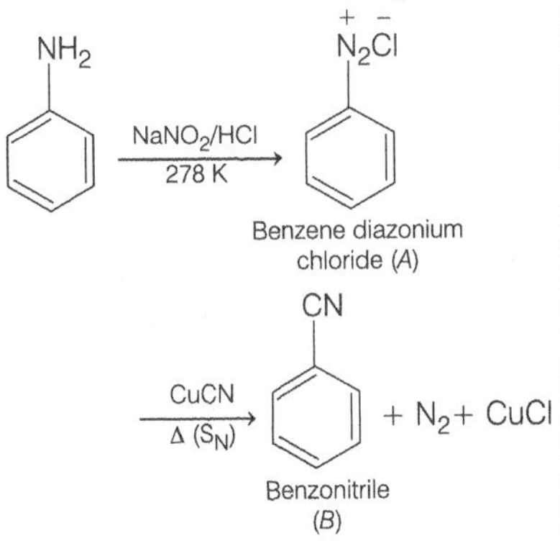

33

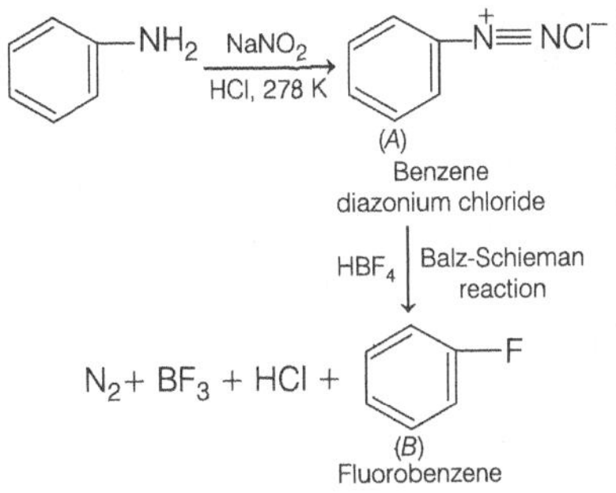

34

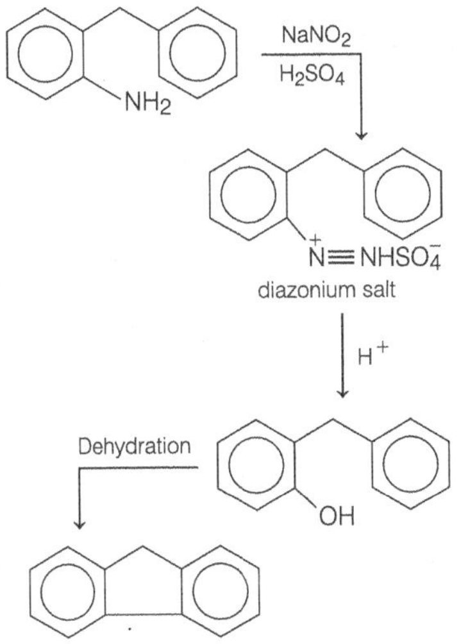

35 **Correct Reason** With Br_2 / NaOH, primary amides are converted into isocyanates which upon alkaline hydrolysis give primary amines.

36 **Correct Reason** Due to delocalisation of the lone pair of electrons of the N-atom over the carbonyl group in the acyl derivative, the electron density on the N-atom decreases to such an extent that it does not act as a nucleophile at all and hence, does not react with another molecule of the acylating agent.

37 It is true that nitrogen atom in amine is sp^3-hybridised but due to the presence of lone pair of electrons, the angle C—N—*E* (where, *E* is C or H) is less than 109.5°. It is 108° in case of trimethyl amine (pyramidal shape).

38 Aromatic amines are less basic than alkyl amines and ammonia due to the electron withdrawing nature of the aryl group. π-electrons are responsible for resonance effect.

39 In strongly acidic medium, aniline gets protonated, so lone pair of electrons are not available to produce mesomeric or electromeric effects. Thus, aniline becomes less reactive.

40 Aniline on reaction with $NaNO_2$ + HCl at 0°C followed by coupling with β-naphthol gives a dark blue coloured precipitate is due to the extended conjugation.

SESSION 2

1 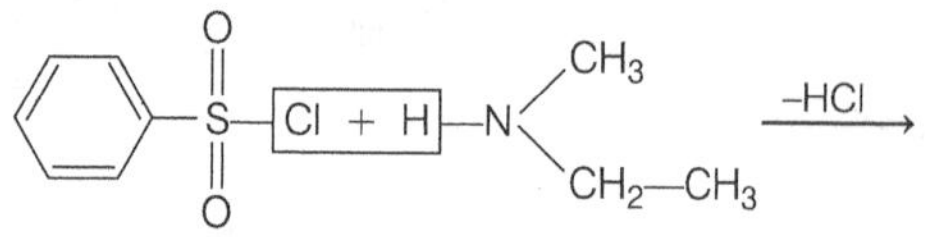

Benzenesulphonyl chloride N-methylmethanamine (2° amine)

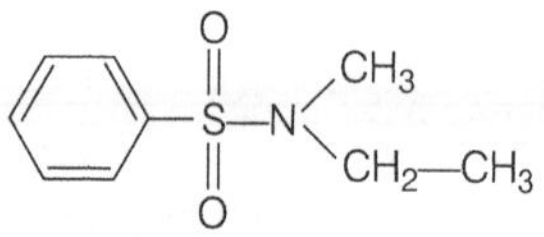

N-ethyl, N-methylbenzene sulphonamide (insoluble in alkali)

2 —NH_2 is more activated than —OH. Hence, —NH_2 is attacked by CH_3COCl.

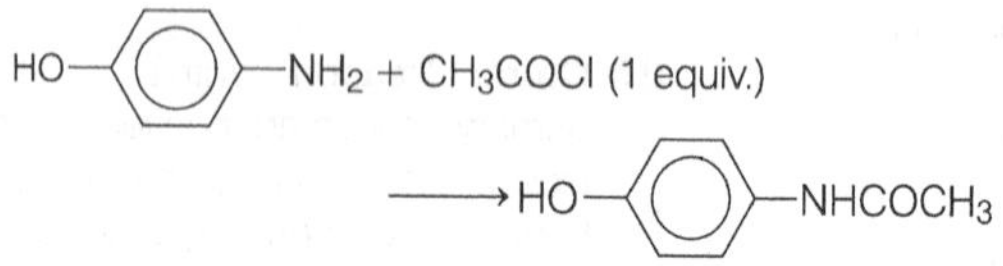

3 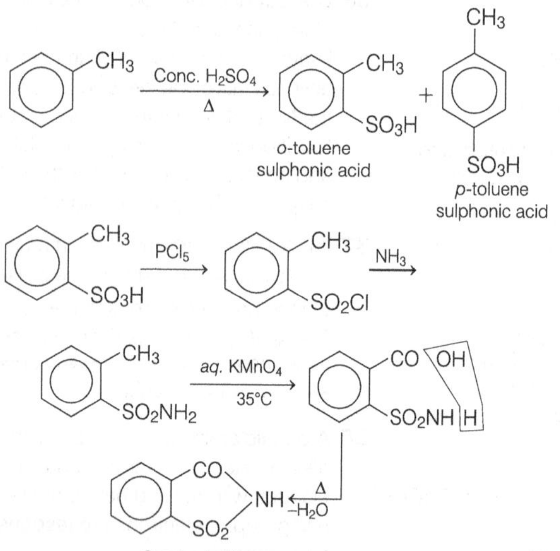

Saccharin (A)

4 Borsch reagent is 2,4-dinitrophenyl hydrazine, thus, it is obtained by treating 2,4-dinitrochlorobenzene with hydrazine hydrate.

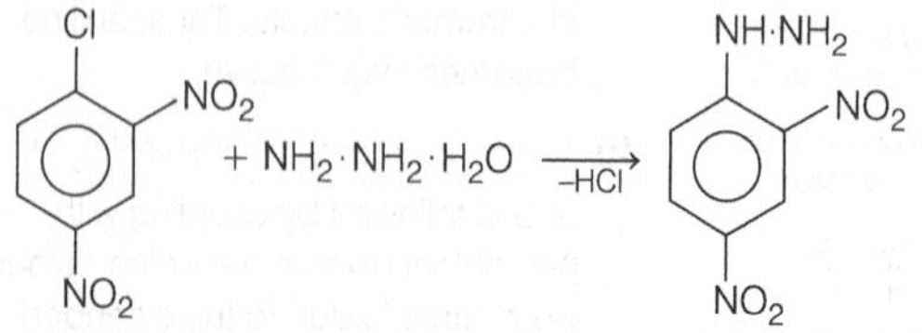

2,4-dinitrochlorobenzene 2,4-dinitrophenyl hydrazine (Borsch reagent)

5. (i) For empirical formula

Element	%	Molar ratio	Simplest ratio
C	61.03	$\frac{61.03}{12} = 5.08$	$\frac{5.08}{1.69} = 3$
H	15.26	$\frac{15.26}{1} = 15.26$	$\frac{15.26}{1.69} = 9$
N	23.71	$\frac{23.71}{14} = 1.69$	$\frac{1.69}{1.69} = 1$

$\therefore$ Empirical formula is C_3H_9N and empirical formula weight = 59

Given, molecular weight = Vapour density × 2 = 29.5 × 2 = 59

$$\therefore \quad n = \frac{59}{59} = 1$$

$\therefore$ Molecular formula is C_3H_9N.

(ii) The amine loses N_2 on treatment with HNO_2 and thus, it is a primary amine. Thus, compound is

$CH_3—CH_2—CH_2—NH_2$ (Propanamine) or $CH_3—CH(NH_2)—CH_3$ (Propan-2-amine)

$(CH_3)_2CHNH_2 + HNO_2 \longrightarrow N_2 + (CH_3)_2CHOH + H_2O$

6 Reaction I

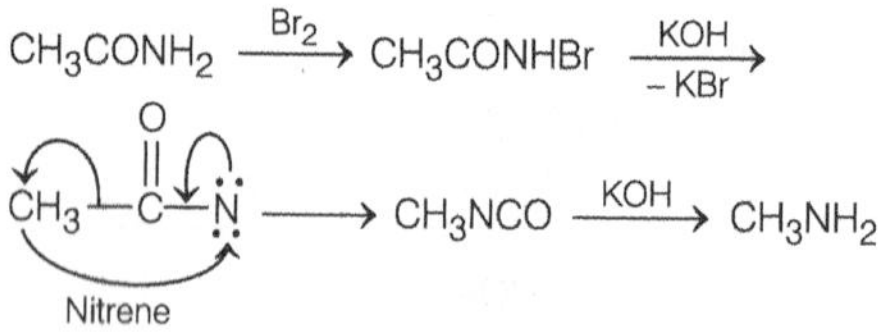

Reaction II

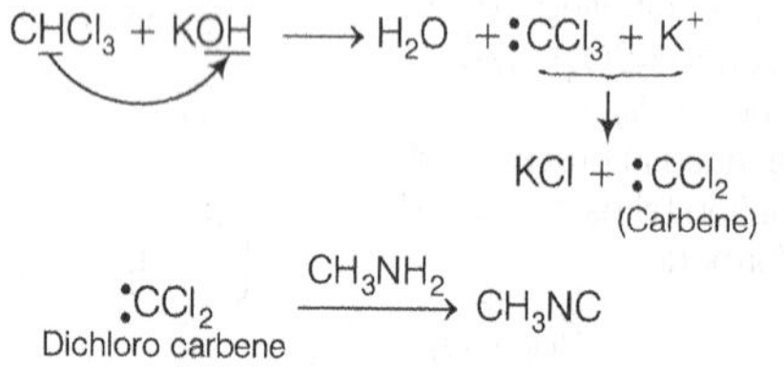

7 Bromination of acetanilide and coupling reactions of diazonium salts both are the examples of electrophilic substitution reaction.

8 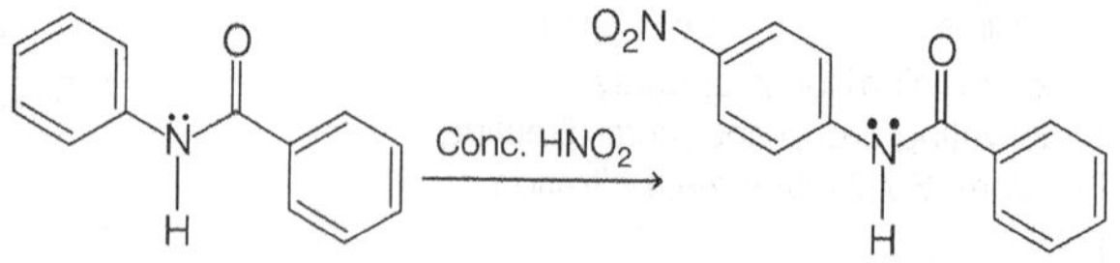

Since, ring 1 is more active, electrophilic substitution takes place over ring 1. Further more —NH—C(=O)—Ph is *ortho-para* directing in nature and *para* product is predominating.

9 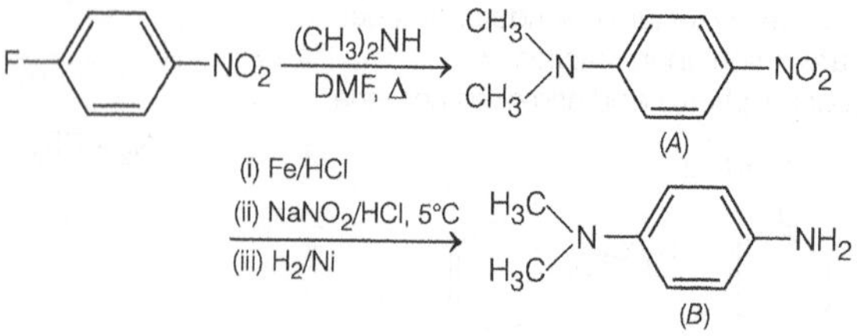

10 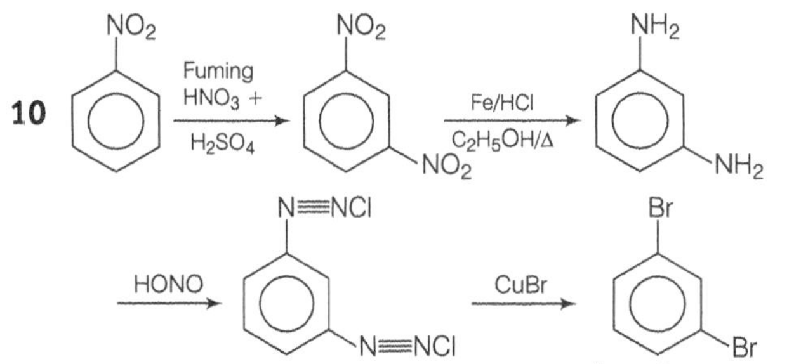

m-dibromobenzene

11 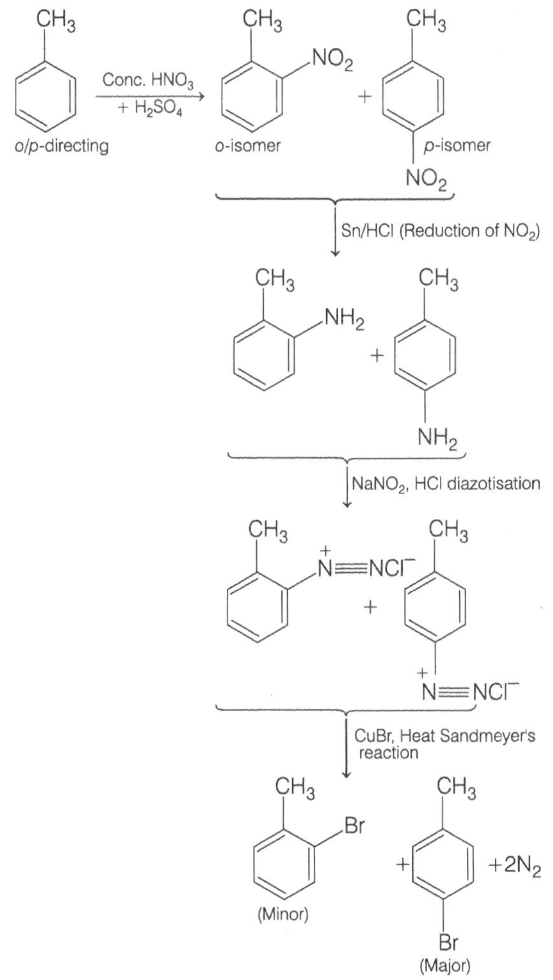

o- and *p*-bromotoluenes

12 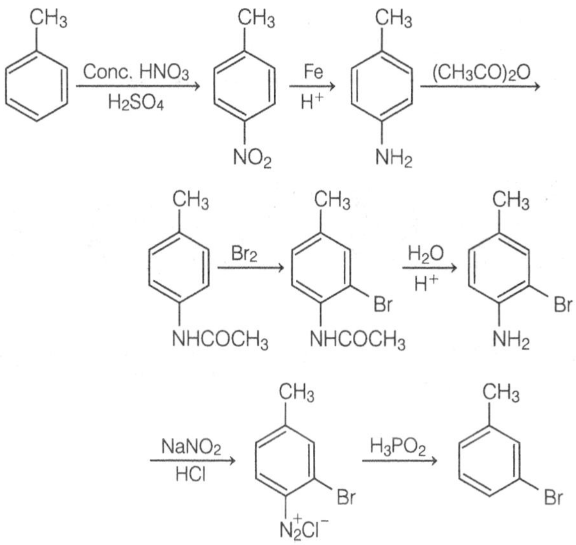

13 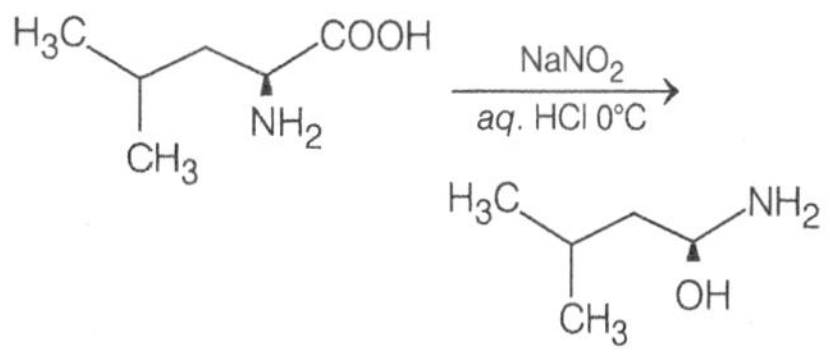

14 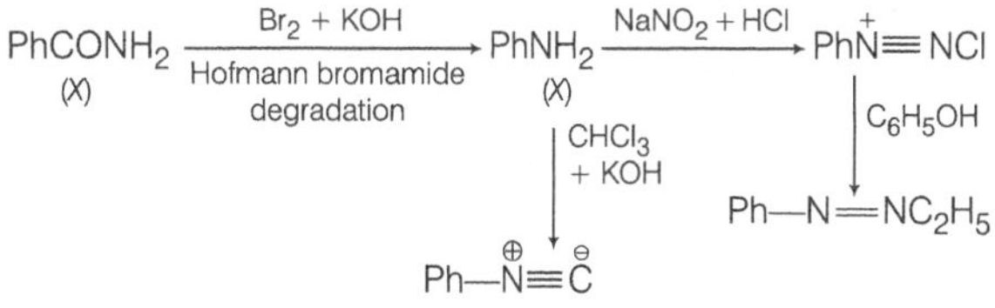

15 CH_3NC (carbylamino methane) when subjected to reduction with Zn-Hg/HCl forms N-methyl methanamine

$$CH_3—NC \xrightarrow[\text{HCl}]{\text{Zn-Hg/}} CH_3NHCH_3$$

16 Electron releasing groups increase the basic strength of amines whereas, electron withdrawing groups decrease it. Therefore, III > II > I > IV, since in $NHCOCH_3$, —NH group is placed between two electron withdrawing groups.

DAY THIRTY TWO

Unit Test 5
(Organic Chemistry-I)

1 Which of the following species is paramagnetic in nature?
(a) Carbonium ion (b) Free radical
(c) Carbene (d) Nitrene

2 Which of the following statement(s) is/are wrong? → NCERT Exemplar
(a) Ozone is not responsible for greenhouse effect
(b) Ozone can oxidise sulphur dioxide present in the atmosphere to sulphur trioxide
(c) Ozone hole is thinning of ozone layer present in stratosphere
(d) Ozone is produced in upper stratosphere by the action of UV-rays on oxygen

3 Among the following, the strongest nucleophile is
(a) C_2H_5SH (b) CH_3COO^-
(c) CH_3NH_2 (d) $NCCH_2^-$

4 Which of the following reactions represent the conversion of CH_4 into CH_3Cl?
(a) Electrophilic substitution (b) Free radical addition
(c) Nucleophilic substitution (d) Free radical substitution

5 In the dehydration reaction,

$CH_3CONH_2 \xrightarrow{P_2O_5} CH_3C{\equiv}N$, the hybridisation state of carbon changes from
(a) sp^3 to sp^2 (b) sp to sp^2
(c) sp^2 to sp (d) sp to sp^3

6 The compound obtained when acetaldehyde reacts with dilute aqueous sodium hydroxide exhibits
(a) geometrical isomerism
(b) optical isomerism
(c) neither optical nor geometrical isomerism
(d) Both optical and geometrical isomerism

7 The most stable free radical is

(a) 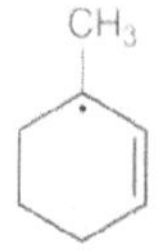
(b) 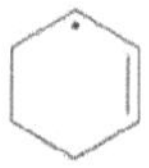
(c) 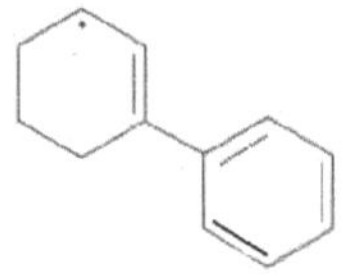
(d) 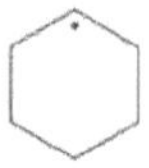

8 Identify Z in the following series.

$$CH_2{=}CH_2 \xrightarrow{HBr} X \xrightarrow{\text{Hydrolysis}} Y \xrightarrow[I_2\ (\text{excess})]{Na_2CO_3} Z$$

(a) C_2H_5I (b) CHI_3
(c) CH_3CHO (d) C_2H_5OH

9 Which of the following structures is enantiomeric with the molecule (*A*) given below. → NCERT Exemplar

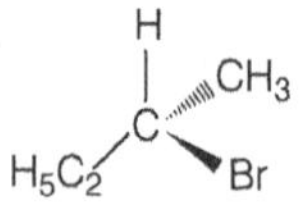
(*A*)

(a) 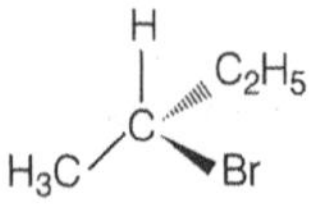
(b) 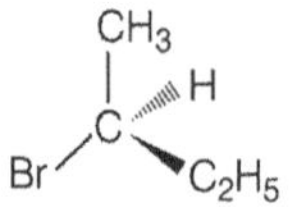
(c) 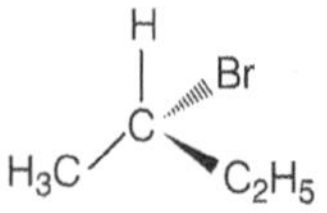
(d) 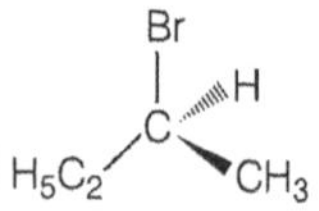

10 Benzene vapours mixed with air when passed over V_2O_5 catalyst at 775 K gives

(a) glyoxal (b) oxalic acid
(c) maleic anhydride (d) fumaric acid

11 Cyclopentene on treatment with alkaline $KMnO_4$ gives

(a) cyclopentanol
(b) *trans*-1, 2-cyclopentanediol
(c) *cis*-1, 2- cyclopentanediol
(d) 1 : 1 mixture of *cis* and *trans*-1, 2-cyclopentanediol

12 $CH_3C \equiv CH \xrightarrow[\text{(ii) } C_2H_5Br]{\text{(i) } NaNH_2} A \xrightarrow[\text{Lindlar's catalyst}]{H_2} B$

What is *B* in the above reaction?

(a) 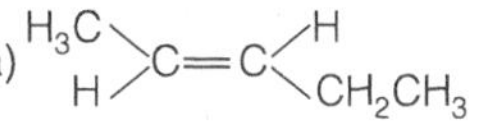

(b) 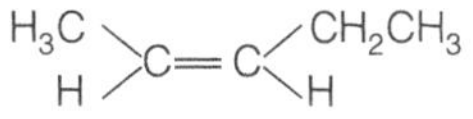

(c) 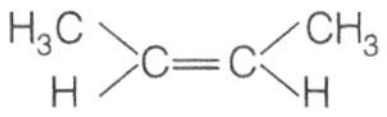

(d) 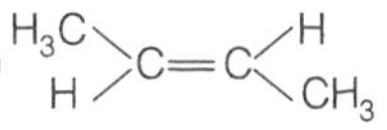

13 $CH_3CH_2CH_2CH_3 \xrightarrow[HCl]{AlCl_3}$ product

Product in the above reaction is

(a) $CH_3—CH(Br)—CH_2—CH_3$ (b) $CH_3—CH(CH_3)—CH_3$
(c) $CH_2(Br)—CH_2—CH_2(Br)$ (d) All of these

14 Identify the reagent *X* in the following reaction.

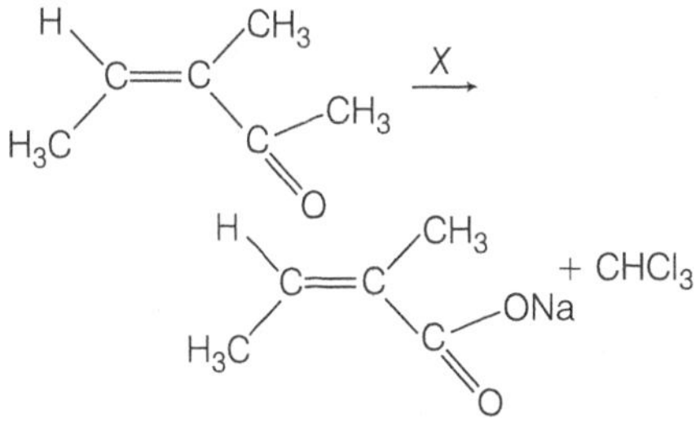

(a) NaCl (b) $SOCl_2$
(c) NaOCl (d) $KMnO_4/H^+$

15 Which of the following statements regarding the effect of trifluoromethyl, $—CF_3$ present on a benzene ring towards electrophilic substitution is correct? [NCERT Exemplar]

(a) $—CF_3$ group activates the ring
(b) $—CF_3$ group renders the ring basic
(c) $—CF_3$ group is a *meta*-directing group, therefore, deactivates the ring
(d) $—CF_3$ group is an *ortho, para*-directing group

16 HBr reacts with $CH_2=CHOCH_3$ under anhydrous conditions at room temperature to give

(a) CH_3CHO and CH_3Br (b) $BrCH_2CHO$ and CH_3OH
(c) $BrCH_2—CH_2—OCH_3$ (d) $H_3C—CHBr—OCH_3$

17 2-methylbutane on reacting with bromine in the presence of sunlight gives mainly

(a) 1-bromo-3-methylbutane
(b) 2-bromo-3-methylbutane
(c) 2-bromo-2-methylbutane
(d) 1-bromo-2- methylbutane

18 Elimination of bromine from 2-bromobutane results in the formation of

(a) predominantly 2-butyne
(b) predominantly 1-butene
(c) predominantly 2-butene
(d) equimolar mixture of 1-butene and 2-butene

19 Reaction of one molecule of HBr with one molecule of 1,3-butadiene at 40°C gives predominantly

(a) 1-bromo-2-butene under kinetically controlled conditions
(b) 3-bromobutene under thermodynamically controlled conditions
(c) 1-bromo-2-butene under thermodynamically controlled conditions
(d) 3-bromobutene under kinetically controlled conditions

20 The compound formed as a result of oxidation of ethyl benzene by $KMnO_4$ is

(a) benzophenone (b) acetophenone
(c) benzoic acid (d) benzyl alcohol

21 Among the following, the one that gives positive iodoform test upon reaction with I_2 and NaOH is

(a) $CH_3CH_2CH(OH)CH_2CH_3$ (b) $C_6H_5CH_2CH_2OH$
(c) 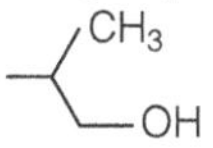 (d) $PhCHOHCH_3$

22 The increasing order of the rate of HCN addition to compound *A-D* is

A. HCHO B. CH_3COCH_3
C. $PhCOCH_3$ D. PhCOPh

(a) *A*<*B*<*C*<*D* (b) *D*<*B*<*C*<*A* (c) *D*<*C*<*B*<*A* (d) *C*<*D*<*B*<*A*

23 Acid catalysed hydration of alkenes except ethene leads to the formation of

(a) mixture of secondary and tertiary alcohols
(b) mixture of primary and secondary alcohols
(c) both secondary or tertiary alcohol
(d) only primary alcohol

24 The best reagent to convert pent-3-en-2-ol into pent-3-en-2-one is

(a) pyridinium chlorochromate
(b) chromic acid in aqueous acetone
(c) acidic dichromate
(d) acidic permanganate

25 Which one of the following undergoes reaction with 50% sodium hydroxide solution to give corresponding alcohols and acids?

(a) Phenol
(b) Benzaldehyde
(c) Butanol
(d) Benzoic acid

26 Ethyl isocyanide on hydrolysis in acidic medium generates

(a) ethylamine salt and methanoic acid
(b) propanoic acid and ammonium salt
(c) ethanoic acid and ammonium salt
(d) methylamine salt and ethanoic acid

27 In the chemical reaction,

$$CH_3CH_2NH_2 + CHCl_3 + 3KOH \longrightarrow A + B + 3H_2O$$

the compounds *A* and *B* respectively are

(a) C_2H_5CN and 3KCl
(b) $CH_3CH_2CONH_2$ and 3KCl
(c) C_2H_5NC and K_2CO_3
(d) C_2H_5NC and 3KCl

28 Examine the following two structures for the anilinium ion and choose the correct statement from the ones given below.

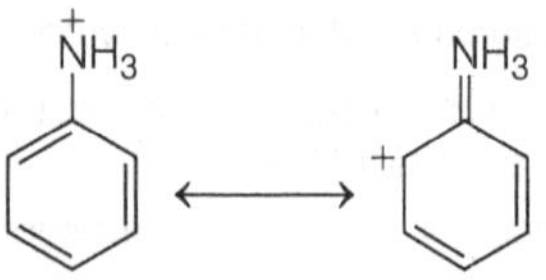

(I) (II)

(a) II is not an acceptable canonical structure because carbonium ions are less stable than ammonium ions
(b) II is not an acceptable canonical structure because it is non-aromatic
(c) II is not an acceptable canonical structure because the nitrogen has 10 valence electrons
(d) II is an acceptable canonical structure

29 The most unlikely representation of resonance structures of *p*-nitrophenoxide ion is

(a) 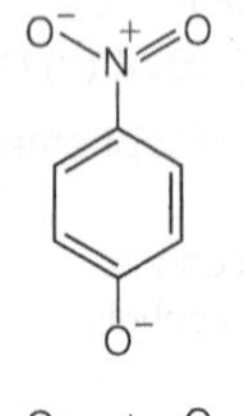

(b) 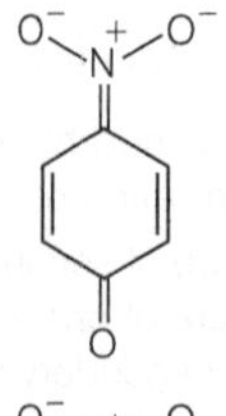

(c) 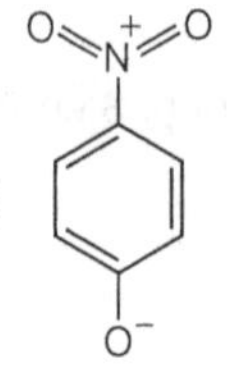

(d) 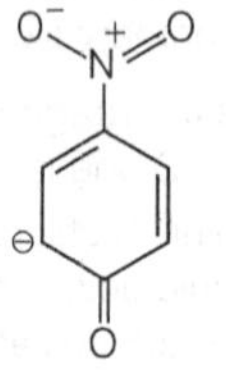

30 The structure of the compound formed, when nitrobenzene is reduced by lithium aluminium hydride ($LiAlH_4$) is

(a) 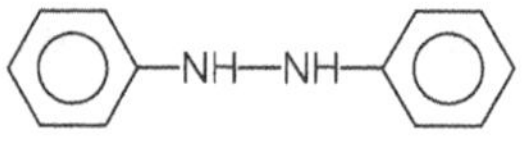

(b) 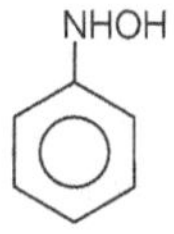

(c) 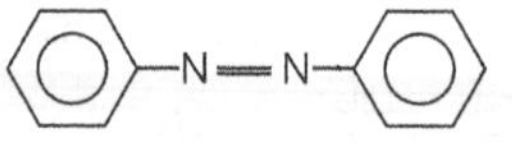

(d) 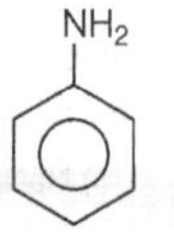

31 Final major product obtained in the following transformation is

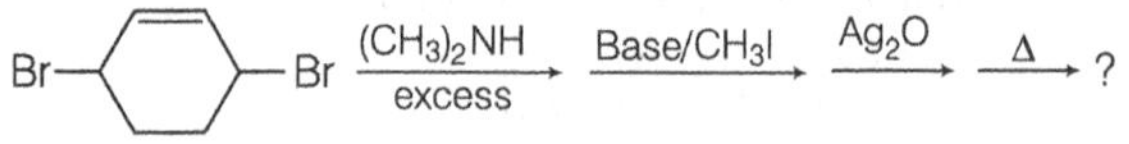

(a) 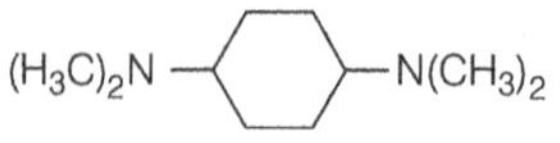

(b) 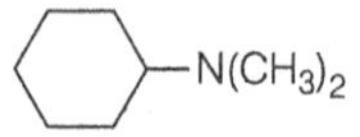

(c) 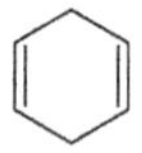

(d) 

32 Cyclobutyl bromide on treatment with magnesium in dry ether forms an organometallic compound *A*. The organometallic compound reacts with ethanal to give an alcohol *B* after mild acidification. Prolonged treatment of alcohol *B* with an equivalent amount of HBr gives 1-bromo-1-methylcyclopentane *C*. Identify the alcohol *B*.

(a) 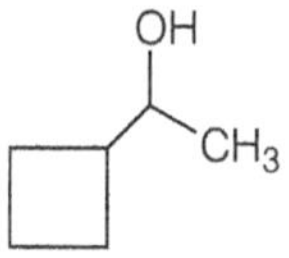

(b) 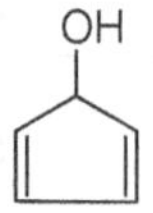

(c) 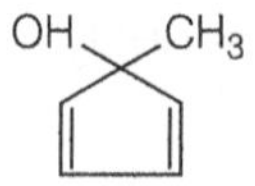

(d) 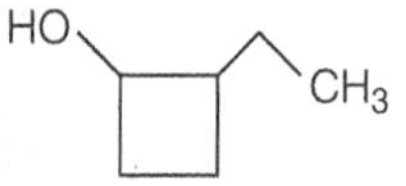

33 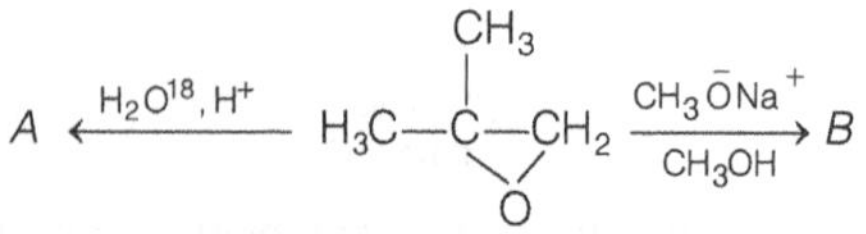

A and *B* respectively are

(a) $(CH_3)_2C(OH)—CH_2\overset{18}{O}H$, $(CH_3)_2C(OH)—CH_2OCH_3$

(b) $(CH_3)_2C(OH)—CH_2\overset{18}{O}H$, $(CH_3)_2C(OCH_3)—CH_2OH$

(c) $(CH_3)_2\underset{|}{\overset{\overset{18}{OH}}{C}}-CH_2OH$, $(CH_3)_2C(OH)-CH_2OCH_3$

(d) $(CH_3)_2C(\overset{18}{OH})-CH_2OCH_3$, $(CH_3)_2C(OCH_3)-CH_2OH$

34 $(CH_3)_3CMgCl$ on reaction with D_2O produces

(a) $(CH_3)_3COD$ (b) $(CD_3)_3CH$
(c) $(CH_3)_3CD$ (d) $(CD_3)_3CD$

35 The reagent(s) for the following conversion, is/are

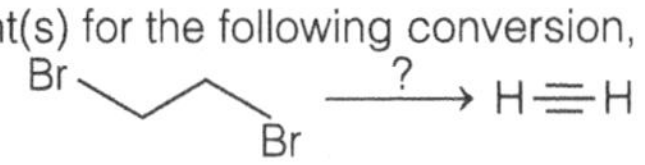

(a) alcoholic KOH → [IIT JEE]
(b) alcoholic KOH followed by $NaNH_2$
(c) aqueous KOH followed by $NaNH_2$
(d) Zn / CH_3OH

36 Identify *B* and *D* in the following sequence of reactions.

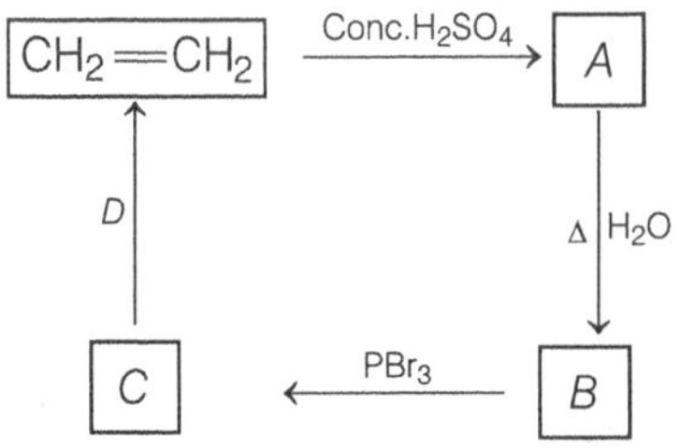

(a) Methanol and bromoethane
(b) Ethyl hydrogen sulphate and alcoholic KOH
(c) Ethyl hydrogen sulphate and aqueous KOH
(d) Ethanol and alcoholic KOH

37 Which compound does not give precipitate with ammoniacal silver nitrate solution?

(a) $C_2H_5-C\equiv CH$ (b) $CH_3-C\equiv C-CH_3$
(c) $CH_3-CH(CH_3)-C\equiv CH$ (d) $Ph-CH_2-C\equiv CH$

38 Thermal decomposition of

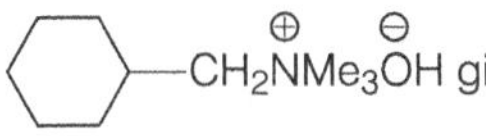

gives

(a) 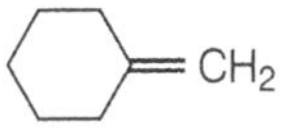

(b) 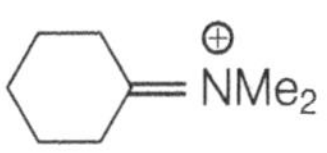

(c) 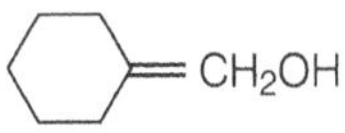

(d) 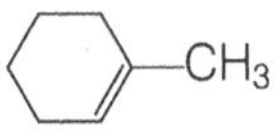

39 The dihalogen derivative *X* of a hydrocarbon with three carbon atoms reacts with alcoholic KOH and produces another hydrocarbon which forms a red precipitate with ammoniacal Cu_2Cl_2. *X* gives an aldehyde on reaction with aqueous KOH. The compound *X* is

(a) 1, 3-dichloropropane (b) 1, 2-dichloropropane
(c) 2, 2-dichloropropane (d) 1, 1-dichloropropane

Direction (Q. Nos. 40-44) *In the following questions, Assertion (A) followed by a Reason (R) is given. Choose the correct answer out of the following choices.*

(a) Both A and R are correct and R is the correct explanation of A
(b) Both A and R are correct but R is not the correct explanation of A
(c) A is correct but R is incorrect
(d) Both A and R are incorrect

40 **Assertion** (A) Phenol forms 2,4,6-tribromophenol on treatment with Br_2 in carbon disulphide at 273 K.

Reason (R) Bromine polarises in carbon disulphide. → [NCERT Exemplar]

41 **Assertion** (A) Pent-1-ene and pent-2-ene are position isomers.

Reason (R) Position isomers differ in the position of functional group or a substituent. → [NCERT Exemplar]

42 **Assertion** (A) The presence of nitro group facilitates nucleophilic substitution reactions in aryl halides.

Reason (R) The intermediate carbanion is stabilised due to presence of nitro group.

43 **Assertion** (A) *trans*-2-butene on reaction with Br_2 gives *meso*-2,3-dibromobutane.

Reason (R) The reaction involves *syn*-addition of bromine.

44 **Assertion** (A) Anilinium chloride is more acidic than ammonium chloride.

Reason (R) Anilinium ion is resonance stabilised.

45 Given,

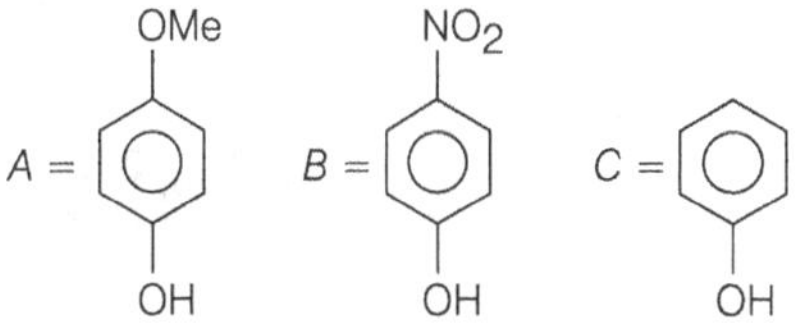

The decreasing order of the acidic character is

(a) $A > B > C$ (b) $B > A > C$
(c) $B > C > A$ (d) $C > B > A$

46 2,3-dimethylpentane on treatment with $Cl_2/h\nu$ produces how many monochloro isomers (only constitutional isomers)?

(a) 4 (b) 5
(c) 6 (d) 8

47 Which one of the following is an example of Hell-Volhard-Zelinsky reaction?

(a) $RCOOH \xrightarrow[\text{(ii) } H_3O^+]{\text{(i) } B_2H_6} RCH_2OH$

(b) $R_2CHCOOH \xrightarrow[\text{(ii) } H_2O]{\text{(i) } Br_2 \text{ (Red P)}} R_2C(Br)COOH$

(c) 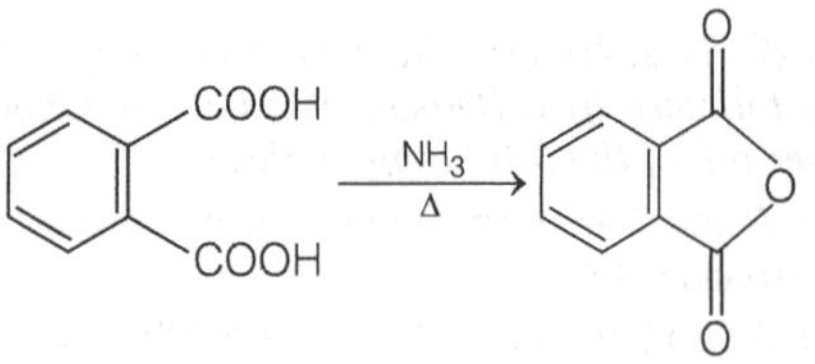

(d) $RCOOH \xrightarrow{PCl_5} RCOCl$

48 Zeisel's method is used to estimate
(a) alcoholic group
(b) amino group
(c) methoxy group
(d) halo group

49 Identify *C*, in the following reaction.

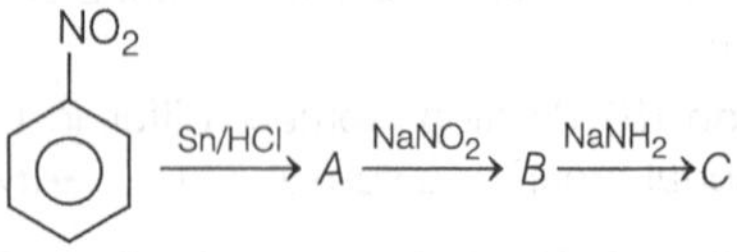

(a) Benzamide (b) Benzoic acid
(c) Chlorobenzene (d) Aniline

50 Match the reactions given in Column I with the names given in Column II.

	Column I	Column II
A.	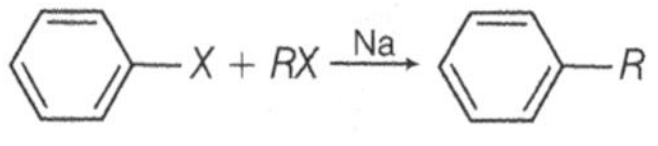	1. Fittig reaction
B.	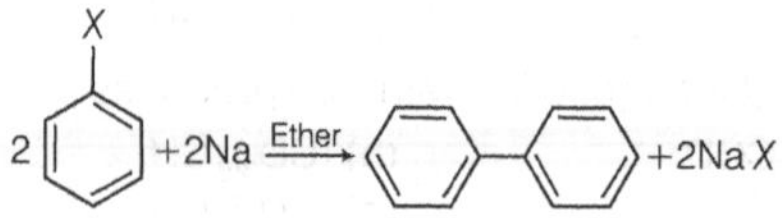	2. Wurtz Fittig reaction
C.	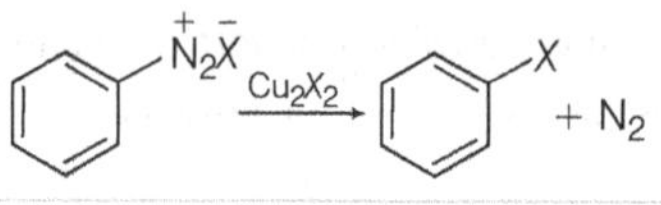	3. Finkelstein reaction
D.	$C_2H_5Cl + NaI \xrightarrow{\text{Dry acetone}} C_2H_5I + NaCl$	4. Sandmeyer reaction

Codes

	A	B	C	D
(a)	1	2	4	3
(b)	2	1	4	3
(c)	1	2	3	4
(d)	2	3	1	4

ANSWERS

1 (b)	**2** (a)	**3** (a)	**4** (d)	**5** (c)	**6** (d)	**7** (c)	**8** (b)	**9** (a)	**10** (c)
11 (c)	**12** (b)	**13** (b)	**14** (c)	**15** (c)	**16** (d)	**17** (c)	**18** (c)	**19** (c)	**20** (c)
21 (d)	**22** (c)	**23** (c)	**24** (b)	**25** (b)	**26** (a)	**27** (d)	**28** (c)	**29** (c)	**30** (c)
31 (d)	**32** (a)	**33** (c)	**34** (c)	**35** (b)	**36** (d)	**37** (b)	**38** (a)	**39** (d)	**40** (d)
41 (a)	**42** (a)	**43** (c)	**44** (c)	**45** (c)	**46** (c)	**47** (b)	**48** (c)	**49** (d)	**50** (b)

Hints and Explanations

1.

Species	Valence electrons	Magnetic behaviour
Carbonium ion	6	Diamagnetic
Free radical	7	Paramagnetic
Carbene	6	Diamagnetic
Nitrene	6	Diamagnetic

2. O_3 is responsible for greenhouse effect. Its contribution is about 8%.

3. Greater the +*I*-effect, greater the nucleophilic power. The +*I*-effect of ethyl is greater than +*I*-effect of methyl group.

$$\underset{+I}{C_2H_5} \rightarrow\!\!\rightarrow S—H,\ \underset{+I}{CH_3} \rightarrow\!\!\rightarrow \overset{O}{\overset{\|}{C}}—O^-,$$

$$CH_3 \rightarrow\!\!\rightarrow NH_2, NC \leftarrow\!\!\leftarrow \bar{C}H_2$$

4. $CH_4 + Cl_2 \xrightarrow{h\nu} CH_3Cl$ (Free radical substitution)

The steps involved in the process are as follows:

$$Cl—Cl \xrightarrow{h\nu} Cl^{\bullet} + Cl^{\bullet}$$

$$CH_4 + Cl^{\bullet} \longrightarrow \dot{C}H_3 + HCl$$

$$\dot{C}H_3 + Cl—Cl \longrightarrow CH_3Cl + Cl^{\bullet}$$

$$Cl^{\bullet} + Cl^{\bullet} \longrightarrow Cl_2$$

5. $$CH_3—\underset{sp^2}{\overset{O}{\overset{\|}{C}}}—NH_2 \xrightarrow{P_2O_5} CH_3\underset{sp}{C} \equiv N + H_2O$$

6. When acetaldehyde is treated with aqueous sodium hydroxide solution, it undergoes aldol condensation because of the presence of α-H-atom.

$$2CH_3CHO \xrightarrow{\text{dil.NaOH}(aq)} CH_3CH(OH)CH_2CHO$$

Aldol shows optical isomerism

$$\xrightarrow{\Delta} CH_3CH{=}CHCHO$$

(Shows geometrical isomerism)

7. Due to resonance stabilisation option (c) forms more stable free radical.

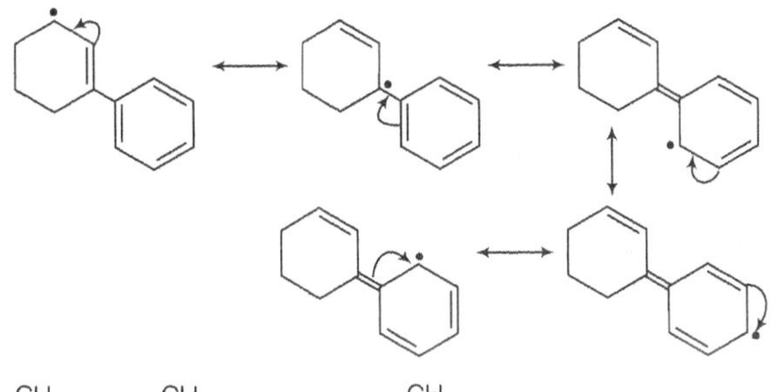

8. $CH_2{=}CH_2 \xrightarrow{HBr} \underset{X}{CH_3CH_2Br} \xrightarrow{\text{Hydrolysis}} \underset{Y}{CH_3CH_2OH} \xrightarrow[I_2\text{(excess)}]{Na_2CO_3} \underset{Z}{CHI_3}$ Iodoform (Yellow ppt.)

9. Compound (a) is an enantiomer of compound (*A*) because in them, the configuration of two groups, i.e. CH_3 and C_2H_5 are reversed at the chiral carbon.

10.

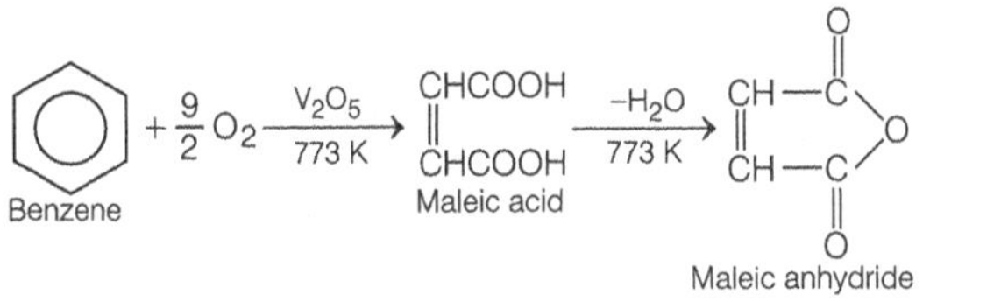

11.

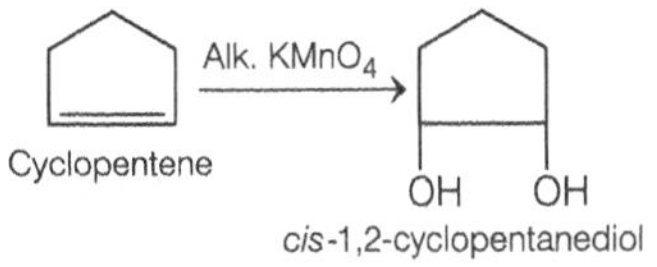

12. $CH_3C{\equiv}CH \xrightarrow[-NH_3]{NaNH_2} CH_3{-}C{\equiv}CNa \xrightarrow[-NaBr]{CH_3CH_2Br}$

$CH_3C{\equiv}C \cdot CH_2CH_3 \xrightarrow[\text{Lindlar's catalyst}]{H_2} (H_3C)(H)C{=}C(CH_2CH_3)(H)$ *cis* alkene

13. When alkane is passed over $AlCl_3$ in presence of HCl, isomerisation takes place.

$$\underset{n\text{-butane}}{CH_3CH_2CH_2CH_3} \xrightarrow[200°C]{AlCl_3/HCl} \underset{iso\text{-butane}}{CH_3{-}CH(CH_3){-}CH_3}$$

14. Methyl ketones are oxidised by sodium hypohalite to sodium salts of corresponding carboxylic acids having one carbon atom less than that of carbonyl compound.

15.

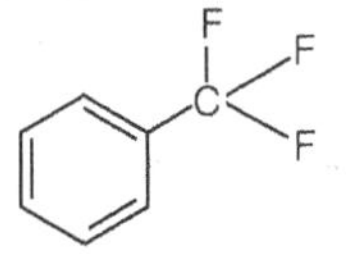

—CF_3 group withdraws electrons from the ring, shows –M effect, makes ring electron deficient, thus deactivates ring for electrophilic substitution.

16. $CH_2{=}CH{-}O{-}CH_3 \xrightarrow{HBr} CH_3{-}CH(Br){-}OCH_3$

First protonation occurs, two possible intermediates are

$\overset{+}{C}H_2 \rightarrow CH(H) \rightarrow OCH_3$ and $CH_3{-}\overset{+}{C}H{-}\ddot{O}CH_3$

(I) (–*I*-effect of OCH_3 destabilises carbocation)

(II) (+*M*- effect stabilises carbocation)

II is more favourable, hence Br^- attacks and product is

$$CH_3{-}CH(Br){-}OCH_3$$

17. $H_3C{-}CH_2{-}CH(CH_3){-}CH_3$ [3° (C — H) bond has minimum bond energy, hence easily gives 2-bromo-2-methylbutane.]

$$\xrightarrow{Br_2} CH_3{-}CH_2{-}C(Br)(CH_3){-}CH_3$$

18. $CH_3CH_2CH(Br)CH_3 \longrightarrow \underset{\text{Carbonium ion}}{CH_3CH_2\overset{\oplus}{C}HCH_3}$

$$\xrightarrow{-H^+} \underset{\text{2-butene}}{CH_3CH{=}CHCH_3} + \underset{\text{1-butene}}{CH_3CH_2CH{=}CH_2}$$

(I) More substituted (II) Less substituted

Stability of I > II, hence I is predominant.

19. $CH_2{=}CHCH{=}CH_2 + HBr \longrightarrow$

$$\underset{\text{1,2-addition product}}{CH_3CH(Br)CH{=}CH_2} + \underset{\text{1,4-addition product}}{CH_3CH{=}CHCH_2Br}$$

Addition is through the formation of allylic carbocation.

$$\underset{2° \text{ allylic (more stable)}}{CH_2{=}CH\overset{\oplus}{C}HCH_3} \longleftrightarrow \underset{1° \text{ allylic (less stable)}}{CH_3CH{=}CH\overset{\oplus}{C}H_2}$$

Under mild conditions (–80°C), kinetic product is the 1,2-addition product and under vigorous conditions (40°C), thermodynamic product is 1,4-addition product. Hence, 1-bromo-2-butene is the major product under given conditions.

20. Any aliphatic carbon with hydrogen attached to it on combination with benzene ring will be oxidised to benzoic acid by $KMnO_4/H^+$.

21. For positive iodoform test, alcohol molecule must have $CH_3{-}CH(OH){-}$ group.

$$Ph{-}CH(OH){-}CH_3 \xrightarrow{I_2+NaOH} CHI_3 + Ph{-}COO^-$$

22. Addition of HCN is nucleophilic reaction. Greater the electron deficiency of carbonyl group, higher the rate of reaction.

$$Ph{-}\overset{O}{\overset{\|}{C}}{-}Ph < Ph{-}\overset{O}{\overset{\|}{C}}{-}CH_3 < CH_3{-}\overset{O}{\overset{\|}{C}}{-}CH_3 < H{-}CHO$$

23. $CH_2{=}CH_2 \xrightarrow{H_2O/H^+} CH_3CH_2OH$

$$CH_3-\underset{CH_3}{CH}-CH{=}CH_2 \xrightarrow{H_2O/H^+}$$

$$CH_3-\underset{CH_3}{CH}-\overset{\oplus}{CH}-CH_3 \xrightarrow{H_2O} 2^\circ \text{ alcohol}$$

1, 2-hydride shift

$$\longrightarrow CH_3-\overset{\oplus}{\underset{CH_3}{C}}-CH_2-CH_3 \xrightarrow{H_2O} 3^\circ \text{ alcohol}$$

24. $CH_3-\underset{OH}{CH}-CH{=}CH-CH_3 \longrightarrow CH_3-\underset{O}{\overset{\|}{C}}-CH{=}CH-CH_3$

Suitable reagent is chromic acid in aqueous acetone solution while others will also affect (C=C) bond.

25. $A + NaOH \longrightarrow$ alcohol + acid (Cannizaro reaction)

A must be aldehyde without H at α-carbon.

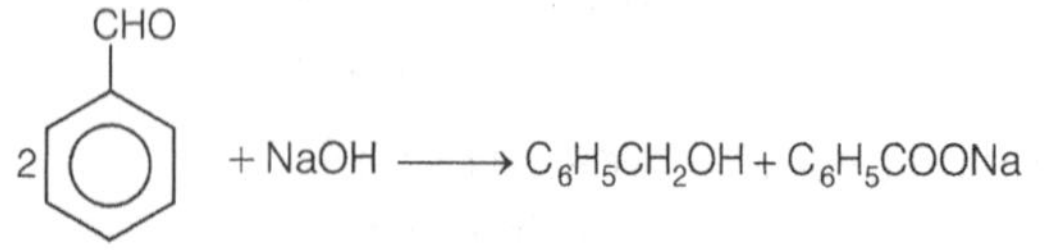

Benzaldehyde

26. $C_2H_5NC + H_2O \xrightarrow{H^+} HCOOH + C_2H_5NH_2$

$C_2H_5NH_2 + H^+ \longrightarrow \underset{\text{Salt}}{C_2H_5NH_3^+}$

27. $CHCl_3 + KOH \longrightarrow\ :CCl_2 + KCl + H_2O$

$CH_3CH_2NH_2 +\ :CCl_2 \longrightarrow \underset{\text{Carbylamine}}{CH_3CH_2NC} + 2HCl$

or $CH_3CH_2NH_2 + CHCl_3 + 3KOH \xrightarrow{\Delta} CH_3CH_2NC + 3KCl + 3H_2O$

28. In the given resonating structures, structure II is not an acceptable canonical structure because the nitrogen has ten valence electrons.

29. In structure (c), N-atom forms five bonds and contains positive charge, so the structure is not possible.

30. Nitrobenzene on reduction with $LiAlH_4$ gives azobenzene.

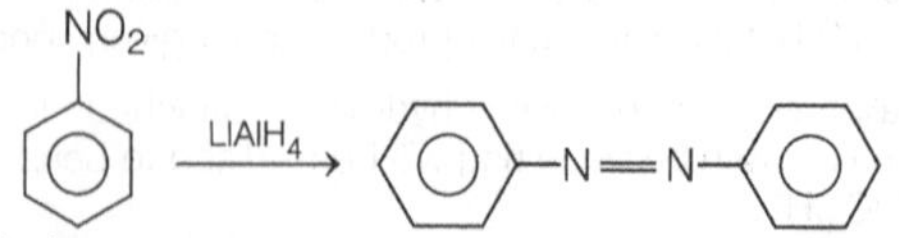

Nitrobenzene Azobenzene

Aromatic nitro compounds on reduction with $LiAlH_4$ gives azo compounds and not primary amines. Aliphatic nitro compounds are reduced to primary amines with $LiAlH_4$.

31.

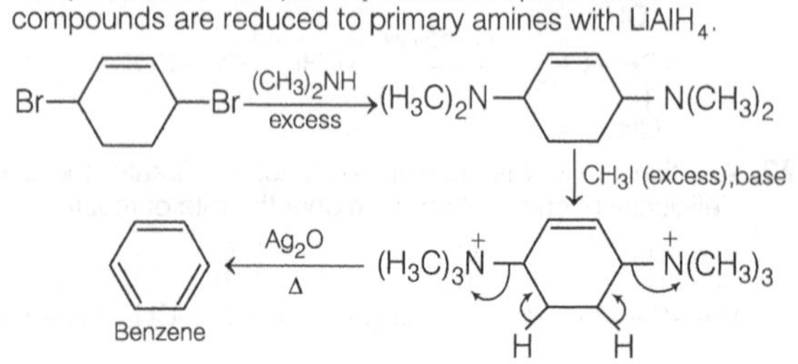

32.

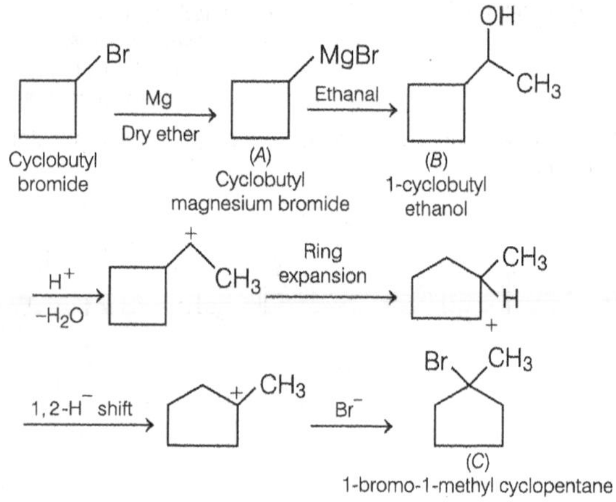

33.

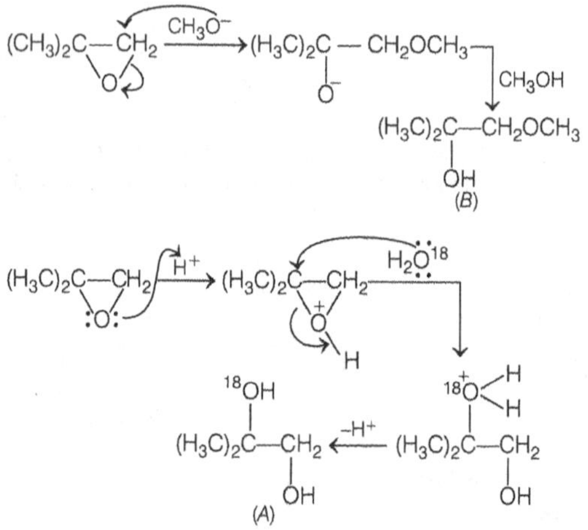

34. $\underset{\text{Grignard reagent}}{(CH_3)_3C{-}MgCl} + D_2O \longrightarrow (CH_3)_3C{\cdot}D + MgCl(OD)$

[Grignard reagent when reacts with a protic solvent, gives alkane.]

35.

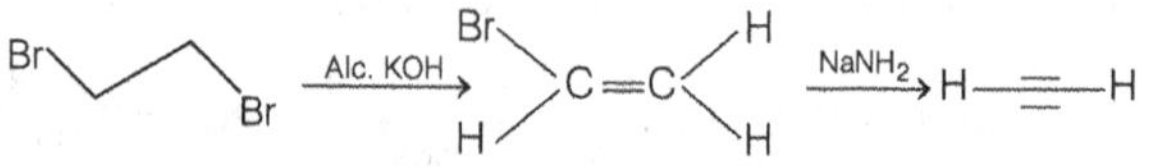

Vinylic bromide is more stable and stronger base ($-NH_2^-$) is required for elimination.

36.

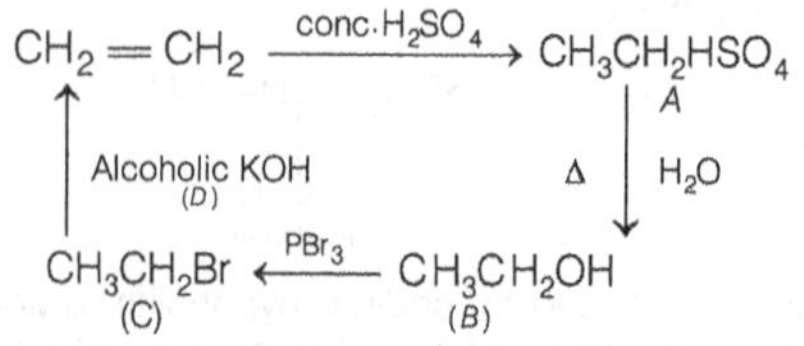

37. Only terminal alkynes give precipitate with ammoniacal silver nitrate solution. Among the given, $CH_3-C\equiv C-CH_3$ is not a terminal alkyne. Thus, it does not give precipitate with ammoniacal $AgNO_3$.

38. The formation of the alkene in an elimination reaction is called Hofmann elimination (thermal decomposition). Elimination of hydrogen occurs from the β-carbon. So,

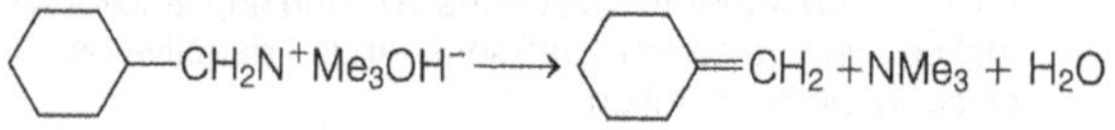

39. X is a three carbon compound with two halogen atom, so its molecular formula is $C_3H_6Cl_2$. Only terminal alkynes give red ppt. with ammoniacal Cu_2Cl_2, so the hydrocarbon produced by the reaction of X with alcoholic KOH, must be a terminal alkyne i.e. $CH_3C \equiv CH$.

$$C_3H_6Cl_2 \xrightarrow{\text{Alc. KOH}} CH_3C \equiv CH \xrightarrow{\text{Ammoniacal } Cu_2Cl_2} \underset{\text{Red ppt.}}{CH_3C \equiv CCu\downarrow}$$

Compound (X) gives an aldehyde when reacts with aqueous KOH. This suggests that both the halogens are present on same terminal carbon atom. Thus, the formula of compound X is

$$\underset{\text{(1, 1-dichloropropane)}}{CH_3 - CH_2 - CHCl_2}$$

and the reactions are as follows :

$$\underset{X}{CH_3CH_2CHCl_2} \xrightarrow{\text{Alc. KOH}}$$

$$CH_3C \equiv CH \xrightarrow{\text{Ammoniacal } Cu_2Cl_2} \underset{\text{Red ppt.}}{CH_3C \equiv CCu\downarrow}$$

$$\underset{\substack{X \\ \text{1, 1-dichloropropane}}}{CH_3CH_2CHCl_2} \xrightarrow{\text{Aq. KOH}}$$

$$\underset{\text{Unstable}}{CH_3CH_2CH(OH)_2} \xrightarrow{-H_2O} \underset{\text{Propanal}}{CH_3CH_2CHO}$$

40. Correct Assertion Phenol forms 2,4,6-tribromophenol on treatment with Br_2 in water.

Correct Reason In H_2O, phenoxide ion is formed which highly activates it towards electrophilic substitution reaction.

41. $\underset{\text{Pent-1-ene}}{\overset{5}{C}H_3 - \overset{4}{C}H_2 - \overset{3}{C}H_2 - H\overset{2}{C} = \overset{1}{C}H_2}$

$\underset{\text{Pent-2-ene}}{\overset{5}{C}H_3 - \overset{4}{C}H_2 - H\overset{3}{C} = \overset{1}{C}H - \overset{1}{C}H_3}$

When two or more compounds differ in the position of substituent atom or functional group on the carbon skeleton then it shows position isomerism. Double bond is a functional group whose position varies.

42. Both Assertion and Reason are correct as presence of nitro group (electron withdrawing group) with respect to halogen activates the aryl halides towards nucleophilic substitution reaction because the carbanions formed in the reaction are stabilised by the NO_2 group.

43. *trans*-2-butene reacts with Br_2 to produce *meso*-2,3- dibromobutane but it does not involve *syn* addition of Br_2.

44. Anilinium chloride is more acidic than NH_4Cl because it forms salt with water and liberate amine when treated with strong base.

$$C_6H_5\overset{+}{N}H_3Cl^- + NaOH \longrightarrow C_6H_5NH_2 + H_2O + NaCl$$

Anilinium ion does not show resonance because charge dispersion at ring may involve pentavalent nitrogen structure.

45. According to Lewis, electron acceptor compounds are called acids. Therefore, compounds having tendency to accept electrons will be more acidic. The correct order of acidic character is as follows :

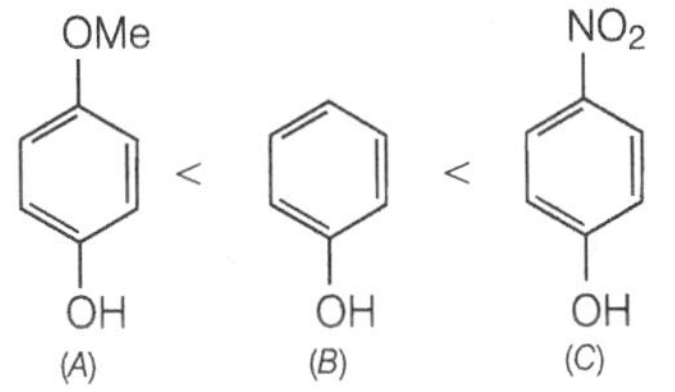

46. Six isomers (only constitutional)

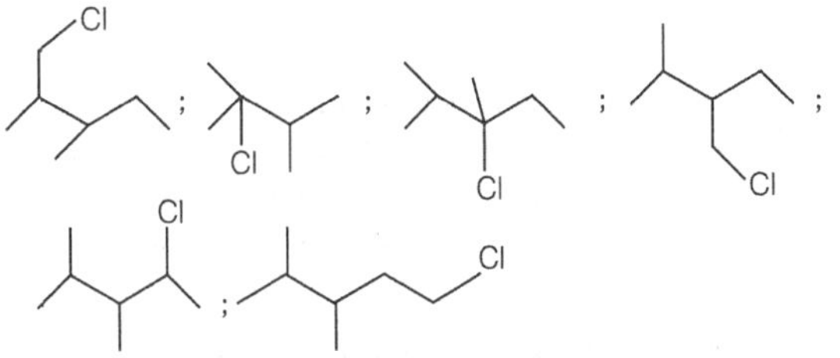

47. When aliphatic carboxylic acids react with bromine in the presence of red phosphorus, they give α-halogenated acids. This reaction is known as Hell-Volhard-Zelinsky reaction.

$$R_2CHCOOH \xrightarrow[\text{(ii) } H_2O]{\text{(i) } Br_2/\text{Red P}} R_2C(Br)-COOH$$

48. Zeisel's method is used to estimate the methoxy group in an organic compound. In this method, organic compound having methoxy group is treated with HI and the alkyl halide thus formed as further treated with $AgNO_3$ to precipitate AgI.

49.

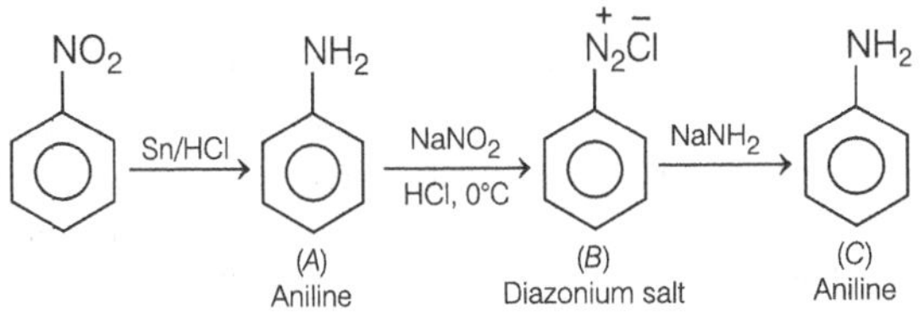

50. A → 2; B → 1; C → 4; D → 3

DAY THIRTY THREE

Polymers

Learning & Revision for the Day

- Classification of Polymers
- Polymerisation
- Some other Important Polymers

Polymers are defined as very large molecules having high molecular mass (10^3-10^7 u). These are formed by joining of repeating structural units, called monomers, on a large scale. The monomer units, are linked to each other by covalent bonds. The process of formation of polymer from respective monomers is called polymerisation.

Classification of Polymers

Some of the common classifications of polymers are :

(i) **On the Basis of Source**

- **Natural Polymers** These polymers are found in plants and animals, e.g. proteins, cellulose, starch, resins, rubber etc.
- Semi-synthetic Polymers These polymers are derived from natural polymers by chemical methods. e.g. cellulose acetate (rayon), cellulose nitrate etc.
- **Synthetic Polymers** These are man-made polymers. e.g. plastic (polythene), synthetic fibres, buna-S etc.

(ii) **On the Basis of Structure of Polymers,**

- **Linear Polymers** These polymers consist of long and straight chains, e.g. high density polythene, polyvinyl chloride etc.
- **Branched Chain Polymers** These polymers consist of linear chains with some branches, e.g. low density polythene.
- **Cross-linked or Network Polymers** These polymers contain strong covalent bonds between various linear polymer chains, e.g. bakelite, melamine etc.

(iii) **On the basis of Mode of Polymerisation**

- **Addition Polymers or Chain Growth Polymers** These are formed by the repeated addition reactions monomer molecules having multiple bonds, e.g. polythene, SBR.
- **Condensation Polymers or Step Growth Polymers** These are formed by repeated condensation process with elimination of small molecules like NH_3, H_2O etc. e.g. terylene, nylon-66, nylon-6 etc.

(iv) **On the Basis of Monomers**

- **Homopolymers** These are formed by polymerisation of single monomeric species. e.g. polythene, polypropene etc.

PREP MIRROR
Your Personal Preparation Indicator

- No. of Questions in Exercises (x)—
- No. of Questions Attempted (y)—
- No. of Correct Questions (z)—
  (Without referring Explanations)

- Accuracy Level ($z/y \times 100$)—
- Prep Level ($z/x \times 100$)—

In order to expect good rank in JEE, your Accuracy Level should be above 85 & Prep Level should be above 75.

- **Copolymers** These are formed by polymerisation of two or more different monomer units.
  e.g. buna-S, buna-N etc.

NOTE Epoxy resins are used in making adhesives such as araldite etc. These are copolymer of epichlorohydrin and bisphenol-A.

(v) **On the Basis of Molecular Forces**

- **Elastomers** These are rubber-like solids with elastic properties. They have weak intermolecular forces. e.g. buna-S, buna-N, neoprene etc.
- **Fibres** These are the thread forming solids and have strong intermolecular hydrogen bonding. e.g. nylon-6,6, polyesters (terylene) etc.
- **Thermoplastic Polymers** These polymers possess intermolecular forces of attraction, intermediate between elastomers and fibres. e.g. polythene, polystyrene, polyvinyls etc.
- **Thermosetting Polymers** These polymers are cross-linked or heavily branched molecules e.g. bakelite, urea-formaldehyde resins etc.
  On long exposure to air and sunlight, thermoplastics become brittle due to the evaporation of plasticiser.

Polymerisation

The process or technique through which monomer units combine to give a polymer is known as polymerisation. The polymerisation reaction can't be controlled easily.

Following are the two principal methods through which monomers combine to give polymers.

1. Addition Polymerisation

It is the process in which molecules of same or different monomers combine together to give a polymer without the elimination of small molecules. It is also known as chain growth polymerisation.

This mode of polymerisation is observed in molecules having multiple bonds, mainly in $CH_2{=}CHY$ type molecules (where, y may be —H, —X, —CO_2R, —CN etc.)

The addition or chain growth polymerisation can proceed by the following two mechanisms:

(i) By the formation of free radicals, i.e. free radical polymerisation.

(ii) By the formation of ionic species, i.e. ionic polymerisation (cationic or anionic).

- **Cationic addition polymerisation** is facilitated in monomers having electron releasing groups such as —CH_3, —C_2H_5 etc. Higher the stability of carbocation intermediate, more is the reactivity of monomer towards cationic addition polymerisation.
- Monomers containing electron withdrawing groups readily undergo **anionic addition polymerisation**.

2. Condensation Polymerisation

It is a combination of molecules of same or different monomers in a controlled stepwise manner with the elimination of small molecules such as water, ammonia, alcohol etc., as by-product.

The process is also known as **step growth polymerisation**

NOTE Other method of polymerisation is copolymerisation. In this method, two or more different monomer units polymerises to form a copolymer which contains multiple units of each monomer used, in the same polymeric chain.

Examples of Addition Homopolymers

(i) **Polymer** Polytetrafluoroethene (teflon)
Monomer Tetrafluoroethene ($CF_2{=}CF_2$)
Uses For making oil seals and gaskets, non-stick surface coated utensils.

(ii) **Polymer** Polythene (or polyethylene)
Monomer Ethene ($CH_2{=}CH_2$)
Uses It is used for making containers, housewares, bottles,toys, electrical insulation etc.

(iii) **Polymer** Polyacrylonitrile (PAN)
Monomer Acrylonitrile ($CH_2{=}CHCN$)
Uses Substitute for wool in making commercial fibres such as orlon or acrilan (or polyproylene).

(iv) **Polymer** Polypropene (or polypropylene)
Monomer Propene
Uses In the manufacture of ropes, toys, pipes, fibres etc.

(v) **Polymer** Polystyrene
Monomer Styrene, ($CH_2{=}CH{-}C_6H_5$)
Uses As an insulator, wrapping material, manufacture of toys, radio and television cabinets, cups for hot drinks.

(vi) **Polymer** Polyvinyl chloride (PVC)
Monomer Vinyl chloride ($CH_2{=}CH{-}Cl$)
Plasticisers are added to plastics to make them soft and rubber like. e.g. addition of dibutyl phthalate to PVC makes it soft and rubber like.
Uses In the manufacture of rain coats, hand bags, vinyl flooring, water pipes etc.

(vii) **Polymer** Natural rubber or polyisoprene (linear polymer)
Monomer Isoprene (2-methyl-1, 3-butadiene)

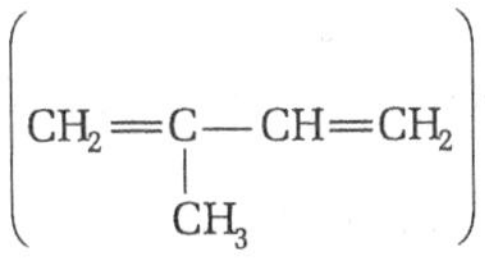

Natural rubber can be stretched like a spring and exhibits elastic properties.

(viii) **Polymer** Neoprene (synthetic rubber)

Monomer Chloroprene (2-chloro-1,3-butadiene)

$$\left(CH_2{=}\underset{\displaystyle Cl}{\underset{|}{C}}-CH{=}CH_2\right)$$

Uses In the manufacture of conveyor belts, gaskets and hoses.

Also, thiokol is a type of synthetic rubber. It is a copolymer of ethylene chloride and sodium tetrasulphide (Na_2S_4).

Examples of Addition Copolymers

(i) **Polymer** Butadiene-styrene (Buna-S)

Monomers 1,3-butadiene ($CH_2{=}CH-CH{=}CH_2$) and styrene ($C_6H_5-CH{=}CH_2$).

Uses It is quite tough and is a good substitute for natural rubber. It is used for the manufacture of auto tyres, floor tiles, footwear components, cable insulation etc.

(ii) **Polymer** Buna-N

Monomers 1,3-butadiene ($CH_2{=}CH-CH{=}CH_2$) and acrylonitrile ($CH_2{=}CHCN$).

Uses It is resistant to the action of petrol, lubricating oil and organic solvents. It is used in making oil seals, tank lining etc.

Examples of Condensation Copolymers

(i) **Polymer** Phenol-formaldehyde polymer (bakelite)

Monomers Phenol (C_6H_5OH) and formaldehyde (HCHO).

Uses It is used for making combs, phonograph records, electrical switches and handles of various utensils.

(ii) **Polymer** Melamine-formaldehyde polymer

Monomer Formaldehyde (HCHO).

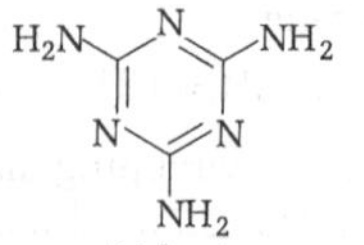

Melamine

Uses In the manufacture of unbreakable crockery.

(iii) **Polymer** Urea-formaldehyde resin

Monomers Urea (NH_2CONH_2) and formaldehyde (HCHO).

Uses For making unbreakable cups and laminated sheets.

Example of Condensation Homopolymers

Polymer Nylon-6 (perlon)

Monomer Caprolactam

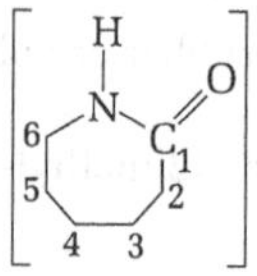

In this reaction caprolactam is first hydrolysed with water to form *E*-amino caproic acid.

Natural and Synthetic Rubber

- Natural rubber is considered as a linear polymer of isoprene (2-methyl-1,3-butadiene) and is also called as *cis*-1,4-polyisoprene.
- It absorbs a large amount of water and has low tensile strength and elasticity. Due to which, vulcanisation of natural rubber is carried out.

Vulcanisation of Rubber

- To improve the undesired properties of natural raw rubber, vulcanisation of rubber is carried out. The process involves the heating of raw rubber with sulphur and appropriate additive (ZnO) at a temperature 373-414 K.
- Sulphur forms cross links at a reactive sites of double bonds and makes the rubber stiffened. Rubber made with 1-3% sulphur is soft and stretchy while rubber made with 3-10% S is more rigid (used in making tyre).

Synthetic Rubber

Synthetic rubber are homopolymers of 1,3-butadiene derivatives or copolymers of 1,3-butadiene or its derivatives with another unsaturated monomer.

Uses Nylon-6 is used for the manufacture of tyre cords, fabrics and ropes.

Some Other Important Polymers

1. **Polythene** is a polymer of ethene. Depending upon the conditions of temperature, pressure and nature of catalyst, polythene are of two types:

(i) **Low Density Polythene** (LDP) is tough, flexible, transparent, chemically inert as well as poor conductor of electricity. It has moderate tensile strength but good tearing strength.

It is used in the insulation of electricity carrying wires and manufacture of squeeze bottles, toys and flexible pipes.

$$n(CH_2{=}CH_2)\xrightarrow[\substack{\text{1000 to 2000 atm}\\ \text{(Traces of oxygen or}\\ \text{a peroxide initiator)}}]{\text{350-570 K}}\underset{\text{LDP}}{-\!\!\left(CH_2-CH_2\right)\!\!-_n}$$

(ii) **High Density Polyethylene** (HDP) has high density due to close packing. It is also chemically inert and more tougher and harder.

$$n(CH_2{=}CH_2)\xrightarrow[\substack{\text{6-7 atm}\\ \text{(Ziegler-Natta}\\ \text{catalyst)}}]{\text{333-343 K}}-\!\!\left(CH_2-CH_2\right)\!\!-_n$$

2. **Polyamides** are polymers having amide linkages. Various types of polyamides are as follows:

- Nylon-6,6 is obtained by the condensation of adipic acid and hexamethylenediamine with the elimination of water molecule.

$$\underset{\text{Hexamethylenediamine}}{nH_2N(CH_2)_6NH_2}+\underset{\text{Adipic acid}}{nHO-\overset{\displaystyle O}{\overset{||}{C}}-(CH_2)_4-\overset{\displaystyle O}{\overset{||}{C}}-OH}$$

$$\xrightarrow[-nH_2O]{\text{553 K, high pressure}}\underset{\text{Nylon-6,6}}{-\!\!\left(\overset{\displaystyle H}{\overset{|}{N}}-(CH_2)_6-\overset{\displaystyle H}{\overset{|}{N}}-\overset{\displaystyle O}{\overset{||}{C}}-(CH_2)_4-\overset{\displaystyle O}{\overset{||}{C}}\right)\!\!-_n}$$

It is a linear polymer and has very high tensile strength. It shows good resistance to abrasion. It is usually fabricated into sheets. It is used in bristles for brushes and in textile industry.

- Nylon-6 is obtained by heating caprolactum with water at a high temperature.

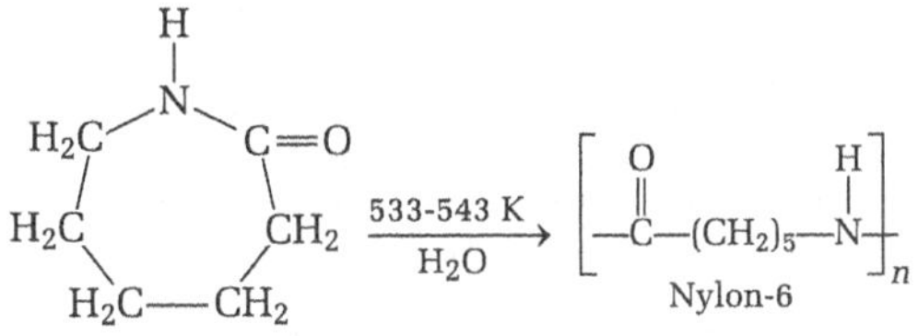

Caprolactam

It is used for the manufacture of tyre cords, fabrics and ropes.

3. **Polyesters** are the polymers which contain an ester linkage. The examples are as follows:

(i) **Polymethylmethacrylate** (PMMA) is prepared by the polymerisation of methylmethacrylate in the presence of suitable organic peroxide.

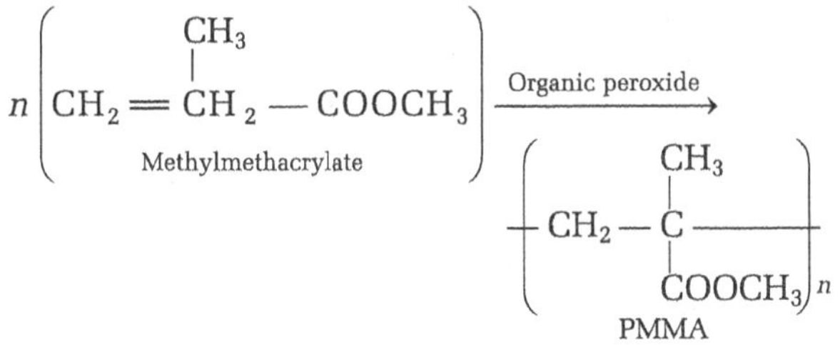

PMMA

The polymer is known by several commercial names such as lucite, acrylite, plexiglass and perspex.

It is a hard and transparent polymer and is quite resistant to the effect of light, heat and ageing. It is used, in the manufacture of unbreakable lights, protective coatings and in making windows for aircrafts.

(ii) **Glyptal** is a polyester having cross-links. It is a thermosetting plastic. It is obtained by condensation of ethylene glycol or glycerol and phthalic acid.

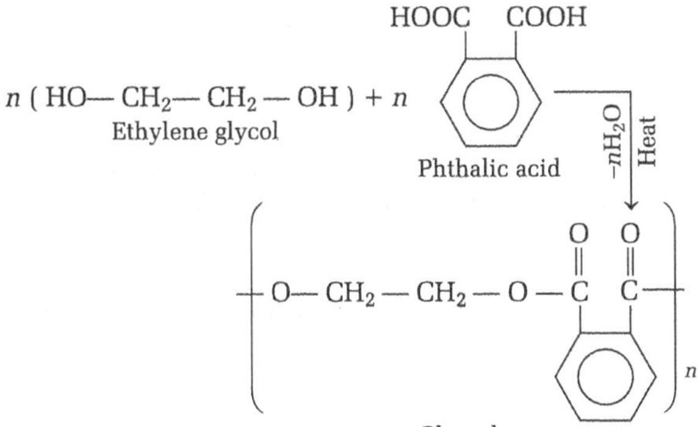

Glyptal

When its solution in a suitable solvent is evaporated, it leaves a tough but non-flexible film. Therefore, it is used in the manufacture of paints and lacquers.

(iii) **Terylene** (Dacron) is a condensation polymer of ethylene glycol and terephthalic acid. Polymerisation is carried out at 420 to 460 K in the presence of catalytic mixture of zinc acetate and antimony trioxide.

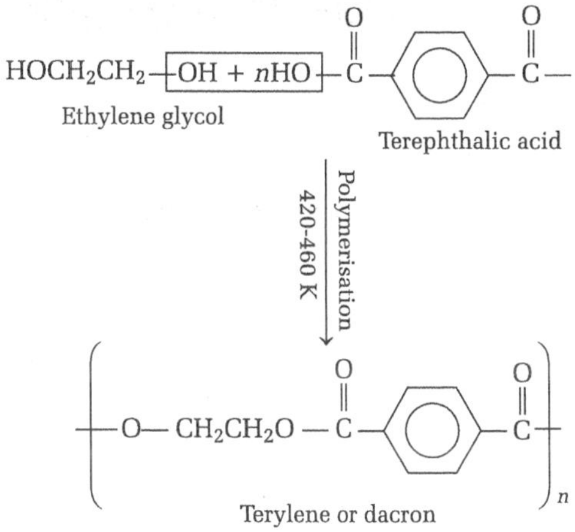

Terylene or dacron

- Terylene is highly resistant to the action of chemicals and biological agents. Its fibres are quite strong and durable. It can also be blended with wool or cotton to obtain fabrics of desired composition.
- It is used in the manufacture of a variety of clothes such as terycot, terywool and terysilk as a result of blending with other yarns. It is also used for preparing magnetic recording tapes, conveyer belts, aprons for industrial workers etc.

4. **Bakelite** is a condensation polymer. The reaction starts with the initial formation of *o*- and *p*-hydroxy methylphenol derivatives which further reacts with phenol to form compounds having rings joined to each other through —CH_2 groups. The initial product formed is novolac which on heating with formaldehyde undergoes cross-linking to form infusible mass called bakelite. It is used in making combs,switches and handles.

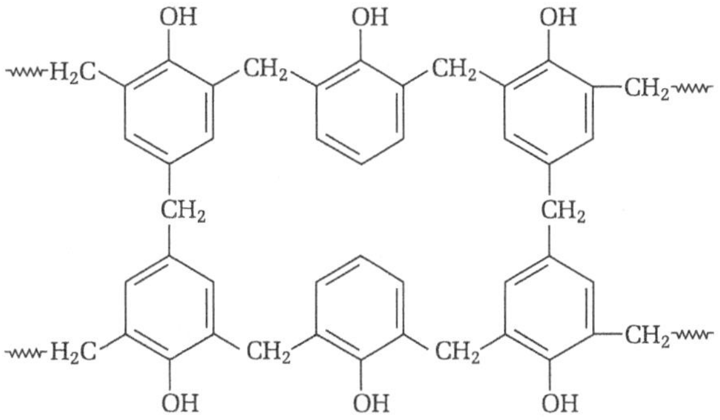

NOTE **Biodegradable polymers** can be degraded by environmental processes like degradation by bacterial enzymes. e.g. poly-β-hydroxy butyrate-co-β-hydroxy valerate (PHBV), nylon-2-nylon-6.

DAY PRACTICE SESSION 1

FOUNDATION QUESTIONS EXERCISE

1 Which of the following is not a semi-synthetic polymer?
(a) *cis*-polyisoprene (b) Cellulose nitrate
(c) Cellulose acetate (d) Vulcanised rubber

2 Which is an example of thermosetting polymer?
(a) Polythene (b) PVC
(c) Neoprene (d) Bakelite

3 The polymer containing strong intermolecular forces e.g. hydrogen bonding, is → AIEEE 2010
(a) teflon (b) nylon-6,6
(c) polystyrene (d) natural rubber

4 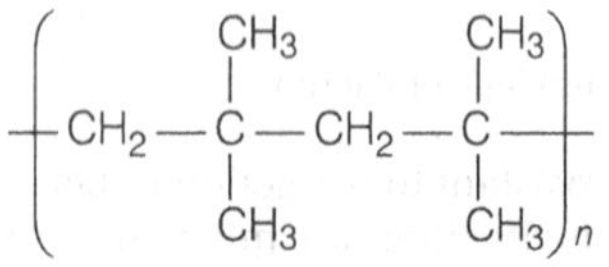

is a polymer having monomer units

(a)  (b) 

(c) 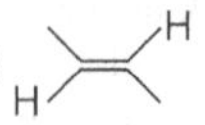 (d) 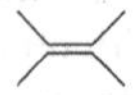

5 Which one of the following can be used as a monomer in a polymerisation reaction?
(a) CH_3CH_2Cl (b) CH_3CH_2OH
(c) C_6H_6 (d) C_3H_6

6 The compound that inhibits the growth of polymer chain during vinyl polymerisation, is
(a) carbon tetrachloride (b) *p*-benzoquinone
(c) benzophenone (d) carbon dioxide

7 Which of the following monomers will react fastest in cationic polymerisation?
$\underset{(I)}{CH_2=CH-CH_3}$, $\underset{(II)}{CH_2=C(CH_3)_2}$, $\underset{(III)}{CH_2=CH-CH(CH_3)_2}$
(a) Only I (b) I, II and III
(c) Only II (d) I and III

8 The species which can best serve as an initiator for the cationic polymerisation is → AIEEE 2012
(a) $LiAlH_4$ (b) HNO_3 (c) $AlCl_3$ (d) BaLi

9 Polymer formation from monomers starts by
(a) condensation reaction between monomers
(b) elimination reaction between monomers
(c) conversion of monomer to monomer ions by protons
(d) hydrolysis of monomers

10 Which of the following is fully fluorinated polymer?
(a) PVC (b) Thiokol
(c) Teflon (d) Neoprene

11 Which of the following contains isoprene units?
(a) Natural rubber (b) Nylon-6,6
(c) Polyethylene (d) Dacron

12 Which of the following monomers gives the polymer neoprene on polymerisation?
(a) $CH_2=CHCl$ (b) $CCl_2=CCl_2$
(c) $CH_2=CCl-CH=CH_2$ (d) $CF_2=CF_2$

13 Buna-N-asynthetic rubber is a copolymer of
(a) $H_2C=CH-\overset{Cl}{\overset{|}{C}}-CH_2$ and $H_2C=CH-CH=CH_2$
(b) $H_2C=CH-CH=CH_2$ and $H_5C_6-CH=CH_2$
(c) $H_2C=CH-CN$ and $H_2C=CH-CH=CH_2$
(d) $H_2C=CH-CN$ and $H_2C-\underset{CH_3}{\underset{|}{C}}=CH_2$

14 Synthetic human hair wigs are made from copolymerisation of vinyl chloride and acrylonitrile and is called
(a) PVC (b) polyacrylonitrile
(c) cellulose (d) dynel

15 Soft drink bottles and baby feeding bottles are generally made up of
(a) polyester (b) polyurea
(c) polyamide (d) polystyrene

16 Polymer obtained by condensation polymerisation is
(a) polythene (b) teflon
(c) phenol-formaldehyde (d) nitrile rubber

17 The catalyst used for the polymerisation of olefins is
(a) *Ziegler-Natta* catalyst (b) Wilkinson's catalyst
(c) Pd-catalyst (d) Zeolite

18 The catalyst used in the manufacture of polyethylene by *Ziegler* method is
(a) titanium tetrachloride and triphenyl aluminium
(b) titanium tetrachloride and triethyl aluminium
(c) titanium dioxide
(d) titanium isopropoxide

19 Which of the following statements about low density polythene is false? → JEE Main 2016
(a) It is a poor conductor of electricity
(b) Its synthesis required dioxygen or a peroxide initiator as a catalyst
(c) It is used in the manufacture of buckets, dustbins etc
(d)Its synthesis requires high pressure

20 Which of the following polymers does not involve cross linkages?

(a) Melmac (b) Bakelite
(c) Polythene (d) Vulcanised rubber

21 Urethane is

(a) $H_2N-C\equiv N$
(b) 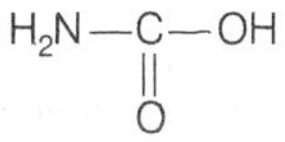
(c) $HO-C\equiv N$
(d) 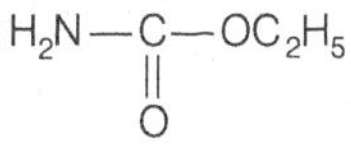

22 Which of the following is currently used as a tyre cord?

(a) Nylon-6
(b) Polyethylene
(c) Polypropylene
(d) Bakelite

23 Which one is classified as a condensation polymer? ➔ JEE Main 2014

(a) Dacron (b) Neoprene
(c) Teflon (d) Acrylonitrile

24 In which of the following polymers, empirical formula resembles with monomer?

(a) Bakelite (b) Teflon
(c) Nylon-6,6 (d) Dacron

25 Which of the following fibres is madeup of polyamides?

(a) Dacron (b) Orlon
(c) Nylon (d) Rayon

26 Which of the following polymer is a polyamide? ➔ JEE Main (Online) 2013

(a) Terylene (b) Nylon-6,6
(c) Rubber (d) Vulcanised rubber

27 The formation of which of the following polymers involves hydrolysis reaction? ➔ JEE Main 2017

(a) Nylon-6 (b) Bakelite
(c) Nylon-6, 6 (d) Terylene

28 Bakelite is obtained from phenol by reaction with ➔ AIEEE 2011, 08

(a) HCHO (b) $(CH_2OH)_2$
(c) CH_3CHO (d) CH_3COCH_3

29 Which of the following contains ester linkage?

(a) Nylon-6,6 (b) Dacron
(c) Bakelite (d) PVC

30 Polymer used in bullet proof glass is

(a) PMMA
(b) lexan
(c) nomex
(d) kevlar

31 The polymer used for optical lenses is ➔ JEE Main (Online) 2013

(a) polypropylene (b) polyvinyl chloride
(c) polythene (d) polymethyl methacrylate

32 Which of the following polymer is used in the manufacture of paints and lacquers? ➔ JEE Main 2015

(a) Bakelite (b) Glyptal
(c) Polypropene (d) Polyvinyl chloride

33 In which of the following polymers, ethylene glycol is one of the monomer units?

(a) 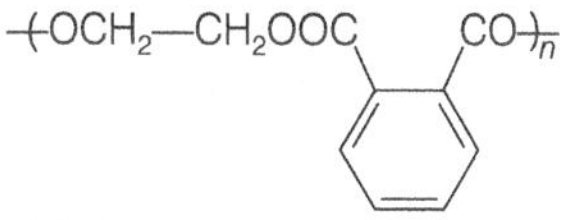
(b) $+CH_2-CH_2+_n$
(c) 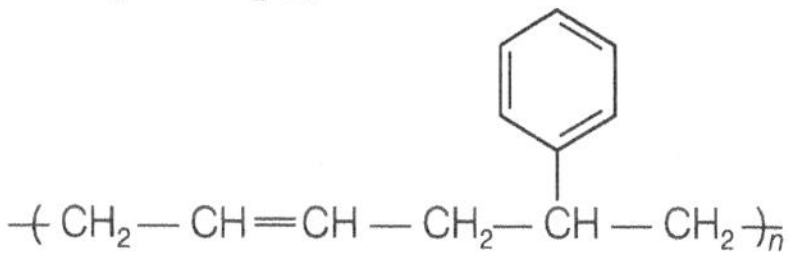
(d) 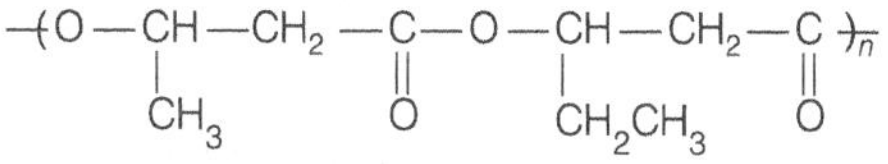

34 Terylene is a condensation polymer of ethylene glycol and

(a) benzoic acid (b) phthalic acid
(c) salicylic acid (d) terephthalic acid

35 Which of the following is not correctly matched?

(a) Neoprene : 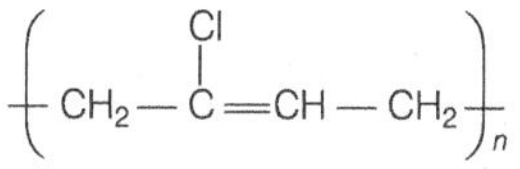
(b) Nylon-6,6 : 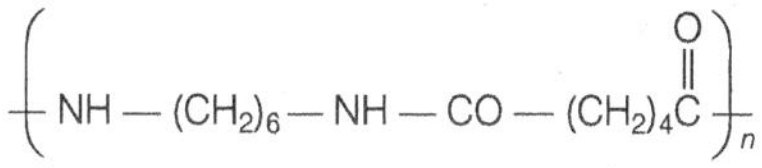
(c) Terylene : 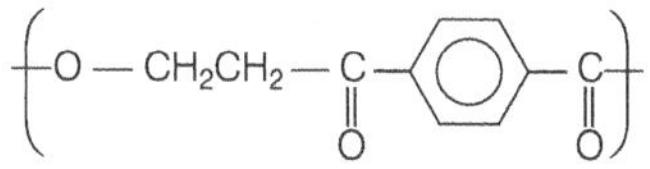
(d) PMMA : 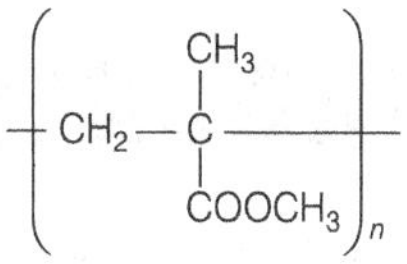

36 Which of the following polymer is biodegradable?

(a) 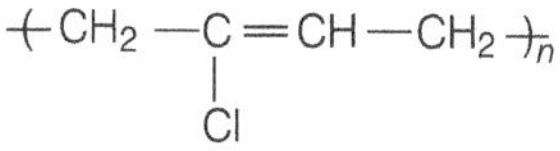
(b) $+CH_2-CH=CH-CH_2-CH_2-CH+_n$
(c) 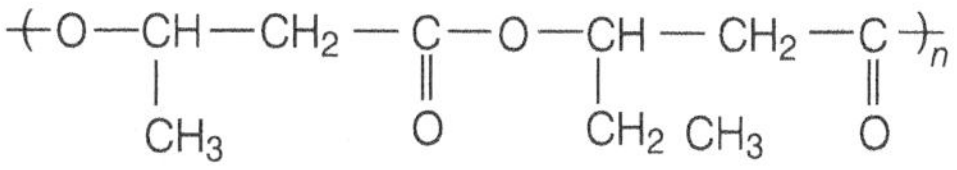
(d) 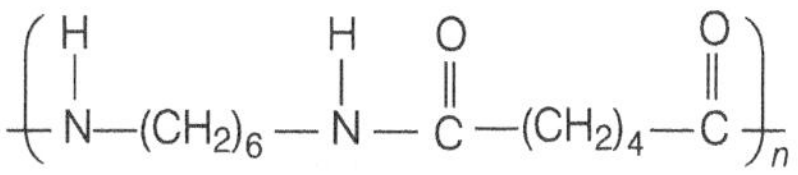

37 An example of biopolymer is

(a) teflon (b) neoprene
(c) nylon-6,6 (d) DNA

38 Match the Columns and choose the correct option.

	Column I		Column II
A.	Nylon-6	1.	Polyvinyl chloride
B.	PVC	2.	Polyacrylonitrile
C.	Acrilan	3.	Polycaprolactum
D.	Natural rubber	4.	*cis*-polyisoprene

Codes

	A	B	C	D		A	B	C	D
(a)	3	1	2	4	(b)	1	2	4	3
(c)	4	1	2	3	(d)	1	4	3	2

39 Number average molecular mass, $\overline{M}_n$ and weight average molecular mass ($\overline{M}_w$) of synthetic polymers are related as

(a) $\overline{M}_n = (\overline{M}_w)^{1/2}$
(b) $\overline{M}_n = \overline{M}_w$
(c) $\overline{M}_w > \overline{M}_n$
(d) $\overline{M}_w < \overline{M}_n$

Direction (Q. Nos 40 and 41) *In the following questions, assertion followed by reason is given. Choose the correct answer out of the following choices.*

(a) Both A and R are true and R is correct explanation of A
(b) Both A and R are true but R is not correct explanation of A
(c) A is true but R is false
(d) Both A and R are false

40 **Assertion** (A) Olefinic monomers undergo addition polymerisation.

Reason (R) Polymerisation of vinyl chloride is initiated by peroxides/persulphates.

41 **Assertion** (A) Most of the synthetic polymers are biodegradable.

Reason (R) Polymerisation process induces toxic character in organic molecules.

Direction (Q. Nos. 42-44) *Each of these questions contains two statements : Statement* I *and Statement* II. *Each of these questions also has four alternative choices, only one of which is the correct answer. You have to select one of the codes (a), (b), (c) and (d) given below :*

(a) Statement I is true, Statement II is true; Statement II is a correct explanation for Statement I
(b) Statement I is true, Statement II is true; Statement II is not a correct explanation for Statement I
(c) Statement I is true; Statement II is false
(d) Statement I is false; Statement II is true

42 **Statement I** Acrilan is a homopolymer.

Statement II Its another name is PAN.

43 **Statement I** Buna-S, buna-N and neoprene rubber are elastomers.

Statement II Polymer chains are held together by weak van der Waals' forces.

44 **Statement I** Phenol-formaldehyde polymer (bakelite) is a thermosetting polymer.

Statement II Bakelite polymer is cross-linked or heavily branched molecules which on heating undergoes extensive cross linking.

DAY PRACTICE SESSION 2

PROGRESSIVE QUESTIONS EXERCISE

1 The chemical name of melamine is

(a) 2, 4- diamino- 1, 3, 5 - triazine
(b) 2- amino -1, 3, 5 - triazine
(c) 2, 4, 6 - triamino –1, 3, 5 - triazine
(d) 1, 3, 5 - triamino –2, 4, 6 - triazine

2 Lactum from which nylon-4 is synthesised is

(a) 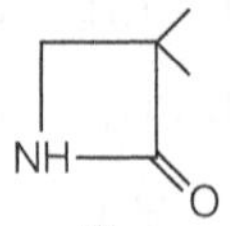

(b) 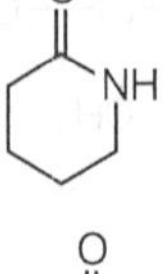

(c) 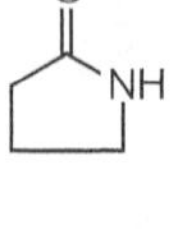

(d) 

3 If N_1, N_2, N_3........ are the number of molecules with moelcular masses M_1, M_2, M_3..... respectively then mass average molar mass is expressed as

(a) $\dfrac{\Sigma N_iM_i^2}{\Sigma N_iM_i}$
(b) $\dfrac{\Sigma N_iM_i^2}{\Sigma N_i}$
(c) $\dfrac{\Sigma M_i^2}{\Sigma N_i}$
(d) $\dfrac{\Sigma N_iM_i^2}{\Sigma M_i}$

4 In the vulcanisation of rubber

(a) sulphur reacts to form a new compound
(b) sulphur cross links are introduced
(c) sulphur forms a very thin protective layer over rubber
(d) All given statements are correct

5 Free radical polymerisation requires a free radical initiator. The most commonly used free radical initiator is
(a) $Ph—CO—O—O—COPh$, benzoylperoxide
(b) $(CH_3)_3C—O—O—C(CH_3)_3$, *tert*-butylperoxide
(c) $C_6H_5—N \longrightarrow O$ (with $C_6H_5—N$ double bonded below), azoxybenzene
(d) CH_2N_2, diazomethane

6 Which of the following is true for polypropylene?
(a) Propylene, condensation polymer
(b) Propylene, addition polymer
(c) propylene, anionic polymers
(d) Propylene, cationic polymers

7 Which of the following sets contain only copolymers?
(a) SBR, glyptal, nylon- 6, 6
(b) Polythene, polyester, PVC
(c) Nylon-6, butyl rubber, neoprene
(d) Melemac, bakelite, teflon

8 A polymeric sample consist of 10% by weight of a macromolecule of molecular weight 10000 and 90% by weight of a macromolecule with molecular weight 100000, then its weight average molecular mass will be
(a) 52000 (b) 85000
(c) 91000 (d) 89000

9 Natural rubber is not used in making footwear for polar regions because
(a) natural rubber becomes soft at temperature lower than 10° C
(b) natural rubber becomes brittle at temperature lower than 10°C
(c) natural rubber melts at temperature lower than 10°C
(d) natural rubber becomes stronger at temperature lower then 10°C

10 Arrange the following in increasing order of their intermolecular forces; Nylon 6,6 (I), Buna-S (II), Polythene (III)
(a) II, I, III (b) III, II, I (c) I, II, III (d) II, III, I

11 A compound (*A*) when reacts with $Ba(CN)_2$ and HCN, addition takes place and cyano derivative (*B*) is formed. On heating compound (*B*) in the presence of $FeSO_4$ and peroxide, orlon polymer is obtained. Compound (*A*) is
(a) $CH_2{=}CHCN$ (b) $CH{\equiv}CH$
(c) $CH_2{=}CHCl$ (d) $H_2C—CH_2$ bridged by O (epoxide)

12 Orlon is a hard, horny and a high melting material. Which of the following represents its structure?
(a) $\left(CH_2—\underset{COOC_2H_5}{\underset{|}{CH}}\right)_n$ (b) $\left(CH_2—\underset{Cl}{\underset{|}{CH}}\right)_n$
(c) $\left(CH_2—\underset{CN}{\underset{|}{CH}}\right)_n$ (d) $\left(CH_2—\overset{CH_3}{\overset{|}{\underset{COOCH_3}{\underset{|}{C}}}}\right)_n$

13 Which of the following pair of monomers form biodegradable polymers?
I. 3-hydroxybutanoic acid + 3-hydroxypentanoic acid
II. Glycine + amino caproic acid
III. Ethylene glycol + phthalic acid
IV. Caprolactam
(a) I, II only (b) II, III only
(c) I, II and III (d) II, III and IV

14 A polymeric sample in which 30% molecules have a molecular mass 20,000, 40% have 30,000 and the rest 30% have 60,000. The $(\overline{M}n)$ and $(\overline{M}w)$ of this sample are,
(a) 36,000, 43,333 (b) 43,333, 36000
(c) 72,000, 86,666 (d) 86,666, 72000

15 Polyethylene is obtained from calcium carbide as
$$CaC_2 + 2H_2O \longrightarrow Ca(OH)_2 + C_2H_2$$
$$C_2H_2 + H_2 \longrightarrow C_2H_4$$
$$nC_2H_4 \longrightarrow (CH_2—CH_2)_n$$
Therefore, the amount of polyethylene obtained from 64 kg CaC_2 is
(a) 7 kg (b) 14 kg (c) 21 kg (d) 28 kg

ANSWERS

SESSION 1	**1** (a)	**2** (d)	**3** (b)	**4** (a)	**5** (d)	**6** (b)	**7** (c)	**8** (c)	**9** (a)	**10** (c)
	11 (a)	**12** (c)	**13** (c)	**14** (d)	**15** (d)	**16** (c)	**17** (a)	**18** (a)	**19** (c)	**20** (c)
	21 (d)	**22** (a)	**23** (a)	**24** (c)	**25** (c)	**26** (b)	**27** (a)	**28** (a)	**29** (b)	**30** (b)
	31 (d)	**32** (b)	**33** (a)	**34** (d)	**35** (c)	**36** (c)	**37** (d)	**38** (a)	**39** (c)	**40** (b)
	41 (d)	**42** (b)	**43** (a)	**44** (a)						
SESSION 2	**1** (c)	**2** (c)	**3** (a)	**4** (b)	**5** (a)	**6** (b)	**7** (a)	**8** (c)	**9** (b)	**10** (d)
	11 (b)	**12** (c)	**13** (a)	**14** (a)	**15** (d)					

Hints and Explanations

SESSION 1

1 *cis*-polyisoprene is a natural polymer.

2 Bakelite is a thermosetting polymer.

3 In nylon-6,6 hydrogen bonds are formed between $-\overset{\overset{O}{\|}}{C}-NH-$ group of successive chains.

4 The repeating structural unit is $-CH_2-\underset{\underset{CH_3}{|}}{\overset{\overset{CH_3}{|}}{C}}-$ and hence, the monomer is isobutylene, 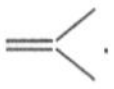 .

5 Only olefins (here, C_3H_6) and dienes undergo addition polymerisation.

6 Certain amines, phenols and quinones are used to inhibit the growth of polymer chain.

7 $CH_2=C(CH_3)_2$ will react faster than the other two because of the formation of a tertiary carbocation (stable) during polymerisation.

8 Electron-deficient species (Lewis acid like $AlCl_3$) is used as an initiator for cationic polymerisation.

9 Polymers are obtained by the condensation reaction between monomers.

10 Teflon is $-(CF_2-CF_2)_n-$. Thus, it is fully fluorinated polymer.

11 Isoprene $\left(CH_2=\underset{\underset{CH_3}{|}}{C}-CH=CH_2\right)$ is the monomer of natural rubber.

$$nCH_2=\underset{\underset{CH_3}{|}}{C}-CH=CH_2 \xrightarrow{\text{Polymerisation}} \left[-CH_2-\underset{\underset{CH_3}{|}}{C}=CHCH_2-\right]_n$$

Natural rubber

12 Chloroprene ($CH_2=CCl-CH=CH_2$) is the monomer of neoprene.

13 Buna-N actually abbreviated as NBR. BR represents butadiene rubber, N represents nitrile (acrylonitrile).

Thus, buna-N is a copolymer of 1, 3-butadiene and acrylonitrile and it usually polymerises in the presence of sodium.

14 Dynel is a copolymer of vinyl chloride and acrylonitrile.

15 Polystyrene is used for making soft drink bottles and baby feeding bottles.

16 Phenol-formaldehyde polymer is obtained by condensation polymerisation.

17 *Ziegler-Natta* catalyst is used for the polymerisation of olefins.

18 Titanium tetrachloride + triethyl aluminium.

19 High density polythene is used in the manufacture of buckets, dustbins etc.

20 Polythene does not involve cross-linkages.

21 Urethane is $H_2N-\underset{\underset{O}{\|}}{C}-OC_2H_5$.

22 Nylon -6 is used as tyre cord.

23 Dacron is a condensation polymer of ethylene glycol and terepthalic acid.

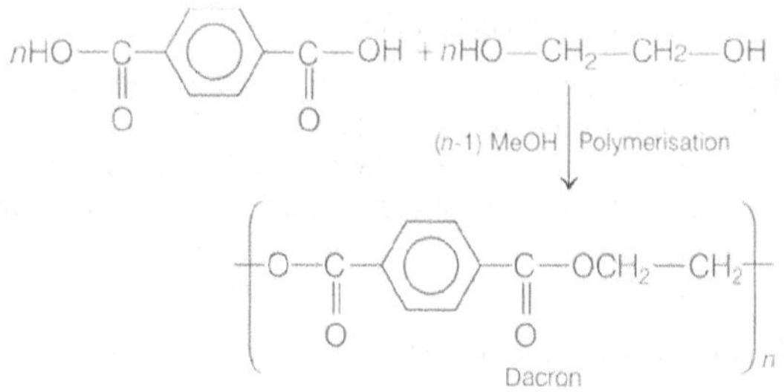

24 Nylon-6,6 is a polyamide of hexamethylene diamine $(CH_2)_6(NH_2)_2$ and adipic acid $(CH_2)_4(COOH)_2$. Each reactant has six carbon chain, hence trade code (6, 6) is used.

25 Nylon fibres is madeup of polyamides.

26 Nylon threads are made up of polyamide, some common are

Nylon-6,6 : $nH_2N(CH_2)_5-\overset{\overset{O}{\|}}{C}-OH \longrightarrow \left[HN-(CH_2)_5-\overset{\overset{O}{\|}}{C}\right]_n$

27 Nylon-6 or perlon is prepared by polymerisation of amino caproic acid at high temperature. Caprolactum is first hydrolysed with water to form amino acid which on heating undergoes polymerisation to give nylon-6.

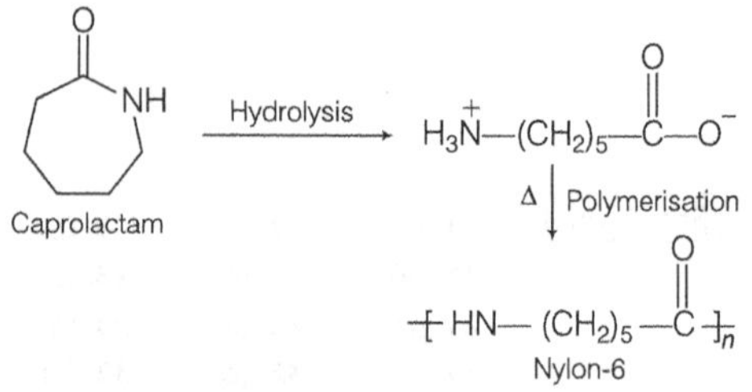

28 Bakelite is obtained from phenol by reacting, with HCHO in acidic or alkaline medium.

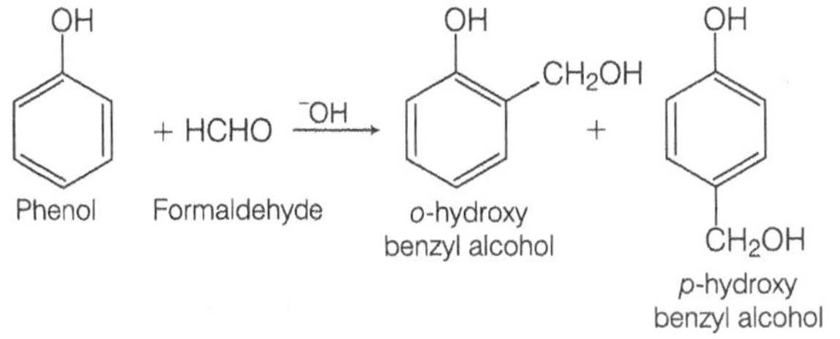

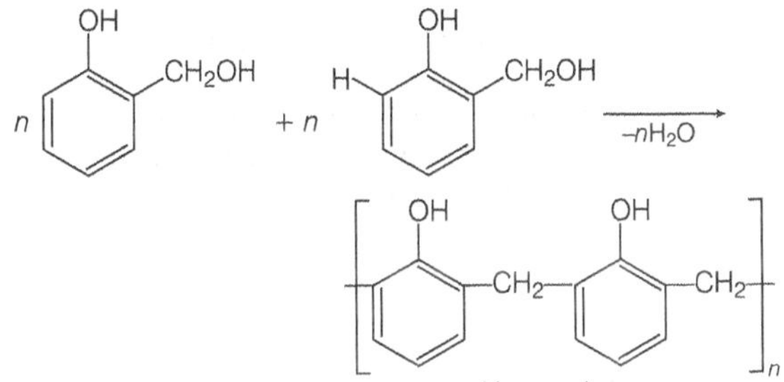

Linear polymer

Here, both the isomers again undergo condensation polymerisation and a highly cross-linked polymer, called 'bakelite' is obtained.

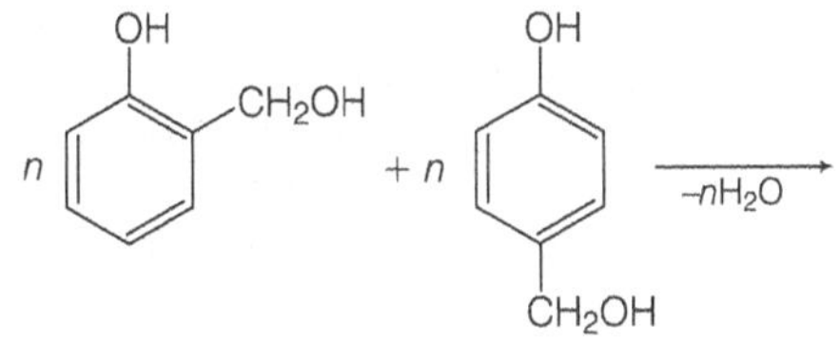

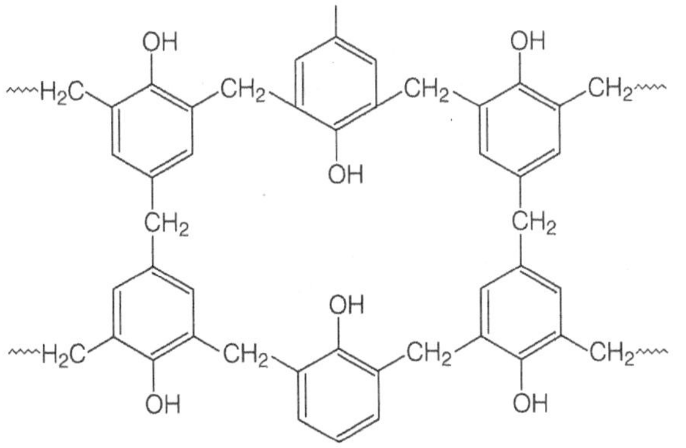

Bakelite

29 When a diacid is condensed with diol, the polymer obtained contains ester linkage, e.g. dacron.

30 Lexan or polycarbonate (polyester) is a condensation copolymer which is used in making bullet proof glass and safety helmets.

31 Polymethyl methacrylate (PMMA or plexi glass) is used as a substitute of glass. Hence, used to prepare optical lenses.

32. (a) Bakelite is used for making gears, protective coating and electrical fittings.

(b) Glyptal is used in the manufacture of paints and lacquers.

(c) Polypropene is used in the manufacture of textile, packaging materials etc.

(d) Poly vinylchloride (PVC) is used in the manufacture of rain coats,hand bags, leather clothes etc.

33 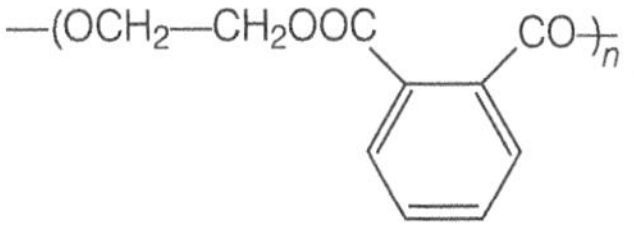 is the strucutre of glyptal whose monomers are are ethylene glycol and phthalic acid.

34 $n\text{HOOC—C}_6\text{H}_4\text{—COOH} + n\text{HOCH}_2\text{—CH}_2\text{OH} \xrightarrow{-\text{H}_2\text{O}}$
Terephthalic acid; Ethylene glycol

$\text{—(OCC}_6\text{H}_4\text{COOCH}_2\text{CH}_2\text{—O)}_n\text{—}$
Terylene or dacron

35 Terylene 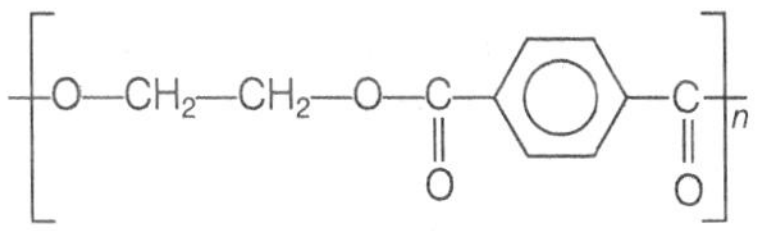 is a polymer of ethylene glycol and terephthalic acid. Thus, it is not correctly matched.

36 PHBV is a biodegradable polymer. Its structure is

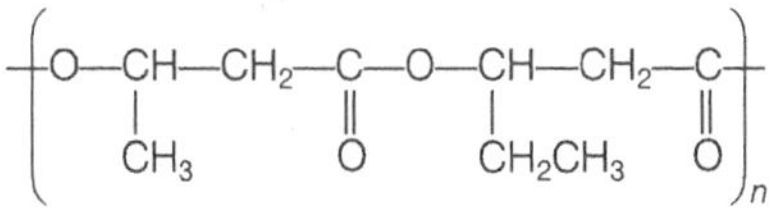

37 DNA is an example of biopolymer.

38 A ⟶ 3; B ⟶ 1; C ⟶ 2; D ⟶ 4

39
$$\text{PDI} = \frac{\overline{M}_w}{\overline{M}_n}$$

For synthetic polymer, PDI > 1

$\therefore \quad \overline{M}_w > \overline{M}_n$

40 **Correct explanation** Due to the presence of double bonds,a large number of olefin molecules simply add to form a high molecular mass molecule.

Thus, both assertion and reason are correct but reason is not the correct explanation of the assertion.

41 **Correct reason** Most of the synthetic polymers are not degraded by enzymatic hydrolysis and environmental oxidation.

42 Both statements are correct but II is not a correct explanation for I. Acrilan is a homopolymer, i.e. it is formed by polymerisation of single monomeric species.

43 Buna-S, buna-N and neoprene rubber are elastomers. Here, polymer chains are held together by weak van der Waal's forces.

44 Phenol-formaldehyde (bakelite) polymer is a thermosetting polymer. It is a cross linked polymer which on heating undergoes extensive cross linking.

SESSION 2

1 Chemical name of melamine is 2, 4, 6-triamino-1, 3, 5-triazine

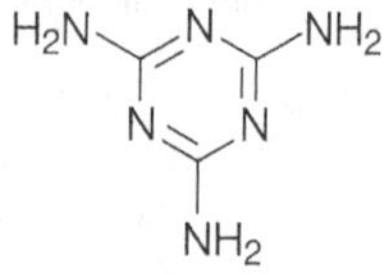

2 Lactum is used in the synthesis of nylon-4.

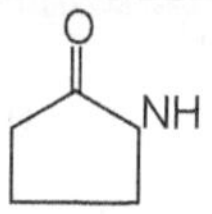

3 The expression for mass-average molar mass is

$$\overline{M}_w = \frac{\Sigma N_i M i^2}{\Sigma N_i M_i}$$

4 In vulcanisation of rubber, sulphur forms cross links at the reactive sites of double bond and makes the rubber stiffened.

5 Most commonly used free radicals generator/initiator are organic and inorganic peroxides like benzoyl peroxide $(PhCOO)_2$.

6 Polypropylene is a thermoplastic addition polymer, formed by monomeric unit, propylene (propene).

7 **Buna-S** (SBR) Monomers- (i) $CH_2=CH-CH-CH_2$ (Buta-1, 3-diene)

(ii) $\underset{\displaystyle C_6H_5}{\underset{|}{CH}}=CH_2$ (Styrene)

Glyptal Monomers (i) $HO-CH_2-CH_2-OH$ (Ethylene glycol)

(ii)

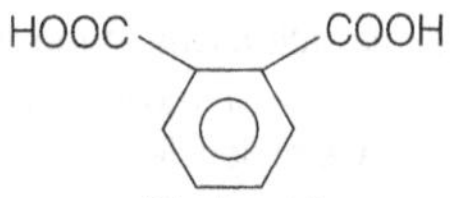

(Pthalic acid)

Nylon-6, 6 Monomers (i) $HO-\overset{O}{\overset{\|}{C}}(CH_2)_4-\overset{O}{\overset{\|}{C}}-OH$ Adipic acid

(ii) $H_2N-(CH_2)_6-NH_2$ Hexamethylene diamine

8 $\overline{M}w = \dfrac{m_1M_1 + m_2M_2 + m_3M_3 + \ldots\ldots}{m_1 + m_2 + m_3 +}$

$$\overline{M}w = \frac{0.1 \times 10000 + 0.9 \times 100000}{(0.1 + 0.9)} \approx 91000$$

9 Natural rubber becomes brittle at temperature lower than 10°C and therefore is not used for making footwears for polar regions.

10 Buna-S is an elastomer, thus has weakest intermolecular forces. Nylon-6,6, is a fibre, thus has strong intermolecular forces like H-bonding. Polythene is a thermoplastic polymers, thus the intermolecular forces present in polythene are in between elastomer and fibres. Thus, the order of intermolecular forces of these polymers is:

Buna-S < Polythene < Nylon-6,6.

11 $\underset{(A)}{CH\equiv CH} \xrightarrow[Ba(CN)_2]{HCN} \underset{\text{Vinyl cyanide } (B)}{CH_2=CHCN} \xrightarrow[FeSO_4,\text{ peroxide}]{\text{Polymerisation}} \underset{\text{Orlon}}{\left[CH_2-\overset{CN}{\overset{|}{C}H}\right]_n}$

12 Orlon is a polymer of vinyl cyanide or acrylonitrile $(CH_2=CHCN)$

13 Monomers 3-hydroxy, butanoic acid and 3-hydroxypentanoic acid react each other and form biodegradable polymer PHBV. Glycine and amino caproic acid also react to form biodegradable polymers nylon-2-nylon-6.

14 $\overline{M}_n = \dfrac{(30 \times 20000) + (40 \times 30000) + (30 \times 60000)}{(30 + 40 + 30)}$

$= 36000$

$$\overline{M}_w = \frac{30(20000)^2 + 40(30000)^2 + 30(60000)^2}{30 \times 20000 + 40 \times 30000 + 30 \times 60000}$$

$= 43333$

15 Moles of $CaC_2 = \dfrac{64 \times 10^3}{64} \approx 1 \times 10^3$ g

∴ From the balanced chemical equation,

Moles of C_2H_4 = moles of C_2H_2

= moles of $CaC_2 = 1 \times 10^3$

∴ Moles of polythene $= \dfrac{1}{n} \times 1 \times 10^3$

∴ Weight of polythene $= \dfrac{1}{n} \times 1 \times 28\, n$ kg $= 28$ kg

DAY THIRTY FOUR

Biomolecules

Learning & Revision for the Day

- Carbohydrates
- Proteins
- Enzymes
- Vitamins
- Nucleic Acids

Biomolecules are the complex lifeless organic chemical substances. These build up living organisms and are required for their growth and maintenance. Carbohydrates, proteins, nucleic acids, lipids etc., are the examples of biomolecules. These biomolecules interact with each other and constitute the molecular logic of life processes. Vitamins and mineral salts also play an important role in functions of organisms.

Carbohydrates

- Carbohydrates are optically active polyhydroxy aldehydes or ketones or substances which produce such units on hydrolysis. Most of them have general formula $C_x(H_2O)_y$. But all the compounds which fits into this formula may not be considered as carbohydrate, e.g. CH_3COOH.
- Carbohydrates are also known as **saccharides** and classified according to their behaviour towards hydrolysis.

Carbohydrates

Depending upon the units produced after hydrolysis

Monosaccharides	Oligosaccharides	Polysaccharides
Simplest saccharides which cannot be hydrolysed to smaller molecules.	(*Oligos*-few) Yield 2-10 monosaccharides on hydrolysis.	Hydrolysed to yield large number of monosaccharides.
Crystalline solids, soluble in water and sweet in taste and collectively called sugars.		Amorphous solids, insoluble in water and tasteless and thus are called **non-sugars**.
Except-sucrose, all are reducing-sugars.		All polysaccharides are non-reducing sugars.

- On the basis of functional group present, monosaccharides are classified as follows:
  (i) **Aldose** mainly contains aldehyde group along with hydroxyl group, e.g. glucose, galactose.
  (ii) **Ketose** mainly contains ketone group along with hydroxyl (—OH) group, e.g. fructose.

PREP MIRROR
Your Personal Preparation Indicator

- No. of Questions in Exercises (x)—
- No. of Questions Attempted (y)—
- No. of Correct Questions (z)— *(Without referring Explanations)*

- Accuracy Level ($z/y \times 100$)—
- Prep Level ($z/x \times 100$)—

In order to expect good rank in JEE, your Accuracy Level should be above 85 & Prep Level should be above 75.

Monosaccharides

A carbohydrate that cannot be hydrolysed further to give simpler units is called **monosaccharide**, e.g. glucose, fructose, galactose are hexose while ribose and arabinose are pentose. All monosaccharides are reducing sugars.

Glucose can be prepared by sucrose and starch. The reactions are as follows:

(i) $\underset{\text{Sucrose}}{C_{12}H_{22}O_{11}} + H_2O \xrightarrow[\text{Or invertase}]{H^+} \underset{\text{Glucose}}{C_6H_{12}O_6} + \underset{\text{Fructose}}{C_6H_{12}O_6}$

(ii) $\underset{\text{Starch}}{(C_6H_{10}O_5)_n} + nH_2O \xrightarrow[\text{393 K, 2-3 atm}]{H^+} \underset{\text{Glucose}}{nC_6H_{12}O_6}$

Some important properties of glucose are as follows:

- Glucose is an aldohexose and is also known as dextrose (Grape sugar). It is the monomer of starch and glucose.
- On prolonged heating with HI, it forms n-hexane, suggesting that all the six carbon atoms are linked in a straight chain.
- Glucose reacts with hydroxylamine to form an oxime and adds a molecule of hydrogen cyanide to give cyanohydrin. These reactions confirm the presence of carbonyl group $>C=O$ in glucose.
- Glucose get oxidised to six carbon carboxylic acid on reaction with a mild oxidising agent like bromine water. This indicates that the carbonyl group is present as an aldehyde group.
- Acetylation of glucose with acetic anhydride gives glucose penta-acetate which confirms the presence of five —OH groups.
- On oxidation with nitric acid, glucose as well as gluconic acid both yield a dicarboxylic acid (saccharic acid). This indicates the presence of primary alcoholic group.
- When glucose is warmed with excess of phenyl hydrazine, crystalline product, glucosazone is formed.

Structure and Isomerism

- Glucose forms a six membered ring in which —OH at C-5 is involved in ring formation. This explains the absence of —CHO group and also existence of glucose in two cyclic forms which are in equilibrium with open structure as (Fischer projection) shown below:

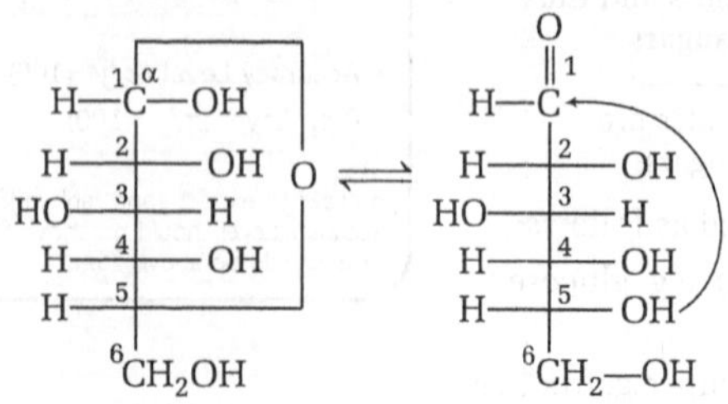

α-D-(+)-glucose

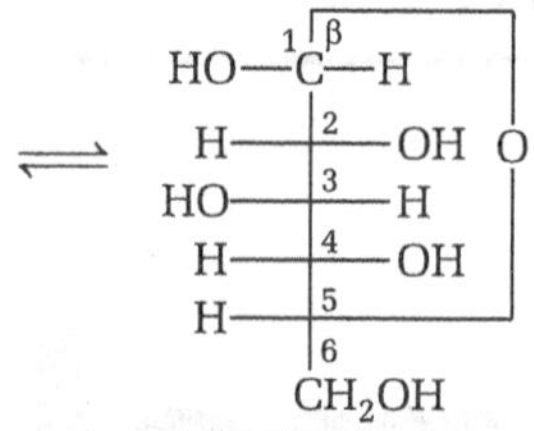

β-D-(+)-glucose

- The cyclic structure of glucose is more correctly represented by **Haworth structure** as given below:

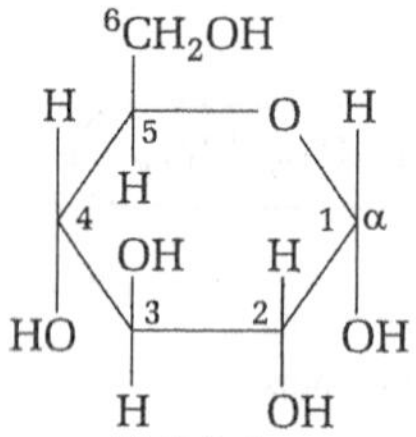

α-D-(+)-glucopyranose

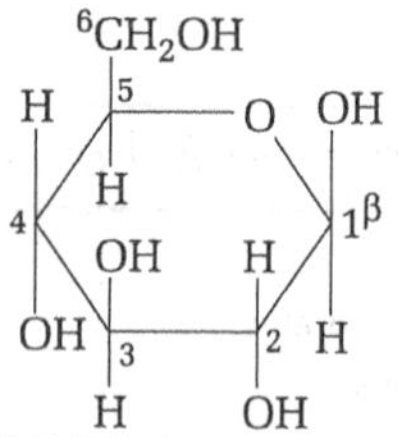

β-D-(+)-glucopyranose

- α and β-D-glucose have different configuration at anomeric (C-1) carbon atom, hence are called **anomers**.
- While the pair of diastereomeric aldoses, e.g. glucose and mannose that differ only in configuration about C-2 are called epimers. Glucose and galactose differ in configuration at C-4 are called C-4-epimers.
- Fructose is a functional isomer of glucose and has ketone group. It is obtained along with glucose by the hydrolysis of disaccharide, sucrose. It exists in two cyclic forms which are obtained by the addition of —OH at C-5 to the $>C=O$ group. The ring formed is a five membered ring and is named as furanose.

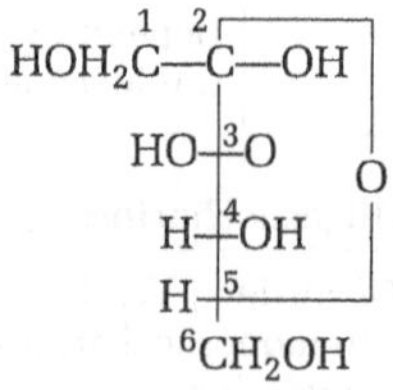

α-D-(−)-fructofuranose

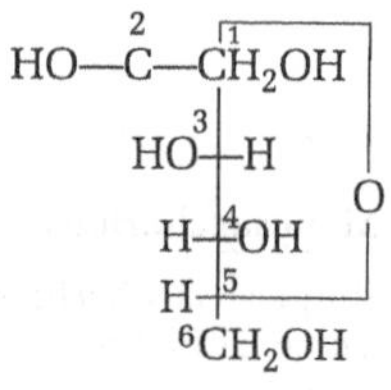

β-D-(−)-fructofuranose

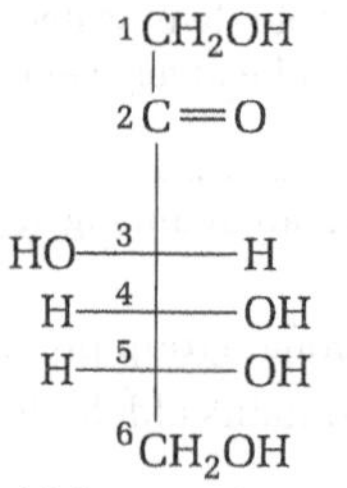

D-(−)-fructose (open structure)

The cyclic structures of two anomers of fructose are represented by Haworth structure as given

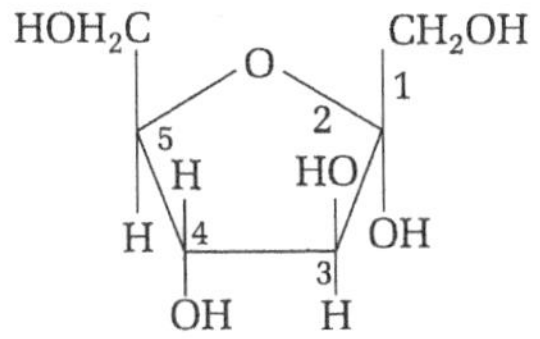

α-D-(−)-fructofuranose

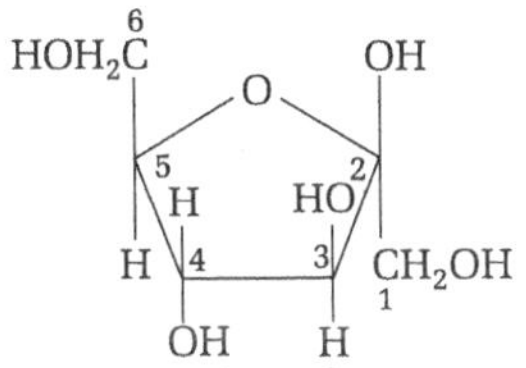

β-D-(−)-fructofuranose

Oligosaccharides

The carbohydrates that yield two to ten monosaccharide units on hydrolysis are known as oligosaccharides. These are further classified as disaccharides, trisaccharides, tetrasaccharides etc., on the basis of number of monosaccharides units obtained on their hydrolysis. e.g. disaccharides : sucrose, maltose, lactose etc.; trisaccharides : raffinose; tetrasaccharides : stachyose.

- Except sucrose all other disaccharides are reducing in nature and hence, are called reducing sugars.
- In disaccharides, the two monosaccharide units are joined together by an oxide linkage formed by the loss of a water molecule, this linkage is known as **glycosidic linkage**.

Various types of disaccharides are as follows:

1. Sucrose

Sucrose is non-reducing due to absence of free aldehyde or ketone group. It is cane sugar or table sugar.

- Sucrose (which is dextrorotatory) is also known as invert sugar. Because on hydrolysis (+) sucrose gets inverted to give a mixture of D-(+) -glucose and D-(−)-fructose.
- In sucrose, free aldehyde or ketone group is absent. It is shown by the facts that it does not form osazone, does not exist in anomeric forms and also does not show mutarotation.

Structure of sucrose is as follows:

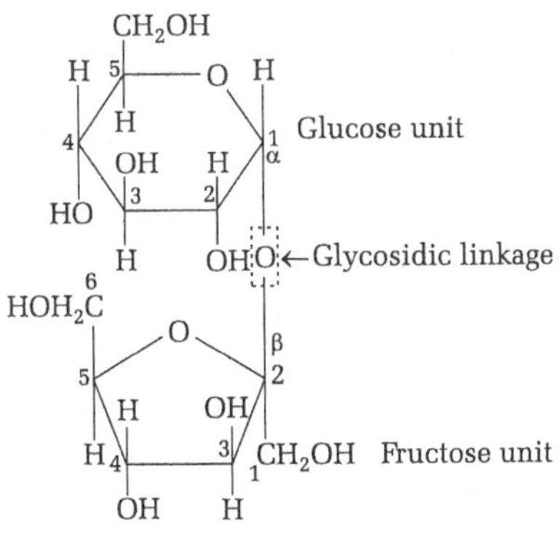

2. Maltose ($C_{12}H_{22}O_{11}$)

It is obtained by partial hydrolysis of starch by diastase enzyme present in malt, i.e. sprouted barley seeds (hence, named maltose or malt sugar).

$$2(C_6H_{10}O_5)_n + nH_2O \xrightarrow{\text{Diastase}} \underset{\text{Maltose}}{nC_{12}H_{22}O_{11}}$$

- It is a white crystalline solid (with melting point 160° − 165°C), soluble in water and dextrorotatory. When it is hydrolysed with dilute acid or by enzyme maltase, yields two molecules of D-(+) -glucose. Hence, maltose is a condensation product of two α-D-glucose units.

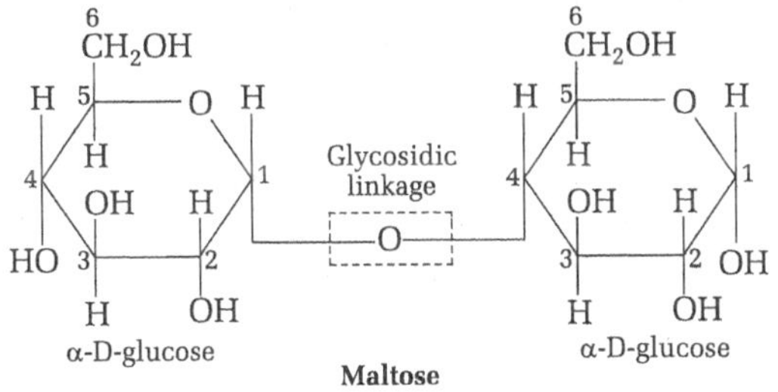

Maltose

- Maltose is a reducing sugar. It reduces Fehling's solution, Tollen's reagent, it forms an oxime and an osazone and undergoes mutarotation. This indicates that at least one aldehyde group is free in maltose.

3. Lactose ($C_{12}H_{22}O_{11}$)

It occurs in the milk of all animals (milk-sugar). It is a white crystalline solid (with melting point 203°C), soluble in water and is dextrorotatory.

- It is hydrolysed by dilute acid or enzyme lactose, to an equimolar mixture of D-(+)-glucose and D-(+)-galactose.
- Lactose is a reducing sugar, forms an oxime and osazone and also undergoes mutarotation.
- It gets hydrolysed by emulsin also, an enzyme which specifically hydrolyses β - glycosidic linkage.

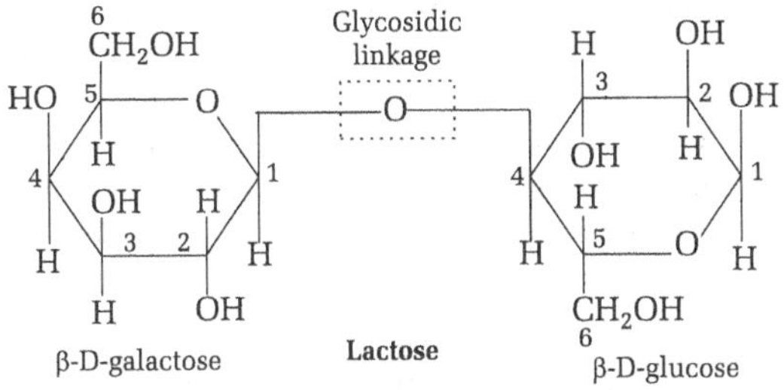

Lactose

Polysaccharides

- Carbohydrates which yield a large number of monosaccharide units on hydrolysis are called polysaccharides. e.g. starch, cellulose, glycogen, gums arabic (acidic) etc.

- Polysaccharides are not sweet in taste, hence they are also called non-sugars.
- Moreover, all polysaccharides are non-reducing due to absence of free —CHO or >CO group.

Some of the important polysaccharides are as follows:

(i) **Starch** is the main storage polysaccharides of plants. It is a polymer of α-D-glucose units and consists of two components-amylose (water soluble) and amylopectin (water insoluble).

(ii) **Cellulose** is a predominant constituent of cell wall of plant cells. It is a straight chain polysaccharide composed only of β-D-glucose units which are joined together by β-1,4-glycosidic linkage, i.e. between C-1 of one glucose and C-4 of the next glucose unit. Cellulose is not digestible by humans due to the absence of enzyme cellulase in digestive system.

(iii) **Glycogen** is the carbohydrate (a condensation polymer of α-D-glucose) which is stored in animal body. When the body needs glucose, enzymes break the glycogen down to glucose.

Proteins

Proteins are large biomolecules that occur in every living organism.

Proteins are polymers of amino acids (the compounds which have both the acid and amino group).

Amino Acids

The total number of amino acids that have been found in proteins are twenty.

On the basis of their synthesis, amino acids are divided into two classes:

1. **Essential amino acids** are the amino acids which cannot be synthesised in the body and must be obtained through diet. e.g. valine, leucine, lysine, isoleucine, arginine etc.
2. **Non-essential amino acids** are the amino acids which can be synthesised in the body. e.g. glycine, alanine, glutamic acid, aspartic acid etc.

   Amino acids behave like salts rather than simple amines or carboxylic acids. This is due to the presence of both acidic and basic groups in the same molecule.

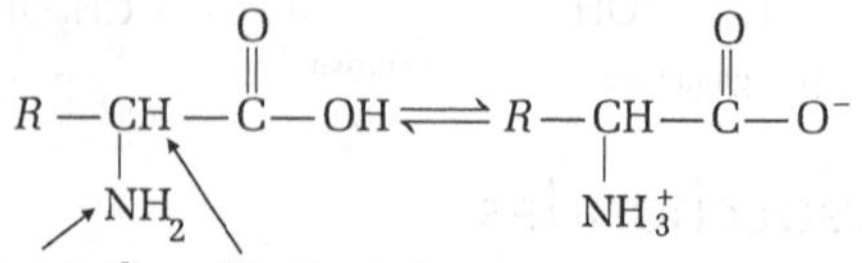

(N-terminal) (Zwitter ion)
(C-terminal) or dipolar ion (neutral)

- At a certain pH of the medium, called the isoelectric point of an amino acid, the structure behaves as a dipolar ion or *Zwitter* ion and does not migrate to any electrode on passing current.
- Proteins give biuret test, Millon's test, Ninhydrin test.

Peptides and Peptide Linkage

The basic structural unit protein is α-amino acid.

- Amino acids may be joined together by an amide linkage called peptide linkage (—CO—NH—).

e.g.

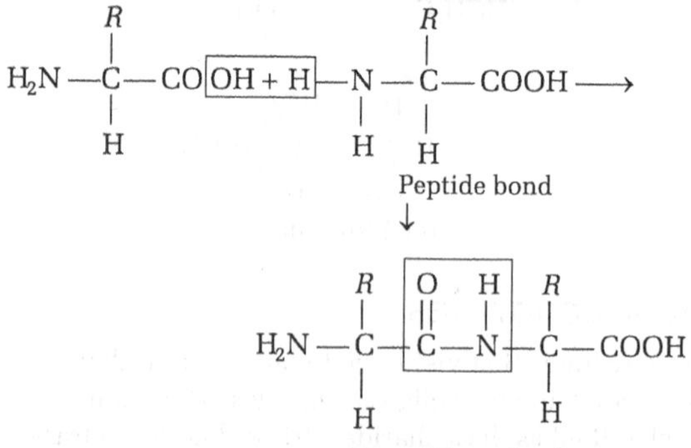

Dipeptide molecule

- The molecule derived from two amino acids containing a single peptide linkage is called a dipeptide, that derived from three amino acids is termed as a tripeptide. The peptides having 2-10 amino acid residues are called oligopeptides while those with greater than 10 amino acid residues are called polypeptides.
- Polypeptide with molecular weight greater than 10,000 u is termed as a protein.

Classification of Proteins

Proteins can be classified into two types on the basis of their molecular shape.

1. **Fibrous Proteins** Polypeptide chains form fibre like structure, e.g. keratin and myosin etc.
2. **Globular Proteins** This structure results when the chains of polypeptides coil around to give a spherical shape. These are usually soluble in water, e.g. insulin and albumins.

Structural Levels of Proteins

Structure and shape of proteins are of four different levels. These are as follows:

(i) **Primary structure** of protein involves the polypeptide that has amino acids linked in specific sequence change in this structure gives a different protein.

(ii) **Secondary structure** of proteins involves the linking of polypeptide chains by hydrogen bonds. They are found to exist in two different types of structures *viz*, α-helix and β-pleated sheet structure.

(iii) **Tertiary structure** of proteins represents overall folding of the polypeptide chains. It involves the polypeptide bonds, hydrogen bonds, disulphide linkages, van der Waals' forces and electrostatic forces of attraction. It gives rise to two major molecular shapes *viz*, fibrous and globular.

(iv) **Quarternary structure** of proteins are composed of two or more polypeptide chains referred to as subunits. The spatial arrangement of these subunits with respect to each other is known as quarternary structure.

Denaturation of Proteins

Disturbance of hydrogen bonds either by acids or alcohols or heat, results in unfolding of globules. Thus, helix get uncoiled and protein loses its biological activity due to change in temperature or pH. During denaturation, secondary and tertiary structures of proteins are destroyed while primary structures remains intact.

Enzymes

- Enzymes are globular protein bodies, which are biological catalysts. Enzyme inhibitors reduce the activity of a particular enzyme. These are mostly inorganic ions or complex organic molecules.
- Almost all the enzymes are globular proteins. These are very specific for a particular reaction and for particular substrate.
- Congenital and albinism diseases are caused by the deficiency of the enzymes phenyl ketone urea and tryosinase respectively.

Vitamins

- Organic compounds required in the diet in small amounts to perform specific biological functions for normal maintenance of optimum growth and health of the organism, are termed as vitamins.
- Provitamins are the biologically inactive compounds that have almost same structure as vitamins and can be converted easily into active vitamins. e.g. β-carotene is a provitamin for vitamin A.

Various types of vitamins are as follows:

(i) **Fat or oil soluble vitamins,** e.g. A D, E and K.

(ii) **Water soluble vitamins,** e.g. B group vitamins and vitamin C.

Some Important Vitamins, Their Sources and Their Deficiency Diseases

Name of Vitamin	Sources	Deficiency Diseases
Vitamin-A (Retinol)	Fish liver oil, carrots, butter and milk	Xerophthalmia (hardening of cornea of eye), night blindness,
Vitamin-B_1 (Thiamine)	Yeast, milk, green vegetables	Beri-beri (loss of appetite)
Vitamin-B_2 (Riboflavin)	Milk, egg white, liver, kidney	Cheilosis (fissuring at corners of mouth and lips)
Vitamin-B_6 (Pyridoxine)	Yeast, milk, egg yolk, cereals	Convulsions, nervousness
Vitamin-B_{12} (Cyanocobalamine)	Meat, fish, egg and curd	Pernicious anaemia (RBC deficient in haemoglobin)
Vitamin-C (Ascorbic acid)	Citrus fruit , amla and green leafy vegetables	Scurvy (bleeding gums)
Vitamin-D	Exposure to sunlight, fish and egg yolk	Rickets and osteomalacia
Vitamin-E	Wheat, germ oil, sunflower oil	Increased fragility of RBC and muscular weakness
Vitamin-K	Green leafy vegetables	Increased blood clotting time.

Nucleic Acids

- These are responsible in the biosynthesis of proteins.
- These are biological polymers. They function as the chemical carriers of cell's genetic information.
- These are mainly of two types the deoxyribonucleic acid (DNA) and ribonucleic acid (RNA).

1. DNA (Deoxyribonucleic Acid)

- DNA is the polymer of nucleotide.
- It is a genetic material.
- It has double helical structure.
- Nucleotide has deoxyribose sugar, phosphate and nitrogenous base.
- Nucleoside has deoxyribose sugar and nitrogenous base. A unit formed by the attachment of a base to 1′ position of sugar is known as nucleoside.
- In nucleosides, the sugar carbons are numbered as 1′, 2′, 3′ etc., in order to distinguish these from the bases. When nucleoside is linked to phosphoric acid at 5′-position of sugar moiety, we get a nucleotide.
- Nucleotides are joined together by phosphodiester linkage between 5′ and 3′ carbon.

- DNA contains four bases *viz* adenine (A), guanine (G), cytosine (C) and thymine (T). Here, adenine and guanine are called purines whereas, thymine and cytosine are called pyrimidine.
- DNA has A = T, C ≡ G

2. RNA (Ribonucleic Acid)

- It is also a polymer of nucleotide units but here, the nucleotide unit contains ribose sugar instead of deoxyribose sugar.
- RNA has adenine (A), guanine (G), cytosine (C) and uracil (U) instead of thymine (T).
- RNA molecules are of three types. These are messenger RNA (*m*RNA), ribosomal RNA (*r*RNA) and transfer RNA (*t*RNA). They perform different functions.

Biological Functions of Nucleic Acids

- DNA has an ability of self duplication during cell division and identical DNA strand are transferred to daughter cell. In this way DNA is responsible for maintaining the identity of different species of organisms over million of years.
- RNA molecules synthesised various types of proteins in the cell but the message for the synthesis of a particular type of protein is present in DNA.
- Genetic messages are encoded in DNA while RNA translated the encoded messages from DNA and thus helps in protein formation.

DAY PRACTICE SESSION 1

FOUNDATION QUESTIONS EXERCISE

1 Which of the following biomolecules contains a non-transition metal ion?

(a) Haemoglobin (b) Chlorophyll
(c) Insulin (d) Vitamin B_{12}

2 Which of the following monosaccharides is a pentose?

(a) Glucose (b) Fructose
(c) Arabinose (d) Galactose

3 D-glucose contains

(a) 50% each of α-D-glucose and β-D-glucose
(b) 64% of α-D-glucose and 36% of β-D-glucose
(c) 36% of α-D-glucose and 64% of β-D-glucose
(d) 33% of each α-D-glucose and β-D-glucose

4 Glucose on prolonged heating with HI gives → JEE Main 2018

(a) *n*-hexane (b) 1-hexene
(c) hexanoic acid (d) 6-iodohexanal

5 The two form of D-glucopyranose obtained from the solution of D-glucose are called

(a) isomer (b) anomer
(c) epimer (d) enantiomer

6 The reagent which forms crystalline osazone derivative when treated with glucose, is

(a) Fehling's solution (b) phenyl hydrazine
(c) Benedict's solution (d) hydroxylamine

7 The reaction with sugars are carried out in neutral or acid medium and not in alkaline medium because in alkaline medium sugars undergoes

(a) decomposition (b) racemisation
(c) inversion (d) rearrangement

8 Optical rotations of some compounds along with their structures are given below which of them have D configuration.

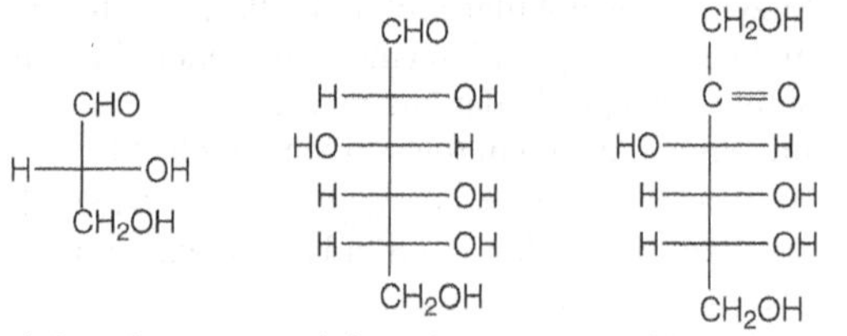

(+) rotation I (+) rotation II (–) rotation III

(a) I, II and III (b) II and III (c) I and II (d) Only III

9 Which of the following pairs give positive Tollen's test?

(a) Glucose, sucrose (b) Glucose, fructose
(c) Hexanal, acetophenone (d) Fructose, sucrose

10 The beta and alpha glucose have different specific rotations. When either is dissolved in water, their rotation changes until the same fixed value results. This is called

(a) epimerisation (b) racemisation
(c) anomerisation (d) mutarotation

11 Which of the following statement is not true about glucose?

(a) It is an aldohexose
(b) On heating with HI it forms *n*-hexane
(c) It is present in furanose form
(d) It does not give 2, 4-DNP test

12 The term invert sugar refers to an equimolar mixture of

(a) D-glucose and D-galactose
(b) D-glucose and D-fructose
(c) D-glucose and D-mannose
(d) D-glucose and D-ribose

13 Biuret test is not given by → AIEEE 2010

(a) carbohydrates (b) polypeptides
(c) urea (d) proteins

14 Which one of the following statements is correct? → AIEEE 2012

(a) All amino acids except lysine are optically active
(b) All amino acids are optically active
(c) All amino acids except glycine are optically active
(d) All amino acids except glutamic acids are optically active

15 The bond that determines the secondary structure of proteins is

(a) coordinate bond (b) covalent bond
(c) hydrogen bond (d) ionic bond

16 Which of the following compounds can be detected by Molisch's test? → AIEEE 2012

(a) Nitro compounds (b) Sugars
(c) Amines (d) Primary alcohols

17 The secondary structure of a protein refers to

(a) α-helical backbone
(b) hydrophobic interaction
(c) sequence of α-amino acids
(d) fixed configuration of the polypeptide back bone

18 Following amino acid has been found in protein prothrombin, but remained undetected due to formation of another common acid (*A*).

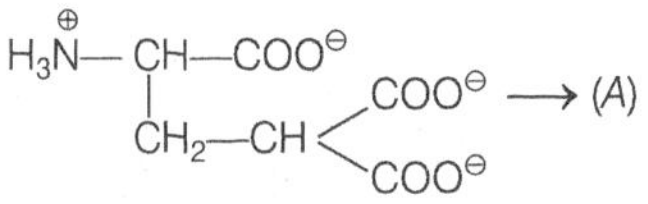

Identify '*A*'

(a) 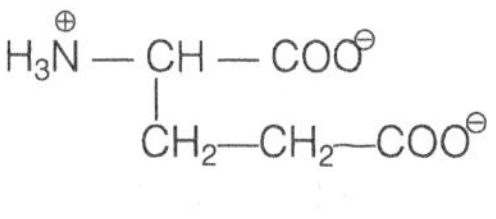 (b) 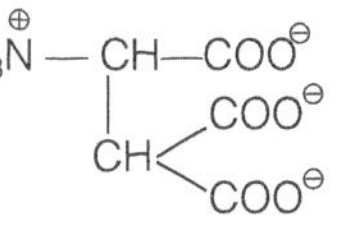

(c) 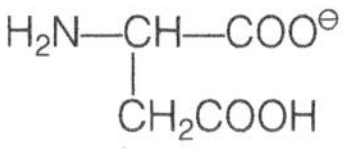 (d) 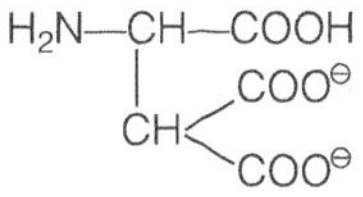

19 On heating with conc. HNO_3 proteins give yellow colour. This test is called

(a) oxidising test (b) xanthoproteic test
(c) Hoppe's test (d) acid-base test

20 Enzyme trypsin converts

(a) proteins into α-amino acids
(b) starch into sugar
(c) glucose into glycogen
(d) α-amino acids into proteins

21 Which of the following enzymes hydrolyses triglycerides into fatty acids and glycerol?

(a) Amylase (b) Maltase (c) Lipase (d) Pepsin

22 Match the following enzymes with the reactions they catalyse.

Enzymes		Reactions
A. Invertase	1.	Conversion of glucose into ethanol
B. Maltase	2.	Hydrolysis of maltose into glucose
C. Pepsin	3.	Hydrolysis of cane suger
D. Zymase	4.	Hydrolysis of proteins into amino acids

Codes

	A	B	C	D		A	B	C	D
(a)	2	1	4	3	(b)	3	2	4	1
(c)	1	4	2	3	(d)	4	2	3	1

23 Which of the vitamins given below is water soluble? → JEE Main 2015

(a) Vitamin C (b) Vitamin D
(c) Vitamin E (d) Vitamin K

24 The vitamin which is water soluble and antioxidant is

(a) Vitamin E (b) Vitamin D
(c) Vitamin C (d) Vitamin B_1

25 Identify the vitamin whose deficiency in our blood decreases reproductive power?

(a) Vitamin E (b) Vitamin D
(c) Vitamin A (d) Vitamin C

26 The chemical name of vitamin B_1 is

(a) ascorbic acid (b) riboflavin
(c) pyridoxine (d) thiamine

27 Which of the following is a fat soluble vitamin?

(a) Vitamin A (b) Riboflavin
(c) Pyridoxine (d) Thiamine

28 Adenosine is an example of

(a) nucleotide (b) nucleoside
(c) purine base (d) pyrimidine base

29 The base adenine occurs in

(a) Only DNA (b) Only RNA
(c) Both DNA and RNA (d) protein

30 The presence or absence of hydroxy group on which carbon atom of sugar differentiates RNA and DNA? → AIEEE 2011

(a) 1st (b) 2nd (c) 3rd (d) 4th

31 Which one of the following bases is not present in DNA? → JEE Main 2014

(a) Quinoline (b) Adenine (c) Cytosine (d) Thymine

Direction (Q.Nos. 32 and 33) *In the following questions. Assertion (A) followed by Reason (R) is given choose the correct answer out of the following choices.*

(a) Assertion and Reason both are correct statements and Reason is the correct explanation of the Assertion
(b) Assertion and Reason both are correct statements but Reason is not the correct explanation of the Assertion
(c) Assertion is correct and Reason is incorrect
(d) Both Assertion and Reason are incorrect

32 **Assertion** (A) All naturally occurring α-amino acids except glycine are optically active.

Reason (R) Most naturally occurring amino acids have L-configuration.

33 **Assertion** (A) Vitamin D can be stored in our body.

Reason (R) Vitamin D is fat soluble vitamin.

DAY PRACTICE SESSION 2

PROGRESSIVE QUESTIONS EXERCISE

1 In the following structure, anomeric carbon is

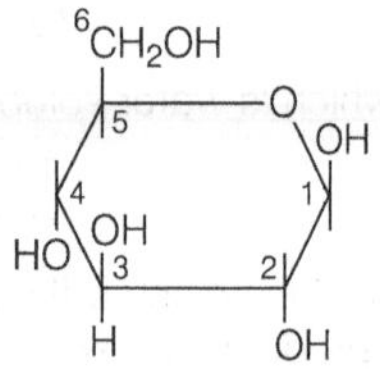

(a) 1 (b) 2 (c) 3 (d) 4

2 How can you say that glucose is cyclic compound?

(a) Glucose undergoes Tollen's reaction
(b) Glucose reacts with phenyl hydrazine
(c) Glucose fails to react with sodium hydrogen sulphate
(d) Glucose react with nitric acid

3 Which of the following reaction establishes difference between glucose and fructose?

(a) Tollen's reagent reaction (b) Phenyl hydrazine
(c) P/HI (d) Conc.HNO_3

4 Which of the following compounds will behave as a reducing sugar in an aqueous KOH solution? → JEE Main 2016

(a)
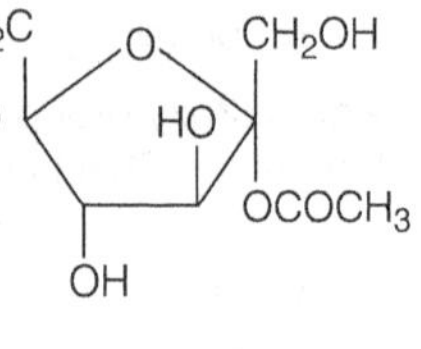

(b)
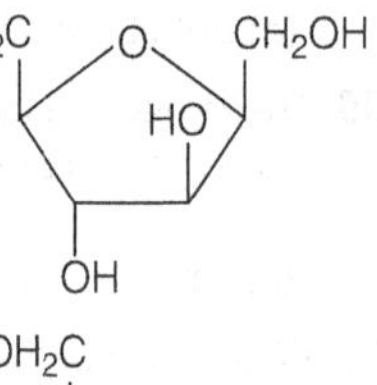

(c)
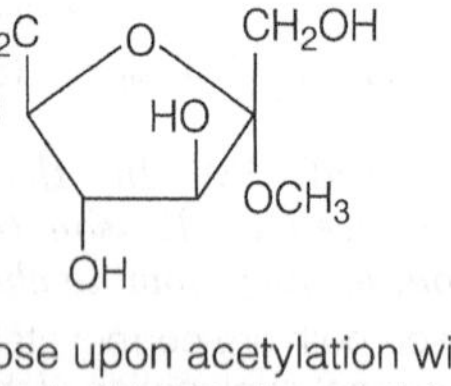

(d)
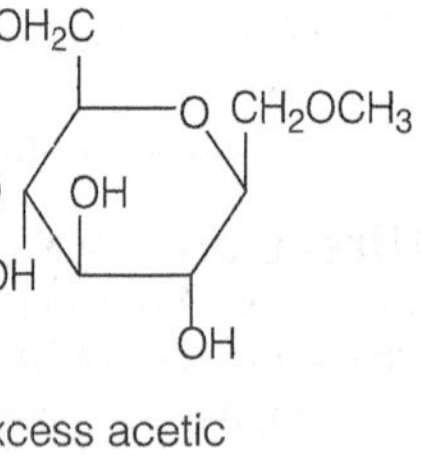

5 Cellulose upon acetylation with excess acetic anhydride / H_2SO_4 (catalytic) gives cellulose triacetate whose structure is

(a)
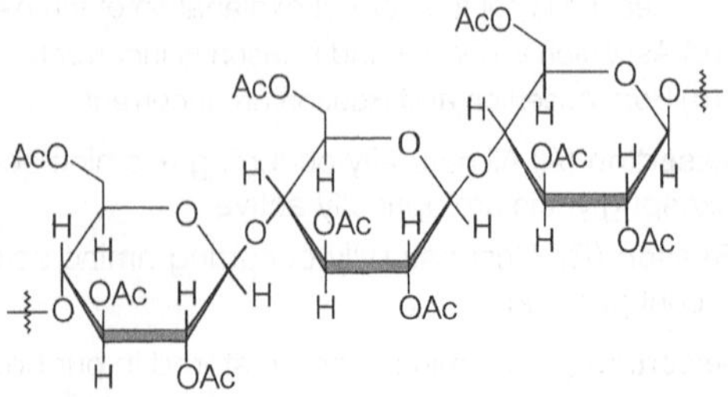

(b)
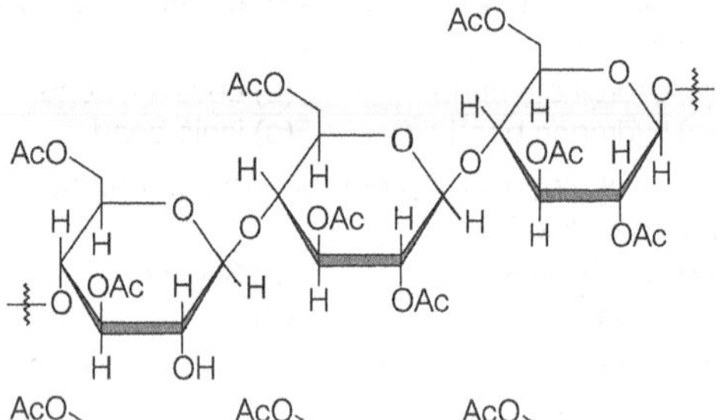

(c)
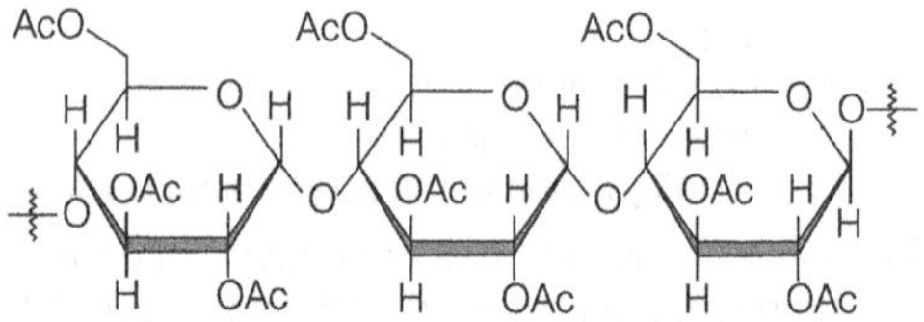

(d)
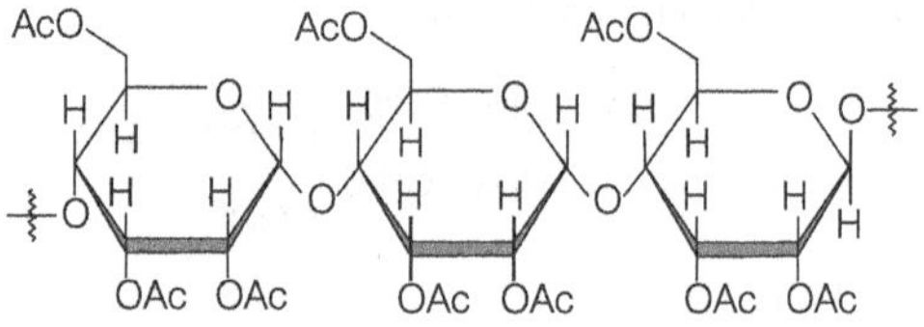

6 In disaccharides, if the reducing groups of monosaccharides, i.e. aldehydic or ketonic groups are bonded, these are non-reducing sugars. Which of the following disaccharide is a non-reducing sugar?

(a)
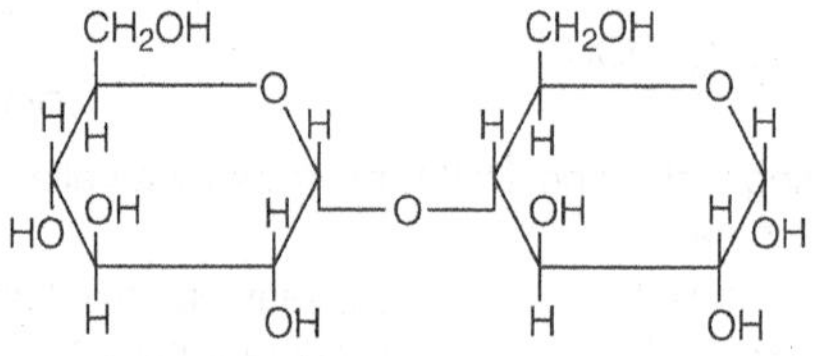

(b)
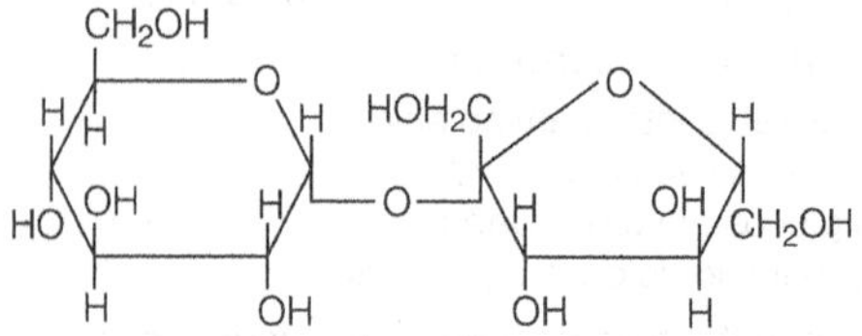

(c)
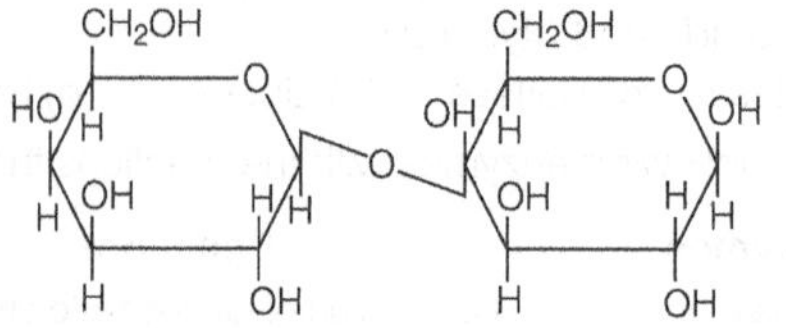

(d) 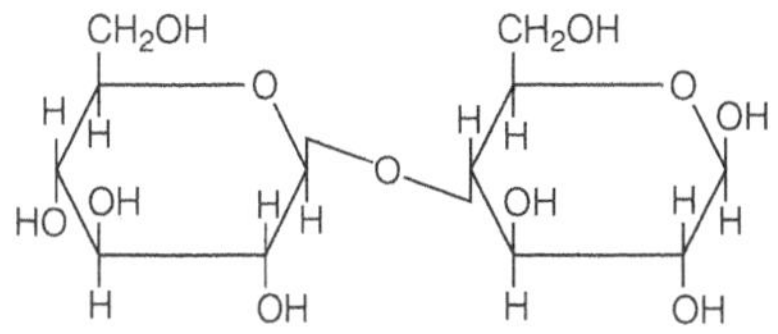

7 D-(+)-glucose reacts with hydroxyl amine and yields an oxime. The structure of the oxime would be.

(a) 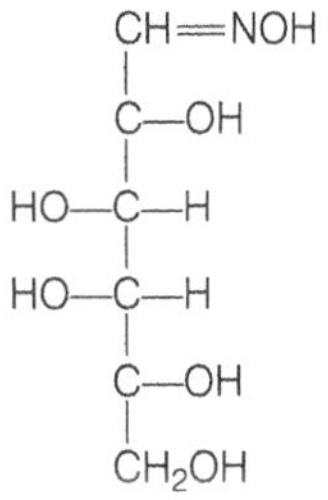

(b) 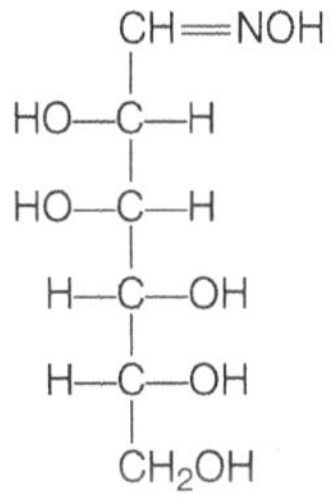

(c) 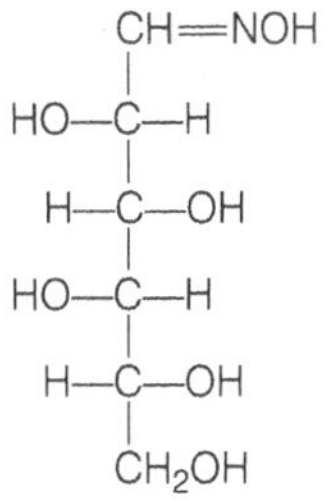

(d) 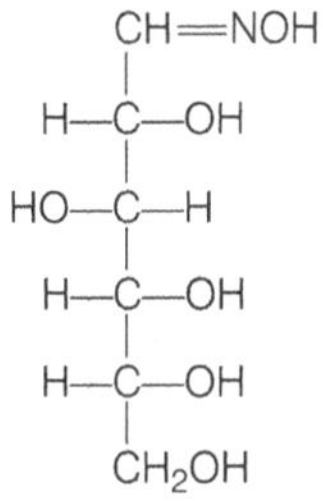

8 Compound *A*, $C_5H_{10}O_5$ gives a tetra acetate with Ac_2O and oxidation of '*A*' with Br_2 / H_2O gives an acid, $C_5H_{10}O_6$. Reduction of '*A*' with HI gives *iso*-pentane. What is the possible structure of *A*?

(a) 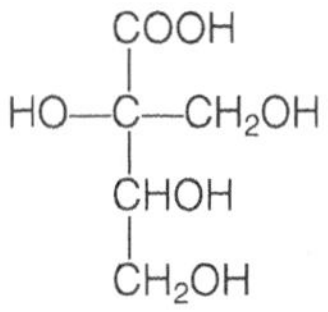

(b) 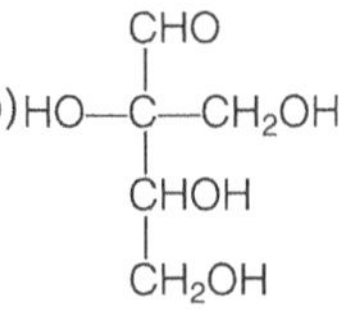

(c) 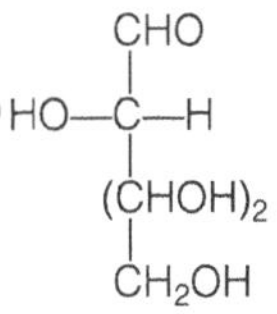

(d) 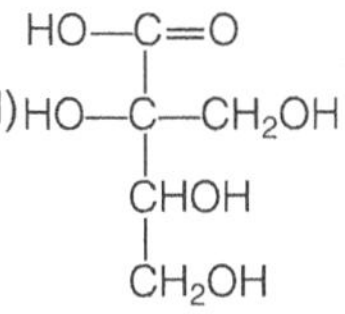

9 The final product of the reaction is

$$\text{D (glucose)} \xrightarrow[\text{HCl}]{\text{EtSH}} ? \xrightarrow[\text{NaOH}]{\text{Me}_2\text{SO}_4} ? \xrightarrow[\text{CdCO}_3]{\text{HgCl}_2} ?$$

(a) 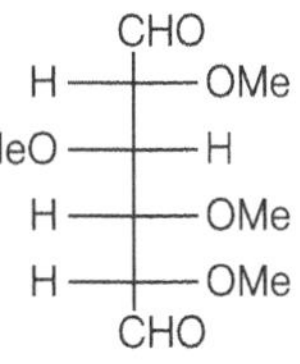

(b) 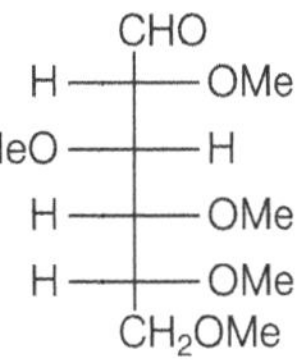

(c) 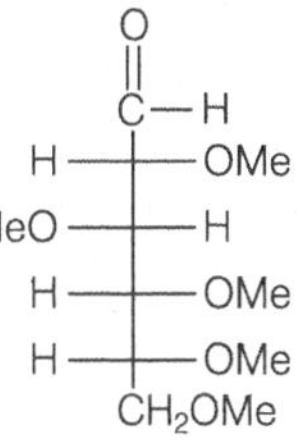

(d) 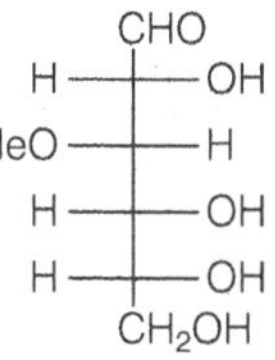

10 A sugar is classified as a D-isomer if the hydroxyl group

(a) on the chiral carbon nearest to the carbonyl point in the left

(b) on the chiral carbon nearest to the carbonyl point on the right

(c) on the chiral carbon farthest from the carbonyl point to the left

(d) on the chiral carbon farthest from the carbonyl point to the right

11 Consider the following reagents.

I. Br_2 water

II. Tollen's reagent

III. Fehling's solution

Which can be used to make distinction between an aldose and a ketose?

(a) I, II and III

(b) Both II and III

(c) Only I

(d) Only II

12 The type of bond that is most important in maintaining secondary structure of a protein in

(a) disulphide bridges

(b) hydrogen bonding within the backbone

(c) hydrogen bonding between *R* groups

(d) salt bridges

13 When a monosaccharide forms a cyclic hemiacetal, the carbon atom that contained the carbonyl group is identified at the anomeric carbon atom because

(a) the carbonyl group is drawn to the right

(b) the carbonyl group is drawn to the left

(c) its substituents can assume α or β-position

(d) it forms bond to an —*OR* and an —*OR'*

14 The backbone of a nucleic acid molecule consists of

(a) alternating sugar and phosphate groups linked by phosphate ester bonds

(b) alternating sugar and nitrogen base groups linked by amide bonds

(c) alternating nitrogen bases and phosphate groups linked by amide bonds and strengthened by hydrogen bonds

(d) sugar molecules bonded from the C-3 of one molecule to the C-5 of the other by glycosidic linkages

15 Which of the following statements about enzymes are true?

I. Enzymes lack in nucleophilic groups.

II. Pepsin is proteolytic enzyme.

III. Enzymes catalyse chemical reactions enhances the rate of reaction by lowering the activation energy.

IV. Enzymes are highly specific both in binding chiral substrates and in catalysing their reactions.

(a) Both I and II

(b) Both I and III

(c) Both I and IV

(d) II, III and IV

16 A decapeptide (mol. wt. 796) on complete hydrolysis gives glycine (mol. wt. 75), alanine and phenylalanine. Glycine contributes 47% to the total weight of the hydrolysed products. The number of glycine units present in the decapeptide is → AIEEE 2011

(a) 3 (b) 4

(c) 5 (d) 6

17 The substituents R_1 and R_2 for nine peptides are listed in the table given below. How many of these peptides are positively charged at pH = 7.0? → AIEEE 2012

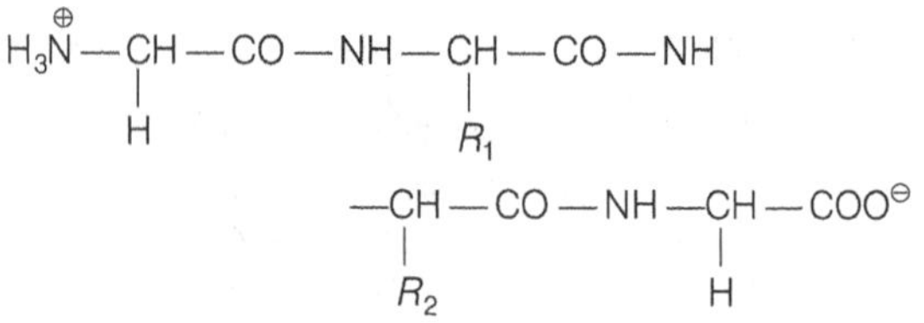

Peptide	R_1	R_2
I	H	H
II	H	CH_3
III	CH_2COOH	H
IV	CH_2CONH_2	$(CH_2)_4NH_2$
V	CH_2CONH_2	CH_2CONH_2
VI	$(CH_2)_4NH_2$	$(CH_2)_4NH_2$
VII	CH_2COOH	CH_2CONH_2
VIII	CH_2OH	$(CH_2)_4NH_2$
IX	$(CH_2)_4NH_2$	CH_3

(a) 2 (b) 4 (c) 6 (d) 8

ANSWERS

SESSION 1	**1** (b)	**2** (c)	**3** (c)	**4** (a)	**5** (b)	**6** (b)	**7** (d)	**8** (a)	**9** (b)	**10** (d)
	11 (c)	**12** (b)	**13** (a)	**14** (c)	**15** (c)	**16** (b)	**17** (d)	**18** (a)	**19** (b)	**20** (a)
	21 (c)	**22** (b)	**23** (a)	**24** (c)	**25** (a)	**26** (d)	**27** (a)	**28** (b)	**29** (c)	**30** (b)
	31 (a)	**32** (b)	**33** (a)							
SESSION 2	**1** (a)	**2** (c)	**3** (d)	**4** (a)	**5** (a)	**6** (b)	**7** (d)	**8** (c)	**9** (b)	**10** (d)
	11 (c)	**12** (b)	**13** (c)	**14** (a)	**15** (d)	**16.** (d)	**17.** (b)			

Hints and Explanations

SESSION 1

1 Chlorophyll is a biomolecules that contains magnesium ion (non-transition metal) that is held at the centre of porphyrin ring in the chlorophyll.

2 Arabinose a monosaccharide is a pentose sugar.

3 D-glucose contains 36% of α-D-glucose and 64% of β-D-glucose.

4 HI is a strong reducing agent. It reduces both primary and secondary alcoholic groups of glucose along with the carbonyl group to produce *n*-hexane as

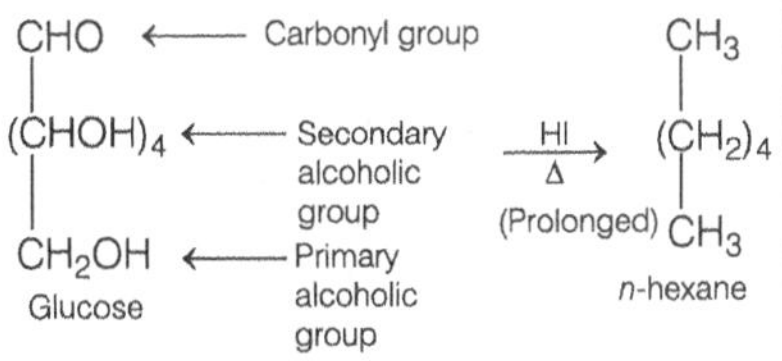

5 Two forms of D-glucopyranose are α and β. These are called **anomers** because they only differ at C_1-configuration.

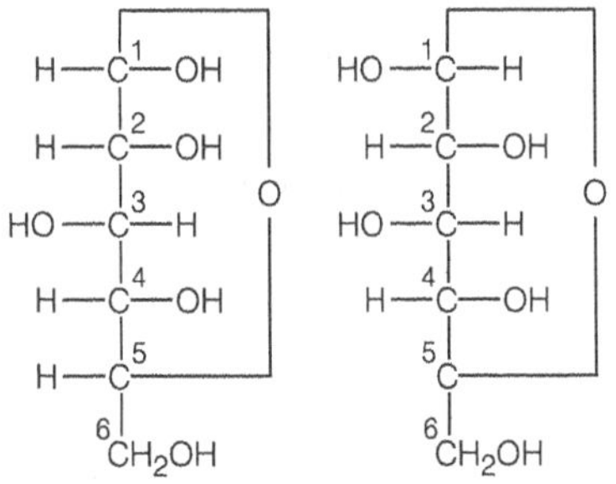

α-D (+)-glucopyranose β-D (+)-glucopyranose

6

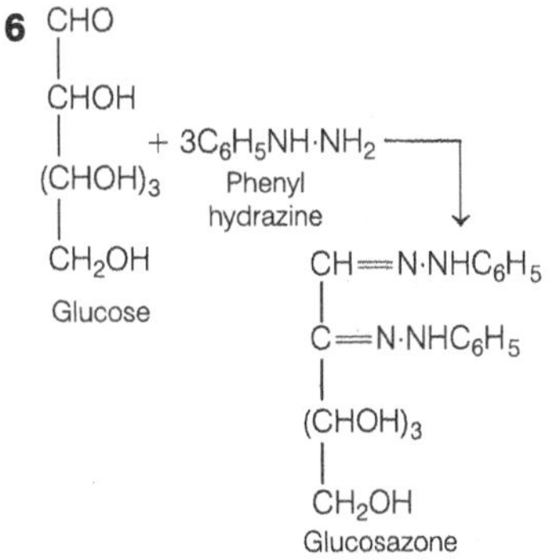

7 In alkaline medium, sugars undergoes rearrangement.

8 In all the three structures (I, II and III), the configuration of OH at lowest assymmetric carbon is towards right and hence, all have D-configuration.

9 Glucose and fructose give positive Tollen's test as they reducing sugar because of the presence of free (— CHO) group.

10 The beta and alpha glucose have different specific rotations. When either is dissolved in water, their rotation changes until the same fixed value results. This is called mutarotation.

11 The cyclic structure of glucose is represented by Haworth projection as a pyranose ring and not in furanose form. Pyranose derived from word pyran which means six membered cyclic ether.

12 D-(+)-sucrose (invert sugar) is an equimolar mixture of D-(+)-glucose and D-(–)-fructose. It is called invert sugar due to the fact that on hydrolysis there is a change in the sign of rotation from positive to negative.

13 Biuret test is characteristically given by the compound having $-\overset{\displaystyle O}{\overset{\|}{C}}-NH-$ functional group.

14 The correct statement is that all amino acids except glycine are optically active. In glycine there is no chiral carbon hence, it is optically inactive.

15 The secondary structure of protein refers to the shape in which long polypeptide chains are held together by means of hydrogen bonding.

16 Carbohydrates (e.g. sugar) can be detected by Molisch's test.

17 Secondary structure involves α-helical and β-pleated sheet like structures. α-helix is formed when the chain of α-amino acid coils as a right handed screw where in β-plated sheet the chains are held together by a large number of H-bonds. Hence, correct statement is (d).

18 Another common acid is

$$H_3\overset{+}{N}-\underset{\displaystyle CH_2-CH_2-COO^{\ominus}}{\underset{|}{CH}}-COO^{\ominus}$$

19 Xanthoproteic test involves heating of a protein with conc. HNO_3 which gives yellow colour.

20 Trypsin converts proteins into α-amino acids.

21 Lipase enzyme hydrolyses triglycerides into fatty acids and glycerol.

22 A → 3; B → 2; C → 4; D → 1

23 Vitamin B and C are water soluble while vitamin A,D,E and K are fat soluble or water insoluble.

24 Vitamin C is water soluble. It also acts as an antioxidant.

25 Deficiency of vitamin E in human being decreases reproductive power.

26 The chemical name of vitamin B_1 is thiamine.

27 Vitamin A, D, E and K are fat soluble while all others are water soluble.

28 Adenosine is a nucleoside while adenosine triphosphate is a nucleotide.

20 Adenine is a purine base which is common in DNA and RNA.

30

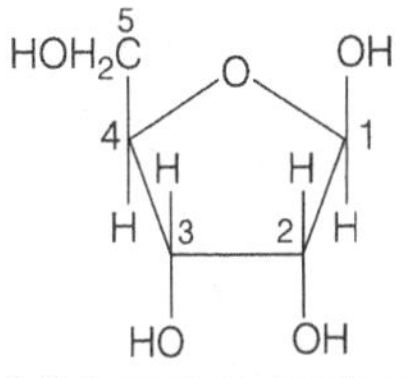

β-D-ribose (sugar base) used in RNA; at 2nd carbon —OH group is present

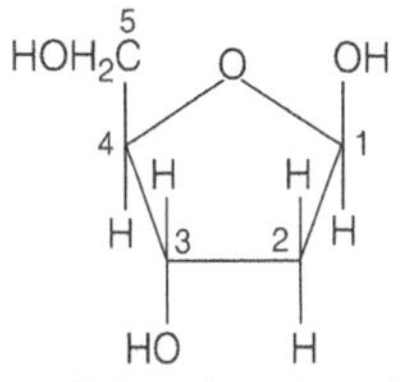

β-D-deoxyribose (sugar base) used in DNA; at 2nd carbon —OH group is missing

31 Quinoline is an alkaloid and is not present in DNA. DNA has four nitrogen bases in adenine, guanine, cytosine and thymine.

32 Correct explanation All α- amino acids except glycine contains at least one chiral carbon.

33 Vitamin D is fat soluble, i.e. water insoluble and hence, can be stored in the body.

SESSION 2

1 In the given structure, anomeric carbon is 1, the Haworth structure showing 2 anomers are given below:

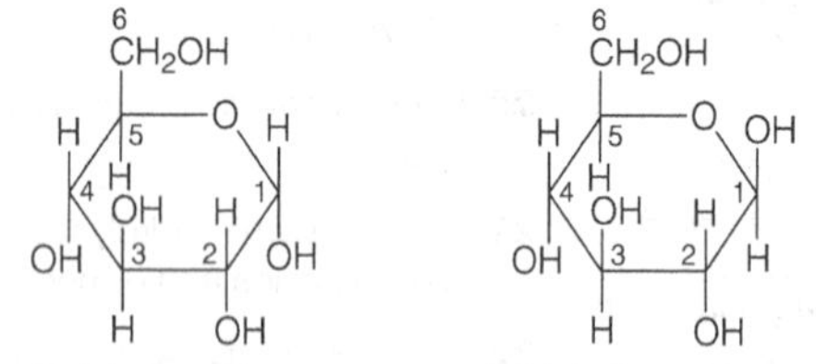

α-D-(+)-glucopyranose β-D(+)-glucopyranose

2 Aldehydic group reacts with hydrogen sulphate to form bisulphite addition product. But in case of glucose, it fails to react with sodium hydrogen sulphate as the aldehydic group is not free this shows that glucose is cyclic compound.

3

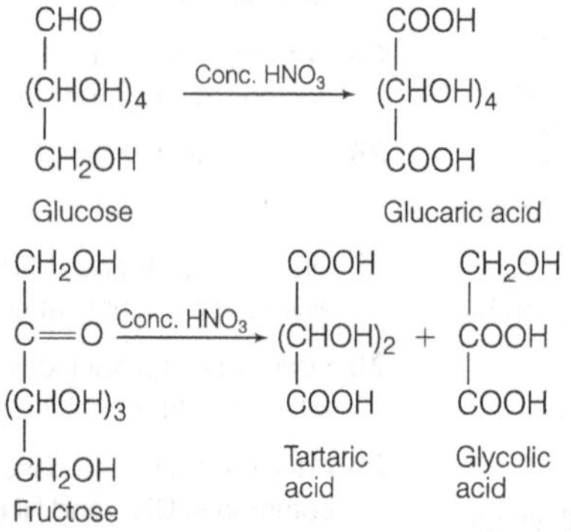

4 Sugars that have free aldehyde, a ketone, a hemiacetal or a hemiketal group is able to reduce an oxidising agent. These sugars are classified as reducing sugars.

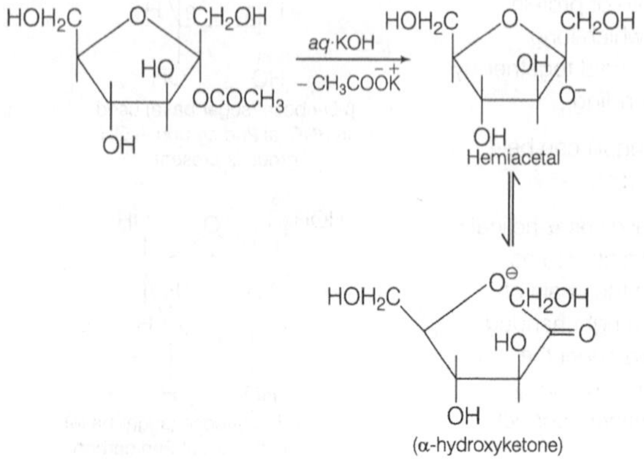

Hemiacetal can be easily reduced by oxidising agent such as Tollen's reagent.

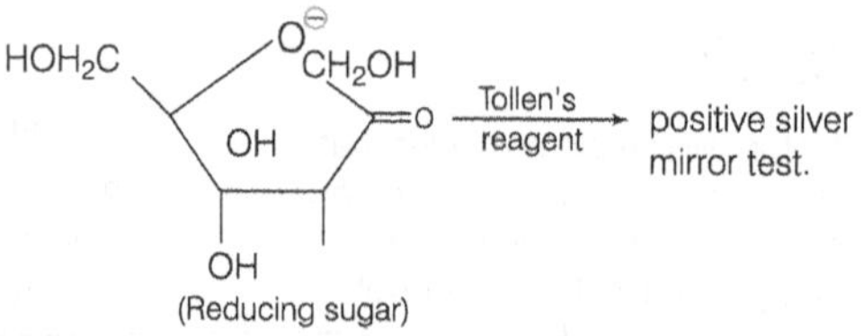

5 Cellulose is a straight chain polysaccharide composed of D-glucose units, which are joined by β-glycosidic linkages between C - 1 of one glucose and C - 4 of the next glucose. In each unit, only three hydroxy groups are free to form acetate, that's why it is called cellulose triacetate. Whose structure is shown in option (a).

6 The structure given in option (b) is of sucrose which is a non-reducing sugar.

7

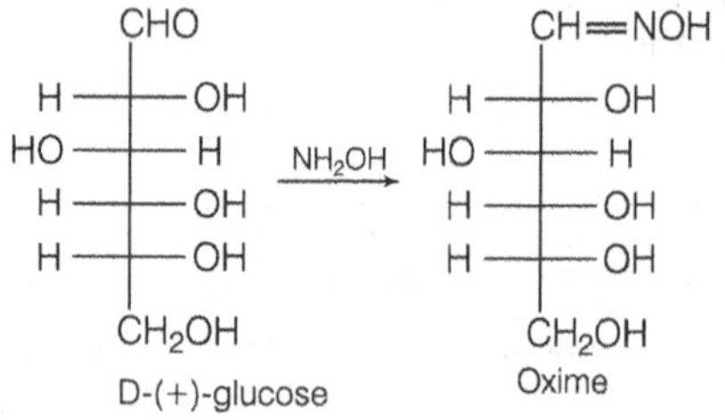

8

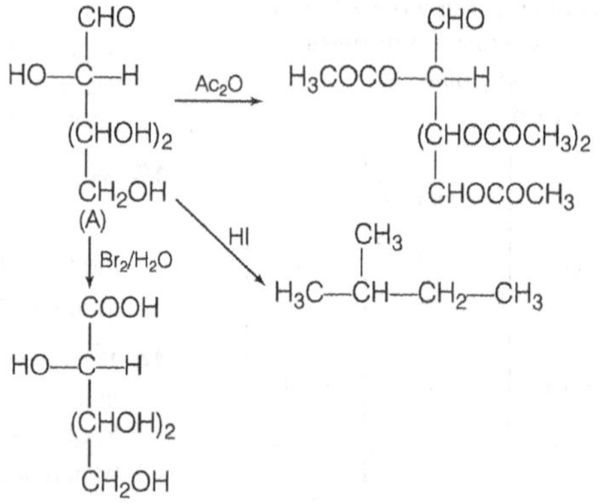

9

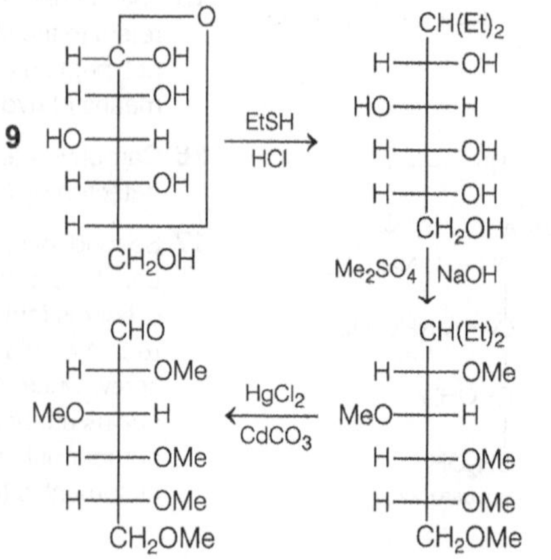

10 A sugar is classified as a D-isomer if the hydroxyl group on the chiral carbon farthest from the carbonyl point to the right. e.g.

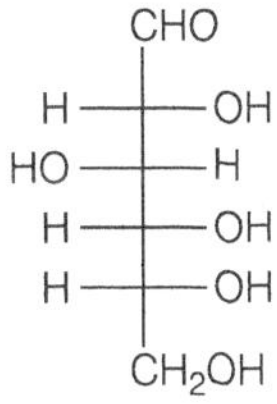

D-glucose

11 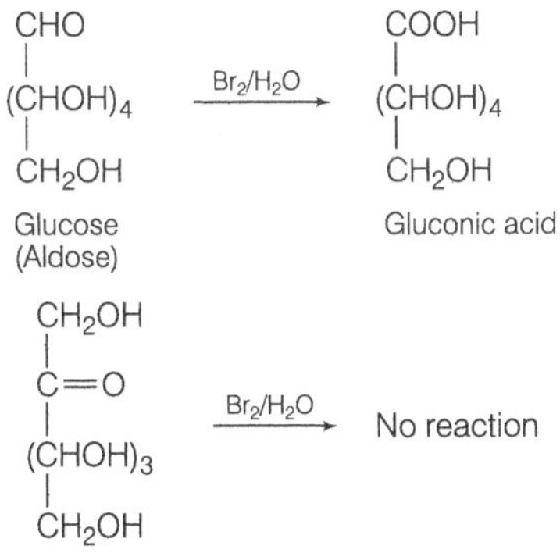

12 The secondary structure of protein refers to the shape in which long polypeptide chains are held together by means of H-bonding within the backbone.

13 When monosaccharide forms a cyclic hemiacetal the carbon atom that contained the carbonyl group is identified at the anomeric carbon because its substitutents can assume α or β-position.

14 The backbone of nucleic acid molecule consists of an alternating sugar and phosphate groups linked by phosphate ester bond.

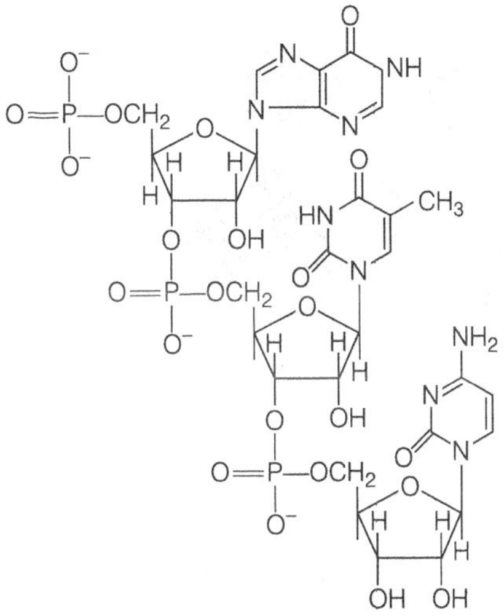

15 Enzymes are globular protein bodies which are biological catalyst. Enzyme catalyse chemical reaction enhances the rate by lowering the activation energy. These are highly specific both in binding chiral substrates and in catalysing their reactions. Pepsin is an example of proteolytic enzyme.

16 The number of glycine units present in the decapeptide is 6. Given, molecular mass of decapeptide = 796. Since, it gives glycine, alanine, phenylalanine on hydrolysis. Thus, it needs 9 molecule of water for hydrolysis.

The mass of product

$$= 796 + 18 \times 9 = 958$$

So, the contribution by glycine

$$= \frac{47}{100} \times 958$$

$$\therefore \text{ Number of glycine unit} = \frac{450}{75} = 6$$

17 The amino acid remain completely in *Zwitter* ionic form at its isoelectric point. Amino acids with additional acidic group have their isoelectric pH less than 7.0 and increasing pH above isoelectric point makes them anionic. On the other hand, amino acids with additional basic group have their isoelectric pH greater than 7.0 and decreasing pH below isoelectric point (by adding acid solution) makes them cationic. The given peptide with followings R_1 and R_2 are basic, will remain protonated (cationic) at pH = 7.0.

Peptide	R_1	R_2
IV	CH_2CONH_2	$(CH_2)_4NH_2$
VI	$(CH_2)_4NH_2$	$(CH_2)_4NH_2$
VIII	CH_2OH	$(CH_2)_4NH_2$
IX	$(CH_2)_4NH_2$	CH_3

DAY THIRTY FIVE

Chemistry in Everyday Life

Learning & Revision for the Day

- Chemicals in Medicines
- Chemicals in Food
- Cleansing Agents

Chemicals in Medicines

Drugs are chemicals of low molecular masses, which interact with macromolecular targets (carbohydrates, proteins and nucleic acids) to produce a biological response. When the biological response is therapeutic and useful, these chemicals are called medicines. The branch of chemistry that deals with the use of such chemicals to cure a disease or to prevent it, is called **chemotherapy**.

Drugs can be classified mainly on criteria outlined as follows:

1. **On the Basis of Pharmacological Effect** It provides the whole range of drugs available for the treatment of a particular type of disease or infection, e.g. analgesics have pain-killer effect while antiseptics kill or arrest the growth of microorganisms.
2. **On the Basis of Drug Action** It is based on the action of a drug on the particular biochemical process, e.g. all antihistamines inhibit the action of histamine which causes inflammation in the body.
3. **On the Basis of Chemical Structure** Drugs classified in this way share common structural features and often have similar pharmacological activity, e.g. all sulphonamide drugs, having the common features as given below are mostly antibacterial.

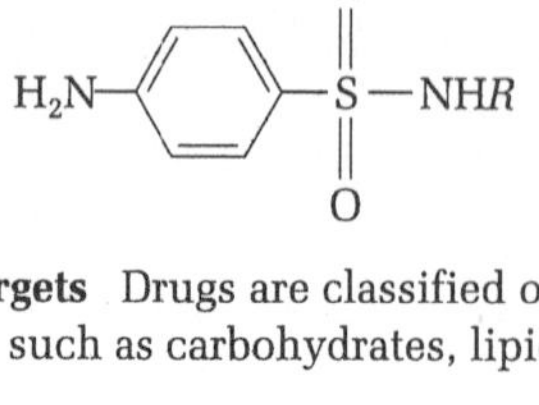

4. **On the Basis of Molecular Targets** Drugs are classified on the basis of their interaction with biomolecules such as carbohydrates, lipids, proteins and nucleic acids.

PREP MIRROR
Your Personal Preparation Indicator

- No. of Questions in Exercises (x)—
- No. of Questions Attempted (y)—
- No. of Correct Questions (z)—
  (Without referring Explanations)

- Accuracy Level ($z/y \times 100$)—
- Prep Level ($z/x \times 100$)—

In order to expect good rank in JEE, your Accuracy Level should be above 85 & Prep Level should be above 75.

- Drugs which can block the binding site of the enzyme and prevent the binding of substrate or can inhibit the catalytic activity of the enzyme are called **enzyme inhibitors**.
- Drugs which compete with the natural substrate for their attachment on the active sites of enzymes are called **competitive inhibitors**.
- If the bond formed between an enzyme and drug (inhibitor) is strong covalent bond then the body degrades the enzyme-drug (inhibitor) complex and synthesises new enzyme.

Therapeutic Action of Different Classes of Drugs

Some important classes of drugs are given below:

1. **Antacids** are used for treatment of acidity. Antacids neutralise the excess acid and raise the pH of stomach to some appropriate level. Sodium hydrogen carbonate, a mixture of aluminium and magnesium hydroxide etc., are the commonly used antacids.
2. **Antihistamines or antiallergic drugs** are used as treatment for allergies. Allergies are caused by an excessive response of the body to allergens (the substances which causes allergies). Excessive secretion of histamines, a chemical secreted by mast cells in body, also cause allergy, e.g. diphenylhydramine, chlorpheniramine, promethazine etc. Some synthetic drugs, e.g. brompheniramine (dimetapp) and terfenadine (seldane), also act as antihistamines.
3. **Neurologically Active Drug** Various types of neurologically active drugs are as follows:

   (i) **Tranquilizers** is a class of chemical compounds used for the treatment of stress, fatigue mild and severe mental diseases. These are commonly called **psychotherapeutic** drugs. These are the essential components of sleeping pills.

   Some examples according to the action of drugs are:

   - Antidepressant drugs, e.g. iproniazid, phenelzine etc., are used to reduce depression.
   - Chlordiazepoxide and meprobamate are relatively mild tranquilizers, suitable for relieving tension. Equanil is used in controlling depression and hypertension.
   - Barbiturates such as seconal, luminal, veronal are hypnotic (sleep producing) drugs while valium and serotonin are non-hypnotic drugs.
   - Reserpine, a powerful tranquiliser is obtained from Indian plant *Rauwolfia serpentina* helps slow down the pulse rate and lowers the blood pressure.
   - Sedatives act as depressant and suppress the activities of the central nervous system. These produce a feeling of calmness, relaxation, drowsiness in the body, e.g. valium (diazepam), calmpose and barbiturates.

   (ii) **Analgesics** reduce or abolish pain. These are classified as follows:

   (a) **Non-narcotic** (non-addictive) analgesics, e.g. aspirin (acetyl salicylic acid) and paracetamol are analgesics (N-acetyl-*p*-aminophenol) as well as antipyretics (fever reducing). Aspirin inhibits the synthesis of chemicals, known as prostaglandins which stimulate inflammation in the tissue and cause pain. Because of its antiblood clotting action, aspirin is also used in the prevention of heart attacks.

   (b) **Narcotic analgesics** These are chiefly used for the relief of post operative pain, cardiac pain and pains of terminal cancer and in child birth e.g. morphine, heroin (morphine diacetate), codeine etc. When administered in medicinal doses, relieve pain and produce sleep.
4. **Antimicrobial drugs** tends to destroy or prevent development or inhibit the pathogenic action of microbes such as bacteria virus or other parasites. Antibiotics, antiseptics and disinfectants are antimicrobial drugs.

   (i) **Antibiotics** are the chemicals synthesised from microbes and have either cidal (killing) effect or a static (inhibitory) effect on microbes. A few examples of the two types of antibiotics are as follows:

   (a) Bactericidal, e.g. penicillin (a narrow spectrum antibiotic), ampicillin and amoxycillin (semisynthetic modification of penicillin), ofloxacin (broad spectrum), aminoglycosides, i.e. streptomycin (broad spectrum) etc.

   (b) Bacteriostatic, e.g. erythromycin, tetracycline, chloramphenicol (a broad spectrum antibiotic) etc.

   (ii) **Antiseptics** are applied to the living tissues such as wounds, cuts, ulcers and diseased skin surfaces, e.g. furacine, soframycin, dettol (a mixture of chloroxylenol and α-terpineol), 0.2 per cent solution of phenol, bithionol is added to soaps, iodine, Boric acid etc.

   (iii) **Disinfectants** are applied to inanimate objects such as floors, drainage system, e.g. 1 per cent solution of phenol, chlorine and SO_2 (in very low concentrations).
5. **Antifertility drugs** have lead to the concept of family planning. Birth control pills essentially contain a mixture of synthetic estrogen and progesterone derivatives, e.g. commonly used pills contain a mixture of norethindrone (progesterone derivative) and ethinylestradiol or novestrol (estrogen).

NOTE The therapeutic properties of a drug depends upon its relative toxicity to the parasite and the host.

$$\text{Therapeutic Index (TI)} = \frac{\text{Maximum Tolerated Dose (MTD)}}{\text{Maximum Curative Dose (MCD)}}$$

The higher the therapeutic index ratio, the safer the drug will be.

Chemicals in Food

Substances which are added to food either to improve its taste, nutritive value and flavour or to preserve it are called food additives. Some important chemicals in food are as follows:

1. **Artificial sweetening agents** controls the intake of calorie and tooth decay. Some examples of artificials weetening agents are saccharin (*ortho*-sulphobenzimide), aspartame (nutra sweet), sucralose, alitame etc. Alitame is high potency sweetener, although it is more stable than aspartame. Sucralose is trichloro derivative of sucrose. Its appearance and taste are sugar like. It is stable at cooking temperature. Hence, its use is of great value to diabetic persons who need to control intake of calories.
2. **Food preservatives** prevent spoilage of food due to microbial growth. e.g. sodium benzoate, salts of sorbic acid and propanoic acid etc.

Cleansing Agents

These are also known as **surfactants** or surface active agents. In fact, those chemicals which concentrate at the surface of the solution or interfaces, form surface films, reduce surface tension of the solution and help in removing dirt and dust by emulsifying grease are known as **surfactants**. Soaps and detergents belong to this class.

Soaps

These are the sodium or potassium salts of higher fatty acids and are prepared by alkaline hydrolysis of fats or oils. Fats or oils are esters of higher fatty acids.

The reaction is known as **saponification**.

Glyceryl ester of stearic acid (fat) + caustic soda $\longrightarrow$ soap + glycerol

Different types of soaps are made by using different raw materials. These are as follows:

1. **Toilet soaps** are prepared by using better grades of fats and oils and care is taken to remove excess alkali.
2. **Transparent soaps** are made by dissolving the soap in ethanol and then evaporating the excess solvent.
3. **Medicated soaps** contain substances of medicinal value.
4. **Shaving soaps** contain glycerol to prevent rapid drying. A gum called rosin is added while making them.
5. **Laundry soaps** contain fillers like sodium rosinate, sodium silicate, borax and sodium carbonate.

Cleansing Action of Soap

- On applying soap to a dirty wet cloth, the hydrocarbon part (non-polar part) of soap dissolves in grease or dust while the polar carboxylate part is directed towards water. Thus, an emulsion is formed between grease particles and water molecules which appears in the form of foam. On washing the cloth with excessive water, these dirt or dust or grease particles are washed away from the surface of cloth along with soap and the cloth becomes clean.
- Soaps do not work in hard water because hard water contains calcium and magnesium ions. These ions form insoluble calcium and magnesium salts of soaps respectively that separate as sum in water.

Synthetic Detergents

These are alkyl benzene sulphonates. These are also called **syndets**. The detergents are classified into following three types on the basis of ionic charge present at the soluble end of their chain. These are as follows:

1. **Anionic detergents** are the sodium salts of sulphonated long chain alcohols or hydrocarbons. Anionic part of these detergents is involved in the cleansing action. These are formed by neutralising alkyl benzene sulphonic acids with alkali.

   e.g. sodiumdodecyl benzene sulphonate Anionic detergents are used in toothpastes and house hold work.
2. **Cationic detergents** are quarternary ammonium salts of amines with acetates, chlorides or bromides as anions. e.g. cetyltrimethyl ammonium bromide
   These are used in hair conditioners and are expensive, therefore these are of limited use.
3. **Non-ionic detergents** do not contain any ion in their constitution. When stearic acid reacts with polyethylene glycol, non-ionic detergent is formed. e.g. ester of stearic acid and polyethylene glycol.

   These are used in liquid dishwashing detergents.

NOTE
- **Detergents** can be used both in soft and hard water as they give foam even in hard water.
- **Straight chain alkyl groups** containing detergents are biodegradable whereas branched chain alkyl groups containing detergents are non-biodegradable.

DAY PRACTICE SESSION 1

FOUNDATION QUESTIONS EXERCISE

1 The pair whose both species are used in antacid medicinal preparation is
(a) $NaHCO_3$ and $Mg(OH)_2$
(b) Na_2CO_3 and $Ca(HCO)_2$
(c) $Ca(HCO_3)_2$ and $Mg(OH)_2$
(d) $Ca(OH)_2$ and $NaHCO_3$

2 Which of the following compounds is not an antacid?
→ JEE Main 2015
(a) Aluminium hydroxide (b) Cimetidine
(c) Phenelzine (d) Ranitidine

3 The drug given during hypertension is
(a) streptomycin (b) chloroxylenol
(c) equanil (d) aspirin

4 The drug used as an antidepressant is
(a) luminal (b) tofranil
(c) mescaline (d) sulphadiazine

5 H_1-receptor antagonists is a term associated with
→ JEE Main (Online) 2013
(a) antiseptics (b) antihistamines
(c) antacids (d) analgesics

6 The local anaesthetic, which is used for small surgical operation is
(a) ether (b) nitrous oxide
(c) cyclopropane (d) procaine

7 Barbituric acid and its derivatives are well known as
(a) tranquilizers
(b) antiseptics
(c) analgesics
(d) antipyretics

8 Tranquilizers are the substances used for the treatment of
(a) cancer (b) AIDS
(c) mental diseases (d) physical disorders

9 Substances used in bringing down the body temperature in high fever are called
(a) antiseptics (b) pyretics
(c) antibiotics (d) antipyretics

10 A drug that is antipyretic as well as analgesic is
(a) chloropromazine hydrochloride
(b) *para*-acetamidophenol
(c) chloroquine
(d) penicillin

11 Compound *A* given below is

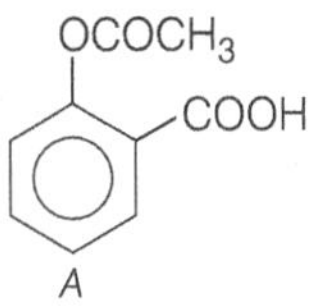

A

(a) antiseptic (b) antibiotic
(c) analgesic (d) pesticide

12 The correct structure of the drug paracetamol is

(a) 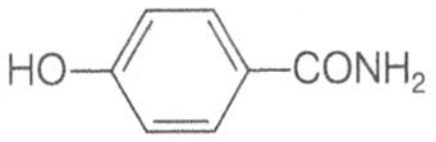

(b) 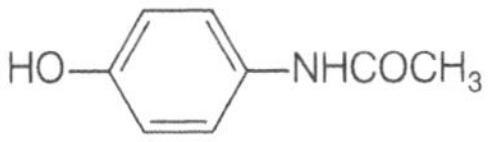

(c) 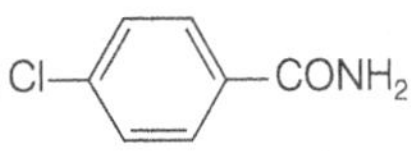

(d) 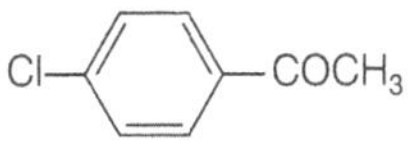

13 Aspirin is known as → AIEEE 2012
(a) acetyl salicylic acid (b) phenyl salicylate
(c) acetyl salicylate (d) methyl salicylic acid

14 A broad spectrum antibiotic is
(a) paracetamol (b) penicillin
(c) aspirin (d) chloramphenicol

15 Which of the following statements about aspirin is not true? → JEE Main (Online) 2013
(a) It is effective in relieving pain
(b) It is a neurologically active drug
(c) It has antiblood clotting action
(d) It belongs to narcotic analgesics

16 An ester used as medicine is
(a) ethyl acetate (b) methyl acetate
(c) methyl salicylate (d) ethyl benzoate

17 One of the most widely used drug in medicine iodex is
(a) methyl salicylate (b) ethyl salicylate
(c) acetyl salicylic acid (d) *o*-hydroxy benzoic acid

18 Streptomycin, a well known antibiotic, is a derivative of
(a) peptides (b) carbohydrates
(c) purines (d) None of these

19 Chloramine-T is a
(a) antiseptic (b) disinfectant
(c) analgesic (d) antipyretics

20 The functional groups present in salol are
(a) $-NH_2$ and $-OR$ (b) $-NH_2$ and $-COOH$
(c) $-OR$ and $-OH$ (d) $-OH$ and $-COOR$

21 Which of the following is used as an antiseptic?
(a) Phenol (b) Benzaldehyde
(c) Benzalamine (d) Maleic anhydride

22 Match the following and choose the correct option.

Column I		Column II	
A.	Ranitidine	1.	Tranquilizer
B.	Furacine	2.	Antibiotic
C.	Phenelzine	3.	Antihistamine
D.	Chloramphenicol	4.	Antiseptic

Codes

	A	B	C	D
(a)	3	4	1	2
(b)	4	1	2	3
(c)	3	1	4	2
(d)	4	2	1	3

23 Which of the following is used as a "morning after pill"?
(a) Mifepristone (b) Ethynylestradiol
(c) Northindrone (d) Bithional

24 Which of the following will not enhance nutritional value of food?
(a) Minerals (b) Artificial sweeteners
(c) Vitamins (d) Amino acids

25 Among the following sweeteners, which one has the lowest sweetness value?
(a) Alitame (b) Aspartame (c) Saccharin (d) Sucralose

26 Which of the following is known as invert soap?
(a) Pentaerythritol monostearate
(b) Sodium stearyl ammonium bromide
(c) Sodium stearyl sulphate
(d) Ethoxylated nonyphenol

27 Which of the following represents soap?
(a) $C_{17}H_{35}COOH$ (b) $C_{17}H_{35}COOK$
(c) $C_{15}H_{31}COOH$ (d) $C_{17}H_{35}(COO)_2Ca$

28 Which of the following is not true for a detergent molecule?
(a) It has a non-polar organic part and a polar group
(b) It is not easily biodegraded
(c) It is sodium salt of a fatty acid
(d) It is a surface active reagent

29 Which of the following is an anionic detergent?
(a) Sodium lauryl sulphate
(b) Cetyltrimethyl ammonium bromide
(c) Glyceryl oleate
(d) Sodium stearate

30 Match the chemicals in List I with their uses in List II.

List I		List II	
A.	Sodium perborate	1.	Disinfectant
B.	Chlorine	2.	Antiseptic
C.	Bithional	3.	Milk bleaching agent
D.	Potassium stearate	4.	Soap

Codes

	A	B	C	D
(a)	2	1	3	4
(b)	3	4	2	1
(c)	3	1	2	4
(d)	2	3	4	1

Direction (Q. Nos. 31-32) *In the following questions a Assertion (A) following by Reason (R) is given. Choose the correct answer out of the following choices.*
(a) Both A and R are true and R is correct explanation of A
(b) Both A and R are true but R is not correct explanation of A
(c) A is true but R is false
(d) Both A and R are false

31 **Assertion** (A) All chemicals added to food items are called food preservatives.
Reason (R) All these chemicals increase the nutritive value of the food.

32 **Assertion** (A) Sodium chloride is added to precipitate soap after saponification.
Reason (R) Hydrolysis of esters of long chain fatty acids by alkali produces soap in colloidal form.

Direction (Q. Nos. 33-35) *Each of these questions contains two statements : Statement I (Assertion) and Statement II (Reason). Each of these questions also has four alternative choices, only one of which is the correct answer. You have to select one of the codes (a), (b), (c) and (d) given below.*
(a) Statement I is true, Statement II is true; Statement II is the correct explanation for Statement I
(b) Statement I is true, Statement II is true; Statement II is not the correct explanation for Statement I
(c) Statement I is true; Statement II is false
(d) Statement I is false; Statement II is true

33 **Statement I** Tranquilizers are used for the treatment of stress and mild or even severe mental diseases.
Statement II Tranquilizers are neurologically active drugs. These affect the message transfer mechanism from nerve to receptor.

34 **Statement I** Aspirin can cause ulcer in stomach when taken empty stomach.
Statement II Aspirin prevents platelet coagulation as it has antiblood clotting action.

35 **Statement I** Tertiary butyl hydroquinone is an antioxidant.
Statement II Antioxidants inhibit free radical reactions.

DAY PRACTICE SESSION 2

PROGRESSIVE QUESTIONS EXERCISE

1 The reagents used in the preparation of aspirin from salicylic acid.

(a) $SOCl_2$, pyridine (b) CH_3COOH, HCl
(c) CH_3Cl, $AlCl_3$ (d) $(CH_3CO)_2O, H^+$

2 Which of the following can possibly be used as analgesic without causing addiction and modification?

(a) Morphine
(b) N-acetyl-*para*-aminophenol
(c) Diazepam
(d) Tetrahydrocatenol

3 The drug *A* is used as a/an

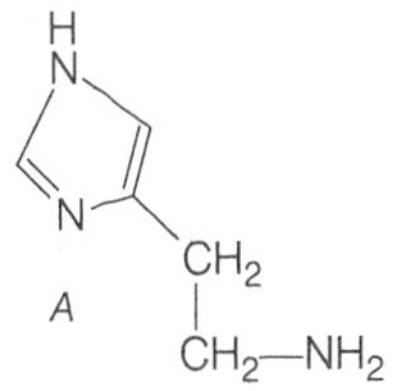

(a) antacid (b) analgesic (c) vasodilator (d) antiseptic

4 Compound which is added to soap to impart antiseptic properties is

(a) sodium lauryl sulphate
(b) sodium dodecylbenzenesulphonate
(c) rosin
(d) bithional

5 Which of the following statements is not correct?

(a) Some antiseptics can be added to soaps
(b) Dilute solutions of some dis-infectants can be used as an antiseptic
(c) Disinfectants are anti-microbial drugs
(d) Antiseptic medicines can be ingested

6 Detergents are known to pollute rivers and water ways. However, detergents can be made biodegradable and pollution free by taking

(a) cyclic hydrocarbon chain
(b) shorter hydrocarbon chain
(c) unbranched hydrocarbon chain
(d) hydrocarbon with more branching

7 The ratio of the therapeutic index is ten means

(a) ten times a dose used for curative purposes would kill the parasite as well as the patient
(b) to cure the patient the dose should be given 10 times to him
(c) 10 times dose given, would not be sufficient to cure the patient
(d) the dose should be given for 10 days

8 Antiseptics and disinfectants either kill or prevent growth of microorganisms. Identify which of the following is not true?

(a) A 0.2% solution of phenol is an antiseptic while 1% solution acts as a disinfectant
(b) Chloride and iodine are used as strong disinfectants
(c) Dilute solutions of boric acid and hydrogen peroxide are strong antiseptics
(d) Disinfectants harm the living tissue

9 Which of the following is not true for antibiotics?

(a) Tetracycline is one of the broad spectrum antibiotic which is effective against a large number of harmful microorganisms
(b) Streptomycin is highly effective against microorganisms which cause tuberculosis
(c) Penicillin has a narrow spectrum and certain persons are sensitive to it
(d) Penicillin may be administered without testing the patients for sensitivity to it

10 The pH value of gastric juice in human stomach is about 1.8 and in the small intestine it is about 7.8. The pK_a value of aspirin is 3.5. Aspirin will be

(a) ionised in the small intestine and almost unionised in the stomach
(b) unionised in the small intestine and in the stomach
(c) completely ionised in the small intestine and in the stomach
(d) ionised in the stomach and almost unionised in the small intestine

ANSWERS

SESSION 1	**1** (a)	**2** (c)	**3** (c)	**4** (a)	**5** (b)	**6** (d)	**7** (a)	**8** (c)	**9** (d)	**10** (b)
	11 (c)	**12** (b)	**13** (a)	**14** (d)	**15** (d)	**16** (c)	**17** (a)	**18** (b)	**19** (b)	**20** (d)
	21 (a)	**22** (a)	**23** (a)	**24** (b)	**25** (b)	**26** (b)	**27** (b)	**28** (c)	**29** (a)	**30** (c)
	31 (d)	**32** (a)	**33** (a)	**34** (c)	**35** (a)					
SESSION 2	**1** (d)	**2** (b)	**3** (c)	**4** (d)	**5** (d)	**6** (c)	**7** (a)	**8** (c)	**9** (d)	**10** (a)

Hints and Explanations

SESSION 1

1 Pair of $NaHCO_3$ and $Mg(OH)_2$ are used in antacid medicinal preparation.

2 The term antihistamine refers only to compounds not have inhibit action at the H_1-receptor so, the term H_1-receptor antagonists is related with antihistamines.

3 Aluminium hydroxide $Al(OH)_3$, cimetidine and ranitidine are antacids while phenelzine is a tranquilizer.

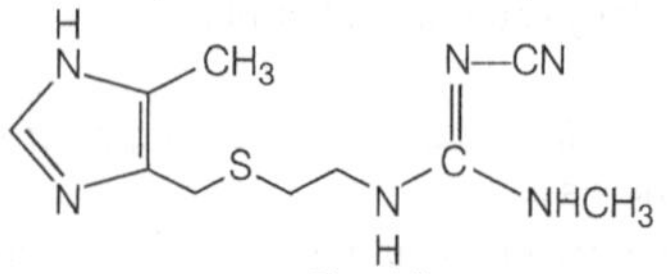

Cimetidine

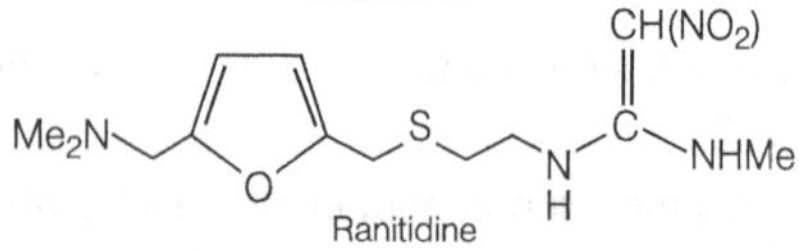

Ranitidine

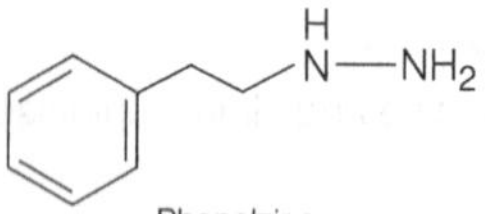

Phenelzine

4 Equanil is a tranquilizer that is widely used during depression and hypertension.

5 Luminal is used as an antidepressant.

6 Procaine is a local anaesthetic, which affects only a part of the body insensitive to pain or feeling. Ether, nitrous oxide and cyclopropane are general anaesthetics.

7 The drugs given to the patients suffering from anxiety and mental tension are called tranquilizers, e.g. barbituric acid and its derivatives.

8 Tranquilizers reduce anxiety and are employed for treatment of mental diseases.

9 The chemical substances used to bring down the body temperature in case of high fever are called antipyretics, e.g. analgin.

10 *para*-acetamidophenol is used as an antipyretic as well as analgesic.

11

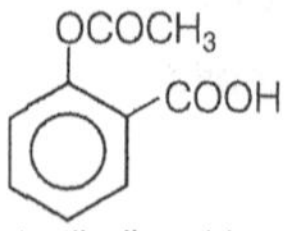

o-acetyl salicylic acid (Aspirin)

It is used as an analgesic.

12

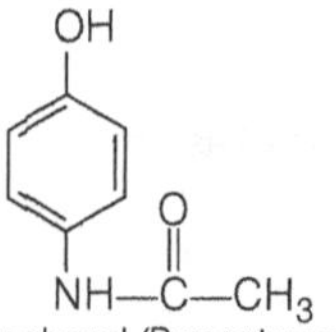

4-acetamidophenol (Paracetamol)

Paracetamol is most popular non-addictive analgesic (pain-killer) and antipyretic (fever reducing).

13 Aspirin is acetyl derivative of salicylic acid.

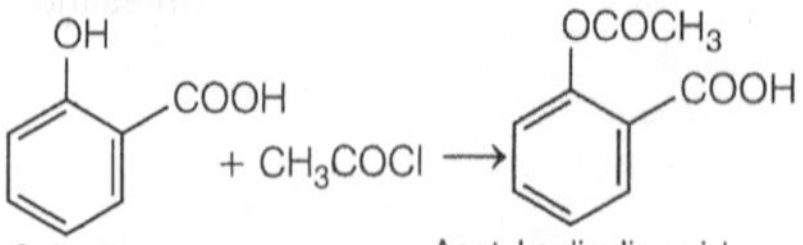

Salicylic acid   Acetyl salicylic acid (Aspirin)

14 Broad spectrum antibiotics are the medicines effective against a large number of harmful microorganisms, e.g. tetracycline, chloramphenicol.

15 Aspirin is a non-narcotic (non-habit forming) analgesic, i.e. give relief from pain and also it has antiblood clotting action. It is a neurologically active drug.

16 An ester used as medicine is methyl salicylate.

17 Methyl salicylate is widely used in iodex. It is also known as oil of winter green.

18 Streptomycin is a carbohydrate derivative. It is highly effective against tuberculosis.

20 In salol (phenyl salicylate), functional groups are —OH and —COO*R*.

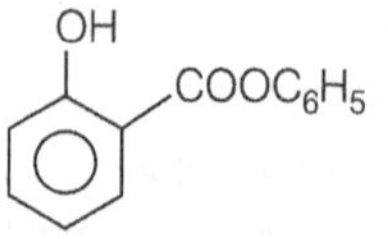

21 0.2 per cent solution of phenol acts as an antiseptic and its 1% solution is a disinfectant.

22 $A \longrightarrow 3; B \longrightarrow 4; C \longrightarrow 1; D \longrightarrow 2$

23 Mifepristone is an anti-fertility drug used as "morning after pill".

24 Artificial sweeteners will not enhance nutritional value of food. It is non-calorific substitute for sugar.

25 (i) Aspartame is 100 times sweeter than sugar (used only in cold foods and soft drinks).

(ii) Saccharin is 550 times sweeter than sugar.

(iii) Sucralose is 600 times sweeter than sugar.

(iv) Alitame is 2000 times sweeter than sugar (the later three are used for cooking and baking).

26 Sodium stearyl ammonium bromide is known as invert soap.

27 Soaps are the sodium or potassium salts of higher fatty acids. e.g. $C_{17}H_{35}COOK$ (potassium stearate). These are obtained by alkaline hydrolysis of oils and fats (saponification).

28 Soap is a sodium or potassium salt of fatty acids while detergent is a sodium salt of benzene sulphonic acid derivatives.

29 Sodium lauryl sulphate $[(CH_3(CH_2)_{10}CH_2OSO_3^-Na^+)]$ is an anionic detergent

Cetyltrimethy ammonium bromide

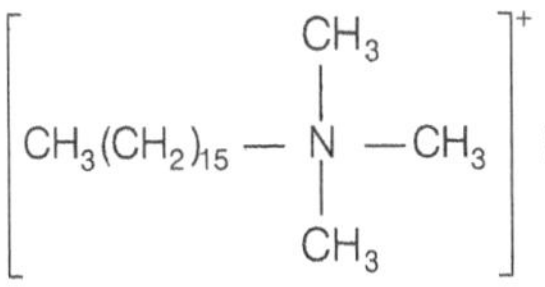

Br⁻ is a cationic detergent.

Glyceryl oleate $[(C_{17}H_{32}COO)_3C_3H_6]$ is a non-ionic detergent.

Sodium stearate $[C_{17}H_{35}COO^-Na^+]$ is a anionic soap.

30 $A \longrightarrow 3; B \longrightarrow 1; C \longrightarrow 2; D \longrightarrow 4$

31 **Correct Assertion** Chemicals which are used to protect food against bacteria, yeasts and moulds are called food preservatives.

Correct Reason Preservatives do not increase the nutritive value of food.

32 Sodium chloride is added to precipitate after saponification because hydrolysis of esters of long chain fatty acids by alkali produces soap in colloidal form.

33 Tranquilizers are also known as psychotherapeutic drugs (neurological active drug) due to their purpose to reduce anxiety and are employed for the treatment of mental diseases by affecting the message transfer mechanism from nerve to receptor.

34 In the stomach, aspirin gets hydrolysed to form salicylic acid which ulcer when taken empty stomach. It prevents heart attack as it has antiblood clotting action.

35 Tertiary butyl hydroquinone is an antioxidant which when added to the fats and fat containing food, prevent their oxidation by interrupting the free radical chain reactions involved in lipid oxidation.

SESSION 2

1

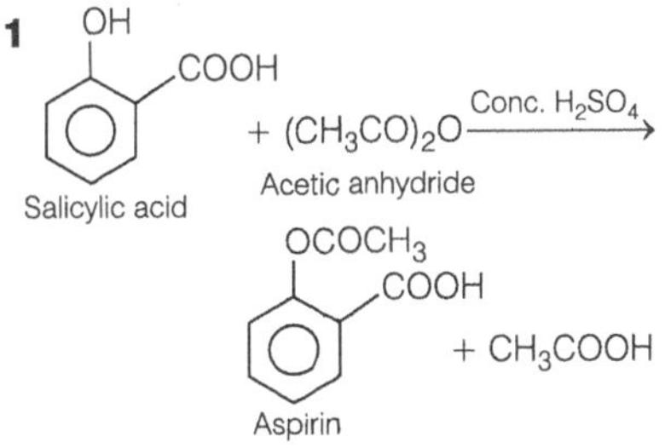

2 N-acetyl para-aminophenol can be used as an analgesic without causing addiction and modification.

3 Drug *A* is a vasodilator that causes vasoditation, a widening (opening) of blood vessels that results from relaxation of the smooth muscle of the vessels.

4 Bithional is an antiseptic which is added to soap to reduce the odour produced by bacterial decomposition of organic matter on the skin.

5 Antiseptic can cause serious poisoning and possibly death if swallowed hence they can't be ingested.

6 Since, unbranched hydrocarbon chain are more prone to attack by bacteria, therefore the detergent can be made biodegradable and pollution free by taking unbranched hydrocarbon chain.

7 $TI = \dfrac{\text{maxium tolerated dose}}{\text{maximum curative dose}}$

If the ratio is ten, then it means ten times a dose used for curative purpose would kill the parasite as well as the patient.

8 Boric acid in form of dilute aqueous solution and hydrogen peroxide are a mild antiseptic used for washing eyes wounds, teeth and ears.

9 Penicillin has a narrow spectrum and is effective only against diseases caused by various cocci and some gram positive bacteria. However, some persons have allergy to penicillin. Hence, it is absolutely essential to test the sensitivity of the patient.

10 As the pK_a value of aspirin is 3.5. So, it will be ionised in small intestine as it is slightly basic and will remain unionised in the stomach due to its acidic medium.

DAY THIRTY SIX

Analytical Chemistry

Learning & Revision for the Day

- Qualitative Inorganic Analysis
- Detection of Functional Group
- Chemistry Involved in Titrimetric Exercises
- Some Important Experiment

Analytical chemistry deals with qualitative and quantitative analysis of the substances.

Qualitative Inorganic Analysis

The qualitative analysis deals with the detection and identification of various constituents present in inorganic salt or a mixture of salts.

1. Inorganic Salts

These are the products of neutralisation reaction of an acid and a base. Therefore, they contain an anion and a cation. By identifying these cations and anions, the salt can be identified.

2. Anions or acid Radicals

These are divided into three groups depending upon their reactions with dilute and concentrated sulphuric acid.

Group I Anions (react with dilute sulphuric acid)

Group I anions	Gases evolved with dil. H_2SO_4
Carbonate (CO_3^{2-})	Brisk effervescence of CO_2
Sulphite (SO_3^{2-})	Colourless gas with odour of burning sulphur (SO_2)
Sulphide (S^{2-})	Colourless gas with smell of rotten eggs (H_2S)
Nitrite (NO_2^-)	Brown fumes with pungent odour (NO_2)

Group II Anions (react with conc. H_2SO_4)

Group II anions	Gases evolved with conc. H_2SO_4
Chloride (Cl^-)	Colourless gas with pungent smell (HCl)
Bromide (Br^-)	Brown fumes (Br_2)
Iodide (I^-)	Violet fumes of (I_2)
Nitrate (NO_3^-)	Pungent, light brown (NO_2) gas

PREP MIRROR

Your Personal Preparation Indicator

- No. of Questions in Exercises (x)—
- No. of Questions Attempted (y)—
- No. of Correct Questions (z)—
  (Without referring Explanations)

- Accuracy Level ($z/y \times 100$)—
- Prep Level ($z/x \times 100$)—

In order to expect good rank in JEE, your Accuracy Level should be above 85 & Prep Level should be above 75.

Group III Anions (not affected by both dil. and conc. H_2SO_4)

Sulphate and phosphate are present in group III anions.

Confirmatory Tests for Group I Anions

Following anions decomposes on reaction with dil.H_2SO_4 to give gases. These gases indicate the nature of acid radical present in the salt.

(i) **Carbonate** (CO_3^{2-}) The evolved gas is carbon dioxide (CO_2) gas. It is recognised by passing it through lime water which turns milky due to formation of $CaCO_3$ and excess of CO_2 form soluble calcium bicarbonate (colourless).

$$Ca(OH)_2(aq) + CO_2(g) \longrightarrow \underset{\text{White}}{CaCO_3(s)\downarrow}$$

$$CaCO_3(s) + H_2O(l) + CO_2(g) \longrightarrow \underset{\text{Colourless}}{Ca(HCO_3)_2(aq)}$$

(ii) **Sulphite** (SO_3^{2-}) The evolved gas is SO_2. This gas turns acidified $K_2Cr_2O_7$, paper green due to reduction of Cr (VI) to Cr (III).

$$K_2Cr_2O_7 + H_2SO_4 + 3SO_2 \longrightarrow K_2SO_4 + \underset{\text{Green}}{Cr_2(SO_4)_3\downarrow} + H_2O$$

Like SO_2, H_2S also turns potassium dichromate paper green. So, before testing SO_3^{2-}, S^{2-} must be tested and if present, it should be removed.

(iii) **Sulphide** (S^{2-}) The evolved gas is $H_2S(g)$. It is tested with lead acetate paper which turns black due to the formation of black lead sulphide.

$$(CH_3COO)_2Pb\downarrow + H_2S(s) \longrightarrow \underset{\text{Black}}{PbS(s)\downarrow} + 2CH_3COOH(aq)$$

H_2S gives violet colour with sodium nitroprusside, $Na_2[Fe(CN)_5NO]$ due to the formation of sodium thio nitroprusside, $Na_4[Fe(CN)_5NOS]$.

(iv) **Nitrite** (NO_2^-) The evolved gas oxidises I^- to I_2 which gives blue colour with starch.

$$2KI + NO(g) + O_2(g) \longrightarrow 2KNO_2(aq) + I_2(aq)$$

$$I_2(aq) + \text{Starch} \longrightarrow \text{Starch iodide complex}$$

Confirmatory Tests for Group II Anions

Following anions decomposes on reaction with conc. H_2SO_4 to give gases.

(i) **Chlorides** (Cl^-) The evolved gas is HCl. It is identified by bringing a rod dipped in ammonium hydroxide near the mouth of the test tube when dense white fumes (NH_4Cl) are produced.

(ii) **Chromyl chloride test** is the confirmatory test for chloride. On heating with solid $K_2Cr_2O_7$ and conc. H_2SO_4, chloride gives orange yellow vapours of chromyl chloride (CrO_2Cl_2) which on passing over NaOH and then on treating with lead acetate forms yellow ppt. of lead chromate.

$$4NaCl + K_2Cr_2O_7(s) + 3H_2SO_4(l) \longrightarrow K_2SO_4 + \underset{\substack{\text{Chromyl chloride}\\\text{(orange yellow fumes)}}}{2CrO_2Cl_2(g)\uparrow} + 3H_2O + 2Na_2SO_4$$

$$CrO_2Cl_2(g) + 4NaOH \longrightarrow Na_2CrO_4(aq) + 2NaCl(aq) + 2H_2O(l)$$

$$(CH_3COO)_2Pb(aq) + Na_2CrO_4(aq) \longrightarrow \underset{\substack{\text{Lead chromate}\\\text{(yellow ppt.)}}}{PbCrO_4(s)\downarrow} + 2CH_3CO\overset{-\ +}{ONa}$$

Chlorides, on treating with HNO_3 + $AgNO_3$ solution, form white precipitate (AgCl) which is soluble in NH_4OH. Chromyl chloride test is not given by chlorides of Hg, Sn, Ag, Pb and Sb.

(iii) **Bromide** (Br^-) The evolved gas is HBr. On treating bromides with HNO_3 and $AgNO_3$ solution, pale yellow precipitate of AgBr is formed which is partially soluble in excess of NH_4OH.

On shaking bromides with chlorine water and chloroform, the chloroform layer turns yellow (or brownish yellow). This is known as **layer test.**

$$2Br^- + Cl_2 \longrightarrow 2Cl^- + \underset{\substack{\text{(Orange yellow)}\\\text{Soluble in } CHCl_3}}{Br_2}$$

(iv) **Iodide** (I^-) The evolved gas is HI along with violet iodine vapours. The obtained I_2 gas is recognised by treating it with starch solution which turns blue.

$$I_2 + \text{Starch} \longrightarrow \underset{\text{(Blue)}}{\text{Starch iodine adsorption complex}\downarrow}$$

If layer test is performed for iodide, the $CHCl_3$ layer turns violet.

(v) **Nitrate** (NO_3^-) On reaction with conc. H_2SO_4 first form colourless HNO_3 which decomposes to give brown fumes of nitrogen dioxide. Presence of nitrate is confirmed by performing ring test. In ring test, the water extract of nitrate salt is treated with freshly prepared $FeSO_4$ solution and then conc. H_2SO_4 is added. Formation of dark brown ring between two layers confirms the presence of nitrate radical.

$$2NaNO_3 + 2H_2SO_4 \longrightarrow 2NaHSO_4 + 2HNO_3$$

$$2HNO_3 + 6FeSO_4 + 3H_2SO_4 \longrightarrow 3Fe_2(SO_4)_3 + 2NO + 4H_2O$$

$$[Fe(H_2O)_6]SO_4 + NO \longrightarrow \underset{\text{Dark brown ring}}{[Fe(H_2O)_5NO]^{2+}SO_4^{2-}} + H_2O$$

Confirmatory Tests for Group III Anions

The following are the steps involved in the confirmatory tests for group III anions.

Sulphate (SO_4^{2-}) $BaCl_2$ test is the confirmatory test for sulphate. The white precipitate of $BaSO_4$ is insoluble in conc. HNO_3.

3. Cations or Basic Radicals

Following dry test methods are used to identify cations.

(i) **Flame test** is used to identify the cations by developing different colours. These are as follows:

Name of radicals	Colour obtained
Sodium (Na^+)	Golden yellow
Lithium (Li^+)	Carmine red
Potassium (K^+)	Violet
Copper (Cu^{2+})	Bluish green
Calcium (Ca^{2+})	Brick red or dull red
Strontium (Sr^{2+})	Crimson red
Barium (Ba^{2+})	Apple green

(ii) **Borax bead test** is used to identify the cation by developing different colour in different flame.

Basic radicals	Oxidising flame (Non-luminous)	Reducing flame (Luminous)
Nickel (Ni)	Light brown	Black
Gold (Au)	Rose-violet	Violet
Chromium (Cr)	Green	Green
Iron (Fe)	Yellow	Bottle green
Copper (Cu)	Light blue or green	Colourless or red

Borax bead test is performed only for coloured substances.

(iii) **Wet Tests** On the basis of solubility product and reagents used, the basic radicals are classified into following six groups.

	Group number (with radicals)	Group reagent
0	(NH_4^+)	NaOH
I	(Pb^{2+}, Ag^+ Hg_2^{2+})	Dil. HCl
II	(Pb^{2+}, Cu^{2+}, Bi^{3+}, Cd^{2+}, Sn^{2+})	H_2S in presence of dil. HCl
III	(Fe^{3+}, Al^{3+}, Cr^{3+})	$NH_4Cl + NH_4OH$
IV	(Ni^{2+}, Zn^{2+}, Mn^{2+}, Co^{2+})	H_2S in presence of NH_4OH
V	(Ba^{2+}, Ca^{2+}, Sr^{2+})	$(NH_4)_2CO_3$ in presence of NH_4Cl and NH_4OH
VI	(Mg^{2+})	Na_2HPO_4

Confirmatory Tests for Zero Group (NH_4^+)

Ammonium ion on heating with NaOH gives NH_3 gas which is recognised by taking a glass rod dipped in HCl above the mouth of test tube.

$$NH_3(g) + HCl(g) \longrightarrow \underset{\text{White fumes}}{NH_4Cl(s)} \uparrow$$

NH_3 gas gives yellowish-brown precipitate with Nessler's reagent.

$$\underset{\text{Nessler's reagent}}{K_2HgI_4} + NH_3 + 3NaOH \longrightarrow$$

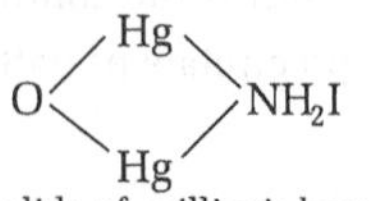

$+ 4KI + 3NaI + 2H_2O$

Iodide of million's base (yellowish-brown ppt.)

Confirmatory Tests for First Group Radicals (Pb^{2+}, Ag^+, Hg_2^{2+})

Lead chloride ($PbCl_2$) is soluble in hot water and gives yellow precipitate with KI and with K_2CrO_4.

$$PbCl_2 + 2KI \longrightarrow \underset{\substack{\text{Lead iodide}\\\text{(yellow ppt.)}}}{PbI_2 \downarrow} + 2KCl$$

$$PbCl_2 + K_2CrO_4 \longrightarrow \underset{\substack{\text{Lead chromate}\\\text{(yellow ppt.)}}}{PbCrO_4 \downarrow} + 2KCl$$

Hg_2Cl_2 is blackened by NH_3 and with K_2CrO_4, it gives red precipitate.

$$Hg_2Cl_2 + 2NH_3 \longrightarrow \underset{\text{Black}}{HgNH_2Cl + Hg} + NH_4Cl$$

$$Hg_2Cl_2 + K_2CrO_4 \longrightarrow \underset{\text{Red}}{Hg_2CrO_4 \downarrow} + 2KCl$$

AgCl gives red precipitate with K_2CrO_4 which is soluble in NH_4OH.

$$2AgCl + K_2CrO_4 \longrightarrow Ag_2CrO_4 + 2KCl$$

For testing sulphate, if Ag and Pb cations are present then Ba $(NO_3)_2$ must be used in place of $BaCl_2$ otherwise chlorides of Ag or Pb may get precipitated.

Confirmatory Tests for Second Group Radicals (Pb^{2+}, Cu^{2+})

On passing H_2S gas, the cations of group II forms precipitates and can be put into two groups on the basis of the colour of their sulphides.

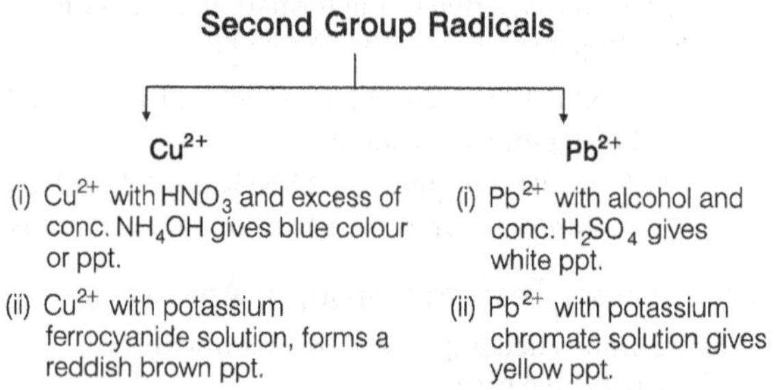

For IInd group cations, H_2S must be passed in hot solution slowly to get more granular and easily filterable precipitates.

The following are some chemical equations which play an important role while confirmation of Cu^{2+} and Pb^{2+} given as:

1. **Copper** (Cu^{2+})

$$\underset{\text{Black}}{3CuS} + 8HNO_3 \longrightarrow 3Cu(NO_3)_2 + 2NO\uparrow + 4H_2O + 3S\downarrow$$

$$Cu(NO_3)_2(aq) + 2NH_4OH(aq) \longrightarrow \underset{\text{Cupric hydroxide}}{Cu(OH)_2(s)} + 2NH_4NO_3$$

$$Cu(OH)_2(s) + 2\,NH_4NO_3(aq) + 2\,NH_4OH \longrightarrow \underset{\text{Deep blue complex}}{[Cu(NH_3)_4](NO_3)_2]} \downarrow + 4H_2O$$

$$2\,Cu(NO_3)_2 + \underset{\substack{\text{Potassium}\\ \text{ferrocyanide}}}{K_4[Fe(CN)_6]} \longrightarrow \underset{\substack{\text{Cupric ferrocyanide}\\ \text{(reddish brown ppt.)}}}{Cu_2[Fe(CN)_6]} \downarrow + 4KNO_3$$

2. **Lead** (Pb^{2+})

$$\underset{\text{Black}}{3\,PbS} + 8\,HNO_3 \longrightarrow 3\,Pb(NO_3)_2 + 2\,NO + 4H_2O \uparrow + 3\,S \downarrow$$

$$Pb(NO_3)_2(aq) + K_2CrO_4(aq) \longrightarrow \underset{\text{Yellow}}{PbCrO_4(s)} \downarrow + 2\,KNO_3(aq)$$

Confirmatory Tests for Third Group Radicals (Fe^{3+}, Al^{3+})

Both are precipitated as hydroxides. These are precipitated by addition of NH_4Cl and NH_4OH.

Third group radicals

Fe^{3+} — Red Brown ppt. $Fe(OH)_3$

(i) The brown ppt is soluble in dil. HCl.

(ii) With $K_4[Fe(CN)_6]$ solution, Fe^{3+} gives a blue colour or ppt.

(iii) With KCNS solution, Fe^{3+} gives a blood red colouration.

Al^{3+} — White gelatinous ppt. $Al(OH)_3$

(i) **Lake test** The white ppt is soluble in NaOH. The obtained solution on treatment with HCl and a few drops of blue litmus solution followed by excess of NH_4OH solution gives a blue lake.

(ii) In charcoal cavity cobalt nitrate test Al^{3+} gives a blue mass.

The following are some chemical equations which play an important role while confirmation of Fe^{3+} and Al^{3+} given as :

(i) **Iron** (Fe^{3+})

$$Fe(OH)_3(s) + 3\,HCl(aq) \longrightarrow FeCl_3(aq) + 3\,H_2O(l)$$

$$FeCl_3 + K_4[Fe(CN)_6] \longrightarrow \underset{\substack{\text{Potassium ferric}\\ \text{ferrocyanide}\\ \text{(Prussian blue)}}}{KFe\,[Fe(CN)_6]} \downarrow + 3\,KCl$$

$$FeCl_3(aq) + 6\,KCNS(aq) \longrightarrow \underset{\text{Blood red colour}}{K_3[Fe\,(CNS)_6]} \downarrow + 3\,KCl$$

(ii) **Aluminium** (Al^{3+})

$$\underset{\substack{\text{Gelatinous}\\ \text{white ppt.}}}{Al(OH)_3} \downarrow + NaOH \longrightarrow NaAlO_2(aq) + 2H_2O(l)$$

$$Al(OH)_3(s) \downarrow + 3\,HCl(aq) \longrightarrow AlCl_3(aq) + 3\,H_2O(l)$$

Addition of ammonium hydroxide precipitates Al as aluminium hydroxide. $Al(OH)_3$ gets adsorbed on the blue colouring matter of litmus to form a complex that floats as blue lake over a colourless solution.

Confirmatory Tests for Fourth Group Radicals (Zn^{2+}, Ni^{2+})

These are precipitaded as sulphides in ammoniacal medium.

Fourth group radicals

Zn^{2+}

(i) **Dirty white ppt.** (ZnS)
In cobalt nitrate test, Zn^{2+} gives green mass.

(ii) The precipitate is soluble in dil. HCl. The obtained solution on treatment with NaOH solution gives white ppt which dissolves in excess of NaOH.

(iii) With potassium ferrocyanide solution, Zn^{2+} gives white ppt.

Ni^{2+}

(i) **Black ppt.** (NiS)
The black ppt is soluble in *aqua-regia*.

(ii) Ni^{2+} with dimethyl glyoxime and excess of NH_4OH gives scarlet red ppt.

In IVth group, H_2S should not be passed continuously, otherwise NiS turns in colloidal state.The following are the some chemical equations which play an important role while confirmation of Zn^{2+} and Ni^{2+}given as:

(i) **Zinc** (Zn^{2+})

$$ZnS(s) + 2HCl(aq) \longrightarrow ZnCl_2(aq) + H_2S(g) \uparrow$$

$$ZnCl_2(aq) + 2NaOH \longrightarrow Zn(OH)_2(s) \downarrow + 2NaCl$$

$$Zn(OH)_2 + \underset{\text{Excess}}{2NaOH\,(aq)} \longrightarrow \underset{\text{Sodium zincate}}{Na_2ZnO_2} + 2H_2O$$

$$2Zn^{2+}(aq) + [Fe(CN)_6]^{4-} \longrightarrow \underset{\substack{\text{Zinc ferrocyanide}\\ \text{(white ppt.)}}}{Zn_2[Fe(CN)_6](s)} \downarrow$$

(ii) **Nickel** (Ni^{2+})

$$NiS + \underset{\textit{Aqua-regia}}{2HCl + NO_3^-} \longrightarrow NiCl_2 + S + NOCl + S \downarrow$$

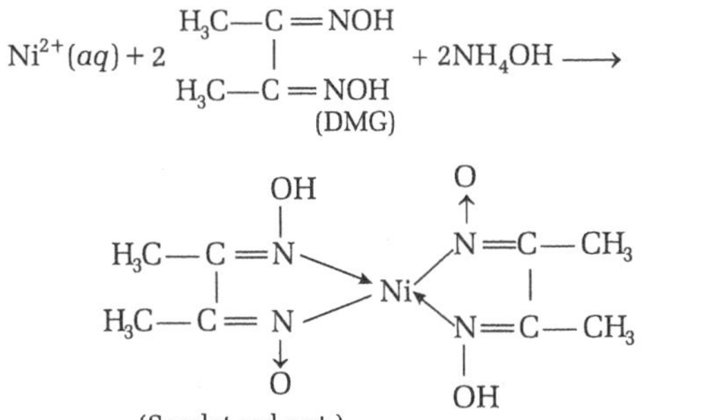

(Scarlet red ppt.)
(Dimethyl glyoximate)

Confirmatory Tests for Fifth Group Radicals

(Ba^{2+}, Ca^{2+}, Sr^{2+}) First boil off H_2S and then add $(NH_4)_2CO_3$ and NH_4OH.

Fifth group radicals

Ba^{2+} White ppt	Ca^{2+} White ppt.
The obtained ppt is soluble in acetic acid.	
(i) With K_2CrO_4 (in excess) solution, Ba^{2+} gives yellow ppt.	(i) With ammonium oxalate solution, Ca^{2+} gives white ppt.
(ii) Ba^{2+} imparts green colour to the flame in flame test.	(ii) Ca^{2+} imparts brick red colour to the flame in flame test.

The following are the some chemical equations which play an important role while confirmation of Ba^{2+} and Ca^{2+} given as:

1. **Barium** (Ba^{2+})

$$\underset{}{BaCl_2} + (NH_4)_2CO_3 \longrightarrow \underset{\text{White ppt.}}{BaCO_3 \downarrow} + 2NH_4Cl$$

$$BaCO_3 + 2CH_3COOH \longrightarrow (CH_3COO)_2Ba + H_2O + CO_2 \uparrow$$

$$(CH_3COO)_2Ba + K_2CrO_4 \longrightarrow \underset{\text{Yellow ppt.}}{BaCrO_4 \downarrow} + 2CH_3COOK$$

$BaCl_2$ is volatile and provides apple green colour to the flame.

2. **Calcium** (Ca^{2+})

$$CaCl_2 + (NH_4)_2CO_3 \longrightarrow \underset{\text{White ppt.}}{CaCO_3 \downarrow} + 2NH_4Cl$$

$$CaCO_3 + 2CH_3COOH \longrightarrow (CH_3COO)_2Ca + H_2O + CO_2 \uparrow$$

$$(CH_3COO)_2Ca + (NH_4)_2C_2O_4 \longrightarrow \underset{\substack{\text{Calcium oxalate}\\ \text{(white ppt.)}}}{CaC_2O_4 \downarrow} + 2CH_3COONH_4$$

3. **Stroncium** (Sr^{2+})

$$Sr^{2+}(aq) + CO_3^{2}(aq) \longrightarrow SrCO_3(s)$$

$$SrCO_3 + 2CH_3COOH \longrightarrow Sr(CH_3COO)_2 + H_2O + CO_2$$

Confirmatory Tests for Sixth Group Radicals (Mg^{2+})

- Mg^{2+} with little NH_4Cl, excess of NH_4OH and sodium phosphate solution, forms white crystalline ppt.
- Mg^{2+} with dil. HCl and a few drops of magneson reagent forms blue colour or ppt.

Chemical Equation Involved in the Confirmation of Mg^{2+}

The following is the chemical equation which plays an important role while confirmation of Mg^{2+} given as:

$$Mg^{2+}(aq) + HPO_4^{2-}(aq) + NH_4^+(aq) \longrightarrow \underset{\substack{\text{Magnesium}\\ \text{ammonium}\\ \text{phosphate (white)}}}{MgNH_4PO_4(s) \downarrow} + H^+$$

Detection of Extra Element in Organic Compounds

Presence of extra elements (i.e. N, S and X) in an organic compound is identified by **Lassaigne's test,** in which the Lassaigne's extract is prepared by fusing the compound with sodium metal. The reason of fusion is to convert covalent compounds into inorganic (ionisable) compounds.

1. **Detection of Nitrogen**

$$\underset{\text{(NaCN)}}{\text{Lassaigne's extract}} + \underset{\text{(Freshly prepared)}}{FeSO_4 + NaOH} \xrightarrow{\text{Boil + Cool}}$$

$$\text{Solution} + \text{dil. } H_2SO_4 + FeCl_3 \longrightarrow \text{Prussian blue colour or greenish blue ppt.}$$

$$\underset{\substack{\text{Ferrous}\\ \text{sulphate}}}{FeSO_4} + 2NaOH \longrightarrow \underset{\substack{\text{Ferrous}\\ \text{hydroxide}}}{Fe(OH)_2 \downarrow} + Na_2SO_4$$

$$Fe(OH)_2 + \underset{\text{Sodium extract}}{6NaCN} \longrightarrow Na_4[Fe(CN)_6] + 2NaOH$$

$$3Na_4[Fe(CN)_6] + 4FeCl_3 \longrightarrow \underset{\substack{\text{Ferric ferrocyanide}\\ \text{(Prussian blue colour)}}}{Fe_4[Fe(CN)_6]_3 \downarrow} + 12NaCl$$

If S is also present alongwith N.

In the test of nitrogen, blood red colour is obtained instead of blue or green.

$$3NaCNS + FeCl_3 \longrightarrow \underset{\substack{\text{Ferric sulpho}\\ \text{cyanide (red)}}}{Fe(SCN)_3} + 3NaCl$$

2. **Detection of Sulphur**

(i) $$\underset{\substack{\text{Sodium}\\ \text{extract}}}{Na_2S} + \underset{\substack{\text{Sodium}\\ \text{nitroprusside}}}{Na_2[Fe(CN)_5 \cdot NO]} \longrightarrow \underset{\substack{\text{Sodium thionitroprusside}\\ \text{(violet colour)}}}{Na_4[Fe(CN)_5 \cdot NOS]}$$

(ii) $$\underset{\substack{\text{Sodium}\\ \text{extract}}}{Na_2S} + \underset{\text{Lead acetate}}{(CH_3COO)_2Pb} \longrightarrow \underset{\text{Black ppt.}}{PbS \downarrow} + 2CH_3COONa$$

3. **Detection of Halogens**

There are two tests involved during the detection of halogens, i.e. $AgNO_3$ test and Beilstein's test.

(i) **$AgNO_3$ Test**

Lassaigne's extract + Conc. HNO_3 + Boil + Cool + $AgNO_3 \longrightarrow$ If

(a) white curdy ppt. is formed which is soluble in NH_4OH, it indicates the presence of Cl^- ion.

$$NaCl + AgNO_3 \longrightarrow \underset{\text{White ppt.}}{AgCl \downarrow} + NaNO_3$$

$$AgCl + 2NH_4OH \longrightarrow [Ag(NH_3)_2]Cl + 2H_2O$$

(b) pale yellow ppt. is obtained which is partially soluble in NH_4OH, it indicates the presence of Br^- ion.

$$NaBr + AgNO_3 \longrightarrow \underset{\substack{\text{Pale}\\ \text{yellow ppt.}}}{AgBr \downarrow} + NaNO_3$$

$$AgBr + 2NH_4OH \longrightarrow [Ag(NH_3)_2]Br + 2H_2O$$

(c) Yellow ppt. is formed which is insoluble in NH_4OH, it indicates the presence of I^- ion.

$$NaI + AgNO_3 \longrightarrow \underset{\text{Yellow ppt.}}{AgI\downarrow} + NaNO_3$$

(ii) **Beilstein's Test**

If a copper wire, dipped in organic compound, on heating gives green or bluish green flame, it shows the presence of halogens.

Several halogen free compounds such as pyridine, purines, urea, thiourea etc., also impart green colour to the flame, so this test is not very reliable. However, production of no green or blue colour confirms the absence of halogen.

Therefore, this test is a confirmative one to show the absence of halogens rather than to show its presence.

Beilstein's test is not given by fluorine because copper fluoride is not volatile.

Detection of Functional Groups

Following methods can be used to detect the functional groups.

1. **Tests for Carboxylic Acid** (—COOH) Carboxylic acids give brisk effervescence of CO_2 with sodium bicarbonate.

   They react with ethyl alcohol and conc. H_2SO_4 to give fruity smell due to formation of ester.
2. **Tests for Alcoholic** (—OH) **Group** Alcohols give red colour with ceric ammonium nitrate. Primary, secondary and tertiary alcohols are distinguished by using Lucas reagent. With Lucas reagent (anhy. $ZnCl_2$ + conc. HCl),
   (i) tertiary alcohols give turbidity immediately.
   (ii) secondary alcohols give turbidity after 5 minutes.
   (iii) primary alcohols do not produce turbidity at room temperature.
3. **Tests for Phenolic** (Ph—OH) **Group** Phenols give characteristic colours (green, blue and violet) with $FeCl_3$ solution due to formation of $[Fe(OC_6H_5)_6]^{3-}$ complex ion. In **Liebermann's nitroso reaction**, phenol gives blue colour (indophenol ion) which turns red (indophenol) on adding NaOH.
   Nitrophenols do not give the above two tests.
4. **Tests for Aldehyde** (—CHO) **Group** Aldehydes give silver mirror with Tollen's reagent, red ppt with Fehling solution, violet colour with Schiff's reagent and red-yellow ppt with Benedict's solution. Except Benedict's solution test, all other are also given by aromatic aldehydes. Therefore, this test is used to differentiate between aliphatic and aromatic aldehydes.
5. **Tests for Ketone Group** $\left(\begin{matrix} R \\ R \end{matrix}\!\!>C=O\right)$

   Ketones give white ppt with sodium bisulphite. With sodium nitroprusside, ketones give red or purple colour.
6. **Tests for Primary Amines** (R—NH_2) On adding acetone and a few drops of sodium nitroprusside to the aliphatic primary amines, violet red colour is obtained.

   Aliphatic primary amines give brisk effervescence with HNO_2 while aromatic primary amines give red dye on treating with HNO_2 and alkaline solution of β-naphthol.

   Tests for Secondary Amines (R—$\overset{R}{\overset{|}{N}}$H) Secondary amines give positive Liebermann's nitroso test. It gives blue colour on heating the mixture with nitrous acids and 1-2 drops of phenol and H_2SO_4. With HNO_2, secondary amines give oily dark coloured liquid.

Chemistry Involved in the Titrimetric Exercises

In titrimetric or volumetric analysis, the amount of a chemical species present in the given unknown solution is determined by measuring the volumes of the solution taking part in the given chemical reactions.

The chemical species react in the ratio of their chemical equivalent masses. The main purpose of this analysis is called titration.

Some important terms used in titrimetric exercises are as follows

(i) **Analyte and Titrant** The substance being analysed is called analyte and that which is added to analyte in a titration is called titrant.

(ii) **Equivalence Point or End Point** It is point at which the reaction between two solutions is just complete. It is generally represented by change in colour, pH, conductivity etc.

(iii) **Standard Solution** Solution of known concentration is called standard solution.

(iv) **Primary Standard Substance** The substance, standard solution of which can be prepared directly by dissolving its definite weight in definite volume of solvent is called primary standard substance, e.g. crystalline oxalic acid, anhydrous Na_2CO_3, Mohr's salt etc. The substance, which occur in pure state, are non-hydroscopic, non-deliquescent, generally behave as primary standard substance.

(v) **Secondary Standard Substance** Their standard solution cannot be prepared directly. e.g. $KMnO_4$, NaOH, KOH etc.

(vi) **Indicator** It shows the end point of a titration.

Titration of Na_2CO_3 *vs* HCl (Acid and Base)

The titration of Na_2CO_3 *vs* HCl is a neutralisation titration (acidimetry and alkalimetry) which involve the neutralisation of an acid with a base e.g.

$$Na_2CO_3 + 2HCl \longrightarrow 2NaCl + H_2O + CO_2\uparrow$$

In it, first the standard solution of Na_2CO_3 $\left(\frac{N}{10}Na_2CO_3\right)$ is prepared and then titrated its 20 mL with HCl solution by adding a few drops of methyl orange indicator. Change in colour shows the end point.

Calculations involved in the titration are as follows:

(i) Weight of Na_2CO_3 dissolved in 250 mL measuring flask $= z = ... g$

Weight of Na_2CO_3 in 100 mL $= .... \times 4$

$= .... g/L$

Normality of Na_2CO_3 (prepared)

$$= \frac{\text{strength (g / L)}}{\text{eq. wt. of oxalic acid}} = \frac{........}{63} N$$

(ii) For the titration using standard Na_2CO_3 solution

$$\underset{\substack{Na_2CO_3\\(\text{known})}}{N_1V_1} = \underset{HCl}{N_2V_2}$$

$N_2 = ... N$

(iii) For the titration using supplied Na_2CO_3 solution

$$\underset{\substack{Na_2CO_3\\(\text{unknown})}}{N_3V_3} = \underset{HCl}{N_4V_4} \qquad [\because N_4 = N_2]$$

$N_3 = ...$

Strength of Na_2CO_3 in g / L $= N_3 \times$ eq. wt. of $Na_2CO_3 = ...$

- In acidimetry and alkalimetry, the choice of indicators mainly depends upon the nature of the acids and alkalies used. Methyl orange, phenolphthalein are some of the important indicators used in these titrations.
- As no indicator gives correct results in the titration of weak acids against weak bases, such titrations are to be avoided.

Use of Indicators

The indicators used in various acid-base titrations are shown below:

Indicators in Acid-base Titrations

Acid-base titrations	Indicators
Strong acid *vs* strong base	Bromothymol blue, phenolphthalein methyl orange, thmyophthalein
Strong base acid *vs* weak base	Methyl orange, methyl red, bromocresol green
Weak acid *vs* strong base	Phenolphthalein, thymophthalein
Weak acid *vs* weak base	Phenol red

Titration of Oxalic Acid *vs* $KMnO_4$

- This is an example of redox titrations, in which a reducing agent (oxalic acid) is estimated by titrating it with a standard solution of oxidising agent ($KMnO_4$). Such reactions are accompanied by the change in valency of ions. In these titrations oxidation and reduction takes place simultaneously, i.e. while one substance is being oxidised, the other one is being reduced.

$$\underset{\text{Oxidising agent}}{2KMnO_4} + 3H_2SO_4 \longrightarrow K_2SO_4 + 2MnSO_4 + 3H_2O + 5[O]$$

$$\underset{\text{Reducing agent}}{5\begin{matrix}COOH\\|\\COOH\end{matrix}} + 5[O] \longrightarrow 5H_2O + 10CO_2\uparrow$$

The last drop of $KMnO_4$ itself acts as an indicator.

- In this titration, first the standard solution of oxalic acid is prepared which is then titrated with $KMnO_4$ solution in the presence of dil. H_2SO_4. The procedure is repeated to obtain a set of concurrent readings.
- Calculations involved in the titration are as follows:

(i) Weight of oxalic acid dissolved in 250 mL measuring flask $= z = ... g$

Weight of oxalic acid in 1000 mL $= \frac{.... \times 1000}{250} = ...$ g/L

Normality of oxalic acid (prepared)

$$= \frac{\text{Strength (g)/ L}}{\text{Eq. wt. of oxalic acid}} = \frac{........}{63.04} N$$

(ii) For the titrations using standard oxalic acid solution

$$\underset{\substack{(\text{Oxalic acid})\\(\text{Known})}}{N_1V_1} = \underset{(KMnO_4)}{N_2V_2}$$

$$\frac{........}{63.04} N \times 20 \text{ mL} = N_2 \times ....$$

(iii) For the titration using supplied oxalic acid solution

$$\underset{\substack{\text{Oxalic acid}\\(\text{unknown})}}{N_3V_3} = N_4V_4 \qquad [\because N_4 = N_2]$$

$N_3 = ... N$

Strength of oxalic acid in g / L $= N_3 \times$ eq. wt. of oxalic acid $= ....$ g/L

- In oxalic acid *vs* H_2SO_4 titration mixture is heated to near about 70°–80°C. Sulphuric acid should be in excess otherwise a brown ppt. due to formation of MnO_2 will be formed. This titration cannot be carried out in the presence of acid like HNO_3 and HCl, because HNO_3 itself is an oxidising agent, so it will interfere with the oxidising action of $KMnO_4$ and HCl reacts chemically with $KMnO_4$ solution.

Titration of Mohr's Salt *vs* $KMnO_4$

This is also an example of redox titrations and work on the same principle as oxalic acid *vs* $KMnO_4$ titration. In this titration, the active constituent of ferrous ammonium sulphate

(Mohr's salt) is ferrous sulphate, which is oxidised to ferric sulphate by acidified potassium permanganate as follows:

$$2KMnO_4 + 3H_2SO_4 \longrightarrow K_2SO_4 + 2MnSO_4 + 3H_2O + 5[O]$$
$$[2FeSO_4 + H_2SO_4 + [O] \longrightarrow Fe_2(SO_4)_3 + H_2O] \times 5$$
$$\overline{2KMnO_4 + 8H_2SO_4 + 10FeSO_4 \longrightarrow K_2SO_4 + 2MnSO_4 + 5Fe_2(SO_4)_3 + 8H_2O}$$

Calculations involved in the titration are as follows:

(i) Weight of ferrous ammonium sulphate dissolved in 250 mL measuring flask = z = ...g

Weight of ferrous ammonium sulphate in 1000 mL

$$= \frac{\ldots \times 1000}{250} = \ldots \text{ g / L}$$

Normality of ferrous ammonium sulphate (prepared)

$$= \frac{\text{strength (g / L)}}{\text{eq. wt. of ferrous ammonium sulphate}} = \frac{\ldots}{63.04} \text{ N}$$

(ii) For the titrations using standard ferrous ammonium sulphate solution

$$\underset{\substack{\text{Mohr's salt}\\ \text{(Known)}}}{N_1V_1} = \underset{KMnO_4}{N_2V_2}$$

$$\frac{\ldots}{63.04} \text{N} \times 20 \text{ mL} = N_2 \times \ldots$$

$$N_2 = \ldots \text{N}$$

(iii) For the titration using supplied ferrous ammonium sulphate solution $N_3V_3 = N_4V_4$

Ferrous ammonium sulphate $KMnO_4$ (unknown) $[\because N_4 = N_2]$

$$N_3 = \ldots N$$

Strength of ferrous ammonium sulphate in g/L = N_3 × eq. wt. of ferrous ammonium sulphate = ...g /L

Chemistry Involved in the Preparation of Compounds

Chemistry involved in the preparation of inorganic and organic compounds are as follows:

Preparation of Inorganic Compounds

- The contents used for the **preparation of Mohr's salt**, i.e. ferrous ammonium sulphate$[FeSO_4 \cdot (NH_4)_2SO_4 \cdot 6H_2O]$.

$$(NH_4)_2SO_4 + FeSO_4 + 6H_2O \longrightarrow \underset{\substack{\text{Mohr's salt}\\ \text{(light green)}}}{FeSO_4 \cdot (NH_4)_2SO_4 \cdot 6H_2O}$$

- **Potash alum** is a double salt having composition $K_2SO_4 \cdot Al_2(SO_4)_3 \cdot 24H_2O$. It is prepared by concentrating a solution containing equimolar quantities of K_2SO_4 and $Al_2(SO_4)_3 \cdot 18H_2O$.

$$K_2SO_4 + Al_2(SO_4)_3 \cdot 18H_2O + 6H_2O \longrightarrow K_2SO_4 \cdot Al_2(SO_4)_3 \cdot 24H_2O$$

Preparation of Organic Compounds

- **Acetanilide** $(C_6H_5NHCOCH_3)$ is obtained from aniline by treating it with acetic anhydride. Since, acetic anhydride is quite costly, it is not used now a days, instead it glacial acetic acid is used.

$$C_6H_5NH_2 + (CH_3CO)_2O \longrightarrow C_6H_5NHCOCH_3 + CH_3COOH$$
$$C_6H_5NH_2 + CH_3COOH \longrightarrow C_6H_5NHCOCH_3 + H_2O$$

- During the preparation of acetanilide, some zinc dust is added to reduce the coloured impurities and also to prevent the oxidation of aniline.
- ***p*-nitro acetanilide** is prepared by the nitration of acetanilide, by using fuming HNO_3 + conc. H_2SO_4.

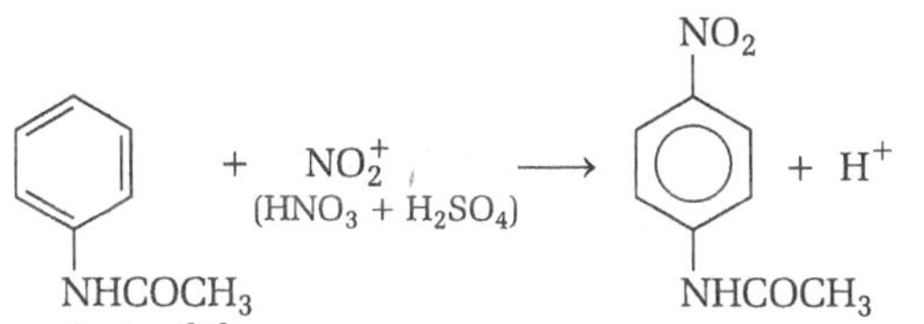

Acetanilide *p*-nitro acetanilide

- During the preparation of *p*-nitro acetanilide from acetanilide, the temperature of the flask should be maintained below 10°C.
- **Iodoform** (CHI_3) is obtained by treating acetone with KI and sodium hypochlorite (NaOCl). The colour of crystals of iodoform is yellow.

$$NaOCl + KI \longrightarrow NaOI + KCl$$
$$CH_3COCH_3 + NaOI \xrightarrow{NaOH} CH_3COONa + \underset{\text{Iodoform}}{CHI_3}$$

- **Aniline yellow**, also called *p*-amino azobenzene, is obtained from diazoamino benzene. It is carcinogenic and basic dye.

Some Important Experiments

1. **Enthalpy of Solution of $CuSO_4$** It is the amount of heat absorbed or evolved on dissolving one mole of it in excess of the solvent at a given temperature and pressure. It is positive when heat is absorbed and negative when heat is given out during dissolution.

   It can be determined by putting a known volume of water in a thermally insulated *vessel* and then finding out its change in temperature on adding a known weight of the substance into it. From the change in temperature, heat absorbed or evolved can be calculated.

   Heat evolved or absorbed = Heat gained or lost

   $$= m \times s \times t$$
   $$= 100 \times 4.2 \times (t_2 - t_1) \text{ J}$$

   Number of moles of $CuSO_4$ dissolved $= \frac{W}{M}\left(\frac{10}{249}\right)$

   Heat of solution of $CuSO_4 \cdot H_2O = 100 + 4.2 \times (t_2 - t_1) \times \frac{249}{10}$

2. **Enthalpy of neutralisation of strong acid and strong base**

It is defined as the heat evolved when one mole of H^+ ions is completely neutralised by a base.

To find this, whole procedure is divided into two step. First step is the determination of water equivalent of the beaker. The second step is the determination of heat of neutralisation by using same beaker.

$$\text{Water equivalent of beaker } (W) = \left[\frac{50 \times 4.2(t_2 - t_3)}{(t_3 - t_1)}\right] - (50 \times 4.2)\ \text{J/°C}$$

t_1 = Initial temperature of cold water

t_2 = Temperature of hot water

t_3 = Temperature of the mixture

Q (heat evolved) $= (100 \times 4.2 + W)(t_5 - t_4)$

t_5 = Temperature of HCl + NaOH

t_4 = Temperature of HCl

3. **Preparation of lyophilic and lyophobic sols** Two methods are employed for preparation of sol. These are as follows

(i) **Condensation method** that involves the increase in size of the solute particle in solution upto the colloidal size. It may be effected by hydrolysis, oxidation, reduction or double decomposition.

(ii) **Dispersion method** that involves crushing of bigger particles in a suspension to the colloidal size. Generally mechanical method is involved to reduce the size of particle.

4. **Kinetic study of reaction of iodide ion with hydrogen peroxide at room temperature.** Here, the rate of reaction is determined by reaction with thiosulphate ions and starch solution.

- The iodine produced in the reaction of iodide and H_2O_2 reacts with thiosulphate and cannot colour starch blue.
- This continues till the whole of thiosulphate has been consumed. After that the solution turns blue.

DAY PRACTICE SESSION 1

FOUNDATION QUESTIONS EXERCISE

1 A substance on treatment with dil. H_2SO_4 liberates a colourless gas which produces (i) turbidity with baryta water and (ii) turns acidified dichromate solution green. The reaction indicates the presence of

(a) CO_3^{2-} (b) S^{2-} (c) SO_3^{2-} (d) NO_2^-

2 The brown ring test for NO_2^- and NO_3^- is due to the formation of complex ion with formula

(a) $[Fe(H_2O)_6]^{2+}$ (b) $[Fe(NO)(CN)_5]^{2-}$

(c) $[Fe(H_2O)_5NO]^{2+}$ (d) $[Fe(H_2O)(NO)_5]^{2+}$

3 The acidic solution of a salt produced a deep blue colour with starch iodide solution. The salt may be

(a) chloride (b) nitrite

(c) acetate (d) bromide

4 Borax when heated on platinum wire forms a glass like bead which is made up of

(a) sodium tetraborate

(b) sodium metaborate

(c) sodium metaborate and boric anhydride

(d) boric anhydride and sodium tetraborate

5 In the borax bead test of Co^{2+}, the blue colour of bead is due to the formation of

(a) B_2O_3 (b) Co_3B_2

(c) $Co(BO_2)_2$ (d) CoO

6 A solution of metal ion when treated with KI gives a red precipitate which dissolves in excess KI to give a colourless solution. Moreover the solution of metal ion on treatment with a solution of cobalt (II) thiocyanate gives rise to a deep blue crystalline precipitate. The metal ion is → AIEEE 2010

(a) Pb^{2+} (b) Hg^{2+} (c) Cu^{2+} (d) Co^{2+}

7 If Fe^{3+} and Cr^{3+} both are present in group III of qualitative analysis, then distinction can be made by

(a) addition of NH_4OH in the presence of NH_4Cl when only $Fe(OH)_3$ is precipitated

(b) addition of NH_4OH in the presence of NH_4Cl when $Cr(OH)_3$ and $Fe(OH)_3$ both are precipitated and on adding Br_2 water and NaOH, $Cr(OH)_3$ dissolves

(c) precipitate of $Cr(OH)_3$ and $Fe(OH)_3$ as obtained in (b) are treated with conc. HCl when only $Fe(OH)_3$ dissolves

(d) Both (b) and (c)

8 Which one among the following pairs of ions cannot be separated by H_2S in dilute hydrochloric acid?

(a) Bi^{3+}, Sn^{4+} (b) Al^{3+}, Hg^{2+}

(c) Zn^{2+}, Cu^{2+} (d) Zn^{2+}, Ni^{2+}

9 The only cations present in a slightly acidic solution are Fe^{3+}, Zn^{2+} and Cu^{2+}. The reagent that when added in excess to this solution would identify and separate Fe^{3+} in one step is

(a) 2 M HCl (b) 6 M NH_3

(c) 6 M NaOH (d) H_2S gas

10 Which one of the following statement is correct?

(a) Fe^{2+} gives brown colour with ammonium thiocyanate
(b) Fe^{2+} gives blue precipitate with potassium ferricyanide
(c) Fe^{3+} gives brown colour with potassium ferricyanide
(d) Fe^{3+} gives red colour with potassium ferrocyanide

11 A metal nitrate reacts with KI to give a black precipitate which on addition of excess of KI convert into orange colour solution. The cation of metal nitrate is

(a) Hg^{2+} (b) Bi^{3+} (c) Pb^{2+} (d) Cu^{+}

12 Sodium carbonate cannot be used in place of $(NH_4)_2CO_3$ for the identification of Ca^{2+}, Ba^{2+} and Sr^{2+} ions (in group V) during mixture analysis because → **JEE Main (Online) 2013**

(a) Mg^{2+} ions will be precipitated
(b) concentration of CO_3^{2-} ions is very low
(c) sodium ions will react with acid radicals
(d) Na^{+} ions will interfere with the detection of Ca^{2+}, Ba^{2+}, Sr^{2+} ions

13 Match the following and choose the correct option.

	Ion+Reagent		Colour
A.	$Fe^{3+} + [Fe(CN)_6]^{4-}$	1.	Cherry red
B.	$Fe^{3+} + CNS^{-}$	2.	Pink
C.	Ni^{2+} + DMG	3.	Blood red
D.	$Mn^{2+} + PbO_2 + H^{+}$	4.	Blue

Codes

	A	B	C	D		A	B	C	D
(a)	1	4	2	3	(b)	4	3	1	2
(c)	3	4	2	1	(d)	3	1	4	3

14 When metal '*M*' is treated with NaOH, a white gelatinous precipitate '*X*' is obtained, which is soluble in excess of NaOH. Compound '*X*' when heated strongly gives an oxide which is used in chromatography as an adsorbent. The metal '*M*' is → **JEE Main 2018**

(a) Zn (b) Ca (c) Al (d) Fe

15 Which of the following gives blood red colour with KCNS?

(a) Cu^{2+} (b) Fe^{3+} (c) Al^{3+} (d) Zn^{2+}

16 In organic analysis, the reagent 2, 4-dinitro phenyl hydrazine is used for the detection of which of the following functional groups?

(a) Alcohol (b) Acid (c) Ketone (d) Amines

17 A compound liberates CO_2 with $NaHCO_3$ and also gives colour with neutral $FeCl_3$ solution. The compound can be

(a) 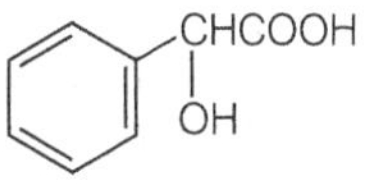

(b) 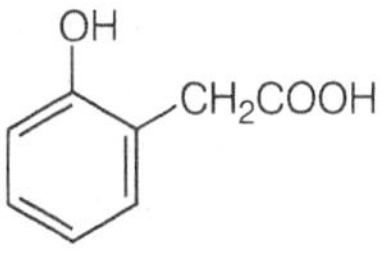

(c) 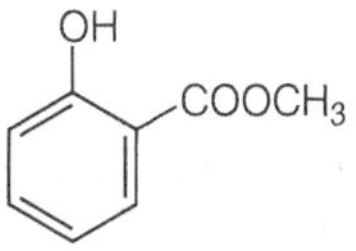

(d) 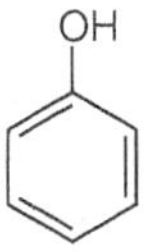

18 Which of the following compounds is not expected to show 'Lassaigne's test for nitrogen? → **JEE Main (Online) 2013**

(a) Propanenitrile
(b) Hydroxylamine hydrochloride
(c) Nitromethane
(d) Ethanamine

19 Which of the following will not give Lassaigne's test for nitrogen ?

(a) $NH_2—NH_2$ (b) $C_6H_5—N=N—C_6H_5$
(c) CH_3CONH_2 (d) $CH_3C\equiv N$

20 Phenolphthalein is an indicator for acid-base titration, it exists as

(a) benzenoid form in acid and quinonoid form in basic solution
(b) quinonoid form in acid and benzenoid form in basic solution
(c) quinonoid form in both
(d) benzenoid form in both

21 The compound formed in the positive test for nitrogen with the Lassaigne solution of an organic compound is

(a) $Fe_4[Fe(CN)_6]_3$
(b) $Na_3[Fe(CN)_6]$
(c) $Fe(CN)_3$
(d) $Na_4[Fe(CN)_5NOS]$

22 In the titration of oxalic acid *vs* potassium permanganate, potassium permanganate acts as

(a) external indicator
(b) self indicator
(c) reductant
(d) Both (b) and (c)

23 Sodium salt of an organic acid '*X*' produces effervescence with conc. H_2SO_4. '*X*' reacts with the acidified aqueous $CaCl_2$ solution to give a white precipitate which decolourises acidic solution of $KMnO_4$. '*X*' is → **JEE Main 2017**

(a) C_6H_5COONa (b) HCOONa
(c) CH_3COONa (d) $Na_2C_2O_4$

24 In the kinetic study of reaction of iodide ion with hydrogen peroxide, a known volume of sodium thiosulphate solution is added to

(a) oxidise iodide ion to iodine
(b) reduce iodine to iodide ion
(c) form a soluble blue complex
(d) induce the reaction rate

25 In the reaction, $2H_2O_2 \xrightarrow{I^-} 2H_2O + O_2$, the rate of reaction

(a) decreases as concentration of I^- ion increases
(b) increases as concentration of I^- ion increases
(c) increases in the presence of UV light
(d) Both (b) and (c)

Direction (Q. Nos. 26-28) *In the following questions, Assertion (A) followed by Reason (R) is given. Choose the correct answer out of the following choices.*

(a) Assertion and Reason both are correct statements and Reason is the correct explanation of the Assertion
(b) Assertion and Reason both are correct statements but Reason is not the correct explanation of the Assertion
(c) Assertion is correct incorrect and Reason is incorrect
(d) Both Assertion and Reason are incorrect

26 **Assertion** (A) NO_2^- is decomposed by urea in the mixture of NO_3^- and NO_2^-.
Reason (R) NO_3^- interfere in the ring test of NO_2^-.

27 **Assertion** (A) Borax bead test is not suitable for Al (III).
Reason (R) Borax bead test is given by cell inorganic salts.

28 **Assertion** (R) Addition of $(NH_4)_2CO_3$ to an aqueous solution of $BaCl_2$ in the presence of NH_4Cl and NH_4OH precipitates $BaCO_3$.
Reason (R) $Ba(OH)_2$ is soluble in water.

DAY PRACTICE SESSION 2

PROGRESSIVE QUESTIONS EXERCISE

1 A metal *X* on heating strongly in the presence of O_2 gives an oxide which is also constituent of a white paint. Metal *X* on treatment with dil. H_2SO_4 evolves the lightest gas *Y* and the resultant solution on crystallisation gives *Z*. The metal is also used as a protective coating on iron. The metal is

(a) Al (b) Cu
(c) Zn (d) Pb

2 A mixed oxide of iron and chromium, $FeO \cdot Cr_2O_3$ is fused with sodium carbonate in the presence of air to form a yellow compound *A*. On acidification, the compound *A* forms an orange compound *B* which is strong oxidising agent. Identify *A* and *B*.

(a) $K_2Cr_2O_7, K_2CrO_4$
(b) $Na_2CrO_4, Na_2Cr_2O_7$
(c) $K_2CrO_4, K_2Cr_2O_7$
(d) $Na_2Cr_2O_7, Na_2CrO_4$

3 A translucent white waxy solid *A* on heating in an inert atmosphere is converted to its allotropic form *B*. Allotrope *A* on reaction with very dilute aqueous KOH liberates a highly poisonous gas *C* having rotten fish smell. With excess of chlorine form *D* which hydrolysis to compound *E*. Identify compound *A*.

(a) White phosphorus
(b) Red phosphours
(c) Phosphine
(d) Phosphorus pentachloride

4 When conc. H_2SO_4 was added to an unknown salt present in a test tube, a brown gas *A* was evolved. The gas intensified when copper turnings were also added into this test tube. On cooling the gas *A* changed into colourless *B*. Identify gases *A* and *B*.

(a) NO_2, N_2O_4 (b) N_2O_4, NO_2
(c) N_2O, NO_2 (d) NO, NO_2

5 A blackish brown solid *A* when fused with alkali metal hydroxide in the presence of air, produces a dark green coloured compound *B* which on electrolytic oxidation in alkaline medium gives a dark purple coloured compound *C*. Identify *A* and *C*.

(a) $MnO_2, KMnO_4$ (b) $K_2MnO_4, KMnO_4$
(c) $KMnO_4, MnO_2$ (d) $KMnO_4, K_2MnO_4$

6 In third group of qualitative analysis, the precipitate employed is a mixture of NH_4Cl and NH_4OH. If NH_4Cl is not available in the laboratory and the mixture does not contain Mn^{2+}, we can use

(a) $(NH_4)_2SO_4$ (b) $(NH_4)_2CO_3$
(c) NH_4NO_3 (d) All of these

7 A metal *X* on heating in nitrogen gas gives *Y*. *Y* on treatment with H_2O gives a colourless gas which when passed through $CuSO_4$ solution gives a blue colour. *Y* is

(a) $Mg(NO_3)_2$ (b) Mg_3N_2 (c) NH_3 (d) MgO

8 Consider the following statements.

(I) *A* black colour compound *B* is formed on passing H_2S through the solution of a compound *A* in NH_4OH
(II) *B* on treatment with HCl and $KClO_3$ gives *A*.
(III) *A* on treatment with KCN given a buff coloured precipitate which dissolves in excess of this reagent forming a compound *C*.
(IV) The compound *C* is changed into compound *D* when its aqueous solution is boiled.
(V) The solution of *A* was treated with excess of $NaHCO_3$ and Br_2 water. On cooling and shaking for sometime a green colour of compound *E* is formed. No change is observed on heating.

The compound *E* is

(a) $Na_4[Fe(CN)_5NOS]$ (b) $Na_3[Co(CO_3)_3]$
(c) $[Cu(NH_3)_4](NO_3)_2$ (d) $Fe_3[Fe(CN)_6]_2$

9 A gaseous mixture containing *X*, *Y* and *Z* gases, when passed into acidified $K_2Cr_2O_7$ solution, gas *X* was absorbed and the solution was turned green. The remainder gas mixture was then pass through lime water, which turns milky by absorbed gas *Y*. The residual gas when passed through alkaline pyrogallol solution, it turned black. The gases *X*, *Y* and *Z* are respectively

(a) SO_2, CO_2, O_2 (b) O_2, CO_2, SO_2
(c) SO_2, O_2, CO_2 (d) CO_2, SO_2, O_2

10 A reddish brown metal *X* when heated in presence of oxygen forms a black compound *Y* which is basic in nature when heated with hydrogen gas gives back *X*. Identify *X* and *Y*.

(a) Zn, ZnO
(b) Cu, CuO
(c) Fe, FeO
(d) Pb, PbO

ANSWERS

SESSION 1	**1** (c)	**2** (c)	**3** (b)	**4** (c)	**5** (c)	**6** (b)	**7** (d)	**8** (d)	**9** (b)	**10** (b)
	11 (b)	**12** (a)	**13** (b)	**14** (c)	**15** (b)	**16** (c)	**17** (b)	**18** (b)	**19** (a)	**20** (a)
	21 (a)	**22** (b)	**23** (d)	**24** (b)	**25** (d)	**26** (c)	**27** (d)	**28** (b)		
SESSION 2	**1** (c)	**2** (b)	**3** (a)	**4** (a)	**5** (a)	**6** (c)	**7** (b)	**8** (b)	**9** (a)	**10** (b)

Hints and Explanations

SESSION 1

1 SO_3^{2-} gives SO_2 with dil. H_2SO_4 which gives turbidity with $Ba(OH)_2$ and turns acidified dichromate solution green due to its reduction to Cr^{3+} ions.

2 The brown ring test for nitrates and nitrites is due to the formation of $[Fe(H_2O)_5NO]^{2+}$ complex ion.

3 Starch iodide solution contains iodide which is oxidised by NO_2^- to I_2 which gives blue colour with starch.

4 Borax on strong heating, first loses its water of crystallisation and then shrinks to form a transparent glassy bead of sodium metaborate and boric anhydride.

$$\underset{\text{Borax}}{Na_2B_4O_7} \longrightarrow \underbrace{\underset{\text{Sodium meta borate}}{2NaBO_2} + \underset{\text{Boric anhydride}}{B_2O_3}}_{\text{transparent glassy bead}}$$

5 $CoO + B_2O_3 \longrightarrow \underset{\text{Blue bead}}{Co(BO_2)_2}$

6 $Hg^{2+} + 2I^- \longrightarrow HgI_2$

$HgI_2 + \underset{\text{Excess}}{2I^-} \longrightarrow [HgI_4]^{2-}$

$Hg^{2+} + Co(SCN)_2 \longrightarrow \underset{\text{Blue crystalline ppt}}{Hg(SCN)_2 \downarrow}$

7 If Fe^{3+} and Cr^{3+} both are present, then very first solid ammonium chloride and ammonium hydroxide is added slowly till the solution gives smell of ammonia. Fe^{3+} and Cr^{3+} precipitates in the hydroxide form.

For identification, precipitate is treated with NaOH and Br_2 water, yellow colouration confirms Cr^{3+} ion.

$$2NaOH + Br_2 \longrightarrow NaBrO + NaBr + H_2O$$

$$NaBrO \longrightarrow NaBr + [O]$$

$$2Cr(OH)_3 + 4\,NaOH + 3\,[O] \longrightarrow 2\,Na_2CrO_4 + 5H_2O$$

Solution is acidified and treated with lead acetate solution.

$$Na_2CrO_4 + Pb(CH_3COO)_2 \longrightarrow \underset{\text{Yellow ppt.}}{PbCrO_4 \downarrow} + 2\,CH_3CO\overset{-}{O}\overset{+}{Na}$$

8 Both Zn^{2+} and Ni^{2+} belong to group IV of qualitative inorganic analysis and will not get precipitated by H_2S.

9 $Fe^{3+} + Zn^{2+} + Cu^{2+} \xrightarrow{6M\ NH_3} \underset{\text{Brown ppt.}}{Fe(OH)_3 \downarrow} + \underset{\text{Soluble}}{[Zn(NH_3)_4]^{2+}} + \underset{\text{Soluble}}{[Cu(NH_3)_4]^{2+}}$

10 The blue precipitate of Fe^{2+} ions with potassium ferricyanide is due to the formation of Turnbull's blue $K\overset{II}{Fe}[\overset{III}{Fe}(CN)_6]$.

$$Fe^{2+} + \underset{\text{Potassium ferricyanide}}{K_3[Fe(CN)_6]} \longrightarrow \underset{\text{Turnbull's blue}}{K\overset{II}{Fe}[\overset{III}{Fe}(CN)_6]} + 2K^+$$

11 Bismuth nitrate, $Bi(NO_3)_3$ (having cation Bi^{3+}), reacts with KI to give a black precipitate of BiI_3 which on addition of excess of KI, dissolved to give orange colour solution of $K[BiI_4]$ complex salt.

$$Bi(NO_3)_3(aq) + 3KI(aq) \longrightarrow \underset{\text{Black ppt.}}{BiI_3(s)\downarrow} + 3KNO_3(aq)$$

$$BiI_3(s) + KI(aq) \longrightarrow \underset{\text{Orange colour}}{K[BiI_4]}$$

12 In fifth group, $(NH_4)_2CO_3$ is used in the presence of NH_4Cl. In presence of NH_4Cl, dissociation of $(NH_4)_2CO_3$ decreases and due to low concentration of CO_3^{2-} ions only V group radical are precipitated out. If Na_2CO_3 is used, concentration of CO_3^{2-} ions will increase and Mg^{2+} ions will also be precipitated.

13 $A \longrightarrow 4; B \longrightarrow 3; C \longrightarrow 1; D \longrightarrow 2$

14 Among the given metals, Al forms white gelatinous ppt. with NaOH. Hence, the probable metal can be Al. This ppt. is dissolved in excess of NaOH due to the formation of sodium metal aluminate. Both the reactions are shown below.

$$Al^{3+} \xrightarrow{NaOH} \underset{\text{White gelatinous ppt. (X) of aluminium hydroxide}}{Al(OH)_3} \xrightarrow{\text{Excess of NaOH}} \underset{\text{Sodium metaaluminate (soluble)}}{NaAlO_2}$$

Aluminium hydroxide on strong heating gives alumina (Al_2O_3) which is used as an adsorbent in chromatography. This reaction can be seen as :

$$2Al(OH)_3 \xrightarrow{\Delta} Al_2O_3 + 3H_2O$$

Thus, metal *M* is Al.

Ca, being below sodium in electrochemical reactivity series, cannot displaces Na from its aqueous solution.

Zn reacts with NaOH to form sodium zincate which is a soluble compound.

Fe reacts with sodium hydroxide to form tetrahydroferrate (II) sodium which is again a soluble complex.

15 Fe^{3+} gives blood red colour with KCNS.

16 The reagent 2, 4-dinitro phenyl hydrazine is used for the detection of carbonyl group, i.e. aldehyde and ketone groups. With carbonyl group, this reagent gives red or yellow ppt due to the formation of respective hydrazones.

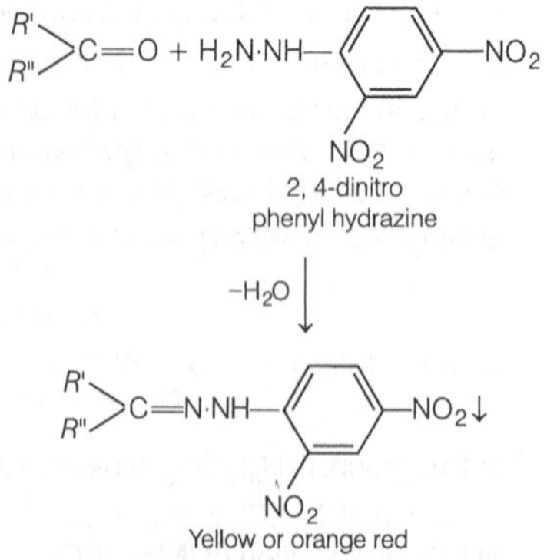

17 The compound liberates CO_2 with $NaHCO_3$, so it contains —COOH group and it also gives colour with neutral $FeCl_3$ solution, so it also contains a —OH group directly attached to the benzene ring (i.e. phenol). Hence, the structure of the compound is

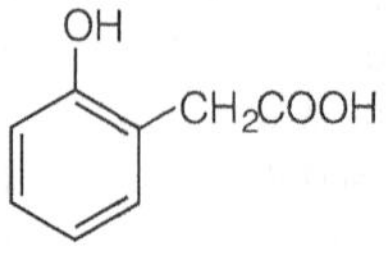

18 Hydroxylamine hydrochloride is not expected to show Lassaigne's test for nitrogen. This is because Lassaigne's test is positive for nitrogen only when the compound contains carbon alongwith nitrogen.

19 $NH_2—NH_2$ will not give Lassaigne's test for nitrogen due to absence carbon alongwith nitrogen.

20 Phenolphthalein is colourless in acid solution (benzenoid form) and pink in alkali (basic) solution (quinonoid form).

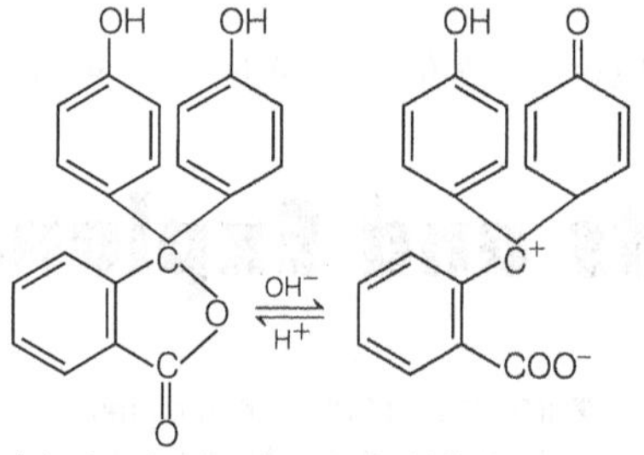

Colourless (acid medium) benzenoid form | Red (alkaline medium) quinonoid form

21 If nitrogen is present in organic compound, then sodium extract contains NaCN.

$$Na + C + N \xrightarrow{Fuse} NaCN$$

$$FeSO_4 + 6NaCN \longrightarrow \underset{(A)}{Na_4[Fe(CN)_6]} + Na_2SO_4$$

A changes to Prussian blue $Fe_4[Fe(CN)_6]_3$ on reaction with $FeCl_3$.

$$4FeCl_3 + 3Na_4[Fe(CN)_6] \longrightarrow Fe_4[Fe(CN)_6]_3 + 12NaCl$$

22 In the titration of oxalic acid *vs* $KMnO_4$, $KMnO_4$ acts as an oxidant as well as a self-indicator.

$$2KMnO_4 + 3H_2SO_4 \longrightarrow K_2SO_4 + 2MnSO_4 + 3H_2O + 5[O]$$

$$5\begin{matrix}COOH\\|\\COOH\end{matrix} + 5[O] \longrightarrow 5H_2O + 10CO_2\uparrow$$

23 The reaction takes place as follows

$$\underset{(X)}{Na_2C_2O_4} + \underset{(conc.)}{H_2SO_4} \longrightarrow Na_2SO_4 + H_2O + CO\uparrow + \underset{\text{Effervescence}}{CO_2\uparrow}$$

$$\underset{(X)}{Na_2C_2O_4} + CaCl_2 \longrightarrow \underset{\text{White ppt.}}{CaC_2O_4} + 2NaCl$$

$$5CaC_2O_4 + \underset{\text{Purple}}{2KMnO_4} + 8H_2SO_4 \longrightarrow K_2SO_4 + 5CaSO_4 + \underset{\text{Colourless}}{2MnSO_4} + 10CO_2 + 8H_2O$$

Hence, *X* is $Na_2C_2O_4$.

24 In the given experiment following reaction occurs

$$H_2O_2 + 2I^- + 2H^+ \longrightarrow 2H_2O + I_2$$

Iodine liberated in this reaction, reacts with sodium thiosulphate solution and is reduced to iodide ions.

$$I_2 + \underset{\text{Thiosulphate ion}}{2S_2O_3^{2-}} \xrightarrow{Fast} \underset{\text{Tetrathionate ion}}{S_4O_6^{2-}} + 2I^-$$

25 The rate of the dissociation of hydrogen peroxide with I^- ion increases by increasing the concentration of I^- ion and also in presence of UV light.

26 Ring test is given by NO_3^- ion. While NO_2^- ion gives thiourea test and sulphanilic acid test.

27 Borax bead test is given by only coloured salts.

28 Barium comes under group V in cation analysis. Addition of $(NH_4)CO_3$ to an aqueous solution of $BaCl_2$ in the presence of NH_4Cl and NH_4OH precipitates $BaCO_3$.

SESSION 2

1 $$\underset{'X'}{Zn} + \frac{1}{2}O_2 \longrightarrow \underset{\text{Zinc oxide (Y)}}{ZnO}$$

$$ZnO + \underset{(Dil.)}{H_2SO_4} \longrightarrow ZnSO_4 + H_2$$

Zinc oxide is used as a constituent in white pain and it evolves H_2 gas on reacting with dilute H_2SO_4.

Zinc is widely used in galvanisation processes to protect steel and iron from rusting. This involves coating the surface of a metal with a thin layer of zinc to create corrosion resistant barrier.

2 $$4FeO\cdot Cr_2O_3 + 8Na_2CO_3 + 7O_2 \longrightarrow \underset{\substack{(A)\\ \text{Sodium chromate}\\ \text{(Yellow compound)}}}{8Na_2CrO_4} + 2Fe_2O_3 + 8CO_2$$

$$\underset{\substack{(A)\\ \text{Yellow compound}}}{2Na_2CrO_4} + H_2SO_4 \longrightarrow \underset{\substack{(B)\\ \text{Orange}}}{Na_2Cr_2O_7} + Na_2SO_4 + H_2O$$

The compound (*B*) is strong oxidising agent.

3 The white waxy solid (*A*) is white phosphorus. When white phosphorus is heated in an inert atmosphere at 573 K, it changes to red phosphorus. (*B*) is red phosphorus.

(*A*) on heating with KOH liberates phosphine (*C*) which is poisonous gas with rotten fish smell.

$$P_4 + 3KOH + 3H_2O \longrightarrow \underset{\text{Phosphine}}{PH_3} + 3KH_2PO_2$$

White phosphorus (P_4) burns with excess of Cl_2 to form phosphorus pentachloride (*D*).

$$P_4 + 10Cl_2 \xrightarrow{\text{Heat}} 4PCl_5$$

Hydrolysis of (*D*) gives phosphoric acid (*E*).

$$PCl_5 + 4H_2O \longrightarrow H_3PO_4 + 5HCl$$

4 The given salt is a nitrate salt which on reaction with conc.H_2SO_4 gives first vapours of HNO_3 which decomposes to give brown gas (NO_2).

$$2NaNO_3 + H_2SO_4 \xrightarrow{\text{Heat}} Na_2SO_4 + \underset{\text{(Colourless)}}{2HNO_3}$$

$$4HNO_3 \xrightarrow{\text{Heat}} \underset{\text{Brown gas}}{4NO_2} + 2H_2O + O_2$$

The gas intensified when copper turnings were added due to reduction of HNO_3 by Cu

$$Cu + 4HNO_3 \xrightarrow{\text{Heat}} Cu(NO_3)_2 + \underset{\text{Brown gas}}{2NO_2\uparrow} + 2H_2O$$

On cooling the gas (*A*) changes into a colourless gas (*B*), N_2O_4.

$$\underset{(A)}{2NO_2} \rightleftharpoons \underset{(B)}{N_2O_4}$$

5 $$\underset{\substack{(A)\\ \text{Blackish brown}\\ \text{compound}}}{2MnO_2} + 4KOH + O_2 \xrightarrow{\text{Fuse}} \underset{\substack{(B)\\ \text{Potassium manganate}\\ \text{(Green coloured)}}}{2K_2MnO_4} + 2H_2O$$

$$\underset{(B)}{2K_2MnO_4} + H_2O + O \xrightarrow[\text{medium}]{\text{Alkaline}} \underset{\substack{(C)\\ \text{Purple coloured}\\ \text{compound}}}{2KMnO_4} + 2KOH$$

6 $(NH_4)SO_4$ will give white ppt. in presence of Ba^{2+}, Sr^{2+}, $(NH_4)_2CO_3$ will give white ppt. in presence of Ba^{2+}, Sr^{2+} and Ca^{2+}. To produce common ion (NH_4^+) and to suppress the ionisation of NH_4OH, NH_4NO_3 can be used in absence of Mn^{2+}.

7 *Y* is Mg_3N_2.

Magnesium react with nitrogen to form magnesium nitride. Magnesium nitride reacts with water to form ammonia and magnesium hydroxide. When ammonia gas is passed through $CuSO_4$ solution, it forms a blue precipitate.

8 The formation of black coloured compound (*B*) by passing H_2S through the alkaline solution of the compound indicate that (*A*) is a salt of the IV group radicals (Co^{2+}, Ni^{2+} or Zn^{2+}). However, the given reactions especially reaction (iii) indicates that compound (*A*) is a cobalt salt ($CoCl_2$) which explains all the given reactions.

(i) $$\underset{(A)}{CoCl_2} + 2NH_4OH + H_2S \longrightarrow \underset{(B)}{CoS} + 2NH_4Cl + 2H_2O$$

(ii) $$CoS + 2HCl + \underset{(\text{From } KClO_3)}{O} \xrightarrow{(B)} CoCl_2 + H_2O + S$$

$$2KClO_3 \longrightarrow 2KCl + 3O_2$$

(iii) $$CoCl_2 + 2KCN \longrightarrow \underset{\text{Buff coloured}}{Co(CN)_2\downarrow} + 2KCl$$

$$Co(CN)_2 + 4KCN \longrightarrow \underset{(C)}{K_4[Co(CN)_6]}$$

(iv) $$2K_4[Co(CN)_6] + O + H_2O \longrightarrow \underset{(D)}{2K_3[Co(CN)_6]} + 2KOH$$

(v) $$CoCl_2 + 6NaHCO_3 \longrightarrow Na_4[Co(CO_3)_3] + 2NaCl + 3CO_2 + 3H_2O$$

$$2Na_4[Co(CO_3)_3] + 2NaHCO_3 + O \longrightarrow \underset{(E)}{2Na_3[Co(CO_3)_3]} + 2Na_2CO_3 + H_2O$$

9 (i) Gas (*X*) is absorbed in acidified $K_2Cr_2O_7$ and the solution turns green, so (*X*) is CO_2.

(ii) Gas (*Y*) is absorbed in lime water turning it white, so *Y* is CO_2.

(iii) Gas (*Z*) is absorbed in pyrogallol, so (*Z*) is O_2.

Reactions

(i) $$\underset{(X)}{3SO_2} + K_2Cr_2O_7 + H_2SO_4 \longrightarrow H_2SO_4 + \underset{\text{Green}}{Cr_2(SO_4)_2} + H_2O$$

(ii) $$Ca(OH)_2 + \underset{(Y)}{CO_2} \longrightarrow \underset{\text{White}}{CaCO_3\downarrow} + H_2O$$

(iii) $$\underset{(Z)}{O_2} + \text{Pyrogallol} \longrightarrow \text{Absorbed.}$$

10 The metal *X* is copper. Copper on being heated in the presence of oxygen forms copper oxide which is basic in nature and black in colour. The chemical for the above reaction is as follows

$$2Cu + O_2 \xrightarrow{\text{Heat}} 2CuO$$

If hydrogen gas is passed over this black coating that is CuO, the black coating turns brown as the reverse reaction takes place and copper is obtained black.

The chemical equation for this reaction is as follows

$$CuO + H_2 \xrightarrow{\text{Heat}} Cu + H_2O$$

Thus, substance *X* is copper (Cu) and *Y* is copper oxide (CuO).

DAY THIRTY SEVEN

Unit Test 6
(Organic Chemistry-II)

1 Identify the correct statement regarding enzymes.
(a) Enzymes are specific biological catalysts that can normally function at very high temperature ($T \sim 1000$ K)
(b) Enzymes are normally heterogeneous catalysts that are very specific in their actions
(c) Enzymes are specific biological catalysts that cannot be poisoned
(d) Enzymes are specific biological catalysts that possess well defined active sites

2 Which of the following is not optically active?
(a) Glycine (b) Alanine
(c) Tyrosine (d) Lysine

3 Isoelectric point is a
(a) specific temperature
(b) suitable concentration of amino acid
(c) hydrogen ion concentration that does not allow migration of amino acid under electric field
(d) melting point of an amino acid under the influence of electric field

4 From the following statements, which one is incorrect?
(a) Albumin is a simple protein
(b) Amino acid lysine contains a basic side chain
(c) Insulin is a hormone
(d) Muscles contain the protein keratin

5 Which of the following statements is not true about glucose? → NCERT Exemplar
(a) It is an aldohexose
(b) On heating with HI, it forms *n*-hexane
(c) It is present in furanose form
(d) It does not give 2,4-DNP test

6 When glucose reacts with bromine water, the main product is
(a) acetic acid (b) saccharic acid
(c) glyceraldehyde (d) gluconic acid

7 Match the following and choose the correct option.

	Column I		Column II
A.	Analgesics	1.	Treatment of stress
B.	Antihistamines	2.	Pain-killing effect
C.	Tranquilizers	3.	Applied to inanimate objects
D.	Disinfectants	4.	Prevents the interaction of histamine with its receptor

Codes

	A	B	C	D		A	B	C	D
(a)	4	3	1	2	(b)	2	4	1	3
(c)	1	2	4	3	(d)	3	4	2	1

8 Amylopectin is a polymer of
(a) α-D-glucose (b) α-D-fructose
(c) lactose (d) amylose

Direction (Q. Nos. 9-10) *In the following questions, Assertion followed by Reason is given. Choose the correct answer out of the following choices.*
(a) Both Assertion and Reason are true and Reason is the correct explanation of Assertion
(b) Both Assertion and Reason are true but Reason is not the correct explanation of Assertion
(c) Assertion is true but Reason is false
(d) Both Assertion and Reason are false

9 **Assertion** (A) β-glycosidic linkage is present in maltose.

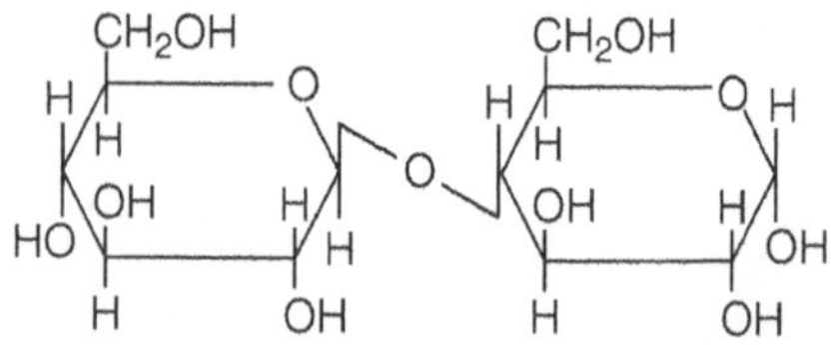

Reason (R) Maltose is composed of two glucose units in which C-1 of one glucose unit is linked to C-2 of another glucose unit.

10 Assertion (A) Polytetrafluoroethene is used in making non-stick cookwares.

Reason (R) Fluorine has highest electronegativity.

→ NCERT Exemplar

11 Food preservatives prevent spoilage of food due to microbial growth. The most commonly used preservatives are

(a) table salt, sugar
(b) vegetable oils and sodium benzoate.
(c) C_6H_5COONa
(d) All of the above

12 Which is not the correct matching of medicine with its disease/activity?

(a) Antihistamines — anti-allergic
(b) Barbiturates — hypnotic
(c) Chloramphenicol — typhoid
(d) 1% phenol — antiseptic

13 Dettol is a mixture of

(a) chloroxylenol and terpeneol
(b) phenol and chlorophenol
(c) phenol and chloroxylenol
(d) chlorophenol and chloroxylenol

14 Sodium benzoate is used as food preservative. It is

(a) metabolised by conversion to hippuric acid and is excreted in the urine
(b) metabolised to benzoic acid and deposited in the bond
(c) decomposed by gastric juice and escapes as CO_2
(d) decomposed by heat of the digestion process and escapes as CO_2

15 Aspartame, an artificial sweetener, is

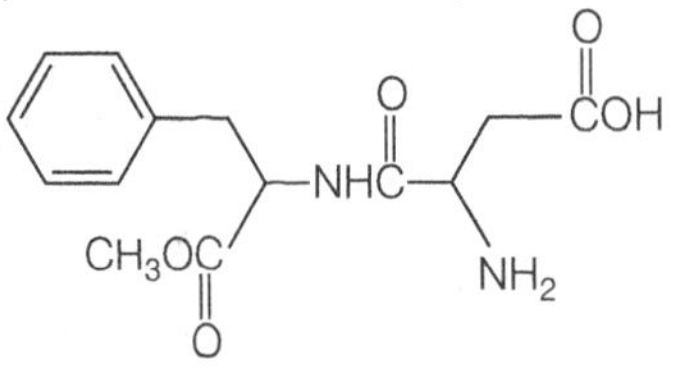

Functional groups, which are not present in aspartame, are

(a) ester, peptide, amino, carboxyl
(b) hydroxyl, keto, methoxy
(c) Both (a) and (b)
(d) None of the above

16 Glyptal polymer is obtained by the reaction of phthalic acid with

(a) glycerol (b) ethylene glycol
(c) acetic acid (d) malonic acid

17 Structurally, the cellulose is a linear polymer of

(a) sucrose molecules (b) β-D-glucose molecules
(c) α-D-glucose molecules (d) fructose molecules

18 The baby feeding bottles are made up of

(a) polyester (b) polyurethane
(c) polystyrene (d) polyamide

19 The monomer of the polymer

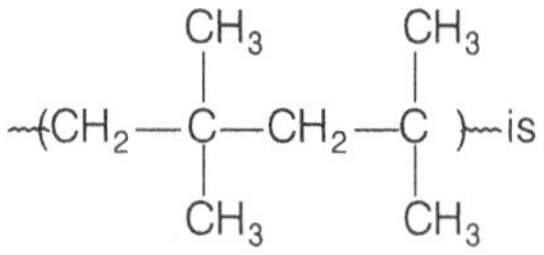

(a) $CH_2=C(CH_3)_2$ (b) $(CH_3)_2C=C(CH_3)_2$
(c) $CH_3CH=CHCH_3$ (d) $CH_3CH=CH_2$

20 The monomers used for the preparation of nylon-2-nylon-6 is/are → [NCERT Exemplar]

(a) caprolactum
(b) alanine and amino caproic acid
(c) glycine and amino caproic acid
(d) hexamethylenediamine and adipic acid

21 Which of the following is a monomer of teflon?

(a) Difluoro ethane (b) Trifluoro ethane
(c) Tetrafluoro ethene (d) None of these

22 A salt is heated first with dil. H_2SO_4 and then with conc. H_2SO_4, no reaction takes place. It may be

(a) nitrate (b) sulphide (c) oxalate (d) sulphate

23 When Cl_2 water is added to a salt solution containing chloroform, chloroform layer turns violet. The salt contains

(a) Cl^- (b) I^- (c) NO_3^- (d) S^{2-}

24 Which will give borax bead test with blue bead?

(a) Cr^{3+} (b) Co^{3+} (c) Ni^{2+} (d) Cd^{2+}

25 Which of the following leaves no residue on heating?

(a) $Pb(NO_3)_2$ (b) NH_4NO_3 (c) $Cu(NO_3)_2$ (d) $NaNO_3$

26 Mark the compound which turns black with NH_4OH.

(a) Lead chloride (b) Mercurous chloride
(c) Mercuric chloride (d) Silver chloride

27 There is foul smell in presence of moisture with

(a) $AlCl_3$ (b) $Al_2(SO_4)_3$ (c) FeS (d) $FeSO_4$

28 A colourless salt changes to yellow on heating. Salt is also soluble in NaOH as well as in dil. HCl. Salt can be

(a) FeO (b) PbO (c) ZnO (d) CdO

29 A glycoside is the carbohydrate form of a/an

(a) ether (b) acetal (c) glycone (d) alcohol

30 Reduction of hexose *A* (mol. formula, $C_6H_{12}O_6$) with $NaBH_4$ gives compounds *B* and *C*. Compound *B* is optically inactive and compound *C* is optically active. Which of the following is compound *A*?

(a) D-fructose (b) D-glucose (c) D-mannose (d) D-psicose

31 Which of the following is true about teflon?

(a) It is linear, unbranched polymer of tetrafluoroethylene
(b) It has very high thermal stability
(c) The polymer molecules are associated by strong dipole-dipole attraction
(d) All of the above

32 Cellulose has very high degree of hydrophilicity because of

(a) its amorphous nature
(b) crystalline nature

(c) presence of excessive voids in solid state
(d) presence of many hydroxyl groups on the polymer backbone

33 Which one among the following is not an analgesic?
(a) Ibuprofen (b) Naproxen
(c) Valium (d) Aspirin

34 Nucleoside involves the combination of
(a) sugar + base + H_3PO_4 (b) sugar + base
(c) sugar + acid (d) sugar + H_3PO_4

35 On hydrolysis of caprolactum, a compound (*B*) is obtained. On polymerisation of (*B*), product (*C*) is formed. (*C*) is
(a) saran (b) nylon-6 (c) terylene (d) bakelite

36 The artificial sweetener aspartame (*A*) is converted to .... on storage for extended period of time in aqueous solution.

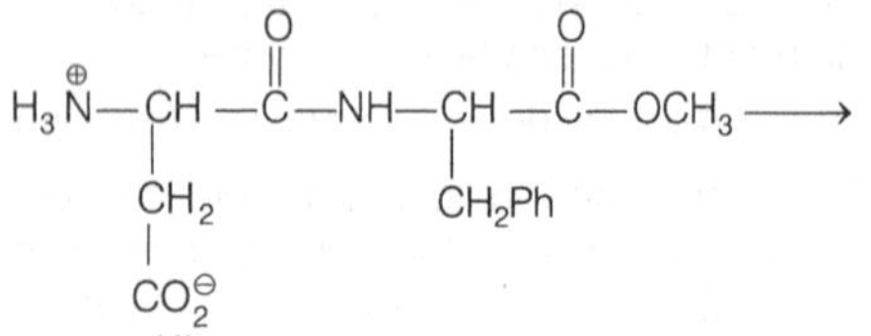

(*A*)

(a) 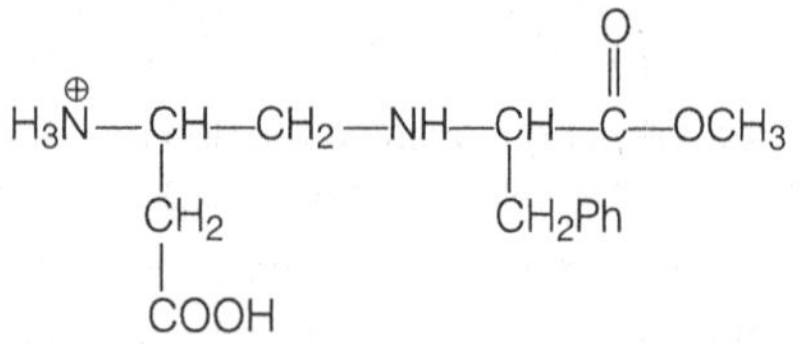

(b) 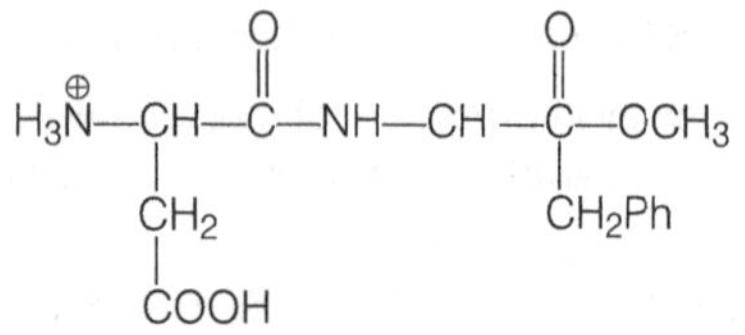

(c) 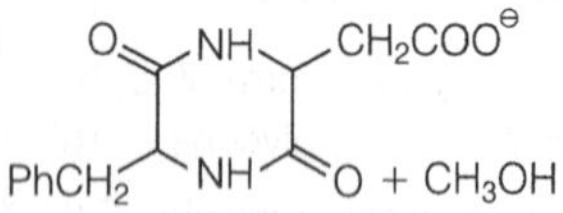

(d) No change, remains as *A*

37 A carbohydrate is treatd with α –naphthol and conc. H_2SO_4. What colour will be formed at the junction of two liquids?
(a) Blood red (b) Violet
(c) Brown (d) Orange

38 (*A*) is a ternary salt with divalent cation. (*A*) gives yellow ppt. with K_2CrO_4 as well as with $AgNO_3$. (*A*) is precipitated by H_2S neither in acidic nor ammoniacal medium, but addition of $(NH_4)_2CO_3$ in NH_4OH gave white ppt. (*B*). 0.297 g of (*A*) gave 0.197 g of (*B*). Identify (*A*).
(a) $BaBr_2$ (b) Cu_2S (c) $FeSO_4$ (d) $ZnCl_2$

39 A shallow eutrophic water lake located in a region where the bedrock and sediments contain limestone, has a pH 7.2 and equilibrium constant $K_a = 4.7 \times 10^{-11}$ for the reaction.

$$H_2O + HCO_3^- \rightleftharpoons H_3O^+ + CO_3^{2-}$$

Concentrations of HCO_3^- and H_3O^+ are respectively 1.06×10^{-3}mol L^{-1} and 6.3×10^{-8}mol L^{-1}. If the concentration of Ca^{2+} is 1.5×10^{-3}mol^{-1} and K_{sp} of $CaCO_3 = 5 \times 10^{-9}$. Would the precipitation of $CaCO_3$ take place?
(a) Precipitation will take place
(b) Precipitation will not take place
(c) It may or may not take place
(d) Cannot be predicted

40 To an aqueous solution containing anions from a few drops of acidified $KMnO_4$ are added. Which one of the following anions, if present will not decolourise the $KMnO_4$ solution?
(a) I^- (b) CO_3^{2-}
(c) S^{2-} (d) NO_2^-

41 In the brown ring test, the brown colour of the ring is due to
(a) ferrous nitrate (b) ferric nitrate
(c) a mixture of NO and NO_2 (d) ferrous nitrososulphate

42 What colour is imparted into the flame when lithium is burnt?
(a) Golden yellow (b) Brick red
(c) Crimson red (d) Grassy green

43 Give the pOH range for the isoelectric point of the amphoteric ion of an amino acid?
(a) 5.5 to 6.3 (b) 2.5 to 5.0
(c) 7.7 to 8.5 (d) 9.0 to 10.7

44 Which of the following describes the overall three dimensional folding of a polypeptide?
(a) Primary structure (b) Secondary structure
(c) Tertiary structure (d) Quaternary structure

Direction (Q. Nos. 45-46) *Each of these questions contains two statements : Statement I (Assertion) and Statement II (Reason). Each of these questions also has four alternative choices, only one of which is the correct answer. You have to select one of the codes (a), (b), (c) and (d) given below:*

(a) Statement I is true, Statement II is true; Statement II is the correct explanation for Statement I
(b) Statement I is true, Statement II is true; Statement II is not the correct explanation for Statement I
(c) Statement I is true; Statement II is false
(d) Statement I is false; Statement II is true

45. Statement I Fructose because of the presence of keto group does not reduce Tollen's reagent.
Statement II Fructose is also called fruit sugar.

46. Statement I MnS is pink in colour which on dissolution in dil. HCl and then heating with NaOH and Br_2 gives pink colour solution changing to brown on heating.
Statement II Mn^{2+} is oxidised to Mn^{4+}.

ANSWERS

1. (d)	**2.** (a)	**3.** (c)	**4.** (d)	**5.** (c)	**6.** (d)	**7.** (b)	**8.** (a)	**9.** (d)	**10.** (c)
11. (d)	**12.** (d)	**13.** (a)	**14.** (a)	**15.** (b)	**16.** (b)	**17.** (b)	**18.** (c)	**19.** (a)	**20.** (c)
21. (c)	**22.** (d)	**23.** (b)	**24.** (b)	**25.** (b)	**26.** (b)	**27.** (c)	**28.** (c)	**29.** (b)	**30.** (d)
31. (d)	**32.** (d)	**33.** (c)	**34.** (b)	**35.** (b)	**36.** (c)	**37.** (b)	**38.** (a)	**39.** (b)	**40.** (b)
41. (d)	**42.** (c)	**43.** (c)	**44.** (c)	**45.** (d)	**46.** (a)				

Hints and Explanations

1. Normal optimum temperature of enzymes is between 25°C to 40°C, hence (a) is false, enzymes have well defined active sites and hence, their actions are specific in nature.

2. Glycine (an amino acid) is optically inactive compound.

3. Isoelectric point is a pH at which *Zwitter* ion does not migrate towards any of the electrodes. Since, amino acids are also *Zwitter* ions, hence, they do not migrate under the influence of electric field at isoelectric point.

4. Muscles contain myosin protein. Keratin is present in hair, wool and silk.

5. Glucose is never present in furanose form.

6. When glucose reacts with Br_2 water, gluconic acid is obtained as main product.

$$\underset{\text{Glucose}}{\begin{matrix} CHO \\ | \\ (CHOH)_4 \\ | \\ CH_2OH \end{matrix}} + [O] \xrightarrow{Br_2/H_2O} \underset{\text{Gluconic acid}}{\begin{matrix} COOH \\ | \\ (CHOH)_4 \\ | \\ CH_2OH \end{matrix}}$$

7.

A.	Analgesics	Pain killing effect
B.	Antihistamines	Prevents the interaction of histamine with its receptor
C.	Tranquilizers	Treatment of stress
D.	Disinfectants	Applied to inanimate objects

8. Amylopectin is a highly branched polymer of α – D glucose units which are joined together through α-glycosidic linkages involving C_1 of one glucose unit with C_4 of the other.

9. Correct Assertion α-glycosidic linkage is present in maltose. Correct reason maltose is a composed of two α-D-glucose units in which C-1 of one glucose unit (I) is linked to C-4 of another glucose unit (II).

10. Correct Explanation Because of great chemical inertness and high thermal stability, teflon (polytetrafluoroethene) is used in making non-stick cookwares.

11. The chemicals which are used to protect food from microbes action are known as food preservatives. Table salt, vegetable oil, sugar, vinegar, sodium benzoate ($C_6H_5COO^-Na^+$) sodium metabisulphite ($Na_2S_2O_5$), vitamin E etc. are food preservatives.

12. 1% phenol is a disinfectant while
0.2% phenol is an antiseptic

13. Dettol is a mixture of chloroxylenol and terpeneol.

14. Sodium benzoate is used as food preservative. It is metabolised to hippuric acid and excreted in urine. It is used in soft drinks and acidic foods.

15. Hydroxyl, keto, methoxy groups are absent in aspartame while ester, peptide amino an carboxyl groups are present.

16. Glyptal is obtained by the reaction of phthalic acid with ethylene glycol.

Structure of glyptals :

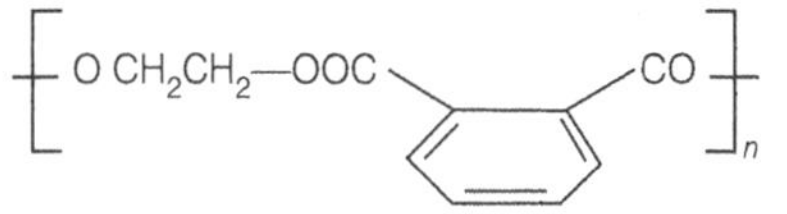

17. Cellulose is a linear condensation polymer of β-D- glucose in which C_1 of one glucose unit is connected to C_4 of the other through β-D-glycosidic linkage.

18. Polystyrene is used for the production of baby feeding bottles.

19. $CH_2{=}C(CH_3)_2$ is the repeating unit of given polymer.

20. $\underset{\text{Glycine}}{nH_2NCH_2COOH} + \underset{\text{Amino caproic acid}}{nH_2N(CH_2)_5COOH} \longrightarrow$

$$\underset{\text{Nylon-2-nylon-6}}{\left[NH - CH_2 - CONH - (CH_2)_5 - \overset{\overset{\displaystyle O}{\|}}{C} \right]}$$

21. Teflon is an addition polymer of tetrafluoroethylene.

$$\underset{\text{Tetrafluoroethylene}}{nF_2C = CF_2} \xrightarrow[\text{Pressure}]{\text{Heat}} \underset{\text{Teflon}}{[F_2C - CF_2]_n}$$

22. SO_4^{2-} does not react with either dilute or conc. H_2SO_4.

23. $2I^- + Cl_2 \longrightarrow I_2 + 2Cl^-$

I_2 gives violet colour to the chloroform layer.

24. Co^{3+} will give blue bead in borax bead test.

25. $NH_4NO_3 \xrightarrow{\Delta} N_2O\uparrow + 2H_2O\uparrow$

26. Mercurous chloride (Hg_2Cl_2) compound turns black with NH_4OH

$$Hg_2Cl_2 + 2NH_4OH \longrightarrow \underset{\text{Black}}{Hg\downarrow + Hg \cdot NH_2Cl} + 2H_2O + NH_4Cl$$

27. $FeS + 2H_2O \longrightarrow Fe(OH)_2 + \underset{\text{Foul smell}}{H_2S}$

28. ZnO is colourless and changes to yellow on heating.

$$ZnO + 2HCl \longrightarrow ZnCl_2 + H_2O$$

$$ZnO + 2NaOH \longrightarrow Na_2ZnO_2 + H_2O$$

29. Acetal form of carbohydrates are known as glycoside.

30.

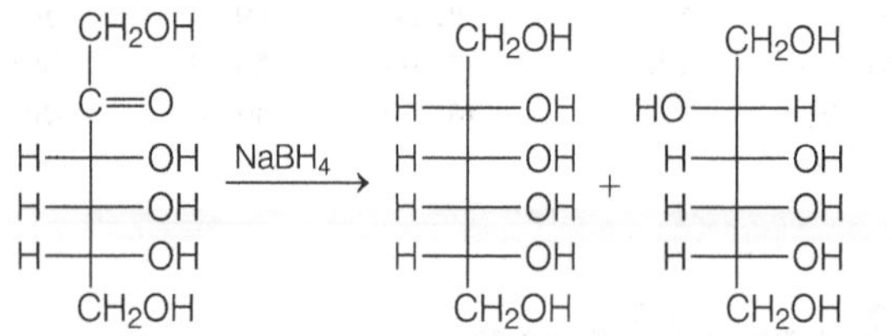

D-psicose (*A*) meso (*B*) Enantiomeric (*C*)

31. Due to symmetry of $F_2C{=}CF_2$ (teflon) and strength of C—F bond, branching doesn't takes place and a linear polymer is formed. High bond energy of C—F bond gives high thermal stability to polymer. Large electronegativity difference of C and F makes C—F bonds highly polar.

32. Cellulose fibres are hydrophilic due to the presence of OH groups at their surfaces.

33. Valium is white all other gives option are the examples of analgesics.

34. A combination of sugar and base is called nucleoside and a combination of sugar, base and phosphate is called nucleotide.

35.

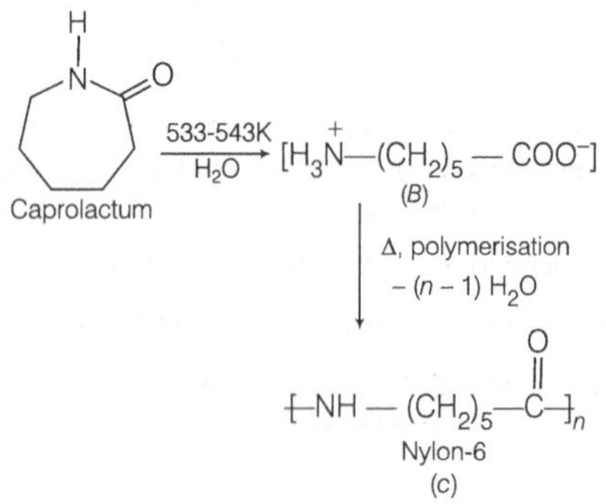

36.

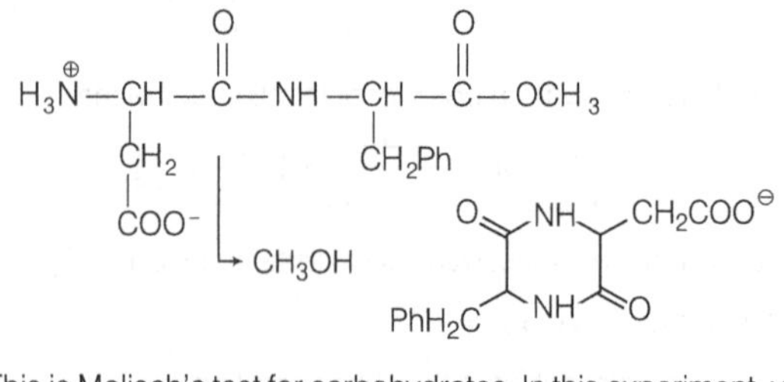

37. This is Molisch's test for carbohydrates. In this experiment, violet ring is formed at the junction of two liquids.

38. (*A*) gives yellow ppt. with K_2CrO_4 as well as with $AgNO_3$. Therefore, (*A*) can have Pb^{2+} or Ba^{2+} (both divalent cations) and Br^-. (*A*) is not precipitated by H_2S in acidic and ammoniacal medium. Hence, (*A*) does not have Pb^{2+} (precipitated as PbS in acidic medium by H_2S gas).

(*A*) contains Ba^{2+} (apple green flame).

$\therefore$ (*A*) is $BaBr_2$.

$$BaBr_2 + 2AgNO_3 \longrightarrow \underset{\text{Yellow}}{2AgBr\downarrow} + Ba(NO_3)_2$$

$$BaBr_2 + K_2CrO_4 \longrightarrow \underset{\text{Yellow}}{BaCrO_4\downarrow} + 2KBr$$

$$\underset{A}{BaBr_2} + (NH_4)_2CO_3 \longrightarrow BaCO_3 + \underset{B}{2NH_4Br}$$

39. Equilibrium constant for reaction:

$$K_a = \frac{[H_3O^+][CO_3^{2-}]}{[HCO_3^-][H_2O]}$$

$$[CO_3^{2-}] = \frac{4.7\times10^{-11}\times1.06\times10^{-3}}{6.3\times10^{-8}}$$

$$= 7.9\times10^{-7}\text{mol L}^{-1}$$

The reaction quotient for the reaction

$Ca^{2+} + CO_3^{2-} \rightleftharpoons CaCO_3$

$$Q = [Ca^{2+}][CO_3^{2-}]$$

$$= 1.5\times10^{-3}\times7.9\times10^{-7}$$

$$= 1.2\times10^{-9} < 5\times10^{-9} = K_{sp}$$

Hence, no precipitation will take place.

40. In CO_3^{2-}, C is present in its highest oxidation state, i.e +4 state, so its further oxidation is not possible, it only undergoes reduction. Acidified $KMnO_4$ is a strong oxidising agent but it cannot oxidise CO_3^{2-}.

41. On adding freshly prepared $FeSO_4$ solution and then conc. H_2SO_4 to water extract of salt mixture carefully by the sides of test-tube, a dark brown ring of ferrous nitrososulphate, $FeSO_4NO$ is formed.

42.

Colour	Cation
Golden yellow	Na^+
Brick red	Ca^{2+}
Crimson red	Li^+
Grassy green	$Cu^{2+}, BO_3^{3-}, Tl^{3+}$

43. The pH at which a particular amino acid does not migrate under the influence of an electric field is called isoelectric point of that amino acid.

The pH range for the isoelectric point is from 5.5 to 6.3 on the pOH range for the isoelectric point is from 7.7 to 8.5.

44. The tertiary structure of proteins refer to three dimensional folding of polymer chain.

45. Fructose although have keto group but it reduces Tollen's reagent to give silver mirror. It is also called fruit sugar.

46. $MnS + \text{dil.}HCl \longrightarrow MnCl_2$

$$MnCl_2 + 2NaOH \longrightarrow Mn(OH)_2 + 2NaCl$$

$$Mn(OH)_2 + [O] \xrightarrow{NaOH + Br_2} \underset{\text{(Brown)}}{MnO_2} + H_2O$$

DAY THIRTY EIGHT

Mock Test 1

(Based on Complete Syllabus)

Instructions

1. This question paper contains of 30 Questions of Chemistry, divided into two Sections :
   Section A Objective Type Questions and **Section B** Numerical Type Questions.
2. Section A contains 20 Objective questions and all Questions are compulsory (**Marking Scheme** : Correct +4, Incorrect −1).
3. Section B contains 10 Numerical value questions out of which only 5 questions are to be attempted (**Marking Scheme** : Correct +4, Incorrect 0).

Section A : Objective Type Questions

1 Electrode potential for Mg electrode varies according to the equation

$$E_{Mg^{2+}/Mg} = E^{\ominus}_{Mg^{2+}/Mg} - \frac{0.059}{2}\log\frac{1}{[Mg^{2+}]}$$

The correct graphical representation is

(a) 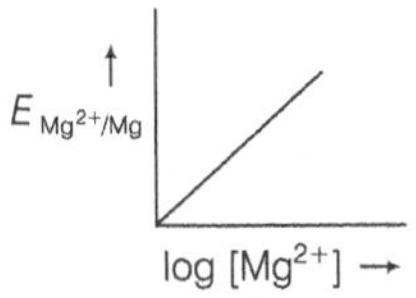

(b) 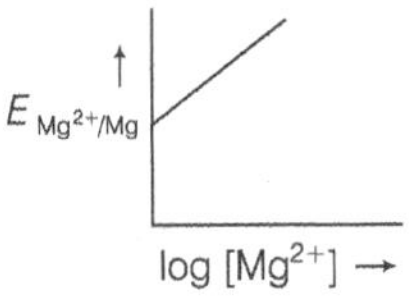

(c) 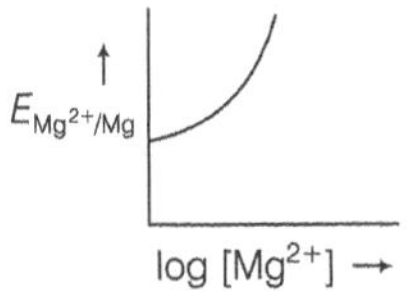

(d) 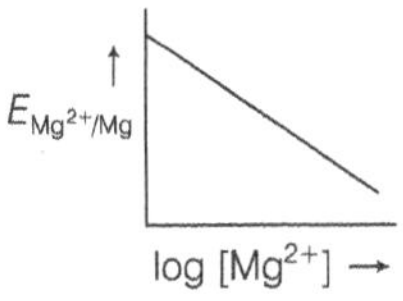

2 Consider the reactions

(i) $CO(g) + H_2O(g) \underset{}{\overset{K_1}{\rightleftharpoons}} CO_2(g) + H_2(g)$

(ii) $CH_4(g) + H_2O(g) \overset{K_2}{\rightleftharpoons} CO(g) + 3H_2(g)$

(iii) $CH_4(g) + 2H_2O(g) \overset{K_3}{\rightleftharpoons} CO_2(g) + 4H_2(g)$

Which of the following is correct relation?

(a) $K_3 = K_1/K_2$ (b) $K_3 = K_1^2/K_2^3$

(c) $K_3 = K_1 \cdot K_2$ (d) $K_3 = K_1 \cdot \sqrt{K_2}$

3 Which of the following will undergo dehydration fast?

(a) 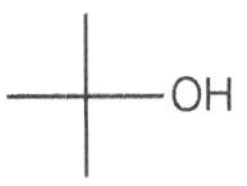

(b) 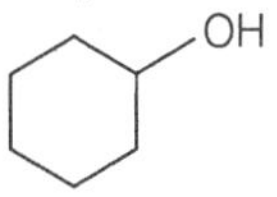

(c) 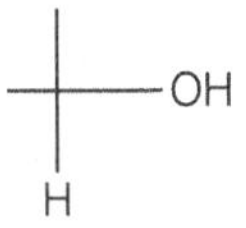

(d) 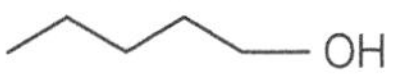

4 The alcohol, having molecular formula C_4H_9OH, when shaken with a mixture of anhydrous and conc. HCl gives an oily layer product after five minutes. The alcohol is

(a) $H_3C—(CH_2)_3—OH$

(b) $(CH_3)_2CHCH_2—OH$

(c) $(CH_3)_3C—OH$

(d) $H_3C—CH(OH)CH_2CH_3$

5 How many moles of $MgIn_2S_4$ can be produced when 1.00 g of magnesium (atomic mass = 24), 1.00 g of indium (atomic mass = 114.8) and 1.00 g of sulphur (atomic mass = 32) react?

(a) 6.74×10^{-4}s (b) 3.1×10^{-2}
(c) 4.17×10^{-2} (d) 8.7×10^{-3}

6 Assertion (A) Deoxyribose, $C_5H_{10}O_4$ is a carbohydrate.
Reason (R) Carbohydrates are hydrates of carbon so compounds which follow $C_x(H_2O)_y$ formula are carbohydrates.

(a) (A) is false but (R) is true.
(b) Both (A) and (R) are true but (R) is not the correct explanation of (A).
(c) (A) is true but (R) is false.
(d) Both (A) and (R) are true and (R) is the correct explanation of (A).

7 Which of the following complexes formed by Cu^{2+} ions is most stable?

(a) $Cu^{2+} + 4NH_3 \rightleftharpoons [Cu(NH_3)_4]^{2+}; \log k = 11.6$
(b) $Cu^{2+} + 4CN^- \rightleftharpoons [Cu(CN)_4]^{2-}; \log k = 27.3$
(c) $Cu^{2+} + 2en \rightleftharpoons [Cu(en)_2]^{2+}; \log k = 15.4$
(d) $Cu^{2+} + 4H_2O \rightleftharpoons [Cu(H_2O)_4]^{2+}; \log k = 8.9$

8 Identify *C* in the following sequence of reactions,

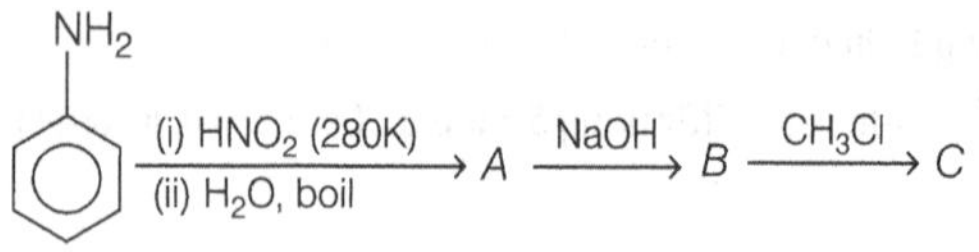

(a) 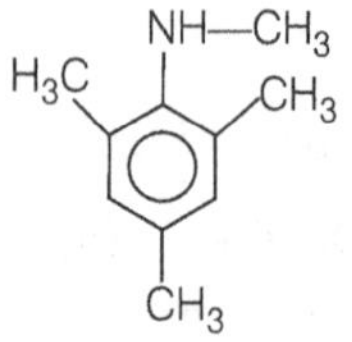 (b) 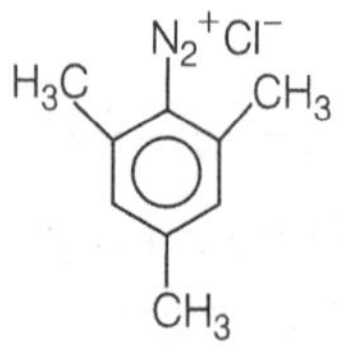

(c) 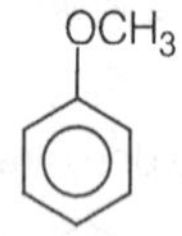 (d) 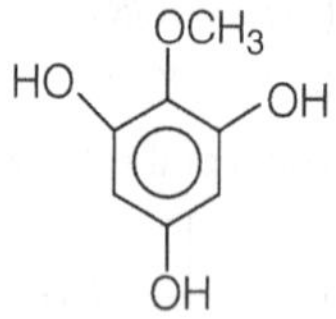

9 Bromine is added to cold dilute aqueous solution of sodium hydroxide. The mixture is boiled. Which of the following statements is not true ?

(a) During the reaction bromine is present in four different oxidation states
(b) The greatest difference between the various oxidation states of bromine is 5
(c) On acidification of the final mixture, bromine is formed
(d) Disproportionation of bromine occurs during the reaction

10 The correct order of increasing N—N bond stability of N_2^{2-}, N_2, $N_2^{\oplus}$, N_2^- is

(a) $N_2^{2-} > N_2 > N_2^{\ominus} > N_2^{\oplus}$ (b) $N_2 > N_2^{\oplus} = N_2^{\ominus} > N_2^{2-}$
(c) $N_2^{2-} > N_2^{\ominus} = N_2^{\oplus} > N_2$ (d) $N_2^{2-} > N_2 > N_2^{\ominus} = N_2^{\oplus}$

11 Given below are two statements.
Statement I Fe^{2+} givens brown colour with ammonium thiocyanate.
Statement II Fe^{3+} gives red colour with potassium ferrocyanide.
In the light of the above statements, choose the most appropriate answer from the options given below.

(a) Statement I is false but Statement II is true.
(b) Statement I is true but Statement II is false.
(c) Both Statement I and Statement II are false.
(d) Both Statement I and Statement II are true.

12 Sulphur is converted into Na_2S in Lassaigne's fusion test. Na_2S can be detected by

(I) $(CH_3COO)_2Pb$ (II) CH_3COOH (III) $Na_2[Fe(CN)_5NO]$

Correct codes are

(a) I and II (b) I and III (c) Only III (d) All of these

13 Which of the following species can act as the strongest base?

(a) ^-OH (b) ^-OR
(c) $^-OC_6H_6$ (d) 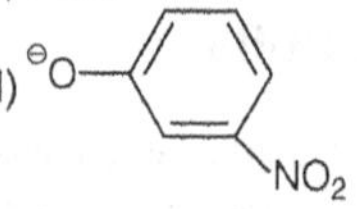

14 An aqueous solution of colourless metal sulphate *M* gives a white precipitate with NH_4OH. This was soluble in excess of NH_4OH. On passing H_2S through this solution a white ppt is formed. The metal *M* in the salt is

(a) Ca (b) Ba (c) Al (d) Zn

15 Match the vitamins given in Column I with the deficiency disease they cause given in Column II.

Column I (Vitamins)	**Column II** (Diseases)
A. Vitamin B_1	1. Muscular weakness
B. Vitamin C	2. Beri-Beri
C. Vitamin E	3. Increased blood clotting time
D. Vitamin K	4. Bleeding gums

Codes

	A	B	C	D		A	B	C	D
(a)	2	1	4	3	(b)	2	4	1	3
(c)	4	1	3	2	(d)	2	3	4	1

16 The emf of the following cell

$Zn | Zn^{2+}(0.004) || Cd^{2+}(0.2) | Cd$ is given by

$[E^\circ_{(Zn^{2+}/Zn)} = -0.763$ V and

$E^\circ_{(Cd^{2+}/Cd)} = -0.403$ V$]$

(a) $E = -0.36 + \frac{0.059}{2} \log 2 \times 10^{-2}$

(b) $E = -0.36 - \frac{0.059}{2} \log 50$

(c) $E = +0.36 - \frac{0.059}{2} \log 2 \times 10^{-2}$

(d) $E = -0.36 + \frac{0.059}{2} \log 50$

17 Which electrolyte is most effective in causing coagulation of ferric hydroxide sol?

(a) KBr (b) K_2SO_4 (c) K_2CrO_4 (d) $K_3[Fe(CN)_6]$

18 Potassium phthalimide on reaction with compound *A* followed by hydrolysis forms *neo*-pentyl amine. Compound *A* is

(a) 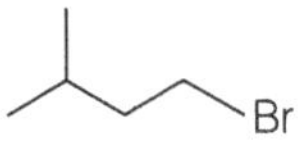 (b) 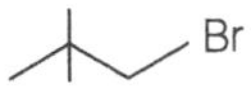

(c) 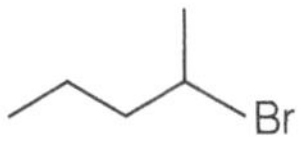 (d) 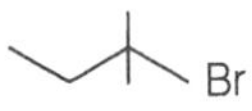

19 Calculate the pressure of O_2 (in atm) over a sample of NiO at 25°C if $\Delta G° = 212$ kJ for the reaction.

$$NiO(s) \rightleftharpoons Ni(s) + \frac{1}{2}O_2(g)$$

(a) 4.9×10^{-75} atm (b) 7×10^{-38} atm
(c) 4.9×10^{-38} atm (d) 7×10^{-75} atm

20 Identify the intermediate for which the rearrangement is not possible.

(a) 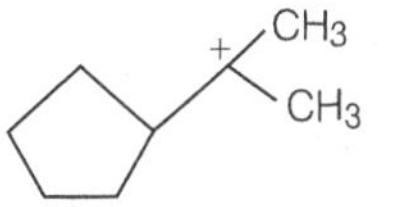

(b) 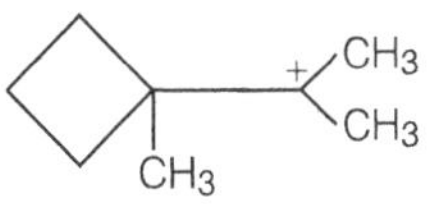

(c) $(CH_3)_3C-\overset{+}{C}H-C(CH_3)_3$

(d) 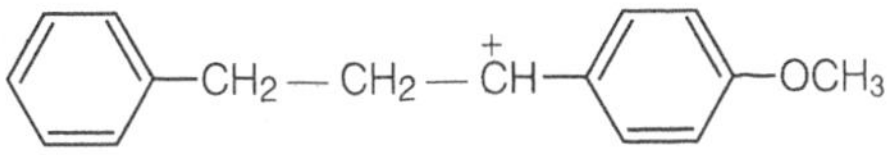

Section B : Numerical Type Questions

21 An ideal solution of benzene and toluene boils at 1 atmospheric pressure at 90°C. At 90°C, benzene has a vapour pressure of 1022 mm and toluene has a vapour pressure of 406 mm. The mole fraction of benzene in the solution is ....

22 If the cell-edge length for CsCl is 0.4123 nm and the ionic radius of a Cl^- ion is 0.181 nm, the ionic radius of a Cs^+ ion is ...... nm.

23 0.0005 mole of strong electrolyte, $Ca(OH)_2$ is dissolved to form 100 mL of a saturated aqueous solution. The pH of this solution is

24 An electron in H-atom in its ground state absorbs 1.50 times as much as energy as the minimum energy required for its escape (13.6 eV) from the atom. Thus, kinetic energy given to emitted electron is ...... eV

25 For a cell reaction, $Zn + Cu^{2+} \rightleftharpoons Cu + Zn^{2+}$ entropy changes $\Delta S°$ is 96.5 $J\,mol^{-1}K^{-1}$ then temperature coefficient of the emf of a cell is ...... $\times 10^{-4}$ VK^{-1}

26 The degree of hydrolysis in hydraulic equilibrium

$$A^-(aq) + H_2O(l) \rightleftharpoons HA(aq) + OH^-(aq)$$

at salt concentration 0.0001 M is ...... $\times 10^{-2}$
$(K_a = 1.0 \times 10^{-6})$

27 The pH of 0.5 L of 1.0 M NaCl after the electrolysis for 965 s using 5.0A current (100 % efficiency) is

28 An organic liquid, *A* is immiscible with water. When boiled together with water, the boiling point is 90°C at which the partial pressure of water is 526 mm Hg. The atmospheric pressure is 736 mm Hg. The weight ratio of the liquid and water collected is 2.5 : 1. The molecular weight of the liquid is ...

29 0.15 mole of CO taken in a 2.5 L flask is maintained at 750 K along with a catalyst so that the following reaction can take place.

$$CO(g) + 2H_2(g) \rightleftharpoons CH_3OH(g)$$

Hydrogen is introduced until the total pressure of the system is 8.5 atm at equilibrium and 0.08 mole of CH_3OH are formed. The value of K_c is ...... $mol^{-2}L^{-2}$ (Assigning ideal behaviour).

30 A reaction proceeds 5 times more at 60°C as it does at 30°C. Its energy of activation is .... Kcal mol^{-1}.

ANSWERS

1 (b)	**2** (c)	**3** (a)	**4** (d)	**5** (d)	**6** (c)	**7** (b)	**8** (c)	**9** (c)	**10** (b)
11 (c)	**12** (b)	**13** (b)	**14** (d)	**15** (b)	**16** (c)	**17** (d)	**18** (b)	**19** (a)	**20** (d)
21 (0.574)	**22** (0.176)	**23** (12)	**24** (20.4)	**25** (5)	**26** (1)	**27** (13)	**28** (112.7)	**29** (187.85)	**30** (10.75)

Hints and Explanations

1. It is known that, $E = E° + \frac{0.059}{2}\log[Mg^{2+}]$. This suggests that the plot of E vs $\log[Mg^{2+}]$ would be linear with positive slope having an intercept of $E°$.

2. For the given reaction,

$$K_1 = \frac{[CO_2]\,[H_2]}{[CO]\,[H_2O]},$$

$$K_2 = \frac{[CO]\,[H_2]^3}{[CH_4]\,[H_2O]}$$

$$K_3 = \frac{[CO_2]\,[H_2]^4}{[CH_4]\,[H_2O]^2}$$

$$\frac{K_1 \times K_2}{K_3} = 1$$

$$K_1 \times K_2 = K_3$$

3. In the dehydration of these compounds, intermediate is a carbocation, thus more stable carbocation (3° carbocation) giving compound undergo dehydration fast, i.e. compound.

4. Secondary alcohol, on reaction with anhydrous $ZnCl_2$ and conc. HCl (Lucas reagent) gives an oil layer product after five minutes.

$$\underset{Sec-\text{ alcohol}}{H_3C{-}CH(OH){-}CH_2CH_3} \xrightarrow{\text{Lucas reagent}} \text{oil layer}$$

product (turbidity) after 5 min

5.

	Mg	+ 2In	+ 4S ⟶ $MgIn_2S_4$
Moles	$\frac{1}{24}$	$\frac{1}{114.8}$	$\frac{1}{32}$
	= 0.0417	= 0.0087	= 0.031

The number of moles of limiting reagent (In) is 0.0087 or 8.7×10^{-3}, hence 8.7×10^{-3} moles of $MgIn_2S_4$ are produced.

6. Correct Assertion Deoxyribose is a carbohydrate.

Correct Reason Carbohydrates are optically active, polyhydroxy aldehydes or polyhydroxy ketones or substances which give these on hydrolysis.

7. Higher the value of log k, larger is the stability of the complex. So, $[Cu(CN)_4]^{2-}$ is the most stable complex.

8.

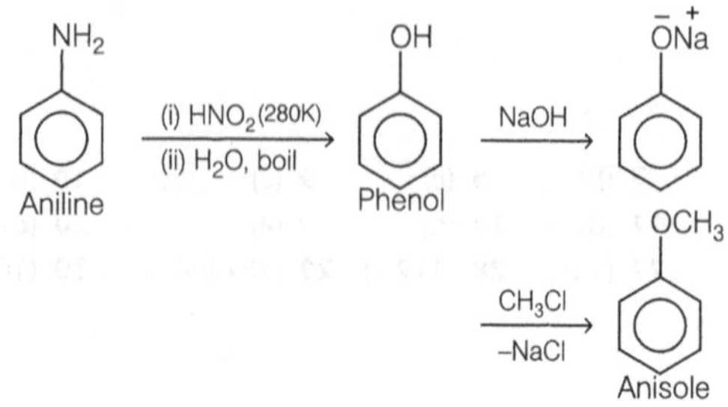

9. Bromine on reaction with cold dilute aq. solution of sodium hydroxide undergoes disproportionation. In this reaction, bromine is present in four oxidation state

$$\overset{0}{Br_2} + NaOH \xrightarrow{\Delta} \overset{-1}{NaBr} + \overset{+1}{NaBrO} + \overset{+5}{NaBrO_3}$$

Reduction (Br_2 → NaBr)

Oxidation (Br_2 → NaBrO)

Oxidation, difference of 5 (greatest)

10. $N_2^{2-}(16): \sigma 1s^2, \overset{*}{\sigma} 1s^2, \sigma 2s^2, \overset{*}{\sigma} 2s^2,$

$\sigma 2p_z^2, \pi 2p_x^2 \approx \pi 2p_y^2, \overset{*}{\pi} 2p_x^1 \approx \overset{*}{\pi} 2p_y^1$

$$\text{Bond order} = \frac{N_b - N_a}{2} = \frac{10-6}{2} = 2$$

In N_2 (14), bond order = 3

In N_2^- (15), bond order = 2.5

In N_2^+ (13), bond order = 2.5

Bond order ∝ stability

Thus, $N_2 > N_2^{\oplus} = N_2^{\ominus} > N_2^{2-}$

11. The blue precipitate of Fe^{2+} ions with potassium ferricyanide is due to the formation of Turnbull's blue $K\overset{II}{Fe}[\overset{III}{Fe}(CN)_6]$.

$$Fe^{2+} + \underset{\text{Potassium ferricyanide}}{K_3[Fe(CN)_6]} \longrightarrow \underset{\text{Turnbull's blue}}{K\overset{II}{Fe}[\overset{III}{Fe}(CN)_6]} + 2K^+$$

Fe^{3+} ion gives white precipitate that will become blue.

$$FeCl_3 + K_4[Fe(CN)_6] \rightarrow KFe[Fe(CN)_6] + 3KCl$$

12. $Na_2S + Na_2[Fe(CN)_5NO] \longrightarrow \underset{\text{Deep violet}}{Na_4[Fe(CN)_5NOS]}$

$$Na_2S + (CH_3COO)_2Pb \xrightarrow{CH_3COOH} \underset{\substack{\text{Black}\\ \text{ppt.}}}{PbS\downarrow} + 2CH_3CO\overset{-}{O}\overset{+}{Na}$$

13. Weakest acid has the strongest conjugate base. Since, ROH is the weakest acid, therefore, RO^- is the strongest base.

14. All the given metals form white ppt with NH_4OH but only ppt of Zn is soluble in excess of NH_4OH and on passing H_2S it gives white ppt of ZnS, so the metal is Zn and reactions takes place are as follows

$$Zn^{2+} + 2NH_4OH \longrightarrow \underset{\text{White ppt.}}{Zn(OH)_2\downarrow} + 2NH_4^+$$

$$Zn(OH)_2 + 2NH_4OH \longrightarrow \underset{\text{Soluble}}{(NH_4)_2ZnO_2} + 2H_2O$$

$$(NH_4)_2ZnO_2 + H_2S \longrightarrow \underset{\text{White ppt.}}{ZnS\downarrow} + 2NH_4OH$$

15. A → 2, B → 4, C → 1, D → 3.

Vitamin-B_1 causes Beri-Beri

Vitamin-C causes bleeding gums

Vitamin-E causes muscular weakness

Vitamin-K causees increased blood clotting time.

16. $E = E°_{cell} - \frac{0.059}{n}\log\frac{[\text{oxidised state}]}{[\text{reduced state}]}$

n = number of electrons taken part in the reaction = 2

$$E = [E^\circ_{Zn/Zn^{2+}} - E^\circ_{Cd^{2+}/Cd}] - \frac{0.059}{2}\log\frac{0.004}{0.2}$$
$$= [0.763 - 0.403] - \frac{0.059}{2}\log\frac{1}{50}$$
$$= 0.36 - \frac{0.059}{2}\log 2\times 10^{-2}$$

17. Ferric hydroxide is a positive sol, thus coagulated by negative ions (like Br^-, SO_4^{2-}, CrO_4^{2-} and $[Fe(CN)_6]^{3-}$). More is the valency of negative ion, more is the coagulating power.

18.

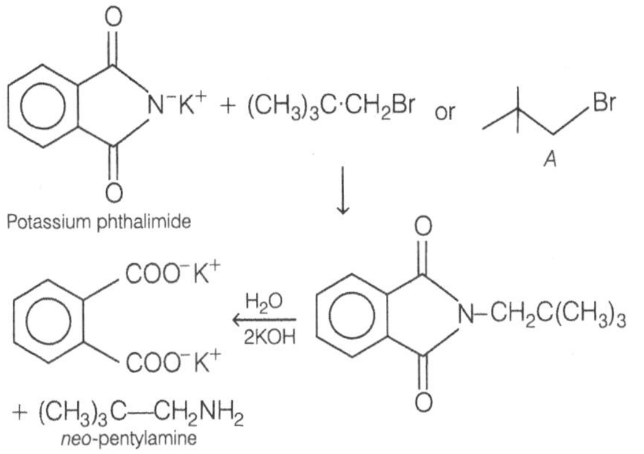

19. $NiO(s) \rightleftharpoons Ni(s) + \frac{1}{2}O_2(g)$

$$k_p = p_{O_2}^{1/2}$$
$$\Delta G^\circ = 212\,kJ = 212000\,J$$
$$\Delta G = -2.303RT\log k_p$$
$$\therefore \quad 212000 = -2.303\times 8.314\times 298\log K_p$$
$$\log k_p = -37.155$$
$$k_p = 7\times 10^{-38}$$
$$\sqrt{p_{O_2}} = 7\times 10^{-38}$$
$$p_{O_2} = 4.9\times 10^{-75}\text{ atm}$$

20. (a)

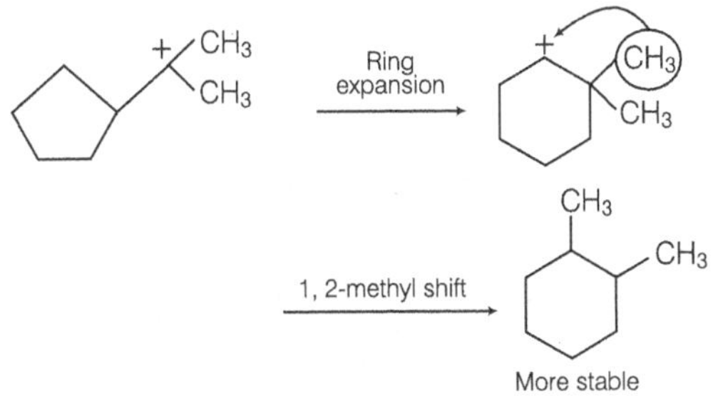

(b)

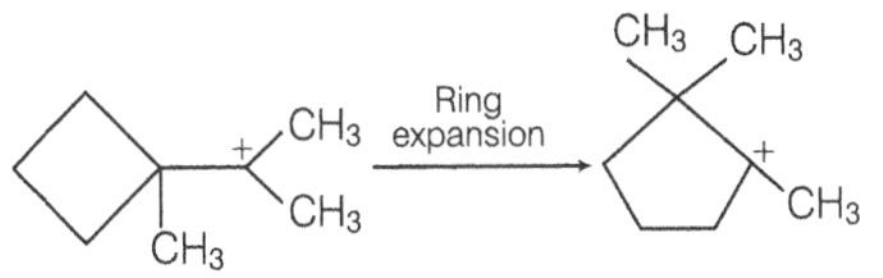

(c)

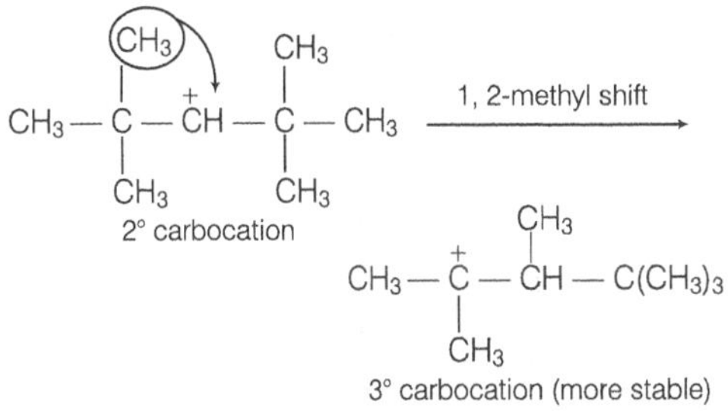

(d)

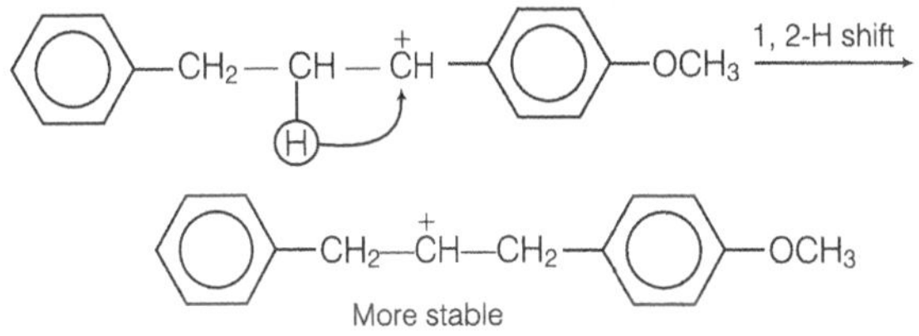

21. $p^\circ_B\chi_B + p^\circ_T\chi_T = p_{total}$

$1022\chi_B + 406(1-\chi_B) = 1$ atm

$= 760$ mm Hg $\qquad [\chi_B + \chi_T = 1]$

$\chi_B = 0.574$

22. CsCl has a bcc lattice, so

$$d = a\sqrt{3}$$
$$= 0.4123\times\sqrt{3}$$
$$= 0.7141\text{ nm}$$
$$r_{Cs^+} + r_{Cl^-} = \frac{d}{2} = \frac{0.7141}{2}$$
$$= 0.3571\text{ nm}$$
$$r_{Cs^+} + 0.181 = 0.3571\text{ nm}$$
$$r_{Cs^+} = 0.176\text{ nm}$$

23. Concentration of $Ca(OH)_2 = \frac{0.0005\times 1000}{100}$

$= 0.005$ mol L^{-1}

	$Ca(OH)_2 \rightleftharpoons$	Ca^{2+} + $2OH^-$
Initially	1	2
	0.005	$0.005\times 2 = 0.01$

$pOH = -\log[OH^-]$

$= -\log[0.01] = 2$

$pH = 14 - 2 = 12$

24. Energy in ground state = 13.6 eV

Energy absorbed = 1.5 × 13.6 = 20.4 eV

Energy in higher level = 34.0 eV

Energy emitted = 34.0 eV − 13.6 eV = 20.4 eV

25. $\Delta G = \Delta H - nFT\left(\frac{dE}{dT}\right)_p \qquad \text{...(i)}$

$\Delta G = \Delta H - T\Delta S \qquad \text{...(ii)}$

On comparing Eq. (i) and (ii)

$$\frac{96.5}{2 \times 96500} = \left(\frac{dE}{dT}\right)_p$$

$$\therefore \quad \left(\frac{dE_{cell}}{dT}\right)_p = \frac{1 \times 10^{-3}}{2} \ VK^{-1} = 5 \times 10^{-4} \ VK^{-1}$$

26. $$K_h = \frac{K_w}{K_a} = \frac{10^{-14}}{10^{-6}} = 10^{-8}$$

$$A^- + H_2O \rightleftharpoons HA + OH^-$$

$$K_h = \frac{[HA][OH^-]}{[A^-]}$$

$$K_h = \frac{(0.0001 \times h)(0.0001 \times h)}{0.0001(1-h)}$$

$$10^{-8} = 0.0001 h^2 \qquad [\because (1-h) \approx 1]$$

$$10^{-4} = h^2$$

$$\therefore \quad h = 10^{-2}$$

27. **At cathode** $2H_2O + 2e^- \longrightarrow H_2 + 2OH^-$

At anode $2Cl^- \longrightarrow Cl_2 + 2e^-$

Moles of OH^- formed $= zit = \frac{1 \times 5 \times 965}{96500}$

$= 0.05$ mol

$$[OH^-] = \frac{0.05}{0.5} = 1 \times 10^{-1}$$

$[H^+] = 1.0 \times 10^{-13}$ and pH = 13

28. Initial pressure of mixture,

$p_{mixture} = 736$mm Hg and at 90°C (boiling point),

$p'_{H_2O} = 526$mm Hg

$p'_{liquid} = 736 - 526$

$= 210$mm Hg

Also, $p' = p_{mixture} \times$ mole fraction in vapour phase .... (i)

Let a g of liquid and water is collected or this is the amount of vapours at equilibrium, thus weight of liquid vapours $= \frac{2.5 \times a}{3.5}$

and weight of water vapours $= \frac{1 \times a}{3.5}$

Now for liquid from Eq. (i)

$$210 = 736 \times \frac{\frac{2.5a}{3.5 \times m}}{\frac{a}{3.5 \times 18} + \frac{2.5a}{3.5 \times m}} \qquad \text{.... (ii)}$$

[here m = molar mass of liquid]

For water, from Eq. (i)

$$526 = 736 \times \frac{\frac{a}{3.5 \times 18}}{\frac{a}{3.5 \times 18} + \frac{2.5a}{3.5 \times m}} \qquad \text{.... (iii)}$$

Thus, from Eqs. (ii) and (iii), we get

$$\frac{210}{526} = \frac{18 \times 2.5}{m}$$

$$\therefore \quad m = 112.7$$

29. $$CO(g) + 2H_2(g) \rightleftharpoons CH_3OH(g)$$

	$CO(g)$	$2H_2(g)$	$CH_3OH(g)$
Initial mol	0.15	a	0
Moles at equilibrium	$(0.15 - x)$	$(a - 2x)$	(x)

Given, $x = 0.08$ mol

$\therefore$ Total moles at equilibrium

$= 0.15 - x + a - 2x + x$

$= a - 0.01$ mol

Total moles at equilibrium from ideal gas equation

$$n = \frac{pV}{RT} = \frac{8.5 \times 2.5}{0.0821 \times 750}$$

$= 0.345$

Hence, $a - 0.01 = 0.345$

or $a = 0.355$

At equilibrium,

moles of CO $= 0.15 - 0.08 = 0.07$

moles of $H_2 = 0.355 - 0.16 = 0.195$

and moles of $CH_3OH = 0.08$

$$\therefore K_C = \frac{[CH_3OH]}{[H_2]^2[CO]} = \frac{0.08/2.5}{[0.195/2.5]^2 \times [0.07/2.5]}$$

$= 187.85 \ mol^{-2} \ L^{-2}$

30. Given, $T_1 = 303K, T_2 = 33K$ and $R = 1.987 \times 10^{-3}$ kcal

$$\because \quad \text{Rate} = k\left[\frac{\text{products}}{\text{reactants}}\right]^n$$

$\therefore \frac{r_2}{r_1} = \frac{k_2}{k_1}$ for a given reaction at different temperature.

$$\because \quad \frac{r_2}{r_1} = 5$$

$$\because \quad \frac{k_2}{k_1} = 5$$

$$\because \quad 2.303 \log \frac{k_2}{k_1} = \frac{E_a}{R}\left[\frac{T_2 - T_1}{T_1 T_2}\right]$$

$$\because \quad 2.303 \log 5 = \frac{E_a}{1.987 \times 10^{-3}}\left[\frac{333 - 303}{333 \times 303}\right]$$

$$\therefore \quad E_a = 10.75 \ kcal \ mol^{-1}$$

DAY THIRTY NINE

Mock Test 2

(Based on Complete Syllabus)

Instructions

1. This question paper contains of 30 Questions of Chemistry, divided into two Sections :
   Section A Objective Type Questions and **Section B** Numerical Type Questions.
2. Section A contains 20 Objective questions and all Questions are compulsory (**Marking Scheme** : Correct +4, Incorrect −1).
3. Section B contains 10 Numerical value questions out of which only 5 questions are to be attempted (**Marking Scheme** : Correct +4, Incorrect 0).

Section A : Objective Type Questions

1 Of the following acids:

I. Hypophosphorus acid

II. Orthophosphorus acid

III. Caro's acid

IV. Glycine

(a) I, II monobasic ; III dibasic acid and IV amphoteric
(b) II monobasic ; I, III dibasic acid and IV amphoteric
(c) I monobasic ; II, III dibasic acid and IV amphoteric
(d) I, II, III dibasic acids and IV amphoteric

2 Magnetic moments of Cr ($Z = 24$), Mn ($Z = 25$) and Fe ($Z = 26$) are x, y and z respectively. Hence,

(a) $x = y = z$ (b) $x = z < y$
(c) $x < y < z$ (d) $x > y > z$

3 **Assertion** (A) Aldehydes and ketones, both react with Tollens' reagent to form silver mirror.

Reason (R) Both, aldehydes and ketones contain a hydrogen attached to a carbonyl group.

(a) Both A and R are true and R is the correct explanation of A
(b) Both A and R are true but R is not the correct explanation of A
(c) A is true but R is false
(d) A and R are false

4 $NaAlH_4$ reduces an ester into HCHO and $(CH_3)_2CHOH$. Thus, ester is

(a) $HCOOCH(CH_3)_2$ (b) $(CH_3)_2CHCOOCH_3$
(c) $HCOOCH_2CH_2CH_3$ (d) $CH_3CH_2COOCH_3$

5 Match the structures given in Column I with the names given in Column II.

	Column I		Column II
A.	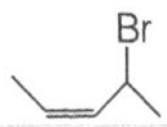	1.	4-bromopent-2-ene
B.	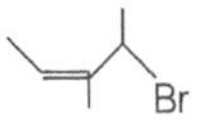	2.	4-bromo-3-methylpent-2-ene
C.	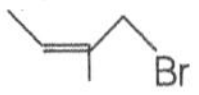	3.	1-bromo-2-methylbut-2-ene
D.	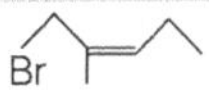	4.	1-bromo-2-methylpent-2-ene

Codes

	A	B	C	D
(a)	4	2	1	3
(b)	1	3	2	4
(c)	1	2	3	4
(d)	2	3	1	4

6 Arrange the following compounds in decreasing order of basicity

I. ethylamine II. 2-amino ethanol
III. 3-amino-1-propanol

(a) I > II > III (b) III > I > II
(c) I > III > II (d) I < II > III

7 In the reaction,

$$CH_3-\underset{\underset{Br}{|}}{CH}-CH_3 \xrightarrow{\text{Alc. KOH}} A \xrightarrow[\text{Peroxide}]{\text{HBr}} B \xrightarrow{CH_3ONa} C, C \text{ is}$$

(a) diethyl ether (b) 1-methoxy propane
(c) isopropyl alcohol (d) propylene glycol.

8 Which one/ones of the following reactions will give 2-propanal? Choose the right answer form (a), (b) and (c) and (d).

(I) $CH_2 = CH - CH_3 + H_2O \xrightarrow{H^+}$

(II) $CH_3\ CHO \xrightarrow[\text{(ii) } H_2O]{\text{(i) } CH_3MgI}$

(III) $CH_2O \xrightarrow[\text{(ii) } H_2O]{\text{(i) } C_2H_5MgI}$

(IV) $CH_2 = CH - CH_3 \xrightarrow{\text{Neutral } KMnO_4}$

(a) I and II (b) II and III
(c) III and I (d) II and IV

9 The order of reactivity of the following al whorls:

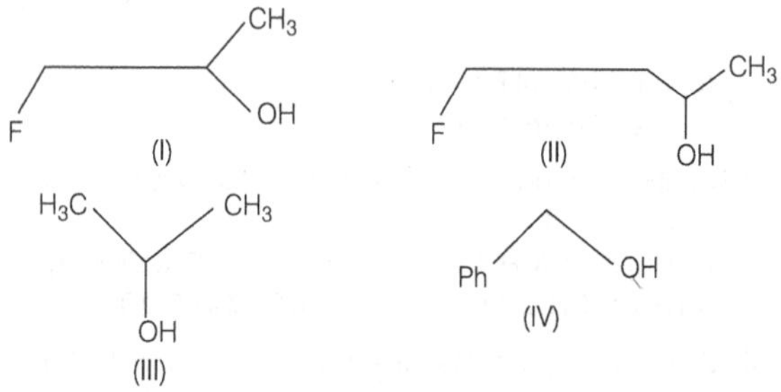

Towards conc. HCl is

(a) I > II > III > IV (b) IV > III > II > I
(c) I > III > IV > II (d) IV > I > II > III

10 **Assertion (A)** Isotonic solutions do not show the phenomenon of osmosis.

Reason (R) Isotonic solutions have different osmotic pressure.

In the light of the above statements, choose the most appropriate answer from the options given below.

(a) (A) is false but (R) is true
(b) (A) is true but (R) is false
(c) Both (A) and (R) are true and (R) is the correct explanation of (A).
(d) Both (A) and (R) are true but (R) is not the correct explanation of (A).

11 **Statement I** All amino acids exist as *Zwitter* ions.

Statement II Amino acids have both $-NH_2$ and $-COOH$ group.

In the light of the above statements, choose the most appropriate answer from the options given below.

(a) Statement I is false but statement II is true.
(b) Statement I is true but statement II is false.
(c) Both statement I and statement II are false.
(d) Both statement I and statement II are true.

12 For the reaction $2Fe^{3+} + Fe \longrightarrow 3Fe^{2+}$, $E° = 1.21$ V, hence

(a) the reaction occurs even in the absence of a second half-cell or a current
(b) in a complete cell, Fe is anode but Pt is the cathode
(c) Fe cannot be used as cathode, because it would directly react with Fe^{3+}, thus short-circuiting the cell
(d) All of the above

13 Reduction of a metal oxide by excess carbon at high temperature is a method for the commercial preparation of some metals. This method can be successfully applied in the case of

(a) BeO and Al_2O_3 (b) ZnO and Fe_2O_3
(c) CaO and Cr_2O_3 (d) BaO and U_3O_8

14 Optical rotations of some compounds along with their structures are given below:

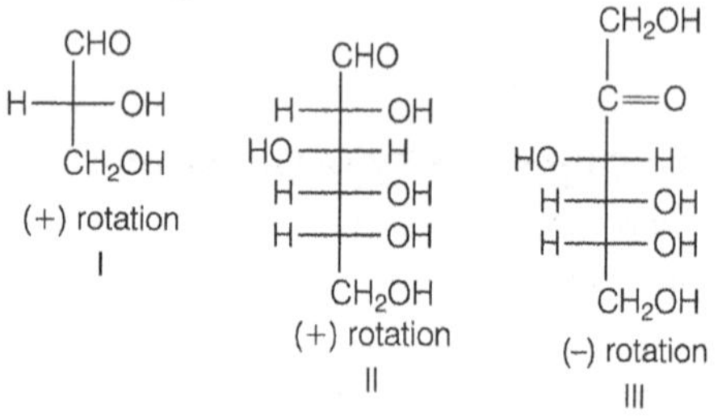

Which of them have D-configuration?

(a) I, II and III (b) II and III
(c) I and II (d) Only III

15 Following is the concentration cell in which electrode is reversible with respect to anion

$$\underset{1\text{ bar}}{Pt(Cl_2)}\,|\,Cl^-(C_1)\,||\,Cl^-(C_2)\,|\,\underset{1\text{ bar}}{Pt(Cl_2)}$$

The cell reaction is spontaneous, if

(a) $C_1 > C_2$ (b) $C_1 < C_2$ (c) $C_1 = C_2$ (d) $C_1 = 0$

16 Consider the following reactions,

(I) $CH_3CH_2\underset{\underset{OH}{|}}{CH}-CH_3 \xrightarrow{H^+} A$ (major)

(II) $CH_3-\underset{\underset{CH_3}{|}}{\overset{\overset{CH_3}{|}}{C}}-\underset{\underset{OH}{|}}{CH}\ CH_3 \xrightarrow{H^+} B$ (major)

A and *B* are respectively

(a) $CH_3CH=CHCH_3$, $CH_3-C(CH_3)_2-CH=CH_2$

(b) $CH_3CH_2CH=CH_2$, $CH_3-C(CH_3)_2-CH=CH_2$

(c) $CH_3CH=CHCH_3$, $CH_3-C(CH_3)=C(CH_3)-CH_3$

(d) $CH_3CH_2CH=CH_2$, $CH_3-C(CH_3)=C(CH_3)-CH_3$

17 The correct name for the complex compound $[Cr(PPh_3)(CO)_5]$ is

(a) pentaphenylpentacarbonylchromium (o)
(b) pentacarbonyltriphenylphosphinechromium (o)
(c) triphenylpentacarbonylphosphinechromium (o)
(d) None of the above

18 A laboratory reagent imparts green colour to the flame. On heating with solid $K_2Cr_2O_7$ and conc. H_2SO_4 it evolves an orange red gas. Identify the reagent.

(a) $CaCl_2$ (b) $BaCl_2$
(c) $CuCl_2$ (d) None of these

19 van't Hoff factors are x, y and z in case of association, ionisation and no change, respectively. Their increasing order is

(a) $x < y < z$
(b) $x = y = z$
(c) $y < x < z$
(d) $x < z < y$

20 Which of the following does not illustrate the anomalous properties of Li?

(a) The melting point and boiling point of Li are comparatively high
(b) Li forms a nitride Li_3N unlike group I metals
(c) Li is much softer than the other group I metals
(d) $LiNO_3$ on decomposition produces its oxide unlike group I metals

Section B : Numerical Type Questions

21 If for a reaction $A \longrightarrow B$, $\Delta H = -10$ kJ mol^{-1} and $E_a = 50$ kJ mol^{-1}, the energy of activation for reaction $B \longrightarrow A$ is ...... kJ mol^{-1}.

22 Relative lowering of vapour pressure of an aqueous glucose dilute solution is found to be 0.018. Hence, elevation in boiling point is ...
(Given, that 1 molal aqueous urea solution boils at 100.54°C at 1 atm pressure)

23 For the following cell reaction,

$$Pb(s) + Hg_2SO_4(s) \rightleftharpoons PbSO_4(s) + 2Hg(l) \quad ;$$

$E^\circ_{cell} = 0.92$ V

$K_{sp}(PbSO_4) = 2 \times 10^{-8}$, $K_{sp}(Hg_2SO_4) = 1 \times 10^{-6}$

Hence, E_{cell} is ...... V.

24 16 g of an ideal gas SO_x occupies 5.6 L at STP. The value of x for this gas is

25 One gram of hydrogen and 112 g of nitrogen are enclosed in two separate containers each of volume 5 L at 27°C. If the pressure of hydrogen is 1 atm, the pressure of nitrogen is ...... atm.

26 Elevation in boiling point of an aqueous urea solution is 0.52° ($K_b = 0.52$°mol^{-1} kg). Hence, mole fraction of urea in this solution is

27 A 100 mL sample is removed from water solution saturated with $CaSO_4$ at 25°C. The water is completely evaporated from the sample and a deposit of 0.24 g $CaSO_4$ is obtained. The K_{sp} of $CaSO_4$ at 25°C is ......$\times 10^4$.

28 A buffer solution contains 100 mL of 0.01 M CH_3COOH and 200 mL of 0.02 M CH_3COONa. 700 mL of water is added to this solution. The pH before and after dilution are ...... (pK_a = 4.74)

29 When FeS_2 is burnt in air, it converts to Fe_2O_3. The change in percentage by weight of iron in the process is ...... % increase (Fe = 56).

30 pK_a of a weak acid (HA) and pK_b of a weak base (BOH) are 3.2 and 3.4, respectively. The pH of their salt (AB) solution is ....

ANSWERS

1 (c)	**2** (d)	**3** (d)	**4** (a)	**5** (c)	**6** (c)	**7** (b)	**8** (a)	**9** (b)	**10** (a)
11 (a)	**12** (d)	**13** (b)	**14** (a)	**15** (a)	**16** (c)	**17** (b)	**18** (b)	**19** (d)	**20** (c)
21 (60)	**22** (0.54)	**23** (0.95)	**24** (2)	**25** (8)	**26** (0.018)	**27** (3.115)	**28** (5.34, 5.34)	**29** (23)	**30** (6.9)

Hints and Explanations

1 I. 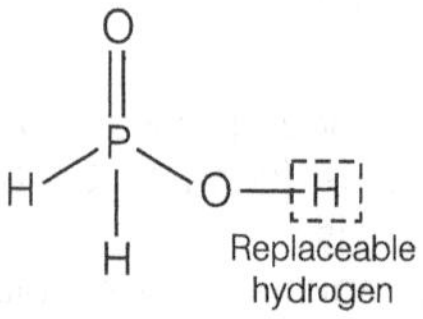

(Hypophosphorous acid)
(monobasic acid)

II. 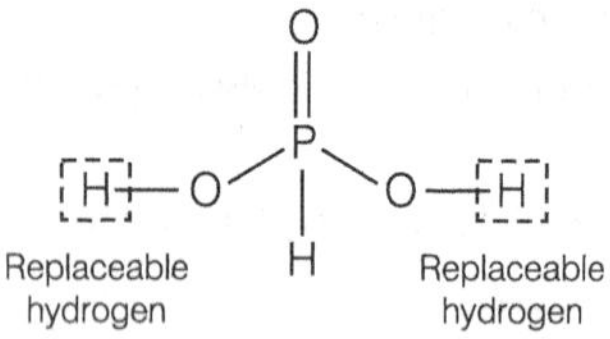

(Orthophosphorous acid)
(dibasic acid)

III. 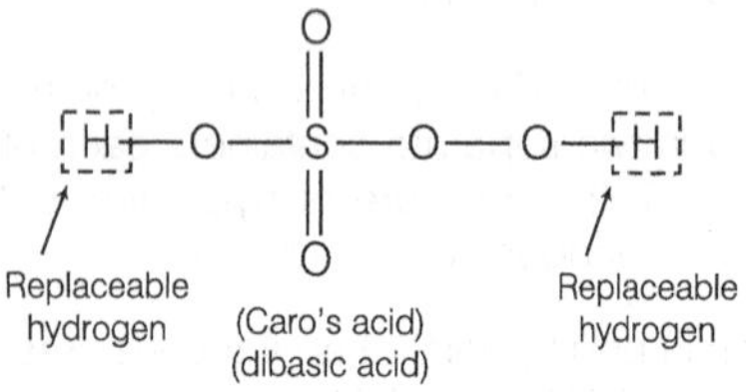

(Caro's acid)
(dibasic acid)

IV. 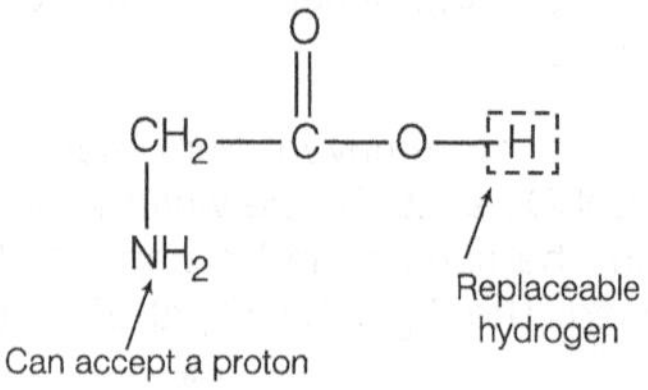

(Glycine) (amphoteric)

2 Fe (26) = $1s^2, 2s^2, 2p^6, 3s^2, 3p^6, 3d^6, 4s^2$

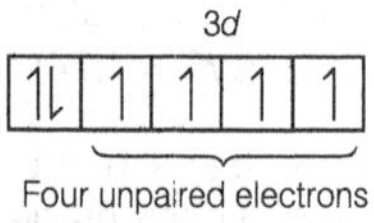

Cr (24) = $1s^2, 2s^2, 2p^6, 3s^2, 3p^6, 3d^5, 4s^1$

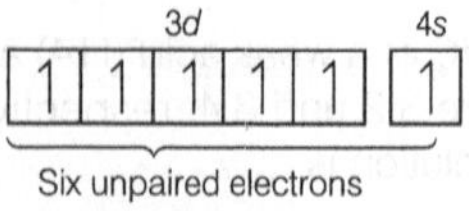

Mn (25) = $1s^2, 2s^2, 2p^6, 3s^2, 3p^6, 3d^5, 4s^2$

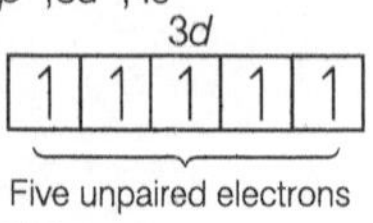

Magnetic moment $(\mu) = \sqrt{n(n+2)}$

where, n = number of unpaired electrons.

Greater the number of unpaired electrons, greater will be the magnetic moment.

Thus, the correct order is $x > y > z$

3 **Correct Assertion** Aldehydes but not ketones react with Tollens' reagent to form silver mirror.

Correct Reason Aldehyde contains a hydrogen attached to carboxyl group while ketone does not.

4 $HCOOCH(CH_3)_2 \xrightarrow{NaAlH_4} HCHO + (CH_3)_2CHOH.$

5 A → 1, B → 2, C → 3, D → 4

6 The electron releasing inductive effect of —OH group decreases the electron density on nitrogen, thus lowers the basicity of amines. This effect diminishes with distance from the amino group. Thus,

ethylamine > 3-amino-1-propanol > 2-amino ethanol.

7 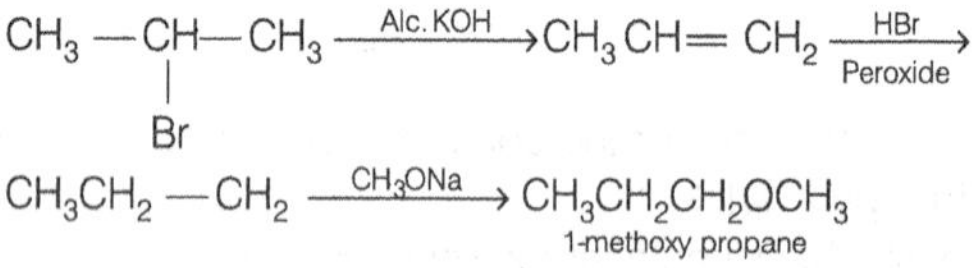

8 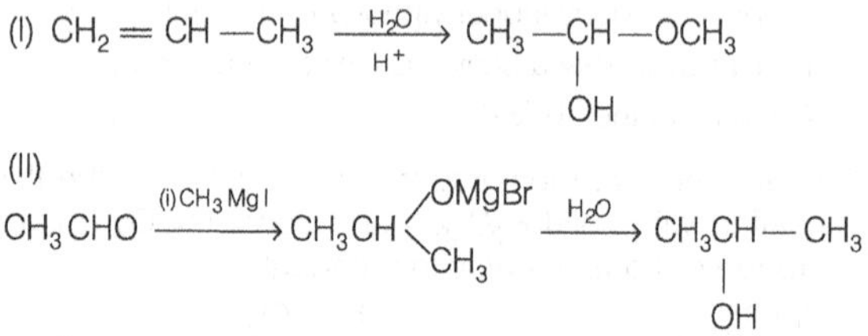

9. The order of reactivity depends upon the stability of carbocations formed. The relative stability of the carbocations follows the order:

$Ph\overset{+}{C}H_2 > CH_2\overset{+}{C}HCH_3 > FCH_2CH_2\overset{+}{C}HCH_3 > FCH_2\overset{+}{C}HCH_3.$

Therefore, the order of reactivity of corresponding alcohols follows the order: IV > III > II > I

10 Isotonic solutions are those solutions which have same osmotic pressure. They do not show the phenomenon of osmosis.

11 All amino acids have amino as well as carboxylic group, $—NH_2$ group is basic while —COOH group is acidic. Hence, they behave as *Zwitter* ion (dipolar ion).

12 $E°$ is positive, thus the reaction occurs even in the absence of half-cell. At anode, oxidation takes place. Thus, all are correct.

13 Least active metals have the most positive standard reduction potentials and are the easiest to reduce, e.g.

$$Fe_2O_3 + 3C \xrightarrow{\text{Blast furnace}} 2Fe + 3CO$$

$$ZnO + C \xrightarrow{1200°C} Zn + CO$$

$$PbO + C \longrightarrow Pb + CO$$

14 In all the given three structures, configuration of OH at C, is towards right, hence all structures possess D-configuration.

15 $Pt(Cl_2)|Cl^-(C_1)||Cl^-(C_2)|Pt(Cl_2)$

1 atm 1 atm

$$E_{cell} = \frac{0.0591}{1} \log \frac{C_1}{C_2}$$

Thus, cell reaction is spontaneous when $C_1 > C_2$.

16 I. $CH_3CH_2CH(OH)—CH_3 \xrightarrow{H^+} CH_3CH=CH—CH_3$ (A)

This is according to Saytzeff's rule.

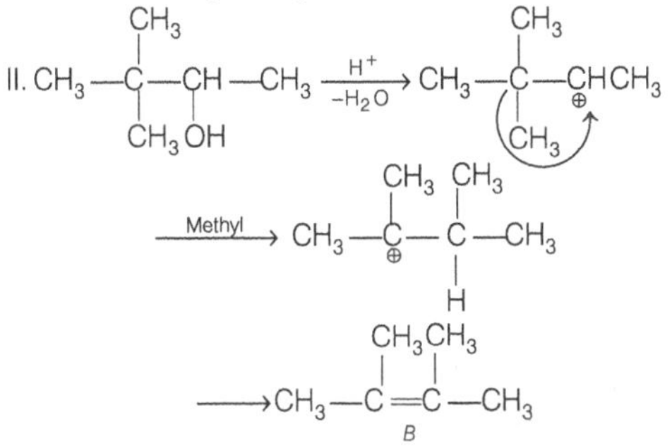

17 The correct name for the given complex compound is pentacarbonyltriphenylphosphinechromium.

18 Ba^{2+} ion imparts green colour to the flame and Cl^- ion forms chromyl chloride (which is orange red in colour) when treated with $K_2Cr_2O_7$ and conc. H_2SO_4. Thus, the reagent is

$$Ba^{2+} + 2Cl^- \longrightarrow BaCl_2$$

$$2BaCl_2 + K_2Cr_2O_7 + 3H_2SO_4 \longrightarrow K_2SO_4 + 2BaSO_4 + \underset{\substack{\text{Chromyl chloride}\\ \text{(orange red gas)}}}{2CrO_2Cl_2\uparrow} + 3H_2O$$

19 In case of ionisation $i > 1$

For no change $i = 0$

For association $i < 1$

Thus, ionisation > no change > association

20 In group I all the metals are soft because of weak metallic bonding and this softness increases down the group.

21 $\Delta H\ (A \to B) = -10$ kJ/mol, i.e. it is an exothermic reaction.

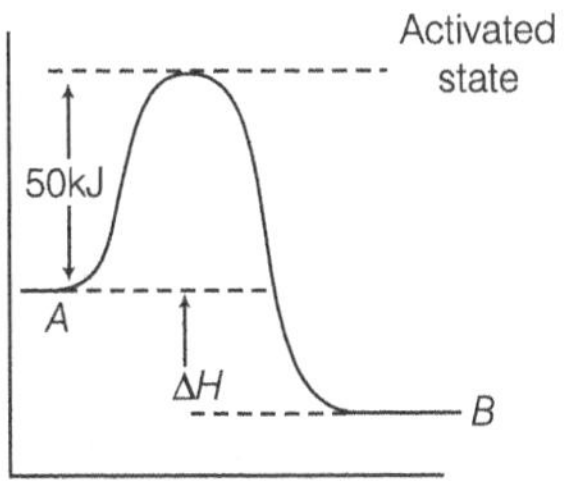

$\therefore$ $\Delta H\ (B \to A) = 10$ kJ/mol

E_a ($A \to$ activated state) = 50 kJ/mol

$\therefore$ E_a ($B \to$ activated state) = 50 + 10 = 60 kJ/mol

22 ΔT_b for urea solution $= 0.54° = K_b \times$ molality

$$K_b = \frac{0.54°}{1} = 0.54° \text{ kg mol}^{-1}$$

$$\frac{\Delta p}{p°} = \frac{w_2 m_1}{mw} \Rightarrow \frac{w_2}{m_2 w_1} = \frac{\Delta p}{p° m_1} = \frac{0.018}{18}$$

$$\Delta T_b \text{ (glucose)} = \frac{1000 K_b w_2}{m_2 w_1} = \frac{1000 \times 0.54 \times 0.018}{18} = 0.54°$$

23 $Q = \dfrac{[Pb^{2+}]}{[Hg_2^{2+}]} = \dfrac{\sqrt{K_{sp}\ (PbSO_4)}}{\sqrt{K_{sp}\ (Hg_2SO_4)}} = \dfrac{\sqrt{2 \times 10^{-8}}}{\sqrt{1 \times 10^{-6}}} = \sqrt{2 \times 10^{-2}} = 0.14$

$$E_{cell} = E°_{cell} - \frac{0.059}{2} \log 0.14 = 0.92 - 0.0295 \log 0.14$$

$$= 0.95 \text{ V}$$

24 1 mole of the gas at STP = 22.4 L

5.6 L at STP = 16 g

Hence, 22.4 L at STP $= \dfrac{16 \times 22.4}{5.6} = 64$ g

Thus, molecular mass of gas = 64 g

Given, gas is SO_x. $32 + 16x = 64$ or $x = 2$

Thus, gas is SO_2.

25 Number of moles of hydrogen $(n_{H_2}) = \dfrac{1}{2}$ mol

Number of moles of nitrogen $(n_{N_2}) = \dfrac{112}{28} = 4$ mol

$$\frac{p_{H_2} V}{p_{N_2} V} = \frac{n_{H_2} RT}{n_{N_2} RT} \Rightarrow \frac{1}{p_{N_2}} = \frac{1}{2 \times 4} \Rightarrow p_{N_2} = 8 \text{ atm}$$

26 $\Delta T_b = m \times K_b \Rightarrow 0.52 = m \times 0.52$

Molality $(m) = 1$ mol kg^{-1}

Moles of urea = 1

Moles of $H_2O = \dfrac{1000}{18} = 55.55 \Rightarrow X_{urea} = \dfrac{1}{56.5} = 0.018$

27 100 mL of saturated $CaSO_4$ solution has 0.24 g of $CaSO_4$.

Thus, solubility of $CaSO_4$ = 2.4 gL^{-1}

$$= \frac{2.4}{136} \text{ mol L}^{-1} = 0.01765$$

$$K_{sp} = S^2 = (0.01765)^2 = 3.115 \times 10^{-4}$$

28 pH (before dilution) $= pK_a + \log \dfrac{[\text{salt}]}{[\text{acid}]} = 4.74 + \log \dfrac{200 \times 0.02}{100 \times 0.01}$

$$= 4.74 + \log 4 = 5.34$$

Dilution has no effect on pH of buffer solution, hence pH after dilution = 5.34

29 $FeS_2 + O_2 \longrightarrow Fe_2O_3 + SO_2$

For FeS_2, Molecular weight $= 56 + 32 \times 2 = 120$

$\therefore$% of Fe $\dfrac{56}{120} \times 100 = 46.67\%$

For Fe_2O_3, Molecular weight $= 56 \times 2 + 3 \times 16 = 160$

$\therefore$ % of Fe $= \dfrac{112}{160} \times 100 = 70\%$

Per cent increase = 70 − 46.67 = 23% (approx.)

30 For a salt of weak acid and weak base,

$$pH = 7 + \frac{1}{2}pK_a - \frac{1}{2}pK_b$$

Given, $pK_a\ (HA) = 3.2, pK_a\ (BOH) = 3.4$

$$\therefore \quad pH = 7 + \frac{1}{2}(3.2), -\frac{1}{2}(3.4)$$

$$= 7 + 1.6 - 1.7 = 6.9$$

DAY FOURTY

Mock Test 3

(Based on Complete Syllabus)

Instructions

1. This question paper contains of 30 Questions of Chemistry, divided into two Sections : **Section A** Objective Type Questions and **Section B** Numerical Type Questions.
2. Section A contains 20 Objective questions and all Questions are compulsory (**Marking Scheme** : Correct +4, Incorrect −1).
3. Section B contains 10 Numerical value questions out of which only 5 questions are to be attempted (**Marking Scheme** : Correct +4, Incorrect 0).

Section A : Objective Type Questions

1 100 mL each of two sols of AgI, one obtained by adding $AgNO_3$, to slight excess of KI and another obtained by adding KI to slight excess of $AgNO_3$ are mixed together. Then,

(a) the two sols will stabilise each other
(b) the sol particles will acquire more electric charge
(c) the sols will coagulate each other mutually
(d) a true solution will be obtained

2 Alkali metals dissolve in liquid NH_3, then which of the following statements about the solution obtained is incorrect?

(a) Solution has high electrical conductivity due to ammoniated electrons
(b) Solution imparts blue colour due to greater polarisation of metal ion
(c) Solution is quite stable, which is considered as a dilute metal
(d) On the addition of substance like iron oxide, solution decomposes and releases hydrogen gas

3 Which of the following does not represent the correct order of properties indicated ?

(a) $NH_3 < PH_3 < AsH_3$ (acidic character)
(b) $Ni^{2+} < Co^{2+} < Fe^{2+} < Mn^{2+}$ (unpaired electrons)
(c) $Al_2O_3 < MgO < Na_2O < K_2O$ (basic nature)
(d) $Li < Be < B < C$ (IE_1)

4 **Assertion** (A) Addition reaction of water to but-1-ene in acidic medium yields butan-1-ol.

Reason (R) Addition of water in acidic medium proceeds through the formation of primary carbocation.

(a) Both Assertion and Reason are true and Reason is the correct explanation of Assertion
(b) Both Assertion and Reason are true but Reason is not the correct explanation of Assertion
(c) Assertion is true but Reason is false
(d) Both Assertion and Reason are false

5 Consider the reduction of 2-butanone.

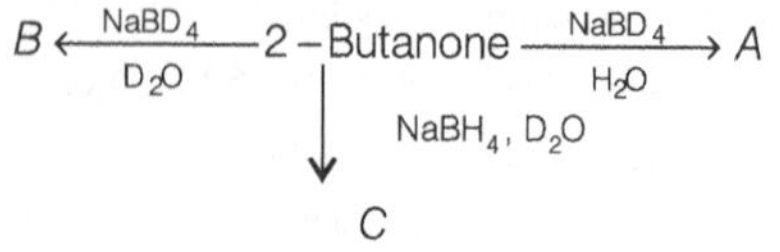

Which of the following statements are true about the products?

(a) Compound (*A*) is $CH_3—C(OH)(D)—CH_2CH_3$

(b) Compound B is 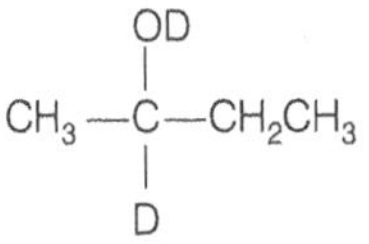

(c) Compound C is 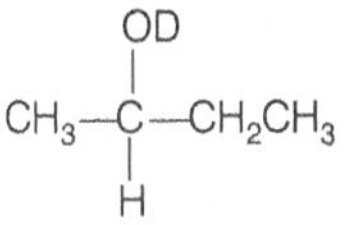

(d) All of the above

6 Which of the following is the strong acid ?

(a) 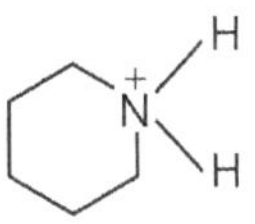 (b) 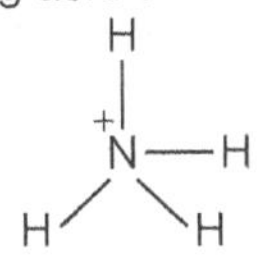

(c) 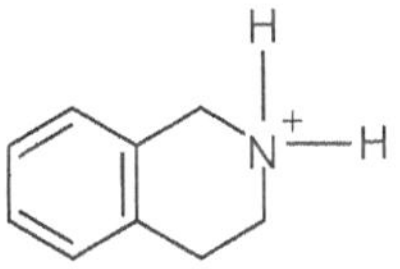 (d) 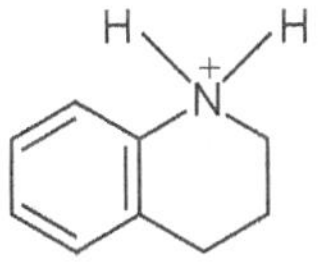

7 'A' is a blue coloured hydrated salt giving white ppt 'B' with $BaCl_2$. 2.495 g of 'A' gave 2.33 g of white ppt 'B'. 'A' gives deep reddish-brown solution on reaction with KI but after addition of hypo, reddish-brown solution appears in the form of white ppt 'C'. 'B' and 'C' respectively are

(a) $BaSO_4, CuCN$ (b) $CuSO_4, Cu_2I_2$
(c) $BaSO_4, Cu_2I_2$ (d) $BaSO_4, CuI_2$

8 Match the polymers (given in Column I) with the type of linkage present in them (given in Column II) and choose the correct option from the codes given below.

Column I	Column II
A. Terylene	1. Glycosidic linkage
B. Nylon	2. Ester linkage
C. Cellulose	3. Phosphodiester linkage
D. RNA	4. Amide linkage

Codes

	A	B	C	D		A	B	C	D
(a)	4	1	3	2	(b)	2	4	1	3
(c)	1	3	2	4	(d)	4	3	1	2

9 A dark brown solid (X) reacts with NH_3 to form a mild explosive which decomposes to give a violet coloured gas. (X) also reacts with H_2 to give an acid (Y). (Y) can also be prepared by heating its salt with H_3PO_4. X and Y are

(a) Cl_2, HCl (b) SO_2, H_2SO_4
(c) Br_2, HBr (d) I_2, HI

10 Compound A and B are treated with dil. HCl separately. The gases liberated are Y and Z respectively. Y turns acidified dichromate paper green while Z turns lead acetate paper black. The compound A and B respectively are

(a) Na_2SO_3 and Na_2S (b) NaCl and Na_2CO_3
(c) Na_2S and Na_2SO_3 (d) Na_2SO_3 and Na_2SO_4

11 Identify A in the following reaction. [NCERT Exemplar]

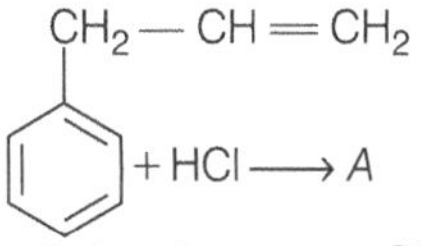

(a) 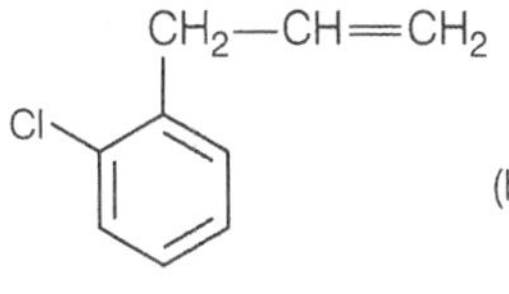 (b) 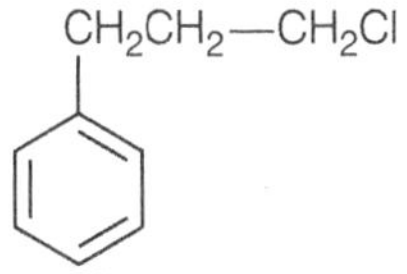

(c) 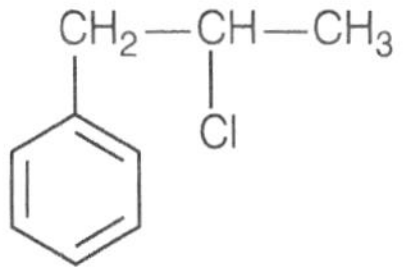 (d) 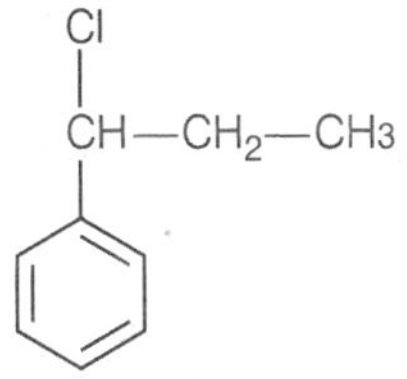

12 Which of the following statements is wrong?

(a) van der Waals' radius of iodine is more than its covalent radius
(b) All isoelectronic ions belong to same period of the periodic table
(c) IE_1 of N is higher than that of O while IE_2 of O is higher than that of N
(d) The electron gain enthalpy of N is zero while that of P is 74.3 kJ mol^{-1}

13 A compound obtained by the hydrolysis of the substance A, on reduction forms 2-hexanol. Hence, the substance A is

(a) 3,3-dichlorohexane (b) 2,3-dichlorohexane
(c) 2,2-dichlorohexane (d) 1,1-dichlorohexane

14 A major alkene 'A' obtained in the following reaction undergoes ozonolysis to give the product.

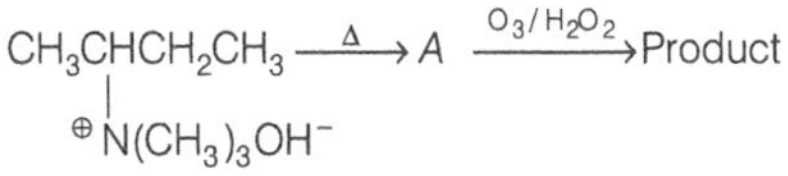

The product obtained is/are identified as

(a) ethanal (b) methanal and propanal
(c) methanal and acetone (d) acetone

15 Which of the following is correct comparison of the stability of the molecules ?

(a) $CN < O_2^+$ (b) $CN^+ = N_2$ (c) $N_2 < O_2$ (d) $H_2^+ > He_2^+$

16 Identify the compound Y in the following reaction.

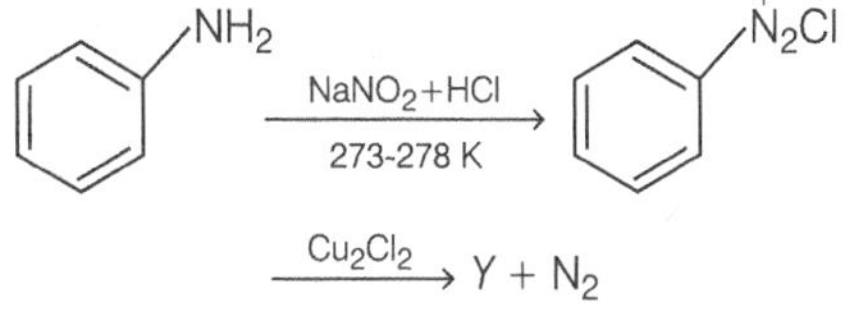

(a) 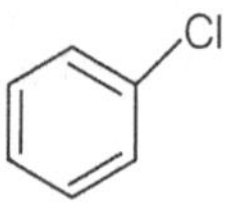
(b) 

(c) 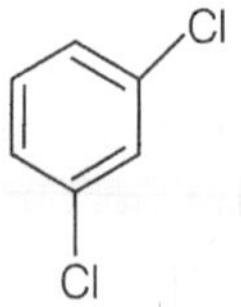
(d) 

17 Assertion (A) Photochemical smog is produced by nitrogen oxides.

Reason (R) Vehicular pollution is a major source of nitrogen oxides.

(a) Both Assertion and Reason are true and Reason is the correct explanation of Assertion
(b) Both Assertion and Reason are true but Reason is not the correct explanation of Assertion
(c) Assertion is true but Reason is false
(d) Both Assertion and Reason are false

18 In a solid 'AB' having the NaCl structure, 'A' atoms occupy the corners of the cubic unit cell. If all the face-centred atoms along one of the axis are removed then the resultant stoichiometry of the solid is

(a) AB_2 (b) A_2B (c) A_3B_4 (d) A_4B_3

19 The hybridisation states of the central atom of $ClF_3, IF_7, PF_4^+, ClO_3^-$ respectively, are

(a) $sp^3d, sp^3d^3, sp^3, sp^3$ (b) $sp^3d, sp^3d^2, sp^3d, sp^3$
(c) $sp^3, sp^3d, sp^3d, sp^3d^2$ (d) $sp^3d^2, sp^3d^3, sp^3, sp^3$

20 Statement I NF_3 is a weaker ligand than N $(CH_3)_3$.
Statement II NF_3 ionises to give F^- ions in aqueous solution.

(a) Statement I is false but Statement II is true.
(b) Statement I is true but Statement II is false.
(c) Both Statement I and Statement II are false.
(d) Both Statement I and Statement II are true.

Section B : Numerical Type Questions

21 One gram of a monobasic acid when dissolved in 100 g of water lowers the freezing point by 0.186°C. Now, 0.25 g of the same acid are dissolved and titrated with 15.1 mL of $N/10$ alkali. The degree of dissociation of the acid is .... $[K_f\,(H_2O) = 1.86]$

22 The radius ratio of Cs^+ to Cl^- is 0.90 Å. If the radius of Cl^- ion is 1.80 Å, the distance between Cs^+ ion and Cl^- ion is ...... Å.

23 The most abundant element dissolved in sea water is chlorine at a concentration of 19 g/L of sea water. The volume of earth's ocean is 1.4×10^{21} L. Gram-atoms of chlorine are potentially available from the oceans are ...... $\times 10^{20}$ (density of sea water is 1 g / cc).

24 When the concentration of 'A' is 0.1M, it decomposes to give 'X' by a first order process with a rate constant of $6.93 \times 10^{-2} \text{min}^{-1}$. The reactant '$A$' in the presence of catalyst gives 'Y' by a second order mechanism with a rate constant of 0.2 min^{-1} M^{-1}. In order that half-life of both the processes is 10 min, one should start with an initial concentration of 'A' as ...... M.

25 4.5 g of PCl_5 was completely vaporised at 252°C, vapour occupying 1700 mL at 1.0 atm. The degree of dissociation is ...... % (Given $V_\infty = 104.175$)

26 When 1 mole of crystalline NaCl is obtained from sodium and chlorine gas, 410 kJ of heat is released. The heat of sublimation of Na metal is 108 kJ mol^{-1} and Cl—Cl bond enthalpy is 242 kJ mol^{-1}. If the ionisation energy of Na is 493.0 kJ mol^{-1} and the electron affinity of chlorine is 368 kJ mol^{-1}, the lattice energy of NaCl is ...... kJ mol^{-1}.

27 A galvanic cell : $Cu|Cu^{2+}||Ag^+|Ag$ initially contains 1 M Ag^+ and 1 M Cu^{2+} ions. The cell potential after the passage of 9.65 A of current for 1 h will be ... V.

28 Given, $Fe^{3+} + e^- \longrightarrow Fe^{2+}, E° = 0.77$ V

$$\frac{1}{2}I_2 + e^- \longrightarrow I^-, \quad E° = 0.54 \text{ V}$$

Thus, equilibrium constant for the reaction in terms of $\log k$ is

$$2Fe^{3+} + 3I^- \rightleftharpoons 2Fe^{2+} + I_3^-$$

29 K_1 and K_2 for carbonic acid (H_2CO_3) are 4.3×10^{-7} and 5.6×10^{-11} respectively. The pH of 0.01 M $NaHCO_3$ solution will be

30 Two first order reactions have half-lives in the ratio 8 : 1. The ratio of time intervals $t_1 : t_2$ is ....
The time t_1 and t_2 are the time period for $(1/4)$th and (3/4)th completion.

ANSWERS

1. (c)	**2.** (b)	**3.** (d)	**4.** (d)	**5.** (d)	**6.** (d)	**7.** (c)	**8.** (b)	**9.** (d)	**10.** (a)
11. (c)	**12.** (b)	**13.** (c)	**14.** (b)	**15.** (d)	**16.** (a)	**17.** (d)	**18.** (c)	**19.** (a)	**20.** (b)
21. (0.67)	**22.** (3.42)	**23.** (7.6)	**24.** (0.5)	**25.** (83)	**26.** (–764)	**27.** (0.01)	**28.** (7.78)	**29.** (9.19)	**30.** (1:0.602)

Hints and Explanations

1. The sols obtained in the two cases will be oppositely charged, thus coagulate each other mutually.

2.
$$\underset{\text{Alkali metal}}{M} + \underset{\text{Solution}}{(x+y)NH_3} \longrightarrow \underset{\text{Ammoniated metal cation}}{[M(NH_3)_x]^+} + \underset{\text{Ammoniated electron (blue)}}{[e(NH_3)_y]^-}$$

It imparts blue colour due to the presence of solvated electrons, not due to polarisation. On addition of iron oxide, solution decomposes as follows

$$2Na + 2NH_3 \xrightarrow{Fe_2O_3} 2NaNH_2 + H_2\uparrow$$

3. Generally, IE increases in a period from left to right but IE of group 2 is greater than group 3 due to stable electronic configuration of group 2.

4. Correct Assertion Addition of water to but-1-ene in acidic medium yields butan-2-ol.

Correct Reason Addition of water in acidic medium proceeds through the formation of secondary carbocation.

5.

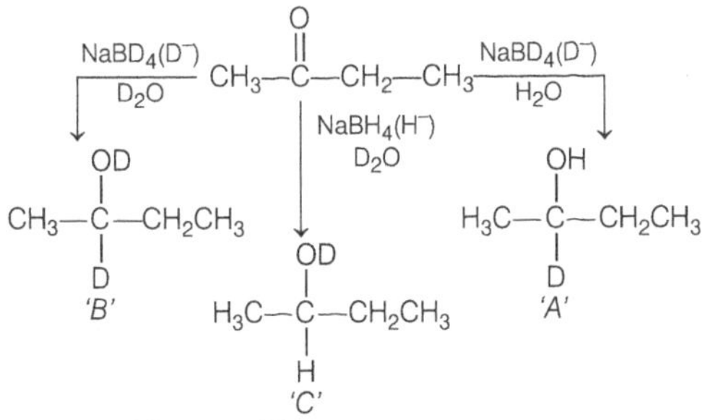

6.

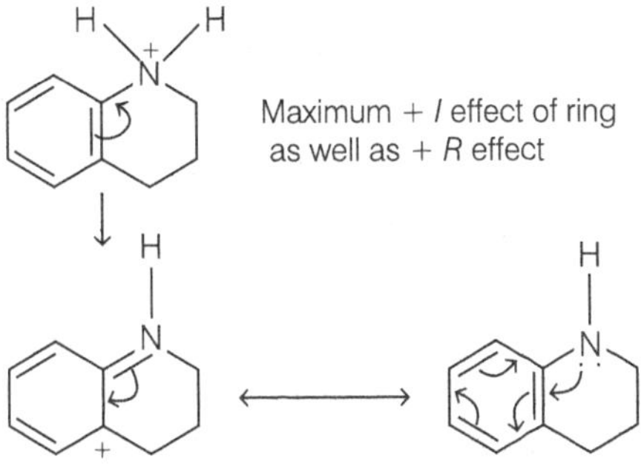

7.
$$\underset{'A'}{CuSO_4\cdot 5H_2O} + BaCl_2 \longrightarrow \underset{'B'}{BaSO_4} + CuCl_2$$

$$2CuSO_4 + 4KI \longrightarrow \underset{\text{White ppt}}{2CuI\downarrow} + I_2$$

$$I_2 + KI \rightleftharpoons \underset{\text{Reddish brown}}{KI_3}$$

On adding hypo, I_2 is reduced to I^- and white ppt of Cu_2I_2 (CuI) appears.

8. $A \rightarrow 2; B \rightarrow 4; C \rightarrow 1; D \rightarrow 3$

9. $X = I_2, Y = HI$

$$\underset{(X)}{3I_2} + 2NH_3 \longrightarrow \underset{\text{(explosive)}}{NH_3\cdot NI_3}$$

$$8NI_3\cdot NH_3 \longrightarrow 5N_2 + I_2 + 6NH_4I$$

$$\underset{(X)}{I_2} + H_2 \longrightarrow \underset{(Y)}{2HI}$$

$$3NaI + H_3PO_4 \xrightarrow{\Delta} Na_3PO_4 + 3HI$$

10. Gas (*Y*) is SO_2 obtained from Na_2SO_3 (*A*) and gas *Z* is H_2S obtained from Na_2S (*B*).

$$\underset{'A'}{Na_2SO_3} + \underset{\text{Dil.}}{2HCl} \longrightarrow 2NaCl + \underset{Y}{SO_2(g)} + H_2O$$

$$\underset{'B'}{Na_2S} + \underset{\text{Dil.}}{2HCl} \longrightarrow 2NaCl + \underset{'Z'}{H_2S(g)}$$

$$K_2Cr_2O_7 + H_2SO_4 + \underset{'Y'}{3SO_2(g)} \longrightarrow K_2SO_4 + \underset{\text{Green}}{Cr_2(SO_4)_3} + H_2O$$

$$(CH_3COO)_2Pb + \underset{'Z'}{H_2S(g)} \longrightarrow \underset{\text{Black}}{PbS\downarrow} + 2CH_3COOH$$

11. Product is in accordance to Markownikoff's rule.

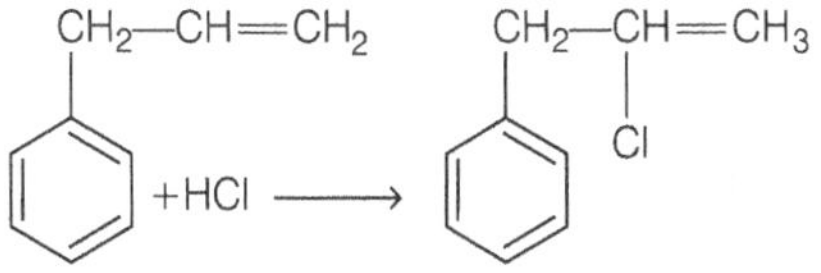

12. In the isoelectronic series, all isoelectronic anions belong to the same period and cations to the next period.

13.
$$\underset{\substack{\text{2,2-dichlorohexane}\\(A)}}{CH_3-CCl_2-CH_2-CH_2-CH_2-CH_3} \xrightarrow[aq.\ KOH]{\text{Hydrolysis}} \underset{\text{2-hexanone}}{CH_3-\overset{O}{\overset{\|}{C}}-CH_2-CH_2CH_2CH_3}$$

$$\xrightarrow[2[H]]{\text{Reduction}} \underset{\text{2-hexanol}}{CH_3-\underset{OH}{\underset{|}{CH}}-CH_2-CH_2-CH_2-CH_3}$$

14.
$$CH_3\underset{\overset{\oplus}{N}(CH_3)_3OH^-}{\underset{|}{C}H}-CH_2CH_3 \xrightarrow{\Delta} \underset{A}{CH_2{=}CHCH_2CH_3} + (CH_3)_3N + H_2O$$

In elimination reaction of alkyl tri-methyl ammonium hydroxide, less substituted alkenes are obtained.

$$CH_2{=}CHCH_2CH_3 \xrightarrow{O_3/H_2O_2} \underset{\text{Methanal}}{HCHO} + \underset{\text{Propanal}}{CH_3CH_2CHO}$$

15. Bond order ∝ stability. When the bond order of two molecules are same, the molecule with least number of antibonding electrons is more stable. Thus,

$$H_2^+(1) = \sigma 1s^1;\ BO = \frac{1-0}{2} = 0.5$$

$He_2^+(3) = \sigma 1s^2, \sigma^* 1s^1,$

$BO = \frac{2-1}{2} = 0.5$

$\therefore$ Stability of H_2^+ > Stability of He_2^+.

16. 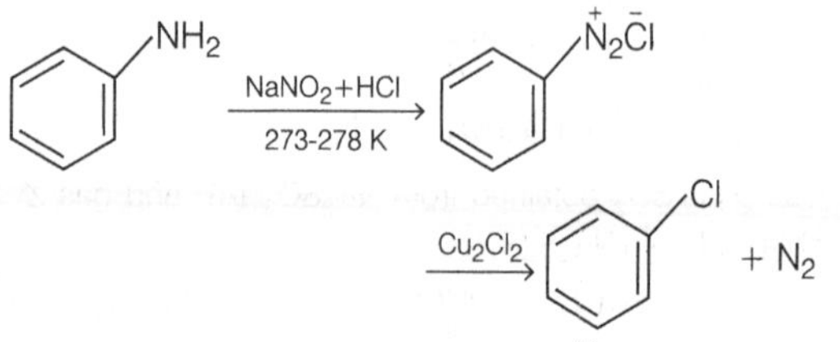

Y

17. Photochemical smog is produced by nitrogen oxides and it is also fact that vehicular pollution is a major source of nitrogen oxide.

18. There are 6 *A* atoms on the face centres. Removing face centred atoms along one of the axis means removal of 2 *A* atoms.

Now, number of *A* atoms per unit cell

$= \underset{\text{(corners)}}{8 \times \frac{1}{8}} + \underset{\text{(face-centred)}}{4 \times \frac{1}{2}} = 3$

Number of *B*-atoms per unit cell

$= \underset{\text{(edge centred)}}{12 \times \frac{1}{4}} + \underset{\text{(body centred)}}{1} = 4$

Hence, the resultant stoichiometry is A_3B_4.

19. Hybridisation in $ClF_3 = sp^3d$

$IF_7 = sp^3d^3$

$PF_4^+ = sp^3$

$ClO_3^- = sp^3$

20. Fluorine is highly electronegative, hence it withdraws electrons from nitrogen atom and therefore, the lone pair of electrons on nitrogen atom cannot be ligated. While $N(CH_3)_3$ is a strong ligand because — CH_3 is a electron releasing group.

21. $w = \frac{ENV}{1000}$ [For monobasic acid $E = M$]

$0.25 = \frac{M \times 1 \times 15.1}{1000 \times 10} \Rightarrow M = 165.56$

Molality $(m) = \frac{1 \times 1000}{165.56 \times 100} = 0.060$

$\Delta T_f = K_f \times m = 1.86 \times 0.060 = 0.1116$

$i = \frac{[\Delta T_f]_{obs}}{[\Delta T_f]_{normal}} = \frac{0.186}{0.1116} = 1.67$

	$HA \rightleftharpoons$	H^+	$+ A^-$
Initial conc.	1	0	0
Conc. at equil.	$1-x$	x	x

$1.67 = \frac{(1-x) + x + x}{1}$

$x = 0.67$

22. $\because (r_{Cs^+}/r_{Cl^-}) = 0.90$ Å,

$\therefore r_{Cs^+} = 0.90 \times r_{Cl^-} = 0.90 \times 1.80 = 1.62$ Å

For precise fitting of the cubic voids, $\frac{r_{Cs^+}}{r_{Cl^-}}$ must be 0.732.

$0.732 < 0.90$

Hence, cubic packing of Cl^- ions is loose to some extent. The distance between Cs^+ and Cl^-

$= r_{Cs^+} + r_{Cl^-} = 1.62$ Å $+ 1.80$ Å $= 3.42$ Å

23. Total weight of ocean = volume of earth ocean × density of sea water

$= 1.4 \times 10^{21} \times 1 \times 10^3 = 1.4 \times 10^{24}$ g water

$[Cl^-] = \frac{19 \text{ g } Cl^-}{1000 \text{ g water}}$

Total weight of chloride

$= 1.4 \times 10^{24}$ g water $\times 19 \times 10^{-3} \frac{gCl^-}{gwater} = 2.7 \times 10^{22}$ g Cl^-

Number of gram atoms $= \frac{2.7 \times 10^{22}}{35.5} = 7.6 \times 10^{20}$ g-atoms

24. $A \xrightarrow{k_1} X; k_1 = 6.93 \times 10^{-2} \text{ min}^{-1}$

$A \xrightarrow{k_2} Y; \ k_2 = 0.2 \text{ min}^{-1} \text{ M}^{-1}$

For I case, $t_{1/2} = \frac{0.693}{k_1}$

$t_{1/2} = \frac{0.693}{6.93 \times 10^{-2}} = 10 \text{ min}$

and independent of initial concentration.

For II case, $t_{1/2} = \frac{1}{k_2 a}; \ a = \frac{1}{k_2 \cdot t_{1/2}} = \frac{1}{0.2 \times 10} = 0.5$ M

25. $PCl_5 \rightleftharpoons PCl_3 + Cl_2$; $pV = \frac{m}{M}RT$

Molecular weight $(M) = \frac{mRT}{pV} = \frac{4.5 \times 0.082 \times (273 + 252)}{1.0 \times 1700/1000}$

$= 113.96$

Observed vapour density $= \frac{113.95}{2} = 56.98$

$\alpha = \frac{V_\infty - V_0}{(n-1)V_0} = \frac{104.175 - 56.98}{(2-1) \times 56.9} = \frac{47.195}{56.98} = 0.828 \approx 0.83$

Degree of dissociation = 83%

26. $Na(s) + \frac{1}{2}Cl_2(g) \longrightarrow NaCl(s), \quad \Delta H = -410$ kJ

$\Delta H_{sub}(Na) + IE(Na(g)) + \frac{1}{2}\Delta H_{Cl-Cl}$

$- EA_{(Cl)} + \Delta H_{lattice}(NaCl) = -410$

$108 + 493 + \frac{1}{2} \times 242 - 368 + \Delta H_{lattice}(NaCl) = -410$

$\Delta H_{lattice}(NaCl) = -764 \text{ kJ mol}^{-1}$

27. $Cu(s) + 2Ag^+(aq) \longrightarrow Cu^{2+}(aq) + 2Ag(s)$

$E_{cell} = E^\circ_{cell} - \frac{0.059}{2} \log \frac{[Cu^{2+}]}{[Ag^+]^2}$

On the passage of 9.65A for 1h (9.65 × 60 × 60 C) externally cell reaction is reversed, now Cu^{2+} will be discharging

(conc. of Cu^{2+} will decrease) and Ag will be ionising (conc. of Ag^+ will increase).

Final mole of $Ag^+ = \dfrac{9.65 \times 60 \times 60}{96500}$

$= 0.36$ g-equivalent $= 0.36$ mol

Final moles of $Cu^{2+} = \dfrac{9.65 \times 60 \times 60}{96500}$

$= 0.36$ g-equivalent $= 0.18$ mol

$[Ag^+] = 1 + 0.36 = 1.36$ M

$[CH^{2+}] = 1 - 0.18 = 0.82$ M

Hence, $E_{cell} = E^\circ_{cell} - \dfrac{0.059}{2} \log \dfrac{[0.82]}{[1.36]^2}$

$E_{cell} = E^\circ_{cell} + 0.010$ V

28. $2\times\ [Fe^{3+} + e^- \longrightarrow Fe^{2+}, E^\circ_{red} = 0.77\text{V}$

$2I^- \longrightarrow I_2 + 2e^-, E_{ox} = -0.54\text{V}$

$I_2 + I^- \rightleftharpoons I_3^-$

$2Fe^{3+} + 3I^- \rightleftharpoons 2Fe^{2+} + I_3^-,\ E^\circ_{cell} = 0.23$ V

$\therefore\ E_{cell} = E^\circ_{cell} - \dfrac{0.0591}{2} \log K_{eq}$

$E_{cell} = 0.00$ V at equilibrium

$K_{eq} = \dfrac{[Fe^{2+}]^2[I_3^-]}{[Fe^{3+}]^2\,[I^-]^3}$

$\therefore\ E^\circ_{cell} = \dfrac{0.0591}{2} \log K_{eq} \Rightarrow \dfrac{0.23 \times 2}{0.0591} = \log K_{eq}$

$\therefore\ \log K_{eq} = 7.78$

29. $NaHCO_3 + H_2O \rightleftharpoons NaOH + H_2CO_3$

or $HCO_3^- + H_2O \rightleftharpoons OH^- + H_2CO_3$

For anion hydrolysis degree of hydrolysis

$K_h = \dfrac{K_w}{K_a} = \dfrac{K_w}{K_1} = \dfrac{1 \times 10^{-14}}{4.3 \times 10^{-7}} = 2.33 \times 10^{-8}$

$[K_2 <<< K_h, [H^+] <<<< [OH^-]]$

$K_h = Ch^2$

$[OH^-] = Ch$

$[OH^-] = \sqrt{C^2h^2} = \sqrt{C \times Ch^2}$

$= \sqrt{C \times K_h} = \sqrt{0.01 \times 2.33 \times 10^{-8}}$

$= 1.53 \times 10^{-5}$

$[H^+] = \dfrac{1 \times 10^{-14}}{1.53 \times 10^{-5}} = 6.53 \times 10^{-10}$

$pH = -\log [H^+]$

$= -\log (6.53 \times 10^{-10}) = 9.19$

30. $t_1 = \dfrac{2.303\,(t_{1/2})_1}{0.693} \log \left(\dfrac{1}{1-(1/4)}\right)$

and $t_2 = \dfrac{2.303\,(t_{1/2})_2}{0.693} \log \left(\dfrac{1}{1-(3/4)}\right)$

$\dfrac{t_1}{t_2} = \dfrac{8}{1} \times \dfrac{\log (4/3)}{\log 4}$

$= \dfrac{8 \times 0.125}{0.602} = 1 : 0.602$

JEE Main 2021

(February, March, July & August Attempt)

Instructions

- This question paper contains of 30 Questions of Chemistry, divided into two Sections : **Section A** Objective Type Questions and **Section B** Numerical Type Questions.
- Section A contains 20 questions and all Questions are compulsory (**Marking Scheme :** Correct + 4, Incorrect −1) .
- Section B contains 10 questions out of which only 5 questions are to be attempted (**Marking Scheme :** Correct + 4, Incorrect 0) .

24 FEBRUARY SHIFT I

Section A : Objective Type Questions

1. Which of the following are isostructural pairs ?

A. SO_4^{2-} and CrO_4^{2-}

B. $SiCl_4$ and $TiCl_4$

C. NH_3 and NO_3^-

D. BCl_3 and $BrCl_3$

a. A and B only **b.** A and C only

c. B and C only **d.** C and D only

2. In Freundlich adsorption isotherm, slope of *AB* line is

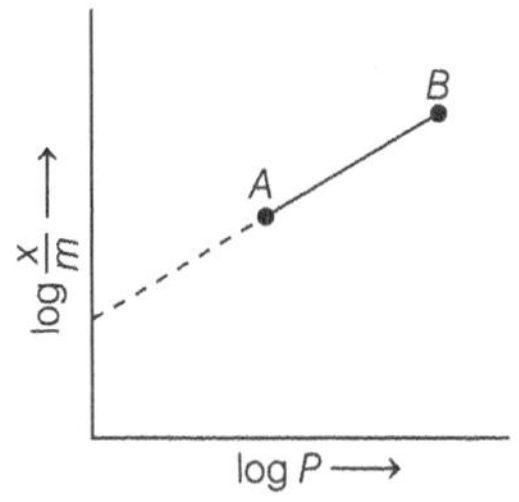

a. n with (n = 0.1 to 0.5) **b.** $\log n$ with ($n > 1$)

c. $\log \frac{1}{n}$ with ($n < 1$) **d.** $\frac{1}{n}$ with $\left(\frac{1}{n} = 0 \text{ to } 1\right)$

3. Consider the elements Mg, Al, S, P and Si, the correct increasing order of their first ionisation enthalpy is

a. Al < Mg < Si < S < P **b.** Mg < Al < Si < P < S

c. Mg < Al < Si < S < P **d.** Al < Mg < S < Si < P

4. Which of the following ore is concentrated using group 1 cyanide salt ?

a. Calamine **b.** Malachite

c. Siderite **d.** Sphalerite

5. (A) $HOCl + H_2O_2 \longrightarrow H_3O^+ + Cl^- + O_2$

(B) $I_2 + H_2O_2 + 2OH^- \longrightarrow 2I^- + 2H_2O + O_2$

Choose the correct option.

a. H_2O_2 acts as oxidising agent in equations (A) and (B)

b. H_2O_2 acts as reducing agent in equations (A) and (B)

c. H_2O_2 act as oxidising and reducing agent respectively in equations (A) and (B)

d. H_2O_2 acts as reducing and oxidising agent respectively in equations (A) and (B).

6. Al_2O_3 was leached with alkali to get *X*. The solution of *X* on passing of gas *Y*, forms *Z*. *X*, *Y* and *Z* respectively are

a. $X = Na[Al(OH)_4]$, $Y = SO_2$, $Z = Al_2O_3$

b. $X = Al(OH)_3$, $Y = SO_2$, $Z = Al_2O_3 \cdot xH_2O$

c. $X = Al(OH)_3$, $Y = CO$, $Z = Al_2O_3$

d. $X = Na[Al(OH)_4]$, $Y = CO_2$, $Z = Al_2O_3 \cdot xH_2O$

7. The electrode potential of M^{2+}/M of 3*d*-series elements shows positive value for

a. Fe **b.** Co

c. Zn **d.** Cu

8. The major components in Gun metal are

a. Cu, Sn and Zn **b.** Cu, Zn and Ni

c. Cu, Ni and Fe **d.** Al, Cu, Mg and Mn

9. The gas released during anaerobic degradation of vegetation may lead to

a. acid rain **b.** global warming and cancer

c. corrosion of metals **d.** ozone hole

10. Which of the following compound gives pink colour on reaction with phthalic anhydride in conc. H_2SO_4 followed by treatment with NaOH?

a.

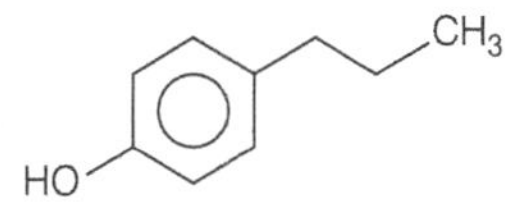

b.

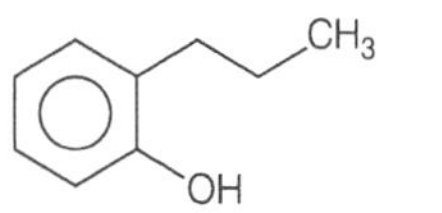

c.

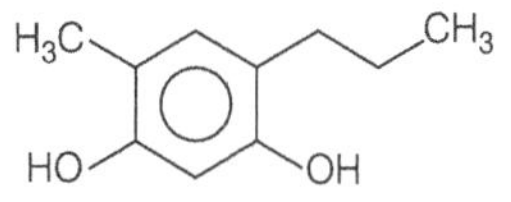

d.

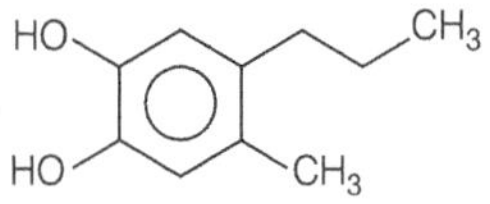

11. What is the major product formed by HI on reaction with

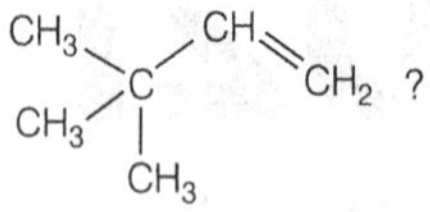 ?

a. 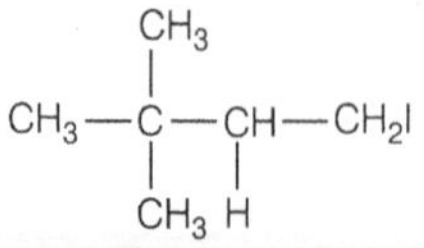 b. 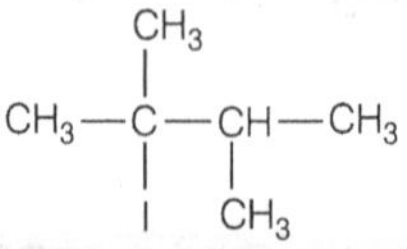

c. 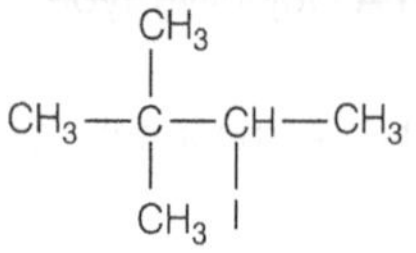 d. 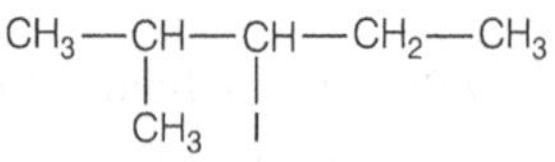

12. Which of the following reagent is used for the following reaction ?

$CH_3CH_2CH_3 \xrightarrow{?} CH_3CH_2CHO$

a. Copper at high temperature and pressure
b. Molybdenum oxide
c. Manganese acetate
d. Potassium permanganate

13. The product formed in the first step of the reaction of

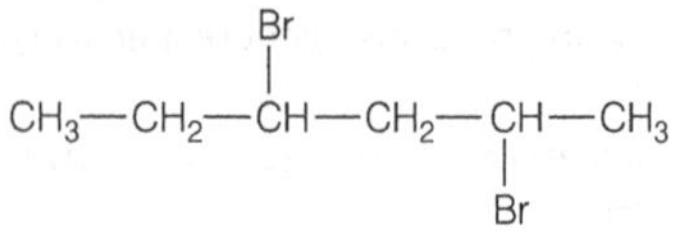 with excess

Mg/Et_2O (Et = C_2H_5) is

a. 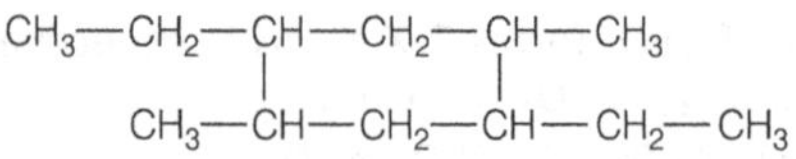

b. 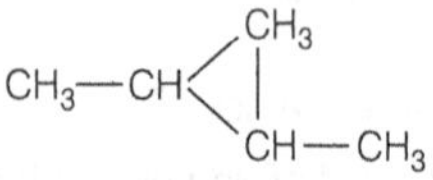

c. 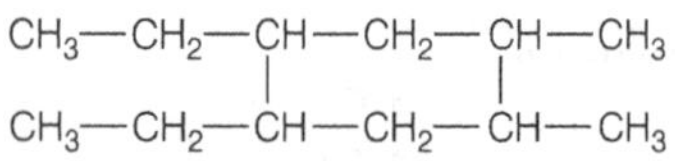

d. 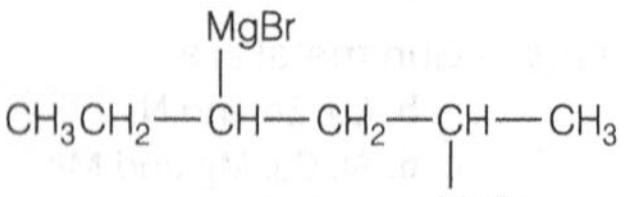

14. What is the final product (major) 'A' in the given reaction ?

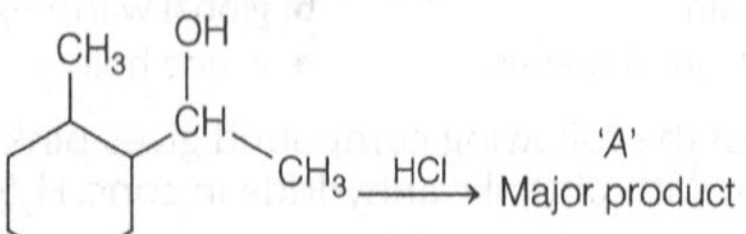

a. 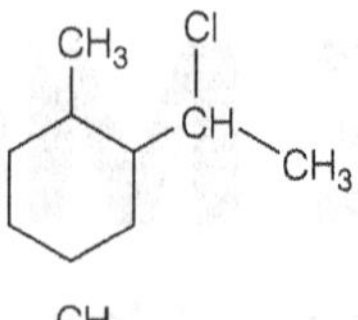 b. 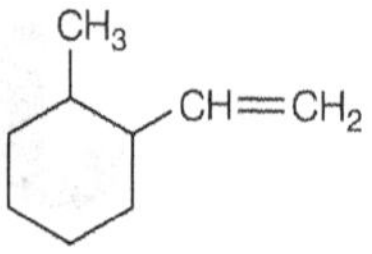

c. 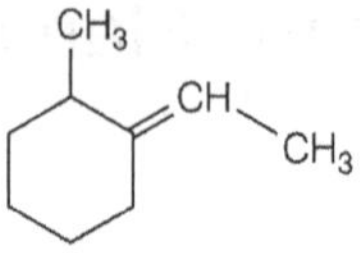 d. 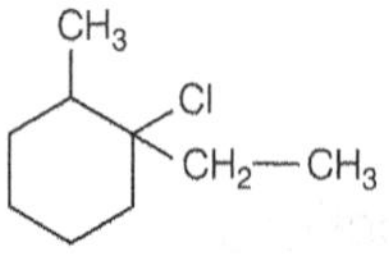

15. Identify products *A* and *B*

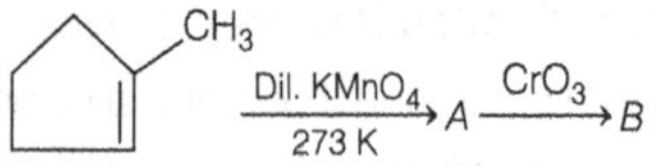

a. A : B : 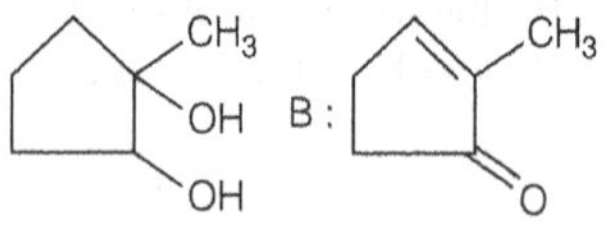

b. A : B : 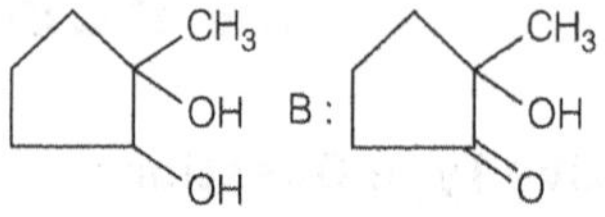

c. A : B : 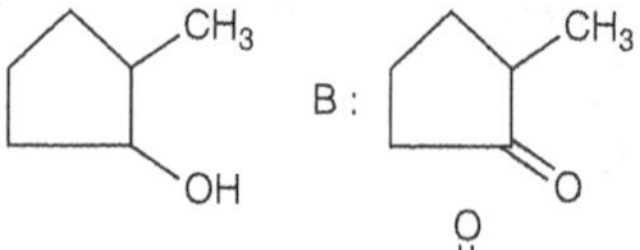

d. A : $OHC-CH_2CH_2CH_2-\overset{O}{\overset{\|}{C}}-CH_3$

B : $HOOC-CH_2CH_2CH_2-\overset{O}{\overset{\|}{C}}-CH_3$

16. In the following reaction the reason why *meta*-nitro product also formed is

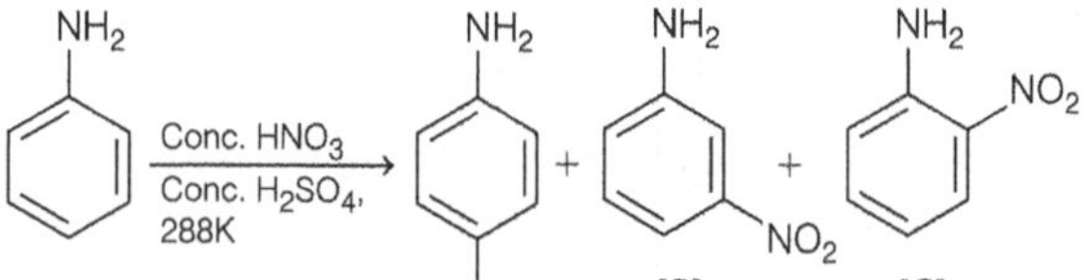

[A] 51% [B] 47% [C] 2%

a. $-NH_2$ group is highly *meta*-directive
b. $-NO_2$ substitution always takes place at *meta*-position
c. Formation of anilinium ion
d. Low temperature

17. 'A' and 'B' in the following reaction are

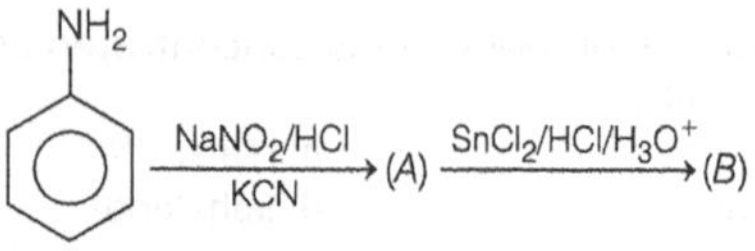

a. (A) 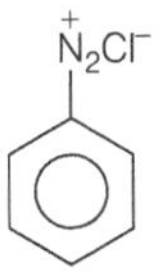 (B) 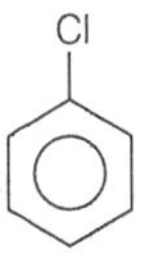

b. (A) 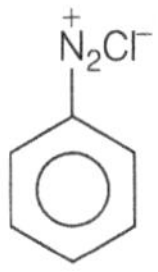 (B) 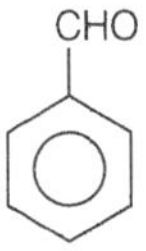

c. (A) 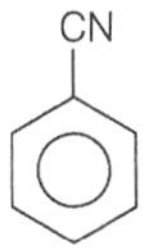 (B) 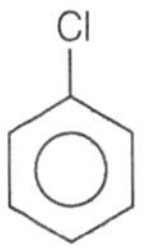

d. (A) 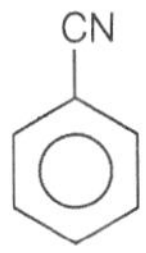 (B) 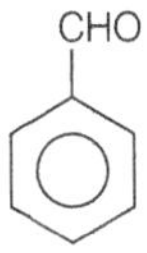

18. Match List-I with List-II.

	List-I (Monomer Unit)		**List-II** (Polymer)
A.	Caprolactum	1.	Natural rubber
B.	2-chloro buta-1,3-diene	2.	Buna-N
C.	Isoprene	3.	Nylon-6
D.	Acrylonitrile	4.	Neoprene

Choose the correct answer from the options given below

	A	B	C	D		A	B	C	D
a.	1	2	3	4	**b.**	4	3	2	1
c.	2	1	4	3	**d.**	3	4	1	2

19. Out of the following, which type of interaction is responsible for the stabilisation of α-helix structure of proteins ?

a. van der Waals' forces **b.** Covalent bonding
c. Ionic bonding **d.** Hydrogen bonding

20. Given below are two statements.

Statement I Colourless cupric metaborate is reduced to cuprous metaborate in a luminous flame.

Statement II Cuprous metaborate is obtained by heating boric anhydride and copper sulphate in a non-luminous flame.

In the light of the above statements, choose the most appropriate answer from the options given below.

a. Both statement I and statement II are true.
b. Both statement I and statement II are false.
c. Statement I is true but statement II is false.
d. Statement I is false but statement II is true.

Section B : Numerical Type Questions

21. 4.5 g of compound *A* (MW = 90) was used to make 250 mL of its aqueous solution. The molarity of the solution in *M* is $x \times 10^{-1}$. The value of *x* is .... (Rounded off to the nearest integer).

22. The coordination number of an atom in a body centered cubic structure is ....

[Assume that the lattice is made up of atoms.]

23. A proton and a Li^{3+} nucleus are accelerated by the same potential. If λ_{Li} and λ_P denote the de-Broglie wavelengths of Li^{3+} and proton respectively, then the value of $\frac{\lambda Li}{\lambda_P}$ is $x \times 10^{-1}$ The value of *x* is ....

(Rounded off to the nearest integer)
(Mass of Li^{3+} = 8.3 mass of proton)

24. For the reaction, $A(g) \longrightarrow B(g)$, the value of the equilibrium constant at 300 K and 1 atm is equal to 100.0. The value of $\Delta_r G$ for the reaction at 300 K and 1 atm in J mol^{-1} is $- xR$, where *x* is .... (Rounded off to the nearest integer)

($R = 8.31$ J mol^{-1} K^{-1} and $\ln 10 = 2.3$)

25. When 9.45 g of $ClCH_2COOH$ is added to 500 mL of water, its freezing point drops by 0.5°C. The dissociation constant of $ClCH_2COOH$ is $x \times 10^{-3}$. The value of *x* is ....
(Rounded off to the nearest integer)
[$K_{f(H_2O)} = 1.86$ K kg mol^{-1}]

26. At 1990 K and 1 atm pressure, there are equal number of Cl_2 molecules and Cl atoms in the reaction mixture. The value of k_p for the reaction $Cl_2(g) \rightleftharpoons 2Cl(g)$ under the above conditions is $x \times 10^{-1}$ The value of *x* is ....
(Rounded off to the nearest integer)

27. The reaction of sulphur in alkaline medium is given below

$$S_8(s) + aOH^-(aq) \longrightarrow bS^{2-}(aq) + cS_2O_3^{2-}(aq) + dH_2O(l)$$

The value of '*a*' is .... (Integer answer)

28. Gaseous cyclobutene isomerises to butadiene in a first order process which has a '*k*' value of 3.3×10^{-4} s^{-1} at 153°C. The time in minutes it takes for the isomerisation to proceed 40 % to completion at this temperature is ....
(Rounded off to the nearest integer)

29. Number of amphoteric compounds among the following is ....

a. BeO **b.** BaO **c.** $Be(OH)_2$ **d.** $Sr(OH)_2$

30. The stepwise formation of $[Cu(NH_3)_4]^{2+}$ is given below

$$Cu^{2+} + NH_3 \xrightleftharpoons{K_1} [Cu(NH_3)]^{2+}$$

$$[Cu(NH_3)]^{2+} + NH_3 \xrightleftharpoons{K_2} [Cu(NH_3)_2]^{2+}$$

$$[Cu(NH_3)_2]^{2+} + NH_3 \xrightleftharpoons{K_3} [Cu(NH_3)]^{2+}$$

$$[Cu(NH_3)_3]^{2+} + NH_3 \xrightleftharpoons{K_4} [Cu(NH_3)_4]^{2+}$$

The value of stability constants K_1, K_2, K_3 and K_4 are 10^4, 1.58×10^3, 5×10^2 and 10^2 respectively. The overall equilibrium constants for dissociation of $[Cu(NH_3)]^{2+}$] is $x \times 10^{-12}$. The value of *x* is .... (Rounded off to the nearest integer)

Answers

For solution scan the QR code

1. *(a)*	2. *(d)*	3. *(a)*	4. *(d)*	5. *(b)*	6. *(d)*	7. *(d)*	8. *(a)*	9. *(b)*	10. *(b)*
11. *(b)*	12. *(b)*	13. *(d)*	14. *(d)*	15. *(d)*	16. *(c)*	17. *(d)*	18. *(d)*	19. *(d)*	20. *(b)*
21. *(2)*	22. *(8)*	23. *(2)*	24. *(1380)*	25. *(34.4)*	26. *(5)*	27. *(12)*	28. *(26)*	29. *(2)*	30. *(1.26)*

24 FEBRUARY SHIFT II

Section A : Objective Type Questions

1. What is the correct sequence of reagents used for converting nitrobenzene into *m*-dibromobenzene?

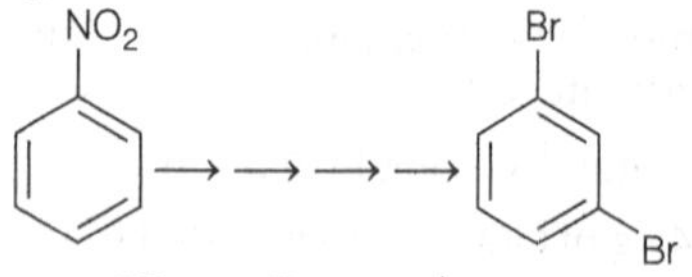

a. $\xrightarrow{NaNO_2}$ / $\xrightarrow{HCl}$ / $\xrightarrow{KBr}$ / $\xrightarrow{H^+}$

b. $\xrightarrow{Br_2/Fe}$ / $\xrightarrow{Sn/HCl}$ / $\xrightarrow{NaNO_2/HCl}$ / $\xrightarrow{CuBr/HBr}$

c. $\xrightarrow{Sn/HCl}$ / $\xrightarrow{KBr}$ / $\xrightarrow{Br_2}$ / $\xrightarrow{H^+}$

d. $\xrightarrow{Sn/HCl}$ / $\xrightarrow{Br_2}$ / $\xrightarrow{NaNO_2}$ / $\xrightarrow{NaBr}$

2. Most suitable salt which can be used for efficient clotting of blood will be

a. $NaHCO_3$ **b.** $FeSO_4$

c. $Mg(HCO_3)_2$ **d.** $FeCl_3$

3. The correct order of the following compounds showing increasing tendency towards nucleophilic substitution reaction is

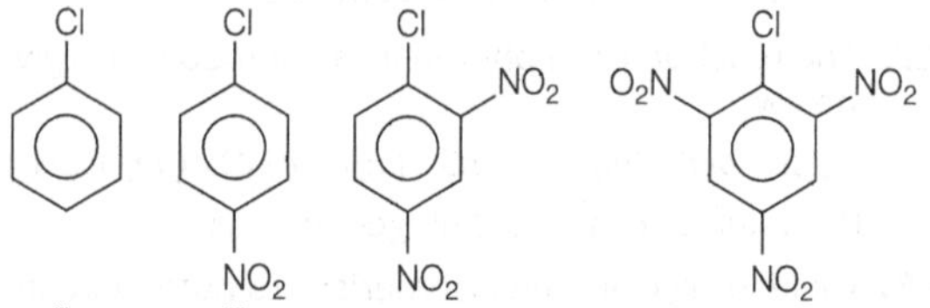

(i) (ii) (iii) (iv)

a. (iv) < (iii) < (ii) < (i) **b.** (iv) < (i) < (ii) < (iii)

c. (iv) < (i) < (iii) < (ii) **d.** (i) < (ii) < (iii) < (iv)

4. According to Bohr's atomic theory,

I. kinetic energy of electron is $\propto \frac{Z^2}{n^2}$

II. the product of velocity (*v*) of electron and principal quantum number (*n*), '*vn*' $\propto Z^2$.

III. frequency of revolution of electron in an orbit is $\propto \frac{Z^3}{n^3}$

IV. coulombic force of attraction on the electron is $\propto \frac{Z^3}{n^4}$

Choose the most appropriate answer from the options given below.

a. Only III

b. Only I

c. I, III and IV

d. I and IV

5. Match List-I with List-II.

	List-I		List-II
A.	$R—\overset{O}{\overset{\|\|}{C}}—Cl \longrightarrow R—CHO$	1.	Br_2/ NaOH
B.	$R—CH_2—COOH \longrightarrow R—\underset{Cl}{\underset{\|}{C}H}—COOH$	2.	H_2/ Pd - $BaSO_4$
C.	$R—\overset{O}{\overset{\|\|}{C}}—NH_2 \longrightarrow R—NH_2$	3.	Zn(Hg) / Conc.HCl
D.	$R—\overset{O}{\overset{\|\|}{C}}—CH_3 \longrightarrow R—CH_2—CH_3$	4.	Cl_2 / Red P, H_2O

Choose the correct answer from the options given below.

	A	B	C	D
a.	2	1	4	3
b.	3	4	1	2
c.	2	4	1	3
d.	3	1	4	2

6. The calculated magnetic moments (spin only value) for species $[FeCl_4]^{2-}$, $[Co(C_2O_4)_3]^{3-}$ and MnO_4^{2-} respectively are

a. 5.82, 0 and 0 BM

b. 4.90, 0 and 1.73 BM

c. 5.92, 4.90 and 0 BM

d. 4.90, 0 and 2.83 BM

7. Match List-I with List-II.

	List-I (Salt)		List-II (Flame colour wavelength)
A.	LiCl	1.	455.5 nm
B.	NaCl	2.	670.8 nm
C.	RbCl	3.	780.0 nm
D.	CsCl	4.	589.2 nm

Choose the correct answer from the options given below.

	A	B	C	D
a.	4	2	3	1
b.	2	1	4	3
c.	1	4	2	3
d.	2	4	3	1

8. Which one of the following carbonyl compounds cannot be prepared by addition of water on an alkyne in the presence of $HgSO_4$ and H_2SO_4?

a. $CH_3—C(=O)—H$

b. 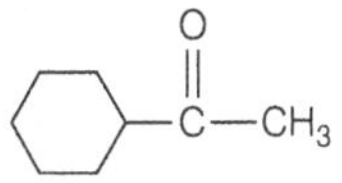

c. $CH_3—CH_2—C(=O)—H$

d. $CH_3—C(=O)—CH_2CH_3$

9. In polymer buna-S: 'S' stands for

a. sulphonation **b.** strength
c. sulphur **d.** styrene

10. 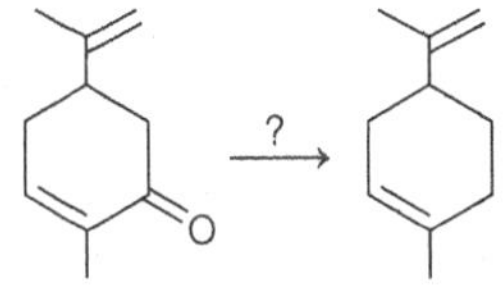

Which of the following reagent is suitable for the preparation of the product in the above reaction?

a. $NaBH_4$ **b.** $NH_2—NH_2 / C_2H_5\overset{\ominus}{O}\overset{\oplus}{Na}$
c. Ni / H_2 **d.** Red P + Cl_2

11. Match List-I and List-II.

	List-I		List-II
A.	Valium	1.	Antifertility drug
B.	Morphine	2.	Pernicious anaemia
C.	Norethindrone	3.	Analgesic
D.	Vitamin B_{12}	4.	Tranquiliser

Choose the correct answer from the option given below.

	A	B	C	D
a.	4	3	2	1
b.	4	3	1	2
c.	2	4	3	1
d.	1	3	4	2

12. Match List-I with List-II.

	List I (Metal)		List II (Ores)
A.	Aluminium	1.	Siderite
B.	Iron	2.	Calamine
C.	Copper	3.	Kaolinite
D.	Zinc	4.	Malachite

Choose the correct answer from the options given below.

	A	B	C	D
a.	4	3	2	1
b.	2	4	1	3
c.	1	2	3	4
d.	3	1	4	2

13. Which one of the following compounds is non-aromatic?

a. 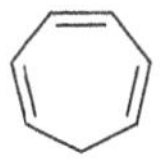

b. 

c. 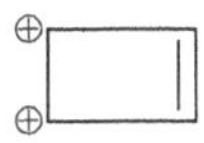

d. 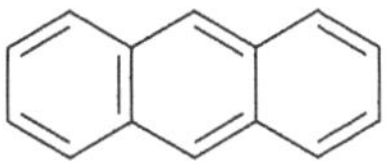

14. What is the correct order of the following elements with respect to their density?

a. Cr < Zn < Co < Cu < Fe **b.** Zn < Cu < Co < Fe < Cr
c. Zn < Cr < Fe < Co < Cu **d.** Cr < Fe < Co < Cu < Zn

15. Given below are two statements.

Statement I The value of the parameter "Biochemical Oxygen Demand (BOD)" is important for survival of aquatic life.

Statement II The optimum value of BOD is 6.5 ppm.

In the light of the above statements, choose the most appropriate answer from the options given below.

a. Statement I is false but statement II is true.
b. Both statements are true.
c. Statement I is true but statement II is false.
d. Both statements are false.

16. The incorrect statement among the following is

a. $VOSO_4$ is a reducing agent.
b. Cr_2O_3 is an amphoteric oxide.
c. RuO_4 is an oxidising agent.
d. Red colour of ruby is due to the presence of Co^{3+}.

17. The correct shape and I—I—I bond angles respectively in I_3^- ion are

a. distorted trigonal planar, 135° and 90°
b. T-shaped, 180° and 90°
c. Trigonal planar, 120°
d. Linear, 180°

18. Given below are two statements.

One is labelled as Assertion A and the other is labelled as Reason R.

Assertion A Hydrogen is the most abundant element in the universe, but it is not the most abundant gas in the troposphere.

Reason R Hydrogen is the lightest element. In the light of the above statements, choose the correct answer from the options given below.

a. A is true but R is false.
b. Both A and R are true and R is the correct explanation of A.
c. A is false but R is true.
d. Both A and R are true but R is not the correct explanation of A.

19. The diazonium salt of which of the following compounds will form a coloured dye on reaction with β-naphthol in NaOH?

a. 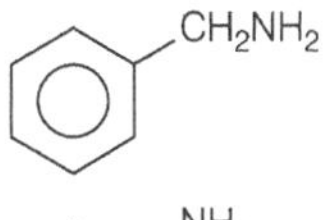

b. 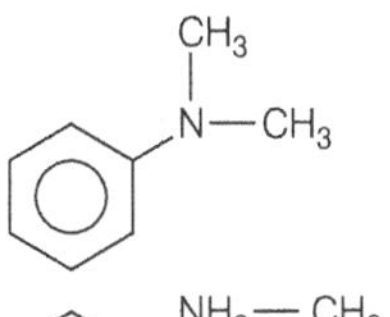

c. 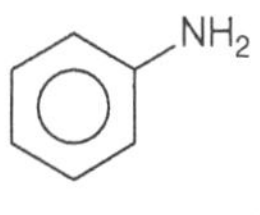

d. 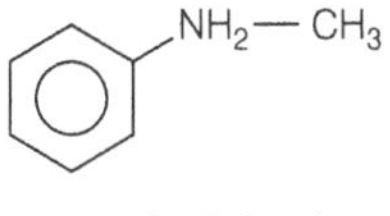

20. The correct set from the following in which both pairs are in correct order of melting point is

a. LiF > LiCl, MgO > NaCl **b.** LiCl > LiF, NaCl > MgO
c. LiF > LiCl, NaCl > MgO **d.** LiCl > LiF, MgO > NaCl

Section B : Numerical Type Questions

21. The total number of amines among the following which can be synthesised by Gabriel synthesis is ....

a. 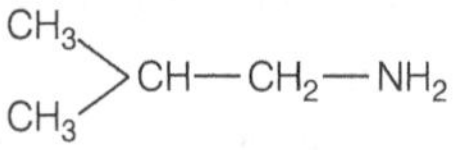

b. $CH_3CH_2NH_2$

c. 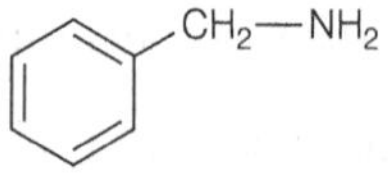

d. 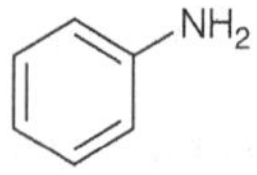

22. Among the following allotropic forms of sulphur, the number of allotropic forms, which will show paramagnetism is ....

a. α-sulphur **b.** β-sulphur **c.** S_2-form

23. The formula of a gaseous hydrocarbon, which requires 6 times of its own volume of O_2 for complete oxidation and produces 4 times its own volume of CO_2 is C_xH_y. The value of y is ....

24. The volume occupied by 4.75 g of acetylene gas at 50°C and 740 mm Hg pressure is ......L (Rounded off to the nearest integer).

[Given, $R = 0.0826$ L atm K^{-1} mol^{-1}]

25. C_6H_6 freezes at 5.5°C. The temperature at which a solution 10 g of C_4H_{10} in 200 g of C_6H_6 freeze is ....°C. (The molal freezing point depression constant of C_6H_6 is 5.12°C/m.)

26. The magnitude of the change in oxidising power of the MnO_4^-/Mn^{2+} couple is $x \times 10^{-4}$ V, if the H^+ concentration is decreased from 1 M to 10^{-4} M at 25°C. (Assume concentration of MnO_4^- and Mn^{2+} to be same on change in H^+ concentration). The value of x is .... (Rounded off to the nearest integer).

$\left[\text{Given, } \frac{2.303\,RT}{F} = 0.059\right]$

27. The solubility product of PbI_2 is 8.0×10^{-9}. The solubility of lead iodide in 0.1 molar solution of lead nitrate is $x \times 10^{-6}$ mol/L. The value of x is .... (Rounded off to the nearest integer).

[Given, : $\sqrt{2} = 1.41$]

28. Sucrose hydrolyses in acid solution into glucose and fructose following first order rate law with a half-life of 3.33 h at 25°C. After 9 h, the fraction of sucrose remaining is f. The value of $\log_{10}(1/f)$ is .... $\times 10^{-2}$ (Rounded off to the nearest integer).

[Assume, ln 10 = 2.303, ln 2 = 0.693]

29. 1.86 g of aniline completely reacts to form acetanilide. 10% of the product is lost during purification. Amount of acetanilide obtained after purification (in g) is .... $\times 10^{-2}$.

30. Assuming ideal behaviour, the magnitude of log K for the following reaction at 25°C is $x \times 10^{-1}$. The value of x is .... (Integer answer)

$$3HC \equiv CH(g) \rightleftharpoons C_6H_6(l)$$

[Given, $\Delta_f G^\circ(HC \equiv CH) = -2.04 \times 10^5$ J mol^{-1},
$\Delta_f G^\circ(C_6H_6) = -1.24 \times 10^5$ J mol^{-1}, $R = 8.314$ J K^{-1} mol^{-1}]

Answers

For solution scan the QR code

1. *(b)*	2. *(d)*	3. *(d)*	4. *(d)*	5. *(c)*	6. *(b)*	7. *(d)*	8. *(c)*	9. *(d)*	10. *(b)*
11. *(b)*	12. *(d)*	13. *(a)*	14. *(c)*	15. *(c)*	16. *(d)*	17. *(d)*	18. *(b)*	19. *(c)*	20. *(a)*
21. *(3)*	22. *(1)*	23. *(8)*	24. *(5)*	25. *(1)*	26. *(3776)*	27. *(141)*	28. *(81)*	29. *(243)*	30. *(855)*

25 FEBRUARY SHIFT I

Section A : Objective Type Questions

1. The plots of radial distribution functions for various orbitals of hydrogen atom against 'r' are given below.

(A) 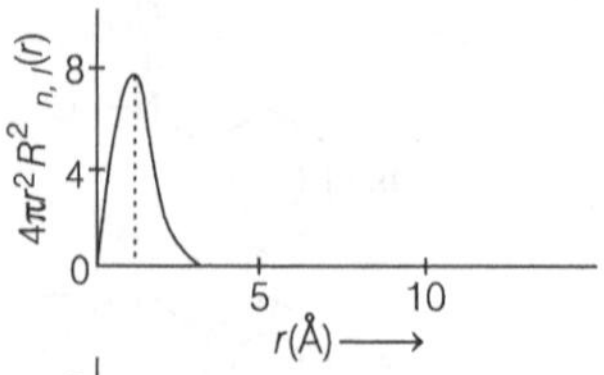

(B) 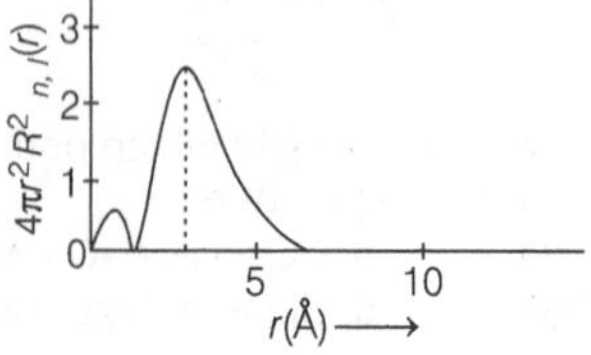

(C) 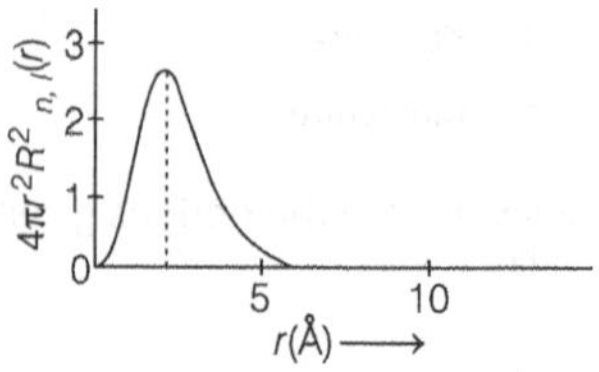

(D) 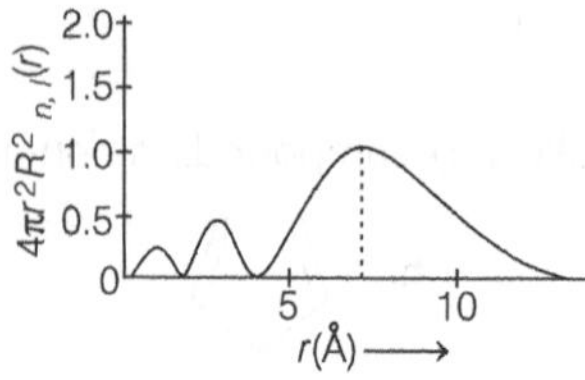

The correct plot for 3s-orbital is

a. (A) **b.** (B) **c.** (C) **d.** (D)

2. According to molecular orbital theory, the species among the following that does not exist is

a. O_2^{2-} **b.** He_2^- **c.** Be_2 **d.** He_2^+

3. The solubility of AgCN in a buffer solution of pH = 3 is x. The value of x is...... .

[Assume : No cyano complex is formed; $K_{sp}(AgCN) = 2.2 \times 10^{-16}$ and $K_a(HCN) = 6.2 \times 10^{-10}$]

a. 0.625×10^{-6} **b.** 1.6×10^{-6}
c. 2.2×10^{-16} **d.** 1.9×10^{-5}

4. In Freundlich adsorption isotherm at moderate pressure, the extent of adsorption $\left(\frac{x}{m}\right)$ is directly proportional to p^x. The value of x is

a. 1 **b.** zero **c.** ∞ **d.** $1/n$

5. Ellingham diagram is a graphical representation of

a. ΔG vs T **b.** ΔH vs T
c. ΔG vs p **d.** $(\Delta G - T\Delta S)$ vs T

6. Which of the following equation depicts the oxidising nature of H_2O_2?

a. $KIO_4 + H_2O_2 \longrightarrow KIO_3 + H_2O + O_2$
b. $I_2 + H_2O_2 + 2OH^- \longrightarrow 2I^- + 2H_2O + O_2$
c. $2I^- + H_2O_2 + 2H^+ \longrightarrow I_2 + 2H_2O$
d. $Cl_2 + H_2O_2 \longrightarrow 2HCl + O_2$

7. The correct statement about B_2H_6 is

(a) all B—H—B angles are of 120°
(b) the two B—H—B bonds are not of same length
(c) terminal B—H bonds have less p-character when compared to bridging bonds
(d) Its fragment, BH_3, behaves as a Lewis base

8. Given below are two statements:

Statement I CeO_2 can be used for oxidation of aldehydes and ketones.

Statement II Aqueous solution of $EuSO_4$ is a strong reducing agent.

In the light of the above statements, choose the correct answer from the options given below.

(a) Both statement I and statement II are true.
(b) Both statement I and statement II are false.
(c) Statement I is true but statement II is false.
(d) Statement I is false but statement II is true.

9. In which of the following pairs, the outer most electronic configuration will be the same?

a. V^{2+} and Cr^+ **b.** Cr^+ and Mn^{2+}
c. Ni^{2+} and Cu^+ **d.** Fe^{2+} and Co^+

10. The hybridisation and magnetic nature of $[Mn(CN)_6]^{4-}$ and $[Fe(CN)_6]^{3-}$, respectively are

a. d^2sp^3 and paramagnetic **b.** sp^3d^2 and diamagnetic
c. d^2sp^3 and diamagnetic **d.** sp^3d^2 and paramagnetic

11. Given below are two statements:

Statement I An allotrope of oxygen is an important intermediate in the formation of reducing smog.

Statement II Gases such as oxides of nitrogen and sulphur present in troposphere contribute to the formation of photochemical smog.

In the light of the above statements, choose the correct answer from the options given below.

(a) Both statement I and statement II are true.
(b) Both statement I and statement II are false.
(c) Statement I is true but statement II is false.
(d) Statement I is false but statement II is true.

12. Complete combustion of 1.80 g of an oxygen containing compound ($C_xH_yO_z$) gave 2.64 g of CO_2 and 1.08 g of H_2O. The percentage of oxygen in the organic compound is

a. 50.33 **b.** 53.33
c. 63.53 **d.** 51.63

13. Identify A in the given chemical reaction.

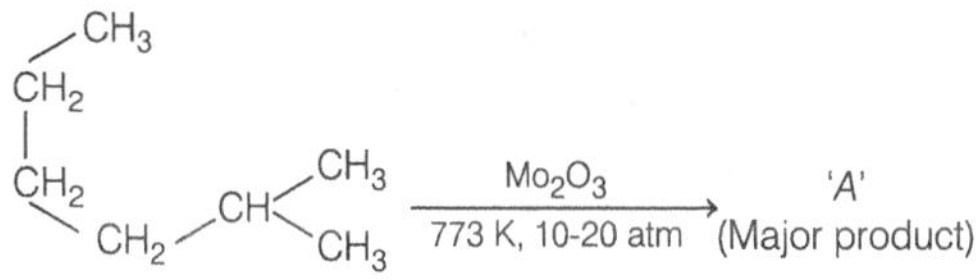

a. 

b. 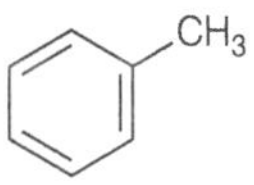

c. 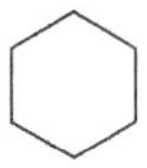

d. 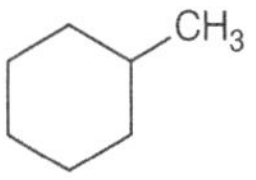

14. Identify A and B in the chemical reaction.

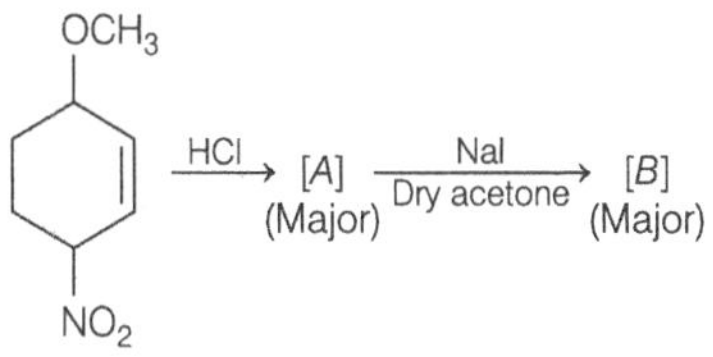

a. $A =$ 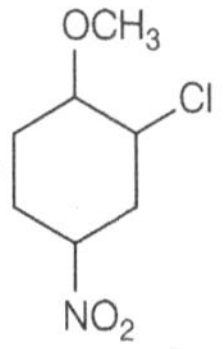

$B =$ 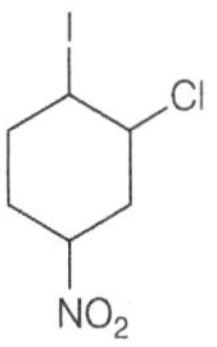

b. $A =$ 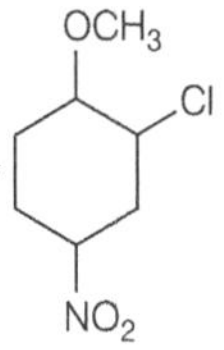

$B =$ 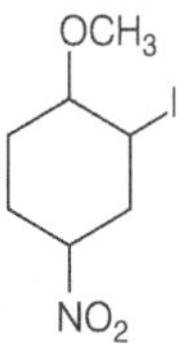

c. $A =$ 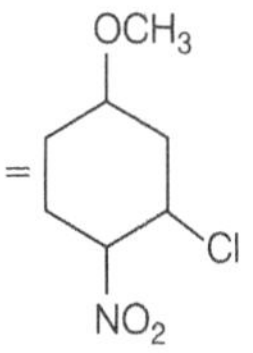

$B =$ 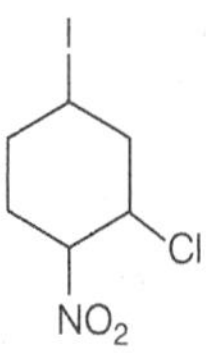

d. $A =$ 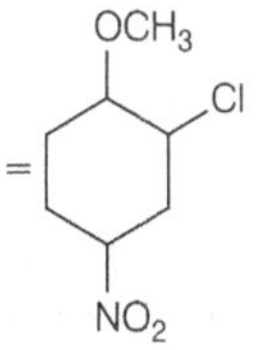

$B =$ 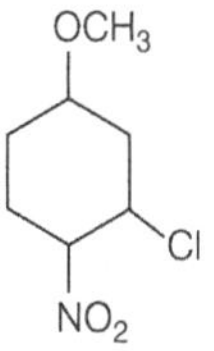

15. The major product of the following chemical reaction is

$$CH_3CH_2CN \xrightarrow[\text{(iii) Pd/BaSO}_4\text{, H}_2]{\text{(i) H}_3\text{O}^+\text{, }\Delta\text{ (ii) SOCl}_2} ?$$

a. $CH_3CH_2CH_2OH$ b. $(CH_3CH_2CO)_2O$
c. $CH_3CH_2CH_3$ d. CH_3CH_2CHO

16. Which one of the following reactions will not form acetaldehyde?

a. $CH_2{=}CH_2 + O_2 \xrightarrow[H_2O]{\text{Pd (II)/Cu (II)}}$ b. $CH_3CH_2OH \xrightarrow{CrO_3\text{-}H_2SO_4}$

c. $CH_3CN \xrightarrow[\text{(ii) H}_2\text{O}]{\text{(i) DIBAL -H}}$ d. $CH_3CH_2OH \xrightarrow[573\text{ K}]{Cu}$

17. Compound(s) which will liberate carbon dioxide with sodium bicarbonate solution is/are

A = 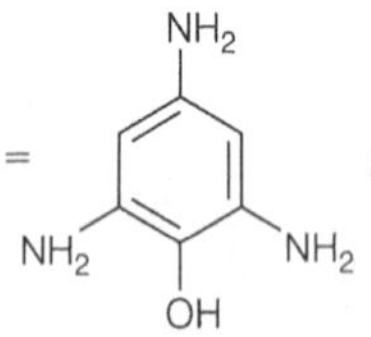

B = 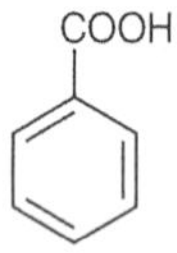

C = 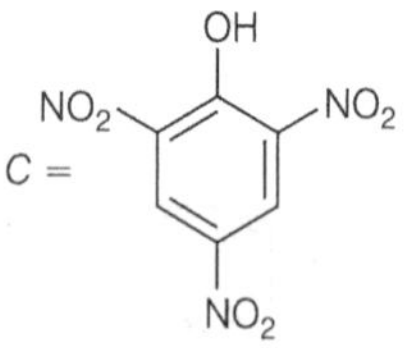

a. *A* and *B* only b. *C* only
c. *B* and *C* only d. *B* only

18. Which of the following reaction(s) will not give *p*-aminoazobenzene?

A = 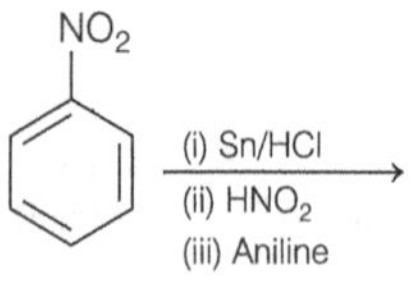

B = 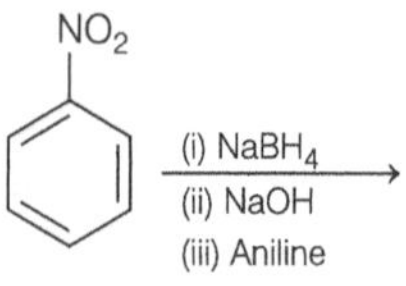

C = 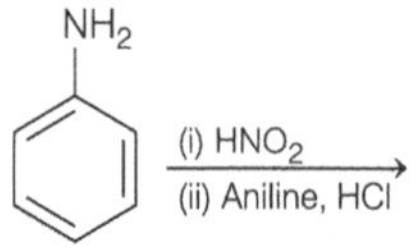

a. *A* only b. *C* only c. *B* only d. *A* and *B*

19. Which statement is correct?

a. Buna-N is a natural polymer.
b. Buna-S is a synthetic and linear thermosetting polymer.
c. Neoprene is an addition copolymer used in plastic bucket manufacturing.
d. Synthesis of buna-S needs nascent oxygen.

20. Which of the glycosidic linkage between galactose and glucose is present in lactose?

a. C-1 of galactose and C-4 of glucose
b. C-1 of galactose and C-6 of glucose
c. C-1 of glucose and C-4 of galactose
d. C-1 of glucose and C-6 of galactose

Section B : Numerical Type Questions

21. 0.4 g mixture of NaOH, Na_2CO_3 and some inert impurities was first titrated with N / 10 HCl using phenolphthalein as an indicator, 17.5 mL of HCl was required at the end point. After this methyl orange was added and titrated. 1.5 mL of same HCl was required for the next end point. The weight percentage of Na_2CO_3 in the mixture is .... (Rounded off to the nearest integer).

22. A car tyre is filled with nitrogen gas at 35 psi at 27°C. It will burst if pressure exceeds 40 psi. The temperature in °C at which the car tyre will burst is .... (Rounded-off to the nearest integer).

23. The reaction of cyanamide, $NH_2CN(s)$ with oxygen was run in a bomb calorimeter and ΔU was found to be -742.24 kJ mol^{-1}. The magnitude of ΔH_{298} for the reaction

$NH_2CN(s) + \frac{3}{2}O_2(g) \longrightarrow N_2(g) + O_2(g) + H_2O(l)$ is .... kJ

(Rounded off to the nearest integer). [Assume ideal gases and $R = 8.314$ J mol^{-1} K^{-1}]

24. 1 molal aqueous solution of an electrolyte A_2B_3 is 60% ionised. The boiling point of the solution at 1 atm is .... K (Rounded off to the nearest integer). [Given, K_b for $(H_2O) = 0.52$ K kg mol^{-1}]

25. In basic medium CrO_4^{2-} oxidises $S_2O_3^{2-}$ to form SO_4^{2-} and itself changes into $Cr(OH)_4^-$. The volume of 0.154 M CrO_4^{2-} required to react with 40 mL of 0.25 M $S_2O_3^{2-}$ is .... mL (Rounded off to the nearest integer).

26. For the reaction, $aA + bB \rightarrow cC + dD$, the plot of $\log k$ vs $\frac{1}{T}$ is given below

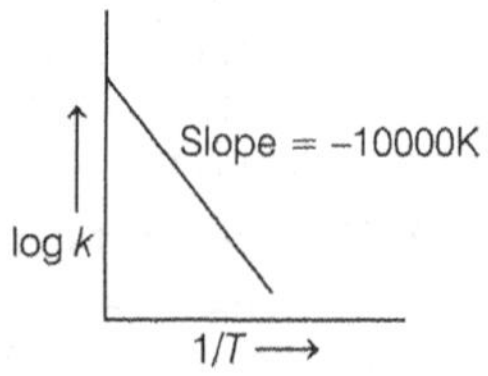

The temperature at which the rate constant of the reaction is $10^{-4}s^{-1}$ is .... K

(Rounded off to the nearest integer).

[Given : The rate constant of the reaction is $10^{-5}s^{-1}$ at 500 K]

27. The ionisation enthalpy of Na^+ formation from $Na(g)$ is 495.8 kJ mol^{-1}, while the electron gain enthalpy of Br is -325.0 kJ mol^{-1}. Given, the lattice enthalpy of NaBr is -728.4 kJ mol^{-1}. The energy for the formation of NaBr ionic solid is (–) .... $\times 10^{-1}$ kJ mol^{-1}.

28. Among the following, the number of halide(s) which is/are inert to hydrolysis is ....

a. BF_3 b. $SiCl_4$ c. PCl_5 d. SF_6

29. Consider the following chemical reaction.

$$CH \equiv CH \xrightarrow[\text{(2) CO, HCl, AlCl}_3]{\text{(1) Red hot Fe tube, 873 K}} \text{Product}$$

The number of sp^2 hybridised carbon atom(s) present in the product is ....

30. Using the provided information in the following paper chromatogram.

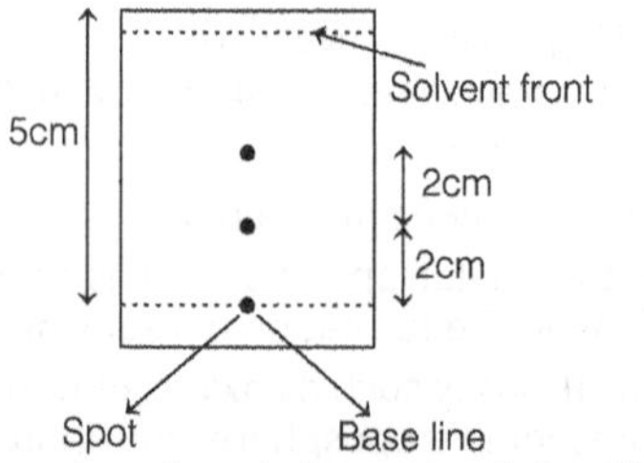

Paper chromatography for compounds *A* and *B*

The calculated R_f value of *A* .... $\times 10^{-1}$.

Answers

For solution scan the QR code

1. *(d)*	2. *(c)*	3. *(d)*	4. *(d)*	5. *(a)*	6. *(c)*	7. *(c)*	8. *(a)*	9. *(b)*	10. *(a)*
11. *(d)*	12. *(b)*	13. *(b)*	14. *(b)*	15. *(d)*	16. *(b)*	17. *(c)*	18. *(c)*	19. *(d)*	20. *(a)*
21. *(4)*	22. *(70)*	23. *(741)*	24. *(375)*	25. *(173)*	26. *(526)*	27. *(5576)*	28. *(1)*	29. *(7)*	30. *(4)*

25 FEBRUARY SHIFT II

Section A : Objective Type Questions

1. Which among the following species has unequal bond lengths ?
a. XeF_4 **b.** SiF_4 **c.** SF_4 **d.** BF_4^-

2. The solubility of $Ca(OH)_2$ in water is

[Given: The solubility product of $Ca(OH)_2$ in water = 5.5×10^{-6}]
a. 1.11×10^{-2} **b.** 1.11×10^{-6}
c. 1.77×10^{-2} **d.** 1.77×10^{-6}

3. Which one of the following statements is false for hydrophilic sols ?
a. They do not require electrolytes for stability.
b. These sols are reversible in nature.
c. Their viscosity is of the order of that of H_2O.
d. The sols cannot be easily coagulated.

4. The correct order of bond dissociation enthalpy of halogens is
a. $F_2 > Cl_2 > Br_2 > I_2$ **b.** $I_2 > Br_2 > Cl_2 > F_2$
c. $Cl_2 > Br_2 > F_2 > I_2$ **d.** $Cl_2 > F_2 > Br_2 > I_2$

5. The method used for the purification of indium is
a. van-Arkel method **b.** liquation
c. zone refining **d.** vapour phase refining

6. Water does not produce CO on reacting with
a. CH_4 **b.** C **c.** CO_2 **d.** C_3H_8

7. Given below are two statements.
Statement I α and β-forms of sulphur can change reversibly between themselves with slow heating or slow cooling.
Statement II At room temperature, the stable crystalline form of sulphur is monoclinic sulphur.
In the light of the above statements, choose the correct answer from the options given below.
a. Both statements I and II are true.
b. Both statements I and II are false.
c. Statement I is true but statement II is false.
d. Statement I is false but statement II is true.

8. The major components of German silver are
a. Cu, Zn and Ag **b.** Cu, Zn and Ni
c. Ge, Cu and Ag **d.** Zn, Ni and Ag

9. In which of the following order the given complex ions are arranged correctly with respect to their decreasing spin only magnetic moment?
(i) $[FeF_6]^{3-}$ (ii) $[Co(NH_3)_6]^{3+}$
(iii) $[NiCl_4]^{2-}$ (iv) $[Cu(NH_3)_4]^{2+}$
a. (i) > (iii) > (iv) > (ii) **b.** (ii) > (iii) > (i) > (iv)
c. (iii) > (iv) > (ii) > (i) **d.** (ii) > (i) > (iii) > (iv)

10. Given below are two statements.
Statement I The pH of rain water is normally ~5.6.
Statement II If the pH of rain water drops below 5.6, it is called acid rain.
In the light of the above statements, choose the correct answer from the options given below.
a. Both statements I and II are true.
b. Both statements I and II are false.
c. Statement I is true but statement II is false.
d. Statement I is false but statement II is true.

11. Which of the following compound is added to the sodium extract before addition of silver nitrate for testing of halogens?
a. Hydrochloric acid **b.** Sodium hydroxide
c. Nitric acid **d.** Ammonia

12. The major product of the following reaction is

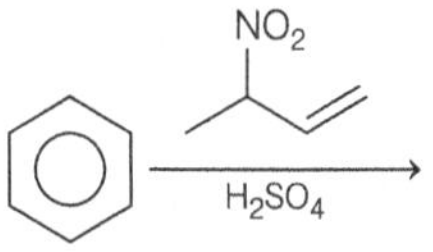

a.
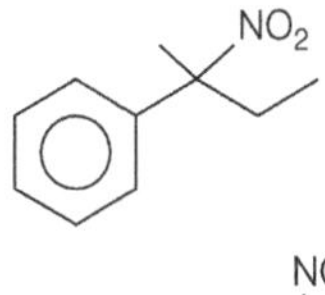

b.
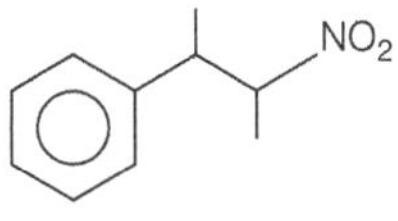

c.
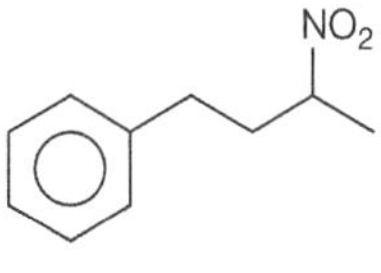

d.
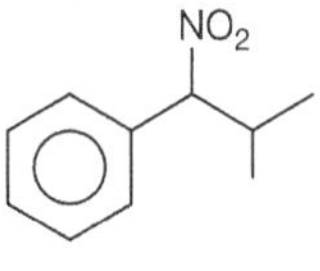

13. The major product of the following reaction is

$$CH_3CH_2CH{=}CH_2 \xrightarrow[\text{Rh catalyst}]{H_2/CO}$$

a. $CH_3CH_2CH_2CH_2CHO$ **b.** $CH_3CH_2CH_2CHO$
c. $CH_3CH_2CH{=}CH{-}CHO$ **d.** $CH_3CH_2\underset{\displaystyle CHO}{\underset{|}{C}}{=}CH_2$

14. The correct sequence of reagents used in the preparation of 4-bromo-2-nitroethyl benzene from benzene is
a. $CH_3COCl / AlCl_3, Br_2 / AlBr_3, HNO_3 / H_2SO_4$, Zn/HCl
b. $CH_3COCl / AlCl_3$, Zn-Hg/HCl, $Br_2 / AlBr_3, HNO_3 / H_2SO_4$
c. $Br_2 / AlBr_3, CH_3COCl / AlCl_3, HNO_3 / H_2SO_4$, Zn/HCl
d. $HNO_3 / H_2SO_4, Br_2 / AlCl_3, CH_3COCl / AlCl_3$, Zn-Hg/HCl

15. The correct order of acid character of the following compounds is

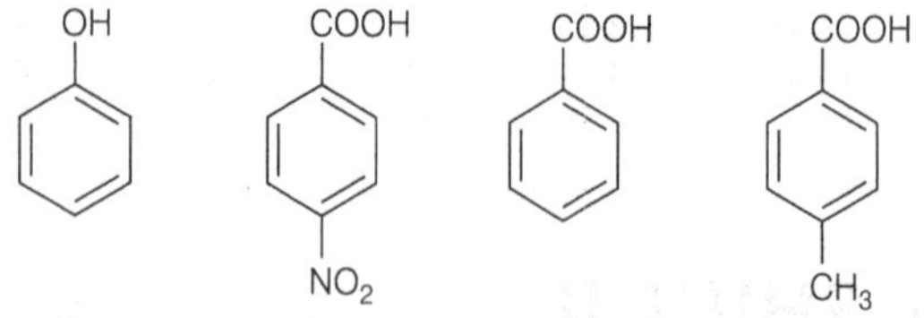

I II III IV

Options

a. I > II > III > IV **b.** III > II > I > IV

c. II > III > IV > I **d.** IV > III > II > I

16. What is '*X*' in the given reaction?

$\underset{CH_2OH}{\overset{CH_2OH}{|}}$ + Oxalic acid $\xrightarrow{210°C}$ X (Major product)

a. $\underset{CHO}{\overset{CHO}{|}}$ **b.** $\underset{CHO}{\overset{CH_2OH}{|}}$ **c.** $\underset{CH_2}{\overset{CH_2}{||}}$ **d.** $\underset{CH_2}{\overset{CH-OH}{||}}$

17.

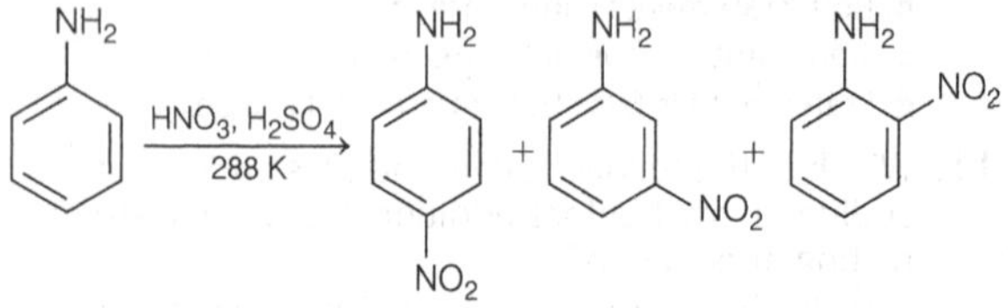

(A) (B) (C)

Correct statement about the given chemical reaction is

a. $-\ddot{N}H_2$ group is *ortho* and *para* directing, so product (*B*) is not possible.

b. reaction is possible and compound (*B*) will be the major product.

c. the reaction will form sulphonated product instead of nitration.

d. reaction is possible and compound (*A*) will be major product.

18. Carbylamine test is used to detect the presence of primary amino group in an organic compound. Which of the following compound is formed when this test is performed with aniline?

a. 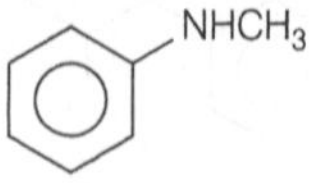 **b.** 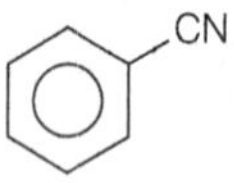

c. 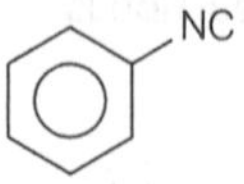 **d.** 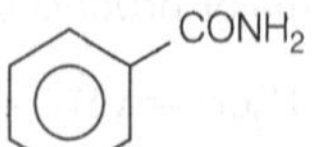

19. Which of the following is correct structure of α-anomer of maltose?

a.

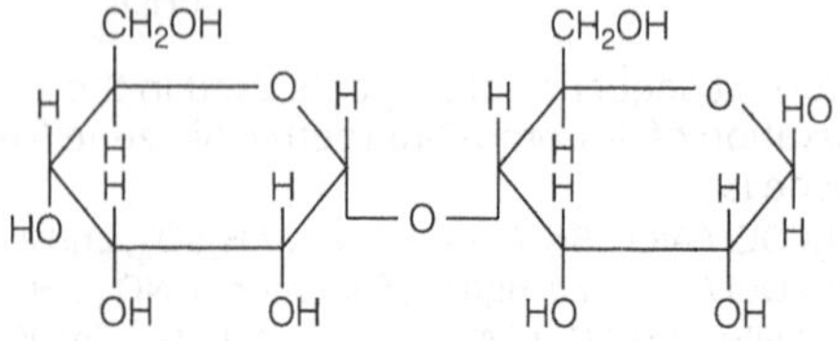

b.

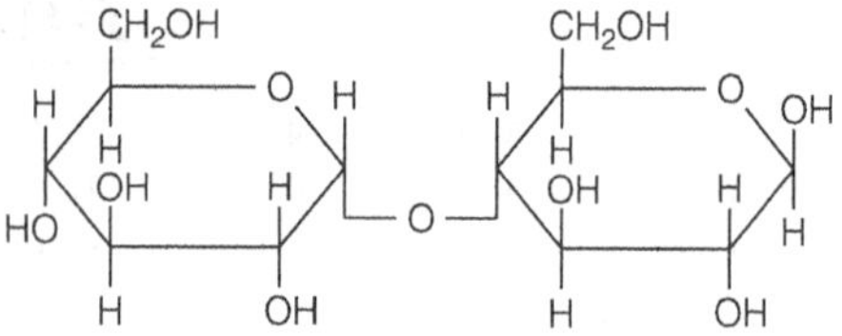

c.

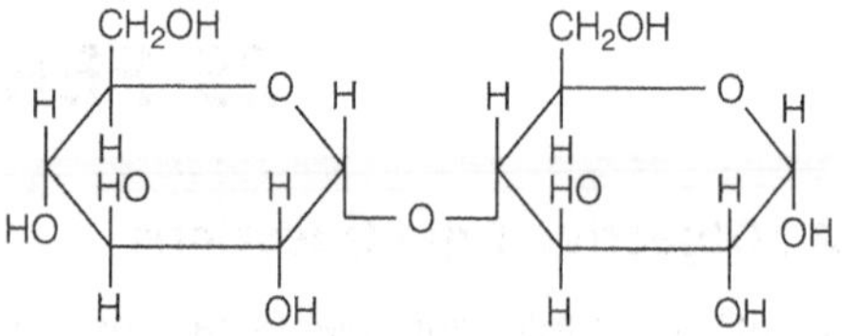

d.

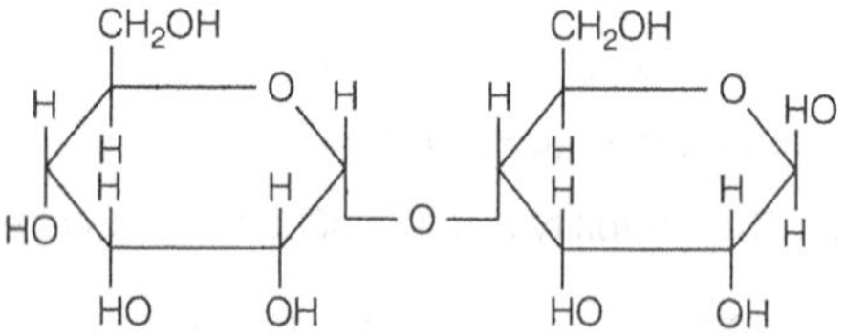

20. Given below are two statements.

Statement I The identification of Ni^{2+} is carried out by dimethyl glyoxime in the presence of NH_4OH.

Statement II The dimethyl glyoxime is a bidentate neutral ligand.

In the light of the above statements, choose the correct answer from the options given below.

a. Both statements I and II are true.

b. Both statements I and II are false.

c. Statement I is true but statement II is false.

d. Statement I is false but statement II is true.

Section B : Numerical Type Questions

21. Consider titration of NaOH solution *versus* 1.25 M oxalic acid solution. At the end point following burette readings were obtained.

(i) 4.5 mL (ii) 4.5 mL

(iii) 4.4 mL (iv) 4.4 mL

(v) 4.4 mL

If the volume of oxalic acid taken was 10.0 mL, then the molarity of the NaOH solution is .... M. (Rounded off to the nearest integer)

22. The unit cell of copper corresponds to a face centered cube of edge length 3.596 Å with one copper atom at each lattice point. The calculated density of copper in kg/m^3 is .... [Molar mass of Cu = 63.54 g ; Avogadro number = 6.022×10^{23}]

23. Electromagnetic radiation of wavelength 663 nm is just sufficient to ionise the atom of metal *A*. The ionisation energy of metal *A* in kJ mol^{-1} is .... (Rounded off to the nearest integer)
[$h = 6.63 \times 10^{-34}$ J-s, $c = 3.00 \times 10^8$ ms^{-1}, $N_A = 6.02 \times 10^{23}$ mol^{-1}]

24. Five moles of an ideal gas at 293 K is expanded isothermally from an initial pressure of 2.1 MPa to 1.3 MPa against at constant external pressure 4.3 MPa. The heat transferred in this process is .... kJ mol^{-1}. (Rounded off to the nearest integer)
[$R = 8.314$ J $mol^{-1}K^{-1}$]

25. If a compound *AB* dissociates to the extent of 75% in an aqueous solution, the molality of the solution which shows a 2.5 K rise in the boiling point of the solution is .... molal. (Rounded off to the nearest integer)
[$K_b = 0.52$ K kg mol^{-1}]

26. Copper reduces NO_3^- into NO and NO_2 depending upon the concentration of HNO_3 in solution. (Assuming fixed [Cu^{2+}] and $p_{NO} = p_{NO_2}$), the HNO_3 concentration at which the thermodynamic tendency for reduction of NO_3^- into NO and NO_2 by copper is same is 10^x M. The value of $2x$ is .... (Rounded off to the nearest integer)
[Given, $E^\circ_{Cu^{2+}/Cu} = 0.34$ V, $E^\circ_{NO_3^-/NO} = 0.96$ V, $E^\circ_{NO_3^-/NO_2} = 0.79$ V and at 298 K, $\frac{RT}{F}(2.303) = 0.059$]

27. The rate constant of a reaction increases by five times on increase in temperature from 27°C to 52°C. The value of activation energy in kJ mol^{-1} is .... (Rounded off to the nearest integer)
[$R = 8.314$ J K^{-1} mol^{-1}]

28. Among the following, number of metal(s) which can be used as electrodes in the photoelectric cell is .... (Integer answer)
(i) Li (ii) Na (ii) Rb (iv) Cs

29. The spin only magnetic moment of a divalent ion in aqueous solution (atomic number = 29) is .... BM.

30. The number of compound(s) given below which contain(s) — COOH group is ....
(i) Sulphanilic acid
(ii) Picric acid
(iii) Aspirin
(iv) Ascorbic acid

Answers

For solution scan the QR code

1. (c)	2. (a)	3. (c)	4. (c)	5. (c)	6. (c)	7. (c)	8. (b)	9. (a)	10. (a)
11. (c)	12. (b)	13. (a)	14. (b)	15. (c)	16. (c)	17. (d)	18. (c)	19. (c)	20. (a)
21. (6)	22. (9077)	23. (180)	24. (15)	25. (3)	26. (4)	27. (52)	28. (1)	29. (2)	30. (1)

26 FEBRUARY SHIFT I

Section A : Objective Type Questions

1. The structure of neoprene is

a. 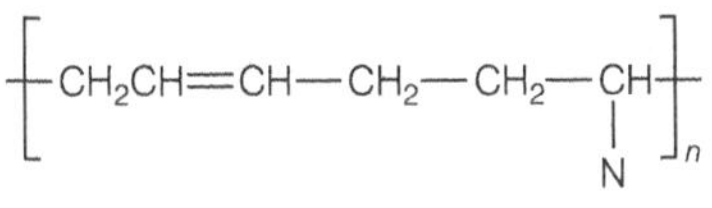

b. 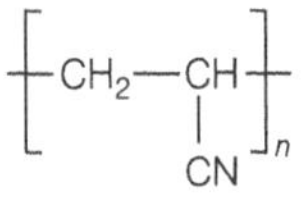

c. 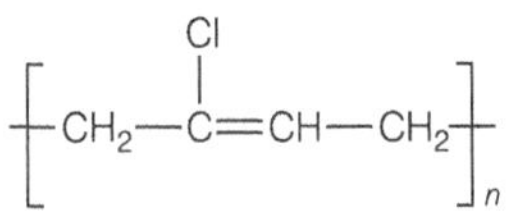

d. 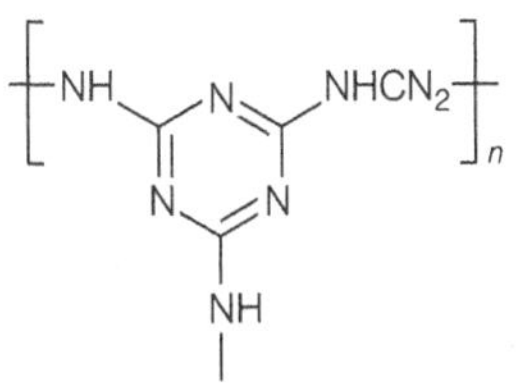

2. Find *A, B* and *C* in the following reactions:
$NH_3 + A + CO_2 \longrightarrow (NH_4)_2CO_3$
$(NH_4)_2CO_3 + H_2O + B \longrightarrow NH_4HCO_3$
$NH_4HCO_3 + NaCl \longrightarrow NH_4Cl + C$

a. $A - O_2$, $B - CO_2$, $C - Na_2CO_3$
b. $A - H_2O$, $B - O_2$, $C - Na_2CO_3$
c. $A - H_2O$, $B - O_2$, $C - NaHCO_3$
d. $A - H_2O$, $B - CO_2$, $C - NaHCO_3$

3. The presence of ozone in troposphere
a. protects us from the UV radiation
b. protects us from the X-ray radiation
c. protects us from green house effect
d. generates photochemical smog

4. Match List-I with List-II.

	List-I (Electronic configuration of elements)		List-II (Δ_i in kJ mol^{-1})
A.	$1s^2 2s^2$	(i)	801
B.	$1s^2 2s^2 2p^4$	(ii)	899
C.	$1s^2 2s^2 2p^3$	(iii)	1314
D.	$1s^2 2s^2 2p^1$	(iv)	1402

Choose the most appropriate answer from the options given below.

	A	B	C	D
a.	(ii)	(iii)	(iv)	(i)
b.	(i)	(iv)	(iii)	(ii)
c.	(i)	(iii)	(iv)	(ii)
d.	(iv)	(i)	(ii)	(iii)

5. Given below are two statements: One is labelled as Assertion (A) and the other is labelled as Reason (R).

Assertion (A) Dipole-dipole interactions are the only non-covalent interactions, resulting in hydrogen bond formation.

Reason (R) Fluorine is the most electronegative element and hydrogen bonds in HF are symmetrical.

In the light of the above statements, choose the most appropriate answer from the options given below.

(a) A is false but R is true.
(b) Both A and R are true and R is the correct explanation of A.
(c) A is true R is false.
(d) Both A and R are true but R is not the correct explanation of A.

6. Statements about heavy water are given below.

A. Heavy water is used in exchange reactions for the study of reaction mechanisms.
B. Heavy water is prepared by exhaustive electrolysis of water.
C. Heavy water has higher boiling point than ordinary water.
D. Viscosity of H_2O is greater than D_2O.

Whic oc the given statement are correct.

a. A, B and C **b.** Only A and B
c. Only A and D **d.** Only A and C

7. The orbital having two radial as well as two angular nodes is

a. 3*p* **b.** 4*f* **c.** 4*d* **d.** 5*d*

8. Match List-I with List-II.

	List-I (Ore)		List-II (Element present)
A.	Kernite	(i)	Tin
B.	Cassiterite	(ii)	Boron
C.	Calamine	(iii)	Fluorine
D.	Cryolite	(iv)	Zinc

Choose the most appropriate answer from the options given below.

	A	B	C	D
a.	(i)	(iii)	(iv)	(ii)
b.	(ii)	(i)	(iv)	(iii)
c.	(ii)	(iv)	(i)	(iii)
d.	(iii)	(i)	(ii)	(iv)

9. Identify the major products *A* and *B* respectively in the following reactions of phenol.

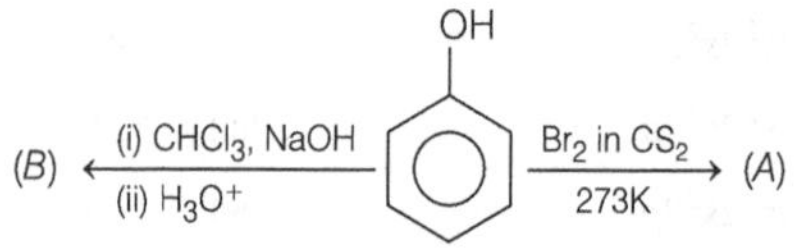

a.
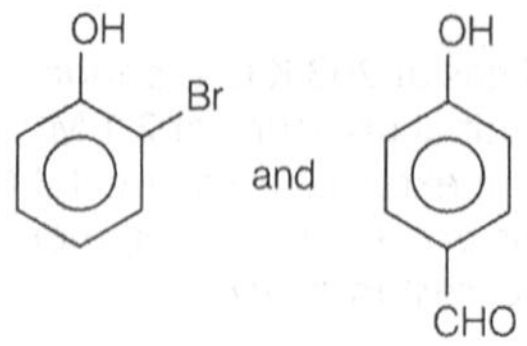

b.
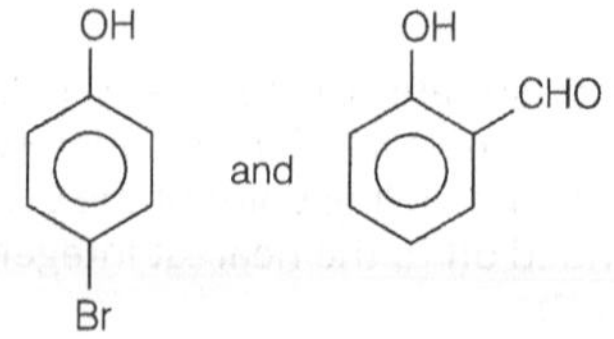

c.
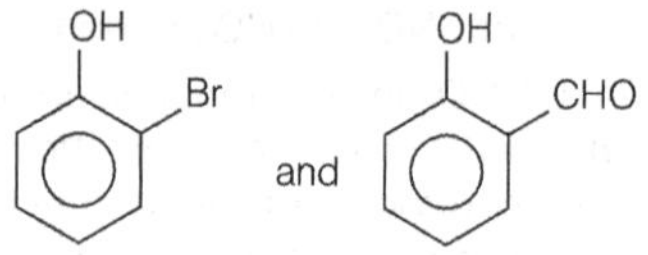

d.
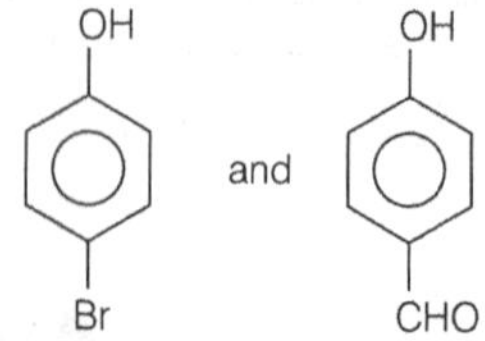

10. Given below are two statements :

Statement I A mixture of chloroform and aniline can be separated by simple distillation.

Statement II When separating aniline from a mixture of aniline and water by steam distillation aniline boils below its boiling point. In the light of the above statements, choose the most appropriate answer from the options given below.

a. Statement I is false but statement II is true
b. Both statement I and statement II are false
c. Statement I is true but statement II is false
d. Both statement I and statement II are true

11. For the given reaction

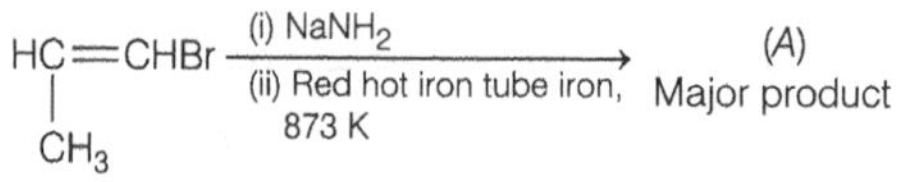

a. $CH_3CH_2CH_2NH_2$ **b.** $\underset{\displaystyle CH_2}{CH}=CH-NH_2$

c.

d.
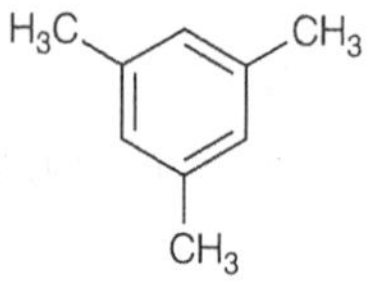

12. On treating a compound with warm dil. H_2SO_4, gas *X* is evolved, which turns $K_2Cr_2O_7$ paper acidified with dil. H_2SO_4 to a green compound *Y*. *X* and *Y* respectively are

a. $X = SO_2, Y = Cr_2O_3$ **b.** $X = SO_3, Y = Cr_2O_3$
c. $X = SO_2, Y = Cr_2(SO_4)_3$ **d.** $X = SO_3, Y = Cr_2(SO_4)_3$

13. Which of the following is a false statement?

(a) Carius tube is used in the estimation of sulphur in an organic compound

(b) Carius method is used for the estimation of nitrogen in an organic compound

(c) Phosphoric acid produced on oxidation of phosphorus present in an organic compound is precipitated as $Mg_2P_2O_7$ by adding magnesia mixture

(d) Kjeldahl's method is used for the estimation of nitrogen in an organic compound

14. Which of the following vitamin is helpful in delaying the blood clotting ?

a. Vitamin C **b.** Vitamin B
c. Vitamin E **d.** Vitamin K

15. $\underset{(C_4H_8Cl_2)}{A} \xrightarrow[373\text{ K}]{\text{Hydrolysis}} \underset{(C_4H_8O)}{B}$

B reacts with hydroxyl amine but does not give Tollen's test. Identify *A* and *B*.

a. 1,1-dichlorobutane and 2-butanone
b. 2,2-dichlorobutane and butanal
c. 1,1-dichlorobutane and butanal
d. 2,2-dichlorobutane and 2-butan-one

16. Compound *A* used as a strong oxidising agent is amphoteric in nature. It is the part of lead storage batteries. Compound *A* is

a. PbO_2 **b.** PbO **c.** $PbSO_4$ **d.** Pb_3O_4

17. Which one of the following lanthanoids does not form MO_2? [*M* is lanthanoid metal]

a. Pr **b.** Dy **c.** Nd **d.** Yb

18. Given below are two statements:

Statement I *o*-nitrophenol is steam volatile due to intramolecular hydrogen bonding.

Statement II *o*-nitrophenol has high melting due to hydrogen bonding.

In the light of the above statements, choose the most appropriate answer from the options given below.

a. Statement I is false but statement II is true
b. Both statement I and statement II are true
c. Both statement I and statement II are false
d. Statement I is true but statement II is false

19. For the given reaction,

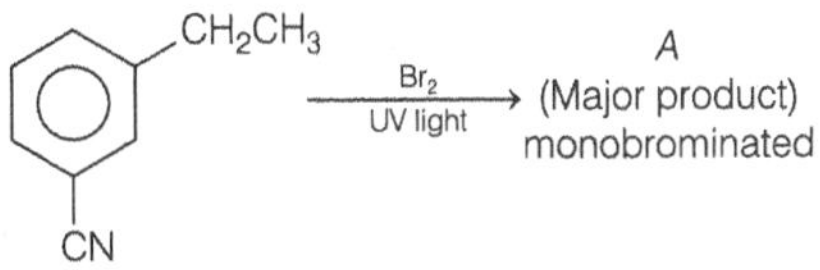

What is *A*?

a. 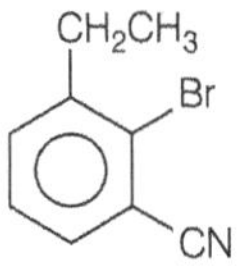

b. 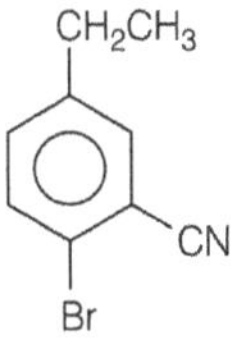

c. 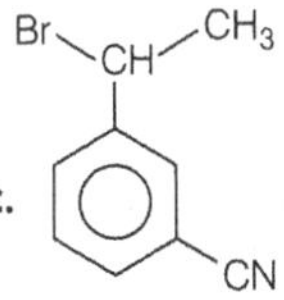

d. 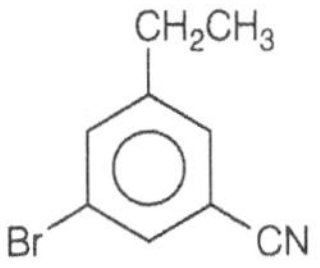

20. An amine on reaction with benzene sulphonyl chloride produces a compound insoluble in alkaline solution. This amine can be prepared by ammonolysis of ethyl chloride. The correct structure of amine is

a. 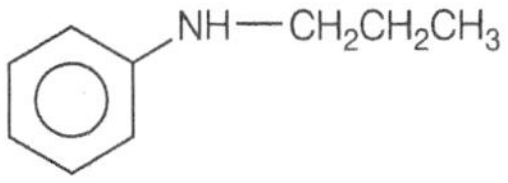

b. $CH_3CH_2NH_2$

c. $CH_3CH_2CH_2NHCH_3$

d. $CH_3CH_2CH_2\overset{H}{N}—CH_2CH_3$

Section B : Numerical Type Questions

21. For a chemical reaction, $A + B \rightleftharpoons C + D$ ($\Delta_r H° = 80$ kJ mol^{-1}) the entropy change $\Delta_r S°$ depends on the temperature *T* (in K) as $\Delta_r S° = 2T$ (JK^{-1} mol^{-1}). Minimum temperature at which it will become spontaneous is .... K.

22. The number of significant figures in 50000.020×10^{-3} is ....

23. An exothermic reaction $X \rightarrow Y$ has an activation energy 30 kJ mol^{-1}. If energy change ΔE during the reaction is –20 kJ mol^{-1}, then the activation energy for the reverse reaction in kJ is ....

24. Consider the following reaction,

$MnO_4^- + 8H^+ + 5e^- \longrightarrow Mn^{2+} + 4H_2O, \quad E° = 1.51$ V.

The quantity of electricity required in Faraday to reduce five moles of MnO_4^- is ....

25. A certain gas obeys $p(V_m - b) = RT$. The value of $\left(\frac{\partial Z}{\partial p}\right)_T$ is xb/RT. The value of *x* is ....
(Z = compressibility factor)

26. A homogeneous ideal gaseous reaction $AB_2(g) \rightleftharpoons A(g) + 2B(g)$ is carried out in a 25 L flask at 27°C. The initial amount of AB_2 was 1 mole and the equilibrium pressure was 1.9 atm. The value of K_p is $x \times 10^{-2}$. The value of *x* is ....

27. Dichromate ion is treated with base, the oxidation number of Cr in the product formed is ....

28. 224 mL of $SO_2(g)$ at 298 K and 1 atm is passed through 100 mL of 0.1 M NaOH solution. The non-volatile solute produced is dissolved in 36 g of water. The lowering of vapour pressure of solution (assuming the solution is dilute), ($p_{(H_2O)}$ = 24 mm of Hg) is $x \times 10^{-2}$ mm of Hg, the value of *x* is ....

29. 3.12 g of oxygen is adsorbed on 1.2 g of platinum metal. The volume of oxygen adsorbed per gram of the adsorbent at 1 atm and 300 K in L is ....
[R = 0.0821 L atm K^{-1} mol^{-1}]

30. Number of bridging CO ligands in $[Mn_2(CO)_{10}]$ is ....

Answers

For solution scan the QR code

1. *(c)*	2. *(d)*	3. *(d)*	4. *(a)*	5. *(a)*	6. *(a)*	7. *(d)*	8. *(b)*	9. *(b)*	10. *(d)*
11. *(d)*	12. *(c)*	13. *(b)*	14. *(d)*	15. *(d)*	16. *(a)*	17. *(d)*	18. *(d)*	19. *(c)*	20. *(d)*
21. *(200)*	22. *(7)*	23. *(50)*	24. *(25)*	25. *(1)*	26. *(73)*	27. *(6)*	28. *(18)*	29. *(2)*	30. *(0)*

26 FEBRUARY SHIFT II

Section A : Objective Type Questions

1. Which of the following forms of hydrogen emits low energy β^- particles ?

a. Deuterium ^2_1H **b.** Tritium ^3_1H
c. Protium ^1_1H **d.** Proton H^+

2. Given below are two statements: one is labelled as Assertion (A) and the other is labelled as Reason (R).

Assertion (A) In TlI_3, isomorphous to CsI_3, the metal is present in +1 oxidation state.

Reason (R) Tl metal has fourteen *f*-electrons in the electronic configuration.

In the light of the above statements, choose the most appropriate answer from the options given below.

a. A is correct but R is not correct.
b. Both A and R are correct and R is the correct explanation of A.
c. A is not correct but R is correct
d. Both A and R are correct but R is not correct explanation of A.

3. Match List-I with List-II.

	List-I		List-II
A.	Sucrose	(i)	β-D-galactose and β-D-glucose
B.	Lactose	(ii)	α-D-glucose and β-D-fructose
C.	Maltose	(iii)	α-D-glucose and α-D-glucose

Choose the correct answer from the options given below.

a. A→ (i), B→ (iii), C→ (ii)
b. A→ (iii), B→ (i), C→ (iii)
c. A→ (ii), B→ (i), C→ (iii)
d. A→ (iii), B→ (ii), C→ (i)

4. A. phenyl methanamine
B. N,N-dimethylaniline
C. N-methyl aniline
D. Benzenamine

Choose the correct order of basic nature of the above amines.

a. A > C > B > D **b.** D > C > B > A
c. D > B > C > A **d.** A > B > C > D

5. The correct order of electron gain enthalpy is

a. S> Se> Te> O **b.** Te > Se > S > O
c. O > S > Se > Te **d.** S > O > Se > Te

6. In $\overset{1}{C}H_2=\overset{2}{C}=\overset{3}{C}H-\overset{4}{C}H_3$ molecule, the hybridisation of carbon 1, 2, 3 and 4 respectively are

a. sp^3, sp, sp^3, sp^3
b. sp^2, sp^2, sp^2, sp^3
c. sp^2, sp, sp^2, sp^3
d. sp^2, sp^3, sp^2, sp^3

7. Seliwanoff test and xanthoproteic test are used for the identification of .... and .... respectively.

a. aldoses, ketoses **b.** proteins, ketoses
c. ketoses, proteins **d.** ketoses, aldoses

8. 2,4-DNP test can be used to identify

a. amine **b.** aldehyde
c. ether **d.** halogens

9. Ceric ammonium nitrate and $CHCl_3$ /alc. KOH are used for the identification of functional groups present in .... and .... respectively.

a. alcohol, phenol
b. amine, alcohol
c. alcohol, amine
d. amine, phenol

10. Which pair of oxides is acidic in nature?

a. B_2O_3, CaO **b.** B_2O_3, SiO_2
c. N_2O, BaO **d.** CaO, SiO_2

11. Identify *A* in the given chemical reaction,

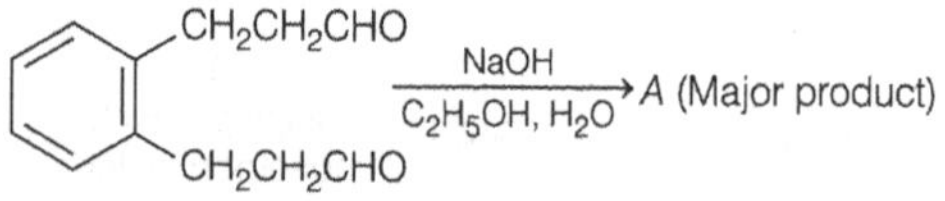

a.
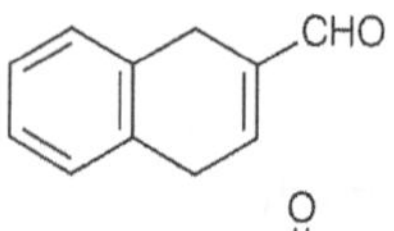

b.
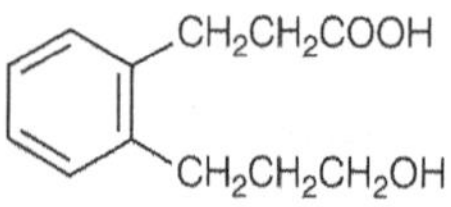

c.
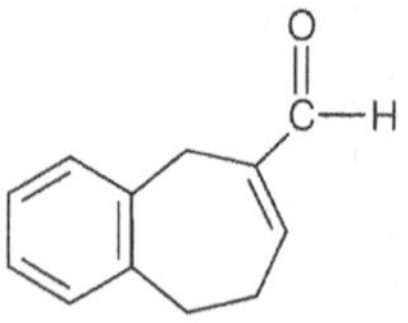

d.
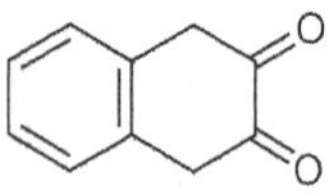

12. Identify *A* in the following chemical reaction.

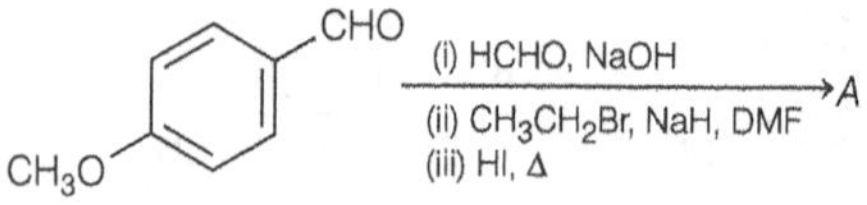

a. 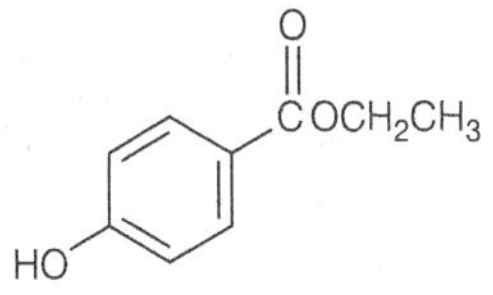
b. 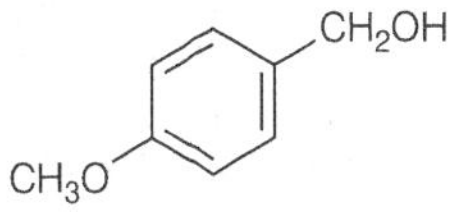
c. 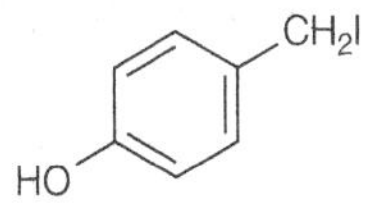
d. 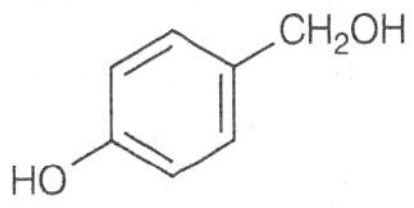

13. Calgon is used for water treatment. Which of the following statement is not true about Calgon?
- **a.** Calgon contains the 2nd most abundant element by weight in the Earth's crust.
- **b.** It is polymeric compound and is water soluble.
- **c.** It is also known as Graham's salt.
- **d.** It does not remove Ca^{2+} ion by precipitation.

14. Match List-I with List-II.

List-I

A. 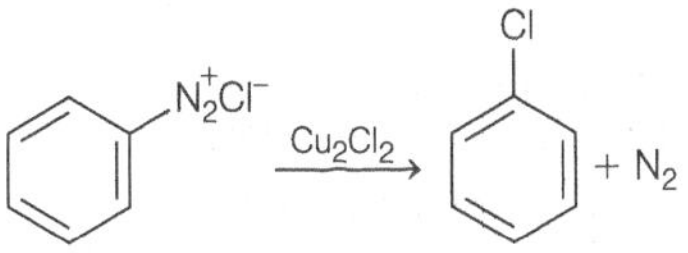

B. 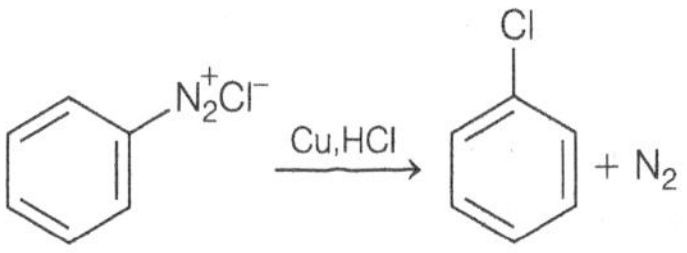

C. $2CH_3CH_2Cl + 2Na \xrightarrow{\text{Ether}} C_2H_5 - C_2H_5 + 2NaCl$

D. $2C_6H_5Cl + 2Na \xrightarrow{\text{Ether}} C_6H_5 - C_6H_5 + 2NaCl$

List-II

(i) Wurtz reaction (ii) Sandmeyer reaction
(iii) Fittig reaction (iv) Gattermann reaction

Choose the correct answer from the options given below.

	A	B	C	D
a.	(iii)	(i)	(iv)	(ii)
b.	(ii)	(i)	(iv)	(iii)
c.	(ii)	(iv)	(i)	(iii)
d.	(iii)	(iv)	(i)	(ii)

15. 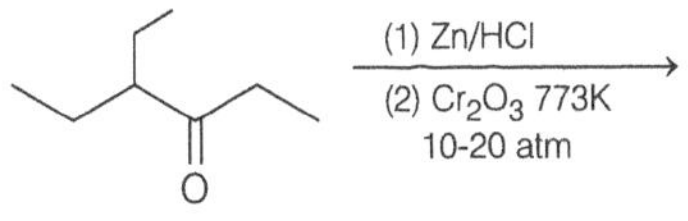

considering the above reaction, the major product among the following is

a. 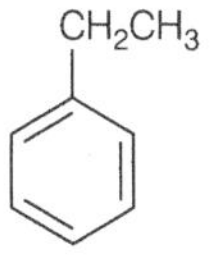
b. 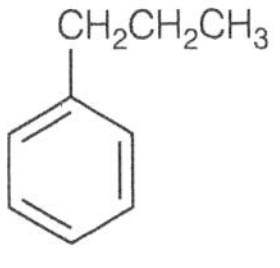
c. 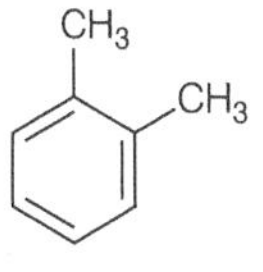
d. 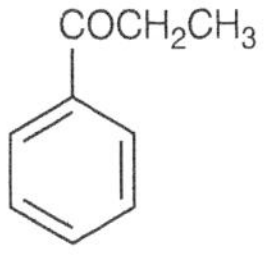

16. Match List-I with List-II

	List-I (Molecule)		List-II (Bond order)
A.	Ne_2	(i)	1
B.	N_2	(ii)	2
C.	F_2	(iii)	0
D.	O_2	(iv)	3

Choose the correct answer from the options given below.

	A	B	C	D
a.	(iii)	(iv)	(i)	(ii)
b.	(i)	(ii)	(iii)	(iv)
c.	(ii)	(i)	(iv)	(iii)
d.	(iv)	(iii)	(ii)	(i)

17. Identify *A* in the given reaction.

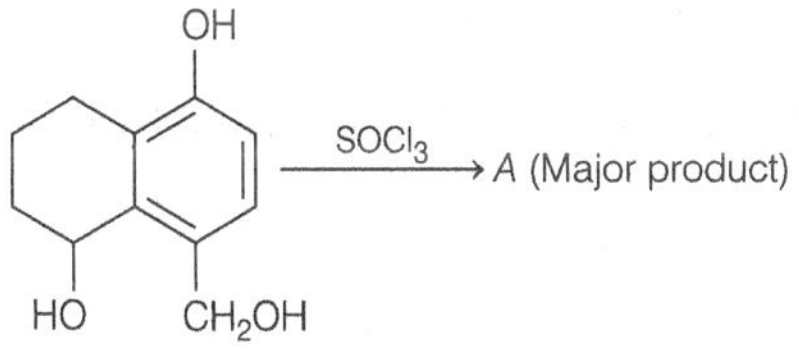

a. 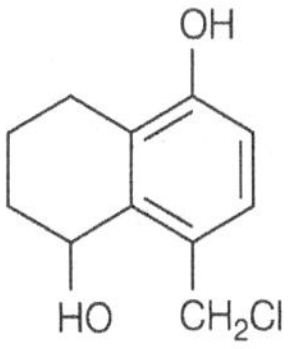
b. 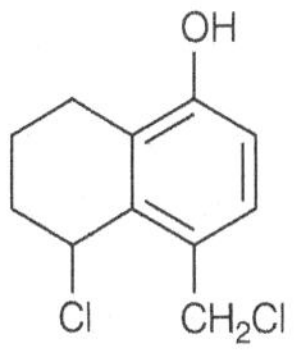
c. 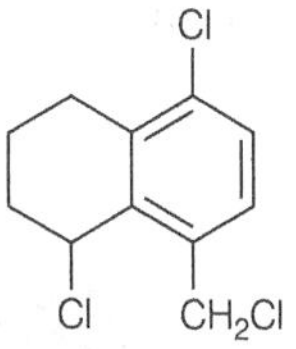
d. 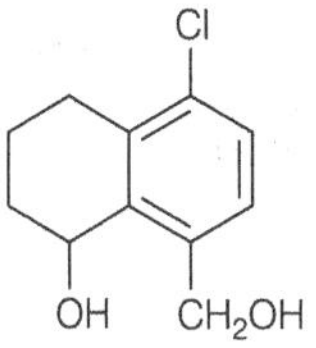

18. Match List-I with List-II.

	List-I		List-II
A.	Siderite	(i)	Cu
B.	Calamine	(ii)	Ca
C.	Malachite	(iii)	Fe
D.	Cryolite	(iv)	Al
		(v)	Zn

Chose the correct answer from the options given below.

	A	B	C	D
a.	(iii)	(i)	(v)	(ii)
b.	(i)	(ii)	(v)	(iii)
c.	(iii)	(v)	(i)	(iv)
d.	(i)	(ii)	(iii)	(iv)

19. The nature of charge on resulting colloidal particles when $FeCl_3$ is added to excess of hot water is
- **a.** Positive
- **b.** Sometimes positive and sometimes negative
- **c.** Neutral
- **d.** Negative

20. Match List-I with List-II

	List-I		List-II
A.	Sodium carbonate	(i)	Deacon
B.	Titanium	(ii)	Castner-Kellner
C.	Chlorine	(iii)	van-Arkel
D.	Sodium hydroxide	(iv)	Solvay

Choose the correct answer form the options given below.

	A	B	C	D
a.	(iv)	(iii)	(i)	(ii)
b.	(i)	(iii)	(iv)	(ii)
c.	(iv)	(i)	(ii)	(iii)
d.	(iii)	(ii)	(i)	(iv)

Section B : Numerical Type Questions

21. The $NaNO_3$ weighed out to make 50 mL of an aqueous solution containing 70.0 mg Na^+ per mL is .... g. (Rounded off to the nearest integer) [Given : Atomic weight in g mol^{-1}, –Na : 23 ; N : 14 ; O : 16].

22. Emf of the following cell at 298 K in V is $x \times 10^{-2}$,
Zn | Zn^{2+} (0.1 M) || Ag^+ (0.01 M) | Ag The value of x is ....
(Rounded off to the nearest integer).
[Given, $E^\circ_{Zn^{2+}/Zn} = -0.76$ V, $E^\circ_{Ag^+/Ag} = +0.80$ V, $\frac{2.303RT}{F} = 0.059$]

23. When 12.2 g of benzoic acid is dissolved in 100 g of water, the freezing point of solution was found to be $-0.93°C$ ($K_f(H_2O) = 1.86$ K kg mol^{-1}). The number (n) of benzoic acid molecules associated (assuming 100% association) is ....

24. The average S—F bond energy in kJ mol^{-1} of SF_6 is .... (Rounded off to the nearest integer)
[Given, the values of standard enthalpy of formation of $SF_6(g)$, $S(g)$ and $F(g)$ are –1100, 275 and 80 kJ mol^{-1} respectively.]

25. A ball weighing 10 g is moving with a velocity of 90 ms^{-1}. If the uncertainity in its velocity is 5%, then the uncertainty in its position is .... $\times 10^{-33}$ m (Rounded off to the nearest integer). [Given, $h = 6.63 \times 10^{-34}$ J-s]

26. The number of octahedral voids per lattice site in a lattice is .... (Rounded off to the nearest integer)

27. In mild alkaline medium, thiosulphate ion is oxidised by MnO_4^- to "A". The oxidation state of sulphur in "A" is ....

28. The number of stereoisomers possible for $[Co(ox)_2(Br)(NH_3)]^{2-}$ is .... [ox = oxalate].

29. If the activation energy of a reaction is 80.9 kJ mol^{-1}, the fraction of molecules at 700 K, having enough energy to react to form products is e^{-x}. The value of x is ....
(Rounded off to the nearest integer)
[Use, $R = 8.31$ JK^{-1} mol^{-1}].

30. The pH of ammonium phosphate solution, if pK_a of phosphoric acid and pK_b of ammonium hydroxide are 5.23 and 4.75 respectively, is ....

Answers

For solution scan the QR code

1. *(b)*	2. *(b)*	3. *(c)*	4. *(d)*	5. *(a)*	6. *(c)*	7. *(c)*	8. *(b)*	9. *(c)*	10. *(b)*
11. *(c)*	12. *(c)*	13. *(a)*	14. *(c)*	15. *(a)*	16. *(a)*	17. *(b)*	18. *(c)*	19. *(a)*	20. *(a)*
21. *(13)*	22. *(147)*	23. *(2)*	24. *(309)*	25. *(1)*	26. *(1)*	27. *(6)*	28. *(3)*	29. *(14)*	30. *(7)*

16 MARCH SHIFT I

Section A : Objective Type Questions

1. Given below are two statements. One is labelled as Assertion (A) and the other is labelled as Reason (R).

Assertion (A) Size of Bk^{3+} ion is less than Np^{3+} ion.

Reason (R) The above is a consequence of the lanthanoid contraction.

In the light of the above statements, choose the correct answer from the options given below

a. A is false but R is true.
b. Both A and R are true but *R* is not the correct explanation of A.
c. Both A and R are true and R is the correct explanation of A.
d. A is true but R is false.

2. Which among the following pair of vitamins is stored in our body relatively for longer duration ?

a. Thiamine and vitamin A **b.** Vitamin A and vitamin D
c. Thiamine and ascorbic acid **d.** Ascorbic acid and vitamin D

3. Given below are two statements.

Statement I Both $CaCl_2 \cdot 6H_2O$ and $MgCl_2 \cdot 8H_2O$ undergo dehydration on heating.

Statement II BeO is amphoteric, whereas the oxides of other elements in the same group are acidic.

In the light of the above statements, choose the correct answer from the options given below.

a. Statement I is false but statement II is true
b. Both statement I and statement II are false
c. Both statement I and statement II are true
d. Statement I is true but statement II is false

4.

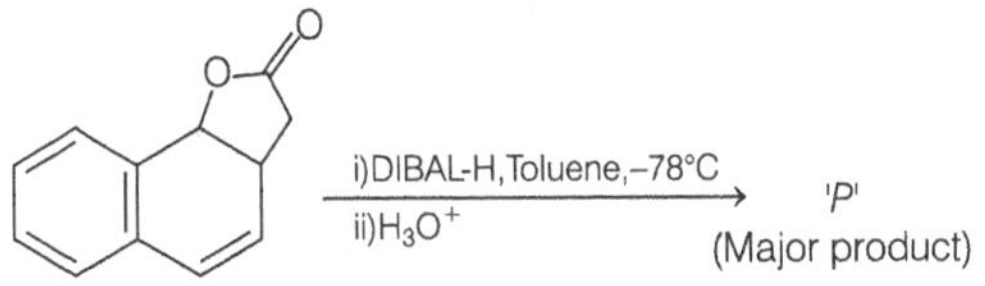

The product *P* in the above reaction is

a.
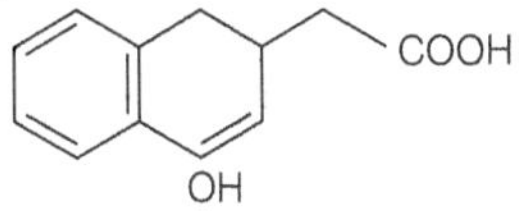

b.
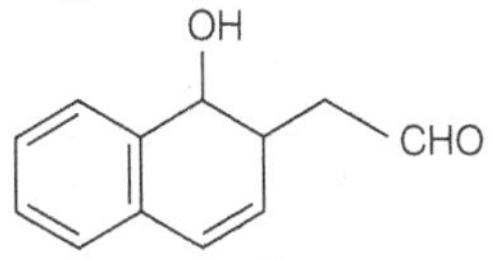

c.
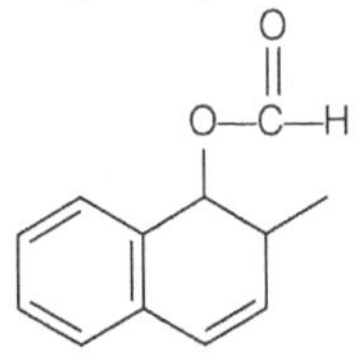

d.
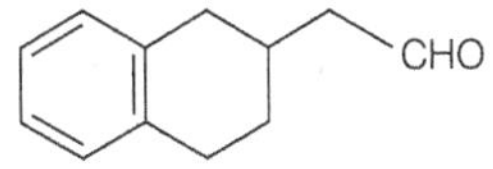

5. Match List -I with List II.

List-I (Industrial process)	List-II (Application)
A. Haber's process	(i) HNO_3 synthesis
B. Ostwald's process	(ii) Aluminium extraction
C. Contact process	(iii) NH_3 synthesis
D. Hall-Heroult process	(iv) H_2SO_4 synthesis

Choose the correct answer from the options given below :

a. A-(ii), B-(iii), C-(iv), D-(i) **b.** A-(iii), B-(iv), C-(i), D-(ii)
c. A-(iii), B-(i), C-(iv), D-(ii) **d.** A-(iv), B-(i), C-(ii), D-(iii)

6. Among the following, the aromatic compounds are

A. 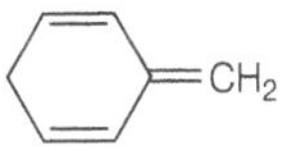

B. 

C. 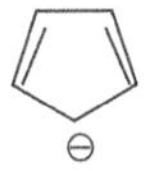
D. 

Choose the correct answer from the following options.

a. Only (A) and (B) **b.** Only (B) and (C)
c. (B), (C) and (D) **d.** (A), (B) and (C)

7.

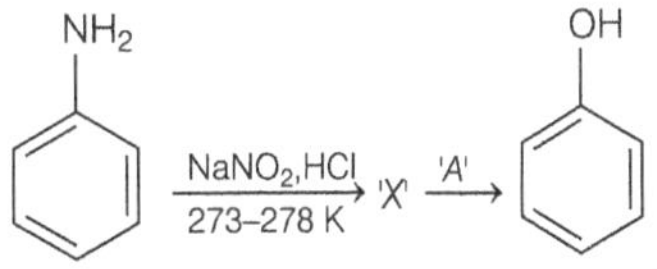

(Major product)

In the above chemical reaction, intermediate *X* and reagent / condition *A* are

a. *X*- 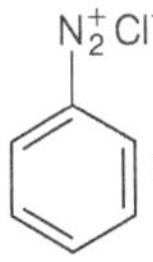

; *A* -H_2O/NaOH

b. *X*- 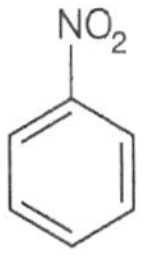

; *A* -H_2O/Δ

c. *X*- 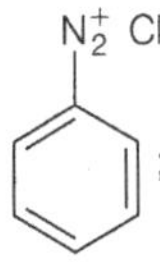

; *A* -H_2O/Δ

d. *X*- 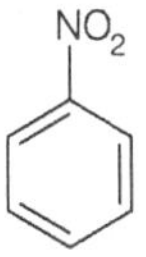

; *A* -H_2O/NaOH

8. Given below are two statements.

Statement I The $E°$ value of Ce^{4+}/Ce^{3+} is $+1.74$ V.

Statement II Ce is more stable in Ce^{4+} state than Ce^{3+} state.

In the light of the above statements, choose the most appropriate answer from the options given below.

a. Both statement I and statement II are correct.
b. Statement I is incorrect but statement II is correct.
c. Both statement I and statement II are incorrect.
d. Statement I is correct but statement II is incorrect.

9. The functions of antihistamine are
a. antiallergic and analgesic
b. antacid and antiallergic
c. analgesic and antacid
d. antiallergic and antidepressant

10. Which of the following is Lindlar catalyst ?
a. Zinc chloride and HCl
b. Cold dilute solution of $KMnO_4$
c. Sodium and liquid NH_3
d. Partially deactivated palladised charcoal

11.

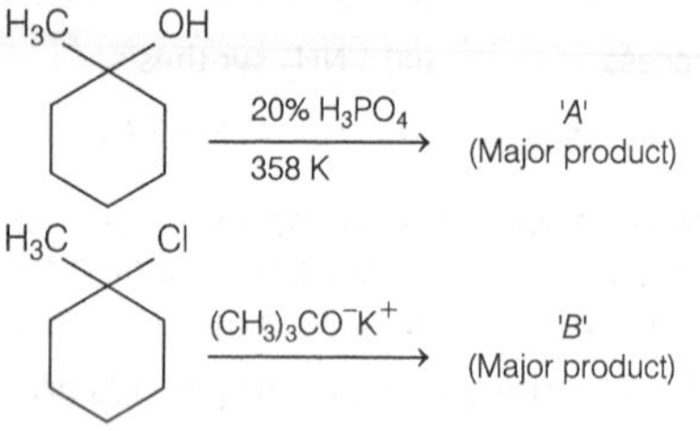

The product *A* and *B* formed in above reactions are

a. 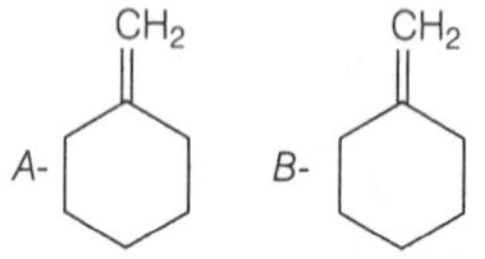

b. 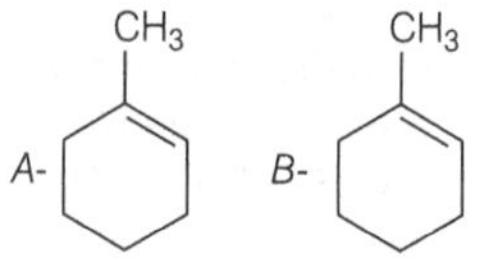

c. 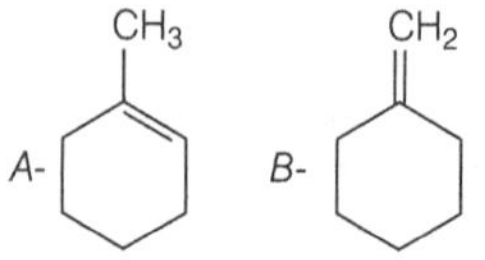

d. 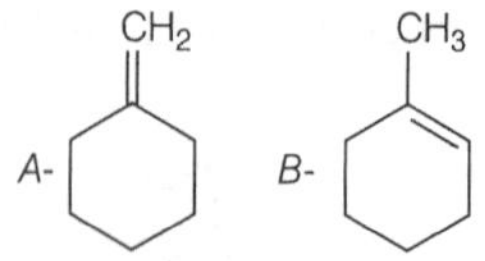

12. Given below are two statements.
Statement I H_2O_2 can act as both oxidising and reducing agent in basic medium.
Statement II In the hydrogen economy, the energy is transmitted in the form of dihydrogen. In the light of the above statements, choose the correct answer from the options given below :
a. Both statement I and statement II are false.
b. Both statement I and statement II are true.
c. Statement I is true but statement II is false.
d. Statement I is false but statement II is true.

13. The type of pollution that gets increased during the day time and in the presence of O_3 is
a. reducing smog **b.** oxidising smog
c. global warming **d.** acid rain

14. Assertion (A) Enol form of acetone $[CH_3COCH_3]$ exists in < 0.1% quantity. However, the enol form of acetyl acetone $[CH_3COCH_2OCCH_3]$ exists in approximately 15% quantity.
Reason (R) Enol form of acetyl acetone is stabilised by intramolecular hydrogen bonding, which is not possible in enol form of acetone.
Choose the correct statement.
a. A is false but R is true.
b. Both A and R are true and R is the correct explanation of A.
c. Both A and R are true but R is not the correct explanation of A.
d. A is true but R is false.

15. Which of the following reaction does not involve Hoffmann bromamide degradation?

a. 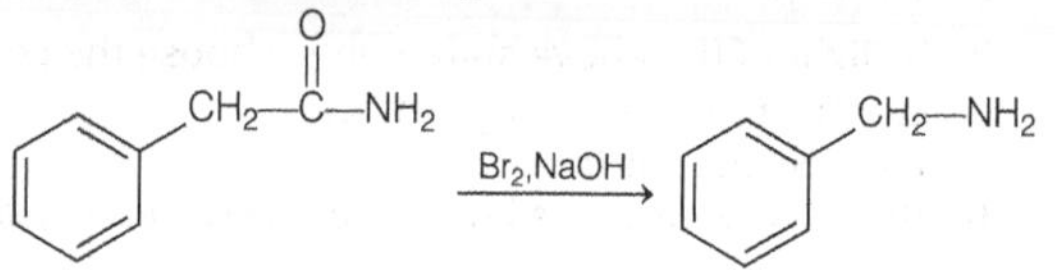

b. 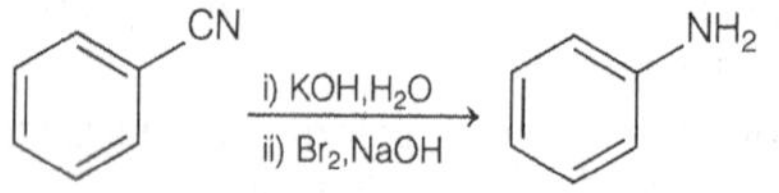

c. 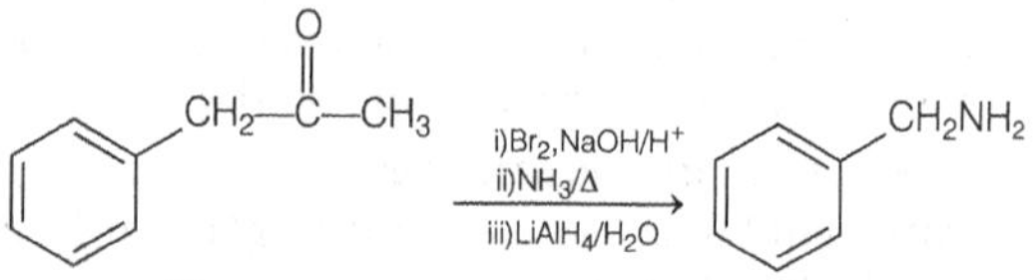

d. 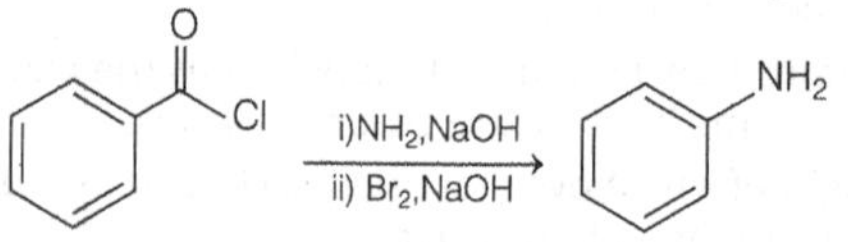

16. The process that involves the removal of sulphur from the ores is
a. smelting **b.** roasting
c. leaching **d.** refining

17. Match List-I with List-II.

List-I (Name of oxo acid)	List-II (Oxidation state of P)
A. Hypophosphorus acid	(i) +5
B. Orthophosphoric acid	(ii) +4
C. Hypophosphoric acid	(iii) +3
D. Orthophosphorus acid	(iv) +2
	(v) +1

Choose the correct answer from the options given below :
a. A-(v), B-(i), C-(ii), D-(iii) **b.** A-(iv), B-(i), C-(ii), D-(iii)
c. A-(iv), B-(v), C-(ii), D-(iii) **d.** A-(v), B-(iv), C-(ii), D-(iii)

18. Given below are two statements : one is labelled as Assertion (A) and the other is labelled as Reason (R).
Assertion (A) The H—O—H bond angle in water molecule is 104.5°.
Reason (R) The lone pair - lone pair repulsion of electrons is higher than the bond pair - bond pair repulsion.
a. A is false but R is true.
b. Both A and R are true, but R is not the correct correct explanation of A.
c. A is true but R is false.
d. Both A and R are true, and R is the correct explanation of A.

19. In chromatography technique, the purification of compound is independent of

a. mobility or flow of solvent system
b. solubility of the compound
c. length of the column or TLC plate
d. physical state of the pure compound

20. A group 15 element, which is a metal and forms a hydride with strongest reducing power among group 15 hydrides. The element is

a. Sb **b.** P **c.** As **d.** Bi

Section B : Numerical Type Questions

21. For the reaction, $A(g) \rightleftharpoons B(g)$ at 495 K,
$\Delta_r G^\circ = -9.478$ kJ mol^{-1}.
If we start the reaction in a closed container at 495 K with 22 millimoles of A, the amount of B is the equilibrium mixture is ....
millimoles (Round off to the nearest integer).
[$R = 8.314$ J mol^{-1} K^{-1}, $\ln 10 = 2.303$]

22. Complete combustion of 750 g of an organic compound provides 420 g of CO_2 and 210 g of H_2O. The percentage composition of carbon and hydrogen in organic compound is 15.3 and ...... respectively (Round off to the nearest integer).

23. $2MnO_4^- + bC_2O_4^{2-} + cH^+ \longrightarrow x\,Mn^{2+} + yCO_2 + zH_2O$

If the above equation is balanced with integer coefficients, the value of c is ....

24. AB_2 is 10% dissociated in water to A^{2+} and B^-. The boiling point of a 10.0 molal aqueous solution of AB_2 is ...... °C (Round off to the nearest integer).
[Given, molal elevation constant of water
$K_b = 0.5$K kg mol^{-1} boiling point of pure water = 100° C]

25. The equivalents of ethylene diamine required to replace the neutral ligands from the coordination sphere of the *trans*-complex of $CoCl_3 \cdot 4NH_3$ is ...... (Round off to the nearest integer).

26. A 6.50 molal solution of KOH (*aq*) has a denisity of 1.89 g cm^{-3}. The molarity of the solution is ...... mol dm^{-3} (Round off to the nearest integer).
[Atomic masses: K : 39.0 u, O : 16.0u, H : 1.0 u]

27. When light of wavelength 248 nm falls on a metal of threshold energy 3.0 eV, the
de- Broglie wavelength of emitted electrons is ...... Å.
[Round off to the nearest integer]
[Use: $\sqrt{3} = 1.73$, $h = 6.63 \times 10^{-34}$ Js $m_e = 9.1 \times 10^{-31}$ kg,
$c = 3.0 \times 10^8$ ms^{-1}, 1 eV $= 1.6 \times 10^{-19}$ J]

28. Two salts A_2X and MX have the same value of solubility product of 4.0×10^{-12}. The ratio of their molar solubilities i.e $\frac{S(A_2X)}{S(MX)} = ...$
(Round off to the nearest integer)

29. A certain element crystallises in a bcc lattice of unit cell edge length 27Å. If the same element under the same conditions crystallises in the fcc lattice, the edge length of the unit cell in Å will be ......
(Round off to the nearest integer)
[Assume each lattice point has a single atom]
[Assume $\sqrt{3} = 1.73$, $\sqrt{2} = 1.41$]

30. The decomposition of formic acid on gold surface follows first order kinetics. If the rate constant at 300K is 1.0×10^{-3} s^{-1} and the activation energy, $E_a = 11.488$ kJ mol^{-1}, the rate constant at 200 K is ...... $\times 10^{-5}$ s^{-1} (Round off to the nearest integer).
[Given, $R = 8.314$ J mol^{-1} K^{-1}]

Answers

For solution scan the QR code

1. (d)	2. (b)	3. (b)	4. (b)	5. (c)	6. (b)	7. (c)	8. (d)	9. (b)	10. (d)
11. (c)	12. (b)	13. (b)	14. (b)	15. (c)	16. (b)	17. (a)	18. (d)	19. (d)	20. (d)
21. (20)	22. (3)	23. (16)	24. (106)	25. (2)	26. (9)	27. (9)	28. (50)	29. (33)	30. (10)

16 MARCH SHIFT II

Section A : Objective Type Questions

1. The green house gas/es is (are)

(A) carbon dioxide (B) oxygen
(C) water vapour (D) methane

Choose the most appropriate answer from the options given below.

a. Only (A) and (C) **b.** Only (A)
c. (A), (C) and (D) **d.** Only (A) and (B)

2. 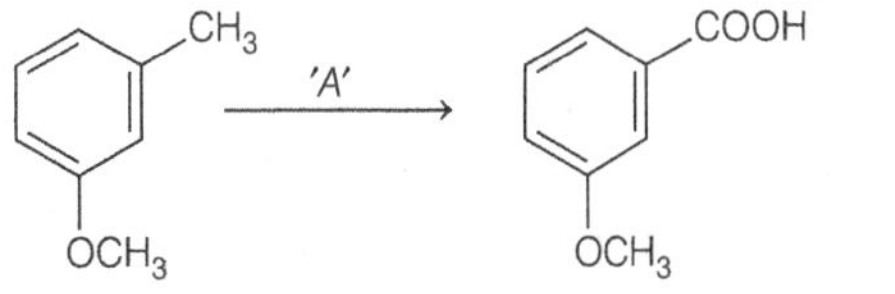

In the above reaction, the reagent 'A' is

a. $NaBH_4$, H_3O^+ **b.** $LiAlH_4$
c. Alkaline $KMNO_4$, H^+ **d.** HCl, Zn - Hg

3. Which of the following reduction reaction cannot be carried out with coke?

a. $Al_2O_3 \longrightarrow Al$ **b.** $ZnO \longrightarrow Zn$
c. $Fe_2O_3 \longrightarrow Fe$ **d.** $Cu_2O \longrightarrow Cu$

4. Identify the elements *X* and *Y* using the ionisation energy values given below.

	Ionisation energy (Ist)	**kJ/mol** (IInd)
X	495	4563
Y	731	1450

a. *X* = Na, *Y* = Mg **b.** *X* = Mg, *Y* = F
c. X = Mg, *Y* = Na **d.** *X* = F, *Y* = Mg

5.

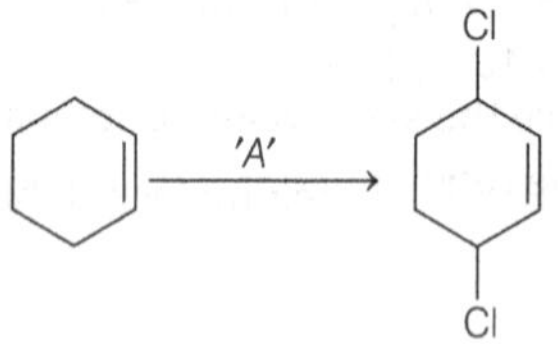

Identify the reagent(s) '*A*' and condition(s) for the reaction:

a. *A* = HCl, anhydrous $AlCl_3$
b. *A* = HCl, $ZnCl_2$
c. *A* = Cl_2, UV light
d. *A* = Cl_2, dark, anhydrous $AlCl_3$

6. The secondary structure of protein is stabilised by

a. peptide bond **b.** glycosidic bond
c. hydrogen bonding **d.** van der Waals' forces

7. Fex_2 and Fey_3 are known when *x* and *y* are

a. *x* = F, Cl, Br, I and *y* = F, Cl and Br
b. *x* = F, Cl, Br and *y* = F, Cl, Br and I
c. *x* = Cl, Br, I and *y* = F, Cl, Br and I
d. *x* = F, Cl, Br, I and *y* = F, Cl, Br and I

8. Which of the following polymer is used in the manufacture of wood laminates ?

a. *cis*-polyisoprene
b. Melamine formaldehyde resin
c. Urea formaldehyde resin
d. Phenol and formaldehyde resin

9. Statement I Sodium hydride can be used as an oxidising agent.

Statement II The lone pair of electrons on nitrogen in pyridine makes it basic.

Choose the correct answer from the options given below.

a. Both statement I and statement II are true
b. Both statement I and statement II are false
c. Statement I is true but statement II is false
d. Statement I is false but statement II is true

10. The incorrect statement regarding the structure of C_{60} is

a. the six-membered rings are fused to both six and five-membered rings
b. each carbon atom forms three sigma bonds
c. the five-membered rings are fused only to six-membered rings
d. it contains 12 six-membered rings and 24 five-membered rings

11. The correct statements about H_2O_2 are

a. used in the treatment of effluents.
b. used as both oxidising and reducing agents.
c. the two hydroxyl groups lie in the same plane.
d. miscible with water.

Choose the correct answer from the options given below.

a. (A), (B), (C) and (D) **b.** (A), (B) and (D)
c. (B), (C) and (D) **d.** (A), (C) and (D)

12. Ammonolysis of alkyl halides followed by the treatment with NaOH solution can be used to prepare primary, secondary and tertiary amines. The purpose of NaOH in the reaction is

a. to remove basic impurities
b. to activate NH_3 used in the reaction
c. to remove acidic impurities
d. to increase the reactivity of alkyl halide

13. An unsaturated hydrocarbon *X* on ozonolysis gives *A*. Compound *A* when warmed with ammonical silver nitrate forms a bright silver mirror along the sides of the test tube. The unsaturated hydrocarbon *X* is

a. $CH_3-\underset{CH_3}{\underset{|}{C}}=\underset{CH_3}{\underset{|}{C}}-CH_3$ **b.** 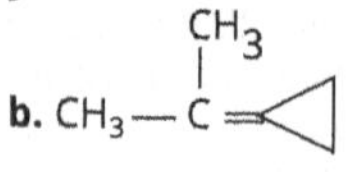

c. $HC\equiv C-CH_2-CH_3$ **d.** $CH_3-C\equiv C-CH_3$

14. Which of the following is least basic?

a. $(CH_3CO)\ddot{N}HC_2H_5$ **b.** $(C_2H_5)_3\ddot{N}$
c. $(CH_3CO)_2\ddot{N}H$ **d.** $(C_2H_5)_2\ddot{N}H$

15. The characteristics of elements *X*, *Y* and *Z* with atomic numbers, respectively, 33, 53 and 83 are

a. *X* and *Y* are metalloids and *Z* is a metal
b. *X* is a metalloid, *Y* is a non-metal and *Z* is a metal
c. *X*, *Y* and *Z* are metals
d. *X* and *Z* are non-metals and *Y* is a metalloid

16. Match List-I with List-II.

	List-I (Test / Reagents / Observation(s))		**List-II** (Species detected)
A.	Lassaigne's test	(i)	Carbon
B.	Cu(II) oxide	(ii)	Sulphur
C.	Silver nitrate	(iii)	N, S, P, and halogen
D.	The sodium fusion extract gives black precipitate with acetic acid and lead acetate	(iv)	Halogen specifically

The correct match is

a. A-(iii), B-(i), C-(ii), D-(iv) **b.** A-(i), B-(iv), C-(iii), D-(ii)
c. A-(iii), B-(i), C-(iv), D-(ii) **d.** A-(i), B-(ii), C-(iv), D-(iii)

17. The incorrect statements below regarding colloidal solutions is

a. a colloidal solution shows colligative properties
b. an ordinary filter paper can stop the flow of colloidal particles

c. the flocculating power of Al^{3+} is more than that of Na^{+}
d. a colloidal solution shows Brownian motion of colloidal particles

18. Arrange the following metal complex/compounds in the increasing order of spin only magnetic moment. Presume all the three, high spin system.
(Atomic numbers Ce = 58, Gd = 64 and Eu = 63.)
A. $(NH_4)_2[Ce(NO_3)_6]$
B. $Gd(NO_3)_3$ and
C. $Eu(NO_3)_3$

a. (B) < (A) < (C) **b.** (C) < (A) < (B)
c. (A) < (B) < (C) **d.** (A) < (C) < (B)

19. The exact volumes of 1 M NaOH solution required to neutralise 50 mL of 1 M H_3PO_3 solution and 100 mL of 2 M H_3PO_2 solution, respectively, are

a. 100 mL and 100 mL **b.** 100 mL and 50 mL
c. 100 mL and 200 mL **d.** 50 mL and 50 mL

20.
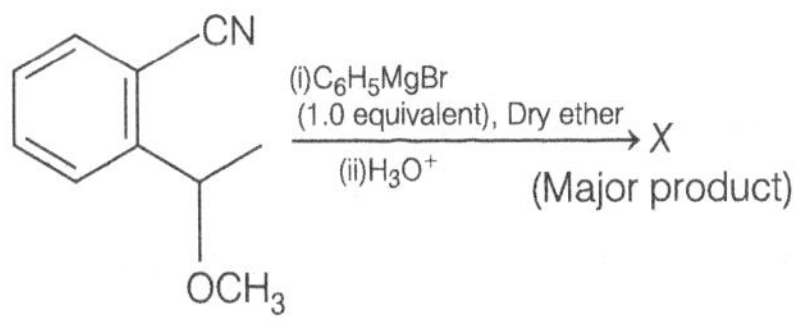

a.
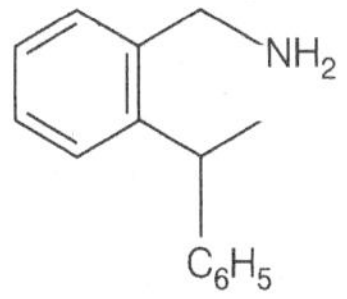

b.
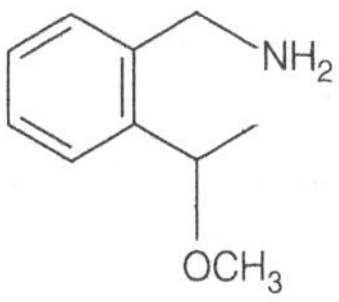

c.
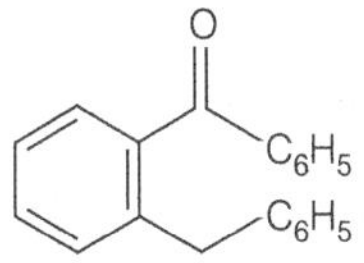

d.
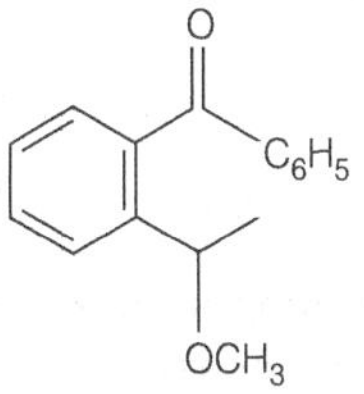

Section B : Numerical Type Questions

21. Ga (atomic mass 70 u) crystallises in a hexagonal close packed structure. The total number of voids in 0.581 g of Ga is ...... $\times 10^{21}$. (Round off to the nearest integer).

22. A 5.0 m mol dm^{-3} aqueous solution of KCl has a conductance of 0.55 mS when measured in a cell constant 1.3 cm^{-1}.
The molar conductivity of this solution is ...... $mSm^2\ mol^{-1}$. (Round off to the nearest integer)

23. *A* and *B* decompose *via* first order kinetics with half-lives 54.0 min and 18.0 min respectively. Starting from an equimolar non-reactive mixture of *A* and *B*, the time taken for the concentration of *A* to become 16 times that of *B* is .... min.
(Round off to the nearest integer).

24. In Duma's method of estimation of nitrogen, 0.1840 g of an organic compound gave 30 mL of nitrogen collected at 287 K and 758 mm of Hg pressure. The percentage composition of nitrogen in the compound is ....
(Round off to the nearest integer).
[Given: Aqueous tension at 287 K = 14 mm of Hg]

25. The number of orbitals with $n = 5$, $m_l = +2$ is ....
(Round off to the nearest integer).

26. At 363 K, the vapour pressure of *A* is 21 kPa and that of *B* is 18 kPa. One mole of *A* and 2 moles of *B* are mixed. Assuming that this solution is ideal, the vapour pressure of the mixture is .... kPa. (Round off to the nearest integer).

27. Sulphurous acid (H_2SO_3) has $K_{a_1} = 1.7 \times 10^{-2}$ and $K_{a_2} = 6.4 \times 10^{-8}$. The pH of 0.588 M is ......
(Round off to the nearest integer)

28. When 35 mL of 0.15 M lead nitrate solution is mixed with 20 mL of 0.12 M chromic sulphate solution, ... $\times 10^{-5}$ moles of lead sulphate precipitate out. (Round off to the nearest integer).

29. At 25°C, 50 g of iron reacts with HCl to form $FeCl_2$. The evolved hydrogen gas expands against a constant pressure of 1 bar. The work done by the gas during this expansion is ...... J. (Round off to the nearest integer)
[Given, $R = 8.314\ J\ mol^{-1}\ K^{-1}$. Assume, hydrogen is an ideal gas]
[Atomic mass off Fe is 55.85 u]

30. $[Ti(H_2O)_6]^{3+}$ absorbs light of wavelength 498 nm during a *d-d* transition. The octahedral splitting energy for the above complex is .... $\times 10^{-9}$ J (Round off to the nearest intger). [$h = 6.626 \times 10^{-34}$ Js, $c = 3 \times 10^{8} ms^{-1}$]

Answers

For solution scan the QR code

1. *(c)*	2. *(c)*	3. *(a)*	4. *(a)*	5. *(c)*	6. *(c)*	7. *(a)*	8. *(c)*	9. *(c)*	10. *(d)*
11. *(b)*	12. *(c)*	13. *(c)*	14. *(c)*	15. *(b)*	16. *(c)*	17. *(b)*	18. *(d)*	19. *(c)*	20. *(d)*
21. *(15)*	22. *(14)*	23. *(108)*	24. *(19)*	25. *(3)*	26. *(19)*	27. *(1)*	28. *(525)*	29. *(2218)*	30. *(4)*

17 MARCH SHIFT I

Section A : Objective Type Questions

1. With respect to drug-enzyme interaction, identify the wrong statement

a. Non-competitive inhibitor binds to the allosteric site.
b. Allosteric inhibitor changes the enzyme's active site.
c. Allosteric inhibitor competes with the enzyme's active site.
d. Competitive inhibitor binds to the enzyme's active site.

2. Which of the following is an aromatic compound?

a. 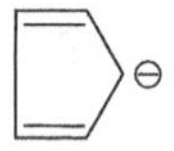

b. 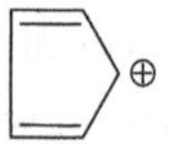

c. 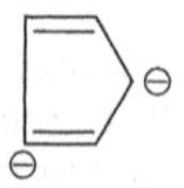

d. 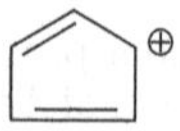

3. 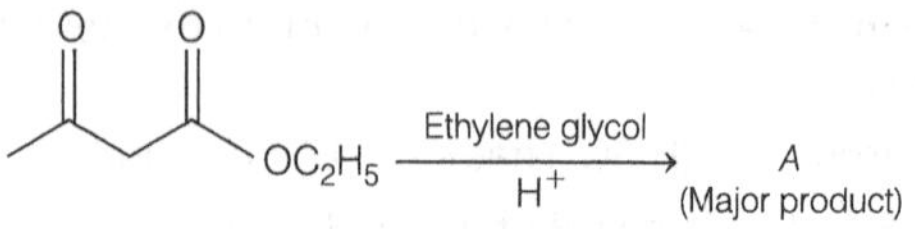

The product *A* in the above reaction is

a. 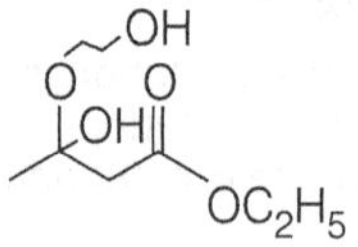

b. 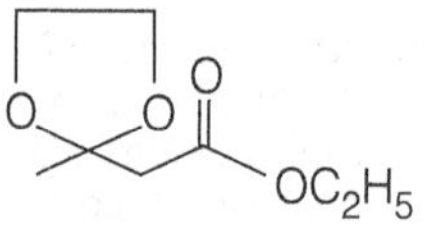

c. 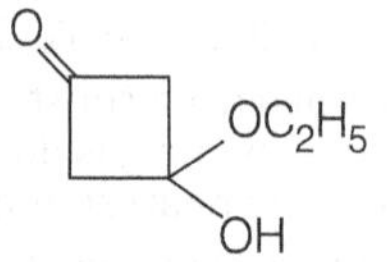

d. 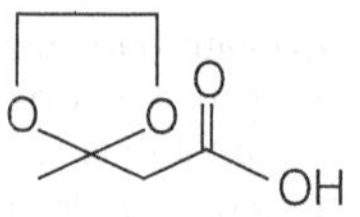

4. A central atom in a molecule has two lone pairs of electrons and forms three single bonds. The shape of this molecule is

a. *see-saw*
b. planar triangular
c. T-shaped
d. trigonal pyramidal

5. Given below are two statements.

Statement I Potassium permanganate on heating at 573 K forms potassium manganate.

Statement II Both potassium permanganate and potassium manganate are tetrahedral and paramagnetic in nature.

In the light of the above statements, choose the most appropriate answer from the options given below

a. Statement I is true but statement II is false
b. Both statement I and statement II are true
c. Statement I is false but statement II is true
d. Both statement I and statement II are false

6. Which of the following is correct structure of tyrosine?

a. 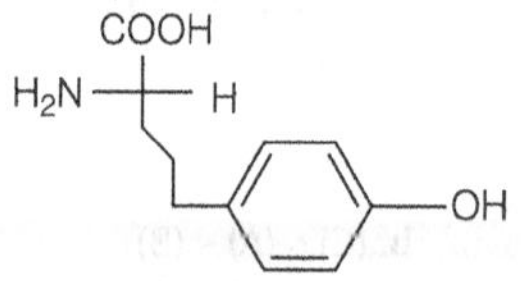

b. 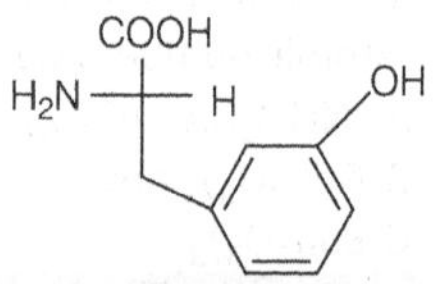

c. 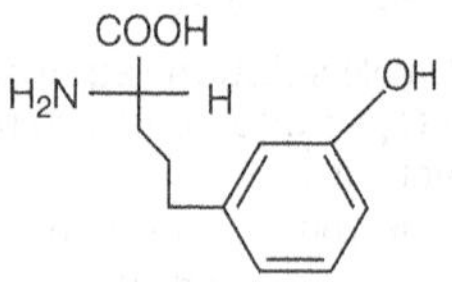

d. 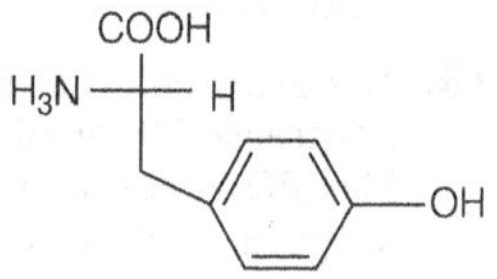

7. 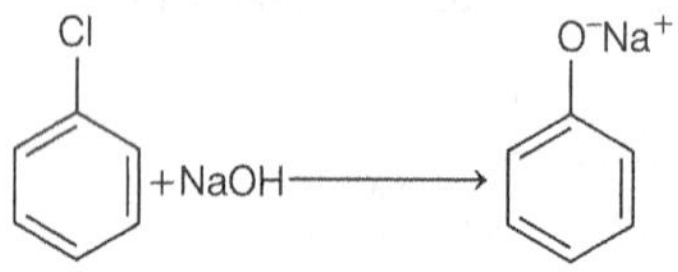

The above reaction requires which of the following reaction conditions?

a. 573 K, Cu, 300 atm
b. 623 K, Cu, 300 atm
c. 573 K, 300 atm
d. 623 K, 300 atm

8. The absolute value of the electron gain enthalpy of halogens satisfies

a. $I > Br > Cl > F$
b. $Cl > Br > F > I$
c. $Cl > F > Br > I$
d. $F > Cl > Br > I$

9. Which of the following compound cannot act as a Lewis base?

a. NF_3 **b.** PCl_5 **c.** SF_4 **d.** ClF_3

10. Reducing smog is a mixture of

a. smoke, fog and O_3
b. smoke, fog and SO_2
c. smoke, fog and $CH_2{=}CH{-}CHO$
d. smoke, fog and N_2O_3

11. Hofmann bromamide degradation of benzamide gives product *A*, which upon heating with $CHCl_3$ and NaOH gives product *B*. The structures of *A* and *B* are

a. $A =$  $B =$ 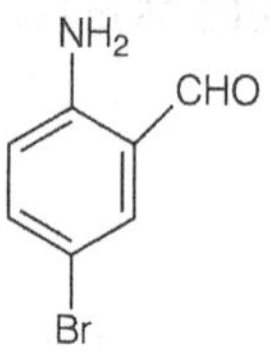

b. $A =$ 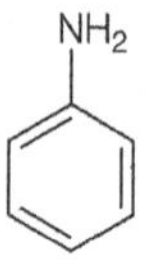 $B =$ 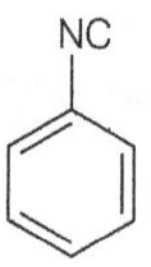

c. $A =$ 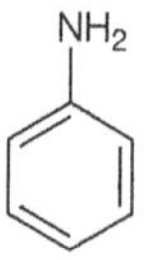 $B =$ 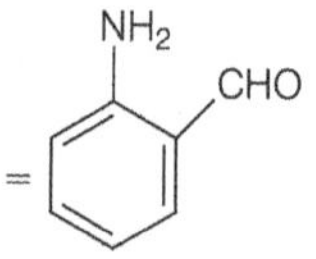

d. $A =$ 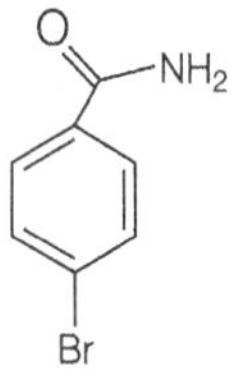 $B =$ 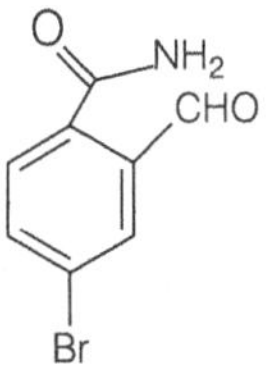

12. Mesityl oxide is a common name of
a. 2,4-dimethyl pentan-3-one
b. 3-methyl cyclohexane carbaldehyde
c. 2-methyl cyclohexanone
d. 4-methyl pent-3-en-2-one

13. Which of the following reaction is an example of ammonolysis?
a. $C_6H_5COCl + C_6H_5NH_2 \longrightarrow C_6H_5CONHC_6H_5$
b. $C_6H_5CH_2CN \xrightarrow{[H]} C_6H_5CH_2CH_2NH_2$
c. $C_6H_5NH_2 \xrightarrow{HCl} C_6H_5\overset{+}{N}H_3Cl^-$
d. $C_6H_5CH_2Cl + NH_3 \longrightarrow C_6H_5CH_2NH_2$

14. 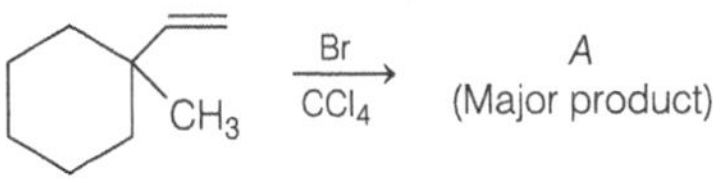

Product A is

a. 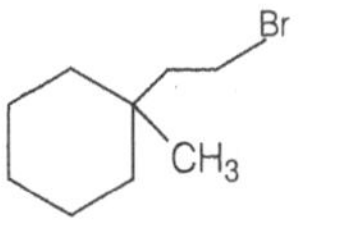 **b.** 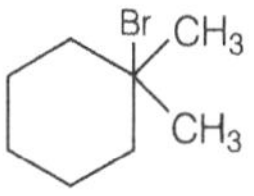

c. 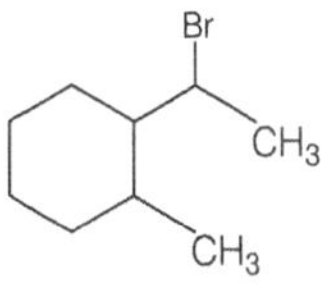 **d.** 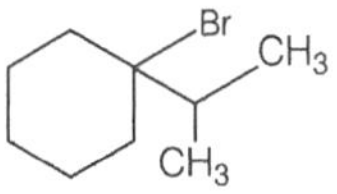

15. A colloidal system consisting of a gas dispersed in a solid is called a/an
a. solid sol **b.** gel
c. aerosol **d.** foam

16. The incorrect statement(s) about heavy water is (are)
(A) used as a moderator in nuclear reactor.
(B) obtained as a by-product in fertiliser industry.
(C) used for the study of reaction mechanism.
(D) has a higher dielectric constant than water.
Choose the correct answer from the options given below :
a. (B) only **b.** (C) only
c. (D) only **d.** (C) and (D) only

17. The correct order of conductivity of ions in water is
a. $Na^+ > K^+ > Rb^+ > Cs^+$ **b.** $Cs^+ > Rb^+ > K^+ > Na^+$
c. $K^+ > Na^+ > Cs^+ > Rb^+$ **d.** $Rb^+ > Na^+ > K^+ > Li^+$

18. What is the spin-only magnetic moment value (BM) of a divalent metal ion with atomic number 25, in it's aqueous solution?
a. 5.92 **b.** 5.0
c. zero **d.** 5.26

19. Given below are two statements.
Statement I Retardation factor (R_f) can be measured in metre/centimetre.
Statement II R_f value of a compound remains constant in all solvents.
Choose the most appropriate answer from the options given below
a. Statement I is true but statement II is false.
b. Both statement I and statement II are true.
c. Both statement I and statement II are false.
d. Statement I is false but statement II is true.

20. The point of intersection and sudden increase in the slope, in the diagram given below, respectively, indicates.

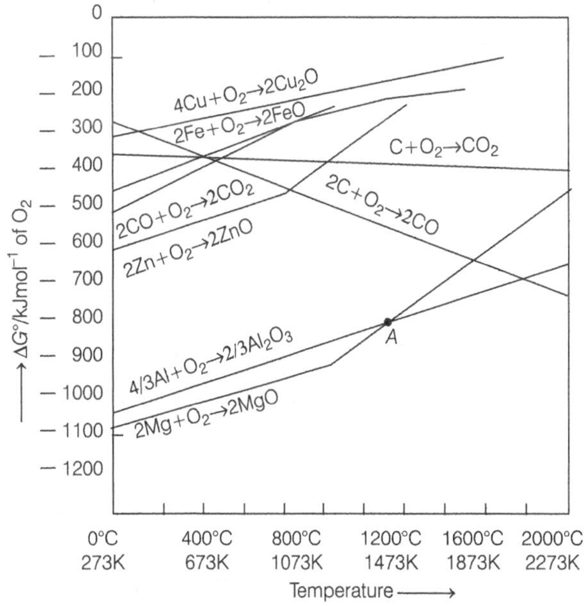

a. $\Delta G = 0$ and melting or boiling point of the metal oxide
b. $\Delta G > 0$ and decomposition of the metal oxide
c. $\Delta G < 0$ and decomposition of the metal oxide
d. $\Delta G = 0$ and reduction of the metal oxide

Section B : Numerical Type Questions

21. The reaction of white phosphorus on boiling with alkali in inert atmosphere resulted in the formation of product A. The reaction of 1 mol of A with excess of $AgNO_3$ in aqueous medium gives ...... mol(s) of Ag (Round off to the nearest integer).

22. 0.01 moles of a weak acid HA ($K_a = 2.0 \times 10^{-6}$) is dissolved in 1.0 L of 0.1 M HCl solution. The degree of dissociation of HA is ...... $\times 10^{-5}$ (Round off to the nearest integer).
[Neglect volume change on adding HA. Assume degree of dissociation << 1]

23. A certain orbital has $n = 4$ and $m_l = -3$. The number of radial nodes in this orbital is ...... (Round off to the nearest integer).

24.

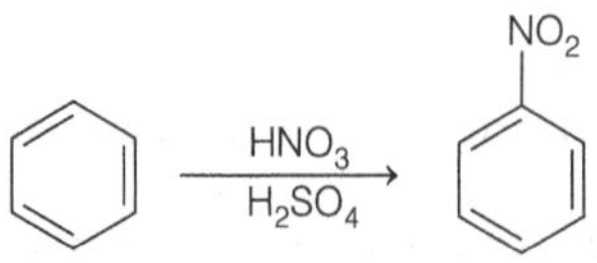

In the above reaction 3.9 g of benzene on nitration gives 4.92 g of nitrobenzene. The percentage yield of nitrobenzene in the above reaction is .... % (Round off to the nearest integer).

(Given, atomic mass C : 12.0 u, H : 1.0 u, O : 16.0 u, N : 14.0 u)

25. The mole fraction of a solute in a 100 molal aqueous solution .... $\times 10^{-2}$

(Round off to the nearest integer).

[Given, atomic masses H : 1.0 u, O : 16.0 u]

26. For a certain first order reaction 32% of the reactant is left after 570 s. The rate constant of this reaction is ...... $\times 10^{-3} s^{-1}$. (Round off to the nearest integer).

[Given, $\log_{10} 2 = 0.301$, $\ln 10 = 2.303$]

27. The standard enthalpies of formation of Al_2O_3 and CaO are -1675 kJ mol^{-1} and -625kJ mol^{-1} respectively.

For the reaction,

$3CaO + 2Al \longrightarrow 3Ca + Al_2O_3$ the standard reaction enthalpy $\Delta_r H^\circ$ .... kJ.

(Round off to the nearest integer).

28. 15 mL of aqueous solution of Fe^{2+} in acidic medium completely reacted with 20 mL of 0.03 M aqueous $Cr_2O_7^{2-}$. The molarity of the Fe^{2+} solution is ...... $\times 10^{-2}$ M (Round off to the nearest integer).

29. The oxygen dissolved in water exerts a partial pressure of 20 kPa in the vapour above water. The molar solubility of oxygen in water is ...... $\times 10^{-5}$ mol dm^{-3}.

(Round off to the nearest integer).

[Given, Henry's law constant $(K_H) = 8.0 \times 10^4$ kPa for O_2, density of water with dissolved oxygen = 1.0 kg dm^{-3}].

30. The pressure exerted by a non-reactive gaseous mixture of 6.4 g of methane and 8.8 g of carbon dioxide in a 10L vessel at 27°C is ...... kPa (Round off to the nearest integer)

(Assume gases are ideal, $R = 8.314$ J mol^{-1} K^{-1} Atomic mass, C : 12.0 u, H : 1.0 u, O : 16.0 u)

Answers

For solution scan the QR code

1. *(c)*	2. *(a)*	3. *(b)*	4. *(c)*	5. *(a)*	6. *(d)*	7. *(d)*	8. *(c)*	9. *(b)*	10. *(b)*
11. *(b)*	12. *(d)*	13. *(d)*	14. *(d)*	15. *(a)*	16. *(c)*	17. *(b)*	18. *(a)*	19. *(c)*	20. *(a)*
21. *(4)*	22. *(2)*	23. *(0)*	24. *(80)*	25. *(64)*	26. *(2)*	27. *(230)*	28. *(24)*	29. *(25)*	30. *(150)*

17 MARCH SHIFT II

Section A : Objective Type Questions

1. Fructose is an example of

a. pyranose **b.** ketohexose
c. aldohexose **d.** heptose

2. The set of elements that differ in mutual relationship from those of the other sets is

a. Li-Mg **b.** B-Si
c. Be-Al **d.** Li-Na

3. The functional groups that are responsible for the ion-exchange property of cation and anion exchange resins, respectively, are

a. $—SO_3H$ and $—NH_2$ **b.** $—SO_3H$ and $—COOH$
c. $—NH_2$ and $—COOH$ **d.** $—NH_2$ and $—SO_3H$

4. Match List-I and List-II.

	List-I		List-II
A.	Haematite	1.	$Al_2O_3 \cdot x H_2O$
B.	Bauxite	2.	Fe_2O_3
C.	Magnetite	3.	$CuCO_3 \cdot Cu(OH)_2$
D.	Malachite	4.	Fe_3O_4

Choose the correct answer from the options given below.

	A	B	C	D		A	B	C	D
(a)	2	3	1	4	(b)	4	1	2	3
(c)	1	3	2	4	(d)	2	1	4	3

5. The correct pair(s) of the ambident nucleophiles is(are)

A. AgCN/KCN B. RCOOAg/RCOOK
C. $AgNO_2/KNO_2$ D. AgI/KI

a. B and C **b.** Only A
c. A and C **d.** Only B

6. The set that represents the pair of neutral oxides of nitrogen is

a. NO and N_2O **b.** N_2O and N_2O_3
c. N_2O and NO_2 **d.** NO and NO_2

7. Match List-I and List-II.

	List-I		List-II
A.	$[Co(NH_3)_6][Cr(CN)_6]$	1.	Linkage isomerism
B.	$[Co(NH_3)_3(NO_2)_3]$	2.	Solvate isomerism
C.	$[Cr(H_2O)_6]Cl_3$	3.	Co-ordination isomerism
D.	*cis*-$[CrCl_2(ox)_2]^{3-}$	4.	Optical isomerism

Choose the correct answer from the options given below.

	A	B	C	D
(a)	3	1	2	4
(b)	4	2	3	1
(c)	2	1	3	4
(d)	1	2	3	4

8. Primary, secondary and tertiary amines can be separated using

a. *para*-toluene sulphonyl chloride
b. chloroform and KOH
c. benzene sulphonic acid
d. acetyl amide

9. The common positive oxidation states for an element with atomic number 24, are

a. + 2 to + 6 **b.** + 1 and + 3 to + 6
c. + 1 and + 3 **d.** + 1 to + 6

10. Match List-I and List-II.

List-I (Chemical compound)	**List-II** (Used as)
A. Sucralose	1. Synthetic detergent
B. Glyceryl ester of stearic acid	2. Artificial sweetener
C. Sodium benzoate	3. Antiseptic
D. Bithionol	4. Food preservative

Choose the correct match.

	A	B	C	D
a.	4	3	2	1
b.	2	1	4	3
c.	3	2	4	1
d.	1	2	4	3

11. Given below are two statements.

Statement-I 2-methylbutane on oxidation with $KMnO_4$ gives 2-methylbutan-2-ol.

Statement-II *n*-alkanes can be easily oxidised to corresponding alcohols with $KMnO_4$.

Choose the correct option.

(a) Both statements I and II are correct.
(b) Both statements I and II are incorrect.
(c) Statement I is correct but statement II is incorrect.
(d) Statement I is incorrect but statement II is correct.

12. Nitrogen can be estimated by Kjeldahl's method for which of the following compound ?

a. 

b. 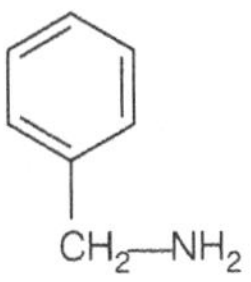

c. 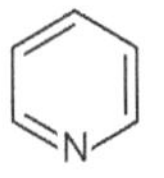

d. 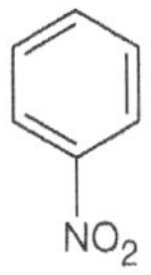

13. Amongst the following, the linear species is

a. NO_2 **b.** Cl_2O
c. O_3 **d.** N_3^-

14. $C_{12}H_{22}O_{11} + H_2O \xrightarrow{\text{Enzyme } A} C_6H_{12}O_6 + C_6H_{12}O_6$
(Sucrose → Glucose + Fructose)

$C_6H_{12}O_6 \xrightarrow{\text{Enzyme } B} 2C_2H_5OH + 2CO_2$
(Glucose → Ethanol)

In the above reactions, the enzyme *A* and enzyme *B* respectively are

a. amylase and invertase **b.** invertase and amylase
c. invertase and zymase **d.** zymase and invertase

15. One of the by-products formed during the recovery of NH_3 from solvay process is

a. $Ca(OH)_2$ **b.** $NaHCO_3$ **c.** $CaCl_2$ **d.** NH_4Cl

16. 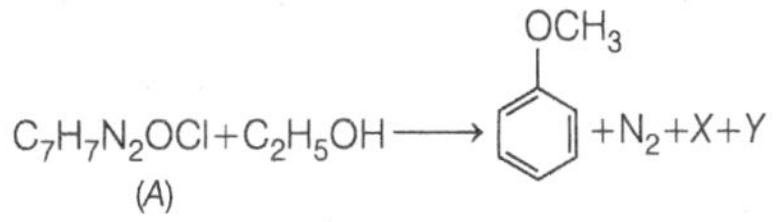

In the above reaction, the structural formula of (*A*), *X* and *Y* respectively are

a. 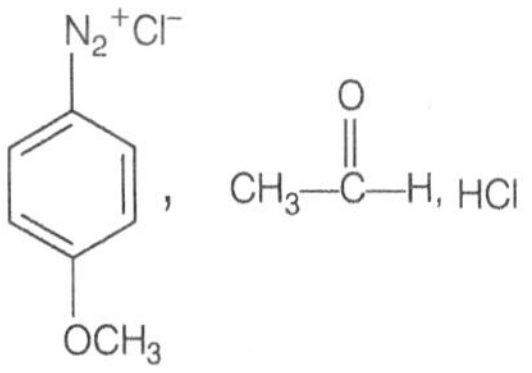

b. 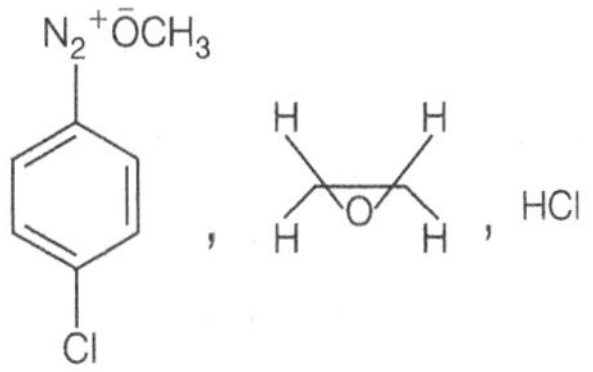

c. 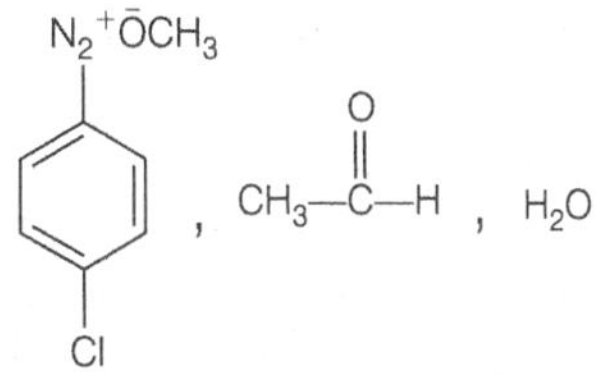

d. 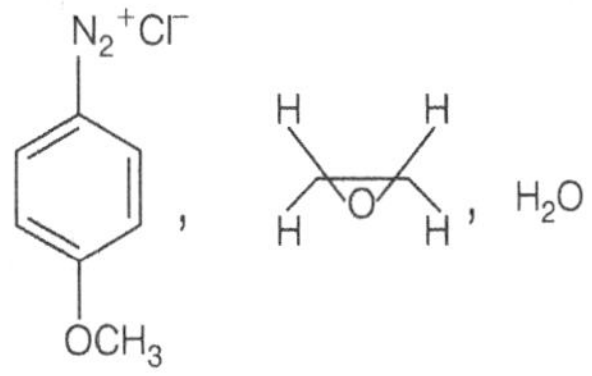

17. For the coagulation of a negative sol, the species below, that has the highest flocculating power is

a. SO_4^{2-} **b.** Ba^{2+} **c.** Na^+ **d.** PO_4^{3-}

18. Which of the following statement(s) is (are) incorrect reason for eutrophication ?

A. Excess usage of fertilisers.
B. Excess usage of detergents.
C. Dense plant population in water bodies.
D. Lack of nutrients in water bodies that prevent plant growth.

Choose the most appropriate answer from the options given below.

a. Only A **b.** Only C **c.** B and D **d.** Only D

19. Choose the correct statement regarding the formation of carbocations *A* and *B*.

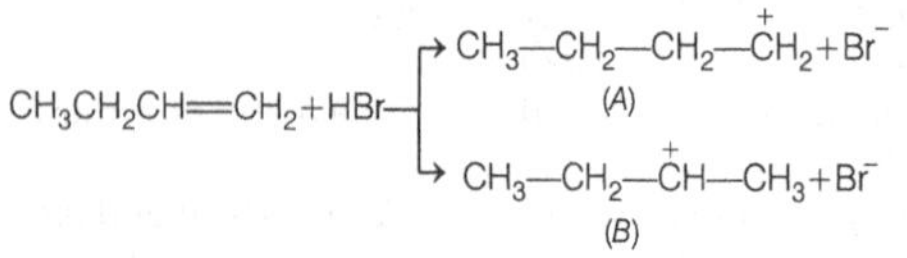

(a) Carbocation *B* is more stable and formed relatively at faster rate.
(b) Carbocation *A* is more stable and formed relatively at slow rate.
(c) Carbocation *B* is more stable and formed relatively at slow rate.
(d) Carbocation *A* is more stable and formed relatively at faster rate.

20. During which of the following processes, does entropy decrease ?

A. Freezing of water to ice at 0°C.
B. Freezing of water to ice at −10°C.
C. $N_2(g) + 3H_2(g) \longrightarrow 2NH_3(g)$
D. Adsorption of $CO(g)$ and lead surface.
E. Dissolution of NaCl in water.

a. A, B, C and D **b.** B and C **c.** A and E **d.** A, C and E

Section B : Numerical Type Questions

21. A KCl solution of conductivity $0.14\ S\ m^{-1}$ shows a resistance of $4.19\ \Omega$ in a conductivity cell. If the same cell is filled with an HCl solution, the resistance drops to $1.03\ \Omega$. The conductivity of the HCl solution is .... $\times 10^{-2}\ S\ m^{-1}$ (Round off to the nearest integer)

22. On complete reaction of $FeCl_3$ with oxalic acid in aqueous solution containing KOH, resulted in the formation of product *A*. The secondary valency of Fe in the product *A* is ....

(Round off to the nearest integer).

23. The reaction $2A + B_2 \longrightarrow 2AB$ is an elementary reaction. For a certain quantity of reactants, if the volume of the reaction vessel is reduced by a factor of 3, the rate of the reaction increases by a factor of .... (Round off to the nearest integer).

24. The total number of C—C sigma bond/s in mesityl oxide ($C_6H_{10}O$) is .... (Round off to the nearest integer).

25. A 1 molal $K_4Fe(CN)_6$ solution has a degree of dissociation of 0.4. Its boiling point is equal to that of another solution which contains 18.1 weight per cent of a non-electrolytic solute *A*. The molar mass of *A* is .... u. (Round off to the nearest integer).
[Density of water = $1.0\ g\ cm^{-3}$]

26. In the ground state of atomic Fe(Z = 26), the spin-only magnetic moment is .... $\times 10^{-1}$ BM. (Round off to the nearest integer).
[Given : $\sqrt{3} = 1.73, \sqrt{2} = 1.41$]

27. The number of chlorine atoms in 20 mL of chlorine gas at STP is _____ 10^{21}. (Round off to the nearest integer).
[Assume chlorine is an ideal gas at STP
$R = 0.083$ L bar $mol^{-1}\ K^{-1}$, $N_A = 6.023 \times 10^{23}$]

28. KBr is doped with 10^{-5} mole per cent of $SrBr_2$. The number of cationic vacancies in 1 g of KBr crystal is .... 10^{14} (Round off to the nearest integer). [Atomic mass : K = 39.1 u, Br = 79.9 u, $N_A = 6.023 \times 10^{23}$]

29. Consider the reaction, $N_2O_4(g) \rightleftharpoons 2NO_2(g)$. The temperature at which $K_C = 20.4$ and $K_p = 600.1$, is .... K. (Round off to the nearest integer).
[Assume all gases are ideal and $R = 0.0831$ L bar, $K^{-1}\ mol^{-1}$].

30.

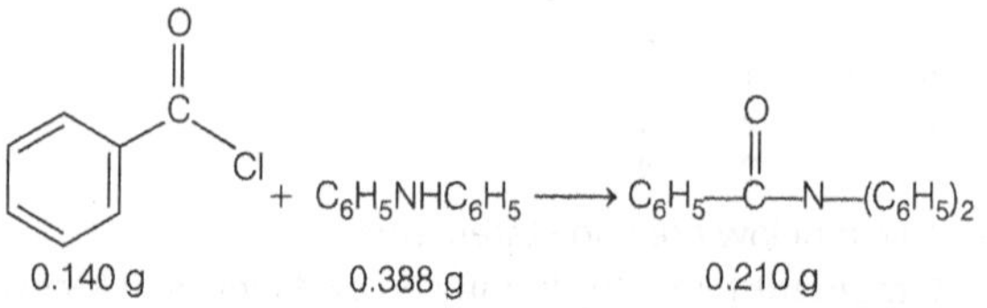

Consider the above reaction. The percentage yield of amide product is .... (Round off to the nearest integer).

(Given : Atomic mass : C : 12.0 u, H : 1.0 u, N : 14.0 u, O : 16.0 u, Cl : 35.5 u)

Answers

For solution scan the QR code

1. (b)	2. (d)	3. (a)	4. (d)	5. (c)	6. (a)	7. (a)	8. (a)	9. (a)	10. (b)
11. (c)	12. (b)	13. (d)	14. (c)	15. (c)	16. (a)	17. (b)	18. (d)	19. (a)	20. (a)
21. (57)	22. (6)	23. (27)	24. (5)	25. (85)	26. (49)	27. (1)	28. (5)	29. (354)	30. (77)

18 MARCH SHIFT I

Section A : Objective Type Questions

1.

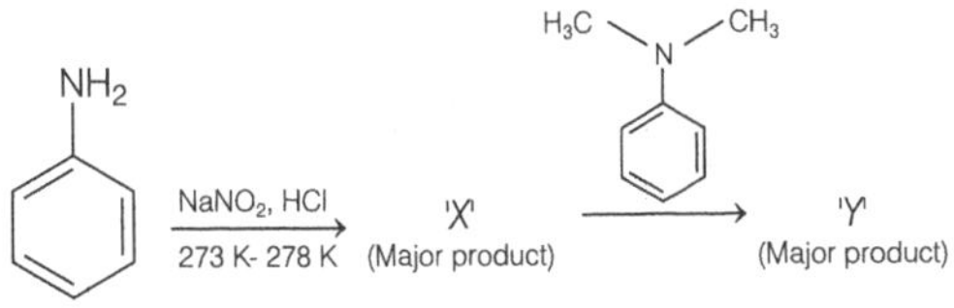

a. 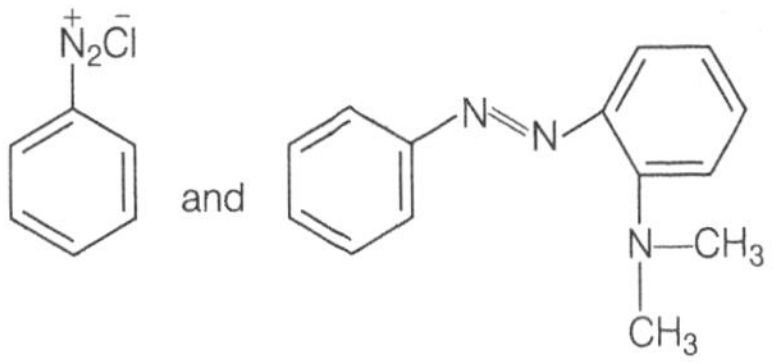

b. 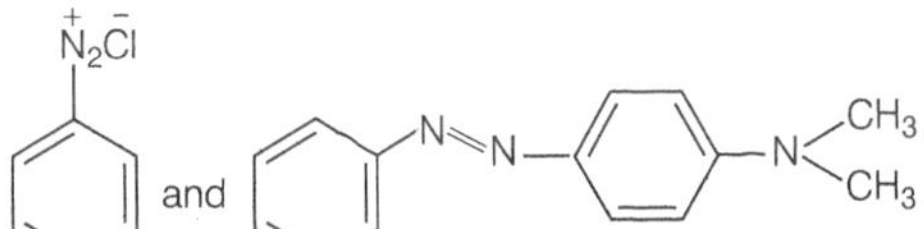

c. 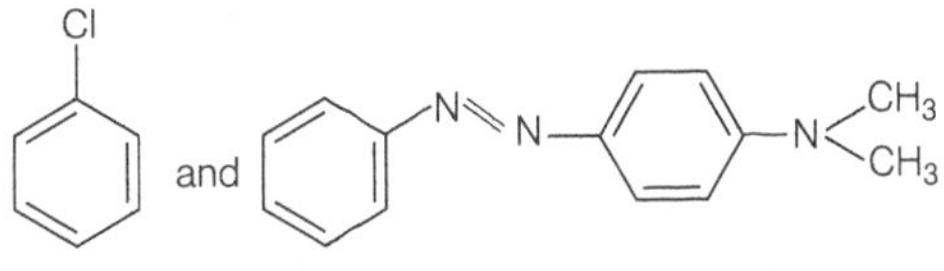

d. 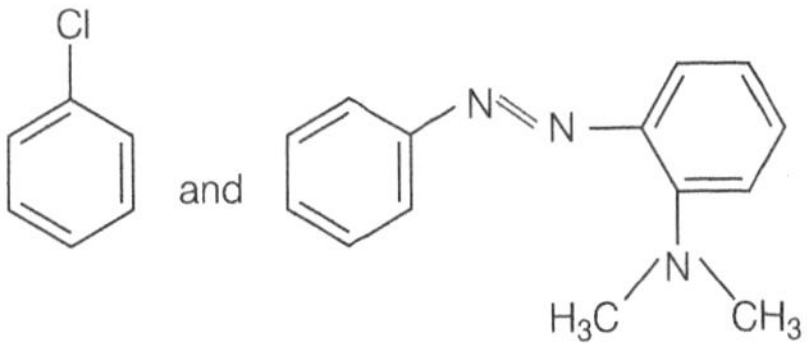

2. The ionic radius of Na^+ ions is 1.02 Å. The ionic radii (in Å) of Mg^{2+} and Al^{3+}, respectively, are

a. 1.05 and 0.99 **b.** 0.72 and 0.54
c. 0.85 and 0.99 **d.** 0.68 and 0.72

3. Reaction of Grignard reagent, C_2H_5MgBr with C_8H_8O followed by hydrolysis gives compound *'A'* which reacts instantly with Lucas reagent to give compound *B*, $C_{10}H_{13}Cl$.

The compound *B* is

a. 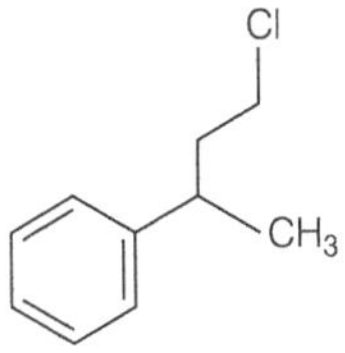

b. 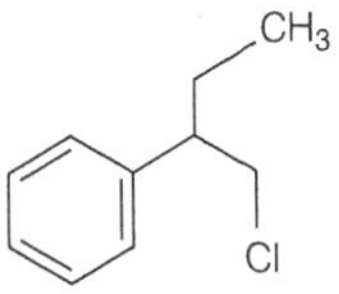

c. 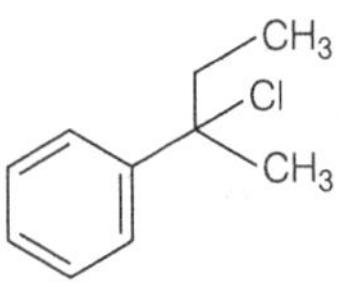

d. 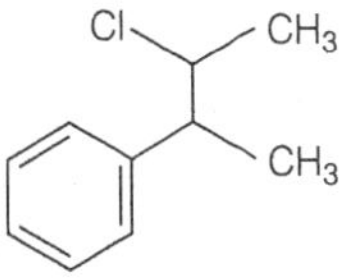

4. Reagent, 1-naphthylamine and sulphanilic acid in acetic acid is used for the detection of

a. N_2O **b.** NO_3^-
c. NO **d.** NO_2^-

5. A non-reducing sugar *"A"* hydrolyses to give two reducing monosaccharides. Sugar *A* is

a. fructose **b.** galactose
c. glucose **d.** sucrose

6. Match the List-I with List- II.

	List-I (Class of drug)		List-II (Example)
(A)	Antacid	(i)	Novestrol
(B)	Artificial sweetener	(ii)	Cimetidine
(C)	Antifertility	(iii)	Valium
(D)	Tranquilizers	(iv)	Alitame

	A	B	C	D
a.	(ii)	(iv)	(i)	(iii)
b.	(iv)	(i)	(ii)	(iii)
c.	(iv)	(iii)	(i)	(ii)
d.	(ii)	(iv)	(iii)	(i)

7.

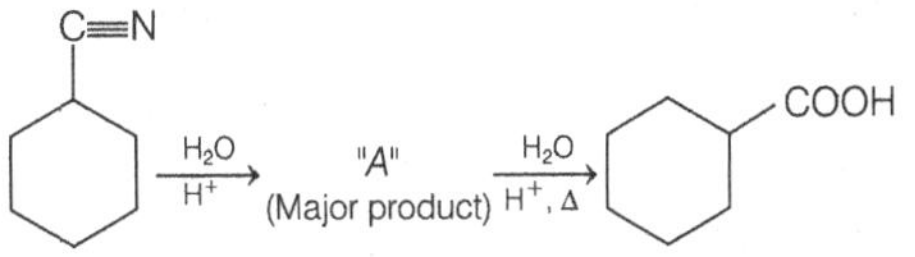

Consider the above chemical reaction and identify product *"A"*.

a. 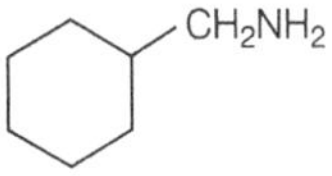

b. 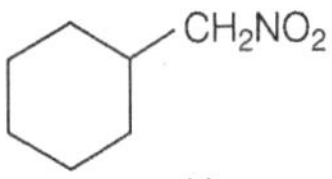

c. 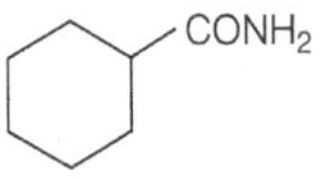

d. 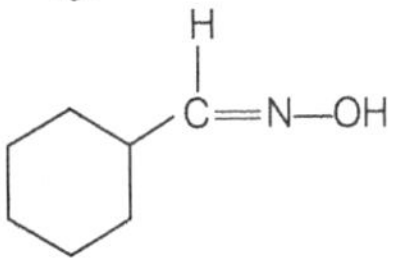

8. Match the list-I with list- II

	List-I		List-II
(A)	Chlorophyll	(i)	Ruthenium
(B)	Vitamin-B_{12}	(ii)	Platinum
(C)	Anticancer drug	(iii)	Cobalt
(D)	Grubbs catalyst	(iv)	Magnesium

Choose the most appropriate answer from the options given below

	A	B	C	D
a.	(iii)	(ii)	(iv)	(i)
b.	(iv)	(iii)	(ii)	(i)
c.	(iv)	(iii)	(i)	(ii)
d.	(iv)	(ii)	(iii)	(i)

9. Match List-I with List-II.

	List-I (Chemicals)		List-II (Use / Preparation / Constituent)
(A)	Alcoholic potassium hydroxide	(i)	Electrodes in batteries
(B)	$Pd/BaSO_4$	(ii)	Obtained by addition reaction
(C)	BHC (Benzene hexachloride)	(iii)	Used for β-elimination reaction
(D)	Polyacetylene	(iv)	Lindlar catalyst

Choose the most appropriate match.

	A	B	C	D
a.	(ii)	(i)	(iv)	(iii)
b.	(iii)	(iv)	(ii)	(i)
c.	(iii)	(i)	(iv)	(ii)
d.	(ii)	(iv)	(i)	(iii)

10. The statements that are true.

(A) Methane leads to both global warming and photochemical smog.
(B) Methane is generated from paddy fields.
(C) Methane is a stronger global warming gas than CO_2.
(D) Methane is a part of reducing smog.

Choose the most appropriate answer from the options given below

a. (A), (B), (C) only **b.** (A) and (B) only
c. (B), (C), (D) only **d.** (A), (B), (D) only

11. Match List-I with List-II.

	List-I		List-II
(A)	$Ca(OCl)_2$	(i)	Antacid
(B)	$CaSO_4 \cdot \frac{1}{2}H_2O$	(ii)	Cement
(C)	CaO	(iii)	Bleach
(D)	$CaCO_3$	(iv)	Plaster of Paris

Choose the most appropriate answer from the options given below.

	A	B	C	D		A	B	C	D
a.	(i)	(iv)	(iii)	(ii)	b.	(iii)	(ii)	(iv)	(i)
c.	(iii)	(iv)	(ii)	(i)	d.	(iii)	(ii)	(i)	(iv)

12. Compound with molecular formula C_3H_6O can show

a. positional isomerism
b. both positional isomerism and metamerism
c. metamerism
d. functional group isomerism

13. The correct structures of *trans*-$[NiBr_2(PPh_3)_2]$ and meridonial-$[Co(NH_3)_3(NO_2)_3]$, respectively are

a. 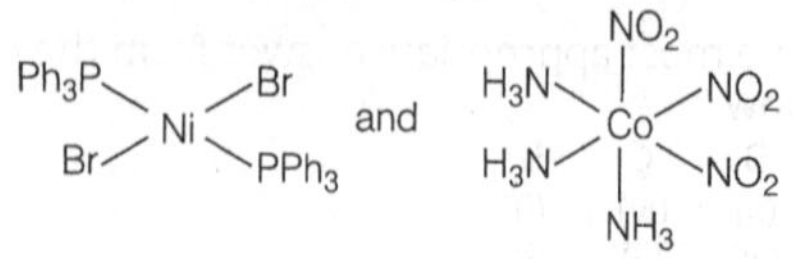

b. 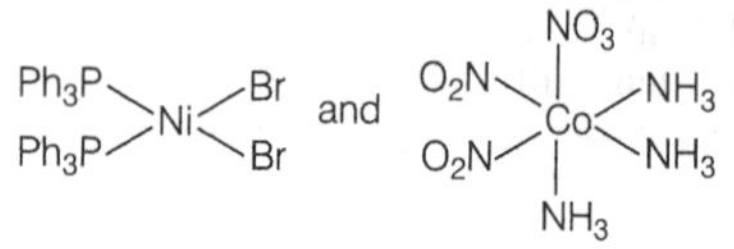

c. 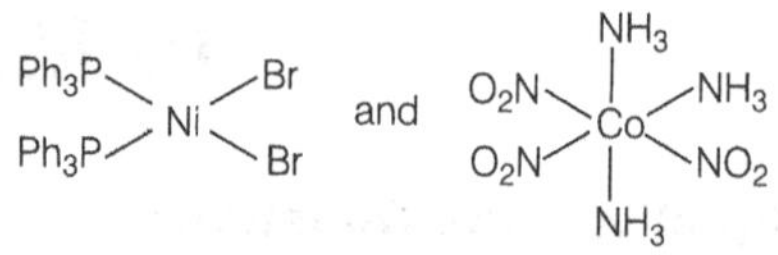

d. 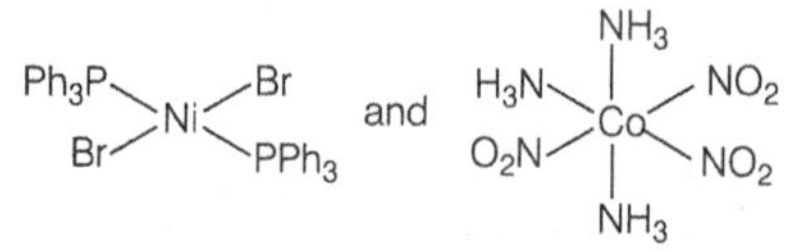

14. A certain orbital has no angular nodes and two radial nodes. The orbital is

a. 2*s* **b.** 3*s*
c. 3*p* **c.** 2*p*

15. 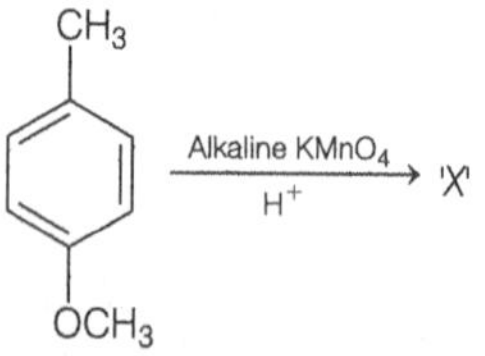

Considering the above chemical reaction, identify the product 'X'.

a. 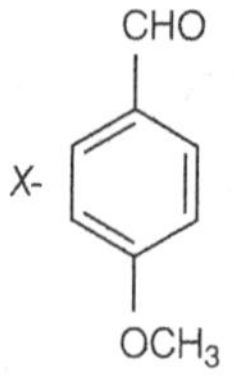 **b.** 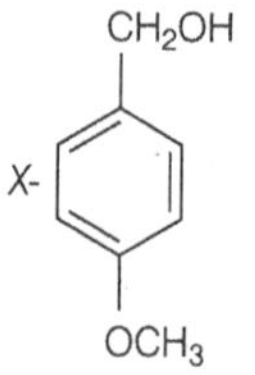

c. 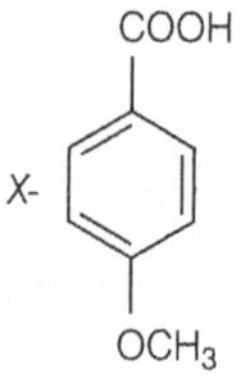 **d.** 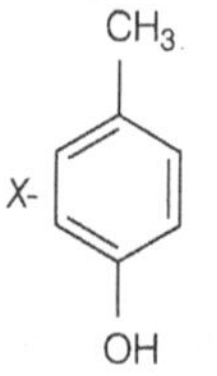

16. Match List-I with List-II.

	List-I (Process)		List-II (Catalyst)
(A)	Deacon's process	(i)	ZSM-5
(B)	Contact process	(ii)	$CuCl_2$
(C)	Cracking of hydrocarbons	(iii)	Particles 'Ni'
(D)	Hydrogenation of vegetable oils	(iv)	V_2O_5

Choose the most appropriate answer from the options given below

	A	B	C	D		A	B	C	D
a.	(ii)	(iv)	(i)	(iii)	b.	(i)	(iii)	(ii)	(iv)
c.	(iii)	(i)	(iv)	(ii)	d.	(iv)	(ii)	(i)	(iii)

17. Given below are two statements : One is labelled as Assertion A and the other labelled as Reason R.

Assertion A During the boiling of water having temporary hardness, $Mg(HCO_3)_2$ is converted to $MgCO_3$.

Reason R The solubility product of $Mg(OH)_2$ is greater than that of $MgCO_3$.

In the light of the above statements, choose the most appropriate answer from the options given below

a. Both A and R are true but R is not the correct explanation of A.
b. A is true but R is false.
c. Both A and R are true and R is the correct explanation of A.
d. A is false but R is true.

18. The number of ionisable hydrogens present in the product obtained from a reaction of phosphorus trichloride and phosphonic acid is

a. 3 b. 0 c. 2 d. 1

19. In a binary compound, atoms of element *A* form a hcp structure and those of element *M* occupy 2/3 of the tetrahedral voids of the hcp structure. The formula of the binary compound is

a. M_2A_3 b. M_4A_3 c. M_4A d. MA_3

20. The chemical that is added to reduce the melting point of the reaction mixture during the extraction of aluminium is

a. cryolite b. bauxite
c. calamine d. kaolite

Section B : Numerical Type Questions

21. *AX* is a covalent diatomic molecule, where *A* and *X* are second row elements of periodic table. Based on molecular orbital theory, the bond order of *AX* is 2.5. The total number of electrons in *AX* is .... (Round off to the nearest integer).

22. In order to prepare a buffer solution of pH 5.74, sodium acetate is added to acetic acid. If the concentration of acetic acid in the buffer is 1.0 M, the concentration of sodium acetate in the buffer is .... M. (Round off to the nearest integer).

[Given : pK_a (acetic acid = 4.74]

23. $2\,NO(g) + Cl_2(g) \rightleftharpoons 2\,NOCl(s)$

This reaction was studied at −10°C and the following data was obtained run

	$[NO]_0$	$[Cl_2]_0$	r_0
1	0.10	0.10	0.18
2	0.10	0.20	0.35
3	0.20	0.20	1.40

$[NO]_0$ and $[Cl_2]_0$ are the initial concentrations and r_0 is the initial reaction rate.

The overall order of the reaction is ....
(Round off to the nearest integer).

24. For the reaction,

$$C_2H_6 \longrightarrow C_2H_4 + H_2$$

the reaction enthalpy $\Delta_r H$ = ... kJ mol^{-1}
(Round off to the nearest integer).
[Given : Bond enthalpies in kJ mol^{-1} :
C—C = 347, C=C = 611; C—H = 414; H—H = 436]

25. .... grams of 3-hydroxy propanal (MW = 74) must be dehydrated to produce 7.8 g of acrolein (MW= 56) (C_3H_4O), if the percentage yield is 64 (Round off to the nearest integer).

[Given : Atomic masses : C = 12.0 u, H = 1.0 u, O = 16.0 u]

26. A reaction of 0.1 mole of benzylamine with bromomethane gave 23 g of benzyl trimethyl ammonium bromide. The number of moles of bromomethane consumed in this reaction are $n \times 10^{-1}$, when n = ... (Round off to the nearest integer).

(Given : Atomic masses : C = 12.0 u, H = 1.0 u, N = 14.0 u, Br = 80.0 u]

27. The total number of unpaired electrons present in the complex K_3 [Cr(oxalate)$_3$] is ....

28. 2 molal solution of a weak acid *HA* has a freezing point of 3.885°C. The degree of dissociation of this acid is _____ $\times 10^{-3}$. (Round off to the nearest integer).

[Given : Molal depression constant of water = 1.85 K kg mol^{-1}, freezing point of pure water = 0°C]

29. For the reaction,

$$2Fe^{3+}(aq) + 2I^-(aq) \longrightarrow 2Fe^{2+}(aq) + I_2(s)$$

The magnitude of the standard molar Gibbs free energy change, $\Delta_r G^\circ_m$ = − .... kJ (Round off to the nearest integer).

$$\left[\begin{array}{ll} E^\circ_{Fe^{2+}/Fe(s)} = -0.440\text{ V}; & E^\circ_{Fe^{3+}/Fe(s)} = -0.036\text{ V} \\ E^\circ_{I_2/2I^-} = 0.539\text{ V}; & F = 96500\text{ C} \end{array}\right]$$

30. Complete combustion of 3 g of ethane gives $x \times 10^{22}$ molecules of water. The value of x is .... (Round off to the nearest integer).

[Use : $N_A = 6.023 \times 10^{23}$; Atomic masses in u : C = 12.0 ; O = 16.0 ; H = 1.0]

Answers

For solution scan the QR code

1. (b)	2. (b)	3. (c)	4. (d)	5. (d)	6. (a)	7. (c)	8. (b)	9. (b)	10. (a)
11. (c)	12. (d)	13. (d)	14. (b)	15. (c)	16. (a)	17. (d)	18. (c)	19. (b)	20. (a)
21. (15)	22. (10)	23. (3)	24. (128)	25. (16)	26. (3)	27. (3)	28. (50)	29. (45)	30. (18)

18 MARCH SHIFT II

Section A : Objective Type Questions

1. The oxidation states of nitrogen in NO, NO_2, N_2O and NO_3^- are in the order of

a. $NO_3^- > NO_2 > NO > N_2O$ **b.** $NO_2 > NO_3^- > NO > N_2O$
c. $N_2O > NO_2 > NO > NO_3^-$ **d.** $NO > NO_2 > N_2O > NO_3^-$

2. In basic medium, H_2O_2 exhibits which of the following reactions ?

A. $Mn^{2+} \rightarrow Mn^{4+}$
B. $I_2 \rightarrow I^-$
C. $PbS \rightarrow PbSO_4$

Choose the most appropriate answer from the options given below.

a. A and C **b.** Only A
c. B Only **d.** A and B

3. In the reaction of hypobromite with amide, the carbonyl carbon is lost as

a. CO_3^{2-} **b.** HCO_3^- **c.** CO_2 **d.** CO

4. The oxide that shows magnetic property is

a. SiO_2 **b.** Mn_3O_4 **c.** Na_2O **d.** MgO

5. Main products formed during a reaction of 1-methoxy naphthalene with hydroiodic acid are

a. 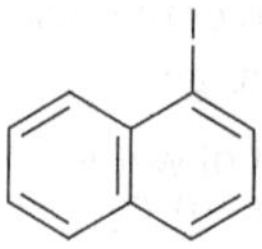 and CH_3OH **b.** 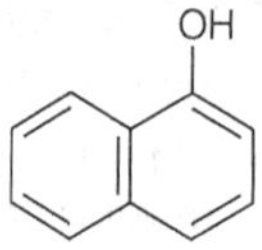 and CH_3I

c. 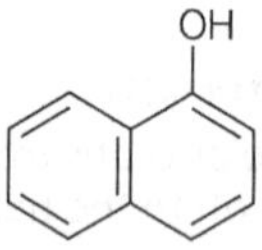 and CH_3OH **d.** 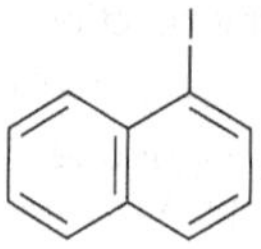 and CH_3I

6. Deficiency of vitamin K causes

a. increase in blood clotting time
b. increase in fragility of RBC's
c. cheilosis
d. decrease in blood clotting time

7. An organic compound "A" on treatment with benzene sulphonyl chloride gives compound *B*. *B* is soluble in dil. NaOH solution. Compound *A* is

a. $C_6H_5-N-(CH_3)_2$ **b.** $C_6H_5-NHCH_2CH_3$
c. $C_6H_5-CH_2NHCH_3$ **d.** $C_6H_5-\underset{\underset{CH_3}{|}}{CH}-NH_2$

8. The first ionisation energy of magnesium is smaller as compared to that of elements *X* and *Y*, but higher than that of *Z*. The elements *X*, *Y* and *Z*, respectively, are

a. chlorine, lithium and sodium
b. argon, lithium and sodium
c. argon, chlorine and sodium
d. neon, sodium and chlorine

9. The secondary valency and the number of hydrogen bonded water molecule(s) in $CuSO_4 \cdot 5H_2O$, respectively, are

a. 6 and 4 **b.** 4 and 1 **c.** 6 and 5 **d.** 5 and 1

10. Given below are two statements.

Statement I Bohr's theory accounts for the stability and line spectrum of Li^+ ion.

Statement II Bohr's theory was unable to explain the splitting of spectral lines in the presence of a magnetic field.

In the light of the above statements, choose the most appropriate answer from the options given below :

a. Both statements I and II are true.
b. Statement I is false but statement II is true.
c. Both statements I and II are false.
d. Statement I is true but statement II is false.

11. 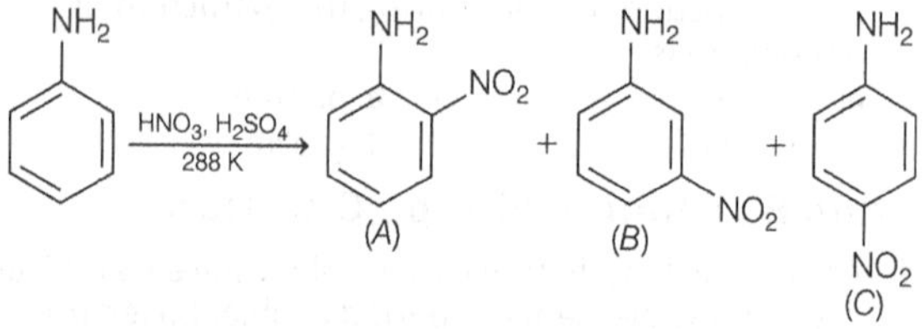

Consider the given reaction, percentage yield of

a. (*C*) > (*A*) > (*B*) **b.** (*B*) > (*C*) > (*A*)
c. (*A*) > (*C*) > (*B*) **d.** (*C*) > (*B*) > (*A*)

12. The charges on the colloidal CdS sol and TiO_2 sol are, respectively

a. positive and positive **b.** positive and negative
c. negative and negative **d.** negative and positive

13. Match List-I with List-II.

	List-I (Class of chemicals)		List-II (Example)
A.	Antifertility drug	1.	Meprobamate
B.	Antibiotic	2.	Alitame
C.	Tranquilizer	3.	Norethindrone
D.	Artificial sweetener	4.	Salvarsan

	A	B	C	D
a.	2	3	4	1
b.	4	3	2	1
c.	3	4	1	2
d.	2	4	1	3

14. 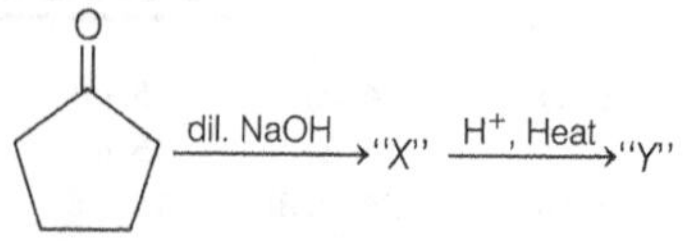

Consider the above reaction, the product '*X*' and '*Y*' respectively are

a. 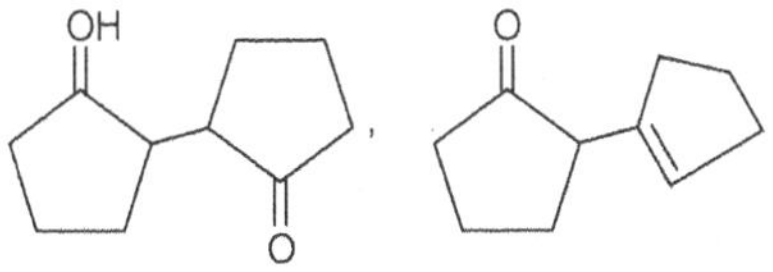

b. 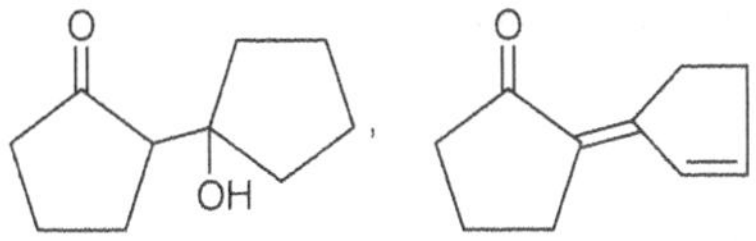

c. 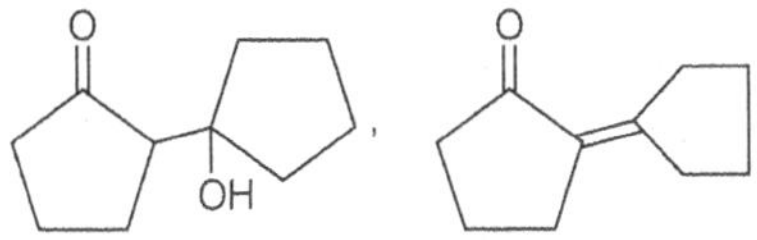

d. 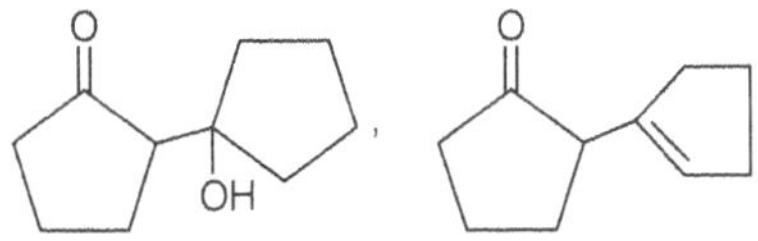

15. Match List-I with List-II

List-I		List-II	
A.	Be	1.	Treatment of cancer
B.	Mg	2.	Extraction of metals
C.	Ca	3.	Incendiary bombs and signals
D.	Ra	4.	Windows of X-ray tubes
		5.	Bearings for motor engines.

Choose the most appropriate answer the option given below.

	A	B	C	D
a.	4	3	1	2
b.	4	3	2	1
c.	3	4	5	2
d.	3	4	2	5

16. Given below are two statements.

Statement I C_2H_5OH and AgCN both can generate nucleophile.

Statement II KCN and AgCN both will generate nitrile nucleophile with all reaction conditions.

Choose the most appropriate option.

a. Statement I is true but statement II is false.
b. Both statements I and II are true.
c. Statement I is false but statement II is true.
d. Both statements I and II are false.

17. Given below are two statements

Statement I Non-biodegradable wastes are generated by the thermal power plants.

Statement II Bio-degradable detergents leads to eutrophication.

In the light of the above statements, choose the most appropriate answer from the option given below.

a. Both statements I and II are false.
b. Statement I is true but statement II is false.
c. Statement I is false but statement II is true.
d. Both statements I and II are true.

18. Match List-I with List-II.

List-I	List-II
A. Mercury	1. Vapour phase refining
B. Copper	2. Distillation refining
C. Silicon	3. Electrolytic refining
D. Nickel	4. Zone refining

Choose the most appropriate answer the option given below

	A	B	C	D
a.	1	4	2	3
b.	2	3	1	4
c.	2	3	4	1
d.	2	4	3	1

19. In the following molecules,

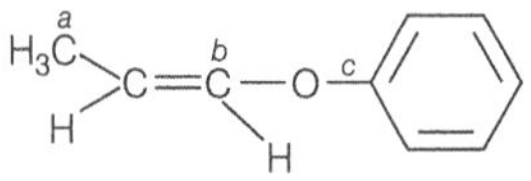

hybridisation of carbon *a*, *b* and *c* respectively are

a. sp^3, sp, sp **b.** sp^3, sp^2, sp
c. sp^3, sp^2, sp^2 **d.** sp^3, sp, sp^2

20. A hard substance melts at high temperature and is an insulator in both solid and in molten state.

This solid is most likely to be a/an

a. ionic solid **b.** molecular solid
c. metallic solid **d.** covalent solid

Numerical Type Questions

21. A reaction has a half-life of 1 min. The time required for 99.9% completion of the reaction is .... min (Round off to the nearest integer).

[Use : $\ln 2 = 0.69$, $\ln 10 = 2.3$]

22. The molar conductivities at infinite dilution of barium chloride, sulphuric acid and hydrochloric acid are 280, 860 and 426 $Scm^2\ mol^{-1}$ respectively. The molar conductivity at infinite dilution of barium sulphate is .... $S\ cm^2\ mol^{-1}$ (Round off to the nearest Integer).

23. The number of species below that have two lone pairs of electrons in their central atom is .... (Round off to the nearest integer)

SF_4, BF_4^-, ClF_3, AsF_3, PCl_5, BrF_5, XeF_4, SF_6

24. A xenon compound '*A*' upon partial hydrolysis gives XeO_2F_2. The number of lone pair of electrons present in compound *A* is .... (Round off to the nearest integer).

25. The gas phase reaction

$$2A(g) \rightleftharpoons A_2(g)$$

at 400 K has $\Delta G° = +25.2$ kJ mol^{-1}.

The equilibrium constant K_C for this reaction is .... $\times 10^{-2}$. (Round off to the nearest integer).
[Use : $R = 8.3\,\text{J mol}^{-1}\,\text{K}^{-1}$, $\ln 10 = 2.3$]
$\log_{10} 2 = 0.30$, 1 atm = 1 bar]
[antilog (– 0.3) = 0.501]

26. In Tollen's test for aldehyde, the overall number of electron(s) transferred to the Tollen's reagent formula $[Ag(NH_3)_2]^+$ per aldehyde group to form silver mirror is .... (Round off to the nearest integer).

27. The solubility of $CdSO_4$ in water is $8.0 \times 10^{-4}\,\text{mol L}^{-1}$. Its solubility in 0.01 M H_2SO_4 solution is .... $\times 10^{-6}\,\text{mol L}^{-1}$. (Round off to the nearest integer) (Assume that, solubility is much less than 0.01 M)

28. A solute a dimerises in water. The boiling point of a 2 molar solution of *A* is 100.52°C.

The percentage association of *A* is .... (Round off to the nearest integer)

[Use : K_b for water = 0.52 K kg mol^{-1}, boiling point of water = 100°C]

29. 10.0 mL of Na_2CO_3 solution is titrated against 0.2 M HCl solution. The following titre values were obtained in 5 readings. 4.8 mL, 4.9 mL, 5.0 mL, 5.0 mL and 5.0 mL based on these readings and convention of titrimetric estimation of concentration of Na_2CO_3 solution is .... mM (Round off to the nearest integer).

30.

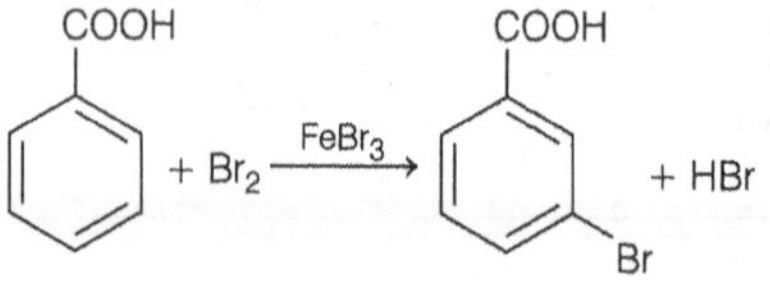

Consider the above reaction where 6.1 g of benzoic acid is used to get 7.8 g of *m*-bromo benzoic acid. The percentage yield of the product is ....
(Round off to the nearest integer).

[Given : Atomic masses : C = 12.0 u, H = 1.0 u, O = 16.0u, Br = 80.0 u]

Answers

For solution scan the QR code

1. (a)	2. (d)	3. (a)	4. (b)	5. (b)	6. (a)	7. (d)	8. (c)	9. (b)	10. (b)
11. (d)	12. (d)	13. (c)	14. (c)	15. (b)	16. (a)	17. (d)	18. (c)	19. (c)	20. (d)
21. (10)	22. (288)	23. (2)	24. (19)	25. (1.66)	26. (2)	27. (64)	28. (100)	29. (50)	30. (78)

20 JULY SHIFT I

Section A : Objective Type Questions

1. According to the valence bond theory the hybridisation of central metal atom is dsp^2 for which one of the following compounds?
   a. $NiCl_2 \cdot 6H_2O$ b. $K_2[Ni(CN)_4]$
   c. $[Ni(CO)_4]$ d. $Na_2[NiCl_4]$

2. The correct structure of Rhumann's Purple, the compound formed in the reaction of ninhydrin with proteins is

a. 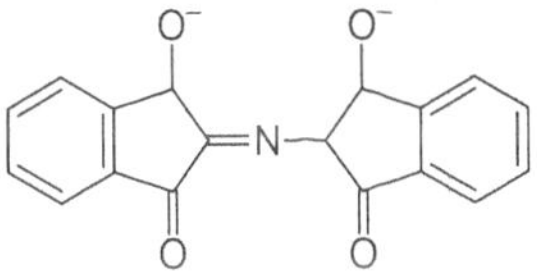

b. 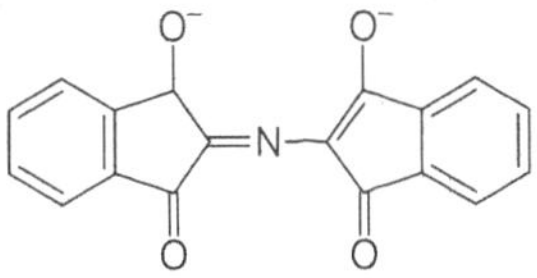

c. 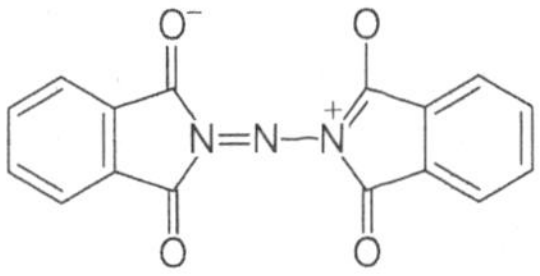

d. 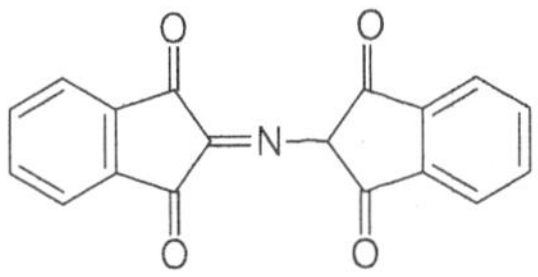

3. Green chemistry in day-to-day life is in the use of
   a. chlorine for bleaching of paper.
   b. large amount of water alone for washing clothes.
   c. Tetrachloroethene for laundry.
   d. Liquified CO_2 for dry cleaning of clothes.

4. The correct order of intensity of colors of the compound is
   a. $[Ni(CN)_4]^{2-} > [NiCl_4]^{2-} > [Ni(H_2O)_6]^{2+}$
   b. $[Ni(H_2O)_6]^{2+} > [NiCl_4]^{2-} > [Ni(CN)_4]^{2-}$
   c. $[NiCl_4]^{2-} > [Ni(H_2O)_6]^{2+} > [Ni(CN)_4]^{2-}$
   d. $[NiCl_4]^{2-} > [Ni(CN)_4]^{2-} > [Ni(H_2O)_6]^{2+}$

5. The set in which compounds have different nature is
   a. $B(OH)_3$ and H_3PO_3 b. $B(OH)_3$ and $Al(OH)_3$
   c. NaOH and $Ca(OH)_2$ d. $Be(OH)_2$ and $Al(OH)_3$

6. The species given below that does not show disproportionation reaction is
   a. BrO_4^- b. BrO^-
   c. BrO_2^- d. BrO_3^-

7. Given below are two statements. One is labelled as Assertion (A) and the other is labelled as Reason (R).
   Assertion (A) Sharp glass edge becomes smooth on heating it upto its melting point.
   Reason (R) The viscosity of glass decreases on melting.
   Choose the most appropriate answer from the options given below.
   a. A is true but R is false
   b. Both A and R are true but R is not the correct explanation of A.
   c. A is false but R is true.
   d. Both A and R are true and R is the correct explanation of A.

8. Orlon fibres are made up of
   a. polyacrylonitrile b. polyester
   c. polyamide d. cellulose

9. Given below are two statements : One is labelled as Assertion A and other is labelled as Reason R.
   Assertion A The dihedral angles in H_2O_2 in gaseous phase is 90.2° and in solid phase is 111.5°.
   Reason R The change in dihedral angle in solid and gaseous phase is due to the difference in the intermolecular forces.
   Choose the most appropriate answer from the options given below for A and R.
   a. A is correct but R is not correct.
   b. Both A and R are correct but R is not the correct explanation of A.
   c. Both A and R are correct and R is the correct explanation of A.
   d. A is not correct but R is correct.

10. Chemical nature of the nitrogen oxide compound obtained from a reaction of concentrated nitric acid and P_4O_{10} (in 4 : 1 ratio) is
    a. acidic b. basic
    c. amphoteric d. neutral

11. An inorganic compound 'X' on treatment with concentrated H_2SO_4 produces brown fumes and gives dark brown ring with $FeSO_4$ in presence of concentrated H_2SO_4. Also compound 'X' gives precipitate 'Y', when its solution in dilute HCl is treated with H_2S gas. The precipitate 'Y' on treatment with concentrated HNO_3 followed by excess of NH_4OH further gives deep blue coloured solution, compound 'X' is
    a. $Co(NO_3)_2$ b. $P(NO_2)_2$
    c. $Cu(NO_3)_2$ d. $Pb(NO_3)_2$

12. 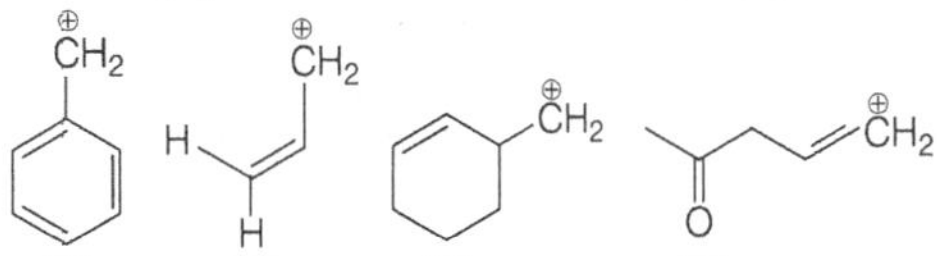

(A) (B) (C) (D)

Among the given species the resonance stabilised carbocations are
    a. (C) and (D) only b. (A), (B) and (D) only
    c. (A) and (B) only d. (A), (B) and (C) only

13. A *s*-block element (*M*) reacts with oxygen to form an oxide of the formula MO_2. The oxide is pale yellow in colour and paramagnetic. The element (*M*) is
    a. Mg b. Na c. Ca d. K

14. In the given reaction, 3-bromo-2, 2-dimethyl butane $\xrightarrow{C_2H_5OH}$ $\underset{\text{Major product}}{A}$.

Product A is

a. 2-ethoxy-3, 3-dimethyl butane
b. 1-ethoxy-3, 3-dimethyl butane
c. 2-ethoxy-2, 3-dimethyl butane
d. 2-hydroxy-3, 3-dimethyl butane

15. The metal that can be purified economically by fractional distillation method is

a. Fe b. Zn c. Cu d. Ni

16. Compound A is converted to B on reaction with $CHCl_3$ and KOH. The compound B is toxic and can be decomposed by C. A, B and C respectively are

a. primary amine, nitrile compound, conc. HCl
b. secondary amine, isonitrile compound, conc. NaOH
c. primary amine, isonitrile compound, conc. HCl
d. secondary amine, nitrile compound, conc. NaOH

17. The conditions given below are in the context of observing Tyndall effect in colloidal solutions

(A) The diameter of the colloidal particles is comparable to the wavelength of light used.
(B) The diameter of the colloidal particles is much smaller than the wavelength of light used.
(C) The diameter of the colloidal particles is much larger than the wavelength of light used.
(D) The refractive indices of the dispersed phase and the dispersion medium are comparable.
(E) The dispersed phase has a very different refractive index from the dispersion medium.

Choose the most appropriate conditions from the options given below.

a. (A) and (E) only b. (C) and (D) only
c. (A) and (D) only d. (B) and (E) only

18. Identify the incorrect statement from the following.

a. Amylose is a branched chain polymer of glucose.
b. Starch is a polymer of α-D-glucose.
c. β-glycosidic linkage makes cellulose polymer.
d. Glycogen is called as animal starch.

19.

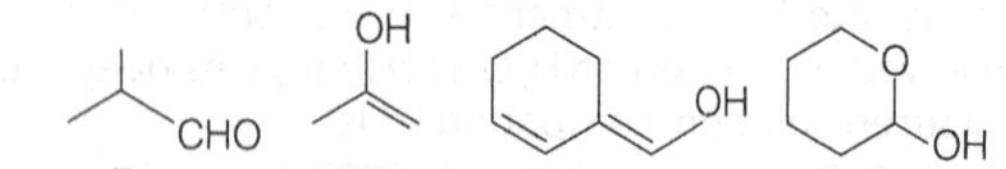

(I) (II) (III) (IV)

Which among the above compound/s does/do not form silver mirror when treated with Tollen's reagent?

a. (I), (III) and (IV) only b. Only (IV)
c. Only (II) d. (III) and (IV) only

20.

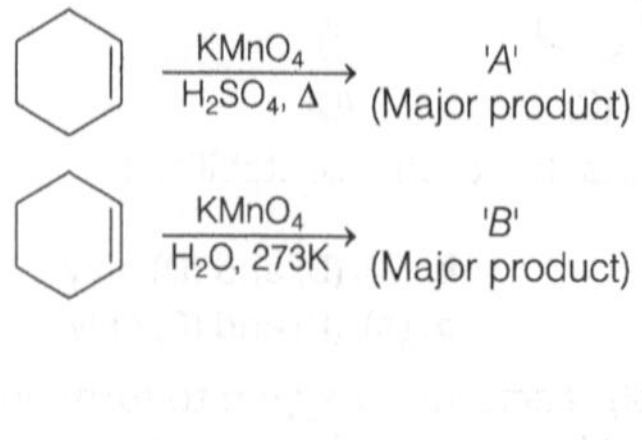

For above chemical reactions, identify the correct statement from the following

a. Both compound 'A' and compound 'B' are dicarboxylic acids.
b. Both compound 'A' and compound 'B' are diols.
c. Compound 'A' is diol and compound 'B' is dicarboxylic acid.
d. Compound 'A' is dicarboxylic acid and compound 'B' is diol.

Section B : Numerical Type Questions

21. The number of lone pairs of electron on the central I atom in I_3^- is ....

22. 250 mL of 0.5 M NaOH was added to 500 mL of 1 M HCl. The number of unreacted HCl molecules in the solution after complete reaction is .... $\times 10^{21}$. (Nearest integer) ($N_A = 6.022 \times 10^{23}$)

23. The azimuthal quantum number for the valence electrons of Ga^+ ion is ....
(Atomic number of Ga = 31)

24. The spin only magnetic moment value for the complex $[Co(CN)_6]^{4-}$ is ...... BM.
[Atomic number of Co = 27]

25. $2SO_2(g) + O_2(g) \rightleftharpoons 2SO_3(g)$

In an equilibrium mixture, the partial pressures are

$p_{SO_3} = 45\text{kPa}$; $p_{O_2} = 530$ Pa and $p_{SO_2} = 45\text{kPa}$. The equilibrium constant $K_p = ...\times 10^{-2}$. (Nearest integer)

26. The number of nitrogen atoms in a semicarbazone molecule of acetone is ....

27. To synthesis 1.0 mole of 2-methylpropan -2-ol from ethylethanoate ...... equivalents of CH_3MgBr reagent will be required. (Integer value)

28. The inactivation rate of a viral preparation is proportional to the amount of virus. In the first minute after preparation, 10% of the virus is inactivated. The rate constant for viral inactivation is ...... $\times 10^{-3}\,\text{min}^{-1}$. (Nearest integer)
[Use: ln 10 = 2.303; $\log_{10} 3 = 0.477$; property of logarithm : $\log x^y = y \log x$]

29. An average person needs about 10000 kJ energy per day. The amount of glucose (molar mass = $180.0\,\text{g mol}^{-1}$) needed to meet this energy requirement is .... g.
[Use $\Delta_c H(\text{glucose}) = 2700\,\text{kJ mol}^{-1}$]

30. At 20°C, the vapour pressure of benzene is 70 torr and that of methyl benzene is 20 torr. The mole fraction of benzene in the vapour phase at 20°C above an equimolar mixture of benzene and methyl benzene is...... $\times 10^{-2}$. (Nearest integer)

Answers

For solution scan the QR code

1. (b)	2. (d)	3. (d)	4. (c)	5. (b)	6. (a)	7. (b)	8. (a)	9. (d)	10. (a)
11. (c)	12. (c)	13. (d)	14. (c)	15. (b)	16. (c)	17. (a)	18. (a)	19. (c)	20. (d)
21. (3)	22. (226)	23. (0)	24. (2)	25. (172)	26. (3)	27. (2)	28. (106)	29. (667)	30. (78)

20 JULY SHIFT II

Section A : Objective Type Questions

1. Which one of the following pairs of isomers is an example of metamerism?

a. $CH_3CH_2CH_2CH_2CH_3$ and 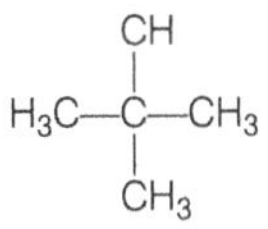

b. 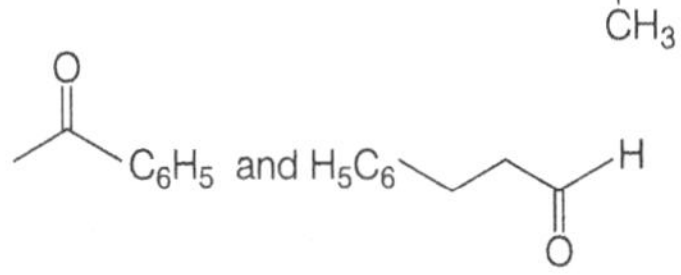

c. 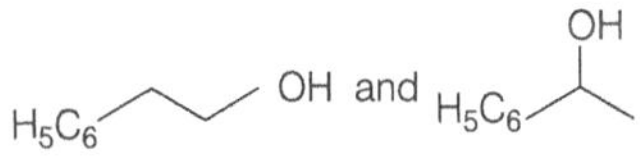

d. 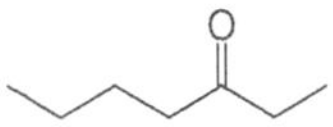 and 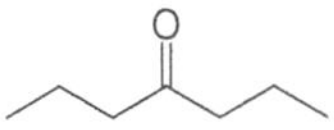

2. 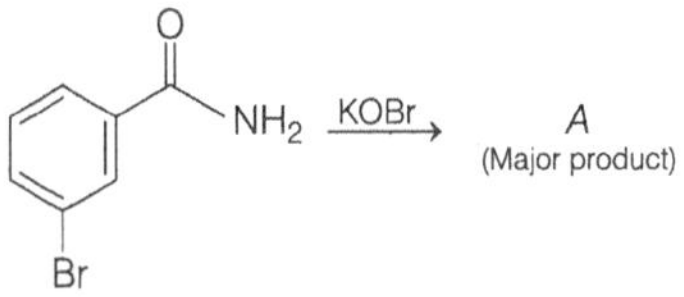

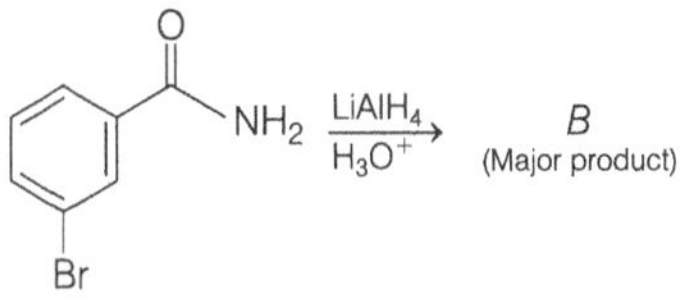

In the above reactions, products *A* and *B* respectively are

a. 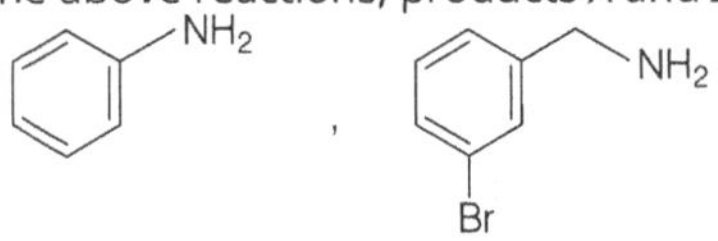

b. 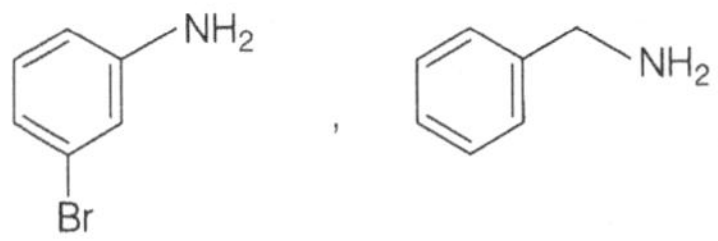

c. 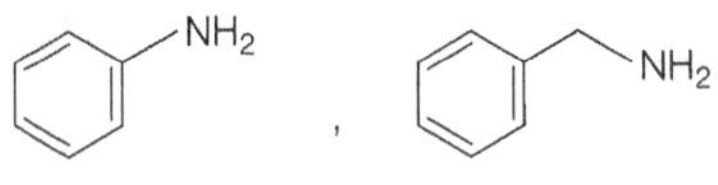

d. 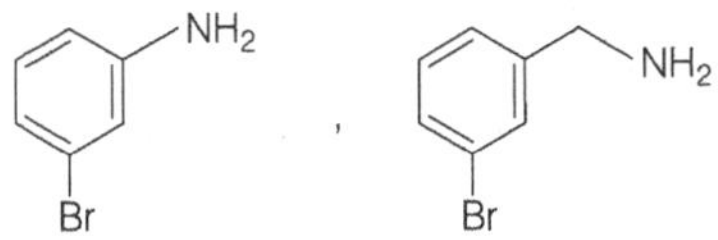

3. The major product *P* in the following reaction is

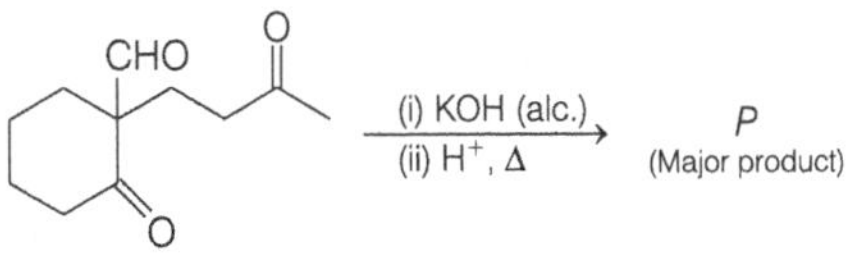

a. 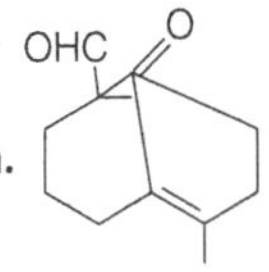 **b.** 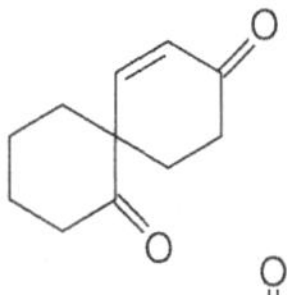

c. 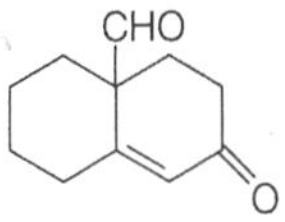 **d.** 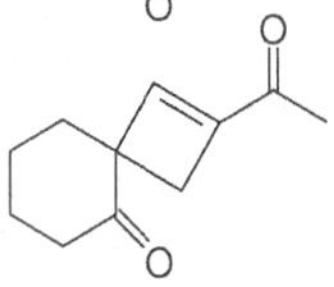

4. The single largest industrial application of dihydrogen is

a. manufacture of metal hydrides
b. rocket fuel in space research
c. in the synthesis of ammonia
d. In the synthesis of nitric acid

5. Consider two chemical reactions (*A*) and (*B*) that take place during metallurgical process :

(*A*) $ZnCO_3(s) \xrightarrow{\Delta} ZnO(s) + CO_2(g)$

(*B*) $2ZnS(s) + 3O_2(g) \xrightarrow{\Delta} 2ZnO(s) + 2SO_2(g)$

The correct option of names given to them respectively is

a. (*A*) is calcination and (*B*) is roasting
b. Both (*A*) and (*B*) are producing same product so both are roasting
c. Both (*A*) and (*B*) are producing same product so both are calcination
d. (*A*) is roasting and (*B*) is calcination

6. A solution is 0.1 M in Cl^- and 0.001 M in CrO_4^{2-}. Solid $AgNO_3$ is gradually added to it.
Assuming that the addition does not change in volume and $K_{sp}(AgCl) = 1.7 \times 10^{-10}\ M^2$ and $K_{sp}(Ag_2CrO_4) = 1.9 \times 10^{-12} M^3$

Select correct statement from the following

a. AgCl precipitates first because its K_{sp} is high.
b. Ag_2CrO_4 precipitates first as its K_{sp} is low.
c. Ag_2CrO_4 precipitates first because the amount of Ag^+ needed is low.
d. AgCl will precipitate first as the amount of Ag^+ needed to precipitate is low.

7. Outermost electronic configuration of a group-13 element *E* is $4s^2\ 4p^1$. The electronic configuration of an element of *p*-block period-five placed diagonally to element, *E* is

a. $[Kr]3d^{10}4s^2 4p^2$ **b.** $[Ar]3d^{10}4s^2 4p^2$
c. $[Xe]5d^{10}6s^2 6p^2$ **d.** $[Kr]4d^{10}5s^2 5p^2$

8. Metallic sodium does not react normally with

a. gaseous ammonia **b.** but-2-yne
c. ethyne **d.** *tert*-butyl alcohol

9. Spin only magnetic moment of an octahedral complex of Fe^{2+} in the presence of a strong field ligand in BM is

a. 4.89 **b.** 2.82
c. 0 **d.** 3.46

10. Which one of the following species doesn't have a magnetic moment of 1.73 BM (spin only value) ?

a. O_2^+ **b.** CuI

c. $[Cu(NH_3)_4]Cl_2$ **d.** O_2^-

11. Which one of the following statements is not true about enzymes ?

a. Enzymes are non-specific for a reaction and substrate.

b. Almost all enzymes are proteins.

c. Enzymes work as catalysts by lowering the activation energy of a biochemical reaction.

d. The action of enzymes is temperature and pH specific.

12. The hybridisations of the atomic orbitals of nitrogen in NO_2^-, NO_2^+ and NH_4^+ respectively are

a. sp^3, sp^2 and sp **b.** sp, sp^2 and sp^3

c. sp^3, sp and sp^2 **d.** sp^2, sp and sp^3

13. Bakelite is a cross-linked polymer of formaldehyde and

a. PHBV **b.** buna-S **c.** novolac **d.** dacron

14. Benzene on nitration gives nitrobenzene in presence of HNO_3 and H_2SO_4 mixutre, where

a. both H_2SO_4 and HNO_3 act as a bases

b. HNO_3 acts as an acid and H_2SO_4 acts as a base

c. both H_2SO_4 and NHO_3 act as an acids

d. HNO_3 acts as a base and H_2SO_4 acts as an acid

15.

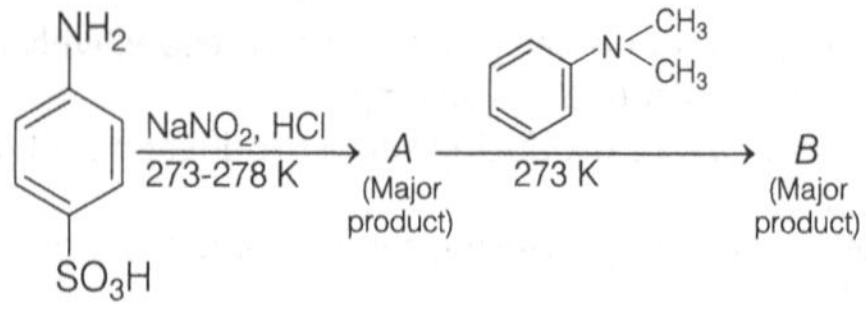

Consider the above reaction, compound *B* is

a.

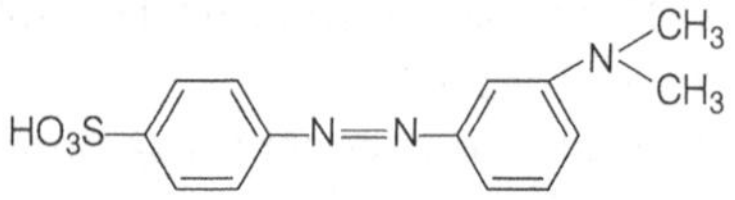

b.

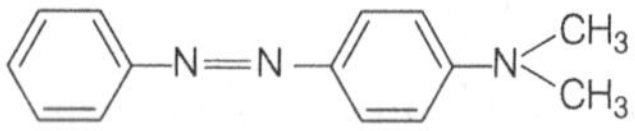

c.

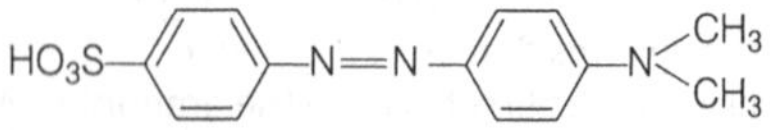

d.

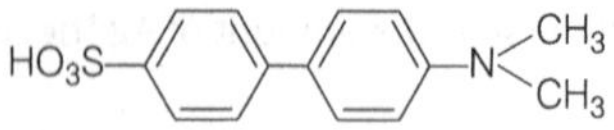

16.

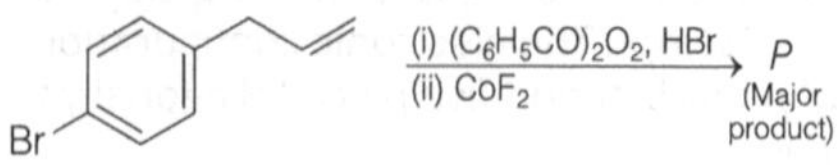

Major product *P* of above reaction is

a.

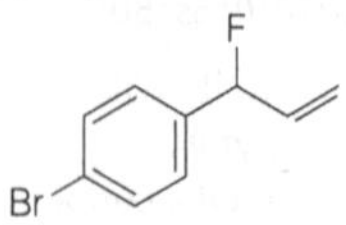

b.

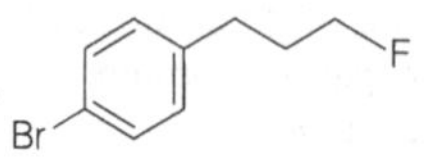

c.

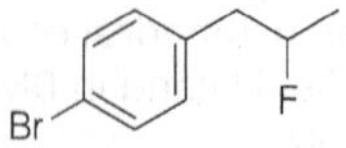

d.

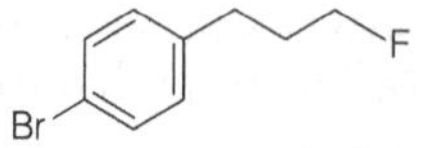

17. Cu^{2+} salt reacts with potassium iodide to give

a. Cu_2I_2 **b.** Cu_2I_3 **c.** CuI **d.** $Cu(I_3)_2$

18. In Carius method, halogen containing organic compound is heated with fuming nitric acid in the presence of

a. HNO_3 **b.** $AgNO_3$ **c.** $CuSO_4$ **d.** $BaSO_4$

19. Which one of the following gases is reported to retard photosynthesis ?

a. CO **b.** CFCs

c. CO_2 **d.** NO_2

20. (A)

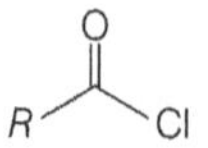

(B)

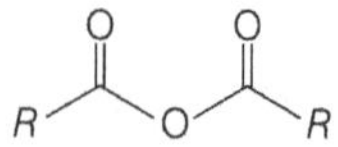

(C)

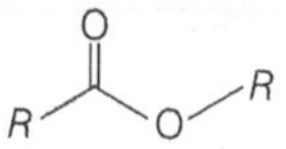

(D)

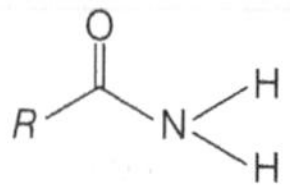

The correct order of their reactivity towards hydrolysis at room temperature is

a. (A) > (B) > (C) > (D) **b.** (D) > (A) > (B) > (C)

c. (D) > (B) > (A) > (C) **d.** (A) > (C) > (B) > (D)

Section B : Numerical Type Questions

21. For a given chemical reaction, $A \longrightarrow B$ at 300 K the free energy change is -49.4 kJ mol^{-1} and the enthalpy of reaction is 51.4 kJ mol^{-1}. The entropy change of the reaction is ...... $JK^{-1}mol^{-1}$.

22. The wavelength of electrons accelerated from rest through a potential difference of 40 kV is $x \times 10^{-12}$ m. The value of *x* is ...... (Nearest integer)

Given : Mass of electron $= 9.1 \times 10^{-31}$ kg

Charge on an electron $= 1.6 \times 10^{-19}$ C

Planck's constant $= 6.63 \times 10^{-34}$ Js

23. The vapour pressures of *A* and *B* at 25°C are 90 mm Hg and 15 mm Hg respectively. If *A* and *B* are mixed such that the mole-fraction of *A* in the mixture is 0.6, then the mole fraction of *B* in the vapour phase is $x \times 10^{-1}$. The value of *x* is .... (Nearest integer)

24. 4g equimolar mixture of NaOH and Na_2CO_3 contains *x* g of NaOH and *y* g of Na_2CO_3. The value of *x* is ...... g. (Nearest integer)

25. When 0.15 g of an organic compound was analysed using Carius method for estimation of bromine, 0.2397 g of AgBr was obtained. The percentage of bromine in the organic compound is .... (Nearest integer)

[Atomic mass : Silver = 108, bromine = 80]

26. 100 mL of 0.0018% (*w*/*v*) solution of Cl^- ion was the minimum concentration of Cl^- required to precipitate a negative sol in one h. The coagulating value of Cl^- ion is .... (Nearest integer)

27. $PCl_5(g) \longrightarrow PCl_3(g) + Cl_2(g)$

In the above first order reaction, the concentration of PCl_5 reduces from initial concentration 50 mol L^{-1} to 10 mol L^{-1} in 120 minutes at 300 K. The rate constant for the reaction at 300 K is $x \times 10^{-2}$ min^{-1}. The value of *x* is ....

[Given, $\log 5 = 0.6989$]

28. Diamond has a three dimensional structure of C atoms formed by covalent bonds. The structure of diamond has face centred cubic lattice, where 50% of the tetrahedral voids are also occupied by carbon atoms. The number of carbon atoms present per unit cell of diamond is ....

29. An aqueous solution of $NiCl_2$ was heated with excess sodium cyanide in presence of strong oxidising agent to form $[Ni(CN)_6]^{2-}$. The total change in number of unpaired electrons on metal centre is ....

30. Potassium chlorate is prepared by electrolysis of KCl in basic solution as shown by following equation

$6OH^- + Cl^- \longrightarrow ClO_3^- + 3H_2O + 6e^-$

A current of xA has to be passed for 10th to produce 10.0 g of potassium chlorate. The value of x is .... (Nearest integer)

(Molar mass of $KClO_3$ =122.6 g mol^{-1}, F= 96500 C)

Answers

For solution scan the QR code

1. (d)	2. (d)	3. (b)	4. (c)	5. (a)	6. (d)	7. (d)	8. (b)	9. (c)	10. (b)
11. (a)	12. (d)	13. (c)	14. (d)	15. (c)	16. (d)	17. (a)	18. (b)	19. (d)	20. (a)
21. (360)	22. (6)	23. (1)	24. (1)	25. (68)	26. (1)	27. (1)	28. (8)	29. (2)	30. (1)

22 JULY SHIFT II

Section A : Objective Type Questions

1. The water having more dissolved O_2 is

a. boiling water **b.** water at 80°C
c. polluted water **d.** water at 4° C

2. Which one of the following statements for D. Mendeleef, is incorrect ?

a. He authored the textbook - principles of chemistry.
b. At the time, he proposed periodic table of elements structure of atom was known.
c. Element with atomic number 101 is named after him.
d. He invented accurate barometer.

3. Which purification technique is used for high boiling organic liquid compound (decomposes near its boiling point)?

a. Simple distillation **b.** Steam distillation
c. Fractional distillation **d.** Reduced pressure distillation

4. Which of the following compounds will provide a tertiary alcohol on reaction with excess of CH_3MgBr followed by hydrolysis?

a. 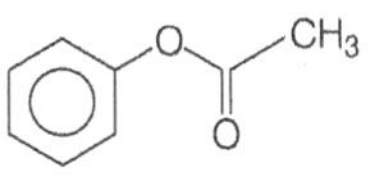

b. 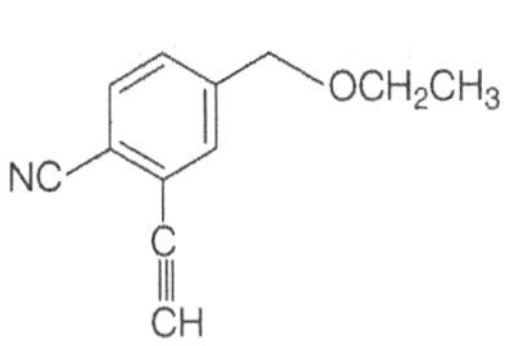

c. 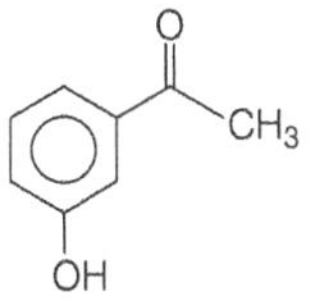

d. 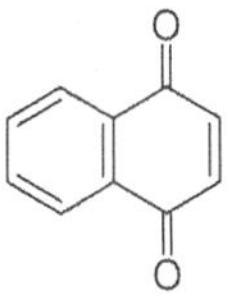

5. Which of the following compounds does not exhibit resonance ?

a. $CH_3CH_2OCH{=}CH_2$

b. 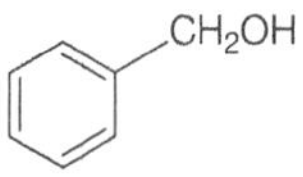

c. $CH_3CH_2CH_2CONH_2$

d. $CH_3CH_2CH{=}CHCH_2NH_2$

6. Match List I with List II.

List-I (Elements)		List-II (Properties)	
A.	Ba	(i)	Organic solvent soluble compounds
B.	Ca	(ii)	Outer electronic configuration $6s^2$
C.	Li	(iii)	Oxalate insoluble in water
D.	Na	(iv)	Formation of very strong monoacidic base

Choose the correct answer from the options given below

	A	B	C	D		A	B	C	D
a.	(ii)	(iii)	(i)	(iv)	**b.**	(iv)	(i)	(ii)	(iii)
c.	(iii)	(ii)	(iv)	(i)	**d.**	(i)	(iv)	(ii)	(iii)

7. 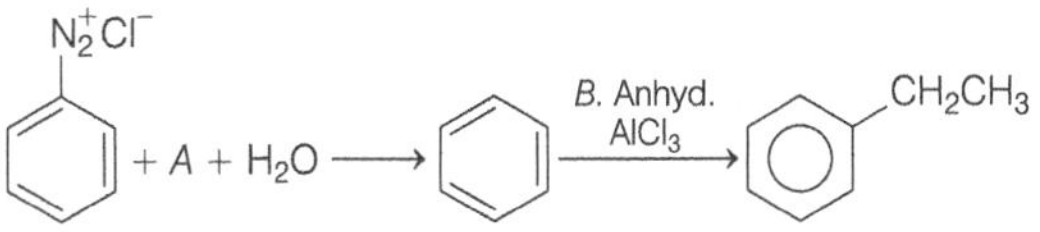

(Major product)

In the chemical reactions given above A and B respectively are

a. H_3PO_2 and CH_3CH_2Cl **b.** CH_3CH_2OH and H_3PO_2
c. H_3PO_2 and CH_3CH_2OH **d.** CH_3CH_2Cl and H_3PO_2

8. Isotope(s) of hydrogen which emits low energy β^- particles with $t_{1/2}$ value > 12 years is/are

a. protium **b.** tritium
c. deuterium **d.** deuterium and tritium

9. Match List I and List II.

List I (Species)		List II (Hybrid orbitals)	
(A)	SF_4	(i)	sp^3d^2
(B)	IF_5	(ii)	d^2sp^3
(C)	NO_2^+	(iii)	sp^3d
(D)	NH_4^+	(iv)	sp^3
		(v)	sp

	A	B	C	D		A	B	C	D
a.	(i)	(ii)	(v)	(iii)	**b.**	(ii)	(i)	(iv)	(v)
c.	(iii)	(i)	(v)	(iv)	**d.**	(iv)	(iii)	(ii)	(v)

10. When silver nitrate solution is added to potassium iodide solution then the sol produced is

a. AgI/I^- **b.** AgI/Ag^+
c. KI/NO_3^- **d.** $AgNO_3/NO_3^-$

11. Which of the following molecules does not show stereoisomerism?

a. 3, 4 - dimethyl hex-3ene **b.** 3- methyl hex-1-ene
c. 3 - ethyl hex-3-ene **d.** 4- methyl hex-1-ene

12. Given below are the statements about diborane.

(A) Diborane is prepared by the oxidation of $NaBH_4$ and I_2.

(B) Each boron atom is in sp^2-hybridised state.

(C) Diborane has one bridged 3 centre -2 - electron bond.

(D) Diborane is a planar molecule.

The option with correct statement(s) is

a. (C) and (D) only **b.** (A) only
c. (C) only **d.** (A) and (B) only

13. Which one of the following group-15 hydride is the strongest reducing agent?

a. AsH_3 **b.** BiH_3 **c.** PH_3 **d.** SbH_3

14. Match List I with List II.

List I		List II	
(A)	Chloroprene	(i)	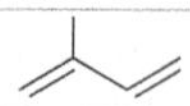
(B)	Neoprene	(ii)	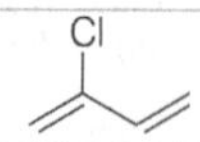
(C)	Acryonitrile	(iii)	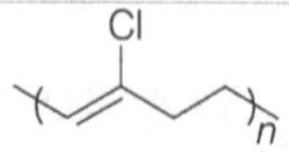
(D)	Isoprene	(iv)	$CH_2=CH—CN$

Choose the correct answer from the options given below.

	A	B	C	D		A	B	C	D
a.	(iii)	(iv)	(ii)	(i)	**b.**	(ii)	(iii)	(iv)	(i)
c.	(ii)	(i)	(iv)	(iii)	**d.**	(iii)	(i)	(iv)	(ii)

15. The set having ions which are coloured and paramagnetic both is

a. Cu^{2+}, Cr^{3+}, Sc^+ **b.** $Cu^{2+}, Zn^{2+}, Mn^{4+}$
c. Sc^{3+}, V^{5+}, Ti^{4+} **d.** $Ni^{2+}, Mn^{7+}, Hg^{2+}$

16. Thiamine and pyridoxine are also known respectively as

a. vitamin B_2 and vitamin E **b.** vitamin E and vitamin B_2
c. vitamin B_6 and vitamin B_2 **d.** vitamin B_1 and vitamin B_6

17. Sulphide ion is soft base and its ores are common for metals

(A) Pb (B) Al (C) Ag (D) Mg

Choose the correct answer from the options given below

a. (A) and (C) only **b.** (A) and (D) only
c. (A) and (B) only **d.** (C) and (D) only

18. An organic compound $A(C_6H_6O)$ gives dark green colouration with ferric chloride. On treatment with $CHCl_3$ and KOH, followed by acidification gives compound *B*. Compound *B* can also be obtained from compound *C* on reaction with pyridinium chlorochromate (PCC). Identify *A*, *B* and C.

a. 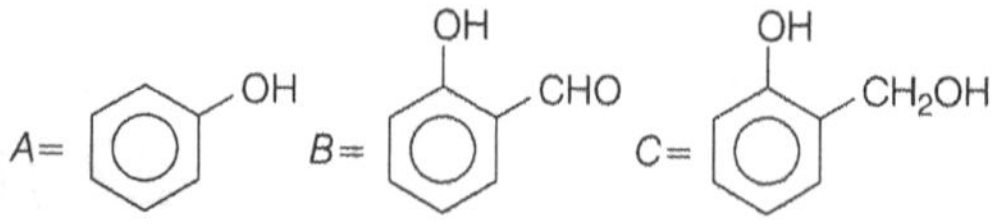

b. 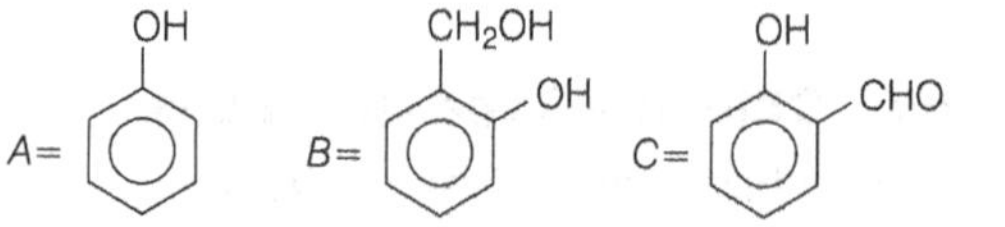

c. 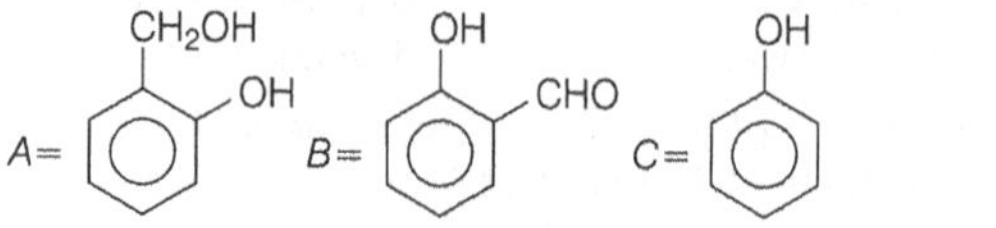

d. 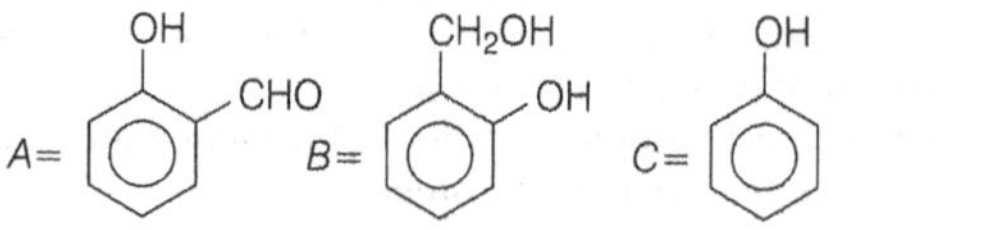

19. Which one of the following reactions does not occur?

a. 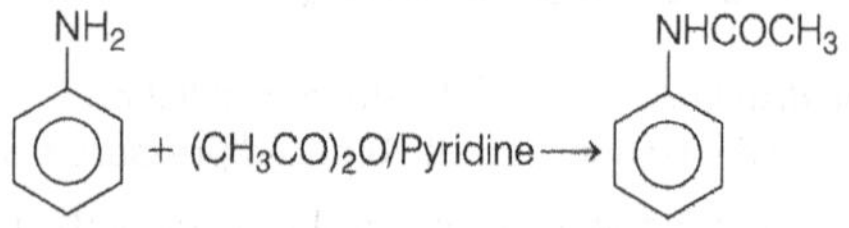

b. 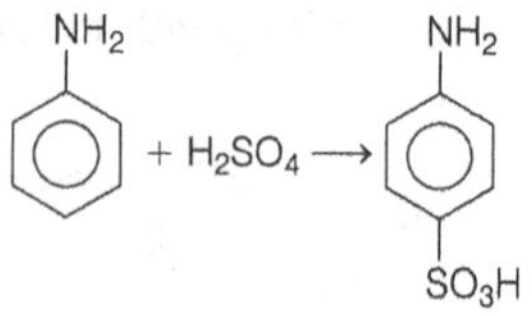

c. 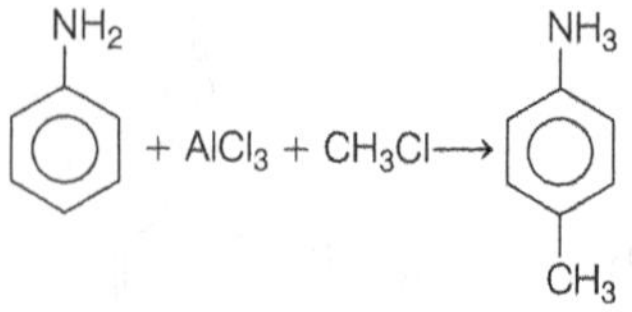

d. 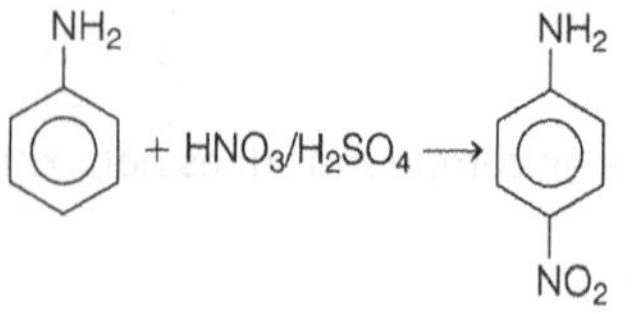

20. Which one of the following 0.06 M aqueous solutions has lowest freezing point?
a. $Al_2(SO_4)_3$ **b.** $C_6H_{12}O_6$
c. KI **d.** K_2SO_4

Section B : Numerical Type Questions

21. The total number of unpaired electrons present in $[Co(NH_3)_6]Cl_2$ and $[Co(NH_3)_6]Cl_3$ is ....

22. Methylation of 10 g of benzene gave 9.2 g of toluene. Calculate the percentage yield of toluene .... (Nearest integer)

23. The number of acyclic structural isomers (including geometrical isomers) for pentene are ....

24. Assume a cell with the following reaction
$Cu(s) + 2Ag^+(1\times10^{-3}M) \longrightarrow Cu^{2+}(0.250M) + 2Ag(s), E^{\ominus}_{cell} = 2.97\,V$
E_{cell} for the above reaction is .... V.
(Nearest integer)
[Given : $\log 2.5 = 0.3979, T = 298$ K]

25. Value of K_p for the equilibrium reaction
$N_2O_4(g) \rightleftharpoons 2NO_2(g)$ at 288 K is 47.9.
The K_C for this reaction at same temperature is ......
(Nearest integer)
($R = 0.083$ L bar $K^{-1}mol^{-1}$)

26. If the standard molar enthalpy change for combustion of graphite powder is -2.48×10^2 kJ mol^{-1}, the amount of heat generated on combustion of 1 g of graphite powder is ...... kJ. (Nearest integer)

27. A copper complex crystallising in a ccp lattice with a cell edge of 0.4518 nm has been revealed by employing X-ray diffraction studies. The density of a copper complex is found to be 7.62 g cm^{-3}.
The molar mass of copper complex is ...... g mol^{-1}. (Nearest integer)
[Given : $N_A = 6.022\times10^{23}\,mol^{-1}$]

28. Number of electrons that vanadium ($Z = 23$) has in *p*-orbitals is equal to ....

29. $N_2O_5(g) \longrightarrow 2NO_2(g) + \frac{1}{2}O_2(g)$
In the above first order reaction the initial concentration of N_2O_5 is 2.40×10^{-2} mol L^{-1} at 318 K. The concentration of N_2O_5 after 1 hour was 1.60×10^{-2} mol L^{-1}. The rate constant of the reaction at 318 K is ...... $\times10^{-3}min^{-1}$. (Nearest integer) [Given : $\log 3 = 0.477, \log 5 = 0.699$]

30. If the concentration of glucose ($C_6H_{12}O_6$) in blood is 0.72 gL^{-1}, the molarity of glucose in blood is ...... $\times10^{-3}$ M. (Nearest integer)
[Given : Atomic mass of C = 12, H = 1, O = 16 u]

Answers

For solution scan the QR code

1. *(d)*	2. *(b)*	3. *(d)*	4. *(a)*	5. *(d)*	6. *(a)*	7. *(a)*	8. *(b)*	9. *(c)*	10. *(a)*
11. *(c)*	12. *(b)*	13. *(b)*	14. *(b)*	15. *(a)*	16. *(d)*	17. *(a)*	18. *(a)*	19. *(c)*	20. *(a)*
21. *(1)*	22. *(78)*	23. *(6)*	24. *(3)*	25. *(2)*	26. *(21)*	27. *(106)*	28. *(12)*	29. *(7)*	30. *(4)*

25 JULY SHIFT I

Section A : Objective Type Questions

1. 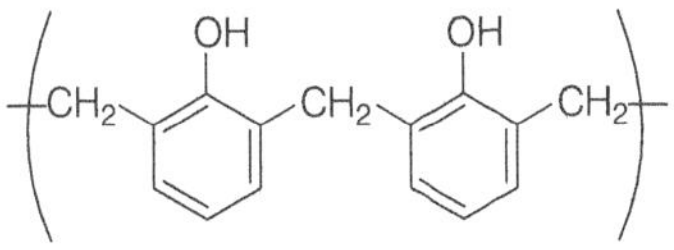

is a repeating unit for
a. novolac **b.** buna-N
c. acrilan **d.** neoprene

2. Which one of the following species responds to an external magnetic field
a. $[Fe(H_2O)_6]^{3+}$ **b.** $[Ni(CN)_4]^{2-}$
c. $[Co(CN)_6]^{3-}$ **d.** $[Ni(CO)_4]$

3. 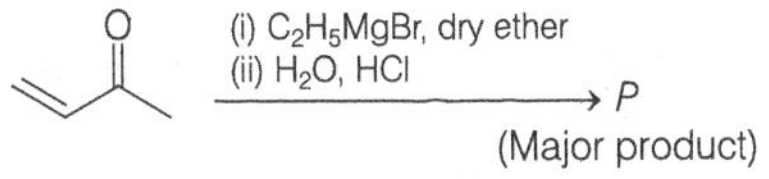

Consider the above reaction, the major product *P* is

a. 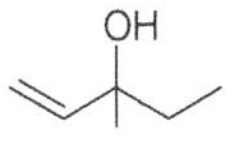

b. 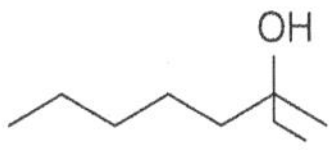

c. 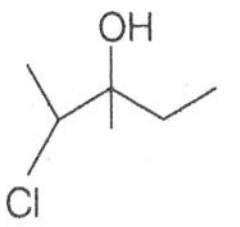

d. 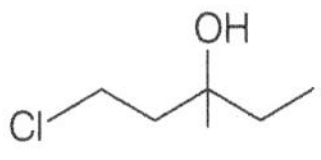

4. Sodium stearate $CH_3(CH_2)_{16}COO^-Na^+$ is an anionic surfactant which forms micelles in oil.
Choose the correct statement for it from the following.
a. It forms spherical micelles with $CH_3(CH_2)_{16}^-$ group pointing towards the centre of sphere.
b. It forms non-spherical micelles with —COO^- group pointing outwards on the surface.
c. It forms spherical micelles with $CH_3(CH_2)_{16}^-$ group pointing outwards on the surface of sphere
d. It forms non-spherical micelles with $CH_3(CH_2)_{16}^-$ group pointing towards the centre.

5. The water soluble protein is

a. fibrin **b.** albumin
c. myosin **d.** collagen

6. At 298.2 K the relationship between enthalpy of bond dissociation (in kJ mol^{-1}) for hydrogen (E_H) and its isotope, deuterium (E_D), is best described by

a. $E_H = \frac{1}{2} E_D$
b. $E_H = E_D$
c. $E_H \simeq E_D - 7.5$
d. $E_H = 2E_D$

7.
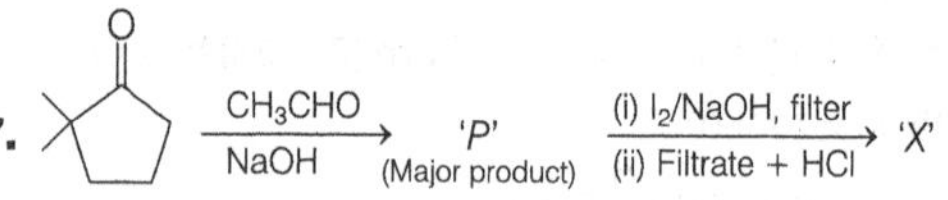

Consider the given reaction, the product X is

a.
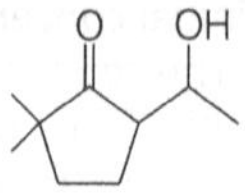

b.
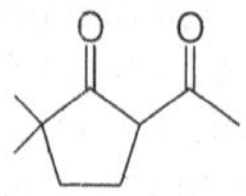

c.
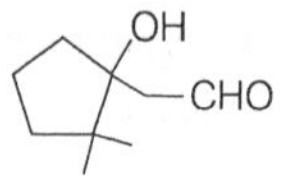

d.
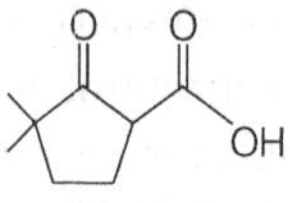

8.
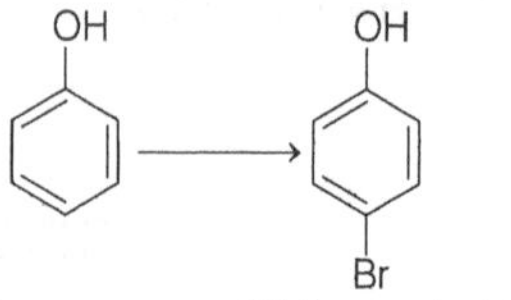

(Major product)

The given reaction can occur in the presence of

(A) bromine water
(B) Br_2 in CS_2, 273 K
(C) Br_2 / $FeBr_3$
(D) Br_2 in $CHCl_3$, 273 K

Choose the correct answer from the options given below

a. (B) and (D) only
b. (A) and (C) only
c. (B), (C) and (D) only
d. (A), (B) and (D) only

9. Given below are two statements, one is labelled as : Assertion (A) and other is labelled as Reason (R).

Assertion (A) Gabriel phthalimide synthesis cannot be used to prepare aromatic primary amines.

Reason (R) Aryl halides do not undergo nucleophilic substitution reaction.

In the light of the above statements, choose the correct answer from the options given below

a. Both (A) and (R) true but (R) is not the correct explanation of (A).
b. (A) is false but (R) is true.
c. Both (A) and (R) true and (R) is correct explanation of (A).
d. (A) is true but (R) is false.

10. For the following graphs

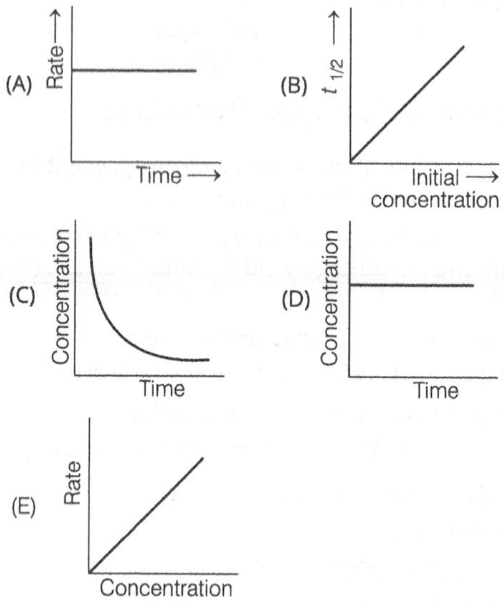

Choose from the options given below, the correct one regarding order of reaction is

a. (B) zero order (C) and (E) first order
b. (A) and (B) zero order (E) first order
c. (B) and (D) zero order (E) first order
d. (A) and (B) zero order (C) and (E) first order

11. Which one of the products of the following reactions does not react with Hinsberg reagent to form sulphonamide?

a.
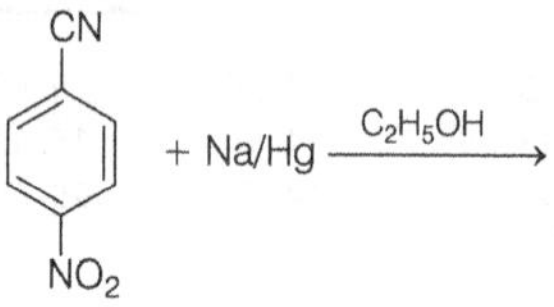

b.
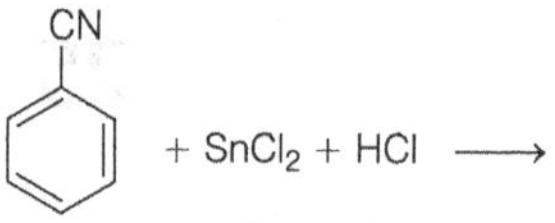

c.
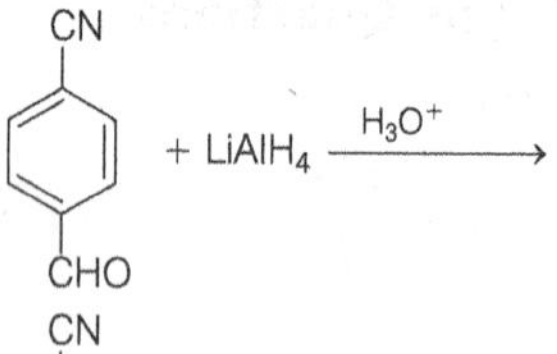

d.
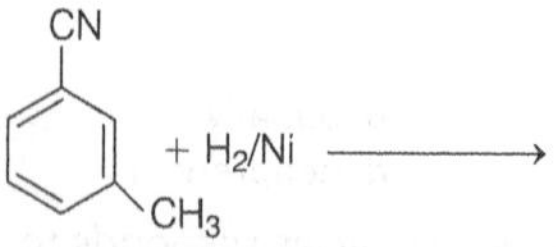

12. The ionic radii of K^+, Na^+, Al^{3+} and Mg^{2+} are in the order

a. $Na^+ < K^+ < Mg^{2+} < Al^{3+}$ **b.** $Al^{3+} < Mg^{2+} < K^+ < Na^+$
c. $Al^{3+} < Mg^{2+} < Na^+ < K^+$ **d.** $K^+ < Al^{3+} < Mg^{2+} < Na^+$

13. Which one of the following compounds of group-14 elements is not known?

a. $[GeCl_6]^{2-}$ **b.** $[Sn(OH)_6]^{2-}$
c. $[SiCl_6]^{2-}$ **d.** $[SiF_6]^{2-}$

14. Which one among the following resonating structures is not correct ?

a. 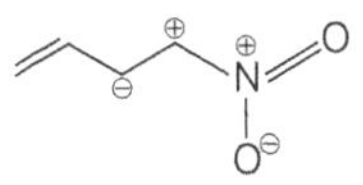
b. 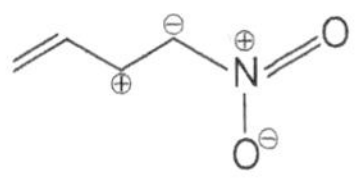
c. 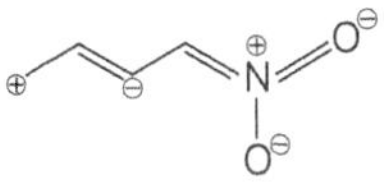
d. 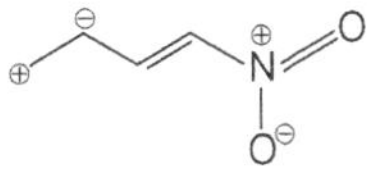

15. Given below are two statements.

Statement I None of the alkaline earth metal hydroxides dissolve in alkali.

Statement II Solubility of alkaline earth metal hydroxides in water increases down the group.

According the above statements, choose the most appropriate answer from the options given below

a. Statement I is correct but statement II is incorrect.
b. Statement I is incorrect but statement II is correct.
c. Statement I and statement II both are incorrect.
d. Statement I and statement II both are correct.

16. The correct order of following 3*d*-metal oxides, according to their oxidation numbers is

(A) CrO_3 (B) Fe_2O_3
(C) MnO_2 (D) V_2O_5
(E) Cu_2O

a. (D) > (A) > (B) > (C) > (E) b. (A) > (C) > (D) > (B) > (E)
c. (A) > (D) > (C) > (B) > (E) d. (C) > (A) > (D) > (E) > (B)

17. Which one of the following chemical agent is not being used for dry cleaning of clothes ?

a. H_2O_2 b. CCl_4 c. Liquid CO_2 d. $Cl_2C=CCl_2$

18. Which one of the following compounds will liberate CO_2, when treated with $NaHCO_3$?

a. $(CH_3)_3\overset{\oplus}{N}H\overset{\ominus}{Cl}$ b. $(CH_3)_3\overset{\oplus\ominus}{NOH}$
c. $CH_3—\overset{O}{\overset{\|}{C}}—NH_2$ d. CH_3NH_2

19. In the leaching of alumina from bauxite, the ore expected to leach out in the process by reacting with NaOH is

a. TiO_2 b. Fe_2O_3 c. ZnO d. SiO_2

20. An organic compound '*A*' C_4H_8 on treatment with $KMnO_4/H^+$ yield's compound '*B*' C_3H_6O.
Compound '*A*' also yields compound '*B*' an ozonolysis. Compound '*A*' is

a. 2-methylpropene b. 1-methylcyclopropane
c. but-2-ene c. cyclobutane

Section B : Numerical Type Questions

21. The number of sigma bonds in

$$H_3C—\underset{\underset{H}{|}}{C}=CH—C\equiv C—H$$ is ....

22. Three moles of AgCl get precipitated when one mole of an octahedral co-ordination compound with empirical formula $CrCl_3 \cdot 3NH_3 \cdot 3H_2O$ reacts with excess of silver nitrate. The number of chloride ions satisfying the secondary valency of the metal ion is ....

23. A source of monochromatic radiation of wavelength 400 nm provides 1000 J of energy in 10 seconds. When this radiation falls on the surface of sodium, $x \times 10^{20}$ electrons are ejected per second. Assume that wavelength 400 nm is sufficient for ejection of electron from the surface of sodium metal. The value of *x* is ....
(Nearest integer)
($h = 6.626 \times 10^{-34}$ Js)

24. CO_2 gas is bubbled through water during a soft drink manufacturing process at 298 K. If CO_2 exerts a partial pressure of 0.835 bar then *x* m mol of CO_2 would dissolve in 0.9 L of water. The value of *x* is ....
(Nearest integer)
(Henry's law constant for CO_2 at 298 K is 1.67×10^3 bar)

25. For the reaction, $A + B \rightleftharpoons 2C$
The value of equilibrium constant is 100 at 298 K. If the initial concentration of all the three species is 1 M each, then the equilibrium concentration of *C* is $x \times 10^{-1}$ M. The value of *x* is .... (Nearest integer)

26. Consider the cell at 25° C
$Zn|Zn^{2+}(aq), (1M)||Fe^{3+}(aq), Fe^{2+}(aq)|Pt(s)$
The fraction of total iron present as Fe^{3+} ion at the cell potential of 1.500 V is $x \times 10^{-2}$. The value of *x* is ....
(Nearest integer)
(Given, $E^\circ_{Fe^{3+}/Fe^{2+}} = 0.77$ V, $E^\circ_{Zn^{2+}/Zn} = -0.76$ V)

27. At 298 K, the enthalpy of fusion of a solid (*X*) is 2.8 kJ mol^{-1} and the enthalpy of vaporisation of the liquid (*X*) is 98.2 kJ mol^{-1}. The enthalpy of sublimation of the substance (*X*) in kJ mol^{-1} is .... (Nearest integer)

28. A home owner uses $4.00 \times 10^3 m^3$ of methane (CH_4) gas, (assume CH_4 is an ideal gas) in a Year to heat his home. Under the pressure of 1.0 atm and 300 K, mass of gas used is $x \times 10^5$ g. The value of *x* is .... (Nearest integer)
(Given, $R = 0.083$ L atm $K^{-1}mol^{-1}$)

29. When 10 mL of an aqueous solution of Fe^{2+} ions was titrated in the presence of dil. H_2SO_4 using diphenylamine indicator, 15 mL of 0.02 M solution of $K_2Cr_2O_7$ was required to get the end point. The molarity of the solution containing Fe^{2+} ions is $x \times 10^{-2}$ M. The value of *x* is .... (Nearest integer)

30. Consider the complete combustion of butane, the amount of butane utilised to produce 72.0 g of water is ...... $\times 10^{-1}$ g. (Nearest integer)

Answers

For solution scan the QR code

1. *(a)*	2. *(a)*	3. *(c)*	4. *(a)*	5. *(b)*	6. *(c)*	7. *(d)*	8. *(c)*	9. *(c)*	10. *(b)*
11. *(b)*	12. *(c)*	13. *(c)*	14. *(a)*	15. *(b)*	16. *(c)*	17. *(b)*	18. *(a)*	19. *(d)*	20. *(a)*
21. *(10)*	22. *(0)*	23. *(2)*	24. *(25)*	25. *(25)*	26. *(24)*	27. *(101)*	28. *(26)*	29. *(18)*	30. *(464)*

25 JULY SHIFT II

Setion A : Objective Type Questions

1. In the following the correct bond order sequence is

a. $O_2^{2-} > O_2^+ > O_2^- > O_2$ **b.** $O_2^+ > O_2^- > O_2^{2-} > O_2$
c. $O_2^+ > O_2 > O_2^- > O_2^{2-}$ **d.** $O_2 > O_2^- > O_2^{2-} > O_2^+$

2. A biodegradable polyamide can be made from

a. glycine and isoprene
b. hexamethylene diamine and adipic acid
c. glycine and aminocaproic acid
d. styrene and caproic acid

3. Match List I with List II.

List I (Elements)			List II (Properties)
(A)	Li	I.	Poor water solubility of I^- salt
(B)	Na	II.	Most abundant element in cell fluid
(C)	K	III.	Bicarbonate salt used in fire extinguisher
(D)	Cs	IV.	Carbonate salt decomposes easily on heating

Choose the correct answer from the options given below.

	A	B	C	D		A	B	C	D
a.	IV	III	II	I	**b.**	I	III	II	IV
c.	IV	II	III	I	**d.**	I	II	III	IV

4. Which one of the following metal complexes is most stable?

a. $[Co(en)(NH_3)_4]Cl_2$ **b.** $[Co(en)_3]Cl_2$
c. $[Co(en)_2(NH_3)_2]Cl_2$ **d.** $[Co(NH_3)_6]Cl_2$

5. Match List I with List II : (Both having metallurgical terms)

List I			List II
(A)	Concentration of Ag ore	I.	Reverberatory furnace
(B)	Blast furnace	II.	Pig iron
(c)	Blister copper	III.	Leaching with dilute NaCN solution
(d)	Froth floatation method	IV.	Sulphide ores

Choose the correct answer from the options given below.

	A	B	C	D		A	B	C	D
a.	III	II	I	IV	**b.**	III	IV	I	II
c.	IV	I	III	II	**d.**	IV	III	II	I

6. The ionic radii of F^- and O^{2-} respectively are 1.33 Å and 1.4 Å while the covalent radius of N is 0.74Å

The correct statement for the ionic radius of N^{3-} from the following is

a. it is smaller than F^- and N
b. it is bigger than O^{2-} and F^-
c. it is bigger than F^- and N, but smaller than of O^{2-}
d. it is smaller than O^{2-} and F^-, but bigger than of N

7. The correct decreasing order of densities of the following compounds is

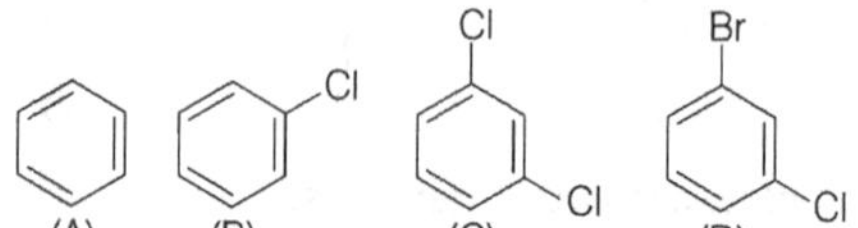

(A) (B) (C) (D)

a. (D) > (C) > (B) > (A) **b.** (C) > (D) > (A) > (B)
c. (C) > (B) > (A) > (D) **d.** (A) > (B) > (C) > (D)

8. $C_6H_5NO_2 \xrightarrow{Sn+HCl} A \xrightarrow[H^+]{C_6H_5N_2^+Cl^-} P$ (Yellow coloured compound)

Consider the above reaction, the product *P* is

a. 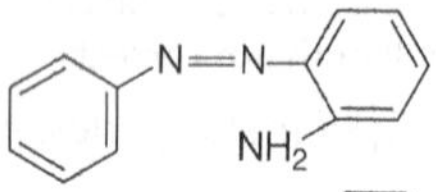

b. 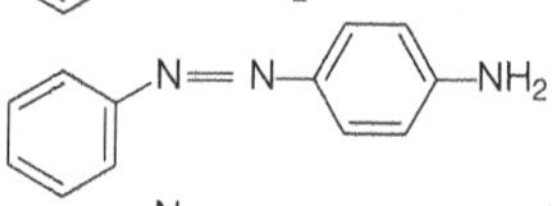

c. 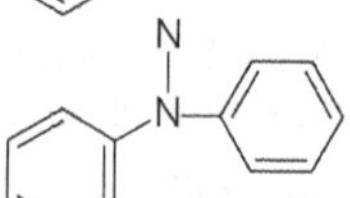

d. 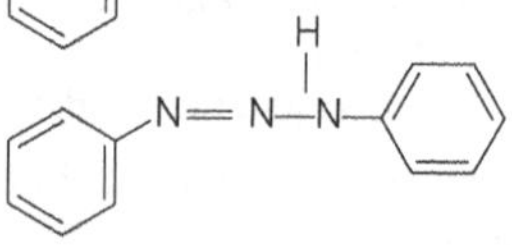

9. A reaction of benzonitrile with one equivalent CH_3MgBr followed by hydrolysis produces a yellow liquid *P*. The compound *P* will give positive

a. iodoform test **b.** Schiff's test
c. ninhydrin's test **d.** Tollen's test

10. The spin only magnetic moments (in BM) for free Ti^{3+}, V^{2+}, and Sc^{3+} ions respectively are
(Atomic number: Sc = 21, Ti = 22, V = 23)

a. 3.87, 1.73, 0 **b.** 1.73, 3.87, 0
c. 1.73, 0, 3.87 **d.** 0, 3.87, 1.73

11. Which one of the following is correct structure for cytosine?

a. 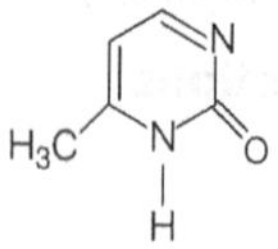

b. 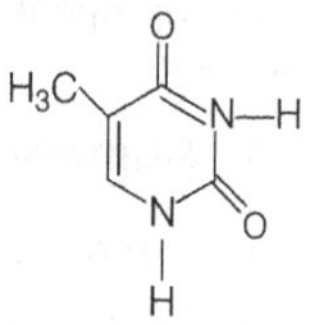

c. 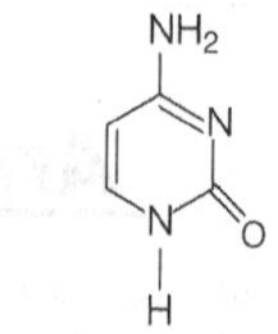

d. 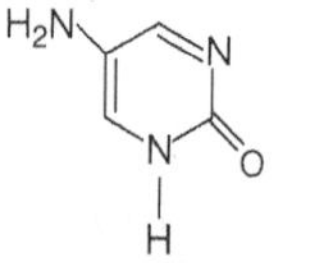

12. Identify the species having one π - bond and maximum number of canonical forms from the following

a. SO_3 **b.** O_2

c. SO_2 **d.** CO_3^{2-}

13. Which one of the following metals forms interstitial hydride easily ?

a. Cr **b.** Fe

c. Mn **d.** Co

14. 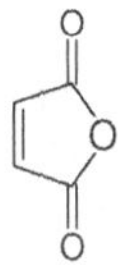

(Maleic anhydride)

Maleic anhydride can be prepared by

a. heating *trans*-but-2-enedioic acid

b. heating *cis*-but-2-enedioic acid

c. treating *cis*-but-2-enedioic acid with alcohol and acid

d. treating *trans*-but-2-enedioic acid with alcohol and acid

15. Given below are two statements :

Statement I Chlorofluoro carbons break down by radiation in the visible energy region and release chlorine gas in the atmosphere which then reacts with stratospheric ozone.

Statement II Atmospheric ozone reacts with nitric oxide to give nitrogen and oxygen gases, which add to the atmosphere.

For the above statements choose the correct answer from the options given below :

a. Statement I is incorrect but statement II is true

b. Both statement I and II are false

c. Statement I is correct but statement II is false

d. Both statement I and II are correct

16. 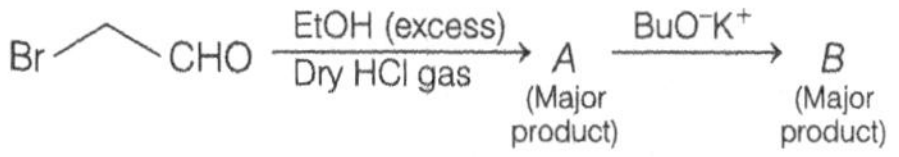

[where, Et = C_2H_5, $^tBu = (CH_3)_3C—$]

Consider the above reaction sequence, product *A* and product *B* formed respectively are

a. 

b. 

c. 

d. 

17. Match List I with List II.

	List I		List II
(A)	Cheese	I.	Dispersion of liquid in liquid
(B)	Pumice stone	II.	Dispersion of liquid in gas
(C)	Hair cream	III.	Dispersion of gas in solid
(D)	Cloud	IV.	Dispersion of liquid in solid

Choose the most appropriate answer from the options given below.

	A	B	C	D
a.	IV	III	II	I
b.	IV	I	III	II
c.	III	IV	I	II
d.	IV	III	I	II

18. What is the major product *P* of the following reaction ?

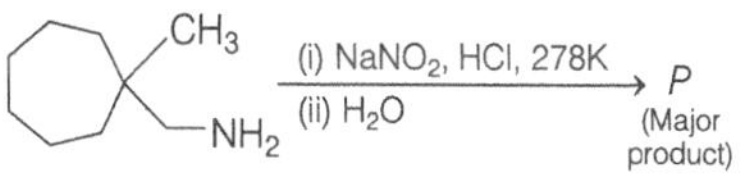

a. 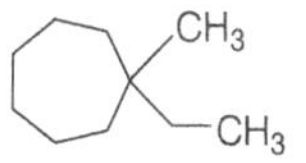

b. 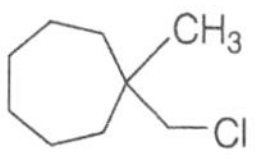

c. 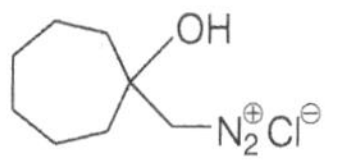

d. 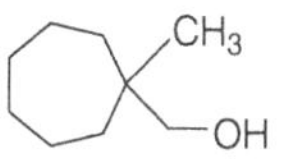

19. Identify the process in which change in the oxidation state is five

a. $Cr_2O_7^{2-} \longrightarrow 2Cr^{3+}$ **b.** $MnO_4^- \longrightarrow Mn^{2+}$

c. $CrO_4^{2-} \longrightarrow Cr^{3+}$ **d.** $C_2O_4^{2-} \longrightarrow 2CO_2$

20. Which among the following is the strongest acid ?

a. $CH_3CH_2CH_2CH_3$ **b.** 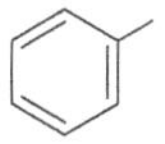

c.  **d.** 

Section B : Numerical Type Questions

21. A system does 200 J of work and at the same time absorbs 150 J of heat. The magnitude of the change in internal energy is .... J. (Nearest integer)

22. An accelerated electron has a speed of $5 \times 10^6 ms^{-1}$ with an uncertainty of 0.02%. The uncertainty in finding its location while in motion is $x \times 10^{-9}$m. The value of *x* is .... (Nearest integer)

[Use mass of electron $= 9.1 \times 10^{-31}$ kg, $h = 6.63 \times 10^{-34}$ Js, $\pi = 3.14$]

23. Number of electrons present in 4*f* orbital of Ho^{3+} ion is .... (Given, atomic number of Ho = 67)

24. 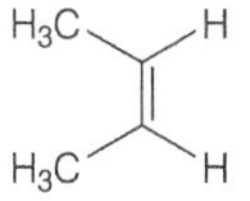 + $Br_2 \xrightarrow{CCl_4}$ Product (*P*)

Consider the above chemical reaction. The total number of stereoisomers possible for product *P* is ......

25. For a chemical reaction $A \longrightarrow B$, it was found that concentration of B is increased by 0.2 mol L^{-1} in 30 min. The average rate of the reaction is ......$\times 10^{-1}$ mol L^{-1} h^{-1}. (Nearest integer)

26. The number of significant figures in 0.00340 is ....

27. Assuming that $Ba(OH)_2$ is completely ionised in aqueous solution under the given conditions the concentration of H_3O^+ ions in 0.005 M aqueous solution of $Ba(OH)_2$ at 298 K is ...... $\times 10^{-12}$mol L^{-1}. (Nearest integer)

28. 0.8 g of an organic compound was analysed by Kjeldahl's method for the estimation of nitrogen. If the percentage of nitrogen in the compound was found to be 42%, then .... mL of 1 M H_2SO_4 would have been neutralised by the ammonia evolved during the analysis.

29. When 3.00 g of a substance 'X' is dissolved in 100 g of CCl_4, it raises the boiling point by 0.60 K. The molar mass of the substance 'X' is .... g mol^{-1}. (Nearest integer)
[Given, K_b for CCl_4 is 5.0 K kg mol^{-1}]

30. An LPG cylinder contains gas at a pressure of 300 kPa at 27°C. The cylinder can withstand the pressure of 12×10^6 Pa. The room in which the cylinder is kept catches fire. The minimum temperature at which the bursting of cylinder will take place is .... °C. (Nearest integer)

Answers

For solution scan the QR code

1. (c)	2. (c)	3. (a)	4. (b)	5. (a)	6. (b)	7. (a)	8. (b)	9. (a)	10. (b)
11. (c)	12. (d)	13. (a)	14. (b)	15. (b)	16. (a)	17. (d)	18. (d)	19. (b)	20. (d)
21. (50)	22. (58)	23. (10)	24. (2)	25. (4)	26. (3)	27. (1)	28. (12)	29. (250)	30. (927)

27 JULY SHIFT I

Section A : Objective Type Questions

1. Which one of the following compounds will give orange precipitate when treated with 2, 4 dinitrophenyl hydrazine?

a. 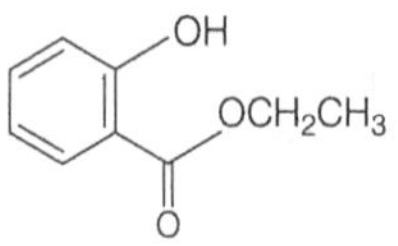

b. 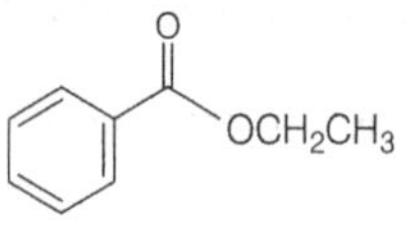

c. 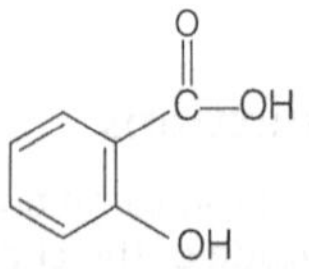

d. 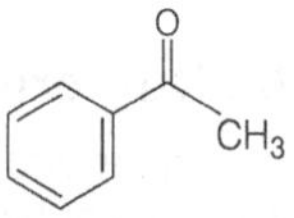

2. The product obtained from the electrolytic oxidation of acidified sulphate solutions, is

a. HSO_4^-
b. HO_3SOOSO_3H
c. HO_2SOSO_2H
d. HO_3SOSO_3H

3. The parameters of the unit cell of a substance are $a = 2.5, b = 3.0, c = 4.0, \alpha = 90°, \beta = 120°, \gamma = 90°$. The crystal system of the substance is

a. hexagonal
b. orthorhombic
c. monoclinic
d. triclinic

4. The oxidation states of 'P' in $H_4P_2O_7$, $H_4P_2O_5$ and $H_4P_2O_6$, respectively are

a. 7, 5 and 6
b. 5, 4 and 3
c. 5, 3 and 4
d. 6, 4 and 5

5. For a reaction of order n, the unit of the rate constant is

a. mol^{1-n} L^{1-n} s
b. mol^{1-n} L^{2n} s^{-1}
c. mol^{1-n} L^{n-1} s^{-1}
d. mol^{1-n} L^{1-n} s^{-1}

6. Given below are two statements.

Statement I Aniline is less basic than acetamide.

Statement II In aniline, the lone pair of electrons on nitrogen atom is delocalised over benzene ring due to resonance and hence less available to a proton.

Choose the most appropriate option :

a. Statement I is true but statement II is false.
b. Statement I is false but statement II is true.
c. Both statement I and statement II are true.
d. Both statement I and statement II are false.

7. The type of hybridisation and magnetic property of the complex $[MnCl_6]^{3-}$, respectively, are

a. sp^3d^2 and diamagnetic
b. d^2sp^3 and diamagnetic
c. d^2sp^3 and paramagnetic
d. sp^3d^2 and paramagnetic

8. The number of geometrical isomers found in the metal complexes $[PtCl_2(NH_3)_2]$, $[Ni(CO)_4]$, $[Ru(H_2O)_3Cl_3]$ and $[CoCl_2(NH_3)_4]^+$ respectively, are

a. 1, 1, 1, 1
b. 2, 1, 2, 2
c. 2, 0, 2, 2
d. 2, 1, 2, 1

9. Which one of the following statements is not correct ?

a. Eutrophication indicates that water body is polluted.
b. The dissolved oxygen concentration below 6 ppm inhibits fish growth.
c. Eutrophication leads to increase in the oxygen level in water.
d. Eutrophication leads to anaerobic conditions.

10. Given below are two statements :

Statement I Rutherford's gold foil experiment cannot explain the line spectrum of hydrogen atom.

Statement II Bohr's model of hydrogen atom contradicts Heisenberg's uncertainty principle.

In the light of the above statements, choose the most appropriate answer from the options given below :

a. Statement I is false but statement II is true.
b. Statement I is true but statement II is false.
c. Both statement I and statement II are false.
d. Both statement I and statement II are true.

11. Presence of which reagent will affect the reversibility of the following reaction, and change it to a irreversible reaction :

$$CH_4 + I_2 \underset{\text{Reversible}}{\overset{h\nu}{\rightleftharpoons}} CH_3{-}I + HI$$

a. HOCl
b. dilute HNO_2
c. Liquid NH_3
d. Concentrated HIO_3

12. Which one among the following chemical tests is used to distinguish monosaccharide from disaccharide ?

a. Seliwanoff's test **b.** Iodine test
c. Barfoed test **d.** Tollen's test

13. Match List-I with List-II.

	List-I (Drug)		**List-II** (Class of drug)
A.	Furacin	I.	Antibiotic
B.	Arsphenamine	II.	Tranquilisers
C.	Dimetone	III.	Antiseptic
D.	Valium	IV.	Synthetic antihistamines

Choose the most appropriate match :

	A	B	C	D
a.	I	III	IV	II
b.	III	IV	II	I
c.	II	I	III	IV
d.	III	I	IV	II

14. The statement that is incorrect about Ellingham diagram is

a. provides idea about the reaction rate.
b. provides idea about free energy change.
c. provides idea about changes in the phases during the reaction.
d. provides idea about reduction of metal oxide.

15. 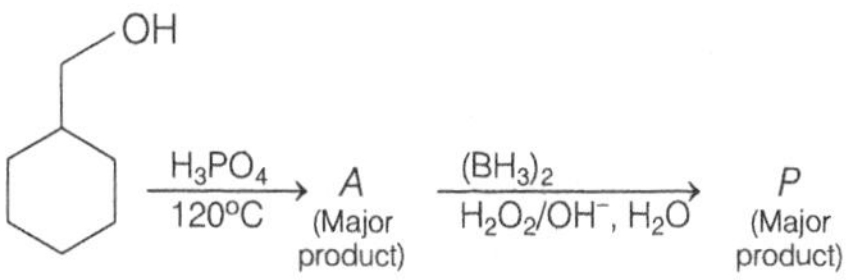

Consider the above reaction and identify the product (*P*).

a. 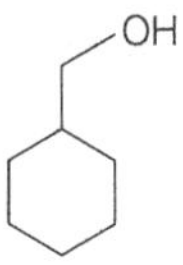

b. 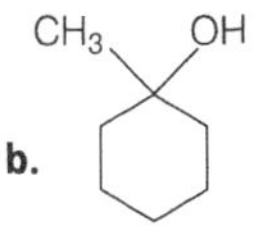

c. 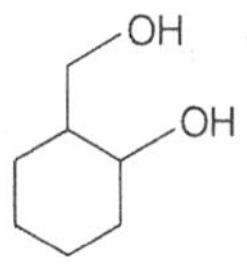

d. 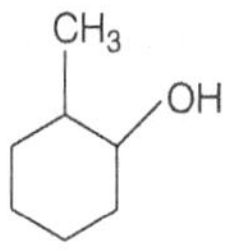

16. 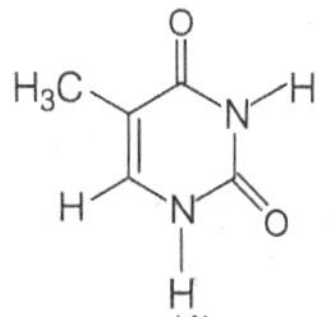

(A)

The compound '*A*' is a complementary base of .... in DNA stands.

a. uracil **b.** guanine
c. adenine **d.** cytosine

17. Staggered and eclipsed conformers of ethane are

a. polymers **b.** rotamers
c. enantiomers **d.** mirror images

18. Match List-I with List-II.

List-I	**List-II**
A. NaOH	I. Acidic
B. $Be(OH)_2$	II. Basic
C. $Ca(OH)_2$	III. Amphoteric
D. $B(OH)_3$	
E. $Al(OH)_3$	

Choose the most appropriate answer from the options given below

	A	B	C	D	E
a.	II	II	III	II	III
b.	II	III	II	I	III
c.	II	II	III	I	III
d.	II	I	II	III	III

19. 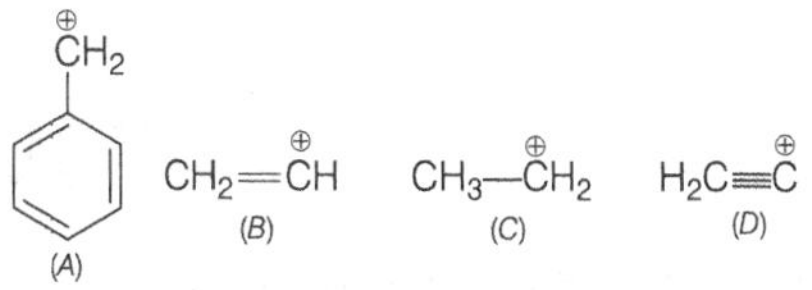

The correct order of stability of given carbocation is

a. $A > C > B > D$ **b.** $D > B > C > A$
c. $D > B > A > C$ **d.** $C > A > D > B$

20. Given below are two statements.

One is labelled as Asseriton A and the other labelled as Reason R.

Assertion A Lithium halides are some what covalent in nature.

Reason R Lithium possess high polarisation capability.

According the above statements, choose the most appropriate answer from the options given below

a. A is true but R is false
b. A is false but R is true
c. Both A and R are true but R is not the correct explanation of A
d. Both A and R are true and R is the correct explanation of A

Section B : Numerical Type Questions

21. The density of NaOH solution is 1.2 g cm^{-3}. The molality of this solution is .... m.

(Round off to the nearest integer)

[Use : Atomic masses : Na = 23.0 u, O = 16.0 u, H = 1.0 u, density of H_2O : 1.0g cm^{-3}]

22. CO_2 gas adsorbs on charcoal following Freundlich adsorption isotherm. For a given amount of charcoal, the mass of CO_2 adsorbed becomes 64 times when the pressure of CO_2 is doubled.

The value of n in the Freundlich isotherm equation is .... $\times 10^{-2}$. (Round off to the nearest integer)

23. The conductivity of a weak acid HA of concentration 0.001 mol L^{-1} is 2.0×10^{-5} S cm^{-1}. If $\Lambda_m^\circ(HA) = 190$ S cm^2 mol^{-1}, the ionisation constant (K_a) of HA is equal to .... $\times 10^{-6}$. (Round off to the nearest integer)

24. 1.46 g of a biopolymer dissolved in a 100 mL water at 300 K exerted an osmotic pressure of 2.42×10^{-3} bar.

The molar mass of the biopolymer is .... $\times 10^4$ g mol^{-1}. (Round off to the nearest integer)

[Use : $R = 0.083$ L bar mol^{-1} K^{-1}]

25. An organic compound is subjected to chlorination to get compound A using 5.0 g of chlorine. When 0.5 g of compound A is reacted with $AgNO_3$ [Carius method], the percentage of chlorine in compound A is .... when it forms 0.3849 g of AgCl. (Round off to the nearest integer)

(Atomic masses of Ag and Cl are 107.87 and 35.5 respectively)

26. The number of geometrical isomers possible in triamminetrinitrocobalt (III) is X and in trioxalatochromate (III) is Y. Then, the value of $X + Y$ is ....

27. In gaseous triethyl amine the "—C—N—C—" bond angle is .... degree.

28. For water at 100°C and 1 bar,

$\Delta_{vap}H - \Delta_{vap}U$ = ... $\times 10^2$ J mol^{-1}.

(Round off to the nearest integer)

[Use : $R = 8.31$ J mol^{-1} K^{-1}]

[Assume volume of $H_2O(l)$ is much smaller than volume of $H_2O(g)$. Assume $H_2O(g)$ treated as an ideal gas]

29. $PCl_5 \rightleftharpoons PCl_3 + Cl_2, K_c = 1.844$

3.0 moles of PCl_5 is introduced in a 1 L closed reaction vessel at 380 K. The number of moles of PCl_5 at equilibrium is .... $\times 10^{-3}$.

(Round off to the nearest integer)

30. The difference between bond orders of CO and $NO^{\oplus}$ is $X/2$. where, x = ...

(Round off to the nearest integer).

Answers

For solution scan the QR code

1. *(d)*	2. *(b)*	3. *(c)*	4. *(c)*	5. *(c)*	6. *(b)*	7. *(d)*	8. *(c)*	9. *(c)*	10. *(d)*
11. *(d)*	12. *(c)*	13. *(d)*	14. *(a)*	15. *(d)*	16. *(c)*	17. *(b)*	18. *(b)*	19. *(a)*	20. *(d)*
21. *(5)*	22. *(17)*	23. *(12)*	24. *(15)*	25. *(19)*	26. *(2)*	27. *(108)*	28. *(31)*	29. *(1400)*	30. *(0)*

27 JULY SHIFT II

Section A : Objective Type Questions

1. Which one of the following set of elements can be detected using sodium fusion extract ?

a. Sulphur, nitrogen, phosphorus, halogens
b. Phosphorus, oxygen, nitrogen, halogens
c. Nitrogen, phosphorus, carbon, sulphur
d. Halogens, nitrogen, oxygen, sulphur

2.

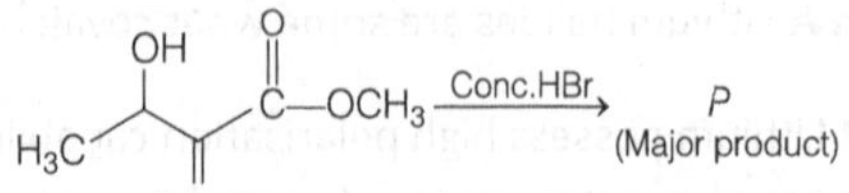

Consider the above reaction, the major product P formed is

a.
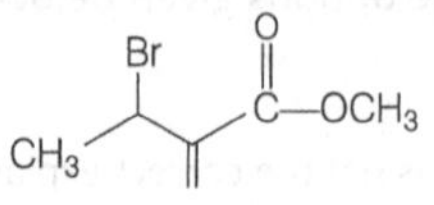

b.
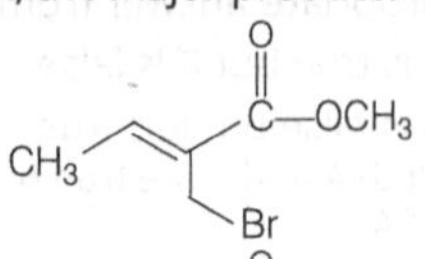

c.
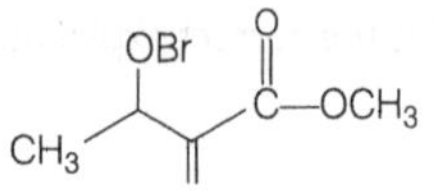

d.
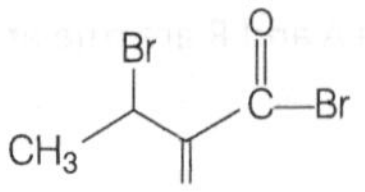

3. The number of neutrons and electrons, respectively, present in the radioactive isotope of hydrogen is

a. 1 and 1 **b.** 3 and 1
c. 2 and 1 **d.** 2 and 2

4. Match List I with List II.

	List I		List II
A.	Li	I.	Photoelectric cell
B.	Na	II.	Absorbent of CO_2
C.	K	III.	Coolant in fast breeder nuclear reactor
D.	Cs	IV.	Treatment of cancer
		V.	Bearings for motor engines

Choose the correct answer from the options given below.

	A	B	C	D
a.	V	I	II	IV
b.	V	II	IV	I
c.	IV	III	I	II
d.	V	III	II	I

5. Given below are two statement : one is labelled as Assertion (A) and the other is labelled as Reason (R).

Assertion (A) $SO_2(g)$ is adsorbed to a large extent than $H_2(g)$ on activated charcoal.

Reason (R) $SO_2(g)$ has a higher critical temperature than $H_2(g)$.

In the light of the above statements, choose the most appropriate answer from the options given below.

a. Both A and R are correct but R is not the correct explanation of A
b. Both A and R are correct and R is the correct explanation of A.
c. A is not correct but R is correct.
d. A is correct but R is not correct.

6. The correct order of first ionisation enthalpy is

a. Mg < S < Al < P **b.** Mg < Al < S < P
c. Al < Mg < S < P **d.** Mg < Al < P < S

7. Given below are two statements.

Statement I Hyperconjugation is a permanent effect.

Statement II Hyperconjugation in ethyl cation $(CH_3—\overset{+}{C}H_2)$ involves the overlapping of $C_{sp^2}—H_{1s}$ bond with empty $2p$ orbital of other carbon.

Choose the correct option :

a. Both statement I and statement II are false.
b. Statement I is incorrect but statement II is true.
c. Statement I is true but statement II is false.
d. Both Statement I and statement II are true.

8. Given below are two statements.

Statement I $[Mn(CN)_6]^{3-}$, $[Fe(CN)_6]^{3-}$ and $[Co(C_2O_4)_3]^{3-}$ are d^2sp^3 hybridised.

Statement II $[MnCl_6]^{3-}$ and $[FeF_6]^{3-}$ are paramagnetic and have 4 and 5 unpaired electrons, respectively.

In the light of the above statements, choose the correct answer from the options given below

a. Statement I is true but statement II is false
b. Both statement I and statement II are false
c. Statement I is false but statement II is true
d. Both statement I and statement II are true

9. To an aqueous solution containing ions such as Al^{3+}, Zn^{2+}, Ca^{2+}, Fe^{3+}, Ni^{2+}, Ba^{2+} and Cu^{2+} conc. HCl, was added followed by H_2S. The total number of cations precipitated during this reaction is/are

a. 1 **b.** 3 **c.** 4 **d.** 2

10. Given below are two statements.

Statement I Penicillin is a bacteriostatic type antibiotic.

Statement II The general structure of penicillin is

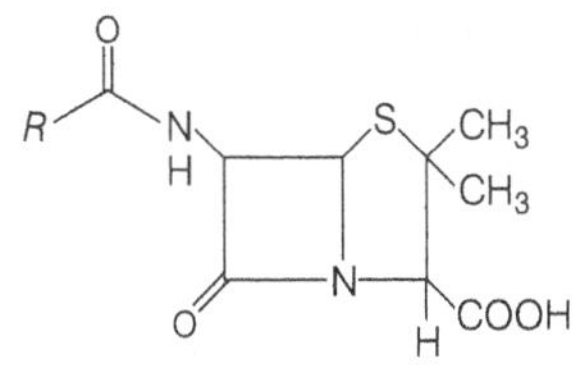

Choose the correct option.

a. Both statement I and statement II are false
b. Statement I is false but statement II is true
c. Both statement I and statement II are true
d. Statement I is true but statement II is false

11. Compound *A* gives D-galactose and D-glucose on hydrolysis. The compound *A* is

a. amylose **b.** sucrose **c.** maltose **d.** lactose

12. $R—CN \xrightarrow[\text{(ii) } H_2O]{\text{(i) DIBAL-H}} R—Y$

Consider the above reaction and identify *Y*.

a. $—CH_2NH_2$ **b.** $—CONH_2$ **c.** $—CHO$ **d.** $—COOH$

13.

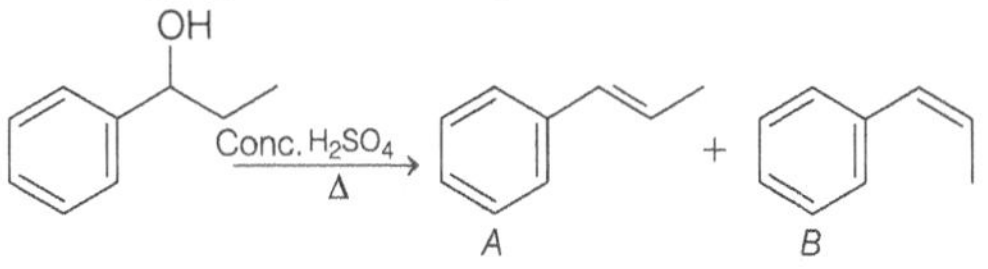

Consider the above reaction, and choose the correct statement.

a. The reaction is not possible in acidic medium.
b. Both compounds *A* and *B* are formed equally.
c. Compound *A* will be the major product.
d. Compound *B* will be the major product.

14. Match List I with List II.

List I (Compound)	List II (Effect/affected species)
A. Carbon monoxide	I. Carcinogenic
B. Sulphur dioxide	II. Metabolised by pyrus plants
C. Polychlorinated biphenyls	III. Haemoglobin
D. Oxides of nitrogen	IV. Stiffness of flower buds

Choose the correct answer from the options given below :

	A	B	C	D		A	B	C	D
a.	III	IV	I	II	**b.**	IV	I	III	II
c.	I	II	III	IV	**d.**	III	IV	II	I

15. If the Thomson model of the atom was correct, then the result of Rutherford's gold foil experiment would have been

a. all of the α-particles pass through the gold foil without decrease in speed.
b. α-particles are deflected over a wide range of angles.
c. all α-particles get bounced back by 180°.
d. α-particles pass through the gold foil deflected by small angles and with reduced speed.

16. Number of $Cl{=}O$ bonds in chlorous acid, chloric acid and perchloric acid respectively are

a. 3, 1 and 1 **b.** 4, 1 and 0 **c.** 1, 1 and 3 **d.** 1, 2 and 3

17. Select the correct statements.

(A) Crystalline solids have long range order.
(B) Crystalline solids are isotropic.
(C) Amorphous solids are sometimes called pseudo solids.
(D) Amorphous solids soften over a range of temperatures.
(E) Amorphous solids have a definite heat of fusion.

Choose the most appropriate answer from the options given below.

a. (A), (B), (E) only **b.** (B), (D) only
c. (C), (D) only **d.** (A), (C), (D) only

18. What is *A* in the following reaction?

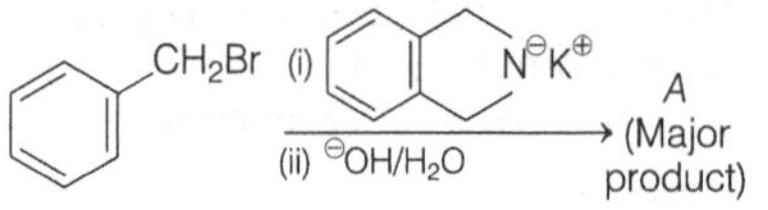

a. 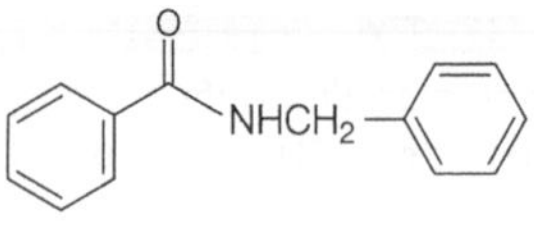

b. 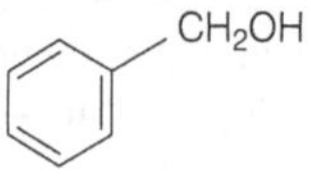

c. 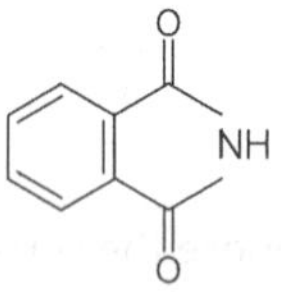

d. 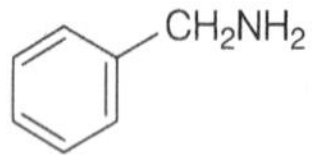

19. The correct sequence of correct reagents for the following transformation is

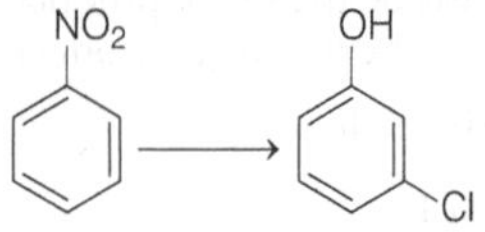

a. (i) Fe, HCl (ii) Cl_2, HCl
(iii) $NaNO_2$, HCl, 0°C (iv) H_2O / H^+
b. (i) Fe, HCl (ii) $NaNO_2$, HCl, 0°C
(iii) H_2O / H^+ (iv) Cl_2, $FeCl_3$
c. (i) Cl_2, $FeCl_3$ (ii) Fe, HCl
(iii) $NaNO_2$, HCl, 0°C (iv) H_2O / H^+
d. (i) Cl_2, $FeCl_3$ (ii) $NaNO_2$, HCl, 0°C
(iii) Fe, HCl (iv) H_2O / H^+

20. The addition of silica during the extraction of copper from its sulphide ore

a. converts copper sulphide into copper silicate
b. converts iron oxide into iron silicate
c. reduces copper sulphide into metallic copper
d. reduces the melting point of the reaction mixture

Section B : Numerical type Questions

21. The equilibrium constant for the reaction

$$A(s) \rightleftharpoons M(s) + \frac{1}{2}O_2(g)$$

is $K_p = 4$. At equilibrium, the partial pressure of O_2 is .... atm. (Round off to the nearest integer)

22. When 400 mL of 0.2 M H_2SO_4 solution is mixed with 600 mL of 0.1 M NaOH solution, the increase in temperature of the final solution is .... $\times 10^{-2}$ K. (Round off to the nearest integer).

[Use : $H^+(aq) + OH^+(aq) \longrightarrow H_2O$; $\Delta_y H = -57.1$ kJ mol^{-1}]
Specific heat of $H_2O = 0.18$ J K^{-1} g^{-1},
density of $H_2O = 1.0$ g cm^{-3}.
Assume no change in volume of solution on mixing.

23. $2SO_2(g) + O_2(g) \longrightarrow 2SO_3(g)$

The above reaction is carried out in a vessel starting with partial pressure $p_{SO_2} = 250$ m bar, $p_{O_2} = 750$ m bar and $p_{SO_3} = 0$. When the reaction is complete, the total pressure in the reaction vessel is .... m bar. (Round off of the nearest integer).

24. 10.0 mL of 0.05 M $KMnO_4$ solution was consumed in a titration with 10.0 mL of given oxalic acid dihydrate solution. The strength of given oxalic acid solution is ...... $\times 10^{-2}$ g/L.
(Round off to the nearest integer)

25. The total number of electrons in all bonding molecular orbitals of O_2^{2-} is ....
(Round off to the nearest integer)

26. 3 moles of metal complex with formula $Co(en)_2Cl_3$ gives 3 moles of silver chloride on treatment with excess of silver nitrate. The secondary valency of Co in the complex is ....
(Round off to the nearest integer)

27. In a solvent 50% of an acid *HA* dimerises and the rest dissociates. The van't Hoff factor of the acid is .... $\times 10^{-2}$.
(Round off to the nearest integer)

28. The dihedral angle in staggered form of Newman projection of 1, 1, 1-trichloro ethane is .... degree.
(Round off to the nearest integer)

29. For the first order reaction, $A \longrightarrow 2B$, 1 mole of reactant *A* gives 0.2 moles of *B* after 100 minutes. The half-life of the reaction is .... min. (Round off to the nearest integer). [Use : ln 2 = 0.69, ln 10 = 2.3
properties of logarithms : $\ln x^y = y \ln x$;
$\ln\left(\frac{x}{y}\right) = \ln x - \ln y$]

30. For the cell
$Cu(s)|Cu^{2+}(aq)$ (0.1) M) || $Ag^+(aq)$ (0.01 M)| Ag(s)
the cell potential, $E_1 = 0.3095$ V
For the cell,
$Cu(s)|Cu^{2+}(aq)$ (0.01 M) || $Ag^+(aq)$ (0.001 M)| Ag(s)
the cell potential =.............. $\times 10^{-2}$ V.
(Round off the nearest integer).
$\left[\text{Use} : \frac{2.303\,RT}{F} = 0.059\right]$

Answers

For solution scan the QR code

1. (a)	2. (b)	3. (c)	4. (d)	5. (b)	6. (c)	7. (c)	8. (d)	9. (a)	10. (b)
11. (d)	12. (c)	13. (c)	14. (a)	15. (d)	16. (d)	17. (d)	18. (d)	19. (c)	20. (b)
21. (16)	22. (82)	23. (875)	24. (1575)	25. (10)	26. (6)	27. (125)	28. (60)	29. (656)	30. (28)

26 AUGUST SHIFT I

Section A : Objective Type Questions

1. Which one of the following complexes is violet in colour?

a. $[Fe(CN)_6]^{4-}$ **b.** $[Fe(SCN)_6]^{4-}$
c. $Fe_4[Fe(CN_6)]_3.H_2O$ **d.** $[Fe(CN)_5NOS]^{4-}$

2. Which one of the following is correct for the adsorption of a gas at a given temperature on a solid surface?

a. $\Delta H > 0, \Delta S > 0$ **b.** $\Delta H > 0, \Delta S < 0$
c. $\Delta H < 0, \Delta S < 0$ **d.** $\Delta H < 0, \Delta S > 0$

3. Which one of the following when dissolved in water gives coloured solution in nitrogen atmosphere?

a. $CuCl_2$ **b.** AgCl
c. $ZnCl_2$ **d.** Cu_2Cl_2

4. The major products formed in the following reaction sequence *A* and *B* are

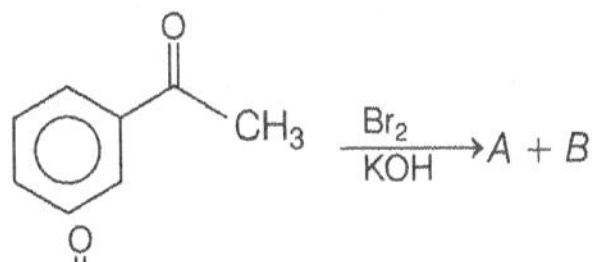

a. 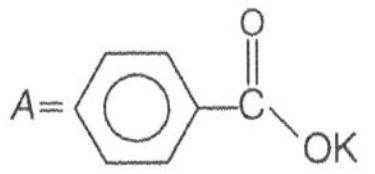

, $B = CHBr_3$

b. 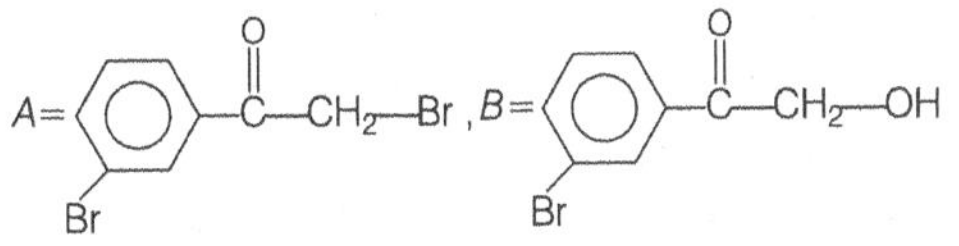

c. 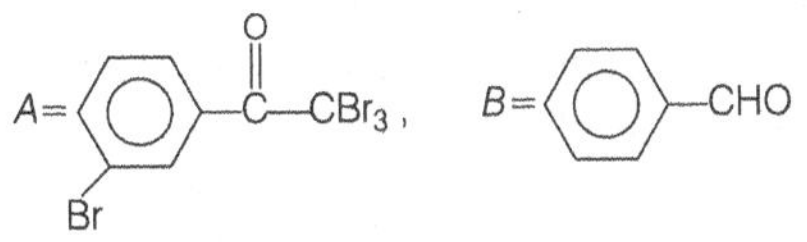

d. 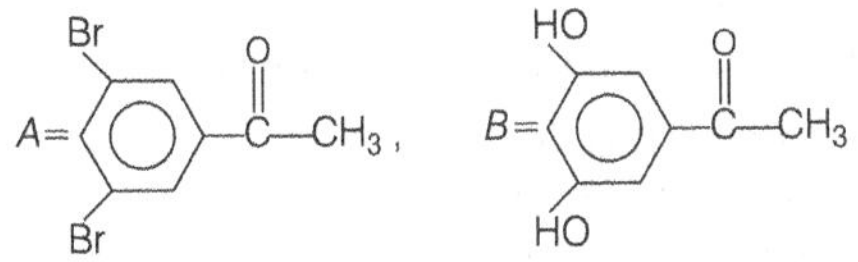

5. The major product formed in the following reaction is

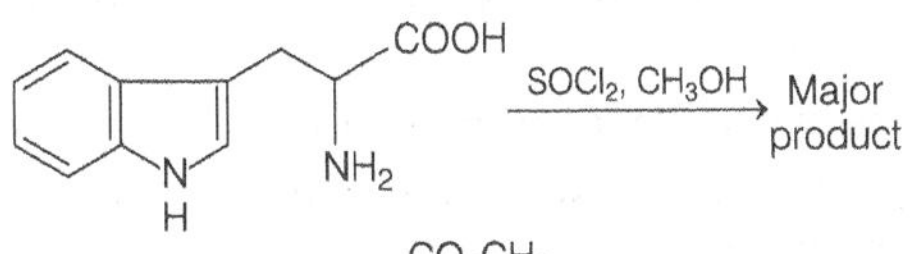

a. 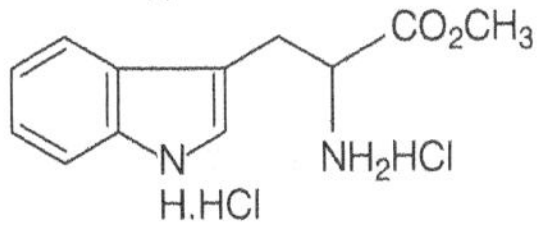

b. 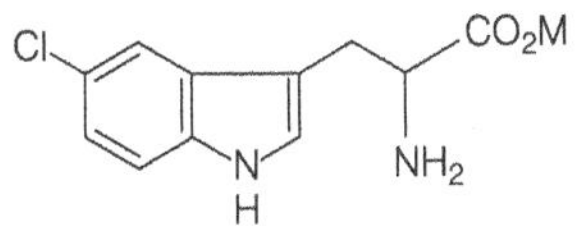

c. 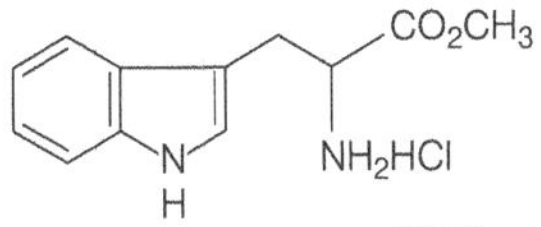

d. 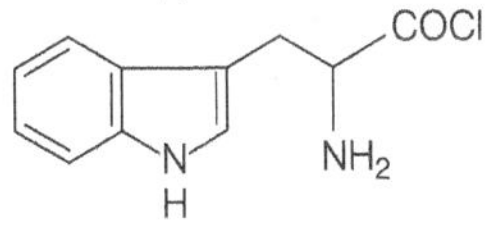

6. The major product formed in the following reaction is

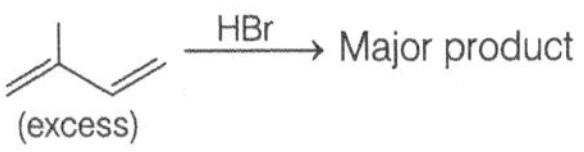

a. 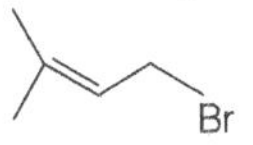

b. 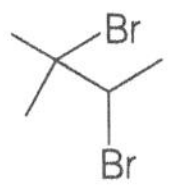

c. 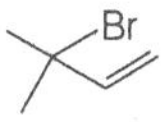

d. 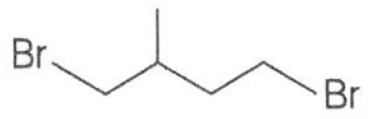

7. The polymer formed on heating novolac with formaldehyde is

a. bakelite **b.** polyester
c. melamine **d.** nylon 6,6

8. Given below are two statements.

Statement I The limiting molar conductivity of KCl (strong electrolyte) is higher compared to that of CH_3COOH (weak electrolyte).

Statement II Molar conductivity decreases with decrease in concentration of electrolyte.

In the light of the above statements, choose the most appropriate answer from the options given below

a. Statement I is true but statement II is false.
b. Statement I is false but statement II is true.
c. Both statement I and statement II are true.
d. Both statement I and statement II are false.

9. The correct options for the products *A* and *B* of the following reactions are

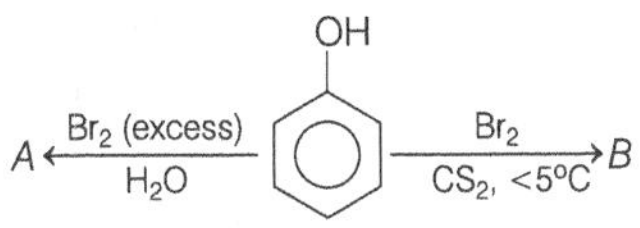

a. 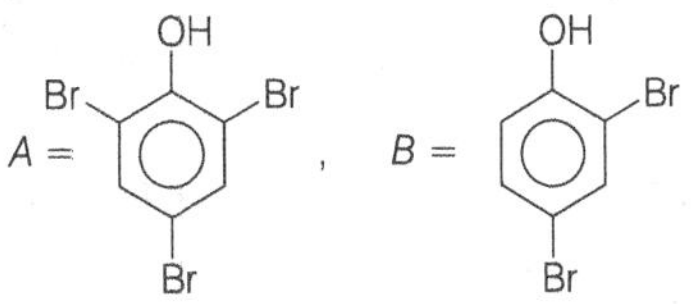

b. 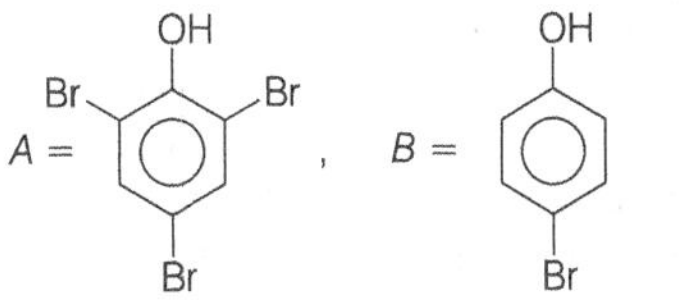

c. $A =$ 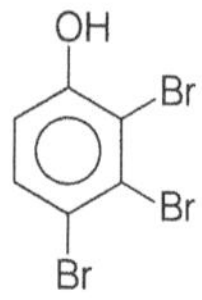 , $B =$ 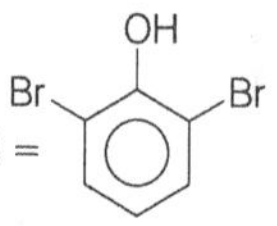

d. $A =$ 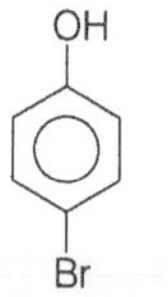 , $B =$ 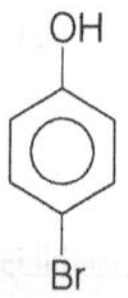

10. The conversion of hydroxyapatite occurs due to presence of F^- ions in water. The correct formula of hydroxyapatite is

a. $[3Ca_3(PO_4)_2Ca(OH)_2]$ **b.** $[3Ca(OH)_2CaF_2]$
c. $[Ca_3(PO_4)_2CaF_2]$ **d.** $[3Ca_3(PO_4)_2CaF_2]$

11. Given below are two statements.

Statement I In the titration between strong acid and weak base methyl orange is suitable as an indicator.

Statement II For titration of acetic acid with NaOH phenolphthalein is not a suitable indicator.

In the light of the above statements, choose the most appropriate answer from the options given below.

a. Statement I is false but statement II is true.
b. Statement I is true but statement II is false.
c. Both statement I and statement II are true.
d. Both statement I and statement II are false.

12. Among the following compounds I-IV, which one forms a yellow precipitate on reacting sequentially with

(i) NaOH (ii) dil. HNO_3
(iii) $AgNO_3$?

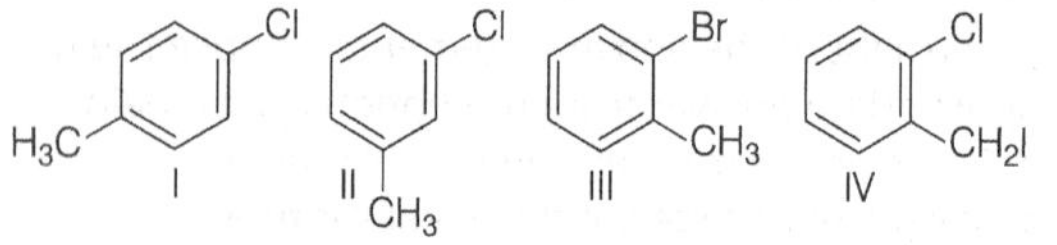

a. II **b.** IV **c.** I **d.** III

13. Which one of the following methods is most suitable for preparing deionised water?

a. Synthetic resin method **b.** Clark's method
c. Calgon's method **d.** Permutit method

14. Given below are two statements.

Statement I The choice of reducing agent for metals extraction can be made by using Ellingham diagram, a plot of ΔG *vs* temperature.

Statement II The value of ΔS increases from left to right in Ellingham diagram.

In the light of the above statements, choose the most appropriate answer from the options given below.

a. Both statement I and statement II are true.
b. Statement I is false but statement II is true.
c. Both statement I and statement II are false.
d. Statement I is true but statement II is false.

15. What are the products formed in sequence when excess of CO_2 is passed in slaked lime?

a. $Ca(HCO_3)_2, CaCO_3$ **b.** $CaCO_3, Ca(HCO_3)_2$
c. $CaO, Ca(HCO_3)_2$ **d.** $CaO, CaCO_3$

16. Given below are two statements.

Statement I According to Bohr's model of an atom, qualitatively the magnitude of velocity of electron increases with decrease in positive charges on the nucleus as there is no strong hold on the electron by the nucleus.

Statement II According to Bohr's model of an atom, qualitatively the magnitude of velocity of electron increases with decrease in principal quantum number.

In the light of the above statements, choose the most appropriate answer from the options given below.

a. Both statement I and statement II are false.
b. Both statement I and statement II are true.
c. Statement I is false but statement II is true.
d. Statement I is true but statement II is false.

17. The correct sequential addition of reagents in the preparation of 3-nitrobenzoic acid from benzene is

a. $Br_2/AlBr_3$, HNO_3/H_2SO_4, Mg/ether, CO_2, H_3O^+
b. $Br_2/AlBr_3$, NaCN, H_3O^+, HNO_3/H_2SO_4
c. $Br_2/AlBr_3$, HNO_3/H_2SO_4, NaCN, H_3O^+
d. HNO_3/H_2SO_4, $Br_2/AlBr_3$, Mg/ether, CO_2, H_3O^+

18. Given below are two statements.

Statement I Frenkel defects are vacancy as well as interstitial defects.

Statement II Frenkel defect leads to colour in ionic solids due to presence of F-centres.

Choose the most appropriate answer for the statements from the options given below.

a. Statement I is false but Statement II is true.
b. Both statement I and statement II are true.
c. Statement I is true but statement II is false.
d. Both Statement I and statement II are false.

19. Choose the incorrect statement.

a. Cl_2 is more reactive than ClF.
b. F_2 is more reactive than ClF.
c. On hydrolysis ClF froms HOCl and HF.
d. F_2 is a stronger oxidising agent than Cl_2 in aqueous solution.

20. Excess of isobutane on reaction with Br_2 in presence of light at 125°C gives which one of the following, as the major product?

a. 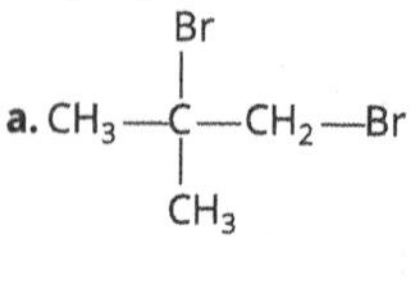 **b.** 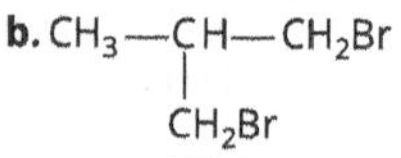

c. 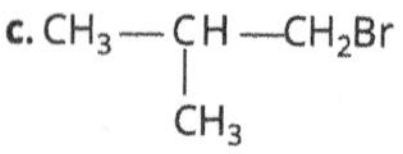 **d.** 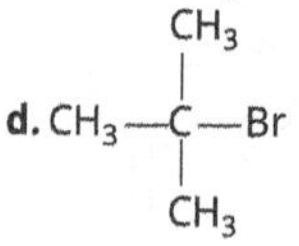

Section B : Numerical Type Questions

21. AB_3 is an interhalogen *T*-shaped molecule. The number of lone pairs of electron on *A* is .... (Integer answer)

22. These are physical properties of an element.

A. Sublimation enthalpy B. Ionisation enthalpy

C. Hydration enthalpy D. Electron gain enthalpy

The total number of above properties that affect the reduction potential is .... (Integer answer)

23. Of the following four aqueous solutions, total number of those solutions whose freezing point is lower than that of 0.10 M C_2H_5OH is .... (Integer answer)

(i) 0.10 M $Ba_3(PO_4)_2$ (ii) 0.10 M Na_2SO_4

(iii) 0.10 M KCl (iv) 0.10 M Li_3PO_4

24. The OH^- concentration in a mixture of 5.0 mL of 0.0504 M NH_4Cl and 2 mL of 0.0210 M NH_3 solution is $x \times 10^{-6}$ M. The value of *x* is .... (Nearest integer)

[Given, $K_w = 1 \times 10^{-14}$ and $K_b = 1.8 \times 10^{-5}$]

25. The number of 4*f* electrons in the ground state electronic configuration of Gd^{2+} is

[Atomic number of Gd is 64.]

26. The ratio of number of water molecules in Mohr's salt and potash alum is .... $\times 10^{-1}$.

(Integer answer)

27. The following data was obtained for chemical reaction given below at 975 K.

$$2NO(g) + 2H_2(g) \longrightarrow N_2(g) + 2H_2O(g)$$

	[NO] mol L^{-1}	$[H_2]$ mol L^{-1}	Rate mol L^{-1}
A.	8×10^{-5}	8×10^{-5}	7×10^{-9}
B.	24×10^{-5}	8×10^{-5}	2.1×10^{-8}
C.	24×10^{-5}	32×10^{-5}	8.4×10^{-8}

The order of the reaction with respect to NO is .... [Integer answer]

28. The Born-Haber cycle for KCl is evaluated with the following data :

$\Delta_f H^\circ$ for KCl = – 436.7 kJ mol^{-1}

$\Delta_{sub} H^\circ$ for K = 89.2 kJ mol^{-1},

$\Delta_{ionisation} H^\circ$ for K = 419.0 kJ mol^{-1}; $\Delta_{electron\ gain} H^{\cup}$ for Cl(g) = – 348.6 kJ mol^{-1}, $\Delta_{bond} H^\circ$ for Cl_2 = 243.0 kJ mol^{-1}

The magnitude of lattice enthalpy of KCl in kJ mol^{-1} is .... (Nearest integer)

29. The total number of negative charge in the tetrapeptide, Gly-Glu-Asp-Tyr, at pH 12.5 will be .... (Integer answer)

30. An aqueous KCl solution of density 1.20gmL^{-1} has a molality of 3.30 mol kg^{-1}. The molarity of the solution in mol L^{-1} is .... (Nearest integer)

[Molar mass of KCl = 74.5]

Answers

For solution scan the QR code

1. (*d*)	2. (*c*)	3. (*a*)	4. (*a*)	5. (*c*)	6. (*a*)	7. (*a*)	8. (*d*)	9. (*b*)	10. (*a*)
11. (*b*)	12. (*b*)	13. (*a*)	14. (*d*)	15. (*b*)	16. (*c*)	17. (*d*)	18. (*c*)	19. (*a*)	20. (*d*)
21. (*2*)	22. (*3*)	23. (*4*)	24. (*3*)	25. (*7*)	26. (*5*)	27. (*1*)	28. (*718*)	29. (*4*)	30. (*3*)

26 AUGUST SHIFT II

Section A : Objective Type Questions

1. Which one of the following phenols does not give colour when condensed with phthalic anhydride in presence of conc. H_2SO_4 ?

a. 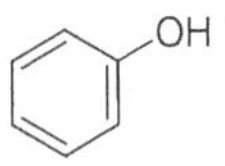

b. 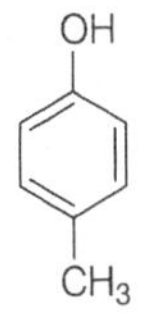

c. 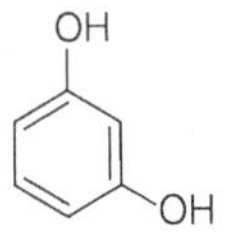

d. 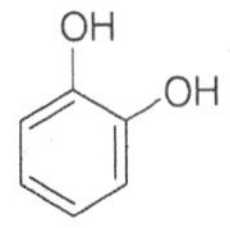

2. Given below are two statements. One is labelled as Assertion (A) and the other is labelled as Reason (R).

Assertion (A) Photochemical smog causes cracking of rubber.

Reason (R) Presence of ozone, nitric oxide, acrolein, formaldehyde and peroxyacetyl nitrate in photochemical smog makes it oxidising.

Choose the most appropriate answer from the options given below.

a. Both (A) and (R) are true but (R) is not the correct explanation of (A).

b. Both (A) and (R) are true and (R) is the correct explanation of (A).

c. (A) is false but (R) is true.

d. (A) is true but (R) is false.

3. The interaction energy of London forces between two particles is proportional to r^x, where r is the distance between the particles. The value of x is

a. 3 b. – 3 c. – 6 d. 6

4. The number of non-ionisable hydrogen atoms present in the final product obtained from the hydrolysis of PCl_5 is

a. 0 b. 2 c. 1 d. 3

5. The bond order and magnetic behaviour of O_2^- ion are, respectively

a. 1.5 and paramagnetic b. 1.5 and diamagnetic
c. 2 and diamagnetic d. 1 and paramagnetic

6. Given below are two statements : One is labelled as Assertion (A) and other is labelled as Reason (R).

Assertion (A) Sucrose is a disaccharide and a non-reducing sugar.

Reason (R) Sucrose involves glycosidic linkage between C_1 of β-glucose and C_2 of α-fructose.

Choose the most appropriate answer from the options given below.

a. Both (A) and (R) are true but (R) is not the correct explanation of (A).
b. (A) is false but (R) is true.
c. (A) is true but (R) is false.
d. Both (A) and (R) are true and (R) is the correct explanation of (A).

7. Match List-I with List-II.

List-I (Chemical reaction)		List-II (Reagent used)
A. $CH_3COOC_2H_5 \longrightarrow C_2H_5OH$	1.	CH_3MgBr / H_3O^+ (1.equivalent)
B. $CH_3COOCH_3 \longrightarrow CH_3CHO$	2.	H_2SO_4 / H_2O
C. $CH_3C{\equiv}N \longrightarrow CH_3CHO$	3.	DIBAL - H / H_2O
D. $CH_3C{\equiv}N \longrightarrow$ 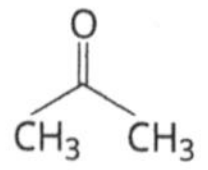	4.	$SnCl_2$, HCl / H_2O

Choose the most appropriate option given below.

	A	B	C	D		A	B	C	D
a.	2	4	3	1	b.	4	2	3	1
c.	2	3	4	1	d.	3	2	1	4

8. Given below are two statements : One is labelled as Assertion (A) and the other is labelled as Reason (R).

Assertion (A) Barium carbonate is insoluble in water and is highly stable.

Reason (R) The thermal stability of the carbonates increases with increasing cationic size.

a. Both (A) and (R) are true and (R) is the correct explanation of (A).
b. (A) is true but (R) is false.
c. Both (A) and (R) are true and (R) is not the true explanation of (A)
d. (A) is false but (R) is true.

9.

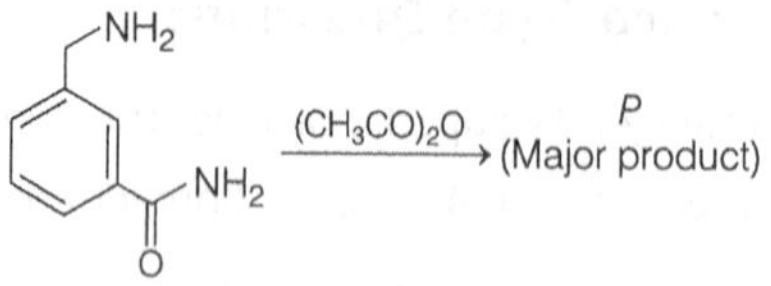

a.

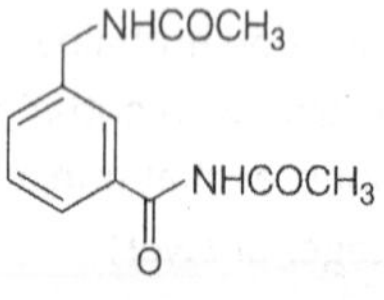

b.

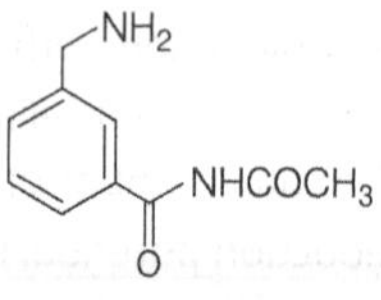

c.

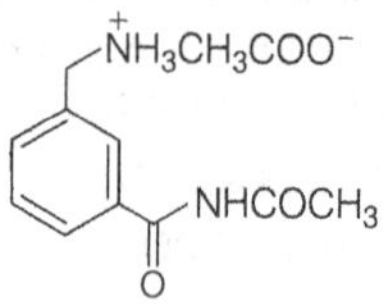

d.

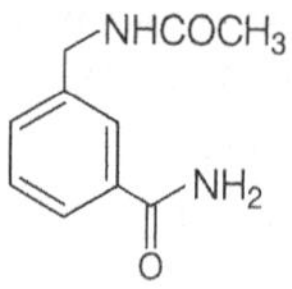

10. Indicate the complex/complex ion which did not show any geometrical isomerism.

a. $[CoCl_2(en)_2]$ b. $[Co(CN)_5(NC)]^{3-}$
c. $[Co(NH_3)_3(NO_2)_3]$ d. $[Co(NH_3)_4Cl_2]^+$

11. The sol given below with negatively charged colloidal particles is

a. $FeCl_3$ added to hot water
b. KI added to $AgNO_3$ solution
c. $AgNO_3$ added to KI solution
d. $Al_2O_3 \cdot xH_2O$ in water

12. Given below are two statements.

Statement I Sphalerite is a sulphide ore of zinc and copper glance is a sulphide ore of copper.

Statement II It is possible to separate two sulphide ores by adjusting proportion of oil to water or by using 'depressants' in a froth flotation method.

Choose the most appropriate answer from the options given below.

a. Statement I is true but statement II is false.
b. Both statement I and statement II are true.
c. Statement I is false but statement II is true.
d. Both statement I and statement II are false.

13. Given below are two statements.
One is labelled as Assertion (A) and the other is labelled as Reason (R).

Assertion (A) Heavy water is used for the study of reaction mechanism.

Reason (R) The rate of reaction for the cleavage of O—H bond is slower than that of O—D bond.

Choose the most appropriate answer from the options given below.

a. Both (A) and (R) are true but (R) is not the corrent explanation of (A).
b. Both (A) and (R) are true and (R) is the correct explanation of (A).
c. (A) is false but (R) is true.
d. (A) is true but (R) is false.

14. Arrange the following cobalt complexes in the order of increasing crystal field stabilisation energy (CFSE) value. Complexes $[CoF_6]^{3-}$, $[Co(H_2O)_6]^{3+}$, $[Co(NH_3)_6]^{3+}$ and $[Co(en)_3]^{3+}$
(A), (B), (C), (D)

Choose the correct option.

a. $A < B < C < D$ **b.** $B < A < C < D$
c. $B < C < D < A$ **d.** $C < D < B < A$

15.

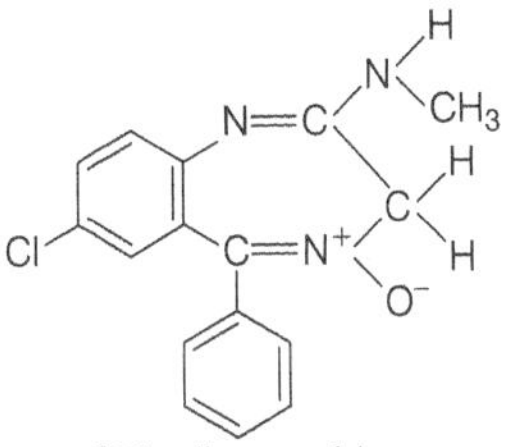

Chlordiazepoxide

The class of drug to which chlordiazepoxide with above structure belongs is

a. antacid **b.** analgesic **c.** tranquilizer **d.** antibiotic

16. Chalcogen group elements are

a. Se, Tb and Pu **b.** Se, Te and Po
c. S, Te and Pm **d.** O, Ti and Po

17. Which one of the following compounds is not aromatic?

a. 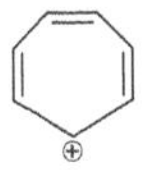

b. 

c. 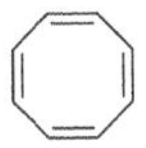

d. 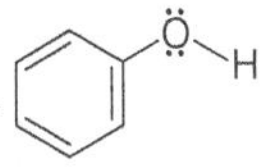

18. The number of stereoisomers possible for 1,2-dimethylcyclopropane is

a. one **b.** four **c.** two **d.** three

19.

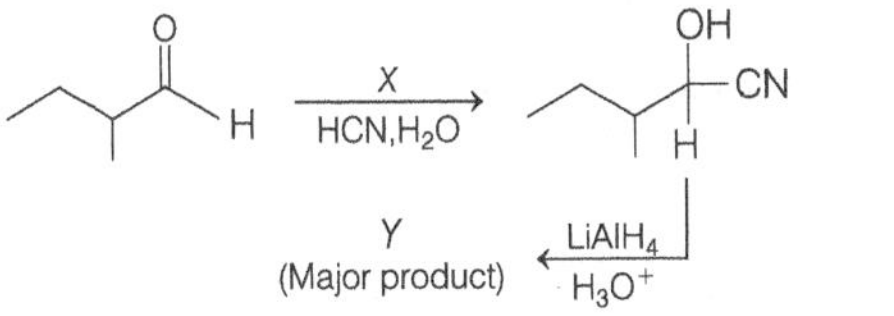

Consider the given reaction, identify X and Y.

a. X = NaOH, Y = 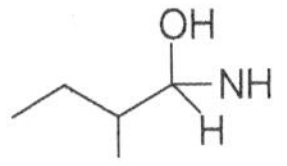

b. $X = HNO_3$, Y = 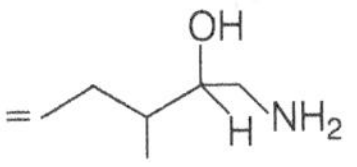

c. X = NaOH, Y = 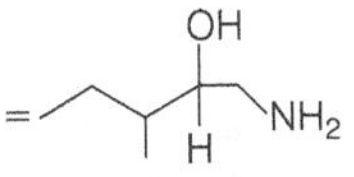

d. $X = HNO_3$, Y = 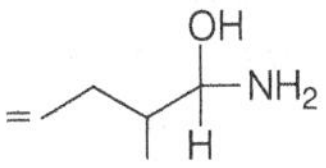

20.

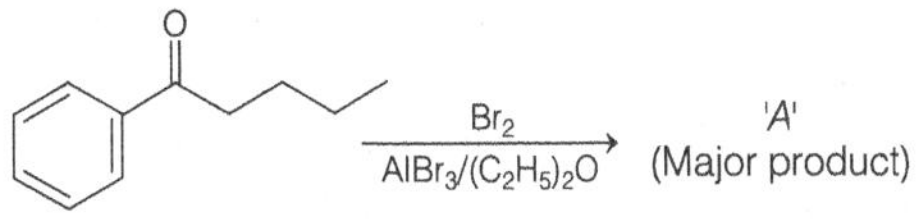

Consider the given reaction, the product A is

a. 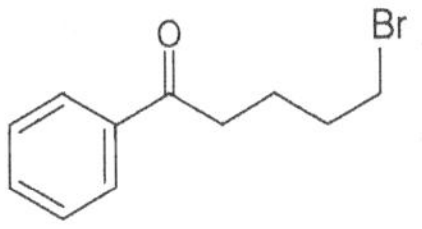

b. 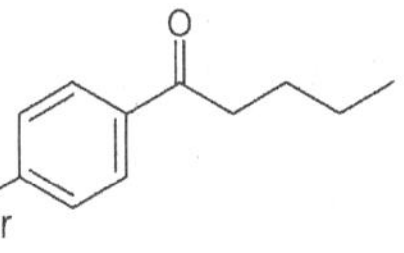

c. 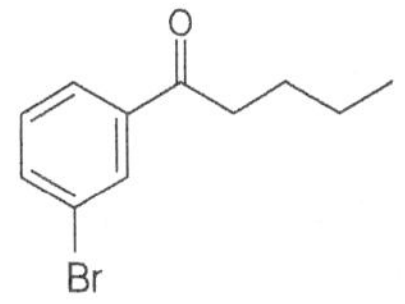

d. 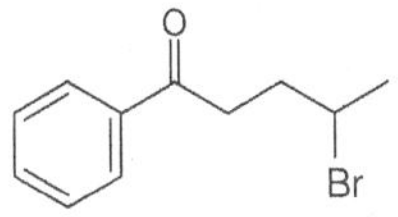

Section B : Numerical Type Questions

21. In the sulphur estimation, 0.471 g of an organic compound gave 1.44 g of barium sulphate. The percentage of sulphur in the compound is ____%. (Nearest integer)
(Atomic mass of Ba = 137 u)

22. The equilibrium constant K_C at 298 K for the reaction $A + B \rightleftharpoons C + D$ is 100. Starting with an equimolar solution with concentrations of A, B, C and D all equal to 1M, the equilibrium concentration of D is .... $\times 10^{-2}$ M. (Nearest integer)

23. For water $\Delta_{vap}H = 41\ kJ\ mol^{-1}$ at 373 K and 1 bar pressure. Assuming that water vapour is an ideal gas that occupies a much larger volume than liquid water, the internal energy change during evaporation of water is .... kJ mol^{-1}.
[Use $R = 8.3\ J\ mol^{-1}\ K^{-1}$]

24. A metal surface is exposed to 500 nm radiation. The threshold frequency of the metal for photoelectric current is 4.3×10^{14} Hz. The velocity of ejected electron is .... $\times 10^5\ ms^{-1}$. (Nearest integer)
[Use $h = 6.63 \times 10^{-34}$ Js, $m_e = 9.0 \times 10^{-31}$ kg]

25. For the galvanic cell,
$Zn(s) + Cu^{2+}(0.02\ M) \longrightarrow Zn^{2+}(0.04M) + Cu(s)$,
E_{Cell} = ... $\times 10^{-2}$ V. (Nearest integer)
[Use $E^{\circ}_{Cu/Cu^{2+}} = -0.34$ V, $E^{\circ}_{Zn/Zn^{2+}} = +0.76$ V,
$\frac{2.303\ RT}{F} = 0.059$ V]

26. 100 mL of Na_3PO_4 solution contains 3.45 g of sodium. The molarity of the solution is ......$\times 10^{-2}$ mol L^{-1}. (Nearest integer)
[Atomic masses - Na = 23.0 u, O = 16.0 u, P = 31.0 u]

27. The overall stability constant of the complex ion $[Cu(NH_3)_4]^{2+}$ is 2.1×10^{13}. The overall dissociations constant is $y \times 10^{-14}$. Then, y is .... (Nearest integer)

28. 83 g of ethylene glycol dissolved in 625 g of water. The freezing point of the solution is ....K. (Nearest integer)

[Use, molal freezing point depression constant of water = $1.86 \text{ K kg mol}^{-1}$,

Freezing point of water = 273 K and

Atomic masses : C = 12.0 u, O = 16.0 u, H = 1.0 u]

29. The reaction rate for the reaction

$$[PtCl_4]^{2-} + H_2O \rightleftharpoons [Pt(H_2O)Cl_3]^- + Cl^-$$

was measured as a function of concentrations of different species. It was observed that

$$\frac{-d[(PtCl_4)^{2-}]}{dt} = 4.8 \times 10^{-5}[(PtCl_4)^{2-}]$$
$$= 2.4 \times 10^{-3}[\{Pt(H_2O)Cl_3\}^-][Cl^-]$$

where, square brackets are used to denote molar concentrations. The equilibrium constant, K_C = ... (Nearest integer)

30. A chloro compound A,

(i) Forms aldehydes on ozonolysis followed by the hydrolysis.

(ii) When vaporised completely, 1.53 g of A gives 448 mL of vapour at STP.

The number of carbon atoms in a molecule of compound A is ....

Answers

For solution scan the QR code

1. *(b)*	2. *(b)*	3. *(c)*	4. *(a)*	5. *(a)*	6. *(c)*	7. *(c)*	8. *(a)*	9. *(d)*	10. *(b)*
11. *(c)*	12. *(b)*	13. *(d)*	14. *(a)*	15. *(c)*	16. *(b)*	17. *(c)*	18. *(d)*	19. *(c)*	20. *(c)*
21. *(42)*	22. *(182)*	23. *(38)*	24. *(5)*	25. *(109)*	26. *(50)*	27. *(5)*	28. *(269)*	29. *(50)*	30. *(3)*

27 AUGUST SHIFT I

Section A : Objective Type Questions

1. In the following sequence of reactions, the final product D is

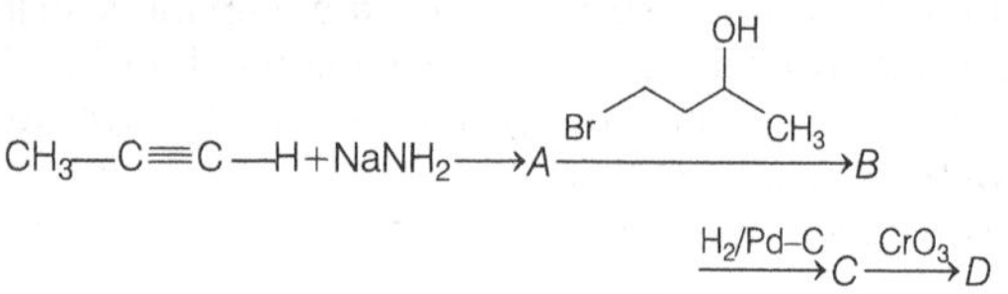

a. $H_3C—CH_2—CH_2—CH_2—CH_2—\overset{O}{\overset{||}{C}}—H$

b. $CH_3—CH{=}CH—CH_2—CH_2—CH_2—COOH$

c. $H_3C—CH{=}CH—CH(OH)—CH_2—CH_2—CH_3$

d. $CH_3—CH_2—CH_2—CH_2—CH_2—\overset{O}{\overset{||}{C}}—CH_3$

2. The structure of the starting compound P used in the reaction given below is

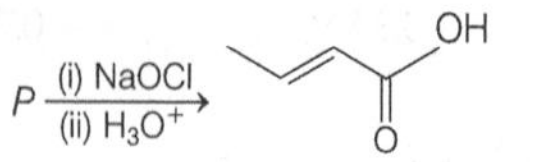

a. 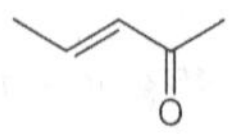

b. 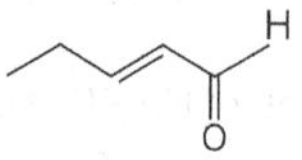

c. 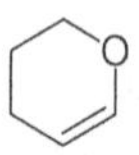

d. 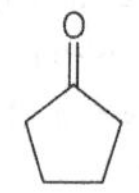

3. Match List-I with List-II.

List-I (Species)	List-II (Number of lone pair of electrons on the central atom)
A. XeF_2	1. 0
B. XeO_2F_2	2. 1
C. XeO_3F_2	3. 2
D. XeF_4	4. 3

Choose the most appropriate answer from the options given below :

	A	B	C	D		A	B	C	D
a.	4	1	2	3	**b.**	3	4	2	1
c.	3	2	4	1	**d.**	4	2	1	3

4. In which one of the following molecules strongest back donation of an electron pair from halide to boron is expected?

a. BCl_3 **b.** BF_3 **c.** BBr_3 **d.** BI_3

5. Deuterium resembles hydrogen in properties but

a. reacts slower than hydrogen

b. reacts vigorously than hydrogen

c. reacts just as hydrogen

d. emits β^+ particles

6. Which refining process is generally used in the purification of low melting metals ?

a. Chromatographic method **b.** Liquation

c. Electrolysis **d.** Zone refining

7. Match List-I with List-II.

	List-I (Property)		List-II (Example)
A.	Diamagnetism	1.	MnO
B.	Ferrimagnetism	2.	O_2
C.	Paramagnetism	3.	NaCl
D.	Antiferromagnetism	4.	Fe_3O_4

Choose the most appropriate answer from the options given below.

	A	B	C	D		A	B	C	D
a.	2	1	3	4	**b.**	1	3	4	2
c.	3	4	2	1	**d.**	4	2	1	3

8. Consider the following structures :

A. 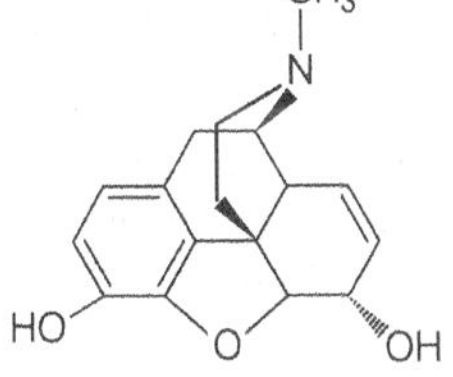

B. 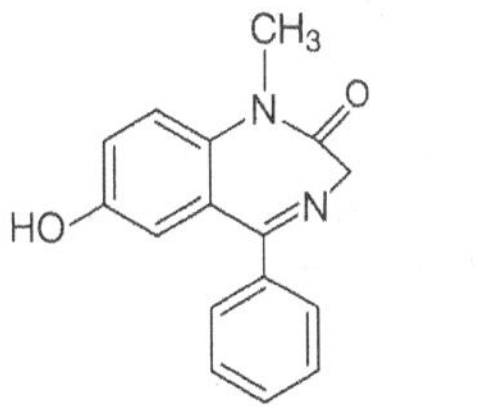

C. 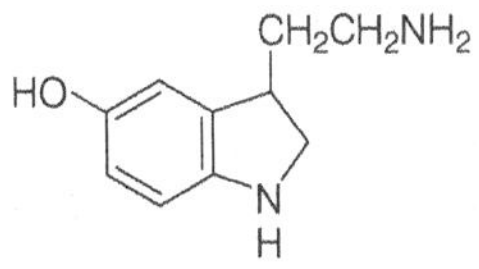

D. 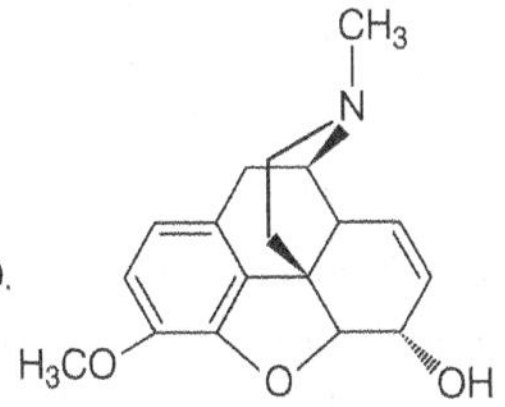

The correct statement about (A), (B), (C) and (D) is

a. (A), (B) and (C) are narcotic analgesics
b. (B), (C) and (D) are tranquillizers
c. (A) and (D) are tranquillizers
d. (B) and (C) are tranquillizers

9. The major product of the following reaction is

$CH_3-CH(CH_3)-CH_2-CH_2-C(=O)-Cl \xrightarrow[\text{(iii) NaNO}_2\text{, HCl; (iv) H}_2\text{O}]{\text{(i) Alcoholic NH}_3\text{; (ii) NaOH, Br}_2}$ Major product

a. $CH_3-CH(CH_3)-CH(Br)-CH_2OH$
b. $CH_3-CH(CH_3)-CH_2-CH_2-CH_2OH$
c. $CH_3-CH(CH_3)-CH_2-CH_2OH$
d. $CH_3-CH(CH_3)-CH_2-CH_2-Cl$

10. Which of the following is not a correct statement for primary aliphatic amines?

a. The intermolecular association in primary amines is less than the intermolecular association in secondary amines.
b. Primary amines on treating with nitrous acid solution form corresponding alcohols except methyl amine.
c. Primary amines are less basic than the secondary amines.
d. Primary amines can be prepared by the gabriel phthalimide synthesis.

11. Acidic ferric chloride solution on treatment with excess of potassium ferrocyanide gives a prussian blue coloured colloidal species. It is

a. $Fe_4[Fe(CN)_6]_3$ **b.** $K_5Fe[Fe(CN)_6]_2$
c. $HFe[Fe(CN)_6]$ **d.** $KFe[Fe(CN)_6]$

12. The gas 'A' is having very low reactivity reaches to stratosphere. It is non-toxic and non-flammable but dissociated by UV-radiations in stratosphere. The intermediates formed initially from the gas 'A' are

a. $Cl\dot{O} + \dot{C}F_2Cl$ **b.** $Cl\dot{O} + \dot{C}H_3$
c. $\dot{C}H_3 + \dot{C}F_2Cl$ **d.** $\dot{C}l + \dot{C}F_2Cl$

13. The number of water molecules in gypsum, dead burnt plaster and plaster of Paris, respectively are

a. 2, 0 and 1 **b.** 0.5, 0 and 2
c. 5, 0 and 0.5 **d.** 2, 0 and 0.5

14. The nature of oxides V_2O_3 and CrO is indexed as 'X' and 'Y' type respectively. The correct set of X and Y is

a. X = basic, Y = amphoteric
b. X = amphoteric, Y = basic
c. X = acidic, Y = acidic
d. X = basic, Y = basic

15. Out of the following isomeric forms of uracil, which one is present in RNA ?

a. 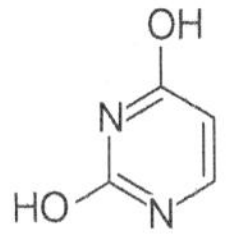

b. 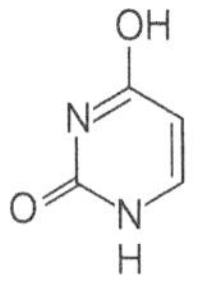

c. 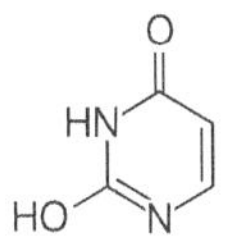

d. 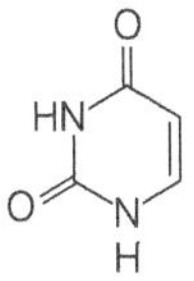

16. Given below are two statements : one is labelled as Assertion (A) and the other is labelled as Reason (R).

Assertion (A) Synthesis of ethyl phenyl ether may be achieved by Williamson synthesis.

Reason (R) Reaction of bromobenzene with sodium ethoxide yields ethyl phenyl ether.

In the light of the above statements, choose the most appropriate answer from the options given below

a. Both (A) and (R) are correct and (R) is the correct explanation of (A)
b. (A) is correct but (R) is incorrect
c. (A) is incorrect but (R) is correct
d. Both (A) and (R) are correct but (R) is not the correct explanation of (A)

17. In the following sequence of reactions the P is

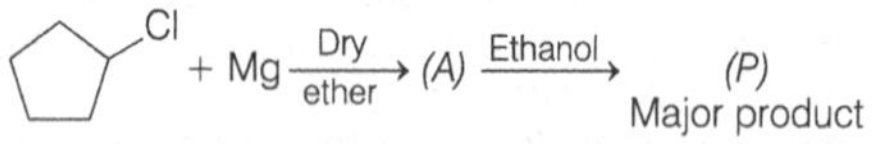

a. 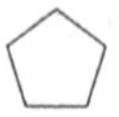

b. 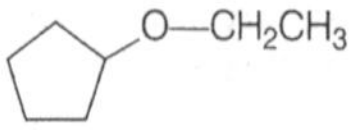

c. 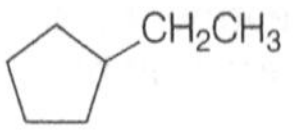

d. 

18. The unit of the van der Waals' gas equation parameter 'a' in $\left(p + \frac{an^2}{V^2}\right)(V - nb) = nRT$ is

a. kg ms^{-2} **b.** dm^3 mol^{-1}
c. kg ms^{-1} **d.** atm dm^{6}mol^{-2}

19. In polythionic acid, $H_2S_xO_6$(x = 3 to 5) the oxidation state(s) of sulphur is/are

a. only + 5 **b.** only + 6
c. + 3 and + 5 **d.** 0 and + 5

20. Tyndall effect is more effectively shown by

a. true solution **b.** lyophilic colloid
c. lyophobic colloid **d.** suspension

Section B : Numerical Type Questions

21. In carius method for estimation of halogens, 0.2 g of an organic compound gave 0.188 g of AgBr. The percentage of bromine in the compound is .... (Nearest integer)
[Atomic mass; Ag = 108, Br = 80]

22. The reaction that occurs in a breath analyser, a device used to determine the alcohol level in a person's blood stream is

$$2K_2Cr_2O_7 + 8H_2SO_4 + 3C_2H_6O \longrightarrow 2Cr_2(SO_4)_3 + 3C_2H_4O_2 + 2K_2SO_4 + 11H_2O$$

If the rate of appearance of $Cr_2(SO_4)_3$ is 2.67 mol min^{-1} at a particular time, the rate of disappearance of C_2H_6O at the same time is .... mol min^{-1}. (Nearest integer)

23. The kinetic energy of an electron in the second Bohr orbit of a hydrogen atom is equal to $\frac{h^2}{xma_0^2}$. The value of $10x$ is ....
(a_0 is radius of Bohr's orbit) (Nearest integer)
[Given, π = 3.14]

24. 1 kg of 0.75 molal aqueous solution of sucrose can be cooled up to – 4°C before freezing. The amount of ice (in g) that will be separated out is .... (Nearest integer)
[Given, $K_f(H_2O)$ = 1.86 K kg mol^{-1}]

25. 1 mol of an octahedral metal complex with formula $MCl_3 \cdot 2L$ on reaction with excess of $AgNO_3$ gives 1 mol of AgCl. The denticity of ligand L is .... (Integer answer)

26. The number of moles of CuO, that will be utilised in Dumas method for estimation nitrogen in a sample of 57.5 g of N, N-dimethylaminopentane is .... $\times 10^{-2}$. (Nearest integer)

27. The number of f-electrons in the ground state electronic configuration of Np (Z = 93) is .... (Nearest integer)

28. 200 mL of 0.2 M HCl is mixed with 300 mL of 0.1 M NaOH. The molar heat of neutralisation of this reaction is – 57.1 kJ. The increase in temperature in °C of the system on mixing is $x \times 10^{-2}$.
The value of x is .... (Nearest integer)
[Given, specific heat of water = 4.18 J g^{-1}K^{-1}
Density of water = 1.00 g cm^{-3}]
(Assume no volume change on mixing)

29. The number of moles of NH_3, that must be added to 2 L of 0.80 M $AgNO_3$ in order to reduce the concentration of Ag^+ ions to 5.0×10^{-8} M ($K_{formation}$ for $[Ag(NH_3)_2]^+ = 1.0 \times 10^8$) is .... (Nearest integer)
[Assume no volume change on adding NH_3]

30. When 10 mL of an aqueous solution of $KMnO_4$ was titrated in acidic medium, equal volume of 0.1 M of an aqueous solution of ferrous sulphate was required for complete discharge of colour.
The strength of $KMnO_4$ in g/L is .... $\times 10^{-2}$. (Nearest integer)
[Atomic mass of K = 39, Mn = 55, O = 16]

Answers

For solution scan the QR code

1. *(d)*	2. *(a)*	3. *(d)*	4. *(b)*	5. *(a)*	6. *(b)*	7. *(c)*	8. *(d)*	9. *(c)*	10. *(a)*
11. *(a)*	12. *(d)*	13. *(d)*	14. *(d)*	15. *(d)*	16. *(b)*	17. *(a)*	18. *(d)*	19. *(d)*	20. *(c)*
21. *(40)*	22. *(4)*	23. *(3155)*	24. *(518)*	25. *(2)*	26. *(1125)*	27. *(18)*	28. *(82)*	29. *(4)*	30. *(316)*

27 AUGUST SHIFT II

Section A : Objective Type Questions

1. Choose the correct statement from the following.
a. The standard enthalpy of formation for alkali metal bromide becomes less negative on descending the group.
b. The low solubility of CsI in water is due to its high lattice enthalpy.
c. Among the alkali metal halides, LiF is least soluble in water.
d. LiF has least negative standard enthalpy of formation among alkali metal fluorides.

2. The addition of dilute NaOH to Cr^{3+} salt solution will give
a. a solution of $[Cr(OH)_4]^-$ **b.** precipitate of $Cr_2O_3(H_2O)_n$
c. precipitate of $[Cr(OH)_6]^{3-}$ **d.** precipitate of $Cr(OH)_3$

3. Given below are two statements.
Statement I Ethyl pent-4-yn-oate on reaction with CH_3MgBr gives a 3° alcohol.
Statement II In this reaction, one mole of ethyl pent-4-yn-oate utilizes two moles of CH_3MgBr.
In the light of the above statements, choose the most appropriate answer from the options given below.
a. Both statement I and statement II are false.
b. Statement I is false but statement II is true.
c. Statement I is true but statement II is false.
d. Both statement I and statement II are true.

4. In stratosphere most of the ozone formation is assisted by
a. cosmic rays **b.** γ-rays
c. ultraviolet radiations **d.** visible radiations

5. The compound/s which will show significant intermolecular H-bonding is/are

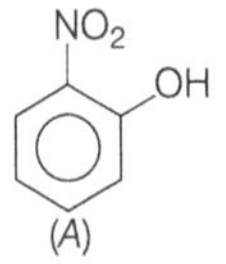

(A)

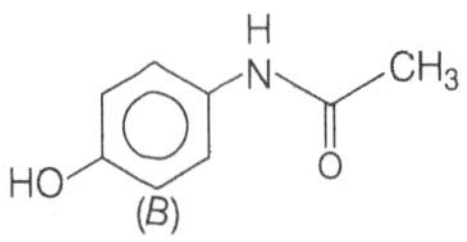

(B)

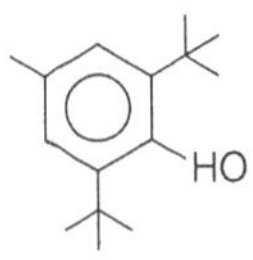

(C)

a. (B) only **b.** (C) only
c. (A) and (B) **d.** (A), (B) and (C)

6. Which one of the following chemicals is responsible for the production of HCl in the stomach leading to irritation and pain?

a. 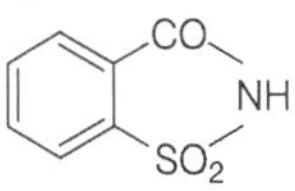

b. 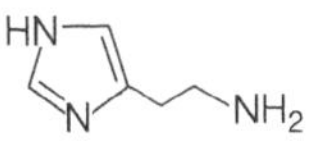

c. 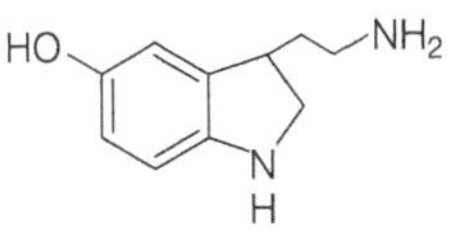

d. 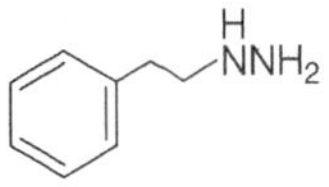

7. The oxide that gives H_2O_2 most readily on treatment with H_2O is
a. PbO_2 **b.** Na_2O_2 **c.** SnO_2 **d.** $BaO_2 \cdot 8H_2O$

8. Which one of the following reactions will not yield propionic acid?
a. $CH_3CH_2COCH_3 + OI^- / H_3O^+$
b. $CH_3CH_2CH_3 + KMnO_4$ (Heat), OH^- / H_3O^+
c. $CH_3CH_2CCl_3 + OH^- / H_3O^+$
d. $CH_3CH_2CH_2Br + Mg, CO_2$, dry ether/$H_3O^+$

9. The correct order of ionic radii for the ions, $P^{3-}, S^{2-}, Ca^{2+}, K^+, Cl^-$ is
a. $P^{3-} > S^{2-} > Cl^- > K^+ > Ca^{2+}$
b. $Cl^- > S^{2-} > P^{3-} > Ca^{2+} > K^+$
c. $P^{3-} > S^{2-} > Cl^- > Ca^{2+} > K^+$
d. $K^+ > Ca^{2+} > P^{3-} > S^{2-} > Cl^-$

10. Which one of the following is the major product of the given reaction?

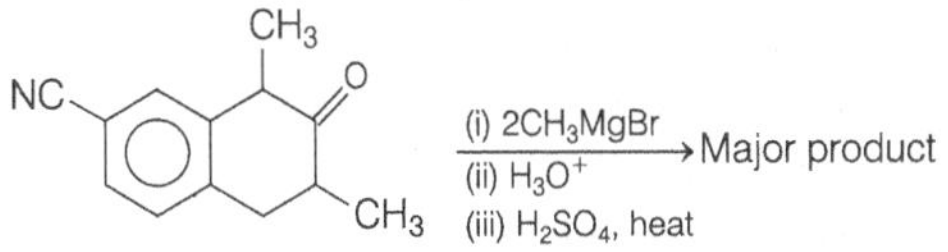

a. 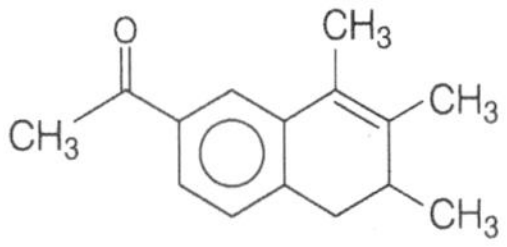

b. 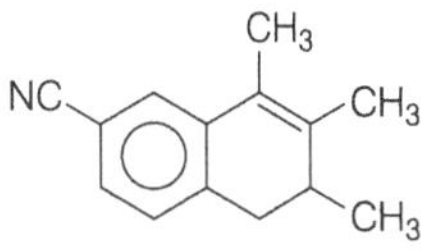

c. 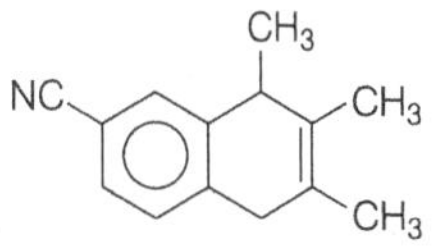

d. 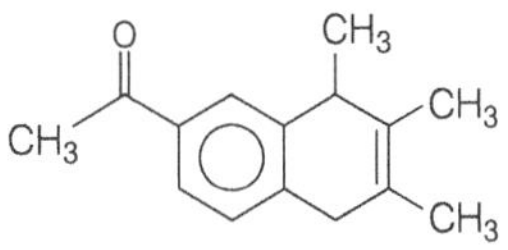

11. The major product (*A*) formed in the reaction given below is

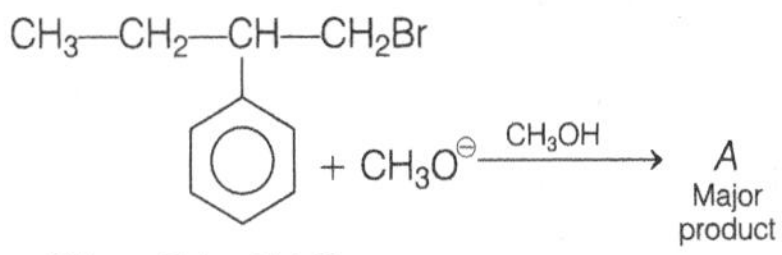

a. 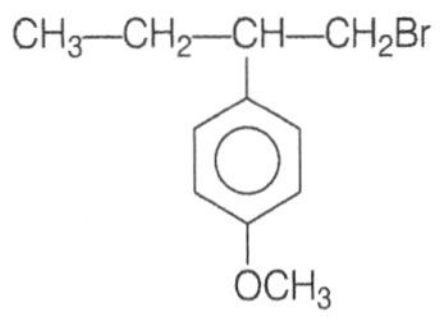

b. 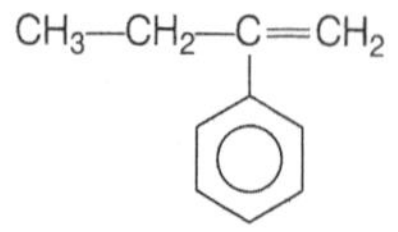

c. 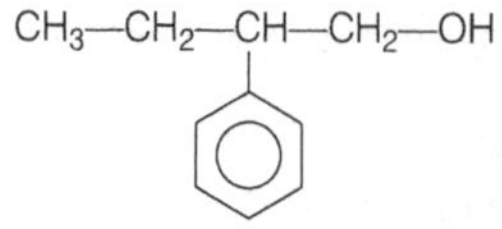

d. 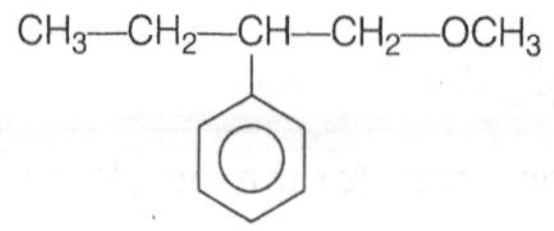

12. Which one of the following is used to remove most of plutonium from spent nuclear fuel?

a. ClF_3 b. O_2F_2 c. I_2O_5 d. BrO_3

13. Lyophilic sols are more stable than lyophobic sols because

a. there is a strong electrostatic repulsion between the negatively charged colloidal particles.
b. the colloidal particles have positive charge.
c. the colloidal particles have no charge.
d. the colloidal particles are solvated.

14. The major product of the following reaction, if it occurs by S_N2 mechanism is

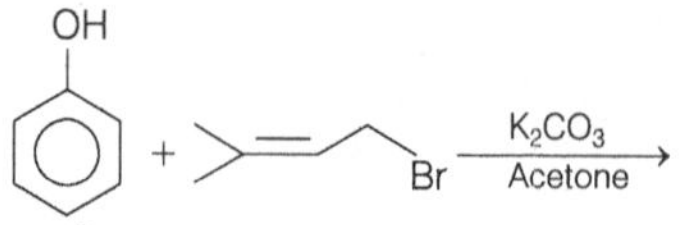

a. 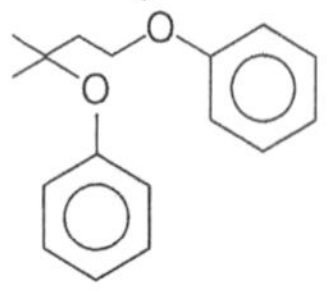
b. 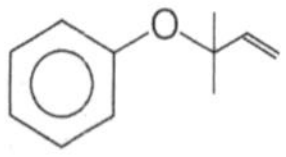

c. 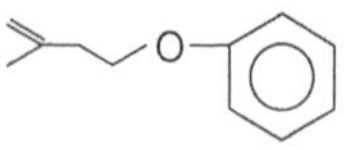
d. 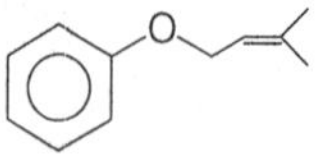

15. Potassium permanganate on heating at 513 K gives a product which is

a. paramagnetic and colourless
b. diamagnetic and green
c. diamagnetic and colourless
d. paramagnetic and green

16. Which one of the following tests used for the identification of functional groups in organic compounds does not use copper reagent ?

a. Barfoed's test b. Seliwanoff's test
c. Benedict's test d. Biuret test for peptide bond

17. Hydrolysis of sucrose gives

a. α-D-(–)-glucose and β-D-(–)-fructose
b. α-D-(+)-glucose and β-D-(+)-fructose
c. α-D-(–)-glucose and β-D-(+)-fructose
d. α-D-(+)-glucose and β-D-(–)-fructose

18. Match List-I with List-II.

	List-I (Name of ore/mineral)		List-I (Chemical formula)
A.	Calamine	1.	ZnS
B.	Malachite	2.	$FeCO_3$
C.	Siderite	3.	$ZnCO_3$
D.	Sphalerite	4.	$CuCO_3 \cdot Cu(OH)_2$

Choose the most appropriate answer from the options given below

	A	B	C	D		A	B	C	D
a.	3	4	2	1	b.	3	4	1	2
c.	4	3	1	2	d.	3	2	4	1

19. Which one of the following is formed (mainly) when red phosphorus is heated in a sealed tube at 803 K ?

a. White phosphorus b. Yellow phosphorus
c. β-black phosphorus d. α-black phosphorus

20. The correct structures of *A* and *B* formed in the following reactions are

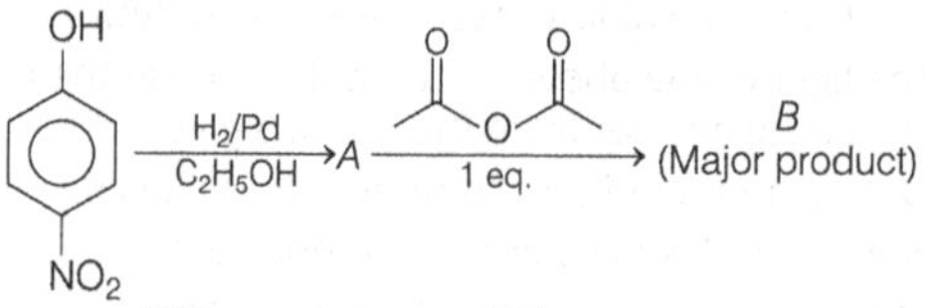

a. 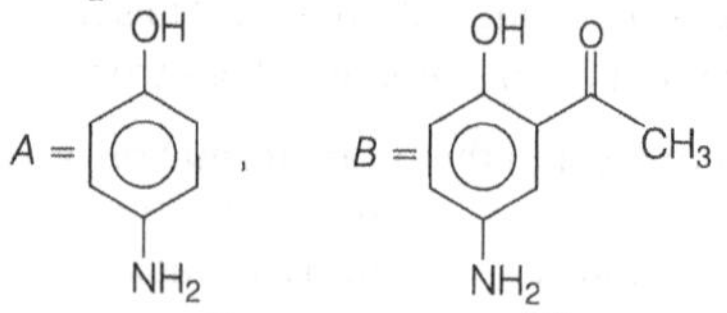

b. 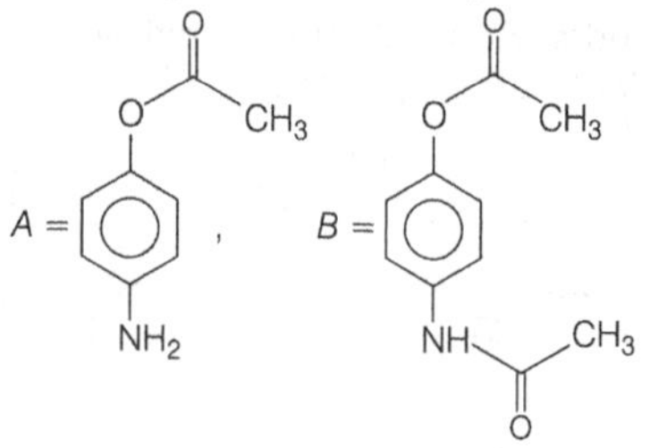

c. 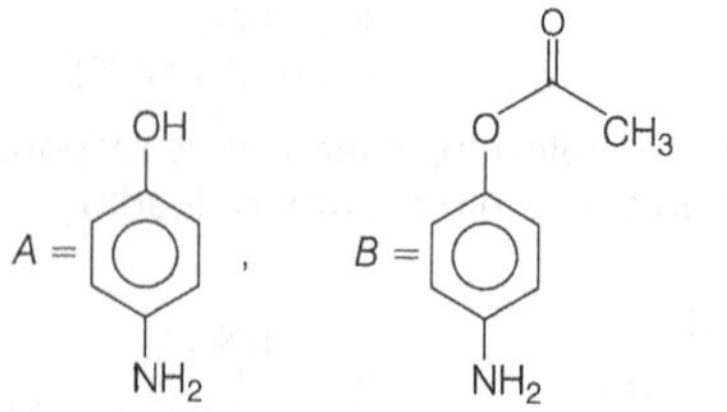

d. 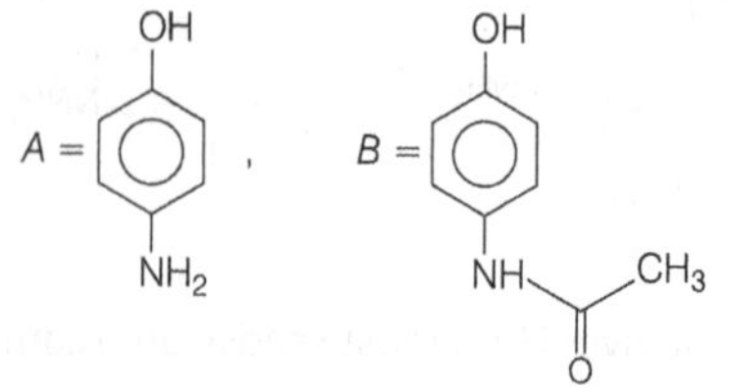

Section B : Numerical Type Questions

21. The first order rate constant for the decomposition of $CaCO_3$ at 700 K is $6.36 \times 10^{-3}\ s^{-1}$ and activation energy is 209 kJ mol^{-1}. Its rate constant (in s^{-1}) at 600 K is $x \times 10^{-6}$. The value of x is .... (Nearest integer)

[Given, $R = 8.31\ JK^{-1}mol^{-1}$, $\log 6.36 \times 10^{-3} = -2.19$, $10^{-4.79} = 1.62 \times 10^{-5}$]

22. The number of optical isomers possible for $[Cr(C_2O_4)_3]^{3-}$ is ...

23. Two flasks I and II shown below are connected by a valve of negligible volume.

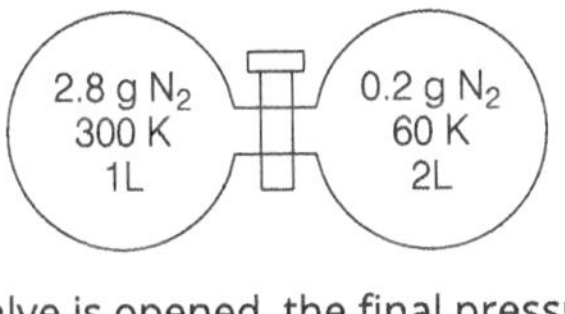

When the valve is opened, the final pressure of the system in bar is $x \times 10^{-2}$. The value of x is .... (Integer answer)

[Assume, Ideal gas, 1 bar = 10^5 Pa, molar mass of $N_2 = 28.0$ g mol^{-1}; $R = 8.31$ J mol^{-1} K^{-1}]

24. 100 g of propane is completely reacted with 1000 g of oxygen. The mole fraction of carbon dioxide in the resulting mixture is $x \times 10^{-2}$. The value of x is .... (Nearest integer)

[Atomic weight : H = 1.008, C = 12.00, O = 16.00]

25. 40 g of glucose (Molar mass = 180) is mixed with 200 mL of water. The freezing point of solution is .... K. (Nearest integer)

[Given, $K_f = 1.86$ K kg mol^{-1}, density of water = 1.00 g cm^{-3}, freezing point of water = 273.15 K]

26. The resistance of a conductivity cell with cell constant 1.14 cm^{-1}, containing 0.001 M KCl at 298 K is 1500 Ω. The molar conductivity of 0.001 M KCl solution at 298 K in S cm^2mol^{-1} is .... (Integer answer)

27. The number of photons emitted by a monochromatic (single frequency) infrared range finder of power 1 mW and wavelength of 1000 nm, in 0.1 second is $x \times 10^{13}$. The value of x is .... (Nearest integer)

($h = 6.63 \times 10^{-34}$ Js, $c = 3.00 \times 10^8\ ms^{-1}$)

28. When 5.1 g of solid NH_4HS is introduced into a two litre evacuated flask at 27°C, 20% of the solid decomposes into gaseous ammonia and hydrogen sulphide. The K_p for the reaction at 27°C is $x \times 10^{-2}$. The value of x is .... (Integer answer)

[Given, $R = 0.082$ L atm K^{-1} mol^{-1}]

29. The number of species having non-pyramidal shape among the following is

(i) SO_3 (ii) NO_3^- (iii) PCl_3 (iv) CO_3^{2-}

30. Data given for the following reaction is as follows.

$FeO(s) + C_{(graphite)} \longrightarrow Fe(s) + CO(g)$

Substance	$\Delta H°$ (kJ mol^{-1})	$\Delta S°$ (J mol^{-1} K^{-1})
FeO(s)	– 266.3	57.49
$C_{(graphite)}$	0	5.74
Fe(s)	0	27.28
CO(g)	– 110.5	197.6

The minimum temperature in K at which the reaction becomes spontaneous is .... (Integer answer)

Answers

For solution scan the QR code

1. (c)	2. (b)	3. (c)	4. (c)	5. (a)	6. (b)	7. (b)	8. (d)	9. (a)	10. (a)
11. (b)	12. (b)	13. (d)	14. (d)	15. (d)	16. (b)	17. (d)	18. (a)	19. (d)	20. (d)
21. (16)	22. (2)	23. (84)	24. (19)	25. (271)	26. (760)	27. (50)	28. (6)	29. (3)	30. (964)

31 AUGUST SHIFT I

Section A : Objective Type Questions

1. The correct order of reactivity of the given chlorides with acetate in acetic acid is

a. 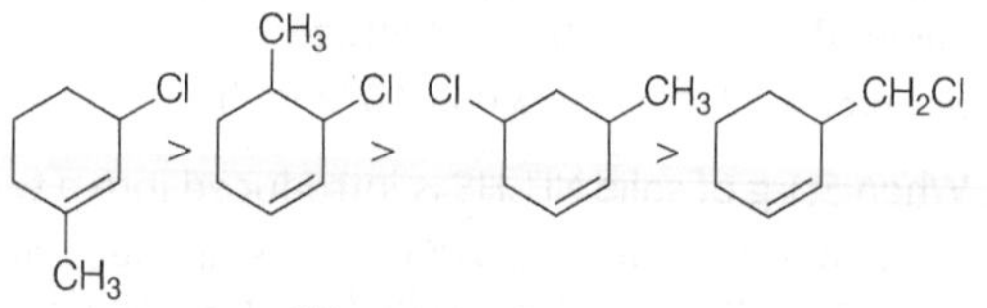

b. 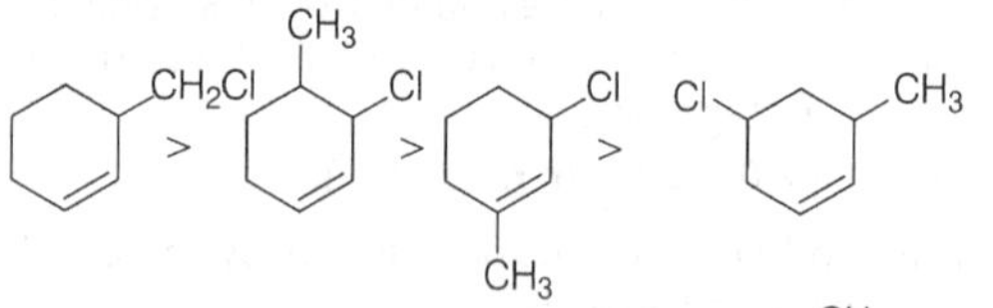

c. 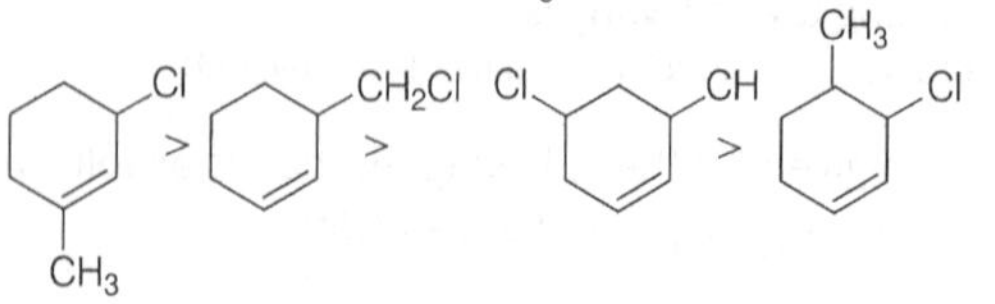

d. 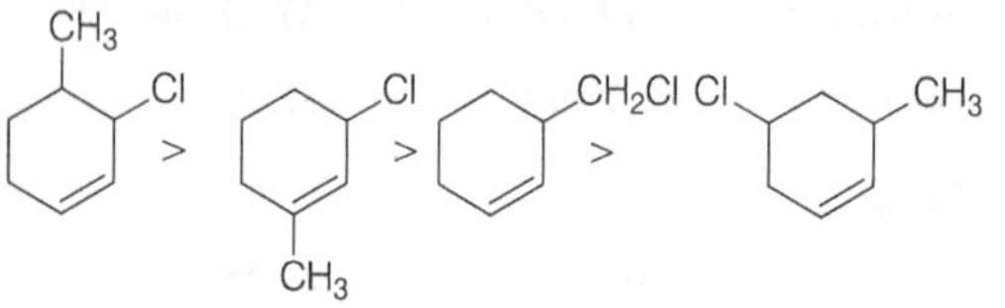

2. Select the graph that correctly describes the adsorption isotherms at two temperatures T_1 and T_2 ($T_1 > T_2$) for a gas. (x = mass of the gas adsorbed ; m = mass of adsorbent ; p = pressure)

a. 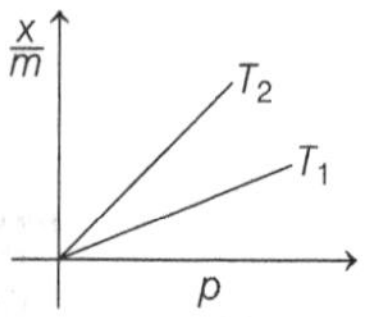

b. 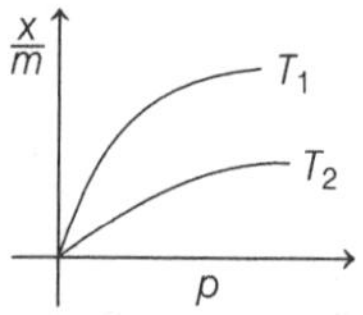

c. 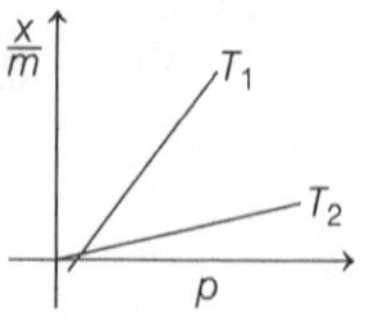

d. 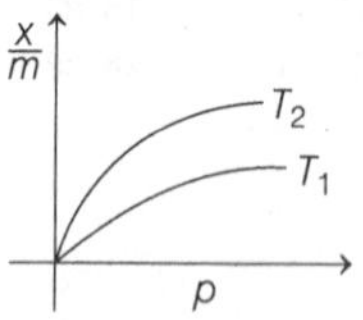

3. The major component/ingredient of Portland cement is

a. tricalcium aluminate
b. tricalcium silicate
c. dicalcium aluminate
d. dicalcium silicate

4. In the structure of the dichromate ion, there is a

a. linear symmetrical Cr—O—Cr bond.
b. non-linear symmetrical Cr—O—Cr bond.
c. linear unsymmetrical Cr—O—Cr bond.
d. non-linear unsymmetrical Cr—O—Cr bond.

5. Which one of the following compounds contain β-C_1—C_4 glycosidic linkage ?

a. Lactose
b. Sucrose
c. Maltose
d. Amylose

6. The major products A and B in the following set of reactions are

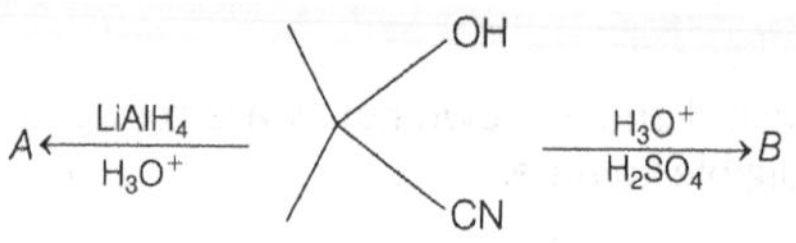

a. A = 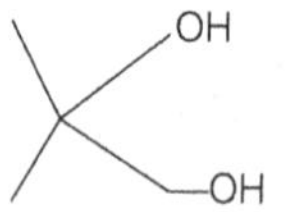 , B = 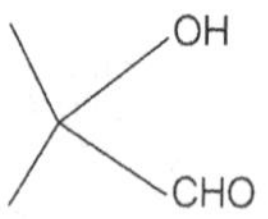

b. A = 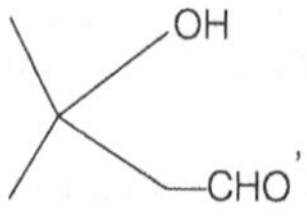 , B = 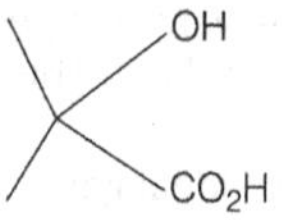

c. A = 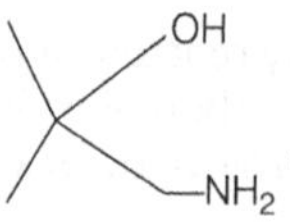 , B = 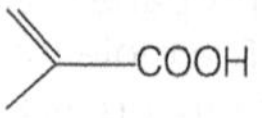

d. A = 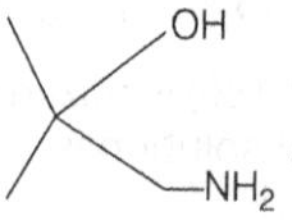 , B = 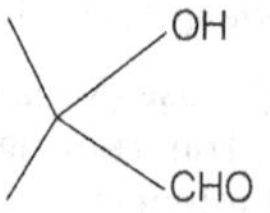

7. Which one of the following lanthanides exhibits + 2 oxidation state with diamagnetic nature ?
(Given, Z for Nd = 60, Yb = 70, La = 57, Ce = 58)

a. Nd
b. Yb
c. La
d. Ce

8. Given below are two statements.
One is labelled as Assertion (A) and the other is labelled as Reason (R).

Assertion (A) Aluminium is extracted from bauxite by the electrolysis of molten mixture of Al_2O_3 with cryolite.

Reason (R) The oxidation state of Al in cryolite is + 3.

In the light of the above statements, choose the most appropriate answer from the options given below.

a. (A) is true but (R) is false
b. (A) is false but (R) is true.
c. Both (A) and (R) are true and (R) is the correct explanation of (A).
d. Both (A) and (R) are true but (R) is not the correct explanation of (A).

9. The major product formed in the following reaction is

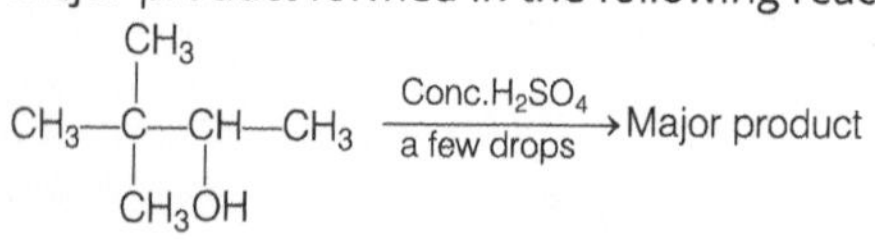

a. 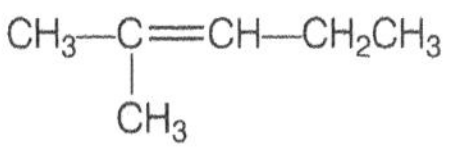 b. 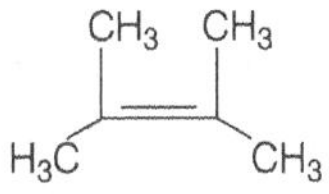

c. 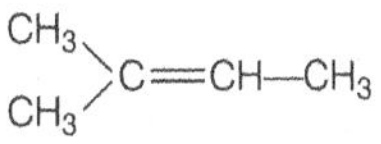 d. 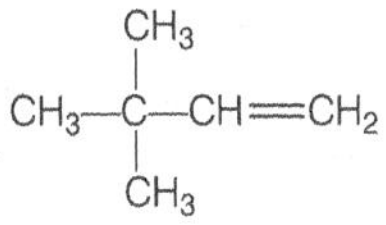

10. Monomer of novolac is

a. 3-hydroxybutanoic acid
b. phenol and melamine
c. *o*-hydroxymethylphenol
d. 1,3-butadiene and styrene

11. Given below are two statements.

Statement I The process of producing *syn*-gas is called gasification of coal.

Statement II The composition of *syn*-gas is $CO + CO_2 + H_2$ (1 : 1 : 1)

In the light of the above statements, choose the most appropriate answer from the options given below.

a. Statement I is false but statement II is true.
b. Statement I is true but statement II is false.
c. Both statement-I and statement II are false.
d. Both statement-I and statement II are true.

12. Given below are two statements.
One is labelled as Assertion (A) and the other is labelled as Reason (R).

Assertion (A) Treatment of bromine water with propene yields 1-bromopropan-2-ol.

Reason (R) Attack of water on bromonium ion follows Markownikoff rule and results in 1-bromopropan-2-ol.

In the light of the above statements, choose the most appropriate answer from the options given below.

a. Both (A) and (R) are true but (R) is not the correct explanation of (A).
b. (A) is false but (R) is true.
c. Both (A) and (R) are true and (R) is the correct explanation of (A).
d. (A) is true but (R) is false

13. The denticity of an organic ligand, biuret is

a. 2 **b.** 4 **c.** 3 **d.** 6

14. Given below are two statements : one is labelled as Assertion (A) and the other is labelled as Reason (R).

Assertion (A) Metallic character decreases and non-metallic character increases on moving from left to right in a period.

Reason (R) It is due to increase in ionisation enthalpy and decrease in electron gain enthalpy, when one moves from left to right in a period.

In the light of the above statements, choose the most appropriate answer from the options given below.

a. (A) is false but (R) is true.
b. (A) is true but (R) is false
c. Both (A) and (R) are correct and (R) is the correct explanation of (A).
d. Both (A) and (R) are correct but (R) is not the correct explanation of (A).

15. Choose the correct name for compound given below

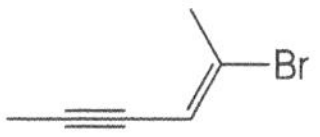

a. (4E)-5-bromohex-4-en-2-yne
b. (2E)-2-bromohex-4-yn-2-ene
c. (2E)-2-bromohex-2-en-4-yne
d. (4E)-5-bromohex-2-en-4-yne

16. Which one of the following is the correct pV *vs* p plot at constant temperature for an ideal gas ? (p and V stand for pressure and volume of the gas respectively)

a. 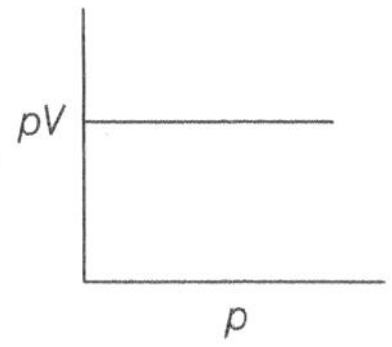

b. 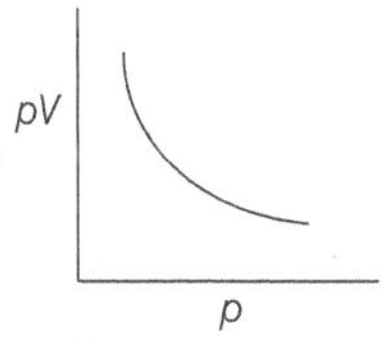

c. 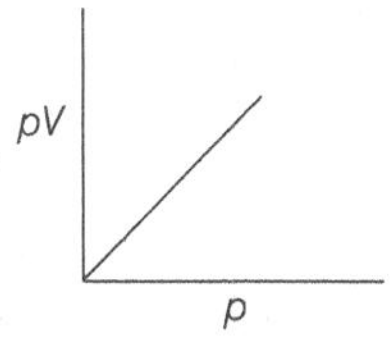

d. 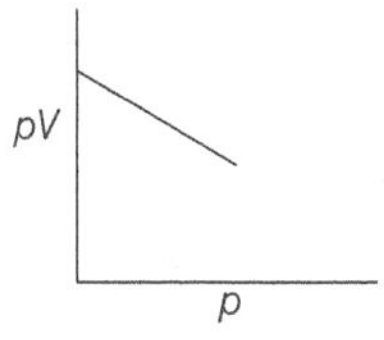

17. Given below are two statements.
One is labelled as Assertion (A) and the other is labelled as Reason (R).

Assertion (A) A simple distillation can be used to separate a mixture of propanol and propanone.

Reason (R) Two liquids with a difference of more than 20°C in their boiling points can be separated by simple distillations.

In the light of the above statements, choose the most appropriate answer from the options given below.

a. (A) is false but (R) is true.
b. Both (A) and (R) are correct but (R) is not the correct explanation of (A).
c. (A) is true but (R) is false
d. Both (A) and (R) are correct and (R) is the correct explanation of (A).

18. Which one of the following 0.10 M aqueous solutions will exhibit the largest freezing point depression ?

a. Hydrazine **b.** Glucose
c. Glycine **d.** $KHSO_4$

19. BOD values (in ppm) for clean water (A) and polluted water (B) are expected respectively

a. $A > 50, B < 27$ **b.** $A > 25, B < 17$
c. $A < 5, B > 17$ **d.** $A > 15, B > 47$

20. The structure of product C, formed by the following sequence of reactions is

$$CH_3COOH + SOCl_2 \longrightarrow A \xrightarrow[AlCl_3]{\text{Benzene}} B \xrightarrow[-OH]{KCN} C$$

a. 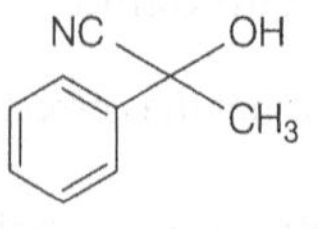 b. 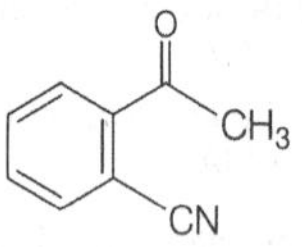

c. 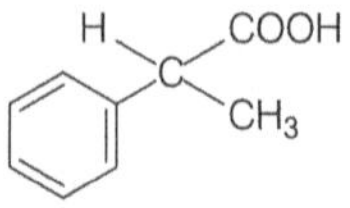 d. 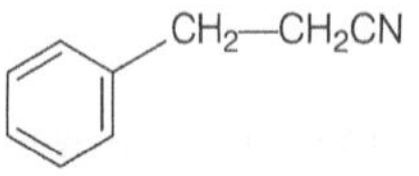

Section B : Numerical Type Questions

21. Consider the following cell reaction,

$$Cd(s) + Hg_2SO_4(s) + \frac{9}{5}H_2O(l) \rightleftharpoons CdSO_4 \cdot \frac{9}{5}H_2O(s) + 2Hg(l)$$

The value of E°_{cell} is 4.315 V at 25°C. If $\Delta H^\circ = -825.2$ kJ mol^{-1}, the standard entropy change ΔS° in J K^{-1} is .... (Nearest integer)

[Given, Faraday constant = 96487 C mol^{-1}]

22. The molarity of the solution prepared by dissolving 6.3 g of oxalic acid $(H_2C_2O_4 \cdot 2H_2O)$ in 250 mL of water in mol L^{-1} is $x \times 10^{-2}$. The value of x is .... (Nearest integer)

[Atomic mass H = 1.0, C = 12.0, O = 16.0]

23. Consider the sulphides HgS, PbS, CuS, Sb_2S_3, As_2S_3 and CdS. Number of these sulphides soluble in 50% HNO_3 is

24. The total number of reagents from those given below, that can convert nitrobenzene into aniline is .... (Integer answer)

I. Sn - HCl II. Sn - NH_4OH
III. Fe - HCl IV. Zn - HCl
V. H_2 - Pd VI. H_2-Raney nickel

25. The number of halogen (s) forming halic (V) acid is

26. For a first order reaction, the ratio of the time for 75% completion of a reaction to the time for 50% completion is .... (Integer answer)

27. The number of hydrogen bonded water molecule(s) associated with stoichiometry $CuSO_4 \cdot 5H_2O$ is/are.

28. According to the following figure, the magnitude of the enthalpy change of the reaction $A + B \rightarrow M + N$ in kJ mol^{-1} is equal to .... (Integer answer)

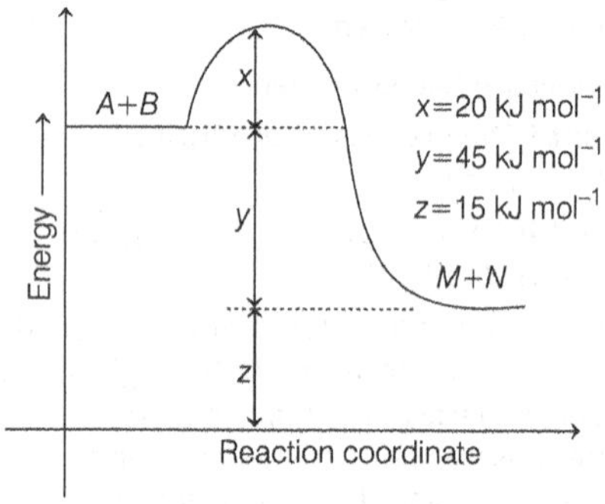

29. Ge(Z = 32) in its ground state electronic configuration has x completely filled orbitals with $m_l = 0$. The value of x is

30. A_3B_2 is a sparingly soluble salt of molar mass M (g mol^{-1}) and solubility x g L^{-1}. The solubility product satisfies $K_{sp} = a\left(\frac{x}{M}\right)^5$. The value of a is .... (Integer answer)

Answers

For solution scan the QR code

1. (a)	2. (d)	3. (b)	4. (b)	5. (a)	6. (c)	7. (b)	8. (d)	9. (b)	10. (c)
11. (b)	12. (c)	13. (a)	14. (b)	15. (c)	16. (a)	17. (d)	18. (d)	19. (c)	20. (a)
21. (25)	22. (20)	23. (4)	24. (5)	25. (3)	26. (2)	27. (1)	28. (45)	29. (7)	30. (108)

31 AUGUST SHIFT II

Section A : Objective Type Questions

1. Arrange the following conformational isomers of *n*-butane in order of their increasing potential energy

I. 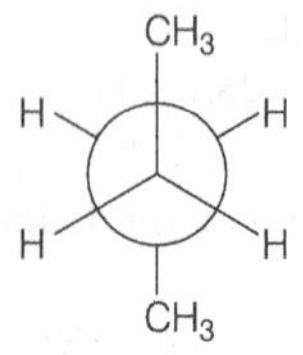 II. 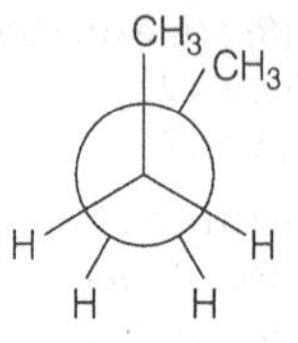

III. 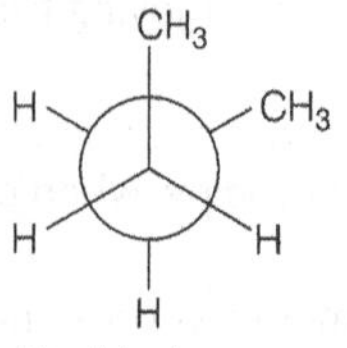 IV. 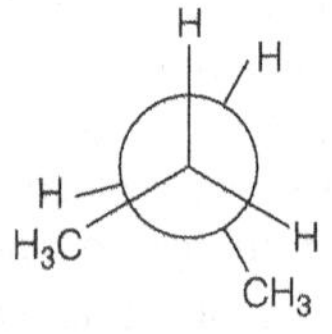

a. II < III < IV < I
b. I < IV < III < II
c. II < IV < III < I
d. I < III < IV < II

2. The Eu^{2+} ion is a strong reducing agent in spite of its ground state electronic configuration (outermost) : [Atomic number of Eu = 63]

a. $4f^7 6s^2$ **b.** $4f^6$ **c.** $4f^7$ **d.** $4f^6 6s^2$

3. The structures of *A* and *B* formed in the following reaction are : [Ph = $-C_6H_5$]

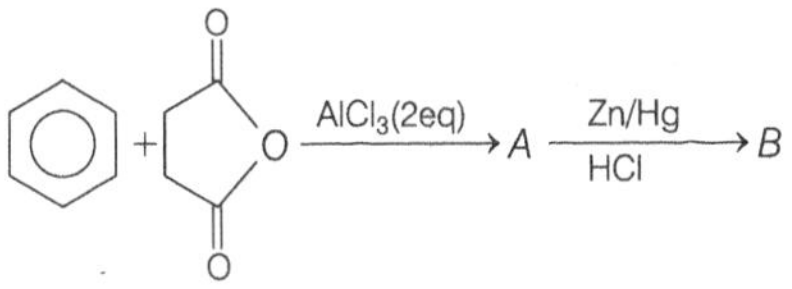

a. $A =$ 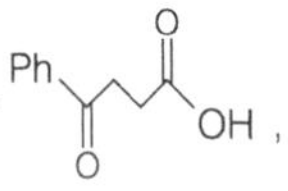 , $B =$ 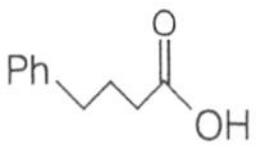

b. $A =$ 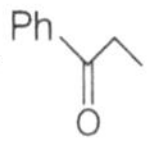 , $B =$ 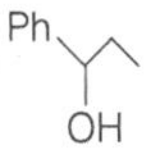

c. $A =$ 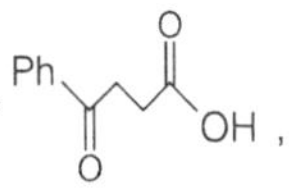 , $B =$ 

d. $A =$ 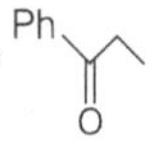 , $B =$ 

4. In which one of the following sets all species show disproportionation reaction

a. ClO_2^-, F_2, MnO_4^{2-} and $Cr_2O_7^{2-}$

b. $Cr_2O_7^{2-}$, MnO_4^-, ClO_2^- and Cl_2

c. ClO_2^-, Cl_2 and Mn^{3+}

d. ClO_4^-, MnO_4^{2-}, ClO_2^- and F_2

5. Match List-I with List-II.

	List-I (Parameter)		**List-II** (Unit)
A.	Cell constant	1.	$S\ cm^2\ mol^{-1}$
B.	Molar conductivity	2.	Dimensionless
C.	Conductivity	3.	m^{-1}
D.	Degree of dissociation of electrolyte	4.	$\Omega^{-1} m^{-1}$

Choose the most appropriate answer from the options given below

	A	B	C	D		A	B	C	D
a.	3	1	4	2	**b.**	3	1	2	4
c.	1	4	3	2	**d.**	2	1	3	4

6. The major products *A* and *B* formed in the following reaction sequence are

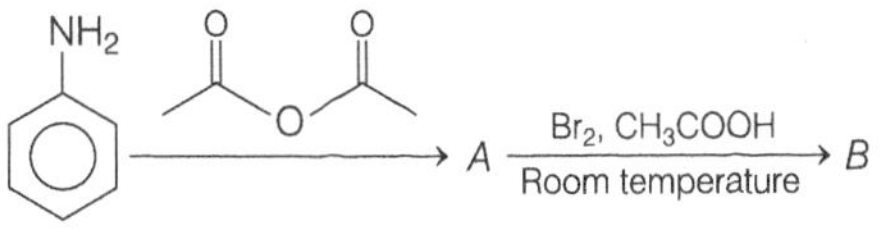

a. $A =$ 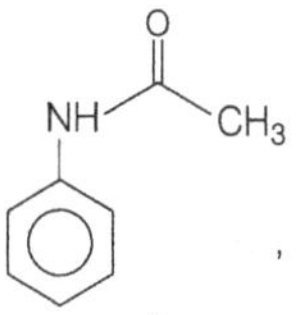 , $B =$ 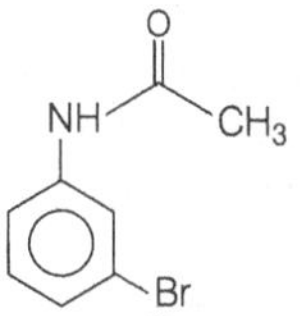

b. $A =$ 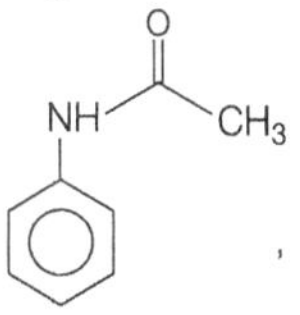 , $B =$ 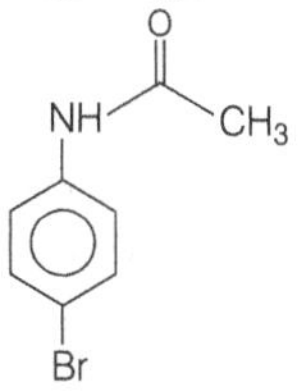

c. $A =$ 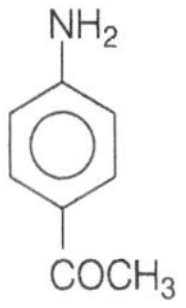 , $B =$ 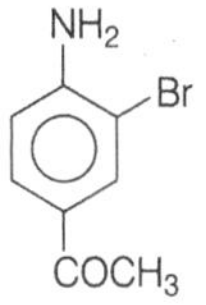

d. $A =$ 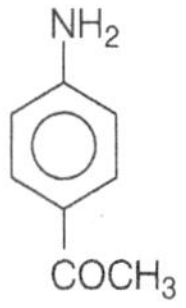 , $B =$ 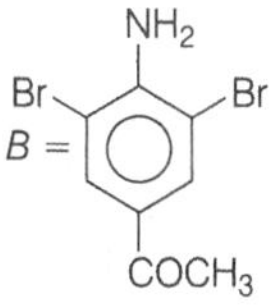

7. Which of the following is not an example of fibrous protein ?

a. Keratin **b.** Albumin

c. Collagen **d.** Myosin

8. The depositions of *X* and *Y* on ground surfaces is referred to as wet and dry depositions, respectively. *X* and *Y* are

a. X = Ammonium salts, $Y = CO_2$

b. $X = SO_2$, Y = Ammonium salts

c. X = Ammonium salts, $Y = SO_2$

d. $X = CO_2$, $Y = SO_2$

9. For the reaction given below.

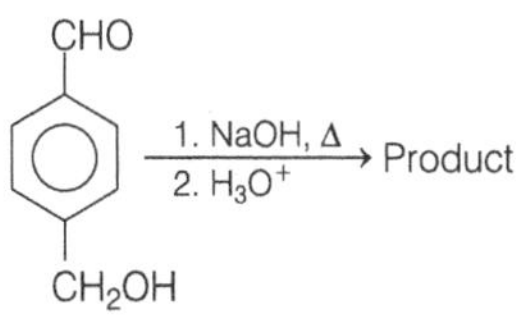

The compound which is not formed as a product in the reaction is a

a. compound with both alcohol and acid functional groups

b. monocarboxylic acid

c. dicarboxylic acid

d. diol

10. Spin only magnetic moment in BM of $[Fe(CO)_4(C_2O_4)]^+$ is

a. 5.92 **b.** 0 **c.** 1 **d.** 1.73

11. Given below are two statements : One is labelled as Assertion (A) and the other is labelled as Reason (R).

Assertion (A) Lithium salts are hydrated.

Reason (R) Lithium has higher polarising power than other alkali metal group members.

In the light of the above statements, choose the most appropriate answer from the options given below

a. Both (A) and (R) are true but (R) is not the correct explanation of (A).

b. (A) is true but (R) is not true

c. (A) is false but (R) is ture.

d. Both (A) and (R) are true (R) is the correct explanation of (A).

12. The incorrect expression among the following is

a. $\frac{\Delta G_{\text{System}}}{\Delta S_{\text{Total}}} = -T$ (at constant p)

b. $\ln k = \frac{\Delta H^\circ - T\Delta S^\circ}{RT}$

c. $k = e^{-\frac{\Delta G^\circ}{RT}}$

d. For isothermal process, $W_{\text{reversible}} = -nRT \ln \frac{V_f}{V_i}$

13. Which one of the following statements is incorrect ?

a. Atomic hydrogen is produced when H_2 molecules at a high temperature are irradiated with UV radiation.

b. At around 2000 K, the dissociation of dihydrogen into its atoms is nearly 8.1%.

c. Bond dissociation enthalpy of H_2 is highest among diatomic gaseous molecules which contain a single bond .

d. Dihydrogen is produced on reacting zinc with HCl as well as NaOH(*aq*).

14. Which among the following is not a polyester ?

a. Novolac **b.** PHBV **c.** Dacron **d.** Glyptal

15. Which one of the following correctly represents the order of stability of oxides, X_2O (X = halogen) ?

a. Br > Cl > I **b.** Br > I > Cl **c.** Cl > I > Br **d.** I > Cl > Br

16. Match List-I with List-II.

List-I (Metal ion)		List-II (Group in qualitative analysis)
A. Mn^{2+}	1.	Group - III
B. As^{3+}	2.	Group - IIA
C. Cu^{2+}	3.	Group - IV
D. Al^{3+}	4.	Group - IIB

Choose the most appropriate answer from the options given below.

	A	B	C	D		A	B	C	D
a.	1	2	3	4	b.	3	4	2	1
c.	1	4	2	3	d.	4	2	3	1

17. The major product of the following reaction is

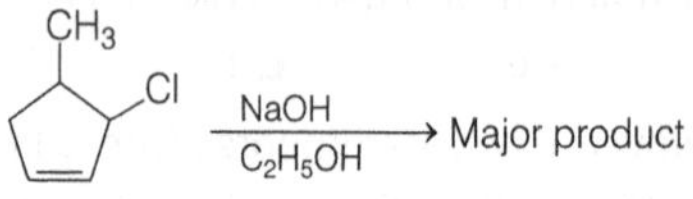

a. 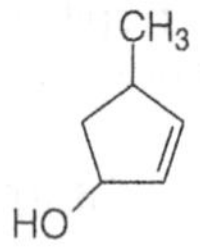

b. 

c. 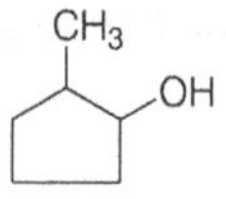

d. 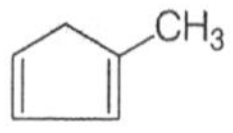

18. For the following

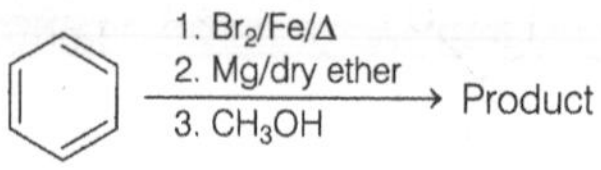

a. 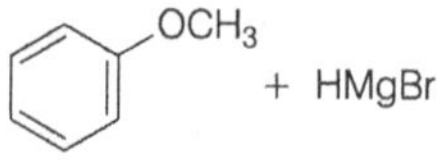

b. 

c. 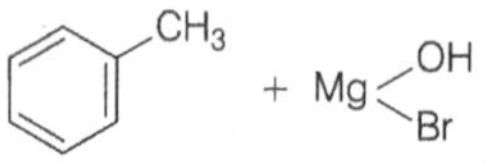

d. 

19. Identify correct *A*, *B* and *C* in the reaction sequence given below.

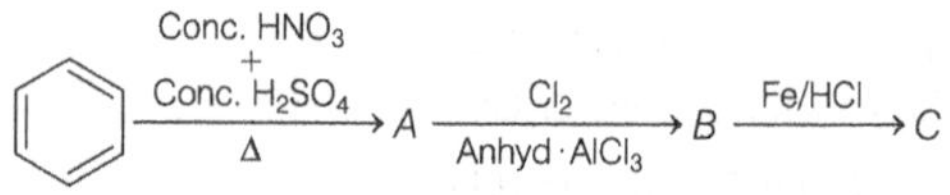

a. 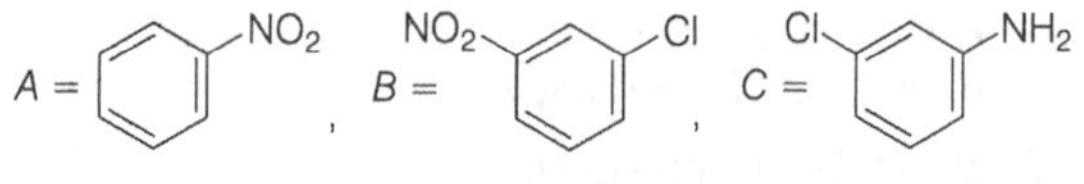

b. 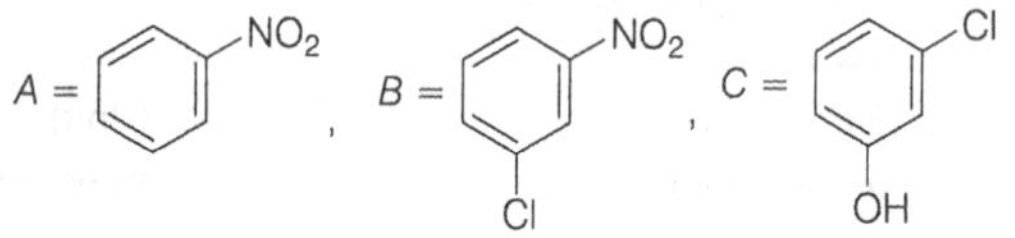

c. 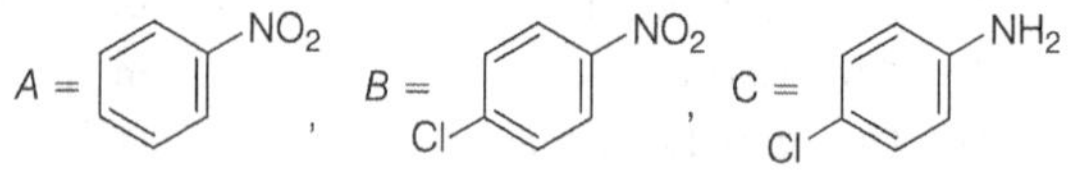

d. 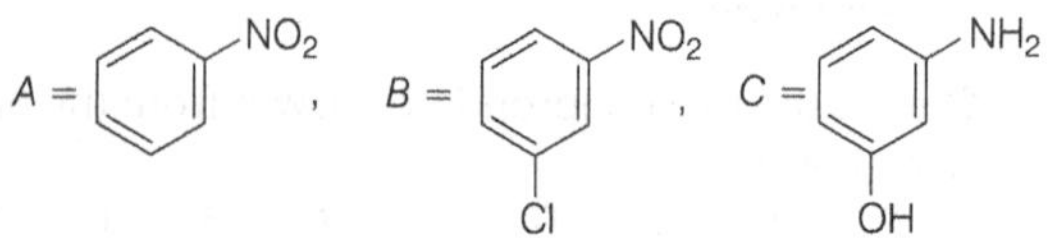

20. The number of S=O bonds present in sulphurous acid, peroxodisulphuric acid and pyrosulphuric acid, respectively are

a. 2, 3 and 4

b. 1, 4 and 3

c. 2, 4 and 3

d. 1, 4 and 4

Section B : Numerical Type Questions

21. CH_4 is adsorbed on 1 g charcoal at 0°C following the Freundlich adsorption isotherm. 10.0 mL of CH_4 is adsorbed at 100 mm of Hg, whereas 15.0 mL is adsorbed at 200 mm of Hg. The volume of CH_4 adsorbed at 300 mm of Hg is 10^x mL. The value of x is .... $\times 10^{-2}$. (Nearest integer)

[Use $\log_{10} 2 = 0.3010$, $\log_{10} 3 = 0.4771$]

22. 1.22 g of an organic acid is separately dissolved in 100 g of benzene ($K_b = 2.6$ K kg mol^{-1}) and 100 g of acetone ($K_b = 1.7$ K kg mol^{-1}). The acid is known to dimerise in benzene but remain as a monomer in acetone. The boiling point of the solution in acetone increases by 0.17°C. The increase in boiling point of solution in benzene in °C is $x \times 10^{-2}$. The value of x is .... (Nearest integer)

[Atomic mass : C = 12.0, H = 1.0, O= 16.0]

23. The value of magnetic quantum number of the outermost electron of Zn^+ ion is

24. The empirical formula for a compound with a cubic close packed arrangement of anions and with cations occupying all the octahedral sites in A_xB. The value of x is .... (Integer answer)

25. In the electrolytic refining of blister copper, the total number of main impurities from the following, removed as anode mud is ......

Pb, Sb, Se, Te, Ru, Ag, Au and Pt

26. The pH of a solution obtained by mixing 50 mL of 1 M HCl and 30 mL of 1 M NaOH is $x \times 10^{-4}$.

The value of x is .... (Nearest integer) [log 2.5 = 0.3979]

27. For the reaction $A \rightarrow B$, the rate constant k (in s^{-1}) is given by $\log_{10} k = 20.35 - \frac{(2.47 \times 10^3)}{T}$.

The energy of activation in kJ mol^{-1} is .... (Nearest integer)

[Given : $R = 8.314$ J K^{-1} mol^{-1}]

28. Sodium oxide reacts with water to produce sodium hydroxide. 20.0 g of sodium oxide is dissolved in 500 mL of water. Neglecting the change in volume, the concentration of the resulting NaOH solution is .... $\times 10^{-1}$ M. (Nearest integer)

[Atomic mass : Na = 23.0, O = 16.0, H = 1.0]

29. According to molecular orbital theory, the number of unpaired electron(s) in O_2^{2-} is ....

30. The transformation occurring in Duma's method is given below

$$C_2H_7N + \left(2x + \frac{y}{2}\right) CuO \longrightarrow xCO_2 + \frac{y}{2} H_2O + \frac{z}{2} N_2 + \left(2x + \frac{y}{2}\right) Cu$$

The value of y is .... (Integer answer)

Answers

For solution scan the QR code

1. (d)	2. (c)	3. (a)	4. (c)	5. (a)	6. (b)	7. (b)	8. (c)	9. (c)	10. (d)
11. (a)	12. (b)	13. (b)	14. (a)	15. (d)	16. (b)	17. (d)	18. (b)	19. (a)	20. (d)
21. (128)	22. (13)	23. (0)	24. (1)	25. (6)	26. (6021)	27. (47)	28. (13)	29. (0)	30. (7)

1 SEPTEMBER SHIFT II

Section A : Objective Type Questions

1. Water sample is called cleanest on the basis of which one of the BOD values given below

a. 11 ppm **b.** 15 ppm **c.** 3 ppm **d.** 21 ppm

2. Calamine and malachite, respectively, are the ores of

a. nickel and aluminium **b.** zinc and copper

c. copper and iron **d.** aluminium and zinc

3. Experimentally reducing a functional group cannot be done by which one of the following reagents ?

a. Pt-C/H_2 **b.** Na / H_2 **c.** Pd-C/H_2 **d.** Zn / H_2O

4. Which one of the following given graphs represents the variation of rate constant (k) with temperature (T) for an endothermic reaction?

a. 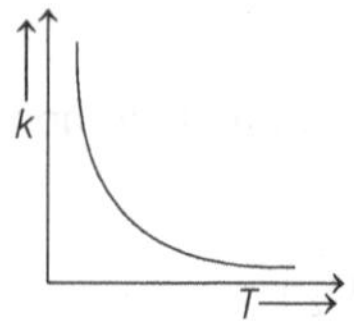

b. 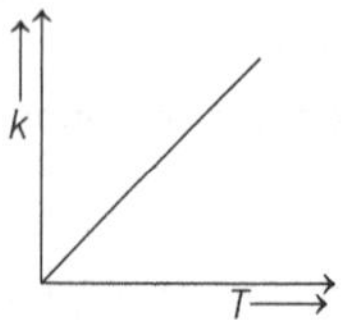

c. 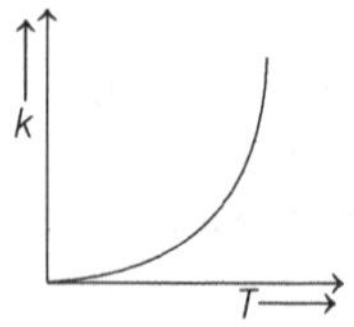

d. 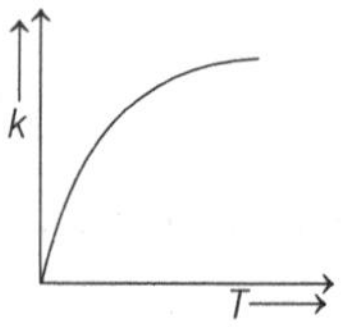

5. Identify *A* in the following reaction.

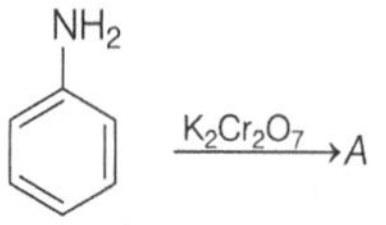

a. 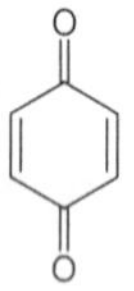

b. 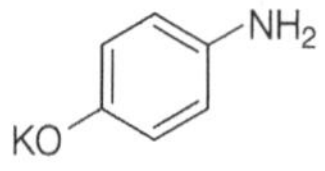

b. 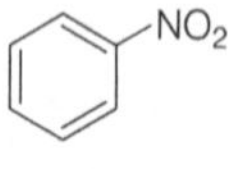

d. 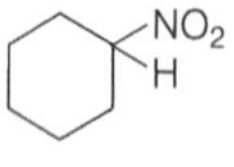

6 In the following sequence of reactions a compound *A*, (molecular formula $C_6H_{12}O_2$) with a straight chain structure gives a C_4 carboxylic acid. *A* is

$$A \xrightarrow[H_3O^+]{LiAlH_4} B \xrightarrow{\text{Oxidation}} C_4 \text{ carboxylic acid}$$

a. $CH_3—CH_2—COO—CH_2—CH_2—CH_3$

b. $CH_3—CH_2—CH(OH)—CH_2—O—CH{=}CH_2$

c. $CH_3—CH_2—CH_2—COO—CH_2—CH_3$

d. $CH_3—CH_2—CH_2—O—CH{=}CH—CH_2—OH$

7. Match List-I with List-II.

List-I (Colloid preparation method)	List-II (Chemical reaction)
A. Hydrolysis	1. $2AuCl_3 + 3HCHO + 3H_2O \rightarrow 2Au(sol) + 3HCOOH + 6HCl$
B. Reduction	2. $As_2O_3 + 3H_2S \rightarrow As_2S_3\ (sol) + 3H_2O$
C. Oxidation	3. $SO_2 + 2H_2S \rightarrow 3S\ (sol) + 2H_2O$
D. Double Decomposition	4. $FeCl_3 + 3H_2O \rightarrow Fe(OH)_3 (sol) + 3HCl$

Choose the most appropriate answer from the options given below.

	A	B	C	D
a.	1	3	2	4
b.	4	1	3	2
c.	4	2	3	1
d.	1	2	4	3

8. The crystal field stabilisation energy (CFSE) and magnetic moment (spin-only) of an octahedral aqua complex of a metal ion (M^+) are $-0.8\ \Delta_o$ and 3.87 BM, respectively. Identify (M^{2+}).

a. V^{3+} **b.** Cr^{3+} **c.** Mn^{4+} **d.** Co^{2+}

9. Monomer units of dacron polymer are

a. ethylene glycol and phthalic acid

b. ethylene glycol and terephthalic acid

c. glycerol and terephthalic acid

d. glycerol and phthalic acid

10. Which one of the following compounds is aromatic in nature ?

a. 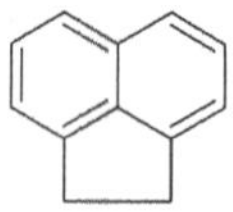

b. 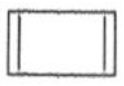

c. 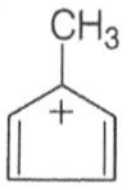

d. 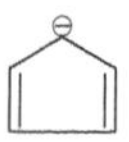

11. In the given chemical reaction, colours of the Fe^{2+} and Fe^{3+} ions, are respectively

$5Fe^{2+} + MnO_4^- + 8H^+ \longrightarrow Mn^{2+} + 4H_2O + 5Fe^{3+}$

a. yellow, orange **b.** yellow, green

c. green, orange **d.** green, yellow

12. The stereoisomers that are formed by electrophilic addition of bromine to *trans*-but-2-ene is/are

a. 2 enantiomers and 2 mesomers

b. 2 identical mesomers

c. 2 enantiomers

d. 1 racemic and 2 enantiomers

13. Hydrogen peroxide reacts with iodine in basic medium to give

a. IO_4^- **b.** IO^- **c.** I^- **d.** IO_3^-

14. In the following sequence of reactions,

$$C_3H_6 \xrightarrow{H^+/H_2O} A \xrightarrow[\text{dil. KOH}]{KIO} B + C$$

The compounds B and C respectively are

a. Cl_3COOK, HCOOH **b.** Cl_3COOK, CH_3I

c. CH_3I, HCOOK **d.** CHI_3, CH_3COOK

15. Given below are two statements.

Statement I The nucleophilic addition of sodium hydrogen sulphite to an aldehyde or a ketone involves proton transfer to form a stable ion.

Statement II The nucleophilic addition of hydrogen cyanide to an aldehyde or a ketone yields amine as final product.

In the light of the above statements, choose the most appropriate answer from the options given below.

a. Both statement I and statement II are true.

b. Statement I is false but statement II is true.

c. Statement I is true but statement II is false.

d. Both statement I and statement II are false.

16. Which one of the following gives the most stable diazonium salt ?

a. $CH_3-CH_2-CH_2-NH_2$

b. 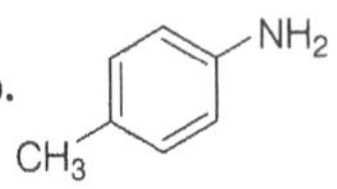

c. $CH_3-\underset{H}{\overset{CH_3}{C}}-NH_2$

d. 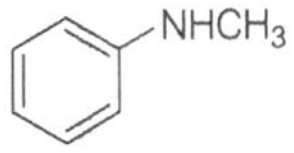

17. The potassium ferrocyanide solution gives a prussian blue colour, when added to

a. $CoCl_3$ **b.** $FeCl_2$ **c.** $CoCl_2$ **d.** $FeCl_3$

18. The oxide without nitrogen-nitrogen bond is

a. N_2O **b.** N_2O_4 **c.** N_2O_3 **d.** N_2O_5

19. Number of paramagnetic oxides among the following given oxides is Li_2O, Na_2O_2, KO_2, HgO and K_2O

a. 1 **b.** 2 **c.** 3 **d.** 0

20. Identify the element for which electronic configuration in +3 oxidation state is [Ar] $3d^5$.

a. Ru **b.** Mn **c.** Co **d.** Fe

Section B : Numerical Type Questions

21. An empty LPG cylinder weight 14.8 kg. When full, it weight 29.0 kg and shows a pressure of 3.47 atm. In the course of use at ambient temperature, the mass of the cylinder is reduced to 23.0 kg. The final pressure inside of the cylinder is .... atm. (Nearest integer)
(Assume LPG of be an ideal gas)

22. The molar solubility of $Zn(OH)_2$ in 0.1 M NaOH solution is $x \times 10^{-18}$ M. The value of x is .... (Nearest integer)
(Given; The solubility product of $Zn(OH)_2$ is 2×10^{-20}).

23. For the reaction, $2NO_2(g) \rightleftharpoons N_2O_4(g)$, when $\Delta S = -176.0\ JK^{-1}$ and $\Delta H = -57.8\ kJ\ mol^{-1}$, the magnitude of ΔG at 298 K for the reaction is .... kJ mol^{-1}. (Nearest integer)

24. The sum of oxidation states of two silver ions in $[Ag(NH_3)_2]\ [Ag(CN)_2]$ complex is ....

25. The number of atoms in 8 g of sodium is $x \times 10^{23}$. The value of x is .... (Nearest integer)
[Given : $N_A = 6.02 \times 10^{23}\ mol^{-1}$, Atomic mass of Na = 23.0 u]

26. If 80 g of copper sulphate $CuSO_4 \cdot 5H_2O$. is dissolved in deionised water to make 5 L of solution. The concentration of the copper sulphate solution is $x \times 10^{-3}$ mol L^{-1}. The value of x is ....
[Atomic masses Cu = 63.54 u, S = 32 u, O = 16 u, H = 1 u]

27. A 50 watt bulb emits monochromatic red light of wavelength of 795 nm. The number of photons emitted per second by the bulb is $x \times 10^{20}$. The value of x is ....
[Given, $h = 6.63 \times 10^{-34}$ Js and $c = 3.0 \times 10^8\ ms^{-1}$]

28. The spin-only magnetic moment value of B_2^+ species is .... $\times 10^{-2}$ BM. (Nearest integer)
[Given, $\sqrt{3} = 1.73$]

29. If the conductivity of mercury at 0°C is $1.07 \times 10^6\ S\ m^{-1}$ and the resistance of a cell containing mercury is 0.243 Ω, then the cell constant of the cell is $x \times 10^4\ m^{-1}$. The value of x is .... (Nearest integer)

30. A peptide synthesised by the reactions of one molecule each of glycine, leucine, aspartic acid and histidine will have .... peptide linkages.

Answers

For solution scan the QR code

1. *(c)*	2. *(b)*	3. *(b)*	4. *(c)*	5. *(a)*	6. *(c)*	7. *(b)*	8. *(d)*	9. *(b)*	10. *(a,d)*
11. *(d)*	12. *(b)*	13. *(c)*	14. *(d)*	15. *(c)*	16. *(b)*	17. *(d)*	18. *(d)*	19. *(a)*	20. *(d)*
21. *(2)*	22. *(2)*	23. *(5)*	24. *(2)*	25. *(2)*	26. *(64)*	27. *(2)*	28. *(173)*	29. *(26)*	30. *(3)*

JEE Main

Solved Papers 2022

ONLINE SOLVED PAPERS

JEE Main 2022

(June Attempt)

24 JUNE 2022, SHIFT-I

Section A : Objective Type Questions

1 If a rocket runs on a fuel ($C_{15}H_{30}$) and liquid oxygen, the weight of oxygen required and CO_2 released for every litre of fuel respectively are (Given, density of the fuel is 0.756 g/mL)

(a) 1188 g and 1296 g (b) 2376 g and 2592 g
(c) 2592g and 2376 g (d) 3429 g and 3142 g

2 Consider the following pairs of electrons :

(A) (a) $n = 3, l = 1, m_1 = -1, m_s = +\frac{1}{2}$

(b) $n = 3, l = 2, m_1 = 1, m_s = +\frac{1}{2}$

(B) (a) $n = 3, l = 2, m_1 = -2, m_s = -\frac{1}{2}$

(b) $n = 3, l = 2, m_1 = -1, m_s = -\frac{1}{2}$

(C) (a) $n = 4, l = 2, m_1 = 2, m_s = +\frac{1}{2}$

(b) $n = 3, l = 2, m_1 = 2, m_s = +\frac{1}{2}$

The pairs of electron present in degenerate orbitals is/are

(a) Only A (b) Only B
(c) Only C (d) Both (B) and (C)

3 Match List-I with List-II.

List-I	List-II
(A) $[PtCl_4]^{2-}$	(I) sp^3d
(B) BrF_5	(II) d^2sp^3
(C) PCl_5	(III) dsp^2
(D) $[Co(NH_3)_6]^{3+}$	(IV) sp^3d^2

Choose the most appropriate answer from the options given below.

(a) (A) $\rightarrow$ (II), (B) $\rightarrow$ (IV), (C) $\rightarrow$ (I), (D) $\rightarrow$ (III)
(b) (A) $\rightarrow$ (III), (B) $\rightarrow$ (IV), (C) $\rightarrow$ (I), (D) $\rightarrow$ (II)
(c) (A) $\rightarrow$ (III), (B) $\rightarrow$ (I), (C) $\rightarrow$ (IV), (D) $\rightarrow$ (II)
(d) (A) $\rightarrow$ (II), (B) $\rightarrow$ (I), (C) $\rightarrow$ (IV), (D) $\rightarrow$ (III)

4 For a reaction at equilibrium,

$$A(g) \rightleftharpoons B(g) + \frac{1}{2}C(g)$$

the relation between dissociation constant (K), degree of dissociation (α) and equilibrium pressure (p) is given by

(a) $K = \dfrac{\alpha^{1/2}p^{3/2}}{\left(1+\dfrac{3}{2}\alpha\right)^{1/2}(1-\alpha)}$ (b) $K = \dfrac{\alpha^{3/2}p^{1/2}}{(2+\alpha)^{1/2}(1-\alpha)}$

(c) $K = \dfrac{(\alpha p)^{3/2}}{\left(1+\dfrac{3}{2}\alpha\right)^{1/2}(1-\alpha)}$ (d) $K = \dfrac{(\alpha\, p)^{3/2}}{(1+\alpha)(1-\alpha)^{1/2}}$

5 Given below are two statements :

Statement I Emulsion of oil in water are unstable and sometimes they separate into two layers on standing.

Statement II For stabilisation of an emulsion, excess of electrolyte is added.

In the light of the above statements, choose the most appropriate answer from the options given below :

(a) Both Statement I and Statement II are correct.
(b) Both Statement I and Statement II are incorrect.
(c) Statement I is correct but Statement II is incorrect.
(d) Statement I is incorrect but Statement II is correct.

6 Given below are the oxides :

$$Na_2O, As_2O_3, N_2O, NO \text{ and } Cl_2O_7$$

Number of amphoteric oxides is

(a) 0 (b) 1
(c) 2 (d) 3

7 Match List-I with List-II.

List-I	List-II
(A) Sphalerite	(I) $FeCO_3$
(B) Calamine	(II) PbS
(C) Galena	(III) $ZnCO_3$
(D) Siderite	(IV) ZnS

Choose the most appropriate answer from the options given below :

(a) (A) - (IV), (B) – (III), (C) – (II), (D) – (I)
(b) (A) - (IV), (B) – (I), (C) – (II), (D) – (III)
(c) (A) - (II), (B) – (III), (C) – (I), (D) – (IV)
(d) (A) - (III), (B) – (IV), (C) – (II), (D) – (I)

8 The highest industrial consumption of molecular hydrogen is to produce compounds of element

(a) carbon (b) nitrogen
(c) oxygen (d) chlorine

9 Which of the following statements are correct ?

(A) Both LiCl and $MgCl_2$ are soluble in ethanol.
(B) The oxides Li_2O and MgO combine with excess of oxygen to give superoxide.
(C) LiF is less soluble in water than other alkali metal fluorides.
(D) Li_2O is more soluble in water than other alkali metal oxides.

Choose the most appropriate answer from the options given below.

(a) (A) and (C) only (b) (A), (C) and (D) only
(c) (B) and (C) only (d) (A) and (D) only

10 Identify the correct statement for B_2H_6 from those given below :

(A) In B_2H_6, all B — H bonds are equivalent.
(B) In B_2H_6 there are four 3-centre-2-electron bonds.
(C) B_2H_6 is a Lewis acid.
(D) B_2H_6 can be synthesised from both BF_3 and $NaBH_4$.
(E) B_2H_6 is a planar molecule.

Choose the most appropriate answer from the options given below :

(a) (A) and (E) only (b) (B), (C) and (E) only
(c) (C) and (D) only (d) (C) and (E) only

11 The most stable trihalide of nitrogen is

(a) NF_3 (b) NCl_3 (c) NBr_3 (d) NI_3

12 Which one of the following elemental forms is not present in the enamel of the teeth?

(a) Ca^{2+} (b) P^{3+} (c) F^- (d) P^{5+}

13 In the given reaction sequence, the major product C is

$$C_8H_{10} \xrightarrow[H_2SO_4]{HNO_3} A \xrightarrow[\Delta]{Br_2} B \xrightarrow[KOH]{Alcoholic} C$$

(a) 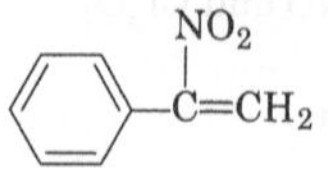

(b) 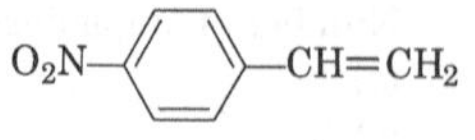

(c) 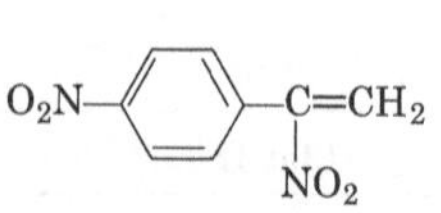

(d) 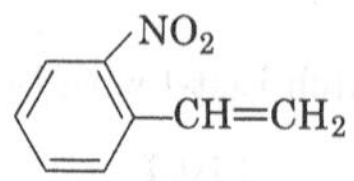

14 Two statements are given below.

Statement I The melting point of monocarboxylic acid with even number of carbon atoms is higher than that of with odd number of carbon atoms acid immediately below and above it in the series.

Statement II The solubility of monocarboxylic acids in water decreases with increase in molar mass.

Choose the most appropriate option.

(a) Both Statement I and Statement II are correct.
(b) Both Statement I and Statement II are incorrect.
(c) Statement I is correct but Statement II is incorrect.
(d) Statement I is incorrect but Statement II is correct.

15 Which of the following is an example of conjugated diketone?

(a) 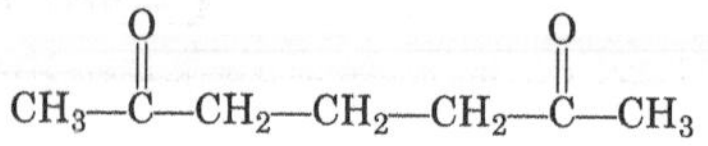

(b) 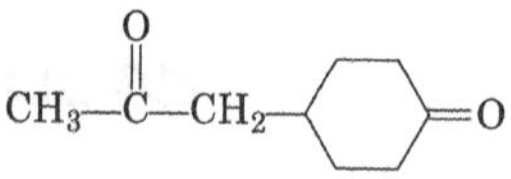

(c) 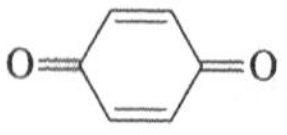

(d) 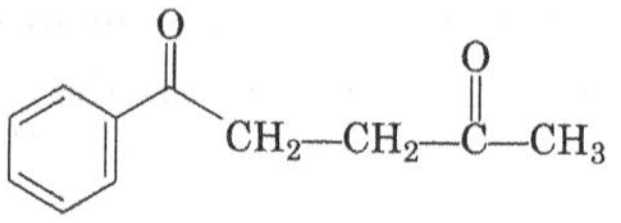

16 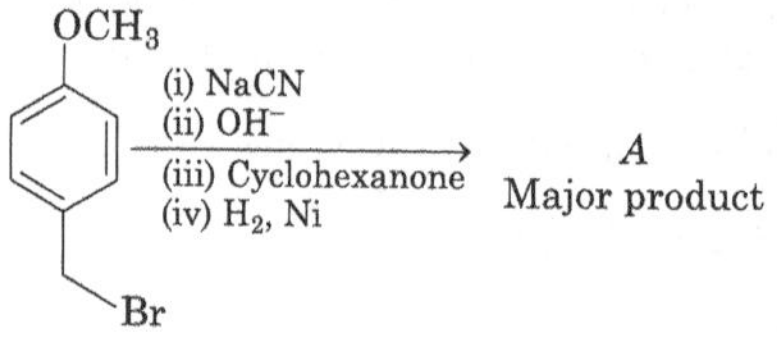

The major product of the above reaction is

(a) 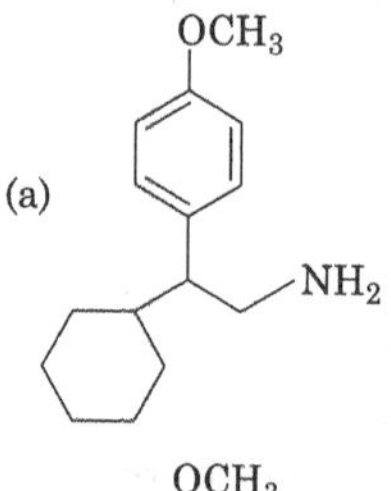

(b) 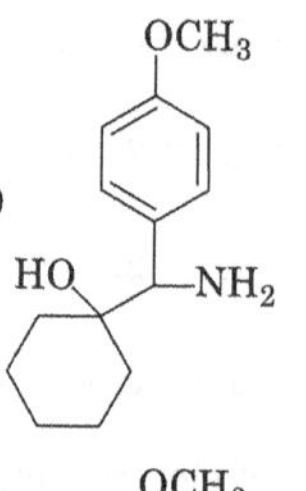

(c) 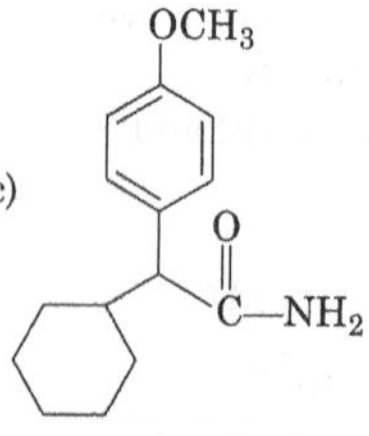

(d) 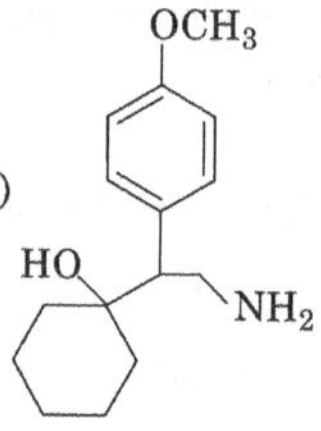

17 Which of the following is an example of polyester?

(a) Butadiene-styrene copolymer
(b) Melamine polymer
(c) Neoprene
(d) Poly-(β)-hydroxybutyrate-co-(β)-hydroxy valerate

18 A polysaccharide 'X' on boiling with dil. H_2SO_4 at 393 K under 2-3 atm pressure yields 'Y'. 'Y' on treatment with bromine water gives gluconic acid. 'X' contains β-glycosidic linkages only. Compound 'X' is

(a) starch (b) cellulose
(c) amylose (d) amylopectin

19 Which of the following is not a broad spectrum antibiotic?
(a) Vancomycin (b) Ampicillin
(c) Ofloxacin (d) Penicillin-G

20 During the qualitative analysis of salt with cation Y^{2+}, addition of a reagent (X) to alkaline solution of the salt gives a bright red precipitate. The reagent (X) and the cation (Y^{2+}) present respectively are
(a) dimethylglyoxime and Ni^{2+}
(b) dimethylglyoxime and Co^{2+}
(c) Nessler's reagent and Hg^{2+}
(d) Nessler's reagent and Ni^{2+}

Section B : Numerical Type Questions

21 Atoms of element X form hcp lattice and those of element Y occupy $\frac{2}{3}$ of its tetrahedral voids. The percentage of element X in the lattice is .... (Nearest integer)

22 At 300 K, ozone is fifty per cent dissociated. The standard free energy change at this temperature and 1 atm pressure is (−) ...... J mol^{-1} (Nearest integer)
[Given, ln 1.35 = 0.3 and $R = 8.3$ JK^{-1} mol^{-1}]

23 The osmotic pressure of blood is 7.47 bar at 300 K. To inject glucose to a patient intravenously, it has to be isotonic with blood. The concentration of glucose solution in gL^{-1} is .... (Molar mass of glucose = 180 g mol^{-1} $R = 0.083$ L bar K^{-1} mol^{-1}) (Nearest integer)

24 The cell potential for the following cell
Pt| $H_2(g)$ | $H^+(aq)$ || Cu^{2+} (0.01 M) | Cu(s)
is 0.576 V at 298 K. The pH of the solution is .... (Nearest integer)

25 The rate constants for decomposition of acetaldehyde have been measured over the temperature range 700-1000 K. The data has been analysed by plotting ln k vs $\frac{10^3}{T}$ graph. The value of activation energy for the reaction is ...... kJ mol^{-1}. [Nearest integer]
(Given, $R = 8.31$ J K^{-1} mol^{-1})

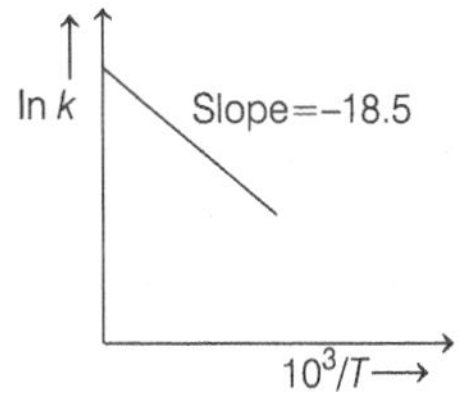

26 The difference in oxidation state of chromium in chromate and dichromate salts is ....

27 In the cobalt-carbonyl complex $[Co_2(CO)_8]$, number of Co — Co bonds is "X" and terminal CO ligands is "Y". $X + Y$ = ...

28 A 0.166 g sample of an organic compound was digested with conc. H_2SO_4 and then distilled with NaOH. The ammonia gas evolved was passed through 50.0 mL of 0.5 N H_2SO_4. The used acid required 30.0 mL of 0.25 N NaOH for complete neutralisation. The mass percentage of nitrogen in the organic compound is ....

29. Number of electrophilic centre in the given compound is ....

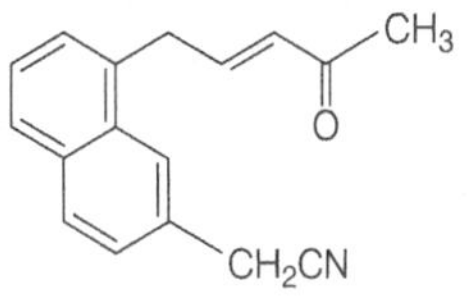

30 The major product 'A' of the following given reaction has ...... sp^2 hybridised carbon atoms.

$$\text{2, 7-dimethyl-2, 6-octadiene} \xrightarrow{H^+} \underset{\text{Major product}}{A}$$

24 JUNE 2022, SHIFT-II

Section A : Objective Type Questions

1 120 g of an organic compound that contains only carbon and hydrogen gives 330 g of CO_2 and 270 g of water on complete combustion. The percentage of carbon and hydrogen, respectively are
(a) 25 and 75 (b) 40 and 60
(c) 60 and 40 (d) 75 and 25

2 The energy of one mole of photons of radiation of wavelength 300 nm is
(Given, $h = 6.63 \times 10^{-34}$ Js, $N_A = 6.02 \times 10^{23}$ mol^{-1}, $c = 3 \times 10^8$ ms^{-1})
(a) 235 kJ mol^{-1} (b) 325 kJ mol^{-1}
(c) 399 kJ mol^{-1} (d) 435 kJ mol^{-1}

3 The correct order of bond orders of C_2^{2-}, N_2^{2-} and O_2^{2-} is, respectively
(a) $C_2^{2-} < N_2^{2-} < O_2^{2-}$ (b) $O_2^{2-} < N_2^{2-} < C_2^{2-}$
(c) $C_2^{2-} < O_2^{2-} < N_2^{2-}$ (d) $N_2^{2-} < C_2^{2-} < O_2^{2-}$

4 At 25°C and 1 atm pressure, the enthalpies of combustion are as given below

Substance	H_2	C(graphite)	$C_2H_6(g)$
$\frac{\Delta_c H^\ominus}{\text{kJ mol}^{-1}}$	−286.0	−394.0	−1560.0

The enthalpy of formation of ethane is
(a) +54.0 kJ mol^{-1} (b) −68.0 kJ mol^{-1}
(c) −86.0 kJ mol^{-1} (d) +97.0 kJ mol^{-1}

5 For a first order reaction, the time required for completion of 90% reaction is 'x' times the half-life of the reaction. The value of 'x' is (Given, $\ln 10 = 2.303$ and $\log 2 = 0.3010$)
(a) 1.12 (b) 2.43 (c) 3.32 (d) 33.31

6 Metals generally melt at very high temperature. Amongst the following, the metal with the highest melting point will be
(a) Hg (b) Ag (c) Ga (d) Cs

7 Which of the following chemical reactions represents Hall-Heroult process?
(a) $Cr_2O_3 + 2Al \longrightarrow Al_2O_3 + 2Cr$
(b) $2Al_2O_3 + 3C \longrightarrow 4Al + 3CO_2$
(c) $FeO + CO \longrightarrow Fe + CO_2$
(d) $2[Au(CN)_2]^-(aq) + Zn(s) \longrightarrow 2Au(s) + [Zn(CN_4)]^{2-}$

8 In the industrial production of which of the following, molecular hydrogen is obtained as a byproduct?
(a) NaOH (b) NaCl (c) Na metal (d) Na_2CO_3

9 Which one of the following compounds is used as a chemical in certain type of fire extinguishers?
(a) Baking soda (b) Soda ash
(c) Washing soda (d) Caustic soda

10 PCl_5 is well known but NCl_5 is not, because
(a) nitrogen is less reactive than phosphorus.
(b) nitrogen doesn't have d-orbitals in its valence shell.
(c) catenation tendency is weaker in nitrogen than phosphorus.
(d) size of phosphorus is larger than nitrogen.

11 Transition metal complex with highest value of crystal field splitting (Δ_o) will be
(a) $[Cr(H_2O)_6]^{3+}$ (b) $[Mo(H_2O)_6]^{3+}$
(c) $[Fe(H_2O)_6]^{3+}$ (d) $[Os(H_2O)_6]^{3+}$

12 Some gases are responsible for heating of atmosphere (green house effect). Identify from the following the gaseous species which does not cause it.
(a) CH_4 (b) O_3 (c) H_2O (d) N_2

13 Arrange the following carbocations in decreasing order of stability.

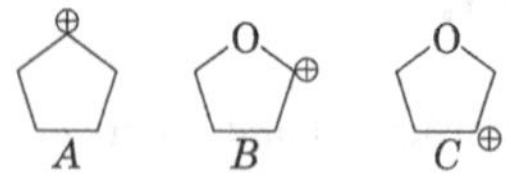

(a) $A > C > B$ (b) $A > B > C$
(c) $C > B > A$ (d) $C > A > B$

14 Given below are two statements.
Statement I The presence of weaker π-bonds make alkenes less stable than alkanes.
Statement II The strength of the double bond is greater than that of carbon-carbon single bond.
In the light of the above statements, choose the correct answer from the options given below.
(a) Both Statement I and Statement II are correct.
(b) Both Statement I and Statement II are incorrect.
(c) Statement I is correct but Statement II is incorrect.
(d) Statement I is incorrect but Statement II is correct.

15 Which of the following reagents/reactions will convert 'A' to 'B'?

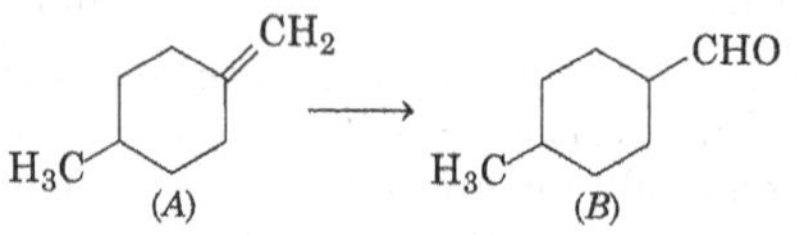

(a) PCC oxidation
(b) Ozonolysis
(c) BH_3, H_2O_2/OH^- followed by PCC oxidation
(d) HBr, hydrolysis followed by oxidation by $K_2Cr_2O_7$

16 Hex-4-ene-2-ol on treatment with PCC gives 'A' on reaction with sodium hypoiodite gives 'B', which on further heating with soda lime gives 'C'.
The compound 'C' is
(a) 2-pentene (b) proponaldehyde
(c) 2-butene (d) 4-methylpent-2-ene

17 The conversion of propan-1-ol to n-butylamine involves the sequential addition of reagents. The correct sequential order of reagents is
(a) (i) $SOCl_2$ (ii) KCN (iii) $H_2/Ni, Na(Hg)/C_2H_5OH$
(b) (i) HCl (ii) H_2/Ni, $Na(Hg)/C_2H_5OH$
(c) (i) $SOCl_2$ (ii) KCN (iii) CH_3NH_2
(d) (i) HCl (ii) CH_3NH_2

18 Which of the following is not an example of a condensation polymer?
(a) Nylon 6,6 (b) Dacron (c) Buna-N (d) Silicon

19 The structure shown below is of which well-known drug molecule?

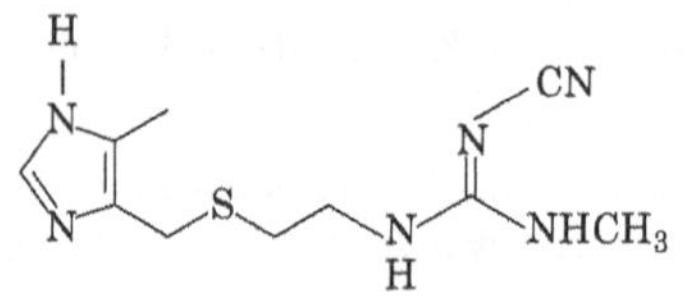

(a) Ranitidine (b) Seldane (c) Cimetidine (d) Codeine

20 In the flame test of a mixture of salts, a green flame with blue centre was observe(d) Which one of the following cations may be present?
(a) Cu^{2+} (b) Sr^{2+} (c) Ba^{2+} (d) Ca^{2+}

Section B : Numerical Type Questions

21 At 300 K, a sample of 3.0 g of gas A occupies the same volume as 0.2 g of hydrogen at 200 K at the same pressure. The molar mass of gas A is ...... g mol^{-1}. (Nearest integer) Assume that, the behaviour of gases as ideal. (Given, the molar mass of hydrogen (H_2) gas is 2.0 g mol^{-1})

22 A company dissolves 'X' amount of CO_2 at 298 K in 1 L of water to prepare soda water, $X = ... \times 10^{-3}$ g. (Nearest integer)
(Given, partial pressure of CO_2 at 298 K= 0.835 bar. Henry's law constant for CO_2 at 298 K = 1.67 kbar. Atomic mass of H,C and O is 1, 12 and 6 g mol^{-1}, respectively).

23 PCl_5 dissociates as

$$PCl_5(g) \rightleftharpoons PCl_3(g) + Cl_2(g)$$

5 moles of PCl_5 are placed in a 200 L vessel which contains 2 moles of N_2 and is maintained at 600 K. The equilibrium pressure is 2.46 atm. The equilibrium constant K_p for the dissociation of PCl_5 is ...... $\times 10^{-3}$.

(Nearest integer)

(Given, $R = 0.082$ L atm K^{-1} mol^{-1}; assume ideal gas behaviour)

24 The resistance of a conductivity cell containing 0.01 M KCl solution at 298 K is 1750 Ω. If the conductivity of 0.01 M KCl solution at 298 K is 0.152×10^{-3} S cm^{-1}, then the cell constant of the conductivity cell is ...... $\times 10^{-3}$ cm^{-1}.

25 When 200 mL of 0.2 M acetic acid is shaken with 0.6 g of wood charcoal, the final concentration of acetic acid after adsorption is 0.1 M. The mass of acetic acid adsorbed per gram of carbon is ...... g.

26 (a) Baryte, (b) Galena, (c) Zinc blende and (d) Copper pyrites. How many of these minerals are sulphide based?

27 Manganese (VI) has ability to disproportionate in acidic solution. The difference in oxidation states of two ions it forms in acidic solution is ....

28 0.2 g of an organic compound was subjected to estimation of nitrogen by Duma's method in which volume of N_2 evolved (at STP) was found to be 22.400 mL. The percentage of nitrogen in the compound is .... [Nearest integer]

(Given molar mass of N_2 is 28 mol^{-1}. Molar volume of N_2 at STP 22.4 L)

29

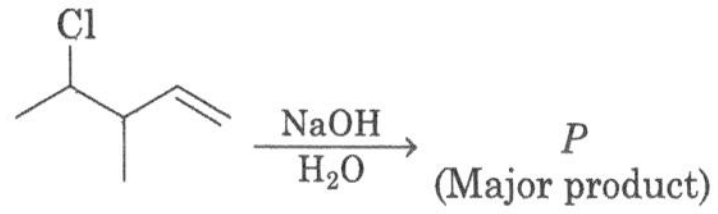

Consider the above reaction. The number of π-electrons present in the product '*P*' is ....

30 In alanylglycyl leucyl alanyl valine, the number of peptide linkages is ....

25 JUNE 2022, SHIFT-I

Section A : Objective Type Questions

1 Bonding in which of the following diatomic molecules becomes stronger, on the basis of MO theory, by removal of an electron?

(A) NO (B) N_2 (C) O_2 (D) C_2 (E) B_2

Choose the most appropriate answer from the options given below

(a) (A), (B), (C) only (b) (B), (C), (E) only
(c) (A), (C) only (d) (D) only

2 Incorrect statement for Tyndall effect is :

(a) The refractive indices of the dispersed phase and the dispersion medium differ greatly in magnitude.
(b) The diameter of the dispersed particles is much smaller than the wavelength of the light used.
(c) During projection of movies in the cinema halls, Tyndall effect is noticed.
(d) It is used to distinguish a true solution from a colloidal solution.

3 The pair, in which ions are isoelectronic with Al^{3+} is

(a) Br^- and Be^{2+} (b) Cl^- and Li^+
(c) S^{2-} and K^+ (d) O^{2-} and Mg^{2+}

4 Leaching of gold with dilute aqueous solution of NaCN in presence of oxygen gives complex [*A*], which on reaction with zinc forms the elemental gold and another complex [*B*]. [*A*] and [*B*], respectively are

(a) $[Au(CN)_4]^-$ and $[Zn(CN)_2(OH)_2]^{2-}$
(b) $[Au(CN)_2]^-$ and $[Zn(OH)_4]^{2-}$
(c) $[Au(CN)_2]^-$ and $[Zn(CN)_4]^{2-}$
(d) $[Au(CN)_4]^{2-}$ and $[Zn(CN)_6]^{4-}$

5 Number of electron deficient molecules among the following

PH_3, B_2H_6, CCl_4 NH_3, LiH and BCl_3 is

(a) 0 (b) 1 (c) 2(d) 3

6 Which one of the following alkaline earth metal ions has the highest ionic mobility in its aqueous solution?

(a) Be^{2+} (b) Mg^{2+}
(c) Ca^{2+} (d) Sr^{2+}

7 White precipitate of AgCl dissolves in aqueous ammonia solution due to formation of

(a) $[Ag(NH_3)_4]Cl_2$ (b) $[Ag(Cl)_2(NH_3)_2]$
(c) $[Ag(NH_3)_2]Cl$ (d) $[Ag(NH_3)Cl]Cl$

8 Cerium (IV) has a noble gas configuration. Which of the following is correct statement about it?

(a) It will not prefer to undergo redox reactions.
(b) It will prefer to gain electron and act as an oxidising agent.
(c) It will prefer to give away an electron and behave as reducing agent.
(d) It acts as both, oxidising and reducing agent.

9 Among the following, which is the strongest oxidising agent?

(a) Mn^{3+} (b) Fe^{3+}
(c) Ti^{3+} (d) Cr^{3+}

10 The eutrophication of water body results in

(a) loss of biodiversity
(b) breakdown of organic matter
(c) increase in biodiversity
(d) decrease in BOD

11 Phenol on reaction with dilute nitric acid, gives two products. Which method will be most effective for large scale separation?
(a) Chromatographic separation
(b) Fractional crystallisation
(c) Steam distillation
(d) Sublimation

12 In the following structures, which one is having staggered conformation with maximum dihedral angle?

(a) 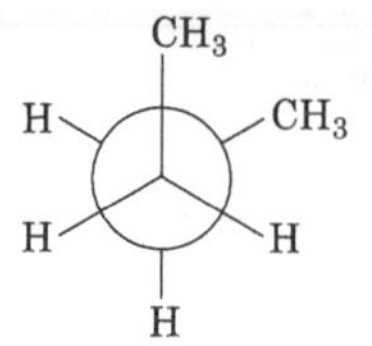
(b) 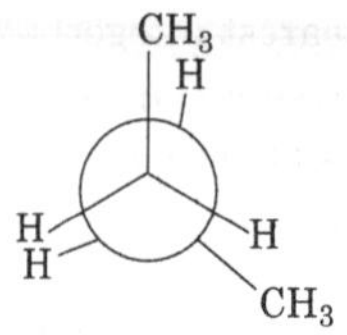
(c) 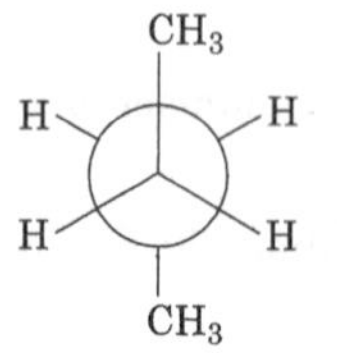
(d) 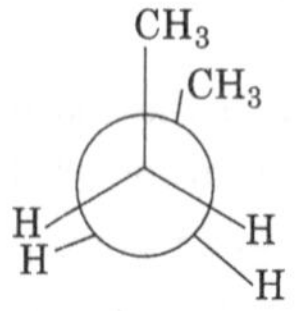

13 The product in the following reaction.

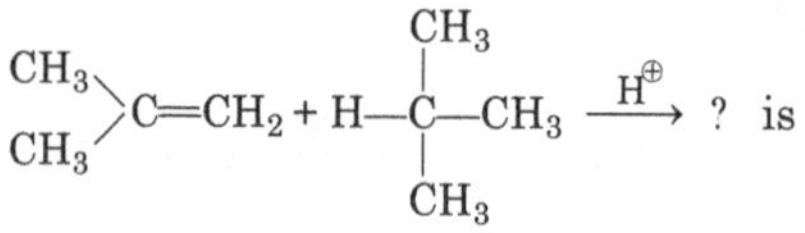

(a) 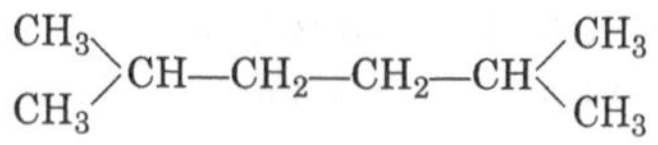
(b) 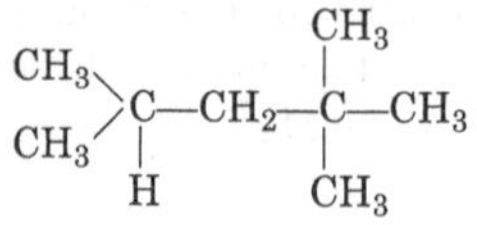
(c) 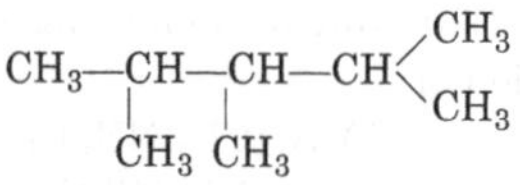
(d) 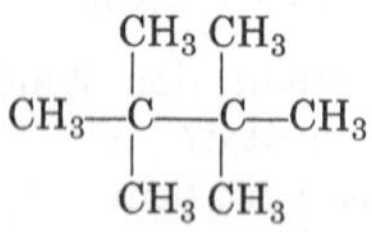

14 The IUPAC name of ethylidene chloride is
(a) 1-chloroethene
(b) 1-chloroethyne
(c) 1, 2 dichloroethane
(d) 1, 1 dichloroethane

15 The major product in the reaction

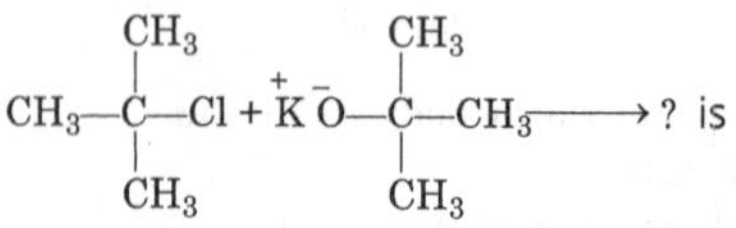

(a) *t*-butyl ethyl ether (b) 2,2 dimethyl butane
(c) 2-methyl pent-1-ene (d) 2-methylprop-1-ene

16 The intermediate *X*, in the reaction

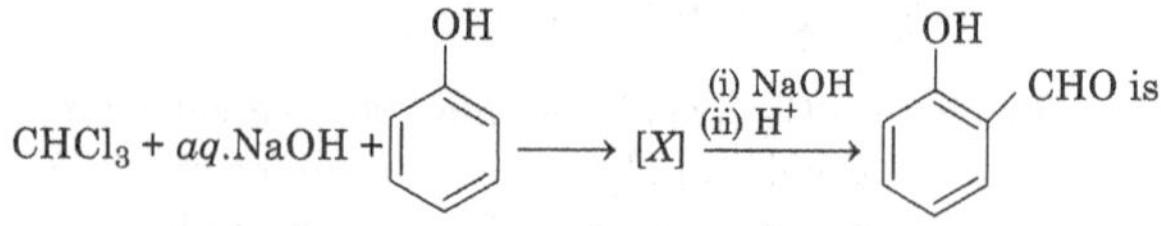

(a) 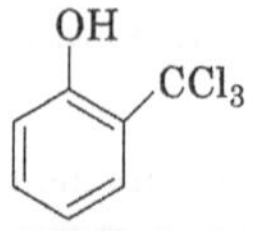
(b) 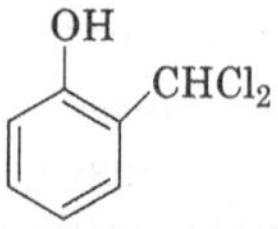
(c) 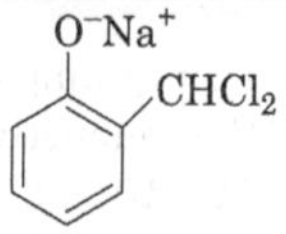
(d) 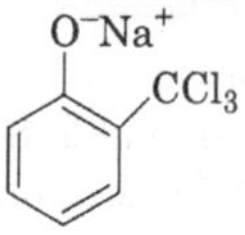

17 In the following reaction,

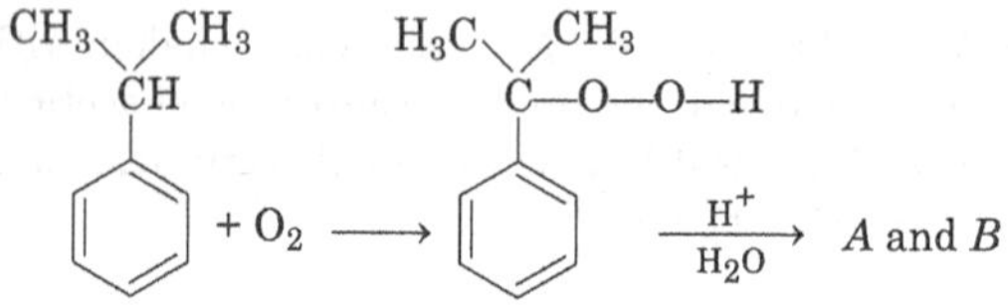

The compounds *A* and *B* respectively are

(a) 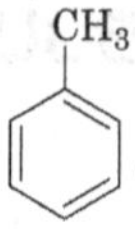, CH_3COOH
(b) 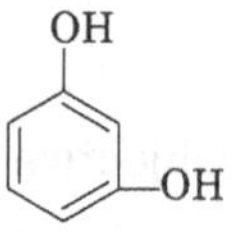, CH_3COOH
(c) 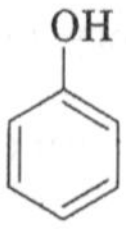, CH_3COCH_3
(d) 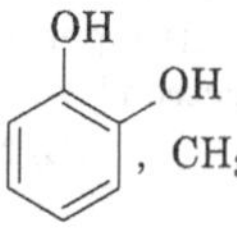, CH_3COCH_3

18 The reaction of $R-\underset{\underset{O}{\|}}{C}-NH_2$ with bromine and KOH gives RNH_2 as the end product. Which one of the following is the intermediate product formed in this reaction?
(a) $R-\underset{\underset{O}{\|}}{C}-NH-Br$ (b) $R-NH-Br$
(c) $R-N=C=O$ (d) $R-\underset{\underset{O}{\|}}{C}-NBr_2$

19 Using very little soap while washing clothes, does not serve the purpose of cleaning of clothes because
(a) soap particles remain floating in water as ions
(b) the hydrophobic part of soap is not able to take away grease
(c) the micelles are not formed due to concentration of soap, below its CMC value
(d) colloidal structure of soap in water is completely disturbed

20 Which one of the following is an example of artificial sweetener?
(a) Bithional (b) Alitame
(c) Salvarsan (d) Lactose

Section B : Numerical Type Questions

21 The number of N atoms is 681 g of $C_7H_5N_3O_6$ is $x \times 10^{21}$. The value of x is ... ($N_A = 6.02 \times 10^{23}$ mol^{-1}) [Nearest integer]

22 The distance between Na^+ and Cl^- ions in solid NaCl of density 43.1 g cm^{-3} is ... $\times 10^{-10}$ m. [Nearest integer]

23 The longest wavelength of light that can be used for the ionisation of lithium atom (Li) in its ground state is $x \times 10^{-8}$ m. The value of x is .... [Nearest integer]

(Given, energy of the electron in the first shell of the hydrogen atom is -2.2×10^{-18} J;

$h = 6.63 \times 10^{-34}$ Js and $c = 3 \times 10^8$ ms^{-1})

24 The standard entropy change for the reaction,

$4Fe(s) + 3O_2(g) \longrightarrow 2Fe_2O_3(s)$ is -550 JK^{-1} at 298 K.

[Given, the standard enthalpy change for the reaction is -165 kJ mol^{-1}]. The temperature in K at which the reaction attains equilibrium is ... [Nearest integer]

25 1 L aqueous solution of H_2SO_4 contains 0.02 m mol H_2SO_4. 50% of this solution is diluted with deionised water to give 1 L solution (*A*). In solution (*A*), 0.01 m mol of H_2SO_4 are added. Total m mol of H_2SO_4 in the final solution is ... $\times 10^3$ m mol.

26 The standard free energy change ($\Delta G°$) for 50% dissociation of N_2O_4 into NO_2 at 27°C and 1 atm pressure is $-x$ J mol^{-1}. The value of x is .... [Nearest integer]

[Given, $R = 8.31$ J K^{-1} mol^{-1}, log 1.33 = 0.1239, In 10 = 2.3]

27 In a cell, the following reactions takes place

$Fe^{2+} \longrightarrow Fe^{3} + e^-$; $\quad E°_{Fe^{3+}/Fe^{2+}} = 0.77$ V

$2I^- \longrightarrow I_2 + 2e^-$; $\quad E°_{I_2/I^-} = 0.54$ V

The standard electrode potential for the spontaneous reaction in the cell is $x \times 10^{-2}$ V 298 K. The value of x is ... [Nearest integer]

28 For a given chemical reaction,

$$\gamma_1 A + \gamma_2 B \longrightarrow \gamma_3 C + \gamma_4 D$$

concentration of *C* changes from 10 m mol dm^{-3} to 20 mmol dm^{-3} in 10 seconds. Rate of appearance of *D* is 1.5 times the rate of disappearance of *B* which is twice the rate of disappearance *A*. The rate of appearance of *D* has been experimentally determined to be 9 m mol dm^{-3} s^{-1}. Therefore, the rate of reaction is ... mmol dm^{-3} s^{-1}. [Nearest integer]

29 If $[Cu(H_2O)_4]^{2+}$ absorbs a light of wavelength 600 nm for *d-d* transition, then the value of octahedral crystal field splitting energy for $[Cu(H_2O)_6]^{2+}$ will be ... $\times 10^{-21}$ J. [Nearest integer]
(Given, $h = 6.63 \times 10^{-34}$ Js and $c = 3.08 \times 10^8$ ms^{-1})

30 Number of grams of bromine that will completely react with 5.0 g of pent-1-ene is ... $\times 10^{-2}$ g. (Atomic mass of Br = 80 g/mol) [Nearest integer]

25 JUNE 2022, SHIFT-II

Section A : Objective Type Questions

1 The minimum energy that must be possessed by photons in order to produce the photoelectric effect with platinum metal is

[Given, the threshold frequency of platinum is 13×10^{15} s^{-1} and $h = 6.6 \times 10^{-34}$ Js]

(a) 3.21×10^{-14} J (b) 6.24×10^{-16} J
(c) 8.58×10^{-19} J (d) 9.76×10^{-20} J

2 At 25°C and 1 atm pressure, the enthalpy of combustion of benzene (*l*) and acetylene (*g*) are -3268 kJ mol^{-1} and -1300 kJ mol^{-1}, respectively.

The change in enthalpy for the reaction

$3C_2H_2(g) \longrightarrow C_6H_6(l)$, is

(a) +324 kJ mol^{-1} (b) +632 kJ mol^{-1}
(c) -632 kJ mol^{-1} (d) -732 kJ mol^{-1}

3 Solute *A* associates in water. When 0.7 g of solute *A* is dissolved in 42.0 g of water, it depresses the freezing point by 0.2°C. The percentage association of solute *A* in water, is

[Given, molar mass of $A = 93$ g mol^{-1}. Molal depression constant of water is 1.86 K kg mol^{-1}]

(a) 50% (b) 60% (c) 70% (d) 80%

4 The K_{sp} for bismuth sulphide (Bi_2S_3) is 1.08×10^{-73}. The solubility of Bi_2S_3 in mol L^{-1} at 298 K is

(a) 1.0×10^{-15} (b) 2.7×10^{-12}
(c) 3.2×10^{-10} (d) 4.2×10^{-8}

5 Match the List I with List II.

	List I		List II
A.	Zymase	I.	Stomach
B.	Diastase	II.	Yeast
C.	Urease	III.	Malt
D.	Pepsin	IV.	Soyabean

Choose the correct answer from the options given below

(a) A-II, B-III, C-I, D-IV
(b) A-II, B-III, C-IV, D-I
(c) A-III, B-II, C-IV, D-I
(d) A-III, B-II, C-I, D-IV

6 The correct order of electron gain enthalpies of Cl, F, Te and Po is

(a) $F < Cl < Te < Po$ (b) $Po < Te < F < Cl$
(c) $Te < Po < Cl < F$ (d) $Cl < F < Te < Po$

7 Given below are two statements.

Statement I During electrolytic refining, blister copper deposits precious metals.

Statement II In the process of obtaining pure copper by electrolysis method, copper blister is used to make the anode.

In the light of the above statements, choose the correct answer from the options given below.

(a) Both Statement I and Statement II are true.
(b) Both Statement I and Statement II are false.
(c) Statement I is true but Statement II is false.
(d) Statement I is false but Statement II is true.

8. Given below are two statements, one is labelled as Assertion A and the other is labelled as Reason R.

Assertion (A) The amphoteric nature of water is explained by using Lewis acid/base concept.

Reason (R) Water acts as an acid with NH_3 and as a base with H_2S.

In the light of the above statements choose the correct answer from the options given below

(a) Both A and R are true and R is the correct explanation of A.
(b) Both A and R are true but R is not the correct explanation of A.
(c) A is true but R is false.
(d) A is false but R is true.

9 The correct order of reduction potentials of the following pairs is

A. Cl_2/Cl^- B. I_2/I^-
C. Ag^+/Ag D. Na^+/Na
E. Li^+/Li

Choose the correct answer from the options given below.

(a) A > C > B > D > E (b) A > B > C > D > E
(c) A > C > B > E > D (d) A > B > C > E > D

10 The number of bridged oxygen atoms present in compound *B* formed from the following reactions is

$$Pb(NO_3)_2 \xrightarrow{673\text{ K}} A + PbO + O_2$$
$$A \xrightarrow{\text{Dimerise}} B$$

(a) 0 (b) 1 (c) 2 (d) 3

11 The metal ion (in gaseous state) with lowest spin (–) only magnetic moment value is

(a) V^{2+} (b) Ni^{2+}
(c) Cr^{2+} (d) Fe^{2+}

12 Given below are two statements.
One is labelled as Assertion A and the other is labelled as Reason R.

Assertion (A) Polluted water may have a value of BOD of the order of 17 ppm.

Reason (R) BOD is a measure of oxygen required to oxidise both the bio-degradable and non-biodegradable organic material in water.

In the light of the above statements, choose the correct answer from the options given below.

(a) Both A and R are correct and R is the correct explanation of A.
(b) Both A and R are correct but R is not the correct explanation of A.
(c) A is correct but R is not correct.
(d) A is not correct but R is correct.

13 Given below are two statements.
One is labelled as Assertion A and the other is labelled as Reason R.

Assertion (A) A mixture contains benzoic acid and naphthalene. The pure benzoic acid can be separated out by the use of benzene.

Reason (R) Benzoic acid is soluble in hot water.

In the light of the above statements, choose the most appropriate answer from the options given below.

(a) Both A and R are true and R is the correct explanation of A.
(b) Both A and R are true but R is not the correct explanation of A.
(c) A is true but R is false.
(d) A is false but R is true.

14 During halogen test, sodium fusion extract is boiled with concentrated HNO_3 to

(a) remove unreacted sodium
(b) decompose cyanide or sulphide of sodium
(c) extract halogen from organic compound
(d) maintain the pH of extract

15 Amongst the following, the major product of the given chemical reaction is

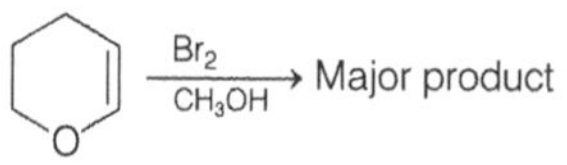

(a) 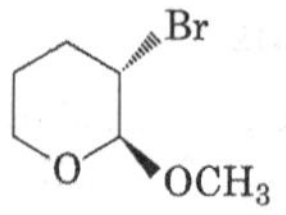

(b) 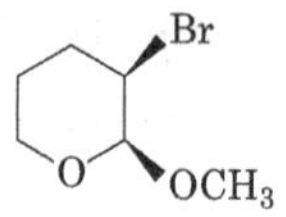

(c) 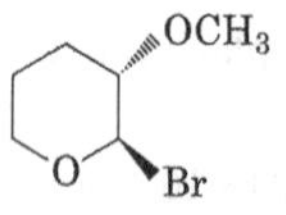

(d) 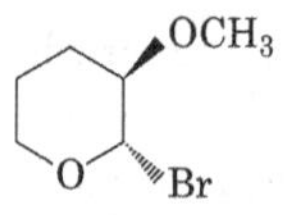

16 In the given reaction,

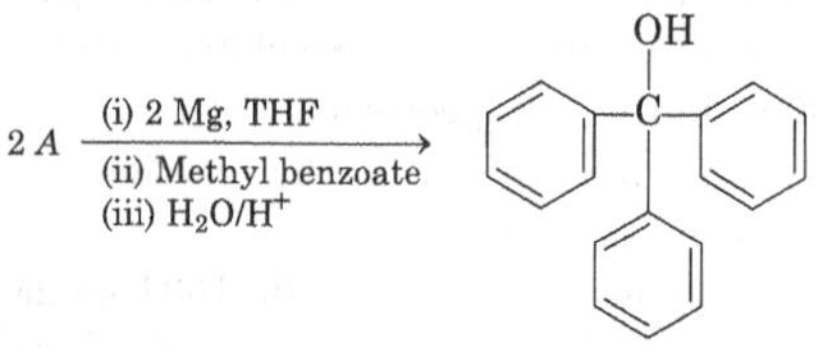

'*A*' can be

(a) benzyl bromide (b) bromobenzene
(c) cyclohexyl bromide (d) methyl bromide

17 Which of the following conditions or reaction sequence will not give acetophenone as the major product ?

(a) I. 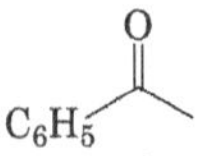 $H + CH_3MgBr$ II. $Na_2Cr_2O_7$, H^+

(b) I. 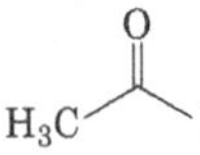 $H + C_6H_5MgBr$ II. PCC, DCM

(c) 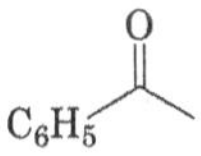 $OC_2H_5 + 2CH_3MgBr$

(d) 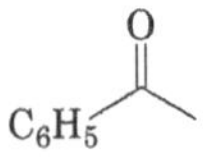 $Cl + CH_3MgBr + CdCl_2$

18 The major product formed in the following reaction, is

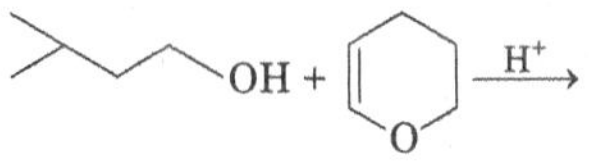

(a) 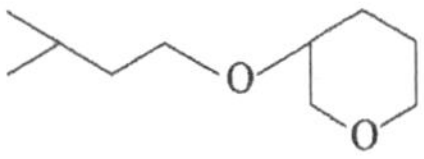

(b) 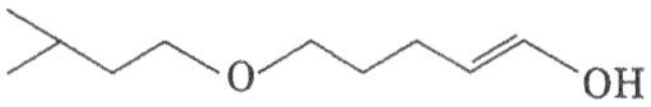

(c) 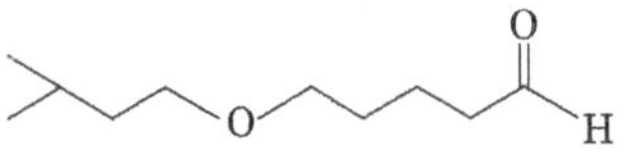

(d) 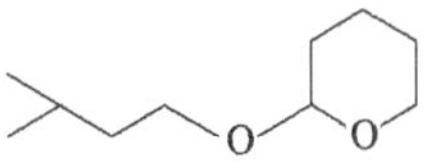

19 Which of the following ketone will not give enamine on treatment with secondary amines?
[where t-Bu is —$C(CH_3)_3$]

(a) 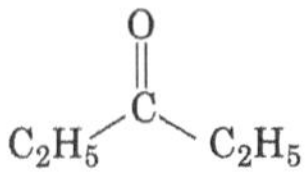
(b) 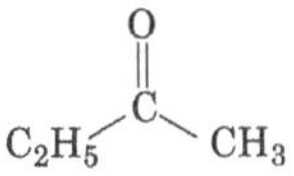
(c) 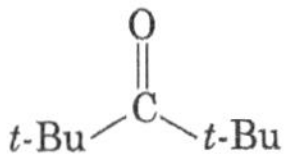
(d) 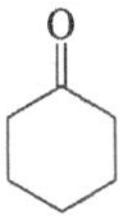

20 An antiseptic dettol is a mixture of two compounds 'A' and 'B', where A has 6π electrons and B has 2π-electrons. What is 'B'?

(a) Bithionol (b) Terpineol
(c) Chloroxylenol (d) Chloramphenicol

Section B : Numerical Type Questions

21 A protein 'A' contains 0.30% of glycine (molecular weight 75). The minimum molar mass of the protein 'A' is .... $\times 10^3$ g mol^{-1}. [Nearest integer]

22 A rigid nitrogen tank stored inside a laboratory has a pressure of 30 atm at 06 : 00 am when the temperature is 27°C. At 03 : 00 pm, when the temperature is 45°C, the pressure in the tank will be .... atm. [Nearest integer]

23 Amongst BeF_2, BF_3, H_2O, NH_3, CCl_4 and HCl, the number of molecules with non-zero net dipole moment is ....

24 At 345 K, the half-life for the decomposition of a sample of a gaseous compound initially at 55.5 kPa was 340 s. When the pressure was 27.8 kPa, the half-life was found to be 170 s. The order of the reaction is .... [Nearest integer]

25 A solution of $Fe_2(SO_4)_3$ is electrolysed for 'x' min with a current of 1.5 A to deposit 0.3482 g of Fe. The value of x is .... [Nearest integer]
Given, 1 F = 96500 C mol^{-1}, atomic mass of Fe = 56 g mol^{-1}

26 Consider the following reactions.

$$PCl_3 + H_2O \longrightarrow A + HCl$$
$$A + H_2O \longrightarrow B + HCl$$

The number of ionisable protons present in the product B is ....

27 Amongst $FeCl_3 \cdot 3H_2O$, $K_3[Fe(CN)_6]$ and $[Co(NH_3)_6]Cl_3$, the spin-only magnetic moment value of the inner-orbital complex that absorbs light at shortest wavelength is .... BM. [Nearest integer]

28. The novolac polymer has mass of 963 g. The number of monomer units present in it are ....

29 How many of the given compounds will give a positive biuret test .... ? Glycine, glycylalanine, tripeptide, biuret.

30 The neutralisation occurs when 10 mL of 0.1 M acid 'A' is allowed to react with 30 mL of 0.05 M base $M(OH)_2$. The basicity of the acid 'A' is .... [M is a metal]

26 JUNE 2022, SHIFT-I

Section A : Objective Type Questions

1 A commercially sold conc. HCl is 35% HCl by mass. If the density of this commercial acid is 1.46 g/mL, the molarity of this solution is
(Atomic mass : Cl = 35.5 amu, H = 1 amu)

(a) 10.2 M (b) 12.5 M (c) 14.0 M (d) 18.2 M

2 An evacuated glass vessel weighs 40.0 g when empty, 135.0 g when filled with a liquid of density 0.95 g mL^{-1} and 40.5 g when filled with an ideal gas at 0.82 atm at 250 K. The molar mass of the gas in g mol^{-1} is
(Given, $R = 0.082$ L atm K^{-1} mol^{-1})

(a) 35 (b) 50 (c) 75 (d) 125

3 If the radius of the 3rd Bohr's orbit of hydrogen atom is r_3 and the radius of 4th Bohr's orbit is r_4. Then

(a) $r_4 = \frac{9}{16} r_3$ (b) $r_4 = \frac{16}{9} r_3$ (c) $r_4 = \frac{3}{4} r_3$ (d) $r_4 = \frac{4}{3} r_3$

4 Consider the ions/molecules O_2^+, O_2, O_2^-, O_2^{2-}

For increasing bond order the correct option is

(a) $O_2^{2-} < O_2^- < O_2 < O_2^+$ (b) $O_2^- < O_2^{2-} < O_2 < O_2^+$

(c) $O_2^- < O_2^{2-} < O_2^+ < O_2$ (d) $O_2^- < O_2^+ < O_2^{2-} < O_2$

5 The $\left(\frac{\partial E}{\partial T}\right)_p$ of different types of half-cells are as follows

A	*B*	*C*	*D*
1×10^{-4}	2×10^{-4}	0.1×10^{-4}	0.2×10^{-4}

(where, *E* is the electromotive force)

Which of the above half-cells would be preferred to be used as reference electrode ?

(a) *A* (b) *B* (c) *C* (d) *D*

6 Choose the correct stability order of group 13 elements in their + 1 oxidation state.

(a) Al < Ga < In < Tl (b) Tl < In < Ga < Al

(c) Al < Ga < Tl < In (d) Al < Tl < Ga < In

7 Given below are two statements :

Statement I According to the Ellingham diagram, any metal oxide with higher $\Delta G°$ is more stable than the one with lower $\Delta G°$.

Statement II The metal involved in the formation of oxide placed lower in the Ellingham diagram can reduce the oxide of a metal placed higher in the diagram.

In the light of the above statements, choose the most appropriate answer from the options given below.

(a) Both Statement I and Statement II are correct.
(b) Both Statement I and Statement II are incorrect.
(c) Statement I is correct but Statement II is incorrect.
(d) Statement I is incorrect but Statement II is correct.

8 Consider the following reaction.

$$2HSO_4^- (aq) \xrightarrow[\text{(ii) Hydrolysis}]{\text{(i) Electrolysis}} 2HSO_4^- + 2H^+ + A$$

The dihedral angle in product *A* in its solid phase at 110 K is

(a) 104° (b) 111.5° (c) 90.2° (d) 111.0°

9 The correct order of melting point is

(a) Be > Mg > Ca > Sr (b) Sr > Ca > Mg > Be

(c) Be > Ca > Mg > Sr (d) Be > Ca > Sr > Mg

10 The correct order of melting point of hydrides of group 16 elements is

(a) $H_2S < H_2Se < H_2Te < H_2O$
(b) $H_2O < H_2S < H_2Se < H_2Te$
(c) $H_2S < H_2Te < H_2Se < H_2O$
(d) $H_2Se < H_2S < H_2Te < H_2O$

11 Consider the following reaction,

A + alkali $\rightarrow B$ (major product)

If *B* is an oxoacid of phosphorus with no P—H bond, then *A* is

(a) white P_4 (b) red P_4 (c) P_2O_3 (d) H_3PO_3

12 Polar stratospheric clouds facilitate the formation of

(a) $ClONO_2$ (b) HOCl

(c) ClO (d) CH_4

13 Given below are two statements :

Statement I In 'Lassaigne's test, when both nitrogen and sulphur are present in an organic compound, sodium thiocyanate is formed.

Statement II If both nitrogen and sulphur are present in an organic compound, then the excess of sodium used in sodium fusion will decompose the sodium thiocyanate formed to give NaCN and Na_2S.

In the light of the above statements, choose the most appropriate answer from the options given below :

(a) Both Statement I and Statement II are correct.
(b) Both Statement I and Statement II are incorrect.
(c) Statement I is correct but Statement II is incorrect.
(d) Statement I is incorrect but Statement II is correct.

14 $(C_7H_5O_2)_2 \xrightarrow{h\nu} [X] + 2C_6H_5 + 2CO_2$

Consider the above reaction and identify the intermediate '*X*'

(a) 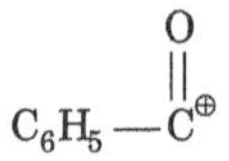

(b) 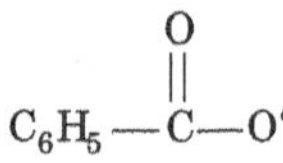

(c) 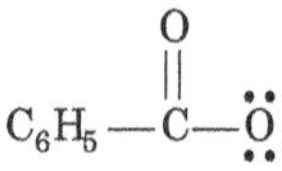

(d) 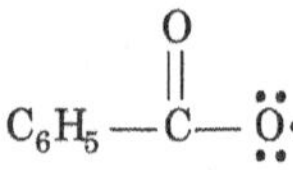

15 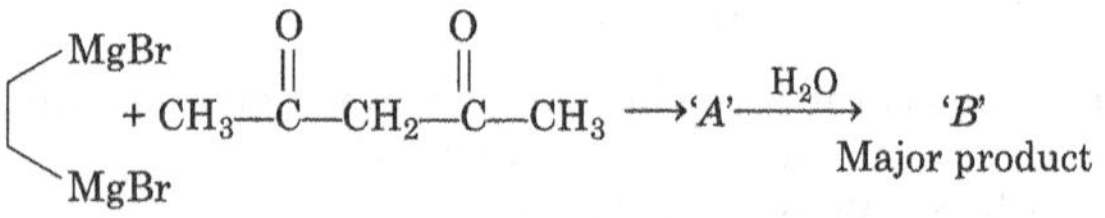

Consider the above reaction sequence and identify the product *B*.

(a) 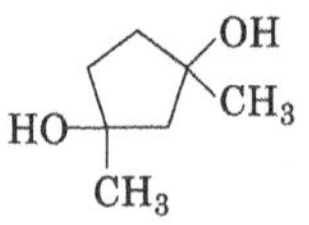

(b) 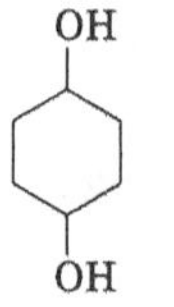

(c) 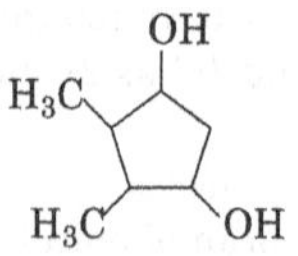

(d) 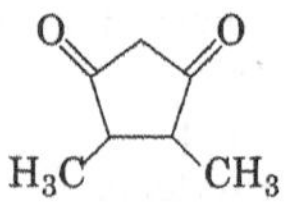

16 Which will have the highest enol content ?

(a) 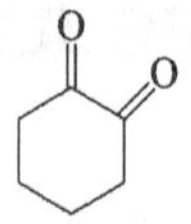

(b) 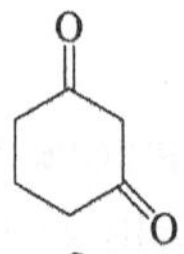

(c) 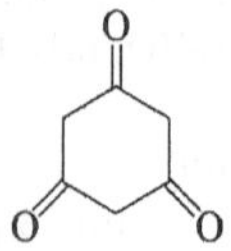

(d) 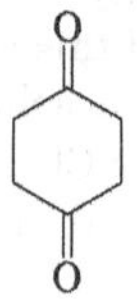

17 Among the following structures, which will show the most stable enamine formation?
(where, Me is —CH_3)

(a) 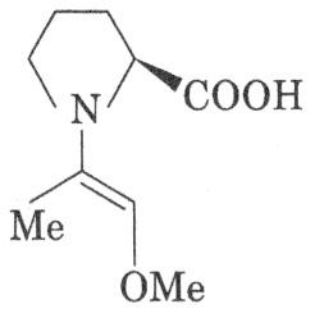

(b) 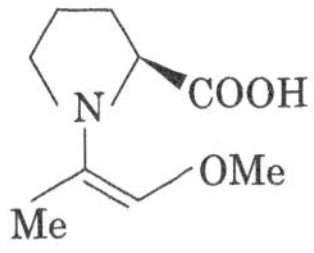

(c) 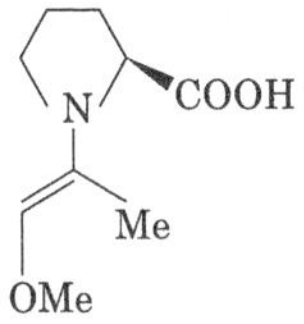

(d) 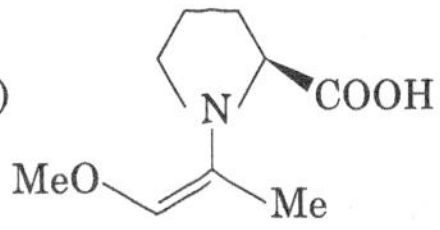

18 Which of the following sets are correct regarding polymer?

(A) Copolymer : Buna-S
(B) Condensation polymer : Nylon-6,6
(C) Fibres : Nylon-6,6
(D) Thermosetting polymer : Terylene
(E) Homopolymer : Buna-N

Choose the correct answer from given options below

(a) (A), (B) and (C) are correct
(b) (B), (C) and (D) are correct
(c) (A), (C) and (E) are correct
(d) (A), (B) and (D) are correct

19 A chemical which stimulates the secretion of pepsin is

(a) anti-histamine (b) cimetidine
(c) histamine (d) zantac

20 Which statement is not true with respect to nitrate ion test ?

(a) A dark brown ring is formed at the junction of two solutions.
(b) Ring is formed due to nitroferrous sulphate complex.
(c) The brown complex is $[Fe(H_2O)_5(NO)]SO_4$.
(d) Heating the nitrate salt with conc. H_2SO_4, light brown fumes are evolved.

Section B : Numerical Type Questions

21 For complete combustion of methanol

$$CH_3OH(l) + \frac{3}{2}O_2(g) \longrightarrow CO_2(g) + 2H_2O(l)$$

The amount of heat produced as measured by bomb calorimeter is 726 kJ mol^{-1} at 27°C. The enthalpy of combustion for the reaction is $-x$ kJ mol^{-1}, where x is .... (Nearest integer)
(Given, $R = 8.3\ JK^{-1}\ mol^{-1}$)

22 A 0.5 per cent solution of potassium chloride was found to freeze at –0.24°C. The percentage dissociation of potassium chloride is .... (Nearest integer)
(Molal depression constant for water is 1.80 K kg mol^{-1} and molar mass of KCl is 74.6 g mol^{-1})

23 50 mL of 0.1 M CH_3COOH is being titrated against 0.1 M NaOH. When 25 mL of NaOH has been added, the pH of the solution will be ...... $\times 10^{-2}$. (Nearest integer)
[Given, pK_a (CH_3COOH) = 4.76]
$\log 2 = 0.30$, $\log 3 = 0.48$, $\log 5 = 0.69$,
$\log 7 = 0.84$, $\log 11 = 1.04$

24 A flask is filled with equal moles of A and B. The half-lives of A and B are 100 s and 50 s respectively and are independent of the initial concentration. The time required for the concentration of A to be four times that of B is ...... s.
(Given, ln 2 = 0.693)

25 2.0 g of H_2 gas is adsorbed on 2.5 g of platinum powder at 300 K and 1 bar pressure. The volume of the gas adsorbed per gram of the adsorbent is ...... mL.
(Given, $R = 0.083$ L bar K^{-1} mol^{-1})

26 The spin–only magnetic moment value of the most basic oxide of vanadium among V_2O_3, V_2O_4 and V_2O_5 is ...... BM. (Nearest integer)

27 The spin–only magnetic moment value of an octahedral complex among $CoCl_3 \cdot 4NH_3$, $NiCl_2 \cdot 6H_2O$ and $PtCl_4 \cdot 2HCl$, which upon reaction with excess of $AgNO_3$ gives 2 moles of AgCl is ...... BM. (Nearest integer)

28 On complete combustion 0.30 g of an organic compound gave 0.20 g of carbon dioxide and 0.10 g of water. The percentage of carbon in the given organic compound is .... (Nearest integer)

29 Compound 'P' on nitration with dil. HNO_3 yields two isomers (A) and (B). Show the intramolecular and intermolecular hydrogen bonding respectively. Compound (P) on reaction with conc. HNO_3 yields a yellow compound 'C', a strong acid. The number of oxygen atoms is present in compound 'C' ....

30 The number of oxygens present in a nucleotide formed from a base, that is present only in RNA is ....

26 JUNE 2022, SHIFT-II

Section A : Objective Type Questions

1 The number of radial and angular nodes in 4*d* orbital are respectively

(a) 1 and 2 (b) 3 and 2
(c) 1 and 0 (d) 2 and 1

2 Match List I with List II.

List-I (Enzyme)		List-II (Conversion of)	
A.	Invertase	I.	Starch into maltose
B.	Zymase	II.	Maltose into glucose
C.	Diastase	III.	Glucose into ethanol
D.	Maltase	IV.	Cane sugar into glucose

Choose the most appropriate answer from the options given below.

	A	B	C	D		A	B	C	D
(a)	III	IV	II	I	(b)	III	II	I	IV
(c)	IV	III	I	II	(d)	IV	II	III	I

3 Which of the following elements is considered as a metalloid?

(a) Sc (b) Pb (c) Bi (d) Te

4 The role of depressants in 'froth floation method' is to

(a) selectively prevent one component of the ore from coming to the froth.
(b) reduce the consumption of oil for froth formation.
(c) stabilise the froth.
(d) enhance non-wettability of the mineral particles.

5 Boiling of hard water is helpful in removing the temporary hardness by converting calcium hydrogen carbonate and magnesium hydrogen carbonate to

(a) $CaCO_3$ and $Mg(OH)_2$ (b) $CaCO_3$ and $MgCO_3$
(c) $Ca(OH)_2$ and $MgCO_3$ (d) $Ca(OH)_2$ and $Mg(OH)_2$

6 *s*-block element which cannot be qualitatively confirmed by the flame test is

(a) Li (b) Na (c) Rb (d) Be

7 The oxide which contains an odd electron at the nitrogen atom is

(a) N_2O (b) NO_2
(c) N_2O_3 (d) N_2O_5

8 Which one of the following is an example of disproportionation reaction?

(a) $3MnO_4^{2-} + 4H^+ \longrightarrow 2MnO_4^- + MnO_2 + 2H_2O$
(b) $MnO_4^- + 4H^+ + 4e^- \longrightarrow MnO_2 + 2H_2O$
(c) $10I^- + 2MnO_4^- + 16H^+ \longrightarrow 2Mn^{2+} + 8H_2O + 5I_2$
(d) $8MnO_4^- + 3S_2O_3^{2-} + H_2O \longrightarrow 8MnO_2 + 6SO_4^{2-} + 2OH^-$

9 The most common oxidation state of lanthanoid elements is + 3. Which of the following is likely to deviate easily from + 3 oxidation state?

(a) Ce (58) (b) La (57)
(c) Lu (71) (d) Gd (64)

10 The measured BOD values for four different water samples (A-D) are as follows
$A = 3$ ppm, $B = 18$ ppm, $C = 21$ ppm,
$D = 4$ ppm. The water samples which can be called as highly polluted with organic wastes, are

(a) *A* and *B* (b) *A* and *D*
(c) *B* and *C* (d) *B* and *D*

11 The correct order of nucleophilicity is

(a) $F^- > OH^-$ (b) $H_2\ddot{O} > OH^-$
(c) $R\ddot{O}H > RO^-$ (d) $NH_2^- > NH_3$

12 Oxidation of toluene to benzaldehyde can be easily carried out with which of the following reagents?

(a) CrO_3/acetic acid, H_3O^+
(b) CrO_3/acetic anhydride, H_3O^+
(c) $KMnO_4$/HCl, H_3O^+
(d) CO/HCl, anhydrous $AlCl_3$

13 The major product in the following reaction

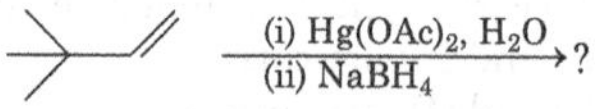

(a) 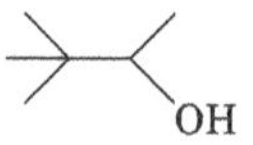 (b) 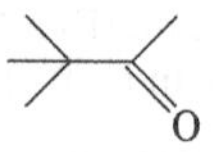

(c) 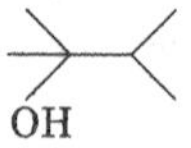 (d) 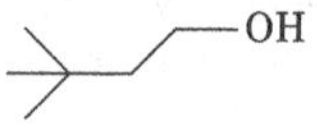

14 Halogenation of which one of the following will yield *m*-substituted product with respect to methyl group as a major product?

(a) 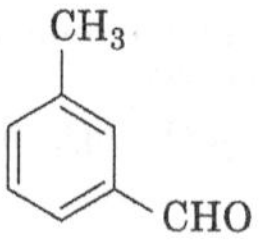 (b) 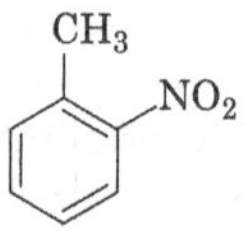

(c) 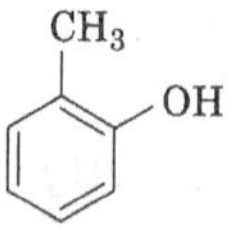 (d) 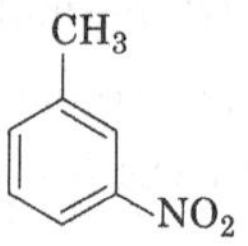

15 The reagent, from the following, which converts benzoic acid to benzaldehyde in one step is

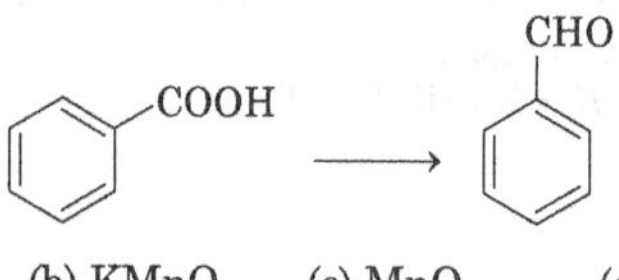

(a) $LiAlH_4$ (b) $KMnO_4$ (c) MnO (d) $NaBH_4$

16 The final product '*A*' in the following reaction sequence

$$CH_3CH_2-\underset{\underset{O}{\|}}{C}-CH_3 \xrightarrow{HCN} ? \xrightarrow[\text{Heat}]{95\%\ H_2SO_4} A$$

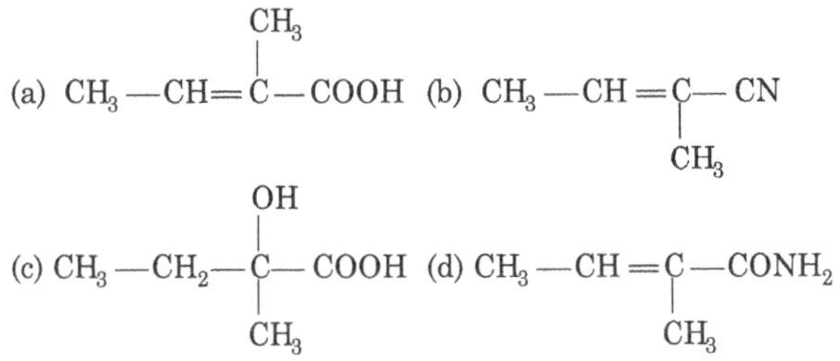

17 Which statement is not correct for *p*-toluene sulphonyl chloride?

(a) It is known as Hinsberg's reagent.
(b) It is used to distinguish primary and secondary amines.
(c) On treatment with secondary amine, it leads to a product, that is soluble in alkali.
(d) It doesn't react with tertiary amines.

18 The final product '*C*' in the following series of reactions.

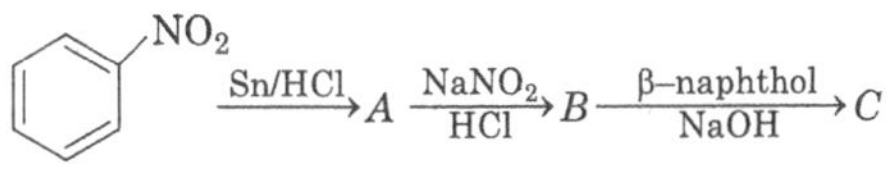

(a)
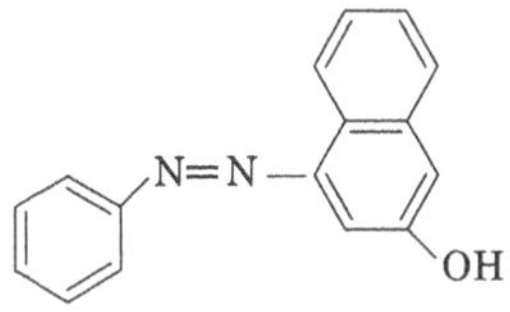

(b)
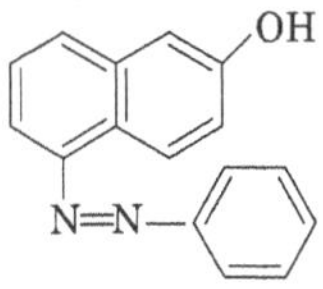

(c)
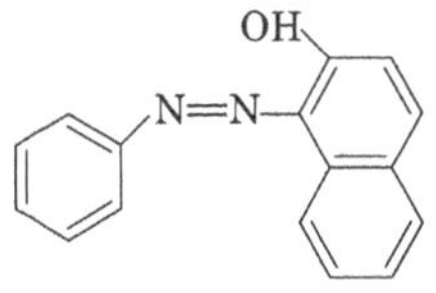

(d)
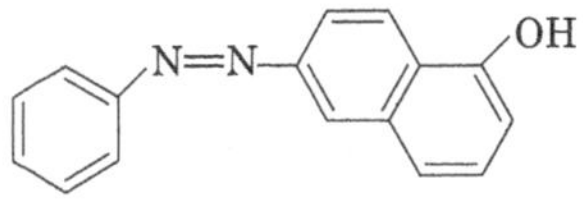

19 Which of the following is not an example of synthetic detergent?

(a)
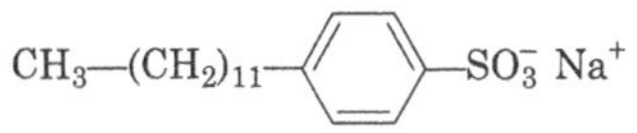

(b) $CH_3-(CH_2)_{16}-COO^-Na^+$

(c)
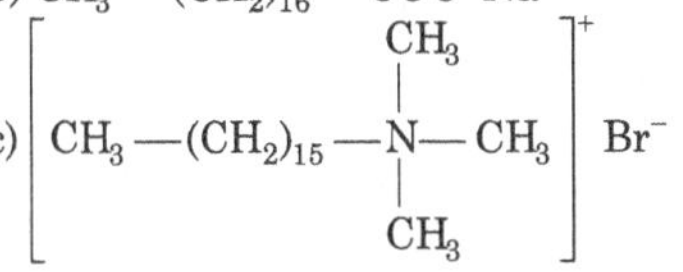

(d) $CH_3(CH_2)_{16}COO(CH_2CH_2O)_nCH_2CH_2OH$

20 Which one of the following is a water soluble vitamin, that is not excreted easily?

(a) Vitamin B_2 (b) Vitamin B_1
(c) Vitamin B_6 (d) Vitamin B_{12}

Section B : Numerical Type Questions

21 CNG is an important transportation fuel. When 100 g CNG is mixed with 208g oxygen in vehicles, it leads to the formation of CO_2 and H_2O and produced large quantity of heat during this combustion, then the amount of carbon dioxide, produced in grams is .... [Nearest integer] [Assume CNG to be methane]

22 In a solid *AB*, *A* atoms are in ccp arrangement and *B* atoms occupy all the octahedral sites. If two atoms from the opposite faces are removed, then the resultant stoichiometry of the compound is A_xB_y. The value of *x* is .... [Nearest integer]

23 Amongst SF_4, XeF_4, CF_4 and H_2O, the number of species with two lone pair of electrons is ....

24 A fish swimming in water body when taken out from the water body is covered with a film of water of weight 36 g. When it is subjected to cooking at 100°C, then the internal energy for vaporisation in kJ mol^{-1} is .... [Nearest integer]

[Assume steam to be an ideal gas. Given, $\Delta_{vap}H^{\ominus}$ for water at 373 K and 1 bar is 41.1 kJ mol^{-1}; $R = 8.31$ JK^{-1} mol^{-1}]

25 The osmotic pressure exerted by a solution prepared by dissolving 2.0 g of protein of molar mass 60 kg mol^{-1} in 200 mL of water at 27°C is ...... Pa. [Integer value] (Use, $R = 0.083$ L bar mol^{-1} K^{-1})

26 40% of HI undergoes decomposition to H_2 and I_2 at 300 K. $\Delta G°$ for this decomposition reaction at one atmosphere pressure is ...... J mol^{-1}. [Nearest integer]

(Use, $R = 8.31$ JK^{-1} mol^{-1}; $\log 2 = 0.3010$, $\ln 10 = 2.3$, $\log 3 = 0.477$)

27 $Cu(s) + Sn^{2+}\ (0.001M) \longrightarrow Cu^{2+}\ (0.01\ M) + Sn(s)$

The Gibb's free energy change for the above reaction at 298 K is $x \times 10^{-1}$ kJ mol^{-1}.

The value of *x* is .... [Nearest integer]

[Given, $E^{\ominus}_{Cu^{2+}/Cu} = 0.34$ V; $E^{\ominus}_{Sn^{2+}/Sn} = -0.14$ V; $F = 96500$ C mol^{-1}].

28 Catalyst *A* reduces the activation energy for a reaction by 10 kJ mol^{-1} at 300 K. The ratio of rate constants, $\frac{K', \text{catalysed}}{K, \text{uncatalysed}}$ is e^x. The value of *x* is .... [Nearest integer]

[Assume that, the pre-exponential factor is same in both the cases. Given, $R = 8.31 JK^{-1}$ mol^{-1}]

29 Reaction of $[Co(H_2O)_6]^{2+}$ with excess ammonia and in the presence of oxygen results into a diamagnetic product. Number of electrons present in t_{2g}-orbitals of the product is ....

30 The moles of methane required to produce 81 g of water after complete combustion is ...... $\times 10^{-2}$ mol. [Nearest integer]

27 JUNE 2022, SHIFT-I

Section A : Objective Type Questions

1 Given below are two statements. One is labelled as Assertion (A) and the other is labelled as Reason (R).

Assertion (A) At 10°C, the density of a 5M solution of KCl [atomic masses of K and Cl are 39 and 35.5 g mol^{-1} respectively], is '*x*' g mL^{-1}. The solution is cooled to – 21°C. The molality of the solution will remain unchanged.

Reason (R) The molality of a solution does not change with temperature as mass remains unaffected with temperature.

In the light of the above statements, choose the correct answer from the options given below.

(a) Both (A) and (R) are true and (R) is the correct explanation of (A)
(b) Both (A) and (R) are true but (R) is not the correct explanation of (A)
(c) (A) is true but (R) is false
(d) (A) is false but (R) is true

2 Based upon VSEPR theory, match the shape (geometry) of the molecules in List-I with the molecules in List-II and select the most appropriate option.

List-I (Shape)		List-II (Molecules)	
A.	T-shaped	I.	XeF_4
B.	Trigonal planar	II.	SF_4
C.	Square planar	III.	ClF_3
D.	*See-saw*	IV.	BF_3

	A	B	C	D
(a)	I	II	III	IV
(b)	III	IV	I	II
(c)	III	IV	II	I
(d)	IV	III	I	II

3 Match List-I with List-II.

List-I		List-II	
A.	Spontaneous process	I.	$\Delta H < 0$
B.	Process with $\Delta p = 0$, $\Delta T = 0$	II.	$\Delta G_{T,p} < 0$
C.	$\Delta H_{reaction}$	III.	Isothermal and isobaric process
D.	Exothermic process	IV.	[Bond energies of molecules in reactants] – [Bond energies of product molecules]

Choose the correct answer from the options given below.

	A	B	C	D
(a)	III	II	IV	I
(b)	II	III	IV	I
(c)	II	III	I	IV
(d)	II	I	III	IV

4 Match List-I with List-II.

List-I		List-II	
A.	Lyophilic colloid	I.	Liquid-liquid colloid
B.	Emulsion	II.	Protective colloid
C.	Positively charged colloid	III.	$FeCl_3$ + NaOH
D.	Negatively charged colloid	IV.	$FeCl_3$ + hot water

Choose the correct answer from the options given below.

	A	B	C	D
(a)	II	I	IV	III
(b)	III	I	IV	II
(c)	II	I	III	IV
(d)	III	II	I	IV

5 Given below are two statements. One is labelled as Assertion (A) and the other is labelled as Reason (R) .

Assertion (A) The ionic radii of O^{2-} and Mg^{2+} are same.

Reason (R) Both O^{2-} and Mg^{2+} are isoelectronic species.

In the light of the above statements, choose the correct answer from the options given below.

(a) Both (A) and (R) are true and (R) is the correct explanation of (A)
(b) Both (A) and (R) are true but (R) is not the correct explanation of (A)
(c) (A) is true but (R) is false
(d) (A) is false but (R) is true

6 Match List-I with List-II.

List-I		List-II	
A.	Concentration of gold ore	I.	Aniline
B.	Leaching of alumina	II.	NaOH
C.	Froth stabiliser	III.	SO_2
D.	Blister copper	IV.	NaCN

Choose the correct answer from the options given below.

	A	B	C	D
(a)	IV	III	II	I
(b)	IV	II	I	III
(c)	III	II	I	IV
(d)	II	IV	III	I

7 Addition of H_2SO_4 to BaO_2 produces

(a) BaO, SO_2 and H_2O
(b) $BaHSO_4$ and O_2
(c) $BaSO_4$, H_2 and O_2
(d) $BaSO_4$ and H_2O_2

8 $BeCl_2$ reacts with $LiAlH_4$ to give

(a) Be + Li[$AlCl_4$] + H_2
(b) Be + AlH_3 + LiCl + HCl
(c) BeH_2 + LiCl + $AlCl_3$
(d) BeH_2 + Li[$AlCl_4$]

9 Match List-I with List-II.

List-I (Si-compounds)		List-II (Si-polymeric/Other products)	
A.	$(CH_3)_4Si$	I.	Chain silicon
B.	$(CH_3)Si(OH)_3$	II.	Dimeric silicon
C.	$(CH_3)_2Si(OH)_2$	III.	Silane
D.	$(CH_3)_3Si(OH)$	IV.	2D–silicon

Choose the correct answer from the options given below

	A	B	C	D		A	B	C	D
(a)	III	II	I	IV	(b)	IV	I	II	III
(c)	II	I	IV	III	(d)	III	IV	I	II

10 Heating white phosphorus with conc. NaOH solution gives mainly

(a) Na_3P and H_2O (b) H_3PO and NaH
(c) $P(OH)_3$ and NaH_2PO_4 (d) PH_3 and NaH_2PO_2

11 Which of the following will have maximum stabilisation due to crystal field?

(a) $[Ti(H_2O)_6]^{3+}$ (b) $[Co(H_2O)_6]^{2+}$
(c) $[Co(CN)_6]^{3-}$ (d) $[Cu(NH_3)_4]^{2+}$

12 Given below are two statements.

Statement I Classical smog occurs in cool humid climate. It is a reducing mixture of smoke, fog and sulphur dioxide.

Statement II Photochemical smog has components, ozone, nitric oxide, acrolein, formaldehyde, PAN etc.

In the light of above statements, choose the correct answer from the options given below.

(a) Both Statement I and Statement II are correct.
(b) Both Statement I and Statement II are incorrect.
(c) Statement I is correct but Statement II is incorrect.
(d) Statement I is incorrect but Statement II is correct.

13 Which of the following is structure of a separating funnel?

(a) 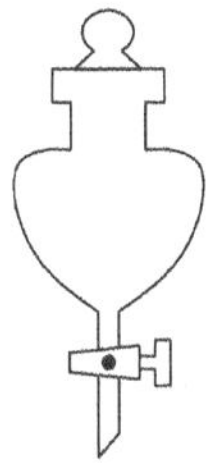
(b) 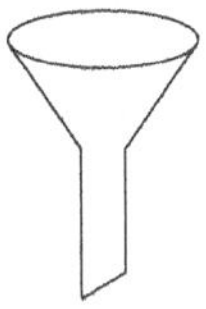
(c) 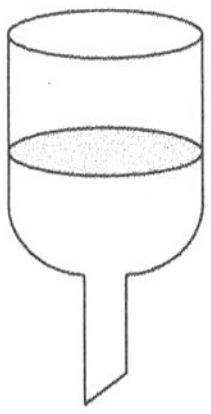
(d) 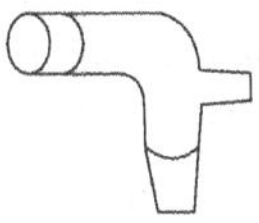

14 '*A*' and '*B*' respectively are

$A \xrightarrow[\text{(ii) Zn-H}_2\text{O}]{\text{(i) O}_3}$ Ethane-1,2-dicarbaldehyde + Glyoxal/Oxaldehyde

$B \xrightarrow[\text{(ii) Zn-H}_2\text{O}]{\text{(i) O}_3}$ 5-oxohexanal

(a) 1-methylcyclohex-1, 3-diene and cyclopentene
(b) Cyclohex-1, 3-diene and cyclopentene
(c) 1-methylcyclohex-1,4-diene and 1-methylcyclopent-1-ene
(d) Cyclohex-1,3-diene and 1-methylcyclopent-1-ene

15 The major product of the following reaction is

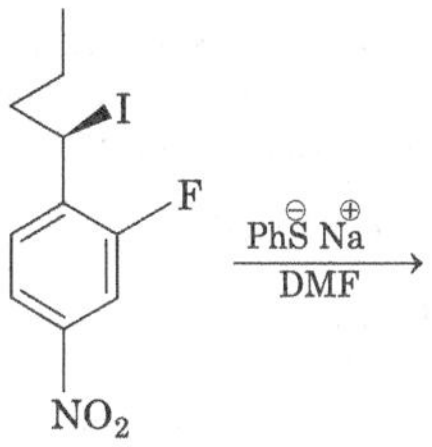

(a) 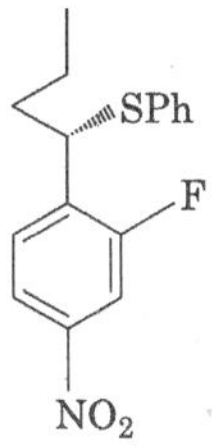

(b) 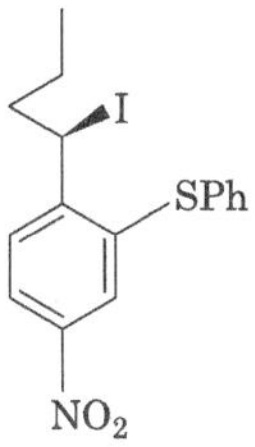

(c) 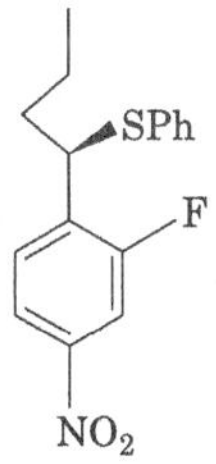

(d) 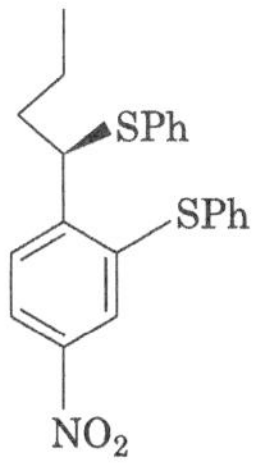

16 Which of the following reactions will yield benzaldehyde as a product?

A. 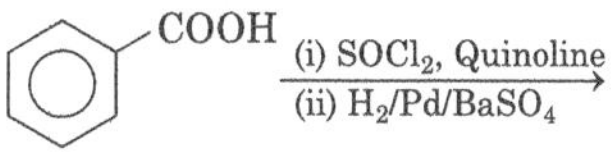

B. 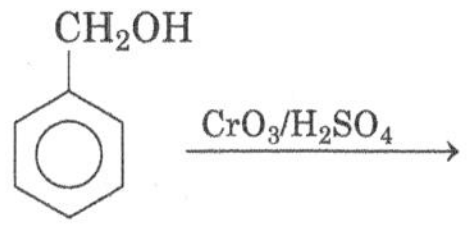

C. 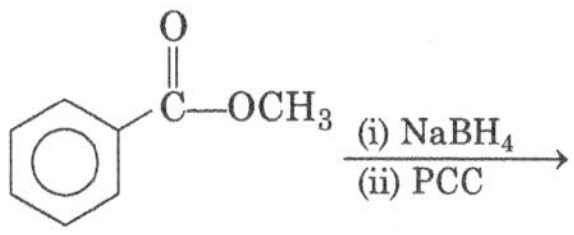

D. 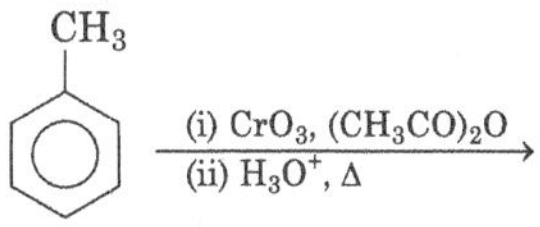

(a) (B) and (C) (b) (C) and (D)
(c) (A) and (D) (d) (A) and (C)

17 Given below are two statements.

Statement I In Hofmann degradation reaction, the migration of only an alkyl group takes place from carbonyl carbon of the amide to the nitrogen atom.

Statement II The group is migrated in Hofmann degradation reaction to electron deficient atom.

In the light of the above statements, choose the correct answer from the options given below.

(a) Both Statement I and Statement II are correct.
(b) Both Statement I and Statement II are incorrect.
(c) Statement I is correct but Statement II is incorrect.
(d) Statement I is incorrect but Statement II is correct.

18 Match List-I with List-II.

List-I (Polymer)		List-II (Used in)	
A.	Bakelite	I.	Radio and television cabinets
B.	Glyptal	II.	Electrical switches
C.	PVC	III.	Paints and Lacquers
D.	Polystyrene	IV.	Water pipes

Choose the correct answer from the options given below.

	A	B	C	D		A	B	C	D
(a)	II	III	IV	I	(b)	I	II	III	IV
(c)	IV	III	II	I	(d)	II	III	I	IV

19 *L*-isomer of a compound (*A*) ($C_4H_8O_4$) gives a positive test with $[Ag(NH_3)_2]^+$. Treatment of *A* with acetic anhydride yields triacetate derivative. Compound (*A*) produces an optically active compound (*B*) and an optically inactive compound (*C*) on treatment with bromine water and HNO_3 respectively. Compound (*A*) is

(a) 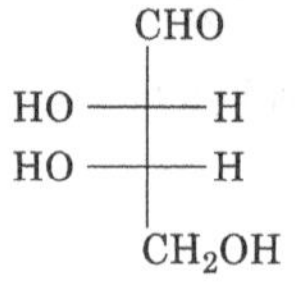

(b) 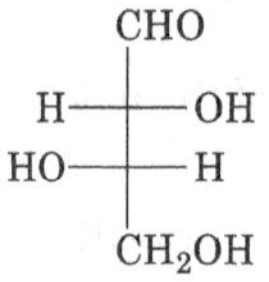

(c) 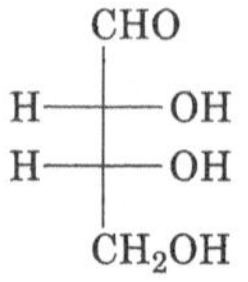

(d) 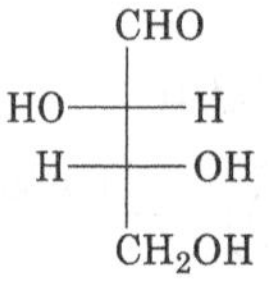

20 Match List-I with List-II.

	List-I		List-II
A.	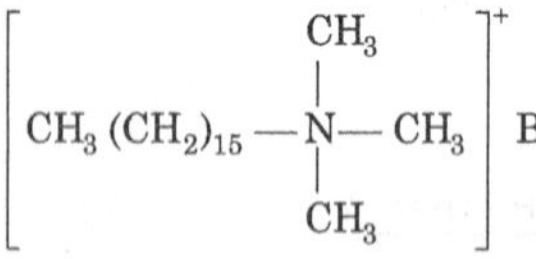	I.	Dishwashing powder
B.	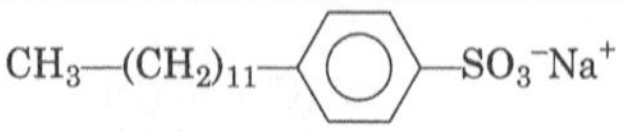	II.	Toothpaste
C.	$C_{17}H_{35}COO^- Na^+ + Na_2CO_3 +$ Rosinate	III.	Laundry soap
D.	$CH_3(CH_2)_{16}COO(CH_2CH_2O)_n CH_2CH_2OH$	IV.	Hair conditioner

Choose the correct answer from the options given below.

	A	B	C	D		A	B	C	D
(a)	III	II	IV	I	(b)	IV	II	III	I
(c)	IV	III	II	I	(d)	III	IV	I	II

Section B : Numerical Type Questions

21 Metal deficiency defect is shown by $Fe_{0.93}O$. In the crystal, some Fe^{2+} cations are missing and loss of positive charge is compensated by the presence of Fe^{3+} ions. The percentage of Fe^{2+} ions in the $Fe_{0.93}O$ crystals is .... [Nearest integer]

22 If the uncertainty in velocity and position of a minute particle in space are, 2.4×10^{-26} (ms^{-1}) and 10^{-7} (m) respectively. The mass of the particle in g is .... [Nearest integer]
(Given, $h = 6.626 \times 10^{-34}$ Js)

23 2 g of a non-volatile non-electrolyte solute is dissolved in 200 g of two different solvents *A* and *B* whose ebullioscopic constants are in the ratio of 1 : 8. The elevation in boiling points of *A* and *B* are in the ratio $\frac{x}{y}$ $(x : y)$.
The value of *y* is ...... (Nearest integer)

24 $2NOCl(g) \rightleftharpoons 2NO(g) + Cl_2(g)$
In an experiment, 2.0 moles of NOCl was placed in a one-litre flask and the concentration of NO after equilibrium established, was found to be 0.4 mol/L. The equilibrium constant at 30°C is ...... $\times 10^{-4}$.

25 The limiting molar conductivities of NaI, $NaNO_3$ and $AgNO_3$ are 12.7, 12.0 and 13.3 mS m^2 mol^{-1}, respectively (all at 25°C). The limiting molar conductivity of AgI at this temperature is ...... mS m^2 mol^{-1}.

26 The rate constant for a first order reaction is given by the following equation

$$\ln k = 33.24 - \frac{2.0 \times 10^4 \text{ K}}{T}$$

The activation energy for the reaction is given by ______ kJ mol^{-1}. [In nearest integer] (Given, $R = 8.3$ J K^{-1} mol^{-1})

27 The number of statement(s) correct from the following for copper (Atomic number 29) is/are ....
(A) Cu(II) complexes are always paramagnetic.
(B) Cu(I) complexes are generally colourless.
(C) Cu(I) is easily oxidised.
(D) In Fehling solution, the active reagent has Cu(I).

28 Acidified potassium permanganate solution oxidises oxalic acid. The spin-only magnetic moment of the manganese product formed from the above reaction is ...... BM. [Nearest integer]

29 Two elements *A* and *B* which form 0.15 moles of A_2B and AB_3 type compounds. If both A_2B and AB_3 weigh equally, then the atomic weight of *A* is ...... times of atomic weight of *B*.

30 Total number of possible stereoisomers of dimethyl cyclopentane is ....

27 JUNE 2022, SHIFT-II

Section A : Objective Type Questions

1 Which amongst the given plots is the correct plot for pressure (p) vs density (d) for an ideal gas?

(a) $T_3 > T_2 > T_1$

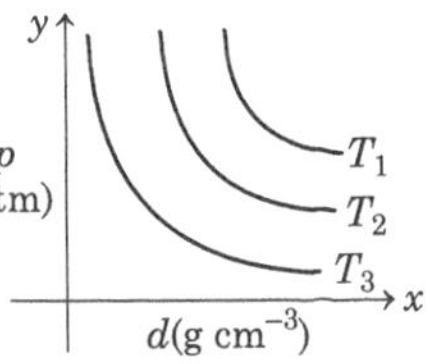

(b) $T_3 > T_2 > T_1$

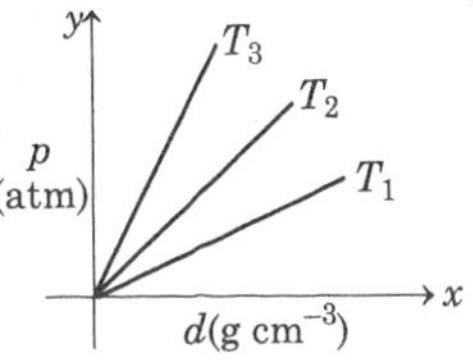

(c) $T_3 > T_2 > T_1$

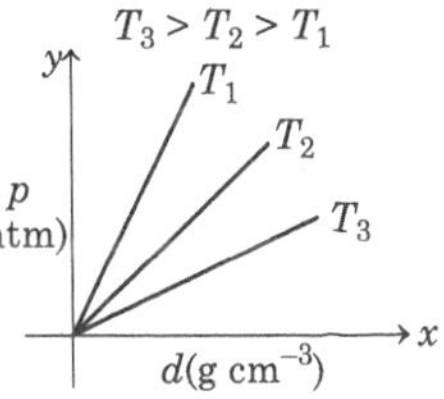

(d) $T_3 > T_2 > T_1$

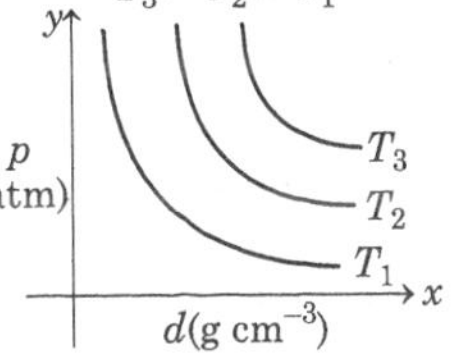

2 Identify the incorrect statement for PCl_5 from the following.

(a) In this molecule, orbitals of phosphorus are assumed to undergo sp^3d hybridisation.
(b) The geometry of PCl_5 is trigonal bipyramidal.
(c) PCl_5 has two axial bonds stronger than three equatorial bonds.
(d) The three equatorial bonds of PCl_5 lie in a plane

3 **Statement I** Leaching of gold with cyanide ion in absence of air / O_2 leads to cyano complex of Au(III).

Statement II Zinc is oxidised during the displacement reaction carried out for gold extraction.

In the light of the above statements, choose the correct answer from the options given below.

(a) Both Statement I and Statement II are correct.
(b) Both Statement I and Statement II are incorrect
(c) Statement I is correct but Statement II is incorrect.
(d) Statement I is incorrect but Statement II is correct.

4 The correct order of increasing intermolecular hydrogen bond strength is

(a) $HCN < H_2O < NH_3$ (b) $HCN < CH_4 < NH_3$
(c) $CH_4 < HCN < NH_3$ (d) $CH_4 < NH_3 < HCN$

5 The correct order of increasing ionic radii is

(a) $Mg^{2+} < Na^+ < F^- < O^{2-} < N^{3-}$
(b) $N^{3-} < O^{2-} < F^- < Na^+ < Mg^{2+}$
(c) $F^- < Na^+ < O^{2-} < Mg^{2+} < N^{3-}$
(d) $Na^+ < F^- < Mg^{2+} < O^{2-} < N^{3-}$

6 The gas produced by treating an aqueous solution of ammonium chloride with sodium nitrite is

(a) NH_3 (b) N_2 (c) N_2O (d) Cl_2

7 Given below are two statements. One is labelled as Assertion A and the other is labelled as Reason R.

Assertion (A) Fluorine forms one oxoacid.

Reason (R) Fluorine has smallest size amongst all halogens and is highly electronegative

In the light of the above statements, choose the most appropriate answer from the options given below.

(a) Both A and R are correct and R is the correct explanation of A.
(b) Both A and R are correct but R is not the correct explanation of A.
(c) A is correct but R is not correct
(d) A is not correct but R is correct

8 In 3d-series, the metal having the highest M^{2+}/M standard electrode potential is

(a) Cr (b) Fe
(c) Cu (d) Zn

9 The 'f' orbitals are half and completely filled, respectively in lanthanide ions

(Given, atomic number : Eu 63; Sm 62; Tm 69; Tb 65; Yb 70; Dy 66]

(a) Eu^{2+} and Tm^{2+} (b) Sm^{2+} and Tm^{3+}
(c) Tb^{4+} and Yb^{2+} (d) Dy^{3+} and Yb^{3+}

10 Arrange the following coordination compounds in the increasing order of magnetic moments. (Atomic numbers : Mn = 25 ; Fe = 26)

(A) $[FeF_6]^{3-}$ (B) $[Fe(CN)_6]^{3-}$
(C) $[MnCl_6]^{3-}$ (high spin) (D) $[Mn(CN)_6]^{3-}$

Choose the correct answer from the options given below :

(a) A < B < D < C (b) B < D < C < A
(c) A < C < D < B (d) B < D < A < C

11 On the surface of polar stratospheric clouds, hydrolysis of chlorine nitrate gives A and B while its reaction with HCl produces B and C. A, B and C are, respectively

(a) HOCl, HNO_3, Cl_2 (b) Cl_2, HNO_3, HOCl
(c) $HClO_2$, HNO_2, HOCl (d) HOCl, HNO_2, Cl_2O

12 Which of the following is most stable?

(a)

(b)

(c)

(d)

13 What will be the major product of following sequence of reactions?

$$n\text{-Bu}-\equiv \xrightarrow[\text{(ii) Lindlar catalysis, } H_2]{\text{(i) } n\text{-BuLi} \quad n\text{-}C_5H_{11}Cl}$$

(a) 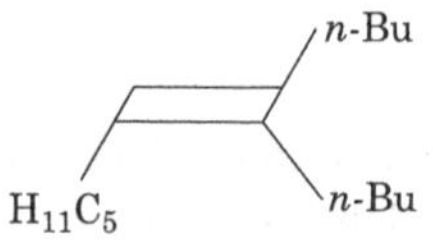

(b) 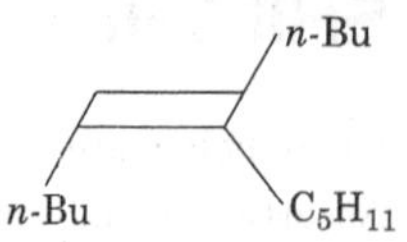

(c) 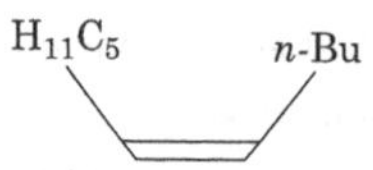

(d) 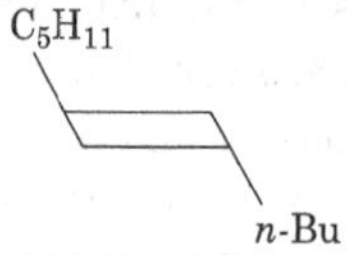

14 Product '*A*' of following sequence of reactions is

$$\text{Ethylbenzene} \xrightarrow[\text{(ii) } Cl_2,\ \Delta \text{ (iii) alc. KOH}]{\text{(i) } Br_2,\ Fe} \text{'}A\text{'} \text{ (Major product)}$$

(a) 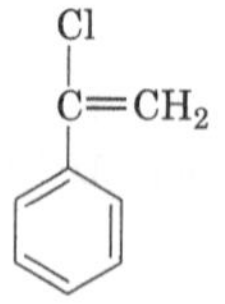

(b) 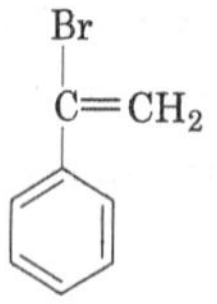

(c) 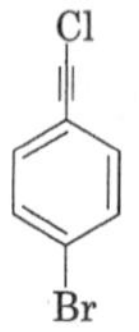

(d) 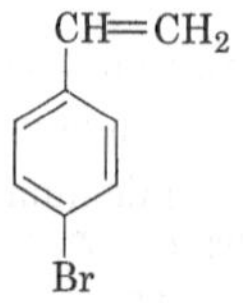

15 Match List I with List II.

	List-I		List-II
A.	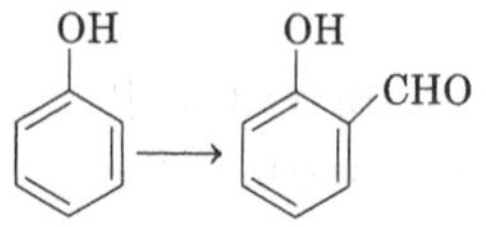	I.	Br_2 in CS_2
B.	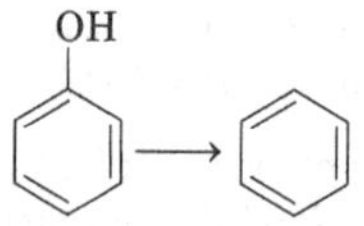	II.	$Na_2Cr_2O_7 / H_2SO_4$
C.	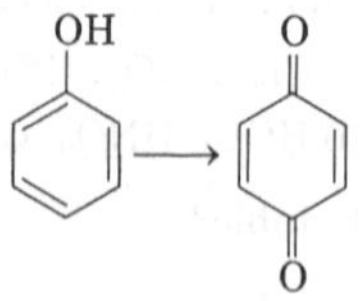	III.	Zn
D.	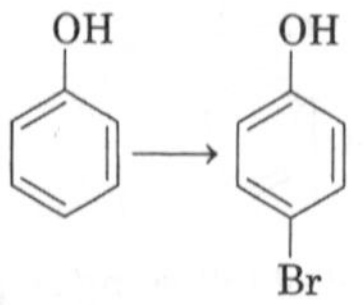	IV.	$CHCl_3 / NaOH$

Choose the correct answer from the options given below.

	A	B	C	D
(a)	IV	III	II	I
(b)	IV	III	I	II
(c)	II	III	I	IV
(d)	IV	II	III	I

16 Decarboxylation of all six possible forms of diaminobenzoic acid $C_6H_3(NH_2)_2COOH$ yields three products *A*, *B* and *C*. Three acids give a product *A*, two acids give a product '*B*' and one acid gives a product *C*. The melting point of product *C* is

(a) 63°C (b) 90°C (c) 104°C (d) 142°C

17 Which is true about buna-N?

(a) It is a linear polymer of 1, 3-butadiene.
(b) It is obtained by co-polymerisation of 1,3-butadiene and styrene.
(c) It is obtained by co-polymerisation of 1,3-butadiene and acrylonitrile.
(d) The suffix N in buna-N stands for its natural occurrence.

18 Given below are two statements.

Statments I Maltose has two α-D-glucose units linked at C_1 and C_4 and is a reducing sugar.

Statement II Maltose has two monosaccharides: α-D-glucose and β-D-glucose linked at C_1 and C_6 and it is a non-reducing sugar.

In the light of the above statements, choose the correct answer from the options given below.

(a) Both Statement I and Statement II are true
(b) Both Statement I and Statement II are false
(c) Statement I is true but Statement II are false
(d) Statement I is false but Statement II is true

19 Match List I with List II.

	List-I		List-II
A.	Antipyretic	I.	Reduces pain
B.	Analgesic	II.	Reduces stress
C.	Tranquiliser	III.	Reduces fever
D.	Antacid	IV.	Reduces acidity (Stomach)

Choose the correct answer from the options given below.

	A	B	C	D
(a)	III	I	II	IV
(b)	III	I	IV	II
(c)	I	IV	II	III
(d)	I	III	II	IV

20 Match List I with List II.

	List-I (Anion)		List-II (Gas evolved on reaction with dil. H_2SO_4)
A.	CO_3^{2-}	I.	Colourless gas which turns lead acetate paper black.
B.	S^{2-}	II.	Colourless gas which turns acidified potassium dichromate solution green.
C.	SO_3^{2-}	III.	Brown fumes which turns acidified KI solution containing starch blue.
D.	NO_2^-	IV.	Colourless gas evolved with brisk effervescence, which turns lime water milky.

Choose the correct answer from the options given below.

	A	B	C	D
(a)	III	I	II	IV
(b)	II	I	IV	III
(c)	IV	I	III	II
(d)	IV	I	II	III

Section B : Numerical Type Questions

21 116 g of a substance upon dissociation reaction, yields 7.5 g of hydrogen, 60g of oxygen and 48.5 g of carbon. Given that, the atomic masses of H, O and C are 1, 16 and 12 respectively. The data agrees with how many formulae of the following?

A. CH_3COOH B. HCHO
C. CH_3OOCH_3 D. CH_3CHO

22 Consider the following set of quantum numbers.

	n	l	m_l		n	l	m_l
A.	3	3	−3	B.	3	2	−2
C.	2	1	+1	D.	2	2	+2

The number of correct sets of quantum numbers is ....

23 BeO reacts with HF in presence of ammonia to give [*A*] which on thermal decomposition produces [*B*] and ammonium fluoride. Oxidation state of Be in [*A*] is ....

24 When 5 moles of He gas expand isothermally and reversibly at 300 K from 10 litre to 20 litre, the magnitude of the maximum work obtained is .... J. [Nearest integer] (Given, $R = 8.3$ J K^{-1} mol^{-1} and $\log 2 = 0.3010$)

25 A solution containing 2.5×10^{-3} kg of a solute dissolved in 75×10^{-3} kg of water boils at 373.535 K. The molar mass of the solute is .... g mol^{-1}. [Nearest integer] (Given, $K_b(H_2O) = 0.52$ K kg mol^{-1} and boiling point of water = 373.15 K).

26 pH value of 0.001 M NaOH solution is ....

27 For the reaction taking place in the cell :
$Pt(s)|H_2(g)|H^+(aq)||Ag^+(aq)|Ag(s)$ $E^\circ_{cell} = +0.5332$ V.
The value of $\Delta_f G^\ominus$ is .... kJ mol^{-1}. [Nearest integer]

28 It has been found that for a chemical reaction with rise in temperature by 9K the rate constant gets doubled. Assuming a reaction to be occurring at 300 K, the value of activation energy is found to be .... kJ mol^{-1}. [Nearest integer]
(Given, $\ln 10 = 2.3$, $R = 8.3$ $JK^{-1}mol^{-1}$, $\log 2 = 0.30$)

29

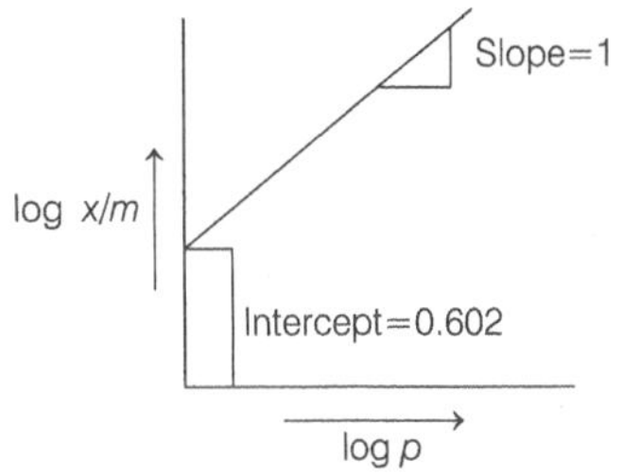

If the initial pressure of a gas is 0.03 atm, the mass of the gas absorbed per gram of the adsorbent is .... $\times 10^{-2}$ g.

30 0.25 g of an organic compound containing chlorine gave 0.40 g of silver chloride in Carius estimation.
The percentage of chlorine present in the compound is .... [Nearest integer]
(Given, Molar mass of Ag is 108 g mol^{-1} and that of Cl is 35.5 g mol^{-1})

28 JUNE 2022, SHIFT-I

Section A : Objective Type Questions

1 The incorrect statement about the imperfections in solids is
(a) Schottky defect decreases the density of the substance.
(b) interstitial defect increases the density of the substance.
(c) Frenkel defect does not alter the density of the substance.
(d) vacancy defect increases the density of the substance.

2 The zeta potential is related to which property of colloids?
(a) Colour
(b) Tyndall effect
(c) Charge on the surface of colloidal particles
(d) Brownian movement

3 Element "*E*" belongs to the period 4 and group 16 of the periodic table. The valence shell electron configuration of the element, which is just above "*E*" in the group is
(a) $3s^2, 3p^4$
(b) $3d^{10}, 4s^2, 4p^4$
(c) $4d^{10}, 5s^2, 5p^4$
(d) $2s^2, 2p^4$

4 Given are two statements one is labelled as : Assertion A and other is labelled as Reason R.

Assertion (A) Magnesium can reduce Al_2O_3 at a temperature below 1350°C, while above 1350°C aluminium can reduce MgO.

Reason (R) The melting and boiling points of magnesium are lower than those of aluminium.

In light of the above statements. Choose most appropriate answer from the options given below
(a) Both A and R are correct and R is correct explanation of A.
(b) Both A and R are correct but R is not the correct explanation of A.
(c) A is correct R is not correct.
(d) A is not correct, R is correct.

5 Dihydrogen reacts with CuO to give
(a) CuH_2 (b) Cu
(c) Cu_2O (d) $Cu(OH)_2$

6 Nitrogen gas is obtained by thermal decomposition of
(a) $Ba(NO_3)_2$ (b) $Ba(N_3)_2$
(c) $NaNO_2$ (d) $NaNO_3$

7 Given below are two statements.

Statement I The pentavalent oxide of group-15 element, E_2O_5, is less acidic than trivalent oxide, E_2O_3 of the same element.

Statement II The acidic character of trivalent oxide of group 15 elements, E_2O_3, decreases down the group.

In light of the above statements, choose most appropriate answer from the options given below

(a) Both Statement I and Statement II are true.
(b) Both Statement I and Statement II are false.
(c) Statement I is true but Statement II is false.
(d) Statement I is false but Statement II is true.

8 Which one of the lanthanoids given below is the most stable in divalent form?

(a) Ce (atomic number 58) (b) Sm (atomic number 62)
(c) Eu (atomic number 63) (d) Yb (atomic number 70)

9 Given below are two statements.

Statement I $[Ni(CN)_4]^{2-}$ is square planar and diamagnetic complex, with dsp^2 hybridisation for Ni but $[Ni(CO)_4]$ is tetrahedral, paramagnetic and with sp^3-hybridisation for Ni.

Statement II $[NiCl_4]^{2-}$ and $[Ni(CO)_4]$ both have same d-electron configuration have same geometry and are paramagnetic.

In light the above statements, choose the correct answer form the options given below

(a) Both Statement I and Statement II are true.
(b) Both Statement I and Statement II are false.
(c) Statement I is true but Statement II is false.
(d) Statement I is false but Statement II is true.

10 Which amongst the following is not a pesticide ?

(a) DDT (b) Organophosphates
(c) Dieldrin (d) Sodium arsenite

11 Which one of the following techniques is not used to spot components of a mixture separated on thin layer chromatographic plate?

(a) I_2 (Solid)
(b) UV light
(c) Visualisation agent as a component of mobile phase
(d) Spraying of an appropriate reagent

12 Which of the following structures are aromatic in nature?

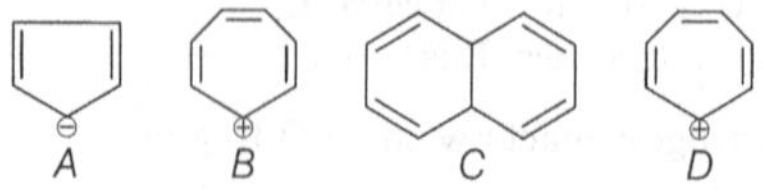

A B C D

(a) A, B, C and D (b) Only A and B
(c) Only A and C (d) Only B, C and D

13 The major product (P) in the reaction.

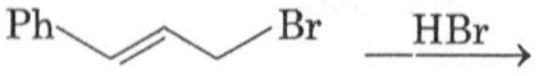

(P) [Ph is $-C_6H_5$] is

(a) 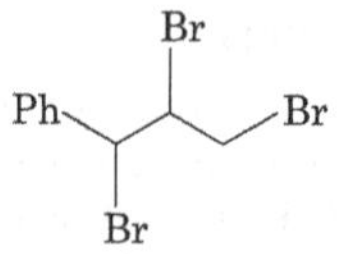

(b) 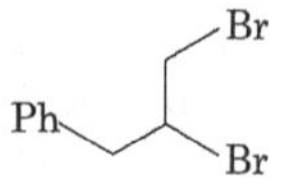

(c) 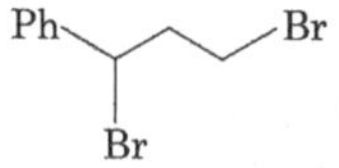

(d) 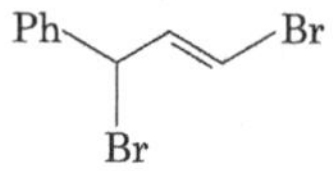

14 The correct structure of product 'A' formed in the following reaction.

$$PhCHO + PhCHO \xrightarrow[\text{in } D_2O]{NaOD} A + Ph-\overset{\overset{O}{\|}}{C}-O^-$$

(Ph is $-C_6H_5$) is

(a) 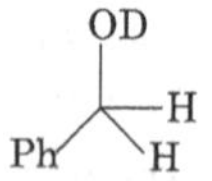

(b) 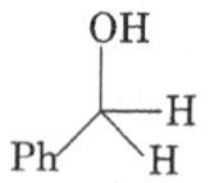

(c) 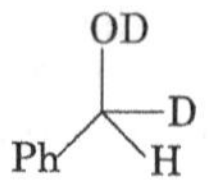

(d) 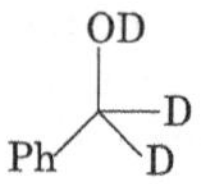

15 Which one of the following compounds is inactive towards S_N1 reaction?

(a) 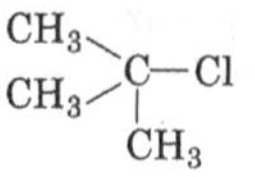

(b) 

(c) 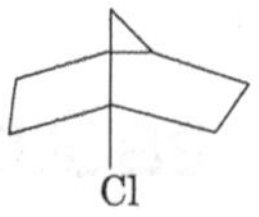

(d) 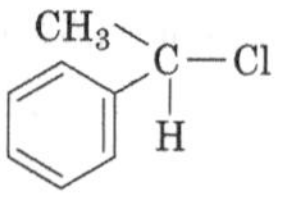

16 Identify the major product formed in the following sequence of reactions.

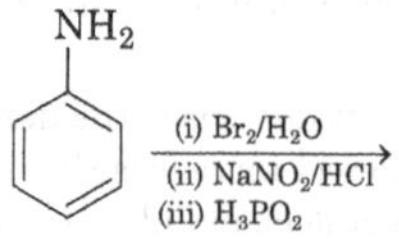

(a) 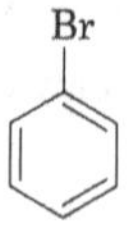

(b) 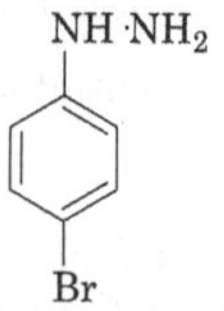

(c) 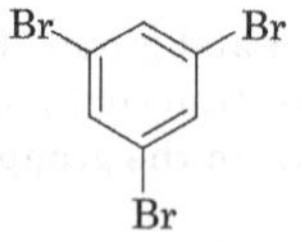

(d) 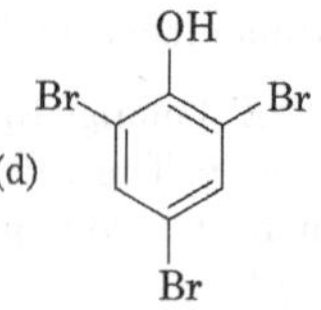

17 A primary aliphatic amine on reaction with nitrous acid in cold (273 K) and there after raising temperature of reaction mixture to room temperature (298 K) gives

(a) nitrile (b) alcohol
(c) diazonium salt (d) secondary amine

18 Which one of the following is not a copolymer ?
(a) Buna-S (b) Neoprene
(c) PHBV (d) Butadiene-styrene

19 Stability of α-helix structure of proteins depends upon
(a) dipolar interaction (b) H-bonding interaction
(c) van der Waals, forces (d) π-stacking interaction

20 The formula of the purple colour formed in Lassaigne's test for sulphur using sodium nitroprusside is
(a) $NaFe[Fe(CN)_6]$ (b) $Na[Cr(NH_3)_2(NCS)_4]$
(c) $Na_2[Fe(CN)_5(NO)]$ (d) $Na_4[Fe(CN)_5(NOS)]$

Section B : Numerical Type Questions

21 A 2.0 g sample containing MnO_2 is treated with HCl liberating Cl_2. The Cl_2 gas is passed into a solution of KI and 60.0 mL of 0.1 M $Na_2S_2O_3$ is required to titrate the liberated iodine. The percentage of MnO_2 in the sample is .... [Nearest integer]
[Atomic masses (in u) Mn = 55; Cl = 35.5; O = 16, I = 127, Na = 23, K = 39, S = 32]

22 If the work function of a metal is 6.63×10^{-19} J, the maximum wavelength of the photon required to remove a photoelectron from the metal is ...... nm. [Nearest integer]
[Given, $h = 6.63 \times 10^{-34}$ J s, and $c = 3 \times 10^8$ m s^{-1}]

23 The hybridisation of P exhibited in PF_5 is $sp^x d^y$. The value of y is ....

24 4.0 L of an ideal gas is allowed to expand isothermally into vacuum until the total volume is 20 L. The amount of heat absorbed in this expansion is ...... L atm.

25 The vapour pressures of two volatile liquids A and B at 25°C are 50 torr and 100 torr, respectively. If the liquid mixture contains 0.3 mole fraction of A, then the mole fraction of liquid B in the vapour phase is $\frac{x}{17}$. The value of x is ....

26 The solubility product of a sparingly soluble salt A_2X_3 is 1.1×10^{-23}. If specific conductance of the solution is 3×10^{-5} S m^{-1}, the limiting molar conductivity of the solution is $x \times 10^{-3}$ S m^2 mol^{-1}. The value of x is ....

27 The quantity of electricity in Faraday needed to reduce 1 mole of $Cr_2O_7^{2-}$ to Cr^{3+} is ....

28 For a first order reaction, $A \rightarrow B$, the rate constant, $k = 5.5 \times 10^{-14}$ s^{-1}. The time required for 67% completion of reaction is $x \times 10^{-1}$ times the half-life of reaction. The value of x is .... [Nearest integer]
(Given, log 3 = 0.4771).

29 Number of complexes which will exhibit synergic bonding amongst, $[Cr(CO)_6]$, $[Mn(CO)_5]$ and $[Mn_2(CO)_{10}]$ is ....

30 In the estimation of bromine, 0.5 g of an organic compound gave 0.40 g of silver bromide. The percentage of bromine in the given compound is ...... %. [Nearest integer] (Relative atomic masses of Ag and Br are 108u and 80u, respectively.)

28 JUNE 2022, SHIFT-II

Section A : Objective Type Questions

1 Compound A contains 8.7% hydrogen, 74% carbon and 17.3% nitrogen. The molecular formula of the compound is (Given, atomic masses of C, H and N are 12, 1 and 14 amu respectively.)
The molar mass of the compound A is 162 g mol^{-1}.
(a) $C_4H_6N_2$ (b) C_2H_3N
(c) C_5H_7N (d) $C_{10}H_{14}N_2$

2 Consider the following statements.
(A) The principal quantum number 'n' is a positive integer with values of 'n' = 1, 2, 3, ...
(B) The azimuthal quantum number 'l' for a given 'n' (principal quantum number) can have values as 'l' = 0, 1, 2,... n.
(C) Magnetic orbital quantum number 'm_l' for a particular 'l' (azimuthal quantum number) has $(2l + 1)$ values.
(D) $\pm 1/2$ are the two possible orientations of electron spin.
(E) For $l = 5$, there will be a total of 9 orbital.

Which of the above statements are correct?
(a) (A), (B) and (C) (b) (A), (C), (D) and (E)
(c) (A), (C) and (D) (d) (A), (B), (C) and (D)

3 In the structure of SF_4, the lone pair of electrons on S is in
(a) equatorial position and there are two lone pair-bond pair repulsions at 90°.
(b) equatorial position and there are three lone pair-bond pair repulsions at 90°.
(c) axial position and there are three lone pair –bond pair repulsion at 90°.
(d) axial position and there are two lone pair–bond pair repulsion at 90°.

4 A student needs to prepare a buffer solution of propanoic acid and its sodium salt with pH 4. The ratio of $\frac{[CH_3CH_2COO^-]}{[CH_3CH_2COOH]}$ required to make buffer is
[Given, K_a $(CH_3CH_2COOH) = 1.3 \times 10^{-5}$]
(a) 0.03
(b) 0.13
(c) 0.23
(d) 0.33

5 Match List-I with List-II.

List I		List II	
A.	Negatively charged sol	(I)	$Fe_2O_3 \cdot xH_2O$
B.	Macromolecular colloid	(II)	CdS sol
C.	Positively charged sol	(III)	Starch
D.	Cheese	(IV)	a gel

Choose the correct answer from the options given below
(a) (A) (II), (B) (III), (C) (IV), (D) (I)
(b) (A) (II), (B) (I), (C) (III), (D) (IV)
(c) (A) (II), (B) (III), (C) (I), (D) (IV)
(d) (A) (I), (B) (III), (C) (II), (D) (IV)

6 Match List I with List II.

List I (Oxide)		List II (Nature)	
A.	Cl_2O_7	(I)	Amphoteric
B.	Na_2O	(II)	Basic
C.	Al_2O_3	(III)	Neutral
D.	N_2O	(IV)	Acidic

Choose the correct answer from the option given below
(a) (A) (IV), (B) (III), (C) (I), (D) (II)
(b) (A) (IV), (B) (II), (C) (I), (D) (III)
(c) (A) (II), (B) (IV), (C) (III), (D) (I)
(d) (A) (I), (B) (II), (C) (III), (D) (IV)

7 In the metallurgical extraction of copper, following reaction is used $FeO + SiO_2 \longrightarrow FeSiO_3$
FeO and $FeSiO_3$ respectively are
(a) gangue and flux (b) flux and slag
(c) slag and flux (d) gangue and slag

8 Hydrogen has three isotopes : protium (^1H), deuterium (^2H or D) and tritium (^3H or T). They have nearly same chemical properties but different physical properties. They differ in
(a) number of protons (b) atomic number
(c) electronic configuration (d) atomic mass

9. Among the following basic oxide is
(a) SO_3 (b) SiO_2
(c) CaO (d) Al_2O_3

10 Among the given oxides of nitrogen; N_2O, N_2O_3, N_2O_4 and N_2O_5, the number of compound(s) having N—N bond is
(a) 1 (b) 2 (c) 3 (d) 4

11 Which of the following oxoacids of sulphur contains "S" in two different oxidation states?
(a) $H_2S_2O_3$ (b) $H_2S_2O_6$
(c) $H_2S_2O_7$ (d) $H_2S_2O_8$

12 Correct statement about photo-chemical smog is
(a) it occurs in humid climate.
(b) it is a mixture of smoke, fog and SO_2.
(c) it is reducing smog.
(d) it results from reaction of unsaturated hydrocarbons.

13 The correct IUPAC name of the following compound is

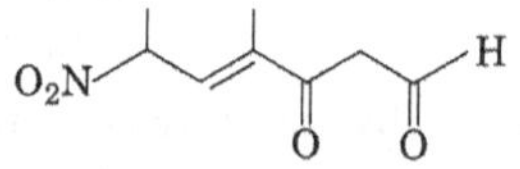

(a) 4-methyl-2-nitro-5-oxohept-3-enal
(b) 4-methyl-5-oxo-2-nitrohept-3-enal
(c) 4-methyl-6-nitro-3-oxohept-4-enal
(d) 6-formyl-4-methyl-2-nitrohex-3-enal

14 The major product (*P*) of the given reaction is (where, Me is —CH_3)

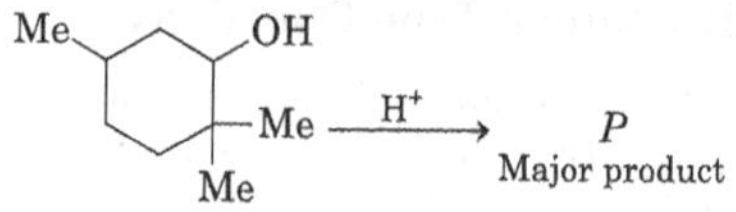

(a) 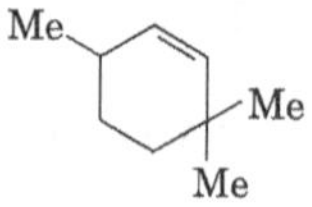

(b) 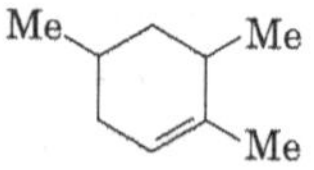

(c) 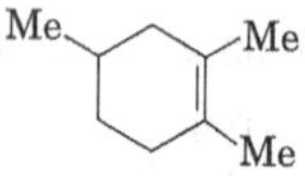

(d) 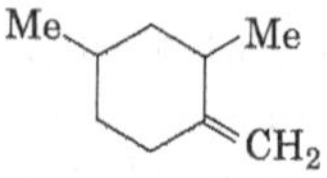

15 $A \xrightarrow[\text{(ii) } CN^- \text{ (iii) } H_2O/H^+]{\text{(i) } Cl_2, \Delta}$ 4-bromophenyl acetic acid.

In the above reaction '*A*' is

(a) 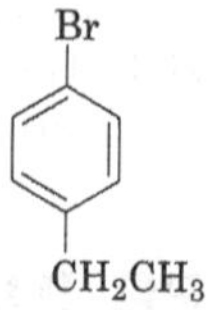

(b) 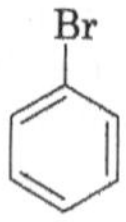

(c) 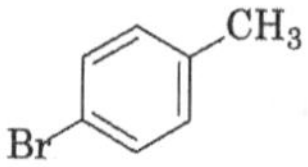

(d) 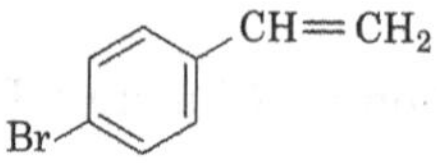

16 Isobutyraldehyde on reaction with formaldehyde and K_2CO_3 gives compound '*A*'. Compound '*A*' reacts with KCN and yields compound '*B*', which on hydrolysis gives a stable compound '*C*'. The compound '*C*' is

(a) $HO—CH_2—\underset{CH_3}{\overset{CH_3}{C}}—\underset{OH}{CH}—COOH$

(b) $HO—CH_2CH_2—\underset{CH_3}{CH}—\underset{OH}{CH}—COOH$

(c) 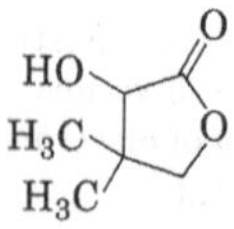

(d) 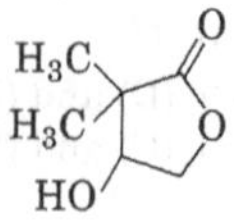

17 With respect to the following reaction, consider the given statements.

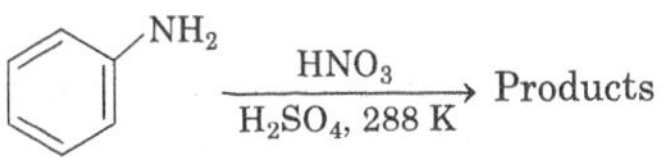

(A) *o*-nitroaniline and *p*-nitroaniline are the predominant products.
(B) *p*-nitroaniline and *m*-nitroaniline are the predominant products.
(C) HNO_3 acts as an acid.
(D) H_2SO_4 acts as an acid.
Choose the correct option.
(a) (A) and (C) are correct statements.
(b) (A) and (D) are correct statements.
(c) (B) and (D) are correct statements.
(d) (B) and (C) are correct statements.

18 Given below are two statements, one is Assertion (A) and other is Reason (R).

Assertion (A) Natural rubber is a linear polymer of isoprene called *cis*-polyisoprene with elastic properties.

Reason (R) The *cis*-polyisoprene molecules consist of various chains held together by strong polar interactions with coiled structure.

In the light of the above statements, choose the correct one from the options given below
(a) Both (A) and (R) are true and (R) is the correct explanation of (A).
(b) Both (A) and (R) are true but (R) is not the correct explanation of (A).
(c) (A) is true but (R) is false.
(d) (A) is false but (R) is true.

19 When sugar '*X*' is boiled with dilute H_2SO_4 in alcoholic solution, two isomers '*A*' and '*B*' are formed. '*A*' on oxidation with HNO_3 yields saccharic acid whereas '*B*' is laevorotatory. The compound '*X*' is
(a) maltose (b) sucrose
(c) lactose (d) starch

20 The drug tegamet is

(a)
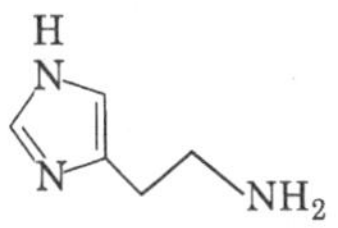

(b)
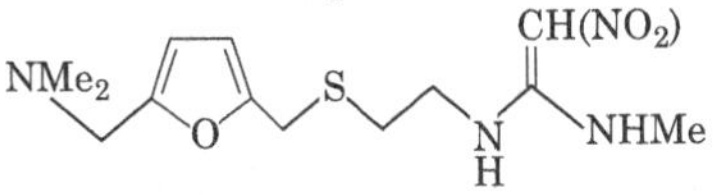

(c)
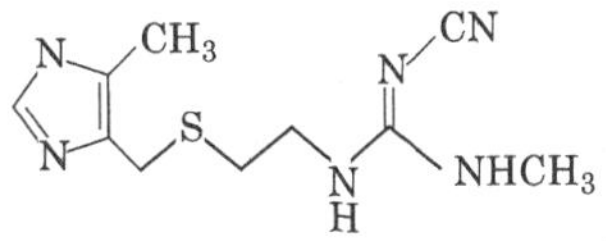

(d)
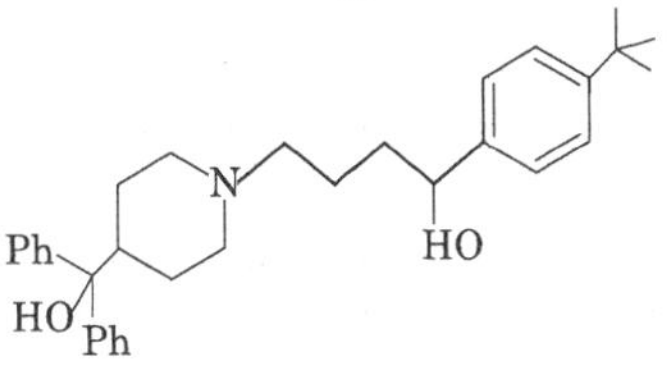

Section B : Numerical Type Questions

21 100 g of an ideal gas is kept in a cylinder of 416 L volume at 27°C under 1.5 bar pressure. The molar mass of the gas is ... g mol^{-1}. (Nearest integer)
(Given, $R = 0.083$ L bar K^{-1} mol^{-1})

22 For combustion of one mole of magnesium in an open container at 300 K and 1 bar pressure, $\Delta_C H^{\ominus} = -601.70$ kJ mol^{-1}, the magnitude of change in internal energy for the reaction is ... kJ. (Nearest integer)
(Given, $R = 8.3$ JK^{-1} mol^{-1})

23 2.5 g of protein containing only glycine ($C_2H_5NO_2$) is dissolved in water to make 500 mL of solution. The osmotic pressure of this solution at 300 K is found to be 5.03×10^{-3} bar. The total number of glycine units present in the protein is ... . (Given $R = 0.083$ L bar K^{-1} mol^{-1})

24 For the given reactions,

$$Sn^{2+} + 2e^- \longrightarrow Sn$$
$$Sn^{4+} + 4e^- \longrightarrow Sn$$

The electrode potentials are $E^{\circ}_{Sn^{2+}/Sn} = -0.140$ V and $E^{\circ}_{Sn^{4+}/Sn} = 0.010$ V. The magnitude of standard electrode potential for Sn^{4+}/Sn^{2+} i.e., $E^{\circ}_{Sn^{4+}/Sn^{2+}}$ is .... $\times 10^{-2}$ V. (Nearest integer)

25 A radioactive element has a half-life of 200 days. The percentage of original activity remaining after 83 days is ... (Nearest integer)
(Given, antilog 0.125 = 1.333, antilog 0.693 = 4.93)

26 $[Fe(CN)_6]^{4-}$, $[Fe(CN)_6]^{3-}$, $[Ti(CN)_6]^{3-}$, $[Ni(CN)_4]^{2-}$ $[Co(CN)_6]^{3-}$
Among the given complexes, number of paramagnetic complexes is ....

27 (i) $CoCl_3 \cdot 4NH_3$, (ii) $CoCl_3 \cdot 5NH_3$, (iii) $CoCl_3 \cdot 6NH_3$ and (iv) $CoCl(NO_3)_2 \cdot 5NH_3$
Number of complex(es) which will exist in *cis-trans* form is/are ....

28 The complete combustion of 0.492 g of an organic compound containing 'C', 'H' and 'O' gives 0.793g of CO_2 and 0.442 g of H_2O. The percentage of oxygen composition in the organic compound is .... (Nearest integer)

29 The major product of the following reaction contains .... bromine atom (s).

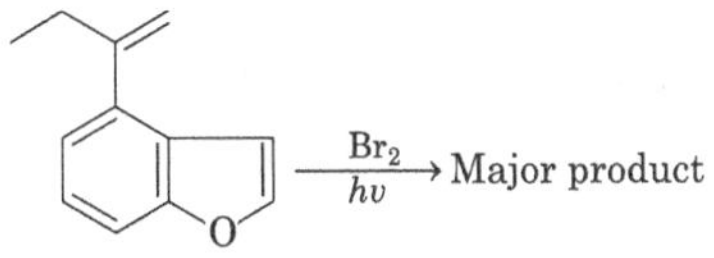

30 0.01 M $KMnO_4$ solution was added to 20.0 mL of 0.05 M Mohr's salt solution through a burette. The initial reading of 50 mL burette is zero. The volume of $KMnO_4$ solution left in burette after the end point is ... mL. (Nearest integer)

29 JUNE 2022, SHIFT-I

Section A : Objective Type Questions

1 Production of iron in blast furnace follows the following equation,

$$Fe_3O_4(s) + 4CO(g) \longrightarrow 3Fe(l) + 4CO_2(g)$$

when 4.640 kg of Fe_3O_4 and 2.520 kg of CO are allowed to react then the amount of iron (in g) produced is

[Given, molar atomic mass (g mol^{-1}) Fe = 56 molar atomic mass (g mol^{-1}) O = 16 and molar atomic mass (g mol^{-1}) C = 12.]

(a) 1400 (b) 2200
(c) 3360 (d) 4200

2 Which of the following statements are correct?

(A) The electronic configuration of Cr is $[Ar]3d^5 4s^1$.
(B) The magnetic quantum number may have a negative value.
(C) In the ground state of an atom, the orbitals are filled in order of their increasing energies.
(D) The total number of nodes are given by $n - 2$.

Choose the most appropriate answer from the options given below

(a) (A), (C) and (D) only (b) (A) and (B) only
(c) (A) and (C) only (d) (A), (B) and (C) only

3 Arrange the following in the decreasing order of their covalent character :

(A) LiCl (B) NaCl (C) KCl (D) CsCl

Choose the most appropriate answer from the options given below

(a) (A) > (C) > (B) > (D) (b) (B) > (A) > (C) > (D)
(c) (A) > (B) > (C) > (D) (d) (A) > (B) > (D) > (C)

4 The solubility of AgCl will be maximum in which of the following ?

(a) 0.01 M KCl (b) 0.01 M HCl
(c) 0.01 M $AgNO_3$ (d) Deionised water

5 Which of the following is a correct statement?

(a) Brownian motion destabilises sols.
(b) Any amount of dispersed phase can be added to emulsion without destabilising it.
(c) Mixing two oppositely charged sols in equal amount neutralises charges and stabilises colloids.
(d) Presence of equal and similar charges on colloidal particles provides stability to the colloidal solution.

6 The electronic configuration of Pt (atomic number 78) is

(a) $[Xe]4f^{14}5d^9 6s^1$ (b) $[Kr]4f^{14}5d^{10}$
(c) $[Xe]4f^{14}5d^{10}$ (d) $[Xe]\,4f^{14}5d^8 6s^2$

7 In isolation of which one of the following metals from their ores, the use of cyanide salt is not commonly involved ?

(a) Zinc (b) Gold (c) Silver (d) Copper

8 Which one of the following reactions indicates the reducing ability of hydrogen peroxide in basic medium ?

(a) $HOCl + H_2O_2 \longrightarrow H_3O^+ + Cl^- + O_2$
(b) $PbS + 4H_2O_2 \longrightarrow PbSO_4 + 4H_2O$
(c) $2MnO_4^- + 3H_2O_2 \longrightarrow 2MnO_2 + 3O_2 + 2H_2O + 2OH^-$
(d) $Mn^{2+} + H_2O_2 \longrightarrow Mn^{4+} + 2OH^-$

9 Match the List-I with List-II.

List-I (Metal)		List-II (Emitted light wavelength (nm))
A. Li	I.	670.8
B. Na	II.	589.2
C. Rb	III.	780.0
D. Cs	IV.	455.5

Choose the most appropriate answer from the options given below.

	A	B	C	D
(a)	I	II	III	IV
(b)	III	II	I	IV
(c)	III	I	II	IV
(d)	IV	II	I	III

10 Match the List-I with List-II.

List-I (Metal)	List-II (Application)
A. Cs	I. High temperature thermometer
B. Ga	II. Water repellent sprays
C. B	III. Photoelectric cells
(d) Si	IV. Bullet proof vest

Choose the most appropriate answer from the options given below.

	A	B	C	D
(a)	III	I	IV	II
(b)	IV	III	II	I
(c)	II	III	IV	I
(d)	I	IV	II	III

11 The oxoacid of phosphorus that is easily obtained from a reaction of alkali and white phosphorus and has two P—H bonds, is

(a) phosphonic acid (b) phosphinic acid
(c) pyrophosphorus acid (d) hypophosphoric acid

12 The acid that is believed to be mainly responsible for the damage of Taj Mahal is

(a) sulphuric acid (b) hydrofluoric acid
(c) phosphoric acid (d) hydrochloric acid

13 Two isomers '*A*' and '*B*' with molecular formula C_4H_8 give different products on oxidation with $KMnO_4/H^+$ results in effervescence of a gas and gives ketone. The compound '*A*' is

(a) but-1-ene (b) *cis*-but-2-ene
(c) *trans*-but-2-ene (d) 2-methyl propene

14 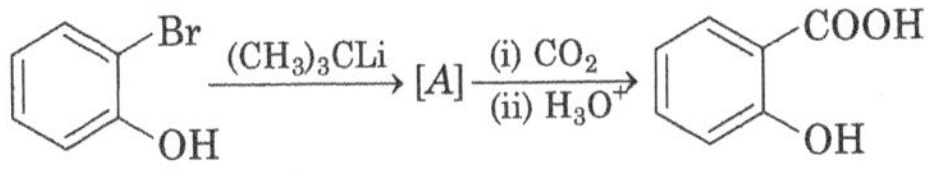

In the given conversion the compound A is

(a) 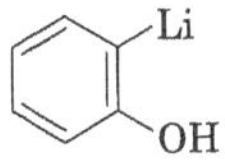

(b) 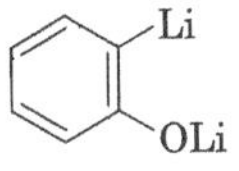

(c) 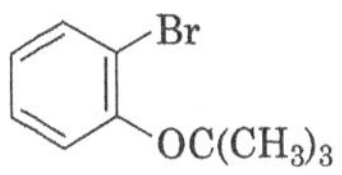

(d) 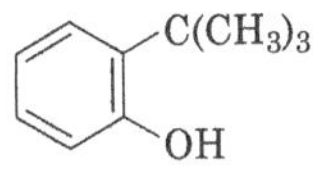

15 Given below are two statements :

Statement I The esterification of carboxylic acid with an alcohol is a nucleophilic acyl substitution.

Statement II Electron withdrawing groups in the carboxylic acid will increase the rate of esterification reaction.

Choose the most appropriate option

(a) Both Statement I and Statement II are correct.
(b) Both Statement I and Statement II are incorrect.
(c) Statement I is correct but Statement II is incorrect.
(d) Statement I is incorrect but Statement II is correct.

16 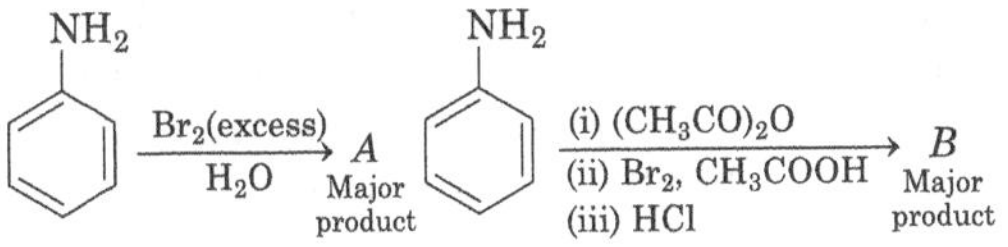

Consider the above reactions, the product A and product B respectively are

(a) 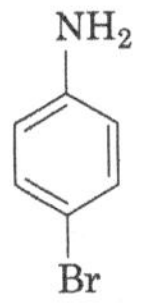 and 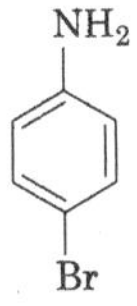

(b) 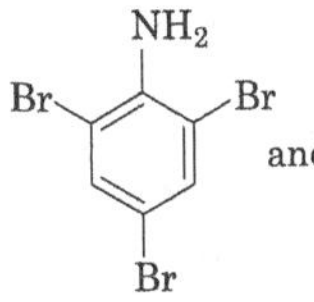 and 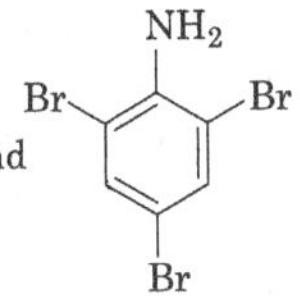

(c) 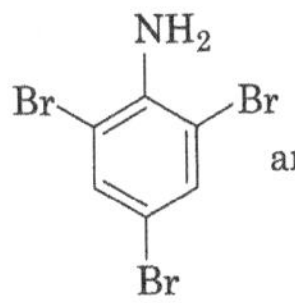 and 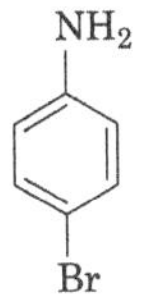

(d) 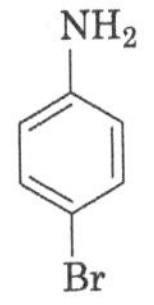 and 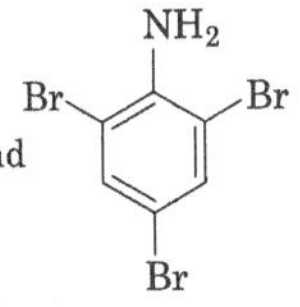

17 The polymer, which can be stretched and retains its original status on releasing the force is

(a) bakelite (b) nylon-6, 6
(c) buna-N (d) terylene

18 Sugar moiety in DNA and RNA molecules respectively are

(a) β-D-2-deoxyribose, β-D-deoxyribose
(b) β-D-2-deoxyribose, β-D-ribose
(c) β-D-ribose, β-D-2-deoxyribose
(d) β-D-deoxyribose, β-D-2-deoxyribose

19. Which of the following compound does not contain sulphur atom?

(a) Cimetidine
(b) Ranitidine
(c) Histamine
(d) Saccharin

20 Given below are two statements.

Statement I Phenols are weakly acidic.

Statement II Therefore they are freely soluble in NaOH solution and are weaker acids than alcohols and water.

Choose the most appropriate option.

(a) Both Statement I and Statement II are correct.
(b) Both Statement I and Statement II are incorrect.
(c) Statement I is correct but Statement II is incorrect.
(d) Statement I is incorrect but Statement II is correct.

Section B : Numerical Type Questions

21 Geraniol, a volatile organic compound, is a component of rose oil. The density of the vapour is 0.46 gL^{-1} at 257°C and 100 mm Hg. The molar mass of geraniol is ...... g mol^{-1}. (Nearest integer)

[Given, $R = 0.082$ L atm K^{-1} mol^{-1}]

22 17.0 g of NH_3 completely vapourises at −33.42°C and 1 bar pressure and the enthalpy change in the process is 23.4 kJ mol^{-1}. The enthalpy change for the vaporisation of 85 g of NH_3 under the same conditions is ...... kJ.

23 1.2 mL of acetic acid is dissolved in water to make 2.0 L of solution. The depression in freezing point observed for this strength of acid is 0.0198°C. The percentage of dissociation of the acid is .... (Nearest integer)

[Given, density of acetic acid is 1.02 g mL^{-1}, molar mass of acetic acid is 60 g mol^{-1} $K_f(H_2O) = 1.85$ K kg mol^{-1}]

24 A dilute solution of sulphuric acid is electrolysed using a current of 0.10 A for 2 hours to produce hydrogen and oxygen gas. The total volume of gases produced at STP is ...... cm^3. (Nearest integer) [Given, Faraday constant, $F = 96500$ C mol^{-1} at STP, molar volume of an ideal gas is 22.7 L mol^{-1}]

25 The activation energy of one of the reactions in a biochemical process is 532611 J mol^{-1}. When the temperature falls from 310 K to 300 K, the change in rate constant observed is $k_{300} = x \times 10^{-3}\ k_{310}$. The value of x is .... [Given, $\ln 10 = 2.3$, $R = 8.3$ J K^{-1} mol^{-1}]

26 The number of terminal oxygen atoms present in the product B obtained from the following reaction is ....

$$FeCr_2O_4 + Na_2CO_3 + O_2 \longrightarrow A + Fe_2O_3 + CO_2$$
$$A + H^+ \longrightarrow B + H_2O + Na^+$$

27 An acidified manganate solution undergoes disproportionation reaction. The spin-only magnetic moment value of the product having manganese in higher oxidation state is ...... BM. (Nearest integer)

28 Kjeldahl's method was used for the estimation of nitrogen in an organic compound. The ammonia evolved from 0.55 g of the compound neutralised 12.5 mL of 1 M H_2SO_4 solution. The percentage of nitrogen in the compound is ...... (Nearest integer)

29 Observe structures of the following compounds.

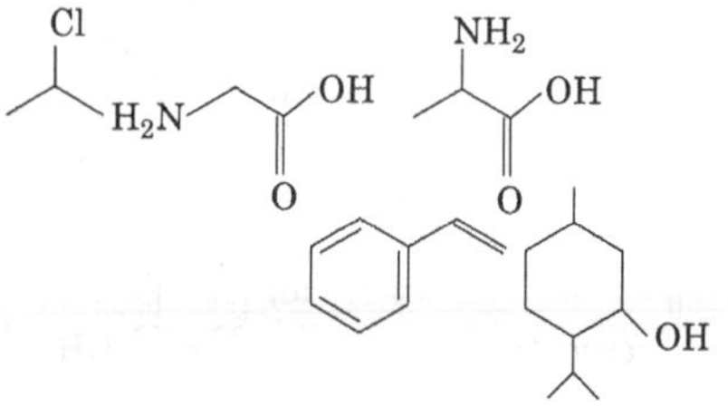

The total number of structures/compounds which possess asymmetric carbon atoms is ....

30 $C_6H_{12}O_6 \xrightarrow{\text{Zymase}} A \xrightarrow[\Delta]{\text{NaOI}} B + CHI_3$

The number of carbon atoms present in the product B is ....

29 JUNE 2022, SHIFT-II

Section A : Objective Type Questions

1 Using the rules for significant figures, the correct answer for the expression

$\dfrac{0.02858 \times 0.112}{0.5702}$ will be

(a) 0.005613 (b) 0.00561 (c) 0.0056 (d) 0.006

2 Which of the following is the correct plot for the probability density $\psi^2(r)$ as a function of distance 'r' of the electron from the nucleus for 2s orbital?

(a)
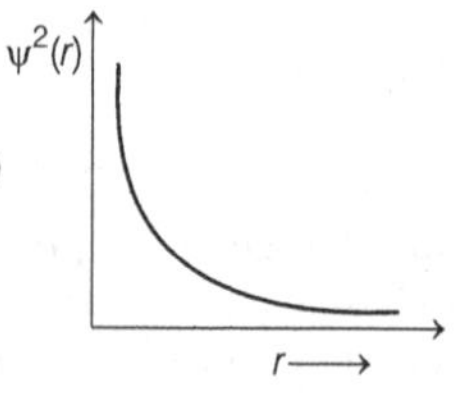

(b)
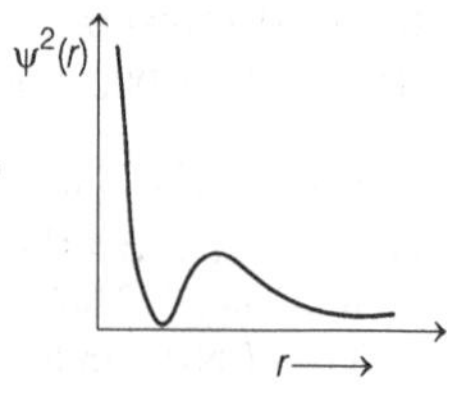

(c)
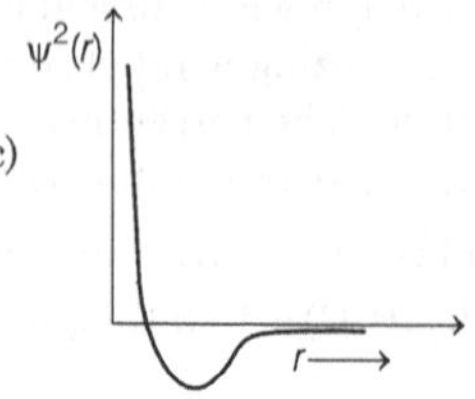

(d)
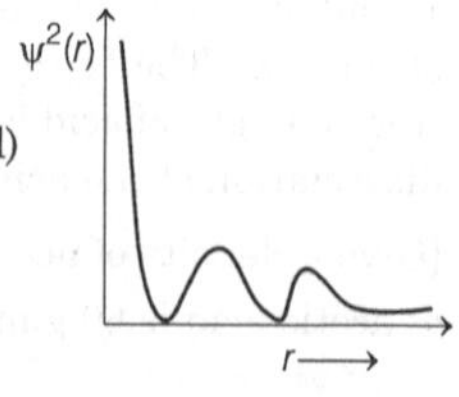

3 Consider the species CH_4, NH_4^+ and BH_4^-. Choose the correct option with respect to these species.

(a) They are isoelectronic and only two have tetrahedral structures.
(b) They are isoelectronic and all have tetrahedral structures.
(c) Only two are isoelectronic and all have tetrahedral structures.
(d) Only two are isoelectronic and only two have tetrahedral structures.

4 4.0 moles of argon and 5.0 moles of PCl_5 are introduced into an evacuated flask of 100 litre capacity at 610 K. The system is allowed to equilibrate. At equilibrium, the total pressure of mixture was found to be 6.0 atm. The K_p for the reaction is [Given, $R = 0.082$ L atm K^{-1} mol^{-1}]

(a) 2.25 (b) 6.24 (c) 12.13 (d) 15.24

5 A 42.12% (w/V) solution of NaCl causes precipitation of a certain sol in 10 hours. The coagulating value of NaCl for the sol is [Given, molar mass : Na = 23.0 g mol^{-1}; Cl = 35.5 g mol^{-1}]

(a) 36 mmol L^{-1} (b) 36 mol L^{-1}
(c) 1440 mol L^{-1} (d) 1440 mmol L^{-1}

6 Given below are two statements. One is labelled as Assertion A and the other is labelled as Reason R.

Assertion (A) The first ionisation enthalpy for oxygen is lower than that of nitrogen.

Reason (R) The four electrons in 2p orbitals of oxygen experience more electron-electron repulsion.

In the light of the above statements, choose the correct answer from the options given below.

(a) Both A and R are correct and R is the correct explanation of A.
(b) Both A and R are correct but R is not the correct explanation of A.
(c) A is correct but R is not correct.
(d) A is not correct but R is correct.

7 Match List I with List II.

	List I (Ore)		List II (Composition)
A.	Siderite	I.	$FeCO_3$
B.	Malachite	II.	$CuCO_3 \cdot Cu(OH)_2$
C.	Sphalerite	III.	ZnS
D.	Calamine	IV.	$ZnCO_3$

Choose the correct answer from the options given below

	A	B	C	D		A	B	C	D
(a)	I	II	III	IV	(b)	III	IV	II	I
(c)	IV	III	I	II	(d)	I	II	IV	III

8 Given below are two statements.

Statement I In $CuSO_4 \cdot 5H_2O$, Cu—O bonds are present.

Statement II In $CuSO_4 \cdot 5H_2O$, ligands coordinating with Cu(II) ion are O-and S-based ligands.

In the light of the above statements, choose the correct answer from the options given below

(a) Both Statement I and Statement II are correct.
(b) Both Statement I and Statement II are incorrect.
(c) Statement I is correct but Statement II is incorrect.
(d) Statement I is incorrect but Statement II is correct.

9 Amongst baking soda, caustic soda and washing soda carbonate anion is present in

(a) washing soda only.
(b) washing soda and caustic soda only.
(c) washing soda and baking soda only.
(d) baking soda, caustic soda and washing soda.

10 Number of lone pair(s) of electrons on central atom and the shape of BrF_3 molecule respectively, are

(a) 0, triangular planar (b) 1, pyramidal
(c) 2, bent T-shape (d) 1, bent T-shape

11 Aqueous solution of which of the following boron compounds will be strongly basic in nature?

(a) $NaBH_4$ (b) $LiBH_4$ (c) B_2H_6 (d) $Na_2B_4O_7$

12 Sulphur dioxide is one of the components of polluted air. SO_2 is also a major contributor to acid rain. The correct and complete reaction to represent acid rain caused by SO_2 is

(a) $2\,SO_2 + O_2 \longrightarrow 2\,SO_3$
(b) $SO_2 + O_3 \longrightarrow SO_3 + O_2$
(c) $SO_2 + H_2O_2 \longrightarrow H_2SO_4$
(d) $2\,SO_2 + O_2 + 2H_2O \longrightarrow 2H_2SO_4$

13 Which of the following carbocations is most stable?

(a) 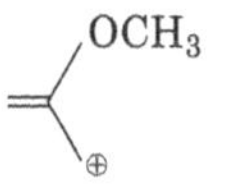
(b) 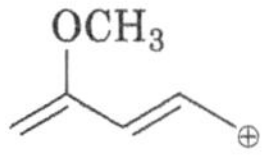
(c) 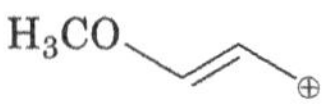
(d) 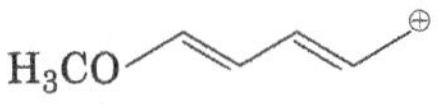

14

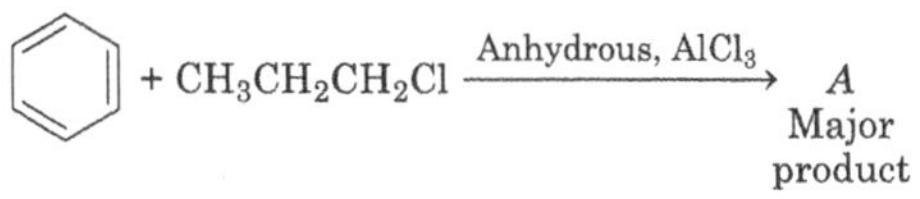

The stable carbocation formed in the above reaction is

(a) $CH_3CH_2\overset{\oplus}{C}H_2$ (b) $CH_3\overset{\oplus}{C}H_2$

(c) $CH_3—\overset{\oplus}{C}H—CH_3$ (d) 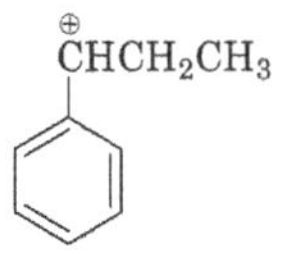

15 Two isomers (*A*) and (*B*) with molar mass 184 g/mol and elemental composition *C*, 52.2%; H, 4.9% and Br, 42.9% gave benzoic acid and *p*-bromobenzoic acid, respectively on oxidation with $KMnO_4$. Isomer *A* is optically active and gives a pale yellow precipitate when warmed with alcoholic $AgNO_3$. Isomers *A* and *B* are, respectively

(a) $H_3C—CHBr—C_6H_5$ and 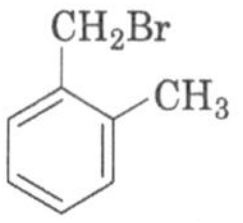

(b) 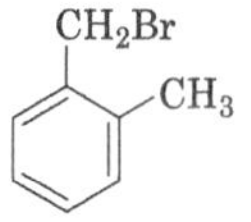 and 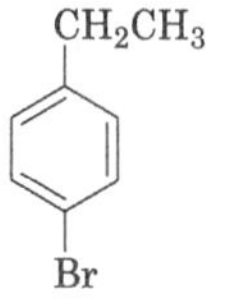

(c) $H_3C—CHBr—C_6H_5$ and 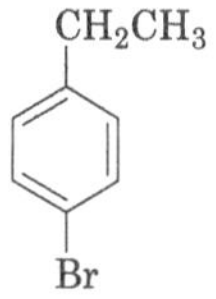

(d) 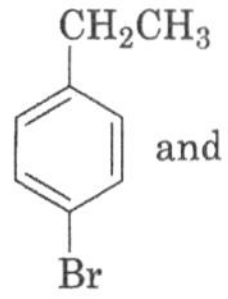 and $H_3C—CHBr—C_6H_5$

16 In Friedel-Crafts alkylation of aniline, one gets

(a) alkylated product with *ortho* and *para* substitution.
(b) secondary amine after acidic treatment.
(c) an amide product.
(d) positively charged nitrogen at benzene ring.

17 Given below are two statements. One is labelled as Assertion A and the other is labelled as Reason R.

Assertion (A) Dacron is an example of polyester polymer.

Reason (R) Dacron is made up of ethylene glycol and terephthalic acid monomers.

In the light of the above statements, choose the most appropriate answer from the options given below.

(a) Both A and R are correct and R is the correct explanation of A.
(b) Both A and R are correct but R is not the correct explanation of A.
(c) A is correct but R is not correct.
(d) A is not correct but R is correct.

18 The structure of protein that is unaffected by heating is

(a) secondary structure
(b) tertiary structure
(c) primary structure
(d) quaternary structure

19 The mixture of chloroxylenol and terpineol is an example of

(a) antiseptic (b) pesticide
(c) disinfectant (d) narcotic analgesic

20 A white precipitate was formed when $BaCl_2$ was added to water extract of an inorganic salt. Further, a gas X with characteristic odour was released when the formed white precipitate was dissolved in dilute HCl. The anion present in the inorganic salt is
(a) I^- (b) SO_3^{2-} (c) S^{2-} (d) NO_2^-

Section B : Numerical Type Questions

21 A box contains 0.90 g of liquid water in equilibrium with water vapour at 27°C. The equilibrium vapour pressure of water at 27°C is 32.0 torr. When the volume of the box is increased, some of the liquid water evaporates to maintain the equilibrium pressure. If all the liquid water evaporates, then the volume of the box must be ... litre. [Nearest integer]
(Given, $R = 0.082$ L atm K^{-1} mol^{-1})
(Ignore the volume of the liquid water and assume water vapours behave as an ideal gas.)

22 2.2 g of nitrous oxide (N_2O) gas is cooled at a constant pressure of 1 atm from 310 K to 270 K causing the compression of the gas from 217.1 mL to 167.75 mL. The change in internal energy of the process, ΔU is '$-x$' J. The value of 'x' is [Nearest integer]
(Given, atomic mass of N = 14 g mol^{-1} and of O = 16 g mol^{-1}. Molar heat capacity of N_2O is 100 JK^{-1} mol^{-1})

23 Elevation in boiling point for 1.5 molal solution of glucose in water is 4K. The depression in freezing point for 4.5 molal solution of glucose in water is 4K. The ratio of molal elevation constant to molal depression constant (K_b/K_f) is...

24 The cell potential for the given cell at 298 K
Pt | H_2 (g, 1 bar) | H^+ (aq) || Cu^{2+} (aq) | Cu (s) is 0.31 V. The pH of the acidic solution is found to be 3, whereas the concentration of Cu^{2+} is 10^{-x} M.
The value of x is ....
(Given, $E^{\ominus}_{Cu^{2+}/Cu} = 0.34$ V and $\frac{2.303RT}{F} = 0.06$ V)

25 The equation, $k = (6.5 \times 10^{12} s^{-1})\, e^{-26000\, K/T}$ is followed for the decomposition of compound *(a)* The activation energy for the reaction is ... kJ mol^{-1}. [Nearest integer]
(Given, $R = 8.314$ J K^{-1} mol^{-1})

26 Spin only magnetic moment of $[MnBr_6]^{4-}$ is ...... BM. (Round off to the closest integer)

27 For the reaction given below
$CoCl_3 \cdot xNH_3 + AgNO_3\ (aq) \longrightarrow$
If two equivalents of AgCl precipitate out, then the value of x will be ....

28 The number of chiral alcohol(s) with molecular formula $C_4H_{10}O$ is...... .

29 In the given reaction,

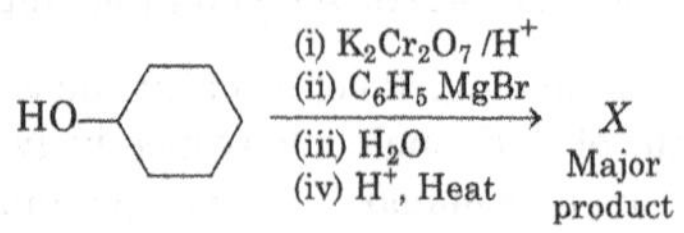

the number of sp^2 hybridised carbon(s) in compound 'X' is ....

30 In the given reaction,

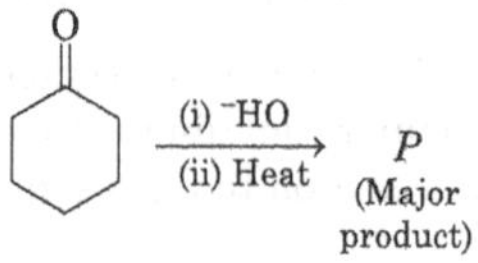

The number of π-electrons present in the product P is ....

Answers

For solutions scan the QR code

24 June 2022, Shift-I

1. *(c)*	2. *(b)*	3. *(b)*	4. *(b)*	5. *(c)*	6. *(b)*	7. *(a)*	8. *(b)*	9. *(a)*	10. *(c)*
11. *(a)*	12. *(b)*	13. *(b)*	14. *(a)*	15. *(c)*	16. *(d)*	17. *(d)*	18. *(b)*	19. *(d)*	20. *(a)*
21. *43*	22. *−747*	23. *54*	24. *5*	25. *154*	26. *0*	27. *7*	28. *63*	29. *3*	30. *2*

24 June 2022, Shift-II

1. *(d)*	2. *(c)*	3. *(b)*	4. *(c)*	5. *(c)*	6. *(b)*	7. *(b)*	8. *(a)*	9. *(a)*	10. *(b)*
11. *(d)*	12. *(d)*	13. *(a)*	14. *(a)*	15. *(c)*	16. *(c)*	17. *(a)*	18. *(c)*	19. *(c)*	20. *(a)*
21. *45*	22. *1221*	23. *1107*	24. *266*	25. *2*	26. *3*	27. *3*	28. *14*	29. *2*	30. *4*

25 June 2022, Shift-I

1. *(c)*	2. *(b)*	3. *(d)*	4. *(c)*	5. *(c)*	6. *(d)*	7. *(c)*	8. *(b)*	9. *(a)*	10. *(a)*
11. *(c)*	12. *(c)*	13. *(b)*	14. *(d)*	15. *(d)*	16. *(c)*	17. *(c)*	18. *(c)*	19. *(c)*	20. *(b)*
21. *5418*	22. *1*	23. *4*	24. *300*	25. *15*	26. *710*	27. *23*	28. *1*	29. *745*	30. *1143*

For solutions scan the QR code

25 June 2022, Shift-II

1. (c)	2. (c)	3. (d)	4. (a)	5. (b)	6. (b)	7. (a)	8. (d)	9. (a)	10. (a)
11. (b)	12. (c)	13. (d)	14. (b)	15. (a)	16. (b)	17. (c)	18. (d)	19. (c)	20. (b)
21. 25	22. 32	23. 3	24. 0	25. 20	26. 2	27. 2	28. 9	29. 2	30. 3

26 June 2022, Shift-I

1. (c)	2. (d)	3. (b)	4. (a)	5. (c)	6. (a)	7. (d)	8. (c)	9. (d)	10. (a)
11. (b)	12. (b)	13. (a)	14. (d)	15. (a)	16. (c)	17. (c)	18. (a)	19. (c)	20. (b)
21. 727	22. 98	23. 476	24. 200	25. 9960	26. 3	27. 3	28. 18	29. 7	30. 9

26 June 2022, Shift-II

1. (a)	2. (c)	3. (d)	4. (a)	5. (a)	6. (d)	7. (b)	8. (a)	9. (a)	10. (c)
11. (d)	12. (b)	13. (a)	14. (c)	15. (c)	16. (a)	17. (c)	18. (c)	19. (b)	20. (d)
21. 143	22. 3	23. 1	24. 38	25. 415	26. 2740	27. 983	28. 4	29. 6	30. 225

27 June 2022, Shift-I

1. (a)	2. (b)	3. (b)	4. (a)	5. (d)	6. (b)	7. (d)	8. (c)	9. (d)	10. (d)
11. (c)	12. (a)	13. (a)	14. (d)	15. (a)	16. (c)	17. (d)	18. (a)	19. (a)	20. (b)
21. 85	22. 22	23. 8	24. 125	25. 14	26. 166	27. 3	28. 6	29. 2	30. (5)

27 June 2022, Shift-II

1. (b)	2. (c)	3. (d)	4. (c)	5. (a)	6. (b)	7. (a)	8. (c)	9. (c)	10. (b)
11. (a)	12. (a)	13. (c)	14. (d)	15. (a)	16. (d)	17. (c)	18. (c)	19. (a)	20. (d)
21. 2	22. 2	23. 2	24. 8630	25. 45	26. 11	27. 51	28. 59	29. 12	30. 40

28 June 2022, Shift-I

1. (d)	2. (c)	3. (a)	4. (b)	5. (b)	6. (b)	7. (d)	8. (c)	9. (b)	10. (d)
11. (c)	12. (b)	13. (c)	14. (a)	15. (c)	16. (c)	17. (b)	18. (b)	19. (b)	20. (d)
21. 13	22. 300	23. 1	24. 0	25. 14	26. 3	27. 6	28. 16	29. 3	30. 34

28 June 2022, Shift-II

1. (d)	2. (c)	3. (a)	4. (b)	5. (c)	6. (b)	7. (d)	8. (d)	9. (c)	10. (c)
11. (a)	12. (d)	13. (c)	14. (c)	15. (c)	16. (c)	17. (c)	18. (c)	19. (b)	20. (c)
21. 4	22. 600	23. 330	24. 16	25. 75	26. 2	27. 1	28. 46	29. 1	30. 30

29 June 2022, Shift-I

1. (c)	2. (d)	3. (c)	4. (d)	5. (d)	6. (a)	7. (d)	8. (c)	9. (a)	10. (a)
11. (b)	12. (a)	13. (d)	14. (b)	15. (a)	16. (c)	17. (c)	18. (b)	19. (c)	20. (c)
21. 152	22. 117	23. 5	24. 127	25. 1	26. 6	27. 0	28. 64	29. 3	30. 1

29 June 2022, Shift-II

1. (b)	2. (b)	3. (b)	4. (a)	5. (d)	6. (b)	7. (a)	8. (c)	9. (a)	10. (c)
11. (d)	12. (d)	13. (d)	14. (c)	15. (c)	16. (d)	17. (a)	18. (c)	19. (a)	20. (b)
21. 29	22. 195	23. 3	24. 7	25. 216	26. 6	27. 5	28. 2	29. 8	30. 4

ONLINE SOLVED PAPERS

JEE Main 2022

(July Attempt)

25 JULY 2022, SHIFT-I

Section A : Objective Type Questions

1 SO_2Cl_2 on reaction with excess of water results into acidic mixture

$$SO_2Cl_2 + 2H_2O \longrightarrow H_2SO_4 + 2HCl$$

16 moles of NaOH is required for the complete neutralisation of the resultant acidic mixture. The number of moles of SO_2Cl_2 used is

(a) 16 (b) 8 (c) 4 (d) 2

2 Which of the following sets of quantum numbers is not allowed?

(a) $n = 3, l = 2, m_l = 0, s = +\frac{1}{2}$

(b) $n = 3, l = 2, m_l = -2, s = +\frac{1}{2}$

(c) $n = 3, l = 3, m_l = -3, s = -\frac{1}{2}$

(d) $n = 3, l = 0, m_l = 0, s = -\frac{1}{2}$

3 The depression in freezing point observed for a formic acid solution of concentration 0.5 mol L^{-1} is 0.0405°C. Density of formic acid is 1.05 g mol L^{-1}. The van't Hoff factor of the formic acid solution is nearly

(Given, for water $K_f = 1.86$ K kg mol^{-1})

(a) 0.8 (b) 1.1 (c) 1.9 (d) 2.4

4 20 mL of 0.1 M NH_4OH is mixed with 40 mL of 0.05 M HCl. The pH of the mixture is nearest to

(Given, $K_b(NH_4OH) = 1 \times 10^{-5}$, $\log 2 = 0.30$, $\log 3 = 0.48$, $\log 5 = 0.69$, $\log 7 = 0.84$, $\log 11 = 1.04$)

(a) 3.2 (b) 4.2 (c) 5.2 (d) 6.2

5. Match List I with List II.

List I	List II
(A) $N_2(g) + 3H_2(g) \longrightarrow 2NH_3(g)$	(I) Cu
(B) $CO(g) + 3H_2(g) \longrightarrow CH_4(g) + H_2O(g)$	(II) $Cu / ZnO—Cr_2O_3$
(C) $CO(g) + H_2(g) \longrightarrow HCHO(g)$	(III) $Fe_xO_y + K_2O + Al_2O_3$
(D) $CO(g) + 2H_2(g) \longrightarrow CH_3OH(g)$	(IV) Ni

Choose the correct answer from the options given below

(a) (A) → (II), (B) → (IV), (C) → (I), (D) → (III)

(b) (A) → (II), (B) → (I), (C) → (IV), (D) → (III)

(c) (A) → (III), (B) → (IV), (C) → (I), (D) → (II)

(d) (A) → (III), (B) → (I), (C) → (IV), (D) → (II)

6 The IUPAC nomenclature of an element with electronic configuration [Rn] $5f^{14}6d^1 7s^2$ is

(a) unnilbium (b) unnilunium

(c) unnilquadium (d) unniltrium

7 The compound(s) that is(are) removed as slag during the extraction of copper is

(A) CaO (B) FeO

(C) Al_2O_3 (D) ZnO

(E) NiO

Choose the correct answer from the options given below

(a) (C), (D) only (b) (A), (B), (E) only

(c) (A), (B) only (d) (B) only

8 The reaction of H_2O_2 with potassium permanganate in acidic medium leads to the formation of mainly

(a) Mn^{2+} (b) Mn^{4+}

(c) Mn^{3+} (d) Mn^{6+}

9 Choose the correct order of density of the alkali metals.

(a) Li < K < Na < Rb < Cs (b) Li < Na < K < Rb < Cs

(c) Cs < Rb < K < Na < Li (d) Li < Na < K < Cs < Rb

10 The geometry around boron in the product '*B*' formed from the following reaction is

$$BF_3 + NaH \xrightarrow{450\text{ K}} A + NaF$$

$$A + NMe_3 \longrightarrow B$$

(a) trigonal planar (b) tetrahedral

(c) pyramidal (d) square planar

11 The interhalogen compound formed from the reaction of bromine with excess of fluorine is a

(a) hypohalite (b) halate

(c) perhalate (d) halite

12 The photochemical smog does not generally contain

(a) NO (b) NO_2

(c) SO_2 (d) HCHO

13 A compound 'A' on reaction with 'X' and 'Y' produces the same major product but different by product 'a' and 'b'. Oxidation of 'a' gives a substance produced by ants.

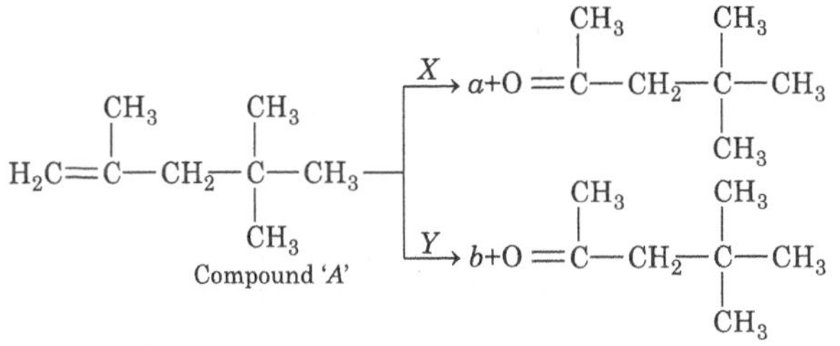

'X' and 'Y' respectively are
(a) $KMnO_4/H^+$ and dil. $KmnO_4$, 273 K
(b) $KMnO_4$(dilute), 273 K and $KMnO_4/H^+$
(c) $KMnO_4/H^+$ and O_3, H_2O/Zn
(d) O_3, H_2O/Zn and $KMnO_4/H^+$

14 Most stable product of the following reaction is

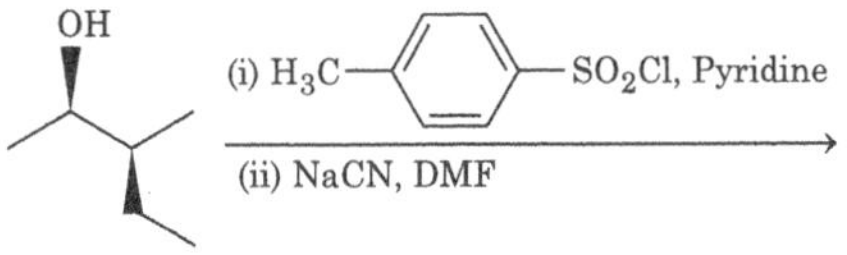

(a) 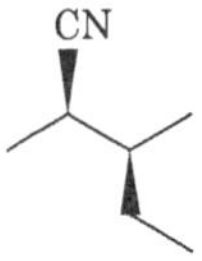
(b) 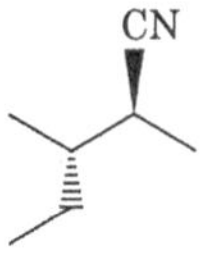
(c) 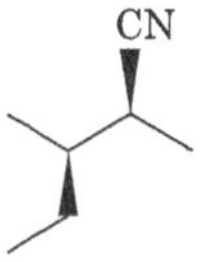
(d) 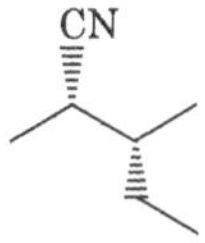

15. Which one of the following reactions does not represent correct combination of substrate and product under the given conditions?

(a) 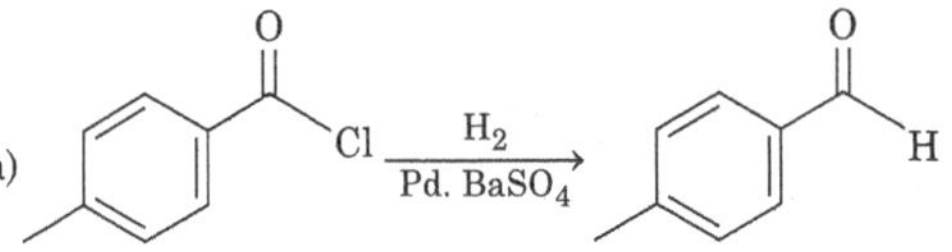
(b) 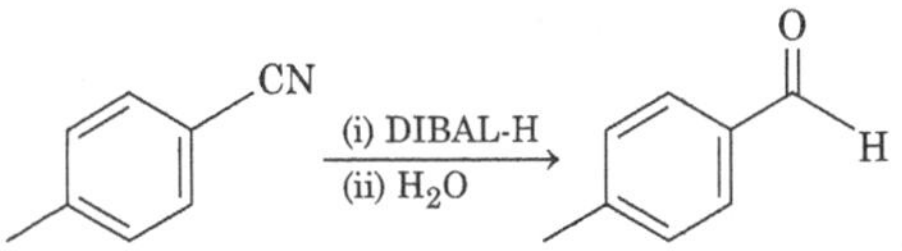
(c) 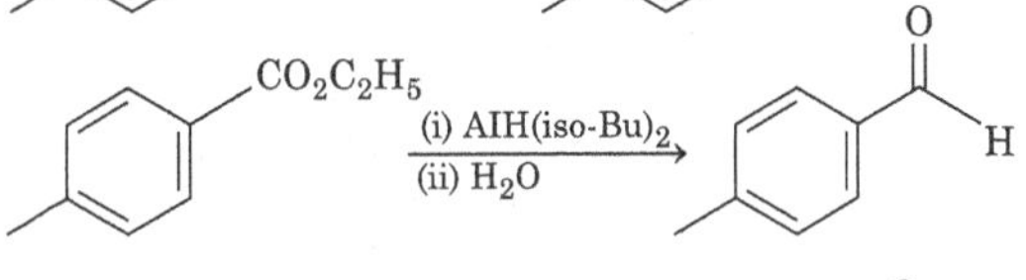
(d) 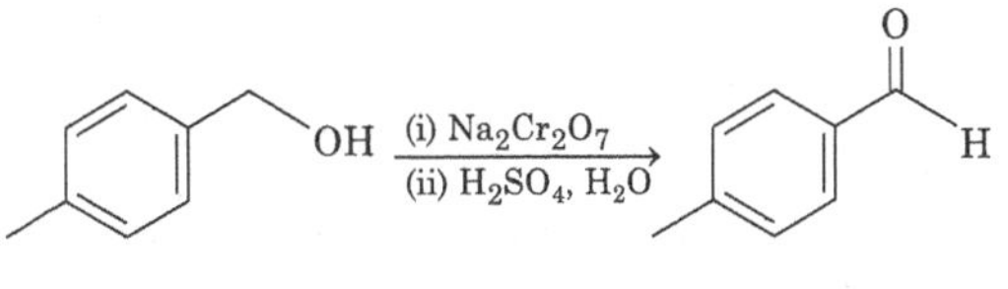

16 An organic compound 'A' on reaction with NH_3 followed by heating gives compound B. Which on further strong heating gives compound $C(C_8H_5NO_2)$. Compound C on sequential reaction with ethanolic KOH, alkyl chloride and hydrolysis with alkali gives a primary amine. The compound A is

(a) 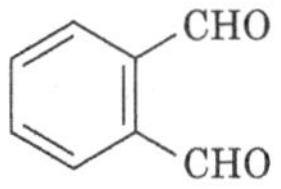
(b) 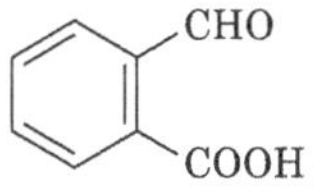
(c) 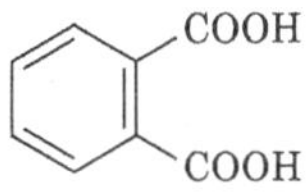
(d) 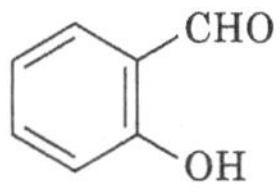

17 Melamine polymer is formed by the condensation of

(a) 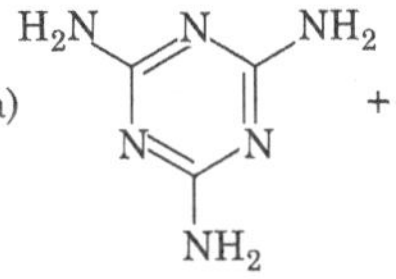 + HCHO
(b) 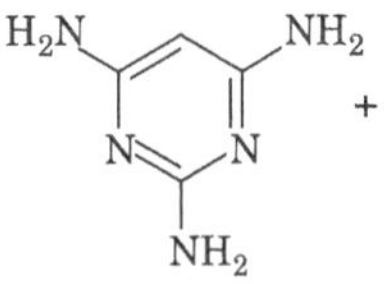 + HCHO
(c) 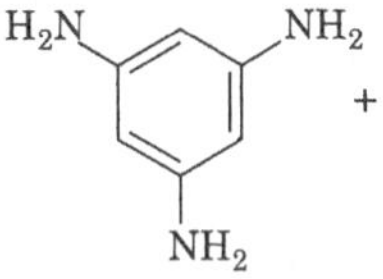 + HCHO
(d) 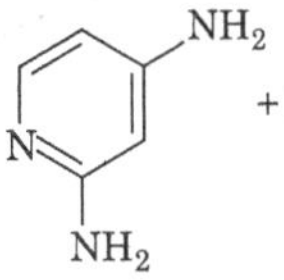 + HCHO

18 During the denaturation of proteins, which of these structures will remain intact?
(a) Primary (b) Secondary
(c) Tertiary (d) Quaternary

19 Drugs used to bind to receptors, inhibiting its natural function and blocking a message are called
(a) agonists (b) antagonists
(c) allosterists (d) anti histamines

20 Given below are two statements.
Statement I On heating with $KHSO_4$, glycerol is dehydrated and acrolein is formed.
Statement II Acrolein has fruity odour and can be used to test glycerol's presence.
Choose the correct option.
(a) Both Statement I and Statement II are correct.
(b) Both Statement I and Statement II are incorrect.
(c) Statement I is correct but Statement II is incorrect.
(d) Statement I is incorrect but Statement II is correct.

Section B : Numerical Type Questions

21 Among the following species $N_2, N_2^+, N_2^-, N_2^{2-}, O_2, O_2^+, O_2^-, O_2^{2-}$, the number of species showing diamagnetism is ....

22 The enthalpy of combustion of propane, graphite and dihydrogen at 298 K are – 2220.0 kJ mol^{-1}, – 393.5 kJ mol^{-1} and –285.8 kJ mol^{-1} respectively. The magnitude of enthalpy of formation of propane (C_3H_8) is .... kJ mol^{-1}. (Nearest integer)

23 The pressure of a moist gas at 27°C is 4 atm. The volume of the container is doubled at the same temperature. The new pressure of the moist gas is .... $\times 10^{-1}$ atm. (Nearest integer)
(Given, the vapour pressure of water at 27°C is 0.4 atm.)

24 The cell potential for $Zn | Zn^{2+}(aq) || Sn^{x+} | Sn$ is 0.801 V at 298 K. The reaction quotient for the above reaction is 10^{-2}. The number of electrons involved in the given electrochemical cell reaction is ....
(Given, $E^\circ_{Zn^{2+}/Zn} = -0.763$ V, $E^\circ_{Sn^{x+}/Sn} = +0.008$ V and $\frac{2.303\,RT}{F} = 0.06$ V)

25 The half-life for the decomposition of gaseous compound A is 240 s when the gaseous pressure was 500 torr initially. When the pressure was 250 torr, the half-life was found to be 4.0 min. The order of the reaction is .... (Nearest integer)

26 Consider the following metal complexes
$[Co(NH_3)_6]^{3+}$; $[CoCl(NH_3)_5]^{2+}$
$[Co(CN)_6]^{3-}$; $[Co(NH_3)_5(H_2O)]^{3+}$
The spin-only magnetic moment value of the complex that absorbs light with shortest wavelength is .... BM. (Nearest integer)

27 Among Co^{3+}, Ti^{2+}, V^{2+} and Cr^{2+} ions, one if used as a reagent cannot liberate H_2 from dilute mineral acid solution, its spin-only magnetic moment in gaseous state is .... BM. (Nearest integer)

28 While estimating the nitrogen present in an organic compound by Kjeldahl's method, the ammonia evolved from 0.25 g of the compound neutralised 2.5 mL of 2 M H_2SO_4. The percentage of nitrogen present in organic compound is ....

29 The number of sp^3 hybridised carbons in an acyclic neutral compound with molecular formula C_4H_5N is ....

30 In the given reaction,

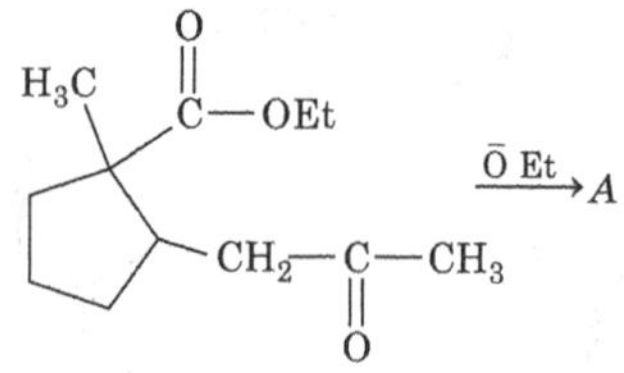

(where, Et is —C_2H_5)
The number of chiral carbon(s) in product A is ....

25 JULY 2022, SHIFT-II

Section A : Objective Type Questions

1 Match List I with List II.

List I (Molecule)			List II (Hybridisation shape)
A.	XeO_3	I	sp^3d; linear
B.	XeF_2	II	sp^3 ; pyramidal
C.	$XeOF_4$	III	sp^3d^3 distorted octahedral
D.	XeF_6	IV	sp^3d^2; square pyramidal

Choose the correct answer from the options given below.
(a) A-II, B-I, C-IV, D-III (b) A-II, B-IV, C-III, D-I
(c) A-IV, B-II, C-III, D-I (d) A-IV, B-II, C-I, D-III

2 Two solutions A and B are prepared by dissolving 1 g of non-volatile solutes X and Y, respectively in 1 kg of water. The ratio of depression in freezing points for A and B is found to be 1 : 4. The ratio of molar masses of X and Y is
(a) 1 : 4 (b) 1 : 0.25
(c) 1 : 0.20 (d) 1 : 5

3 K_{a_1}, K_{a_2} and K_{a_3} are the respective ionisation constants for the following reactions A, B, and C.
A. $H_2C_2O_4 \rightleftharpoons H^+ + HC_2O_4^-$
B. $HC_2O_4^- \rightleftharpoons H^+ + C_2O_4^{2-}$
C. $H_2C_2O_4 \rightleftharpoons 2H^+ + C_2O_4^{2-}$
The relationship between K_{a_1}, K_{a_2} and K_{a_3} is given as
(a) $K_{a_3} = K_{a_1} + K_{a_2}$ (b) $K_{a_3} = K_{a_1} - K_{a_2}$
(c) $K_{a_3} = K_{a_1} / K_{a_2}$ (d) $K_{a_3} = K_{a_1} \times K_{a_2}$

4 The molar conductivity of a conductivity cell filled with 10 moles of 20 mL NaCl solution is Λ_{M_1} and that of 20 moles another identical cell having 80 mL NaCl solution is Λ_{M_2}. The conductivities exhibited by these two cells are same. The relationship between Λ_{M_2} and Λ_{M_1} is
(a) $\Lambda_{M_2} = 2\Lambda_{M_1}$ (b) $\Lambda_{M_2} = \Lambda_{M_1}/2$
(c) $\Lambda_{M_2} = \Lambda_{M_1}$ (d) $\Lambda_{M_2} = 4\Lambda_{M_1}$

5 For micelle formation, which of the following statements are correct?
A. Micelle formation is an exothermic process.
B. Micelle formation is an endothermic process.

C. The entropy change is positive.
D. The entropy change is negative.
(a) A and D only (b) A and C only
(c) B and C only (d) B and D only

6 The first ionisation enthalpies of Be, B, N and O follow the order
(a) O < N < B < Be (b) Be < B < N < O
(c) B < Be < N < O (d) B < Be < O < N

7 Given below are two statements.
Statement I Pig iron is obtained by heating cast iron with scrap iron.
Statement II Pig iron has a relatively lower carbon content than that of cast iron.
In the light of the above statements, choose the correct answer from the options given below.
(a) Both Statement I and Statement II are correct.
(b) Both Statement I and Statement II are not correct.
(c) Statement I is correct but Statement II is not correct.
(d) Statement I is not correct but Statement II is correct.

8 High purity (>99.95%) dihydrogen is obtained by
(a) reaction of zinc with aqueous alkali
(b) electrolysis of acidified water using platinum electrodes
(c) electrolysis of warm aqueous barium hydroxide solution between nickel electrodes
(d) reaction of zinc with dilute acid

9 The correct order of density is
(a) Be > Mg > Ca > Sr (b) Sr > Ca > Mg > Be
(c) Sr > Be > Mg > Ca (d) Be > Sr > Mg > Ca

10 The total number of acidic oxides from the following list is
$NO, N_2O, B_2O_3, N_2O_5, CO, SO_3, P_4O_{10}$
(a) 3 (b) 4 (c) 5 (d) 6

11 The correct order of energy of absorption for the following metal complexes is
A. $[Ni(en)_3]^{2+}$ B. $[Ni(NH_3)_6]^{2+}$
C. $[Ni(H_2O)_6]^{2+}$
(a) C < B < A (b) B < C < A
(c) C < A < B (d) A < C < B

12 Match List I with List II.

	List I		List II
A	Sulphate	I	Pesticide
B	Fluoride	II	Bending of bones
C	Nicotine	III	Laxative effect
D	Sodium arsinite	IV	Herbicide

Choose the correct answer from the options given below.
(a) A-II, B-III, C-IV, D-I (b) A-IV, B-III, C-II, D-I
(c) A-III, B-II, C-I, D-IV (d) A-III, B-II, C-IV, D-I

13 Major product of the following reaction is

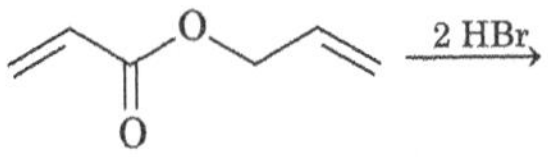

(a) 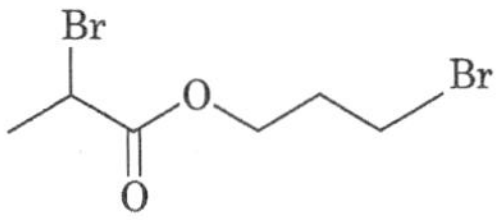

(b) 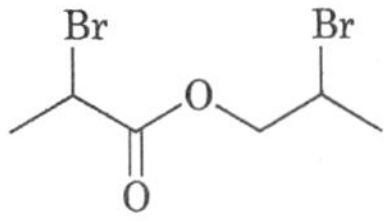

(c) 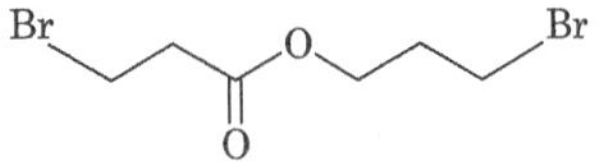

(d) 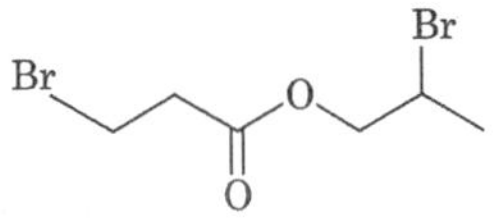

14 What is the major product of the following reaction?

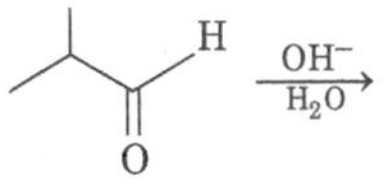

(a) 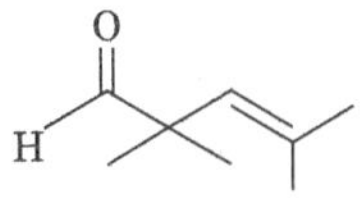
(b) 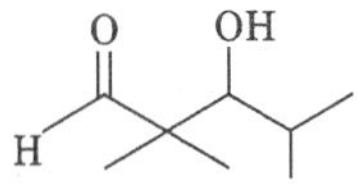

(c) 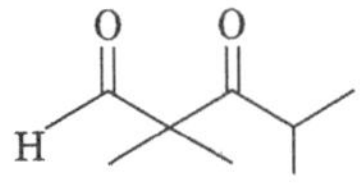
(d) 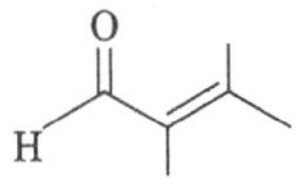

15 Arrange the following in decreasing acidic strength.

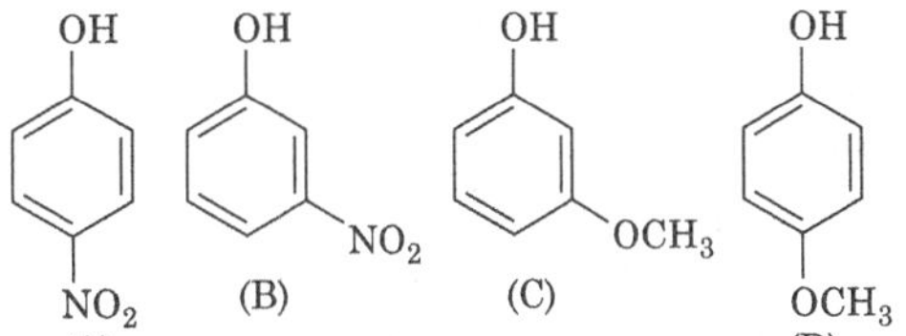

(A) (B) (C) (D)

(a) A > B > C > D (b) B > A > C > D
(c) D > C > A > B (d) D > C > B > A

16 $CH_3-CH_2\equiv CN \xrightarrow[\text{Ether}]{CH_3MgBr} A \xrightarrow{H_3O^+} B \xrightarrow[\text{HCl}]{\text{Zn - Hg}} C$

The correct structure of C is
(a) $CH_3-CH_2-CH_2-CH_3$ (b) $CH_3-CH_2-\overset{O}{\overset{\|}{C}}-CH_3$
(c) $CH_3-CH_2-\overset{OH}{\overset{|}{CH}}-CH_3$ (d) $CH_3-CH_2-CH=CH_2$

17 Match List I with List II.

List I (Polymer)		List II (Used for items)	
A.	Nylon 6, 6	I	Buckets
B.	Low density polythene	II.	Non-stick utensils
C.	High density polythene	III.	Bristles of brushes
D.	Teflon	IV.	Toys

Choose the correct answer from the options given below.

(a) A-III, B-I, C-IV, D-II (b) A-III, B-IV, C-I, D-II
(c) A-II, B-I, C-IV, D-III (d) A-II, B-IV, C-I, D-III

18 Glycosidic linkage between C_1 of α-glucose and C_2 of β-fructose is found in

(a) maltose (b) sucrose
(c) lactose (d) amylose

19 Some drugs bind to a site other than the active site of an enzyme. This site is known as

(a) non-active site (b) allosteric site
(c) competitive site (d) therapeutic site

20 In base *vs* acid titration, at the end point methyl orange is present as

(a) quinonoid form (b) heterocyclic form
(c) phenolic form (d) benzenoid form

Section B : Numerical Type Questions

21 56.0 L of nitrogen gas is mixed with excess of hydrogen gas and it is found that 20 L of ammonia gas is produced. The volume of unused nitrogen gas is found to be .... L.

22 A sealed flask with a capacity of 2 dm^3 contains 11 g of propane gas. The flask is so weak that it will burst if the pressure becomes 2 MPa. The minimum temperature at which the flask will burst is .... °C. [Nearest integer]

(Given, $R = 8.3 JK^{-1} mol^{-1}$, Atomic masses of C and H are 12 u and 1 u, respectively.) (Assume that, propane behaves as an ideal gas.)

23 When the excited electron of a H-atom from $n = 5$ drops to the ground state, the maximum number of emission lines observed are ....

24 While performing a thermodynamics experiment, a student made the following observations.

$$HCl + NaOH \longrightarrow NaCl + H_2O \,;\, \Delta H = -57.3 \text{ kJ mol}^{-1}$$

$$CH_3COOH + NaOH \longrightarrow CH_3COONa + H_2O; \quad \Delta H = -55.3 \text{ kJ mol}^{-1}$$

The enthalpy of inoisation of CH_3COOH as calculated by the student is .... kJ mol^{-1}. [nearest integer]

25 For the decomposition of azomethane, $CH_3N_2CH_3(g) \longrightarrow CH_3CH_3(g) + N_2(g)$, a first order reaction, the variation in partial pressure with time at 600 K is given as

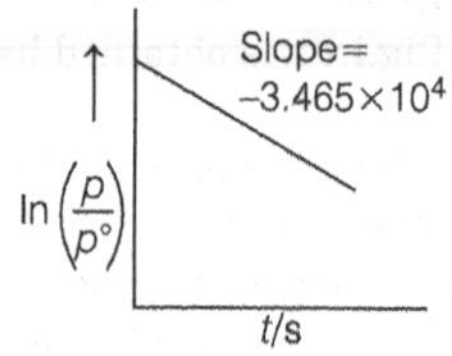

The half-life of the reaction is ...... $\times 10^{-5}$ s.

26 The sum of number of lone pairs of electrons present on the central atoms of XeO_3, $XeOF_4$ and XeF_6, is ....

27 The spin - only magnetic moment value of M^{3+} ion (in gaseous state) from the pairs Cr^{3+}/Cr^{2+}, Mn^{3+}/Mn^{2+}, Fe^{3+}/Fe^{2+} and Co^{3+}/Co^{2+} that has negative standard electrode potential, is .... BM. [nearest integer]

28 A sample of 4.5 mg of an unknown monohydric alcohol, *R*–OH was added to methylmagnesium iodide. A gas is evolved and is collected and its volume measured to be 3.1 mL. The molecular weight of the unknown alcohol is .... g/mol. [nearest integer]

29 The separation of two coloured substances was done by paper chromatography. The distances travelled by solvent front, substance *A* and substance *B* from the base line are 3.25 cm, 2.08 cm and 1.05 cm, respectively. The ratio of R_f values of *A* to *B* is......... .

30 The total number of monobromo derivatives formed by the alkanes with molecular formula C_5H_{12} is (excluding stereo isomers)........... .

26 JULY 2022, SHIFT-I

Section A : Objective Type Questions

1 Match List I with List II.

List I (Compound)		List II (Shape)	
(A)	BrF_5	(I)	bent
(B)	$[CrF_6]^{3-}$	(II)	square pyramidal
(C)	O_3	(III)	trigonal bipyramidal
(D)	PCl_5	(IV)	octahedral

Choose the correct answer from the options given below :

	A	B	C	D		A	B	C	D
(a)	I	II	III	IV	(b)	IV	III	II	I
(c)	II	IV	I	III	(d)	III	IV	II	I

2 Match List I with List II.

	List I (Processes / Reactions)		List II (Catalyst)
A.	$2SO_2(g) + O_2(g) \longrightarrow 2SO_3(g)$	I.	Fe(*s*)
B.	$4NH_3(g) + 5O_2(g) \longrightarrow 4NO(g) + 6H_2O(g)$	II.	Pt(*s*)—Rh(*s*)
C.	$N_2(g) + 3H_2(g) \longrightarrow 2NH_3(g)$	III.	V_2O_5
D.	Vegetable oil (*l*) + $H_2 \longrightarrow$ Vegetable ghee (*s*)	IV.	Ni(*s*)

Choose the correct answer from the options given below :

	A	B	C	D		A	B	C	D
(a)	III	I	II	IV	(b)	III	II	I	IV
(c)	IV	III	I	II	(d)	IV	II	III	I

3 Given two statements below.

Statement I In Cl_2 molecule the covalent radius is double of the atomic radius of chlorine.

Statement II Radius of anionic species is always greater than their parent atomic radius.

Choose the most appropriate answer from options given below

(a) Both Statement I and Statement II are correct.
(b) Both Statement I and Statement II are incorrect.
(c) Statement I is correct but Statement II is incorrect.
(d) Statement I is incorrect but Statement II is correct.

4 Refining using liquation method is the most suitable for metals with

(a) low melting point
(b) high boiling point
(c) high electrical conductivity
(d) less tendency to be soluble in melts than impurities

5 Which of the following can be used to prevent the decomposition of H_2O_2 ?

(a) Urea (b) Formaldehyde
(c) Formic acid (d) Ethanol

6 Reaction of $BeCl_2$ with $LiAlH_4$ gives

(A) $AlCl_3$ (B) BeH_2 (C) LiH (D) LiCl
(E) $BeAlH_4$

Choose the correct answer from options given below

(a) (A), (D) and (E) (b) (A), (B) and (D)
(c) (D) and (E) (d) (B), (C) and (D)

7 Borazine, also known as inorganic benzene, can be prepared by the reaction of 3-equivalents of "X" with 6-equivalents of "Y". "X" and "Y", respectively are

(a) $B(OH)_3$ and NH_3 (b) B_2H_6 and NH_3
(c) B_2H_6 and NH_3 (d) NH_3 and B_2O_3

8 Which of the given reactions is not an example of disproportionation reaction ?

(a) $2H_2O_2 \longrightarrow 2H_2O + O_2$
(b) $2NO_2 + H_2O \longrightarrow HNO_3 + HNO_2$
(c) $MnO_4^- + 4H^+ + 3e^- \longrightarrow MnO_2 + 2H_2O$
(d) $3MnO_4^{2-} + 4H^+ \longrightarrow 2MnO_4^- + MnO_2 + 2H_2O$

9 The dark purple colour of $KMnO_4$ disappears in the titration with oxalic acid in acidic medium. The overall change in the oxidation number of manganese in the reaction is

(a) 5 (b) 1 (c) 7 (d) 2

10 $\dot{C}l + CH_4 \longrightarrow A + B$

A and *B* in the above atmospheric reaction step are

(a) C_2H_6 and Cl_2 (b) $\dot{C}HCl_2$ and H_2
(c) $\dot{C}H_3$ and HCl (d) C_2H_6 and HCl

11 Which technique among the following, is most appropriate in separation of a mixture of 100 mg of *p*-nitrophenol and picric acid?

(a) Steam distillation
(b) 2-5 ft long column of silica gel
(c) Sublimation
(d) Preparative TLC (Thin Layer Chromatography)

12 The difference in the reaction of phenol with bromine in chloroform and bromine in water medium is due to

(a) hyperconjugation in substrate
(b) polarity of solvent
(c) free radical formation
(d) electromeric effect of substrate

13 Which of the following compounds is not aromatic?

(a) 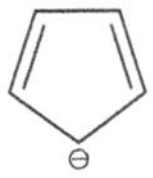
(b) 
(c) 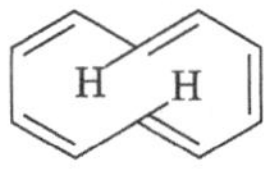
(d) 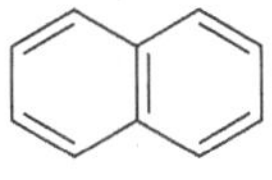

14 The products formed in the following reaction, *A* and *B* are

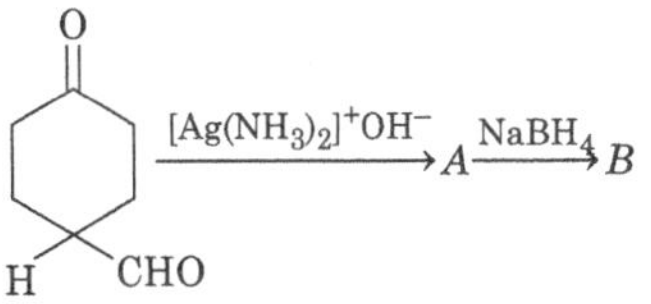

(a) *A* = 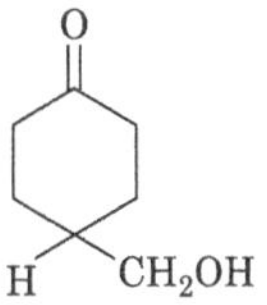 *B* = 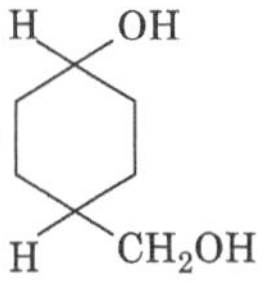

(b) *A* = 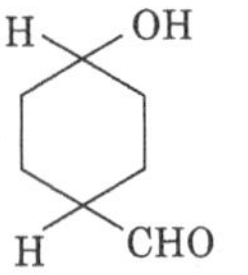 *B* = 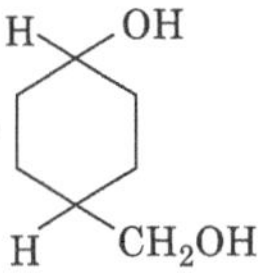

(c) *A* = 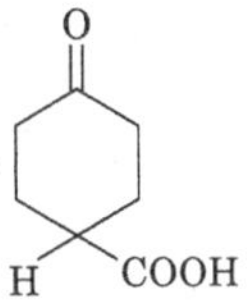 *B* = 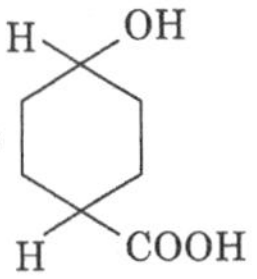

(d) *A* = 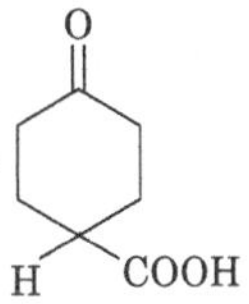 *B* = 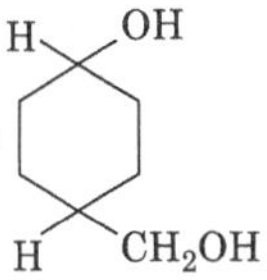

15 Which reactant will give the following alcohol on reaction with one mole of phenyl magnesium bromide (PhMgBr) followed by acidic hydrolysis?

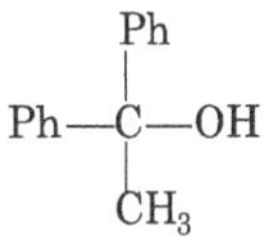

(a) $CH_3—C\equiv N$ (b) $Ph—C\equiv N$
(c) $CH_3—C(=O)—O—Ph$ (d) $Ph—C(=O)—CH_3$

16 The major product of the following reaction is

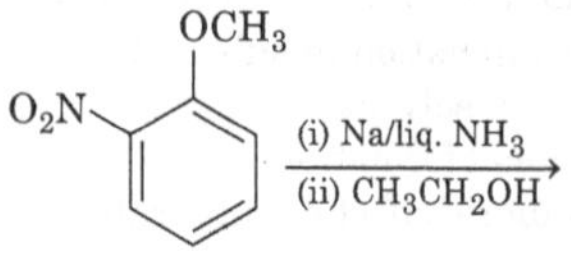

(a) 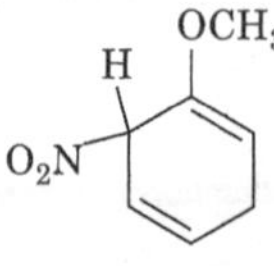
(b) 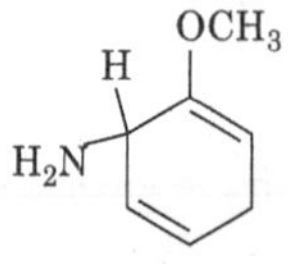
(c) 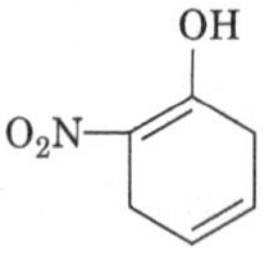
(d) 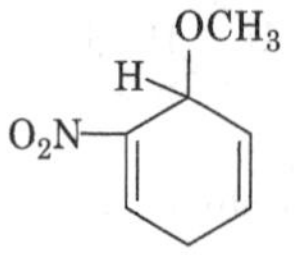

17 The correct stability order of the following diazonium salt is

(A) 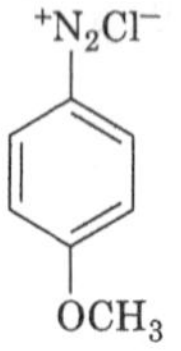
(B) 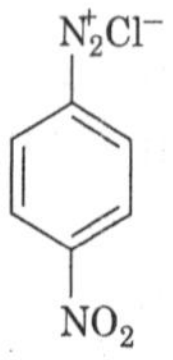
(C) 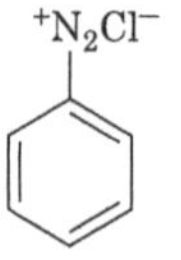
(D) 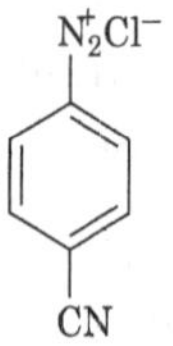

(a) (A) > (B) > (C) > (D)
(b) (A) > (C) > (D) > (B)
(c) (C) > (A) > (D) > (B)
(d) (C) > (D) > (B) > (A)

18 Stearic acid and polyethylene glycol react to form which one of the following soap/s detergents?

(a) Cationic detergent
(b) Soap
(c) Anionic detergent
(d) Non-ionic detergent

19 Which one of the following is a reducing sugar?

(a) 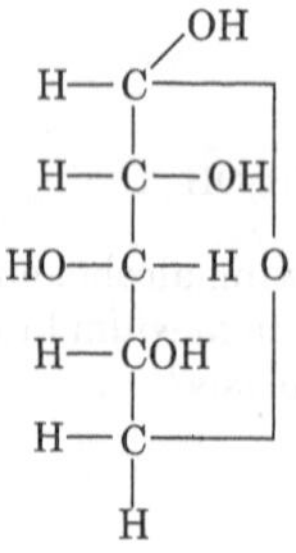
(b) 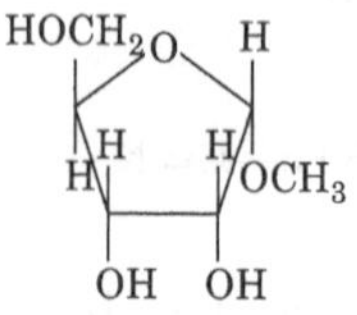
(c) 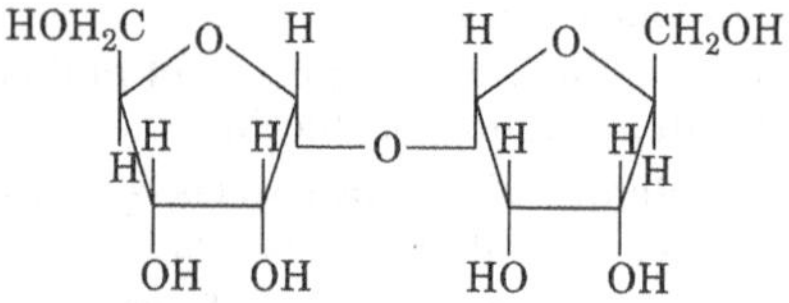
(d) 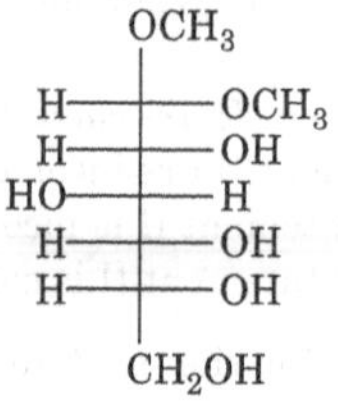

20 Given below are two statements, one is labelled as Assertion (A) and the other is labelled as Reason (R).

Assertion (A) Experimental reaction of CH_3Cl with aniline and anhydrous $AlCl_3$ does not give *o* and *p*-methylaniline.

Reason (R) The —NH_2 group of aniline becomes deactivating because of salt formation with anhydrous $AlCl_3$ and hence yields *m*-methyl aniline as the product.

In the light of the above statements, choose the most appropriate answer from the options given below

(a) Both (A) and (R) are true and (R) is the correct explanation of (A).
(b) Both (A) and (R) are true but (R) is not the correct explanation of (A).
(c) (A) is true, but (R) is false.
(d) (A) is false, but (R) is true.

Section B : Numerical Type Questions

21 Chlorophyll extracted from the crushed green leaves was dissolved in water to make 2 L solution of Mg of concentration 48 ppm. The number of atoms of Mg in this solution is $x \times 10^{20}$ atoms. The value of x is .... (Nearest integer)

(Given, atomic mass of Mg is 24 g mol^{-1}; $N_A = 6.02 \times 10^{23} mol^{-1}$)

22 A mixture of hydrogen and oxygen contains 40% hydrogen by mass when the pressure is 22 bar. The partial pressure of hydrogen is .... bar. (Nearest integer)

23 The wavelength of an electron and a neutron will become equal when the velocity of the electron is x times the velocity of neutron. The value of x is .... (Nearest integer)

(Mass of electron is 9.1×10^{-31} kg and mass of neutron is 1.6×10^{-27} kg)

24 2.4 g coal is burnt in a bomb calorimeter in excess of oxygen at 298 K and 1 atm pressure. The temperature of the calorimeter rises from 298 K to 300 K. The enthalpy change during the combustion of coal is $-x$ kJ mol^{-1}. The value of x is ....

(Nearest integer)

(Given, heat capacity of bomb calorimeter is 20.0 kJ K^{-1}. Assume coal to be pure carbon.)

25 When 800 mL of 0.5 M nitric acid is heated in a beaker, its volume is reduced to half and 11.5 g of nitric acid is evaporated. The molarity of the remaining nitric acid solution is $x \times 10^{-2}$ M. (Nearest integer)
(Molar mass of nitric acid is 63 g mol^{-1})

26 At 298 K, the equilibrium constant is 2×10^{15} for the reaction

$$Cu(s) + 2Ag^+(aq) \rightleftharpoons Cu^{2+}(aq) + 2Ag(s)$$

The equilibrium constant for the reaction

$\frac{1}{2}Cu^{2+}(aq) + Ag(s) \rightleftharpoons \frac{1}{2}Cu(s) + Ag^+(aq)$ is $x \times 10^{-8}$. The value of x is .... (nearest integer)

27 The amount of charge in F (Faraday) required to obtain one mole of iron from Fe_3O_4 is .... (Nearest integer)

28 For a reaction, $A \longrightarrow 2B + C$ the half-lifes are 100 s and 50 s when the concentration of reactant A is 0.5 and 1.0 mol L^{-1} respectively. The order of the reaction is .... (Nearest integer)

29 The difference between spin only magnetic moment value of $[Co(H_2O)_6]Cl_2$ and $[Cr(H_2O)_6]Cl_3$ is ....

30 In the presence of sunlight, benzene reacts with Cl_2 to give product X. The number of hydrogens in X is ....

26 JULY 2022, SHIFT-II

Section A : Objective Type Questions

1 Haemoglobin contains 0.34% of iron by mass. The number of Fe-atoms in 3.3 g of haemoglobin is (Given, atomic mass of Fe is 56 u, $N_A = 6.022 \times 10^{23}$ mol^{-1})

(a) 1.21×10^5 (b) 12.0×10^{16}
(c) 1.21×10^{20} (d) 3.4×10^{22}

2 Arrange the following in increasing order of their covalent character.

A. CaF_2 B. $CaCl_2$ C. $CaBr_2$ D. CaI_2

Choose the correct answer from the option given below.

(a) B < A < C < D (b) A < B < C < D
(c) A < B < D < C (d) A < C < B < D

3 Class XII students were asked to prepare one litre of buffer solution of pH 8.26 by their chemistry teacher. The amount of ammonium chloride to be dissolved by the student in 0.2 M ammonia solution to make one litre of the buffer is
(Given, pK_b $(NH_3) = 4.74$, Molar mass of $NH_3 = 17$ $gmol^{-1}$, Molar mass of $NH_4Cl = 53.5$ g mol^{-1})

(a) 53.5 g (b) 72.3 g (c) 107.0 g (d) 126.0 g

4 At 30°C, the half-life for the decomposition of AB_2 is 200 s and is independent of the initial concentration of AB_2. The time required for 80% of the AB_2 to decompose is
(Given, log 2 = 0.30, log 3 = 0.48)

(a) 200 s (b) 323 s (c) 467 s (d) 532 s

5 Given below are two statements.
One is labelled as Assertion A and other is labelled as Reason R.

Assertion (A) Finest gold is red in colour, as the size of the particles increases, it appears purple then blue and finally gold.

Reason (R) The colour of the colloidal solution depends on the wavelength of light scattered by the dispersed particles.

In the light of the above statements, choose the most appropriate answer from the options given below.

(a) Both A and R are true and R is the correct explanation of A.
(b) Both A and R are true but R is not the correct explanation of A.
(c) A is true but R is false.
(d) A is false but R is true.

6 The metal that has very low melting point and its periodic position is closer to a metalloid is

(a) Al (b) Ga (c) Se (d) In

7 The metal that is not extracted from its sulphide ore is

(a) aluminium (b) iron
(c) lead (d) zinc

8 The products obtained from a reaction of hydrogen peroxide and acidified potassium permanganate are

(a) Mn^{4+}, H_2O only (b) Mn^{2+}, H_2O only
(c) Mn^{4+}, H_2O, O_2 only (d) Mn^{2+}, H_2O, O_2 only

9 Given below are two statements.
One is labelled as Assertion A and the other is labelled as Reason R.

Assertion (A) LiF is sparingly soluble in water.

Reason (R) The ionic radius of Li^+ ion is smallest among its group members, hence has least hydration enthalpy.

In the light of the above statements, choose the most appropriate answer from the options given below.

(a) Both A and R are true and R is the correct explanation of A.
(b) Both A and R are true but R is not the correct explanation of A.
(c) A is true but R is false.
(d) A is false but R is true.

10 Given below are two statements.
One is labelled as Assertion A and the other is labelled as Reason R.

Assertion (A) Boric acid is a weak acid.

Reason (R) Boric acid is not able to release H^+ ion on its own. It receives OH^- ion from water and releases H^+ ion.

In the light of the above statements, choose the most appropriate answer from the options given below.
(a) Both A and R are correct and R is the correct explanation of A.
(b) Both A and R are correct but R is not the correct explanation of A.
(c) A is correct but R is not correct.
(d) A is not correct but R is correct.

11 The metal complex that is diamagnetic is (Atomic number Fe, 26; Cu, 29)
(a) $K_3[Cu(CN)_4]$ (b) $K_2[Cu(CN)_4]$
(c) $K_3[Fe(CN)_6]$ (d) $K_4[FeCl_6]$

12 Match List I with List II.

List I (Pollutant)		List II (Source)	
A.	Microorganisms	I.	Strip mining
B.	Plant nutrients	II.	Domestic sewage
(c)	Toxic heavy metals	III.	Chemical fertiliser
D.	Sediment	IV.	Chemical factory

Choose the correct answer from the options given below.
(a) A-II, B-III, C-IV, D-I (b) A-II, B-I, C-IV, D-III
(c) A-I, B-IV, C-II, D-III (d) A-I, B-IV, C-III, D-II

13 The correct decreasing order of priority of functional groups in naming an organic compound as per IUPAC system of nomenclature is
(a) $—COOH > —CONH_2 > —COCl > —CHO$
(b) $—SO_3H > —COCl > —CONH_2 > —CN$
(c) $—COOR > —COCl > —NH_2 > \rangle C{=}O$
(d) $—COOH > —COOR > —CONH_2 > —COCl$

14 Which of the following is not an example of benzenoid compound?

(a) 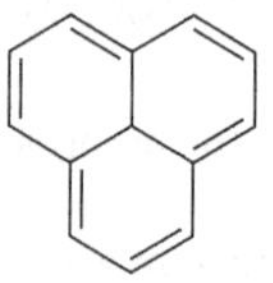
(b) 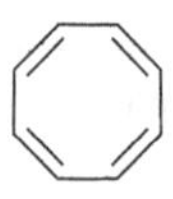
(c) 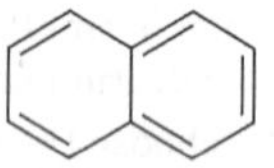
(d) 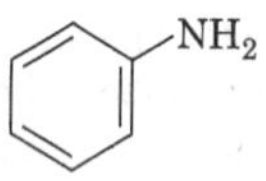

15 Hydrolysis of which compound will give carbolic acid?
(a) Cumene
(b) Benzene diazonium chloride
(c) Benzal chloride
(d) Ethylene glycol ketal

16 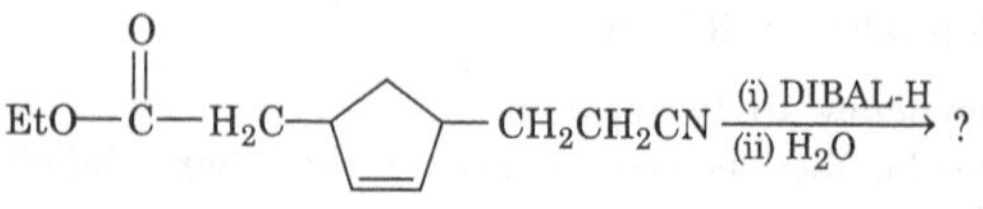

[Et is $—C_2H_5$]
Consider the above reaction and predict the major product.

(a) 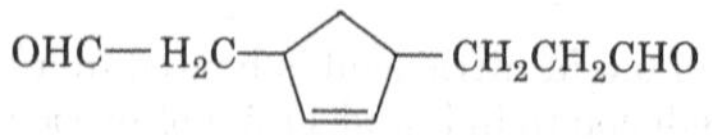
(b) 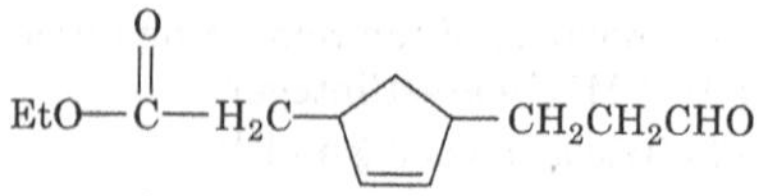
(c) 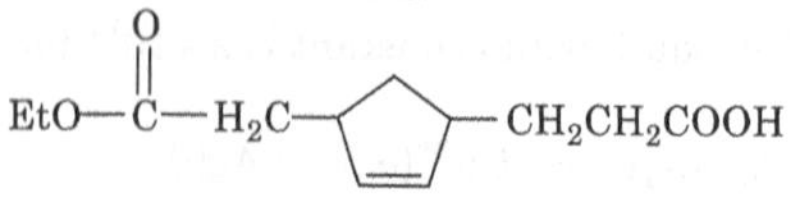
(d) 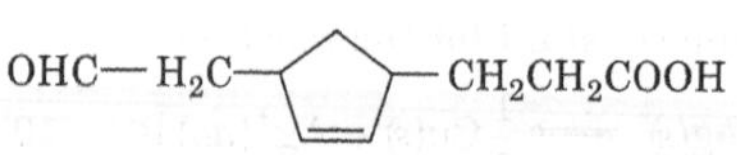

17 The correct sequential order of the reagents for the given reaction is

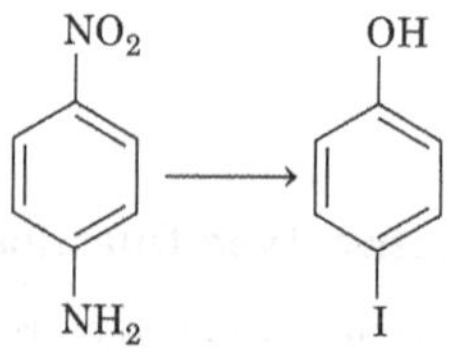

(a) HNO_2, Fe/H^+, HNO_2, KI, H_2O/H^+
(b) HNO_2, KI, Fe/H^+, HNO_2, H_2O/warm
(c) HNO_2, KI, HNO_2, Fe/H^+, H_2O/H^+
(d) HNO_2, Fe/H^+, KI, HNO_2, H_2O/warm

18 Vulcanisation of rubber is carried out by heating a mixture of
(a) isoprene and styrene (b) neoprene and sulphur
(c) isoprene and sulphur (d) neoprene and styrene

19 Animal starch is the other name of
(a) amylose (b) maltose
(c) glycogen (d) amylopectin

20 Given below are two statements.
One is labelled as Assertion A and the other is labelled as Reason R.

Assertion (A) Phenolphthalein is a pH dependent indicator, remains colourless in acidic solution and gives pink colour in basic medium.

Reason (R) Phenolphthalein is a weak aci(d) It doesn't dissociate in basic medium.
In the light of the above statements, choose the most appropriate answer from the options given below.
(a) Both A and R are true and R is the correct explanation of A.
(b) Both A and R are true but R is not the correct explanation of A.
(c) A is true but R is false.
(d) A is false but R is true.

Section B : Numerical Type Questions

21 A 10 g mixture of hydrogen and helium is contained in a vessel of capacity $0.0125\ m^3$ at 6 bar and 27°C. The mass of helium in the mixture is .... g. (nearest integer)
Given : $R = 8.3\ JK^{-1}\ mol^{-1}$
(Atomic masses of H and He are 1 u and 4 u, respectively)

22 Consider an imaginary ion $^{48}_{22}X^{3-}$. The nucleus contains 'a'% more neutrons than the number of electrons in the ion. The value of 'a' is .... [nearest integer]

23 For the reaction,

$$H_2F_2(g) \longrightarrow H_2(g) + F_2(g)$$

$\Delta U = -59.6 \text{ kJ mol}^{-1}$ at 27° C.

The enthalpy change for the above reaction is (–) .... kJ mol^{-1} [nearest integer]

Given : $R = 8.314 \text{ JK}^{-1}\text{mol}^{-1}$

24 The elevation in boiling point for 1 molal solution of non-volatile solute A is 3 K. The depression in freezing point for 2 molal solution of A in the same solvent is 6 K. The ratio of K_b and K_f i.e., K_b/K_f is 1 : x. The value of x is......... . [nearest integer]

25 20 mL of 0.02 M hypo solution is used for the titration of 10 mL of copper sulphate solution, in the presence of excess of KI using starch as an indicator. The molarity of Cu^{2+} is found to be .... $\times 10^{-2}$ M. [nearest integer]

Given : $2Cu^{2+} + 4I^- \longrightarrow Cu_2I_2 + I_2$

$$I_2 + 2S_2O_3^{2-} \longrightarrow 2I^- + S_4O_6^{2-}$$

26 The number of non-ionisable protons present in the product B obtained from the following reactions is ....

$C_2H_5OH + PCl_3 \longrightarrow C_2H_5Cl + A \qquad A + PCl_3 \longrightarrow B$

27 The spin-only magnetic moment value of the compound with strongest oxidising ability among MnF_4, MnF_3 and MnF_2 is .... BM. [nearest integer]

28 Total number of isomers (including stereoisomers) obtained on monochlorination of methylcyclohexane is ....

29 A 100 mL solution of CH_3CH_2MgBr on treatment with methanol produces 2.24 mL of a gas at STP. The weight of gas produced is .... mg. [nearest integer]

30 How many of the following drugs is/are example(s) of broad-spectrum antibiotics? Ofloxacin, Penicillin-G, Terpineol, Salvarsan

27 JULY 2022, SHIFT-I

Section A : Objective Type Questions

1 250 g solution of D-glucose in water contains 10.8% of carbon by weight. The molality of the solution is nearest to (Given: Atomic weights are, H, 1u; C, 12u; O, 16u)

(a) 1.03 (b) 2.06 (c) 3.09 (d) 5.40

2 Given below are two statements.

Statement I O_2, Cu^{2+} and Fe^{3+} are weakly attracted by magnetic field and are magnetised in the same direction as magnetic field.

Statement II NaCl and H_2O are weakly magnetised in opposite direction to magnetic field.

In the light of the above statements, choose the most appropriate answer from the options given below.

(a) Both Statement I and Statement II are correct.
(b) Both Statement I and Statement II are incorrect.
(c) Statement I is correct but Statement II is incorrect.
(d) Statement I is incorrect but Statement II is correct.

3 Given below are two statements. One is labelled as Assertion (A) and the other is labelled as Reason (R).

Assertion (A) Energy of 2s-orbital of hydrogen atom is greater than that of 2s-orbital of lithium.

Reason (R) Energies of the orbitals in the same subshell decrease with increase in the atomic number.

In the light of the above statements, choose the correct answer from the options given below.

(a) Both A and R are true and R is the correct explanation of A.
(b) Both A and R are true but R is not the correct explanation of A.
(c) A is true but R is false.
(d) A is false but R is true.

4 Given below are two statements. One is labelled as Assertion (A) and the other is labelled as Reason (R).

Assertion (A) Activated charcoal adsorbs SO_2 more efficiently than CH_4.

Reason (R) Gases with lower critical temperatures are readily adsorbed by activated charcoal.

In the light of the above statements, choose the correct answer from the options given below.

(a) Both A and R are correct and R is the correct explanation of A.
(b) Both A and R are correct but R is NOT the correct explanation of A.
(c) A is correct but R is not correct.
(d) A is not correct but R is correct.

5 Boiling point of a 2% aqueous solution of a non-volatile solute A is equal to the boiling point of 8% aqueous solution of a non-volatile solute B. The relation between molecular weights of A and B is

(a) $M_A = 4M_B$
(b) $M_B = 4M_A$
(c) $M_A = 8M_B$
(d) $M_B = 8M_A$

6 The incorrect statement is

(a) the first ionisation enthalpy of K is less than that of Na and Li.
(b) Xe does not have the lowest first ionisation enthalpy in its group.
(c) the first ionisation enthalpy of element with atomic number 37 is lower than that of the element with atomic number 38.
(d) the first ionisation enthalpy of Ga is higher than that of the d-block element with atomic number 30.

7 Which of the following methods are not used to refine any metal?

A. Liquation B. Calcination
C. Electrolysis D. Leaching
E. Distillation

Choose the correct answer from the options given below
(a) B and D only
(b) A, B, D and E only
(c) B, D and E only
(d) A, C and E only

8 Given below are two statements.

Statement I Hydrogen peroxide can act as an oxidising agent in both acidic and basic conditions.

Statement II Density of hydrogen peroxide at 298 K is lower than that of D_2O.

In the light of the above statements, choose the correct answer from the options given below.
(a) Both Statement I and Statement II are true.
(b) Both Statement I and Statement II are false.
(c) Statement I is true but Statement II is false.
(d) Statement I is false but Statement II is true.

9 Given below are two statements.

Statement I The chlorides of Be and Al have Cl-bridged structure. Both are soluble in organic solvents and act as Lewis bases.

Statement II Hydroxides of Be and Al dissolve in excess alkali to give beryllate and aluminate ions.

In the light of the above statements, choose the correct answer from the options given below.
(a) Both Statement I and Statement II are true.
(b) Both Statement I and Statement II are false.
(c) Statement I is true but Statement II is false.
(d) Statement I is false but Statement II is true.

10 Which oxoacid of phosphorus has the highest number of oxygen atoms present in its chemical formula?
(a) Pyrophosphorus acid
(b) Hypophosphoric acid
(c) Phosphoric acid
(d) Pyrophosphoric acid

11 Given below are two statements.

Statement I Iron (III) catalyst, acidified $K_2Cr_2O_7$ and neutral $KMnO_4$ have the ability to oxidise I^- to I_2 independently.

Statement II Manganate ion is paramagnetic in nature and involves $p\pi - p\pi$ bonding.

In the light of the above statements, choose the correct answer from the options given below.
(a) Both Statement I and Statement II are true.
(b) Both Statement I and Statement II are false.
(c) Statement I is true but Statement II is false.
(d) Statement I is false but Statement II is true.

12 The total number of Mn=O bonds in Mn_2O_7 is
(a) 4 (b) 5
(c) 6 (d) 3

13 Match List I with List II.

List I (Pollutant)	List II (Disease / sickness)
A. Sulphate (> 500 ppm)	I. Methemoglobinemia
B. Nitrate (> 50 ppm)	II. Brown mottling of teeth
C. Lead (> 50 ppb)	III. Laxative effect
D. Fluoride (> 2 ppm)	IV. Kidney damage

Choose, the correct answer from the options given below.
(a) A-IV, B-I, C-II, D-III (b) A-III, B-I, C-IV, D-II
(c) A-II, B-IV, C-I, D-III (d) A-II, B-IV, C-III, D-I

14 Given below are two statements, one is labelled as Assertion (A) and, the other is labelled as Reason (R).

Assertion (A) [6] Annulene, [8] Annulene and *cis*-[10] Annulene, are respectively aromatic, not aromatic and aromatic.

[6] Annulene 

[8] Annulene 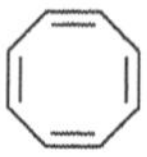

cis-[10] Annulene 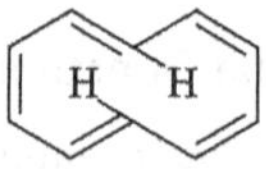

Reason (R) Planarity is one, of the requirements of aromatic systems.

In the light of the above statements, choose the most appropriate answer from the options given below.
(a) Both A and R are correct and R is the correct explanation of A.
(b) Both A and R are correct but R is not the correct explanation of A.
(c) A is correct but R is not correct.
(d) A is not correct but R is correct.

15 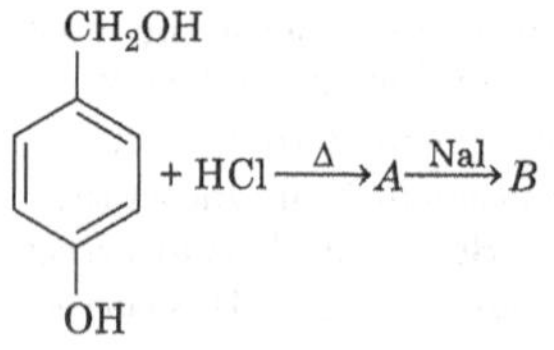

In the above reaction product *B* is

(a) 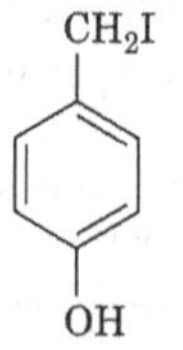

(b) 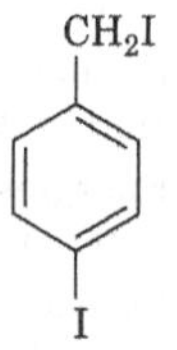

(c) 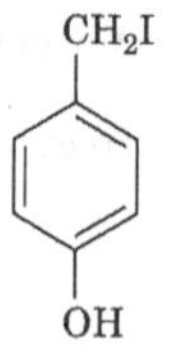

(d) 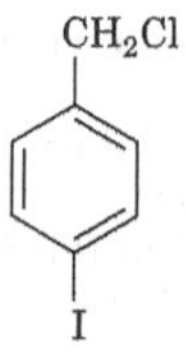

16 Match List I with List II.

List I (Polymers)	List II (Commercial names)
A. Phenol formaldehyde resin	I. Glyptal
B. Copolymer of 1,3-butadiene and styrene	II. Novolac
C. Polyester of glycol and phthalic acid	III. Buna-S
D. Polyester of glycol and terephthalic acid	IV. Dacron

Choose the correct answer from the options give below.

(a) A-II, B-III, C-IV, D-I (b) A-II, B-III, C-I, D-IV
(c) A-II, B-I, C-III, D-IV (d) A-III, B-II, C-IV, D- I

17 A sugar '*X*' dehydrates very slowly under acidic condition to give furfural which on further reaction with resorcinol gives the coloured product after sometime. Sugar '*X*' is

(a) aldopentose (b) aldotetrose
(c) oxalic acid (d) ketotetrose

18 Match List I and List II.

List I	List II
A. 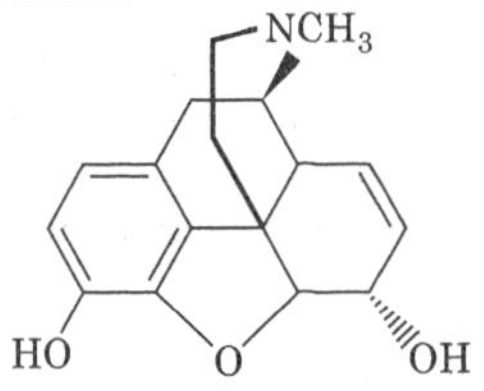	I. Antidepressant
B. 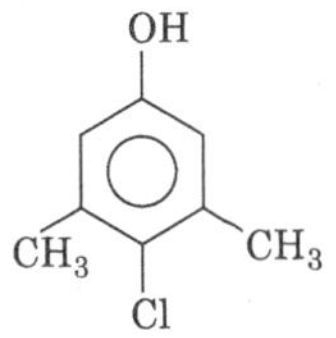	II. 550 times sweeter than cane sugar
C. 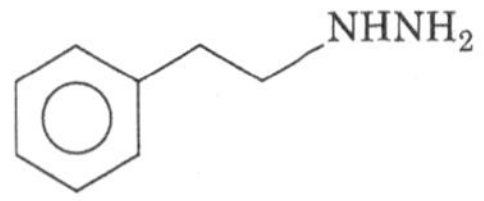	III. Narcotic analgesic
D. 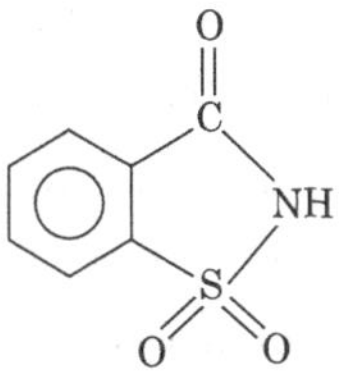	IV. Antiseptic

Choose the correct answer from the options given below.

(a) A-IV, B-III, C-II, D-I (b) A-III, B-I, C-II, D-IV
(c) A-III, B-IV, C-I, D-II (d) A-III, B-I, C-IV, D-II

19 In Carius method of estimation of halogen, 0.45 g of an organic compound gave 0.36 g of AgBr. Find out the percentage of bromine in the compound. (Molar masses $AgBr = 188\ g\ mol^{-1}$; $Br = 80\ g\ mol^{-1}$)

(a) 34.04% (b) 40.04%
(c) 36.03% (d) 38.04%

20 Match list I with List II .

List I	List II
A. Benzene sulphonyl chloride	I. Test for primary amines
B. Hofmann bromamide reaction	II. Anti saytzeff
C. Carbylamine reaction	III. Hinsberg reagent
D. Hofmann orientation	IV. Known reaction of isocyanates

Choose the correct answer from the options given below.

(a) A-IV, B-III, C-II, D-I
(b) A-IV, B-II, C-I, D-II
(c) A-III, B-IV, C-I, D-II
(d) A-IV, B-III, C-I, D-II

Section B : Numerical Type Questions

21 20 mL of 0.02 M $K_2Cr_2O_7$ solution is used for the titration of 10 mL of Fe^{2+} solution in the acidic medium. The molarity of Fe^{2+} solution is ...... $\times 10^{-2}$ M (Nearest integer)

22 $2NO + 2H_2 \longrightarrow N_2 + 2H_2O$

The above reaction has been studied at 800° C. The related date are given in the table below.

Reaction serial number	Initial pressure of H_2/kPa	Initial pressure of NO/kPa	Initial rate $\left(\frac{-dp}{dt}\right)$/ (kPa/s)
1	65.6	40.0	0.135
2	65.6	20.1	0.033
3	38.6	65.6	0.214
4	19.2	65.6	0.106

The order of the reaction with respect to NO is...... .

23 Amongst the following, the number of oxide(s) which are paramagnetic in nature is

Na_2O, KO_2, NO_2, N_2O, ClO_2, NO, SO_2, Cl_2O

24 The molar heat capacity for an ideal gas at constant pressure is 20.785 J K^{-1} mol^{-1}. The change in internal energy is 5000 J upon heating it from 300 K to 500 K. The number of moles of the gas at constant volume is .... (Nearest integer)
(Given, $R = 8.314\ JK^{-1}\ mol^{-1}$)

25 According to MO (molecular orbital) theory, number of species/ions from the following having identical bond order is ...... CN^-, NO^+, O_2, O_2^+, O_2^{2+}

26 At 310 K, the solubility of CaF_2 in water is 2.34×10^{-3} g/100 mL. The solubility product of CaF_2 is ...... $\times 10^{-8} (mol/L)^3$.
(Given, molar mass of $CaF_2 = 78\ g\ mol^{-1}$)

27 The conductivity of a solution of complex with formula $CoCl_3(NH_3)_4$ corresponds to 1 : 1 electrolyte, then the primary valency of central metal ion is ....

28 In the titration of $KMnO_4$ and oxalic acid in acidic medium, the change in oxidation number of carbon at the end point is ....

29 Optical activity of an enantiomeric mixture is +12.6° and the specific rotation of (+) isomer is +30°. The optical purity is ...... %.

30 In the following reaction,

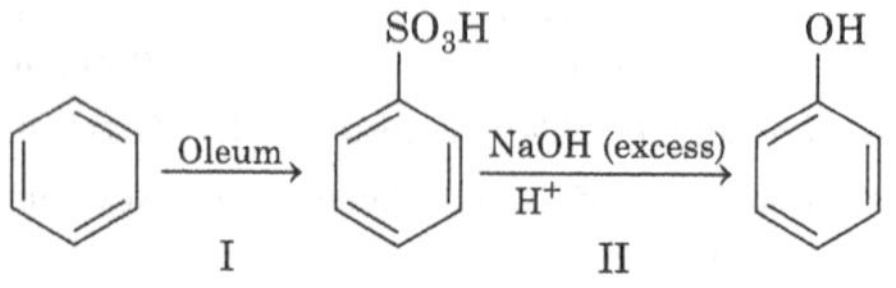

the % yield for reaction I is 60% and that of reaction II is 50%. The overall yield of the complete reaction is ______%. [Nearest integer]

27 JULY 2022, SHIFT-II

Section A : Objective Type Questions

1 The correct decreasing order of energy for the orbitals having, following set of quantum numbers

(A) $n = 3, l = 0, m = 0$; (B) $n = 4, l = 0, m = 0$

(C) $n = 3, l = 1, m = 0$; (D) $n = 3, l = 2, m = 1$

(a) (D) > (B) > (C) > (A) (b) (B) > (D) > (C) > (A)

(c) (C) > (B) > (D) > (A) (d) (B) > (C) > (D) > (A)

2 Match List I with List II.

	List-I		List-II
(A)	$\psi_{MO} = \psi_A - \psi_B$	(I)	Dipole moment
(B)	$\mu = Q \times r$	(II)	Bonding molecular orbital
(C)	$\frac{N_b - N_a}{2}$	(III)	Anti-bonding molecular orbital
(D)	$\psi_{MO} = \psi_A + \psi_B$	(IV)	Bond order

Choose the correct answer from options given below:

(a) (A)-(II), (B)-(I), (C)-(IV), (D)-(III)

(b) (A)-(III), (B)-(IV), (C)-(I), (D)-(II)

(c) (A)-(III), (B)-(I), (C)-(IV), (D)-(II)

(d) (A)-(III), (B)-(IV), (C)-(II), (D)-(I)

3 The plot of pH-metric titration of weak base NH_4OH *vs* strong acid HCl looks like

(a)
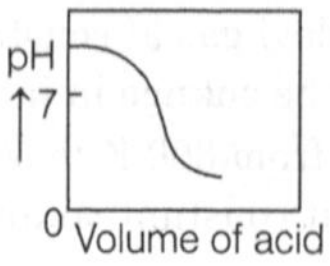

(b)
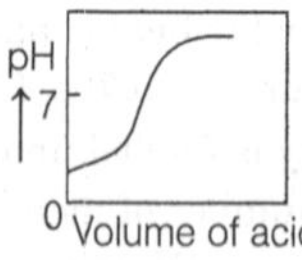

(c)
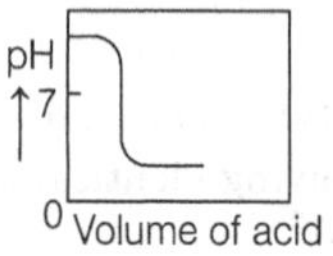

(d)
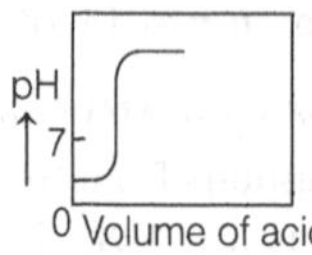

4 Given below are two statements.

Statement I For KI, molar conductivity increases steeply with dilution.

Statement II For carbonic acid, molar conductivity increases slowly with dilution.

In the light of the above statements, choose the correct answer from the options given below.

(a) Both Statement I and Statement II are true.

(b) Both Statement I and Statement II are false.

(c) Statement I is true but Statement II is false.

(d) Statement I is false but Statement II is true.

5 Given below are two statements, one is labelled as Assertion (A) and the other is labelled as Reason (R).

Assertion (A) Dissolved substances can be removed from a colloidal solution by diffusion through a parchment paper.

Reason (R) Particles in a true solution cannot pass through parchment paper but the colloidal particles can pass through the parchment paper.

In the light of the above statements, choose the correct answer from the options given below.

(a) Both (A) and (R) are correct and (R) is the correct explanation of (A).

(b) Both (A) and (R) are correct but (R) is not the correct explanation of (A).

(c) (A) is correct but (R) is not correct.

(d) (A) is not correct but (R) is correct.

6 Outermost electronic configurations of four elements A, B, C, D are given below.

(A) $3s^2$ (B) $3s^2 3p^1$ (C) $3s^2 3p^3$ (D) $3s^2 3p^4$

The correct order of fist ionisation enthalpy for them is

(a) (A) < (B) < (C) < (D) (b) (B) < (A) < (D) < (C)

(c) (B) < (D) < (A) < (C) (d) (B) < (A) < (C) < (D)

7 An element A of group 1 shows similarity to an element B belonging to group 2. If A has maximum hydration enthalpy in group 1 then B is

(a) Mg (b) Be (c) Ca (d) Sr

8 Given below are two statements, one is labelled as Assertion (A) and the other is labelled as Reason (R).

Assertion (A) Boron is unable to form BF_6^{3-}.

Reason (R) Size of B is very small.

In the light of the above statements, choose the correct answer from the options given below.

(a) Both (A) and (R) are true and (R) is the correct explanation of (A).
(b) Both (A) and (R) are true but (R) is not the correct explanation of (A).
(c) (A) is true but (R) is false.
(d) (A) is false but (R) is true.

9 In neutral or alkaline solution, MnO_4^- oxidises thiosulphate to

(a) $S_2O_7^{2-}$ (b) $S_2O_8^{2-}$
(c) SO_3^{2-} (d) SO_4^{2-}

10 Low oxidation state of metals in their complexes are common when ligands

(a) have good π-accepting character
(b) have good σ-donor character
(c) are having good π-donating ability
(d) are having poor σ-donating ability

11 Given below are two statements:

Statement I The non bio-degradable fly ash and slag from steel industry can be used by cement industry.

Statement II The fuel obtained from plastic waste is lead free.

In the light of the above statements, choose the most appropriate answer from the options given below.

(a) Both Statement I and Statement II are correct.
(b) Both Statement I and Statement II are incorrect.
(c) Statement I is correct but Statement II is incorrect.
(d) Statement I is incorrect but Statement II is correct.

12 The structure of A in the given reaction is

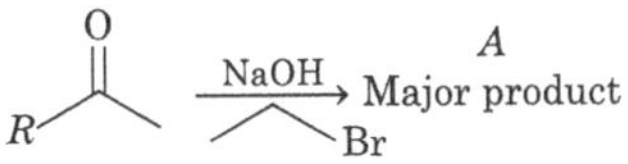

(a) 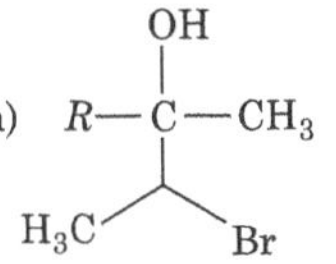

(b) 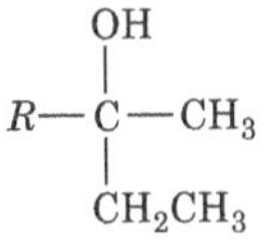

(c) 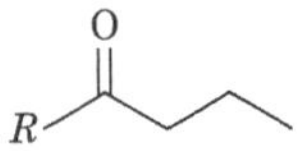

(d) 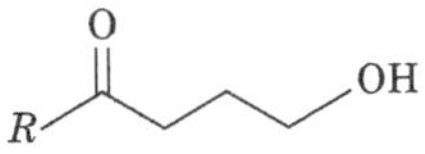

13 Major product 'B' of the following reaction sequence is

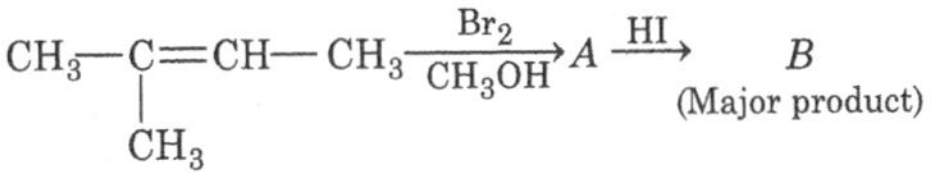

(a) 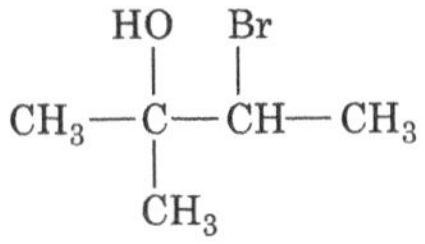

(b) 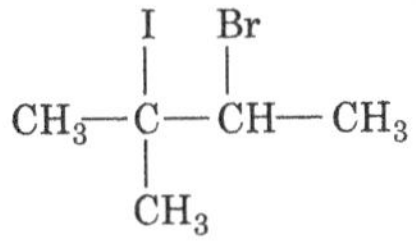

(c) 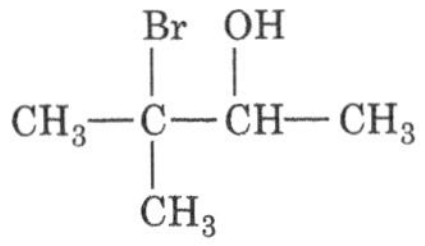

(d) 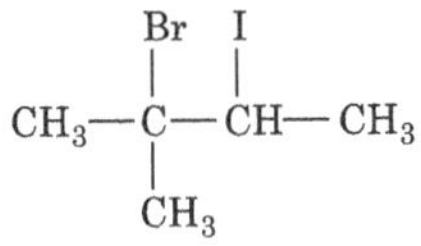

14 Match List I with List II.

	List I	List II
(A)	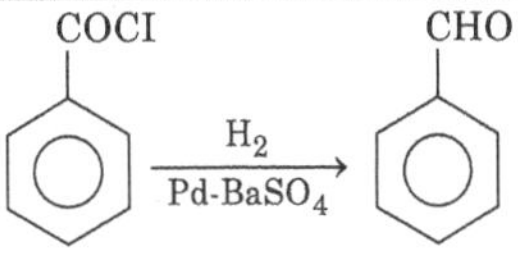	(I) Gattermann Koch reaction
(B)	$CH_3-CN \xrightarrow[H_3O^+]{SnCl_2/HCl} CH_3-CHO$	(II) Etard reaction
(C)	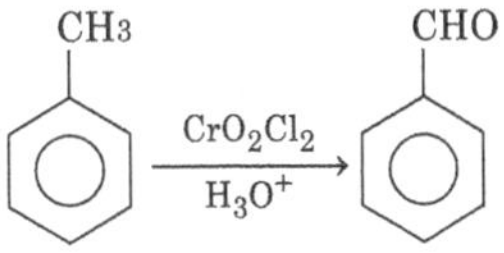	(III) Stephen reaction
(D)	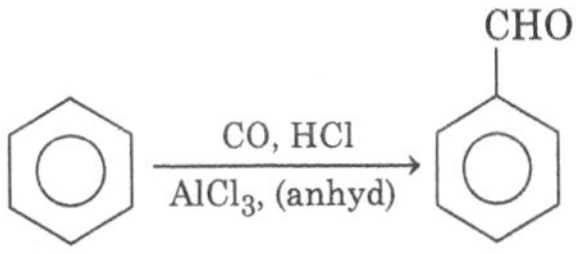	(IV) Rosenmund reaction

Choose the correct answer from the options given below.

(a) (A) → (IV), (B) → (III), (C) → (II), (D) → (I)
(b) (A) → (I), (B) → (II), (C) → (III), (D) → (IV)
(c) (A) → (II), (B) → (III), (C) → (IV), (D) → (I)
(d) (A) → (III), (B) → (II), (C) → (I), (D) → (IV)

15 Match List I with List II.

	List I (Polymer)		List II (Monomer)
(A)	Neoprene	(I)	Acrylonitrile
(B)	Teflon	(II)	Chloroprene
(C)	Acrilan	(III)	Tetrafluoroethene
(D)	Natural rubber	(IV)	Isoprene

Choose the correct answer from the options given below.

(a) (A) → (II), (B) → (III), (C) → (I), (D) → (IV)
(b) (A) → (II), (B) → (I), (C) → (III), (D) → (IV)
(c) (A) → (II), (B) → (I), (C) → (IV), (D) → (III)
(d) (A) → (I), (B) → (II), (C) → (III), (D) → (IV)

16 An organic compound 'A' contains nitrogen and chlorine. It dissolves readily in water to give a solution that turns litmus red. Titration of compound 'A' with standard base indicates that the molecular weight of 'A' is 131 ± 2. When a sample of 'A' is treated with aq. NaOH, a liquid separates which contains N but not Cl. Treatment of the obtained liquid with nitrous acid followed by phenol gives orange precipitate. The compound A is

(a) 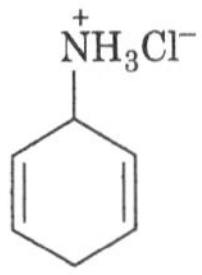

(b) 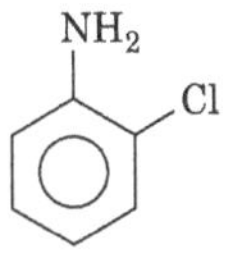

(c) 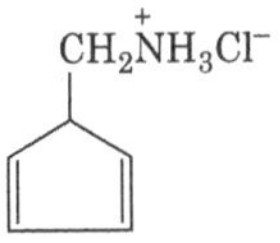

(d) 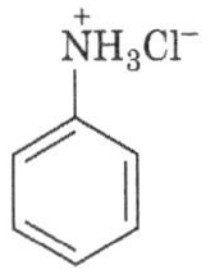

17 Match List I with List II.

	List I		List II
(A)	Glucose + HI	(I)	Gluconic acid
(B)	Glucose + Br_2 water	(II)	Glucose pentacetate
(C)	Glucose + acetic anhydride	(III)	Saccharic acid
(D)	Glucose + HNO_3	(IV)	Hexane

Choose the correct answer from the options given below.

(a) (A)-(IV), (B)-(I), (C)-(II), (D)-(III)
(b) (A)-(IV), (B)-(III), (C)-(II), (D)-(I)
(c) (A)-(III), (B)-(I), (C)-(IV), (D)-(II)
(d) (A)-(I), (B)-(III), (C)-(IV), (D)-(II)

18 Which of the following enhances the lathering property of soap?

(a) Sodium stearate (b) Sodium carbonate
(c) Sodium rosinate (d) Trisodium phosphate

19 Match List I with List II.

	List I (Mixture)		List II (Purification process)
(A)	Chloroform and aniline	(I)	Steam distillation
(B)	Benzoic acid and naphthalene	(II)	Sublimation
(C)	Water and aniline	(III)	Distillation
(D)	Naphthalene and sodium chloride	(IV)	Crystallisation

Choose the correct answer from the options given below

(a) (A) $\rightarrow$ (IV), (B) $\rightarrow$ (III), (C) $\rightarrow$ (I), (D) $\rightarrow$ (II)
(b) (A) $\rightarrow$ (III), (B) $\rightarrow$ (I), (C) $\rightarrow$ (IV), (D) $\rightarrow$ (II)
(c) (A) $\rightarrow$ (III), (B) $\rightarrow$ (IV), (C) $\rightarrow$ (II), (D) $\rightarrow$ (I)
(d) (A) $\rightarrow$ (III), (B) $\rightarrow$ (IV), (C) $\rightarrow$ (I), (D) $\rightarrow$ (II)

20 Fe^{3+} cation gives a prussian blue precipitate on addition of potassium ferrocyanide solution due to the formation of

(a) $[Fe(H_2O)_6]_2[Fe(CN)_6]$ (b) $Fe_2[Fe(CN)_6]_2$
(c) $Fe_3[Fe(OH)_2(CN)_4]_2$ (d) $Fe_4[Fe(CN)_6]_3$

Section B : Numerical Type Questions

21 The normality of H_2SO_4 in the solution obtained on mixing 100 mL of 0.1 M H_2SO_4 with 50 mL of 0.1 M NaOH is .... $\times 10^{-1}$N. (nearest integer)

22 For a real gas at 25°C temperature and high pressure (99 bar) the value of compressibility factor is 2, so the value of van der Waals' constant 'b' should be .... $\times 10^{-2}$ L mol^{-1}. (nearest integer)

(Given, $R = 0.083$ L bar K^{-1} mol^{-1}).

23 A gas (molar mass = 280 g mol^{-1}) was burnt in excess O_2 in a constant volume calorimeter and during combustion the temperature of calorimeter increased from 298.0 K to 298.45 K. If the heat capacity of calorimeter is 2.5 kJ K^{-1} and enthalpy of combustion of gas is 9 kJ mol^{-1} then amount of gas burnt is .... g. (nearest integer)

24 When a certain amount of solid A is dissolved in 100 g of water at 25°C to make a dilute solution, the vapour pressure of the solution is reduced to one-half of that of pure water. The vapour pressure of pure water is 23.76 mmHg. The number of moles of solute A added is .... (nearest integer)

25 $[A] \longrightarrow [B]$

Reactant Product

If formation of compound $[B]$ follows the first order of kinetics and after 70 minutes the concentration of $[A]$ was found to be half of its initial concentration. Then the rate constant of the reaction is $x \times 10^{-6} s^{-1}$. The value of x is .... (nearest integer)

26 Among the following ores Bauxite, Siderite, Cuprite, Calamine, Haematite, Kaolinite, Malachite, Magnetite, Sphalerite, Limonite, Cryolite.

The number of principal ores of iron is ....

27 The oxidation state of manganese in the product obtained in a reaction of potassium permanganate and hydrogen peroxide in basic medium is ....

28 The number of molecule(s) or ion(s) from the following having non-planar structure is ....

NO_3^-, H_2O_2, BF_3, PCl_3, XeF_4, SF_4, XeO_3, PH_4^+, SO_3, $[Al(OH)_4]^-$

29 The spin only magnetic moment of the complex present in Fehling's reagent is .... BM. (nearest integer).

30

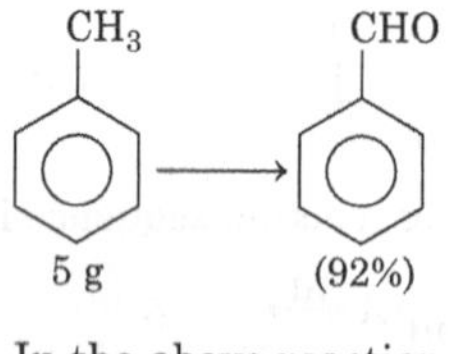

In the above reaction, 5 g of toluene is converted into benzaldehyde with 92% yield. The amount of benzaldehyde produced is .... $\times 10^{-2}$ g. (nearest integer).

28 JULY 2022, SHIFT-I

Section A : Objective Type Questions

1 Identify the incorrect statement from the following.

(a) A circular path around the nucleus in which an electron moves is proposed as Bohr's orbit.

(b) An orbital is the one electron wave function (ψ) in an atom.

(c) The existence of Bohr's orbits is supported by hydrogen spectrum.

(d) Atomic orbital is characterised by the quantum numbers n and l only.

2 Which of the following relation is not correct?

(a) $\Delta H = \Delta U - p\Delta V$ (b) $\Delta U = q + W$

(c) $\Delta S_{say} + \Delta S_{surr} \geq 0$ (d) $\Delta G = \Delta H - T\Delta S$

3 Match List I with List II.

List I	List II
(A) $Cd(s) + 2Ni(OH)_3(s) \longrightarrow CdO(s) + 2Ni(OH)_2(s) + H_2O(l)$	(I) Primary battery
(B) $Zn(Hg) + HgO(s) \longrightarrow ZnO(s) + Hg(l)$	(II) Discharging of secondary battery
(C) $2PbSO_4(s) + 2H_2O(l) \longrightarrow Pb(s) + PbO_2(s) + 2H_2SO_4(aq)$	(III) Fuel cell
(D) $2H_2(g) + O_2(g) \longrightarrow 2H_2O(l)$	(IV) Charging of secondary battery

Choose the correct answer from the options given below.

(a) (A) → (I), (B) → (II), (C) → (III), (D) → (IV)
(b) (A) → (IV), (B) → (I), (C) → (II), (D) → (III)
(c) (A) → (II), (B) → (I), (C) → (IV), (D) → (III)
(d) (A) → (II), (B) → (I), (C) → (III), (D) → (IV)

4 Match List I with List II.

List I (Reaction)	List II (Catalyst)
(A) $4NH_3(g) + 5O_2(g) \longrightarrow 4NO(g) + 6H_2O(g)$	(I) $NO(g)$
(B) $N_2(g) + 3H_2(g) \longrightarrow 2NH_3(g)$	(II) $H_2SO_4(l)$
(C) $C_{12}H_{22}O_{11}(aq) + H_2O(l) \longrightarrow \underset{\text{Glucose}}{C_6H_{12}O_6} + \underset{\text{Fructose}}{C_6H_{12}O_6}$	(III) $Pt(s)$
(D) $2SO_2(g) + O_2(g) \longrightarrow 2SO_3(g)$	(IV) $Fe(s)$

Choose the correct answer from the options given below.

(a) (A) → (II), (B) → (III), (C) → (I), (D) → (IV)
(b) (A) → (III), (B) → (II), (C) → (I), (D) → (IV)
(c) (A) → (III), (B) → (IV), (C) → (II), (D) → (I)
(d) (A) → (III), (B) → (II), (C) → (IV), (D) → (I)

5 In which of the following pairs, electron gain enthalpies of constituent elements are nearly the same or identical?

(A) Rb and Cs (B) Na and K
(C) Ar and Kr (D) I and At

Choose the correct answer from the options given below.

(a) (A) and (B) only (b) (B) and (C) only
(c) (A) and (C) only (d) (C) and (D) only

6 Which of the reaction is suitable for concentrating ore by leaching process?

(a) $2Cu_2S + 3O_2 \longrightarrow 2Cu_2O + 2SO_2$
(b) $Fe_3O_4 + CO \longrightarrow 3FeO + CO_2$
(c) $Al_2O_3 + 2NaOH + 3H_2O \longrightarrow 2Na[Al(OH)_4]$
(d) $Al_2O_3 + 6Mg \longrightarrow 6MgO + 4Al$

7 The metal salts formed during softening of hardwater using Clark's method are

(a) $Ca(OH)_2$ and $Mg(OH)_2$ (b) $CaCO_3$ and $Mg(OH)_2$
(c) $Ca(OH)_2$ and $MgCO_3$ (d) $CaCO_3$ and $MgCO_3$

8 Which of the following statement is incorrect?

(a) Low solubility of LiF in water is due to its small hydration enthalpy.

(b) KO_2 is paramagnetic.

(c) Solution of sodium in liquid ammonia is conducting in nature.

(d) Sodium metal has higher density than potassium metal.

9 Match List I with List II, match the gas evolved during each reaction.

List I	List II
(A) $(NH_4)_2Cr_2O_7 \xrightarrow{\Delta}$	(I) H_2
(B) $KMnO_4 + HCl \longrightarrow$	(II) N_2
(C) $Al + NaOH + H_2O \longrightarrow$	(III) O_2
(D) $NaNO_3 \longrightarrow$	(IV) Cl_2

Choose the correct answer from the options given below.

(a) (A) → (II), (B) → (III), (C) → (I), (D) → (IV)
(b) (A) → (III), (B) → (I), (C) → (IV), (D) → (II)
(c) (A) → (II), (B) → (IV), (C) → (I), (D) → (III)
(d) (A) → (III), (B) → (IV), (C) → (I), (D) → (II)

10 Which of the following has least tendency to liberate H_2 from mineral acids?

(a) Cu (b) Mn (c) Ni (d) Zn

11 Given below are two statements.

Statement I In polluted water values of both dissolved oxygen and BOD are very low.

Statement II Eutrophication results in decrease in the amount of dissolved oxygen.

In the light of the above statements, choose the most appropriate answer from the options given below.

(a) Both Statement I and Statement II are true
(b) Both Statement I and Statement II are false
(c) Statement I is true but Statement II is false
(d) Statement I is false but Statement II is true

12 Match List I with List II.

	List I		List II
(A)	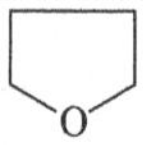	(I)	Spiro compound
(B)	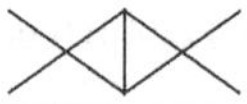	(II)	Aromatic compound
(C)	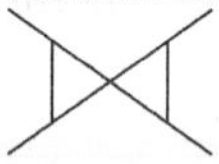	(III)	Non-planar heterocyclic compound
(D)	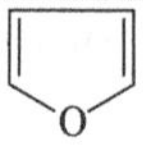	(IV)	Bicyclo compound

(a) (A) → (II), (B) → (I), (C) → (IV), (D) → (III)
(b) (A) → (IV), (B) → (III), (C) → (I), (D) → (II)
(c) (A) → (III), (B) → (IV), (C) → (I), (D) → (II)
(d) (A) → (IV), (B) → (III), (C) → (II), (D) → (I)

13 Choose the correct option for the following reactions.

$$B \xleftarrow[H_2O_2/OH^-]{(BH_3)_2} H_3C-\underset{CH_3}{\overset{CH_3}{C}}-CH=CH_2 \xrightarrow[NaBH_4]{Hg(OAc)_2/H_2O} A$$

(a) '*A*' and '*B*' are both Markownikoff addition products.
(b) '*A*' is Markownikoff product and '*B*' is anti-Markownikoff product.
(c) '*A*' and '*B*' are both anti-Markownikoff products.
(d) '*B*' is Markownikoff and '*A*' is anti-Markownikoff product.

14 Among the following marked proton of which compound shows lowest pK_a value?

(a) $H_2\overset{H}{C}-COOH$

(b) $H_2\overset{H}{C}-\overset{O}{\overset{\|}{C}}-CH_3$

(c) 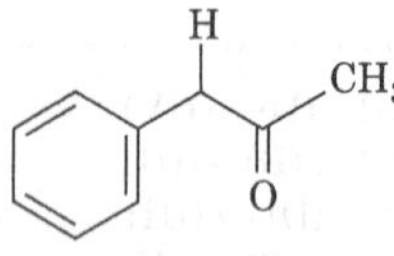

(d) 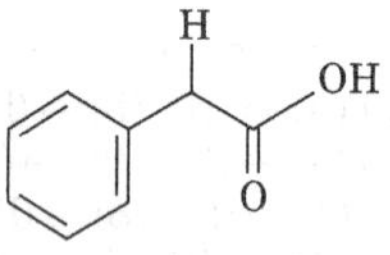

15 Identify the major products *A* and *B* for the below given reaction sequence.

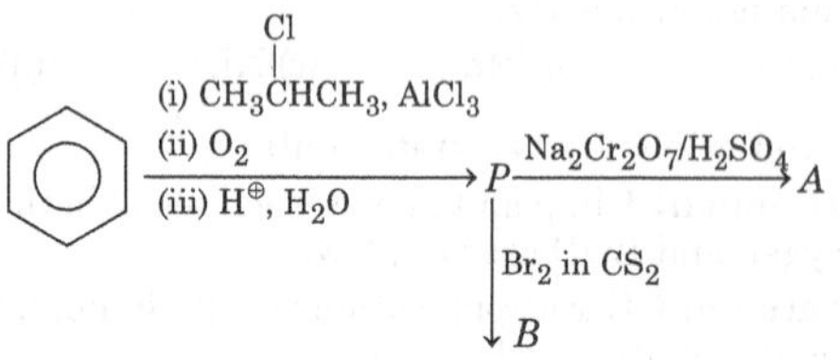

	Ⓐ		Ⓑ
(a)	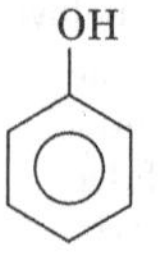	and	
(b)	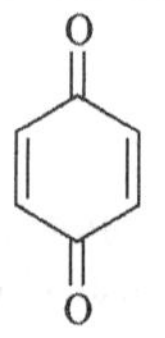	and	
(c)	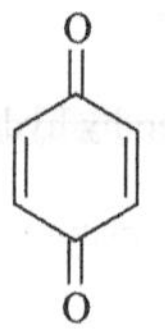	and	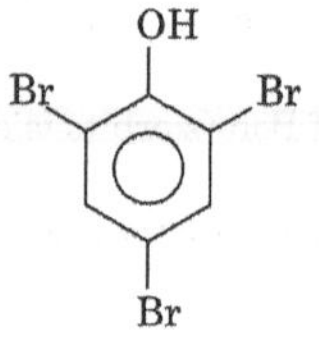
(d)	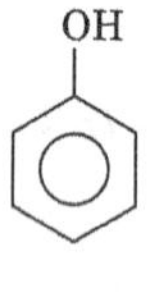	and	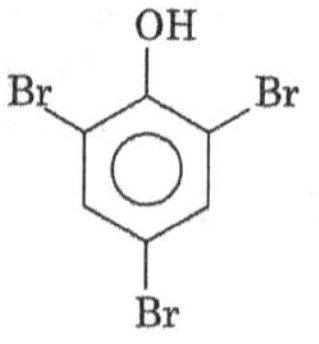

16 Identify the correct statement for the below given transformation.

$$CH_3-CH_2-CH_2-\underset{\overset{+}{N}(CH_3)_3}{CH}-CH_3 \xrightarrow[C_2H_5OH]{C_2H_5ONa} \underset{(Major)}{A} + \underset{(Minor)}{B}$$

(a) *A*- $CH_3CH_2CH=CH-CH_3$,
B- $CH_3CH_2CH_2CH=CH_2$ Saytzeff products
(b) *A*- $CH_3CH_2CH=CH-CH_3$,
B- $CH_3CH_2CH_2CH=CH_2$ Hofmann products
(c) *A*- $CH_3CH_2CH_2CH=CH_2$,
B- $CH_3CH_2CH=CH-CH_2$ Hofmann poducts
(d) *A*- $CH_3CH_2CH_2CH=CH_2$,
B- $CH_3CH_2CH=CH-CH_3$ Saytzeff products

17 Terylene polymer is obtained by condensation of
(a) ethane-1, 2-diol and benzene-1, 3-dicarboxylic acid
(b) propane-1, 2-diol and benzene-1, 4 dicarboxylic acid
(c) ethane-1, 2-diol and benzene-1, 4-dicarboxylic acid
(d) ethane-1, 2-diol and benzene-1, 2-dicarboxylic acid

18 For the below given cyclic hemiacetal (*X*), the correct pyranose structure is

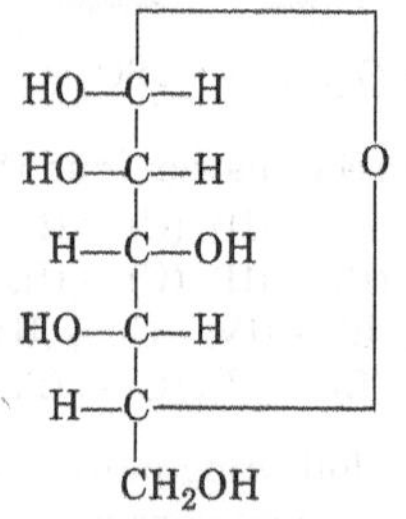

(*X*)

(a) 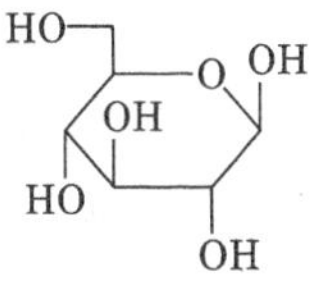

(b) 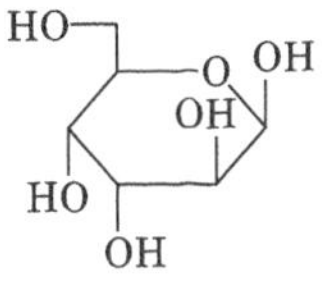

(c) 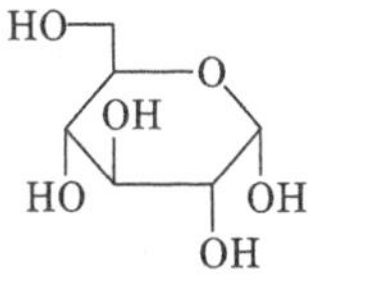

(d) 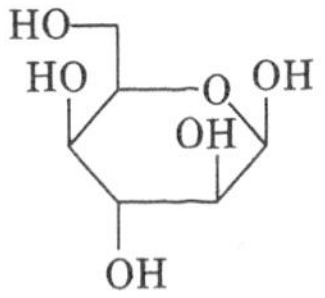

19 Statements about enzyme inhibitor drugs are given below.

(A) There are competitive and non-competitive inhibitor drugs.

(B) These can bind at the active sites and allosteric sites.

(C) Competitive drugs are allosteric site blocking drugs.

(D) Non-competitive drugs are active site blocking drugs.

Choose the correct answer from the options given below

(a) (A) (D) only (b) (A), (C) only

(c) (A), (B) only (d) (A), (B), (C) only

20 For kinetic study of the reaction of iodide ion with H_2O_2 at room temperature :

(A) Always use freshly prepared starch solution.

(B) Always keep the concentration of sodium thiosulphate solution less than that of KI solution.

(C) Record the time immediately after the appearance of blue colour.

(D) Record the time immediately before the appearance of blue colour.

(E) Always keep the concentration of sodium thiosulphate solution more than that of KI solution.

Choose the correct answer from the options given below.

(a) (A), (B), (C) only (b) (A), (D), (E) only

(c) (D), (E) only (d) (A), (B), (E) only

Section B : Numerical Type Questions

21 In the given reaction,

$$X + Y + 3Z \rightleftharpoons XYZ_3$$

if one mole of each of X and Y with 0.05 mole of Z gives compound XYZ_3. (Given : Atomic masses of X, Y and Z are 10, 20 and 30 amu, respectively.) the yield of XYZ_3 is ...... g. (Nearest integer)

22 An element M crystallises in a body centred cubic unit cell with a cell edge of 300 pm. The density of the element is 6.0 g cm^{-3}. The number of atoms present in 180 g of the element is ....$\times 10^{23}$. (Nearest integer)

23 The number of paramagnetic species among the following is ....

B_2, Li_2, C_2, C_2^-, O_2^{2-}, O_2^+ and He_2^+

24 150 g of acetic acid was contaminated with 10.2 g ascorbic acid ($C_6H_8O_6$) to lower down its freezing point by $(x \times 10^{-1})$°C. The value of x is .... (Nearest integer)

(Given : K_f = 3.9K kg mol^{-1}; molar mass of ascorbic acid = 176g mol^{-1})

25 K_a for butyric acid (C_3H_7COOH) is 2×10^{-5}. The pH of 0.2 M solution of butyric acid is ....$\times 10^{-1}$. (Nearest integer) (Given, $\log 2 = 0.30$)

26 For the given first order reaction $A \rightarrow B$, the half-life of the reaction is 0.3010 min. The ratio of the initial concentration of reactant to the concentration of reactant at time 2.0 min will be equal to .... (Nearest integer)

27 The number of interhalogens from the following having square pyramidal structure is

ClF_3, IF_7, BrF_5, BrF_3, I_2Cl_6, IF_5, ClF, ClF_5

28 The disproportionation of MnO_4^{2-} in acidic medium resulted in the formation of two manganese compounds A and B. If the oxidation state of Mn in B is smaller than that of A, then the spin-only magnetic moment (μ) value of B in BM is .... (Nearest integer)

29 Total number of relatively more stable isomer(s) possible for octahedral complex $[Cu(en)_2(SCN)_2]$ will be ....

30 On complete combustion of 0.492 g of an organic compound containing C, H and O, 0.7938 g of CO_2 and 0.4428 g of H_2O was produced. The % composition of oxygen in the compound is ....

28 JULY 2022, SHIFT-II

Section A : Objective Type Questions

1 Given below are two statements, one is labelled as Assertion A and the other is labelled as Reason R.

Assertion (A) Zero orbital overlap is an out of phase overlap.

Reason (R) It results due to different orientation/direction of approach of orbitals.

In the light of the above statements, choose the correct answer from the options given below

(a) Both A and R are true and R is the correct explanation of A

(b) Both A and R are true but R is not the correct explanation of A

(c) A is true but R is false

(d) A is false but R is true

2 The correct decreasing order for metallic character is

(a) Na > Mg > Be > Si > P (b) P > Si > Be > Mg > Na

(c) Si > P > Be > Na > Mg (d) Be > Na > Mg > Si > P

3 Given below are two statements, one is labelled as Assertion A and the other is labelled as Reason R.

Assertion (A) The reduction of a metal oxide is easier if the metal formed is in liquid state than solid state.

Reason (R) The value of $\Delta G^{\ominus}$ becomes more on negative side as entropy is higher in liquid state than solid state.

In the light of the above statements, choose the most appropriate answer from the options given below

(a) Both A and R are correct and R is the correct explanation of A
(b) Both A and R are correct but R is not the correct explanation of A
(c) A is correct but R is not correct
(d) A is not correct but R is correct

4 The products obtained during treatment of hard water using Clark's method are

(a) $CaCO_3$ and $MgCO_3$ (b) $Ca(OH)_2$ and $Mg(OH)_2$
(c) $CaCO_3$ and $Mg(OH)_2$ (d) $Ca(OH)_2$ and $MgCO_3$

5 **Statement I** An alloy of lithium and magnesium is used to make aircraft plates.

Statement II The magnesium ions are important for cell-membrane integrity.

In the light of the above statements, choose the correct answer from the options given below

(a) Both Statement-I and Statement-II are true
(b) Both Statement-I and Statement-II are false
(c) Statement-I is true but Statement-II is false
(d) Statement-I is false but Statement-II is true

6 White phosphorus reacts with thionyl chloride to give?

(a) PCl_5, SO_2 and S_2Cl_2
(b) PCl_3, SO_2 and S_2Cl_2
(c) PCl_3, SO_2 and Cl_2
(d) PCl_5, SO_2 and Cl_2

7 Concentrated HNO_3 reacts with Iodine to give

(a) HI, NO_2 and H_2O
(b) HIO_2, N_2O and H_2O
(c) HIO_3, NO_2 and H_2O
(d) HIO_4, N_2O and H_2O

8 Which of the following pair is not isoelectronic species? (Atomic number : Sm62; Er68; Yb70; Lu71; Eu63; Tb65; Tm69)

(a) Sm^{2+} and Er^{3+} (b) Yb^{2+} and Lu^{3+}
(c) Eu^{2+} and Tb^{4+} (d) Tb^{2+} and Tm^{4+}

9 Given below are two statements, one is labelled as Assertion A and the other is labelled as Reason R.

Assertion (A) Permanganate titrations are not performed in presence of hydrochloric acid.

Reason (R) Chlorine is formed as a consequence of oxidation of hydrochloric acid.

In the light of the above statements, choose the correct answer from the options given below.

(a) Both A and R are true and R is the correct explanation of A
(b) Both A and R are true but R is not the correct explanation of A
(c) A is true but R is false
(d) A is false but R is true

10 Match List I with List II.

List I (Complex)		List II (Hybridisation)
A. $Ni(CO)_4$	I.	sp^3
B. $[Ni(CN)_4]^{2-}$	II.	sp^3d^2
C. $[Co(CN)_6]^{3-}$	III.	d^2sp^3
D. $[CoF_6]^{3-}$	IV.	dsp^2

Choose the correct answer from the options given below

(a) A-IV, B-I, C-III, D-II (b) A-I, B-IV, C-III, D-II
(c) A-I, B-IV, C-II, D-III (d) A-IV, B-I, C-II, D-III

11 Dinitrogen and dioxygen, the main constituents of air do not react with each other in atmosphere to form oxides of nitrogen because

(a) N_2 is unreactive in the condition of atmosphere
(b) oxides of nitrogen are unstable
(c) reaction between them can occur in the presence of a catalyst
(d) the reaction is endothermic and require very high temperature

12 The major product in the given reaction is

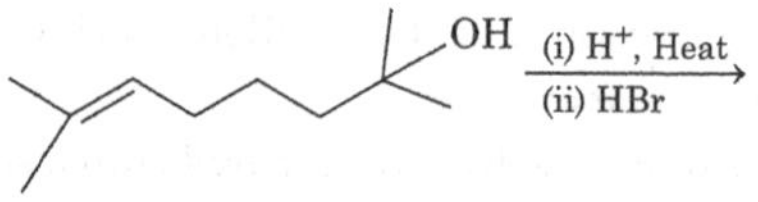

(a)
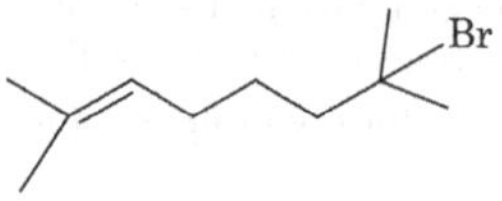

(b)
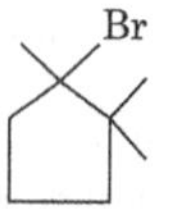

(c)
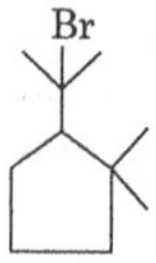

(d)
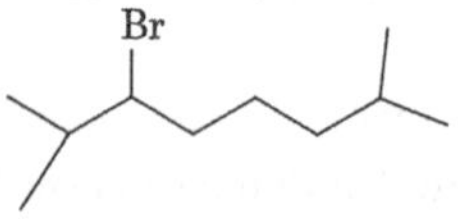

13 Arrange the following in increasing order of reactivity towards nitration.

A. *p*-xylene B. Bromobenzene
C. Mesitylene D. Nitrobenzene
E. Benzene

(a) C < D < E < A < B (b) D < B < E < A < C
(c) D < C < E < A < B (d) C < D < E < B < A

14 Compound I is heated with conc. HI to give a hydroxy compound *A* which is further heated with Zn dust to give compound *B*. Identify *A* and *B*.

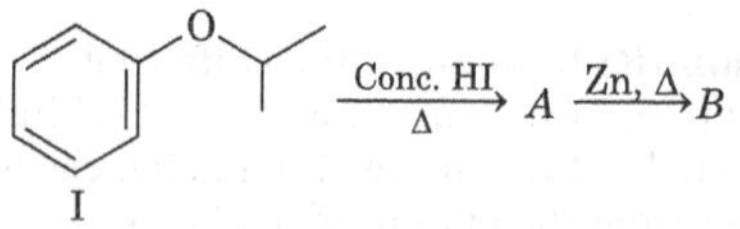

I

(a) $A =$ 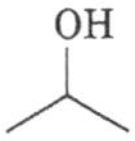 $B =$ 

(b) $A =$ 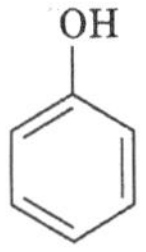 $B =$ 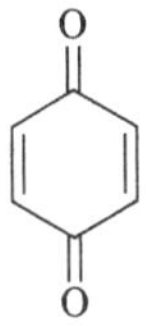

(c) $A =$ 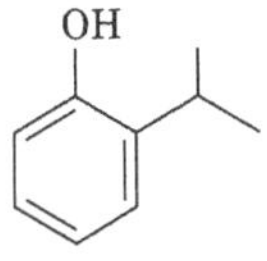 $B =$ 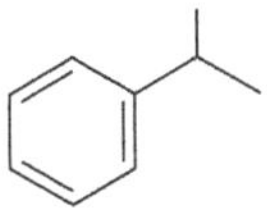

(d) $A =$ 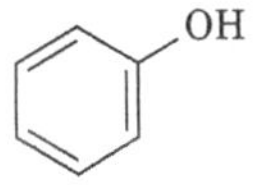 $B =$ 

15 Given below are two statements, one is labelled as Assertion A and the other is labelled as Reason R.

Assertion (A) Aniline on nitration yields *ortho*, *meta* and *para* nitro derivatives of aniline.

Reason (R) Nitrating mixture is a strong acidic mixture.

In the light of the above statements, choose the correct answer from the options given below.

(a) Both A and R are true and R is the correct explanation of A
(b) Both A and R are true but R is not the correct explanation of A
(c) A is true but R is false
(d) A is false but R is true

16 Match List I with List II.

List I (Polymer)	List II (Nature)
A. $-\!\!\left[CH_2-\underset{Cl}{\underset{\mid}{C}}=CH-CH_2\right]\!\!-_n$	I. Thermosetting polymer
B. $\left[\overset{H}{\overset{\mid}{N}}-(CH_2)_6-\overset{H}{\overset{\mid}{N}}-\overset{O}{\overset{\parallel}{C}}-(CH_2)_4-\overset{O}{\overset{\parallel}{C}}\right]_n$	II. Fibers
C. $-\!\!\left[CH_2-\overset{Cl}{\overset{\mid}{CH}}\right]\!\!-_n$	III. Elastomer
D. 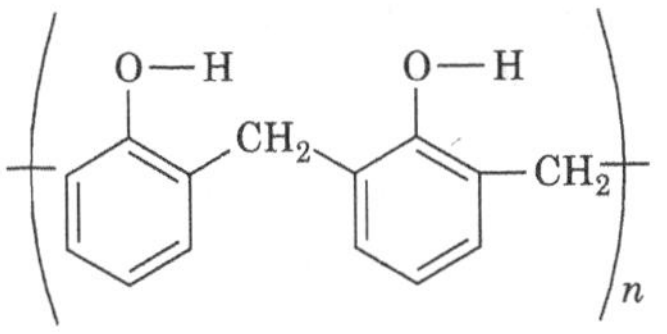	IV. Thermoplastic polymer

Choose the correct answer from the options given below

(a) A-II, B-III, C-IV, D-I
(b) A-III, B-II, C-IV, D-I
(c) A-III, B-I, C-IV, D-II
(d) A-I, B-III, C-IV, D-II

17 Two statements in respect of drug-enzyme interaction are given below

Statement I Action of an enzyme can be blocked only when an inhibitor blocks the active site of the enzyme.

Statement II An inhibitor can form a strong covalent bond with the enzyme.

In the light of the above statements, choose the correct answer from the options given below

(a) Both Statement I and Statement II are true
(b) Both Statement I and Statement II are false
(c) Statement I is true but Statement II is false
(d) Statement I is false but Statement II is true

18 Given below are two statements, one is labelled as Assertion A and the other is labelled as Reason R.

Assertion (A) Thin layer chromatography is an adsorption chromatography.

Reason (R) A thin layer of silica gel is spread over a glass plate of suitable size in thin layer chromatography which acts as an adsorbent.

In the light of the above statements, choose the correct answer from the options given below

(a) Both A and R are true and R is the correct explanation of A.
(b) Both A and R are true but R is not the correct explanation of A
(c) A is true but R is false
(d) A is false but R is true

19 The formulas of A and B for the following reaction sequence

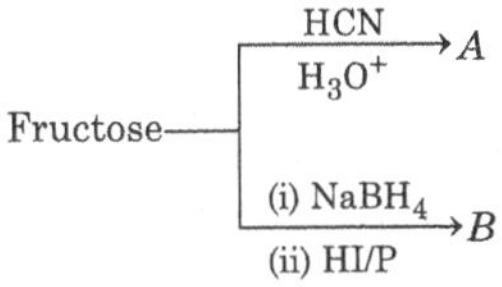

(a) $A = C_7H_{14}O_8$, $B = C_6H_{14}$
(b) $A = C_7H_{13}O_7$, $B = C_7H_{14}O$
(c) $A = C_7H_{12}O_8$, $B = C_6H_{14}$
(d) $A = C_7H_{14}O_8$, $B = C_6H_{14}O_6$

20 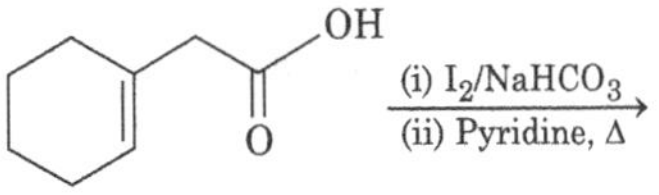

Find out the major product for the above reaction.

(a) 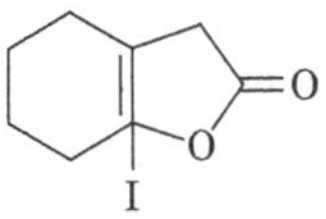
(b) 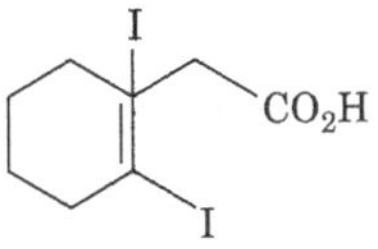
(c) 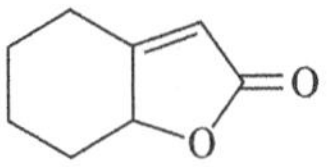
(d) 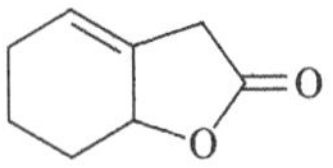

Section B : Numerical Type Questions

21 2 L of 0.2 M H_2SO_4 is reacted with 2 L of 0.1 M NaOH solution, the molarity of the resulting product Na_2SO_4 in the solution is .... millimolar. (Nearest integer)

22 Metal M cystallises into a fcc lattice with the edge length of 4.0×10^{-8} cm. The atomic mass of the metal is .... g/mol. (Nearest integer)
(Use, $N_A = 6.02 \times 10^{23}$ mol^{-1},
density of metal, $M = 9.03$ g cm^{-3})

23 If the wavelength for an electron emitted from H-atom is 3.3×10^{-10} m, then energy absorbed by the electron in its ground state compared to minimum energy required for its escape from the atom, is .... times. (Nearest integer)
[Given, $h = 6.626 \times 10^{-34}$ Js]
[Mass of electron $= 9.1 \times 10^{-31}$ kg]

24 A gaseous mixture of two substances A and B, under a total pressure of 0.8 atm is in equilibrium with an ideal liquid solution. The mole fraction of substance A is 0.5 in the vapour phase and 0.2 in the liquid phase. The vapour pressure of pure liquid A is .... atm. (Nearest integer)

25 At 600 K, 2 moles of NO are mixed with 1 mole of O_2.

$$2NO(g) + O_2(g) \rightleftharpoons 2NO_2(g)$$

The reaction occurring as above comes to equilibrium under a total pressure of 1 atm. Analysis of the system shows that 0.6 mole of oxygen are present at equilibrium. The equilibrium constant for the reaction is .... (Nearest integer)

26 A sample of 0.125 g of an organic compound when analysed by Duma's method yields 22.78 mL of nitrogen gas collected over KOH solution at 280 K and 759 mm Hg. The percentage of nitrogen in the given organic compound is ....(Nearest integer)
Given,
(i) The vapour pressure of water of 280 K is 14.2 mm Hg.
(ii) $R = 0.082$ L atm K^{-1}mol^{-1}.

27 On reaction with stronger oxidising agent like KIO_4, hydrogen peroxide oxidises with the evolution of O_2. The oxidation number of I in KIO_4 changes to ....

28 For a reaction, given below is the graph of ln k vs $1/T$. The activation energy for the reaction is equal to .... cal mol^{-1}. (Nearest integer)
(Given, $R = 2$ cal K^{-1} mol^{-1})

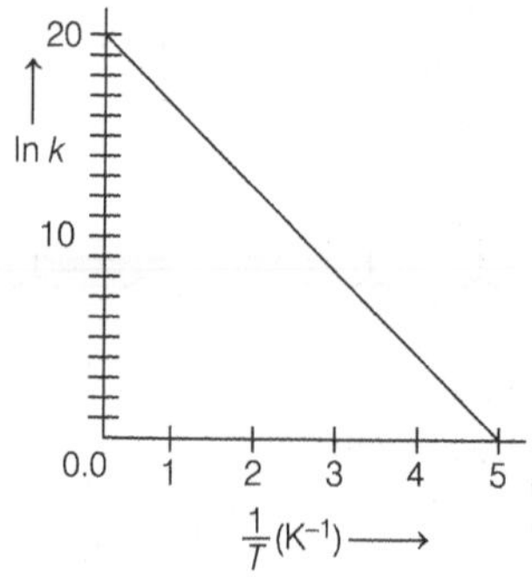

29 Among the following the number of curves not in accordance with Freundlich adsorption isotherm is ....

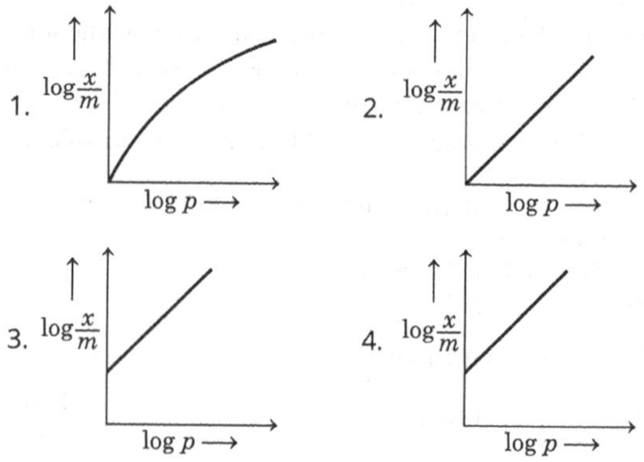

30 Among the following the number of state variables is ....
Internal energy (U)
Volume (V)
Heat (q)
Enthalpy (H)

29 JULY 2022, SHIFT-I

Section A : Objective Type Questions

1 Which of the following pair of molecules contain odd electron molecule and an expanded octet molecule?
(a) BCl_3 and SF_6 (b) NO and H_2SO_4
(c) SF_6 and H_2SO_4 (d) BCl_3 and NO

2 $\underset{(20\text{ g})}{N_2(g)} + \underset{(5\text{ g})}{3H_2(g)} \rightleftharpoons 2NH_3(g)$

Consider the above reaction, the limiting reagent of the reaction and number of moles of NH_3 formed respectively are
(a) H_2, 1.42 mol (b) H_2, 0.71 mol
(c) N_2, 1.42 mol (d) N_2, 0.71 mol

3 100 mL of 5% (w/V) solution of NaCl in water was prepared in 250 mL beaker. Albumin from the egg was poured into NaCl solution and stirred well. This resulted in a/an
(a) lyophilic sol (b) lyophobic sol
(c) emulsion (d) precipitate

4 The first ionisation enthalpy of Na, Mg and Si, respectively, are 496, 737 and 786 kJ mol^{-1}. The first ionisation enthalpy (kJ mol^{-1}) of Al is
(a) 487 (b) 768
(c) 577 (d) 856

5 In metallurgy the term "gangue" is used for
(a) contamination of undesired earthy materials.
(b) contamination of metals, other than desired metal.
(c) minerals which are naturally occurring in pure form.
(d) magnetic impurities in an ore.

6 The reaction of zinc with excess of aqueous alkali, evolves hydrogen gas and gives
(a) $Zn(OH)_2$ (b) ZnO
(c) $[Zn(OH)_4]^{2-}$ (d) $[ZnO_2]^{2-}$

7 Lithium nitrate and sodium nitrate, when heated separately, respectively, give
(a) $LiNO_2$ and $NaNO_2$
(b) Li_2O and Na_2O
(c) Li_2O and $NaNO_2$
(d) $LiNO_2$ and Na_2O

8 Number of lone pairs of electrons in the central atom of SCl_2, O_3, ClF_3 and SF_6, respectively, are
(a) 0, 1, 2 and 2 (b) 2, 1, 2 and 0
(c) 1, 2, 2 and 0 (d) 2, 2, 2 and 0

9 In following pairs, the one in which both transition metal ions are colourless is
(a) Sc^{3+}, Zn^{2+} (b) Ti^{4+}, Cu^{2+}
(c) V^{2+}, Ti^{3+} (d) Zn^{2+}, Mn^{2+}

10 In neutral or faintly alkaline medium, $KMnO_4$ being a powerful oxidant can oxidise, thiosulphate almost quantitatively, to sulphate. In this reaction overall change in oxidation state of manganese will be
(a) 5 (b) 1
(c) 0 (d) 3

11 Which among the following pairs has only herbicides ?
(a) Aldrin and dieldrin
(b) Sodium chlorate and aldrin
(c) Sodium arsinate and dieldrin
(d) Sodium chlorate and sodium arsinite

12 Which among the following is the strongest Bronsted base?

(a) 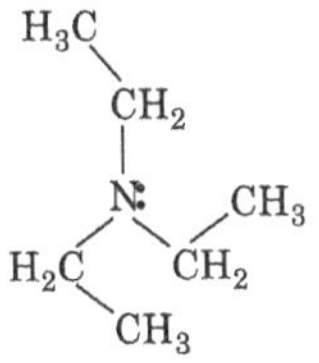

(b) 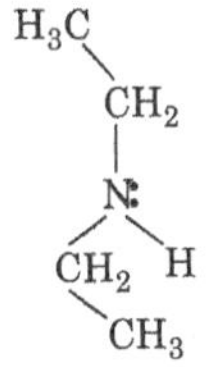

(c) 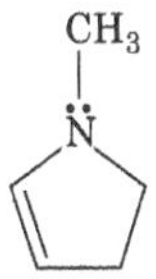

(d) 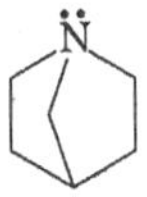

13 Which among the following pairs of the structures will give different products on ozonolysis? (Consider the double bonds in the structures are rigid and not delocalised.)

(a) 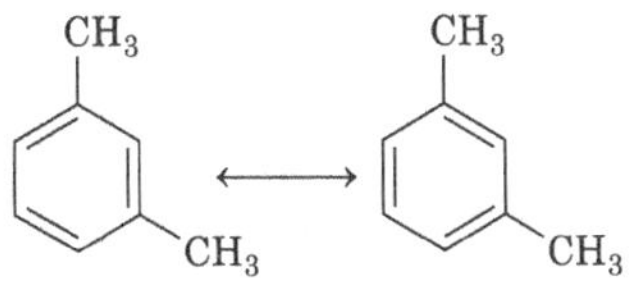

(b) 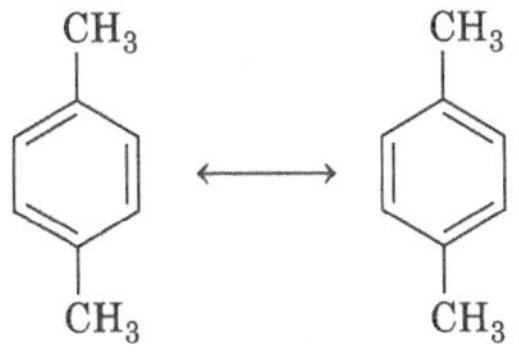

(c) 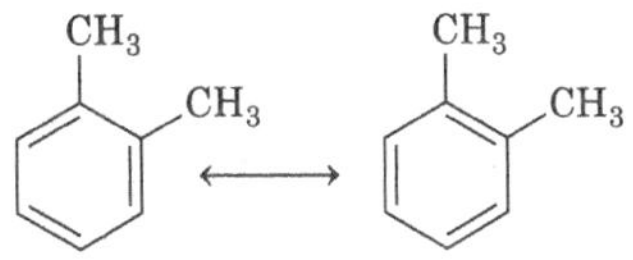

(d) 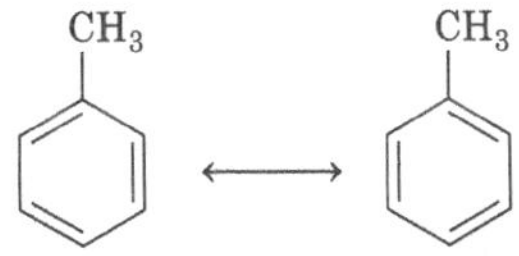

14 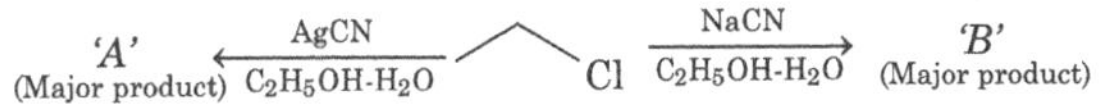

Considering the above reactions, the compound '*A*' and compound '*B*' respectively are

(a) 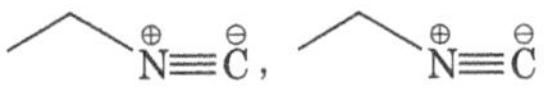

(b) 

(c) 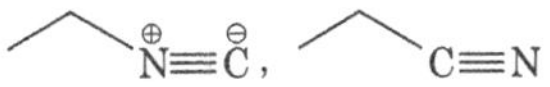

(d) 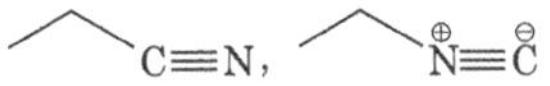

15 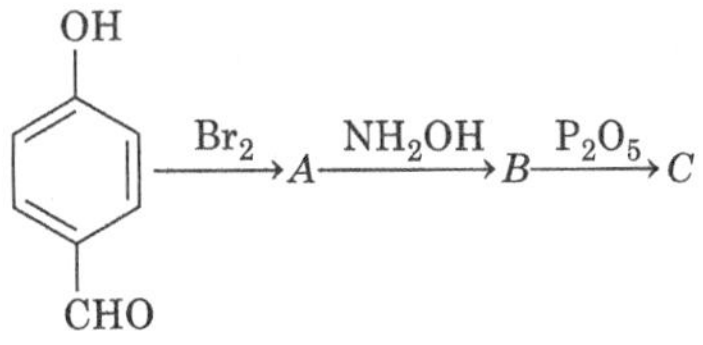

Consider the above reaction sequence, the product '*C*' is

(a) 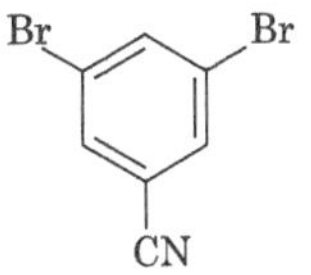

(b) 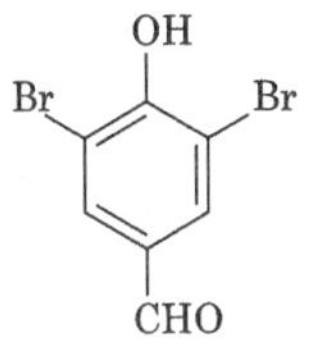

(c) 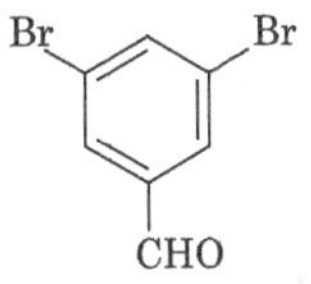

(d) 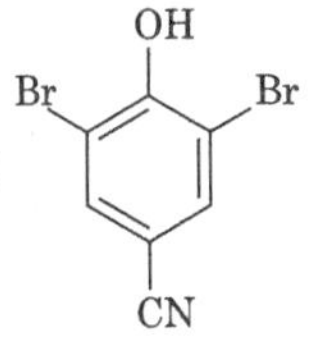

16 'A' ($C_8H_6Cl_2O$) $\xrightarrow{NH_3}$ C_8H_8ClNO $\xrightarrow[NaOH]{Br_2}$ 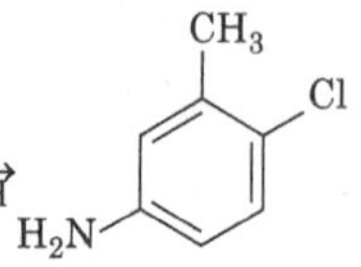

Consider the above reaction, the compound 'A' is

(a) 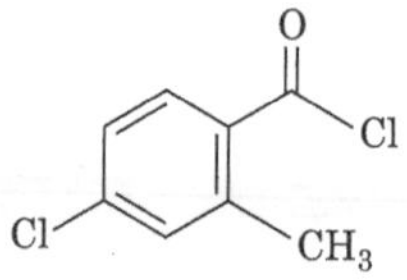

(b) 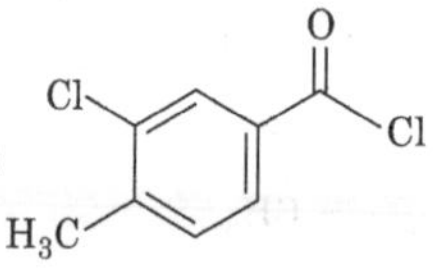

(c) 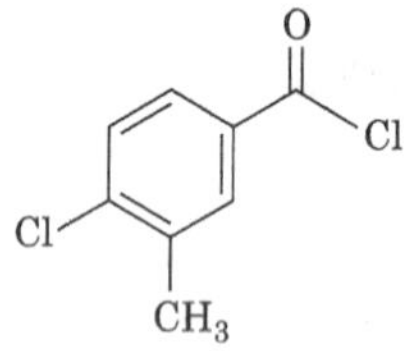

(d) 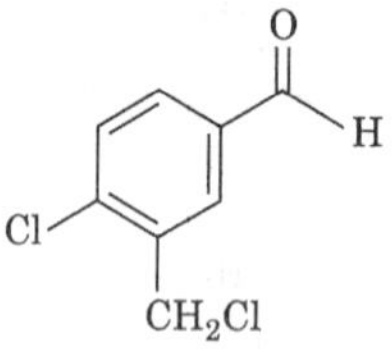

17 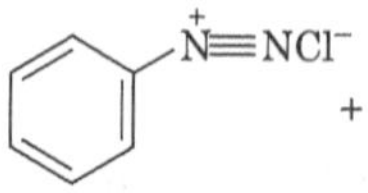

+ 'A' $\xrightarrow{NaOH}$ Product (Orange-red dye)

Which among the following represent reagent 'A'?

(a) 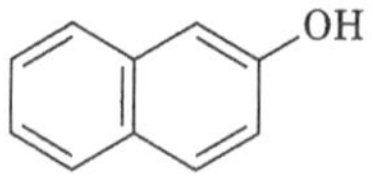

(b) 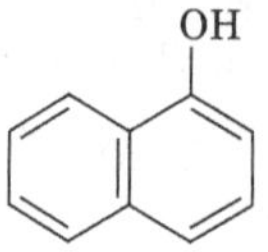

(c) 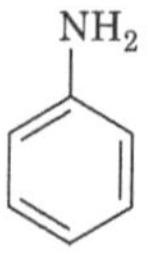

(d) 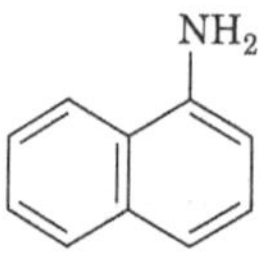

18 Consider the following reaction sequence

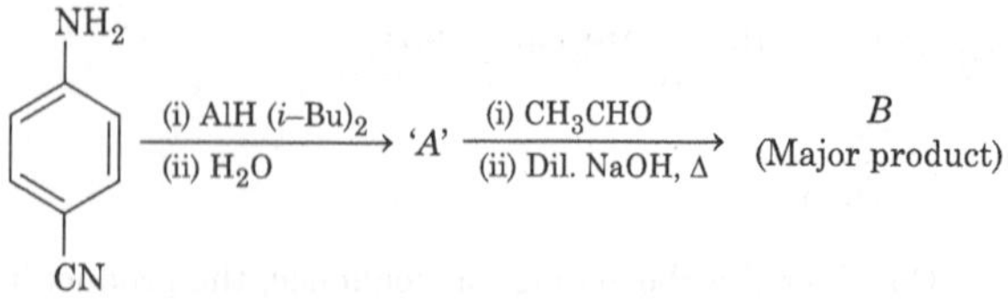

The product 'B' is

(a) 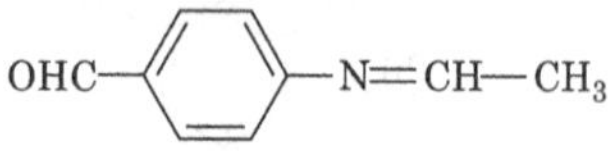

(b) 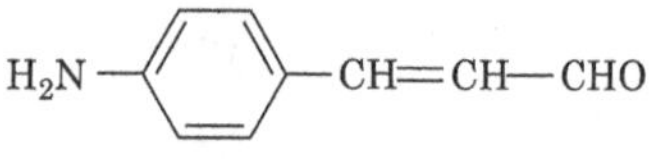

(c) 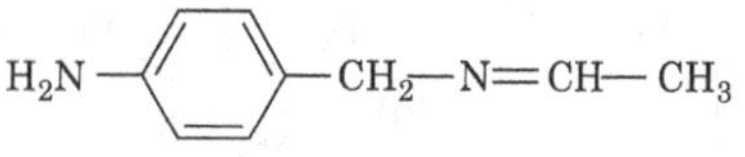

(d) 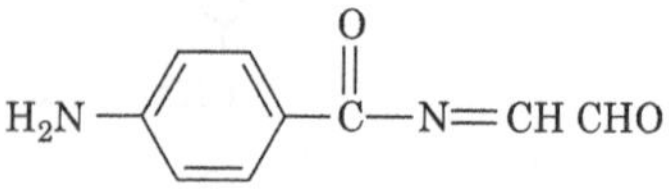

19 Which of the following compounds is an example of hypnotic drug ?

(a) Seldane (b) Amytal
(c) Aspartame (d) Prontosil

20 A compound 'X' is acidic and it is soluble in NaOH solution, but insoluble in $NaHCO_3$ solution. Compound 'X' also gives violet colour with neutral $FeCl_3$ solution. The compound 'X' is

(a) 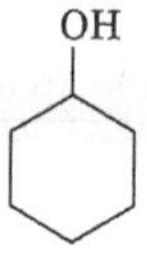

(b) 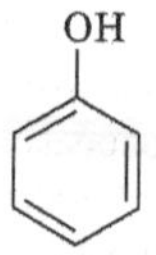

(c) 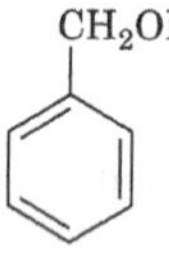

(d) 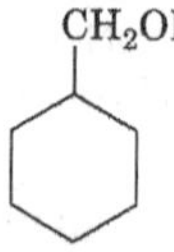

Section B : Numerical Type Questions

21 Resistance of a conductivity cell (cell constant 129 m^{-1}) filled with 74.5 ppm solution of KCl is 100 Ω (labelled as solution 1). When the same cell is filled with KCl solution of 149 ppm, the resistance is 50 Ω (labelled as solution 2). The ratio of molar conductivity of solution 1 and solution 2 is i.e. $\frac{\lambda_1}{\lambda_2} = x \times 10^{-3}$. The value of x is .... (Nearest integer)

22 Ionic radii of cation A^+ and anion B^- are 102 and 181 pm respectively. These ions are allowed to crystallise into an ionic solid. This crystal has cubic close packing for B^-. A^+ is present in all octahedral voids. The edge length of the unit cell of the crystal AB is .... pm. (Nearest integer)

23 The minimum uncertainty in the speed of an electron in an one dimensional region of length $2a_0$ (where, a_0 = Bohr radius 52.9 pm) is ...... kms^{-1}.

(Given, mass of electron = 9.1×10^{-31} kg. Planck's constant, $h = 6.63 \times 10^{-34}$ Js)

24 When 600 mL of 0.2 M HNO_3 is mixed with 400 mL of 0.1 M NaOH solution in a flask, the rise in temperature of the flask is ...... $\times 10^{-2}$°C.

(Enthalpy of neutralisation = 57 kJ mol^{-1} and specific heat of water = 4. 2JK^{-1} g^{-1})

(Neglect heat capacity of flask)

25 If O_2 gas is bubbled through water at 303 K, the number of millimoles of O_2 gas that dissolve in 1 litre of water is .... (Nearest integer).

(Given, Henry's law constant for O_2 at 303 K is 46.82 k bar and partial pressure of O_2 = 0.920 bar)

(Assume, solubility of O_2 in water is too small, nearly negligible.)

26 If the solubility product of PbS is 8×10^{-28}, then the solubility of PbS in pure water at 298 K is $x \times 10^{-16}$ mol L^{-1}. The value of x is ...... (Nearest integer)

[Given, $\sqrt{2} = 1.41$]

27 The reaction between X and Y is first order with respect to X and zero order with respect to Y.

Experiment	$[X]$ (mol L^{-1})	$[Y]$ (mol L^{-1})	Initial rate (mol L^{-1} min^{-1})
I.	0.1	0.1	2×10^{-3}
II.	L	0.2	4×10^{-3}
III.	0.4	0.4	$M \times 10^{-3}$
IV.	0.1	0.2	2×10^{-3}

Examine the data of table and calculate ratio of numerical values of M and L. (Nearest integer)

28 In a linear tetrapeptide (Constituted with different amino acids), (number of amino acids) – (number of peptide bonds) is ....

29 In bromination of propyne, with bromine 1, 1, 2, 2-tetrabromopropane is obtained in 27% yield. The amount of 1, 1, 2, 2 tetrabromopropane obtained from 1 g of bromine in this reaction is .... $\times 10^{-1}$ g. (Nearest integer)

(Molar mass of bromine = 80 g/mol)

30 $[Fe(CN)_6]^{3-}$ should be an inner orbital complex. Ignoring the pairing energy, the value of crystal field stabilisation energy for this complex is (–) .... Δ_o. (Nearest integer)

29 JULY 2022, SHIFT-II

Section A : Objective Type Questions

1 Consider the reaction,

$$4HNO_3(l) + 3KCl(s) \longrightarrow Cl_2(g) + NOCl(g) + 2H_2O(g) + 3KNO_3(s)$$

The amount of HNO_3 required to produce 110.0 g of KNO_3 is

(Given, atomic masses of H, O, N and K are 1, 16, 14 and 39 respectively.)

(a) 32.2 g (b) 69.4 g
(c) 91.5 g (d) 162.5 g

2 Given below are the quantum numbers for 4 electrons.

A. $n = 3, l = 2, m_1 = 1, m_s = +1/2$
B. $n = 4, l = 1, m_1 = 0, m_s = +1/2$
C. $n = 4, l = 2, m_1 = -2, m_s = -1/2$
(d) $n = 3, l = 1, m_1 = -1, m_s = +1/2$

The correct order of increasing energy is

(a) D < B < A < C
(b) D < A < B < C
(c) B < D < A < C
(d) B < D < C < A

3 $C(s) + O_2(s) \longrightarrow CO_2(g) + 400$ kJ

$C(s) + \frac{1}{2}O_2(g) \longrightarrow CO(g) + 100$ kJ

When coal of purity 60% is allowed to burn in presence of insufficient oxygen, 60% of carbon is converted into 'CO' and the remaining is converted into 'CO_2'. The heat generated when 0.6 kg of coal is burnt is ....

(a) 1600 kJ (b) 3200 kJ
(c) 4400 kJ (d) 6600 kJ

4 200 mL of 0.01 M HCl is mixed with 400 mL of 0.01 M H_2SO_4. The pH of the mixture is ....

(a) 1.14 (b) 1.78
(c) 2.34 (d) 3.02

5 Given below are the critical temperatures of some of the gases.

Gas	Critical temperature (K)
He	5.2
CH_4	190
CO_2	304.2
NH_3	405.5

The gas showing least adsorption on a definite amount of charcoal is

(a) He (b) CH_4 (c) CO_2 (d) NH_3

6 In liquation process used for tin (Sn), the metal

(a) is reacted with acid
(b) is dissolved in water
(c) is brought to molten form which is made to flow on a slope
(d) is fused with NaOH

7 Given below are two statements.

Statement I Stannane is an example of a molecular hydride.

Statement II Stannane is a planar molecule.

In the light of the above statements, choose the most appropriate answer from the options given below

(a) Both Statement I and Statement II are true.
(b) Both Statement I and Statement II are false.
(c) Statement I is true but Statement II is false.
(d) Statement I is false but Statement II is true.

8 Portland cement contains 'X' to enhance the setting time. What is 'X'?

(a) $CaSO_4 \cdot \frac{1}{2}H_2O$ (b) $CaSO_4 \cdot 2H_2O$
(c) $CaSO_4$ (d) $CaCO_3$

9 When borax is heated with CoO on a platinum loop, blue coloured bead formed is largely due to

(a) B_2O_3 (b) $Co(BO_2)_2$
(c) CoB_4O_7 (d) $Co[B_4O_5(OH)_4]$

10 Which of the following 3*d*-metal ion will give the lowest enthalpy of hydration ($\Delta_{hyd} H$), when dissolved in water ?
(a) Cr^{2+} (b) Mn^{2+}
(c) Fe^{2+} (d) Co^{2+}

11 Octahedral complexes of copper (II) undergo structural distortion (Jahn-Teller). Which one of the given copper (II) complexes will show the maximum structural distortion ?
(en–ethylenediamine; $H_2N—CH_2—CH_2—NH_2$)
(a) $[Cu(H_2O)_6]SO_4$ (b) $[Cu(en)(H_2O)_4]SO_4$
(c) *cis*-$[Cu(en)_2Cl_2]$ (d) *trans*-$[Cu(en)_2Cl_2]$

12 Dinitrogen is a robust compound, but reacts at high altitude to form oxides. The oxide of nitrogen that can damage plant leaves and retard photosynthesis is
(a) NO (b) NO_3^- (c) NO_2 (d) NO_2^-

13 Correct structure of γ-methylcyclohexane carbaldehyde is

(a) 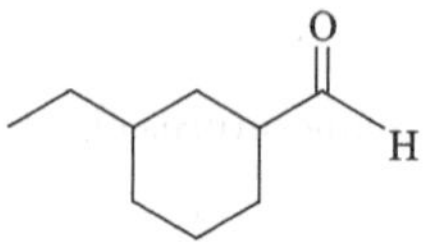
(b) 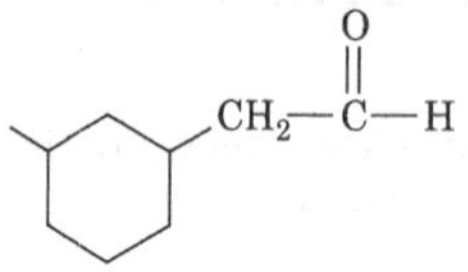
(c) 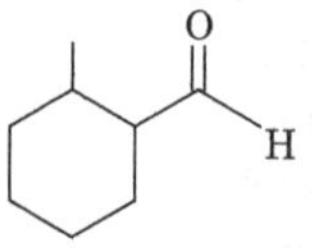
(d) 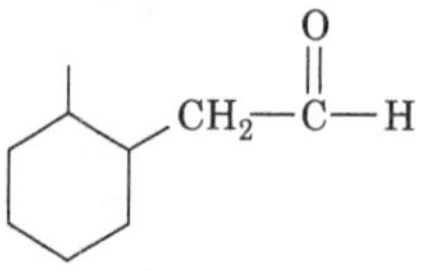

14 Compound '*A*' undergoes following sequence of reactions to give compound '*B*'. The correct structure and chirality of compound '*B*' is
[where, Et is $—C_2H_5$]

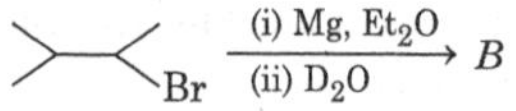

Compound '*A*'

(a) 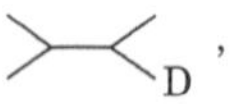 , Achiral
(b) 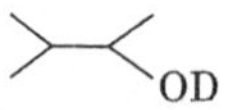 , Chiral
(c) 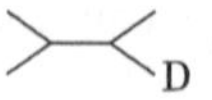 , Chiral
(d)  , Achiral

15 Given below are two statements.
Statement I The compound A.

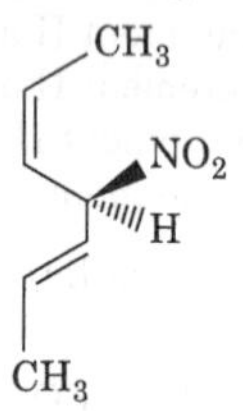

is optically active.

Statement II 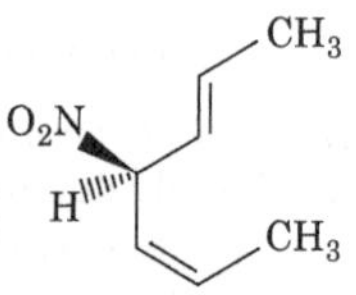 is mirror image of above compound A.

In the light of the above statements, choose the most appropriate answer from the options given below.
(a) Both statement I and statement II are correct.
(b) Both statement I and statement II are incorrect.
(c) Statement I is correct but statement II is incorrect.
(d) Statement I is incorrect but statement II is correct.

16 When ethanol is heated with conc. H_2SO_4, a gas is produced. The compound formed, when this gas is treated with cold dilute aqueous solution of Baeyer's reagent, is
(a) formaldehyde (b) formic acid
(c) glycol (d) ethanoic acid

17 The Hinsberg reagent is

(a) 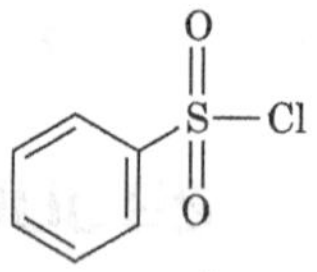
(b) 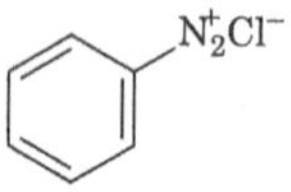
(c) 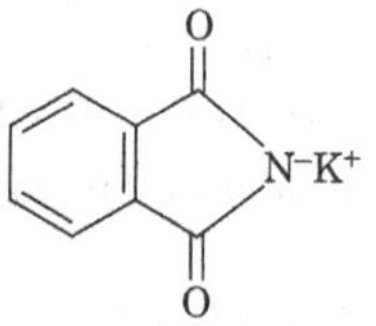
(d) 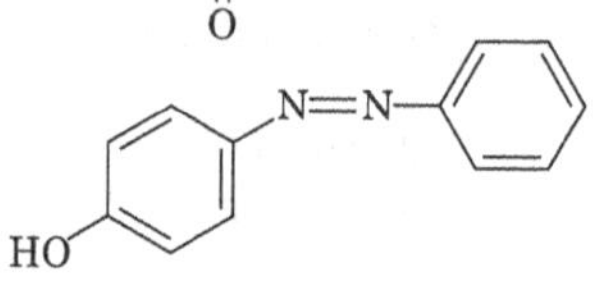

18 Which of the following is not a natural polymer?
(a) Protein (b) Starch
(c) Rubber (d) Rayon

19 Given below are two statements. One is labelled as Assertion (A) and the other is labelled as Reason (R).
Assertion (A) Amylose is insoluble in water.
Reason (R) Amylose is a long linear molecule with more than 200 glucose units.
In the light of the above statements, choose the correct answer from the options given below.
(a) Both A and R are correct and R is the correct explanation of A.
(b) Both A and R are correct and R is NOT the correct explanation of A.
(c) A is correct but R is not correct.
(d) A is not correct but R is correct.

20 A compound '*X*' is a weak acid and it exhibits colour change at pH close to the equivalence point during neutralisation of NaOH with CH_3COOH. Compound '*X*' exists in ionised form in basic medium. The compound '*X*' is
(a) methyl orange
(b) methyl red
(c) phenolphthalein
(d) erichrome black T

Section B : Numerical Type Questions

21 'x' g of molecular oxygen (O_2) is mixed with 200 g of neon (Ne). The total pressure of the non-reactive mixture of O_2 and Ne in the cylinder is 25 bar. The partial pressure of Ne is 20 bar at the same temperature and volume. The value of 'x' is ....

[Given : Molar mass of O_2 = 32 g mol^{-1}.

Molar mass of Ne = 20 g mol^{-1}]

22 Consider, PF_5, BrF_5, PCl_3, SF_6, $[ICl_4]^-$, ClF_3 and IF_5.

Amongst the above molecule(s)/ion(s), the number of molecule(s)/ion(s) having sp^3d^2 hybridisation is ....

23 1.80 g of solute A was dissolved in 62.5 cm^3 of ethanol and freezing point of the solution was found to be 155.1 K. The molar mass of solute A is .... g mol^{-1}.

[Given : Freezing point of ethanol is 156.0 K.

Density of ethanol is 080 g cm^{-3}.

Freezing point depression constant of ethanol is 2.00 K kg mol^{-1}]

24 For a cell, Cu (s) | Cu^{2+} (0.001 M) || Ag^+ (0.01 M | Ag(s)

the cell potential is found to be 0.43 V at 298 K. The magnitude of standard electrode potential for Cu^{2+} /Cu is .... $\times 10^{-2}$ V.

[Given : $E^\circ_{Ag^+/Ag}$ = 0.80 V and $\frac{2.303\,RT}{F}$ = 0.06 V]

25 Assuming 1 μ g of trace radioactive element X with a half-life of 30 years is absorbed by a growing tree. The amount of X remaining in the tree after 100 years is .... $\times 10^{-1}$μ g.

[Given : ln 10 = 2.303; log 2 = 0.30]

26 Sum of oxidation state (magnitude) and coordination number of cobalt in Na[Co(bpy)Cl_4] is ....

(Given, bpy = 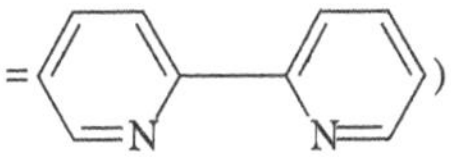)

27 Consider the following sulphur based oxoacids.

H_2SO_3, H_2SO_4, $H_2S_2O_8$ and $H_2S_2O_7$.

Amongst these oxoacids, the number of those with peroxo (O — O) bond is ....

28 A 1.84 mg sample of polyhydric alcoholic compound 'X' of molar mass 92.0 g/mol gave 1.344 mL of H_2 gas at STP. The number of alcoholic hydrogens present in compound 'X' is ....

29 The number of stereoisomers formed in a reaction of (±) Ph(C=O)C(OH)(CN)Ph with HCN is ....

30 The number of chlorine atoms in Bithional is ....

Answers

For solutions scan the QR code

25 July 2022, Shift-I

1. (c)	2. (c)	3. (c)	4. (c)	5. (c)	6. (d)	7. (d)	8. (a)	9. (a)	10. (b)
11. (b)	12. (c)	13. (d)	14. (b)	15. (d)	16. (c)	17. (a)	18. (a)	19. (b)	20. (c)
21. 2	22. 104	23. 22	24. 4	25. 1	26. 0	27. 5	28. 56	29. 1	30. 2

25 July 2022, Shift-II

1. (a)	2. (b)	3. (d)	4. (a)	5. (c)	6. (d)	7. (b)	8. (c)	9. (c)	10. (b)
11. (a)	12. (c)	13. (d)	14. (b)	15. (a)	16. (a)	17. (b)	18. (b)	19. (b)	20. (a)
21. 46	22. 1655	23. 4	24. 10	25. 2	26. 3	27. 4	28. 33	29. 2	30. 8

26 July 2022, Shift-I

1. (c)	2. (b)	3. (d)	4. (a)	5. (a)	6. (b)	7. (b)	8. (c)	9. (a)	10. (c)
11. (d)	12. (b)	13. (c)	14. (c)	15. (d)	16. (a)	17. (b)	18. (d)	19. (a)	20. (c)
21. 24	22. 2	23. 1758	24. 200	25. 54	26. 2	27. 3	28. 2	29. 0	30. 6

26 July 2022, Shift-II

1. (c)	2. (b)	3. (c)	4. (c)	5. (a)	6. (b)	7. (a)	8. (d)	9. (c)	10. (a)
11. (a)	12. (a)	13. (b)	14. (b)	15. (b)	16. (a)	17. (b)	18. (c)	19. (c)	20. (c)
21. 8	22. 4	23. 57	24. 1	25. 4	26. 2	27. 5	28. 12	29. 3	30. 1

27 July 2022, Shift-I

1. (b)	2. (a)	3. (a)	4. (c)	5. (b)	6. (d)	7. (a)	8. (c)	9. (d)	10. (d)
11. (b)	12. (c)	13. (b)	14. (d)	15. (a)	16. (b)	17. (a)	18. (c)	19. (a)	20. (c)
21. 24	22. 2	23. 4	24. 2	25. 3	26. 0	27. 3	28. 1	29. 42	30. 30

For solutions scan the QR code

27 July 2022, Shift-II

1. *(a)*	2. *(c)*	3. *(a)*	4. *(b)*	5. *(c)*	6. *(b)*	7. *(a)*	8. *(b)*	9. *(d)*	10. *(a)*
11. *(a)*	12. *(c)*	13. *(b)*	14. *(a)*	15. *(a)*	16. *(d)*	17. *(a)*	18. *(c)*	19. *(d)*	20. *(d)*
21. *1*	22. *25*	23. *35*	24. *6*	25. *165*	6. *4*	27. *4*	28. *6*	29. *2*	30. *530*

28 July 2022, Shift-I

1. *(d)*	2. *(a)*	3. *(c)*	4. *(c)*	5. *(c)*	6. *(c)*	7. *(b)*	8. *(a)*	9. *(c)*	10. *(a)*
11. *(d)*	12. *(c)*	13. *(b)*	14. *(c)*	15. *(b)*	16. *(c)*	17. *(c)*	18. *(d)*	19. *(c)*	20. *(a)*
21. *2*	22. *22*	23. *4*	24. *15*	25. *27*	26. *100*	27. *3*	28. *4*	29. *3*	30. *46*

28 July 2022, Shift-II

1. *(a)*	2. *(a)*	3. *(a)*	4. *(c)*	5. *(d)*	6. *(b)*	7. *(c)*	8. *(a,d)*	9. *(a)*	10. *(b)*
11. *(d)*	12. *(c)*	13. *(b)*	14. *(d)*	15. *(a)*	16. *(b)*	17. *(d)*	18. *(a)*	19. *(a)*	20. *(c)*
21. *25*	22. *87*	23. *2*	24. *2*	25. *2*	26. *22*	27. *5*	28. *8*	29. *3*	30. *3*

29 July 2022, Shift-I

1. *(b)*	2. *(c)*	3. *(a)*	4. *(c)*	5. *(a)*	6. *(d)*	7. *(c)*	8. *(b)*	9. *(a)*	10. *(d)*
11. *(d)*	12. *(d)*	13. *(c)*	14. *(c)*	15. *(d)*	16. *(c)*	17. *(a)*	18. *(b)*	19. *(b)*	20. *(b)*
21. *1000*	22. *566*	23. *548*	24. *54*	25. *1*	26. *282*	27. *40*	28. *1*	29. *3*	30. *2*

29 July 2022, Shift-II

1. *(c)*	2. *(b)*	3. *(d)*	4. *(b)*	5. *(a)*	6. *(c)*	7. *(c)*	8. *(b)*	9. *(b)*	10. *(b)*
11. *(a)*	12. *(c)*	13. *(a)*	14. *(c)*	15. *(c)*	16. *(c)*	17. *(a)*	18. *(d)*	19. *(d)*	20. *(c)*
21. *80*	22. *4*	23. *80*	24. *34*	25. *1*	26. *9*	27. *1*	28. *3*	29. *3*	30. *4*

ONLINE SOLVED PAPERS

JEE Main 2023

(January Attempt)

24 JANUARY 2023, SHIFT-I

Section A : Objective Type Questions

1 Which of the phosphorus oxoacid can create silver mirror from $AgNO_3$ solution?

(a) $H_4P_2O_6$ (b) $H_4P_2O_7$ (c) $(HPO_3)_n$ (d) $H_4P_2O_5$

2 Order of covalent bond;

A. KF > KI; LiF > KF
B. KF < KI; LiF > KF
C. $SnCl_4 > SnCl_2$; CuCl > NaCl
D. LiF > KF; CuCl < NaCl
E. KF< KI; CuCl > NaCl

Choose the correct answer from the options given below.

(a) B, C only (b) A, B only (c) C, E only (d) B,C,E only

3 In the depression of freezing point experiment.

A. Vapour pressure of the solution is less than that of pure solvent
B. Vapour pressure of the solution is more than that of pure solvent
C. Only solute molecules solidify at the freezing point
D. Only solvent molecules solidify at the freezing point

Choose the most appropriate answer from the options given below.

(a) B and C only (b) A and D only
(c) A and C only (d) A only

4 The primary and secondary valencies of cobalt respectively in $[Co(NH_3)_5Cl]Cl_2$ are

(a) 3 and 6 (b) 2 and 6 (c) 3 and 5 (d) 2 and 8

5 Increasing order of stability of the resonance structures is

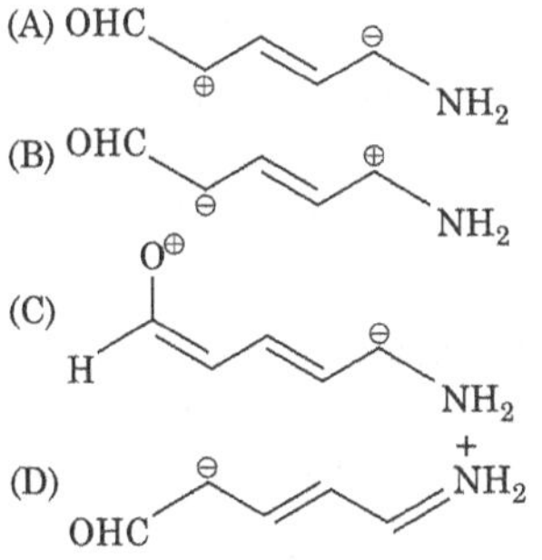

Choose the correct answer from the options given below.

(a) D, C, B, A (b) D, C, A, B
(c) C, D, B, A (d) C, D, A, B

6 'A' and 'B' formed in the following set of reactions are

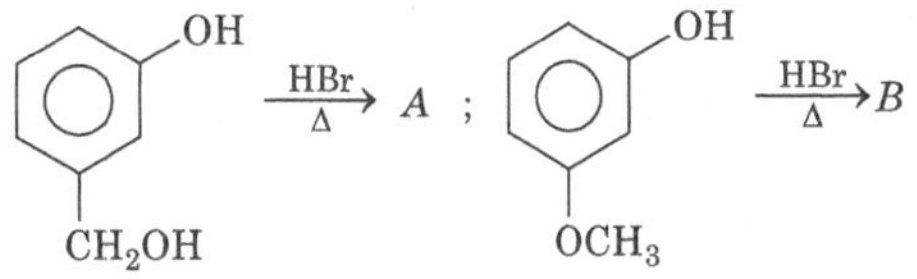

(a) 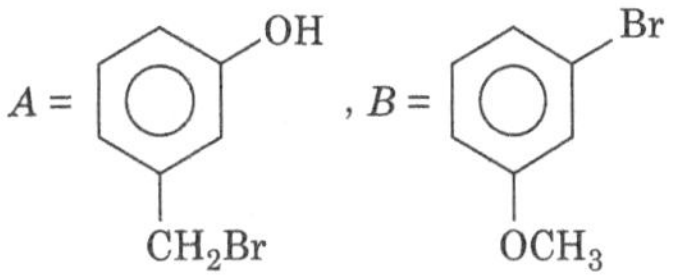

(b) 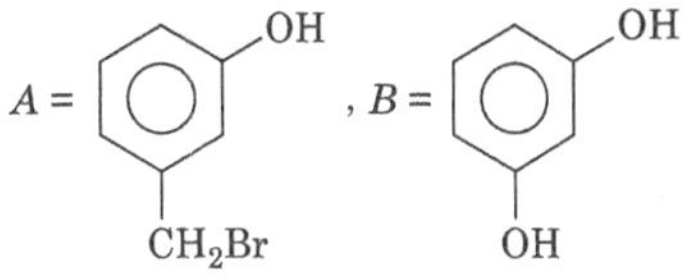

(c) 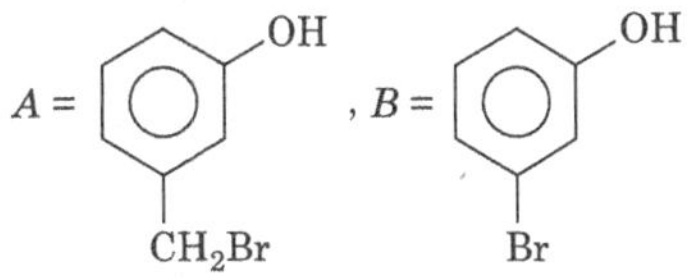

(d) 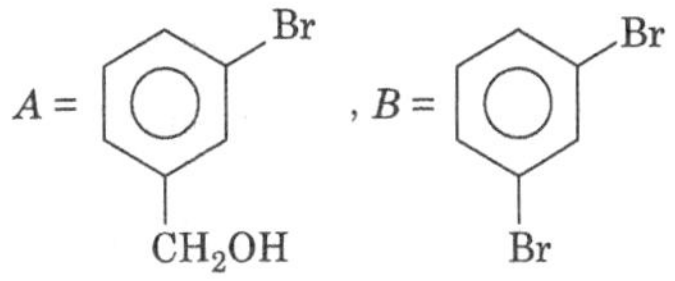

7 Reaction of BeO with ammonia and hydrogen fluoride gives A which on thermal decomposition gives BeF_2 and NH_4F. What is 'A'?

(a) $(NH_4)_2BeF_4$ (b) $(NH_4)BeF_3$
(c) H_3NBeF_3 (d) $(NH_4)Be_2F_5$

8 An ammoniacal metal salt solution gives a brilliant red precipitate on addition of dimethylglyoxime. The metal ion is

(a) Ni^{2+} (b) Cu^{2+} (c) Co^{2+} (d) Fe^{2+}

9 **Statement I** For colloidal particles, the values of colligative properties are of small order as compared to values shown by true solutions at same concentration.

Statement II For colloidal particles, the potential difference between the fixed layer and the diffused layer of same charges is called the electrokinetic potential or zeta potential.

In the light of the above statements, choose the correct answer from the options given below.

(a) Statement I is false but Statement II is true.
(b) Statement I is true but Statement II is false.
(c) Both Statement I and Statement II are false.
(d) Both Statement I and Statement II are true.

10 Decreasing order of the hydrogen bonding in following forms of water is correctly represented by

A. liquid water B. ice C. impure water

Choose the correct answer from the options given below.

(a) A > B > C (b) C > B > A
(c) B > A > C (d) A = B > C

11 **Assertion (A)** Hydrolysis of an alkyl chloride is a slow reaction but in the presence of NaI, the rate of the hydrolysis increases.

Reason (R) I^- is a good nucleophile as well as a good leaving group.

In the light of the above statements, choose the correct answer from the options given below.

(a) Both A and R are true and R is the correct explanation of A.
(b) Both A and R are true but R is not the correct explanation of A.
(c) A is false but R is true.
(d) A is true but R is false.

12 Match List-I with List-II.

List-I		List-II	
A.	Chlorophyll	I.	Na_2CO_3
B.	Soda ash	II.	$CaSO_4$
C.	Dentistry, ornamental work	III.	Mg^{2+}
D.	Used in white washing	IV.	$Ca(OH)_2$

Choose the correct answer from the options given below.

(a) A-II, B-I, C-III, D-IV (b) A-III, B-I, C-II, D-IV
(c) A -II, B-III, C-IV, D-I (d) A-III, B-IV, C-I, D-II

13 It is observed that characteristic X-ray spectra of elements show regularity. When frequency to the power 'n' i.e. v^n of X-rays emitted is plotted against atomic number 'Z', following graph is obtained.

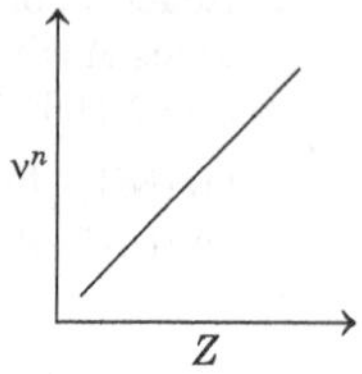

The value of 'n' is

(a) 1 (b) 3 (c) 1/2 (d) 2

14 Match List-I with List-II.

List-I		List-II	
A.	Reverberatory furnace	I.	Pig iron
B.	Electrolytic cell	II.	Aluminium
C.	Blast furnace	III.	Silicon
D.	Zone-refining furnace	IV.	Copper

Choose the correct answer from the options given below.

(a) A-I, B-III, C-II, D-IV (b) A-I, B-IV, C-II, D-III
(c) A -IV, B-II, C-I, D-III (d) A-III, B-IV, C-I, D-II

15 The magnetic moment of a transition metal compound has been calculated to be 3.87 BM. The metal ion is

(a) Ti^{2+} (b) Cr^{2+}
(c) Mn^{2+} (d) V^{2+}

16 Compound (X) undergoes following sequence of reactions to give the lactone (Y).

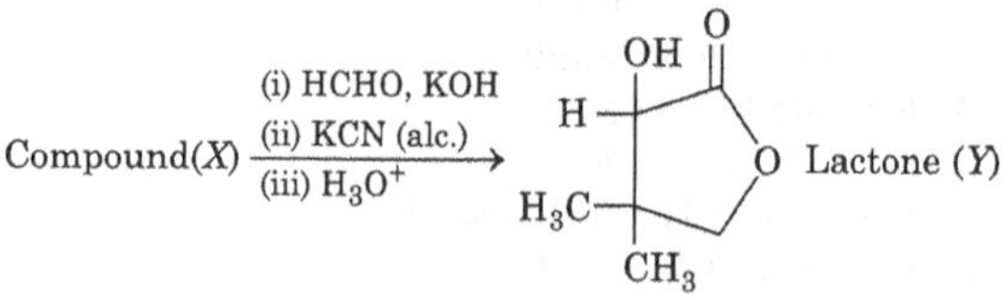

Compound (X) is

(a) $HOH_2C—C(CH_3)_2—CHO$ (b) $H_2C(OH)—CH_2CH_2CH_2CHO$

(c) $H—C(CH_3)_2—CHO$ (d) $H_2C(CH_3)—CH_2—CHO$

17 In the following given reaction, 'A' is

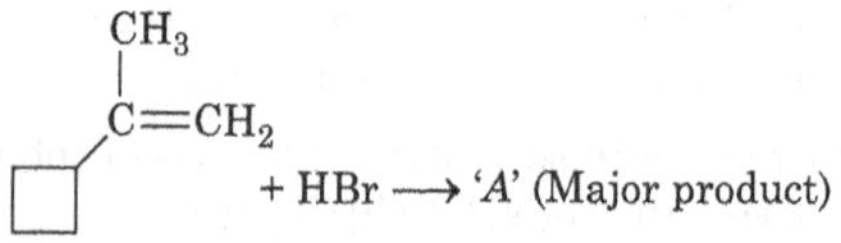

(a) 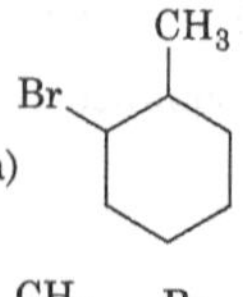

(b) 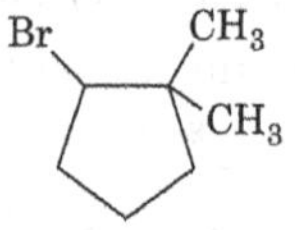

(c) 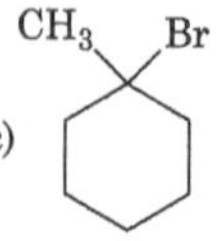

(d) 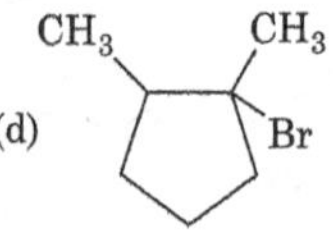

18 Given below are two statements.

Statement I Noradrenaline is a neurotransmitter.

Statement II Low level of noradrenaline is not the cause of depression in human.

In the light of the above statements, choose the correct answer from the options given below.

(a) Statement I is false but Statement II is true.
(b) Both Statement I and Statement II are false.
(c) Statement I is true but Statement II is false.
(d) Both Statement I and Statement II are true.

19 'R' formed in the following sequence of reaction is

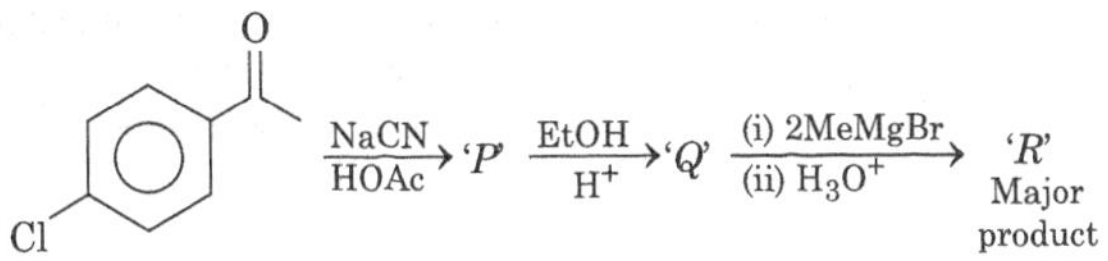

(a) 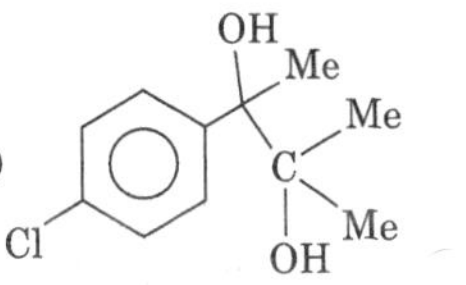 (b) 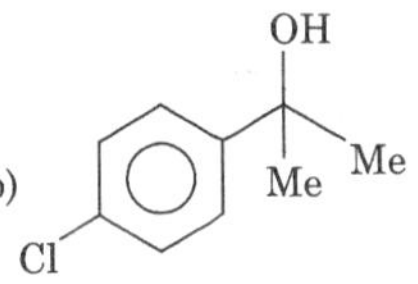

(c) 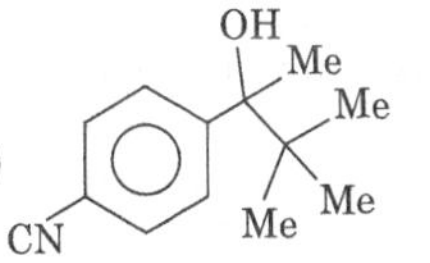 (d) 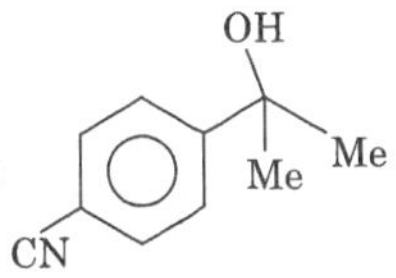

20 Which of the following is true about freons?

(a) All radicals are called freons
(b) These are radicals of chlorine and chlorine monoxide
(c) These are chemicals causing skin cancer
(d) These are chlorofluorocarbon compounds

Section B : Numerical Value Type Questions

21 If wavelength of the first line of the Paschen series of hydrogen atom is 720 nm, then the wavelength of the second line of this series is .... nm. (Nearest integer)

22 The d-electronic configuration of $[CoCl_4]^{2-}$ in tetrahedral crystal field is $e^m t_2^n$. Sum of "m" and "number of unpaired electrons" is ....

23 For independent processes at 300 K.

Process	ΔH/kJ mol^{-1}	ΔS/J K^{-1}
A	− 25	− 80
B	− 22	40
C	25	− 50
D	22	20

The number of non-spontaneous processes from the given data is ....

> **Key Idea** In this question, to find out the number of non-spontaneous processes, first find the value of Gibbs free energy by using the equation $\Delta G = \Delta H - T\Delta S$. If the value of ΔG is negative, then the process is spontaneous, while if ΔG is positive then the value is non-spontaneous.

24 The dissociation constant of acetic acid is $x \times 10^{-5}$. When 25 mL of 0.2 M CH_3COONa solution is mixed with 25 mL of 0.02 M CH_3COOH solution, the pH of the resultant solution is found to be equal to 5. The value of x is ....

25 The number of correct statements from the following is ....

(A) larger the activation energy, smaller is the value of the rate constant.
(B) the higher is the activation energy, higher is the value of the temperature coefficient.
(C) at lower temperature, increase in temperature causes more change in the value of k than at higher temperature.
(D) a plot of ln k vs $\frac{1}{T}$ is a straight line with slope equal to $-\frac{E_a}{R}$

26 At 298 K, a 1 litre solution containing 10 m mol of $Cr_2O_7^{2-}$ and 100 m mol of Cr^{3+} shows a pH of 3.0.

[Given : $Cr_2O_7^{2-} \longrightarrow Cr^{3+}$; $E° = 1.330$ V and $\frac{2.303RT}{F} = 0.059$ V]

The potential for the half-cell reaction is $x \times 10^{-3}$ V. The value of x is ....

> **Key Idea** To calculate the potential for the half-cell reaction. First of all, write the balanced chemical equation. Then calculate the value of n from it. After that use Nernst equation to calculate potential for half-cell reaction.

27 Number of moles of AgCl formed in the following reaction is ....

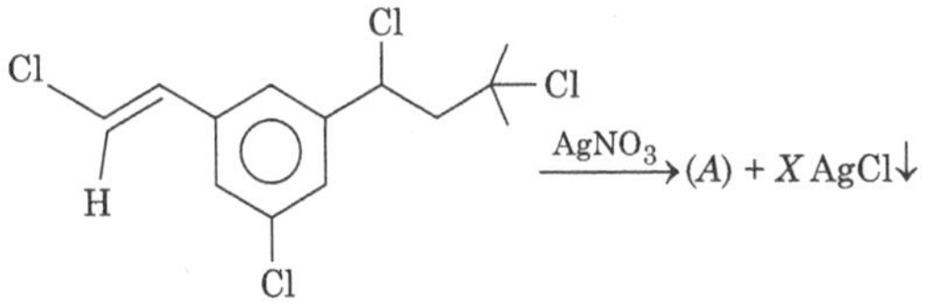

28 When $Fe_{0.93}O$ is heated in presence of oxygen, it converts to Fe_2O_3. The number of correct statements from the following is ....

(A) the equivalent weight of $Fe_{0.93}O$ is molecular weight 0.79
(B) the number of moles of Fe^{2+} and Fe^{3+} in 1 mole of $Fe_{0.93}O$ is 0.79 and 0.14 respectively
(C) $Fe_{0.93}O$ is metal deficient with lattice comprising of cubic closed packed arrangement of O^{2-} ions
(D) The % composition of Fe^{2+} and Fe^{3+} in $Fe_{0.93}O$ is 85% and 15% respectively

29 5 g of NaOH was dissolved in deionised water to prepare a 450 mL stock solution. What volume (in mL) of this solution would be required to prepare 500 mL of 0.1 M solution?

[Given : molar mass of Na, O and H is 23, 16 and 1 g mol^{-1} respectively.]

30 Uracil is a base present in RNA with the following structure, % of N in uracil is ......

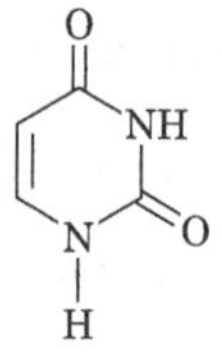

[Given : Molar mass of $N = 14 \text{ g mol}^{-1}$

$O = 16 \text{ g mol}^{-1}$

$C = 12 \text{ g mol}^{-1}$

$H = 1 \text{ g mol}^{-1}$]

> **Key Idea** In this question, to calculate the percentage of N in uracil. First of all count the number of nitrogen and find its molecular mass then calculate the molecular mass of uracil. After that find the percentage, by using formula $\frac{\text{molecular mass of nitrogen}}{\text{molecular mass of uracil}} \times 100$

24 JANUARY 2023, SHIFT-II

Section A : Objective Type Questions

1 $K_2Cr_2O_7$ paper acidified with dilute H_2SO_4 turns green when exposed to

(a) sulphur trioxide (b) hydrogen sulphide
(c) carbon dioxide (d) sulphur dioxide

2 The number of *s*-electrons present in an ion with 55 protons in its unipositive state is

(a) 9 (b) 8
(c) 10 (d) 12

> **Key Idea** In this question, to calculate the number of *s*-electrons present in an ion with given number or protons. First of all note that atomic number will be equal to number of protons. Then write down the electronic configuration of element with given atomic number. Then count the number of *s*-electrons with the help of electronic configuration.

3 Which one amongst the following are good oxidising agents?

A. Sm^{2+} B. Ce^{2+}
C. Ce^{4+} D. Tb^{4+}

Choose the most appropriate answer from the options given below.

(a) C and D (b) A and B
(c) Only C (d) Only D

4 Given below are two statements.

Statement I 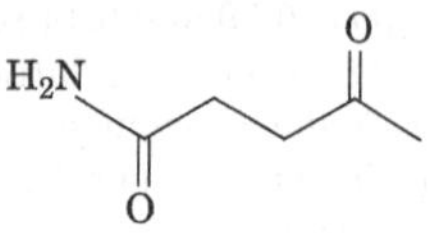 under Clemmensen reduction conditions will give 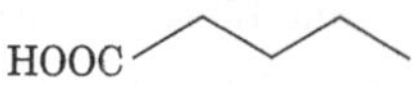

Statement II 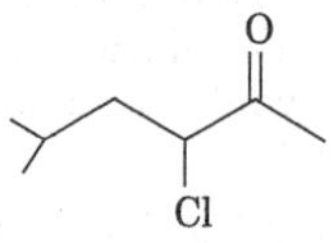 under Wolff-Kishner reduction condition will give 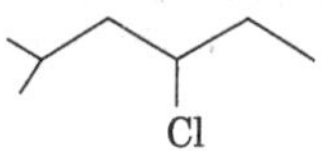

In the light of the above statements, choose the correct answer from the options given below.

(a) Statement I is false but Statement II is true.
(b) Both Statement I and Statement II are true.
(c) Both Statement I and Statement II are false.
(d) Statement I is true but Statement II is false.

5 Given below are two statements :

Statement I Pure aniline and other arylamines are usually colourless.

Statement II Arylamines get coloured on storage due to atmospheric reduction.

In the light of the above statements, choose the most appropriate answer form the options given below.

(a) Both Statement I and Statement II are incorrect.
(b) Both Statement I and Statement II are correct.
(c) Statement I is correct but Statement II is incorrect.
(d) Statement I is incorrect but Statement II is correct.

6 What is the number of unpaired electrons in the highest occupied molecular orbital of the following species N_2; N_2^+; O_2; O_2^+ ?

(a) 2, 1, 2, 1 (b) 0, 1, 0, 1
(c) 2, 1, 0, 1 (d) 0, 1, 2, 1

7 Which of the following cannot be explained by crystal field theory?

(a) The order of spectrochemical series
(b) Magnetic properties of transition metal complexes
(c) Stability of metal complexes
(d) Colour of metal complexes

8 Identify the correct statements about alkali metals.

A. The order of standard reduction potential $(M^+ \mid M)$ for alkali metal ions is Na > Rb > Li.
B. CsI is highly soluble in water.
C. Lithium carbonate is highly stable to heat.
D. Potassium dissolved in concentrated liquid ammonia is blue in colour and paramagnetic.
E. All the alkali metal hydrides are ionic solids.

Choose the correct answer from the options given below.

(a) C and E (b) A and E
(c) A, B and D (d) A, B and E

9 Choose the correct representation of conductometric titration of benzoic acid *vs* sodium hydroxide.

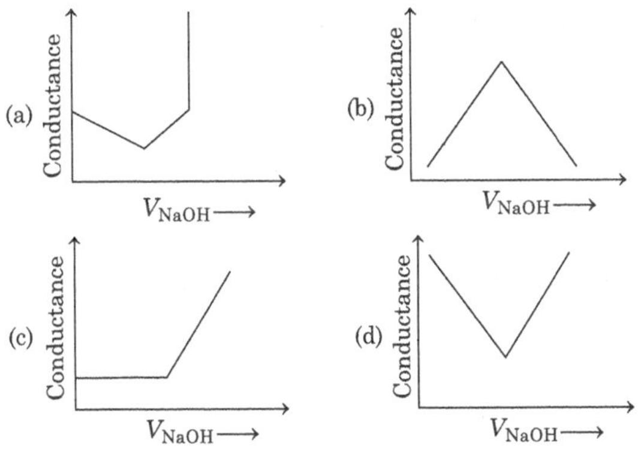

10 The metal which is extracted by oxidation and subsequent reduction from its ore is
(a) Fe (b) Al (c) Cu (d) Ag

11 A student has studied the decomposition of a gas AB_3 at 25°C. He obtained the following data.

p (mm Hg)	50	100	200	400
Relative $t_{1/2}$ (s)	4	2	1	0.5

The order of the reaction is
(a) 1 (b) 0 (zero) (c) 0.5 (d) 2

12 Given below are two statements, one is labelled as Assertion (A) and the other is labelled as Reason (R).

Assertion (A) Benzene is more stable than hypothetical cyclohexatriene.

Reason (R) The delocalised π-electron cloud is attracted more strongly by nuclei of carbon atoms.

In the light of the above statements, choose the correct answer from the options given below.
(a) Both A and R are correct and R is the correct explanation of A.
(b) A is false but R is true.
(c) A is true but R is false.
(d) Both A and R are correct but R is not the correct explanation of A.

13 Which will undergo deprotonation most readily in basic medium?

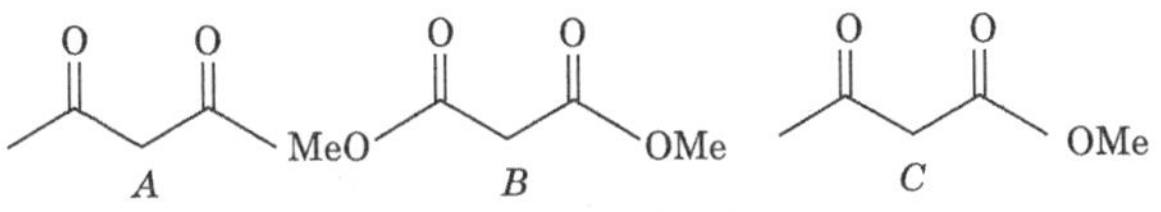

(a) *C* only (b) *B* only
(c) Both *A* and *C* (d) *A* only

14 The hybridisation and magnetic behaviour of cobalt ion in $[Co(NH_3)_6]^{3+}$ complex, respectively is
(a) d^2sp^3 and diamagnetic (b) d^2sp^3 and paramagnetic
(c) sp^3d^2 and diamagnetic (d) sp^3d^2 and paramagnetic

15 In which of the following reactions the hydrogen peroxide acts as a reducing agent?
(a) $Mn^{2+} + H_2O_2 \longrightarrow Mn^{4+} + 2OH^-$
(b) $HOCl + H_2O_2 \longrightarrow H_3O^+ + Cl^- + O_2$
(c) $PbS + 4H_2O_2 \longrightarrow PbSO_4 + 4H_2O$
(d) $2Fe^{2+} + H_2O_2 \longrightarrow 2Fe^{3+} + 2OH^-$

16 Given below are two statements, one is labelled as Assertion (A) and the other is labelled as Reason (R).

Assertion (A) Beryllium has less negative value of reduction potential compared to the other alkaline earth metals.

Reason (R) Beryllium has large hydration energy due to small size of Be^{2+} but relatively large value of atomisation enthalpy.

In the light of the above statements, choose the most appropriate answer from the options given below.
(a) A is correct but R is not correct.
(b) Both A and R are correct but R is not the correct explanation of A.
(c) Both A and R are correct and R is the correct explanation of A.
(d) A is not correct but R is correct

17 Correct statement is
(a) an average human being consumes equal amount of food and air.
(b) an average human being consumes 100 times more air than food.
(c) an average human being consumes more food than air.
(d) an average human being consumes nearly 15 times more air than food.

18 Match List-I with List-II.

	List-I (Type)		**List-II** (Name)
A.	Antifertility drug	I.	Norethindrone
B.	Tranquilizer	II.	Meprobomate
C.	Antihistamine	III.	Seldane
D.	Antibiotic	IV.	Ampicillin

Choose the correct answer from the options given below.
(a) A-I, B-II, C-III, D-IV (b) A-I, B-III, C-II, D-IV
(c) A-II, B-I, C-III, D-IV (d) A-IV, B-III, C-II, D-I

19 Choose the correct colour of the product for the following reaction.

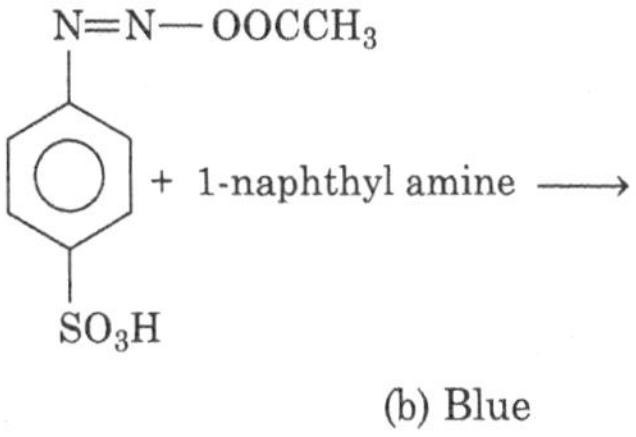

(a) White (b) Blue
(c) Yellow (d) Red

20 Find out the major products from the following reactions.

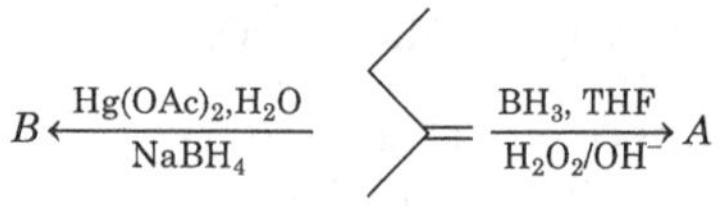

(a) $A =$ 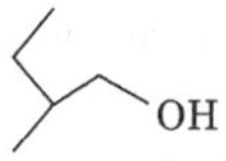 , $B =$ 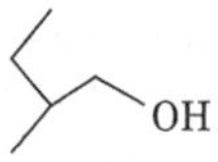

(b) $A =$ 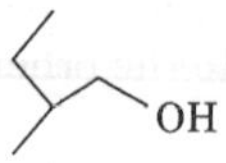 , $B =$ 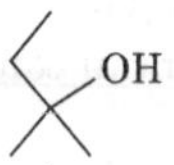

(c) $A =$  , $B =$ 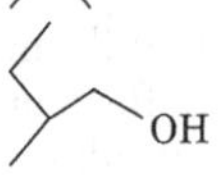

(d) $A =$  , $B =$ 

Section B : Numerical Value Type Questions

21 Sum of π-bonds present in peroxodisulphuric acid and pyrosulphuric acid is ....

22 The number of statements which are the characteristics of physisorption is ....

A. it is highly specific in nature
B. enthalpy of adsorption is high
C. it decreases with increase in temperature
D. it results into unimolecular layer
E. no activation energy is needed

23 Maximum number of isomeric monochloro derivatives which can be obtained from 2,2,5,5-tetramethylhexane by chlorination is ....

24 The number of units, which are used to express concentration of solutions from the following is ....
mass per cent, mole, mole-fraction, molarity, ppm, molality

25 The total pressure observed by mixing two liquids A and B is 350 mm Hg when their mole fractions are 0.7 and 0.3 respectively.
The total pressure becomes 410 mm Hg if the mole fractions are changed to 0.2 and 0.8 respectively for A and B. The vapour pressure of pure A is .... mm Hg. (Nearest integer) Consider the liquids and solutions behave ideally.

26 Following figure shows spectrum of an ideal black body at four different temperatures. The number of correct statements from the following is ....

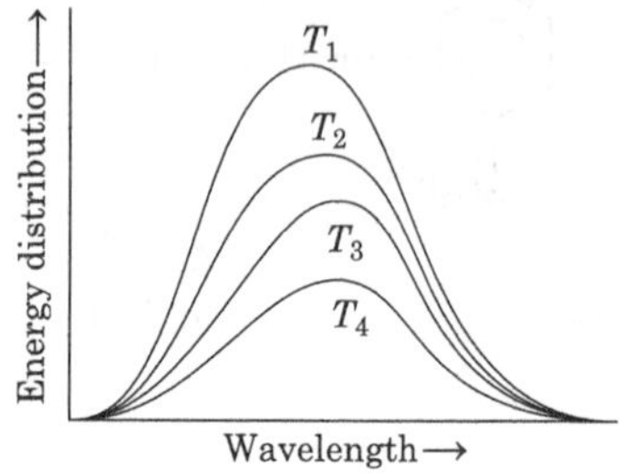

A. $T_4 > T_3 > T_2 > T_1$
B. the black body consists of particles performing simple harmonic motion.
C. the peak of the spectrum shifts to shorter wavelength as temperature increases.
D. $\frac{T_1}{\nu_1} = \frac{T_2}{\nu_2} = \frac{T_3}{\nu_3} \neq$ constant
E. the given spectrum could be explained using quantisation of energy.

27 If the pK_b of lactic acid is 5, then the pH of 0.005 M calcium lactate solution at 25°C is .... $\times 10^{-1}$. (Nearest integer)

Lactic acid $CH_3—C(H)(OH)—COOH$

28 Total number of tripeptides possible by mixing of valine and proline is ....

29 One mole of an ideal monoatomic gas is subjected to changes as shown in the graph. The magnitude of the work done (by the system or on the system) is .... J. (Nearest integer)

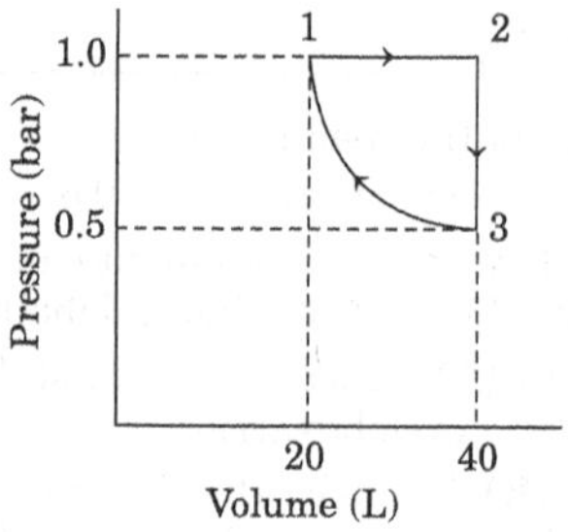

Given : $\log 2 = 0.3$
$\ln 10 = 2.3$

30 The number of statements, which are correct with respect to the compression of carbon dioxide from point (a) in the Andrews isotherm from the following is ....

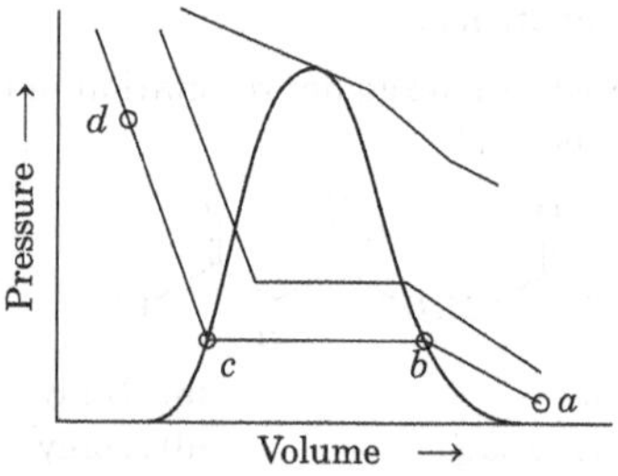

A. carbon dioxide remains as a gas upto point (b)
B. liquid carbon dioxide appears at point (c)
C. liquid and gaseous carbon dioxide coexist between points (b) and (c)
D. as the volume decreases from (b) to (c), the amount of liquid decreases

25 JANUARY 2023, SHIFT-I

Section A : Objective Type Questions

1 A cubic solid is made up of two elements X and Y. Atoms of X are present on every alternate corner and one at the center of cube. Y is at $\frac{1}{3}$rd of the total faces. The empirical formula of the compound is

(a) $X_{2.5}Y$ (b) $X_2Y_{1.5}$
(c) $X_{1.5}Y_2$ (d) $XY_{2.5}$

2 Match List-I with List-II.

List-I (Element)		List-II (Colour imparted to the flame)	
A.	K	I.	Brick red
B.	Ca	II.	Violet
C.	Sr	III.	Apple green
D.	Ba	IV.	Crimson red

Choose the correct answer from the options given below:

(a) A-II, B-I, C-IV, D-III
(b) A-II, B-IV, C-I, D-III
(c) A-IV, B-III, C-II, D-I
(d) A-II, B-I, C-III, D-IV

3 Inert gases have positive electron gain enthalpy. Its correct order is

(a) Xe < Kr < Ne < He (b) He < Ne < Kr < Xe
(c) He < Kr < Xe < Ne (d) He < Xe < Kr < Ne

4 The correct order in aqueous medium of basic strength in case of methyl substituted amines is

(a) $Me_2NH > Me_3N > MeNH_2 > NH_3$
(b) $NH_3 > Me_3N > MeNH_2 > Me_2NH$
(c) $Me_2NH > MeNH_2 > Me_3N > NH_3$
(d) $Me_3N > Me_2NH > MeNH_2 > NH_3$

5 Which of the following conformations will be the most stable?

(a)
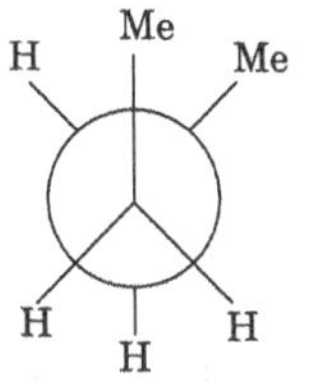

(b)
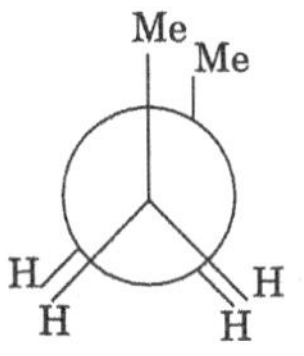

(c)
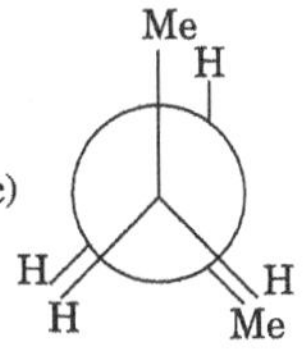

(d)
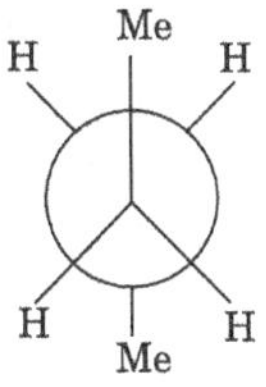

6 Identify the products formed (A and E).

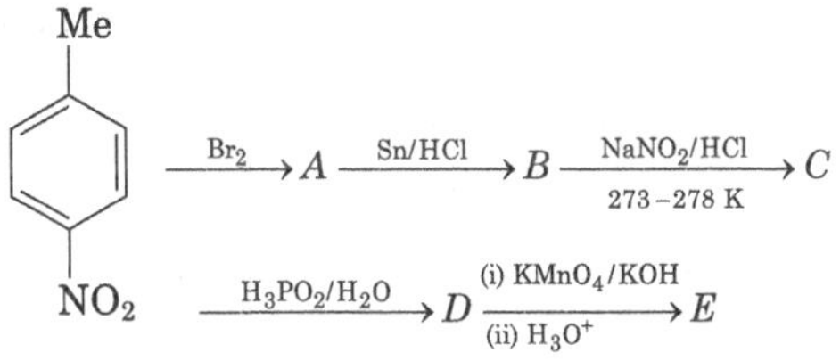

(a)
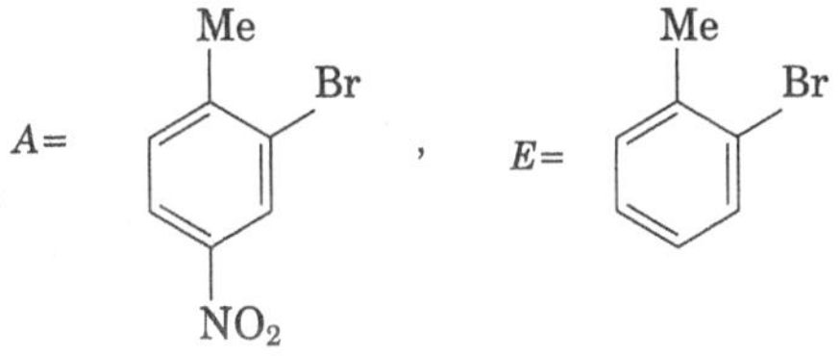

(b)
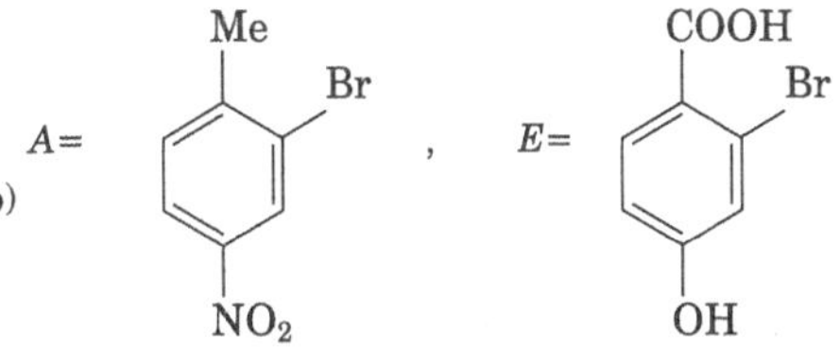

(c)
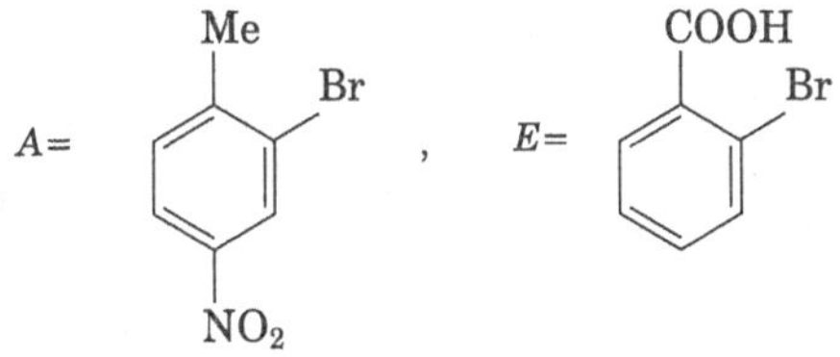

(d)
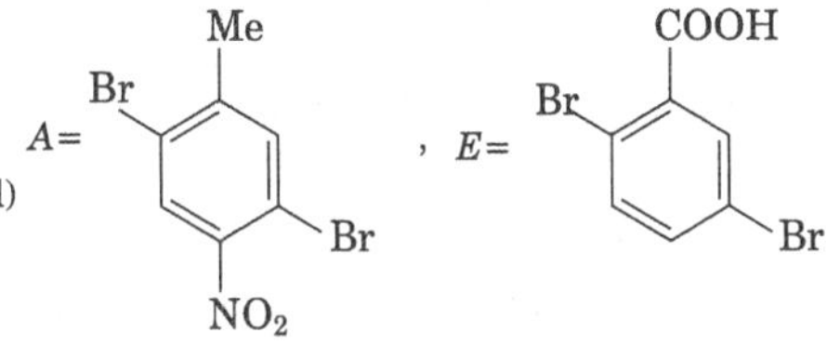

7 The variation of the rate of an enzyme catalysed reaction with substrate concentration is correctly represented by graph

(A)
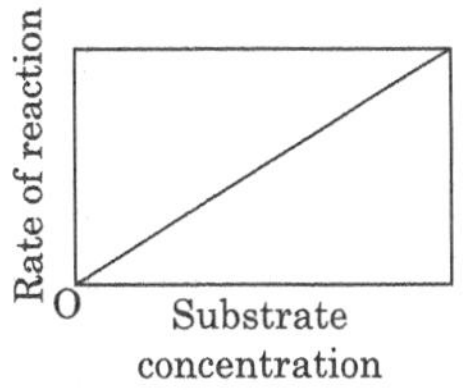

(B)
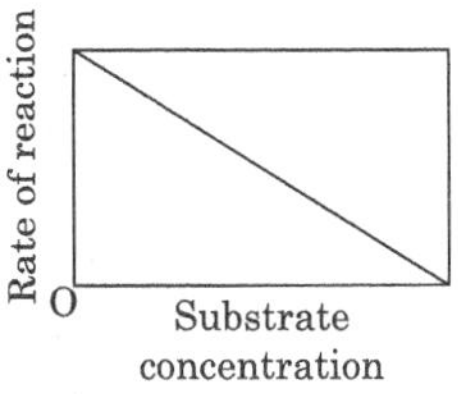

(C)
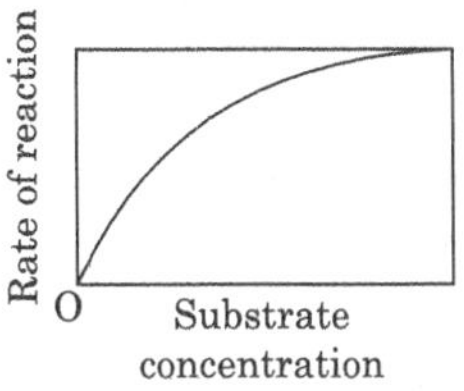

(D)
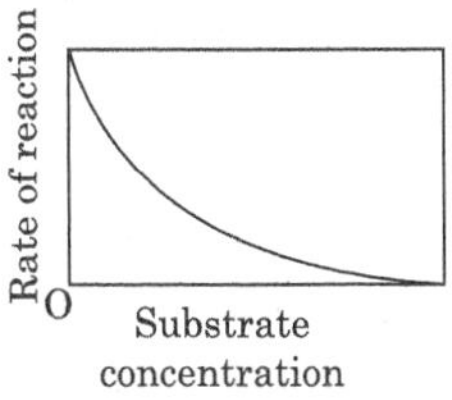

(a) B (b) A (c) C (d) D

8 Given below are two statements, one is labelled as Assertion (A) and the other is labelled as Reason (R).

Assertion (A) Acetal/ketal is stable in basic medium.

Reason (R) The high leaving tendency of alkoxide ion gives the stability to acetal/ketal in basic medium.

In the light of the above statements, choose the correct answer from the options given below :

(a) A is true but R is false.
(b) A is false but R is true.
(c) Both A and R are true but R is not the correct explanation of A.
(d) Both A and R are true and R is the correct explanation of A.

9 '25 volume' hydrogen peroxide means

(a) 1 L marketed solution contains 75 g of H_2O_2.
(b) 1 L marketed solution contains 250 g of H_2O_2.
(c) 100 mL marketed solution contains 25 g of H_2O_2.
(d) 1 L marketed solution contains 25 g of H_2O_2.

10  $\xrightarrow{?}$ PhCOOH + $PhCH_2OH$

P Q R

The correct sequence of reagents for the preparation of *Q* and *R* is

(a) (i) Mo_2O_3, Δ; (ii) CrO_2Cl_2, H_3O^+; (iii) NaOH; (iv) H_3O^+
(b) (i) CrO_2Cl_2, H_3O^+; (ii) Cr_2O_3, 770 K, 20 atm; (iii) NaOH; (iv) H_3O^+
(c) (i) $KMnO_4$, OH^-; (ii) Mo_2O_3, Δ; (iii) NaOH; (iv) H_3O^+
(d) (i) Cr_2O_3, 770 K, 20 atm; (ii) CrO_2Cl_2,H_3O^+ ; (iii) NaOH; (iv) H_3O^+

11 Reaction of thionyl chloride with white phosphorus forms a compound [*A*], which on hydrolysis gives [*B*], a dibasic acid. [*A*] and [*B*] are respectively

(a) PCl_3 and H_3PO_3 (b) PCl_5 and H_3PO_4
(c) $POCl_3$ and H_3PO_4 (d) P_4O_6 and H_3PO_3

12 Which of the following statements is incorrect for antibiotics?

(a) An antibiotic should promote the growth or survival of microorganisms.
(b) An antibiotic is a synthetic substance produced as a structural analogue of naturally occurring antibiotic.
(c) An antibiotic should be effective in low concentrations.
(d) An antibiotic must be a product of metabolism.

13 Compound *A* reacts with NH_4Cl and forms a compound *B*. Compound *B* reacts with H_2O and excess of CO_2 to form compound *C* which on passing through or reaction with saturated NaCl solution forms sodium hydrogen carbonate. Compounds *A*, *B* and *C* are respectively

(a) $CaCl_2$, $NH_4^{\oplus}$, $(NH_4)_2CO_3$
(b) $Ca(OH)_2$, NH_3, NH_4HCO_3
(c) $CaCl_2$, NH_3,NH_4HCO_3
(d) $Ca(OH)_2$, $NH_4^{\oplus}$, $(NH_4)_2CO_3$

14 Match List-I with List-II.

List-I (Cations)		List-II (Group reagents)
A. Pb^{2+}, Cu^{2+}	I.	H_2S gas in presence of dilute HCl
B. Al^{3+}, Fe^{3+}	II.	$(NH_4)_2CO_3$ in presence of NH_4OH
C. Co^{2+}, Ni^{2+}	III.	NH_4OH in presence of NH_4Cl
D. Ba^{2+}, Ca^{2+}	IV.	H_2S in presence of NH_4OH

Correct match is.

(a) A-III, B-I, C-IV, D-II (b) A-I, B-III, C-IV, D-II
(c) A -I, B-III, C-II, D-IV (d) A-IV, B-II, C-III, D-I

15 Some reactions of NO_2 relevant to photochemical smog formation are

$$NO_2 \xrightarrow{\text{Sunlight}} X + Y$$
$$X \xrightarrow{A} B$$

Identify *A*, *B*, *X* and *Y*.

(a) $X = \frac{1}{2}O_2$, $Y = NO_2$, $A = O_3$, $B = O_2$
(b) $X = N_2O$, $Y = [O]$, $A = O_3$; $B = NO$
(c) $X = NO$, $Y = [O]$, $A = O_2$, $B = N_2O_3$
(d) $X = [O]$, $Y = NO$, $A = O_2$, $B = O_3$

16 Match items of Row-I with those of Row-II.

Row I

(A) 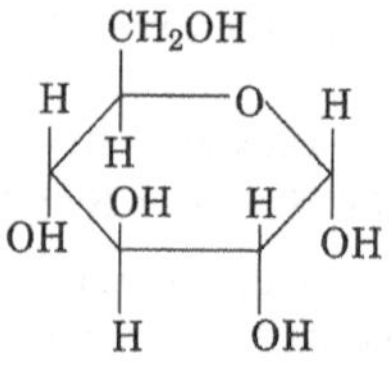

(B) 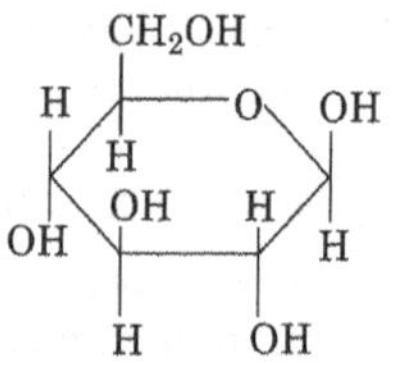

(C) 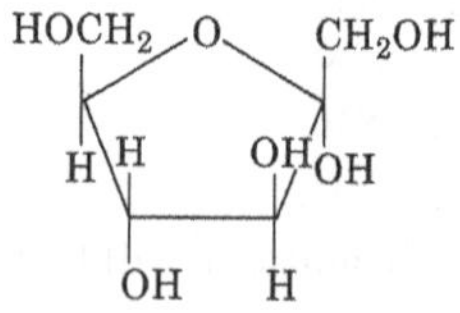

(D) 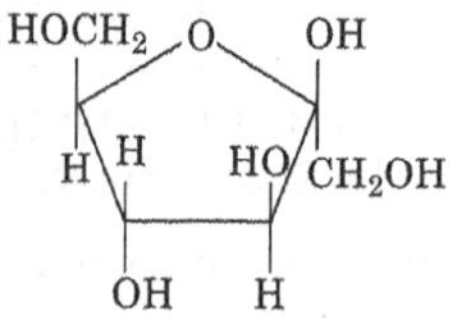

Row II

(i) α-D-(–)-fructofuranose,
(ii) β-D-(–)-fructofuranose
(iii) α-D-(–) glucopyranose,
(iv) β-D-(–)-glucopyranose

Correct match is

(a) A-(iv), B-(iii), C-(i), D-(ii) (b) A-(iii), B-(iv), C-(i), D-(ii)
(c) A-(i), B-(ii), C-(iii), D-(iv) (d) A-(iii), B-(iv), C-(ii), D-(i)

17 Which one of the following reactions does not occur during extraction of copper?

(a) $CaO + SiO_2 \longrightarrow CaSiO_3$
(b) $FeO + SiO_2 \longrightarrow FeSiO_3$
(c) $2Cu_2S + 3O_2 \longrightarrow 2Cu_2O + 2SO_2$
(d) $2FeS + 3O_2 \longrightarrow 2FeO + 2SO_2$

18 The radius of the 2nd orbit of Li^{2+} is x. The expected radius of the 3rd orbit of Be^{3+} is

(a) $\frac{4}{9}x$ (b) $\frac{27}{16}x$ (c) $\frac{9}{4}x$ (d) $\frac{16}{27}x$

19 The compound which will have the lowest rate towards nucleophilic aromatic substitution on treatment with OH^- is

(a) 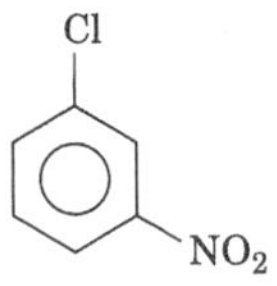

(b) 

(c) 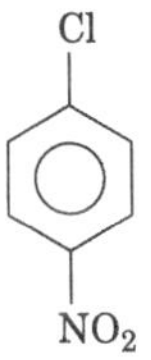

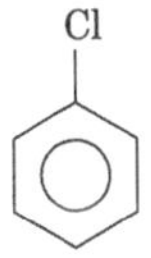

20 In the cumene to phenol preparation in presence of air, the intermediate is

(a) 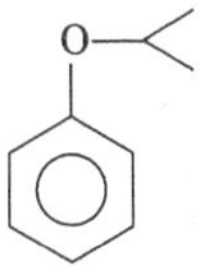

(b) 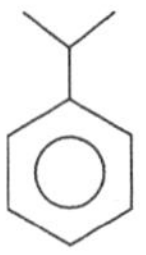

(c) 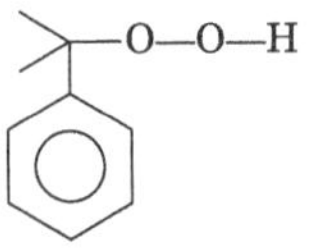

(d) 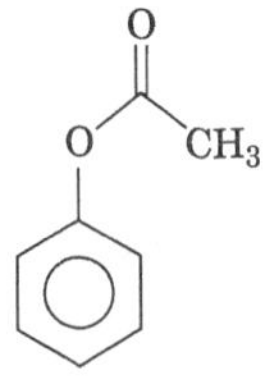

Section B : Numerical Value Type Questions

21 The total number of lone pairs of electrons on oxygen atoms of ozone is ....

22 An athlete is given 100 g of glucose ($C_6H_{12}O_6$) for energy. This is equivalent to 1800 kJ of energy. The 50% of this energy gained is utilised by the athlete for sports activities at the event. In order to avoid storage of energy, the weight of extra water he would need to perspire is .... g. (Nearest integer)

Assume that there is no other way of consuming stored energy.

[Given, the enthalpy of evaporation of water is 45 kJ mol^{-1}

Molar mass of C, H and O are 12, 1 and 16 g mol^{-1} respectively]

23 A litre of buffer solution contains 0.1 mole of each of NH_3 and NH_4Cl. On the addition of 0.02 mole of HCl by dissolving/ gaseous HCl, the pH of the solution is found to be ...... $\times 10^{-3}$ (Nearest integer)

[Given : $pK_b(NH_3) = 4.745$

$\log 2 = 0.301$, $\log 3 = 0.477$, $T = 298$ K]

24 In sulphur estimation, 0.471 g of an organic compound gave 1.4439 g of barium sulphate. The percentage of sulphur in the compound is .... (Nearest integer)

(Given : atomic mass Ba = 137 u, S = 32 u, O = 16 u)

25 For the first order reaction $A \rightarrow B$, the half-life is 30 min. The time taken for 75% completion of the reaction is .... min. (Nearest integer).

[Given : $\log 2 = 0.3010$, $\log 3 = 0.4771$, $\log 5 = 0.6989$]

26 The number of paramagnetic species from the following is ....

$[Ni(CN)_4]^{2-}$, $[Ni(CO)_4]$, $[NiCl_4]^{2-}$,

$[Fe(CN)_6]^{4-}$, $[Cu(NH_3)_4]^{2+}$

$[Fe(CN)_6]^{3-}$ and $[Fe(H_2O)_6]^{2+}$

27 How many of the following metal ions have similar value of spin only magnetic moment in gaseous state? ....

[Given : atomic number V = 23; Cr = 24; Fe = 26; Ni = 28]

V^{3+}, Cr^{3+}, Fe^{2+}, Ni^{3+}

Key Idea Note that, in this type of questions, the metal ions with same number of unpaired electron will have similar value of spin only magnetic moment.

28 The osmotic pressure of solutions of PVC in cyclohexanone at 300K are plotted on the graph.

The molar mass of PVC is .... g mol^{-1}. (Nearest integer)

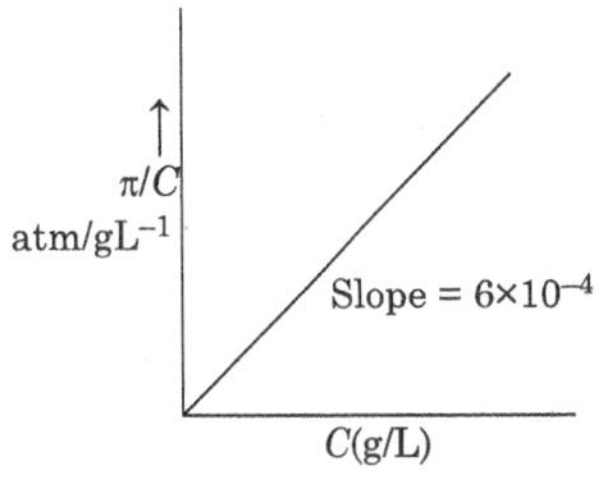

(Given : $R = 0.083$ L atm K^{-1} mol^{-1}).

(Given : $R = 0.083$ L atm K^{-1} mol^{-1}).

Key Idea To find out the volume of solution required, first of all calculate the molarity. It will be equal to the normality of solution due to its monobasic nature. After that use the formula $N_1V_1 = N_2V_2$

29 The density of a monobasic strong acid (molar mass 24.2 g/mol) is 1.21 kg/L. The volume of its solution required for the complete neutralisation of 25 mL of 0.24 M NaOH is .... $\times 10^{-2}$ mL. (Nearest integer)

30 Consider the cell

Pt(s) | $H_2(g)$ (1 atm) $H^+(aq.\ [H^+]=1)$ || $Fe^{3+}(aq)$, $Fe^{2+}(aq)$ | Pt(s)

Given, $E^\circ_{Fe^{3+}/Fe^{2+}} = 0.771$ V and $E^\circ_{H^+/1/2H_2} = 0$ V, $T = 298$ K

If the potential of the cell is 0.712 V, the ratio of concentration of Fe^{2+} to Fe^{3+} is .... (Nearest integer)

> **Key Idea** To calculate the ratio of concentration of Fe^{2+} to Fe^{3+}, Nernst equation is used. First determine the value of *n*, from balanced chemical equation and then put this value in Nernst equation.

25 JANUARY 2023, SHIFT-II

Section A : Objective Type Questions

1 Given below are two statements, one is labelled as Assertion (A) and the other is labelled as Reason (R).

Assertion (A) The alkali metals and their salts impart characteristic colour to reducing flame.

Reason (R) Alkali metals can be detected using flame tests.

In the light of the above statements, choose the most appropriate answer from the options given below.

(a) A is not correct but R is correct
(b) A is correct but R is not correct
(c) Both A and R are correct but R is not the correct explanation of A
(d) Both A and R are correct and R is the correct explanation of A

2 A. Ammonium salts produce haze in atmosphere.

B. Ozone gets produced when atmospheric oxygen reacts with chlorine radicals.

C. Polychlorinated biphenyls act as cleansing solvents.

D. 'Blue baby' syndrome occurs due to the presence of excess of sulphate ions in water.

Choose the correct answer from the options given below.

(a) A and D (b) A, B and C
(c) B and C (d) A and C

3 Match List-I with List-II.

List-I	List-II
A. Cobalt catalyst	I. $(H_2 + Cl_2)$ production
B. *Syn* gas	II. Water gas production
C. Nickel catalyst	III. Coal gasification
D. Brine solution	IV. Methanol production

Choose the correct answer from the options given below.

(a) A-IV, B-I, C-II, D-III (b) A-II, B-III, C-IV, D-I
(c) A-IV, B-III, C-I, D-II (d) A-IV, B-III, C-II, D-I

4 Match List-I with List-II.

List-I (Amines)		List-II (pK_b)
A. Aniline	I.	3.25
B. Ethanamine	II.	3.00
C. N-ethylethanamine	III.	9.38
D. N, N-diethylethanamine	IV.	3.29

Choose the correct answer from the options given below.

(a) A-III, B-II, C-I, D-IV (b) A-III, B-IV, C-II, D-I
(c) A-I, B-IV, C-II, D-III (d) A-III, B-II, C-IV, D-I

5 Potassium dichromate acts as a strong oxidising agent in acidic solution. During this process, the oxidation state changes from

(a) + 6 to + 3 (b) + 2 to + 1
(c) + 6 to + 2 (d) + 3 to + 1

6 '*A*' in the given reaction is

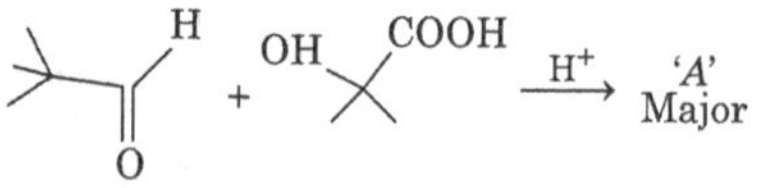

(a)
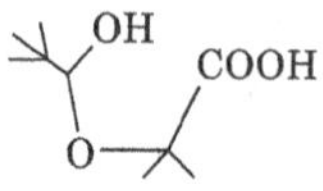

(b)
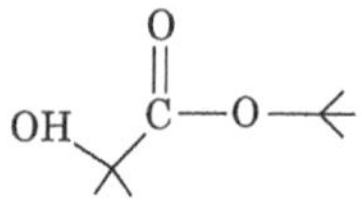

(c)
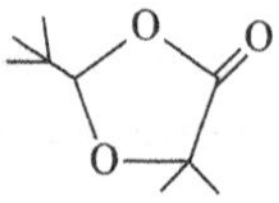

(d)
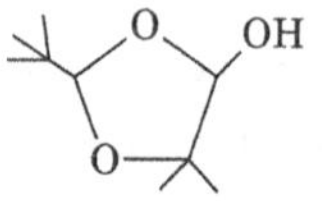

7 What is the mass ratio of ethylene glycol ($C_2H_6O_2$, molar mass = 62 g/mol) required for making 500 g of 0.25 molal aqueous solution and 250 mL of 0.25 molal aqueous solution?

(a) 1 : 2 (b) 1 : 1 (c) 3 : 1 (d) 2 : 1

> **Key Idea** To calculate the mass ratio of ethylene glycol, use the formula of molalty. i.e.,
>
> $$\text{Molality} = \frac{\text{mass of solute} \times 1000}{\text{molar mass} \times \text{mass of solvent (in kg)}}$$
>
> From this formula mass of solute can be calculated and then we can find the ratio.

8 Given below are two statements, one is labelled as Assertion (A) and the other is labelled as Reason (R).

Assertion (A) Carbon forms two important oxides —CO and CO_2, CO is neutral whereas CO_2 is acidic in nature.

Reason (R) CO_2 can combine with water in a limited way to form carbonic acid, while CO is sparingly soluble in water.

In the light of the above statements, choose the most appropriate answer from the options given below.

(a) Both A and R are correct but R is not the correct explanation of A
(b) Both A and R are correct and R is the correct explanation of A
(c) A is correct but R is not correct
(d) A is not correct but R is correct

9 Find out the major product from the following reaction.

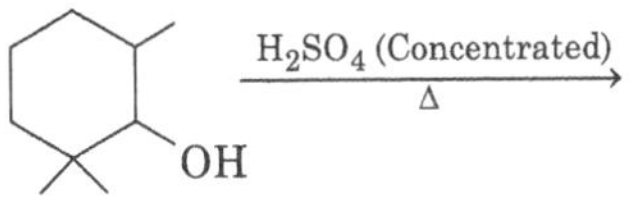

(a) 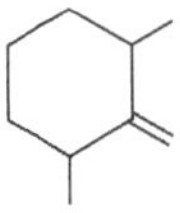 (b) 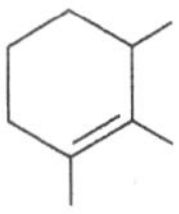 (c) 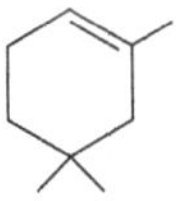 (d) 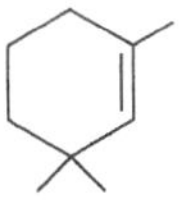

10 Match List-I with List-II.

List-I (Coordination entity)	List-II (Wavelength of light absorbed in nm)
A. $[CoCl(NH_3)_5]^{2+}$	I. 310
B. $[Co(NH_3)_6]^{3+}$	II. 475
C. $[Co(CN)_6]^{3-}$	III. 535
D. $[Cu(H_2O)_4]^{2+}$	IV. 600

Choose the correct answer from the options given below.

(a) A-II, B-III, C-IV, D-I (b) A-III, B-II, C-I, D-IV
(c) A-IV, B-I, C-III, D-II (d) A-III, B-I, C-II, D-IV

11 When the hydrogen ion concentration $[H^+]$ changes by a factor of 1000, the value of pH of the solution

(a) decreases by 3 units (b) increases by 2 units
(c) decreases by 2 units (d) increases by 1000 units

12 A chloride salt solution acidified with dil. HNO_3 gives a curdy white precipitate [*A*], on addition of $AgNO_3$. [*A*] on treatment with NH_4OH gives a clear solution *B*. *A* and *B* are respectively

(a) $H[AgCl_3]$ and $(NH_4)[Ag(OH)_2]$
(b) AgCl and $[Ag(NH_3)_2]Cl$
(c) $H[AgCl_3]$ and $[Ag(NH_3)_2]Cl$
(d) AgCl and $(NH_4)[Ag(OH)_2]$

13 The isomeric deuterated bromide with molecular formula C_4H_8DBr having two chiral carbon atoms is

(a) 2-bromo-3-deuterobutane
(b) 2-bromo-1-deuterobutane
(c) 2-bromo-1-deutero-2-methylpropane
(d) 2-bromo-2-deuterobutane

14 Which of the following represents the correct order of metallic character of the given elements?

(a) Be < Si < K < Mg (b) Be < Si < Mg < K
(c) Si < Be < Mg < K (d) K < Mg < Be < Si

15 Match List-I with List-II.

List-I (Name of polymer)	List-II (Uses)
A. Glyptal	I. Flexible pipes
B. Neoprene	II. Synthetic wool
C. Acrilan	III. Paints and Lacquers
D. LDP	IV. Gaskets

Choose the correct answer from the options given below.

(a) A-III, B-IV, C-I, D-II (b) A-III, B-II, C-IV, D-I
(c) A-III, B-IV, C-II, D-I (d) A-III, B-I, C-IV, D-II

16 Given below are two statements.

Statement I In froth floatation method a rotating paddle agitates the mixture to drive air out of it.

Statement II Iron pyrites are generally avoided for extraction of iron due to environmental reasons.

In the light of the above statements, choose the correct answer from the options given below.

(a) Both Statement I and Statement II are incorrect
(b) Statement I is correct but Statement II is incorrect
(c) Statement I is incorrect but Statement II is correct
(d) Both Statement I and Statement II are correct

17 Given below are two statements.

Statement I Dipole moment is a vector quantity and by convention it is depicted by a small arrow with tail on the negative centre and head pointing towards the positive centre.

Statement II The crossed arrow of the dipole moment symbolises the direction of the shift of charges in the molecules.

In the light of the above statements, choose the most appropriate answer from the options given below.

(a) Both Statement I and Statement II are incorrect
(b) Statement I is correct but Statement II is incorrect
(c) Statement I is incorrect but Statement II is correct
(d) Both Statement I and Statement II are correct

18 Match List-I with List-II.

List-I (Isomeric pairs)	List-II (Type of isomers)
A. Propanamine and N-methylethanamine	I. Metamers
B. Hexan-2-one and hexan-3-one	II. Positional isomers
C. Ethanamide and hydroxyethanimine	III. Functional isomers
D. *o*-nitrophenol and *p*-nitrophenol	IV. Tautomers

Choose the correct answer from the options given below.

(a) A-II, B-III, C-I, D-IV (b) A-III, B-I, C-IV, D-II

(c) A-III, B-IV, C-I, D-II (d) A-IV, B-III, C-I, D-II

19 Which one among the following metals is the weakest reducing agent?

(a) Na (b) K (c) Li (d) Rb

20 Given below are two statements, one is labelled as Assertion (A) and the other is labelled as Reason (R).

Assertion (A) Butylated hydroxy anisole when added to butter increases its shelf life.

Reason (R) Butylated hydroxy anisole is more reactive towards oxygen than food.

In the light of the above statements, choose the most appropriate answer from the options given below.

(a) A is correct but R is not correct

(b) Both A and R are correct and R is the correct explanation of A

(c) A is not correct but R is correct

(d) Both A and R are correct but R is not the correct explanation of A

Section B : Numerical Value Type Questions

21 The number of incorrect statements from the following is/are ....

A. Water vapours are adsorbed by anhydrous calcium chloride.

B. There is a decrease in surface energy during adsorption.

C. As the adsorption proceeds, ΔH becomes more and more negative.

D. Adsorption is accompanied by decrease in entropy of the system.

22 Based on the given figure, the number of correct statements is/are ....

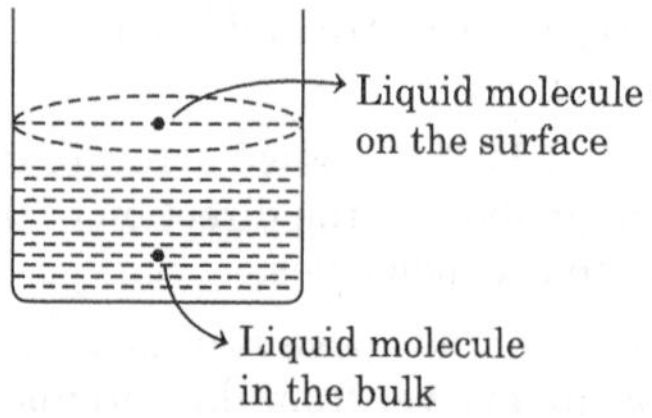

A. Surface tension is the outcome of equal attractive and repulsive forces acting on the liquid molecule in bulk.

B. Surface tension is due to uneven forces acting on the molecules present on the surface.

C. The molecule in the bulk can never come to the liquid surface.

D. The molecules on the surface are responsible for vapour pressure if the system is a closed system.

23 Total number of moles of AgCl precipitated on addition of excess of $AgNO_3$ to one mole each of the following complexes $[Co(NH_3)_4Cl_2]Cl$, $[Ni(H_2O_6]Cl_2$, $[Pt(NH_3)_2Cl_2]$ and $[Pd(NH_3)_4]Cl_2$ is ....

24 Number of hydrogen atoms per molecule of a hydrocarbon A having 85.8 % carbon is .... [Given, molar mass of $A = 84$ g mol^{-1}]

25 Number of compounds giving (i) red colouration with ceric ammonium nitrate and also (ii) positive iodoform test from the following is ....

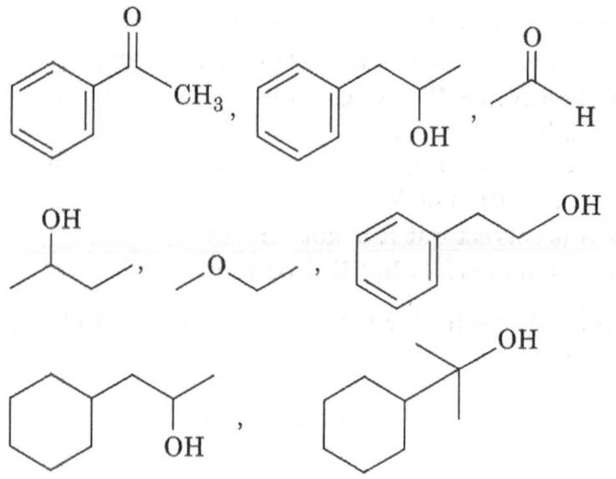

26 Pt(s) | $H_2(g)$ (1 bar) | $H^+(aq)$ (1 M) | $M^{3+}(aq)$ | $M^+(aq)$ | Pt(s)

The E_{cell} for the given cell is 0.1115 V at 298 K when $\dfrac{[M^+(aq)]}{[M^{3+}(aq)]} = 10^a$

The value of a is ....

[Given, $E^\circ_{M^{3+}/M^+} = 0.2$ V

$\dfrac{2.303RT}{F} = 0.059$ V]

> **Key Idea** To calculate the value of a, Nernst equation is used. First determine the value of E_{cell} from given $E^\circ_{X^{2+}/X}$ and $E^\circ_{Y^{2+}/Y}$ value. Then, put this value in Nernst equation.

27 The number of pairs of the solutions having the same value of the osmotic pressure from the following is .... (Assume 100% ionisation)

A. 0.500 M $C_2H_5OH(aq)$ and 0.25 M $KBr(aq)$

B. 0.100 M $K_4[Fe(CN)_6]$ (aq) and 0.100 M $FeSO_4(NH_4)_2SO_4(aq)$

C. 0.05 M $K_4[Fe(CN)_6]$ (aq) and 0.25 M $NaCl(aq)$

D. 0.15 M $NaCl(aq)$ and 0.1 M $BaCl_2(aq)$

E. 0.02 M $KCl \cdot MgCl_2 \cdot 6H_2O(aq)$ and 0.05 M $KCl(aq)$

28 28.0 L of CO_2 is produced on complete combustion of 16.8 L gaseous mixture of ethene and methane at 25°C and 1 atm. Heat evolved during the combustion process is .... kJ.

[Given, $\Delta H_C(CH_4) = -900$ kJ mol^{-1}

$\Delta H_C(C_2H_4) = -1400$ kJ mol^{-1}]

> **Key Idea** This problem is based on the concept of stoichiometry and mole concept. The step involved to solve this question are, firstly write the balanced chemical equation, from that calculate the value of volume. After that calculate the No. of moles, using the value of No. of moles, calculate the heat evolved.

29 The number of given orbitals which have electron density along the axis is ....

$p_x, p_y, p_z, d_{xy}, d_{yz}, d_{z^2}, d_{x^2-y^2}$

30 A first order reaction has the rate constant, $k = 4.6 \times 10^{-3}\ s^{-1}$. The number of correct statements from the following is/are ....

[Given, log 3 = 0.48]

A. reaction completes in 1000 s.
B. the reaction has a half-life of 500 s.
C. the time required for 10% completion is 25 times the time required for 90% completion.
D. the degree of dissociation is equal to $(1 - e^{-kt})$
E. the rate and the rate constant have the same unit.

29 JANUARY 2023, SHIFT-I

Section A : Objective Type Questions

1 The correct order of hydration enthalpies is

(A) K^+ (B) Rb^+ (C) Mg^{2+} (D) Cs^+ (E) Ca^{2+}

Choose the correct answer from the options given below.

(a) C > E > A > B > D
(b) C > A > E > B > D
(c) E > C > A > B > D
(d) C > E > A > D > B

2 Match List-I with List-II.

List-I (Antimicrobials)	List-II (Names)
A. Narrow spectrum antibiotic	I. Furacin
B. Antiseptic	II. Sulphur dioxide
C. Disinfectants	III. Penicillin G
D. Broad spectrum antibiotic	IV. Chloramphenicol

Choose the correct answer from the options given below.

(a) A-III, B-I, C-IV, D-II
(b) A-I, B-II, C-IV, D-III
(c) A-III, B-I, C-II, D-IV
(d) A-II, B-I, C-IV, D-III

3 For 1 mol of gas, the plot of pV *vs* p is shown below. p is the pressure and V is the volume of the gas.

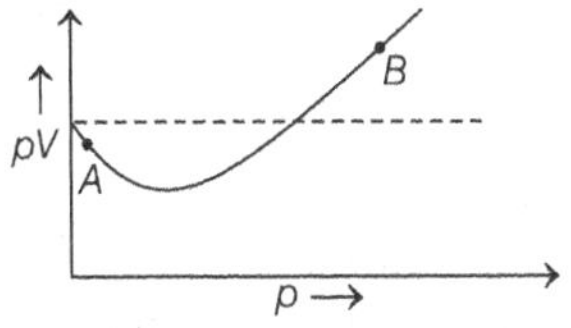

What is the value of compressibility factor at point A?

(a) $1 - \frac{a}{RTV}$
(b) $1 - \frac{b}{V}$
(c) $1 + \frac{b}{V}$
(d) $1 + \frac{a}{RTV}$

4 The increasing order of pK_a for the following phenols is

(A) 2, 4 - dinitrophenol
(B) 4 - nitrophenol
(C) 2, 4, 5 - trimethylphenol
(D) phenol
(E) 3-chlorophenol

Choose the correct answer from the option given below.

(a) A, E, B, D, C
(b) A, B, E, D, C
(c) C, D, E, B, A
(d) C, E, D, B, A

5 The major product '*P*' for the following sequence of reactions is

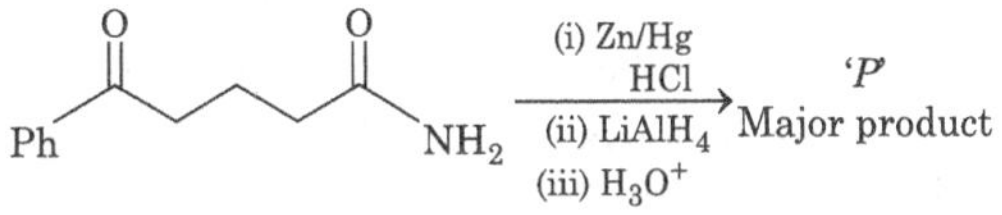

(a) 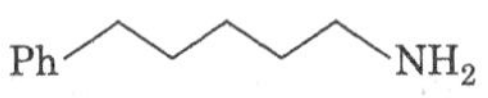

(b) 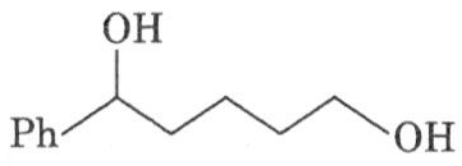

(c) 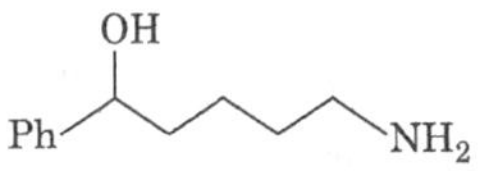

(d) 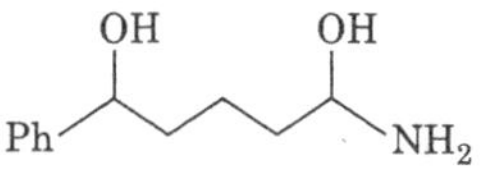

6 Correct statement about smog is

(a) both NO_2 and SO_2 are present in classical smog.
(b) classical smog also has high concentration of oxidising agents.
(c) NO_2 is present in classical smog.
(d) photochemical smog has high concentration of oxidising agents.

7 The bond dissociation energy is highest for

(a) Br_2 (b) F_2 (c) Cl_2 (d) I_2

8 Identify the correct order for the given property for following compounds.

(A) Boiling point 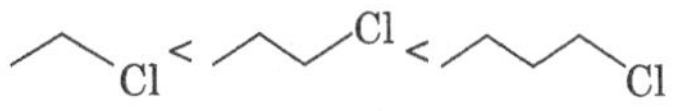

(B) Density 

(C) Boiling point 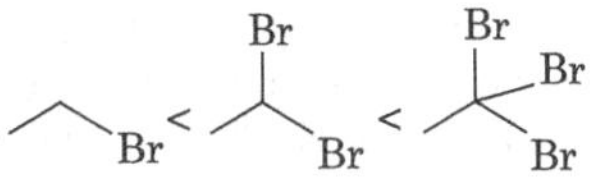

(D) Density 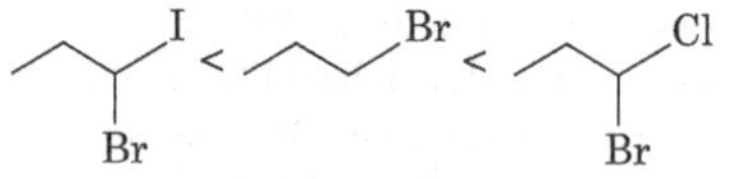

(E) Boiling point 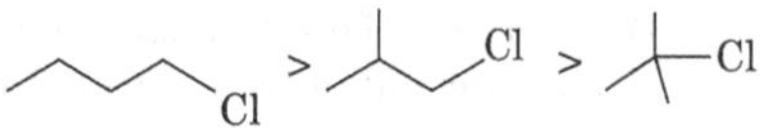

Choose the correct answer from the option given below.

(a) (B), (C) and (D)
(b) (A), (B) and (E)
(c) (A), (C) and (D)
(d) (A), (C) and (E)

9 The magnetic behaviour of Li_2O, Na_2O_2 and KO_2 respectively, are

(a) paramagnetic, diamagnetic and paramagnetic
(b) diamagnetic, paramagnetic and diamagnetic
(c) diamagnetic, diamagnetic and paramagnetic
(d) paramagnetic, paramagnetic and diamagnetic

10 The standard electrode potential (M^{3+}/M^{2+}) for V, Cr, Mn and Co are -0.26 V, -0.41 V, $+1.57$ V and $+1.97$ V, respectively. The metal ions which can liberate H_2 from a dilute acid are

(a) Mn^{2+} and Co^{2+}
(b) V^{2+} and Mn^{2+}
(c) V^{2+} and Cr^{2+}
(d) Cr^{2+} and Co^{2+}

11 During the borax bead test with $CuSO_4$, a blue green colour of the bead was observed in oxidising flame due to the formation of

(a) CuO
(b) $Cu(BO_2)_2$
(c) Cu
(d) Cu_3B_2

12 Number of cyclic tripeptides formed with 2 amino acids *A* and *B* is

(a) 3 (b) 4 (c) 2 (d) 5

13 Chiral complex from the following is

(Here : en = ethylene diamine)

(a) *cis*-$[PtCl_2(NH_3)_2]$
(b) *cis*-$[PtCl_2(en)_2]^{2+}$
(c) *trans*-$[Co(NH_3)_4Cl_2]^+$
(d) *trans*-$[PtCl_2(en)_2]^{2+}$

14 Compound that will give positive Lassaigne's test for both nitrogen and halogen is

(a) $NH_2OH \cdot HCl$
(b) NH_4Cl
(c) $N_2H_4 \cdot HCl$
(d) $CH_3NH_2 \cdot HCl$

15 The reaction representing the Mond process for metal refining is

(a) $2K[Au(CN)_2] + Zn \xrightarrow{\Delta} K_2[Zn(CN)_4] + 2Au$
(b) $ZnO + C \xrightarrow{\Delta} Zn + CO$
(c) $Zr + 2I_2 \xrightarrow{\Delta} ZrI_4$
(d) $Ni + 4CO \xrightarrow{\Delta} Ni(CO)_4$

16 *A* obtained by Ostwald's method involving air oxidation of NH_3, upon further air oxidation produce *B*. *B* on hydration forms an oxoacid of nitrogen along with evolution of *A*. The oxoacid also produces *A* and gives positive brown ring test.

Identify *A* and *B*, respectively

(a) NO_2, N_2O_5
(b) NO_2, N_2O_4
(c) N_2O_3, NO_2
(d) NO, NO_2

17 Which of the given compounds can enhance the efficiency of hydrogen storage tank?

(a) Di-isobutylaluminium hydride
(b) SiH_4
(c) $NaNi_5$
(d) Li/P_4

18 Which of the following salt solutions would coagulate the colloid solution formed when $FeCl_3$ is added to NaOH solution, at the fastest rate?

(a) 10 mL of 0.2 mol dm^{-3} $AlCl_3$
(b) 10 mL of 0.1 mol dm^{-3} $Ca_3(PO_4)_2$
(c) 10 mL of 0.15 mol dm^{-3} $CaCl_2$
(d) 10 mL of 0.1 mol dm^{-3} Na_2SO_4

Key Idea In this question, the concept of Hardy-Schulze rule is used. According to which, more the charge on the electrolyte, more will be the coagulation.

19 The shortest wavelength of hydrogen atom in Lyman series is λ. The longest wavelength in Balmer series of He^+ is

(a) $\frac{5}{9\lambda}$
(b) $\frac{9\lambda}{5}$
(c) $\frac{5\lambda}{9}$
(d) $\frac{36\lambda}{5}$

20 Match List-I with List-II

List-I (Reaction)		List-II (Reagents)
A. Hoffmann degradation	I.	Conc. KOH, Δ
B. Clemmensen reduction	II.	$CHCl_3$, $NaOH/H_3O^+$
C. Cannizzaro reaction	III.	Br_2, NaOH
D. Reimer-Tiemann reaction	IV.	Zn-Hg/HCl

Choose the correct answer from the options given below.

(a) A-II, B-I, C-III, D-IV
(b) A-II, B-IV, C-I, D-III
(c) A-III, B-IV, C-I, D-II
(d) A-III, B-IV, C-II, D-I

Section B : Numerical Value Type Questions

21 Consider the following reaction approaching equilibrium at 27°C and 1 atm pressure.

$$A + B \underset{K_b = 10^2}{\overset{K_f = 10^3}{\rightleftharpoons}} C + D$$

The standard Gibb's energy change $(\Delta_r G^\circ)$ at 27°C is (–) .... kJ mol^{-1}. (Nearest integer).

(Given: $R = 8.3$ J K^{-1} mol^{-1} and ln 10 = 2.3)

Key Idea To calculate the value of Gibbs free energy, first of all calculate K_{eq} by using the formula $K_{eq} = \frac{K_f}{K_b}$. After that substitute the value obtained, in the equation $\Delta G^\circ = -RT \ln K_{eq}$

22 The number of molecules or ions from the following, which do not have odd number of electrons are ....

(A) NO_2 (B) ICl_4^- (C) BrF_3 (D) ClO_2
(E) NO_2^+ (F) NO

23 For certain chemical reaction $X \rightarrow Y$, the rate of formation of product is plotted against the time as shown in the figure. The number of correct statements from the following is ....

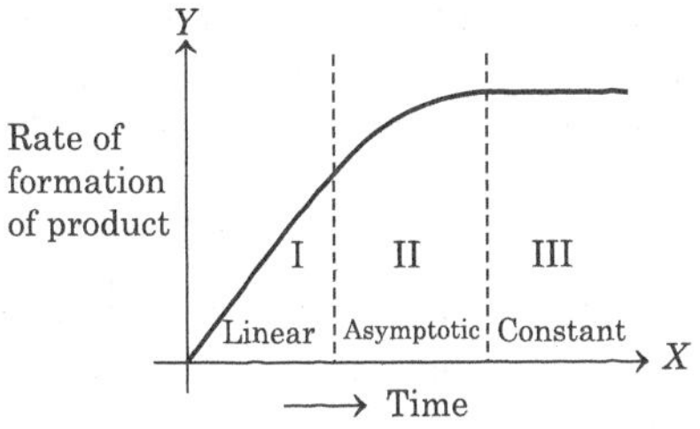

(A) over all order of this reaction is one
(B) order of this reaction can't be determined
(C) in region I and III, the reaction is of first and zero order respectively
(D) in region-II, the reaction is of first order
(E) in region-II, the order of reaction is in the range of 0.1 to 0.9.

24 Following figure shows dependence of molar conductance of two electrolytes on concentration. Λ_m° is the limiting molar conductivity.

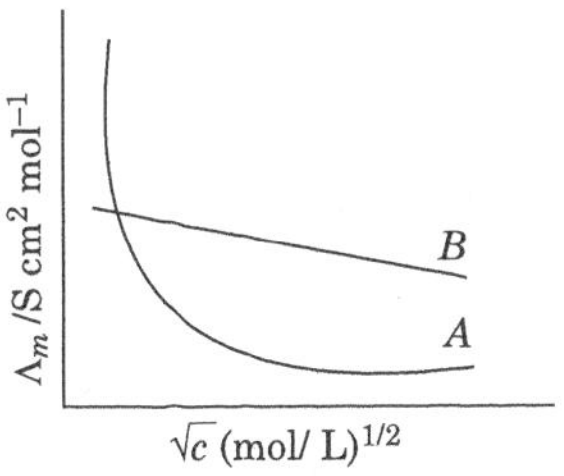

The number of incorrect statements from the following is ....
(A) Λ_m° for electrolyte A is obtained by extrapolation
(B) For electrolyte B, Λ_m *vs* $\sqrt{c}$ graph is a straight line with intercept equal to Λ_m°
(C) At infinite dilution, the value of degree of dissociation approaches zero for electrolyte B.
(D) Λ_m° for any electrolyte A or B can be calculated using λ° for individual ions

25 The sum of bridging carbonyls in $W(CO)_6$ and $Mn_2(CO)_{10}$ is ....

26 Water decomposes at 2300 K

$$H_2O(g) \longrightarrow H_2(g) + \frac{1}{2}O_2(g)$$

The percent of water decomposing at 2300 K and 1 bar is ...... (Nearest integer).

Equilibrium constant for the reaction is 2×10^{-3} at 2300 K.

27 17 mg of a hydrocarbon (molecular formula $C_{10}H_{16}$) takes up 8.40 mL of the H_2 gas measured at 0° C and 760 mm of Hg. Ozonolysis of the same hydrocarbon yields

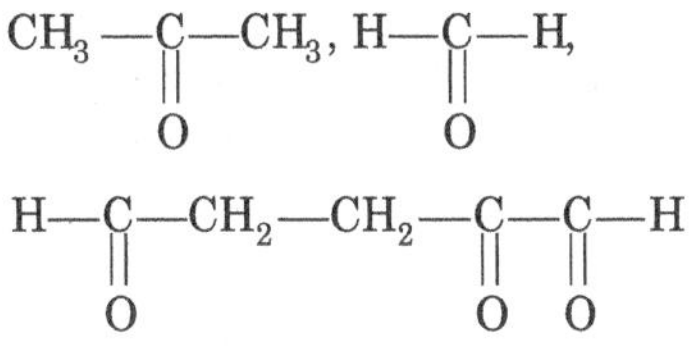

The number of double bonds present in the hydrocarbon is ....

28 Solid lead nitrate is dissolved in 1 litre of water. The solution was found to boil at 100.15° C. When 0.2 mol of NaCl is added to the resulting solution, it was observed that the solution froze at – 0.8°C. The solubility product of $PbCl_2$ formed is .... $\times 10^{-6}$ at 298 K. (Nearest integer)

[Given $K_b = 0.5$ K kg mol^{-1} and $K_f = 1.8$ K kg mol^{-1}. Assume molality to be equal to molarity in all cases.]

29 Following chromatogram was developed by adsorption of compound 'A' on a 6 cm TLC glass plate. Retardation factor of the compound 'A' is .... $\times 10^{-1}$.

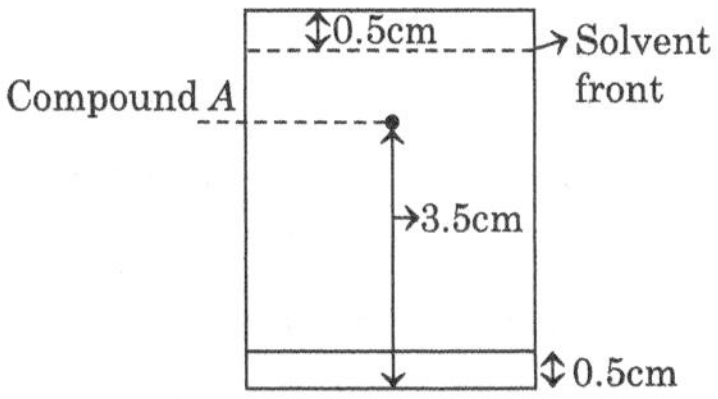

Key Idea In this question, the concept of thin layer chromatography is used to calculate the retardation factor, use the following formula,

$$R_f = \frac{\text{Distance moved by the substance from base line}}{\text{Distance moved by the solvent from base line}}$$

30 Millimoles of calcium hydroxide required to produce 100 mL of the aqueous solution of pH 12 is $x \times 10^{-1}$. The value of x is .... (Nearest integer).

Assume complete dissociation.

29 JANUARY 2023, SHIFT-II

Section A : Objective Type Questions

1 Match List-I with List-II.

List-I		List-II	
A.	van't Hoff factor, i	I.	Cryoscopic constant
B.	k_f	II.	Isotonic solutions
C.	Solutions with same osmotic pressure	III.	$\frac{\text{Normal molar mass}}{\text{Abnormal molar mass}}$
D.	Azeotropes	IV.	Solutions with same composition of vapour above it

Choose the correct answer from the options given below :

(a) A-III, B-I, C-IV, D-II (b) A-I, B-III, C-II, D-IV
(c) A-III, B-I, C-II, D-IV (d) A-III, B-II, C-I, D-IV

2 When a hydrocarbon A undergoes combustion in the presence of air, it requires 9.5 equivalents of oxygen and produces 3 equivalents of water. What is the molecular formula of A?

(a) C_9H_9 (b) C_6H_6 (c) C_8H_6 (d) C_9H_6

> **Key Idea** While solving this problem, students are advised to use the concept of mole concept and stoichiometry. For this, write the chemical equation of the combustion reaction of hydrocarbon. Then equate the stoichiometric coefficients of reactant and product with given values. Finally use this information to complete the question.

3 Match List-I with List-II.

List-I		List-II	
A.	Elastomeric polymer	I.	Urea formaldehyde resin
B.	Fibre polymer	II.	Polystyrene
C.	Thermosetting polymer	III.	Polyester
D.	Thermoplastic polymer	IV.	Neoprene

Choose the correct answer from the options given below :

(a) A-II, B-I, C-IV, D-III (b) A-IV, B-III, C-I, D-II
(c) A-IV, B-I, C-III, D-II (d) A-II, B-III, C-I, D-IV

4 The set of correct statements is

(i) Manganese exhibits + 7 oxidation state in its oxide.
(ii) Ruthenium and osmium exhibit + 8 oxidation in their oxides.
(iii) Sc shows + 4 oxidation state which is oxidising in nature.
(iv) Cr shows oxidising nature in + 6 oxidation state.

(a) (ii), (iii) and (iv) (b) (ii) and (iii)
(c) (i) and (iii) (d) (i), (ii) and (iv)

5 The one giving maximum number of isomeric alkenes on dehydrohalogenation reaction is (excluding rearrangement)

(a) l-bromo-2-methylbutane
(b) 2-bromopentane
(c) 2-bromo-3 ,3-dimethylpentane
(d) 2-bromopropane

6 The concentration of dissolved oxygen in water for growth of fish should be more than X ppm and biochemical oxygen demand in clean water should be less than Y ppm. X and Y in ppm are, respectively

(a) X 6, Y 5 (b) X 4, Y 15 (c) X 4, Y 8 (d) X 6, Y 12

7 According to MO theory, the bond orders for O_2^{2-}, CO and NO^+ respectively, are

(a) 2, 3 and 3 (b) 1, 2 and 3
(c) 1, 3 and 3 (d) 1, 3 and 2

8 Find out the major products from the following reaction sequence.

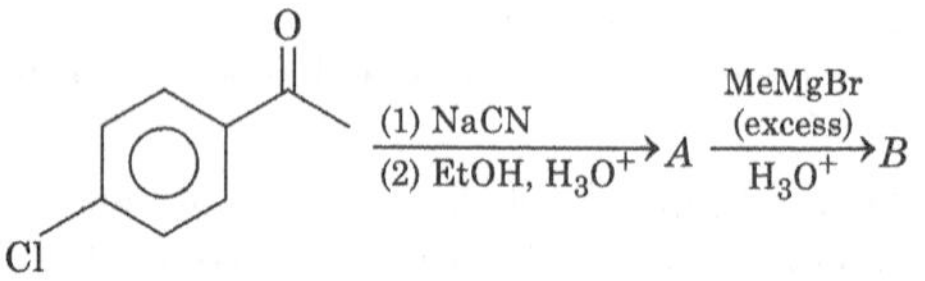

(a) A = 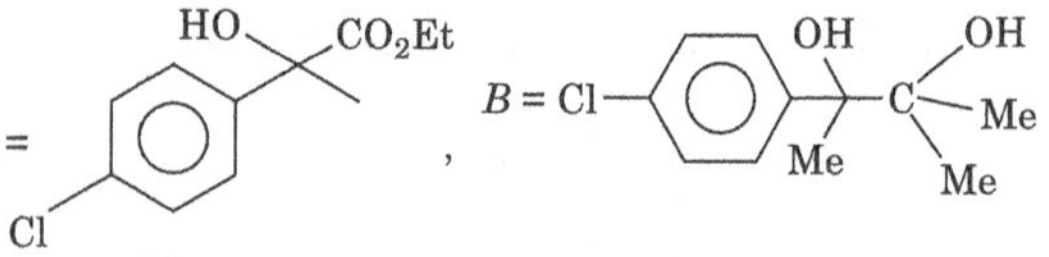

(b) A = 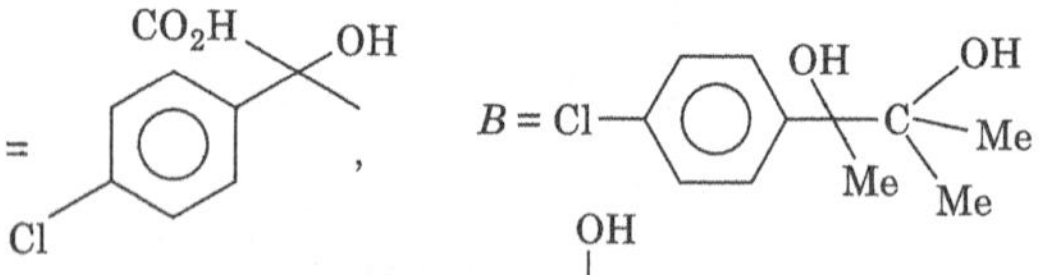

(c) A = 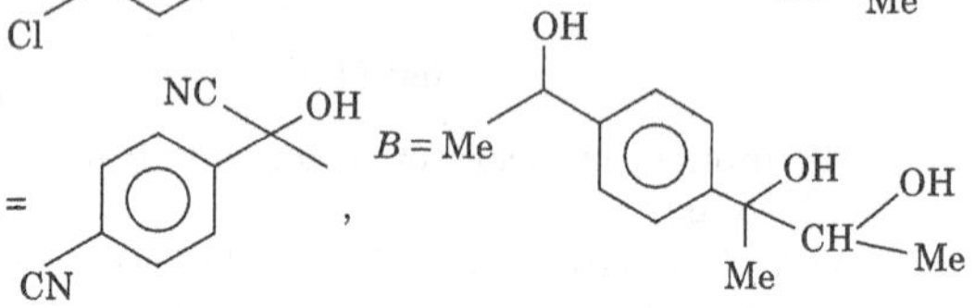

(d) A = 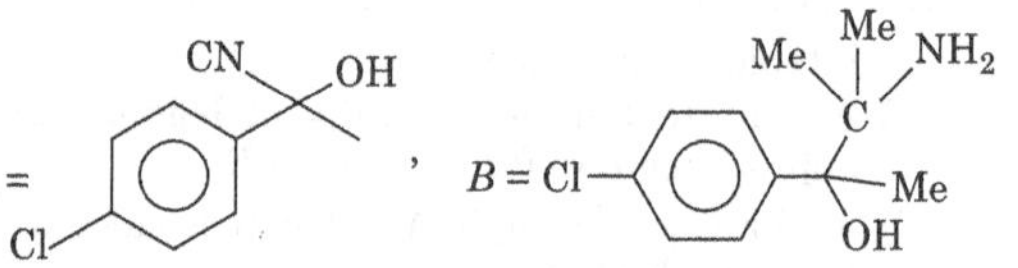

9 Which of the following relations are correct?

A. $\Delta U = q + p\Delta V$ B. $\Delta G = \Delta H - T\Delta S$
C. $\Delta S = \frac{q_{rev}}{T}$ D. $\Delta H = \Delta U - \Delta nRT$

Choose the most appropriate answer from the options given below.

(a) A and B only (b) C and D only
(c) B and C only (d) B and D only

10 Find out the major product for the following reaction.

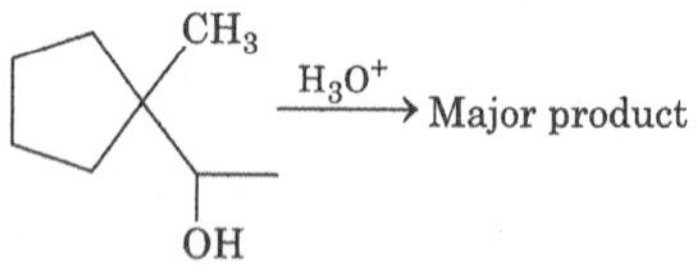

(a) 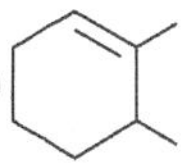
(b) 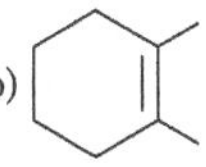
(c) 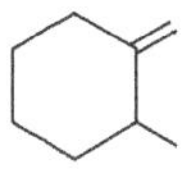
(d) 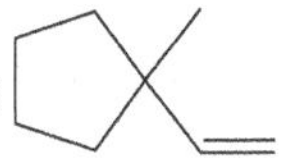

11 An indicator '*X*' is used for studying the effect of variation in concentration of iodide on the rate of reaction of iodide ion with H_2O_2 at room temperature. The indicator '*X*' forms blue coloured complex with compound
'*A*' present in the solution. The indicator '*X*' and compound '*A*' respectively are

(a) methyl orange and iodine (b) starch and H_2O_2
(c) starch and iodine (d) methyl orange and H_2O_2

12 Reaction of propanamide with Br_2 / KOH(*aq*) produces

(a) propyl amine (b) ethylamine
(c) ethylnitrile (d) propanenitrile

13 The major component of which of the following ore is sulphide based mineral?

(a) Siderite (b) Sphalerite
(c) Malachite (d) Calamine

14 Match List-I with List-II.

List-I		List-II
A. Osmosis	I.	Solvent molecules pass through semi-permeable membrane towards solvent side.
B. Reverse osmosis	II.	Movement of charged colloidal particles under the influence of applied electric potential towards oppositely charged electrodes.
C. Electro-osmosis	III.	Solvent molecules pass through semi-permeable membrane towards solution side.
D. Electrophoresis	IV.	Dispersion medium moves in an electric field.

Choose the correct answer from the options given below.

(a) A-I, B-III, C-II, D-IV (b) A-III, B-I, C-IV, D-II
(c) A-III, B-I, C-II, D-IV (d) A-I, B-III, C-IV, D-II

15 A doctor prescribed the drug, equanil to a patient. The patient was likely to have symptoms of which disease?

(a) Stomach ulcers
(b) Depression and hypertension
(c) Hyperacidity
(d) Anxiety and stress

16 Given below are two statements.

Statement I The decrease in first ionisation enthalpy from B to Al is much larger than that from Al to Ga.

Statement II The *d*-orbitals in Ga are completely filled.

In the light of the above statements, choose the most appropriate answer from the options given below.

(a) Both the Statements I and II are incorrect.
(b) Statement I is correct but Statement II is incorrect.
(c) Both the Statements I and II are correct.
(d) Statement I is incorrect but Statement II is correct.

17 Correct order of spin only magnetic moment of the following complex ions is
(Given, atomic number Fe = 26, Co = 27)

(a) $[CoF_6]^{3-} > [FeF_6]^{3-} > [Co(C_2O_4)_3]^{3-}$
(b) $[Co(C_2O_4)_3]^{3-} > [CoF_6]^{3-} > [FeF_6]^{3-}$
(c) $[FeF_6]^{3-} > [CoF_6]^{3-} > [Co(C_2O_4)_3]^{3-}$
(d) $[FeF_6]^{3-} > [Co(C_2O_4)_3]^{3-} > [CoF_6]^{3-}$

18 A solution of CrO_5 in amyl alcohol has a .... colour.

(a) orange-red (b) yellow
(c) green (d) blue

19 Following tetrapeptide can be represented as

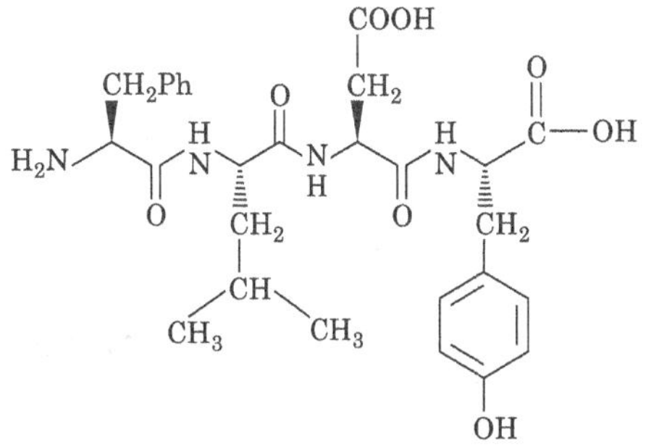

(*F*, *L*, *D*, *Y*, *I*, *Q*, *P* are one letter codes for amino acids)

(a) *FIQY* (b) *YQLF* (c) *FLDY* (d) *PLDY*

20 Given below are two statements :

Statement I Nickel is being used as the catalyst for producing *syn* gas and edible fats.

Statement II Silicon forms both electron rich and electron deficient hydrides.

In the light of the above statements, choose the most appropriate answer from the options given below;

(a) Statement I is correct but Statement II is incorrect
(b) Both the Statements I and II are correct
(c) Both the Statements I and II are incorrect
(d) Statement I is incorrect but Statement II is correct

Section B : Numerical Value Type Questions

21 The equilibrium constant for the reaction
$Zn(s) + Sn^{2+}(aq) \rightleftharpoons Zn^{2+}(aq) + Sn(s)$ is 1×10^{20} at 298 K. The magnitude of standard electrode potential of Sn/Sn^{2+} if $E^{\circ}_{Zn^{2+}/Zn} = -0.76$ V is .... $\times 10^{-2}$ V.
(Nearest integer).
Given, $\frac{2.303RT}{F} = 0.059$ V

22 A metal *M* forms hexagonal close-packed structure. The total number of voids in 0.02 mol of it is .... $\times 10^{21}$. (Nearest integer). (Given, $N_A = 6.02 \times 10^{23}$)

23 The denticity of the ligand present in the Fehling's reagent is ....

24 For conversion of compound $A \rightarrow B$, the rate constant of the reaction was found to be 4.6×10^{-5} L mol^{-1} s^{-1}. The order of the reaction is ....

> **Key Idea** To approach this type of questions, simply solve the unit of rate constant by using the general expression $[mol\ L^{-1}]^{1-n}\ s^{-1}$.

25 At 298 K

$N_2(g) + 3H_2(g) \rightleftharpoons 2NH_3(g);\ K_1 = 4 \times 10^5$

$N_2(g) + O_2(g) \rightleftharpoons 2NO(g);\ K_2 = 1.6 \times 10^{12}$

$H_2(g) + \frac{1}{2}O_2(g) \rightleftharpoons H_2O(g);\ K_3 = 1.0 \times 10^{-13}$

Based on above equilibria, the equilibrium constant of the reaction,

$2NH_3(g) + \frac{5}{2}O_2(g) \rightleftharpoons 2NO(g) + 3H_2O(g)$ is ..... $\times 10^{-33}$. (Nearest integer)

26 Total number of acidic oxides among N_2O_3, NO_2, N_2O, Cl_2O_7, SO_2, CO, CaO, Na_2O and NO is ....

27 The volume of HCl, containing 73 g L^{-1}, required to completely neutralise NaOH obtained by reacting 0.69 g of metallic sodium with water, is .... mL. (Nearest integer)

(Given; molar masses of Na, Cl, O and H, are 23, 35.5, 16 and 1 g mol^{-1} respectively)

28 Assume that, the radius of the first Bohr orbit of hydrogen atom is 0.6 Å. The radius of the third Bohr orbit of He^+ is .... picometre. (Nearest integer)

> **Key Idea** This question includes the concept of calculating the radius of hydrogen or hydrogen like species i.e., He^+. For this, we will use the formula $r_n = r_0\left(\frac{n^2}{Z}\right)$ Å

29 On heating, $LiNO_3$ gives how many compounds among the following?

Li_2O, N_2, O_2, $LiNO_2$, NO_2

30 When 0.01 mol of an organic compound containing 60% carbon was burnt completely, 4.4 g of CO_2 was produced. The molar mass of compound is ......g mol^{-1}. (Nearest integer)

30 JANUARY 2023, SHIFT-I

Section A : Objective Type Questions

1 Given below are two statements.

One is labelled as Assertion (A) and the other is labelled as Reason (R).

Assertion (A) In expensive scientific instruments, silica gel is kept in watch-glasses or in semipermeable membrane bags.

Reason (R) Silica gel adsorbs moisture from air *via* adsorption, thus protects the instrument from water corrosion (rusting) and/or prevents malfunctioning.

In the light of the above statements, choose the correct answer from the options given below.

(a) Both A and R are true but R is not the correct explanation of A
(b) A is false but R is true
(c) Both A and R are true and R is the correct explanation of A
(d) A is true but R is false

2 Caprolactam when heated at high temperature in presence of water, gives

(a) nylon 6,6
(b) nylon 6
(c) dacron
(d) teflon

3 Match List-I with List-II.

List-I (Molecules/ions)		List-II (No. of lone pairs of e^- on central atom)	
A.	IF_7	I.	Three
B.	ICl_4^-	II.	One
C.	XeF_6	III.	Two
D.	XeF_2	IV.	Zero

Choose the **correct** answer from the options given below.

(a) A-II, B-III, C-IV, D-I (b) A-IV, B-I, C-II, D-III
(c) A-II, B-I, C-IV, D-III (d) A-IV, B-III, C-II, D-I

4 Lithium aluminium hydride can be prepared from the reaction of

(a) LiCl and Al_2H_6 (b) LiCl, Al and H_2
(c) LiH and $Al(OH)_3$ (d) LiH and Al_2Cl_6

5 Benzyl isocyanide can be obtained by

A. 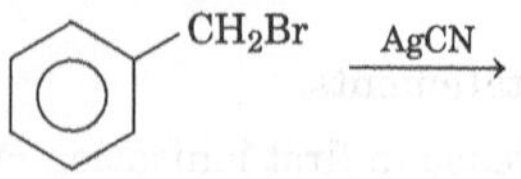

B. 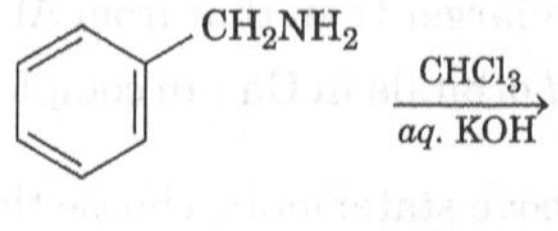

C. 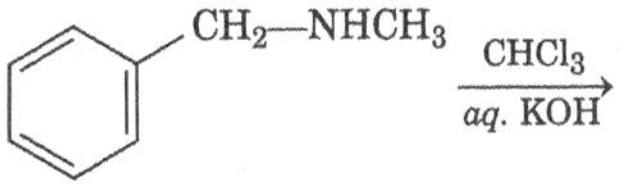

D. 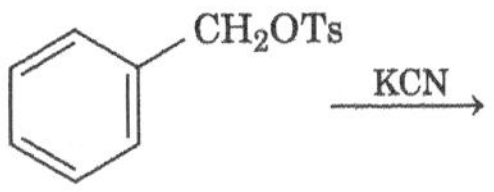

Choose the correct answer from the options given below.

(a) B and C (b) A and B (c) Only B (d) A and D

6 Amongst the following compounds, which one is an antacid?

(a) Ranitidine (b) Meprobamate
(c) Terfenadine (d) Brompheniramine

7 To inhibit the growth of tumours, identify the compounds used from the following

A. EDTA
B. coordination compounds of Pt
C. *d*-penicillamine
D. *cis*-platin

Choose the correct answer from the options given below.

(a) A and C (b) B and D (c) A and B (d) C and D

8 During the qualitative analysis of SO_3^{2-} using dilute H_2SO_4, SO_2 gas is evolved which turns $K_2Cr_2O_7$ solution (acidified with dilute H_2SO_4)

(a) blue (b) red (c) black (d) green

9 In the wet tests for identification of various cations by precipitation, which transition element cation doesn't belong to group IV in qualitative inorganic analysis?

(a) Ni^{2+} (b) Co^{2+} (c) Fe^{3+} (d) Zn^{2+}

10 The major products '*A*' and '*B*' respectively, are

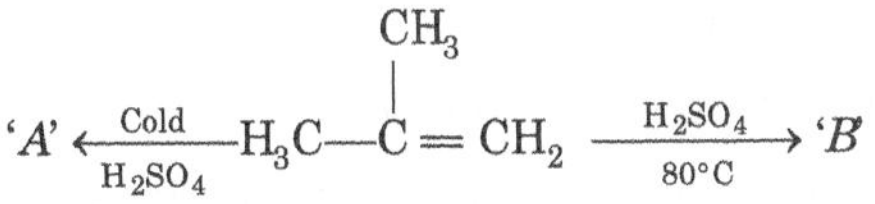

(a) 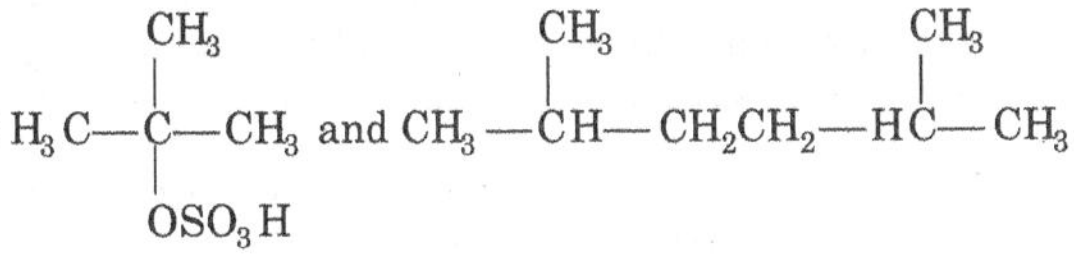

(b) 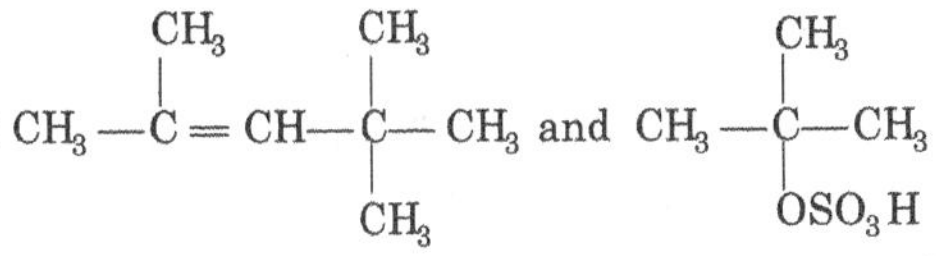

(c) 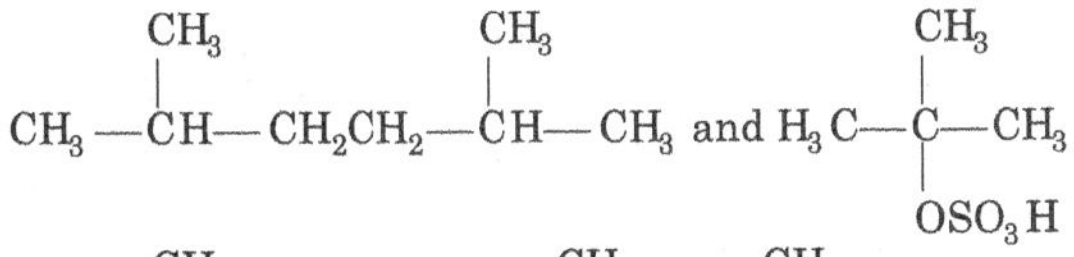

(d) 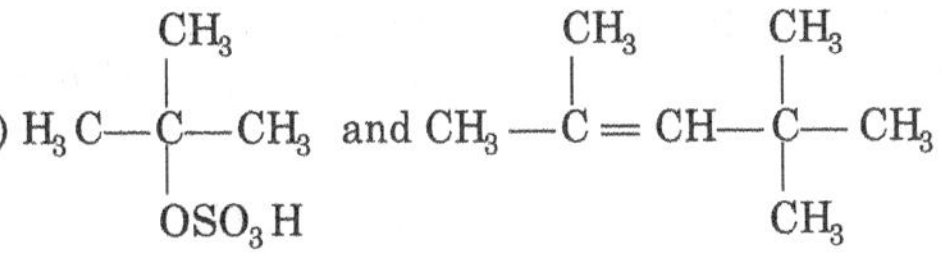

11 For OF_2 molecule, consider the following :

A. Number of lone pairs on oxygen is 2.
B. FOF angle is less than 104.5°.
C. Oxidation state of O is – 2.
D. Molecule is bent 'V' shaped.
E. Molecular geometry is linear.

Correct options are

(a) A, B and D only (b) B, E and A only
(c) A, C and D only (d) C, D and E only

12 In the extraction of copper, its sulphide ore is heated in a reverberatory furnace after mixing with silica to:

(a) separate CuO as $CuSiO_3$
(b) decrease the temperature needed for roasting of Cu_2S
(c) remove FeO as $FeSiO_3$
(d) remove calcium as $CaSiO_3$

13 Match List-I with List-II.

List-I (Atomic number)		List-II (Block of periodic table)	
A.	37	I.	*p*-block
B.	78	II.	*d*-block
C.	52	III.	*f*-block
D.	65	IV.	*s*-block

Choose the correct answer from the options given below.

(a) A-II, B-IV, C-I, D-III (b) A-I, B-III, C-IV, D-II
(c) A-IV, B-III, C-II, D-I (d) A-IV, B-II, C-I, D-III

Key Idea To solve this type of questions, simply see that valence electrons fills in which orbital i.e., if valence electrons are filled in *s*, *p*, *d* or *f*-orbital then the element belongs to *s*, *p*, *d* or *f*-block, respectively.

14 What is the correct order of acidity of the protons marked *A*-*D* in the given compounds?

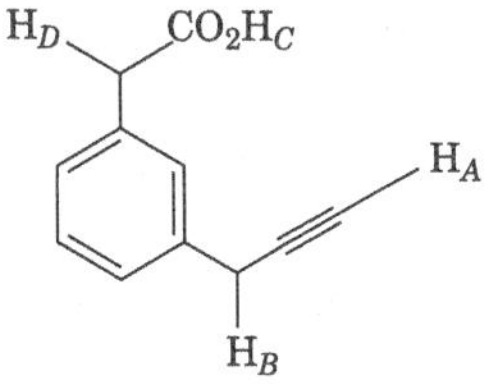

(a) $H_C > H_D > H_B > H_A$ (b) $H_C > H_D > H_A > H_B$
(c) $H_D > H_C > H_B > H_A$ (d) $H_C > H_A > H_D > H_B$

15 Formation of photochemical smog involves the following reaction in which *A*, *B* and *C* are respectively.

(i) $NO_2 \xrightarrow{h\nu} A + B$
(ii) $B + O_2 \longrightarrow C$
(iii) $A + C \longrightarrow NO_2 + O_2$

Choose the correct answer from the options given below.

(a) NO, O and O_3 (b) O, N_2O and NO
(c) O, NO and NO_3^- (d) N, O_2 and O_3

16 Which of the following is correct order of ligand field strength?

(a) $NH_3 < en < CO < S^{2-} < C_2O_4^{2-}$

(b) $CO < en < NH_3 < C_2O_4^{2-} < S^{2-}$

(c) $S^{2-} < C_2O_4^{2-} < NH_3 < en < CO$

(d) $S^{2-} < NH_3 < en < CO < C_2O_4^{2-}$

17 Which of the following compounds would give the following set of qualitative analysis?

(i) Fehling's test: positive
(ii) Na fusion extract upon treatment with sodium nitroprusside gives a blood red colour but not prussian blue.

(a) 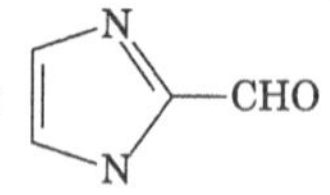

(b) 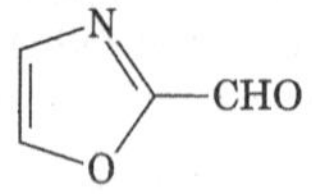

(c) 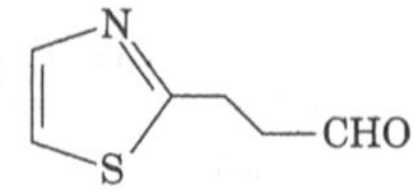

(d) 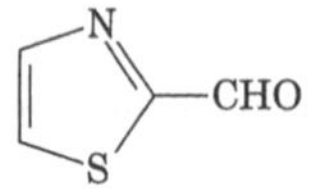

18 Match List-I with List-II.

	List-I		List-II
A.	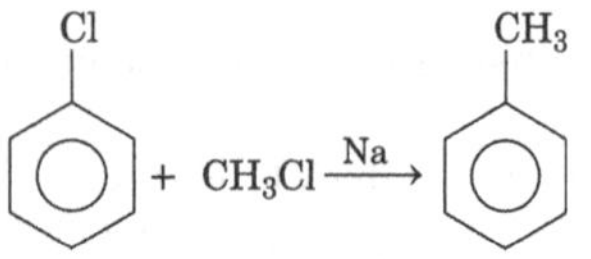	I.	Fittig reaction
B.	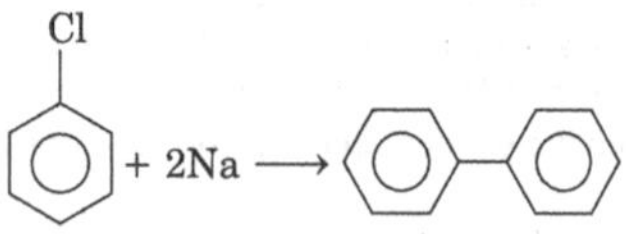	II.	Wurtz-Fittig reaction
C.	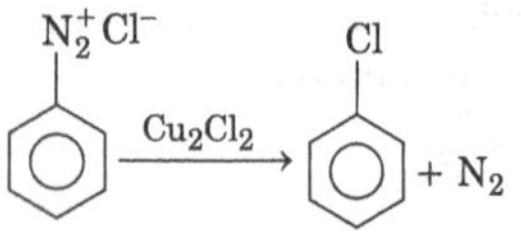	III.	Finkelstein reaction
D.	$C_2H_5Cl + NaI \longrightarrow C_2H_5I + NaCl$	IV.	Sandmeyer reaction

Choose the correct answer from the options given below.

(a) A-II, B-I, C-IV, D-III
(b) A-III, B-II, C-IV, D-I
(c) A-II, B-I, C-III, D-IV
(d) A-IV, B-II, C-III, D-I

19 The alkaline earth metal sulphate(s) which are readily soluble in water is/are

A. $BeSO_4$
B. $MgSO_4$
C. $CaSO_4$
D. $SrSO_4$
E. $BaSO_4$

Choose the correct answer from the options given below.

(a) A only
(b) B only
(c) A and B
(d) B and C

20 Given below are two statements.

One is labelled as Assertion (A) and the other is labelled as Reason (R).

Assertion (A) Ketoses give Seliwanoff's test faster than aldoses.

Reason (R) Ketoses undergo β-elimination followed by formation of furfural.

In the light of the above statements, choose the correct answer from the options given below.

(a) A is false but R is true
(b) A is true but R is false
(c) Both A and R are true but R is not the correct explanation of A
(d) Both A and R are true and R is the correct explanation of A

Section B : Numerical Value Type Questions

21 A solution containing 2 g of a non-volatile solute in 20 g of water boils at 373.52 K. The molecular mass of the solute is .... g mol^{-1}. (Nearest integer)

[Given : water boils at 373 K, K_b for water $= 0.52$ K kg mol^{-1}]

22 The number of electrons involved in the reduction of permanganate to manganese dioxide in acidic medium is ....

23 A trisubstituted compound '*A*', $C_{10}H_{12}O_2$ gives neutral $FeCl_3$ test positive. Treatment of compound '*A*' with NaOH and CH_3Br gives $C_{11}H_{14}O_2$, with hydroiodic acid gives methyl iodide and with hot conc. NaOH gives a compound *B*, $C_{10}H_{12}O_2$. Compound '*A*' also decolourises alkaline $KMnO_4$. The number of π bonds present in the compound '*A*' is ....

24 Some amount of dichloromethane (CH_2Cl_2) is added to 671.141 mL of chloroform ($CHCl_3$) to prepare 2.6×10^{-3} M solution of CH_2Cl_2 (DCM). The concentration of DCM is .... ppm (by mass).

[Given, atomic mass C = 12, H = 1, Cl = 35.5, density of $CHCl_3 = 1.49$ g cm^{-3}]

25 When 2 litre of ideal gas expands isothermally into vacuum to a total volume of 6 litre, the change in internal energy is .... J. (Nearest integer)

26 600 mL of 0.01 M HCl is mixed with 400 mL of 0.01 M H_2SO_4. The pH of the mixture is .... $\times 10^{-2}$. (Nearest integer)

[Given, log 2 = 0.30, log 3 = 0.48, log 5 = 0.69, log 7 = 0.84, log 11 = 1.04]

> **Key Idea** To calculate the pH of the mixture, first of all calculate the total moles of H^+ by using the formula $N_1V_1 = N_2V_2$. After that, use the formula of pH i.e., $pH = -\log[H^+]$.

27 The energy of one mole of photons of radiation of frequency 2×10^{12} Hz in J mol^{-1} is .... (Nearest integer)

[Given, $h = 6.626 \times 10^{-34}$ Js, $N_A = 6.022 \times 10^{23}$ mol^{-1}]

28 A 300 mL bottle of soft drink has 0.2 M CO_2 dissolved in it. Assuming CO_2 behaves as an ideal gas, the volume of the dissolved CO_2 at STP is .... mL. (Nearest integer)

[Given : at STP, molar volume of an ideal gas is 22.7 L mol^{-1}]

> **Key Idea** In this type of questions, where ideal behaviour of any gas is given, use the formula $v = 22.7 \times n$, to calculate the volume of gas in litre.

29 Consider the cell

$Pt(s) \mid H_2\,(g, 1\text{ atm}) \mid H^+\,(aq, 1\text{ M}) \mid\mid Fe^{3+}(aq),$
$Fe^{2+}(aq) \mid\mid Pt(s)$

When the potential of the cell is 0.712 V at 298 K, the ratio $[Fe^{2+}]/[Fe^{3+}]$ is .... (Nearest integer)

[Given : $Fe^{3+} + e^- \rightleftharpoons Fe^{2+}$, $E^\circ_{Fe^{3+}, Fe^{2+}/Pt} = 0.771$ V

$\frac{2.303RT}{F} = 0.06$ V]

30 If compound A reacts with B following first order kinetics with rate constant $2.011 \times 10^{-3} s^{-1}$. The time taken by A (in seconds) to reduce from 7 g to 2 g will be .... (Nearest integer) [log 5 = 0.698, log 7 = 0.845, log 2 = 0.301]

30 JANUARY 2023, SHIFT-II

Section A : Objective Type Questions

1 Bond dissociation energy of "E—H" bond of the "H_2E" hydrides of group 16 elements (given below), follows order.

A. O B. S C. Se D. Te

Choose the correct from the options given below:

(a) A > B > C > D (b) D > C > B > A
(c) B > A > C > D (d) A > B > D > C

2 The wave function (ψ) of 2s is given by

$$\psi_{2s} = \frac{1}{2\sqrt{2\pi}}\left(\frac{1}{a_0}\right)^{1/2}\left(2 - \frac{r}{a_0}\right)e^{-\frac{r}{2a_0}}$$

At $r = r_0$, radial node is formed. Thus, r_0 in terms of a_0

(a) $r_0 = a_0$ (b) $r_0 = 2a_0$
(c) $r_0 = \frac{a_0}{2}$ (d) $r_0 = 4a_0$

3 The water quality of a pond was analysed and its BOD was found to be 4. The pond has

(a) slightly polluted water
(b) very clean water
(c) water has high amount of fluoride compounds
(d) highly polluted water

4 Given below are two statements.

One is labelled as Assertion (A) and the other is labelled as Reason (R).

Assertion (A) Antihistamines do not affect the secretion of acid in stomach.

Reason (R) Antiallergic and antacid drugs work on different receptors.

In the light of the above statements, choose the correct answer from the options given below.

(a) Both A and R are true but R is not the correct explanation of A
(b) Both A and R are true and R is the correct explanation of A
(c) A is true but R is false
(d) A is false but R is true

5

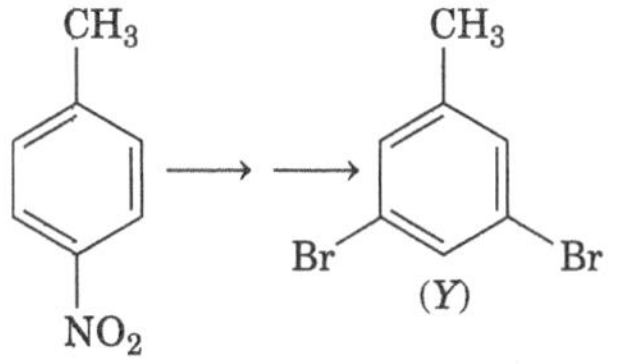

(X)

In the above conversion of compound (X) to product (Y), the sequence of reagents to be used will be

(a) (i) Fe, H^+ (ii) $Br_2(aq)$ (iii) HNO_2 (iv) CuBr
(b) (i) Fe, H^+ (ii) $Br_2(aq)$ (iii) HNO_2 (iv) H_3PO_2
(c) (i) Br_2, Fe (ii) Fe , H^+ (iii) $LiAlH_4$
(d) (i) $Br_2(aq)$ (ii) $LiAlH_4$ (iii) H_3O^+

6 The Cl—Co—Cl bond angle values in a fac—$[Co(NH_3)_3Cl_3]$ complex is/are:

(a) 90° and 120° (b) 180°
(c) 90° and 180° (d) 90°

7 Decreasing order towards S_N1 reaction for the following compounds is

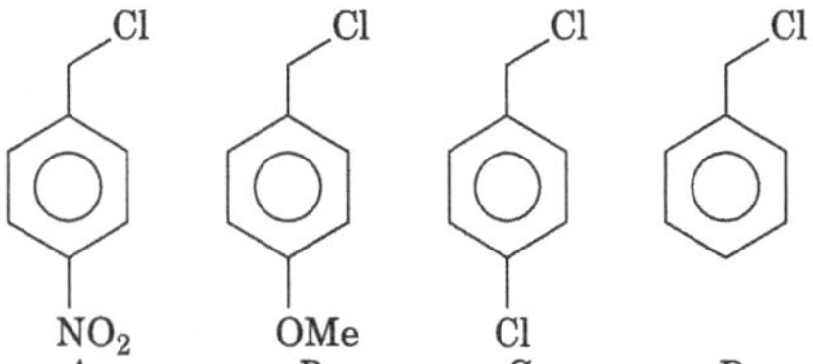

A B C D

(a) $A > B > C > D$ (b) $B > D > C > A$
(c) $A > C > D > B$ (d) $D > B > C > A$

> **Key Idea** This problem is based upon the aliphatic nucleophilic substitution. The benzene ring shown helps in the stabilisation of intermediate formed in the reaction.

8 Maximum number of electrons that can be accommodated in shell with $n = 4$ are

(a) 72 (b) 32
(c) 16 (d) 50

9 Formula for Nessler's reagent is

(a) $KHgI_3$ (b) HgI_2
(c) K_2HgI_4 (d) KHg_2I_2

10 Given below are two statements.

Statement I During electrolytic refining, the pure metal is made to act as anode and its impure metallic form is used as cathode.

Statement II During the Hall-Heroult electrolysis process, purified Al_2O_3 is mixed with Na_3AlF_6 to lower the melting point of the mixture.

In the light of the above statements, choose the most appropriate answer from the options given below.

(a) Statement I is correct but Statement II is incorrect
(b) Both Statement I and Statement II are correct
(c) Statement I is incorrect but Statement II is correct
(d) Both Statement I and Statement II are incorrect

11 Chlorides of which metal are soluble in organic solvents?

(a) Be (b) Mg
(c) Ca (d) K

12 Match List-I with List-II.

List-I (Complexes)	List-II (Hybridisation)
A. $[Ni(CO)_4]$	I. sp^3
B. $[Cu(NH_3)_4]^{2+}$	II. dsp^2
C. $[Fe(NH_3)_6]^{2+}$	III. sp^3d^2
D. $[Fe(H_2O)_6]^{2+}$	IV. d^2sp^3

(a) A-II, B-I, C-III, D-IV (b) A-II, B-I, C-IV, D-III
(c) A-I, B-II, C-IV, D-III (d) A-I, B-II, C-III, D-IV

13 The most stable carbocation for the following is

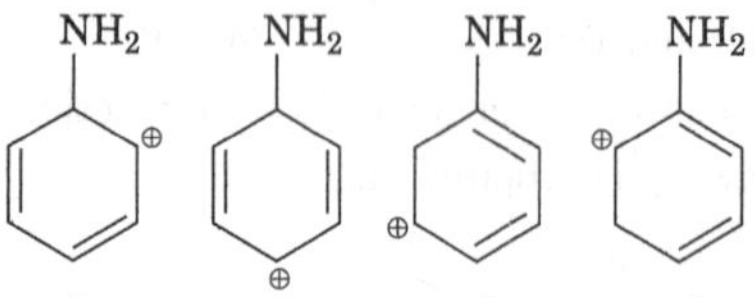

A B C D

(a) *D* (b) *C*
(c) *B* (d) *A*

Key Idea Use the concept of stability of carbocation to solve this problem. This stability can be judged on the basis of resonance and hyperconjugation.

14 $KMnO_4$ oxidises I^- in acidic and neutral/faintly alkaline solution, respectively, to

(a) IO_3^- and IO_3^- (b) I_2 and IO_3^-
(c) I_2 and I_2 (d) IO_3^- and I_2

15 The correct order of pK_a values for the following compounds is

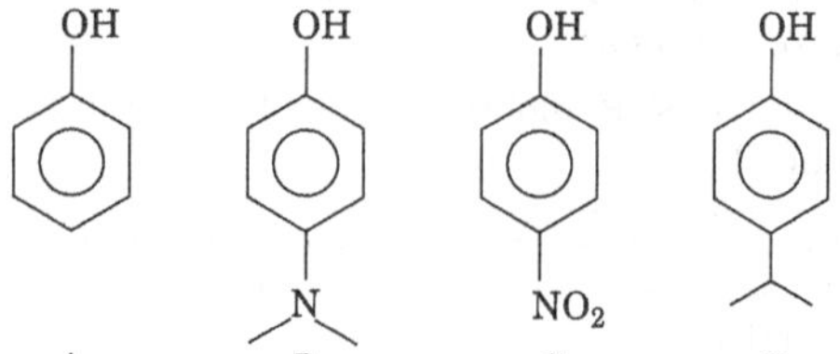

A B C D

(a) $B > D > A > C$ (b) $B > A > D > C$
(c) $C > A > D > B$ (d) $A > B > C > D$

16 Given below are two statements.

One is labelled as Assertion (A) and the other is labelled as Reason (R).

Assertion (A) 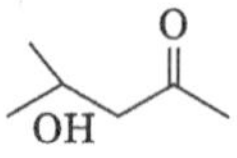 can be easily reduced using Zn-Hg / HCl to 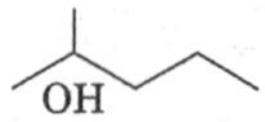

Reason (R) Zn-Hg/HCl is used to reduce carbonyl group to $—CH_2—$ group.

In the light of the above statements, choose the correct answer from the options given below.

(a) A is false but R is true
(b) Both A and R are true but R is not the correct explanation of A
(c) Both A and R are true and R is the correct explanation of A
(d) A is true but R is false

17 Boric acid is solid, whereas BF_3 is a gas at room temperature because of

(a) strong ionic bond in boric acid
(b) strong covalent bond in BF_3
(c) strong hydrogen bond in boric acid
(d) strong van der Waals' interaction in boric acid

18 Which of the following reaction is correct?

(a) $2LiNO_3 \xrightarrow{\Delta} 2NaNO_2 + O_2$
(b) $2LiNO_3 \xrightarrow{\Delta} 2Li + 2NO_2 + O_2$
(c) $4LiNO_3 \xrightarrow{\Delta} 2Li_2O + 4NO_2 + O_2$
(d) $4LiNO_3 \xrightarrow{\Delta} 2Li_2O + 2N_2O_4 + O_2$

19 1 L, 0.02 M solution of $[Co(NH_3)_5SO_4]Br$ is mixed with 1 L, 0.02 M solution of $[Co(NH_3)_5Br]SO_4$.

The resulting solution is divided into two equal parts (*X*) and treated with excess of $AgNO_3$ solution and $BaCl_2$ solution respectively as shown below.

1 L solution (*X*) + $AgNO_3$ solution (excess) $\longrightarrow$ *Y*
1 L solution (*X*) + $BaCl_2$ solution (excess) $\longrightarrow$ *Z*

The number of moles of *Y* and *Z* respectively are

(a) 0.02, 0.01 (b) 0.01, 0.02 (c) 0.02, 0.02 (d) 0.01, 0.01

Key Idea This problem of mole concept is based on the fact that mixing and division of solutions of equal molarity with different anions give same result.

20 Match List-I with List-II.

	List-I (Mixture)		List-II (Separation technique)
A.	$CHCl_3 + C_6H_5NH_2$	I.	Steam distillation
B.	$C_6H_{14} + C_5H_{12}$	II.	Differential extraction
C.	$C_6H_5NH_2 + H_2O$	III.	Distillation
D.	Organic compound in H_2O	IV.	Fractional distillation

(a) A-II, B-I, C-III, D-IV (b) A-IV, B-I, C-III, D-II
(c) A-III, B-I, C-IV, D-II (d) A-III, B-IV, C-I, D-II

Section B : Numerical Value Type Questions

21 The strength of 50 volume solution of hydrogen peroxide is .... g/L. (Nearest integer).
[Given : Molar mass of H_2O_2 is 34 g mol^{-1}
Molar volume of gas at STP = 22.7 L.]

22 A short peptide on complete hydrolysis produces 3 moles of glycine (*G*), two moles of leucine (*L*) and two moles of valine (*V*) per mole of peptide. The number of peptide linkages in it are ....

23 Lead storage battery contains 38% by weight solution of H_2SO_4. The van't Hoff factor is 2.67 at this concentration. The temperature in Kelvin at which the solution in the battery will freeze is .... (Nearest integer).

[Given : K_f = 1.8 K kg mol^{-1}]

> **Key Idea** Consider the percentage concentration in grams and use the formula $\Delta T = i_{max} \times K_f \times$ molality.

24 The graph of $\log \frac{x}{m}$ *vs* $\log p$ for an adsorption process is a straight line inclined at an angle of 45° with intercept equal to 0.6020. The mass of gas adsorbed per unit mass of adsorbent at the pressure of 0.4 atm is
.... $\times 10^{-1}$. (Nearest integer).

[Given : log 2 = 0.3010]

25 1 mole of ideal gas is allowed to expand reversibly and adiabatically from a temperature of 27°C. The work done is
3 kJ mol^{-1}. The final temperature of the gas is .... K. (Nearest integer).
[Given : C_V = 20 J mol^{-1} K^{-1}.]

26 The electrode potential of the following half cell at 298 K $X \mid X^{2+}$ (0.001 M) $\|$ Y^{2+} (0.01 M) $\mid Y$ is
.... $\times 10^{-2}$ V. (Nearest integer).

[Given : $E^\circ_{X^{2+}/X} = -2.36$ V
$E^\circ_{Y^{2+}/Y} = +0.36$ V
$\frac{2.303RT}{F} = 0.06$ V]

27 Iron oxide FeO, crystallises in a cubic lattice with a unit cell edge length of 5.0 Å. If density of the FeO in the crystal is 4.0 g cm^{-3}, then the number of FeO units present per unit cell is .... (Nearest integer)

[Given : molar mass of Fe and O is 56 and 16 g mol^{-1} respectively. $N_A = 6.0 \times 10^{23}$ mol^{-1}]

28 Consider the following equation:

$2SO_2(g) + O_2(g) \rightleftharpoons 2SO_3(g), \Delta H = -190$ kJ

The number of factors which will increase the yield of SO_3 at equilibrium from the following is ....
A. increasing temperature
B. increasing pressure
C. adding more SO_2
D. adding more O_2
E. addition of catalyst

29 An organic compound undergoes first order decomposition. If the time taken for the 60% decomposition is 540 s, then the time required for 90% decomposition will be is .... s. (Nearest integer)

[Given : ln 10 = 2.3; log 2 = 0.3]

30 Number of compounds from the following which will not dissolve in cold $NaHCO_3$ and NaOH solutions but will dissolve in hot NaOH solution is ....

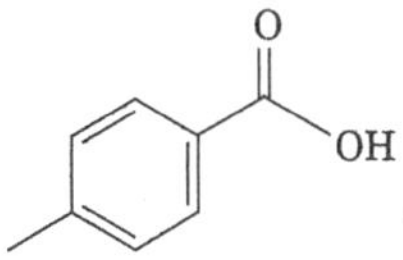

,
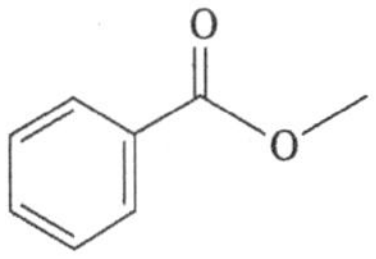

,
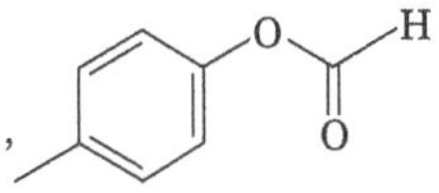

,
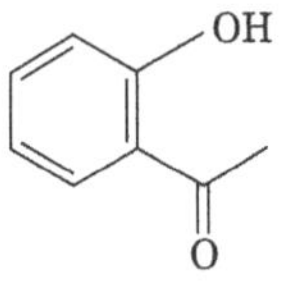

,
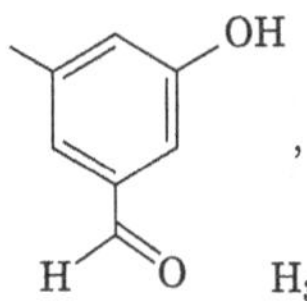

,
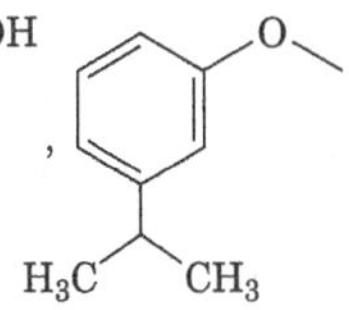

,
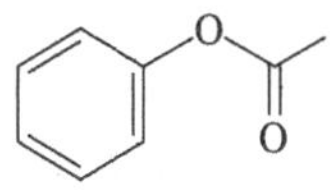

31 JANUARY 2023, SHIFT-I

Section A : Objective Type Questions

1 An organic compound '*A*' with empirical formula C_6H_6O gives sooty flame on burning. Its reaction with bromine solution in low polarity solvent results in high yield of *B*. *B* is

(a) 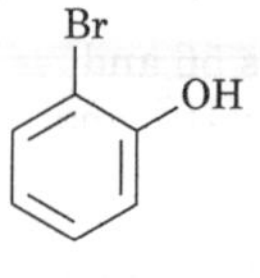 (b) 

(c) 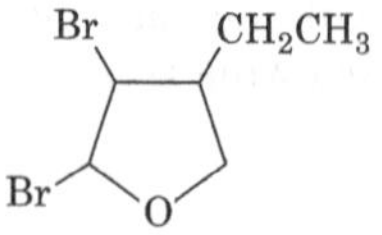 (d) 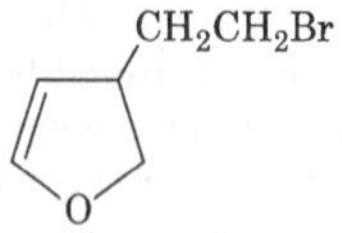

2 Nd^{2+} =........

(a) $4f^3$ (b) $4f^4$ (c) $4f^4 6s^2$ (d) $4f^2 6s^2$

3 Choose the correct set of reagents for the following conversion.

trans (Ph—CH= CH—CH_3) ⟶ *cis* (Ph—CH == CH—CH_3)

(a) Br_2, aq · KOH, $NaNH_2$, H_2 Lindlar catalyst
(b) Br_2, alc · KOH, $NaNH_2$, H_2 Lindlar catalyst
(c) Br_2, aq · KOH, $NaNH_2$, Na(liq. NH_3)
(d) Br_2 , alc · KOH, $NaNH_2$, Na(liq. NH_3)

4 Consider the following reaction.

$$\text{Propanal + Methanal} \xrightarrow[\text{(ii) }\Delta \text{ (iii) NaCN (iv) } H_3O^+]{\text{(i) dil. NaOH}} \text{Product } B\ (C_5H_8O_3)$$

The correct statement for product *B* is
(a) optically active and adds one mole of bromine
(b) optically active alcohol and is neutral
(c) racemic mixture and gives a gas with saturated $NaHCO_3$ solution
(d) racemic mixture and is neutral

5 The correct order of basicity of oxides of vanadium is
(a) $V_2O_3 > V_2O_4 > V_2O_5$
(b) $V_2O_4 > V_2O_3 > V_2O_5$
(c) $V_2O_5 > V_2O_4 > V_2O_3$
(d) $V_2O_3 > V_2O_5 > V_2O_4$

6 The methods not involved in concentration of ore are
A. liquation B. leaching
C. electrolysis D. hydraulic washing
E. froth floatation

Choose the correct answer from the options given below :
(a) B, D and E (b) A and C
(c) C, D and E (d) B, D and C

7 Which of the following artificial sweeteners has the highest sweetness value in comparison to cane sugar?
(a) Alitame (b) Saccharin
(c) Aspartame (d) Sucralose

8 Match List-I with List-II.

	List-I		List-II
A.	XeF_4	I.	*See-saw*
B.	SF_4	II.	Square planar
C.	NH_4^+	III.	Bent T-shaped
D.	BrF_3	IV.	Tetrahedral

Choose the correct answer from the options given below :
(a) A-II, B-I, C-IV, D-III (b) A-IV, B-I, C-II, D-III
(c) A -IV, B-III, C-II, D-I (d) A-II, B-I, C-III, D-IV

9 The correct increasing order of the ionic radii is
(a) $Ca^{2+} < K^+ < Cl^- < S^{2-}$ (b) $S^{2-} < Cl^- < Ca^{2+} < K^+$
(c) $K^+ < S^{2-} < Ca^{2+} < Cl^-$ (d) $Cl^- < Ca^{2+} < K^+ < S^{2-}$

10 Which transition in the hydrogen spectrum would have the same wavelength as the Balmer type transition from $n = 4$ to $n = 2$ of He^+ spectrum
(a) $n = 1$ to $n = 3$ (b) $n = 1$ to $n = 2$
(c) $n = 3$ to $n = 4$ (d) $n = 2$ to $n = 1$

11 Match List-I with List-II.

	List-I (Mixture of compounds)		List-II (Separation technique)
A.	H_2O/CH_2Cl_2	I.	Crystallisation
B.	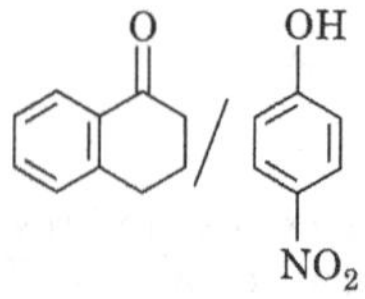	II.	Differential solvent extraction
C.	Kerosene / naphthalene	III.	Column chromatography
D.	$C_6H_{12}O_6$/NaCl	IV.	Fractional distillation

Correct match is
(a) A-I, B-III, C-II, D-IV (b) A-II, B-IV, C-I, D-III
(c) A -III, B-IV, C-II, D-I (d) A-II, B-III, C-IV, D-I

12 H_2O_2 acts as a reducing agent in
(a) $Na_2S + 4H_2O_2 \longrightarrow Na_2SO_4 + 4H_2O$
(b) $2Fe^{2+} + 2H^+ + H_2O_2 \longrightarrow 2Fe^{3+} + 2H_2O$
(c) $Mn^{2+} + 2H_2O_2 \longrightarrow MnO_2 + 2H_2O$
(d) $2NaOCl + H_2O_2 \longrightarrow 2NaCl + H_2O + O_2$

13 A protein '*X*' with molecular weight of 70,000 u, on hydrolysis gives amino acids. One of these amino acid is

(a) 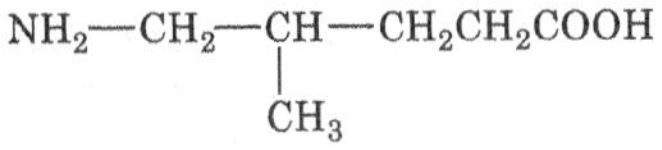

(b) 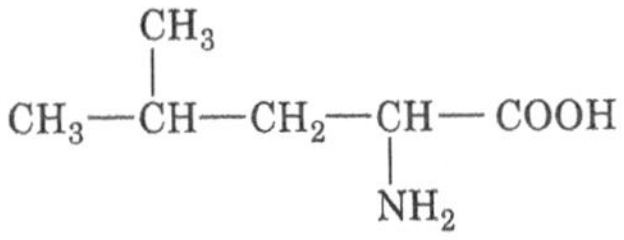

(c) 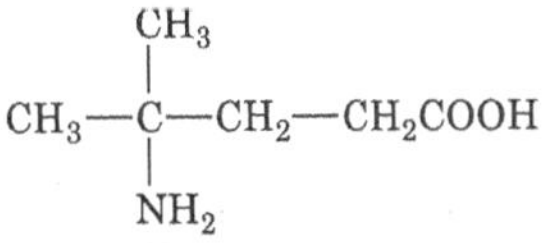

(d) 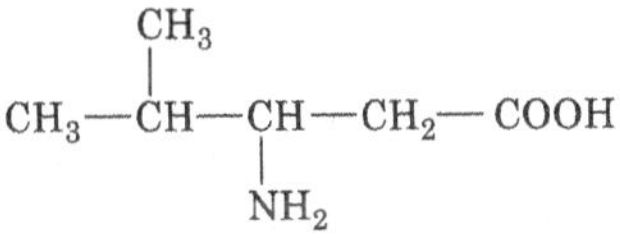

14 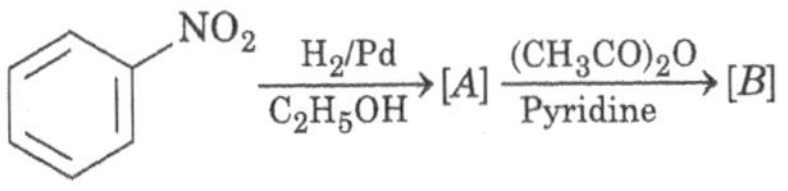

Consider the above reaction and identify the product *B*.

(a) 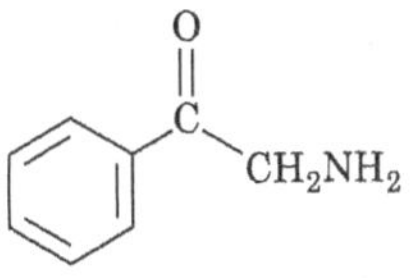
(b) 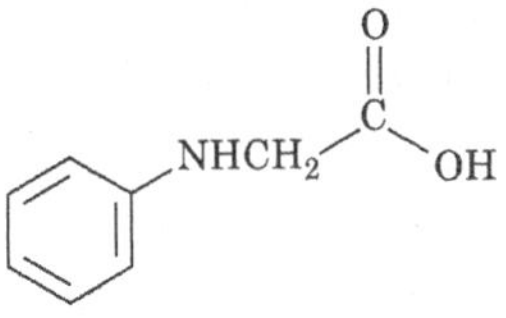

(c) 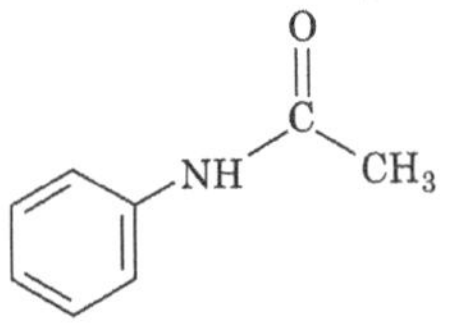
(d) 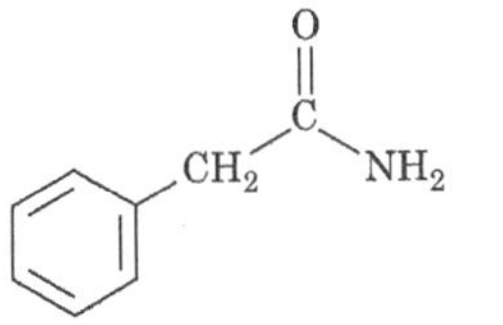

15 The correct order of melting points of dichlorobenzenes is

(a) 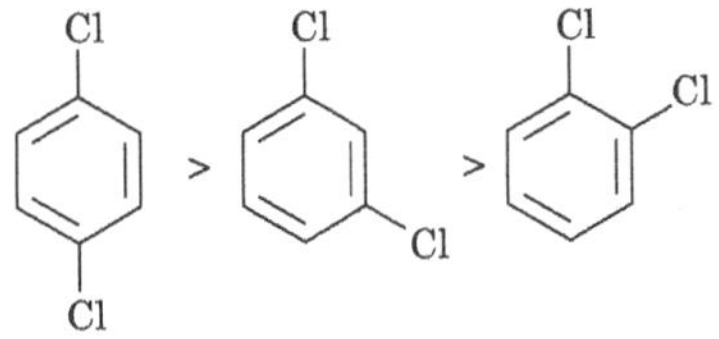

(b) 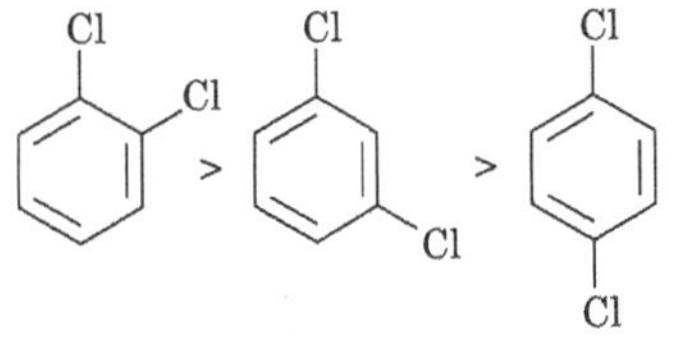

(c) 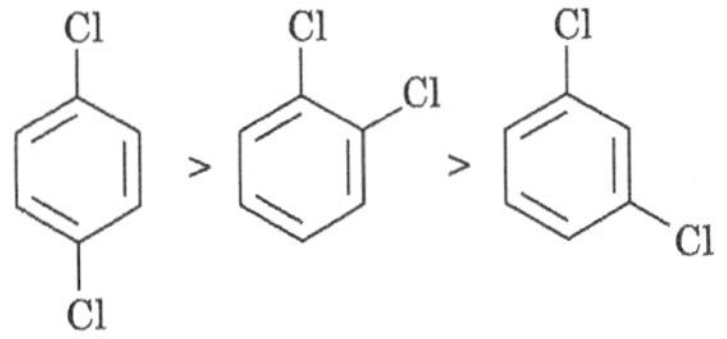

(d) 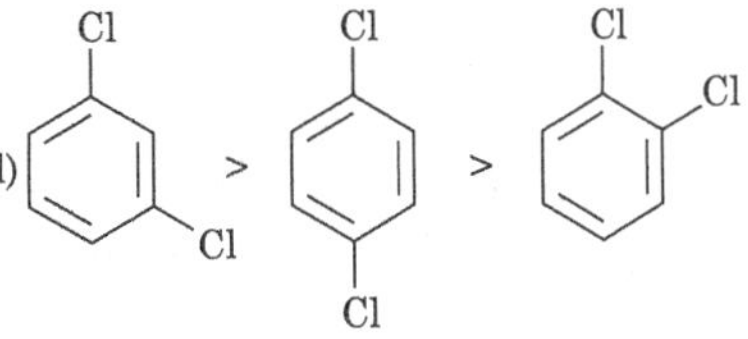

16 Cobalt chloride when dissolved in water forms pink coloured complex $\underline{X}$ which has octahedral geometry. This solution on treating with conc. HCl forms deep blue complex, $\underline{Y}$ which has a $\underline{Z}$ geometry. *X*, *Y* and *Z*, respectively, are

(a) $X = [Co(H_2O)_4Cl_2]^+, Y = [CoCl_4]^{2-}, Z =$ tetrahedral
(b) $X = [Co(H_2O)_6]^{2+}, Y = [CoCl_6]^{3-}, Z =$ octahedral
(c) $X = [Co(H_2O)_6]^{2+}, Y = [CoCl_4]^{2-}, Z =$ tetrahedral
(d) $X = [Co(H_2O)_6]^{3+}, Y = [CoCl_6]^{3+}, Z =$ octahedral

17 Adding surfactants in non-polar solvent, the micelles structure will look like

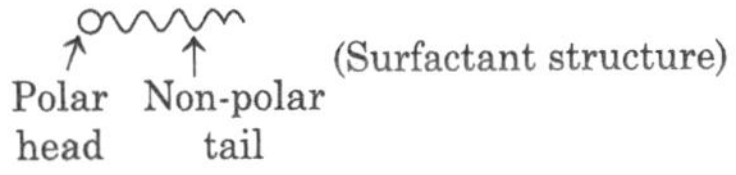

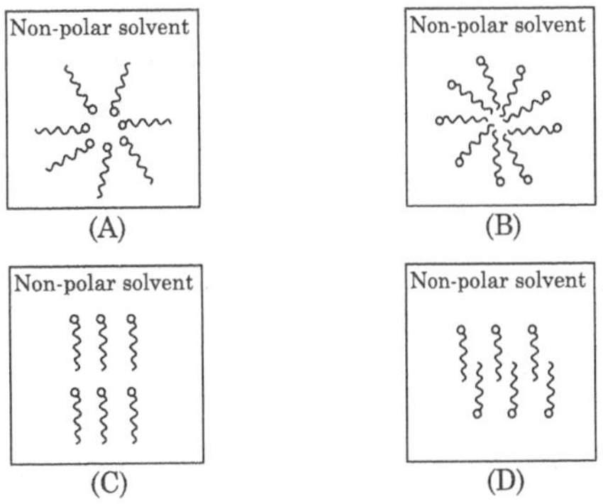

(a) B (b) C (c) A (d) D

18 When Cu^{2+} ion is treated with KI, a white precipitate, *X* appears in solution. The solution is titrated with sodium thiosulphate, the compound *Y* is formed. *X* and *Y* respectively are

(a) $X = Cu_2I_2 \quad Y = Na_2S_4O_5$ (b) $X = CuI_2 \quad Y = Na_2S_2O_3$
(c) $X = CuI_2 \quad Y = Na_2S_4O_6$ (d) $X = Cu_2I_2 \quad Y = Na_2S_4O_6$

19 Which one of the following statements is correct for electrolysis of brine solution?

(a) Cl_2 is formed at cathode. (b) OH^- is formed at cathode.
(c) O_2 is formed at cathode. (d) H_2 is formed at anode.

20 Identify *X*, *Y* and *Z* in the following reaction. (Equation not balanced)

$$Cl\dot{O} + NO_2 \longrightarrow \underline{X} \xrightarrow{H_2O} \underline{Y} + \underline{Z}$$

(a) $X = ClONO_2, Y = HOCl, Z = NO_2$
(b) $X = ClNO_2, Y = HCl, Z = HNO_3$
(c) $X = ClNO_3, Y = Cl_2, Z = NO_2$
(d) $X = ClONO_2, Y = HOCl, Z = HNO_3$

Section B : Numerical Value Type Questions

21 The total pressure of a mixture of non-reacting gases X(0.6 g) and Y(0.45 g) in a vessel is 740 mm of Hg. The partial pressure of the gas X is .... mm of Hg. (Nearest integer)

(Given : molar mass $X = 20$ and $Y = 45$ g mol^{-1})

22 At 27°C, a solution containing 2.5 g of solute in 250.0 mL of solution exerts an osmotic pressure of 400 Pa. The molar mass of the solute is .... g mol^{-1}. (Nearest integer) (Given : $R = 0.083$ L bar K^{-1} mol^{-1})

23 For reaction,

$$SO_2(g) + \frac{1}{2}O_2(g) \rightleftharpoons SO_3(g)$$

$K_p = 2 \times 10^{12}$ at 27°C and 1 atm pressure. The K_C for the same reaction is .... $\times 10^{13}$. (Nearest integer)

(Given : $R = 0.082$ L atm K^{-1} mol^{-1})

> **Key Idea** To calculate the molar mass of solute, use the formula of osmotic pressure i.e., $\pi = CRT$.
> Here we will express C in terms of molar mass, which is as follows
> $C = \frac{W_2}{M_2 V}$
> Where, W_2 = mass of solute, M_2 = molar mass of solute
> V = volume of solution

24 The logarithm of equilibrium constant for the reaction $Pd^{2+} + 4Cl^- \rightleftharpoons PdCl_4^{2-}$ is .... (Nearest integer)

Given : $\frac{2.303RT}{F} = 0.06$ V

$Pd^{2+}(aq) + 2e^- \rightleftharpoons Pd(s)$, $E^\ominus = 0.83$ V

$PdCl_4^{2-}(aq) + 2e^- \rightleftharpoons Pd(s) + 4Cl^-(aq)$, $E^\ominus = 0.65$ V

> **Key Idea** To find out the value of $\log K$, first of all we will calculate the value of n and $E°$ from the given equation. Then we will substitute these values in the equation $\Delta G° = -RT \ln K$.

25 $A \rightarrow B$

The rate constants of the above reaction at 200 K and 300 K are 0.03 min^{-1} and 0.05 min^{-1} respectively. The activation energy for the reaction is .... J. (Nearest integer)

[Given : $\ln 10 = 2.3$, $R = 8.3$J K^{-1} mol^{-1}

$\log 5 = 0.70$, $\log 3 = 0.48$, $\log 2 = 0.30$]

26 On complete combustion, 0.492 g of an organic compound gave 0.792 g of CO_2. The % of carbon in the organic compound is .... (Nearest integer)

27 The enthalpy change for the conversion of $\frac{1}{2}Cl_2(g)$ to $Cl^-(aq)$ is (–) .... kJ mol^{-1} (Nearest integer)

[Given, $\Delta_{dis}H^\ominus_{Cl_{2(g)}} = 240$ kJ mol^{-1},

$\Delta_{eg}H^\ominus_{Cl(g)} = -350$ kJ mol^{-1},

$\Delta_{hyd}H^\ominus_{Cl^-_{(g)}} = -380$ kJ mol^{-1}]

28 Zinc reacts with hydrochloric acid to give hydrogen and zinc chloride. The volume of hydrogen gas produced at STP from the reaction of 11.5 g of zinc with excess HCl is .... L. (Nearest integer)

[Given molar mass of Zn is 65.4 g mol^{-1} and molar volume of H_2 at STP = 22.7 L]

29 The oxidation state of phosphorus in hypophosphoric acid is + ...

30 How many of the transformations given below would result in aromatic amines?

(a) 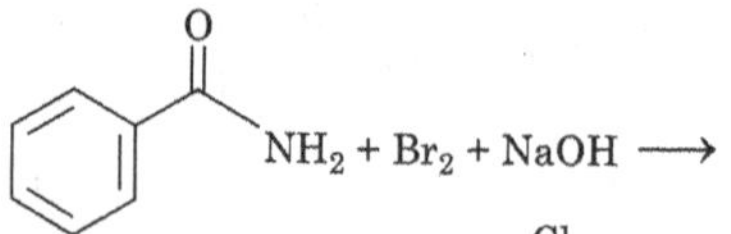

(b) 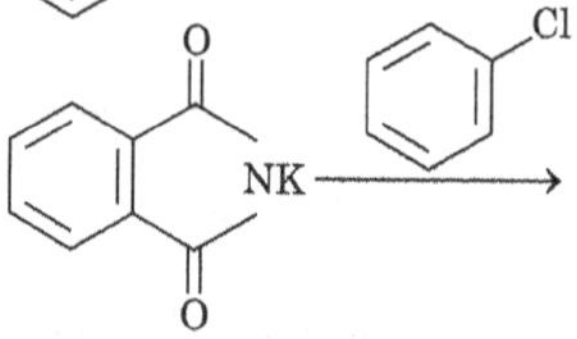

(c) 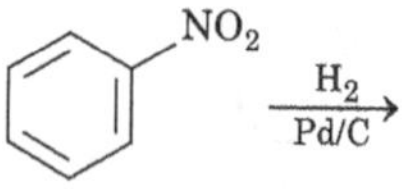

(d) 

31 JANUARY 2023, SHIFT-II

Section A : Objective Type Questions

1 In the following halogenated organic compounds the one with maximum number of chlorine atoms in its structure is

(a) gammaxene (b) chloropicrin
(c) freon-12 (d) chloral

2 Cyclohexylamine when treated with nitrous acid yields (P). On treating (P) with PCC results in (Q). When (Q) is heated with dil. NaOH we get (R). The final product (R) is

(a) 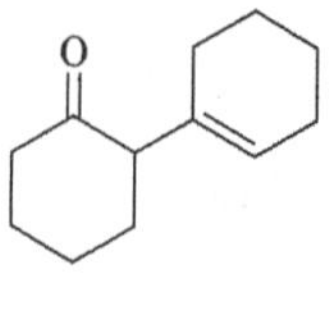

(b) 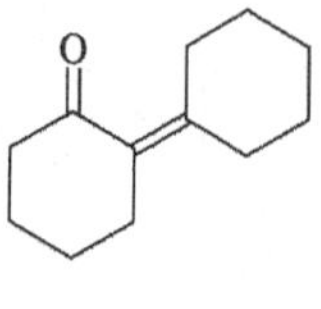

(c) 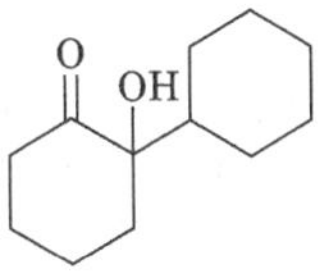

(d) 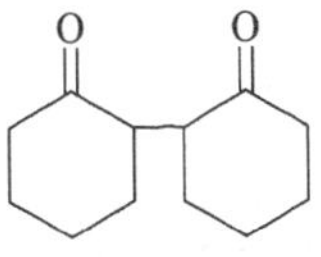

3 Match List-I with List-II.

List-I	List-II
A. Physisorption	I. Single layer adsorption
B. Chemisorption	II. 20 - 40 kJ mol^{-1}
C. $N_2(g) + 3H_2(g) \xrightarrow{Fe(s)} 2NH_3(g)$	III. Chromatography
D. Analytical application or adsorption	IV. Heterogeneous catalysis

Choose the correct answer from the options given below.

(a) A-II, B-III, C-I, D-IV (b) A-II, B-I, C-IV, D-III
(c) A-III, B-IV, C-I, D-II (d) A-IV, B-II, C-III, D-I

4 The Lewis acid character of boron trihalides follows the order

(a) $BBr_3 > BI_3 > BCl_3 > BF_3$ (b) $BF_3 > BCl_3 > BBr_3 > BI_3$
(c) $BI_3 > BBr_3 > BCl_3 > BF_3$ (d) $BCl_3 > BF_3 > BBr_3 > BI_3$

5 Incorrect statement for the use of indicators in acid-base titration is

(a) methyl orange may be used for a weak acid *vs* weak base titration.
(b) phenolphthalein is a suitable indicator for a weak acid *vs* strong base titration.
(c) phenolphthalein may be used for a strong acid *vs* strong base titration.
(d) methyl orange is a suitable indicator for a strong acid *vs* weak base titration.

6 The element playing significant role in neuromuscular function and interneuronal transmission is

(a) Mg (b) Ca (c) Li (d) Be

7 Compound *A*, $C_5H_{10}O_5$, given a tetraacetate with Ac_2O and oxidation of *A* with $Br_2—H_2O$ gives an acid, $C_5H_{10}O_6$. Reduction of *A* with HI gives isopentane. The possible structure of *A* is

(a) 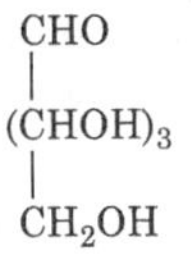

(b) 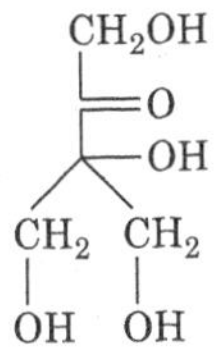

(c) 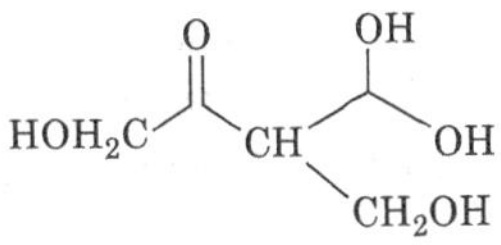

(d) 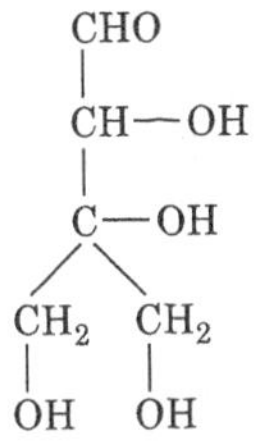

8 In Duma's method for the estimation of N_2, the sample is heated with copper oxide and the gas evolved is passed over

(a) copper gauze (b) Pd
(c) Ni (d) copper oxide

9 When a hydrocarbon *A* undergoes complete combustion it requires 11 equivalents of oxygen and produces 4 equivalents of water. What is the molecular formula of *A*?

(a) C_5H_8 (b) $C_{11}H_8$ (c) C_9H_8 (d) $C_{11}H_4$

10 Given below are two statements.

Statement I Upon heating a borax bead dipped in cupric sulphate in a luminous flame, the colour of the bead becomes green.

Statement II The green colour observed is due to the formation of copper (I) metaborate.

In the light of the above statements, choose the most appropriate answer from the options given below.

(a) Statement I is true but Statement II is false
(b) Statement I is false but Statement II is true
(c) Both Statement I and Statement II are true
(d) Both Statement I and Statement II are false

11 The normal rain water is slightly acidic and its pH value is 5.6 because of which one of the following?

(a) $4NO_2 + O_2 + 2H_2O \longrightarrow 4HNO_3$
(b) $N_2O_5 + H_2O \longrightarrow 2HNO_3$
(c) $CO_2 + H_2O \longrightarrow H_2CO_3$
(d) $2SO_2 + O_2 + 2H_2O \longrightarrow 2H_2SO_4$

12 Given below are two statements.

One is labelled as Assertion (A) and the other is labelled as Reason (R).

Assertion (A) The first ionisation enthalpy of 3*d* series elements is more than that of group 2 metals.

Reason (R) In 3*d* series of elements successive filling of *d*-orbitals takes place.

In the light of the above statements, choose the correct answer from the options given below.

(a) (A) is true but (R) is false
(b) (A) is false but (R) is true
(c) Both (A) and (R) are true and (R) is the correct explanation of (A)
(d) Both (A) and (R) are true but (R) is not the correct explanation of (A)

13 Which one of the following statements is incorrect?

(a) Cast iron is obtained by melting pig iron with scrap iron and coke using hot air blast.
(b) The malleable iron is prepared from cast iron by oxidising impurities in a reverberatory furnace.
(c) van Arkel method is used to purify tungsten.
(d) Boron and indium can be purified by zone refining method.

14 Arrange the following orbitals in decreasing order of energy.

A. $n = 3, l = 0, m = 0$ B. $n = 4, l = 0, m = 0$
C. $n = 3, l = 1, m = 0$ D. $n = 3, l = 2, m = 1$

The correct option for the order is

(a) D > B > A > C (b) A > C > B > D
(c) B > D > C > A (d) D > B > C > A

15 Which of the following compounds are not used as disinfectants?

A. Chloroxylenol
B. Bithional
C. Veronal
D. Prontosil
E. Terpineol

Choose the correct answer from the options given below.

(a) A and B (b) A, B and E
(c) C and D (d) B, D and E

16 Which of the following elements have half-filled *f*-orbitals in their ground state?

(Given: atomic number Sm = 62; Eu = 63; Tb = 65; Gd = 64, Pm = 61)

A. Sm
B. Eu
C. Tb
D. Gd
E. Pm

Choose the correct answer from the options given below.

(a) C and D (b) A and E
(c) B and D (d) A and B

17 Evaluate the following statements for their correctness.

A. The elevation in boiling point temperature of water will be same for 0.1 M NaCl and 0.1 M urea.
B. Azeotropic mixtures boil without change in their composition.
C. Osmosis always takes place from hypertonic to hypotonic solution.
D. The density of 32% H_2SO_4 solution having molarity 4.09 M is approximately 1.26 g mL^{-1}.
E. A negatively charged sol is obtained when KI solution is added to silver nitrate solution.

Choose the correct answer from the options given below.

(a) B and D (b) A and C
(c) A, B and D (d) B, D and E

18 A hydrocarbon '*X*' with formula C_6H_8 uses two moles of H_2 on catalytic hydrogenation of its one mole. On ozonolysis, '*X*' yields two moles of methane dicarbaldehyde. The hydrocarbon '*X*' is

(a) 1-methylcyclopenta-l, 4-diene
(b) cyclohexa - 1, 3 - diene
(c) hexa-1, 3, 5-triene
(d) cyclohexa-1, 4-diene

19 Given below are two statements.

Statement I H_2O_2 is used in the synthesis of cephalosporin.

Statement II H_2O_2 is used for the restoration of aerobic conditions to sewage wastes.

In the light of the above statements, choose the most appropriate answer from the options given below.

(a) Both Statement I and Statement II are correct
(b) Statement I is incorrect but Statement II is correct
(c) Both Statement I and Statement II are incorrect
(d) Statement I is correct but Statement II is incorrect

20 An organic compound [*A*] ($C_4H_{11}N$), shows optical activity and gives N_2 gas on treatment with HNO_2. The compound [*A*] reacts with $PhSO_2Cl$ producing a compound which is soluble in KOH. The structure of *A* is

(a) 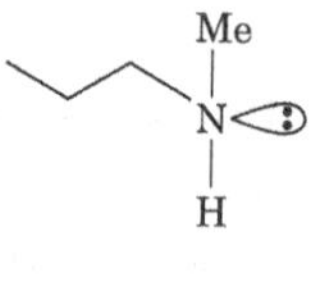
(b) 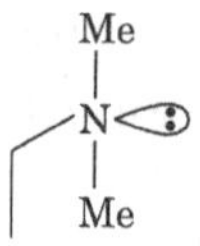
(c) 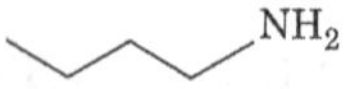
(d) 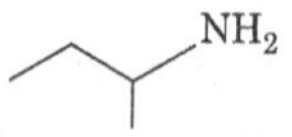

Section B : Numerical Value Type Questions

21 A sample of metal oxide has formula $M_{0.83}O_{100}$. The metal *M* can exist in two oxidation states + 2 and + 3. In the sample of $M_{0.83}O_{1.00}$, the percentage of metal ions existing in + 2 oxidation state is ....%. (Nearest integer)

22 Assume carbon burns according to following equation.

$$2C(s) + O_2(g) \longrightarrow 2CO(g)$$

When 12 g carbon is burnt in 48 g of oxygen, the volume of carbon monoxide produced is .... $\times 10^{-1}$ L at STP. [Nearest integer]

[Given, assume CO as ideal gas, mass of C is 12 g mol^{-1}, mass of O is 16 g mol^{-1} and molar volume of an ideal gas at STP is 22.7 L mol^{-1}]

23 At 298 K, the solubility of silver chloride in water is 1.434×10^{-3} g L^{-1}. The value of $-\log K_{sp}$ for silver chloride is ....

[Given, mass of Ag is 107.9 g mol^{-1} and mass of Cl is 35.5 g mol^{-1}]

Key Idea To solve this question, the concept of solubility can be used. For this, write the equation involved in the reaction. Then use the formula of K_{sp} in terms of solubility *S*. Then solve the required value i. e., $\log K_{sp}$.

24 Amongst the following, the number of species having the linear shape is ....

XeF_2, I_3^+, C_3O_2, I_3^-, CO_2, SO_2, $BeCl_2$ and $BCl_2^{\ominus}$

25 The rate constant for a first order reaction is 20 min^{-1}. The time required for the initial concentration of the reactant to reduce to its $\frac{1}{32}$ level is .... $\times 10^{-2}$ min. (Nearest integer)

[Given, ln 10 = 2.303, log 2 = 0.3010]

26 The number of molecules which gives haloform test among the following molecules is ....

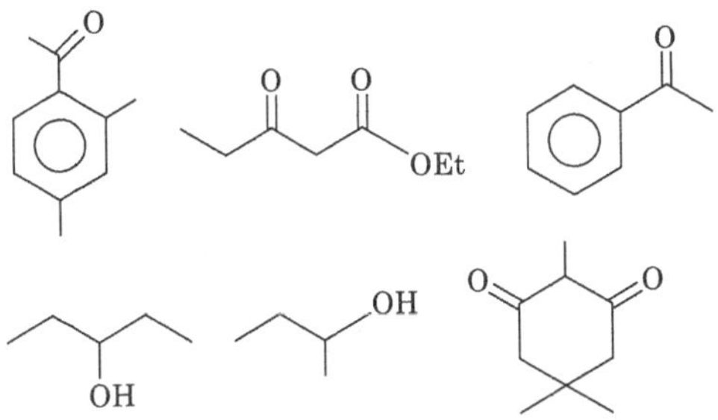

27 Enthalpies of formation of $CCl_4(g)$, $H_2O(g)$, $CO_2(g)$ and $HCl(g)$ are -105, -242, -394 and -92 kJ mol^{-1} respectively. The magnitude of enthalpy of the reaction given below is .... kJ mol^{-1}. (Nearest integer)

$$CCl_4(g) + 2H_2O(g) \longrightarrow CO_2(g) + 4HCl(g)$$

28 The number of alkali metals, from Li, K, Cs, Rb having ionisation enthalpy greater than 400 kJ mol^{-1} and forming stable super oxide is ....

29 If the CFSE of $[Ti(H_2O)_6]^{3+}$ is -96.0 kJ/mol, this complex will absorb maximum at wavelength .... nm. (Nearest integer)

[Assume Planck's constant $(h) = 6.4 \times 10^{-34}$ Js, speed of light $(c) = 3.0 \times 10^8$ m/s and Avogadro's constant $(N_A) = 6 \times 10^{23}$/mol.]

30 The resistivity of a 0.8 M solution of an electrolyte is $5 \times 10^{-3} \Omega$ cm. Its molar conductivity is .... $\times 10^4$ $\Omega^{-1}cm^2 mol^{-1}$.

(Nearest integer)

> **Key Idea** This question can be solved by using the concept of molar conductivity for this, use the formula $\Lambda_m = \frac{\kappa \times 1000}{M}$.
> where κ can be expressed in terms of resistivity as $\frac{1}{\rho}$.

1 FEBRUARY 2023, SHIFT-I

Section A : Objective Type Questions

1 But-2-yne is reacted separately with one mole of hydrogen as shown below :

$$\underline{B} \xleftarrow[\text{liq. } NH_3]{Na} CH_3—C{\equiv}C—CH_3 + H_2 \xrightarrow[\Delta]{Pd/C} \underline{A}$$

A. A is more soluble than B.

B. The boiling point and melting point of A are higher and lower than B respectively.

C. A is more polar than B because dipole moment of A is zero.

D. Br_2 adds easily to B than A.

Identify the incorrect statements from the options given below.

(a) A, C and D only (b) B, C and D only
(c) B and C only (d) A and B only

2 Choose the correct statements.

A. Beryllium oxide is purely acidic in nature.

B. Beryllium carbonate is kept in the atmosphere of CO_2.

C. Beryllium sulphate is readily soluble in water.

D. Beryllium shows anomalous behaviour.

(a) A, B and C only (b) B, C and D only
(c) A and B only (d) A only

3 Given below are two statements, one is labelled as Assertion (A) and the other is labelled as Reason (R).

Assertion (A) Amongst He, Ne, Ar and Kr, 1g of activated charcoal adsorbs more of Kr.

Reason (R) The critical volume V_c (cm^3 mol^{-1}) and critical pressure p_c (atm) is highest for krypton but the compressibility factor at critical point Z_c is lowest for krypton.

In the light of the above statements, choose the correct answer from the options given below.

(a) A is true but R is false.
(b) Both A and R are true and R is the correct explanation of A.
(c) Both A and R are true but R is not the correct explanation of A.
(d) A is false but R is true.

4 Which of the following are the example of double salt?

A. $FeSO_4 \cdot (NH_4)_2SO_4 \cdot 6H_2O$

B. $CuSO_4 \cdot 4NH_3 \cdot H_2O$

C. $K_2SO_4 \cdot Al_2(SO_4)_3 \cdot 24H_2O$

D. $Fe(CN)_2 \cdot 4KCN$

Choose the correct answer given below.

(a) A and B only (b) A, B and D only
(c) A and C only (d) B and D only

5 Which of the following represents the lattice structure of $A_{0.95}O$ containing A^{2+}, A^{3+} and O^{2-} ions?

□ A^{2+} △ A^{3+} ○ O^{2-}

(A) 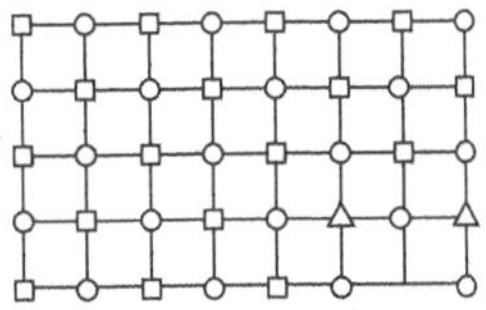

(B) 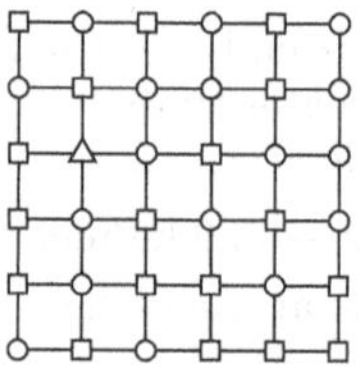

(C) 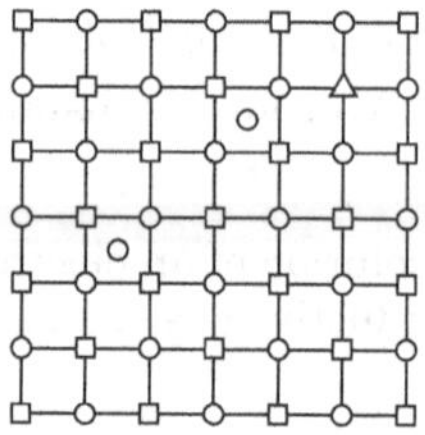

(a) B only (b) A and B only
(c) B and C only (d) A only

6 Identify the incorrect option from the following :

(a) 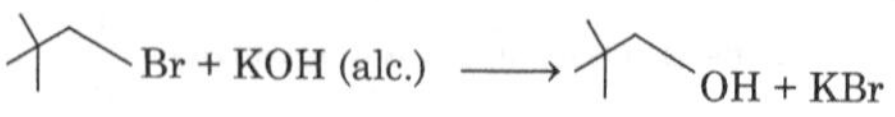

(b) 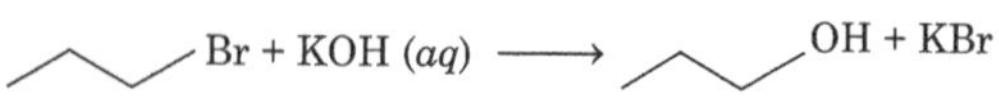

(c) 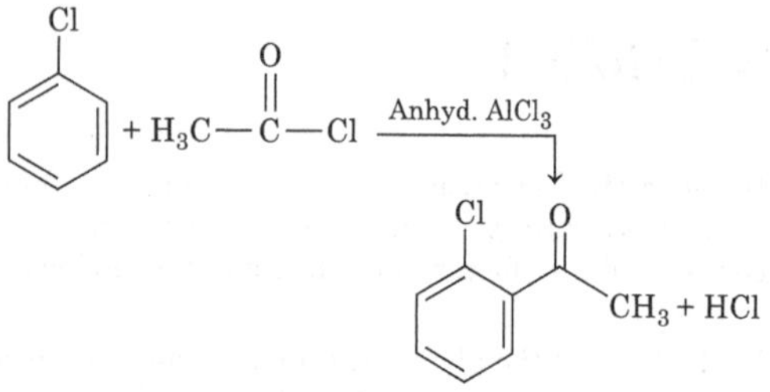

(d) 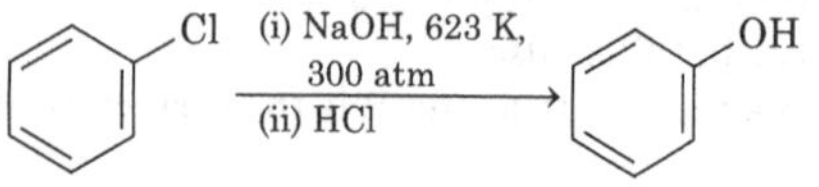

7 Match List-I with List-II.

List-I		List-II
A. Slaked lime	I.	NaOH
B. Dead burnt plaster	II.	$Ca(OH)_2$
C. Caustic soda	III.	$Na_2CO_3 \cdot 10H_2O$
D. Washing soda	IV.	$CaSO_4$

Choose the correct answer from the options given below.
(a) A-III, B-II, C-IV, D-I (b) A-I, B-IV, C-II, D-III
(c) A-III, B-IV, C-II, D-I (d) A-II, B-IV, C-I, D-III

8 Given below are two statements, one is labelled as Assertion (A) and the other is labelled as Reason (R).

Assertion (A) Hydrogen is an environment friendly fuel.

Reason (R) Atomic number of hydrogen is 1 and it is a very light element.

In the light of the above statements, choose the correct answer from the options given below.
(a) A is false but R is true.
(b) Both A and R are true and R is the correct explanation of A.
(c) A is true but R is false.
(d) Both A and R are true but R is not the correct explanation of A.

9 Match List-I with List-II.

List-I (Test)		List-II (Functional group/Class of Compound)
A. Molisch's test	I.	Peptide
B. Biuret test	II.	Carbohydrate
C. Carbylamine test	III.	Primary amine
D. Schiff's test	IV.	Aldehyde

Choose the correct answer from the options given below.
(a) A-III, B-IV, C-II, D-I (b) A-II, B-I, C-III, D-IV
(c) A-III, B-IV, C-I, D-II (d) A-I, B-II, C-III, D-IV

10 A solution of $FeCl_3$ when treated with $K_4[Fe(CN)_6]$ gives a prussian blue precipitate due to the formation of
(a) $K[Fe_2(CN)_6]$ (b) $Fe_4[Fe(CN)_6]_3$
(c) $Fe_3[Fe(CN)_6]_2$ (d) $Fe[Fe(CN)_6]$

11 Given below are two statements, one is labelled as Assertion (A) and the other is labelled as Reason (R).

Assertion (A) In an Ellingham diagram, the oxidation of carbon to carbon monoxide shows a negative slope with respect to temperature.

Reason (R) CO tends to get decomposed at higher temperature.

In the light of the above statements, choose the correct answer from the options given below
(a) Both A and R are true but R is not the correct explanation of A.
(b) Both A and R are true and R is the correct explanation of A.
(c) A is not correct but R is correct.
(d) A is correct but R is not correct.

12 Highest oxidation state of Mn is exhibited in Mn_2O_7. The correct statements about Mn_2O_7 are
A. Mn is tetrahedrally surrounded by oxygen atoms.
B. Mn is octahedrally surrounded by oxygen atoms.
C. contains Mn—O—Mn bridge.
D. contains Mn—Mn bond.

Choose the correct answer from the options given below.
(a) B and D only (b) B and C only
(c) A and C only (d) A and D only

13 Resonance in carbonate ion (CO_3^{2-}) is

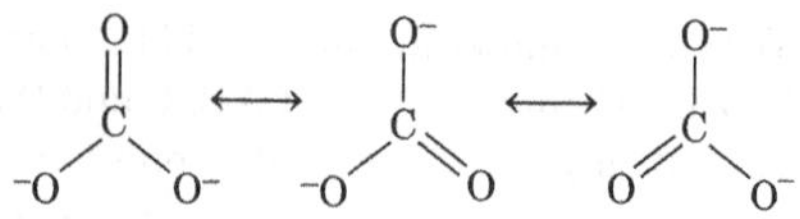

Which of the following is true?
(a) Each structure exists for equal amount of time.
(b) It is possible to identify each structure individually by some physical or chemical method.

(c) All these structures are in dynamic equilibrium with each other.

(d) CO_3^{2-} has a single structure i.e., resonance hybrid of the above three structures.

14 The correct representation in six membered pyranose form for the following sugar [*X*] is

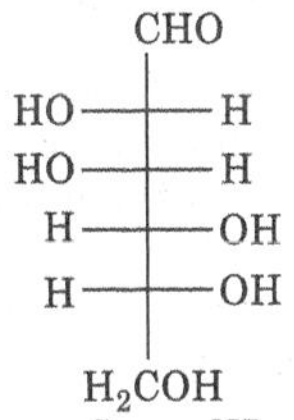

Sugar [*X*]

(a) 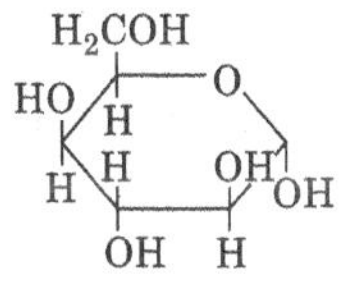

(b) 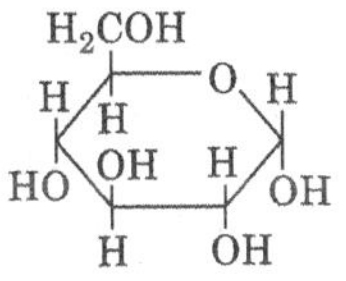

(c) 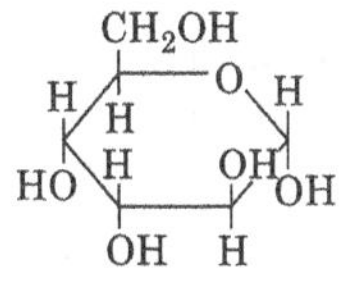

(d) 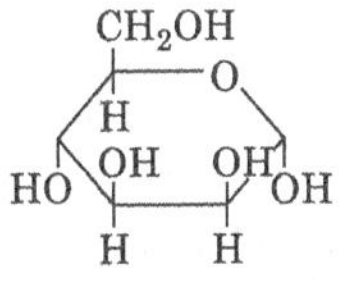

15 Decreasing order of dehydration of the following alcohols is

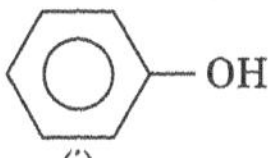

(i)

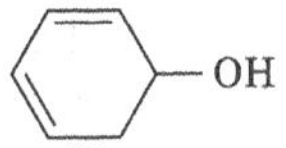

(ii)

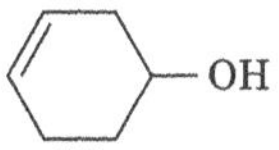

(iii)

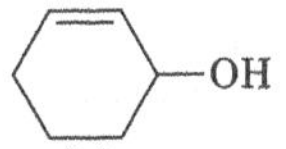

(iv)

(a) (iv) > (ii) > (iii) > (i)
(b) (i) > (iv) > (ii) > (iii)
(c) (ii) > (i) > (iv) > (iii)
(d) (ii) > (iv) > (iii) > (i)

16 Which of the following complexes will show largest splitting of *d*-orbitals?

(a) $[Fe(CN)_6]^{3-}$
(b) $[Fe(NH_3)_6]^{3+}$
(c) $[FeF_6]^{3-}$
(d) $[Fe(C_2O_4)_3]^{3-}$

> **Key Idea** The complex containing strongest field ligand in the given options will show largest splitting of *d*-orbitals.

17 How can photochemical smog be controlled?

(a) By using catalyst.
(b) By using tall chimneys.
(c) By complete combustion of fuel.
(d) By using catalytic converters in the automobiles/industry.

18 In the following reaction, '*A*' is

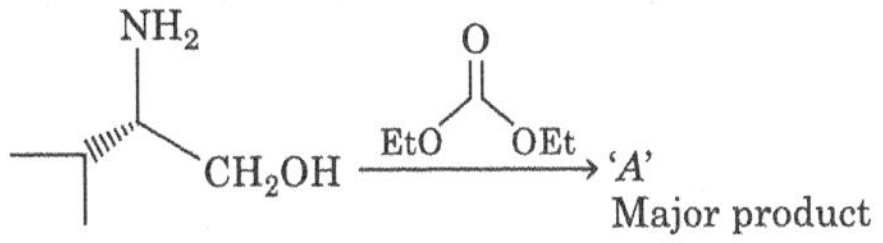

(a) 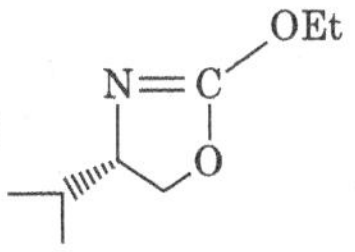

(b) 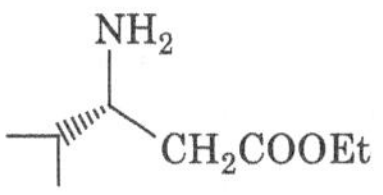

(c) 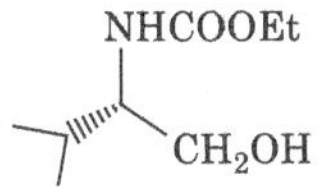

(d) 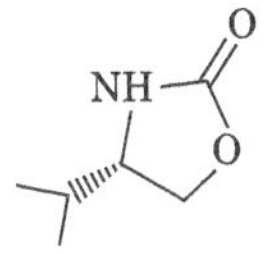

19 Given below are two statements.

Statement I Chlorine can easily combine with oxygen to form oxides and the product has a tendency to explode.

Statement II Chemical reactivity of an element can be determined by its reaction with oxygen and halogens.

In the light of the above statements, choose the correct answer from the options given below.

(a) Both the Statements I and II are true
(b) Statement I is true but Statement II is false
(c) Both the Statements I and II are false
(d) Statement I is false but Statement II is true

20 Match List-I with List-II.

	List-I		List-II
A.	Tranquilizers	I.	Anti blood clotting
B.	Aspirin	II.	Salvarsan
C.	Antibiotic	III.	Antidepressant drugs
D.	Antiseptic	IV.	Soframicine

Choose the correct answer from the options given below.

(a) A-IV, B-II, C-I, D-III
(b) A-III, B-I, C-II, D-IV
(c) A-II, B-IV, C-I, D-III
(d) A-II, B-I, C-III, D-IV

Section B : Numerical Value Type Questions

21 At what pH, given half-cell MnO_4^- (0.1 M) | Mn^{2+} (0.001 M) will have electrode potential of 1.282 V? .... (Nearest integer)

Given, $E^\circ_{MnO_4/Mn^{2+}} = 1.54\text{ V}, \dfrac{2.303RT}{F} = 0.059\text{ V}$

22 The total number of chiral compounds from the following is ....

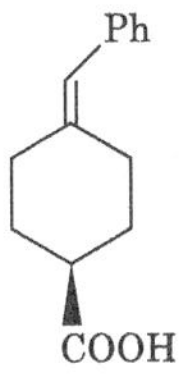

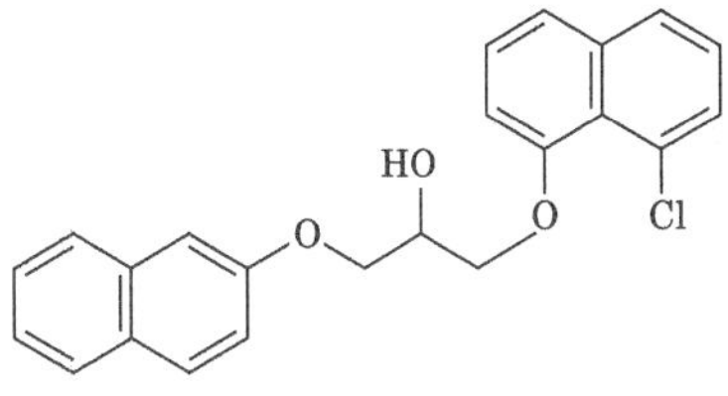

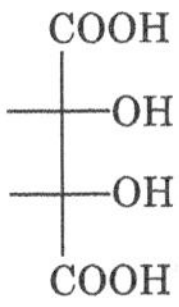

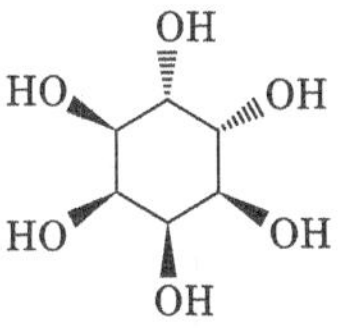

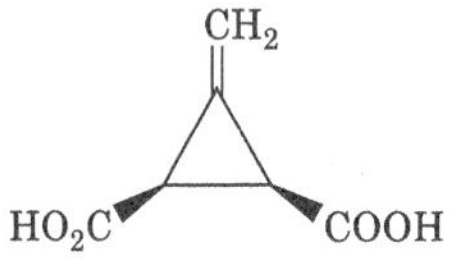

23 Electrons in a cathode ray tube have been emitted with a velocity of 1000 m s^{-1}. The number of following statements which is/are true about the emitted radiation is ....
[Given, $h = 6 \times 10^{-34}$ J s, $m_e = 9 \times 10^{-31}$ kg.]

(A) The de-Broglie wavelength of the electron emitted is 666.67 nm.
(B) The characteristic of electrons emitted depend upon the material of the electrodes of the cathode ray tube.
(C) The cathode rays start from cathode and move towards anode.
(D) The nature of the emitted electrons depends on the nature of the gas present in cathode ray tube.

24 (i) $X(g) \rightleftharpoons Y(g) + Z(g), K_{p_1} = 3$
(ii) $A(g) \rightleftharpoons 2B(g), K_{p_2} = 1$
If the degree of dissociation and initial concentration of both the reactants $X(g)$ and $A(g)$ are equal, then the ratio of the total pressure at equilibrium $\left(\frac{p_1}{p_2}\right)$ is equal to $x:1$. The value of x is .... (Nearest integer)

25 At 25°C, the enthalpy of the following processes are given

$$H_2(g) + O_2(g) \longrightarrow 2OH(g), \quad \Delta H° = 78 \text{ kJ mol}^{-1}$$
$$H_2(g) + 1/2O_2(g) \longrightarrow H_2O(g), \quad \Delta H° = -242 \text{ kJ mol}^{-1}$$
$$H_2(g) \longrightarrow 2H(g), \quad \Delta H° = 436 \text{ kJ mol}^{-1}$$
$$1/2O_2(g) \longrightarrow [O](g), \Delta H° = 249 \text{ kJ mol}^{-1}$$

What would be the value of X for the following reaction ....? (Nearest integer)
$$H_2O(g) \longrightarrow H(g) + OH(g), \Delta H° = X \text{ kJ mol}^{-1}$$

26 The density of 3 M solution of NaCl is 1.0 g mL^{-1}. Molality of the solution is .... $\times 10^{-2}$ m. (Nearest integer).
[Given, molar mass of Na and Cl is 23 and 35.5 g mol^{-1} respectively.]

> **Key Idea** In this type of questions, whenever the density and molarity is given, then use this formula of molality i.e.,
> $$m = \frac{1000 \times M}{1000d - M \times \text{molecular weight}}$$

27 A and B are two substances undergoing radioactive decay in a container. The half-life of A is 15 min and that of B is 5 min. If the initial concentration of B is 4 times that of A and they both start decaying at the same time, how much time will it take for the concentration of both of them to be same? .... min.

28 Sum of oxidation states of bromine in bromic acid and perbromic acid is ....

29 Number of isomeric compounds with molecular formula $C_9H_{10}O$ which
(i) do not dissolve in NaOH
(ii) do not dissolve in HCl
(iii) do not give orange precipitate with 2, 4-DNP
(iv) on hydrogenation give identical compound with molecular formula $C_9H_{12}O$ is ....

30 25 mL of an aqueous solution of KCl was found to require 20 mL of 1 M $AgNO_3$ solution when titrated using K_2CrO_4 as an indicator. What is the depression in freezing point of KCl solution of the given concentration? .... (Nearest integer)
(Given, $K_f = 2.0$ K kg mol^{-1})

Assume
1. 100% ionisation and
2. density of the aqueous solution as 1 g mL^{-1}

> **Key Idea** To calculate the value of depression in freezing point, we need to know the value of molality and van't Hoff factor. So, first of all calculate the above two values. then, substitute the obtained value in the equation $\Delta T_f = i \times K_f \times m$

Answers

For solutions scan the QR code

24 Jan 2023, Shift-I

1. (d)	2. (d)	3. (b)	4. (a)	5. (d)	6. (b)	7. (a)	8. (a)	9. (b)	10. (c)
11. (a)	12. (b)	13. (c)	14. (c)	15. (d)	16. (c)	17. (d)	18. (c)	19. (a)	20. (d)
21. 492	22. 7	23. 2	24. 10	25. 3	26. 917	27. 2	28. 4	29. 180	30. 25

24 Jan 2023, Shift-II

1. (d)	2. (c)	3. (a)	4. (d)	5. (c)	6. (d)	7. (a)	8. (b)	9. (a)	10. (d)
11. (d)	12. (a)	13. (d)	14. (a)	15. (b)	16. (c)	17. (d)	18. (a)	19. (d)	20. (b)
21. 8	22. 2	23. 3	24. 5	25. 314	26. 2	27. 85	28. 8	29. 620	30. 2

25 Jan 2023, Shift-I

1. (*)	2. (a)	3. (d)	4. (c)	5. (d)	6. (c)	7. (c)	8. (a)	9. (a)	10. (d)
11. (a)	12. (a)	13. (b)	14. (b)	15. (d)	16. (b)	17. (a)	18. (b)	19. (a)	20. (c)
21. 6	22. 360	23. 9079	24. 42	25. 60	26. 4	27. 2	28. 41500	29. 12	30. 10

25 Jan 2023, Shift-II

1. (a)	2. (d)	3. (d)	4. (b)	5. (a)	6. (c)	7. (d)	8. (b)	9. (b)	10. (b)
11. (a)	12. (b)	13. (a)	14. (c)	15. (c)	16. (c)	17. (b)	18. (b)	19. (a)	20. (b)
21. 2	22. 2	23. 5	24. 12	25. 3	26. 3	27. 4	28. 925	29. 5	30. 2

29 Jan 2023, Shift-I

1. (a)	2. (c)	3. (a)	4. (b)	5. (a)	6. (d)	7. (c)	8. (d)	9. (c)	10. (c)
11. (b)	12. (b)	13. (b)	14. (d)	15. (d)	16. (d)	17. (c)	18. (a)	19. (c)	20. (c)
21. 6	22. 3	23. 2	24. 2	25. 0	26. 2	27. 3	28. 13	29. 6	30. 5

29 Jan 2023, Shift-II

1. (c)	2. (c)	3. (b)	4. (d)	5. (b)	6. (a)	7. (c)	8. (a)	9. (*)	10. (b)
11. (c)	12. (b)	13. (b)	14. (b)	15. (b)	16. (c)	17. (c)	18. (d)	19. (c)	20. (a)
21. 17	22. 36	23. 4	24. 2	25. 4	26. 4	27. 15	28. 270	29. 3	30. 200

30 Jan 2023, Shift-I

1. (c)	2. (b)	3. (d)	4. (d)	5. (b)	6. (a)	7. (b)	8. (d)	9. (c)	10. (d)
11. (a)	12. (c)	13. (d)	14. (b)	15. (a)	16. (c)	17. (c)	18. (a)	19. (c)	20. (b)
21. 100	22. 3	23. 4	24. 148	25. 0	26. 186	27. 798	28. 1362	29. 10	30. 623

30 Jan 2023, Shift-II

1. (a)	2. (b)	3. (b)	4. (b)	5. (b)	6. (d)	7. (b)	8. (b)	9. (c)	10. (c)
11. (a)	12. (c)	13. (b)	14. (b)	15. (a)	16. (a)	17. (c)	18. (c)	19. (d)	20. (d)
21. 150	22. 6	23. 243	24. 16	25. 150	26. 275	27. 4	28. 3	29. 1350	30. 3

31 Jan 2023, Shift-I

1. (b)	2. (b)	3. (b)	4. (c)	5. (a)	6. (b)	7. (a)	8. (a)	9. (a)	10. (d)
11. (d)	12. (d)	13. (b)	14. (c)	15. (c)	16. (c)	17. (c)	18. (d)	19. (b)	20. (d)
21. 555	22. 62250	23. 1	24. 6	25. 2520	26. 44	27. 610	28. 4	29. 4	30. 3

31 Jan 2023, Shift-II

1. (a)	2. (b)	3. (b)	4. (c)	5. (a)	6. (b)	7. (d)	8. (a)	9. (c)	10. (d)
11. (c)	12. (c)	13. (c)	14. (d)	15. (c)	16. (c)	17. (a)	18. (d)	19. (a)	20. (d)
21. 59	22. 227	23. 10	24. 5	25. 17	26. 3	27. 173	28. 3	29. 480	30. 25

1 Feb 2023, Shift-I

1. (*)	2. (b)	3. (a)	4. (c)	5. (d)	6. (a)	7. (d)	8. (d)	9. (b)	10. (b)
11. (d)	12. (c)	13. (d)	14. (d)	15. (d)	16. (a)	17. (d)	18. (d)	19. (a)	20. (b)
21. 3	22. 2	23. 2	24. 12	25. 499	26. 364	27. 15	28. 12	29. 2	30. 3

ONLINE SOLVED PAPERS

JEE Main 2023

(April Attempt)

6 APRIL 2023, SHIFT-I

Section A : Objective Type Questions

1 The major products A and B from the following reactions are

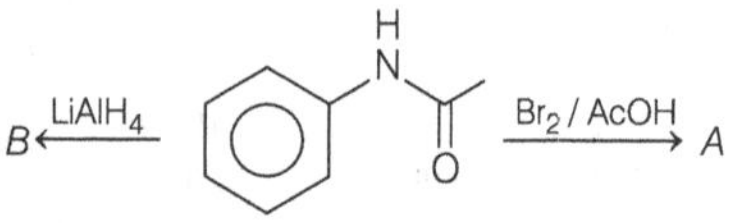

(a) 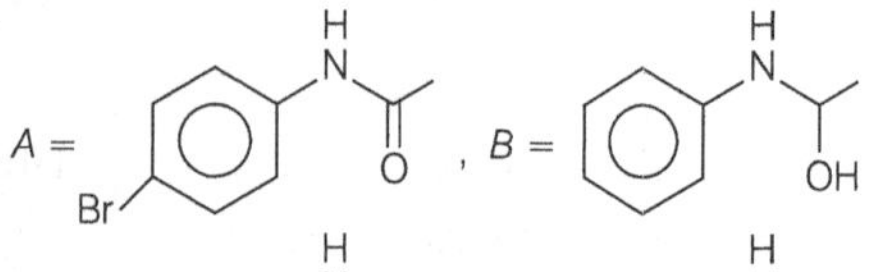

(b) 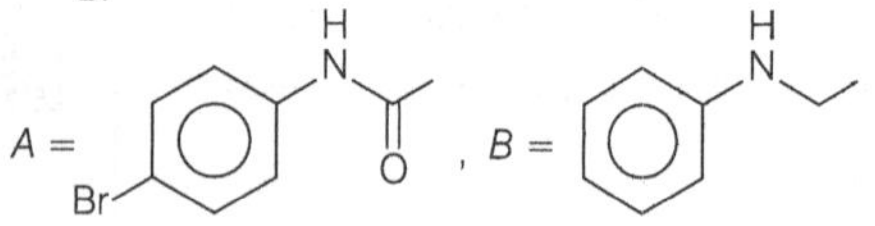

(c) 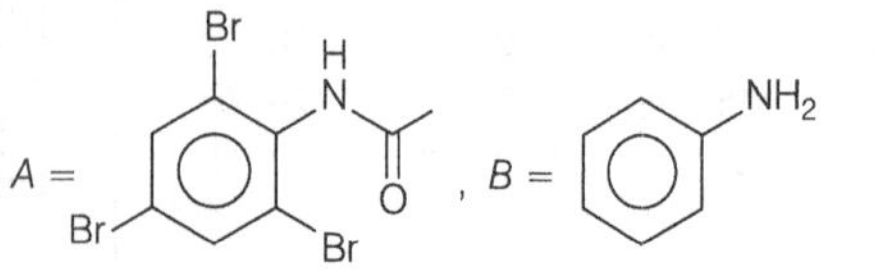

(d) 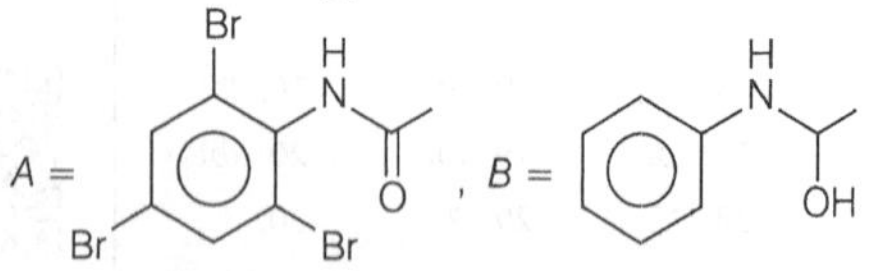

2 A compound is formed by two elements X and Y. The element Y forms cubic close packed arrangement and those of element X occupy one third of the tetrahedral voids. What is the formula of the compound ?

(a) X_3Y_2 (b) XY_3 (c) X_3Y (d) X_2Y_3

3 Given below are two statements, one is labelled as Assertion (A) and the other is labelled as Reason (R).

Assertion (A) Loss of electron from hydrogen atom results in nucleus of $\sim 1.5 \times 10^{-3}$ pm size.

Reason (R) Proton (H^+) always exists in combined form.

In the light of the above statements, choose the most appropriate answer from the options given below.

(a) Both A and R are correct and R is the correct explanation of A.
(b) Both A and R are correct but R is not the correct explanation of A.
(c) A is correct but R is not correct.
(d) A is not correct but R is correct.

4 Match List-I with List-II.

List-I (Enzymatic reaction)	List-II (Enzyme)
A. Sucrose → Glucose and fructose	I. Zymase
B. Glucose → Ethyl alcohol and CO_2	II. Pepsin
C. Starch → Maltose	III. Invertase
D. Proteins → Amino acids	IV. Diastase

Choose the correct answer from the options given below.

(a) A-I, B-IV, C-III, D-II
(b) A-I, B-II, C-IV, D-III
(c) A-III, B-I, C-II, D-IV
(d) A-III, B-I, C-IV, D-II

5 Given below are two statements, one is labelled as Assertion (A) and the other is labelled as Reason (R).

Assertion (A) The spin only magnetic moment value for $(Fe(CN)_6]^{3-}$ is 1.74 BM, whereas for $[Fe(H_2O_6)]^{3+}$ is 5.92 BM.

Reason (R) In both complexes, Fe is present in + 3 oxidation state.

In the light of the above statements, choose the correct answer from the options given below.

(a) Both A and R are true and R is the correct explanation of A.
(b) Both A and R are true but R is not the correct explanation of A.
(c) A is true but R is false.
(d) A is false but R is true.

6 The setting time of cement is increased by adding

(a) gypsum (b) limestone
(c) silica (d) clay

7 Match List-I with List-II.

List-I (Element detected)		List-II (Reagent used/Product formed)	
A.	Nitrogen	I.	$Na_2[Fe(CN)_5NO]$
B.	Sulphur	II.	$AgNO_3$
C.	Phosphorus	III.	$Fe_4[Fe(CN)_6]_3$
D.	Halogen	IV.	$(NH_4)_2MoO_4$

Choose the correct answer from the options given below.

(a) A-II, B-IV, C-I, D-III (b) A-III, B-I, C-IV, D-II
(c) A-IV, B-II, C-I, D-III (d) A-II, B-I, C-IV, D-III

8 Match List-I with List-II.

List-I (Oxide)	List-II (Type of bond)
A. N_2O_4	I. 1 N=O bond
B. NO_2	II. 1 N—O—N bond
C. N_2O_5	III. 1 N—N bond
D. N_2O	IV. 1 N= N / N≡ N bond

Choose the correct answer from the options given below.

(a) A-III, B-I, C-IV, D-II (b) A-II, B-I, C-III, D-IV
(c) A-II, B-IV, C-III, D-I (d) A-III, B-I, C-II, D-IV

9 For the reaction,

$$RCH_2\,Br + I^- \xrightarrow{\text{Acetone}} \underset{\text{(Major)}}{R\,CH_2I} + Br^-$$

The correct statement is

(a) Br^- can act as competing nucleophile.
(b) The solvent used in the reaction solvates the ions formed in rate determining step.
(c) The transition state formed in the above reaction is less polar than the localised anion.
(d) The reaction can occur in acetic acid also.

10 Compound (P) $\xrightarrow{HCl,\,\Delta}$ Filter → Residue (Q); → Filtrate $\xrightarrow{NaOH}$ Oily liquid (R)

(M.F. $C_{14}H_{13}ON$)
M.F. = Molecular formula

Compound (P) is neutral, Q gives effervescence with $NaHCO_3$ while R reacts with Hinsbergs reagent to give solid soluble in NaOH. Compound (P) is

(a) 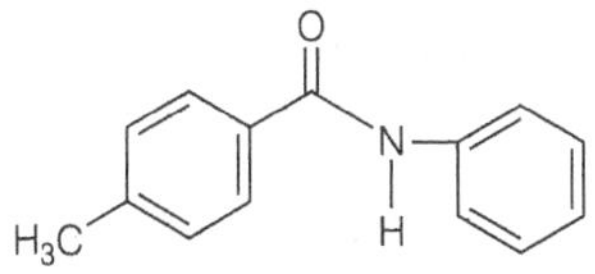

(b) 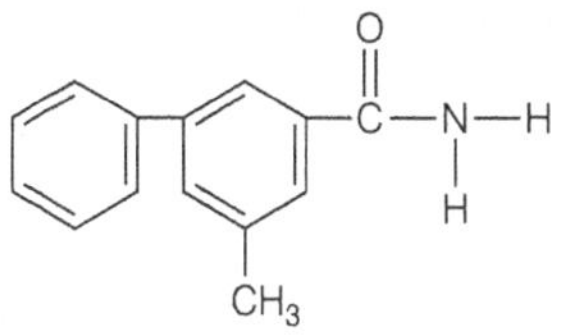

(c) 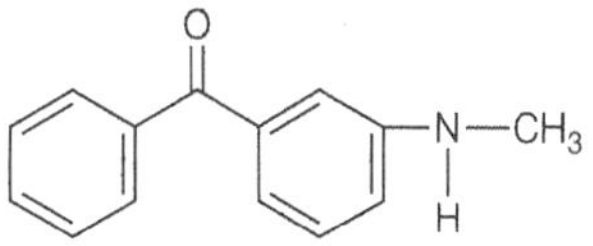

(d) 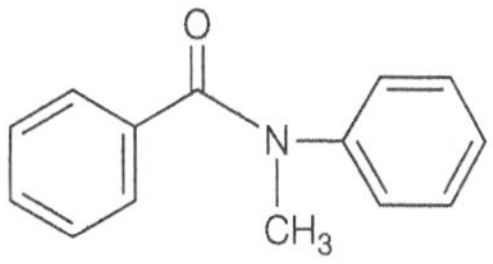

11 Match List-I with List-II.

List-I (Name of reaction)		List-II (Reagent used)	
A.	Hell-Volhard-Zelinsky reaction	I.	$NaOH + I_2$
B.	Iodoform reaction	II.	(i) CrO_2Cl_2, CS_2 (ii) H_2O
C.	Etard reaction	III.	(i) Br_2/red phosphorus (ii) H_2O
D.	Gattermann-Koch reaction	IV.	CO, HCl, anhyd. $AlCl_3$

Choose the correct answer from the options given below.

(a) A-I, B-II, C-III, D-IV
(b) A-III, B-I, C-II, D-IV
(c) A-III, B-II, C-I, D-IV
(d) A-III, B-I, C-IV, D-II

12 The difference between electron gain enthalpies will be maximum between

(a) Ar and Cl (b) Ar and F
(c) Ne and Cl (d) Ne and F

13 The standard electrode potential of M^+/M in aqueous solution does not depend on

(a) ionisation of a solid metal atom
(b) ionisation of a gaseous metal atom
(c) hydration of a gaseous metal ion
(d) sublimation of a solid metal

14 The possibility of photochemical smog formation is more at

(a) the places with healthy vegetation
(b) industrial areas
(c) marshy lands
(d) himalayan villages in winter

15 For a concentrated solution of a weak electrolyte (K_{eq} = equilibrium constant) A_2B_3 of concentration c, the degree of dissociation α is

(a) $\left(\frac{K_{eq}}{108c^4}\right)^{\frac{1}{5}}$ (b) $\left(\frac{K_{eq}}{25c^2}\right)^{\frac{1}{5}}$

(c) $\left(\frac{K_{eq}}{5c^4}\right)^{\frac{1}{5}}$ (d) $\left(\frac{K_{eq}}{6c^5}\right)^{\frac{1}{5}}$

16 Match List-I with List-II.

List-I (Vitamin)		List-II (Deficiency disease)	
A.	Vitamin A	I.	Beri-Beri
B.	Thiamine	II.	Cheilosis
C.	Ascorbic acid	III.	Xerophthalmia
D.	Riboflavin	IV.	Scurvy

Choose the correct answer from the options given below.
(a) A-III, B-II, C-IV, D-I (b) A-IV, B-II, C-III, D-I
(c) A-III, B-I, C-IV, D-II (d) A-IV, B-I, C-III, D-II

17 Strong reducing and oxidising agents among the following, respectively, are
(a) Eu^{2+} and Ce^{4+} (b) Ce^{4+} and Tb^{4+}
(c) Ce^{3+} and Ce^{4+} (d) Ce^{4+} and Eu^{2+}

18 Polymer used in orlon is
(a) polyacrylonitrile (b) polycarbonate
(c) polyamide (d) polyethene

19 Which of the following options are correct for the reaction?

$$2[Au(CN)_2]^-(aq) + Zn(s) \longrightarrow 2Au(s) + [Zn(CN)_4]^{2-}(aq)$$

A. Redox reaction
B. Displacement reaction
C. Decomposition reaction
D. Combination reaction

Choose the correct answer from the options given below:
(a) C and D only (b) A and B only
(c) A and D only (d) A only

20 The major product formed in the following reaction is

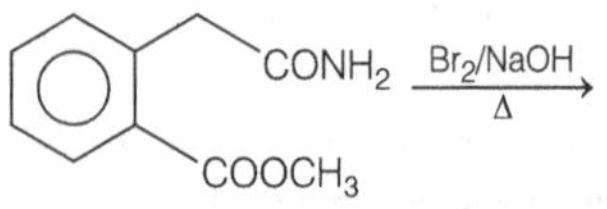

(a)
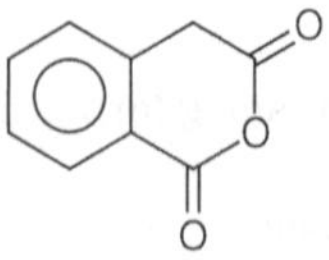

(b)
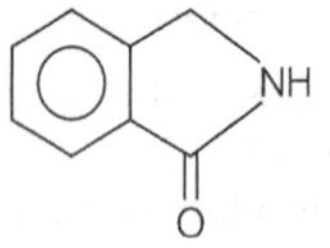

(c)
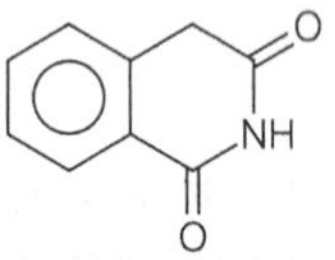

(d)
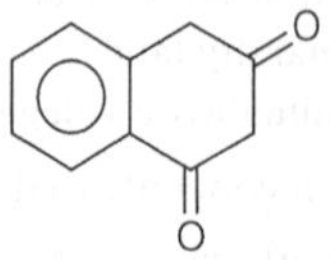

Section B : Numerical Value Type Questions

21 In ammonium - phosphomolybdate, the oxidation state of Mo is + ...

22 The value of $\log K$ for the reaction $A \rightleftharpoons B$ at 298 K is .... (Nearest integer)

Given, $\Delta H^\circ = -54.07\,kJ\,mol^{-1}$, $\Delta S^\circ = 10\,JK^{-1}\,mol^{-1}$

(Take, $2.303 \times 8.314 \times 298 = 5705$)

23 The wavelength of an electron of kinetic energy 4.50×10^{-29} J is .... $\times 10^{-5}$ m. (Nearest integer)

24 The number of species from the following which have square pyramidal structure is .... PF_5, BrF_4^-, IF_5, BrF_5, $XeOF_4$, ICl_4^-

25 Mass of urea (NH_2CONH_2) required to be dissolved in 1000 g of water in order to reduce the vapour pressure of water by 25% is .... g. (Nearest integer)

Given, molar mass of N, C, O and H are 14, 12, 16 and 1g mol^{-1}, respectively.

26 Consider the graph of Gibbs free energy G vs extent of reaction. The number of statement(s) from the following which are true with respect to points (*a*), (*b*) and (*c*) is ....

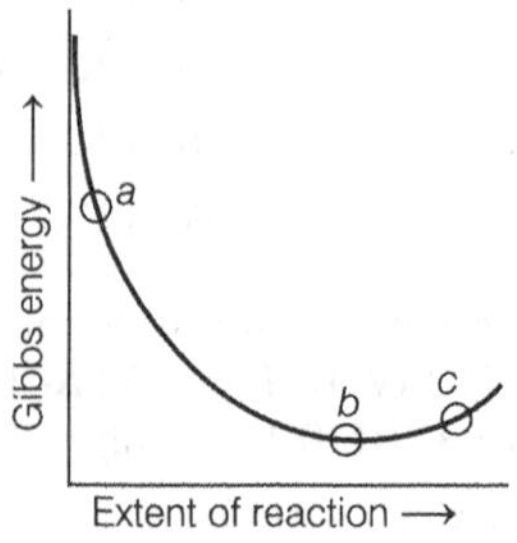

A. Reaction is spontaneous at (*a*) and (*b*).
B. Reaction is at equilibrium at point (*b*) and non-spontaneous at point (*c*).
C. Reaction is spontaneous at (*a*) and non-spontaneous at (*c*).
D. Reaction is non-spontaneous at (*a*) and (*b*).

27 Number of ambidentate ligands in a representative metal complex $[M(en)(SCN)_4]$ is .... [en = ethylenediamine]

28 For the adsorption of hydrogen on platinum, the activation energy is 30 kJ mol^{-1} and for the adsorption of hydrogen on nickel, the activation energy is 41.4 kJ mol^{-1}. The logarithm of the ratio of the rates of chemisorption on equal areas of the metals at 300 K is .... (Nearest integer)
($R = 8.3\,JK^{-1}\,mol^{-1}$)

29 Number of bromo derivatives obtained on treating ethane with excess of Br_2, in diffused sunlight is ....

30 If 5 moles of $BaCl_2$, is mixed with 2 moles of Na_3PO_4, the maximum number of moles of $Ba_3(PO_4)_2$ formed is .... (Nearest integer)

6 APRIL 2023, SHIFT-II

Section A : Objective Type Questions

1 Element not present in Nessler's reagent is

(a) N (b) Hg (c) K (d) I

2 The IUPAC name of $K_3[Co(C_2O_4)_3]$ is

(a) Potassium trioxalatocobaltate (III)
(b) Potassium tris(oxalato)cobalt (III)
(c) Potassium tris(oxalato)cobaltate (III)
(d) Potassium trioxalatocobalt (III)

3 From the figure of column chromatography given below, identify incorrect statements.

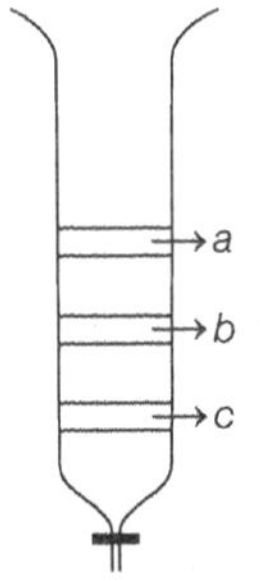

A. Compound '*c*' is more polar than '*a*' and '*b*'.
B. Compound '*a*' is least polar.
C. Compound '*b*' comes out of the column before '*c*' and after '*a*'.
D. Compound '*a*' spends more time in the column.

Choose the correct answer from the options given below.

(a) B, C and D only (b) A, B and D only
(c) A, B and C only (d) B and D only

4 Formation of which complex, among the following, is not a confirmatory test of Pb^{2+} ions

(a) lead chromate (b) lead nitrate
(c) lead sulphate (d) lead iodide

5 Structures of $BeCl_2$ in solid state, vapour phase and at very high temperature respectively are

(a) polymeric, monomeric, dimeric
(b) monomeric, dimeric, polymeric
(c) polymeric, dimeric, monomeric
(d) dimeric, polymeric, monomeric

6 Find out the major product from the following reaction.

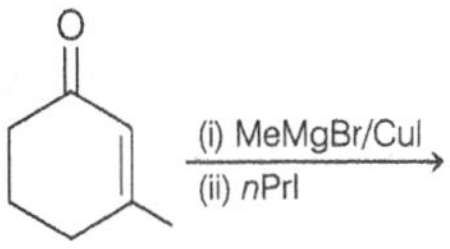

(a) 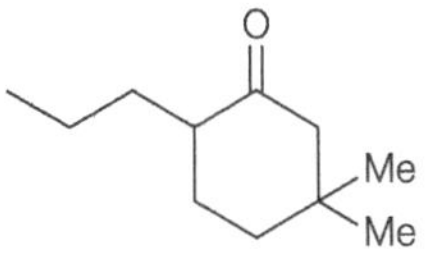

(b) 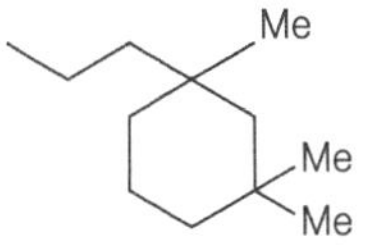

(c) 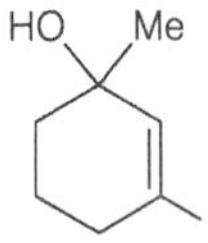

(d) 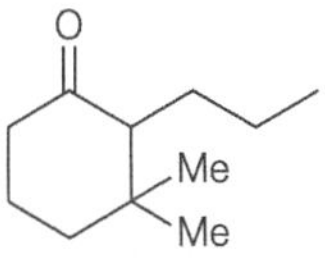

7 If the radius of the first orbit of hydrogen atom is α_0, then de-Broglie's wavelength of electron in 3rd orbit is

(a) $3\pi a$ (b) $\frac{\pi a}{3}$ (c) $6\pi a$ (d) $\frac{\pi a}{6}$

8 The volume of 0.02 M aqueous HBr required to neutralise 10.0 mL of 0.01 M aqueous $Ba(OH)_2$ is (assume complete neutralisation)

(a) 5.0 mL (b) 2.5 mL (c) 7.5 mL (d) 10.0 mL

9 The strongest acid from the following is

(a) 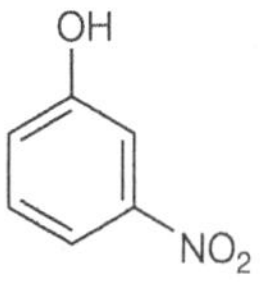

(b) 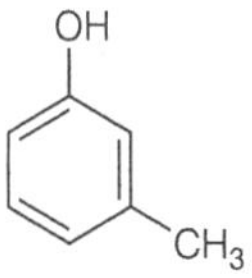

(c) 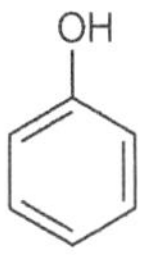

(d) 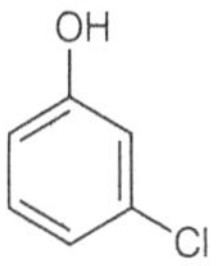

10 The group of chemicals used as pesticide is

(a) sodium chlorate, DDT, PAN
(b) DDT, aldrin
(c) dieldrin, sodium arsinite, tetrachloroethene
(d) aldrin, sodium chlorate, sodium arsinite

11 In the following reaction, '*B*' is

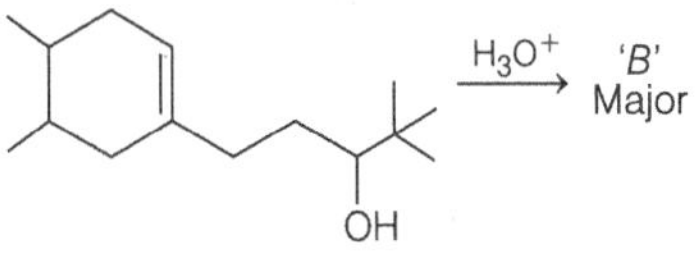

(a) 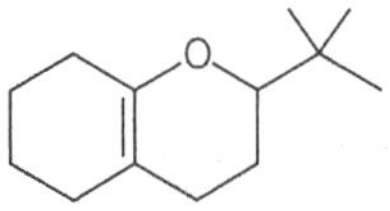

(b) 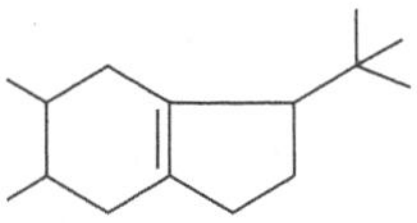

(c) 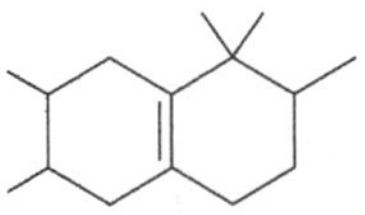

(d) 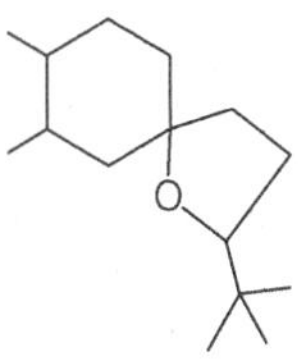

12 Consider the following reaction that goes from A to B in three steps as shown below

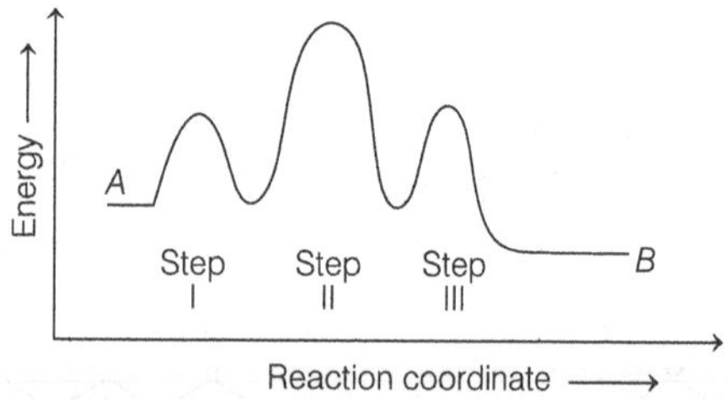

Choose the correct option.

	Number of intermediates	Number of activated complexes	Rate determining step
(a)	2	3	III
(b)	3	2	II
(c)	2	3	II
(d)	2	3	I

13 Given below are two statements: one is labelled as Assertion (A) and the other is labelled as Reason (R).

Assertion (A) In the complex $Ni(CO)_4$ and $Fe(CO)_5$, the metals have zero oxidation state.

Reason (R) Low oxidation states are found when a complex has ligands capable of π-donor character in addition to the σ-bonding.

In the light of the above statements, choose the most appropriate answer from the options given below.
(a) Both A and R are correct but R is not the correct explanation of A.
(b) Both A and R are correct and R is the correct explanation of A.
(c) A is correct but R is not correct.
(d) A is not correct but R is correct.

14 During the reaction of permanganate with thiosulphate, the change in oxidation of manganese occurs by value of 3. Identify which of the below medium will favour the reaction.
(a) Both aqueous acidic and faintly alkaline
(b) Aqueous acidic
(c) Aqueous neutral
(d) Both aqueous acidic and neutral

15 Given below are two statements:

Statement I Morphine is a narcotic analgesic. It helps in relieving pain without producing sleep.

Statement II Morphine and its derivatives are obtained from opium poppy.

In the light of the above statements, choose the correct answer from the options given below.
(a) Statement I is false but Statement II is true.
(b) Statement I is true but Statement II is false.
(c) Both Statement I and Statement II are false.
(d) Both Statement I and Statement II are true.

16 Group-13 elements react with O_2 in amorphous form to form oxides of type M_2O_3 (M = element). Which among the following is the most basic oxide?
(a) Ga_2O_3 (b) B_2O_3 (c) Al_2O_3 (d) Tl_2O_3

17 Which one of the following elements will remain as liquid inside pure boiling water?
(a) Ga (b) Li (c) Cs (d) Br

18 Match List-I with List-II.

List-I (Natural amino acid)	List-II (One letter code)
A. Arginine	I. D
B. Aspartic acid	II. N
C. Asparagine	III. A
D. Alanine	IV. R

Choose the correct answer from the options given below
(a) A-III, B-I, C-II, D-IV (b) A-IV, B-I, C-III, D-II
(c) A-I, B-III, C-IV, D-II (d) A-IV, B-I, C-II, D-III

19 Ion having highest hydration enthalpy among the given alkaline earth metal ions is
(a) Sr^{2+} (b) Ba^{2+} (c) Be^{2+} (d) Ca^{2+}

20 The product, which is not obtained during the electrolysis of brine solution is
(a) NaOH (b) Cl_2 (c) H_2 (d) HCl

Section B : Numerical Value Type Questions

21 Number of crystal systems from the following where body centred unit cell can be found, is

Cubic, tetragonal, orthorhombic, hexagonal, rhombohedral, monoclinic, triclinic.

22 The standard reduction potentials at 298 K for the following half-cells are given below:

$NO_3^- + 4H^+ + 3e^- \longrightarrow NO(g) + 2H_2O, E^\circ = 0.97\,V$

$V^{2+}(aq) + 2e^- \longrightarrow V \quad E^\circ = -1.19\,V$

$Fe^{3+}(aq) + 3e^- \longrightarrow Fe \quad E^\circ = -0.04\,V$

$Ag^+(aq) + e^- \longrightarrow Ag(s) \quad E^\circ = 0.80\,V$

$Au^{3+}(aq) + 3e^- \longrightarrow Au(s) \quad E^\circ = 1.40\,V$

The number of metal(s) which will be oxidised by NO_3^- in aqueous solution is

23 The equilibrium composition for the reaction $PCl_3 + Cl_2 \rightleftharpoons PCl_5$ at 298 K is given below
$[PCl_3]_{eq} = 0.2$ mol L^{-1}, $[Cl_2]_{eq} = 0.1$ mol L^{-1}, $[PCl_5]_{eq} = 0.40$ mol L^{-1}
If 0.2 mol of Cl_2 is added at the same temperature, the equilibrium concentrations of PCl_5 is .... $\times 10^{-2}$ molL^{-1}.
Given, K_C for the reaction at 298 K is 20.

24 Among the following, the number of compounds which will give positive iodoform reaction is
(A) 1-phenylbutan-2-one (B) 2-methylbutan-2-ol
(C) 3-methylbutan-2-ol (D) 1-phenylethanol
(E) 3,3-dimethylbutan-2-one
(F) 1-phenylpropan-2-ol

25 Consider the following pairs of solution which will be isotonic at the same temperature. The number of pairs of solutions is/are

A. 1 M aq. NaCl and 2 M aq. urea
B. 1 M aq. $CaCl_2$ and 1.5 M aq. KCl
C 1.5 M aq. $AlCl_3$ and 2 M aq. Na_2SO_4
D. 2.5 M aq. KCl and 1 M aq. $Al_2(SO_4)_3$

26 The number of species having a square planar shape from the following is

XeF_4, SF_4, SiF_4, BF_4^-, BrF_4^-, $[Cu(NH_3)_4]^{2+}$, $[FeCl_4)]^{2-}$, $[PtCl_4]^{2-}$

27 Consider the following data.

Heat of combustion of $H_2(g) = -241.8$ kJ mol^{-1}
Heat of combustion of $C(s) = -393.5$ kJ mol^{-1}
Heat of combustion of $C_2H_5OH(l) = -1234.7$ kJ mol^{-1}

The heat of formation of $C_2H_5OH(l)$ is (–) .... kJ mol^{-1}. (Nearest integer)

28 In an ice crystal, each water molecule is hydrogen bonded to .... neighbouring molecules.

29 Number of isomeric aromatic amines with molecular formula $C_8H_{11}N$, which can be synthesised by Gabriel phthalimide synthesis is

30 The number of colloidal systems from the following, which will have 'liquid' as the dispersion medium, is .... Gem stones, paints, smoke, cheese, milk, hair cream, insecticide sprays, froth, soap lather.

8 APRIL 2023, SHIFT-I

Section A : Objective Type Questions

1 In chromyl chloride, the number of *d*-electrons present on chromium is same as in (Given, atomic number of Ti : 22, V : 23, Cr : 24, Mn : 25, Fe : 26)

(a) V (IV) (b) Ti (III) (c) Mn (VII) (d) Fe (III)

2 Choose the halogen which is most reactive towards S_N1 reaction in the given compounds (A, B, C and D).

(A) 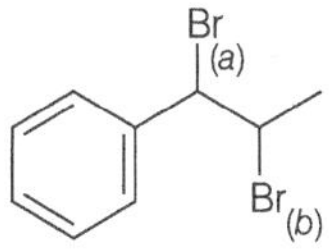

(B) 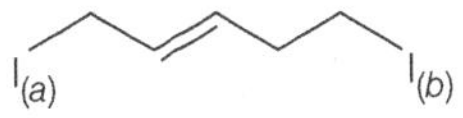

(C) 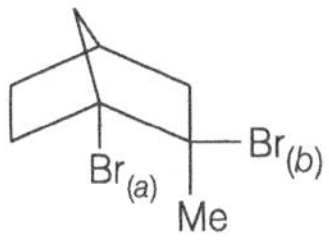

(D) 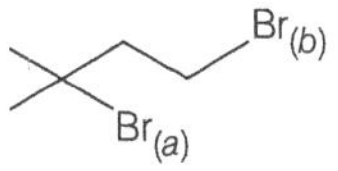

(a) A-$Br_{(a)}$; B-$I_{(a)}$; C-$Br_{(b)}$; D-$Br_{(a)}$
(b) A-$Br_{(b)}$; B-$I_{(b)}$; C-$Br_{(b)}$; D-$Br_{(b)}$
(c) A-$Br_{(a)}$; B-$I_{(a)}$; C-$Br_{(a)}$; D-$Br_{(a)}$
(d) A-$Br_{(a)}$; B-$I_{(a)}$; C-$Br_{(a)}$; D-$Br_{(a)}$

3 What is the purpose of adding gypsum to cement?

(a) To slow down the process of setting
(b) To speed up the process of setting
(c) To give a hard mass
(d) To facilitate the hydration of cement

4 Match List-I with List-II.

List-I (Reagents used)		List-II (Compound with functional group detected)
A. Alkaline solution of copper sulphate and sodium citrate	I.	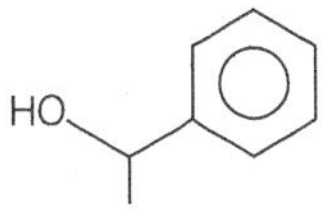
B. Neutral $FeCl_3$ solution	II.	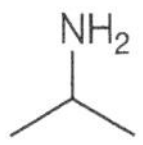
C. Alkaline chloroform solution	III.	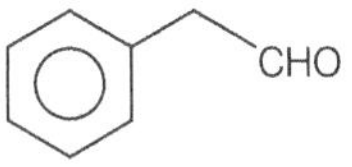
D. Potassium iodide and sodium hypochlorite	IV.	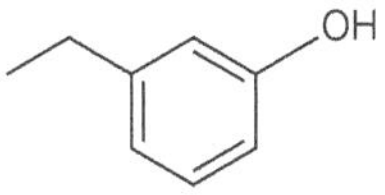

Choose the correct answer from the options given below:

(a) A-III, B-IV, C-I, D-II (b) A-II, B-IV, C-III, D-I
(c) A-IV, B-I, C-II, D-III (d) A-III, B-IV, C-II, D-I

5 Given below are two statements :

Statement I Lithium and magnesium do not form superoxide.

Statement II The ionic radius of Li^+ is larger than ionic radius of Mg^{2+}.

In the light of the above statements, choose the **most appropriate** answer from the options given below

(a) Statement I is incorrect but Statement II is correct.
(b) Both Statement I and Statement II are incorrect.
(c) Statement I is correct but Statement II is incorrect.
(d) Both Statement I and Statement II are correct.

6 Which halogen is known to cause the reaction given below?

$$2Cu^{2+} + 4X^- \longrightarrow Cu_2X_{2(s)} + X_2$$

(a) All halogens (b) Only chlorine
(c) Only iodine (d) Only bromine

7 Match List-I with List-II.

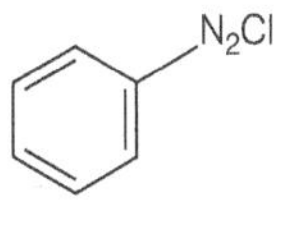

is reacted with reagents in List-I to form products in List-II.

	List-I (Reagent)		List-II (Product)
A.	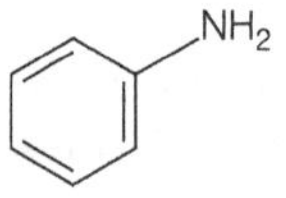	I.	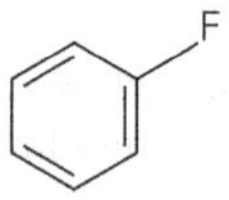
B.	HBF_4, Δ	II.	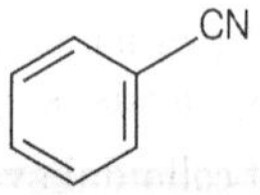
C.	Cu, HCl	III.	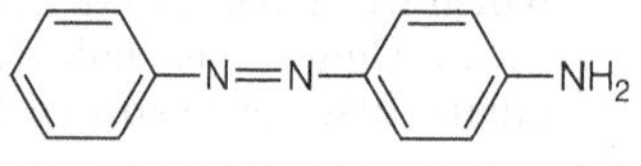
D.	CuCN/KCN	IV.	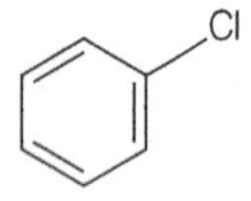

Choose the **correct** answer from the options given below :

(a) A-IV.,B-III, C-II, D-I (b) A-III, B-I, C-IV, D-II

(c) A-III, B-I, C-II, D-IV (d) A-I, B-III, C-IV, D-II

8 The reaction

$$\frac{1}{2}H_2(g) + AgCl(s) \rightleftharpoons H^+(aq) + Cl^-(aq) + Ag(s)$$

occurs in which of the given galvanic cell.

(a) $Pt | H_2(g) | HCl(sol^n) | AgCl(s) | Ag$

(b) $Pt | H_2(g) | KCl(sol^n) | AgCl(s) | Ag$

(c) $Ag | AgCl(s) | KCl(sol^n) | AgNO_3 | Ag$

(d) $Pt | H_2(g) | HCl(sol^n) | AgNO_3(sol^n) | Ag$

9 Sulphur (S) containing amino acids from the following are

(A) isoleucine (B) cysteine

(C) lysine (D) methionine

(E) glutamic acid

(a) A, B, C (b) B, C, E

(c) B, D (d) A, D

10 The water gas on reacting with cobalt as a catalyst forms

(a) methanoic acid (b) ethanol

(c) methanal (d) methanol

11 The major product formed in the following reaction is

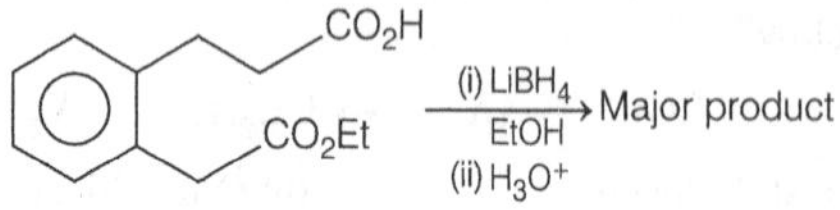

(a) 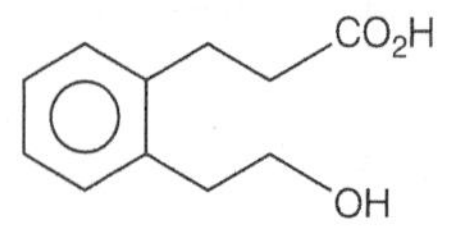

(b) 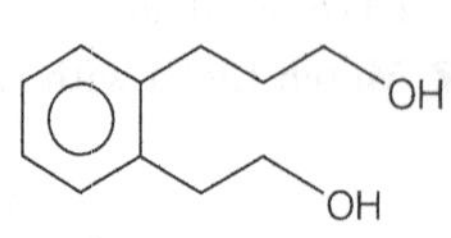

(c) 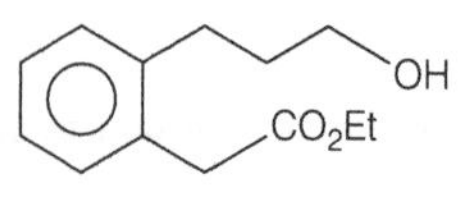

(d) 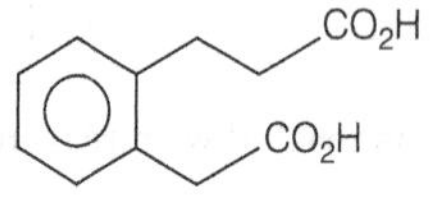

12 Which of the following complex is octahedral, diamagnetic and the most stable?

(a) $[Ni(NH_3)_6]Cl_2$ (b) $[Co(H_2O)_6]Cl_2$

(c) $Na_3[CoCl_6]$ (d) $K_3[Co(CN)_6]$

13 Which of the following represent the Freundlich adsorption isotherms?

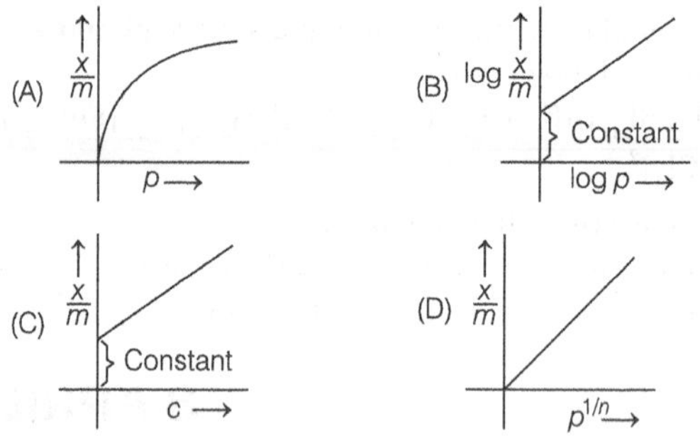

Choose the correct answer from the options given below

(a) A, C, D only (b) B, C, D only

(c) A, B, D only (d) A, B only

14 Match List-I with List-II.

	List-I (Species)		List-II (Maximum allowed concentration in ppm in drinking water)
A.	F^-	I.	< 50 ppm
B.	SO_4^{2-}	II.	< 5 ppm
C.	NO_3^-	III.	< 2 ppm
D.	Zn	IV.	< 500 ppm

Choose the correct answer from the options given below

(a) A-IV, B-III, C-II, D-I (b) A-II, B-I, C-III, D-IV

(c) A-III, B-II, C-I, D-IV (d) A-I, B-II, C-III, D-IV

15 Match List-I with List-II.

	List-I		List-II
A.	Saccharin	I.	High potency sweetener
B.	Aspartame	II.	First artificial sweetening agent
C.	Alitame	III.	Stable at cooking temperature
D.	Sucralose	IV.	Unstable at cooking temperature

Choose the **correct** answer from the options given below

(a) A-IV, B-III, C-I, D-II

(b) A-II, B-III, C-IV, D-I

(c) A-II, B-IV, C-III, D-I

(d) A-II, B-IV, C-I, D-III

16 $2IO_3^- + xI^- + 12H^+ \longrightarrow 6I_2 + 6H_2O$

What is the value of x?

(a) 6 (b) 2 (c) 10 (d) 12

17 The correct order of electronegativity for given elements is

(a) P > Br > C > At (b) C > P > At > Br

(c) Br > P > At > C (d) Br > C > At > P

18 The correct order of spin only magnetic moments for the following complex ions is

(a) $[Fe(CN)_6]^{3-} < [CoF_6]^{3-} < [MnBr_4]^{2-} < [Mn(CN)_6]^{3-}$
(b) $[Fe(CN)_6]^{3-} < [Mn(CN)_6]^{3-} < [CoF_6]^{3-} < [MnBr_4]^{2-}$
(c) $[MnBr_4]^{2-} < [CoF_6]^{3-} < [Fe(CN)_6]^{3-} < [Mn(CN)_6]^{3-}$
(d) $[CoF_6]^{3-} < [MnBr_4]^{2-} < [Fe(CN)_6]^{3-} < [Mn(CN)_6]^{3-}$

19 Given below are two statements: One is labelled as Assertion A and the other is labelled as Reason R.

Assertion (A) Butan-1-ol has higher boiling point than ethoxyethane.

Reason (R) Extensive hydrogen bonding leads to stronger association of molecules.

In the light of the above statements, choose the **correct** answer from the options given below

(d) A is true but R is false.
(b) Both A and R are true but R is not the correct explanation of A.
(c) Both A and R are true and R is the correct explanation of A.
(d) A is false but R is true.

20 Which of the following metals can be extracted through alkali leaching technique?

(a) Au (b) Pb (c) Cu (d) Sn

Section B : Numerical Value Type Questions

21 The vapour pressure *vs* temperature curve for a solution solvent system is shown below.

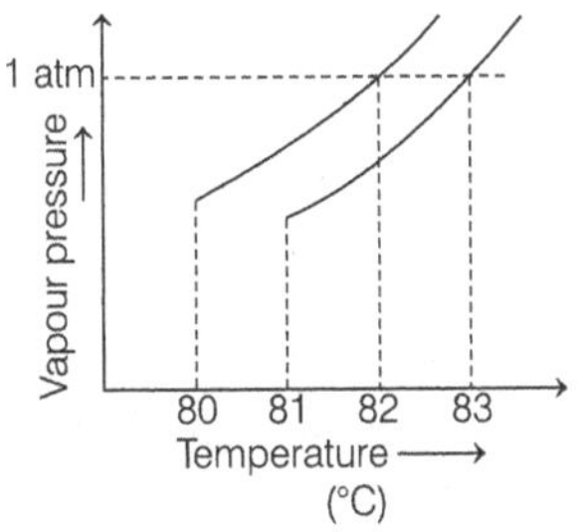

The boiling point of the solvent is .... °C.

22 Molar mass of the hydrocarbon (X) which on ozonolysis consumes one mole of O_3 per mole of (X) and gives one mole each of ethanal and propanone is .... g mol^{-1}. (Molar mass of C : 12 g mol^{-1}, H : 1 g mol^{-1})

23 When a 60 W electric heater is immersed in a gas for 100s in a constant volume container with adiabatic walls, the temperature of the gas rises by 5°C. The heat capacity of the given gas is .... JK^{-1}. (Nearest integer)

24 The number of given statement/s which is/are correct is:

(A) The stronger the temperature dependence of the rate constant, the higher is the activation energy.
(B) If a reaction has zero activation energy, its rate is independent of temperature.
(C) The stronger the temperature dependence of the rate constant, the smaller is the activation energy.
(D) If there is no correlation between the temperature and the rate constant then it means that the reaction has negative activation energy.

25

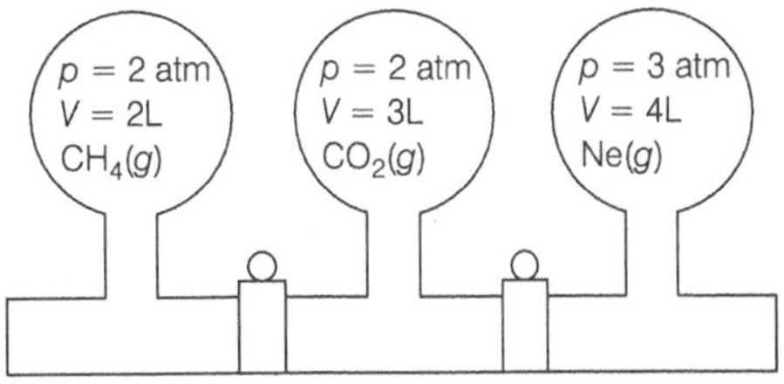

Three bulbs are filled with CH_4, CO_2 and Ne as shown in the picture. The bulbs are connected through pipes of zero volume. When the stopcocks are opened and the temperature is kept constant throughout, the pressure of the system is found to be ...... atm. (Nearest integer)

26 0.5 g of an organic compound (X) with 60% carbon will produce ..... $\times 10^{-1}$ g of CO_2 on complete combustion.

27 XeF_4 reacts with SbF_5 to form $[XeF_m]^{n+} [SbF_y]^{z-}$

$m + n + y + z = ...$

28 The number of following factors which affect the per cent covalent character of the ionic bond is ....

(A) polarising power of cation
(B) extent of distortion of anion
(C) polarisability of the anion
(D) polarising power of anion

29 The number of following statement/s which is/are incorrect is

(A) line emission spectra are used to study the electronic structure.
(B) the emission spectra of atoms in the gas phase show a continuous spread of wavelength from red to violet.
(C) an absorption spectrum is like the photographic negative of an emission spectrum.
(D) the element helium was discovered in the sun by spectroscopic method.

30 The titration curve of weak acid *vs* strong base with phenolphthalein (as indictor) is shown below. The $K_{Phenolphthalein} = 4 \times 10^{-10}$.

(Given, log 2 = 0.3)

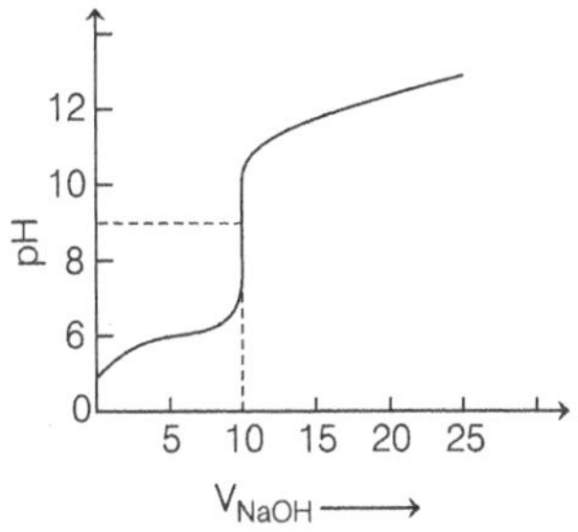

The number of following statement/s which is/are correct about phenolphthalein is ....

A. it can be used as an indicator for the titration of weak acid with weak base.
B. it begins to change colour at pH = 8.4
C. it is a weak organic base
D. it is colourless in acidic medium

8 APRIL 2023, SHIFT-II

Section A : Objective Type Questions

1 A compound X when treated with phthalic anhydride in presence of concentrated H_2SO_4 yields Y. Y is used as an acid-base indicator. X and Y are respectively

(a) carbolic acid, phenolphthalein
(b) salicylaldehyde, phenolphthalein
(c) toludine, phenolphthalein
(d) anisole, methyl orange

2 For a good quality cement, the ratio of lime to the total of the oxides of Si, Al and Fe should be as close as to

(a) 4 (b) 1 (c) 2 (d) 3

3 Match List-I with List-II.

	List-I (Natural amino acid)		List-II (One letter code)
A.	Glutamic acid	I.	Q
B.	Glutamine	II.	W
C.	Tyrosine	III.	E
D.	Tryptophan	IV.	Y

Choose the correct answer from the options given below:

(a) A-III, B-IV, C-I, D-II (b) A-III, B-I, C-IV, D-II
(c) A-II, B-I, C-IV, D-III (d) A-IV, B-III. C-I.D-II

4 Given below are two statements :

Statement I Methyl orange is a weak acid.

Statement II The benzenoid form of methyl orange is more intense/deeply coloured than the quinonoid form.

In the light of the above statement, choose the most appropriate answer from the options given below

(a) Both Statement I and Statement II are correct.
(b) Statement I is incorrect but Statement II is correct.
(c) Both Statement I and Statement II are incorrect.
(d) Statement I is correct but Statement II is incorrect.

5 Match List-I with List-II

	List-I (Coordination complex)		List-II (Number of unpaired electrons)
A.	$[Cr(CN)_6]^{3-}$	I.	0
B.	$[Fe(H_2O)_6]^{2+}$	II.	3
C.	$[Co(NH_3)_6]^{3+}$	III.	2
D.	$[Ni(NH_3)_6]^{2+}$	IV.	4

Choose the correct answer from the options given below:

(a) A-II, B-IV, C-I, D-III
(b) A-III, B-IV, C-I, D-II
(c) A-II, B-I, C-IV, D-III
(d) A-IV, B-III, C-II, D-I

6 Henry Moseley studied characteristic X-ray spectra of elements. The graph which represents his observation correctly is
(Given, ν = frequency of X-ray emitted,
Z = atomic number)

7 Which of the following can reduce decomposition of H_2O_2 on exposure to light

(a) glass containers (b) dust
(c) urea (d) alkali

8 The statements which are true about antagonists from the following is/are

A. They bind to the receptor site.
B. Get transferred inside the cell for their action.
C. Inhibit the natural communication of the body.
D. Mimic the natural messenger.

Choose the correct answer from the options given below :

(a) A and C (b) B only
(c) A and B (d) A, C and D

9 Which of these reactions is not a part of breakdown of ozone in stratosphere?

(a) $2Cl\dot{O} \longrightarrow ClO_2(g) + \dot{Cl}(g)$

(b) $Cl\dot{O}(g) + O(g) \longrightarrow \dot{Cl}(g) + O_2(g)$

(c) $CF_2Cl_2(g) \xrightarrow{UV} \dot{C}l(g) + \dot{C}F_2Cl(g)$

(d) $\dot{Cl}(g) + O_3(g) \longrightarrow Cl\dot{O}(g) + O_2(g)$

10 The correct order of reactivity of following haloarenes towards nucleophilic substitution with aqueous NaOH is

(A) 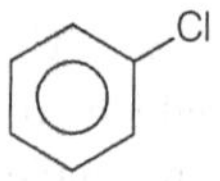

(B) 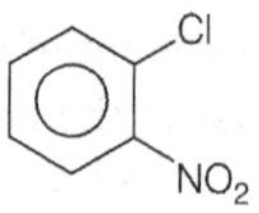

(C) 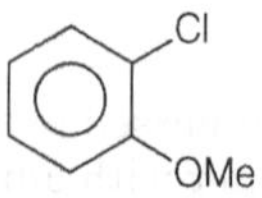

(D) 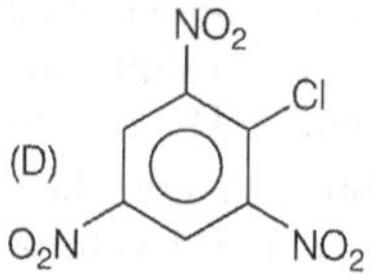

Choose the correct answer from the options given below

(a) C > A > D > B (b) A > B > D > C
(c) D > B > A > C (d) D > C > B > A

11 The descending order of acidity for the following carboxylic acid is

A. CH_3COOH B. F_3C—COOH
C. $ClCH_2$—COOH D. FCH_2—COOH
E. $BrCH_2$—COOH

Choose the correct answer from the options given below:

(a) D > B > A > E > C (b) B > D > C > E > A
(c) E > D > B > A > C (d) B > C > D > E > A

12 The correct reaction profile diagram for a positive catalyst reaction is

(a)
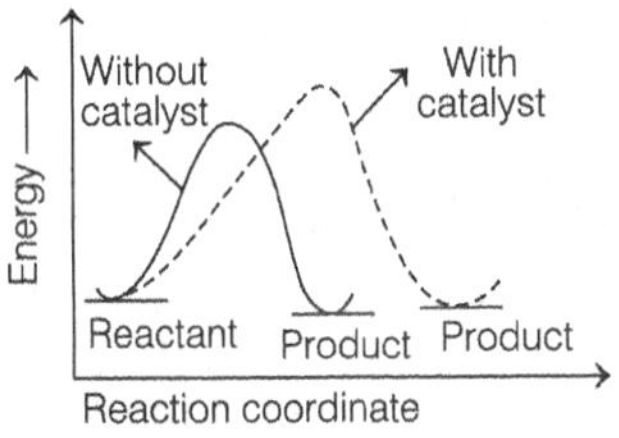

(b)
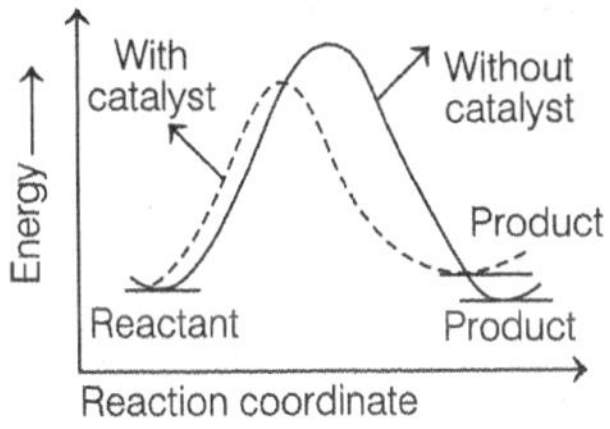

(c)
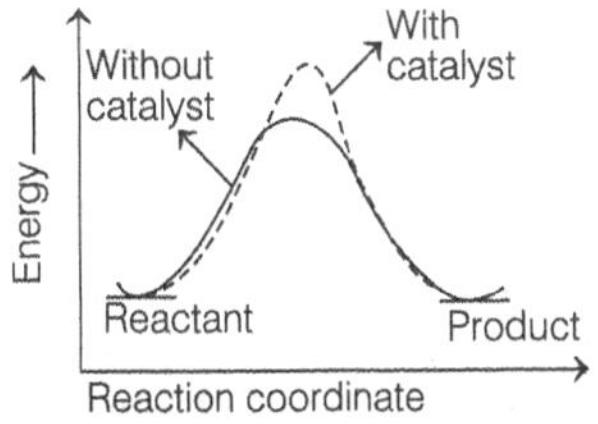

(d)
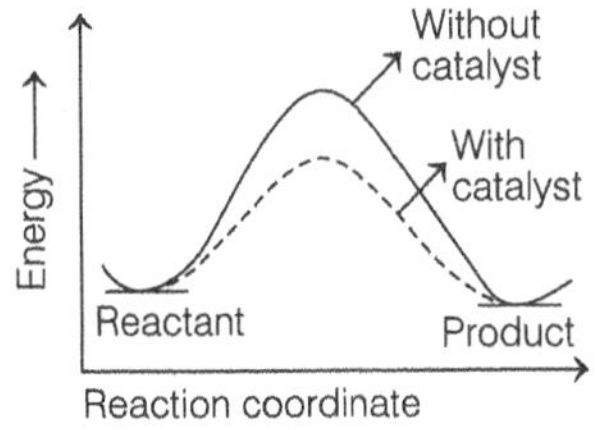

13 Major product *P* formed in the following reaction is

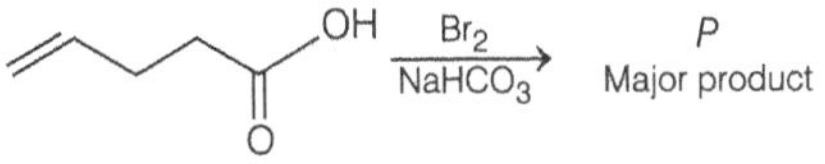

(a) 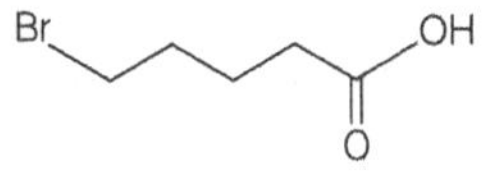 (b) 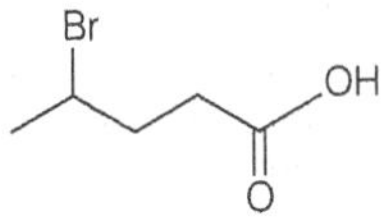

(c) 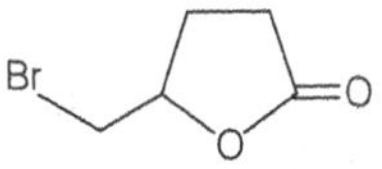 (d) 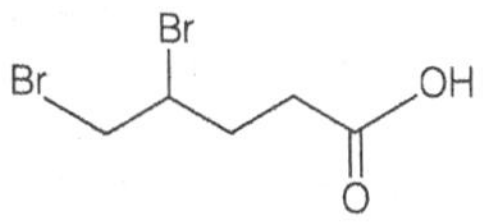

14 Given below are two statements: One is labelled as Assertion (A) and the other is labelled as Reason (R).

Assertion (A) Sodium is about 30 times as abundant as potassium in the oceans.

Reason (R) Potassium is bigger in size than sodium.

In the light of the above statements, choose the correct answer from the options given below

(a) A is true but R is false.
(b) A is false but R is true.
(c) Both A and R are true and R is the correct explanation of A.
(d) Both A and R are true but R is not the correct explanation of A.

15 Arrange the following gases in increasing order of van der Waals' constant *a*.

A. Ar B. CH_4
C. H_2O D. C_6H_6

Choose the correct option from the following.

(a) D, C, B and A (b) C, D, B and A
(c) B, C, D and A (d) A, B, C and D

16 Which of the following have same number of significant figures?

A. 0.00253 B. 1.0003
C. 15.0 D. 163

Choose the correct answer from the options given below

(a) A, C and D only (b) A, B and C only
(c) C and D only (d) B and C only

17 The product (*P*) formed from the following multistep reaction is

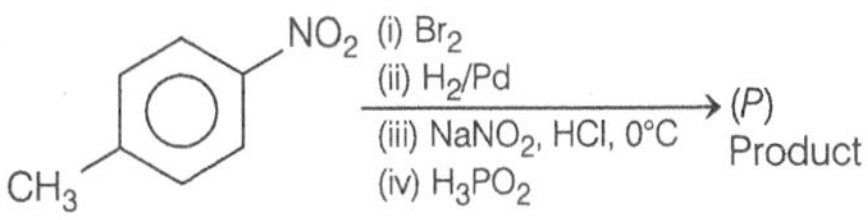

(a) 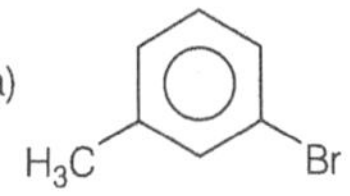 (b) 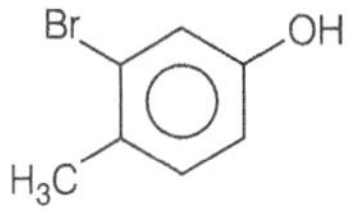

(c) 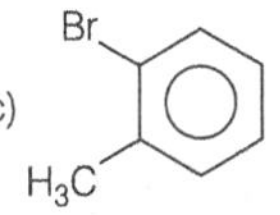 (d) 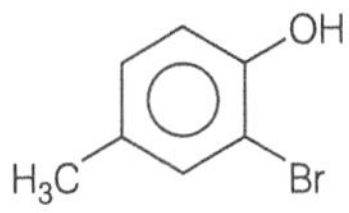

18 The correct IUPAC nomenclature for the following compound is

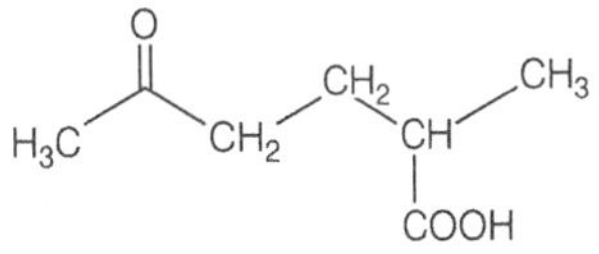

(a) 5-formyl -2-methylhexanoic acid
(b) 5-methyl-2-oxohexan-6-oic acid
(c) 2-methyl-5-oxohexanoic acid
(d) 2-formyl -5-methylhexan-6-oic acid

19 Given below are two statements:

Statement I In redox titration, the indicators used are sensitive to change in pH of the solution.

Statement II In acid-base titration, the indicators used are sensitive to change in oxidation potential.

In the light of the above statements, choose the most appropriate answer from the options given below

(a) Both Statement I and Statement II are correct.
(b) Both Statement I and Statement II are incorrect.
(c) Statement I is incorrect but Statement II is correct.
(d) Statement I is correct but Statement II is incorrect.

20 In Hall - Heroult process, the following is used for reducing Al_2O_3

(a) Na_3AlF_6 (b) graphite
(c) magnesium (d) CaF_2

Section B : Numerical Value Type Questions

21 The observed magnetic moment of the complex $[Mn(NCS)_6]^{x-}$ is 6.06 BM. The numerical value of x is ....

22 The number of species from the following carrying a single lone pair on central atom xenon is ....

XeF_5^+, XeO_3, XeO_2F_2, XeF_5^-, XeO_3F_2, $XeOF_4$, XeF_4

23 The number of incorrect statements from the following is ....

A. The electrical work that a reaction can perform at constant pressure and temperature is equal to the reaction Gibbs energy.

B. E°_{cell} is dependent on the pressure.

C. $\dfrac{dE^\circ_{cell}}{dT} = \dfrac{\Delta_r S^\circ}{nF}$

D. A cell is operating reversibly if the cell potential is exactly balanced by an opposing source of potential difference.

24 The sum of oxidation state of the metals in $Fe(CO)_5$, VO^{2+} and WO_3 is ....

25 The ratio of σ and π bonds present in pyrophosphoric acid is ....

26 Coagulating value of the electrolytes $AlCl_3$ and NaCl for As_2S_3 are 0.09 and 50.04 respectively. The coagulating power of $AlCl_3$ is x times the coagulating power of NaCl. The value of x is ....

27 The number of atomic orbitals from the following having 5 radial nodes is 7*s*, 7*p*, 6*s*, 8*p*, 8*d*

28 If the boiling points of two solvents X and Y (having same molecular weights) are in the ratio 2:1 and their enthalpy of vaporisations are in the ratio 1:2, then the boiling point elevation constant of X is m times the boiling point elevation constant of Y. The value of m is .... (nearest integer).

29 The solubility product of $BaSO_4$ is 1×10^{-10} at 298K. The solubility of $BaSO_4$ in 0.1 M K_2SO_4 (*aq*) solution is ...... $\times 10^{-9}$ gL^{-1} (nearest integer).

Given, molar mass of $BaSO_4$ is 233 g mol^{-1}

30 For complete combustion of ethene,

$$C_2H_4(g) + 3O_2(g) \longrightarrow 2CO_2(g) + 2H_2O(l)$$

the amount of heat produced as measured in bomb calorimeter is 1406 kJ mol^{-1} at 300K. The minimum value of $T\Delta S$ needed to reach equilibrium is (–) .... kJ. (Nearest integer)

(Given, $R = 8.3$ J K^{-1} mol^{-1})

10 APRIL 2023, SHIFT-I

Section A : Objective Type Questions

1 Match List-I with List-II.

List-I (Polymer)		List-II (Type/ Class)
A. Nylon-2-nylon-6	I.	Thermosetting polymer
B. Buna-N	II.	Biodegradable polymer
C. Urea-formaldehyde resin	III.	Synthetic rubber
D. Dacron	IV.	Polyester

Choose the correct answer from the options given below:

(a) (A)-(IV), (B)-(I), (C)-(III), (D)-(II)
(b) (A)-(IV), (B)-(III), (C)-(I), (D)-(II)
(c) (A)-(II), (B)-(III), (C)-(I), (D)-(IV)
(d) (A)-(II), (B)-(I), (C)-(IV), (D)-(II

2 Lime reacts exothermally with water to give '*A*' which has low solubility in water. Aqueous solution of '*A*' is often used for the test of CO_2, a test in which insoluble *B* is formed. If *B* is further reacted with CO_2, then soluble compound is formed. '*A*' is

(a) white lime (b) lime water
(c) slaked lime (d) quick lime

3 Given below are two statements :

Statement I Aqueous solution of $K_2Cr_2O_7$ is preferred as a primary standard in volumetric analysis over $Na_2Cr_2O_7$ aqueous solution.

Statement II $K_2Cr_2O_7$ has a higher solubility in water than $Na_2Cr_2O_7$.

In the light of the above statements, choose the correct answer from the options given below :

(a) Both Statement I and Statement II are true.
(b) Statement I is false but Statement II is true.
(c) Statement I is true but Statement II is false.
(d) Both Statement I and Statement II are false.

4 The number of molecules and moles in 2.8375 litres of O_2 at STP are respectively

(a) 7.527×10^{23} and 0.125 mol
(b) 1.505×10^{23} and 0.250 mol
(c) 7.527×10^{22} and 0.250 mol
(d) 7.527×10^{22} and 0.125 mol

5 Identify the correct order of reactivity for the following pairs towards the respective mechanism

(A) S_N2
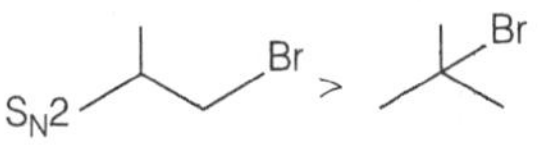

(B) S_N1
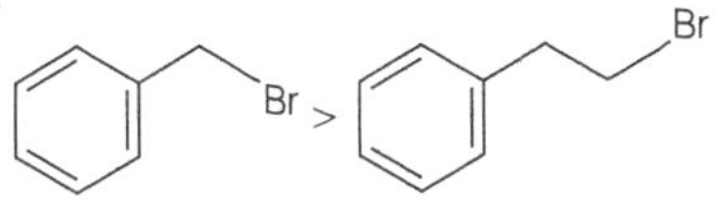

(C) Electrophilic substitution
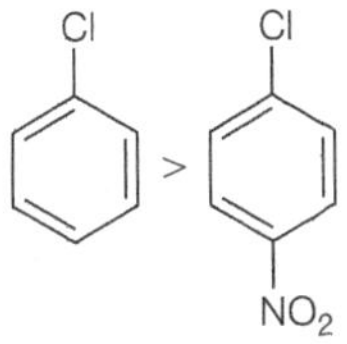

(D) Nucleophilic substitution
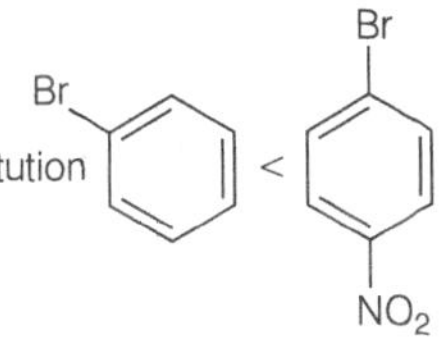

Choose the correct answer from the options given below
(a) (A), (C) and (D) only (b) (A), (B), (C) and (D)
(c) (A), (B) and (D) only (d) (B), (C) and (D) only

6 Isomeric amines with molecular formula $C_8H_{11}N$ give the following tests.
Isomer (*P*) ⇒ can be prepared by Gabriel phthalimide synthesis.
Isomer (*Q*) ⇒ reacts with Hinsberg's reagent to give solid insoluble in NaOH.
Isomer (R) ⇒ reacts with HONO followed by β-naphthol in NaOH to give red dye.
Isomers (*P*), (*Q*) and (*R*) respectively are

P Q R

(a)
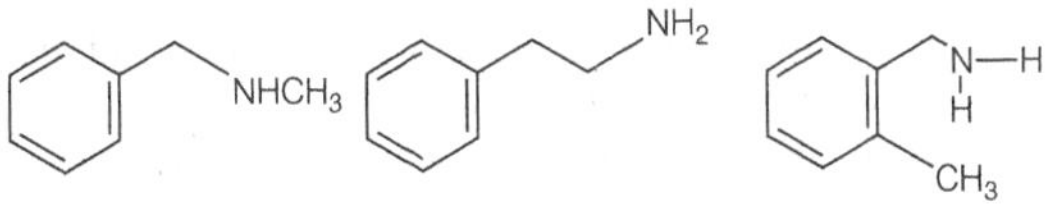

(b)
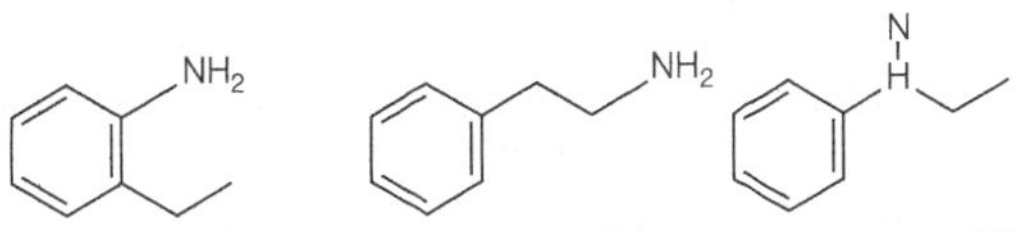

(c)
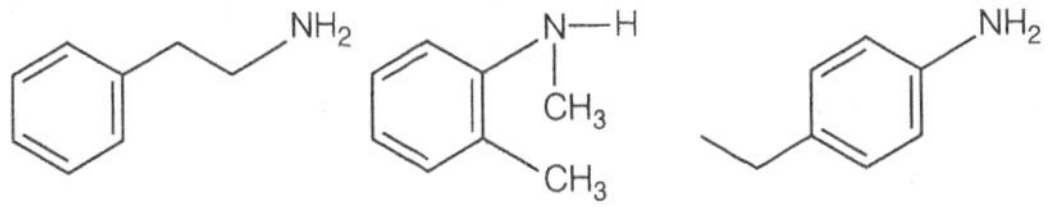

(d)
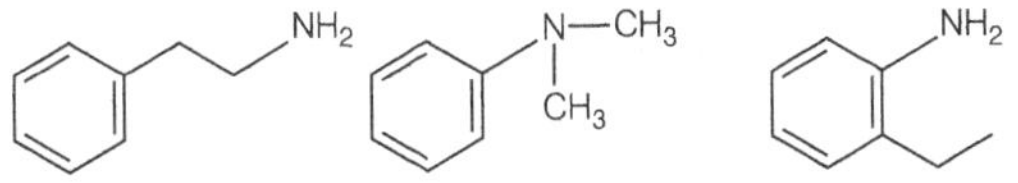

7 The major product *P* formed in the given reaction is

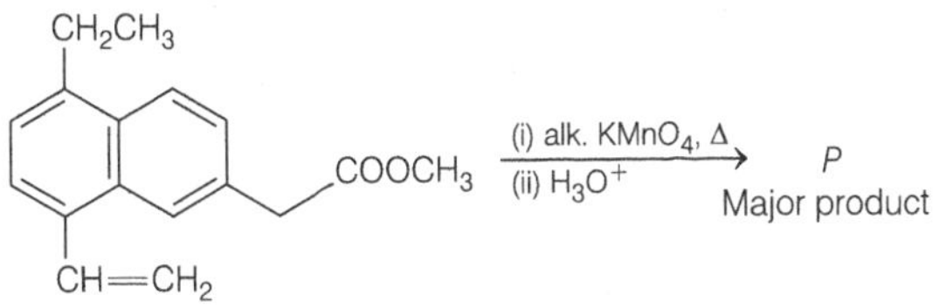

(a)
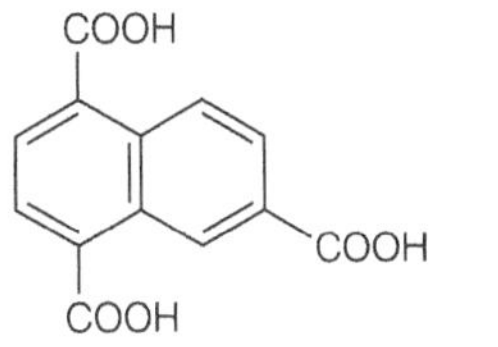

(b)
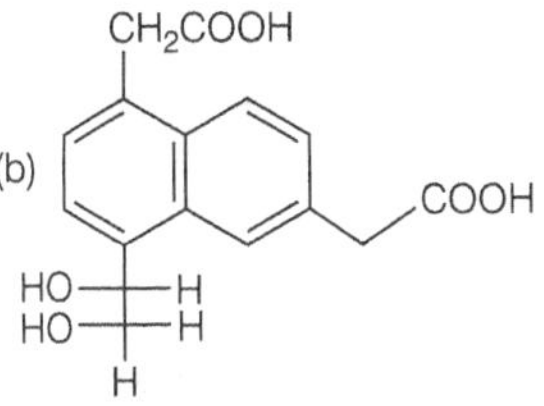

(c)
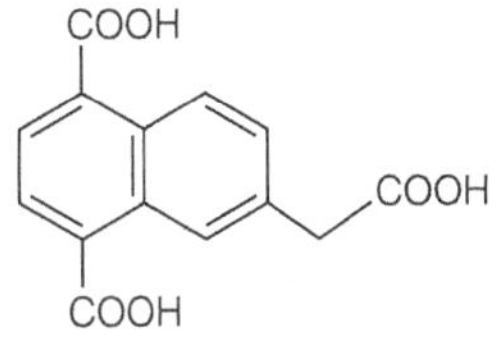

(d)
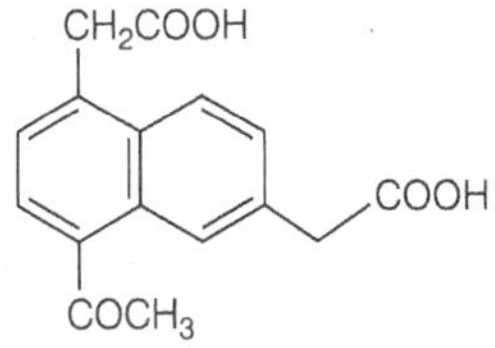

8 The enthalpy change for the adsorption process and micelle formation respectively are
(a) $\Delta H_{ads} > 0$ and $\Delta H_{mix} < 0$ (b) $\Delta H_{ads} < 0$ and $\Delta H_{mix} > 0$
(c) $\Delta H_{ads} > 0$ and $\Delta H_{mix} > 0$ (d) $\Delta H_{ads} < 0$ and $\Delta H_{mix} < 0$

9 Match List-I with List-II.

List-I (Industry)		List-II (Waste generated)	
A.	Steel plants	I.	Gypsum
B.	Thermal power plants	II.	Fly ash
C.	Fertiliser industries	III.	Slag
D.	Paper mills	IV.	Bio-degradable wastes

Choose the correct answer from the options given below
(a) (A)-(III), (B)-(IV), (C)-(I), (D)-(II)
(b) (A)-(II), (B)-(III), (C)-(IV), (D)-(I)
(c) (A)-(IV), (B)-(I), (C)-(II), (D)-(III)
(d) (A)-(III), (B)-(II), (C)-(I), (D)-(IV)

10 The one that does not stabilise 2° and 3° structures of proteins is
(a) H-bonding (b) —S—S—linkage
(c) van der Waals' forces (d) —O—O— linkage

11 The octahedral diamagnetic low spin complex among the following is
(a) $[CoF_6]^{3-}$ (b) $[Co(NH_3)_6]^{3+}$
(c) $[NiCl_4]^{2-}$ (d) $[CoCl_6]^{3-}$

12 Prolonged heating is avoided during the preparation of ferrous ammonium sulphate to
(a) prevent reduction (b) prevent hydrolysis
(c) prevent oxidation (d) prevent breaking

13 Suitable reaction condition for preparation of methylphenylether is
(a) Ph$\overset{\ominus}{O}$Na$^{\oplus}$, MeOH (b) Ph—Br, Me$\overset{\ominus}{O}$Na$^{\oplus}$
(c) Benzene, MeBr (d) Ph$\overset{\ominus}{O}$Na$^{\oplus}$, MeBr

14 The pair from the following pairs having both compounds with net non-zero dipole moment is
(a) 1, 4-dichlorobenzene, 1,3-dichlorobenzene
(b) CH_2Cl_2, $CHCl_3$
(c) Benzene, anisidine
(d) *cis*-butene, *trans*-butene

15 Using column chromatography, mixture of two compounds A and B was separated.

A eluted first, this indicates 'B' has

(a) low R_f, weaker adsorption
(b) low R_f, stronger adsorption
(c) high R_f, weaker adsorption
(d) high R_f, stronger adsorption

16 Given,

(A) $2CO(g) + O_2(g) \longrightarrow 2CO_2(g)\ \Delta H_1^\circ = -x \text{ kJ mol}^{-1}$

(B) $C\text{ (graphite)} + O_2(g) \longrightarrow CO_2(g)\ \Delta H_2^\circ = -y \text{ kJ mol}^{-1}$

The ΔH° for the reaction,

$C\text{ (graphite)} + \frac{1}{2}O_2(g) \longrightarrow CO(g)$ is

(a) $\frac{x+2y}{2}$ (b) $\frac{x-2y}{2}$
(c) $\frac{2x-y}{2}$ (d) $2y - x$

17 Which of the following statements are correct?

(A) The M^{3+}/M^{2+} reduction potential for iron is greater than manganese.
(B) The higher oxidation states of first row d-block elements get stabilised by oxide ion.
(C) Aqueous solution of Cr^{2+} can liberate hydrogen from dilute acid.
(D) Magnetic moment of V^{2+} is observed between 4.4 - 5.2 BM.

Choose the correct answer from the options given below:

(a) (A), (B), (D) only (b) (A), (B) only
(c) (B), (C) only (d) (C), (D) only

18 Which of the following is used as a stabiliser during the concentration of sulphide ores?

(a) Fatty acids (b) Xanthates
(c) Pine oils (d) Cresols

19 The compound which does not exist is

(a) $PbEt_4$ (b) NaO_2
(c) BeH_2 (d) $(NH_4)_2BeF_4$

20 Given below are two reactions, involved in the commercial production of dihydrogen (H_2). The two reactions are carried out at temperature "T_1" and "T_2", respectively

$C(s) + H_2O(g) \xrightarrow{T_1} CO(g) + H_2(g)$

$CO(g) + H_2O(g) \xrightarrow[\text{Catalyst}]{T_2} CO_2(g) + H_2(g)$

The temperatures T_1 and T_2 are correctly related as

(a) $T_1 < T_2$
(b) $T_1 = 100$ K, $T_2 = 1270$ K
(c) $T_1 = T_2$
(d) $T_1 > T_2$

Section B : Numerical Value Type Questions

21 The number of correct statement/s involving equilibria in physical processes from the following is

(A) Equilibrium is possible only in a closed system at a given temperature.
(B) Both the opposing processes occur at the same rate.
(C) When equilibrium is attained at a given temperature, the value of all its parameters became equal.
(D) For dissolution of solids in liquids, the solubility is constant at a given temperature.

22 If the degree of dissociation of aqueous solution of weak monobasic acid is determined to be 0.3, then the observed freezing point will be ....% higher than the expected/theoretical freezing point. (Nearest integer)

23 At constant temperature, a gas is at a pressure of 940.3 mm Hg. The pressure at which its volume decreases by 40% is ______ mm Hg. (Nearest integer)

24 The sum of lone pairs present on the central atom of the interhalogen IF_5 and IF_7 is ....

25 In the following reactions, the total number of oxygen atoms in X and Y is ....

$$Na_2O + H_2O \longrightarrow 2X$$
$$Cl_2O_2 + H_2O \longrightarrow 2Y$$

26 A molecule undergoes two independent first order reactions whose respective half-lifes are 12 min and 3 min. If both the reactions are occurring then the time taken for the 50% consumption of the reactant is min. (Nearest integer)

27 The number of incorrect statement/s about the black body from the following is

(A) Emit or absorb energy in the form of electromagnetic radiation.
(B) Frequency distribution of the emitted radiation depends on temperature.
(C) At a given temperature, intensity *vs* frequency curve passes through a maximum value.
(D) The maximum of the intensity *vs* frequency curve is at a higher frequency at higher temperature compared to that at lower temperature.

28 In potassium ferrocyanide, there are .... pairs of electrons in the t_{2g} set of orbitals.

29 $FeO_4^{2-} \xrightarrow{+2.2\text{ V}} Fe^{3+} \xrightarrow{+0.70\text{ V}} Fe^{2+} \xrightarrow{-0.045\text{ V}} Fe^0$

$E^\circ_{FeO_4^{2-}/Fe^{2+}}$ is $x \times 10^{-3}$ V.

The value of x is ....

30 The number of bent-shaped molecule/s from the following is

$$N_3^-, NO_2^-, I_3^-, O_3, SO_2$$

10 APRIL 2023, SHIFT-II

Section A : Objective Type Questions

1 The correct relationships between unit cell edge length 'a' and radius of sphere 'r' for face-centred and body-centred cubic structures respectively are

(a) $r = 2\sqrt{2}a$ and $\sqrt{3}r = 4a$
(b) $r = 2\sqrt{2}a$ and $4r = \sqrt{3}a$
(c) $2\sqrt{2}r = a$ and $4r = \sqrt{3}a$
(d) $2\sqrt{2}r = a$ and $\sqrt{3}r = 4a$

2 Incorrect method of preparation for alcohols from the following is

(a) reaction of ketone with RMgBr followed by hydrolysis.
(b) hydroboration-oxidation of alkene.
(c) ozonolysis of alkene.
(d) reaction of alkyl halide with aqueous NaOH.

3 The reaction used for preparation of soap from fat is

(a) an oxidation reaction
(b) reduction reaction
(c) an addition reaction
(d) alkaline hydrolysis reaction

4 Given below are two statements: one is labelled as Assertion (A) and the other is labelled as Reason (R).

Assertion (A) 3.1500 g of hydrated oxalic acid dissolved in water to make 250.0 mL solution will result in 0.1 M oxalic acid solution.

Reason (R) Molar mass of hydrated oxalic acid is 126 g mol^{-1}.

In the light of the above statements, choose the correct answer from the options given below.

(a) A is true but R is false.
(b) Both A and R are true and R is the correct explanation of A.
(c) A is false but R is true.
(d) Both A and R are true but R is not the correct explanation of A.

5 Buna-S can be represented as

(a) 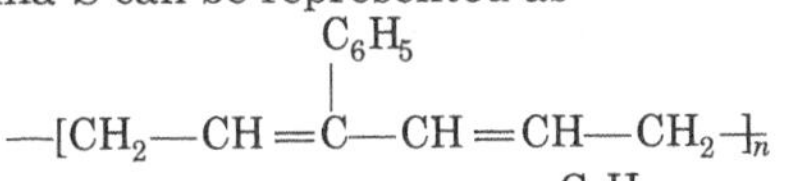

(b) 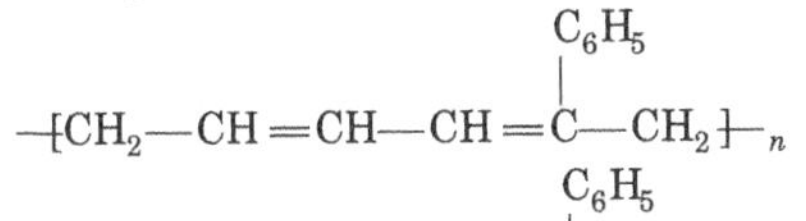

(c) 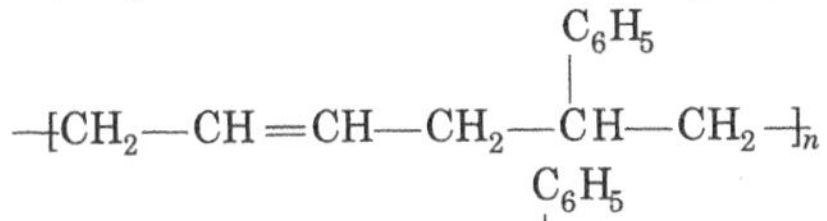

(d) 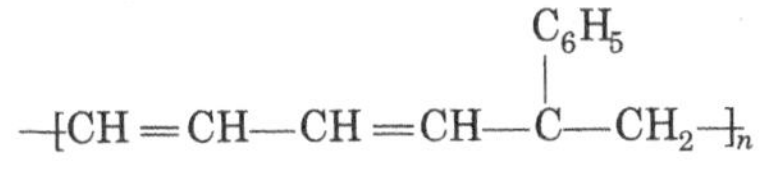

6 The correct order of metallic character is

(a) K > Be > Ca
(b) K > Ca > Be
(c) Be > Ca > K
(d) Ca > K > Be

7 Ferric chloride is applied to stop bleeding because

(a) Cl^- ions cause coagulation of blood.
(b) blood absorbs $FeCl_3$ and forms a complex.
(c) Fe^{3+} ions coagulate blood which is a negatively charged sol.
(d) $FeCl_3$ reacts with the constituents of blood which is a positively charged sol.

8 Given below are two statements: one is labelled as Assertion (A) and the other is labelled as Reason (R).

Assertion (A) Physical properties of isotopes of hydrogen are different.

Reason (R) Mass difference between isotopes of hydrogen is very large.

In the light of the above statements, choose the correct answer from the options given below.

(a) A is false but R is true.
(b) Both A and R are true but R is not the correct explanation of A.
(c) A is true but R is false.
(d) Both A and R are true and R is the correct explanation of A.

9 Match List-I with List-II.

	List-I Complex		List-II Crystal field splitting energy (Δ_0)
A.	$[Ti(H_2O)_6]^{2+}$	I.	− 1.2
B.	$[V(H_2O)_6]^{2+}$	II.	− 0.6
C.	$[Mn(H_2O)_6]^{3+}$	III.	0
D.	$[Fe(H_2O)_6]^{3+}$	IV.	− 0.8

Choose the correct answer from the options given below.

(a) A-II, B-IV, C-I, D-III
(b) A-IV, B-I, C-III, D-II
(c) A-IV, B-I, C-II, D-III
(d) A-II, B-IV, C-III, D-I

10 In Carius tube, an organic compound X is treated with sodium peroxide to form a mineral acid Y. The solution of $BaCl_2$ is added to Y to form a precipitate Z. Z is used for the quantitative estimation of an extra element.

X could be

(a) cytosine
(b) methionine
(c) a nucleotide
(d) chloroxylenol

11 Given below are two statements: one is labelled as Assertion (A) and the other is labelled as Reason (R).

Assertion (A) The energy required to form Mg^{2+} from Mg is much higher than that required to produce Mg^+.

Reason (R) Mg^{2+} is small ion and carry more charge than Mg^+.

In the light of the above statements, choose the correct answer from the options given below.
(a) A is true but R is false.
(b) A is false but R is true.
(c) Both A and R are true and R is the correct explanation of A.
(d) Both A and R are true but R is not the correct explanation of A.

12 The major product '*P*' formed in the given reaction is

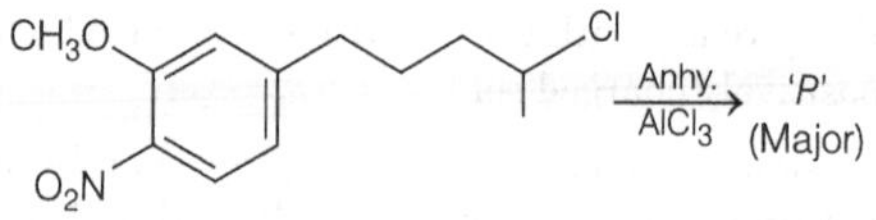

(a)
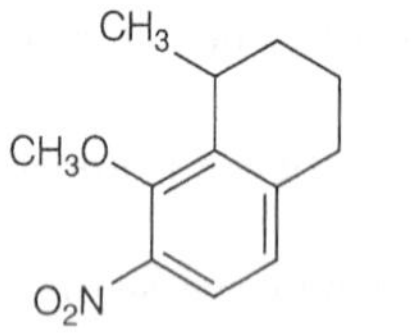

(b)
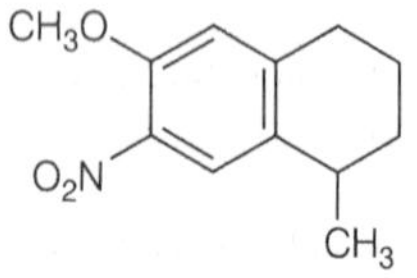

(c)
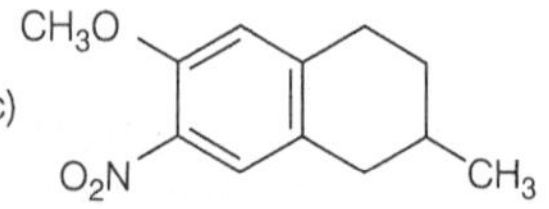

(d)
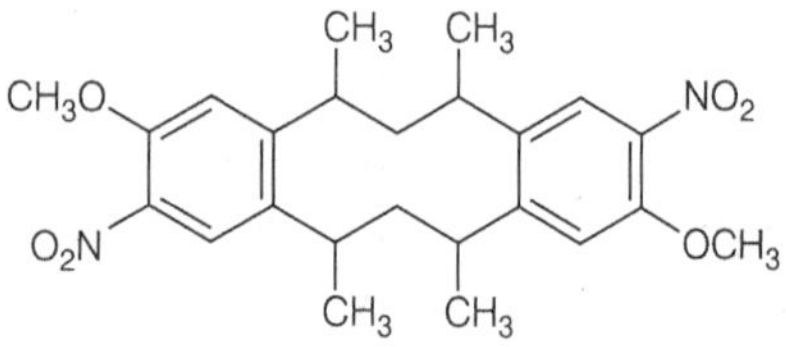

13 The delicate balance of CO_2 and O_2 is not disturbed by
(a) burning of coal (b) respiration
(c) deforestation (d) burning of petroleum

14 The decreasing order of hydride affinity for following carbocations is

A. $CH_2 = CH-\overset{+}{C}(CH_3)-CH_3$ B. $C_6H_5-\overset{+}{C}(C_6H_5)-C_6H_5$

C. $H_3C-\overset{+}{C}(CH_3)-CH_3$ D.
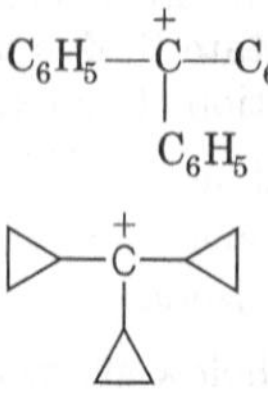

Choose the correct answer from the options given below
(a) C, A, B, D (b) C, A, D, B
(c) A, C, D, B (d) A, C, B, D

15 The correct order of the number of unpaired electrons in the given complexes is
A. $[Fe(CN)_6]^{3-}$ B. $[FeF_6]^{3-}$
C. $[CoF_6]^{3-}$ D. $[Cr(oxalate)_3]^{3-}$
E. $[Ni(CO)_4]$
Choose the correct answer from the options given below:
(a) E < A < B < D < C (b) A < E < C < B < D
(c) A < E < D < C < B (d) E < A < D < C < B

16 Gibbs energy *vs* *T* plot for the formation of oxides is given below

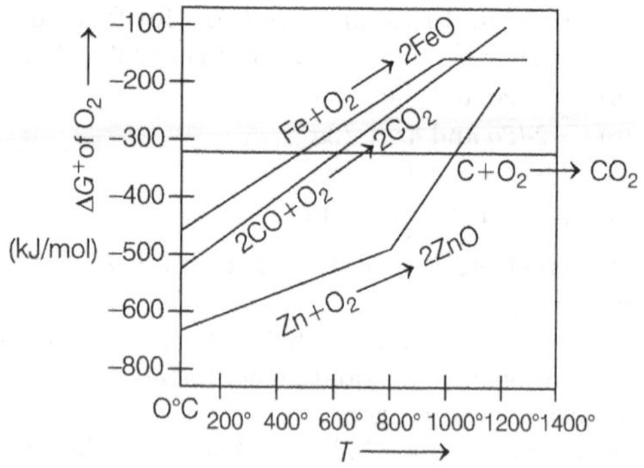

For the given diagram, the correct statement is
(a) at 600 °C, C can reduce FeO
(b) at 600 °C. CO cannot reduce FeO
(c) at 600 °C,C can reduce ZnO
(d) at 600 °C, CO can reduce ZnO

17 Match List-I with List-II.

	List-I		List-II
A.	16 g of $CH_4(g)$	I.	Weighs 28 g
B.	1 g of $H_2(g)$	II.	60.2×10^{23} electrons
C.	1 mole of $N_2(g)$	III.	Weighs 32 g
D.	0.5 mol of $SO_2(g)$	IV.	Occupies 11.4 L volume at STP

Choose the correct answer from the options given below
(a) A-I, B-III, C-II, D-IV (b) A-II, B-IV, C-I, D-III
(c) A-II. B-III, C-IV, D-I (d) A-II, B-IV, C-III, D-I

18 Number of water molecules in washing soda and soda ash respectively are
(a) 10 and 0 (b) 1 and 0
(c) 1 and 10 (d) 10 and 1

19 In the reaction given below

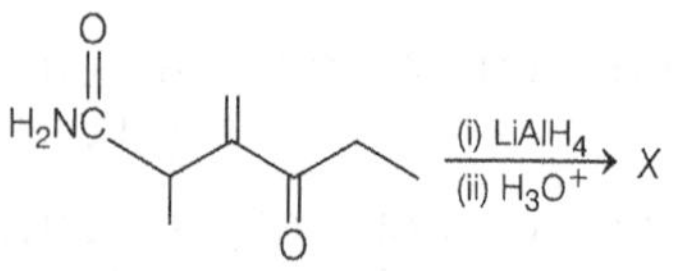

The product *X* is

(a)
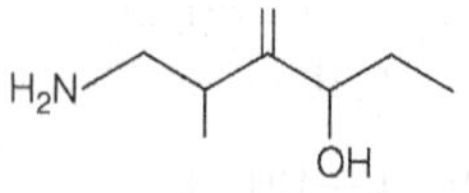

(b)
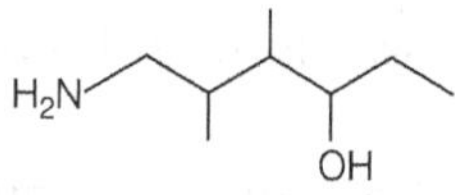

(c)
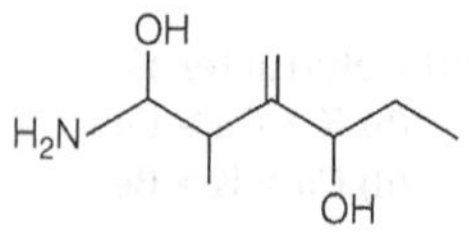

(d)
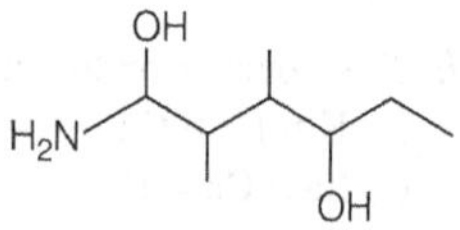

20 The correct order for acidity of the following hydroxyl compound is

A. CH_3OH B. $(CH_3)_3COH$

C. 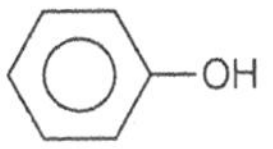
D. 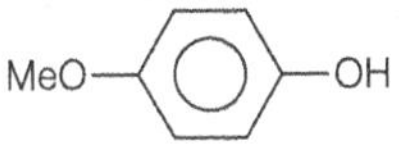
E. 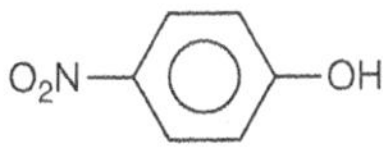

Choose the correct answer from the options given below

(a) C > E > D > B > A (b) E > D > C > B > A
(c) E > C > D > A > B (d) D > E > C > A > B

Section B : Numerical Value Type Questions

21 The number of endothermic processes from the following is ....

A. $I_2(g) \longrightarrow 2I(g)$ B. $HCl(g) \longrightarrow H(g) + Cl(g)$
C. $H_2O(l) \longrightarrow H_2O(g)$ D. $C(s) + O_2(g) \longrightarrow CO_2(g)$
E. Dissolution of ammonium chloride in water

22 For a metal ion, the calculated magnetic moment is 4.90 BM. This metal ion has .... number of unpaired electrons.

23 The electron in the nth orbit of Li^{2+} is excited to $(n+1)$ orbit using the radiation of energy 1.47×10^{-17} J (as shown in the diagram). The value of n is ....
Given, $R_H = 2.18 \times 10^{-18}$ J

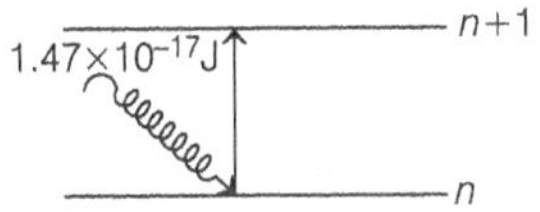

24 $A(g) \rightleftharpoons 2B(g) + C(g))$

For the given reaction, if the initial pressure is 450 mm Hg and the pressure at time t is 720 mm Hg at a constant temperature T and constant volume V. The fraction of $A(g)$ decomposed under these conditions is $x \times 10^{-1}$. The value of x is .... (nearest integer)

25 The number of incorrect statements from the following is

A. The successive half-lifes of zero order reactions decreases with time.
B. A substance appearing as reactant in the chemical equation may not affect the rate of reaction.
C. Order and molecularity of a chemical reaction can be a fractional number.
D. The rate constant units of zero and second order reaction are mol $L^{-1}s^{-1}$ and $mol^{-1}Ls^{-1}$ respectively.

26 The difference in the oxidation state of Xe between the oxidised product of Xe formed on complete hydrolysis of XeF_4 and XeF_4 is

27 The number of molecules from the following which contain only two lone pair of electrons is

H_2O, N_2, CO, XeF_4, NH_3, NO, CO_2, F_2

28 In alkaline medium, the reduction of permanganate anion involves a gain of ....electrons.

29 The specific conductance of 0.0025 M acetic acid is 5×10^{-5} S cm^{-1} at a certain temperature. The dissociation constant of acetic acid is ....$\times 10^{-7}$ (Nearest integer)

Consider limiting molar conductivity of CH_3COOH as 400 S cm^2 mol^{-1}.

30 An aqueous solution of volume 300 cm^3 contains 0.63 g of protein. The osmotic pressure of the solution at 300 K is 1.29 m bar. The molar mass of the protein is ...... g mol^{-1}
[Given, $R = 0.083$ L bar K^{-1} mol^{-1}]

11 APRIL 2023, SHIFT-I

Section A : Objective Type Questions

1 Match List-I with List-II.

	List-I		List-II
A.	K	I.	Thermonuclear reactions
B.	KCl	II.	Fertiliser
C.	KOH	III.	Sodium potassium pump
D.	Li	IV.	Absorbent of CO_2

Choose the correct answer from the options given below.

(a) A-III, B-IV, C-II, D-I (b) A-IV, B-I, C-III, D-II
(c) A-III, B-II, C-IV, D-I (d) A-IV, B-III, C-I, D-II

2 Given below are two statements, one is labelled as Assertion (A) and the other is labelled as Reason (R).

Assertion (A) In the photoelectric effect, the electrons are ejected from the metal surface as soon as the beam of light of frequency greater than threshold frequency strikes the surface.

Reason (R) When the photon of any energy strikes an electron in the atom, transfer of energy from the photon to the electron takes place.

In the light of the above statements, choose the most appropriate answer from the options given below.

(a) Both A and R are correct but R is not the correct explanation of A.
(b) Both A and R are correct and R is the correct explanation of A.
(c) A is correct but R is not correct.
(d) A is not correct but R is correct.

3 In the extraction process of copper, the product obtained after carrying out the reactions

(i) $2Cu_2S + 3O_2 \longrightarrow 2Cu_2O + 2SO_2$

(ii) $2Cu_2O + Cu_2S \longrightarrow 6Cu + SO_2$ is called

(a) copper scrap (b) blister copper
(c) reduced copper (d) copper matte

4 Thin layer chromatography of a mixture shows the following observation.

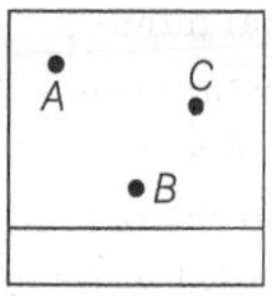

The correct order of elution in the silica gel column chromatography is

(a) *B, C, A* (b) *A, C, B*
(c) *C, A, B* (d) *B, A, C*

5 25 mL of silver nitrate solution (1M) is added dropwise to 25 mL of potassium iodide (1.05 M) solution. The ion(s) present in very small quantity in the solution is/are

(a) K^+ only (b) NO_3^- only
(c) Ag^+ and I^- both (d) I^- only

6 Match List-I with List-II.

List-I (Species)	List-II (Geometry/Shape)
A. H_3O^+	I. Tetrahedral
B. Acetylide anion	II. Linear
C. NH_4^+	III. Pyramidal
D. ClO_2^-	IV. Bent

Choose the correct answer from the options given below.

(a) A-III, B-IV, C-II, D-I (b) A-III, B-I, C-II, D-IV
(c) A-III, B-IV, C-I, D-II (d) A-III, B-II, C-I, D-IV

7 The polymer *X*-consists of linear molecules and is closely packed. It is prepared in the presence of triethylaluminium and titanium tetrachloride under low pressure. The polymer *X* is

(a) low density polythene
(b) high density polythene
(c) polyacrylonitrile
(d) polytetrafluoroethane

8

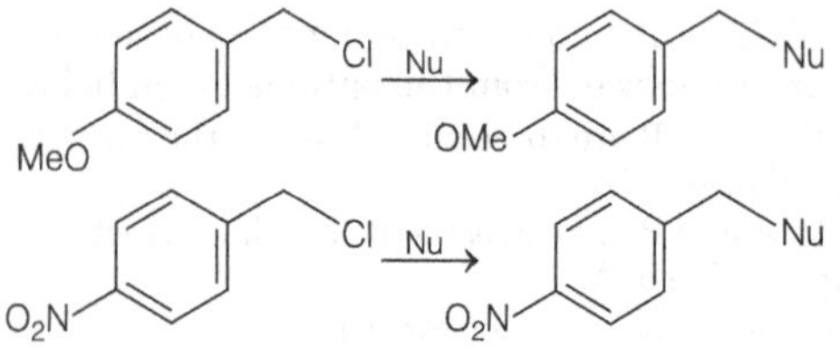

....(I)
....(II)

(where, Nu = nucleophile)

Find out the correct statement from the options given below for the above 2 reactions.

(a) Reactions (I) and (II) both are of 1st order.
(b) Reactions (I) and (II) both are of 2nd order.
(c) Reaction (I) is of 2nd order and reaction (II) is of 1st order.
(d) Reaction (I) is of 1st order and reaction (II) is of 2nd order.

9 For elements B, C, N, Li, Be, O and F, the correct order of first ionisation enthalpy is

(a) B > Li > Be > C > N > O > F
(b) Li < B < Be < C < O < N < F
(c) Li < Be < B < C < O < N < F
(d) Li < Be < B < C < N < O < F

10 *o*-phenylenediamine $\xrightarrow{HNO_2}$ *X*
(Major product)

X is

(a)
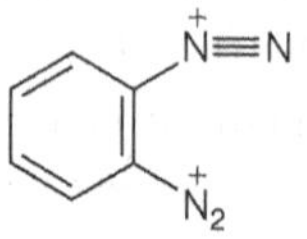

(b)
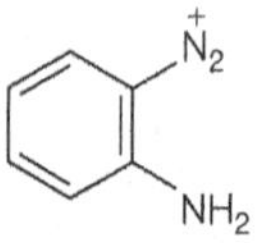

(c)
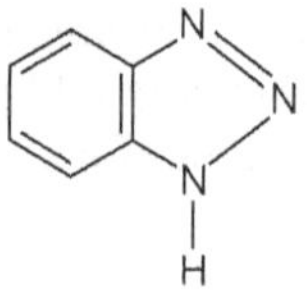

(d)
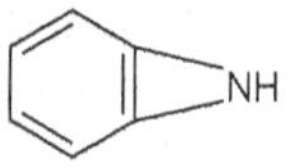

11 The complex that dissolves in water is

(a) $[Fe_3(OH)_2(OAc)_6]Cl$
(b) $Fe_4[Fe(CN)_6]_3$
(c) $K_3[Co(NO_2)_6]$
(d) $(NH_4)_3[As(Mo_3O_{10})_4]$

12 Given below are two statements :

Statement I If BOD is 4 ppm and dissolved oxygen is 8 ppm, then it is a good quality water.

Statement II If the concentration of zinc and nitrate salts are 5 ppm each, then it can be a good quality water.

In the light of the above statements, choose the most appropriate answer from the options given below.

(a) Both the statements I and II are correct
(b) Both the statements I and II are incorrect.
(c) Statement I is incorrect but statement II is correct.
(d) Statement I is correct but statement II is incorrect.

13 Arrange the following compounds in increasing order of rate of aromatic electrophilic substitution reaction

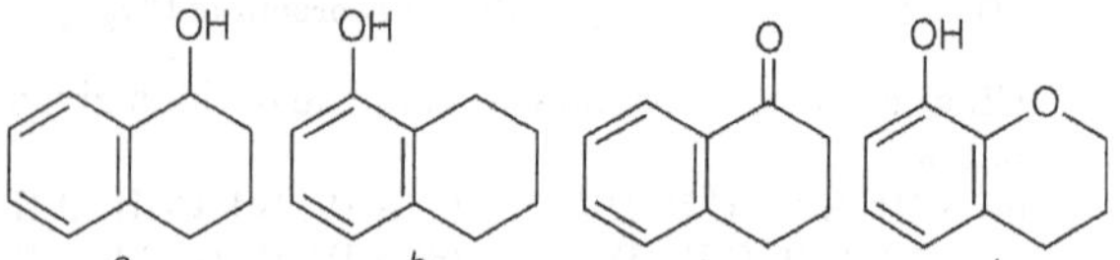

a b c d

(a) *d, b, c, a* (b) *b, c, a, d*
(c) *c, a, b, d* (d) *d, b, a, c*

14 The set which does not have ambidentate ligand(s) is
(a) NO_2^-, $C_2O_4^{2-}$, $EDTA^{4-}$
(b) $EDTA^{4-}$, NCS^-, $C_2O_4^{2-}$
(c) $C_2O_4^{2-}$, NO_2^-, NCS^-
(d) $C_2O_4^{2-}$, ethylene diammine, H_2O

15 *L*-isomer of tetrose $X(C_4H_8O_4)$ gives positive Schiff's test and has two chiral carbons. On acetylation, '*X*' yields triacetate. '*X*' also undergoes following reactions

$$'A' \xleftarrow{HNO_3} 'X' \xrightarrow{NaBH_4} 'B'$$
Chiral compound

'*X*' is

(a) 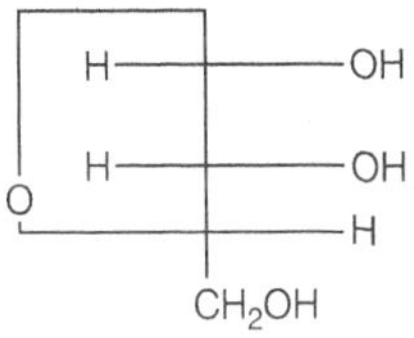
(b) 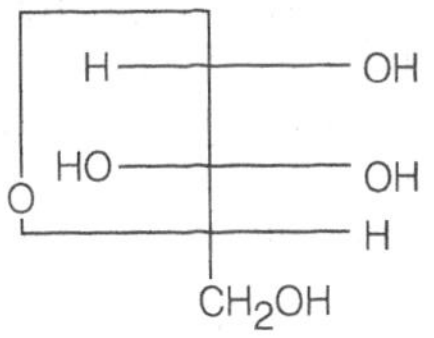
(c) 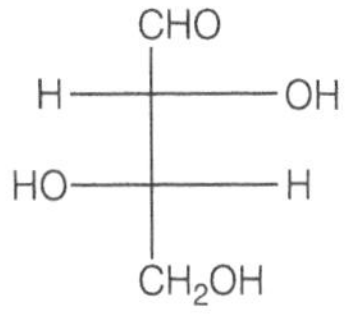
(d) 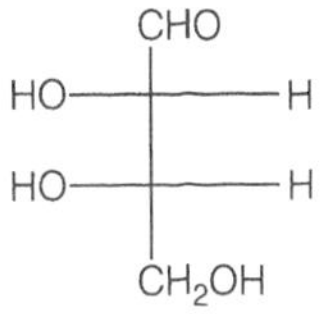

16 Which of the following complex has a possibility to exist as meridional isomer?
(a) $[Pt(NH_3)_2Cl_2]$ (b) $[Co(NH_3)_3(NO_2)_3]$
(c) $[Co(en)_2Cl_2]$ (d) $[Co(en)_3]$

17 For compound having the formula $GaAlCl_4$, the correct option from the following is
(a) Ga is more electronegative than Al and is present as a cationic part of the salt $GaAlCl_4$.
(b) Ga is coordinated with Cl in $GaAlCl_4$.
(c) Cl forms bond with both Al and Ga in $GaAlCl_4$.
(d) Oxidation state of Ga in the salt $GaAlCl_4$ is + 3.

18 Given below are two statements.

Statement I Methane and steam passed over a heated Ni catalyst produces hydrogen gas.

Statement II Sodium nitrite reacts with NH_4Cl to give H_2O, N_2 and NaCl.

In the light of the above statements, choose the most appropriate answer from the options given below
(a) Statement I is correct but Statement II is incorrect.
(b) Statement I is incorrect but Statement II is correct.
(c) Both the statements I and II are correct.
(d) Both the statements I and II are incorrect.

19 When a solution of mixture having two inorganic salts was treated with freshly prepared ferrous sulphate in acidic medium, a dark brown ring was formed whereas on treatment with neutral $FeCl_3$, it gave deep red colour which disppeared on boiling and a brown red ppt. was formed. The mixture contains
(a) CH_3COO^- and NO_3^- (b) $C_2O_4^{2-}$ and NO_3^-
(c) SO_3^{2-} and $C_2O_4^{2-}$ (d) SO_3^{2-} and CH_3COO^-

20 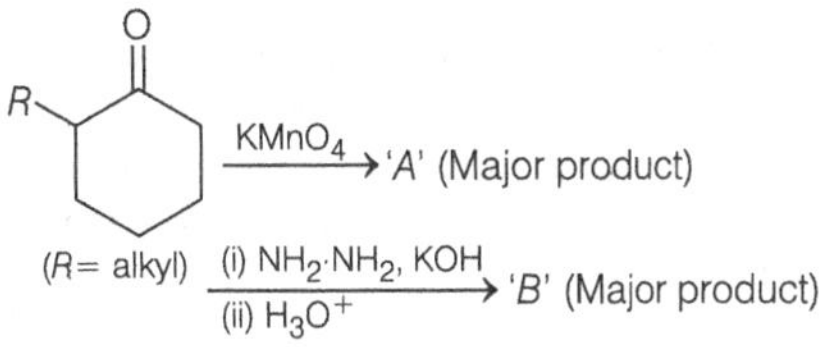

'*A*' and '*B*' in the above reactions are

(a) 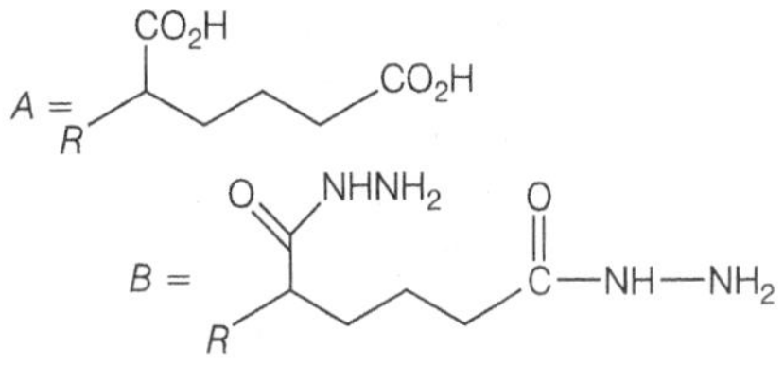
(b) 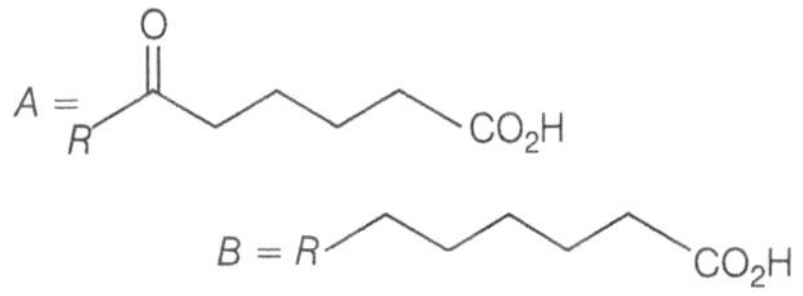
(c) 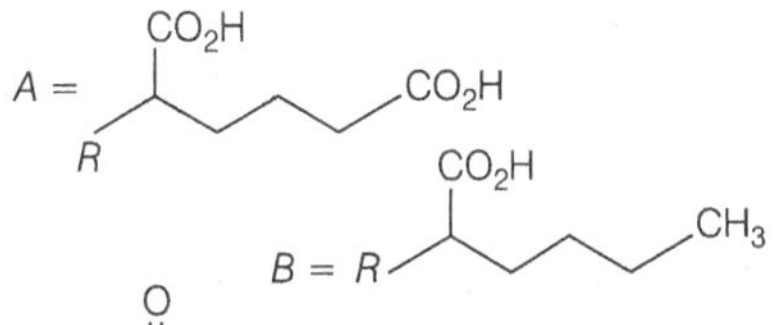
(d) 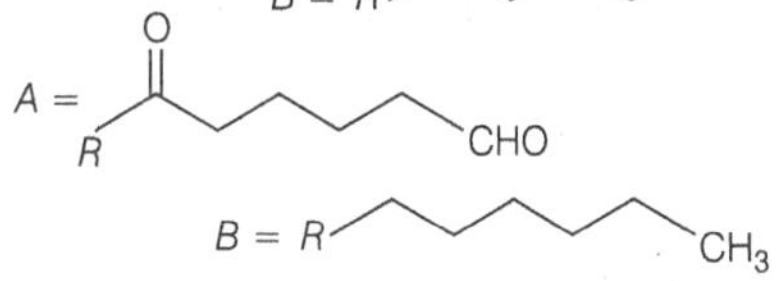

Section B : Numerical Value Type Questions

21 A mixture of 1 mole of H_2O and 1 mole of CO is taken in a 10 litre container and heated to 725 K. At equilibrium 40% of water by mass reacts with carbon monoxide according to the equation
$CO(g) + H_2O(g) \rightleftharpoons CO_2(g) + H_2(g)$.
The equilibrium constant $K_C \times 10^2$ for the reaction is .... (Nearest integer)

22 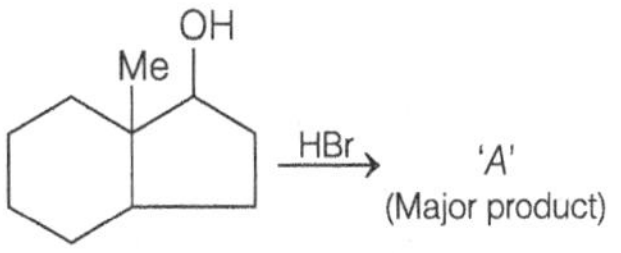

The number of hyperconjugation structure involved to stablise carbocation formed in the above reaction is ....

23 In an electrochemical reaction of lead, at standard temperature, if $E^\circ_{(Pb^{2+}/Pb)} = m$ volt and $E^\circ_{(Pb^{4+}/Pb)} = n$ volt, then the value of $E^\circ_{(Pb^{2+}/Pb^{4+})}$ is given by $m - xn$. Then, value of x is .... (Nearest integer).

24 $KClO_3 + 6FeSO_4 + 3H_2SO_4 \longrightarrow$
$KCl + 3Fe_2(SO_4)_3 + 3H_2O$

The above reaction was studied at 300 K by monitoring the concentration of $FeSO_4$ in which initial concentration was 10 M and after half an hour became 8.8 M. The rate of production of $Fe_2(SO_4)_3$ is ………… $\times 10^{-6}$ mol $L^{-1}s^{-1}$.

(Nearest integer)

25 0.004 M K_2SO_4 solution is isotonic with 0.01 M glucose solution. Percentage dissociation of K_2SO_4 is ………… (Nearest integer)

26 The ratio of spin-only magnetic moment values $\mu_{eff}[Cr(CN)_6]^{3-} / \mu_{eff}[Cr(H_2O_6]^{3+}$ is ………… .

27 An atomic substance *A* of molar mass 12 g mol^{-1} has a cubic crystal structure with edge length of 300 pm. The number of atoms present in one unit cell of *A* is ………… . (Nearest integer)

(Given, the density of *A* is 3.0 g mL^{-1} and $N_A = 6.02 \times 10^{23} mol^{-1}$)

28

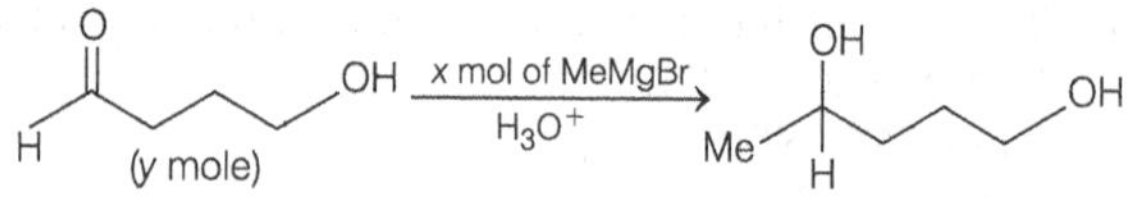

The ratio x/y on completion of the above reaction is ………… .

29 A solution of sugar is obtained by mixing 200 g of its 25% solution and 500 g of its 40% solution (both by mass). The mass percentage of the resulting sugar solution is ………… (Nearest integer)

30 Solid fuel used in rocket is a mixture of Fe_2O_3 and Al (in ratio 1 : 2). The heat evolved (kJ) per gram of the mixture is ………… (Nearest integer)

(Given, $\Delta H_f^\circ (Al_2O_3) = -1700$ kJ mol^{-1}

$\Delta H_f^\circ (Fe_2O_3) = -840$ kJ mol^{-1}

Molar mass of Fe, Al and O are 56, 27 and 16 g mol^{-1} respectively).

11 APRIL 2023, SHIFT-II

Section A : Objective Type Questions

1 A solution is prepared by adding 2 g of "X" to 1 mole of water. Mass per cent of "X" in the solution is

(a) 20% (b) 10% (c) 5%(d) 2%

2 Given below are two statements :

Statement I Ethene at 333 to 343 K and 6-7 atm pressure in the presence of $AlEt_3$ and $TiCl_4$ undergoes addition polymerisation to give LDP.

Statement II Caprolactam at 533-543 K in H_2O through step growth polymerises to give nylon 6.

In the light of the above statements, choose the correct answer from the options given below.

(a) Statement I is false but Statement II is true.
(b) Both Statement I and Statement II are true.
(c) Statement I is true but Statement II is false.
(d) Both Statement I and Statement II are false.

3 What weight of glucose must be dissolved in 100 g of water to lower the vapour pressure by 0.20 mm Hg ?

(Assume, dilute solution is being formed.)

(Given; vapour pressure of pure water is 54.2 mm Hg at room temperature. Molar mass of glucose is 180 g mol^{-1}.)

(a) 3.69 g (b) 3.59 g (c) 2.59 g (d) 4.69 g

4 Given below are two statements, one is labelled as Assertion (A) and the other is labelled as Reason (R).

Assertion (A) $[CoCl(NH_3)_5]^{2+}$ absorbs at lower wavelength of light with respect to $[Co(NH_3)_5(H_2O)]^{3+}$.

Reason (R) It is because the wavelength of the light absorbed depends on the oxidation state of the metal ion.

In the light of the above statements, choose the most appropriate answer from the options given below.

(a) Both A and R are true and R is the correct explanation of A.
(b) A is true but R is false.
(c) Both A and R are true but R is not the correct explanation of A.
(d) A is false but R is true.

5 Compound from the following that will not produce precipitate on reaction with $AgNO_3$ is

(a) 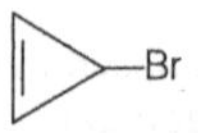

(b) 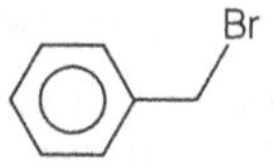

(c) 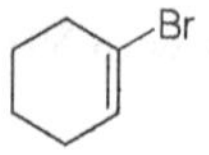

(d) 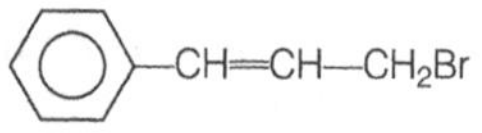

6 Alkali metal from the following with least melting point is

(a) K (b) Cs (c) Rb (d) Na

7 Which of the following compounds is an example of freon?

(a) C_2HF_3 (b) C_2F_4
(c) $C_2Cl_2F_2$ (d) $C_2H_2F_2$

8 Given below are two statements, one is labelled as Assertion (A) and the other is labelled as Reason (R).

Assertion (A) 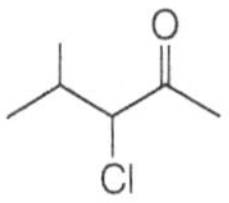 can be subjected to

Wolff-Kishner reduction to give 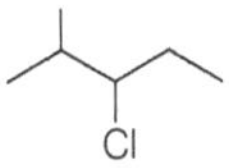

Reason (R) Wolff-Kishner reduction is used to

convert 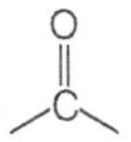 into ⁄CH_2⟍.

In the light of the above statements, choose the most appropriate answer from the options given below.

(a) Both A and R are true and R is the correct explanation of A.
(b) A is true but R is false.
(c) A is false but R is true.
(d) Both A and R are true but R is not the correct explanation of A.

9 The magnetic moment is measured in Bohr Magneton (BM). Spin only magnetic moment of Fe in $[Fe(H_2O)_6]^{3+}$ and $[Fe(CN)_6]^{3-}$ complexes respectively is

(a) 6.92 BM in both (b) 4.89 BM and 6.92 BM
(c) 5.92 BM and 1.732 BM (d) 3.87 BM and 1.732 BM

10 Which one of the following pairs is an example of polar molecular solids?

(a) $MgO(s)$, $SO_2(s)$ (b) $HCl(s)$, $AlN(s)$
(c) $SO_2(s)$, $CO_2(s)$ (d) $SO_2(s)$, $NH_3(s)$

11 Match List-I with List-II.

List-I (Complex)		List-II (Colour)
A. $Mg(NH_4)PO_4$	I.	Brown
B. $K_3[Co(NO_2)_6]$	II.	White
C. $MnO[OH]_2$	III.	Yellow
D. $Fe_4[Fe(CN)_6]_3$	IV.	Blue

Choose the correct answer from the options given below.

(a) A-III, B-IV, C-II, D-I (b) A-II, B-III, C-IV, D-I
(c) A-II, B-IV, C-I, D-III (d) A-II, B-III, C-I, D-IV

12 $H_3C—CH_2—CH(OH)—CH_3 \xrightarrow[\text{(iii) } D_2O]{\text{(i) NaI, } H_3PO_4 \text{ (ii) Mg, Dry ether}} [X]$ Product

Product $[X]$ formed in the above reaction is

(a) $H_3C—CH=CH—CH_3$ (b) $H_3C—CH_2—CH=CH_2$
(c) $H_3C—CH_2—CH(D)—CH_3$ (d) $H_3C—CH_2—C(H)(OH)—CH_3$

13 Compound 'B' is

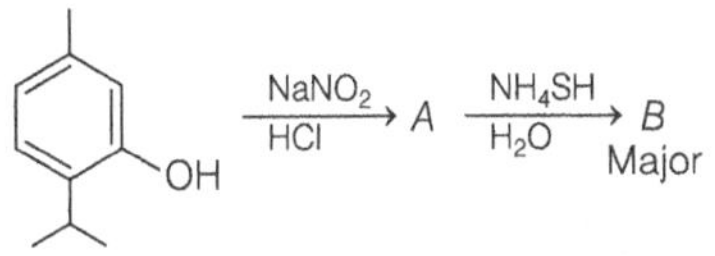

(a) 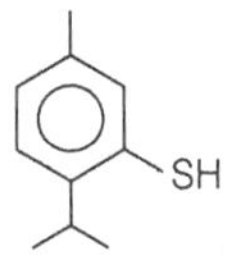

(b) 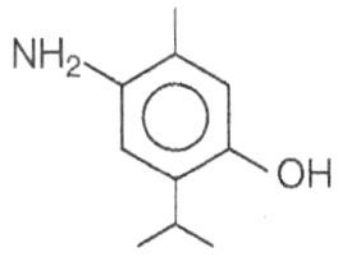

(c) 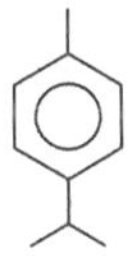

(d) 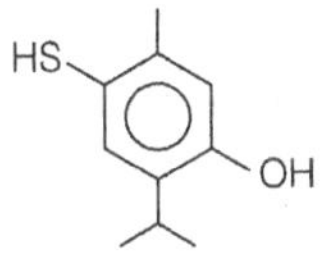

14 Given below are two statements :

Statement I In the metallurgy process, sulphide ore is converted to oxide before reduction.

Statement II Oxide ores in general are easier to reduce.

In the light of the above statements, choose the most appropriate answer from the options given below.

(a) Statement I is correct but Statement II is incorrect.
(b) Both Statement I and Statement II are correct.
(c) Statement I is incorrect but Statement II is correct.
(d) Both Statement I and Statement II are incorrect.

15 If Ni^{2+} is replaced by Pt^{2+} in the complex $[NiCl_2Br_2]^{2-}$, which of the following properties are expected to get changed?

A. Geometry
B. Geometrical isomerism
C. Optical isomerism
D. Magnetic properties

(a) A, B and D (b) A and D
(c) A, B and C (d) B and C

16 For a chemical reaction, $A + B \longrightarrow$ Product, the order is 1 with respect to A and B.

Rate (mol L^{-1} s^{-1})	$[A]$ (mol L^{-1})	$[B]$ (mol L^{-1})
0.10	20	0.5
0.40	x	0.5
0.80	40	y

What is the value of x and y?

(a) 80 and 4 (b) 80 and 2 (c) 40 and 4 (d) 160 and 4

17 The major product formed in the following reaction is

$C_6H_5—CH(OH)—CH(CH_3)—CH_2CHO \xrightarrow[\Delta]{Zn(Hg)/HCl}$ Major products

(A) $C_6H_5—CH(OH)—CH(CH_3)—C_2H_5$ (B) $C_6H_5—CH{=}C(CH_3)—C_2H_5$

(C) $C_6H_5—C(CH_3){=}CH—C_2H_5$ (D) 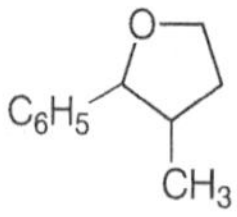

Choose the correct answer from the options given below

(a) A only (b) D only (c) B only (d) C only

18 Given below are two statements, one is labelled as Assertion (A) and the other is labelled as Reason (R).

Assertion (A) A solution of the product obtained by heating a mole of glycine with a mole of chlorine in presence of red phosphorus generates chiral carbon atom.

Reason (R) A molecule with 2 chiral carbons is always optically active.

In the light of the above statements, choose the most appropriate answer from the options given below.

(a) Both A and R are true and R is the correct explanation of A.
(b) Both A and R are true but R is not the correct explanation of A.
(c) A is false but R is true.
(d) A is true but R is false.

19 One mole of P_4 reacts with 8 moles of $SOCl_2$ to give 4 mole of A, x mole of SO_2 and 2 moles of B. A, B and x respectively are

(a) $POCl_3$, S_2Cl_2 and 4. (b) PCl_3, S_2Cl_2 and 4.
(c) PCl_3, S_2Cl_2 and 2. (d) $POCl_3$, S_2Cl_2 and 2.

20 Which hydride among the following is less stable?

(a) HF (b) BeH_2 (c) LiH (d) NH_3

Section B : Numerical Value Type Questions

21 $Mg(NO_3)_2 \cdot XH_2O$ and $Ba(NO_3)_2 \cdot YH_2O$, represent formula of the crystalline forms of nitrate salts. Sum of X and Y is

22 The total number of intensive properties from the following isvolume, molar heat capacity, molarity, E°_{cell}, Gibbs free energy change, molar mass, mole.

23 The volume of hydrogen liberated at STP by treating 2.4 g of magnesium with excess of hydrochloric acid is .... $\times 10^{-2}$ L

Given, molar volume of gas is 22.4L at STP.
Molar mass of magnesium is 24 g mol^{-1}.

24 The number of possible isomeric products formed when 3-chloro-1-butene reacts with HCl through carbocation formation is

25 The number of correct statements about modern adsorption theory of heterogeneous catalysis from the following is

A. The catalyst is diffused over the surface of reactants.
B. Reactants are adsorbed on the surface of the catalyst.
C. Occurrence of chemical reaction on the catalyst's surface through formation of an intermediate.
D. It is a combination of intermediate compound formation theory and the old adsorption theory.
E. It explains the action of the catalyst as well as those of catalytic promoters and poisons.

26 Number of compounds from the following which will not produce orange-red precipitate with Benedict solution is ....

Glucose, maltose, sucrose, ribose, 2-deoxyribose, amylose, lactose

27 The number of correct statements from the following is ......

A. E_{cell} is an intensive parameter.
B. A negative $E^\ominus$ means that the redox couple is a stronger reducing agent than the H^+/H_2 couple.
C. The amount of electricity required for oxidation or reduction depends on the stoichiometry of the electrode reaction.
D. The amount of chemical reaction which occurs at any electrode during electrolysis by a current is proportional to the quantity of electricity passed through the electrolyte.

28 The number of correct statements from the following is ....

A. For 1s orbital, the probability density is maximum at the nucleus.
B. For 2s orbital, the probability density first increases to maximum and then decreases sharply to zero.
C. Boundary surface diagrams of the orbitals encloses a region of 100% probability of finding the electron.
D. p and d-orbitals have 1 and 2 angular nodes respectively.
E. Probability density of p-orbital is zero at the nucleus.

29 4.5 moles each of hydrogen and iodine is heated in a sealed ten litre vessel. At equilibrium, 3 moles of HI were found. The equilibrium constant for $H_2(g) + I_2(g) \rightleftharpoons 2HI(g)$ is ....

30 The maximum number of lone pairs of electrons on the central atom from the following species is .... ClO_3^-, XeF_4, SF_4 and I_3^-

12 APRIL 2023, SHIFT-I

Section A : Objective Type Questions

1 Given below are two statements, one is labelled as Assertion (A) and the other is labelled as Reason (R).

Assertion (A) In the Ellingham diagram, a sharp change in slope of the line is observed for $Mg \longrightarrow MgO$ at ~1120°C.

Reason (R) There is a large change of entropy associated with the change of state.

In the light of the above statements, choose the most appropriate answer from the options given below.

(a) Both A and R are true and R is the correct explanation of A.
(b) A is false but R is true.
(c) A is true but R is false.
(d) Both A and R are true but R is not the correct explanation of A.

2 Given below are two statements :

Statement I Boron is extremely hard indicating its high lattice energy.

Statement II Boron has highest melting and boiling point compared to its other group members.

In the light of the above statements, choose the most appropriate answer from the options given below.

(a) Both Statement I and Statement II are correct.
(b) Statement I is incorrect but Statement II is correct.
(c) Statement I is correct but Statement II is incorrect.
(d) Both Statement I and Statement II are incorrect.

3 Match List-I with List-II.

List-I		List-II	
A.	Nitrogen oxides in air	I.	Eutrophication
B.	Methane in air	II.	pH of rain water becomes 5.6
C.	Carbon dioxide	III.	Global warming
D.	Phosphate fertilisers in water	IV.	Acid rain

Choose the correct answer from the options given below.

(a) A-IV, B-III, C-II, D-I
(b) A-I, B-II, C-III, D-IV
(c) A-II, B-III, C-I, D-IV
(d) A-IV, B-II, C-III, D-I

4 Correct statements for the given reaction are

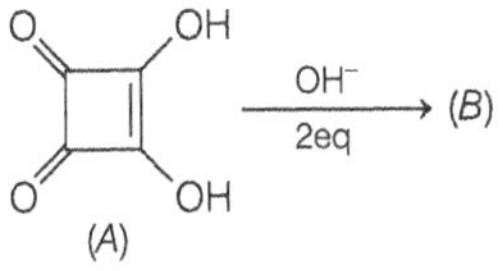

A. Compound '*B*' is aromatic.
B. The completion of above reaction is very slow.
C. '*A*' shows tautomerism.
D. The bond lengths of C—C in compound *B* are found to be same.

Choose the correct answer from the options given below.

(a) B, C and D only
(b) A, C and D only
(c) A, B and D only
(d) A, B and C only

5 The bond order and magnetic property of acetylide ion are same as that of

(a) O_2^+
(b) N_2^+
(c) O_2^-
(d) NO^+

6 The incorrect statement regarding the reaction given below is

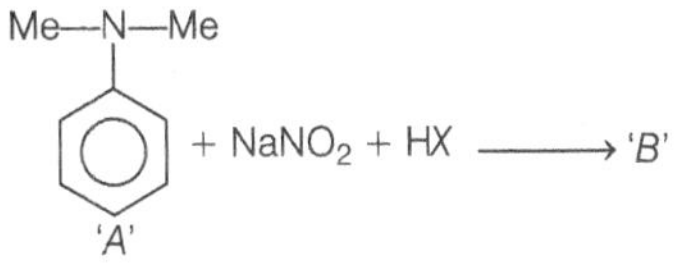

(a) The electrophile involved in the reaction is NO^+.
(b) '*B*' is N-nitroso ammonium compound.
(c) The product '*B*' formed in the above reaction is *p*-nitroso compound at low temperature.
(d) The reaction occurs at low temperature.

7 The major product '*P*' formed in the following sequence of reactions is

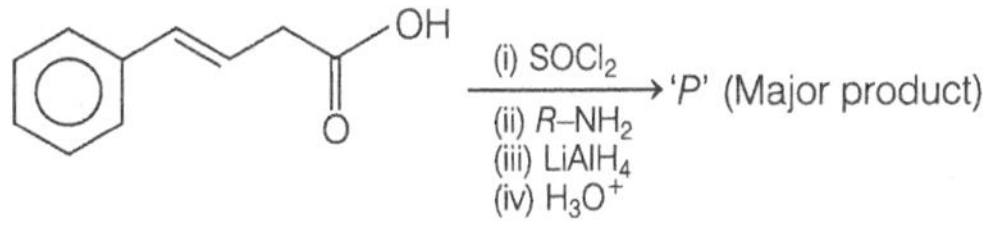

(a) 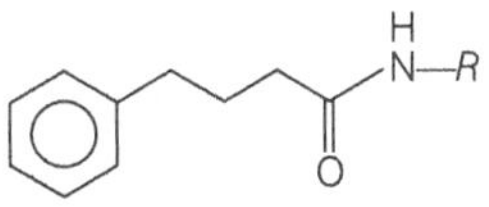

(b) 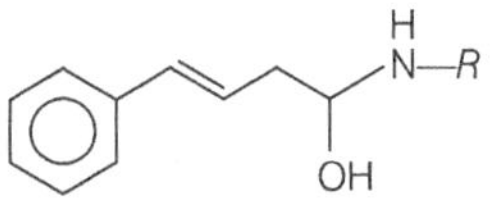

(c) 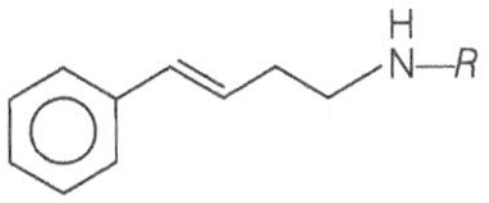

(d) 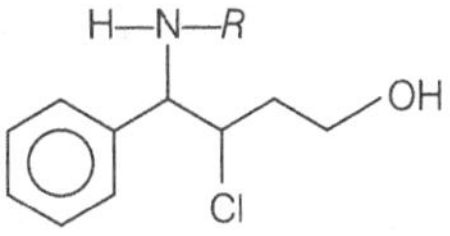

8 Given below are two statements :

Statement I $SbCl_5$ is more covalent than $SbCl_3$.

Statement II The higher oxides of halogens also tend to be more stable than the lower ones.

In the light of the above statements, choose the correct answer from the options given below.

(a) Statement I is incorrect but Statement II is correct.
(b) Both statement I and Statement II are correct.
(c) Statement I is correct but Statement II is incorrect.
(d) Both Statement I and Statement II are incorrect.

9 In the following reaction,

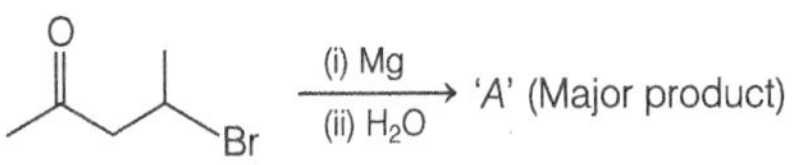

A is

(a) 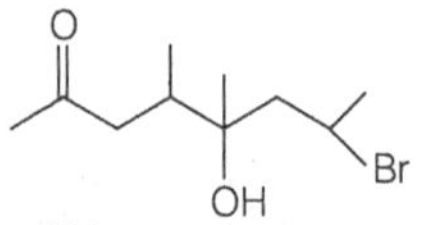

(b) 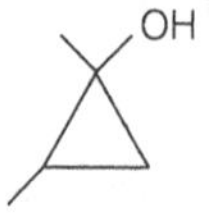

(c) 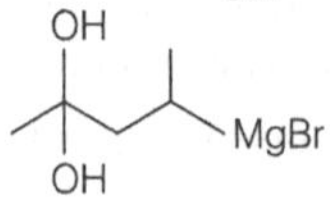

(d) 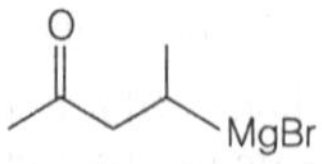

10 Match List-I with List-II.

List-I (Complex)	List-II (CFSE (Δ_0))
A. $[Cu(NH_3)_6]^{2+}$	I. -0.6
B. $[Ti(H_2O)_6]^{3+}$	II. -2.0
C. $[Fe(CN)_6]^{3-}$	III. -1.2
D. $[NiF_6]^{4-}$	IV. -0.4

Choose the correct answer from the options given below.

(a) A-I, B-II, C-IV, D-III (b) A-II, B-III, C-I, D-IV
(c) A-I, B-IV, C-II, D-III (d) A-III, B-IV, C-I, D-II

11 Match List-I with List-II

List-I (Type of hydride)	List-II (Example)
A. Electron deficient hydride	I. MgH_2
B. Electron rich hydride	II. HF
C. Electron precise hydride	III. B_2H_6
D. Saline hydride	IV. CH_4

Choose the correct answer from the options given below.

(a) A-II, B-III, C-IV, D-I (b) A-III, B-II, C-IV, D-I
(c) A-II, B-III, C-I, D-IV (d) A-III, B-II, C-I, D-IV

12 For lead storage battery pick the correct statements.

A. During charging of battery, $PbSO_4$ on anode is converted into PbO_2.
B. During charging of battery, $PbSO_4$ on cathode is converted into PbO_2.
C. Lead storage battery consists of grid of lead packed with PbO_2 as anode.
D. Lead storage battery has ~ 38% solution of sulphuric acid as an electrolyte.

Choose the correct answer from the options given below.

(a) B, D only (b) B, C only
(c) B, C, D only (d) A, B, D only

13 A metal chloride contains 55.0% of chlorine by weight. 100 mL vapours of the metal chloride at STP weigh 0.57 g. The molecular formula of the metal chloride is

(Given, atomic mass of chlorine is 35.5 u)

(a) $MnCl_4$ (b) MCl_2 (c) MCl (d) MCl_3

14 2-hexene $\xrightarrow[\text{(ii) } H_2O]{\text{(i) } O_3}$ Products

The two products formed in above reaction are

(a) butanoic acid and acetaldehyde
(b) butanal and acetic acid
(c) butanal and acetaldehyde
(d) butanoic acid and acetic acid

15 In the given reaction cycle,

$CaCl_2 + Na_2CO_3 \longrightarrow \underline{X} + \underline{Y}$ (Y $\xrightarrow{Z}$ back to $CaCl_2$)

X, *Y* and *Z* respectively are

	X	Y	Z
(a)	$CaCO_3$	NaCl	HCl
(b)	CaO	NaCl + CO_2	NaCl
(c)	$CaCO_3$	NaCl	KCl
(d)	CaO	NaCl + CO_2	KCl

16 Four gases *A*, *B*, *C* and *D* have critical temperatures 5.3, 33.2, 126.0 and 154.3K respectively. For their adsorption on a fixed amount of charcoal, the correct order is

(a) D > C > B > A (b) D > C > A > B
(c) C > D > B > A (d) C > B > D > A

17 Match List-I with List-II.

List-I (Examples)	List-II (Type)
A. 2-chloro-1, 3-butadiene	I. Biodegradable polymer
B. Nylon 2-nylon 6	II. Synthetic rubber
C. Polyacrylonitrile	III. Polyester
D. Dacron	IV. Addition polymer

Choose the correct answer from the options given below.

(a) A-IV, B-III, C-I, D-II (b) A-II, B-I, C-IV, D-III
(c) A-IV, B-I, C-III, D-II (d) A-II, B-IV, C-I, D-III

18 The density of alkali metals is in the order

(a) Na < K < Cs < Rb (b) K < Cs < Na < Rb
(c) Na < Rb < K < Cs (d) K < Na < Rb < Cs

19

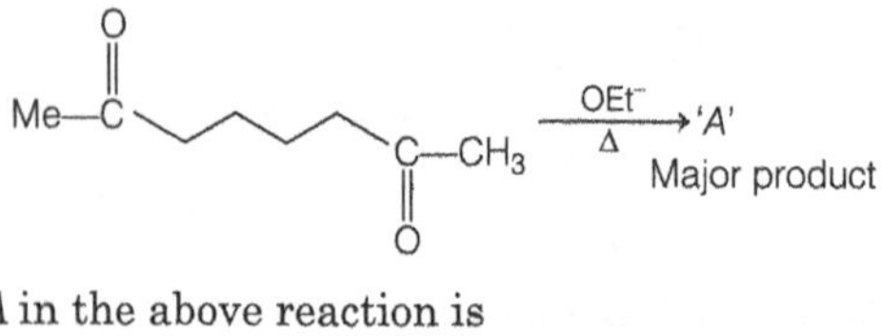

A in the above reaction is

(a) 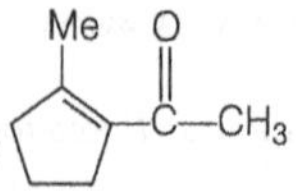

(b) 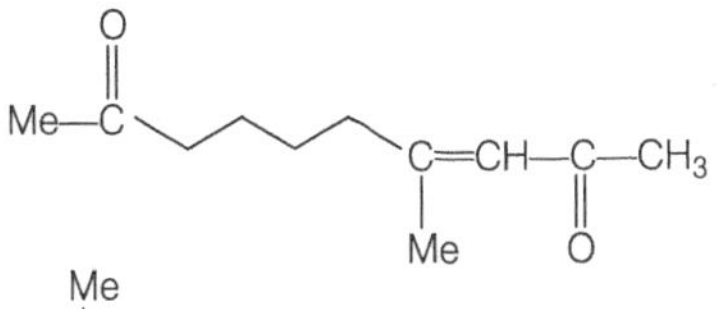

(c) 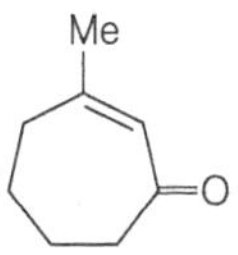

(d) 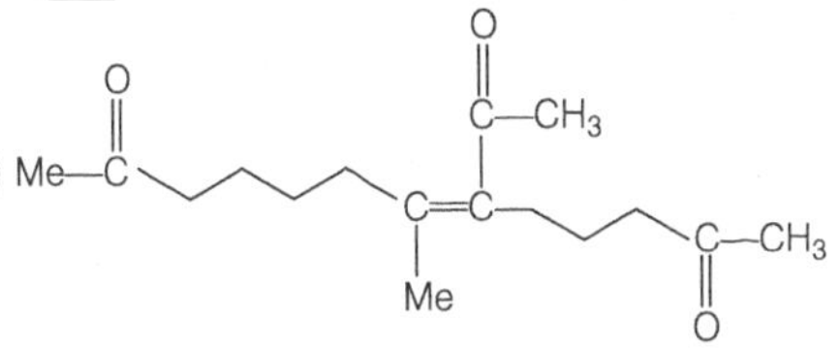

20 Given below are two statement : One is labelled as Assertion (A) and the other is labelled as Reason (R).

Assertion (A) $5f$ electrons can participate in bonding to a far greater extent than $4f$ electrons.

Reason (R) $5f$ orbitals are not as buried as $4f$ orbitals.

In the light of the above statements, choose the correct answer from the options given below.

(a) A is true but R is false.
(b) Both A and R are true and R is the correct explanation of A.
(c) Both A and R are true but R is not the correct explanation of A.
(d) A is false but R is true.

Section B : Numerical Value Type Questions

21 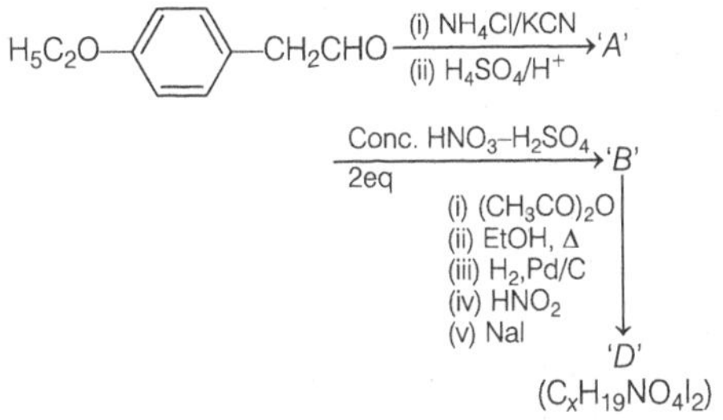

The value of x in compound 'D' is ....

22 In an oligopeptide named alanylglycylphenyl alanyl isoleucine, the number of sp^2 hybridised carbons is ....

23 Values of work function (W_0) for a few metals are given below

Metal	Li	Na	K	Mg	Cu	Ag
W_0 / eV	2.42	2.3	2.25	3.7	4.8	4.3

The number of metals which will show photoelectric effect when light of wavelength 400nm falls on it is ....

Given, $h = 6.6 \times 10^{-34}$ Js

$c = 3 \times 10^8$ ms^{-1}

$e = 1.6 \times 10^{-19}$C

24 80 mole per cent of $MgCl_2$ is dissociated in aqueous solution. The vapour pressure of 1.0 molal aqueous solution of $MgCl_2$ at 38°C is .... mm Hg. (Nearest integer)

(Given : Vapour pressure of water at 38°C is 50 mm Hg)

25 The mass of NH_3 produced when 131.8 kg of cyclohexanecarbaldehyde undergoes Tollen's test is .... kg. (Nearest integer)

Molar mass of C = 12 g/mol
N = 14 g/mol
O = 16 g/mol

26 At 600K, the root mean square (rms) speed of gas X (molar mass = 40) is equal to the most probable speed of gas Y at 90K. The molar mass of the gas Y is .... g mol^{-1}. (Nearest integer)

27 One mole of an ideal gas at 350 K is in a 2.0 L vessel of thermally conducting walls, which are in contact with the surroundings. It undergoes isothermal reversible expansion from 2.0 L to 3.0 L against a constant pressure of 4 atm. The change in entropy of the surroundings (ΔS) is .... J K^{-1}. (Nearest integer)

Given, $R = 8.314$ J K^{-1} mol^{-1}

28 Three organic compounds A, B and C were allowed to run in thin layer chromatography using hexane and gave the following result (see figure). The R_f value of the most polar compound is .... $\times 10^{-2}$.

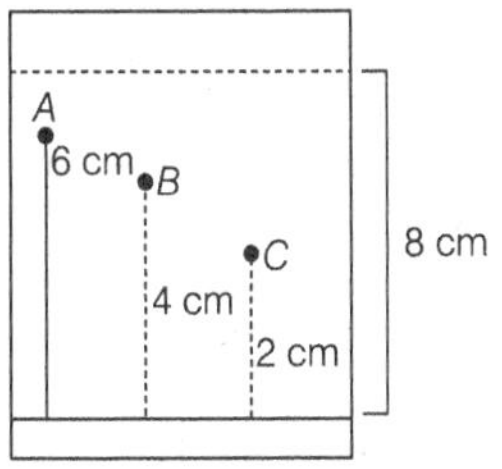

29 The reaction $2NO + Br_2 \longrightarrow 2NOBr$ takes places through the mechanism given below

$NO + Br_2 \longrightarrow NOBr_2$ (fast)

$NOBr_2 + NO \longrightarrow 2NOBr$ (Slow)

The overall order of the reaction is ....

30 An analyst wants to convert 1L HCl of pH = 1 to a solution of HCl of pH 2. The volume of water needed to do this dilution is .... mL. (Nearest integer)

13 APRIL 2023, SHIFT-I

Section A : Objective Type Questions

1 The incorrect statement from the following for borazine is
(a) it is a cyclic compound.
(b) it contains banana bonds.
(c) it has electronic delocalisation.
(d) it can react with water.

2 D-(+)-glyceraldehyde $\xrightarrow[\text{(ii) } H_2O / H^+ \text{ (iii) } HNO_3]{\text{(i) HCN}}$

The products formed in the above reaction are
(a) one optically inactive products
(b) two optically active products
(c) two optically inactive products
(d) one optically active and one meso product

3 Among the following compounds, the one which shows highest dipole moment is

(a) 
(b) 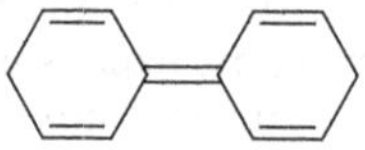
(c) 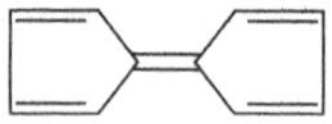
(d) 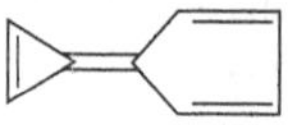

4 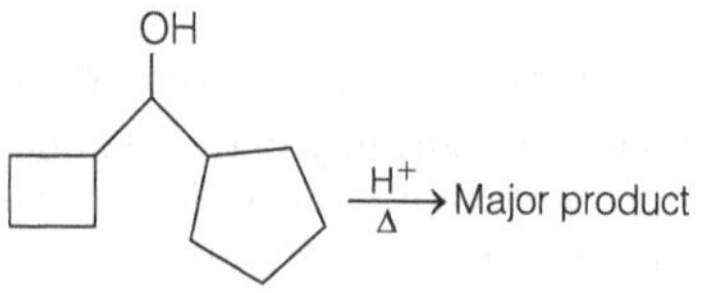

In the above reaction, left hand side and right hand side rings are named as '*A*' and '*B*' respectively. They undergo ring expansion. The correct statement for this process is
(a) only *A* will become 6 membered.
(b) finally both rings will become six membered each.
(c) finally both rings will become five membered each.
(d) ring expansion can go upto seven membered rings.

5 Which of the following statements are not correct ?
A. The electron gain enthalpy of F is more negative than that of Cl.
B. Ionisation enthalpy decreases in a group of periodic table.
C. The electronegativity of an atom depends upon the atoms bonded to it.
D. Al_2O_3 and NO are examples of amphoteric oxides.

Choose the most appropriate answer from the options given below
(a) B and D only
(b) A, B, C and D
(c) A, B and D only
(d) A, C and D only

6 The pair of lanthanides in which both elements have high third - ionisation energy
(a) Lu, Yb (b) Dy, Gd (c) Eu, Yb (d) Eu, Gd

7 The radical which mainly causes ozone depletion in the presence of UV radiations is
(a) $\dot{O}H$ (b) $Cl^\bullet$ (c) $NO^\bullet$ (d) $CH_3^\bullet$

8 In the following reaction '*X*' is

$$CH_3(CH_2)_4CH_3 \xrightarrow[\text{HCl}, \Delta]{\text{Anhy. AlCl}_3} \underset{\text{Major product}}{'X'}$$

(a) $CH_3(CH_2)_4CH_2Cl$
(b) $CH_3CH(CH_3)—(CH_2)_2CH_3$
(c) $Cl—CH_2—(CH_2)_4—CH_2Cl$
(d) 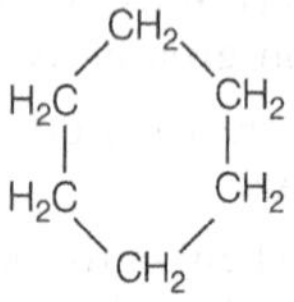

9 $Be(OH)_2$ reacts with $Sr(OH)_2$ to yield an ionic salt. Choose the incorrect option related to this reaction from the following.
(a) The element Be is present in the cationic part of the ionic salt.
(b) Both Sr and Be elements are present in the ionic salt.
(c) Be is tetrahedrally coordinated in the ionic salt.
(d) The reaction is an example of acid-base neutralisation reaction.

10 Which one of the following is most likely a mismatch?
(a) Zinc - Liquation
(b) Copper - Electrolysis
(c) Titanium - van Arkel method
(d) Nickel - Mond process

11 The mismatched combinations are
A. Chlorophyll - Co
B. Water hardness - EDTA
C. Photography -$[Ag(CN)_2]^-$
D. Wilkinson catalyst - $[(Ph_3P)_3RhCl]$
E. Chelating ligand -D-penicillamine

Choose the correct answer from the options given below
(a) A and C only (b) D and E only
(c) A, C, and E only (d) A and E only

12 Match List-I with List-II.

List-I	List-II
A. Nylon-6	I. Natural rubber
B. Vulcanised rubber	II. Cross-linked
C. *cis*-1, 4-polyisoprene	III. Caprolactam
D. Polychloroprene	IV. Neoprene

Choose the correct answer from the options given below.

(a) A-II, B-III, C-IV, D-I

(b) A-III, B-IV, C-I, D-II

(c) A-III, B-II, C-I, D-IV

(d) A-IV, B-III, C-II, D-I

13 The energy of an electron in the first Bohr orbit of hydrogen atom is -2.18×10^{-18} J. Its energy in the third Bohr orbit is ....

(a) three times of this value (b) $\frac{1}{9}$th of this value

(c) one third of this value (d) $\frac{1}{27}$ of this value

14 In the reaction given below

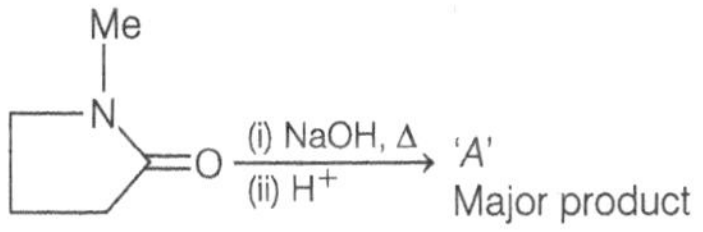

'*A*' is

(a)
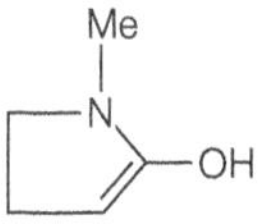

(b)
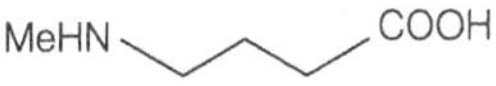

(c)
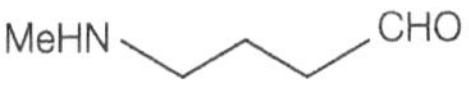

(d)
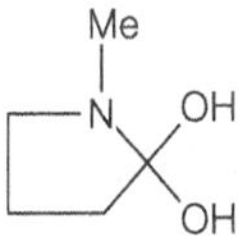

15 What happens when a lyophilic sol is added to a lyophobic sol?

(a) Film of lyophobic sol is formed over lyophilic sol.

(b) Lyophobic sol is coagulated

(c) Lyophilic sol is dispersed in lyophobic sol

(d) Film of lyophilic sol is formed over lyophobic sol

16 In the reaction given below

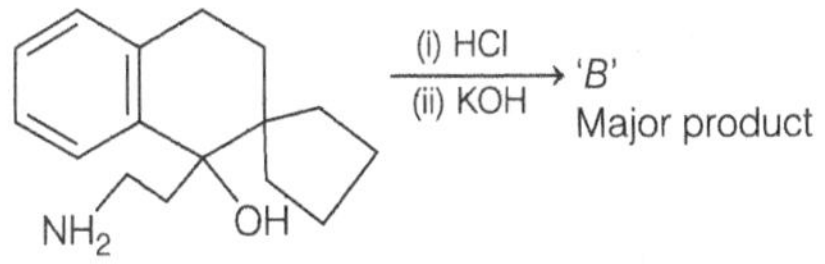

'*B*' is

(a)
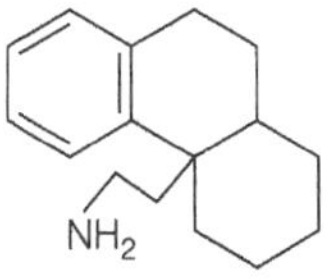

(b)
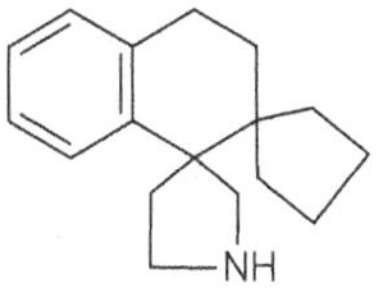

(c)
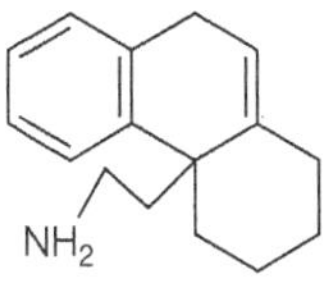

(d)
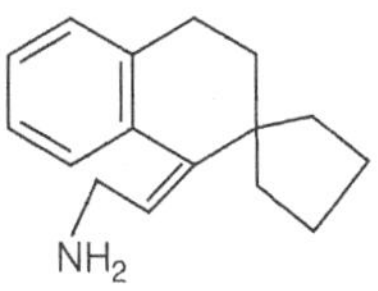

17 In which of the following processes, the bond order increases and paramagnetic character changes to diamagnetic one?

(a) $O_2 \longrightarrow O_2^+$ (b) $NO \longrightarrow NO^+$

(c) $O_2 \longrightarrow O_2^{2-}$ (d) $N_2 \longrightarrow N_2^+$

18 Given below are two statements :

Statement I Permutit process is more efficient compared to the synthetic resin method for the softening of water.

Statement II Synthetic resin method results in the formation of soluble sodium salts.

In the light of above statements, choose the most appropriate answer from the options given below.

(a) Statement I is correct but Statement II is incorrect.

(b) Both the Statement I and II are incorrect.

(c) Both the Statement I and II are correct.

(d) Statement I is incorrect but Statement II is correct.

19 ClF_5 at room temperature is a

(a) colourless gas with trigonal bipyramidal geometry.

(b) colourless liquid with square pyramidal geometry.

(c) colourless gas with square pyramidal geometry.

(d) colourless liquid with trigonal bipyramidal geometry.

20 2-methyl propyl bromide reacts with $C_2H_5O^-$ and gives *A* whereas on reaction with C_2H_5OH it gives *B*. The mechanism followed in these reactions and the products '*A*' and '*B*' respectively are

(a) S_N1, *A* = tert-butyl ethyl ether; S_N2, *B* = *iso*-butyl ethyl ether

(b) S_N2, *A* = 2-butyl ethyl ether; S_N2, *B* = *iso*-butyl ethyl ether

(c) S_N2, *A* = *iso*-butyl ethyl ether; S_N1, *B* = *tert*-butyl ethyl ether

(d) S_N1, *A* = *iso*-butyl ethyl ether; S_N1, *B* = 2-butyl ethyl ether

Section B : Numerical Value Type Questions

21 A metal surface of 100 cm^2 area has to be coated with nickel layer of thickness 0.001 mm. A current of 2A was passed through a solution of $Ni(NO_3)_2$ for *x* seconds to coat the desired layer. The value of *x* is .... (Nearest integer) (ρ_{Ni} (density of nickel) is 10 g mL^{-1}. Molar mass of nickel is 60 g mol^{-1}, $F = 96500$ C mol^{-1})

22 $t_{87.5}$ is the time required for the reaction to undergo 87.5% completion and t_{50} is the time required for the reaction to undergo 50% completion. The relation between $t_{87.5}$ and t_{50} for a first order reaction is

$t_{87.5} = x \times t_{50}$

The value of *x* is .... (Nearest integer)

23 For the given reaction

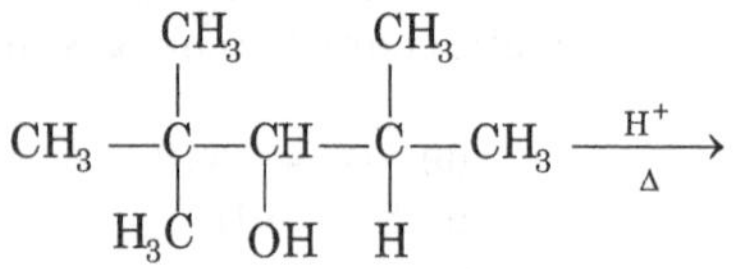

The total number of possible products formed by tertiary carbocation of A is ....

24 25.0 mL of 0.050 M $Ba(NO_3)_2$ is mixed with 25.0 mL of 0.020 M NaF. K_{sp} of BaF_2 is 0.5×10^{-6} at 298 K. The ratio of $[Ba^{2+}][F^-]^2$ and K_{sp} is ....

(Nearest integer)

25 $KMnO_4$ is titrated with ferrous ammonium sulphate hexahydrate in presence of dilute H_2SO_4. Number of water molecules produced for 2 molecules of $KMnO_4$ is ....

26 An organic compound gives 0.220 g of CO_2 and 0.126 g of H_2O on complete combustion. If the % of carbon is 24 then the % of hydrogen is .... $\times 10^{-1}$. (Nearest integer)

27 Solution of 12 g of non-electrolyte (A) prepared by dissolving it in 1000 mL of water exerts the same osmotic pressure as that of 0.05 M glucose solution at the same temperature. The empirical formula of A is CH_2O. The molecular mass of A is .... g. (Nearest integer)

28 20 mL of calcium hydroxide was consumed when it was reacted with 10 mL of unknown solution of H_2SO_4. Also 20 mL standard solution of 0.5 M HCl containing 2 drops of phenolphthalein was titrated with calcium hydroxide, the mixture showed pink colour when burette displayed the value of 35.5 mL whereas the burette showed 25.5 mL initially. The concentration of H_2SO_4 is .... M. (Nearest integer)

29 $A_2 + B_2 \longrightarrow 2AB$. $\Delta H_f^\circ = -200 \text{ kJ mol}^{-1}$

AB, A_2 and B_2 are diatomic molecules. If the bond enthalpies of A_2, B_2 and AB are in the ratio 1 : 0.5 : 1, then the bond enthalpy of A_2 is .... kJ mol^{-1}

(Nearest integer)

30 A certain quantity of real occupies a volume of 0.15 dm^3 at 100 atm and 500 K when its compressibility factor is 1.07. Its volume at 300 atm and 300 K (when its compressibility factor is 1.4) is .... $\times 10^{-4} \text{ dm}^3$. (Nearest integer)

13 APRIL 2023, SHIFT-II

Section A : Objective Type Questions

1 Identify the correct order of standard enthalpy of formation of sodium halides.

(a) NaI < NaBr < NaF < NaCl
(b) NaCl < NaF < NaBr < NaI
(c) NaF < NaCl < NaBr < NaI
(d) NaI < NaBr < NaCl < NaF

2 Which of the following complexes will exhibit maximum attraction to an applied magnetic field?

(a) $[Co(H_2O)_6]^{2+}$ (b) $[Co(en)_3]^{3+}$
(c) $[Ni(H_2O)_6]^{2+}$ (d) $[Zn(H_2O)_6]^{2+}$

3 Match List-I with List-II.

1-bromopropane is reacted with reagents in List-I to give product in List-II.

List-I (Reagent)		List-II (Product)
A. KOH (alc)	I.	Nitrite
B. KCN (alc)	II.	Ester
C. $AgNO_2$	III.	Alkene
D. $H_3CCOOAg$	IV.	Nitroalkane

Choose the correct answer from the options given below.

(a) A-I, B-II, C-III, D-IV (b) A-III, B-I, C-IV, D-II
(c) A-IV, B-III, C-II, D-I (d) A-I, B-III, C-IV, D-II

4 Better method for preparation of BeF_2, among the following is

(a) $BeO + C + F_2 \xrightarrow{\Delta} BeF_2$ (b) $Be + F_2 \xrightarrow{\Delta} BeF_2$
(c) $(NH_4)_2BeF_4 \xrightarrow{\Delta} BeF_2$ (d) $BeH_2 + F_2 \xrightarrow{\Delta} BeF_2$

5 The total number of stereoisomers for the complex $[Co(ox)_2ClBr]^{3-}$ (where, ox = oxalate) is

(a) 1 (b) 3 (c) 2 (d) 4

6 Given below are two statements.

Statement I Tropolone is an aromatic compound and has 8 π electrons.

Statement II π electrons of $>C=O$ group in tropolone is involved in aromaticity.

In the light of the above statements, choose the correct answer from the options given below.

(a) Both Statement I and Statement II are true.
(b) Statement I is true but Statement II is false.
(c) Statement I is false but Statement II is true.
(d) Both Statement I and Statement II are false.

7 Given below are two statements, one is labelled as Assertion (A) and the other is labelled as Reason (R).

Assertion (A) Isotopes of hydrogen have almost same chemical properties but difference in their rates of reaction.

Reason (R) Isotopes of hydrogen have different enthalpy of bond dissociation.

In the light of the above statements, choose the most appropriate answer from the options given below.
(a) Both A and R are correct but R is not the correct explanation of A.
(b) Both A and R are correct and R is the correct explanation of A.
(c) A is not correct but R is correct.
(d) A is correct but R is not correct.

8 In the wet tests for detection of various cations by precipitation, Ba^{2+} cations are detected by obtaining precipitate of
(a) Ba(ox) : Barium oxalate (b) $BaSO_4$
(c) $Ba(OAc)_2$ (d) $BaCO_3$

9 Match List-I with List-II.

	List-I		List-II
A.	Weak intermolecular forces of attraction	I.	Hexamethylene diamine + adipic acid
B.	Hydrogen bonding	II.	$AlEt_3$ + $TiCl_4$
C.	Heavily branched polymer	III.	2-chloro-1, 3-butandiene
D.	High density polymer	IV.	Phenol + formaldehyde

Choose the correct answer from the options given below.
(a) A-II, B-IV, C-I, D-III
(b) A-III, B-I, C-IV, D-II
(c) A-IV, B-II, C-III, D-I
(d) A-IV, B-I, C-III, D-II

10 The naturally occurring amino acid that contains only one basic functional group in its chemical structure is
(a) histidine (b) asparagine
(c) lysine (d) arginine

11 Given below are two statements related to Ellingham diagram.

Statement I Ellingham diagrams can be constructed for formation of oxides, sulphides and halides of metals.

Statement II It consists plots of $\Delta_f H^\circ$ *vs* T for formation of oxides of elements.

In the light of the above statements, choose the correct answer from the options given below.
(a) Both Statement I and Statement II are correct.
(b) Statement I is correct but Statement II is incorrect.
(c) Both Statement I and Statement II are incorrect.
(d) Statement I is incorrect but Statement II is correct.

12 The correct group of halide ions which can be oxidised by oxygen in acidic medium is
(a) I^- only
(b) Br^- only
(c) Cl^-, Br^- and I^- only
(d) Br^- and I^- only

13 Compound *A* from the following reaction sequence is

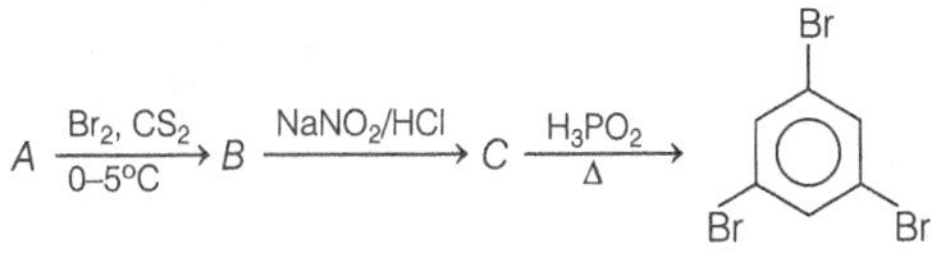

(a) aniline (b) benzoic acid
(c) salicylic acid (d) phenol

14 The covalency and oxidation state respectively of boron in $[BF_4]^-$, are
(a) 4 and 3 (b) 3 and 5
(c) 3 and 4 (d) 4 and 4

15 What happens when methane undergoes combustion in systems *A* and *B* respectively?

Adiabatic system	Diathermic container
System *A*	System *B*

(a)

System *A*	System *B*
Temperature remains same	Temperature rises

(b)

System *A*	System *B*
Temperature falls	Temperature remains same

(c)

System *A*	System *B*
Temperature falls	Temperature rises

(d)

System *A*	System *B*
Temperature rises	Temperature remains same

16 Given below are two statements, one is labelled as Assertion (A) and the other is labelled as Reason (R).

Assertion (A) The diameter of colloidal particles in solution should not be much smaller than wavelength of light to show Tyndall effect.

Reason (R) The light scatters in all direction when the size of particles is large enough.

In the light of the above statements, choose the most appropriate answer from the options given below.
(a) A is false but R is true.
(b) Both A and R are correct but R is not the correct explanation of A.
(c) A is true but R is false.
(d) Both A and R are correct and R is the correct explanation of A.

17 Given below are two statements.

Statement I SO_2 and H_2O both possess V-shaped structure.

Statement II The bond angle of SO_2 is less than that of H_2O.

In the light of the above statements, choose the most appropriate answer from the options given below.

(a) Both Statement I and Statement II are incorrect.
(b) Statement I is correct but Statement II is incorrect.
(c) Both Statement I and Statement II are correct.
(d) Statement I is incorrect but Statement II is correct.

18 The major product for the following reaction is

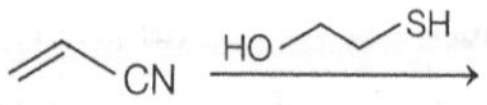

(a) 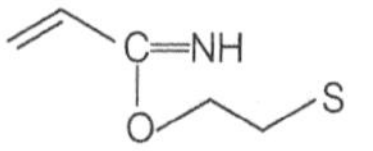

(b) 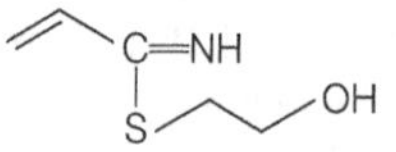

(c) 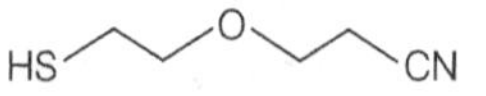

(d) 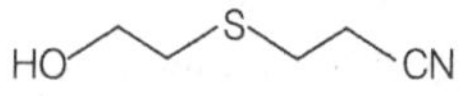

19 Which of the following are the green house gases?

A. Water vapour
B. Ozone
C. I_2
D. Molecular hydrogen

Choose the most appropriate answer from the options given below.

(a) B and C only (b) A and B only
(c) C and D only (d) A and D only

20 Given below are two statements, one is labelled as Assertion (A) and the other is labelled as Reason (R).

Assertion (A) Order of acidic nature of the following compounds is $A > B > C$.

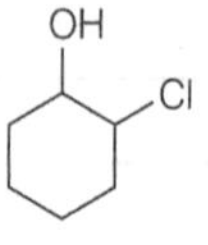

(A)

(B)

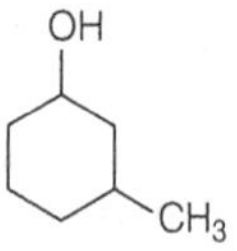

(C)

Reason (R) Fluoro is a stronger electron withdrawing group than chloro group.

In the light of the above statements, choose the most appropriate correct answer from the options given below.

(a) A is true but R is false.
(b) Both A and R are correct and R is the correct explanation of A.
(c) Both A and R are correct but R is not the correct explanation of A.
(d) A is false but R is true.

Section B : Numerical Value Type Questions

21 $A(g) \longrightarrow 2B(g) + C(g)$ is a first order reaction. The initial pressure of the system was found to be 800 mm Hg which increased to 1600 mm Hg after 10 min. The total pressure of the system after 30 min will be .... mm Hg. (Nearest integer)

22 At 298 K, the standard reduction potential for Cu^{2+}/Cu electrode is 0.34 V.

Given : $K_{sp} Cu(OH)_2 = 1 \times 10^{-20}$

The reduction potential at pH = 14 for the above couple is $(-)x \times 10^{-2}$ V.

The value of x is ....

$$\left[\text{Take } \frac{2.303\,RT}{F} = 0.059 \text{ V}\right]$$

23 Sea water contains 29.25% NaCl and 19% $MgCl_2$ by weight of solution. The normal boiling point of the sea water is .... °C (Nearest integer)

Assume 100% ionisation for both NaCl and $MgCl_2$.

[Given : $K_b(H_2O) = 0.52$ K kg mol^{-1}

Molar mass of NaCl and $MgCl_2$ is 58.5 and 95 g mol^{-1} respectively.]

24 If the formula of borax is $Na_2B_4O_x(OH)_y \cdot zH_2O$, then $x + y + z =$ ....

25 The orbital angular momentum of an electron in 3s orbital is $\frac{xh}{2\pi}$. The value of x is .... (nearest integer)

26 1g of a carbonate (M_2CO_3) on treatment with excess HCl products 0.01 mol of CO_2. The molar mass of M_2CO_3 is .... g mol^{-1}. (Nearest integer)

27 20 mL of 0.1 M NaOH is added to 50 mL of 0.1 M acetic acid solution. The pH of the resulting solution is .... $\times 10^{-2}$. (Nearest integer)

[Given : $pK_a(CH_3COOH) = 4.76$
$\log 2 = 0.30$
$\log 3 = 0.48$]

28 0.400 g of an organic compound (X) gave 0.376 g of AgBr in Carius method for estimation of bromine. % of bromine in the compound (X) is ....

(Given : Molar mass AgBr = 188 g mol^{-1}
Br = 80 g mol^{-1})

29 See the following chemical reaction.

$$Cr_2O_7^{2-} + XH^+ + 6Fe^{2+} \longrightarrow YCr^{3+} + 6Fe^{3+} + ZH_2O$$

The sum of X, Y and Z is ....

30 Sodium metal crystallise in a body centered cubic lattice with unit cell edge length of 4 Å. The radius of sodium atom is .... $\times 10^{-1}$ Å. (Nearest integer)

Answers

For solutions scan the QR code

6 April Shift-I

1. (b)	2. (d)	3. (b)	4. (d)	5. (b)	6. (a)	7. (b)	8. (d)	9. (c)	10. (a)
11. (b)	12. (c)	13. (a)	14. (b)	15. (a)	16. (c)	17. (a)	18. (a)	19. (b)	20. (b)
21. 6	22. 10	23. 7	24. 3	25. 1111	26. 2	27. 4	28. 2	29. 9	30. 1

6 April Shift-II

1. (a)	2. (a)	3. (c)	4. (b)	5. (c)	6. (d)	7. (c)	8. (d)	9. (a)	10. (b)
11. (c)	12. (c)	13. (c)	14. (c)	15. (a)	16. (d)	17. (a)	18. (d)	19. (c)	20. (d)
21. 3	22. 3	23. 48	24. 4	25. 4	26. 4	27. 278	28. 4	29. 5	30. 5

8 April Shift-I

1. (c)	2. (a)	3. (a)	4. (d)	5. (d)	6. (c)	7. (b)	8. (a)	9. (c)	10. (d)
11. (a)	12. (d)	13. (c)	14. (c)	15. (d)	16. (c)	17. (d)	18. (b)	19. (c)	20. (a)
21. 82	22. 70	23. 1200	24. 2	25. 3	26. 11	27. 11	28. 3	29. 1	30. 2

8 April Shift-II

1. (a)	2. (c)	3. (b)	4. (d)	5. (a)	6. (a)	7. (c)	8. (a)	9. (a)	10. (c)
11. (b)	12. (d)	13. (c)	14. (c)	15. (d)	16. (a)	17. (c)	18. (c)	19. (b)	20. (b)
21. 4	22. 4	23. 1	24. 10	25. 6	26. 556	27. 3	28. 8	29. 233	30. 1411

10 April Shift-I

1. (c)	2. (c)	3. (c)	4. (d)	5. (b)	6. (c)	7. (a)	8. (b)	9. (d)	10. (d)
11. (b)	12. (c)	13. (d)	14. (b)	15. (b)	16. (b)	17. (c)	18. (d)	19. (b)	20. (d)
21. 3	22. 30	23. 1567	24. 1	25. 5	26. 2.4	27. 0	28. 3	29. 1825	30. 3

10 April Shift-II

1. (c)	2. (c)	3. (d)	4. (b)	5. (c)	6. (b)	7. (c)	8. (d)	9. (c)	10. (b)
11. (c)	12. (b)	13. (b)	14. (a)	15. (d)	16. (a)	17. (b)	18. (a)	19. (a)	20. (c)
21. 4	22. 4	23. 1	24. 3	25. 1	26. 2	27. 4	28. 3	29. 66	30. 40535

11 April Shift-I

1. (a)	2. (c)	3. (b)	4. (b)	5. (c)	6. (d)	7. (b)	8. (d)	9. (b)	10. (c)
11. (a)	12. (a)	13. (c)	14. (d)	15. (c)	16. (b)	17. (a)	18. (c)	19. (a)	20. (b)
21. 44	22. 7	23. 2	24. 333	25. 75	26. 1	27. 4	28. 2	29. 36	30. 4

11 April Shift-II

1. (b)	2. (a)	3. (a)	4. (d)	5. (c)	6. (b)	7. (c)	8. (c)	9. (c)	10. (d)
11. (d)	12. (c)	13. (b)	14. (b)	15. (a)	16. (b)	17. (c)	18. (d)	19. (b)	20. (b)
21. 6	22. 4	23. 224	24. 4	25. 3	26. 2	27. 4	28. 3	29. 1	30. 3

12 April Shift-I

1. (a)	2. (a)	3. (a)	4. (b)	5. (d)	6. (b)	7. (c)	8. (b)	9. (a)	10. (c)
11. (b)	12. (a)	13. (b)	14. (d)	15. (a)	16. (a)	17. (b)	18. (d)	19. (a)	20. (b)
21. 15	22. 10	23. 3	24. 48	25. 60	26. 4	27. 3	28. 25	29. 3	30. 9000

13 April Shift-I

1. (b)	2. (d)	3. (d)	4. (b)	5. (d)	6. (c)	7. (b)	8. (b)	9. (a)	10. (a)
11. (d)	12. (c)	13. (b)	14. (b)	15. (d)	16. (c)	17. (b)	18. (b)	19. (b)	20. (c)
21. 161	22. 3	23. 4	24. 5	25. 68	26. 56	27. 240	28. 1	29. 800	30. 392

13 April Shift-II

1. (d)	2. (a)	3. (b)	4. (c)	5. (b)	6. (b)	7. (b)	8. (d)	9. (b)	10. (b)
11. (b)	12. (a)	13. (a)	14. (a)	15. (d)	16. (d)	17. (b)	18. (d)	19. (b)	20. (c)
21. 2200	22. 25	23. 116	24. 17	25. 0	26. 100	27. 458	28. 40	29. 23	30. 17